Applied and Numerical Harmonic Analysis

Series Editor
John J. Benedetto
University of Maryland
College Park, MD, USA

Editorial Advisory Board

Akram Aldroubi
Vanderbilt University
Nashville, TN, USA

Douglas Cochran
Arizona State University
Phoenix, AZ, USA

Hans G. Feichtinger
University of Vienna
Vienna, Austria

Christopher Heil
Georgia Institute of Technology
Atlanta, GA, USA

Stéphane Jaffard
University of Paris XII
Paris, France

Jelena Kovačević
Carnegie Mellon University
Pittsburgh, PA, USA

Gitta Kutyniok
Technische Universität Berlin
Berlin, Germany

Mauro Maggioni
Duke University
Durham, NC, USA

Zuowei Shen
National University of Singapore
Singapore, Singapore

Thomas Strohmer
University of California
Davis, CA, USA

Yang Wang
Michigan State University
East Lansing, MI, USA

More information about this series at http://www.springer.com/series/4968

Birkhäuser

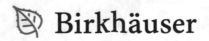

Bin Han

Framelets and Wavelets

Algorithms, Analysis, and Applications

Birkhäuser

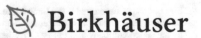

Bin Han
Department of Mathematical and Statistical
Sciences
University of Alberta
Edmonton, Alberta, Canada

ISSN 2296-5009 ISSN 2296-5017 (electronic)
Applied and Numerical Harmonic Analysis
ISBN 978-3-319-88617-6 ISBN 978-3-319-68530-4 (eBook)
https://doi.org/10.1007/978-3-319-68530-4

Mathematics Subject Classification (2010): 42C40, 42C15, 41A05, 41A30, 65T60

Softcover re-print of the Hardcover 1st edition 2017

Printed on acid-free paper

This book is published under the trade name Birkhäuser, www.birkhauser-science.com
The registered company is Springer International Publishing AG
The registered company address is: Gewerbestrasse 11, 6330 Cham, Switzerland

To my wife Shuang, daughter Rachel, son James, and parents

ANHA Series Preface

The *Applied and Numerical Harmonic Analysis* (*ANHA*) book series aims to provide the engineering, mathematical, and scientific communities with significant developments in harmonic analysis, ranging from abstract harmonic analysis to basic applications. The title of the series reflects the importance of applications and numerical implementation, but richness and relevance of applications and implementation depend fundamentally on the structure and depth of theoretical underpinnings. Thus, from our point of view, the interleaving of theory and applications and their creative symbiotic evolution is axiomatic.

Harmonic analysis is a wellspring of ideas and applicability that has flourished, developed, and deepened over time within many disciplines and by means of creative cross-fertilization with diverse areas. The intricate and fundamental relationship between harmonic analysis and fields such as signal processing, partial differential equations (PDEs), and image processing is reflected in our state-of-the-art *ANHA* series.

Our vision of modem harmonic analysis includes mathematical areas such as wavelet theory, Banach algebras, classical Fourier analysis, time-frequency analysis, and fractal geometry, as well as the diverse topics that impinge on them.

For example, wavelet theory can be considered an appropriate tool to deal with some basic problems in digital signal processing, speech and image processing, geophysics, pattern recognition, biomedical engineering, and turbulence. These areas implement the latest technology from sampling methods on surfaces to fast algorithms and computer vision methods. The underlying mathematics of wavelet theory depends not only on classical Fourier analysis but also on ideas from abstract harmonic analysis, including von Neumann algebras and the affine group. This leads to a study of the Heisenberg group and its relationship to Gabor systems, and of the metaplectic group for a meaningful interaction of signal decomposition methods. The unifying influence of wavelet theory in the aforementioned topics illustrates the justification for providing a means for centralizing and disseminating information from the broader, but still focused, area of harmonic analysis. This will be a key role of *ANHA*. We intend to publish with the scope and interaction that such a host of issues demands.

Along with our commitment to publish mathematically significant works at the frontiers of harmonic analysis, we have a comparably strong commitment to publish major advances in the following applicable topics in which harmonic analysis plays a substantial role:

Antenna theory	*Prediction theory*
Biomedical signal processing	*Radar applications*
Digital signal processing	*Sampling theory*
Fast algorithms	*Spectral estimation*
Gabor theory and applications	*Speech processing*
Image processing	*Time-frequency and*
Numerical partial differential equations	*time-scale analysis*
	Wavelet theory

The above point of view for the *ANHA* book series is inspired by the history of Fourier analysis itself, whose tentacles reach into so many fields.

In the last two centuries Fourier analysis has had a major impact on the development of mathematics, on the understanding of many engineering and scientific phenomena, and on the solution of some of the most important problems in mathematics and the sciences. Historically, Fourier series were developed in the analysis of some of the classical PDEs of mathematical physics; these series were used to solve such equations. In order to understand Fourier series and the kinds of solutions they could represent, some of the most basic notions of analysis were defined, e.g., the concept of "function." Since the coefficients of Fourier series are integrals, it is no surprise that Riemann integrals were conceived to deal with uniqueness properties of trigonometric series. Cantor's set theory was also developed because of such uniqueness questions.

A basic problem in Fourier analysis is to show how complicated phenomena, such as sound waves, can be described in terms of elementary harmonics. There are two aspects of this problem: first, to find, or even define properly, the harmonics or spectrum of a given phenomenon, e.g., the spectroscopy problem in optics; second, to determine which phenomena can be constructed from given classes of harmonics, as done, for example, by the mechanical synthesizers in tidal analysis.

Fourier analysis is also the natural setting for many other problems in engineering, mathematics, and the sciences. For example, Wiener's Tauberian theorem in Fourier analysis not only characterizes the behavior of the prime numbers but also provides the proper notion of spectrum for phenomena such as white light; this latter process leads to the Fourier analysis associated with correlation functions in filtering and prediction problems, and these problems, in turn, deal naturally with Hardy spaces in the theory of complex variables.

Nowadays, some of the theory of PDEs has given way to the study of Fourier integral operators. Problems in antenna theory are studied in terms of unimodular trigonometric polynomials. Applications of Fourier analysis abound in signal processing, whether with the fast Fourier transform (FFT), or filter design, or the adaptive modeling inherent in time-frequency-scale methods such as wavelet theory. The coherent states of mathematical physics are translated and modulated Fourier transforms, and these are used, in conjunction with the uncertainty principle, for dealing with signal reconstruction in communications theory. We are back to the raison d'être of the *ANHA* series!

University of Maryland
College Park, MD, USA

John J. Benedetto
Series Editor

Preface

As a rapidly growing, multidisciplinary field of mathematics, wavelet theory provides the major mathematical multiscale representation for analyzing functions/data and has undergone extensive development for more than 30 years, including numerous applications in science and industry. This book introduces readers to the mathematical theory of wavelets from the perspective of framelets (i.e., frame wavelets) and discrete framelet transforms by systematically covering various topics. It can serve not only as a textbook for graduate students, senior undergraduate students, and beginning researchers, offering them self-contained elementary proofs and supplementary exercises, but also as an advanced reference guide for experienced researchers in mathematics, physics, and engineering by systematically addressing a wide range of topics on wavelet theory in depth. The book can also be used to teach or study some specific topics in approximation theory, Fourier analysis, applied harmonic analysis, functional analysis, and wavelet-based signal/image processing. It consists of seven chapters and an appendix with 362 exercise problems and 72 figures.

The classical part of wavelet theory, in particular, orthogonal wavelets and biorthogonal wavelets in the square integrable function space $L_2(\mathbb{R})$ with filter banks, multiresolution analysis, and homogeneous affine systems:

$$\mathsf{AS}(\Psi) := \{\psi_{2^j;k} := 2^{j/2}\psi(2^j \cdot -k) \ : \ j,k \in \mathbb{Z}, \psi \in \Psi\},$$

where $\Psi \subseteq L_2(\mathbb{R})$, has been extensively addressed in many books under the framework of the space $L_2(\mathbb{R})$. A closely related but much less studied system is a nonhomogeneous affine system:

$$\mathsf{AS}_J(\Phi;\Psi) := \{\phi_{2^J;k} \ : \ k \in \mathbb{Z}, \phi \in \Phi\} \cup \{\psi_{2^j;k} \ : \ k \in \mathbb{Z}, j \geqslant J, \psi \in \Psi\}, \quad J \in \mathbb{Z}$$

with $\Phi, \Psi \subseteq L_2(\mathbb{R})$. The word "framelet" is formed by combining "frame wavelet." For the purpose of this book, a framelet (or a wavelet) is a set of functions whose generated affine system is a frame (or a basis), where a frame is a generalization of a basis by allowing additional/extra elements. Hence, a wavelet is merely a special framelet but without redundancy. This book was largely inspired by recent findings on the framelet aspects of wavelet theory, in particular, (a) the theory and applications of framelets for image processing; (b) framelets and wavelets in Sobolev spaces beyond the classical platform $L_2(\mathbb{R})$; (c) the algorithmic approach to framelets and wavelets in the discrete setting; and (d) nonhomogeneous affine systems and frequency-based dual framelets.

Most wavelet-related books introduce orthogonal wavelets first, focus on multiresolution analysis (MRA) and homogeneous affine systems, and then derive a discrete wavelet transform as a byproduct of MRA. By taking a different approach, this book systematically introduces the theory of framelets and wavelets from the perspective of framelets and discrete transforms. As we shall see, multiresolution analysis and homogeneous affine systems play only marginal roles in the theory of framelets, largely because a wavelet has no redundancy while a framelet may have redundancy. In fact, the theory of framelets and nonhomogeneous affine systems is not only richer than the classical theory of wavelets and homogeneous affine systems, but also reveals many new interesting features. In the classical theory of wavelets, a refinable function $\phi \in L_2(\mathbb{R})$, satisfying $\widehat{\phi}(2\xi) = \widehat{a}(\xi)\widehat{\phi}(\xi)$ with Fourier transform $\widehat{\phi}(\xi) := \int_{\mathbb{R}} \phi(x)e^{-ix\xi}dx, \xi \in \mathbb{R}$ and Fourier series $\widehat{a}(\xi) := \sum_{k\in\mathbb{Z}} a(k)e^{-ik\xi}$ for a sequence/filter $a = \{a(k)\}_{k\in\mathbb{Z}} \in l_2(\mathbb{Z})$, is the key ingredient in a multiresolution analysis for constructing wavelets. One may ask

Where do these magic refinable functions ϕ come from?

This question can easily be answered in the theory of framelets and nonhomogeneous affine systems. We provide an example here to illustrate why this book takes a different approach to the theory of framelets and wavelets, in contrast to most available wavelet-related books. Let $\Phi = \{\phi^1, \dots, \phi^r\}$ and $\Psi = \{\psi^1, \dots, \psi^s\}$ be finite subsets of $L_2(\mathbb{R})$ such that $\{\Phi; \Psi\}$ is a tight framelet for $L_2(\mathbb{R})$; that is, $\mathsf{AS}_0(\Phi; \Psi)$ is a (normalized) tight frame of $L_2(\mathbb{R})$ satisfying

$$\|f\|^2_{L_2(\mathbb{R})} = \sum_{h\in\mathsf{AS}_0(\Phi;\Psi)} |\langle f, h\rangle|^2, \qquad \forall f \in L_2(\mathbb{R}).$$

Then the following statements must hold:

(1) $\mathsf{AS}_J(\Phi; \Psi)$ is a tight frame of $L_2(\mathbb{R})$ for every $J \in \mathbb{Z}$.
(2) As the limiting system of $\mathsf{AS}_J(\Phi; \Psi)$ when $J \to -\infty$, the homogeneous affine system $\mathsf{AS}(\Psi)$ is a tight frame of $L_2(\mathbb{R})$; that is, Ψ is a homogeneous tight framelet for $L_2(\mathbb{R})$.
(3) There exist sequences/filters $a \in (l_2(\mathbb{Z}))^{r\times r}$ and $b \in (l_2(\mathbb{Z}))^{s\times r}$ such that the refinable structure holds:

$$\widehat{\phi}(2\xi) = \widehat{a}(\xi)\widehat{\phi}(\xi), \qquad \widehat{\psi}(2\xi) = \widehat{b}(\xi)\widehat{\phi}(\xi), \qquad a.e.\, \xi \in \mathbb{R},$$

where $\phi := (\phi^1, \dots, \phi^r)^\mathsf{T}$ is a refinable vector function and $\psi := (\psi^1, \dots, \psi^s)^\mathsf{T}$ is a tight framelet.

(4) The set $\{a; b\}$ of filters is a generalized tight framelet filter bank (see Theorem 4.5.4 for its definition).
(5) $\mathscr{V}_j + \mathscr{W}_j = \mathscr{V}_{j+1}$ for all $j \in \mathbb{Z}$, where $\mathscr{V}_j$ and $\mathscr{W}_j$ are the closed linear spans of $\{\phi^\ell(2^j \cdot -k) \; : \; k \in \mathbb{Z}, \ell = 1, \dots, r\}$ and of $\{\psi^\ell(2^j \cdot -k) \; : \; k \in \mathbb{Z}, \ell = 1, \dots, s\}$, respectively.
(6) $\{\mathscr{V}_j\}_{j\in\mathbb{Z}}$ forms a multiresolution analysis (MRA): (i) $\mathscr{V}_0$ is shift-invariant (i.e., $f(\cdot - k) \in \mathscr{V}_0$ for all $f \in \mathscr{V}_0$ and $k \in \mathbb{Z}$) and $\mathscr{V}_j = \{f(2^j\cdot) \; : \; f \in \mathscr{V}_0\}$ for all $j \in \mathbb{Z}$; (ii) $\mathscr{V}_j \subseteq \mathscr{V}_{j+1}$ for all $j \in \mathbb{Z}$; (iii) $\cap_{j\in\mathbb{Z}}\mathscr{V}_j = \{0\}$ and $\cup_{j\in\mathbb{Z}}\mathscr{V}_j$ is dense in $L_2(\mathbb{R})$.

A scalar framelet/wavelet refers to the special case $r = 1$; otherwise, it is called a multiframelet/multiwavelet with multiplicity r. If $\mathsf{AS}_0(\Phi; \Psi)$ is an orthonormal basis (a special case of a tight frame) of $L_2(\mathbb{R})$, then Ψ is an orthogonal wavelet and item (5) can be strengthened as $\mathscr{V}_j \oplus \mathscr{W}_j = \mathscr{V}_{j+1}$ and $\mathscr{V}_j \cap \mathscr{W}_j = \{0\}$ for all $j \in \mathbb{Z}$, leading to the important MRA decomposition $L_2(\mathbb{R}) = \oplus_{j\in\mathbb{Z}}\mathscr{W}_j$. However, in the setting of framelets with $s > r$, we almost surely have $\mathscr{V}_j \cap \mathscr{W}_j \neq \{0\}$ (and even $\mathscr{W}_j = \mathscr{V}_{j+1}$ if $s \geqslant 2r$). Consequently, the popular MRA decomposition technique of $L_2(\mathbb{R})$ for classical wavelets is no longer applicable.

Tight framelets are useful in applications, largely due to their desirable features of redundancy and flexibility over orthogonal wavelets. For example, the undecimated wavelet transform using an orthogonal wavelet can be used to effectively remove white Gaussian noise in signals and images; interestingly, its underlying system is in fact a tight framelet with high redundancy (that is, large numbers r and s of the generators). To capture anisotropic structures such as edge singularities in multidimensional data sets, directional representation systems are highly desirable, but the desired level of directionality can only be achieved by redundant systems such as (tight) framelets. For example, the directional tensor product complex tight framelets discussed in Sect. 7.4 offer significantly better performance for image denoising/inpainting than orthogonal wavelets and their undecimated versions. To reflect recent developments in wavelet theory, a book that systematically introduces and studies the framelet aspect of wavelet theory, which is often ignored or barely touched on in most wavelet-related books, is called for.

Assume that $\{\Phi; \Psi\}$ is a tight framelet for $L_2(\mathbb{R})$ with $\Phi = \{\phi^1, \dots, \phi^r\}$ and $\Psi = \{\psi^1, \dots, \psi^s\}$. According to item (1), every $f \in L_2(\mathbb{R})$ has the following framelet/wavelet representation:

$$f = \sum_{\ell=1}^{r} \sum_{k\in\mathbb{Z}} \langle f, \phi^\ell_{2^J;k}\rangle \phi^\ell_{2^J;k} + \sum_{j=J}^{\infty} \sum_{\ell=1}^{s} \sum_{k\in\mathbb{Z}} \langle f, \psi^\ell_{2^j;k}\rangle \psi^\ell_{2^j;k}.$$

Considering the difference of the above identities between $J - 1$ and J, we have the cascade structure

$$\sum_{\ell=1}^{r} \sum_{k\in\mathbb{Z}} \langle f, \phi^\ell_{2^J;k}\rangle \phi^\ell_{2^J;k} = \sum_{\ell=1}^{r} \sum_{k\in\mathbb{Z}} \langle f, \phi^\ell_{2^{J-1};k}\rangle \phi^\ell_{2^{J-1};k} + \sum_{\ell=1}^{s} \sum_{k\in\mathbb{Z}} \langle f, \psi^\ell_{2^{J-1};k}\rangle \psi^\ell_{2^{J-1};k}.$$

Items (3)–(6) largely follow from the above cascade structure, which is also the key to introducing a discrete wavelet transform in the classical theory of wavelets. Choose $J \in \mathbb{Z}$ large enough so that the left-hand side of the above identity approximates the function f well. More precisely, for every $J \in \mathbb{Z}$, define a discretization/sampling operator $\mathbb{D}_J$ by

$$\mathbb{D}_J : L_2(\mathbb{R}) \to (l_2(\mathbb{Z}))^r \quad \text{with} \quad \mathbb{D}_J f := \{\langle f, \phi^\ell_{2^J;k}\rangle\}_{k\in\mathbb{Z},\ell=1,\dots,r}.$$

Then $\mathbb{D}_J f$ is a discrete version of f at the resolution level J. Now the filters a and b in item (3) can be employed to compute the coefficients $\{\langle f, \phi^\ell_{2^{J-1};k}\rangle\}_{k\in\mathbb{Z},\ell=1,\dots,r}$ (i.e., $\mathbb{D}_{J-1}f$) and $\{\langle f, \psi^\ell_{2^{J-1};k}\rangle\}_{k\in\mathbb{Z},\ell=1,\dots,s}$ at the level $J-1$ from $\mathbb{D}_J f$ at the level J recursively (see Sect. 4.1.4 for details).

Now let us explain our motivations a bit more, based on the example above. For the sake of simplicity, from now on we will only consider the scalar case $r = 1$; for this scalar case, $a \in l_2(\mathbb{Z})$ and $b = (b_1, \dots, b_s)^\mathsf{T} \in (l_2(\mathbb{Z}))^s$. In reality, a discrete signal $v \in l_2(\mathbb{Z})$ may be given in advance. If the sampling operator $\mathbb{D}_J$ is onto (which is indeed the case for classical wavelets), for every $v \in l_2(\mathbb{Z})$, there exists $f \in L_2(\mathbb{R})$ satisfying $v = \mathbb{D}_J f$ and hence, the above described discrete transform can work. However, in the setting of framelets (e.g., the directional complex tight framelets in Sect. 7.4 for image processing), $\mathbb{D}_J L_2(\mathbb{R}) \subsetneq l_2(\mathbb{Z})$ indeed can happen and consequently, a discrete framelet transform is no longer a trivial byproduct of a framelet transform with MRA and filter banks in the function setting. In fact, the sampling operator $\mathbb{D}_J$ is onto if and only if $[\widehat{\phi}, \widehat{\phi}](\xi) := \sum_{k\in\mathbb{Z}} |\widehat{\phi}(\xi + 2\pi k)|^2 \ne 0$ for almost every $\xi \in \mathbb{R}$; for this particular case, a generalized (scalar) tight framelet filter bank $\{a; b_1, \dots, b_s\}$ in item (4) must be a (standard scalar) tight framelet filter bank satisfying

$$|\widehat{a}(\xi)|^2 + \sum_{\ell=1}^{s} |\widehat{b_\ell}(\xi)|^2 = 1, \quad \widehat{a}(\xi)\overline{\widehat{a}(\xi+\pi)} + \sum_{\ell=1}^{s} \widehat{b_\ell}(\xi)\overline{\widehat{b_\ell}(\xi+\pi)} = 0, \quad a.e.\ \xi \in \mathbb{R}.$$

Though every function in $L_2(\mathbb{R})$ has a framelet representation under the tight frame $\mathsf{AS}_J(\phi; \psi^1, \dots, \psi^s)$, the above discussion shows that the generalized tight framelet filter bank $\{a; b_1, \dots, b_s\}$ in item (4) with the MRA structure in item (6) does not necessarily lead to a discrete framelet transform for every input from $l_2(\mathbb{Z})$. More surprisingly, we shall prove in Sect. 4.1.3 that there exist two auxiliary filters $b_{s+1}, b_{s+2} \in l_2(\mathbb{Z})$ such that by adding b_{s+1}, b_{s+2} to the generalized tight framelet filter bank $\{a; b_1, \dots, b_s\}$ in item (4), $\{a; b_1, \dots, b_{s+2}\}$ forms a tight framelet filter bank, based on which we have a discrete framelet transform for every input from $l_2(\mathbb{Z})$. However, their associated wavelet functions ψ^{s+1} and ψ^{s+2} defined by

$$\widehat{\psi^{s+1}}(\xi) := \widehat{b_{s+1}}(\xi/2)\widehat{\phi}(\xi/2), \qquad \widehat{\psi^{s+2}}(\xi) := \widehat{b_{s+2}}(\xi/2)\widehat{\phi}(\xi/2), \qquad \xi \in \mathbb{R}$$

are identically zero (that is, $\psi^{s+1} = \psi^{s+2} = 0$) and hence $\mathsf{AS}_J(\phi; \psi^1, \dots, \psi^{s+2}) = \mathsf{AS}_J(\phi; \psi^1, \dots, \psi^s)$. That is, the underlying tight frame is unchanged.

Even if the sampling operator $\mathbb{D}_J$ is onto (that is, $[\widehat{\phi}, \widehat{\phi}](\xi) \neq 0$ for a.e. $\xi \in \mathbb{R}$), some obstacles still remain to transferring important properties of an affine system $\mathsf{AS}_J(\phi; \psi^1, \dots, \psi^s)$ in the function setting to its discrete version. For example, as we shall see in Chap. 6, the stability of a discrete affine system of $l_2(\mathbb{Z})$ induced by a filter bank $\{a; b_1, \dots, b_s\}$ can only be guaranteed by the stability of its underlying affine system $\mathsf{AS}_0(\phi; \psi^1, \dots, \psi^s)$ in $L_2(\mathbb{R})$, provided that $\{\phi(\cdot - k) \ : \ k \in \mathbb{Z}\}$ is a Riesz sequence in $L_2(\mathbb{R})$ (which is required by classical wavelets). Note that $\{\phi(\cdot - k) \ : \ k \in \mathbb{Z}\}$ is a Riesz sequence in $L_2(\mathbb{R})$ if and only if there exists $C > 0$ such that $C^{-1} \leqslant [\widehat{\phi}, \widehat{\phi}](\xi) \leqslant C$ for a.e. $\xi \in \mathbb{R}$ (which is further equivalent to saying that $\mathbb{D}_J : \mathscr{V}_J \to l_2(\mathbb{Z})$ is an isomorphism). However, such a condition often fails in the setting of framelets. For example, for the refinable functions ϕ in the directional complex tight framelets in Sect. 7.4 for image processing, $[\widehat{\phi}, \widehat{\phi}](\xi) = 0$ on a set of positive measure. Hence, the sampling operator $\mathbb{D}_J$ is not even surjective. The common and popular approach in the literature is to introduce classical wavelets in the function setting first with MRA and filter banks, and then derive a discrete wavelet transform from the function setting as a byproduct of MRA. This classical approach now proves to be problematic in the setting of framelets.

Since discrete transforms using framelets and wavelets are of paramount importance in applications, this motivates us to directly take an algorithmic approach to studying framelets and wavelets in Chap. 1 using only filters and sequences, without involving any functions from $L_2(\mathbb{R})$. Without requiring advanced mathematics, the algorithmic approach in Chap. 1 for scalar framelets/wavelets not only yields a self-contained approach to studying discrete framelet transforms but also provides a faithful analysis and precise understanding of the widely used discrete framelet/wavelet transforms. The discrete transforms associated with multiframelets/multiwavelets with multiplicity $r > 1$ take vector-valued inputs from $(l_2(\mathbb{Z}))^r$ and have further related issues to be addressed such as the balanced approximation property. One can only clearly resolve the balanced property of multiframelets/multiwavelets from the discrete setting. See Sect. 7.6 for details. Moreover, item (3) requires a thorough study of refinable vector functions ϕ and multiframelets/multiwavelets ψ, which are only briefly discussed in many wavelet-related books due to their highly involved mathematical complexity.

This book not only introduces

(1) a comprehensive treatment on the theory of framelets (including wavelets as special cases);
(2) an algorithmic approach to the theory of scalar framelets and wavelets in the discrete world;
(3) recent algorithms for constructing real- or complex-valued scalar wavelet or framelet filter banks;
(4) the approach of frequency-based nonhomogeneous affine systems and frequency-based dual framelets naturally linking most aspects of wavelet theory together,

but also provides

(5) a detailed and deep analysis of refinable vector functions, multiframelets, and multiwavelets;
(6) a comprehensive study on linear independence, stability, approximation orders, shift-invariant spaces, and refinable Hermite interpolants, which are also important topics in approximation theory, applied harmonic analysis, and computational mathematics;
(7) advances in subdivision schemes for computer graphics and directional complex tight framelets for image processing;
(8) a step-by-step introduction to wavelet theory with self-contained (simplest possible) proofs and supplementary exercises (a total of 362 exercise problems).

This book mainly focuses on an introduction to and recent developments in one-dimensional (dyadic) wavelet theory and its applications from the perspective of framelets and discrete transforms. To the best of our knowledge, most topics/approaches regarding the above items (1)–(6) are barely covered in any other wavelet-related books. Though this book discusses many topics in one-dimensional wavelet theory in depth, several important topics on wavelet theory are only briefly touched on and not fully addressed: for example, one-dimensional M-band framelets and wavelets, the construction of one-dimensional matrix-valued filter banks for multiframelets/multiwavelets, and multidimensional aspects of wavelet theory and its applications. However, most proofs in this book are presented in a way that they can be easily modified for the corresponding results in multiple dimensions.

For the convenience of students and beginning/junior researchers, all the results in the book are proved in a self-contained (and as elementary as possible) fashion. Experienced researchers may realize that many results in this book are not covered in other books or even articles. For some well-known results such as linear independence of the integer shifts of functions, the proofs are (greatly) simplified with elementary proofs. This book was purposefully written using the language of Fourier analysis. We believe that, in order to better understand framelets and wavelets, it is crucial to understand their frequency aspects and their behaviors in the frequency domain via the Fourier transform. As such, readers do not need any background in Fourier analysis. In fact, we provide a self-contained brief introduction to real analysis and Fourier analysis in Appendix A. But only a few basic facts from Appendix A are needed in order to read and benefit from this book. The well-known fundamental results on Banach spaces and Hilbert spaces in Appendix A are only needed for Sect. 4.2 on frames and bases in Hilbert spaces.

Students and researchers without any background in wavelet theory can use this book as an introductory textbook with self-contained proofs and supplementary exercises, while seasoned researchers can use it as a reference guide to recent advances in and further research on wavelet theory. The book essentially consists of three parts: I) discrete framelet transforms and scalar filter banks in Chaps. 1–3; II) analysis of affine systems and refinable vector functions in Chaps. 4–6; and III) selected applications of framelets and wavelets such as directional complex tight framelets for image processing in Chap. 7. Exception for Chaps. 6 and 7, which draw on results from Chaps. 4 and 5, all the chapters are effectively self-contained.

More than 72 numerical examples of framelets and wavelets are provided in this book for two purposes: (1) so that the readers can apply the algorithms in this book to reproduce these examples as exercises; (2) so that they can apply these examples to their own work and further research. For all numerical examples in this book, we use at least 12 decimal places for the filters (which are sufficiently accurate for their applications) and use 6 decimal places for their associated mathematical quantities. Sections marked with $*$ often have a higher level of technicality and are mainly intended for research purposes. These sections marked with $*$ can be initially skipped by students and beginners. Exercise problems, with varying difficulty levels, are provided at the end of each chapter. All the examples in this book can be reproduced using the `maple` (a computer mathematics software) routines posted at

http://www.ualberta.ca/~bhan/bookfw.html

We will now explain how to use this book for the purpose of teaching or research on wavelet theory and its applications. Chapters 1–3 in Part I are intended for readers who are mainly interested in fast framelet/wavelet transforms and their algorithmic implementations for applications. For mathematics-oriented readers and researchers, Chaps. 4–6 in Part II provide a systematic introduction to and mathematical treatment of the theory of framelets and wavelets in the function setting. Chapter 6 in Part III is mainly for experienced researchers looking for recent advances in and applications of framelets and wavelets. We will now elaborate on these three parts one by one.

Without requiring mathematical knowledge beyond calculus and trigonometric/ Laurent polynomials with complex coefficients in $\mathbb{C}$, Chaps. 1–3 in Part I are written at the level of senior undergraduate students and provide a self-contained comprehensive theory of framelets and wavelets in the discrete setting. These chapters are ideally suited for those who want to learn about the algorithmic aspects of scalar framelets and wavelets (but with much less demanding mathematics) so that discrete framelet/wavelet transforms employing various types of filter banks can be implemented and used in applications. Chapter 1 can be used for a short course on discrete framelet/wavelet transforms or can be combined with Chaps. 2 and/or 3 as a one-semester course for senior level undergraduate students in applied mathematics, computer science, or engineering. Depending on the length of the planned course and its difficulty level, there are three potential options: (Basic) Sects. 1.1–1.3, 2.1, 2.2, 2.6, 2.7, 3.1–3.3; (Extended): Sects. 1.1–1.4, 2.1–2.3, 2.6, 2.7, 3.1–3.4; and (Advanced): Sects. 1.1–1.5, 2.1–2.7, 3.1–3.5.

For experienced researchers, the stability of discrete affine systems in $l_2(\mathbb{Z})$, covered in Sect. 1.3, has to be further studied and established (currently, there are virtually no results on this topic). The algorithms for constructing various scalar wavelet filter banks in Chap. 2 are essentially complete. One remaining question is whether there exist finitely supported real-valued orthogonal wavelet filter banks with arbitrarily high orders of linear-phase moments. Such wavelets with high orders of linear-phase moments are of particular interest in numerical mathematics. The algorithms in Chap. 3 for constructing scalar framelet filter banks $\{a; b_1, \dots, b_s\}$ are essentially satisfactory for $s = 2$ or $s = 3$ with or without symmetry. As we shall see in Sect. 7.5, wavelets and framelets with symmetry are of particular interest for constructing wavelets/framelets on the interval $[0, 1]$. Most of the results in Chap. 1 can be generalized to multidimensional multiframelets/multiwavelets. However, currently there are very few general algorithms for constructing M-wavelet filter banks and M-framelet filter banks with a general $d \times d$ dilation matrix M and with multiplicity $r \geqslant 1$, even for dimension one with either $\mathsf{M} > 2$ or $r > 1$. Chapters 1–3 of this book only deal with $\mathsf{M} = 2$ (dyadic) and $r = 1$ (scalar).

Chapters 4–6 introduce the mathematical foundation of wavelets and framelets for mathematics-oriented junior or experienced researchers, in particular, graduate students or researchers from mathematics, physics, or engineering. Some working knowledge of Fourier analysis and distribution theory will be quite helpful (see Appendix A for a brief self-contained introduction to these topics). Chapter 4 can be used for a one-semester course on affine systems, shift-invariant spaces, and multiresolution analysis; Chap. 5 can be used for a one-semester course on refinable vector functions, vector cascade algorithms, stability and linear independence of generators of shift-invariant spaces. Chapters 4 and 5, together with part or all of Chap. 6, can be used for a two-semester graduate course on the function aspects of wavelet theory. Depending on the length of the respective course and its difficulty level, there are several potential options: (Basic) Sects. 4.2–4.5, 5.1, 5.6–5.8, 6.2, 6.5; (Extended): Sects. 4.1–4.5, 5.1–5.3, 5.6–5.8, 6.2–6.5; (Advanced): Sects. 4.1–4.6, 5.1–5.8, 6.1–6.5, 6.7; and (Research level): Sects. 4.1–4.6, 4.8, 4.9, 5.1–5.9, 6.1–6.8. The mathematical study of refinable vector functions and multiframelets/multiwavelets in Chaps. 5 and 6 is often much more involved than their scalar counterparts. To the best of our knowledge, this book is likely the first one to provide a comprehensive treatment of refinable vector functions and their associated multiframelets/multiwavelets. Our treatments and proofs for these topics, which may appear to be technical and demanding at first glance, are in fact the least complicated ones in the literature, thanks to the notion we introduce of the normal form of matrix-valued filters. For those who are only interested in the theory of scalar framelets and wavelets, the complexity associated with refinable vector functions and their multiframelets/multiwavelets can be easily avoided, since we provide separate simple treatments for several key results on scalar refinable functions and their scalar framelets/wavelets.

For researchers, most results in Chaps. 4 and 5 can be generalized to multiple dimensions with a general (or isotropic) dilation matrix (see Sect. 7.1). However, there are hardly any results on affine systems beyond Sobolev spaces such as a general Besov space. The convergence of a (vector) cascade algorithm in a general Besov space has not yet well studied extensively and is important for subdivision schemes in computer graphics. The study of compactly supported refinable (vector) functions and cascade algorithms in the Sobolev space $W_p^\tau(\mathbb{R})$ with $1 \leqslant p \leqslant \infty$ and $\tau \geqslant 0$ has not yet been done extensively, either. Except in Sect. 6.6, in this book we do not discuss one very important aspect of wavelet theory: refinable vector functions and multiframelets/multiwavelets that are not compactly supported. As we shall see in Sect. 6.6, the study of such topics is often much more demanding and technical than their compactly supported counterparts; such topics are very important for studying compactly supported Riesz multiwavelet bases, for constructing multiframelets/multiwavelets on a finite interval, and for employing new types of framelets/wavelets in applications. However, so far there are very few known results on these topics.

Chapter 7 is mainly intended for practitioners and researchers who are interested in the applications of wavelets and framelets. Parts of Chap. 7 can be combined with other chapters such as Chaps. 1 or 6 in a course to illustrate the practical applications of framelets and wavelets. The directional tensor product complex tight framelets for image processing covered in Sect. 7.4 have only been developed in the past few years, and the general method for constructing multiframelets/multiwavelets on $[0, 1]$ from symmetric multiframelets/multiwavelets on $\mathbb{R}$ in Sect. 7.5 is, to the best of our knowledge, new. There are many problems to be studied in order to further improve these results or their performance in applications. For example, fast multiframelet/multiwavelet transforms have to be better understood and developed to reveal their potential advantages over the commonly used (scalar) fast framelet/wavelet transforms.

Various sections of this book can also be used to teach special topics in applied harmonic analysis, approximation theory, or wavelet theory. For example, Sects. 4.2, 4.4, 4.5, 5.2–5.5 on approximation in shift-invariant spaces; Sects. 5.1–5.3, 5.6–5.8, 6.1, 6.2 on refinable vector functions; Sects. 3.2, 3.3, 5.1, 6.2–6.5, 7.4 on wavelet-based numerical algorithms; Sects. 1.1–1.3, 3.3–3.5, 4.1, 4.5, 7.2 on tight framelets and their applications in image processing; Sects. 5.6–5.9, 7.1, 7.2 on subdivision schemes in computer graphics; and Sects. A.6, 4.6, 6.3, 6.4 on distribution theory, Sobolev function spaces and wavelets/framelets in such spaces.

This book is built on many excellent works, papers, and books written by pioneers and numerous researchers in the areas of wavelet theory, applied harmonic analysis, and approximation theory. The author greatly benefitted from these references on wavelet theory while writing this book, and wishes to acknowledge their valuable contributions, as well as those who have influenced the author regarding wavelet theory in various ways. The author would like to thank all his teachers and senior coauthors (in alphabetical order): Wolfgang Dahmen, Ingrid Daubechies, Zeev Ditzian, Rong-Qing Jia, Rui-Lin Long, Sherman D. Riemenschneider, and Di-Rong Chen, Charles K. Chui, Qingtang Jiang, Amos Ron, and Zuowei Shen, Ding-Xuan Zhou, etc., from whom the author has learned a great deal about wavelet theory and mathematics. Though this book introduces and covers many results on one-dimensional dyadic wavelet theory with self-contained proofs, it is not exhaustive with regard to wavelet theory, since as we mentioned before many important topics even on one-dimensional wavelet theory are not discussed. Due to the vast literature and so many contributors on wavelet theory, only a very tiny relevant portion of references and books are listed at the end of this book. To preserve the integrity and simplicity of this book and to avoid the highly disputable task of correctly tracing credits of numerous contributors to wavelet theory, we do not plan to elaborate on the origin of each result in the book (e.g., which paper first obtained/stated it and by whom); instead, we only provide brief discussions on references for well-known important results/concepts and provide some remarks/acknowledgments for some key results/approaches in the Notes and Acknowledgments at the end of the book. In particular, closely related results in published works, which are employed by or appear in this book, have been acknowledged in the Notes and Acknowledgments. Any mistakes, disputable remarks, and missing important/relevant references in this book are surely the author's responsibility and are due to my limited knowledge and personal biases on the vast multidisciplinary area of wavelet theory and its applications.

Due to different/inconsistent notation/symbol systems existing in the literature on wavelet-related papers and books, the writing of this book started in 2008 by first spending half a year on designing a comprehensive system of notations and symbols. The book is a product of this notation/symbol system, which is still far from perfect. Some symbols and notations of this book may seem complicated to readers at first glance, but they are indeed very helpful for dealing with a wide range of complicated topics in wavelet theory. For example, the classical notation $\psi_{j,k}$ for $2^{j/2}\psi(2^j \cdot -k)$ is replaced by $\psi_{2^j;k}$, which is a special case of the operator $\psi_{c;k} := |c|^{1/2}\psi(c \cdot -k)$ acting on ψ with dilation $c \in \mathbb{R}$ and translation $k \in \mathbb{R}$. The notation $\psi_{c;k}$ is very handy for studying nonstationary framelets and shift-invariant spaces, while the classical notation $\psi_{j,k}$ is obscure, as it does not explicitly specify its underlying dilation (see Sect. 4.8).

For readers' convenience, a list of symbols is presented at the beginning of the book, while a list of indices is provided at the end.

Parts of this book manuscript were taught as two graduate courses at the University of Alberta in 2013 and 2017, as two short courses in the summer school at the University of Alberta in the summer of 2011, and as a short course at the Chinese Academy of Sciences in the spring of 2014. Special thanks go to my former students, postdoctoral fellows, and visitors (in alphabetical order): Yi Shen, Zhenpeng Zhao, and Xiaosheng Zhuang for plotting the graphs and the numerical experiments on image processing in Sect. 7.3, and to Elmira Ashpazzadeh, Menglu Che, Chenzhe Diao, Rejoyce Gavhi, Jaewon Jung, Rongrong Lin, Ran Lu, Michelle Michelle, Qun Mo, and Jie Zhou for pointing out some typos and mistakes in the manuscript. I am also grateful to the reviewers of this book, whose suggestions and comments have greatly improved it. The author would like to thank the Department of Mathematical and Statistical Sciences at the University of Alberta, and the University of Alberta in general, for providing an excellent working environment during the writing of this book. The author also wishes to thank the Natural Sciences and Engineering Research Council of Canada for their generous financial support of his research activities. The author is very grateful to all his family members, to whom this book is dedicated. Last but not least, the author would like to thank Springer for publishing this book.

Edmonton, Canada
2017

Bin Han

Contents

List of Symbols

$\mathbb{C}$	Complex numbers
$\mathbb{N}$	Natural numbers $\{1, 2, 3, \ldots\}$
$\mathbb{N}_0$	Whole numbers $\mathbb{N} \cup \{0\} := \{0, 1, 2, \ldots\}$
$\mathbb{Q}$	Rational numbers
$\mathbb{R}$	Real numbers
$\mathbb{T}$	The unit circle $\mathbb{T} := \{z \in \mathbb{C} : \lvert z\rvert = 1\}$ in the complex plane $\mathbb{C}$
$\mathbb{Z}$	Integers
$\mathbb{P}$, $\mathbb{P}_m$	Polynomials, Polynomials of degree no more than m, page 13
$\mathsf{P}_{m,n}$	Special polynomials $\mathsf{P}_{m,n}(x) := \sum_{j=0}^{n-1}(-1)^j\binom{-m}{j}x^j$, see (2.1.4)
a.e.	Almost everywhere
:	Set separator for sets
$\sim$	Bijection between primal and dual systems, see (4.1.3)
$\boldsymbol{\delta}$	Dirac/Kronecker sequence: $\boldsymbol{\delta}(0) = 1$ and $\boldsymbol{\delta}(k) = 0$, $k \neq 0$, see (1.1.11)
$\operatorname{sgn}(z)$	Sign/Phase $\operatorname{sgn}(z) := \frac{z}{\lvert z\rvert}$ of a complex number z, see (3.1.14)
$\operatorname{odd}(n)$	Odd function $\operatorname{odd}(n) = 1$ for odd n, and 0 for even n, see (2.5.10)
$\mathscr{O}(\lvert\xi - \xi_0\rvert^m)$	Big O notation, see equation (1.2.8)
$u * v$	Convolution $u * v := \sum_{k\in\mathbb{Z}} u(k)v(\cdot - k)$, see (1.1.27), and (1.2.5) for $\mathsf{p} * u$, (5.2.1) for $v * \phi$
$f', f^{(j)}$	Classical derivatives, see (6.1.2)
Df, D^jf	Distributional derivatives, page 688
$\operatorname{vec}(A)$	Column vector by stacking columns of a matrix A, see (5.8.13)
$A \otimes B$	Right Kronecker product $A \otimes B$ of matrices A, B, see (5.8.12)
$\rho(A)$	Spectral radius of a matrix/operator A, see (5.7.5)
$\operatorname{spec}(A)$	Spectrum set of a matrix or operator A, see (5.6.32)
Ω_j	Basic sets of shifts for periodic framelets/wavelets, see (4.9.1)
$\Omega_{\mathsf{M}}, \Gamma_{\mathsf{M}}$	Complete sets of representatives of distinct cosets $[(\mathsf{M}^{\mathsf{T}})^{-1}\mathbb{Z}^d]/\mathbb{Z}$ and $\mathbb{Z}^d/[\mathsf{M}\mathbb{Z}^d]$, see (7.1.3)
$\widehat{v}$	Fourier series $\widehat{v}(\xi) := \sum_{k\in\mathbb{Z}} v(k)e^{-ik\xi}$ for a filter/sequence $v = \{v(k)\}_{k\in\mathbb{Z}} : \mathbb{Z} \to \mathbb{C}$ on $\mathbb{Z}$, see (1.1.1)

$\{\widehat{f}(k)\}_{k\in\mathbb{Z}}$	Fourier coefficients $\widehat{f}(k) := \frac{1}{2\pi}\int_{-\pi}^{\pi} f(t)e^{-ikt}dt, f \in L_1(\mathbb{T})$, see (A.4)
$\{\widehat{u}(k)\}_{k=0}^{N-1}$	N-point discrete Fourier transform of $\{u(j)\}_{j=0}^{N-1}$, see (A.8)
$\mathscr{F}f, \widehat{f}$	Fourier transform $\widehat{f} := \int_{\mathbb{R}} f(x)e^{-ix\cdot}dx$, see (A.9), or $\langle \widehat{f}; \varphi\rangle := \langle f; \widehat{\varphi}\rangle$ for tempered distributions f
$f_{\lambda;k,n}$	Dilation/translation/modulation $f_{\lambda;k,n} := f_{[\![\lambda;k,n]\!]} = \|\lambda\|^{\frac{1}{2}} e^{-in\lambda\cdot} f(\lambda \cdot -k)$, see (4.0.1), (7.1.7)
$f_{\lambda;k}, f_{k,n}$	Dilation/translation/modulation $f_{\lambda;k} := \|\lambda\|^{\frac{1}{2}} f(\lambda \cdot -k), f_{k,n} := e^{-in\cdot} f(\cdot - k)$, see (4.0.2), (7.1.8)
$\mathrm{supp}(\mathbf{f})$	Support of a measurable function **f**, see (4.1.14)
Θ_n	Special moment correcting filters, page 181
υ	Moment matching filters υ for matrix-valued filters, see (5.5.9)
u	Laurent polynomial $\mathsf{u}(z) := \sum_{k\in\mathbb{Z}} u(k)z^k$ for $u = \{u(k)\}_{k\in\mathbb{Z}}$, see (2.7.1)
$\otimes^d u$	Tensor product filters, see (7.1.18)
$u^{[\gamma]}, a^{[\gamma]}$	γ-coset sequence $u^{[\gamma]} := \{u(\gamma + 2k)\}_{k\in\mathbb{Z}}$ of u, see (1.1.30), (5.5.12)
$u^{[r]}, u^{[i]}$	Real and imaginary parts of a complex-valued sequence u, see (2.5.1)
$u^\star, \mathsf{u}^\star$	Reflected conjugate filter $u^\star(k) := \overline{u(-k)}^{\mathsf{T}}$, see (2.7.2), (2.7.3), (5.6.12)
$\mathrm{fsupp}(u)$	Filter support $\mathrm{fsupp}(u) := [m, n]$ if $u(m)u(n) \neq 0$ and $u(k) = 0$ for all $k \in \mathbb{Z}\backslash[m, n]$, page 10
a_m^B	B-spline filter of order m, see (1.2.24)
a_m^D	Daubechies orthogonal wavelet filter of order m, see (2.2.4)
a_m^H	Complex orthogonal wavelet filters with linear-phase moments and symmetry, see (2.5.21)
$a_{2rm}^{H_r}$	Hermite interpolatory filters, page 505
a_{2m}^I	Interpolatory filters with shortest supports, see (2.1.6)
a_m^{IS}	Interpolatory splitting filters, see (2.6.4)
a_m^{S}	Complex orthogonal wavelet filters with symmetry, see (2.4.7)
$a_{2m-1,2n}$	Filters with linear-phase moments, see (2.1.12)
$a_{2m,2n}$	Filters with linear-phase moments, see (2.1.11)
Pu, Pf	Projected filter and projected function, see (7.1.20), page 588
$\mathrm{ao}(\phi)$	Accuracy order of ϕ, page 405
$\mathrm{jsr}_p(\mathcal{A})$	p-norm joint spectral radius of matrices/operators, see (5.7.1)
$\mathrm{lpm}(u)$	Order of linear-phase moments of a scalar filter u, page 69
$\mathrm{sr}(a), \mathrm{sr}(u)$	Order of sum rules of filters, see (5.5.9), (5.5.13), (5.6.28), (7.2.1)
$\mathrm{sr}(a\\|\upsilon)$	Order of sum rules with a matching filter υ, see (5.5.9), page 406
$\mathrm{sr}(a\|\upsilon)$	Order of general sum rules with a matching filter υ, see (7.6.11)
$\mathrm{sm}_p(a)$	L_p smoothness exponent of filters, $\mathrm{sm}(a) := \mathrm{sm}_2(a)$, see (5.6.44), (5.6.56), (2.0.7), (6.6.5), (7.2.2)
$\mathrm{sm}_p(\phi)$	L_p smoothness exponent, $\mathrm{sm}(\phi) := \mathrm{sm}_2(\phi)$, see (5.8.1), (6.3.1)
$\mathrm{vm}(u)$	Order of vanishing moments of filters or functions, see (6.4.3), page 69
$\rho(\mathcal{T}_{\widehat{a}})$	Spectral radius associated with $\mathcal{T}_{\|\widehat{a}\|^2}$, see (6.6.5)
$\rho_m(a, \upsilon)_p$	Spectral radius associated with a filter a, see (5.6.41), (5.6.55), (7.2.3)
$\mathrm{DS}(a)$	All distributional solutions of refinement equation with a filter/mask a

$\mathrm{E}(u)$	Expectation of a scalar filter u, see (2.0.8)
$\mathrm{Fsi}(u)$	Frequency separation indicator of a filter u, see (2.0.11)
$\mathrm{Fsi}(u, \upsilon)$	Frequency separation indicator of filters u, υ, see (2.0.10)
$\mathrm{Fsp}(b)$	Frequency separation of a filter b, see (7.4.7)
$\mathrm{Ofi}(u)$	Orthogonal wavelet filter indicator, see (2.6.1)
$\mathrm{Ofi}(u, \upsilon)$	Perpendicular filter indicator of filters u and υ, see (2.6.2)
$\mathrm{Var}(u)$	Variance of a scalar filter u, see (2.0.9)
$\uparrow \mathsf{d}, \downarrow \mathsf{d}$	Downsampling and upsampling operators with a factor d, see (1.1.28)
$\mathring{E}$	Standard vector conversion operator for sequences, see (7.6.4)
$F_{c,\epsilon}(f)$	Folding operator of functions, see (7.5.3)
$\mathcal{M}_{a,\tilde{a},\Theta}$	2×2 matrix associated with filters $a, \tilde{a}, \Theta$, see (3.0.2)
$\mathcal{N}_{a,\tilde{a},\Theta}$	2×2 matrix associated with filters $a, \tilde{a}, \Theta$, $\mathcal{N}_{a,\Theta} := \mathcal{N}_{a,a,\Theta}$
$\mathcal{R}_a$	Refinement/cascade operator, see (5.6.1), (7.2.4)
$\mathcal{S}_u, \mathcal{S}_{u,\mathsf{d}}$	Subdivision operator, see (1.1.2), (1.3.12),(2.7.7), (5.6.30)
$\mathcal{S}_{u,\mathsf{M}}$	Subdivision operator with a dilation matrix M, see (7.1.1)
$\mathcal{T}_u, \mathcal{T}_{u,\mathsf{d}}$	Transition operator, see (1.1.3), (1.3.13),(2.7.7), (5.6.31)
$\mathcal{T}_\gamma, \mathring{\mathcal{T}}_\gamma$	Shifted transition operators, see (5.7.10), (5.7.11)
$\mathcal{T}_{u,\mathsf{M}}$	Transition operator with a dilation matrix M, see (7.1.2)
$\mathcal{T}_{a,b}$	Convolved transition operators using matrix filters a and b, see (5.8.15)
$\mathsf{S}\widehat{u}, \mathsf{Sp}$	Symmetry operator, see (1.2.18), (3.1.1)
$\mathbb{S}\widehat{u}, \mathbb{S}\mathsf{p}$	Complex symmetry operator, see (1.2.20), (3.1.2)
$\mathcal{F}$	Frame operator in Hilbert spaces, see (4.2.4)
$\mathcal{V}, \widetilde{\mathcal{V}}$	Synthesis/reconstruction operator, see (1.1.8), (1.1.20), (1.3.5), (4.2.3)
$\mathcal{W}, \widetilde{\mathcal{W}}$	Analysis/decomposition operator, see (1.1.7), (1.1.19), (1.3.4), (4.2.2)
$\mathcal{V}^\star, \mathcal{W}^\star$	Adjoint operators of $\mathcal{V}$ and $\mathcal{W}$, see (1.1.22), (1.1.21)
$\mathcal{W}_\psi f$	Continuous wavelet transform of f, see (4.3.26)
$l(\mathbb{Z})$	Sequences $\upsilon : \mathbb{Z} \to \mathbb{C}$ on $\mathbb{Z}$
$l_0(\mathbb{Z})$	Finitely supported sequences $u : \mathbb{Z} \to \mathbb{C}$ on $\mathbb{Z}$
$l_2(\phi)$	Subspace of $l_2(\mathbb{Z})$ associated with ϕ, see (4.4.5)
$l_2^\tau(\mathbb{Z})$	Sobolev sequence spaces, page 356
$(l_p(\mathbb{Z}))^{r\times s}$	Sequences $\upsilon : \mathbb{Z} \to \mathbb{C}^{r\times s}$ in $l_p(\mathbb{Z})$, page 421
$\mathscr{B}_{m,\upsilon}$	Generating sets of the sequence space $\mathscr{V}_{m,\upsilon}$, see (5.6.23)
$\mathscr{F}_{m,\upsilon,p}$	Initial vector functions in L_p associated with a matching filter υ, see (5.6.16), (5.6.53)
$\mathscr{P}_{m,\upsilon}$	Vector polynomials with a matching filter υ, see (2.1.4), (5.6.52)
$\mathscr{V}_{m,\upsilon}$	Vector sequences with a matching filter υ, see (5.6.15), (5.6.52)
B_m	B-spline function B_m of order m, see (5.4.5), (6.1.1)
$\chi_{[c_l,c_r];\varepsilon_l,\varepsilon_r}$	Bump functions, see (4.6.21)
φ^a	Frequency-based scalar refinable function, see (4.1.32)
ϕ^a	Refinable function $\phi^a = 2\sum_{k\in\mathbb{Z}} a(k)\phi^a(2\cdot -k)$, page 68
ψ^a	Wavelet function $\psi^a := 2\sum_{k\in\mathbb{Z}}(-1)^{1-k}\overline{a(1-k)}\phi^a(2\cdot -k)$, page 68
$\psi^{a,b}$	Wavelet functions $\psi^{a,b} := 2\sum_{k\in\mathbb{Z}} b(k)\phi^a(2\cdot -k)$, page 68
$\dim_{\mathsf{S}(\Phi\|L_2(\mathbb{R}))}$	Dimension function of a shift-invariant space, see (4.4.9)
$\mathrm{len}(\mathsf{S}(\Phi\|L_2))$	Length of a shift-invariant space, see (4.4.9)

$\langle f; g\rangle$	Pairing or linear functional $\langle f; g\rangle := \int_{\mathbb{R}} f(x)g(x)^{\mathsf{T}}dx$, see (4.1.2), page 686
$\langle f, g\rangle$	Matrix inner product $\langle f, g\rangle := \int_{\mathbb{R}} f(x)\overline{g(x)}^{\mathsf{T}} dx$, see (4.1.2), (4.4.14)
$\langle\cdot,\cdot\rangle_{H^\tau(\mathbb{R})}$	Inner product in Sobolev space $H^\tau(\mathbb{R})$, page 326
$\langle\cdot,\cdot\rangle_{l_2^\tau(\mathbb{Z})}$	Inner product in Sobolev sequence spaces $l_2^\tau(\mathbb{Z})$, page 356
$[f, g]$	Bracket product $[f, g] := \sum_{k\in\mathbb{Z}} f(\cdot+2\pi k)\overline{g(\cdot+2\pi k)}^{\mathsf{T}}$, see (4.4.14)
$[f, g]_\tau$	$[\mathbf{f},\mathbf{g}]_\tau := \sum_{k\in\mathbb{Z}} \mathbf{f}(\cdot+2\pi k)\overline{\mathbf{g}(\cdot+2\pi k)}(1+\lvert\cdot+2\pi k\rvert^2)^\tau$, see (4.6.14)
$[\mathbf{f}, \boldsymbol{\psi}]_{2^j\mathbb{Z}}$	Bracket product over $2^j\mathbb{Z}$ for sequences, see (4.9.2)
σ_φ	Support of $[\varphi,\varphi] := \sum_{k\in\mathbb{Z}} \lvert\varphi(\cdot+2\pi k)$, i.e., $\sigma_\varphi := \mathrm{supp}([\varphi,\varphi])$, page 261
$\mathsf{Z}(\mathsf{p}, z_0)$	Multiplicity of zeros of $\mathsf{p}(z)$ at $z = z_0$, , page 155
$\mathcal{I}_{\boldsymbol{\Psi}}^k$	$\mathcal{I}_{\boldsymbol{\Psi}}^k := \sum_{\psi\in\boldsymbol{\Psi}} \tilde{\boldsymbol{\psi}}(\cdot)\overline{\boldsymbol{\psi}(\cdot+2\pi k)}$ if $k\in\mathbb{Z}$, and 0 if $k \in \mathbb{R}\backslash\mathbb{Z}$, see (4.1.20)
$\mathsf{AS}(\Psi)$	Homogeneous affine system, see (4.3.2)
$\mathsf{AS}^{\mathsf{M}}(\Psi)$	Homogeneous affine system, see (7.1.10)
$\mathsf{AS}_J(\Phi;\Psi)$	Nonhomogeneous affine system, see (4.3.1)
$\mathsf{AS}_J^{\mathsf{M}}(\Phi;\Psi)$	Nonhomogeneous affine system with a dilation matrix M, see (7.1.9)
$\mathsf{AS}_J^\tau(\Phi;\Psi)$	Nonhomogeneous affine system (scaled), see (4.6.6)
FAS_J	Frequency-based nonhomogeneous affine system, see (4.1.1)
FPAS_J	Frequency-based periodic affine system , page 351
$\mathcal{A}_n$	Approximation operators, see (4.7.1)
$\mathcal{Q}_\lambda$	Quasi-projection operators, $\mathcal{Q} := \mathcal{Q}_1$, see (4.7.2), (5.4.9), (6.2.2)
$\mathscr{P}f, f^{per}$	Periodization of a function f on $\mathbb{R}$, see (4.9.27)
$\mathsf{p}(\cdot - i\frac{d}{d\xi})$	Polynomial differentiation operator, see equation (1.2.1)
$\nabla_k v, \nabla_t f$	Difference operators on sequences v or functions f, see (5.4.2)
$\eta_\lambda^{hard}, \eta_\lambda^{soft}$	Hard and soft thresholding operators, see (1.3.2)
$\omega_m(f,\lambda)_p$	mth modulus of smoothness, $\omega(f,\lambda)_p = \omega_1(f,\lambda)_p$, see (5.4.3)
$A(\mathbb{T})$	Functions having absolutely convergent Fourier series, page 678
$\mathscr{D}(\mathbb{R})$	Compactly supported $\mathscr{C}^\infty(\mathbb{R})$ test functions, page 686
$\mathscr{D}'(\mathbb{R})$	Distributions, page 686
$\mathscr{S}(\mathbb{R})$	Functions of Schwartz class, page 686
$\mathscr{S}'(\mathbb{R})$	Tempered distributions, page 686
$\mathtt{S}_\lambda(\Phi\vert L_2)$	Shift-invariant subspace generated by a set of functions $\Phi\subseteq L_2(\mathbb{R})$ at the scale λ, $\mathtt{S}(\Phi\vert L_2(\mathbb{R})) := \mathtt{S}_1(\Phi\vert L_2(\mathbb{R}))$, see (4.4.17)
$\mathtt{S}(\Phi)$	Shift-invariant space generated by compactly supported functions in Φ, see (5.2.3)
$\mathscr{V}_n(\Psi)$	Subspace of $L_2(\mathbb{R})$ generated by Ψ, see (4.5.17)
L_p	L_p spaces, $0 < p \leqslant \infty$, page 670
$L_p(\mathbb{T})$	2π-periodic functions in L_p, page 674
$\mathcal{L}_p(\mathbb{R})$	Special subspace of L_p, see (5.3.2)
$L_p^{loc}(\mathbb{R})$	Locally L_p functions, page 246
$W_p^m(\mathbb{R})$	L_p Sobolev space with exponent $m\in\mathbb{N}_0$, see (5.4.1)
$H^\tau(\mathbb{R})$	L_2 Sobolev space with exponent τ, see (4.6.1)
$L_{2,1,\tau}(\mathbb{R})$	Weighted subspaces of $L_2(\mathbb{R})$, see (6.6.27)
CBC	Coset by coset algorithm, see Algorithms 2.6.2 and 6.5.2, , page 126

CWT	Continuous wavelet transform, see (4.3.26), page 288
DAS_J	Discrete affine system (J-level), see (1.3.19), page 31
DFrT	Discrete framelet transform, page 1
FFrT	Fast framelet transform, page 26
OEP	Oblique extension principle, page 35
TPCTF	Directional tensor product complex tight framelets, page 621

Chapter 1
Discrete Framelet Transforms

Discrete wavelet/framelet transforms are the backbone of wavelet theory for its applications in a wide scope of areas. In this chapter we study algorithmic aspects and key properties of wavelets and framelets in the discrete setting. First, we introduce a standard (both one-level and multilevel) discrete framelet transform and filter banks. Then we investigate three fundamental properties of a standard discrete framelet transform: perfect reconstruction, sparsity, and stability; these properties are very much desired and crucial in successful applications of wavelets and framelets. Furthermore, we discuss several variants of a standard discrete framelet transform such as nonstationary discrete framelet transforms and undecimated discrete framelet transforms. We fully analyze, in the discrete setting, basic properties related to sparsity such as vanishing moments, sum rules, polynomial reproduction, linear-phase moments, and symmetry. Next, we introduce a general discrete framelet transform that is based on the oblique extension principle, which allows us to increase vanishing moments of high-pass filters in a filter bank. Finally, we describe in detail several algorithms to concretely implement a discrete framelet transform and its variants for processing signals on a bounded interval. We also discuss such algorithms implemented equivalently and completely in the frequency domain using the discrete Fourier transform.

1.1 Perfect Reconstruction of Discrete Framelet Transforms

In this section we introduce one-level (standard) discrete framelet transforms (DFrT). There are three fundamental properties of a discrete framelet transform: perfect reconstruction, sparsity, and stability. In this section we study the perfect reconstruction property.

B. Han, *Framelets and Wavelets*, Applied and Numerical Harmonic Analysis,
https://doi.org/10.1007/978-3-319-68530-4_1

1.1.1 One-Level Standard Discrete Framelet Transforms

To introduce a discrete framelet transform, we need some definitions and notation. By $l(\mathbb{Z})$ we denote the linear space of all sequences $v = \{v(k)\}_{k\in\mathbb{Z}} : \mathbb{Z} \to \mathbb{C}$ of complex numbers on $\mathbb{Z}$. A one-dimensional discrete input signal is often regarded as an element in $l(\mathbb{Z})$. Similarly, by $l_0(\mathbb{Z})$ we denote the linear space of all sequences $u = \{u(k)\}_{k\in\mathbb{Z}} : \mathbb{Z} \to \mathbb{C}$ on $\mathbb{Z}$ such that $u(k) \neq 0$ only for finitely many $k \in \mathbb{Z}$. An element in $l_0(\mathbb{Z})$ is often regarded as a finitely supported filter or mask in the literature of wavelet analysis (which is also called a finite-impulse-response filter in engineering). In this book we often use u for a general filter and v for a general signal. It is often convenient to use the formal Fourier series (or symbol) $\widehat{v}$ of a sequence $v = \{v(k)\}_{k\in\mathbb{Z}}$, which is defined to be

$$\widehat{v}(\xi) := \sum_{k\in\mathbb{Z}} v(k)e^{-ik\xi}, \qquad \xi \in \mathbb{R}, \tag{1.1.1}$$

where i in this book always denotes the imaginary unit. For $v \in l_0(\mathbb{Z})$, the Fourier series $\widehat{v}$ is a 2π-periodic trigonometric polynomial. See Appendix A for a brief introduction to Fourier series.

A discrete framelet transform can be described using two linear operators—the subdivision operator and the transition operator. For a filter $u \in l_0(\mathbb{Z})$ and a sequence $v \in l(\mathbb{Z})$, the *subdivision operator* $\mathcal{S}_u : l(\mathbb{Z}) \to l(\mathbb{Z})$ is defined to be

$$[\mathcal{S}_u v](n) := 2\sum_{k\in\mathbb{Z}} v(k)u(n-2k), \qquad n \in \mathbb{Z} \tag{1.1.2}$$

and the *transition operator* $\mathcal{T}_u : l(\mathbb{Z}) \to l(\mathbb{Z})$ is defined to be

$$[\mathcal{T}_u v](n) := 2\sum_{k\in\mathbb{Z}} v(k)\overline{u(k-2n)}, \qquad n \in \mathbb{Z}. \tag{1.1.3}$$

The transition operator plays the role of coarsening and frequency-separating a signal to lower resolution levels; while the subdivision operator plays the role of refining and predicting a signal to higher resolution levels.

In terms of Fourier series, the subdivision operator $\mathcal{S}_u$ in (1.1.2) and the transition operator $\mathcal{T}_u$ in (1.1.3) can be equivalently rewritten as

$$\widehat{\mathcal{S}_u v}(\xi) = 2\widehat{v}(2\xi)\widehat{u}(\xi), \qquad \xi \in \mathbb{R} \tag{1.1.4}$$

and

$$\widehat{\mathcal{T}_u v}(\xi) = \widehat{v}(\xi/2)\overline{\widehat{u}(\xi/2)} + \widehat{v}(\xi/2+\pi)\overline{\widehat{u}(\xi/2+\pi)}, \qquad \xi \in \mathbb{R} \tag{1.1.5}$$

for $u, v \in l_0(\mathbb{Z})$, where $\overline{c}$ denotes the complex conjugate of a complex number $c \in \mathbb{C}$. Though most results in this book can be stated and proofs can be carried out equivalently in the space/time domain, to understand wavelets and framelets better, we shall take a frequency/Fourier based approach as the main theme of this book.

A one-level standard discrete framelet transform has two parts: a one-level discrete framelet decomposition and a one-level discrete framelet reconstruction. A set $\{\tilde{u}_0, \ldots, \tilde{u}_s\}$ of filters $\tilde{u}_0, \ldots, \tilde{u}_s \in l_0(\mathbb{Z})$ forms *a filter bank* for decomposition. For a given signal $v \in l(\mathbb{Z})$, *a one-level discrete framelet decomposition* employing the filter bank $\{\tilde{u}_0, \ldots, \tilde{u}_s\}$ is

$$w_\ell := \tfrac{\sqrt{2}}{2}\, \mathcal{T}_{\tilde{u}_\ell} v, \qquad \ell = 0, \ldots, s, \tag{1.1.6}$$

where w_ℓ are called sequences of *framelet coefficients* of the input signal v. We can group all sequences of framelet coefficients in (1.1.6) together and define *a discrete framelet analysis (or decomposition) operator* $\widetilde{\mathcal{W}} : l(\mathbb{Z}) \to (l(\mathbb{Z}))^{1\times(s+1)}$ employing the filter bank $\{\tilde{u}_0, \ldots, \tilde{u}_s\}$ as follows:

$$\widetilde{\mathcal{W}} v := \tfrac{\sqrt{2}}{2}(\mathcal{T}_{\tilde{u}_0} v, \ldots, \mathcal{T}_{\tilde{u}_s} v), \qquad v \in l(\mathbb{Z}). \tag{1.1.7}$$

Let $\{u_0, \ldots, u_s\}$ with $u_0, \ldots, u_s \in l_0(\mathbb{Z})$ be a filter bank for reconstruction. *A one-level discrete framelet reconstruction* employing the filter bank $\{u_0, \ldots, u_s\}$ can be described by *a discrete framelet synthesis (or reconstruction) operator* $\mathcal{V} : (l(\mathbb{Z}))^{1\times(s+1)} \to l(\mathbb{Z})$ which is defined to be

$$\mathcal{V}(w_0, \ldots, w_s) := \frac{\sqrt{2}}{2} \sum_{\ell=0}^{s} \mathcal{S}_{u_\ell} w_\ell, \qquad w_0, \ldots, w_s \in l(\mathbb{Z}). \tag{1.1.8}$$

Throughout the book we denote a discrete framelet analysis operator employing the filter bank $\{u_0, \ldots, u_s\}$ by $\mathcal{W}$ and similarly, a discrete framelet synthesis operator employing the filter bank $\{\tilde{u}_0, \ldots, \tilde{u}_s\}$ by $\widetilde{\mathcal{V}}$. See Fig. 1.1 for a diagram of a one-level discrete framelet transform using a filter bank $\{\tilde{u}_0, \ldots, \tilde{u}_s\}$ for decomposition and a filter bank $\{u_0, \ldots, u_s\}$ for reconstruction.

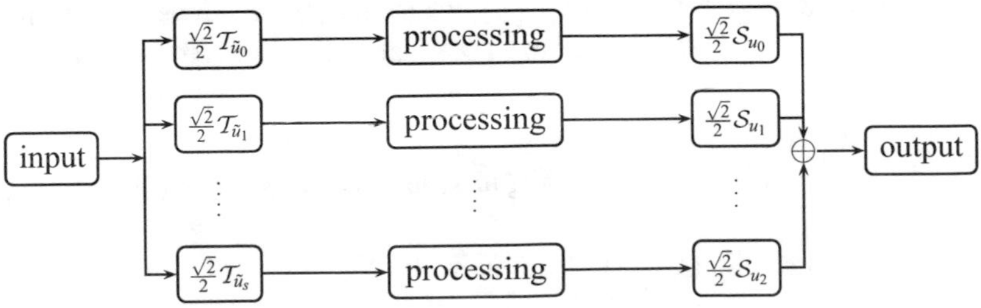

Fig. 1.1 Diagram of a one-level discrete framelet transform using a filter bank $\{\tilde{u}_0, \ldots, \tilde{u}_s\}$ for decomposition and a filter bank $\{u_0, \ldots, u_s\}$ for reconstruction

1.1.2 Perfect Reconstruction of Discrete Framelet Transforms

We say that a filter bank $(\{\tilde{u}_0, \dots, \tilde{u}_s\}, \{u_0, \dots, u_s\})$ (or more precisely, its associated discrete framelet transform) has the *perfect reconstruction* property if $\mathcal{V}\widetilde{\mathcal{W}}v = v$ for all input signals $v \in l(\mathbb{Z})$. A necessary and sufficient condition for the perfect reconstruction property of a general one-level discrete framelet transform is as follows:

Theorem 1.1.1 *Let $\tilde{u}_0, \dots, \tilde{u}_s, u_0, \dots, u_s \in l_0(\mathbb{Z})$. Then the following are equivalent:*

(i) The filter bank $(\{\tilde{u}_0, \dots, \tilde{u}_s\}, \{u_0, \dots, u_s\})$ has the following perfect reconstruction property: for all $v \in l(\mathbb{Z})$,

$$v = \mathcal{V}\widetilde{\mathcal{W}}v = \frac{1}{2}\sum_{\ell=0}^{s} \mathcal{S}_{u_\ell}\mathcal{T}_{\tilde{u}_\ell}v, \tag{1.1.9}$$

where $\widetilde{\mathcal{W}}$ and $\mathcal{V}$ are defined in (1.1.7) *and* (1.1.8)*, respectively.*

(ii) The identity in (1.1.9) *holds for all $v \in l_0(\mathbb{Z})$.*

(iii) (1.1.9) *holds for the two sequences $v = \boldsymbol{\delta}$ and $\boldsymbol{\delta}(\cdot - 1)$, more explicitly,*

$$\sum_{\ell=0}^{s}\sum_{k\in\mathbb{Z}} \overline{\tilde{u}_\ell(\gamma + 2k)}u_\ell(n+\gamma+2k) = \frac{1}{2}\boldsymbol{\delta}(n), \qquad \forall\, \gamma \in \{0,1\},\ n \in \mathbb{Z}, \tag{1.1.10}$$

where $\boldsymbol{\delta}$ is the Dirac (or Kronecker) sequence *such that*

$$\boldsymbol{\delta}(0) = 1 \qquad \text{and} \qquad \boldsymbol{\delta}(k) = 0, \qquad \forall\, k \neq 0. \tag{1.1.11}$$

(iv) The following perfect reconstruction condition holds: for all $\xi \in \mathbb{R}$,

$$\widehat{\tilde{u}_0}(\xi)\overline{\widehat{u_0}(\xi)} + \widehat{\tilde{u}_1}(\xi)\overline{\widehat{u_1}(\xi)} + \cdots + \widehat{\tilde{u}_s}(\xi)\overline{\widehat{u_s}(\xi)} = 1, \tag{1.1.12}$$

$$\widehat{\tilde{u}_0}(\xi)\overline{\widehat{u_0}(\xi+\pi)} + \widehat{\tilde{u}_1}(\xi)\overline{\widehat{u_1}(\xi+\pi)} + \cdots + \widehat{\tilde{u}_s}(\xi)\overline{\widehat{u_s}(\xi+\pi)} = 0. \tag{1.1.13}$$

Proof Since $\{\boldsymbol{\delta}, \boldsymbol{\delta}(\cdot - 1)\} \subseteq l_0(\mathbb{Z}) \subseteq l(\mathbb{Z})$, we trivially have (i)$\Longrightarrow$(ii)$\Longrightarrow$(iii). By (1.1.2) and (1.1.3), it is straightforward to see that (1.1.9) holds for $v = \boldsymbol{\delta}, \boldsymbol{\delta}(\cdot - 1)$ if and only if (1.1.10) holds. Using (1.1.4) and (1.1.5) for $v \in l_0(\mathbb{Z})$, we have

$$\frac{1}{2}\widehat{[\mathcal{S}_{u_\ell}\mathcal{T}_{\tilde{u}_\ell}v]}(\xi) = \widehat{\mathcal{T}_{\tilde{u}_\ell}v}(2\xi)\widehat{u_\ell}(\xi) = \widehat{v}(\xi)\overline{\widehat{\tilde{u}_\ell}(\xi)}\widehat{u_\ell}(\xi) + \widehat{v}(\xi+\pi)\overline{\widehat{\tilde{u}_\ell}(\xi+\pi)}\widehat{u_\ell}(\xi).$$

Therefore, for $v \in l_0(\mathbb{Z})$, (1.1.9) holds if and only if

$$\overline{\widehat{v}(\xi)} = \overline{\widehat{v}(\xi)}\sum_{\ell=0}^{s}\widehat{\tilde{u}_\ell}(\xi)\overline{\widehat{u_\ell}(\xi)} + \overline{\widehat{v}(\xi+\pi)}\sum_{\ell=0}^{s}\widehat{\tilde{u}_\ell}(\xi+\pi)\overline{\widehat{u_\ell}(\xi)}. \tag{1.1.14}$$

To prove (iii)$\Longrightarrow$(iv), plugging $v = \boldsymbol{\delta}$ into (1.1.14) and noting $\widehat{\boldsymbol{\delta}}(\xi) = 1$, we see that (1.1.14) becomes

$$1 = \sum_{\ell=0}^{s} \widehat{\tilde{u}_\ell}(\xi)\overline{\widehat{u_\ell}(\xi)} + \sum_{\ell=0}^{s} \widehat{\tilde{u}_\ell}(\xi+\pi)\overline{\widehat{u_\ell}(\xi)}.$$

Plugging $v = \boldsymbol{\delta}(\cdot - 1)$ into (1.1.14) and noting $\widehat{\boldsymbol{\delta}(\cdot - 1)}(\xi) = e^{-i\xi}$, we conclude from (1.1.14) that

$$1 = \sum_{\ell=0}^{s} \widehat{\tilde{u}_\ell}(\xi)\overline{\widehat{u_\ell}(\xi)} - \sum_{\ell=0}^{s} \widehat{\tilde{u}_\ell}(\xi+\pi)\overline{\widehat{u_\ell}(\xi)}.$$

From these two identities, by adding or subtracting one from the other, we conclude that (1.1.12) and (1.1.13) must hold. Therefore, (iii)$\Longrightarrow$(iv).

If (1.1.12) and (1.1.13) are satisfied, then it is straightforward to see that (1.1.14) holds for all $v \in l_0(\mathbb{Z})$. That is, we proved (iv)$\Longrightarrow$(ii).

To complete the proof, we prove (ii)$\Longrightarrow$(i) by using the locality of the subdivision and transition operators. Let $v \in l(\mathbb{Z})$. Since all the filters are finitely supported, there exists a positive integer N such that all filters $u_0, \dots, u_s, \tilde{u}_0, \dots, \tilde{u}_s$ are supported inside $[-N, N]$. Let $n \in \mathbb{Z}$ be fixed. Define a finitely supported sequence $v_n \in l_0(\mathbb{Z})$ by $v_n(k) := v(k)$ for all $k \in \mathbb{Z} \cap [n-2N, n+2N]$, and $v_n(k) = 0$ otherwise. For all $k \in \mathbb{Z} \cap [\frac{n-N}{2}, \frac{n+N}{2}]$, since all involved filters are supported inside $[-N, N]$, we have

$$\begin{aligned}
[\mathcal{T}_{\tilde{u}_\ell} v](k) &= 2\sum_{j\in\mathbb{Z}} v(j)\overline{\tilde{u}_\ell(j-2k)} = 2\sum_{j=n-2N}^{n+2N} v(j)\overline{\tilde{u}_\ell(j-2k)}\\
&= 2\sum_{j=n-2N}^{n+2N} v_n(j)\overline{\tilde{u}_\ell(j-2k)} = [\mathcal{T}_{\tilde{u}_\ell} v_n](k).
\end{aligned}$$

Therefore, we deduce that

$$\begin{aligned}
\frac{1}{2}\sum_{\ell=0}^{s}[\mathcal{S}_{u_\ell}\mathcal{T}_{\tilde{u}_\ell} v](n) &= \sum_{\ell=0}^{s}\sum_{k\in\mathbb{Z}\cap[\frac{n-N}{2},\frac{n+N}{2}]} [\mathcal{T}_{\tilde{u}_\ell} v](k)u_\ell(n-2k)\\
&= \sum_{\ell=0}^{s}\sum_{k\in\mathbb{Z}\cap[\frac{n-N}{2},\frac{n+N}{2}]} [\mathcal{T}_{\tilde{u}_\ell} v_n](k)u_\ell(n-2k)\\
&= \frac{1}{2}\sum_{\ell=0}^{s}[\mathcal{S}_{u_\ell}\mathcal{T}_{\tilde{u}_\ell} v_n](n) = v_n(n) = v(n),
\end{aligned}$$

where we used (ii) in the second-to-last identity. Hence, (ii)$\Longrightarrow$(i). $\square$

The equivalence between items (ii) and (iii) of Theorem 1.1.1 can be easily understood through the following simple relation: for $m \in \mathbb{Z}$,

$$\frac{1}{2}\sum_{\ell=0}^{s} \mathcal{S}_{u_\ell}\mathcal{T}_{\tilde{u}_\ell}(v(\cdot-2m)) = \frac{1}{2}\sum_{\ell=0}^{s} \mathcal{S}_{u_\ell}([\mathcal{T}_{\tilde{u}_\ell}v](\cdot-m)) = \left[\frac{1}{2}\sum_{\ell=0}^{s}(\mathcal{S}_{u_\ell}\mathcal{T}_{\tilde{u}_\ell}v)\right](\cdot-2m).$$

Therefore, if the identity in (1.1.9) holds for a particular sequence v, then it also holds for all $v(\cdot-2m), m \in \mathbb{Z}$. Note that the space $l_0(\mathbb{Z})$ is generated by the finite linear combinations of $\boldsymbol{\delta}(\cdot-k), k \in \mathbb{Z}$. Now it is not surprising to see the equivalence between items (ii) and (iii) of Theorem 1.1.1.

The perfect reconstruction condition in (1.1.12) and (1.1.13) can be equivalently rewritten into the following matrix form:

$$\begin{bmatrix} \widehat{\tilde{u}_0}(\xi) & \cdots & \widehat{\tilde{u}_s}(\xi) \\ \widehat{\tilde{u}_0}(\xi+\pi) & \cdots & \widehat{\tilde{u}_s}(\xi+\pi) \end{bmatrix} \begin{bmatrix} \widehat{u_0}(\xi) & \cdots & \widehat{u_s}(\xi) \\ \widehat{u_0}(\xi+\pi) & \cdots & \widehat{u_s}(\xi+\pi) \end{bmatrix}^{\star} = I_2, \qquad (1.1.15)$$

where I_2 denotes the 2×2 identity matrix and $A^\star$ denotes the transpose of the complex conjugate of a matrix A, that is, $A^\star := \overline{A}^{\mathsf{T}}$.

A filter bank satisfying the perfect reconstruction condition in (1.1.15) is called *a dual framelet filter bank*. It is trivial from (1.1.15) that $(\{\tilde{u}_0,\dots,\tilde{u}_s\},\{u_0,\dots,u_s\})$ is a dual framelet filter bank if and only if $(\{u_0,\dots,u_s\},\{\tilde{u}_0,\dots,\tilde{u}_s\})$ is a dual framelet filter bank. In other words, $\mathcal{V}\widetilde{\mathcal{W}} = \mathrm{Id}_{l(\mathbb{Z})}$ if and only if $\widetilde{\mathcal{V}}\mathcal{W} = \mathrm{Id}_{l(\mathbb{Z})}$.

In particular, a dual framelet filter bank with $s = 1$ is called *a biorthogonal wavelet filter bank* which, by the following result, is a nonredundant filter bank.

Proposition 1.1.2 *Let* $(\{\tilde{u}_0,\dots,\tilde{u}_s\},\{u_0,\dots,u_s\})$ *be a dual framelet filter bank. Let the discrete framelet analysis operator* $\widetilde{\mathcal{W}} : l(\mathbb{Z}) \to (l(\mathbb{Z}))^{1\times(s+1)}$ *and the discrete framelet synthesis operator* $\mathcal{V} : (l(\mathbb{Z}))^{1\times(s+1)} \to l(\mathbb{Z})$ *be defined in* (1.1.7) *and* (1.1.8), *respectively. Then the following statements are equivalent:*

(i) $\widetilde{\mathcal{W}}$ *is onto.*
(ii) $\mathcal{V}$ *is one-to-one.*
(iii) $\mathcal{V}\widetilde{\mathcal{W}} = \mathrm{Id}_{l(\mathbb{Z})}$ *and* $\widetilde{\mathcal{W}}\mathcal{V} = \mathrm{Id}_{(l(\mathbb{Z}))^{1\times(s+1)}}$, *that is,* $\mathcal{V}^{-1} = \widetilde{\mathcal{W}}$ *and* $\widetilde{\mathcal{W}}^{-1} = \mathcal{V}$.
(iv) $s = 1$.

Proof It is trivial that (iii) implies both (i) and (ii). Note that $\mathcal{V}\widetilde{\mathcal{W}} = \mathrm{Id}_{l(\mathbb{Z})}$ follows directly from the perfect reconstruction property. If (i) holds, by $\widetilde{\mathcal{W}}\mathcal{V}\widetilde{\mathcal{W}}v = \widetilde{\mathcal{W}}v$ and $\widetilde{\mathcal{W}}(l(\mathbb{Z}))^{1\times(s+1)} = l(\mathbb{Z})$, we must have $\widetilde{\mathcal{W}}\mathcal{V} = \mathrm{Id}_{(l(\mathbb{Z}))^{1\times(s+1)}}$. Thus, (i)$\Longrightarrow$(iii). By $\mathcal{V}\widetilde{\mathcal{W}} = \mathrm{Id}_{l(\mathbb{Z})}$, we have $\mathcal{V}\widetilde{\mathcal{W}}\mathcal{V} = \mathcal{V}$ and therefore, $\mathcal{V}(\widetilde{\mathcal{W}}\mathcal{V} - \mathrm{Id}_{(l(\mathbb{Z}))^{1\times(s+1)}}) = 0$. Since $\mathcal{V}$ is one-to-one by item (ii), we must have $\widetilde{\mathcal{W}}\mathcal{V} = \mathrm{Id}_{(l(\mathbb{Z}))^{1\times(s+1)}}$ and hence, (ii)$\Longrightarrow$(iii).

We now prove (iii) $\Longleftrightarrow$ (iv). If (iii) holds, then $\widetilde{\mathcal{W}}\mathcal{V} = \mathrm{Id}_{(l(\mathbb{Z}))^{1\times(s+1)}}$ implies

$$\frac{1}{2}\mathcal{T}_{\tilde{u}_\ell}\Big(\mathcal{S}_{u_0}w_0+\cdots+\mathcal{S}_{u_s}w_s\Big) = w_\ell, \ \forall\, w_0,\dots,w_s \in l_0(\mathbb{Z}),\ \ell = 0,\dots,s. \qquad (1.1.16)$$

Taking Fourier series on both sides of (1.1.16), we see that (1.1.16) is equivalent to

$$\begin{aligned}&[\widehat{u_0}(\xi/2)\overline{\widehat{\tilde{u}_\ell}(\xi/2)} + \widehat{u_0}(\xi/2+\pi)\overline{\widehat{\tilde{u}_\ell}(\xi/2+\pi)}]\widehat{w_0}(\xi) + \cdots \\ &\quad + [\widehat{u_s}(\xi/2)\overline{\widehat{\tilde{u}_\ell}(\xi/2)} + \widehat{u_s}(\xi/2+\pi)\overline{\widehat{\tilde{u}_\ell}(\xi/2+\pi)}]\widehat{w_s}(\xi) = \widehat{w_\ell}(\xi)\end{aligned}$$

for all $\ell = 0, \dots, s$. It is trivial to see that the above identities hold if and only if

$$\overline{\widehat{u_m}(\xi/2)}\widehat{\tilde{u}_\ell}(\xi/2) + \overline{\widehat{u_m}(\xi/2+\pi)}\widehat{\tilde{u}_\ell}(\xi/2+\pi) = \boldsymbol{\delta}(\ell - m), \qquad \ell, m = 0, \dots, s,$$

where $\boldsymbol{\delta}$ is the Dirac sequence defined in (1.1.11). Consequently, we can rewrite the above identities into the following matrix form:

$$\begin{bmatrix} \widehat{u_0}(\xi) & \cdots & \widehat{u_s}(\xi) \\ \widehat{u_0}(\xi+\pi) & \cdots & \widehat{u_s}(\xi+\pi) \end{bmatrix}^\star \begin{bmatrix} \widehat{\tilde{u}_0}(\xi) & \cdots & \widehat{\tilde{u}_s}(\xi) \\ \widehat{\tilde{u}_0}(\xi+\pi) & \cdots & \widehat{\tilde{u}_s}(\xi+\pi) \end{bmatrix} = I_{s+1}, \tag{1.1.17}$$

where I_{s+1} denotes the $(s+1) \times (s+1)$ identity matrix. Noting that the traces of the matrices on the left-hand sides of (1.1.15) and (1.1.17) must be the same, we conclude from (1.1.15) and (1.1.17) that $s = 1$. Therefore, (iii)$\Longrightarrow$(iv).

Conversely, if $s = 1$, then the two matrices in (1.1.15) are square matrices and hence (1.1.15) directly implies (1.1.17). By the above argument, (1.1.16) must hold. Using the locality of the subdivision and transition operators as in the proof of Theorem 1.1.1, we see that (1.1.16) holds for all $w_0, \dots, w_s \in l(\mathbb{Z})$. Thus, (iii) must hold. ☐

Consequently, under a biorthogonal wavelet filter bank, any input signal $v \in l(\mathbb{Z})$ has a nonredundant representation $v = \mathcal{V}w$ with the unique choice $w = \widetilde{\mathcal{W}}v$; while under a dual framelet filter bank with $s > 1$, an input signal v can be represented as $v = \mathcal{V}w$ from infinitely many $w \in (l(\mathbb{Z}))^{1\times(s+1)}$ of framelet coefficients.

Quite often, one only needs to deal with v in the space $l_2(\mathbb{Z})$, which is equipped with the following inner product:

$$\langle v, w \rangle := \sum_{k\in\mathbb{Z}} v(k)\overline{w(k)}, \qquad v, w \in l_2(\mathbb{Z})$$

and $\|v\|^2_{l_2(\mathbb{Z})} := \langle v, v\rangle < \infty$. For $v \in l_2(\mathbb{Z})$, its Fourier series $\widehat{v}$ is a 2π-periodic square integrable function on $\mathbb{R}$ satisfying $\widehat{v}(\xi + 2\pi) = \widehat{v}(\xi)$ and $\frac{1}{2\pi}\int_{-\pi}^{\pi} |\widehat{v}(\xi)|^2 d\xi = \|v\|^2_{l_2(\mathbb{Z})} = \sum_{k\in\mathbb{Z}} |v(k)|^2$. See Appendix A for a brief introduction to Fourier series.

In the following we explain the role played by the factor $\frac{\sqrt{2}}{2}$ in (1.1.7) and (1.1.8). To do so, we need the following duality relation between the subdivision operator $\mathcal{S}_u$ and the transition operator $\mathcal{T}_u$ acting on the space $l_2(\mathbb{Z})$.

Lemma 1.1.3 *Let $u \in l_0(\mathbb{Z})$ be a finitely supported filter on $\mathbb{Z}$. Then $\mathcal{S}_u : l_2(\mathbb{Z}) \to l_2(\mathbb{Z})$ is the adjoint operator of $\mathcal{T}_u : l_2(\mathbb{Z}) \to l_2(\mathbb{Z})$, that is, $\mathcal{T}_u^\star = \mathcal{S}_u$:*

$$\langle \mathcal{S}_u v, w \rangle = \langle \mathcal{T}_u^\star v, w \rangle := \langle v, \mathcal{T}_u w \rangle, \qquad \forall\, v, w \in l_2(\mathbb{Z}). \tag{1.1.18}$$

Proof Applying (1.1.4) and (1.1.18), we have

$$\begin{aligned}\langle \mathcal{S}_u v, w \rangle &= \frac{1}{\pi} \int_{-\pi}^{\pi} \widehat{v}(2\xi)\widehat{u}(\xi)\overline{\widehat{w}(\xi)} d\xi = \frac{1}{2\pi} \int_{-2\pi}^{2\pi} \widehat{v}(\xi)\widehat{u}(\xi/2)\overline{\widehat{w}(\xi/2)} d\xi \\ &= \frac{1}{2\pi} \int_{-\pi}^{\pi} \widehat{v}(\xi)\overline{\widehat{\mathcal{T}_u w}(\xi)} d\xi = \langle v, \mathcal{T}_u w \rangle = \langle \mathcal{T}_u^\star v, w \rangle.\end{aligned}$$

Hence, (1.1.18) holds. □

Note that the space $(l_2(\mathbb{Z}))^{1\times(s+1)}$ is equipped with the following inner product:

$$\langle (w_0, \dots, w_s), (\tilde{w}_0, \dots, \tilde{w}_s) \rangle := \langle w_0, \tilde{w}_0 \rangle + \cdots + \langle w_s, \tilde{w}_s \rangle, \qquad w_0, \dots, w_s, \tilde{w}_0, \dots, \tilde{w}_s \in l_2(\mathbb{Z})$$

and

$$\|(w_0, \dots, w_s)\|^2_{(l_2(\mathbb{Z}))^{1\times(s+1)}} := \|w_0\|^2_{l_2(\mathbb{Z})} + \cdots + \|w_s\|^2_{l_2(\mathbb{Z})}.$$

For a filter bank $\{u_0, \dots, u_s\}$, recall that

$$\mathcal{W} : l_2(\mathbb{Z}) \to (l_2(\mathbb{Z}))^{1\times(s+1)}, \quad \mathcal{W} v := \tfrac{\sqrt{2}}{2}(\mathcal{T}_{u_0} v, \dots, \mathcal{T}_{u_s} v), \quad v \in l_2(\mathbb{Z}) \tag{1.1.19}$$

and

$$\begin{aligned}&\mathcal{V} : (l_2(\mathbb{Z}))^{1\times(s+1)} \to l_2(\mathbb{Z}) \quad \text{with} \\ &\mathcal{V}(w_0, \dots, w_s) := \tfrac{\sqrt{2}}{2} \sum_{\ell=0}^{s} \mathcal{S}_{u_\ell} w_\ell, \quad w_0, \dots, w_s \in l_2(\mathbb{Z}).\end{aligned} \tag{1.1.20}$$

The *adjoint operators* of $\mathcal{W}$ and $\mathcal{V}$ are defined to be

$$\mathcal{W}^\star : (l_2(\mathbb{Z}))^{1\times(s+1)} \to l_2(\mathbb{Z}) \qquad \text{through} \qquad \langle v, \mathcal{W}^\star w \rangle := \langle \mathcal{W} v, w \rangle \tag{1.1.21}$$

and

$$\mathcal{V}^\star : l_2(\mathbb{Z}) \to (l_2(\mathbb{Z}))^{1\times(s+1)} \qquad \text{through} \qquad \langle \mathcal{V}^\star v, w \rangle := \langle v, \mathcal{V} w \rangle \tag{1.1.22}$$

for all $v \in l_2(\mathbb{Z})$ and $w \in (l_2(\mathbb{Z}))^{1\times(s+1)}$. By Lemma 1.1.3, it is easy to directly check that $\mathcal{W}^\star = \mathcal{V}$ and $\mathcal{V}^\star = \mathcal{W}$.

The role played by $\frac{\sqrt{2}}{2}$ in (1.1.7) and (1.1.8) is explained by the following result:

Theorem 1.1.4 *Let $u_0, \dots, u_s \in l_0(\mathbb{Z})$ be finitely supported sequences on $\mathbb{Z}$. Let $\mathcal{W} : l_2(\mathbb{Z}) \to (l_2(\mathbb{Z}))^{1\times(s+1)}$ be defined in* (1.1.19). *Then the following are equivalent:*

(i) $\|\mathcal{W}v\|^2_{(l_2(\mathbb{Z}))^{1\times(s+1)}} = \|v\|^2_{l_2(\mathbb{Z})}$ *for all $v \in l_2(\mathbb{Z})$, that is,*

$$\|\mathcal{T}_{u_0} v\|^2_{l_2(\mathbb{Z})} + \cdots + \|\mathcal{T}_{u_s} v\|^2_{l_2(\mathbb{Z})} = 2\|v\|^2_{l_2(\mathbb{Z})}, \qquad \forall\, v \in l_2(\mathbb{Z}).$$

(ii) $\langle \mathcal{W}v, \mathcal{W}\tilde{v}\rangle = \langle v, \tilde{v}\rangle$ *for all $v, \tilde{v} \in l_2(\mathbb{Z})$.*
(iii) $\mathcal{W}^\star\mathcal{W} = \mathrm{Id}_{l_2(\mathbb{Z})}$*, that is, $\mathcal{W}^\star\mathcal{W}v = v$ for all $v \in l_2(\mathbb{Z})$.*
(iv) The filter bank $\{u_0, \dots, u_s\}$ satisfies the perfect reconstruction condition:

$$\begin{bmatrix} \widehat{u_0}(\xi) & \cdots & \widehat{u_s}(\xi) \\ \widehat{u_0}(\xi+\pi) & \cdots & \widehat{u_s}(\xi+\pi) \end{bmatrix} \begin{bmatrix} \widehat{u_0}(\xi) & \cdots & \widehat{u_s}(\xi) \\ \widehat{u_0}(\xi+\pi) & \cdots & \widehat{u_s}(\xi+\pi) \end{bmatrix}^\star = I_2, \quad \xi \in \mathbb{R}. \tag{1.1.23}$$

Proof Obviously, (ii)$\Longrightarrow$(i). Note that (i) implies $\langle \mathcal{W}^\star\mathcal{W}v, v\rangle = \langle \mathcal{W}v, \mathcal{W}v\rangle = \langle v, v\rangle$. Using the well-known polarization identity in Exercise 1.2, it is straightforward to see that $\langle \mathcal{W}v, \mathcal{W}\tilde{v}\rangle = \langle \mathcal{W}^\star\mathcal{W}v, \tilde{v}\rangle = \langle v, \tilde{v}\rangle$. Hence, (i)$\Longrightarrow$(ii). The equivalence between (ii) and (iii) is trivial. Note that $\mathcal{W}^\star = \mathcal{V}$. The equivalence between (iii) and (iv) follows directly from Theorem 1.1.1. □

A filter bank $\{u_0, \dots, u_s\}$ satisfying the perfect reconstruction condition in (1.1.23) is called *a tight framelet filter bank*. In particular, a tight framelet filter bank with $s = 1$ is called *an orthogonal wavelet filter bank*. By Theorem 1.1.4, if (1.1.23) is satisfied, then the energy is preserved after a framelet decomposition: $\sum_{\ell=0}^{s} \|w_\ell\|^2_{l_2(\mathbb{Z})} = \|\mathcal{W}v\|^2_{l_2(\mathbb{Z})} = \|v\|^2_{l_2(\mathbb{Z})}$ for all $v \in l_2(\mathbb{Z})$, where $(w_0, \dots, w_s) := \mathcal{W}v$ is the sequence of framelet coefficients.

By Proposition 1.1.2 and Theorem 1.1.4, we have the following result on orthogonal wavelet filter banks.

Proposition 1.1.5 *Let $\{u_0, \dots, u_s\}$ be a tight framelet filter bank. Define $\mathcal{W}$ as in* (1.1.19) *and $\mathcal{V}$ as in* (1.1.20). *Then the following are equivalent:*

(1) $\mathcal{W}$ is an invertible orthogonal mapping satisfying $\langle \mathcal{W}v, \mathcal{W}\tilde{v}\rangle = \langle v, \tilde{v}\rangle$, $v, \tilde{v} \in l_2(\mathbb{Z})$.
(2) $\mathcal{V}$ is an invertible orthogonal mapping such that for all $w_0, \dots, w_s, \tilde{w}_0, \dots, \tilde{w}_s \in l_2(\mathbb{Z})$, $\langle \mathcal{V}(w_0, \dots, w_s), \mathcal{V}(\tilde{w}_0, \dots, \tilde{w}_s)\rangle = \langle (w_0, \dots, w_s), (\tilde{w}_0, \dots, \tilde{w}_s)\rangle$.
(3) $\mathcal{W}^\star\mathcal{W} = \mathrm{Id}_{l_2(\mathbb{Z})}$ and $\mathcal{W}\mathcal{W}^\star = \mathrm{Id}_{(l_2(\mathbb{Z}))^{1\times(s+1)}}$.
(4) $s = 1$.

We shall discuss how to design orthogonal or biorthogonal wavelet filter banks in Chap. 2, and tight or dual framelet filter banks in Chap. 3.

1.1.3 Some Examples of Wavelet or Framelet Filter Banks

In the following, let us provide a few examples to illustrate various types of filter banks. For a filter $u = \{u(k)\}_{k\in\mathbb{Z}}$ such that $u(k) = 0$ for all $k \in \mathbb{Z}\backslash[m,n]$ and $u(m)u(n) \neq 0$, we denote by $\mathrm{fsupp}(u) := [m,n]$ as its *filter support*. To list the filter u, we shall adopt the following notation throughout the book:

$$u = \{u(m), u(m+1), \dots, u(-1), \underline{\mathbf{u(0)}}, u(1), \dots, u(n-1), u(n)\}_{[m,n]}, \qquad (1.1.24)$$

where we underlined and boldfaced the number $u(0)$ to indicate its position at 0.

Example 1.1.1 $\{u_0, u_1\}$ is an orthogonal wavelet filter bank (called the Haar orthogonal wavelet filter bank), where

$$u_0 = \{\underline{\tfrac{\mathbf{1}}{\mathbf{2}}}, \tfrac{1}{2}\}_{[0,1]}, \quad u_1 = \{\underline{-\tfrac{\mathbf{1}}{\mathbf{2}}}, \tfrac{1}{2}\}_{[0,1]}. \qquad (1.1.25)$$

Example 1.1.2 $(\{\tilde{u}_0, \tilde{u}_1\}, \{u_0, u_1\})$ is a biorthogonal wavelet filter bank, where

$$\tilde{u}_0 = \{-\tfrac{1}{8}, \tfrac{1}{4}, \underline{\tfrac{\mathbf{3}}{\mathbf{4}}}, \tfrac{1}{4}, -\tfrac{1}{8}\}_{[-2,2]}, \qquad \tilde{u}_1 = \{\underline{-\tfrac{\mathbf{1}}{\mathbf{4}}}, \tfrac{1}{2}, -\tfrac{1}{4}\}_{[0,2]},$$

$$u_0 = \{\tfrac{1}{4}, \underline{\tfrac{\mathbf{1}}{\mathbf{2}}}, \tfrac{1}{4}\}_{[-1,1]}, \qquad u_1 = \{-\tfrac{1}{8}, \underline{-\tfrac{\mathbf{1}}{\mathbf{4}}}, \tfrac{3}{4}, -\tfrac{1}{4}, -\tfrac{1}{8}\}_{[-1,3]}.$$

Example 1.1.3 $\{u_0, u_1, u_2\}$ is a tight framelet filter bank, where

$$u_0 = \{\tfrac{1}{4}, \underline{\tfrac{\mathbf{1}}{\mathbf{2}}}, \tfrac{1}{4}\}_{[-1,1]}, \quad u_1 = \{-\tfrac{\sqrt{2}}{4}, \underline{\mathbf{0}}, \tfrac{\sqrt{2}}{4}\}_{[-1,1]}, \quad u_2 = \{-\tfrac{1}{4}, \underline{\tfrac{\mathbf{1}}{\mathbf{2}}}, -\tfrac{1}{4}\}_{[-1,1]}.$$

Example 1.1.4 $(\{\tilde{u}_0, \tilde{u}_1, \tilde{u}_2\}, \{u_0, u_1, u_2\})$ is a dual framelet filter bank, where

$$\tilde{u}_0 = \{\underline{\tfrac{\mathbf{1}}{\mathbf{2}}}, \tfrac{1}{2}\}_{[0,1]}, \qquad \tilde{u}_1 = \{-\tfrac{1}{2}, \underline{\tfrac{\mathbf{1}}{\mathbf{2}}}\}_{[-1,0]}, \qquad \tilde{u}_2 = \{\underline{-\tfrac{\mathbf{1}}{\mathbf{2}}}, \tfrac{1}{2}\}_{[0,1]},$$

$$u_0 = \{\tfrac{1}{8}, \underline{\tfrac{\mathbf{3}}{\mathbf{8}}}, \tfrac{3}{8}, \tfrac{1}{8}\}_{[-1,2]}, \qquad u_1 = \{-\tfrac{1}{4}, \underline{\tfrac{\mathbf{1}}{\mathbf{4}}}\}_{[-1,0]}, \qquad u_2 = \{-\tfrac{1}{8}, \underline{-\tfrac{\mathbf{3}}{\mathbf{8}}}, \tfrac{3}{8}, \tfrac{1}{8}\}_{[-1,2]}.$$

At the end of this section, we illustrate a one-level discrete framelet transform using the Haar orthogonal wavelet filter bank in (1.1.25). Let

$$v = \{-21, -22, -23, -23, -25, 38, 36, 34\}_{[0,7]} \qquad (1.1.26)$$

be a test input signal. Note that

$$[\mathcal{T}_{u_0} v](n) = v(2n+1) + v(2n), \qquad [\mathcal{T}_{u_1} v](n) = v(2n+1) - v(2n), \qquad n \in \mathbb{Z}.$$

Therefore, we have the wavelet coefficients:

$$w_0 = \tfrac{\sqrt{2}}{2}\{-43, -46, 13, 70\}_{[0,3]}, \qquad w_1 = \tfrac{\sqrt{2}}{2}\{-1, 0, 63, -2\}_{[0,3]}.$$

On the other hand, we have

$$[\mathcal{S}_{u_0} w_0](2n) = w_0(n), \qquad [\mathcal{S}_{u_0} w_0](2n+1) = w_0(n), \quad n \in \mathbb{Z},$$
$$[\mathcal{S}_{u_1} w_1](2n) = -w_1(n), \qquad [\mathcal{S}_{u_1} w_1](2n+1) = w_1(n), \quad n \in \mathbb{Z}.$$

Hence, we have

$$\tfrac{\sqrt{2}}{2}\mathcal{S}_{u_0} w_0 = \tfrac{1}{2}\{-43, -43, -46, -46, 13, 13, 70, 70\}_{[0,7]},$$
$$\tfrac{\sqrt{2}}{2}\mathcal{S}_{u_1} w_1 = \tfrac{1}{2}\{1, -1, 0, 0, -63, 63, 2, -2\}_{[0,7]}.$$

Clearly, we have the perfect reconstruction of the original input signal v:

$$\tfrac{\sqrt{2}}{2}\mathcal{S}_{u_0} w_0 + \tfrac{\sqrt{2}}{2}\mathcal{S}_{u_1} w_1 = \{-21, -22, -23, -23, -25, 38, 36, 34\}_{[0,7]} = v$$

and the following energy-preserving identity

$$\|w_0\|^2_{l_2(\mathbb{Z})} + \|w_1\|^2_{l_2(\mathbb{Z})} = 4517 + 1987 = 6504 = \|v\|^2_{l_2(\mathbb{Z})}.$$

The subdivision operator and the transition operator in applications are often implemented through the widely used convolution operation in mathematics and engineering. For $u \in l_0(\mathbb{Z})$ and $v \in l(\mathbb{Z})$, the convolution $u * v$ is defined to be

$$[u * v](n) := \sum_{k\in\mathbb{Z}} u(k)v(n-k), \qquad n \in \mathbb{Z}. \tag{1.1.27}$$

By the definition of the convolution in (1.1.27), we note that $\widehat{u * v}(\xi) = \widehat{u}(\xi)\widehat{v}(\xi)$. To implement the subdivision and transition operators using the convolution operation, we also need the upsampling and downsampling operators on sequences in $l(\mathbb{Z})$. The *downsampling (or decimation) operator* $\downarrow \mathsf{d} : l(\mathbb{Z}) \to l(\mathbb{Z})$ and the *upsampling operator* $\uparrow \mathsf{d} : l(\mathbb{Z}) \to l(\mathbb{Z})$ with a sampling factor $\mathsf{d} \in \mathbb{Z}\backslash\{0\}$ are given by

$$[v \downarrow \mathsf{d}](n) := v(\mathsf{d}n) \quad \text{and} \quad [v \uparrow \mathsf{d}](n) := \begin{cases} v(n/\mathsf{d}), & \text{if } n/\mathsf{d} \text{ is an integer,} \\ 0, & \text{otherwise,} \end{cases} \tag{1.1.28}$$

for $n \in \mathbb{Z}$. For a sequence $v = \{v(k)\}_{k\in\mathbb{Z}}$, we denote its complex conjugate sequence reflected about the origin by $v^\star$, which is defined to be

$$v^\star(k) := \overline{v(-k)}, \qquad k \in \mathbb{Z}.$$

Note that $\widehat{v^\star}(\xi) = \overline{\widehat{v}(\xi)}$. Now the subdivision operator $\mathcal{S}_u$ in (1.1.2) and the transition operator $\mathcal{T}_u$ in (1.1.3) can be equivalently expressed as follows:

$$\mathcal{S}_u v = 2(v \uparrow 2) * u \quad \text{and} \quad \mathcal{T}_u v = 2(v * u^\star) \downarrow 2. \tag{1.1.29}$$

For $u = \{u(k)\}_{k\in\mathbb{Z}}$ and $\gamma \in \mathbb{Z}$, we define the associated *coset sequence* $u^{[\gamma]}$ of u at the coset $\gamma + 2\mathbb{Z}$ by

$$\widehat{u^{[\gamma]}}(\xi) := \sum_{k\in\mathbb{Z}} u(\gamma + 2k)e^{-ik\xi}, \text{ i.e., } u^{[\gamma]} = u(\gamma + \cdot) \downarrow 2 = \{u(\gamma + 2k)\}_{k\in\mathbb{Z}}. \tag{1.1.30}$$

Using the coset sequences of u, we can rewrite (1.1.29) as

$$[\mathcal{S}_u v]^{[0]} = 2v * u^{[0]}, \quad [\mathcal{S}_u v]^{[1]} = 2v * u^{[1]}, \quad \mathcal{T}_u v = 2\big(v^{[0]} * (u^{[0]})^\star + v^{[1]} * (u^{[1]})^\star\big).$$

1.2 Sparsity of Discrete Framelet Transforms

Sparse representation for smooth or piecewise smooth signals is a highly desired property of a discrete transform in applications. To achieve sparsity, it is desirable to have as many as possible negligible framelet coefficients for smooth signals. In this section, we study several basic mathematical properties that are closely related to sparsity of a discrete framelet transform in the discrete setting, in particular, properties such as vanishing moments, sum rules, polynomial reproduction, linear-phase moments, and symmetry. For the convenience of the reader, basic definitions such as vanishing moments, sum rules, linear-phase moments, and symmetry will be repeated at the beginning of Chap. 2.

1.2.1 Convolution and Transition Operators on Polynomial Spaces

Smooth signals are theoretically modeled by polynomials of various degrees. Let $\mathsf{p} : \mathbb{R} \to \mathbb{C}$ be a polynomial, that is, $\mathsf{p}(x) = \sum_{j=0}^m p_j x^j$ with $p_0, \ldots, p_m \in \mathbb{C}$ and a nonnegative integer m; if the leading coefficient $p_m \ne 0$, then we define $\deg(\mathsf{p}) = m$, which is the degree of the polynomial p. For the zero polynomial, we use the convention $\deg(0) = -\infty$. Sampling a polynomial p on the integer lattice $\mathbb{Z}$, we have a polynomial sequence $\mathsf{p}|_{\mathbb{Z}} : \mathbb{Z} \to \mathbb{C}$ which is given by $[\mathsf{p}|_{\mathbb{Z}}](k) = \mathsf{p}(k), k \in \mathbb{Z}$. If a sequence $v = \{v(k)\}_{k\in\mathbb{Z}}$ is a polynomial sequence, then a polynomial p, satisfying $v(k) = \mathsf{p}(k)$ for all $k \in \mathbb{Z}$, is uniquely determined. Therefore, for simplicity of presentation, we shall use p to denote both a polynomial function p on $\mathbb{R}$ and its induced polynomial sequence $\mathsf{p}|_{\mathbb{Z}}$ on $\mathbb{Z}$. One can easily tell them apart from the context. In case of confusion, we explicitly use $\mathsf{p}|_{\mathbb{Z}}$ instead of p.

Define $\mathbb{N}_0 := \mathbb{N} \cup \{0\}$, the set of all nonnegative integers. For $m \in \mathbb{N}_0$, $\mathbb{P}_m$ denotes the space of all polynomials of degree no more than m. In particular, $\mathbb{P} := \cup_{m=0}^{\infty}\mathbb{P}_m$ denotes the space of all polynomials on $\mathbb{R}$. For a polynomial $\mathsf{p}(x) = \sum_{j=0}^{\infty} p_j x^j \in \mathbb{P}$ and a smooth function $\mathbf{f}(\xi)$, $\mathsf{p}^{(n)}$ is the nth derivative of p and we use the following polynomial differentiation operator:

$$\mathsf{p}(x - i\tfrac{d}{d\xi})\mathbf{f}(\xi) := \sum_{j=0}^{\infty} p_j (x - i\tfrac{d}{d\xi})^j \mathbf{f}(\xi). \tag{1.2.1}$$

By the definition of $\mathsf{p}(x - i\frac{d}{d\xi})$ in (1.2.1) and the Taylor expansion of $\mathsf{p}(y+z)$ at the point y, we deduce $\mathsf{p}(y+z) = \sum_{j=0}^{\infty} \mathsf{p}^{(j)}(y)\frac{z^j}{j!}$ and hence,

$$\mathsf{p}(x - i\tfrac{d}{d\xi})\mathbf{f}(\xi) = \sum_{j=0}^{\infty} \frac{(-i)^j}{j!}\mathsf{p}^{(j)}(x)\mathbf{f}^{(j)}(\xi) = \sum_{j=0}^{\infty} \frac{x^j}{j!}\mathsf{p}^{(j)}(-i\tfrac{d}{d\xi})\mathbf{f}(\xi). \tag{1.2.2}$$

Using the Leibniz differentiation formula and (1.2.2), we have the following generalized product rule for differentiation:

$$\mathsf{p}(x - i\tfrac{d}{d\xi})(\mathbf{g}(\xi)\mathbf{f}(\xi)) = \sum_{j=0}^{\infty} \frac{(-i)^j}{j!}\mathbf{g}^{(j)}(\xi)\mathsf{p}^{(j)}(x - i\tfrac{d}{d\xi})\mathbf{f}(\xi). \tag{1.2.3}$$

It follows directly from (1.2.2) and (1.2.3) that

$$\left[\mathsf{p}(-i\tfrac{d}{d\xi})(e^{ix\xi}\mathbf{f}(\xi))\right]\Big|_{\xi=0} = \left[\mathsf{p}(x - i\tfrac{d}{d\xi})\mathbf{f}(\xi)\right]\Big|_{\xi=0}. \tag{1.2.4}$$

To study sparsity of a discrete framelet transform, we have to understand how the subdivision operator and the transition operator act on polynomial spaces. Because the subdivision and transition operators can be expressed via the convolution operation, in the following we first study the convolution operation acting on polynomial spaces.

Lemma 1.2.1 *Let* $u = \{u(k)\}_{k\in\mathbb{Z}} \in l_0(\mathbb{Z})$ *be a finitely supported sequence on* $\mathbb{Z}$ *and* $\mathsf{p} \in \mathbb{P}$. *Then* $\mathsf{p} * u$ *is a polynomial sequence satisfying* $\deg(\mathsf{p} * u) \leqslant \deg(\mathsf{p})$ *and*

$$\begin{aligned} \mathsf{p} * u(x) &:= \sum_{k\in\mathbb{Z}} \mathsf{p}(x-k)u(k) = \left[\mathsf{p}(x - i\tfrac{d}{d\xi})\widehat{u}(\xi)\right]\Big|_{\xi=0} \\ &= \sum_{j=0}^{\infty} \frac{(-i)^j}{j!}\mathsf{p}^{(j)}(x)\widehat{u}^{(j)}(0) = \sum_{j=0}^{\infty} \frac{x^j}{j!}\left[\mathsf{p}^{(j)}(-i\tfrac{d}{d\xi})\widehat{u}(\xi)\right]\Big|_{\xi=0}. \end{aligned} \tag{1.2.5}$$

Moreover, $\mathsf{p} * (u \uparrow 2) = [\mathsf{p}(2\cdot) * u](2^{-1}\cdot)$,

$$\mathsf{p}^{(j)} * u = [\mathsf{p} * u]^{(j)}, \quad \forall\, j \in \mathbb{N}_0 \quad \textit{and} \quad \mathsf{p}(\cdot - y) * u = [\mathsf{p} * u](\cdot - y), \quad \forall\, y \in \mathbb{R}. \tag{1.2.6}$$

Proof By the Taylor expansion $\mathsf{p}(x-k) = \sum_{j=0}^{\infty} \mathsf{p}^{(j)}(x) \frac{(-k)^j}{j!}$, we have

$$[\mathsf{p} * u](x) = \sum_{k \in \mathbb{Z}} \mathsf{p}(x-k) u(k) = \sum_{k \in \mathbb{Z}} \sum_{j=0}^{\infty} \mathsf{p}^{(j)}(x) u(k) \frac{(-k)^j}{j!} = \sum_{j=0}^{\infty} \mathsf{p}^{(j)}(x) \sum_{k \in \mathbb{Z}} u(k) \frac{(-k)^j}{j!}.$$

By $\widehat{u}(\xi) = \sum_{k \in \mathbb{Z}} u(k) e^{-ik\xi}$, we have $\widehat{u}^{(j)}(0) = \sum_{k \in \mathbb{Z}} u(k)(-ik)^j = i^j \sum_{k \in \mathbb{Z}} u(k)(-k)^j$. Now we conclude that

$$[\mathsf{p} * u](x) = \sum_{j=0}^{\infty} \frac{(-i)^j}{j!} \mathsf{p}^{(j)}(x) \widehat{u}^{(j)}(0). \tag{1.2.7}$$

Therefore, (1.2.5) and (1.2.6) follow directly from (1.2.2) and (1.2.7). □

For smooth functions **f** and **g**, we shall use the following big $\mathscr{O}$ notation:

$$\mathbf{f}(\xi) = \mathbf{g}(\xi) + \mathscr{O}(|\xi - \xi_0|^m), \qquad \xi \to \xi_0 \tag{1.2.8}$$

to mean that the derivatives of **f** and **g** at $\xi = \xi_0$ agree to the orders up to $m-1$:

$$\mathbf{f}^{(j)}(\xi_0) = \mathbf{g}^{(j)}(\xi_0), \qquad \forall\, j = 0, \ldots, m-1.$$

For a polynomial $\mathsf{p} \in \mathbb{P}_{m-1}$ of degree less than m, by (1.2.5), it is evident that the polynomial $\mathsf{p} * u$ depends only on the values $\widehat{u}(0), \widehat{u}'(0), \ldots, \widehat{u}^{(m-1)}(0)$ of $\widehat{u}$ at the origin. Consequently, if two sequences $u, v \in l_0(\mathbb{Z})$ satisfy $\widehat{u}(\xi) = \widehat{v}(\xi) + \mathscr{O}(|\xi|^m)$ as $\xi \to 0$, then $\mathsf{p} * u = \mathsf{p} * v$ for all $\mathsf{p} \in \mathbb{P}_{m-1}$. For simplicity, in this book we shall frequently use the big $\mathscr{O}$ notation in (1.2.8).

The action of the transition operator on polynomial spaces is as follows.

Theorem 1.2.2 *Let $u \in l_0(\mathbb{Z})$ be a finitely supported sequence on $\mathbb{Z}$. For $\mathsf{p} \in \mathbb{P}$,*

$$\mathcal{T}_u \mathsf{p} = 2[\mathsf{p} * u^\star](2\cdot) = 2\mathsf{p}(2\cdot) * u_{\mathsf{p}} = \sum_{j=0}^{\infty} \frac{2(-i)^j}{j!} \mathsf{p}^{(j)}(2\cdot) \overline{\widehat{u}^{(j)}(0)}, \tag{1.2.9}$$

where $u_{\mathsf{p}} \in l_0(\mathbb{Z})$ is any finitely supported sequence on $\mathbb{Z}$ such that

$$\widehat{u_{\mathsf{p}}}(\xi) = \overline{\widehat{u}(\xi/2)} + \mathscr{O}(|\xi|^{\deg(\mathsf{p})+1}), \qquad \xi \to 0.$$

In particular, for any positive integer $m \in \mathbb{N}$, the following are equivalent:

(1) $\mathcal{T}_u \mathsf{p} = 0$ *for all polynomial sequences* $\mathsf{p} \in \mathbb{P}_{m-1}$.
(2) $\mathcal{T}_u \mathsf{q} = 0$ *for some polynomial sequence* q *with* $\deg(\mathsf{q}) = m-1$.
(3) $\widehat{u}(\xi) = \mathscr{O}(|\xi|^m)$ *as* $\xi \to 0$*, that is,* $\widehat{u}^{(j)}(0) = 0$ *for all* $j = 0, \dots, m-1$.
(4) $\widehat{u}(\xi) = (1 - e^{-i\xi})^m \mathbf{Q}(\xi)$ *for some* 2π*-periodic trigonometric polynomial* $\mathbf{Q}$.

Proof Since $\mathcal{T}_u \mathsf{p} = 2(\mathsf{p} * u^\star) \downarrow 2 = 2[\mathsf{p} * u^\star](2\cdot)$, by Lemma 1.2.1, we see that $\mathcal{T}_u \mathsf{p}$ is a polynomial sequence and

$$2[\mathsf{p} * u^\star](2\cdot) = \sum_{j=0}^{\infty} \frac{2(-i)^j}{j!} \mathsf{p}^{(j)}(2\cdot)\overline{\widehat{u}^{(j)}(0)} = \sum_{j=0}^{\infty} \frac{(-i)^j}{j!} [\mathsf{p}(2\cdot)]^{(j)} 2^{1-j}\overline{\widehat{u}^{(j)}(0)}.$$

By $\widehat{u_{\mathsf{p}}}^{(j)}(0) = 2^{-j}\overline{\widehat{u}^{(j)}(0)}$ for all $j = 0, \dots, \deg(\mathsf{p})$, the identities in (1.2.9) follow directly from (1.2.5). From (1.2.5), we see that

$$\mathcal{T}_u \mathsf{p}^{(j)} = 2^{j+1} \mathsf{p}^{(j)}(2\cdot) * \mathring{u}, \qquad \forall\, \mathsf{p} \in \mathbb{P}_{m-1},\ j \in \mathbb{N}_0, \tag{1.2.10}$$

where $\mathring{u}$ is any finitely supported sequence satisfying $\widehat{\mathring{u}}(\xi) = \overline{\widehat{u}(\xi/2)} + \mathscr{O}(|\xi|^m)$ as $\xi \to 0$. For a polynomial q with $\deg(\mathsf{q}) = m-1$, the set $\{\mathsf{q}, \mathsf{q}', \dots, \mathsf{q}^{(m-1)}\}$ is a basis for $\mathbb{P}_{m-1}$. Now the equivalence among items (1)–(4) is a direct consequence of (1.2.9) and (1.2.10). □

We say that a filter u (or its Fourier series $\widehat{u}$) has *m vanishing moments* if any of items (1)–(4) in Theorem 1.2.2 holds. The notion of vanishing moments is important for sparse framelet expansions, since most framelet coefficients are identically zero for any input signal which is a polynomial to certain degree. More precisely, suppose that u has m vanishing moments. For a signal v, if v agrees with some polynomial of degree less than m on the support of $u(\cdot - 2n)$, then by the definition of the transition operator and the definition of vanishing moments, we have $[\mathcal{T}_u v](n) = 0$.

Note that $\mathcal{T}_u \mathbb{P}_{m-1} \subseteq \mathbb{P}_{m-1}$ for all $m \in \mathbb{N}$. In particular, $\mathcal{T}_u \mathbb{P}_{m-1} = \mathbb{P}_{m-1}$ if $\widehat{u}(0) \ne 0$. Moreover, all the eigenvalues of $\mathcal{T}_u|_{\mathbb{P}_{m-1}}$ are $2\overline{\widehat{u}(0)}, \dots, 2^m\overline{\widehat{u}(0)}$ (see Exercise 1.15).

1.2.2 Subdivision Operator on Polynomial Spaces

We now investigate the subdivision operator acting on polynomial spaces. In contrast to the case of the transition operator, $\mathcal{S}_u \mathsf{p}$ is not always a polynomial sequence for an input polynomial sequence p. A simple example is $\mathsf{p} = 1$ and $u = \{1\}_{[0,0]}$ (that is, $u = \boldsymbol{\delta}$). Then $[\mathcal{S}_u \mathsf{p}]^{[0]} := [\mathcal{S}_u \mathsf{p}](2\cdot) = 2$ and $[\mathcal{S}_u \mathsf{p}]^{[1]} := [\mathcal{S}_u \mathsf{p}](2\cdot +1) = 0$.

Lemma 1.2.3 *Let* $u = \{u(k)\}_{k\in\mathbb{Z}} \in l_0(\mathbb{Z})$ *and* q *be a polynomial. Then the following are equivalent:*

(i) $\sum_{k\in\mathbb{Z}} \mathsf{q}(-\frac{1}{2}-k)u(1+2k) = \sum_{k\in\mathbb{Z}} \mathsf{q}(-k)u(2k)$, *i.e.*, $(\mathsf{q} * u^{[1]})(-\frac{1}{2}) = (\mathsf{q} * u^{[0]})(0)$.

(ii) $[\mathsf{q}(-i\frac{d}{d\xi})(e^{-i\xi/2}\widehat{u^{[1]}}(\xi))]|_{\xi=0} = [\mathsf{q}(-i\frac{d}{d\xi})\widehat{u^{[0]}}(\xi)]|_{\xi=0}$.

(iii) $[\mathsf{q}(-\frac{i}{2}\frac{d}{d\xi})\widehat{u}(\xi)]|_{\xi=\pi} = 0$.

Proof (i) $\iff$ (ii) follows directly from

$$\big[\mathsf{q}(-i\tfrac{d}{d\xi})(e^{-i\xi\gamma/2}\widehat{u^{[\gamma]}}(\xi))\big]\big|_{\xi=0} = \big[\mathsf{q}(-\tfrac{\gamma}{2}-i\tfrac{d}{d\xi})\widehat{u^{[\gamma]}}(\xi)\big]\big|_{\xi=0} = [\mathsf{q} * u^{[\gamma]}](-\tfrac{\gamma}{2})$$
$$= \sum_{k\in\mathbb{Z}} \mathsf{q}(-\tfrac{\gamma}{2}-k)u(\gamma+2k)$$

for $\gamma \in \mathbb{Z}$, where we used (1.2.4) and (1.2.5). By $\widehat{u}(\xi) = \widehat{u^{[0]}}(2\xi) + e^{-i\xi}\widehat{u^{[1]}}(2\xi)$, we have $\widehat{u}(2^{-1}\xi+\pi) = \widehat{u^{[0]}}(\xi) - e^{-i\xi/2}\widehat{u^{[1]}}(\xi)$. Now item (ii) is equivalent to $[\mathsf{q}(-i\frac{d}{d\xi})\widehat{u}(2^{-1}\xi+\pi)]|_{\xi=0} = 0$, which is simply item (iii). □

A necessary and sufficient condition for $\mathcal{S}_u\mathsf{p} \in \mathbb{P}$ is as follows.

Theorem 1.2.4 *Let $u = \{u(k)\}_{k\in\mathbb{Z}} \in l_0(\mathbb{Z})$ be a finitely supported sequence on $\mathbb{Z}$ and $\mathsf{p} \in \mathbb{P}$ be a polynomial. Then the following are equivalent:*

(1) $\mathcal{S}_u\mathsf{p}$ is a polynomial sequence, i.e., $\mathcal{S}_u\mathsf{p} \in \mathbb{P}$.

(2) $\sum_{k\in\mathbb{Z}} \mathsf{p}^{(j)}(-\frac{1}{2}-k)u(1+2k) = \sum_{k\in\mathbb{Z}} \mathsf{p}^{(j)}(-k)u(2k)$ for all $j \in \mathbb{N}_0$.

(3) $[\mathsf{p}^{(j)}(-i\frac{d}{d\xi})(e^{-i\xi/2}\widehat{u^{[1]}}(\xi))]|_{\xi=0} = [\mathsf{p}^{(j)}(-i\frac{d}{d\xi})\widehat{u^{[0]}}(\xi)]|_{\xi=0}$ for all $j \in \mathbb{N}_0$.

(4) $[\mathsf{p}^{(j)}(-\frac{1}{2}-i\frac{d}{d\xi})\widehat{u^{[1]}}(\xi)]|_{\xi=0} = [\mathsf{p}^{(j)}(-i\frac{d}{d\xi})\widehat{u^{[0]}}(\xi)]|_{\xi=0}$ for all $j \in \mathbb{N}_0$.

(5) $[\mathsf{p}^{(j)}(-\frac{i}{2}\frac{d}{d\xi})\widehat{u}(\xi)]|_{\xi=\pi} = 0$ for all nonnegative integers $j \in \mathbb{N}_0$.

Moreover, if any of the above items (1)–(5) holds, then $\deg(\mathcal{S}_u\mathsf{p}) \leqslant \deg(\mathsf{p})$,

$$\mathcal{S}_u\mathsf{p} = \big(\mathsf{p}(2^{-1}\cdot)\big) * u = \sum_{j=0}^{\infty} \frac{(-i)^j}{2^j j!}\mathsf{p}^{(j)}(2^{-1}\cdot)\widehat{u}^{(j)}(0), \tag{1.2.11}$$

and

$$\mathcal{S}_u(\mathsf{p}^{(j)}) = \mathsf{p}^{(j)}(2^{-1}\cdot) * u = 2^j[\mathcal{S}_u\mathsf{p}]^{(j)}, \qquad j \in \mathbb{N}_0,$$
$$\mathcal{S}_u(\mathsf{p}(\cdot - y)) = \mathsf{p}(2^{-1}\cdot - y) * u = [\mathcal{S}_u\mathsf{p}](\cdot - 2y), \qquad y \in \mathbb{R}.$$

Proof By the definition of the subdivision operator $\mathcal{S}_u$ in (1.1.2), for $n, \gamma \in \mathbb{Z}$,

$$[\mathcal{S}_u\mathsf{p}](\gamma+2n) = 2\sum_{m\in\mathbb{Z}} \mathsf{p}(m)u(\gamma+2n-2m) = 2\sum_{k\in\mathbb{Z}} \mathsf{p}(2^{-1}(\gamma+2n)-\tfrac{\gamma}{2}-k)u(\gamma+2k).$$

Hence, $[\mathcal{S}_u \mathsf{p}](\gamma + 2\cdot)$ is a polynomial sequence on each coset for every $\gamma \in \mathbb{Z}$. Now it is easy to see that $\mathcal{S}_u \mathsf{p}$ is a polynomial sequence if and only if $\sum_{k\in\mathbb{Z}} \mathsf{p}(\cdot - \frac{\gamma}{2} - k)u(\gamma + 2k)$ is independent of γ. Using the Taylor expansion of p, we have

$$\begin{aligned}\sum_{k\in\mathbb{Z}} \mathsf{p}(x - \tfrac{\gamma}{2} - k)u(\gamma + 2k) &= \sum_{k\in\mathbb{Z}} \sum_{j=0}^{\infty} \frac{x^j}{j!} \mathsf{p}^{(j)}(-\tfrac{\gamma}{2} - k)u(\gamma + 2k)\\ &= \sum_{j=0}^{\infty} \frac{x^j}{j!} \sum_{k\in\mathbb{Z}} \mathsf{p}^{(j)}(-\tfrac{\gamma}{2} - k)u(\gamma + 2k).\end{aligned}$$

Hence, the sequence $\sum_{k\in\mathbb{Z}} \mathsf{p}(\cdot - \frac{\gamma}{2} - k)u(\gamma + 2k)$ is independent of γ if and only if all $\sum_{k\in\mathbb{Z}} \mathsf{p}^{(j)}(-\frac{\gamma}{2} - k)u(\gamma + 2k), j \in \mathbb{N}_0$ are independent of γ, which are obviously equivalent to the conditions in item (2). Thus, we proved (1) $\iff$ (2). Moreover, when $\mathcal{S}_u \mathsf{p} \in \mathbb{P}$, the above argument also yields

$$\mathcal{S}_u \mathsf{p} = 2\sum_{k\in\mathbb{Z}} \mathsf{p}(2^{-1} \cdot - \tfrac{\gamma}{2} - k)u(\gamma + 2k) = \sum_{k\in\mathbb{Z}} \mathsf{p}(2^{-1}(\cdot - k))u(k), \qquad \forall\, \gamma \in \mathbb{Z},$$

from which we see that (1.2.11) holds. The equivalence among (2)–(5) follows directly from Lemma 1.2.3. □

For the subdivision operator acting on polynomial spaces, we have

Theorem 1.2.5 *Let $u = \{u(k)\}_{k\in\mathbb{Z}}$. For $m \in \mathbb{N}$, the following are equivalent:*

(1) $\mathcal{S}_u \mathbb{P}_{m-1} \subseteq \mathbb{P}$.
(2) $\mathcal{S}_u \mathsf{q} \in \mathbb{P}$ for some polynomial $\mathsf{q} \in \mathbb{P}$ with $\deg(\mathsf{q}) = m - 1$.
(3) $\mathcal{S}_u \mathbb{P}_{m-1} \subseteq \mathbb{P}_{m-1}$.
(4) $\widehat{u}^{(j)}(\pi) = 0$ for all $j = 0, \dots, m-1$, in other words,

$$\widehat{u}(\xi + \pi) = \mathscr{O}(|\xi|^m), \qquad \xi \to 0. \tag{1.2.12}$$

(5) $\widehat{u}(\xi) = (1 + e^{-i\xi})^m \mathbf{Q}(\xi)$ for some 2π-periodic trigonometric polynomial $\mathbf{Q}$.
(6) $[e^{-i\xi/2}\widehat{u^{[1]}}(\xi)]^{(j)}(0) = [\widehat{u^{[0]}}(\xi)]^{(j)}(0)$ for all $j = 0, \dots, m-1$, that is,

$$e^{-i\xi/2}\widehat{u^{[1]}}(\xi) = \widehat{u^{[0]}}(\xi) + \mathscr{O}(|\xi|^m), \qquad \xi \to 0, \tag{1.2.13}$$

or its equivalent form in the space/time domain:

$$\sum_{k\in\mathbb{Z}} u(1 + 2k)(1 + 2k)^j = \sum_{k\in\mathbb{Z}} u(2k)(2k)^j, \qquad \forall\, j = 0, \dots, m-1.$$

In particular, if (1.2.12) *holds, then for all* $\mathsf{p} \in \mathbb{P}_{m-1}$ *and* $v \in l_0(\mathbb{Z})$,

$$\mathcal{S}_u(\mathsf{p} * v) = 2^{-1}\mathsf{p}(2^{-1}\cdot) * [\mathcal{S}_u v] = \sum_{j=0}^{\infty} \frac{(-i)^j}{2^j j!} \mathsf{p}^{(j)}(2^{-1}\cdot)[\widehat{u}\,\widehat{v}(2\cdot)]^{(j)}(0), \qquad (1.2.14)$$

and furthermore, $\mathcal{S}_u\mathbb{P}_{m-1} = \mathbb{P}_{m-1}$ *if* $\widehat{u}(0) \neq 0$.

Proof (1)$\Longrightarrow$(2) is obvious. By Theorem 1.2.4, if $\mathcal{S}_u\mathsf{p} \in \mathbb{P}$, then $\mathcal{S}_u\mathsf{p}^{(j)} \in \mathbb{P}$ for all $j \in \mathbb{N}_0$. Since $\{\mathsf{q}, \mathsf{q}', \ldots, \mathsf{q}^{(m-1)}\}$ is a basis for $\mathbb{P}_{m-1}$, we now see that (2)$\Longrightarrow$(1). The equivalence between (1) and (3) follows from Theorem 1.2.4.

Applying Theorem 1.2.4 with $\mathsf{p} \in \{1, x, \ldots, x^{m-1}\}$ which is a basis for $\mathbb{P}_{m-1}$, we see that (3) $\Longleftrightarrow$ (4). (4) $\Longleftrightarrow$ (5) is trivial. By Theorem 1.2.4 or a direct proof, we have (5) $\Longleftrightarrow$ (6).

By (1.2.11), it is straightforward to see that (1.2.14) holds. □

We say that a filter u (or its Fourier series $\widehat{u}$) has *m sum rules* if any of items (1)–(6) in Theorem 1.2.5 is satisfied. If u has m sum rules, then $\mathcal{S}_u\mathbb{P}_{m-1} \subseteq \mathbb{P}_{m-1}$ and all the eigenvalues of $\mathcal{S}_u|_{\mathbb{P}_{m-1}}$ are $\widehat{u}(0), 2^{-1}\widehat{u}(0), \ldots, 2^{1-m}\widehat{u}(0)$ (see Exercise 1.16).

1.2.3 Linear-Phase Moments and Symmetry Property of Filters

For certain applications, the image of a polynomial under a convolution operation is required to be exactly itself or its translated version. For this purpose, we have the following result:

Lemma 1.2.6 *Let* $u \in l_0(\mathbb{Z})$ *be a finitely supported sequence on* $\mathbb{Z}$. *Let* p *be a polynomial and define* $m := \deg(\mathsf{p}) + 1$. *For a real number* $c \in \mathbb{R}$, *the identity* $\mathsf{p} * u = \mathsf{p}(\cdot - c)$ *holds if and only if* u *has* m linear-phase moments with phase c:

$$\widehat{u}(\xi) = e^{-ic\xi} + \mathscr{O}(|\xi|^m), \qquad \xi \to 0. \qquad (1.2.15)$$

Proof By Lemma 1.2.1, we have (1.2.7). On the other hand, using the Taylor expansion of p, we have

$$\mathsf{p}(x - c) = \sum_{j=0}^{m-1} \mathsf{p}^{(j)}(x)\frac{(-c)^j}{j!} = \sum_{j=0}^{m-1} \frac{(-i)^j}{j!}\mathsf{p}^{(j)}(x)(-ic)^j. \qquad (1.2.16)$$

Comparing the coefficients of $\mathsf{p}^{(j)}, j = 0, \ldots, m-1$ in both (1.2.7) and (1.2.16), we see that $\mathsf{p} * u = \mathsf{p}(\cdot - c)$ if and only if $\widehat{u}^{(j)}(0) = (-ic)^j$ for all $j = 0, \ldots, m-1$, which can be equivalently rewritten as (1.2.15). □

If a filter has linear-phase moments, then the action of the subdivision operator and the transition operator on polynomial spaces has some particular structure.

Proposition 1.2.7 *Let $u \in l_0(\mathbb{Z})$ and $c \in \mathbb{R}$. Then u has m linear-phase moments with phase c if and only if $\mathcal{T}_u \mathsf{p} = 2\mathsf{p}(2\cdot + c)$ for all $\mathsf{p} \in \mathbb{P}_{m-1}$ (or for some polynomial p with $\deg(\mathsf{p}) = m - 1$). Similarly, u has m sum rules and m linear-phase moments with phase c if and only if $\mathcal{S}_u \mathsf{p} = \mathsf{p}(2^{-1}(\cdot - c))$ for all $\mathsf{p} \in \mathbb{P}_{m-1}$ (or for some polynomial p with $\deg(\mathsf{p}) = m - 1$).*

Proof The first part is a direct consequence of (1.2.9) in Theorem 1.2.2 and Lemma 1.2.6. The second part is a direct consequence of (1.2.11) and Theorem 1.2.5. □

We now discuss symmetry property which is desirable in many applications. We say that a filter or a sequence $u = \{u(k)\}_{k\in\mathbb{Z}} : \mathbb{Z} \to \mathbb{C}$ has *symmetry* if

$$u(c - k) = \epsilon u(k), \qquad \forall\, k \in \mathbb{Z} \tag{1.2.17}$$

with $c \in \mathbb{Z}$ and $\epsilon \in \{-1, 1\}$. A filter u is *symmetric* about the point $\frac{c}{2}$ if (1.2.17) holds with $\epsilon = 1$, and *antisymmetric* about the point $\frac{c}{2}$ if (1.2.17) holds with $\epsilon = -1$. We call $\frac{c}{2}$ the *symmetry center* of the filter u, which is simply the center of its filter support $\mathrm{fsupp}(u)$. Recall that $\mathrm{fsupp}(u) = [m, n]$ if u vanishes outside $[m, n]$ and $u(m)u(n) \neq 0$. It is often convenient to use a *symmetry operator* S to record the symmetry type of a filter having symmetry. For this purpose, we define

$$[\mathsf{S}\widehat{u}](\xi) := \frac{\widehat{u}(\xi)}{\widehat{u}(-\xi)}, \qquad \xi \in \mathbb{R}. \tag{1.2.18}$$

Now it is straightforward to see that (1.2.17) holds if and only if $[\mathsf{S}\widehat{u}](\xi) = \epsilon e^{-ic\xi}$. It is easy to see that $[\mathsf{S}\, \widehat{u(\cdot - m)}](\xi) = [\mathsf{S}\widehat{u}](\xi)e^{-i2m\xi}$ for any integer m. Consequently, up to an integer shift, there are essentially four types of symmetries $\mathsf{S}\widehat{u}(\xi) = \epsilon e^{-ic\xi}$ with $c \in \{0, 1\}$ and $\epsilon \in \{-1, 1\}$.

Since in this book we address both real-valued and complex-valued filters, there is a closely related notion of symmetry for complex-valued filters. We say that a filter or a sequence $u = \{u(k)\}_{k\in\mathbb{Z}} : \mathbb{Z} \to \mathbb{C}$ has *complex symmetry* if

$$u(c - k) = \epsilon \overline{u(k)}, \qquad \forall\, k \in \mathbb{Z} \tag{1.2.19}$$

with $c \in \mathbb{Z}$ and $\epsilon \in \{-1, 1\}$. That is, $u^\star(k) = \epsilon u(c + k)$ for all $k \in \mathbb{Z}$. Define a *complex symmetry operator* $\mathbb{S}$ by

$$[\mathbb{S}\widehat{u}](\xi) := \frac{\widehat{u}(\xi)}{\overline{\widehat{u}(\xi)}}, \qquad \xi \in \mathbb{R}. \tag{1.2.20}$$

Then a filter u has complex symmetry in (1.2.19) if and only if $[\mathbb{S}\widehat{u}](\xi) = \epsilon e^{-ic\xi}$. It is trivial to see that a filter u is real-valued if and only if $\overline{\widehat{u}(\xi)} = \widehat{u}(-\xi)$. Therefore, for a real-valued filter u, there is no difference between symmetry and complex symmetry since $\mathsf{S}\widehat{u} = \mathbb{S}\widehat{u}$. We say that u has *essential complex symmetry* if (1.2.19) holds with $c \in \mathbb{Z}$ and $\epsilon \in \mathbb{T} := \{\zeta \in \mathbb{C} : |\zeta| = 1\}$.

In the following, we make some remarks on the relation between linear-phase moments and symmetry. Note that a filter u has one linear-phase moment is equivalent to saying that $\widehat{u}(0) = 1$.

Proposition 1.2.8 *Suppose that $u \in l_0(\mathbb{Z})$ has m but not $m + 1$ linear-phase moments with phase $c \in \mathbb{R}$. If $m > 1$, then the phase c is uniquely determined by u through*

$$c = i\widehat{u}'(0) = \sum_{k\in\mathbb{Z}} u(k)k. \tag{1.2.21}$$

Moreover,

(i) if u has symmetry: $u(c_u - k) = u(k)$ for all $k \in \mathbb{Z}$ for some $c_u \in \mathbb{Z}$, then $c = c_u/2$ (that is, the phase c agrees with the symmetry center $c_u/2$ of u) and m must be an even integer;

(ii) if u has complex symmetry: $u(c_u - k) = \overline{u(k)}$ for all $k \in \mathbb{Z}$ for some $c_u \in \mathbb{Z}$, then $c = c_u/2$.

Proof (1.2.21) follows directly from the definition of linear-phase moments in (1.2.15).

For (i), we have $[\mathsf{S}\widehat{u}](\xi) = e^{-ic_u\xi}$. Then it follows from (1.2.15) that we must have $e^{-ic_u\xi} = \frac{\widehat{u}(\xi)}{\widehat{u}(-\xi)} = e^{-i2c\xi} + \mathscr{O}(|\xi|^m)$ as $\xi \to 0$. Since $m > 1$, we must have $c = c_u/2$. Note that $\mathsf{S}\widehat{u}(\xi) = e^{-ic_u\xi}$ and $c_u = 2c$ imply $\widehat{u}(\xi)e^{ic\xi} = \widehat{u}(-\xi)e^{-ic\xi}$, from which we see that

$$[\widehat{u}(\cdot)e^{ic\cdot}]^{(j)}(0) = 0, \qquad \text{for all positive odd integers } j. \tag{1.2.22}$$

On the other hand, the definition of linear-phase moments in (1.2.15) is equivalent to

$$\widehat{u}(\xi)e^{ic\xi} = 1 + \mathscr{O}(|\xi|^m), \qquad \xi \to 0.$$

Since u has m but not $m+1$ linear-phase moments with phase c, it now follows from (1.2.22) that m must be an even integer.

For (ii), we have $[\mathbb{S}\widehat{u}](\xi) = e^{-ic_u\xi}$. Then it follows from (1.2.15) that we must have $e^{-ic_u\xi} = \frac{\widehat{u}(\xi)}{\overline{\widehat{u}(\xi)}} = e^{-i2c\xi} + \mathscr{O}(|\xi|^m)$ as $\xi \to 0$. Thus, $c = c_u/2$ holds. □

In the following we explain the relation between complex symmetry and linear phase of a filter.

Theorem 1.2.9 *Let $u \in l_0(\mathbb{Z})$ and $\xi_0 \in (-\pi, \pi)$ such that $\widehat{u}(\xi_0) \ne 0$. Write $\widehat{u}(\xi_0) = M_0 e^{-i\theta_0}$ for some $M_0, \theta_0 \in \mathbb{R}$. Then there exist unique real-valued continuous functions $M, \theta : (-\pi, \pi) \to \mathbb{R}$ such that*

$$\widehat{u}(\xi) = M(\xi)e^{-i\theta(\xi)} \quad \forall\, \xi \in (-\pi, \pi) \quad \textit{with} \quad M(\xi_0) = M_0,\ \theta(\xi_0) = \theta_0. \tag{1.2.23}$$

Moreover, the filter u has essential complex symmetry (that is, $e^{id}u$ has complex symmetry for some $d \in \mathbb{R}$) if and only if $\theta(\xi) = c\xi + \lambda, \xi \in (-\pi, \pi)$, where $\lambda \in \mathbb{R}$ and $c = phase(u) := Re(\sum_{k\in\mathbb{Z}} u(k)k)$. In addition, if $\widehat{u}(0) = 1$, then $d = 0$; if u is real-valued, then $d \in \pi\mathbb{Z}$.

Proof We first define a function $m : (-\pi, \pi) \to \mathbb{N}_0$ such that $m(\xi)$ denotes the number of all zeros, counting multiplicity, of $\widehat{u}$ on the open interval between ξ and ξ_0. Define $M : (-\pi, \pi) \to \mathbb{R}$ by

$$M(\xi) = |\widehat{u}(\xi)|(-1)^{m(\xi)}\frac{M_0}{|M_0|}.$$

Then it is pretty straightforward to conclude that M is a real-valued continuous function. Moreover, $\frac{\widehat{u}(\xi)}{M(\xi)}$ is a continuous function on $(-\pi, \pi)$ with all singularities removable. This can be seen as follows. Considering the Taylor expansion of $\widehat{u}$ near ξ_1, we have

$$\widehat{u}(\xi) = C_{\xi_1}(\xi - \xi_1)^n + \mathscr{O}(|\xi - \xi_1|^{n+1}), \qquad \xi \to \xi_1$$

for some $C_{\xi_1} \neq 0$ and $n \in \mathbb{N}_0$. Therefore,

$$\begin{aligned}\frac{\widehat{u}(\xi)}{M(\xi)} &= \frac{|M_0|C_{\xi_1}(\xi - \xi_1)^n}{M_0\,|C_{\xi_1}|\,|\xi - \xi_1|^n(-1)^{m(\xi)}} + \mathscr{O}(|\xi - \xi_1|) \\ &= \frac{C_{\xi_1}}{|C_{\xi_1}|}\frac{|M_0|}{M_0}\frac{(\xi - \xi_1)^n}{|\xi - \xi_1|^n(-1)^{m(\xi)}} + \mathscr{O}(|\xi - \xi_1|),\ \xi \to 0.\end{aligned}$$

By the definition of the function $m(\xi)$, we see that $\frac{(\xi-\xi_1)^n}{|\xi-\xi_1|^n(-1)^{m(\xi)}}$ is a constant function in a neighborhood of ξ_1. Therefore, $\frac{\widehat{u}(\xi)}{M(\xi)}$ is a continuous function at $\xi = \xi_1$ (that is, the singularity at ξ_1 is removable). Consequently, by $|\frac{\widehat{u}(\xi)}{M(\xi)}| = 1$, define $\theta(\xi) = i \ln \frac{\widehat{u}(\xi)}{M(\xi)}, \xi \in (-\pi, \pi)$, where ln denotes a branch of the natural log function with $\theta(\xi_0) = \theta_0$. Therefore, θ is a real-valued continuous function such that (1.2.23) holds.

We now show that a filter u has essential complex symmetry if and only if it has linear phase. Suppose that $e^{id}u$ has complex symmetry for some $d \in \mathbb{R}$, that is, $[\mathbb{S}(e^{id}\widehat{u})](\xi) = \epsilon e^{-i2c\xi}$ for some $\epsilon \in \{-1, 1\}$ and $c \in \frac{1}{2}\mathbb{Z}$. Then $\widehat{u}(\xi)e^{i(c\xi+d)} = \epsilon\overline{\widehat{u}(\xi)e^{i(c\xi+d)}}$. If $\epsilon = 1$, then $\widehat{u}(\xi)e^{i(c\xi+d)} \in \mathbb{R}$ for all $\xi \in (-\pi, \pi)$. Consequently, we must have $M(\xi) = \widehat{u}(\xi)e^{i(c\xi+d)}$ or $-\widehat{u}(\xi)e^{i(c\xi+d)}$ for all $\xi \in (-\pi, \pi)$. This implies that we must have $\theta(\xi) = c\xi + d + \kappa\pi$ for all $\xi \in (-\pi, \pi)$ for some integer $\kappa \in \mathbb{Z}$. Hence, θ is a linear function on $(-\pi, \pi)$. If $\epsilon = -1$, then $i\widehat{u}(\xi)e^{i(c\xi+d)} \in \mathbb{R}$ and we have $M(\xi) = i\widehat{u}(\xi)e^{i(c\xi+d)}$ or $-i\widehat{u}(\xi)e^{i(c\xi+d)}$ for all $\xi \in (-\pi, \pi)$. A similar argument shows that $\theta(\xi) = c\xi + d + \frac{\pi}{2} + \kappa\pi$ for all $\xi \in (-\pi, \pi)$ for some integer κ. Hence,

θ is a linear function on $(-\pi, \pi)$. Therefore, if a filter u has essential complex symmetry, then it must have linear phase on $(-\pi, \pi)$.

Conversely, suppose that θ is a linear function. Then $\theta(\xi) = c\xi + \lambda$ for some $c, \lambda \in \mathbb{R}$. Since both M and θ are real-valued, we deduce that

$$\mathbb{S}\widehat{u}(\xi) = \frac{\widehat{u}(\xi)}{\overline{\widehat{u}(\xi)}} = \frac{M(\xi)e^{-i\theta(\xi)}}{M(\xi)e^{i\theta(\xi)}} = e^{-i2\theta(\xi)} = e^{-i2(c\xi+\lambda)}.$$

The above identity implies that $e^{i\lambda}u$ has complex symmetry $\mathbb{S}(e^{i\lambda}\widehat{u}) = e^{-i2c\xi}$. Thus, if the phase θ of u is a linear function on $(-\pi, \pi)$, then the filter u must have essential complex symmetry. □

If $M_0 > 0$ and $\widehat{u}(\xi) \neq 0$ for all $\xi \in (-\pi, \pi)$, from the proof of Theorem 1.2.9 we see that $M(\xi) = |\widehat{u}(\xi)|$ for all $\xi \in (-\pi, \pi)$. For filters u such that $\widehat{u}(0) \in \mathbb{R}\backslash\{0\}$, without further mention in this book, we always take $\xi_0 = 0, M_0 = \widehat{u}(0)$, and $\theta_0 = 0$ in Theorem 1.2.9. We call $M(\xi)$ the default magnitude function of $\widehat{u}$ and $\theta(\xi)$ the default phase function of $\widehat{u}$. For this case, u has complex symmetry if and only if u has linear phase $\theta(\xi) = c\xi$ for $\xi \in (-\pi, \pi)$.

1.2.4 An Example

We complete this section by presenting an example. According to item (5) of Theorem 1.2.5, a natural filter having m sum rules and the shortest possible filter support is

$$\widehat{a_m^B}(\xi) := 2^{-m}(1 + e^{-i\xi})^m, \qquad m \in \mathbb{N}, \tag{1.2.24}$$

which is called the *B-spline filter (or mask)* of order m in the literature of wavelet analysis and approximation theory. Note that $\widehat{a_m^B}(\xi) = 2^{-m}\sum_{j=0}^{m}\binom{m}{j}e^{-ij\xi}$ and a_m^B has the symmetry type $[\mathsf{S}\,\widehat{a_m^B}](\xi) = e^{-im\xi}$, where

$$\binom{m}{0} := 1, \qquad \binom{m}{j} := \frac{m!}{j!(m-j)!} \quad \text{with} \quad j! := 1\cdot 2\cdots(j-1)j. \tag{1.2.25}$$

Hence, we have

$$\widehat{a_m^B}^{(n)}(0) = (-i)^n 2^{-m}\sum_{j=0}^{m}\binom{m}{j}j^n, \qquad n \in \mathbb{N}_0.$$

Let us consider $a_4^B = \{\underline{\tfrac{1}{16}}, \tfrac{1}{4}, \tfrac{3}{8}, \tfrac{1}{4}, \tfrac{1}{16}\}_{[0,4]}$. Then

$$\widehat{a_4^B}(0) = 1, \quad \widehat{a_4^B}'(0) = -2i, \quad \widehat{a_4^B}''(0) = -5, \quad \widehat{a_4^B}'''(0) = 14i.$$

For any $\mathsf{p} \in \mathbb{P}_3$, by Lemma 1.2.1 and (1.2.5), we have

$$[\mathsf{p} * a_4^B](x) = \mathsf{p}(x) - 2\mathsf{p}'(x) + \frac{5}{2}\mathsf{p}''(x) - \frac{7}{3}\mathsf{p}'''(x).$$

By Theorem 1.2.2,

$$[\mathcal{T}_{a_4^B}\mathsf{p}](x) = 2\mathsf{p}(2x) + 4\mathsf{p}'(2x) + 5\mathsf{p}''(2x) + \frac{14}{3}\mathsf{p}'''(2x).$$

By Theorem 1.2.4 and (1.2.11),

$$[\mathcal{S}_{a_4^B}\mathsf{p}](x) = \mathsf{p}(x/2) - \mathsf{p}'(x/2) + \frac{5}{8}\mathsf{p}''(x/2) - \frac{7}{24}\mathsf{p}'''(x/2).$$

Let $c := \sum_{k\in\mathbb{Z}} a_4^B(k)k = 2$, which is also the symmetry center of a_4^B. Then a_4^B has no more than two linear-phase moments with the phase $c = 2$. Moreover, for $\mathsf{p}(x) = t_0 + t_1 x$ with $t_0, t_1 \in \mathbb{C}$, we have

$$[\mathsf{p} * a_4^B](x) = (t_0 + t_1 x) - 2t_1 = (t_0 - 2t_1) + t_1 x = \mathsf{p}(\cdot - 2),$$
$$[\mathcal{T}_{a_4^B}\mathsf{p}](x) = 2(t_0 + 2t_1 x) + 4t_1 = (2t_0 + 4t_1) + 4t_1 x = 2\mathsf{p}(2x + 2),$$
$$[\mathcal{S}_{a_4^B}\mathsf{p}](x) = (t_0 + t_1 x/2) - t_1 = (t_0 - t_1) + t_1 x/2 = \mathsf{p}(2^{-1}(x-2)) = \mathsf{p}(2^{-1}x - 1).$$

1.3 Multilevel Discrete Framelet Transforms and Stability

To extract the multiscale structure embedded in signals, a multilevel discrete framelet transform is used in applications by recursively applying one-level discrete framelet transforms on selected sequences of framelet coefficients at the immediate higher scale level. In this section we discuss a (standard) multilevel discrete framelet transform, study its stability in the space $l_2(\mathbb{Z})$, and introduce the notion of discrete affine systems in $l_2(\mathbb{Z})$.

1.3.1 *Multilevel Discrete Framelet Transforms*

A standard multilevel discrete framelet transform is obtained by recursively performing one-level discrete framelet transforms on only one selected sequence of framelet coefficients. Certainly one may select several or even all the sequences of framelet coefficients for further decomposition, but we have more or less the same algorithm as we shall see in Sect. 1.3.4. The framelet coefficients and their associated filters in such selected sequences for further decomposition are called *parent (or low-pass) framelet coefficients* and *parent (or low-pass) filters or masks* (since they are often low-pass filters), respectively. In this book we use a or its indexed version to denote a low-pass (or parent) filter and use v or its indexed version to denote low-pass (or parent) framelet coefficients. The framelet coefficients and their associated filters in other not-selected sequences for decomposition are called *child (or high-pass) framelet coefficients* and *child (or high-pass) filters* (since they are often high-pass filters), respectively. In this book we use b or its indexed version to denote a high-pass (or child) filter and use w or its indexed version to denote high-pass (or child) framelet coefficients.

Readers may notice that our definitions of low-pass and high-pass filters are different from those used in the literature of engineering, where a low-pass filter u means $\widehat{u}(0) \neq 0$ and $\widehat{u}(\pi) = 0$ while a high-pass filer v means $\widehat{v}(0) = 0$ and $\widehat{v}(\pi) \neq 0$. Such a purposely misuse of the notion of low-pass and high-pass filters in this book is not serious but convenient for our discussion, since quite often only wavelet coefficients associated with low-pass filters in the sense of engineering are selected for further decomposition.

Let $\tilde{a}, \tilde{b}_1, \ldots, \tilde{b}_s$ be filters for decomposition. For a positive integer J, *a J-level discrete framelet decomposition* is given by

$$v_j := \tfrac{\sqrt{2}}{2}\mathcal{T}_{\tilde{a}}v_{j-1}, \quad w_{\ell,j} := \tfrac{\sqrt{2}}{2}\mathcal{T}_{\tilde{b}_\ell}v_{j-1}, \qquad \ell = 1,\ldots,s, \quad j = 1,\ldots,J, \tag{1.3.1}$$

where $v_0 : \mathbb{Z} \to \mathbb{C}$ is an input signal. The filter $\tilde{a}$ is often called a dual low-pass filter and the filters $\tilde{b}_1, \ldots, \tilde{b}_s$ are called dual high-pass filters. After a J-level discrete framelet decomposition, the original input signal v_0 is decomposed into one sequence v_J of low-pass framelet coefficients and sJ sequences $w_{\ell,j}$ of high-pass framelet coefficients for $\ell = 1, \ldots, s$ and $j = 1, \ldots, J$. Such framelet coefficients are often processed for various purposes. One of the most commonly employed operations is thresholding so that the low-pass framelet coefficients v_J and high-pass framelet coefficients $w_{\ell,j}$ become $\mathring{v}_J$ and $\mathring{w}_{\ell,j}$, respectively. More precisely, $\mathring{w}_{\ell,j}(k) = \eta(w_{\ell,j}(k)), k \in \mathbb{Z}$, where $\eta : \mathbb{C} \to \mathbb{C}$ is a thresholding function. For example, for a given threshold value $\lambda > 0$, the hard thresholding function η_λ^{hard} and soft-thresholding function η_λ^{soft} are defined to be

$$\eta_\lambda^{hard}(z) = \begin{cases} z, & \text{if } |z| \geqslant \lambda; \\ 0, & \text{otherwise} \end{cases} \quad \text{and} \quad \eta_\lambda^{soft}(z) = \begin{cases} z - \lambda\frac{z}{|z|}, & \text{if } |z| \geqslant \lambda; \\ 0, & \text{otherwise.} \end{cases} \tag{1.3.2}$$

Quantization is another commonly employed operation after or without thresholding. For example, for a given quantization level $q > 0$, the quantization function

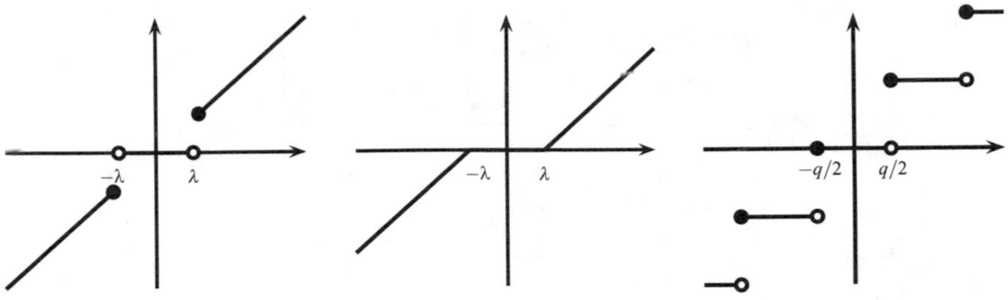

Fig. 1.2 The hard thresholding function η_λ^{hard}, the soft thresholding function η_λ^{soft}, and the quantization function, respectively. Both thresholding and quantization operations are often used to process framelet coefficients in a discrete framelet transform

$\mathcal{Q} : \mathbb{R} \to q\mathbb{Z}$ is defined to be $\mathcal{Q}(x) := q\lfloor \frac{x}{q} + \frac{1}{2} \rfloor$, $x \in \mathbb{R}$, where $\lfloor \cdot \rfloor$ is the floor function such that $\lfloor x \rfloor = n$ if $n \leqslant x < n+1$ for an integer n. See Fig. 1.2 for illustration.

Let $a, b_1, \dots, b_s$ be filters for reconstruction. Now *a J-level discrete framelet reconstruction* is

$$\mathring{v}_{j-1} := \frac{\sqrt{2}}{2}\mathcal{S}_a \mathring{v}_j + \frac{\sqrt{2}}{2}\sum_{\ell=1}^{s} \mathcal{S}_{b_\ell} \mathring{w}_{\ell,j}, \qquad j = J, \dots, 1. \tag{1.3.3}$$

The filter a is often called a primal low-pass filter and the filters $b_1, \dots, b_s$ are called primal high-pass filters. To analyze a multilevel discrete framelet transform, we rewrite the J-level discrete framelet decomposition employing the filter bank $\{\tilde{a}; \tilde{b}_1, \dots, \tilde{b}_s\}$ by using *a J-level discrete framelet analysis operator* $\widetilde{\mathcal{W}}_J : l(\mathbb{Z}) \to (l(\mathbb{Z}))^{1\times(sJ+1)}$ as follows:

$$\widetilde{\mathcal{W}}_J v_0 := (w_{1,1}, \dots, w_{s,1}, \dots, w_{1,J}, \dots, w_{s,J}, v_J), \tag{1.3.4}$$

where $w_{\ell,j}$ and v_J are defined in (1.3.1). Similarly, *a J-level discrete framelet synthesis operator* $\mathcal{V}_J : (l(\mathbb{Z}))^{1\times(sJ+1)} \to l(\mathbb{Z})$ employing the filter bank $\{a; b_1, \dots, b_s\}$ is defined by

$$\mathcal{V}_J(\mathring{w}_{1,1}, \dots, \mathring{w}_{s,1}, \dots, \mathring{w}_{1,J}, \dots, \mathring{w}_{s,J}, \mathring{v}_J) = \mathring{v}_0, \tag{1.3.5}$$

where $\mathring{v}_0$ is computed via the recursive formulas in (1.3.3). Note that

$$\widetilde{\mathcal{W}}_J = (\mathrm{Id}_{(l(\mathbb{Z}))^{1\times s(J-1)}} \otimes \widetilde{\mathcal{W}}) \cdots (\mathrm{Id}_{(l(\mathbb{Z}))^{1\times s}} \otimes \widetilde{\mathcal{W}})\widetilde{\mathcal{W}}$$

and

$$\mathcal{V}_J = \mathcal{V}(\mathrm{Id}_{(l(\mathbb{Z}))^{1\times s}} \otimes \mathcal{V}) \cdots (\mathrm{Id}_{(l(\mathbb{Z}))^{1\times s(J-1)}} \otimes \mathcal{V}).$$

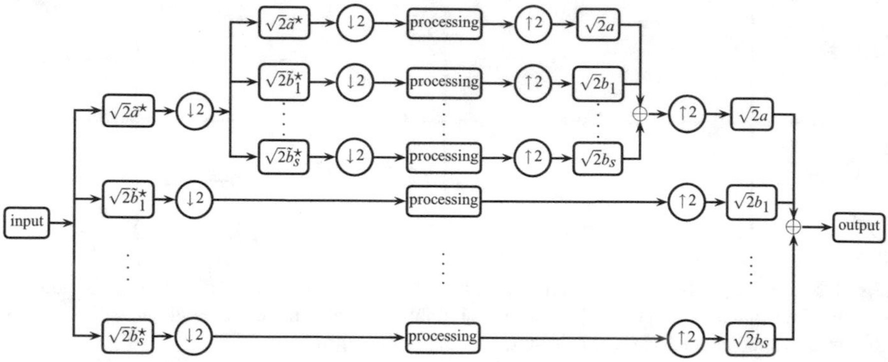

Fig. 1.3 Diagram of a two-level discrete framelet transform employing filter banks $\{\tilde{a}; \tilde{b}_1, \ldots, \tilde{b}_s\}$ and $\{a; b_1, \ldots, b_s\}$. Note that $\frac{\sqrt{2}}{2}\mathcal{T}_{\tilde{b}_\ell} v = \sqrt{2}(v * \tilde{b}_\ell^\star) \downarrow 2$ and $\frac{\sqrt{2}}{2}\mathcal{S}_{b_\ell} v = \sqrt{2}(v \uparrow 2) * b_\ell$ for $\ell = 1, \ldots, s$

Due to (1.3.1) and (1.3.3), a multilevel discrete framelet transform implemented using the recursive cascade structure is often called a *fast framelet transform (FFrT)*. Due to Proposition 1.1.2, a fast framelet transform with $s = 1$ is called a *fast wavelet transform (FWT)*. We shall denote a J-level discrete framelet analysis operator employing the filter bank $\{a; b_1, \ldots, b_s\}$ by $\mathcal{W}_J$ and a J-level discrete framelet synthesis operator employing the filter bank $\{\tilde{a}; \tilde{b}_1, \ldots, \tilde{b}_s\}$ by $\widetilde{\mathcal{V}}_J$. When $J = 1$, these operators become the analysis $\mathcal{W}$ and synthesis operators $\mathcal{V}$ that we discussed in Sect. 1.1.

If $(\{\tilde{a}; \tilde{b}_1, \ldots, \tilde{b}_s\}, \{a; b_1, \ldots, b_s\})$ is a dual framelet filter bank, i.e., it satisfies the perfect reconstruction condition:

$$\begin{bmatrix} \widehat{\tilde{a}}(\xi) & \widehat{\tilde{b}_1}(\xi) & \cdots & \widehat{\tilde{b}_s}(\xi) \\ \widehat{\tilde{a}}(\xi+\pi) & \widehat{\tilde{b}_1}(\xi+\pi) & \cdots & \widehat{\tilde{b}_s}(\xi+\pi) \end{bmatrix} \begin{bmatrix} \widehat{a}(\xi) & \widehat{b_1}(\xi) & \cdots & \widehat{b_s}(\xi) \\ \widehat{a}(\xi+\pi) & \widehat{b_1}(\xi+\pi) & \cdots & \widehat{b_s}(\xi+\pi) \end{bmatrix}^\star = I_2, \tag{1.3.6}$$

then Theorem 1.1.1 tells us that $\mathcal{V}_J \widetilde{\mathcal{W}}_J = \mathrm{Id}_{l(\mathbb{Z})}$ for all $J \in \mathbb{N}$, that is, all the J-level discrete framelet transforms have the perfect reconstruction property. Observe that $\mathcal{W}_J^\star = \mathcal{V}_J$ and $\widetilde{\mathcal{W}}_J^\star = \widetilde{\mathcal{V}}_J$. Thus $\mathcal{V}_J \widetilde{\mathcal{W}}_J = \mathrm{Id}_{l(\mathbb{Z})}$ if and only if $\widetilde{\mathcal{V}}_J \mathcal{W}_J = \mathrm{Id}_{l(\mathbb{Z})}$. See Fig. 1.3 for a diagram of a 2-level discrete framelet transform with a pair of filter banks $\{\tilde{a}; \tilde{b}_1, \ldots, \tilde{b}_s\}$ and $\{a; b_1, \ldots, b_s\}$.

1.3.2 Stability of Multilevel Discrete Framelet Transforms

In this section all our input signals and domains of the analysis/synthesis operators are from the space $l_2(\mathbb{Z})$. A key property of a multilevel framelet transform is its stability. A filter bank $\{a; b_1, \ldots, b_s\}$ is said to have *stability* in $l_2(\mathbb{Z})$ if there exist positive constants C_1 and C_2 such that

$$C_1\|v\|^2_{l_2(\mathbb{Z})} \leqslant \|\mathcal{W}_J v\|^2_{(l_2(\mathbb{Z}))^{1\times(sJ+1)}} \leqslant C_2\|v\|^2_{l_2(\mathbb{Z})}, \qquad \forall\, v \in l_2(\mathbb{Z}), J \in \mathbb{N}. \tag{1.3.7}$$

A filter bank $\{a; b_1, \dots, b_s\}$ having stability in $l_2(\mathbb{Z})$ is called *a framelet filter bank* in $l_2(\mathbb{Z})$. The inequalities (1.3.7) imply $\|\mathcal{W}_J\|^2 \leqslant C_2$ for all $J \in \mathbb{N}$. By (1.3.7), the l_2-norm of framelet coefficients provides an equivalent norm for the sequence space $l_2(\mathbb{Z})$.

For stability of a multilevel discrete framelet transform, we have

Theorem 1.3.1 *Let $(\{\tilde{a}; \tilde{b}_1, \dots, \tilde{b}_s\}, \{a; b_1, \dots, b_s\})$ be a dual framelet filter bank. Let $\mathcal{W}_J, \widetilde{\mathcal{W}}_J$ be its associated J-level discrete framelet analysis operators and $\mathcal{V}_J, \widetilde{\mathcal{V}}_J$ be its associated J-level discrete framelet synthesis operators. Let C_1, C_2 be positive numbers. Then the following statements are equivalent:*

(1) Both filter banks $\{\tilde{a}; \tilde{b}_1, \dots, \tilde{b}_s\}$ and $\{a; b_1, \dots, b_s\}$ have stability in the space $l_2(\mathbb{Z})$ satisfying (1.3.7) *and*

$$C_2^{-1}\|v\|^2_{l_2(\mathbb{Z})} \leqslant \|\widetilde{\mathcal{W}}_J v\|^2_{(l_2(\mathbb{Z}))^{1\times(sJ+1)}} \leqslant C_1^{-1}\|v\|^2_{l_2(\mathbb{Z})}, \qquad \forall\, v \in l_2(\mathbb{Z}), J \in \mathbb{N}. \tag{1.3.8}$$

(2) $\|\mathcal{W}_J\|^2 \leqslant C_2$ and $\|\widetilde{\mathcal{W}}_J\|^2 \leqslant C_1^{-1}$ for all $J \in \mathbb{N}$.
(3) $\|\mathcal{V}_J\|^2 \leqslant C_2$ and $\|\widetilde{\mathcal{V}}_J\|^2 \leqslant C_1^{-1}$ for all $J \in \mathbb{N}$.
(4) $\|\mathcal{V}_J\|^2 \leqslant C_2$ and $\|\widetilde{\mathcal{W}}_J\|^2 \leqslant C_1^{-1}$ for all $J \in \mathbb{N}$.
(5) $\|\mathcal{W}_J\|^2 \leqslant C_2$ and $\|\widetilde{\mathcal{V}}_J\|^2 \leqslant C_1^{-1}$ for all $J \in \mathbb{N}$.

If in addition $s = 1$, then each of the above statements is further equivalent to

(6) For all $w \in (l_2(\mathbb{Z}))^{1\times(sJ+1)}$ and $J \in \mathbb{N}$,

$$C_1\|w\|^2_{(l_2(\mathbb{Z}))^{1\times(sJ+1)}} \leqslant \|\mathcal{V}_J w\|^2_{l_2(\mathbb{Z})} \leqslant C_2\|w\|^2_{(l_2(\mathbb{Z}))^{1\times(sJ+1)}}, \tag{1.3.9}$$

$$C_2^{-1}\|w\|^2_{(l_2(\mathbb{Z}))^{1\times(sJ+1)}} \leqslant \|\widetilde{\mathcal{V}}_J w\|^2_{l_2(\mathbb{Z})} \leqslant C_1^{-1}\|w\|^2_{(l_2(\mathbb{Z}))^{1\times(sJ+1)}}. \tag{1.3.10}$$

Proof Note that

$$\mathcal{V}_J = \mathcal{W}_J^\star, \quad \widetilde{\mathcal{V}}_J = \widetilde{\mathcal{W}}_J^\star, \quad \|\mathcal{W}_J^\star\| = \|\mathcal{W}_J\|, \quad \|\widetilde{\mathcal{W}}_J^\star\| = \|\widetilde{\mathcal{W}}_J\|.$$

We trivially have (1)$\Longrightarrow$(2) $\iff$ (3) $\iff$ (4) $\iff$ (5). We now prove that (2) and (3) together imply (1). Since $(\{\tilde{a}; \tilde{b}_1, \dots, \tilde{b}_s\}, \{a; b_1, \dots, b_s\})$ is a dual framelet filter bank, we have $\widetilde{\mathcal{V}}_J\mathcal{W}_J = \mathcal{V}_J\widetilde{\mathcal{W}}_J = \mathrm{Id}_{l_2(\mathbb{Z})}$. By item (3),

$$\|v\|^2_{l_2(\mathbb{Z})} = \|\widetilde{\mathcal{V}}_J\mathcal{W}_J v\|^2_{l_2(\mathbb{Z})} \leqslant \|\widetilde{\mathcal{V}}_J\|^2\|\mathcal{W}_J v\|^2_{l_2(\mathbb{Z})} \leqslant C_1^{-1}\|\mathcal{W}_J v\|^2_{(l_2(\mathbb{Z}))^{1\times(sJ+1)}},$$

which is simply the left-hand inequality of (1.3.7). By item (2), (1.3.7) holds. The inequalities in (1.3.8) can be proved similarly.

Since $v = \mathcal{V}_J\widetilde{\mathcal{W}}_J v$, replacing $\|v\|_{l_2(\mathbb{Z})}$ in (1.3.8) by $\|\mathcal{V}_J\widetilde{\mathcal{W}}_J v\|_{l_2(\mathbb{Z})}$, we deduce that

$$C_1\|\widetilde{\mathcal{W}}_J v\|^2_{(l_2(\mathbb{Z}))^{1\times(sJ+1)}} \leqslant \|\mathcal{V}_J\widetilde{\mathcal{W}}_J v\|^2_{l_2(\mathbb{Z})} \leqslant C_2\|\widetilde{\mathcal{W}}_J v\|^2_{(l_2(\mathbb{Z}))^{1\times(sJ+1)}}. \qquad (1.3.11)$$

If $s = 1$, by Proposition 1.1.2, then $\widetilde{\mathcal{W}}_J$ is onto and hence, (1.3.9) follows directly from (1.3.11). The inequalities in (1.3.10) can be proved similarly. □

For $\mathcal{V}_J$ and $\widetilde{\mathcal{V}}_J$, generally we can only have (1.3.11) and its duality part by replacing $\widetilde{\mathcal{W}}_J$ and $\mathcal{V}_J$ in (1.3.11) with $\mathcal{W}_J$ and $\widetilde{\mathcal{V}}_J$, respectively. For $s > 1$, both (1.3.9) and (1.3.10) cannot hold, since by Proposition 1.1.2, there exists $w \in l_0(\mathbb{Z})\backslash\{0\}$ such that $\mathcal{V}_J w = 0$. The stability of a multilevel discrete framelet transform implies that a small change of an input signal v induces a small change of all framelet coefficients, and a small perturbation of all framelet coefficients results in a small perturbation of a reconstructed signal. The notion of stability of a multilevel discrete framelet transform can be extended to other (weighted) sequence spaces and is closely related to refinable functions. The study of stability of multilevel discrete framelet transforms is an important part of mathematical analysis of wavelets and framelets. We shall devote Chaps. 5 and 6 of this book to address such issues.

1.3.3 *Discrete Affine Systems in $l_2(\mathbb{Z})$*

In this section we shall introduce the notion of discrete affine systems. A multilevel discrete framelet transform can be fully expressed through discrete affine systems. To do so, let us first generalize the definition of the subdivision operator and the transition operator. For a nonzero integer d and a finitely supported sequence u, the subdivision operator $\mathcal{S}_{u,\mathsf{d}} : l(\mathbb{Z}) \to l(\mathbb{Z})$ and the transition operator $\mathcal{T}_{u,\mathsf{d}} : l(\mathbb{Z}) \to l(\mathbb{Z})$ are defined to be

$$[\mathcal{S}_{u,\mathsf{d}}v](n) := |\mathsf{d}|\sum_{k\in\mathbb{Z}} v(k)u(n-\mathsf{d}k), \qquad n \in \mathbb{Z}, \qquad (1.3.12)$$

$$[\mathcal{T}_{u,\mathsf{d}}v](n) := |\mathsf{d}|\sum_{k\in\mathbb{Z}} v(k)\overline{u(k-\mathsf{d}n)} = |\mathsf{d}|\langle v, u(\cdot-\mathsf{d}n)\rangle, \qquad n \in \mathbb{Z} \qquad (1.3.13)$$

for $v \in l(\mathbb{Z})$. For $v \in l_0(\mathbb{Z})$, one can check that $\widehat{\mathcal{S}_{u,\mathsf{d}}v}(\xi) = |\mathsf{d}|\widehat{v}(\mathsf{d}\xi)\widehat{u}(\xi)$ and

$$\widehat{\mathcal{T}_{u,\mathsf{d}}v}(\xi) = \sum_{\gamma=0}^{|\mathsf{d}|-1} \widehat{v}(\tfrac{\xi+2\pi\gamma}{\mathsf{d}})\overline{\widehat{u}(\tfrac{\xi+2\pi\gamma}{\mathsf{d}})}.$$

Moreover, $\mathcal{S}_{u,\mathsf{d}}v = |\mathsf{d}|u * (v \uparrow \mathsf{d})$, $\mathcal{T}_{u,\mathsf{d}}v = |\mathsf{d}|(u^\star * v) \downarrow \mathsf{d}$, $\mathcal{S}_{u,\mathsf{d}}^\star = \mathcal{T}_{u,\mathsf{d}}$, and

$$\langle \mathcal{S}_{u,\mathsf{d}}v, w\rangle = \langle \mathcal{T}_{u,\mathsf{d}}^\star v, w\rangle = \langle v, \mathcal{T}_{u,\mathsf{d}}w\rangle, \qquad \forall\, v, w \in l_2(\mathbb{Z}). \tag{1.3.14}$$

To understand the operators $\widetilde{\mathcal{W}}_J$ and $\mathcal{V}_J$ in a J-level discrete framelet transform, we need the following auxiliary result.

Lemma 1.3.2 *For* $\mathsf{d}_1, \mathsf{d}_2 \in \mathbb{Z}\backslash\{0\}$ *and* $u_1, u_2 \in l_0(\mathbb{Z})$,

$$\mathcal{S}_{u_1,\mathsf{d}_1}\mathcal{S}_{u_2,\mathsf{d}_2}v = \mathcal{S}_{u_1*(u_2\uparrow \mathsf{d}_1),\mathsf{d}_1\mathsf{d}_2}v = |\mathsf{d}_1\mathsf{d}_2|u_1 * (u_2 \uparrow \mathsf{d}_1) * (v \uparrow \mathsf{d}_1\mathsf{d}_2) \tag{1.3.15}$$

and

$$\mathcal{T}_{u_2,\mathsf{d}_2}\mathcal{T}_{u_1,\mathsf{d}_1}v = \mathcal{T}_{u_1*(u_2\uparrow \mathsf{d}_1),\mathsf{d}_1\mathsf{d}_2}v = |\mathsf{d}_1\mathsf{d}_2|(u_1^\star * (u_2^\star \uparrow \mathsf{d}_1) * v) \downarrow \mathsf{d}_1\mathsf{d}_2. \tag{1.3.16}$$

Proof By $\widehat{\mathcal{S}_{u,\mathsf{d}}v}(\xi) = |\mathsf{d}|\widehat{v}(\mathsf{d}\xi)\widehat{u}(\xi)$, the Fourier series of the sequence $\mathcal{S}_{u_1,\mathsf{d}_1}\mathcal{S}_{u_2,\mathsf{d}_2}v$ is

$$|\mathsf{d}_1|\widehat{u_1}(\xi)\widehat{\mathcal{S}_{u_2,\mathsf{d}_2}v}(\mathsf{d}_1\xi) = |\mathsf{d}_1\mathsf{d}_2|\widehat{u_1}(\xi)\widehat{u_2}(\mathsf{d}_1\xi)\widehat{v}(\mathsf{d}_1\mathsf{d}_2\xi) = |\mathsf{d}_2|\widehat{\mathcal{S}_{u_1,\mathsf{d}_1}u_2}(\xi)\widehat{v}(\mathsf{d}_1\mathsf{d}_2\xi).$$

Therefore, (1.3.15) holds. By duality in (1.3.14) and (1.3.15), we have

$$\begin{aligned}\langle w, \mathcal{T}_{u_2,\mathsf{d}_2}\mathcal{T}_{u_1,\mathsf{d}_1}v\rangle &= \langle \mathcal{S}_{u_2,\mathsf{d}_2}w, \mathcal{T}_{u_1,\mathsf{d}_1}v\rangle = \langle \mathcal{S}_{u_1,\mathsf{d}_1}\mathcal{S}_{u_2,\mathsf{d}_2}w, v\rangle\\ &= \langle \mathcal{S}_{u_1*(u_2\uparrow \mathsf{d}_1),\mathsf{d}_1\mathsf{d}_2}w, v\rangle = \langle w, \mathcal{T}_{u_1*(u_2\uparrow \mathsf{d}_1),\mathsf{d}_1\mathsf{d}_2}v\rangle,\end{aligned}$$

from which we see that (1.3.16) holds. The identities in (1.3.15) and (1.3.16) can also be seen as follows:

$$\begin{aligned}\mathcal{S}_{u_1,\mathsf{d}_1}\mathcal{S}_{u_2,\mathsf{d}_2}v &= |\mathsf{d}_1\mathsf{d}_2|u_1 * ((u_2 * (v \uparrow \mathsf{d}_2)) \uparrow \mathsf{d}_1) = |\mathsf{d}_1\mathsf{d}_2|u_1 * (u_2 \uparrow \mathsf{d}_1) * (v \uparrow \mathsf{d}_1\mathsf{d}_2),\\ \mathcal{T}_{u_2,\mathsf{d}_2}\mathcal{T}_{u_1,\mathsf{d}_1}v &= |\mathsf{d}_1\mathsf{d}_2|(u_2^\star * ((u_1^\star * v) \downarrow \mathsf{d}_1)) \downarrow \mathsf{d}_2 = |\mathsf{d}_1\mathsf{d}_2|(u_1^\star * (u_2 \uparrow \mathsf{d}_1) * v) \downarrow \mathsf{d}_1\mathsf{d}_2,\end{aligned}$$

where we used $(u * v) \uparrow \mathsf{d} = (u \uparrow \mathsf{d}) * (v \uparrow \mathsf{d})$ and $u * (v \downarrow \mathsf{d}) = ((u \uparrow \mathsf{d}) * v) \downarrow \mathsf{d}$. □

Define filters $a_j, \tilde{a}_j, b_{\ell,j}, \tilde{b}_{\ell,j}$ with $j \in \mathbb{N}_0$ by

$$\widehat{a_j}(\xi) := \widehat{a}(\xi)\widehat{a}(2\xi)\cdots\widehat{a}(2^{j-1}\xi), \qquad \widehat{\tilde{a}_j}(\xi) := \widehat{\tilde{a}}(\xi)\widehat{\tilde{a}}(2\xi)\cdots\widehat{\tilde{a}}(2^{j-1}\xi), \tag{1.3.17}$$

$$\begin{aligned}\widehat{b_{\ell,j}}(\xi) &:= \widehat{a}(\xi)\widehat{a}(2\xi)\cdots\widehat{a}(2^{j-2}\xi)\widehat{b_\ell}(2^{j-1}\xi),\\ \widehat{\tilde{b}_{\ell,j}}(\xi) &:= \widehat{\tilde{a}}(\xi)\widehat{\tilde{a}}(2\xi)\cdots\widehat{\tilde{a}}(2^{j-2}\xi)\widehat{\tilde{b}_\ell}(2^{j-1}\xi)\end{aligned}$$

with the convention that $a_0 = \tilde{a}_0 = b_{\ell,0} = \tilde{b}_{\ell,0} := \boldsymbol{\delta}$. In other words,

$$a_j = a * (a \uparrow 2) * \cdots * (a \uparrow 2^{j-1}) \quad \text{and} \quad \tilde{a}_j := \tilde{a} * (\tilde{a} \uparrow 2) * \cdots * (\tilde{a} \uparrow 2^{j-1}).$$

From the definition of framelet coefficients $w_{\ell,j}$ in (1.3.1), noting that $\mathcal{T}_{\tilde{u}} = \mathcal{T}_{\tilde{u},2}$ and $v_0 = v$, we see that

$$v_j = \tfrac{\sqrt{2}}{2}\mathcal{T}_{\tilde{a}} v_{j-1} = \cdots = (\tfrac{\sqrt{2}}{2})^j \mathcal{T}_{\tilde{a}}^j v_0 = (\tfrac{\sqrt{2}}{2})^j \mathcal{T}_{\tilde{a}*(\tilde{a}\uparrow 2)*\cdots*(\tilde{a}\uparrow 2^{j-1}),2^j} v = \langle v, \tilde{a}_{j;\cdot}\rangle$$

and

$$\begin{aligned} w_{\ell,j} &= \tfrac{\sqrt{2}}{2}\mathcal{T}_{\tilde{b}_\ell} v_{j-1} = (\tfrac{\sqrt{2}}{2})^j \mathcal{T}_{\tilde{b}_\ell} \mathcal{T}_{\tilde{a}}^{j-1} v_0 \\ &= (\tfrac{\sqrt{2}}{2})^j \mathcal{T}_{\tilde{a}*(\tilde{a}\uparrow 2)*\cdots*(\tilde{a}\uparrow 2^{j-2})*(\tilde{b}_\ell \uparrow 2^{j-1}),2^j} v = \langle v, \tilde{b}_{\ell,j;\cdot}\rangle, \end{aligned}$$

where $\tilde{a}_{j;k}$ and $\tilde{b}_{\ell,j;k}$ are defined to be

$$\tilde{a}_{j;k} := 2^{j/2}\tilde{a}_j(\cdot - 2^j k), \quad \tilde{b}_{\ell,j;k} := 2^{j/2}\tilde{b}_{\ell,j}(\cdot - 2^j k), \qquad j \in \mathbb{N}_0, k \in \mathbb{Z}.$$

Similarly, we deduce that

$$\mathcal{V}_J(0,\ldots,0,v_J) = (\tfrac{\sqrt{2}}{2})^J \mathcal{S}_a^J v_J = (\tfrac{\sqrt{2}}{2})^J \mathcal{S}_{a*(a\uparrow 2)*\cdots*(a\uparrow 2^{J-1}),2^J} v_J = \sum_{k\in\mathbb{Z}} v_J(k) a_{J;k}$$

and

$$\begin{aligned} \mathcal{V}_J(0,\ldots,0,w_{\ell,j},0,\ldots,0) &= (\tfrac{\sqrt{2}}{2})^j \mathcal{S}_a^{j-1}\mathcal{S}_{b_\ell} w_{\ell,j} \\ &= (\tfrac{\sqrt{2}}{2})^j \mathcal{S}_{a*(\tilde{a}\uparrow 2)*\cdots*(a\uparrow 2^{j-2})*(b_\ell\uparrow 2^{j-1}),2^j} w_{\ell,j} = \sum_{k\in\mathbb{Z}} w_{\ell,j}(k) b_{\ell,j;k}, \end{aligned}$$

where $a_{j;k}$ and $b_{\ell,j;k}$ are defined to be

$$a_{j;k} := 2^{j/2}a_j(\cdot - 2^j k), \quad b_{\ell,j;k} := 2^{j/2}b_{\ell,j}(\cdot - 2^j k), \qquad j \in \mathbb{N}_0, k \in \mathbb{Z}.$$

Now a J-level discrete framelet transform employing a dual framelet filter bank $(\{\tilde{a}; \tilde{b}_1,\ldots,\tilde{b}_s\}, \{a; b_1,\ldots,b_s\})$ can be equivalently rewritten as

$$v = \sum_{k\in\mathbb{Z}} \langle v, \tilde{a}_{J;k}\rangle a_{J;k} + \sum_{j=1}^{J}\sum_{\ell=1}^{s}\sum_{k\in\mathbb{Z}} \langle v, \tilde{b}_{\ell,j;k}\rangle b_{\ell,j;k}. \tag{1.3.18}$$

By employing the dilation factor 2, a multilevel discrete framelet transform provides a multiscale representation of a signal, which is the key to extract the multiscale structure in a signal. The representation in (1.3.18) also shows that the stability of a multilevel discrete framelet transform in the space $l_2(\mathbb{Z})$ is closely related to the asymptotic behavior of the sequences a_J (and $\tilde{a}_J$) in (1.3.17) as $J \to \infty$, which is in turn closely related to the behavior of the frequency-based refinable function $\varphi^a(\xi) := \prod_{j=1}^{\infty} \widehat{a}(2^{-j}\xi)$ for $\xi \in \mathbb{R}$. Roughly speaking, $2^J a_J(k) =$

$\mathcal{S}_a^J\delta(k) \approx \phi^a(2^{-J}k), k \in \mathbb{Z}$ as $J \to \infty$, where $\phi^a(x) := \frac{1}{2\pi}\int_{\mathbb{R}} \varphi^a(\xi)e^{i\xi x}d\xi$ is the inverse Fourier transform of φ^a. We shall address the stability issue and refinable functions in Chaps. 5 and 6 of this book.

The above discussion motivates us to define *discrete affine systems* as follows:

$$\begin{aligned}\mathsf{DAS}_J(\{a; b_1, \ldots, b_s\}) :=& \{a_{J;k} \; : \; k \in \mathbb{Z}\} \\ & \cup \{b_{\ell,j;k} \; : \; \ell = 1, \ldots, s, j = 1, \ldots, J, k \in \mathbb{Z}\}\end{aligned} \tag{1.3.19}$$

and similarly

$$\mathsf{DAS}_J(\{\tilde{a}; \tilde{b}_1, \ldots, \tilde{b}_s\}) := \{\tilde{a}_{J;k} \; : \; k \in \mathbb{Z}\} \cup \{\tilde{b}_{\ell,j;k} \; : \; \ell = 1, \ldots, s, j = 1, \ldots, J, k \in \mathbb{Z}\}.$$

Under the convention that

$$\sim : \mathsf{DAS}_J(\{a; b_1, \ldots, b_s\}) \to \mathsf{DAS}_J(\{\tilde{a}; \tilde{b}_1, \ldots, \tilde{b}_s\}) \quad \text{with} \quad u \mapsto \tilde{u},$$

that is, $(u, \tilde{u})$ is always regarded as a pair together, the representation of $v \in l_2(\mathbb{Z})$ in (1.3.18) can be rewritten as

$$v = \sum_{u \in \mathsf{DAS}_J(\{a; b_1, \ldots, b_s\})} \langle v, \tilde{u}\rangle u, \qquad v \in l_2(\mathbb{Z}), J \in \mathbb{N}. \tag{1.3.20}$$

Therefore, the stability of a filter bank $\{a; b_1, \ldots, b_s\}$ in $l_2(\mathbb{Z})$ as defined in (1.3.7) simply means

$$C_1\|v\|^2_{l_2(\mathbb{Z})} \leqslant \sum_{u \in \mathsf{DAS}_J(\{a; b_1, \ldots, b_s\})} |\langle v, u\rangle|^2 \leqslant C_2\|v\|^2_{l_2(\mathbb{Z})}, \qquad \forall\, v \in l_2(\mathbb{Z}) \tag{1.3.21}$$

for all $J \in \mathbb{N}$. This is equivalent to saying that $\mathsf{DAS}_J(\{a; b_1, \ldots, b_s\})$ is a frame in $l_2(\mathbb{Z})$ with uniform lower and upper frame bounds for all $J \in \mathbb{N}$. It is also easy to prove that $\{a; b_1, \ldots, b_s\}$ is a tight framelet filter bank if and only if (1.3.21) holds with $C_1 = C_2 = 1$. Furthermore, $\{a; b\}$ is an orthogonal wavelet filter bank if and only if $\mathsf{DAS}_J(\{a; b\})$ is an orthonormal basis for $l_2(\mathbb{Z})$ for every $J \in \mathbb{N}$ (see Exercises 1.24–1.27).

We complete this section on stability of a multilevel discrete framelet transform by presenting an example to illustrate the elements in a discrete affine system. Let $\{a; b\}$ be an orthogonal wavelet filter bank given by

$$\begin{aligned} a &= \{\tfrac{1+\sqrt{3}}{8}, \underline{\tfrac{3+\sqrt{3}}{8}}, \tfrac{3-\sqrt{3}}{8}, \tfrac{1-\sqrt{3}}{8}\}_{[-1,2]}, \\ b &= \{\tfrac{1-\sqrt{3}}{8}, \underline{\tfrac{\sqrt{3}-3}{8}}, \tfrac{3+\sqrt{3}}{8}, -\tfrac{1+\sqrt{3}}{8}\}_{[-1,2]}. \end{aligned} \tag{1.3.22}$$

Note that each $\mathsf{DAS}_J(\{a; b\})$ is an orthonormal basis of $l_2(\mathbb{Z})$. Some generators of the discrete affine system $\mathsf{DAS}_J(\{a; b\})$ are presented in Fig. 1.4.

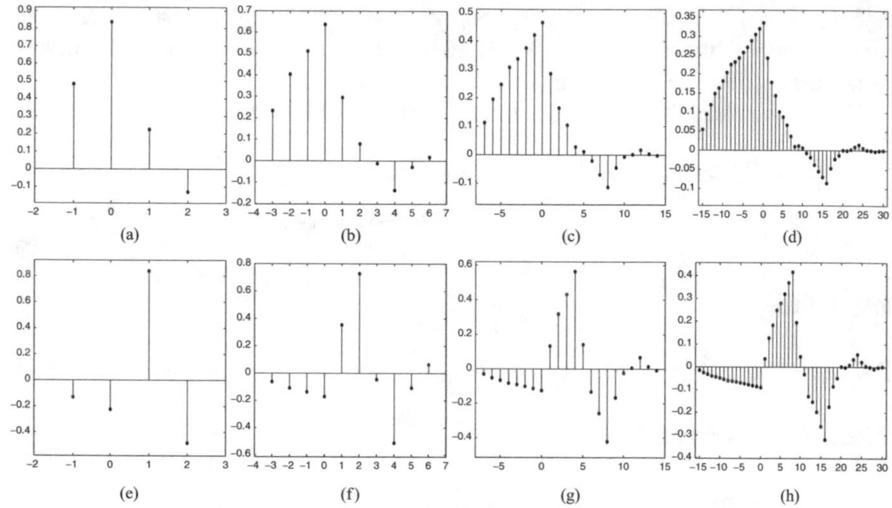

Fig. 1.4 $\{a; b\}$ is the orthogonal wavelet filter bank given in (1.3.22). Some generators of the discrete affine systems $\mathsf{DAS}_J(\{a; b\})$, which are orthonormal bases of $l_2(\mathbb{Z})$. **(a)** $a_{1;0}$. **(b)** $a_{2;0}$. **(c)** $a_{3;0}$. **(d)** $a_{4;0}$. **(e)** $b_{1;0}$. **(f)** $b_{2;0}$. **(g)** $b_{3;0}$. **(h)** $b_{4;0}$

1.3.4 Nonstationary and Undecimated Discrete Framelet Transforms

For some applications, a standard discrete framelet transform is often modified to achieve better performance. Here we discuss some of them such as nonstationary multilevel discrete framelet transforms, framelet packets, and undecimated discrete framelet transforms.

Let us first discuss a nonstationary multilevel discrete framelet transform which includes wavelet packets and undecimated discrete framelet transforms as special cases. The key idea of a nonstationary multilevel discrete framelet transform is to use possibly different filter banks at every scale level. Let

$$\left(\{\tilde{a}_{j,1},\ldots,\tilde{a}_{j,r_j};\tilde{b}_{j,1},\ldots,\tilde{b}_{j,s_j}\},\{a_{j,1},\ldots,a_{j,r_j};b_{j,1},\ldots,b_{j,s_j}\}\right),\ 1\leqslant j\leqslant J \tag{1.3.23}$$

be a sequence of filter banks. *A J-level nonstationary discrete framelet decomposition* with the J-level nonstationary filter bank in (1.3.23) is given by

$$v_{j:k_1,k_2,\ldots,k_j} := \frac{\sqrt{2}}{2}\mathcal{T}_{\tilde{a}_{j,k_j}} v_{j-1;k_1,\ldots,k_{j-1}}, \qquad k_j = 1,\ldots,r_j,$$

$$w_{\ell,j:k_1,k_2,\ldots,k_j} := \frac{\sqrt{2}}{2}\mathcal{T}_{\tilde{b}_{j,\ell}} v_{j-1:k_1,\ldots,k_{j-1}}, \qquad \ell = 1,\ldots,s_j \quad \text{and} \quad k_j = 1,\ldots,r_j,$$

for $j = 1,\ldots,J$, where $v_{0;}$ (i.e., v_0) is an input signal and we used the convention that the subscript index chain $k_m,\ldots,k_n$ is empty if $m > n$. *A J-level nonstationary discrete framelet reconstruction* with the J-level nonstationary filter bank in (1.3.23) is given by

$$\mathring{v}_{j-1:k_1,k_2,\dots,k_{j-1}} := \frac{\sqrt{2}}{2}\sum_{k_j=1}^{r_j} \mathcal{S}_{a_{j,k_j}} v_{j:k_1,k_2,\dots,k_j} + \frac{\sqrt{2}}{2}\sum_{\ell=1}^{s_j} \mathcal{S}_{b_{j,\ell}} w_{\ell,j:k_1,k_2,\dots,k_j}$$

for $j = J, \dots, 1$. If all the pairs in (1.3.23) are dual framelet filter banks, then the above nonstationary J-level discrete framelet transform has the perfect reconstruction property. The word *nonstationary* refers to the fact that the filter bank at the scale level j in (1.3.23) depends on the scale level j. A standard multilevel discrete framelet transform in Sect. 1.3.1 uses the same filter bank for all scale levels with $r_j = 1$ and therefore, it is formally called a (stationary) discrete framelet transform. If all s_j are zero (that is, no child filters), then it corresponds to *a (nonstationary) framelet packet*. Furthermore, if $r_j = 2$ and $s_j = 0$, then it is called *a (nonstationary) wavelet packet*.

Observe that

$$\begin{aligned} &\mathcal{S}_u(v(\cdot - n)) = [\mathcal{S}_u v](\cdot - 2n), \qquad &&\mathcal{S}_{u(\cdot-n)} v = [\mathcal{S}_u v](\cdot - n), \qquad && n \in \mathbb{Z},\\ &\mathcal{T}_u(v(\cdot - 2n)) = [\mathcal{T}_u v](\cdot - n), \qquad &&\mathcal{T}_{u(\cdot+2n)} v = [\mathcal{T}_u v](\cdot - n), \qquad && n \in \mathbb{Z}. \end{aligned} \tag{1.3.24}$$

Hence, if we shift an input signal v or a filter u by an integer, then its output under the subdivision operator is a shifted version of $\mathcal{S}_u v$. But for the transition operator, $\mathcal{T}_u(v(\cdot - n))$ or $\mathcal{T}_{u(\cdot+n)} v$ is generally no longer a shifted version of $\mathcal{T}_u v$ for an odd integer n. This shift sensitivity of framelet coefficients with respect to a shift of an input signal is not desirable in some applications such as signal denoising, since a simple shift of a noise wouldn't change the characteristics of a noise. To overcome this difficulty, a simple solution is to consider both sequences $\mathcal{T}_u v$ and $\mathcal{T}_u(v(\cdot - 1))$. Since $\mathcal{T}_u(v(\cdot - 1)) = \mathcal{T}_{u(\cdot+1)} v$, we end up with a discrete framelet transform by considering two sequences $\mathcal{T}_u v$ and $\mathcal{T}_{u(\cdot+1)} v$ instead of just one sequence $\mathcal{T}_u v$ of framelet coefficients.

Suppose that we have a J-level nonstationary filter bank in (1.3.23). To achieve shift invariance of framelet coefficients, we end up with another J-level nonstationary discrete framelet transform with nonstationary filter banks

$$\begin{aligned} &\Big(\tfrac{\sqrt{2}}{2}\{\tilde{a}_{j,1}, \tilde{a}_{j,1}(\cdot + 1), \dots, \tilde{a}_{j,r_j}, \tilde{a}_{j,r_j}(\cdot + 1); \tilde{b}_{j,1}, \tilde{b}_{j,1}(\cdot + 1), \dots, \tilde{b}_{j,s_j}, \tilde{b}_{j,s_j}(\cdot + 1)\},\\ &\tfrac{\sqrt{2}}{2}\{a_{j,1}, a_{j,1}(\cdot + 1), \dots, a_{j,r_j}, a_{j,r_j}(\cdot + 1); b_{j,1}, b_{j,1}(\cdot + 1), \dots, b_{j,s_j}, b_{j,s_j}(\cdot + 1)\}\Big) \end{aligned} \tag{1.3.25}$$

for $j = 1, \dots, J$. Simply speaking, the above new nonstationary filter banks are obtained by replacing each filter $\tilde{u}$ in (1.3.23) with $\frac{\sqrt{2}}{2}\tilde{u}$ and $\frac{\sqrt{2}}{2}\tilde{u}(\cdot + 1)$. Note that

$$\mathcal{T}_{\frac{\sqrt{2}}{2}\tilde{u}} v = \sqrt{2}(\tilde{u}^\star * v)(2\cdot) \quad \text{and} \quad \mathcal{T}_{\frac{\sqrt{2}}{2}\tilde{u}(\cdot+1)} v = \sqrt{2}(\tilde{u}^\star * v)(2\cdot -1).$$

Consequently, the two sequences $\mathcal{T}_{\frac{\sqrt{2}}{2}\tilde{u}}v$ and $\mathcal{T}_{\frac{\sqrt{2}}{2}\tilde{u}(\cdot+1)}v$ putting together in a disjoint way are simply the sequence $\sqrt{2}\tilde{u}^\star * v$. Similarly, it is easy to verify that

$$\mathcal{S}_{\frac{\sqrt{2}}{2}u}(w(2\cdot)) + \mathcal{S}_{\frac{\sqrt{2}}{2}u(\cdot+1)}(w(2\cdot -1)) = \sqrt{2}u * w. \tag{1.3.26}$$

In other words, the new nonstationary discrete framelet transform is undecimated by removing the downsampling (that is, decimation) and upsampling operations in the original discrete framelet transform. Consequently, the J-level nonstationary discrete framelet transform with the new filter bank in (1.3.25) is called a J-level undecimated discrete nonstationary framelet transform employing the filter bank in (1.3.23). By the above discussion, the seemingly complicated undecimated nonstationary discrete framelet transform with the filter bank in (1.3.23) in fact has a very simple structure as follows. *A J-level undecimated nonstationary discrete framelet decomposition* with the filter bank in (1.3.23) becomes

$$v_{j:k_1,k_2,\ldots,k_j} := (\tilde{a}^\star_{j,k_j} \uparrow 2^j) * v_{j-1:k_1,\ldots,k_{j-1}}, \qquad k_j = 1,\ldots,r_j \quad \text{and} \quad j = 1,\ldots,J,$$

$$w_{\ell,j:k_1,k_2,\ldots,k_j} := (\tilde{b}^\star_{j,\ell} \uparrow 2^j) * v_{j-1:k_1,\ldots,k_{j-1}}, \qquad \ell = 1,\ldots,s_j \quad \text{and} \quad j = 1,\ldots,J.$$

A J-level undecimated nonstationary discrete framelet reconstruction with the filter bank in (1.3.23) becomes

$$\mathring{v}_{j-1:k_1,k_2,\ldots,k_{j-1}} := \sum_{k_j=1}^{r_j}(a_{j,k_j} \uparrow 2^j) * \mathring{v}_{j:k_1,k_2,\ldots,k_j} + \sum_{\ell=1}^{s_j}(b_{j,\ell} \uparrow 2^j) * \mathring{w}_{\ell,j:k_1,k_2,\ldots,k_j}$$

for $j = J,\ldots,1$. To illustrate, we present a J-level undecimated (stationary) discrete framelet transform employing a stationary filter bank $(\{\tilde{a};\tilde{b}_1,\ldots,\tilde{b}_s\}, \{a;b_1,\ldots,b_s\})$. *A J-level undecimated (stationary) discrete framelet decomposition* is given by

$$v_j := (\tilde{a}^\star \uparrow 2^j) * v_{j-1}, \qquad w_{\ell,j} := (\tilde{b}^\star_\ell \uparrow 2^j) * v_{j-1}, \qquad \ell = 1,\ldots,s, \quad j = 1,\ldots,J,$$

where $v_0 : \mathbb{Z} \to \mathbb{C}$ is an input signal. *A J-level undecimated (stationary) discrete framelet reconstruction* is given by

$$\mathring{v}_{j-1} := (a \uparrow 2^j) * \mathring{v}_j + \sum_{\ell=1}^{s}(b_\ell \uparrow 2^j) * \mathring{w}_{\ell,j}, \qquad j = J,\ldots,1.$$

We observe that the above J-level undecimated (stationary) discrete framelet transform has the perfect reconstruction property if and only if

$$\widehat{\tilde{a}}(\xi)\overline{\widehat{a}(\xi)} + \widehat{\tilde{b}_1}(\xi)\overline{\widehat{b_1}(\xi)} + \cdots + \widehat{\tilde{b}_s}(\xi)\overline{\widehat{b_s}(\xi)} = 1. \tag{1.3.27}$$

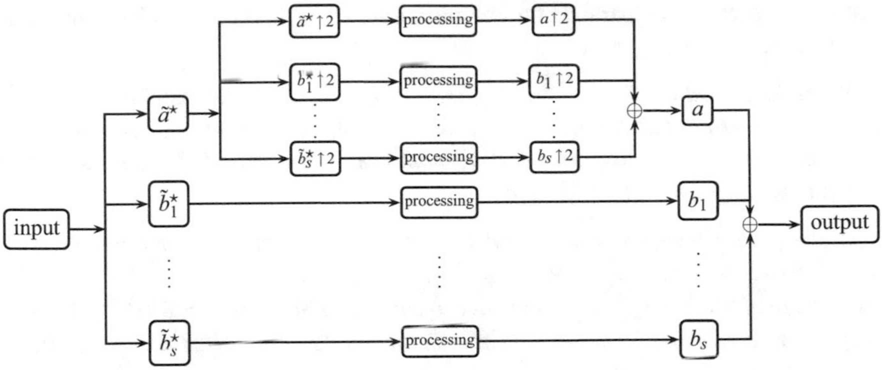

Fig. 1.5 Diagram of a two-level discrete undecimated framelet transform using filter banks $\{\tilde{a}; \tilde{b}_1, \ldots, \tilde{b}_s\}$ and $\{a; b_1, \ldots, b_s\}$

If (1.3.27) holds, then we have the following signal representation similar to (1.3.18):

$$v = \sum_{k\in\mathbb{Z}} \langle v, \tilde{a}_J(\cdot - k)\rangle a_J(\cdot - k) + \sum_{j=0}^{J-1} \sum_{\ell=1}^{s} \sum_{k\in\mathbb{Z}} \langle v, \tilde{b}_{\ell,j}(\cdot - k)\rangle b_{\ell,j}(\cdot - k)$$

whose underlying discrete affine system for reconstruction is

$$\{a_J(\cdot - k) \ : \ k \in \mathbb{Z}\} \cup \{b_{\ell,j}(\cdot - k) \ : \ j = 1, \ldots, J, \ell = 1, \ldots, s, k \in \mathbb{Z}\}.$$

See Fig. 1.5 for a diagram of a two-level undecimated (stationary) discrete framelet transform.

1.4 The Oblique Extension Principle (OEP)

In this section we introduce a generalized discrete framelet transform based on the oblique extension principle (OEP). Let us first explain our motivation for generalizing the standard discrete framelet transform described in Sects. 1.1 and 1.3.

1.4.1 Oblique Extension Principle

As we discussed in Sect. 1.2, to have sparse representations for smooth signals, it is important for high-pass filters to possess high vanishing moments and for low-pass filters to have high sum rules. However, vanishing moments of high-pass filters put some necessary constraints on low-pass filters in a dual framelet filter bank, as shown by the following result.

Lemma 1.4.1 *Let* $(\{\tilde{a};\tilde{b}_1,\ldots,\tilde{b}_s\},\{a;b_1,\ldots,b_s\})$ *be a dual framelet filter bank. If all primal high-pass filters* $b_1,\ldots,b_s$ *have* $\tilde{m}$ *vanishing moments and all dual high-pass filters* $\tilde{b}_1,\ldots,\tilde{b}_s$ *have* m *vanishing moments, where* m *and* $\tilde{m}$ *are nonnegative integers satisfying* $m+\tilde{m}>0$*, then*

(i) the primal low-pass filter a *must have* m *sum rules:* $\widehat{a}(\xi+\pi)=\mathscr{O}(|\xi|^m)$, $\xi\to 0$;
(ii) the dual low-pass filter $\tilde{a}$ *must have* $\tilde{m}$ *sum rules:* $\widehat{\tilde{a}}(\xi+\pi)=\mathscr{O}(|\xi|^{\tilde{m}})$, $\xi\to 0$;
(iii) $\widehat{\tilde{a}}\overline{\widehat{a}}$ *(or equivalently* $\tilde{a}*a^\star$*) has* $m+\tilde{m}$ *linear-phase moments with phase* 0*:*

$$1-\widehat{\tilde{a}}(\xi)\overline{\widehat{a}(\xi)}=\mathscr{O}(|\xi|^{m+\tilde{m}}),\qquad \xi\to 0. \tag{1.4.1}$$

Proof Since $\widehat{b_\ell}(\xi)=\mathscr{O}(|\xi|^{\tilde{m}})$ and $\widehat{\tilde{b}_\ell}(\xi)=\mathscr{O}(|\xi|^m)$ as $\xi\to 0$ for $\ell=1,\ldots,s$, it is straightforward to deduce from (1.3.6) that (1.4.1) holds. Thus, item (iii) holds. In particular, by $m+\tilde{m}>0$, we have $\widehat{\tilde{a}}(0)\overline{\widehat{a}(0)}=1$. By (1.3.6), we have

$$\widehat{\tilde{a}}(\xi)\overline{\widehat{a}(\xi+\pi)}=\widehat{\tilde{b}_1}(\xi)\overline{\widehat{b_1}(\xi+\pi)}+\cdots+\widehat{\tilde{b}_s}(\xi)\overline{\widehat{b_s}(\xi+\pi)}=\mathscr{O}(|\xi|^m),\qquad \xi\to 0.$$

Since $\widehat{\tilde{a}}(0)\neq 0$, we must have $\widehat{a}(\xi+\pi)=\mathscr{O}(|\xi|^m)$ as $\xi\to 0$. That is, item (i) holds. By the same argument, we see that item (ii) holds. □

Since (1.4.1) implies $\widehat{\tilde{a}}(0)\overline{\widehat{a}(0)}=1$, we often normalize a low-pass filter a by $\widehat{a}(0)=1$. Let $\widehat{a}(\xi)=\widehat{a_m^B}(\xi)=2^{-m}(1+e^{-i\xi})^m$ and $\widehat{\tilde{a}}(\xi)=\widehat{a_{\tilde{m}}^B}(\xi)=2^{-\tilde{m}}(1+e^{-i\xi})^{\tilde{m}}$ be two B-spline low-pass filters in (1.2.24). Then $\widehat{\tilde{a}}(\xi)\overline{\widehat{a}(\xi)}=2^{-m-\tilde{m}}(1+e^{-i\xi})^{\tilde{m}}(1+e^{i\xi})^m$ and it is not difficult to check that

$$\widehat{\tilde{a}}(0)\overline{\widehat{a}(0)}=1,\quad \big[\widehat{\tilde{a}}\overline{\widehat{a}}\big]'(0)=\tfrac{i(m-\tilde{m})}{2},\quad \big[\widehat{\tilde{a}}\overline{\widehat{a}}\big]''(0)=\tfrac{(m-\tilde{m})^2+m+\tilde{m}}{4}.$$

Note that $\big[\widehat{\tilde{a}}\overline{\widehat{a}}\big]'(0)=0$ if and only if $m=\tilde{m}$. Regardless of the choices of the positive integers m and $\tilde{m}$, the above identities imply that (1.4.1) cannot be true if $m+\tilde{m}>2$. More generally, let $n,\tilde{n}$ be any integers and define $\widehat{a}(\xi):=e^{-in\xi}\widehat{a_m^B}(\xi)$ and $\widehat{\tilde{a}}(\xi):=e^{-i\tilde{n}\xi}\widehat{a_{\tilde{m}}^B}(\xi)$. Then we can show (see Exercise 1.28) that the relation $\widehat{\tilde{a}}(\xi)\overline{\widehat{a}(\xi)}=1+O(|\xi|^3),\xi\to 0$ can never be true, regardless of the choices of $m,n,\tilde{m},\tilde{n}$. Consequently, for any dual framelet filter bank $(\{\tilde{a};\tilde{b}_1,\ldots,\tilde{b}_s\},\{a;b_1,\ldots,b_s\})$ with B-spline low-pass filters a and $\tilde{a}$, some primal high-pass filters $b_1,\ldots,b_s$ and some dual high-pass filters $\tilde{b}_1,\ldots,\tilde{b}_s$ must have no more than one vanishing moment. Therefore, a more general filter bank is needed in order to improve vanishing moments of high-pass filters derived from B-spline low-pass filters.

The main goal of the following oblique extension principle (OEP) is to increase vanishing moments of high-pass filters derived from a given pair of low-pass filters.

Theorem 1.4.2 (Oblique Extension Principle) *Let $\Theta, \tilde{a}, \tilde{b}_1, \dots, \tilde{b}_s, a, b_1, \dots, b_s \in l_0(\mathbb{Z})$ be finitely supported sequences on $\mathbb{Z}$. Then the following statements are equivalent:*

(i) The filter bank $(\{\tilde{a}; \tilde{b}_1, \dots, \tilde{b}_s\}, \{a; b_1, \dots, b_s\})_\Theta$ has the following generalized perfect reconstruction property: for all $v \in l(\mathbb{Z})$,

$$\Theta^\star * v = \frac{1}{2}\mathcal{S}_a(\Theta^\star * \mathcal{T}_{\tilde{a}} v) + \frac{1}{2}\sum_{\ell=1}^{s} \mathcal{S}_{b_\ell}\mathcal{T}_{\tilde{b}_\ell} v. \tag{1.4.2}$$

(ii) The identity in (1.4.2) holds for all $v \in l_0(\mathbb{Z})$.
(iii) The identity in (1.4.2) holds for the two particular sequences $v = \boldsymbol{\delta}$ and $\boldsymbol{\delta}(\cdot - 1)$.
(iv) The following perfect reconstruction condition holds: for all $\xi \in \mathbb{R}$,

$$\widehat{\Theta}(2\xi)\widehat{\tilde{a}}(\xi)\overline{\widehat{a}(\xi)} + \widehat{\tilde{b}_1}(\xi)\overline{\widehat{b_1}(\xi)} + \cdots + \widehat{\tilde{b}_s}(\xi)\overline{\widehat{b_s}(\xi)} = \widehat{\Theta}(\xi), \tag{1.4.3}$$

$$\widehat{\Theta}(2\xi)\widehat{\tilde{a}}(\xi)\overline{\widehat{a}(\xi + \pi)} + \widehat{\tilde{b}_1}(\xi)\overline{\widehat{b_1}(\xi + \pi)} + \cdots + \widehat{\tilde{b}_s}(\xi)\overline{\widehat{b_s}(\xi + \pi)} = 0. \tag{1.4.4}$$

Proof Taking Fourier series on both sides of (1.4.2), we see that (1.4.2) is equivalent to

$$\overline{\widehat{\Theta}(\xi)\widehat{v}(\xi)} = \overline{\widehat{v}(\xi)}\Big[\overline{\widehat{\Theta}(2\xi)}\,\overline{\widehat{\tilde{a}}(\xi)}\widehat{a}(\xi) + \sum_{\ell=1}^{s} \overline{\widehat{\tilde{b}_\ell}(\xi)}\widehat{b_\ell}(\xi)\Big]$$

$$+ \overline{\widehat{v}(\xi + \pi)}\Big[\overline{\widehat{\Theta}(2\xi)}\,\overline{\widehat{\tilde{a}}(\xi + \pi)}\widehat{a}(\xi) + \sum_{\ell=1}^{s} \overline{\widehat{\tilde{b}_\ell}(\xi + \pi)}\widehat{b_\ell}(\xi)\Big].$$

All the claims follow from the same argument as in the proof of Theorem 1.1.1. □

1.4.2 OEP-Based Tight Framelet Filter Banks

For OEP-based filter banks, we have the following result generalizing Theorem 1.1.4:

Theorem 1.4.3 *Let $\theta, a, b_1, \dots, b_s \in l_0(\mathbb{Z})$ be sequences on $\mathbb{Z}$. Then*

$$\|\theta * \mathcal{T}_a v\|^2_{l_2(\mathbb{Z})} + \|\mathcal{T}_{b_1} v\|^2_{l_2(\mathbb{Z})} + \cdots + \|\mathcal{T}_{b_s} v\|^2_{l_2(\mathbb{Z})} = 2\|\theta * v\|^2_{l_2(\mathbb{Z})}, \qquad \forall\, v \in l_2(\mathbb{Z}),$$

if and only if the filter bank $\{a; b_1, \ldots, b_s\}_\Theta$ *satisfies*

$$\begin{bmatrix} \widehat{b_1}(\xi) & \cdots & \widehat{b_s}(\xi) \\ \widehat{b_1}(\xi+\pi) & \cdots & \widehat{b_s}(\xi+\pi) \end{bmatrix} \begin{bmatrix} \widehat{b_1}(\xi) & \cdots & \widehat{b_s}(\xi) \\ \widehat{b_1}(\xi+\pi) & \cdots & \widehat{b_s}(\xi+\pi) \end{bmatrix}^\star = \mathcal{M}_{a,\Theta}(\xi), \qquad (1.4.5)$$

where

$$\mathcal{M}_{a,\Theta}(\xi) := \begin{bmatrix} \widehat{\Theta}(\xi) - \widehat{\Theta}(2\xi)|\widehat{a}(\xi)|^2 & -\widehat{\Theta}(2\xi)\widehat{a}(\xi)\overline{\widehat{a}(\xi+\pi)} \\ -\widehat{\Theta}(2\xi)\widehat{a}(\xi+\pi)\overline{\widehat{a}(\xi)} & \widehat{\Theta}(\xi+\pi) - \widehat{\Theta}(2\xi)|\widehat{a}(\xi+\pi)|^2 \end{bmatrix} \qquad (1.4.6)$$

and

$$\Theta := \theta * \theta^\star, \quad \textit{that is,} \quad \widehat{\Theta}(\xi) := |\widehat{\theta}(\xi)|^2. \qquad (1.4.7)$$

Proof Note that $\Theta^\star = \Theta$ and $\|\theta * v\|^2_{l_2(\mathbb{Z})} = \frac{1}{2\pi}\int_{-\pi}^{\pi} |\widehat{\theta}(\xi)|^2 |\widehat{v}(\xi)|^2 d\xi = \langle \Theta^\star * v, v\rangle$. By the relation of $\mathcal{S}_a$ and $\mathcal{T}_a$ in Lemma 1.1.3,

$$\|\theta * \mathcal{T}_a v\|^2_{l_2(\mathbb{Z})} = \langle \theta * \mathcal{T}_a v, \theta * \mathcal{T}_a v\rangle = \langle \Theta^\star * \mathcal{T}_a v, \mathcal{T}_a v\rangle = \langle \mathcal{S}_a(\Theta^\star * \mathcal{T}_a v), v\rangle.$$

All claims follow from the same proof as in Theorem 1.1.4. □

A filter bank $(\{\tilde{a}; \tilde{b}_1, \ldots, \tilde{b}_s\}, \{a; b_1, \ldots, b_s\})_\Theta$ satisfying the perfect reconstruction condition in (1.4.3) and (1.4.4) is called *an (OEP-based) dual framelet filter bank*. Similarly, an OEP-based filter bank $\{a; b_1, \ldots, b_s\}_\Theta$ satisfying the perfect reconstruction condition in (1.4.5) is called *a tight framelet filter bank*. From the perfect reconstruction condition in (1.4.3) and (1.4.4), it is straightforward to see that $(\{\tilde{a}; \tilde{b}_1, \ldots, \tilde{b}_s\}, \{a; b_1, \ldots, b_s\})_\Theta$ is a dual framelet filter bank if and only if $(\{a; b_1, \ldots, b_s\}, \{\tilde{a}; \tilde{b}_1, \ldots, \tilde{b}_s\})_{\Theta^\star}$ is a dual framelet filter bank.

Here is an example using B-spline filters. We shall provide many examples in Chap. 3 to illustrate how the oblique extension principle can be used to improve vanishing moments of high-pass filters.

Example 1.4.1 $\{a; b_1, b_2\}_\Theta$ is a tight framelet filter bank, where

$$a = \{\tfrac{1}{4}, \underline{\tfrac{1}{2}}, \tfrac{1}{4}\}_{[-1,1]}, \qquad b_1 = \{-\tfrac{1}{4}, \underline{\tfrac{1}{2}}, -\tfrac{1}{4}\}_{[-1,1]}, \qquad b_2 = \tfrac{\sqrt{6}}{24}\{1, \underline{2}, -6, 2, 1\}_{[-1,3]},$$
$$\Theta = \{-\tfrac{1}{6}, \underline{\tfrac{4}{3}}, -\tfrac{1}{6}\}_{[-1,1]}.$$

In comparison with Example 1.1.3 where one of the two high-pass filters has only one vanishing moment, both high-pass filters b_1 and b_2 here have two vanishing moments.

The following well-known *Fejér-Riesz Lemma* will be needed later.

Lemma 1.4.4 (the Fejér-Riesz Lemma) *Let* $\widehat{\Theta}$ *be a* 2π*-periodic trigonometric polynomial with real coefficients (or with complex coefficients) such that* $\widehat{\Theta}(\xi) \geqslant 0$

for all $\xi \in \mathbb{R}$. *Then there exists a* 2π*-periodic trigonometric polynomial* $\widehat{\theta}$ *with real coefficients (or with complex coefficients) such that* $|\widehat{\theta}(\xi)|^2 = \widehat{\Theta}(\xi)$ *for all* $\xi \in \mathbb{R}$. *Moreover, we can further require* $\widehat{\theta}(0) = \sqrt{\widehat{\Theta}(0)}$.

Proof Since $\widehat{\Theta}(\xi) = \sum_{k\in\mathbb{Z}} \Theta(k)e^{-ik\xi}$, we define a Laurent polynomial $\Theta(z) := \sum_{k\in\mathbb{Z}} \Theta(k)z^k, z \in \mathbb{C}\backslash\{0\}$. Let Z be the set of all the roots, counting multiplicity, of $\Theta(z) = 0, z \in \mathbb{C}\backslash\{0\}$. Since $\overline{\widehat{\Theta}(\xi)} = \widehat{\Theta}(\xi)$ implies $\overline{\Theta(z)} = \Theta(\bar{z}^{-1})$, we see that Z is invariant under the mapping $z \mapsto \bar{z}^{-1}$. Denote $\mathbb{T} := \{z \in \mathbb{C} : |z| = 1\}$. Then $\mathbb{T}$ is the invariant set of the mapping $z \mapsto \bar{z}^{-1}$. Since $\Theta(z) \geqslant 0$ for all $z \in \mathbb{T}$, we see that any point in $\mathbb{T} \cap Z$ has an even multiplicity in Z. Consequently, Z can be described as $\{\zeta, \bar{\zeta}^{-1} : \zeta \in Y\}$ for a unique subset Y of $Z \cap \{z \in \mathbb{C} : |z| \leqslant 1\}$. Define $\boldsymbol{\theta}(z) := \prod_{\zeta\in Y}(z - \zeta)$. Then $\Theta(z) = c^2\boldsymbol{\theta}(z)\boldsymbol{\theta}(\bar{z}^{-1})$ for some $c > 0$. Set $\widehat{\theta}(\xi) := c\boldsymbol{\theta}(e^{-i\xi})$. Now it is straightforward to check that $|\widehat{\theta}(\xi)|^2 = \widehat{\Theta}(\xi)$ for all $\xi \in \mathbb{R}$.

We now show that if $\widehat{\Theta}$ has real coefficients, then so does $\widehat{\theta}$. Since $\widehat{\Theta}$ has real coefficients, we have $\overline{\Theta(z)} = \Theta(\bar{z})$, from which we see that Z is invariant under the mapping $z \mapsto \bar{z}$. Note that the real line $\mathbb{R}$ is the invariant set of the mapping $z \mapsto \bar{z}$. Now we see that the set Y can be described as $\{x, \bar{x} : x \in Y, \mathrm{Im}(x) > 0\} \cup \{y \in Y : \mathrm{Im}(y) = 0\}$. From the definition of $\boldsymbol{\theta}$, now we see that $\boldsymbol{\theta}$ must have real coefficients and therefore, $\widehat{\theta}$ has real coefficients. □

Up to a factor $e^{i(n\xi+c)}$ for some integer n and $c \in \mathbb{R}$, the 2π-periodic trigonometric polynomial $\widehat{\theta}$ constructed in the proof of the above Fejér-Riesz Lemma is unique and there are many numerical algorithms in the literature to compute $\widehat{\theta}$. However, other than the choice in the proof of Lemma 1.4.4, there are many other choices of pairing the roots of $\Theta(z)$ which lead to different 2π-periodic trigonometric polynomials $\widehat{\theta}$ (with real coefficients) satisfying $|\widehat{\theta}(\xi)|^2 = \widehat{\Theta}(\xi)$. For this reason, $\widehat{\theta}$ constructed in the proof of Lemma 1.4.4 is often called the canonical choice.

We now show that a filter Θ in every tight framelet filter bank must take the special form in (1.4.7). Recall that an $r \times r$ matrix U of complex numbers is *positive semidefinite*, denoted by $U \geqslant 0$, if $\bar{x}^{\mathsf{T}}Ux \geqslant 0$ for all $x \in \mathbb{C}^r$.

Lemma 1.4.5 *Let* $a, \Theta \in l_0(\mathbb{Z})\backslash\{0\}$ *and* $\mathcal{M}_{a,\Theta}$ *be defined in* (1.4.6). *Then*

$$\mathcal{M}_{a,\Theta}(\xi) \geqslant 0 \qquad \forall\, \xi \in \mathbb{R}, \tag{1.4.8}$$

if and only if for all $\xi \in \mathbb{R}$, $\widehat{\Theta}(\xi) \geqslant 0$ *and*

$$\begin{aligned}\det(\mathcal{M}_{a,\Theta}(\xi)) =&\widehat{\Theta}(\xi)\widehat{\Theta}(\xi+\pi)\\ &- \widehat{\Theta}(2\xi)\big[\widehat{\Theta}(\xi+\pi)|\widehat{a}(\xi)|^2 + \widehat{\Theta}(\xi)|\widehat{a}(\xi+\pi)|^2\big] \geqslant 0.\end{aligned} \tag{1.4.9}$$

Proof The inequality (1.4.8) implies $\mathcal{M}^{\star}_{a,\Theta}(\xi) = \mathcal{M}_{a,\Theta}(\xi)$, $\det(\mathcal{M}_{a,\Theta}(\xi)) \geqslant 0$, and

$$\widehat{\Theta}(\xi) - \widehat{\Theta}(2\xi)|\widehat{a}(\xi)|^2 \geqslant 0 \qquad \forall\, \xi \in \mathbb{R}, \tag{1.4.10}$$

which is the $(1,1)$-entry of $\mathcal{M}_{a,\Theta}$. By $\overline{[\mathcal{M}_{a,\Theta}(\xi)]_{1,2}} = [\mathcal{M}_{a,\Theta}(\xi)]_{2,1}$, we must have $\overline{\widehat{\Theta}(\xi)} = \widehat{\Theta}(\xi)$ since $\widehat{a}$ is not identically zero. Hence, $\widehat{\Theta}(\xi) \in \mathbb{R}$ for all $\xi \in \mathbb{R}$.

Suppose $\widehat{\Theta}(\xi) < 0$ for $\xi \in (c, d)$ for some $c < d$. Then (1.4.10) implies $\widehat{\Theta}(\xi) < 0$ for all $\xi \in (2^j c, 2^j d)$ and $j \in \mathbb{N}$. For $j \in \mathbb{N}$ such that $2^j(d-c) > 2\pi$, since $\widehat{\Theta}$ is 2π-periodic, this leads to $\widehat{\Theta}(\xi) < 0$ for all $\xi \in \mathbb{R}$. Now by (1.4.9) and $\widehat{\Theta}(\xi) < 0$,

$$[\widehat{\Theta}(\xi) - \widehat{\Theta}(2\xi)|\widehat{a}(\xi)|^2]\widehat{\Theta}(\xi+\pi) \geqslant \widehat{\Theta}(2\xi)\widehat{\Theta}(\xi)|\widehat{a}(\xi+\pi)|^2 \geqslant 0 \tag{1.4.11}$$

for all $\xi \in \mathbb{R}$. Since $\widehat{\Theta}(\xi) < 0$ for all $\xi \in \mathbb{R}$, the above inequality and (1.4.10) imply

$$\widehat{\Theta}(\xi) - \widehat{\Theta}(2\xi)|\widehat{a}(\xi)|^2 = 0 \qquad \forall\, \xi \in \mathbb{R}. \tag{1.4.12}$$

From (1.4.11) and (1.4.12), we must have $\widehat{\Theta}(2\xi)\widehat{\Theta}(\xi)|\widehat{a}(\xi+\pi)|^2 = 0$ contradicting our assumption $a, \Theta \in l_0(\mathbb{Z})\backslash\{0\}$. This proves that $\widehat{\Theta}(\xi) \geqslant 0$ for all $\xi \in \mathbb{R}$.

Conversely, if $\widehat{\Theta}(\xi) \geqslant 0$ and (1.4.9) holds, then (1.4.11) holds. By $\widehat{\Theta}(\xi) \geqslant 0$, we conclude that (1.4.10) holds. That is, the $(1,1)$-entry of $\mathcal{M}_{a,\Theta}$ must be nonnegative. Since (1.4.9) holds, by a standard result from linear algebra, (1.4.8) must hold. □

As a direct consequence of Lemma 1.4.5, we have

Corollary 1.4.6 *If $\{a; b_1, \dots, b_s\}_\Theta$ is a tight framelet filter bank, then $\widehat{\Theta}(\xi) \geqslant 0$ for all $\xi \in \mathbb{R}$ and consequently, there exists $\theta \in l_0(\mathbb{Z})$ such that $|\widehat{\theta}(\xi)|^2 = \widehat{\Theta}(\xi)$ and $\widehat{\theta}(0) = \sqrt{\widehat{\Theta}(0)}$.*

Proof By (1.4.5), we have $\mathcal{M}_{a,\Theta}(\xi) \geqslant 0$ for all $\xi \in \mathbb{R}$. If $a = 0$, then $\mathcal{M}_{a,\Theta}(\xi) \geqslant 0$ directly implies $\widehat{\Theta}(\xi) \geqslant 0$. If $a \in l_0(\mathbb{Z})\backslash\{0\}$, then it follows from Lemma 1.4.5 that $\widehat{\Theta}(\xi) \geqslant 0$. By the Fejér-Riesz Lemma in Lemma 1.4.4, (1.4.7) holds for some $\theta \in l_0(\mathbb{Z})$ with $\widehat{\theta}(0) = \sqrt{\widehat{\Theta}(0)}$. □

1.4.3 OEP-Based Filter Banks with One Pair of High-Pass Filters

Let $u \in l_0(\mathbb{Z})$ be a finitely supported sequence. Recall that $\mathrm{fsupp}(u) = [m, n]$ is the filter support of u if u vanishes outside $[m, n]$ and $u(m)u(n) \neq 0$. If u is the zero sequence, by default $\mathrm{fsupp}(u) = \emptyset$, the empty set. For simplicity, we also set $\mathrm{fsupp}(\widehat{u}) = \mathrm{fsupp}(u)$ and $\mathrm{len}(u) := n - m$.

For some applications, the number of high-pass filters is preferred to be as small as possible. As demonstrated by the following result, the number of high-pass filters in an (OEP-based) dual framelet filter bank can seldom be $s = 1$; otherwise, it is essentially a usual biorthogonal wavelet filter bank.

Theorem 1.4.7 *Let $(\{\tilde{a}; \tilde{b}\}, \{a; b\})_\Theta$ be a dual framelet filter bank such that Θ is not identically zero. Then there exists a nonzero number $\lambda \in \mathbb{C}\backslash\{0\}$ such that*

$$\widehat{\Theta}(2\xi) = \lambda \widehat{\Theta}(\xi)\widehat{\Theta}(\xi + \pi), \qquad \forall\, \xi \in \mathbb{R} \tag{1.4.13}$$

and

$$\begin{bmatrix} \widehat{\mathring{a}}(\xi) & \widehat{\mathring{b}}(\xi) \\ \widehat{\mathring{a}}(\xi+\pi) & \widehat{\mathring{b}}(\xi+\pi) \end{bmatrix} \begin{bmatrix} \widehat{a}(\xi) & \widehat{b}(\xi) \\ \widehat{a}(\xi+\pi) & \widehat{b}(\xi+\pi) \end{bmatrix}^\star = \begin{bmatrix} 1 & 0 \\ 0 & 1 \end{bmatrix}, \tag{1.4.14}$$

where all the above filters are finitely supported and are given by

$$\widehat{\mathring{a}}(\xi) := \widehat{\tilde{a}}(\xi)\lambda\widehat{\Theta}(\xi+\pi) \qquad \text{and} \qquad \widehat{\mathring{b}}(\xi) := \widehat{\tilde{b}}(\xi)/\widehat{\Theta}(\xi). \tag{1.4.15}$$

That is, $(\{\mathring{a}; \mathring{b}\}, \{a; b\})$ is a biorthogonal wavelet filter bank. Moreover, (1.4.14) implies that

$$\widehat{\mathring{b}}(\xi) = \overline{c^{-1}} e^{i(2n-1)\xi} \overline{\widehat{a}(\xi+\pi)}, \quad \widehat{b}(\xi) = c e^{i(2n-1)\xi} \overline{\widehat{\mathring{a}}(\xi+\pi)}, \\ \text{for some } \; c \in \mathbb{C}\backslash\{0\},\; n \in \mathbb{Z}. \tag{1.4.16}$$

If $\{a; b\}_\Theta$ is a tight framelet filter bank, then $\Theta = \theta * \theta^\star$ for some $\theta \in l_0(\mathbb{Z})$ and $\{\breve{a}; \breve{b}\}$ is an orthogonal wavelet filter bank, where $\breve{a}, \breve{b} \in l_0(\mathbb{Z})$ are given by

$$\widehat{\breve{a}}(\xi) := \widehat{a}(\xi)\sqrt{\lambda}\widehat{\theta}(\xi+\pi), \qquad \widehat{\breve{b}}(\xi) := \widehat{b}(\xi)/\widehat{\theta}(\xi). \tag{1.4.17}$$

Proof (1.4.3) and (1.4.4) with $s = 1$ can be rewritten as

$$\begin{bmatrix} \widehat{\tilde{a}}(\xi) & \widehat{\tilde{b}}(\xi) \\ \widehat{\tilde{a}}(\xi+\pi) & \widehat{\tilde{b}}(\xi+\pi) \end{bmatrix} \begin{bmatrix} \widehat{\Theta}(2\xi) & 0 \\ 0 & 1 \end{bmatrix} \begin{bmatrix} \widehat{a}(\xi) & \widehat{b}(\xi) \\ \widehat{a}(\xi+\pi) & \widehat{b}(\xi+\pi) \end{bmatrix}^\star = \begin{bmatrix} \widehat{\Theta}(\xi) & 0 \\ 0 & \widehat{\Theta}(\xi+\pi) \end{bmatrix}. \tag{1.4.18}$$

Taking determinant on both sides of (1.4.18), we have

$$\widehat{\Theta}(2\xi)\tilde{\boldsymbol{\theta}}(\xi)\overline{\boldsymbol{\theta}(\xi)} = \widehat{\Theta}(\xi)\widehat{\Theta}(\xi+\pi),$$

where

$$\tilde{\boldsymbol{\theta}}(\xi) := e^{i\xi}\big(\widehat{\tilde{a}}(\xi)\widehat{\tilde{b}}(\xi+\pi) - \widehat{\tilde{a}}(\xi+\pi)\widehat{\tilde{b}}(\xi)\big),$$
$$\boldsymbol{\theta}(\xi) := e^{i\xi}\big(\widehat{a}(\xi)\widehat{b}(\xi+\pi) - \widehat{a}(\xi+\pi)\widehat{b}(\xi)\big).$$

Since $\widehat{\Theta}$ is not identically zero and $\operatorname{len}(\widehat{\Theta}(2\cdot)) = 2\operatorname{len}(\widehat{\Theta}) = \operatorname{len}(\widehat{\Theta}\widehat{\Theta}(\cdot+\pi))$, we see that (1.4.13) must hold for some $\lambda \in \mathbb{C}\backslash\{0\}$. Thus, $\lambda\tilde{\boldsymbol{\theta}}(\xi)\overline{\boldsymbol{\theta}(\xi)} = 1$. Since $\boldsymbol{\theta}$ and $\tilde{\boldsymbol{\theta}}$ are π-periodic trigonometric polynomials, this identity forces $\boldsymbol{\theta}(\xi) = -ce^{i2n\xi}$ for some $c \in \mathbb{C}\backslash\{0\}$ and some $n \in \mathbb{Z}$. By (1.4.18), a direct calculation shows that (1.4.14) must hold with $\widehat{\mathring{a}}$ and $\widehat{\mathring{b}}$ in (1.4.15). By (1.4.14), we deduce that

$$\begin{bmatrix} \widehat{\mathring{a}}(\xi) & \widehat{\mathring{b}}(\xi) \\ \widehat{\mathring{a}}(\xi+\pi) & \widehat{\mathring{b}}(\xi+\pi) \end{bmatrix} = \begin{bmatrix} \overline{\widehat{a}(\xi)} & \overline{\widehat{a}(\xi+\pi)} \\ \overline{\widehat{b}(\xi)} & \overline{\widehat{b}(\xi+\pi)} \end{bmatrix}^{-1} = \frac{1}{\overline{e^{i\xi}\boldsymbol{\theta}(\xi)}} \begin{bmatrix} \overline{\widehat{b}(\xi+\pi)} & -\overline{\widehat{a}(\xi+\pi)} \\ -\overline{\widehat{b}(\xi)} & \overline{\widehat{a}(\xi)} \end{bmatrix}.$$

Plugging $\boldsymbol{\theta}(\xi) = -ce^{i2n\xi}$ into the above identity and comparing the entries of the matrices on both sides, we conclude that (1.4.16) holds. Consequently, $\mathring{b}$ must be a finitely supported sequence. By (1.4.15), $\mathring{a}$ is a finitely supported sequence.

Suppose that $\tilde{a} = a$ and $\tilde{b} = b$. By Lemma 1.4.5 and (1.4.13), we see that $\lambda > 0$ and $\Theta = \theta * \theta^\star$ for some $\theta \in l_0(\mathbb{Z})$. It is obvious from (1.4.15) that $\breve{b} \in l_0(\mathbb{Z})$. Using (1.4.13) and (1.4.17), we can directly check that $\{\breve{a}; \breve{b}\}$ is an orthogonal wavelet filter bank. □

1.4.4 OEP-Based Multilevel Discrete Framelet Transforms

We now discuss a multilevel discrete framelet transform employing a dual framelet filter bank $(\{\tilde{a}; \tilde{b}_1, \dots, \tilde{b}_s\}, \{a; b_1, \dots, b_s\})_\Theta$. A J-level discrete framelet decomposition is exactly the same as the one in (1.3.1). A J-level discrete framelet reconstruction, which will be described as follows, is a slight modification to the J-level discrete framelet reconstruction in (1.3.3). For given low-pass framelet coefficients $\mathring{v}_J$ and high-pass framelet coefficients $\mathring{w}_{\ell,j}, \ell = 1, \dots, s$ and $j = 1, \dots, J$, *a J-level discrete framelet reconstruction* is

$$\breve{v}_J := \Theta^\star * \mathring{v}_J, \tag{1.4.19}$$

$$\breve{v}_{j-1} := \frac{\sqrt{2}}{2}\mathcal{S}_a \breve{v}_j + \frac{\sqrt{2}}{2}\sum_{\ell=1}^{s} \mathcal{S}_{b_\ell} \mathring{w}_{\ell,j}, \qquad j = J, \dots, 1, \tag{1.4.20}$$

recover $\mathring{v}_0$ from $\breve{v}_0$ via the relation $\mathring{v}_0 = \Theta^\star * \breve{v}_0$. (1.4.21)

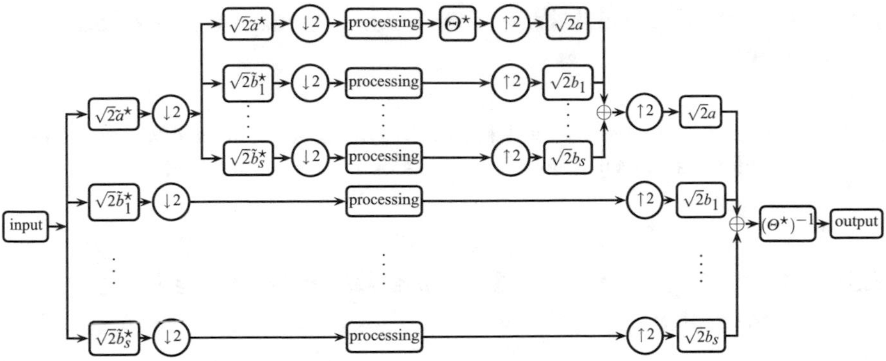

Fig. 1.6 Diagram of a two-level discrete framelet transform implemented using (1.1.29) employing a dual framelet filter bank $(\{\tilde{a}; \tilde{b}_1, \dots, \tilde{b}_s\}, \{a; b_1, \dots, b_s\})_\Theta$

If $(\{\tilde{a}; \tilde{b}_1, \dots, \tilde{b}_s\}, \{a; b_1, \dots, b_s\})_\Theta$ is a dual framelet filter bank satisfying (1.4.3) and (1.4.4), then Theorem 1.4.2 tells us that its associated J-level discrete framelet transform has the perfect reconstruction property. See Fig. 1.6 for a diagram of a two-level discrete framelet transform employing a dual framelet filter bank $(\{\tilde{a}; \tilde{b}_1, \dots, \tilde{b}_s\}, \{a; b_1, \dots, b_s\})_\Theta$. Using discrete affine systems, a J-level discrete framelet transform using an OEP-based dual framelet filter bank $(\{\tilde{a}; \tilde{b}_1, \dots, \tilde{b}_s\}, \{a; b_1, \dots, b_s\})_\Theta$ can be equivalently expressed as

$$v * \Theta^\star = \sum_{k\in\mathbb{Z}} \Big(\sum_{m\in\mathbb{Z}} \langle v, \tilde{a}_{J;m} \rangle \Theta^\star(k-m) \Big) a_{J;k} + \sum_{j=1}^{J} \sum_{\ell=1}^{s} \sum_{k\in\mathbb{Z}} \langle v, \tilde{b}_{\ell,j;k} \rangle b_{\ell,j;k}.$$

In the above multilevel discrete framelet transform, there is a deconvolution in (1.4.21) to recover $\mathring{v}_0$ from $\breve{v}_0$ if $\widehat{\Theta}$ is not a nonzero monomial. We can easily avoid this troubling de-convolution by the following argument. Let

$$(\{\tilde{a}_m; \tilde{b}_{m,1}, \dots, \tilde{b}_{m,s_m}\}, \{a_m; b_{m,1}, \dots, b_{m,s_m}\})_{\Theta_m},$$

$m = 1, \dots, n$ be a family of dual framelet filter banks. For any input signal v, let $\breve{v}_{0,m}$ be the reconstructed signal using the dual framelet filter bank $(\{\tilde{a}_m; \tilde{b}_{m,1}, \dots, \tilde{b}_{m,s_m}\}, \{a_m; b_{m,1}, \dots, b_{m,s_m}\})_{\Theta_m}$. Suppose that there exist $\tilde{\Theta}_1, \dots, \tilde{\Theta}_n \in l_0(\mathbb{Z})$ such that

$$\widehat{\tilde{\Theta}_1}(\xi)\overline{\widehat{\Theta_1}(\xi)} + \dots + \widehat{\tilde{\Theta}_n}(\xi)\overline{\widehat{\Theta_n}(\xi)} = 1. \tag{1.4.22}$$

Avoiding de-convolution in (1.4.21), we can recover $\mathring{v}_0$ via the following formula:

$$\mathring{v}_0 = \tilde{\Theta}_1 * \breve{v}_{0,1} + \dots + \tilde{\Theta}_n * \breve{v}_{0,n}.$$

As shown by many examples in Chap. 3, quite often $n = 2$ is sufficient to achieve (1.4.22).

1.5 Discrete Framelet Transforms for Signals on Bounded Intervals

In this section we present several algorithms to implement a discrete framelet transform and its variants to deal with signals on a bounded interval.

1.5.1 Boundary Effect in a Standard Discrete Framelet Transform

The input signals to discrete framelet transforms described in Sect. 1.1 have support on the integer lattice $\mathbb{Z}$, which has infinite length and no boundaries. However, signals in applications often have finite length and can be modeled by $v^b = \{v^b(k)\}_{k=0}^{N-1} : [0, N-1] \cap \mathbb{Z} \to \mathbb{C}$, where the superscript b over v is used to emphasize that v^b is a signal on a bounded interval. To apply a discrete framelet transform discussed in Sects. 1.1 and 1.3, we have to extend signals from the bounded interval $[0, N-1]$ to the integer lattice $\mathbb{Z}$.

Since a J-level discrete framelet transform employs one-level discrete framelet transforms recursively, it suffices for us to discuss how to implement one-level discrete framelet transforms for signals on the interval $[0, N-1]$. Let $(\{\tilde{u}_0, \dots, \tilde{u}_s\}, \{u_0, \dots, u_s\})$ be a dual framelet filter bank satisfying the perfect reconstruction condition in (1.1.12) and (1.1.13). In order to modify a one-level discrete framelet transform for signals on a bounded interval, one often first shifts the filters in a given dual framelet filter bank $(\{\tilde{u}_0, \dots, \tilde{u}_s\}, \{u_0, \dots, u_s\})$ properly so that the data structure for storing framelet coefficients is simple. This can be easily done, according to the following result.

Proposition 1.5.1 *Let* $(\{\tilde{u}_0, \dots, \tilde{u}_s\}, \{u_0, \dots, u_s\})$ *be a dual framelet filter bank. Suppose that* $n_0, \dots, n_s$ *are integers such that*

$$n_1 - n_0, \ \dots, \ n_s - n_0 \quad \text{are even integers.} \tag{1.5.1}$$

Then the shifted filter bank $(\{\tilde{u}_0(\cdot - n_0), \dots, \tilde{u}_s(\cdot - n_s)\}, \{u_0(\cdot - n_0), \dots, u_s(\cdot - n_s)\})$ *is also a dual framelet filter bank.*

Proof Since $\widehat{[u_\ell(\cdot - n_\ell)]}(\xi) = e^{-in_\ell\xi}\widehat{u_\ell}(\xi)$ for $\ell = 1, \dots, s$ and (1.5.1) holds, for $\omega \in \{0, 1\}$, we have

$$\begin{aligned}\widehat{[\tilde{u}_\ell(\cdot - n_\ell)]}(\xi)\overline{\widehat{[u_\ell(\cdot - n_\ell)]}(\xi + \pi\omega)} &= e^{i\pi n_\ell\omega}\widehat{\tilde{u}_\ell}(\xi)\overline{\widehat{u_\ell}(\xi + \pi\omega)} \\ &= e^{i\pi n_0\omega}\widehat{\tilde{u}_\ell}(\xi)\overline{\widehat{u_\ell}(\xi + \pi\omega)},\end{aligned}$$

where we used (1.5.1) in the last identity. We conclude that (1.1.12) and (1.1.13) are satisfied if $\tilde{u}_0, \ldots, \tilde{u}_s, u_0, \ldots, u_s$ are replaced by $\tilde{u}_0(\cdot - n_0), \ldots, \tilde{u}_s(\cdot - n_s), u_0(\cdot - n_0), \ldots, u_s(\cdot - n_s)$, respectively. By Theorem 1.1.1, the shifted filter bank $(\{\tilde{u}_0(\cdot - n_0), \ldots, \tilde{u}_s(\cdot - n_s)\}, \{u_0(\cdot - n_0), \ldots, u_s(\cdot - n_s)\})$ also satisfies the perfect reconstruction condition and therefore is a dual framelet filter bank. □

For a finitely supported filter/sequence $u \in l_0(\mathbb{Z})$, recall that its filter support $[n_-, n_+]$ is defined to be $u(k) = 0$ for all $k \in \mathbb{Z}\backslash[n_-, n_+]$, but $u(n_-)u(n_+) \neq 0$. That is, the filter support of a sequence u is the smallest interval (with integer endpoints) outside which u vanishes. Obviously, if all $n_0, \ldots, n_s$ are even integers, then the condition in (1.5.1) is automatically satisfied. Due to (1.3.24), for a pair $(\tilde{u} := \tilde{u}_\ell, u := u_\ell)$ in practical implementation, we often replace $(\tilde{u}, u)$ by $(\tilde{u}(\cdot - 2n), u(\cdot - 2n))$, where n is an appropriate integer chosen in such a way that the filter support $[n_-, n_+]$ of the filter $u(\cdot - 2n)$ contains 0; moreover, in implementation one often further assumes that its middle point $(n_- + n_+)/2$ is the smallest in modulus. In this section, for a filter bank $(\{\tilde{u}_0, \ldots, \tilde{u}_s\}, \{u_0, \ldots, u_s\})$, we always assume that the filter support $[n_-, n_+]$ of u_ℓ satisfies $n_- \leqslant 0$ and $n_+ \geqslant 0$ for every filter u_ℓ, $\ell = 0, \ldots, s$.

We now discuss how to modify a discrete framelet transform for handling signals on $\mathbb{Z}$ into a framelet algorithm for handling a signal $v^b = \{v^b(k)\}_{k=0}^{N-1}$ with N being a positive integer. First, one extends v^b from $[0, N-1] \cap \mathbb{Z}$ to a sequence v (which can be also explicitly denoted by v^e) on $\mathbb{Z}$ by any method that the reader prefers. For example, the signal v^b can be extended from the interval $[0, N-1] \cap \mathbb{Z}$ to $\mathbb{Z}$ by the simple zero-padding extension: $v(k) = v^b(k)$ for $k = 0, \ldots, N-1$, and $v(k) = 0$ for $k \in \mathbb{Z}\backslash[0, N-1]$. To preserve the perfect reconstruction property for a given signal v^b, a framelet algorithm shall be able to reconstruct all the original values $v^b(0), \ldots, v^b(N-1)$, while the artificial values outside the interval $[0, N-1]$ may or may not be preserved after reconstruction. This implies that we have to calculate $[\mathcal{S}_{u_\ell}\mathcal{T}_{\tilde{u}_\ell} v](n)$ for all $n = 0, \ldots, N-1$ and all the values $[\mathcal{T}_{\tilde{u}_\ell} v](k)$, which are involved in the calculation of $[\mathcal{S}_{u_\ell}\mathcal{T}_{\tilde{u}_\ell} v](n)$ for $n = 0, \ldots, N-1$, must be retained.

In the following, let us look at one typical pair $\tilde{u} := \tilde{u}_\ell$ and $u := u_\ell$ such that the filter support $[n_-, n_+]$ of u satisfies $n_- \leqslant 0$ and $n_+ \geqslant 0$. By the definition of the subdivision operator in (1.1.2), we have

$$\frac{1}{2}[\mathcal{S}_u\mathcal{T}_{\tilde{u}} v](n) = \sum_{k\in\mathbb{Z}} [\mathcal{T}_{\tilde{u}} v](k)u(n-2k) = \sum_{k=\lceil (n-n_+)/2\rceil}^{\lfloor (n-n_-)/2\rfloor} [\mathcal{T}_{\tilde{u}} v](k)u(n-2k), \qquad n \in \mathbb{Z},$$

where $\lfloor x \rfloor = m$ is the floor function for $m \leqslant x < m+1$ and $\lceil y \rceil = m$ is the ceiling function for $m - 1 < y \leqslant m$ with $m \in \mathbb{Z}$. Thus, to calculate $[\mathcal{S}_u\mathcal{T}_{\tilde{u}} v](n)$ for $n = 0, \ldots, N-1$, we have to record all the framelet coefficients:

$$[\mathcal{T}_{\tilde{u}} v](k), \qquad k = \lceil \tfrac{-n_+}{2} \rceil, \ldots, \lfloor \tfrac{N-1-n_-}{2} \rfloor. \tag{1.5.2}$$

Assuming that N is a positive even integer, by $n_- \leqslant 0$ and $n_+ \geqslant 0$, we always have $\lceil \frac{-n_+}{2} \rceil \leqslant 0$ and $\lfloor \frac{N-1-n_-}{2} \rfloor \geqslant \frac{N}{2} - 1$. In other words, regardless of the filter support

$[n_-, n_+]$ of the filter u, the framelet coefficients $\{[\mathcal{T}_{\tilde{u}}v](k)\}_{k=0}^{\frac{N}{2}-1}$ must be recorded. Note that the must-be-recorded $\{[\mathcal{T}_{\tilde{u}}v](k)\}_{k=0}^{\frac{N}{2}-1}$ has exactly half of the length N of the original signal v. This is the ideal situation, since the total number of framelet coefficients to be recorded will be $(s+1)N/2$; in particular, for the wavelet case $s = 1$, one is using a biorthogonal wavelet filter bank and the total number of recorded wavelet coefficients will be the same as the length of the original signal. This ideal situation is very convenient from the viewpoint of data structure for practical programming.

Now the extra work of a framelet algorithm for an arbitrary bounded signal v^b on $[0, N-1]$ is to record the extra framelet coefficients:

$$[\mathcal{T}_{\tilde{u}}v](k), \qquad k = \lceil \tfrac{-n_+}{2} \rceil, \dots, -1 \quad \text{and} \quad k = \tfrac{N}{2}, \dots, \lfloor \tfrac{N-1-n_-}{2} \rfloor. \tag{1.5.3}$$

The ideal situation, that there are no extra framelet coefficients to be recorded, can happen if and only if $0 \leqslant n_+ \leqslant 1$ and $n_- = 0$; in other words, u has a very short support and the only possible nonzero values of u are $u(0)$ and $u(1)$. Obviously, many filters have a much longer filter support than $[0, 1]$ and we have to find a way to avoid directly recording the extra framelet coefficients in (1.5.3).

The main idea in an efficient framelet algorithm for signals on a bounded interval to handle the extra framelet coefficients in (1.5.3) is to use correlation: one hopes that the extra framelet coefficients in (1.5.3) are linked in some simple way to the must-be-recorded framelet coefficients $[\mathcal{T}_{\tilde{u}}v](k), k = 0, \dots, \frac{N}{2}-1$. In other words, if all the extra framelet coefficients in (1.5.3) are completely determined by the must-be-recorded framelet coefficients $[\mathcal{T}_{\tilde{u}}v](k), k = 0, \dots, \frac{N}{2}-1$, then there is no need to record the extra framelet coefficients explicitly, since they can be recovered from the must-be-recorded framelet coefficients. In the following, let us present two possible ways of achieving this goal: one is to explore the periodic structure, and the other is to take advantage of symmetries of filters.

For $0 < p < \infty$, we denote by $l_p(\mathbb{Z})$ the space of all sequences $v = \{v(k)\}_{k\in\mathbb{Z}} \in l(\mathbb{Z})$ such that

$$\|v\|_{l_p(\mathbb{Z})}^p := \sum_{k\in\mathbb{Z}} |v(k)|^p < \infty.$$

For $p = \infty$, we use $\|v\|_{l_\infty(\mathbb{Z})} := \sup_{k\in\mathbb{Z}} |v(k)| < \infty$.

For the convenience of discussion on frequency-based framelet algorithms in the next section, from now on in this section, we assume that all filters $u, \tilde{u} \in l_1(\mathbb{Z})$ instead of its subspace $l_0(\mathbb{Z})$. Since $v^b = \{v^b(k)\}_{k=0}^{N-1}$ has only finitely many values, it is natural for us to assume that its extended sequence v belongs to $l_\infty(\mathbb{Z})$ and consequently, it is natural to require that $u \in l_1(\mathbb{Z})$ so that the convolution $u * v$ is well defined.

1.5.2 Discrete Framelet Transforms Using Periodic Extension

We first explore the periodic structure by the following result.

Proposition 1.5.2 *Let $u \in l_1(\mathbb{Z})$ be a filter and $v^b = \{v^b(k)\}_{k=0}^{N-1}$ be an arbitrary input signal. Extend v^b into an N-periodic sequence v on $\mathbb{Z}$ as follows:*

$$v(Nn+k) := v^b(k), \qquad k = 0, \ldots, N-1 \quad \text{and} \quad n \in \mathbb{Z}. \tag{1.5.4}$$

Then the following properties hold:

(i) $u * v$ is an N-periodic sequence on $\mathbb{Z}$;
(ii) $\mathcal{S}_u v$ is a $2N$-periodic sequence on $\mathbb{Z}$;
(iii) If N is a positive even integer, then $\mathcal{T}_u v$ is an $\frac{N}{2}$-periodic sequence on $\mathbb{Z}$;
(iv) If N is a positive odd integer, then $\mathcal{T}_u v$ is an N-periodic sequence on $\mathbb{Z}$ and is given by

$$[\mathcal{T}_u v](k) = 2(u^\star * v)(2k), \qquad k = 0, \ldots, N-1.$$

That is, $\mathcal{T}_u v$ is a simple rearrangement of the N-periodic sequence $2u^\star * v$.

Proof Since v is N-periodic, we have $v(N+k) = v(k)$ and

$$[u * v](N+n) = \sum_{k\in\mathbb{Z}} u(k)v(N+n-k) = \sum_{k\in\mathbb{Z}} u(k)v(n-k) = [u * v](n), \qquad n \in \mathbb{Z}.$$

Hence, $u * v$ is also N-periodic and item (i) holds.

It is not difficult to see that $v \uparrow 2$ is the same sequence as the sequence obtained by first upsampling v^b then extending it into a $2N$-periodic sequence. Now by $\mathcal{S}_u v = 2u * (v \uparrow 2)$, we see that $\mathcal{S}_u v$ is a $2N$-periodic sequence.

Since $\mathcal{T}_u v = 2(u^\star * v) \downarrow 2$ and N is a positive even integer, it is straightforward to check that $\mathcal{T}_u v$ is an $\frac{N}{2}$-periodic sequence on $\mathbb{Z}$. Item (iv) is left as Exercise 1.34. □

In the following we describe a one-level periodic discrete framelet transform for signals $v^b = \{v^b(k)\}_{k=0}^{N-1}$ with N being a positive even integer. Let $\tilde{u}_0, \ldots, \tilde{u}_s \in l_1(\mathbb{Z})$ be filters for decomposition. *A one-level periodic discrete framelet decomposition* is

$$w_\ell^b = \left\{ w_\ell^b(k) := \tfrac{\sqrt{2}}{2}[\mathcal{T}_{\tilde{u}_\ell} v](k) \right\}_{k=0}^{\frac{N}{2}-1}, \qquad \ell = 0, \ldots, s, \tag{1.5.5}$$

where v is the N-periodic extension of v^b given in (1.5.4). Grouping all framelet coefficients in (1.5.5) together, we can define a periodic discrete framelet analysis operator $\widetilde{\mathcal{W}}^{per}$ employing the filter bank $\{\tilde{u}_0, \ldots, \tilde{u}_s\}$ as follows:

$$\begin{aligned} \widetilde{\mathcal{W}}^{per}(v^b) &= (w_0^b, w_1^b, \ldots, w_s^b)^\mathsf{T} \\ &= (w_0^b(0), \ldots, w_0^b(\tfrac{N}{2}-1), \ldots, w_s^b(0), \ldots, w_s^b(\tfrac{N}{2}-1))^\mathsf{T}. \end{aligned} \tag{1.5.6}$$

In other words, if we regard v^b as an $N \times 1$ column vector, then the coefficient matrix, still denoted by $\widetilde{\mathcal{W}}^{per}$, of the linear operator $\widetilde{\mathcal{W}}^{per}$ is an $N(\frac{s+1}{2}) \times N$ matrix and a one-level periodic discrete framelet decomposition in (1.5.5) simply becomes $\widetilde{\mathcal{W}}^{per} v^b$, where v^b here is regarded as an $N \times 1$ column vector. We shall use $\mathcal{W}^{per}$ to denote the associated periodic discrete framelet analysis operator employing the filter bank $\{u_0, \dots, u_s\}$.

Let $u_0, \dots, u_s \in l_1(\mathbb{Z})$ be filters for reconstruction. For given sequences of framelet coefficients $w_0^b = \{w_0^b(k)\}_{k=0}^{\frac{N}{2}-1}, \dots, w_s^b = \{w_s^b(k)\}_{k=0}^{\frac{N}{2}-1}$, *a one-level periodic discrete framelet reconstruction* is

$$v^b = \mathcal{V}^{per}(w_0^b, \dots, w_s^b) := \left\{ v^b(k) := \frac{\sqrt{2}}{2} \sum_{\ell=0}^{s} [\mathcal{S}_{u_\ell} w_\ell](k) \right\}_{k=0}^{N-1}, \tag{1.5.7}$$

where w_ℓ is the $\frac{N}{2}$-periodic extension of w_ℓ^b. If we still denote by $\mathcal{V}^{per}$ the $N \times N(\frac{s+1}{2})$ coefficient matrix of the linear operator $\mathcal{V}^{per}$, then a one-level periodic discrete framelet synthesis simply becomes $\mathcal{V}^{per}(w_0^b, \dots, w_s^b)^{\mathsf{T}}$. We shall use $\widetilde{\mathcal{V}}^{per}$ to denote the associated periodic discrete framelet reconstruction operator employing the filter bank $\{\tilde{u}_0, \dots, \tilde{u}_s\}$. For a dual framelet filter bank $(\{\tilde{u}_0, \dots, \tilde{u}_s\}, \{u_0, \dots, u_s\})$, we have $\mathcal{V}^{per}\widetilde{\mathcal{W}}^{per} = \widetilde{\mathcal{V}}^{per}\mathcal{W}^{per} = I_N$. In case that $\tilde{u}_0 = u_0, \dots, \tilde{u}_s = u_s$ (that is, a tight framelet filter bank), it is easy to see that $\mathcal{V}^{per} = (\mathcal{W}^{per})^\star = (\overline{\mathcal{W}^{per}})^{\mathsf{T}}$. In particular, for an orthogonal wavelet filter bank $\{u_0, u_1\}$, $\mathcal{V}^{per}$ and $\mathcal{W}^{per}$ are $N \times N$ unitary matrices satisfying $(\mathcal{V}^{per})^\star \mathcal{V}^{per} = (\mathcal{W}^{per})^\star \mathcal{W}^{per} = I_N$ and $\mathcal{V}^{per} = (\mathcal{W}^{per})^\star = (\mathcal{W}^{per})^{-1}$.

It is easy to see that the above described periodic discrete framelet transform can be straightforwardly modified for a dual framelet filter bank $(\{\tilde{a}; \tilde{b}_1, \dots, \tilde{b}_s\}, \{a; b_1, \dots, b_s\})_\Theta$ such that the one-level periodic discrete framelet decomposition is the same as (1.5.5) and the one-level periodic discrete framelet reconstruction is

$$\breve{v}^b = \left\{ \breve{v}^b(k) := \frac{\sqrt{2}}{2}[\mathcal{S}_a(\Theta^\star * w_0)](k) + \frac{\sqrt{2}}{2} \sum_{\ell=1}^{s} [\mathcal{S}_{b_\ell} w_\ell](k) \right\}_{k=0}^{N-1}$$

and the reconstructed signal $v^b := \{v(k)\}_{k=0}^{N-1}$ is obtained from $\breve{v}^b$ via $v = \Theta^\star * \breve{v}$, where $\breve{v}$ is the N-periodic extension of $\breve{v}^b$. As discussed at the end of Sect. 1.4, the deconvolution here can be easily avoided by combining more than one discrete framelet transforms employing several dual framelet filter banks.

We present an example to illustrate the periodic discrete framelet transform.

Example 1.5.1 Let v^b be the same input signal as in (1.1.26), that is,

$$v^b = \{\underline{\mathbf{-21}}, -22, -23, -23, -25, 38, 36, 34\}_{[0,7]}. \tag{1.5.8}$$

We apply the tight framelet filter bank in Example 1.1.3 to v^b in (1.5.8). We extend v^b to an 8-periodic sequence v on $\mathbb{Z}$, given by

$$v = \{\dots, -25, 38, 36, 34, \underline{\mathbf{-21}, -22, -23, -23, -25, 38, 36, 34}, -21, -22, -23, \dots\}.$$

Then all sequences $\mathcal{T}_{u_0}v$, $\mathcal{T}_{u_1}v$, $\mathcal{T}_{u_2}v$ are 4-periodic and

$$w_0 = \tfrac{\sqrt{2}}{2}\mathcal{T}_{u_0}v = \tfrac{\sqrt{2}}{2}\{\dots, -15, -\tfrac{91}{2}, -\tfrac{35}{2}, 72, \underline{\mathbf{-15}, -\tfrac{91}{2}, -\tfrac{35}{2}, 72}, -15, -\tfrac{91}{2}, -\tfrac{35}{2}, \dots\},$$

$$w_1 = \tfrac{\sqrt{2}}{2}\mathcal{T}_{u_1}v = \{\dots, -28, -\tfrac{1}{2}, \tfrac{61}{2}, -2, \underline{\mathbf{-28}, -\tfrac{1}{2}, \tfrac{61}{2}, -2}, -28, -\tfrac{1}{2}, \tfrac{61}{2}, -2, \dots\},$$

$$w_2 = \tfrac{\sqrt{2}}{2}\mathcal{T}_{u_2}v = \tfrac{\sqrt{2}}{2}\{\dots, -27, -\tfrac{1}{2}, -\tfrac{65}{2}, 0, \underline{\mathbf{-27}, -\tfrac{1}{2}, -\tfrac{65}{2}, 0}, -27, -\tfrac{1}{2}, -\tfrac{65}{2}, 0, \dots\}.$$

It is also easy to directly check that $\frac{\sqrt{2}}{2}(\mathcal{S}_{u_0}w_0 + \mathcal{S}_{u_1}w_1 + \mathcal{S}_{u_2}w_2) = v$.

1.5.3 Discrete Framelet Transforms Using Symmetric Extension

The periodic extension in Proposition 1.6.2 is often used for a filter without any symmetry. Taking advantages of symmetry of a dual framelet filter bank, we now discuss a symmetric discrete framelet transform for signals on a bounded interval $[0, N-1]$. Generally, we need to adapt a symmetric discrete framelet transform for different types of symmetries.

Recall that $\mathsf{S}\widehat{u}(\xi) = \widehat{u}(\xi)/\widehat{u}(-\xi)$ records the symmetry type of a filter u and $\mathbb{S}\widehat{u}(\xi) = \widehat{u}(\xi)/\overline{\widehat{u}(\xi)}$ records the complex symmetry type of u. The following result can be easily verified and will be needed later.

Proposition 1.5.3 *If sequences u and v have symmetry such that*

$$[\mathsf{S}\widehat{u}](\xi) = \epsilon_u e^{-ic_u\xi}, \quad [\mathsf{S}\widehat{v}](\xi) = \epsilon_v e^{-ic_v\xi} \quad \textit{for some} \quad c_u, c_v \in \mathbb{Z}, \epsilon_u, \epsilon_v \in \{-1, 1\},$$

then both $u * v$ and $\mathcal{S}_u v$ have symmetry satisfying

$$[\mathsf{S}(\widehat{u * v})](\xi) = [\mathsf{S}\widehat{u}](\xi)[\mathsf{S}\widehat{v}](\xi) = \epsilon_u\epsilon_v e^{-i(c_u+c_v)\xi}$$

and

$$[\mathsf{S}\,\widehat{\mathcal{S}_u v}](\xi) = [\mathsf{S}\widehat{u}](\xi)[\mathsf{S}\widehat{v}](2\xi) = \epsilon_u\epsilon_v e^{-i(c_u+2c_v)\xi}.$$

If in addition $c_v - c_u$ is an even integer, then $\mathcal{T}_u v$ also has symmetry satisfying

$$[\mathsf{S}(\widehat{\mathcal{T}_u v})](\xi) = \overline{[\mathsf{S}\widehat{u}](\xi/2)}[\mathsf{S}\widehat{v}](\xi/2) = \epsilon_u\epsilon_v e^{-i\xi(c_v-c_u)/2}.$$

Moreover, all the identities and conclusions still hold if the symmetry operator S *is replaced by the complex symmetry operator* $\mathbb{S}$.

Symmetry of a filter bank is not only very much desired for better visual quality of reconstructed signals in a lot of applications, but also plays a critical role in the implementation of a symmetric discrete framelet transform for signals on a bounded interval, which we shall address in detail here. Due to the same behavior of the symmetry operator S and the complex symmetry operator $\mathbb{S}$, in this section we only consider filters and signals with symmetry; the closely related case for complex symmetry can be deduced similarly and easily.

First, we discuss some natural conditions that we will put on a given dual framelet filter bank $(\{\tilde{u}_0, \ldots, \tilde{u}_s\}, \{u_0, \ldots, u_s\})$ with each filter having symmetry. Let v, w be two arbitrary sequences with symmetry. It is pretty easy to see from Proposition 1.5.3 that $v * w$ also has symmetry and $\mathsf{S}\widehat{v * w} = \mathsf{S}\widehat{v}\mathsf{S}\widehat{w}$. However, the sum $v + w$ generally does not have any symmetry. If $\mathsf{S}\widehat{v} = \mathsf{S}\widehat{w}$, that is, both v and w have the same symmetry type, then indeed $v + w$ has symmetry and $\mathsf{S}(\widehat{v + w}) = \mathsf{S}\widehat{v} = \mathsf{S}\widehat{w}$. Now assume that we have a dual framelet filter bank $(\{\tilde{u}_0, \ldots, \tilde{u}_s\}, \{u_0, \ldots, u_s\})$ with each filter having symmetry such that

$$\mathsf{S}\widehat{u_\ell} = \epsilon_{u_\ell} e^{-ic_{u_\ell}\xi}, \quad \mathsf{S}\widehat{\tilde{u}_\ell} = \epsilon_{\tilde{u}_\ell} e^{-ic_{\tilde{u}_\ell}\xi}, \text{ where } \epsilon_{u_\ell}, \epsilon_{\tilde{u}_\ell} \in \{-1, 1\}, \quad c_{u_\ell}, c_{\tilde{u}_\ell} \in \mathbb{Z}, \tag{1.5.9}$$

for $\ell = 0, \ldots, s$. By (1.1.12) and the above discussion on compatibility of symmetry types, it is natural for us to assume that

$$\mathsf{S}\widehat{\tilde{u}_0}\mathsf{S}\overline{\widehat{u_0}} = \cdots = \mathsf{S}\widehat{\tilde{u}_s}\mathsf{S}\overline{\widehat{u_s}} = \mathsf{S}1 = 1.$$

Since $\mathsf{S}\overline{\widehat{u_\ell}} = \overline{\mathsf{S}\widehat{u_\ell}} = (\mathsf{S}\widehat{u_\ell})^{-1}$, by (1.5.9), the above relation is equivalent to assuming that

$$\epsilon_{\tilde{u}_\ell} = \epsilon_{u_\ell}, \quad c_{\tilde{u}_\ell} = c_{u_\ell}, \qquad \ell = 0, \ldots, s. \tag{1.5.10}$$

Similarly, by (1.1.13), it is natural for us to assume that

$$\mathsf{S}\widehat{\tilde{u}_0}\overline{\mathsf{S}\widehat{u_0}(\cdot + \pi)} = \cdots = \mathsf{S}\widehat{\tilde{u}_s}\overline{\mathsf{S}\widehat{u_s}(\cdot + \pi)}.$$

Since $\mathsf{S}(\widehat{u_\ell}(\xi + \pi)) = [\mathsf{S}\widehat{u_\ell}](\xi + \pi)$, by (1.5.9), the above relation is equivalent to assuming that

$$\epsilon_{u_\ell}\epsilon_{\tilde{u}_\ell} e^{-i\pi c_{u_\ell}} = \epsilon_{u_0}\epsilon_{\tilde{u}_0} e^{-i\pi c_{u_0}} \quad \text{and} \quad c_{\tilde{u}_\ell} - c_{u_\ell} = c_{\tilde{u}_0} - c_{u_0}, \qquad \ell = 0, \ldots, s.$$

For a dual framelet filter bank $(\{\tilde{u}_0, \ldots, \tilde{u}_s\}, \{u_0, \ldots, u_s\})$ with each filter having symmetry in (1.5.9), taking into account of (1.5.10), we always assume the following natural condition:

$$\epsilon_{\tilde{u}_\ell} = \epsilon_{u_\ell}, \quad c_{\tilde{u}_\ell} = c_{u_\ell}, \quad \text{and} \quad c_{u_\ell} - c_{u_0} \in 2\mathbb{Z}, \qquad \ell = 0, \ldots, s. \tag{1.5.11}$$

After shifting the filters by even integers according to Proposition 1.5.1, we can further assume that $(\{\tilde{u}_0, \dots, \tilde{u}_s\}, \{u_0, \dots, u_s\})$ with each filter having symmetry in (1.5.9) is normalized so that (1.5.11) is satisfied with all $c_{u_0}, \dots, c_{u_s} \in \{-1, 0, 1, 2\}$. By (1.5.11), we see that either $c_{u_0}, \dots, c_{u_s} \in \{0, 2\}$ or $c_{u_0}, \dots, c_{u_s} \in \{-1, 1\}$.

For a dual framelet filter bank $(\{\tilde{a}; \tilde{b}_1, \dots, \tilde{b}_s\}, \{a; b_1, \dots, b_s\})_\Theta$ with all involved filters having symmetry in (1.5.9) and $\mathsf{S}\widehat{\Theta}(\xi) = \epsilon_\Theta e^{-ic_\Theta \xi}$, by the same argument, we have

$$\epsilon_{\tilde{a}} = \epsilon_a, \ \epsilon_{\tilde{b}_\ell} = \epsilon_\Theta \epsilon_{b_\ell}, \ c_{\tilde{a}} = c_a - c_\Theta, \ c_{\tilde{b}_\ell} = c_{b_\ell} + c_\Theta, \ c_{b_\ell} - c_a \in 2\mathbb{Z}, \quad \ell = 1, \dots, s.$$

We say that an interval $[m, n]$ is *a control interval* of $v \in l_0(\mathbb{Z})$ if v is uniquely determined by $\{v(k)\}_{k=m}^n$ through the periodicity and/or symmetry of v.

Building on Propositions 1.5.2 and 1.5.3, we have the following result on a symmetric discrete framelet decomposition for signals on a bounded interval.

Proposition 1.5.4 *Let $u \in l_1(\mathbb{Z})$ be a decomposition filter such that $\mathsf{S}\widehat{u}(\xi) = \epsilon e^{-ic\xi}$ for some $\epsilon \in \{-1, 1\}$ and $c \in \mathbb{Z}$, that is,*

$$u(c - k) = \epsilon u(k) \qquad \forall\, k \in \mathbb{Z}. \tag{1.5.12}$$

Let $v^b = \{v^b(k)\}_{k=0}^{N-1}$ be an arbitrary input signal. Extend v^b, with both endpoints non-repeated (EN), into a $(2N-2)$-periodic sequence v on $\mathbb{Z}$ by

$$\begin{aligned} &v(k) = v^b(k), \qquad k = 0, \dots, N-1 \quad \text{and} \\ &v(k) = v^b(2N-2-k), \qquad k = N, \dots, 2N-3. \end{aligned} \tag{1.5.13}$$

(i) Then $u^\star * v$ is a $(2N-2)$-periodic sequence, has the following symmetries:

$$[u^\star * v](-c-k) = [u^\star * v](2N-2-c-k) = \epsilon [u^\star * v](k), \qquad \forall\, k \in \mathbb{Z}, \tag{1.5.14}$$

and $[-\lfloor \frac{c}{2} \rfloor, N-1-\lceil \frac{c}{2} \rceil]$ is a control interval of $u^\star * v$.

(ii) If c is an even integer, then $\mathcal{T}_u v$ is an $(N-1)$-periodic sequence, has the following symmetries:

$$[\mathcal{T}_u v](-\tfrac{c}{2} - k) = [\mathcal{T}_u v](N - 1 - \tfrac{c}{2} - k) = \epsilon [\mathcal{T}_u v](k) \qquad \forall\, k \in \mathbb{Z}, \tag{1.5.15}$$

and $[\lceil -\frac{c}{4} \rceil, \lfloor \frac{N-1}{2} - \frac{c}{4} \rfloor]$ is a control interval of $\mathcal{T}_u v$.

Proof Since v is $(2N-2)$-periodic, we see that v is symmetric about the points 0 and $N-1$:

$$v(-k) = v(2N-2-k) = v(k) \qquad \forall\, k \in \mathbb{Z}. \tag{1.5.16}$$

Applying Proposition 1.5.2, we deduce that $u^\star * v$ is $(2N-2)$-periodic and $\mathcal{T}_u v$ is $(N-1)$-periodic. On the other hand, by Proposition 1.5.3 and (1.5.16), we have $\mathsf{S}\widehat{u^\star * v}(\xi) = \epsilon e^{-i\xi(c_v - c)}$ and $\mathsf{S}\widehat{\mathcal{T}_u v}(\xi) = \epsilon e^{-i\xi(c_v-c)/2}$ for $c_v = 0, N-1$. Hence, (1.5.14) and (1.5.15) hold true. □

Proposition 1.5.5 *Let $u \in l_1(\mathbb{Z})$ be a decomposition filter such that* (1.5.12) *holds for some $\epsilon \in \{-1,1\}$ and $c \in \mathbb{Z}$. Let $v^b = \{v^b(k)\}_{k=0}^{N-1}$ be an input signal. Extend v^b, with both endpoints repeated (ER), into a $2N$-periodic sequence v on $\mathbb{Z}$ by*

$$\begin{aligned} &v(k) = v^b(k), \qquad k = 0,\dots,N-1 \quad \textit{and} \\ &v(k) = v^b(2N-1-k), \qquad k = N,\dots,2N-1. \end{aligned} \tag{1.5.17}$$

(i) Then $u^\star * v$ is $2N$-periodic, has the following symmetries:

$$[u^\star * v](-1-c-k) = [u^\star * v](2N-1-c-k) = \epsilon[u^\star * v](k), \qquad \forall\, k \in \mathbb{Z},$$

and $[-\lfloor \frac{1+c}{2} \rfloor, N - \lceil \frac{1+c}{2} \rceil]$ is a control interval of $u^\star * v$.

(ii) If c is an odd integer, then $\mathcal{T}_u v$ is an N-periodic sequence, has the following symmetries:

$$[\mathcal{T}_u v](-\tfrac{1+c}{2} - k) = [\mathcal{T}_u v](N - \tfrac{1+c}{2} - k) = \epsilon[\mathcal{T}_u v](k) \qquad \forall\, k \in \mathbb{Z},$$

and $[\lceil -\frac{1+c}{4} \rceil, \lfloor \frac{N}{2} - \frac{1+c}{4} \rfloor]$ is a control interval of $\mathcal{T}_u v$.

Proof Since v is $2N$-periodic, (1.5.17) implies that v is symmetric about the points $-\frac{1}{2}$ and $N - \frac{1}{2}$:

$$v(-1-k) = v(2N-1-k) = v(k) \qquad \forall\, k \in \mathbb{Z}.$$

Now the claims can be verified by a similar argument as in the proof of Proposition 1.5.4. □

For the convenience of the reader, the results in Propositions 1.5.4 and 1.5.5 are summarized in Tables 1.1 and 1.2.

For an input signal $v^b = \{v^b(k)\}_{k=0}^{N-1}$ with an even integer N, by Tables 1.1 and 1.2, except the two cases for $c = -1$, we only need to record $\{\mathcal{T}_u v(k)\}_{k=0}^{\frac{N}{2}-1}$, which has exactly half of the length of v^b. For the particular case that $c = -1$ and $\epsilon = 1$, we have to record $\{\mathcal{T}_u v(k)\}_{k=0}^{\frac{N}{2}}$, which has $\frac{N}{2}+1$ coefficients; while for the particular case that $c = -1$ and $\epsilon = -1$, we only have to record $\{\mathcal{T}_u v(k)\}_{k=1}^{\frac{N}{2}-1}$, which has $\frac{N}{2}-1$ coefficients. If the case $c = 1$ does not appear and we have only the case $c = -1$ in a dual framelet filter bank $(\{\tilde{u}_0,\dots,\tilde{u}_s\},\{u_0,\dots,u_s\})$, then by Proposition 1.5.1 we may use the shifted dual framelet filter bank $(\{\tilde{u}_0(\cdot+1),\dots,\tilde{u}_s(\cdot+1)\},\{u_0(\cdot+1),\dots,u_s(\cdot+1)\})$. Note that all the filters in the shifted dual framelet filter bank have the symmetry center $1/2$.

Table 1.1 The analysis/decomposition filter u has symmetry $\mathsf{S}\widehat{u}(\xi) = \epsilon e^{-ic\xi}$, where $\epsilon \in \{-1, 1\}$ and $c \in \{0, 2\}$. v is a symmetric extension with both endpoints non-repeated (EN) from an input signal $v^b = \{v^b(k)\}_{k=0}^{N-1}$ in (1.5.13). For the control interval of $\mathcal{T}_u v$, we assumed that N is an even integer

Filter u	$u^\star * v$ with v extended by EN	$\mathcal{T}_u v$ with v extended by EN
$c = 0$, $\epsilon = 1$	$(2N-2)$-periodic, symmetric about 0 and $N-1$, a control interval $[0, N-1]$	$(N-1)$-periodic, symmetric about 0 and $\frac{N-1}{2}$, a control interval $[0, \frac{N}{2}-1]$
$c = 0$, $\epsilon = -1$	$(2N-2)$-periodic, antisymmetric about 0 and $N-1$, a control interval $[0, N-1]$, $[u^\star * v](0) = [u^\star * v](N-1) = 0$	$(N-1)$-periodic, antisymmetric about 0 and $\frac{N-1}{2}$, a control interval $[0, \frac{N}{2}-1]$, $[\mathcal{T}_u v](0) = 0$
$c = 2$, $\epsilon = 1$	$(2N-2)$-periodic, symmetric about -1 and $N-2$, a control interval $[-1, N-2]$	$(N-1)$-periodic, symmetric about $-\frac{1}{2}$ and $\frac{N}{2}-1$, a control interval $[0, \frac{N}{2}-1]$
$c = 2$, $\epsilon = -1$	$(2N-2)$-periodic, antisymmetric about -1 and $N-2$, a control interval $[-1, N-2]$, $[u^\star * v](-1) = [u^\star * v](N-2) = 0$	$(N-1)$-periodic, antisymmetric about $-\frac{1}{2}$ and $\frac{N}{2}-1$, a control interval $[0, \frac{N}{2}-1]$, $[\mathcal{T}_u v](\frac{N}{2}-1) = 0$

Table 1.2 The analysis/decomposition filter u has symmetry $\mathsf{S}\widehat{u}(\xi) = \epsilon e^{-ic\xi}$, where $\epsilon \in \{-1, 1\}$ and $c \in \{-1, 1\}$. v is a symmetric extension with both endpoints repeated (ER) from an input signal $v^b = \{v^b(k)\}_{k=0}^{N-1}$ in (1.5.17). For the control interval of $\mathcal{T}_u v$, we assumed that N is an even integer

Filter u	$u^\star * v$ with v extended by ER	$\mathcal{T}_u v$ with v extended by ER
$c = 1$, $\epsilon = 1$	$2N$-periodic, symmetric about -1 and $N-1$, a control interval $[-1, N-1]$	N-periodic, symmetric about $-\frac{1}{2}$ and $\frac{N-1}{2}$, a control interval $[0, \frac{N}{2}-1]$
$c = 1$, $\epsilon = -1$	$2N$-periodic, antisymmetric about -1 and $N-1$, a control interval $[-1, N-1]$, $[u^\star * v](-1) = [u^\star * v](N-1) = 0$	N-periodic, antisymmetric about $-\frac{1}{2}$ and $\frac{N-1}{2}$, a control interval $[0, \frac{N}{2}-1]$
$c = -1$, $\epsilon = 1$	$2N$-periodic, symmetric about 0 and N, a control interval $[0, N]$	N-periodic, symmetric about 0 and $\frac{N}{2}$, a control interval $[0, \frac{N}{2}]$
$c = -1$, $\epsilon = -1$	$2N$-periodic, antisymmetric about 0 and N, a control interval $[0, N]$, $[u^\star * v](0) = [u^\star * v](N) = 0$	N-periodic, antisymmetric about 0 and $\frac{N}{2}$, a control interval $[0, \frac{N}{2}]$, $[\mathcal{T}_u v](0) = [\mathcal{T}_u v](\frac{N}{2}) = 0$

We now discuss a symmetric discrete framelet reconstruction for signals on a bounded interval. Since the reconstruction filter u has the same symmetry type as the decomposition filter $\tilde{u}$, for recorded framelet coefficients $\{w^b(k)\}_{k=0}^{\frac{N}{2}-1}$, according to the cases c and ϵ, we extend w^b into a sequence w on $\mathbb{Z}$ according to the symmetries and periodicity of $\mathcal{T}_u v$ in Tables 1.1 and 1.2. Then the reconstructed sequence $\mathcal{S}_u w$ will have the corresponding same symmetry property as v in Tables 1.1 and 1.2, which can be easily verified by Propositions 1.5.2 and 1.5.3.

We present a few examples to illustrate the symmetric discrete framelet transforms.

Example 1.5.2 We apply the biorthogonal wavelet filter bank in Example 1.1.2 to v^b in (1.5.8). Since $\mathsf{S}\widehat{\tilde{u}_0} = 1$ and $\mathsf{S}\widehat{\tilde{u}_1} = e^{-i2\xi}$, we extend v^b according to Table 1.1 by both endpoints non-repeated (EN): v is 14-periodic, is symmetric about the points 0 and 7, and is given by

$$v = \{\dots, -25, -23 - 23, -22, \underline{\mathbf{-21}, -22, -23, -23, -25, 38, 36, 34}, 36, 38, -25, \dots\}. \tag{1.5.18}$$

Then $\mathcal{T}_{\tilde{u}_0} v$ is 7-periodic and is symmetric about the points 0 and $7/2$:

$$w_0 = \tfrac{\sqrt{2}}{2}\mathcal{T}_{\tilde{u}_0} v = \tfrac{\sqrt{2}}{2}\{\dots, -\tfrac{133}{4}, -\tfrac{91}{2}, \underline{\mathbf{-42}, -\tfrac{91}{2}, -\tfrac{133}{4}, \tfrac{349}{4}}, \tfrac{349}{4}, -\tfrac{133}{4}, -\tfrac{91}{2}, -42, \dots\},$$

and $\mathcal{T}_{\tilde{u}_1} v$ is 7-periodic and is symmetric about the points $-\frac{1}{2}$ and 3:

$$w_1 = \tfrac{\sqrt{2}}{2}\mathcal{T}_{\tilde{u}_1} v = \tfrac{\sqrt{2}}{2}\{\dots, -2, \tfrac{65}{2}, 1, 0, \underline{\mathbf{0}, 1, \tfrac{65}{2}, -2}, \tfrac{65}{2}, 1, 0, 0, 1, \tfrac{65}{2}, \dots\}.$$

Both the control intervals of $\mathcal{T}_{\tilde{u}_0} v$ and $\mathcal{T}_{\tilde{u}_1} v$ are underlined and have 4 coefficients. It is easy to directly check that $\frac{\sqrt{2}}{2}(\mathcal{S}_{u_0} w_0 + \mathcal{S}_{u_1} w_1) = v$.

Example 1.5.3 We apply the tight framelet filter bank in Example 1.1.3 to v^b in (1.5.8). Since $\mathsf{S}\widehat{u_0} = \mathsf{S}\widehat{u_2} = 1$ and $\mathsf{S}\widehat{u_1} = -1$, we extend v^b according to Table 1.1 with both endpoints non-repeated (EN): v is 14-periodic, is symmetric about both 0 and 7, and is given in (1.5.18). Then both $\mathcal{T}_{u_0} v$ and $\mathcal{T}_{u_2} v$ are 7-periodic and symmetric about the points 0 and $7/2$, and $\mathcal{T}_{u_1} v$ is 7-periodic and antisymmetric about the points 0 and $7/2$:

$$w_0 = \tfrac{\sqrt{2}}{2}\mathcal{T}_{u_0} v = \tfrac{\sqrt{2}}{2}\{\dots, 72, -\tfrac{35}{2}, -\tfrac{91}{2}, \underline{\mathbf{-43}, -\tfrac{91}{2}, -\tfrac{35}{2}, 72}, 72, -\tfrac{35}{2}, -\tfrac{91}{2}, -43, \dots\},$$

$$w_1 = \tfrac{\sqrt{2}}{2}\mathcal{T}_{u_1} v = \{\dots, 2, -\tfrac{61}{2}, \tfrac{1}{2}, \underline{\mathbf{0}, -\tfrac{1}{2}, \tfrac{61}{2}, -2}, 2, -\tfrac{61}{2}, \tfrac{1}{2}, 0, -\tfrac{1}{2}, \tfrac{61}{2}, \dots\},$$

$$w_2 = \tfrac{\sqrt{2}}{2}\mathcal{T}_{u_2} v = \tfrac{\sqrt{2}}{2}\{\dots, 0, -\tfrac{65}{2}, -\tfrac{1}{2}, \underline{\mathbf{1}, -\tfrac{1}{2}, -\tfrac{65}{2}, 0}, 0, -\tfrac{65}{2}, -\tfrac{1}{2}, 1, -\tfrac{1}{2}, \dots\}.$$

It is easy to directly check that $\frac{\sqrt{2}}{2}(\mathcal{S}_{u_0} w_0 + \mathcal{S}_{u_1} w_1 + \mathcal{S}_{u_2} w_2) = v$.

Example 1.5.4 We apply the dual framelet filter bank in Example 1.1.4 to v^b in (1.5.8). Since $\mathsf{S}\widehat{\tilde{u}_0} = e^{-i\xi}$, we extend v^b according to Table 1.2 by both endpoints repeated (ER): v is 16-periodic, is symmetric about the points $-\frac{1}{2}$ and $7\frac{1}{2}$, and is given by

$$v = \{\ldots, -23, -23, -22, -21, \underline{\mathbf{-21}, -22, -23, -23, -25, 38, 36, 34}, 34, 36, 38, \ldots\}.$$

Then all $\mathcal{T}_{\tilde{u}_0} v$, $\mathcal{T}_{\tilde{u}_1} v$ and $\mathcal{T}_{\tilde{u}_2} v$ are 8-periodic. $\mathcal{T}_{\tilde{u}_0} v$ is symmetric about the points $-\frac{1}{2}$ and $\frac{7}{2}$, $\mathcal{T}_{\tilde{u}_1} v$ is antisymmetric about the points 0 and 4, and $\mathcal{T}_{\tilde{u}_2} v$ is antisymmetric about the points $-\frac{1}{2}$ and $\frac{7}{2}$:

$$w_0 = \tfrac{\sqrt{2}}{2}\mathcal{T}_{\tilde{u}_0} v = \tfrac{\sqrt{2}}{2}\{\ldots, 70, 13, -46, -43, \underline{\mathbf{-43}, -46, 13, 70}, 70, 13, -46, -43, \ldots\},$$

$$w_1 = \tfrac{\sqrt{2}}{2}\mathcal{T}_{\tilde{u}_1} v = \tfrac{\sqrt{2}}{2}\{\ldots, 0, 2, 2, 1, \underline{\mathbf{0}, -1, -2, -2}, 0, 2, 2, 1, 0, \ldots\},$$

$$w_2 = \tfrac{\sqrt{2}}{2}\mathcal{T}_{\tilde{u}_2} v = \tfrac{\sqrt{2}}{2}\{\ldots, 2, -63, 0, 1, \underline{\mathbf{-1}, 0, 63, -2}, 2, -63, 0, 1, -1, 0, \ldots\}.$$

It is easy to directly check that $\frac{\sqrt{2}}{2}(\mathcal{S}_{u_0} w_0 + \mathcal{S}_{u_1} w_1 + \mathcal{S}_{u_2} w_2) = v$.

1.5.4 Symmetric Extension for Filter Banks Without Symmetry

On one hand, the implementation of a discrete framelet transform using symmetry extension in Sect. 1.5.3 is complicated by the many different symmetry patterns of the filters. On the other hand, many filter banks do not have any symmetry at all. In order to reduce artificial jumps near boundaries induced by periodic extension, if storage of framelet coefficients is not an issue, there are two ways to employ symmetric extension for general filter banks without symmetry. The first way is to extend the input signal by either EN or ER (or any other extension method) and then directly record all the framelet coefficients in (1.5.2) including the extra framelet coefficients in (1.5.3). The second way is to use the following simple algorithm.

Algorithm 1.5.6 *Let $v^b = \{v^b(k)\}_{k=0}^{N-1}$ be an arbitrary input signal.*

(S1) (Pre-processing) Extend the right-hand endpoint of v^b by ER (often used) or EN to obtain another signal $\mathring{v}^b$. More precisely, for ER extension, $\mathring{v}^b(k) = \mathring{v}^b(2N-1-k) := v^b(k)$, $k = 0, \ldots, N-1$, that is,

$$\mathring{v}^b = \{v^b(0), \ldots, v^b(N-2), v^b(N-1), v^b(N-1), v^b(N-2), \ldots, v^b(0)\}_{[0,2N-1]};$$

for EN extension, $\mathring{v}^b(k) = \mathring{v}^b(2N-2-k) := v^b(k)$, $k = 0, \ldots, N-1$, that is,

$$\mathring{v}^b = \{v^b(0), \ldots, v^b(N-2), v^b(N-1), v^b(N-2), \ldots, v^b(1)\}_{[0,2N-3]}.$$

(S2) Apply a periodic discrete framelet transform to $\mathring{v}^b$. Denote the reconstructed signal of $\mathring{v}^b$ by $\mathring{v}^r$.

(S3) *(After-processing) If ER is used in (S1), a reconstructed signal v^r of v^b is obtained by $v^r(k) = \frac{\mathring{v}^r(k)+\mathring{v}^r(2N-1-k)}{2}$, $k = 0, \dots, N-1$. If EN is used in (S1), a reconstructed signal v^r of v^b is obtained by $v^r(k) = \frac{\mathring{v}^r(k)+\mathring{v}^r(2N-2-k)}{2}$, $k = 0, \dots, N-1$.*

1.6 Discrete Framelet Transforms Implemented in the Frequency Domain*

In this section we provide an equivalent implementation in the frequency domain of the framelet algorithms for signals on bounded intervals in Sect. 1.5.

Filters having infinite support are also used in applications. Such filters $u = \{u(k)\}_{k\in\mathbb{Z}}$ are often given in the frequency domain such that $\widehat{u}$ has an explicit expression, while its time domain form $\{u(k)\}_{k\in\mathbb{Z}}$ is only implicitly given by $u(k) = \frac{1}{2\pi}\int_{-\pi}^{\pi}\widehat{u}(\xi)e^{-ik\xi}d\xi, k \in \mathbb{Z}$ and lacks an explicit expression. On the other hand, for certain circumstances, it is easier to design a filter in the frequency domain, that is, to design $\widehat{u}$, rather than its time domain form $\{u(k)\}_{k\in\mathbb{Z}}$. This can be seen from the perfect reconstruction condition in (1.1.12) and (1.1.13) which are expressed in the frequency domain. As a consequence, it is important to have an equivalent implementation in the frequency domain of the framelet algorithms described in Sect. 1.5 for signals on bounded intervals.

A periodic discrete framelet transform can be implemented using discrete Fourier transform (DFT). For $v^b = \{v^b(k)\}_{k=0}^{N-1}$, its N-point *discrete Fourier transform* is another N-point sequence $\{\widehat{v^b}(\frac{2\pi n}{N})\}_{n=0}^{N-1}$ on the interval $[0, N-1] \cap \mathbb{Z}$, where $\widehat{v^b}(\xi) := \sum_{k=0}^{N-1} v^b(k)e^{-ik\xi}$ for $\xi \in \mathbb{R}$. That is, if we regard v^b as a sequence on $\mathbb{Z}$ by the simple zero-padding extension, then the N-point discrete Fourier transform of $v^b = \{v^b(k)\}_{k=0}^{N-1}$ is obtained by sampling the Fourier series $\widehat{v^b}(\xi)$ at $\xi = \frac{2\pi n}{N}$ for $n = 0, \dots, N-1$. It is well known that the original signal $v^b = \{v^b(k)\}_{k=0}^{N-1}$ can be recovered from its N-point discrete Fourier transform via the *inverse discrete Fourier transform*:

$$v^b(k) = \frac{1}{N}\sum_{n=0}^{N-1}\widehat{v^b}(\tfrac{2\pi n}{N})e^{i\frac{2\pi nk}{N}}, \qquad k = 0, \dots, N-1. \tag{1.6.1}$$

For N-point signals, both the discrete Fourier transform and its inverse can be implemented by fast Fourier transform (FFT) with computational complexity $\mathscr{O}(N \log N)$. See Appendix A for basic properties of the discrete Fourier transform.

The periodic discrete framelet transform in Proposition 1.5.2 can be implemented in the frequency domain using DFT as follows.

Proposition 1.6.1 *Let $u \in l_1(\mathbb{Z})$ be a filter and $v^b = \{v^b(k)\}_{k=0}^{N-1}$ be an arbitrary input signal. Extend v^b into an N-periodic sequence v on $\mathbb{Z}$ as in (1.5.4). Then $u * v$ is N-periodic and the following properties hold:*

(i) the N-point discrete Fourier transform of $\{[u*v](k)\}_{k=0}^{N-1}$ is $\{\widehat{u}(\frac{2\pi n}{N})\widehat{v^b}(\frac{2\pi n}{N})\}_{n=0}^{N-1}$. Therefore, $\{[u * v](k)\}_{k=0}^{N-1}$ can be obtained by the inverse discrete Fourier transform (see (1.6.1)) of $\{\widehat{u}(\frac{2\pi n}{N})\widehat{v^b}(\frac{2\pi n}{N})\}_{n=0}^{N-1}$ as follows:

$$[u * v](k) = \frac{1}{N}\sum_{n=0}^{N-1}\widehat{u}(\tfrac{2\pi n}{N})\widehat{v^b}(\tfrac{2\pi n}{N})e^{i\frac{2\pi nk}{N}}, \qquad k \in \mathbb{Z};$$

(ii) the $2N$-point discrete Fourier transform of $\{[\mathcal{S}_u v](k)\}_{k=0}^{2N-1}$ is given by

$$\sum_{n=0}^{2N-1}[\mathcal{S}_u v](k)e^{-i\frac{2\pi kn}{2N}} = 2\widehat{u}(\tfrac{\pi n}{N})\widehat{v^b}(\tfrac{2\pi n}{N}), \qquad n \in \mathbb{Z}; \tag{1.6.2}$$

(iii) if N is a positive even integer, then the $\frac{N}{2}$-point discrete Fourier transform of $\{[\mathcal{T}_u v](k)\}_{k=0}^{N/2-1}$ is given by: for $n = 0, \dots, N/2 - 1$,

$$\sum_{k=0}^{\frac{N}{2}-1}[\mathcal{T}_u v](k)e^{-i\frac{2\pi kn}{N/2}} = \overline{\widehat{u}(\tfrac{2\pi n}{N})}\widehat{v^b}(\tfrac{2\pi n}{N}) + \overline{\widehat{u}(\tfrac{2\pi n}{N}+\pi)}\widehat{v^b}(\tfrac{2\pi n}{N}+\pi). \tag{1.6.3}$$

Proof By Proposition 1.5.2, $u * v$ is N-periodic. Item (i) is a basic property of discrete Fourier transform, see Appendix A for details. By $\mathcal{S}_u v = 2u * (v \uparrow 2)$ and item (i), we see that $\mathcal{S}_u v$ is $(2N)$-periodic and (1.6.2) holds. By item (i), we have

$$[u^\star * v](k) = \frac{1}{N}\sum_{n=0}^{N-1}\overline{\widehat{u}(\tfrac{2\pi n}{N})}\widehat{v^b}(\tfrac{2\pi n}{N})e^{i\frac{2\pi nk}{N}}, \qquad k = 0, \dots, N-1.$$

Since $\mathcal{T}_u v = 2(u^\star * v) \downarrow 2$ and N is even, it is straightforward to check that $\mathcal{T}_u v$ is $\frac{N}{2}$-periodic and

$$\sum_{k=0}^{\frac{N}{2}-1}[\mathcal{T}_u v](k)e^{-i\frac{2\pi kn}{N/2}} = \sum_{k=0}^{\frac{N}{2}-1}2(u^\star * v)(2k)e^{-i\frac{4\pi kn}{N}}$$

$$= 2\sum_{k=0}^{\frac{N}{2}-1}\frac{1}{N}\sum_{m=0}^{N-1}\overline{\widehat{u}(\tfrac{2\pi m}{N})}\widehat{v^b}(\tfrac{2\pi m}{N})e^{i\frac{4\pi(m-n)k}{N}} = \frac{2}{N}\sum_{m=0}^{N-1}\overline{\widehat{u}(\tfrac{2\pi m}{N})}\widehat{v^b}(\tfrac{2\pi m}{N})\sum_{k=0}^{\frac{N}{2}-1}e^{i\frac{4\pi(m-n)k}{N}}.$$

Note that for $n = 0, \dots, \frac{N}{2} - 1$, we have

$$\sum_{k=0}^{\frac{N}{2}-1} e^{i\frac{4\pi(m-n)k}{N}} = \begin{cases} \frac{N}{2}, & m = n \quad \text{or} \quad n + \frac{N}{2}, \\ 0, & m \in \{0, \dots, N-1\} \backslash \{n, n + \frac{N}{2}\}. \end{cases}$$

Now we can easily see that (1.6.3) holds. □

As a direct consequence of Propositions 1.1.1 and 1.5.2, we have the following result on the perfect reconstruction property of a periodic discrete framelet transform for signals on a bounded interval stated right after Proposition 1.5.2.

Proposition 1.6.2 *Let $N \in 2\mathbb{N}$. The one-level periodic discrete framelet transform has the perfect reconstruction property (i.e., $\mathcal{V}^{per}\widetilde{\mathcal{W}}^{per}v^b = v^b$ for all $v^b = \{v^b(k)\}_{k=0}^{N-1}$) if and only if* (1.1.12) *and* (1.1.13) *hold for all $\xi = 0, \frac{2\pi}{N}, \dots, \frac{2\pi(N-1)}{N}$.*

Due to many different types of symmetries, it is a little bit more complicated to implement a symmetric discrete framelet transform in the frequency domain. Since both the transition operator and subdivision operator use convolution operation as the core operation, it is essential for us to discuss the convolution operation in a symmetric discrete framelet transform.

Let $v^b = \{v^b(k)\}_{k=0}^{N-1}$ be an input signal. Let $u \in l_1(\mathbb{Z})$ be a filter with $\mathsf{S}\widehat{u}(\xi) = \varepsilon e^{ic\xi}$ for $\varepsilon \in \{-1, 1\}$ and $c \in \mathbb{Z}$. Define $\mathbf{u}_c(\xi) := e^{ic\xi/2}\widehat{u}(\xi)$. Then $\mathbf{u}_c(\xi) = \varepsilon\mathbf{u}_c(-\xi)$. Moreover, if the filter u is real-valued, then $\sqrt{\varepsilon}\mathbf{u}_c(\xi)$ is real-valued for all $\xi \in \mathbb{R}$.

Let v be the sequence in (1.5.13) extending v^b with both endpoints non-repeated (EN). Using $(2N-2)$-point discrete Fourier transform of the signal $\{v(k)\}_{k=2-N}^{N-1}$ and its inverse Fourier transform, the $(2N-2)$-periodic sequence $u * v$ is computed via

$$[u * v](k) = \frac{1}{2N-2} \sum_{n=2-N}^{N-1} \widehat{u}(\tfrac{\pi n}{N-1}) \Big(\sum_{m=2-N}^{N-1} v(m) e^{-i\frac{\pi nm}{N-1}} \Big) e^{i\frac{\pi kn}{N-1}}, \quad k \in \mathbb{Z}. \tag{1.6.4}$$

To compute (1.6.4) efficiently, we use variants of discrete cosine transforms (DCTs) and discrete sine transforms (DSTs). Since v extends v^b as in (1.5.13), the DCT_I (Discrete Cosine Transform of Type I) of the N-point $\{v^b(k)\}_{k=0}^{N-1}$ is defined by

$$\widehat{v}^{DCT_I}(n) := \frac{1}{2} \sum_{m=2-N}^{N-1} v(m) e^{-i\frac{\pi nm}{N-1}} = \frac{1}{2} v^b(0) + \frac{(-1)^n}{2} v^b(N-1) + \sum_{m=1}^{N-2} v^b(m) \cos(\tfrac{\pi nm}{N-1}).$$

Note that $\widehat{v}^{DCT_I}(-n) = \widehat{v}^{DCT_I}(n)$ for all $n \in \mathbb{Z}$. If $\varepsilon = 1$, the identity (1.6.4) becomes

$$[u * v](k) = \frac{1}{N-1} \sum_{n=2-N}^{N-1} \mathbf{u}_c(\tfrac{\pi n}{N-1}) \widehat{v}^{DCT_I}(n) e^{i\frac{\pi(k-c/2)n}{N-1}} = \frac{1}{N-1} \mathbf{u}_c(0) \widehat{v}^{DCT_I}(0)$$

$$+ \frac{e^{i\pi(k-c/2)}}{N-1} \mathbf{u}_c(\pi) \widehat{v}^{DCT_I}(N-1) + \frac{2}{N-1} \sum_{n=1}^{N-2} \mathbf{u}_c(\tfrac{\pi n}{N-1}) \widehat{v}^{DCT_I}(n) \cos(\tfrac{\pi(k-c/2)n}{N-1}),$$

which is the N-point DCT_I of $\{\frac{2}{N-1}\mathbf{u}_c(\frac{\pi n}{N-1})\widehat{v}^{DCT_I}(n)\}_{n=0}^{N-1}$ if $c=0$. The above is the $(N-1)$-point DCT_{III} of $\{\frac{2}{N-1}\mathbf{u}_c(\frac{\pi n}{N-1})\widehat{v}^{DCT_I}(n)\}_{n=0}^{N-2}$ if $c=-1$, since $\mathbf{u}_c(\pi)=0$ if $\varepsilon=1$ and c is odd. If $\varepsilon=-1$, then we have $\mathbf{u}_c(0)=0$ and similarly (1.6.4) becomes

$$[u*v](k)=\frac{e^{i\pi(k-\frac{c}{2})}}{N-1}\mathbf{u}_c(\pi)\widehat{v}^{DCT_I}(N-1)+\frac{2}{N-1}\sum_{n=1}^{N-2}i\mathbf{u}_c(\tfrac{\pi n}{N-1})\widehat{v}^{DCT_I}(n)\sin(\tfrac{\pi(k-\frac{c}{2})n}{N-1}),$$

which is linked to variants of discrete sine transforms.

Similarly, let v be the sequence in (1.5.17) extending v^b with both endpoints repeated (ER). Using $2N$-point discrete Fourier transform of $\{v(k)\}_{k=-N}^{N-1}$ and its inverse, the $2N$-periodic sequence $u*v$ can be computed via

$$[u*v](k)=\frac{1}{2N}\sum_{n=-N}^{N-1}\widehat{u}(\tfrac{\pi n}{N})\Big(\sum_{m=-N}^{N-1}v(m)e^{-i\frac{\pi nm}{N}}\Big)e^{i\frac{\pi kn}{N}},\qquad k\in\mathbb{Z}. \tag{1.6.5}$$

By (1.5.17), the (widely used) DCT_{II} of the N-point $\{v^b(k)\}_{k=0}^{N-1}$ is defined by

$$\widehat{v}^{DCT_{II}}(n):=\frac{e^{-i\frac{n\pi}{2N}}}{2}\sum_{m=-N}^{N-1}v(m)e^{-i\frac{\pi nm}{N}}=\sum_{m=0}^{N-1}v^b(m)\cos(\tfrac{\pi n(m+1/2)}{N}).$$

Note that $\widehat{v}^{DCT_{II}}(-n)=\widehat{v}^{DCT_{II}}(n)$ for all $n\in\mathbb{Z}$ and $\widehat{v}^{DCT_{II}}(N)=0$. If $\varepsilon=1$, then the identity (1.6.5) becomes

$$\begin{aligned}[u*v](k)&=\frac{1}{N}\sum_{n=-N}^{N-1}\mathbf{u}_c(\tfrac{\pi n}{N})\widehat{v}^{DCT_{II}}(n)e^{i\frac{\pi(k+\frac{1-c}{2})n}{N}}\\&=\frac{1}{N}\mathbf{u}_c(0)\widehat{v}^{DCT_{II}}(0)+\frac{2}{N}\sum_{n=1}^{N-1}\mathbf{u}_c(\tfrac{\pi n}{N})\widehat{v}^{DCT_{II}}(n)\cos(\tfrac{\pi(k+\frac{1-c}{2})n}{N}),\end{aligned}$$

which is the N-point DCT_{III} of $\{\frac{2}{N}\widehat{u}(\frac{\pi n}{N})\widehat{v}^{DCT_{II}}(n)\}_{n=0}^{N-1}$ if $c=0$. If $\varepsilon=1$ and c is odd, then $\mathbf{u}_c(\pi)=0$ and the above is the $(N+1)$-point DCT_I of $\{\frac{2}{N}\widehat{u}(\frac{\pi n}{N})\widehat{v}^{DCT_{II}}(n)\}_{n=0}^{N}$ if $c=1$. If $\varepsilon=-1$, then we have $\mathbf{u}_c(0)=0$ and similarly (1.6.4) becomes

$$[u*v](k)=\frac{2}{N}\sum_{n=1}^{N-1}i\mathbf{u}_c(\tfrac{\pi n}{N})\widehat{v}^{DCT_{II}}(n)\sin(\tfrac{\pi(k+\frac{1-c}{2})n}{N}),$$

which is linked to variants of discrete sine transforms.

1.7 Exercises

1.1. Prove the following identities: For $u, v, w \in l_2(\mathbb{Z})$ and $n \in \mathbb{Z}$, $\langle v, w\rangle = [v * w^\star](0)$, $[v * w](n) = \langle v, w^\star(\cdot - n)\rangle$, $u * v = v * u$, $\langle u * v, w\rangle = \langle v, u^\star * w\rangle$, $\langle v \uparrow \mathsf{d}, w\rangle = \langle v, w \downarrow \mathsf{d}\rangle$, where $\mathsf{d} \in \mathbb{Z}\backslash\{0\}$ is a sampling factor.

1.2. Let $(\mathcal{H}, \langle\cdot,\cdot\rangle)$ be an inner product space over the complex field $\mathbb{C}$, e.g., $\mathcal{H} = l_2(\mathbb{Z})$. Let $T : \mathcal{H} \to \mathcal{H}$ be a linear mapping. Prove the polarization identity:

$$\begin{aligned}\langle Tv, w\rangle =& \frac{1}{4}\big[\langle T(v+w), v+w\rangle - \langle T(v-w), v-w\rangle \\ &+ i\langle T(v+iw), v+iw\rangle - i\langle T(v-iw), v-iw\rangle\big], \quad v, w \in \mathcal{H}.\end{aligned}$$

1.3. Prove Proposition 1.1.5.

1.4. Prove that the perfect reconstruction condition in (1.3.6) is equivalent to

$$\begin{bmatrix} \widehat{\tilde{a}^{[0]}}(\xi) & \widehat{\tilde{b}_1^{[0]}}(\xi) & \cdots & \widehat{\tilde{b}_s^{[0]}}(\xi) \\ \widehat{\tilde{a}^{[1]}}(\xi) & \widehat{\tilde{b}_1^{[1]}}(\xi) & \cdots & \widehat{\tilde{b}_s^{[1]}}(\xi) \end{bmatrix} \begin{bmatrix} \widehat{a^{[0]}}(\xi) & \widehat{b_1^{[0]}}(\xi) & \cdots & \widehat{b_s^{[0]}}(\xi) \\ \widehat{a^{[1]}}(\xi) & \widehat{b_1^{[1]}}(\xi) & \cdots & \widehat{b_s^{[1]}}(\xi) \end{bmatrix}^\star = \frac{1}{2} I_2.$$

1.5. Prove the Leibniz differentiation formula: $[\mathbf{fg}]^{(n)} = \sum_{j=0}^n \frac{n!}{j!(n-j)!} \mathbf{f}^{(j)}(\cdot)\mathbf{g}^{(n-j)}(\cdot)$.

1.6. Let $\partial_1 := \frac{\partial}{\partial \xi_1}$ and $\partial_2 := \frac{\partial}{\partial \xi_2}$. Using directional derivatives to prove $[\mathbf{f}(\xi)\mathbf{g}(\xi)]^{(n)} = [(\partial_1 + \partial_2)^n(\mathbf{f}(\xi_1)\mathbf{g}(\xi_2))]|_{\xi_1=\xi,\xi_2=\xi}$ and use it to prove Exercise 1.5.

1.7. Prove the generalized product rule for differentiation in (1.2.3).

1.8. Prove the identity in (1.2.4).

1.9. For $u = \{u(k)\}_{k\in\mathbb{Z}} \in l_0(\mathbb{Z})$, (1.2.15) holds $\iff \sum_{k\in\mathbb{Z}} u(k)k^j = c^j$ for all $j = 0, \dots, m \iff \sum_{k\in\mathbb{Z}} u(k)(k-c)^j = \boldsymbol{\delta}(j), j = 0, \dots, m$.

1.10. For $0 < p \leqslant \infty$ and $v \in l(\mathbb{Z})$, define $\|v\|_{l_p(\mathbb{Z})} := (\sum_{k\in\mathbb{Z}} |v(k)|^p)^{1/p}$. Prove that $\|u+v\|_{l_p(\mathbb{Z})}^{\min(p,1)} \leqslant \|u\|_{l_p(\mathbb{Z})}^{\min(p,1)} + \|v\|_{l_p(\mathbb{Z})}^{\min(p,1)}$ and $\|v\|_{l_p(\mathbb{Z})} \leqslant \|v\|_{l_q(\mathbb{Z})}$ $\forall 0 < q \leqslant p \leqslant \infty$.

1.11. For a linear operator $T : l_p(\mathbb{Z}) \to l_p(\mathbb{Z})$, its operator norm is defined to be $\|T\| := \sup\{\|Tv\|_{l_p(\mathbb{Z})} : \|v\|_{l_p(\mathbb{Z})} \leqslant 1\}$. For a filter $u \in l_0(\mathbb{Z})$, prove that all the linear operators $u* : l_p(\mathbb{Z}) \to l_p(\mathbb{Z})$, $\mathcal{S}_u : l_p(\mathbb{Z}) \to l_p(\mathbb{Z})$, $\mathcal{T}_u : l_p(\mathbb{Z}) \to l_p(\mathbb{Z})$ are well defined and bounded for all $0 < p \leqslant \infty$. In particular, with $q := \min(p, 1)$,

$$\begin{aligned} &\|u * v\|_{l_p(\mathbb{Z})} \leqslant \|u\|_{l_q(\mathbb{Z})}\|v\|_{l_p(\mathbb{Z})}, \qquad \|\mathcal{T}_u v\|_{l_p(\mathbb{Z})} \leqslant \|u\|_{l_q(\mathbb{Z})}\|v\|_{l_p(\mathbb{Z})}, \\ &\|\mathcal{S}_u v\|_{l_p(\mathbb{Z})} \leqslant \|v\|_{l_p(\mathbb{Z})} \max(\|u^{[0]}\|_{l_q(\mathbb{Z})}, \|u^{[1]}\|_{l_q(\mathbb{Z})}). \end{aligned}$$

1.12. Prove that Proposition 1.1.2 is still true if $l(\mathbb{Z})$ is replaced by $l_p(\mathbb{Z})$, $0 < p \leqslant \infty$.

1.13. Show that $\mathcal{W} : l_p(\mathbb{Z}) \to (l_p(\mathbb{Z}))^{1\times(s+1)}$ in (1.1.19) and $\mathcal{V} : (l_p(\mathbb{Z}))^{1\times(s+1)} \to l_p(\mathbb{Z})$ in (1.1.20) are well-defined bounded linear operators for $0 < p \leqslant \infty$.

1.14. Let m be a nonnegative integer. Let $\mathbf{u}, \mathbf{v}$ be functions which are m-times differentiable at the origin and satisfy $\mathbf{u}(0) = \mathbf{v}(0) \ne 0$. Suppose that d and λ are real numbers such that $\mathsf{d}^n\mathbf{u}(0) \ne \lambda^n\mathbf{v}(0)$ for all $n = 1, \dots, m$. Show that there exists a finitely supported sequence $\theta \in l_0(\mathbb{Z})$ satisfying

$$\widehat{\theta}(0) = 1 \quad \text{and} \quad \widehat{\theta}(\mathsf{d}\xi)\mathbf{u}(\xi) = \widehat{\theta}(\lambda\xi)\mathbf{v}(\xi) + \mathscr{O}(|\xi|^{m+1}), \quad \xi \to 0.$$

More precisely, $\widehat{\theta}^{(j)}(0), 0 \leqslant j \leqslant m$ are uniquely determined by $\mathbf{u}^{(j)}(0)$ and $\mathbf{v}^{(j)}(0), 0 \leqslant j \leqslant m$ via the following recursive formula: For $n = 1, \dots, m$,

$$\widehat{\theta}(0) = 1 \quad \text{and} \quad \widehat{\theta}^{(n)}(0) = \sum_{j=0}^{n-1} \frac{n!}{j!(n-j)!} \frac{\lambda^j\mathbf{v}^{(n-j)}(0) - \mathsf{d}^j\mathbf{u}^{(n-j)}(0)}{\mathsf{d}^n\mathbf{u}(0) - \lambda^n\mathbf{v}(0)} \widehat{\theta}^{(j)}(0).$$

1.15. Let $u \in l_0(\mathbb{Z})$ and m be a positive integer. Show that the coefficient matrix of $\mathcal{T}_u|_{\mathbb{P}_{m-1}}$ under the basis $\{1, x, \dots, x^{m-1}\}$ of $\mathbb{P}_{m-1}$ is a lower triangular matrix with its diagonal entries being $2\overline{\widehat{u}(0)}, 2^2\overline{\widehat{u}(0)}, \dots, 2^m\overline{\widehat{u}(0)}$. Moreover, if $\widehat{u}(0) \ne 0$, then

$$\mathcal{T}_u[(\cdot)^j * \theta] = 2^{j+1}\overline{\widehat{u}(0)}[(\cdot)^j * \theta], \qquad j = 0, \dots, m-1,$$

that is, $(\cdot)^j * \theta \in \mathbb{P}_{m-1}$ is a nonzero eigenvector of $\mathcal{T}_u : \mathbb{P}_{m-1} \to \mathbb{P}_{m-1}$ corresponding to the eigenvalue $2^{n+1}\overline{\widehat{u}(0)}$ for all $n = 0, \dots, m-1$, where $(\cdot)^j$ is the polynomial sequence induced by x^j and $\theta \in l_0(\mathbb{Z})$ satisfies

$$\widehat{\theta}(0) = 1 \quad \text{and} \quad \widehat{\theta}(2\xi)\overline{\widehat{u}(0)} = \widehat{\theta}(\xi)\overline{\widehat{u}(\xi)} + \mathscr{O}(|\xi|^m), \qquad \xi \to 0.$$

1.16. For $\mathsf{p} \in \mathbb{P}_{m-1}$ and $\widehat{v}(\xi) = \mathscr{O}(|\xi|^m)$ as $\xi \to 0$, show that

$$\langle \mathcal{S}_u\mathsf{p}, v\rangle = \left[\mathsf{p}\left(-i\frac{d}{d\xi}\right)\left(\widehat{u}(\xi/2+\pi)\overline{\widehat{v}(\xi/2+\pi)}\right)\right]\Big|_{\xi=0} = \left[\mathsf{p}\left(-\frac{i}{2}\frac{d}{d\xi}\right)\left(\widehat{u}(\xi)\overline{\widehat{v}(\xi)}\right)\right]\Big|_{\xi=\pi}.$$

Then use it to prove the equivalence between items (1) and (5) of Theorem 1.2.4.

1.17. Prove the identity in (1.3.26).

1.18. Let $u, \tilde{u} \in l_0(\mathbb{Z})$ and $m \in \mathbb{N}$. Prove that $\frac{1}{2}\mathcal{S}_u\mathcal{T}_{\tilde{u}}\mathsf{p} = \mathsf{p}$ for all $\mathsf{p} \in \mathbb{P}_{m-1}$ if and only if u has m sum rules and $\overline{\widehat{u}(\xi)}\widehat{\tilde{u}}(\xi) = 1 + \mathscr{O}(|\xi|^m)$ as $\xi \to 0$, that is, $u^\star * \tilde{u}$ has m linear-phase moments with phase 0.

1.19. Let $u \in l_0(\mathbb{Z})$.

a. Show that there exists a nontrivial sequence $v \in l_0(\mathbb{Z})$ such that $\mathcal{T}_u v = 0$.
b. If $\mathcal{S}_u v = 0$ for some $v \in l_0(\mathbb{Z})$, prove that $v(k) = 0$ for all $k \in \mathbb{Z}$.
c. For $u = \{\frac{1}{2}, \underline{\mathbf{0}}, \frac{1}{2}\}_{[-1,1]}$, find a nontrivial sequence $v \in l(\mathbb{Z})$ such that $\mathcal{S}_u v = 0$.

1.20. Let $m \in \mathbb{N}$ and $u \in l_0(\mathbb{Z})$ such that u has m sum rules. Show that the coefficient matrix of $\mathcal{S}_u$ under the basis $\{1, x, \dots, x^{m-1}\}$ of $\mathbb{P}_{m-1}$ is a lower triangular matrix with its diagonal entries being $\widehat{u}(0), 2^{-1}\widehat{u}(0), \dots, 2^{1-m}\widehat{u}(0)$. Moreover, if $\widehat{u}(0) \neq 0$, then $\mathcal{S}_u[(\cdot)^j * \vartheta] = 2^{-j}\widehat{u}(0)[(\cdot)^j * \vartheta], j = 0, \dots, m-1$, where $\vartheta \in l_0(\mathbb{Z})$ satisfies $\widehat{\vartheta}(0) = 1$ and $\widehat{\vartheta}(2\xi)\widehat{u}(\xi) = \widehat{\vartheta}(\xi)\widehat{u}(0) + \mathcal{O}(|\xi|^m)$, $\xi \to 0$.

1.21. Let $\mathsf{p} \in \mathbb{P}$ be a polynomial and $v \in l(\mathbb{Z})$ such that $v(n) = \mathsf{p}(n)$ for all $n \geqslant M$ for some $M \in \mathbb{N}$. Let $u \in l_0(\mathbb{Z})$. If there exist $N \in \mathbb{N}$ and $\mathsf{q} \in \mathbb{P}$ such that $\mathcal{S}_u v(n) = \mathsf{q}(n)$ for all $n \geqslant N$, prove that $\mathcal{S}_u \mathsf{p}(n) = \mathsf{q}(n)$ for all $n \in \mathbb{Z}$.

1.22. Let $(\{\tilde{u}_0, \dots, \tilde{u}_s\}, \{u_0, \dots, u_s\})$ be a dual framelet filter bank. If $s > 1$, by Proposition 1.1.2, then $\mathcal{V}$ is not one-to-one. Explicitly construct $w \in (l_0(\mathbb{Z}))^{1\times(s+1)}$ such that w is not identically zero but $\mathcal{V}w = 0$, where $\mathcal{V}$ is the discrete framelet synthesis operator defined in (1.1.8).

1.23. Let the J-level discrete framelet analysis operator $\mathcal{W}_J$ and the J-level discrete framelet synthesis operator $\mathcal{V}_J$ employing a filter bank $\{a; b_1, \dots, b_s\}$ be defined in Sect. 1.3. Show that $\mathcal{W}_J^\star = \mathcal{V}_J$ and $\mathcal{V}_J^\star = \mathcal{W}_J$.

1.24. Prove that $\{a; b_1, \dots, b_s\}$ is a tight framelet filter bank if and only if the discrete affine system $\mathsf{DAS}_J(\{a; b_1, \dots, b_s\})$ is a (normalized) tight frame for the space $l_2(\mathbb{Z})$ for every $J \in \mathbb{N}$, that is,

$$\|v\|^2_{l_2(\mathbb{Z})} = \sum_{u \in \mathsf{DAS}_J(\{a;b_1,\dots,b_s\})} |\langle v, u\rangle|^2, \qquad \forall\, v \in l_2(\mathbb{Z}).$$

1.25. Prove that $\{a; b\}$ is an orthogonal wavelet filter bank if and only if $\mathsf{DAS}_J(\{a; b\})$ is an orthonormal basis for $l_2(\mathbb{Z})$ for every $J \in \mathbb{N}$.

1.26. Prove that $(\{\tilde{a}; \tilde{b}\}, \{a; b\})$ is a biorthogonal wavelet filter bank if and only if $\mathsf{DAS}_J(\{\tilde{a}; \tilde{b}\})$ and $\mathsf{DAS}_J(\{a; b\})$ are biorthogonal to each other in the space $l_2(\mathbb{Z})$ for every $J \in \mathbb{N}$:

$$\langle u, v\rangle = \begin{cases} 1, & \text{if } \tilde{u} = v \\ 0, & \text{if } \tilde{u} \neq v, \end{cases} \qquad \forall\, u \in \mathsf{DAS}_J(\{a; b\}), \quad v \in \mathsf{DAS}_J(\{\tilde{a}; \tilde{b}\}).$$

1.27. Prove that $(\{\tilde{a}; \tilde{b}_1, \dots, \tilde{b}_s\}, \{a; b_1, \dots, b_s\})$ is a dual framelet filter bank if and only if $(\mathsf{DAS}_J(\{\tilde{a}; \tilde{b}_1, \dots, \tilde{b}_s\}), \mathsf{DAS}(\{a; b_1, \dots, b_s\}))$ is a dual frame in $l_0(\mathbb{Z})$, that is,

$$\langle v, w\rangle = \sum_{u \in \mathsf{DAS}_J(\{a;b_1,\dots,b_s\})} \langle v, \tilde{u}\rangle \langle u, w\rangle, \qquad \forall\, v, w \in l_0(\mathbb{Z}).$$

Note that the above summation is in fact finite since $v, w \in l_0(\mathbb{Z})$.

1.28. Let $m, \tilde{m}, n, \tilde{n} \in \mathbb{Z}$. Define $\widehat{a}(\xi) := e^{-in\xi}\widehat{a^B_m}(\xi)$ and $\widehat{\tilde{a}}(\xi) := e^{-i\tilde{n}\xi}\widehat{a^B_{\tilde{m}}}(\xi)$. Show that $\widehat{\tilde{a}}(\xi)\overline{\widehat{a}(\xi)} = 1 + O(|\xi|^3), \xi \to 0$ cannot hold.

1.29. Suppose that $\{a; b_1, \dots, b_s\}_\Theta$ is a tight framelet filter bank. Show that all the high-pass filters $b_1, \dots, b_s$ have m vanishing moments if and only if

$$\widehat{\Theta}(2\xi) - \widehat{\Theta}(\xi)|\widehat{a}(\xi)|^2 = \mathscr{O}(|\xi|^{2m}), \qquad \xi \to 0.$$

If in addition $\widehat{\Theta}(0) \neq 0$, then a must have m sum rules.

1.30. Define $A(\xi) := \prod_{j=1}^{N} \frac{e^{-i\xi} - e^{it_j}}{1 - e^{it_j}}$, $\xi \in \mathbb{R}$ with $t_1, \dots, t_N \in \mathbb{R}\backslash[2\pi\mathbb{Z}]$. Prove that $A''(0) - [A'(0)]^2 = -\sum_{j=1}^{N} \frac{1}{2(1-\cos(t_j))} \leqslant 0$ and the equality holds if and only if $A(\xi) = 1$.

1.31. Let $a, b_1, \dots, b_s, \Theta \in l_0(\mathbb{Z})$ with $\widehat{a}(0) = 1$ and $\widehat{\Theta}(0) = 1$. Suppose $\{a; b_1, \dots, b_s\}_\Theta$ is a tight framelet filter bank. If all the roots of the Laurent polynomial $\sum_{k \in \mathbb{Z}} a(k) z^k$ lie on the unit circle, prove that one of the high-pass filters $b_1, \dots, b_s$ must have at most one vanishing moment.

1.32. Let $(\{\tilde{a}; \tilde{b}_1, \dots, \tilde{b}_s\}, \{a; b_1, \dots, b_s\})_\Theta$ be a dual framelet filter bank. For $\lambda_0, \dots, \lambda_s \in \mathbb{T}$ and $n_\Theta, n_0, \dots, n_s \in \mathbb{Z}$ such that $\tilde{n}_0 = n_0 - n_\Theta, \tilde{n}_\ell = n_\ell + n_\Theta, n_\ell - n_0 \in 2\mathbb{Z}$ for all $\ell = 1, \dots, s$, show that $(\{\lambda_0 \tilde{a}(\cdot - n_0); \lambda_1 \tilde{b}_1(\cdot - n_1), \dots, \lambda_s \tilde{b}_s(\cdot - n_s)\}, \{\lambda_0 a(\cdot - n_0); \lambda_1 b_1(\cdot - n_1), \dots, \lambda_s b_s(\cdot - n_s)\})_{\Theta(\cdot - n_\Theta)}$ is also a dual framelet filter bank.

1.33. Prove that the perfect reconstruction condition in (1.4.3) and (1.4.4) for a dual framelet filter bank $(\{\tilde{a}; \tilde{b}_1, \dots, \tilde{b}_s\}, \{a; b_1, \dots, b_s\})_\Theta$ is equivalent to

$$\begin{bmatrix} \widehat{\tilde{a}^{[0]}}(\xi) & \widehat{\tilde{b}_1^{[0]}}(\xi) & \cdots & \widehat{\tilde{b}_s^{[0]}}(\xi) \\ \widehat{\tilde{a}^{[1]}}(\xi) & \widehat{\tilde{b}_1^{[1]}}(\xi) & \cdots & \widehat{\tilde{b}_s^{[1]}}(\xi) \end{bmatrix} \begin{bmatrix} \widehat{\Theta}(2\xi) & 0 \\ 0 & 1 \end{bmatrix}$$

$$\times \begin{bmatrix} \widehat{a^{[0]}}(\xi) & \widehat{b_1^{[0]}}(\xi) & \cdots & \widehat{b_s^{[0]}}(\xi) \\ \widehat{a^{[1]}}(\xi) & \widehat{b_1^{[1]}}(\xi) & \cdots & \widehat{b_s^{[1]}}(\xi) \end{bmatrix}^\star = \frac{1}{2} \begin{bmatrix} \widehat{\Theta^{[0]}}(\xi) & 0 \\ 0 & \widehat{\Theta^{[1]}}(\xi) \end{bmatrix}.$$

1.34. Prove item (iv) of Proposition 1.5.2.

1.35. Let $u_0, \dots, u_s \in l_0(\mathbb{Z})$. Let $\mathcal{W}^{per}$ be the coefficient matrix of the periodic discrete framelet analysis operator defined in (1.5.6) and let $\mathcal{V}^{per}$ be the coefficient matrix of the periodic discrete framelet synthesis operator defined in (1.5.7) but with $\tilde{u}_0, \dots, \tilde{u}_s$ being replaced by $u_0, \dots, u_s$. Show that $[\mathcal{W}^{per}]^\star = \mathcal{V}^{per}$, that is, $\mathcal{V}^{per}$ is the complex conjugate of the transpose of $\mathcal{W}^{per}$.

1.36. Let $\{a; b_1, \dots, b_s\}$ be a filter bank. Suppose that there exists a positive constant C such that $\|\mathcal{W}_J v\|^2_{(l_2(\mathbb{Z}))^{1 \times (sJ+1)}} \leqslant C \|v\|^2_{l_2(\mathbb{Z})}$ for all $v \in l_2(\mathbb{Z})$ and $J \in \mathbb{N}$. Prove $|\widehat{a}(0)| \leqslant 1$. If in addition $|\widehat{a}(0)| = 1$, then $\widehat{b_1}(0) = \cdots \widehat{b_s}(0) = 0$.

1.37. Let $\zeta \in \mathbb{C}$ and $\lambda > 0$. Prove (i) $\eta_\lambda^{soft}(\zeta) = \operatorname{argmin}_{z \in \mathbb{C}} \frac{1}{2}|z - \zeta|^2 + \lambda |z|$, where the soft thresholding function η_λ^{soft} is defined in (1.3.2); (ii) $\eta_\lambda^{Hard}(\zeta) = \operatorname{argmin}_{z \in \mathbb{C}} \frac{1}{2}|z - \zeta|^2 + \lambda |z|_0$, where $|0|_0 := 0$ and $|z|_0 := 1$ for $z \in \mathbb{C}\backslash\{0\}$, and $\eta_\lambda^{Hard}(\zeta) = \{\zeta\}$ for $|\zeta| > \lambda$, $\eta_\lambda^{Hard}(\zeta) = \{0, \zeta\}$ if $|\zeta| = \lambda$, and $\eta_\lambda^{Hard}(\zeta) := 0$ for $|\zeta| < \lambda$.

The sampling factor used in this Chapter is 2. In fact, there are more general discrete framelet transforms and filter banks using a general sampling factor d, where d is a nonzero integer. For simplicity, here we only consider a positive sampling factor d with $\mathsf{d} \geqslant 1$. Define $\mathcal{S}_{u,\mathsf{d}}$ and $\mathcal{T}_{u,\mathsf{d}}$ as in (1.3.12) and (1.3.13). For a filter bank $(\{\tilde{u}_0, \ldots, \tilde{u}_s\}, \{u_0, \ldots, u_s\})$, a one-level discrete d-framelet decomposition is

$$\mathcal{W}v := (w_0, \ldots, w_s) \quad \text{with} \quad w_\ell := \mathsf{d}^{-1/2}\mathcal{T}_{\tilde{u}_\ell,\mathsf{d}}v, \quad \ell = 0, \ldots, s, \; v \in l(\mathbb{Z})$$

and a one-level discrete d-framelet reconstruction is

$$\mathring{v} = \mathcal{V}(\mathring{w}_0, \ldots, \mathring{w}_s) = \mathsf{d}^{-1/2} \sum_{\ell=1}^{s} \mathcal{S}_{u_\ell,\mathsf{d}}\mathring{w}_\ell, \qquad \mathring{w}_0, \ldots, \mathring{w}_s \in l(\mathbb{Z}).$$

$(\{\tilde{u}_0, \ldots, \tilde{u}_s\}, \{u_0, \ldots, u_s\})$ is called a dual d-framelet filter bank if it has the perfect reconstruction property, that is, $\sum_{\ell=0}^{s} \mathcal{S}_{u_\ell,\mathsf{d}}\mathcal{T}_{\tilde{u}_\ell,\mathsf{d}}v = \mathsf{d}v$ for all $v \in l(\mathbb{Z})$. $\{u_0, \ldots, u_s\}$ is called a tight d-framelet filter bank if $(\{u_0, \ldots, u_s\}, \{u_0, \ldots, u_s\})$ is a dual d-framelet filter bank. A dual d-framelet filter bank with $s = |\mathsf{d}| - 1$ is called a biorthogonal d-wavelet filter bank and a tight d-framelet filter bank with $s = |\mathsf{d}| - 1$ is called an orthogonal d-wavelet filter bank. The coset sequences $u^{[\gamma:\mathsf{d}]}$ are

$$\widehat{u^{[\gamma:\mathsf{d}]}}(\xi) := \sum_{k\in\mathbb{Z}} u(\gamma+\mathsf{d}k)e^{-ik\xi}, \quad \text{that is,} \quad u^{[\gamma:\mathsf{d}]} = u(\gamma+\cdot)\downarrow\mathsf{d} = \{u(\gamma+\mathsf{d}k)\}_{k\in\mathbb{Z}}.$$

1.38. Prove that $(\{\tilde{u}_0, \ldots, \tilde{u}_s\}, \{u_0, \ldots, u_s\})$ is a dual d-framelet filter bank if and only if

$$\widehat{\tilde{u}_0}(\xi)\overline{\widehat{u_0}(\xi + 2\pi\gamma/\mathsf{d})} + \cdots + \widehat{\tilde{u}_s}(\xi)\overline{\widehat{u_s}(\xi + 2\pi\gamma/\mathsf{d})} = \boldsymbol{\delta}(\gamma), \qquad \gamma = 0, \ldots, \mathsf{d}-1.$$

1.39. Prove $(\{\tilde{u}_0, \ldots, \tilde{u}_s\}, \{u_0, \ldots, u_s\})$ is a dual d-framelet filter bank if and only if

$$\begin{bmatrix} \widehat{\tilde{u}_0^{[0:\mathsf{d}]}}(\xi) & \widehat{\tilde{u}_1^{[0:\mathsf{d}]}}(\xi) & \cdots & \widehat{\tilde{u}_s^{[0:\mathsf{d}]}}(\xi) \\ \widehat{\tilde{u}_0^{[1:\mathsf{d}]}}(\xi) & \widehat{\tilde{u}_1^{[1:\mathsf{d}]}}(\xi) & \cdots & \widehat{\tilde{u}_s^{[1:\mathsf{d}]}}(\xi) \\ \vdots & \vdots & \ddots & \vdots \\ \widehat{\tilde{u}_0^{[\mathsf{d}-1:\mathsf{d}]}}(\xi) & \widehat{\tilde{u}_1^{[\mathsf{d}-1:\mathsf{d}]}}(\xi) & \cdots & \widehat{\tilde{u}_s^{[\mathsf{d}-1:\mathsf{d}]}}(\xi) \end{bmatrix} \begin{bmatrix} \widehat{u_0^{[0:\mathsf{d}]}}(\xi) & \widehat{u_1^{[0:\mathsf{d}]}}(\xi) & \cdots & \widehat{u_s^{[0:\mathsf{d}]}}(\xi) \\ \widehat{u_0^{[1:\mathsf{d}]}}(\xi) & \widehat{u_1^{[1:\mathsf{d}]}}(\xi) & \cdots & \widehat{u_s^{[1:\mathsf{d}]}}(\xi) \\ \vdots & \vdots & \ddots & \vdots \\ \widehat{u_0^{[\mathsf{d}-1:\mathsf{d}]}}(\xi) & \widehat{u_1^{[\mathsf{d}-1:\mathsf{d}]}}(\xi) & \cdots & \widehat{u_s^{[\mathsf{d}-1:\mathsf{d}]}}(\xi) \end{bmatrix}^{\star} = \mathsf{d}^{-1}I_{\mathsf{d}}.$$

1.40. Let $(\{\tilde{u}_0, \ldots, \tilde{u}_s\}, \{u_0, \ldots, u_s\})$ be a dual d-framelet filter bank. Prove that the following statements are equivalent:

a. $\widetilde{\mathcal{W}}$ is onto or $\mathcal{V}$ is one-to-one.
b. $\mathcal{V}\widetilde{\mathcal{W}} = \mathrm{Id}_{l(\mathbb{Z})}$ and $\widetilde{\mathcal{W}}\mathcal{V} = \mathrm{Id}_{(l(\mathbb{Z}))^{1\times(s+1)}}$, that is, $\mathcal{V}^{-1} = \widetilde{\mathcal{W}}$ and $\widetilde{\mathcal{W}}^{-1} = \mathcal{V}$.
c. $s = |\mathsf{d}| - 1$.

1.41. Prove that $\{u_0, \ldots, u_s\}$ is a tight d-framelet filter bank if and only if

$$\|\mathcal{T}_{u_0,\mathsf{d}}v\|^2_{l_2(\mathbb{Z})} + \cdots + \|\mathcal{T}_{u_s,\mathsf{d}}v\|^2_{l_2(\mathbb{Z})} = \mathsf{d}\|v\|^2_{l_2(\mathbb{Z})}, \qquad \forall\, v \in l_2(\mathbb{Z}).$$

1.42. Prove that $\mathcal{T}_{u,\mathsf{d}}\mathsf{p} = \mathsf{d}[\mathsf{p} * u^\star](\mathsf{d}\cdot)$ for any polynomial p.

1.43. Let $u = \{u(k)\}_{k\in\mathbb{Z}} \in l_0(\mathbb{Z})$ be a finitely supported sequence on $\mathbb{Z}$ and $\mathsf{p} \in \mathbb{P}$ be a polynomial. Show that the following statements are equivalent:

a. $\mathcal{S}_{u,\mathsf{d}}\mathsf{p}$ is a polynomial sequence;
b. $\sum_{k\in\mathbb{Z}} \mathsf{p}^{(j)}(-\mathsf{d}^{-1}\gamma - k)u(\gamma + \mathsf{d}k) = \sum_{k\in\mathbb{Z}} \mathsf{p}^{(j)}(-k)u(\mathsf{d}k)$ for all $j \in \mathbb{N}_0$ and $\gamma = 0, \ldots, \mathsf{d}-1$;
c. For all $j \in \mathbb{N}_0$, $[\mathsf{p}^{(j)}(-\frac{0}{\mathsf{d}} - i\frac{d}{d\xi})\widehat{u^{[0:\mathsf{d}]}}(\xi)]|_{\xi=0} = [\mathsf{p}^{(j)}(-\frac{1}{\mathsf{d}} - i\frac{d}{d\xi})\widehat{u^{[1:\mathsf{d}]}}(\xi)]|_{\xi=0}$ $= \cdots = [\mathsf{p}^{(j)}(-\frac{\mathsf{d}-1}{\mathsf{d}} - i\frac{d}{d\xi})\widehat{u^{[\mathsf{d}-1:\mathsf{d}]}}(\xi)]|_{\xi=0}$;
d. $[\mathsf{p}^{(j)}(-\frac{i}{\mathsf{d}}\frac{d}{d\xi})\widehat{u}(\xi)]|_{\xi=\pi} = \cdots = [\mathsf{p}^{(j)}(-i\frac{\mathsf{d}-1}{\mathsf{d}}\frac{d}{d\xi})\widehat{u}(\xi)]|_{\xi=\pi} = 0$ for all $j \in \mathbb{N}_0$.

Moreover, if any of the above items holds, then $\deg(\mathcal{S}_{u,\mathsf{d}}\mathsf{p}) \leqslant \deg(\mathsf{p})$,

$$\mathcal{S}_{u,\mathsf{d}}\mathsf{p} = \mathsf{p}(\mathsf{d}^{-1}\cdot) * u = \sum_{j=0}^{\infty} \frac{(-i)^j}{\mathsf{d}^j j!}\mathsf{p}^{(j)}(\mathsf{d}^{-1}\cdot)\widehat{u}^{(j)}(0).$$

1.44. For any positive integer $m \in \mathbb{N}$, the following statements are equivalent

a. $\mathcal{S}_{u,\mathsf{d}}\mathsf{q} \in \mathbb{P}$ for some polynomial $\mathsf{q} \in \mathbb{P}$ with $\deg(\mathsf{q}) = m-1$;
b. $\mathcal{S}_{u,\mathsf{d}}\mathbb{P}_{m-1} \subseteq \mathbb{P}_{m-1}$;
c. $\widehat{u}^{(j)}(\pi\gamma/\mathsf{d}) = 0$ for all $0 \leqslant j < m$ and $1 \leqslant \gamma < \mathsf{d}$, i.e., $\widehat{u}(\xi + \pi\gamma/\mathsf{d}) = \mathscr{O}(|\xi|^m)$ as $\xi \to 0$ for $1 \leqslant \gamma < \mathsf{d}$;
d. $\widehat{u}(\xi) = (1 + e^{-i\xi} + \cdots + e^{-i(\mathsf{d}-1)\xi})^m\widehat{v}(\xi)$ for some $v \in l_0(\mathbb{Z})$;
e. $\sum_{k\in\mathbb{Z}} u(\gamma + \mathsf{d}k)(\frac{\gamma}{\mathsf{d}} + k)^j = \sum_{k\in\mathbb{Z}} u(\mathsf{d}k)k^j$ for all $0 \leqslant j < m$ and $0 \leqslant \gamma < \mathsf{d}$.

1.45. Show that $(\{\tilde{a}; \tilde{b}_1, \ldots, \tilde{b}_s\}, \{a; b_1, \ldots, b_s\})_\Theta$ is a dual d-framelet filter bank, that is, it has the following perfect reconstruction property:

$$\Theta^\star * v = \frac{1}{\mathsf{d}}\mathcal{S}_{a,\mathsf{d}}(\Theta^\star * \mathcal{T}_{\tilde{a},\mathsf{d}}v) + \frac{1}{\mathsf{d}}\sum_{\ell=1}^{s} \mathcal{S}_{b_\ell,\mathsf{d}}\mathcal{T}_{\tilde{b}_\ell,\mathsf{d}}v, \qquad \forall\, v \in l(\mathbb{Z}),$$

if and only if for all $\gamma = 0, \ldots, \mathsf{d}-1$ and for all $\xi \in \mathbb{R}$,

$$\widehat{\Theta}(\mathsf{d}\xi)\widehat{a}(\xi)\overline{\widehat{a}(\xi + \tfrac{2\pi\gamma}{\mathsf{d}})} + \widehat{b_1}(\xi)\overline{\widehat{b_1}(\xi + \tfrac{2\pi\gamma}{\mathsf{d}})} + \cdots + \widehat{b_s}(\xi)\overline{\widehat{b_s}(\xi + \tfrac{2\pi\gamma}{\mathsf{d}})} = \boldsymbol{\delta}(\gamma)\widehat{\Theta}(\xi).$$

1.46. Suppose that $\{a; b_1, \ldots, b_s\}_\Theta$ is a tight d-framelet filter bank. Prove (i) $\widehat{\Theta}(\xi) \geqslant 0$ for all $\xi \in \mathbb{R}$; (ii) All the high-pass filters $b_1, \ldots, b_s$ have m vanishing moments if and only if $\widehat{\Theta}(\mathsf{d}\xi) - \widehat{\Theta}(\xi)|\widehat{a}(\xi)|^2 = \mathscr{O}(|\xi|^{2m})$ as $\xi \to 0$.

Chapter 2
Wavelet Filter Banks

Wavelet filter banks are the indispensable key part in any discrete wavelet transform and are one of the major topics in the classical theory of wavelets. In this chapter we discuss how to systematically design orthogonal and biorthogonal wavelet filter banks such that they have some desirable properties such as sum rules, vanishing moments, linear-phase moments, and symmetry property.

Since an orthogonal wavelet filter bank is a special case of a biorthogonal wavelet filter bank, we recall from Sect. 1.1 that $(\{\tilde{a};\tilde{b}\},\{a;b\})$ with filters $\tilde{a},\tilde{b},a,b \in l_0(\mathbb{Z})$ is *a biorthogonal wavelet filter bank* if it satisfies the perfect reconstruction condition:

$$\begin{bmatrix} \widehat{\tilde{a}}(\xi) & \widehat{\tilde{b}}(\xi) \\ \widehat{\tilde{a}}(\xi+\pi) & \widehat{\tilde{b}}(\xi+\pi) \end{bmatrix} \begin{bmatrix} \widehat{a}(\xi) & \widehat{b}(\xi) \\ \widehat{a}(\xi+\pi) & \widehat{b}(\xi+\pi) \end{bmatrix}^{\star} = I_2, \tag{2.0.1}$$

for all $\xi \in \mathbb{R}$, where $\widehat{u}(\xi) := \sum_{k\in\mathbb{Z}} u(k)e^{-ik\xi}$ for a filter $u = \{u(k)\}_{k\in\mathbb{Z}}$, and $A^{\star} := \overline{A}^{\mathsf{T}}$ is the transpose of the complex conjugate of a matrix A. By Theorem 1.4.7, we must have

$$\widehat{\tilde{b}}(\xi) = ce^{i(2n-1)\xi}\overline{\widehat{a}(\xi+\pi)}, \qquad \widehat{b}(\xi) = \tfrac{1}{c}e^{i(2n-1)\xi}\overline{\widehat{\tilde{a}}(\xi+\pi)}, \tag{2.0.2}$$

for some $c \in \mathbb{C}\backslash\{0\}$ and $n \in \mathbb{Z}$. With the form of $\widehat{\tilde{b}}, \widehat{b}$ in (2.0.2), it is now pretty easy to verify that (2.0.1) holds if and only if for all $\xi \in \mathbb{R}$,

$$\widehat{\tilde{a}}(\xi)\overline{\widehat{a}(\xi)} + \widehat{\tilde{a}}(\xi+\pi)\overline{\widehat{a}(\xi+\pi)} = 1. \tag{2.0.3}$$

If (2.0.3) holds, then $(\tilde{a}, a)$ is called *a pair of biorthogonal wavelet filters*, or simply, $\tilde{a}$ is *a dual (wavelet) filter* of a, and vice versa. Without loss of generality, we often take $c = 1$ and $n = 0$ in (2.0.2) and use the following standard primal high-pass filter b and standard dual high-pass filter $\tilde{b}$ derived from a pair of biorthogonal

B. Han, *Framelets and Wavelets*, Applied and Numerical Harmonic Analysis,
https://doi.org/10.1007/978-3-319-68530-4_2

wavelet filters $(\tilde{a}, a)$:

$$\widehat{\tilde{b}}(\xi) = e^{-i\xi}\overline{\widehat{a}(\xi+\pi)} \quad \text{and} \quad \widehat{b}(\xi) = e^{-i\xi}\overline{\widehat{\tilde{a}}(\xi+\pi)}. \tag{2.0.4}$$

In the space/time domain, (2.0.4) simply becomes

$$\tilde{b}(k) = (-1)^{1-k}\overline{a(1-k)} \quad \text{and} \quad b(k) = (-1)^{1-k}\overline{\tilde{a}(1-k)}, \qquad k \in \mathbb{Z}.$$

If (2.0.3) holds with $\tilde{a} = a$, more explicitly,

$$|\widehat{a}(\xi)|^2 + |\widehat{a}(\xi+\pi)|^2 = 1, \tag{2.0.5}$$

then a is called *an orthogonal wavelet filter* and the standard high-pass filter b is derived from a via

$$\widehat{b}(\xi) = e^{-i\xi}\overline{\widehat{a}(\xi+\pi)}. \tag{2.0.6}$$

For a low-pass filter $a = \{a(k)\}_{k\in\mathbb{Z}}$ satisfying $\widehat{a}(0) = \sum_{k\in\mathbb{Z}} a(k) = 1$, we can define a frequency-based refinable function φ^a by $\varphi^a(\xi) := \prod_{j=1}^{\infty} \widehat{a}(2^{-j}\xi), \xi \in \mathbb{R}$. As we shall discuss in Chaps. 4 and 5, we can define a compactly supported function/distribution ϕ^a, called the standard (time-domain) refinable function associated with the filter a, through the inverse Fourier transform of φ^a, such that $\phi^a(x) := \frac{1}{2\pi}\langle \varphi^a; e^{ix\cdot}\rangle = \frac{1}{2\pi}\int_{\mathbb{R}} \varphi^a(\xi)e^{i\xi x}d\xi$ (or equivalently $\widehat{\phi^a} = \varphi^a$) and ϕ^a satisfies a time-domain refinement equation:

$$\phi^a = 2\sum_{k\in\mathbb{Z}} a(k)\phi^a(2\cdot -k).$$

For a filter $b = \{b(k)\}_{k\in\mathbb{Z}}$, we shall define the following function $\psi^{a,b}$ by

$$\psi^{a,b} := 2\sum_{k\in\mathbb{Z}} b(k)\phi^a(2\cdot -k).$$

If b is the standard high-pass filter derived from a as in (2.0.6), then we define

$$\psi^a := 2\sum_{k\in\mathbb{Z}} (-1)^{1-k}\overline{a(1-k)}\phi^a(2\cdot -k).$$

The functions ϕ^a and ψ^a are called the *standard refinable function* and *standard wavelet function* associated with the filter a, respectively. As we shall see in Chaps. 5 and 6, the properties of ϕ^a and ψ^a are closely related to the stability of a discrete wavelet transform. In this chapter we only plot graphs of ϕ^a, $\psi^{a,b}$ without any discussion on their properties. See Sect. 6.2.3 for plotting ϕ^a and $\psi^{a,b}$.

In this chapter we shall present several ways of constructing wavelet filter banks with various basic properties. We first recall some basic quantities of a filter $u \in l_0(\mathbb{Z})$:

(1) u has m *vanishing moments* if $\widehat{u}(\xi) = \mathscr{O}(|\xi|^m)$ as $\xi \to 0$. If u has m but not $m+1$ vanishing moments, then we define $\operatorname{vm}(u) := m$, the highest order of vanishing moments satisfied by u, and we say that u has the vanishing moments of order m;
(2) Filter u has m *sum rules* if $\widehat{u}(\xi + \pi) = \mathscr{O}(|\xi|^m)$ as $\xi \to 0$. If u has m but not $m+1$ sum rules, then we define $\operatorname{sr}(u) := m$, the highest order of sum rules satisfied by u, and we say that u has the sum rules of order m;
(3) Filter u has m *linear-phase moments with phase* $c \in \mathbb{R}$ if $\widehat{u}(\xi) = e^{-ic\xi} + \mathscr{O}(|\xi|^m)$ as $\xi \to 0$. By Proposition 1.2.8, if $m > 1$, then the phase c is uniquely determined by u through the identity $c = \sum_{k\in\mathbb{Z}} u(k)k$. Hence, we denote by $\operatorname{phase}(u) := \operatorname{Re}(\sum_{k\in\mathbb{Z}} u(k)k)$ as the default phase of $\widehat{u}$. Consequently, we simply say that u has m *linear-phase moments* if $\widehat{u}(\xi) = e^{-ic\xi} + \mathscr{O}(|\xi|^m)$ as $\xi \to 0$ with $c = \operatorname{phase}(u) \in \mathbb{R}$. If u has m but not $m+1$ linear-phase moments, then we define $\operatorname{lpm}(u) := m$, the highest order of linear-phase moments satisfied by u, and we say that u has the linear-phase moments of order m (with default phase $c = \operatorname{phase}(u)$);
(4) u has *symmetry* if $u(c_u - k) = \epsilon u(k)\ \forall\, k \in \mathbb{Z}$ for some $c_u \in \mathbb{Z}$ and $\epsilon \in \{-1, 1\}$. The symmetry type of u is often recorded by $[\mathsf{S}\widehat{u}](\xi) = \epsilon e^{-ic_u\xi}$ using the symmetry operator $[\mathsf{S}\widehat{u}](\xi) := \widehat{u}(\xi)/\widehat{u}(-\xi)$. Similarly, u has *complex symmetry* if $u(c_u - k) = \epsilon\overline{u(k)}\ \forall\, k \in \mathbb{Z}$ for some $c_u \in \mathbb{Z}$ and $\epsilon \in \{-1, 1\}$. The complex symmetry type of u is often recorded by $[\mathbb{S}\widehat{u}](\xi) = \epsilon e^{-ic_u\xi}$ using the complex symmetry operator $[\mathbb{S}\widehat{u}](\xi) := \widehat{u}(\xi)/\overline{\widehat{u}(\xi)}$. As proved in Proposition 1.2.8, if u has symmetry or complex symmetry with the symmetry center $c_u/2$, then $\operatorname{phase}(u) = c_u/2$. When u is the zero sequence, $\mathsf{S}\widehat{u}$ and $\mathbb{S}\widehat{u}$ can be assigned any symmetry types;
(5) If $u(m)u(n) \neq 0$ and $u(k) = 0$ for all $k \in \mathbb{Z}\backslash[m, n]$, where m and n are integers satisfying $m \leqslant n$, then we define $\operatorname{fsupp}(u) := [m, n]$, called the *filter support* of u. Moreover, the length of the filter u is $\operatorname{len}(u) := |\operatorname{fsupp}(u)| = n - m$, which is the length of the filter support of u. If u is the zero sequence, then by default $\operatorname{fsupp}(u) = \emptyset$ is the empty set and $\operatorname{len}(u) = -\infty$. Note that $\operatorname{fsupp}(u)$ is always an interval and may not agree with $\operatorname{supp}(u) := \{k \in \mathbb{Z} \ :\ u(k) \neq 0\}$, the support of u.
(6) The *smoothness exponent* $\operatorname{sm}(u)$ of a filter u is closely related to the stability of a multilevel discrete framelet transform and to the smoothness of the associated refinable function. To define $\operatorname{sm}(u)$, we first write $\widehat{u}(\xi) = (1 + e^{-i\xi})^m \widehat{v}(\xi)$ for a nonnegative integer m and a finitely supported sequence v such that $\widehat{v}(\pi) \neq 0$. Then we define the *smoothness exponent* of the filter u by

$$\operatorname{sm}(u) := -1/2 - \log_2 \sqrt{\rho(u)}, \tag{2.0.7}$$

where $\rho(u)$ denotes the spectral radius, the largest of the modulus of all the eigenvalues, of the square matrix $(w(2j-k))_{-K\leqslant j,k\leqslant K}$, where w is determined by $\sum_{k=-K}^{K} w(k)e^{-ik\xi} := |\widehat{v}(\xi)|^2$.

As we shall see in Chap. 6, the smoothness exponent $\mathrm{sm}(a)$ is tightly linked to the stability of a multilevel discrete framelet transform employing a filter bank $\{a; b_1, \dots, b_s\}$. Roughly speaking, the larger the quantity $\mathrm{sm}(a)$, the wider family of sequence spaces in which a multilevel discrete framelet transform has stability. The connections between the smoothness quantity and the stability of a multilevel discrete framelet transform will become clear in Chap. 6 of this book.

In the context of filter design, there are a few statistics-related quantities that are of interest in applications. For $u \in l_0(\mathbb{Z})$, we define its expectation/mean $\mathrm{E}(u)$ by

$$\mathrm{E}(u) := \frac{\sum_{k\in\mathbb{Z}} |u(k)|^2 k}{\|u\|_{l_2(\mathbb{Z})}^2}, \qquad \text{where} \quad \|u\|_{l_2(\mathbb{Z})}^2 := \sum_{k\in\mathbb{Z}} |u(k)|^2. \tag{2.0.8}$$

We define (normalized) variance $\mathrm{Var}(u)$ by

$$\mathrm{Var}(u) := \frac{\sum_{k\in\mathbb{Z}} |u(k)|^2 (k-\mathrm{E}(u))^2}{\|u\|_{l_2(\mathbb{Z})}^2}. \tag{2.0.9}$$

Note that $\mathrm{Var}(u) = \min_{c\in\mathbb{R}} \sum_{k\in\mathbb{Z}} |u(k)|^2 (k-c)^2/\|u\|_{l_2(\mathbb{Z})}^2$, with the minimum value achieved at $c = \mathrm{E}(u)$. For an orthogonal wavelet filter u, we always have $\|u\|_{l_2(\mathbb{Z})}^2 = 1/2$ (see Exercise 2.2).

In many practical applications, small variance $\mathrm{Var}(u)$ is desirable so that most significant coefficients of the filter u concentrate around the point $\mathrm{E}(u)$. The frequency separation of two filters u and v is also of interest in filter design. In other words, we are interested in how well two 2π-periodic functions $|\widehat{u}|$ and $|\widehat{v}|$ are separated from each other in the frequency domain. For two sequences u and v, we define their *frequency separation indicator* $\mathrm{Fsi}(u,v)$ to be

$$\mathrm{Fsi}(u,v) := \frac{\int_{-\pi}^{\pi} |\widehat{u}(\xi)|^2 |\widehat{v}(\xi)|^2 d\xi}{\sqrt{\int_{-\pi}^{\pi} |\widehat{u}(\xi)|^4 d\xi}\sqrt{\int_{-\pi}^{\pi} |\widehat{v}(\xi)|^4 d\xi}}. \tag{2.0.10}$$

It is trivial to see that $0 \leqslant \mathrm{Fsi}(u,v) \leqslant 1$. The smaller the quantity $\mathrm{Fsi}(u,v)$, the better separation between u and v in the frequency domain. For example, $\mathrm{Fsi}(u,v) = 0$ if and only if $\widehat{u}(\xi)\widehat{v}(\xi) = 0$ for all $\xi \in \mathbb{R}$; $\mathrm{Fsi}(u,v) = 1$ if and only if for some nonnegative constant λ, either $|\widehat{u}(\xi)| = \lambda|\widehat{v}(\xi)|$ or $|\widehat{v}(\xi)| = \lambda|\widehat{u}(\xi)|$ for almost every

$\xi \in \mathbb{R}$. The quantity $\operatorname{Fsi}(u, v)$ can be easily computed by the following formula:

$$\operatorname{Fsi}(u, v) = \frac{\sum_{k\in\mathbb{Z}} \mathring{u}(k)\mathring{v}(-k)}{\sqrt{\sum_{k\in\mathbb{Z}} |\mathring{u}(k)|^2}\sqrt{\sum_{k\in\mathbb{Z}} |\mathring{v}(k)|^2}},$$

where the sequences $\mathring{u}$ and $\mathring{v}$ are determined by $\sum_{k\in\mathbb{Z}} \mathring{u}(k)e^{-ik\xi} := |\widehat{u}(\xi)|^2$ and $\sum_{k\in\mathbb{Z}} \mathring{v}(k)e^{-ik\xi} := |\widehat{v}(\xi)|^2$. In particular, we define

$$\operatorname{Fsi}(u) := \operatorname{Fsi}(u, v) \qquad \text{with} \quad \widehat{v}(\xi) := e^{-i\xi}\overline{\widehat{u}(\xi + \pi)}. \tag{2.0.11}$$

It follows from (2.0.2) that for a biorthogonal wavelet filter bank $(\{\tilde{a}; \tilde{b}\}, \{a; b\})$, we always have $\operatorname{vm}(\tilde{b}) = \operatorname{sr}(a)$ and $\operatorname{vm}(b) = \operatorname{sr}(\tilde{a})$. Therefore, to design a biorthogonal wavelet filter bank $(\{\tilde{a}; \tilde{b}\}, \{a; b\})$ with its high-pass filters $\tilde{b}, b$ having high vanishing moments, it is necessary and sufficient to design a pair of biorthogonal wavelet filters $(\tilde{a}, a)$ with high sum rules. Thus, for a wavelet filter bank, the orders of sum rules for low-pass filters and the orders of vanishing moments for high-pass filters are closely related to each other.

To design a filter $u = \{u(k)\}_{k\in\mathbb{Z}}$ with some desired basic properties, it is convenient to use the coset sequences of a filter u. Recall that the *coset sequence* of u at γ is $u^{[\gamma]} := \{u(\gamma + 2k)\}_{k\in\mathbb{Z}}$, that is,

$$\widehat{u^{[\gamma]}}(\xi) = \sum_{k\in\mathbb{Z}} u(\gamma + 2k)e^{-ik\xi} = 2^{-1}e^{i\xi\gamma/2}[\widehat{u}(\xi/2) + e^{-i\gamma\pi}\widehat{u}(\xi/2 + \pi)], \quad \gamma \in \mathbb{Z}.$$

In terms of coset sequences of filters, $(\tilde{a}, a)$ is a pair of biorthogonal wavelet filters satisfying (2.0.3) if and only if

$$\widehat{\tilde{a}^{[0]}}(\xi)\overline{\widehat{a^{[0]}}(\xi)} + \widehat{\tilde{a}^{[1]}}(\xi)\overline{\widehat{a^{[1]}}(\xi)} = 1/2. \tag{2.0.12}$$

2.1 Interpolatory Filters and Filters with Linear-Phase Moments

In this section we shall study and construct interpolatory filters and filters with linear-phase moments. Both types of filters are closely related to orthogonal wavelet filters and biorthogonal wavelet filters. As we shall see in this chapter, all the orthogonal wavelet filters are derived from interpolatory filters having nonnegative Fourier series through the Fejér-Riesz Lemma in Lemma 1.4.4. Interpolatory filters are also of great interest in the study of subdivision curves in computer graphics, as we shall discuss in Sect. 7.3.

2.1.1 Interpolatory Filters with Sum Rules and Minimum Supports

Let $(\tilde{a}, a)$ be a pair of biorthogonal wavelet filters satisfying (2.0.3). The correlation filter u of filters $\tilde{a}$ and a is defined to be $u := \tilde{a} * a^\star$, that is, $\widehat{u}(\xi) = \widehat{\tilde{a}}(\xi)\overline{\widehat{a}(\xi)}$. Therefore, $(\tilde{a}, a)$ is a pair of biorthogonal wavelet filters if and only if their correlation filter u satisfies

$$\widehat{u}(\xi) + \widehat{u}(\xi + \pi) = 1. \tag{2.1.1}$$

If (2.1.1) holds true, then we say that u is *an interpolatory filter*. Note that (2.1.1) is equivalent to

$$\widehat{u^{[0]}}(\xi) = 1/2, \quad \text{or equivalently, } u(0) = 1/2, \quad u(2k) = 0 \quad \forall\, k \in \mathbb{Z}\backslash\{0\}.$$

That u is an interpolatory filter can be also equivalently interpreted as the condition that $(u, \boldsymbol{\delta})$ is a pair of biorthogonal wavelet filters, where $\boldsymbol{\delta}$ is the *Dirac sequence* such that

$$\boldsymbol{\delta}(0) = 1 \quad \text{and} \quad \boldsymbol{\delta}(k) = 0, \qquad \forall\, k \in \mathbb{Z}\backslash\{0\}.$$

Hence, an interpolatory filter is simply a dual filter of the Dirac filter $\boldsymbol{\delta}$.

The name of an interpolatory filter is due to the following interpolation property.

Proposition 2.1.1 *A filter $u \in l_0(\mathbb{Z})$ is an interpolatory filter if and only if*

$$(\mathcal{S}_u v)\downarrow 2 = v, \quad \textit{that is,} \quad [\mathcal{S}_u v](2k) = v(k), \qquad \forall\, k \in \mathbb{Z} \quad \textit{and} \quad v \in l(\mathbb{Z}).$$

Proof By (1.1.4) and (1.1.29), we see that $(\mathcal{S}_u v)\downarrow 2 = v$ becomes

$$\widehat{(\mathcal{S}_u v)\downarrow 2}(\xi) = [\widehat{u}(\xi/2) + \widehat{u}(\xi/2 + \pi)]\widehat{v}(\xi) = \widehat{v}(\xi).$$

Therefore, $(\mathcal{S}_u v)\downarrow 2 = v$ if and only if $\widehat{u}(\xi/2) + \widehat{u}(\xi/2 + \pi) = 1$. □

To construct an interpolatory filter u with some desirable properties, we only need to design the odd coset $u^{[1]}$ of u such that $u^{[1]}$ satisfies certain conditions. For this purpose, we need the following simple fact.

Lemma 2.1.2 *Let $m \in \mathbb{N}$ and $\mathbf{v}$ be a smooth function. For a subset $\Lambda \subseteq \mathbb{R}$ with cardinality m (that is, Λ has m distinct points), then there is a unique solution $\{c_\lambda\}_{\lambda\in\Lambda}$ to the system of linear equations induced by*

$$\sum_{\lambda\in\Lambda} c_\lambda e^{-i\lambda\xi} = \mathbf{v}(\xi) + \mathscr{O}(|\xi|^m), \qquad \xi \to 0. \tag{2.1.2}$$

If in addition all $i^j\mathbf{v}^{(j)}(0) \in \mathbb{R}$ for $j = 0, \ldots, m-1$, then all $c_\lambda \in \mathbb{R}$ for $\lambda \in \Lambda$.

Proof The system induced by (2.1.2) is $\sum_{\lambda\in\Lambda} c_\lambda \lambda^j = i^j \mathbf{v}^{(j)}(0), j = 0, \ldots, m-1$. Since its coefficient matrix $(\lambda^j)_{\lambda\in\Lambda, 0\leqslant j<m}$ is a Vandermonde matrix which is invertible, there is a unique solution $\{c_\lambda\}_{\lambda\in\Lambda}$ to the linear system in (2.1.2). □

For any real number m and any nonnegative integer n, we define $\mathsf{P}_{m,n}$ to be the $(n-1)$th Taylor polynomial of $(1-x)^{-m}$ at $x = 0$. That is, $\mathsf{P}_{m,n}$ is the unique polynomial of degree $n-1$ and satisfies

$$\mathsf{P}_{m,n}(x) = (1-x)^{-m} + \mathscr{O}(x^n), \qquad x \to 0. \tag{2.1.3}$$

More explicitly,

$$\mathsf{P}_{m,n}(x) := \sum_{j=0}^{n-1} (-1)^j \binom{-m}{j} x^j = \sum_{j=0}^{n-1} \binom{m+j-1}{j} x^j, \tag{2.1.4}$$

where $\binom{m}{j}$ is defined in (1.2.25) and we used $(-1)^j\binom{-m}{j} = \binom{m+j-1}{j}$. Moreover,

$$(1-x)^m \mathsf{P}_{m,m}(x) + x^m \mathsf{P}_{m,m}(1-x) = 1 \qquad \forall\, x \in \mathbb{R}, m \in \mathbb{N}. \tag{2.1.5}$$

Indeed, expand and write $(x+y)^{2m-1} = x^m \mathsf{P}(x,y) + y^m \mathsf{P}(y,x)$ with $\mathsf{P}(y,x) := \sum_{j=0}^{m-1} \binom{2m-1}{j} x^j y^{m-j-1}$. Set $y = 1-x$. Then $\deg(\mathsf{P}(1-x,x)) \leqslant m-1$ and $x^m \mathsf{P}(x, 1-x) + (1-x)^m \mathsf{P}(1-x,x) = 1$, from which we further deduce that

$$\begin{aligned} \mathsf{P}(1-x,x) &= (1-x)^{-m}[(1-x)^m \mathsf{P}(1-x,x)] = (1-x)^{-m}[1 - x^m \mathsf{P}(x, 1-x)] \\ &= (1-x)^{-m} + \mathscr{O}(x^m), \ x \to 0. \end{aligned}$$

By the uniqueness of the polynomial $\mathsf{P}_{m,m}$, we must have $\mathsf{P}(1-x,x) = \mathsf{P}_{m,m}(x)$ and therefore, (2.1.5) holds. The identity in (2.1.5) has a simple explanation in probability theory. Suppose that we are doing an experiment by tossing a coin sequentially until one of the head or tail appears exactly m times. Assume $0 < x < 1$ and the probability for the head of the coin showing up is x. Then it is easy to see that in the process the tail appears exactly m times and the head appears j times with $j < m$ is $(1-x)^m \binom{m+j-1}{j} x^j$. Consequently, the probability that the tail appears exactly m times while the head appears less than m times is $(1-x)^m \mathsf{P}_{m,m}(x)$. Similarly, the probability that the head appears exactly m times while the tail appears less than m times is $x^m \mathsf{P}_{m,m}(1-x)$. As a result, the identity (2.1.5) is now obviously true.

For a family of filters having property P, in this book we shall use the notation a_m^P or $a_{m,n}^P$ to index them, where m and n refer to the highest order of sum rules and the highest order of linear-phase moments satisfied by the filter, respectively.

We have the following family of interpolatory filters a_{2m}^I (I=Interpolatory) for $m \in \mathbb{N}$, which have the shortest filter supports with respect to their orders of sum rules.

Theorem 2.1.3 *For every positive integer m, there exists a unique interpolatory filter a^I_{2m} such that a^I_{2m} has 2m sum rules and vanishes outside* $[1-2m, 2m-1]$. *Moreover, a^I_{2m} is a real-valued filter, symmetric about the origin, and given by*

$$\widehat{a^I_{2m}}(\xi) = \cos^{2m}(\xi/2)\mathsf{P}_{m,m}(\sin^2(\xi/2)). \tag{2.1.6}$$

Proof Note that a filter u is an interpolatory filter if and only if $\widehat{u^{[0]}}(\xi) = 1/2$. Therefore, to design an interpolatory filter u satisfying $2m$ sum rules and vanishing outside $[1-2m, 2m-1]$, by (1.2.13) of Theorem 1.2.5, we only need to design the odd coset $u^{[1]} = \{u(1+2k)\}_{k\in\Lambda}$ with $\Lambda := \{-m, \dots, m-1\}$ such that

$$\widehat{u^{[1]}}(\xi) = e^{i\xi/2}\widehat{u^{[0]}}(\xi) + \mathscr{O}(|\xi|^{2m}) = \tfrac{1}{2}e^{i\xi/2} + \mathscr{O}(|\xi|^{2m}), \qquad \xi \to 0. \tag{2.1.7}$$

That is, $2\widehat{u^{[1]}}$ has $2m$ linear-phase moments with phase $-1/2$. Since $\widehat{u^{[1]}}(\xi) = \sum_{k\in\Lambda} u(1+2k)e^{-ik\xi}$ and the cardinality of Λ is $2m$, by Lemma 2.1.2, there is a unique solution $\{u(1+2k)\}_{k\in\Lambda}$ to the system of linear equations induced by (2.1.7). So, there is a unique interpolatory filter u such that u is real-valued, has $2m$ sum rules, and vanishes outside $[1-2m, 2m-1]$. Note that $u(-\cdot)$ is also an interpolatory filter which has $2m$ sum rules and vanishes outside $[1-2m, 2m-1]$. By the uniqueness of such a filter, we must have $u(-k) = u(k)$ for all $k \in \mathbb{Z}$. That is, u must be symmetric about the origin.

On the other hand, it is easy to check that the filter a^I_{2m} defined via (2.1.6) has $2m$ sum rules and is supported inside $[1-2m, 2m-1]$. By the identity in (2.1.5), it is also trivial to verify that the filter a^I_{2m}, defined via (2.1.6), satisfies $\widehat{a^I_{2m}}(\xi)+\widehat{a^I_{2m}}(\xi+\pi) = 1$. Consequently, the filter a^I_{2m} must be interpolatory. Now by the uniqueness of such a filter u, we conclude that $u = a^I_{2m}$. Therefore, (2.1.6) must hold. □

Using the notation in (1.1.24) for representing filters, in the following we explicitly list the filters a^I_{2m} for $m = 1, \dots, 8$ as follows:

$$a^I_2 = \{\tfrac{1}{4}, \underline{\tfrac{1}{2}}, \tfrac{1}{4}\}_{[-1,1]},$$

$$a^I_4 = \{-\tfrac{1}{32}, 0, \tfrac{9}{32}, \underline{\tfrac{1}{2}}, \tfrac{9}{32}, 0, -\tfrac{1}{32}\}_{[-3,3]},$$

$$a^I_6 = \{\tfrac{3}{512}, 0, -\tfrac{25}{512}, 0, \tfrac{75}{256}, \underline{\tfrac{1}{2}}, \tfrac{75}{256}, 0, -\tfrac{25}{512}, 0, \tfrac{3}{512}\}_{[-5,5]},$$

$$a^I_8 = \{-\tfrac{5}{4096}, 0, \tfrac{49}{4096}, 0, -\tfrac{245}{4096}, 0, \tfrac{1225}{4096}, \underline{\tfrac{1}{2}}, \tfrac{1225}{4096}, 0, -\tfrac{245}{4096}, 0, \tfrac{49}{4096}, 0, -\tfrac{5}{4096}\}_{[-7,7]},$$

$$a^I_{10}|_{[0,9]} = \{\underline{\tfrac{1}{2}}, \tfrac{19845}{65536}, 0, -\tfrac{2205}{32768}, 0, \tfrac{567}{32768}, 0, -\tfrac{405}{131072}, 0, \tfrac{35}{131072}\}_{[0,9]},$$

$$a^I_{12}|_{[0,11]} = \{\underline{\tfrac{1}{2}}, \tfrac{160083}{524288}, 0, -\tfrac{38115}{524288}, 0, \tfrac{22869}{1048576}, 0, -\tfrac{5445}{1048576}, 0, \tfrac{847}{1048576}, 0, -\tfrac{63}{1048576}\}_{[0,11]},$$

$$a^I_{14}|_{[0,13]} = \{\underline{\tfrac{1}{2}}, \tfrac{1288287}{4194304}, 0, -\tfrac{1288287}{16777216}, 0, \tfrac{429429}{16777216}, 0, -\tfrac{61347}{8388608}, 0, \tfrac{13013}{8388608}, 0,$$
$$-\tfrac{3549}{16777216}, 0, \tfrac{231}{16777216}\}_{[0,13]},$$

$$a_{16}^{I}|_{[0,15]} = \{\underline{\tfrac{1}{2}}, \tfrac{41409225}{134217728}, 0, -\tfrac{10735725}{134217728}, 0, \tfrac{3864861}{134217728}, 0, -\tfrac{1254825}{134217728}, 0, \tfrac{325325}{134217728}, 0,$$
$$-\tfrac{61425}{134217728}, 0, \tfrac{7425}{134217728}, 0, -\tfrac{429}{134217728}\}_{[0,15]}.$$

Note that $\mathrm{E}(a_{2m}^{I}) = 0$ and $\widehat{a_{2m}^{I}}(\xi) \geqslant 0$ for all $m \in \mathbb{N}$. The smoothness exponents $\mathrm{sm}(a_{2m}^{I})$, variances $\mathrm{Var}(a_{2m}^{I})$, and frequency separation indicators $\mathrm{Fsi}(a_{2m}^{I})$ are presented in Table 2.1 and the graphs of $\phi^{a_{2m}^{I}}$ are given in Fig. 2.1 for $1 \leqslant m \leqslant 8$.

Table 2.1 The smoothness exponents $\mathrm{sm}(a_{2m}^{I})$, $\|a_{2m}^{I}\|_{l_2(\mathbb{Z})}^2$, variances $\mathrm{Var}(a_{2m}^{I})$, and frequency separation indicators $\mathrm{Fsi}(a_{2m}^{I})$ for $1 \leqslant m \leqslant 8$, where the interpolatory filter a_{2m}^{I} is defined in (2.1.6)

m	1	2	3	4
$\mathrm{sm}(a_{2m}^{I})$	1.5	2.440765	3.175132	3.793134
$\|a_{2m}^{I}\|_{l_2(\mathbb{Z})}^2$	0.375	0.410156	0.426498	0.436333
$\mathrm{Var}(a_{2m}^{I})$	0.333333	0.428571	0.507137	0.574308
$\mathrm{Fsi}(a_{2m}^{I})$	0.085714	0.048316	0.035932	0.029520
m	5	6	7	8
$\mathrm{sm}(a_{2m}^{I})$	4.344084	4.8620120	5.362830	5.852926
$\|a_{2m}^{I}\|_{l_2(\mathbb{Z})}^2$	0.443063	0.448035	0.451899	0.455014
$\mathrm{Var}(a_{2m}^{I})$	0.633798	0.687718	0.737374	0.783634
$\mathrm{Fsi}(a_{2m}^{I})$	0.025506	0.022714	0.020637	0.019019

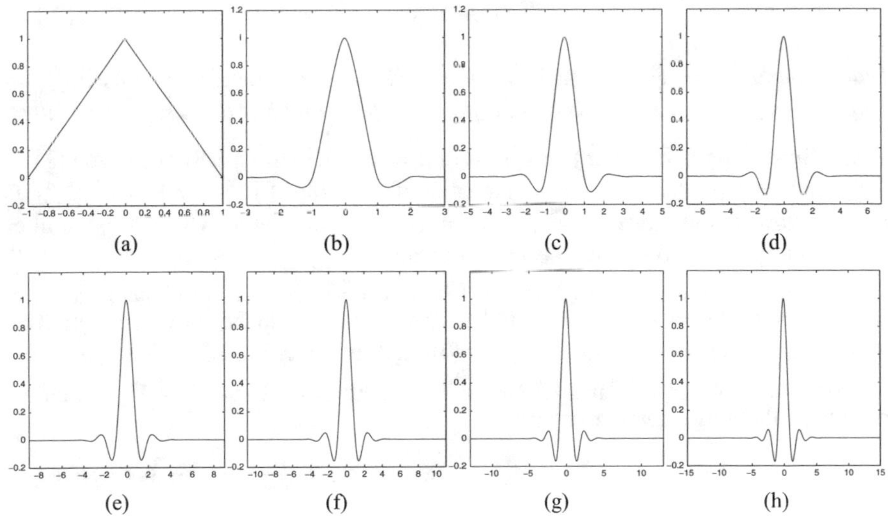

Fig. 2.1 The graphs of the standard refinable functions $\phi^{a_{2m}^{I}}$ associated with the interpolatory filters a_{2m}^{I}, $m = 1, \ldots, 8$. Note that the continuous function $\phi^{a_{2m}^{I}}$ is supported inside $[1-2m, 2m-1]$ and has the interpolation property: $\phi^{a_{2m}^{I}}(0) = 1$ and $\phi^{a_{2m}^{I}}(k) = 0$ for all $k \in \mathbb{Z}\backslash\{0\}$. **(a)** $\phi^{a_2^{I}}$. **(b)** $\phi^{a_4^{I}}$. **(c)** $\phi^{a_6^{I}}$. **(d)** $\phi^{a_8^{I}}$. **(e)** $\phi^{a_{10}^{I}}$. **(f)** $\phi^{a_{12}^{I}}$. **(g)** $\phi^{a_{14}^{I}}$. **(h)** $\phi^{a_{16}^{I}}$

The following result shows that the only meaningful symmetry type of an interpolatory filter u is symmetric about the origin, that is, $\mathsf{S}\widehat{u} = 1$.

Lemma 2.1.4 *If u is an interpolatory filter such that $\mathsf{S}\widehat{u}(\xi) = \epsilon e^{-ic\xi}$ (or $\mathbb{S}\widehat{u}(\xi) = \epsilon e^{-ic\xi}$) for some $\epsilon \in \{-1, 1\}$ and $c \in \mathbb{Z}$, then either $\epsilon = 1$ and $c = 0$, or $\widehat{u}(\xi) = (1 + \epsilon e^{-ic\xi})/2$ and c is an odd integer.*

Proof If c is even, since u is interpolatory, then it follows directly from $u(c-k) = \epsilon u(k)$ for all $k \in \mathbb{Z}$ that we must have $c = 0$ and $\epsilon = 1$. If c is odd, then $\widehat{u^{[1]}}(\xi) = \varepsilon e^{-i\xi(c-1)/2}\widehat{u^{[0]}}(\xi)$ and $\widehat{u}(\xi) = \widehat{u^{[0]}}(2\xi) + e^{-i\xi}\widehat{u^{[1]}}(2\xi) = (1 + \epsilon e^{-ic\xi})/2$. □

2.1.2 Interpolatory Filters Constructed by Convolution Method

Note that $\cos^2(\xi/2) = \widehat{a_2^I}(\xi)$. The identity in (2.1.5) and the expression in (2.1.6) also motivate us another way of constructing interpolatory filters.

Theorem 2.1.5 (Convolution Method) *Let a be an interpolatory filter having n sum rules. Let m be a positive integer and P be a polynomial satisfying*

$$(1-x)^m \mathsf{P}(x) + x^m \mathsf{P}(1-x) = 1 \qquad (2.1.8)$$

(For example, $\mathsf{P} = \mathsf{P}_{m,m}$ in (2.1.4) satisfies (2.1.8).) Define a filter u by $\widehat{u}(\xi) := (\widehat{a}(\xi))^m \mathsf{P}(\widehat{a}(\xi + \pi))$. Then u is an interpolatory filter and has at least mn sum rules.

Proof Since $\widehat{a}(\xi) + \widehat{a}(\xi + \pi) = 1$, setting $x = 1 - \widehat{a}(\xi)$ in (2.1.8), we have $\widehat{u}(\xi) + \widehat{u}(\xi + \pi) = 1$. Note that a has n sum rules if and only if $\widehat{a}(\xi + \pi) = \mathscr{O}(|\xi|^n)$ as $\xi \to 0$. Now it is obvious that $\widehat{u}(\xi + \pi) = \mathscr{O}(|\xi|^{mn})$ as $\xi \to 0$. Therefore, the filter u is an interpolatory filter and has mn sum rules. □

If we plug the interpolatory filter $\widehat{a_2^I}(\xi) = \cos^2(\xi/2)$ as $\widehat{a}$ into Theorem 2.1.5, then the interpolatory filter constructed in Theorem 2.1.5 via the convolution method is the same interpolatory filter a_{2m}^I constructed in Theorem 2.1.3. If we plug the interpolatory filter $\widehat{a_4^I}$ as $\widehat{a}$ into Theorem 2.1.5 with $m = 2$ and $\mathsf{P} = \mathsf{P}_{2,2}$, then we have the following interpolatory filter:

$$\{\tfrac{1}{16384}, 0, -\tfrac{27}{16384}, 0, \tfrac{27}{2048}, 0, -\tfrac{63}{1024}, 0, \tfrac{2457}{8192}, \underline{\tfrac{1}{2}}, \tfrac{2457}{8192}, 0, -\tfrac{63}{1024}, 0, \tfrac{27}{2048}, 0, -\tfrac{27}{16384}, 0, \tfrac{1}{16384}\}_{[-9,9]}.$$

In Table 2.2, we present the smoothness exponents of the interpolatory filters constructed by the convolution method in Theorem 2.1.5 using $a = a_4^I$ and $\mathsf{P} = \mathsf{P}_{m,m}$. See Fig. 2.2 for the graphs of some interpolatory refinable functions.

Table 2.2 The smoothness exponents $\mathrm{sm}(a_{4m}^{CM})$, $\|a_{4m}^{CM}\|_{l_2(\mathbb{Z})}^2$, variances $\mathrm{Var}(a_{4m}^{CM})$, and frequency separation indicators $\mathrm{Fsi}(a_{4m}^{CM})$ for $m = 1, \dots, 8$, where the interpolatory filter a_{4m}^{CM} is constructed by the convolution method in Theorem 2.1.5 using $a = a_4^I$ and $\mathsf{P} = \mathsf{P}_{m,m}$. Note that $\mathrm{E}(a_{4m}^{CM}) = 0$ and $\mathrm{fsupp}(a_{4m}^{CM}) = [3 - 6m, 6m - 3]$ while $\mathrm{fsupp}(a_{4m}^I) = [1 - 4m, 4m - 1]$

m	1	2	3	4
$\mathrm{sm}(a_{4m}^{CM})$	2.440765	3.899101	5.068043	6.157088
$\|a_{4m}^{CM}\|_{l_2(\mathbb{Z})}^2$	0.410156	0.437835	0.449789	0.456780
$\mathrm{Var}(a_{4m}^{CM})$	0.428571	0.586982	0.709939	0.813479
$\mathrm{Fsi}(a_{4m}^{CM})$	0.048316	0.028563	0.021735	0.018104
m	5	6	7	8
$\mathrm{sm}(a_{4m}^{CM})$	7.212459	8.246510	9.265337	10.272686
$\|a_{4m}^{CM}\|_{l_2(\mathbb{Z})}^2$	0.461492	0.464941	0.467605	0.469742
$\mathrm{Var}(a_{4m}^{CM})$	0.904531	0.986718	1.06220	1.13238
$\mathrm{Fsi}(a_{4m}^{CM})$	0.015788	0.014156	0.012930	0.011966

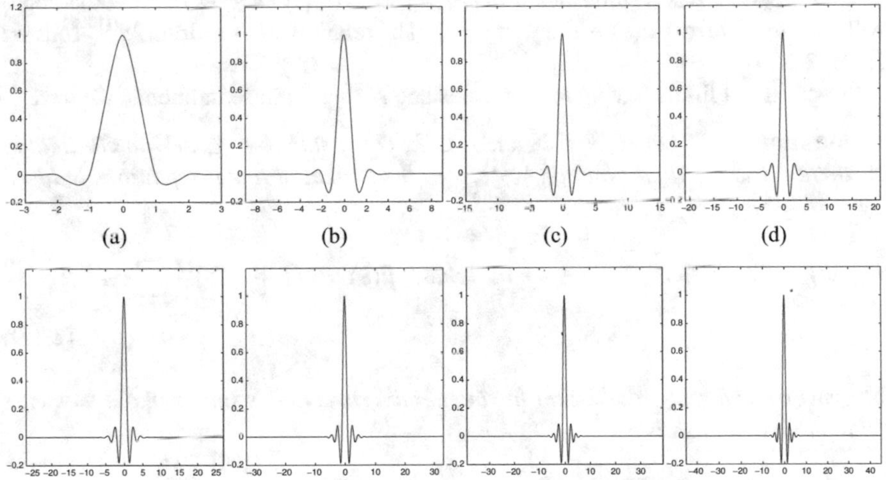

Fig. 2.2 The graphs of the standard refinable functions $\phi^{a_{4m}^{CM}}$ associated with the interpolatory filters a_{4m}^{CM}, $m = 1, \dots, 8$. Note that $\phi^{a_{4m}^{CM}}$ is supported inside $[3 - 6m, 6m - 3]$ and has the interpolation property: $\phi^{a_{4m}^I}(0) = 1$ and $\phi^{a_{4m}^I}(k) = 0$ for all $k \in \mathbb{Z}\backslash\{0\}$. Moreover, $\widehat{a_{4m}^{CM}}(\xi) \geqslant 0$ for $\xi \in \mathbb{R}$. **(a)** $\phi^{a_4^{CM}}$. **(b)** $\phi^{a_8^{CM}}$. **(c)** $\phi^{a_{12}^{CM}}$. **(d)** $\phi^{a_{16}^{CM}}$. **(e)** $\phi^{a_{20}^{CM}}$. **(f)** $\phi^{a_{24}^{CM}}$. **(g)** $\phi^{a_{28}^{CM}}$. **(h)** $\phi^{a_{32}^{CM}}$

2.1.3 Filters Having Linear-Phase Moments

As we have seen in Proposition 1.2.7, filters having linear-phase moments possess the desirable almost-interpolation property: up to shifts and dilations, polynomials to certain degrees can be exactly reproduced/preserved by their associated subdivision or transition operators. Also, as we will see in Chap. 3, filters having linear-phase moments are of importance in building dual or tight framelet filter banks with vanishing moments.

For interpolatory filters, there are some close connections between sum rules and linear-phase moments.

Proposition 2.1.6 *For an interpolatory filter u, u has n sum rules $\iff$ u has n linear-phase moments with phase 0 $\iff$ $2u^{[1]}$ has n linear-phase moments with phase $-1/2$. That is,* $\operatorname{sr}(u) = \operatorname{lpm}(u) = \operatorname{lpm}(2u^{[1]})$ *for any interpolatory filter u.*

Proof Recall that u has n linear-phase moments with phase 0 if $\widehat{u}(\xi) = 1 + \mathscr{O}(|\xi|^n)$ as $\xi \to 0$, and u has n sum rules if $\widehat{u}(\xi + \pi) = \mathscr{O}(|\xi|^n)$ as $\xi \to 0$. Now the claim follows directly from $\widehat{u}(\xi) + \widehat{u}(\xi + \pi) = 1$. The relation $\operatorname{sr}(u) = \operatorname{lpm}(2u^{[1]})$ follows from (2.1.7). □

To construct filters having M sum rules and N linear-phase moments, we have

Proposition 2.1.7 *Let $M, N \in \mathbb{N}$ and $c \in \mathbb{R}$. For a subset $\Lambda \subset \mathbb{Z}$ with cardinality N, there exists a unique solution $\{c_k\}_{k\in\Lambda}$ to the system of linear equations induced by*

$$\widehat{u}(\xi) = e^{-ic\xi} + \mathscr{O}(|\xi|^N),\ \xi \to 0, \ \ \textit{where}\ \ \widehat{u}(\xi) := (1 + e^{-i\xi})^M \sum_{k\in\Lambda} c_k e^{-ik\xi}. \tag{2.1.9}$$

Moreover, u is a real-valued filter having M sum rules and N linear-phase moments with phase c, and satisfies

$$\mathcal{S}_u \mathsf{p} = \mathsf{p}(2^{-1}(\cdot - c)) \qquad \forall\, \mathsf{p} \in \mathbb{P}_{\min(M,N)-1}. \tag{2.1.10}$$

Proof Equation (2.1.9) is equivalent to $\sum_{k\in\Lambda} c_k e^{-ik\xi} = e^{-ic\xi}(1+e^{-i\xi})^{-M} + \mathscr{O}(|\xi|^N)$ as $\xi \to 0$. By Lemma 2.1.2, there is a unique real-valued solution $\{c_k\}_{k\in\Lambda}$ to (2.1.9). (2.1.10) is a direct consequence of Theorem 1.2.5 and Proposition 1.2.7. □

Example 2.1.1 Taking $N = 1$ and $M = m$ in Proposition 2.1.7, we have the B-spline filter a_m^B in (1.2.24). Since $\widehat{a_m^B}(\xi) = 2^{-m}(1 + e^{-i\xi})^m$, it is trivial to see that $\operatorname{E}(a_m^B) = m/2$ and $\rho(a_m^B) = 2^{-2m}$. Therefore, we have $\operatorname{sm}(a_m^B) = m - 1/2$. See

Fig. 2.3 for some graphs of the refinable functions $\phi^{a_m^B}$ and see Table 2.3 for some statistics quantities of a_m^B for $m = 1, \ldots, 8$. For example,

$$a_1^B = \{\tfrac{\mathbf{1}}{\underline{\mathbf{2}}}, \tfrac{1}{2}\}_{[0,1]}, \qquad a_2^B = \{\tfrac{\mathbf{1}}{\underline{\mathbf{4}}}, \tfrac{1}{2}, \tfrac{1}{4}\}_{[0,2]}, \qquad a_3^B = \{\tfrac{\mathbf{1}}{\underline{\mathbf{8}}}, \tfrac{3}{8}, \tfrac{3}{8}, \tfrac{1}{8}\}_{[0,3]},$$

$$a_4^B = \{\tfrac{\mathbf{1}}{\underline{\mathbf{16}}}, \tfrac{1}{4}, \tfrac{3}{8}, \tfrac{1}{4}, \tfrac{1}{16}\}_{[0,4]}, \qquad a_5^B = \{\tfrac{\mathbf{1}}{\underline{\mathbf{32}}}, \tfrac{5}{32}, \tfrac{5}{16}, \tfrac{5}{16}, \tfrac{15}{32}, \tfrac{1}{32}\}_{[0,5]},$$

$$a_6^B = \{\tfrac{\mathbf{1}}{\underline{\mathbf{64}}}, \tfrac{3}{32}, \tfrac{15}{64}, \tfrac{5}{16}, \tfrac{15}{64}, \tfrac{3}{32}, \tfrac{1}{64}\}_{[0,6]}, \quad a_7^B = \{\tfrac{\mathbf{1}}{\underline{\mathbf{128}}}, \tfrac{7}{128}, \tfrac{21}{128}, \tfrac{35}{128}, \tfrac{35}{128}, \tfrac{21}{128}, \tfrac{7}{128}, \tfrac{1}{128}\}_{[0,7]},$$

$$a_8^B = \{\tfrac{\mathbf{1}}{\underline{\mathbf{256}}}, \tfrac{1}{32}, \tfrac{7}{64}, \tfrac{7}{32}, \tfrac{35}{128}, \tfrac{7}{32}, \tfrac{7}{64}, \tfrac{1}{32}, \tfrac{1}{256}\}_{[0,8]}.$$

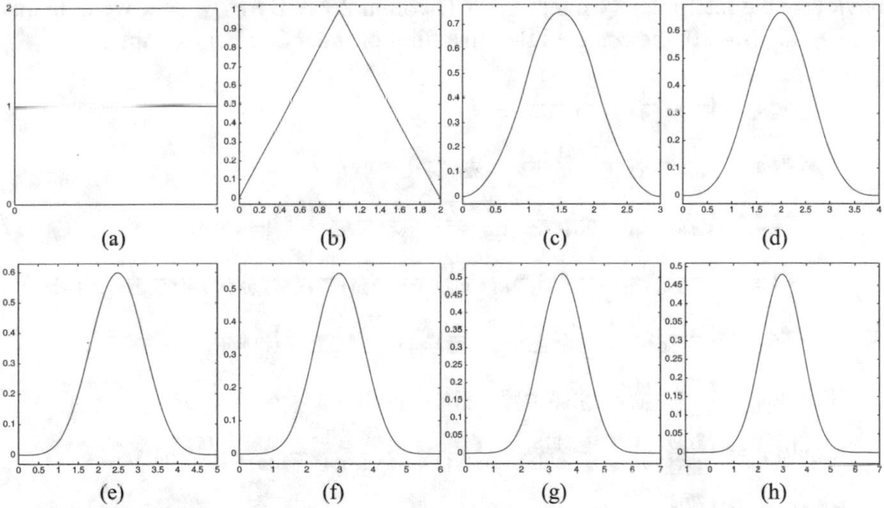

Fig. 2.3 The graphs of the standard refinable functions $\phi^{a_m^B}$ associated with the *B*-spline filters a_m^B (also see (1.2.24)) with $m = 1, \ldots, 8$ in Example 2.1.1. **(a)** $\phi^{a_1^B}$. **(b)** $\phi^{a_2^B}$. **(c)** $\phi^{a_3^B}$. **(d)** $\phi^{a_4^B}$. **(e)** $\phi^{a_5^B}$. **(f)** $\phi^{a_6^B}$. **(g)** $\phi^{a_7^B}$. **(h)** $\phi^{a_8^B}$

Table 2.3 The smoothness exponents $\mathrm{sm}(a_m^B)$, $\|a_m^B\|^2_{l_2(\mathbb{Z})}$, variances $\mathrm{Var}(a_m^B)$, and frequency separation indicators $\mathrm{Fsi}(a_m^B)$ for $m = 1, \ldots, 8$

m	1	2	3	4
$\mathrm{sm}(a_m^B)$	0.5	1.5	2.5	3.5
$\|a_m^B\|^2_{l_2(\mathbb{Z})}$	0.5	0.375	0.3125	0.273438
$\mathrm{Var}(a_m^B)$	0.25	0.333333	0.45	0.571429
$\mathrm{Fsi}(a_m^B)$	0.333333	0.085714	0.021645	0.005439
m	5	6	7	8
$\mathrm{sm}(a_m^B)$	4.5	5.5	6.5	7.5
$\|a_m^B\|^2_{l_2(\mathbb{Z})}$	0.246094	0.225586	0.209473	0.196381
$\mathrm{Var}(a_m^B)$	0.694444	0.818182	0.942308	1.06667
$\mathrm{Fsi}(a_m^B)$	0.001364	0.000342	0.000086	0.000021

Using Proposition 2.1.7, we present two families of filters with linear-phase moments and sum rules.

Example 2.1.2 For any positive integers m and n, set $c = 0$, $M = 2m$, $N = 2n - 1$, and $\Lambda = \{1 - n - m, \dots, n - m - 1\}$ in Theorem 2.1.7. Then the unique filter in Proposition 2.1.7, denoted by $a_{2m,2n}$, must take the form

$$\widehat{a_{2m,2n}}(\xi) = \cos^{2m}(\xi/2)\mathsf{P}_{m,n}(\sin^2(\xi/2)), \tag{2.1.11}$$

where $\mathsf{P}_{m,n}$ is the polynomial defined in (2.1.4). The filter $a_{2m,2n}$ has $2m$ sum rules, $2n$ linear-phase moments with phase 0, is symmetric about the origin, and has filter support $[1 - m - n, m + n - 1]$. Moreover, when $n = m$, the above filter $a_{2m,2m}$ is obviously the interpolatory filter a^I_{2m} in Theorem 2.1.3. By $\mathsf{P}_{m,1} = 1$, we note that $a_{2m,2} = a^B_{2m}(\cdot - m)$, the centered B-spline filter of order $2m$. For example,

$$\begin{aligned}
a_{2,4} &= \{-\tfrac{1}{16}, \tfrac{1}{4}, \underline{\tfrac{\mathbf{5}}{\mathbf{8}}}, \tfrac{1}{4}, -\tfrac{1}{16}\}_{[-2,2]},\\
a_{2,6} &= \{\tfrac{1}{64}, -\tfrac{3}{32}, \tfrac{15}{64}, \underline{\tfrac{\mathbf{11}}{\mathbf{16}}}, \tfrac{15}{64}, -\tfrac{3}{32}, \tfrac{1}{64}\}_{[-3,3]},\\
a_{4,6} &= \{\tfrac{3}{256}, -\tfrac{1}{32}, -\tfrac{3}{64}, \tfrac{9}{32}, \underline{\tfrac{\mathbf{73}}{\mathbf{128}}}, \tfrac{9}{32}, -\tfrac{3}{64}, -\tfrac{1}{32}, \tfrac{3}{256}\}_{[-4,4]},\\
a_{4,8} &= \{-\tfrac{1}{256}, \tfrac{5}{256}, -\tfrac{5}{256}, -\tfrac{5}{64}, \tfrac{35}{128}, \underline{\tfrac{\mathbf{79}}{\mathbf{128}}}, \tfrac{35}{128}, -\tfrac{5}{64}, -\tfrac{5}{256}, \tfrac{5}{256}, -\tfrac{1}{256}\}_{[-5,5]},\\
a_{6,4} &= \{-\tfrac{3}{256}, -\tfrac{1}{32}, \tfrac{3}{64}, \tfrac{9}{32}, \underline{\tfrac{\mathbf{55}}{\mathbf{128}}}, \tfrac{9}{32}, \tfrac{3}{64}, -\tfrac{1}{32}, -\tfrac{3}{256}\}_{[-4,4]},\\
a_{6,8}|_{[0,6]} &= \{\underline{\tfrac{\mathbf{281}}{\mathbf{512}}}, \tfrac{75}{256}, -\tfrac{75}{2048}, -\tfrac{25}{512}, \tfrac{15}{1024}, \tfrac{3}{512}, -\tfrac{5}{2048}\}_{[0,6]},\\
a_{6,10}|_{[0,7]} &= \{\underline{\tfrac{\mathbf{1199}}{\mathbf{2048}}}, \tfrac{4725}{16384}, -\tfrac{525}{8192}, -\tfrac{665}{16384}, \tfrac{105}{4096}, \tfrac{21}{16384}, -\tfrac{35}{8192}, \tfrac{15}{16384}\}_{[0,7]},\\
a_{6,12}|_{[0,8]} &= \{\underline{\tfrac{\mathbf{20129}}{\mathbf{32768}}}, \tfrac{1155}{4096}, -\tfrac{693}{8192}, -\tfrac{119}{4096}, \tfrac{525}{16384}, -\tfrac{21}{4096}, -\tfrac{35}{8192}, \tfrac{9}{4096}, -\tfrac{21}{65536}\}_{[0,8]},\\
a_{8,4} &= \{-\tfrac{1}{256}, -\tfrac{5}{256}, -\tfrac{5}{256}, \tfrac{5}{64}, \tfrac{35}{128}, \underline{\tfrac{\mathbf{49}}{\mathbf{128}}}, \tfrac{35}{128}, \tfrac{5}{64}, -\tfrac{5}{256}, -\tfrac{5}{256}, -\tfrac{1}{256}\}_{[-5,5]},\\
a_{8,6}|_{[0,6]} &= \{\underline{\tfrac{\mathbf{231}}{\mathbf{512}}}, \tfrac{75}{256}, \tfrac{75}{2048}, -\tfrac{25}{512}, -\tfrac{15}{1024}, \tfrac{3}{512}, \tfrac{5}{2048}\}_{[-6,6]}.
\end{aligned}$$

We present the smoothness exponents, variances, and frequency separation indicators of the filters $a_{2m,2n}$ in Table 2.4 and the graphs of $\phi^{a_{2m,2n}}$ in Fig. 2.4.

Example 2.1.3 For $m, n \in \mathbb{N}$, set $c = 1/2$, $M = 2m - 1$, $N = 2n - 1$, and $\Lambda = \{2 - n - m, \dots, n - m\}$ in Theorem 2.1.7. Then the unique filter in Theorem 2.1.7 is

$$\widehat{a_{2m-1,2n}}(\xi) = 2^{-1}(1 + e^{-i\xi}) \cos^{2m-2}(\xi/2)\mathsf{P}_{m-1/2,n}(\sin^2(\xi/2)), \tag{2.1.12}$$

Table 2.4 The smoothness exponents $\mathrm{sm}(a_{2m,2n})$, $\|a_{2m,2n}\|^2_{l_2(\mathbb{Z})}$, variances $\mathrm{Var}(a_{2m,2n})$, and frequency separation indicators $\mathrm{Fsi}(a_{2m,2n})$, where the real-valued filters $a_{2m,2n}$ are defined in (2.1.11). Note that $\mathrm{E}(a_{2m,2n}) = 0$ and all the filters $a_{2m,2n}$ are symmetric about the origin with sum rule order $\mathrm{sr}(a_{2m,2n}) = 2m$ and linear-phase moment order $\mathrm{lpm}(a_{2m,2n}) = 2n$

$(2m, 2n)$	(2,4)	(2,6)	(4,6)	(4,8)	(4,10)	(6,4)	(6,8)	(6,10)
$\mathrm{sm}(a_{2m,2n})$	0.885296	0.557291	1.793801	1.351645	1.030888	4.098191	2.509852	2.006574
$\|a_{2m,2n}\|^2_{l_2(\mathbb{Z})}$	0.523438	0.600586	0.490082	0.544220	0.584049	0.349457	0.480834	0.521954
$\mathrm{Var}(a_{2m,2n})$	0.298507	0.307317	0.403512	0.400942	0.405972	0.565889	0.487315	0.481568
$\mathrm{Fsi}(a_{2m,2n})$	0.263077	0.415829	0.137202	0.242003	0.337543	0.007165	0.0928606	0.167347
$(2m, 2n)$	(6,12)	(6,14)	(8,4)	(8,6)	(8,10)	(8,12)	(10,4)	(10,6)
$\mathrm{sm}(a_{2m,2n})$	1.61616	1.308201	5.820374	4.659613	3.119511	2.584188	7.586972	6.219362
$\|a_{2m,2n}\|^2_{l_2(\mathbb{Z})}$	0.554531	0.581190	0.309845	0.383178	0.477388	0.510428	0.281315	0.351056
$\mathrm{Var}(a_{2m,2n})$	0.482287	0.486388	0.704225	0.611512	0.557726	0.550928	0.842854	0.716001
$\mathrm{Fsi}(a_{2m,2n})$	0.244506	0.314823	0.000916	0.007757	0.070689	0.126938	0.000105	0.001436

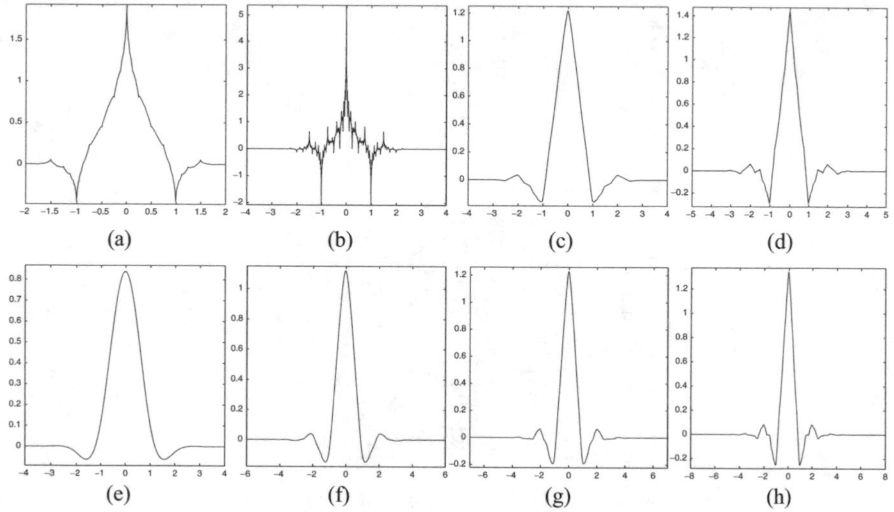

Fig. 2.4 The graphs of the standard refinable functions $\phi^{a_{2m,2n}}$ associated with the filters $a_{2m,2n}$ in (2.1.11). Note that $\mathrm{E}(a_{2m,2n}) = 0$ and both $a_{2m,2n}$ and $\phi^{a_{2m,2n}}$ are symmetric about the origin. (**a**) $\phi^{a_{2,4}}$. (**b**) $\phi^{a_{2,6}}$. (**c**) $\phi^{a_{4,6}}$. (**d**) $\phi^{a_{4,8}}$. (**e**) $\phi^{a_{6,4}}$. (**f**) $\phi^{a_{6,8}}$. (**g**) $\phi^{a_{6,10}}$. (**h**) $\phi^{a_{6,12}}$

where $\mathsf{P}_{m-1/2,n}$ is the polynomial defined in (2.1.4). The filter $a_{2m-1,2n}$ has $2m-1$ sum rules, $2n$ linear-phase moments with phase $1/2$, is symmetric about the point $1/2$, and has filter support $[2-m-n, m+n-1]$. By $\mathsf{P}_{2m-1,1} = 1$, we see that $a_{2m-1,2} = a^B_{2m-1}(\cdot - m + 1)$, the shifted B-spline filter of order $2m-1$. For example,

$$a_{3,4} = \{-\tfrac{3}{64}, \tfrac{5}{64}, \underline{\tfrac{\mathbf{15}}{\mathbf{32}}}, \tfrac{15}{32}, \tfrac{5}{64}, -\tfrac{3}{64}\}_{[-2,3]},$$

$$a_{3,6} = \{\tfrac{15}{1024}, -\tfrac{63}{1024}, \tfrac{35}{1024}, \underline{\tfrac{\mathbf{525}}{\mathbf{1024}}}, \tfrac{525}{1024}, \tfrac{35}{1024}, -\tfrac{63}{1024}, \tfrac{15}{1024}\}_{[-3,4]},$$

$$a_{3,8} = \{-\tfrac{35}{8192}, \tfrac{225}{8192}, -\tfrac{63}{1024}, 0, \underline{\tfrac{\mathbf{2205}}{\mathbf{4096}}}, \tfrac{2205}{4096}, 0, -\tfrac{63}{1024}, \tfrac{225}{8192}, -\tfrac{35}{8192}\}_{[-4,5]},$$

$$a_{5,4} = \{-\tfrac{5}{256}, -\tfrac{7}{256}, \tfrac{35}{256}, \underline{\tfrac{\mathbf{105}}{\mathbf{256}}}, \tfrac{105}{256}, \tfrac{35}{256}, -\tfrac{7}{256}, -\tfrac{5}{256}\}_{[-3,4]},$$

$$a_{5,6} = \{\tfrac{35}{4096}, -\tfrac{45}{4096}, -\tfrac{63}{1024}, \tfrac{105}{1024}, \underline{\tfrac{\mathbf{945}}{\mathbf{2048}}}, \tfrac{945}{2048}, \tfrac{105}{1024}, -\tfrac{63}{1024}, -\tfrac{45}{4096}, \tfrac{35}{4096}\}_{[-4,5]},$$

$$a_{5,8}|_{[1,6]} = \{\tfrac{8085}{16384}, \tfrac{1155}{16384}, -\tfrac{2541}{32768}, \tfrac{165}{32768}, \tfrac{385}{32768}, -\tfrac{105}{32768}\}_{[1,6]},$$

$$a_{5,10}|_{[1,7]} = \{\tfrac{135135}{262144}, \tfrac{45045}{1048576}, -\tfrac{87087}{1048576}, \tfrac{10725}{524288}, \tfrac{5005}{524288}, -\tfrac{6825}{1048576}, \tfrac{1155}{1048576}\}_{[1,7]},$$

$$a_{7,4} = \{-\tfrac{7}{1024}, -\tfrac{27}{1024}, 0, \tfrac{21}{128}, \underline{\tfrac{\mathbf{189}}{\mathbf{512}}}, \tfrac{189}{512}, \tfrac{21}{128}, 0, -\tfrac{27}{1024}, -\tfrac{7}{1024}\}_{[-4,5]}.$$

We present the smoothness exponents, variances, and frequency separation indicators of the filters $a_{2m-1,2n}$ in Table 2.5 and the graphs of $\phi^{a_{2m-1,2n}}$ in Fig. 2.5.

Table 2.5 The smoothness exponents $\mathrm{sm}(a_{2m-1,2n})$, $\|a_{2m-1,2n}\|^2_{l_2(\mathbb{Z})}$ variances $\mathrm{Var}(a_{2m-1,2n})$, and frequency separation indicators $\mathrm{Fsi}(a_{2m-1,2n})$, where the filters $a_{2m-1,2n}$ are defined in (2.1.12)

$(2m-1, 2n)$	(3,4)	(3,6)	(3,8)	(3,10)	(5,4)	(5,6)	(5,8)	(5,10)
$\mathrm{sm}(a_{2m-1,2n})$	1.646884	1.154269	0.829750	0.600824	3.259609	2.469332	1.913082	1.500069
$\|a_{2m-1,2n}\|^2_{l_2(\mathbb{Z})}$	0.456055	0.536049	0.588714	0.626740	0.376099	0.454814	0.509337	0.550071
$\mathrm{Var}(a_{2m-1,2n})$	0.361349	0.353059	0.359147	0.369643	0.496998	0.455130	0.443946	0.443587
$\mathrm{Fsi}(a_{2m-1,2n})$	0.116398	0.246172	0.363284	0.454085	0.019019	0.072124	0.153618	0.242329
$(2m-1, 2n)$	(5,12)	(7,4)	(7,6)	(7,8)	(7,10)	(7,12)	(9,4)	(9,6)
$\mathrm{sm}(a_{2m-1,2n})$	1.184771	4.952718	3.906495	3.137799	2.547492	2.083505	6.699013	5.431405
$\|a_{2m-1,2n}\|^2_{l_2(\mathbb{Z})}$	0.582033	0.327847	0.403048	0.456881	0.498069	0.530974	0.294540	0.366042
$\mathrm{Var}(a_{2m-1,2n})$	0.448050	0.634999	0.559294	0.530799	0.519645	0.516612	0.773519	0.663753
$\mathrm{Fsi}(a_{2m-1,2n})$	0.323278	0.002602	0.017058	0.0535398	0.110987	0.179160	0.000314	0.003395

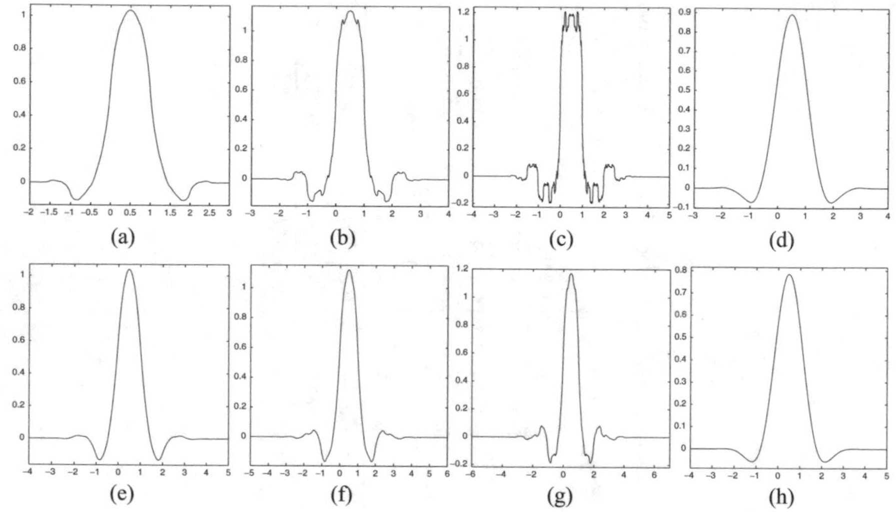

Fig. 2.5 The graphs of the standard refinable functions $\phi^{a_{2m-1,2n}}$ associated with the filters $a_{2m-1,2n}$ in (2.1.12). Note that both $a_{2m-1,2n}$ and $\phi^{a_{2m-1,2n}}$ are symmetric about $1/2$. **(a)** $\phi^{a_{3,4}}$. **(b)** $\phi^{a_{3,6}}$. **(c)** $\phi^{a_{3,8}}$. **(d)** $\phi^{a_{5,4}}$. **(e)** $\phi^{a_{5,6}}$. **(f)** $\phi^{a_{5,8}}$. **(g)** $\phi^{a_{5,10}}$. **(h)** $\phi^{a_{7,4}}$

2.2 Real Orthogonal Wavelet Filter Banks with Minimal Supports

For an orthogonal wavelet filter bank $\{a; b\}$, as we discussed at the beginning of this chapter, the high-pass filter b is determined by the low-pass filter a and we often take the standard high-pass filter b in (2.0.6). Therefore, to construct an orthogonal wavelet filter bank, it suffices to design a low-pass orthogonal wavelet filter a. In this section, we discuss how to systematically design real-valued low-pass orthogonal wavelet filters with high sum rules and minimal filter supports.

As we discussed before, there is a close relation between interpolatory filters and orthogonal low-pass filters. This relation is stated in the following result.

Theorem 2.2.1 *If $a \in l_0(\mathbb{Z})$ is a real-valued orthogonal wavelet filter and has m sum rules, then*

$$|\widehat{a}(\xi)|^2 = \cos^{2m}(\xi/2)\mathsf{P}(\sin^2(\xi/2)) \tag{2.2.1}$$

for some polynomial P with real coefficients such that (2.1.8) *holds and*

$$\mathsf{P}(x) \geqslant 0, \qquad \forall\, x \in [0, 1]. \tag{2.2.2}$$

Conversely, if (2.1.8) *and* (2.2.2) *are satisfied for a polynomial P with real coefficients and if $m \geqslant 1$, then there exists a real-valued orthogonal wavelet filter a with $\widehat{a}(0) = 1$ such that a has m sum rules and* (2.2.1) *holds.*

Proof Suppose that a is a real-valued orthogonal wavelet filter and has m sum rules. Then we can write $\widehat{a}(\xi) = 2^{-m}(1+e^{-i\xi})^m\widehat{u}(\xi)$ for some 2π periodic trigonometric polynomial $\widehat{u}$ with real coefficients. Then there is a polynomial P with real coefficients such that $|\widehat{u}(\xi)|^2 = \mathsf{P}(\sin^2(\xi/2))$ (see Exercise 2.8). Therefore, both (2.2.1) and (2.2.2) hold. By the orthogonality condition in (2.0.5),

$$\cos^{2m}(\xi/2)\mathsf{P}(\sin^2(\xi/2)) + \sin^{2m}(\xi/2)\mathsf{P}(\cos^2(\xi/2)) = |\widehat{a}(\xi)|^2 + |\widehat{a}(\xi+\pi)|^2 = 1.$$

Setting $x = \sin^2(\xi/2)$ in the above identity, we see that (2.1.8) holds.

Conversely, suppose that (2.1.8) and (2.2.2) are satisfied. Then $\mathsf{P}(\sin^2(\xi/2)) \geqslant 0$ for all $\xi \in \mathbb{R}$. By (2.1.8) and $m \geqslant 1$, we must have $\mathsf{P}(0) = 1$. Now by the Fejér-Riesz Lemma in Lemma 1.4.4, there exists a 2π-periodic trigonometric polynomial $\mathbf{Q}$ with real coefficients such that $|\mathbf{Q}(\xi)|^2 = \mathsf{P}(\sin^2(\xi/2))$ and $\mathbf{Q}(0) = \sqrt{\mathsf{P}(0)} = 1$. Define $\widehat{a}(\xi) := 2^{-m}(1+e^{-i\xi})^m\mathbf{Q}(\xi)$. Then a is a finitely supported real-valued orthogonal wavelet filter with $\widehat{a}(0) = 1$ and has m sum rules. □

In the following we provide an algorithm to derive a 2π-periodic trigonometric polynomial $\mathbf{Q}$ from P (assuming $\mathsf{P}(0) = 1$) such that $|\mathbf{Q}(\xi)|^2 = \mathsf{P}(\sin^2(\xi/2))$ and $\mathbf{Q}(0) = 1$.

Algorithm 2.2.2 *Input a polynomial* P *with real coefficients such that* $\mathsf{P}(x) \geqslant 0$ *for all* $x \in [0,1]$ *and* $\mathsf{P}(0) = 1$.

(1) Factorize the polynomial P *as a product of linear factors and irreducible quadratic factors with real coefficients:*

$$\mathsf{P}(x) = \left(\prod_{j=1}^{J_0}(1-t_jx)^{m_j}\right)\left(\prod_{j=J_0+1}^{J}(1-t_jx)^{m_j}\right)\left(\prod_{k=1}^{K}(1+x_kx+y_kx^2)^{n_k}\right),$$

where all x_k, y_k *are real numbers such that* $x_k^2 - 4y_k < 0$ *(irreducible), and* $t_1,\ldots,t_{J_0} \in [1,\infty)$, $t_{J_0+1},\ldots,t_J \in (-\infty,1)\backslash\{0\}$ *are distinct. Since* $\mathsf{P}(x) \geqslant 0$ *for all* $[0,1]$, *we see that the multiplicities* $m_j, j = 1,\ldots,J_0$ *must be even integers;*

(2) Note $\sin^2(\xi/2) = \frac{1}{2} - \frac{e^{-i\xi}+e^{i\xi}}{4}$. *Define* $\zeta := \frac{1}{2} - \frac{z+z^{-1}}{4}$. *For* $j = J_0+1,\ldots,J$, *since* $t_j \in (-\infty,1)\backslash\{0\}$, *factorize* $1 - t_j\zeta = \frac{(z-c_j)(z^{-1}-c_j)}{(1-c_j)^2}$ *for some real number* $c_j \neq 1$;

(3) For $k = 1,\ldots,K$, *factorize* $1 + x_k\zeta + y_k\zeta^2 = (z^2+\tilde{x}_kz+\tilde{y}_k)(z^{-2}+\tilde{x}_kz^{-1}+\tilde{y}_k)/(1+\tilde{x}_k+\tilde{y}_k)^2$ *for some real numbers* $\tilde{x}_k, \tilde{y}_k$ *with* $1+\tilde{x}_k+\tilde{y}_k \neq 0$;

(4) Define $\mathbf{Q}(\xi) := \tilde{\mathbf{Q}}(\xi)/\tilde{\mathbf{Q}}(0)$ *with*

$$\tilde{\mathbf{Q}}(\xi) := \prod_{j=1}^{J_0}(1-t_j\sin^2(\xi/2))^{m_j/2}\prod_{j=J_0+1}^{J}(e^{-i\xi}-c_j)^{m_j}\prod_{k=1}^{K}(e^{-2i\xi}+\tilde{x}_ke^{-i\xi}+y_k)^{n_k}.$$

Output a 2π*-periodic trigonometric polynomial* $\mathbf{Q}$ *with real coefficients such that* $|\mathbf{Q}(\xi)|^2 = \mathsf{P}(\sin^2(\xi/2))$ *and* $\mathbf{Q}(0) = 1$.

Since both (2.1.8) and (2.2.2) hold for all $\mathsf{P} = \mathsf{P}_{m,m}$ in (2.1.4), by $\mathsf{P}_{m,m}(0) = 1$, it follows from Theorem 2.2.1 that there is a real-valued filter $\mathcal{Q}_m^D \in l_0(\mathbb{Z})$ such that

$$|\widehat{\mathcal{Q}_m^D}(\xi)|^2 = \mathsf{P}_{m,m}(\sin^2(\xi/2)), \quad \widehat{\mathcal{Q}_m^D}(0) = 1, \quad \text{fsupp}(\mathcal{Q}_m^D) = [0, m-1]. \tag{2.2.3}$$

Define

$$\widehat{a_m^D}(\xi) := 2^{-m} e^{i(m-1)\xi}(1 + e^{-i\xi})^m \widehat{\mathcal{Q}_m^D}(\xi), \qquad m \in \mathbb{N}. \tag{2.2.4}$$

Then a_m^D is a finitely supported real-valued orthogonal wavelet filter such that the filter support of a_m^D is $[1-m, m]$ and a_m^D satisfies

$$|\widehat{a_m^D}(\xi)|^2 = \widehat{a_{2m}^I}(\xi) = \cos^{2m}(\xi/2)\mathsf{P}_{m,m}(\sin^2(\xi/2)) \quad \text{with} \quad \widehat{a_m^D}(0) = 1. \tag{2.2.5}$$

The above orthogonal wavelet filters a_m^D in (2.2.5) were constructed by I. Daubechies in 1988 and is called a *Daubechies orthogonal wavelet filter* of order m. Note that a_m^D has m sum rules, i.e., $\text{sr}(a_m^D) = m$. Up to trivial variants such as a shifted version $a_m^D(\cdot - n)$ for some $n \in \mathbb{Z}$ or a flipped version $a_m^D(-\cdot)$, the filter a_m^D is unique for $m = 1, 2, 3$, since $\mathsf{P}_{m,m}$ has only one unique factor with real coefficients. Explicitly,

$$\mathcal{Q}_1^D = \{\underline{1}\}_{[0,0]}, \qquad \mathcal{Q}_2^D = \{\underline{\tfrac{1+\sqrt{3}}{2}}, \tfrac{1-\sqrt{3}}{2}\}_{[0,1]},$$

$$\mathcal{Q}_3^D = \{\underline{\tfrac{1+\sqrt{10}+\sqrt{5+2\sqrt{10}}}{4}}, \tfrac{1-\sqrt{10}}{2}, \tfrac{1+\sqrt{10}-\sqrt{5+2\sqrt{10}}}{4}\}_{[0,2]}.$$

Therefore, we have

$$a_1^D = \{\underline{\tfrac{1}{2}}, \tfrac{1}{2}\}_{[0,1]}, \qquad a_2^D = \{\tfrac{1+\sqrt{3}}{8}, \underline{\tfrac{3+\sqrt{3}}{8}}, \tfrac{3-\sqrt{3}}{8}, \tfrac{1-\sqrt{3}}{8}\}_{[-1,2]}$$

$$a_3^D = \{\tfrac{1+\sqrt{10}+\sqrt{5+2\sqrt{10}}}{32}, \tfrac{5+\sqrt{10}+3\sqrt{5+2\sqrt{10}}}{32}, \underline{\tfrac{5-\sqrt{10}+\sqrt{5+2\sqrt{10}}}{16}}, \tfrac{5-\sqrt{10}-\sqrt{5+2\sqrt{10}}}{16},$$

$$\tfrac{5+\sqrt{10}-3\sqrt{5+2\sqrt{10}}}{32}, \tfrac{1+\sqrt{10}-\sqrt{5+2\sqrt{10}}}{32}\}_{[-2,3]}.$$

Since $\deg(\mathsf{P}_{m,m}) = m-1$, we have closed-form expressions in radicals for all filters $\mathcal{Q}_m^D$ and a_m^D for $m = 1, \ldots, 5$. For example, when $m = 4$, we have

$$\mathsf{P}(x) = 20x^3 + 10x^2 + 4x + 1 = (x - t)[20x^2 + (20t + 10)x + (20t^2 + 10t + 4)],$$

where $t := \frac{7}{6}(350+105\sqrt{15})^{-1/3} - \frac{1}{30}(350+105\sqrt{15})^{1/3} - \frac{1}{6} \approx -0.342384094858$.

We now present a few numerical examples of a_m^D by listing an associated filter Q_m^D such that $\mathrm{Var}(a_m^D)$ is the smallest among all possible choices in (2.2.3):

$$Q_4^D = \{\underline{\mathbf{-0.857191211347}}, 3.093477124385, -1.60084868, 0.364562767168\}_{[0,3]},$$

$$Q_5^D = \{\underline{\mathbf{0.618476735277}}, -2.424433845637, 5.051894897560, -2.688052234523, 0.442114447325\}_{[0,4]},$$

$$Q_6^D = \{\underline{\mathbf{0.697110410451}}, -4.024690866806, 8.351866150884, -5.869382002997, 2.198116068769, -0.35301976030\}_{[0,5]},$$

$$Q_7^D = \{\underline{\mathbf{1.087511610137}}, -6.054604296959, 13.66968441474, -12.40850877024, 6.247686022060, -1.749202110010, 0.207433130274\}_{[0,6]},$$

$$Q_8^D = \{\underline{\mathbf{-0.612282689473}}, 4.80012508106, -15.519661593664, 25.41872339414, -20.685506683565, 10.04825984189, -2.79177490467, 0.342117554\}_{[0,7]}.$$

The orthogonal wavelet filters $a_m^D, m = 4, \ldots, 8$ are listed explicitly as follows:

$$a_4^D = \{-0.0535744507091, -0.0209554825625, 0.351869534328, \underline{\mathbf{0.568329121704}}, 0.210617267102, -0.0701588120893, -0.00891235072084, 0.0227851729480\}_{[-3,4]},$$

$$a_5^D = \{0.0193273979774, 0.0208734322107, -0.0276720930583, 0.140995348427, \underline{\mathbf{0.511526483447}}, 0.448290824190, 0.0117394615681, -0.123975681306, -0.0149212499343, 0.0138160764789\}_{[-4,5]},$$

$$a_6^D = \{0.0108923501633, 0.00246830618592, -0.0834316077061, -0.0341615607933, 0.347228986479, \underline{\mathbf{0.556946391963}}, 0.238952185666, -0.0513624849308, -0.0148918756493, 0.0316252813300, 0.00124996104640, -0.00551593375469\}_{[-5,6]},$$

$$a_7^D = \{0.00849618445420, 0.0121716951094, -0.0458968894615, -0.0453476699090, 0.254712916415, 0.552902060973, \underline{\mathbf{0.341964557935}}, -0.0401668308104, -0.0714255071200, 0.0316376187153, 0.0144703799497, -0.0128174454083, -0.00232164217260, 0.00162057133027\}_{[-6,7]},$$

$$a_8^D = \{-0.00239172925575, -0.000383345448116, 0.0224118115218,$$
$$0.00537930587528, -0.101324327643, -0.0433268077029,$$
$$0.340372673595, \underline{\mathbf{0.549553315268}}, 0.257699335187,$$
$$-0.0367312543805, -0.0192467606317, 0.0347452329557,$$
$$0.00269319437688, -0.0105728432642, -0.000214197150122,$$
$$0.00133639669641\}_{[-7,8]}.$$

Note that both $\mathrm{sm}(a_m^D)$ and $\mathrm{sr}(a_m^D)$ are independent of the choice Q_m^D and are determined by a_{2m}^I. The smoothness exponents $\mathrm{sm}(a_m^D)$, expectations $\mathrm{E}(a_m^D)$, variances $\mathrm{Var}(a_m^D)$, and frequency separation indicators $\mathrm{Fsi}(a_m^D)$ are given in Table 2.6 and graphs of $\phi^{a_m^D}$ and $\psi^{a_m^D}$ are given in Figs. 2.6 and 2.7 for $m = 1, \dots, 8$.

Though the family of real-valued Daubechies orthogonal wavelet filters can achieve increasingly high orders of sum rules, as demonstrated by the following result, they have two shortcomings: they lack symmetry and linear-phase moments.

Proposition 2.2.3 *Let $a \in l_0(\mathbb{Z})$ be a finitely supported complex-valued orthogonal wavelet filter satisfying* (2.0.5). *If the filter a has complex symmetry $\mathbb{S}\widehat{a}(\xi) = \epsilon e^{-ic\xi}$ for some $\epsilon \in \{-1, 1\}$ and $c \in \mathbb{Z}$, then*

(i) for the case of an even integer c, the filter a must take the form

$$\widehat{a}(\xi) = \tfrac{\sqrt{2}}{2} e^{i\lambda} e^{-in\xi}; \qquad (2.2.6)$$

(ii) for the case of an odd integer c, the filter a must take the form

$$\widehat{a}(\xi) = \tfrac{1}{2}(e^{i\lambda} e^{i2m\xi} + \epsilon e^{-i\lambda} e^{-i(2m+c)\xi}), \qquad (2.2.7)$$

for some real number λ and some integers $m, n \in \mathbb{Z}$.

Table 2.6 The smoothness exponents $\mathrm{sm}(a_m^D)$, expectations $\mathrm{E}(a_m^D)$, variances $\mathrm{Var}(a_m^D)$, and frequency separation indicators $\mathrm{Fsi}(a_m^D)$, where a_m^D in (2.2.4) is a Daubechies orthogonal wavelet filter of order m (with the smallest variance). Note that $\|a_m^D\|^2_{l_2(\mathbb{Z})} = 1/2$ and $\mathrm{sr}(a_m^D) = m$

m	1	2	3	4
$\mathrm{sm}(a_m^D)$	0.5	1.0	1.415037	1.775565
$\mathrm{E}(a_m^D)$	0.5	−0.149518	−0.835864	−0.153565
$\mathrm{Var}(a_m^D)$	0.25	0.328124	0.453684	0.425360
$\mathrm{Fsi}(a_m^D)$	0.333333	0.219047	0.172336	0.145914
m	5	6	7	8
$\mathrm{sm}(a_m^D)$	2.096787	2.388374	2.658660	2.914722
$\mathrm{E}(a_m^D)$	0.449967	−0.154343	−0.869398	−0.154614
$\mathrm{Var}(a_m^D)$	0.559572	0.531640	0.569226	0.631786
$\mathrm{Fsi}(a_m^D)$	0.128509	0.115986	0.106442	0.098868

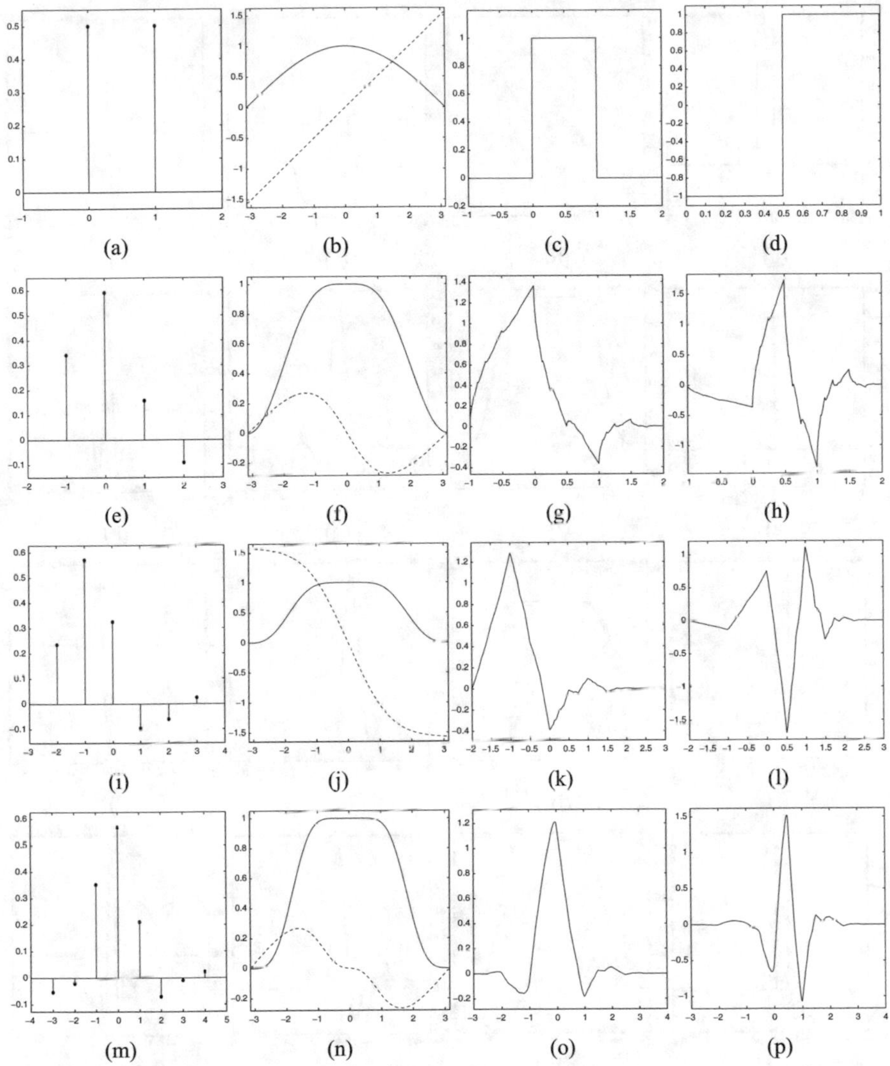

Fig. 2.6 The orthogonal wavelet filter a_m^D, magnitude and phase of $\widehat{a_m^D}$ (see Theorem 1.2.9), and the graphs of the standard orthogonal refinable function $\phi^{a_m^D}$ and its associated standard orthogonal wavelet function $\psi^{a_m^D}$, where $m = 1, \ldots, 4$. **(a)** Filter a_1^D. **(b)** $\widehat{a_1^D}$. **(c)** $\phi^{a_1^D}$. **(d)** $\psi^{a_1^D}$. **(e)** Filter a_2^D. **(f)** $\widehat{a_2^D}$. **(g)** $\phi^{a_2^D}$. **(h)** $\psi^{a_2^D}$. **(i)** Filter a_3^D. **(j)** $\widehat{a_3^D}$. **(k)** $\phi^{a_3^D}$. **(l)** $\psi^{a_3^D}$. **(m)** Filter a_4^D. **(n)** $\widehat{a_4^D}$. **(o)**$\phi^{a_4^D}$. **(p)** $\psi^{a_4^D}$

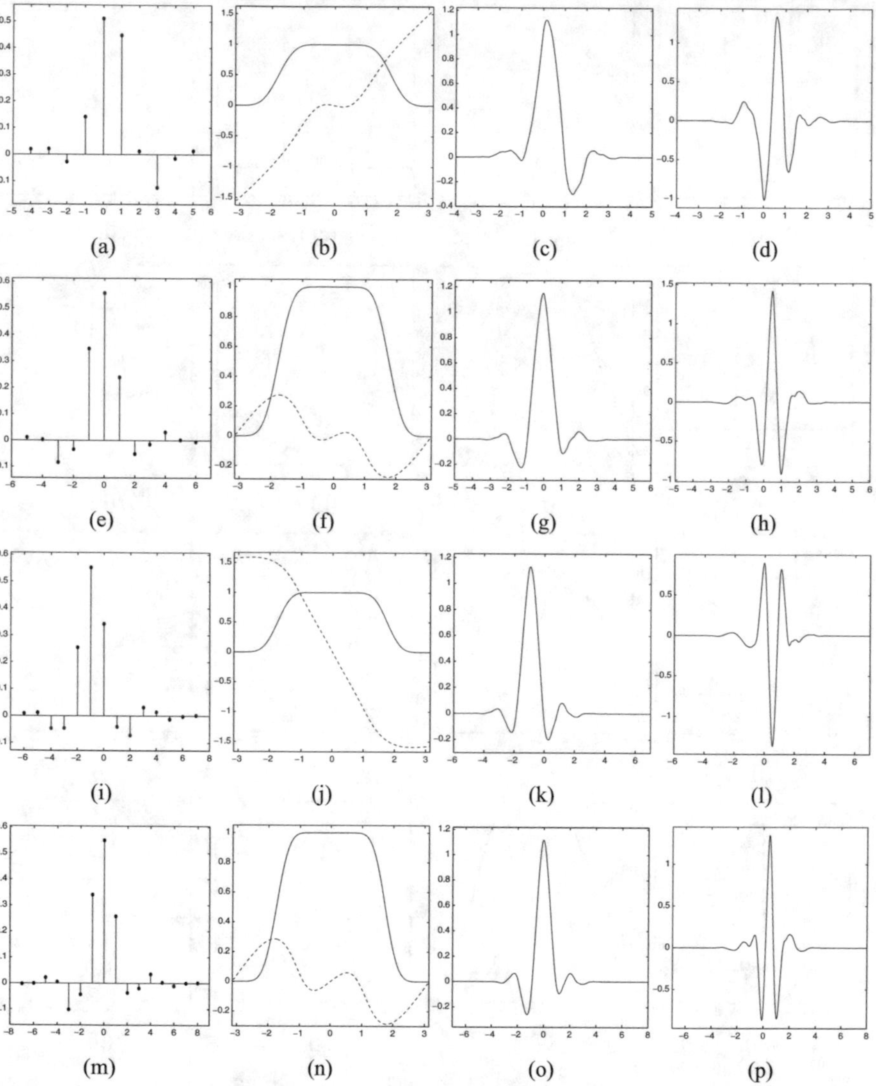

Fig. 2.7 The orthogonal wavelet filter a_m^D, magnitude and phase of $\widehat{a_m^D}$, and the graphs of the standard orthogonal refinable function $\phi^{a_m^D}$ and its associated standard orthogonal wavelet function $\psi^{a_m^D}$, where $m = 5, \ldots, 8$. **(a)** Filter a_5^D. **(b)** $\widehat{a_5^D}$. **(c)** $\phi^{a_5^D}$. **(d)** $\psi^{a_5^D}$. **(e)** Filter a_6^D. **(f)** $\widehat{a_6^D}$. **(g)** $\phi^{a_6^D}$. **(h)** $\psi^{a_6^D}$. **(i)** Filter a_7^D. **(j)** $\widehat{a_7^D}$. **(k)** $\phi^{a_7^D}$. **(l)** $\psi^{a_7^D}$. **(m)** Filter a_8^D. **(n)** $\widehat{a_8^D}$. **(o)** $\phi^{a_8^D}$. **(p)** $\psi^{a_8^D}$

If the complex symmetry operator $\mathbb{S}$ *is replaced by the symmetry operator* S*, then the claim in item (i) still holds, but item (ii) only holds under the additional assumption that a is real-valued.*

Proof Suppose that c is an even integer. Then a is supported inside $[c-n, n]$ for some integer n such that $a(c-n) = \epsilon \overline{a(n)} \neq 0$. It is easy to see that its autocorrelation filter $u = a * a^\star$ is supported inside $[c-2n, 2n-c]$ and $u(2n-c) = a(n)\overline{a(c-n)} \neq 0$. Since u is an interpolatory filter and c is even, we must have $2n - c = 0$, that is, $c = 2n$. But a is supported inside $[c-n, n]$. We conclude that $a = C\boldsymbol{\delta}(\cdot - n)$ for some number C. Now by the orthogonality condition $|\widehat{a}(\xi)|^2 + |\widehat{a}(\xi+\pi)|^2 = 1$, we deduce $2|C|^2 = 1$, i.e., $C = \frac{\sqrt{2}}{2}e^{i\lambda}$ for some real number λ. Hence, item (i) is verified.

Suppose that c is an odd integer. Then $c = 2n+1$ for some integer n. Since $a(2n+1-k) = \varepsilon\overline{a(k)}$ for all $k \in \mathbb{Z}$, we have $\widehat{a^{[1]}}(\xi) = \epsilon e^{-in\xi}\overline{\widehat{a^{[0]}}(\xi)}$. Note that the condition in (2.0.5) for an orthogonal wavelet filter is equivalent to $|\widehat{a^{[0]}}(\xi)|^2 + |\widehat{a^{[1]}}(\xi)|^2 = 1/2$. Therefore, by $\widehat{a^{[1]}}(\xi) = \epsilon e^{-in\xi}\overline{\widehat{a^{[0]}}(\xi)}$, we must have $|\widehat{a^{[0]}}(\xi)|^2 = 1/4$. Since $\widehat{a^{[0]}}$ is a trigonometric polynomial, we deduce that $\widehat{a^{[0]}}(\xi) = \frac{1}{2}e^{i\lambda}e^{im\xi}$ for some $\lambda \in \mathbb{R}$ and $m \in \mathbb{Z}$. Now we conclude that

$$\widehat{a}(\xi) = \widehat{a^{[0]}}(2\xi) + e^{-i\xi}\widehat{a^{[1]}}(2\xi) = \frac{1}{2}e^{i\lambda}e^{i2m\xi} + \frac{1}{2}\epsilon e^{-i\lambda}e^{-i(2m+2n+1)\xi}.$$

That is, we see that (2.2.6) must hold.

When $\mathbb{S}$ is replaced by S, the proof of item (i) is the same. If a is real-valued, then $\mathsf{S}\widehat{a} = \mathbb{S}\widehat{a}$, from which we see that item (ii) must hold (with $\lambda \in \pi\mathbb{Z}$). □

Therefore, beyond the two trivial cases in (2.2.6) and (2.2.7), there is no finitely supported (not necessarily low-pass) complex-valued orthogonal wavelet filter with complex symmetry and there is no finitely supported real-valued orthogonal wavelet filter with symmetry. To achieve symmetry for orthogonal wavelet filters, Proposition 2.2.3 tells us that the only possibility left is to consider complex-valued orthogonal wavelet filters having symmetry with the symmetry center being a half integer.

Another shortcoming of the orthogonal wavelet filters a_m^D is that they lack linear-phase moments, which we shall address in the next section.

2.3 Real Orthogonal Wavelet Filter Banks with Linear-Phase Moments

Having the almost-interpolation property in Proposition 1.2.7, real-valued orthogonal wavelet filter banks with linear moments are of particular interest in applications of orthogonal wavelets in numerical algorithms and computational mathematics. In this section we study and construction real-valued orthogonal wavelet filter banks with linear-phase moments.

For any real-valued filter u with $\widehat{u}(0) = 1$, using the Taylor expansion $\widehat{u}(\xi) = \widehat{u}(0) + \widehat{u}'(0)\xi + \mathscr{O}(|\xi|^2)$ as $\xi \to 0$, we can easily see that u always has 2 linear-phase moments with phase $i\widehat{u}'(0)$. Note that the phase $i\widehat{u}'(0) = \sum_{k\in\mathbb{Z}} u(k)k$ is a real number, since u is real-valued. Regardless of the fact that a^I_{2m} in (2.1.6) has $2m$ linear-phase moments, by the above argument, a derived Daubechies orthogonal wavelet filter a^D_m from a^I_{2m} has 2 linear-phase moments. By the following result, we now show that as a real-valued orthogonal wavelet filter, $a^D_m, m \geqslant 2$ always has 3 linear-phase moments, but generally no linear-phase moments higher than order 3.

Proposition 2.3.1 *Let $a \in l_0(\mathbb{Z})$ be a finitely supported real-valued orthogonal wavelet filter such that $\widehat{a}(0) = 1$. Then one of the following two cases must hold:*

(i) $\operatorname{lpm}(a) = 2\operatorname{sr}(a)$ *and* $\operatorname{lpm}(a)$ *is an even integer;*
(ii) $\operatorname{lpm}(a) < 2\operatorname{sr}(a)$ *and* $\operatorname{lpm}(a)$ *must be an odd integer.*

Proof Denote $m = \operatorname{sr}(a)$ and $n = \operatorname{lpm}(a)$. By $\widehat{a}(0) = 1$, we must have $n \geqslant 1$. Since a is an orthogonal wavelet filter and has at least n linear-phase moments, we have $|\widehat{a}(\xi)|^2 = 1 + \mathscr{O}(|\xi|^n)$ as $\xi \to 0$ and $|\widehat{a}(\xi+\pi)|^2 = 1 - |\widehat{a}(\xi)|^2 = \mathscr{O}(|\xi|^n)$ as $\xi \to 0$. By $m = \operatorname{sr}(a)$, we deduce that $m \geqslant n/2$, that is, $n \leqslant 2m$. So, we have two cases.

Case 1: $n = 2m$. Then it is obvious that item (i) holds.

Case 2: $n < 2m$. We use proof by contradiction to show that n must be an odd integer. Suppose n is even. Since a has n linear-phase moments with phase $c \in \mathbb{R}$,

$$\mathbf{f}(\xi) = 1 + \mathscr{O}(|\xi|^n), \qquad \xi \to 0 \qquad \text{with} \quad \mathbf{f}(\xi) := e^{ic\xi}\widehat{a}(\xi). \tag{2.3.1}$$

Because a is a real-valued sequence, we have $\overline{\widehat{a}(\xi)} = \widehat{a}(-\xi)$. Hence, $\mathbf{f}(\xi)\mathbf{f}(-\xi) = |\widehat{a}(\xi)|^2$. Since a is an orthogonal wavelet filter and has at least m sum rules,

$$\mathbf{f}(\xi)\mathbf{f}(-\xi) = |\widehat{a}(\xi)|^2 = 1 - |\widehat{a}(\xi+\pi)|^2 = 1 + \mathscr{O}(|\xi|^{2m}), \qquad \xi \to 0.$$

That is, we have

$$\mathbf{f}(\xi)\mathbf{f}(-\xi) = 1 + \mathscr{O}(|\xi|^{2m}), \qquad \xi \to 0. \tag{2.3.2}$$

Since (2.3.1) holds and n is an even integer, by $n < 2m$, we deduce from (2.3.2) that

$$2\mathbf{f}^{(n)}(0) = \mathbf{f}^{(n)}(0) + (-1)^n\mathbf{f}^{(n)}(0) = [\mathbf{f}(\cdot)\mathbf{f}(-\cdot)]^{(n)}(0) = 0.$$

Hence $\mathbf{f}^{(n)}(0) = 0$ and consequently, (2.3.1) must hold with n being replaced by $n+1$. That is, $\operatorname{lpm}(a) \geqslant n+1$, which is a contradiction to $\operatorname{lpm}(a) = n$. Thus, $n = \operatorname{lpm}(a)$ must be an odd integer and item (ii) holds. □

Corollary 2.3.2 *If $a \in l_0(\mathbb{Z})$ is a real-valued orthogonal wavelet filter with $\widehat{a}(0) = 1$ and* $\operatorname{sr}(a) \geqslant 2$, *then* $\operatorname{lpm}(a) \geqslant 3$.

Proof Since a is real-valued and $\widehat{a}(0) = 1$, we must have $\operatorname{lpm}(a) \geqslant 2$. Suppose that $\operatorname{lpm}(a) < 3$. Then we must have $\operatorname{lpm}(a) = 2$. Therefore, by $\operatorname{sr}(a) \geqslant 2$, we have $\operatorname{lpm}(a) < 2\operatorname{sr}(a)$ and consequently, item (ii) of Proposition 2.3.1 must be true. In particular, $\operatorname{lpm}(a)$ must be an odd integer, which is a contradiction to $\operatorname{lpm}(a) = 2$. Hence, we must have $\operatorname{lpm}(a) \geqslant 3$. □

Currently there is no systematic construction or existence result for a family of finitely supported real-valued orthogonal wavelet filters a such that $\widehat{a}(0) = 1$ and a has arbitrarily preassigned orders of linear-phase moments. In the rest of this section, by solving nonlinear systems of quadratic algebraic equations, we present an algorithm to construct real-valued low-pass orthogonal wavelet filters with linear-phase moments.

Algorithm 2.3.3 *A filter $a \in l_0(\mathbb{Z})$ is a real-valued orthogonal wavelet filter having m sum rules and n linear-phase moments with phase $c \in \mathbb{R}$, if and only if,*

$$\widehat{a}(\xi) = e^{-i\beta\xi}2^{-m}(1+e^{-i\xi})^m(\boldsymbol{\theta}_{m,n}(\xi) + (1-e^{-i\xi})^n\boldsymbol{\theta}(\xi)), \tag{2.3.3}$$

where $\beta \in \mathbb{Z}$ is the left-hand endpoint of the filter support of the filter a and

(1) $\boldsymbol{\theta}_{m,n}(\xi) = \sum_{j=0}^{n-1}\lambda_j e^{-ij\xi}$ and its real-valued coefficients $\lambda_0,\ldots,\lambda_{n-1}$ are uniquely determined by the system of linear equations induced by

$$\boldsymbol{\theta}_{m,n}(\xi) = e^{-i\mathring{c}\xi}2^m(1+e^{-i\xi})^{-m} + \mathscr{O}(|\xi|^n), \qquad \xi \to 0,$$

where $\mathring{c} := c - \beta$. Note that $\lambda_0,\ldots,\lambda_{n-1}$ are polynomials in $\mathring{c}$ of degree less than n with real coefficients;

(2) $\boldsymbol{\theta}(\xi) = \sum_{j=0}^{\ell-1} t_j e^{-ij\xi}$ for some $\ell \in \mathbb{N}$, where the real-valued unknown coefficients $t_0,\ldots,t_{\ell-1}$ and the unknown shifted phase $\mathring{c}$ are to be determined by solving the system of nonlinear equations induced by the orthogonality condition in (2.0.5).

Proof The sufficiency part is trivial. We now prove the necessity part. Denote $[\beta,\beta'] := \operatorname{fsupp}(a)$, that is, a vanishes outside $[\beta,\beta']$ and $a(\beta)a(\beta') \neq 0$. Define

$$\widehat{a_0}(\xi) = e^{-i\beta\xi}2^{-m}(1+e^{-i\xi})^m\boldsymbol{\theta}_{m,n}(\xi) \quad \text{and} \quad \widehat{a_1}(\xi) = \widehat{a}(\xi) - \widehat{a_0}(\xi).$$

Since both a and a_0 have m sum rules and n linear-phase moments with phase c,

$$\widehat{a_1}(\xi+\pi) = \mathscr{O}(|\xi|^m) \quad \text{and} \quad \widehat{a_1}(\xi) = \mathscr{O}(|\xi|^n), \qquad \xi \to 0.$$

Note that $\operatorname{len}(a_0) \leqslant \operatorname{len}(a)$, because a_0 is the shortest filter having m sum rules and n linear-phase moments with phase c. Since $\operatorname{fsupp}(a) = [\beta,\beta']$ and $\operatorname{fsupp}(a_1) \subseteq [\beta,\infty)$, we must have $\widehat{a_1}(\xi) = e^{-i\beta\xi}2^{-m}(1+e^{-i\xi})^m(1-e^{-i\xi})^n\boldsymbol{\theta}(\xi)$ for $\boldsymbol{\theta}(\xi) = \sum_{j=0}^{\ell-1} t_j e^{-ij\xi}$ with $\ell := \beta' - \beta + 1 - m - n$ for some real numbers $t_0,\ldots,t_{\ell-1}$. Hence, (2.3.3) holds. □

Note that the filter a in (2.3.3) has the filter support $[\beta, m+n+\ell-1+\beta]$. In Algorithm 2.3.3, the unknowns c and β can be arbitrary as long as $\beta \in \mathbb{Z}$ and $c - \beta = \mathring{c}$; we often choose c and β in such a way that $\|\theta\|_{L_\infty(\mathbb{R})}$ (or simply $|c|$) is small, where θ is the phase function of $\widehat{a}$ defined in Proposition 1.2.9.

Define a sequence $v \in l_0(\mathbb{Z})$ by $\widehat{v}(\xi) := |\widehat{a}(\xi/2)|^2 + |\widehat{a}(\xi/2+\pi)|^2$. Then v is symmetric about the origin and is supported inside $[-\tilde{n}, \tilde{n}]$ where $\tilde{n} := \lfloor \frac{m+n+\ell-1}{2} \rfloor$. Since a has m sum rules and n linear-phase moments, the filter v must have at least $\min(2m, 2\lceil \frac{n}{2} \rceil)$ linear-phase moments. Consequently, there are essentially $\tilde{n}+1-\min(m, \lceil \frac{n}{2} \rceil)$ constraints induced by the orthogonality condition $\widehat{v}(\xi) = 1$. On the other hand, there are $\ell+1$ unknowns $t_0, \ldots, t_{\ell-1}$ and $\mathring{c}$. Consequently, we have

$$N_{m,n,\ell} := \ell + \min(m, \lceil \tfrac{n}{2} \rceil) - \lfloor \tfrac{m+n+\ell-1}{2} \rfloor$$

free parameters/coefficients in finding the unknown coefficients $t_0, \ldots, t_{\ell-1}$ and the unknown shifted phase $\mathring{c}$. We often pick ℓ to be the smallest integer so that $N_{m,n,\ell} = 0$ (no free parameter), or $N_{m,n,\ell} = 1$ (one free parameter $\mathring{c}$ to optimize $\operatorname{Var}(a)$ or $\operatorname{sm}(a)$). By Corollary 2.3.2, Algorithm 2.3.3 can be used to construct all real-valued orthogonal wavelet filters with at least 2 sum rules if we set $n = 3$ in Algorithm 2.3.3.

In the following we shall use $a^O_{m,n}$ (O=Orthogonal) to denote a real-valued orthogonal wavelet filter a constructed in (2.3.3) of Algorithm 2.3.3 with m sum rules and n linear-phase moments. Due to Corollary 2.3.2, we consider $n \geqslant 3$ and choose β so that $|\operatorname{E}(a^O_{m,n})|$ is small. By Proposition 2.3.1, we only consider odd integers n.

Example 2.3.1 Let $m = 2, n = 3$ in Algorithm 2.3.3. Then

$$\boldsymbol{\theta}_{2,3}(\xi) = (\tfrac{1}{2}\mathring{c}^2 - \tfrac{5}{2}\mathring{c} + \tfrac{11}{4}) + (-\mathring{c}^2 + 4\mathring{c} - \tfrac{5}{2})e^{-i\xi} + (\tfrac{1}{2}\mathring{c}^2 - \tfrac{3}{2}\mathring{c} + \tfrac{3}{4})e^{-i2\xi}.$$

$\ell = 1$ is the smallest integer satisfying $N_{2,3,\ell} = 1$ and there is one free parameter $\mathring{c}$. Set $\mathring{c} = 1$. Then $t_0 = \frac{3-\sqrt{15}}{8}$. For $c = 0$ and $\beta = -1$ satisfying $c - \beta = \mathring{c} = 1$, the real-valued orthogonal wavelet filter $a^O_{2,3}$ is supported inside $[-1, 4]$ with $\operatorname{sr}(a^O_{2,3}) = 2$ and $\operatorname{lpm}(a^O_{2,3}) = 3$ with phase $c = 0$. By calculation, $\operatorname{E}(a^O_{2,3}) \approx 0.38393$, $\operatorname{Var}(a^O_{2,3}) \approx 0.465706$, $\operatorname{Fsi}(a^O_{2,3}) \approx 0.184371$, and $\operatorname{sm}(a^O_{2,3}) \approx 1.232138$. The orthogonal wavelet filter $a^O_{2,3}$ is given by

$$\{0.160219270431, \underline{\mathbf{0.527280729569}}, 0.429561459138, -0.0545614591380,$$
$$-0.0897807295690, 0.0272807295690\}_{[-1,4]}.$$

If $\mathring{c} = 4/5$ with $\beta = -1$ and $c = -1/5$, then $t_0 = \frac{63-\sqrt{6751}}{200}$. By calculation, $\operatorname{E}(a^O_{2,3}) \approx 0.13926$, $\operatorname{Var}(a^O_{2,3}) \approx 0.447488$, $\operatorname{Fsi}(a^O_{2,3}) \approx 0.172542$, $\operatorname{sm}(a^O_{2,3}) \approx$

1.410020, and $\operatorname{fsupp}(a^O_{2,3}) = [-1,4]$. The orthogonal wavelet filter $a^O_{2,3}$ is given by

$$\{0.243544413492, \underline{\mathbf{0.573955586508}}, 0.312911173016, -0.0979111730165, \\ -0.0564555865082, 0.0239555865082\}_{[-1,4]}.$$

Example 2.3.2 Let $m = 3, n = 3$ in Algorithm 2.3.3. Then

$$\boldsymbol{\theta}_{3,3}(\xi) = (\tfrac{1}{2}\mathring{c}^2 - 3\mathring{c} + 4) + (-\mathring{c}^2 + 5\mathring{c} - \tfrac{9}{2})e^{-i\xi} + (\tfrac{1}{2}\mathring{c}^2 - 2\mathring{c} + \tfrac{3}{2})e^{-i2\xi}.$$

$\ell = 2$ is the smallest integer satisfying $N_{3,3,\ell} = 1$ and there is one free parameter $\mathring{c}$. Set $\mathring{c} = 1$ with $\beta = -1$ and $c = 0$. Then $t_0 = \frac{3}{8} - \frac{3\sqrt{1495}}{208}, t_1 = -\frac{5}{16} + \frac{\sqrt{1495}}{104}$. The real-valued orthogonal wavelet filter $a^O_{3,3}$ is supported inside $[-1,6]$ with $\operatorname{sr}(a^O_{3,3}) = 3$ and $\operatorname{lpm}(a^O_{3,3}) = 3$ with phase $c = 0$. By calculation, $\operatorname{E}(a^O_{3,3}) \approx 0.45390$, $\operatorname{Var}(a^O_{3,3}) \approx 0.600262$, $\operatorname{Fsi}(a^O_{3,3}) \approx 0.145924$, $\operatorname{sm}(a^O_{3,3}) \approx 1.775280$, and $\operatorname{fsupp}(a^O_{3,3}) = [-1,6]$. The orthogonal wavelet filter $a^O_{3,3}$ is given by

$$\{0.164666051938, \underline{\mathbf{0.507410132041}}, 0.443501844186, -0.0222303961239, \\ -0.131001844186, 0.0222303961239, 0.02283394806, -0.0074101320413\}_{[-1,6]}.$$

Set $\mathring{c} = 3$ with $\beta = -3$ and $c = 0$. Then $t_0 = \frac{4-\sqrt{7}}{16}, t_1 = -\frac{3}{16}$. The orthogonal wavelet filter $a^O_{3,3}$ is supported inside $[-3,4]$ with $\operatorname{sr}(a^O_{3,3}) = 3$ and $\operatorname{lpm}(a^O_{3,3}) = 3$ with phase $c = 0$. Then $\operatorname{E}(a^O_{3,3}) \approx -0.133052$, $\operatorname{Var}(a^O_{3,3}) \approx 0.426144$, $\operatorname{Fsi}(a^O_{3,3}) \approx 0.14599$, and $\operatorname{sm}(a^O_{3,3}) \approx 1.773409$. The orthogonal wavelet filter $a^O_{3,3}$ is given by

$$\{-0.0519199321177, -0.0234375, 0.343259796353, \underline{\mathbf{0.5703125}}, 0.219240203647, \\ -0.0703125, -0.0105800678823, 0.0234375\}_{[-3,4]}.$$

See Fig. 2.8 for the graphs of their orthogonal refinable and wavelet functions.

Example 2.3.3 Let $m = 4, n = 5$ in Algorithm 2.3.3. Then

$$\boldsymbol{\theta}_{4,5}(\xi) = (\tfrac{1}{24}\mathring{c}^4 - \tfrac{3}{4}\mathring{c}^3 + \tfrac{113}{24}\mathring{c}^2 - 12\mathring{c} + \tfrac{163}{16}) + (-\tfrac{1}{6}\mathring{c}^4 + \tfrac{17}{6}\mathring{c}^3 - \tfrac{49}{3}\mathring{c}^2 + \tfrac{217}{6}\mathring{c} - \tfrac{93}{4})e^{-i\xi} \\ + (\tfrac{1}{4}\mathring{c}^4 - 4\mathring{c}^3 + \tfrac{85}{4}\mathring{c}^2 - 42\mathring{c} + \tfrac{185}{8})e^{-i2\xi} + (-\tfrac{1}{6}\mathring{c}^4 + \tfrac{5}{2}\mathring{c}^3 - \tfrac{37}{3}\mathring{c}^2 + \tfrac{45}{2}\mathring{c} - \tfrac{45}{4})e^{-i3\xi} \\ + (\tfrac{1}{24}\mathring{c}^4 - \tfrac{7}{12}\mathring{c}^3 + \tfrac{65}{24}\mathring{c}^2 - \tfrac{14}{3}\mathring{c} + \tfrac{35}{16})e^{-i4\xi}.$$

$\ell = 1$ is the smallest integer satisfying $N_{4,5,\ell} = 0$. Then $\mathring{c} \approx 2.45227510835$ is a real root of a polynomial and $t_0 \approx 0.019975982075$ is a polynomial of $\mathring{c}$. Let $\beta = -3$ and $c = \mathring{c} + \beta \approx -0.54772489165$. The real-valued orthogonal wavelet filter $a^O_{4,5}$ is supported inside $[-3,6]$ with $\operatorname{sr}(a^O_{4,5}) = 4$ and $\operatorname{lpm}(a^O_{4,5}) = 5$ with

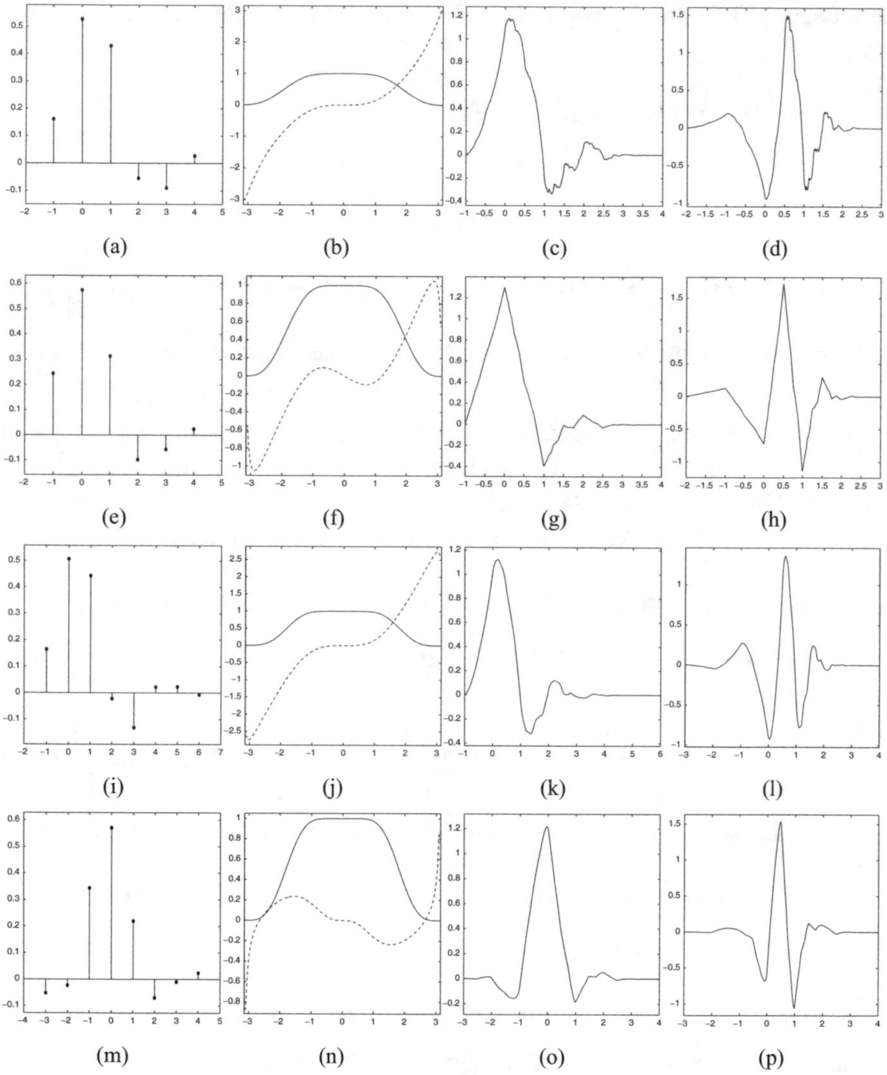

Fig. 2.8 The filter $a^{O}_{m,n}$, magnitude and phase of $\widehat{a^{O}_{m,n}}$, and the graphs of the standard orthogonal refinable function $\phi^{a^{O}_{m,n}}$ and its associated standard orthogonal wavelet function $\psi^{a^{O}_{m,n}}$. The first row is for Example 2.3.1 with $\beta = -1$ and $c = 0$. The second row is for Example 2.3.1 with $\beta = -1$ and $c = -1/5$. The third row is for Example 2.3.2 with $\beta = -1$ and $c = 0$. The fourth row is for Example 2.3.2 with $\beta = -3$ and $c = 0$. **(a)** Filter $a^{O}_{2,3}, c = 0$. **(b)** $\widehat{a^{O}_{2,3}}, c = 0$. **(c)** $\phi^{a^{O}_{2,3}}, c = 0$. **(d)** $\psi^{a^{O}_{2,3}}, c = 0$. **(e)** Filter $a^{O}_{2,3}, c = -\frac{1}{5}$. **(f)** $\widehat{a^{O}_{2,3}}, c = -\frac{1}{5}$. **(g)** $\phi^{a^{O}_{2,3}}, c = -\frac{1}{5}$. **(h)** $\psi^{a^{O}_{2,3}}, c = -\frac{1}{5}$. **(i)** Filter $a^{O}_{3,3}, \beta = -1$. **(j)** $\widehat{a^{O}_{3,3}}, \beta = -1$. **(k)** $\phi^{a^{O}_{3,3}}, \beta = -1$. **(l)** $\psi^{a^{O}_{3,3}}, \beta = -1$. **(m)** Filter $a^{O}_{3,3}, \beta = -3$. **(n)** $\widehat{a^{O}_{3,3}}, \beta = -3$. **(o)** $\phi^{a^{O}_{3,3}}, \beta = -3$. **(p)** $\psi^{a^{O}_{3,3}}, \beta = -3$

phase $c \approx -0.54772489165$. Then $\mathrm{E}(a^O_{4,5}) \approx -0.308736$, $\mathrm{Var}(a^O_{4,5}) \approx 0.488352$, $\mathrm{Fsi}(a^O_{4,5}) \approx 0.143608$, and $\mathrm{sm}(a^O_{4,5}) \approx 1.806529$. The orthogonal wavelet filter is given by

$$\{-0.0286910093834, 0.0648570567748, 0.432076687413, \underline{\mathbf{0.530824322734}},$$
$$0.106028599696, -0.120537091758, -0.00659200118465, 0.026104211129,$$
$$-0.00282227654116, -0.00124849887972\}_{[-3,6]}.$$

$\ell = 3$ is the smallest integer satisfying $N_{4,5,\ell} = 1$ with one free parameter $\mathring{c}$. If $\mathring{c} = 83/16$, then t_1 is a real root of a polynomial and t_0, t_2 are rational polynomials of t_1:

$$t_0 \approx 0.219980402824, \quad t_1 \approx 0.156474926633, \quad t_2 \approx 0.190848744653.$$

Let $\beta = -5$ and $c = \mathring{c} + \beta = 3/16$. The real-valued orthogonal wavelet filter $a^O_{4,5}$ is supported inside $[-5, 6]$ with $\mathrm{sr}(a^O_{4,5}) = 4$ and $\mathrm{lpm}(a^O_{4,5}) = 5$ with phase $c = 3/16$. By calculation, $\mathrm{E}(a^O_{4,5}) = -0.159240$, $\mathrm{Var}(a^O_{4,5}) \approx 0.692656$, $\mathrm{Fsi}(a^O_{4,5}) \approx 0.0942839$, and $\mathrm{sm}(a^O_{4,5}) \approx 2.413420$. The orthogonal wavelet filter $a^O_{4,5}$ is given by

$$\{0.0209579569620, 0.00377474318732, -0.109950370960, -0.0405234182158,$$
$$0.344777273107, \underline{\mathbf{0.542734607689}}, 0.261738102445, -0.0378009800214,$$
$$-0.01967132518, 0.0437430939, 0.00214836362626, -0.01192804654\}_{[-5,6]}.$$

If $\mathring{c} = 5$, then $t_1 \approx 0.153004834376$, $t_0 \approx 0.228811811631$ and $t_2 \approx 0.129435659085$. Let $\beta = -5$ and $c = 0$. The real-valued orthogonal wavelet filter $a^O_{4,5}$ is supported inside $[-5, 6]$ such that $\mathrm{sr}(a^O_{4,5}) = 4$ and $\mathrm{lpm}(a^L_{4,5}) = 5$ with phase $c = 0$. By calculation, $\mathrm{E}(a^O_{4,5}) \approx -0.445580$, $\mathrm{Var}(a^O_{4,5}) \approx 0.659709$, $\mathrm{Fsi}(a^O_{4,5}) \approx 0.106290$, and $\mathrm{sm}(a^O_{4,5}) \approx 2.173867$. The orthogonal wavelet filter $a^O_{4,5}$ is given by

$$\{0.0260194882268, -0.00473793607836, -0.136801026363, 0.0108620156207,$$
$$0.443259223184, \underline{\mathbf{0.503931298301}}, 0.168333606358, -0.0295866278433,$$
$$0.0006617820496, 0.02762097869, -0.0014730734557, -0.00808972869\}_{[-5,6]}.$$

Example 2.3.4 Let $m = 5, n = 5$ in Algorithm 2.3.3. Then

$$\boldsymbol{\theta}_{5,5}(\xi) = (\tfrac{1}{24}\mathring{c}^4 - \tfrac{5}{6}\mathring{c}^3 + \tfrac{35}{6}\mathring{c}^2 - \tfrac{50}{3}\mathring{c} + 16) + (-\tfrac{1}{6}\mathring{c}^4 + \tfrac{19}{6}\mathring{c}^3 - \tfrac{247}{12}\mathring{c}^2$$
$$+ \tfrac{157}{3}\mathring{c} - \tfrac{325}{8})e^{-i\xi} + (\tfrac{1}{4}\mathring{c}^4 - \tfrac{9}{2}\mathring{c}^3 + \tfrac{109}{4}\mathring{c}^2 - 63c + \tfrac{345}{8})e^{-2\xi}$$

$$+(-\tfrac{1}{6}\mathring{c}^4+\tfrac{17}{6}\mathring{c}^3-\tfrac{193}{12}\mathring{c}^2+\tfrac{104}{3}\mathring{c}-\tfrac{175}{8})e^{-i3\xi}$$
$$+(\tfrac{1}{24}\mathring{c}^4-\tfrac{2}{3}\mathring{c}^3+\tfrac{43}{12}\mathring{c}^2-\tfrac{22}{3}\mathring{c}+\tfrac{35}{8})e^{-i4\xi}.$$

$\ell = 2$ is the smallest integer satisfying $N_{4,5,\ell} = 0$ and there is no free parameter in finding the two unknowns t_0 and $\mathring{c}$. By calculation, $\mathring{c} \approx 3.39365335163$ is a real root of a polynomial, $t_0 \approx 0.264214310980$ and $t_1 \approx -0.0695388971119$ are polynomials of $\mathring{c}$. Let $\beta = -4$ and $c = \mathring{c} + \beta \approx -0.60634664837$. The real-valued orthogonal wavelet filter $a^O_{5,5}$ is supported inside $[-4, 7]$ such that $\mathrm{sr}(a^O_{5,5}) = 5$ and $\mathrm{lpm}(a^O_{5,5}) = 5$. Then $\mathrm{E}(a^O_{5,5}) \approx -0.229750$, $\mathrm{Var}(a^O_{5,5}) \approx 0.675247$, $\mathrm{Fsi}(a^O_{5,5}) \approx 0.126290$, and $\mathrm{sm}(a^O_{5,5}) \approx 2.137549$. The orthogonal wavelet filter $a^O_{5,5}$ is given by

$$\{-0.00495331804585, -0.0188202224784, 0.0873861361603, 0.415764463514,$$
$$\underline{\mathbf{0.526176706277}}, 0.127106694105, -0.151368583976, -0.0224661362097,$$
$$0.0510157568, -0.003757889465, -0.0082566972, 0.00217309053475\}_{[-4,7]}.$$

$\ell = 4$ is the smallest integer satisfying $N_{5,5,\ell} = 1$ with one free parameter $\mathring{c}$. Set $\mathring{c} = 5$. Then $t_2 \approx 0.0118500404042$ is a root of a polynomial and

$$t_0 = -\tfrac{95569}{9840}t_2 - \tfrac{65}{2048} + \tfrac{14272}{41}t_2^2 - \tfrac{69632}{75}t_2^3 \approx -0.0994932509701,$$
$$t_1 = -\tfrac{5}{4}t_2 + \tfrac{205}{1024} \approx 0.185382761995,$$
$$t_3 = -\tfrac{1}{4}t_2 - \tfrac{15}{1024} \approx -0.0176109476010.$$

Let $\beta = -5$ and $c = \mathring{c} - \beta = 0$. The real-valued orthogonal wavelet filter $a^O_{5,5}$ is supported inside $[-5, 8]$ such that $\mathrm{sr}(a^O_{5,5}) = 5$ and $\mathrm{lpm}(a^O_{5,5}) = 5$. By calculation, $\mathrm{E}(a^O_{5,5}) \approx -0.0500912$, $\mathrm{Var}(a^O_{5,5}) \approx 0.530668$, $\mathrm{Fsi}(a^O_{5,5}) \approx 0.113206$, and $\mathrm{sm}(a^O_{5,5}) \approx 2.449284$. The orthogonal wavelet filter $a^O_{5,5}$ is given by

$$\{0.00860958590718, 0.00579321131234, -0.0622088657733, -0.0295163986742,$$
$$0.318619290259, \underline{\mathbf{0.560683823686}}, 0.269169778554, -0.0634355342487,$$
$$-0.0387802080904, 0.034469477687, 0.00496073290597, -0.008544921875,$$
$$-0.000370313762630, 0.000550342112533\}_{[-5,8]}.$$

See Fig. 2.9 for the graphs of $\phi^{a^O_{m,n}}$ and $\psi^{a^O_{m,n}}$ with $(m, n) = (4, 5), (5, 5)$.

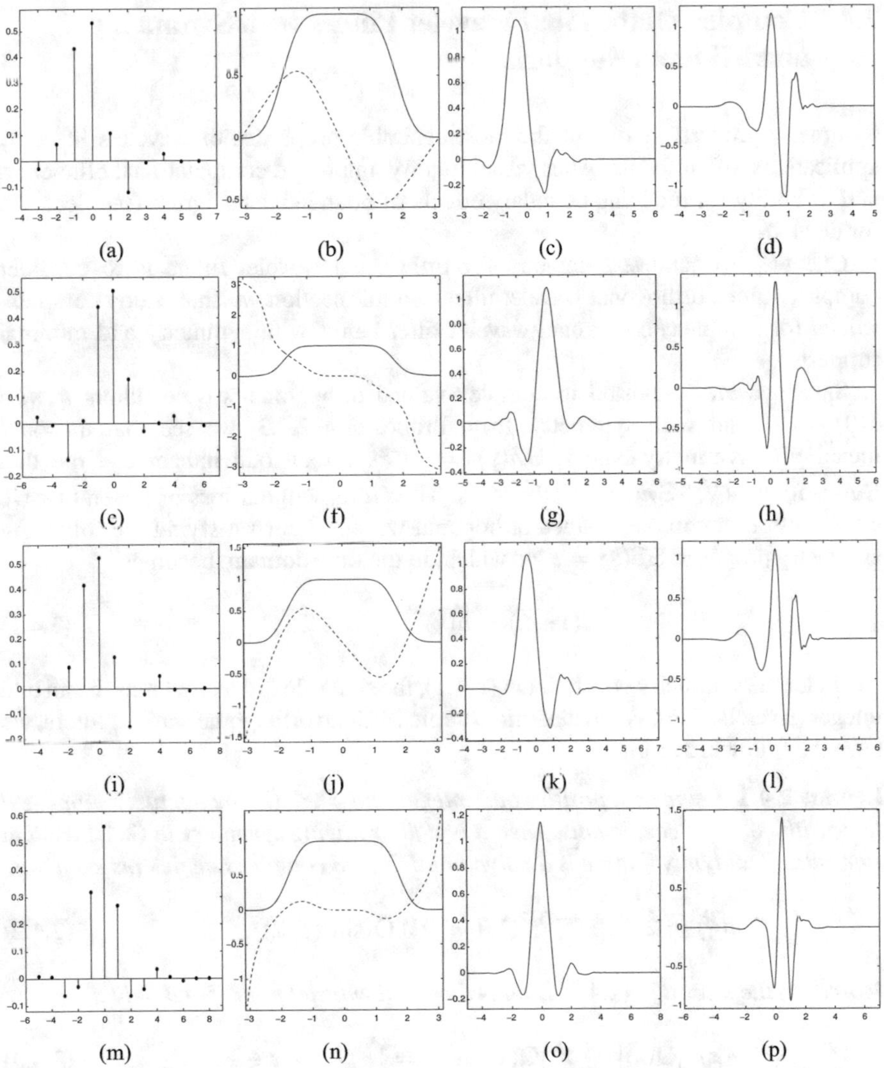

Fig. 2.9 The filter $a^O_{m,n}$, magnitude and phase of $\widehat{a^O_{m,n}}$, and the graphs of the standard orthogonal refinable functions $\phi^{a^O_{m,n}}$ and its standard orthogonal wavelet function $\psi^{a^O_{m,n}}$. The first row is for Example 2.3.3 with $\ell = 1$. The second row is for Example 2.3.3 with $\ell = 3$ and $c = 0$. The third row is for Example 2.3.4 with $\ell = 2$. The fourth row is for Example 2.3.4 with $\ell = 4$ and $c = 0$. **(a)** Filter $a^O_{4,5}, \ell = 1$. **(b)** $\widehat{a^O_{4,5}}, \ell = 1$. **(c)** $\phi^{a^O_{4,5}}, \ell = 1$. **(d)** $\psi^{a^O_{4,5}}, \ell = 1$. **(e)** Filter $a^O_{4,5}, c = 0$. **(f)** $\widehat{a^O_{4,5}}, c = 0$. **(g)** $\phi^{a^O_{4,5}}, c = 0$. **(h)** $\psi^{a^O_{4,5}}, c = 0$. **(i)** Filter $a^O_{5,5}, \ell = 2$. **(j)** $\widehat{a^O_{5,5}}, \ell = 2$. **(k)** $\phi^{a^O_{5,5}}, \ell = 2$. **(l)** $\psi^{a^O_{5,5}}, \ell = 2$. **(m)** Filter $a^O_{5,5}, \ell = 4$. **(n)** $\widehat{a^O_{5,5}}, \ell = 4$. **(o)** $\phi^{a^O_{5,5}}, \ell = 4$. **(p)** $\psi^{a^O_{5,5}}, \ell = 4$

2.4 Complex Orthogonal Wavelet Filters with Symmetry and Minimal Supports

Symmetry property is one of the most desirable properties of wavelets in many applications, often for its better visual quality, improved computational efficiency, and its ability for reducing boundary effects of bounded data/signals (see Sect. 1.5 for details).

One way to achieve symmetry for orthogonal wavelet filters is to consider complex-valued orthogonal wavelet filters. In this section we shall study complex-valued (or complex) orthogonal wavelet filter banks with symmetry and minimal supports.

Since we are interested in complex-valued orthogonal wavelet filters a with $\widehat{a}(0) = 1$ and with symmetry, from Proposition 2.2.3, we see that the only meaningful symmetry type is $\mathsf{S}\widehat{a}(\xi) = e^{-ic\xi}$ for an odd integer c. Note that $\mathsf{S}\widehat{a(\cdot - n)} = e^{-i2n\xi}\mathsf{S}\widehat{a}(\xi)$ for all $n \in \mathbb{Z}$. Therefore, without loss of generality, we only consider a complex-valued orthogonal wavelet filter satisfying the following symmetry property: $\mathsf{S}\widehat{a}(\xi) = e^{-i\xi}$, which, in the time domain, becomes

$$a(1-k) = a(k), \qquad k \in \mathbb{Z}. \tag{2.4.1}$$

Under the symmetry constraint in (2.4.1), the sum rule order $\operatorname{sr}(a)$ must be an odd integer (Exercise 2.6). For symmetric complex-valued orthogonal wavelet filters, we have the following result.

Lemma 2.4.1 *Let m be a positive odd integer and $a \in l_0(\mathbb{Z})$ be a finitely supported filter with complex coefficients. Then a is a filter having symmetry in* (2.4.1) *and m sum rules if and only if there is a polynomial Q with complex coefficients such that*

$$\widehat{a}(\xi) = 2^{-m} e^{i\xi(m-1)/2}(1 + e^{-i\xi})^m \mathsf{Q}(\sin^2(\xi/2)). \tag{2.4.2}$$

Moreover, the filter a in (2.4.2) *is an orthogonal wavelet filter if and only if*

$$(1-x)^m|\mathsf{Q}(x)|^2 + x^m|\mathsf{Q}(1-x)|^2 = 1, \qquad \forall\, x \in \mathbb{R}. \tag{2.4.3}$$

Proof Since a has m sum rules and m is an odd integer, we have $\widehat{a}(\xi) = 2^{-m} e^{i\xi(m-1)/2}(1 + e^{-i\xi})^m \widehat{u}(\xi)$ for some 2π-periodic trigonometric polynomial $\widehat{u}$. Due to the symmetry property in (2.4.1), we see that $\mathsf{S}\widehat{u} = 1$. By Exercise 2.8, $\widehat{u}(\xi) = \mathsf{Q}(\sin^2(\xi/2))$ for some polynomial Q with complex coefficients. Consequently, a is a filter having symmetry in (2.4.1) and m sum rules if and only if (2.4.2) holds. Note

$$|\widehat{a}(\xi)|^2 = \cos^{2m}(\xi/2)|\widehat{u}(\xi)|^2 = \cos^{2m}(\xi/2)|\mathsf{Q}(\sin^2(\xi/2))|^2.$$

Setting $x := \sin^2(\xi/2)$ and using the above identity, now we can easily deduce that a is an orthogonal wavelet filter satisfying $|\widehat{a}(\xi)|^2 + |\widehat{a}(\xi+\pi)|^2 = 1$ if and only if

(2.4.3) holds for all $x \in [0,1]$. Consequently, (2.4.3) holds for all $x \in [0,1]$ if and only if it holds for all $x \in \mathbb{R}$, since the left-hand side of (2.4.3) is a polynomial by $|\mathsf{Q}(x)|^2 = \mathsf{Q}(x)\overline{\mathsf{Q}(x)}$ and $\bar{x} = x$ for $x \in \mathbb{R}$. □

The following simple lemma plays a critical role in the study of complex-valued orthogonal wavelet filters with symmetry.

Lemma 2.4.2 *Let* P *be a polynomial with* $\mathsf{P}(0) \neq 0$. *Then* $\mathsf{P}(x) = |\mathsf{Q}(x)|^2, x \in \mathbb{R}$ *for some polynomial* Q *with complex coefficients and* $\mathsf{Q}(0) = \sqrt{\mathsf{P}(0)}$ *if and only if the polynomial* P *has real coefficients and is nonnegative on the real line:*

$$\mathsf{P}(x) \geqslant 0, \qquad \forall\, x \in \mathbb{R}. \tag{2.4.4}$$

Proof Necessity ($\Rightarrow$). It follows directly from $\mathsf{P}(x) = |\mathsf{Q}(x)|^2$ that (2.4.4) holds and P has real coefficients.

Sufficiency ($\Leftarrow$). Suppose that (2.4.4) holds. By (2.4.4), we see that every real root of P has even multiplicity. Let X denote the set of all the roots, counting multiplicity, of P in the complex plane. Since P has real coefficients, we see that $\mathsf{P}(z_0) = 0$ implies $\mathsf{P}(\overline{z_0}) = 0$. Therefore, since every real root of P has even multiplicity, there is a subset Y of X such that $X = \{y, \bar{y} \;:\; y \in Y\}$. Note that $0 \notin Y$ since $\mathsf{P}(0) \neq 0$. Define a polynomial Q with complex coefficients by

$$\mathsf{Q}(x) = \sqrt{\mathsf{P}(0)} \prod_{y \in Y} \left(1 - \frac{x}{y}\right). \tag{2.4.5}$$

By the choice of the subset Y, we must have $|\mathsf{Q}(x)|^2 = \mathsf{P}(x)$ and $\mathsf{Q}(0) = \sqrt{\mathsf{P}(0)}$. □

The polynomial Q in Lemma 2.4.2 is often not unique, due to many different choices of a subset Y in the proof of Lemma 2.4.2. The corresponding polynomial Q in (2.4.5) is called the canonical choice if $Y \subset \{z \in \mathbb{C} \;:\; \mathrm{Im}(z) \geqslant 0\}$. Also note that even though Q has complex coefficients, the polynomial $\mathsf{P}(x) := |\mathsf{Q}(x)|^2, x \in \mathbb{R}$ always has real coefficients.

For complex-valued orthogonal wavelet filters with symmetry in (2.4.1), we have

Theorem 2.4.3 *Let* m *be a positive odd integer and* $a \in l_0(\mathbb{Z})$ *be a finitely supported filter with complex coefficients and* $\widehat{a}(0) \neq 0$. *If* a *is an orthogonal wavelet filter having symmetry in* (2.4.1) *and* m *sum rules, then there is a polynomial* P *with real coefficients such that all* (2.2.1), (2.1.8), (2.4.4) *are satisfied. Conversely, if* (2.1.8) *and* (2.4.4) *hold for a polynomial* P *with real coefficients, then there exists a finitely supported complex-valued orthogonal wavelet filter* a, *having symmetry in* (2.4.1) *and* m *sum rules, such that* (2.2.1) *holds and* $\widehat{a}(0) = 1$.

Proof By Lemma 2.4.1, there is a polynomial Q such that (2.4.2) and (2.4.3) holds. Define $\mathsf{P}(x) := |\mathsf{Q}(x)|^2, x \in \mathbb{R}$. It is trivial to see that P is a polynomial with real coefficients and satisfies the condition in (2.4.4). It is also straightforward to deduce from (2.4.3) that (2.1.8) holds.

We now prove the converse direction. By (2.1.8) and $m \geqslant 1$, we have $\mathsf{P}(0) = 1$. By (2.4.4) and Lemma 2.4.2, there is a polynomial Q such that $|\mathsf{Q}(x)|^2 = \mathsf{P}(x)$ for all $x \in \mathbb{R}$ and $\mathsf{Q}(0) = \sqrt{\mathsf{P}(0)} = 1$. Define a filter a through (2.4.2). Then a is a desired orthogonal wavelet filter satisfying all the requirements. □

Comparing with Theorem 2.2.1 for real-valued orthogonal wavelet filters, we see that the nonnegativity condition in (2.4.4) is stronger than (2.2.2). By Theorem 2.4.3, in a certain sense the set of all complex orthogonal filters with symmetry in (2.4.1) is a proper subset of all real-valued orthogonal wavelet filters.

To present a family of complex-valued orthogonal wavelet filters with symmetry and minimal supports, we need the following result.

Lemma 2.4.4 *For any positive odd integer m, the polynomial* $\mathsf{P}_{m,m}$, *defined in* (2.1.4), *satisfies* $\mathsf{P}_{m,m}(x) > 0$ *for all* $x \in \mathbb{R}$.

Proof By (2.1.4), we see that all the coefficients of $\mathsf{P}_{m,m}$ are nonnegative. Consequently, by $\mathsf{P}(0) = 1$, it is trivial that $\mathsf{P}_{m,m}(x) \geqslant 1 > 0$ for all $x \geqslant 0$. On the other hand, by the identity in (2.1.5), noting that m is an odd integer and $1 - x > 0$ whenever $x < 0$, we have

$$(1-x)^m \mathsf{P}_{m,m}(x) = 1 - x^m \mathsf{P}_{m,m}(1-x) \geqslant 1, \qquad \forall\, x < 0,$$

from which we have $\mathsf{P}_{m,m}(x) > 0$ for all $x < 0$. Hence, we proved that $\mathsf{P}_{m,m}(x) > 0$ for all $x \in \mathbb{R}$ and for all positive odd integers m. □

As a direct consequence of Theorem 2.4.3 and Lemma 2.4.4, for every positive odd integer m, there is a subset Y_m^{S} (S=Symmetry) of $\mathbb{C}$ such that

$$\Big|\prod_{y \in Y_m^{\mathsf{S}}} (1 + yx)\Big|^2 = \mathsf{P}_{m,m}(x). \tag{2.4.6}$$

That is, Y_m^{S} is a subset of $\mathbb{C}$ such that $\{-y^{-1}, -\bar{y}^{-1} \;:\; y \in Y_m^{\mathsf{S}}\}$ is the set of all complex zeros, counting multiplicity, of $\mathsf{P}_{m,m}$. For every positive odd integer m, define

$$\widehat{a_m^{\mathsf{S}}}(\xi) := 2^{-m} e^{i\xi(m-1)/2} (1 + e^{-i\xi})^m \prod_{y \in Y_m^{\mathsf{S}}} (1 + y \sin^2(\xi/2)). \tag{2.4.7}$$

Then

(1) a_m^{S} is a complex-valued orthogonal wavelet filter satisfying $a_m^{\mathsf{S}}(1-k) = a_m^{\mathsf{S}}(k)$ for all $k \in \mathbb{Z}$;
(2) $|\widehat{a_m^{\mathsf{S}}}(\xi)|^2 = |\widehat{a_m^{D}}(\xi)|^2 = \widehat{a_{2m}^{I}}(\xi)$ and $\widehat{a_m^{\mathsf{S}}}(0) = 1$;
(3) $\operatorname{sr}(a_m^{\mathsf{S}}) = m$ and the filter support of a_m^{S} is $[1-m, m]$.

In the following we present a few examples of a_m^{S} by listing Y_m^{S} such that $\mathrm{Var}(a_m^{\mathsf{S}})$ is the smallest among all possible choices of Y_m^{S} in (2.4.6). Note that $a_1^{\mathsf{S}} = a_1^D$.

$$Y_3^{\mathsf{S}} = \{\tfrac{3-\sqrt{15}i}{2}\},$$

$$Y_5^{\mathsf{S}} = \{2.67984516848 + 1.60066496071i, -0.179845168483 - 2.67428235144i\},$$

$$Y_7^{\mathsf{S}} = \{3.2285892049 + 1.30036579467i, 1.301987015 + 2.95813693603i,$$
$$- 1.03057621991 - 2.49790321151i\},$$

$$Y_9^{\mathsf{S}} = \{3.54409519917 - 1.08582974979i, 2.21753883450 + 2.76788363446i,$$
$$0.208274762391 + 3.2108725564i, -1.46990879606 - 2.24325452765i\},$$

$$Y_{11}^{\mathsf{S}} = \{3.74979134181 - 0.929822989894i, 2.79293701197 - 2.50485336419i,$$
$$1.19309156479 + 3.31669500563i, -0.51543192051 + 3.12940759622i,$$
$$- 1.72038799806 - 2.01935282657i\},$$

$$Y_{13}^{\mathsf{S}} = \{3.89511452442 + 0.812322557651i, 3.17592039995 - 2.25749176419i,$$
$$1.90740348671 - 3.20548726789i, 0.392414210452 + 3.45278767653i,$$
$$- 0.996034279170 + 2.95594908817i, -1.87481834237 - 1.83491941i\},$$

$$Y_{15}^{\mathsf{S}} = \{4.00362108613 + 0.720949682904i, 3.44463249605 + 2.04259015503i,$$
$$2.42618732008 - 3.02431803222i, 1.13094137196 - 3.50542114323i,$$
$$- 0.20465820086 + 3.41178928525i, -1.3249952648 + 2.7675829609i,$$
$$- 1.97572880855 - 1.68364140435i\},$$

$$Y_{17}^{\mathsf{S}} = \{4.08796398218 - 0.647971916690i, 3.64155893176 - 1.85947337418i,$$
$$2.81078377777 - 2.82909042709i, 1.71169364447 + 3.43214725089i,$$
$$0.499626068616 - 3.593556312i, -0.64958161881 + 3.29767061055i,$$
$$- 1.55744482356 + 2.5889940048i, -2.0445999624 - 1.5583140929i\}.$$

Explicitly, we have

$$a_3^{\mathsf{S}} = \{\tfrac{\sqrt{15}i-3}{64}, \tfrac{\sqrt{15}i+5}{64}, \tfrac{\mathbf{15}-\sqrt{\mathbf{15}}\mathbf{i}}{\mathbf{32}}, \tfrac{15-\sqrt{15}i}{32}, \tfrac{\sqrt{15}i+5}{64}, \tfrac{\sqrt{15}i-3}{64}\}_{[-2,3]},$$

$$\begin{aligned} a_5^{\mathsf{S}}|_{[1,5]} = \{&0.454671947451 - 0.129296005476i,\\ &0.107041618366 + 0.0666261866324i,\\ &- 0.0570208816340 + 0.0834014583626i,\\ &- 0.0121119670915 - 0.00617200182672i,\\ &0.00741928290851 - 0.0145596376918i\}_{[1,5]}, \end{aligned}$$

$$
\begin{aligned}
a_7^{\mathsf{S}}|_{[1,7]} = \{&0.480049233235 - 0.0128267889802i,\\
&0.0947790667727 + 0.0359219721071i,\\
&-0.0847256809483 + 0.0193061206810i,\\
&-0.00915454265935 - 0.0241545526008i,\\
&0.0227305055174 - 0.0191146817061i,\\
&-0.000295066186610 - 0.000655386356666i,\\
&-0.00338351573081 + 0.00152331685558i\}_{[1,7]},\\
a_9^{\mathsf{S}}|_{[1,9]} = \{&0.475746199486 + 0.0529154250772i,\\
&0.0980402300290 + 0.0168568487986i,\\
&-0.0884664036553 - 0.0246025530855i,\\
&-0.00825675975262 - 0.0342006104779i,\\
&0.0315874621271 - 0.0149154632742i,\\
&-0.00117727923225 + 0.000922959609887i,\\
&-0.00836058017750 + 0.00258220363043i,\\
&0.0000225539929357 + 0.000481672933847i,\\
&0.000864577183062 - 0.0000404832123811i\}_{[1,9]},\\
a_{11}^{\mathsf{S}}|_{[1,11]} = \{&0.475946940575 + 0.0132276341809i,\\
&0.105613540552 + 0.0239501047203i,\\
&-0.0927497518692 - 0.00126103671116i,\\
&-0.0161032009435 - 0.0306157082521i,\\
&0.0352301638497 - 0.0177206438108i,\\
&0.00106588512721 + 0.00581121511828i,\\
&-0.0111739127826 + 0.00707913328213i,\\
&0.000147091920955 + 0.000364784856321i,\\
&0.00223155015595 - 0.000854349189803i,\\
&-0.0000119656551977 - 0.0000399465211857i,\\
&-0.000196340930399 + 0.0000588123271966i\}_{[1,11]}.
\end{aligned}
$$

We present smoothness exponents, variances, and frequency separation indicators of the complex symmetric orthogonal wavelet filters a_m^{S} in Table 2.7 and graphs of $\phi^{a_m^{\mathsf{S}}}$ and $\psi^{a_m^{\mathsf{S}}}$ in Figs. 2.10 and 2.11. Note that the symmetry in (2.4.1)

Table 2.7 The smoothness exponents, variances, and frequency separation indicators of a_m^{S}, where a_m^{S} is defined in (2.4.7). Note that $\|a_m^{\mathsf{S}}\|_{l_2(\mathbb{Z})}^2 = \mathrm{E}(a_m^{\mathsf{S}}) = 1/2$, and $\mathrm{sm}(a_m^{\mathsf{S}}) = \mathrm{sm}(a_m^{D})$

m	3	5	7	9
$\mathrm{sm}(a_m^{\mathsf{S}})$	1.415037	2.096787	2.658660	3.161667
$\mathrm{Var}(a_m^{\mathsf{S}})$	0.468752	0.652384	0.618386	0.701961
$\mathrm{Fsi}(a_m^{\mathsf{S}})$	0.172338	0.128507	0.106441	0.0926729
m	11	13	15	17
$\mathrm{sm}(a_m^{\mathsf{S}})$	3.639798	4.106047	4.565135	5.019141
$\mathrm{Var}(a_m^{\mathsf{S}})$	0.767464	0.839722	0.893810	0.937622
$\mathrm{Fsi}(a_m^{\mathsf{S}})$	0.0830663	0.0758893	0.0702697	0.0657165

implies $\sum_{k\in\mathbb{Z}} a(k)k = \frac{1}{2}\widehat{a}(0)$. Therefore, all filters a_m^{S} have at least two linear-phase moments. However, the filter a_m^{S} in general has no more than two linear-phase moments. For example, $[\widehat{a_3^{\mathsf{S}}}]'''(0) = -\frac{1+\sqrt{15}i}{4} \neq -\frac{1}{4}$ and hence, $\mathrm{lpm}(a_3^{\mathsf{S}}) = 2$. In the next section we study symmetric complex-valued orthogonal wavelet filters with linear-phase moments.

2.5 Complex Orthogonal Wavelet Filters with Symmetry and Linear-Phase Moments

In this section we present a family of symmetric complex-valued orthogonal low-pass filters with increasing orders of linear-phase moments and sum rules.

2.5.1 Properties of Complex-Valued Orthogonal Wavelet Filters

For a complex number $c = c^{[r]} + ic^{[i]}$ with $c^{[r]}, c^{[i]} \in \mathbb{R}$, recall that $\mathrm{Re}(c) := c^{[r]}$ and $\mathrm{Im}(c) := c^{[i]}$ denote the real and imaginary parts of the complex number c, respectively. Separating the real and imaginary parts of a complex-valued sequence u, we can uniquely write $u = u^{[r]} + iu^{[i]}$, where

$$
\begin{aligned}
\widehat{u^{[r]}}(\xi) &:= [\widehat{u}(\xi) + \overline{\widehat{u}(-\xi)}]/2 = \sum_{k\in\mathbb{Z}} \mathrm{Re}(u(k))e^{-ik\xi},\\
\widehat{u^{[i]}}(\xi) &:= i[\overline{\widehat{u}(-\xi)} - \widehat{u}(\xi)]/2 = \sum_{k\in\mathbb{Z}} \mathrm{Im}(u(k))e^{-ik\xi}.
\end{aligned}
\tag{2.5.1}
$$

Similarly, for a polynomial P with complex coefficients, we write $\mathsf{P} = \mathsf{P}^{[r]} + i\mathsf{P}^{[i]}$, where $\mathsf{P}^{[r]}$ and $\mathsf{P}^{[i]}$ are polynomials with real coefficients. If u has symmetry satisfying $\mathsf{S}\widehat{u}(\xi) = \epsilon e^{-ic\xi}$ for some $\epsilon \in \{-1, 1\}$ and $c \in \mathbb{Z}$, then $\mathsf{S}\widehat{u^{[r]}}(\xi) =$

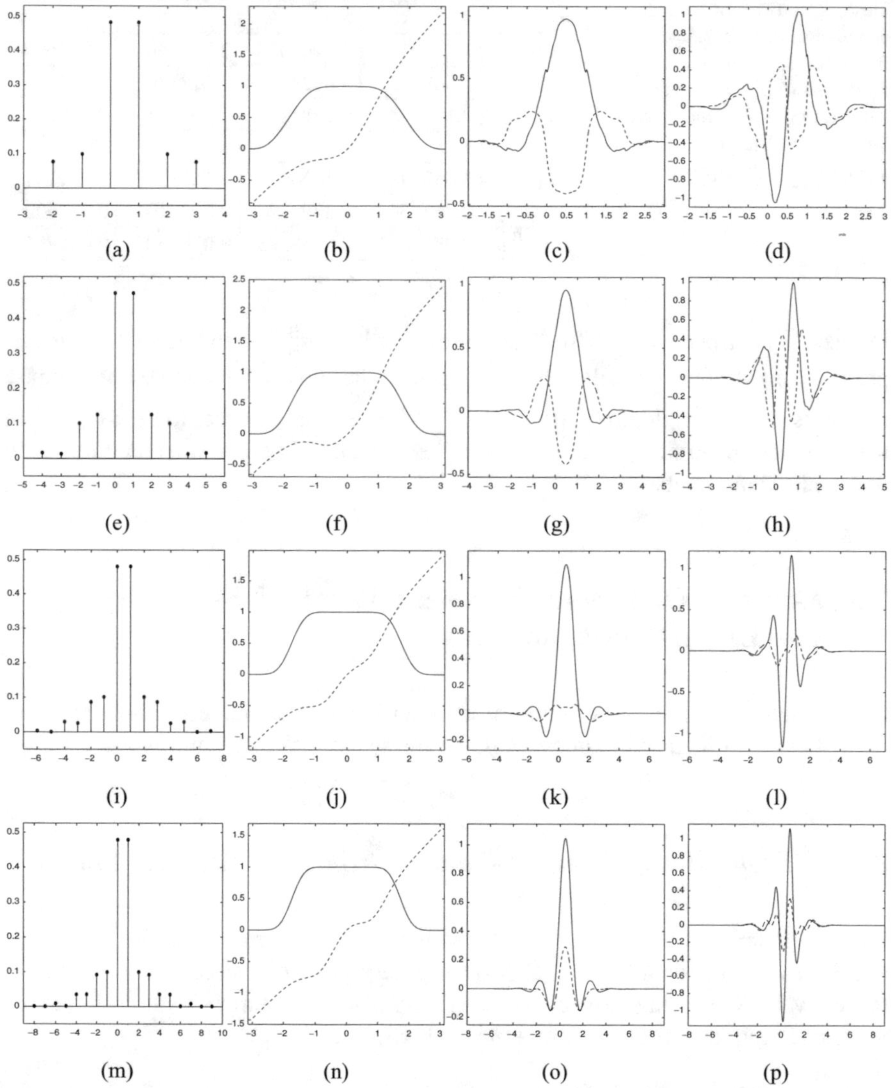

Fig. 2.10 The complex-valued orthogonal wavelet filter a_m^{S} with symmetry, magnitude and phase of $\widehat{a_m^{\mathsf{S}}}$, and the graphs of the standard orthogonal refinable function $\phi^{a_m^{\mathsf{S}}}$ and its associated standard orthogonal wavelet function $\psi^{a_m^{\mathsf{S}}}$ (solid line for the real part and dashed line for the imaginary part), where $m = 3, 5, 7, 9$. Note that $|a_m^{\mathsf{S}}| = \{|a_m^{\mathsf{S}}(k)|\}_{k\in\mathbb{Z}}$. **(a)** Filter $|a_3^{\mathsf{S}}|$. **(b)** $\widehat{a_3^{\mathsf{S}}}$. **(c)** $\phi^{a_3^{\mathsf{S}}}$. **(d)** $\psi^{a_3^{\mathsf{S}}}$. **(e)** Filter $|a_5^{\mathsf{S}}|$. **(f)** $\widehat{a_5^{\mathsf{S}}}$. **(g)** $\phi^{a_5^{\mathsf{S}}}$. **(h)** $\psi^{a_5^{\mathsf{S}}}$. **(i)** Filter $|a_7^{\mathsf{S}}|$. **(j)** $\widehat{a_7^{\mathsf{S}}}$. **(k)** $\phi^{a_7^{\mathsf{S}}}$. **(l)** $\psi^{a_7^{\mathsf{S}}}$. **(m)** Filter $|a_9^{\mathsf{S}}|$. **(n)** $\widehat{a_9^{\mathsf{S}}}$. **(o)** $\phi^{a_9^{\mathsf{S}}}$. **(p)** $\psi^{a_9^{\mathsf{S}}}$

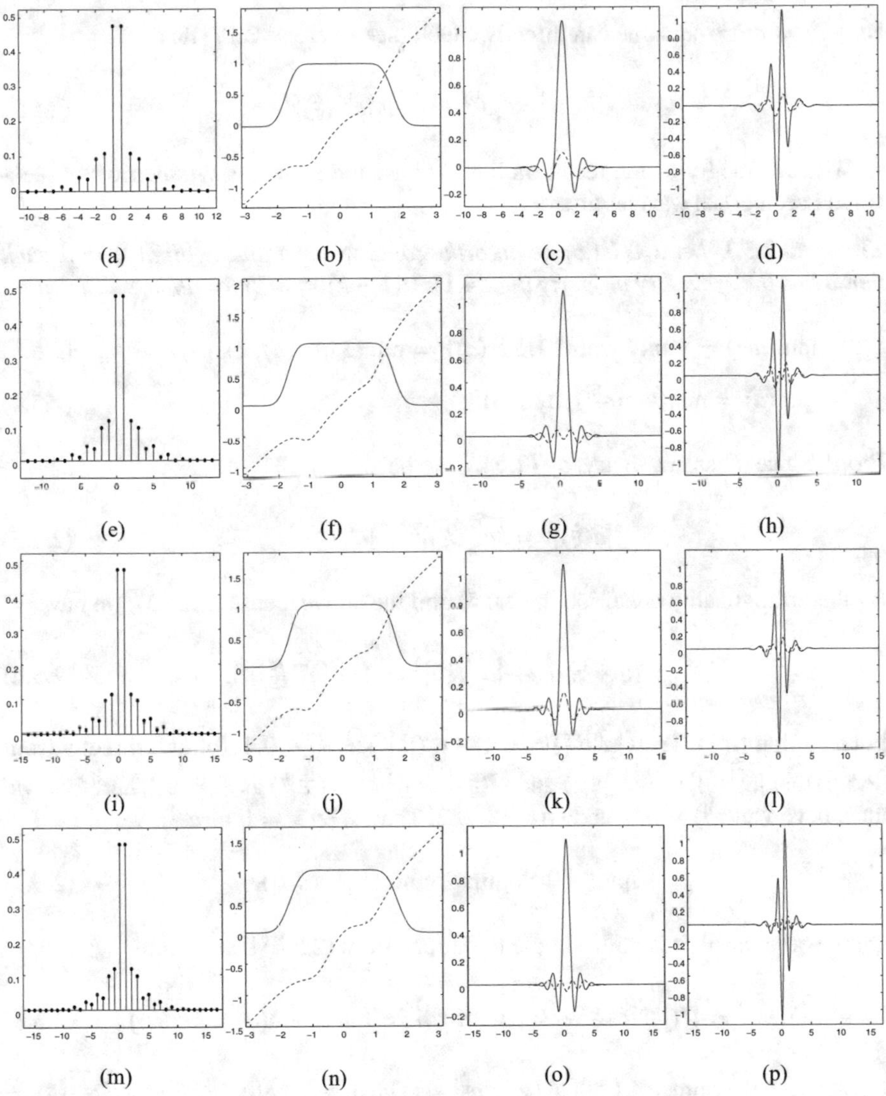

Fig. 2.11 The complex-valued orthogonal wavelet filter a_m^{S} with symmetry, magnitude and phase of $\widehat{a_m^{\mathsf{S}}}$, and the graphs of the standard orthogonal refinable function $\phi^{a_m^{\mathsf{S}}}$ and its associated standard orthogonal wavelet function $\psi^{a_m^{\mathsf{S}}}$ (solid line for the real part and dashed line for the imaginary part), where $m = 11, 13, 15, 17$. Note that $|a_m^{\mathsf{S}}| = \{|a_m^{\mathsf{S}}(k)|\}_{k\in\mathbb{Z}}$. **(a)** Filter $|a_{11}^{\mathsf{S}}|$. **(b)** $\widehat{a_{11}^{\mathsf{S}}}$. **(c)** $\phi^{a_{11}^{\mathsf{S}}}$. **(d)** $\psi^{a_{11}^{\mathsf{S}}}$. **(e)** Filter $|a_{13}^{\mathsf{S}}|$. **(f)** $\widehat{a_{13}^{\mathsf{S}}}$. **(g)** $\phi^{a_{13}^{\mathsf{S}}}$. **(h)** $\psi^{a_{13}^{\mathsf{S}}}$. **(i)** Filter $|a_{15}^{\mathsf{S}}|$. **(j)** $\widehat{a_{15}^{\mathsf{S}}}$. **(k)** $\phi^{a_{15}^{\mathsf{S}}}$. **(l)** $\psi^{a_{15}^{\mathsf{S}}}$. **(m)** Filter $|a_{17}^{\mathsf{S}}|$. **(n)** $\widehat{a_{17}^{\mathsf{S}}}$. **(o)** $\phi^{a_{17}^{\mathsf{S}}}$. **(p)** $\psi^{a_{17}^{\mathsf{S}}}$

$\widehat{\mathsf{S}u^{[i]}}(\xi) = \epsilon e^{-ic\xi}$ and one can directly check (see Exercise 2.17) that

$$|\widehat{u}(\xi)|^2 = |\widehat{u^{[r]}}(\xi)|^2 + |\widehat{u^{[i]}}(\xi)|^2. \tag{2.5.2}$$

We have the following result on linear-phase moments for symmetric complex-valued orthogonal wavelet filters.

Theorem 2.5.1 *Let $a \in l_0(\mathbb{Z})$ be an orthogonal wavelet filter with $\widehat{a}(0) = 1$ such that a has the symmetry property in* (2.4.1)*: $a(1-k) = a(k)$ for all $k \in \mathbb{Z}$. Then*

$$\operatorname{lpm}(a^{[r]}) = \min(2\operatorname{vm}(a^{[i]}), 2\operatorname{sr}(a)) = \min(2\operatorname{lpm}(a), 2\operatorname{sr}(a)), \tag{2.5.3}$$

$$\operatorname{lpm}(a) = \min(\operatorname{vm}(a^{[i]}), 2\operatorname{sr}(a)). \tag{2.5.4}$$

Proof Since a has symmetry, by (2.5.2), we have

$$|\widehat{a}(\xi)|^2 = |\widehat{a^{[r]}}(\xi)|^2 + |\widehat{a^{[i]}}(\xi)|^2. \tag{2.5.5}$$

By the orthogonality condition in (2.0.5) and the above identity (2.5.5), we have

$$|\widehat{a}(\xi+\pi)|^2 + |\widehat{a^{[i]}}(\xi)|^2 = 1 - |\widehat{a^{[r]}}(\xi)|^2. \tag{2.5.6}$$

Let $n := \operatorname{lpm}(a^{[r]})$. Then $|\widehat{a^{[r]}}(\xi)|^2 = 1 + \mathscr{O}(|\xi|^n)$ as $\xi \to 0$. It follows directly from (2.5.6) that $|\widehat{a^{[i]}}(\xi)|^2 = \mathscr{O}(|\xi|^n)$ and $|\widehat{a}(\xi+\pi)|^2 = \mathscr{O}(|\xi|^n)$ as $\xi \to 0$. Therefore, we must have $\operatorname{vm}(a^{[i]}) \geqslant n/2$ and $\operatorname{sr}(a) \geqslant n/2$. That is, by $n = \operatorname{lpm}(a^{[r]})$, we proved

$$\operatorname{lpm}(a^{[r]}) \leqslant \min(2\operatorname{vm}(a^{[i]}), 2\operatorname{sr}(a)). \tag{2.5.7}$$

On the other hand, it is quite simple to deduce from (2.5.6) that

$$1 - |\widehat{a^{[r]}}(\xi)|^2 = \mathscr{O}(|\xi|^{\tilde{n}}),\ \xi \to 0 \quad \text{with} \quad \tilde{n} := \min(2\operatorname{vm}(a^{[i]}), 2\operatorname{sr}(a)). \tag{2.5.8}$$

Because a has symmetry, both $a^{[r]}$ and $a^{[i]}$ also have symmetry satisfying $\widehat{\mathsf{S}a^{[r]}}(\xi) = \widehat{\mathsf{S}a^{[i]}}(\xi) = \mathsf{S}\widehat{a}(\xi) = e^{-i\xi}$. Thus, $|\widehat{a^{[r]}}(\xi)|^2 = [e^{i\xi/2}\widehat{a^{[r]}}(\xi)]^2$. Now (2.5.8) becomes

$$(1 - e^{i\xi/2}\widehat{a^{[r]}}(\xi))(1 + e^{i\xi/2}\widehat{a^{[r]}}(\xi)) = \mathscr{O}(|\xi|^{\tilde{n}}), \qquad \xi \to 0. \tag{2.5.9}$$

Since $\widehat{a^{[r]}}(0) = \operatorname{Re}(\widehat{a}(0)) = 1$, we have $(1 + e^{i\xi}\widehat{a^{[r]}}(\xi))|_{\xi=0} = 2 \neq 0$. By (2.5.9), we get $1 - e^{i\xi}\widehat{a^{[r]}}(\xi) = \mathscr{O}(|\xi|^{\tilde{n}})$, $\xi \to 0$, i.e., $\widehat{a^{[r]}}(\xi) = e^{-i\xi/2} + \mathscr{O}(|\xi|^{\tilde{n}})$ as $\xi \to 0$. Thus,

$$\operatorname{lpm}(a^{[r]}) \geqslant \tilde{n} = \min(2\operatorname{vm}(a^{[i]}), 2\operatorname{sr}(a)).$$

Combining the above identity with (2.5.7), we see that the first identity in (2.5.3) must hold true. Next, we prove (2.5.4). By Exercise 2.19, we have $\operatorname{lpm}(a) = \min(\operatorname{lpm}(a^{[r]}), \operatorname{vm}(a^{[i]}))$. It follows from the proved first identity in (2.5.3) that

$$\begin{aligned}\operatorname{lpm}(a) &= \min(\operatorname{lpm}(a^{[r]}), \operatorname{vm}(a^{[i]})) = \min(\min(2\operatorname{vm}(a^{[i]}), 2\operatorname{sr}(a)), \operatorname{vm}(a^{[i]}))\\ &= \min(2\operatorname{vm}(a^{[i]}), 2\operatorname{sr}(a), \operatorname{vm}(a^{[i]})) = \min(\operatorname{vm}(a^{[i]}), 2\operatorname{sr}(a)).\end{aligned}$$

Hence, (2.5.4) holds. Now it follows from the first identity (2.5.3) and (2.5.4) that

$$\begin{aligned}\min(2\operatorname{lpm}(a), 2\operatorname{sr}(a)) &= \min(\min(2\operatorname{vm}(a^{[i]}), 4\operatorname{sr}(a)), 2\operatorname{sr}(a))\\ &= \min(2\operatorname{vm}(a^{[i]}), 2\operatorname{sr}(a)) = \operatorname{lpm}(a^{[r]}).\end{aligned}$$

Therefore, the second identity in (2.5.3) holds. □

For an integer n, we define

$$\operatorname{odd}(n) := \begin{cases} 1, & \text{if } n \text{ is odd},\\ 0, & \text{if } n \text{ is even},\end{cases} \quad \text{or equivalently,} \quad \operatorname{odd}(n) := \frac{1-(-1)^n}{2}. \tag{2.5.10}$$

We now present a family of symmetric complex-valued orthogonal wavelet filters with arbitrarily preassigned orders of linear-phase moments and sum rules.

Theorem 2.5.2 *Let m and n be positive integers such that $n \leqslant m$. Then*

$$H_{m,n}(x) := 1 - x^{2m-1}[\mathsf{P}_{m-1/2,n}(1-x)]^2 - (1-x)^{2m-1}[\mathsf{P}_{m-1/2,n}(x)]^2 \geqslant 0 \tag{2.5.11}$$

for all $x \in [0,1]$, and $x^n(1-x)^n \mid H_{m,n}(x)$, where the polynomial $\mathsf{P}_{m-1/2,n}$ is defined in (2.1.4). In other words,

$$H_{m,n}(x) = x^n(1-x)^n R_{m,n}(4x(1-x)) \quad \text{and} \quad R_{m,n}(x) \geqslant 0, \qquad \forall\, x \in [0,1] \tag{2.5.12}$$

for a polynomial $R_{m,n}$ with real coefficients and with $\deg(R_{m,n}) \leqslant m-2$. Define

$$\widehat{a_{m;n}}(\xi) := \widehat{a^{[r]}_{m;n}}(\xi) + i\widehat{a^{[i]}_{m;n}}(\xi), \tag{2.5.13}$$

$$\widehat{a^{[r]}_{m;n}}(\xi) := 2^{1-2m} e^{i(m-1)\xi}(1+e^{-i\xi})^{2m-1}\mathsf{P}_{m-1/2,n}(\sin^2(\xi/2)), \tag{2.5.14}$$

$$\widehat{a^{[i]}_{m;n}}(\xi) := 2^{-2n-1}(e^{i\xi} - e^{-i\xi})^n[\widehat{Q_{m,n}}(2\xi) + (-1)^n e^{-i\xi}\widehat{Q_{m,n}}(-2\xi)], \tag{2.5.15}$$

where $Q_{m,n}$ is a finitely supported real-valued sequence obtained via the Fejér-Riesz Lemma from $R_{m,n}$ such that $|\widehat{Q_{m,n}}(\xi)|^2 = R_{m,n}(\sin^2(\xi/2))$. Then $a_{m;n}$ is a complex-valued orthogonal wavelet filter with $\widehat{a_{m;n}}(0) = 1$ and $\mathsf{S}\widehat{a_{m;n}}(\xi) = e^{-i\xi}$, has at least

$n+\mathrm{odd}(n)$ linear-phase moments (with phase $1/2$), and at least $n+1-\mathrm{odd}(n)$ sum rules. Moreover, if the filter support of $Q_{m,n}$ is contained inside $[1-\frac{m}{2}+\frac{\mathrm{odd}(m)}{2}, \frac{m}{2}-1+\frac{\mathrm{odd}(m)}{2}]$, then the filter support of $a_{m;n}$ is $[2-m-n, m+n-1]$.

Proof It is trivial to see that $\mathsf{S}\widehat{a_{m;n}}(\xi) = \mathsf{S}\widehat{a^{[r]}_{m;n}}(\xi) = \mathsf{S}\widehat{a^{[i]}_{m;n}}(\xi) = e^{-i\xi}$. We first prove (2.5.12). By the definition of $\mathsf{P}_{m-1/2,n}$, it follows from (2.1.3) that

$$[\mathsf{P}_{m-1/2,n}(x)]^2 = (1-x)^{1-2m} + \mathscr{O}(x^n), \qquad x \to 0.$$

On the other hand, we have

$$\begin{aligned}(1-x)^{1/2-m} &= \mathsf{P}_{m-1/2,2m-1}(x) + \mathscr{O}(x^{2m-1})\\ &= \mathsf{P}_{m-1/2,n}(x) + \sum_{j=n}^{2m-2}(-1)^j\binom{1/2-m}{j}x^j + \mathscr{O}(x^{2m-1}), \quad x \to 0.\end{aligned}$$

Squaring both sides, we deduce from the above relation that

$$\begin{aligned}(1-x)^{1-2m} &= \mathsf{P}_{2m-1,2m-1}(x) + \mathscr{O}(x^{2m-1})\\ &= [\mathsf{P}_{m-1/2,n}(x)]^2 + S_{m,n}(x) + \mathscr{O}(x^{2m-1}),\ x \to 0,\end{aligned} \tag{2.5.16}$$

where $S_{m,n}$ is a unique polynomial of degree at most $2m-2$. Since $\deg(\mathsf{P}_{m-1/2,n}) = n-1$ and $n \leqslant m$, we see that $\deg([\mathsf{P}_{m-1/2,n}]^2) = 2(n-1) \leqslant 2m-2$. Since $\mathsf{P}_{m-1/2,n}(x) = (1-x)^{1/2-m} + \mathscr{O}(x^n)$ as $x \to 0$, we have $[\mathsf{P}_{m-1/2,n}(x)]^2 = (1-x)^{1-2m} + \mathscr{O}(x^n)$ as $x \to 0$. Consequently, it follows from (2.5.16) that $x^n \mid S_{m,n}(x)$. In other words, from (2.5.16), we have the relation:

$$\mathsf{P}_{2m-1,2m-1}(x) = [\mathsf{P}_{m-1/2,n}(x)]^2 + S_{m,n}(x) \quad \text{and} \quad x^n \mid S_{m,n}(x). \tag{2.5.17}$$

Now by (2.1.5), we observe that

$$\begin{aligned}1 &= (1-x)^{2m-1}\mathsf{P}_{2m-1,2m-1}(x) + (1-x)^{2m-1}\mathsf{P}_{2m-1,2m-1}(1-x)\\ &= (1-x)^{2m-1}[\mathsf{P}_{m-1/2,n}(x)]^2 + x^{2m-1}[\mathsf{P}_{m-1/2,n}(1-x)]^2\\ &\quad + (1-x)^{2m-1}S_{m,n}(x) + x^{2m-1}S_{m,n}(1-x).\end{aligned}$$

Because $(-1)^j\binom{1/2-m}{j} = \frac{(m-1/2)(m+1/2)\cdots(m+j-3/2)}{j!} \geqslant 0$ for all $j \in \mathbb{N}$, it is easy to see that all the coefficients of $S_{m,n}$ are nonnegative. Therefore, we deduce from the above identity that for all $x \in [0,1]$,

$$\begin{aligned}H_{m,n}(x) &:= 1 - x^{2m-1}[\mathsf{P}_{m-1/2,n}(1-x)]^2 - (1-x)^{2m-1}[\mathsf{P}_{m-1/2,n}(x)]^2\\ &= (1-x)^{2m-1}S_{m,n}(x) + x^{2m-1}S_{m,n}(1-x) \geqslant 0.\end{aligned}$$

Moreover, by (2.5.17) and $n \leqslant m$, it is evident that $2m-1 \geqslant n$ and

$$x^n(1-x)^n \mid (1-x)^{2m-1} S_{m,n}(x) \quad \text{and} \quad x^n(1-x)^n \mid x^{2m-1} S_{m,n}(1-x).$$

Therefore, we proved (2.5.12).

We now prove that $a_{m;n}$ in (2.5.13) is an orthogonal wavelet filter. By $\mathsf{S}\widehat{a_{m;n}}(\xi) = e^{-i\xi}$ and Exercise 2.17,

$$|\widehat{a_{m;n}}(\xi)|^2 = |\widehat{a_{m;n}^{[r]}}(\xi)|^2 + |\widehat{a_{m;n}^{[i]}}(\xi)|^2. \tag{2.5.18}$$

By (2.5.14) and (2.5.15), we observe that

$$|\widehat{a_{m;n}^{[r]}}(\xi)|^2 = \cos^{4m-2}(\xi/2)\mathsf{P}_{m-1/2,n}(\sin^2(\xi/2)), \tag{2.5.19}$$

$$\begin{aligned} |\widehat{a_{m;n}^{[i]}}(\xi)|^2 =& 2^{-1}\cos^{2n}(\tfrac{\xi}{2})\sin^{2n}(\tfrac{\xi}{2})\big(|\widehat{Q_{m,n}}(2\xi)|^2 \\ &+ (-1)^n \operatorname{Re}(\widehat{Q_{m,n}}(2\xi)\overline{\widehat{Q_{m,n}}(-2\xi)}e^{-i\xi})\big). \end{aligned} \tag{2.5.20}$$

Set $x = \sin^2(\xi/2)$. Then

$$\begin{aligned} |\widehat{a_{m;n}^{[r]}}(\xi)|^2 + |\widehat{a_{m;n}^{[r]}}(\xi+\pi)|^2 &= (1-x)^{2m-1}[\mathsf{P}_{m-1/2,n}(x)]^2 + x^{2m-1}[\mathsf{P}_{m-1/2,n}(1-x)]^2 \\ &= 1 - H_{m,n}(x). \end{aligned}$$

By (2.5.20) and $\sin^2\xi = 4\sin^2(\xi/2)\cos^2(\xi/2) = 4x(1-x)$,

$$|\widehat{a_{m;n}^{[i]}}(\xi)|^2 + |\widehat{a_{m;n}^{[i]}}(\xi+\pi)|^2 = x^n(1-x)^n|\widehat{Q_{m,n}}(2\xi)|^2 = x^n(1-x)^n R_{m,n}(4x(1-x)).$$

Now by (2.5.19) and (2.5.12), it follows from the above identity and (2.5.18) that

$$|\widehat{a_{m;n}}(\xi)|^2 + |\widehat{a_{m;n}}(\xi+\pi)|^2 = 1 - H_{m,n}(x) + x^n(1-x)^n R_{m,n}(4x(1-x)) = 1.$$

Thus, $a_{m;n}$ is an orthogonal wavelet filter.

From (2.5.14) and (2.5.15), by $n \leqslant m$, it is obvious to see that both $a_{m;n}^{[r]}$ and $a_{m;n}^{[i]}$ have n sum rules. Therefore, $a_{m;n}$ has at least n sum rules. From (2.5.15), it is also trivial to see that $a_{m;n}^{[i]}$ has n vanishing moments. By Theorem 2.5.1, we conclude that the orthogonal wavelet filter $a_{m;n}$ has at least n linear-phase moments. □

2.5.2 Complex Orthogonal Wavelet Filters with Linear-Phase Moments

For every positive integer m, letting $n = m$ in Theorem 2.5.2, we denote the filter $a_{m;m}$ constructed in Theorem 2.5.2 by a_m^H, that is,

$$\begin{aligned}\widehat{a_m^H}(\xi) :=& 2^{1-2m} e^{i(m-1)\xi}(1+e^{-i\xi})^{2m-1} \mathsf{P}_{m-1/2,m}(\sin^2(\xi/2)) \\ &+ i2^{-2m-1}(e^{i\xi}-e^{-i\xi})^m [\widehat{Q_m^H}(2\xi) + (-1)^m e^{-i\xi}\widehat{Q_m^H}(-2\xi)],\end{aligned} \tag{2.5.21}$$

where $Q_m^H := Q_{m,m}$ is a finitely supported sequence which satisfies $|\widehat{Q_m^H}(\xi)|^2 = R_{m,m}(\sin^2(\xi/2))$ and has its filter support contained inside $[1 - \frac{m}{2} + \frac{\text{odd}(m)}{2}, \frac{m}{2} - 1 + \frac{\text{odd}(m)}{2}]$. Note that the filter support of a_m^H is contained inside $[2-2m, 2m-1]$.

In the following we present a few examples of a_m^H. Since the choice of Q_m^H is not unique and $\text{sm}(a_m^H)$ depends on Q_m^H, we shall pick Q_m^H such that $\text{sm}(a_m^H)$ is the largest. Because the real part $a_{m;m}^{[r]}$ and imaginary part $a_{m;m}^{[i]}$ of a_m^H are explicitly given in (2.5.14), in the following we only need to state the coefficients of Q_m^H explicitly.

$$Q_2^H = \{\underline{\tfrac{\sqrt{15}}{2}}\}_{[0,0]}, \qquad Q_3^H = \{\underline{\mathbf{0.383876505437}}, -4.00672069200\}_{[0,1]},$$

$$Q_4^H = \{-7.93455097214, \underline{\mathbf{1.20587188834}}, -0.121275295606\}_{[-1,1]},$$

$$Q_5^H = \{0.370734915266, \underline{\mathbf{-1.12662000411}}, 15.7161958860, -1.88630270865\}_{[-1,2]},$$

$$Q_6^H = \{-0.537573675927, 7.07825179788, \underline{\mathbf{-30.4983020577}}, -0.150565196702, -1.02128030839\}_{[-2,2]},$$

$$Q_7^H = \{0.172874353501, -2.85241492229, \underline{\mathbf{19.5216175954}}, -58.3747657116, -4.41135751405, -2.61351669354\}_{[-2,3]},$$

$$Q_8^H = \{0.00325757829328, -0.0534725690374, 0.425909831939, \underline{\mathbf{-2.22794916183}}, 8.87097217020, -30.6291371050, 117.822391376\}_{[-3,3]},$$

$$Q_9^H = \{0.159837839290, -2.31096472400, 18.3445994658, \underline{\mathbf{-80.2118913613}}, 226.056354889, 14.8365505888, 8.59683655672, -2.08707485227\}_{[-3,4]}.$$

Note that $a_2^H = a_3^{\mathsf{S}}$ and

$$\begin{aligned}
a_3^H|_{[1,5]} = \{&0.461425781250 + 0.102904621815i,\\
&0.102539062500 - 0.0402996110023i,\\
&-0.0615234375000 - 0.0969065514171i,\\
&-0.0109863281250 + 0.00299903519872i,\\
&0.00854492187500 + 0.0313025054061i\}_{[1,5]},\\
a_4^H|_{[1,7]} = \{&0.458221435547 + 0.0770674539082i,\\
&0.114555358887 - 0.0263392388655i,\\
&-0.0687332153320 - 0.102640759144i,\\
&-0.0196380615234 + 0.00330268177883i,\\
&0.0152740478516 + 0.0643438980017i,\\
&0.00208282470703 - 0.000236865811731i,\\
&-0.00176239013672 - 0.0154971698675i\}_{[1,7]},\\
a_5^H|_{[1,9]} = \{&0.456431508064 + 0.0877506577252i,\\
&0.121715068817 - 0.0466019684241i,\\
&-0.0730290412903 - 0.0888812601966i,\\
&-0.0260818004608 + 0.0122346899703i,\\
&0.0202858448029 + 0.0481301887308i,\\
&0.00474214553833 - 0.00145522196311i,\\
&-0.00401258468628 - 0.0122791549948i,\\
&-0.000434696674347 + 0.000181022907845i,\\
&0.000383555889130 + 0.000921046244453i\}_{[1,9]},\\
a_6^H|_{[1,11]} = \{&0.455290429294 + 0.0882854570406i,\\
&0.126469563693 - 0.0625963764673i,\\
&-0.0758817382157 - 0.0741236646838i,\\
&-0.0309721380472 + 0.0254193350050i,\\
&0.0240894407034 + 0.0365923459028i,\\
&0.00739107839763 - 0.00548268011514i,\\
&-0.00625398941338 - 0.00989153051561i,
\end{aligned}$$

$$- 0.00120447203517 + 0.000729628497755i,$$
$$0.00106276944280 + 0.00125777512860i,$$
$$0.0000950898975134 - 0.000124668006395i,$$
$$- 0.0000860337167978 - 0.0000656217866122i\}_{[1,11]}.$$

We present smoothness exponents, variances, and frequency separation indicators of a_m^H in Table 2.8 and graphs of $\phi^{a_m^H}$ and $\psi^{a_m^H}$ in Figs. 2.12 and 2.13.

2.5.3 Algorithm for Symmetric Complex Orthogonal Wavelet Filters with Linear-Phase Moments and Minimum Supports

The condition $n \leqslant m$ in Theorem 2.5.2 is used to guarantee (2.5.11), that is, the nonnegativity of $H_{m,n}$ on the interval $[0, 1]$. In fact, from the proof of Theorem 2.5.11, we see that if (2.5.11) holds, as long as $n < 2m$ (which is a natural constraint, see Exercise 2.21), then all the claims of Theorem 2.5.2 still hold. Though symmetric complex-valued orthogonal wavelet filters with high linear-phase moments can be easily derived from Theorem 2.5.2, the constructed filters $a_{m;n}$ generally do not have the shortest possible filter supports with respect to their orders of linear-phase moments. Fortunately, Theorem 2.5.2 can be easily modified into the following algorithm, whose proof follows the same line as Theorem 2.5.2.

Algorithm 2.5.3 *Let ℓ, m, n be nonnegative integers such that $n < 2m$. Define*

$$\mathsf{P}_{m-1/2,n;\ell}(x) := \mathsf{P}_{m-1/2,n}(x) + \sum_{k=0}^{\ell-1} t_k x^{k+n},$$

where $t_0, \ldots, t_{\ell-1} \in \mathbb{R}$ are ℓ real-valued free parameters chosen in such a way that

$$H_{\ell,m,n}(x) := 1 - x^{2m-1}[\mathsf{P}_{m-1/2,n;\ell}(1-x)]^2 - (1-x)^{2m-1}[\mathsf{P}_{m-1/2,n;\ell}(x)]^2 \geqslant 0$$

Table 2.8 The smoothness exponents, variances, and frequency separation indicators of a_m^H, where a_m^H is defined in (2.5.21). Note that $\|a_m^H\|^2_{l_2(\mathbb{Z})} = \mathrm{E}(a_m^H) = 1/2$

m	2	3	4	5
$\mathrm{sm}(a_m^H)$	1.415037	1.72798	1.95884	2.61918
$\mathrm{Var}(a_m^H)$	0.468752	0.753788	1.13706	0.992888
$\mathrm{Fsi}(a_m^H)$	0.172338	0.112118	0.0932405	0.0817623
m	6	7	8	9
$\mathrm{sm}(a_m^H)$	3.43832	3.71051	3.89248	4.23551
$\mathrm{Var}(a_m^H)$	0.944372	1.03059	2.74195	1.16315
$\mathrm{Fsi}(a_m^H)$	0.0869154	0.0801046	0.101746	0.0731063

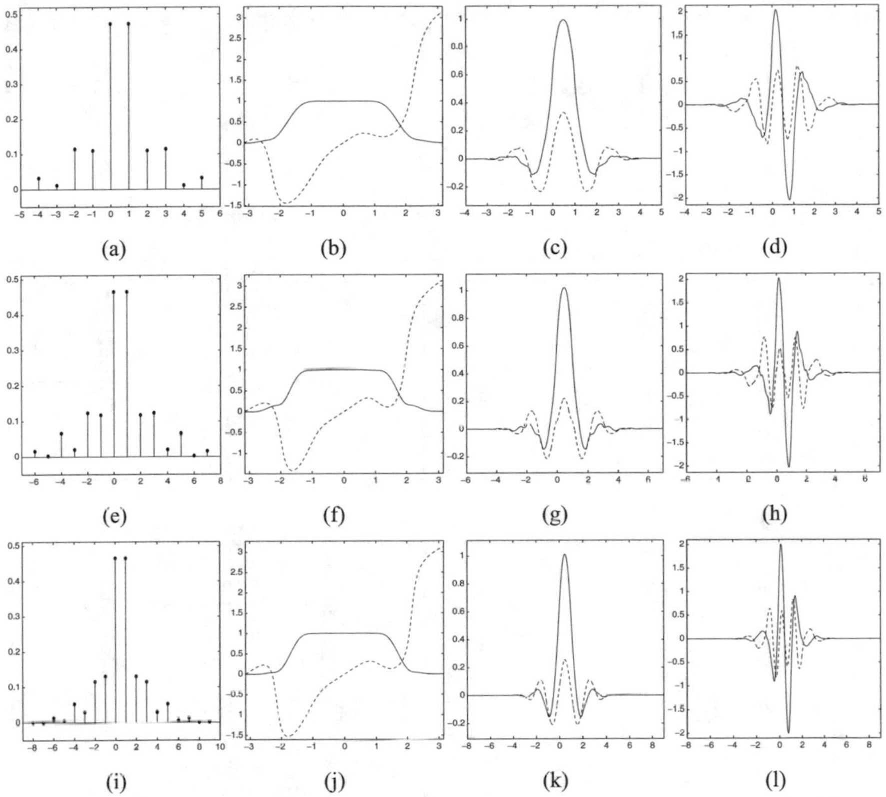

Fig. 2.12 The complex-valued orthogonal wavelet filter a^H_m with symmetry and linear-phase moments, magnitude and phase of $\widehat{a^H_m}$, and the graphs of the standard orthogonal refinable function $\phi^{a^H_m}$ and its associated standard orthogonal wavelet function $\psi^{a^H_m}$ (solid line for the real part and dashed line for the imaginary part), where $m = 3, 4, 5$. Note that $|a^H_m| = \{|a^H_m(k)|\}_{k\in\mathbb{Z}}$. **(a)** Filter $|a^H_3|$. **(b)** $\widehat{a^H_3}$. **(c)** $\phi^{a^H_3}$. **(d)** $\psi^{a^H_3}$. **(e)** Filter $|a^H_4|$. **(f)** $\widehat{a^H_4}$. **(g)** $\phi^{a^H_4}$. **(h)** $\psi^{a^H_4}$. **(i)** Filter $|a^H_5|$. **(j)** $\widehat{a^H_5}$. **(k)** $\phi^{a^H_5}$. **(l)** $\psi^{a^H_5}$

for all $x \in [0, 1]$. *Consequently,*

$$H_{\ell,m,n}(x) = x^n(1-x)^n R_{\ell,m,n}(4x(1-x)) \quad \textit{and} \quad R_{\ell,m,n}(x) \geqslant 0, \ \forall\, x \in [0,1] \tag{2.5.22}$$

for a polynomial $R_{\ell,m,n}$ *with real coefficients and* $\deg(R_{\ell,m,n}) \leqslant \ell + m - 2$. *Define*

$$\widehat{a_{\ell,m;n}}(\xi) := \widehat{a^{[r]}_{\ell,m;n}}(\xi) + i\widehat{a^{[i]}_{\ell,m;n}}(\xi),$$

$$\widehat{a^{[r]}_{\ell,m;n}}(\xi) := 2^{1-2m} e^{i(m-1)\xi}(1+e^{-i\xi})^{2m-1} \mathsf{P}_{\ell,m-1/2,n}(\sin^2(\xi/2)),$$

$$\widehat{a^{[i]}_{\ell,m;n}}(\xi) := 2^{-2n-1}(e^{i\xi} - e^{-i\xi})^n [\widehat{Q_{\ell,m,n}}(2\xi) + (-1)^n e^{-i\xi}\widehat{Q_{\ell,m,n}}(-2\xi)],$$

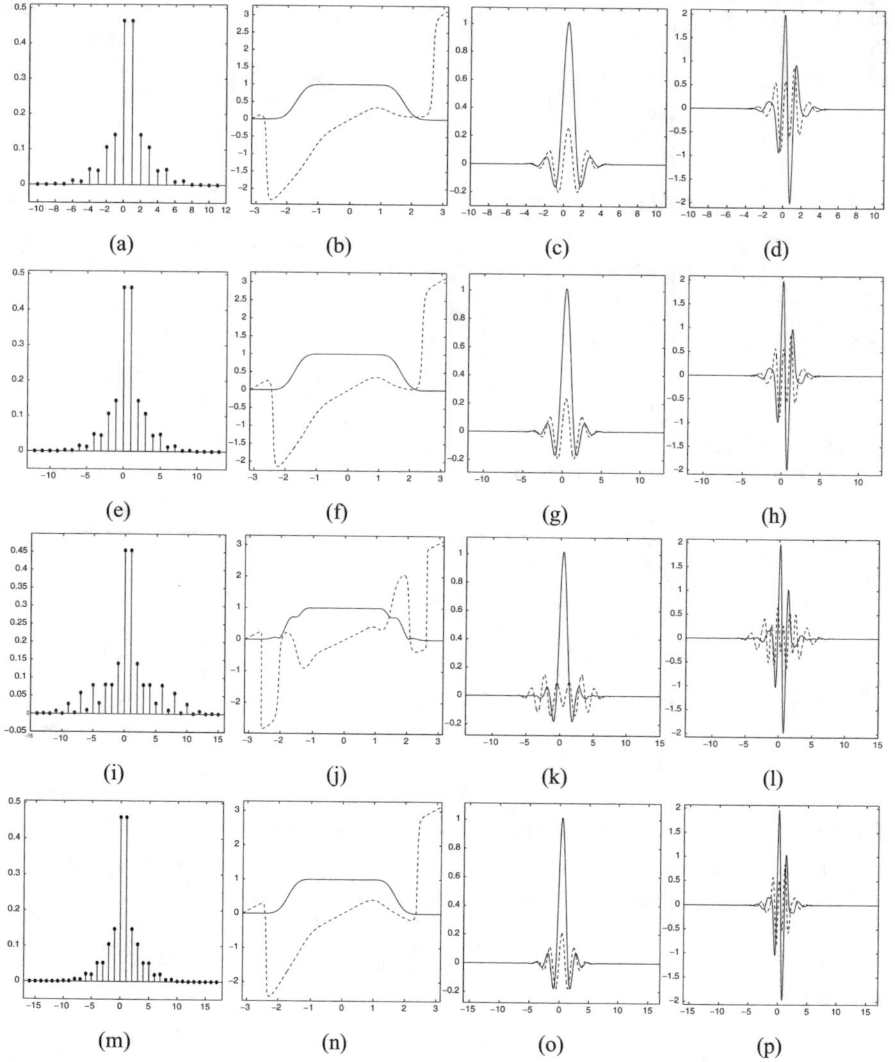

Fig. 2.13 The complex-valued orthogonal wavelet filter a^H_m with symmetry and linear-phase moments, magnitude and phase of $\widehat{a^H_m}$, and the graphs of the standard orthogonal refinable function $\phi^{a^H_m}$ and its associated standard orthogonal wavelet function $\psi^{a^H_m}$ (solid line for the real part and dashed line for the imaginary part), where $m = 6, 7, 8, 9$. Note that $|a^H_m| = \{|a^H_m(k)|\}_{k\in\mathbb{Z}}$. **(a)** Filter $|a^H_6|$. **(b)** $\widehat{a^H_6}$. **(c)** $\phi^{a^H_6}$. **(d)** $\psi^{a^H_6}$. **(e)** Filter $|a^H_7|$. **(f)** $\widehat{a^H_7}$. **(g)** $\phi^{a^H_7}$. **(h)** $\psi^{a^H_7}$. **(i)** Filter $|a^H_8|$. **(j)** $\widehat{a^H_8}$. **(k)** $\phi^{a^H_8}$. **(l)** $\psi^{a^H_8}$. **(m)** Filter $|a^H_9|$. **(n)** $\widehat{a^H_9}$. **(o)** $\phi^{a^H_9}$. **(p)** $\psi^{a^H_9}$

where $Q_{\ell,m,n} \in l_0(\mathbb{Z})$ *has real coefficients and is obtained via the Féjer-Riesz Lemma through* $|\widehat{Q_{\ell,m,n}}(\xi)|^2 = R_{\ell,m,n}(\sin^2(\xi/2))$. *Then* $a_{\ell,m;n}$ *is a complex-valued orthogonal wavelet filter with* $\widehat{a_{\ell,m;n}}(0) = 1$ *such that*

(i) $a_{\ell,m;n}$ *is symmetric about the point* $1/2$: $\mathsf{S}\widehat{a_{\ell,m;n}}(\xi) = e^{-i\xi}$;
(ii) $a_{\ell,m;n}$ *has at least* $n + \operatorname{odd}(n)$ *linear-phase moments (with phase* $1/2$*) and at least* $n + 1 - \operatorname{odd}(n)$ *sum rules.*

In fact, any complex-valued orthogonal wavelet filter satisfying the above items (i) and (ii) can be obtained by the above procedure with $m = \frac{n-\operatorname{odd}(n)}{2} + 1$ for some parameters $t_0, \dots, t_{\ell-1}$.

Clearly, Theorem 2.5.2 is a special case of Algorithm 2.5.3 with $\ell = 0$ and $n \leqslant m$. The freedom in the parameters $t_0, \dots, t_{\ell-1}$ can be used to maximize $\operatorname{sm}(a_{\ell,m;n})$ or to minimize $\|\widehat{a^{[i]}_{\ell,m;n}}\|_{L_\infty(\mathbb{T})}$. In the following, we present a few examples with $\ell = 1$ or 2.

Example 2.5.1 Let $\ell = 1, m = 2, n = 3$ in Algorithm 2.5.3. Then $\mathsf{P}_{3/2,3;1}(x) = 1 + \frac{3}{2}x + \frac{15}{8}x^2 + t_0x^3$. When $t_0 = -35/8$, we have $\mathsf{P}_{3/2,3;1}(1) = 0$ and therefore, the constructed filter in Algorithm 2.5.3 is a^H_3, since $\mathsf{P}_{3/2,3;1}(x) = (1-x)\mathsf{P}_{5/2,3}(x)$.

When $t_0 = -7/4$, we have $H_{1,2,3}(1/2) = 0$ (that is, $R_{1,2,3}(1) = 0$) and the nonnegativity condition in (2.5.22) holds. We have $Q_{1,2,3}(\xi) = -3\sqrt{7}(1+e^{-i\xi})/16$. $a_{1,2;3}$ is a symmetric orthogonal wavelet filter with filter support $[-4, 5]$, has 3 sum rules and 4 linear-phase moments. Then $\operatorname{Var}(a_{1,2;3}) \approx 0.376834$, $\operatorname{Fsi}(a_{1,2;3}) \approx 0.165140$, and $\operatorname{sm}(a_{1,2;3}) \approx 1.476203$. The orthogonal wavelet filter $a_{1,2;3}$ is given by

$$a_{1,2;3} = \{\tfrac{7+3\sqrt{7}i}{2048}, \tfrac{9-3\sqrt{7}i}{2048}, -\tfrac{63-3\sqrt{7}i}{1024}, \tfrac{63+3\sqrt{7}i}{1024}, \underline{\mathbf{\tfrac{63}{128}}}, \tfrac{63}{128}, \tfrac{63+3\sqrt{7}i}{1024}, -\tfrac{63-3\sqrt{7}i}{1024},$$
$$\tfrac{9-3\sqrt{7}i}{2048}, \tfrac{7+3\sqrt{7}i}{2048}\}_{[-4,5]}.$$

Example 2.5.2 Let $\ell = 1, m = 3, n = 4$ in Algorithm 2.5.3. Then

$$\mathsf{P}_{5/2,4;1}(x) = 1 + \tfrac{5}{2}x + \tfrac{35}{8}x^2 + + \tfrac{105}{16}x^3 + t_0x^4.$$

When $t_0 = -231/16$, we have $\mathsf{P}_{5/2,4;1}(1) = 0$ and therefore, the constructed filter in Algorithm 2.5.3 is a^H_4, since $\mathsf{P}_{5/2,3;1}(x) = (1-x)\mathsf{P}_{5/2,4}(x)$.

When $t_0 = -21/8$, we have $H_{1,3,4}(1/2) = 0$ (that is, $R_{1,3,4}(1) = 0$) and the nonnegativity condition in (2.5.22) holds. We have $Q_{1,3,4}(\xi) = 2.48903989648e^{-i\xi} + 2.41334182205 - 0.0756980744309e^{i\xi}$. $a_{1,3;4}$ is a symmetric orthogonal wavelet filter with filter support $[-6, 7]$, has 5 sum rules and 4 linear-phase moments. By calculation, $\operatorname{Var}(a_{1,3;4}) \approx 0.472287$, $\operatorname{Fsi}(a_{1,3;4}) \approx 0.132626$, and $\operatorname{sm}(a_{1,3;4}) \approx$

2.027941. The orthogonal wavelet filter $a_{1,3;4}$ is given by

$$\begin{aligned} a_{1,3;4}|_{[1,7]} = \{&0.487060546875 + 0.00942711649234i,\\ &0.0785064697266 + 0.0101663555004i,\\ &-0.0759429931641 - 0.0148799137466i,\\ &0.000549316406250 - 0.0147320659450i,\\ &0.0123901367188 + 0.00530494945266i,\\ &-0.00224304199219 + 0.00486140604779i,\\ &-0.000320434570312 - 0.000147847801622i\}_{[1,7]}. \end{aligned}$$

Example 2.5.3 Let $\ell = 2, m = 3, n = 5$ in Algorithm 2.5.3. Then

$$\mathsf{P}_{5/2,5;2}(x) = 1 + \tfrac{5}{2}x + \tfrac{35}{8}x^2 + \tfrac{105}{16}x^3 + \tfrac{1155}{128}x^4 + t_0x^5 + t_1y^6.$$

When $t_0 = -4719/64$ and $t_1 = 6435/128$, we have $\mathsf{P}_{5/2,5;2}(1) = \mathsf{P}'_{5/2,5;2}(1) = 0$ and the constructed filter in Algorithm 2.5.3 is a_4^H, since $\mathsf{P}_{5/2,5;2}(x) = (1-x)^2\mathsf{P}_{5-1/2,5}(x)$.

When $t_0 = -1491/64$ and $t_1 = 0$, we have $H_{2,3,5}(1/2) = 0$ (that is, $R_{2,3,5}(1) = 0$) and the nonnegativity condition in (2.5.22) holds. We have

$$Q_{2,3,5}(\xi) = 5.51150309451e^{i\xi} + 4.18412986761 - 1.32737322690e^{-i\xi}.$$

The filter $a_{2,3;5}$ is a symmetric orthogonal wavelet filter with filter support $[-7, 8]$, has 5 sum rules and 6 linear-phase moments. By calculation, $\mathrm{Var}(a_{2,3;5}) \approx 0.554505$, $\mathrm{Fsi}(a_{2,3;5}) \approx 0.112273$, and $\mathrm{sm}(a_{2,3;5}) \approx 2.488790$. See the graphs of $\phi^{a_{\ell,m;n}}$ and $\psi^{a_{\ell,m;n}}$ in Fig. 2.14. The orthogonal wavelet filter $_{2,3;5}$ is given by

$$\begin{aligned} a_{2,3;5}|_{[1,8]} = \{&0.483505725861 + 0.00324065729224i,\\ &0.0891709327698 + 0.0134558181018i,\\ &-0.0837635993958 - 0.0160483439356i,\\ &-0.00442743301392 - 0.0140053117737i,\\ &0.0202107429504 + 0.0114127859399i,\\ &-0.00295400619507 + 0.00528368945416i,\\ &-0.00245332717896 - 0.00269116362037i,\\ &0.000710964202881 - 0.000648131458448i\}_{[1,8]}. \end{aligned}$$

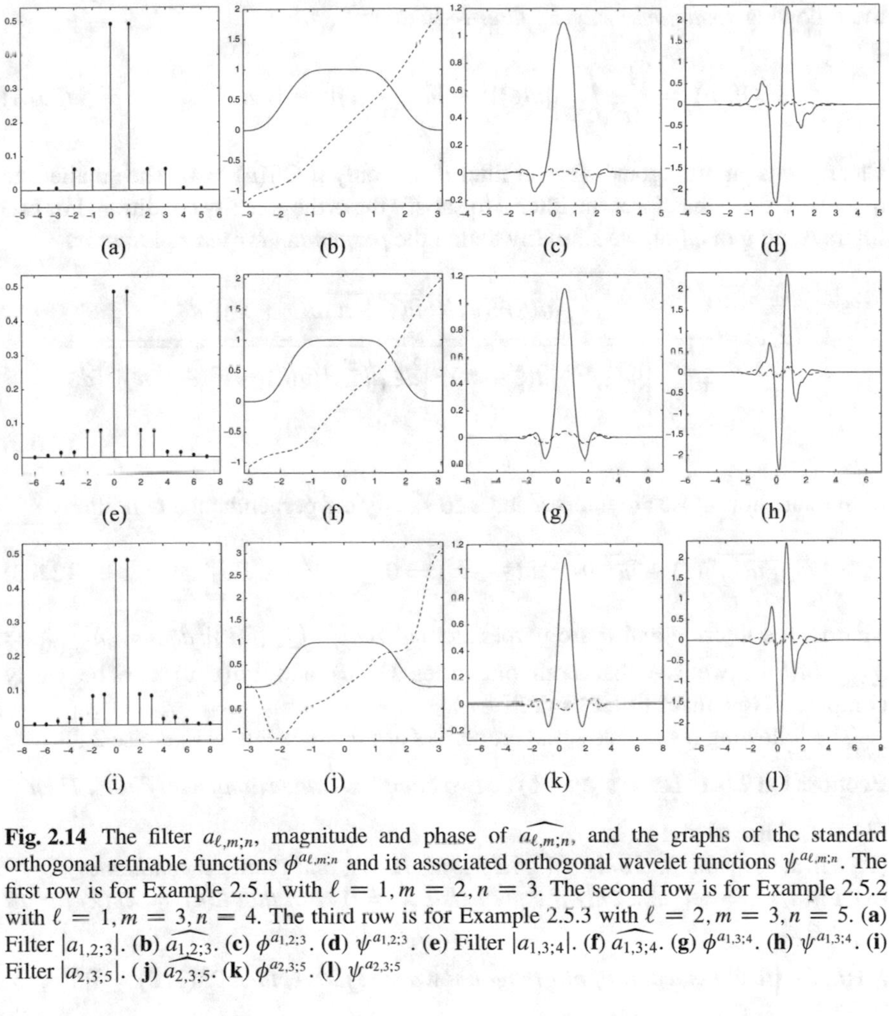

Fig. 2.14 The filter $a_{\ell,m;n}$, magnitude and phase of $\widehat{a_{\ell,m;n}}$, and the graphs of the standard orthogonal refinable functions $\phi^{a_{\ell,m;n}}$ and its associated orthogonal wavelet functions $\psi^{a_{\ell,m;n}}$. The first row is for Example 2.5.1 with $\ell = 1, m = 2, n = 3$. The second row is for Example 2.5.2 with $\ell = 1, m = 3, n = 4$. The third row is for Example 2.5.3 with $\ell = 2, m = 3, n = 5$. **(a)** Filter $|a_{1,2;3}|$. **(b)** $\widehat{a_{1,2;3}}$. **(c)** $\phi^{a_{1,2;3}}$. **(d)** $\psi^{a_{1,2;3}}$. **(e)** Filter $|a_{1,3;4}|$. **(f)** $\widehat{a_{1,3;4}}$. **(g)** $\phi^{a_{1,3;4}}$. **(h)** $\psi^{a_{1,3;4}}$. **(i)** Filter $|a_{2,3;5}|$. **(j)** $\widehat{a_{2,3;5}}$. **(k)** $\phi^{a_{2,3;5}}$. **(l)** $\psi^{a_{2,3;5}}$

2.6 Biorthogonal Wavelet Filter Banks by CBC (Coset by Coset) Algorithm

Due to Proposition 2.2.3, except the very special filters in (2.2.7), a finitely supported real-valued orthogonal wavelet filter cannot have symmetry. Another way to achieve symmetry is to consider biorthogonal wavelet filter banks. In this section we discuss systematic ways of constructing biorthogonal wavelet filter banks with or without symmetry.

It is of interest to measure how close is a filter u to almost satisfy the orthogonality condition $|\widehat{u}(\xi)|^2 + |\widehat{u}(\xi + \pi)|^2 = 1$. For this purpose, we introduce the following *orthogonal wavelet filter indicator*:

$$\mathrm{Ofi}(u) := \frac{1}{2\pi}\int_{-\pi}^{\pi}\Big| |\widehat{u}(\xi)|^2 + |\widehat{u}(\xi+\pi)|^2 - 1\Big|^2 d\xi. \tag{2.6.1}$$

Clearly, u is an orthogonal wavelet filter if and only if $\mathrm{Ofi}(u) = 0$. The smaller the quantity $\mathrm{Ofi}(u)$, the closer the filter u to satisfy the orthogonality condition. For two filters u and v in $l_0(\mathbb{Z})$, we similarly define the *perpendicular filter indicator*:

$$\mathrm{Pfi}(u,v) := \frac{\int_{-\pi}^{\pi}\Big|\overline{\widehat{u}(\xi)}\widehat{v}(\xi) + \overline{\widehat{u}(\xi+\pi)}\widehat{v}(\xi+\pi)\Big|^2 d\xi}{\sqrt{\int_{-\pi}^{\pi}\Big||\widehat{u}(\xi)|^2 + |\widehat{u}(\xi+\pi)|^2\Big|^2 d\xi}\sqrt{\int_{-\pi}^{\pi}\Big||\widehat{v}(\xi)|^2 + |\widehat{v}(\xi+\pi)|^2\Big|^2 d\xi}}. \tag{2.6.2}$$

to measure how close are filters u and v to satisfy the perpendicular condition

$$\overline{\widehat{u}(\xi)}\widehat{v}(\xi) + \overline{\widehat{u}(\xi+\pi)}\widehat{v}(\xi+\pi) = 0, \qquad \forall\, \xi \in \mathbb{R}. \tag{2.6.3}$$

That is, the filters u and v are perpendicular. By $\frac{1}{2\pi}\int_{-\pi}^{\pi}|\widehat{u}(\xi)|^2 d\xi = \|u\|^2_{l_2(\mathbb{Z})} = \sum_{k\in\mathbb{Z}}|u(k)|^2$, we see that both quantities $\mathrm{Ofi}(u)$ and $\mathrm{Pfi}(u,v)$ can be easily computed. Note that $\mathrm{Pfi}(u,v) = \mathrm{Pfi}(v,u)$.

The following result lists some properties of Pfi and is left as Exercise 2.23.

Proposition 2.6.1 *Let $u, v \in l_0(\mathbb{Z})$ be two complex-valued nontrivial filters. Then*

(1) $0 \leqslant \mathrm{Pfi}(u,v) \leqslant 1$*;*
(2) $\mathrm{Pfi}(u,v) = 0$ *if and only if* (2.6.3) *holds (i.e., u and v are perpendicular);*
(3) $\mathrm{Pfi}(u,v) = 1$ *if and only if there exists $\lambda \neq 0$ such that $u(k) = \lambda v(k)$ for all $k \in \mathbb{Z}$.*

If $(\{\tilde{u},\tilde{v}\},\{u,v\})$ is a pair of biorthogonal wavelet filters, then $\mathrm{Pfi}(\tilde{u},\tilde{v}) = \mathrm{Pfi}(u,v)$.

2.6.1 Biorthogonal Wavelet Filters by Splitting Interpolatory Filters

As discussed in Sect. 2.1, $(\tilde{a}, a)$ is a pair of biorthogonal wavelet filters if and only if their correlation filter $u := \tilde{a} * a^\star$ is an interpolatory filter. Therefore, a simple way of obtaining pairs of biorthogonal wavelet filters is to split the Fourier series of interpolatory filters. In the following let us present a few examples by splitting the Fourier series of interpolatory filters a^I_{2m} in (2.1.6). More precisely, since $\widehat{a^I_{2m}}(\xi) = \cos^{2m}(\xi/2)\mathsf{P}_{m,m}(\sin^2(\xi/2))$, we factorize the polynomial $\mathsf{P}_{m,m}$ as $\mathsf{P}_{m,m}(x) = \mathsf{P}(x)\tilde{\mathsf{P}}(x)$, where P and $\tilde{\mathsf{P}}$ are polynomials with real coefficients and

$\mathsf{P}(0) = \tilde{\mathsf{P}}(0) = 1$. Then we define a pair $(\tilde{a}_m^{IS}, a_m^{IS})$ (IS=Interpolatory Splitting) of biorthogonal wavelet filters by splitting $\widehat{a_{2m}^I}$ as follows:

$$\widehat{a_m^{IS}}(\xi) = 2^{-m} e^{i\xi\lfloor m/2 \rfloor}(1 + e^{-i\xi})^m \mathsf{P}(\sin^2(\xi/2)),$$
$$\widehat{\tilde{a}_m^{IS}}(\xi) = 2^{-m} e^{i\xi\lfloor m/2 \rfloor}(1 + e^{-i\xi})^m \tilde{\mathsf{P}}(\sin^2(\xi/2)). \tag{2.6.4}$$

Their primal high-pass filter b_m^{IS} and dual high-pass filter $\tilde{b}_m^{IS}$ are defined to be

$$\widehat{b_m^{IS}}(\xi) := e^{-i\xi}\overline{\widehat{\tilde{a}_m^{IS}}(\xi + \pi)} \qquad \text{and} \qquad \widehat{\tilde{b}_m^{IS}}(\xi) := e^{-i\xi}\overline{\widehat{a_m^{IS}}(\xi + \pi)}.$$

Note that $\mathsf{S}\widehat{a_m^{IS}}(\xi) = \mathsf{S}\widehat{\tilde{a}_m^{IS}}(\xi) = e^{-i\xi \operatorname{odd}(m)}$ and $\operatorname{sr}(a_m^{IS}) = \operatorname{sr}(\tilde{a}_m^{IS}) = m$. Generally, we pick up a factorization $\mathsf{P}_{m,m} = \mathsf{P}\tilde{\mathsf{P}}$ such that $\operatorname{Var}(a_m^{IS}) + \operatorname{Var}(\tilde{a}_m^{IS})$ is the smallest.

Example 2.6.1 For $m = 2$, we have $\mathsf{P}_{2,2}(x) = 1 + 2x$. Choose $\mathsf{P}(x) = 1$ and $\tilde{\mathsf{P}}(x) = 1 + 2x$ so that $\mathsf{P}(x)\tilde{\mathsf{P}}(x) = \mathsf{P}_{2,2}(x)$. By (2.6.4) with $m = 2$, the filters a_2^{IS} and $\tilde{a}_2^{IS}$ are given as follows (also see Example 1.1.2):

$$a_2^{IS} = u_0 = \{\tfrac{1}{4}, \underline{\mathbf{\tfrac{1}{2}}}, \tfrac{1}{4}\}_{[-1,1]}, \qquad \tilde{a}_2^{IS} = \tilde{u}_0 = \{-\tfrac{1}{8}, \tfrac{1}{4}, \underline{\mathbf{\tfrac{3}{4}}}, \tfrac{1}{4}, -\tfrac{1}{8}\}_{[-2,2]}.$$

Example 2.6.2 For $m = 3$, we have $\mathsf{P}_{3,3}(x) = 1 + 3x + 6x^2$. Choose $\mathsf{P}(x) = 1$ and $\tilde{\mathsf{P}}(x) = 1 + 3x + 6x^2$ so that $\mathsf{P}(x)\tilde{\mathsf{P}}(x) = \mathsf{P}_{3,3}(x)$. By (2.6.4) with $m = 3$, we have

$$a_3^{IS} = \{\tfrac{1}{8}, \underline{\mathbf{\tfrac{3}{8}}}, \tfrac{3}{8}, \tfrac{1}{8}\}_{[-1,2]}, \qquad \tilde{a}_3^{IS} = \{\tfrac{3}{64}, -\tfrac{9}{64}, -\tfrac{7}{64}, \underline{\mathbf{\tfrac{45}{64}}}, \tfrac{45}{64}, -\tfrac{7}{64}, -\tfrac{9}{64}, \tfrac{3}{64}\}_{[-3,4]}.$$

Example 2.6.3 For $m = 4$, we have $\mathsf{P}_{4,4}(x) = 1+4x+10x^2+20x^3$. Choose $\mathsf{P}(x) = 1 + tx$ and $\tilde{\mathsf{P}}(x) = 1 + (4-t)x + (10-4t+t^2)x^2$ so that $\mathsf{P}(x)\tilde{\mathsf{P}}(x) = \mathsf{P}_{4,4}(x)$, where $t := \frac{30t_0}{t_0^2+5t_0-35} \approx 2.92069641964$ with $t_0 := (350 + 105\sqrt{15})^{1/3}$, or equivalently, t is a real number satisfying $\mathsf{P}_{4,4}(-1/t) = 0$. By (2.6.4) with $m = 4$, we have

$$a_4^{IS} = \{-\tfrac{t}{64}, \tfrac{2-t}{32}, \tfrac{16+t}{64}, \underline{\mathbf{\tfrac{6+t}{16}}}, \tfrac{16+t}{64}, \tfrac{2-t}{32}, -\tfrac{t}{64}\}_{[-3,3]},$$

$$\tilde{a}_4^{IS} = \{\tfrac{t^2-4t+10}{256}, \tfrac{t-4}{64}, \tfrac{-t^2+6t-14}{64}, \tfrac{20-t}{64}, \underline{\mathbf{\tfrac{3t^2-20t+110}{128}}}, \tfrac{20-t}{64}, \tfrac{-t^2+6t-14}{64}, \tfrac{t-4}{64}, \tfrac{t^2-4t+10}{256}\}_{[-4,4]}.$$

Numerically,

$$a_4^{IS}|_{[0,3]} = \{\underline{\mathbf{0.557543526229}}, 0.295635881557, -0.0287717631143, -0.0456358815571\}_{[0,3]},$$

$$\tilde{a}_4^{IS}|_{[0,4]} = \{\underline{\mathbf{0.602949018236}}, 0.266864118443, -0.0782232665290, -0.0168641184429, 0.0267487574108\}_{[0,4]}.$$

Example 2.6.4 For $m = 5$, $\mathsf{P}_{5,5}(x) = 1 + 5x + 15x^2 + 35x^3 + 70x^4 = \mathsf{P}(x)\tilde{\mathsf{P}}(x)$, where

$$\mathsf{P}(x) = 1 - 0.359690336966x + 7.18413037985x^2,$$
$$\tilde{\mathsf{P}}(x) = 1 + 5.35969033697x + 9.74369844348x^2.$$

Numerically,

$$a_5^{IS}|_{[1,5]} = \{0.382638624101, 0.102934062165, -0.0164457763200, 0.0168415854057, 0.0140315046482\}_{[1,5]},$$
$$\tilde{a}_5^{IS}|_{[1,5]} = \{0.636046869922, 0.0382547751527, -0.170490386362, -0.0228419197351, 0.0190306610224\}_{[1,5]}.$$

Example 2.6.5 For $m = 6$, we have $\mathsf{P}_{6,6} = 1+6x+21x^2+56x^3+126x^4+252x^5 = \mathsf{P}(x)\tilde{\mathsf{P}}(x)$, where

$$\mathsf{P}(x) = 1 + 3.99545808068x + 10.2689777690x^2,$$
$$\tilde{\mathsf{P}}(x) = (1 - 1.35941315184x + 7.29496378734x^2)(1 + 3.36395507115x).$$

Numerically,

$$a_6^{IS}|_{[0,5]} = \{\underline{\mathbf{0.588912164508}}, 0.316860629716, -0.0489054213313, -0.0768889283185, 0.00444933907740, 0.0100282986026\}_{[0,5]},$$
$$\tilde{a}_6^{IS}|_{[0,6]} = \{\underline{\mathbf{0.542524255039}}, 0.271012293703, -0.0487041816195, -0.0236704568097, 0.0334332480516, 0.00265816310684, -0.00599119395171\}_{[0,6]}.$$

Example 2.6.6 For $m = 7$, we have $\mathsf{P}_{7,7} = 1+7x+28x^2+84x^3+210x^4+462x^5+924x^6 = \mathsf{P}(x)\tilde{\mathsf{P}}(x)$, where

$$\mathsf{P}(x) = 1 + 6.45717840982x + 12.1147394540x^2,$$
$$\tilde{\mathsf{P}}(x) = (1 - 2.06115243982x + 7.30160779913x^2) \times (1 + 2.60397403000x + 10.4457443195x^2).$$

Numerically,

$$a_7^{IS}|_{[1,6]} = \{0.532816573879, 0.128570099256, -0.111275314017, -0.0611612830323, 0.00513452379038, 0.00591540012401\}_{[1,6]},$$

$$\begin{aligned}\tilde{a}_7^{IS}|_{[1,8]} = \{&0.448231035722, 0.0509354602345, -0.0645597678838,\\ &0.0435025013662, 0.0315244335194, -0.00994092300497,\\ &-0.00202033785938, 0.00232759790622\}_{[1,8]}.\end{aligned}$$

Example 2.6.7 For $m = 8$, we have $\mathsf{P}_{8,8} = 1 + 8x + 36x^2 + 120x^3 + 330x^4 + 792x^5 + 1716x^6 + 3432x^7 = \mathsf{P}(x)\tilde{\mathsf{P}}(x)$, where

$$\begin{aligned}\mathsf{P}(x) &= (1 + 3.62942129189x)(1 + 1.40307962889x + 10.4447693621x^2),\\ \tilde{\mathsf{P}}(x) &= (1 + 5.53389375876x + 12.4722589554x^2)\\ &\quad\times (1 - 2.56639467955x + 7.25881825429x^2).\end{aligned}$$

Numerically,

$$\begin{aligned}a_8^{IS}|_{[0,7]} = \{&\underline{\mathbf{0.609805812159}}, 0.329468439571, -0.0638390774673,\\ &-0.0989818698434, 0.00977042375954, 0.0218271795586,\\ &-0.000834252371686, -0.00231374928660\}_{[0,7]},\\ \tilde{a}_8^{IS}|_{[0,8]} = \{&\underline{\mathbf{0.508999816067}}, 0.272610427155, -0.0290591010958,\\ &-0.0271041968160, 0.0338756084605, 0.00499186475425,\\ &-0.0106978524861, -0.000498095093275, 0.00138143708767\}_{[0,8]}.\end{aligned}$$

Example 2.6.8 For $m = 9$, we have $\mathsf{P}_{9,9} = 1 + 9x + 45x^2 + 165x^3 + 495x^4 + 1287x^5 + 3003x^6 + 6435x^7 + 12870x^8 = \mathsf{P}(x)\tilde{\mathsf{P}}(x)$, where

$$\begin{aligned}\mathsf{P}(x) &= (1 + 7.08819039834x + 13.7396370263x^2)\\ &\quad\times (1 - 2.93981759213x + 7.19282274456x^2),\\ \tilde{\mathsf{P}}(x) &= (1 + 4.43507766900x + 12.5786582964x^2)\\ &\quad\times (1 + 0.416549524783x + 10.3530809501x^2).\end{aligned}$$

Numerically,

$$\begin{aligned}a_9^{IS}|_{[1,9]} = \{&0.398908152483, 0.125807683403, -0.0273888459628,\\ &0.00105833058506, 0.0144501406957, -0.00703008551184,\\ &-0.00699010824798, 0.000430744108599, 0.000753988446842\}_{[1,9]},\end{aligned}$$

$$\tilde{a}_9^{IS}|_{[1,9]} = \{0.594284585805, 0.0563844965548, -0.183679754506, \\ -0.0251769191496, 0.0578787495629, 0.00949007145245, \\ -0.00960718130476, -0.000567608195994, 0.00099355978078\}_{[1,9]}.$$

We present smoothness exponents, variances, frequency separation indicators, and orthogonal/perpendicular filter indicators of a_m^{IS} and $\tilde{a}_m^{IS}$ in Table 2.9 and graphs of the pair of biorthogonal refinable functions $\phi^{a_m^{IS}}, \phi^{\tilde{a}_m^{IS}}$ and the pair of biorthogonal wavelet functions $\psi^{a_m^{IS}, b_m^{IS}}, \psi^{\tilde{a}_m^{IS}, \tilde{b}_m^{IS}}$ in Figs. 2.15 and 2.16.

2.6.2 Biorthogonal Wavelet Filters by CBC Algorithm

However, splitting the Fourier series $\widehat{u}$ of an interpolatory filter u into a product $\widehat{\tilde{a}a}$ greatly restricts the choices of a (the filter for reconstruction), whose shape and smoothness exponent play an important role for a practical satisfactory appli-

Table 2.9 The smoothness exponents, squared norms, variances, frequency separation indicators, and orthogonal/perpendicular filter indicators of a_m^{IS} and $\tilde{a}_m^{IS}$ in (2.6.4). Note that $\mathrm{E}(a_m^{IS}) = \mathrm{E}(\tilde{a}_m^{IS}) = \mathrm{odd}(m)/2$, $\mathrm{Fsi}(a_m^{IS}, b_m^{IS}) = \mathrm{Fsi}(\tilde{a}_m^{IS}, \tilde{b}_m^{IS})$, and $\mathrm{Pfi}(a_m^{IS}, b_m^{IS}) = \mathrm{Pfi}(\tilde{a}_m^{IS}, \tilde{b}_m^{IS})$

m	2	3	4	5
$\mathrm{sm}(a_m^{IS})$	1.5	2.5	2.122644	2.662135
$\mathrm{sm}(\tilde{a}_m^{IS})$	0.440765	0.175132	1.409968	1.530850
$\\|a_m^{IS}\\|^2_{l_2(\mathbb{Z})}$	0.375	0.3125	0.491478	0.315518
$\\|\tilde{a}_m^{IS}\\|^2_{l_2(\mathbb{Z})}$	0.718750	1.05664	0.520217	0.871940
$\mathrm{Var}(a_m^{IS})$	0.333333	0.45	0.445416	0.441148
$\mathrm{Var}(\tilde{a}_m^{IS})$	0.347826	0.569778	0.421747	0.687719
$\mathrm{Fsi}(a_m^{IS}, b_m^{IS})$	0.215436	0.159414	0.152107	0.107347
$\mathrm{Ofi}(a_m^{IS})$	0.09375	0.210938	0.0122206	0.177758
$\mathrm{Ofi}(\tilde{a}_m^{IS})$	0.318359	2.36290	0.0192038	0.805697
$\mathrm{Pfi}(a_m^{IS}, b_m^{IS})$	0.013692	0.030800	0.0079638	0.0000386
m	6	7	8	9
$\mathrm{sm}(a_m^{IS})$	2.638762	3.268119	3.096332	3.675177
$\mathrm{sm}(\tilde{a}_m^{IS})$	2.127349	2.018217	2.7293236	2.648124
$\\|a_m^{IS}\\|^2_{l_2(\mathbb{Z})}$	0.564468	0.633216	0.617861	0.352030
$\\|\tilde{a}_m^{IS}\\|^2_{l_2(\mathbb{Z})}$	0.449413	0.421335	0.413449	0.788519
$\mathrm{Var}(a_m^{IS})$	0.588186	0.735834	0.733994	0.499588
$\mathrm{Var}(\tilde{a}_m^{IS})$	0.477656	0.611810	0.520229	0.985709
$\mathrm{Fsi}(a_m^{IS}, b_m^{IS})$	0.120248	0.121768	0.100035	0.0767112
$\mathrm{Ofi}(a_m^{IS})$	0.0279181	0.134896	0.0814951	0.136545
$\mathrm{Ofi}(\tilde{a}_m^{IS})$	0.0170676	0.0566292	0.0412137	0.590129
$\mathrm{Pfi}(a_m^{IS}, b_m^{IS})$	0.0058367	0.020734	0.0046650	0.00004888

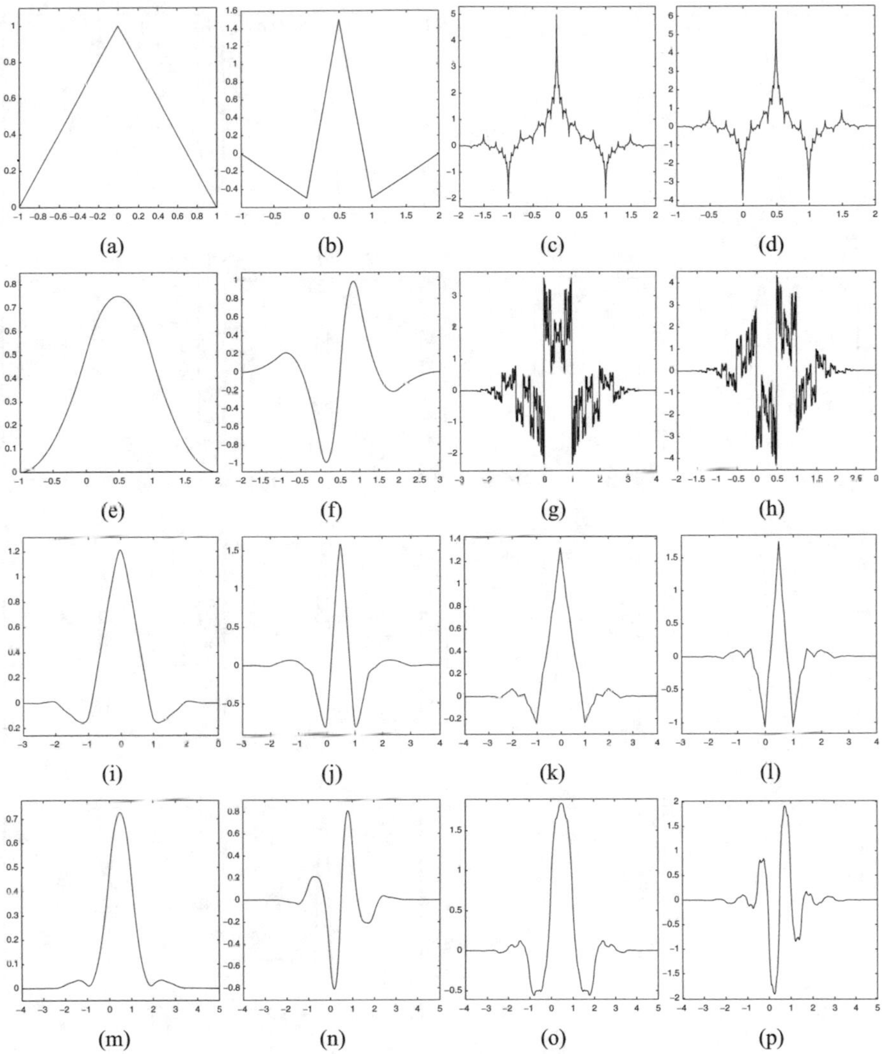

Fig. 2.15 The graphs of the standard refinable function $\phi^{a_m^{IS}}$, its associated wavelet function $\psi^{a_m^{IS},b_m^{IS}}$, the standard dual refinable function $\phi^{\tilde{a}_m^{IS}}$, and its associated dual wavelet function $\psi^{\tilde{a}_m^{IS},\tilde{b}_m^{IS}}$, where $m = 2, 3, 4, 5$. **(a)** $\phi^{a_2^{IS}}$. **(b)** $\psi^{a_2^{IS},b_2^{IS}}$. **(c)** $\phi^{\tilde{a}_2^{IS}}$. **(d)** $\psi^{\tilde{a}_2^{IS},\tilde{b}_2^{IS}}$. **(e)** $\phi^{a_3^{IS}}$. **(f)** $\psi^{a_3^{IS},b_3^{IS}}$. **(g)** $\phi^{\tilde{a}_3^{IS}}$. **(h)** $\psi^{\tilde{a}_3^{IS},\tilde{b}_3^{IS}}$. **(i)** $\phi^{a_4^{IS}}$. **(j)** $\psi^{a_4^{IS},b_4^{IS}}$. **(k)** $\phi^{\tilde{a}_4^{IS}}$. **(l)** $\psi^{\tilde{a}_4^{IS},\tilde{b}_4^{IS}}$. **(m)** $\phi^{a_5^{IS}}$. **(n)** $\psi^{a_5^{IS},b_5^{IS}}$. **(o)** $\phi^{\tilde{a}_5^{IS}}$. **(p)** $\psi^{\tilde{a}_5^{IS},\tilde{b}_5^{IS}}$

cation of a discrete wavelet transform using a biorthogonal wavelet filter bank $(\{\tilde{a}; \tilde{b}\}, \{a; b\})$. In the rest of this section we present a systematic way of constructing dual filters $\tilde{a} \in l_0(\mathbb{Z})$ of a given filter $a \in l_0(\mathbb{Z})$ such that $\tilde{a}$ has a preassigned order of sum rules. For a set Λ, we define $\#\Lambda$ to be the cardinality of the set Λ. Recall that $a^{[\gamma]} = \{a(\gamma + 2k)\}_{k \in \mathbb{Z}}$ is the γ-coset sequence of the filter a, where $\gamma \in \mathbb{Z}$.

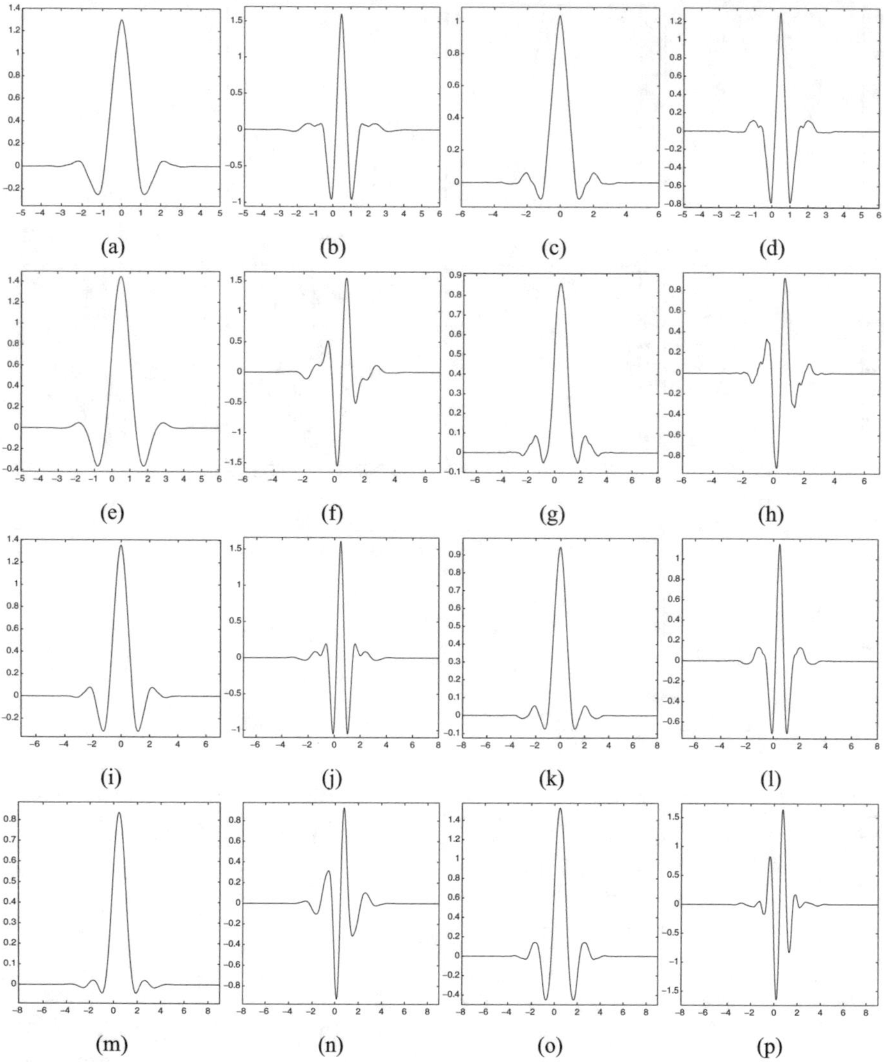

Fig. 2.16 The graphs of the standard refinable function $\phi^{a_m^{IS}}$, its associated wavelet function $\psi^{a_m^{IS},b_m^{IS}}$, the standard dual refinable function $\phi^{\tilde{a}_m^{IS}}$, and its associated dual wavelet function $\psi^{\tilde{a}_m^{IS},\tilde{b}_m^{IS}}$, where $m = 6, 7, 8, 9$. **(a)** $\phi^{a_6^{IS}}$. **(b)** $\psi^{a_6^{IS},b_6^{IS}}$. **(c)** $\phi^{\tilde{a}_6^{IS}}$. **(d)** $\psi^{\tilde{a}_6^{IS},\tilde{b}_6^{IS}}$. **(e)** $\phi^{a_7^{IS}}$. **(f)** $\psi^{a_7^{IS},b_7^{IS}}$. **(g)** $\phi^{\tilde{a}_7^{IS}}$. **(h)** $\psi^{\tilde{a}_7^{IS},\tilde{b}_7^{IS}}$. **(i)** $\phi^{a_8^{IS}}$. **(j)** $\psi^{a_8^{IS},b_8^{IS}}$. **(k)** $\phi^{\tilde{a}_8^{IS}}$. **(l)** $\psi^{\tilde{a}_8^{IS},\tilde{b}_8^{IS}}$. **(m)** $\phi^{a_9^{IS}}$. **(n)** $\psi^{a_9^{IS},b_9^{IS}}$. **(o)** $\phi^{\tilde{a}_9^{IS}}$. **(p)** $\psi^{\tilde{a}_9^{IS},\tilde{b}_9^{IS}}$

We now state the CBC (coset by coset) algorithm for constructing dual filters with sum rules of a given filter.

Algorithm 2.6.2 (CBC Algorithm) *Let $a \in l_0(\mathbb{Z})$ be a finitely supported filter such that $\widehat{a}(0) \neq 0$ and a has at least one finitely supported dual filter $\mathring{a}$. For every $m \in \mathbb{N}$,*

(S1) choose a subset Λ *of* $\mathbb{Z}$ *such that* $\#\Lambda \geqslant m$. *Then there exists a solution* $\{c_n\}_{n\in\Lambda}$ *with* $(\#\Lambda - m)$ *free parameters to the system of linear equations induced by*

$$\sum_{n\in\Lambda} c_n e^{-in\xi} = \frac{e^{i\xi/2}\widehat{\mathring{a}}(\xi/2+\pi)}{\overline{\widehat{a}(\xi/2)}} + \mathscr{O}(|\xi|^m), \quad \xi \to 0. \tag{2.6.5}$$

In particular, if $\#\Lambda = m$, *then the solution* $\{c_n\}_{n\in\Lambda}$ *to* (2.6.5) *is unique;*

(S2) construct a filter $\tilde{a}$ *coset by coset as follows:*

$$\begin{aligned} \tilde{a}(2k) &= \mathring{a}(2k) - \sum_{n\in\Lambda} c_n\, \overline{a(1+2n-2k)}, \qquad k \in \mathbb{Z}, \\ \tilde{a}(1+2k) &= \mathring{a}(1+2k) + \sum_{n\in\Lambda} c_n\, \overline{a(2n-2k)}, \qquad k \in \mathbb{Z}. \end{aligned} \tag{2.6.6}$$

Then $\tilde{a}$ *is a dual filter of the given filter* a *and* $\tilde{a}$ *has at least* m *sum rules. Moreover, if the filters* a *and* $\mathring{a}$ *have real coefficients, then so does* $\tilde{a}$.

Proof Note that the coset by coset construction in (2.6.6) is equivalent to

$$\widehat{\tilde{a}^{[0]}}(\xi) = \widehat{\mathring{a}^{[0]}}(\xi) - \theta(\xi)\overline{\widehat{a^{[1]}}(\xi)} \quad \text{and} \quad \widehat{\tilde{a}^{[1]}}(\xi) = \widehat{\mathring{a}^{[1]}}(\xi) + \theta(\xi)\overline{\widehat{a^{[0]}}(\xi)}, \tag{2.6.7}$$

where $\theta(\xi) := \sum_{n\in\Lambda} c_n e^{-in\xi}$. By the definition of coset sequences, we have $\widehat{\tilde{a}}(\xi) = \widehat{\tilde{a}^{[0]}}(2\xi) + e^{-i\xi}\widehat{\tilde{a}^{[1]}}(2\xi)$. Now it is not difficult to check that (2.6.7) is equivalent to

$$\widehat{\tilde{a}}(\xi) = \widehat{\mathring{a}}(\xi) + \theta(2\xi)e^{-i\xi}\overline{\widehat{a}(\xi+\pi)}. \tag{2.6.8}$$

Since $\mathring{a}$ is a dual filter of a, by (2.6.8), it is straightforward to check that $\tilde{a}$ is a dual filter of a. Moreover, by (2.6.5) and (2.6.8), we deduce that

$$\begin{aligned} \widehat{\tilde{a}}(\xi+\pi) &= \widehat{\mathring{a}}(\xi+\pi) - \theta(2\xi)e^{-i\xi}\overline{\widehat{a}(\xi)} \\ &= \widehat{\mathring{a}}(\xi+\pi) - \frac{e^{i\xi}\widehat{\mathring{a}}(\xi+\pi)}{\overline{\widehat{a}(\xi)}}e^{-i\xi}\overline{\widehat{a}(\xi)} + \mathscr{O}(|\xi|^m) = \mathscr{O}(|\xi|^m), \quad \xi \to 0. \end{aligned}$$

Therefore, a has at least m sum rules. □

For an interpolatory filter a (that is, $\widehat{a^{[0]}}(\xi) = 1/2$), we can take $\mathring{a} = \boldsymbol{\delta}$ in Algorithm 2.6.2. Then (2.6.6) becomes

$$\tilde{a}(2k) = \boldsymbol{\delta}(k) - \sum_{n\in\Lambda} c_n\overline{a(1+2n-2k)}, \quad k \in \mathbb{Z} \quad \text{and} \quad \tilde{a}(1+2k) = c_k, \quad k \in \Lambda.$$

That is, a dual filter $\tilde{a}$ is obtained coset by coset: the odd coset $\{\tilde{a}(1+2k)\}_{k\in\mathbb{Z}}$ is designed such that the moment conditions in (2.6.5) are satisfied, and the even coset $\{\tilde{a}(2k)\}_{k\in\mathbb{Z}}$ is uniquely determined by the biorthogonality relation in (2.0.3).

If a given filter a has symmetry, we often require that a constructed dual filter $\tilde{a}$ should have symmetry too. This can be easily achieved by the following simple fact.

Lemma 2.6.3 *Let $(\tilde{a}, a)$ be a pair of biorthogonal wavelet filters.*

(i) If a has symmetry: $\mathsf{S}\widehat{a}(\xi) = \epsilon e^{-ic\xi}$ for some $\epsilon \in \{-1, 1\}$ and $c \in \mathbb{Z}$, define

$$\widehat{\tilde{a}_{\mathsf{S}}}(\xi) := (\widehat{\tilde{a}}(\xi) + \epsilon e^{-ic\xi}\widehat{\tilde{a}}(-\xi))/2, \tag{2.6.9}$$

then $\tilde{a}_{\mathsf{S}}$ is a dual filter of a and $\tilde{a}_{\mathsf{S}}$ has the same symmetry type as a: $\mathsf{S}\widehat{\tilde{a}_{\mathsf{S}}} = \mathsf{S}\widehat{a}$.

(ii) If a has complex symmetry: $\mathbb{S}\widehat{a}(\xi) = \epsilon e^{-ic\xi}$ for $\epsilon \in \{-1, 1\}$ and $c \in \mathbb{Z}$, define

$$\widehat{\tilde{a}_{\mathbb{S}}}(\xi) := (\widehat{\tilde{a}}(\xi) + \epsilon e^{-ic\xi}\overline{\widehat{\tilde{a}}(\xi)})/2, \tag{2.6.10}$$

then $\tilde{a}_{\mathbb{S}}$ is a dual filter of a and $\tilde{a}_{\mathbb{S}}$ has the same symmetry type as a: $\mathbb{S}\widehat{\tilde{a}_{\mathbb{S}}} = \mathbb{S}\widehat{a}$.

In fact, for a pair $(\tilde{a}, a)$ of biorthogonal wavelet filters such that both $\tilde{a}$ and a have symmetry or complex symmetry (but their symmetry types may be different), it is very natural and almost surely to assume that both $\tilde{a}$ and a have the same symmetry type. In fact, note that $u := \tilde{a} * a^\star$ is interpolatory. According to Lemma 2.1.4, except the trivial case $\widehat{\tilde{a}}(\xi)\overline{\widehat{a}(\xi)} = (1 + \epsilon e^{-ic\xi})/2$ for some $\epsilon \in \{-1, 1\}$ and an odd integer c (this case cannot happen if $\operatorname{sr}(\tilde{a}) + \operatorname{sr}(a) \geqslant 2$), we must have $\mathsf{S}\widehat{\tilde{a}} = \mathsf{S}\widehat{a}$ or $\mathbb{S}\widehat{\tilde{a}} = \mathbb{S}\widehat{a}$, that is, both $\tilde{a}$ and a have the same symmetry type.

Symmetry can be also directly incorporated into the CBC algorithm. By the following fact, we only need to consider filters which are symmetric about the origin, though we can have a CBC algorithm to handle other cases directly.

Lemma 2.6.4 *$(\tilde{a}, a)$ is a pair of biorthogonal wavelet filters such that $\mathsf{S}\widehat{\tilde{a}}(\xi) = \mathsf{S}\widehat{a}(\xi) = e^{-ic\xi}$ for an odd integer c if and only if $(\tilde{a}^{new}, a^{new})$ is a pair of biorthogonal wavelet filters such that $\mathsf{S}\widehat{\tilde{a}^{new}}(\xi) = \mathsf{S}\widehat{a^{new}}(\xi) = 1$, $\operatorname{sr}(a^{new}) = \operatorname{sr}(a) - 1$, and $\operatorname{sr}(\tilde{a}^{new}) = \operatorname{sr}(\tilde{a}) + 1$, where*

$$\widehat{a^{new}}(\xi) := 2e^{i\xi(c-1)/2}(1 + e^{-i\xi})^{-1}\widehat{a}(\xi), \tag{2.6.11}$$

$$\widehat{\tilde{a}^{new}}(\xi) := 2^{-1}e^{i(c-1)/2}(1 + e^{i\xi})\widehat{\tilde{a}}(\xi). \tag{2.6.12}$$

Proof If a filter a is symmetric about the point $c/2$ for an odd integer c, then $\widehat{a}$ must contain the factor $1 + e^{-i\xi}$. Now all the claims can be directly verified. □

By Exercise 2.6, if a filter $\tilde{a}$ is symmetric about the origin, then $\operatorname{sr}(\tilde{a})$ must be an even integer.

Algorithm 2.6.5 (CBC Algorithm with Symmetry) *Let $(\mathring{a}, a)$ be a pair of biorthogonal wavelet filters such that $\widehat{a}(0) \neq 0$ and $\mathsf{S}\widehat{\mathring{a}}(\xi) = \mathsf{S}\widehat{a}(\xi) = 1$ (that is, both $\mathring{a}$ and a are symmetric about the origin). For any positive integer m,*

(S0) compute the numbers $h_{a,\mathring{a}}(j)$ by

$$h_{a,\mathring{a}}(j) := \frac{(-1)^j}{2} \frac{d^{2j}}{d\xi^{2j}} \Big[\frac{\widehat{\mathring{a}}(\xi/2+\pi)}{\widehat{a}(\xi/2)} \Big]\Big|_{\xi=0}, \qquad j = 0, \ldots, m-1; \qquad (2.6.13)$$

(S1) choose a subset Λ_m of $\mathbb{Z} \cap [0, \infty)$ with $\#\Lambda_m \geqslant m$. Then there exists a solution $\{t_n\}_{n\in\Lambda_m}$ with $(\#\Lambda_m - m)$ free parameters to the system of linear equations induced by

$$\sum_{n\in\Lambda_m} t_n(1/2+n)^{2j} = h_{a,\mathring{a}}(j), \qquad j = 0, \ldots, m-1; \qquad (2.6.14)$$

(S2) construct a filter $\tilde{a}$ coset by coset as follows: For $k \in \mathbb{Z}$,

$$\tilde{a}(2k) = \mathring{a}(2k) - \sum_{n\in\Lambda_m} t_n \left(\overline{a(1+2n-2k)} + \overline{a(-1-2n-2k)}\right),$$

$$\tilde{a}(1+2k) = \mathring{a}(1+2k) + \sum_{n\in\Lambda_m} t_n \left(\overline{a(2n-2k)} + \overline{a(-2-2n-2k)}\right).$$

Then $\tilde{a}$ is a dual filter of the given filter a such that $\tilde{a}$ is symmetric about the origin and $\tilde{a}$ has at least $2m$ sum rules. Moreover, if the filters a and $\mathring{a}$ have real coefficients, then so does $\tilde{a}$.

Proof We take $\Lambda := \{n, -1-n \ : \ n \in \Lambda\}$ and $c_n = c_{-1-n} = t_n, n \in \Lambda$ in Algorithm 2.6.2. In other words, we use $\theta(\xi) = \sum_{k\in\Lambda_m} t_n(e^{-ik\xi} + e^{i(k+1)\xi})$ in (2.6.8). It is straightforward to see that (2.6.5) with m being replaced by $2m$ is equivalent to (2.6.14) with $h_{a,\mathring{a}}(j)$ defined in (2.6.13). □

The initial dual filter $\mathring{a}$ can be obtained by long division from a given filter a and we shall address this issue in Sect. 2.8. Generally we can assume that the filter support of $\mathring{a}$ is contained inside that of a (see Proposition 2.7.2 and Theorem 2.7.4). We often take $\Lambda_m = \{0, \ldots, m-1\}$ in Algorithm 2.6.5 so that the constructed unique symmetric dual filter $\tilde{a}$ in Algorithm 2.6.5 has the shortest filter support among all symmetric dual filters having $2m$ sum rules.

Many examples can be easily obtained by the CBC algorithms. Here we provide some examples to illustrate the CBC algorithm stated in Algorithm 2.6.5.

Example 2.6.9 Let $a = a_4^I = \{-\frac{1}{32}, 0, \frac{9}{32}, \underline{\mathbf{\frac{1}{2}}}, \frac{9}{32}, 0, -\frac{1}{32}\}_{[-3,3]}$. Then $\mathring{a} = \boldsymbol{\delta}$ is a dual filter of a such that $\mathsf{S}\widehat{a}(\xi) = \mathsf{S}\widehat{\mathring{a}}(\xi) = 1$. By (2.6.13), we have

$$h_{a,\mathring{a}}(0) = \tfrac{1}{2}, \quad h_{a,\mathring{a}}(1) = 0, \quad h_{a,\mathring{a}}(2) = \tfrac{9}{64}, \quad h_{a,\mathring{a}}(3) = \tfrac{45}{128}, \quad h_{a,\mathring{a}}(4) = \tfrac{1827}{512}.$$

For $m = 1$ and $\Lambda_m = \{0\}$, then $t_0 = \frac{1}{2}$ is a solution to (2.6.14) and the dual filter $\tilde{a}$ of the given primal filter a is given by

$$\tilde{a} = \{\tfrac{1}{64}, 0, -\tfrac{1}{8}, \tfrac{1}{4}, \underline{\mathbf{\tfrac{23}{32}}}, \tfrac{1}{4}, -\tfrac{1}{8}, 0, \tfrac{1}{64}\}_{[-4,4]}.$$

For $m = 2$ and $\Lambda_m = \{0, 1\}$, then $\{t_0 = \frac{9}{16}, t_1 = -\frac{1}{16}\}$ is a solution to (2.6.14) and the dual filter $\tilde{a}$ of the given primal filter a is given by

$$\tilde{a} = \{-\tfrac{1}{512}, 0, \tfrac{9}{256}, -\tfrac{1}{32}, -\tfrac{63}{512}, \tfrac{9}{32}, \underline{\mathbf{\tfrac{87}{128}}}, \tfrac{9}{32}, -\tfrac{63}{512}, -\tfrac{1}{32}, \tfrac{9}{256}, 0, -\tfrac{1}{512}\}_{[-6,6]}.$$

Example 2.6.10 Let $a = a_{6,4} = \{-\frac{3}{256}, -\frac{1}{32}, \frac{3}{64}, \frac{9}{32}, \underline{\mathbf{\frac{55}{128}}}, \frac{9}{32}, \frac{3}{64}, -\frac{1}{32}, -\frac{3}{256}\}_{[-4,4]}$. Then $\mathring{a} = \{\frac{9}{80}, -\frac{3}{10}, \frac{3}{16}, \underline{\mathbf{1}}, \frac{3}{16}, -\frac{3}{10}, \frac{9}{80}\}_{[-3,3]}$ is a dual filter of a (see Proposition 2.7.2) such that $\mathsf{S}\widehat{a}(\xi) = \mathsf{S}\widehat{\mathring{a}}(\xi) = 1$. By (2.6.13), we have

$$h_{a,\mathring{a}}(0) = -\tfrac{1}{10}, \quad h_{a,\mathring{a}}(1) = -\tfrac{3}{5}, \quad h_{a,\mathring{a}}(2) = -\tfrac{15}{16}, \quad h_{a,\mathring{a}}(3) = -\tfrac{4389}{640}, \quad h_{a,\mathring{a}}(4) = -\tfrac{78117}{1024}.$$

For $m = 2$ and $\Lambda_m = \{0, 1\}$, then $\{t_0 = \frac{3}{16}, t_1 = -\frac{23}{80}\}$ is a solution to (2.6.14) and the dual filter $\tilde{a}$ of the given primal filter a is given by

$$\begin{aligned}\tilde{a} = \{&\tfrac{69}{20480}, -\tfrac{23}{2560}, -\tfrac{321}{20480}, \tfrac{111}{1280}, -\tfrac{91}{20480}, -\tfrac{681}{2560}, \tfrac{5463}{20480}, \underline{\mathbf{\tfrac{561}{640}}}, \tfrac{5463}{20480}, -\tfrac{681}{2560},\\ &-\tfrac{91}{20480}, \tfrac{111}{1280}, -\tfrac{321}{20480}, -\tfrac{23}{2560}, \tfrac{69}{20480}\}_{[-7,7]}.\end{aligned}$$

For $m = 3$ and $\Lambda_m = \{0, 1, 2\}$, then $\{t_0 = \frac{147}{640}, t_1 = -\frac{449}{1280}, t_2 = \frac{27}{1280}\}$ is a solution to (2.6.14) and the dual filter $\tilde{a}$ of the given primal filter a is given by

$$\begin{aligned}\tilde{a} = \{&-\tfrac{81}{327680}, \tfrac{27}{40960}, \tfrac{1671}{327680}, -\tfrac{173}{10240}, -\tfrac{165}{16384}, \tfrac{1023}{10240}, -\tfrac{2389}{81920}, -\tfrac{2643}{10240}, \tfrac{46593}{163840}, \underline{\mathbf{\tfrac{3477}{4096}}},\\ &\tfrac{46593}{163840}, -\tfrac{2643}{10240}, -\tfrac{2389}{81920}, \tfrac{1023}{10240}, -\tfrac{165}{16384}, -\tfrac{173}{10240}, \tfrac{1671}{327680}, \tfrac{27}{40960}, -\tfrac{81}{327680}\}_{[-9,9]}.\end{aligned}$$

Example 2.6.11 Let $a = a_{5,4} = \{-\frac{5}{256}, -\frac{7}{256}, \frac{35}{256}, \underline{\mathbf{\frac{105}{256}}}, \frac{105}{256}, \frac{35}{256}, -\frac{7}{256}, -\frac{5}{256}\}_{[-3,4]}$. Since $\mathsf{S}\widehat{a}(\xi) = e^{-i\xi}$, we consider $a^{new} = \{-\frac{5}{128}, -\frac{1}{64}, \frac{37}{128}, \underline{\mathbf{\frac{17}{32}}}, \frac{37}{128}, -\frac{1}{64}, -\frac{5}{128}\}_{[-3,3]}$ in (2.6.11). Then $\mathring{a} = \{\frac{5}{144}, -\frac{1}{72}, \underline{\mathbf{\frac{23}{24}}}, -\frac{1}{72}, \frac{5}{144}\}_{[-2,2]}$ is a dual filter of a^{new} (see Proposition 2.7.2) such that $\mathsf{S}\widehat{a^{new}}(\xi) = \mathsf{S}\widehat{\mathring{a}}(\xi) = 1$. By (2.6.13), we have

$$h_{a,\mathring{a}}(0) = \tfrac{19}{36}, \quad h_{a,\mathring{a}}(1) = \tfrac{41}{576}, \quad h_{a,\mathring{a}}(2) = \tfrac{1237}{4608}, \quad h_{a,\mathring{a}}(3) = \tfrac{87373}{73728}, \quad h_{a,\mathring{a}}(4) = \tfrac{3638893}{294912}.$$

For $m = 2$ and $\Lambda_m = \{0, 1\}$, then $\{t_0 = \frac{643}{1152}, t_1 = -\frac{35}{1152}\}$ is a solution to (2.6.14) with a being replaced by a^{new} and a dual filter $\tilde{a}^{new}$ is given by

$$\begin{aligned}\tilde{a}^{new} = \{&-\tfrac{175}{147456}, \tfrac{35}{73728}, \tfrac{2255}{73728}, -\tfrac{611}{24576}, -\tfrac{14161}{147456}, \tfrac{10115}{36864}, \underline{\mathbf{\tfrac{23345}{36864}}}, \tfrac{10115}{36864}, -\tfrac{14161}{147456},\\ &-\tfrac{611}{24576}, \tfrac{2255}{73728}, \tfrac{35}{73728}, -\tfrac{175}{147456}\}_{[-6,6]}.\end{aligned}$$

A dual filter $\tilde{a}$ of the given primal filter a is uniquely determined by the relation in (2.6.12) with $c = 1$ and is given by

$$\tilde{a} = \{-\tfrac{175}{73728}, \tfrac{245}{73728}, \tfrac{4265}{73728}, -\tfrac{7931}{73728}, -\tfrac{3115}{36864}, \underline{\mathbf{\tfrac{23345}{36864}}}, \tfrac{23345}{36864}, -\tfrac{3115}{36864}, -\tfrac{7931}{73728}, \tfrac{4265}{73728}, \tfrac{245}{73728}, -\tfrac{175}{73728}\}_{[-5,6]}.$$

Note that $\mathsf{S}\widehat{\tilde{a}}(\xi) = \mathsf{S}\widehat{a}(\xi) = e^{-i\xi}$ and $\operatorname{sr}(\tilde{a}) = 3$.

For $m = 3$ and $\Lambda_m = \{0, 1, 2\}$, then $\{t_0 = \frac{10883}{18432}, t_1 = -\frac{2905}{36864}, t_2 = \frac{595}{36864}\}$ is a solution to (2.6.14) with a being replaced by a^{new} and a dual filter $\tilde{a}^{new}$ is given by

$$\tilde{a}^{new} = \{\tfrac{2975}{4718592}, -\tfrac{595}{2359296}, -\tfrac{1015}{131072}, \tfrac{23135}{2359296}, \tfrac{48575}{1179648}, -\tfrac{13459}{262144}, -\tfrac{105553}{1179648}, \tfrac{688415}{2359296}, \underline{\mathbf{\tfrac{160125}{262144}}}, \tfrac{688415}{2359296}, -\tfrac{105553}{1179648}, -\tfrac{13459}{262144}, \tfrac{48575}{1179648}, \tfrac{23135}{2359296}, -\tfrac{1015}{131072}, -\tfrac{595}{2359296}, \tfrac{2975}{4718592}\}_{[-8,8]}.$$

A dual filter $\tilde{a}$ of a is uniquely determined by (2.6.12) with $c = 1$ and is given by

$$\tilde{a} = \{\tfrac{2975}{2359296}, -\tfrac{4165}{2359296}, -\tfrac{32375}{2359296}, \tfrac{26215}{786432}, \tfrac{115655}{2359296}, -\tfrac{357917}{2359296}, -\tfrac{64295}{2359296}, \underline{\mathbf{\tfrac{160125}{262144}}}, \tfrac{160125}{262144}, -\tfrac{64295}{2359296}, -\tfrac{357917}{2359296}, \tfrac{115655}{2359296}, \tfrac{26215}{786432}, -\tfrac{32375}{2359296}, -\tfrac{4165}{2359296}, \tfrac{2975}{2359296}\}_{[-7,8]}.$$

Note that $\mathsf{S}\widehat{\tilde{a}}(\xi) = \mathsf{S}\widehat{a}(\xi) = e^{-i\xi}$ and $\operatorname{sr}(\tilde{a}) = 5$.

See Table 2.10 for their smoothness exponents and statistics-related quantities. See Figs. 2.17 and 2.18 for their associated refinable and wavelet functions.

Example 2.6.12 Let $a = a_3^B(\cdot - 1) = \{\frac{1}{8}, \underline{\mathbf{\frac{3}{8}}}, \frac{3}{8}, \frac{1}{8}\}_{[-1,2]}$. All dual filters having symmetry and the filter support $[-3, 4]$ is given by

$$\tilde{a} = \{t, -3t, 3t - \tfrac{1}{4}, -t + \tfrac{3}{4}, \underline{\mathbf{-t + \tfrac{3}{4}}}, 3t - \tfrac{1}{4}, -3t, t\}_{[-3,4]}.$$

When $t = 0$, $\tilde{a}$ is the unique dual filter of a such that $\tilde{a}$ has the shortest possible filter support. In this case, we have $\operatorname{sm}(\tilde{a}) \approx -0.559235$, $\operatorname{sr}(\tilde{a}) = 1$, and $\operatorname{fsupp}(\tilde{a}) = [-1, 2]$. Hence, to find a dual filter $\tilde{a}$ with $\operatorname{sm}(\tilde{a}) > 0$, the shortest possible filter support is $[-3, 4]$. When $t = \frac{3}{64}$, $\operatorname{sm}(\tilde{a}) \approx 0.175132$ and $\operatorname{sr}(\tilde{a}) = 3$, which is the same as Example 2.6.2. By calculation, $\operatorname{sm}(\tilde{a})$ almost achieves its maximum value at $t = \frac{29}{512}$ with $\operatorname{sm}(\tilde{a}) \approx 0.330582$ and $\operatorname{sr}(\tilde{a}) = 1$.

Example 2.6.13 Let $a = a_4^B(\cdot - 2) = \{\frac{1}{16}, \frac{1}{4}, \underline{\mathbf{\frac{3}{8}}}, \frac{1}{4}, \frac{1}{16}\}_{[-2,2]}$. A dual filter having the shortest filter support is unique and is given by $\tilde{a} = \{-\frac{1}{2}, \underline{\mathbf{2}}, -\frac{1}{2}\}_{[-1,1]}$ with $\operatorname{sm}(\tilde{a}) \approx -1.5592345$ and $\operatorname{sr}(\tilde{a}) = 0$. All dual filters $\tilde{a}$ having symmetry, $\operatorname{sr}(\tilde{a}) > 0$, and the filter support $[-5, 5]$ is given by

$$\tilde{a} = \{t, -4t, 5t + \tfrac{3}{32}, -\tfrac{3}{8}, -6t + \tfrac{5}{32}, \underline{\mathbf{8t + \tfrac{5}{4}}}, -6t + \tfrac{5}{32}, -\tfrac{3}{8}, 5t + \tfrac{3}{32}, -4t, t\}_{[-5,5]}.$$

Table 2.10 The smoothness exponents, squared norms, variances, frequency separation indicators, and orthogonal/perpendicular filter indicators of a and $\tilde{a}$ in Examples 2.6.9, 2.6.10, and 2.6.11. Note that $\mathrm{Pfi}(a, b) = \mathrm{Pfi}(\tilde{a}, \tilde{b})$ and $\mathrm{Fsi}(a, b) = \mathrm{Fsi}(\tilde{a}, \tilde{b})$

CBC Algorithm	sr	$\|\cdot\|^2_{l_2(\mathbb{Z})}$	Var	sm	Ofi	$\mathrm{Pfi}(a, b)$	$\mathrm{Fsi}(a, b)$
$a = a^I_4$	4	0.410156	0.428571	2.440765	0.0650482		
$\tilde{a}$ with $m = 1$	2	0.673340	0.382886	0.593223	0.249223	0.0339170	0.188788
$\tilde{a}$ with $m = 2$	4	0.654892	0.514178	1.179370	0.218591	0.0101275	0.141911
$a = a_{6,4}$	6	0.349457	0.565889	4.098191	0.176972		
$\tilde{a}$ with $m = 2$	4	1.06795	0.906996	0.349587	3.29133	0.0463589	0.135131
$\tilde{a}$ with $m = 3$	6	1.03806	1.01883	0.649332	3.08117	0.0320641	0.112711
$a = a_{5,4}$	5	0.376099	0.496998	3.259609	0.122794		
$\tilde{a}$ with $m = 2$	3	0.846221	0.543673	0.346291	1.07161	0.0566240	0.183283
$\tilde{a}$ with $m = 3$	5	0.801151	0.740546	1.042980	0.904897	0.0193244	0.121141

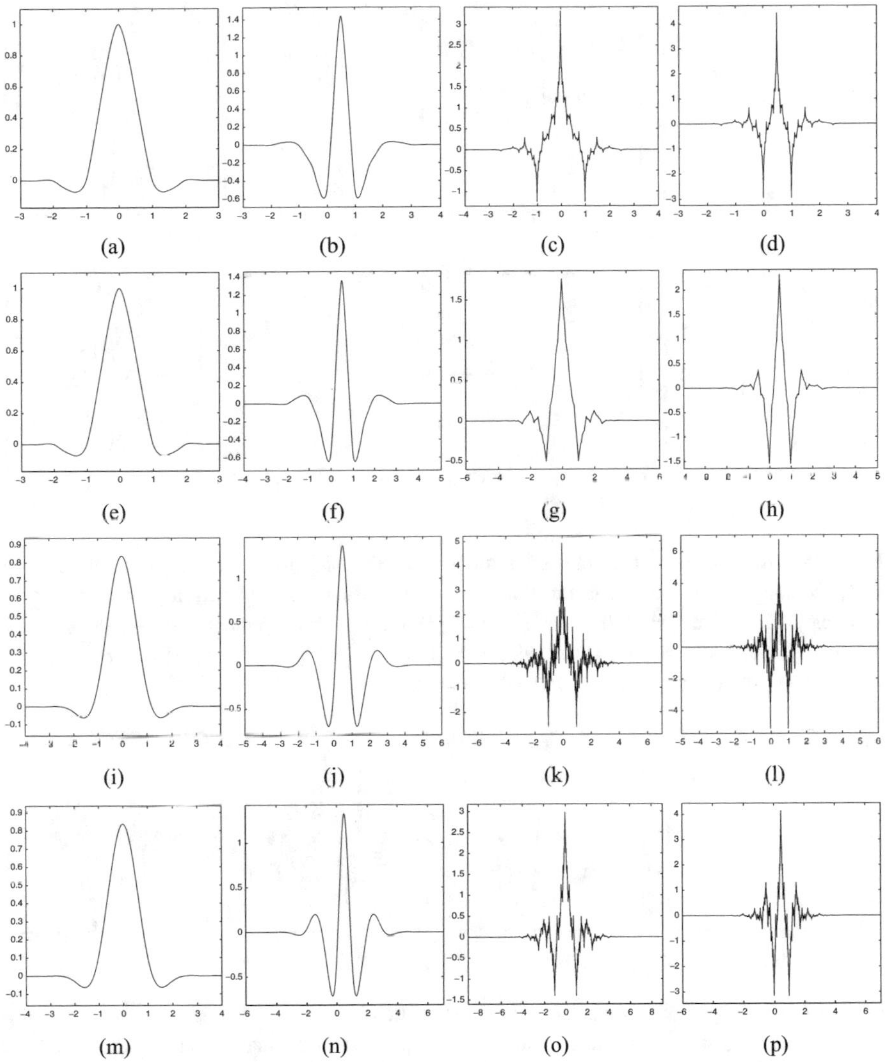

Fig. 2.17 The graphs of the standard refinable function ϕ^a, its associated wavelet function $\psi^{a,b}$, the standard dual refinable function $\phi^{\tilde{a}}$, and its associated dual wavelet function $\psi^{\tilde{a},\tilde{b}}$ in Examples 2.6.9 and 2.6.10. The pair $(\{\phi^a; \psi^{a,b}\}, \{\phi^{\tilde{a}}; \psi^{\tilde{a},\tilde{b}}\})$ forms a biorthogonal wavelet. **(a)** ϕ^a with $a = a_4^I$. **(b)** $\psi^{a,b}$ with $m = 1$. **(c)** $\phi^{\tilde{a}}$ with $m = 1$. **(d)** $\psi^{\tilde{a},\tilde{b}}$ with $m = 1$. **(e)** ϕ^a with $a = a_4^I$. **(f)** $\psi^{a,b}$ with $m = 2$. **(g)** $\phi^{\tilde{a}}$ with $m = 2$. **(h)** $\psi^{\tilde{a},\tilde{b}}$ with $m = 2$. **(i)** ϕ^a with $a = a_{6,4}$. **(j)** $\psi^{a,b}$ with $m = 2$. **(k)** $\phi^{\tilde{a}}$ with $m = 2$. **(l)** $\psi^{\tilde{a},\tilde{b}}$ with $m = 2$. **(m)** ϕ^a with $a = a_{6,4}$. **(n)** $\psi^{a,b}$ with $m = 3$. **(o)** $\phi^{\tilde{a}}$ with $m = 3$. **(p)** $\psi^{\tilde{a},\tilde{b}}$ with $m = 3$

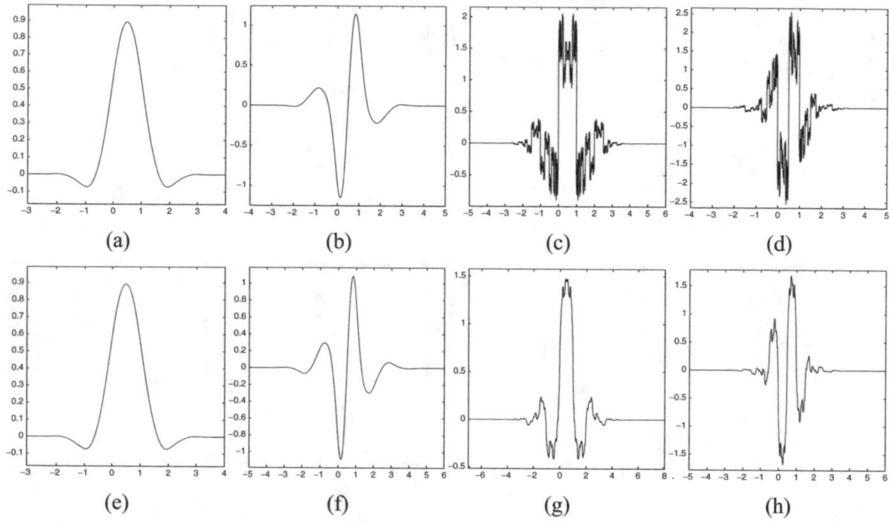

Fig. 2.18 The graphs of the standard refinable function ϕ^a, its associated wavelet function $\psi^{a,b}$, the standard dual refinable function $\phi^{\tilde{a}}$, and its associated dual wavelet function $\psi^{\tilde{a},\tilde{b}}$ in Example 2.6.11. The pair $(\{\phi^a; \psi^{a,b}\}, \{\phi^{\tilde{a}}; \psi^{\tilde{a},\tilde{b}}\})$ forms a biorthogonal wavelet. (**a**) ϕ^a with $a = a_{5,4}$. (**b**) $\psi^{a,b}$ with $m = 2$. (**c**) $\phi^{\tilde{a}}$ with $m = 2$. (**d**) $\psi^{\tilde{a},\tilde{b}}$ with $m = 2$. (**e**) ϕ^a with $a = a_{5,4}$. (**f**) $\psi^{a,b}$ with $m = 3$. (**g**) $\phi^{\tilde{a}}$ with $m = 3$. (**h**) $\psi^{\tilde{a},\tilde{b}}$ with $m = 3$

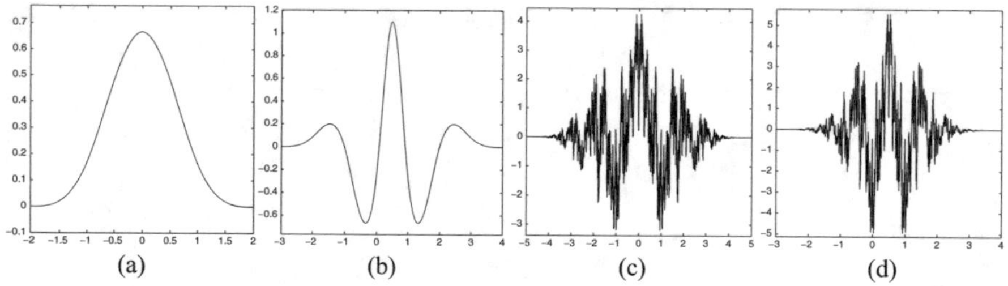

Fig. 2.19 The graphs of the standard refinable function ϕ^a, and its associated wavelet function $\psi^{a,b}$, the standard dual refinable function $\phi^{\tilde{a}}$, and its associated dual wavelet function $\psi^{\tilde{a},\tilde{b}}$ in Example 2.6.13 with $t = -\frac{1}{32}$. The pair $(\{\phi^a; \psi^{a,b}\}, \{\phi^{\tilde{a}}; \psi^{\tilde{a},\tilde{b}}\})$ forms a biorthogonal wavelet. (**a**) ϕ^a. (**b**) $\psi^{a,b}$. (**c**) $\phi^{\tilde{a}}$. (**d**) $\psi^{\tilde{a},\tilde{b}}$

When $t = 0$, $\tilde{a}$ is the unique dual filter of a such that $\tilde{a}$ has symmetry, the shortest possible filter support, and $\mathrm{sr}(\tilde{a}) > 0$. In this case, we have $\mathrm{sm}(\tilde{a}) \approx -0.824868$, $\mathrm{sr}(\tilde{a}) = 2$, and $\mathrm{fsupp}(\tilde{a}) = [-3, 3]$. Hence, to find a dual filter $\tilde{a}$ with $\mathrm{sm}(\tilde{a}) > 0$, the shortest possible filter support is $[-5, 5]$. When $t = -\frac{5}{256}$, $\mathrm{sm}(\tilde{a}) \approx -0.206866$ and $\mathrm{sr}(\tilde{a}) = 4$. By calculation, $\mathrm{sm}(\tilde{a})$ almost achieves its maximum value at $t = -\frac{1}{32}$ with $\mathrm{sm}(\tilde{a}) \approx 0.129764$ and $\mathrm{sr}(\tilde{a}) = 2$. See Fig. 2.19 for their associated refinable and wavelet functions.

2.6.3 Particular Biorthogonal Wavelet Filters Having Short Supports

For practical applications, it is also interesting to consider all possible pairs $(\tilde{a}, a)$ of biorthogonal wavelet filters such that the filter supports of $\tilde{a}$ and a are preassigned. We present here three parameterized pairs $(\tilde{a}, a)$ of biorthogonal wavelet filters with short support and symmetry such that $\mathrm{Ofi}(\tilde{a}) + \mathrm{Ofi}(a)$ is nearly the smallest. We shall label them as $\tilde{a}_{[m:n]}$ and $a_{[m:n]}$ referring to $\mathrm{len}(a_{[m:n]}) = m$ and $\mathrm{len}(\tilde{a}_{[m:n]}) = n$.

Example 2.6.14 Let $(a_{[6:4]}, \tilde{a}_{[6:4]})$ be a pair of biorthogonal wavelet filters given by

$$\widehat{a_{[6:4]}}(\xi) = 2^{-2}e^{i\xi}(1 + e^{-i\xi})^2[1 - 2t_1 - 2t_2 + t_1(e^{i\xi} + e^{-i\xi}) + t_2(e^{i2\xi} + e^{-i2\xi})],$$

$$\widehat{\tilde{a}_{[6:4]}}(\xi) = 2^{-2}e^{i\xi}(1 + e^{-i\xi})^2[1 - 2t + t(e^{i\xi} + e^{-i\xi})],$$

where t is a free parameter and $t_1 = \frac{(2t+1)^2}{4t-2}$, $t_2 = -\frac{2t^2+t}{4t-2}$. When $t = -27/128$, the quantity $\mathrm{Ofi}(a_{[6:4]}) + \mathrm{Ofi}(\tilde{a}_{[6:4]})$ is nearly the smallest and

$$a_{[6:4]} = \{-\tfrac{999}{93184}, -\tfrac{37}{728}, \tfrac{24295}{93184}, \underline{\mathbf{\tfrac{219}{364}}}, \tfrac{24295}{93184}, -\tfrac{37}{728}, -\tfrac{999}{93184}\}_{[-3,3]},$$

$$\tilde{a}_{[6:4]} = \{-\tfrac{27}{512}, \tfrac{1}{4}, \underline{\mathbf{\tfrac{155}{256}}}, \tfrac{1}{4}, -\tfrac{27}{512}\}_{[-2,2]}.$$

Example 2.6.15 Let $(a_{[7:7]}, \tilde{a}_{[7:7]})$ be a pair of biorthogonal wavelet filters given by

$$\widehat{a_{[7:7]}}(\xi) = 2^{-3}e^{i\xi}(1 + e^{-i\xi})^3(1 - 2t_3 - 2t_2 + t_3(e^{i\xi} + e^{-i\xi}) + t(e^{-i2\xi} + e^{i2\xi})),$$

$$\widehat{\tilde{a}_{[7:7]}}(\xi) = 2^{-3}e^{i\xi}(1 + e^{-i\xi})^3(1 - 2t_1 - 2t_2 + t_1(e^{i\xi} + e^{-i\xi}) + t_2(e^{-i2\xi} + e^{i2\xi})),$$

where t is a free parameter, t_3 is a real root of

$$8t_3^3 + (80t + 6)t_3^2 + (256t^2 + 20t + 3)t_3 + 256t^3 = 0,$$

$t_2 = t_3^2 + (8t + \frac{3}{4})t_3 + 16t^2 + 2t + \frac{3}{8}$, and $t_1 = -t_3 - 4t_2 - 4t - \frac{3}{4}$. The quantity $\mathrm{Ofi}(a_{[7:7]}) + \mathrm{Ofi}(\tilde{a}_{[7:7]})$ is nearly the smallest when $t = -1/8$. When $t = -1/8$, we have $t_3 \approx 0.120972063761$ is a real root of $8t_3^3 - 4t_3^2 + \frac{9}{2}t_3 - \frac{1}{2}$, and $t_1 = -4t_3^2 - \frac{7}{4} \approx -1.80853696084$, $t_2 = t_3^2 - \frac{1}{4}t_3 + \frac{3}{8} \approx 0.359391224271$. The filters are numerically given by

$$\begin{aligned} a_{[7:7]} = \{&-0.015625, -0.0317534920299, 0.124496507970, \underline{\mathbf{0.422881984060}},\\ &0.422881984060, 0.124496507970, -0.03175349203, -0.015625\}_{[-3,4]},\end{aligned}$$

$$\begin{aligned} \tilde{a}_{[7:7]} = \{&0.0449239030338, -0.091295411004, -0.056143217072,\\ &\underline{\mathbf{0.602514725044}}, 0.602514725045, -0.056143217072,\\ &-0.091295411004, 0.0449239030338\}_{[-3,4]}.\end{aligned}$$

Example 2.6.16 Let $(a_{[6:8]}, \tilde{a}_{[6:8]})$ be a pair of biorthogonal wavelet filters given by

$$\widehat{a_{[6:8]}}(\xi) = 2^{-4}e^{i2\xi}(1+e^{-i\xi})^4(1-2t+t(e^{i\xi}+e^{-i\xi})),$$
$$\widehat{\tilde{a}_{[6:8]}}(\xi) = 2^{-2}e^{i\xi}(1+e^{-i\xi})^2(1-2t_1-2t_2-2t_3+t_1(e^{i\xi}+e^{-i\xi}) + t_2(e^{i2\xi}+e^{-i2\xi})+t_3(e^{i3\xi}+e^{-i3\xi})),$$

where t is a free parameter and

$$t_1 = \frac{56t^3+58t^2+41t+18}{16t-8}, \quad t_2 = -\frac{32t^3+32t^2+18t+3}{16t-8}, \quad t_3 = \frac{8t^3+6t^2+3t}{16t-8}.$$

When $t = -13/16$, the quantity $\operatorname{Ofi}(a_{[6:8]}) + \operatorname{Ofi}(\tilde{a}_{[6:8]})$ is nearly the smallest and the filters are given by

$$a_{[6:8]} = \{-\tfrac{13}{256}, -\tfrac{5}{128}, \tfrac{77}{256}, \underline{\mathbf{\tfrac{37}{64}}}, \tfrac{77}{256}, -\tfrac{5}{128}, -\tfrac{13}{256}\}_{[-3,3]},$$
$$\tilde{a}_{[6:8]} = \{\tfrac{1417}{43008}, -\tfrac{545}{21504}, -\tfrac{11}{168}, \tfrac{5921}{21504}, \underline{\mathbf{\tfrac{12151}{21504}}}, \tfrac{5921}{21504}, -\tfrac{11}{168}, -\tfrac{545}{21504}, \tfrac{1417}{43008}\}_{[-4,4]}.$$

See Table 2.11 for the smoothness exponents, variances, frequency separation indicators, and orthogonal wavelet filter indicators of the pairs of biorthogonal wavelet filters $a_{[m;n]}$ and $\tilde{a}_{[m;n]}$. See Fig. 2.20 for their associated refinable and wavelet functions.

2.7 Polyphase Matrix and Chain Structure of Biorthogonal Wavelet Filters

Though it is very convenient to use Fourier series (also called symbols) of filters for analyzing and understanding various mathematical properties of wavelets and framelets, representing filters using Laurent polynomials has the main advantage of simplicity and convenience for constructing wavelet or framelet filter banks. Consequently, in the literature of engineering, filters are often represented using Laurent polynomials (called the z-transform in the engineering literature), instead of the Fourier series in (1.1.1), for the purpose of filter design.

In this section we discuss z-transform and Laurent polynomial representation for filters and we shall translate several important definitions using the language of Laurent polynomials instead of Fourier series. Then we address chain structure and polyphase matrix of biorthogonal wavelet filter banks.

For a sequence $u = \{u(k)\}_{k\in\mathbb{Z}} \in l_0(\mathbb{Z})$, its *z-transform* is a Laurent polynomial defined by

$$\mathsf{u}(z) := \sum_{k\in\mathbb{Z}} u(k)z^k, \qquad z \in \mathbb{C}\backslash\{0\}. \tag{2.7.1}$$

Table 2.11 The smoothness exponents, variances, frequency separation indicators, and orthogonal wavelet filter indicators of a and $\tilde{a}$ in Example 2.6.14 with $t = -27/128$, Example 2.6.15 with $t = -1/8$, and Example 2.6.16 with $t = -13/16$. Note that $\mathrm{Pfi}(\tilde{a}, \tilde{b}) = \mathrm{Pfi}(a, b)$ and $\mathrm{Fsi}(\tilde{a}, \tilde{b}) = \mathrm{Fsi}(a, b)$

	sr	$\|\cdot\|^2_{l_2(\mathbb{Z})}$	Var	sm	Ofi	$\mathrm{Pfi}(a, b)$	$\mathrm{Fsi}(a, b)$
$a_{[6:4]}$	2	0.5033327	0.315270	0.967876	0.000128808	0.000844269	0.215553
$\tilde{a}_{[6:4]}$	2	0.497154	0.296182	1.074442	0.000109112	0.000844269	0.215553
$a_{[7:7]}$	3	0.391166	0.454411	2.539580	0.0954151	0.0448151	0.189071
$\tilde{a}_{[7:7]}$	3	0.753063	0.463882	0.480803	0.556954	0.0448151	0.189071
$a_{[6:8]}$	4	0.523376	0.457726	2.023656	0.010191	0.00258807	0.137094
$\tilde{a}_{[6:8]}$	2	0.482949	0.480847	1.533249	0.00952426	0.00258807	0.137094

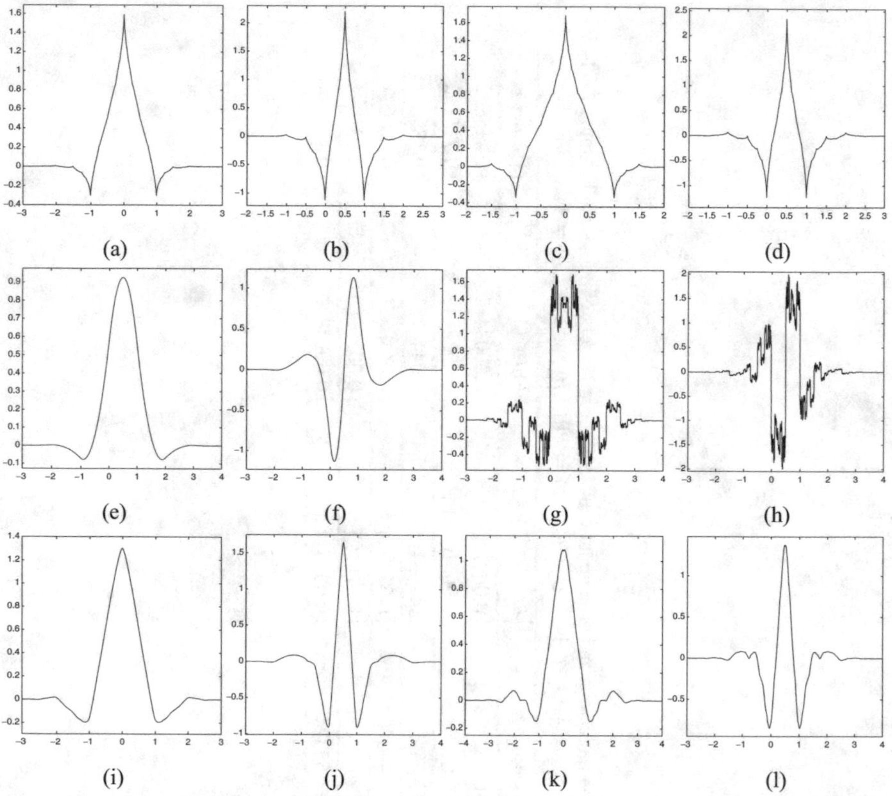

Fig. 2.20 The graphs of the standard refinable function $\phi^{a_{[m:n]}}$, its associated wavelet function $\psi^{a_{[m:n]},b_{[m:n]}}$, the standard dual refinable function $\phi^{\tilde{a}_{[m:n]}}$, and its associated dual wavelet function $\psi^{\tilde{a}_{[m:n]},\tilde{b}_{[m:n]}}$, where $(m,n) = (6,4), (7,7), (6,8)$. **(a)** $\phi^{a_{[6:4]}}$. **(b)** $\psi^{a_{[6:4]},b_{[6:4]}}$. **(c)** $\phi^{\tilde{a}_{[6:4]}}$. **(d)** $\psi^{\tilde{a}_{[6:4]},\tilde{b}_{[4:6]}}$. **(e)** $\phi^{a_{[7:7]}}$. **(f)** $\psi^{a_{[7:7]},b_{[7:7]}}$. **(g)** $\phi^{\tilde{a}_{[7:7]}}$. **(h)** $\psi^{\tilde{a}_{[7:7]},\tilde{b}_{[7:7]}}$. **(i)** $\phi^{a_{[6:8]}}$. **(j)** $\psi^{a_{[6:8]},b_{[6:8]}}$. **(k)** $\phi^{\tilde{a}_{[6:8]}}$. **(l)** $\psi^{\tilde{a}_{[6:8]},\tilde{b}_{[6:8]}}$

Let $u : \mathbb{Z} \to \mathbb{C}^{r\times s}$ be a sequence of $r \times s$ matrices. We define $u^\star$ to be its associated adjoint sequence as follows:

$$u^\star(k) := \overline{u(-k)}^{\mathsf{T}}, \qquad k \in \mathbb{Z}. \tag{2.7.2}$$

In terms of Fourier series, we have $\widehat{u^\star}(\xi) = \overline{\widehat{u}(\xi)}^{\mathsf{T}}$ and $\widehat{u}(\xi) = \mathsf{u}(e^{-i\xi})$. Using the notation of Laurent polynomials, the definition in (2.7.2) becomes

$$\mathsf{u}^\star(z) := [\mathsf{u}(z)]^\star := \sum_{k\in\mathbb{Z}} \overline{u(k)}^{\mathsf{T}} z^{-k}, \qquad z \in \mathbb{C}\backslash\{0\}. \tag{2.7.3}$$

The following fact can be directly checked.

Lemma 2.7.1 *Let* $u = \{u(k)\}_{k\in\mathbb{Z}} : \mathbb{Z} \to \mathbb{C}^{r\times s}$. *For any* $c \in \mathbb{C}\backslash\{0\}$ *and* $m \in \mathbb{Z}$, $[\mathsf{u}(cz^m)]^\star = \mathsf{u}^\star(\bar{c}^{-1}z^m)$. *Moreover, if* u *is real-valued, then* $\mathbf{u}^\star(z) = [\mathbf{u}(z^{-1})]^\mathsf{T}$.

Using notation of Laurent polynomials, we can rewrite the definitions of sum rules, vanishing moments, linear-phase moments, and symmetry. It is quite trivial to see that a filter u has m sum rules if and only if $(z+1)^m \mid \mathsf{u}(z)$, that is, the Laurent polynomial $\mathsf{u}(z)$ contains the polynomial factor $(z+1)^m$, or equivalently, $\mathsf{u}(-1) = \mathsf{u}'(-1) = \cdots = \mathsf{u}^{(m-1)}(-1) = 0$, or in short, $\mathsf{u}(z) = \mathscr{O}(|z+1|^m)$ as $z \to -1$. Similarly, a filter u has m vanishing moments if and only if $(z-1)^m \mid \mathsf{u}(z)$, that is, $\mathsf{u}(z)$ contains the polynomial $(z-1)^m$, or equivalently, $\mathsf{u}(1) = \mathsf{u}'(1) = \cdots = \mathsf{u}^{(m-1)}(1) = 0$, or in short, $\mathsf{u}(z) = \mathscr{O}(|z-1|^m)$ as $z \to 1$.

For a filter u having symmetry $u(k) = \epsilon u(c-k)$ for all $k \in \mathbb{Z}$, where $\epsilon \in \{-1, 1\}$ and $c \in \mathbb{Z}$, it is also trivial to see that

$$\mathsf{Su}(z) := \frac{\mathsf{u}(z)}{\mathsf{u}(z^{-1})} = \epsilon z^c. \tag{2.7.4}$$

For a filter u having complex symmetry $u(k) = \epsilon\overline{u(c-k)}$ for all $k \in \mathbb{Z}$, where $\epsilon \in \{-1, 1\}$ and $c \in \mathbb{Z}$, it is also trivial to see that

$$\mathbb{S}\mathsf{u}(z) := \frac{\mathsf{u}(z)}{\mathsf{u}^\star(z)} = \epsilon z^c. \tag{2.7.5}$$

Recall that a filter u has m linear-phase moments with phase $c \in \mathbb{R}$ if $\widehat{u}(\xi) = e^{-ic\xi} + \mathscr{O}(|\xi|^m)$ as $\xi \to 0$. Note that $e^{-i\xi}|_{\xi=0} = 1$. Using $z = e^{-i\xi}$, we see that u has m linear-phase moments with phase c if and only if

$$\mathsf{u}(z) = z^c + \mathscr{O}(|z-1|^m), \qquad z \to 1. \tag{2.7.6}$$

Note that z^c with $c \in \mathbb{R}$ is well defined for z in a neighborhood of 1. Moreover, it is trivial to see that (2.7.6) is equivalent to

$$\mathsf{u}(1) = 1 \quad \text{and} \quad \mathsf{u}^{(j)}(1) = c(c-1)\cdots(c-j+1), \qquad j = 1, \ldots, m-1.$$

Also note that the subdivision operator $\mathcal{S}_u$ in (1.1.4) and transition operator $\mathcal{T}_u$ in (1.1.5) can be equivalently expressed using Laurent polynomials as follows:

$$[\mathcal{S}_\mathsf{u}\mathsf{v}](z) := 2\mathsf{v}(z^2)\mathsf{u}(z), \qquad [\mathcal{T}_\mathsf{u}\mathsf{v}](z^2) = \mathsf{v}(z)\mathsf{u}^\star(z) + \mathsf{v}(-z)\mathsf{u}^\star(-z). \tag{2.7.7}$$

Using Laurent polynomials, the biorthogonality condition in (2.0.3) becomes

$$\tilde{\mathsf{a}}(z)\mathsf{a}^\star(z) + \tilde{\mathsf{a}}(-z)\mathsf{a}^\star(-z) = 1, \qquad z \in \mathbb{C}\backslash\{0\}. \tag{2.7.8}$$

It is trivial to observe that $\tilde{a}$ is a dual filter of a (i.e., (2.7.8) holds) if and only if $\mathcal{T}_a\tilde{a} = \boldsymbol{\delta}$, i.e., $\mathcal{T}_\mathsf{a}\tilde{\mathsf{a}} = 1$.

The following result gives a necessary and sufficient condition for the existence of a finitely supported dual filter of a given finitely supported filter.

Proposition 2.7.2 *A filter $a \in l_0(\mathbb{Z})$ has at least one finitely supported dual filter if and only if* $\mathsf{a}(z)$ *and* $\mathsf{a}(-z)$ *have no common zeros in* $\mathbb{C}\backslash\{0\}$: $\gcd(\mathsf{a}(z), \mathsf{a}(-z)) = 1$.

Proof Suppose that a has a finitely supported dual filter $\tilde{a}$. Then (2.7.8) holds, from which we see that $\mathsf{a}(z)$ and $\mathsf{a}(-z)$ have no common zeros in $\mathbb{C}\backslash\{0\}$, i.e., $\gcd(\mathsf{a}(z), \mathsf{a}(-z)) = 1$.

Conversely, suppose that $\mathsf{a}(z)$ and $\mathsf{a}(-z)$ have no common zeros in $\mathbb{C}\backslash\{0\}$. Assume that the filter support of a is $[m, n]$. Then $z^{-m}\mathsf{a}(z)$ and $(-z)^{-m}\mathsf{a}(-z)$ are polynomials having no common zeros in $\mathbb{C}$. Hence, by Euclidean division algorithm (see Sect. 3.1.1 on greatest common divisors and Euclidean division algorithm), there exist polynomials $\mathsf{u}(z)$ and $\mathsf{v}(z)$ such that $\max(\deg(\mathsf{u}), \deg(\mathsf{v})) < n - m$ and

$$z^{-m}\mathsf{a}(z)\mathsf{u}(z) + (-z)^{-m}\mathsf{a}(-z)\mathsf{v}(z) = 1.$$

Define $\tilde{\mathsf{a}}(z) := z^m(\mathsf{u}^\star(z) + \mathsf{v}^\star(-z))/2$. Then $\tilde{a}$ is a dual filter of a and $\mathrm{len}(\tilde{a}) < \mathrm{len}(a)$. This completes the proof. □

The following result can be directly verified.

Proposition 2.7.3 *Let $\tilde{a}, a \in l_0(\mathbb{Z})$ such that $\tilde{a}$ is a dual filter of a. Define*

$$\tilde{\mathsf{b}}(z) := z\mathsf{a}^\star(-z) \quad \text{and} \quad \mathsf{b}(z) := z\tilde{\mathsf{a}}^\star(-z). \tag{2.7.9}$$

If $\tilde{a}^{new} \in l_0(\mathbb{Z})$ is also a dual filter of a, then

$$\tilde{\mathsf{a}}^{new}(z) = \tilde{\mathsf{a}}(z) + \Theta(z^2)\tilde{\mathsf{b}}(z), \tag{2.7.10}$$

where

$$\Theta(z^2) := (\tilde{\mathsf{a}}^{new}(z) - \tilde{\mathsf{a}}(z))\mathsf{b}^\star(z) + (\tilde{\mathsf{a}}^{new}(-z) - \tilde{\mathsf{a}}(-z))\mathsf{b}^\star(-z).$$

Conversely, if $\tilde{\mathsf{a}}^{new}$ is defined in (2.7.10) *for an arbitrary Laurent polynomial Θ, then $\tilde{a}^{new}$ is a dual filter of a.*

Namely, if we have one dual filter $\tilde{a}$ of a given filter a, then all dual filters $\tilde{a}^{new}$ of a must take the form in (2.7.10). Suppose that $\tilde{a}^{new}$ is defined in (2.7.10). Define

$$\mathsf{b}^{new}(z) := z[\tilde{\mathsf{a}}^{new}(-z)]^\star = \mathsf{b}(z) - \Theta^\star(z^2)\mathsf{a}(z).$$

In the following we study the relations between the discrete wavelet transform using the biorthogonal wavelet filter bank $(\{\tilde{a}; \tilde{b}\}, \{a; b\})$ and the discrete wavelet transform using the new biorthogonal wavelet filter bank $(\{\tilde{a}^{new}; \tilde{b}\}, \{a; b^{new}\})$. By the definition of the transition operator in (1.1.3) and the subdivision operator in (1.1.2), from (2.7.9) and (2.7.10), it is trivial to see that

$$\mathcal{T}_{\tilde{a}^{new}} v = \mathcal{T}_{\tilde{a}} v + \Theta^\star * \mathcal{T}_{\tilde{b}} v \quad \text{and} \quad \mathcal{S}_{b^{new}} w = \mathcal{S}_b w - \mathcal{S}_a(\Theta^\star * w). \tag{2.7.11}$$

The implementation of the discrete wavelet transform using (2.7.11) is called the *lifting scheme* in the literature of engineering. Generally, the filters in the original biorthogonal wavelet filter bank $(\{\tilde{a}; \tilde{b}\}, \{a; b\})$ have shorter support than those in the new one. Therefore, the use of the lifting scheme in (2.7.11) can improve the efficiency of the discrete wavelet transform which employs the new biorthogonal wavelet filter bank $(\{\tilde{a}^{new}; \tilde{b}\}, \{a; b^{new}\})$.

We now discuss the chain structure of a biorthogonal wavelet filter bank.

Theorem 2.7.4 *Let $a \in l_0(\mathbb{Z})$ be a filter such that a has at least one finitely supported dual filter $\tilde{a}$. Assume that the filter support of a is $[m, n]$ for some integers m and n. If $\operatorname{len}(a) = n - m > 0$, then there exists a unique dual filter $\mathring{a}$ of a such that $\operatorname{fsupp}(\mathring{a}) \subseteq [m, n-1]$. More precisely,*

(i) if $\operatorname{len}(a)$ is a positive even integer, then there exists a unique dual filter $\mathring{a}$ of a such that $\operatorname{fsupp}(\mathring{a}) \subseteq [m+1, n-1]$;
(ii) if $\operatorname{len}(a)$ is an odd integer greater than 1, then there exists a unique dual filter $\mathring{a}$ of a such that $\operatorname{fsupp}(\mathring{a}) \subseteq [m, n-2]$. Similarly, if $\operatorname{len}(a)$ is an odd integer greater than 1, then there exists a unique dual filter $\breve{a}$ of a such that $\operatorname{fsupp}(\breve{a}) \subseteq [m+2, n]$;
(iii) if $\operatorname{len}(a) = 1$, then both $\mathring{a} = \frac{1}{2a(m)}\delta(\cdot - m)$ and $\mathring{a} = \frac{1}{2a(n)}\delta(\cdot - n)$ (that is, $\widehat{\mathring{a}}(\xi) = \frac{1}{2a(m)}e^{-im\xi}$ or $\frac{1}{2a(n)}e^{-in\xi}$) are all the possible dual filters of a such that $\operatorname{len}(\mathring{a}) = 0$.

Moreover, if a is real-valued, then $\mathring{a}$ is also real-valued.

Proof Assume that the filter support of $\tilde{a}$ is $[\tilde{m}, \tilde{n}]$. Since $\tilde{a}$ is a dual filter of a, $\tilde{a} * a^\star$ is an interpolatory filter and $\operatorname{fsupp}(\tilde{a} * a^\star) = [\tilde{m} - n, \tilde{n} - m]$. Thus, either $\tilde{m} = n$ or $\tilde{m} - n$ must be an odd integer. Similarly, either $\tilde{n} = m$ or $\tilde{n} - m$ must be odd.

We first prove that if $\operatorname{len}(a) = n - m \geqslant 1$, then there is a unique dual filter $\mathring{a}$ of a such that $\operatorname{fsupp}(\mathring{a}) \subseteq [m, n-1]$. Let $\tilde{b}$ be the high-pass filter as defined in (2.7.9). Then it is obvious that

$$\mathsf{a}^\star(z)\tilde{\mathsf{b}}(z) + \mathsf{a}^\star(-z)\tilde{\mathsf{b}}(-z) = 0. \tag{2.7.12}$$

Note that $\operatorname{len}(a) = n - m \geqslant 1$ implies $m < n$. Suppose that $\tilde{n} \geqslant n$. Then we must have $\tilde{n} \neq m$; otherwise, $\tilde{n} = m < n \leqslant \tilde{n}$, a contradiction. Therefore, by what has been proved at the beginning of the proof, $\tilde{n} - m$ must be an odd integer. We consider another filter

$$\tilde{\mathsf{a}}_1(z) := \tilde{\mathsf{a}}(z) + \frac{\tilde{a}(\tilde{n})}{a(m)}(-1)^m z^{\tilde{n}+m-1}\tilde{\mathsf{b}}(z).$$

Since $\tilde{a}$ is a dual filter of a and $\tilde{n} + m - 1$ is an even integer, by Proposition 2.7.3, it follows from (2.7.12) that $\tilde{a}_1$ is also a dual filter of a such that $\operatorname{fsupp}(\tilde{a}_1) \subseteq [\min(\tilde{m}, \tilde{n} - n + m), \tilde{n} - 2] \subseteq [\min(\tilde{m}, m), \tilde{n} - 2]$ by $\tilde{n} \geqslant n$. If necessary, repeating this procedure finitely many times, we now can assume that a has a dual filter $\tilde{a}$ such that $\operatorname{fsupp}(\tilde{a}) = [\tilde{m}, \tilde{n}]$ with $\tilde{m} \leqslant m$ and $\tilde{n} \leqslant n-1$.

Suppose that $\tilde{m} < m$. Then we must have $\tilde{m} \neq n$; otherwise, $n = \tilde{m} \leqslant \tilde{n} \leqslant n-1$, a contradiction. Therefore, $\tilde{m}-n$ must be an odd integer. We take a similar procedure and define

$$\tilde{\mathsf{a}}_2(z) := \tilde{\mathsf{a}}(z) + \frac{\tilde{a}(\tilde{m})}{\overline{a(n)}}(-1)^n z^{n+\tilde{m}-1}\tilde{\mathsf{b}}(z).$$

Since $n + \tilde{m} - 1$ is an even integer, we see that $\tilde{a}_2$ is a dual filter of a such that $\operatorname{fsupp}(\tilde{a}_2) \subseteq [\tilde{m}-2, \max(\tilde{n}, n+\tilde{m}-m)] \subseteq [\tilde{m}-2, n-1]$ by $\tilde{m} < m$ and $\tilde{n} \leqslant n-1$. If necessary, we repeat this procedure finitely many times. In other words, we proved that there is a dual filter $\mathring{a}$ of a such that $\operatorname{fsupp}(\mathring{a}) \subseteq [m, n-1]$.

We now prove that such a filter $\mathring{a}$ is unique. Otherwise, suppose that $\mathring{a}_1$ is another such dual filter. Consider $\breve{a} := \mathring{a} - \mathring{a}_1$. Then $\mathsf{a}^\star(z)\breve{\mathsf{a}}(z) - \mathsf{a}^\star(-z)\breve{\mathsf{a}}(-z) = 0$ and $\operatorname{fsupp}(\breve{a}) \subseteq [m, n-1]$. By Proposition 2.7.2, $\mathsf{a}^\star(z)$ and $\mathsf{a}^\star(-z)$ have no common zeros in $\mathbb{C}\backslash\{0\}$. Therefore, $\mathsf{a}^\star(z) \mid \breve{\mathsf{a}}(-z)$. Since $\operatorname{len}(a) = n - m > \operatorname{len}(\breve{a})$, we must have $\breve{\mathsf{a}} = 0$. That is, $\mathring{a} = \mathring{a}_1$. Hence, such a dual filter $\mathring{a}$ is unique.

For case (i), since $\operatorname{len}(a) = n - m$ is an even integer and since $\mathring{a}$ is a dual filter of a, we must have $\mathring{a}(m) = 0$. Therefore, by $\operatorname{fsupp}(\mathring{a}) \subseteq [m, n-1]$, we deduce that $\operatorname{fsupp}(\mathring{a}) \subseteq [m+1, n-1]$.

For case (ii), since $\operatorname{len}(a) = n - m$ is an odd integer and since $\mathring{a}$ is a dual filter of a, we must have $\mathring{a}(n-1) = 0$. Therefore, by $\operatorname{fsupp}(\mathring{a}) \subseteq [m, n-1]$, we conclude that $\operatorname{fsupp}(\mathring{a}) \subseteq [m, n-2]$.

The second case in (ii) can be proved similarly. Case (iii) is straightforward. □

For filters having symmetry, we have the following result.

Corollary 2.7.5 *Let $a \in l_0(\mathbb{Z})$ be a finitely supported filter such that a has at least one finitely supported dual filter and a has symmetry $\mathsf{Sa}(z) = \epsilon z^c$ for some $\epsilon \in \{-1, 1\}$ and $c \in \mathbb{Z}$. Suppose that $\operatorname{len}(a) \geqslant 1$.*

(i) If c is an even integer, then there exists a unique dual filter $\mathring{a}$ of a such that $\operatorname{fsupp}(\mathring{a})$ is strictly contained inside $\operatorname{fsupp}(a)$, $\operatorname{len}(\mathring{a}) \leqslant \operatorname{len}(a) - 2$, and $\mathring{a}$ has the same symmetry type as the filter a: $\mathsf{S}\mathring{\mathsf{a}}(z) = \epsilon z^c$;

(ii) If c is an odd integer, then there exists a dual filter $\mathring{a}$ of a such that $\operatorname{fsupp}(\mathring{a}) \subseteq \operatorname{fsupp}(a)$ and $\mathring{a}$ has the same symmetry type as the filter a: $\mathsf{S}\mathring{\mathsf{a}}(z) = \epsilon z^c$.

The same conclusion holds if S is replaced by the complex symmetry operator $\mathbb{S}$.

Proof Suppose that $\operatorname{fsupp}(a) = [c-n, n]$ for some integer n. If c is an even integer, then $\operatorname{len}(a) = 2n - c$ is an even integer. By Theorem 2.7.4, there exists a dual filter $\tilde{a}$ of a such that $\operatorname{fsupp}(\tilde{a}) \subseteq [c-n+1, n-1]$. Define a symmetrized filter $\tilde{a}_{\mathsf{S}}$ as in (2.6.9). Then $\tilde{a}_{\mathsf{S}}$ is a dual filter of a, $\operatorname{fsupp}(\tilde{a}_{\mathsf{S}}) \subseteq [c-n+1, n-1]$, and $\mathsf{S}\tilde{\mathsf{a}}_{\mathsf{S}}(z) = \epsilon z^c$. That is, we can take $\mathring{a} = \tilde{a}_{\mathsf{S}}$ and all the claims in (i) hold. The claim in (ii) follows directly from Theorem 2.7.4 and the above argument. For complex symmetry, we simply take a symmetrized filter $\tilde{a}_{\mathbb{S}}$ as in (2.6.10) and all the claims hold. □

However, as shown by the following two examples, if c is an odd integer in Corollary 2.7.5, then both $\text{len}(\mathring{a}) < \text{len}(a)$ and $\text{len}(\mathring{a}) = \text{len}(a)$ indeed can happen. So, the filter support of a symmetric dual filter cannot be reduced for the later case.

Example 2.7.1 Let $\mathsf{a}(z) = z^{-1}(1+z)^3(-z^{-1}+4-z)/16$ and $\tilde{\mathsf{a}}(z) = (1+z)/2$. Then it is trivial to check that both a and $\tilde{a}$ are symmetric about the point $1/2$ and $\tilde{a}$ is a dual filter of a (in fact, $\tilde{\mathsf{a}}(z)\mathsf{a}^\star(z) = \mathsf{a}_4^I(z)$, where a_4^I is the interpolatory filter with filter support $[-3,3]$). Obviously, $1 = \text{len}(\tilde{a}) < \text{len}(a) = 5$.

Example 2.7.2 A simple example is $\mathsf{a}(z) = (1+z)/2$. If $\tilde{a}$ is a dual filter of a and $\tilde{a}$ is symmetric about the point $1/2$, then we must have $[0,1] \subseteq \text{fsupp}(\tilde{a})$. A little bit more complicated example is as follows. Let $\mathsf{a}(z) = z^{-1}(1+z)^3/8$. Then $\mathsf{Sa}(z) = z$. By direct calculation, if $\tilde{a}$ is a dual filter of a such that $\tilde{a}$ is symmetric about the point $1/2$ and $\text{fsupp}(\tilde{a}) \subseteq [-1,2]$, then $\tilde{a}$ is unique and is given by $\tilde{\mathsf{a}}(z) = -\frac{1}{4}z^{-1} + \frac{3}{4} + \frac{3}{4}z - \frac{1}{4}z^2$. In other words, a does not have a dual filter whose filter support is strictly smaller than $[-1,2]$ and which is symmetric about the point $1/2$.

However, this shortcoming for the symmetry center $\frac{c}{2}$ being a half integer in Corollary 2.7.5 is not serious and can be easily overcome by the simple fact in Lemma 2.6.4. Let (a_1, a_2) be a pair of biorthogonal wavelet filters. By Theorem 2.7.4, we have a sequence $a_3, \ldots, a_r$ of filters:

$$a_1 \longrightarrow a_2 \longrightarrow a_3 \longrightarrow \cdots \longrightarrow a_r \tag{2.7.13}$$

such that

(1) $\text{len}(a_{r-1}) = \text{len}(a_r) = 0$;
(2) a_j is a dual filter of a_{j-1} and $\text{len}(a_j) \leqslant \text{len}(a_{j-1}) - 2$ for all $j = 3, \ldots, r-1$.

Then we have a sequence of pairs of biorthogonal wavelet filters:

$$(a_1, a_2) \longrightarrow (a_2, a_3) \longrightarrow \cdots \longrightarrow (a_{r-1}, a_r). \tag{2.7.14}$$

We call the sequence in (2.7.14) satisfying items (1) and (2) the *chain structure* of the pair (a_1, a_2). If both a_1 and a_2 have symmetry with the same symmetry type such that the symmetry center $\frac{c}{2}$ is an integer, by Corollary 2.7.5, we can further require that all $a_1, a_2, \ldots, a_r$ in the chain structure of (a_1, a_2) have the same symmetry type.

Define $\mathsf{b}_j(z) := z\mathsf{a}_j^\star(-z)$, $j = 1, \ldots, r$. Consequently, we have a sequence of biorthogonal wavelet filter banks:

$$(\{a_1, b_2\}, \{a_2, b_1\}) \longrightarrow (\{a_2, b_3\}, \{a_3, b_2\}) \longrightarrow \cdots \longrightarrow (\{a_{r-1}, b_r\}, \{a_r, b_{r-1}\}).$$

By Proposition 2.7.3, we have the following decomposition:

$$\begin{bmatrix} \mathsf{a}_j(z) & \mathsf{b}_{j+1}(z) \\ \mathsf{a}_j(-z) & \mathsf{b}_{j+1}(-z) \end{bmatrix} = \begin{bmatrix} \mathsf{a}_{j+2}(z) & \mathsf{b}_{j+1}(z) \\ \mathsf{a}_{j+2}(-z) & \mathsf{b}_{j+1}(-z) \end{bmatrix} \begin{bmatrix} 1 & 0 \\ \Theta_{j,j+1,j+2}(z^2) & 1 \end{bmatrix}, \tag{2.7.15}$$

where

$$\Theta_{j,j+1,j+2}(z^2) := (\mathsf{a}_j(z) - \mathsf{a}_{j+2}(z))\mathsf{b}^{\star}_{j+2}(z) + (\mathsf{a}_j(-z) - \mathsf{a}_{j+2}(-z))\mathsf{b}^{\star}_{j+2}(-z). \tag{2.7.16}$$

Similarly,

$$\begin{bmatrix} \mathsf{a}_{j+1}(z) & \mathsf{b}_j(z) \\ \mathsf{a}_{j+1}(-z) & \mathsf{b}_j(-z) \end{bmatrix} = \begin{bmatrix} \mathsf{a}_{j+1}(z) & \mathsf{b}_{j+2}(z) \\ \mathsf{a}_{j+1}(-z) & \mathsf{b}_{j+2}(-z) \end{bmatrix} \begin{bmatrix} 1 & -\Theta^{\star}_{j,j+1,j+2}(z^2) \\ 0 & 1 \end{bmatrix}. \tag{2.7.17}$$

By the definition of $\Theta_{j,j+1,j+2}$, we have the following relation:

$$\mathsf{a}_j(z) = \mathsf{a}_{j+2}(z) + \Theta_{j,j+1,j+2}(z^2)\mathsf{b}_{j+2}(z),$$
$$\mathsf{b}_j(z) = \mathsf{b}_{j+2}(z) - \Theta^{\star}_{j,j+1,j+2}(z^2)\mathsf{a}_{j+1}(z).$$

Therefore, the following matrices

$$\begin{bmatrix} \mathsf{a}_1(z) & \mathsf{b}_2(z) \\ \mathsf{a}_1(-z) & \mathsf{b}_2(-z) \end{bmatrix} \quad \text{and} \quad \begin{bmatrix} \mathsf{a}_2(z) & \mathsf{b}_1(z) \\ \mathsf{a}_2(-z) & \mathsf{b}_1(-z) \end{bmatrix}$$

can be decomposed into a product of the elementary matrices, which are the last matrices in (2.7.15) and (2.7.17), in an alternate way.

According to Corollary 2.7.5, if both $\mathsf{Sa}_1(z) = \mathsf{Sa}_2(z) = \epsilon z^c$ for some $\epsilon \in \{-1, 1\}$ and $c \in 2\mathbb{Z}$, then we can further require that all the filters a_j in the chain (2.7.13) have the symmetry type $\mathsf{Sa}_j(z) = \epsilon z^c$ for all $j = 1, \dots, r$. Consequently, $\Theta_{j,j+1,j+2}$ in (2.7.16) also have the symmetry type $\mathsf{S}\Theta_{j,j+1,j+2}(z) = z^2$. That is, $\Theta_{j,j+1,j+2}$ is symmetric about the point 1.

We complete this section by rewriting the perfect reconstruction condition in (1.1.15) in terms of Laurent polynomials and polyphase matrices.

Using the notation of Laurent polynomials, we see that the perfect reconstruction condition in (1.1.15) becomes

$$\begin{bmatrix} \tilde{\mathsf{u}}_0(z) & \cdots & \tilde{\mathsf{u}}_s(z) \\ \tilde{\mathsf{u}}_0(-z) & \cdots & \tilde{\mathsf{u}}_s(-z) \end{bmatrix} \begin{bmatrix} \mathsf{u}_0(z) & \cdots & \mathsf{u}_s(z) \\ \mathsf{u}_0(-z) & \cdots & \mathsf{u}_s(-z) \end{bmatrix}^{\star} = I_2, \qquad z \in \mathbb{C}\backslash\{0\}. \tag{2.7.18}$$

Recall that $u^{[\gamma]} = \{u(\gamma + 2k)\}_{k\in\mathbb{Z}}$ is the γ-coset sequence of u. It is trivial to see that $\mathsf{u}(z) = \mathsf{u}^{[0]}(z^2) + z\mathsf{u}^{[1]}(z^2)$. Consequently, we have

$$\begin{bmatrix} \mathsf{u}_0(z) & \cdots & \mathsf{u}_s(z) \\ \mathsf{u}_0(-z) & \cdots & \mathsf{u}_s(-z) \end{bmatrix} = \begin{bmatrix} 1 & z \\ 1 & -z \end{bmatrix} \begin{bmatrix} \mathsf{u}_0^{[0]}(z^2) & \cdots & \mathsf{u}_s^{[0]}(z^2) \\ \mathsf{u}_0^{[1]}(z^2) & \cdots & \mathsf{u}_s^{[1]}(z^2) \end{bmatrix}.$$

Note that the inverse matrix of $\begin{bmatrix} 1 & z \\ 1 & -z \end{bmatrix}$ is $\frac{1}{2}\begin{bmatrix} 1 & 1 \\ z^{-1} & z^{-1} \end{bmatrix}$. We see that (2.7.18) becomes

$$\begin{bmatrix} \tilde{u}_0^{[0]}(z^2) \cdots \tilde{u}_s^{[0]}(z^2) \\ \tilde{u}_0^{[1]}(z^2) \cdots \tilde{u}_s^{[1]}(z^2) \end{bmatrix} \begin{bmatrix} u_0^{[0]}(z^2) \cdots u_s^{[0]}(z^2) \\ u_0^{[1]}(z^2) \cdots u_s^{[1]}(z^2) \end{bmatrix}^\star = \frac{1}{4}\begin{bmatrix} 1 & 1 \\ z^{-1} & -z^{-1} \end{bmatrix}\begin{bmatrix} 1 & 1 \\ z^{-1} & -z^{-1} \end{bmatrix}^\star = \frac{1}{2}I_2.$$

Therefore, the perfect reconstruction condition in (2.7.18) is equivalent to

$$\begin{bmatrix} \tilde{u}_0^{[0]}(z) \cdots \tilde{u}_s^{[0]}(z) \\ \tilde{u}_0^{[1]}(z) \cdots \tilde{u}_s^{[1]}(z) \end{bmatrix} \begin{bmatrix} u_0^{[0]}(z) \cdots u_s^{[0]}(z) \\ u_0^{[1]}(z) \cdots u_s^{[1]}(z) \end{bmatrix}^\star = \frac{1}{2}I_2,$$

where the first matrix on the left-hand side is called the *polyphase matrix* of the filter bank $\{\tilde{u}_0, \dots, \tilde{u}_s\}$.

Similarly, the perfect reconstruction condition in (1.4.3) and (1.4.4) for an OEP-based dual framelet filter bank becomes

$$\begin{aligned} &\begin{bmatrix} \tilde{a}(z) & \tilde{b}_1(z) & \cdots & \tilde{b}_s(z) \\ \tilde{a}(-z) & \tilde{b}_1(-z) & \cdots & \tilde{b}_s(-z) \end{bmatrix} \begin{bmatrix} \Theta(z^2) & 0 \\ 0 & 1 \end{bmatrix} \begin{bmatrix} a(z) & b_1(z) & \cdots & b_s(z) \\ a(-z) & b_1(-z) & \cdots & b_s(-z) \end{bmatrix}^\star \\ &\qquad = \begin{bmatrix} \Theta(z) & 0 \\ 0 & \Theta(-z) \end{bmatrix}. \end{aligned} \tag{2.7.19}$$

In terms of polyphase matrices, the equation (2.7.19) can be equivalently rewritten as

$$\begin{aligned} &\begin{bmatrix} \tilde{a}^{[0]}(z) \, \tilde{b}_1^{[0]}(z) \cdots \tilde{b}_s^{[0]}(z) \\ \tilde{a}^{[1]}(z) \, \tilde{b}_1^{[1]}(z) \cdots \tilde{b}_s^{[1]}(z) \end{bmatrix} \begin{bmatrix} \Theta(z) & 0 \\ 0 & 1 \end{bmatrix} \begin{bmatrix} a^{[0]}(z) \, b_1^{[0]}(z) \cdots b_s^{[0]}(z) \\ a^{[1]}(z) \, b_1^{[1]}(z) \cdots b_s^{[1]}(z) \end{bmatrix}^\star \\ &\qquad = \frac{1}{2}\begin{bmatrix} \Theta^{[0]}(z) & z\Theta^{[1]}(z) \\ \Theta^{[1]}(z) & \Theta^{[0]}(z) \end{bmatrix}. \end{aligned} \tag{2.7.20}$$

Note that $\Theta^{[1]}(z) = z^{-1}\Theta^{[-1]}(z)$ in the matrix on the right-hand side of (2.7.20).

2.8 Exercises

2.1. Let $P_{m,n}$ be the polynomial defined in (2.1.4). Prove that $P'_{m,n}(x) = \frac{m}{1-x}(P_{m,n}(x) - P_{m,n}(1)x^{n-1})$.

2.2. For any orthogonal wavelet filter $a \in l_0(\mathbb{Z})$, prove that $\|a\|^2_{l_2(\mathbb{Z})} = 1/2$.

2.3. Prove that $a \in l_0(\mathbb{Z})$ is an orthogonal wavelet filter if and only if (2.0.12) holds.

2.4. Suppose that $u = \{u(k)\}_{k\in\mathbb{Z}} \in l_0(\mathbb{Z})$ has symmetry or complex symmetry with symmetry center c_u. If $\sum_{k\in\mathbb{Z}} u(k) = 1$, prove that $\mathrm{E}(u) = c_u$.

2.5. Let $a \in l_0(\mathbb{Z})$ be an interpolatory filter such that a is also an orthogonal wavelet filter. Prove that a must take the form $\widehat{a}(\xi) = (1 + \lambda e^{-i(2k+1)\xi})/2$ for some $k \in \mathbb{Z}$ and $\lambda \in \mathbb{T}$.

2.6. Suppose that a filter a has symmetry: $\mathsf{S}\widehat{a}(\xi) = \epsilon e^{-ic\xi}$ for some $\epsilon \in \{-1, 1\}$ and $c \in \mathbb{Z}$. Prove that $\mathrm{sr}(a)$ must be an even integer if c is even; and $\mathrm{sr}(a)$ must be an odd integer if c is odd. Present examples to show that the same conclusion fails if symmetry is replaced by complex symmetry.

2.7. For every $n \in \mathbb{N}$, prove that $T_n(x) := \cos(n \arccos x)$ is a polynomial of degree n with rational coefficients (called the Chebychev polynomial of degree n).

2.8. Let $\boldsymbol{\Theta}$ be a 2π-periodic trigonometric polynomial with complex coefficients such that $\boldsymbol{\Theta}(-\xi) = \boldsymbol{\Theta}(\xi)$. Then there exists a polynomial P such that $\boldsymbol{\Theta}(\xi) = \mathsf{P}(\sin^2(\xi/2))$. If $\boldsymbol{\Theta}$ has real coefficients, then so does P. Hint: use Exercise 2.7.

2.9. Let $\boldsymbol{\Theta}$ be a 2π-periodic trigonometric polynomial. Show that $\overline{\boldsymbol{\Theta}(\xi)} = \boldsymbol{\Theta}(\xi)$ if and only if there exist polynomials P, Q with real coefficients such that

$$\boldsymbol{\Theta}(\xi) = \mathsf{P}(\sin^2(\xi/2)) + \sin(\xi)\mathsf{Q}(\sin^2(\xi/2)).$$

2.10. Let m be a nonnegative integer. Show that a polynomial Q satisfies $(1-x)^m\mathsf{Q}(x)+x^m\mathsf{Q}(1-x) = 0$ for all $x \in \mathbb{R}$ if and only if $\mathsf{Q}(x) = x^m(\frac{1}{2}-x)\mathsf{R}((\frac{1}{2}-x)^2)$, where R is some polynomial. Therefore, a polynomial P satisfies (2.1.8) if and only if P must take the form $\mathsf{P}(x) = \mathsf{P}_{m,m}(x) + x^m(\frac{1}{2} - x)\mathsf{R}((\frac{1}{2} - x)^2)$, where $\mathsf{P}_{m,m}$ is the polynomial defined in (2.1.6) and R is any polynomial.

2.11. For any nonnegative integer m, show that a finitely supported filter u is an interpolatory complex-valued filter, $u^\star = u$, and has $2m$ sum rules if and only if there are polynomials P and Q with real coefficients such that (2.1.8) holds and

$$\widehat{u}(\xi) = \cos^{2m}(\xi/2)\mathsf{P}(\sin^2(\xi/2)) + \sin^{2m+1}(\xi)\mathsf{Q}(\sin^2(\xi)).$$

2.12. Let $u, v \in l_0(\mathbb{Z})$ be real-valued filters such that $\widehat{u}(0) = 1$ and $u(1-k) = u(k)$ for all $k \in \mathbb{Z}$. If $|\widehat{u}(\xi)|^2 + |\widehat{v}(\xi)|^2 = 1$ for all $\xi \in \mathbb{R}$ and v has symmetry, prove that

$$\widehat{u}(\xi) = 2^{-1}(e^{ij\xi} + e^{-i(j+1)\xi}), \quad \widehat{v}(\xi) = 2^{-1}e^{-ik\xi}(e^{ij\xi} - e^{-i(j+1)\xi}),$$

for some $j, k \in \mathbb{Z}$.

2.13. Show that $\mathrm{fsupp}(a^I_{2m}) = [1-2m, 2m-1]$ and $\mathrm{sr}(a^I_{2m}) = 2m$.

2.14. Prove that $\mathrm{fsupp}(a_{2m,2n}) = [1-m-n, m+n-1]$, $\mathrm{sr}(a_{2m,2n}) = 2m$, and $\mathrm{lpm}(a_{2m,2n}) = 2n$; Show that $\mathrm{fsupp}(a_{2m-1,2n}) = [2-m-n, m+n-1]$, $\mathrm{sr}(a_{2m-1,2n}) = 2m-1$, and $\mathrm{lpm}(a_{2m-1,2n}) = 2n$.

2.15. Prove that $\widehat{a^I_{2m}}(\xi) = 1 - \frac{\int_0^\xi (\sin t)^{2m-1} dt}{\int_0^\pi (\sin t)^{2m-1} dt}$ for every $m \in \mathbb{N}$. *Hint:* use the uniqueness result in Theorem 2.1.3.

2.16. Prove that

$$\lim_{m\to\infty} \widehat{a^I_{2m}}(\xi) = \begin{cases} 1, & \text{if } \xi \in (-\pi/2, \pi/2); \\ \frac{1}{2}, & \text{if } \xi = \pm\pi/2; \\ 0, & \text{if } \xi \in [-\pi, \pi]\backslash[-\pi/2, \pi/2]. \end{cases}$$

The above limiting filter a defined on the right-hand side in the frequency domain is called the Shannon orthogonal wavelet filter, which is the refinement filter for the sinc function $\operatorname{sinc}(x) := \frac{\sin(\pi x)}{\pi x}$ with $\operatorname{sinc}(0) := 1$.

2.17. Suppose that a filter $u \in l_0(\mathbb{Z})$ has symmetry: $\mathsf{S}\widehat{u}(\xi) = \epsilon e^{-ic\xi}$ for some $\epsilon \in \{-1, 1\}$ and some $c \in \mathbb{Z}$ (that is, $u(c-k) = \epsilon u(k)$ for all $k \in \mathbb{Z}$). Prove (i) $\mathsf{S}\widehat{u^{[r]}}(\xi) = \mathsf{S}\widehat{u^{[i]}}(\xi) = \mathsf{S}\widehat{u}(\xi)$ and (ii) the identity in (2.5.2), where $u^{[r]}$ and $u^{[i]}$ are the real and imaginary parts of the sequence u defined in (2.5.1).

2.18. Suppose that a filter $u \in l_0(\mathbb{Z})$ has complex symmetry: $\mathbb{S}\overline{\widehat{u}(\xi)} = \epsilon e^{-ic\xi}$ for some $\epsilon \in \{-1, 1\}$ and some $c \in \mathbb{Z}$ (that is, $u(c-k) = \epsilon \overline{u(k)}$ for all $k \in \mathbb{Z}$). Prove that (i) $\mathbb{S}\widehat{u^{[r]}}(\xi) = \mathbb{S}\widehat{u^{[i]}}(\xi) = \mathbb{S}\widehat{u}(\xi)$ and (ii) (2.5.2) holds if and only if one of $u^{[r]}$ and $u^{[i]}$ must be identically zero.

2.19. For a complex-valued filter u, show that

$$\operatorname{sr}(u) = \min(\operatorname{sr}(u^{[r]}), \operatorname{sr}(u^{[i]})) \quad \text{and} \quad \operatorname{lpm}(u) = \min(\operatorname{lpm}(u^{[r]}), \operatorname{lpm}(u^{[i]})).$$

2.20. Let $a \in l_0(\mathbb{Z})$ with $\widehat{a}(0) = 1$. Define $\widehat{\phi}(\xi) := \prod_{j=1}^\infty \widehat{a}(2^{-j}\xi)$. We say that ϕ has m linear-phase moments with phase $c_\phi \in \mathbb{R}$ if $\widehat{\phi}(\xi) = e^{-ic_\phi\xi} + \mathscr{O}(|\xi|^m)$ as $\xi \to 0$. We define $\operatorname{lpm}(\phi) := m$ with m being the largest nonnegative integer. Prove that

$$\operatorname{lpm}(\phi) = \operatorname{lpm}(a), \quad \operatorname{vm}(\phi^{[i]}) = \operatorname{vm}(a^{[i]}),$$

$$\min(\operatorname{lpm}(\phi^{[r]}), 2\operatorname{vm}(a^{[i]})) = \min(\operatorname{lpm}(a^{[r]}), 2\operatorname{vm}(a^{[i]}),$$

where the real-valued functions $\phi^{[r]}$ and $\phi^{[i]}$ are defined by $\widehat{\phi^{[r]}}(\xi) := [\widehat{\phi}(\xi) + \overline{\widehat{\phi}(-\xi)}]/2$ and $\widehat{\phi^{[i]}}(\xi) := i[\overline{\widehat{\phi}(-\xi)} - \widehat{\phi}(\xi)]/2$.

2.21. Suppose that a filter $a \in l_0(\mathbb{Z})$ has the symmetry $a(c-k) = a(k)$ for all $k \in \mathbb{Z}$ and for some $c \in \mathbb{Z}$. If $|\widehat{a}(\xi)| \leqslant 1$ for all ξ in some neighborhood of 0, prove that

$$\operatorname{lpm}(a^{[r]}) \leqslant 2\operatorname{vm}(a^{[i]}), \quad \operatorname{lpm}(\phi^{[r]}) \leqslant 2\operatorname{vm}(a^{[i]}), \quad \operatorname{lpm}(\phi^{[r]}) = \operatorname{lpm}(a^{[r]}).$$

2.22. Let $a \in l_0(\mathbb{Z})$ be given by $\widehat{a}(\xi) = 1 + i(e^{i\xi} - e^{-i\xi})^2$. Then a is symmetric about the origin and $\widehat{a}(0) = 1$. Prove that $\operatorname{lpm}(a^{[r]}) = \infty$ and $\operatorname{lpm}(\phi^{[r]}) = 4$.

Therefore, $\operatorname{lpm}(\phi^{[r]}) \neq \operatorname{lpm}(a^{[r]})$ (Note that $|\widehat{a}(\xi)| \leqslant 1$ fails in a neighborhood of 0).

2.23. Prove the properties of Pfi in Proposition 2.23.

2.24. Let $(\{\tilde{a}; \tilde{b}\}, \{a; b\})$ be a biorthogonal wavelet filter bank. Prove that $\operatorname{Fsi}(a, b) = \operatorname{Fsi}(\tilde{a}, \tilde{b})$.

2.25. Prove Proposition 2.7.3

2.26. Let $a \in l_0(\mathbb{Z})$ be a filter such that a has at least one finitely supported dual filter. For any integers $\tilde{m}$ and $\tilde{n}$ such that $\tilde{n} - \tilde{m} = \operatorname{len}(a) - 1$, prove that there is a unique dual filter $\tilde{a}$ of a such that $\operatorname{fsupp}(\tilde{a}) \subseteq [\tilde{m}, \tilde{n}]$.

2.27. Let $a \in l_0(\mathbb{Z})$. Prove that

a. $z^k \gcd(\mathsf{a}(z), \mathsf{a}(-z)) = \mathsf{u}(z^2)$ for some $k \in \{0, 1\}$ and $u \in l_0(\mathbb{Z})$. Therefore, we can assume that $\gcd(\mathsf{a}(z), \mathsf{a}(-z)) = \mathsf{u}(z^2)$ always holds.

b. $\mathcal{T}_a l_0(\mathbb{Z}) = \{u * v \ : \ v \in l_0(\mathbb{Z})\}$, where $\mathsf{u}(z^2) := \gcd(\mathsf{a}(z), \mathsf{a}(-z))$.

c. $\mathcal{T}_a l_0(\mathbb{Z}) = l_0(\mathbb{Z})$ if and only if $\mathsf{a}(z)$ and $\mathsf{a}(-z)$ have no common zeros in $\mathbb{C}\backslash\{0\}$.

d. $\mathcal{T}_a l(\mathbb{Z}) = l(\mathbb{Z})$ if and only if $\mathsf{a}(z)$ and $\mathsf{a}(-z)$ have no common zeros in $\mathbb{C}\backslash\{0\}$.

2.28. For $a \in l_0(\mathbb{Z})$ such that $\gcd(\mathsf{a}(z), \mathsf{a}(-z)) = 1$, if $\mathcal{S}_a v = 0$ for some $v \in l(\mathbb{Z})$, prove that $v = 0$. Consequently, $\mathcal{T}_a : l_0(\mathbb{Z}) \to l_0(\mathbb{Z})$ is a bijection provided that $\mathsf{a}(z)$ and $\mathsf{a}(-z))$ have no common zeros in $\mathbb{C}\backslash\{0\}$.

2.29. For $a = \{\frac{1}{2}, \underline{\mathbf{0}}, \frac{1}{2}\}_{[-1,1]}$, find a nontrivial sequence $v \in l(\mathbb{Z})$ such that $\mathcal{S}_a v = 0$. Generally, if $\mathsf{u}(z^2) = \gcd(\mathsf{a}(z), \mathsf{a}(-z))$ is not a monomial, construct a nontrivial sequence $v \in l(\mathbb{Z})$ such that $\mathcal{S}_a v = 0$.

2.30. For $a \in l_0(\mathbb{Z})$ such that $\gcd(\mathsf{a}(z), \mathsf{a}(-z)) = 1$, prove that $\ker(\mathcal{T}_a) := \{v \in l_0(\mathbb{Z}) \ : \ \mathcal{T}_a v = 0\} = \{b * (u(2\cdot)) \ : \ u \in l_0(\mathbb{Z})\}$, where $\mathsf{b}(z) := z\mathsf{a}^\star(-z)$.

2.31. Let $z_0 \in \mathbb{C}\backslash\{0\}$. Define a filter $u := \{\underline{\mathbf{z_0}}, -1\}_{[0,1]}$ and a sequence $v \in l(\mathbb{Z})$ by $v(k) := z_0^k$ for $k \in \mathbb{Z}$. Prove that $u * v = 0$.

2.32. Let $u \in l_0(\mathbb{Z})$ such that $\mathsf{u}(z)$ is not a monomial (i.e., $\operatorname{len}(u) > 0$). It is easy to deduce that if $u * v = 0$ for some $v \in l_0(\mathbb{Z})$, then we must have $v = 0$. Prove that there exists a nontrivial sequence $v \in l(\mathbb{Z})$ such that $u * v = 0$.

2.33. If $a \in l_0(\mathbb{Z})$ is a filter such that a is an interpolatory wavelet filter and an orthogonal wavelet filter, prove that $\widehat{a}(\xi) = \frac{1}{2} + \frac{1}{2}\lambda e^{-i(2j+1)\xi}$ for some $j \in \mathbb{Z}$ and $\lambda \in \mathbb{C}$ with $|\lambda| = 1$.

For all exercises below, we assume that $\mathsf{d} > 1$ is a positive integer. We say that a filter u is *a* d*-interpolatory filter* if $u(0) = \mathsf{d}^{-1}$ and $u(\mathsf{d}k) = 0$ for all $k \in \mathbb{Z}\backslash\{0\}$. A filter u is an orthogonal d-wavelet filter if $u * u^\star$ is a d-interpolatory filter. Similarly, we say that $(\tilde{u}, u)$ is *a pair of biorthogonal* d*-wavelet filters* if $\tilde{u} * u^\star$ is a d-interpolatory filter. Recall that $a \in l_0(\mathbb{Z})$ has m sum rules with respect to d if $\widehat{a}(\xi) = (1 + e^{-i\xi} + \cdots + e^{-i(\mathsf{d}-1)\xi})^m \mathbf{Q}(\xi)$ for some 2π-periodic trigonometric polynomial $\mathbf{Q}$. Define $\operatorname{sr}(a, \mathsf{d}) := m$ with m being the largest such integer m.

2.34. For nonnegative integers m and n, prove that there exists a unique interpolatory filter a such that a has $m+n$ sum rules and vanishes outside $[1-\mathsf{d}m, \mathsf{d}n-1]$.

2.35. Denote by $a_{2m}^{I,\mathsf{d}}$ the unique interpolatory filter which has $2m$ sum rules and vanishes outside $[1-\mathsf{d}m, \mathsf{d}m-1]$. Prove that

$$\widehat{a_{2m}^{I,\mathsf{d}}}(\xi) = \mathsf{d}^{-2m}|1+e^{-i\xi}+\cdots+e^{-i(\mathsf{d}-1)\xi}|^{2m}\mathsf{P}_{\mathsf{d};m}(\sin^2(\xi/2)),$$

where

$$\mathsf{P}_{\mathsf{d};n}(x) := \sum_{j=0}^{n-1}\sum_{j_1+\cdots+j_\mathsf{d}=j}\prod_{k=1}^{\mathsf{d}-1}\frac{(n-1+j_k)!}{j_k!(n-1)!}\left(\sin\frac{k\pi}{\mathsf{d}}\right)^{-2j_k}, \qquad n\in\mathbb{N}.$$

2.36. A filter $a\in l_0(\mathbb{Z})$ is an orthogonal d-wavelet filter if $\sum_{\omega=0}^{\mathsf{d}-1}|\widehat{a}(\xi+\frac{2\pi\omega}{\mathsf{d}})|^2=1$. Let $\mathbf{Q}$ be a 2π-periodic trigonometric polynomial such that $\mathbf{Q}(0)=1$ and $|\mathbf{Q}(\xi)|^2=\mathsf{P}_{\mathsf{d};m}(\sin^2(\xi/2))$. Prove that $a_m^{D,\mathsf{d}}$ is a orthogonal d-wavelet filter, where

$$\widehat{a_m^{D,\mathsf{d}}}(\xi) := \mathsf{d}^{-m}(1+e^{-i\xi}+\cdots+e^{-i(\mathsf{d}-1)\xi})^m\mathbf{Q}(\xi).$$

2.37. Define $\widehat{\phi}(\xi) := \prod_{j=1}^{\infty}\widehat{a_m^{D,\mathsf{d}}}(\mathsf{d}^{-j}\xi)$, $\xi\in\mathbb{R}$. Prove the following statements:

a. There exists a positive constant C such that $|\widehat{\phi}(\xi)|\geqslant C$ for all $\xi\in[-\pi,\pi]$;
b. Define $\widehat{\phi_n}(\xi) := \chi_{(-\pi,\pi]}(\mathsf{d}^{-n}\xi)\prod_{j=1}^{n}\widehat{a_m^{D,\mathsf{d}}}(\mathsf{d}^{-j}\xi)$ for $n\in\mathbb{N}$. Prove that $\{\phi_n(\cdot-k) : k\in\mathbb{Z}\}$ is an orthonormal system in $L_2(\mathbb{R})$.
c. Prove that $\lim_{n\to\infty}\|\widehat{\phi_n}-\widehat{\phi}\|_{L_2(\mathbb{R})}=0$ and $\{\phi(\cdot-k) : k\in\mathbb{Z}\}$ is an orthonormal system in $L_2(\mathbb{R})$.

2.38. Let h be a polynomial such that $h(0)=1$ and all the roots of h in the complex plane lie on the interval $(0,\infty)$. Let $H_n(x)$ be the $(n-1)$th-degree Taylor polynomial of $h(x)^{-1}$ at $x=0$, that is, $H_n(x)=h(x)^{-1}+\mathscr{O}(x^n), x\to 0$. Prove that

$$(-1)^{n-1}(h(x)H_n(x)-1)>0, \qquad \forall\, x<0,\ n\in\mathbb{N}.$$

2.39. Prove the following three identities: (i) $\prod_{j=1}^{\mathsf{d}-1}|1-e^{i\frac{2\pi j}{\mathsf{d}}}|^2=\mathsf{d}^2$; (ii) $\sin^2\frac{\pi j}{\mathsf{d}}-\sin^2\frac{\xi}{2}=\frac{1}{4}e^{i\xi}(e^{-i\xi}-e^{-i\frac{2\pi j}{\mathsf{d}}})(e^{-i\xi}-e^{i\frac{2\pi j}{\mathsf{d}}})$; (iii)

$$\left|\frac{1+e^{-i\xi}+\cdots+e^{-i(\mathsf{d}-1)\xi}}{\mathsf{d}}\right|^2=\mathsf{d}^{-2}\prod_{j=1}^{\mathsf{d}-1}|1-e^{i\xi}e^{i\frac{2\pi j}{\mathsf{d}}}|^2=\prod_{j=1}^{\mathsf{d}-1}\left(1-\frac{\sin^2\frac{\xi}{2}}{\sin^2\frac{j\pi}{\mathsf{d}}}\right).$$

2.40. Prove that $\mathsf{P}_{\mathsf{d};2n-1}(x) > 0$ for all $x \in \mathbb{R}$ and $n \in \mathbb{N}$. Therefore, there exist polynomials $\mathsf{P}^{[r]}_{\mathsf{d};2n-1}$ and $\mathsf{P}^{[i]}_{\mathsf{d};2n-1}$ with real coefficients such that $[\mathsf{P}^{[r]}_{\mathsf{d};2n-1}(x)]^2 + [\mathsf{P}^{[i]}_{\mathsf{d};2n-1}(x)]^2 = \mathsf{P}(x)$. Define a symmetric filter $a^{\mathsf{S},\mathsf{d}}_{2n-1}$ by

$$\widehat{a^{\mathsf{S},\mathsf{d}}_{2n-1}}(\xi) := \mathsf{d}^{1-2n}(1 + e^{-i\xi} + \cdots + e^{-i(\mathsf{d}-1)\xi})^{2n-1} \times \left[\mathsf{P}^{[r]}_{\mathsf{d};2n-1}(\sin^2(\xi/2)) + i\mathsf{P}^{[i]}_{\mathsf{d};2n-1}(\sin^2(\xi/2))\right].$$

Then $a^{\mathsf{S},\mathsf{d}}_{2n-1}$ is an orthogonal d-wavelet filter and has $2n-1$ sum rules.

2.41. For a biorthogonal d-wavelet filter bank $(\{\tilde{a}; \tilde{b}_1, \ldots, \tilde{b}_{\mathsf{d}-1}\}, \{a; b_1, \ldots, b_{\mathsf{d}-1}\})$, prove $\operatorname{sr}(a) = \min(\operatorname{vm}(\tilde{b}_1), \ldots, \operatorname{vm}(\tilde{b}_{\mathsf{d}-1}))$ and $\operatorname{sr}(\tilde{a}) = \min(\operatorname{vm}(b_1), \ldots, \operatorname{vm}(b_{\mathsf{d}-1}))$.

2.42. Let $(a, \tilde{a})$ be a pair of biorthogonal d-wavelet filters with $\tilde{a}, a \in l_0(\mathbb{Z})$. Define $[m, n] := \operatorname{fsupp}(a)$ and assume $\operatorname{len}(a) = n - m \geqslant 1$. Prove that there exists a filter $\mathring{a} \in l_0(\mathbb{Z})$ such that $(a, \mathring{a})$ is a pair of biorthogonal d-wavelet filters and $\operatorname{fsupp}(\mathring{a}) \subseteq [m+1, n]$ or $\operatorname{fsupp}(\mathring{a}) \subseteq [m, n-1]$.

2.43. Let $(a, \tilde{a})$ be a pair of biorthogonal d-wavelet filter with $a, \tilde{a} \in l_0(\mathbb{Z})$. Prove that there exists a sequence of finitely supported filters $a_0, a_1, \ldots, a_r \in l_0(\mathbb{Z})$ such that $a_0 \to a_1 \cdots \to a_{r-1} \to a_r$ is a chain satisfying

a. $(a_0, a_1) = (a, \tilde{a})$ and $a_r = t\boldsymbol{\delta}(\cdot - k)$ for some $t \in \mathbb{C}$ and $k \in \mathbb{Z}$;
b. every (a_{j-1}, a_j) is a pair of biorthogonal d-wavelet filters for all $j = 1, \ldots, r$;
c. $\operatorname{fsupp}(a_{j+1}) \subsetneq \operatorname{fsupp}(a_j)$ for all $j = 1, \ldots, r-1$.

A biorthogonal d-wavelet filter bank $(\{a; b_1, \ldots, b_{\mathsf{d}-1}\}, \{\tilde{a}; \tilde{b}_1, \ldots, \tilde{b}_{\mathsf{d}-1}\})$ can be constructed from a given pair $(a, \tilde{a})$ by reversing the chain with an initial d-biorthogonal wavelet filter bank easily derived from the pair (a_{r-1}, a_r).

2.44. Let $(a, \tilde{a})$ be a pair of biorthogonal d-wavelet filters with $a, \tilde{a} \in l_0(\mathbb{Z})$ such that $\mathsf{S}a(z) = \mathsf{S}\tilde{a}(z) = \epsilon z^c$ for some $\epsilon \in \{-1, 1\}$ and $c \in \mathbb{Z}$. Define $[m, n] := \operatorname{fsupp}(a)$ and assume that at least two cosets $a^{[\gamma:\mathsf{d}]}$ are not identically zero for $\gamma = 0, \ldots, \mathsf{d}-1$. Prove that there exists a filter $\mathring{a} \in l_0(\mathbb{Z})$ such that $(a, \mathring{a})$ is a pair of biorthogonal d-wavelet filters, $\mathsf{S}\mathring{a}(z) = \epsilon z^c$, and $\operatorname{fsupp}(\mathring{a}) \subseteq [m+1, n-1]$.

2.45. Let $(a, \tilde{a})$ be a pair of biorthogonal d-wavelet filters with $a, \tilde{a} \in l_0(\mathbb{Z})$ such that $\mathsf{S}a(z) = \mathsf{S}\tilde{a}(z) = \epsilon z^c$ for some $\epsilon \in \{-1, 1\}$ and $c \in \mathbb{Z}$. Prove that there exists a sequence of filters $a_0, a_1, \ldots, a_r \in l_0(\mathbb{Z})$ such that $a_0 \to a_1 \cdots \to a_{r-1} \to a_r$ is a symmetric chain satisfying

a. $\mathsf{S}a_j(z) = \epsilon z^c$ for all $j = 0, \ldots, r$;
b. $(a_0, a_1) = (a, \tilde{a})$ and the filter a_r has no more than two nontrivial coset sequences $a_r^{[\gamma:\mathsf{d}]} := \{a_r(\gamma + \mathsf{d}k)\}_{k \in \mathbb{Z}}$ for $\gamma = 0, \ldots, \mathsf{d}-1$;
c. every (a_{j-1}, a_j) is a pair of biorthogonal d-wavelet filters for all $j = 1, \ldots, r$;
d. $\operatorname{fsupp}(a_{j+1}) \subsetneq \operatorname{fsupp}(a_j)$ for all $j = 1, \ldots, r-1$.

A biorthogonal d-wavelet filter bank $(\{a; b_1, \ldots, b_{\mathsf{d}-1}\}, \{\tilde{a}; \tilde{b}_1, \ldots, \tilde{b}_{\mathsf{d}-1}\})$ with each filter being symmetric or antisymmetric can be constructed from a given symmetric pair $(a, \tilde{a})$ by reversing the chain with an initial biorthogonal d-wavelet filter bank having symmetry easily derived from the pair (a_{r-1}, a_r).

Chapter 3
Framelet Filter Banks

Framelet filter banks are employed in a discrete framelet transform and are more general and flexible than wavelet filter banks. In this chapter we discuss how to systematically design dual framelet filter banks $(\{\tilde{a}; \tilde{b}_1, \tilde{b}_2\}, \{a; b_1, b_2\})_\Theta$ and tight framelet filter banks $\{a; b_1, \ldots, b_s\}_\Theta$ with $s = 2$ or 3 having some desirable properties such as vanishing moments and symmetry. As we shall see in Sect. 7.5, tight or dual framelet filter banks with symmetry property are of particular interest for constructing framelets on the interval $[0, 1]$.

The design of a framelet filter bank is quite different in nature to the construction of a wavelet filter bank. As we have seen in Chap. 2, the key issue for the design of an orthogonal or biorthogonal wavelet filter bank is the construction of a low-pass filter satisfying (2.0.5) or a pair of low-pass filters satisfying (2.0.3); while up to an integer shift and a multiplicative constant (see (2.0.2)), the high-pass filters are uniquely derived via (2.0.6) for an orthogonal wavelet filter bank or via (2.0.4) for a biorthogonal wavelet filter bank. Due to the unavoidable orthogonality/biorthogonality constraints on the low-pass filters in a wavelet filter bank, not all low-pass filters can be used in a wavelet filter bank. For example, the B-spline filter a_m^B in (1.2.24) with $m > 1$ does not satisfy (2.0.5) and therefore, cannot be used in an orthogonal wavelet filter bank.

As a generalization of a wavelet filter bank by using more than one high-pass filter, a framelet filter bank allows us to relax such constraints in a wavelet filter bank and consequently more low-pass filters can be used in a framelet filter bank. To form a tight or dual framelet filter bank, quite often the low-pass filters are (arbitrarily) given in advance, and one has to construct the associated high-pass filters with some desirable properties. In this chapter we shall mainly use the Laurent polynomial representation of filters, because we frequently need the root information of a Laurent polynomial on the complex plane $\mathbb{C}$.

Since all framelet filter banks can be obtained via the oblique extension principle (OEP) described in Sect. 1.4, we first revisit here the oblique extension principle from Sect. 1.4. Recall from (2.7.4) and (2.7.5) that $\mathsf{a}(z) := \sum_{k\in\mathbb{Z}} a(k)z^k$ for

B. Han, *Framelets and Wavelets*, Applied and Numerical Harmonic Analysis,
https://doi.org/10.1007/978-3-319-68530-4_3

$a = \{a(k)\}_{k\in\mathbb{Z}} \in l_0(\mathbb{Z})$. For $\tilde{a}, \tilde{b}_1, \dots, \tilde{b}_s, a, b_1, \dots, b_s \in l_0(\mathbb{Z})$, we say that $(\{\tilde{a}; \tilde{b}_1, \dots, \tilde{b}_s\}, \{a; b_1, \dots, b_s\})_\Theta$ is *a dual framelet filter bank* if the following perfect reconstruction condition, in terms of Laurent polynomial representation, holds:

$$\begin{bmatrix} \tilde{\mathsf{b}}_1(z) & \cdots & \tilde{\mathsf{b}}_s(z) \\ \tilde{\mathsf{b}}_1(-z) & \cdots & \tilde{\mathsf{b}}_s(-z) \end{bmatrix} \begin{bmatrix} \mathsf{b}_1(z) & \cdots & \mathsf{b}_s(z) \\ \mathsf{b}_1(-z) & \cdots & \mathsf{b}_s(-z) \end{bmatrix}^\star = \mathcal{M}_{a,\tilde{a},\Theta}(z), \tag{3.0.1}$$

where

$$\mathcal{M}_{a,\tilde{a},\Theta}(z) := \begin{bmatrix} \Theta(z) - \Theta(z^2)\tilde{\mathsf{a}}(z)\mathsf{a}^\star(z) & -\Theta(z^2)\tilde{\mathsf{a}}(z)\mathsf{a}^\star(-z) \\ -\Theta(z^2)\tilde{\mathsf{a}}(-z)\mathsf{a}^\star(z) & \Theta(-z) - \Theta(z^2)\tilde{\mathsf{a}}(-z)\mathsf{a}^\star(-z) \end{bmatrix} \tag{3.0.2}$$

and $\mathsf{P}^\star(z) := \sum_{k\in\mathbb{Z}} \overline{P_k}^\mathsf{T} z^{-k}$ for a matrix $\mathsf{P}(z) = \sum_{k\in\mathbb{Z}} P_k z^k$ of Laurent polynomials.

The low-pass filters a and $\tilde{a}$ are often given in advance. As we shall see in this chapter, one often designs a moment correcting filter Θ with some desirable properties first. Then the matrix $\mathcal{M}_{a,\tilde{a},\Theta}$ is given and the construction of high-pass filters $b_1, \dots, b_s, \tilde{b}_1, \dots, \tilde{b}_s$ now becomes: how to properly factorize a given matrix $\mathcal{M}_{a,\tilde{a},\Theta}(z)$ of Laurent polynomials in (3.0.2) so that (3.0.1) holds.

If $(\{a; b_1, \dots, b_s\},\{a; b_1, \dots, b_s\})_\Theta$ is a dual framelet filter bank, then we call $\{a; b_1, \dots, b_s\}_\Theta$ *a tight framelet filter bank*. To reduce computational complexity in the implementation of a framelet filter bank, we prefer a small number s of high-pass filters. As shown in Theorem 1.4.7, it is often necessary that $s > 1$. So, we shall consider the case of either $s = 2$ or $s = 3$ for a framelet filter bank in this chapter.

3.1 Properties of Laurent Polynomials with Symmetry

To construct framelet filter banks in this chapter, we first study some basic properties of Laurent polynomials.

3.1.1 GCD of Laurent Polynomials with Symmetry

As we discussed in Sect. 1.2.3, there are two related but different notions of symmetry. Recall that a Laurent polynomial p has *symmetry* if

$$\mathsf{S}\mathsf{p}(z) := \frac{\mathsf{p}(z)}{\mathsf{p}(z^{-1})} = \epsilon z^c \quad \forall\, z \in \mathbb{C}\backslash\{0\} \quad \text{with} \quad \epsilon \in \{-1, 1\},\ c \in \mathbb{Z}. \tag{3.1.1}$$

Similarly, a Laurent polynomial p has *complex symmetry* if

$$\mathbb{S}\mathsf{p}(z) := \frac{\mathsf{p}(z)}{\mathsf{p}^\star(z)} = \epsilon z^c \quad \forall\, z \in \mathbb{C}\backslash\{0\} \quad \text{with} \quad \epsilon \in \{-1, 1\},\ c \in \mathbb{Z}. \tag{3.1.2}$$

For a Laurent polynomial p, we observe that $\mathsf{S}[\lambda \mathsf{p}](z) = \mathsf{Sp}(z)$ and $\mathbb{S}[\lambda \mathsf{p}](z) = \frac{\lambda}{\bar{\lambda}}\mathbb{S}\mathsf{p}(z)$ for all $\lambda \in \mathbb{C}\backslash\{0\}$. Hence, we say that p has *essential complex symmetry* if (3.1.2) holds with $\epsilon \in \mathbb{T} := \{\zeta \in \mathbb{C} \ : \ |\zeta| = 1\}$ instead of $\epsilon \in \{-1, 1\}$. It is straightforward to see that p has essential complex symmetry if and only if $\lambda\mathsf{p}$ has complex symmetry for some $\lambda \in \mathbb{T}$. Note that a Laurent polynomial p has real coefficients if and only if $\mathsf{p}^\star(z) = \mathsf{p}(z^{-1})$. Therefore, for a Laurent polynomial p having real coefficients, p has complex symmetry $\iff$ p has essential complex symmetry $\iff$ p has symmetry.

We say that p is a nontrivial Laurent polynomial if it is not identically zero. Particularly, $\mathsf{S}0$ (or $\mathbb{S}0$) can be assigned any choice of ϵz^c for $\epsilon \in \{-1, 1\}$ and $c \in \mathbb{Z}$. Consequently, if $\mathsf{S}0$ (or $\mathbb{S}0$) appears in an identity, it is conventionally understood that it takes the proper choice of ϵz^c so that the identity is satisfied. We shall frequently use the following two obvious facts:

$$\mathsf{p}^\star(z) = \overline{\mathsf{p}(\bar{z}^{-1})}, \quad z \in \mathbb{C}\backslash\{0\}, \tag{3.1.3}$$

and $z = \bar{z}^{-1}$ if and only if $|z| = 1$, i.e., $z \in \mathbb{T}$. For a Laurent polynomial p and $z_0 \in \mathbb{C}\backslash\{0\}$, by $\mathsf{Z}(\mathsf{p}, z_0)$ we denote the multiplicity of zeros of $\mathsf{p}(z)$ at the point $z = z_0$. We shall use [complex] symmetry to handle both symmetry and complex symmetry simultaneously.

We have the following basic result on [complex] symmetry of Laurent polynomials, whose proof is left as Exercise 3.6.

Proposition 3.1.1 *Let p be a nontrivial Laurent polynomial.*

(i) The Laurent polynomial p has symmetry if and only if $\mathsf{Z}(\mathsf{p}, z) = \mathsf{Z}(\mathsf{p}, z^{-1})$ for all $z \in \mathbb{C}\backslash\{0\}$.

(ii) The Laurent polynomial p has essential complex symmetry if and only if

$$\mathsf{Z}(\mathsf{p}, z) = \mathsf{Z}(\mathsf{p}, \bar{z}^{-1}) \quad \forall\, z \in \mathbb{C}\backslash\{0\}. \tag{3.1.4}$$

(iii) Either $\mathsf{p}(z) \geqslant 0$ or $\mathsf{p}(z) \leqslant 0$ for all $z \in \mathbb{T}$ if and only if $\mathsf{p}^\star(z) = \mathsf{p}(z)$ and

$$\mathsf{Z}(\mathsf{p}, z) \quad \text{is an even integer for every} \quad z \in \mathbb{T}. \tag{3.1.5}$$

(iv) If both (3.1.4) and (3.1.5) are satisfied, then there exist $k \in \mathbb{Z}$ and $\lambda \in \mathbb{T}$ ($\lambda \in \{-1, 1\}$ if p has real coefficients) such that $\lambda z^k \mathsf{p}(z) \geqslant 0$ for all $z \in \mathbb{T}$.

Lemma 3.1.2 *If a Laurent polynomial p has symmetry and $\mathsf{p}(z) \geqslant 0$ for all $z \in \mathbb{T}$, then all the coefficients of p must be real numbers.*

Proof By $\mathsf{p}(z) \geqslant 0$ for all $z \in \mathbb{T}$, we have $\mathsf{p}^\star = \mathsf{p}$ and $\mathbb{S}\mathsf{p} = 1$. Since p also has symmetry, by $\mathbb{S}\mathsf{p} = 1$, we must have $\mathsf{Sp}(z) = \epsilon$ for some $\epsilon \in \{-1, 1\}$. Write $\mathsf{p}(z) = (z-1)^{\mathsf{Z}(\mathsf{p},1)}\mathsf{q}(z)$ for some Laurent polynomial q. Then

$$\epsilon = \mathsf{Sp}(z) = \frac{\mathsf{p}(z)}{\mathsf{p}(z^{-1})} = \frac{(z-1)^{\mathsf{Z}(\mathsf{p},1)}\mathsf{q}(z)}{(z^{-1}-1)^{\mathsf{Z}(\mathsf{p},1)}\mathsf{q}(z^{-1})} = (-z)^{\mathsf{Z}(\mathsf{p},1)}\frac{\mathsf{q}(z)}{\mathsf{q}(z^{-1})}.$$

By item (iii) of Proposition 3.1.1, the inequality $\mathsf{p}(z) \geqslant 0$ for all $z \in \mathbb{T}$ implies that $\mathsf{Z}(\mathsf{p}, 1)$ must be an even integer. Since $\mathsf{q}(1) \neq 0$, plugging $z = 1$ into the above identity, we deduce that $\epsilon = 1$. Hence, $\mathsf{Sp} = 1$ and $\mathsf{p}^\star(z) = \mathsf{p}(z) = \mathsf{p}(z^{-1})$ from which we can easily conclude that all the coefficients of p must be real numbers. □

We now discuss greatest common divisors of Laurent polynomials. For two nontrivial Laurent polynomials p_1 and p_2, we say that a Laurent polynomial p is a greatest common divisor of p_1 and p_2 if

(i) p is a common divisor/factor of p_1 and p_2, that is, $\mathsf{p} \mid \mathsf{p}_1$ and $\mathsf{p} \mid \mathsf{p}_2$;
(ii) any common divisor/factor of p_1 and p_2 must be a factor of p.

Using the notation $\mathsf{Z}(\mathsf{p}, z)$, we see that p is a greatest common divisor of p_1 and p_2 if and only if $\mathsf{Z}(\mathsf{p}, z) = \min(\mathsf{Z}(\mathsf{p}_1, z), \mathsf{Z}(\mathsf{p}_2, z))$ for all $z \in \mathbb{C}\backslash\{0\}$. Up to a multiplicative monomial λz^k for some $\lambda \in \mathbb{C}\backslash\{0\}$ and $k \in \mathbb{Z}$, a greatest common divisor of two Laurent polynomials p_1 and p_2 is unique and can be found by the following well-known algorithm using Euclidean long division for Laurent polynomials.

Algorithm 3.1.3 *Let p_1 and p_2 be nontrivial Laurent polynomials. Set $j = 1$.*

(S1) Apply the long division to obtain Laurent polynomials q_{j+2} and p_{j+2} by

$$\mathsf{p}_j(z) = \mathsf{p}_{j+1}(z)\mathsf{q}_{j+2}(z) + \mathsf{p}_{j+2}(z) \quad \textit{satisfying} \quad \operatorname{len}(\mathsf{p}_{j+2}) < \operatorname{len}(\mathsf{p}_{j+1}).$$

(S2) If p_{j+2} is not a monomial and not identically zero, then increase j by one and go to step (S1); otherwise, set $\mathsf{p} := 1$ if p_{j+2} is a nonzero monomial or $\mathsf{p} := \mathsf{p}_{j+1}$ if p_{j+2} is identically zero.

Then p is a greatest common divisor of p_1 and p_2.

If p_1 and p_2 have real coefficients, then so does p. Due to Proposition 3.1.1, if both p_1 and p_2 have [complex] symmetry (and/or both $\mathsf{p}_1(z) \geqslant 0$ and $\mathsf{p}_2(z) \geqslant 0$ for all $z \in \mathbb{T}$), then there exist $\lambda \in \mathbb{T}$ and $k \in \mathbb{Z}$ such that $\lambda z^k \mathsf{p}(z)$ has [complex] symmetry (and/or $\lambda z^k \mathsf{p}(z) \geqslant 0$ for all $z \in \mathbb{T}$). We define such normalized Laurent polynomial $\lambda z^k \mathsf{p}(z)$ as the greatest common divisor $\gcd(\mathsf{p}_1, \mathsf{p}_2)$ of p_1 and p_2.

3.1.2 *Sum of Squares of Laurent Polynomials with Symmetry Property*

To investigate tight framelet filter banks with [complex] symmetry, we have to study a closely related problem about how to split a Laurent polynomial into a sum of squares of no more than two Laurent polynomials with [complex] symmetry.

For $\epsilon \in \{-1, 1\}$ and $c \in \mathbb{Z}$, we say that a Laurent polynomial p has *the complex SOS (sum of squares) property with respect to the symmetry type* ϵz^c if there exist two Laurent polynomials p_1 and p_2 having complex symmetry such that

$$\mathsf{p}_1(z)\mathsf{p}_1^\star(z) + \mathsf{p}_2(z)\mathsf{p}_2^\star(z) = \mathsf{p}(z) \quad \text{and} \quad \frac{\mathbb{S}\mathsf{p}_1(z)}{\mathbb{S}\mathsf{p}_2(z)} = \epsilon z^c. \tag{3.1.6}$$

Similarly, we say that a Laurent polynomial p has *the real SOS (sum of squares) property with respect to the symmetry type* ϵz^c if there exist two Laurent polynomials p_1 and p_2 having symmetry and complex coefficients such that

$$\mathsf{p}_1(z)\mathsf{p}_1^\star(z) + \mathsf{p}_2(z)\mathsf{p}_2^\star(z) = \mathsf{p}(z) \quad \text{and} \quad \frac{\mathsf{S}\mathsf{p}_1(z)}{\mathsf{S}\mathsf{p}_2(z)} = \epsilon z^c. \tag{3.1.7}$$

As we shall see in Theorem 3.1.6, if p has the real SOS property with respect to the symmetry type ϵz^c, then we can always construct two Laurent polynomials p_1 and p_2 having symmetry and *real coefficients* such that (3.1.7) is satisfied.

To study the SOS properties, we need the following result, which can be directly verified and is left as Exercise 3.7.

Lemma 3.1.4 *Let* $\mathsf{p}_1, \mathsf{p}_2, \mathsf{p}_3, \mathsf{p}_4$ *be Laurent polynomials with [complex] symmetry such that* $\frac{\mathsf{S}\mathsf{p}_1(z)}{\mathsf{S}\mathsf{p}_2(z)} = \frac{\mathsf{S}\mathsf{p}_3(z)}{\mathsf{S}\mathsf{p}_4(z)}$ *(replace* S *by* $\mathbb{S}$ *throughout for complex symmetry). Then*

$$\mathsf{p}_5(z) := \mathsf{p}_1(z)\mathsf{p}_3(z) + (\mathsf{S}\mathsf{p}_1(z))\mathsf{p}_2^\star(z)\mathsf{p}_4(z), \quad \mathsf{p}_6(z) := \mathsf{p}_2(z)\mathsf{p}_3(z) - (\mathsf{S}\mathsf{p}_1(z))\mathsf{p}_1^\star(z)\mathsf{p}_4(z)$$

have [complex] symmetry and satisfy $\frac{\mathsf{S}\mathsf{p}_5(z)}{\mathsf{S}\mathsf{p}_6(z)} = \frac{\mathsf{S}\mathsf{p}_1(z)}{\mathsf{S}\mathsf{p}_2(z)}$ *and*

$$\mathsf{p}_5(z)\mathsf{p}_5^\star(z) + \mathsf{p}_6(z)\mathsf{p}_6^\star(z) = [\mathsf{p}_1(z)\mathsf{p}_1^\star(z) + \mathsf{p}_2(z)\mathsf{p}_2^\star(z)][\mathsf{p}_3(z)\mathsf{p}_3^\star(z) + \mathsf{p}_4(z)\mathsf{p}_4^\star(z)].$$

For Laurent polynomials having the complex SOS property, we have

Theorem 3.1.5 *Let* $\epsilon \in \{-1, 1\}$ *and* $c \in \mathbb{Z}$. *A Laurent polynomial* p *has the complex SOS property with respect to the symmetry type* ϵz^c *if and only if* $\mathsf{p}(z) \geqslant 0$, $\forall\, z \in \mathbb{T}$.

Proof The necessity part ($\Rightarrow$) is trivial, since (3.1.6) implies $\mathsf{p}(z) \geqslant 0$ for all $z \in \mathbb{T}$.

We now prove the sufficiency part ($\Leftarrow$). If (3.1.6) holds, then for $k \in \mathbb{Z}$,

$$\mathsf{p}_1(z)\mathsf{p}_1^\star(z) + [iz^k\mathsf{p}_2(z)][iz^k\mathsf{p}_2(z)]^\star = \mathsf{p}(z) \quad \text{and} \quad \frac{\mathbb{S}\mathsf{p}_1(z)}{\mathbb{S}(iz^k\mathsf{p}_2(z))} = -\epsilon z^{c-2k}.$$

Therefore, it suffices to prove (3.1.6) for the cases $\epsilon = -1$ and $c \in \{0, 1\}$.

Since $\mathsf{p}(z) \geqslant 0$ for all $z \in \mathbb{T}$, by the Fejér-Riesz Lemma in Lemma 1.4.4, there exists a Laurent polynomial u such that $\mathsf{u}(z)\mathsf{u}^\star(z) = \mathsf{p}(z)$. Define

$$\mathsf{p}_1(z) := [\mathsf{u}(z) + \mathsf{u}^\star(z)]/2, \qquad \mathsf{p}_2(z) := [\mathsf{u}(z) - \mathsf{u}^\star(z)]/2. \tag{3.1.8}$$

Then (3.1.6) must hold with $\epsilon = -1$ and $c = 0$ by $\mathbb{S}\mathsf{p}_1(z) = 1$ and $\mathbb{S}\mathsf{p}_2(z) = -1$.

We now deal with the case $\epsilon = -1$ and $c = 1$. Since $\mathsf{p}(z) \geqslant 0$ for all $z \in \mathbb{T}$, by Proposition 3.1.1, (3.1.4) and (3.1.5) must hold. Therefore, there is a positive constant λ such that $\lambda\mathsf{p}$ is a product of the factors $(z-z_0)(z^{-1}-\overline{z_0})$ with $z_0 \in \mathbb{C}\backslash\{0\}$. Define

$$\tilde{\mathsf{p}}_1(z) := 1 - |z_0|, \qquad \tilde{\mathsf{p}}_2(z) := \overline{\sqrt{z_0}} - \sqrt{z_0}z^{-1},$$

where $\sqrt{z_0}$ denotes a complex number such that $(\sqrt{z_0})^2 = z_0$ (more precisely, write $z_0 = re^{i\theta}$ with $r \geqslant 0$ and $\theta \in \mathbb{R}$, then $\sqrt{z_0} := \sqrt{r}e^{i\theta/2}$ or $\sqrt{z_0} := -\sqrt{r}e^{i\theta/2}$). Then

$$\tilde{\mathsf{p}}_1(z)\tilde{\mathsf{p}}_1^\star(z) + \tilde{\mathsf{p}}_2(z)\tilde{\mathsf{p}}_2^\star(z) = (z - z_0)(z^{-1} - \overline{z_0})$$

and $\frac{\mathbb{S}\tilde{\mathsf{p}}_1(z)}{\mathbb{S}\tilde{\mathsf{p}}_2(z)} = -z$ by $\mathbb{S}\tilde{\mathsf{p}}_1(z) = 1$ and $\mathbb{S}\tilde{\mathsf{p}}_2(z) = -z^{-1}$. Now it follows from Lemma 3.1.4 that the claim in (3.1.6) holds with $\epsilon = -1$ and $c = 1$. ☐

For Laurent polynomials having the real SOS property, the situation is more complicated and we have the following result.

Theorem 3.1.6 *Let $\epsilon \in \{-1, 1\}$ and $c \in \mathbb{Z}$. A Laurent polynomial p has the real SOS property with respect to the symmetry type ϵz^c if and only if*

(i) the Laurent polynomial p has real coefficients and $\mathsf{p}(z) \geqslant 0$ for all $z \in \mathbb{T}$;
(ii) the Laurent polynomial p satisfies the root condition for the real SOS property with respect to the symmetry type ϵz^c, that is, one of the following four cases must hold:

(1) if $\epsilon = -1$ and c is an even integer, then there is no condition;
(2) if $\epsilon = -1$ and c is an odd integer, then $\mathsf{Z}(\mathsf{p}, x) \in 2\mathbb{Z}$ for all $x \in (-1, 0)$;
(3) if $\epsilon = 1$ and c is an even integer, then $\mathsf{Z}(\mathsf{p}, x) \in 2\mathbb{Z}$ for all $x \in (-1, 0) \cup (0, 1)$;
(4) if $\epsilon = 1$ and c is an odd integer, then $\mathsf{Z}(\mathsf{p}, x) \in 2\mathbb{Z}$ for all $x \in (0, 1)$.

Moreover, if items (i) and (ii) are satisfied, then there exist Laurent polynomials p_1 and p_2 with real coefficients and symmetry such that (3.1.7) *holds.*

Proof Necessity ($\Rightarrow$). (3.1.7) trivially implies $\mathsf{p}(z) \geqslant 0$ for all $z \in \mathbb{T}$. Because p_1 and p_2 have symmetry, by (3.1.7), the Laurent polynomial p has symmetry. It now follows from Lemma 3.1.2 that p must have real coefficients. Therefore, item (i) holds.

By the symmetry property of p_1 and p_2, we have $\mathsf{p}_1^\star(z) = \overline{\mathsf{p}_1(\overline{z}^{-1})} = \overline{\mathsf{p}_1(\overline{z})}/\mathsf{S}\mathsf{p}_1(z)$ and hence (3.1.7) implies

$$\mathsf{p}(x) = [|\mathsf{p}_1(x)|^2 + \epsilon x^c|\mathsf{p}_2(x)|^2]/\mathsf{S}\mathsf{p}_1(x), \qquad x \in \mathbb{R}\backslash\{0\}. \tag{3.1.9}$$

When $\epsilon = 1$ and c is an even integer, we have $\epsilon x^c > 0$ for all $x \in \mathbb{R}\backslash\{0\}$ and hence, one can easily deduce from (3.1.9) that $\mathsf{Z}(\mathsf{p}, x) = 2\min(\mathsf{Z}(\mathsf{p}_1, x), \mathsf{Z}(\mathsf{p}_2, x))$ for all $x \in \mathbb{R}\backslash\{0\}$. Hence, item (3) holds. Items (2) and (4) can be proved similarly.

Sufficiency ($\Leftarrow$). Note that (3.1.7) implies $\mathsf{p}_1(z)\mathsf{p}_1^\star(z) + [z^k\mathsf{p}_2(z)][z^k\mathsf{p}_2(z)]^\star = \mathsf{p}(z)$ and $\frac{\mathsf{S}\mathsf{p}_1(z)}{\mathsf{S}(z^k\mathsf{p}_2(z))} = \epsilon z^{c-2k}$ for $k \in \mathbb{Z}$. Hence, it suffices to prove (3.1.7) for $c \in \{0, 1\}$ and $\epsilon \in \{-1, 1\}$. Since $\mathsf{p}(z) \geqslant 0$ for all $z \in \mathbb{T}$ and p has real coefficients, by the Fejér-Riesz Lemma in Lemma 1.4.4, there exists a Laurent polynomial u with real coefficients such that $\mathsf{u}(z)\mathsf{u}^\star(z) = \mathsf{p}(z)$. Define p_1 and p_2 as in the equation (3.1.8). Since u has real coefficients, we have $\mathsf{u}^\star(z) = \mathsf{u}(z^{-1})$ and consequently, the identity (3.1.7) is satisfied with $\epsilon = -1$ and $c = 0$ by $\mathsf{S}\mathsf{p}_1(z) = 1$ and $\mathsf{S}\mathsf{p}_2(z) = -1$. This proves the case $\epsilon = -1$ and $c = 0$ in item (1).

We now prove the other three cases. Since $\mathsf{p}(z) \geqslant 0$ for all $z \in \mathbb{T}$, we see that (3.1.4) and (3.1.5) must hold. On the other hand, since p has real coefficients, we have $\mathsf{p}^\star(z) = \mathsf{p}(z^{-1})$. Therefore, we deduce from $\mathsf{p}(z) = \mathsf{p}^\star(z) = \mathsf{p}(z^{-1})$ that

$$\mathsf{Z}(\mathsf{p}, z) = \mathsf{Z}(\mathsf{p}, z^{-1}), \qquad \forall\, z \in \mathbb{C}\backslash\{0\}. \tag{3.1.10}$$

Now (3.1.4), (3.1.5) and (3.1.10) together imply that there is a positive number λ such that $\lambda\mathsf{p}$ can be written as a product of the following two types of factors

$$\tilde{\mathsf{p}}(z|z_0)\tilde{\mathsf{p}}(z|\overline{z_0}), \qquad z_0 \in \mathbb{C} \text{ with } 0 < |z_0| \leqslant 1,\ \mathrm{Im}(z_0) > 0 \tag{3.1.11}$$

or

$$\tilde{\mathsf{p}}(z|x_0), \qquad x_0 \in [-1, 1]\backslash\{0\}, \tag{3.1.12}$$

where $\tilde{\mathsf{p}}(z|z_0) := (z - z_0)(z^{-1} - \overline{z_0})$ for $z, z_0 \in \mathbb{C}\backslash\{0\}$. For the factors in (3.1.11),

$$\begin{aligned}\tilde{\mathsf{p}}(z|z_0)\tilde{\mathsf{p}}(z|\overline{z_0}) =& |z_0|^2 z^{-2} - 2\,\mathrm{Re}(z_0)(1 + |z_0|^2)z^{-1} + 1 + 4(\mathrm{Re}(z_0))^2 + |z_0|^4 \\ & - 2\,\mathrm{Re}(z_0)(1 + |z_0|^2)z + |z_0|^2 z^2\end{aligned}$$

and the following identity holds

$$\tilde{\mathsf{p}}_1(z)\tilde{\mathsf{p}}_1^\star(z) + \tilde{\mathsf{p}}_2(z)\tilde{\mathsf{p}}_2^\star(z) = \tilde{\mathsf{p}}(z|z_0)\tilde{\mathsf{p}}(z|\overline{z_0}) \quad \text{and} \quad \frac{\mathsf{S}\tilde{\mathsf{p}}_1(z)}{\mathsf{S}\tilde{\mathsf{p}}_2(z)} = \epsilon z^c,$$

where the two Laurent polynomials $\tilde{\mathsf{p}}_1$ and $\tilde{\mathsf{p}}_2$ with real coefficients are defined according to the following three cases. For $\epsilon = -1$ and $c = 1$,

$$\tilde{\mathsf{p}}_1(z) := |z_0|(z + z^{-1}) + (1 - |z_0|)^2 - 2\,\mathrm{Re}(z_0),$$

$$\tilde{\mathsf{p}}_2(z) := (1 - |z_0|)\sqrt{2|z_0| + 2\,\mathrm{Re}(z_0)}(z^{-1} - 1),$$

with $\frac{\mathsf{S}\tilde{\mathsf{p}}_1(z)}{\mathsf{S}\tilde{\mathsf{p}}_2(z)} = \frac{1}{-z^{-1}} = -z$. For $\epsilon = 1$ and $c = 0$,

$$\tilde{\mathsf{p}}_1(z) := -\mathrm{Re}(z_0)(z^{-1} + z) - (\mathrm{Im}(z_0))^2 + (\mathrm{Re}(z_0))^2 + 1,$$

$$\tilde{\mathsf{p}}_2(z) := \mathrm{Im}(z_0)(z^{-1} + z) - 2\,\mathrm{Re}(z_0)\,\mathrm{Im}(z_0),$$

with $\frac{\mathsf{S}\tilde{\mathsf{p}}_1(z)}{\mathsf{S}\tilde{\mathsf{p}}_2(z)} = \frac{1}{1} = 1$. For $\epsilon = 1$ and $c = 1$,

$$\tilde{\mathsf{p}}_1(z) := -|z_0|(z^{-1} + z) + (|z_0| - 1)^2 + 2\,\mathrm{Re}(z_0),$$
$$\tilde{\mathsf{p}}_2(z) := (1 - |z_0|)\sqrt{2|z_0| - 2\,\mathrm{Re}(z_0)}(z^{-1} + 1),$$

with $\frac{\mathsf{S}\tilde{\mathsf{p}}_1(z)}{\mathsf{S}\tilde{\mathsf{p}}_2(z)} = \frac{1}{z^{-1}} = z$.

We now deal with the factors in (3.1.12). Note that both $\mathsf{Z}(\mathsf{p}, 1)$ and $\mathsf{Z}(\mathsf{p}, -1)$ are even integers by $-1, 1 \in \mathbb{T}$. If $\mathsf{Z}(\mathsf{p}, x_0)$ is an even integer, then it is trivial to have

$$\mathring{\mathsf{p}}_1(z)\mathring{\mathsf{p}}_1^\star(z) + \mathring{\mathsf{p}}_2(z)\mathring{\mathsf{p}}_2^\star(z) = [\tilde{\mathsf{p}}(z|x_0)]^{\mathsf{Z}(\mathsf{p},x_0)} \quad \text{and} \quad \frac{\mathsf{S}\mathring{\mathsf{p}}_1(z)}{\mathsf{S}\mathring{\mathsf{p}}_2(z)} = \epsilon z^c,$$

for $\epsilon \in \{-1, 1\}$ and $c \in \{0, 1\}$ with the trivial choice $\mathring{\mathsf{p}}_1(z) := [\tilde{\mathsf{p}}(z|x_0)]^{\mathsf{Z}(\mathsf{p},x_0)/2}$ and $\mathring{\mathsf{p}}_2(z) := 0$. Note that $\mathsf{S}0$ can be assigned any symmetry type. By (3) in item (ii), this proves the claim for the case $\epsilon = 1$ and $c = 0$ in item (3).

We now handle the factor $\tilde{\mathsf{p}}(z|x_0) = 1 + x_0^2 - x_0(z^{-1} + z)$, $x_0 \in (-1, 1)\backslash\{0\}$ such that $\mathsf{Z}(\mathsf{p}, x_0)$ is an odd integer. For the case $\epsilon = -1$ and $c = 1$, since $\mathsf{Z}(\mathsf{p}, x_0)$ is odd, by item (2), we must have $x_0 \in (0, 1)$. Therefore,

$$\mathring{\mathsf{p}}_1(z)\mathring{\mathsf{p}}_1^\star(z) + \mathring{\mathsf{p}}_2(z)\mathring{\mathsf{p}}_2^\star(z) = \tilde{\mathsf{p}}(z|x_0) \quad \text{with} \quad \frac{\mathsf{S}\mathring{\mathsf{p}}_1(z)}{\mathsf{S}\mathring{\mathsf{p}}_2(z)} = \epsilon z^c \tag{3.1.13}$$

holds with $\mathring{\mathsf{p}}_1(z) = 1 - x_0$ and $\mathring{\mathsf{p}}_2(z) = \sqrt{x_0}(z^{-1} - 1)$ by $\mathsf{S}\mathring{\mathsf{p}}_1 = 1$ and $\mathsf{S}\mathring{\mathsf{p}}_2 = -z^{-1}$.

For $\epsilon = 1$ and $c = 1$, since $\mathsf{Z}(\mathsf{p}, x_0)$ is odd, by item (4), we must have $x_0 \in (-1, 0)$. Thus, (3.1.13) holds with $\mathring{\mathsf{p}}_1(z) = 1 + x_0$ and $\mathring{\mathsf{p}}_2(z) = \sqrt{-x_0}(z^{-1} + 1)$ by $\mathsf{S}\mathring{\mathsf{p}}_1 = 1$ and $\mathsf{S}\mathring{\mathsf{p}}_2 = z^{-1}$. Now it follows from Lemma 3.1.4 that the claim in (3.1.7) holds for all the cases $\epsilon \in \{-1, 1\}$ and $c \in \{0, 1\}$. This also proves that if items (i) and (ii) hold, then there exist Laurent polynomials p_1 and p_2 with real coefficients and symmetry such that (3.1.7) is satisfied. □

3.1.3 Splitting Nonnegative Laurent Polynomials with Symmetry

Let us now consider the problem on sum of squares using only one Laurent polynomial with [complex] symmetry. In other words, we study the Fejér-Riesz Lemma under the symmetry property constraint.

For a complex number $z \in \mathbb{C}$, its sign is defined to be

$$\mathrm{sgn}(z) := \begin{cases} \frac{z}{|z|}, & \text{if } z \in \mathbb{C}\backslash\{0\}; \\ 1, & \text{if } z = 0. \end{cases} \tag{3.1.14}$$

For a real number $x \in \mathbb{R}$, we have $\mathrm{sgn}(x) = 1$ if $x \geqslant 0$ and $\mathrm{sgn}(x) = -1$ if $x < 0$.

For the case of complex symmetry, we have

Theorem 3.1.7 *Let* $\mathsf{p}(z) = p_0 + \sum_{k=1}^{N}(p_k z^k + \overline{p_k} z^{-k})$ *with* $p_N \neq 0, p_0 \in \mathbb{R}$ *and* $p_1, \ldots, p_N \in \mathbb{C}$. *Then the following statements are equivalent:*

(1) There exists a Laurent polynomial q *having complex symmetry and satisfying* $\mathsf{p}(z) = \mathsf{q}(z)\mathsf{q}^\star(z)$ *for all* $z \in \mathbb{C}\backslash\{0\}$.

(2) $\mathsf{p}(z) \geqslant 0$ *for all* $z \in \mathbb{T}$ *and the following relation holds:*

$$\mathsf{Z}(\mathsf{p}, z) \in 2\mathbb{Z} \qquad \forall\, z \in \mathbb{C}\backslash\{0\}. \tag{3.1.15}$$

(3) There exists a nonzero number $\lambda \in \mathbb{C}$ *such that the Laurent polynomial* $\lambda\mathsf{q}$ *has complex symmetry and* $[\lambda\mathsf{q}(z)][\lambda\mathsf{q}(z)]^\star = \mathsf{p}(z)$ *for all* $z \in \mathbb{C}\backslash\{0\}$, *where*

$$\mathsf{q}(z) := \prod_{z_0 \in \mathbb{C}\backslash\{0\}} (z - z_0)^{\mathsf{Z}(\mathsf{p}, z_0)/2}, \qquad z \in \mathbb{C}\backslash\{0\}. \tag{3.1.16}$$

Moreover, if p *has real coefficients, then* $\lambda\mathsf{q}$ *can also have real coefficients.*

(4) $\mathsf{p}(z) = \mathsf{q}_{\mathsf{p}}(z)\mathsf{q}_{\mathsf{p}}^\star(z)$ *for all* $z \in \mathbb{C}\backslash\{0\}$, *where the Laurent polynomial* q_{p} *is defined by one of the following two cases:*

Case 1: If $N = 2n$ *for some* $n \in \mathbb{N}_0$ *(i.e., N is an even integer), define* $\mathsf{q}_{\mathsf{p}}(z) := \sqrt{|p_N|}\Big[\frac{e^{i\alpha}t_0 + e^{-i\alpha}\overline{t_0}}{2} + \sum_{k=1}^{n}(e^{i\alpha}t_k z^k + e^{-i\alpha}\overline{t_k}z^{-k})\Big]$.

Case 2: If $N = 2n+1$ *for* $n \in \mathbb{N}_0$, *define* $\mathsf{q}_{\mathsf{p}}(z) := \sqrt{|p_N|}\sum_{k=0}^{n}(e^{i\alpha}t_k z^k + e^{-i\alpha}\overline{t_k}z^{-1-k})$,

where α *is a real number satisfying* $e^{i\alpha} := \mathrm{sgn}(\sqrt{p_N})$, $t_n := 1$ *and recursively*

$$t_{n-j} := \frac{1}{2}\Big[\frac{p_{N-j}}{p_N} - \sum_{k=n-j+1}^{n-1} t_k t_{2n-j-k}\Big], \qquad j = 1, \ldots, n. \tag{3.1.17}$$

If in addition the Laurent polynomial p *has real coefficients, replace* q_{p} *by* $i\mathsf{q}_{\mathsf{p}}$ *if* $p_N < 0$, *then* q_{p} *also has real coefficients.*

Proof (1)$\Longrightarrow$(2). Since q has complex symmetry, we have $\mathsf{p}(z) = \mathsf{q}(z)\mathsf{q}^\star(z) = [\mathsf{q}(z)]^2/\mathbb{S}\mathsf{q}(z)$. So, $\mathsf{Z}(\mathsf{p}, z) = 2\mathsf{Z}(\mathsf{q}, z) \in 2\mathbb{Z}$ for all $z \in \mathbb{C}\backslash\{0\}$ and $\mathsf{p}(z) \geqslant 0$, $\forall\, z \in \mathbb{T}$.

(2)$\Longrightarrow$(3). By (3.1.15), the Laurent polynomial q in (3.1.16) is a well-defined polynomial. Since $\mathsf{p}^\star = \mathsf{p}$, we see that (3.1.4) holds. Consequently, by the definition of q in (3.1.16), we see that $\mathsf{Z}(\mathsf{q}, z) = \mathsf{Z}(\mathsf{q}, \overline{z}^{-1})$ for all $z \in \mathbb{C}\backslash\{0\}$. Note that if p has real coefficients, then q also has real coefficients. By Proposition 3.1.1, there is $\tilde{\lambda} \in \mathbb{T}$ ($\tilde{\lambda} = 1$ if q has real coefficients) such that $\tilde{\lambda}\mathsf{q}$ has complex symmetry. It is also easy to see that there exists a positive number ρ such that $\rho[\tilde{\lambda}\mathsf{q}(z)][\tilde{\lambda}\mathsf{q}(z)]^\star = \mathsf{p}(z)$ for all $z \in \mathbb{C}\backslash\{0\}$. Define $\lambda := \tilde{\lambda}\sqrt{\rho}$. Then item (3) holds.

(3)$\Longrightarrow$(1) is trivial. Item (4) is an explicit way of finding a particular q in item (1) and the solution in (3.1.17) is obtained by directly comparing the first $n+1$ highest terms in the Laurent polynomials $\mathsf{q}_{\mathsf{p}}(z)\mathsf{q}_{\mathsf{p}}^{\star}(z)$ and $\mathsf{p}(z)$. □

By item (3) of Theorem 3.1.7, up to a multiplicative monomial λz^k with $\lambda \in \{\pm 1, \pm i\}$ and $k \in \mathbb{Z}$, a Laurent polynomial q, having complex symmetry and satisfying $\mathsf{p}(z) = \mathsf{q}(z)\mathsf{q}^{\star}(z)$ in Theorem 3.1.7, is unique.

For a Laurent polynomial p having symmetry, we have

Theorem 3.1.8 *Let $\mathsf{p}(z) = p_0 + \sum_{k=1}^{N} p_k(z^k + z^{-k})$ with $p_N \ne 0$ be a Laurent polynomial with complex coefficients. Then the following statements are equivalent:*

(1) $\mathsf{p}(z) = \mathsf{q}(z)\mathsf{q}^{\star}(z)$ *for some Laurent polynomial* q *having symmetry.*

(2) $\mathsf{p}(z) = \mathsf{q}^{[r]}(z)(\mathsf{q}^{[r]}(z))^{\star} + \mathsf{q}^{[i]}(z)(\mathsf{q}^{[i]}(z))^{\star}$ *for some Laurent polynomials* $\mathsf{q}^{[r]}$ *and* $\mathsf{q}^{[i]}$ *having real coefficients and symmetry such that* $\mathsf{S}\mathsf{q}^{[r]}(z) = \mathsf{S}\mathsf{q}^{[i]}(z)$.

(3) All $p_0, \ldots, p_N \in \mathbb{R}$, $\mathsf{p}(z) \geqslant 0$ *for all* $z \in \mathbb{T}$, *and* $\mathsf{Z}(\mathsf{p}, x) \in 2\mathbb{Z}$ *for all* $x \in (-1, 0) \cup (0, 1)$ *(the last condition can be replaced by* $\mathsf{Z}(\mathsf{p}, x) \in 2\mathbb{Z}$ *for all* $x \in \mathbb{R}\backslash\{0\}$*).*

(4) $\mathsf{Z}(\mathsf{p}, 1)$ *and* $\mathsf{Z}(\mathsf{p}, -1)$ *are even integers and* $\mathsf{p}(z) = \zeta^{\mathsf{Z}(\mathsf{p},1)/2}(1-\zeta)^{\mathsf{Z}(\mathsf{p},-1)/2}\mathsf{P}(\zeta)$ *with* $\zeta := \frac{1}{2} - \frac{1}{4}z - \frac{1}{4}z^{-1}$, *i.e.,* $\mathsf{p}(e^{-i\xi}) = \sin^{\mathsf{Z}(\mathsf{p},1)}(\frac{\xi}{2})\cos^{\mathsf{Z}(\mathsf{p},-1)}(\frac{\xi}{2})\mathsf{P}(\sin^2(\xi/2))$, *where* P *is a polynomial having real coefficients and satisfying* $\mathsf{P}(x) \geqslant 0$, $\forall\, x \in \mathbb{R}$.

(5) All coefficients $p_0, \ldots, p_N \in \mathbb{R}$ *and* $\mathsf{p}(z) = \mathsf{q}_{\mathsf{p}}(z)\mathsf{q}_{\mathsf{p}}^{\star}(z)$, *where the Laurent polynomial* q_{p} *is defined by one of the following two cases:*

Case 1: If $N = 2n$ *for some* $n \in \mathbb{N}_0$ *(i.e., N is an even integer), define* $\mathsf{q}_{\mathsf{p}}(z) := \sqrt{|p_N|}\Big[\frac{1+sgn(p_N)}{2}t_0 + \sum_{k=1}^{n} t_k(z^k + sgn(p_N)z^{-k})\Big]$;

Case 2: If $N = 2n+1$ *for* $n \in \mathbb{N}_0$, *define* $\mathsf{q}_{\mathsf{p}}(z) := \sqrt{|p_N|}\sum_{k=0}^{n} t_k(z^k + sgn(p_N)z^{-1-k})$,

where $t_n := 1$ *and*

$$\operatorname{Re}(t_{n-j}) := \frac{1}{2}\Big[\frac{p_{N-j}}{p_N} - \sum_{k=n-j+1}^{n-1} \operatorname{Re}(t_k\overline{t_{2n-j-k}})\Big], \qquad j = 1, \ldots, n. \tag{3.1.18}$$

Proof (1)$\Longrightarrow$(2). Write $\mathsf{q} = \mathsf{q}^{[r]} + i\mathsf{q}^{[i]}$, where $\mathsf{q}^{[r]}$ and $\mathsf{q}^{[i]}$ are Laurent polynomials having real coefficients. Since q has symmetry, we have $\mathsf{S}\mathsf{q}^{[r]}(z) = \mathsf{S}\mathsf{q}^{[i]}(z) = \mathsf{S}\mathsf{q}(z)$ and $\mathsf{q}^{[r]}(z)(\mathsf{q}^{[r]}(z))^{\star} + \mathsf{q}^{[i]}(z)(\mathsf{q}^{[i]}(z))^{\star} = \mathsf{q}(z)\mathsf{q}^{\star}(z) = \mathsf{p}(z)$. So, (1)$\Longrightarrow$(2). Conversely, we take $\mathsf{q} = \mathsf{q}^{[r]} + i\mathsf{q}^{[i]}$ and therefore, (2)$\Longrightarrow$(1).

The equivalence between items (2) and (3) is proved in Theorem 3.1.6. We now prove (3)$\Longrightarrow$(4). Since $\mathsf{S}\mathsf{p} = 1$ and p has real coefficients, by Exercise 2.8, we can always write $\mathsf{p}(z) = \zeta^{\mathsf{Z}(\mathsf{p},1)/2}(1-\zeta)^{\mathsf{Z}(\mathsf{p},-1)/2}\mathsf{P}(\zeta)$ with $\zeta = \frac{1}{2} - \frac{1}{4}z - \frac{1}{4}z^{-1}$, where P is a polynomial with real coefficients. Consider the map $\eta : \mathbb{C}\backslash\{0\} \to \mathbb{C}$ with $\eta(z) = \frac{1}{2} - \frac{1}{4}z - \frac{1}{4}z^{-1}$. Then η is a bijection from $\{z \in \mathbb{T} : \operatorname{Im}(z) > 0\}$ to $(0, 1)$ and from $(-1, 0) \cup (0, 1)$ to $\mathbb{R}\backslash[0, 1]$. Note that $\mathsf{Z}(\mathsf{P}, 0) = \mathsf{Z}(\mathsf{P}, 1) = 0$. By item (3), we must have $\mathsf{Z}(\mathsf{P}, x) \in 2\mathbb{Z}$ for all $x \in \mathbb{R}$. Since P has real coefficients, either $\mathsf{P}(x) \geqslant 0$

or $\mathsf{P}(x) \leqslant 0$ for all $x \in \mathbb{R}$. Since $\sin^{\mathsf{Z}(\mathsf{p},1)}(\xi/2)\cos^{\mathsf{Z}(\mathsf{p},-1)}(\xi/2)\mathsf{P}(\sin^2(\xi/2)) = \mathsf{p}(e^{-i\xi}) \geqslant 0$ for all $\xi \in \mathbb{R}$ and since both $\mathsf{Z}(\mathsf{p},1)$ and $\mathsf{Z}(\mathsf{p},-1)$ are even, we must have $\mathsf{P}(\sin^2(\xi/2)) \geqslant 0$ for all $\xi \in \mathbb{R}$, i.e., $\mathsf{P}(x) \geqslant 0$ for all $x \in [0,1]$. Consequently, by $\mathsf{P}(0) \neq 0$, we must have $\mathsf{P}(0) \geqslant 0$ and hence $\mathsf{P}(x) \geqslant 0$ for all $x \in \mathbb{R}$.

(4)$\Longrightarrow$(2) follows directly from Lemma 2.4.2. Item (5) is an explicit way of finding a particular Laurent polynomial q in item (1) and (3.1.18) is obtained by directly comparing the first $n+1$ highest terms in $\mathsf{q}_\mathsf{p}(z)\mathsf{q}_\mathsf{p}^\star(z)$ and $\mathsf{p}(z)$. □

There are often finitely many but essentially different solutions q having symmetry and satisfying $\mathsf{p}(z) = \mathsf{q}(z)\mathsf{q}^\star(z)$. Nevertheless, up to a factor z^{2k} with $k \in \mathbb{Z}$, its symmetry type Sq is uniquely determined by p. This fact is left as Exercise 3.27.

3.2 Dual Framelet Filter Banks with Symmetry and Two High-Pass Filters

In this section, we shall present an algorithm to construct dual framelet filter banks $(\{\tilde{a};\tilde{b}_1,\tilde{b}_2\},\{a;b_1,b_2\})_\Theta$ having the shortest possible filter supports with or without [complex] symmetry.

For a general dual framelet filter bank $(\{\tilde{a};\tilde{b}_1,\ldots,\tilde{b}_s\},\{a;b_1,\ldots,b_s\})_\Theta$, it is desirable for the high-pass filters to possess certain numbers of vanishing moments:

$$\mathsf{b}_\ell(z) = (1-z^{-1})^{n_b}\mathring{\mathsf{b}}_\ell(z), \qquad \tilde{\mathsf{b}}_\ell(z) = (1-z^{-1})^{n_{\tilde{b}}}\mathring{\tilde{\mathsf{b}}}_\ell(z), \qquad \ell = 1,\ldots,s,$$

where n_b and $n_{\tilde{b}}$ are nonnegative integers. That is, all the primal high-pass filters $b_1,\ldots,b_s$ have at least n_b vanishing moments, while all the dual high-pass filters $\tilde{b}_1,\ldots,\tilde{b}_s$ have at least $n_{\tilde{b}}$ vanishing moments. Now the perfect reconstruction condition in (3.0.1) for a general dual framelet filter bank $(\{\tilde{a};\tilde{b}_1,\ldots,\tilde{b}_s\},\{a;b_1,\ldots,b_s\})_\Theta$ can be equivalently expressed as

$$\begin{bmatrix} \mathring{\tilde{\mathsf{b}}}_1(z) & \cdots & \mathring{\tilde{\mathsf{b}}}_s(z) \\ \mathring{\tilde{\mathsf{b}}}_1(-z) & \cdots & \mathring{\tilde{\mathsf{b}}}_s(-z) \end{bmatrix} \begin{bmatrix} \mathring{\mathsf{b}}_1(z) & \cdots & \mathring{\mathsf{b}}_s(z) \\ \mathring{\mathsf{b}}_1(-z) & \cdots & \mathring{\mathsf{b}}_s(-z) \end{bmatrix}^\star = \mathcal{M}_{a,\tilde{a},\Theta|n_b,n_{\tilde{b}}}(z), \tag{3.2.1}$$

where

$$\mathcal{M}_{a,\tilde{a},\Theta|n_b,n_{\tilde{b}}}(z) := \begin{bmatrix} \mathsf{A}(z) & \mathsf{B}(z) \\ \mathsf{B}(-z) & \mathsf{A}(-z) \end{bmatrix} \tag{3.2.2}$$

with

$$\mathsf{A}(z) := \frac{\Theta(z) - \Theta(z^2)\tilde{\mathsf{a}}(z)\mathsf{a}^\star(z)}{(1-z)^{n_b}(1-z^{-1})^{n_{\tilde{b}}}}, \quad \mathsf{B}(z) := -\Theta(z^2)\frac{\tilde{\mathsf{a}}(z)}{(1+z)^{n_b}}\frac{\mathsf{a}^\star(-z)}{(1-z^{-1})^{n_{\tilde{b}}}}. \tag{3.2.3}$$

Now the construction of a dual framelet filter bank with preassigned orders of vanishing moments is simply to properly factorize the matrix $\mathcal{M}_{a,\tilde{a},\Theta|n_b,n_{\tilde{b}}}$ in (3.2.2) so that (3.2.1) is satisfied.

As shown in Theorem 1.4.7, it is often necessary that $s > 1$. Hence, to reduce computational complexity, we only consider the case $s = 2$ for a dual framelet filter bank. For the case $s = 2$, the equation (3.2.1) is equivalent to

$$\begin{bmatrix} \mathring{\mathsf{b}}_1(z) & \mathring{\mathsf{b}}_2(z) \\ \mathring{\mathsf{b}}_1(-z) & \mathring{\mathsf{b}}_2(-z) \end{bmatrix} \begin{bmatrix} \mathring{\tilde{\mathsf{b}}}_1(z) & \mathring{\tilde{\mathsf{b}}}_2(z) \\ \mathring{\tilde{\mathsf{b}}}_1(-z) & \mathring{\tilde{\mathsf{b}}}_2(-z) \end{bmatrix}^\star = \begin{bmatrix} \mathsf{A}^\star(z) & \mathsf{B}^\star(-z) \\ \mathsf{B}^\star(z) & \mathsf{A}^\star(-z) \end{bmatrix}. \tag{3.2.4}$$

It is easy to find particular solutions to (3.2.4) by choosing $\mathring{\mathsf{b}}_1$ and $\mathring{\mathsf{b}}_2$ in such a way that the determinant of the first 2×2 matrix on the left-hand side of (3.2.4) is a nonzero monomial, e.g., $(\{\tilde{a}; \tilde{b}_1, \tilde{b}_2\}, \{a; b_1, b_2\})_\Theta$ is a dual framelet filter bank if

$$\begin{aligned} \mathring{\mathsf{b}}_1(z) &= 1, & \mathring{\mathsf{b}}_2(z) &= z, \\ \mathring{\tilde{\mathsf{b}}}_1(z) &= [\mathsf{A}(z) + \mathsf{B}(z)]/2, & \mathring{\tilde{\mathsf{b}}}_2(z) &= z[\mathsf{A}(z) - \mathsf{B}(z)]/2 \end{aligned} \tag{3.2.5}$$

or

$$\begin{aligned} \mathring{\mathsf{b}}_1(z) &= (1+z)/2, & \mathring{\mathsf{b}}_2(z) &= (1-z)/2, \\ \mathring{\tilde{\mathsf{b}}}_1(z) &= \mathring{\mathsf{b}}_1(z)\mathsf{A}(z) + \mathring{\mathsf{b}}_2(z)\mathsf{B}(z), & \mathring{\tilde{\mathsf{b}}}_2(z) &= \mathring{\mathsf{b}}_2(z)\mathsf{A}(z) + \mathring{\mathsf{b}}_1(z)\mathsf{B}(z). \end{aligned} \tag{3.2.6}$$

We now discuss dual framelet filter banks with [complex] symmetry. Assume that all the filters $\Theta, a, \tilde{a}$ have the following symmetry (or complex symmetry by replacing S with $\mathbb{S}$):

$$\mathsf{S}\Theta(z) = \epsilon_\Theta z^{c_\Theta}, \quad \mathsf{S}\tilde{\mathsf{a}}(z) = \epsilon z^{c}, \quad \mathsf{S}\tilde{\mathsf{a}}(z) = \tilde{\epsilon} z^{\tilde{c}}. \tag{3.2.7}$$

In order for the Laurent polynomial A to have [complex] symmetry, by Exercise 3.13, it is natural to assume that

$$\tilde{c} = c - c_\Theta \quad \text{and} \quad \tilde{\epsilon} = \epsilon. \tag{3.2.8}$$

Then A and B have the following [complex] symmetry:

$$\mathsf{SA}(z) = (-1)^{n_{\tilde{b}}+n_b} \epsilon_\Theta z^{c_\Theta + n_{\tilde{b}} - n_b}, \quad \mathsf{SB}(z) = (-1)^{c+n_{\tilde{b}}} \epsilon_\Theta z^{c_\Theta + n_{\tilde{b}} - n_b}.$$

Note that the filters $\mathring{\tilde{b}}_1$ and $\mathring{\tilde{b}}_2$ in (3.2.5) have symmetry if $c + n_b$ is even (but often lose symmetry if $c + n_b$ is odd), while the filters $\mathring{\tilde{b}}_1$ and $\mathring{\tilde{b}}_2$ in (3.2.6) have symmetry if $c + n_b$ is odd (but often lose symmetry if $c + n_b$ is even).

If all the filters $\tilde{a}, \tilde{b}_1, \tilde{b}_2, a, b_1, b_2, \Theta$ are required to have symmetry, by Exercise 3.13, it is natural to require $\mathsf{S}(\tilde{\mathsf{b}}_1(z)\mathsf{b}_1^\star(-z)) = \mathsf{S}(\tilde{\mathsf{b}}_2(z)\mathsf{b}_2^\star(-z))$. Therefore, from the perfect reconstruction condition in (3.0.1), we must have the following relation on the lengths of filter supports of the high-pass filters:

$$\max(\mathrm{len}(b_1) + \mathrm{len}(\tilde{b}_1), \mathrm{len}(b_2) + \mathrm{len}(\tilde{b}_2)) = \mathrm{len}(a) + \mathrm{len}(\tilde{a}) + 2\,\mathrm{len}(\Theta) + 2\epsilon_{\mathrm{len}}$$

with $\epsilon_{\mathrm{len}} \in \mathbb{N}_0$. Our goal in this section is to find all possible dual framelet filter banks $(\{\tilde{a}; \tilde{b}_1, \tilde{b}_2\}, \{a; b_1, b_2\})_\Theta$ having the shortest possible filter supports from any given filters $a, \tilde{a}$ and Θ with or without [complex] symmetry.

To guarantee that both A and B defined in (3.2.3) are Laurent polynomials, under the natural assumption that $\Theta(1)\tilde{\mathsf{a}}(1)\mathsf{a}(1) \neq 0$, it is necessary and sufficient to assume (see Lemma 1.4.1 and Exercise 3.10) that

$$\begin{aligned} &0 \leqslant n_b \leqslant \mathrm{sr}(\tilde{a}), \quad 0 \leqslant n_{\tilde{b}} \leqslant \mathrm{sr}(a), \\ &\Theta(z) - \Theta(z^2)\tilde{\mathsf{a}}(z)\mathsf{a}^\star(z) = \mathscr{O}(|1-z|^{n_b+n_{\tilde{b}}}), \quad z \to 1. \end{aligned} \tag{3.2.9}$$

Due to the role of Θ in (3.2.9) to achieve the high vanishing moment order $\mathrm{vm}(\Theta(z) - \Theta(z^2)\tilde{\mathsf{a}}(z)\mathsf{a}^\star(z))$, we call such a filter Θ *a moment correcting filter*. Recall that the $\mathrm{odd}(n)$ function is defined in (2.5.10).

We now present an algorithm to construct all possible dual framelet filter banks having symmetry or complex symmetry and having real coefficients or complex coefficients with short filter supports.

Algorithm 3.2.1 *Let $a, \tilde{a}, \Theta \in l_0(\mathbb{Z})$ be given filters having [complex] symmetry (and real coefficients) satisfying* (3.2.7), (3.2.8), *and* (3.2.9) *for some $c, \tilde{c}, c_\Theta \in \mathbb{Z}$, $\epsilon, \tilde{\epsilon}, \epsilon_\Theta \in \{-1, 1\}$, and $n_b, n_{\tilde{b}} \in \mathbb{N}_0$. Assume that* $\mathrm{len}(a) + \mathrm{len}(\tilde{a}) + \mathrm{len}(\Theta) > 0$ *(that is, $\mathsf{a}, \tilde{\mathsf{a}}, \Theta$ cannot be simultaneously monomials). For the case of complex symmetry, replace the symmetry operator S by the complex symmetry operator $\mathbb{S}$ throughout.*

(S1) Define Laurent polynomials A and B as in (3.2.3), *and*

$$\begin{aligned} &\mathsf{p}(z^2) := \gcd(\mathsf{A}(z), \mathsf{A}(-z), \mathsf{B}(z), \mathsf{B}(-z)), \\ &\mathring{\mathsf{A}}(z) := \mathsf{A}(z)/\mathsf{p}(z^2), \quad \mathring{\mathsf{B}}(z) := \mathsf{B}(z)/\mathsf{p}(z^2). \end{aligned} \tag{3.2.10}$$

Then the Laurent polynomials $\mathsf{p}, \mathsf{A}, \mathsf{B}, \mathring{\mathsf{A}}, \mathring{\mathsf{B}}$ have [complex] symmetry (and real coefficients). Define ϵ_0, c_0, n_0 by

$$\epsilon_0 z^{c_0} := \mathsf{S}\mathring{\mathsf{A}}(z) \quad \textit{and} \quad [c_0 - n_0, n_0] := \mathrm{fsupp}(\mathring{\mathsf{A}});$$

(S2) Select $\mathsf{d}, c_1, \epsilon_1, n_1, n_2, \epsilon_{\mathrm{len}}$ as follows:

(1) Select a Laurent polynomial d with [complex] symmetry (and real coefficients) satisfying

$$\mathsf{d}(z) \mid \mathsf{D}(z) \quad \textit{with} \quad \mathsf{D}(z^2) := [\mathring{\mathsf{A}}(z)\mathring{\mathsf{A}}(-z) - \mathring{\mathsf{B}}(z)\mathring{\mathsf{B}}(-z)]^\star. \tag{3.2.11}$$

Define $\epsilon_{\mathsf{d}}, c_{\mathsf{d}}, n_{\mathsf{d}}$ *by* $\epsilon_{\mathsf{d}} z^{c_{\mathsf{d}}} := \mathsf{Sd}(z)$, $[c_{\mathsf{d}} - n_{\mathsf{d}}, n_{\mathsf{d}}] := \mathrm{fsupp}(\mathsf{d})$. *Without loss of generality, we often restrict* $c_{\mathsf{d}} \in \{0,1\}$ *by multiplying* d *with a monomial;*

(2) *Select* $c_1 \in \{\mathrm{odd}(c+n_b), \mathrm{odd}(c+n_b)+2\}$. *Define* $c_2 := 2c_{\mathsf{d}} + 2 - c_1$;

(3) *Set* $\epsilon_1 = 1$ *if* $(-1)^{c_1}\epsilon_{\mathsf{d}} = -1$; *otherwise,* $\epsilon_1 \in \{-1,1\}$. *Define* $\epsilon_2 := (-1)^{c_1}\epsilon_{\mathsf{d}}\epsilon_1$;

(4) *Set* $\epsilon_{\mathrm{len}} = 0$ *for the shortest possible filter support; otherwise, select* $\epsilon_{\mathrm{len}} = 1$;

(5) *Select* $n_1 \in \mathbb{Z}$ *satisfying* $\frac{c_1}{2} \leqslant n_1 \leqslant \frac{c_1 - c_0}{2} + n_0 + \epsilon_{\mathrm{len}}$;

(6) *Select* $n_2 \in \mathbb{Z}$ *satisfying* $\max(\frac{c_2}{2}, 2n_{\mathsf{d}} + 1 - n_1) \leqslant n_2 \leqslant \frac{c_2 - c_0}{2} + n_0 + \epsilon_{\mathrm{len}}$;

(S3) *Parameterize a filter* $\mathring{\mathsf{b}}_1$ *such that* $\mathsf{S}\mathring{\mathsf{b}}_1(z) = \epsilon_1 z^{c_1}$ *and* $\mathrm{fsupp}(\mathring{\mathsf{b}}_1) = [c_1 - n_1, n_1]$. *Find the unknown coefficients of* $\mathring{\mathsf{b}}_1$ *by solving a system* X_1 *of linear equations induced by* $\mathcal{R}_1(z) = 0$ *(i.e., all the coefficients of* $\mathcal{R}_1$ *are zero) and*

$$coeff(\mathring{\tilde{\mathsf{b}}}_2^{\star}, z, j) = 0, \quad n_0 - n_2 - c_0 + 1 + \epsilon_{\mathrm{len}} \leqslant j \leqslant n_0 + n_1 - c_0 - 2n_{\mathsf{d}} - 1, \tag{3.2.12}$$

where $\mathcal{R}_1$ *and* $\mathring{\tilde{\mathsf{b}}}_2^{\star}$ *are Laurent polynomials uniquely determined, through long division using the divisor* $\mathsf{d}(z^2)$, *by* $\mathrm{fsupp}(\mathcal{R}_1) \subseteq [2(c_{\mathsf{d}} - n_{\mathsf{d}}), 2n_{\mathsf{d}} - 1]$ *and*

$$\mathring{\mathsf{B}}^{\star}(z)\mathring{\mathsf{b}}_1(z) - \mathring{\mathsf{A}}^{\star}(z)\mathring{\mathsf{b}}_1(-z) = \mathsf{d}(z^2) z \mathring{\tilde{\mathsf{b}}}_2^{\star}(z) + \mathcal{R}_1(z). \tag{3.2.13}$$

Note that $coeff(\mathring{\tilde{\mathsf{b}}}_2^{\star}, z, j)$ *stands for the coefficient of* z^j *in the Laurent polynomial* $\mathring{\tilde{\mathsf{b}}}_2^{\star}$. *If* X_1 *has no nontrivial solution, then restart the algorithm from (S2) by selecting other choices of* $\mathsf{d}, c_1, \epsilon_1, n_1, n_2, \epsilon_{\mathrm{len}}$;

(S4) *Parameterize a filter* $\mathring{\mathsf{b}}_2$ *such that* $\mathsf{S}\mathring{\mathsf{b}}_2(z) = \epsilon_2 z^{c_2}$ *and* $\mathrm{fsupp}(\mathring{\mathsf{b}}_2) = [c_2 - n_2, n_2]$. *Find the unknown coefficients of the filter* $\mathring{\mathsf{b}}_2$ *by solving a system* X_2 *of linear equations induced by* $\mathcal{R}_2(z) = 0$ *(i.e., all the coefficients of* $\mathcal{R}_2$ *are zero) and*

$$coeff(\mathring{\tilde{\mathsf{b}}}_1^{\star}, z, j) = 0, \quad n_0 - n_1 - c_0 + 1 + \epsilon_{\mathrm{len}} \leqslant j \leqslant n_0 + n_2 - c_0 - 2n_{\mathsf{d}} - 1, \tag{3.2.14}$$

where $\mathcal{R}_2$ *and* $\mathring{\tilde{\mathsf{b}}}_1^{\star}$ *are Laurent polynomials uniquely determined, through long division using the divisor* $\mathsf{d}(z^2)$, *by* $\mathrm{fsupp}(\mathcal{R}_2) \subseteq [2(c_{\mathsf{d}} - n_{\mathsf{d}}), 2n_{\mathsf{d}} - 1]$ *and*

$$\mathring{\mathsf{B}}^{\star}(z)\mathring{\mathsf{b}}_2(z) - \mathring{\mathsf{A}}^{\star}(z)\mathring{\mathsf{b}}_2(-z) = -\mathsf{d}(z^2) z \mathring{\tilde{\mathsf{b}}}_1^{\star}(z) + \mathcal{R}_2(z). \tag{3.2.15}$$

If X_2 *has no nontrivial solution, then restart the algorithm from (S2) by selecting other choices of* $\mathsf{d}, c_1, \epsilon_1, n_1, n_2, \epsilon_{\mathrm{len}}$;

(S5) There must exist a complex number $\lambda \in \mathbb{C}$ such that

$$\lambda \mathsf{d}(z^2) = z^{-1}[\mathring{\mathsf{b}}_1(z)\mathring{\mathsf{b}}_2(-z) - \mathring{\mathsf{b}}_1(-z)\mathring{\mathsf{b}}_2(z)]. \tag{3.2.16}$$

If $\lambda = 0$, then restart from (S2) by selecting other choices of d, c_1, ϵ_1, $n_1, n_2, \epsilon_{\text{len}}$; Otherwise, for $\lambda \neq 0$, replace $\mathring{\tilde{\mathsf{b}}}_1, \mathring{\tilde{\mathsf{b}}}_2$ by $\bar{\lambda}^{-1}\mathring{\tilde{\mathsf{b}}}_1, \bar{\lambda}^{-1}\mathring{\tilde{\mathsf{b}}}_2$, respectively. Moreover,

$$\mathsf{S}\mathring{\tilde{\mathsf{b}}}_1(z) = \epsilon_0\epsilon_1 z^{c_0+c_1}, \qquad \mathsf{S}\mathring{\tilde{\mathsf{b}}}_2(z) = \epsilon_0\epsilon_2 z^{c_0+c_2}; \tag{3.2.17}$$

(S6) Find Laurent polynomials q and $\tilde{\mathsf{q}}$ having [complex] symmetry (and real coefficients) such that $\mathsf{p}(z) = \tilde{\mathsf{q}}(z)\mathsf{q}^\star(z)$. Define

$$\mathsf{b}_1(z) := (1-z^{-1})^{n_b}\mathring{\mathsf{b}}_1(z)\mathsf{q}(z^2), \qquad \mathsf{b}_2(z) := (1-z^{-1})^{n_b}\mathring{\mathsf{b}}_2(z)\mathsf{q}(z^2), \tag{3.2.18}$$

$$\tilde{\mathsf{b}}_1(z) := (1-z^{-1})^{n_{\tilde{b}}}\mathring{\tilde{\mathsf{b}}}_1(z)\tilde{\mathsf{q}}(z^2), \qquad \tilde{\mathsf{b}}_2(z) := (1-z^{-1})^{n_{\tilde{b}}}\mathring{\tilde{\mathsf{b}}}_2(z)\tilde{\mathsf{q}}(z^2). \tag{3.2.19}$$

Then $(\{\tilde{a}; \tilde{b}_1, \tilde{b}_2\}, \{a; b_1, b_2\})_\Theta$ is a dual framelet filter bank having [complex] symmetry (and real coefficients) such that $\operatorname{vm}(b_1) \geqslant n_b$, $\operatorname{vm}(b_2) \geqslant n_b$, $\operatorname{vm}(\tilde{b}_1) \geqslant n_{\tilde{b}}$, $\operatorname{vm}(\tilde{b}_2) \geqslant n_{\tilde{b}}$, *and*

$$\max(\operatorname{len}(b_1)+\operatorname{len}(\tilde{b}_1), \operatorname{len}(b_2)+\operatorname{len}(\tilde{b}_2)) \leqslant \operatorname{len}(a)+\operatorname{len}(\tilde{a})+2\operatorname{len}(\Theta)+2\epsilon_{\text{len}}. \tag{3.2.20}$$

Proof We first look at the symmetry property and filter supports of $\mathring{\mathsf{A}}$ and $\mathring{\mathsf{B}}$. By our assumption in (3.2.8), we have $\epsilon\tilde{\epsilon} = 1$, $c_\Theta = c - \tilde{c}$, and

$$\mathsf{S}(\Theta(z^2)\tilde{\mathsf{a}}(z)\mathsf{a}^\star(z)) = \epsilon_\Theta\epsilon\tilde{\epsilon} z^{2c_\Theta+\tilde{c}-c} = \epsilon_\Theta z^{c_\Theta} = \mathsf{S}\Theta(z). \tag{3.2.21}$$

Hence, both A and B have symmetry. Since $\operatorname{len}(a) + \operatorname{len}(\tilde{a}) + \operatorname{len}(\Theta) > 0$, we have

$$\operatorname{len}(\Theta) < \operatorname{len}(a) + \operatorname{len}(\tilde{a}) + 2\operatorname{len}(\Theta) = \operatorname{len}(\Theta(z^2)\tilde{\mathsf{a}}(z)\mathsf{a}^\star(z)).$$

From the definition of A and B in (3.2.3), it follows from the above relation and (3.2.21) that $\operatorname{fsupp}(\mathsf{A}) = \operatorname{fsupp}(\mathsf{B})$. Since both A and B have symmetry, the Laurent polynomial p has symmetry too. Define $\epsilon_{\mathsf{p}} z^{c_{\mathsf{p}}} := \mathsf{Sp}(z)$ to be the symmetry type of the Laurent polynomial p. By the definition of $\mathring{\mathsf{A}}$ and $\mathring{\mathsf{B}}$ in (3.2.10), we conclude that

$$\operatorname{fsupp}(\mathring{\mathsf{A}}) = \operatorname{fsupp}(\mathring{\mathsf{B}}) = [c_0 - n_0, n_0], \quad \mathsf{S}\mathring{\mathsf{A}}(z) = \epsilon_0 z^{c_0}, \quad \mathsf{S}\mathring{\mathsf{B}}(z) = \epsilon_{\mathring{\mathsf{B}}} z^{c_0} \tag{3.2.22}$$

with

$$\epsilon_0 = (-1)^{n_{\tilde{b}}+n_b}\epsilon_\Theta\epsilon_{\mathsf{p}}, \quad \epsilon_{\mathring{\mathsf{B}}} = \epsilon_0(-1)^{c+n_b}, \quad c_0 = c_\Theta + n_{\tilde{b}} - n_b - 2c_{\mathsf{p}}. \tag{3.2.23}$$

By (3.2.23) and $\mathsf{S}\mathring{\mathsf{b}}_1(z) = \epsilon_1 z^{c_1}$, we have

$$\mathsf{S}(\mathring{\mathsf{B}}^\star(z)\mathring{\mathsf{b}}_1(z)) = \epsilon_{\mathring{\mathsf{B}}}\epsilon_1 z^{-c_0}z^{c_1} = (-1)^{c+n_b}\epsilon_0\epsilon_1 z^{c_1-c_0}$$

and

$$\mathsf{S}(\mathring{\mathsf{A}}^\star(z)\mathring{\mathsf{b}}_1(-z)) = \epsilon_0\epsilon_1 z^{-c_0}(-z)^{c_1} = (-1)^{c_1}\epsilon_0\epsilon_1 z^{c_1-c_0}.$$

By item (2) of (S2), we have $(-1)^{c_1} = (-1)^{c+n_b}$. Therefore,

$$\mathsf{S}(\mathring{\mathsf{B}}^\star(z)\mathring{\mathsf{b}}_1(z)) = (-1)^{c+n_b}\epsilon_0\epsilon_1 z^{c_1-c_0} = (-1)^{c_1}\epsilon_0\epsilon_1 z^{c_1-c_0} = \mathsf{S}(\mathring{\mathsf{A}}^\star(z)\mathring{\mathsf{b}}_1(-z)).$$

Consequently, it follows from (3.2.13) and $\mathcal{R}_1 = 0$ that

$$\mathsf{S}(\mathsf{d}(z^2)z\mathring{\tilde{\mathsf{b}}}_2^\star(z)) = \mathsf{S}(\mathring{\mathsf{A}}^\star(z)\mathring{\mathsf{b}}_1(-z)) = (-1)^{c_1}\epsilon_0\epsilon_1 z^{c_1-c_0},$$

from which we conclude that $\mathring{\tilde{\mathsf{b}}}_2$ has symmetry such that

$$\begin{aligned}\mathsf{S}\mathring{\tilde{\mathsf{b}}}_2(z) &= \frac{\mathsf{S}(\mathsf{d}(z^2)z)}{\mathsf{S}(\mathring{\mathsf{A}}^\star(z)\mathring{\mathsf{b}}_1(-z))} = \frac{\epsilon_{\mathsf{d}} z^{2c_{\mathsf{d}}+2}}{(-1)^{c_1}\epsilon_0\epsilon_1 z^{c_1-c_0}}\\ &= (-1)^{c_1}\epsilon_{\mathsf{d}}\epsilon_1\epsilon_0 z^{2c_{\mathsf{d}}+2-c_1+c_0} = \epsilon_0\epsilon_2 z^{c_0+c_2},\end{aligned}$$

where we used the definition of c_2 and ϵ_2 in (S2). Hence, the second identity in (3.2.17) holds.

By (3.2.22) and (3.2.23), since $\operatorname{fsupp}(\mathring{\mathsf{A}}) = \operatorname{fsupp}(\mathring{\mathsf{B}})$, we obtain

$$\begin{aligned}\operatorname{fsupp}(\mathring{\mathsf{B}}^\star(z)\mathring{\mathsf{b}}_1(z)) &= \operatorname{fsupp}(\mathring{\mathsf{b}}_1) - \operatorname{fsupp}(\mathring{\mathsf{B}}) = \operatorname{fsupp}(\mathring{\mathsf{b}}_1) - \operatorname{fsupp}(\mathring{\mathsf{A}})\\ &= \operatorname{fsupp}(\mathring{\mathsf{A}}^\star(z)\mathring{\mathsf{b}}_1(-z)).\end{aligned}$$

Hence, by $\operatorname{fsupp}(\mathsf{d}) = [c_{\mathsf{d}} - n_{\mathsf{d}}, n_{\mathsf{d}}]$ and $\operatorname{fsupp}(\mathring{\mathsf{b}}_1) = [c_1 - n_1, n_1]$, we deduce from (3.2.13) and $\mathcal{R}_1 = 0$ that

$$\operatorname{fsupp}(\mathring{\tilde{\mathsf{b}}}_2^\star) \subseteq [c_1 - n_0 - n_1 + 2n_{\mathsf{d}} - 2c_{\mathsf{d}} - 1, n_0 + n_1 - c_0 - 2n_{\mathsf{d}} - 1].$$

By the proved symmetry property $\mathsf{S}\mathring{\tilde{\mathsf{b}}}_2(z) = \epsilon_0\epsilon_2 z^{c_0+c_2}$ and (3.2.12), using the definition $c_2 = 2c_{\mathsf{d}} + 2 - c_1$, we obtain

$$\operatorname{fsupp}(\mathring{\tilde{\mathsf{b}}}_2) \subseteq [c_0 - n_0 + n_2 - \epsilon_{\mathrm{len}}, c_2 + n_0 - n_2 + \epsilon_{\mathrm{len}}]. \tag{3.2.24}$$

By a similar argument and using (S4) instead of (S3), we can check that the first identity in (3.2.17) holds and

$$\operatorname{fsupp}(\mathring{\tilde{\mathsf{b}}}_1) \subseteq [c_0 - n_0 + n_1 - \epsilon_{\text{len}}, c_1 + n_0 - n_1 + \epsilon_{\text{len}}]. \tag{3.2.25}$$

Since $\mathcal{R}_1 = \mathcal{R}_2 = 0$, the identities (3.2.13) and (3.2.15) together imply

$$\mathsf{d}(z^2)\begin{bmatrix} z\mathring{\tilde{\mathsf{b}}}_1^\star(z) \\ z\mathring{\tilde{\mathsf{b}}}_2^\star(z) \end{bmatrix} = \begin{bmatrix} \mathring{\mathsf{b}}_2(-z) & -\mathring{\mathsf{b}}_2(z) \\ -\mathring{\mathsf{b}}_1(-z) & \mathring{\mathsf{b}}_1(z) \end{bmatrix}\begin{bmatrix} \mathring{\mathsf{A}}^\star(z) \\ \mathring{\mathsf{B}}^\star(z) \end{bmatrix}. \tag{3.2.26}$$

Multiplying $\begin{bmatrix} \mathring{\mathsf{b}}_1(z) & \mathring{\mathsf{b}}_2(z) \\ \mathring{\mathsf{b}}_1(-z) & \mathring{\mathsf{b}}_2(-z) \end{bmatrix}$ from the left on both sides of (3.2.26), we have

$$\mathsf{d}(z^2)\begin{bmatrix} \mathring{\mathsf{b}}_1(z) & \mathring{\mathsf{b}}_2(z) \\ \mathring{\mathsf{b}}_1(-z) & \mathring{\mathsf{b}}_2(-z) \end{bmatrix}\begin{bmatrix} \mathring{\tilde{\mathsf{b}}}_1^\star(z) \\ \mathring{\tilde{\mathsf{b}}}_2^\star(z) \end{bmatrix} = \mathsf{D}_{\mathring{\mathsf{b}}}(z^2)\begin{bmatrix} \mathring{\mathsf{A}}^\star(z) \\ \mathring{\mathsf{B}}^\star(z) \end{bmatrix},$$

where $\mathsf{D}_{\mathring{\mathsf{b}}}(z^2) := z^{-1}[\mathring{\mathsf{b}}_1(z)\mathring{\mathsf{b}}_2(-z) - \mathring{\mathsf{b}}_1(-z)\mathring{\mathsf{b}}_2(z)]$. From the above identity we further deduce that

$$\begin{bmatrix} \mathring{\mathsf{b}}_1(z) & \mathring{\mathsf{b}}_2(z) \\ \mathring{\mathsf{b}}_1(-z) & \mathring{\mathsf{b}}_2(-z) \end{bmatrix}\begin{bmatrix} \mathring{\tilde{\mathsf{b}}}_1(z) & \mathring{\tilde{\mathsf{b}}}_2(z) \\ \mathring{\tilde{\mathsf{b}}}_1(-z) & \mathring{\tilde{\mathsf{b}}}_2(-z) \end{bmatrix}^\star = \frac{\mathsf{D}_{\mathring{\mathsf{b}}}(z^2)}{\mathsf{d}(z^2)}\begin{bmatrix} \mathring{\mathsf{A}}^\star(z) & \mathring{\mathsf{B}}^\star(-z) \\ \mathring{\mathsf{B}}^\star(z) & \mathring{\mathsf{A}}^\star(-z) \end{bmatrix}. \tag{3.2.27}$$

Since $\gcd(\mathring{\mathsf{A}}(z), \mathring{\mathsf{A}}(-z), \mathring{\mathsf{B}}(z), \mathring{\mathsf{B}}(-z)) = 1$ by (S1), we obtain

$$\gcd(\mathring{\mathsf{A}}^\star(z), \mathring{\mathsf{A}}^\star(-z), \mathring{\mathsf{B}}^\star(z), \mathring{\mathsf{B}}^\star(-z)) = 1.$$

Therefore, we must have $\mathsf{d}(z^2) \mid \mathsf{D}_{\mathring{\mathsf{b}}}(z^2)$. Thus, the above relation in (3.2.27) particularly implies

$$\mathring{\mathsf{b}}_1(z)\mathring{\tilde{\mathsf{b}}}_1^\star(z) + \mathring{\mathsf{b}}_2(z)\mathring{\tilde{\mathsf{b}}}_2^\star(z) = \frac{\mathsf{D}_{\mathring{\mathsf{b}}}(z^2)}{\mathsf{d}(z^2)}\mathring{\mathsf{A}}^\star(z). \tag{3.2.28}$$

Since $\operatorname{fsupp}(\mathring{b}_1) \subseteq [c_1 - n_1, n_1]$ and $\operatorname{fsupp}(\mathring{b}_2) \subseteq [c_2 - n_2, n_2]$, by (3.2.24) and (3.2.25), we must have

$$\begin{aligned} &\operatorname{fsupp}(\mathring{\mathsf{b}}_1(z)\mathring{\tilde{\mathsf{b}}}_1^\star(z)) \subseteq [-n_0 - \epsilon_{\text{len}}, n_0 - c_0 + \epsilon_{\text{len}}], \\ &\operatorname{fsupp}(\mathring{\mathsf{b}}_2(z)\mathring{\tilde{\mathsf{b}}}_2^\star(z)) \subseteq [-n_0 - \epsilon_{\text{len}}, n_0 - c_0 + \epsilon_{\text{len}}]. \end{aligned} \tag{3.2.29}$$

From (3.2.28) and $\mathrm{fsupp}(\mathring{\mathsf{A}}^\star) = [-n_0, n_0 - c_0]$, we have $\mathrm{fsupp}(\mathsf{D}_{\mathring{\mathsf{b}}}(z^2)/\mathsf{d}(z^2)) \subseteq [-\epsilon_{\mathrm{len}}, \epsilon_{\mathrm{len}}]$. Since $\epsilon_{\mathrm{len}} \in \{0, 1\}$, this forces $\mathrm{fsupp}(\mathsf{D}_{\mathring{\mathsf{b}}}(z^2)/\mathsf{d}(z^2)) \subseteq \{0\}$. That is, $\lambda := \mathsf{D}_{\mathring{\mathsf{b}}}(z^2)/\mathsf{d}(z^2)$ must be a constant. In other words, $\mathsf{D}_{\mathring{\mathsf{b}}}(z) = \lambda \mathsf{d}(z)$. By our assumption $\lambda \neq 0$, after replacing $\mathring{\tilde{\mathsf{b}}}_1, \mathring{\tilde{\mathsf{b}}}_2$ by $\bar{\lambda}^{-1}\mathring{\tilde{\mathsf{b}}}_1, \bar{\lambda}^{-1}\mathring{\tilde{\mathsf{b}}}_2$, respectively, we deduce from (3.2.27) that

$$\begin{bmatrix} \mathring{\tilde{\mathsf{b}}}_1(z) & \mathring{\tilde{\mathsf{b}}}_2(z) \\ \mathring{\tilde{\mathsf{b}}}_1(-z) & \mathring{\tilde{\mathsf{b}}}_2(-z) \end{bmatrix} \begin{bmatrix} \mathring{\mathsf{b}}_1(z) & \mathring{\mathsf{b}}_2(z) \\ \mathring{\mathsf{b}}_1(-z) & \mathring{\mathsf{b}}_2(-z) \end{bmatrix}^\star = \begin{bmatrix} \mathring{\mathsf{A}}(z) & \mathring{\mathsf{B}}(z) \\ \mathring{\mathsf{B}}(-z) & \mathring{\mathsf{A}}(-z) \end{bmatrix}.$$

For the case of complex symmetry, since $\lambda = \mathsf{D}_{\mathring{\mathsf{b}}}(z^2)/\mathsf{d}(z^2)$ and since both $\mathsf{D}_{\mathring{\mathsf{b}}}$ and d have complex symmetry, the constant λ must have complex symmetry. This is only possible for $\lambda \in \mathbb{R}$ or $i\lambda \in \mathbb{R}$. Hence, complex symmetry will be preserved after replacing $\mathring{\tilde{\mathsf{b}}}_1, \mathring{\tilde{\mathsf{b}}}_2$ by $\bar{\lambda}^{-1}\mathring{\tilde{\mathsf{b}}}_1, \bar{\lambda}^{-1}\mathring{\tilde{\mathsf{b}}}_2$, respectively. If all filters have real coefficients, then λ must be a real number and therefore, all the constructed high-pass filters must have real coefficients too.

Now it is straightforward to check that $(\{\tilde{a}; \tilde{b}_1, \tilde{b}_2\}, \{a; b_1, b_2\})_\Theta$ is a dual framelet filter bank with [complex] symmetry (and real coefficients).

By (3.2.22) and (3.2.23), we have

$$\mathrm{len}(\mathring{\mathsf{A}}) = \mathrm{len}(\mathring{\mathsf{B}}) = 2n_0 - c_0 = 2n_0 + n_b - n_{\tilde{b}} + 2c_{\mathsf{p}} - c_\Theta.$$

To prove (3.2.20), it follows from (3.2.18), (3.2.19), and (3.2.29) that

$$\begin{aligned} \mathrm{len}(b_1) + \mathrm{len}(\tilde{b}_1) &= n_b + n_{\tilde{b}} + 2\,\mathrm{len}(\mathsf{p}) + \mathrm{len}(\mathring{b}_1) + \mathrm{len}(\mathring{\tilde{b}}_1) \\ &\leqslant n_b + n_{\tilde{b}} + 2\,\mathrm{len}(\mathsf{p}) + 2n_0 - c_0 + 2\epsilon_{\mathrm{len}}. \end{aligned}$$

By the definition of B in (3.2.3), we have

$$\begin{aligned} 2\,\mathrm{len}(\mathsf{p}) + 2n_0 - c_0 &= 2\,\mathrm{len}(\mathsf{p}) + \mathrm{len}(\mathring{\mathsf{B}}) = \mathrm{len}(\mathsf{B}) \\ &= \mathrm{len}(a) + \mathrm{len}(\tilde{a}) + 2\,\mathrm{len}(\Theta) - n_b - n_{\tilde{b}}. \end{aligned}$$

We deduce from the above inequalities that $\mathrm{len}(b_1) + \mathrm{len}(\tilde{b}_1) \leqslant \mathrm{len}(a) + \mathrm{len}(\tilde{a}) + 2\,\mathrm{len}(\Theta) + 2\epsilon_{\mathrm{len}}$. Similarly, we can verify $\mathrm{len}(b_2) + \mathrm{len}(\tilde{b}_2) \leqslant \mathrm{len}(a) + \mathrm{len}(\tilde{a}) + 2\,\mathrm{len}(\Theta) + 2\epsilon_{\mathrm{len}}$. Hence, the inequality (3.2.20) holds. □

If $c + n_b$ is an even integer, then the particular construction of a dual framelet filter bank in (3.2.5) can be recovered by Algorithm 3.2.1; If $c + n_b$ is an odd integer, then the particular construction in (3.2.6) can be recovered by Algorithm 3.2.1. By the discussion before Algorithm 3.2.1, almost all dual framelet filter banks with [complex] symmetry satisfying (3.2.20) can be constructed via Algorithm 3.2.1.

For a given pair of filters a and $\tilde{a}$, to construct a dual framelet filter bank $(\{\tilde{a}; \tilde{b}_1, \tilde{b}_2\}, \{a; b_1, b_2\})_\Theta$ having high vanishing moments, it is important to first construct a moment correcting filter Θ satisfying the last equation in (3.2.9). The following result guarantees the existence of such a desired moment correcting filter having the shortest filter support.

Lemma 3.2.2 *Let u be a filter having symmetry $\mathsf{S}\mathsf{u}(z) = z^c$ (or complex symmetry $\mathbb{S}\mathsf{u}(z) = z^c$) for some $c \in \mathbb{Z}$ and satisfying $\mathsf{u}(1) = 1$. For any nonnegative integer n, there exists a finitely supported filter $\Theta \in l_0(\mathbb{Z})$ such that*

$$\Theta(1) = 1 \qquad \text{and} \qquad \Theta(z) - \Theta(z^2)\mathsf{u}(z) = \mathscr{O}(|1-z|^n), \qquad z \to 1, \qquad (3.2.30)$$

$\mathsf{S}\Theta(z) = z^{-c}$ *(or $\mathbb{S}\Theta(z) = z^{-c}$), and* $\operatorname{fsupp}(\Theta) \subseteq [-c-m, m]$ *with $m := \lceil \frac{n-c-1}{2} \rceil$. Moreover, if u has real coefficients, then Θ has real coefficients.*

Proof By Exercise 1.14, there always exists a unique filter Θ with $\operatorname{fsupp}(\Theta) \subseteq [m-n+1, m]$ such that (3.2.30) holds. For the case of symmetry, we replace Θ by $[\Theta(z) + z^{-c}\Theta(z^{-1})]/2$; for the case of complex symmetry, we replace Θ by $[\Theta(z) + z^{-c}\Theta^\star(z)]/2$. Noting that $-c-m \leqslant m-n+1$, we see that Θ is a desired moment correcting filter satisfying all the requirements. □

When a moment correcting filter Θ is given in advance, we can also design a filter $\tilde{a}$ derived from a given low-pass filter a such that (3.2.30) holds. Since a general moment correcting filter does not introduce any additional difficulty, here we only discuss the commonly used case $\Theta = 1$. The following result can be proved in the same way as in Proposition 2.1.7 or Lemma 3.2.2.

Proposition 3.2.3 *Let $M, N \in \mathbb{N}$ be positive integers and $a \in l_0(\mathbb{Z})$ be a filter with $\mathsf{a}(1) = 1$. For any subset Λ of $\mathbb{Z}$ such that the cardinality of Λ is N, there exists a unique solution $\{t_k\}_{k\in\Lambda}$ to the system of linear equations induced by*

$$\tilde{\mathsf{a}}(z)\mathsf{a}^\star(z) = 1 + \mathscr{O}(|z-1|^N), \qquad z \to 1, \quad \text{with} \quad \tilde{\mathsf{a}}(z) := (1+z)^M \sum_{k\in\Lambda} t_k z^k.$$

If the filter a is real-valued, then so is the filter $\tilde{a}$. If in addition a has symmetry $\mathsf{S}\mathsf{a}(z) = z^c$ (or complex symmetry $\mathbb{S}\mathsf{a}(z) = z^c$) for some $c \in \mathbb{Z}$, take $\Lambda = \{\lceil \frac{c-M+1-N}{2} \rceil, \ldots, \lfloor \frac{c-M-1+N}{2} \rfloor\}$ provided that $c+N+M$ is an odd integer (this requirement can be dropped if N is even and either a is real-valued or a has symmetry), then $\mathsf{S}\tilde{\mathsf{a}}(z) = z^c$ (or $\mathbb{S}\tilde{\mathsf{a}}(z) = z^c$).

Since Algorithm 3.2.1 only involves systems of linear equations, dual framelet filter banks with symmetry can be easily obtained via Algorithm 3.2.1. Using Lemma 3.2.2 or Proposition 3.2.3, here we present several examples of dual framelet filter banks to illustrate Algorithm 3.2.1.

Example 3.2.1 Let $a = a_3^B(\cdot - 1) = \{\frac{1}{8}, \underline{\frac{3}{8}}, \frac{3}{8}, \frac{1}{8}\}_{[-1,2]}$ and $\Theta = \boldsymbol{\delta}$. Setting $M = 3$ and $N = 4$ in Proposition 3.2.3, we have a filter $\tilde{a}$, with $\operatorname{sm}(\tilde{a}) \approx 1.0981905$, given by

$$\tilde{a} = \{-\tfrac{3}{32}, \tfrac{1}{32}, \underline{\tfrac{9}{16}}, \tfrac{9}{16}, \tfrac{1}{32}, -\tfrac{3}{32}\}_{[-2,3]}.$$

If $n_b = 1$ and $n_{\tilde{b}} = 3$, then $\mathsf{p}(z) = 1$ and $\mathsf{D}(z) = -\frac{3}{128}z^{-1}$. Taking $\mathsf{d}(z) = 1, c_1 = 0, \epsilon_1 = 1, n_1 = 2, n_2 = 2$ in Algorithm 3.2.1, we have

$$\tilde{b}_1 = \tfrac{3}{32}\{1, -3, \underline{\mathbf{3}}, -1\}_{[-2,1]}, \qquad \tilde{b}_2 = \tfrac{1}{32}\{3, -1, \underline{\mathbf{-12}}, 12, 1, -3\}_{[-2,3]},$$

$$b_1 = \tfrac{1}{2}\{-1, \underline{\mathbf{1}}\}_{[-1,0]}, \qquad b_2 = \tfrac{1}{8}\{-1, \underline{\mathbf{-3}}, 3, 1\}_{[-1,2]}.$$

Then $(\{\tilde{a}; \tilde{b}_1, \tilde{b}_2\}, \{a; b_1, b_2\})$ is a real-valued dual framelet filter bank with symmetry such that $\operatorname{vm}(b_1) = \operatorname{vm}(b_2) = 1$ and $\operatorname{vm}(\tilde{b}_1) = \operatorname{vm}(\tilde{b}_2) = 3$.

By Lemma 3.2.2 with $n = 6$, consider $\Theta = \{\frac{1}{40}, -\frac{1}{10}, \underline{\mathbf{\frac{23}{20}}}, -\frac{1}{10}, \frac{1}{40}\}_{[-2,2]}$. If $n_b = n_{\tilde{b}} = 3$, then $\mathsf{p}(z) = 1$. Taking $\mathsf{d}(z) = z$, $c_1 = 2, \epsilon_1 = 1, n_1 = 4, n_2 = 5$ in Algorithm 3.2.1, we have

$$\tilde{b}_1 = \tfrac{1}{640}\{6, 1, -63, 142, \underline{\mathbf{-142}}, 63, -1, -6\}_{[-4,3]},$$

$$\tilde{b}_2 = \tfrac{1}{640}\{-1, 7, -61, \underline{\mathbf{155}}, -155, 61, -7, 1\}_{[-3,4]},$$

$$b_1 = \tfrac{1}{32}\{-1, -3, 0, 8, 6, \underline{\mathbf{-6}}, -8, 0, 3, 1\}_{[-5,4]},$$

$$b_2 = \tfrac{1}{32}\{-3, -9, 7, \underline{\mathbf{45}}, -45, -7, 9, 3\}_{[-3,4]}.$$

Then $(\{\tilde{a}; \tilde{b}_1, \tilde{b}_2\}, \{a; b_1, b_2\})_\Theta$ is a real-valued dual framelet filter bank with symmetry such that $\operatorname{vm}(b_1) = \operatorname{vm}(b_2) = 3$ and $\operatorname{vm}(\tilde{b}_1) = \operatorname{vm}(\tilde{b}_2) = 3$. See Fig. 3.1 for their associated refinable functions and framelet functions.

Example 3.2.2 Let $a = a_{2,4} = \{-\frac{1}{16}, \frac{1}{4}, \underline{\mathbf{\frac{5}{8}}}, \frac{1}{4}, -\frac{1}{16}\}_{[-2,2]}$ and $\Theta = \boldsymbol{\delta}$ with $\operatorname{sm}(a) \approx 0.885296$. Let $\tilde{a} = a$. If $n_b = 2$ and $n_{\tilde{b}} = 2$, then $\mathsf{p}(z) = 1$ and $\mathsf{D}(z) = -\frac{1}{128}$. Taking $\mathsf{d}(z) = z, c_1 = 2, \epsilon_1 = 1, n_1 = 2, n_2 = 3$ in Algorithm 3.2.1, we have

$$\tilde{b}_1 = \{\tfrac{1}{16}, -\tfrac{1}{4}, \underline{\mathbf{\tfrac{3}{8}}}, -\tfrac{1}{4}, \tfrac{1}{16}\}_{[-2,2]}, \qquad \tilde{b}_2 = \{\underline{\mathbf{-\tfrac{\sqrt{2}}{4}}}, \tfrac{\sqrt{2}}{2}, -\tfrac{\sqrt{2}}{4}\}_{[0,2]},$$

$$b_1 = \{-\tfrac{1}{16}, \tfrac{1}{4}, \underline{\mathbf{-\tfrac{3}{8}}}, \tfrac{1}{4}, -\tfrac{1}{16}\}_{[-2,2]}, \qquad b_2 = \{\underline{\mathbf{-\tfrac{\sqrt{2}}{4}}}, \tfrac{\sqrt{2}}{2}, -\tfrac{\sqrt{2}}{4}\}_{[0,2]}.$$

Then $(\{\tilde{a}; \tilde{b}_1, \tilde{b}_2\}, \{a; b_1, b_2\})$ is a real-valued dual framelet filter bank with $\operatorname{vm}(b_1) = \operatorname{vm}(\tilde{b}_1) = 4$ and $\operatorname{vm}(b_2) = \operatorname{vm}(\tilde{b}_2) = 2$. Note that $\tilde{a} = a$, $\tilde{b}_1 = -b_1$, and $\tilde{b}_2 = b_2$. In other words, the constructed dual framelet filter bank $(\{a; -b_1, b_2\}, \{a; b_1, b_2\})$ appears to be almost a tight framelet filter bank and we call it *a quasi-tight framelet filter bank*.

Let $\tilde{a} = a_2^B(\cdot - 1) = \{\frac{1}{4}, \underline{\mathbf{\frac{1}{2}}}, \frac{1}{4}\}_{[-1,1]}$ and $\Theta = \boldsymbol{\delta}$. If $n_b = 1$ and $n_{\tilde{b}} = 1$, then $\mathsf{p}(z) = 1$ and $\mathsf{D}(z) = \frac{1}{16}$. Taking $\mathsf{d}(z) = 1, c_1 = 1, \epsilon_1 = 1, n_1 = 2, n_2 = 2$ in Algorithm 3.2.1, we have

$$\tilde{b}_1 = \{-\tfrac{1}{4}, \underline{\mathbf{0}}, -\tfrac{1}{4}\}_{[-1,1]}, \qquad \tilde{b}_2 = \{-\tfrac{1}{8}, \underline{\mathbf{\tfrac{1}{4}}}, -\tfrac{1}{8}\}_{[-1,1]},$$

$$b_1 = \{\tfrac{1}{8}, -\tfrac{1}{2}, \underline{\mathbf{0}}, \tfrac{1}{2}, -\tfrac{1}{8}\}_{[-2,2]}, \qquad b_2 = \{\tfrac{1}{8}, -\tfrac{1}{2}, \underline{\mathbf{\tfrac{3}{4}}}, -\tfrac{1}{2}, \tfrac{1}{8}\}_{[-2,2]}.$$

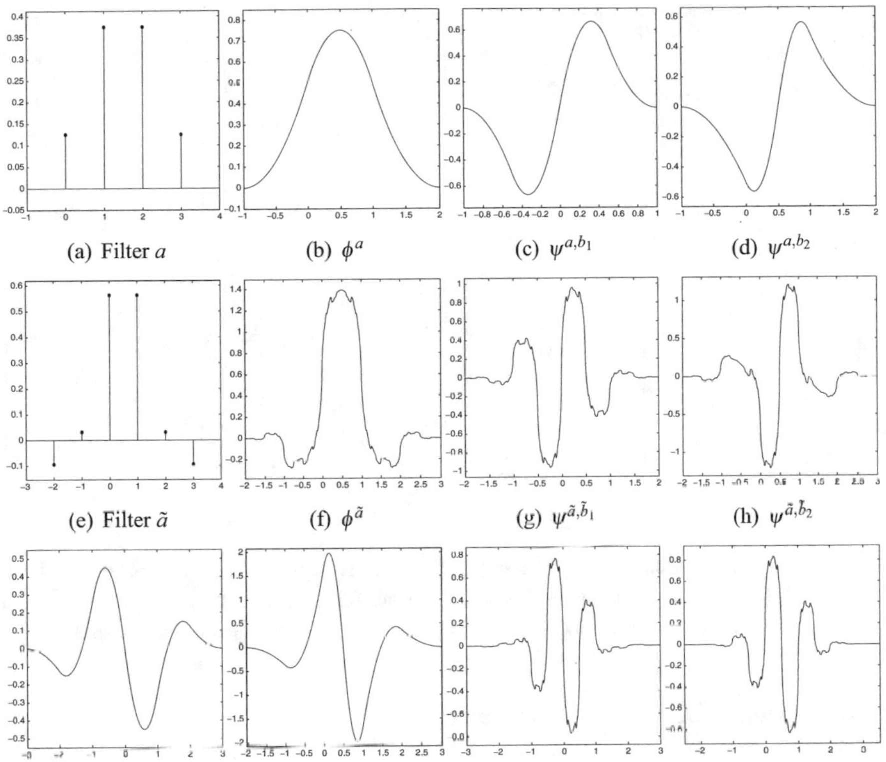

(a) Filter a (b) ϕ^a (c) ψ^{a,b_1} (d) ψ^{a,b_2}

(e) Filter $\tilde{a}$ (f) $\phi^{\tilde{a}}$ (g) $\psi^{\tilde{a},\tilde{b}_1}$ (h) $\psi^{\tilde{a},\tilde{b}_2}$

(i) ψ^{a,b_1} (j) ψ^{a,b_2} (k) $\psi^{\tilde{a},\tilde{b}_1}$ (l) $\psi^{\tilde{a},\tilde{b}_2}$

Fig. 3.1 Low-pass filters $a, \tilde{a}$ and functions $\phi^a, \psi^{a,b_1}, \psi^{a,b_2}, \phi^{\tilde{a}}, \psi^{\tilde{a},\tilde{b}_1}, \psi^{\tilde{a},\tilde{b}_2}$ in **(a)**–**(h)** are associated with the dual framelet filter bank $(\{\tilde{a}; \tilde{b}_1, \tilde{b}_2\}, \{a; b_1, b_2\})_\Theta$ in Example 3.2.1 with $\Theta = \boldsymbol{\delta}$. **(i)**–**(l)** are graphs of $\psi^{a,b_1}, \psi^{a,b_2}, \psi^{\tilde{a},\tilde{b}_1}, \psi^{\tilde{a},\tilde{b}_2}$ with $\Theta = \{\frac{1}{40}, -\frac{1}{10}, \underline{\mathbf{\frac{23}{20}}}, -\frac{1}{10}, \frac{1}{40}\}_{[-2,2]}$

Then $(\{\tilde{a}; \tilde{b}_1, \tilde{b}_2\}, \{a; b_1, b_2\})$ is a dual framelet filter bank with $\text{vm}(b_1) = \text{vm}(\tilde{b}_1) = 1$, $\text{vm}(\tilde{b}_2) = 2$, and $\text{vm}(b_2) = 4$. See Fig. 3.2 for their framelet functions.

Example 3.2.3 Let $a = \tilde{a} = a_4^I = \{-\frac{1}{32}, 0, \frac{9}{32}, \underline{\mathbf{\frac{1}{2}}}, \frac{9}{32}, 0, -\frac{1}{32}\}_{[-3,3]}$. If $\Theta = \boldsymbol{\delta}$ and $n_b = n_{\tilde{b}} = 2$, then $\mathsf{p}(z) = 1$ and $\mathsf{D}(z) = \frac{7}{256} - \frac{1}{512}(z + z^{-1})$. Taking $\mathsf{d}(z) = 1, c_1 = 0, \epsilon_1 = 1, n_1 = 2, n_2 = 3$ in Algorithm 3.2.1, we have

$$\tilde{b}_1 = \tfrac{1}{8}\{-1, -2, 6, \underline{\mathbf{-2}}, -1\}_{[-3,1]}, \qquad \tilde{b}_2 = \tfrac{1}{32}\{1, 0, -9, \underline{\mathbf{16}}, -9, 0, 1\}_{[-3,3]},$$

$$b_1 = \tfrac{1}{8}\{-1, 0, 2, \underline{\mathbf{0}}, -1\}_{[-3,1]}, \qquad b_2 = \tfrac{1}{32}\{-1, 0, -7, \underline{\mathbf{16}}, -7, 0, -1\}_{[-3,3]}.$$

Then $(\{\tilde{a}; \tilde{b}_1, \tilde{b}_2\}, \{a; b_1, b_2\})$ is a real-valued dual framelet filter bank with symmetry such that $\text{vm}(b_1) = \text{vm}(b_2) = \text{vm}(\tilde{b}_1) = 2$ and $\text{vm}(\tilde{b}_2) = 4$.

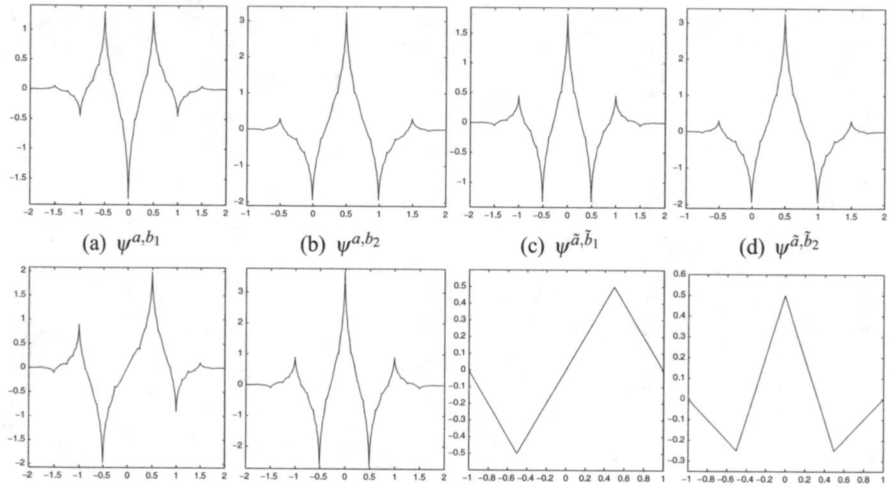

(a) ψ^{a,b_1} (b) ψ^{a,b_2} (c) $\psi^{\tilde{a},\tilde{b}_1}$ (d) $\psi^{\tilde{a},\tilde{b}_2}$

(e) ψ^{a,b_1} (f) ψ^{a,b_2} (g) $\psi^{\tilde{a},\tilde{b}_1}$ (h) $\psi^{\tilde{a},\tilde{b}_2}$

Fig. 3.2 Wavelet functions $\psi^{a,b_1}, \psi^{a,b_2}, \psi^{\tilde{a},\tilde{b}_1}, \psi^{\tilde{a},\tilde{b}_2}$ associated with the dual framelet filter bank $(\{\tilde{a}; \tilde{b}_1, \tilde{b}_2\}, \{a; b_1, b_2\})_\Theta$ in Example 3.2.2. **(a)**–**(d)** are for $\tilde{a} = a$ with $\psi^{\tilde{a},\tilde{b}_1} = -\psi^{a,b_1}$ and $\psi^{\tilde{a},\tilde{b}_2} = \psi^{a,b_2}$. **(i)**–**(l)** are for $\tilde{a} = a_2^B(\cdot - 1)$. The graph ϕ^a with $a = a_{2,4}$ is given in Fig. 2.4

By Lemma 3.2.2 with $n = 8$, we can take

$$\Theta = \{-\tfrac{11}{5040}, \tfrac{4}{105}, -\tfrac{223}{1680}, \underline{\mathbf{\tfrac{376}{315}}}, -\tfrac{223}{1680}, \tfrac{4}{105}, -\tfrac{11}{5040}\}_{[-3,3]}.$$

If $n_b = n_{\tilde{b}} = 4$, then $\mathsf{p}(z) = 1$. Taking $\mathsf{d}(z) = z, c_1 = 2, \epsilon_1 = 1, n_1 = 5, n_2 = 6$ in Algorithm 3.2.1, we have

$$\tilde{b}_1 = \tfrac{1}{40320}\{-44, 0, 933, 896, -4668, -2736, 11238, \underline{\mathbf{-2736}}, -4668, 896, 933, 0, -44\}_{[-7,5]},$$

$$\tilde{b}_2 = \tfrac{1}{13440}\{-55, -64, 1489, -112, -10554, \underline{\mathbf{18592}}, -10554, -112, 1489, -64, -55\}_{[-5,5]}$$

and

$$b_1 = \tfrac{1}{512}\{-1, 0, 18, 16, -63, -144, 348, \underline{\mathbf{-144}}, -63, 16, 18, 0, -1\}_{[-7,5]},$$

$$b_2 = \tfrac{1}{512}\{-1, 0, 3, 16, -66, \underline{\mathbf{96}}, -66, 16, 3, 0, -1\}_{[-5,5]}.$$

Then $(\{\tilde{a}; \tilde{b}_1, \tilde{b}_2\}, \{a; b_1, b_2\})_\Theta$ is a real-valued dual framelet filter bank with symmetry such that $\mathrm{vm}(b_1) = \mathrm{vm}(b_2) = \mathrm{vm}(\tilde{b}_1) = \mathrm{vm}(\tilde{b}_2) = 4$.

If $\Theta = \boldsymbol{\delta}$ and $n_b = n_{\tilde{b}} = 2$, setting $M = 2$ and $N = 4$ in Proposition 3.2.3, we obtain

$$\tilde{a} = \{-\tfrac{1}{16}, \tfrac{1}{4}, \underline{\mathbf{\tfrac{5}{8}}}, \tfrac{1}{4}, -\tfrac{1}{16}\}_{[-2,2]} \tag{3.2.31}$$

with $\mathrm{sm}(\tilde{a}) \approx 0.885296$ and $\mathrm{sr}(\tilde{a}) = 2$. Then $\mathsf{p}(z) = 1$ and $\mathsf{D}(z) = \frac{1}{64}$. Taking $\mathsf{d}(z) = 1, c_1 = 0, \epsilon_1 = 1, n_1 = 2, n_2 = 3$ in Algorithm 3.2.1, we have

$$\tilde{b}_1 = \tfrac{1}{4}\{-1, 2, \underline{\mathbf{-1}}\}_{[-2,0]}, \qquad \tilde{b}_2 = \tfrac{1}{16}\{1, -4, \underline{\mathbf{6}}, -4, 1\}_{[-2,2]},$$

$$b_1 = \tfrac{1}{4}\{-1, 0, 2, \underline{\mathbf{0}}, -1\}_{[-3,1]}, \qquad b_2 = \tfrac{1}{32}\{-1, 0, -7, \underline{\mathbf{16}}, -7, 0, -1\}_{[-3,3]}.$$

Then $(\{\tilde{a}; \tilde{b}_1, \tilde{b}_2\}, \{a; b_1, b_2\})$ is a real-valued dual framelet filter bank with symmetry such that $\mathrm{vm}(\tilde{b}_1) = \mathrm{vm}(b_1) = \mathrm{vm}(b_2) = 2$ and $\mathrm{vm}(\tilde{b}_2) = 4$. See Fig. 3.3 for their associated refinable functions and framelet functions.

Without the symmetry property, we now present an algorithm to construct all dual framelet filter banks $(\{\tilde{a}; \tilde{b}_1, \tilde{b}_2\}, \{a; b_1, b_2\})_\Theta$ having the shortest filter support.

Algorithm 3.2.4 *Let $a, \tilde{a}, \Theta \in l_0(\mathbb{Z})$ be given filters and $n_b, n_{\tilde{b}} \in \mathbb{N}_0$ satisfy* (3.2.9).

(S1) Define A *and* B *in* (3.2.3) *and* $\mathsf{p}, \mathring{\mathsf{A}}, \mathring{\mathsf{B}}$ *in* (3.2.10). *Define* $[m_0, n_0] := \mathrm{fsupp}(\mathring{\mathsf{B}}^\star)$.

(S2) Select $\epsilon_{\mathrm{len}}, s_1, s_2 \in \{0, 1\}$ *and* $\ell_1, \ell_2 \in \mathbb{N}_0$ *such that* $\max(\ell_1, \ell_2) \leqslant n_0 - m_0 + \epsilon_{\mathrm{len}}$. *Select a Laurent polynomial* d *satisfying* (3.2.11) *and* $\lceil \frac{s_1+s_2-1}{2} \rceil \leqslant m_{\mathsf{d}} \leqslant n_{\mathsf{d}} \leqslant \lfloor \frac{s_1+s_2+\ell_1+\ell_2-1}{2} \rfloor$, *where* $[m_{\mathsf{d}}, n_{\mathsf{d}}] := \mathrm{fsupp}(\mathsf{d})$;

(S3) Parameterize a filter $\mathring{\mathsf{b}}_1$ *by* $\mathring{\mathsf{b}}_1(z) := z^{s_1} \sum_{j=0}^{\ell_1} t_j z^j$. *Find the unknown coefficients* $\{t_0, \ldots, t_{\ell_1}\}$ *by solving a system* X_1 *of linear equations induced by* $\mathcal{R}_1(z) = 0$ *and*

$$\mathit{coeff}(\mathring{\tilde{\mathsf{b}}}_2^\star, z, j) = 0, \quad j = m_0 + s_1 - 2m_{\mathsf{d}} - 1, \ldots, m_0 - s_2 - \epsilon_{\mathrm{len}} - 1 \ \mathit{and}$$
$$j = n_0 - s_2 - \ell_2 + \epsilon_{\mathrm{len}} + 1, \ldots, n_0 + s_1 + \ell_1 - 2n_{\mathsf{d}} - 1,$$

where $\mathcal{R}_1$ *and* $\mathring{\tilde{\mathsf{b}}}_2^\star$ *are Laurent polynomials uniquely determined by* $\mathrm{fsupp}(\mathcal{R}_1) \subseteq [2m_{\mathsf{d}}, 2n_{\mathsf{d}} - 1]$ *and* (3.2.13);

(S4) Parameterize a filter $\mathring{\mathsf{b}}_2$ *by* $\mathring{\mathsf{b}}_2(z) := z^{s_2} \sum_{j=0}^{\ell_2} \tilde{t}_j z^j$. *Find the unknown coefficients* $\{\tilde{t}_0, \ldots, \tilde{t}_{\ell_2}\}$ *by solving a system* X_2 *of linear equations induced by* $\mathcal{R}_2(z) = 0$ *and*

$$\mathit{coeff}(\mathring{\tilde{\mathsf{b}}}_1^\star, z, j) = 0, \quad j = m_0 + s_2 - 2m_{\mathsf{d}} - 1, \ldots, m_0 - s_1 - \epsilon_{\mathrm{len}} - 1 \ \mathit{and}$$
$$j = n_0 - s_1 - \ell_1 + \epsilon_{\mathrm{len}} + 1, \ldots, n_0 + s_2 + \ell_2 - 2n_{\mathsf{d}} - 1,$$

where $\mathcal{R}_2$ *and* $\mathring{\tilde{\mathsf{b}}}_1^\star$ *are Laurent polynomials uniquely determined by* $\mathrm{fsupp}(\mathcal{R}_2) \subseteq [2m_{\mathsf{d}}, 2n_{\mathsf{d}} - 1]$ *and* (3.2.15). *If either X_1 or X_2 has only the trivial solution, then restart the algorithm from (S2) by selecting other choices of* $\mathsf{d}, \epsilon_{\mathrm{len}}, s_1, s_2, \ell_1, \ell_2$;

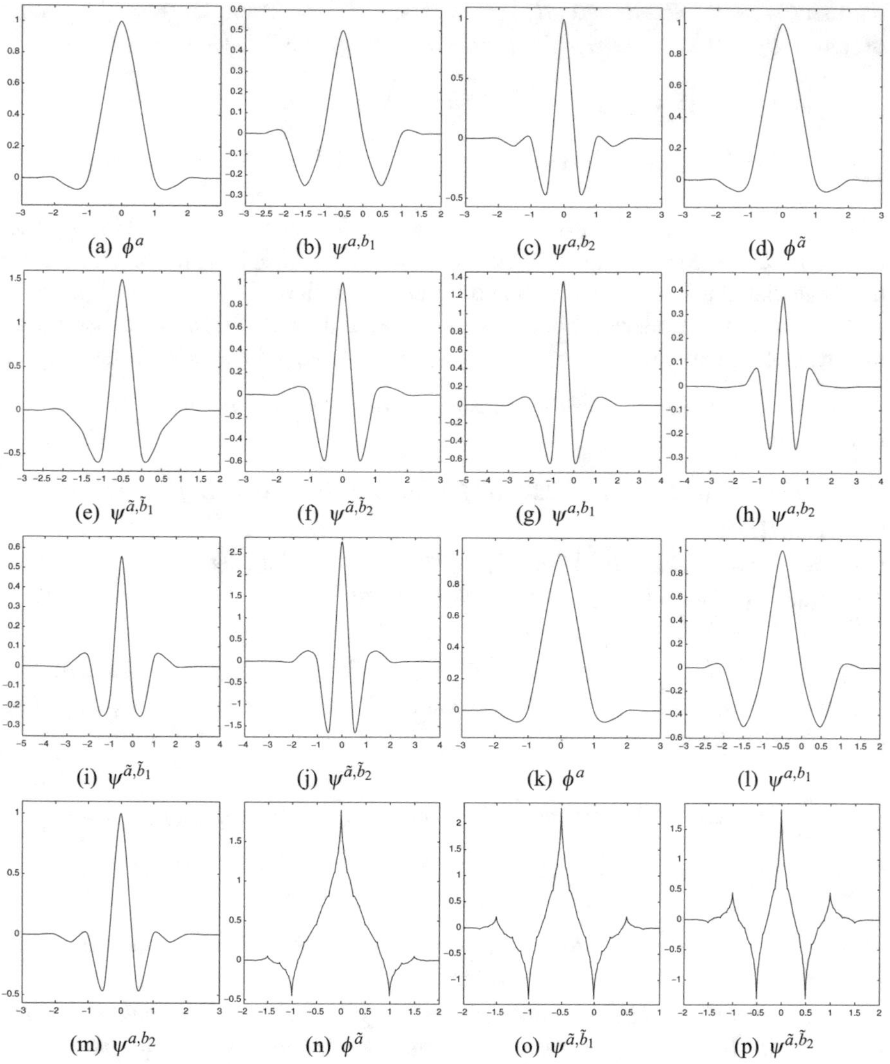

(a) ϕ^a (b) ψ^{a,b_1} (c) ψ^{a,b_2} (d) $\phi^{\tilde{a}}$

(e) $\psi^{\tilde{a},\tilde{b}_1}$ (f) $\psi^{\tilde{a},\tilde{b}_2}$ (g) ψ^{a,b_1} (h) ψ^{a,b_2}

(i) $\psi^{\tilde{a},\tilde{b}_1}$ (j) $\psi^{\tilde{a},\tilde{b}_2}$ (k) ϕ^a (l) ψ^{a,b_1}

(m) ψ^{a,b_2} (n) $\phi^{\tilde{a}}$ (o) $\psi^{\tilde{a},\tilde{b}_1}$ (p) $\psi^{\tilde{a},\tilde{b}_2}$

Fig. 3.3 Functions $\phi^a, \psi^{a,b_1}, \psi^{a,b_2}, \phi^{\tilde{a}}, \psi^{\tilde{a},\tilde{b}_1}, \psi^{\tilde{a},\tilde{b}_2}$ of the dual framelet filter bank $(\{\tilde{a};\tilde{b}_1,\tilde{b}_2\},\{a;b_1,b_2\})_\Theta$ in Example 3.2.3. (**a**)–(**f**) are for $\Theta = \boldsymbol{\delta}$ with $\phi^{\tilde{a}} = \phi^a$. (**g**)–(**j**) are $\psi^{a,b_1}, \psi^{a,b_2}, \phi^{\tilde{a}}, \psi^{\tilde{a},\tilde{b}_1}, \psi^{\tilde{a},\tilde{b}_2}$ with $\Theta = \{-\frac{11}{5040}, \frac{4}{105}, -\frac{223}{1680}, \underline{\mathbf{\frac{376}{315}}}, -\frac{223}{1680}, \frac{4}{105}, -\frac{11}{5040}\}_{[-3,3]}$. (**k**)–(**p**) are graphs of $\phi^a, \psi^{a,b_1}, \psi^{a,b_2}, \phi^{\tilde{a}}, \psi^{\tilde{a},\tilde{b}_1}, \psi^{\tilde{a},\tilde{b}_2}$ for $\Theta = \boldsymbol{\delta}$ and $\tilde{a}$ in (3.2.31)

(S5) There must exist $\lambda \in \mathbb{C}$ *such that* (3.2.16) *holds. If* $\lambda = 0$*, then restart the algorithm from (S2) by selecting other choices of* $\mathsf{d}, \epsilon_{\text{len}}, s_1, s_2, \ell_1, \ell_2$*. Otherwise, for* $\lambda \neq 0$*, replace* $\mathring{\mathsf{b}}_1, \mathring{\mathsf{b}}_2$ *by* $\bar{\lambda}^{-1}\mathring{\mathsf{b}}_1, \bar{\lambda}^{-1}\mathring{\mathsf{b}}_2$*, respectively.*

(S6) Find Laurent polynomials q *and* $\tilde{\mathsf{q}}$ *such that* $\mathsf{p}(z) = \tilde{\mathsf{q}}(z)\mathsf{q}^\star(z)$*. Define* $\mathsf{b}_1, \mathsf{b}_2$ *as in* (3.2.18) *and* $\tilde{\mathsf{b}}_1, \tilde{\mathsf{b}}_2$ *as in* (3.2.19)*.*

Then $(\{\tilde{a}; \tilde{b}_1, \tilde{b}_2\}, \{a; b_1, b_2\})_\Theta$ *is a dual framelet filter bank satisfying* (3.2.20).

Proof Note that $\operatorname{fsupp}(\mathring{b}_1) \subseteq [s_1, s_1 + \ell_1]$ and $\operatorname{fsupp}(\mathring{b}_2) \subseteq [s_2, s_2 + \ell_2]$. The equation (3.2.15) implies that $\operatorname{fsupp}(\mathring{\tilde{b}}_1^\star) \subseteq [m_0 + s_2 - 2m_{\mathsf{d}} - 1, n_0 + s_2 + \ell_2 - 2n_{\mathsf{d}} - 1]$ and (3.2.13) implies that $\operatorname{fsupp}(\mathring{\tilde{b}}_2^\star) \subseteq [m_0 + s_1 - 2m_{\mathsf{d}} - 1, n_0 + s_1 + \ell_1 - 2n_{\mathsf{d}} - 1]$. The constraint on the coefficients of $\mathring{\tilde{b}}_1^\star$ in (S4) implies $\operatorname{fsupp}(\mathring{\tilde{b}}_1^\star) \subseteq [m_0 - s_1 - \epsilon_{\text{len}}, n_0 - s_1 - \ell_1 + \epsilon_{\text{len}}]$, and the constraint on the coefficients of $\mathring{\tilde{b}}_2^\star$ in (S3) implies $\operatorname{fsupp}(\mathring{\tilde{b}}_2^\star) \subseteq [m_0 - s_2 - \epsilon_{\text{len}}, n_0 - s_2 - \ell_2 + \epsilon_{\text{len}}]$. Therefore, $\operatorname{fsupp}(\mathring{b}_1(z)\mathring{\tilde{b}}_1^\star(-z)) \subseteq [m_0 - \epsilon_{\text{len}}, n_0 + \epsilon_{\text{len}}]$ and $\operatorname{fsupp}(\mathring{b}_2(z)\mathring{\tilde{b}}_2^\star(-z)) \subseteq [m_0 - \epsilon_{\text{len}}, n_0 + \epsilon_{\text{len}}]$. By the same proof of Algorithm 3.2.1, we see that (3.2.27) holds. In particular, we have $\mathring{\mathsf{b}}_1(-z)\mathring{\tilde{\mathsf{b}}}_1^\star(z) + \mathring{\mathsf{b}}_2(-z)\mathring{\tilde{\mathsf{b}}}_2^\star(z) = \frac{\mathsf{D}_{\mathring{\mathsf{b}}}(z^2)}{\mathsf{d}(z^2)}\mathring{\mathsf{B}}^\star(z)$. This identity forces $\operatorname{fsupp}(\mathsf{D}_{\mathring{\mathsf{b}}}(z^2)/\mathsf{d}(z^2)) \subseteq [-\epsilon_{\text{len}}, \epsilon_{\text{len}}]$ by $\operatorname{fsupp}(\mathring{\mathsf{B}}^\star) = [m_0, n_0]$. Since $\epsilon_{\text{len}} \in \{0, 1\}$, $\mathsf{D}_{\mathring{\mathsf{b}}}(z^2)/\mathsf{d}(z^2)$ must be a constant. By a similar argument as in Algorithm 3.2.1, then $(\{\tilde{a}; \tilde{b}_1, \tilde{b}_2\}, \{a; b_1, b_2\})_\Theta$ is a dual framelet filter bank satisfying (3.2.20). □

Because Algorithm 3.2.4 only involves linear equations, dual framelet filter banks $(\{\tilde{a}; \tilde{b}_1, \tilde{b}_2\}, \{a, b_1, b_2\})_\Theta$ without symmetry can be easily obtained from any pair of low-pass filters $(\tilde{a}, a)$ and a moment correcting filter Θ. Since Algorithm 3.2.4 is similar to Algorithm 3.2.1, for simplicity, we do not present any examples here.

3.3 Tight Framelet Filter Banks with Symmetry and Two High-Pass Filters

Tight framelet filter banks are special cases of dual framelet filter banks and are widely used in applications (e.g., the directional complex tight framelets for image processing in Sect. 7.4), largely due to their desired energy preservation property stated in item (i) of Theorem 1.1.4.

In this section we discuss how to systematically construct all tight framelet filter banks $\{a; b_1, b_2\}_\Theta$ with the symmetry property and short support.

3.3.1 Vanishing Moments of Tight Framelet Filter Banks

Before discussing how to construct a tight framelet filter bank with the symmetry property, let us first study several basic properties of a general tight framelet filter bank. We first investigate vanishing moments of a general tight framelet filter bank. For convenience, we shall use the following slightly abused notation: For $u \in l_0(\mathbb{Z})$,

$$\operatorname{vm}(\mathsf{u}) := \operatorname{vm}(\widehat{u}) := \operatorname{vm}(u), \quad \operatorname{sr}(\mathsf{u}) := \operatorname{sr}(\widehat{u}) := \operatorname{sr}(u), \quad \operatorname{lpm}(\mathsf{u}) := \operatorname{lpm}(\widehat{u}) := \operatorname{lpm}(u).$$

Using a similar argument as in the proof of Lemma 1.4.1, we have

Proposition 3.3.1 *Let $\{a; b_1, \ldots, b_s\}_\Theta$ be a tight framelet filter bank with all filters from $l_0(\mathbb{Z})$. If $\Theta(1)\mathsf{a}(1) \neq 0$, then*

$$\min\big(\operatorname{vm}(b_1), \ldots, \operatorname{vm}(b_s)\big) = \min\big(\operatorname{sr}(a), \tfrac{1}{2}\operatorname{vm}(\Theta(z) - \Theta(z^2)\mathsf{a}(z)\mathsf{a}^\star(z))\big). \tag{3.3.1}$$

In particular, if $\Theta(z) = 1$ (that is, $\Theta = \boldsymbol{\delta}$) and $\mathsf{a}(1) \neq 0$, then

$$\min\big(\operatorname{vm}(b_1), \ldots, \operatorname{vm}(b_s)\big) = \min\big(\operatorname{sr}(a), \tfrac{1}{2}\operatorname{lpm}(a * a^\star)\big).$$

If in addition a has complex symmetry and $\mathsf{a}(1) \neq -1$, then $\operatorname{lpm}(a * a^\star) = \operatorname{lpm}(a)$.

Proof For a tight framelet filter bank $\{a; b_1, \ldots, b_s\}_\Theta$, we have

$$|\widehat{b_1}(\xi)|^2 + \cdots + |\widehat{b_s}(\xi)|^2 = \Theta(\xi) - \widehat{\Theta}(2\xi)|\widehat{a}(\xi)|^2, \tag{3.3.2}$$

$$\overline{\widehat{b_1}(\xi + \pi)}\widehat{b_1}(\xi) + \cdots + \overline{\widehat{b_s}(\xi + \pi)}\widehat{b_s}(\xi) = -\Theta(2\xi)\overline{\widehat{a}(\xi + \pi)}\widehat{a}(\xi). \tag{3.3.3}$$

From (3.3.2), it is easy to deduce that

$$2\min\big(\operatorname{vm}(b_1), \ldots, \operatorname{vm}(b_s)\big) = \operatorname{vm}(\widehat{\Theta}(\xi) - \widehat{\Theta}(2\xi)|\widehat{a}(\xi)|^2). \tag{3.3.4}$$

By $\widehat{\Theta}(0)\widehat{a}(0) \neq 0$, we deduce from (3.3.3) that $\min\big(\operatorname{vm}(b_1), \ldots, \operatorname{vm}(b_s)\big) \leqslant \operatorname{sr}(a)$. This inequality together with (3.3.4) implies (3.3.1).

By the definition of linear-phase moments, we can directly check $\operatorname{lpm}(a * a^\star) \geqslant \operatorname{lpm}(a)$. Suppose that a has complex symmetry and $\widehat{a}(0) \neq -1$. Then we must have $\overline{\widehat{a}(\xi)} = \widehat{a}(\xi)e^{ic\xi}$ for some $c \in \mathbb{Z}$. Thus $|\widehat{a}(\xi)|^2 = [e^{ic\xi/2}\widehat{a}(\xi)]^2$. Let $n := \operatorname{lpm}(a * a^\star)$. By $\widehat{a * a^\star}(\xi) = |\widehat{a}(\xi)|^2$, we must have $[e^{ic\xi/2}\widehat{a}(\xi)]^2 = \widehat{a}(\xi)\overline{\widehat{a}(\xi)} = 1 + \mathscr{O}(|\xi|^n)$ as $\xi \to 0$, which can be rewritten as

$$[1 - e^{ic\xi/2}\widehat{a}(\xi)][1 + e^{ic\xi/2}\widehat{a}(\xi)] = 1 - [e^{ic\xi/2}\widehat{a}(\xi)]^2 = \mathscr{O}(|\xi|^n), \qquad \xi \to 0.$$

Since $[1 + e^{ic\xi/2}\widehat{a}(\xi)]|_{\xi=0} \neq 0$, the above identity forces $1 - e^{ic\xi/2}\widehat{a}(\xi) = \mathscr{O}(|\xi|^n)$ as $\xi \to 0$. That is, we proved $\operatorname{lpm}(a) \geqslant n = \operatorname{lpm}(a * a^\star)$. □

The claim $\operatorname{lpm}(a * a^\star) = \operatorname{lpm}(a)$ is not necessarily true even if $\mathsf{a}(1) = 1$ and a has symmetry (but not complex symmetry). See Exercise 3.8 for details.

Due to Proposition 3.3.1, for $n_b \in \mathbb{N}_0$ satisfying

$$0 \leqslant n_b \leqslant \min\big(\operatorname{sr}(a), \tfrac{1}{2}\operatorname{vm}(\Theta(z) - \Theta(z^2)\mathsf{a}(z)\mathsf{a}^\star(z))\big), \tag{3.3.5}$$

we often write

$$\mathsf{b}_1(z) = (1 - z^{-1})^{n_b}\mathring{\mathsf{b}}_1(z), \quad \ldots, \quad \mathsf{b}_s(z) = (1 - z^{-1})^{n_b}\mathring{\mathsf{b}}_s(z).$$

The perfect reconstruction condition for a tight framelet filter bank $\{a; b_1, \ldots, b_s\}_\Theta$ becomes

$$\begin{bmatrix} \mathring{\mathsf{b}}_1(z) & \cdots & \mathring{\mathsf{b}}_s(z) \\ \mathring{\mathsf{b}}_1(-z) & \cdots & \mathring{\mathsf{b}}_s(-z) \end{bmatrix} \begin{bmatrix} \mathring{\mathsf{b}}_1(z) & \cdots & \mathring{\mathsf{b}}_s(z) \\ \mathring{\mathsf{b}}_1(-z) & \cdots & \mathring{\mathsf{b}}_s(-z) \end{bmatrix}^\star = \mathcal{M}_{a,\Theta|n_b}(z), \tag{3.3.6}$$

where

$$\mathcal{M}_{a,\Theta|n_b}(z) := \begin{bmatrix} \mathsf{A}(z) & \mathsf{B}(z) \\ \mathsf{B}(-z) & \mathsf{A}(-z) \end{bmatrix}, \qquad \mathcal{M}_{a,\Theta}(z) := \mathcal{M}_{a,\Theta|0}(z) \tag{3.3.7}$$

with A and B being well-defined Laurent polynomials given by

$$\mathsf{A}(z) := \frac{\Theta(z) - \Theta(z^2)\mathsf{a}(z)\mathsf{a}^\star(z)}{(1-z)^{n_b}(1-z^{-1})^{n_b}}, \quad \mathsf{B}(z) := -\Theta(z^2)\frac{\mathsf{a}(z)}{(1+z)^{n_b}}\frac{\mathsf{a}^\star(-z)}{(1-z^{-1})^{n_b}}. \tag{3.3.8}$$

Recall that $\mathsf{b}^{[\gamma]}(z) := \sum_{k\in\mathbb{Z}} b(\gamma+2k)z^k$, $\gamma \in \mathbb{Z}$ for $b = \{b(k)\}_{k\in\mathbb{Z}} \in l_0(\mathbb{Z})$. In terms of polyphase matrices and coset sequences, (3.3.6) can be further rewritten as

$$\begin{bmatrix} \mathring{\mathsf{b}}_1^{[0]}(z) & \cdots & \mathring{\mathsf{b}}_s^{[0]}(z) \\ \mathring{\mathsf{b}}_1^{[1]}(z) & \cdots & \mathring{\mathsf{b}}_s^{[1]}(z) \end{bmatrix} \begin{bmatrix} \mathring{\mathsf{b}}_1^{[0]}(z) & \cdots & \mathring{\mathsf{b}}_s^{[0]}(z) \\ \mathring{\mathsf{b}}_1^{[1]}(z) & \cdots & \mathring{\mathsf{b}}_s^{[1]}(z) \end{bmatrix}^\star = \mathcal{N}_{a,\Theta|n_b}(z), \tag{3.3.9}$$

where

$$\mathcal{N}_{a,\Theta|n_b}(z) := \frac{1}{2}\begin{bmatrix} \mathsf{A}^{[0]}(z) + \mathsf{B}^{[0]}(z) & z(\mathsf{A}^{[1]}(z) - \mathsf{B}^{[1]}(z)) \\ \mathsf{A}^{[1]}(z) + \mathsf{B}^{[1]}(z) & \mathsf{A}^{[0]}(z) - \mathsf{B}^{[0]}(z) \end{bmatrix}. \tag{3.3.10}$$

Recall that an $r \times r$ matrix U of complex numbers is *positive semidefinite*, denoted by $U \geqslant 0$, if $\bar{x}^\mathsf{T} U x \geqslant 0$ for all $x \in \mathbb{C}^r$. If $U \geqslant 0$, then $U^\star = U$ (see Exercise 3.16). Obviously, by (3.3.6) a necessary condition for constructing a tight framelet filter bank is $\mathcal{M}_{a,\Theta|n_b}(z) \geqslant 0$ for all $z \in \mathbb{T}$ (or equivalently $\mathcal{N}_{a,\Theta|n_b}(z) \geqslant 0$ for all $z \in \mathbb{T}$ by (3.3.9)). This necessary condition can be equivalently expressed as follows:

Lemma 3.3.2 *Let* $\mathcal{M}_{a,\Theta}(z) := \mathcal{M}_{a,\Theta|0}(z)$ *be defined in* (3.3.7) *with* $n_b = 0$. *Then the following statements are equivalent:*

(i) $\mathcal{M}_{a,\Theta}(z) \geqslant 0$ *for all* $z \in \mathbb{T}$.
(ii) $\mathcal{N}_{a,\Theta}(z) \geqslant 0$ *for all* $z \in \mathbb{T}$, *where* $\mathcal{N}_{a,\Theta}(z) := \mathcal{N}_{a,\Theta|0}(z)$ *is defined in* (3.3.10).
(iii) The following two conditions hold: For all $z \in \mathbb{T}$,

$$\Theta(z) \geqslant 0 \tag{3.3.11}$$

and $\det(\mathcal{M}_{a,\Theta}(z)) \geqslant 0$, *i.e.*,

$$\Theta(z)\Theta(-z) - \Theta(z^2)\big[\Theta(-z)\mathsf{a}(z)\mathsf{a}^\star(z) + \Theta(z)\mathsf{a}(-z)\mathsf{a}^\star(-z)\big] \geqslant 0. \tag{3.3.12}$$

Proof From the definition of $\mathcal{N}_{a,\Theta} := \mathcal{N}_{a,\Theta|0}(z)$ in (3.3.10) with $n_b = 0$ and

$$\mathcal{M}_{a,\Theta}(z) = \begin{bmatrix} 1 & z \\ 1 & -z \end{bmatrix} \mathcal{N}_{a,\Theta}(z^2) \begin{bmatrix} 1 & z \\ 1 & -z \end{bmatrix}^\star,$$

it is straightforward to see that (i) is equivalent to (ii). Note that $\det(\mathcal{M}_{a,\Theta}(z))$ is equal to the left-hand side of (3.3.12). The equivalence between (i) and (iii) has been established in Lemma 1.4.5 of Chap. 1. □

To construct a tight framelet filter bank $\{a; b_1, \dots, b_s\}_\Theta$, according to Lemma 3.3.2 and the perfect reconstruction condition (3.3.6), the condition $\mathcal{M}_{a,\Theta}(z) \geqslant 0$ for all $z \in \mathbb{T}$ (which is equivalent to (3.3.11) and (3.3.12)) has to be satisfied. For $\Theta = \boldsymbol{\delta}$, such a necessary condition in item (iii) of Lemma 3.3.2 simply becomes

$$\mathsf{a}(z)\mathsf{a}^\star(z) + \mathsf{a}(-z)\mathsf{a}^\star(-z) \leqslant 1, \qquad \forall\, z \in \mathbb{T}. \tag{3.3.13}$$

To construct tight framelet filter banks with symmetry and vanishing moments, there are two approaches. The first approach is to use $\Theta = \boldsymbol{\delta}$ and then construct low-pass filters $a \in l_0(\mathbb{Z})$ satisfying (3.3.13) and having large $\min(\operatorname{sr}(a), \frac{1}{2}\operatorname{lpm}(a * a^\star))$. A family of such low-pass filters is the symmetric low-pass filters $a_{m,n}$, with m sum rules and n linear-phase moments, defined in (2.1.11) and (2.1.12).

The second approach is to construct a moment correcting filter Θ for a given low-pass filter a. Since we often require $\mathsf{a}(1) = 1$, it is natural to require $\Theta(1) = 1$ as well. For a given low-pass filter $a \in l_0(\mathbb{Z})$, a natural question for constructing a tight framelet filter bank is whether there exists a desirable moment correcting filter Θ satisfying

$$\Theta(1) = 1 \quad \text{and} \quad \Theta(z) - \Theta(z^2)\mathsf{a}(z)\mathsf{a}^\star(z) = \mathscr{O}(|z-1|^{2n}), \qquad z \to 1, \tag{3.3.14}$$

and the necessary conditions in (3.3.11) and (3.3.12); if so, how to construct such a moment correcting filter Θ. From the expression of $\det(\mathcal{M}_{a,\Theta}(z))$ in (3.3.12), we observe that

$$\frac{\det(\mathcal{M}_{a,\Theta}(z))}{\Theta(z)\Theta(-z)\Theta(z^2)} = \frac{1}{\Theta(z^2)} - \Big[\mathsf{a}(z)\mathsf{a}^\star(z)\frac{1}{\Theta(z)} + \mathsf{a}(-z)\mathsf{a}^\star(-z)\frac{1}{\Theta(-z)}\Big].$$

The existence of such a desirable moment correcting filter is guaranteed by the following result, whose proof will be developed in Exercises 3.21–3.23.

Theorem 3.3.3 *Let $a \in l_0(\mathbb{Z})$ such that $\mathsf{a}(1) = 1$. Define a filter u by $u := a * a^\star$; that is, $\mathsf{u}(z) := \mathsf{a}(z)\mathsf{a}^\star(z)$. Suppose that there exists $v \in l_0(\mathbb{Z})$ such that*

$$\mathcal{T}_u v = 2v \quad \text{and} \quad \mathsf{v}(z) > 0 \qquad \forall\, z \in \mathbb{T}. \tag{3.3.15}$$

Then for any positive integer n, there exists $\Theta \in l_0(\mathbb{Z})$ *such that* (3.3.14) *holds,* $\Theta(z) > 0$ *and* $\det(\mathcal{M}_{a,\Theta}(z)) \geqslant 0$ *for all* $z \in \mathbb{T}$.

In the following we present a heuristic way of constructing a moment correcting filter Θ with the shortest possible filter support.

Proposition 3.3.4 *Let* $a \in l_0(\mathbb{Z})$ *such that* $\mathsf{a}(1) = 1$. *For any positive integer n, there exists a unique moment correcting filter* Θ *such that* $\operatorname{fsupp}(\Theta) \subseteq [-n, n-1]$ *and* (3.3.14) *is satisfied. In addition,*

(i) if a is real-valued or a has symmetry, then $\Theta_n := \Theta$ *is real-valued,* $\operatorname{fsupp}(\Theta_n) \subseteq [1-n, n-1]$, $\mathsf{S}\Theta_n(z) = 1$, *and* (3.3.14) *holds with* Θ *being replaced by* Θ_n;
(ii) if a is complex-valued, set $\Theta_n(z) := [\Theta(z) + \Theta^\star(z)]/2$, *then* $\operatorname{fsupp}(\Theta_n) \subseteq [-n, n]$, $\mathbb{S}\Theta_n(z) = 1$, *and* (3.3.14) *holds with* Θ *being replaced by* Θ_n.

Proof By Exercise 1.14 with $\mathsf{d} = 1$, $\lambda = 2$, $\mathbf{u}(\xi) = 1$ and $\mathbf{v}(\xi) = |\widehat{a}(\xi)|^2$, there exists a unique moment correcting filter Θ such that $\operatorname{fsupp}(\Theta) \subseteq [-n, n-1]$ and (3.3.14) is satisfied. In fact, such Θ can be easily obtained by solving

$$\sum_{k=-n}^{n-1} \Theta(k) k^j = i^j h_a(j), \qquad j = 0, \ldots, 2n-1, \tag{3.3.16}$$

where the values $h_a(j), j \in \mathbb{N}_0$ are real numbers given recursively by

$$h_a(0) = 1, \quad h_a(j) = \frac{1}{1-2^j} \sum_{m=0}^{j-1} \frac{j! 2^m}{m!(j-m)!} [\widehat{a}\overline{\widehat{a}}]^{(j-m)}(0) h_a(m), \quad j \in \mathbb{N}. \tag{3.3.17}$$

Define $u \in l_0(\mathbb{Z})$ by $\widehat{u}(\xi) := |\widehat{a}(\xi)|^2$. If a has symmetry, then u also has symmetry. Since $\mathsf{u}(z) \geqslant 0$ for all $z \in \mathbb{T}$, by Lemma 3.1.2, the sequence u must have real coefficients. If a is real-valued, then u must have real coefficients. In other words, $\widehat{u}(\xi) = \widehat{u}(-\xi)$ and consequently,

$$[\widehat{a}\overline{\widehat{a}}]^{(j)}(0) = [\widehat{u}]^{(j)}(0) = 0, \qquad \text{for all positive odd integers } j.$$

Now we can deduce from (3.3.17) that $h_a(j) = 0$ for all positive odd integers j and $h_a(j)$ are real numbers for all even integers j. There is a unique solution $\{t_0, \ldots, t_n\}$ of real numbers to

$$\sum_{k=0}^{n-1} t_k k^{2j} = \frac{(-1)^j}{2} h_a(2j), \qquad j = 0, \ldots, n-1.$$

Define $\widehat{\Theta}(\xi) := \sum_{k=0}^{n-1} t_k (e^{ik\xi} + e^{-ik\xi})$. Using the fact that $\widehat{\Theta}^{(j)}(0) = 0$ for all positive odd integers j, we see that Θ must be the unique solution to (3.3.16), since $\widehat{\Theta}^{(j)}(0) = h_a(j)$ for all $j = 0, \ldots, 2n-1$ and $\operatorname{fsupp}(\Theta) \subseteq [1-n, n-1]$.

Therefore, Θ is real-valued, $\overline{\widehat{\Theta}(\xi)} = \widehat{\Theta}(\xi)$, and $\mathrm{fsupp}(\Theta) \subseteq [1-n, n-1]$. Item (ii) is trivial. □

The moment correcting filters Θ_n in Proposition 3.3.4 often satisfy the necessary conditions in (3.3.11) and (3.3.12) for constructing tight framelet filter banks.

3.3.2 Algorithm for Tight Framelet Filter Banks with Symmetry

Since a tight framelet filter bank is a special case of a dual framelet filter bank by using the same set of filters for both analysis and synthesis, algorithms developed in Sect. 3.2 for dual framelet filter banks allow us to obtain tight framelet filter banks as special cases.

To present the algorithm, we first show that the symmetry types of the high-pass filters b_1 and b_2 in a tight framelet filter bank $\{a; b_1, b_2\}_\Theta$ with [complex] symmetry are essentially uniquely determined by the filters a and Θ.

Theorem 3.3.5 *Let $\{a; b_1, b_2\}_\Theta$ be a tight framelet filter bank such that all the filters a, b_1, b_2, Θ are not identically zero and have the following symmetry property:*

$$\mathsf{S}\Theta(z) = 1, \qquad \mathsf{S}\mathsf{a}(z) = \epsilon z^c, \qquad \mathsf{S}\mathsf{b}_1(z) = \epsilon_1 z^{c_1}, \qquad \mathsf{S}\mathsf{b}_2(z) = \epsilon_2 z^{c_2} \tag{3.3.18}$$

[or (3.3.18) holds with S being replaced by $\mathbb{S}$]. If

$$\mathrm{len}(b_1) \leqslant \mathrm{len}(b_2) \leqslant \mathrm{len}(a) + \mathrm{len}(\Theta) \neq 0, \tag{3.3.19}$$

then up to a trivial switch of b_1 and b_2 for the case $\mathrm{len}(b_1) = \mathrm{len}(b_2)$, the symmetry centers $\frac{c_1}{2}$ and $\frac{c_2}{2}$ are essentially uniquely determined by

$$\tfrac{c_2}{2} - (\tfrac{c}{2} - n_\Theta) \in 2\mathbb{Z}, \quad \tfrac{c_1}{2} - (n_{\mathcal{M}} + 1 - \tfrac{c_2}{2}) \in 2\mathbb{Z}, \tag{3.3.20}$$

and additionally for the case of symmetry, ϵ_1 and ϵ_2 are uniquely determined by

$$\epsilon_2 = -\epsilon sgn(\Theta(n_\Theta)), \quad \epsilon_1 = (-1)^c \epsilon_2 sgn(\lambda), \tag{3.3.21}$$

where $\Theta(n_\Theta)$ is the leading coefficient of Θ (that is, $\Theta(n_\Theta) \neq 0$ and $\Theta(k) = 0$ for all $k > n_\Theta$), $\lambda z^{2n_{\mathcal{M}}}$ is the leading term of the Laurent polynomial $\det(\mathcal{M}_{a,\Theta}(z))$. Moreover, if $c + n_{\mathcal{M}}$ is an even integer, then $\mathrm{len}(b_1) < \mathrm{len}(b_2) = \mathrm{len}(a) + \mathrm{len}(\Theta)$.

Proof Note that the second equation of the perfect reconstruction condition of a tight framelet filter bank $\{a; b_1, b_2\}_\Theta$ is

$$\Theta(z^2)\mathsf{a}(z)\mathsf{a}^\star(-z) + \mathsf{b}_1(z)\mathsf{b}_1^\star(-z) + \mathsf{b}_2(z)\mathsf{b}_2^\star(-z) = 0. \tag{3.3.22}$$

The symmetry types of each term in the above identity are

$$\mathsf{S}(\Theta(z^2)\mathsf{a}(z)\mathsf{a}^\star(-z)) = \mathsf{S}(\Theta(z^2))\mathsf{S}(\mathsf{a}(z))\mathsf{S}(\mathsf{a}^\star(-z)) = \epsilon z^c \epsilon(-z)^{-c} = (-1)^c,$$

$$\mathsf{S}(\mathsf{b}_1(z)\mathsf{b}_1^\star(-z)) = \mathsf{S}\mathsf{b}_1(z)(\mathsf{S}\mathsf{b}_1(-z))^\star = (-1)^{c_1}, \qquad \mathsf{S}(\mathsf{b}_2(z)\mathsf{b}_2^\star(-z)) = (-1)^{c_2}.$$

Because all c, c_1, c_2 are integers, two of the three terms $(-1)^c, (-1)^{c_1}, (-1)^{c_2}$ must be the same. In other words, at least two terms in (3.3.22) must have the same symmetry type. Consequently, (3.3.22) forces $(-1)^c = (-1)^{c_1} = (-1)^{c_2}$, i.e.,

$$c_1 - c \in 2\mathbb{Z} \quad \text{and} \quad c_2 - c \in 2\mathbb{Z}. \tag{3.3.23}$$

To prove (3.3.20) and (3.3.21), we compare the leading coefficients in the equations of the perfect reconstruction condition. Since $\Theta(n_\Theta)$ is the leading coefficients of Θ and $\mathsf{S}\Theta(z) = 1$, we have $\operatorname{fsupp}(\Theta) = [-n_\Theta, n_\Theta]$. Define $n, n_1, n_2 \in \mathbb{Z}$ by

$$[c - n, n] := \operatorname{fsupp}(a), \quad [c_1 - n_1, n_1] := \operatorname{fsupp}(b_1), \quad [c_2 - n_2, n_2] := \operatorname{fsupp}(b_2).$$

For the case of complex symmetry, we define

$$\lambda_0 := \epsilon\Theta(n_\Theta)(a(n))^2, \quad \lambda_1 := \epsilon_1(b_1(n_1))^2, \quad \lambda_2 := \epsilon_2(b_2(n_2))^2.$$

For the case of symmetry or real coefficients, we define

$$\lambda_0 := \epsilon\Theta(n_\Theta)|a(n)|^2, \quad \lambda_1 := \epsilon_1|b_1(n_1)|^2, \quad \lambda_2 := \epsilon_2|b_2(n_2)|^2. \tag{3.3.24}$$

The leading terms of each addent in the first equation of the perfect reconstruction condition:

$$\Theta(z^2)\mathsf{a}(z)\mathsf{a}^\star(z) + \mathsf{b}_1(z)\mathsf{b}_1^\star(z) + \mathsf{b}_2(z)\mathsf{b}_2^\star(z) = \Theta(z) \tag{3.3.25}$$

are

$$\lambda_0 z^{2n_\Theta+2n-c}, \qquad \lambda_1 z^{2n_1-c_1}, \quad \lambda_2 z^{2n_2-c_2}, \quad \Theta(n_\Theta)z^{n_\Theta},$$

and the leading terms of each addent in the second equation (3.3.22) of the perfect reconstruction condition are

$$(-1)^{n-c}\lambda_0 z^{2n_\Theta+2n-c}, \quad (-1)^{n_1-c_1}\lambda_1 z^{2n_1-c_1}, \quad (-1)^{n_2-c_2}\lambda_2 z^{2n_2-c_2}, \tag{3.3.26}$$

respectively. Note that all $\lambda_0, \lambda_1, \lambda_2$ are nonzero. Our assumption in (3.3.19) becomes $2n_1 - c_1 \leqslant 2n_2 - c_2 \leqslant 2n - c + 2n_\Theta \neq 0$, from the last relation we must have $n_\Theta < 2n - c + 2n_\Theta$ (otherwise, $\operatorname{len}(a) = \operatorname{len}(\Theta) = 0$). Since the perfect reconstruction condition in (3.3.25) and (3.3.22) must hold, by (3.3.23), we consider two cases.

Case 1: $2n_1 - c_1 = 2n_2 - c_2 = 2n - c + 2n_\Theta$. By (3.3.23) and comparing leading terms in (3.3.26), the following two equations must hold:

$$\lambda_0 + \lambda_1 + \lambda_2 = 0, \qquad \lambda_0 + (-1)^{n_1-n}\lambda_1 + (-1)^{n_2-n}\lambda_2 = 0, \tag{3.3.27}$$

from which we deduce that $(-1)^{n_1-n} = (-1)^{n_2-n} = 1$ (otherwise, the above two equations will force at least one of $\lambda_0, \lambda_1, \lambda_2$ to be zero). Hence, $n_1 - n \in 2\mathbb{Z}$ and $n_2 - n \in 2\mathbb{Z}$. Now we deduce from $2n_2 - c_2 = 2n - c + 2n_\Theta$ that

$$c_2 = c + 2n_2 - 2n - 2n_\Theta = c - 2n_\Theta + 4k \quad \text{with} \quad k := (n_2 - n)/2 \in \mathbb{Z}. \tag{3.3.28}$$

Hence, $c_2 - (c - 2n_\Theta) \in 4\mathbb{Z}$. Similarly, we also have $c_1 - (c - 2n_\Theta) \in 4\mathbb{Z}$. Since $\{a; b_1, b_2\}_\Theta$ is a tight framelet filter bank, we must have $\Theta(z) \geqslant 0$ for all $z \in \mathbb{T}$. Consequently, $\mathbb{S}\Theta(z) = 1$. If all filters have real coefficients or have symmetry, by Lemma 3.1.2, we conclude that Θ must have real coefficients. Therefore, by (3.3.24), all $\lambda_0, \lambda_1, \lambda_2$ are real numbers. Now it follows from the first identity of (3.3.27) that at least one of the signs of λ_1 and λ_2 must be different to that of λ_0. Without loss of generality, we assume $\lambda_0\lambda_2 < 0$. That is, by (3.3.24), we must have $\epsilon_2 = \operatorname{sgn}(\lambda_2) = -\operatorname{sgn}(\lambda_0) = -\epsilon \operatorname{sgn}(\Theta(n_\Theta))$.

Case 2: $2n_1 - c_1 < 2n_2 - c_2 = 2n - c + 2n_\Theta$. By (3.3.23) and comparing leading terms in (3.3.26), the following two equations must hold:

$$\lambda_0 + \lambda_2 = 0, \quad \lambda_0 + (-1)^{n_2-n}\lambda_2 = 0, \tag{3.3.29}$$

from which we must have $(-1)^{n_2-n} = 1$, that is, $n_2 - n \in 2\mathbb{Z}$. We deduce from $2n_2 - c_2 = 2n - c + 2n_\Theta$ that (3.3.28) holds. Hence, we also have $c_2 - (c - 2n_\Theta) \in 4\mathbb{Z}$. If all the filters have symmetry or real coefficients, then it follows from the first equation of (3.3.29) that $\lambda_2 = -\lambda_0$, that is, we must have $\epsilon_2 = -\epsilon \operatorname{sgn}(\Theta(n_\Theta))$.

We now investigate the property of c_1 and ϵ_1. By the perfect reconstruction condition in (3.3.6) with $s = 2$ and $n_b = 0$, we have $\mathsf{D}(z^2)\mathsf{D}^\star(z^2) = \det(\mathcal{M}_{a,\Theta}(z))$, where $\mathsf{D}(z^2) := z^{-1}[\mathsf{b}_1(z)\mathsf{b}_2(-z) - \mathsf{b}_1(-z)\mathsf{b}_2(z)]$. By (3.3.23), we have $\mathsf{S}(\mathsf{b}_1(z)\mathsf{b}_2(-z)) = \epsilon_1\epsilon_2(-1)^c z^{c_1+c_2} = \mathsf{S}(\mathsf{b}_1(-z)\mathsf{b}_2(z))$. Hence, by the definition of D, we have $\mathsf{SD}(z) = \epsilon_1\epsilon_2(-1)^c z^{\frac{c_1+c_2}{2}-1}$, where we used (3.3.23). Let the leading term of $\mathsf{D}(z)$ be $\lambda_{\mathsf{D}} z^{n_{\mathsf{D}}}$. Then the leading term of $\mathsf{D}(z^2)\mathsf{D}^\star(z^2)$ must be $|\lambda_0|^2\epsilon_1\epsilon_2(-1)^c z^{4n_{\mathsf{D}}-(c_1+c_2-2)}$ for symmetry and $\lambda_0^2\epsilon_1\epsilon_2(-1)^c z^{4n_{\mathsf{D}}-(c_1+c_2-2)}$ for complex symmetry. Comparing the leading terms in the identity $\mathsf{D}(z^2)\mathsf{D}^\star(z^2) = \det(\mathcal{M}_{a,\Theta}(z))$, we must have $c_1 + c_2 - 2 = 4n_{\mathsf{D}} - 2n_{\mathcal{M}}$ from which we have the second relation in (3.3.20). For symmetry, we further have $\epsilon_1\epsilon_2(-1)^c = \operatorname{sgn}(\lambda)$ and hence $\epsilon_1 = (-1)^c\epsilon_2 \operatorname{sgn}(\lambda)$.

If $c + n_{\mathcal{M}}$ is an even integer, by (3.3.23), we conclude that

$$\tfrac{c_1}{2} - (\tfrac{c}{2} - n_\Theta) = n_{\mathcal{M}} - c + 1 + [\tfrac{c_1}{2} - (n_{\mathcal{M}} + 1 - \tfrac{c_2}{2})] + 2n_\Theta - [\tfrac{c_2}{2} - (\tfrac{c}{2} - n_\Theta)] \notin 2\mathbb{Z}.$$

Therefore, Case 1 cannot happen and we must have Case 2, that is, $\mathrm{len}(b_1) < \mathrm{len}(b_2) = \mathrm{len}(a) + \mathrm{len}(\Theta)$. $\sqcap$

We now derive from Algorithm 3.2.1 a simple algorithm for constructing tight framelet filter banks $\{a; b_1, b_2\}_\Theta$ with the shortest possible filter supports and with the symmetry property.

Algorithm 3.3.6 *Let $a, \Theta \in l_0(\mathbb{Z})$ be filters having [complex] symmetry (and real coefficients) such that $\mathsf{S}\mathsf{a}(z) = \epsilon z^c$ with $\epsilon \in \{-1, 1\}$ and $c \in \mathbb{Z}$, $\mathsf{S}\Theta(z) = 1$, and $\mathcal{M}_{a,\Theta}(z) \geqslant 0$ for all $z \in \mathbb{T}$. Choose $n_b \in \mathbb{N}_0$ satisfying* (3.3.5)*. For the case of complex symmetry, replace S by $\mathbb{S}$ throughout.*

(S1) Define Laurent polynomials A and B as in (3.3.8) *and Laurent polynomials $\mathsf{p}, \mathring{\mathsf{A}}, \mathring{\mathsf{B}}$ as in* (3.2.10).

(S2) Select a Laurent polynomial d with [complex] symmetry (and real coefficients) such that $\mathsf{d}(z)\mathsf{d}^\star(z) = \mathsf{D}(z)$, where D is defined in (3.2.11)*. Define $\epsilon_{\mathsf{d}} z^{c_{\mathsf{d}}} := \mathsf{S}\mathsf{d}(z)$, $[c_{\mathsf{d}} - n_{\mathsf{d}}, n_{\mathsf{d}}] :=$ fsupp(d), $[-n_0, n_0] :=$ fsupp$(\mathring{\mathsf{B}})$, and*

$$c_2 := c - 2n_\Theta + n_b, \quad \epsilon_2 := \epsilon(-1)^{n_b+1} sgn(\Theta(n_\Theta)), \quad n_2 := \tfrac{n_0+c_2}{2},$$

where n_Θ is defined in Theorem 3.3.5. If $\Theta(n_\Theta)$ is not real, define $\epsilon_2 := 1$;

(S3) Parameterize a filter $\mathring{\mathsf{b}}_2$ such that $\mathsf{S}\mathring{\mathsf{b}}_2(z) = \epsilon_2 z^{c_2}$ and fsupp$(\mathring{\mathsf{b}}_2) = [c_2 - n_2, n_2]$. Find the unknown coefficients of $\mathring{\mathsf{b}}_2$ by solving a system X of linear equations induced by $\mathcal{R}(z) = 0$ (i.e., all the coefficients of $\mathcal{R}$ are zero) and

$$coeff(\mathring{\mathsf{b}}_1^\star, z, j) = 0, \quad j = n_2 - c_{\mathsf{d}}, \ldots, n_0 + n_2 - 2n_{\mathsf{d}} - 1, \tag{3.3.30}$$

with $coeff(\mathring{\mathsf{b}}_1^\star, z, j)$ being the coefficient of z^j in the Laurent polynomial $\mathring{\mathsf{b}}_1^\star$, where $\mathcal{R}$ and $\mathring{\mathsf{b}}_1^\star$ are uniquely determined by fsupp$(\mathcal{R}) \subseteq [2(c_{\mathsf{d}} - n_{\mathsf{d}}), 2n_{\mathsf{d}} - 1]$ *and*

$$\mathring{\mathsf{B}}^\star(z)\mathring{\mathsf{b}}_2(z) - \mathring{\mathsf{A}}^\star(z)\mathring{\mathsf{b}}_2(-z) = -\mathsf{d}(z^2)z\mathring{\mathsf{b}}_1^\star(z) + \mathcal{R}(z); \tag{3.3.31}$$

(S4) For any nontrivial solution to the homogeneous system X in (S3), there must exist $\lambda > 0$ such that (3.2.16) *holds. Replace $\mathring{\mathsf{b}}_1, \mathring{\mathsf{b}}_2$ by $\lambda^{-1/2}\mathring{\mathsf{b}}_1, \lambda^{-1/2}\mathring{\mathsf{b}}_2$, respectively;*

(S5) Find two Laurent polynomials $\mathsf{q}_1, \mathsf{q}_2$ with [complex] symmetry (and real coefficients) such that

$$\mathsf{q}_1(z)\mathsf{q}_1^\star(z) + \mathsf{q}_2(z)\mathsf{q}_2^\star(z) = \mathsf{p}(z) \quad \text{and} \quad \frac{\mathsf{S}\mathsf{q}_1(z)}{\mathsf{S}\mathsf{q}_2(z)} = (-1)^{c_2}\epsilon_{\mathsf{d}} z^{c_2 - c_{\mathsf{d}} - 1}. \tag{3.3.32}$$

For any choice of $s_1, s_2 \in \mathbb{Z}$, *define*

$$\begin{aligned} \mathsf{b}_1(z) &:= z^{2s_1}(1-z^{-1})^{n_b}[\mathring{\mathsf{b}}_1(z)\mathsf{q}_1(z^2) + \mathring{\mathsf{b}}_2(z)\mathsf{q}_2(z^2)], \\ \mathsf{b}_2(z) &:= z^{2s_2}(1-z^{-1})^{n_b}[\mathring{\mathsf{b}}_2(z)\mathsf{q}_1^\star(z^2) - \mathring{\mathsf{b}}_1(z)\mathsf{q}_2^\star(z^2)]. \end{aligned} \tag{3.3.33}$$

Then $\{a; b_1, b_2\}_\Theta$ *is a tight framelet filter bank having [complex] symmetry (and real coefficients) such that* $\operatorname{vm}(b_1) \geqslant n_b$, $\operatorname{vm}(b_2) \geqslant n_b$,

$$\mathsf{S}b_1(z) = (-1)^{n_b+c_2}\epsilon_2\epsilon_{\mathsf{d}}\epsilon_{\mathsf{q}} z^{2c_{\mathsf{d}}+2c_{\mathsf{q}}+2-c_2-n_b+4s_1}, \quad \mathsf{S}b_2(z) = (-1)^{n_b}\epsilon_2\epsilon_{\mathsf{q}} z^{c_2-n_b-2c_{\mathsf{q}}+4s_2}$$

with $\epsilon_{\mathsf{q}} z^{c_{\mathsf{q}}} := \mathsf{S}\mathsf{q}_1(z)$, *and* $\max(\operatorname{len}(b_1), \operatorname{len}(b_2)) \leqslant \operatorname{len}(a) + \operatorname{len}(\Theta)$.

Proof By the definition of $\mathring{\mathsf{B}}$ in (3.3.8), we see that $n_0 = 2n_\Theta + \operatorname{len}(a) - n_b$ and $n_2 := \frac{n_0+c_2}{2} = \frac{c+\operatorname{len}(a)}{2}$ must be an integer. We deduce from (3.3.31) with $\mathcal{R} = 0$ that

$$\operatorname{fsupp}(\mathring{\mathsf{b}}_1^\star) \subseteq [c_2 - n_0 - n_2 + 2n_{\mathsf{d}} - 2c_{\mathsf{d}} - 1, n_0 + n_2 - 2n_{\mathsf{d}} - 1]$$

and $\mathsf{S}\mathring{\mathsf{b}}_1(z) = \epsilon_1 z^{c_1}$ with $c_1 := 2c_{\mathsf{d}} + 2 - c_2$ and $\epsilon_1 := (-1)^{c_2}\epsilon_{\mathsf{d}}\epsilon_2$. It follows from (3.3.30) that $\operatorname{fsupp}(\mathring{\mathsf{b}}_1^\star) \subseteq [-c_1 - (n_2 - c_{\mathsf{d}} - 1), n_2 - c_{\mathsf{d}} - 1]$. Hence, $\operatorname{len}(\mathring{\mathsf{b}}_1) \leqslant 2n_2 + c_1 - 2c_{\mathsf{d}} - 2 = n_0$. By the definition of n_2, we also have $\operatorname{len}(\mathring{b}_2) \leqslant 2n_2 - c_2 = n_0$.

Since $\mathcal{R} = 0$, we deduce from (3.3.31) that $\mathring{\mathsf{b}}_1(z)\mathsf{d}^\star(z^2)z^{-1} = \mathring{\mathsf{A}}(z)\mathring{\mathsf{b}}_2^\star(-z) - \mathring{\mathsf{B}}(z)\mathring{\mathsf{b}}_2^\star(z)$, from which we see that

$$\begin{aligned} &[\mathring{\mathsf{B}}^\star(z)\mathring{\mathsf{b}}_1(z) - \mathring{\mathsf{A}}^\star(z)\mathring{\mathsf{b}}_1(-z)]\mathsf{d}^\star(z^2)z^{-1} \\ &\quad= \mathring{\mathsf{B}}^\star(z)[\mathring{\mathsf{b}}_1(z)\mathsf{d}^\star(z^2)z^{-1}] + \mathring{\mathsf{A}}^\star(z)[\mathring{\mathsf{b}}_1(-z)\mathsf{d}^\star(z^2)(-z)^{-1}] \\ &\quad= \mathring{\mathsf{B}}^\star(z)[\mathring{\mathsf{A}}(z)\mathring{\mathsf{b}}_2^\star(-z) - \mathring{\mathsf{B}}(z)\mathring{\mathsf{b}}_2^\star(z)] + \mathring{\mathsf{A}}^\star(z)[\mathring{\mathsf{A}}(-z)\mathring{\mathsf{b}}_2^\star(z) - \mathring{\mathsf{B}}(-z)\mathring{\mathsf{b}}_2^\star(-z)] \\ &\quad= [\mathring{\mathsf{A}}^\star(z)\mathring{\mathsf{A}}(-z) - \mathring{\mathsf{B}}^\star(z)\mathring{\mathsf{B}}(z)]\mathring{\mathsf{b}}_2^\star(z) + [\mathring{\mathsf{B}}^\star(z)\mathring{\mathsf{A}}(z) - \mathring{\mathsf{A}}^\star(z)\mathring{\mathsf{B}}(-z)]\mathring{\mathsf{b}}_2^\star(-z). \end{aligned}$$

By the definition of $\mathring{\mathsf{A}}$ and $\mathring{\mathsf{B}}$ in (3.3.8), we have $\mathring{\mathsf{A}}^\star(z) = \mathring{\mathsf{A}}(z)$ and $\mathring{\mathsf{B}}^\star(z) = \mathring{\mathsf{B}}(-z)$. Hence, $\mathring{\mathsf{B}}^\star(z)\mathring{\mathsf{A}}(z) - \mathring{\mathsf{A}}^\star(z)\mathring{\mathsf{B}}(-z) = 0$. By $\mathsf{d}(z^2)\mathsf{d}^\star(z^2) = \mathsf{D}^\star(z^2) = \mathring{\mathsf{A}}(z)\mathring{\mathsf{A}}(-z) - \mathring{\mathsf{B}}(z)\mathring{\mathsf{B}}(-z)$, we deduce from the above identities that

$$[\mathring{\mathsf{B}}^\star(z)\mathring{\mathsf{b}}_1(z) - \mathring{\mathsf{A}}^\star(z)\mathring{\mathsf{b}}_1(-z)]\mathsf{d}^\star(z^2)z^{-1} = \mathsf{d}(z^2)\mathsf{d}^\star(z^2)\mathring{\mathsf{b}}_2^\star(z).$$

Since the Laurent polynomial d is not identically zero, the above identity implies

$$\mathring{\mathsf{B}}^\star(z)\mathring{\mathsf{b}}_1(z) - \mathring{\mathsf{A}}^\star(z)\mathring{\mathsf{b}}_1(-z) = \mathsf{d}(z^2)z\mathring{\mathsf{b}}_2^\star(z).$$

Now the same proof to establish (3.2.27) in Algorithm 3.2.1 shows that (3.2.27) is satisfied with $\mathring{\mathring{b}}_1 = \mathring{b}_1, \mathring{\mathring{b}}_2 = \mathring{b}_2$, and $D_{\mathring{b}}(z^2) := z^{-1}[\mathring{b}_1(z)\mathring{b}_2(-z) - \mathring{b}_1(-z)\mathring{b}_2(z)]$. In particular, we have

$$\mathring{b}_1(-z)\mathring{b}_1^\star(z) + \mathring{b}_2(-z)\mathring{b}_2^\star(z) = \frac{D_{\mathring{b}}(z^2)}{d(z^2)}\mathring{B}^\star(z). \tag{3.3.34}$$

By $\max(\operatorname{len}(\mathring{b}_1), \operatorname{len}(\mathring{b}_2)) \leqslant n_0$, $\operatorname{fsupp}(\mathring{b}_1(-z)\mathring{b}_1^\star(z) + \mathring{b}_2(-z)\mathring{b}_2^\star(z)) \subseteq [-n_0, n_0] = \operatorname{fsupp}(\mathring{B}^\star)$. Hence, the above identity forces that $\frac{D_{\mathring{b}}(z^2)}{d(z^2)}$ must be a nonnegative number, since $D(z) = d(z)d^\star(z) \geqslant 0$ for all $z \in \mathbb{T}$. If $\lambda = 0$, then (3.3.34) implies $\mathring{b}_1(z)\mathring{b}_1^\star(z) + \mathring{b}_2(z)\mathring{b}_2^\star(z) = 0$, which is only possible when $\mathring{b}_1 = \mathring{b}_2 = 0$, a contradiction to our assumption on a nontrivial solution to the system X. Hence, $\lambda > 0$. After replacing $\mathring{b}_1, \mathring{b}_2$ by $\lambda^{-1/2}\mathring{b}_1, \lambda^{-1/2}\mathring{b}_2$, respectively, we deduce from (3.2.27) that

$$\begin{bmatrix} \mathring{b}_1(z) & \mathring{b}_2(z) \\ \mathring{b}_1(-z) & \mathring{b}_2(-z) \end{bmatrix} \begin{bmatrix} \mathring{b}_1(z) & \mathring{b}_2(z) \\ \mathring{b}_1(-z) & \mathring{b}_2(-z) \end{bmatrix}^\star = \begin{bmatrix} A(z) & B(z) \\ B(-z) & A(-z) \end{bmatrix}. \tag{3.3.35}$$

Note that

$$\begin{bmatrix} b_1(z) & b_2(z) \\ b_1(-z) & b_2(-z) \end{bmatrix} = \begin{bmatrix} z^{2s_1}(1-z^{-1})^{n_b} & \\ & z^{2s_2}(1+z^{-1})^{n_b} \end{bmatrix} \begin{bmatrix} \mathring{b}_1(z) & \mathring{b}_2(z) \\ \mathring{b}_1(-z) & \mathring{b}_2(-z) \end{bmatrix} \begin{bmatrix} q_1(z^2) & -q_2^\star(z^2) \\ q_2(z^2) & q_1^\star(z^2) \end{bmatrix}.$$

Using (3.3.32), we can directly check that $\{a; b_1, b_2\}_\Theta$ is a tight framelet filter bank having [complex] symmetry (and real coefficients). □

To have a tight framelet filter bank $\{a; b_1, b_2\}_\Theta$ with [complex] symmetry (and real coefficients), by (3.3.6) with $s = 2$, it is necessary that there exists a Laurent polynomial d with [complex] symmetry (and real coefficients) in item (S2) of Algorithm 3.3.6 such that $d(z)d^\star(z) = D(z)$. According to Theorem 3.1.7, up to a multiplicative monomial, such d with complex symmetry is essentially unique. Hence, it suffices to use only one particular choice of d in Algorithm 3.3.6 for complex symmetry. A moment correcting filter Θ with [complex] symmetry can be obtained by Proposition 3.3.4 with $n = n_b$. But item (S2) of Algorithm 3.3.6 often fails for such Θ. Thus, one has to solve nonlinear equations to obtain a moment correcting filter Θ satisfying both (3.3.14) and the condition in item (S2) of Algorithm 3.3.6.

The equation (3.3.32) in item (S5) of Algorithm 3.3.6 belongs to the SOS problem with symmetry, which is well studied in Theorems 3.1.5 and 3.1.6. In Algorithm 3.3.6, we often have $p = 1$, which must hold if $\Theta = \boldsymbol{\delta}$. Hence, for $p = 1$, we always take $q_1 = 1$ and $q_2 = 0$ in this section as the default solution to (3.3.32) for $p = 1$. We often use the choices of $s_1, s_2 \in \mathbb{Z}$ in (3.3.33) so that the symmetry centers of the constructed high-pass filters b_1, b_2 are near the origin. By the following result, the existence of Laurent polynomials $\mathsf{d}, \mathsf{q}_1, \mathsf{q}_2$ with [complex] symmetry in items (S2) and (S5) of Algorithm 3.3.6 is a necessary and sufficient condition for having a tight framelet filter bank $\{a; b_1, b_2\}_\Theta$ with symmetry property.

Theorem 3.3.7 *Let $a, \Theta \in l_0(\mathbb{Z})$ have [complex] symmetry (and real coefficients) such that $\mathsf{S}\Theta(z) = 1$ and $\mathsf{S}a(z) = \epsilon z^c$. Let $n_b \in \mathbb{N}_0$ satisfy (3.3.5). Then there exists a tight framelet filter bank $\{a; b_1, b_2\}_\Theta$ for some $b_1, b_2 \in l_0(\mathbb{Z})$ having [complex] symmetry (and real coefficients) if and only if*

(i) $\Theta(z) \geqslant 0$ for all $z \in \mathbb{T}$, and $\det(\mathcal{N}_{a,\Theta|n_b}(z)) = \mathsf{d}_{n_b}(z)\mathsf{d}^\star_{n_b}(z)$ for a Laurent polynomial d_{n_b} having [complex] symmetry (and real coefficients), where $\mathcal{N}_{a,\Theta|n_b}$ is defined in (3.3.10);
(ii) $\mathsf{p}(z) := \gcd([\mathcal{N}_{a,\Theta|n_b}(z)]_{1,1}, [\mathcal{N}_{a,\Theta|n_b}(z)]_{1,2}, [\mathcal{N}_{a,\Theta|n_b}(z)]_{2,1}, [\mathcal{N}_{a,\Theta|n_b}(z)]_{2,2})$ has the real SOS property with respect to the symmetry type $(-1)^{c+n_b} z^{\operatorname{odd}(c+n_b)-1} \mathsf{S}\mathsf{d}_{n_b}(z)$.

Moreover, if all involved Laurent polynomials and filters have complex symmetry (that is, all filters a, b_1, b_2, Θ and d_{n_b}) instead of symmetry, replace the symmetry operator S by the complex symmetry operator $\mathbb{S}$, then the same necessary and sufficient condition, with item (ii) removed, still holds.

If $\Theta = \boldsymbol{\delta}$, then $\mathsf{p} = 1$ and item (ii) is automatically satisfied in Theorem 3.3.7. For different choices of n_b in Theorem 3.3.7, the Laurent polynomial p in Theorem 3.3.7 differs only by a factor $(1-z)^{2m}$ for some nonnegative integer m and the symmetry type $(-1)^{c+n_b} z^{\operatorname{odd}(c+n_b)-1} \mathsf{S}\mathsf{d}_{n_b}(z)$ only differs by a factor of z^{2k} for some $k \in \mathbb{Z}$. Consequently, it doesn't matter which $n_b \in \mathbb{N}$ satisfying (3.3.5) in Theorem 3.3.7 is chosen. For simplicity, we often set $n_b = 0$ in Theorem 3.3.7.

We shall prove Theorem 3.3.7 in Sect. 3.6 by generalizing the results in Sect. 3.4.2 under the symmetry property. Currently, we are unable to prove the existence of a tight framelet filter bank with symmetry in Algorithm 3.3.6, largely because the linear system X in item (S3) of Algorithm 3.3.6 is overcomplete, i.e., there are more linear equations than the number of free unknowns. If only numerical (not exact) solutions of d with [complex] symmetry satisfying $\mathsf{d}(z)\mathsf{d}^\star(z) = \mathsf{D}(z)$ are available, then the overcompleteness of X often leads to the trivial zero solution. To overcome this numerical issue, we form a new linear system $\tilde{X}$ by adding an auxiliary equation $\mathring{b}_2(n_2) = 1$ to X. Then we find the least square solution to $\tilde{X}$ and check whether such a solution is indeed a nontrivial solution to X or not.

We now present some examples of tight framelet filter banks $\{a; b_1, b_2\}_\Theta$ with symmetry property to illustrate Algorithm 3.3.6.

Example 3.3.1 Let $a = a_2^B = \{\frac{1}{4}, \frac{1}{2}, \frac{1}{4}\}_{[0,2]}$ be the B-spline filter of order 2. Then $\epsilon z^c := \mathsf{S}a(z) = z^2$. Setting $n_b = 1$ in Algorithms 3.3.6 with $\Theta = \boldsymbol{\delta}$, we have

$$\mathsf{p}(z) = 1, \quad \mathsf{D}(z) = 1/8, \quad c_2 = 3, \quad \epsilon_2 = 1, \quad n_0 = 1, \quad n_2 = 2.$$

Taking $\mathsf{d}(z) = \frac{\sqrt{2}}{4}z$ with $\mathsf{Sd}(z) = z^2$ and $\mathsf{d}(z)\mathsf{d}^\star(z) = \mathsf{D}(z)$ in Algorithm 3.3.6, we have

$$\mathsf{b}_1(z) = -\tfrac{1}{4}(z-1)^2 = \{\underline{-\tfrac{1}{4}}, \tfrac{1}{2}, -\tfrac{1}{4}\}_{[0,2]},$$
$$\mathsf{b}_2(z) = \tfrac{\sqrt{2}}{4}(z^2-1) = \{\underline{-\tfrac{\sqrt{2}}{4}}, 0, \tfrac{\sqrt{2}}{4}\}_{[0,2]},$$

with $s_1 = s_2 = 0$. Then $\{a; b_1, b_2\}$ is a tight framelet filter bank with $\mathrm{vm}(b_1) = 2$, $\mathrm{vm}(b_2) = 1$ and $\mathsf{Sb}_1(z) = z^2$, $\mathsf{Sb}_2(z) = -z^2$. By calculation, $\mathrm{Var}(b_1) = \frac{1}{3}$, $\mathrm{Var}(b_2) = 1$, and $\mathrm{Fsi}(a, b_1) \approx 0.0857143$, $\mathrm{Fsi}(a, b_2) \approx 0.487950$, $\mathrm{Fsi}(b_1, b_2) \approx 0.487950$. Setting $n_b = 2$ in Algorithm 3.3.6 with $\Theta = \{-\frac{1}{6}, \underline{\frac{4}{3}}, -\frac{1}{6}\}_{[-1,1]}$, we have

$$\mathsf{p}(z) = 1, \quad \mathsf{D}(z) = 1/24, \quad n_\Theta = 1, \quad c_2 = 2, \quad \epsilon_2 = 1, \quad n_0 = 2, \quad n_2 = 2.$$

Taking $\mathsf{d}(z) = \frac{\sqrt{6}}{12}z$ with $\mathsf{Sd}(z) = z^2$ and $\mathsf{d}(z)\mathsf{d}^\star(z) = \mathsf{D}(z)$ in Algorithm 3.3.6, we have

$$\mathsf{b}_1(z) = -\tfrac{1}{4}(z-1)^2 = \{\underline{-\tfrac{1}{4}}, \tfrac{1}{2}, -\tfrac{1}{4}\}_{[0,2]},$$
$$\mathsf{b}_2(z) = -\tfrac{\sqrt{6}}{24}(z-1)^2(z^{-2}+4z^{-1}+1) = \tfrac{\sqrt{6}}{24}\{-1, -2, \underline{\mathbf{6}}, -2, -1\}_{[-2,2]},$$

with $s_1 = s_2 = 0$. Then $\{a; b_1, b_2\}_\Theta$ is a tight framelet filter bank with $\mathrm{vm}(b_1) = \mathrm{vm}(b_2) = 2$ and $\mathsf{Sb}_1(z) = z^2$, $\mathsf{Sb}_2(z) = z^4$. By calculation, $\mathrm{Var}(b_1) = \frac{1}{3}$, $\mathrm{Var}(b_2) = \frac{8}{23}$ and $\mathrm{Fsi}(a, b_1) \approx 0.0857143$, $\mathrm{Fsi}(a, b_2) \approx 0.215435$, $\mathrm{Fsi}(b_1, b_2) \approx 0.904828$. See Fig. 3.4 for the graphs of their associated refinable and framelet functions.

Example 3.3.2 Let $a = a_3^B = \{\underline{\frac{1}{8}}, \frac{3}{8}, \frac{3}{8}, \frac{1}{8}\}_{[0,3]}$ be the B-spline filter of order 3. Then $\epsilon z^c := \mathsf{Sa}(z) = z^3$. Setting $n_b = 1$ in Algorithm 3.3.6 with $\Theta = \boldsymbol{\delta}$, we have

$$\mathsf{p}(z) = 1, \quad \mathsf{D}(z) = 3/16, \quad c_2 = 4, \quad \epsilon_2 = 1, \quad n_0 = 2, \quad n_2 = 3.$$

Taking $\mathsf{d}(z) = \frac{\sqrt{3}}{4}z$ with $\mathsf{Sd}(z) = z^2$ and $\mathsf{d}(z)\mathsf{d}^\star(z) = \mathsf{D}(z)$ in Algorithm 3.3.6, we have

$$\mathsf{b}_1(z) = \tfrac{\sqrt{3}}{4}(z-1) = \tfrac{\sqrt{3}}{4}\{\underline{-\mathbf{1}}, 1\}_{[0,1]},$$
$$\mathsf{b}_2(z) = \tfrac{1}{8}(z-1)(1+4z+z^2) = \{\underline{-\tfrac{1}{8}}, -\tfrac{3}{8}, \tfrac{3}{8}, \tfrac{1}{8}\}_{[0,3]}$$

with $s_1 = s_2 = 0$. Then $\{a; b_1, b_2\}$ is a tight framelet filter bank with $\mathrm{vm}(b_1) = \mathrm{vm}(b_2) = 1$ and $\mathsf{Sb}_1(z) = -z$, $\mathsf{Sb}_2(z) = -z^3$. By calculation, $\mathrm{Var}(b_1) = \frac{1}{4}$, $\mathrm{Var}(b_2) = \frac{9}{20}$ and $\mathrm{Fsi}(a, b_1) \approx 0.134304$, $\mathrm{Fsi}(a, b_2) \approx 0.350020$, $\mathrm{Fsi}(b_1, b_2) \approx 0.846642$.

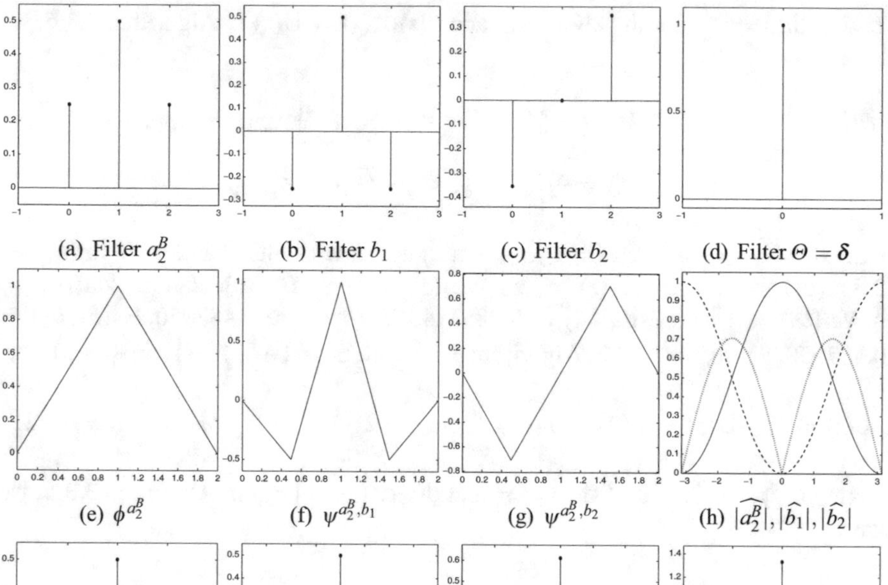

(a) Filter a_2^B (b) Filter b_1 (c) Filter b_2 (d) Filter $\Theta = \boldsymbol{\delta}$

(e) $\phi^{a_2^B}$ (f) $\psi^{a_2^B, b_1}$ (g) $\psi^{a_2^B, b_2}$ (h) $|\widehat{a_2^B}|, |\widehat{b_1}|, |\widehat{b_2}|$

(i) Filter a_2^B (j) Filter b_1 (k) Filter b_2 (l) Filter Θ

(m) $\phi^{a_2^B}$ (n) $\psi^{a_2^B, b_1}$ (o) $\psi^{a_2^B, b_2}$ (p) $|\widehat{a_2^B}|, |\widehat{b_1}|, |\widehat{b_2}|$

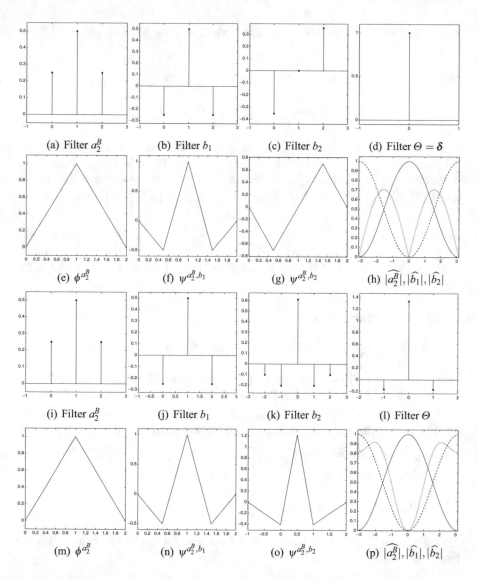

Fig. 3.4 The first two rows are for the tight framelet filter bank $\{a_2^B; b_1, b_2\}$ with symmetry constructed in Example 3.3.1 with $\Theta = \boldsymbol{\delta}$. (**a**), (**b**), (**c**), (**d**) are the graphs of the filters $a_2^B, b_1, b_2, \Theta = \boldsymbol{\delta}$, respectively. (**e**), (**f**), (**g**) are the graphs of the refinable function $\phi^{a_2^B}$ and framelet functions $\psi^{a_2^B, b_1}$, $\psi^{a_2^B, b_2}$, respectively. (**h**) is the magnitudes of $\widehat{a_2^B}$ (in solid line), $\widehat{b_1}$ (in dashed line), and $\widehat{b_2}$ (in dotted line) on the interval $[-\pi, \pi]$. The last two rows are for the tight framelet filter bank $\{a_2^B; b_1, b_2\}_\Theta$ with symmetry constructed in Example 3.3.1 with $\Theta = \{-\frac{1}{6}, \underline{\frac{4}{3}}, -\frac{1}{6}\}_{[-1,1]}$

To achieve 3 vanishing moments, we consider $\Theta(z) := \Theta_3(z) + (2 - z^{-1} - z)^3(\lambda_1 z^{-1} + \lambda_0 + \lambda_1 z)$ with $\Theta_3 = \{\frac{13}{240}, -\frac{7}{15}, \underline{\mathbf{\frac{73}{40}}}, -\frac{7}{15}, \frac{13}{240}\}_{[-2,2]}$. The condition in Theorem 3.3.7 is satisfied with $\lambda_0 \approx 0.014844579427$ and $\lambda_1 \approx -0.0148559654632$. Set $n_b = 3$. Then $n_\Theta = 4$, $c_2 = -2$, $\epsilon_2 = 1$, $n_0 = 8, n_2 = 3$, and we pick

$$\mathsf{d}(z) = 0.119556925161+0.123189685153(z^{-1}+z)+0.00946857807736(z^{-2}+z^2).$$

Then Algorithm 3.3.6 yields

$$\begin{aligned}\mathsf{b}_1(z) &= (1-z^{-1})^3[0.012317964188(z^{-1}+z^5)+0.073907785129(1+z^4)\\&\quad+0.193590774860(z+z^3)-0.0114508083672z^2],\\ \mathsf{b}_2(z) &= (1-z^{-1})^3[0.015235631275(z^{-1}+z^7)+0.091413787653(1+z^6)\\&\quad+0.215942972647(z+z^5)+0.229163646602(z^2+z^4)+0.062720194480z^3],\end{aligned}$$

with $s_1 = 0$ and $s_2 = 4$. Then $\{a; b_1, b_2\}_\Theta$ is a tight framelet filter bank with $\mathrm{vm}(b_1) = \mathrm{vm}(b_2) = 3$ such that $\mathsf{Sb}_1(z) = -z$ and $\mathsf{Sb}_2(z) = -z^3$. Note that $\mathrm{fsupp}(b_1) = [-4, 5]$ and $\mathrm{fsupp}(b_2) = [-4, 7]$. By calculation, $\mathrm{Var}(b_1) \approx 0.704865$, $\mathrm{Var}(b_1) \approx 0.970973$ and $\mathrm{Fsi}(a, b_1) \approx 0.0093227$, $\mathrm{Fsi}(a, b_2) \approx 0.067623$, $\mathrm{Fsi}(b_1, b_2) \approx 0.77172$. See Fig. 3.5 for the graphs of their associated refinable and framelet functions.

Example 3.3.3 Let $a = a_7^B = \frac{1}{128}\{\underline{\mathbf{1}}, 7, 21, 35, 35, 21, 7, 1\}_{[0,7]}$ be the B-spline filter of order 7. Then $\epsilon z^c := \mathsf{Sa}(z) = z^7$. Setting $n_b = 1$ in Algorithm 3.3.6 with $\Theta = \boldsymbol{\delta}$, we have

$$\mathsf{p}(z) = 1, \quad \mathsf{D}(z) = \tfrac{7}{4096}(z^{-1} + 14 + z)^2, \quad c_2 = 8, \quad \epsilon_2 = 1, \quad n_0 = 6, \quad n_2 = 7.$$

Taking $\mathsf{d}(z) = \frac{\sqrt{7}}{64}(z^{-1} + 14 + z)$ with $\mathsf{Sd}(z) = 1$ and $\mathsf{d}(z)\mathsf{d}^\star(z) = \mathsf{D}(z)$ in Algorithm 3.3.6, we have

$$\begin{aligned}\mathsf{b}_1(z) &= \tfrac{\sqrt{7}}{64}z^{-1}(1-z^{-1})(1+z)^2(1+6z+z^2) = \tfrac{\sqrt{7}}{64}\{-1,-7,\underline{\mathbf{-6}},6,7,1\}_{[-2,3]},\\ \mathsf{b}_2(z) &= \tfrac{1}{128}z^{-2}(1-z)^3(1+10z+34z^2+10z^3+z^4)\\ &= \tfrac{1}{128}\{1,7,\underline{\mathbf{7}},-63,63,-7,-7,-1\}_{[-2,5]},\end{aligned}$$

with $s_1 = 2$ and $s_2 = -1$. Then $\{a; b_1, b_2\}$ is a tight framelet filter bank with $\mathrm{vm}(b_1) = 1, \mathrm{vm}(b_2) = 3$ and $\mathsf{Sb}_1(z) = -z, \mathsf{Sb}_2(z) = -z^3$. By calculation, $\mathrm{Var}(b_1) = \frac{251}{174}$, $\mathrm{Var}(b_2) = \frac{1421}{4068}$ and $\mathrm{Fsi}(a, b_1) \approx 0.363244$, $\mathrm{Fsi}(a, b_2) \approx 0.0140206$, $\mathrm{Fsi}(b_1, b_2) \approx 0.289827$. See Fig. 3.6 for the graphs of their associated framelet functions.

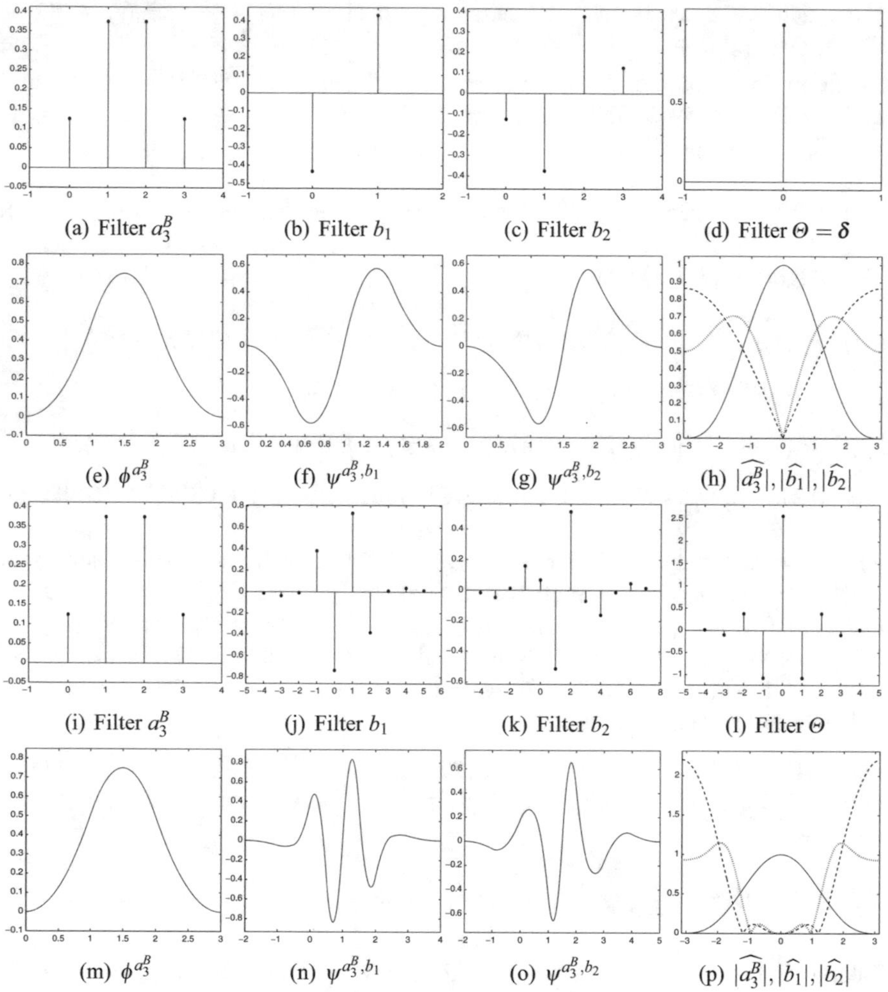

(a) Filter a_3^B (b) Filter b_1 (c) Filter b_2 (d) Filter $\Theta = \boldsymbol{\delta}$

(e) $\phi^{a_3^B}$ (f) $\psi^{a_3^B, b_1}$ (g) $\psi^{a_3^B, b_2}$ (h) $|\widehat{a_3^B}|, |\widehat{b_1}|, |\widehat{b_2}|$

(i) Filter a_3^B (j) Filter b_1 (k) Filter b_2 (l) Filter Θ

(m) $\phi^{a_3^B}$ (n) $\psi^{a_3^B, b_1}$ (o) $\psi^{a_3^B, b_2}$ (p) $|\widehat{a_3^B}|, |\widehat{b_1}|, |\widehat{b_2}|$

Fig. 3.5 The first two rows are for the tight framelet filter bank $\{a_3^B; b_1, b_2\}$ with symmetry and one vanishing moment constructed in Example 3.3.2. (**a**), (**b**), (**c**), (**d**) are the graphs of the filters $a_3^B, b_1, b_2, \Theta = \boldsymbol{\delta}$, respectively. (**e**), (**f**), (**g**) are the graphs of the refinable function $\phi^{a_3^B}$ and framelet functions $\psi^{a_3^B, b_1}$, $\psi^{a_3^B, b_2}$, respectively. (**h**) is the magnitudes of $\widehat{a_3^B}$ (in solid line), $\widehat{b_1}$ (in dashed line), and $\widehat{b_2}$ (in dotted line) on the interval $[-\pi, \pi]$. The last two rows are for the tight framelet filter bank $\{a_3^B; b_1, b_2\}_\Theta$ with symmetry and 3 vanishing moments constructed in Example 3.3.1

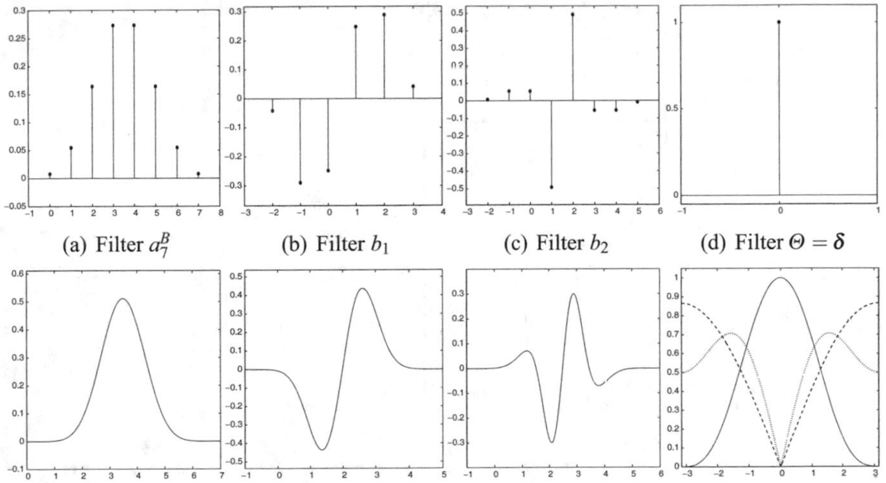

(a) Filter a_7^B (b) Filter b_1 (c) Filter b_2 (d) Filter $\Theta = \boldsymbol{\delta}$

(e) $\phi^{a_7^B}$ (f) $\psi^{a_7^B, b_1}$ (g) $\psi^{a_7^B, b_2}$ (h) $|\widehat{a_7^B}|, |\widehat{b_1}|, |\widehat{b_2}|$

Fig. 3.6 The tight framelet filter bank $\{a_7^B; b_1, b_2\}$ with symmetry and one vanishing moment constructed in Example 3.3.3. (**a**), (**b**), (**c**), (**d**) are the graphs of the filters $a_7^B, b_1, b_2, \Theta = \boldsymbol{\delta}$, respectively. (**e**), (**f**), (**g**) are the graphs of the refinable function $\phi^{a_7^B}$ and framelet functions $\psi^{a_7^B, b_1}$, $\psi^{a_7^B, b_2}$, respectively. (**h**) is the magnitudes of $\widehat{a_7^B}$ (in solid line), $\widehat{b_1}$ (in dashed line), and $\widehat{b_2}$ (in dotted line) on the interval $[-\pi, \pi]$

Example 3.3.4 Let $a = a_{3,4} = \{-\frac{3}{64}, \frac{5}{64}, \underline{\mathbf{\frac{15}{32}}}, \frac{15}{32}, \frac{5}{64}, -\frac{3}{64}\}_{[-2,3]}$ (see (2.1.12)). Then $\epsilon z^c := \mathsf{S}a(z) = z$. Setting $n_b = 2$ in Algorithm 3.3.6 with $\Theta = \boldsymbol{\delta}$, we have

$$\mathsf{p}(z) = 1, \quad \mathsf{D}(z) = 15/1024, \quad c_2 = 3, \quad \epsilon_2 = -1, \quad n_0 = 3, \quad n_2 = 3.$$

Taking $\mathsf{d}(z) = \frac{\sqrt{15}}{32}$ with $\mathsf{S}\mathsf{d}(z) = 1$ and $\mathsf{d}(z)\mathsf{d}^\star(z) = \mathsf{D}(z)$ in Algorithm 3.3.6, we have

$$\mathsf{b}_1(z) = -\tfrac{\sqrt{15}}{64}(1 - z^{-1})^2(1 + z)(3 - 2z + 3z^2) = \tfrac{\sqrt{15}}{64}\{-3, 5, \underline{\mathbf{-2}}, -2, 5, -3\}_{[-2,3]},$$
$$\mathsf{b}_2(z) = \tfrac{1}{16}z^{-3}(z - 1)^3(3z^{-1} + 4 + 3z) = \tfrac{1}{16}\{-3, 5, \underline{\mathbf{0}}, 0, -5, 3\}_{[-2,3]},$$

with $s_1 = 1$ and $s_2 = 0$. Then $\{a; b_1, b_2\}$ is a tight framelet filter bank with $\operatorname{vm}(b_1) = 2$, $\operatorname{vm}(b_2) = 3$ and $\mathsf{S}b_1(z) = z, \mathsf{S}b_2(z) = -z$. By calculation, $\operatorname{Var}(b_1) = \frac{227}{76}$, $\operatorname{Var}(b_2) = \frac{225}{68}$ and $\operatorname{Fsi}(a, b_1) \approx 0.0748685$, $\operatorname{Fsi}(a, b_2) \approx 0.253998$, $\operatorname{Fsi}(b_1, b_2) \approx 0.344246$. See Fig. 3.7 for the graphs of their associated refinable and framelet functions.

For $a = a_6^I$, $a = a_{7,6}$ or $a = a_{7,8}$ with $\Theta = \boldsymbol{\delta}$, item (ii) of Theorem 3.3.7 fails for complex symmetry, but items (i) and (ii) of Theorem 3.3.7 are satisfied

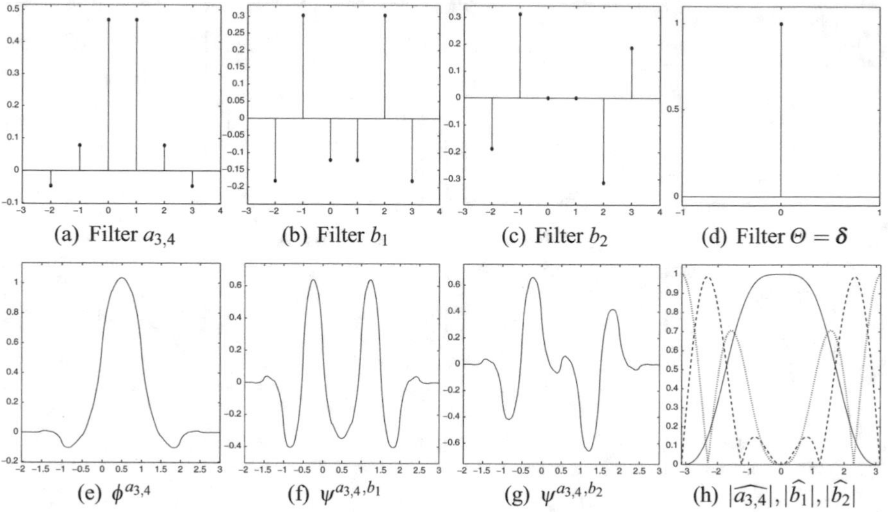
(a) Filter $a_{3,4}$
(b) Filter b_1
(c) Filter b_2
(d) Filter $\Theta = \boldsymbol{\delta}$
(e) $\phi^{a_{3,4}}$
(f) $\psi^{a_{3,4},b_1}$
(g) $\psi^{a_{3,4},b_2}$
(h) $|\widehat{a_{3,4}}|, |\widehat{b_1}|, |\widehat{b_2}|$

Fig. 3.7 The tight framelet filter bank $\{a_{3,4}; b_1, b_2\}$ with symmetry and 2 vanishing moments constructed in Example 3.3.4. (**a**), (**b**), (**c**), (**d**) are the graphs of the filters $a_{3,4}, b_1, b_2, \Theta = \boldsymbol{\delta}$, respectively. (**e**), (**f**), (**g**) are the graphs of the refinable function $\phi^{a_{3,4}}$ and the framelet functions $\psi^{a_{3,4},b_1}$, $\psi^{a_{3,4},b_2}$, respectively. (**h**) is the magnitudes of $\widehat{a_{3,4}}$ (in solid line), $\widehat{b_1}$ (in dashed line), and $\widehat{b_2}$ (in dotted line) on the interval $[-\pi, \pi]$

for symmetry. Therefore, complex-valued tight framelet filter banks $\{a; b_1, b_2\}$ with symmetry can be easily constructed by Algorithm 3.3.6 from such low-pass filters. See Example 3.5.5 for a tight framelet filter bank $\{a_6^I; b_1, b_2\}$ with symmetry, which is constructed by a different algorithm with the additional interpolation property.

To construct tight framelet filter banks $\{a; b_1, b_2\}$ with [complex] symmetry and n_b vanishing moments, we may also construct low-pass filters a such that all the necessary conditions in Algorithm 3.3.6 or Theorem 3.3.7 with $\Theta = \boldsymbol{\delta}$ are satisfied.

Algorithm 3.3.8 *Let $n_b, N \in \mathbb{N}$. Parameterize a filter $\mathring{a} \in l_0(\mathbb{Z})$ with complex [or real] coefficients such that $\mathring{\mathsf{a}}(1) = 1$, $\operatorname{fsupp}(\mathring{a}) = [0, N - n_b]$, and $\mathring{a}$ is [complex] symmetric about the point $(N - n_b)/2$. Define $\mathsf{a}(z) := 2^{-n_b}(1 + z)^{n_b}\mathring{\mathsf{a}}(z)$.*

(S1) For complex symmetry (or real coefficients), solve linear equations induced by

$$1 - \mathsf{a}(z) = \mathscr{O}(|z - 1|^{2n_b}), \quad z \to 1. \tag{3.3.36}$$

Otherwise, solve equations induced by $1 - \mathsf{a}(z)\mathsf{a}^\star(z) = \mathscr{O}(|z-1|^{2n_b})$ as $z \to 1$.

(S2) Define $\mathsf{p}(z^2) := 1 - \mathsf{a}(z)\mathsf{a}^\star(z) - \mathsf{a}(-z)\mathsf{a}^\star(-z)$ and obtain a Laurent polynomial q_p by item (4) of Theorem 3.1.7 for complex symmetry or by item (5) of Theorem 3.1.8 for symmetry. Solve the nonlinear equations induced by $\mathsf{q}_\mathsf{p}(z)\mathsf{q}_\mathsf{p}^\star(z) = \mathsf{p}(z)$ to determine the coefficients of the filter a.

(S3) Apply Algorithm 3.3.6 with $\Theta = \boldsymbol{\delta}$ to derive high-pass filters b_1 and b_2 having [complex] symmetry (and real coefficients).

Then $\{a; b_1, b_2\}$ is a tight framelet filter bank with [complex] symmetry (and real coefficients) having n_b vanishing moments.

Here we provide some examples of real-valued tight framelet filter banks $\{a; b_1, b_2\}$ with symmetry to illustrate Algorithm 3.3.8.

Example 3.3.5 Let $n_b = 2$ and $N = 6$ in Algorithm 3.3.8. Solving the linear equations in (3.3.36) of item (S1), we have a low-pass filter a parameterized by a parameter $\lambda \in \mathbb{R}$ as follows:

$$a = \{\underline{\boldsymbol{\lambda}}, -\tfrac{1}{16} - 2\lambda, \tfrac{1}{4} - \lambda, \tfrac{5}{8} + 4\lambda, \tfrac{1}{4} - \lambda, -\tfrac{1}{16} - 2\lambda, \lambda\}_{[0,6]}.$$

By item (4) of Theorem 3.1.8, we have $\mathsf{q_p}(z) = \frac{\sqrt{2}}{512}(1-z)[512\lambda z^{-1} + (\lambda^{-1} + 192 + 1024\lambda) + 512\lambda z]$. Solving the nonlinear equations induced by $\mathsf{q_p}(z)\mathsf{q_p^\star}(z) = \mathsf{p}(z)$ in (S2), we get $\lambda = -\frac{3}{64} \pm \frac{\sqrt{7}}{64}$. Applying Algorithm 3.3.6 with $\Theta = \boldsymbol{\delta}$ and $n_b = 2$, we obtain high-pass filters b_1 and b_2 such that $\operatorname{fsupp}(b_1) = \operatorname{fsupp}(b_2) = [0, 6]$ and

$$\mathsf{b}_1(z) = \tfrac{7-3t}{448}(z-1)^2[7(1+z^4) + (42+14t)(z+z^3) + (14+4t)z^2],$$

$$\mathsf{b}_2(z) = \tfrac{(3-t)\sqrt{2}}{32}(z-1)^3(z+1)[1 + (6+2t)z + z^2],$$

with $t = \pm\sqrt{7}$. Then $\{a; b_1, b_2\}$ is a real-valued tight framelet filter bank with $\operatorname{vm}(b_1) = 2$, $\operatorname{vm}(b_2) = 3$ and $\mathsf{S}\mathsf{b}_1(z) = z^6$, $\mathsf{S}\mathsf{b}_2(z) = -z^6$.

When $t = \sqrt{7}$, we have $\operatorname{sm}(a) \approx 1.023927066$, $\operatorname{Var}(a) \approx 0.304992$, $\operatorname{Var}(b_1) = 1.27283$, $\operatorname{Var}(b_2) = 1.48267$ and $\operatorname{Fsi}(a, b_1) \approx 0.0480952$, $\operatorname{Fsi}(a, b_2) \approx 0.0118341$, $\operatorname{Fsi}(b_1, b_2) \approx 0.351069$. When $t = -\sqrt{7}$, we have $\operatorname{sm}(a) \approx 1.22062838$, $\operatorname{Var}(a) \approx 1.37265$, $\operatorname{Var}(b_1) = 5.65925$, $\operatorname{Var}(b_2) = 5.57583$ as well as $\operatorname{Fsi}(a, b_1) \approx 0.0837388$, $\operatorname{Fsi}(a, b_2) \approx 0.225926$, $\operatorname{Fsi}(b_1, b_2) \approx 0.341421$. See Fig. 3.8 for the graphs of their associated refinable functions and framelet functions.

Example 3.3.6 Let $n_b = 3$ and $N = 9$ in Algorithm 3.3.8. Solving the linear equations in (3.3.36) of item (S1), we have a low-pass filter a parameterized by a parameter $\lambda \in \mathbb{R}$ as follows:

$$a = \{\underline{\boldsymbol{\lambda}}, \tfrac{15}{1024} - 3\lambda, -\tfrac{63}{1024}, \tfrac{35}{1024} + 8\lambda, \tfrac{525}{1024} - 6\lambda, \tfrac{525}{1024} - 6\lambda,$$
$$\tfrac{35}{1024} + 8\lambda, -\tfrac{63}{1024}, \tfrac{15}{1024} - 3\lambda, \lambda\}_{[0,9]}.$$

By item (4) of Theorem 3.1.8 and solving the nonlinear equations induced by $\mathsf{q_p}(z)\mathsf{q_p^\star}(z) = \mathsf{p}(z)$ in item (S2), we get $\lambda = \frac{5}{1024}, \frac{7}{2048}$, or $\frac{43+2\sqrt{226}}{4096}$.

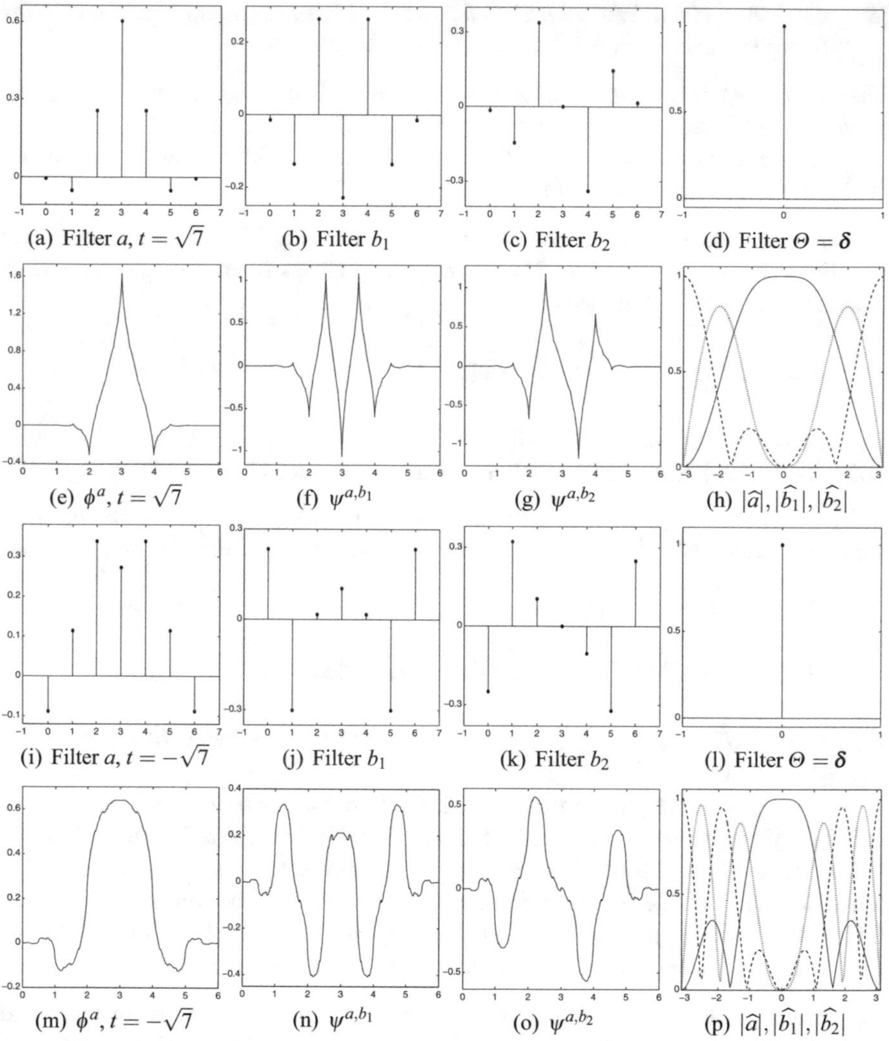

(a) Filter a, $t = \sqrt{7}$ (b) Filter b_1 (c) Filter b_2 (d) Filter $\Theta = \boldsymbol{\delta}$

(e) ϕ^a, $t = \sqrt{7}$ (f) ψ^{a,b_1} (g) ψ^{a,b_2} (h) $|\widehat{a}|, |\widehat{b_1}|, |\widehat{b_2}|$

(i) Filter a, $t = -\sqrt{7}$ (j) Filter b_1 (k) Filter b_2 (l) Filter $\Theta = \boldsymbol{\delta}$

(m) ϕ^a, $t = -\sqrt{7}$ (n) ψ^{a,b_1} (o) ψ^{a,b_2} (p) $|\widehat{a}|, |\widehat{b_1}|, |\widehat{b_2}|$

Fig. 3.8 Tight framelet filter banks $\{a; b_1, b_2\}$ with symmetry and 2 vanishing moments constructed in Example 3.3.5. (**a**)–(**h**) are for $t = \sqrt{7}$, while (**i**)–(**p**) are for $t = -\sqrt{7}$. (**h**) and (**p**) are for the magnitudes of $\widehat{a}$ (in solid line), $\widehat{b_1}$ (in dashed line), and $\widehat{b_2}$ (in dotted line) on $[-\pi, \pi]$

Applying $\lambda = \frac{5}{1024}$ and Algorithm 3.3.6 with $\Theta = \boldsymbol{\delta}$ and $n_b = 3$, we obtain

$$a = \tfrac{1}{1024}\{\underline{\mathbf{5}}, 0, -63, 75, 495, 495, 75, -63, 0, 5\}_{[0,9]},$$
$$b_1 = \tfrac{\sqrt{15}}{512}\{\underline{\mathbf{-3}}, 0, 22, -45, 45, -22, 0, 3\}_{[0,7]},$$
$$b_2 = \tfrac{1}{1024}\{\underline{\mathbf{-5}}, 0, 117, -75, -315, 315, 75, -117, 0, 5\}_{[0,9]}.$$

Then $\{a; b_1, b_2\}$ is a real-valued tight framelet filter bank with $\mathrm{vm}(b_1) = \mathrm{vm}(b_2) = 3$ and $\mathsf{S}b_1(z) = -z^7, \mathsf{S}b_2(z) = -z^9$. By calculation, $\mathrm{sm}(a) \approx 1.67850576$, $\mathrm{Var}(a) \approx 0.389663$, $\mathrm{Var}(b_1) = 0.677324$, $\mathrm{Var}(b_2) = 1.04185$ and $\mathrm{Fsi}(a, b_1) \approx 0.0282147$, $\mathrm{Fsi}(a, b_2) \approx 0.30197$, $\mathrm{Fsi}(b_1, b_2) \approx 0.478219$.

Applying $\lambda = \frac{7}{2048}$ and Algorithm 3.3.6 with $\Theta = \boldsymbol{\delta}$ and $n_b = 3$, we obtain

$$a = \tfrac{1}{2048}\{\underline{\mathbf{7}}, 9, -126, 126, 1008, 1008, 126, 9, 7\}_{[0,9]},$$
$$b_1 = \tfrac{1}{1024}\{\underline{\mathbf{7}}, 9, -126126, 126, -126, -126, 126, -9, -7\}_{[0,9]},$$
$$b_2 = \tfrac{\sqrt{7}}{1024}\{\underline{\mathbf{-21}}, -27, 186, -186, 27, 21\}_{[0,5]}.$$

Then $\{a; b_1, b_2\}$ is a real-valued tight framelet filter bank with $\mathrm{vm}(b_1) = \mathrm{vm}(b_2) = 3$ and $\mathsf{S}b_1(z) = -z^9, \mathsf{S}b_2(z) = -z^5$. By calculation, $\mathrm{sm}(a) \approx 1.4932156$, $\mathrm{Var}(a) \approx 0.373060$, $\mathrm{Var}(b_1) = 2.95029$, $\mathrm{Var}(b_2) = 0.364745$ and $\mathrm{Fsi}(a, b_1) \approx 0.363628$, $\mathrm{Fsi}(a, b_2) \approx 0.150445$, $\mathrm{Fsi}(b_1, b_2) \approx 0.603034$.

Applying $\lambda = \frac{43+2\sqrt{226}}{4096}$ and Algorithm 3.3.6 with $\Theta = \boldsymbol{\delta}$ and $n_b = 3$, we obtain

$$a = \tfrac{1}{4096}\{43 + 2t, -69 - 6t, -252, 484 + 16t, 1842 - 12t, 1842 - 12t,$$
$$484 + 16t, -252, -69 - 6t, 43 + 2t\}_{[0,9]},$$
$$b_1 = \tfrac{1}{1024}\{-76 - 5t, 164 + 11t, -24 - 6t, -24 - 6t, -272 + 16t, 272 - 16t,$$
$$24 + 6t, 24 + 6t, -164 - 11t, 76 + 5t\}_{[0,9]},$$
$$\mathsf{b}_2(z) = \tfrac{(19741+1066t)\sqrt{24056071+1734104t}}{54432\times 10^7}(z-1)^4(z+1)[-315z^4 + (40t - 860)z^3$$
$$+ (318 - 48t)z^2 + (40t - 860)z - 315],$$

with $t := \sqrt{226}$. Then $\{a; b_1, b_2\}$ is a real-valued tight framelet filter bank with $\mathrm{vm}(b_1) = 4$, $\mathrm{vm}(b_2) = 3$ and $\mathsf{S}b_1(z) = z^9, \mathsf{S}b_2(z) = -z^9$. By calculation, we have $\mathrm{sm}(a) \approx 1.8212795$, $\mathrm{Var}(a) \approx 0.79492$, $\mathrm{Var}(b_1) = 12.1545$, $\mathrm{Var}(b_2) = 12.0119$ and $\mathrm{Fsi}(a, b_1) \approx 0.0554945$, $\mathrm{Fsi}(a, b_2) \approx 0.146920$, $\mathrm{Fsi}(b_1, b_2) \approx 0.334758$. See Fig. 3.9 for the graphs of their associated refinable and framelet functions.

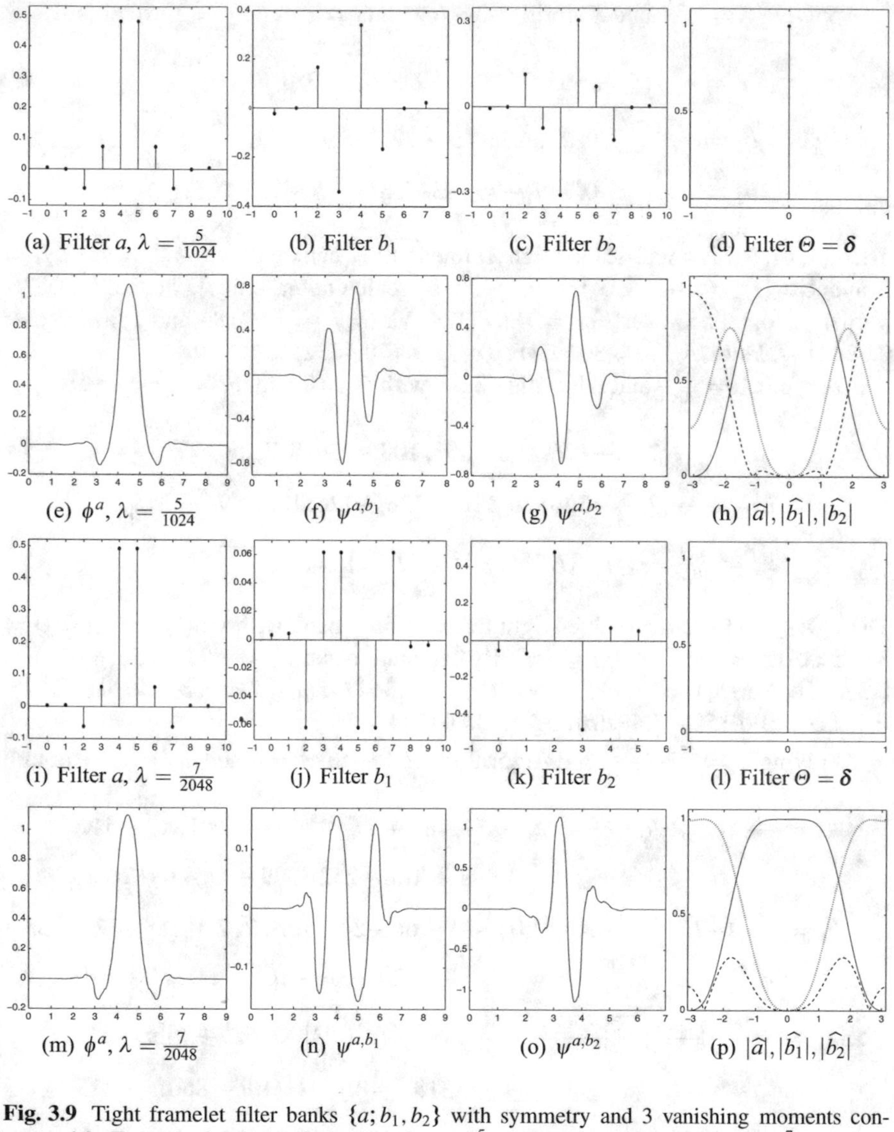

Fig. 3.9 Tight framelet filter banks $\{a; b_1, b_2\}$ with symmetry and 3 vanishing moments constructed in Example 3.3.6. (**a**)–(**h**) are for $\lambda = \frac{5}{1024}$, while (**i**)–(**p**) are for $\lambda = \frac{7}{2048}$. (**h**) and (**p**) are for the magnitudes of $\widehat{a}$ (in solid line), $\widehat{b_1}$ (in dashed line), and $\widehat{b_2}$ (in dotted line) on $[-\pi, \pi]$

3.4 Tight Framelet Filter Banks with Two High-Pass Filters

As we have seen in Sect. 3.3, to construct a tight framelet filter bank $\{a; b_1, b_2\}_\Theta$ with [complex] symmetry, the necessary and sufficient condition in Theorem 3.3.7 has to be satisfied. However, the condition in Theorem 3.3.7 is often too restricted to be satisfied by many low-pass filters a and moment correcting filters Θ. Therefore, to obtain a wide class of tight framelet filter banks, it is natural to drop the requirement for symmetry property of a tight framelet filter bank.

In this section we discuss how to systematically construct all tight framelet filter banks $\{a; b_1, b_2\}_\Theta$ with short support but without the symmetry property.

3.4.1 Algorithm and Examples of Tight Framelet Filter Banks

Without the symmetry constraint, based on Algorithm 3.2.1 and similar to Algorithm 3.3.6, we now present an algorithm to construct all possible tight framelet filter banks $\{a; b_1, b_2\}_\Theta$ having the shortest filter support and derived from any given filters a and Θ satisfying the necessary condition $\mathcal{M}_{a,\Theta}(z) \geqslant 0$ for all $z \in \mathbb{T}$.

Algorithm 3.4.1 *Let $a, \Theta \in l_0(\mathbb{Z})$ be filters satisfying* (3.3.11) *and* (3.3.12). *Let $n_b \in \mathbb{N}_0$ be a nonnegative integer satisfying* (3.3.5). *Assume* $\mathrm{len}(a) + \mathrm{len}(\Theta) > 0$.

(S1) Define A *and* B *as in* (3.3.8) *and* $\mathsf{p}, \mathring{\mathsf{A}}, \mathring{\mathsf{B}}$ *as in* (3.2.10). *Define* $[-n_0, n_0] := \mathrm{fsupp}(\mathring{\mathsf{B}})$ *and* D *as in* (3.2.11);

(S2) Select $s, \epsilon_{\mathrm{len}} \in \{0, 1\}$. *Find a Laurent polynomial* d *such that* $\mathsf{d}(z)\mathsf{d}^\star(z) = \mathsf{D}(z)$. *Define* $[m_\mathsf{d}, n_\mathsf{d}] := \mathrm{fsupp}(\mathsf{d})$. *Without loss of generality, we often set* $m_\mathsf{d} = 0$;

(S3) Parameterize a filter $\mathring{\mathsf{b}}_2$ *by* $\mathring{\mathsf{b}}_2(z) = z^s \sum_{j=0}^{n_0+\epsilon_{\mathrm{len}}} t_j z^j$. *Let* $\mathcal{R}$ *and* $\mathring{\mathsf{b}}_1^\star$ *be uniquely determined by* (3.3.31) *and* $\mathrm{fsupp}(\mathcal{R}) \subseteq [2m_\mathsf{d}, 2n_\mathsf{d} - 1]$. *Find the unknown coefficients* $\{t_0, \dots, t_{n_0+\epsilon_{\mathrm{len}}}\}$ *by solving a system X of linear equations induced by* $\mathcal{R}(z) = 0$ *(i.e., all the coefficients of* $\mathcal{R}$ *are zero) and*

$$\mathit{coeff}(\mathring{\mathsf{b}}_1^\star, z, j) = 0, \quad j = s - n_0 - 2m_\mathsf{d} - 1, \dots, s + n_0 - 2n_\mathsf{d} - 3,$$
$$\text{and} \quad j = s + \epsilon_{\mathrm{len}} + 1 - 2m_\mathsf{d}, \dots, s + 2n_0 + \epsilon_{\mathrm{len}} - 2n_\mathsf{d} - 1; \tag{3.4.1}$$

(S4) For any nontrivial solution to the homogeneous system X in (S3), there must exist $\lambda > 0$ *such that* (3.2.16) *holds. Replace* $\mathring{\mathsf{b}}_1, \mathring{\mathsf{b}}_2$ *by* $\lambda^{-1/2}\mathring{\mathsf{b}}_1, \lambda^{-1/2}\mathring{\mathsf{b}}_2$, *respectively;*

(S5) Find Laurent polynomials $\mathsf{q}_1, \mathsf{q}_2$ *such that* $\mathsf{q}_1(z)\mathsf{q}_1^\star(z) + \mathsf{q}_2(z)\mathsf{q}_2^\star(z) = \mathsf{p}(z)$.

For any choice of $s_1, s_2 \in \mathbb{Z}$, *define* b_1 *and* b_2 *as in* (3.3.33). *Then* $\{a; b_1, b_2\}_\Theta$ *is a tight framelet filter bank satisfying* $\max(\mathrm{len}(b_1), \mathrm{len}(b_2)) \leqslant \mathrm{len}(a) + \mathrm{len}(\Theta) + \epsilon_{\mathrm{len}}$.

Proof Note that $\mathrm{fsupp}(\mathring{b}_2) = [s, s + n_0 + \epsilon_{\mathrm{len}}]$ and $\mathrm{len}(\mathring{b}_2) \leqslant n_0 + \epsilon_{\mathrm{len}}$. By (3.3.31) with $\mathcal{R} = 0$, we deduce that

$$\mathrm{fsupp}(\mathring{b}_1^\star) \subseteq [s - n_0 - 2m_\mathsf{d} - 1, s + 2n_0 + \epsilon_{\mathrm{len}} - 2n_\mathsf{d} - 1].$$

It follows from (3.4.1) that $\operatorname{fsupp}(\mathring{b}_1) = [2m_{\mathsf{d}} - s - \epsilon_{\text{len}}, 2n_d + 2 - s - n_{\mathsf{o}}]$. Since $\operatorname{len}(\mathsf{d}) = n_{\mathsf{d}} - m_{\mathsf{d}} \leqslant n_{\mathsf{o}} - 1$, we conclude that $\operatorname{len}(\mathring{b}_1) \leqslant n_{\mathsf{o}} + \epsilon_{\text{len}}$. By the proofs of Algorithms 3.3.6 and 3.2.1, we see that (3.2.27) and (3.3.34) are satisfied with $\tilde{\mathring{\mathsf{b}}}_1 = \mathring{\mathsf{b}}_1, \tilde{\mathring{\mathsf{b}}}_2 = \mathring{\mathsf{b}}_2$, and $\mathsf{D}_{\mathring{\mathsf{b}}}(z^2) := z^{-1}[\mathring{\mathsf{b}}_1(z)\mathring{\mathsf{b}}_2(-z) - \mathring{\mathsf{b}}_1(-z)\mathring{\mathsf{b}}_2(z)]$. Now the same proof of Algorithm 3.3.6 shows that $\{a; b_1, b_2\}_\Theta$ is a tight framelet filter bank. □

For simplicity of presentation, in the following we only provide the constructed tight framelet filter banks $\{a; b_1, b_2\}_\Theta$ without the intermediate steps.

Example 3.4.1 Let $a = a_2^B = \{\underline{\tfrac{1}{4}}, \tfrac{1}{2}, \tfrac{1}{4}\}_{[0,2]}$ be the *B*-spline filter of order 2. Setting $n_b = 1$ in Algorithm 3.4.1 with $\Theta = \boldsymbol{\delta}$, we have $\mathcal{M}_{a,\boldsymbol{\delta}}(z) \geqslant 0$ for all $z \in \mathbb{T}$ and

$$\mathsf{b}_1(z) = \tfrac{\sqrt{6}}{6}(z-1) = \tfrac{\sqrt{6}}{6}\{\underline{-1}, 1\}_{[0,1]}, \ \mathsf{b}_2(z) = \tfrac{\sqrt{3}}{12}(z-1)(3z+1) = \tfrac{\sqrt{3}}{12}\{\underline{-1}, -2, 3\}_{[0,2]}.$$

Then $\{a; b_1, b_2\}$ is a tight framelet filter bank with $\operatorname{vm}(b_1) = \operatorname{vm}(b_2) = 1$ and $\operatorname{fsupp}(b_1) = [0,1], \operatorname{fsupp}(b_2) = [0,2]$. Moreover, $\operatorname{Var}(b_1) = \frac{1}{4}, \operatorname{Var}(b_2) = \frac{19}{49}$, and $\operatorname{Fsi}(a, b_1) \approx 0.19518, \operatorname{Fsi}(a, b_2) \approx 0.350543, \operatorname{Fsi}(b_1, b_2) \approx 0.93704$. See Fig. 3.10 for the graphs of their associated refinable functions and framelet functions.

Note that the necessary and sufficient condition in Theorem 3.3.7 is indeed satisfied for $a = a_2^B$ and $\Theta = \boldsymbol{\delta}$. A real-valued tight framelet filter bank $\{a_2^B; b_1, b_2\}$ with symmetry is presented in Example 3.3.1.

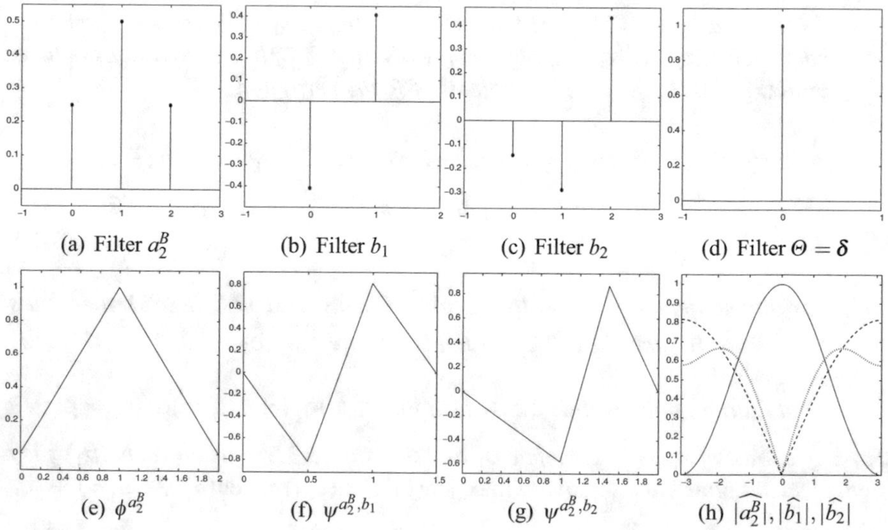

(a) Filter a_2^B (b) Filter b_1 (c) Filter b_2 (d) Filter $\Theta = \boldsymbol{\delta}$

(e) $\phi^{a_2^B}$ (f) $\psi^{a_2^B, b_1}$ (g) $\psi^{a_2^B, b_2}$ (h) $|\widehat{a_2^B}|, |\widehat{b_1}|, |\widehat{b_2}|$

Fig. 3.10 $\{a_2^B; b_1, b_2\}$ is the tight framelet filter bank constructed in Example 3.4.1. **(a)**, **(b)**, **(c)**, **(d)** are the graphs of the filters $a_2^B, b_1, b_2, \Theta = \boldsymbol{\delta}$, respectively. **(e)**, **(f)**, **(g)** are the graphs of the refinable function $\phi^{a_2^B}$ and framelet functions $\psi^{a_2^B, b_1}$, $\psi^{a_2^B, b_2}$, respectively. **(h)** is the magnitudes of $\widehat{a_2^B}$ (in solid line), $\widehat{b_1}$ (in dashed line) and $\widehat{b_2}$ (in dotted line) on the interval $[-\pi, \pi]$

Example 3.4.2 Let $a = a_3^B = \{\frac{1}{8}, \frac{3}{8}, \frac{3}{8}, \frac{1}{8}\}_{[0,3]}$ be the *B*-spline filter of order 3. Setting $n_b = 3$ in Algorithm 3.4.1 with

$$\Theta = \{\tfrac{13}{240}, -\tfrac{7}{15}, \tfrac{73}{40}, -\tfrac{7}{15}, \tfrac{13}{240}\}_{[-2,2]}$$

constructed by Proposition 3.3.4 with $n = 3$, we have $\mathcal{M}_{a,\Theta}(z) \geqslant 0$ for all $z \in \mathbb{T}$ and

$$\mathsf{b}_1(z) = z^2(1-z^{-1})^3 \frac{\sqrt{\lambda}}{1248\sqrt{1805+64\lambda}}[416\,z^{-1} + 2496 + (5607-\lambda)z],$$

$$\mathsf{b}_2(z) = z^2(1-z^{-1})^3 \frac{\sqrt{5}}{1920\sqrt{1805+64\lambda}}[(3\lambda-741)z^{-1}$$

$$+ (18\lambda - 4446) + (38\lambda - 10906)z + (18\lambda - 13566)z^2 + (3\lambda - 2261)z^3],$$

where $\lambda := 2719 + 4\sqrt{458247}$. Then $\{a; b_1, b_2\}_\Theta$ is a tight framelet filter bank with $\operatorname{vm}(b_1) = \operatorname{vm}(b_2) = 3$ and $\operatorname{fsupp}(b_1) = [-2, 3], \operatorname{fsupp}(b_2) = [-2, 5]$. Moreover, $\operatorname{Var}(b_1) \approx 0.386160$, $\operatorname{Var}(b_2) \approx 0.503131$ and $\operatorname{Fsi}(a, b_1) \approx 0.0342814$, $\operatorname{Fsi}(a, b_2) \approx 0.144436$, $\operatorname{Fsi}(b_1, b_2) \approx 0.818812$. See Fig. 3.11 for the graphs of their associated refinable functions and framelet functions.

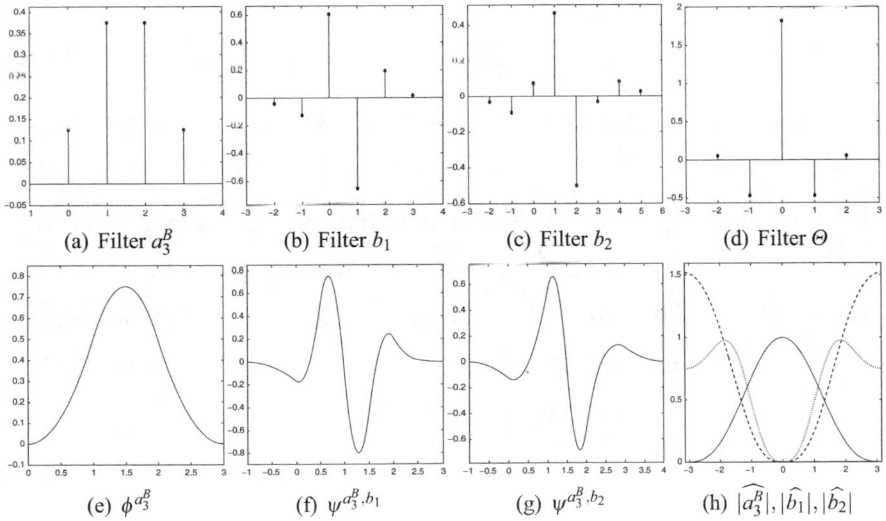

Fig. 3.11 $\{a_3^B; b_1, b_2\}_\Theta$ is a tight framelet filter bank constructed in Example 3.4.2. (**a**), (**b**), (**c**), (**d**) are the graphs of the filters a_3^B, b_1, b_2, Θ, respectively. (**e**), (**f**), (**g**) are the graphs of the refinable function $\phi^{a_3^B}$ and framelet functions $\psi^{a_3^B, b_1}$, $\psi^{a_3^B, b_2}$, respectively. (**h**) is the magnitudes of $\widehat{a_3^B}$ (in solid line), $\widehat{b_1}$ (in dashed line), and $\widehat{b_2}$ (in dotted line) on the interval $[-\pi, \pi]$

Note that the necessary and sufficient condition in Theorem 3.3.7 is indeed satisfied for $a = a_3^B$ and $\Theta = \boldsymbol{\delta}$. A real-valued tight framelet filter bank $\{a_3^B; b_1, b_2\}$ with symmetry is presented in Example 3.3.2.

Example 3.4.3 Let $a = a_4^B = \{\frac{1}{16}, \frac{1}{4}, \frac{3}{8}, \frac{1}{4}, \frac{1}{16}\}_{[0,4]}$ be the B-spline filter of order 4. Setting $n_b = 1$ in Algorithm 3.4.1 with $\Theta = \boldsymbol{\delta}$, we have $\mathcal{M}_{a,\boldsymbol{\delta}}(z) \geqslant 0 \;\forall\, z \in \mathbb{T}$ and

$$\mathsf{b}_1(z) = \frac{1}{20\sqrt{\lambda}\sqrt{1362-364\sqrt{14}}} z^2(1-z^{-1})[(91\lambda-3)z^{-1}+10\lambda],$$

$$\mathsf{b}_2(z) = \frac{\sqrt{2}}{320\sqrt{1362-364\sqrt{14}}}(1-z^{-1})[(3-41\lambda)z^{-1}+(15-205\lambda) + (25-75\lambda)z+(5-15\lambda)z^2],$$

with $\lambda := 15-4\sqrt{14}$. Then $\{a; b_1, b_2\}$ is a tight framelet filter bank with $\mathrm{vm}(b_1) = \mathrm{vm}(b_2) = 1$ and $\mathrm{fsupp}(b_1) = [0,2]$, $\mathrm{fsupp}(b_2) = [-2,2]$. Moreover, $\mathrm{Var}(b_1) \approx 0.261154$, $\mathrm{Var}(b_2) \approx 0.487093$ and $\mathrm{Fsi}(a, b_1) \approx 0.130064$, $\mathrm{Fsi}(a, b_2) \approx 0.321036$, $\mathrm{Fsi}(b_1, b_2) \approx 0.884775$.

Note that the necessary and sufficient condition in Theorem 3.3.7 fails for $a = a_4^B$ and $\Theta = \boldsymbol{\delta}$. Consequently, there does not exist a finitely supported tight framelet filter bank $\{a_4^B; b_1, b_2\}$ with the symmetry property.

Setting $n_b = \mathrm{sr}(a) = 4$ in Algorithm 3.4.1 with

$$\Theta = \{-\frac{311}{15120}, \frac{22}{105}, -\frac{1657}{1680}, \frac{2452}{945}, -\frac{1657}{1680}, \frac{22}{105}, -\frac{311}{15120}\}_{[-3,3]}, \tag{3.4.2}$$

constructed by Proposition 3.3.4 with $n = 4$, we have $\mathcal{M}_{a,\Theta}(z) \geqslant 0$ for all $z \in \mathbb{T}$ and

$$\mathsf{b}_1(z) = (1-z^{-1})^4[0.00590078608420 + 0.0472062886736z + 0.152616280681z^2 + 0.271002525136z^3 + 0.0338753156420z^4],$$

$$\mathsf{b}_2(z) = (1-z^{-1})^4[0.00541138817875 + 0.0432911054300z + 0.150624304062z^2 + 0.295881218467z^3 + 0.348787397939z^4 + 0.118781807859z^5 + 0.0148477259824z^6].$$

Then $\{a; b_1, b_2\}_\Theta$ is a tight framelet filter bank with $\mathrm{vm}(b_1) = \mathrm{vm}(b_2) = 4$ and $\mathrm{fsupp}(b_1) = [-4,4]$, $\mathrm{fsupp}(b_2) = [-4,6]$. Moreover, $\mathrm{Var}(b_1) \approx 0.404568$, $\mathrm{Var}(b_2) \approx 0.493907$ and $\mathrm{Fsi}(a, b_1) \approx 0.0201925$, $\mathrm{Fsi}(a, b_2) \approx 0.0735715$, $\mathrm{Fsi}(b_1, b_2) \approx 0.92617$. See Fig. 3.12 for the graphs of their associated refinable and framelet functions.

(a) Filter a_4^B (b) Filter b_1 (c) Filter b_2 (d) Filter $\Theta = \delta$

(e) $\phi^{a_4^B}$ (f) $\psi^{a_4^B,b_1}$ (g) $\psi^{a_4^B,b_2}$ (h) $|\widehat{a_4^B}|, |\widehat{b_1}|, |\widehat{b_2}|$

(i) Filter a_4^B (j) Filter b_1 (k) Filter b_2 (l) Filter Θ

(m) $\phi^{a_4^B}$ (n) $\psi^{a_4^B,b_1}$ (o) $\psi^{a_4^B,b_2}$ (p) $|\widehat{a_4^B}|, |\widehat{b_1}|, |\widehat{b_2}|$

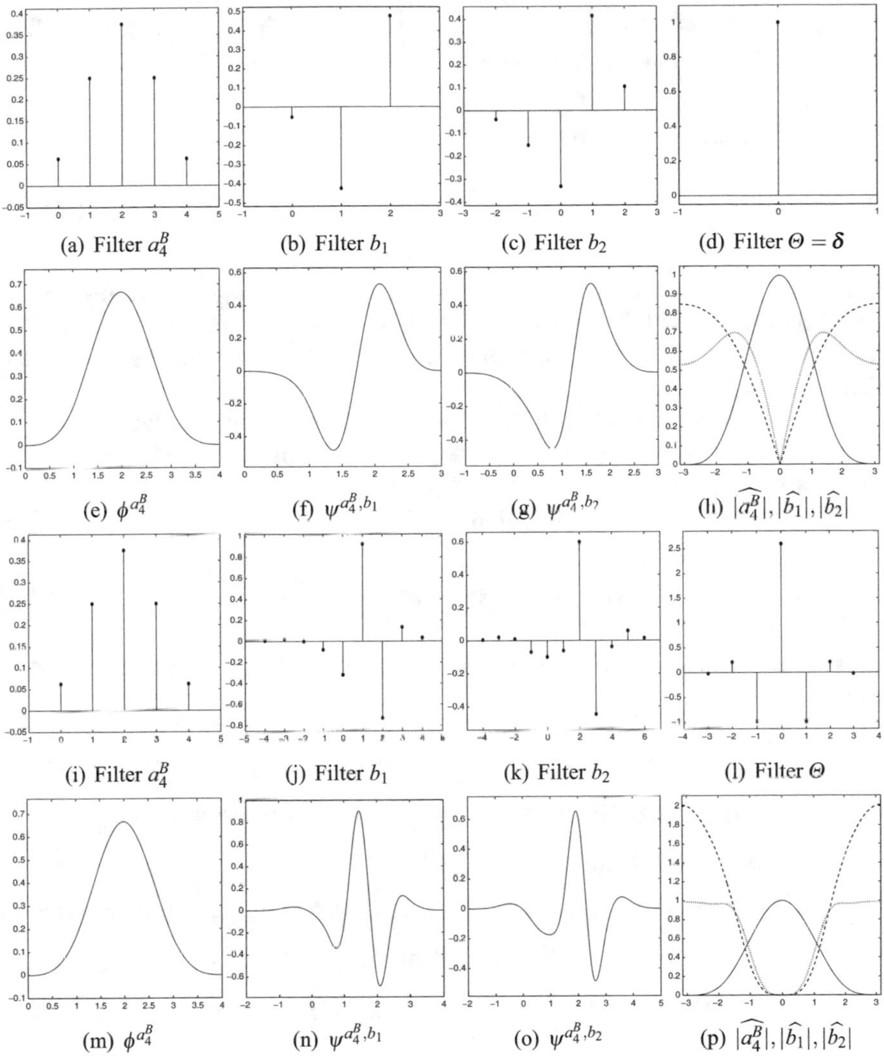

Fig. 3.12 The first two rows are for $\{a_3^B; b_1, b_2\}$ which is a tight framelet filter bank constructed in Example 3.4.3. (**a**), (**b**), (**c**), (**d**) are the graphs of the filters $a_4^B, b_1, b_2, \Theta = \delta$. (**e**), (**f**), (**g**) are the graphs of the refinable function $\phi^{a_4^B}$ and framelet functions $\psi^{a_4^B,b_1}, \psi^{a_4^B,b_2}$, respectively. (**h**) is the magnitudes of $\widehat{a_4^B}$ (in solid line), $\widehat{b_1}$ (in dashed line), and $\widehat{b_2}$ (in dotted line) on $[-\pi, \pi]$. The last two rows are for a tight framelet filter bank $\{a_3^B; b_1, b_2\}_\Theta$ in Example 3.4.3 with Θ in (3.4.2)

Example 3.4.4 Let $a = a_4^I = \{-\frac{1}{32}, 0, \frac{9}{32}, \underline{\mathbf{\frac{1}{2}}}, \frac{9}{32}, 0, -\frac{1}{32}\}_{[-3,3]}$. Setting $n_b = 2$ in Algorithm 3.4.1 with $\Theta = \boldsymbol{\delta}$, we have $\mathcal{M}_{a,\boldsymbol{\delta}}(z) \geqslant 0$ for all $z \in \mathbb{T}$ and

$$\mathsf{b}_1(z) = \frac{\sqrt{2}}{8\sqrt{9-4\sqrt{3}}} z^2(1-z^{-1})^2(z^{-1}-\sqrt{3})(z+2-\sqrt{3}),$$

$$\mathsf{b}_2(z) = \frac{2\sqrt{3}+1}{352\sqrt{9-4\sqrt{3}}}(1-z^{-1})^2(x+2-\sqrt{3})[(1-2\sqrt{3})z^{-1}+(6-\sqrt{3})+33\,z+11\sqrt{3}z^2].$$

Then $\{a; b_1, b_2\}$ is a tight framelet filter bank with $\mathrm{vm}(b_1) = \mathrm{vm}(b_2) = 2$ and $\mathrm{fsupp}(b_1) = [-1, 3], \mathrm{fsupp}(b_2) = [-3, 3]$. Moreover, $\mathrm{Var}(b_1) \approx 0.374694$, $\mathrm{Var}(b_2) \approx 0.935396$ and $\mathrm{Fsi}(a, b_1) \approx 0.0612604, \mathrm{Fsi}(a, b_2) \approx 0.349023$, $\mathrm{Fsi}(b_1, b_2) \approx 0.541160$.

Note that the necessary and sufficient condition in Theorem 3.3.7 fails for $a = a_4^I$ and $\Theta = \boldsymbol{\delta}$. Consequently, there does not exist a finitely supported tight framelet filter bank $\{a_4^I; b_1, b_2\}$ with the symmetry property.

Setting $n_b = \mathrm{sr}(a) = 4$ in Algorithm 3.4.1 with

$$\Theta = \{-\tfrac{11}{5040}, \tfrac{4}{105}, -\tfrac{223}{1680}, \underline{\mathbf{\tfrac{376}{315}}}, -\tfrac{223}{1680}, \tfrac{4}{105}, -\tfrac{11}{5040}\}_{[-3,3]}, \tag{3.4.3}$$

constructed by Proposition 3.3.4 with $n = 4$, we have $\mathcal{M}_{a,\Theta}(z) \geqslant 0$ for all $z \in \mathbb{T}$ and

$$\begin{aligned}\mathsf{b}_1(z) = (1-z^{-1})^4[&0.0312508038066z^{-1} + 0.125003215226 + 0.0652710805584z\\ &+ 0.0209620231613z^2 + 0.00232937166827z^3 - 0.00129383738758z^4\\ &- 0.000323459346894z^5],\\ \mathsf{b}_2(z) = (1-z^{-1})^4[&0.0200800818532z^{-1} + 0.0803203274128 + 0.201908825322z\\ &+ 0.0847523545721z^2 - 0.00470092293218z^3 - 0.0126384333272z^4\\ &- 0.00220430955269z^5 + 0.000424577235155z^6 + 0.00010614430879z^7].\end{aligned}$$

Then $\{a; b_1, b_2\}_\Theta$ is a tight framelet filter bank with $\mathrm{vm}(b_1) = \mathrm{vm}(b_2) = 4$ and $\mathrm{fsupp}(b_1) = [-5, 5], \mathrm{fsupp}(b_2) = [-5, 7]$. Moreover, $\mathrm{Var}(b_1) \approx 0.389998$, $\mathrm{Var}(b_2) \approx 0.486932$ and $\mathrm{Fsi}(a, b_1) \approx 0.0681027$, $\mathrm{Fsi}(a, b_2) \approx 0.099625$, $\mathrm{Fsi}(b_1, b_2) \approx 0.988415$. See Fig. 3.13 for the graphs of their associated refinable and framelet functions.

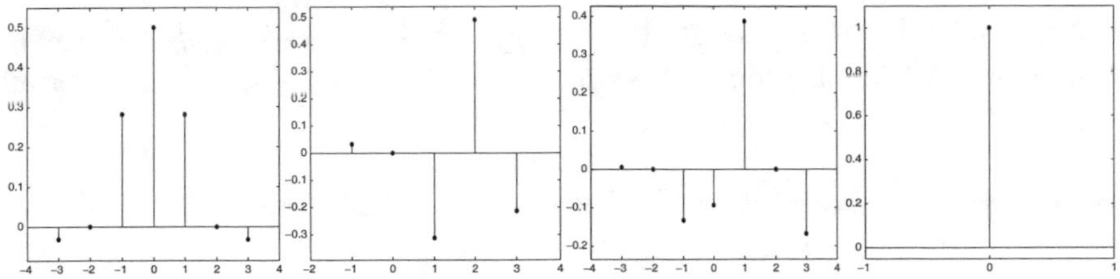

(a) Filter a_4^I (b) Filter b_1 (c) Filter b_2 (d) Filter $\Theta = \boldsymbol{\delta}$

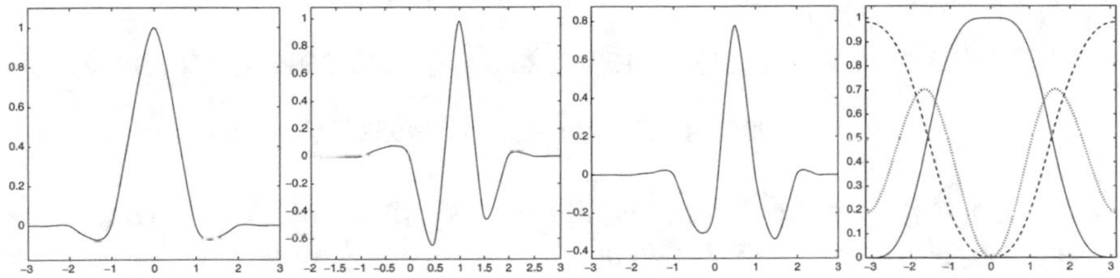

(e) $\phi^{a_4^I}$ (f) $\psi^{a_4^I,b_1}$ (g) $\psi^{a_4^I,b_2}$ (h) $|\widehat{a_4^I}|, |\widehat{b_1}|, |\widehat{b_2}|$

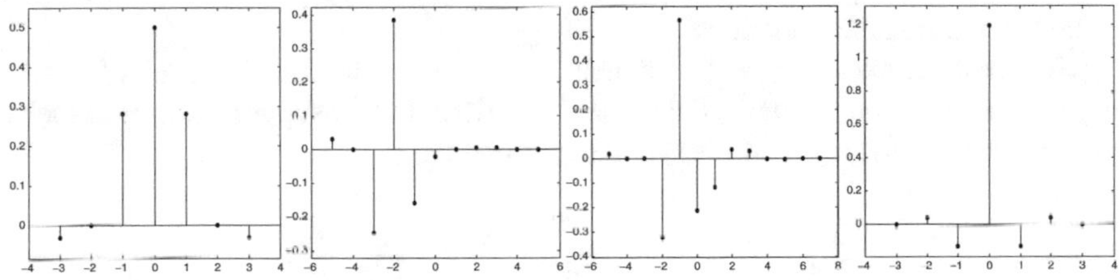

(i) Filter a_4^I (j) Filter b_1 (k) Filter b_2 (l) Filter Θ

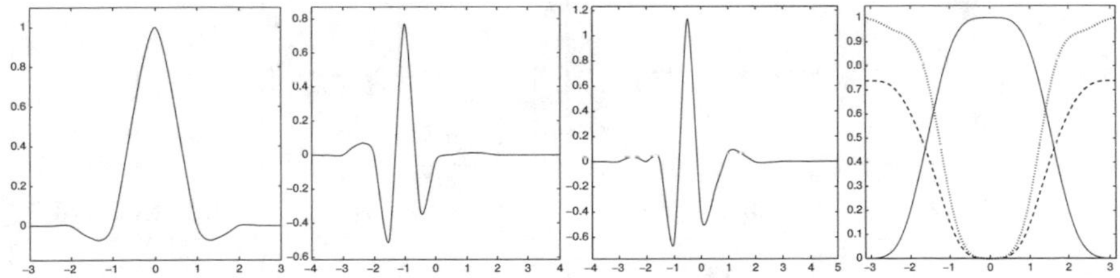

(m) $\phi^{a_4^I}$ (n) $\psi^{a_4^I,b_1}$ (o) $\psi^{a_4^I,b_2}$ (p) $|\widehat{a_4^I}|, |\widehat{b_1}|, |\widehat{b_2}|$

Fig. 3.13 The first two rows are for the tight framelet filter bank $\{a_4^I; b_1, b_2\}$ constructed in Example 3.4.4. (**a**), (**b**), (**c**), (**d**) are the graphs of the filters $a_4^I, b_1, b_2, \Theta = \boldsymbol{\delta}$. (**e**), (**f**), (**g**) are the graphs of the refinable function $\phi^{a_4^I}$ and framelet functions $\psi^{a_4^I,b_1}$, $\psi^{a_4^I,b_2}$. (**h**) is the magnitudes of $\widehat{a_4^I}$ (in solid line), $\widehat{b_1}$ (in dashed line), and $\widehat{b_2}$ (in dotted line) on the interval $[-\pi, \pi]$. The last two rows are for a tight framelet filter bank $\{a_4^I; b_1, b_2\}_\Theta$ constructed in Example 3.4.4 with Θ in (3.4.3)

Example 3.4.5 Let $a = a_{5,4} = \{-\frac{5}{256}, -\frac{7}{256}, \frac{35}{256}, \frac{\mathbf{105}}{\mathbf{256}}, \frac{105}{256}, \frac{35}{256}, -\frac{7}{256}, -\frac{5}{256}\}_{[-3,4]}$. Setting $n_b = 2$ in Algorithm 3.4.1 with $\Theta = \boldsymbol{\delta}$, we have $\mathcal{M}_{a,\boldsymbol{\delta}}(z) \geqslant 0$ for all $z \in \mathbb{T}$ and

$$\mathsf{b}_1(z) = \tfrac{(65+38\sqrt{5})\sqrt{440899}\sqrt{7}}{8451152032}(1-z^{-1})^2[(4940\sqrt{5}-11445)z^{-1} + (16796\sqrt{5}-38913) - 10183\,z - 2995\,z^2],$$

$$\mathsf{b}_2(z) = \tfrac{(968+315\sqrt{5})\sqrt{440899}}{49764333619456}(1-z^{-1})^2[(7165745-3049200\sqrt{5})z^{-1} + (24363533-10367280\sqrt{5}) + (47675506-22305360\sqrt{5})z - (709006+6560400\sqrt{5})z^2 - 7495283z^3 - 2204495z^4].$$

Then $\{a; b_1, b_2\}$ is a tight framelet filter bank with $\mathrm{vm}(b_1) = \mathrm{vm}(b_2) = 2$ and $\mathrm{fsupp}(b_1) = [-3, 2], \mathrm{fsupp}(b_2) = [-3, 4]$. Moreover, $\mathrm{Var}(b_1) \approx 0.335466, \mathrm{Var}(b_2) \approx 0.423234$ and $\mathrm{Fsi}(a, b_1) \approx 0.122346, \mathrm{Fsi}(a, b_2) \approx 0.214975, \mathrm{Fsi}(b_1, b_2) \approx 0.962760$. See Fig. 3.14 for the graphs of their associated refinable and framelet functions.

Note that the necessary and sufficient condition in Theorem 3.3.7 fails for $a = a_{5,4}$ and $\Theta = \boldsymbol{\delta}$. Consequently, there does not exist a finitely supported tight framelet filter bank $\{a_{5,4}; b_1, b_2\}$ with the symmetry property.

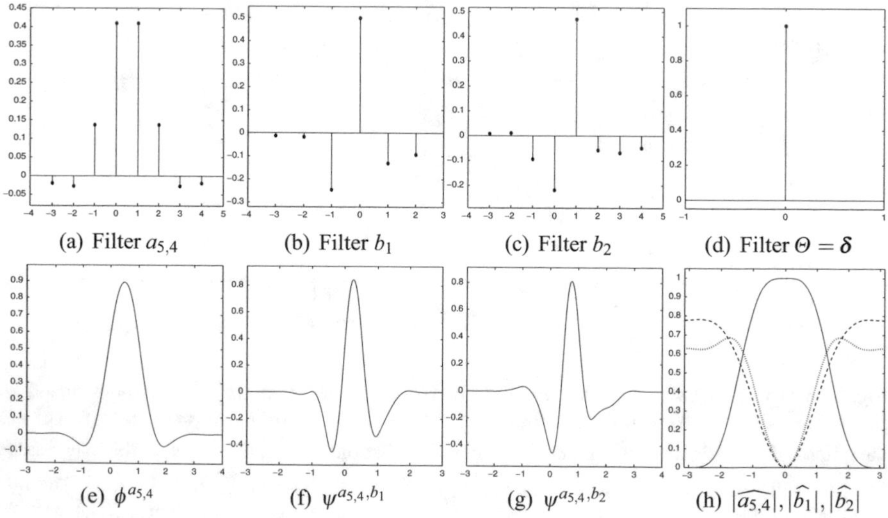
(a) Filter $a_{5,4}$
(b) Filter b_1
(c) Filter b_2
(d) Filter $\Theta = \boldsymbol{\delta}$
(e) $\phi^{a_{5,4}}$
(f) $\psi^{a_{5,4},b_1}$
(g) $\psi^{a_{5,4},b_2}$
(h) $|\widehat{a_{5,4}}|, |\widehat{b_1}|, |\widehat{b_2}|$

Fig. 3.14 The tight framelet filter bank $\{a_{5,4}; b_1, b_2\}$ is constructed in Example 3.4.5. (**a**), (**b**), (**c**), (**d**) are the graphs of the filters $a_{5,4}, b_1, b_2, \Theta = \boldsymbol{\delta}$, respectively. (**e**), (**f**), (**g**) are the graphs of the refinable function $\phi^{a_{5,4}}$ and framelet functions $\psi^{a_{5,4},b_1}$, $\psi^{a_{5,4},b_2}$, respectively. (**h**) is the magnitudes of $\widehat{a_{5,4}}$ (in solid line), $\widehat{b_1}$ (in dashed line), and $\widehat{b_2}$ (in dotted line) on the interval $[-\pi, \pi]$

3.4.2 Existence of Tight Framelet Filter Banks Without Symmetry

Using the Fejér-Riesz Lemma and solving linear equations, under the necessary condition $\mathcal{M}_{a,\Theta}(z) \geqslant 0$ for all $z \in \mathbb{T}$, Algorithm 3.4.1 can construct all possible tight framelet filter banks $\{a; b_1, b_2\}_\Theta$ with the shortest possible filter supports. However, we haven't proved yet whether this necessary condition always guarantees the existence of a tight framelet filter bank in Algorithm 3.4.1. In this subsection, we establish such existence. To do so, we first study how to split/factorize a positive semidefinite 2×2 matrix of Laurent polynomials.

To split a positive semidefinite matrix of Laurent polynomials, we need the following known matrix-valued Fejér-Riesz lemma.

Lemma 3.4.2 (Matrix-valued Fejér-Riesz Lemma) *Let $r \in \mathbb{N}$ and $\mathcal{N}(z)$ be an $r \times r$ matrix of Laurent polynomials with complex (or real) coefficients such that $\mathcal{N}(z) \geqslant 0$ for all $z \in \mathbb{T}$ and the filter supports of all entries of $\mathcal{N}$ are contained inside $[-n, n]$. Then there exists an $r \times r$ matrix $\mathcal{U}$ of Laurent polynomials with complex (or real) coefficients such that $\mathcal{U}(z)\mathcal{U}^\star(z) = \mathcal{N}(z)$ for all $z \in \mathbb{C}\backslash\{0\}$ and the filter supports of all entries in $\mathcal{U}$ are contained inside $[0, n]$.*

In this book we only need the case $r = 2$, for which we provide a constructive/algorithmic proof for the special case $r = 2$ in Lemma 3.4.2.

Algorithm 3.4.3 *Let $\mathcal{N}$ be a 2×2 matrix of Laurent polynomials such that*

$$\mathcal{N}(z) := \begin{bmatrix} \mathcal{N}_{1,1}(z) & \mathcal{N}_{1,2}(z) \\ \mathcal{N}_{2,1}(z) & \mathcal{N}_{2,2}(z) \end{bmatrix} \geqslant 0 \qquad \forall\, z \in \mathbb{T}$$

and $\mathcal{N}$ is not identically zero. We further assume that $\mathcal{N}_{1,1}$ is not identically zero.

(S1) Define $\mathsf{q}(z) := \mathsf{q}_1(z)/\mathsf{q}_2(z)$, where $\mathsf{q}_1(z) := \gcd(\mathcal{N}_{1,1}(z), \mathcal{N}_{1,2}(z))$ and q_2 is a Laurent polynomial obtained via the Fejér-Riesz Lemma (see Lemma 1.4.4) satisfying $\mathsf{q}_2(z)\mathsf{q}_2^\star(z) = \mathsf{q}_1(z)\mathsf{q}_1^\star(z)/\breve{\mathsf{q}}(z)$ with $\breve{\mathsf{q}}(z) := \gcd(\mathcal{N}_{1,1}(z), \mathcal{N}_{1,2}(z)\mathcal{N}_{1,2}^\star(z))$. Then q is a well-defined Laurent polynomial satisfying

$$\mathsf{q}(z)\mathsf{q}^\star(z) = \gcd(\mathcal{N}_{1,1}(z), \mathcal{N}_{1,2}(z)\mathcal{N}_{1,2}^\star(z)) \quad \text{and} \quad \mathsf{q} \mid \mathcal{N}_{1,2}. \tag{3.4.4}$$

Define $\mathring{\mathcal{N}}$ by

$$\mathring{\mathcal{N}}(z) := \begin{bmatrix} \mathring{\mathcal{N}}_{1,1}(z) & \mathring{\mathcal{N}}_{1,2}(z) \\ \mathring{\mathcal{N}}_{2,1}(z) & \mathring{\mathcal{N}}_{2,2}(z) \end{bmatrix} := \begin{bmatrix} \frac{\mathcal{N}_{1,1}(z)}{\mathsf{q}(z)\mathsf{q}^\star(z)} & \frac{\mathcal{N}_{1,2}(z)}{\mathsf{q}(z)} \\ \frac{\mathcal{N}_{2,1}(z)}{\mathsf{q}^\star(z)} & \mathcal{N}_{2,2}(z) \end{bmatrix}. \tag{3.4.5}$$

Then $\mathring{\mathcal{N}}_{1,1}$ and $\mathring{\mathcal{N}}_{1,2}$ have no common zeros in $\mathbb{C}\backslash\{0\}$;

(S2) Define $[-\mathring{n}, \mathring{n}] := \mathrm{fsupp}(\mathring{\mathcal{N}}_{1,1})$. *Use the Fejér-Riesz Lemma (see Lemma 1.4.4) to find a Laurent polynomial* $\mathring{\mathsf{d}}$ *satisfying* $\mathring{\mathsf{d}}(z)\mathring{\mathsf{d}}^\star(z) = \det(\mathring{\mathcal{N}}(z))$.

(S3) Write $\mathring{\mathcal{U}}_{1,1}(z) = \sum_{j=0}^{\mathring{n}} t_j z^j$ *and* $\mathring{\mathcal{U}}_{1,2}(z) = \sum_{j=0}^{\mathring{n}} \tilde{t}_j z^j$, *where* $\{t_0, \ldots, t_{\mathring{n}}, \tilde{t}_0, \ldots, \tilde{t}_{\mathring{n}}\}$ *is a nontrivial solution to the homogeneous system X of* $2\mathring{n}$ *linear equations induced by* $\mathring{\mathcal{R}}(z) = 0$, *where* $\mathring{\mathcal{R}}$ *and* $\mathring{\mathcal{U}}_{2,1}$ *are uniquely determined through long division using* $\mathring{\mathcal{N}}_{1,1}$ *by*

$$\mathring{\mathcal{N}}_{2,1}(z)\mathring{\mathcal{U}}_{1,1}(z) - \mathring{\mathsf{d}}(z)\mathring{\mathcal{U}}_{1,2}^\star(z) = \mathring{\mathcal{N}}_{1,1}(z)\mathring{\mathcal{U}}_{2,1}(z) + \mathring{\mathcal{R}}(z) \quad \text{with} \quad \mathrm{fsupp}(\mathring{\mathcal{R}}) \subseteq [-\mathring{n}, \mathring{n}-1]. \tag{3.4.6}$$

Then the space of all solutions to X has dimension at least two and every nontrivial solution to X must satisfy $|\mathring{\mathcal{U}}_{1,1}(1)|^2 + |\mathring{\mathcal{U}}_{1,2}(1)|^2 \neq 0$;

(S4) Normalize $\{\mathring{\mathcal{U}}_{1,1}, \mathring{\mathcal{U}}_{1,2}\}$ *in (S3) by multiplying them with the positive number* $\sqrt{\frac{\mathring{\mathcal{N}}_{1,1}(1)}{|\mathring{\mathcal{U}}_{1,1}(1)|^2+|\mathring{\mathcal{U}}_{1,2}(1)|^2}}$.

(S5) Re-define (more precisely, re-normalize) $\mathring{\mathcal{U}}_{2,1}$ *in (S3) and define* $\mathring{\mathcal{U}}_{2,2}$ *as follows:*

$$\mathring{\mathcal{U}}_{2,1}(z) := \frac{\mathring{\mathcal{N}}_{2,1}(z)\mathring{\mathcal{U}}_{1,1}(z) - \mathring{\mathsf{d}}(z)\mathring{\mathcal{U}}_{1,2}^\star(z)}{\mathring{\mathcal{N}}_{1,1}(z)} \tag{3.4.7}$$

and

$$\mathring{\mathcal{U}}_{2,2}(z) := \frac{\mathring{\mathcal{N}}_{2,1}(z)\mathring{\mathcal{U}}_{1,2}(z) + \mathring{\mathsf{d}}(z)\mathring{\mathcal{U}}_{1,1}^\star(z)}{\mathring{\mathcal{N}}_{1,1}(z)}. \tag{3.4.8}$$

Then $\mathring{\mathcal{U}}_{2,1}$ *and* $\mathring{\mathcal{U}}_{2,2}$ *are well-defined Laurent polynomials;*

(S6) Define a 2×2 *matrix* $\mathcal{U}$ *of Laurent polynomials by*

$$\mathcal{U}(z) := \begin{bmatrix} \mathcal{U}_{1,1}(z) & \mathcal{U}_{1,2}(z) \\ \mathcal{U}_{2,1}(z) & \mathcal{U}_{2,2}(z) \end{bmatrix} := \begin{bmatrix} \mathsf{q}(z) & 0 \\ 0 & 1 \end{bmatrix} \mathring{\mathcal{U}}(z) = \begin{bmatrix} \mathsf{q}(z)\mathring{\mathcal{U}}_{1,1}(z) & \mathsf{q}(z)\mathring{\mathcal{U}}_{1,2}(z) \\ \mathring{\mathcal{U}}_{2,1}(z) & \mathring{\mathcal{U}}_{2,2}(z) \end{bmatrix}.$$

Then $\mathcal{U}(z)\mathcal{U}^\star(z) = \mathcal{N}(z)$. *If all Laurent polynomials in* $\mathcal{N}$ *have real coefficients, then all Laurent polynomials in* $\mathcal{U}$ *can have real coefficients.*

Proof We first prove that q_2 and q are well-defined Laurent polynomials. Since $\mathcal{N}_{1,1}(z) \geqslant 0$ for all $z \in \mathbb{T}$, we have $\mathcal{N}_{1,1}^\star(z) = \mathcal{N}_{1,1}(z)$ and $\breve{\mathsf{q}}(z) \geqslant 0$ for all $z \in \mathbb{T}$. By the definition of the greatest common divisors and the following simple inequality

$$\min(\mathsf{Z}(\mathcal{N}_{1,1}, z), \mathsf{Z}(\mathcal{N}_{1,2}, z)) + \min(\mathsf{Z}(\mathcal{N}_{1,1}, z), \mathsf{Z}(\mathcal{N}_{1,2}^\star, z)) \geqslant \min(\mathsf{Z}(\mathcal{N}_{1,1}, z), \mathsf{Z}(\mathcal{N}_{1,2}, z) + \mathsf{Z}(\mathcal{N}_{1,2}^\star, z))$$

for all $z \in \mathbb{C}\backslash\{0\}$, we conclude that $\mathsf{q}_1(z)\mathsf{q}_1^\star(z)/\breve{\mathsf{q}}(z)$ is a well-defined Laurent polynomial and is nonnegative for all $z \in \mathbb{T}$. This proves the existence of a Laurent polynomial q_2 satisfying $\mathsf{q}_2(z)\mathsf{q}_2^\star(z) = \mathsf{q}_1(z)\mathsf{q}_1^\star(z)/\breve{\mathsf{q}}(z)$. On the other hand, by $\mathsf{q}_1(z)\mathsf{q}_1^\star(z) = \breve{\mathsf{q}}(z)\frac{\mathsf{q}_1(z)\mathsf{q}_1^\star(z)}{\breve{\mathsf{q}}(z)} = \breve{\mathsf{q}}(z)\mathsf{q}_2(z)\mathsf{q}_2^\star(z)$, we see that $\mathsf{q}(z) = \frac{\mathsf{q}_1(z)}{\mathsf{q}_2(z)} = \mathsf{q}_2^\star(z)\frac{\breve{\mathsf{q}}(z)}{\mathsf{q}_1^\star(z)}$ is a well-defined Laurent polynomial, since $\mathsf{q}_1^\star \mid \breve{\mathsf{q}}$.

By the definition of q_2, we see that $\mathsf{q}(z)\mathsf{q}^\star(z) = \frac{\mathsf{q}_1(z)\mathsf{q}_1^\star(z)}{\mathsf{q}_2(z)\mathsf{q}_2^\star(z)} = \breve{\mathsf{q}}(z)$. Since $\mathsf{q}_1 \mid \mathcal{N}_{1,2}$ and $\mathsf{q} \mid \mathsf{q}_1$, we trivially have $\mathsf{q} \mid \mathcal{N}_{1,2}$. This proves (3.4.4). Moreover, by $\mathsf{q}(z)\mathsf{q}^\star(z) = \breve{\mathsf{q}}(z)$, we see that $\gcd(\mathring{\mathcal{N}}_{1,1}(z), \mathring{\mathcal{N}}_{1,2}(z)) = \gcd(\frac{\mathcal{N}_{1,1}(z)}{\breve{\mathsf{q}}(z)}, \frac{\mathcal{N}_{1,2}(z)}{\mathsf{q}(z)})$ must be a factor of $\gcd(\frac{\mathcal{N}_{1,1}(z)}{\breve{\mathsf{q}}(z)}, \frac{\mathcal{N}_{1,2}(z)\mathcal{N}_{1,2}^\star(z)}{\breve{\mathsf{q}}(z)}) = 1$, where we used the definition of $\breve{\mathsf{q}}$. Hence, we conclude that $\mathring{\mathcal{N}}_{1,1}$ and $\mathring{\mathcal{N}}_{1,2}$ have no common zeros in $\mathbb{C}\backslash\{0\}$. This proves (S1).

By $\operatorname{fsupp}(\mathring{\mathcal{R}}) \subseteq [-\mathring{n}, \mathring{n}-1]$, there are no more than $2\mathring{n}$ homogeneous linear equations in X, while there are $2\mathring{n}+2$ unknowns. Therefore, the space of all solutions to X must have dimension at least two.

Since $\mathring{\mathcal{R}} = 0, \mathring{\mathcal{U}}_{2,1}$ in (3.4.7) is well defined. Next, we show that $\mathring{\mathcal{U}}_{2,2}$ in (3.4.8) is a well-defined Laurent polynomial. Since $\mathring{\mathsf{d}}(z)\mathring{\mathsf{d}}^\star(z) = \det(\mathring{\mathcal{N}}(z)) = \mathring{\mathcal{N}}_{1,1}(z)\mathring{\mathcal{N}}_{2,2}(z) - \mathring{\mathcal{N}}_{1,2}(z)\mathring{\mathcal{N}}_{2,1}(z)$, we have

$$
\begin{aligned}
&\mathring{\mathcal{N}}_{1,2}(z)[\mathring{\mathcal{N}}_{2,1}(z)\mathring{\mathcal{U}}_{1,2}(z) + \mathring{\mathsf{d}}(z)\mathring{\mathcal{U}}_{1,1}^\star(z)]\\
&\quad = \mathring{\mathcal{N}}_{1,2}(z)\mathring{\mathcal{N}}_{2,1}(z)\mathring{\mathcal{U}}_{1,2}(z) + \mathring{\mathsf{d}}(z)\mathring{\mathcal{N}}_{1,2}(z)\mathring{\mathcal{U}}_{1,1}^\star(z)\\
&\quad = [\mathring{\mathcal{N}}_{1,1}(z)\mathring{\mathcal{N}}_{2,2}(z) - \mathring{\mathsf{d}}(z)\mathring{\mathsf{d}}^\star(z)]\mathring{\mathcal{U}}_{1,2}(z) + \mathring{\mathsf{d}}(z)\mathring{\mathcal{N}}_{1,2}(z)\mathring{\mathcal{U}}_{1,1}^\star(z)\\
&\quad = \mathring{\mathcal{N}}_{1,1}(z)\mathring{\mathcal{N}}_{2,2}(z)\mathring{\mathcal{U}}_{1,2}(z) + \mathring{\mathsf{d}}(z)[\mathring{\mathcal{N}}_{1,2}(z)\mathring{\mathcal{U}}_{1,1}^\star(z) - \mathring{\mathsf{d}}^\star(z)\mathring{\mathcal{U}}_{1,2}(z)]\\
&\quad = \mathring{\mathcal{N}}_{1,1}(z)\mathring{\mathcal{N}}_{2,2}(z)\mathring{\mathcal{U}}_{1,2}(z) + \mathring{\mathsf{d}}(z)[\mathring{\mathcal{N}}_{1,2}^\star(z)\mathring{\mathcal{U}}_{1,1}(z) - \mathring{\mathsf{d}}(z)\mathring{\mathcal{U}}_{1,2}^\star(z)]^\star.
\end{aligned}
$$

By $\mathring{\mathcal{N}}(z) \geqslant 0$ for all $z \in \mathbb{T}$, we have $\mathring{\mathcal{N}}_{1,1}^\star(z) = \mathring{\mathcal{N}}_{1,1}(z)$ and $\mathring{\mathcal{N}}_{1,2}^\star(z) = \mathring{\mathcal{N}}_{2,1}(z)$. Using (3.4.7), we deduce from the above identity that

$$
\begin{aligned}
\mathring{\mathcal{N}}_{1,2}(z)[\mathring{\mathcal{N}}_{2,1}(z)\mathring{\mathcal{U}}_{1,2}(z) + \mathring{\mathsf{d}}(z)\mathring{\mathcal{U}}_{1,1}^\star(z)] &= \mathring{\mathcal{N}}_{1,1}(z)\mathring{\mathcal{N}}_{2,2}(z)\mathring{\mathcal{U}}_{1,2}(z) + \mathring{\mathsf{d}}(z)\mathring{\mathcal{N}}_{1,1}^\star(z)\mathring{\mathcal{U}}_{2,1}^\star(z)\\
&= \mathring{\mathcal{N}}_{1,1}(z)[\mathring{\mathcal{N}}_{2,2}(z)\mathring{\mathcal{U}}_{1,2}(z) + \mathring{\mathsf{d}}(z)\mathring{\mathcal{U}}_{2,1}^\star(z)].
\end{aligned}
$$

Since $\mathring{\mathcal{N}}_{1,1}$ and $\mathring{\mathcal{N}}_{1,2}$ have no common zeros in $\mathbb{C}\backslash\{0\}$, we conclude from the above identity that $\mathring{\mathcal{N}}_{1,1}(z) \mid [\mathring{\mathcal{N}}_{2,1}(z)\mathring{\mathcal{U}}_{1,2}(z) + \mathring{\mathsf{d}}(z)\mathring{\mathcal{U}}_{1,1}^\star(z)]$. That is, we proved that $\mathring{\mathcal{U}}_{2,2}$ in (3.4.8) is a well-defined Laurent polynomial.

Now (3.4.7) and (3.4.8) together imply

$$
\begin{bmatrix} \mathring{\mathcal{U}}_{2,2}(z) & -\mathring{\mathcal{U}}_{1,2}(z)\\ -\mathring{\mathcal{U}}_{2,1}(z) & \mathring{\mathcal{U}}_{1,1}(z) \end{bmatrix}
\begin{bmatrix} \mathring{\mathcal{N}}_{1,1}(z)\\ \mathring{\mathcal{N}}_{2,1}(z) \end{bmatrix}
= \mathring{\mathsf{d}}(z)\begin{bmatrix} \mathring{\mathcal{U}}_{1,1}^\star(z)\\ \mathring{\mathcal{U}}_{1,2}^\star(z) \end{bmatrix}. \tag{3.4.9}
$$

Multiplying $[-\mathring{\mathcal{U}}^{\star}_{1,2}(z), \mathring{\mathcal{U}}^{\star}_{1,1}(z)]$ from the left on both sides of (3.4.9), we have

$$-[\mathring{\mathcal{U}}_{2,2}(z)\mathring{\mathcal{U}}^{\star}_{1,2}(z)+\mathring{\mathcal{U}}_{2,1}(z)\mathring{\mathcal{U}}^{\star}_{1,1}(z)]\mathring{\mathcal{N}}_{1,1}(z)\\ +[\mathring{\mathcal{U}}_{1,1}(z)\mathring{\mathcal{U}}^{\star}_{1,1}(z)+\mathring{\mathcal{U}}_{1,2}(z)\mathring{\mathcal{U}}^{\star}_{1,2}(z)]\mathring{\mathcal{N}}_{2,1}(z)=0.$$

That is,

$$[\mathring{\mathcal{U}}_{1,1}(z)\mathring{\mathcal{U}}^{\star}_{1,1}(z)+\mathring{\mathcal{U}}_{1,2}(z)\mathring{\mathcal{U}}^{\star}_{1,2}(z)]\mathring{\mathcal{N}}_{2,1}(z)\\ =[\mathring{\mathcal{U}}_{2,2}(z)\mathring{\mathcal{U}}^{\star}_{1,2}(z)+\mathring{\mathcal{U}}_{2,1}(z)\mathring{\mathcal{U}}^{\star}_{1,1}(z)]\mathring{\mathcal{N}}_{1,1}(z). \tag{3.4.10}$$

Since the solution $\{\mathring{\mathcal{U}}_{1,1}, \mathring{\mathcal{U}}_{1,2}\}$ to the system X of linear equations is nontrivial, we see that

$$\mathring{\mathcal{U}}_{1,1}(z)\mathring{\mathcal{U}}^{\star}_{1,1}(z)+\mathring{\mathcal{U}}_{1,2}(z)\mathring{\mathcal{U}}^{\star}_{1,2}(z) \quad \text{must be nontrivial.} \tag{3.4.11}$$

Because $\mathring{\mathcal{U}}_{1,1}$ and $\mathring{\mathcal{U}}_{1,2}$ are polynomials of degree at most $\mathring{n}$, the filter support of the Laurent polynomial on the left-hand side of (3.4.11) is contained inside $[-\mathring{n}, \mathring{n}]$. Note that $\mathring{\mathcal{N}}_{1,1} = \mathring{\mathcal{N}}^{\star}_{1,1}$ and $\mathring{\mathcal{N}}_{2,1} = \mathring{\mathcal{N}}^{\star}_{1,2}$. Since $\mathring{\mathcal{N}}_{1,1}$ and $\mathring{\mathcal{N}}_{1,2}$ have no common zeros in $\mathbb{C}\backslash\{0\}$, we see that $\mathring{\mathcal{N}}_{1,1}$ and $\mathring{\mathcal{N}}_{2,1}$ also have no common zeros in $\mathbb{C}\backslash\{0\}$. By (3.4.11) and noting that the filter support of the Laurent polynomial in (3.4.11) is contained inside $[-\mathring{n}, \mathring{n}] = \operatorname{fsupp}(\mathring{\mathcal{N}}_{1,1})$, from (3.4.10), we must have

$$\mathring{\mathcal{U}}_{1,1}(z)\mathring{\mathcal{U}}^{\star}_{1,1}(z)+\mathring{\mathcal{U}}_{1,2}(z)\mathring{\mathcal{U}}^{\star}_{1,2}(z)=\lambda^{-1}\mathring{\mathcal{N}}_{1,1}(z) \qquad \text{for some } \lambda \in \mathbb{C}\backslash\{0\}. \tag{3.4.12}$$

We now show that

$$\lambda = \frac{\mathring{\mathcal{N}}_{1,1}(1)}{|\mathring{\mathcal{U}}_{1,1}(1)|^2+|\mathring{\mathcal{U}}_{1,2}(1)|^2} > 0. \tag{3.4.13}$$

Since $\mathring{\mathcal{N}}_{1,1}$ and $\mathring{\mathcal{N}}_{2,1}$ have no common zeros in $\mathbb{C}\backslash\{0\}$ and since $\mathring{\mathcal{N}}(z) \geqslant 0$ for all $z \in \mathbb{T}$, we see that $\mathring{\mathcal{N}}_{1,1}(z) > 0$ for all $z \in \mathbb{T}$ (if $\mathring{\mathcal{N}}_{1,1}(z_0) = 0$ for some $z_0 \in \mathbb{T}$, then $\mathring{\mathcal{N}}_{1,2}(z_0) = 0$ by $\mathring{\mathcal{N}}(z_0) \geqslant 0$). In particular, $\mathring{\mathcal{N}}_{1,1}(1) > 0$. Consequently, (3.4.11) and (3.4.12) imply that $|\mathring{\mathcal{U}}_{1,1}(1)|^2 + |\mathring{\mathcal{U}}_{1,2}(1)|^2 > 0$. Hence, (3.4.13) holds. Normalizing the solution $\{\mathring{\mathcal{U}}_{1,1}, \mathring{\mathcal{U}}_{1,2}\}$ by multiplying them with the factor $\sqrt{\lambda}$, we have

$$\mathring{\mathcal{U}}_{1,1}(z)\mathring{\mathcal{U}}^{\star}_{1,1}(z)+\mathring{\mathcal{U}}_{1,2}(z)\mathring{\mathcal{U}}^{\star}_{1,2}(z)=\mathring{\mathcal{N}}_{1,1}(z). \tag{3.4.14}$$

Now by (3.4.10) and (3.4.14), we further have

$$\mathring{\mathcal{U}}_{2,2}(z)\mathring{\mathcal{U}}^\star_{1,2}(z) + \mathring{\mathcal{U}}_{2,1}(z)\mathring{\mathcal{U}}^\star_{1,1}(z) = \mathring{\mathcal{N}}_{2,1}(z). \tag{3.4.15}$$

Multiplying $[\mathring{\mathcal{U}}_{1,1}(z), \mathring{\mathcal{U}}_{1,2}(z)]$ from the left on both sides of (3.4.9), we have

$$[\mathring{\mathcal{U}}_{1,1}(z)\mathring{\mathcal{U}}_{2,2}(z) - \mathring{\mathcal{U}}_{1,2}(z)\mathring{\mathcal{U}}_{2,1}(z)]\mathring{\mathcal{N}}_{1,1}(z) = \mathring{\mathsf{d}}(z)[\mathring{\mathcal{U}}_{1,1}(z)\mathring{\mathcal{U}}^\star_{1,1}(z) + \mathring{\mathcal{U}}_{1,2}(z)\mathring{\mathcal{U}}^\star_{1,2}(z)].$$

Combining the above identity with (3.4.14), we conclude that

$$\det(\mathring{\mathcal{U}}(z)) = \mathring{\mathcal{U}}_{1,1}(z)\mathring{\mathcal{U}}_{2,2}(z) - \mathring{\mathcal{U}}_{1,2}(z)\mathring{\mathcal{U}}_{2,1}(z) = \mathring{\mathsf{d}}(z).$$

Multiplying $[\mathring{\mathcal{U}}^\star_{2,2}(z), -\mathring{\mathcal{U}}^\star_{2,1}(z)]$ from the left on both sides of (3.4.9), by (3.4.15) and the above identity, we have

$$\begin{aligned}[]
[\mathring{\mathcal{U}}_{2,2}(z)\mathring{\mathcal{U}}^\star_{2,2}(z)+\mathring{\mathcal{U}}_{2,1}(z)\mathring{\mathcal{U}}^\star_{2,1}(z)]\mathring{\mathcal{N}}_{1,1}(z) - \mathring{\mathcal{N}}_{2,1}(z)\mathring{\mathcal{N}}^\star_{2,1}(z) &= \mathring{\mathsf{d}}(z)\mathring{\mathsf{d}}^\star(z)\\
&= \det(\mathring{\mathcal{N}}(z)) = \mathring{\mathcal{N}}_{2,2}(z)\mathring{\mathcal{N}}_{1,1}(z) - \mathring{\mathcal{N}}_{2,1}(z)\mathring{\mathcal{N}}_{1,2}(z).
\end{aligned}$$

Since $\mathring{\mathcal{N}}^\star(z) = \mathring{\mathcal{N}}(z)$, we have $\mathring{\mathcal{N}}_{1,2}(z) = \mathring{\mathcal{N}}^\star_{2,1}(z)$ and therefore,

$$[\mathring{\mathcal{U}}_{2,2}(z)\mathring{\mathcal{U}}^\star_{2,2}(z) + \mathring{\mathcal{U}}_{2,1}(z)\mathring{\mathcal{U}}^\star_{2,1}(z)]\mathring{\mathcal{N}}_{1,1}(z) = \mathring{\mathcal{N}}_{2,2}(z)\mathring{\mathcal{N}}_{1,1}(z),$$

from which we deduce that

$$\mathring{\mathcal{U}}_{2,2}(z)\mathring{\mathcal{U}}^\star_{2,2}(z) + \mathring{\mathcal{U}}_{2,1}(z)\mathring{\mathcal{U}}^\star_{2,1}(z) = \mathring{\mathcal{N}}_{2,2}(z). \tag{3.4.16}$$

Now (3.4.14), (3.4.15), and (3.4.16) together imply that $\mathring{\mathcal{U}}(z)\mathring{\mathcal{U}}^\star(z) = \mathring{\mathcal{N}}(z)$. It is trivial to verify that $\mathcal{U}(z)\mathcal{U}^\star(z) = \mathcal{N}(z)$. $\square$

Using long division in Algorithm 3.1.3, we often add a preprocessing step as (S0) in Algorithm 3.4.3 to $\mathcal{N}$ as well as $\mathring{\mathcal{N}}$ such that

$$\breve{\mathcal{N}}(z) := \begin{bmatrix} \breve{\mathcal{N}}_{1,1}(z) & \breve{\mathcal{N}}_{1,2}(z) \\ \breve{\mathcal{N}}_{2,1}(z) & \breve{\mathcal{N}}_{2,2} \end{bmatrix} = \mathcal{R}(z)\mathcal{N}(z)\mathcal{R}^\star(z)$$

satisfies $\operatorname{fsupp}(\breve{\mathcal{N}}_{1,2}) \subsetneq \operatorname{fsupp}(\breve{\mathcal{N}}_{1,1})$ and $\operatorname{fsupp}(\breve{\mathcal{N}}_{1,2}) \subsetneq \operatorname{fsupp}(\breve{\mathcal{N}}_{2,2})$, where $\mathcal{R}$ is a 2×2 matrix of Laurent polynomials such that $\det(\mathcal{R}(z)) = 1$. This can improve the efficiency of Algorithm 3.4.3.

Using Algorithm 3.4.3 for splitting a 2×2 matrix of Laurent polynomials, we now discuss another way to construct two high-pass filters b_1, b_2 in a tight framelet filter bank $\{a, b_1, b_2\}_\Theta$ from a low-pass filter a and a moment correcting filter Θ.

Algorithm 3.4.4 *Let $a = \{a(k)\}_{k\in\mathbb{Z}} \in l_0(\mathbb{Z})$. Choose $n \in \mathbb{N}$ such that $n \leqslant \operatorname{sr}(a)$.*

(S1) Construct a moment correcting filter $\Theta \in l_0(\mathbb{Z})$ such that (3.3.11), (3.3.12), *and* (3.3.14) *are satisfied;*

(S2) Choose a nonnegative integer n_b such that (3.3.5) *holds. Calculate the 2×2 matrix $\mathcal{N}(z) := \mathcal{N}_{a,\Theta|n_b}(z)$ as in* (3.3.10).

(S3) Apply Algorithm 3.4.3 to $\mathcal{N}$ to find a 2×2 matrix $\mathcal{U}$ of Laurent polynomials such that $\mathcal{U}(z)\mathcal{U}^\star(z) = \mathcal{N}(z) = \mathcal{N}_{a,\Theta|n_b}(z)$ for all $z \in \mathbb{C}\backslash\{0\}$;

(S4) Define two finitely supported filters b_1 and b_2 by

$$\begin{aligned} \mathsf{b}_1(z) &:= (1-z^{-1})^{n_b}[\mathcal{U}_{1,1}(z^2) + z\mathcal{U}_{2,1}(z^2)], \\ \mathsf{b}_2(z) &:= (1-z^{-1})^{n_b}[\mathcal{U}_{1,2}(z^2) + z\mathcal{U}_{2,2}(z^2)]. \end{aligned} \tag{3.4.17}$$

Then $\{a; b_1, b_2\}_\Theta$ is a tight framelet filter bank such that

$$\min\big(\operatorname{vm}(b_1), \operatorname{vm}(b_2)\big) = \min\big(\operatorname{sr}(a), \tfrac{1}{2}\operatorname{vm}(\Theta(z) - \Theta(z^2)\mathsf{a}(z)\mathsf{a}^\star(z))\big) \geqslant n.$$

Summarizing the above results, we have

Theorem 3.4.5 *Let $a, \Theta \in l_0(\mathbb{Z})$ with complex (or real) coefficients and $\mathsf{a}(1) = \Theta(1) = 1$. If there is a tight framelet filter bank $\{a; b_1, \dots, b_s\}_\Theta$ for some $b_1, \dots, b_s \in l_0(\mathbb{Z})$, then* (3.3.11) *and* (3.3.12) *must hold. Conversely, if* (3.3.11) *and* (3.3.12) *are satisfied, then there is a tight framelet filter bank $\{a; b_1, b_2\}_\Theta$ for some $b_1, b_2 \in l_0(\mathbb{Z})$ with complex (or real) coefficients, which can be constructed by Algorithm 3.4.4.*

Proof Suppose that there is a tight framelet filter bank $\{a; b_1, \dots, b_s\}_\Theta$ for some $b_1, \dots, b_s \in l_0(\mathbb{Z})$. By (3.0.1), it is straightforward to conclude that $\mathcal{M}_{a,\Theta}(z) \geqslant 0$ for all $z \in \mathbb{T}$. By Lemma 3.3.2, we see that (3.3.11) and (3.3.12) must hold.

The converse part follows directly from Algorithm 3.4.4. □

3.4.3 Symmetrize Tight Framelet Filter Banks

Let $a, \Theta \in l_0(\mathbb{Z})$ such that a has [complex] symmetry and $\mathcal{M}_{a,\Theta}(z) \geqslant 0$ for all $z \in \mathbb{T}$. A tight framelet filter bank $\{a; b_1, b_2\}_\Theta$ can be easily constructed by Algorithm 3.4.1, but its high-pass filters b_1, b_2 may not possess any symmetry property at all. By doubling the number of high-pass filters, a tight framelet filter bank with [complex] symmetry can be trivially derived as follows:

Proposition 3.4.6 *Let $\{a; b_1, \dots, b_s\}_\Theta$ be a tight framelet filter bank such that either $\mathsf{Sa}(z) = \epsilon z^c$ or $\mathbb{S}\mathsf{a}(z) = \epsilon z^c$ with $\epsilon \in \{-1, 1\}$ and $c \in \mathbb{Z}$. Let c_ℓ be integers satisfying $c_\ell - c \in 2\mathbb{Z}$ for all $\ell = 1, \dots, s$.*

(i) If a has complex symmetry, define

$$\mathsf{b}_\ell^{\mathbb{S}} := [\mathsf{b}_\ell(z) + z^{c_\ell}\mathsf{b}_\ell^\star(z)]/2, \quad \mathsf{b}_\ell^{\mathbb{A}}(z) := [\mathsf{b}_\ell(z) - z^{c_\ell}\mathsf{b}_\ell^\star(z)]/2, \quad \ell = 1,\dots,s,$$

then $\{a; b_1^{\mathbb{S}}, b_1^{\mathbb{A}}, \dots, b_s^{\mathbb{S}}, b_s^{\mathbb{A}}\}_\Theta$ *is a tight framelet filter bank with complex symmetry.*

(ii) If a has symmetry and Θ *also has symmetry, define*

$$\mathsf{b}_\ell^{\mathsf{S}} := [\mathsf{b}_\ell(z)+z^{c_\ell}\mathsf{b}_\ell(z^{-1})]/2, \quad \mathsf{b}_\ell^{\mathsf{A}}(z) := [\mathsf{b}_\ell(z)-z^{c_\ell}\mathsf{b}_\ell(z^{-1})]/2, \quad \ell = 1,\dots,s,$$

then $\{a; b_1^{\mathsf{S}}, b_1^{\mathsf{A}}, \dots, b_s^{\mathsf{S}}, b_s^{\mathsf{A}}\}_\Theta$ *is a tight framelet filter bank with symmetry.*

Proof Since $\{a; b_1, \dots, b_s\}_\Theta$ is a tight framelet filter bank, by Lemma 3.3.2, we always have $\Theta(z) \geqslant 0$ for all $z \in \mathbb{T}$. Hence, $\mathbb{S}\Theta = 1$. For (i), by calculation, we have

$$\mathsf{b}_\ell^{\mathbb{S}}(z)(\mathsf{b}_\ell^{\mathbb{S}}(z))^\star + \mathsf{b}_\ell^{\mathbb{A}}(z)(\mathsf{b}_\ell^{\mathbb{A}}(z))^\star = \mathsf{b}_\ell(z)\mathsf{b}_\ell^\star(z)$$

and

$$\mathsf{b}_\ell^{\mathbb{S}}(z)(\mathsf{b}_\ell^{\mathbb{S}}(-z))^\star + \mathsf{b}_\ell^{\mathbb{A}}(z)(\mathsf{b}_\ell^{\mathbb{A}}(-z))^\star = \tfrac{1}{2}\mathsf{b}_\ell(z)\mathsf{b}_\ell^\star(-z) + \tfrac{1}{2}(-1)^{c_\ell}\mathsf{b}_\ell^\star(z)\mathsf{b}_\ell(-z).$$

Since $\Theta(z^2)\mathsf{a}^\star(z)\mathsf{a}(-z) = (-1)^c\Theta(z^2)\mathsf{a}(z)\mathsf{a}^\star(-z)$ and $(-1)^{c_\ell} = (-1)^c$, we can directly check that $\{a; b_1^{\mathbb{S}}, b_1^{\mathbb{A}}, \dots, b_s^{\mathbb{S}}, b_s^{\mathbb{A}}\}_\Theta$ is a tight framelet filter bank with complex symmetry.

For (ii), by Lemma 3.1.2, Θ must have real coefficients and $\mathbb{S}\Theta = \mathsf{S}\Theta = 1$. By direct calculation, we have

$$\mathsf{b}_\ell^{\mathsf{S}}(z)(\mathsf{b}_\ell^{\mathsf{S}}(z))^\star + \mathsf{b}_\ell^{\mathsf{A}}(z)(\mathsf{b}_\ell^{\mathsf{A}}(z))^\star = \tfrac{1}{2}\mathsf{b}_\ell(z)\mathsf{b}_\ell^\star(z) + \tfrac{1}{2}\mathsf{b}_\ell(z^{-1})\mathsf{b}_\ell^\star(z^{-1})$$

and

$$\mathsf{b}_\ell^{\mathsf{S}}(z)(\mathsf{b}_\ell^{\mathsf{S}}(-z))^\star + \mathsf{b}_\ell^{\mathsf{A}}(z)(\mathsf{b}_\ell^{\mathsf{A}}(-z))^\star = \tfrac{1}{2}\mathsf{b}_\ell(z)\mathsf{b}_\ell^\star(-z) + \tfrac{1}{2}(-1)^{c_\ell}\mathsf{b}_\ell(z^{-1})\mathsf{b}_\ell^\star(-z^{-1}).$$

By our assumption $\mathsf{S}\mathsf{a}(z) = \epsilon z^c$, we have $\mathsf{a}(z) = \epsilon z^c\mathsf{a}(z^{-1})$. Therefore, we deduce that $\Theta(z^2)\mathsf{a}(z^{-1})\mathsf{a}^\star(z^{-1}) = \Theta(z^2)\mathsf{a}(z)\mathsf{a}^\star(z)$ and

$$\Theta(z^{-2})\mathsf{a}(z^{-1})\mathsf{a}^\star(-z^{-1}) = (-1)^c\Theta(z^2)\mathsf{a}(z)\mathsf{a}^\star(-z).$$

Since $(-1)^{c_\ell} = (-1)^c$ for all $\ell = 1,\dots,s$, by direction calculation we can verify that $\{a; b_1^{\mathsf{S}}, b_1^{\mathsf{A}}, \dots, b_s^{\mathsf{S}}, b_s^{\mathsf{A}}\}_\Theta$ is a tight framelet filter bank with symmetry. □

3.5 Tight Framelet Filter Banks with Symmetry and Three High-Pass Filters

As we have seen in Sect. 3.3, to construct a tight framelet filter bank $\{a; b_1, b_2\}_\Theta$ with [complex] symmetry, the low-pass filter a and the moment correcting filter Θ have to satisfy the necessary and sufficient condition in Theorem 3.3.7, which is often too restrictive to be satisfied by many low-pass filters and moment correcting filters. For example, for $a = a_4^I$ and $\Theta = \boldsymbol{\delta}$, the necessary and sufficient condition in Theorem 3.3.7 fails for both symmetry and complex symmetry. Therefore, it is natural to use more than two high-pass filters. The main purpose of this section is to construct tight framelet filter banks $\{a; b_1, b_2, b_3\}_\Theta$ with [complex] symmetry and with the shortest possible filter supports, when filters a and Θ are given in advance.

3.5.1 *Severable and Non-Severable Tight Framelet Filter Banks*

According to the following result on the symmetry centers, there are two types of tight framelet filter banks $\{a; b_1, b_2, b_3\}_\Theta$ with [complex] symmetry.

Proposition 3.5.1 *Let $\{a; b_1, b_2, b_3\}_\Theta$ be a tight framelet filter bank such that all the filters $\Theta, a, b_1, b_2, b_3 \in l_0(\mathbb{Z})$ are not identically zero and have symmetry (or complex symmetry by replacing S with $\mathbb{S}$ below):*

$$\mathsf{S}\Theta(z) = 1, \quad \mathsf{S}a(z) = \epsilon z^c, \quad \mathsf{S}b_1(z) = \epsilon_1 z^{c_1}, \quad \mathsf{S}b_2(z) = \epsilon_2 z^{c_2}, \quad \mathsf{S}b_3(z) = \epsilon_3 z^{c_3} \tag{3.5.1}$$

for some $\epsilon, \epsilon_1, \epsilon_2, \epsilon_3 \in \{-1, 1\}$ and $c, c_1, c_2, c_3 \in \mathbb{Z}$. Up to reordering of the high-pass filters b_1, b_2, b_3, one of the following must hold:

(i) $c_3 - c$ is even and $c_1 - c$, $c_2 - c$ are odd. Moreover, the following identities hold:

$$\Theta(z^2)a(z)a^\star(-z) + b_3(z)b_3^\star(-z) = 0, \quad b_1(z)b_1^\star(-z) + b_2(z)b_2^\star(-z) = 0. \tag{3.5.2}$$

(ii) all $c_3 - c$, $c_1 - c$, $c_2 - c$ are even integers.

Proof By the definition of a tight framelet filter bank $\{a; b_1, b_2, b_3\}_\Theta$, its perfect reconstruction condition is

$$\Theta(z^2)a(z)a^\star(z) + b_1(z)b_1^\star(z) + b_2(z)b_2^\star(z) + b_3(z)b_3^\star(z) = \Theta(z) \tag{3.5.3}$$

and

$$\Theta(z^2)a(z)a^\star(-z) + b_1(z)b_1^\star(-z) + b_2(z)b_2^\star(-z) + b_3(z)b_3^\star(-z) = 0. \tag{3.5.4}$$

It is trivial to deduce the identity $\mathsf{S}(\mathsf{u}(z)\mathsf{v}(z)) = \mathsf{S}\mathsf{u}(z)\mathsf{S}\mathsf{v}(z)$. Therefore, we conclude from (3.5.1) that

$$\mathsf{S}(\Theta(z^2)\mathsf{a}(z)\mathsf{a}^\star(-z)) = (-1)^c, \qquad \mathsf{S}(\mathsf{b}_\ell(z)\mathsf{b}_\ell^\star(-z)) = (-1)^{c_\ell}, \qquad \ell = 1,2,3.$$

If $(-1)^{c_\ell} \neq (-1)^c$ for all $\ell = 1,2,3$, moving all the terms involving $\mathsf{b}_1, \mathsf{b}_2, \mathsf{b}_3$ to the right-hand side of (3.5.4), then we must have $\Theta(z^2)\mathsf{a}(z)\mathsf{a}^\star(-z) = 0$, which contradicts our assumption. By the same argument, we see that there are exactly either one or three of $(-1)^{c_1}, (-1)^{c_2}, (-1)^{c_3}$ having the same value as $(-1)^c$. Without loss of generality, we assume $(-1)^{c_3} = (-1)^c \neq (-1)^{c_1} = (-1)^{c_2}$ which is item (i), or $(-1)^{c_1} = (-1)^{c_2} = (-1)^{c_3} = (-1)^c$ which is item (ii). For the symmetry centers satisfying item (i), by (3.5.4), since the two sides of

$$\Theta(z^2)\mathsf{a}(z)\mathsf{a}^\star(-z) + \mathsf{b}_3(z)\mathsf{b}_3^\star(-z) = -\mathsf{b}_1(z)\mathsf{b}_1^\star(-z) - \mathsf{b}_2(z)\mathsf{b}_2^\star(-z)$$

have different symmetry patterns $(-1)^c$ and $(-1)^{c+1}$, both sides of the above identity must vanish and therefore, (3.5.2) must hold. □

A tight framelet filter bank $\{a; b_1, b_2, b_3\}_\Theta$ is called *severable* if the equations in (3.5.2) are satisfied. Note that tight framelet filter banks $\{a; b_1, b_2, b_3\}_\Theta$ satisfying item (ii) of Proposition 3.5.1 can also be severable. Severable tight framelet filter banks $\{a; b_1, b_2, b_3\}_\Theta$ must have the following severable structure.

Theorem 3.5.2 *Let $a, \Theta, b_1, b_2, b_3 \in l_0(\mathbb{Z})$ be filters which are not identically zero. Then $\{a; b_1, b_2, b_3\}_\Theta$ is a tight framelet filter bank with [complex] symmetry (and real coefficients) satisfying (3.5.2) (i.e., severable) if and only if*

$$\Theta(z^2) = \boldsymbol{\theta}(z)\boldsymbol{\theta}^\star(-z), \quad \mathsf{a}(z) = \mathsf{d_a}(z)\mathring{\mathsf{a}}(z), \quad \mathsf{b}_3(z) = \mathsf{d_a}(z)\boldsymbol{\theta}(z)z\mathring{\mathsf{a}}^\star(-z) \tag{3.5.5}$$

and

$$\mathsf{b}_1(z)\mathsf{b}_1^\star(z) + \mathsf{b}_2(z)\mathsf{b}_2^\star(z) = \mathsf{p}(z), \qquad \mathsf{b}_1(z)\mathsf{b}_1^\star(-z) + \mathsf{b}_2(z)\mathsf{b}_2^\star(-z) = 0, \tag{3.5.6}$$

where $\boldsymbol{\theta}, \mathsf{d_a}, \mathring{\mathsf{a}}$ are Laurent polynomials with [complex] symmetry (and real coefficients) and the Laurent polynomial p is defined to be

$$\mathsf{p}(z) := \Theta(z) - \mathsf{d_a}(z)\mathsf{d_a^\star}(z)\boldsymbol{\theta}(z)[\mathring{\mathsf{a}}(z)\mathring{\mathsf{a}}^\star(z)\boldsymbol{\theta}^\star(-z) + \mathring{\mathsf{a}}(-z)\mathring{\mathsf{a}}^\star(-z)\boldsymbol{\theta}^\star(z)]. \tag{3.5.7}$$

Proof Sufficiency ($\Leftarrow$). By (3.5.5), we have

$$\Theta(z^2)\mathsf{a}(z)\mathsf{a}^\star(-z) = \boldsymbol{\theta}(z)\boldsymbol{\theta}^\star(-z)\mathsf{d_a}(z)\mathsf{d_a^\star}(-z)\mathring{\mathsf{a}}(z)\mathring{\mathsf{a}}^\star(-z) = -\mathsf{b}_3(z)\mathsf{b}_3^\star(-z).$$

Hence, the first identity in (3.5.2) holds. By the second identity in (3.5.6), we conclude that both (3.5.2) and (3.5.4) are satisfied. By (3.5.5), we also have

$$\Theta(z^2)\mathsf{a}(z)\mathsf{a}^\star(z) = \boldsymbol{\theta}(z)\boldsymbol{\theta}^\star(-z)\mathsf{d_a}(z)\mathsf{d_a^\star}(z)\mathring{\mathsf{a}}(z)\mathring{\mathsf{a}}^\star(z),$$

$$\mathsf{b}_3(z)\mathsf{b}_3^\star(z) = \boldsymbol{\theta}(z)\boldsymbol{\theta}^\star(z)\mathsf{d_a}(z)\mathsf{d_a^\star}(z)\mathring{\mathsf{a}}(-z)\mathring{\mathsf{a}}^\star(-z).$$

By the definition of p in (3.5.7), we conclude that $\Theta(z^2)\mathsf{a}(z)\mathsf{a}^\star(z) + \mathsf{b}_3(z)\mathsf{b}_3^\star(z) = \Theta(z) - \mathsf{p}(z)$. Hence, using the first identity in (3.5.6), we see that (3.5.3) must hold. Therefore, $\{a; b_1, b_2, b_3\}_\Theta$ is a tight framelet filter bank satisfying (3.5.2).

Necessity ($\Rightarrow$). Define $\mathsf{d_a} := \gcd(\mathsf{a}, \mathsf{b}_3)$. Since both a and b_3 have [complex] symmetry (and real coefficients), so does $\mathsf{d_a}$. Then we can write $\mathsf{a}(z) = \mathsf{d_a}(z)\mathring{\mathsf{a}}(z)$ and $\mathsf{b}_3(z) = \mathsf{d_a}(z)\mathring{\mathsf{b}}_3(z)$ for some Laurent polynomials $\mathring{\mathsf{a}}$ and $\mathring{\mathsf{b}}_3$ with [complex] symmetry (and real coefficients). Now the first equation in (3.5.2) holds if and only if

$$\Theta(z^2)\mathring{\mathsf{a}}(z)\mathring{\mathsf{a}}^\star(-z) + \mathring{\mathsf{b}}_3(z)\mathring{\mathsf{b}}_3^\star(-z) = 0. \tag{3.5.8}$$

Since $\gcd(\mathring{\mathsf{a}}, \mathring{\mathsf{b}}_3) = 1$, the above identity implies $\mathring{\mathsf{b}}_3(z) \mid \Theta(z^2)\mathring{\mathsf{a}}^\star(-z)$, that is,

$$\Theta(z^2)\mathring{\mathsf{a}}^\star(-z) = z^{-1}\boldsymbol{\theta}^\star(-z)\mathring{\mathsf{b}}_3(z) \tag{3.5.9}$$

for some Laurent polynomial $\boldsymbol{\theta}$ with [complex] symmetry (and real coefficients). Plugging (3.5.9) back to (3.5.8), we deduce that $\mathring{\mathsf{b}}_3(z)[z^{-1}\boldsymbol{\theta}^\star(-z)\mathring{\mathsf{a}}(z) + \mathring{\mathsf{b}}_3^\star(-z)] = 0$, from which we must have $\mathring{\mathsf{b}}_3^\star(-z) = -z^{-1}\boldsymbol{\theta}^\star(-z)\mathring{\mathsf{a}}(z)$, i.e., $\mathring{\mathsf{b}}_3(z) = z\boldsymbol{\theta}(z)\mathring{\mathsf{a}}^\star(-z)$. Therefore, $\mathsf{b}_3(z) = \mathsf{d_a}(z)\mathring{\mathsf{b}}_3(z) = \mathsf{d_a}(z)\boldsymbol{\theta}(z)z\mathring{\mathsf{a}}^\star(-z)$. Moreover, by (3.5.9),

$$\Theta(z^2)\mathring{\mathsf{a}}^\star(-z) = z^{-1}\boldsymbol{\theta}^\star(-z)\mathring{\mathsf{b}}_3(z) = z^{-1}\boldsymbol{\theta}^\star(-z)z\boldsymbol{\theta}(z)\mathring{\mathsf{a}}^\star(-z) = \boldsymbol{\theta}(z)\boldsymbol{\theta}^\star(-z)\mathring{\mathsf{a}}^\star(-z),$$

which forces $\Theta(z^2) = \boldsymbol{\theta}(z)\boldsymbol{\theta}^\star(-z)$ because a is not identically zero. This proves (3.5.5). We proved in the sufficiency part that (3.5.5) implies the first identity in (3.5.2) and $\Theta(z^2)\mathsf{a}(z)\mathsf{a}^\star(z) + \mathsf{b}_3(z)\mathsf{b}_3^\star(z) = \Theta(z) - \mathsf{p}(z)$. Now (3.5.6) follows directly from the perfect reconstruction conditions in (3.5.3) and (3.5.4). □

Due to Theorem 3.5.2, severable tight framelet filter banks $\{a;, b_1, b_2, b_3\}_\Theta$ can be easily constructed by the following algorithm.

Algorithm 3.5.3 *Let $a, \Theta \in l_0(\mathbb{Z})$ be filters having [complex] symmetry (and real coefficients) such that $\mathcal{M}_{a,\Theta}(z) \geqslant 0$ for all $z \in \mathbb{T}$, $\mathsf{S}\mathsf{a}(z) = \epsilon z^c$ and $\mathsf{S}\Theta = 1$, where $\epsilon \in \{-1, 1\}$ and $c \in \mathbb{Z}$. Replace S by $\mathbb{S}$ for complex symmetry.*

(S1) Construct a Laurent polynomial $\boldsymbol{\theta}$ with [complex] symmetry (and real coefficients) such that

$$\Theta(z^2) = \boldsymbol{\theta}(z)\boldsymbol{\theta}^\star(-z), \qquad \forall\, z \in \mathbb{C}\backslash\{0\}. \tag{3.5.10}$$

(S2) Write $\mathsf{a}(z) = \mathsf{d_a}(z)\mathring{\mathsf{a}}(z)$ by selecting a Laurent polynomial $\mathsf{d_a}$ with [complex] symmetry (and real coefficients) such that $\mathsf{d_a}(z) \mid \mathsf{a}(z)$.

(S3) Let p *be the Laurent polynomial in* (3.5.7) *and define* $\mathsf{q}(z^2) := \gcd(\mathsf{p}(z), \mathsf{p}(-z))$. *Find Laurent polynomials* $\mathsf{d_p}$ *and* b *with [complex] symmetry (and real coefficients) satisfying*

$$\mathsf{d_p}(z)\mathsf{d_p^\star}(z) = \frac{\mathsf{p}(z)}{\mathsf{q}(z^2)} \tag{3.5.11}$$

and

$$\mathsf{b}(z)\mathsf{b}^\star(z) + \mathsf{b}(-z)\mathsf{b}^\star(-z) = \mathsf{q}(z^2). \tag{3.5.12}$$

If there are no solutions of such desired Laurent polynomials $\mathsf{d_p}$ *and* b*, then restart the algorithm from (S1) by selecting other choices of* $\boldsymbol{\theta}$ *and* $\mathsf{d_a}$.

Define high-pass filters $b_1, b_2, b_3 \in l_0(\mathbb{Z})$ *by*

$$\mathsf{b}_1(z) = \mathsf{d_p}(z)\mathsf{b}(z), \quad \mathsf{b}_2(z) = \mathsf{d_p}(z)z\mathsf{b}^\star(-z), \quad \mathsf{b}_3(z) = \mathsf{d_a}(z)\boldsymbol{\theta}(z)z\mathring{\mathsf{a}}^\star(-z). \tag{3.5.13}$$

Then $\{a; b_1, b_2, b_3\}_\Theta$ *is a tight framelet filter bank with [complex] symmetry (and real coefficients) satisfying* (3.5.2) *(i.e., severable).*

Proof Obviously, by the definition in (3.5.13), all high-pass filters b_1, b_2, b_3 have [complex] symmetry (and real coefficients). By the first two identities in (3.5.13), it follows from (3.5.11) and (3.5.12) that

$$\mathsf{b}_1(z)\mathsf{b}_1^\star(z) + \mathsf{b}_2(z)\mathsf{b}_2^\star(z) = \mathsf{d_p}(z)\mathsf{d_p^\star}(z)[\mathsf{b}(z)\mathsf{b}^\star(z) + \mathsf{b}(-z)\mathsf{b}^\star(-z)] = \mathsf{p}(z)$$

and

$$\mathsf{b}_1(z)\mathsf{b}_1^\star(-z) + \mathsf{b}_2(z)\mathsf{b}_2^\star(-z) = \mathsf{d_p}(z)\mathsf{d_p^\star}(-z)[\mathsf{b}(z)\mathsf{b}^\star(-z) - \mathsf{b}(z)\mathsf{b}^\star(-z)] = 0.$$

That is, (3.5.6) holds. By definition, (3.5.5) trivially holds. By Theorem 3.5.2, we conclude that $\{a; b_1, b_2, b_3\}_\Theta$ is a tight framelet filter bank satisfying (3.5.2). □

The problem in (3.5.11) is well studied in Theorems 3.1.7 and 3.1.8. We often have either $\mathsf{p}(z) = \mathsf{q}(z^2)$ (e.g., this must hold if $\Theta(z) = \mathsf{d_a}(z) = 1$) and hence we can set $\mathsf{d_p}(z) = 1$, or $\mathsf{p}(z) = 1$ and hence, we can set $\mathsf{b}(z) = \sqrt{2}/2$. If $\Theta(z) = 1$, then we often set $\boldsymbol{\theta}(z) = 1$ as the default solution to $\Theta(z^2) = \boldsymbol{\theta}(z)\boldsymbol{\theta}^\star(-z)$.

Using the coset notation of the sequence $\mathsf{b}(z) = \mathsf{b}^{[0]}(z^2) + z\mathsf{b}^{[1]}(z^2)$ and noting

$$\mathsf{b}(z)\mathsf{b}^\star(z) + \mathsf{b}(-z)\mathsf{b}^\star(-z) = 2\mathsf{b}^{[0]}(z^2)(\mathsf{b}^{[0]}(z^2))^\star + 2\mathsf{b}^{[1]}(z^2)(\mathsf{b}^{[1]}(z^2))^\star,$$

the problem in (3.5.12) is equivalent to the sum of squares (SOS) problem:

$$\mathsf{b}^{[0]}(z)(\mathsf{b}^{[0]}(z))^\star + \mathsf{b}^{[1]}(z)(\mathsf{b}^{[1]}(z))^\star = \mathsf{q}(z)/2. \tag{3.5.14}$$

Suppose that b has [complex] symmetry with $\mathsf{Sb}(z) = \epsilon_{\mathsf{b}} z^{c_{\mathsf{b}}}$, where $\epsilon_{\mathsf{b}} \in \{-1, 1\}$ and $c_{\mathsf{b}} \in \mathbb{Z}$. If c_{b} is an even integer, then we can directly check that both $\mathsf{b}^{[0]}$ and $\mathsf{b}^{[1]}$ have [complex] symmetry; consequently, the SOS problem with symmetry in (3.5.14) has been well studied in Theorems 3.1.5 and 3.1.6. If c_{b} is an odd integer, then $\mathsf{b}^{[1]}(z) = \epsilon_{\mathsf{b}} z^{(c_{\mathsf{b}}-1)/2}(\mathsf{b}^{[0]}(z))^\star$ and all the solutions to (3.5.12) are given by

$$\mathsf{b}(z) = [\mathsf{u}(z^2) + \epsilon_{\mathsf{b}} z^{c_{\mathsf{b}}} \mathsf{u}^\star(z^2)]/2 \quad \text{or} \quad \mathsf{b}(z) = [\mathsf{u}(z^2) + \epsilon_{\mathsf{b}} z^{c_{\mathsf{b}}} \mathsf{u}(z^{-2})]/2 \tag{3.5.15}$$

with the first for complex symmetry and the second for symmetry, where u is any Laurent polynomial obtained via the Fejér-Riesz Lemma such that $\mathsf{u}(z)\mathsf{u}^\star(z) = \mathsf{q}(z)$. Usually, (3.5.12) has infinitely many solutions and we often choose a particular solution b of the equation in (3.5.12) with the shortest possible support.

The following result provides a necessary and sufficient condition for the existence of a Laurent polynomial solution $\boldsymbol{\theta}$ to the equation in (3.5.10).

Theorem 3.5.4 *Let Θ be a nontrivial Laurent polynomial.*

(i) If (3.5.10) *has a Laurent polynomial solution $\boldsymbol{\theta}$, then $\Theta^\star = \Theta$. Conversely, if $\Theta^\star = \Theta$, then* (3.5.10) *has a Laurent polynomial solution $\boldsymbol{\theta}$ with complex symmetry.*

(ii) If (3.5.10) *has a Laurent polynomial solution $\boldsymbol{\theta}$ with symmetry, then*

(1) $\Theta^\star = \Theta$ and Θ has real coefficients;
(2) $\mathsf{Z}(\Theta, x)$ is an even integer for every $-1 \leqslant x < 0$.

Moreover, if items (1) and (2) are satisfied, then (3.5.10) *must have a Laurent polynomial solution $\boldsymbol{\theta}$ with symmetry and real coefficients.*

For both (i) and (ii), the symmetry center of $\boldsymbol{\theta}$ satisfying (3.5.10) *must be an integer.*

Proof If $\Theta(z^2) = \boldsymbol{\theta}(z)\boldsymbol{\theta}^\star(-z)$, replacing z by $-z$, then $\Theta^\star(z^2) = \Theta^\star((-z)^2) = \boldsymbol{\theta}^\star(-z)\boldsymbol{\theta}(z) = \Theta(z^2)$. Hence, $\Theta^\star = \Theta$. Suppose that $\mathbb{S}\boldsymbol{\theta}(z) = \epsilon_{\boldsymbol{\theta}} z^{c_{\boldsymbol{\theta}}}$. Then (3.5.10) and $\mathbb{S}\Theta(z) = 1$ imply $(-1)^{c_{\boldsymbol{\theta}}} = \mathbb{S}\boldsymbol{\theta}(z)\mathbb{S}\boldsymbol{\theta}^\star(-z) = \mathbb{S}\Theta(z) = 1$, from which we have $c_{\boldsymbol{\theta}} \in 2\mathbb{Z}$. The proof of $c_{\boldsymbol{\theta}} \in 2\mathbb{Z}$ for the case of symmetry is the same.

Define a finite set $\mathsf{Z}_\Theta := \{z \in \mathbb{C}\backslash\{0\} \ : \ \Theta(z) = 0\}$. To prove item (i), we define

$$\boldsymbol{\theta}(z) := \left(\prod_{\zeta \in \mathsf{Z}_\Theta \cap \mathbb{T}} (z - \sqrt{\zeta})^{\mathsf{Z}(\Theta,\zeta)} \right) \left(\prod_{\zeta \in \mathsf{Z}_\Theta, 0<|\zeta|<1} \left[(z - \sqrt{\zeta})(z^{-1} - \overline{\sqrt{\zeta}}) \right]^{\mathsf{Z}(\Theta,\zeta)} \right),$$

where $\sqrt{\zeta} \in \mathbb{C}$ is a solution to $(\sqrt{\zeta})^2 = \zeta$. By item (ii) of Proposition 3.1.1, we conclude from the above definition of $\boldsymbol{\theta}$ that the Laurent polynomial $\boldsymbol{\theta}$ has essential complex symmetry. Therefore, multiplying some number from $\mathbb{T}$ with $\boldsymbol{\theta}$, we can assume that $\boldsymbol{\theta}$ has complex symmetry. Since $\Theta^\star = \Theta$, by item (ii) of Proposition 3.1.1 and the construction of the Laurent polynomial $\boldsymbol{\theta}$, the two Laurent polynomials $\Theta(z^2)$ and $\boldsymbol{\theta}(z)\boldsymbol{\theta}^\star(-z)$ must have the same zeros on $\mathbb{C}\backslash\{0\}$ and $\mathbb{S}\Theta(z) = \mathbb{S}(\boldsymbol{\theta}(z)\boldsymbol{\theta}^\star(-z)) = 1$, where we used $c_{\boldsymbol{\theta}} \in 2\mathbb{Z}$. Hence,

$\Theta(z^2) = \lambda\boldsymbol{\theta}(z)\boldsymbol{\theta}^\star(-z)$ for some constant $\lambda \in \mathbb{R}\backslash\{0\}$. If $\lambda > 0$, replace $\boldsymbol{\theta}$ by $\sqrt{\lambda}\boldsymbol{\theta}$; if $\lambda < 0$, replace $\boldsymbol{\theta}$ by $\sqrt{|\lambda|}z\boldsymbol{\theta}(z)$. Now it is straightforward to check that $\Theta(z^2) = \boldsymbol{\theta}(z)\boldsymbol{\theta}^\star(-z)$ and $\boldsymbol{\theta}$ has complex symmetry.

Necessity of item (ii) ($\Rightarrow$). We proved that $\Theta^\star = \Theta$ and hence $\mathbb{S}\Theta = 1$. Since $\boldsymbol{\theta}$ has symmetry, say, $\mathsf{S}\boldsymbol{\theta}(z) = \epsilon_{\boldsymbol{\theta}} z^{c_{\boldsymbol{\theta}}}$, then Θ must have symmetry $\mathsf{S}\Theta(z^2) = \mathsf{S}\boldsymbol{\theta}(z)\mathsf{S}\boldsymbol{\theta}^\star(-z) = (-1)^{c_{\boldsymbol{\theta}}}$. Since we proved that $c_{\boldsymbol{\theta}}$ must be an even integer, we have $\mathsf{S}\Theta = \mathbb{S}\Theta = 1$. Therefore, the Laurent polynomial Θ must have real coefficients.

Plugging $z = ix, x \in \mathbb{R}\backslash\{0\}$ into $\Theta(z^2) = \boldsymbol{\theta}(z)\boldsymbol{\theta}^\star(-z) = \boldsymbol{\theta}(z)\overline{\boldsymbol{\theta}(-\bar{z}^{-1})}$, we have

$$\Theta(-x^2) = \boldsymbol{\theta}(ix)\overline{\boldsymbol{\theta}((ix)^{-1})} = \boldsymbol{\theta}(ix)\overline{\boldsymbol{\theta}(ix)}/\overline{\mathsf{S}\boldsymbol{\theta}(ix)} = |\boldsymbol{\theta}(ix)|^2/\overline{\mathsf{S}\boldsymbol{\theta}(ix)}.$$

Hence, $\mathsf{Z}(\Theta, -x^2) \in 2\mathbb{N}_0$ for all $x > 0$. This proves the necessity part of item (ii).

Sufficiency of item (ii) ($\Leftarrow$). We define $\mathsf{Z}_{\mathbb{T}} := \mathbb{T}\backslash\{-1, 1\}$ and

$$\boldsymbol{\theta}(z) := \boldsymbol{\theta}_1(z)\boldsymbol{\theta}_2(z)\boldsymbol{\theta}_3(z)\boldsymbol{\theta}_4(z)\left(\prod_{\zeta \in \mathsf{Z}_\Theta \cap \mathsf{Z}_{\mathbb{T}}} \left[(z - \sqrt{\zeta})(z - \overline{\sqrt{\zeta}})\right]^{\mathsf{Z}(\Theta,\zeta)}\right),$$

where $\boldsymbol{\theta}_1(z) := (z-1)^{\mathsf{Z}(\Theta,1)}(z^2+1)^{\mathsf{Z}(\Theta,-1)/2}$ and

$$\boldsymbol{\theta}_2(z) := \prod_{\zeta \in \mathsf{Z}_\Theta, 0<\zeta<1} \left[(z - \sqrt{\zeta})(z^{-1} - \sqrt{\zeta})\right]^{\mathsf{Z}(\Theta,\zeta)},$$

$$\boldsymbol{\theta}_3(z) := \prod_{\zeta \in \mathsf{Z}_\Theta, -1<\zeta<0} \left[(z^2 - \zeta)(z^{-2} - \zeta)\right]^{\mathsf{Z}(\Theta,\zeta)/2},$$

$$\boldsymbol{\theta}_4(z) := \prod_{\zeta \in \mathsf{Z}_\Theta, 0<|\zeta|<1} \left[(z^2 - 2\operatorname{Re}(\sqrt{\zeta})z + |\zeta|)(z^{-2} - 2\operatorname{Re}(\sqrt{\zeta})z^{-1} + |\zeta|)\right]^{\mathsf{Z}(\Theta,\zeta)}.$$

Since $\mathsf{Z}(\Theta, x) \in 2\mathbb{N}_0$ for all $x \in [-1, 0)$, the above $\boldsymbol{\theta}$ is a well-defined Laurent polynomial with real coefficients. By $\Theta^\star = \Theta$ and Θ has real coefficients, we conclude from item (i) of Proposition 3.1.1 that $\boldsymbol{\theta}$ must have symmetry. By the same argument as in item (i), we see that $\Theta(z^2) = \lambda\boldsymbol{\theta}(z)\boldsymbol{\theta}^\star(-z)$ for some constant $\lambda \in \mathbb{R}\backslash\{0\}$. After multiplying a monomial with $\boldsymbol{\theta}$, we have $\Theta(z^2) = \boldsymbol{\theta}(z)\boldsymbol{\theta}^\star(-z)$. ☐

Though severable tight framelet filter banks $\{a; b_1, b_2, b_3\}_\Theta$ with [complex] symmetry can be easily obtained by Algorithm 3.5.3, without the extra severable condition in (3.5.2), the construction of non-severable tight framelet filter banks is often much more difficult. Here we present a heuristic algorithm to construct all tight framelet filter banks $\{a; b_1, b_2, b_3\}_\Theta$ with [complex] symmetry.

Algorithm 3.5.5 *Let $a, \Theta \in l_0(\mathbb{Z})$ be filters having [complex] symmetry (and real coefficients) such that $\mathcal{M}_{a,\Theta}(z) \geqslant 0$ for all $z \in \mathbb{T}$, $\mathsf{S}a(z) = \epsilon z^c$ and $\mathsf{S}\Theta(z) = 1$,* *where* $\epsilon \in \{-1, 1\}$ *and* $c \in \mathbb{Z}$ *[Replace* S *by* $\mathbb{S}$ *for complex symmetry]. Choose a nonnegative integer* n_b *such that* (3.3.5) *holds.*

(S1) Select $c_3 \in \{c, c+1, c+2, c+3\}$ *and* $\epsilon_3 \in \{-1, 1\}$.

(S2) Parameterize a filter $\mathring{b}_3$ *such that* $\mathsf{S}\mathring{\mathsf{b}}_3(z) = (-1)^{n_b}\epsilon_3 z^{c_3+n_b}$ *and* $\operatorname{len}(\mathring{b}_3) \leqslant \operatorname{len}(a) + \operatorname{len}(\Theta) - n_b$. *Define*

$$\mathsf{A}(z) := \frac{\Theta(z) - \Theta(z^2)\mathsf{a}(z)\mathsf{a}^\star(z)}{(1-z)^{n_b}(1-z^{-1})^{n_b}} - \mathring{\mathsf{b}}_3(z)\mathring{\mathsf{b}}_3^\star(z),$$

$$\mathsf{B}(z) := -\Theta(z^2)\frac{\mathsf{a}(z)}{(1+z)^{n_b}}\frac{\mathsf{a}^\star(-z)}{(1-z^{-1})^{n_b}} - \mathring{\mathsf{b}}_3(z)\mathring{\mathsf{b}}_3^\star(-z).$$

(S3) Define $\mathsf{p}(z^2) := [\mathsf{A}(z)\mathsf{A}(-z) - \mathsf{B}(z)\mathsf{B}(-z)]$ *and obtain a Laurent polynomial* q_p *by item (4) of Theorem 3.1.7 for complex symmetry or by item (5) of Theorem 3.1.8 for symmetry. Solve the nonlinear equations induced by* $\mathsf{q}_\mathsf{p}(z)\mathsf{q}_\mathsf{p}^\star(z) = \mathsf{p}(z)$ *to determine the coefficients of the filter* $\mathring{b}_3$.

(S4) Apply Algorithm 3.3.6 to derive high-pass filters b_1, b_2 *with [complex] symmetry satisfying* (3.3.35) *and* (3.3.33).

Then $\{a; b_1, b_2, b_3\}_\Theta$ *is a tight framelet filter bank with [complex] symmetry (and real coefficients), where* $\mathsf{b}_3(z) := (1 - z^{-1})^{n_b}\mathring{\mathsf{b}}_3(z)$.

3.5.2 Examples of (Non)severable Tight Framelet Filter Banks

For any given filters $a, \Theta \in l_0(\mathbb{Z})$ with [complex] symmetry such that $\mathcal{M}_{a,\Theta}(z) \geqslant 0$ for all $z \in \mathbb{T}$, severable tight framelet filter banks $\{a; b_1, b_2, b_3\}_\Theta$ with [complex] symmetry can be easily constructed by Algorithm 3.5.3. By solving nonlinear equations, non-severable tight framelet filter banks $\{a; b_1, b_2, b_3\}_\Theta$ with [complex] symmetry can be also constructed by Algorithm 3.5.5. We now provide a few examples to illustrate Algorithms 3.5.3 and 3.5.5.

Example 3.5.1 Let $a = a_2^B = \{\underline{\frac{1}{4}}, \frac{1}{2}, \frac{1}{4}\}_{[0,2]}$ be the B-spline filter of order 2. Applying Algorithm 3.5.3 with $\Theta = \boldsymbol{\delta}$ and setting $\boldsymbol{\theta} = -1$ and $\mathsf{d_a}(z) = 1$, we have $\mathsf{p}(z) = \mathsf{q}(z^2)$ with $\mathsf{q}(z) := (1 - z^{-1})(1 - z)/8$. Selecting $\mathsf{d_p} = 1$ and $\mathsf{b}(z) = (z^{-1} - z)/4$, we have

$$\mathsf{b}_1(z) = \mathsf{d_p}(z)\mathsf{b}(z) = \tfrac{1}{4}(1-z)(z^{-1}+1) = \{\tfrac{1}{4}, \underline{\mathbf{0}}, -\tfrac{1}{4}\}_{[-1,1]},$$

$$\mathsf{b}_2(z) = \mathsf{d_p}(z)z\mathsf{b}^\star(-z) = z\mathsf{b}_1(z) = \tfrac{1}{4}(1-z)(1+z) = \{\underline{\tfrac{1}{4}}, 0, -\tfrac{1}{4}\}_{[0,2]},$$

$$\mathsf{b}_3(z) = \mathsf{d_a}(z)\boldsymbol{\theta}(z)z\mathring{\mathsf{a}}^\star(-z) = \tfrac{1}{4}(1-z^{-1})(1-z) = \{-\tfrac{1}{4}, \underline{\tfrac{1}{2}}, -\tfrac{1}{4}\}_{[-1,1]}.$$

Then $\{a; b_1, b_2, b_3\}$ is a real-valued severable tight framelet filter bank with $\mathrm{vm}(b_1) = \mathrm{vm}(b_2) = 1$, $\mathrm{vm}(b_3) = 2$ and $\mathsf{Sb}_1(z) = -1$, $\mathsf{Sb}_2(z) = -z^2$, $\mathsf{Sb}_3(z) = 1$. By calculation, we have $\mathrm{Var}(b_1) = \mathrm{Var}(b_2) = 1$, $\mathrm{Var}(b_3) = \frac{1}{3}$ and $\mathrm{Fsi}(a, b_1) = \mathrm{Fsi}(a, b_2) \approx 0.48795$, $\mathrm{Fsi}(a, b_3) \approx 0.00857$, $\mathrm{Fsi}(b_1, b_2) = 1$, $\mathrm{Fsi}(b_1, b_3) = \mathrm{Fsi}(b_2, b_3) \approx 0.48795$.

We can also apply (3.5.15) using the Fejér-Riesz Lemma to choose $\mathsf{b}(z) = -\frac{\sqrt{2}}{8}(z^{-1} + 1)(1 - z)^2$, leading to high-pass filters with longer supports:

$$b_1 = \tfrac{\sqrt{2}}{8}\{-1, \underline{\mathbf{1}}, 1, -1\}_{[-1,2]}, \quad b_2 = \tfrac{\sqrt{2}}{8}\{-1, \underline{\mathbf{-1}}, 1, 1\}_{[-1,2]}, \quad b_3 = \tfrac{1}{4}\{-1, \underline{\mathbf{2}}, -1\}_{[-1,1]}.$$

Take $\mathsf{d_a}(z) = 1 + z$ instead in Algorithm 3.5.3. Then $\mathring{\mathsf{a}}(z) = \frac{1}{4}(1 + z)$, $\mathsf{q}(z) = 1$, and $\mathsf{p}(z) = \frac{1}{4}(1 - z^{-1})(1 - z)$. Selecting $\mathsf{d_p}(z) = \frac{1}{2}(z^{-1} - 1)$ and $\mathsf{b}(z) = \frac{\sqrt{2}}{2}$, we have

$$\mathsf{b}_1(z) = \mathsf{d_p}(z)\mathsf{b}(z) = \tfrac{\sqrt{2}}{4}(z^{-1} - 1) = \tfrac{\sqrt{2}}{4}\{1, \underline{\mathbf{-1}}\}_{[-1,0]},$$

$$\mathsf{b}_2(z) = \mathsf{d_p}(z)z\mathsf{b}^\star(-z) = z\mathsf{b}_1(z) = \tfrac{\sqrt{2}}{4}(z - 1) = \tfrac{\sqrt{2}}{4}\{\underline{\mathbf{1}}, -1\}_{[0,1]},$$

$$\mathsf{b}_3(z) = \mathsf{d_a}(z)\boldsymbol{\theta}(z)z\mathring{\mathsf{a}}^\star(-z) = \tfrac{1}{4}(z^2 - 1) = \{\underline{\mathbf{\tfrac{1}{4}}}, 0, -\tfrac{1}{4}\}_{[0,2]}.$$

Then $\{a; b_1, b_2, b_3\}$ is a severable tight framelet filter bank with $\mathrm{vm}(b_1) = \mathrm{vm}(b_2) = \mathrm{vm}(b_3) = 1$ and $\mathsf{Sb}_1(z) = -z^{-1}$, $\mathsf{Sb}_2(z) = -z$, $\mathsf{Sb}_3(z) = -z^2$. By calculation, we have $\mathrm{Var}(b_1) = \mathrm{Var}(b_2) = \frac{1}{4}$, $\mathrm{Var}(b_3) = 1$ and $\mathrm{Fsi}(a, b_1) = \mathrm{Fsi}(a, b_2) \approx 0.1952$, $\mathrm{Fsi}(a, b_3) \approx 0.488$, $\mathrm{Fsi}(b_1, b_2) = 1$, $\mathrm{Fsi}(b_1, b_3) = \mathrm{Fsi}(b_2, b_3) \approx 0.66667$.

However, applying Algorithm 3.5.3 with $\Theta = \{-\frac{1}{6}, \underline{\mathbf{\frac{4}{3}}}, -\frac{1}{6}\}_{[-1,1]}$, though the necessary condition $\mathcal{M}_{a,\Theta}(z) \geqslant 0$ for all $z \in \mathbb{T}$ is satisfied, Algorithm 3.5.3 is unable to find any solution regardless of the choices of $\boldsymbol{\theta}(z) = \frac{\sqrt{15}}{3} \pm \frac{\sqrt{6}}{6}(z^{-1} + z)$ (or $\boldsymbol{\theta}(z) = -\frac{\sqrt{15}}{3} \pm \frac{\sqrt{6}}{6}(z^{-1} + z)$) and $\mathsf{d_a}(z) \mid \mathsf{a}(z)$ (because (3.5.11) always fails). That is, a severable tight framelet filter bank $\{a; b_1, b_2, b_3\}_\Theta$ with either symmetry or complex symmetry cannot be derived from the filters a and Θ.

For $\Theta = \{-\frac{1}{6}, \underline{\mathbf{\frac{4}{3}}}, -\frac{1}{6}\}_{[-1,1]}$, we now employ Algorithm 3.5.5 to construct a real-valued non-severable tight framelet filter bank $\{a; b_1, b_2, b_3\}_\Theta$ with symmetry. Considering $n_b = 2$ and $\mathring{\mathsf{b}}_3(z) = tz$ in Algorithm 3.5.5 with $\Theta = \{-\frac{1}{6}, \underline{\mathbf{\frac{4}{3}}}, -\frac{1}{6}\}_{[-1,1]}$, we find that the condition in item (3) of Algorithm 3.5.5 is satisfied with $t = \pm\frac{\sqrt{10}}{10}$. Taking $t = -\frac{\sqrt{10}}{10}$ and applying Algorithm 3.5.5, we have $b_3 = \frac{\sqrt{10}}{10}\{-1, \underline{\mathbf{2}}, -1\}_{[-1,1]}$ and

$$b_1 = \tfrac{\sqrt{15}}{120}\{-5, 2, \underline{\mathbf{6}}, 2, -5\}_{[-2,2]}, \qquad b_2 = \tfrac{1}{8}\{1, -2, \underline{\mathbf{0}}, 2, -1\}_{[-2,2]}.$$

Then $\{a; b_1, b_2, b_3\}_\Theta$ is a non-severable tight framelet filter bank with $\mathrm{vm}(b_1) = \mathrm{vm}(b_3) = 2$, $\mathrm{vm}(b_2) = 3$ and $\mathsf{Sb}_1(z) = \mathsf{Sb}_3(z) = 1$, $\mathsf{Sb}_2(z) = -1$. By calculation, $\mathrm{Var}(b_1) \approx 2.212765$, $\mathrm{Var}(b_2) \approx 1.6$, $\mathrm{Var}(b_3) \approx 0.33333$ and $\mathrm{Fsi}(a, b_1) \approx$

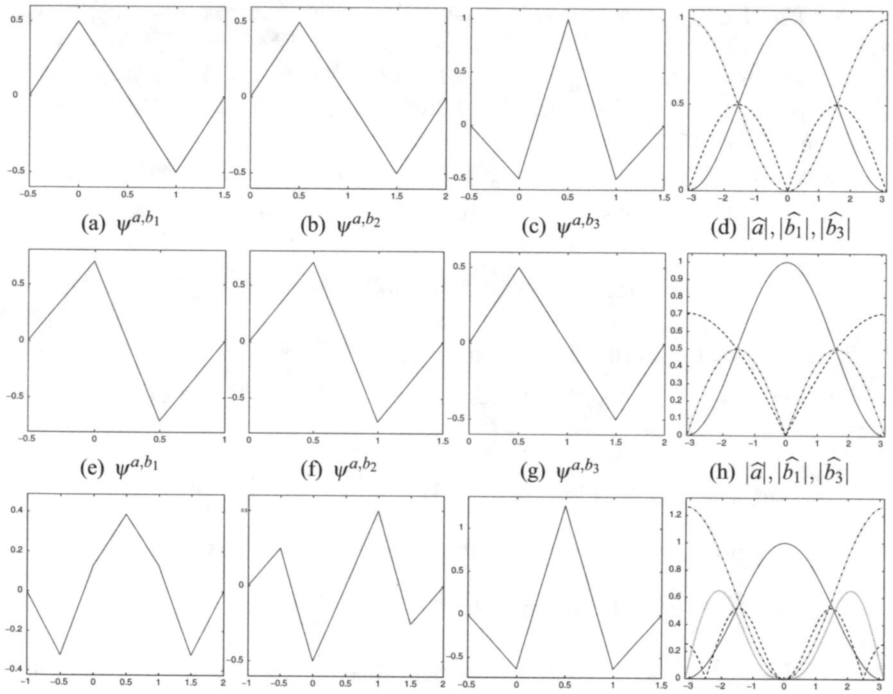

(a) ψ^{a,b_1} (b) ψ^{a,b_2} (c) ψ^{a,b_3} (d) $|\widehat{a}|, |\widehat{b_1}|, |\widehat{b_3}|$

(e) ψ^{a,b_1} (f) ψ^{a,b_2} (g) ψ^{a,b_3} (h) $|\widehat{a}|, |\widehat{b_1}|, |\widehat{b_3}|$

(i) ψ^{a,b_1} (j) ψ^{a,b_2} (k) ψ^{a,b_3} (l) $|\widehat{a}|, |\widehat{b_1}|, |\widehat{b_2}|, |\widehat{b_3}|$

Fig. 3.15 Tight framelet filter banks $\{a; b_1, b_2, b_3\}_\Theta$ derived from $a = a_2^B$ and Θ in Example 3.5.1. (**a**)–(**d**) are for $\Theta = \boldsymbol{\delta}$ and $\mathsf{d_a}(z) = 1$; (**e**)–(**h**) are for $\Theta = \boldsymbol{\delta}$ and $\mathsf{d_a}(z) = 1 + z$; (**i**)–(**l**) are for $\Theta = \{-\frac{1}{6}, \underline{\mathbf{\tfrac{4}{3}}}, -\frac{1}{6}\}_{[-1,1]}$ by Algorithm 3.5.5. (**d**), (**h**), and (**l**) are the magnitudes of $\widehat{a}$ (in solid line), $\widehat{b_1}$ (in dashed line), $\widehat{b_2}$ (in dotted line) and $\widehat{b_3}$ (in dashed-dotted line) on the interval $[-\pi, \pi]$. Note that $|\widehat{b_2}| = |\widehat{b_1}|$ for (**d**) and (**h**)

0.441223, $\operatorname{Fsi}(a, b_2) \approx 0.169882$, $\operatorname{Fsi}(a, b_3) \approx 0.0857143$, $\operatorname{Fsi}(b_1, b_2) \approx 0.621175$, $\operatorname{Fsi}(b_1, b_3) \approx 0.382888$, $\operatorname{Fsi}(b_2, b_3) \approx 0.713506$. See Fig. 3.15 for their framelet functions.

Example 3.5.2 Let $a = a_3^B = \{\underline{\mathbf{\tfrac{1}{8}}}, \frac{3}{8}, \frac{3}{8}, \frac{1}{8}\}_{[0,3]}$ be the B-spline filter of order 3. Applying Algorithm 3.5.3 with $\Theta = \boldsymbol{\delta}$ and setting $\boldsymbol{\theta}(z) = 1$ and $\mathsf{d_a}(z) = 1$, we have $\mathsf{p}(z) = \mathsf{q}(z^2)$ with $\mathsf{q}(z) := \frac{3}{16}(1 - z^{-1})(1 - z)$. Selecting $\mathsf{d_p}(z) = 1$ and $\mathsf{b}(z) = \frac{\sqrt{6}}{8}(z - z^{-1})$, we have

$$\mathsf{b}_1(z) = \mathsf{d_p}(z)\mathsf{b}(z) = \frac{\sqrt{6}}{8}(z - z^{-1}) = \frac{\sqrt{6}}{8}\{-1, \underline{\mathbf{0}}, 1\}_{[-1,1]},$$

$$\mathsf{b}_2(z) = \mathsf{d_p}(z)z\mathsf{b}^\star(-z) = z\mathsf{b}_1(z) = \frac{\sqrt{6}}{8}(z^2 - 1) = \frac{\sqrt{6}}{8}\{\underline{\mathbf{-1}}, 0, 1\}_{[0,2]},$$

$$\mathsf{b}_3(z) = \mathsf{d_a}(z)\boldsymbol{\theta}(z)z\mathring{\mathsf{a}}^\star(-z) = \frac{1}{8}z(1 - z^{-1})^3 = \{-\frac{1}{8}, \frac{3}{8}, \underline{\mathbf{-\tfrac{3}{8}}}, \frac{1}{8}\}_{[-2,1]}.$$

Then $\{a; b_1, b_2, b_3\}$ is a severable tight framelet filter bank with $\operatorname{vm}(b_1) = \operatorname{vm}(b_2) = 1$, $\operatorname{vm}(b_3) = 3$ and $\mathsf{Sb}_1(z) = -1, \mathsf{Sb}_2(z) = -z^2, \mathsf{Sb}_3(z) = -z^{-1}$. By calculation, we have $\operatorname{Var}(b_1) = \operatorname{Var}(b_2) = 1, \operatorname{Var}(b_3) = \frac{9}{20}$ and $\operatorname{Fsi}(a, b_1) = \operatorname{Fsi}(a, b_2) \approx 0.376051$, $\operatorname{Fsi}(a, b_3) \approx 0.021645$, $\operatorname{Fsi}(b_1, b_2) = 1$, $\operatorname{Fsi}(b_1, b_3) = \operatorname{Fsi}(b_2, b_3) \approx 0.376051$.

Take $\mathsf{d_a}(z) = 1 + z$ and $\boldsymbol{\theta} = -1$ in Algorithm 3.5.3. Then $\mathring{\mathsf{a}}(z) = (1 + z)^2/8$, $\mathsf{p}(z) = \frac{1}{32}(1 - z^{-1})(1 - z)(z^{-2} + 4z^{-1} + 14 + 4z + z^2)$ and $\mathsf{q}(z) = 1$. The condition in (3.5.11) can only hold for symmetry (not for complex symmetry). Selecting $\mathsf{d_p}(z) = \frac{\sqrt{2}}{8}(1 - z^{-1})[z^{-1} + (2 + 2\sqrt{2}i) + z]$ and $\mathsf{b}(z) = \frac{\sqrt{2}}{2}$, we have

$$\mathsf{b}_1(z) = \mathsf{d_p}(z)\mathsf{b}(z) = \{-\tfrac{1}{8}, \underline{-\tfrac{\mathbf{1+2\sqrt{2}i}}{\mathbf{8}}}, \tfrac{1+2\sqrt{2}i}{8}, \tfrac{1}{8}\}_{[-2,1]},$$

$$\mathsf{b}_2(z) = \mathsf{d_p}(z)z\mathsf{b}^\star(-z) = z\mathsf{b}_1(z) = \{-\tfrac{1}{8}, \underline{-\tfrac{\mathbf{1+2\sqrt{2}i}}{\mathbf{8}}}, \tfrac{1+2\sqrt{2}i}{8}, \tfrac{1}{8}\}_{[-1,2]},$$

$$\mathsf{b}_3(z) = \mathsf{d_a}(z)\boldsymbol{\theta}(z)z\mathring{\mathsf{a}}^\star(-z) = \{\tfrac{1}{8}, \underline{-\tfrac{\mathbf{1}}{\mathbf{8}}}, -\tfrac{1}{8}, \tfrac{1}{8}\}_{[-1,2]}.$$

Then $\{a; b_1, b_2, b_3\}$ is a severable tight framelet filter bank with $\operatorname{vm}(b_1) = \operatorname{vm}(b_2) = 1$, $\operatorname{vm}(b_3) = 2$ and $\mathsf{Sb}_1(z) = -z^{-1}, \mathsf{Sb}_2(z) = -z, \mathsf{Sb}_3(z) = z$. By calculation, we have $\operatorname{Var}(b_1) = \operatorname{Var}(b_2) = \frac{9}{20}, \operatorname{Var}(b_3) = \frac{5}{4}$ and $\operatorname{Fsi}(a, b_1) = \operatorname{Fsi}(a, b_2) \approx 0.239373$, $\operatorname{Fsi}(a, b_3) \approx 0.174078$, $\operatorname{Fsi}(b_1, b_2) = 1$, $\operatorname{Fsi}(b_1, b_3) = \operatorname{Fsi}(b_2, b_3) \approx 0.838473$.

Let $\Theta = \{-\frac{1}{4}, \underline{\frac{\mathbf{3}}{\mathbf{2}}}, -\frac{1}{4}\}_{[-1,1]}$. Then $\boldsymbol{\theta}(z) = -\sqrt{2} - (z^{-1} + z)/2$ satisfies $\Theta(z^2) = \boldsymbol{\theta}(z)\boldsymbol{\theta}^\star(-z)$. Take $\mathsf{d_a}(z) = 1 + z$ in Algorithm 3.5.3. Then $\mathring{\mathsf{a}}(z) = (1 + z)^2/8$ and

$$\mathsf{p}(z) = \tfrac{2-\sqrt{2}}{64}(1 - z^{-1})^2(1 - z)^2[(z^{-2} + z^2) + 6(z^{-1} + z) + (18 + 8\sqrt{2})], \quad \mathsf{q}(z) = 1.$$

The condition in (3.5.11) can only hold for symmetry (not for complex symmetry). Selecting $\mathsf{b}(z) = \sqrt{2}/2$ and

$$\mathsf{d_p}(z) = \tfrac{\sqrt{4+2\sqrt{2}}}{16}(1 - z^{-1})^2[(\sqrt{2} - 2)(1 + z^2) + \sqrt{2} - 4 - \sqrt{14 - 8\sqrt{2}}\, i],$$

Algorithm 3.5.3 yields $\mathsf{b}_2(z) = \mathsf{d_p}(z)z\mathsf{b}^\star(-z) = z\mathsf{b}_1(z)$,

$$\mathsf{b}_1(z) = \mathsf{d_p}(z)\mathsf{b}(z) = \tfrac{\sqrt{4+2\sqrt{2}}}{16}\{1 - \sqrt{2}, -1 - \sqrt{7 - 4\sqrt{2}}\, i, \underline{\mathbf{2\sqrt{2} + 2\sqrt{7 - 4\sqrt{2}}}},$$
$$-1 - \sqrt{7 - 4\sqrt{2}}\, i, 1 - \sqrt{2}\}_{[-2,2]},$$

$$\mathsf{b}_3(z) = \mathsf{d_a}(z)\boldsymbol{\theta}(z)z\mathring{\mathsf{a}}^\star(-z) = \{-\tfrac{1}{16}, \tfrac{1-2\sqrt{2}}{16}, \underline{\tfrac{\mathbf{\sqrt{2}}}{\mathbf{8}}}, \tfrac{\sqrt{2}}{8}, \tfrac{1-2\sqrt{2}}{16}, -\tfrac{1}{16}\}_{[-2,3]}.$$

Then $\{a; b_1, b_2, b_3\}_\Theta$ is a severable tight framelet filter bank with $\operatorname{vm}(b_1) = \operatorname{vm}(b_2) = \operatorname{vm}(b_3) = 2$ and $\mathsf{Sb}_1(z) = 1, \mathsf{Sb}_2(z) = z^2, \mathsf{Sb}_3(z) = z$. By calculation, $\operatorname{Var}(b_1) = \operatorname{Var}(b_2) \approx 0.329252$, $\operatorname{Var}(b_3) = 1.27780$ and $\operatorname{Fsi}(a, b_1) = \operatorname{Fsi}(a, b_2) \approx$

0.0894569, $\mathrm{Fsi}(a, b_3) \approx 0.324434$, $\mathrm{Fsi}(b_1, b_2) = 1$, $\mathrm{Fsi}(b_1, b_3) = \mathrm{Fsi}(b_2, b_3) \approx$ 0.531843.

Considering $n_b = 2$ and $\mathring{\mathsf{b}}_3(z) = tz(1+z)$ in Algorithm 3.5.5 with $\Theta = \{-\frac{1}{4}, \underline{\frac{3}{2}}, -\frac{1}{4}\}_{[-1,1]}$, we find that the condition in item (3) of Algorithm 3.5.5 is satisfied with $t = \pm\sqrt{782-544\sqrt{2}}/136$. Taking $t = -\sqrt{782-544\sqrt{2}}/136$ and applying Algorithm 3.5.5, we have

$$\mathsf{b}_1(z) = \tfrac{2-\sqrt{2}}{16} z^{-2}(z-1)^3[1+(4+2\sqrt{2})z+z^2],$$

$$\mathsf{b}_2(z) = -\frac{\sqrt{391-272\sqrt{2}}(5+2\sqrt{2})}{4624} z^{-2}(z-1)^2(z+1)[17+(32+28\sqrt{2})z+17z^2],$$

$$\mathsf{b}_3(z) = -\frac{\sqrt{782-544\sqrt{2}}}{136} z^{-1}(z-1)^2(z+1).$$

Then $\{a; b_1, b_2, b_3\}_\Theta$ is a real-valued non-severable tight framelet filter bank with $\mathrm{vm}(b_1) = 3$, $\mathrm{vm}(b_2) = \mathrm{vm}(b_3) = 2$ and $\mathsf{Sb}_1(z) = -z$, $\mathsf{Sb}_2(z) = \mathsf{Sb}_3(z) = z$. Note $\mathrm{fsupp}(b_1) = \mathrm{fsupp}(b_2) = [-2, 3]$ and $\mathrm{fsupp}(b_3) = [-1, 2]$. By calculation, we have $\mathrm{Var}(b_1) \approx 0.3728640$, $\mathrm{Var}(b_2) \approx 1.16659$, $\mathrm{Var}(b_3) \approx 1.25$ and $\mathrm{Fsi}(a, b_1) \approx$ 0.009749, $\mathrm{Fsi}(a, b_2) \approx 0.03404$, $\mathrm{Fsi}(a, b_3) \approx 0.174078$, $\mathrm{Fsi}(b_1, b_2) \approx 0.2211$, $\mathrm{Fsi}(b_1, b_3) \approx 0.1482$, $\mathrm{Fsi}(b_2, b_3) \approx 0.08977$. See Fig. 3.16 for their framelet functions.

Example 3.5.3 Let $a = a_4^B = \{\underline{\frac{1}{16}}, \frac{1}{4}, \frac{3}{8}, \frac{1}{4}, \frac{1}{16}\}_{[0,4]}$ be the B-spline filter of order 4. Let $\Theta = \boldsymbol{\delta}$. According to Theorem 3.3.7, it is impossible to construct a tight framelet filter bank $\{a, b_1, b_2\}$ with [complex] symmetry since there does not exist a desired Laurent polynomial d_{n_b} with [complex] symmetry in item (i) of Theorem 3.3.7. Applying Algorithm 3.5.3 with $\Theta = \boldsymbol{\delta}$ and setting $\boldsymbol{\theta}(z) = 1$ and $\mathsf{d_a}(z) = 1$, we have $\mathsf{p}(z) = \mathsf{q}(z^2)$ with $\mathsf{q}(z) := (1-z^{-1})(1-z)(z^{-1}+30+z)/128$. Select $\mathsf{d_p}(z) = 1$ and $\mathsf{b}(z) = (1-z^2)(1+2\sqrt{7}z+z^2)/16$, where we directly solved the SOS problem (3.5.14) with symmetry instead of using the Fejér-Riesz Lemma and (3.5.15). Then

$$\mathsf{b}_1(z) = \mathsf{d_p}(z)\mathsf{b}(z) = \mathsf{b}(z) = \{\underline{\tfrac{1}{16}}, \tfrac{\sqrt{7}}{8}, 0, -\tfrac{\sqrt{7}}{8}, -\tfrac{1}{16}\}_{[0,4]},$$

$$\mathsf{b}_2(z) = \mathsf{d_p}(z)z\mathsf{b}^\star(-z) = z\mathsf{b}^\star(-z) = \{-\tfrac{1}{16}, \tfrac{\sqrt{7}}{8}, 0, \underline{-\tfrac{\sqrt{7}}{8}}, \tfrac{1}{16}\}_{[-3,1]},$$

$$\mathsf{b}_3(z) = \mathsf{d_a}(z)\boldsymbol{\theta}(z)z\mathring{\mathsf{a}}^\star(-z) = \tfrac{1}{16}z^{-3}(1-z)^4 = \{\tfrac{1}{16}, -\tfrac{1}{4}, \tfrac{3}{8}, \underline{-\tfrac{1}{4}}, \tfrac{1}{16}\}_{[-3,1]}.$$

Then $\{a; b_1, b_2, b_3\}$ is a severable tight framelet filter bank with $\mathrm{vm}(b_1) = \mathrm{vm}(b_2) = 1$, $\mathrm{vm}(b_3) = 4$ and $\mathsf{Sb}_1(z) = -z^4$, $\mathsf{Sb}_2(z) = -z^{-2}$, $\mathsf{Sb}_3(z) = z^2$. By calculation, we have $\mathrm{Var}(b_1) = \mathrm{Var}(b_2) = \frac{32}{29}$, $\mathrm{Var}(b_3) = \frac{4}{7}$ and $\mathrm{Fsi}(a, b_1) = \mathrm{Fsi}(b_2, b_3) \approx 0.421257$, $\mathrm{Fsi}(a, b_2) = \mathrm{Fsi}(b_1, b_3) \approx 0.177111$, $\mathrm{Fsi}(a, b_3) \approx$ 0.005439, $\mathrm{Fsi}(b_1, b_2) \approx 0.833519$.

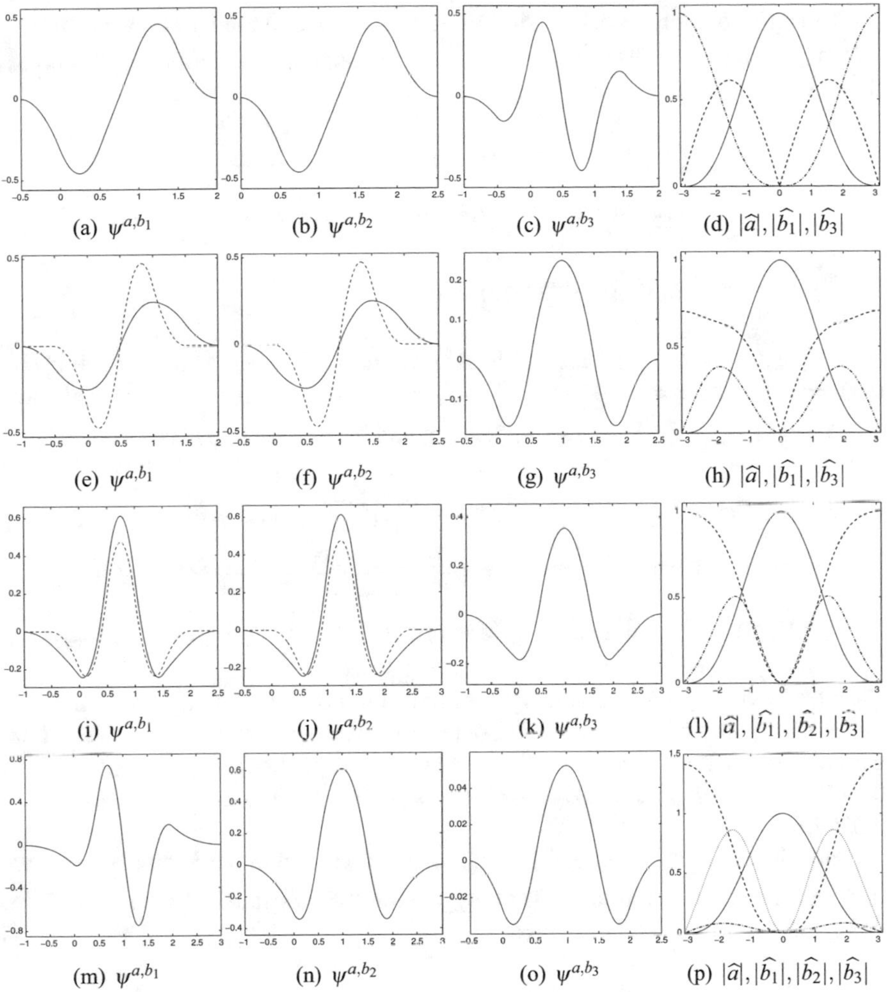

Fig. 3.16 Tight framelet filter banks $\{a; b_1, b_2, b_3\}_\Theta$ derived from $a = a_3^B$ and Θ in Example 3.5.2. **(a)**–**(d)** are for $\Theta = \boldsymbol{\delta}$ and $\mathsf{d_a}(z) = 1$; **(e)**–**(h)** are for $\Theta = \boldsymbol{\delta}$ and $\mathsf{d_a}(z) = 1+z$; **(i)**–**(l)** are for $\Theta = \{-\frac{1}{4}, \underline{\frac{3}{2}}, -\frac{1}{4}\}_{[-1,1]}$ and $\mathsf{d_a}(z) = 1 + z$ by Algorithm 3.5.3; **(m)**–**(p)** are for $\Theta = \{-\frac{1}{4}, \underline{\frac{3}{2}}, -\frac{1}{4}\}_{[-1,1]}$ by Algorithm 3.5.5. **(d)**, **(h)**, **(l)** and **(p)** are the magnitudes of $\widehat{a}$ (in solid line), $\widehat{b_1}$ (in dashed line), $\widehat{b_2}$ (in dotted line) and $\widehat{b_3}$ (in dashed-dotted line) on the interval $[-\pi, \pi]$. $|\widehat{b_2}| = |\widehat{b_1}|$ for **(d)** and **(h)**

We can also apply (3.5.15) using the Fejér-Riesz Lemma by choosing $\mathsf{b}(z) = \frac{\sqrt{14}-4}{32}(z-1)(1+z^{-1})^2[1+(14+4\sqrt{14})z+z^2]$, leading to the following high-pass filters with longer supports:

$$b_1 = \{\tfrac{4-\sqrt{14}}{32}, \tfrac{4+\sqrt{14}}{32}, \underline{\tfrac{\sqrt{14}}{16}}, -\tfrac{\sqrt{14}}{16}, -\tfrac{4+\sqrt{14}}{32}, -\tfrac{4-\sqrt{14}}{32}\}_{[-2,3]},$$

$$b_2 = \{\tfrac{4-\sqrt{14}}{32}, -\tfrac{4+\sqrt{14}}{32}, \underline{\tfrac{\sqrt{14}}{16}}, \tfrac{\sqrt{14}}{16}, -\tfrac{4+\sqrt{14}}{32}, \tfrac{4-\sqrt{14}}{32}\}_{[-2,3]},$$

$$b_3 = \{\tfrac{1}{16}, -\tfrac{1}{4}, \tfrac{3}{8}, \underline{-\tfrac{1}{4}}, \tfrac{1}{16}\}_{[-3,1]}.$$

Take $\mathsf{d_a}(z) = (1+z)^3$ instead in Algorithm 3.5.3. Then $\mathring{\mathsf{a}}(z) = (1+z)/16$, $\mathsf{q}(z) = 1$, and $\mathsf{p}(z) = \frac{1}{64}(1-z^{-1})(1-z)(z^{-2}+8z^{-1}+30+8z+z^2)$. Selecting $\mathsf{d_p}(z) = \frac{1}{8}(1-z^{-1})[z^{-1}+(4+2\sqrt{3}i)+z]$ and $\mathsf{b}(z) = \frac{\sqrt{2}}{2}$, we have

$$\mathsf{b}_1(z) = \mathsf{d_p}(z)\mathsf{b}(z) = \{\tfrac{\sqrt{2}}{16}, \tfrac{3\sqrt{2}+2\sqrt{6}i}{16}, \underline{-\tfrac{3\sqrt{2}+2\sqrt{6}i}{16}}, -\tfrac{\sqrt{2}}{16}\}_{[-2,1]},$$

$$\mathsf{b}_2(z) = \mathsf{d_p}(z)z\mathsf{b}^\star(-z) = z\mathsf{b}_1(z) = \{\tfrac{\sqrt{2}}{16}, \underline{\tfrac{3\sqrt{2}+2\sqrt{6}i}{16}}, -\tfrac{3\sqrt{2}+2\sqrt{6}i}{16}, -\tfrac{\sqrt{2}}{16}\}_{[-1,2]},$$

$$\mathsf{b}_3(z) = \mathsf{d_a}(z)\boldsymbol{\theta}(z)z\mathring{\mathsf{a}}^\star(-z) = \tfrac{1}{16}(z-1)(z+1)^3 = \{\underline{-\tfrac{1}{16}}, -\tfrac{1}{8}, 0, \tfrac{1}{8}, \tfrac{1}{16}\}_{[0,4]}.$$

Then $\{a; b_1, b_2, b_3\}$ is a severable tight framelet filter bank with $\operatorname{vm}(b_1) = \operatorname{vm}(b_2) = \operatorname{vm}(b_3) = 1$ and $\mathsf{Sb}_1(z) = -z^{-1}, \mathsf{Sb}_2(z) = -z, \mathsf{Sb}_3(z) = -z^4$. By calculation, we have $\operatorname{Var}(b_1) = \operatorname{Var}(b_2) = \frac{15}{44}, \operatorname{Var}(b_3) = \frac{8}{5}$ and $\operatorname{Fsi}(a, b_1) = \operatorname{Fsi}(a, b_2) \approx 0.186254$, $\operatorname{Fsi}(a, b_3) \approx 0.537484$, $\operatorname{Fsi}(b_1, b_2) = 1$, $\operatorname{Fsi}(b_1, b_3) = \operatorname{Fsi}(b_2, b_3) \approx 0.538751$.

Considering $n_b = 1$ and $\mathring{\mathsf{b}}_3(z) = t(1+z)$ in Algorithm 3.5.5 with $\Theta = \boldsymbol{\delta}$, we find that the condition in item (3) of Algorithm 3.5.5 is satisfied with $t = \frac{3-\sqrt{7}}{8}$. Applying Algorithm 3.5.5, we have

$$\mathsf{b}_1(z) = \tfrac{2-\sqrt{7}}{48}(1-z^{-1})^2[3+(14+4\sqrt{7})z+z^2] = \{\tfrac{2-\sqrt{7}}{16}, -\tfrac{1}{4}, \underline{\tfrac{2+\sqrt{7}}{8}}, -\tfrac{1}{4}, \tfrac{2-\sqrt{7}}{16}\}_{[-2,2]},$$

$$\mathsf{b}_2(z) = t(1-z^{-1})(1+z)[z^{-1}+(2+\sqrt{7})+z] = t\{1, -2-\sqrt{7}, \underline{\mathbf{0}}, 2+\sqrt{7}, 1\}_{[-2,2]},$$

$$\mathsf{b}_3(z) = t(z-z^{-1}) = t\{-1, \underline{\mathbf{0}}, 1\}_{[-1,1]}.$$

Then $\{a; b_1, b_2, b_3\}$ is a non-severable tight framelet filter bank with $\operatorname{vm}(b_1) = 2, \operatorname{vm}(b_2) = \operatorname{vm}(b_3) = 1$ and $\mathsf{Sb}_1(z) = 1, \mathsf{Sb}_2(z) = \mathsf{Sb}_3(z) = -1$. By calculation, $\operatorname{Var}(b_1) \approx 0.296527$, $\operatorname{Var}(b_2) \approx 1.13285$, $\operatorname{Var}(b_3) \approx 1$ and $\operatorname{Fsi}(a, b_1) \approx 0.0515092$, $\operatorname{Fsi}(a, b_2) \approx 0.434991$, $\operatorname{Fsi}(a, b_3) \approx 0.302283$, $\operatorname{Fsi}(b_1, b_2) \approx 0.430157$, $\operatorname{Fsi}(b_1, b_3) \approx 0.608831$, $\operatorname{Fsi}(b_2, b_3) \approx 0.946017$. See Fig. 3.17 for their framelet functions.

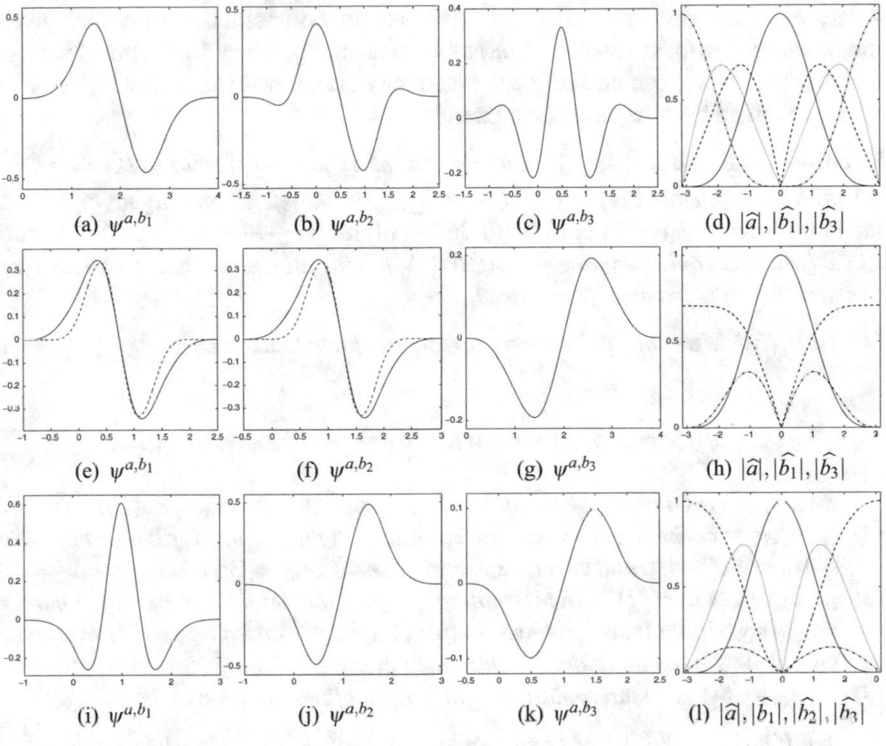

Fig. 3.17 Tight framelet filter banks $\{a; b_1, b_2, b_3\}$ derived from $a = a_4^B$ and $\Theta = \boldsymbol{\delta}$ in Example 3.5.3. **(a)**–**(d)** are for $\mathsf{d_a}(z) = 1$; **(e)**–**(h)** are for $\mathsf{d_a}(z) = (1+z)^3$; **(i)**–**(l)** are obtained by Algorithm 3.5.5. **(d)**, **(h)**, and **(l)** are the magnitudes of $\widehat{a}$ (in solid line), $\widehat{b_1}$ (in dashed line), $\widehat{b_2}$ (in dotted line) and $\widehat{b_3}$ (in dashed-dotted line) on the interval $[-\pi, \pi]$. $|\widehat{b_2}| = |\widehat{b_1}|$ for **(h)**

3.5.3 Interpolatory Tight Framelet Filter Banks

Recall that $a \in l_0(\mathbb{Z})$ is *interpolatory* if $a(2k) = 0$ for all $k \in \mathbb{Z}\backslash\{0\}$. Interpolatory filters form an important family of filters in wavelet theory and are of particular interest in certain applications such as computer graphics and image processing. In this section, we investigate tight framelet filter banks with the interpolation property.

For $a, b_1, \ldots, b_s \in l_0(\mathbb{Z})$, we say that $\{a; b_1, \ldots, b_s\}$ is *an interpolatory tight framelet filter bank* if it is a tight framelet filter bank such that all the filters $a, b_1, \ldots, b_s$ are interpolatory. In this subsection we only consider the cases $s = 2$ or 3 and we often write

$$\mathsf{b}_1(z) = \tfrac{1}{2}c_1 + z\mathsf{d}_1(z^2), \quad \mathsf{b}_2(z) = \tfrac{1}{2}c_2 + z\mathsf{d}_2(z^2), \quad \mathsf{b}_3(z) = \tfrac{1}{2}c_3 + z\mathsf{d}_3(z^2), \tag{3.5.16}$$

where $c_1, c_2, c_3 \in \mathbb{C}$ and $\mathsf{d}_1, \mathsf{d}_2, \mathsf{d}_3$ are Laurent polynomials. From any given interpolatory low-pass filter a, it turns out that interpolatory tight framelet filter banks $\{a; b_1, b_2, b_3\}$ can be easily constructed by the following result, which can be directly verified by calculation (see Exercise 3.41).

Theorem 3.5.6 *Let $a \in l_0(\mathbb{Z})$ be an interpolatory filter such that $a(2k) = \frac{1}{2}\delta(k)$ for all $k \in \mathbb{Z}$. Define $\mathsf{p}(z) := \mathsf{a}^{[1]}(z) = \sum_{k\in\mathbb{Z}} a(2k+1)z^k$ and define high-pass filters $b_1, b_2, b_3 \in l_0(\mathbb{Z})$ as in (3.5.16). If $1 - \mathsf{a}(z)\mathsf{a}^\star(z) - \mathsf{a}(-z)\mathsf{a}^\star(-z) \geqslant 0$ for all $z \in \mathbb{T}$ (which is equivalent to $\frac{1}{2} - 2\mathsf{p}(z)\mathsf{p}^\star(z) \geqslant 0$ for all $z \in \mathbb{T}$ and is necessary for constructing tight framelet filter banks), then*

(1) $\{a; b_1, b_2\}$ is an interpolatory tight framelet filter bank with $|c_1|^2 + |c_2|^2 = 1$ and

$$\mathsf{d}_1(z) := \overline{c_2}\mathsf{p}_0(z) - c_1\mathsf{p}(z), \quad \mathsf{d}_2(z) := -\overline{c_1}\mathsf{p}_0(z) - c_2\mathsf{p}(z),$$

where p_0 is a Laurent polynomial satisfying $\mathsf{p}_0(z)\mathsf{p}_0^\star(z) = \frac{1}{2} - 2\mathsf{p}(z)\mathsf{p}^\star(z)$.

(2) $\{a; b_1, b_2\}$ in item (1) is an interpolatory tight framelet filter bank with symmetry, if both p and p_0 have symmetry, and if $\mathsf{S}\mathsf{p}_0 = \mathsf{S}\mathsf{p}$ whenever $c_1c_2 \neq 0$.

(3) $\{a; b_1, b_2\}$ in item (1) is an interpolatory tight framelet filter bank with complex symmetry, if both p and p_0 have complex symmetry, and if $\mathbb{S}\mathsf{p}_0 = \mathbb{S}\mathsf{p}$ whenever $c_1c_2 \neq 0$ with the additional requirement $c_1, c_2 \in \mathbb{R}$.

(4) $\{a; b_1, b_2, b_3\}$ is an interpolatory tight framelet filter bank which is defined by

(i) If $|c_1| < 1$, under the condition $|c_1|^2 + |c_2|^2 + |c_3|^2 = 1$, then we define

$$\mathsf{d}_1(z) = \sqrt{1-|c_1|^2}\,\mathsf{p}_1(z) - c_1\mathsf{p}(z),$$
$$\mathsf{d}_2(z) = \frac{\overline{c_3}}{\sqrt{1-|c_1|^2}}\mathsf{p}_2(z) - \frac{\overline{c_1}c_2}{\sqrt{1-|c_1|^2}}\mathsf{p}_1(z) - c_2\mathsf{p}(z),$$
$$\mathsf{d}_3(z) = -\frac{\overline{c_1}c_3}{\sqrt{1-|c_1|^2}}\mathsf{p}_1(z) - \frac{\overline{c_2}}{\sqrt{1-|c_1|^2}}\mathsf{p}_2(z) - c_3\mathsf{p}(z).$$

(ii) If $|c_1| = 1$, then we define $\mathsf{d}_1(z) = -c_1\mathsf{p}(z)$, $\mathsf{d}_2(z) = \mathsf{p}_1(z)$ and $\mathsf{d}_3(z) = \mathsf{p}_2(z)$,

where p_1 and p_2 are Laurent polynomials satisfying

$$\mathsf{p}_1(z)\mathsf{p}_1^\star(z) + \mathsf{p}_2(z)\mathsf{p}_2^\star(z) = \frac{1}{2} - 2\mathsf{p}(z)\mathsf{p}^\star(z). \tag{3.5.17}$$

(5) $\{a; b_1, b_2, b_3\}$ in item (4) is an interpolatory tight framelet filter bank with symmetry if all $\mathsf{p}, \mathsf{p}_1, \mathsf{p}_2$ have symmetry, and if $\mathsf{S}\mathsf{p}_1 = \mathsf{S}\mathsf{p}_2 = \mathsf{S}\mathsf{p}$ when $c_1c_2c_3 \neq 0$, or $\mathsf{S}\mathsf{p}_1 = \mathsf{S}\mathsf{p}$ when only c_2 or c_3 is 0 and $|c_1| < 1$, or $\mathsf{S}\mathsf{p}_2 = \mathsf{S}\mathsf{p}$ when only $c_1 = 0$.

(6) $\{a; b_1, b_2, b_3\}$ *in item (4) is an interpolatory tight framelet filter bank with complex symmetry if all* $\mathsf{p}, \mathsf{p}_1, \mathsf{p}_2$ *have complex symmetry, and if* $\mathbb{S}\mathsf{p}_1 = \mathbb{S}\mathsf{p}_2 = \mathbb{S}\mathsf{p}$ *when* $c_1c_2c_3 \ne 0$*, or* $\mathbb{S}\mathsf{p}_1 = \mathbb{S}\mathsf{p}$ *when only* c_2 *or* c_3 *is 0 and* $|c_1| < 1$*, or* $\mathbb{S}\mathsf{p}_2 = \mathbb{S}\mathsf{p}$ *when only* $c_1 = 0$ *with the additional requirement* $c_1, c_2, c_3 \in \mathbb{R}$.

Many interpolatory tight framelet filter banks can be easily derived via Theorem 3.5.6 from interpolatory low-pass filters. Here we provide two examples.

Example 3.5.4 Let $a = a_4^I = \{-\frac{1}{32}, 0, \frac{9}{32}, \underline{\frac{1}{2}}, \frac{9}{32}, 0, -\frac{1}{32}\}_{[-3,3]}$ be an interpolatory filter with $\operatorname{sr}(a) = 4$ and $\operatorname{sm}(a_4^I) \approx 2.440765$. Then $\mathsf{p}(z) = \frac{1}{32}(1 + z^{-1})(-z^{-1} + 10 - z)$ and

$$\tfrac{1}{2} - 2\mathsf{p}(z)\mathsf{p}^\star(z) = (1-z)^2(1-1/z)^2(14 - z^{-1} - z)/512.$$

Note that $7 - 4\sqrt{3}$ lies on the interval $(0, 1)$ and is a simple root of $\mathsf{P}(z) := \frac{1}{2} - 2\mathsf{p}(z)\mathsf{p}^\star(z)$, that is, $\mathsf{Z}(\mathsf{P}, 7 - 4\sqrt{3}) = 1$. Consequently, both item (2) of Theorem 3.1.7 and item (3) of Theorem 3.1.8 fail. Therefore, according to Theorems 3.1.7 and 3.1.8, there does not exist a Laurent polynomial d with [complex] symmetry such that $\mathsf{d}(z)\mathsf{d}^\star(z) = \mathsf{P}(z)$. Hence, by $\mathcal{N}_{a,\Theta\,|\,0}(z) = \frac{1}{4}\mathsf{P}(z)$, we conclude that the second condition in item (i) of Theorem 3.3.7 with $n_b = 0$ cannot hold. Consequently, by Theorem 3.3.7, there does not exist a finitely supported tight framelet filter bank $\{a; b_1, b_2\}$ with [complex] symmetry. Taking $\mathsf{p}_0(z) = (1 - z^{-1})^2[(\sqrt{6} - 2\sqrt{2}) + (\sqrt{6} + 2\sqrt{2})z]/32$ and $c_1 = 1, c_2 = 0$, Theorem 3.5.6 yields $\mathsf{d}_1(z) = -\mathsf{p}(z)$, $\mathsf{d}_2(z) = -\mathsf{p}_0(z)$ and

$$b_1 = \{\tfrac{1}{32}, 0, -\tfrac{9}{32}, \underline{\tfrac{1}{2}}, -\tfrac{9}{32}, 0, \tfrac{1}{32}\}_{[-3,3]},$$

$$b_2 = \{\tfrac{2\sqrt{2}-\sqrt{6}}{32}, 0, \tfrac{\sqrt{6}-6\sqrt{2}}{32}, 0, \tfrac{\sqrt{6}+6\sqrt{2}}{32}, 0, -\tfrac{2\sqrt{2}+\sqrt{6}}{32}\}_{[-3,3]}.$$

Then $\{a; b_1, b_2\}$ is a real-valued interpolatory tight framelet filter bank with $\operatorname{vm}(b_1) = 4$ and $\operatorname{vm}(b_2) = 2$. By calculation, we have $\operatorname{Var}(b_1) = \frac{3}{7}$, $\operatorname{Var}(b_2) = \frac{741}{529}$ and $\operatorname{Fsi}(a, b_1) \approx 0.0483164$, $\operatorname{Fsi}(a, b_2) = \operatorname{Fsi}(b_1, b_2) \approx 0.387387$.

Since both a and b_1 have symmetry, by Proposition 3.4.6, $\{a; b_1, (b_2 + b_2^\star)/2, (b_2 - b_2^\star)/2\}$ is a real-valued interpolatory tight framelet filter bank with symmetry. This example can be recovered by Theorem 3.5.6 by selecting $c_1 = 1, c_2 = 0, c_3 = 0$ and

$$\mathsf{p}_1(z) = -\tfrac{\sqrt{6}}{32}(1 - z^{-1})^2(1 + z), \quad \mathsf{p}_2(z) = \tfrac{\sqrt{2}}{16} z^{-2}(z - 1)^3.$$

Then p_1 and p_2 satisfy (3.5.17) with symmetry $\mathsf{S}\mathsf{p}_1(z) = z^{-1}$ and $\mathsf{S}\mathsf{p}_2(z) = -z^{-1}$. Hence, Theorem 3.5.6 yields $\mathsf{d}_1(z) = -\mathsf{p}(z)$, $\mathsf{d}_2(z) = \mathsf{p}_1(z)$, $\mathsf{d}_3(z) = \mathsf{p}_2(z)$ and

$$b_1 = \tfrac{1}{32}\{1, 0, -9, \underline{\mathbf{16}}, -9, 0, 1\}_{[-3,3]}, \qquad b_2 = \tfrac{\sqrt{6}}{32}\{-1, 0, 1, \underline{\mathbf{0}}, 1, 0, -1\}_{[-3,3]},$$

$$b_3 = \tfrac{\sqrt{2}}{16}\{-1, 0, 3, \underline{\mathbf{0}}, -3, 0, 1\}_{[-3,3]}.$$

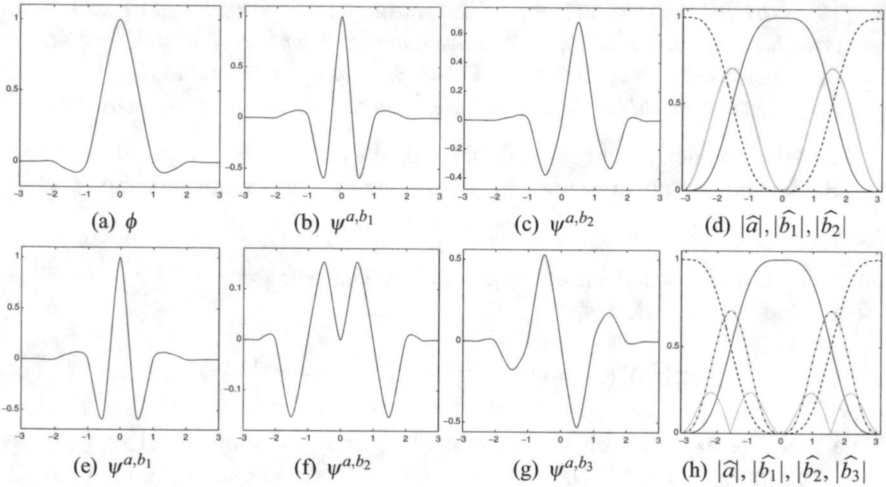

Fig. 3.18 (**a**)–(**d**) are for the real-valued interpolatory tight framelet filter bank $\{a; b_1, b_2\}$ with $a = a_4^I$ without symmetry in Example 3.5.4. (**e**)–(**h**) are for the real-valued interpolatory tight framelet filter bank $\{a; b_1, b_2, b_3\}$ with symmetry. (**d**) and (**h**) are the magnitudes of $\widehat{a}$ (in solid line), $\widehat{b_1}$ (in dashed line), $\widehat{b_2}$ (in dotted line) and $\widehat{b_3}$ (in dashed-dotted line) on the interval $[-\pi, \pi]$

Then $\{a; b_1, b_2, b_3\}$ is a real-valued interpolatory tight framelet filter bank with $\mathrm{vm}(b_1) = 4$, $\mathrm{vm}(b_2) = 2$, $\mathrm{vm}(b_3) = 3$ and $\mathsf{Sb}_1(z) = \mathsf{Sb}_2(z) = 1$, $\mathsf{Sb}_3(z) = -1$. By calculation, $\mathrm{Var}(b_1) = \frac{3}{7}$, $\mathrm{Var}(b_2) = 5$, $\mathrm{Var}(b_3) = \frac{9}{5}$ and $\mathrm{Fsi}(a, b_1) \approx 0.0483164$, $\mathrm{Fsi}(a, b_2) = \mathrm{Fsi}(b_1, b_2) \approx 0.517887$, $\mathrm{Fsi}(a, b_3) = \mathrm{Fsi}(b_1, b_3) \approx 0.348427$, $\mathrm{Fsi}(b_2, b_3) \approx 0.522233$. See Fig. 3.18 for their associated refinable and framelet functions.

Example 3.5.5 Let $a = a_6^I = \{\frac{3}{512}, 0, -\frac{25}{512}, 0, \frac{75}{256}, \underline{\frac{1}{2}}, \frac{75}{256}, 0, -\frac{25}{512}, 0, \frac{3}{512}\}_{[-5,5]}$ be an interpolatory filter with $\mathrm{sr}(a) = 6$ and $\mathrm{sm}(a_6^I) \approx 3.175132$. Then $\mathsf{p}(z) = (1 + z^{-1})(3z^{-2} - 28z^{-1} + 178 - 28z + 3z^2)/512$ and

$$\tfrac{1}{2} - 2\mathsf{p}(z)\mathsf{p}^\star(z) = (1-z)^3(1-z^{-1})^3(9z^{-2} - 96z^{-1} + 814 - 96z + 9z^2)/131072.$$

Taking $\mathsf{p}_0(z) = \frac{\sqrt{2}}{512}(1-z^{-1})^3[(8+2\sqrt{10}+\lambda)+(4\sqrt{10}-16)z+(8+2\sqrt{10}-\lambda)z^2]$ and $c_1 = 1, c_2 = 0$ in Theorem 3.5.6, we have $\mathsf{d}_1(z) = -\mathsf{p}(z)$, $\mathsf{d}_2(z) = -\mathsf{p}_0(z)$ and

$$b_1 = \{-\tfrac{3}{512}, 0, \tfrac{25}{512}, 0, -\tfrac{75}{256}, \underline{\tfrac{1}{2}}, -\tfrac{75}{256}, 0, \tfrac{25}{512}, 0 - \tfrac{3}{512}\}_{[-5,5]},$$

$$b_2 = \tfrac{\sqrt{2}}{512}\{8 + 2\sqrt{10} + \lambda, 0, -40 - 2\sqrt{10} - 3\lambda, 0, 80 - 4\sqrt{10} + 2\lambda, \underline{\mathbf{0}},$$
$$-80 + 4\sqrt{10} + 2\lambda, 0, 40 + 2\sqrt{10} - 3\lambda, 0, -8 - 2\sqrt{10} + \lambda\}_{[-5,5]},$$

with $\lambda := \sqrt{95 + 32\sqrt{10}}$. Then $\{a; b_1, b_2\}$ is a real-valued interpolatory tight framelet filter bank with $\mathrm{vm}(b_1) = 6$ and $\mathrm{vm}(b_2) = 3$. By calculation,

$\mathrm{Var}(b_1) = 0.507137$, $\mathrm{Var}(b_2) = 1.96382$ and $\mathrm{Fsi}(a, b_1) \approx 0.035932, \mathrm{Fsi}(a, b_2) = \mathrm{Fsi}(b_1, b_2) \approx 0.339939$. Since both a and b_1 have symmetry, by Proposition 3.4.6, $\{a; b_1, (b_2 + b_2^\star)/2, (b_2 - b_2^\star)/2\}$ is a real-valued interpolatory tight framelet filter bank with symmetry.

By Theorems 3.1.7 and 3.1.8, there is a Laurent polynomial p_0 with symmetry:

$$\mathsf{p}_0(z) = \tfrac{\sqrt{2}}{512}(1 - z^{-1})^3[3 + (-16 + 6\sqrt{15}\,i)z + 3z^2],$$

(but not possible for complex symmetry) satisfying $\mathsf{p}_0(z)\mathsf{p}_0^\star(z) = \frac{1}{2} - 2\mathsf{p}(z)\mathsf{p}^\star(z)$. Taking $c_1 = 1, c_2 = 0$, Theorem 3.5.6 yields $\mathsf{d}_1(z) = -\mathsf{p}(z)$, $\mathsf{d}_2(z) = -\mathsf{p}_0(z)$ and

$$b_1 = \{-\tfrac{3}{512}, 0, \tfrac{25}{512}, 0, -\tfrac{75}{256}, \underline{\mathbf{\tfrac{1}{2}}}, -\tfrac{75}{256}, 0, \tfrac{25}{512}, 0, -\tfrac{3}{512}\}_{[-5,5]},$$

$$b_2 = \tfrac{\sqrt{2}}{512}\{3, 0, 6\sqrt{15}\,i - 25, 0, 60 - 18\sqrt{15}\,i, \underline{\mathbf{0}}, 18\sqrt{15}\,i - 60, 0, 25 - 6\sqrt{15}\,i, 0, -3\}_{[-5,5]}.$$

Then $\{a; b_1, b_2\}$ is an interpolatory tight framelet filter bank with $\mathrm{vm}(b_1) = 6$, $\mathrm{vm}(b_2) = 3$ and $\mathsf{S}b_1(z) = 1$, $\mathsf{S}b_2(z) = -1$. By calculation, $\mathrm{Var}(b_1) = 0.507137$, $\mathrm{Var}(b_2) = 1.98983$ and $\mathrm{Fsi}(a, b_1) \approx 0.035932, \mathrm{Fsi}(a, b_2) = \mathrm{Fsi}(b_1, b_2) \approx 0.339939$. Moreover, $\{a; b_1, b_2^{[r]}, b_2^{[i]}\}$ is a real-valued interpolatory tight framelet filter bank with symmetry $\mathsf{S}b_1(z) = 1$ and $\mathsf{S}b_2^{[r]}(z) = \mathsf{S}b_2^{[i]}(z) = -1$, where $b_2^{[r]}$ and $b_2^{[i]}$ are the real and imaginary parts of b_2. See Fig. 3.19 for their framelet functions.

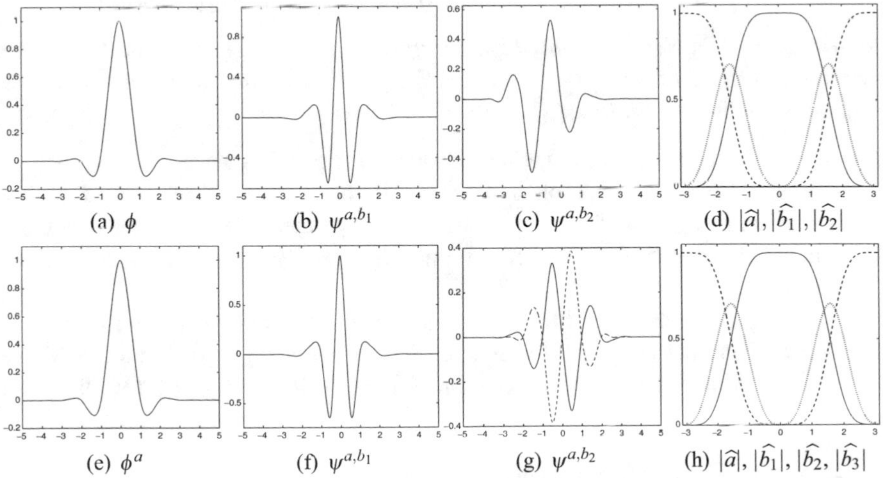

Fig. 3.19 (**a**)–(**d**) are for the real-valued interpolatory tight framelet filter bank $\{a; b_1, b_2\}$ with $a = a_6^I$ without symmetry in Example 3.5.5. (**e**)–(**h**) are for the complex-valued interpolatory tight framelet filter bank $\{a; b_1, b_2\}$ with symmetry. Note that $\{a; b_1, b_2^{[r]}, b_2^{[i]}\}$ is a real-valued interpolatory tight framelet filter bank with symmetry (The dotted line in (**g**) for $\psi^{a,b_2^{[i]}}$). (**d**) and (**h**) are the magnitudes of $\widehat{a}$ (in solid line), $\widehat{b_1}$ (in dashed line) and $\widehat{b_2}$ (in dotted line) on the interval $[-\pi, \pi]$

3.6 Existence of Tight Framelet Filter Banks with Symmetry*

Employing a similar idea as in Sect. 3.4.2 but under the symmetry property, we now prove Theorem 3.3.7 which provides a necessary and sufficient condition for the existence of a tight framelet filter bank $\{a; b_1, b_2\}_\Theta$ with [complex] symmetry.

To prove Theorem 3.3.7, we first present a general result on splitting a 2×2 matrix of Laurent polynomials under the symmetry constraint.

Theorem 3.6.1 *Let $\mathcal{N}$ be a 2×2 matrix of Laurent polynomials such that all its four entries $\mathcal{N}_{1,1}, \mathcal{N}_{1,2}, \mathcal{N}_{2,1}, \mathcal{N}_{2,2}$ have symmetry (and real coefficients) and $\mathcal{N}$ is not identically zero. Then there exist Laurent polynomials $\mathcal{U}_{1,1}, \mathcal{U}_{1,2}, \mathcal{U}_{2,1}, \mathcal{U}_{2,2}$ having symmetry (and real coefficients) such that*

$$\mathcal{U}(z)\mathcal{U}^\star(z) = \mathcal{N}(z), \quad z \in \mathbb{C}\backslash\{0\},$$
$$\frac{\mathsf{S}\mathcal{U}_{1,1}(z)}{\mathsf{S}\mathcal{U}_{2,1}(z)} = \frac{\mathsf{S}\mathcal{U}_{1,2}(z)}{\mathsf{S}\mathcal{U}_{2,2}(z)} \quad \text{with} \quad \mathcal{U}(z) := \begin{bmatrix} \mathcal{U}_{1,1}(z) & \mathcal{U}_{1,2}(z) \\ \mathcal{U}_{2,1}(z) & \mathcal{U}_{2,2}(z) \end{bmatrix} \tag{3.6.1}$$

if and only if

(i) $\mathcal{N}(z) \geqslant 0$ for all $z \in \mathbb{T}$ and $\det(\mathcal{N}(z)) = \mathsf{d}(z)\mathsf{d}^\star(z)$ for some Laurent polynomial d with symmetry (and real coefficients);
(ii) the Laurent polynomial $\mathsf{p}(z) := \gcd(\mathcal{N}_{1,1}(z), \mathcal{N}_{1,2}(z), \mathcal{N}_{2,1}(z), \mathcal{N}_{2,2}(z))$ has the real SOS property with respect to the symmetry type $[\mathsf{S}\mathcal{N}_{2,1}(z)][\mathsf{S}\mathsf{d}(z)]$.

Moreover, if all involved Laurent polynomials (that is, all $\mathcal{N}, \mathcal{U}$ and d) have complex symmetry instead of symmetry, replace S by $\mathbb{S}$, then the same necessary and sufficient condition, with item (ii) removed, still holds.

Proof The sufficiency part ($\Leftarrow$) will be proved in Algorithm 3.6.2.

Necessity ($\Rightarrow$). (3.6.1) implies $\mathcal{N}(z) \geqslant 0$ for all $z \in \mathbb{T}$ and $\mathsf{d}(z)\mathsf{d}^\star(z) = \det(\mathcal{N}(z))$ with $\mathsf{d}(z) := \det(\mathcal{U}(z))$. Due to the symmetry relation in (3.6.1), the Laurent polynomial d has symmetry with $\mathsf{S}\mathsf{d}(z) = [\mathsf{S}\mathcal{U}_{1,1}(z)][\mathsf{S}\mathcal{U}_{2,2}(z)]$. Hence, item (i) holds.

We now prove item (ii) for the case of symmetry. We first prove that $\mathcal{N}(z) \geqslant 0$ for all $z \in \mathbb{T}$ implies $\mathsf{p}(z) \geqslant 0$ for all $z \in \mathbb{T}$. Because $\mathcal{N}(z) \geqslant 0$ for all $z \in \mathbb{T}$, we have $\mathcal{N}^\star = \mathcal{N}$ and thus, $\mathcal{N}_{1,2}^\star = \mathcal{N}_{2,1}$ and $\mathcal{N}_{1,1}(z) \geqslant 0, \mathcal{N}_{2,2}(z) \geqslant 0$ for all $z \in \mathbb{T}$. Since

$$\mathsf{Z}(\mathsf{p}, z) = \min(\mathsf{Z}(\mathcal{N}_{1,1}, z), \mathsf{Z}(\mathcal{N}_{2,2}, z), \mathsf{Z}(\mathcal{N}_{1,2}, z), \mathsf{Z}(\mathcal{N}_{1,2}^\star, z)), \quad \forall\, z \in \mathbb{C}\backslash\{0\}, \tag{3.6.2}$$

we observe that (3.1.4) must hold. By $\mathcal{N}_{1,2}^\star(z) = \overline{\mathcal{N}_{1,2}(\overline{z}^{-1})} = \overline{\mathcal{N}_{1,2}(z)}$ for all $z \in \mathbb{T}$ and $\det(\mathcal{N}(z)) \geqslant 0$ for all $z \in \mathbb{T}$, we have

$$0 \leqslant |\mathcal{N}_{1,2}(z)|^2 = \mathcal{N}_{1,2}(z)\mathcal{N}_{1,2}^\star(z) = \mathcal{N}_{1,2}(z)\mathcal{N}_{2,1}(z) \leqslant \mathcal{N}_{1,1}(z)\mathcal{N}_{2,2}(z), \quad \forall\, z \in \mathbb{T},$$

from which we deduce that $2\mathsf{Z}(\mathcal{N}_{1,2}, z) \geqslant \mathsf{Z}(\mathcal{N}_{1,1}, z) + \mathsf{Z}(\mathcal{N}_{2,2}, z)$ for all $z \in \mathbb{T}$. By item (iii) of Proposition 3.1.1, since $\mathcal{N}_{1,1}(z) \geqslant 0$ and $\mathcal{N}_{2,2}(z) \geqslant 0$ for all $z \in \mathbb{T}$, we conclude from (3.6.2) that (3.1.5) holds. Now it follows from item (iv) of Proposition 3.1.1 that $\mathsf{p}(z) \geqslant 0$ for all $z \in \mathbb{T}$. Since p also has symmetry, by Lemma 3.1.2, p must have real coefficients. Since $\mathcal{N}$ is not identically zero, so is p. Define

$$\breve{\mathcal{N}}(z) := \begin{bmatrix} \breve{\mathcal{N}}_{1,1}(z) & \breve{\mathcal{N}}_{1,2}(z) \\ \breve{\mathcal{N}}_{2,1}(z) & \breve{\mathcal{N}}_{2,2}(z) \end{bmatrix} := \begin{bmatrix} \frac{\mathcal{N}_{1,1}(z)}{\mathsf{p}(z)} & \frac{\mathcal{N}_{1,2}(z)}{\mathsf{p}(z)} \\ \frac{\mathcal{N}_{2,1}(z)}{\mathsf{p}(z)} & \frac{\mathcal{N}_{2,2}(z)}{\mathsf{p}(z)} \end{bmatrix} = \frac{1}{\mathsf{p}(z)}\mathcal{N}(z). \tag{3.6.3}$$

Since $\mathsf{p}(z) \geqslant 0$ and $\mathcal{N}(z) \geqslant 0$ for all $z \in \mathbb{T}$, it is not difficult to see that $\breve{\mathcal{N}}(z) \geqslant 0$ for all $z \in \mathbb{T}$. Since all entries of $\mathcal{N}$ have symmetry and $\mathsf{Sp}(z) = 1$, all the entries of $\breve{\mathcal{N}}$ have symmetry. We now show that $\det(\breve{\mathcal{N}}(z)) = \breve{\mathsf{d}}(z)\breve{\mathsf{d}}^\star(z)$ for some Laurent polynomial $\breve{\mathsf{d}}$ with symmetry. By Theorem 3.1.8 and $[\mathsf{p}(z)]^2 \det(\breve{\mathcal{N}}(z)) = \det(\mathcal{N}(z)) = \mathsf{d}(z)\mathsf{d}^\star(z)$, it is trivial to see that item (3) of Theorem 3.1.8 with p being replaced by $\det(\breve{\mathcal{N}}(z))$ is satisfied. Therefore, there exists a Laurent polynomial $\breve{\mathsf{d}}$ with symmetry such that $\det(\breve{\mathcal{N}}(z)) = \breve{\mathsf{d}}(z)\breve{\mathsf{d}}^\star(z)$.

Note that $\gcd(\breve{\mathcal{N}}_{1,1}, \breve{\mathcal{N}}_{1,2}, \breve{\mathcal{N}}_{2,1}, \breve{\mathcal{N}}_{2,2}) = 1$. Therefore, items (i) and (ii) in Theorem 3.6.1 with $\mathcal{N}$ being replaced by $\breve{\mathcal{N}}$ are satisfied. By the sufficiency part of Theorem 3.6.1 (see Algorithm 3.6.2), there exists a 2×2 matrix $\breve{\mathcal{U}}$ of Laurent polynomials having symmetry (and real coefficients) such that $\breve{\mathcal{U}}(z)\breve{\mathcal{U}}^\star(z) = \breve{\mathcal{N}}(z)$ for all $z \in \mathbb{C}\backslash\{0\}$ and $\frac{\mathsf{S}\breve{\mathcal{U}}_{1,1}(z)}{\mathsf{S}\breve{\mathcal{U}}_{2,1}(z)} = \frac{\mathsf{S}\breve{\mathcal{U}}_{1,2}(z)}{\mathsf{S}\breve{\mathcal{U}}_{2,2}(z)}$, from which and the identity $\breve{\mathcal{N}}_{2,1}(z) = \mathcal{N}_{2,1}(z)/\mathsf{p}(z)$ and $\mathsf{Sp}(z) = 1$ we deduce that

$$\frac{\mathsf{S}\breve{\mathcal{U}}_{2,1}(z)}{\mathsf{S}\breve{\mathcal{U}}_{1,1}(z)} = \frac{\mathsf{S}\breve{\mathcal{U}}_{2,2}(z)}{\mathsf{S}\breve{\mathcal{U}}_{1,2}(z)} = \mathsf{S}\breve{\mathcal{N}}_{2,1}(z) = \mathsf{S}\mathcal{N}_{2,1}(z) = \frac{\mathsf{S}\mathcal{U}_{2,1}(z)}{\mathsf{S}\mathcal{U}_{1,1}(z)} = \frac{\mathsf{S}\mathcal{U}_{2,2}(z)}{\mathsf{S}\mathcal{U}_{1,2}(z)}, \tag{3.6.4}$$

where we used (3.6.1). Consequently, from $\mathcal{U}(z)\mathcal{U}^\star(z) = \mathcal{N}(z) = \mathsf{p}(z)\breve{\mathcal{N}}(z) = \mathsf{p}(z)\breve{\mathcal{U}}(z)\breve{\mathcal{U}}^\star(z)$, we must have

$$\mathcal{Q}(z)\mathcal{Q}^\star(z) = \breve{\mathsf{d}}(z)\breve{\mathsf{d}}^\star(z)\mathsf{p}(z)I_2, \tag{3.6.5}$$

where

$$\mathcal{Q}(z) := \begin{bmatrix} \mathsf{q}_1(z) & \mathsf{q}_2(z) \\ \mathsf{q}_3(z) & \mathsf{q}_4(z) \end{bmatrix} = \operatorname{adj}(\breve{\mathcal{U}}(z))\mathcal{U}(z) = \begin{bmatrix} \breve{\mathcal{U}}_{2,2}(z) & -\breve{\mathcal{U}}_{1,2}(z) \\ -\breve{\mathcal{U}}_{2,1}(z) & \breve{\mathcal{U}}_{1,1}(z) \end{bmatrix} \begin{bmatrix} \mathcal{U}_{1,1}(z) & \mathcal{U}_{1,2}(z) \\ \mathcal{U}_{2,1}(z) & \mathcal{U}_{2,2}(z) \end{bmatrix}.$$

By the symmetry relations in (3.6.4), we can check that all the entries of $\mathcal{Q}$ have symmetry and

$$\begin{aligned} \frac{\mathsf{Sq}_1(z)}{\mathsf{Sq}_2(z)} &= \frac{\mathsf{S}\breve{\mathcal{U}}_{2,2}(z)\mathsf{S}\mathcal{U}_{1,1}(z)}{\mathsf{S}\breve{\mathcal{U}}_{2,2}(z)\mathsf{S}\mathcal{U}_{1,2}(z)} = \frac{\mathsf{S}\mathcal{U}_{1,1}(z)}{\mathsf{S}\mathcal{U}_{1,2}(z)} \\ &= \frac{\mathsf{S}\mathcal{U}_{2,2}(z)}{\mathsf{S}\mathcal{U}_{1,2}(z)} \frac{\mathsf{S}\mathcal{U}_{1,1}(z)}{\mathsf{S}\mathcal{U}_{2,2}(z)} = z^{2k}[\mathsf{S}\mathcal{U}_{2,2}(z)]^{-2}\mathsf{S}\mathcal{N}_{2,1}(z)\mathsf{Sd}(z), \end{aligned}$$

where we used the fact that $\mathsf{Sd}(z) = z^{-2k}\mathsf{S}\mathcal{U}_{1,1}(z)\mathsf{S}\mathcal{U}_{2,2}(z)$ for some $k \in \mathbb{Z}$. Note that (3.6.5) implies $\mathsf{q}_1(z)\mathsf{q}_1^\star(z) + \mathsf{q}_2(z)\mathsf{q}_2^\star(z) = \breve{\mathsf{d}}(z)\breve{\mathsf{d}}^\star(z)\mathsf{p}(z)$. Since q_1 and q_2 have symmetry, by Theorem 3.1.6, $\breve{\mathsf{d}}(z)\breve{\mathsf{d}}^\star(z)\mathsf{p}(z)$ must satisfy the root condition for the real SOS property with respect to $z^{2k}[\mathsf{S}\mathcal{U}_{2,2}(z)]^{-2}[\mathsf{S}\mathcal{N}_{2,1}(z)][\mathsf{Sd}(z)]$. Because $\breve{\mathsf{d}}$ has symmetry, we have $\mathsf{Z}(\breve{\mathsf{d}}^\star, x) = \mathsf{Z}(\breve{\mathsf{d}}, \bar{x}^{-1}) = \mathsf{Z}(\breve{\mathsf{d}}, \bar{x}) = \mathsf{Z}(\breve{\mathsf{d}}, x)$ for all $x \in \mathbb{R}\backslash\{0\}$. We conclude that p must satisfy the root condition for the real SOS property with respect to $[\mathsf{S}\mathcal{N}_{2,1}(z)][\mathsf{Sd}(z)]$. Since $\mathsf{p}(z) \geqslant 0$ for all $z \in \mathbb{T}$ and p has real coefficients, it follows from Theorem 3.1.6 that item (ii) must hold for the case of symmetry. □

We now prove the sufficiency part of Theorem 3.6.1 by the following algorithm:

Algorithm 3.6.2 *Let $\mathcal{N}$ be a 2×2 matrix of Laurent polynomials such that all the entries of $\mathcal{N}$ have [complex] symmetry (and real coefficients) and $\mathcal{N}$ is not identically zero. Assume that items (i) and (ii) of Theorem 3.6.1 are satisfied [for the case of complex symmetry, remove item (ii) and replace S by $\mathbb{S}$].*

(S1) Define a 2×2 matrix $\breve{\mathcal{N}}$ as in (3.6.3). Define $\breve{\mathsf{q}}(z) := \gcd(\breve{\mathcal{N}}_{1,1}(z), \breve{\mathcal{N}}_{1,2}(z)\breve{\mathcal{N}}_{1,2}^\star(z))$ and $\mathsf{q}_1(z) := \gcd(\breve{\mathcal{N}}_{1,1}(z), \breve{\mathcal{N}}_{1,2}(z))$. Then there exists a Laurent polynomial q_2 with [complex] symmetry (and real coefficients) such that q_2 satisfies $\mathsf{q}_2(z)\mathsf{q}_2^\star(z) = \mathsf{q}_1(z)\mathsf{q}_1^\star(z)/\breve{\mathsf{q}}(z)$. Define $\mathring{\mathcal{N}}$ as in (3.4.5) with $\mathcal{N}$ being replaced by $\breve{\mathcal{N}}$ and $\mathsf{q}(z) := \mathsf{q}_1(z)/\mathsf{q}_2(z)$. Then all entries of $\mathring{\mathcal{N}}$ must have [complex] symmetry (and real coefficients), and $\mathring{\mathcal{N}}_{1,1}$ and $\mathring{\mathcal{N}}_{1,2}$ have no common zeros in $\mathbb{C}\backslash\{0\}$;

(S2) There exists a Laurent polynomial $\tilde{\mathsf{d}}$ with [complex] symmetry (and real coefficients) such that $\tilde{\mathsf{d}}(z)\tilde{\mathsf{d}}^\star(z) = \det(\mathring{\mathcal{N}}(z))$. Define $[-\mathring{n}, \mathring{n}] := \mathrm{fsupp}(\mathring{\mathcal{N}}_{1,1})$, $\mathring{\epsilon}z^{\mathring{c}} := \mathsf{S}\tilde{\mathsf{d}}(z)$, $\epsilon_{\mathring{\mathcal{N}}_{2,1}} z^{c_{\mathring{\mathcal{N}}_{2,1}}} := \mathsf{S}\mathring{\mathcal{N}}_{2,1}(z)$, and

$$c_{\mathrm{odd}} := \mathrm{odd}(c_{\mathring{\mathcal{N}}_{2,1}} - \mathring{c}), \qquad \mathring{\mathsf{d}}(z) := z^{\mathring{n}+(c_{\mathring{\mathcal{N}}_{2,1}}-\mathring{c}-c_{\mathrm{odd}})/2}\tilde{\mathsf{d}}(z). \tag{3.6.6}$$

Then $\mathring{\mathsf{d}}$ has [complex] symmetry $\mathsf{S}\mathring{\mathsf{d}}(z) = \mathring{\epsilon}z^{2\mathring{n}+c_{\mathring{\mathcal{N}}_{2,1}}-c_{\mathrm{odd}}}$ (and real coefficients);

(S3) Write $\mathring{\mathcal{U}}_{1,1}(z) = \sum_{j=0}^{\mathring{n}-c_{\mathrm{odd}}} t_j z^j$ and $\mathring{\mathcal{U}}_{1,2}(z) = \sum_{j=0}^{\mathring{n}} \tilde{t}_j z^j$ (when $\mathring{n} < c_{\mathrm{odd}}$, set $\mathring{\mathcal{U}}_{1,1}(z) = 0$), where $\{t_0, \ldots, t_{\mathring{n}-c_{\mathrm{odd}}}, \tilde{t}_0, \ldots, \tilde{t}_{\mathring{n}}\}$ is a nontrivial solution to the homogeneous system X of $2\mathring{n}$ linear equations induced by $\mathring{\mathcal{R}}(z) = 0$, where $\mathring{\mathcal{R}}$ and $\mathring{\mathcal{U}}_{2,1}$ are uniquely determined by (3.4.6). Then the space of all solutions to X has dimension at least one. For the case of symmetry, replace $\mathring{\mathcal{U}}_{1,1}$ and $\mathring{\mathcal{U}}_{1,2}$ by

$$[\mathring{\mathcal{U}}_{1,1}(z) + \epsilon\epsilon_{\mathring{\mathcal{N}}_{2,1}} z^{\mathring{n}-c_{\mathrm{odd}}}\mathring{\mathcal{U}}_{1,1}(z^{-1})]/2, \quad [\mathring{\mathcal{U}}_{1,2}(z) + \epsilon\mathring{\epsilon}z^{\mathring{n}}\mathring{\mathcal{U}}_{1,2}(z^{-1})]/2. \tag{3.6.7}$$

For the case of complex symmetry, replace $\mathring{\mathcal{U}}_{1,1}$ *and* $\mathring{\mathcal{U}}_{1,2}$ *by*

$$[\breve{\mathcal{U}}_{1,1}(z) + \epsilon\epsilon_{\mathring{\mathcal{N}}_{2,1}} z^{\mathring{n}-c_{\text{odd}}}\mathring{\mathcal{U}}^{\star}_{1,1}(z)]/2, \qquad [\mathring{\mathcal{U}}_{1,2}(z) + \epsilon\mathring{\epsilon} z^{\mathring{n}}\mathring{\mathcal{U}}^{\star}_{1,2}(z)]/2, \tag{3.6.8}$$

where $\epsilon \in \{-1, 1\}$ *such that the Laurent polynomials in* (3.6.7) *or* (3.6.8) *are not simultaneously identically zero. Then the symmetrized pair* $\{\mathring{\mathcal{U}}_{1,1}, \mathring{\mathcal{U}}_{1,2}\}$ *satisfies*

$$\mathring{\mathcal{N}}_{1,1}(z) \mid [\mathring{\mathcal{N}}_{2,1}(z)\mathring{\mathcal{U}}_{1,1}(z) - \mathring{\mathsf{d}}(z)\mathring{\mathcal{U}}^{\star}_{1,2}(z)] \tag{3.6.9}$$

with $\mathsf{S}\mathring{\mathcal{U}}_{1,1}(z) = \epsilon\epsilon_{\mathring{\mathcal{N}}_{2,1}} z^{\mathring{n}-c_{\text{odd}}}$ *and* $\mathsf{S}\mathring{\mathcal{U}}_{1,2}(z) = \epsilon\mathring{\epsilon} z^{\mathring{n}}$*;*

(S4) Multiply $\{\mathring{\mathcal{U}}_{1,1}, \mathring{\mathcal{U}}_{1,2}\}$ *in (S3) with the well-defined positive number* $\sqrt{\frac{\mathring{\mathcal{N}}_{1,1}(1)}{|\mathring{\mathcal{U}}_{1,1}(1)|^2+|\mathring{\mathcal{U}}_{1,2}(1)|^2}}$*. Define* $\mathring{\mathcal{U}}_{2,1}$ *and* $\mathring{\mathcal{U}}_{2,2}$ *as in* (3.4.7) *and* (3.4.8)*. Then* $\mathring{\mathcal{U}}_{2,1}$ *and* $\mathring{\mathcal{U}}_{2,2}$ *are well-defined Laurent polynomials having [complex] symmetry (and real coefficients)* $\mathsf{S}\mathring{\mathcal{U}}_{2,1}(z) = \epsilon z^{\mathring{n}+c_{\mathring{\mathcal{N}}_{2,1}}-c_{\text{odd}}}$ *and* $\mathsf{S}\mathring{\mathcal{U}}_{2,2}(z) = \epsilon\mathring{\epsilon}\epsilon_{\mathring{\mathcal{N}}_{2,1}} z^{\mathring{n}+c_{\mathring{\mathcal{N}}_{2,1}}}$*. Moreover,*

$$\mathring{\mathcal{U}}(z)\mathring{\mathcal{U}}^{\star}(z) = \mathring{\mathcal{N}}(z) \quad \text{and} \quad \frac{\mathsf{S}\mathring{\mathcal{U}}_{1,2}(z)}{\mathsf{S}\mathring{\mathcal{U}}_{1,1}(z)} = \frac{\mathsf{S}\mathring{\mathcal{U}}_{2,2}(z)}{\mathsf{S}\mathring{\mathcal{U}}_{2,1}(z)} = \mathring{\epsilon}\epsilon_{\mathring{\mathcal{N}}_{2,1}} z^{c_{\text{odd}}};$$

(S5) By Theorem 3.1.6 and item (ii) for the case of symmetry (or by Theorem 3.1.5 for complex symmetry), there exist Laurent polynomials p_1 *and* p_2 *having [complex] symmetry (and real coefficients) such that* (3.1.7) *(or* (3.1.6)*) holds with respect to the symmetry type* $\mathring{\epsilon}\epsilon_{\mathring{\mathcal{N}}_{2,1}} z^{c_{\text{odd}}}$*, which equals* $z^{2k}\mathsf{S}\mathcal{N}_{2,1}(z)\mathsf{Sd}(z)$ *for some* $k \in \mathbb{Z}$*;*

(S6) Define a 2×2 *matrix* $\mathcal{U}$ *of Laurent polynomials by*

$$\mathcal{U}(z) := \begin{bmatrix} \mathcal{U}_{1,1}(z) & \mathcal{U}_{1,2}(z) \\ \mathcal{U}_{2,1}(z) & \mathcal{U}_{2,2}(z) \end{bmatrix} := \begin{bmatrix} \mathsf{q}(z) & 0 \\ 0 & 1 \end{bmatrix} \mathring{\mathcal{U}}(z) \begin{bmatrix} \mathsf{p}_1(z) & -\mathsf{p}^{\star}_2(z) \\ \mathsf{p}_2(z) & \mathsf{p}^{\star}_1(z) \end{bmatrix}.$$

Then $\mathcal{U}$ *has [complex] symmetry (and real coefficients) and satisfies* (3.6.1).

Proof By Theorem 3.1.8 and item (ii) of Theorem 3.6.1, $\mathsf{p}(z) \geqslant 0$ for all $z \in \mathbb{T}$ and p has real coefficients. Hence, $\gcd(\breve{\mathcal{N}}_{1,1}, \breve{\mathcal{N}}_{1,2}, \breve{\mathcal{N}}_{2,1}, \breve{\mathcal{N}}_{2,2}) = 1$ and $\breve{\mathcal{N}}(z) \geqslant 0$ for all $z \in \mathbb{T}$. Since $\breve{\mathcal{N}}_{1,1}(z) \geqslant 0$ for all $z \in \mathbb{T}$, we see that $\breve{\mathsf{p}}(z) \geqslant 0$ for all $z \in \mathbb{T}$. Note that

$$\mathsf{d}(z)\mathsf{d}^{\star}(z) = (\mathsf{p}(z))^2[\breve{\mathcal{N}}_{1,1}(z)\breve{\mathcal{N}}_{2,2}(z) - \breve{\mathcal{N}}_{1,2}(z)\breve{\mathcal{N}}_{2,1}(z)]. \tag{3.6.10}$$

For the case of complex symmetry, since d has complex symmetry (due to item (i)), we have $\mathsf{Z}(\mathsf{dd}^{\star}, z) \in 2\mathbb{Z}$ for all $z \in \mathbb{C}\backslash\{0\}$. Hence, it follows from (3.6.10) that

$$\mathsf{Z}(\det(\breve{\mathcal{N}}), z) = \mathsf{Z}(\breve{\mathcal{N}}_{1,1}\breve{\mathcal{N}}_{2,2} - \breve{\mathcal{N}}_{1,2}\breve{\mathcal{N}}_{2,1}, z) \in 2\mathbb{Z} \tag{3.6.11}$$

for all $z \in \mathbb{C}\backslash\{0\}$. Since $\breve{\mathcal{N}}_{2,1} = \breve{\mathcal{N}}_{1,2}^{\star} = \breve{\mathcal{N}}_{1,2}/\mathbb{S}\breve{\mathcal{N}}_{1,2}$, we have $\mathsf{Z}(\breve{\mathcal{N}}_{1,2}\breve{\mathcal{N}}_{2,1}, z) \in 2\mathbb{Z}$. If $\mathsf{Z}(\breve{\mathsf{q}}, z_0) = 2m-1$ for some $m \in \mathbb{N}$ and $z_0 \in \mathbb{C}\backslash\{0\}$, by $\breve{\mathsf{q}} = \gcd(\breve{\mathcal{N}}_{1,1}, \breve{\mathcal{N}}_{1,2}\breve{\mathcal{N}}_{1,2}^{\star})$, then $\mathsf{Z}(\breve{\mathcal{N}}_{1,2}\breve{\mathcal{N}}_{2,1}, z_0) \geqslant 2m$ and $\mathsf{Z}(\breve{\mathcal{N}}_{1,1}, z_0) = 2m-1$. Now it follows from (3.6.11) that $\mathsf{Z}(\breve{\mathcal{N}}_{2,2}, z_0)$ must be at least one. However, this implies $\breve{\mathcal{N}}_{1,1}(z_0) = \breve{\mathcal{N}}_{2,2}(z_0) = \breve{\mathcal{N}}_{1,2}(z_0) = \breve{\mathcal{N}}_{2,1}(z_0) = 0$, contradicting $\gcd(\breve{\mathcal{N}}_{1,1}, \breve{\mathcal{N}}_{1,2}, \breve{\mathcal{N}}_{2,1}, \breve{\mathcal{N}}_{2,2}) = 1$. So, $\mathsf{Z}(\breve{\mathsf{q}}, z) \in 2\mathbb{Z}$ for all $z \in \mathbb{C}\backslash\{0\}$. Since q_1 has complex symmetry, $\mathsf{Z}(\mathsf{q}_1\mathsf{q}_1^{\star}/\breve{\mathsf{q}}, z) \in 2\mathbb{Z}$ for all $z \in \mathbb{C}\backslash\{0\}$. By Theorem 3.1.7 and $\breve{\mathsf{q}}(z) \geqslant 0$, there exists a Laurent polynomial q_2 with complex symmetry such that $\mathsf{q}_2(z)\mathsf{q}_2^{\star}(z) = \mathsf{q}_1(z)\mathsf{q}_1^{\star}(z)/\breve{\mathsf{q}}(z)$. Since $\det(\breve{\mathcal{N}}(z)) = \mathsf{q}(z)\mathsf{q}^{\star}(z)\det(\mathring{\mathcal{N}}(z))$ and (3.6.11) holds for all $z \in \mathbb{C}\backslash\{0\}$, we conclude from Theorem 3.1.7 that there exists a desired Laurent polynomial $\tilde{\mathsf{d}}$ in (S2).

For the case of symmetry, since d and $\breve{\mathcal{N}}_{1,2}$ have symmetry, by Theorem 3.1.8, we have $\mathsf{Z}(\mathsf{d}\mathsf{d}^{\star}, z) \in 2\mathbb{Z}$ for all $z \in \mathbb{R}\backslash\{0\}$. Therefore, it follows from (3.6.10) that (3.6.11) holds for all $z \in \mathbb{R}\backslash\{0\}$. Now using the same proof by contradiction for complex symmetry by considering only $z_0 \in \mathbb{R}\backslash\{0\}$, we conclude that $\mathsf{Z}(\breve{\mathsf{q}}, x) \in 2\mathbb{Z}$ for all $x \in \mathbb{R}\backslash\{0\}$ and consequently, $\mathsf{Z}(\mathsf{q}_1\mathsf{q}_1^{\star}/\breve{\mathsf{q}}, x) \in 2\mathbb{Z}$ for all $x \in \mathbb{R}\backslash\{0\}$. Since $\mathsf{q}_1(z)\mathsf{q}_1^{\star}(z)/\breve{\mathsf{q}}(z) \geqslant 0$ for all $z \in \mathbb{T}$ and has real coefficients, by Theorem 3.1.8, there exists a Laurent polynomial q_2 with symmetry and real coefficients such that $\mathsf{q}_2(z)\mathsf{q}_2^{\star}(z) = \mathsf{q}_1(z)\mathsf{q}_1^{\star}(z)/\breve{\mathsf{q}}(z)$. Now by the same argument as in Algorithm 3.4.3, all the claims in (S1) hold with $\mathsf{q}(z)\mathsf{q}^{\star}(z) = \breve{\mathsf{q}}(z)$ and $\mathsf{q} \mid \breve{\mathcal{N}}_{1,2}$. Since $\det(\breve{\mathcal{N}}(z)) = \mathsf{q}(z)\mathsf{q}^{\star}(z)\det(\mathring{\mathcal{N}}(z))$ and (3.6.11) holds for all $z \in \mathbb{R}\backslash\{0\}$, we deduce from Theorem 3.1.8 that there exists a desired Laurent polynomial $\tilde{\mathsf{d}}$ in (S2).

For the case of complex symmetry, since $\{\mathring{\mathcal{U}}_{1,1}, \mathring{\mathcal{U}}_{1,2}\}$ obtained in (S3) before symmetrization is a solution to (3.6.9), then

$$\epsilon z^{\mathring{n}+c_{\mathring{\mathcal{N}}_{2,1}}-c_{\text{odd}}}[\mathring{\mathcal{N}}_{2,1}^{\star}(z)\mathring{\mathcal{U}}_{1,1}^{\star}(z) - \mathring{\mathsf{d}}^{\star}(z)\mathring{\mathcal{U}}_{1,2}(z)] = \epsilon z^{\mathring{n}+c_{\mathring{\mathcal{N}}_{2,1}}-c_{\text{odd}}}\mathring{\mathcal{N}}_{1,1}^{\star}(z)\mathring{\mathcal{U}}_{2,1}^{\star}(z).$$

Note that $\mathring{\mathcal{N}}_{1,1}^{\star}(z) = \mathring{\mathcal{N}}_{1,1}(z)$ and $\mathring{\mathcal{N}}_{2,1}^{\star}(z) = \epsilon_{\mathring{\mathcal{N}}_{2,1}} z^{-c_{\mathring{\mathcal{N}}_{2,1}}}\mathring{\mathcal{N}}_{2,1}(z)$. We deduce from the above identity and $\mathring{\mathsf{d}}^{\star}(z) = \mathring{\epsilon} z^{-2\mathring{n}-c_{\mathring{\mathcal{N}}_{2,1}}+c_{\text{odd}}}\mathring{\mathsf{d}}(z)$ that

$$\mathring{\mathcal{N}}_{2,1}(z)\epsilon\epsilon_{\mathring{\mathcal{N}}_{2,1}} z^{\mathring{n}-c_{\text{odd}}}\mathring{\mathcal{U}}_{1,1}^{\star}(z) - \mathring{\mathsf{d}}(z)\epsilon\mathring{\epsilon} z^{-\mathring{n}}\mathring{\mathcal{U}}_{1,2}(z) = \mathring{\mathcal{N}}_{1,1}(z)\epsilon z^{\mathring{n}+c_{\mathring{\mathcal{N}}_{2,1}}-c_{\text{odd}}}\mathring{\mathcal{U}}_{2,1}^{\star}(z).$$

Now the above identity and (3.6.9) together imply that the symmetrized pair $\{\mathring{\mathcal{U}}_{1,1}, \mathring{\mathcal{U}}_{1,2}\}$ defined in (3.6.8) is also a solution to (3.6.9) with $\text{fsupp}(\mathring{\mathcal{U}}_{1,1}) \subseteq [0, \mathring{n} - c_{\text{odd}}] \subseteq [0, \mathring{n}]$ and $\text{fsupp}(\mathring{\mathcal{U}}_{1,2}) \subseteq [0, \mathring{n}]$.

For the case of symmetry, since $\{\mathring{\mathcal{U}}_{1,1}, \mathring{\mathcal{U}}_{1,2}\}$ is a solution to (3.6.9),

$$\begin{aligned}\epsilon z^{\mathring{n}+c_{\mathring{\mathcal{N}}_{2,1}}-c_{\text{odd}}}&[\mathring{\mathcal{N}}_{2,1}(z^{-1})\mathring{\mathcal{U}}_{1,1}(z^{-1}) - \mathring{\mathsf{d}}(z^{-1})\mathring{\mathcal{U}}_{1,2}^{\star}(z^{-1})]\\ &= \epsilon z^{\mathring{n}+c_{\mathring{\mathcal{N}}_{2,1}}-c_{\text{odd}}}\mathring{\mathcal{N}}_{1,1}(z^{-1})\mathring{\mathcal{U}}_{2,1}(z^{-1}).\end{aligned}$$

Since $\mathring{\mathcal{N}}_{1,1}$ has symmetry and $\mathring{\mathcal{N}}_{1,1}(z) \geqslant 0$ for all $z \in \mathbb{T}$, by Lemma 3.1.2, $\mathring{\mathcal{N}}_{1,1}$ must have real coefficients and $\mathsf{S}\mathring{\mathcal{N}}_{1,1}(z) = \mathbb{S}\mathring{\mathcal{N}}_{1,1}(z) = 1$. That is, $\mathring{\mathcal{N}}_{1,1}(z^{-1}) = \mathring{\mathcal{N}}_{1,1}(z)$. By $\mathring{\mathcal{N}}_{2,1}(z^{-1}) = \epsilon_{\mathring{\mathcal{N}}_{2,1}} z^{-c_{\mathring{\mathcal{N}}_{2,1}}} \mathring{\mathcal{N}}_{2,1}(z)$ and $\mathring{\mathsf{d}}(z^{-1}) = \mathring{\epsilon} z^{-2\mathring{n}-c_{\mathring{\mathcal{N}}_{2,1}}+c_{\text{odd}}} \mathring{\mathsf{d}}(z)$,

$$\mathring{\mathcal{N}}_{2,1}(z)\epsilon\epsilon_{\mathring{\mathcal{N}}_{2,1}} z^{\mathring{n}-c_{\text{odd}}} \mathring{\mathcal{U}}_{1,1}(z^{-1}) - \mathring{\mathsf{d}}(z)\epsilon\mathring{\epsilon} z^{-n} \mathring{\mathcal{U}}^{\star}_{1,2}(z^{-1})$$
$$= \mathring{\mathcal{N}}_{1,1}(z)\epsilon z^{\mathring{n}+c_{\mathring{\mathcal{N}}_{2,1}}-c_{\text{odd}}} \mathring{\mathcal{U}}_{2,1}(z^{-1}).$$

Now the above identity and (3.6.9) together imply that the symmetrized pair $\{\mathring{\mathcal{U}}_{1,1}, \mathring{\mathcal{U}}_{1,2}\}$ defined in (3.6.7) is also a solution to (3.6.9), and $\operatorname{fsupp}(\mathring{\mathcal{U}}_{1,1}) \subseteq [0, \mathring{n} - c_{\text{odd}}] \subseteq [0, \mathring{n}]$ and $\operatorname{fsupp}(\mathring{\mathcal{U}}_{1,2}) \subseteq [0, \mathring{n}]$.

By $\operatorname{fsupp}(\mathring{\mathcal{R}}) \subseteq [-\mathring{n}, \mathring{n} - 1]$, there are no more than $2\mathring{n}$ homogeneous linear equations in X, while there are $2\mathring{n} + 2 - c_{\text{odd}} \geqslant 2\mathring{n} + 1$ unknowns. Hence, the space of all solutions to X must have dimension at least one. Therefore, we must have a nontrivial symmetrized solution $\{\mathring{\mathcal{U}}_{1,1}, \mathring{\mathcal{U}}_{1,2}\}$ to X in (S3). By $\mathring{\mathcal{R}}(z) = 0$ and $\mathsf{S}\mathring{\mathcal{N}}_{2,1}(z)\mathsf{S}\mathring{\mathcal{U}}_{1,1}(z) = \mathsf{S}\mathring{\mathsf{d}}(z)\mathsf{S}\mathring{\mathcal{U}}^{\star}_{1,2}(z)$, we see that $\mathring{\mathcal{U}}_{2,1}$ defined in (3.4.7) is a well-defined Laurent polynomial with [complex] symmetry. This proves step (S3).

By $\mathcal{N}_{2,1}(z) = \mathsf{p}(z)\mathsf{q}^{\star}(z)\mathring{\mathcal{N}}_{2,1}(z)$ and $\mathsf{Sd}(z) = \mathring{\epsilon} z^{\mathring{c}+2j}\mathsf{Sq}(z)$ for some $j \in \mathbb{Z}$, by (3.6.6) we see that

$$\mathsf{S}\mathcal{N}_{2,1}(z)\mathsf{Sd}(z) = \mathring{\epsilon}\epsilon_{\mathring{\mathcal{N}}_{2,1}} z^{c_{\mathring{\mathcal{N}}_{2,1}}+\mathring{c}+2j} = \mathring{\epsilon}\epsilon_{\mathring{\mathcal{N}}_{2,1}} z^{c_{\text{odd}}} z^{2k},$$

where $k := j + (c_{\mathring{\mathcal{N}}_{2,1}} + \mathring{c} - c_{\text{odd}})/2 \in \mathbb{Z}$. Now by Theorem 3.1.6 or Theorem 3.1.5, the existence of p_1 and p_2 in (S5) is justified.

The rest of claims can be proved by the same argument as in Algorithm 3.4.3. □

We finish this section by proving Theorem 3.3.7.

Proof (of Theorem 3.3.7) Necessity (⇒). Suppose that there is a tight framelet filter bank $\{a; b_1, b_2\}_\Theta$ such that a, b_1, b_2, Θ have [complex] symmetry in (3.3.18) for some $\epsilon, \epsilon_1, \epsilon_2 \in \{-1, 1\}$ and $c, c_1, c_2 \in \mathbb{Z}$. By Lemma 3.3.2, $\mathcal{N}_{a,\Theta}(z) \geqslant 0$ for all $z \in \mathbb{T}$. By the relation between $\mathcal{N}_{a,\Theta}$ and $\mathcal{N}_{a,\Theta|n_b}$, we have $\mathcal{N}_{a,\Theta|n_b}(z) \geqslant 0$ for all $z \in \mathbb{T}$ and $\mathbb{S}\Theta = 1$. Define

$$\mathcal{N}(z) := \begin{bmatrix} \mathcal{N}_{1,1}(z) & \mathcal{N}_{1,2}(z) \\ \mathcal{N}_{2,1}(z) & \mathcal{N}_{2,2}(z) \end{bmatrix} = \begin{cases} \mathcal{N}_{a,\Theta|n_b}(z), & \text{if } c + n_b \text{ is even}, \\ P_{k_a}(z)\mathcal{N}_{a,\Theta|n_b}(z)P^{\star}_{k_a}(z), & \text{if } c + n_b \text{ is odd}, \end{cases} \tag{3.6.12}$$

where $\mathcal{N}_{a,\Theta|n_b}$ is defined in (3.3.10) and

$$P_{k_a}(z) := \frac{1}{\sqrt{2}} \begin{bmatrix} 1 & z^{k_a} \\ 1 & -z^{k_a} \end{bmatrix} \tag{3.6.13}$$

with $k_a := 0$. Trivially, $\det(\mathcal{N}(z)) = \det(\mathcal{N}_{a,\Theta|n_b}(z))$ and $\mathcal{N}^\star = \mathcal{N}$. We now show that all the entries of $\mathcal{N}$ have [complex] symmetry. By the definition of A, B in (3.3.8), we have

$$\mathsf{SA}(z) = 1, \quad \mathsf{SB}(z) = (-1)^{c+n_b}. \tag{3.6.14}$$

By (3.6.14), we see that

$$\mathsf{SA}^{[0]}(z)=1,\ \mathsf{SA}^{[1]}(z)=z^{-1},\ \mathsf{SB}^{[0]}(z)=(-1)^{c+n_b},\ \mathsf{SB}^{[1]}(z)=(-1)^{c+n_b}z^{-1}. \tag{3.6.15}$$

When $c + n_b$ is even, by (3.6.12) and the definition of $\mathcal{N}_{a,\Theta|n_b}(z)$ in (3.3.10) and by (3.6.15), we see that $\mathsf{S}\mathcal{N}_{1,1}(z) = \mathsf{S}\mathcal{N}_{2,2}(z) = 1$, $\mathsf{S}\mathcal{N}_{1,2}(z) = z$, and $\mathsf{S}\mathcal{N}_{2,1}(z) = z^{-1}$.

When $c + n_b$ is odd, though some entries of $\mathcal{N}_{a,\Theta|n_b}$ may no longer have any symmetry property, since $\mathcal{N}(z) = P_{k_a}(z)\mathcal{N}_{a,\Theta|n_b}(z)P^\star_{k_a}(z)$, we have

$$\begin{aligned}
[\mathcal{N}(z)]_{1,1} &= \tfrac{1}{2}\big([\mathcal{N}_{a,\Theta|n_b}(z)]_{1,1} + [\mathcal{N}_{a,\Theta|n_b}(z)]_{2,2} + z^{k_a}[\mathcal{N}_{a,\Theta|n_b}(z)]_{2,1} + z^{-k_a}[\mathcal{N}_{a,\Theta|n_b}(z)]_{1,2}\big)\\
&= \tfrac{1}{4}\big(2\mathsf{A}^{[0]}(z) + (z^{1-k_a} + z^{k_a})\mathsf{A}^{[1]}(z) + (z^{1-k_a} - z^{k_a})\mathsf{B}^{[1]}(z)\big),\\
[\mathcal{N}(z)]_{1,2} &= \tfrac{1}{2}\big([\mathcal{N}_{a,\Theta|n_b}(z)]_{1,1} - [\mathcal{N}_{a,\Theta|n_b}(z)]_{2,2} + z^{k_a}[\mathcal{N}_{a,\Theta|n_b}(z)]_{2,1} - z^{-k_a}[\mathcal{N}_{a,\Theta|n_b}(z)]_{1,2}\big)\\
&= \tfrac{1}{4}\big(2\mathsf{B}^{[0]}(z) - (z^{1-k_a} - z^{k_a})\mathsf{A}^{[1]}(z) - (z^{1-k_a} + z^{k_a})\mathsf{B}^{[1]}(z)\big),\\
[\mathcal{N}(z)]_{2,1} &= \tfrac{1}{2}\big([\mathcal{N}_{a,\Theta|n_b}(z)]_{1,1} - [\mathcal{N}_{a,\Theta|n_b}(z)]_{2,2} - z^{k_a}[\mathcal{N}_{a,\Theta|n_b}(z)]_{2,1} + z^{-k_a}[\mathcal{N}_{a,\Theta|n_b}(z)]_{1,2}\big)\\
&= \tfrac{1}{4}\big(2\mathsf{B}^{[0]}(z) + (z^{1-k_a} - z^{k_a})\mathsf{A}^{[1]}(z) + (z^{1-k_a} + z^{k_a})\mathsf{B}^{[1]}(z)\big),\\
[\mathcal{N}(z)]_{2,2} &= \tfrac{1}{2}\big([\mathcal{N}_{a,\Theta|n_b}(z)]_{1,1} + [\mathcal{N}_{a,\Theta|n_b}(z)]_{2,2} - z^{k_a}[\mathcal{N}_{a,\Theta|n_b}(z)]_{2,1} - z^{-k_a}[\mathcal{N}_{a,\Theta|n_b}(z)]_{1,2}\big)\\
&= \tfrac{1}{4}\big(2\mathsf{A}^{[0]}(z) - (z^{1-k_a} + z^{k_a})\mathsf{A}^{[1]}(z) - (z^{1-k_a} - z^{k_a})\mathsf{B}^{[1]}(z)\big).
\end{aligned}$$

Since $c+n_b$ is odd, by (3.6.15), we have $\mathsf{S}\mathcal{N}_{1,1}(z) = \mathsf{S}\mathcal{N}_{2,2}(z) = 1$ and $\mathsf{S}\mathcal{N}_{1,2}(z) = \mathsf{S}\mathcal{N}_{2,1}(z) = -1$.

Hence, for both cases we have $\mathsf{S}\mathcal{N}_{2,1}(z) = (-1)^{c+n_b}z^{\operatorname{odd}(c+n_b)-1}$. We proved in Theorem 3.3.5 that (3.3.23) holds. Suppose that $c+n_b$ is an even integer. By (3.3.23) we see that both $c_1 + n_b$ and $c_2 + n_b$ are even integers. Since $\mathcal{N} = \mathcal{N}_{a,\Theta}$, setting

$$\mathcal{U}(z) := \begin{bmatrix} \mathcal{U}_{1,1}(z) & \mathcal{U}_{1,2}(z)\\ \mathcal{U}_{2,1}(z) & \mathcal{U}_{2,2}(z)\end{bmatrix} := \begin{bmatrix} \mathring{\mathsf{b}}_1^{[0]}(z) & \mathring{\mathsf{b}}_2^{[0]}(z)\\ \mathring{\mathsf{b}}_1^{[1]}(z) & \mathring{\mathsf{b}}_2^{[1]}(z)\end{bmatrix},$$

we deduce that all the entries in $\mathcal{U}$ have symmetry, $\mathcal{U}(z)\mathcal{U}^\star(z) = \mathcal{N}(z)$, and

$$\frac{\mathsf{S}\mathcal{U}_{1,1}(z)}{\mathsf{S}\mathcal{U}_{2,1}(z)} = \frac{\mathsf{S}\mathring{\mathsf{b}}_1^{[0]}(z)}{\mathsf{S}\mathring{\mathsf{b}}_1^{[1]}(z)} = \frac{(-1)^{n_b}\epsilon_1 z^{\frac{c_1+n_b}{2}}}{(-1)^{n_b}\epsilon_1 z^{\frac{c_1+n_b}{2}-1}} = z$$
$$= \frac{(-1)^{n_b}\epsilon_2 z^{\frac{c_2+n_b}{2}}}{(-1)^{n_b}\epsilon_2 z^{\frac{c_2+n_b}{2}-1}} = \frac{\mathsf{S}\mathring{\mathsf{b}}_2^{[0]}(z)}{\mathsf{S}\mathring{\mathsf{b}}_2^{[1]}(z)} = \frac{\mathsf{S}\mathcal{U}_{1,2}(z)}{\mathsf{S}\mathcal{U}_{2,2}(z)}.$$

Suppose that $c + n_b$ is an odd integer. By (3.3.23) we see that both $c_1 + n_b$ and $c_2 + n_b$ are odd integers. Define

$$\mathcal{U}(z) := \begin{bmatrix} \mathcal{U}_{1,1}(z) & \mathcal{U}_{1,2}(z) \\ \mathcal{U}_{2,1}(z) & \mathcal{U}_{2,2}(z) \end{bmatrix} := P_{k_a}(z) \begin{bmatrix} \mathring{\mathsf{b}}_1^{[0]}(z) & \mathring{\mathsf{b}}_2^{[0]}(z) \\ \mathring{\mathsf{b}}_1^{[1]}(z) & \mathring{\mathsf{b}}_2^{[1]}(z) \end{bmatrix}.$$

We can directly verify that all the entries in $\mathcal{U}$ have symmetry, $\mathcal{U}(z)\mathcal{U}^\star(z) = \mathcal{N}(z)$, and

$$\frac{\mathsf{S}\mathcal{U}_{1,1}(z)}{\mathsf{S}\mathcal{U}_{2,1}(z)} = \frac{(-1)^{n_b}\epsilon_1 z^{(c_1+n_b-1)/2}}{(-1)^{n_b+1}\epsilon_1 z^{(c_1+n_b-1)/2}} = -1 = \frac{(-1)^{n_b}\epsilon_2 z^{(c_2+n_b-1)/2}}{(-1)^{n_b+1}\epsilon_2 z^{(c_2+n_b-1)/2}} = \frac{\mathsf{S}\mathcal{U}_{1,2}(z)}{\mathsf{S}\mathcal{U}_{2,2}(z)}.$$

Observe that $\mathsf{S}\mathcal{N}_{2,1}(z)\mathsf{Sd}_{n_b}(z) = (-1)^{c+n_b} z^{\operatorname{odd}(c+n_b)-1}\mathsf{Sd}_{n_b}(z)$. By Theorem 3.6.1, we conclude that items (i) and (ii) must hold. Hence, we verified the necessity part. The proof for the case of complex symmetry is the same by replacing S with $\mathbb{S}$.

Sufficiency ($\Leftarrow$). Define $\mathcal{N}$ as in (3.6.12), where P_{k_a} is defined in (3.6.13) and $k_a := \lfloor \frac{n_+ + n_-}{2} \rfloor$ with $[n_-, n_+] := \operatorname{fsupp}([\mathcal{N}_{a,\Theta|n_b}]_{1,2})$.

We now prove that items (i) and (ii) of Theorem 3.6.1 are satisfied. Note that item (i) in Theorem 3.3.7 implies that $\det(\mathcal{N}_{a,\Theta}(z)) = |1 - z^{-2}|^{n_b} |\mathsf{d}_{n_b}(z)|^2 \geqslant 0$ for all $z \in \mathbb{T}$. Together with item (i) in Theorem 3.3.7, we must have $\mathcal{N}_{a,\Theta}(z) \geqslant 0$ for all $z \in \mathbb{T}$. Now it is not difficult to see that $\mathcal{N}_{a,\Theta|n_b}(z) \geqslant 0$ for all $z \in \mathbb{T}$. By the definition of $\mathcal{N}$ in (3.6.12), it is now trivial to see that $\mathcal{N}(z) \geqslant 0$ for all $z \in \mathbb{T}$. As we argued in the proof of Theorem 3.3.7, $\mathsf{S}\mathcal{N}_{2,1}(z) = (-1)^{c+n_b} z^{\operatorname{odd}(c+n_b)-1}$. Since $\det(\mathcal{N}_{a,\Theta|n_b}(z)) = \det(\mathcal{N}(z))$ and $\gcd(\mathcal{N}_{1,1}, \mathcal{N}_{1,2}, \mathcal{N}_{2,1}, \mathcal{N}_{2,2}) = \mathsf{p}$, items (i) and (ii) of Theorem 3.6.1 simply correspond to items (i) and (ii) of Theorem 3.3.7.

By Theorem 3.6.1 (more precisely, Algorithm 3.6.2), we can factorize $\mathcal{N}$ such that (3.6.1) is satisfied. If $c + n_b$ is even, define b_1 and b_2 as in (3.4.17). If $c + n_b$ is odd, define

$$\begin{aligned} \mathsf{b}_1(z) &:= (1 - z^{-1})^{n_b} \left[\frac{1 + z^{1-2k_a}}{\sqrt{2}} \mathcal{U}_{1,1}(z^2) + \frac{1 - z^{1-2k_a}}{\sqrt{2}} \mathcal{U}_{2,1}(z^2) \right], \\ \mathsf{b}_2(z) &:= (1 - z^{-1})^{n_b} \left[\frac{1 + z^{1-2k_a}}{\sqrt{2}} \mathcal{U}_{1,2}(z^2) + \frac{1 - z^{1-2k_a}}{\sqrt{2}} \mathcal{U}_{2,2}(z^2) \right]. \end{aligned} \tag{3.6.16}$$

By the definition of b_1 and b_2 in (3.6.16) and by the definition of $\mathcal{N}$ in (3.6.12), we can directly check that $\{a; b_1, b_2\}_\Theta$ is a tight framelet filter bank with [complex] symmetry (and real coefficients). □

3.7 Exercises

3.1. Let $a \in l_0(\mathbb{Z})$ such that $\mathsf{Sa}(z) = \epsilon z^c$ with $\epsilon \in \{-1, 1\}$ and $c \in \mathbb{Z}$. Prove that (i) if c is an even integer, then $\mathsf{a}^{[0]}(z) = \epsilon z^{\frac{c}{2}} \mathsf{a}^{[0]}(z^{-1})$ and $\mathsf{a}^{[1]}(z) = \epsilon z^{\frac{c}{2}-1} \mathsf{a}^{[1]}(z^{-1})$; (ii) if c is an odd integer, then $\mathsf{a}^{[1]}(z) = \epsilon z^{(c-1)/2} \mathsf{a}^{[0]}(z^{-1})$.

3.2. Let $a \in l_0(\mathbb{Z})$ such that $\mathbb{S}\mathsf{a}(z) = \epsilon z^c$ with $\epsilon \in \{-1, 1\}$ and $c \in \mathbb{Z}$. Prove that (i) if c is an even integer, then $\mathsf{a}^{[0]}(z) = \epsilon z^{\frac{c}{2}} (\mathsf{a}^{[0]}(z))^\star$ and $\mathsf{a}^{[1]}(z) = \epsilon z^{\frac{c}{2}-1} (\mathsf{a}^{[1]}(z))^\star$; (ii) if c is an odd integer, then $\mathsf{a}^{[1]}(z) = \epsilon z^{(c-1)/2} (\mathsf{a}^{[0]}(z))^\star$.

3.3. Let $u \in l_0(\mathbb{Z})$ such that u has both symmetry and complex symmetry. Prove that either u or iu must be real-valued.

3.4. If a Laurent polynomial p has symmetry $\mathsf{Sp}(z) = \epsilon z^c$, prove that $\mathsf{Sp}^\star(z) = \epsilon z^{-c}$.

3.5. For a Laurent polynomial p, prove that $\gcd(\mathsf{p}, \mathsf{p}^\star)$ has complex symmetry, i.e., if q is a greatest common divisor of p and $\mathsf{p}^\star$, then there exist $k \in \mathbb{Z}$ and $\lambda \in \mathbb{T}$ such that $\lambda z^k \mathsf{q}(z)$ has complex symmetry.

3.6. Prove Proposition 3.1.1.

3.7. Prove Lemma 3.1.4.

3.8. Let $a = a_3^{\mathsf{S}}$ be the low-pass filter defined in (2.4.7). Then $\mathsf{Sa}(z) = z$ and $a * a^\star = a_6^I$. Prove that $\operatorname{lpm}(a_3^{\mathsf{S}}) = 2$ and $\operatorname{lpm}(a_6^I) = 6$.

3.9. Prove that $(\{\tilde{a}; \tilde{b}_1, \ldots, \tilde{b}_s\}, \{a; b_1, \ldots, b_s\})_\Theta$ is a dual framelet filter bank if and only if

$$\begin{bmatrix} \tilde{\mathsf{b}}_1^{[0]}(z) \cdots \tilde{\mathsf{b}}_s^{[0]}(z) \\ \tilde{\mathsf{b}}_1^{[1]}(z) \cdots \tilde{\mathsf{b}}_s^{[1]}(z) \end{bmatrix} \begin{bmatrix} \mathsf{b}_1^{[0]}(z) \cdots \mathsf{b}_s^{[0]}(z) \\ \mathsf{b}_1^{[1]}(z) \cdots \mathsf{b}_s^{[1]}(z) \end{bmatrix}^\star = \mathcal{N}_{a,\tilde{a},\Theta}(z),$$

where

$$\mathcal{N}_{a,\tilde{a},\Theta}(z) := \begin{bmatrix} \frac{1}{2}\Theta^{[0]}(z) - \Theta(z)\tilde{\mathsf{a}}^{[0]}(z)(\mathsf{a}^{[0]}(z))^\star & \frac{1}{2}z\Theta^{[1]}(z) - \Theta(z)\tilde{\mathsf{a}}^{[0]}(z)(\mathsf{a}^{[1]}(z))^\star \\ \frac{1}{2}\Theta^{[1]}(z) - \Theta(z)\tilde{\mathsf{a}}^{[1]}(z)(\mathsf{a}^{[0]}(z))^\star & \frac{1}{2}\Theta^{[0]}(z) - \Theta(z)\tilde{\mathsf{a}}^{[1]}(z)(\mathsf{a}^{[1]}(z))^\star \end{bmatrix}.$$

3.10. Let $(\{\tilde{a}; \tilde{b}_1, \ldots, \tilde{b}_s\}, \{a; b_1, \ldots, b_s\})_\Theta$ be a dual framelet filter bank such that $\mathsf{a}(1)\tilde{\mathsf{a}}(1)\Theta(1) \neq 0$. Suppose that all dual high-pass filters $\tilde{b}_1, \ldots, \tilde{b}_s$ have $n_{\tilde{b}}$ vanishing moments and all primal high-pass filters $b_1, \ldots, b_s$ have n_b vanishing moments. Prove that $\operatorname{sr}(a) \geqslant n_{\tilde{b}}$, $\operatorname{sr}(\tilde{a}) \geqslant n_b$, and $\Theta(z) - \Theta(z^2)\tilde{\mathsf{a}}(z)\mathsf{a}^\star(z) = \mathscr{O}(|1 - z|^{n_b + n_{\tilde{b}}})$ as $z \to 1$. *Hint:* cf. Lemma 1.4.1.

3.11. If $(\{\tilde{u}_0, \ldots, \tilde{u}_s\}, \{u_0, \ldots, u_s\})$ is a dual framelet filter bank, prove that

$$(\{\tilde{u}_0^{[r]}, \ldots, \tilde{u}_s^{[r]}, \tilde{u}_0^{[i]}, \ldots, \tilde{u}_s^{[i]}\}, \{u_0^{[r]}, \ldots, u_s^{[r]}, u_0^{[i]}, \ldots, u_s^{[i]}\})$$

is a real-valued dual framelet filter bank, where $u^{[r]}, u^{[i]}$ are the real and imaginary parts of $u \in l_0(\mathbb{Z})$. In particular, if $\{u_0, \dots, u_s\}$ is a tight framelet filter bank, then $\{u_0^{[r]}, \dots, u_s^{[r]}, u_0^{[i]}, \dots, u_s^{[i]}\}$ is a real-valued tight framelet filter bank.

3.12. If $(\{\tilde{a}; \tilde{b}_1, \dots, \tilde{b}_s\}, \{a; b_1, \dots, b_s\})_\Theta$ is a dual framelet filter bank with real-valued $a, \tilde{a}, \Theta$, prove that $(\{\tilde{a}; \tilde{b}_1^{[r]}, \dots, \tilde{b}_s^{[r]}, \tilde{b}_1^{[i]}, \dots, \tilde{b}_s^{[i]}\}, \{a; b_1^{[r]}, \dots, b_s^{[r]}, b_1^{[i]}, \dots, b_s^{[i]}\})_\Theta$ is a real-valued dual framelet filter bank. In particular, if $\{a; b_1, \dots, b_s\}_\Theta$ is a tight framelet filter bank with real-valued a and Θ, then $\{a; b_1^{[r]}, \dots, b_s^{[r]}, b_1^{[i]}, \dots, b_s^{[i]}\}_\Theta$ is a real-valued tight framelet filter bank.

3.13. Let $u, v \in l_0(\mathbb{Z})$ be nontrivial filters having symmetry $\mathsf{S}\mathsf{u}(z) = \epsilon_u z^{c_u}$ and $\mathsf{S}\mathsf{v}(z) = \epsilon_v z^{c_v}$ (or complex symmetry $\mathbb{S}\mathsf{u}(z) = \epsilon_u z^{c_u}$ and $\mathbb{S}\mathsf{v}(z) = \epsilon_v z^{c_v}$) for some $\epsilon_u, \epsilon_v \in \{-1, 1\}$ and $c_u, c_v \in \mathbb{Z}$. If $u + v$ also has symmetry $\mathsf{S}(\mathsf{u} + \mathsf{v}) = \epsilon z^c$ (or complex symmetry $\mathbb{S}(\mathsf{u} + \mathsf{v}) = \epsilon z^c$), then one of the following two cases must hold:

(i) $\epsilon_u = \epsilon_v$ and $c_u = c_v$, that is, u and v have the same [complex] symmetry type.

(ii) $\epsilon\epsilon_u z^{c-c_u} \ne 1$, $\epsilon\epsilon_v z^{c-c_v} \ne 1$, and $\mathsf{u}(z) = \mathsf{w}(z)\mathring{\mathsf{u}}(z), \mathsf{v}(z) = \mathsf{w}(z)\mathring{\mathsf{v}}(z)$ with $\mathsf{w} := \gcd(\mathsf{u}, \mathsf{v})$, $\mathring{\mathsf{u}}(z) := \frac{1-\epsilon\epsilon_v z^{c-c_v}}{\mathsf{q}(z)}$, $\mathring{\mathsf{v}}(z) := \frac{\epsilon\epsilon_u z^{c-c_u}-1}{\mathsf{q}(z)}$, where q is a Laurent polynomial with [complex] symmetry such that $\mathsf{q}(z) \mid \gcd(1 - \epsilon\epsilon_u z^{c-c_u}, 1 - \epsilon\epsilon_v z^{c-c_v})$.

3.14. Let $(\{\tilde{a}; \tilde{b}_1, \tilde{b}_2\}, \{a; b_1, b_2\})_\Theta$ be a dual framelet filter bank with complex symmetry: $\mathbb{S}\Theta(z) = \epsilon_\Theta z^{c_\Theta}$, $\mathbb{S}\mathsf{a}(z) = \epsilon z^c$, $\mathbb{S}\tilde{\mathsf{a}}(z) = \tilde{\epsilon} z^{\tilde{c}}$, $\mathbb{S}\mathsf{b}_1(z) = \epsilon_1 z^{c_1}$, $\mathbb{S}\mathsf{b}_2(z) = \epsilon_2 z^{c_2}$, $\mathbb{S}\tilde{\mathsf{b}}_1(z) = \tilde{\epsilon}_1(z) z^{\tilde{c}_1}$, $\mathbb{S}\tilde{\mathsf{b}}_2(z) = \tilde{\epsilon}_2(z) z^{\tilde{c}_2}$ such that $\tilde{c} = c - c_\Theta$, $\tilde{c}_1 = c_1 + c_\Theta$, and $\tilde{c}_2 = c_2 + c_\Theta$. Prove that $\epsilon_\Theta \epsilon\tilde{\epsilon} = \epsilon_1\tilde{\epsilon}_1 = \epsilon_2\tilde{\epsilon}_2$ and $(-1)^c = (-1)^{c_1} = (-1)^{c_2}$.

3.15. Let $U = U_1 + iU_2$ such that U_1 and U_2 are $n \times n$ matrices of real numbers. Suppose that $x^\mathsf{T} U x \in \mathbb{R}$ for all $x \in \mathbb{R}^n$. Prove that $U_1^\mathsf{T} = U_1$ and $U_2^\mathsf{T} = -U_2$.

3.16. Let U be an $n \times n$ matrix in $\mathbb{C}$. If $\bar{x}^\mathsf{T} U x \in \mathbb{R}$ for all $x \in \mathbb{C}^n$, prove that $\overline{U}^\mathsf{T} = U$.

3.17. Suppose that p_1 and p_2 are Laurent polynomials having symmetry such that (3.1.7) holds with c being an odd integer. Prove that $\max(\operatorname{len}(\mathsf{p}_1), \operatorname{len}(\mathsf{p}_2)) \leqslant \operatorname{len}(\mathsf{p})/2$. Also prove that the same conclusion holds for complex symmetry.

3.18. Let p_1 and p_2 be two nontrivial Laurent polynomials such that $\mathsf{p}_1^\star = \mathsf{p}_1$. Show that p_1 and p_2 have no common zeros in $\mathbb{C}\backslash\{0\}$ if and only if $\gcd(\mathsf{p}_1, \mathsf{p}_2\mathsf{p}_2^\star) = 1$.

3.19. Let p_1 and p_2 be two nontrivial Laurent polynomials such that $\mathsf{p}_1(z) \geqslant 0$ for all $z \in \mathbb{T}$. Define $\mathsf{q}_1 := \gcd(\mathsf{p}_1, \mathsf{p}_2)$ and $\check{\mathsf{q}} := \gcd(\mathsf{p}_1, \mathsf{p}_2\mathsf{p}_2^\star)$.

a. If p_2 has complex symmetry, prove that there exists a Laurent polynomial q_2 with complex symmetry satisfying $\mathsf{q}_2(z)\mathsf{q}_2^\star(z) = \mathsf{p}_1(z)\mathsf{p}_1^\star(z)/\check{\mathsf{q}}(z)$ if and only if $\mathsf{Z}(\check{\mathsf{q}}, z) \in 2\mathbb{Z}$ for all $z \in \mathbb{C}\backslash\{0\}$.

b. If both p_1 and p_2 have symmetry, prove that there exists a Laurent polynomial q_2 with symmetry and real coefficients satisfying $\mathsf{q}_2(z)\mathsf{q}_2^\star(z) = \mathsf{p}_1(z)\mathsf{p}_1^\star(z)/\check{\mathsf{q}}(z)$ if and only if $\mathsf{Z}(\check{\mathsf{q}}, x) \in 2\mathbb{Z}$ for all $x \in (-1, 0) \cup (0, 1)$.

Define $\mathsf{q}(z) := \mathsf{q}_1(z)/\mathsf{q}_2(z)$, prove that q is well defined, has [complex] symmetry, $\mathsf{q}\mathsf{q}^\star = \breve{\mathsf{q}}$, $\mathsf{q} \mid \mathsf{p}_2$, and $\gcd(\mathsf{p}_1/\breve{\mathsf{q}}, \mathsf{p}_2/\mathsf{q}) = 1$.

3.20. Let $u \in l_0(\mathbb{Z})$ be a finitely supported filter such that $\widehat{u}(0) = 1$. Let $\Theta \in l_0(\mathbb{Z})$ be a finitely supported sequence. Prove that (3.2.30) holds if and only if $\widehat{\Theta}(\xi) = \frac{1}{\varphi(\xi)} + \mathscr{O}(|\xi|^n)$ as $\xi \to 0$, where $\varphi(\xi) := \prod_{j=1}^\infty \widehat{u}(2^{-j}\xi)$.

3.21. Let $\mathbf{f} \in C^\infty(\mathbb{T})$ and $\mathbf{g} \in C(\mathbb{T})$ such that $\mathbf{g} - \mathbf{f} = \mathbf{q}\mathbf{h}$ for some 2π-periodic trigonometric polynomial $\mathbf{q}$ and $\mathbf{h} \in C(\mathbb{T})$. If $\mathbf{q}(\xi) \geqslant 0$ and $\mathbf{h}(\xi) > 0$ for all $\xi \in \mathbb{R}$, prove that there exists a 2π-periodic trigonometric polynomial $\mathbf{p}$ such that $\mathbf{f}(\xi) \leqslant \mathbf{p}(\xi) \leqslant \mathbf{g}(\xi)$ for all $\xi \in \mathbb{R}$.

3.22. Let $\widehat{a}(\xi) = (1 + e^{-i\xi})^m \widehat{b}(\xi)$ for some $b \in l_0(\mathbb{Z})$. For $\mathbf{f} \in C^\infty(\mathbb{T})$ and $m \in \mathbb{N}$, define

$$H_{a,m}(\mathbf{f})(\xi) := \big(\mathbf{f}(\xi) - [|\widehat{a}(\xi/2)|^2 \mathbf{f}(\xi/2) + |\widehat{a}(\xi/2+\pi)|^2 \mathbf{f}(\xi/2+\pi)]\big)/\sin^{2m}(\xi/2).$$

Assume that (i) $\mathbf{f}(0) = 1$ and $\mathbf{f}(\xi) > 0$ for all $\xi \in \mathbb{R}$; (ii) $H_{a,m}(\mathbf{f}) \in C(\mathbb{T})$ and $d_{\mathbf{f}} := \min_{\xi \in \mathbb{R}} H_{a,m}(\mathbf{f})(\xi) > 0$. For any $\mathbf{g} \in C^\infty(\mathbb{T})$ such that $|\frac{\mathbf{g}(\xi) - \mathbf{f}(\xi)}{\sin^{2m}(\xi/2)}| < \frac{d_{\mathbf{f}}}{d_b}$ for all $\xi \in \mathbb{R}$ with $d_b := \max_{\xi \in \mathbb{R}}(1 + |\widehat{b}(\xi)|^2 + |\widehat{b}(\xi + \pi)|^2)$, then $\mathbf{g}(0) = 1$, $H_{a,m}(\mathbf{g}) \in C(\mathbb{T})$ and $H_{a,m}(\mathbf{g})(\xi) > 0$ for all $\xi \in \mathbb{R}$.

3.23. Let $a \in l_0(\mathbb{Z})$ such that $\widehat{a}(\xi) = (1 + e^{-i\xi})^m \widehat{b}(\xi)$ for some $b \in l_0(\mathbb{Z})$. If (3.3.15) holds, then there exists a 2π-periodic trigonometric polynomial $\boldsymbol{\theta}_1$ such that (i) $\boldsymbol{\theta}_1(0) = 1$ and $H_{a,m}(\frac{1}{\boldsymbol{\theta}_1}) \in C(\mathbb{T})$; (ii) $\boldsymbol{\theta}_1(\xi) > 0$ and $H_{a,m}(\frac{1}{\boldsymbol{\theta}_1})(\xi) > 0$, $\forall\, \xi \in \mathbb{R}$.

3.24. Let Θ_n be the moment correcting filter in Proposition 3.3.4 for the B-spline filter a_m^B with $m \in \mathbb{N}$. (i) Prove that $\widehat{\Theta_n}(\xi) = \mathsf{p}_{m,n}^B(\sin^2(\xi/2)) \geqslant 0$ for all $\xi \in \mathbb{R}$, where $\mathsf{p}_{m,n}^B$ is the unique polynomial of degree at most $n - 1$ such that $\left(1 + \sum_{j=1}^{n-1} \frac{(2j-1)!!}{(2j)!!(2j+1)} x^j\right)^{2m} = \mathsf{p}_{m,n}^B(x) + \mathscr{O}(x^n)$ as $x \to 0$. (ii) Prove that $\widehat{\Theta_n}(\xi) - \widehat{\Theta_n}(2\xi)|\widehat{a_m^B}(\xi)|^2 \geqslant 0$ for all $\xi \in \mathbb{R}$ and $1 \leqslant n \leqslant m$.

3.25. Prove Proposition 3.3.1.

3.26. Suppose that q_1 and q_2 are two nontrivial Laurent polynomials with complex symmetry and they satisfy $\mathsf{q}_1(z)\mathsf{q}_1^\star(z) = \mathsf{q}_2(z)\mathsf{q}_2^\star(z)$. Prove that there exist an integer $k \in \mathbb{Z}$ and $\lambda \in \{\pm 1, \pm i\}$ such that $\mathsf{q}_1(z) = \lambda z^k \mathsf{q}_2(z)$.

3.27. Suppose that q_1 and q_2 are two nontrivial Laurent polynomials with symmetry and satisfy $\mathsf{q}_1(z)\mathsf{q}_1^\star(z) = \mathsf{q}_2(z)\mathsf{q}_2^\star(z)$. Prove that there exists an integer k such that $\mathsf{S}\mathsf{q}_1(z) = z^{2k}\mathsf{S}\mathsf{q}_2(z)$. That is, the symmetry type $\mathsf{S}\mathsf{q}_1$ is essentially unique.

3.28. Let $\mathsf{d}_1(z) := z + z^{-1} + i$ and $\mathsf{d}_2(z) := z + z^{-1} + 2i$. Define $\mathsf{q}_1 := \mathsf{d}_1\mathsf{d}_2$ and $\mathsf{q}_2 = \mathsf{d}_1\mathsf{d}_2^\star$. Prove (i) q_1 and q_2 have symmetry and $\mathsf{q}_1(z)\mathsf{q}_1^\star(z) = \mathsf{q}_2(z)\mathsf{q}_2^\star(z)$; (ii) both $\mathsf{q}_1^\star/\mathsf{q}_2$ and $\mathsf{q}_1/\mathsf{q}_2$ are not monomials. Hence, q_1 and q_2 are essentially different.

3.29. Prove Proposition 3.2.3

3.30. Let $a, \Theta \in l_0(\mathbb{Z})$. Define $\mathcal{M}_{a,\Theta} := \mathcal{M}_{a,\Theta|0}$ as in (3.3.7). If $\det(\mathcal{M}_{a,\Theta}(z)) = 0$ for all $z \in \mathbb{C}\backslash\{0\}$, prove that there exists $b \in l_0(\mathbb{Z})$ such that $\{a; b\}_\Theta$ is a tight framelet filter bank. Then explicitly construct such a filter $b \in l_0(\mathbb{Z})$.

3.31. Let $\mathcal{M}_{a,\tilde{a},\Theta}$ be defined in (3.0.2). Suppose that $\det(\mathcal{M}_{a,\tilde{a},\Theta}(z)) = 0$ for all $z \in \mathbb{C}\backslash\{0\}$. Prove that there exist $b, \tilde{b} \in l_0(\mathbb{Z})$ such that $(\{\tilde{a}; \tilde{b}\}, \{a; b\})_\Theta$ is a dual framelet filter bank. Then explicitly construct such filters $b, \tilde{b}$. According to Theorem 1.4.7, from the filters $a, \tilde{a}, \Theta, b, \tilde{b}$, explicitly construct filters $\mathring{a}, \mathring{b}, \mathring{\tilde{a}}, \mathring{\tilde{b}}$ such that $(\{\mathring{\tilde{a}}; \mathring{\tilde{b}}\}, \{a; b\})$ and $(\{\tilde{a}; \tilde{b}\}, \{\mathring{a}; \mathring{b}\})$ are biorthogonal wavelet filter banks.

3.32. Suppose that $\{a; b_1, \ldots, b_s\}_\Theta$ is a tight framelet filter bank. Let U be an $s \times s$ paraunitary matrix of Laurent polynomials, that is, $\mathsf{U}^\star(z)\mathsf{U}(z) = I_s$. Define $[\mathring{\mathsf{b}}_1(z), \ldots, \mathring{\mathsf{b}}_s(z)] := [\mathsf{b}_1(z), \ldots, \mathsf{b}_s(z)]\mathsf{U}(z^2)$. Show that $\{a; \mathring{b}_1, \ldots, \mathring{b}_s\}_\Theta$ is also a tight framelet filter bank.

3.33. We say that a tight framelet filter bank $\{a; b_1, \ldots, b_s\}_\Theta$ is *support-irreducible* if there doesn't exist a paraunitary matrix U of Laurent polynomials such that $\operatorname{len}(\mathring{b}_1) \leqslant \operatorname{len}(b_1), \ldots, \operatorname{len}(\mathring{b}_s) \leqslant \operatorname{len}(b_s)$ and $\operatorname{len}(\mathring{b}_1) + \cdots + \operatorname{len}(\mathring{b}_s) < \operatorname{len}(b_1) + \cdots + \operatorname{len}(b_s)$, where $[\mathring{\mathsf{b}}_1(z), \ldots, \mathring{\mathsf{b}}_s(z)] := [\mathsf{b}_1(z), \ldots, \mathsf{b}_s(z)]\mathsf{U}(z^2)$. When all filters have [complex] symmetry (or real coefficients), we also require that all $\mathring{b}_1, \ldots, \mathring{b}_s$ have [complex] symmetry (or real coefficients) in the definition of support-irreducibility of a tight framelet filter bank. Suppose that $\{a; b_1, \ldots, b_s\}_\Theta$ is a tight framelet filter bank such that all $\Theta, a, b_1, \ldots, b_s$ have symmetry (and real coefficients). If there exist $1 \leqslant m \neq n \leqslant s$ such that $\mathsf{Sb}_m(z) = \mathsf{Sb}_n(z)$ and $\operatorname{len}(b_m) = \operatorname{len}(b_n)$, show that $\{a; b_1, \ldots, b_s\}_\Theta$ is support-reducible.

3.34. Let $\mathsf{a}, \mathsf{b}_1, \mathsf{b}_2, \mathsf{b}_3, \Theta$ be nontrivial Laurent polynomial such that

$$\mathsf{S}\Theta(z) = 1, \quad \mathsf{Sa}(z) = \epsilon z^c, \quad \mathsf{Sb}_1(z) = \epsilon_1 z^{c_1}, \quad \mathsf{Sb}_2(z) = \epsilon_2 z^{c_2}, \quad \mathsf{Sb}_3(z) = \epsilon_3 z^{c_3}$$

with $\epsilon, \epsilon_1, \epsilon_2, \epsilon_3 \in \{-1, 1\}$ and $c, c_1, c_2, c_3 \in \mathbb{Z}$. If $\{a; b_1, b_2, b_3\}_\Theta$ is a tight framelet filter bank satisfying $\max(\operatorname{len}(b_1), \operatorname{len}(b_2), \operatorname{len}(b_3)) \leqslant \operatorname{len}(a) + \operatorname{len}(\Theta) \neq 0$, prove that up to reordering of b_1, b_2, b_3, one of the following four cases must hold:

a. $\max(\operatorname{len}(b_1), \operatorname{len}(b_2)) < \operatorname{len}(b_3)$ with $\frac{c_3}{2} - (\frac{c}{2} + n_\Theta) \in 2\mathbb{Z}$;
b. $\operatorname{len}(b_1) < \operatorname{len}(b_2) = \operatorname{len}(b_3)$ with $\frac{c_\ell}{2} - (\frac{c}{2} + n_\Theta) \in 2\mathbb{Z}$ for $\ell = 2, 3$;
c. $\operatorname{len}(b_1) = \operatorname{len}(b_2) = \operatorname{len}(b_3)$ with $\frac{c_\ell}{2} - (\frac{c}{2} + n_\Theta) \in 2\mathbb{Z}$ for $\ell = 1, 2, 3$;
d. $\operatorname{len}(b_1) = \operatorname{len}(b_2) = \operatorname{len}(b_3)$ with $\frac{c_3}{2} - (\frac{c}{2} + n_\Theta) \in 2\mathbb{Z}$ and $\frac{c_\ell}{2} - (\frac{c+1}{2} + n_\Theta) \in 2\mathbb{Z}$ for $\ell = 1, 2$.

Moreover, $\epsilon_3 = -\epsilon \operatorname{sgn}(\Theta(n_\Theta))$ and for item (4) we additionally have $\epsilon_1\epsilon_2 = -1$, where $\Theta(n_\Theta)$ is the leading coefficient of Θ.

3.35. Under the same conditions as in Exercise 3.34, if $\{a; b_1, b_2, b_3\}_\Theta$ is support-irreducible, prove that $\operatorname{len}(b_1) < \operatorname{len}(a) + \operatorname{len}(\Theta)$.

3.36. Let $a \in l_0(\mathbb{Z})$ be a real-valued filter having symmetry. Assume that $\mathsf{p}(z) := 1 - \mathsf{a}(z)\mathsf{a}^\star(z) - \mathsf{a}(-z)\mathsf{a}^\star(-z) \geqslant 0$ for all $z \in \mathbb{T}$. If $\operatorname{len}(a)$ is an odd integer or $\mathsf{Z}(\mathsf{p}, x) \in 2\mathbb{Z}$ for all $x \in (0, 1)$, prove that there is a real-valued tight framelet filter bank $\{a; b_1, b_2, b_3\}$ with symmetry such that $\max(\operatorname{len}(b_1), \operatorname{len}(b_2), \operatorname{len}(b_3)) \leqslant \operatorname{len}(a)$.

3.37. For any B-spline filter $a = a_m^B$ of order $m \in \mathbb{N}$, prove that $\mathsf{p}(z) \geqslant 0$ for all $z \in \mathbb{T}$ and $\mathsf{Z}(\mathsf{p}, x) = 0$ for all $x \in (0, 1)$, where $\mathsf{p}(z) := 1 - \mathsf{a}(z)\mathsf{a}^\star(z) - \mathsf{a}(-z)\mathsf{a}^\star(-z)$.

3.38. Let $a \in l_0(\mathbb{Z})$ be a filter having complex symmetry and $\mathsf{a}(1) = 1$. Prove that there exists a tight framelet filter bank $\{a; b_1, b_2, b_3\}$ having complex symmetry and satisfying $\max(\operatorname{len}(b_1), \operatorname{len}(b_2), \operatorname{len}(b_3)) \leqslant \operatorname{len}(a)$ if and only if $\mathsf{a}(z)\mathsf{a}^\star(z) + \mathsf{a}(-z)\mathsf{a}^\star(-z) \leqslant 1$ for all $z \in \mathbb{T}$.

3.39. Let p be a Laurent polynomial and define $\mathsf{q}(z^2) := \gcd(\mathsf{p}(z), \mathsf{p}(-z))$. Prove that there exist Laurent polynomials $\mathsf{b}_1, \mathsf{b}_2$ with symmetry satisfying (3.5.6) if and only if (i) p has real coefficients and $\mathsf{p}(z) \geqslant 0$ for all $z \in \mathbb{T}$. (ii) $\frac{\mathsf{p}(z)}{\mathsf{q}(z^2)} = \mathsf{d}_\mathsf{p}(z)\mathsf{d}_\mathsf{p}^\star(z)$ for some Laurent polynomial d_p with symmetry.

3.40. Let $t \approx -0.7025$ be a root of $4t^7 - 65t^6 + 252t^5 + 402t^4 - 2420t^3 - 225t^2 + 872t - 162 = 0$. Define $\mathsf{a}(z) = z^{-1}(1 + z)^2(-tz^{-1} + 2t + 2 - tz)/8$ and $\Theta(z) := \boldsymbol{\theta}(z)\boldsymbol{\theta}(z^2)$ with $\boldsymbol{\theta}(z) := 1 + \frac{1-2t}{30}(2 - z - z^{-1})$. Then $\operatorname{sm}(a) \approx 1.9888$ and $\operatorname{sr}(a) = 2$. Prove (1) $\mathcal{M}_{a,\Theta}(z) \geqslant 0$ for all $z \in \mathbb{T}$; (2) There exists a Laurent polynomial d_1 with symmetry and $\operatorname{fsupp}(\mathsf{d}_1) = [-1, 1]$ such that $\det(\mathcal{N}_{a,\Theta|1}(z)) = \mathsf{d}_1(z)\mathsf{d}_1^\star(z)$; (3) $\mathsf{p} = \boldsymbol{\theta}$, where p is defined in item (iii) of Theorem 3.3.7; (4) p does not satisfy the root condition for the real SOS property in Theorem 3.1.8.

3.41. Prove Theorem 3.5.6.

3.42. Prove that the sufficient conditions for interpolatory tight framelet filter banks in Theorem 3.5.6 are also necessary.

Chapter 4
Analysis of Affine Systems and Dual Framelets

One of the major tasks in mathematics and its applications is to develop representation systems and mathematical transforms with desirable properties so that various types of functions can be effectively represented and efficiently transformed for the purpose of further study and analysis. The wavelet theory employs affine systems which are generated from a set of functions (often from the space $L_2(\mathbb{R})$ of square integrable functions on the real line $\mathbb{R}$) through dilation and translation on $\mathbb{R}$. More precisely, in this chapter, we shall study the following affine systems

$$\mathsf{AS}_J(\Phi;\Psi) := \{\phi_{2^J;k} \ : \ k \in \mathbb{Z}, \phi \in \Phi\} \cup \{\psi_{2^j;k} \ : \ k \in \mathbb{Z}, j \geqslant J, \psi \in \Psi\}, \quad J \in \mathbb{Z},$$

where $\psi_{2^j;k} := 2^{j/2}\psi(2^j \cdot -k)$ and Φ, Ψ consist of Lebesgue measurable functions or distributions on the real line. For example, if $\mathsf{AS}_J(\Phi;\Psi)$ is a (normalized) tight frame of $L_2(\mathbb{R})$, i.e., $\|f\|^2_{L_2(\mathbb{R})} = \sum_{h\in \mathsf{AS}_J(\Phi;\Psi)} |\langle f, h\rangle|^2$ for all $f \in L_2(\mathbb{R})$. Then every function $f \in L_2(\mathbb{R})$ has the following representation:

$$f = \sum_{\phi\in\Phi}\sum_{k\in\mathbb{Z}} \langle f, \phi_{2^J;k}\rangle \phi_{2^J;k} + \sum_{j=J}^{\infty}\sum_{\psi\in\Psi}\sum_{k\in\mathbb{Z}} \langle f, \psi_{2^j;k}\rangle \psi_{2^j;k}$$

with the series converging in $L_2(\mathbb{R})$. Then properties of a function f can be analyzed through its coefficients $\langle f, \phi_{2^J;k}\rangle$ and $\langle f, \psi_{2^j;k}\rangle$. The affine systems $\mathsf{AS}_J(\Phi;\Psi)$ are widely known as the leading multiscale representation systems in mathematics having many desirable properties such as sparse multiscale representation, simultaneous good time and frequency localization, fast transforms with filter banks, high approximation ability, and etc. One of the major mathematical tasks of wavelet theory is to investigate such affine systems and their mathematical properties.

We now introduce in this chapter the basic mathematical theory of framelets and wavelets in the function setting by studying framelets and wavelets in the continuum domain $\mathbb{R}$. As we discussed in Sect. 1.3 of Chap. 1, a filter bank $\{a; b_1, \dots, b_s\}$ induces a sequence of discrete affine systems $\mathsf{DAS}_J(\{a; b_1, \dots, b_s\})$ for $J \in \mathbb{N}$,

B. Han, *Framelets and Wavelets*, Applied and Numerical Harmonic Analysis,
https://doi.org/10.1007/978-3-319-68530-4_4

which are defined in (1.3.19) in the discrete domain $\mathbb{Z}$. To understand various properties of discrete affine systems, it is important to study their asymptotic behavior as $J \to \infty$. This naturally motivates us to define affine systems on the continuum domain $\mathbb{R}$, which in turn are closely related to discrete affine systems on the discrete domain $\mathbb{Z}$.

In this chapter we first introduce the notion of frequency-based affine systems and frequency-based dual framelets in the continuum domain $\mathbb{R}$. Then we shall provide a complete characterization of frequency-based dual framelets. This characterization enables us to establish a natural connection between a dual framelet filter bank and a frequency-based dual framelet. The classical theory of framelets and wavelets is often developed for the square integrable function space $L_2(\mathbb{R})$ which is a particular example of Hilbert spaces. We shall study frames and bases in a general Hilbert space. Next we shall investigate nonhomogeneous and homogeneous affine systems in $L_2(\mathbb{R})$. We shall show that a nonhomogeneous affine system in $L_2(\mathbb{R})$, which is either nonredundant or forms a tight frame, naturally leads to the notion of refinable structure and multiresolution analysis. In order to study some properties of a multiresolution analysis, we shall investigate some basic properties of shift-invariant subspaces of $L_2(\mathbb{R})$. Then we shall further study framelets and wavelets in Sobolev spaces and discuss the approximation property of dual framelets in Sobolev spaces. Also, we shall address frequency-based framelets and wavelets in the nonstationary setting. As a byproduct development we shall establish at the end of this chapter a comprehensive theory for periodic framelets and wavelets on the unit circle $\mathbb{T}$.

In this book we often use dilation, translation, and modulation of functions on $\mathbb{R}$. For a function $f : \mathbb{R} \to \mathbb{C}$, we shall adopt the following notation: For $x, \lambda, k, n \in \mathbb{R}$,

$$f_{\lambda;k,n}(x) := f_{[\![\lambda;k,n]\!]}(x) := [\![\lambda; k, n]\!] f(x) := |\lambda|^{1/2} e^{-in\lambda x} f(\lambda x - k). \tag{4.0.1}$$

In particular, we shall use the abbreviated notation: $[\![\lambda; k]\!] := [\![\lambda; k, 0]\!]$ and $[\![k, n]\!] := [\![1; k, n]\!]$, that is,

$$f_{\lambda;k} := f_{[\![\lambda;k]\!]} := |\lambda|^{1/2} f(\lambda \cdot -k), \quad f_{k,n} := f_{[\![k,n]\!]} := f_{1;k,n} = e^{-in\cdot} f(\cdot - k). \tag{4.0.2}$$

By $\mathscr{D}(\mathbb{R})$ we denote the linear space of all $\mathscr{C}^\infty(\mathbb{R})$ functions having compact support in $\mathbb{R}$. In this chapter we shall discuss frequency-based affine systems and frequency-based dual framelets within the framework of the locally square integrable function space $L_2^{loc}(\mathbb{R})$, where $\mathbf{f} \in L_p^{loc}(\mathbb{R})$ with $1 \leqslant p < \infty$ means that $\int_K |\mathbf{f}(\xi)|^p d\xi < \infty$ for every bounded interval K in $\mathbb{R}$. We remind the reader that throughout the book bold-face letters often indicate functions or sets of functions in the frequency domain. The discussions on frequency-based dual framelets can take place in a wider framework of the distribution space $\mathscr{D}'(\mathbb{R})$, that is, the dual space of $\mathscr{D}(\mathbb{R})$. Note that $L_2^{loc}(\mathbb{R}) \subseteq \mathscr{D}'(\mathbb{R})$ and the definition in (4.0.1) can be easily extended to distributions in $\mathscr{D}'(\mathbb{R})$ through duality. The Fourier transform used in this book is $\widehat{f}(\xi) := \int_{\mathbb{R}} f(x) e^{-ix\xi} dx$ for $f \in L_1(\mathbb{R})$ and is naturally extended to square integrable functions and tempered distributions. For a short self-contained introduction to Fourier transform on $\mathbb{R}$ and Fourier series on $\mathbb{T}$, see Appendix A. The following basic identities can be directly verified: For $\lambda \in \mathbb{R}\backslash\{0\}$ and $k, n \in \mathbb{R}$,

$$\widehat{f_{\lambda;k,n}} = e^{-ikn}\widehat{f}_{\lambda^{-1};-n,k}, \qquad \widehat{f_{\lambda;k}} = \widehat{f}_{\lambda^{-1};0,k}, \qquad \widehat{f_{k,n}} = e^{-ikn}\widehat{f}_{-n,k}. \tag{4.0.3}$$

4.1 Frequency-Based Dual Framelets and Connections to Filter Banks

To study the affine systems $\mathsf{AS}_J(\Phi;\Psi)$, let us first investigate their associated frequency-based affine systems, which are the images of $\mathsf{AS}_J(\Phi;\Psi)$ under the Fourier transform. In this section we introduce and characterize frequency-based dual framelets. Then we establish a natural one-to-one correspondence between dual framelet filter banks and frequency-based dual framelets. We also discuss connections between framelet transforms in the function setting and discrete framelet transforms.

4.1.1 *Frequency-Based Affine Systems and Dual Framelets*

Let $\boldsymbol{\Phi}$ and $\boldsymbol{\Psi}$ be subsets of $L_2^{loc}(\mathbb{R})$ (or $\mathscr{D}'(\mathbb{R})$). For every integer $J \in \mathbb{Z}$, *a frequency-based (nonhomogeneous) affine system* is defined to be

$$\mathsf{FAS}_J(\boldsymbol{\Phi};\boldsymbol{\Psi}) := \{\boldsymbol{\varphi}_{2^{-J};0,k} : k \in \mathbb{Z}, \boldsymbol{\varphi} \in \boldsymbol{\Phi}\} \cup \{\boldsymbol{\psi}_{2^{-j};0,k} : j \geqslant J, k \in \mathbb{Z}, \boldsymbol{\psi} \in \boldsymbol{\Psi}\}. \tag{4.1.1}$$

It is important to point out here that the elements in a set $\boldsymbol{\Psi}$ of this book are not necessarily distinct or finite and $\boldsymbol{\psi} \in \boldsymbol{\Psi}$ in a summation means that $\boldsymbol{\psi}$ visits every element (with multiplicity) in $\boldsymbol{\Psi}$ once and only once. For example, for $\boldsymbol{\Psi} = \{\boldsymbol{\psi}^1, \dots, \boldsymbol{\psi}^s\}$, all the functions $\boldsymbol{\psi}^1, \dots, \boldsymbol{\psi}^s$ are not necessarily distinct and $\boldsymbol{\psi} \in \boldsymbol{\Psi}$ means $\boldsymbol{\psi} = \boldsymbol{\psi}^1, \dots, \boldsymbol{\psi}^s$.

For measurable functions $\mathbf{f}, \boldsymbol{\psi} : \mathbb{R} \to \mathbb{C}$ with $\mathbf{f}\boldsymbol{\psi} \in L_1(\mathbb{R})$, we use the pairing:

$$\langle \mathbf{f}, \boldsymbol{\psi} \rangle := \int_{\mathbb{R}} \mathbf{f}(\xi)\overline{\boldsymbol{\psi}(\xi)}d\xi \quad \text{and} \quad \langle \mathbf{f}; \boldsymbol{\psi} \rangle := \int_{\mathbb{R}} \mathbf{f}(\xi)\boldsymbol{\psi}(\xi)d\xi. \tag{4.1.2}$$

We are ready to introduce frequency-based dual framelets. Throughout this chapter, we always assume that $\boldsymbol{\Phi}, \tilde{\boldsymbol{\Phi}}, \boldsymbol{\Psi}, \tilde{\boldsymbol{\Psi}}$ are (countable) subsets of $L_2^{loc}(\mathbb{R})$ (or $\mathscr{D}'(\mathbb{R})$) such that there is a bijection between $\boldsymbol{\Phi}$ and $\tilde{\boldsymbol{\Phi}}$ (and between $\boldsymbol{\Psi}$ and $\tilde{\boldsymbol{\Psi}}$):

$$\sim : \boldsymbol{\Phi} \to \tilde{\boldsymbol{\Phi}}, \; \boldsymbol{\varphi} \mapsto \tilde{\boldsymbol{\varphi}} \qquad \text{and} \qquad \sim : \boldsymbol{\Psi} \to \tilde{\boldsymbol{\Psi}}, \; \boldsymbol{\psi} \mapsto \tilde{\boldsymbol{\psi}}. \tag{4.1.3}$$

More explicitly, $\boldsymbol{\Phi} = \{\boldsymbol{\varphi}^\ell : \ell \in \Lambda_{\boldsymbol{\Phi}}\}$ and $\tilde{\boldsymbol{\Phi}} = \{\tilde{\boldsymbol{\varphi}}^\ell : \ell \in \Lambda_{\boldsymbol{\Phi}}\}$ for some index set $\Lambda_{\boldsymbol{\Phi}}$ so that $(\boldsymbol{\varphi}^\ell, \tilde{\boldsymbol{\varphi}}^\ell)$ is always regarded as a pair together. Similarly, $\boldsymbol{\Psi} = \{\boldsymbol{\psi}^\ell : \ell \in \Lambda_{\boldsymbol{\Psi}}\}$ and $\tilde{\boldsymbol{\Psi}} = \{\tilde{\boldsymbol{\psi}}^\ell : \ell \in \Lambda_{\boldsymbol{\Psi}}\}$ for some index set $\Lambda_{\boldsymbol{\Psi}}$ so that $(\boldsymbol{\psi}^\ell, \tilde{\boldsymbol{\psi}}^\ell)$ is always a pair.

For $J \in \mathbb{Z}$ and $\mathsf{FAS}_J(\boldsymbol{\Phi}; \boldsymbol{\Psi})$ as in (4.1.1), we say that $(\mathsf{FAS}_J(\tilde{\boldsymbol{\Phi}}; \tilde{\boldsymbol{\Psi}}), \mathsf{FAS}_J(\boldsymbol{\Phi}; \boldsymbol{\Psi}))$ is *a pair of frequency-based dual frames* if the following identity holds:

$$\sum_{\boldsymbol{\varphi} \in \boldsymbol{\Phi}} \sum_{k \in \mathbb{Z}} \langle \mathbf{f}, \tilde{\boldsymbol{\varphi}}_{2^{-J};0,k} \rangle \langle \boldsymbol{\varphi}_{2^{-J};0,k}, \mathbf{g} \rangle + \sum_{j=J}^{\infty} \sum_{\boldsymbol{\psi} \in \boldsymbol{\Psi}} \sum_{k \in \mathbb{Z}} \langle \mathbf{f}, \tilde{\boldsymbol{\psi}}_{2^{-j};0,k} \rangle \langle \boldsymbol{\psi}_{2^{-j};0,k}, \mathbf{g} \rangle = 2\pi \langle \mathbf{f}, \mathbf{g} \rangle$$

for all $\mathbf{f}, \mathbf{g} \in \mathscr{D}(\mathbb{R})$, where the above infinite series converge in the following sense:

(i) For every $\mathbf{f}, \mathbf{g} \in \mathscr{D}(\mathbb{R})$, the following series

$$\sum_{\boldsymbol{\varphi} \in \boldsymbol{\Phi}} \sum_{k \in \mathbb{Z}} \langle \mathbf{f}, \tilde{\boldsymbol{\varphi}}_{2^{-J};0,k} \rangle \langle \boldsymbol{\varphi}_{2^{-J};0,k}, \mathbf{g} \rangle \quad \text{and} \quad \sum_{\boldsymbol{\psi} \in \boldsymbol{\Psi}} \sum_{k \in \mathbb{Z}} \langle \mathbf{f}, \tilde{\boldsymbol{\psi}}_{2^{-j};0,k} \rangle \langle \boldsymbol{\psi}_{2^{-j};0,k}, \mathbf{g} \rangle$$

converge absolutely for all integers $j \geqslant J$;

(ii) For every $\mathbf{f}, \mathbf{g} \in \mathscr{D}(\mathbb{R})$, the following limit exists and

$$\lim_{J' \to \infty} \Big(\sum_{\boldsymbol{\varphi} \in \boldsymbol{\Phi}} \sum_{k \in \mathbb{Z}} \langle \mathbf{f}, \tilde{\boldsymbol{\varphi}}_{2^{-J};0,k} \rangle \langle \boldsymbol{\varphi}_{2^{-J};0,k}, \mathbf{g} \rangle + \sum_{j=J}^{J'-1} \sum_{\boldsymbol{\psi} \in \boldsymbol{\Psi}} \sum_{k \in \mathbb{Z}} \langle \mathbf{f}, \tilde{\boldsymbol{\psi}}_{2^{-j};0,k} \rangle \langle \boldsymbol{\psi}_{2^{-j};0,k}, \mathbf{g} \rangle \Big) = 2\pi \langle \mathbf{f}, \mathbf{g} \rangle. \tag{4.1.4}$$

For two measurable functions $\mathbf{f}, \mathbf{g} : \mathbb{R} \to \mathbb{C}$, the *bracket product* is defined to be

$$[\mathbf{f}, \mathbf{g}](\xi) := \sum_{k \in \mathbb{Z}} \mathbf{f}(\xi + 2\pi k) \overline{\mathbf{g}(\xi + 2\pi k)}, \qquad \xi \in \mathbb{R}, \tag{4.1.5}$$

provided that the series converges absolutely for almost every $\xi \in \mathbb{R}$. By the Cauchy-Schwarz inequality, we see that $|[\mathbf{f}, \mathbf{g}](\xi)|^2 \leqslant [\mathbf{f}, \mathbf{f}](\xi)[\mathbf{g}, \mathbf{g}](\xi)$. The bracket product links functions on the real line with 2π-periodic functions. If both $\boldsymbol{\Phi}$ and $\boldsymbol{\Psi}$ are finite subsets of $L_2^{loc}(\mathbb{R})$, we now show that item (i) above is automatically satisfied.

Lemma 4.1.1 *Let λ be a nonzero real number. Let $\mathbf{f}, \mathbf{g}$ be (Lebesgue) measurable functions on $\mathbb{R}$ and $\boldsymbol{\Psi}, \tilde{\boldsymbol{\Psi}}$ be sets of measurable functions on $\mathbb{R}$ such that*

$$\sum_{\boldsymbol{\psi} \in \boldsymbol{\Psi}} \sum_{k \in \mathbb{Z}} \int_{\mathbb{R}} |\mathbf{f}(\xi) \mathbf{f}(\xi + 2\pi\lambda^{-1}k) \boldsymbol{\psi}(\lambda\xi) \boldsymbol{\psi}(\lambda\xi + 2\pi k)| d\xi < \infty, \tag{4.1.6}$$

and

$$\sum_{\tilde{\psi}\in\tilde{\Psi}}\sum_{k\in\mathbb{Z}}\int_{\mathbb{R}}|\mathbf{g}(\xi)\mathbf{g}(\xi+2\pi\lambda^{-1}k)\tilde{\psi}(\lambda\xi)\tilde{\psi}(\lambda\xi+2\pi k)|d\xi<\infty.$$

Then $\sum_{\psi\in\Psi}\sum_{k\in\mathbb{Z}}|\langle\mathbf{f},\psi_{\lambda;0,k}\rangle|^2<\infty$, $\sum_{\tilde{\psi}\in\tilde{\Psi}}\sum_{k\in\mathbb{Z}}|\langle\mathbf{g},\tilde{\psi}_{\lambda;0,k}\rangle|^2<\infty$, *and*

$$\begin{aligned}\sum_{\psi\in\Psi}\sum_{k\in\mathbb{Z}}&\langle\mathbf{f},\psi_{\lambda;0,k}\rangle\langle\tilde{\psi}_{\lambda;0,k},\mathbf{g}\rangle\\&=2\pi|\lambda|^{-1}\int_{-\pi}^{\pi}\sum_{\psi\in\Psi}[\mathbf{f}(\lambda^{-1}\cdot),\psi](\xi)[\tilde{\psi},\mathbf{g}(\lambda^{-1}\cdot)](\xi)d\xi\\&=2\pi\int_{\mathbb{R}}\sum_{k\in\mathbb{Z}}\mathbf{f}(\xi)\overline{\mathbf{g}(\xi+2\pi\lambda^{-1}k)}\sum_{\psi\in\Psi}\overline{\psi(\lambda\xi)}\tilde{\psi}(\lambda\xi+2\pi k)d\xi\end{aligned}\tag{4.1.7}$$

with the series on the left-hand side converging absolutely and all other series converging absolutely for almost every $\xi\in\mathbb{R}$. *In particular, the condition in* (4.1.6) *is satisfied if one of the following two cases holds:*

(i) $\mathbf{f}\in\mathscr{D}(\mathbb{R})$ *and* $\sum_{\psi\in\Psi}|\psi(\cdot)|^2\in L_1^{loc}(\mathbb{R})$ *(which holds if* $\Psi\subseteq L_2^{loc}(\mathbb{R})$ *is finite).*
(ii) $\mathbf{f}\in L_2(\mathbb{R})$ *and* $\sum_{\psi\in\Psi}[\psi,\psi]\in L_\infty(\mathbb{T})$.

Proof By calculation, we have

$$\begin{aligned}\sum_{\psi\in\Psi}&|\lambda|^{-1}\int_{-\pi}^{\pi}|\,[|\mathbf{f}(\lambda^{-1}\cdot)|,|\psi|](\xi)\,|^2d\xi\\&=|\lambda|^{-1}\sum_{\psi\in\Psi}\int_{\mathbb{R}}|\mathbf{f}(\lambda^{-1}\xi)\psi(\xi)|\,[|\mathbf{f}(\lambda^{-1}\cdot)|,|\psi|](\xi)d\xi\\&=\sum_{\psi\in\Psi}\sum_{k\in\mathbb{Z}}|\lambda|^{-1}\int_{\mathbb{R}}|\mathbf{f}(\lambda^{-1}\xi)\psi(\xi)||\mathbf{f}(\lambda^{-1}(\xi+2\pi k))\psi(\xi+2\pi k)|d\xi\\&=\sum_{\psi\in\Psi}\sum_{k\in\mathbb{Z}}\int_{\mathbb{R}}|\mathbf{f}(\xi)\mathbf{f}(\xi+2\pi\lambda^{-1}k)\psi(\lambda\xi)\psi(\lambda\xi+2\pi k)|d\xi<\infty.\end{aligned}$$

Thus, all $[\mathbf{f}(\lambda^{-1}\cdot),\psi]$ are well-defined functions in $L_2(\mathbb{T})$. Similarly, we have $\sum_{\tilde{\psi}\in\tilde{\Psi}}|\lambda|^{-1}\int_{-\pi}^{\pi}|\,[|\mathbf{g}(\lambda^{-1}\cdot)|,|\tilde{\psi}|](\xi)\,|^2d\xi<\infty$ and all $[\mathbf{g}(\lambda^{-1}\cdot),\tilde{\psi}]$ are well-defined functions in $L_2(\mathbb{T})$. By calculation, we have

$$\int_{-\pi}^{\pi}[\mathbf{f}(\lambda^{-1}\cdot),\psi](\xi)e^{ik\xi}d\xi=\int_{\mathbb{R}}\mathbf{f}(\lambda^{-1}\xi)\overline{\psi(\xi)}e^{ik\xi}d\xi=|\lambda|^{1/2}\langle\mathbf{f},\psi_{\lambda;0,k}\rangle,\quad k\in\mathbb{Z}.$$

By Parseval's identity (see (A.6) in Appendix A), we conclude that

$$\sum_{\psi\in\Psi}\sum_{k\in\mathbb{Z}}|\langle \mathbf{f}, \boldsymbol{\psi}_{\lambda;0,k}\rangle|^2 = 2\pi|\lambda|^{-1}\int_{-\pi}^{\pi}\sum_{\psi\in\Psi}|[\mathbf{f}(\lambda^{-1}\cdot),\boldsymbol{\psi}](\xi)|^2 d\xi < \infty.$$

The claim $\sum_{\tilde{\psi}\in\tilde{\Psi}}\sum_{k\in\mathbb{Z}}|\langle \mathbf{g}, \tilde{\boldsymbol{\psi}}_{\lambda;0,k}\rangle|^2 < \infty$ can be proved similarly. The identities in (4.1.7) also follow directly from Parseval's identity.

If item (i) holds, then there exists $c > 0$ such that $\mathbf{f}(\xi) = 0$ for all $\xi \in \mathbb{R}\backslash[-c, c]$. By the Cauchy-Schwarz inequality and because $[-c, c]\cap[-c-2\pi\lambda^{-1}k, c-2\pi\lambda^{-1}k]$ has measure zero for all $k \in \mathbb{Z}\backslash[-|\lambda|c/\pi, |\lambda|c/\pi]$, we have

$$\begin{aligned}
&\sum_{\psi\in\Psi}\sum_{k\in\mathbb{Z}}\int_{\mathbb{R}}|\mathbf{f}(\xi)\mathbf{f}(\xi+2\pi\lambda^{-1}k)\boldsymbol{\psi}(\lambda\xi)\boldsymbol{\psi}(\lambda\xi+2\pi k)|d\xi\\
&\leqslant \|\mathbf{f}\|^2_{L_\infty(\mathbb{R})}\sum_{k\in\mathbb{Z}}\sum_{\psi\in\Psi}\int_{[-c,c]\cap[-c-2\pi\lambda^{-1}k,c-2\pi\lambda^{-1}k]}|\boldsymbol{\psi}(\lambda\xi)\boldsymbol{\psi}(\lambda\xi+2\pi k)|d\xi\\
&\leqslant \|\mathbf{f}\|^2_{L_\infty(\mathbb{R})}\sum_{k\in\mathbb{Z}\cap[-|\lambda|c/\pi,|\lambda|c/\pi]}\sum_{\psi\in\Psi}\int_{[-c,c]}|\boldsymbol{\psi}(\lambda\xi)|^2 d\xi < \infty.
\end{aligned}$$

If item (ii) holds, then by $([|\mathbf{f}(\lambda^{-1}\cdot)|, |\boldsymbol{\psi}|](\xi))^2 \leqslant [\mathbf{f}(\lambda^{-1}\cdot), \mathbf{f}(\lambda^{-1}\cdot)](\xi)[\boldsymbol{\psi},\boldsymbol{\psi}](\xi)$,

$$\begin{aligned}
&\sum_{\psi\in\Psi}\sum_{k\in\mathbb{Z}}\int_{\mathbb{R}}|\mathbf{f}(\xi)\mathbf{f}(\xi+2\pi\lambda^{-1}k)\boldsymbol{\psi}(\lambda\xi)\boldsymbol{\psi}(\lambda\xi+2\pi k)|d\xi\\
&= \sum_{\psi\in\Psi}|\lambda|^{-1}\int_{-\pi}^{\pi}|\,[|\mathbf{f}(\lambda^{-1}\cdot)|, |\boldsymbol{\psi}|](\xi)\,|^2 d\xi\\
&\leqslant \sum_{\psi\in\Psi}|\lambda|^{-1}\int_{-\pi}^{\pi}[\mathbf{f}(\lambda^{-1}\cdot), \mathbf{f}(\lambda^{-1}\cdot)](\xi)[\boldsymbol{\psi},\boldsymbol{\psi}](\xi)d\xi\\
&= |\lambda|^{-1}\int_{-\pi}^{\pi}[\mathbf{f}(\lambda^{-1}\cdot), \mathbf{f}(\lambda^{-1}\cdot)](\xi)\sum_{\psi\in\Psi}[\boldsymbol{\psi},\boldsymbol{\psi}](\xi)d\xi\\
&\leqslant \Big\|\sum_{\psi\in\Psi}[\boldsymbol{\psi},\boldsymbol{\psi}]\Big\|_{L_\infty(\mathbb{T})}|\lambda|^{-1}\int_{-\pi}^{\pi}[\mathbf{f}(\lambda^{-1}\cdot), \mathbf{f}(\lambda^{-1}\cdot)](\xi)d\xi\\
&= \Big\|\sum_{\psi\in\Psi}[\boldsymbol{\psi},\boldsymbol{\psi}]\Big\|_{L_\infty(\mathbb{T})}\|\mathbf{f}\|^2_{L_2(\mathbb{R})} < \infty.
\end{aligned}$$

This completes the proof. □

Note that if the functions $\mathbf{f}, \mathbf{g}$ are compactly supported, then the sum $\sum_{k\in\mathbb{Z}}$ on the right-hand side of (4.1.7) is in fact finite for every $\xi \in \mathbb{R}$.

We now study a basic property of a pair of frequency-based dual frames.

Proposition 4.1.2 $(\mathsf{FAS}_J(\tilde{\boldsymbol{\Phi}};\tilde{\boldsymbol{\Psi}}), \mathsf{FAS}_J(\boldsymbol{\Phi};\boldsymbol{\Psi}))$ *is a pair of frequency-based dual frames for some* $J \in \mathbb{Z}$ *if and only if* $(\mathsf{FAS}_J(\tilde{\boldsymbol{\Phi}};\tilde{\boldsymbol{\Psi}}), \mathsf{FAS}_J(\boldsymbol{\Phi};\boldsymbol{\Psi}))$ *is a pair of frequency-based dual frames for every* $J \in \mathbb{Z}$.

Proof If (4.1.4) holds for a particular integer J, for any given integer m, replacing $\mathbf{f}$ and $\mathbf{g}$ by $\mathbf{f}_{2^m;0,0}$ and $\mathbf{g}_{2^m;0,0}$, respectively, we have

$$\langle \mathbf{f}_{2^m;0,0}, \tilde{\boldsymbol{\psi}}_{2^{-j};0,k}\rangle\langle \boldsymbol{\psi}_{2^{-j};0,k}, \mathbf{g}_{2^m;0,0}\rangle = \langle \mathbf{f}, \tilde{\boldsymbol{\psi}}_{2^{-j-m};0,k}\rangle\langle \boldsymbol{\psi}_{2^{-j-m};0,k}, \mathbf{g}\rangle \tag{4.1.8}$$

and therefore, the identity (4.1.4) still holds with J being replaced by $J + m$. □

We call $(\{\tilde{\boldsymbol{\Phi}};\tilde{\boldsymbol{\Psi}}\}, \{\boldsymbol{\Phi};\boldsymbol{\Psi}\})$ *a frequency-based dual framelet* if $(\mathsf{FAS}_0(\tilde{\boldsymbol{\Phi}};\tilde{\boldsymbol{\Psi}}), \mathsf{FAS}_0(\boldsymbol{\Phi};\boldsymbol{\Psi}))$ is a pair of frequency-based dual frames. By Proposition 4.1.2, a frequency-based dual framelet $(\{\tilde{\boldsymbol{\Phi}};\tilde{\boldsymbol{\Psi}}\}, \{\boldsymbol{\Phi};\boldsymbol{\Psi}\})$ induces a sequence of pairs of frequency-based dual frames $(\mathsf{FAS}_J(\tilde{\boldsymbol{\Phi}};\tilde{\boldsymbol{\Psi}}), \mathsf{FAS}_J(\boldsymbol{\Phi};\boldsymbol{\Psi}))$ at every scale level $J \in \mathbb{Z}$.

4.1.2 Characterization of Frequency-Based Dual Framelets

We next characterize a frequency-based dual framelet. To do so, we need some auxiliary results.

Lemma 4.1.3 *Let* $\boldsymbol{\Phi}, \tilde{\boldsymbol{\Phi}} \subseteq L_2^{loc}(\mathbb{R})$ *such that*

$$\sum_{\varphi \in \boldsymbol{\Phi}} (|\varphi(\cdot)|^2 + |\tilde{\varphi}(\cdot)|^2) \in L_1^{loc}(\mathbb{R}). \tag{4.1.9}$$

Then

$$\lim_{j\to\infty} \sum_{\varphi\in\boldsymbol{\Phi}} \sum_{k\in\mathbb{Z}} \langle \mathbf{f}, \tilde{\varphi}_{2^{-j};0,k}\rangle\langle \varphi_{2^{-j};0,k}, \mathbf{g}\rangle = 2\pi\langle \mathbf{f}, \mathbf{g}\rangle \qquad \forall\, \mathbf{f}, \mathbf{g} \in \mathscr{D}(\mathbb{R}) \tag{4.1.10}$$

if and only if

$$\lim_{j\to\infty} \Big\langle \sum_{\varphi\in\boldsymbol{\Phi}} \tilde{\varphi}(2^{-j}\cdot)\overline{\varphi(2^{-j}\cdot)}, \mathbf{h}\Big\rangle = \langle 1, \mathbf{h}\rangle \qquad \forall\, \mathbf{h} \in \mathscr{D}(\mathbb{R}). \tag{4.1.11}$$

Proof By Lemma 4.1.1, we have

$$\sum_{\varphi\in\boldsymbol{\Phi}} \sum_{k\in\mathbb{Z}} \langle \mathbf{f}, \tilde{\varphi}_{2^{-j};0,k}\rangle\langle \varphi_{2^{-j};0,k}, \mathbf{g}\rangle$$

$$= 2\pi \int_{\mathbb{R}} \sum_{k\in\mathbb{Z}} \mathbf{f}(\xi)\overline{\mathbf{g}(\xi + 2\pi 2^j k)} \sum_{\varphi\in\boldsymbol{\Phi}} \overline{\tilde{\varphi}(2^{-j}\xi)}\varphi(2^{-j}\xi + 2\pi k)d\xi.$$

Since $\mathbf{f}, \mathbf{g} \in \mathscr{D}(\mathbb{R})$ are compactly supported, there exists an integer $J_{\mathbf{f},\mathbf{g}}$ such that $\mathbf{f}(\xi)\overline{\mathbf{g}(\xi + 2\pi 2^j k)} = 0$ for all $\xi \in \mathbb{R}$, $k \in \mathbb{Z}\backslash\{0\}$, and $j \geqslant J_{\mathbf{f},\mathbf{g}}$. That is, for $j \geqslant J_{\mathbf{f},\mathbf{g}}$, the above identity becomes

$$\sum_{\varphi\in\Phi}\sum_{k\in\mathbb{Z}}\langle \mathbf{f}, \tilde{\varphi}_{2^{-j};0,k}\rangle\langle \varphi_{2^{-j};0,k}, \mathbf{g}\rangle = 2\pi\int_{\mathbb{R}} \mathbf{f}(\xi)\overline{\mathbf{g}(\xi)}\sum_{\varphi\in\Phi}\overline{\tilde{\varphi}(2^{-j}\xi)}\varphi(2^{-j}\xi)d\xi. \tag{4.1.12}$$

Note that $\mathbf{f}\overline{\mathbf{g}} \in \mathscr{D}(\mathbb{R})$. If (4.1.11) holds, then it follows directly from (4.1.12) that (4.1.10) holds. Conversely, we can select $\mathbf{g} \in \mathscr{D}(\mathbb{R})$ such that $\mathbf{g}$ takes value one on the support of $\mathbf{f}$, now it follows from (4.1.10) and (4.1.12) that

$$\begin{aligned}\lim_{j\to\infty}\int_{\mathbb{R}} \mathbf{f}(\xi)\sum_{\varphi\in\Phi}\overline{\tilde{\varphi}(2^{-j}\xi)}\varphi(2^{-j}\xi)d\xi &= \lim_{j\to\infty}\frac{1}{2\pi}\sum_{\varphi\in\Phi}\sum_{k\in\mathbb{Z}}\langle \mathbf{f}, \tilde{\varphi}_{2^{-j};0,k}\rangle\langle \varphi_{2^{-j};0,k}, \mathbf{g}\rangle \\ &= \langle \mathbf{f}, \mathbf{g}\rangle = \langle \mathbf{f}, 1\rangle.\end{aligned}$$

Hence, the identity (4.1.11) holds. □

Under some mild condition, we show that (4.1.11) holds if and only if $\lim_{j\to\infty} \sum_{\varphi\in\Phi} \tilde{\varphi}(2^{-j}\xi)\overline{\varphi(2^{-j}\xi)} = 1$ for almost every $\xi \in \mathbb{R}$.

Lemma 4.1.4 *Let* $\mathbf{f}$ *be a measurable function on* $\mathbb{R}$ *such that* $\mathbf{F}(\xi) := \lim_{j\to\infty} \mathbf{f}(2^{-j}\xi)$ *exists for almost every* $\xi \in \mathbb{R}$ *and there exist positive numbers* τ *and* C *such that* $|\mathbf{f}(\xi)| \leqslant C$ *for almost every* $\xi \in [-\tau, \tau]$. *Then the following identity holds*

$$\lim_{j\to\infty}\langle \mathbf{f}(2^{-j}\cdot), \mathbf{h}\rangle = \langle 1, \mathbf{h}\rangle, \qquad \forall\, \mathbf{h} \in \mathscr{D}(\mathbb{R}) \tag{4.1.13}$$

if and only if $\mathbf{F}(\xi) = 1$ *for almost every* $\xi \in \mathbb{R}$.

Proof Note that $\mathbf{h} \in \mathscr{D}(\mathbb{R})$ vanishes outside $[-N, N]$ for some $N \in \mathbb{N}$. For all $j \geqslant \log_2 \frac{N}{\tau}$ and $\xi \in [-N, N]$, we have $|2^{-j}\xi| \leqslant \tau$. By $|f(\xi)| \leqslant C$ for a.e. $\xi \in [-\tau, \tau]$,

$$|\mathbf{f}(2^{-j}\xi)\overline{\mathbf{h}(\xi)}| \leqslant C|\mathbf{h}(\xi)| \in L_1(\mathbb{R}) \qquad \forall\, j \geqslant \log_2 \tfrac{N}{\tau}.$$

By the Dominated Convergence Theorem, we have

$$\lim_{j\to\infty}\langle \mathbf{f}(2^{-j}\cdot), \mathbf{h}\rangle = \int_{\mathbb{R}}\lim_{j\to\infty}\mathbf{f}(2^{-j}\xi)\overline{\mathbf{h}(\xi)}d\xi = \int_{\mathbb{R}}\mathbf{F}(\xi)\overline{\mathbf{h}(\xi)}d\xi.$$

Note that $|\mathbf{F}(\xi)| \leqslant C$ for almost every $\xi \in \mathbb{R}$. Therefore, (4.1.13) holds if and only if $\mathbf{F}(\xi) = 1$ for almost every $\xi \in \mathbb{R}$. □

For a function $\mathbf{f}: \mathbb{R} \to \mathbb{C}$, its support supp(**f**) in this chapter is simply defined by

$$\operatorname{supp}(\mathbf{f}) := \{\xi \in \mathbb{R} \ : \ \mathbf{f}(\xi) \neq 0\}. \tag{4.1.14}$$

Note that supp(**f**) may not be a closed subset of $\mathbb{R}$. See Exercise 4.1 for a related notion of support of a measurable function using closed subsets of $\mathbb{R}$.

Lemma 4.1.5 *Let λ be a nonzero real number. Let $\boldsymbol{\Phi}, \tilde{\boldsymbol{\Phi}}, \boldsymbol{\Psi}, \tilde{\boldsymbol{\Psi}}, \mathbf{H}, \tilde{\mathbf{H}} \subseteq L_2^{loc}(\mathbb{R})$ such that* (4.1.9) *is satisfied,*

$$\sum_{\boldsymbol{\psi}\in\boldsymbol{\Psi}} (|\boldsymbol{\psi}(\cdot)|^2 + |\tilde{\boldsymbol{\psi}}(\cdot)|^2) \in L_1^{loc}(\mathbb{R}) \tag{4.1.15}$$

and $\sum_{\eta\in\mathbf{H}} (|\boldsymbol{\eta}(\cdot)|^2 + |\tilde{\boldsymbol{\eta}}(\cdot)|^2) \in L_1^{loc}(\mathbb{R})$. *Then*

$$\begin{aligned}\sum_{\varphi\in\boldsymbol{\Phi}}\sum_{k\in\mathbb{Z}} \langle \mathbf{f}, \tilde{\boldsymbol{\varphi}}_{1;0,k}\rangle \langle \boldsymbol{\varphi}_{1;0,k}, \mathbf{g}\rangle + \sum_{\psi\in\boldsymbol{\Psi}}\sum_{k\in\mathbb{Z}} \langle \mathbf{f}, \tilde{\boldsymbol{\psi}}_{1;0,k}\rangle \langle \boldsymbol{\psi}_{1;0,k}, \mathbf{g}\rangle \\ = \sum_{\eta\in\mathbf{H}}\sum_{k\in\mathbb{Z}} \langle \mathbf{f}, \tilde{\boldsymbol{\eta}}_{\lambda;0,k}\rangle \langle \boldsymbol{\eta}_{\lambda;0,k}, \mathbf{g}\rangle \quad \forall\, \mathbf{f}, \mathbf{g} \in \mathscr{D}(\mathbb{R})\end{aligned} \tag{4.1.16}$$

if and only if

$$\mathcal{I}_{\boldsymbol{\Phi}}^k(\xi) + \mathcal{I}_{\boldsymbol{\Psi}}^k(\xi) = \mathcal{I}_{\mathbf{H}}^{\lambda k}(\lambda\xi), \qquad a.e.\, \xi \in \mathbb{R}, \ \forall\, k \in \mathbb{Z} \cap [\lambda^{-1}\mathbb{Z}], \tag{4.1.17}$$

$$\mathcal{I}_{\boldsymbol{\Phi}}^k(\xi) + \mathcal{I}_{\boldsymbol{\Psi}}^k(\xi) = 0, \qquad a.e.\, \xi \in \mathbb{R}, \ \forall\, k \in \mathbb{Z}\backslash[\lambda^{-1}\mathbb{Z}], \tag{4.1.18}$$

$$\mathcal{I}_{\mathbf{H}}^{\lambda k}(\xi) = 0, \qquad a.e.\, \xi \in \mathbb{R}, \ \forall\, k \in [\lambda^{-1}\mathbb{Z}]\backslash\mathbb{Z}, \tag{4.1.19}$$

where $\mathcal{I}_{\mathbf{H}}^{\lambda k}(\xi) := \sum_{\eta\in\mathbf{H}} \tilde{\boldsymbol{\eta}}(\xi)\overline{\boldsymbol{\eta}(\xi + 2\pi\lambda k)}$ *for* $k \in \lambda^{-1}\mathbb{Z}$, *and*

$$\mathcal{I}_{\boldsymbol{\Phi}}^k(\xi) := \sum_{\varphi\in\boldsymbol{\Phi}} \tilde{\boldsymbol{\varphi}}(\xi)\overline{\boldsymbol{\varphi}(\xi + 2\pi k)}, \quad \mathcal{I}_{\boldsymbol{\Psi}}^k(\xi) := \sum_{\psi\in\boldsymbol{\Psi}} \tilde{\boldsymbol{\psi}}(\xi)\overline{\boldsymbol{\psi}(\xi + 2\pi k)}, \tag{4.1.20}$$

for $k \in \mathbb{Z}$, where all the above series converge absolutely for almost every $\xi \in \mathbb{R}$.

Proof By Lemma 4.1.1, all the infinite series in (4.1.16) converge absolutely and (4.1.16) is equivalent to

$$\int_{\mathbb{R}} \sum_{k\in\mathbb{Z}} \mathbf{f}(\xi)\overline{\mathbf{g}(\xi + 2\pi k)}\Big(\overline{\mathcal{I}_{\boldsymbol{\Phi}}^k(\xi)} + \overline{\mathcal{I}_{\boldsymbol{\Psi}}^k(\xi)}\Big) d\xi = \int_{\mathbb{R}} \sum_{k\in[\lambda^{-1}\mathbb{Z}]} \mathbf{f}(\xi)\overline{\mathbf{g}(\xi + 2\pi k)\mathcal{I}_{\mathbf{H}}^{\lambda k}(\lambda\xi)} d\xi,$$

which can be easily rewritten as

$$\int_{\mathbb{R}} \sum_{k\in\mathbb{Z}\cap[\lambda^{-1}\mathbb{Z}]} \overline{\mathbf{f}(\xi)}\mathbf{g}(\xi+2\pi k)\Big(\mathcal{I}^k_{\boldsymbol{\Phi}}(\xi)+\mathcal{I}^k_{\boldsymbol{\Psi}}(\xi)-\mathcal{I}^{\lambda k}_{\mathbf{H}}(\lambda\xi)\Big)d\xi+$$
$$\int_{\mathbb{R}} \sum_{k\in\mathbb{Z}\backslash[\lambda^{-1}\mathbb{Z}]} \overline{\mathbf{f}(\xi)}\mathbf{g}(\xi+2\pi k)\Big(\mathcal{I}^k_{\boldsymbol{\Phi}}(\xi)+\mathcal{I}^k_{\boldsymbol{\Psi}}(\xi)\Big)d\xi \tag{4.1.21}$$
$$=\int_{\mathbb{R}} \sum_{k\in[\lambda^{-1}\mathbb{Z}]\backslash\mathbb{Z}} \overline{\mathbf{f}(\xi)}\mathbf{g}(\xi+2\pi k)\mathcal{I}^{\lambda k}_{\mathbf{H}}(\lambda\xi)d\xi.$$

Sufficiency ($\Leftarrow$). If all the identities (4.1.17), (4.1.18), and (4.1.19) are satisfied, then it is obvious that (4.1.21) is true and therefore, the identity (4.1.16) holds.

Necessity ($\Rightarrow$). Denote $\Lambda := \mathbb{Z}\cup[\lambda^{-1}\mathbb{Z}]$. For a point $x\in\mathbb{R}$, we define $\operatorname{dist}(x,\Lambda) := \inf_{y\in\Lambda}|x-y|$. By (4.1.16), then (4.1.21) holds. Let $k_0\in\mathbb{Z}\cap[\lambda^{-1}\mathbb{Z}]$ and $\xi_0\in\mathbb{R}$ be temporarily fixed. Then $\varepsilon := \frac{\pi}{2}\operatorname{dist}(k_0,\Lambda\backslash\{k_0\})>0$. Consider all $\mathbf{f},\mathbf{g}\in\mathscr{D}(\mathbb{R})$ such that the support of $\mathbf{f}$ is contained inside $(\xi_0-\varepsilon,\xi_0+\varepsilon)$ and the support of $\mathbf{g}$ is contained inside $(\xi_0-2\pi k_0-\varepsilon,\xi_0-2\pi k_0+\varepsilon)$. Then it is not difficult to verify that $\overline{\mathbf{f}(\xi)}\mathbf{g}(\xi+2\pi k)=0$ for all $\xi\in\mathbb{R}$ and $k\in\Lambda\backslash\{k_0\}$, from which we see that (4.1.21) becomes

$$\int_{\mathbb{R}} \overline{\mathbf{f}(\xi)}\mathbf{g}(\xi+2\pi k_0)(\mathcal{I}^{k_0}_{\boldsymbol{\Phi}}(\xi)+\mathcal{I}^{k_0}_{\boldsymbol{\Psi}}(\xi)-\mathcal{I}^{\lambda k_0}_{\mathbf{H}}(\lambda\xi))d\xi=0 \tag{4.1.22}$$

for all $\mathbf{f},\mathbf{g}\in\mathscr{D}(\mathbb{R})$ such that $\operatorname{supp}(\mathbf{f})\subseteq(\xi_0-\varepsilon,\xi_0+\varepsilon)$ and $\operatorname{supp}(\mathbf{g})\subseteq(\xi_0-2\pi k_0-\varepsilon,\xi_0-2\pi k_0+\varepsilon)$. From (4.1.22), we must have $\mathcal{I}^{k_0}_{\boldsymbol{\Phi}}(\xi)+\mathcal{I}^{k_0}_{\boldsymbol{\Psi}}(\xi)-\mathcal{I}^{\lambda k_0}_{\mathbf{H}}(\lambda\xi)=0$ for a.e. $\xi\in(\xi_0-\varepsilon,\xi_0+\varepsilon)$. Since ξ_0 is arbitrarily fixed, we see that (4.1.17) must be true. Equations (4.1.18) and (4.1.19) can be proved by the same argument. □

We now completely characterize a frequency-based dual framelet.

Theorem 4.1.6 *Let* $\boldsymbol{\Phi},\tilde{\boldsymbol{\Phi}},\boldsymbol{\Psi},\tilde{\boldsymbol{\Psi}}\subseteq L_2^{loc}(\mathbb{R})$ *such that* $\sum_{\boldsymbol{\varphi}\in\boldsymbol{\Phi}}(|\boldsymbol{\varphi}(\cdot)|^2+|\tilde{\boldsymbol{\varphi}}(\cdot)|^2)\in L_1^{loc}(\mathbb{R})$ *and* $\sum_{\boldsymbol{\psi}\in\boldsymbol{\Psi}}(|\boldsymbol{\psi}(\cdot)|^2+|\tilde{\boldsymbol{\psi}}(\cdot)|^2)\in L_1^{loc}(\mathbb{R})$. *Then the following are equivalent:*

(i) $(\{\tilde{\boldsymbol{\Phi}};\tilde{\boldsymbol{\Psi}}\},\{\boldsymbol{\Phi};\boldsymbol{\Psi}\})$ *is a frequency-based dual framelet.*
(ii) The identity (4.1.10) *holds and for all* $\mathbf{f},\mathbf{g}\in\mathscr{D}(\mathbb{R})$,

$$\sum_{\boldsymbol{\varphi}\in\boldsymbol{\Phi}}\sum_{k\in\mathbb{Z}}\langle\mathbf{f},\tilde{\boldsymbol{\varphi}}_{1;0,k}\rangle\langle\boldsymbol{\varphi}_{1;0,k},\mathbf{g}\rangle+\sum_{\boldsymbol{\psi}\in\boldsymbol{\Psi}}\sum_{k\in\mathbb{Z}}\langle\mathbf{f},\tilde{\boldsymbol{\psi}}_{1;0,k}\rangle\langle\boldsymbol{\psi}_{1;0,k},\mathbf{g}\rangle$$
$$=\sum_{\boldsymbol{\varphi}\in\boldsymbol{\Phi}}\sum_{k\in\mathbb{Z}}\langle\mathbf{f},\tilde{\boldsymbol{\varphi}}_{2^{-1};0,k}\rangle\langle\boldsymbol{\varphi}_{2^{-1};0,k},\mathbf{g}\rangle. \tag{4.1.23}$$

(iii) The identity (4.1.11) *holds and for almost every* $\xi \in \mathbb{R}$,

$$\begin{aligned}\sum_{\varphi\in\Phi}\tilde{\varphi}(2\xi)\overline{\varphi(2(\xi+2\pi k))}+\sum_{\psi\in\Psi}\tilde{\psi}(2\xi)\overline{\psi(2(\xi+2\pi k))}\\ =\sum_{\varphi\in\Phi}\tilde{\varphi}(\xi)\overline{\varphi(\xi+2\pi k)},\quad \forall\, k\in\mathbb{Z},\end{aligned}\tag{4.1.24}$$

$$\sum_{\varphi\in\Phi}\tilde{\varphi}(\xi)\overline{\varphi(\xi+2\pi k_0)}+\sum_{\psi\in\Psi}\tilde{\psi}(\xi)\overline{\psi(\xi+2\pi k_0)}=0,\ \forall\, k_0\in\mathbb{Z}\backslash[2\mathbb{Z}].\tag{4.1.25}$$

Proof We shall prove (i)$\Longrightarrow$(ii)$\Longrightarrow$(iii)$\Longrightarrow$(i). Suppose that item (i) holds. By Proposition 4.1.2, we see that $(\mathsf{FAS}_J(\tilde{\Phi};\tilde{\Psi}),\mathsf{FAS}_J(\Phi;\Psi))$ is a pair of frequency-based dual frames for every $J\in\mathbb{Z}$. Considering the difference between the frequency-based affine systems at two consecutive levels $J=j$ and $j+1$, we can easily deduce that for all $\mathbf{f},\mathbf{g}\in\mathscr{D}(\mathbb{R})$ and for all $j\in\mathbb{Z}$,

$$\begin{aligned}\sum_{\varphi\in\Phi}\sum_{k\in\mathbb{Z}}\langle\mathbf{f},\tilde{\varphi}_{2^{-j};0,k}\rangle\langle\varphi_{2^{-j};0,k},\mathbf{g}\rangle+\sum_{\psi\in\Psi}\sum_{k\in\mathbb{Z}}\langle\mathbf{f},\tilde{\psi}_{2^{-j};0,k}\rangle\langle\psi_{2^{-j};0,k},\mathbf{g}\rangle\\ =\sum_{\varphi\in\Phi}\sum_{k\in\mathbb{Z}}\langle\mathbf{f},\tilde{\varphi}_{2^{-1-j};0,k}\rangle\langle\varphi_{2^{-1-j};0,k},\mathbf{g}\rangle.\end{aligned}\tag{4.1.26}$$

When $j=0$, the above equation (4.1.26) becomes (4.1.23). By (4.1.26), we observe

$$\begin{aligned}\sum_{\varphi\in\Phi}\sum_{k\in\mathbb{Z}}\langle\mathbf{f},\tilde{\varphi}_{2^{-J};0,k}\rangle\langle\varphi_{2^{-J};0,k},\mathbf{g}\rangle+\sum_{j=J}^{J'-1}\sum_{\psi\in\Psi}\sum_{k\in\mathbb{Z}}\langle\mathbf{f},\tilde{\psi}_{2^{-j};0,k}\rangle\langle\psi_{2^{-j};0,k},\mathbf{g}\rangle\\ =\sum_{\varphi\in\Phi}\sum_{k\in\mathbb{Z}}\langle\mathbf{f},\tilde{\varphi}_{2^{-J'};0,k}\rangle\langle\varphi_{2^{-J'};0,k},\mathbf{g}\rangle.\end{aligned}\tag{4.1.27}$$

(4.1.10) now follows directly from the above relation and (4.1.4). Hence, (i)$\Longrightarrow$(ii).

By Lemma 4.1.5, we see that (4.1.23) is equivalent to (4.1.24) and (4.1.25). By Lemma 4.1.3, we see that (4.1.10) implies (4.1.11). Hence, (ii)$\Longrightarrow$(iii).

Suppose that item (iii) holds. By Lemma 4.1.3, we see that (4.1.10) holds. By Lemma 4.1.5, (4.1.24) and (4.1.25) imply (4.1.23). By the relation in (4.1.8), we see that (4.1.26) must hold for all $j\in\mathbb{Z}$. Consequently, (4.1.27) holds. Now it follows from the assumption in (4.1.10) that (4.1.4) must hold. Hence, $(\mathsf{FAS}_J(\tilde{\Phi};\tilde{\Psi}),\mathsf{FAS}_J(\Phi;\Psi))$ must be a pair of frequency-based dual frames for every integer J. Thus, $(\{\tilde{\Phi};\tilde{\Psi}\},\{\Phi;\Psi\})$ is a frequency-based dual framelet and item (i) holds. □

Since (4.1.23) is equivalent to (4.1.26) which essentially reveals the multiscale relation embedded in a frequency-based affine system, we call (4.1.23) or (4.1.26) the *cascade structure* of a frequency-based dual framelet $(\{\tilde{\Phi};\tilde{\Psi}\},\{\Phi;\Psi\})$.

When all the generators in the sets $\boldsymbol{\Phi}, \tilde{\boldsymbol{\Phi}}, \boldsymbol{\Psi}, \tilde{\boldsymbol{\Psi}}$ are nonnegative, Theorem 4.1.6 reduces to the following simple result, whose proof is left as Exercise 4.4.

Corollary 4.1.7 *Let* $\boldsymbol{\Phi}, \tilde{\boldsymbol{\Phi}}, \boldsymbol{\Psi}, \tilde{\boldsymbol{\Psi}} \subseteq L_2^{loc}(\mathbb{R})$ *such that* (4.1.9) *and* (4.1.15) *are satisfied. If* $(\{\tilde{\boldsymbol{\Phi}}; \tilde{\boldsymbol{\Psi}}\}, \{\boldsymbol{\Phi}; \boldsymbol{\Psi}\})$ *is a frequency-based dual framelet and* $\mathbf{h}(\xi) \geqslant 0$ *for almost every* $\xi \in \mathbb{R}$ *for all* $\mathbf{h} \in \boldsymbol{\Phi} \cup \tilde{\boldsymbol{\Phi}} \cup \boldsymbol{\Psi} \cup \tilde{\boldsymbol{\Psi}}$, *then* (4.1.11) *holds,*

$$\tilde{\varphi}(\xi)\overline{\varphi(\xi + 2\pi k)} = 0, \quad \tilde{\psi}(\xi)\overline{\psi(\xi + 2\pi k)} = 0, \qquad a.e.\, \xi \in \mathbb{R}, \quad \forall\, k \in \mathbb{Z}\backslash\{0\},\, \varphi \in \boldsymbol{\Phi},\, \psi \in \boldsymbol{\Psi} \tag{4.1.28}$$

and

$$\sum_{\varphi\in\boldsymbol{\Phi}} \tilde{\varphi}(2\xi)\overline{\varphi(2\xi)} + \sum_{\psi\in\boldsymbol{\Psi}} \tilde{\psi}(2\xi)\overline{\psi(2\xi)} = \sum_{\varphi\in\boldsymbol{\Phi}} \tilde{\varphi}(\xi)\overline{\varphi(\xi)}, \qquad a.e.\, \xi \in \mathbb{R}. \tag{4.1.29}$$

Conversely, if all the three conditions (4.1.11), (4.1.28), *and* (4.1.29) *are satisfied, then* $(\{\tilde{\boldsymbol{\Phi}}; \tilde{\boldsymbol{\Psi}}\}, \{\boldsymbol{\Phi}; \boldsymbol{\Psi}\})$ *is a frequency-based dual framelet.*

4.1.3 Connections to Dual Framelet Filter Banks

We now show that under some natural condition there is a natural one-to-one correspondence between dual framelet filter banks and frequency-based dual framelets.

To establish the correspondence between frequency-based dual framelets and dual framelet filter banks, let us first study the existence of a frequency-based scalar refinable function associated with a low-pass filter. As a special case of Theorem 5.1.2, we have the following result.

Lemma 4.1.8 *Let* $\widehat{a} : \mathbb{R} \to \mathbb{C}$ *be a* 2π*-periodic measurable function such that there exist positive constants* C_0 *and* τ *satisfying*

$$|1 - \widehat{a}(\xi)| \leqslant C_0|\xi|^\tau, \qquad \forall\, \xi \in [-\pi, \pi]. \tag{4.1.30}$$

Then

$$\left|\prod_{j=1}^{n} \widehat{a}(2^{-j}\xi)\right| \leqslant C(\xi) \qquad \forall\, \xi \in \mathbb{R}, n \in \mathbb{N} \tag{4.1.31}$$

with $C(\xi) := e^{\frac{C_0\pi^\tau}{1-2^{-\tau}}} \max\left(1, (\frac{|\xi|}{\pi})^{\log_2 \|\widehat{a}\|_{L_\infty(\mathbb{R})}}\right)$ *and*

$$\varphi^a(\xi) := \prod_{j=1}^{\infty} \widehat{a}(2^{-j}\xi) := \lim_{n\to\infty} \prod_{j=1}^{n} \widehat{a}(2^{-j}\xi) \tag{4.1.32}$$

is well defined with the infinite product uniformly converging for ξ in every bounded interval. Moreover, for all $\xi \in \mathbb{R}$,

$$\Big|\prod_{j=1}^{m}\widehat{a}(2^{-j}\xi)-\prod_{j=1}^{n}\widehat{a}(2^{-j}\xi)\Big| \leqslant \frac{2^{-\tau(m+1)}}{1-2^{-\tau}}C_0[C(\xi)]^2|\xi|^\tau, \quad \forall\, 0\leqslant m<n\leqslant\infty \tag{4.1.33}$$

with the convention $\prod_{j=n}^{m} := 1$ for $m < n$.

Proof By (4.1.30), since $\widehat{a}$ is 2π-periodic, we have $1 \leqslant \|\widehat{a}\|_{L_\infty(\mathbb{R})} \leqslant 1 + C_0\pi^\tau$ and (4.1.30) must hold for all $\xi \in \mathbb{R}$. By $|z| \leqslant 1 + |1 - z| \leqslant e^{|1-z|}$ for all $z \in \mathbb{C}$,

$$\Big|\prod_{j=1}^{n}\widehat{a}(2^{-j}\xi)\Big| \leqslant e^{\sum_{j=1}^{n}|1-\widehat{a}(2^{-j}\xi)|} \leqslant e^{C_0\sum_{j=1}^{n}2^{-\tau j}|\xi|^\tau} \leqslant e^{C_0|\xi|^\tau\sum_{j=1}^{\infty}2^{-\tau j}} = e^{\frac{C_0|\xi/2|^\tau}{1-2^{-\tau}}}.$$

Hence, for $|\xi| \leqslant 2\pi$, we have $|\prod_{j=1}^{n}\widehat{a}(2^{-j}\xi)| \leqslant C(0)$. For $2^m\pi < |\xi| \leqslant 2^{m+1}\pi$ with $m \in \mathbb{N}$, noting that $|2^{-m}\xi| \leqslant 2\pi$ and $m < \log_2(|\xi|/\pi)$, we deduce

$$\begin{aligned}\Big|\prod_{j=1}^{n}\widehat{a}(2^{-j}\xi)\Big| &= \prod_{j=1}^{\min(m,n)}|\widehat{a}(2^{-j}\xi)|\prod_{j=1}^{\max(0,n-m)}|\widehat{a}(2^{-j}2^{-m}\xi)| \leqslant C(0)\prod_{j=1}^{\min(m,n)}|\widehat{a}(2^{-j}\xi)|\\ &\leqslant C(0)\|\widehat{a}\|_{L_\infty(\mathbb{R})}^{\min(m,n)} \leqslant C(0)\|\widehat{a}\|_{L_\infty(\mathbb{R})}^{m} \leqslant C(0)\|\widehat{a}\|_{L_\infty(\mathbb{R})}^{\log_2(|\xi|/\pi)} = C(\xi).\end{aligned}$$

Therefore, we proved (4.1.31). For $0 \leqslant m < n < \infty$, we have

$$\begin{aligned}\Big|\prod_{j=1}^{m}\widehat{a}(2^{-j}\xi)-\prod_{j=1}^{n}\widehat{a}(2^{-j}\xi)\Big| &= \Big|\Big[\prod_{j=1}^{m}\widehat{a}(2^{-j}\xi)\Big]\Big[\sum_{k=m+1}^{n}(1-\widehat{a}(2^{-k}\xi))\prod_{j=k+1}^{n}\widehat{a}(2^{-j}\xi)\Big]\Big|\\ &\leqslant C(\xi)\sum_{k=m+1}^{n}|1-\widehat{a}(2^{-k}\xi)|C(2^{-k}\xi) \leqslant C_0C(\xi)\sum_{k=m+1}^{n}|2^{-k}\xi|^\tau C(2^{-k}\xi)\\ &\leqslant C_0[C(\xi)]^2|\xi|^\tau\sum_{k=m+1}^{\infty}2^{-\tau k} = \frac{2^{-\tau(m+1)}}{1-2^{-\tau}}C_0[C(\xi)]^2|\xi|^\tau,\end{aligned}$$

since $C(\xi)$ is increasing on $[0,\infty)$. Hence, (4.1.33) is verified. For every fixed ξ, (4.1.33) implies that $\{\prod_{j=1}^{n}\widehat{a}(2^{-j}\xi)\}_{n\in\mathbb{N}}$ is a Cauchy sequence and so, $\boldsymbol{\varphi}^a$ in (4.1.32) is well defined. Setting $n = \infty$ in (4.1.33), we see that $\prod_{j=1}^{m}\widehat{a}(2^{-j}\xi)$ converges uniformly to $\boldsymbol{\varphi}^a(\xi)$ as $m \to \infty$ for ξ in every bounded interval. □

Let $a = \{a(k)\}_{k\in\mathbb{Z}} \in l_0(\mathbb{Z})$ be a low-pass filter such that $\widehat{a}(0) = 1$, where we recall that $\widehat{a}(\xi) := \sum_{k\in\mathbb{Z}} a(k)e^{-ik\xi}$. Since $|1-\widehat{a}(\xi)| = |\int_0^\xi \widehat{a}'(\xi)d\xi| \leqslant \|\widehat{a}'\|_{L_\infty(\mathbb{R})}|\xi|$ for all $\xi \in [-\pi,\pi]$, by Lemma 4.1.8, (4.1.32) defines a well-defined continuous function $\boldsymbol{\varphi}^a$, which is called *a frequency-based refinable function* associated with the filter/mask a and φ^a satisfies the frequency-based refinement equation $\varphi^a(2\xi) = \widehat{a}(\xi)\varphi^a(\xi)$ for all $\xi \in \mathbb{R}$. By (4.1.31), we conclude that φ^a must be a tempered distribution and thus $\varphi^a = \widehat{\phi^a}$ for some tempered distribution ϕ^a, which is called the (spatial) refinable function/distrubition. If in addition $a \in l_0(\mathbb{Z})$, then ϕ^a must be compactly supported with $\mathrm{supp}(\phi^a) \subseteq \mathrm{fsupp}(a)$ and $\phi^a = 2\sum_{k\in\mathbb{Z}} a(k)\phi^a(2\cdot -k)$, see Exercise 4.5.

Recall that $(\{\tilde{a}; \tilde{b}_1, \ldots, \tilde{b}_s\}, \{a; b_1, \ldots, b_s\})_\Theta$ is *a dual framelet filter bank* if

$$\widehat{\Theta}(2\xi)\widehat{\tilde{a}}(\xi)\overline{\widehat{a}(\xi)} + \widehat{\tilde{b}_1}(\xi)\overline{\widehat{b_1}(\xi)} + \cdots + \widehat{\tilde{b}_s}(\xi)\overline{\widehat{b_s}(\xi)} = \widehat{\Theta}(\xi), \tag{4.1.34}$$

$$\widehat{\Theta}(2\xi)\widehat{\tilde{a}}(\xi)\overline{\widehat{a}(\xi+\pi)} + \widehat{\tilde{b}_1}(\xi)\overline{\widehat{b_1}(\xi+\pi)} + \cdots + \widehat{\tilde{b}_s}(\xi)\overline{\widehat{b_s}(\xi+\pi)} = 0. \tag{4.1.35}$$

To be consistent with the notation of a dual framelet filter bank, if all the elements in the sets $\boldsymbol{\Phi}, \tilde{\boldsymbol{\Phi}}, \boldsymbol{\Psi}, \tilde{\boldsymbol{\Psi}}$ are explicitly listed as $\boldsymbol{\Phi} = \{\varphi^1, \ldots, \varphi^r\}$, $\boldsymbol{\Psi} = \{\psi^1, \ldots, \psi^s\}$, and $\tilde{\boldsymbol{\Phi}} = \{\tilde{\varphi}^1, \ldots, \tilde{\varphi}^r\}$, $\tilde{\boldsymbol{\Psi}} = \{\tilde{\psi}^1, \ldots, \tilde{\psi}^s\}$, then we use the notation $(\{\tilde{\varphi}^1, \ldots, \tilde{\varphi}^r; \tilde{\psi}^1, \ldots, \tilde{\psi}^s\}, \{\varphi^1, \ldots, \varphi^r; \psi^1, \ldots, \psi^s\})$ or $(\{\tilde{\varphi}; \tilde{\psi}\}, \{\varphi; \psi\})$ to stand for $(\{\tilde{\boldsymbol{\Phi}}; \tilde{\boldsymbol{\Psi}}\}, \{\boldsymbol{\Phi}, \boldsymbol{\Psi}\})$ with $\varphi := (\varphi^1, \ldots, \varphi^r)^\mathsf{T}$ and $\psi := (\psi^1, \ldots, \psi^s)^\mathsf{T}$.

The following result establishes a natural correspondence between a dual framelet filter bank and a frequency-based dual framelet.

Theorem 4.1.9 *Let* $\theta, a, b_1, \ldots, b_s, \tilde{\theta}, \tilde{a}, \tilde{b}_1, \ldots, \tilde{b}_s \in l_0(\mathbb{Z})$ *such that* $\widehat{a}(0) = \widehat{\tilde{a}}(0) = 1$. *Define two frequency-based refinable functions* $\varphi := \varphi^a$ *and* $\tilde{\varphi} := \varphi^{\tilde{a}}$ *as in* (4.1.32) *associated with the filters* a *and* $\tilde{a}$*, respectively. Define functions* $\eta, \tilde{\eta}$ *by*

$$\eta(\xi) := \widehat{\theta}(\xi)\varphi(\xi), \qquad \tilde{\eta}(\xi) := \widehat{\tilde{\theta}}(\xi)\tilde{\varphi}(\xi), \qquad \xi \in \mathbb{R}, \tag{4.1.36}$$

and

$$\psi^\ell(\xi) := \widehat{b_\ell}(\xi/2)\varphi(\xi/2), \quad \tilde{\psi}^\ell(\xi) := \widehat{\tilde{b}_\ell}(\xi/2)\tilde{\varphi}(\xi/2), \qquad \ell = 1, \ldots, s. \tag{4.1.37}$$

Then $(\{\tilde{\eta}; \tilde{\psi}^1, \ldots, \tilde{\psi}^s\}, \{\eta; \psi^1, \ldots, \psi^s\})$ *is a frequency-based dual framelet if and only if* $(\{\tilde{a}; \tilde{b}_1, \ldots, \tilde{b}_s\}, \{a; b_1, \ldots, b_s\})_\Theta$ *is a dual framelet filter bank and* $\widehat{\Theta}(0) = 1$*, where* $\widehat{\Theta}(\xi) := \widehat{\tilde{\theta}}(\xi)\overline{\widehat{\theta}(\xi)}$.

Proof By Lemma 4.1.8, both φ and $\tilde{\varphi}$ are well-defined continuous functions. Hence, all functions $\eta, \psi^1, \ldots, \psi^s, \tilde{\eta}, \tilde{\psi}^1, \ldots, \tilde{\psi}^s \in L_2^{loc}(\mathbb{R})$. We now apply Theorem 4.1.6 with $\boldsymbol{\Phi} := \{\eta\}$ and $\tilde{\boldsymbol{\Phi}} := \{\tilde{\eta}\}$ to prove our claim.

By (4.1.32), we have $\varphi(2\xi) = \widehat{a}(\xi)\varphi(\xi)$ and $\tilde{\varphi}(2\xi) = \widehat{\tilde{a}}(\xi)\tilde{\varphi}(\xi)$ for all $\xi \in \mathbb{R}$. Using the definition of Θ and the functions $\eta, \tilde{\eta}$ in (4.1.36), we have

$$\begin{aligned}\tilde{\eta}(2\xi)\overline{\eta(2(\xi+2\pi k))} &= \tilde{\varphi}(\xi)\overline{\varphi(\xi+2\pi k)}\widehat{\tilde{\theta}}(2\xi)\overline{\widehat{\theta}(2\xi)}\widehat{\tilde{a}}(\xi)\overline{\widehat{a}(\xi)} \\ &= \tilde{\varphi}(\xi)\overline{\varphi(\xi+2\pi k)}\widehat{\Theta}(2\xi)\widehat{\tilde{a}}(\xi)\overline{\widehat{a}(\xi)}\end{aligned}$$

and similarly by (4.1.37),

$$\sum_{\ell=1}^{s} \widehat{\tilde{\psi}^{\ell}}(2\xi)\overline{\widehat{\psi^{\ell}}(2(\xi+2\pi k))} = \widehat{\tilde{\varphi}}(\xi)\overline{\widehat{\varphi}(\xi+2\pi k)}\sum_{\ell=1}^{s}\widehat{\tilde{b}_\ell}(\xi)\overline{\widehat{b_\ell}(\xi)}$$

for all integers $k \in \mathbb{Z}$. Now the equation (4.1.24) in Theorem 4.1.6 with $\boldsymbol{\Phi} = \{\boldsymbol{\eta}\}$ and $\tilde{\boldsymbol{\Phi}} = \{\tilde{\boldsymbol{\eta}}\}$ is equivalent to

$$\begin{aligned} \widehat{\tilde{\varphi}}(\xi)\overline{\widehat{\varphi}(\xi+2\pi k)}\Big(\widehat{\Theta}(2\xi)\widehat{\tilde{a}}(\xi)\overline{\widehat{a}(\xi)} + \sum_{\ell=1}^{s}\widehat{\tilde{b}_\ell}(\xi)\overline{\widehat{b_\ell}(\xi)}\Big) \\ = \widehat{\tilde{\varphi}}(\xi)\overline{\widehat{\varphi}(\xi+2\pi k)}\widehat{\Theta}(\xi) \end{aligned} \tag{4.1.38}$$

for all $k \in \mathbb{Z}$. Since $\widehat{a}$ is a 2π-periodic trigonometric polynomial, the set of all zeros of $\widehat{a}$ is countable. Therefore, by Lemma 4.1.8 and $\widehat{\varphi}(\xi) = \prod_{j=1}^{\infty}\widehat{a}(2^{-j}\xi)$ for $\xi \in \mathbb{R}$, the set $\{\xi \in \mathbb{R} \; : \; \widehat{\varphi}(\xi) = 0\}$ is countable and hence $\widehat{\varphi}(\xi) \neq 0$ for almost every $\xi \in \mathbb{R}$. Similarly, $\widehat{\tilde{\varphi}}(\xi) \neq 0$ for almost every $\xi \in \mathbb{R}$. Therefore, it is straightforward to deduce that (4.1.38) is equivalent to (4.1.34).

For $k_0 = 2k+1$ with $k \in \mathbb{Z}$, replacing ξ in (4.1.25) by 2ξ, by the same argument for (4.1.38), we see that (4.1.25) with $\boldsymbol{\Phi} = \{\boldsymbol{\eta}\}$ and $\tilde{\boldsymbol{\Phi}} = \{\tilde{\boldsymbol{\eta}}\}$ is equivalent to

$$\widehat{\tilde{\varphi}}(\xi)\overline{\widehat{\varphi}(\xi+\pi+2\pi k)}\Big(\widehat{\Theta}(2\xi)\widehat{\tilde{a}}(\xi)\overline{\widehat{a}(\xi+\pi)} + \sum_{\ell=1}^{s}\widehat{\tilde{b}_\ell}(\xi)\overline{\widehat{b_\ell}(\xi+\pi)}\Big) = 0$$

for all $k \in \mathbb{Z}$. It is trivial that the above identity is equivalent to (4.1.35). Hence, we proved that $(\{\tilde{a}; \tilde{b}_1, \ldots, \tilde{b}_s\}, \{a; b_1, \ldots, b_s\})_\Theta$ is a dual framelet filter bank if and only if (4.1.24) and (4.1.25) of Theorem 4.1.6 are satisfied with $\boldsymbol{\Phi} = \{\boldsymbol{\eta}\}$ and $\tilde{\boldsymbol{\Phi}} = \{\tilde{\boldsymbol{\eta}}\}$.

To complete the proof, by Theorem 4.1.6 we prove that $\widehat{\Theta}(0) = 1$ if and only if

$$\lim_{j\to\infty}\langle \tilde{\eta}(2^{-j}\cdot)\overline{\eta(2^{-j}\cdot)}, \mathbf{h}\rangle = \langle 1, \mathbf{h}\rangle, \qquad \forall\, \mathbf{h} \in \mathscr{D}(\mathbb{R}). \tag{4.1.39}$$

Note that both $\widehat{\varphi}$ and $\widehat{\tilde{\varphi}}$ are continuous functions with $\widehat{\varphi}(0) = \widehat{\tilde{\varphi}}(0) = 1$. By the definition of η and $\tilde{\eta}$ in (4.1.36), we see that

$$\tilde{\eta}(2^{-j}\xi)\overline{\eta(2^{-j}\xi)} = \widehat{\tilde{\varphi}}(2^{-j}\xi)\overline{\widehat{\varphi}(2^{-j}\xi)}\widehat{\Theta}(2^{-j}\xi).$$

Therefore, we have $|\tilde{\eta}(\xi)\overline{\eta(\xi)}| \leqslant \max_{\xi\in[-1,1]} |\widehat{\tilde{\varphi}}(\xi)\widehat{\varphi}(\xi)\widehat{\Theta}(\xi)| < \infty$ for all $\xi \in [-1, 1]$ and

$$\lim_{j\to\infty}\tilde{\eta}(2^{-j}\xi)\overline{\eta(2^{-j}\xi)} = \lim_{j\to\infty}\widehat{\tilde{\varphi}}(2^{-j}\xi)\overline{\widehat{\varphi}(2^{-j}\xi)}\widehat{\Theta}(2^{-j}\xi) = \widehat{\Theta}(0).$$

By Lemma 4.1.4, the identity (4.1.39) holds if and only if $\widehat{\Theta}(0) = 1$. ∎

Furthermore we have the following more general result.

Theorem 4.1.10 *Let $\varphi^1, \dots, \varphi^r, \tilde{\varphi}^1, \dots, \tilde{\varphi}^r \in L_2^{loc}(\mathbb{R})$ such that there exist $r \times r$ matrices $\widehat{a}$, $\widehat{\tilde{a}}$ of 2π-periodic measurable functions satisfying*

$$\boldsymbol{\varphi}(2\xi) = \widehat{a}(\xi)\boldsymbol{\varphi}(\xi) \quad \text{and} \quad \tilde{\boldsymbol{\varphi}}(2\xi) = \widehat{\tilde{a}}(\xi)\tilde{\boldsymbol{\varphi}}(\xi), \qquad a.e.\, \xi \in \mathbb{R}, \tag{4.1.40}$$

where $\boldsymbol{\varphi} := (\varphi^1, \dots, \varphi^r)^\mathsf{T}$ and $\tilde{\boldsymbol{\varphi}} := (\tilde{\varphi}^1, \dots, \tilde{\varphi}^r)^\mathsf{T}$. For $r \times r$ matrices $\widehat{\theta}, \widehat{\tilde{\theta}}$ and $s \times r$ matrices $\widehat{b}, \widehat{\tilde{b}}$ of 2π-periodic measurable functions on $\mathbb{R}$, define $\boldsymbol{\eta}, \tilde{\boldsymbol{\eta}}, \boldsymbol{\psi}, \tilde{\boldsymbol{\psi}}$ by

$$\boldsymbol{\eta}(\xi) := \widehat{\theta}(\xi)\boldsymbol{\varphi}(\xi), \quad \tilde{\boldsymbol{\eta}}(\xi) := \widehat{\tilde{\theta}}(\xi)\tilde{\boldsymbol{\varphi}}(\xi), \quad \boldsymbol{\psi}(\xi) := \widehat{b}(\tfrac{\xi}{2})\boldsymbol{\varphi}(\tfrac{\xi}{2}), \quad \tilde{\boldsymbol{\psi}}(\xi) := \widehat{\tilde{b}}(\tfrac{\xi}{2})\tilde{\boldsymbol{\varphi}}(\tfrac{\xi}{2}).$$

Define $\widehat{\Theta}(\xi) := \widehat{\tilde{\theta}}(\xi)^\mathsf{T}\overline{\widehat{\theta}(\xi)}$. Suppose that all the entries in $\boldsymbol{\eta}, \tilde{\boldsymbol{\eta}}, \boldsymbol{\psi}, \tilde{\boldsymbol{\psi}}$ belong to $L_2^{loc}(\mathbb{R})$. Then $(\{\tilde{\boldsymbol{\eta}}; \tilde{\boldsymbol{\psi}}\}, \{\boldsymbol{\eta}; \boldsymbol{\psi}\})$ is a frequency-based dual framelet if and only if

$$\lim_{j\to\infty} \langle \tilde{\boldsymbol{\varphi}}(2^{-j}\cdot)^\mathsf{T}\widehat{\Theta}(2^{-j}\cdot)\overline{\boldsymbol{\varphi}(2^{-j}\cdot)}, \mathbf{h}\rangle = \langle 1, \mathbf{h}\rangle, \qquad \forall\, \mathbf{h} \in \mathscr{D}(\mathbb{R}) \tag{4.1.41}$$

and $(\{\widehat{\tilde{a}}; \widehat{\tilde{b}}\}, \{\widehat{a}; \widehat{b}\})_{\widehat{\Theta}}$ *is a* (frequency-based) generalized dual framelet (matrix-valued) filter bank, *that is, for all $k \in \mathbb{Z}$ and for almost every $\xi \in \mathbb{R}$,*

$$\tilde{\boldsymbol{\varphi}}(\xi)^\mathsf{T}\Big[\widehat{\tilde{a}}(\xi)^\mathsf{T}\widehat{\Theta}(2\xi)\overline{\widehat{a}(\xi)} + \widehat{\tilde{b}}(\xi)^\mathsf{T}\overline{\widehat{b}(\xi)} - \widehat{\Theta}(\xi)\Big]\overline{\boldsymbol{\varphi}(\xi + 2\pi k)} = 0, \tag{4.1.42}$$

$$\tilde{\boldsymbol{\varphi}}(\xi)^\mathsf{T}\Big[\widehat{\tilde{a}}(\xi)^\mathsf{T}\widehat{\Theta}(2\xi)\overline{\widehat{a}(\xi + \pi)} + \widehat{\tilde{b}}(\xi)^\mathsf{T}\overline{\widehat{b}(\xi + \pi)}\Big]\overline{\boldsymbol{\varphi}(\xi + 2\pi k)} = 0. \tag{4.1.43}$$

Proof The argument is the same as in Theorem 4.1.9. Indeed, for $k \in \mathbb{Z}$, we have

$$\begin{aligned}\sum_{\ell=1}^{r} \tilde{\boldsymbol{\eta}}^\ell(2\xi)\overline{\boldsymbol{\eta}^\ell(2(\xi + 2\pi k))} &= \tilde{\boldsymbol{\varphi}}(2\xi)^\mathsf{T}\widehat{\tilde{\theta}}(2\xi)^\mathsf{T}\overline{\widehat{\theta}(2\xi)}\overline{\boldsymbol{\varphi}(2(\xi + 2\pi k))}\\ &= \tilde{\boldsymbol{\varphi}}(\xi)^\mathsf{T}\widehat{\tilde{a}}(\xi)^\mathsf{T}\widehat{\Theta}(2\xi)\overline{\widehat{a}(\xi)}\,\overline{\boldsymbol{\varphi}(\xi + 2\pi k)}\end{aligned}$$

and $\sum_{\ell=1}^{s} \tilde{\boldsymbol{\psi}}^\ell(2\xi)\overline{\boldsymbol{\psi}^\ell(2(\xi + 2\pi k))} = \tilde{\boldsymbol{\varphi}}(\xi)^\mathsf{T}\widehat{\tilde{b}}(\xi)^\mathsf{T}\overline{\widehat{b}(\xi)}\overline{\boldsymbol{\varphi}(\xi + 2\pi k)}$. Replacing $\boldsymbol{\Phi}$ and $\tilde{\boldsymbol{\Phi}}$ in Theorem 4.1.6 by $\boldsymbol{\eta}$ and $\tilde{\boldsymbol{\eta}}$ (as sets instead of vectors) respectively, we see that (4.1.24) is equivalent to (4.1.42) and (4.1.25) is equivalent to (4.1.43). Now the claim follows directly from Theorem 4.1.6. □

For the scalar case $r = 1$ in Theorem 4.1.10, (4.1.42) and (4.1.43) become

$$\widehat{\Theta}(2\xi)\widehat{\tilde{a}}(\xi)\overline{\widehat{a}(\xi)} + \sum_{\ell=1}^{s} \widehat{\tilde{b}_\ell}(\xi)\overline{\widehat{b_\ell}(\xi)} = \widehat{\Theta}(\xi), \quad a.e.\, \xi \in \sigma_{\boldsymbol{\varphi}} \cap \sigma_{\tilde{\boldsymbol{\varphi}}}, \tag{4.1.44}$$

$$\widehat{\Theta}(2\xi)\widehat{\tilde{a}}(\xi)\overline{\widehat{a}(\xi + \pi)} + \sum_{\ell=1}^{s} \widehat{\tilde{b}_\ell}(\xi)\overline{\widehat{b_\ell}(\xi + \pi)} = 0, \quad a.e.\, \xi \in \sigma_{\boldsymbol{\varphi}} \cap (\sigma_{\tilde{\boldsymbol{\varphi}}} - \pi), \tag{4.1.45}$$

where $(\widehat{b_1}, \ldots, \widehat{b_s}) := \widehat{b}^{\mathsf{T}}$, $(\widehat{\tilde{b}_1}, \ldots, \widehat{\tilde{b}_s}) := \widehat{\tilde{b}}^{\mathsf{T}}$, and

$$\sigma_{\varphi} := \operatorname{supp}([\varphi, \varphi]), \qquad \sigma_{\tilde{\varphi}} := \operatorname{supp}([\tilde{\varphi}, \tilde{\varphi}]). \tag{4.1.46}$$

Comparing with a standard dual framelet filter bank, the two equations in (4.1.44) and (4.1.45) may not hold for almost every $\xi \in \mathbb{R}$ for a generalized dual framelet filter bank. We now show that a generalized dual framelet filter bank can be modified into a standard dual framelet filter bank without changing its underlying frequency-based dual framelet $(\{\tilde{\eta}; \tilde{\boldsymbol{\psi}}\}, \{\eta; \boldsymbol{\psi}\})$ in Theorem 4.1.10 with $r = 1$. Define

$$\boldsymbol{\theta}(\xi) := \begin{cases} \widehat{\theta}(\xi), & \xi \in \sigma_{\varphi}, \\ 1, & \xi \in \mathbb{R}\backslash\sigma_{\varphi}, \end{cases} \qquad \tilde{\boldsymbol{\theta}}(\xi) := \begin{cases} \widehat{\tilde{\theta}}(\xi), & \xi \in \sigma_{\tilde{\varphi}}, \\ 1, & \xi \in \mathbb{R}\backslash\sigma_{\tilde{\varphi}}, \end{cases}$$

where σ_{φ} and $\sigma_{\tilde{\varphi}}$ are defined in (4.1.46). Define $\boldsymbol{\Theta}(\xi) := \tilde{\boldsymbol{\theta}}(\xi)\overline{\boldsymbol{\theta}(\xi)}$ and

$$\mathbf{a}(\xi) := \widehat{a}(\xi)\chi_{\sigma_{\varphi}}(\xi), \quad \tilde{\mathbf{a}}(\xi) := \widehat{\tilde{a}}(\xi)\chi_{\sigma_{\tilde{\varphi}}}(\xi),$$

$$\mathbf{b}_{\ell}(\xi) := \widehat{b_{\ell}}(\xi)\chi_{\sigma_{\varphi}}(\xi), \quad \tilde{\mathbf{b}}_{\ell}(\xi) := \widehat{\tilde{b}_{\ell}}(\xi)\chi_{\sigma_{\tilde{\varphi}}}(\xi), \quad \ell = 1, \ldots, s.$$

Then it is straightforward to directly check that all the identities in (4.1.40), (4.1.36), (4.1.37), and (4.1.41) are satisfied and consequently, by Theorem 4.1.10, both (4.1.44) and (4.1.45) are also satisfied when $\widehat{a}, \widehat{b_1}, \ldots, \widehat{b_s}, \widehat{\tilde{a}}, \widehat{\tilde{b}_1}, \ldots, \widehat{\tilde{b}_s}, \widehat{\theta}, \widehat{\tilde{\theta}}, \widehat{\Theta}$ are replaced by $\mathbf{a}, \mathbf{b}_1, \ldots, \mathbf{b}_s, \tilde{\mathbf{a}}, \tilde{\mathbf{b}}_1, \ldots, \tilde{\mathbf{b}}_s, \boldsymbol{\theta}, \tilde{\boldsymbol{\theta}}, \boldsymbol{\Theta}$, respectively. Define 2π-periodic measurable functions as follows: For $\xi \in [-\pi, \pi)$,

$$\mathbf{b}_{s+1}(\xi) := \boldsymbol{\theta}(\xi)\chi_{[-\pi,0]\backslash\sigma_{\varphi}}(\xi), \quad \mathbf{b}_{s+2}(\xi) := \boldsymbol{\theta}(\xi)\chi_{[0,\pi]\backslash\sigma_{\varphi}}(\xi),$$

$$\tilde{\mathbf{b}}_{s+1}(\xi) := \tilde{\boldsymbol{\theta}}(\xi)\chi_{[-\pi,0]\backslash\sigma_{\tilde{\varphi}}}(\xi), \quad \tilde{\mathbf{b}}_{s+2}(\xi) := \tilde{\boldsymbol{\theta}}(\xi)\chi_{[0,\pi]\backslash\sigma_{\tilde{\varphi}}}(\xi).$$

If in addition $(\mathbb{R}\backslash(\sigma_{\varphi} \cap \sigma_{\tilde{\varphi}})) \cap (\mathbb{R}\backslash(\sigma_{\varphi} \cap \sigma_{\tilde{\varphi}}) - \pi)$ has measure zero, then we can simply take $\mathbf{b}_{s+1}(\xi) := \boldsymbol{\theta}(\xi)\chi_{\mathbb{R}\backslash(\sigma_{\varphi}\cap\sigma_{\tilde{\varphi}})}(\xi)$ and $\tilde{\mathbf{b}}_{s+1}(\xi) := \tilde{\boldsymbol{\theta}}(\xi)\chi_{\mathbb{R}\backslash(\sigma_{\varphi}\cap\sigma_{\tilde{\varphi}})}(\xi)$ for $\xi \in [-\pi, \pi)$, and $\mathbf{b}_{s+2} = \tilde{\mathbf{b}}_{s+2} = 0$. Then $(\{\tilde{\mathbf{a}}; \tilde{\mathbf{b}}_1, \ldots, \tilde{\mathbf{b}}_{s+2}\}, \{\mathbf{a}; \mathbf{b}_1, \ldots, \mathbf{b}_{s+2}\})_{\boldsymbol{\Theta}}$ is a (frequency-based) dual framelet filter bank. If we define $\boldsymbol{\psi}^{\ell}$ and $\tilde{\boldsymbol{\psi}}^{\ell}$ as in (4.1.37) for all $\ell = 1, \ldots, s+2$, then we have $\boldsymbol{\psi}^{s+1} = \boldsymbol{\psi}^{s+2} = \tilde{\boldsymbol{\psi}}^{s+1} = \tilde{\boldsymbol{\psi}}^{s+2} = 0$ and all other functions $\tilde{\eta}, \tilde{\boldsymbol{\psi}}^{1}, \ldots \tilde{\boldsymbol{\psi}}^{s}, \eta, \boldsymbol{\psi}^{1}, \ldots, \boldsymbol{\psi}^{s}$ are the same/unchanged as before.

4.1.4 Framelet Transforms Versus Discrete Framelet Transforms

Based on Theorems 4.1.9 and 4.1.10, we now discuss the connections between framelet transforms in the function setting and discrete framelet transforms in the discrete setting induced by a dual framelet filter bank. Since we described the discrete framelet transforms in the time/space domain, we shall discuss the connections between framelet transforms and discrete framelet transforms using Theorem 4.1.9.

Assume that $(\{\tilde{a}; \tilde{b}_1, \ldots, \tilde{b}_s\}, \{a; b_1, \ldots, b_s\})_\Theta$ is a dual framelet filter bank and $\widehat{\Theta}(0) = 1$, where $\widehat{\Theta}(\xi) := \widehat{\overline{\theta}}(\xi)\overline{\widehat{\theta}(\xi)}$. Let $\boldsymbol{\varphi}, \boldsymbol{\eta}, \boldsymbol{\psi}^1, \ldots, \boldsymbol{\psi}^s, \tilde{\boldsymbol{\varphi}}, \tilde{\boldsymbol{\eta}}, \tilde{\boldsymbol{\psi}}^1, \ldots, \tilde{\boldsymbol{\psi}}^s$ be defined in Theorem 4.1.9. Then by Theorem 4.1.9, $(\{\tilde{\boldsymbol{\eta}}; \tilde{\boldsymbol{\psi}}^1, \ldots, \tilde{\boldsymbol{\psi}}^s\}, \{\boldsymbol{\eta}; \boldsymbol{\psi}^1, \ldots, \boldsymbol{\psi}^s\})$ is a frequency-based dual framelet. Consequently, for every $\mathbf{f} \in \mathscr{D}(\mathbb{R})$, we have the following representation:

$$\mathbf{f} = \frac{1}{2\pi} \sum_{k \in \mathbb{Z}} \langle \mathbf{f}, \tilde{\boldsymbol{\eta}}_{1;0,k} \rangle \boldsymbol{\eta}_{1;0,k} + \frac{1}{2\pi} \sum_{j=0}^{\infty} \sum_{\ell=1}^{s} \sum_{k \in \mathbb{Z}} \langle \mathbf{f}, \tilde{\boldsymbol{\psi}}^\ell_{2^{-j};0,k} \rangle \boldsymbol{\psi}^\ell_{2^{-j};0,k} \tag{4.1.47}$$

in the sense of distributions as described in (4.1.4). To discuss framelet transforms in the function setting, we need the following trivial fact.

Lemma 4.1.11 *Let* $\boldsymbol{\eta}, \tilde{\boldsymbol{\eta}}$ *as in* (4.1.36) *such that* $\boldsymbol{\eta}, \tilde{\boldsymbol{\eta}}, \boldsymbol{\varphi}, \tilde{\boldsymbol{\varphi}}, \mathring{\boldsymbol{\varphi}}, \mathring{\tilde{\boldsymbol{\varphi}}} \in L_2^{loc}(\mathbb{R})$, *where* $\mathring{\boldsymbol{\varphi}}(\xi) := \overline{\widehat{\Theta}(\xi)}\boldsymbol{\varphi}(\xi)$ *and* $\mathring{\tilde{\boldsymbol{\varphi}}}(\xi) := \widehat{\Theta}(\xi)\tilde{\boldsymbol{\varphi}}(\xi)$ *with* $\widehat{\Theta}(\xi) := \widehat{\overline{\theta}}(\xi)\overline{\widehat{\theta}(\xi)}$. *Then*

$$\begin{aligned} \sum_{k \in \mathbb{Z}} \langle \mathbf{f}, \tilde{\boldsymbol{\eta}}_{2^{-j};0,k} \rangle \langle \boldsymbol{\eta}_{2^{-j};0,k}, \mathbf{g} \rangle &= \sum_{k \in \mathbb{Z}} \langle \mathbf{f}, \tilde{\boldsymbol{\varphi}}_{2^{-j};0,k} \rangle \langle \mathring{\boldsymbol{\varphi}}_{2^{-j};0,k}, \mathbf{g} \rangle \\ &= \sum_{k \in \mathbb{Z}} \langle \mathbf{f}, \mathring{\tilde{\boldsymbol{\varphi}}}_{2^{-j};0,k} \rangle \langle \boldsymbol{\varphi}_{2^{-j};0,k}, \mathbf{g} \rangle, \end{aligned} \tag{4.1.48}$$

for all $\mathbf{f}, \mathbf{g} \in \mathscr{D}(\mathbb{R})$ *and for all* $j \in \mathbb{Z}$.

Proof By Lemma 4.1.1 and the definition of $\boldsymbol{\eta}, \tilde{\boldsymbol{\eta}}$ in (4.1.36), we have

$$\begin{aligned} \sum_{k \in \mathbb{Z}} \langle \mathbf{f}, \tilde{\boldsymbol{\eta}}_{2^{-j};0,k} \rangle \langle \boldsymbol{\eta}_{2^{-j};0,k}, \mathbf{g} \rangle &= 2\pi \int_{\mathbb{R}} \sum_{k \in \mathbb{Z}} \mathbf{f}(\xi) \overline{\mathbf{g}(\xi + 2\pi 2^j k)}\, \overline{\tilde{\boldsymbol{\eta}}(2^{-j}\xi)} \boldsymbol{\eta}(2^{-j}\xi + 2\pi k) d\xi \\ &= 2\pi \int_{\mathbb{R}} \sum_{k \in \mathbb{Z}} \mathbf{f}(\xi) \overline{\mathbf{g}(\xi + 2\pi 2^j k)}\, \overline{\widehat{\Theta}(2^{-j}\xi)} \overline{\tilde{\boldsymbol{\varphi}}(2^{-j}\xi)} \boldsymbol{\varphi}(2^{-j}\xi + 2\pi k) d\xi. \end{aligned}$$

Using Lemma 4.1.1 again, we see that (4.1.48) holds. □

Consequently, the framelet representation in (4.1.47) is equivalent to

$$\mathbf{f} = \frac{1}{2\pi}\sum_{k\in\mathbb{Z}}\langle \mathbf{f}, \tilde{\boldsymbol{\varphi}}_{1;0,k}\rangle \mathring{\boldsymbol{\varphi}}_{1;0,k} + \frac{1}{2\pi}\sum_{j=0}^{\infty}\sum_{\ell=1}^{s}\sum_{k\in\mathbb{Z}}\langle \mathbf{f}, \tilde{\boldsymbol{\psi}}^{\ell}_{2^{-j};0,k}\rangle \boldsymbol{\psi}^{\ell}_{2^{-j};0,k}.$$

We now discuss *framelet transforms* for functions. Define framelet coefficients by

$$v^j(k) := \tfrac{1}{2\pi}\langle \mathbf{f}, \tilde{\boldsymbol{\varphi}}_{2^{-j};0,k}\rangle, \quad w^{\ell,j}(k) := \tfrac{1}{2\pi}\langle \mathbf{f}, \tilde{\boldsymbol{\psi}}^{\ell}_{2^{-j};0,k}\rangle, \ j,k\in\mathbb{Z}, \ell = 1,\dots,s.$$

If there exist tempered distributions (or functions) $f, \tilde{\phi}, \tilde{\psi}^{\ell}$ such that $\widehat{f} = \mathbf{f}, \widehat{\tilde{\phi}} = \tilde{\boldsymbol{\varphi}}$ and $\widehat{\tilde{\psi}^{\ell}} = \tilde{\boldsymbol{\psi}}^{\ell}, \ell = 1,\dots,s$, then it follows from Plancherel's Theorem in Theorem A.5.6 that $v^j(k) = \langle f, \tilde{\phi}_{2^j;k}\rangle$ and $w^{\ell,j}(k) = \langle f, \tilde{\psi}^{\ell}_{2^j;k}\rangle$. Note that both

$$(\{\tilde{\boldsymbol{\varphi}}; \tilde{\boldsymbol{\psi}}^1,\dots,\tilde{\boldsymbol{\psi}}^s\}, \{\mathring{\boldsymbol{\varphi}}; \boldsymbol{\psi}^1,\dots,\boldsymbol{\psi}^s\}) \quad \text{and} \quad (\{\tilde{\mathring{\boldsymbol{\varphi}}}; \tilde{\boldsymbol{\psi}}^1,\dots,\tilde{\boldsymbol{\psi}}^s\}, \{\boldsymbol{\varphi}; \boldsymbol{\psi}^1,\dots,\boldsymbol{\psi}^s\})$$

are frequency-based dual framelets. Hence, $\lim_{J\to\infty}\langle \mathbf{f}_J, \mathbf{g}\rangle = \langle \mathbf{f}, \mathbf{g}\rangle$ for all $\mathbf{f}, \mathbf{g} \in \mathscr{D}(\mathbb{R})$, that is, $\mathbf{f} = \lim_{J\to\infty}\mathbf{f}_J$ in the sense of distributions, where

$$\mathbf{f}_j := \frac{1}{2\pi}\sum_{k\in\mathbb{Z}}\langle \mathbf{f}, \tilde{\boldsymbol{\varphi}}_{2^{-j};0,k}\rangle \mathring{\boldsymbol{\varphi}}_{2^{-j};0,k} = \sum_{k\in\mathbb{Z}} v^j(k)\mathring{\boldsymbol{\varphi}}_{2^{-j};0,k}, \qquad j\in\mathbb{Z}.$$

As long as J is large enough, the function $\mathbf{f}_J$ approximates the function $\mathbf{f}$ well enough. In certain sense, the sequence v^J can be regarded as a digitalized/discretized version on the discrete domain $\mathbb{Z}$ (more precisely, at the resolution level $2^{-J}\mathbb{Z}$) of the function $\mathbf{f}$ on the continuum domain $\mathbb{R}$. We now discuss how to efficiently compute the framelet coefficients $w^{\ell,j}$ and $v^j, j = 0,\dots,J-1$ from v^J by a discrete framelet transform. Note that the cascade structure in (4.1.26) simply means

$$\mathbf{f}_j = \mathbf{f}_{j-1} + \frac{1}{2\pi}\sum_{\ell=1}^{s}\sum_{k\in\mathbb{Z}}\langle \mathbf{f}, \tilde{\boldsymbol{\psi}}^{\ell}_{2^{1-j};0,k}\rangle \boldsymbol{\psi}^{\ell}_{2^{1-j};0,k}, \qquad j\in\mathbb{Z}. \tag{4.1.49}$$

Consequently, we have the following multiscale representation of $\mathbf{f}_J$:

$$\begin{aligned}\mathbf{f}_J &= \frac{1}{2\pi}\sum_{k\in\mathbb{Z}}\langle \mathbf{f}, \tilde{\boldsymbol{\varphi}}_{2^{-J};0,k}\rangle \mathring{\boldsymbol{\varphi}}_{2^{-J};0,k} \\ &= \frac{1}{2\pi}\sum_{j=0}^{J-1}\sum_{\ell=1}^{s}\sum_{k\in\mathbb{Z}}\langle \mathbf{f}, \tilde{\boldsymbol{\psi}}^{\ell}_{2^{-j};0,k}\rangle \boldsymbol{\psi}^{\ell}_{2^{-j};0,k} + \frac{1}{2\pi}\sum_{k\in\mathbb{Z}}\langle \mathbf{f}, \tilde{\boldsymbol{\varphi}}_{1;0,k}\rangle \mathring{\boldsymbol{\varphi}}_{1;0,k}.\end{aligned} \tag{4.1.50}$$

The discrete framelet transform is built on the cascade structure in (4.1.49). Since $\tilde{\varphi}(2\xi) = \widehat{\tilde{a}}(\xi)\tilde{\varphi}(\xi)$ and $\tilde{\psi}^\ell(2\xi) = \widehat{\tilde{b}_\ell}(\xi)\tilde{\varphi}(\xi)$, by $\widehat{\tilde{b}_\ell}(\xi) = \sum_{k\in\mathbb{Z}} \tilde{b}_\ell(k)e^{-ik\xi}$, we have

$$
\begin{aligned}
w^{\ell,j-1}(k) &:= \frac{1}{2\pi}\langle \mathbf{f}, \tilde{\psi}^\ell_{2^{1-j};0,k}\rangle = \frac{2^{(1-j)/2}}{2\pi}\int_{\mathbb{R}} \mathbf{f}(\xi)e^{i2^{1-j}k\xi}\overline{\tilde{\psi}^\ell(2^{1-j}\xi)}d\xi \\
&= \frac{2^{(1-j)/2}}{2\pi}\int_{\mathbb{R}} \overline{\widehat{\tilde{b}_\ell}(2^{-j}\xi)}\mathbf{f}(\xi)e^{i2^{1-j}k\xi}\overline{\tilde{\varphi}(2^{-j}\xi)}d\xi \\
&= \sum_{n\in\mathbb{Z}} \overline{\tilde{b}_\ell(n)}\frac{2^{(1-j)/2}}{2\pi}\int_{\mathbb{R}} \mathbf{f}(\xi)e^{i2^{-j}(2k+n)\xi}\overline{\tilde{\varphi}(2^{-j}\xi)}d\xi \\
&= \sqrt{2}\sum_{n\in\mathbb{Z}} \overline{\tilde{b}_\ell(n)}\frac{1}{2\pi}\langle \mathbf{f}, \tilde{\varphi}_{2^{-j};0,n+2k}\rangle \\
&= \sqrt{2}\sum_{n\in\mathbb{Z}} v^j(n+2k)\overline{\tilde{b}_\ell(n)} \\
&= \tfrac{\sqrt{2}}{2}[\mathcal{T}_{\tilde{b}_\ell} v^j](k),
\end{aligned}
$$

where $[\mathcal{T}_u v](n) := 2\sum_{k\in\mathbb{Z}} v(k)\overline{u(k-2n)}$, see (1.1.3). Similarly, we have $v^{j-1}(k) = \frac{\sqrt{2}}{2}[\mathcal{T}_{\tilde{a}} v^j](k)$. Hence, we verified that all the framelet coefficients $w^{\ell,j}, v^j, j < J$ can be computed from v^J via the J-level discrete framelet decomposition as described in (1.3.1) with $v_j := v^{J-j}$ and $w_{\ell,j} := w^{\ell,J-j}$. In other words, the J-level discrete framelet decomposition in (1.3.1) can be used to compute the multiscale representation of $\mathbf{f}_J$ in (4.1.50) on the right-hand side from the left-hand side of (4.1.50).

We now discuss the framelet reconstruction. By Lemma 4.1.11, the identity (4.1.48) holds. Define $\mathring{v}^j(k) := \frac{1}{2\pi}\langle \mathbf{f}, \mathring{\tilde{\varphi}}_{2^{-j};0,k}\rangle$ for $j,k\in\mathbb{Z}$. Since $\mathring{\tilde{\varphi}}(\xi) := \widehat{\Theta}(\xi)\tilde{\varphi}(\xi)$,

$$
\mathring{v}^j = \Theta^\star * v^j, \qquad j \in \mathbb{Z}. \tag{4.1.51}
$$

By the relation $\psi^\ell(2\xi) = \widehat{b_\ell}(\xi)\varphi(\xi)$ and $\widehat{b_\ell}(\xi) = \sum_{k\in\mathbb{Z}} b_\ell(k)e^{-ik\xi}$, we have

$$
\begin{aligned}
\sum_{k\in\mathbb{Z}} w^{\ell,j-1}(k)\psi^\ell_{2^{1-j};0,k}(\xi) &= \sum_{k\in\mathbb{Z}} w^{\ell,j-1}(k)2^{(1-j)/2}e^{-i2^{1-j}k\xi}\widehat{b_\ell}(2^{-j}\xi)\varphi(2^{-j}\xi) \\
&= \sqrt{2}\sum_{k\in\mathbb{Z}}\sum_{n\in\mathbb{Z}} w^{\ell,j-1}(k)b_\ell(n)\varphi_{2^{-j};0,n+2k}(\xi) \\
&= \sum_{n\in\mathbb{Z}} \varphi_{2^{-j};0,n}(\xi)\tfrac{\sqrt{2}}{2}[\mathcal{S}_{b_\ell} w^{\ell,j-1}](n),
\end{aligned}
$$

where $[\mathcal{S}_u v](n) := 2\sum_{k\in\mathbb{Z}} v(k)u(n-2k)$, see (1.1.2). Similarly, by $\varphi(2\xi) := \widehat{a}(\xi)\varphi(\xi)$,

$$f_{j-1} = \sum_{k\in\mathbb{Z}} \breve{v}^{j-1}(k)\varphi_{2^{1-j};0,k}(\xi) = \sum_{n\in\mathbb{Z}} \varphi_{2^{-j};0,n}(\xi)\tfrac{\sqrt{2}}{2}[\mathcal{S}_a v^{j-1}](n).$$

By (4.1.49), we deduce that

$$\sum_{k\in\mathbb{Z}} \breve{v}^j(k)\varphi_{2^{-j};0,k} = f_j = \sum_{k\in\mathbb{Z}} \Big[\frac{\sqrt{2}}{2}\mathcal{S}_a\breve{v}^{j-1} + \frac{\sqrt{2}}{2}\sum_{\ell=1}^{s}\mathcal{S}_{b_\ell}w^{\ell,j-1}\Big](k)\varphi_{2^{-j};0,k}.$$

Since $[\varphi,\varphi](\xi) \ne 0$ for a.e. $\xi \in \mathbb{R}$ and $\varphi_{2^{-j};0,k} = 2^{-j/2}e^{-i2^{-j}k\xi}\varphi(2^{-j}\xi)$, we have

$$\sum_{k\in\mathbb{Z}} \breve{v}^j(k)e^{-ik\xi} = \sum_{k\in\mathbb{Z}} \Big[\frac{\sqrt{2}}{2}\mathcal{S}_a\breve{v}^{j-1} + \frac{\sqrt{2}}{2}\sum_{\ell=1}^{s}\mathcal{S}_{b_\ell}w^{\ell,j-1}\Big](k)e^{-ik\xi} \tag{4.1.52}$$

for almost every $\xi \in \mathbb{R}$. Consequently, comparing coefficients in (4.1.52), we have

$$\breve{v}^j = \frac{\sqrt{2}}{2}\mathcal{S}_a\breve{v}^{j-1} + \frac{\sqrt{2}}{2}\sum_{\ell=1}^{s}\mathcal{S}_{b_\ell}w^{\ell,j-1}, \qquad j \in \mathbb{Z}. \tag{4.1.53}$$

If a generalized dual framelet filter bank satisfying (4.1.44) and (4.1.45) is used, (4.1.52) only holds for almost every $\xi \in \operatorname{supp}([\varphi,\varphi])$ and hence, (4.1.53) may fail.

By Lemma 4.1.11, we see that $\mathbf{f}_J = \sum_{k\in\mathbb{Z}} \breve{v}^J(k)\varphi_{2^{-J};0,k} = \sum_{k\in\mathbb{Z}} v^J(k)\mathring{\varphi}_{2^{-J};0,k}$. Now by the relation in (4.1.51), we can obtain v^J from $\breve{v}^J$ through deconvolution $\Theta^\star * v^J = \breve{v}^J$. In the function setting, the deconvolution can be avoided by directly using the expression $\mathbf{f}_J = \sum_{k\in\mathbb{Z}} \breve{v}^J(k)\varphi_{2^{-J};0,k}$. The above algorithm is exactly the J-level discrete framelet reconstruction as described in (1.4.19), (1.4.20), and (1.4.21) of Chap. 1 with the notation change $v_j := v^{J-j}$ and $w_{\ell,j} := w^{\ell,J-j}$. When $\Theta = \boldsymbol{\delta}$, as we explained in Sect. 1.3 of Chap. 1 that a J-level discrete framelet transform can be equivalently expressed as in (1.3.20) using a discrete affine system. Therefore, in certain sense a frequency-based affine system $\mathsf{FAS}_J(\varphi;\psi^1,\dots,\psi^s)$ is naturally linked to a discrete affine system $\mathsf{DAS}_J(\{a;b_1,\dots,b_s\})$.

As a conclusion, a discrete framelet transform discussed in Chap. 1 can be used to compute a framelet transform of the multiscale representation of $\mathbf{f}_J$ ($\approx \mathbf{f}$) in (4.1.50) in the continuum domain $\mathbb{R}$. Further connections between dual framelet filter banks and frequency-based dual framelets will be addressed in Sect. 4.5.

4.2 Frames and Bases in Hilbert Spaces

Classical theory of framelets and wavelets is often developed for the square integrable function space $L_2(\mathbb{R})$, which is a particular example of separable Hilbert spaces. Before we discuss framelets and wavelets in $L_2(\mathbb{R})$ and Sobolev spaces, in this section we introduce and study frames and bases in a general separable Hilbert space. For definitions and some basic properties of a Hilbert space, see Appendix A.

Let $(\mathcal{H}, \langle\cdot,\cdot\rangle)$ be a separable Hilbert space equipped with an inner product $\langle\cdot,\cdot\rangle$. The induced norm on $\mathcal{H}$ is $\|h\| := \sqrt{\langle h, h\rangle}$. Let Λ be a countable index set. Throughout this section $\{\Lambda_j\}_{j\in\mathbb{N}}$ always denotes a sequence of finite subsets Λ_j of Λ such that $\Lambda_j \subseteq \Lambda_{j+1}$ for all $j \in \mathbb{N}$ and $\cup_{j=1}^{\infty}\Lambda_j = \Lambda$. For $\{f_k\}_{k\in\Lambda}$ in $\mathcal{H}$, we say that $\sum_{k\in\Lambda} f_k$ *converges unconditionally* in $\mathcal{H}$ if for every choice of $\{\Lambda_j\}_{j\in\mathbb{N}}$, the limit $\lim_{j\to\infty}\sum_{k\in\Lambda_j} f_k$ exists in $\mathcal{H}$.

Let $\{h_k\}_{k\in\Lambda}$ be a sequence in a Hilbert space $\mathcal{H}$. We say that $\{h_k\}_{k\in\Lambda}$ is *a Bessel sequence* in $\mathcal{H}$ if there exists a positive constant C such that for all $h \in \mathcal{H}$,

$$\sum_{k\in\Lambda} |\langle h, h_k\rangle|^2 \leqslant C\|h\|^2. \tag{4.2.1}$$

To study frames and bases in a Hilbert space $\mathcal{H}$, we now introduce the following operators. The *analysis operator* $\mathcal{W}$ associated with $\{h_k\}_{k\in\Lambda}$ is defined to be

$$\mathcal{W}: \mathcal{H} \to l_2(\Lambda) \quad \text{with} \quad \mathcal{W}(h) := \{\langle h, h_k\rangle\}_{k\in\Lambda} \tag{4.2.2}$$

and the *synthesis operator* $\mathcal{V}$ associated with $\{h_k\}_{k\in\Lambda}$ is defined to be

$$\mathcal{V}: l_2(\Lambda) \to \mathcal{H} \quad \text{with} \quad \mathcal{V}(\{c_k\}_{k\in\Lambda}) := \sum_{k\in\Lambda} c_k h_k. \tag{4.2.3}$$

The *frame operator* $\mathcal{F}$ associated with $\{h_k\}_{k\in\Lambda}$ is defined to be

$$\mathcal{F}: \mathcal{H} \to \mathcal{H} \quad \text{with} \quad \mathcal{F}(h) := \sum_{k\in\Lambda} \langle h, h_k\rangle h_k. \tag{4.2.4}$$

Though the above operators may not be well defined in general, by the following result, all of them are indeed well-defined bounded linear operators if $\{h_k\}_{k\in\Lambda}$ is a Bessel sequence in $\mathcal{H}$.

Proposition 4.2.1 *Let $\{h_k\}_{k\in\Lambda}$ be a sequence in a Hilbert space $\mathcal{H}$. Then the following statements are equivalent:*

(1) Inequality (4.2.1) *holds for all h in a dense subset $D_{\mathcal{H}}$ of $\mathcal{H}$ for some $C > 0$.*
(2) $\{h_k\}_{k\in\Lambda}$ is a Bessel sequence in $\mathcal{H}$: (4.2.1) *holds for all $h \in \mathcal{H}$ for some $C > 0$.*
(3) The analysis operator $\mathcal{W}$ in (4.2.2) *is well defined: $\sum_{k\in\Lambda} |\langle h, h_k\rangle|^2 < \infty$ $\forall h \in \mathcal{H}$.*

(4) *The analysis operator $\mathcal{W}$ in* (4.2.2) *is a well-defined bounded linear operator.*
(5) *There exists a positive constant C such that*

$$\Big\| \sum_{k\in\Lambda} c_k h_k \Big\|^2 \leqslant C \sum_{k\in\Lambda} |c_k|^2 \tag{4.2.5}$$

for all finitely supported sequences $\{c_k\}_{k\in\Lambda}$.
(6) *The series $\sum_{k\in\Lambda} c_k h_k$ converges unconditionally in $\mathcal{H}$ for every $\{c_k\}_{k\in\Lambda} \in l_2(\Lambda)$.*
(7) *The synthesis operator $\mathcal{V}$ in* (4.2.3) *is well-defined, that is, there exists $\{\Lambda_j\}_{j\in\mathbb{N}}$ such that $\lim_{j\to\infty} \sum_{k\in\Lambda_j} c_k h_k$ exists in $\mathcal{H}$ for every $\{c_k\}_{k\in\Lambda} \in l_2(\Lambda)$.*
(8) *The synthesis operator $\mathcal{V}$ in* (4.2.3) *is a well-defined bounded linear operator.*
(9) *The series $\sum_{k\in\Lambda} \langle h, h_k\rangle h_k$ converges unconditionally in $\mathcal{H}$ for every $h \in \mathcal{H}$.*
(10) *The frame operator $\mathcal{F}$ in* (4.2.4) *is well defined, that is, there exists $\{\Lambda_j\}_{j\in\mathbb{N}}$ such that $\lim_{j\to\infty} \sum_{k\in\Lambda_j} \langle h, h_k\rangle h_k$ exists in $\mathcal{H}$ for every $h \in \mathcal{H}$.*
(11) *The frame operator $\mathcal{F}$ in* (4.2.4) *is a well-defined bounded linear operator.*

Moreover, any of the above statements implies that $\mathcal{V} = \mathcal{W}^\star, \mathcal{W} = \mathcal{V}^\star$, $\mathcal{F} = \mathcal{V}\mathcal{W}$, and $\mathcal{F}$ is self-adjoint satisfying $\mathcal{F}^\star = \mathcal{F}$, where $\mathcal{F}^\star$ is the adjoint operator of $\mathcal{F}$ (see Theorem A.1.5 for definition).

Proof (1)$\Longrightarrow$(2). Since $D_{\mathcal{H}}$ is dense in $\mathcal{H}$, for any $h \in \mathcal{H}$, there exists $\{f_n\}_{n\in\mathbb{N}}$ in $D_{\mathcal{H}}$ such that $\lim_{n\to\infty} \|f_n - h\| = 0$. By our assumption in item (1), we have

$$\sum_{k\in\Lambda_j} |\langle f_n, h_k\rangle|^2 \leqslant C\|f_n\|^2 \qquad \forall\, j, n \in \mathbb{N}.$$

Since Λ_j is a finite set, taking $n \to \infty$, we see that $\sum_{k\in\Lambda_j} |\langle h, h_k\rangle|^2 \leqslant C\|h\|^2$. Now taking $j \to \infty$, we see that (4.2.1) holds. Hence, (1)$\Longrightarrow$(2).

(2)$\Longrightarrow$(3) is trivial. (3)$\Longrightarrow$(4). Define $\mathcal{W}_j : \mathcal{H} \to l_2(\Lambda)$ by $[\mathcal{W}_j(h)](k) := \langle h, h_k\rangle$ for $k \in \Lambda_j$ and $[\mathcal{W}_j(h)](k) := 0$ for all $k \in \Lambda\backslash\Lambda_j$. Since Λ_j is a finite set, it is obvious that all operators $\mathcal{W}_j$ are well-defined bounded linear operators. The condition in item (3) is equivalent to $\sup_{j\in\mathbb{N}} \|\mathcal{W}_j h\|^2_{l_2(\Lambda)} < \infty$ for all $h \in \mathcal{H}$. Now by the Uniform Boundedness Principle (see Theorem A.1.2), we conclude that there exists a positive constant C such that $\|\mathcal{W}_j\| \leqslant C$ for all $j \in \mathbb{N}$. By $\lim_{j\to\infty} \|\mathcal{W}_j h\|_{l_2(\Lambda)} = \|\mathcal{W}h\|_{l_2(\Lambda)}$, it is trivial to see that $\mathcal{W}$ is a well-defined bounded linear operator with $\|\mathcal{W}\| \leqslant C$.

(4)$\Longrightarrow$(5). By the Cauchy-Schwarz inequality and item (4), we have

$$\Big|\Big\langle h, \sum_{k\in\Lambda} c_k h_k \Big\rangle\Big|^2 = \Big| \sum_{k\in\Lambda} \langle h, h_k\rangle \overline{c_k} \Big|^2 \leqslant \Big(\sum_{k\in\Lambda} |\langle h, h_k\rangle|^2 \Big)\Big(\sum_{k\in\Lambda} |c_k|^2 \Big) \leqslant \|\mathcal{W}\|^2 \|h\|^2 \sum_{k\in\Lambda} |c_k|^2.$$

Consequently,

$$\Big\| \sum_{k\in\Lambda} c_k h_k \Big\|^2 = \sup_{h\in\mathcal{H}, \|h\|\leqslant 1} \Big|\Big\langle h, \sum_{k\in\Lambda} c_k h_k \Big\rangle\Big|^2 \leqslant \|\mathcal{W}\|^2 \sum_{k\in\Lambda} |c_k|^2.$$

(5)$\Longrightarrow$(6). By (4.2.5) and $\sum_{k\in\Lambda}|c_k|^2<\infty$, we have

$$\Big\|\sum_{k\in\Lambda_n\backslash\Lambda_m} c_k h_k\Big\|^2 \leqslant C\sum_{k\in\Lambda_n\backslash\Lambda_m}|c_k|^2\to 0,\quad \text{as } m,n\to\infty.$$

It follows directly from the above inequality that $\{\sum_{k\in\Lambda_j} c_k h_k\}_{j\in\mathbb{N}}$ is a Cauchy sequence in $\mathcal{H}$ and therefore, $\lim_{j\to\infty}\sum_{k\in\Lambda_j} c_k h_k$ exists in $\mathcal{H}$. Hence, the series $\sum_{k\in\Lambda} c_k h_k$ converges unconditionally in $\mathcal{H}$.

(6)$\Longrightarrow$(7) is trivial. (7)$\Longrightarrow$(8) can be proved by a similar argument as in the proof of (3)$\Longrightarrow$(4).

(8)$\Longrightarrow$(9). By item (8), we have

$$\begin{aligned}\sum_{k\in\Lambda}|\langle h,h_k\rangle|^2 = \|\mathcal{W}h\|^2_{l_2(\Lambda)} &= \sup_{\sum_{k\in\Lambda}|c_k|^2\leqslant 1}|\langle\{c_k\}_{k\in\Lambda},\mathcal{W}h\rangle_{l_2(\Lambda)}|^2\\ &= \sup_{\sum_{k\in\Lambda}|c_k|^2\leqslant 1}|\langle\mathcal{V}(\{c_k\}_{k\in\Lambda}),h\rangle|^2\leqslant\|\mathcal{V}\|^2\|h\|^2<\infty.\end{aligned}$$

Now item (9) follows from item (8), since

$$\Big\|\sum_{k\in\Lambda_n\backslash\Lambda_m}\langle h,h_k\rangle h_k\Big\|^2\leqslant\|\mathcal{V}\|^2\sum_{k\in\Lambda_n\backslash\Lambda_m}|\langle h,h_k\rangle|^2\to 0\quad\text{as}\quad m,n\to\infty.$$

(9)$\Longrightarrow$(10) is trivial. (10)$\Longrightarrow$(11) can be proved by a similar argument as in the proof of (3)$\Longrightarrow$(4). (11)$\Longrightarrow$(1) follows from $\sum_{k\in\Lambda}|\langle h,h_k\rangle|^2=\langle\mathcal{F}h,h\rangle\leqslant\|\mathcal{F}h\|\|h\|\leqslant\|\mathcal{F}\|\|h\|^2$. For $h\in\mathcal{H}$ and $\{c_k\}_{k\in\Lambda}\in l_2(\Lambda)$,

$$\langle\mathcal{W}h,\{c_k\}_{k\in\Lambda}\rangle=\sum_{k\in\Lambda}[\mathcal{W}h](k)\overline{c_k}=\sum_{k\in\Lambda}\langle h,h_k\rangle\overline{c_k}=\Big\langle h,\sum_{k\in\Lambda}c_k h_k\Big\rangle=\langle h,\mathcal{V}(\{c_k\}_{k\in\Lambda})\rangle.$$

Therefore, we obtain $\mathcal{V}=\mathcal{W}^\star$ and $\mathcal{W}=\mathcal{V}^\star$. Since $\mathcal{F}=\mathcal{V}\mathcal{W}$, we see that $\mathcal{F}^\star=\mathcal{W}^\star\mathcal{V}^\star=\mathcal{V}\mathcal{W}=\mathcal{F}$, that is, $\mathcal{F}$ is self-adjoint. □

Proposition 4.2.2 *Let $\mathcal{H}$ be a Hilbert space and $\mathcal{W}:\mathcal{H}\to l_2(\Lambda)$ be a bounded linear operator. Then there exists a Bessel sequence $\{h_k\}_{k\in\Lambda}$ in $\mathcal{H}$ such that $\mathcal{W}(h)=\{\langle h,h_k\rangle\}_{k\in\Lambda}$, i.e., $\mathcal{W}$ is the analysis operator associated with $\{h_k\}_{k\in\Lambda}$.*

Proof For each $k\in\Lambda$, consider the linear functional $\ell_k:\mathcal{H}\to\mathbb{C}$ with $\ell_k(h):=[\mathcal{W}(h)](k)$, the kth component of the sequence $\mathcal{W}(h)$. It is evident that $|\ell_k(h)|\leqslant\|\mathcal{W}(h)\|_{l_2(\Lambda)}\leqslant\|\mathcal{W}\|\|h\|$. Hence, ℓ_k is a continuous linear functional on $\mathcal{H}$. Therefore, by the Riesz Representation Theorem in Theorem A.1.4, there exists $h_k\in\mathcal{H}$ such that $\ell_k(h)=\langle h,h_k\rangle$ for all $h\in\mathcal{H}$. Since $\{\langle h,h_k\rangle\}_{k\in\Lambda}=\mathcal{W}(h)$, it is obvious that $\{h_k\}_{k\in\Lambda}$ is a Bessel sequence in $\mathcal{H}$ and $\mathcal{W}$ agrees with the analysis operator associated with $\{h_k\}_{k\in\Lambda}$. □

In fact, there is a bijection between Bessel sequences in $\mathcal{H}$ and bounded linear operators from $\mathcal{H}$ to $l_2(\Lambda)$, see Exercise 4.8 for details.

We say that $\{h_k\}_{k\in\Lambda}$ is *a frame for* $\mathcal{H}$ if there exist positive constants C_1 and C_2 such that for all $h \in \mathcal{H}$,

$$C_1\|h\|^2 \leqslant \sum_{k\in\Lambda} |\langle h, h_k\rangle|^2 \leqslant C_2\|h\|^2. \tag{4.2.6}$$

The positive constants C_1 and C_2 are called a lower frame bound and an upper frame bound, respectively. Moreover, if C_1 is the largest possible constant and C_2 is the smallest possible constant such that (4.2.6) holds, then the optimal constants C_1 and C_2 are called *the lower frame bound* and *the upper frame bound* of the frame $\{h_k\}_{k\in\Lambda}$, respectively.

If (4.2.6) is satisfied with $C_1 = C_2 = 1$, that is,

$$\sum_{k\in\Lambda} |\langle h, h_k\rangle|^2 = \|h\|^2, \qquad \forall\, h \in \mathcal{H}, \tag{4.2.7}$$

then we say that $\{h_k\}_{k\in\Lambda}$ is *a (normalized) tight frame for* $\mathcal{H}$. If (4.2.6) is satisfied with $C_1 = C_2$, then we say that $\{h_k\}_{k\in\Lambda}$ is *a general tight frame* or *a* C_1*-tight frame* for $\mathcal{H}$..

To study a frame in a Hilbert pace, we need the following auxiliary result.

Theorem 4.2.3 *Let $\mathcal{H}$ and $\mathcal{K}$ be two Hilbert spaces and $T : \mathcal{H} \to \mathcal{K}$ be a linear operator. Then there exist two positive constants C_1 and C_2 such that*

$$C_1\|h\|^2_{\mathcal{H}} \leqslant \|Th\|^2_{\mathcal{K}} \leqslant C_2\|h\|^2_{\mathcal{H}} \qquad \forall\, h \in \mathcal{H} \tag{4.2.8}$$

if and only if T is an injective bounded linear operator such that $\operatorname{ran}(T) := \{Th : h \in \mathcal{H}\}$ *is a closed linear space.*

Proof Necessity ($\Rightarrow$). It follows directly from the right-hand side inequality in (4.2.8) that $\|T\| \leqslant \sqrt{C_2}$. If $Th = 0$, then the left-hand side inequality in (4.2.8) implies $C_1\|h\|^2_{\mathcal{H}} \leqslant \|Th\|^2_{\mathcal{K}} = 0$ and hence $h = 0$. So, T is an injective bounded linear operator. Since T is linear, it is obvious that $\operatorname{ran}(T)$ is a linear space. We now prove that $\operatorname{ran}(T)$ is closed. Indeed, for f in the closure of $\operatorname{ran}(T)$, there exists a sequence $\{h_n\}_{n\in\mathbb{N}}$ in $\mathcal{H}$ such that $\lim_{n\to\infty} \|Th_n - f\|_{\mathcal{K}} = 0$. Therefore, the sequence $\{Th_n\}_{n\in\mathbb{N}}$ is a Cauchy sequence in $\mathcal{K}$. By the left-hand side inequality in (4.2.8), we have $C_1\|h_n - h_m\|^2_{\mathcal{H}} \leqslant \|Th_n - Th_m\|^2_{\mathcal{K}} \to 0$ as $m, n \to \infty$. Hence, the sequence $\{h_n\}_{n\in\mathbb{N}}$ must be a Cauchy sequence in the Hilbert space $\mathcal{H}$ and therefore, there exists $h \in \mathcal{H}$ such that $\lim_{n\to\infty} \|h_n - h\|_{\mathcal{H}} = 0$. Since T is bounded, we conclude that $\lim_{n\to\infty} \|Th_n - Th\|_{\mathcal{K}} = 0$. By our assumption $\lim_{n\to\infty} \|Th_n - f\|_{\mathcal{K}} = 0$, we must have $f = Th \in \operatorname{ran}(T)$. Thus, we conclude that $\operatorname{ran}(T)$ is a closed linear space.

Sufficiency ($\Leftarrow$). We consider the operator $\mathring{T} : \mathcal{H} \to \operatorname{ran}(T)$ with $\mathring{T}h := Th$. By our assumption, it is trivial to see that $\mathring{T}$ is a bijective bounded linear operator. Since $\operatorname{ran}(T)$ is a closed linear subspace of a Hilbert space $\mathcal{K}$, the space $\operatorname{ran}(T)$ itself is a

Hilbert space. By the Open Mapping Theorem (see Theorem A.1.1), the operator $\mathring{T}$ has a bounded inverse linear operator $\mathring{T}^{-1}$. Now it is easy to see that (4.2.8) holds with $C_1 = \|\mathring{T}^{-1}\|^{-2}$ and $C_2 = \|\mathring{T}\|^2$, since $\|h\|_{\mathcal{H}} = \|\mathring{T}^{-1}\mathring{T}h\|_{\mathcal{H}} \leqslant \|\mathring{T}^{-1}\|\|\mathring{T}h\|_{\mathcal{K}}$ and $\|\mathring{T}h\|_{\mathcal{K}} \leqslant \|\mathring{T}\|\|h\|_{\mathcal{H}}$. □

Some equivalent conditions about a frame for $\mathcal{H}$ are as follows:

Theorem 4.2.4 *Let $\{h_k\}_{k\in\Lambda}$ be a sequence in a Hilbert space $\mathcal{H}$. Then the following statements are equivalent:*

(1) The condition in (4.2.6) *holds for all h in a dense subset $D_{\mathcal{H}}$ of $\mathcal{H}$ with $C_1, C_2 > 0$.*
(2) $\{h_k\}_{k\in\Lambda}$ is a frame for $\mathcal{H}$, that is, (4.2.6) *holds for all $h \in \mathcal{H}$ with $C_1, C_2 > 0$.*
(3) The frame operator $\mathcal{F}$ in (4.2.4) *is a well-defined, bounded, self-adjoint, and bijective linear operator with a bounded inverse operator $\mathcal{F}^{-1}$.*
(4) The synthesis operator $\mathcal{V}$ in (4.2.3) *is a well-defined, bounded, and surjective linear operator.*
(5) The analysis operator $\mathcal{W}$ in (4.2.2) *is a well-defined, bounded, injective linear operator and its range* $\operatorname{ran}(\mathcal{W}) := \{\mathcal{W}h \;:\; h \in \mathcal{H}\}$ *is a closed subspace of* $l_2(\Lambda)$.

Moreover, any of the above statements implies that $\mathcal{V}|_{\operatorname{ran}(\mathcal{W})} : \operatorname{ran}(\mathcal{W}) \to \mathcal{H}, f \mapsto \mathcal{V}f$ is a well-defined bounded and bijective linear operator with a bounded inverse.

Proof (1)$\Longrightarrow$(2). By Proposition 4.2.1, we have $\|\mathcal{W}h\|^2_{l_2(\Lambda)} \leqslant C_2\|h\|^2$ for all $h \in \mathcal{H}$ and hence $\mathcal{W}$ is bounded. Let $h \in \mathcal{H}$. Since $D_{\mathcal{H}}$ is dense in $\mathcal{H}$, there exists $\{f_n\}_{n\in\mathbb{N}}$ in $D_{\mathcal{H}}$ such that $\lim_{n\to\infty} \|f_n - h\| = 0$. Since $\mathcal{W}$ is bounded, we have $\lim_{n\to\infty} \|\mathcal{W}f_n - \mathcal{W}h\| = 0$. By item (1), we have $C_1\|f_n\|^2 \leqslant \|\mathcal{W}f_n\|^2_{l_2(\Lambda)}$. Taking $n \to \infty$ on both sides, we have $C_1\|h\|^2 \leqslant \lim_{n\to\infty} \|\mathcal{W}f_n\|^2_{l_2(\Lambda)} = \|\lim_{n\to\infty} \mathcal{W}f_n\|^2_{l_2(\Lambda)} = \|\mathcal{W}h\|^2_{l_2(\Lambda)}$. Therefore, item (2) holds.

(2)$\Longrightarrow$(3). By Proposition 4.2.1, we see that $\mathcal{F}$ is well defined, bounded, and self-adjoint. We now prove that $\mathcal{F}$ is bijective. Note that (4.2.6) is equivalent to

$$C_1\langle h, h\rangle \leqslant \langle \mathcal{F}h, h\rangle \leqslant C_2\langle h, h\rangle, \qquad \forall\, h \in \mathcal{H}, \tag{4.2.9}$$

from which we deduce that $C_1\|h\|^2 \leqslant \langle \mathcal{F}h, h\rangle \leqslant \|\mathcal{F}h\|\|h\|$. Hence, we have $C_1^2\|h\|^2 \leqslant \|\mathcal{F}h\|^2 \leqslant \|\mathcal{F}\|^2\|h\|^2$ for all $h \in \mathcal{H}$. By Theorem 4.2.3, the frame operator $\mathcal{F}$ is injective and $\operatorname{ran}(\mathcal{F})$ is closed. Since $\mathcal{F}$ is self-adjoint, by Theorem A.1.5, we have $\operatorname{ran}(\mathcal{F})^\perp = \ker(\mathcal{F}^\star) = \ker(\mathcal{F}) = \{0\}$. Therefore, the space $\operatorname{ran}(\mathcal{F})$ is dense in $\mathcal{H}$. Since $\operatorname{ran}(\mathcal{F})$ is closed, we must have $\operatorname{ran}(\mathcal{F}) = \mathcal{H}$. Hence, the frame operator $\mathcal{F}$ is onto. Since $\mathcal{F}$ is bounded and bijective, by the Open Mapping Theorem (see Theorem A.1.1), its inverse operator $\mathcal{F}^{-1}$ is also bounded.

(3)$\Longrightarrow$(4). By Proposition 4.2.1, the synthesis operator $\mathcal{V}$ is a well-defined bounded linear operator. Since $\mathcal{F} = \mathcal{V}\mathcal{W}$ and $\mathcal{F}$ has a bounded inverse operator $\mathcal{F}^{-1}$, we must have $\mathcal{V}\mathcal{W}\mathcal{F}^{-1} = \mathcal{F}\mathcal{F}^{-1} = \mathrm{Id}_{\mathcal{H}}$. Hence, the synthesis operator $\mathcal{V}$ is onto.

(4)$\Longrightarrow$(5). By Proposition 4.2.1, the analysis operator $\mathcal{W}$ is a well-defined bounded linear operator and $\mathcal{W} = \mathcal{V}^\star$. If $\mathcal{W}h = 0$ for some $h \in \mathcal{H}$, then for all $\vec{c} \in l_2(\Lambda)$, $0 = \langle \mathcal{W}h, \vec{c}\rangle = \langle h, \mathcal{V}\vec{c}\rangle$. Since $\mathcal{V}$ is onto, we must have $h = 0$. Therefore, the analysis operator $\mathcal{W}$ is injective. Obviously, the set ran($\mathcal{W}$) is a linear space since $\mathcal{W}$ is a linear operator. We now prove that ran($\mathcal{W}$) must be closed. Define $\mathring{\mathcal{V}} : \ker(\mathcal{V})^\perp \to \mathcal{H}$ with $\mathring{\mathcal{V}}\vec{c} = \mathcal{V}\vec{c}$. Clearly, the space $\ker(\mathcal{V})^\perp$ is a closed subspace of a Hilbert space $\mathcal{H}$ and therefore, it is a Hilbert space. By item (4), it is also trivial to see that $\mathring{\mathcal{V}}$ is a bounded bijection. By the Open Mapping Theorem (see Theorem A.1.1), the operator $\mathring{\mathcal{V}}$ has a bounded inverse operator $\mathring{\mathcal{V}}^{-1}$. Note that $\mathcal{V}\mathring{\mathcal{V}}^{-1} = \mathrm{Id}_\mathcal{H}$. Hence, we have $(\mathring{\mathcal{V}}^{-1})^\star\mathcal{V}^\star = \mathrm{Id}_\mathcal{H}$, that is, $(\mathring{\mathcal{V}}^{-1})^\star\mathcal{W} = \mathrm{Id}_\mathcal{H}$. Therefore, we have $\|h\|^2 = \|(\mathring{\mathcal{V}}^{-1})^\star\mathcal{W}h\|^2 \leqslant \|(\mathring{\mathcal{W}}^{-1})^\star\|^2\|\mathcal{W}h\|^2$, from which we obtain $\|(\mathring{\mathcal{W}}^{-1})^\star\|^{-2}\|h\|^2 \leqslant \|\mathcal{W}h\|^2_{l_2(\Lambda)} \leqslant \|\mathcal{W}\|^2\|h\|^2$ for all $h \in \mathcal{H}$. By Theorem 4.2.3, item (5) holds.

(5)$\Longrightarrow$(1) follows directly from Theorem 4.2.3 and $\|\mathcal{W}h\|^2_{l_2(\Lambda)} = \sum_{k\in\Lambda} |\langle h, h_k\rangle|^2$.

By (1)–(5), the space ran($\mathcal{W}$) is closed. Therefore, by $\mathcal{W} = \mathcal{V}^\star$, we conclude that $\ker(\mathcal{V})^\perp = \mathrm{ran}(\mathcal{W})$. We proved in (4)$\Longrightarrow$(5) that $\mathcal{V}|_{\mathrm{ran}(\mathcal{W})}$ is a well-defined bijective linear operator with a bounded inverse operator. □

For two sequences $\{\tilde{h}_k\}_{k\in\Lambda}$ and $\{h_k\}_{k\in\Lambda}$ in $\mathcal{H}$, we say that $(\{\tilde{h}_k\}_{k\in\Lambda}, \{h_k\}_{k\in\Lambda})$ is *a pair of dual frames* for $\mathcal{H}$, or simply $\{\tilde{h}_k\}_{k\in\Lambda}$ is *a dual frame* of $\{h_k\}_{k\in\Lambda}$, if each of $\{\tilde{h}_k\}_{k\in\Lambda}$ and $\{h_k\}_{k\in\Lambda}$ is a frame for $\mathcal{H}$ and the following identity holds: for all $f, g \in \mathcal{H}$,

$$\langle f, g\rangle = \sum_{k\in\Lambda} \langle f, \tilde{h}_k\rangle\langle h_k, g\rangle \tag{4.2.10}$$

with the series converging absolutely. If $(\{\tilde{h}_k\}_{k\in\Lambda}, \{h_k\}_{k\in\Lambda})$ is a pair of dual frames for $\mathcal{H}$, from (4.2.10) we see that every $f \in \mathcal{H}$ has the following representations:

$$f = \sum_{k\in\Lambda} \langle f, \tilde{h}_k\rangle h_k = \sum_{k\in\Lambda} \langle f, h_k\rangle \tilde{h}_k \tag{4.2.11}$$

with the series on the right-hand side converging unconditionally in $\mathcal{H}$. For a pair of dual frames, we have the following basic result.

Theorem 4.2.5 *Let $\{\tilde{h}_k\}_{k\in\Lambda}$ and $\{h_k\}_{k\in\Lambda}$ be two sequences in a Hilbert space $\mathcal{H}$. Let $D_\mathcal{H}$ be a dense subset of $\mathcal{H}$. Then $(\{\tilde{h}_k\}_{k\in\Lambda}, \{h_k\}_{k\in\Lambda})$ is a pair of dual frames for $\mathcal{H}$ if and only if there exists a positive constant C such that*

$$\sum_{k\in\Lambda} (|\langle h, h_k\rangle|^2 + |\langle h, \tilde{h}_k\rangle|^2) \leqslant C\|h\|^2, \qquad \forall\, h \in D_\mathcal{H} \tag{4.2.12}$$

and the identity (4.2.10) *holds for all $f, g \in D_\mathcal{H}$.*

Proof Necessity is trivial. We now prove sufficiency. Let $\mathcal{W}$ and $\widetilde{\mathcal{W}}$ be the analysis operators associated with $\{h_k\}_{k\in\Lambda}$ and $\{\tilde{h}\}_{k\in\Lambda}$, respectively. Since (4.2.10) and (4.2.12) hold for all $f, g, h \in D_{\mathcal{H}}$, using the Cauchy-Schwarz inequality, we have

$$|\langle f, g\rangle|^2 = |\langle \widetilde{\mathcal{W}} f, \mathcal{W} g\rangle_{l_2(\Lambda)}|^2 \leqslant \|\widetilde{\mathcal{W}} f\|^2_{l_2(\Lambda)} \|\mathcal{W} g\|^2_{l_2(\Lambda)} \leqslant C\|f\|^2\|\mathcal{W} g\|^2_{l_2(\Lambda)}.$$

Therefore, we deduce that $\|g\|^2 = \sup_{f\in D_{\mathcal{H}}, \|f\|\leqslant 1} |\langle f, g\rangle|^2 \leqslant C\|\mathcal{W} g\|^2_{l_2(\Lambda)}$, from which we have $C^{-1}\|g\|^2 \leqslant \|\mathcal{W} g\|^2_{l_2(\Lambda)} \leqslant C\|g\|^2$ for all $g \in D_{\mathcal{H}}$. Since $D_{\mathcal{H}}$ is dense in $\mathcal{H}$, by Theorem 4.2.4, we see that $\{h_k\}_{k\in\Lambda}$ is a frame for $\mathcal{H}$. By the same argument, $\{\tilde{h}_k\}_{k\in\Lambda}$ is also a frame for $\mathcal{H}$. For $f, g \in \mathcal{H}$, since $D_{\mathcal{H}}$ is dense in $\mathcal{H}$, there exist $\{f_n\}_{n\in\mathbb{N}}$ and $\{g_n\}_{n\in\mathbb{N}}$ in $\mathcal{H}$ such that $\lim_{n\to\infty} \|f_n - f\| = 0$ and $\lim_{n\to\infty} \|g_n - g\| = 0$. By (4.2.10), we have $\langle f_n, g_n\rangle = \langle \widetilde{\mathcal{W}} f_n, \mathcal{W} g_n\rangle_{l_2(\Lambda)}$. Since both $\widetilde{\mathcal{W}}$ and $\mathcal{W}$ are bounded, we have $\langle f, g\rangle = \lim_{n\to\infty}\langle f_n, g_n\rangle = \lim_{n\to\infty}\langle \widetilde{\mathcal{W}} f_n, \mathcal{W} g_n\rangle_{l_2(\Lambda)} = \langle \widetilde{\mathcal{W}} f, \mathcal{W} g\rangle_{l_2(\Lambda)}$. This shows that (4.2.10) holds for all $f, g \in \mathcal{H}$. This proves that $(\{\tilde{h}_k\}_{k\in\Lambda}, \{h_k\}_{k\in\Lambda})$ is a pair of dual frames for $\mathcal{H}$. □

The following result shows that there exists a special dual frame to a given frame.

Proposition 4.2.6 *Let $\{h_k\}_{k\in\Lambda}$ be a frame for a Hilbert space $\mathcal{H}$ such that* (4.2.6) *holds. Then the following statements hold:*

(i) $\{\mathcal{F}^{-1}h_k\}_{k\in\Lambda}$ is a dual frame of $\{h_k\}_{k\in\Lambda}$, called the canonical dual frame *of the frame $\{h_k\}_{k\in\Lambda}$, such that*

$$C_2^{-1}\|h\|^2 \leqslant \sum_{k\in\Lambda} |\langle h, \mathcal{F}^{-1}h_k\rangle|^2 \leqslant C_1^{-1}\|h\|^2, \qquad \forall\, h \in \mathcal{H}; \tag{4.2.13}$$

(ii) The frame operator associated with the frame $\{\mathcal{F}^{-1}h_k\}_{k\in\Lambda}$ is $\mathcal{F}^{-1}$;

(iii) For every $f \in \mathcal{H}$ and for all sequences $\{c_k\}_{k\in\Lambda} \in l_2(\Lambda)$ satisfying $f = \sum_{k\in\Lambda} c_k h_k$,

$$\sum_{k\in\Lambda} |\langle f, \mathcal{F}^{-1}h_k\rangle|^2 \leqslant \sum_{k\in\Lambda} |c_k|^2, \tag{4.2.14}$$

and the equality in (4.2.14) *holds if and only if $c_k = \langle f, \mathcal{F}^{-1}h_k\rangle$ for all $k \in \Lambda$.*

(iv) $\{\tilde{h}_k\}_{k\in\Lambda}$ is a dual frame of $\{h_k\}_{k\in\Lambda}$ if and only if the linear operator $U : \mathcal{H} \to \ker(\mathcal{V})$ with $U(h) := \{\langle h, \mathcal{F}^{-1}h_k - \tilde{h}_k\rangle\}_{k\in\Lambda}$ is well defined, where $\mathcal{V}$ is the synthesis operator in (4.2.3) *associated with $\{h_k\}_{k\in\Lambda}$.*

Proof By Theorem 4.2.4, the frame operator $\mathcal{F}$ is self-adjoint and bijective with a bounded inverse operator $\mathcal{F}^{-1}$. Note that (4.2.6) is equivalent to (4.2.9). Replacing h by $\mathcal{F}^{-1}h$ in (4.2.9), we have $C_1\|\mathcal{F}^{-1}h\|^2 \leqslant \langle \mathcal{F}^{-1}h, h\rangle \leqslant \|\mathcal{F}^{-1}h\|\|h\|$. Hence, we have $\|\mathcal{F}^{-1}h\| \leqslant C_1^{-1}\|h\|$, i.e., $\|\mathcal{F}^{-1}\| \leqslant C_1^{-1}$.

Since $\mathcal{F}$ is self-adjoint, by the definition of $\mathcal{F}g := \sum_{k\in\Lambda}\langle g, h_k\rangle h_k$, we have

$$\langle f, g\rangle = \langle \mathcal{F}^{-1}f, \mathcal{F}g\rangle = \sum_{k\in\Lambda}\langle \mathcal{F}^{-1}f, h_k\rangle\langle h_k, g\rangle = \sum_{k\in\Lambda}\langle f, \mathcal{F}^{-1}h_k\rangle\langle h_k, g\rangle. \tag{4.2.15}$$

Taking $f = h$ and $g = \mathcal{F}^{-1}h$ in (4.2.15), we have $\sum_{k\in\Lambda} |\langle h, \mathcal{F}^{-1}h_k\rangle|^2 = \langle h, \mathcal{F}^{-1}h\rangle \leqslant \|\mathcal{F}^{-1}\|\|h\|^2 \leqslant C_1^{-1}\|h\|^2$. Now it follows from (4.2.15) with $f = g = h$ that

$$\|h\|^4 = |\langle h, h\rangle|^2 \leqslant \Big(\sum_{k\in\Lambda} |\langle h, \mathcal{F}^{-1}h_k\rangle|^2\Big)\Big(\sum_{k\in\Lambda} |\langle h_k, h\rangle|^2\Big) \leqslant C_2\|h\|^2 \sum_{k\in\Lambda} |\langle h, \mathcal{F}^{-1}h_k\rangle|^2,$$

from which we see that (4.2.13) holds. Therefore, the sequence $\{\mathcal{F}^{-1}h_k\}_{k\in\Lambda}$ is a frame for $\mathcal{H}$. By (4.2.15), we conclude that $\{\mathcal{F}^{-1}h_k\}_{k\in\Lambda}$ is a dual frame of $\{h_k\}_{k\in\Lambda}$. Hence, item (i) holds. Item (ii) follows from (4.2.15) by taking $g = \mathcal{F}^{-1}h$.

Define $d_k := \langle f, \mathcal{F}^{-1}h_k\rangle, k \in \Lambda$. By item (i), we have $f = \sum_{k\in\Lambda} d_k h_k$. By our assumption in item (iii), we have $f = \sum_{k\in\Lambda} c_k h_k$. Hence, $\sum_{k\in\Lambda}(c_k - d_k)h_k = 0$ and

$$0 = \sum_{k\in\Lambda}(c_k - d_k)\langle h_k, \mathcal{F}^{-1}f\rangle = \sum_{k\in\Lambda}(c_k - d_k)\langle \mathcal{F}^{-1}h_k, f\rangle = \sum_{k\in\Lambda}(c_k - d_k)\overline{d_k}.$$

So, $\{c_k - d_k\}_{k\in\Lambda}$ is perpendicular to $\{d_k\}_{k\in\Lambda}$ in $l_2(\Lambda)$. By the Pythagorean theorem,

$$\sum_{k\in\Lambda} |c_k|^2 = \sum_{k\in\Lambda} |d_k + (c_k - d_k)|^2 = \sum_{k\in\Lambda} |d_k|^2 + \sum_{k\in\Lambda} |c_k - d_k|^2 \geqslant \sum_{k\in\Lambda} |d_k|^2.$$

Obviously, from the above identity, the equality sign in (4.2.14) holds if and only if $c_k = d_k$ for all $k \in \Lambda$.

Suppose that $\{\tilde{h}_k\}_{k\in\Lambda}$ is a dual frame of $\{h_k\}_{k\in\Lambda}$. Then it is trivial to verify that $\{\mathcal{F}^{-1}h_k - \tilde{h}_k\}_{k\in\Lambda}$ is a Bessel sequence in $\mathcal{H}$. Therefore, $U(h) \in l_2(\Lambda)$. On the other hand, since both $\{\mathcal{F}^{-1}h_k\}_{k\in\Lambda}$ and $\{\tilde{h}_k\}_{k\in\Lambda}$ are dual frames of $\{h_k\}_{k\in\Lambda}$, we have

$$\sum_{k\in\Lambda} \langle h, \mathcal{F}^{-1}h_k - \tilde{h}_k\rangle\langle h_k, f\rangle = 0, \qquad \forall\, f, h \in \mathcal{H}. \tag{4.2.16}$$

That is, $U(h) \in \operatorname{ran}(\mathcal{W})^{\perp} = \ker(\mathcal{V})$, where the last identity follows from the relation $\mathcal{W} = \mathcal{V}^{\star}$. Therefore, the mapping $U : \mathcal{H} \to \ker(\mathcal{V})$ is well defined.

Conversely, if the mapping U is well defined. By Proposition 4.2.1, the sequence $\{\mathcal{F}^{-1}h_k - \tilde{h}_k\}_{k\in\Lambda}$ is a Bessel sequence in $\mathcal{H}$. Consequently, $\{\tilde{h}_k\}_{k\in\Lambda}$ is a Bessel sequence in $\mathcal{H}$. By the relation $\operatorname{ran}(\mathcal{W})^{\perp} = \ker(\mathcal{V})$, we see that (4.2.16) holds, from which and item (i) we see that (4.2.10) holds. By Theorem 4.2.5, we conclude that $\{\tilde{h}_k\}_{k\in\Lambda}$ is a dual frame of $\{h_k\}_{k\in\Lambda}$. □

We now study bases in a Hilbert space. We say that $\{h_k\}_{k\in\Lambda}$ is *a Riesz sequence in* $\mathcal{H}$ if there exist positive constants C_3 and C_4 such that

$$C_3 \sum_{k\in\Lambda} |c_k|^2 \leqslant \Big\|\sum_{k\in\Lambda} c_k h_k\Big\|^2 \leqslant C_4 \sum_{k\in\Lambda} |c_k|^2 \tag{4.2.17}$$

for all finitely supported sequences $\{c_k\}_{k\in\Lambda}$. If in addition the linear span of $\{h_k\}_{k\in\Lambda}$ is dense in $\mathcal{H}$, then $\{h_k\}_{k\in\Lambda}$ is called *a Riesz basis for* $\mathcal{H}$. The optimal constants

C_3 and C_4 in (4.2.17) are called *the lower Riesz bound* and *the upper Riesz bound*, respectively. By Proposition 4.2.1, it is easy to see that (4.2.17) holds for all finitely supported sequences $\{c_k\}_{k\in\Lambda}$ if and only if $\sum_{k\in\Lambda} c_k h_k$ converges unconditionally and (4.2.17) holds for all $\{c_k\}_{k\in\Lambda} \in l_2(\Lambda)$.

Let $\{h_k\}_{k\in\Lambda}$ and $\{\tilde{h}_k\}_{k\in\Lambda}$ be two sequences in a Hilbert space $\mathcal{H}$. We say that $(\{\tilde{h}_k\}_{k\in\Lambda}, \{h_k\}_{k\in\Lambda})$ is *a pair of biorthogonal bases* for $\mathcal{H}$, or $\{\tilde{h}_k\}_{k\in\Lambda}$ is *a dual Riesz basis* of $\{h_k\}_{k\in\Lambda}$, if each of $\{h_k\}_{k\in\Lambda}$ and $\{\tilde{h}_k\}_{k\in\Lambda}$ is a Riesz basis for $\mathcal{H}$ and

$$\langle \tilde{h}_j, h_k\rangle = \boldsymbol{\delta}_{j,k}, \qquad \forall\, j,k \in \Lambda, \tag{4.2.18}$$

where $\boldsymbol{\delta}_{j,k} = 1$ if $j = k$, and $\boldsymbol{\delta}_{j,k} = 0$ if $j \ne k \in \Lambda$.

We say that $\{h_k\}_{k\in\Lambda}$ is *an orthonormal basis* for $\mathcal{H}$ if $\langle h_j, h_k\rangle = \boldsymbol{\delta}_{j,k}$ for all $j,k \in \Lambda$ and the linear span of $\{h_k\}_{k\in\Lambda}$ is dense in $\mathcal{H}$. It is straightforward to see that an orthonormal basis for $\mathcal{H}$ is also a Riesz basis for $\mathcal{H}$.

In the following we shall characterize Riesz bases in a Hilbert space $\mathcal{H}$ and we shall see that a Riesz basis for $\mathcal{H}$ must be a frame for $\mathcal{H}$. Some basic facts about a Riesz basis for $\mathcal{H}$ are as follows:

Theorem 4.2.7 *Let $\{h_k\}_{k\in\Lambda}$ be a sequence in a Hilbert space $\mathcal{H}$. Then the following statements are equivalent:*

(1) The sequence $\{h_k\}_{k\in\Lambda}$ is a Riesz basis for $\mathcal{H}$.
(2) The synthesis operator $\mathcal{V}$ in (4.2.3) *is a well-defined, bounded, and bijective linear operator with a bounded inverse operator.*
(3) The analysis operator $\mathcal{W}$ in (4.2.2) *is a well-defined, bounded, and bijective linear operator with a bounded inverse operator.*
(4) The sequence $\{h_k\}_{k\in\Lambda}$ is a frame for $\mathcal{H}$ and $\langle \mathcal{F}^{-1}h_j, h_k\rangle = \boldsymbol{\delta}_{j,k}$ for all $j,k \in \Lambda$.
(5) The sequence $\{h_k\}_{k\in\Lambda}$ is a frame for $\mathcal{H}$ and $\{h_k\}_{k\in\Lambda}$ has a unique dual frame.
(6) The sequence $\{h_k\}_{k\in\Lambda}$ is a frame for $\mathcal{H}$ and is l_2-linearly independent, that is,

$$\sum_{k\in\Lambda} c_k h_k = 0 \quad \textit{with} \quad \sum_{k\in\Lambda} |c_k|^2 < \infty \implies c_k = 0 \quad \forall\, k \in \Lambda. \tag{4.2.19}$$

(7) The sequence $\{h_k\}_{k\in\Lambda}$ is a frame for $\mathcal{H}$ and $\{h_k\}_{k\in\Lambda}$ is minimal, that is, every element $h_j \notin \overline{\mathrm{span}\{h_k : k \in \Lambda\backslash\{j\}\}}$ for all $j \in \Lambda$.

Moreover, any of the above statements implies that $\{\mathcal{F}^{-1}h_k\}_{k\in\Lambda}$ is a dual Riesz basis of $\{h_k\}_{k\in\Lambda}$.

Proof (1)$\Longrightarrow$(2). By Proposition 4.2.1, we see that $\mathcal{V}$ is a well-defined bounded linear operator, and (4.2.17) holds for all $\vec{c} = \{c_k\}_{k\in\Lambda} \in l_2(\Lambda)$, i.e., $C_3\|\vec{c}\|^2_{l_2(\Lambda)} \leqslant \|\mathcal{V}\vec{c}\|^2 \leqslant C_4\|\vec{c}\|^2_{l_2(\Lambda)}$ for all $\vec{c} \in l_2(\Lambda)$. By Theorem 4.2.3, the synthesis operator $\mathcal{V}$ is injective and $\mathrm{ran}(\mathcal{V})$ is closed. Since the linear span of $\{h_k\}_{k\in\Lambda}$ is dense in $\mathcal{H}$, we see that $\mathrm{ran}(\mathcal{V})$ is dense in $\mathcal{H}$. Consequently, we have $\mathrm{ran}(\mathcal{V}) = \mathcal{H}$ and $\mathcal{V}$ is onto. Hence, the synthesis operator $\mathcal{V}$ is a bijection. Since $\mathcal{V}$ is bounded, by the Open Mapping Theorem, we conclude that $\mathcal{V}$ has a bounded inverse operator.

(2) $\iff$ (3) follows trivially from the fact that $\mathcal{W} = \mathcal{V}^\star$ and $\mathcal{V} = \mathcal{W}^\star$.

(3)$\Longrightarrow$(4). By Theorem 4.2.4, $\{h_k\}_{k\in\Lambda}$ is a frame for $\mathcal{H}$. By Proposition 4.2.6,

$$h_j = \sum_{k\in\Lambda} \langle h_j, \mathcal{F}^{-1}h_k\rangle h_k, \qquad \forall\, j,k \in \Lambda.$$

Since (3)$\Longrightarrow$(2), the synthesis operator $\mathcal{V}$ must be injective. By $\{\langle h_j, \mathcal{F}^{-1}h_k\rangle\}_{k\in\Lambda} \in l_2(\Lambda)$, now the above identity will force $\langle h_j, \mathcal{F}^{-1}h_k\rangle = \delta_{j,k}$ for all $j,k \in \Lambda$.

(4)$\Longrightarrow$(5). Suppose that $\{\tilde{h}_k\}_{k\in\Lambda}$ is another dual frame of $\{h_k\}_{k\in\Lambda}$. Then

$$\mathcal{F}^{-1}h_j = \sum_{k\in\Lambda} \langle \mathcal{F}^{-1}h_j, h_k\rangle \tilde{h}_k, \qquad \forall\, j \in \Lambda.$$

Using the biorthogonality relation in item (4), we deduce from the above identity that $\mathcal{F}^{-1}h_j = \tilde{h}_j$ for all $j \in \Lambda$. Hence, $\{h_k\}_{k\in\Lambda}$ must have a unique dual frame.

(5)$\Longrightarrow$(6). Since $\{h_k\}_{k\in\Lambda}$ is a frame, items (3) and (4) of Theorem 4.2.4 hold. It suffices to prove (4.2.19). We use proof by contradiction. Suppose that (4.2.19) fails. Then $\ker(\mathcal{V}) \neq \{0\}$. Let $g \in \ker(\mathcal{V})\backslash\{0\}$ and $f \in \mathcal{H}\backslash\{0\}$. Therefore, there exists a nontrivial bounded linear operator $U : \mathcal{H} \to \ker(\mathcal{V})$ defined by $U(f) = g$ and $U(h) = 0$ for all $h \in \mathcal{H}$ satisfying $\langle h, f\rangle = 0$. By Proposition 4.2.2, there is a nontrivial Bessel sequence $\{g_k\}_{k\in\Lambda}$ in $\mathcal{H}$ such that $\langle h, g_k\rangle = [Uh](k)$ for all $k \in \Lambda$. Define $\tilde{h}_k := \mathcal{F}^{-1}h_k - g_k$ for $k \in \Lambda$. By item (iv) of Proposition 4.2.6, $\{\tilde{h}_k\}_{k\in\Lambda}$ is a dual frame of $\{h_k\}_{k\in\Lambda}$. As a consequence, $\{h_k\}_{k\in\Lambda}$ has at least two different dual frames, which is a contradiction to item (5). Hence, (4.2.19) must hold.

(6)$\Longrightarrow$(1). Since $\{h_k\}_{k\in\Lambda}$ is a frame for $\mathcal{H}$, by Theorem 4.2.4, the synthesis operator $\mathcal{V}$ is a bounded surjective linear operator. In particular, the linear span of $h_k, k \in \Lambda$ is dense in $\mathcal{H}$. Note that (4.2.19) is equivalent to $\ker(\mathcal{V}) = \{0\}$. Hence, $\mathcal{V}$ is a bounded bijection. By the Open Mapping Theorem, $\mathcal{V}$ has a bounded inverse operator. Now it is straightforward to see that (4.2.17) holds with $C_3 = \|\mathcal{V}^{-1}\|^{-2}$ and $C_4 = \|\mathcal{V}\|^2$.

(7)$\Longrightarrow$(6) is trivial. We prove (4)$\Longrightarrow$(7) using proof by contradiction. Suppose that there exists $j \in \Lambda$ such that $h_j \in \overline{\text{span}}\{h_k : k \in \Lambda\backslash\{j\}\}$, i.e., there is a sequence $\{f_n\}_{n=1}^\infty$ in $\text{span}\{h_k : k \in \Lambda\backslash\{j\}\}$ such that $\lim_{n\to\infty} \|f_n - h_j\| = 0$. By item (4), we get $1 = \langle \mathcal{F}^{-1}h_j, h_j\rangle = \lim_{n\to\infty}\langle \mathcal{F}^{-1}h_j, f_n\rangle = 0$, a contradiction. Hence, (4)$\Longrightarrow$(7).

That $\{\mathcal{F}^{-1}h_k\}_{k\in\Lambda}$ is a dual Riesz basis of $\{h_k\}_{k\in\Lambda}$ follows directly from item (4) and Proposition 4.2.6. □

Theorem 4.2.7 shows that a Riesz basis for a Hilbert space $\mathcal{H}$ is just a frame for $\mathcal{H}$ but without redundancy in the sense of (4.2.19). Therefore, a general frame may have redundancy, which is a desirable property in some applications. Moreover, Exercise 4.14 shows that a frame in a Hilbert space $\mathcal{H}$ is just the image of a Riesz basis in a larger Hilbert space $\mathcal{K}$ under the orthogonal projection from $\mathcal{K}$ to $\mathcal{H}$.

Let $\{h_k\}_{k\in\Lambda}$ be a sequence in a Hilbert space $\mathcal{K}$. Then $\mathcal{H} := \overline{\text{span}}\{h_k : k \in \Lambda\}$ is a closed subspace of $\mathcal{K}$. Now it is trivial to see that $\{h_k\}_{k\in\Lambda}$ is a Riesz sequence in $\mathcal{K}$ if and only if $\{h_k\}_{k\in\Lambda}$ is a Riesz basis for the subspace $\mathcal{H}$. Similarly, we say that $\{h_k\}_{k\in\Lambda}$ is *a frame sequence* in $\mathcal{K}$ if there exist positive constants C_1 and C_2 such that (4.2.6) holds for all $h \in \mathcal{H}$. Hence, $\{h_k\}_{k\in\Lambda}$ is a frame sequence in $\mathcal{K}$ if and only if it is a frame for $\overline{\text{span}}\{h_k : k \in \Lambda\}$. Now it is trivial to see that $\{h_k\}_{k\in\Lambda}$ is a Riesz sequence in $\mathcal{K}$ is equivalent to each of items (2)–(7) in Theorem 4.2.7.

The following results are direct consequences of Theorem 4.2.7.

Corollary 4.2.8 *Let $\{\tilde{h}_k\}_{k\in\Lambda}$ and $\{h_k\}_{k\in\Lambda}$ be two sequences in a Hilbert space $\mathcal{H}$. Then $(\{\tilde{h}_k\}_{k\in\Lambda}, \{h_k\}_{k\in\Lambda})$ is a pair of biorthogonal bases for $\mathcal{H}$ if and only if $(\{\tilde{h}_k\}_{k\in\Lambda}, \{h_k\}_{k\in\Lambda})$ is a pair of dual frames for $\mathcal{H}$ and the biorthogonality relation in* (4.2.18) *is satisfied (or any of items (1)–(7) in Theorem 4.2.7 is satisfied).*

Corollary 4.2.9 *Let $\{h_k\}_{k\in\Lambda}$ be a sequence in a Hilbert space $\mathcal{H}$. Then $\{h_k\}_{k\in\Lambda}$ is an orthonormal basis for $\mathcal{H}$ if and only if $\{h_k\}_{k\in\Lambda}$ is a (normalized) tight frame for $\mathcal{H}$ (i.e.,* (4.2.7) *is satisfied) and $\|h_k\| = 1$ for all $k \in \Lambda$.*

Therefore, an orthonormal basis for a Hilbert space $\mathcal{H}$ is just a special tight frame for $\mathcal{H}$ with all its elements having norm one.

4.3 Nonhomogeneous and Homogeneous Affine Systems in $L_2(\mathbb{R})$

Classical theory of wavelets and framelets is often developed for the square integrable function space $L_2(\mathbb{R})$. In this section we systematically study nonhomogeneous and homogeneous affine systems in $L_2(\mathbb{R})$.

4.3.1 Nonhomogeneous and Homogeneous Framelets in $L_2(\mathbb{R})$

Let us first introduce nonhomogeneous and homogeneous affine systems in the time/space domain. Let Φ and Ψ be two subsets of $L_2(\mathbb{R})$. For every integer $J \in \mathbb{Z}$, *a (nonhomogeneous) affine system* $\mathsf{AS}_J(\Phi; \Psi)$ is defined to be

$$\mathsf{AS}_J(\Phi; \Psi) := \{\phi_{2^J;k} : k \in \mathbb{Z}, \phi \in \Phi\} \cup \{\psi_{2^j;k} : j \geqslant J, k \in \mathbb{Z}, \psi \in \Psi\} \tag{4.3.1}$$

and *a homogeneous affine system* $\mathsf{AS}(\Psi)$ is defined to be

$$\mathsf{AS}(\Psi) := \{\psi_{2^j;k} : j \in \mathbb{Z}, k \in \mathbb{Z}, \psi \in \Psi\}, \tag{4.3.2}$$

where $f_{\lambda;k} := |\lambda|^{1/2} f(\lambda \cdot -k)$ is defined in (4.0.2). By the basic identity in (4.0.3), we see that $\mathsf{FAS}_J(\widehat{\Phi}; \widehat{\Psi}) = \{\widehat{f} : f \in \mathsf{AS}_J(\Phi; \Psi)\}$ is the image of $\mathsf{AS}_J(\Phi; \Psi)$ under the Fourier transform, where $\widehat{\Phi} := \{\widehat{\phi} : \phi \in \Phi\}$ and $\widehat{\Psi} := \{\widehat{\psi} : \psi \in \Psi\}$.

To study homogeneous affine systems, we need the following estimate later:

Lemma 4.3.1 *For $f, \psi \in L_2(\mathbb{R})$ and $j \in \mathbb{Z}$,*

$$\sum_{k\in\mathbb{Z}} |\langle f, \psi_{2^j;k}\rangle|^2 \leqslant \max(1, 2^j)\|[\widehat{f},\widehat{f}]\|_{L_\infty(\mathbb{R})}\frac{1}{2\pi}\int_{2^{-j}\operatorname{supp}(\widehat{f})} |\widehat{\psi}(\xi)|^2 d\xi.$$

Proof By Plancherel's Theorem, we have $\langle f, \psi_{2^j;k}\rangle = \frac{1}{2\pi}\langle \widehat{f}, \widehat{\psi}_{2^{-j};0,k}\rangle$. Let $K := \operatorname{supp}(\widehat{f}) := \{x \in \mathbb{R} : \widehat{f}(\xi) \neq 0\}$. By Lemma 4.1.1, we have

$$\begin{aligned}
\sum_{k\in\mathbb{Z}} |\langle f, \psi_{2^j;k}\rangle|^2 &= \frac{1}{4\pi^2}\sum_{k\in\mathbb{Z}} |\langle \widehat{f}, \widehat{\psi}_{2^{-j};0,k}\rangle|^2 = \frac{2^j}{2\pi}\int_{-\pi}^{\pi} |[\widehat{f}(2^j\cdot), \widehat{\psi}](\xi)|^2 d\xi \\
&\leqslant 2^j\|[\widehat{f}(2^j\cdot),\widehat{f}(2^j\cdot)]\|_{L_\infty(\mathbb{R})}\frac{1}{2\pi}\int_{-\pi}^{\pi} |[\chi_K(2^j\cdot)\widehat{\psi}, \chi_K(2^j\cdot)\widehat{\psi}](\xi)|^2 d\xi \\
&\leqslant 2^j \max(2^{-j}, 1)\|[\widehat{f},\widehat{f}]\|_{L_\infty(\mathbb{R})}\frac{1}{2\pi}\int_{-\pi}^{\pi} \sum_{k\in\mathbb{Z}} \chi_K(2^j(\xi + 2\pi k))|\widehat{\psi}(\xi + 2\pi k)|^2 d\xi \\
&= \max(1, 2^j)\|[\widehat{f},\widehat{f}]\|_{L_\infty(\mathbb{R})}\frac{1}{2\pi}\int_{2^{-j}K} |\widehat{\psi}(\xi)|^2 d\xi,
\end{aligned}$$

where we used

$$\|[\widehat{f}(2^j\cdot),\widehat{f}(2^j\cdot)]\|_{L_\infty(\mathbb{R})} = \|\sum_{k\in\mathbb{Z}} |\widehat{f}(2^j\cdot + 2\pi 2^j k)|^2\|_{L_\infty(\mathbb{R})} \leqslant \max(2^{-j}, 1)\|[\widehat{f},\widehat{f}]\|_{L_\infty(\mathbb{R})}.$$

This completes the proof. □

As an application of Lemma 4.3.1, we have the following auxiliary result.

Lemma 4.3.2 *Let $\Phi \subseteq L_2(\mathbb{R})$ such that $\sum_{\phi\in\Phi} \|\phi\|^2_{L_2(\mathbb{R})} < \infty$ and there exists $C > 0$ satisfying*

$$\sum_{\phi\in\Phi}\sum_{k\in\mathbb{Z}} |\langle g, \phi(\cdot - k)\rangle|^2 \leqslant C\|g\|^2_{L_2(\mathbb{R})} \qquad \forall\, g \in L_2(\mathbb{R}). \tag{4.3.3}$$

Then for all $f \in L_2(\mathbb{R})$,

$$\lim_{j\to-\infty}\sum_{\phi\in\Phi}\sum_{k\in\mathbb{Z}} |\langle f, \phi_{2^j;k}\rangle|^2 = 0. \tag{4.3.4}$$

Proof Define a dense subset $L^{cpt}_\infty(\mathbb{R}\backslash\{0\})$ of $L_2(\mathbb{R})$ as follows:

$$L^{cpt}_\infty(\mathbb{R}\backslash\{0\}) := \{\mathbf{f} \in L_\infty(\mathbb{R}) : \mathbf{f}(\xi) = 0 \;\; \forall\, \xi \in \mathbb{R}\backslash K \text{ for some compact set } K \subseteq \mathbb{R}\backslash\{0\}\}.$$

Let $f \in L_\infty^{cpt}(\mathbb{R}\backslash\{0\})$. By Lemma 4.3.1, we deduce that for all $j \leqslant 0$,

$$E_j(f) := \sum_{\phi\in\Phi}\sum_{k\in\mathbb{Z}} |\langle f, \phi_{2^j;k}\rangle|^2 \leqslant \|[\widehat{f},\widehat{f}]\|_{L_\infty(\mathbb{R})}\frac{1}{2\pi}\int_{2^{-j}\operatorname{supp}(\widehat{f})}\sum_{\phi\in\Phi}|\widehat{\phi}(\xi)|^2 d\xi.$$

Since $\int_{\mathbb{R}}\sum_{\phi\in\Phi}|\widehat{\phi}(\xi)|^2 d\xi = 2\pi\sum_{\phi\in\Phi}\|\phi\|^2_{L_2(\mathbb{R})} < \infty$ and $\operatorname{supp}(\widehat{f})$ is outside some neighborhood of the origin, we have $\lim_{j\to-\infty}\int_{2^{-j}\operatorname{supp}(\widehat{f})}\sum_{\phi\in\Phi}|\widehat{\phi}(\xi)|^2 d\xi = 0$. Therefore, (4.3.4) holds for all $f \in L_\infty^{cpt}(\mathbb{R}\backslash\{0\})$.

Note that (4.3.3) implies $E_j(f) \leqslant C\|f\|^2_{L_2(\mathbb{R})}$ for all $f \in L_2(\mathbb{R})$. Since $L_\infty^{cpt}(\mathbb{R}\backslash\{0\})$ is dense in $L_2(\mathbb{R})$, for $f \in L_2(\mathbb{R})$ and $\varepsilon > 0$, there exists $g \in L_\infty^{cpt}(\mathbb{R}\backslash\{0\})$ such that $\|g - f\|_{L_2(\mathbb{R})} \leqslant \varepsilon$. By what has been proved, there exists $J \in \mathbb{Z}$ such that $E_j(g) \leqslant \varepsilon^2$ for all $j \leqslant J$. Hence, we deduce that

$$\sqrt{E_j(f)} \leqslant \sqrt{E_j(f-g)} + \sqrt{E_j(g)} \leqslant \sqrt{C}\|f-g\|_{L_2(\mathbb{R})} + \sqrt{E_j(g)} \leqslant (\sqrt{C}+1)\varepsilon,$$

for all $j \leqslant J$. This completes the proof of the identity in (4.3.4). □

We now show that a nonhomogeneous affine system in $L_2(\mathbb{R})$ naturally leads to a sequence of nonhomogeneous affine systems at every scale level J and a homogeneous affine system in $L_2(\mathbb{R})$ as its limiting system when $J \to -\infty$.

Theorem 4.3.3 *Let Φ and Ψ be subsets of $L_2(\mathbb{R})$. Suppose that $\mathsf{AS}_J(\Phi;\Psi)$ is a frame for $L_2(\mathbb{R})$ for $J \in \mathbb{Z}$, that is, there exist positive constants C_1 and C_2 such that*

$$C_1\|f\|^2_{L_2(\mathbb{R})} \leqslant \sum_{\phi\in\Phi}\sum_{k\in\mathbb{Z}}|\langle f, \phi_{2^J;k}\rangle|^2 + \sum_{j=J}^{\infty}\sum_{\psi\in\Psi}\sum_{k\in\mathbb{Z}}|\langle f, \psi_{2^j;k}\rangle|^2 \leqslant C_2\|f\|^2_{L_2(\mathbb{R})}, \tag{4.3.5}$$

for all $f \in L_2(\mathbb{R})$. Then (4.3.5) *holds for all integers J, in other words, $\mathsf{AS}_J(\Phi;\Psi)$ is a frame for $L_2(\mathbb{R})$ with the same lower and upper frame bounds for every integer $J \in \mathbb{Z}$. If in addition $\sum_{\phi\in\Phi}\|\phi\|^2_{L_2(\mathbb{R})} < \infty$, then the homogeneous affine system $\mathsf{AS}(\Psi)$ is a frame for $L_2(\mathbb{R})$ with the same lower and upper frame bounds, that is,*

$$C_1\|f\|^2_{L_2(\mathbb{R})} \leqslant \sum_{j\in\mathbb{Z}}\sum_{\psi\in\Psi}\sum_{k\in\mathbb{Z}}|\langle f, \psi_{2^j;k}\rangle|^2 \leqslant C_2\|f\|^2_{L_2(\mathbb{R})}, \qquad \forall f \in L_2(\mathbb{R}). \tag{4.3.6}$$

Proof By $\|f_{2^n;0}\|_{L_2(\mathbb{R})} = \|f\|_{L_2(\mathbb{R})}$ and $\langle f_{2^n;0}, \psi_{2^j;k}\rangle = \langle f, \psi_{2^{j-n};k}\rangle$, it is straightforward to see that if (4.3.5) holds for one integer J, then (4.3.5) holds for all integers J. We now prove (4.3.6). Since $\sum_{\phi\in\Phi}\|\phi\|^2_{L_2(\mathbb{R})} < \infty$ and (4.3.3) holds with $C = C_2$, by Lemma 4.3.2, (4.3.4) holds. For $f \in L_2(\mathbb{R})$, it follows trivially from (4.3.5) that

$$C_1\|f\|^2_{L_2(\mathbb{R})} - \sum_{\phi\in\Phi}\sum_{k\in\mathbb{Z}}|\langle f, \phi_{2^J;k}\rangle|^2 \leqslant \sum_{j=J}^{\infty}\sum_{\psi\in\Psi}\sum_{k\in\mathbb{Z}}|\langle f, \psi_{2^j;k}\rangle|^2 \leqslant C_2\|f\|^2_{L_2(\mathbb{R})}$$

for all integers $J \in \mathbb{Z}$. Using (4.3.4) and taking $J \to -\infty$ in the above two inequalities, we see that (4.3.6) holds for all $f \in L_2(\mathbb{R})$. □

Due to Theorem 4.3.3, we say that $\{\Phi; \Psi\}$ is *a framelet in* $L_2(\mathbb{R})$ if $\mathsf{AS}_0(\Phi; \Psi)$ is a frame for $L_2(\mathbb{R})$. In particular, the set $\{\Phi; \Psi\}$ of generators is called *a tight framelet in* $L_2(\mathbb{R})$ if $\mathsf{AS}_0(\Phi; \Psi)$ is a (normalized) tight frame for $L_2(\mathbb{R})$, that is, (4.3.5) holds with $C_1 = C_2 = 1$ and $J = 0$. Similarly, we say that Ψ is *a homogeneous framelet in* $L_2(\mathbb{R})$ if $\mathsf{AS}(\Psi)$ is a frame for $L_2(\mathbb{R})$. We call Ψ *a homogeneous tight framelet in* $L_2(\mathbb{R})$ if (4.3.6) holds with $C_1 = C_2 = 1$.

Theorem 4.3.3 shows that a framelet in $L_2(\mathbb{R})$ naturally leads to a homogeneous framelet in $L_2(\mathbb{R})$. The converse direction is also true.

Theorem 4.3.4 *Let* $\Psi \subseteq L_2(\mathbb{R})$ *such that* $\sum_{\psi\in\Psi} \|\psi\|^2_{L_2(\mathbb{R})} < \infty$. *Then* Ψ *is a homogeneous framelet in* $L_2(\mathbb{R})$ *if and only if* $\{\Phi; \Psi\}$ *is a framelet in* $L_2(\mathbb{R})$ *having the same lower and upper frame bounds, where* $\Phi := \{2^{-j}\psi(2^{-j}\cdot) : \psi \in \Psi, j \in \mathbb{N}\}$.

Proof Suppose that $\{\Phi; \Psi\}$ is a framelet in $L_2(\mathbb{R})$ satisfying (4.3.5) with $J = 0$. By the definition of Φ and $\|2^{-j}\psi(2^{-j}\cdot)\|^2_{L_2(\mathbb{R})} = 2^{-j}\|\psi\|^2_{L_2(\mathbb{R})}$, we have

$$\sum_{\phi\in\Phi} \|\phi\|^2_{L_2(\mathbb{R})} = \sum_{\psi\in\Psi}\sum_{j=1}^{\infty} 2^{-j}\|\psi\|^2_{L_2(\mathbb{R})} = \sum_{\psi\in\Psi} \|\psi\|^2_{L_2(\mathbb{R})} < \infty. \tag{4.3.7}$$

By Theorem 4.3.3, we conclude that $\mathsf{AS}(\Psi)$ is a frame for $L_2(\mathbb{R})$ satisfying (4.3.6).

Conversely, suppose that Ψ is a homogeneous framelet in $L_2(\mathbb{R})$ satisfying (4.3.6). Define

$$E^{\Psi}_j(f) := \sum_{\psi\in\Psi}\sum_{k\in\mathbb{Z}} |\langle f, \psi_{2^j;k}\rangle|^2, \qquad j \in \mathbb{Z}.$$

Let $J \in \mathbb{N}_0$. By $\langle f(\cdot+\gamma), \psi_{2^j;k}\rangle = \langle f, \psi_{2^j;k+2^j\gamma}\rangle$, it is straightforward to verify that

$$\sum_{\psi\in\Psi}\sum_{k\in\mathbb{Z}} |\langle f(\cdot+\gamma), \psi_{2^j;k}\rangle|^2 = \sum_{\psi\in\Psi}\sum_{k\in\mathbb{Z}} |\langle f, \psi_{2^j;k+2^j\gamma}\rangle|^2 = \sum_{\psi\in\Psi}\sum_{k\in\mathbb{Z}} |\langle f, \psi_{2^j;k}\rangle|^2,$$

for all $j \in \mathbb{N}_0 := \mathbb{N} \cup \{0\}$ and $\gamma \in \mathbb{Z}$. For all $J \leqslant j < 0$, noting that $\gamma \in \{0, \dots, 2^{-J}-1\}$ can be uniquely written as $\gamma = \gamma' + 2^{-j}m$ with $\gamma' \in \{0, \dots, 2^{-j}-1\}$ and $m \in \{0, \dots, 2^{j-J}-1\}$, we have

$$\begin{aligned}
&2^J \sum_{\gamma=0}^{2^{-J}-1} \sum_{\psi\in\Psi}\sum_{k\in\mathbb{Z}} |\langle f(\cdot+\gamma), \psi_{2^j;k}\rangle|^2\\
&\quad = \sum_{\psi\in\Psi}\sum_{\gamma'=0}^{2^{-j}-1}\sum_{m=0}^{2^{j-J}-1}\sum_{k\in\mathbb{Z}} 2^J |\langle f, 2^{j/2}\psi(2^j(\cdot-\gamma'-2^{-j}m-2^{-j}k))\rangle|^2
\end{aligned}$$

$$= \sum_{\psi\in\Psi}\sum_{\gamma'=0}^{2^{-j}-1}\sum_{m=0}^{2^{j-J}-1}\sum_{k\in\mathbb{Z}} 2^J |\langle f, 2^{j/2}\psi(2^j(\cdot-\gamma'-2^{-j}k))\rangle|^2$$

$$= \sum_{\psi\in\Psi}\sum_{\gamma'=0}^{2^{-j}-1}\sum_{k\in\mathbb{Z}} 2^{j-J}2^J |\langle f, 2^{j/2}\psi(2^j(\cdot-\gamma'-2^{-j}k))\rangle|^2$$

$$= \sum_{\psi\in\Psi}\sum_{k\in\mathbb{Z}} |\langle f, 2^j\psi(2^j(\cdot-k))\rangle|^2.$$

By (4.3.6) and the above identities, we deduce that for all $J < 0$ and $f \in L_2(\mathbb{R})$,

$$\begin{aligned} C_1\|f\|^2_{L_2(\mathbb{R})} &\leqslant 2^J \sum_{\gamma=0}^{2^{-J}-1}\sum_{j\in\mathbb{Z}}\sum_{\psi\in\Psi}\sum_{k\in\mathbb{Z}} |\langle f(\cdot+\gamma), \psi_{2^j;k}\rangle|^2 \\ &= 2^J \sum_{\gamma=0}^{2^{-J}-1}\sum_{j\in\mathbb{Z}} E_j^{\Psi}(f(\cdot+\gamma)) \\ &= R_J^{\Psi}(f) + \sum_{j=J}^{-1}\sum_{\psi\in\Psi}\sum_{k\in\mathbb{Z}} |\langle f, 2^j\psi(2^j(\cdot-k))\rangle|^2 + \sum_{j=0}^{\infty} E_j^{\Psi}(f) \\ &\leqslant C_2\|f\|^2_{L_2(\mathbb{R})}, \end{aligned} \tag{4.3.8}$$

where we used $\mathbb{Z} = \{j \geqslant 0\} \cup \{J \leqslant j < 0\} \cup \{j < J\}$ and we define

$$R_J^{\Psi}(f) := 2^J \sum_{\gamma=0}^{2^{-J}-1}\sum_{j<J} E_j^{\Psi}(f(\cdot+\gamma)).$$

It follows directly from (4.3.8) that

$$C_1\|f\|^2_{L_2(\mathbb{R})} - R_J^{\Psi}(f) \leqslant \sum_{j=J}^{-1}\sum_{\psi\in\Psi}\sum_{k\in\mathbb{Z}} |\langle f, 2^j\psi(2^j(\cdot-k))\rangle|^2 + \sum_{j=0}^{\infty} E_j^{\Psi}(f) \leqslant C_2\|f\|^2_{L_2(\mathbb{R})}.$$

Note that $\sum_{j<0}\sum_{\psi\in\Psi}\sum_{k\in\mathbb{Z}} |\langle f, 2^j\psi(2^j(\cdot-k))\rangle|^2 = E_0^{\Phi}(f)$. If we can prove that

$$\lim_{J\to-\infty} R_J^{\Psi}(f) = 0, \qquad \forall\, f \in L_\infty^{cpt}(\mathbb{R}\backslash\{0\}), \tag{4.3.9}$$

by a similar argument as in the proof of Theorem 4.3.3, taking $J \to -\infty$ in the inequalities right above (4.3.9), we see that (4.3.5) holds for all $f \in L_\infty^{cpt}(\mathbb{R}\backslash\{0\})$. Since $L_\infty^{cpt}(\mathbb{R}\backslash\{0\})$ is dense in $L_2(\mathbb{R})$, applying Theorem 4.2.4, we conclude that (4.3.5) holds for all $f \in L_2(\mathbb{R})$.

We now prove (4.3.9). Let $f \in L_\infty^{cpt}(\mathbb{R}\backslash\{0\})$ and define $C := (2\pi)^{-1}\|[\widehat{f},\widehat{f}]\|_{L_\infty(\mathbb{R})} < \infty$. By Lemma 4.3.1, we have

$$R_J^\Psi(f) \leqslant 2^J \sum_{\gamma=0}^{2^{-J}-1} \sum_{j<J} C \int_{2^{-j}\operatorname{supp}(\widehat{f})} \sum_{\psi\in\Psi} |\widehat{\psi}(\xi)|^2 d\xi = C \int_{\mathbb{R}} H_J(\xi) \sum_{\psi\in\Psi} |\widehat{\psi}(\xi)|^2 d\xi,$$

where $H_J(\xi) := \sum_{j<J} \chi_{2^{-j}\operatorname{supp}(\widehat{f})}(\xi)$. By $f \in L_\infty^{cpt}(\mathbb{R}\backslash\{0\})$, $\operatorname{supp}(\widehat{f})$ is bounded and is outside some neighborhood of the origin. Therefore, there exists a positive constant C_0 depending only on $\operatorname{supp}(\widehat{f})$ such that $|H_J(\xi)| \leqslant C_0$ for all $\xi \in \mathbb{R}$ and for all negative integers J. By $\sum_{\psi\in\Psi} \|\widehat{\psi}\|_{L_2(\mathbb{R})}^2 = 2\pi \sum_{\psi\in\Psi} \|\psi\|_{L_2(\mathbb{R})}^2 < \infty$, we see that $|H_J(\xi)| \sum_{\psi\in\Psi} |\widehat{\psi}(\xi)|^2 \leqslant C_0 \sum_{\psi\in\Psi} |\widehat{\psi}(\xi)|^2 \in L_1(\mathbb{R})$ for all negative integers J. By the Dominated Convergence Theorem and $\lim_{J\to-\infty} H_J(\xi) = 0$, we conclude that

$$0 \leqslant \limsup_{J\to-\infty} R_J^\Psi(f) \leqslant \lim_{J\to-\infty} C \int_{\mathbb{R}} H_J(\xi) \sum_{\psi\in\Psi} |\widehat{\psi}(\xi)|^2 d\xi = 0.$$

This completes the proof of (4.3.9). ☐

We now discuss affine dual frames in $L_2(\mathbb{R})$. Let $\Phi, \tilde{\Phi}, \Psi, \tilde{\Psi} \subseteq L_2(\mathbb{R})$. As for the frequency case in (4.1.3), we always assume that there is a bijection between Φ and $\tilde{\Phi}$ (and between Ψ and $\tilde{\Psi}$), that is, $(\phi, \tilde{\phi})$ as well as $(\psi, \tilde{\psi})$ is always regarded as a pair together. We say that $(\mathsf{AS}_J(\tilde{\Phi}; \tilde{\Psi}), \mathsf{AS}_J(\Phi; \Psi))$ is a pair of dual frames for $L_2(\mathbb{R})$ if each of $\mathsf{AS}_J(\tilde{\Phi}; \tilde{\Psi})$ and $\mathsf{AS}_J(\Phi; \Psi)$ is a frame for $L_2(\mathbb{R})$ and

$$\langle f, g\rangle = \sum_{\phi\in\Phi} \sum_{k\in\mathbb{Z}} \langle f, \tilde{\phi}_{2^J;k}\rangle \langle \phi_{2^J;k}, g\rangle + \sum_{j=J}^{\infty} \sum_{\psi\in\Psi} \sum_{k\in\mathbb{Z}} \langle f, \tilde{\psi}_{2^j;k}\rangle \langle \psi_{2^j;k}, g\rangle \qquad (4.3.10)$$

for all $f, g \in L_2(\mathbb{R})$ with the series converging absolutely.

Theorem 4.3.5 *Let $\Phi, \tilde{\Phi}, \Psi, \tilde{\Psi} \subseteq L_2(\mathbb{R})$. Suppose that $(\mathsf{AS}_J(\tilde{\Phi}; \tilde{\Psi}), \mathsf{AS}_J(\Phi; \Psi))$ is a pair of dual frames for $L_2(\mathbb{R})$ for some integer J. Then it is a pair of dual frames for $L_2(\mathbb{R})$ for all integers J. If in addition $\sum_{\phi\in\Phi} (\|\phi\|_{L_2(\mathbb{R})}^2 + \|\tilde{\phi}\|_{L_2(\mathbb{R})}^2) < \infty$, then the pair $(\mathsf{AS}(\tilde{\Psi}), \mathsf{AS}(\Psi))$ is a pair of homogeneous dual frames for $L_2(\mathbb{R})$, that is, each of $\mathsf{AS}(\tilde{\Psi})$ and $\mathsf{AS}(\Psi)$ is a frame for $L_2(\mathbb{R})$ and the following identity holds:*

$$\langle f, g\rangle = \sum_{j\in\mathbb{Z}} \sum_{\psi\in\Psi} \sum_{k\in\mathbb{Z}} \langle f, \tilde{\psi}_{2^j;k}\rangle \langle \psi_{2^j;k}, g\rangle, \qquad \forall\, f, g \in L_2(\mathbb{R}) \qquad (4.3.11)$$

with the series converging absolutely.

Proof By $\langle f_{2^n;0}, g_{2^n;0}\rangle = \langle f, g\rangle$ and $\langle f_{2^n;0}, \tilde{\psi}_{2^j;k}\rangle = \langle f, \tilde{\psi}_{2^{j-n};k}\rangle$, we see that (4.3.10) holds for all integers J. By Theorem 4.3.3, we conclude that $(\mathsf{AS}_J(\tilde{\Phi};\tilde{\Psi}), \mathsf{AS}_J(\Phi;\Psi))$ is a pair of dual frames for $L_2(\mathbb{R})$ for all integers J. By Theorem 4.3.3 again, each of $\mathsf{AS}(\tilde{\Psi})$ and $\mathsf{AS}(\Psi)$ is a frame for $L_2(\mathbb{R})$. The identity in (4.3.11) can be proved similarly as in the proof of Theorem 4.3.3 for all $f, g \in L_2(\mathbb{R})$, since by Lemma 4.3.2,

$$\Big|\sum_{\phi\in\Phi}\sum_{k\in\mathbb{Z}}\langle f, \tilde{\phi}_{2^J;k}\rangle\langle \phi_{2^J;k}, g\rangle\Big| \leqslant \sqrt{E_J^{\tilde{\Phi}}(f)E_J^{\Phi}(g)} \to 0 \quad \text{as} \quad J \to -\infty.$$

This completes the proof. □

Due to Theorem 4.3.5, we call $(\{\tilde{\Phi};\tilde{\Psi}\}, \{\Phi;\Psi\})$ *a dual framelet in* $L_2(\mathbb{R})$ if $(\mathsf{AS}_0(\tilde{\Phi};\tilde{\Psi}), \mathsf{AS}_0(\Phi;\Psi))$ is a pair of dual frames for $L_2(\mathbb{R})$. Similarly, we call $(\tilde{\Psi}, \Psi)$ *a homogeneous dual framelet in* $L_2(\mathbb{R})$ if $(\mathsf{AS}(\tilde{\Psi}), \mathsf{AS}(\Psi))$ is a pair of homogeneous dual frames for $L_2(\mathbb{R})$. For a dual framelet $(\{\tilde{\Phi};\tilde{\Psi}\}, \{\Phi;\Psi\})$ in $L_2(\mathbb{R})$ and a homogeneous dual framelet $(\tilde{\Psi}, \Psi)$ in $L_2(\mathbb{R})$, every function $f \in L_2(\mathbb{R})$ has the following representations: For every $J \in \mathbb{Z}$,

$$\begin{aligned} f &= \sum_{\phi\in\Phi}\sum_{k\in\mathbb{Z}}\langle f, \tilde{\phi}_{2^J;k}\rangle\phi_{2^J;k} + \sum_{j=J}^{\infty}\sum_{\psi\in\Psi}\sum_{k\in\mathbb{Z}}\langle f, \tilde{\psi}_{2^j;k}\rangle\psi_{2^j;k} \\ &= \sum_{\phi\in\Phi}\sum_{k\in\mathbb{Z}}\langle f, \phi_{2^J;k}\rangle\tilde{\phi}_{2^J;k} + \sum_{j=J}^{\infty}\sum_{\psi\in\Psi}\sum_{k\in\mathbb{Z}}\langle f, \psi_{2^j;k}\rangle\tilde{\psi}_{2^j;k} \end{aligned} \tag{4.3.12}$$

and

$$f = \sum_{j\in\mathbb{Z}}\sum_{\psi\in\Psi}\sum_{k\in\mathbb{Z}}\langle f, \tilde{\psi}_{2^j;k}\rangle\psi_{2^j;k} = \sum_{j\in\mathbb{Z}}\sum_{\psi\in\Psi}\sum_{k\in\mathbb{Z}}\langle f, \psi_{2^j;k}\rangle\tilde{\psi}_{2^j;k} \tag{4.3.13}$$

with all series converging unconditionally in $L_2(\mathbb{R})$.

Theorem 4.3.5 shows that a dual framelet in $L_2(\mathbb{R})$ naturally leads to a homogeneous dual framelet in $L_2(\mathbb{R})$. The converse direction is also true.

Theorem 4.3.6 *Let* $\Psi, \tilde{\Psi} \subseteq L_2(\mathbb{R})$ *such that* $\sum_{\psi\in\Psi}(\|\psi\|^2_{L_2(\mathbb{R})} + \|\tilde{\psi}\|^2_{L_2(\mathbb{R})}) < \infty$. *Then* $(\tilde{\Psi}, \Psi)$ *is a homogeneous dual framelet in* $L_2(\mathbb{R})$ *if and only if* $(\{\tilde{\Phi};\tilde{\Psi}\}, \{\Phi;\Psi\})$ *is a dual framelet in* $L_2(\mathbb{R})$, *where*

$$\Phi := \{2^{-j}\psi(2^{-j}\cdot) \;:\; \psi \in \Psi, j \in \mathbb{N}\}, \; \tilde{\Phi} := \{2^{-j}\tilde{\psi}(2^{-j}\cdot) \;:\; \tilde{\psi} \in \tilde{\Psi}, j \in \mathbb{N}\}. \tag{4.3.14}$$

Proof As proved in Theorem 4.3.4, for $f, g \in L_\infty^{cpt}(\mathbb{R}\backslash\{0\})$, we have $\lim_{J\to-\infty} R_J^{\tilde{\Psi}}(f) = 0$ and $\lim_{J\to-\infty} R_J^{\Psi}(g) = 0$. Hence, by the Cauchy-Schwarz inequality,

$$\left|2^J \sum_{\gamma=0}^{2^{-J}-1} \sum_{\psi\in\Psi} \sum_{j<J} \langle f, \tilde{\psi}_{2^j;k}\rangle \langle \psi_{2^j;k}, g\rangle\right| \leqslant \sqrt{R_J^{\tilde{\Psi}}(f) R_J^{\Psi}(g)} \to 0, \quad \text{as } J \to -\infty.$$

Now all the claims follow directly from the same argument as in Theorem 4.3.4. □

Under certain conditions, the sets $\Phi, \tilde{\Phi}$ in (4.3.14) with infinitely many generators in Theorems 4.3.4 and 4.3.6 can be replaced by sets with a single generator. See Theorems 4.5.12 and 4.5.15 for details.

4.3.2 Dual Framelets and Homogeneous Dual Framelets in $L_2(\mathbb{R})$

We now characterize dual framelets and homogeneous dual framelets in $L_2(\mathbb{R})$.

Theorem 4.3.7 *Let $\Phi, \tilde{\Phi}, \Psi, \tilde{\Psi} \subseteq L_2(\mathbb{R})$. Then $(\{\tilde{\Phi}; \tilde{\Psi}\}, \{\Phi; \Psi\})$ is a dual framelet in $L_2(\mathbb{R})$ if and only if*

(i) For a dense subset D of $L_2(\mathbb{R})$, there is a constant $C > 0$ such that for all $f \in D$,

$$\sum_{\phi\in\Phi} \sum_{k\in\mathbb{Z}} (|\langle f, \phi_{1;k}\rangle|^2 + |\langle f, \tilde{\phi}_{1;k}\rangle|^2)$$
$$+ \sum_{j=0}^{\infty} \sum_{\psi\in\Psi} \sum_{k\in\mathbb{Z}} (|\langle f, \psi_{2^j;k}\rangle|^2 + |\langle f, \tilde{\psi}_{2^j;k}\rangle|^2) \leqslant C\|f\|_{L_2(\mathbb{R})}^2;$$

(ii) $(\{\tilde{\boldsymbol{\Phi}}; \tilde{\boldsymbol{\Psi}}\}, \{\boldsymbol{\Phi}; \boldsymbol{\Psi}\})$ is a frequency-based dual framelet, where

$$\begin{aligned} \boldsymbol{\Phi} &:= \{\widehat{\phi} \,:\, \phi \in \Phi\}, \quad & \tilde{\boldsymbol{\Phi}} &:= \{\widehat{\tilde{\phi}} \,:\, \tilde{\phi} \in \tilde{\Phi}\}, \\ \boldsymbol{\Psi} &:= \{\widehat{\psi} \,:\, \psi \in \Psi\}, \quad & \tilde{\boldsymbol{\Psi}} &:= \{\widehat{\tilde{\psi}} \,:\, \tilde{\psi} \in \tilde{\Psi}\}. \end{aligned} \tag{4.3.15}$$

Proof If $(\{\tilde{\Phi}; \tilde{\Psi}\}, \{\Phi; \Psi\})$ is a dual framelet in $L_2(\mathbb{R})$, then items (i) and (ii) are obviously true. Since both D and $\mathscr{D}(\mathbb{R})$ are dense in $L_2(\mathbb{R})$, the converse direction follows directly from Proposition 4.2.1 and Theorem 4.2.5 as well as the definition of a frequency-based dual framelet. □

Note that under the assumption $\sum_{\phi\in\Phi}(|\widehat{\phi}(\cdot)|^2 + |\widehat{\tilde{\phi}}(\cdot)|^2) \in L_1^{loc}(\mathbb{R})$ and $\sum_{\psi\in\Psi}(|\widehat{\psi}(\cdot)|^2 + |\widehat{\tilde{\psi}}(\cdot)|^2) \in L_1^{loc}(\mathbb{R})$, we completely characterized a frequency-based dual framelet $(\{\widehat{\tilde{\Phi}}; \widehat{\tilde{\Psi}}\}, \{\widehat{\Phi}; \widehat{\Psi}\})$ in Theorem 4.1.6. In particular, we have a complete characterization of a tight framelet in $L_2(\mathbb{R})$.

Theorem 4.3.8 *Let $\boldsymbol{\Phi}, \boldsymbol{\Psi} \subseteq L_2^{loc}(\mathbb{R})$ such that $\sum_{\boldsymbol{\varphi}\in\boldsymbol{\Phi}} |\boldsymbol{\varphi}(\cdot)|^2 + \sum_{\boldsymbol{\psi}\in\boldsymbol{\Psi}} |\boldsymbol{\psi}(\cdot)|^2 \in L_1^{loc}(\mathbb{R})$. Then the following statements are equivalent:*

(i) $\{\boldsymbol{\Phi}; \boldsymbol{\Psi}\}$ *is* a frequency-based tight framelet *in $L_2(\mathbb{R})$, that is, $\boldsymbol{\Phi}, \boldsymbol{\Psi} \subseteq L_2(\mathbb{R})$ and*

$$2\pi \|\mathbf{f}\|^2_{L_2(\mathbb{R})} = \sum_{\boldsymbol{\varphi}\in\boldsymbol{\Phi}} \sum_{k\in\mathbb{Z}} |\langle \mathbf{f}, \boldsymbol{\varphi}_{2^{-J};0,k}\rangle|^2 + \sum_{j=J}^{\infty} \sum_{\boldsymbol{\psi}\in\boldsymbol{\Psi}} \sum_{k\in\mathbb{Z}} |\langle \mathbf{f}, \boldsymbol{\psi}_{2^{-j};0,k}\rangle|^2 \qquad (4.3.16)$$

for all $\mathbf{f} \in L_2(\mathbb{R})$ and for every (or for some) $J \in \mathbb{Z}$.

(ii) $(\{\boldsymbol{\Phi}; \boldsymbol{\Psi}\}, \{\boldsymbol{\Phi}; \boldsymbol{\Psi}\})$ *is a frequency-based dual framelet.*

(iii) $\lim_{j\to\infty} \langle \sum_{\boldsymbol{\varphi}\in\boldsymbol{\Phi}} |\boldsymbol{\varphi}(2^{-j}\cdot)|^2, \mathbf{h}\rangle = \langle 1, \mathbf{h}\rangle, \ \forall\, \mathbf{h} \in \mathscr{D}(\mathbb{R})$, *and for almost every $\xi \in \mathbb{R}$,*

$$\sum_{\boldsymbol{\varphi}\in\boldsymbol{\Phi}} \boldsymbol{\varphi}(2\xi)\overline{\boldsymbol{\varphi}(2(\xi + 2\pi k))} + \sum_{\boldsymbol{\psi}\in\boldsymbol{\Psi}} \boldsymbol{\psi}(2\xi)\overline{\boldsymbol{\psi}(2(\xi + 2\pi k))} = \sum_{\boldsymbol{\varphi}\in\boldsymbol{\Phi}} \boldsymbol{\varphi}(\xi)\overline{\boldsymbol{\varphi}(\xi + 2\pi k)}, \quad \forall\, k \in \mathbb{Z},$$

$$\sum_{\boldsymbol{\varphi}\in\boldsymbol{\Phi}} \boldsymbol{\varphi}(\xi)\overline{\boldsymbol{\varphi}(\xi + 2\pi k_0)} + \sum_{\boldsymbol{\psi}\in\boldsymbol{\Psi}} \boldsymbol{\psi}(\xi)\overline{\boldsymbol{\psi}(\xi + 2\pi k_0)} = 0, \quad \forall\, k_0 \in \mathbb{Z}\backslash[2\mathbb{Z}].$$

(iv) There exist $\Phi, \Psi \subseteq L_2(\mathbb{R})$ such that the relation in (4.3.15) *holds and $\{\Phi; \Psi\}$ is a tight framelet in $L_2(\mathbb{R})$.*

Proof By definition, it is trivial to see that (i)$\Longrightarrow$(ii) and (i) $\Longleftrightarrow$ (iv). By Theorem 4.1.6, we see that (ii) $\Longleftrightarrow$ (iii). To complete the proof, we show that (ii)$\Longrightarrow$(i). By definition and item (ii), we see that (4.3.16) holds for all $\mathbf{f} \in \mathscr{D}(\mathbb{R})$. In particular, we have $|\langle \mathbf{f}, \boldsymbol{\varphi}_{2^{-J};0,0}\rangle|^2 \leqslant 2\pi \|\mathbf{f}\|^2_{L_2(\mathbb{R})}$ for all $\mathbf{f} \in \mathscr{D}(\mathbb{R})$. Since $\mathscr{D}(\mathbb{R})$ is dense in $L_2(\mathbb{R})$, we conclude that $\|\boldsymbol{\varphi}_{2^{-J};0,0}\|^2_{L_2(\mathbb{R})} = \sup_{\mathbf{f}\in\mathscr{D}(\mathbb{R}), \|\mathbf{f}\|_{L_2(\mathbb{R})}\leqslant 1} |\langle \mathbf{f}, \boldsymbol{\varphi}_{2^{-J};0,0}\rangle|^2 \leqslant 2\pi$, from which we have $\boldsymbol{\varphi} \in L_2(\mathbb{R})$ with $\|\boldsymbol{\varphi}\|^2_{L_2(\mathbb{R})} \leqslant 2\pi$. Consequently, we proved $\boldsymbol{\Phi} \subseteq L_2(\mathbb{R})$. Similarly, we can prove that $\boldsymbol{\Psi} \subseteq L_2(\mathbb{R})$. Since (4.3.16) holds for all $\mathbf{f} \in \mathscr{D}(\mathbb{R})$ and the space $\mathscr{D}(\mathbb{R})$ is dense in $L_2(\mathbb{R})$, we conclude (see Theorem 4.2.5) that (4.3.16) holds for all $\mathbf{f} \in L_2(\mathbb{R})$. □

As a direct consequence of Theorem 4.3.8, we have

Corollary 4.3.9 *Let $\Phi, \Psi \subseteq L_2(\mathbb{R})$ such that $\sum_{\phi\in\Phi} |\widehat{\phi}(\cdot)|^2 \in L_1^{loc}(\mathbb{R})$ and $\sum_{\psi\in\Psi} |\widehat{\psi}(\cdot)|^2 \in L_1^{loc}(\mathbb{R})$. Then $\{\Phi; \Psi\}$ is a tight framelet in $L_2(\mathbb{R})$ if and only if item (iii) in Theorem 4.3.8 holds with $\boldsymbol{\Phi} := \{\widehat{\phi} : \phi \in \Phi\}$ and $\boldsymbol{\Psi} := \{\widehat{\psi} : \psi \in \Psi\}$.*

To characterize homogeneous dual framelets, we need the following result.

Lemma 4.3.10 *Let $\Psi \subseteq L_2(\mathbb{R})$ such that $\sum_{\psi\in\Psi} |\widehat{\psi}(\cdot)|^2 \in L_1^{loc}(\mathbb{R})$. If there exists a positive constant C_2 such that the right-hand inequality of* (4.3.6) *holds for all f in a dense subset of $L_2(\mathbb{R})$, then $\sum_{j\in\mathbb{Z}} \sum_{\psi\in\Psi} |\widehat{\psi}(2^j\xi)|^2 \leqslant C_2$ for almost every $\xi \in \mathbb{R}$.*

Proof By Proposition 4.2.1, the right-hand inequality of (4.3.6) holds for all $f \in L_2(\mathbb{R})$. Let $J \in \mathbb{Z}$ and $\xi_0 \in \mathbb{R}$. For $f \in L_2(\mathbb{R})$ such that $\widehat{f}$ vanishes outside $(\xi_0 - 2^J\pi, \xi_0 + 2^J\pi)$, then $\widehat{f}(\xi)\overline{\widehat{f}(\xi + 2\pi 2^j k)} = 0$ for all $j \geqslant J, k \in \mathbb{Z}\backslash\{0\}$ and $\xi \in \mathbb{R}$. Now by Lemma 4.1.1 and $\|\widehat{f}\|^2_{L_2(\mathbb{R})} = 2\pi \|f\|^2_{L_2(\mathbb{R})}$,

$$
\begin{aligned}
C_2\|\widehat{f}\|^2_{L_2(\mathbb{R})} &\geqslant 2\pi \sum_{j=J}^{\infty}\sum_{\psi\in\Psi}\sum_{k\in\mathbb{Z}} |\langle f, \psi_{2^j;k}\rangle|^2 = \frac{1}{2\pi}\sum_{j=J}^{\infty}\sum_{\psi\in\Psi}\sum_{k\in\mathbb{Z}} |\langle \widehat{f}, \widehat{\psi}_{2^{-j};0,k}\rangle|^2 \\
&= \sum_{j=J}^{\infty} \int_{\mathbb{R}} \sum_{k\in\mathbb{Z}} \widehat{f}(\xi)\overline{\widehat{f}(\xi + 2\pi 2^j k)} \sum_{\psi\in\Psi} \overline{\widehat{\psi}(2^{-j}\xi)}\widehat{\psi}(2^{-j}\xi + 2\pi k) d\xi \\
&= \int_{\mathbb{R}} |\widehat{f}(\xi)|^2 \sum_{j=J}^{\infty}\sum_{\psi\in\Psi} |\widehat{\psi}(2^{-j}\xi)|^2 d\xi,
\end{aligned}
$$

from which we deduce that $\sum_{j=J}^{\infty}\sum_{\psi\in\Psi}|\widehat{\psi}(2^{-j}\xi)|^2 \leqslant C_2$ for almost every $\xi \in (\xi_0 - 2^J\pi, \xi_0 + 2^J\pi)$. Since both ξ_0 and $J \in \mathbb{Z}$ are arbitrary, we conclude that $\sum_{j\in\mathbb{Z}}\sum_{\psi\in\Psi}|\widehat{\psi}(2^j\xi)|^2 \leqslant C_2$ for almost every $\xi \in \mathbb{R}$. □

The following result characterizes homogeneous dual framelets in $L_2(\mathbb{R})$.

Theorem 4.3.11 *Let $\Psi, \tilde{\Psi} \subseteq L_2(\mathbb{R})$ such that $\sum_{\psi\in\Psi}(\|\psi\|^2_{L_2(\mathbb{R})} + \|\tilde{\psi}\|^2_{L_2(\mathbb{R})}) < \infty$. Then $(\tilde{\Psi}, \Psi)$ is a homogeneous dual framelet in $L_2(\mathbb{R})$ if and only if*

(i) For a dense subset D of $L_2(\mathbb{R})$, there is a constant $C > 0$ such that for all $f \in D$,

$$
\sum_{j\in\mathbb{Z}}\sum_{\psi\in\Psi}\sum_{k\in\mathbb{Z}}(|\langle f, \psi_{2^j;k}\rangle|^2 + |\langle f, \tilde{\psi}_{2^j;k}\rangle|^2) \leqslant C\|f\|^2_{L_2(\mathbb{R})};
$$

(ii) The following identities are satisfied:

$$
\sum_{j\in\mathbb{Z}}\sum_{\psi\in\Psi} \widehat{\tilde{\psi}}(2^j\xi)\overline{\widehat{\psi}(2^j\xi)} = 1, \quad a.e.\ \xi \in \mathbb{R} \tag{4.3.17}
$$

$$
\sum_{j=0}^{\infty}\sum_{\psi\in\Psi} \widehat{\tilde{\psi}}(2^j\xi)\overline{\widehat{\psi}(2^j(\xi + 2\pi k_0))} = 0, \quad a.e.\ \xi \in \mathbb{R},\ \forall\, k_0 \in \mathbb{Z}\backslash[2\mathbb{Z}] \tag{4.3.18}
$$

with all series converging absolutely for almost every $\xi \in \mathbb{R}$.

Proof Let Φ and $\tilde{\Phi}$ as in (4.3.14), $\boldsymbol{\Phi} = \widehat{\Phi}$ and $\tilde{\boldsymbol{\Phi}} = \widehat{\tilde{\Phi}}$. Theorem 4.3.6 says that $(\tilde{\Psi}, \Psi)$ is a homogeneous dual framelet in $L_2(\mathbb{R})$ if and only if $(\{\tilde{\Phi}; \tilde{\Psi}\}, \{\Phi; \Psi\})$ is a dual framelet in $L_2(\mathbb{R})$. Since $\sum_{\psi\in\Psi}\|\psi\|^2_{L_2(\mathbb{R})} < \infty$, (4.3.7) holds and hence, $\sum_{\phi\in\Phi}|\widehat{\phi}(\cdot)|^2 \in L_1(\mathbb{R}) \subseteq L_1^{loc}(\mathbb{R})$. Now all the claims follow directly from Theorem 4.3.7. Indeed, by the definition of Φ and $\tilde{\Phi}$, (4.1.24) is automatically true and (4.1.25) is equivalent to (4.3.18). By item (i) and Lemma 4.3.10, we have

$$\sum_{j\in\mathbb{Z}}\sum_{\psi\in\Psi}(|\widehat{\psi}(2^j\xi)|^2+|\widehat{\tilde{\psi}}(2^j\xi)|^2)\leqslant C, \qquad a.e.\ \xi\in\mathbb{R}. \tag{4.3.19}$$

Under the condition in (4.3.19), we now prove that (4.3.17) is equivalent to (4.1.11). Note that $\boldsymbol{\Phi}=\{\widehat{\psi}(2^m\cdot)\ :\ m\in\mathbb{N}, \psi\in\Psi\}$. Hence,

$$H_j(\xi):=\sum_{\varphi\in\boldsymbol{\Phi}}\tilde{\varphi}(2^{-j}\xi)\overline{\varphi(2^{-j}\xi)}=\sum_{m=1-j}^{\infty}\sum_{\psi\in\Psi}\widehat{\tilde{\psi}}(2^m\xi)\overline{\widehat{\psi}(2^m\xi)}.$$

By (4.3.19) and the Cauchy-Schwarz inequality, we see that $|H_j(\xi)|\leqslant C$ for a.e. $\xi\in\mathbb{R}$ for all $j\in\mathbb{N}$. Moreover, $\lim_{j\to\infty}H_j(\xi)=\sum_{m\in\mathbb{Z}}\sum_{\psi\in\Psi}\widehat{\tilde{\psi}}(2^m\xi)\overline{\widehat{\psi}(2^m\xi)}$ exists for almost every $\xi\in\mathbb{R}$. Now by Lemma 4.1.4, (4.3.17) is equivalent to (4.1.11). □

As a consequence of Theorem 4.3.11, we characterize homogeneous tight framelets in $L_2(\mathbb{R})$ as follows.

Corollary 4.3.12 *Let $\Psi\subseteq L_2(\mathbb{R})$ such that $\sum_{\psi\in\Psi}\|\psi\|^2_{L_2(\mathbb{R})}<\infty$. Then Ψ is a homogeneous tight framelet in $L_2(\mathbb{R})$ if and only if*

$$\sum_{j\in\mathbb{Z}}\sum_{\psi\in\Psi}|\widehat{\psi}(2^j\xi)|^2=1, \quad a.e.\ \xi\in\mathbb{R}, \tag{4.3.20}$$

$$\sum_{j=0}^{\infty}\sum_{\psi\in\Psi}\widehat{\psi}(2^j\xi)\overline{\widehat{\psi}(2^j(\xi+2\pi k_0))}=0, \quad a.e.\ \xi\in\mathbb{R},\ \forall\, k_0\in\mathbb{Z}\backslash[2\mathbb{Z}] \tag{4.3.21}$$

with all series converging absolutely for almost every $\xi\in\mathbb{R}$.

Proof The necessity part (⇒) follows directly from Theorem 4.3.11.

Sufficiency (⇐). Let $\tilde{\Phi}=\Phi:=\{2^{-j}\psi(2^{-j}\cdot)\ :\ \psi\in\Psi, j\in\mathbb{N}\}$ and $\tilde{\boldsymbol{\Phi}}=\boldsymbol{\Phi}:=\widehat{\Phi}$. Note that (4.3.20) directly implies (4.3.19) with $\tilde{\Psi}:=\Psi$ and $C=2$. As we proved in the proof of Theorem 4.3.11, the equations (4.3.20) and (4.3.21) are equivalent to item (iii) of Theorem 4.1.6 with $\tilde{\boldsymbol{\Psi}}:=\boldsymbol{\Psi}:=\widehat{\Psi}$, from which we deduce that $(\{\boldsymbol{\Phi};\boldsymbol{\Psi}\},\{\boldsymbol{\Phi};\boldsymbol{\Psi}\})$ is a frequency-based dual framelet. By Theorems 4.3.8 and 4.3.3, we conclude that Ψ is a homogeneous tight framelet in $L_2(\mathbb{R})$. □

Since a frequency-based dual framelet has the intrinsic cascade structure as in (4.1.26), a dual framelet $(\{\tilde{\Phi};\tilde{\Psi}\},\{\Phi;\Psi\})$ in $L_2(\mathbb{R})$ also has the *cascade structure*:

$$\begin{aligned}\sum_{\phi\in\Phi}\sum_{k\in\mathbb{Z}}\langle f,\tilde{\phi}_{2^j;k}\rangle\langle\phi_{2^j;k},g\rangle+\sum_{\psi\in\Psi}\sum_{k\in\mathbb{Z}}\langle f,\tilde{\psi}_{2^j;k}\rangle\langle\psi_{2^j;k},g\rangle\\=\sum_{\phi\in\Phi}\sum_{k\in\mathbb{Z}}\langle f,\tilde{\phi}_{2^{j+1};k}\rangle\langle\phi_{2^{j+1};k},g\rangle,\end{aligned} \tag{4.3.22}$$

for all $j \in \mathbb{Z}$ and $f, g \in L_2(\mathbb{R})$. By (4.3.22), Theorem 4.3.5 shows that a homogeneous dual framelet $(\tilde{\Psi}, \Psi)$ in $L_2(\mathbb{R})$ also has the following cascade structure:

$$\sum_{m=0}^{\infty}\sum_{k\in\mathbb{Z}}\sum_{\psi\in\Psi}(-1)^k\langle f, 2^{-m}\tilde{\psi}(2^{-m}(\cdot-k/2))\rangle\langle 2^{-m}\psi(2^{-m}(\cdot-k/2)), g\rangle = 0 \tag{4.3.23}$$

for all $f, g \in L_2(\mathbb{R})$, which is equivalent to (4.3.18).

4.3.3 Nonhomogeneous and Homogeneous Wavelets in $L_2(\mathbb{R})$

We now study affine systems which are Riesz bases in $L_2(\mathbb{R})$. Let Φ and Ψ be subsets of $L_2(\mathbb{R})$. Recall that $\mathsf{AS}_J(\Phi;\Psi)$ is a Riesz basis for $L_2(\mathbb{R})$ if the linear span of $\mathsf{AS}_J(\Phi;\Psi)$ is dense in $L_2(\mathbb{R})$ and there exist two positive constants C_3 and C_4 such that

$$\begin{aligned}C_3\Big[\sum_{\phi\in\Phi}\sum_{k\in\mathbb{Z}}|v_{k,\phi}|^2+\sum_{j=J}^{\infty}\sum_{\psi\in\Psi}\sum_{k\in\mathbb{Z}}|w_{j;k,\psi}|^2\Big] &\leqslant \Big\|\sum_{\phi\in\Phi}\sum_{k\in\mathbb{Z}}v_{k,\phi}\phi_{2^J;k}+\sum_{j=J}^{\infty}\sum_{\psi\in\Psi}\sum_{k\in\mathbb{Z}}w_{j;k,\psi}\psi_{2^j;k}\Big\|^2_{L_2(\mathbb{R})}\\ &\leqslant C_4\Big[\sum_{\phi\in\Phi}\sum_{k\in\mathbb{Z}}|v_{k,\phi}|^2+\sum_{j=J}^{\infty}\sum_{\psi\in\Psi}\sum_{k\in\mathbb{Z}}|w_{j;k,\psi}|^2\Big]\end{aligned} \tag{4.3.24}$$

for all finitely supported sequences $\{v_{k,\phi}\}_{k\in\mathbb{Z},\phi\in\Phi}$ and $\{w_{j;k,\psi}\}_{j\geqslant J,k\in\mathbb{Z},\psi\in\Psi}$, where the best possible constants C_3 and C_4 are called the lower and upper Riesz bounds of $\mathsf{AS}_J(\Phi;\Psi)$, respectively.

For an affine system which is a Riesz basis, the following result is similar to Theorem 4.3.3.

Theorem 4.3.13 *Let Φ and Ψ be subsets of $L_2(\mathbb{R})$. Suppose that $\mathsf{AS}_J(\Phi;\Psi)$ is a Riesz basis for $L_2(\mathbb{R})$ satisfying (4.3.24) for some integer J. Then $\mathsf{AS}_J(\Phi;\Psi)$ is a Riesz basis for $L_2(\mathbb{R})$ with the same lower and upper Riesz bounds for every integer $J\in\mathbb{Z}$. If in addition $\sum_{\phi\in\Phi}\|\phi\|^2_{L_2(\mathbb{R})}<\infty$, then $\mathsf{AS}(\Psi)$ is a homogeneous Riesz basis for $L_2(\mathbb{R})$ with the same lower and upper Riesz bounds, that is, the linear span of $\mathsf{AS}(\Psi)$ is dense in $L_2(\mathbb{R})$ and*

$$\begin{aligned}C_3\sum_{j\in\mathbb{Z}}\sum_{\psi\in\Psi}\sum_{k\in\mathbb{Z}}|w_{j;k,\psi}|^2 &\leqslant \Big\|\sum_{j\in\mathbb{Z}}\sum_{\psi\in\Psi}\sum_{k\in\mathbb{Z}}w_{j;k,\psi}\psi_{2^j;k}\Big\|^2_{L_2(\mathbb{R})}\\ &\leqslant C_4\sum_{j\in\mathbb{Z}}\sum_{\psi\in\Psi}\sum_{k\in\mathbb{Z}}|w_{j;k,\psi}|^2\end{aligned} \tag{4.3.25}$$

for all finitely supported sequences $\{w_{j;k,\psi}\}_{j\in\mathbb{Z},k\in\mathbb{Z},\psi\in\Psi}$.

Proof By $\|f_{2^j;k}\|_{L_2(\mathbb{R})} = \|f\|_{L_2(\mathbb{R})}$, if (4.3.24) holds for some $J \in \mathbb{Z}$, then it is trivial to see that (4.3.24) holds for all integers J. If the linear span of $\mathsf{AS}_J(\Phi; \Psi)$ is dense in $L_2(\mathbb{R})$, then it is true for every integer J by a simple scaling. Setting all $v_{k,\phi} = 0$ in (4.3.24), we see that (4.3.25) holds. By Theorem 4.3.3, $\mathsf{AS}(\Psi)$ must be a frame for $L_2(\mathbb{R})$ from which we see that the linear span of $\mathsf{AS}(\Psi)$ is dense in $L_2(\mathbb{R})$. □

Corollary 4.3.14 *Let $\Phi, \tilde{\Phi}, \Psi, \tilde{\Psi} \subseteq L_2(\mathbb{R})$. If $(\mathsf{AS}_J(\tilde{\Phi}; \tilde{\Psi}), \mathsf{AS}_J(\Phi; \Psi))$ is a pair of biorthogonal bases for $L_2(\mathbb{R})$ for some $J \in \mathbb{Z}$, then $(\mathsf{AS}_J(\tilde{\Phi}; \tilde{\Psi}), \mathsf{AS}_J(\Phi; \Psi))$ is a pair of biorthogonal bases for $L_2(\mathbb{R})$ for every integer $J \in \mathbb{Z}$. If in addition $\sum_{\phi \in \Phi}(\|\phi\|^2_{L_2(\mathbb{R})} + \|\tilde{\phi}\|^2_{L_2(\mathbb{R})}) < \infty$, then $(\mathsf{AS}(\tilde{\Psi}), \mathsf{AS}(\Psi))$ is a pair of biorthogonal bases for $L_2(\mathbb{R})$.*

Proof By Theorem 4.3.13, each of $\mathsf{AS}(\tilde{\Psi})$ and $\mathsf{AS}(\Psi)$ is a Riesz basis for $L_2(\mathbb{R})$. The biorthogonality of $(\mathsf{AS}(\tilde{\Psi}), \mathsf{AS}(\Psi))$ follows trivially from the biorthogonality of $(\mathsf{AS}_J(\tilde{\Phi}; \tilde{\Psi}), \mathsf{AS}_J(\Phi; \Psi))$. □

Due to Theorem 4.3.13 and Corollary 4.3.14, we call $\{\Phi; \Psi\}$ *a Riesz wavelet in $L_2(\mathbb{R})$* if $\mathsf{AS}_0(\Phi; \Psi)$ is a Riesz basis for $L_2(\mathbb{R})$, and we call $(\{\tilde{\Phi}; \tilde{\Psi}\}, \{\Phi; \Psi\})$ *a biorthogonal wavelet in $L_2(\mathbb{R})$* if $(\mathsf{AS}_0(\tilde{\Phi}; \tilde{\Psi}), \mathsf{AS}_0(\Phi; \Psi))$ is a pair of biorthogonal bases for $L_2(\mathbb{R})$. In particular, we call $\{\Phi; \Psi\}$ *an orthogonal wavelet in $L_2(\mathbb{R})$* if $\mathsf{AS}_0(\Phi; \Psi)$ is an orthonormal basis for $L_2(\mathbb{R})$. Similarly, we say that Ψ is *a homogeneous Riesz wavelet in $L_2(\mathbb{R})$* if $\mathsf{AS}(\Phi)$ is a Riesz basis for $L_2(\mathbb{R})$, and we call $(\tilde{\Psi}, \Psi)$ *a homogeneous biorthogonal wavelet in $L_2(\mathbb{R})$* if $(\mathsf{AS}(\tilde{\Psi}), \mathsf{AS}(\Psi))$ is a pair of biorthogonal bases for $L_2(\mathbb{R})$. We call Ψ *a homogeneous orthogonal wavelet in $L_2(\mathbb{R})$* if $\mathsf{AS}(\Psi)$ is an orthonormal basis for $L_2(\mathbb{R})$.

As we shall see in Sect. 4.5.5, the converse direction in Theorem 4.3.13 and Corollary 4.3.14 only holds under certain conditions.

4.3.4 Continuous Wavelet Transform

Homogeneous framelets and wavelets originated from discretizing *continuous wavelet transform (CWT)*. For $f, \psi \in L_2(\mathbb{R})$ and $\alpha \in \mathbb{R}\backslash\{0\}, \beta \in \mathbb{R}$, define

$$\mathcal{W}_\psi f(\alpha, \beta) := \langle f, \psi_{\alpha^{-1};\alpha^{-1}\beta} \rangle = \int_{\mathbb{R}} f(x)|\alpha|^{-1/2}\overline{\psi\Big(\frac{x-\beta}{\alpha}\Big)}dx. \tag{4.3.26}$$

Theorem 4.3.15 *Let $\psi, \eta \in L_2(\mathbb{R})$ such that*

$$C_\psi := \int_{\mathbb{R}} \frac{|\widehat{\psi}(\xi)|^2}{|\xi|} d\xi < \infty, \qquad C_\eta := \int_{\mathbb{R}} \frac{|\widehat{\eta}(\xi)|^2}{|\xi|} d\xi < \infty.$$

Then $\|\mathcal{W}_\psi f\|^2 := \langle \mathcal{W}_\psi f, \mathcal{W}_\psi f\rangle = C_\psi \|f\|^2_{L_2(\mathbb{R})}$ *and for all* $f, g \in L_2(\mathbb{R})$,

$$\langle \mathcal{W}_\psi f, \mathcal{W}_\eta g\rangle := \int_{\mathbb{R}}\int_{\mathbb{R}} \mathcal{W}_\psi f(\alpha,\beta)\overline{\mathcal{W}_\eta g(\alpha,\beta)}d\beta\frac{d\alpha}{\alpha^2} = C_{\psi,\eta}\langle f, g\rangle_{L_2(\mathbb{R})},$$

where $C_{\psi,\eta} := \int_{\mathbb{R}} \frac{\overline{\widehat{\psi}(\xi)}\widehat{\eta}(\xi)}{|\xi|}d\xi < \infty$. *In particular, if* $C_{\psi,\eta} \neq 0$, *then*

$$f(\cdot) = \frac{1}{C_{\psi,\eta}}\int_{\mathbb{R}}\int_{\mathbb{R}} \mathcal{W}_\psi f(\alpha,\beta)\eta_{\alpha^{-1};\alpha^{-1}\beta}(\cdot)d\beta\frac{d\alpha}{\alpha^2} \tag{4.3.27}$$

holds in the weak sense, that is,

$$\langle f, g\rangle = \frac{1}{C_{\psi,\eta}}\int_{\mathbb{R}}\int_{\mathbb{R}} \mathcal{W}_\psi f(\alpha,\beta)\langle \eta_{\alpha^{-1};\alpha^{-1}\beta}, g\rangle d\beta\frac{d\alpha}{\alpha^2}, \qquad \forall\, g \in L_2(\mathbb{R}).$$

Proof By Plancherel's Theorem, we have

$$\mathcal{W}_\psi f(\alpha,\beta) = \frac{1}{2\pi}\langle \widehat{f}, \widehat{\psi}_{\alpha;0,\alpha^{-1}\beta}\rangle = \frac{\sqrt{\alpha}}{2\pi}\int_{\mathbb{R}} \widehat{f}(\xi)\overline{\widehat{\psi}(\alpha\xi)}e^{i\beta\xi}d\xi = |\alpha|^{1/2}(\widehat{f}\overline{\widehat{\psi}(\alpha\cdot)})^\vee(\beta),$$

where $\vee$ is the inverse Fourier transform. Consequently,

$$\begin{aligned}\|\mathcal{W}_\psi f\|^2 &= \int_{\mathbb{R}} \|\mathcal{W}_\psi f(\alpha,\cdot)\|^2_{L_2(\mathbb{R})}\frac{d\alpha}{\alpha^2} = \frac{1}{2\pi}\int_{\mathbb{R}}\int_{\mathbb{R}} |\widehat{f}(\xi)\overline{\widehat{\psi}(\alpha\xi)}|^2 d\xi\frac{d\alpha}{|\alpha|}\\ &= \frac{1}{2\pi}\int_{\mathbb{R}}\int_{\mathbb{R}} |\widehat{f}(\xi)|^2|\widehat{\psi}(\alpha\xi)|^2\frac{d\alpha}{|\alpha|}d\xi = \frac{C_\psi}{2\pi}\int_{\mathbb{R}} |\widehat{f}(\xi)|^2 d\xi = C_\psi\|f\|^2_{L_2(\mathbb{R})} < \infty,\end{aligned}$$

where we used the Fubini-Tonelli Theorem in the third identity. Therefore,

$$\begin{aligned}\langle \mathcal{W}_\psi f, \mathcal{W}_\eta g\rangle &= \int_{\mathbb{R}} \langle \mathcal{W}_\psi f(\alpha,\cdot), \mathcal{W}_\eta g(\alpha,\cdot)\rangle\frac{d\alpha}{\alpha^2} = \frac{1}{2\pi}\int_{\mathbb{R}} \langle \widehat{f}\,\overline{\widehat{\psi}(\alpha\cdot)}, \widehat{g}\,\overline{\widehat{\eta}(\alpha\cdot)}\rangle\frac{d\alpha}{|\alpha|}\\ &= \frac{1}{2\pi}\int_{\mathbb{R}}\int_{\mathbb{R}} \widehat{f}(\xi)\overline{\widehat{g}(\xi)}\,\overline{\widehat{\psi}(\alpha\xi)}\widehat{\eta}(\alpha\xi)d\xi\frac{d\alpha}{|\alpha|}\\ &= \frac{1}{2\pi}\int_{\mathbb{R}}\int_{\mathbb{R}} \widehat{f}(\xi)\overline{\widehat{g}(\xi)}\,\overline{\widehat{\psi}(\alpha\xi)}\widehat{\eta}(\alpha\xi)\frac{d\alpha}{|\alpha|}d\xi\\ &= \frac{C_{\psi,\eta}}{2\pi}\int_{\mathbb{R}} \widehat{f}(\xi)\overline{\widehat{g}(\xi)}d\xi = C_{\psi,\eta}\langle f, g\rangle,\end{aligned}$$

where we used the Fubini-Tonelli Theorem again in the fourth identity. Therefore, we proved (4.3.27). □

For example, the condition $C_\psi < \infty$ holds if $\psi(x) = G''(x) = \frac{1}{\sqrt{2\pi}}(x^2 - 1)e^{-x^2/2}$, the second-order derivative of the standard Gaussian function $G(x) = \frac{1}{\sqrt{2\pi}}e^{-x^2/2}$.

4.4 Shift-Invariant Subspaces of $L_2(\mathbb{R})$

As we have seen before, an affine system is generated by dyadic dilations and integer shifts of generating functions. Therefore, it is important to separately investigate a subspace generated by integer shifts of functions from a subset of $L_2(\mathbb{R})$.

In this section we introduce shift-invariant subspaces of $L_2(\mathbb{R})$ and study their basic properties. The results developed in this section for shift-invariant spaces of $L_2(\mathbb{R})$ are mainly for the study of the well-known concept of a multiresolution analysis of $L_2(\mathbb{R})$ in Sect. 4.5.3.

A closed subspace $\mathscr{V}$ of $L_2(\mathbb{R})$ is said to be *shift-invariant* if $f(\cdot - k) \in \mathscr{V}$ for all $f \in \mathscr{V}$ and $k \in \mathbb{Z}$. For a subset $\Phi \subseteq L_2(\mathbb{R})$, we define the shift-invariant subspace $\mathsf{S}(\Phi|L_2(\mathbb{R}))$ generated by Φ as follows:

$$\mathsf{S}(\Phi|L_2(\mathbb{R})) := \overline{\operatorname{span}\{\phi(\cdot - k) \; : \; k \in \mathbb{Z}, \phi \in \Phi\}}, \qquad (4.4.1)$$

where the above bar refers to the closure of the set $\operatorname{span}\{\phi(\cdot - k) \; : \; k \in \mathbb{Z}, \phi \in \Phi\}$ in $L_2(\mathbb{R})$ and span refers to all finite linear combinations of elements in the set. It is easy to see that $\mathsf{S}(\Phi|L_2(\mathbb{R}))$ is the smallest shift-invariant space containing Φ. If $\Phi = \{\phi\}$ is a singleton, then we simply use $\mathsf{S}(\phi|L_2(\mathbb{R}))$ for $\mathsf{S}(\{\phi\}|L_2(\mathbb{R}))$ in (4.4.1).

The following properties of the bracket product in (4.1.5) will be needed later.

Lemma 4.4.1 *For $f, g \in L_2(\mathbb{R})$, the bracket product $[\widehat{f}, \widehat{g}] \in L_1(\mathbb{T})$ and its Fourier series is $\sum_{k\in\mathbb{Z}} \langle f, g(\cdot - k)\rangle e^{-ik\xi}$. In particular,*

(i) $\langle f, g(\cdot - k)\rangle = 0$ for all $k \in \mathbb{Z}$ if and only if $[\widehat{f}, \widehat{g}](\xi) = 0$ for a.e. $\xi \in \mathbb{R}$.
(ii) $\langle f, g(\cdot - k)\rangle = \boldsymbol{\delta}(k)$ for all $k \in \mathbb{Z}$ if and only if $[\widehat{f}, \widehat{g}](\xi) = 1$ for a.e. $\xi \in \mathbb{R}$.
(iii) If in addition $[\widehat{f}, \widehat{f}] \in L_\infty(\mathbb{T})$, then $\sum_{k\in\mathbb{Z}} |\langle f, g(\cdot - k)\rangle|^2 = \frac{1}{2\pi}\int_{-\pi}^{\pi} |[\widehat{f}, \widehat{g}](\xi)|^2 d\xi$.

Proof Since $f, g \in L_2(\mathbb{R})$, by $|[\widehat{f}, \widehat{g}](\xi)|^2 \leqslant [\widehat{f}, \widehat{f}](\xi)[\widehat{g}, \widehat{g}](\xi)$, we have

$$\int_{-\pi}^{\pi} |[\widehat{f}, \widehat{g}](\xi)| d\xi \leqslant \Big(\int_{-\pi}^{\pi} [\widehat{f}, \widehat{f}](\xi) d\xi\Big)^{1/2} \Big(\int_{-\pi}^{\pi} [\widehat{g}, \widehat{g}](\xi) d\xi\Big)^{1/2} = \|\widehat{f}\|_{L_2(\mathbb{R})} \|\widehat{g}\|_{L_2(\mathbb{R})}.$$

Therefore, $[\widehat{f}, \widehat{g}] \in L_1(\mathbb{T})$. We now calculate its Fourier coefficients:

$$\frac{1}{2\pi}\int_{-\pi}^{\pi} [\widehat{f}, \widehat{g}](\xi) e^{ik\xi} d\xi = \frac{1}{2\pi}\int_{\mathbb{R}} \widehat{f}(\xi)\overline{\widehat{g}(\xi)} e^{ik\xi} d\xi = \frac{1}{2\pi}\langle \widehat{f}, \widehat{g(\cdot - k)}\rangle = \langle f, g(\cdot - k)\rangle,$$

where we used $\widehat{g(\cdot - k)}(\xi) = \overline{\widehat{g}(\xi)}e^{ik\xi}$ and Plancherel's Theorem. Because $[\widehat{f}, \widehat{g}]$ has the Fourier series $\sum_{k\in\mathbb{Z}}\langle f, g(\cdot - k)\rangle e^{-ik\xi}$, items (i) and (ii) hold. If $[\widehat{f}, \widehat{f}] \in L_\infty(\mathbb{T})$, then $[\widehat{f}, \widehat{g}] \in L_2(\mathbb{T})$. Item (iii) is a direct consequence of Parseval's identity. □

4.4.1 Principal Shift-Invariant Spaces

A shift-invariant space $S(\phi|L_2(\mathbb{R}))$ with a single generator is called *a principal shift-invariant space*. We now study some basic properties of principal shift-invariant subspaces of $L_2(\mathbb{R})$.

Theorem 4.4.2 *Let $\phi \in L_2(\mathbb{R})$ and $\mathcal{P} : L_2(\mathbb{R}) \to S(\phi|L_2(\mathbb{R}))$ be the orthogonal projection from $L_2(\mathbb{R})$ to $S(\phi|L_2(\mathbb{R}))$. Then $\widehat{\mathcal{P}f}(\xi) = \mathbf{u}_f(\xi)\widehat{\phi}(\xi)$ for almost every $\xi \in \mathbb{R}$, where $\mathbf{u}_f : \mathbb{R} \to \mathbb{C}$ is a 2π-periodic measurable function defined by*

$$\mathbf{u}_f(\xi) := \frac{[\widehat{f}, \widehat{\phi}](\xi)}{[\widehat{\phi}, \widehat{\phi}](\xi)}, \quad 0 < [\widehat{\phi}, \widehat{\phi}](\xi) < \infty \quad \textit{and} \quad \mathbf{u}_f(\xi) = 0, \quad \textit{otherwise}. \tag{4.4.2}$$

Proof By $|[\widehat{f}, \widehat{\phi}]|^2 \leqslant [\widehat{f}, \widehat{f}][\widehat{\phi}, \widehat{\phi}]$, we have $|\mathbf{u}_f(\xi)|^2[\widehat{\phi}, \widehat{\phi}](\xi) \leqslant [\widehat{f}, \widehat{f}](\xi)$ and

$$\begin{aligned}\int_{\mathbb{R}} |\mathbf{u}_f(\xi)\widehat{\phi}(\xi)|^2 d\xi &= \int_{-\pi}^{\pi} |\mathbf{u}_f(\xi)|^2[\widehat{\phi}, \widehat{\phi}](\xi) d\xi \leqslant \int_{-\pi}^{\pi} [\widehat{f}, \widehat{f}](\xi) d\xi \\ &= \int_{\mathbb{R}} |\widehat{f}(\xi)|^2 d\xi = 2\pi \|f\|^2_{L_2(\mathbb{R})} < \infty.\end{aligned}$$

Hence, there exists $g \in L_2(\mathbb{R})$ such that $\widehat{g} = \mathbf{u}_f\widehat{\phi}$. To prove $g \in S(\phi|L_2(\mathbb{R}))$, it suffices to prove $\langle g, h\rangle = 0$ for all $h \in S(\phi|L_2(\mathbb{R}))^\perp$. By Lemma 4.4.1, $h \in S(\phi|L_2(\mathbb{R}))^\perp$ if and only if $[\widehat{\phi}, \widehat{h}](\xi) = 0$ for a.e. $\xi \in \mathbb{R}$. Thus, $[\widehat{g}, \widehat{h}] = [\mathbf{u}_f\widehat{\phi}, \widehat{h}] = \mathbf{u}_f[\widehat{\phi}, \widehat{h}] = 0$. By Lemma 4.4.1, $\langle g, h\rangle = 0$. Hence, $g \in S(\phi|L_2(\mathbb{R}))$. Next we show $\langle f - g, \phi(\cdot - k)\rangle = 0$ for all $k \in \mathbb{Z}$. By Lemma 4.4.1, it suffices to prove $[\widehat{f} - \widehat{g}, \widehat{\phi}] = 0$. Note that

$$[\widehat{f} - \widehat{g}, \widehat{\phi}] = [\widehat{f}, \widehat{\phi}] - [\widehat{g}, \widehat{\phi}] = [\widehat{f}, \widehat{\phi}] - |\mathbf{u}_f|^2[\widehat{\phi}, \widehat{\phi}].$$

Note that $[\widehat{\phi}, \widehat{\phi}](\xi) < \infty$ for a.e. $\xi \in \mathbb{R}$ by $[\widehat{\phi}, \widehat{\phi}] \in L_1(\mathbb{T})$. If $[\widehat{\phi}, \widehat{\phi}](\xi) \neq 0$, by the definition of $\mathbf{u}_f(\xi)$, we have

$$[\widehat{f} - \widehat{g}, \widehat{\phi}](\xi) = [\widehat{f}, \widehat{\phi}](\xi) - |\mathbf{u}_f(\xi)|^2[\widehat{\phi}, \widehat{\phi}](\xi) = 0.$$

If $[\widehat{\phi}, \widehat{\phi}](\xi) = 0$ and $[\widehat{f}, \widehat{f}](\xi) < \infty$, by $|[\widehat{f}, \widehat{\phi}](\xi)|^2 \leqslant [\widehat{f}, \widehat{f}](\xi)[\widehat{\phi}, \widehat{\phi}](\xi)$, we must have $[\widehat{f}, \widehat{\phi}](\xi) = 0$. Since $\int_{-\pi}^{\pi} [\widehat{f}, \widehat{f}](\xi) d\xi = \int_{\mathbb{R}} |\widehat{f}(\xi)|^2 d\xi < \infty$, we have $[\widehat{f}, \widehat{f}](\xi) <$

∞ for a.e. $\xi \in \mathbb{R}$. So, $[\widehat{f}-\widehat{g},\widehat{\phi}](\xi) = 0$ a.e. $\xi \in \mathbb{R}$. By Lemma 4.4.1, $\langle f-g, \phi(\cdot - k)\rangle = 0$ for all $k \in \mathbb{Z}$. Therefore, $(f-g) \perp \mathrm{S}(\phi|L_2(\mathbb{R}))$. This proves $\mathcal{P}f = g$. $\square$

We now characterize a principal shift-invariant space $\mathrm{S}(\phi|L_2(\mathbb{R}))$.

Theorem 4.4.3 *Let $\phi \in L_2(\mathbb{R})$. Then $f \in \mathrm{S}(\phi|L_2(\mathbb{R}))$ if and only if $f \in L_2(\mathbb{R})$ and there exists a 2π-periodic measurable function $\mathbf{u}: \mathbb{R} \to \mathbb{C}$ such that $\widehat{f}(\xi) = \mathbf{u}(\xi)\widehat{\phi}(\xi)$ for almost every $\xi \in \mathbb{R}$.*

Proof If $f \in \mathrm{S}(\phi|L_2(\mathbb{R}))$, then $\mathcal{P}f = f$ and by Theorem 4.4.2, we have $\widehat{f} = \widehat{\mathcal{P}f} = \mathbf{u}_f\widehat{\phi}$, where $\mathbf{u}_f$ is a 2π-periodic measurable function defined in (4.4.2). Conversely, suppose that $f \in L_2(\mathbb{R})$ and $\widehat{f} = \mathbf{u}\widehat{\phi}$ for some 2π-periodic measurable function $\mathbf{u}$. For $0 < [\widehat{\phi},\widehat{\phi}](\xi) < \infty$, by Theorem 4.4.2 and $\widehat{f} = \mathbf{u}\widehat{\phi}$, we have

$$\mathbf{u}_f(\xi) = \frac{[\widehat{f},\widehat{\phi}](\xi)}{[\widehat{\phi},\widehat{\phi}](\xi)} = \frac{[\mathbf{u}\widehat{\phi},\widehat{\phi}](\xi)}{[\widehat{\phi},\widehat{\phi}](\xi)} = \frac{\mathbf{u}(\xi)[\widehat{\phi},\widehat{\phi}](\xi)}{[\widehat{\phi},\widehat{\phi}](\xi)} = \mathbf{u}(\xi)$$

and $\widehat{\mathcal{P}f}(\xi) = \mathbf{u}_f(\xi)\widehat{\phi}(\xi) = \mathbf{u}(\xi)\widehat{\phi}(\xi) = \widehat{f}(\xi)$. When $[\widehat{\phi},\widehat{\phi}](\xi) = 0$, it is trivial to see that the above identities still hold. Since $[\widehat{\phi},\widehat{\phi}] \in L_1(\mathbb{T})$, we have $[\widehat{\phi},\widehat{\phi}] < \infty$ a.e. Hence, $\widehat{\mathcal{P}f}(\xi) = \widehat{f}(\xi)$ for a.e. $\xi \in \mathbb{R}$ and consequently $f = \mathcal{P}f \in \mathrm{S}(\phi|L_2(\mathbb{R}))$. $\square$

To study frames and bases in a principal shift-invariant space, we need the following result.

Proposition 4.4.4 *Let $\phi \in L_2(\mathbb{R})$ and C be a positive constant.*

(i) $C[\widehat{\phi},\widehat{\phi}](\xi) \leqslant |[\widehat{\phi},\widehat{\phi}](\xi)|^2$ for a.e. $\xi \in \mathbb{R}$ (i.e., $C \leqslant [\widehat{\phi},\widehat{\phi}](\xi)$ for a.e. $\xi \in \operatorname{supp}([\widehat{\phi},\widehat{\phi}])$) if and only if $C\|f\|^2_{L_2(\mathbb{R})} \leqslant \sum_{k\in\mathbb{Z}} |\langle f, \phi(\cdot - k)\rangle|^2$ for all $f \in \mathrm{S}(\phi|L_2(\mathbb{R}))$.

(ii) $[\widehat{\phi},\widehat{\phi}](\xi) \leqslant C$ for a.e. $\xi \in \mathbb{R}$ if and only if $\sum_{k\in\mathbb{Z}} |\langle f, \phi(\cdot - k)\rangle|^2 \leqslant C\|f\|^2_{L_2(\mathbb{R})}$ for all $f \in \mathrm{S}(\phi|L_2(\mathbb{R}))$ (or for all $f \in L_2(\mathbb{R})$).

Proof By Theorem 4.4.3, we see that $f \in \mathrm{S}(\phi|L_2(\mathbb{R}))$ if and only if $f \in L_2(\mathbb{R})$ and $\widehat{f} = \mathbf{u}\widehat{\phi}$ for some 2π-periodic measurable function $\mathbf{u}$. By Lemma 4.4.1,

$$2\pi \sum_{k\in\mathbb{Z}} |\langle f, \phi(\cdot - k)\rangle|^2 = \int_{-\pi}^{\pi} |[\widehat{f},\widehat{\phi}](\xi)|^2 d\xi = \int_{-\pi}^{\pi} |\mathbf{u}(\xi)|^2 |[\widehat{\phi},\widehat{\phi}](\xi)|^2 d\xi. \tag{4.4.3}$$

(i) $\Rightarrow$. By (4.4.3) and $|[\widehat{\phi},\widehat{\phi}](\xi)|^2 \geqslant C[\widehat{\phi},\widehat{\phi}](\xi)$, we have

$$\begin{aligned}\sum_{k\in\mathbb{Z}} |\langle f, \phi(\cdot - k)\rangle|^2 &= \frac{1}{2\pi}\int_{-\pi}^{\pi} |\mathbf{u}(\xi)|^2 |[\widehat{\phi},\widehat{\phi}](\xi)|^2 d\xi \\ &\geqslant \frac{C}{2\pi}\int_{-\pi}^{\pi} |\mathbf{u}(\xi)|^2 [\widehat{\phi},\widehat{\phi}](\xi) d\xi = \frac{C}{2\pi}\int_{\mathbb{R}} |\widehat{f}(\xi)|^2 d\xi = C\|f\|^2_{L_2(\mathbb{R})}.\end{aligned}$$

(i) $\Leftarrow$. Suppose that the claim $C[\widehat{\phi},\widehat{\phi}](\xi) \leqslant |[\widehat{\phi},\widehat{\phi}](\xi)|^2$ for a.e. $\xi \in \mathbb{R}$ fails. Then there exist $0 < C' < C$ and a measurable set $E \subseteq (-\pi,\pi]$ such that E has a positive Lebesgue measure and $0 < [\widehat{\phi},\widehat{\phi}](\xi) \leqslant C'$ for a.e. $\xi \in E$. Take $\widehat{f} := \mathbf{u}\widehat{\phi}$ with a 2π-periodic function $\mathbf{u}$ being defined by $\mathbf{u}(\xi) := \chi_E(\xi), \xi \in [-\pi,\pi)$. Then $\|\widehat{f}\|^2_{L_2(\mathbb{R})} = \int_E |[\widehat{\phi},\widehat{\phi}](\xi)|^2 d\xi > 0$ and by (4.4.3),

$$\sum_{k\in\mathbb{Z}} |\langle f, \phi(\cdot - k)\rangle|^2 = \frac{1}{2\pi}\int_{-\pi}^{\pi} |\mathbf{u}(\xi)|^2 |[\widehat{\phi},\widehat{\phi}](\xi)|^2 d\xi$$

$$\leqslant \frac{C'}{2\pi}\int_{-\pi}^{\pi} \chi_E(\xi)[\widehat{\phi},\widehat{\phi}](\xi) = C'\|f\|^2_{L_2(\mathbb{R})} < C\|f\|^2_{L_2(\mathbb{R})},$$

which is a contradiction to our assumption. Hence, we must have $C[\widehat{\phi},\widehat{\phi}](\xi) \leqslant |[\widehat{\phi},\widehat{\phi}](\xi)|^2$ for a.e. $\xi \in \mathbb{R}$.

The claim in item (ii) can be proved similarly. □

We characterize frames and bases in principal shift-invariant subspaces of $L_2(\mathbb{R})$.

Theorem 4.4.5 *Let $\phi \in L_2(\mathbb{R})$. Then the following statements are equivalent:*

(1) $\{\phi(\cdot - k) : k \in \mathbb{Z}\}$ is a frame for $\mathrm{S}(\phi|L_2(\mathbb{R}))$ (or equivalently, $\{\phi(\cdot - k) : k \in \mathbb{Z}\}$ is a frame sequence in $L_2(\mathbb{R})$), i.e., there exist positive constants C_1, C_2 such that

$$C_1\|f\|^2_{L_2(\mathbb{R})} \leqslant \sum_{k\in\mathbb{Z}} |\langle f, \phi(\cdot - k)\rangle|^2 \leqslant C_2\|f\|^2_{L_2(\mathbb{R})}, \quad \forall\, f \in \mathrm{S}(\phi|L_2(\mathbb{R})). \tag{4.4.4}$$

(2) $C_1 \leqslant [\widehat{\phi},\widehat{\phi}](\xi) \leqslant C_2$ for a.e. $\xi \in \operatorname{supp}([\widehat{\phi},\widehat{\phi}])$, or equivalently, $C_1[\widehat{\phi},\widehat{\phi}](\xi) \leqslant |[\widehat{\phi},\widehat{\phi}](\xi)|^2 \leqslant C_2[\widehat{\phi},\widehat{\phi}](\xi)$ for a.e. $\xi \in \mathbb{R}$.

(3) The operator $\mathcal{V} : l_2(\phi) \to \mathrm{S}(\phi|L_2(\mathbb{R}))$ with $\{v(k)\}_{k\in\mathbb{Z}} \mapsto \sum_{k\in\mathbb{Z}} v(k)\phi(\cdot - k)$ is a well-defined bounded and bijective linear operator with a bounded inverse, where

$$l_2(\phi) := \{v \in l_2(\mathbb{Z}) : \widehat{v}(\xi) = 0 \ \ for \ \ a.e.\, \xi \in \mathbb{R}\backslash \operatorname{supp}([\widehat{\phi},\widehat{\phi}])\}. \tag{4.4.5}$$

(4) The operator $\mathcal{W} : \mathrm{S}(\phi|L_2(\mathbb{R})) \to l_2(\phi)$ with $f \mapsto \{\langle f, \phi(\cdot - k)\rangle\}_{k\in\mathbb{Z}}$ is a well-defined bounded and bijective linear operator with a bounded inverse.

Proof (1) $\Longleftrightarrow$ (2) has been proved in Proposition 4.4.4.

(2)$\Longrightarrow$(3). Since $\{\phi(\cdot - k) : k \in \mathbb{Z}\}$ is a frame for $\mathrm{S}(\phi|L_2(\mathbb{R}))$, it follows from Theorem 4.2.4 that $\mathcal{V}$ is a well-defined bounded linear operator. We now prove that $\mathcal{V}$ is bijective. Suppose that $\mathcal{V}u = \mathcal{V}v$ for $u, v \in l_2(\phi)$. Then $\widehat{u}(\xi)\widehat{\phi}(\xi) = \widehat{v}(\xi)\widehat{\phi}(\xi)$ for almost every $\xi \in \mathbb{R}$. Hence, $\widehat{u}(\xi)[\widehat{\phi},\widehat{\phi}](\xi) = [\widehat{u}\widehat{\phi},\widehat{\phi}](\xi) = [\widehat{v}\widehat{\phi},\widehat{\phi}] = \widehat{v}(\xi)[\widehat{\phi},\widehat{\phi}](\xi)$ for almost every $\xi \in \mathbb{R}$ from which we conclude that $\widehat{u}(\xi) = \widehat{v}(\xi)$ for a.e. $\xi \in \operatorname{supp}([\widehat{\phi},\widehat{\phi}])$. Since $u, v \in l_2(\phi)$, we must have $u = v$. Thus, $\mathcal{V}$ is one-to-one. For $f \in \mathrm{S}(\phi|L_2(\mathbb{R}))$, by Theorem 4.4.3 and Proposition 4.4.2, we have

$\widehat{f} = \mathbf{u}_f \widehat{\phi}$ for some 2π-periodic measurable function $\mathbf{u}_f$ such that $\mathbf{u}_f(\xi) = 0$ for a.e. $\xi \in \mathbb{R} \setminus \operatorname{supp}([\widehat{\phi}, \widehat{\phi}])$. Since $C_1 \leqslant [\widehat{\phi}, \widehat{\phi}](\xi) \leqslant C_2$ for almost every $\xi \in \operatorname{supp}([\widehat{\phi}, \widehat{\phi}])$, by $\mathbf{u}_f(\xi) = \frac{[\widehat{f}, \widehat{\phi}](\xi)}{[\widehat{\phi}, \widehat{\phi}](\xi)}$ for a.e. $\xi \in \operatorname{supp}([\widehat{\phi}, \widehat{\phi}])$, we have

$$|\mathbf{u}_f(\xi)|^2 \leqslant C_1^{-2} |[\widehat{f}, \widehat{\phi}](\xi)|^2 \leqslant C_1^{-2} [\widehat{f}, \widehat{f}](\xi) [\widehat{\phi}, \widehat{\phi}](\xi) \leqslant C_1^{-2} C_2 [\widehat{f}, \widehat{f}](\xi).$$

So, $\int_{-\pi}^{\pi} |\mathbf{u}_f(\xi)|^2 d\xi \leqslant C_1^{-2} C_2 \int_{-\pi}^{\pi} [\widehat{f}, \widehat{f}](\xi) = C_1^{-2} C_2 \int_{\mathbb{R}} |\widehat{f}(\xi)|^2 d\xi < \infty$. This shows that $\mathbf{u}_f \in L_2(\mathbb{T})$. Therefore, there exists a unique sequence $u \in l_2(\phi)$ such that $\widehat{u} = \mathbf{u}$. This proves that $\mathcal{V}$ is onto. Since $l_2(\phi)$ is a closed subspace of $l_2(\mathbb{Z})$ and $\mathcal{V}$ is a bijective bounded linear operator, by the Open Mapping Theorem (see Theorem A.1.1), the linear operator $\mathcal{V}$ has a bounded inverse.

(3) $\iff$ (4) follows directly from $\mathcal{W} = \mathcal{V}^\star$. To prove (4)$\Longrightarrow$(1), it is trivial to see from item (4) that (4.4.4) holds with $C_1 = \|\mathcal{W}^{-1}\|^{-2}$ and $C_2 = \|\mathcal{W}\|^2$. □

Theorem 4.4.6 *Let $\phi \in L_2(\mathbb{R})$. Then the following statements are equivalent:*

(1) $\{\phi(\cdot - k) : k \in \mathbb{Z}\}$ is a Riesz basis for $\mathrm{S}(\phi|L_2(\mathbb{R}))$ (namely, $\{\phi(\cdot - k) : k \in \mathbb{Z}\}$ is a Riesz sequence in $L_2(\mathbb{R})$): there exist positive constants C_1 and C_2 such that

$$C_1 \sum_{k\in\mathbb{Z}} |v(k)|^2 \leqslant \Big\| \sum_{k\in\mathbb{Z}} v(k)\phi(\cdot - k) \Big\|_{L_2(\mathbb{R})}^2 \leqslant C_2 \sum_{k\in\mathbb{Z}} |v(k)|^2 \qquad (4.4.6)$$

for all finitely supported sequences $v \in l_0(\mathbb{Z})$.

(2) $C_1 \leqslant [\widehat{\phi}, \widehat{\phi}](\xi) \leqslant C_2$ for almost every $\xi \in \mathbb{R}$.

(3) The synthesis operator $\mathcal{V} : l_2(\mathbb{Z}) \to \mathrm{S}(\phi|L_2(\mathbb{R}))$ with $\{v(k)\}_{k\in\mathbb{Z}} \mapsto \sum_{k\in\mathbb{Z}} v(k)\phi(\cdot - k)$ is a well-defined bounded and bijective operator with a bounded inverse.

(4) The analysis operator $\mathcal{W} : \mathrm{S}(\phi|L_2(\mathbb{R})) \to l_2(\mathbb{Z})$ with $f \mapsto \{\langle f, \phi(\cdot - k)\rangle\}_{k\in\mathbb{Z}}$ is a well-defined bounded and bijective linear operator with a bounded inverse.

Proof By the definition of $\mathrm{S}(\phi|L_2(\mathbb{R}))$, the set of all finite linear combinations of $\phi(\cdot - k), k \in \mathbb{Z}$ is dense in $\mathrm{S}(\phi|L_2(\mathbb{R}))$. Hence, if (4.4.6) is satisfied, then $\{\phi(\cdot - k) : k \in \mathbb{Z}\}$ is a Riesz basis for $\mathrm{S}(\phi|L_2(\mathbb{R}))$. By Theorem 4.2.7, $\{\phi(\cdot - k) : k \in \mathbb{Z}\}$ is a Riesz basis for $\mathrm{S}(\phi|L_2(\mathbb{R}))$ if and only if $\{\phi(\cdot - k) : k \in \mathbb{Z}\}$ is a frame for $\mathrm{S}(\phi|L_2(\mathbb{R}))$ and $\mathcal{W} : \mathrm{S}(\phi|L_2(\mathbb{R})) \to l_2(\mathbb{Z})$ is a well-defined bounded and bijective linear operator. By Theorem 4.4.5, these two conditions are equivalent to that $\{\phi(\cdot - k) : k \in \mathbb{Z}\}$ is a frame for $\mathrm{S}(\phi|L_2(\mathbb{R}))$ and $l_2(\phi) = l_2(\mathbb{Z})$, where the latter condition is further equivalent to the condition that $\mathbb{R} \setminus \operatorname{supp}([\widehat{\phi}, \widehat{\phi}])$ has measure zero. Now all the claims follow directly from Theorem 4.4.5. □

4.4.2 Finitely Generated Shift-Invariant Spaces

We now study finitely or countably generated shift-invariant spaces $\mathrm{S}(\Phi|L_2(\mathbb{R}))$ of $L_2(\mathbb{R})$. For two measurable functions f and g, if $g(\xi) \in \{0, \pm\infty\}$, without further mention we shall use the convention $(f/g)(\xi) := 0$ in this section. To facilitate the study of $\mathrm{S}(\Phi|L_2(\mathbb{R}))$, we obtain another generating set with some desired properties for $\mathrm{S}(\Phi|L_2(\mathbb{R}))$ through a standard procedure of orthogonalization.

Proposition 4.4.7 *Let $\Phi = \{\phi^1, \dots, \phi^r\}$ be a countable subset of $L_2(\mathbb{R})$. Define $\varphi^1, \dots, \varphi^r$ recursively by*

$$\widehat{\varphi^\ell}(\xi) := \frac{\widehat{\mathring{\phi}^\ell}(\xi)}{\sqrt{[\widehat{\mathring{\phi}^\ell}, \widehat{\mathring{\phi}^\ell}](\xi)}} \quad \text{with} \quad \widehat{\mathring{\phi}^\ell}(\xi) := \widehat{\phi^\ell}(\xi) - \sum_{j=1}^{\ell-1} [\widehat{\phi^\ell}, \widehat{\varphi^j}](\xi)\widehat{\varphi^j}(\xi), \tag{4.4.7}$$

for $\ell = 1, \dots, r$. Then $\varphi^1, \dots, \varphi^r \in L_2(\mathbb{R})$. Moreover,

(i) There exist $r \times r$ lower triangular matrices $\mathbf{u}, \mathbf{v} : \mathbb{R} \to \mathbb{C}^{r\times r}$ of 2π-periodic measurable functions such that $\widehat{\phi}(\xi) = \mathbf{u}(\xi)\widehat{\varphi}(\xi)$ and $\widehat{\varphi}(\xi) = \mathbf{v}(\xi)\widehat{\phi}(\xi)$ for a.e. $\xi \in \mathbb{R}$, where $\phi := (\phi^1, \dots, \phi^r)^\mathsf{T}$ and $\varphi := (\varphi^1, \dots, \varphi^r)^\mathsf{T}$.

(ii) $\|\varphi^j\|_{L_2(\mathbb{R})} \leqslant 1$, $[\widehat{\varphi^j}, \widehat{\varphi^j}] = \chi_{\mathrm{supp}([\widehat{\varphi^j}, \widehat{\varphi^j}])}$, and $[\widehat{\varphi^j}, \widehat{\varphi^\ell}] = 0$ for all $1 \leqslant j \neq \ell \leqslant r$.

(iii) $\mathrm{S}(\Phi|L_2(\mathbb{R})) = \mathrm{S}(\varphi^1|L_2(\mathbb{R})) \oplus \cdots \oplus \mathrm{S}(\varphi^r|L_2(\mathbb{R}))$, where $\oplus$ means the orthogonal sum in $L_2(\mathbb{R})$.

(iv) $\{\varphi^1(\cdot - k), \dots, \varphi^r(\cdot - k) \; : \; k \in \mathbb{Z}\}$ is a (normalized) tight frame for $\mathrm{S}(\Phi|L_2(\mathbb{R}))$.

Proof Note that $\mathrm{supp}([\widehat{\varphi^j}, \widehat{\varphi^j}]) = \mathrm{supp}([\widehat{\mathring{\phi}^j}, \widehat{\mathring{\phi}^j}])$ for all $j = 1, \dots, r$. We prove the claims by induction on $r \in \mathbb{N}$. Suppose $r = 1$. It is trivial that item (i) holds. Note that $[\widehat{\varphi^1}, \widehat{\varphi^1}] = \chi_{\mathrm{supp}([\widehat{\varphi^1}, \widehat{\varphi^1}])} \leqslant 1$. Hence, $\|\widehat{\varphi^1}\|^2_{L_2(\mathbb{R})} = \int_{-\pi}^{\pi} [\widehat{\varphi^1}, \widehat{\varphi^1}](\xi) d\xi \leqslant 2\pi$ and $\|\varphi^1\|^2_{L_2(\mathbb{R})} = \frac{1}{2\pi}\|\widehat{\varphi^1}\|^2_{L_2(\mathbb{R})} \leqslant 1$. So, item (ii) holds. Item (iii) is a direct consequence of Theorem 4.4.3. By $[\widehat{\varphi^1}, \widehat{\varphi^1}] = \chi_{\mathrm{supp}([\widehat{\varphi^1}, \widehat{\varphi^1}])}$ and Theorem 4.4.5, we conclude that $\{\varphi^1(\cdot - k) \; : \; k \in \mathbb{Z}\}$ is a tight frame for $\mathrm{S}(\phi^1|L_2(\mathbb{R}))$. Hence, item (iv) holds.

Suppose that items (i)–(iv) hold for $r = \ell - 1$. We now prove the claims for $r = \ell$. By the definition of $\widehat{\varphi^\ell}$ in (4.4.7), we observe that

$$\widehat{\phi^\ell}(\xi) = \sqrt{[\widehat{\mathring{\phi}^\ell}, \widehat{\mathring{\phi}^\ell}](\xi)}\widehat{\varphi^\ell} + \sum_{j=1}^{\ell-1} [\widehat{\phi^\ell}, \widehat{\varphi^j}](\xi)\widehat{\varphi^j}(\xi)$$

and

$$\widehat{\varphi^\ell}(\xi) = \frac{1}{\sqrt{[\widehat{\mathring{\phi}^\ell}, \widehat{\mathring{\phi}^\ell}](\xi)}}\widehat{\phi^\ell}(\xi) - \sum_{j=1}^{\ell-1} \frac{[\widehat{\phi^\ell}, \widehat{\varphi^j}](\xi)}{\sqrt{[\widehat{\mathring{\phi}^\ell}, \widehat{\mathring{\phi}^\ell}](\xi)}}\widehat{\varphi^j}(\xi).$$

By the induction hypothesis for $r = \ell - 1$ and the above two identities, it is straightforward to see that item (i) holds for $r = \ell$.

By Theorem 4.4.2 and $[\widehat{\varphi^j}, \widehat{\varphi^j}] = \chi_{\operatorname{supp}([\widehat{\varphi^j},\widehat{\varphi^j}])}$ for $j = 1, \dots, \ell - 1$, we see that $[\widehat{\phi^\ell}, \widehat{\varphi^j}](\xi)\widehat{\varphi^j}(\xi) = \widehat{\mathcal{P}_j\phi^\ell}(\xi)$ for all $j = 1, \dots, \ell - 1$, where $\mathcal{P}_j$ is the orthogonal projection from $L_2(\mathbb{R})$ to $\mathsf{S}(\varphi^j|L_2(\mathbb{R}))$. Hence, (4.4.7) implies

$$\mathring{\phi}^\ell = \phi^\ell - \sum_{j=1}^{\ell-1} \mathcal{P}_j\phi^\ell \in L_2(\mathbb{R}) \tag{4.4.8}$$

and we can directly check that $[\widehat{\mathring{\phi}^\ell}, \widehat{\varphi^j}] = 0$ for all $j = 1, \dots, \ell - 1$. Thus, we have $[\widehat{\varphi^\ell}, \widehat{\varphi^j}] = 0$ for all $j = 1, \dots, \ell - 1$ and $[\widehat{\varphi^\ell}, \widehat{\varphi^\ell}] = \chi_{\operatorname{supp}([\widehat{\mathring{\phi}^\ell},\widehat{\mathring{\phi}^\ell}])}$. Therefore, $[\widehat{\varphi^\ell}, \widehat{\varphi^\ell}] = \chi_{\operatorname{supp}([\widehat{\varphi^\ell},\widehat{\varphi^\ell}])}$. This proves item (ii).

By item (ii) and Lemma 4.4.1, we have $\mathsf{S}(\varphi^j|L_2(\mathbb{R})) \perp \mathsf{S}(\varphi^\ell|L_2(\mathbb{R}))$ for all $j \ne \ell$. By Theorem 4.4.3, we have $\mathsf{S}(\mathring{\phi}^\ell|L_2(\mathbb{R})) = \mathsf{S}(\varphi^\ell|L_2(\mathbb{R}))$. Now it follows from (4.4.8) that $\phi^\ell = \mathring{\phi}^\ell + \sum_{j=1}^{\ell-1} \mathcal{P}_j\phi^\ell \in \mathsf{S}(\varphi^1|L_2(\mathbb{R})) \oplus \cdots \oplus \mathsf{S}(\varphi^\ell|L_2(\mathbb{R}))$. Hence, we have $\mathsf{S}(\{\phi^1, \dots, \phi^\ell\}|L_2(\mathbb{R})) \subseteq \mathsf{S}(\varphi^1|L_2(\mathbb{R})) \oplus \cdots \oplus \mathsf{S}(\varphi^\ell|L_2(\mathbb{R}))$. Conversely, by (4.4.8) and

$$\mathcal{P}_j\phi^\ell \in \mathsf{S}(\varphi^1|L_2(\mathbb{R})) \oplus \cdots \oplus \mathsf{S}(\varphi^{\ell-1}|L_2(\mathbb{R})) = \mathsf{S}(\{\phi^1, \dots, \phi^{\ell-1}\}|L_2(\mathbb{R}))$$

(the induction hypothesis), we have $\mathring{\phi}^\ell = \phi^\ell - \sum_{j=1}^{\ell-1} \mathcal{P}_j\phi^\ell \in \mathsf{S}(\{\phi^1, \dots, \phi^\ell\}|L_2(\mathbb{R}))$. Consequently, $\mathsf{S}(\varphi^\ell|L_2(\mathbb{R})) = \mathsf{S}(\mathring{\phi}^\ell|L_2(\mathbb{R})) \subseteq \mathsf{S}(\{\phi^1, \dots, \phi^\ell\}|L_2(\mathbb{R}))$. Therefore, $\mathsf{S}(\varphi^1|L_2(\mathbb{R})) \oplus \cdots \oplus \mathsf{S}(\varphi^\ell|L_2(\mathbb{R})) \subseteq \mathsf{S}(\{\phi^1, \dots, \phi^\ell\}|L_2(\mathbb{R}))$. This proves item (iii). By Theorem 4.4.5, we see that $\{\varphi^\ell(\cdot - k) \; : \; k \in \mathbb{Z}\}$ is a tight frame for $\mathsf{S}(\varphi^\ell|L_2(\mathbb{R}))$. By the induction hypothesis, $\{\varphi^1(\cdot - k), \dots, \varphi^{\ell-1}(\cdot - k) \; : \; k \in \mathbb{Z}\}$ is a tight frame for $\mathsf{S}(\{\varphi^1, \dots, \varphi^{\ell-1}\}|L_2(\mathbb{R}))$. Since $\mathsf{S}(\{\varphi^1, \dots, \varphi^{\ell-1}\}|L_2(\mathbb{R})) \perp \mathsf{S}(\varphi^\ell|L_2(\mathbb{R}))$, we conclude that item (iv) holds. The proof is now completed by induction. □

Let $\varphi^1, \dots, \varphi^r$ be defined in (4.4.7) and be derived from $\Phi = \{\phi^1, \dots, \phi^r\}$. We define the *dimension function* and *length* of a shift-invariant space $\mathsf{S}(\Phi|L_2(\mathbb{R}))$ to be

$$\begin{aligned} \dim_{\mathsf{S}(\Phi|L_2(\mathbb{R}))}(\xi) &:= \dim_\Phi(\xi) := \sum_{\ell=1}^{r} [\widehat{\varphi^\ell}, \widehat{\varphi^\ell}](\xi), \\ \operatorname{len}(\mathsf{S}(\Phi|L_2(\mathbb{R}))) &:= \| \dim_\Phi(\cdot)\|_{L_\infty(\mathbb{T})}. \end{aligned} \tag{4.4.9}$$

For a shift-invariant subspace $\mathscr{V}$ of $L_2(\mathbb{R})$, since $L_2(\mathbb{R})$ is separable, there exists a countable subset $\Phi = \{\phi^1, \dots, \phi^r\} \subseteq L_2(\mathbb{R})$ such that $\mathscr{V} = \mathsf{S}(\Phi|L_2(\mathbb{R}))$. The following result shows that the definition of the dimension function $\dim_{\mathsf{S}(\Phi|L_2(\mathbb{R}))}$ is independent of the choice of a tight frame for $\mathsf{S}(\Phi|L_2(\mathbb{R}))$.

Proposition 4.4.8 *Let Φ and H be countable subsets of $L_2(\mathbb{R})$. Suppose that $\{\eta(\cdot - k) : k \in \mathbb{Z}, \eta \in H\}$ is a tight frame for $\mathrm{S}(\Phi|L_2(\mathbb{R}))$. Then $\dim_{\mathrm{S}(\Phi|L_2(\mathbb{R}))}(\xi) = \sum_{\eta\in H}[\widehat{\eta},\widehat{\eta}](\xi)$ for a.e. $\xi \in \mathbb{R}$.*

Proof Let $\mathring{\Phi}$ be obtained in (4.4.7) from Φ such that $\{\varphi(\cdot - k) : k \in \mathbb{Z}, \varphi \in \mathring{\Phi}\}$ is a tight frame for $\mathrm{S}(\Phi|L_2(\mathbb{R}))$ and $\mathrm{S}(\Phi|L_2(\mathbb{R})) = \oplus_{\varphi\in\mathring{\Phi}}\mathrm{S}(\varphi|L_2(\mathbb{R}))$. Since $\{\eta(\cdot - k) : k \in \mathbb{Z}, \eta \in H\}$ is a tight frame for $\mathrm{S}(\Phi|L_2(\mathbb{R}))$, for every $\varphi \in \mathring{\Phi}$, we have $\varphi = \sum_{\eta\in H}\sum_{k\in\mathbb{Z}}\langle\varphi, \eta(\cdot-k)\rangle\eta(\cdot-k)$ with the series converging unconditionally in $L_2(\mathbb{R})$. In terms of Fourier transform, this identity becomes $\widehat{\varphi} = \sum_{\eta\in H}[\widehat{\varphi},\widehat{\eta}]\widehat{\eta}$. Similarly, we have $\widehat{\eta} = \sum_{\varphi\in\mathring{\Phi}}[\widehat{\eta},\widehat{\varphi}]\widehat{\varphi}$. So, $[\widehat{\varphi},\widehat{\varphi}] = \sum_{\eta\in H}|[\widehat{\varphi},\widehat{\eta}]|^2$ and $[\widehat{\eta},\widehat{\eta}] = \sum_{\varphi\in\mathring{\Phi}}|[\widehat{\eta},\widehat{\varphi}]|^2$. Hence,

$$\sum_{\eta\in H}[\widehat{\eta},\widehat{\eta}](\xi) = \sum_{\eta\in H}\sum_{\varphi\in\mathring{\Phi}}|[\widehat{\eta},\widehat{\varphi}](\xi)|^2 = \sum_{\varphi\in\mathring{\Phi}}\sum_{\eta\in H}|[\widehat{\eta},\widehat{\varphi}](\xi)|^2 = \sum_{\varphi\in\mathring{\Phi}}|[\widehat{\varphi},\widehat{\varphi}](\xi)|^2,$$

which is equal to the dimension function $\dim_{\mathrm{S}(\Phi|L_2(\mathbb{R}))}(\xi)$. □

The generating set for $\mathrm{S}(\Phi|L_2(\mathbb{R}))$ obtained in Proposition 4.4.7 can be further improved with the extra nested property of supports as follows:

Proposition 4.4.9 *Let $\Phi = \{\phi^1, \dots, \phi^r\}$ be a countable subset of $L_2(\mathbb{R})$. Then there exist $\varphi^1, \dots, \varphi^r \in L_2(\mathbb{R})$ such that all items (i)–(iv) of Proposition 4.4.7 are satisfied and the following additional property holds:*

$$\mathrm{supp}([\widehat{\varphi^1},\widehat{\varphi^1}]) \supseteq \mathrm{supp}([\widehat{\varphi^2},\widehat{\varphi^2}]) \supseteq \cdots \supseteq \mathrm{supp}([\widehat{\varphi^r},\widehat{\varphi^r}]). \qquad (4.4.10)$$

Moreover, $\mathrm{supp}([\widehat{\varphi^j},\widehat{\varphi^j}]) = \{\xi \in \mathbb{R} : \dim_{\mathrm{S}(\Phi|L_2(\mathbb{R}))}(\xi) \geqslant j\}$ *for all $j = 1, \dots, r$, and consequently $\varphi^\ell = 0$ for all* $\mathrm{len}(\mathrm{S}(\Phi|L_2(\mathbb{R}))) < \ell \leqslant r$.

Proof We use the cut-and-paste technique to prove the claim. Let $\varphi^1, \dots, \varphi^r$ be given in Proposition 4.4.7. Define 2π-periodic sets $E_j := \mathrm{supp}([\widehat{\varphi^j},\widehat{\varphi^j}])$ for $j = 1, \dots, r$. Since E_j and $\mathbb{R}\backslash E_j$ form a disjoint partition of $\mathbb{R}$, overlaying all E_j and $\mathbb{R}\backslash E_j$ to cut the real line $\mathbb{R}$, we see that there exists a countable collection $\{F_k\}_{k\in\Lambda_0}$ of (smaller) measurable subsets such that (i) all χ_{F_k} are 2π-periodic, (ii) $\{F_k\}_{k\in\Lambda_0}$ forms a disjoint partition (up to a set of measure zero of their intersection sets) of $\mathbb{R}$, and (iii) for each $1 \leqslant j \leqslant r$, there is a subset Λ_j of Λ_0 such that E_j is a disjoint union of $F_k, k \in \Lambda_j$. Define $\varphi^{j,k}$ by $\widehat{\varphi^{j,k}} := \chi_{F_k}\widehat{\varphi^j}$ for $k \in \Lambda_j$. By Theorem 4.4.3 it is trivial to verify that $\mathrm{S}(\varphi^j|L_2(\mathbb{R})) = \oplus_{k\in\Lambda_j}\mathrm{S}(\varphi^{j,k}|L_2(\mathbb{R}))$. Consequently, we have $\mathrm{S}(\Phi|L_2(\mathbb{R})) = \oplus_{n\in\Lambda}\mathrm{S}(\mathring{\varphi}^n|L_2(\mathbb{R}))$ with Λ being a countable index set such that $[\widehat{\mathring{\varphi}^m},\widehat{\mathring{\varphi}^n}] = 0$ for all $m \neq n$. Define $\sigma_n := [\widehat{\mathring{\varphi}^n},\widehat{\mathring{\varphi}^n}]$ for $n \in \Lambda$. Then $\sigma_n \in \{F_k\}_{k\in\Lambda_0}$.

We now construct the desired new functions $\varphi^1, \dots, \varphi^r$ by induction. Pick a subset S_1 of Λ such that $\{\sigma_n\}_{n\in S_1}$ forms a disjoint partition of the measurable set $H_1 := \cup_{n\in\Lambda}\sigma_n$. If H_1 has measure zero, then we redefine $\varphi^1 = 0$; otherwise, redefine $\varphi^1 := \sum_{n\in S_1}\mathring{\varphi}^n$. Similarly, the function φ^ℓ is redefined in the same way as the new φ^1 by replacing the index set Λ with $\Lambda\backslash(\cup_{j=1}^{\ell-1}S_j)$. Now we can directly verify that all items (i)–(iv) of Proposition 4.4.7 and (4.4.10) are satisfied. The relation $\operatorname{supp}([\widehat{\varphi^j},\widehat{\varphi^j}]) = \{\xi \in \mathbb{R} : \dim_{\mathsf{S}(\Phi|L_2(\mathbb{R}))}(\xi) \geqslant j\}$ follows directly from the above construction. Hence, we have $\varphi^\ell = 0$ for all $\operatorname{len}(\mathsf{S}(\Phi|L_2(\mathbb{R}))) < \ell \leqslant r$. ☐

As a consequence of Proposition 4.4.7, we have the following result on the dimension function of a countably generated shift-invariant space $\mathsf{S}(\Phi|L_2(\mathbb{R}))$.

Corollary 4.4.10 *Let Φ be a countable subset of $L_2(\mathbb{R})$. Then*

$$\dim_{\mathsf{S}(\Phi|L_2(\mathbb{R}))}(\xi) = \dim(span\{\{\widehat{\phi}(\xi+2\pi k)\}_{k\in\mathbb{Z}} : \phi \in \Phi\}), \quad a.e. \xi \in \mathbb{R}. \tag{4.4.11}$$

Proof Let $\varphi^1,\dots,\varphi^r$ be defined as in Proposition 4.4.7. Then $\dim_{\mathsf{S}(\Phi|L_2(\mathbb{R}))}(\xi) = \sum_{\ell=1}^r[\widehat{\varphi^\ell},\widehat{\varphi^\ell}](\xi)$. By item (iii) of Proposition 4.4.7 and $[\widehat{\varphi^\ell},\widehat{\varphi^\ell}] = \chi_{\operatorname{supp}([\widehat{\varphi^\ell},\widehat{\varphi^\ell}])}$, we observe that $\|\{\widehat{\varphi^\ell}(\xi+2\pi k)\}_{k\in\mathbb{Z}}\|_{l_2(\mathbb{Z})} \in \{0,1\}$ and $\{\widehat{\varphi^\ell}(\xi+2\pi k)\}_{k\in\mathbb{Z}} \perp \{\widehat{\varphi^j}(\xi+2\pi k)\}_{k\in\mathbb{Z}}$ for a.e. $\xi\in\mathbb{R}$ for all $1 \leqslant \ell \neq j \leqslant r$. Therefore,

$$\dim_{\mathsf{S}(\Phi|L_2(\mathbb{R}))}(\xi) = \sum_{\ell=1}^r[\widehat{\varphi^\ell},\widehat{\varphi^\ell}](\xi) = \dim(\operatorname{span}\{\{\widehat{\varphi^\ell}(\xi+2\pi k)\}_{k\in\mathbb{Z}} : \ell = 1,\dots,r\}).$$

By item (i) of Proposition 4.4.7, we can deduce that the right-hand side of the above identity is equal to the right-hand side of (4.4.11) for almost every $\xi \in \mathbb{R}$. ☐

We now characterize a finitely generated shift-invariant space $\mathsf{S}(\Phi|L_2(\mathbb{R}))$.

Theorem 4.4.11 *Let $\Phi = \{\phi^1,\dots,\phi^r\}$ be a finite subset of $L_2(\mathbb{R})$. Then $f \in \mathsf{S}(\Phi|L_2(\mathbb{R}))$ if and only if $f \in L_2(\mathbb{R})$ and there exist 2π-periodic measurable functions $\mathbf{u}_1,\dots,\mathbf{u}_r : \mathbb{R} \to \mathbb{C}$ such that*

$$\widehat{f}(\xi) = \mathbf{u}_1(\xi)\widehat{\phi^1}(\xi) + \cdots + \mathbf{u}_r(\xi)\widehat{\phi^r}(\xi), \qquad a.e.\ \xi \in \mathbb{R}. \tag{4.4.12}$$

Proof Let $\varphi^1,\dots,\varphi^r,\mathcal{P}_1,\dots,\mathcal{P}_r$ as in Proposition 4.4.7. Let $\mathcal{P}$ be the orthogonal projection from $L_2(\mathbb{R})$ to $\mathsf{S}(\Phi|L_2(\mathbb{R}))$. By Proposition 4.4.7, $\mathcal{P} = \mathcal{P}_1 + \cdots + \mathcal{P}_r$.

Necessity ($\Rightarrow$). $f \in \mathsf{S}(\Phi|L_2(\mathbb{R}))$ if and only if $f = \mathcal{P}f = \mathcal{P}_1 f + \cdots + \mathcal{P}_r f$. By Theorem 4.4.2, $\widehat{\mathcal{P}_j f}(\xi) = \mathbf{v}_j(\xi)\widehat{\varphi^j}(\xi)$ for a.e. $\xi \in \mathbb{R}$ for some 2π-periodic measurable function $\mathbf{v}_j$. Therefore,

$$\widehat{f}(\xi) = \mathbf{v}_1(\xi)\widehat{\varphi^1}(\xi) + \cdots + \mathbf{v}_r(\xi)\widehat{\varphi^r}(\xi). \tag{4.4.13}$$

By item (i) of Proposition 4.4.7, the identity (4.4.12) holds.

Sufficiency ($\Leftarrow$). If (4.4.12) holds, by item (i) of Proposition 4.4.7, we see that (4.4.13) holds for some 2π-periodic measurable functions $\mathbf{v}_1,\dots,\mathbf{v}_r$. By Item (ii) of Proposition 4.4.7 and $f \in L_2(\mathbb{R})$, we have

$$[\widehat{f},\widehat{f}](\xi) = |\mathbf{v}_1(\xi)|^2[\widehat{\varphi^1},\widehat{\varphi^1}](\xi) + \cdots + |\mathbf{v}_r(\xi)|^2[\widehat{\varphi^r},\widehat{\varphi^r}](\xi).$$

Let $\widehat{g^j}(\xi) := \mathbf{v}_j(\xi)\widehat{\varphi^j}(\xi)$ for $j = 1, \dots, r$. Then

$$\int_{\mathbb{R}} |\widehat{g^j}(\xi)|^2 d\xi = \int_{-\pi}^{\pi} |\mathbf{v}_j(\xi)|^2 [\widehat{\varphi^j}, \widehat{\varphi^j}](\xi) d\xi \leqslant \int_{-\pi}^{\pi} [\widehat{f}, \widehat{f}](\xi) d\xi = \|\widehat{f}\|^2_{L_2(\mathbb{R})} < \infty.$$

Hence, the function g^j belongs to $L_2(\mathbb{R})$ and by Theorem 4.4.3 we have $g^j \in \mathsf{S}(\varphi^j|L_2(\mathbb{R}))$. Therefore, by (4.4.13) and item (iii) of Proposition 4.4.7, we conclude that $f = g^1 + \cdots + g^r \in \mathsf{S}(\varphi^1|L_2(\mathbb{R})) \oplus \cdots \oplus \mathsf{S}(\varphi^r|L_2(\mathbb{R})) = \mathsf{S}(\Phi|L_2(\mathbb{R}))$. $\square$

For a matrix M, we denote by $[\mathsf{M}]_{m,n}$ its (m,n)-entry. By $v \in (l_p(\mathbb{Z}))^{r\times s}$ we mean that $v : \mathbb{Z} \to \mathbb{C}^{r\times s}$ is a sequence of $r \times s$ matrices such that $\|v\|^p_{(l_p(\mathbb{Z}))^{r\times s}} := \sum_{k\in\mathbb{Z}} \sum_{m=1}^{r} \sum_{n=1}^{s} |[v(k)]_{m,n}|^p < \infty$. Similarly, we can define the space $(L_p(\mathbb{R}))^{r\times s}$. Let $f \in (L_2(\mathbb{R}))^{r\times s}$ and $g \in (L_2(\mathbb{R}))^{t\times s}$, for $\xi \in \mathbb{R}$, we define

$$\langle f, g\rangle := \int_{\mathbb{R}} f(x)\overline{g(x)}^{\mathsf{T}} dx, \quad [f, g](\xi) := \sum_{k\in\mathbb{Z}} f(\xi + 2\pi k)\overline{g(\xi + 2\pi k)}^{\mathsf{T}}. \qquad (4.4.14)$$

Note that $\langle f, g\rangle$ is an $r \times t$ matrix of complex numbers and $[f, g]$ is an $r \times t$ matrix of 2π-periodic functions. By Lemma 4.4.1, $\{\phi^1(\cdot - k), \dots, \phi^r(\cdot - k) \ : \ k \in \mathbb{Z}\}$ is an orthonormal basis of $\mathsf{S}(\{\phi^1, \dots, \phi^r\})$ if and only if $[\widehat{\phi}, \widehat{\phi}](\xi) = I_r$ a.e. $\xi \in \mathbb{R}$, where $\phi := (\phi^1, \dots, \phi^r)^{\mathsf{T}}$.

For two $r \times r$ matrices E and F such that $E^\star = E$ and $F^\star = F$, we say that $E \leqslant F$ if $\bar{x}^{\mathsf{T}}(F - E)x \geqslant 0$ for all $x \in \mathbb{C}^r$. By $\rho_{\min}(E)$ and $\rho_{\max}(E)$ we denote the smallest and largest eigenvalues of E, respectively. Note that $\rho_{\min}(E) = \inf_{x\in\mathbb{C}^r, \bar{x}^{\mathsf{T}}x=1} \bar{x}^{\mathsf{T}}Ex$ and $\rho_{\max}(E) = \sup_{x\in\mathbb{C}^r, \bar{x}^{\mathsf{T}}x=1} \bar{x}^{\mathsf{T}}Ex$ if $E^\star = E$.

We now characterize bases in a finitely generated shift-invariant space.

Theorem 4.4.12 *Let $\Phi = \{\phi^1, \dots, \phi^r\}$ be a finite subset of $L_2(\mathbb{R})$. Then the following statements are equivalent:*

(1) $\{\phi^\ell(\cdot - k) \ : \ k \in \mathbb{Z}, \ell = 1, \dots, r\}$ is a Riesz basis for $\mathsf{S}(\Phi|L_2(\mathbb{R}))$, that is, there exist positive constants C_1 and C_2 such that

$$C_1 \sum_{\ell=1}^{r} \sum_{k\in\mathbb{Z}} |v_\ell(k)|^2 \leqslant \Big\| \sum_{\ell=1}^{r} \sum_{k\in\mathbb{Z}} v_\ell(k)\phi^\ell(\cdot - k) \Big\|^2_{L_2(\mathbb{R})} \leqslant C_2 \sum_{\ell=1}^{r} \sum_{k\in\mathbb{Z}} |v_\ell(k)|^2 \qquad (4.4.15)$$

for all finitely supported sequences $v_\ell \in l_0(\mathbb{Z}), \ell = 1, \dots, r$.

(2) $C_1 I_r \leqslant [\widehat{\phi}, \widehat{\phi}](\xi) \leqslant C_2 I_r$ for almost every $\xi \in \mathbb{R}$, where $\phi := (\phi^1, \dots, \phi^r)^{\mathsf{T}}$.

(3) $C_1 \leqslant \rho_{\min}([\widehat{\phi}, \widehat{\phi}](\xi)) \leqslant \rho_{\max}([\widehat{\phi}, \widehat{\phi}](\xi)) \leqslant C_2$ for almost every $\xi \in \mathbb{R}$.

Proof The equivalence between (2) and (3) is well known in linear algebra. For $\widehat{v}(\xi) = (\widehat{v_1}(\xi), \dots, \widehat{v_r}(\xi))$, we have $\sum_{\ell=1}^{r} \sum_{k\in\mathbb{Z}} |v_\ell(k)|^2 = \frac{1}{2\pi} \int_{-\pi}^{\pi} \widehat{v}(\xi)\widehat{v}^\star(\xi) d\xi$ and

$$\Big\| \sum_{\ell=1}^{r} \sum_{k\in\mathbb{Z}} v_\ell(k)\phi^\ell(\cdot - k) \Big\|^2_{L_2(\mathbb{R})} = \frac{1}{2\pi} \int_{-\pi}^{\pi} \widehat{v}(\xi)[\widehat{\phi}, \widehat{\phi}](\xi)\widehat{v}^\star(\xi) d\xi.$$

(2)$\Longrightarrow$(1). If item (2) holds, then (4.4.15) follows directly from the above identities and

$$C_1\widehat{v}(\xi)\widehat{v}^\star(\xi) \leqslant \widehat{v}(\xi)[\widehat{\phi},\widehat{\phi}](\xi)\widehat{v}^\star(\xi) \leqslant C_2\widehat{v}(\xi)\widehat{v}^\star(\xi).$$

(1)$\Longrightarrow$(3). We first show that the functions $\rho_{min}([\widehat{\phi},\widehat{\phi}])$ and $\rho_{max}([\widehat{\phi},\widehat{\phi}])$ are measurable. By Theorem 4.4.6, we see that all the entries of $[\widehat{\phi},\widehat{\phi}]$ belong to $L_\infty(\mathbb{T})$. By basic knowledge from real analysis, there exists a sequence $\{M_n(\xi)\}_{n\in\mathbb{N}}$ of $r\times r$ matrices of 2π-periodic simple functions such that $M_n^\star(\xi) = M_n(\xi)$ and $\lim_{n\to\infty}\|M_n - [\widehat{\phi},\widehat{\phi}]\|_{(L_\infty(\mathbb{T}))^{r\times r}} = 0$. By $\rho_{\min}(M_n(\xi)) = \inf_{\|x\|_{l_2}=1}\overline{x}^\mathsf{T}M_n(\xi)x$, it is quite straightforward to see that

$$|\rho_{\min}(M_n(\xi)) - \rho_{\min}([\widehat{\phi},\widehat{\phi}](\xi))| \leqslant \sqrt{r}\|M_n - [\widehat{\phi},\widehat{\phi}]\|_{(L_\infty(\mathbb{T}))^{r\times r}},\ a.e.\,\xi\in\mathbb{R}. \tag{4.4.16}$$

Since all $\rho_{\min}(M_n)$ are simple functions, we conclude that their limit $\rho_{\min}([\widehat{\phi},\widehat{\phi}])$ must be measurable. Similarly, (4.4.16) holds if $\rho_{\min}$ is replaced by $\rho_{\max}$ and therefore, the function $\rho_{\max}([\widehat{\phi},\widehat{\phi}])$ is also measurable.

We now prove (1)$\Longrightarrow$(3) using proof by contradiction. If the left-hand inequality in item (3) fails, then there exist $0<\varepsilon<C_1/2$ and a measurable set $E\subseteq[-\pi,\pi)$ with a positive measure such that $\rho_{\min}([\widehat{\phi},\widehat{\phi}](\xi)) \leqslant C_1 - 4\varepsilon$ for all $\xi\in E$. By (4.4.16), there exists a sufficiently large n such that $|\rho_{\min}(M_n(\xi)) - \rho_{\min}([\widehat{\phi},\widehat{\phi}](\xi))| < \varepsilon$ for almost every $\xi\in\mathbb{R}$ and $\|M_n - [\widehat{\phi},\widehat{\phi}]\|_{(L_\infty(\mathbb{T}))^{r\times r}} \leqslant \varepsilon/r$. Hence, $\rho_{\min}(M_n(\xi)) \leqslant C_1 - 3\varepsilon$ for all $\xi\in E$. Since M_n is a matrix of simple functions, there exist a measurable subset $F\subseteq E$ and a vector $x\in\mathbb{C}^n$ with $\overline{x}^\mathsf{T}x = 1$ such that F has a positive measure and $\overline{x}^\mathsf{T}M_n(\xi)x \leqslant C_1 - 2\varepsilon$ for all $\xi\in F$. Hence, $\overline{x}^\mathsf{T}[\widehat{\phi},\widehat{\phi}](\xi)x \leqslant C_1-\varepsilon$ for $\xi\in F$ by $\|M_n - [\widehat{\phi},\widehat{\phi}]\|_{(L_\infty(\mathbb{T}))^{r\times r}} \leqslant \varepsilon/r$. Define a 2π-periodic function $\widehat{v}$ by $\widehat{v}(\xi) = \overline{x}^\mathsf{T}\chi_F(\xi)$ for $\xi\in[-\pi,\pi)$. Then $\int_{-\pi}^{\pi}\widehat{v}(\xi)\widehat{v}^\star(\xi)d\xi = \int_F 1d\xi > 0$ but

$$\int_{-\pi}^{\pi}\widehat{v}(\xi)[\widehat{\phi},\widehat{\phi}](\xi)\widehat{v}^\star(\xi)d\xi \leqslant \int_F(C_1-\varepsilon)d\xi = (C_1-\varepsilon)\int_{-\pi}^{\pi}\widehat{v}(\xi)\widehat{v}^\star(\xi)d\xi,$$

contradicting our assumption in item (1). Hence, the left-hand inequality in item (3) must hold. The right-hand inequality in item (3) can be proved similarly. □

By a similar argument as in Theorem 4.4.12, we characterize frames in a finitely generated shift-invariant space in the following result, whose proof is left as Exercise 4.34.

Theorem 4.4.13 *Let $\Phi = \{\phi^1,\ldots,\phi^r\}$ be a finite subset of $L_2(\mathbb{R})$ and define $\phi := (\phi^1,\ldots,\phi^r)^\mathsf{T}$. Then the following statements are equivalent:*

(1) $\{\phi^\ell(\cdot-k) : k\in\mathbb{Z}, \ell=1,\ldots,r\}$ is a frame for $\mathsf{S}(\Phi|L_2(\mathbb{R}))$, that is, there exist positive constants C_1 and C_2 such that

$$C_1\|f\|^2_{L_2(\mathbb{R})} \leqslant \sum_{\ell=1}^{r}\sum_{k\in\mathbb{Z}}|\langle f,\phi^\ell(\cdot-k)\rangle|^2 \leqslant C_2\|f\|^2_{L_2(\mathbb{R})}, \qquad \forall f\in\mathsf{S}(\Phi|L_2(\mathbb{R})).$$

(2) $C_1[\widehat{\phi},\widehat{\phi}](\xi) \leqslant [\widehat{\phi},\widehat{\phi}](\xi)[\widehat{\phi},\widehat{\phi}]^\star(\xi) \leqslant C_2[\widehat{\phi},\widehat{\phi}](\xi)$ *for almost every* $\xi \in \mathbb{R}$.
(3) $C_1 \leqslant \mathring{\rho}_{\min}([\widehat{\phi},\widehat{\phi}](\xi)) \leqslant \rho_{\max}([\widehat{\phi},\widehat{\phi}](\xi)) \leqslant C_2$ *for almost every* $\xi \in \operatorname{supp}([\widehat{\phi},\widehat{\phi}])$, *where* $\mathring{\rho}_{\min}([\widehat{\phi},\widehat{\phi}](\xi))$ *is the smallest nonzero eigenvalue of* $[\widehat{\phi},\widehat{\phi}](\xi)$.

By Theorem 4.4.13, $\{\phi^1(\cdot-k),\dots,\phi^r(\cdot-k)\}$ is a tight frame of the shift-invariant space $S(\{\phi^1,\dots,\phi^r\}\mid L_2(\mathbb{R}))$ if and only if $[\widehat{\phi},\widehat{\phi}](\xi)[\widehat{\phi},\widehat{\phi}](\xi) = [\widehat{\phi},\widehat{\phi}](\xi)$ a.e., i.e., the matrix $[\widehat{\phi},\widehat{\phi}](\xi)$ has only eigenvalues 0 or 1 for almost every $\xi \in \mathbb{R}$.

4.4.3 Sampling Theorems in Shift-Invariant Spaces

We now discuss reproducing kernel Hilbert spaces and sampling theorems in shift-invariant spaces.

A reproducing kernel Hilbert space (RKHS) is a Hilbert space of functions on a set X in which point evaluation is a continuous linear functional.

By the Riesz representation theorem, for every $x \in X$, the continuous point-evaluation linear functional K_x on $\mathcal{H}$ with $K_x(f) := f(x) = \langle f, K_x\rangle$ is an element in $\mathcal{H}$. Hence, we can define a reproducing kernel $K(y,x) := K_x(y) = \langle K_x, K_y\rangle$ for $x, y \in X$. Then $K(x,x) = \langle K_x, K_x\rangle = \|K_x\|^2 \geqslant 0$, K is symmetric

$$K(y,x) = \langle K_x, K_y\rangle = \overline{\langle K_y, K_x\rangle} = \overline{K(x,y)}$$

and K is positive semi-definite: the matrix $(K(x_j,x_k))_{1\leqslant j,k\leqslant n} \geqslant 0$ for all points $x_1,\dots,x_n \in X$ and $n \in \mathbb{N}$.

Theorem 4.4.14 **(Moore–Aronszajn Theorem)** *If* $K : X \times X \to \mathbb{F}$ *with* $\mathbb{F} = \mathbb{R}$ *or* $\mathbb{C}$ *is a symmetric and positive semi-definite kernel on a set* X*, then there is a unique Hilbert space* $\mathcal{H}$ *of functions on* X *for which* K *is the reproducing kernel for* $\mathcal{H}$.

In fact, the obtained Hilbert space is the completion of the linear space spanned by $\{K(\cdot,x) : x \in X\}$ under the inner product

$$\langle \sum_{j=1}^{m} \alpha_j K_{x_j}, \sum_{k=1}^{n} \beta_k K_{y_k}\rangle = \sum_{j=1}^{m}\sum_{k=1}^{n} \alpha_j \overline{\beta_k} K(y_k, x_j).$$

For a subset $\Phi \subseteq L_2(\mathbb{R})$ and a nonzero scaling factor $\lambda \in \mathbb{R}\backslash\{0\}$, we now generalize the notion of shift-invariant spaces by defining

$$S_\lambda(\Phi \mid L_2(\mathbb{R})) := \overline{\operatorname{span}\{\phi(\lambda\cdot -k) : k \in \mathbb{Z}, \phi \in \Phi\}}, \qquad \lambda \in \mathbb{R}\backslash\{0\}. \qquad (4.4.17)$$

That is, $S_\lambda(\Phi \mid L_2(\mathbb{R})) = \{f(\lambda\cdot) : f \in S(\Phi|L_2(\mathbb{R}))\}$. If $\Phi = \{\phi\}$ is a singleton, we use the notation $S(\phi|L_2(\mathbb{R}))$ for $S(\{\phi\}|L_2(\mathbb{R}))$ and $S_\lambda(\phi|L_2(\mathbb{R}))$ for $S_\lambda(\{\phi\}|L_2(\mathbb{R}))$.

Many shift-invariant spaces of $L_2(\mathbb{R})$ are reproducing kernel Hilbert spaces.

Theorem 4.4.15 *Let ϕ be a continuous square integrable function on $\mathbb{R}$. If $\{\phi(\cdot - k) : k \in \mathbb{Z}\}$ is a frame for $\mathrm{S}(\phi|L_2(\mathbb{R}))$ and $\sum_{k\in\mathbb{Z}} |\phi(\cdot - k)|^2 \in L_\infty(\mathbb{R})$, then*

$$\mathrm{S}_\lambda(\phi|L_2(\mathbb{R})) = \Big\{ \sum_{k\in\mathbb{Z}} v(k)\phi(\lambda \cdot - k) \ : \ v \in l_2(\mathbb{Z}) \Big\} \subset \mathscr{C}(\mathbb{R}), \qquad \forall\, \lambda > 0$$

with the series converging absolutely and uniformly on $\mathbb{R}$, and $\mathrm{S}_\lambda(\phi|L_2(\mathbb{R}))$ is a reproducing kernel Hilbert space with the reproducing kernel

$$K_\lambda(x, y) := \lambda \sum_{k\in\mathbb{Z}} \varphi(\lambda x - k)\overline{\varphi(\lambda y - k)},$$

where $\varphi \in \mathrm{S}(\phi|L_2(\mathbb{R}))$ is given by $\widehat{\varphi}(\xi) := \frac{\widehat{\phi}(\xi)}{\sqrt{[\widehat{\phi},\widehat{\phi}](\xi)}} \chi_{\mathrm{supp}([\widehat{\phi},\widehat{\phi}])}(\xi)$ and φ satisfies $\sup_{x\in\mathbb{R}} \sum_{k\in\mathbb{Z}} |\varphi(x - k)|^2 < \infty$.

Proof It suffices to prove the claim for $\lambda = 1$. Let $C := \| \sum_{k\in\mathbb{Z}} |\phi(\cdot - k)|^2 \|_{L_\infty(\mathbb{R})}$. Then $\sum_{|k|<N} |\phi(x - k)|^2 \leqslant C$ for almost every $x \in \mathbb{R}$. Since ϕ is continuous, it is trivial that $\sum_{|k|<N} |\phi(x - k)|^2 \leqslant C$ for all $x \in \mathbb{R}$. Hence, $\sum_{k\in\mathbb{Z}} |\phi(x - k)|^2 \leqslant C$ for all $x \in \mathbb{R}$. Since $\{\phi(\cdot - k) : k \in \mathbb{Z}\}$ is a frame for $\mathrm{S}(\phi|L_2(\mathbb{R}))$, by Theorem 4.4.5, $\mathcal{V} : l_2(\phi) \to \mathrm{S}(\phi|L_2(\mathbb{R}))$ is a bounded bijective operator. For $f \in \mathrm{S}(\phi|L_2(\mathbb{R}))$, there exists $v \in l_2(\phi)$ such that $f = \sum_{k\in\mathbb{Z}} v(k)\phi(\cdot - k)$. Since $\sum_{k\in\mathbb{Z}} |\phi(x - k)|^2 \leqslant C$ for all $x \in \mathbb{R}$, by the Cauchy-Schwarz inequality, we deduce from

$$\Big(\sum_{|k|>N} |v(k)\phi(x - k)| \Big)^2 \leqslant \Big(\sum_{|k|>N} |v(k)|^2 \Big)\Big(\sum_{|k|>N} |\phi(x - k)|^2 \Big) \leqslant C \sum_{|k|>N} |v(k)|^2$$

that $\sum_{k\in\mathbb{Z}} v(k)\phi(\cdot - k)$ converges absolutely and uniformly on $\mathbb{R}$. Hence, f is continuous and by $v = \mathcal{V}^{-1}f$,

$$|f(x)| = \Big| \sum_{k\in\mathbb{Z}} v(k)\phi(x - k) \Big| \leqslant \|v\|_{l_2(\mathbb{Z})} \sqrt{C} \leqslant \|\mathcal{V}^{-1}\| \sqrt{C} \|f\|_{L_2(\mathbb{R})}.$$

Therefore, all the point-evaluation functionals are continuous and $\mathrm{S}(\phi|L_2(\mathbb{R}))$ is a reproducing kernel Hilbert space with a reproducing kernel $K : \mathbb{R}^2 \to \mathbb{C}$ satisfying $\|K_x\|_{L_2(\mathbb{R})} \leqslant \|\mathcal{V}^{-1}\| \sqrt{C}$ for all $x \in \mathbb{R}$ with $K_x := K(\cdot, x)$. Note that $\varphi \in \mathrm{S}(\phi \mid L_2(\mathbb{R}))$ is continuous and $\{\varphi(\cdot - k) : k \in \mathbb{Z}\}$ is a tight frame for $\mathrm{S}(\phi|L_2(\mathbb{R}))$ by $[\widehat{\varphi}, \widehat{\varphi}] = \chi_{\mathrm{supp}([\widehat{\varphi},\widehat{\varphi}])}$. Hence, by $K_x \in \mathrm{S}(\phi|L_2(\mathbb{R}))$,

$$K_x = \sum_{k\in\mathbb{Z}} \langle K_x, \varphi(\cdot - k)\rangle \varphi(\cdot - k) \qquad \text{in } L_2(\mathbb{R}), \tag{4.4.18}$$

from which we particularly have $\|K_x\|^2_{L_2(\mathbb{R})} = \sum_{k\in\mathbb{Z}} |\langle K_x, \varphi(\cdot - k)\rangle|^2 = \sum_{k\in\mathbb{Z}} |\varphi(x - k)|^2$. This proves that $\sum_{k\in\mathbb{Z}} |\varphi(x - k)|^2 = \|K_x\|^2_{L_2(\mathbb{R})} \leqslant \|\mathcal{V}^{-1}\|^2 C$ for all $x \in \mathbb{R}$. Since

$\sum_{k\in\mathbb{Z}} |\varphi(\cdot - k)|^2 \in L_\infty(\mathbb{R})$ and $\{\langle K_x, \varphi(\cdot - k)\rangle\}_{k\in\mathbb{Z}} \in l_2(\mathbb{Z})$, as we proved before, the series in (4.4.18) converges absolutely and uniformly. We deduce from (4.4.18) that

$$K(y,x) = K_x(y) = \sum_{k\in\mathbb{Z}} \langle K_x, \varphi(\cdot - k)\rangle \varphi(y-k) = \sum_{k\in\mathbb{Z}} \overline{\varphi(x-k)}\varphi(y-k).$$

This completes the proof. □

Interpolating functions play an indispensable role in sampling theorems in shift-invariant spaces. A continuous function $\varphi : \mathbb{R} \to \mathbb{C}$ is called *interpolating* if $\varphi(0) = 1$ and $\varphi(k) = 0$ for all $k \in \mathbb{Z}\backslash\{0\}$. Interpolating functions can be characterized by its Fourier transform as follows:

Lemma 4.4.16 *If φ is a continuous square integrable function on $\mathbb{R}$ and $\widehat{\varphi} \in L_1(\mathbb{R})$, then φ is interpolating if and only if $\sum_{k\in\mathbb{Z}} \widehat{\varphi}(\xi + 2\pi k) = 1$ a.e. $\xi \in \mathbb{R}$.*

Proof Define $g := \sum_{k\in\mathbb{Z}} \widehat{\varphi}(\cdot + 2\pi k)$. Since $\widehat{\varphi} \in L_1(\mathbb{R})$ and φ is continuous, we have $\int_{-\pi}^{\pi} \sum_{k\in\mathbb{Z}} |\widehat{\varphi}(\xi + 2\pi k)| d\xi = \|\widehat{\varphi}\|_{L_1(\mathbb{R})} < \infty$ and $\varphi(x) = \frac{1}{2\pi}\int_{\mathbb{R}} \widehat{\varphi}(\xi)e^{ix\xi} d\xi$. Hence, $g \in L_1(\mathbb{T})$ and for $n \in \mathbb{Z}$,

$$\widehat{g}(n) := \frac{1}{2\pi}\int_{\mathbb{T}} \sum_{k\in\mathbb{Z}} \widehat{\varphi}(\xi + 2\pi k)e^{-in\xi} d\xi = \frac{1}{2\pi}\int_{\mathbb{R}} \widehat{\varphi}(\xi)e^{-in\xi} d\xi = \varphi(-n).$$

Therefore, $\varphi(0) = 1$ and $\varphi(k) = 0$ for all $k \in \mathbb{Z}\backslash\{0\}$ if and only if $g(\xi) = 1$ for almost every $\xi \in \mathbb{R}$. □

Lemma 4.4.17 *Let $\widehat{\varphi} \in L_2(\mathbb{R})$ such that*

$$\sum_{k\in\mathbb{Z}} |\widehat{\varphi}(\cdot + 2\pi k)| \in L_2(\mathbb{T}). \tag{4.4.19}$$

Let φ be the inverse Fourier transform of $\widehat{\varphi}$ given by $\varphi(x) := \frac{1}{2\pi}\int_{\mathbb{R}} \widehat{\varphi}(\xi)e^{ix\xi} d\xi$. Then φ is a continuous function satisfying

$$\sum_{n\in\mathbb{Z}} |\varphi(x-n)|^2 \leqslant \Big\| \sum_{k\in\mathbb{Z}} |\widehat{\varphi}(\cdot + 2\pi k)| \Big\|_{L_2(\mathbb{T})}^2 < \infty, \qquad \forall\, x \in \mathbb{R}. \tag{4.4.20}$$

Proof By $L_2(\mathbb{T}) \subset L_1(\mathbb{T})$, we conclude from (4.4.19) that $\sum_{k\in\mathbb{Z}} |\widehat{\varphi}(\cdot + 2\pi k)| \in L_1(\mathbb{T})$ and therefore, $\widehat{\varphi} \in L_1(\mathbb{R})$. Consequently, as the inverse Fourier transform of $\widehat{\varphi}$, the function φ is continuous.

For each fixed $x \in \mathbb{R}$, by (4.4.19), we observe that the 2π-periodic function $g_x(\xi) := \sum_{k\in\mathbb{Z}} \widehat{\varphi}(\xi + 2\pi k)e^{i(\xi+2\pi k)x}$ is a well-defined function in $L_1(\mathbb{T})$ and its Fourier coefficients are

$$\widehat{g_x}(n) = \frac{1}{2\pi}\int_{-\pi}^{\pi} g_x(\xi)e^{-in\xi} d\xi = \frac{1}{2\pi}\int_{\mathbb{R}} \widehat{\varphi}(\xi)e^{i(x-n)\xi} d\xi = \varphi(x-n), \qquad n \in \mathbb{Z}.$$

By Parseval's identity and (4.4.19), we conclude that

$$\sum_{n\in\mathbb{Z}} |\varphi(x-n)|^2 = \frac{1}{2\pi}\int_{-\pi}^{\pi} |g_x(\xi)|^2 d\xi \leqslant \frac{1}{2\pi}\int_{-\pi}^{\pi} \Big(\sum_{k\in\mathbb{Z}} |\widehat{\varphi}(\xi+2\pi k)|\Big)^2 d\xi$$
$$= \Big\| \sum_{k\in\mathbb{Z}} |\widehat{\varphi}(\cdot+2\pi k)| \Big\|^2_{L_2(\mathbb{T})} < \infty.$$

This proves (4.4.20). □

The sampling theorem in a general shift-invariant space is as follows.

Theorem 4.4.18 *Let ϕ be a continuous function in $L_2(\mathbb{R})$ such that $\sum_{k\in\mathbb{Z}} |\widehat{\phi}(\cdot + 2\pi k)| \in L_2(\mathbb{T})$ and there exist positive constants C_1 and C_2 satisfying*

$$C_1 \leqslant |[\widehat{\phi}, 1](\xi)| \leqslant C_2 \quad \text{and} \quad C_1 \leqslant [\widehat{\phi}, \widehat{\phi}](\xi) \leqslant C_2, \qquad a.e.\, \xi \in \mathbb{R}, \tag{4.4.21}$$

where $[\widehat{\phi}, 1](\xi) := \sum_{k\in\mathbb{Z}} \widehat{\phi}(\xi+2\pi k)$ and $[\widehat{\phi}, \widehat{\phi}](\xi) := \sum_{k\in\mathbb{Z}} |\widehat{\phi}(\xi+2\pi k)|^2$. Define

$$\varphi := \sum_{k\in\mathbb{Z}} v_\phi(k)\phi(\cdot - k) \qquad \text{with} \quad \widehat{v_\phi}(\xi) = \sum_{k\in\mathbb{Z}} v_\phi(k)e^{-ik\xi} := \frac{1}{[\widehat{\phi}, 1](\xi)}. \tag{4.4.22}$$

Then φ is a well-defined continuous interpolating function in $L_2(\mathbb{R})$, $\mathsf{S}_\lambda(\phi|L_2(\mathbb{R})) = \mathsf{S}_\lambda(\varphi|L_2(\mathbb{R}))$ is a reproducing kernel Hilbert space for every $\lambda > 0$, and the sampling formula

$$f(x) = \sum_{k\in\mathbb{Z}} f(k/\lambda)\varphi(\lambda x - k), \qquad \forall f \in \mathsf{S}_\lambda(\varphi|L_2(\mathbb{R})), \lambda > 0, x \in \mathbb{R} \tag{4.4.23}$$

holds with the series converging absolutely and uniformly on $\mathbb{R}$.

Proof By (4.4.19), both $[\widehat{\phi}, 1](\xi)$ and $[\widehat{\phi}, \widehat{\phi}](\xi)$ are well defined for almost every $\xi \in \mathbb{R}$. Note that (4.4.22) is equivalent to $\widehat{\varphi}(\xi) = \widehat{v_\phi}(\xi)\widehat{\phi}(\xi) = \widehat{\phi}(\xi)/[\widehat{\phi}, 1](\xi)$. By the first inequality in (4.4.21), we have $\frac{1}{C_2}|\widehat{\phi}(\xi)| \leqslant |\widehat{\varphi}(\xi)| \leqslant \frac{1}{C_1}|\widehat{\phi}(\xi)|$. Consequently, (4.4.19) holds by $\sum_{k\in\mathbb{Z}} |\widehat{\varphi}(\cdot + 2\pi k)| \leqslant \frac{1}{C_1}\sum_{k\in\mathbb{Z}} |\widehat{\phi}(\cdot + 2\pi k)| \in L_2(\mathbb{T})$. By Lemma 4.4.17, φ is a well-defined continuous function in $L_2(\mathbb{R})$ and (4.4.20) holds.

By the first inequality in (4.4.21) and Theorem 4.4.3, we conclude from $\widehat{\varphi} = \widehat{\phi}/[\widehat{\phi}, 1] \in L_2(\mathbb{R})$ and $\widehat{\phi} = [\widehat{\phi}, 1]\widehat{\varphi} \in L_2(\mathbb{R})$ that $\mathsf{S}(\phi|L_2(\mathbb{R})) = \mathsf{S}(\varphi|L_2(\mathbb{R}))$.

By Theorem 4.4.6, it follows from the second inequality in (4.4.21) that $\{\varphi(\cdot - k) : k \in \mathbb{Z}\}$ is a Riesz basis and thus, a frame for $\mathsf{S}(\varphi|L_2(\mathbb{R}))$. By Theorem 4.4.15, the space $\mathsf{S}_\lambda(\varphi|L_2(\mathbb{R}))$ is a reproducing kernel Hilbert space for every $\lambda > 0$.

For $f \in \mathsf{S}_\lambda(\varphi|L_2(\mathbb{R}))$, by Theorem 4.4.15, there exists $v \in l_2(\mathbb{Z})$ such that $f(x) = \sum_{k\in\mathbb{Z}} v(k)\varphi(\lambda x - k)$ with the series converging absolutely and uniformly on $\mathbb{R}$. It is trivial that $\sum_{k\in\mathbb{Z}} \widehat{\varphi}(\xi + 2\pi k) = 1$. Since $\widehat{\varphi} \in L_1(\mathbb{R})$, Lemma 4.4.16 guarantees that φ is interpolating. Therefore, for $n \in \mathbb{Z}$, $f(n/\lambda) = \sum_{k\in\mathbb{Z}} v(k)\varphi(n-k) = v(n)$. This proves the sampling formula (4.4.23). □

The conditions in (4.4.19) and (4.4.21) are often satisfied by many generators of shift-invariant spaces.

Example 4.4.1 The Paley-Wiener space PW_λ with bandwidth $\lambda > 0$ consists of all functions $f \in L_2(\mathbb{R})$ such that $\widehat{f}(\xi) = 0$ for all $\xi \in \mathbb{R}\backslash[-\lambda, \lambda]$. We first claim that $\mathrm{PW}_{\lambda\pi} = \mathrm{S}_\lambda(\mathrm{sinc}\,|L_2(\mathbb{R}))$, where the sinc function is defined to be

$$\mathrm{sinc}(0) := 1 \quad \text{and} \quad \mathrm{sinc}(x) := \frac{\sin(\pi x)}{\pi x}, \qquad x \in \mathbb{R}\backslash\{0\}. \tag{4.4.24}$$

Noting that $\widehat{\chi_{(-\pi,\pi]}}(\xi) = \frac{2\sin(\pi\xi)}{\xi}$, we have $\widehat{\mathrm{sinc}} = \chi_{(-\pi,\pi]}$. Hence, $\mathrm{sinc} \in \mathrm{PW}_\pi$ and $\mathrm{S}(\mathrm{sinc}\,|L_2(\mathbb{R})) \subseteq \mathrm{PW}_\pi$. Conversely, for any $f \in \mathrm{PW}_\pi$. We define a 2π-periodic function $\widehat{v}$ by $\widehat{v}(\xi) := \widehat{f}(\xi)$ for all $\xi \in (-\pi, \pi]$. Then $\widehat{f} = \widehat{v}\,\widehat{\mathrm{sinc}}$ and by Theorem 4.4.3, we have $f \in \mathrm{S}(\mathrm{sinc}\,|L_2(\mathbb{R}))$. Therefore, $\mathrm{PW}_\pi \subseteq \mathrm{S}(\mathrm{sinc}\,|L_2(\mathbb{R}))$. This proves $\mathrm{PW}_\pi = \mathrm{S}(\mathrm{sinc}\,|L_2(\mathbb{R}))$ and consequently, $\mathrm{PW}_{\lambda\pi} = \mathrm{S}_\lambda(\mathrm{sinc}\,|L_2(\mathbb{R}))$.

By $\widehat{\mathrm{sinc}} = \chi_{(-\pi,\pi]}$, we have $[\widehat{\mathrm{sinc}}, 1] = 1$ and $[\widehat{\mathrm{sinc}}, \widehat{\mathrm{sinc}}] = 1$. Therefore, $\{\mathrm{sinc}(\cdot - k) \, : \, k \in \mathbb{Z}\}$ is an orthonormal basis of $\mathrm{S}(\mathrm{sinc}\,|L_2(\mathbb{R}))$, and the conditions in (4.4.19) and (4.4.21) are satisfied with $\phi = \mathrm{sinc}$. Since φ in (4.4.22) is just the sinc function itself by $[\widehat{\mathrm{sinc}}, 1] = 1$, it follows trivially from Theorem 4.4.18 that for $\lambda > 0$, every function f in the reproducing kernel Hilbert space $\mathrm{PW}_{\lambda\pi}$ is uniquely determined by its samples $\{f(k/\lambda)\}_{k\in\mathbb{Z}}$ through the following sampling formula:

$$f(x) = \sum_{k\in\mathbb{Z}} f(k/\lambda)\,\mathrm{sinc}(\lambda x - k), \qquad x \in \mathbb{R}, f \in \mathrm{PW}_{\lambda\pi} = \mathrm{S}_\lambda(\mathrm{sinc}\,|L_2(\mathbb{R})) \tag{4.4.25}$$

with the series converging absolutely and uniformly on $\mathbb{R}$. The formula in (4.4.25) is called the Whittaker-Nyquist-Kotelnikov-Shannon Sampling Theorem in the literature.

Example 4.4.2 For $m \in \mathbb{N}$, the *B-spline function* B_m *of order* m is defined to be

$$B_1 := \chi_{(0,1]} \quad \text{and} \quad B_m := B_{m-1} * B_1 = \int_0^1 B_{m-1}(\cdot - t)dt. \tag{4.4.26}$$

Then $\mathrm{supp}(B_m) = [0, m]$ and $B_m \in \mathscr{C}^{m-2}(\mathbb{R})$. The centered B-spline is defined to be $\phi_m := B_m(\cdot + m/2)$. Since $\widehat{\phi_1}(\xi) = \widehat{\chi_{(-\frac{1}{2},\frac{1}{2}]}}(\xi) = \frac{\sin(\xi/2)}{\xi/2}$, it is evident that $\widehat{\phi_m}(\xi) = (\frac{\sin(\xi/2)}{\xi/2})^m$ for all $\xi \in \mathbb{R}$. We now show that all ϕ_m with $m \geqslant 2$ satisfies the conditions in (4.4.19) and (4.4.21). Clearly, ϕ_m with $m \geqslant 2$ is continuous. Since $|\widehat{\phi_m}(\xi)| \leqslant \max(1, |\xi|^{-m}) \leqslant \max(1, |\xi|^{-2})$ by $m \geqslant 2$, the series $\sum_{k\in\mathbb{Z}} |\widehat{\phi_m}(\xi + 2\pi k)|$ converges uniformly to a continuous function on $\mathbb{R}$. Hence, (4.4.19) holds and $[\widehat{\phi_m}, 1](\xi) := \sum_{k\in\mathbb{Z}} \widehat{\phi_m}(\xi + 2\pi k)$ is well defined and continuous. We now prove that $[\widehat{\phi_m}, 1](\xi) \neq 0$ for all $\xi \in \mathbb{R}$. If m is even, then $\widehat{\phi_m}(\xi) \geqslant 0$ for all $\xi \in \mathbb{R}$

and consequently, $[\widehat{\phi_m}, 1](\xi) \geqslant \widehat{\phi_m}(\xi) > 0$ for all $\xi \in [0, 2\pi)$. We now suppose that m is odd. Note that $\widehat{\phi_m}(0) = 1$ and $\widehat{\phi_m}(2\pi k) = 0$ for all $k \in \mathbb{Z}\backslash\{0\}$. Hence, $[\widehat{\phi_m}, 1](0) = 1 \neq 0$. For $\xi \in (0, 2\pi)$, since m is odd and $\widehat{\phi_m}(\xi) = (\frac{\sin(\xi/2)}{\xi/2})^m$, we have

$$
\begin{aligned}
[\widehat{\phi_m}, 1](\xi) &= \sum_{k\in\mathbb{Z}} \frac{(-1)^{km}\sin^m(\xi/2)}{(\xi/2+k\pi)^m}\\
&= \sin^m(\xi/2)\left(\sum_{k=0}^{\infty}\frac{(-1)^k}{(\xi/2+k\pi)^m} + \sum_{k=-\infty}^{-1}\frac{(-1)^k}{(\xi/2+k\pi)^m}\right)\\
&= \sin^m(\xi/2)\left(\sum_{k=0}^{\infty}\frac{(-1)^k}{(\xi/2+k\pi)^m} + \sum_{k=0}^{\infty}\frac{(-1)^{-1-k}}{(\xi/2+(-1-k)\pi)^m}\right)\\
&= \sin^m(\xi/2)\sum_{k=0}^{\infty}(-1)^k\left(\frac{1}{(k\pi+\xi/2)^m} + \frac{1}{(k\pi+\pi-\xi/2)^m}\right)\\
&\geqslant \sin^m(\xi/2)\left[\left(\frac{1}{(\xi/2)^m} + \frac{1}{(\pi-\xi/2)^m}\right) - \left(\frac{1}{(\pi+\xi/2)^m} + \frac{1}{(2\pi-\xi/2)^m}\right)\right]\\
&= \sin^m(\xi/2)\left[\left(\frac{1}{(\xi/2)^m} - \frac{1}{(\pi+\xi/2)^m}\right) + \left(\frac{1}{(\pi-\xi/2)^m} - \frac{1}{(2\pi-\xi/2)^m}\right)\right] > 0.
\end{aligned}
$$

Since $[\widehat{\phi_m}, 1]$ is continuous, 2π-periodic, and does not vanish everywhere, we conclude that the first inequality in (4.4.21) holds for some positive constants C_1 and C_2. Since $(2/\pi)^m \leqslant |\widehat{\phi_m}(\xi)|^2 \leqslant [\widehat{\phi_m}, \widehat{\phi_m}] \leqslant [\widehat{\phi_m}, 1](\xi)$ for all $\xi \in (-\pi, \pi]$, we conclude that the second inequality in (4.4.21) must hold as well. By Theorem 4.4.18, the function φ defined in (4.4.22) with $\phi = \phi_m$ is a well-defined interpolating function (which is called the fundamental interpolating spline of order m). Moreover, the sampling formula in (4.4.23) holds in the reproducing kernel Hilbert space $\mathrm{S}_\lambda(B_m|L_2(\mathbb{R}))$.

If $m \geqslant 2$ is an odd integer, then $[\widehat{B_m}, 1](\pi) = 0$. Therefore, the not-centered spline $\phi = B_m$ with an odd integer m does not satisfy the first inequality in (4.4.21).

Finally, we discuss sampling theorems in shift-invariant spaces generated by a family of compactly supported interpolating functions.

Example 4.4.3 For $m \in \mathbb{N}$, recall that the interpolatory filter a^I_{2m} is defined in (2.1.6). Define $\widehat{\phi}(\xi) := \prod_{j=1}^{\infty} \widehat{a^I_{2m}}(2^{-j}\xi)$. We now prove that ϕ satisfies the conditions in (4.4.19) and (4.4.21). Define $f_n(\xi) := \chi_{(-2^n\pi, 2^n\pi]}(\xi)\prod_{j=1}^{n}\widehat{a^I_{2m}}(2^{-j}\xi)$ for $n \in \mathbb{N}\cup\{0\}$. Then $\lim_{n\to\infty} f_n(\xi) = \widehat{\phi}(\xi)$ for every $\xi \in \mathbb{R}$. Since $f_0 = \chi_{(-\pi,\pi]}$, we trivially have $[f_0, 1](\xi) = \sum_{k\in\mathbb{Z}} f_0(\xi + 2\pi k) = 1$. Suppose that $[f_{n-1}, 1] = 1$.

Then by $f_n(\xi) = \widehat{a^I_{2m}}(\xi/2) f_{n-1}(\xi/2)$, we have

$$
\begin{aligned}
[f_n, 1](\xi) &= \sum_{k\in\mathbb{Z}} \widehat{a^I_{2m}}(\xi/2 + k\pi) f_{n-1}(\xi/2 + k\pi) \\
&= \sum_{k\in\mathbb{Z}} \widehat{a^I_{2m}}(\xi/2) f_{n-1}(\xi/2 + 2\pi k) + \sum_{k\in\mathbb{Z}} \widehat{a^I_{2m}}(\xi/2 + \pi) f_{n-1}(\xi/2 + 2\pi k + \pi) \\
&= \widehat{a^I_{2m}}(\xi/2) + \widehat{a^I_{2m}}(\xi/2 + \pi) = 1,
\end{aligned}
$$

where we used the identity $\widehat{a^I_{2m}}(\xi) + \widehat{a^I_{2m}}(\xi + \pi) = 1$ and the induction hypothesis $[f_{n-1}, 1] = 1$. By induction, we have $[f_n, 1] = 1$ for all $n \in \mathbb{N} \cup \{0\}$. Consequently, $\int_{\mathbb{R}} f_n(\xi) d\xi = \int_{-\pi}^{\pi} [f_n, 1](\xi) d\xi = 2\pi$. Since $f_n(\xi) \geqslant 0$ for all $\xi \in \mathbb{R}$ due to $\widehat{a^I_{2m}}(\xi) \geqslant 0$, by Fatou's Lemma,

$$
\int_{\mathbb{R}} \widehat{\phi}(\xi) d\xi = \int_{\mathbb{R}} \lim_{n\to\infty} f_n(\xi) d\xi \leqslant \liminf_{n\to\infty} \int_{\mathbb{R}} f_n(\xi) d\xi = 2\pi.
$$

This proves $\widehat{\phi} \in L_1(\mathbb{R})$ by $\widehat{\phi}(\xi) \geqslant 0$. By $0 \leqslant \widehat{a^I_{2m}}(\xi) \leqslant 1$, we have $0 \leqslant \widehat{\phi}(\xi) \leqslant 1$. Hence, $\int_{\mathbb{R}} |\widehat{\phi}(\xi)|^2 d\xi \leqslant \int_{\mathbb{R}} \widehat{\phi}(\xi) d\xi \leqslant 2\pi$, which proves $\phi \in L_2(\mathbb{R})$.

By $\widehat{a^I_{2m}}(\xi) > 0$ for all $\xi \in (-\pi, \pi)$, since $\widehat{\phi}$ is continuous, we have $c := \inf_{\xi\in[-\pi,\pi]} \widehat{\phi}(\xi) > 0$ and hence $0 \leqslant f_n(\xi) \leqslant c^{-1}\widehat{\phi}(\xi) \in L_1(\mathbb{R})$. By the Dominated Convergence Theorem, $\int_{\mathbb{R}} \widehat{\phi}(\xi) d\xi = \lim_{n\to\infty} \int_{\mathbb{R}} f_n(\xi) d\xi = 2\pi$. Since $0 \leqslant \widehat{\phi} \leqslant 1$, we have $0 \leqslant \widehat{\phi}\chi_{(-2^n\pi, 2^n\pi]} \leqslant f_n$. Hence,

$$
\sum_{k\in\mathbb{Z}} \widehat{\phi}(\xi + 2\pi k) \chi_{(-2^n\pi, 2^n\pi]}(\xi + 2\pi k) \leqslant [f_n, 1](\xi) = 1.
$$

Taking $n \to \infty$, we conclude from the above inequality that $[\widehat{\phi}, 1] \leqslant 1$. Since $\int_{\mathbb{R}} \widehat{\phi}(\xi) d\xi = 2\pi$, we have $2\pi = \int_{\mathbb{R}} \widehat{\phi}(\xi) d\xi = \int_{-\pi}^{\pi} [\widehat{\phi}, 1](\xi) d\xi \leqslant 2\pi$, from which we have $[\widehat{\phi}, 1](\xi) = 1$ a.e. $\xi \in \mathbb{R}$. Hence, by $\widehat{\phi}(\xi) \geqslant 0$ for all $\xi \in \mathbb{R}$, the condition in (4.4.19) and the first inequality in (4.4.21) are satisfied. Moreover, we have $0 < c^2 \leqslant [\widehat{\phi}, \widehat{\phi}](\xi) \leqslant 1$ for all $\xi \in \mathbb{R}$. This proves the second inequality in (4.4.21).

Since $[\widehat{\phi}, 1] = 1$, we have $\varphi = \phi$ in Theorem 4.4.18. Therefore, ϕ is a well-defined continuous interpolating function and the sampling formula in the reproducing kernel Hilbert space $\mathsf{S}_\lambda(\phi|L_2(\mathbb{R}))$ holds:

$$
f(x) = \sum_{k\in\mathbb{Z}} f(k/\lambda)\phi(\lambda x - k), \qquad \forall f \in \mathsf{S}_\lambda(\phi|L_2(\mathbb{R})), \lambda > 0, x \in \mathbb{R}
$$

with the series converging absolutely and uniformly on $\mathbb{R}$. In fact, ϕ has compact support and the above series is finite for every $x \in \mathbb{R}$.

4.5 Refinable Structure and Multiresolution Analysis

A major problem in wavelet theory is about how to construct subsets Φ and Ψ in $L_2(\mathbb{R})$ such that the affine system $\mathsf{AS}_0(\Phi;\Psi)$ is a frame or a Riesz basis for $L_2(\mathbb{R})$. Though various characterizations have been obtained in Sect. 4.3 in terms of Φ and Ψ, the theoretical results in Sect. 4.3 are still far from easy for the purpose of constructing such desirable subsets Φ and Ψ.

Fortunately, in this section we show that there are an embedded refinable structure and a multiresolution analysis underlying a nonhomogeneous affine system in $L_2(\mathbb{R})$ which is either nonredundant or a tight frame. Such a refinable structure links an affine system to filter banks and consequently leads to a very efficient way for constructing framelets and wavelets in $L_2(\mathbb{R})$ from filter banks through the refinable structure and refinable functions.

4.5.1 *Biorthogonal Wavelets and Refinable Structure*

We now study refinable structure and multiresolution analysis underlying nonredundant nonhomogeneous affine systems in $L_2(\mathbb{R})$. For simplicity of discussion, from now on we assume that

$$\begin{aligned} \Phi &= \{\phi^1,\dots,\phi^r\}, \quad & \tilde{\Phi} &= \{\tilde{\phi}^1,\dots,\tilde{\phi}^r\}, \\ \Psi &= \{\psi^1,\dots,\psi^s\}, \quad & \tilde{\Psi} &= \{\tilde{\psi}^1,\dots,\tilde{\psi}^s\}, \end{aligned} \tag{4.5.1}$$

are finite sets (i.e., $r,s\in\mathbb{N}_0$), though the results can be extended to $r,s=\infty$.

The following result characterizes a biorthogonal wavelet in $L_2(\mathbb{R})$ through filter banks and the refinable structure.

Theorem 4.5.1 *Let* $\Phi,\tilde{\Phi},\Psi,\tilde{\Psi}\subseteq L_2(\mathbb{R})$ *be as in* (4.5.1). *Then* $(\{\tilde{\Phi};\tilde{\Psi}\},\{\Phi;\Psi\})$ *is a biorthogonal wavelet in* $L_2(\mathbb{R})$ *if and only if*

(1) $\lim_{j\to\infty}\langle\sum_{\ell=1}^r \widehat{\tilde{\phi}^\ell}(2^{-j}\cdot)\overline{\widehat{\phi^\ell}(2^{-j}\cdot)},\mathbf{h}\rangle=\langle 1,\mathbf{h}\rangle$ *for all* $\mathbf{h}\in\mathscr{D}(\mathbb{R})$.

(2) ϕ *and* $\tilde{\phi}$ *satisfy the biorthogonality condition:* $\langle\tilde{\phi},\phi(\cdot-k)\rangle=\boldsymbol{\delta}(k)I_r$ *for all* $k\in\mathbb{Z}$, *where the vector functions* $\phi,\tilde{\phi}:\mathbb{R}\to\mathbb{C}^{r\times 1}$ *and* $\psi,\tilde{\psi}:\mathbb{R}\to\mathbb{C}^{s\times 1}$ *are defined by*

$$\begin{aligned} \phi &:= (\phi^1,\dots,\phi^r)^{\mathsf{T}}, \quad & \psi &:= (\psi^1,\dots,\psi^s)^{\mathsf{T}}, \\ \tilde{\phi} &:= (\tilde{\phi}^1,\dots,\tilde{\phi}^r)^{\mathsf{T}}, \quad & \tilde{\psi} &:= (\tilde{\psi}^1,\dots,\tilde{\psi}^s)^{\mathsf{T}}. \end{aligned} \tag{4.5.2}$$

(3) There exist $a, \tilde{a} \in (l_2(\mathbb{Z}))^{r\times r}$ *and* $b, \tilde{b} \in (l_2(\mathbb{Z}))^{s\times r}$ *such that*

$$\widehat{\phi}(2\xi) = \widehat{a}(\xi)\widehat{\phi}(\xi), \qquad \widehat{\psi}(2\xi) = \widehat{b}(\xi)\widehat{\phi}(\xi), \qquad a.e.\ \xi \in \mathbb{R}, \tag{4.5.3}$$

$$\widehat{\tilde{\phi}}(2\xi) = \widehat{\tilde{a}}(\xi)\widehat{\tilde{\phi}}(\xi), \qquad \widehat{\tilde{\psi}}(2\xi) = \widehat{\tilde{b}}(\xi)\widehat{\tilde{\phi}}(\xi), \qquad a.e.\ \xi \in \mathbb{R}, \tag{4.5.4}$$

and $(\{\tilde{a}; \tilde{b}\}, \{a; b\})$ *is a biorthogonal wavelet (matrix) filter bank, i.e.,* $s = r$ *and*

$$\begin{bmatrix} \widehat{\tilde{a}}(\xi) & \widehat{\tilde{a}}(\xi+\pi) \\ \widehat{\tilde{b}}(\xi) & \widehat{\tilde{b}}(\xi+\pi) \end{bmatrix} \begin{bmatrix} \overline{\widehat{a}(\xi)}^{\mathsf{T}} & \overline{\widehat{b}(\xi)}^{\mathsf{T}} \\ \overline{\widehat{a}(\xi+\pi)}^{\mathsf{T}} & \overline{\widehat{b}(\xi+\pi)}^{\mathsf{T}} \end{bmatrix} = I_{2r}, \qquad a.e.\ \xi \in \mathbb{R}. \tag{4.5.5}$$

(4) $\mathsf{AS}_0(\Phi; \Psi)$ *and* $\mathsf{AS}_0(\tilde{\Phi}; \tilde{\Psi})$ *are Bessel sequences in* $L_2(\mathbb{R})$.

Proof Necessity ($\Rightarrow$). By definition, items (2) and (4) are obvious. By Corollary 4.2.8, the pair $(\{\tilde{\Phi}; \tilde{\Psi}\}, \{\Phi; \Psi\})$ is a dual framelet in $L_2(\mathbb{R})$. Now item (1) follows directly from Theorem 4.1.6.

We now prove item (3). By Corollary 4.3.14, $(\mathsf{AS}_J(\tilde{\Phi}; \tilde{\Psi}), \mathsf{AS}_J(\Phi; \Psi))$ is a pair of biorthogonal bases for $L_2(\mathbb{R})$ for all $J \in \mathbb{Z}$. To prove (4.5.3), we consider the expansions of $\phi(2^{-1}\cdot)$ and $\psi(2^{-1}\cdot)$ under the Riesz basis $\mathsf{AS}_0(\Phi; \Psi)$. By biorthogonality of $(\mathsf{AS}_0(\tilde{\Phi}; \tilde{\Psi}), \mathsf{AS}_0(\Phi; \Psi))$, we get from their representations in (4.3.12) that

$$\begin{aligned} \phi(2^{-1}\cdot) &= \sum_{k\in\mathbb{Z}} \langle \phi(2^{-1}\cdot), \tilde{\phi}(\cdot - k)\rangle \phi(\cdot - k), \\ \psi(2^{-1}\cdot) &= \sum_{k\in\mathbb{Z}} \langle \psi(2^{-1}\cdot), \tilde{\phi}(\cdot - k)\rangle \phi(\cdot - k), \end{aligned} \tag{4.5.6}$$

with the series in (4.5.6) converging in $L_2(\mathbb{R})$. Define

$$a(k) := 2^{-1}\langle \phi(2^{-1}\cdot), \tilde{\phi}(\cdot - k)\rangle, \quad b(k) := 2^{-1}\langle \psi(2^{-1}\cdot), \tilde{\phi}(\cdot - k)\rangle, \quad k \in \mathbb{Z}.$$

Then $a \in (l_2(\mathbb{Z}))^{r\times r}$, $b \in (l_2(\mathbb{Z}))^{s\times r}$, and (4.5.3) follows directly from (4.5.6). Similarly, we see that (4.5.4) holds with

$$\tilde{a}(k) := 2^{-1}\langle \tilde{\phi}(2^{-1}\cdot), \phi(\cdot - k)\rangle, \quad \tilde{b}(k) := 2^{-1}\langle \tilde{\psi}(2^{-1}\cdot), \phi(\cdot - k)\rangle, \quad k \in \mathbb{Z}.$$

From the biorthogonality relations, we have

$$\langle \tilde{\phi}, \phi(\cdot - k)\rangle = \boldsymbol{\delta}(k)I_r, \ \langle \tilde{\phi}, \psi(\cdot - k)\rangle = 0, \ \langle \tilde{\psi}, \phi(\cdot - k)\rangle = 0, \ \langle \tilde{\psi}, \psi(\cdot - k)\rangle = \boldsymbol{\delta}(k)I_s$$

for all $k \in \mathbb{Z}$. By Lemma 4.4.1, the above identities are equivalent to

$$[\widehat{\tilde{\phi}}, \widehat{\phi}](\xi) = I_r, \quad [\widehat{\tilde{\phi}}, \widehat{\psi}](\xi) = 0, \quad [\widehat{\tilde{\psi}}, \widehat{\phi}](\xi) = 0, \quad [\widehat{\tilde{\psi}}, \widehat{\psi}](\xi) = I_r, \quad a.e.\ \xi \in \mathbb{R}.$$

Using (4.5.3) and (4.5.4), we deduce from the above identities that

$$
\begin{aligned}
I_r &= [\widehat{\tilde{\phi}}, \widehat{\phi}](2\xi) = \widehat{\tilde{a}}(\xi)[\widehat{\tilde{\phi}}, \widehat{\phi}](\xi)\overline{\widehat{a}(\xi)}^{\mathsf{T}} + \widehat{\tilde{a}}(\xi+\pi)[\widehat{\tilde{\phi}}, \widehat{\phi}](\xi+\pi)\overline{\widehat{a}(\xi+\pi)}^{\mathsf{T}}\\
&= \widehat{\tilde{a}}(\xi)\overline{\widehat{a}(\xi)}^{\mathsf{T}} + \widehat{\tilde{a}}(\xi+\pi)\overline{\widehat{a}(\xi+\pi)}^{\mathsf{T}},\\
0 &= [\widehat{\tilde{\phi}}, \widehat{\psi}](2\xi) = \widehat{\tilde{a}}(\xi)[\widehat{\tilde{\phi}}, \widehat{\phi}](\xi)\overline{\widehat{b}(\xi)}^{\mathsf{T}} + \widehat{\tilde{a}}(\xi+\pi)[\widehat{\tilde{\phi}}, \widehat{\phi}](\xi+\pi)\overline{\widehat{b}(\xi+\pi)}^{\mathsf{T}}\\
&= \widehat{\tilde{a}}(\xi)\overline{\widehat{b}(\xi)}^{\mathsf{T}} + \widehat{\tilde{a}}(\xi+\pi)\overline{\widehat{b}(\xi+\pi)}^{\mathsf{T}},
\end{aligned}
$$

and by the same argument we have

$$
\begin{aligned}
I_s &= [\widehat{\tilde{\psi}}, \widehat{\psi}](2\xi) = \widehat{\tilde{b}}(\xi)\overline{\widehat{b}(\xi)}^{\mathsf{T}} + \widehat{\tilde{b}}(\xi+\pi)\overline{\widehat{b}(\xi+\pi)}^{\mathsf{T}},\\
0 &= [\widehat{\tilde{\psi}}, \widehat{\phi}](2\xi) = \widehat{\tilde{b}}(\xi)\overline{\widehat{a}(\xi)}^{\mathsf{T}} + \widehat{\tilde{b}}(\xi+\pi)\overline{\widehat{a}(\xi+\pi)}^{\mathsf{T}}.
\end{aligned}
$$

The above four identities can be rewritten into the matrix form in (4.5.5) but with I_{2r} on the right-hand side being replaced by I_{r+s}. To complete the proof, we have to prove $s = r$. Define $H := \{\eta^1, \dots .\eta^{2s}\}$, where $\eta^\ell := \phi^\ell_{2;0}$ and $\eta^{\ell+s} := \phi^\ell_{2;1}$, $\ell = 1, \dots, s$. Representing ϕ^ℓ, ψ^m under the pair $(\mathsf{AS}_1(\tilde{\Phi}; \tilde{\Psi}), \mathsf{AS}_1(\Phi; \Psi))$ of biorthogonal bases for $L_2(\mathbb{R})$ (c.f. (4.5.6)), we can easily deduce that $\phi^\ell, \psi^m \in \mathsf{S}(H|L_2(\mathbb{R}))$ for all $\ell = 1, \dots, r$ and $m = 1, \dots, s$. Similarly, representing η^ℓ under the pair $(\mathsf{AS}_0(\tilde{\Phi}; \tilde{\Psi}), \mathsf{AS}_0(\Phi; \Psi))$ of biorthogonal bases for $L_2(\mathbb{R})$, we see that $\eta^\ell \in \mathsf{S}(\Phi \cup \Psi|L_2(\mathbb{R}))$ for all $\ell = 1, \dots, 2s$. Therefore, $\mathsf{S}(\Phi \cup \Psi|L_2(\mathbb{R})) = \mathsf{S}(H|L_2(\mathbb{R})) =: \mathscr{V}$. Consequently, both $\{\phi^\ell(\cdot - k), \psi^m(\cdot - k) \; : \; \ell = 1, \dots, r, m = 1, \dots, s, k \in \mathbb{Z}\}$ and $\{\eta^\ell(\cdot - k) \; : \; \ell = 1, \dots, 2s, k \in \mathbb{Z}\}$ are Riesz bases for the same space $\mathscr{V}$. Define $\eta := (\eta^1, \dots, \eta^{2s})^{\mathsf{T}}$ and $\theta := (\phi^1, \dots, \phi^r, \psi^1, \dots, \psi^s)^{\mathsf{T}}$. By Theorems 4.4.11,

$$
\widehat{\eta}(\xi) = \mathbf{u}(\xi)\widehat{\theta}(\xi) \quad \text{and} \quad \widehat{\theta}(\xi) = \mathbf{v}(\xi)\widehat{\eta}(\xi), \qquad a.e.\, \xi \in \mathbb{R}, \tag{4.5.7}
$$

where $\mathbf{u}$ and $\mathbf{v}$ are $(2s)\times(r+s)$ and $(r+s)\times(2s)$ matrices of 2π-periodic measurable functions. It follows from (4.5.7) that $\widehat{\eta}(\xi) = \mathbf{U}(\xi)\widehat{\eta}(\xi)$ with $\mathbf{U} = \mathbf{uv}$ and therefore, $[\widehat{\eta}, \widehat{\eta}](\xi) = \mathbf{U}(\xi)[\widehat{\eta}, \widehat{\eta}](\xi)\mathbf{U}(\xi)^\star$. By Theorem 4.4.12, $\det([\widehat{\eta}, \widehat{\eta}](\xi)) \ne 0$ for almost every $\xi \in \mathbb{R}$. Consequently, $\det(\mathbf{U}(\xi)) \ne 0$ a.e. $\xi \in \mathbb{R}$. By $\mathbf{U} = \mathbf{uv}$, we conclude that $r + s \geqslant 2s$, that is, $r \geqslant s$. We can similarly prove that $s \geqslant r$ by considering $\widehat{\theta}(\xi) = \mathbf{V}(\xi)\widehat{\theta}(\xi)$ with $\mathbf{V} = \mathbf{vu}$. This proves $s = r$. Hence, item (3) holds.

Sufficiency ($\Leftarrow$). By Theorems 4.3.7 and 4.1.10 with $\widehat{\theta} = \widehat{\tilde{\theta}} = I_r$, we deduce from items (1), (3), and (4) that $(\{\tilde{\Phi}; \tilde{\Psi}\}, \{\Phi; \Psi\})$ is a dual framelet in $L_2(\mathbb{R})$. By items (2) and (3), noting $\mathsf{S}(\Phi \cup \Psi|L_2(\mathbb{R})) \subseteq \{f(2\cdot) \; : \; f \in \mathsf{S}(\Phi|L_2(\mathbb{R}))\}$ and $\mathsf{S}(\tilde{\Phi} \cup \tilde{\Psi}|L_2(\mathbb{R})) \subseteq \{f(2\cdot) \; : \; f \in \mathsf{S}(\tilde{\Phi}|L_2(\mathbb{R}))\}$, we deduce that $\mathsf{AS}_0(\tilde{\Phi}; \tilde{\Psi})$ and $\mathsf{AS}_0(\Phi; \Psi)$ are biorthogonal to each other. Hence, we conclude that $(\mathsf{AS}_0(\tilde{\Phi}; \tilde{\Psi}), \mathsf{AS}_0(\Phi; \Psi))$ must be a pair of biorthogonal bases for $L_2(\mathbb{R})$. □

We call the relations in (4.5.3) *the refinable structure of* ϕ *and* ψ. The Bessel property in item (4) of Theorem 4.5.1 will be studied in Corollary 4.6.6.

As a special case of Theorem 4.5.1, an orthogonal wavelet in $L_2(\mathbb{R})$ can be characterized as follows:

Corollary 4.5.2 *Let* $\Phi = \{\phi^1, \dots, \phi^r\} \subseteq L_2(\mathbb{R})$ *and* $\Psi = \{\psi^1, \dots, \psi^s\} \subseteq L_2(\mathbb{R})$. *Then* $\{\Phi; \Psi\}$ *is an orthogonal wavelet in* $L_2(\mathbb{R})$ *if and only if all items (1)–(3) of Theorem 4.5.1 are satisfied with* $\tilde{\phi} = \phi, \tilde{\psi} = \psi$ *and* $\tilde{a} = a, \tilde{b} = b$.

4.5.2 Tight Framelets and Refinable Structure

In this subsection we show that the cascade structure in a tight framelet leads to the refinable structure.

Lemma 4.5.3 *Let* Φ, Ψ *be subsets of* $L_2(\mathbb{R})$. *If* $\{\Phi; \Psi\}$ *is a tight framelet in* $L_2(\mathbb{R})$, *then* $\mathsf{S}(\Phi \cup \Psi | L_2(\mathbb{R})) = \mathsf{S}_2(\Phi | L_2(\mathbb{R})) := \overline{\text{span}\{\phi(2\cdot - k) \ : \ k \in \mathbb{Z}, \phi \in \Phi\}}$.

Proof Since $\{\Phi; \Psi\}$ is a tight framelet in $L_2(\mathbb{R})$, the following cascade structure holds (see (4.3.22)):

$$\sum_{\phi \in \Phi} \sum_{k \in \mathbb{Z}} |\langle f, \phi_{2^j;k} \rangle|^2 + \sum_{\psi \in \Psi} \sum_{k \in \mathbb{Z}} |\langle f, \psi_{2^j;k} \rangle|^2 = \sum_{\phi \in \Phi} \sum_{k \in \mathbb{Z}} |\langle f, \phi_{2^{j+1};k} \rangle|^2, \tag{4.5.8}$$

for all $f \in L_2(\mathbb{R})$ and $j \in \mathbb{Z}$. It follows trivially from (4.5.8) with $j = 0$ that $\langle f, \phi_{2;k} \rangle = 0$ for all $k \in \mathbb{Z}$ and $\phi \in \Phi$ if and only if $\langle f, \phi(\cdot - k) \rangle = \langle f, \psi(\cdot - k) \rangle = 0$ for all $k \in \mathbb{Z}$ and $\phi \in \Phi, \psi \in \Psi$. That is, we proved $\mathsf{S}_2(\Phi | L_2(\mathbb{R}))^\perp = \mathsf{S}(\Phi \cup \Psi | L_2(\mathbb{R}))^\perp$. Therefore, we must have $\mathsf{S}(\Phi \cup \Psi | L_2(\mathbb{R})) = \mathsf{S}_2(\Phi | L_2(\mathbb{R}))$. □

We now characterize tight framelets through filter banks and refinable structure in the following result, which is more general than Corollary 4.5.2 for orthogonal wavelets.

Theorem 4.5.4 *Let* $\Phi = \{\phi^1, \dots, \phi^r\}$ *and* $\Psi = \{\psi^1, \dots, \psi^s\}$ *be finite subsets of tempered distributions such that all* $\widehat{\phi^1}, \dots, \widehat{\phi^r}, \widehat{\psi^1}, \dots, \widehat{\psi^s}$ *belong to* $L_2^{loc}(\mathbb{R})$. *Define vector functions* $\phi := (\phi^1, \dots, \phi^r)^\mathsf{T}$ *and* $\psi := (\psi^1, \dots, \psi^s)^\mathsf{T}$. *Then* $\Phi \cup \Psi \subset L_2(\mathbb{R})$ *and* $\{\Phi; \Psi\}$ *is a tight framelet in* $L_2(\mathbb{R})$ *if and only if*

(1) $\lim_{j \to \infty} \langle \widehat{\phi}(2^{-j}\cdot)^\mathsf{T} \overline{\widehat{\phi}(2^{-j}\cdot)}, \mathbf{h} \rangle = \langle 1, \mathbf{h} \rangle$ *for all* $\mathbf{h} \in \mathscr{D}(\mathbb{R})$.

(2) *There exist an* $r \times r$ *matrix* $\widehat{a}$ *and an* $s \times r$ *matrix* $\widehat{b}$ *of* 2π*-periodic measurable functions such that* (4.5.3) *holds and* $\{a; b\}$ *is* a generalized tight framelet filter bank *satisfying*

$$\widehat{\phi}(\xi)^\mathsf{T} \Big(\widehat{a}(\xi)^\mathsf{T} \overline{\widehat{a}(\xi)} + \widehat{b}(\xi)^\mathsf{T} \overline{\widehat{b}(\xi)} - I_r \Big) \overline{\widehat{\phi}(\xi + 2\pi k)} = 0,$$

$$\widehat{\phi}(\xi)^\mathsf{T} \Big(\widehat{a}(\xi)^\mathsf{T} \overline{\widehat{a}(\xi + \pi)} + \widehat{b}(\xi)^\mathsf{T} \overline{\widehat{b}(\xi + \pi)} \Big) \overline{\widehat{\phi}(\xi + 2\pi k)} = 0,$$

for almost every $\xi \in \mathbb{R}$ *and for all* $k \in \mathbb{Z}$.

Proof Sufficiency ($\Leftarrow$). By Theorem 4.1.10 with $\widehat{\widetilde{\theta}} = \widehat{\theta} = I_r$, items (1) and (2) imply that $(\{\widehat{\Phi}; \widehat{\Psi}\}, \{\widehat{\Phi}; \widehat{\Psi}\})$ is a frequency-based dual framelet. By Theorem 4.3.8, we conclude that $\{\Phi; \Psi\}$ is a tight framelet in $L_2(\mathbb{R})$.

Necessity ($\Rightarrow$). The existence of $\widehat{a}$ and $\widehat{b}$ satisfying (4.5.3) is guaranteed by Lemma 4.5.3 and Theorem 4.4.11. Now items (1) and (2) follows directly from Theorems 4.3.8 and 4.1.10. □

Note that $\widehat{\phi}(\xi)^{\mathsf{T}}\overline{\widehat{\phi}(\xi)} = \|\widehat{\phi}(\xi)\|_{l_2}^2$. If $\|\widehat{\phi}(\xi)\|_{l_2}^2$ is continuous at the point $\xi = 0$, then Lemma 4.1.4 tells us that item (1) of Theorem 4.5.4 is equivalent to the simple normalization condition $\|\widehat{\phi}(0)\|_{l_2} = 1$.

4.5.3 *Multiresolution Analysis and Orthogonal Wavelets in $L_2(\mathbb{R})$*

A refinable structure is closely related to a multiresolution analysis of $L_2(\mathbb{R})$. Though a multiresolution analysis plays a critical role in the classical theory of wavelets, we do not need the results in this section on a multiresolution analysis in this book; we provide this section to study a multiresolution analysis mainly for the purpose of completeness of the wavelet theory.

We say that a sequence $\{\mathscr{V}_j\}_{j\in\mathbb{Z}}$ of closed subspaces in $L_2(\mathbb{R})$ forms *a (wavelet) multiresolution analysis (MRA)* of $L_2(\mathbb{R})$ if

(i) $\mathscr{V}_j = \{f(2^j\cdot) \; : \; f \in \mathscr{V}_0\}$ and $\mathscr{V}_j \subseteq \mathscr{V}_{j+1}$ for all integers $j \in \mathbb{Z}$;
(ii) $\overline{\cup_{j\in\mathbb{Z}}\mathscr{V}_j} = L_2(\mathbb{R})$ (that is, $\cup_{j\in\mathbb{Z}}\mathscr{V}_j$ is dense in $L_2(\mathbb{R})$) and $\cap_{j\in\mathbb{Z}}\mathscr{V}_j = \{0\}$;
(iii) there exists a finite subset $\Phi \subseteq L_2(\mathbb{R})$ such that $\{\phi(\cdot - k) \; : \; k \in \mathbb{Z}, \phi \in \Phi\}$ is a Riesz basis for $\mathscr{V}_0$.

Note that item (i) can be replaced by $\mathscr{V}_0 \subseteq \mathscr{V}_1$ and $\mathscr{V}_j = \{f(2^j\cdot) \; : \; f \in \mathscr{V}_0\}$ for all integers j. Clearly, a multiresolution analysis $\{\mathscr{V}_j\}_{j\in\mathbb{Z}}$ is completely determined by Φ in item (iii), since $\mathscr{V}_j = \mathsf{S}_{2^j}(\Phi|L_2(\mathbb{R}))$.

Suppose that $\{\Phi; \Psi\}$ is an orthogonal wavelet in $L_2(\mathbb{R})$. Define $\mathscr{V}_j := \mathsf{S}_{2^j}(\Phi|L_2(\mathbb{R}))$ and $\mathscr{W}_j := \mathsf{S}_{2^j}(\Psi|L_2(\mathbb{R}))$ for all $j \in \mathbb{Z}$. By Corollary 4.5.2, the sequence $\{\mathscr{V}_j\}_{j\in\mathbb{Z}}$ of subspaces of $L_2(\mathbb{R})$ forms a multiresolution analysis of $L_2(\mathbb{R})$ and

$$\mathscr{V}_{J+1} = \mathscr{V}_J \oplus \mathscr{W}_J \quad \text{and} \quad L_2(\mathbb{R}) = \mathscr{V}_J \oplus_{j=J}^{\infty} \mathscr{W}_j = \oplus_{j\in\mathbb{Z}}\mathscr{W}_j, \qquad \forall\, J \in \mathbb{Z},$$

where $\oplus$ means the orthogonal sum of closed subspaces in $L_2(\mathbb{R})$. Note that $\{\phi_{2^j;k} \; : \; k \in \mathbb{Z}, \phi \in \Phi\}$ and $\{\psi_{2^j;k} \; : \; k \in \mathbb{Z}, \psi \in \Psi\}$ are orthonormal bases for $\mathscr{V}_j$ and $\mathscr{W}_j$, respectively.

To have a more general notion of a (framelet) multiresolution analysis, item (iii) in the definition of a multiresolution analysis may be replaced by the existence of a finite subset $\Phi \subseteq L_2(\mathbb{R})$ such that $\mathscr{V}_0 = \mathsf{S}(\Phi|L_2(\mathbb{R}))$. As we shall see in Theorem 6.8.3 of Chap. 6, the Riesz basis property in item (iii) will play an indispensable role in linking dual framelets in the function setting to discrete framelet transforms in the discrete domain.

We now study the first condition $\overline{\cup_{j\in\mathbb{Z}}\mathscr{V}_j} = L_2(\mathbb{R})$ of item (ii) in the definition of a multiresolution analysis.

Theorem 4.5.5 *Let $\Phi \subseteq L_2(\mathbb{R})$ and $\mathscr{V}_j := \mathsf{S}_{2^j}(\Phi|L_2(\mathbb{R}))$ for all $j \in \mathbb{Z}$. If $\mathscr{V}_0 \subseteq \mathscr{V}_1$, then $\overline{\cup_{j\in\mathbb{Z}}\mathscr{V}_j} = L_2(\mathbb{R})$ if and only if $\mathbb{R}\backslash(\cup_{j\in\mathbb{Z}}\cup_{\phi\in\Phi} 2^j\,\mathrm{supp}(\widehat{\phi}))$ has measure zero.*

Proof Necessity ($\Rightarrow$). We use proof by contradiction. Let $X := \cup_{j\in\mathbb{Z}}\cup_{\phi\in\Phi} 2^j\,\mathrm{supp}(\widehat{\phi})$. If $\mathbb{R}\backslash X$ has a positive measure, then there is a measurable set $E \subseteq \mathbb{R}\backslash X$ such that E has a finite positive measure. Define f by $\widehat{f} := \chi_E$. Then $\|f\|_{L_2(\mathbb{R})} > 0$. Since $E \subseteq \mathbb{R}\backslash X$, we must have $2\pi\langle f, \phi_{2^j;k}\rangle = \langle \widehat{f}, \widehat{\phi}_{2^{-j};0,k}\rangle = 0$ for all $j, k \in \mathbb{Z}$ and $\phi \in \Phi$. Hence, we have $f \perp \mathscr{V}_j$ for all $j \in \mathbb{Z}$, from which we conclude that $f \perp \overline{U}$, where $U := \cup_{j\in\mathbb{Z}}\mathscr{V}_j$. This shows that $\overline{U} \ne L_2(\mathbb{R})$, a contradiction to our assumption. Hence, $\overline{\cup_{j\in\mathbb{Z}}\mathscr{V}_j} = L_2(\mathbb{R})$ must imply that $\mathbb{R}\backslash X$ has measure zero.

Sufficiency ($\Leftarrow$). We first prove that $\overline{U}$ is translation-invariant, that is, $f(\cdot - t) \in \overline{U}$ for all $f \in \overline{U}$ and $t \in \mathbb{R}$. Let $\varepsilon > 0$. For $f \in \overline{U}$, there exists $g \in U$ such that $\|f - g\|_{L_2(\mathbb{R})} < \varepsilon/2$. By the definition of U, there exists $J \in \mathbb{Z}$ such that $g \in \mathscr{V}_J$. Note that $\mathscr{V}_0 \subseteq \mathscr{V}_1$ implies $\mathscr{V}_j \subseteq \mathscr{V}_{j+1}$ for all $j \in \mathbb{Z}$. Consequently, $g \in \mathscr{V}_j$ for all $j \geqslant J$. By the fact $\lim_{y\to 0}\|g - g(\cdot - y)\|_{L_2(\mathbb{R})} = 0$, there exists $\delta > 0$ such that $\|g - g(\cdot - y)\|_{L_2(\mathbb{R})} < \varepsilon/2$ for all $|y| < \delta$. For sufficiently large $j \geqslant J$, we have $|t - 2^{-j}k| < \delta$ for some $k \in \mathbb{Z}$. Since $g \in \mathscr{V}_j$, we have $g(\cdot - 2^{-j}k) \in \mathscr{V}_j$. In summary, we have

$$\begin{aligned}\|f(\cdot - t) - g(\cdot - 2^{-j}k)\|_{L_2(\mathbb{R})} \leqslant& \|f(\cdot - t) - g(\cdot - t)\|_{L_2(\mathbb{R})} \\ &+ \|g(\cdot - t) - g(\cdot - 2^{-j}k)\|_{L_2(\mathbb{R})} < \tfrac{\varepsilon}{2} + \tfrac{\varepsilon}{2} = \varepsilon.\end{aligned}$$

Since $g(\cdot - 2^{-j}k) \in U$ and ε is arbitrary, we proved $f(\cdot - t) \in \overline{U}$. Hence, the linear space $\overline{U}$ is translation-invariant. We use proof by contradiction to show that $\overline{U} = L_2(\mathbb{R})$. Suppose not. Then there exists $f \in L_2(\mathbb{R})$ such that $\|f\|_{L_2(\mathbb{R})} > 0$ and $f \perp \overline{U}$. For each $\phi \in \Phi$, since $\overline{U}$ is translation-invariant, we have $\phi_{2^{-j};t} \in \overline{U}$. Therefore, $f \perp \phi_{2^{-j};t}$ for all $\phi \in \Phi$, $t \in \mathbb{R}$ and $j \in \mathbb{Z}$. In other words,

$$0 = 2\pi\langle f, \phi_{2^{-j};t}\rangle = \langle \widehat{f}, \widehat{\phi}_{2^j;0,t}\rangle = 2^{j/2}\int_{\mathbb{R}} \widehat{f}(\xi)\overline{\widehat{\phi}(2^j\xi)}e^{i2^jt\xi}\,d\xi, \qquad \forall\, t \in \mathbb{R}.$$

Hence, $\widehat{f}(\xi)\overline{\widehat{\phi}(2^j\xi)} = 0$ for a.e. $\xi \in \mathbb{R}$ for all $j \in \mathbb{Z}$. This shows that $\mathrm{supp}(\widehat{f})$ is essentially contained inside the null set $\mathbb{R}\backslash X$. Hence, $\mathrm{supp}(\widehat{f})$ must have measure zero, which is a contradiction to $\|f\|_{L_2(\mathbb{R})} > 0$. This proves $\overline{U} = L_2(\mathbb{R})$. ☐

The following result shows that the second condition $\cap_{j\in\mathbb{Z}}\mathscr{V}_j = \{0\}$ in item (ii) is a direct consequence of item (iii) in the definition of a multiresolution analysis.

Proposition 4.5.6 *For a finite subset $\Phi \subseteq L_2(\mathbb{R})$, define $\mathscr{V}_j := \mathsf{S}_{2^j}(\Phi|L_2(\mathbb{R}))$ for $j \in \mathbb{Z}$. Then $\cap_{j\in\mathbb{Z}}\mathscr{V}_j = \{0\}$.*

Proof Since Φ is finite, we can write $\Phi = \{\phi^1, \dots, \phi^r\}$. Define functions $\varphi^1, \dots, \varphi^r \in L_2(\mathbb{R})$ as in (4.4.7). By Proposition 4.4.7, we see that the orthogonal projection operator $\mathcal{P}_j : L_2(\mathbb{R}) \to \mathsf{S}_{2^j}(\Phi|L_2(\mathbb{R}))$ can be expressed as

$$\mathcal{P}_j f = \sum_{\ell=1}^{r} \sum_{k\in\mathbb{Z}} \langle f, \varphi^{\ell}_{2^j;k}\rangle \varphi^{\ell}_{2^j;k}, \qquad f \in L_2(\mathbb{R}).$$

By $[\widehat{\varphi^\ell}, \widehat{\varphi^\ell}](\xi) \leqslant 1$ and Lemma 4.3.2, the limit in (4.3.4) holds with $\Phi = \{\varphi^1, \dots, \varphi^r\}$. Let $f \in \cap_{j\in\mathbb{Z}} \mathscr{V}_j$. Then $\mathcal{P}_j f = f$ for all $j \in \mathbb{Z}$. Now it follows directly from (4.3.4) with $\Phi = \{\varphi^1, \dots, \varphi^r\}$ that

$$\|f\|^2_{L_2(\mathbb{R})} = \langle \mathcal{P}_j f, f\rangle = \sum_{\ell=1}^{r} \sum_{k\in\mathbb{Z}} |\langle f, \varphi^{\ell}_{2^j;k}\rangle|^2 \to 0 \quad \text{as } j \to -\infty.$$

Hence, $\|f\|_{L_2(\mathbb{R})} = 0$ which implies $f = 0$. So, we must have $\cap_{j\in\mathbb{Z}} \mathscr{V}_j = \{0\}$. ☐

The following is a direct consequence of Theorem 4.4.11.

Proposition 4.5.7 *Let $\Phi = \{\phi^1, \dots, \phi^r\}$ be a finite subset of $L_2(\mathbb{R})$. Then $\mathsf{S}(\Phi|L_2(\mathbb{R})) \subseteq \mathsf{S}_2(\Phi|L_2(\mathbb{R}))$ if and only if there exists an $r \times r$ matrix $\mathbf{a}$ of 2π-periodic measurable functions such that $\widehat{\phi}(2\xi) = \mathbf{a}(\xi)\widehat{\phi}(\xi)$ for a.e. $\xi \in \mathbb{R}$, where $\phi := (\phi^1, \dots, \phi^r)^{\mathsf{T}}$.*

The following result characterizes a multiresolution analysis generated by a finite subset Φ of $L_2(\mathbb{R})$.

Theorem 4.5.8 *Let $\Phi = \{\phi^1, \dots, \phi^r\}$ be a finite subset of $L_2(\mathbb{R})$. Define $\mathscr{V}_j := \mathsf{S}_{2^j}(\Phi|L_2(\mathbb{R}))$ for all $j \in \mathbb{Z}$ and $\phi := (\phi^1, \dots, \phi^r)^{\mathsf{T}}$. Then $\{\mathscr{V}_j\}_{j\in\mathbb{Z}}$ forms a (wavelet) multiresolution analysis of $L_2(\mathbb{R})$ if and only if*

(i) $\mathbb{R}\backslash(\cup_{\ell=1}^{r} \cup_{j\in\mathbb{Z}} 2^j \operatorname{supp}(\widehat{\phi^\ell}))$ *has measure zero;*
(ii) There exists an $r \times r$ matrix $\mathbf{a}$ of 2π-periodic measurable functions such that $\widehat{\phi}(2\xi) = \mathbf{a}(\xi)\widehat{\phi}(\xi)$ for almost every $\xi \in \mathbb{R}$;
(iii) $\dim_{\mathsf{S}(\Phi|L_2(\mathbb{R}))}(\xi) = \operatorname{len}(\mathsf{S}(\Phi|L_2(\mathbb{R})))$ *for almost every $\xi \in \mathbb{R}$.*

Proof Since $\mathscr{V}_j = \mathsf{S}_{2^j}(\Phi|L_2(\mathbb{R}))$, we see that $\mathscr{V}_j \subseteq \mathscr{V}_{j+1}$ for all $j \in \mathbb{Z}$ if and only if $\mathscr{V}_0 \subseteq \mathscr{V}_1$. By Proposition 4.5.7, we see that $\mathscr{V}_0 \subseteq \mathscr{V}_1$ if and only if item (ii) holds. By Theorem 4.5.5 and $\mathscr{V}_0 \subseteq \mathscr{V}_1$, we see that $\cup_{j\in\mathbb{Z}} \mathscr{V}_j$ is dense in $L_2(\mathbb{R})$ if and only if item (i) holds. By Proposition 4.5.6, we automatically have $\cap_{j\in\mathbb{Z}} \mathscr{V}_j = \{0\}$.

Define $s := \operatorname{len}(\mathsf{S}(\Phi|L_2(\mathbb{R})))$. Let $\varphi^1, \dots, \varphi^r$ be defined in Proposition 4.4.9. If item (iii) holds, then we must have $\varphi^\ell = 0$ for all $s < \ell \leqslant r$ and $\mathbb{R}\backslash \operatorname{supp}([\widehat{\varphi^\ell}, \widehat{\varphi^\ell}])$ has measure zero for all $\ell = 1, \dots, s$. By item (ii) of Proposition 4.4.9, we must have $[\widehat{\varphi^\ell}, \widehat{\varphi^{\ell'}}] = \boldsymbol{\delta}(\ell - \ell')$ a.e. for all $\ell, \ell' = 1, \dots, s$. By Lemma 4.4.1, the set $\{\varphi^\ell(\cdot - k) : k \in \mathbb{Z}, \ell = 1, \dots, s\}$ is an orthonormal (and hence Riesz) basis for $\mathscr{V}_0$. Conversely, suppose that there exist $\varphi^1, \dots, \varphi^s \in \mathscr{V}_0$ such that $\{\varphi^\ell(\cdot - k) : k \in \mathbb{Z}, \ell = 1, \dots, s\}$ is a Riesz basis for $\mathscr{V}_0$. By Theorem 4.4.12, we have $[\widehat{\varphi},\widehat{\varphi}] > 0$ a.e. with $\varphi := (\varphi^1,\dots,\varphi^s)^\mathsf{T}$. In particular, this implies $\dim(\mathrm{span}\{\{\widehat{\varphi^\ell}(\xi+2\pi k)\}_{k\in\mathbb{Z}} : \ell = 1,\dots,s\}) = s$ for a.e. $\xi \in \mathbb{R}$. We conclude from Corollary 4.4.10 that $\dim_{\mathsf{S}(\Phi|L_2(\mathbb{R}))}(\xi) = s$ for a.e. $\xi \in \mathbb{R}$, which also implies $\mathrm{len}(\mathsf{S}(\Phi|L_2(\mathbb{R}))) = s$ by (4.4.9). □

Orthogonal wavelets can be easily derived from a multiresolution analysis.

Theorem 4.5.9 *Let $\Phi = \{\phi^1,\dots,\phi^r\}$ be a finite subset of $L_2(\mathbb{R})$ such that $\{\phi(\cdot - k) : k \in \mathbb{Z}, \phi \in \Phi\}$ is a Riesz basis for $\mathscr{V}_0 := \mathsf{S}(\Phi|L_2(\mathbb{R}))$ and $\{\mathscr{V}_j\}_{j\in\mathbb{Z}}$ forms a multiresolution analysis of $L_2(\mathbb{R})$ with $\mathscr{V}_j := \{f(2^j\cdot) : f \in \mathscr{V}_0\}$. Define $\phi := (\phi^1,\dots,\phi^r)^\mathsf{T}$ and*

$$\widehat{\varphi}(\xi) := \Theta(\xi)^{-1}\widehat{\phi}(\xi),$$

where Θ is an $r\times r$ matrix of 2π-periodic measurable functions satisfying $\Theta(\xi)\overline{\widehat{\Theta}(\xi)}^\mathsf{T} = [\widehat{\phi},\widehat{\phi}](\xi)$. Then there exists $a \in (l_2(\mathbb{Z}))^{r\times r}$ such that $\widehat{\varphi}(2\xi) = \widehat{a}(\xi)\widehat{\varphi}(\xi)$ with $\widehat{a}(\xi) := \sum_{k\in\mathbb{Z}} a(k)e^{-ik\xi}$, and a is an orthogonal low-pass filter satisfying

$$\widehat{a}(\xi)\overline{\widehat{a}(\xi)}^\mathsf{T} + \widehat{a}(\xi+\pi)\overline{\widehat{a}(\xi+\pi)}^\mathsf{T} = I_r, \qquad a.e.\, \xi \in \mathbb{R}. \tag{4.5.9}$$

If there exists $b \in (l_2(\mathbb{Z}))^{r\times r}$ such that $\{a; b\}$ is an orthogonal wavelet filter bank satisfying

$$\begin{bmatrix} \widehat{a}(\xi) & \widehat{a}(\xi+\pi) \\ \widehat{b}(\xi) & \widehat{b}(\xi+\pi) \end{bmatrix} \begin{bmatrix} \overline{\widehat{a}(\xi)}^\mathsf{T} & \overline{\widehat{b}(\xi)}^\mathsf{T} \\ \overline{\widehat{a}(\xi+\pi)}^\mathsf{T} & \overline{\widehat{b}(\xi+\pi)}^\mathsf{T} \end{bmatrix} = I_{2r}, \qquad a.e.\, \xi \in \mathbb{R}. \tag{4.5.10}$$

Define $\widehat{\psi}(2\xi) = \widehat{b}(\xi)\widehat{\varphi}(\xi)$. Then $\{\varphi; \psi\}$ is an orthogonal wavelet in $L_2(\mathbb{R})$.

Proof Since $\{\phi(\cdot - k) : k \in \mathbb{Z}, \phi \in \Phi\}$ is a Riesz basis for $\mathsf{S}(\Phi|L_2(\mathbb{R}))$, by Theorem 4.4.12, there exist $C_1, C_2 > 0$ such that $C_1 I_r \leqslant [\widehat{\phi},\widehat{\phi}](\xi) \leqslant C_2 I_r$ for almost every $\xi \in \mathbb{R}$. Therefore, $\det(\Theta(\xi)) \neq 0$ a.e. $\xi \in \mathbb{R}$ by $\Theta(\xi)\overline{\widehat{\Theta}(\xi)}^\mathsf{T} = [\widehat{\phi},\widehat{\phi}](\xi)$. Hence, $\widehat{\varphi}$ is well defined and $\mathsf{S}(\varphi|L_2(\mathbb{R})) = \mathsf{S}(\phi|L_2(\mathbb{R})) = \mathscr{V}_0$. Moreover,

$$[\widehat{\varphi},\widehat{\varphi}](\xi) = \Theta(\xi)^{-1}[\widehat{\phi},\widehat{\phi}](\xi)\overline{\Theta(\xi)}^\mathsf{T} = I_r, \qquad a.e.\, \xi \in \mathbb{R},$$

from which we conclude that each entry in $\varphi = (\varphi^1,\dots,\varphi^r)^\mathsf{T}$ belongs to $L_2(\mathbb{R})$ and $\{\varphi^1(\cdot - k),\dots,\varphi^r(\cdot - k) : k \in \mathbb{Z}\}$ is an orthonormal basis of $\mathscr{V}_0$. By Theorem 4.5.8, there exists an $r\times r$ matrix $\widehat{a}$ of 2π-periodic measurable functions such that $\widehat{\varphi}(2\xi) = \widehat{a}(\xi)\widehat{\varphi}(\xi)$. Since $\{\mathscr{V}_j\}_{j\in\mathbb{Z}}$ forms a multiresolution analysis of $L_2(\mathbb{R})$ and $\{\varphi^\ell(\cdot - k) : k \in \mathbb{Z}, \ell = 1,\dots,r\}$ is an orthonormal basis of $\mathscr{V}_0$, we see that $\mathcal{P}_j f := \sum_{\ell=1}^r \sum_{k\in\mathbb{Z}} \langle f, \varphi^\ell_{2^j;k}\rangle \varphi^\ell_{2^j;k}$ converges to f in $L_2(\mathbb{R})$. By Lemma 4.1.3, this implies $\lim_{j\to\infty}\langle\sum_{\ell=1}^r|\widehat{\varphi^\ell}(2^j\cdot)|^2,\mathbf{h}\rangle=\langle 1,\mathbf{h}\rangle$ for all $h\in\mathscr{D}(\mathbb{R})$. By Corollary 4.5.2, we conclude that $\{\varphi;\psi\}$ is an orthogonal wavelet in $L_2(\mathbb{R})$. □

For the particular case $r=1$ in Theorem 4.5.9, we have

Corollary 4.5.10 *Let $\phi\in L_2(\mathbb{R})$ such that*

(i) $\mathbb{R}\backslash(\cup_{j\in\mathbb{Z}}[2^j\operatorname{supp}(\widehat{\phi})])$ has measure zero;
(ii) $\{\phi(\cdot-k)\ :\ k\in\mathbb{Z}\}$ is a Riesz basis of $\mathscr{V}_0:=\mathrm{S}(\phi\,|\,L_2(\mathbb{R}))$;
(iii) There exists a 2π-periodic measurable function $\widehat{u}$ such that $\widehat{\phi}(2\xi)=\widehat{u}(\xi)\widehat{\phi}(\xi)$ a.e. $\xi\in\mathbb{R}$.

Let Θ be a 2π-periodic measurable function such that $|\Theta(\xi)|^2=[\widehat{\phi},\widehat{\phi}](\xi)$. Define

$$\widehat{\varphi}(\xi):=\frac{\widehat{\phi}(\xi)}{\Theta(\xi)},\quad \widehat{a}(\xi):=\frac{\Theta(2\xi)}{\Theta(\xi)}\widehat{u}(\xi),\qquad \widehat{b}(\xi):=e^{-i\xi}\overline{\widehat{a}(\xi+\pi)}.$$

Then $\widehat{\varphi}(2\xi)=\widehat{a}(\xi)\widehat{\varphi}(\xi)$ and $\{a;b\}$ is an orthogonal wavelet filter bank. Define $\widehat{\psi}(2\xi):=\widehat{b}(\xi)\widehat{\varphi}(\xi)$ and $\mathscr{V}_j:=\{f(2^j\cdot)\ :\ f\in\mathscr{V}_0\}$ for $j\in\mathbb{Z}$. Then $\{\mathscr{V}_j\}_{j\in\mathbb{Z}}$ forms a multiresolution analysis of $L_2(\mathbb{R})$ and $\{\varphi;\psi\}$ is an orthogonal wavelet in $L_2(\mathbb{R})$.

Proof By Theorem 4.5.8, $\{\mathscr{V}_j\}_{j\in\mathbb{Z}}$ forms a multiresolution analysis of $L_2(\mathbb{R})$. The rest of the claim follows directly from Theorem 4.5.9. □

We now present several examples of orthogonal wavelets in $L_2(\mathbb{R})$.

Example 4.5.1 Let $\phi=\operatorname{sinc}$. By $\widehat{\operatorname{sinc}}=\chi_{(-\pi,\pi]}$, we see that all the conditions in Corollary 4.5.10 are satisfied. Note that $\varphi=\phi=\operatorname{sinc}$ in Corollary 4.5.10 and the 2π-periodic measurable functions $\widehat{a}$ and $\widehat{b}$ are given by

$$\widehat{a}(\xi)=\chi_{(-\pi/2,\pi/2]}(\xi),\quad \widehat{b}(\xi)=e^{-i\xi}\chi_{[-\pi,-\pi/2)\cup(\pi/2,\pi]}(\xi),\qquad \xi\in(-\pi,\pi].$$

Then by Corollary 4.5.10, $\{\phi;\psi\}$ is an orthogonal wavelet in $L_2(\mathbb{R})$ (called the *Shannon wavelet*), where $\widehat{\psi}(2\xi):=\widehat{b}(\xi)\widehat{\varphi}(\xi)$.

Example 4.5.2 Let $\phi=B_m$ be the B-spline of order m with $m\in\mathbb{N}$. By $\widehat{B_m}(\xi)=(\frac{1-e^{-i\xi}}{i\xi})^m$, we have $\widehat{B_m}(2\xi)=\widehat{a_m^B}(\xi)\widehat{B_m}(\xi)$ with $\widehat{a_m^B}(\xi):=2^{-m}(1+e^{-i\xi})^m$. Since $\widehat{B_m}(0)=1$ and $\widehat{B_m}$ is continuous, item (i) in Corollary 4.5.10 holds. We proved in Example 4.4.2 that $C_1\leqslant[\widehat{B_m},\widehat{B_m}](\xi)\leqslant C_2$ a.e. $\xi\in\mathbb{R}$ for some positive constants C_1 and C_2. Hence, item (ii) in Corollary 4.5.10 holds. Therefore, all the conditions in Corollary 4.5.10 are satisfied with $\widehat{u}(\xi)=\widehat{a_m^B}(\xi)$. By Corollary 4.5.10, we conclude that $\{\varphi;\psi\}$ is an (spline) orthogonal wavelet in $L_2(\mathbb{R})$. In particular, when $m=1$, we have $\Theta=1$ and therefore, $\phi=B_1=\chi_{[0,1]}$ and $\psi=\chi_{[1/2,1]}-\chi_{[0,1/2]}$. This orthogonal wavelet $\{\phi;\psi\}$ is called the *Haar orthogonal wavelet*.

Example 4.5.3 Let $\widehat{a^I_{2m}}$ be defined in (2.1.6) with $m \in \mathbb{N}$. Since $\widehat{a^I_{2m}}(\xi) \geqslant 0$ for all $\xi \in \mathbb{R}$ and $\widehat{a^I_{2m}}(0) = 1$, by the Fejér-Riesz Lemma in Lemma 1.4.4, there exists a 2π-periodic trigonometric polynomial $\widehat{a^D_m}$ such that $|\widehat{a^D_m}(\xi)|^2 = \widehat{a^I_{2m}}(\xi)$ and $\widehat{a^D_m}(0) = 1$ (see Sect. 2.2 for details). Define $\widehat{\varphi}(\xi) := \prod_{j=1}^\infty \widehat{a^D_m}(2^{-j}\xi)$ and $\widehat{\phi}(\xi) := \prod_{j=1}^\infty \widehat{a^I_{2m}}(2^{-j}\xi)$. Note that $|\widehat{\varphi}(\xi)|^2 = \widehat{\phi}(\xi)$. Since we proved in Example 4.4.3 that $[\widehat{\phi}, 1] = 1$, we trivially have $[\widehat{\varphi}, \widehat{\varphi}] = [\widehat{\phi}, 1] = 1$. Therefore, $\varphi \in L_2(\mathbb{R})$ and $\{\varphi(\cdot - k) : k \in \mathbb{Z}\}$ is an orthonormal basis of $\mathscr{V}_0 := \mathrm{S}(\varphi|L_2(\mathbb{R}))$. Moreover, it is trivial to see that $\widehat{\varphi}(2\xi) = \widehat{a^D_m}(\xi)\widehat{\varphi}(\xi)$. Hence, all the conditions in Corollary 4.5.10 are satisfied. Note that $\Theta := [\widehat{\varphi}, \widehat{\varphi}] = 1$ and define

$$\widehat{b^D_m}(\xi) := e^{-i\xi}\overline{\widehat{a^D_m}(\xi + \pi)}, \quad \widehat{\psi}(\xi) := \widehat{b^D_m}(\xi/2)\widehat{\varphi}(\xi/2).$$

Then $\{\varphi; \psi\}$ is an orthonormal wavelet in $L_2(\mathbb{R})$. Such orthogonal wavelets are called the *Daubechies orthogonal wavelets* in the literature and $\{a^D_m; b^D_m\}$ is an orthogonal wavelet filter bank. See Sect. 2.2 for details.

4.5.4 Homogeneous Framelets in $L_2(\mathbb{R})$ with Refinable Structure

To study and construct homogeneous framelets in $L_2(\mathbb{R})$, we link them to nonhomogeneous framelets with the refinable structure. In this subsection we show that the infinite set Φ in Theorems 4.3.4 and 4.3.6 can be reduced to a single generator if a homogeneous framelet has refinable structure.

The following simple result reduces a set of infinitely many generators from a principal shift-invariant space into a single generator while keeps their energy preserved.

Lemma 4.5.11 *Let $\phi \in L_2(\mathbb{R})$ and $H \subseteq L_2(\mathbb{R})$ such that $\sum_{h\in H} \|h\|^2_{L_2(\mathbb{R})} < \infty$ and $\widehat{h}(\xi) = \widehat{b_h}(\xi)\widehat{\phi}(\xi)$ for almost every $\xi \in \mathbb{R}$ for some 2π-periodic measurable functions $\widehat{b_h}$, $h \in H$ (the last condition implies $H \subseteq \mathrm{S}(\phi|L_2(\mathbb{R}))$). Define*

$$\widehat{\eta}(\xi) := \widehat{\theta}(\xi)\widehat{\phi}(\xi) \quad \text{with} \quad \widehat{\theta}(\xi) := \chi_{\mathrm{supp}([\widehat{\phi},\widehat{\phi}])}(\xi)\sqrt{\sum_{h\in H} |\widehat{b_h}(\xi)|^2}. \tag{4.5.11}$$

Then $\eta \in L_2(\mathbb{R})$, $|\widehat{\theta}(\xi)| < \infty$ for almost every $\xi \in \mathbb{R}$, and

$$\sum_{h\in H}\sum_{k\in\mathbb{Z}} |\langle f, h(\cdot - k)\rangle|^2 = \sum_{k\in\mathbb{Z}} |\langle f, \eta(\cdot - k)\rangle|^2, \qquad \forall f \in \check{\mathscr{D}}(\mathbb{R}) := \{f : \widehat{f} \in \mathscr{D}(\mathbb{R})\}.$$

Proof By $\widehat{h}(\xi) = \widehat{b_h}(\xi)\widehat{\phi}(\xi)$ and (4.5.11), we get $[\widehat{\eta}, \widehat{\eta}](\xi) = |\widehat{\theta}(\xi)|^2[\widehat{\phi}, \widehat{\phi}](\xi)$ and

$$\int_{\mathbb{R}} |\widehat{\eta}(\xi)|^2 d\xi = \int_{-\pi}^{\pi} \sum_{h\in H} |\widehat{b_h}(\xi)|^2[\widehat{\phi}, \widehat{\phi}](\xi) d\xi = \int_{\mathbb{R}} \sum_{h\in H} |\widehat{b_h}(\xi)|^2 |\widehat{\phi}(\xi)|^2 d\xi$$
$$= \sum_{h\in H} \int_{\mathbb{R}} |\widehat{h}(\xi)|^2 d\xi = 2\pi \sum_{h\in H} \|h\|^2_{L_2(\mathbb{R})} < \infty.$$

Hence, we have $\widehat{\eta} \in L_2(\mathbb{R})$ and thus, $\eta \in L_2(\mathbb{R})$. The above identity also implies $\sum_{h\in H} |\widehat{b_h}(\xi)|^2[\widehat{\phi}, \widehat{\phi}] < \infty$ a.e.. Since the 2π-periodic function $\widehat{\theta}$ vanishes outside $\operatorname{supp}([\widehat{\phi}, \widehat{\phi}])$ and $[\widehat{\phi}, \widehat{\phi}](\xi) < \infty$ for a.e. $\xi \in \mathbb{R}$, we conclude that $|\widehat{\theta}(\xi)| < \infty$ for a.e. $\xi \in \mathbb{R}$. Since $f \in \mathscr{D}(\mathbb{R})$ and $\sum_{h\in H} \|h\|^2_{L_2(\mathbb{R})} < \infty$, by Lemma 4.1.1, we have

$$\sum_{h\in H} \sum_{k\in\mathbb{Z}} |\langle f, h(\cdot - k)\rangle|^2 = \frac{1}{(2\pi)^2} \sum_{h\in H} \sum_{k\in\mathbb{Z}} |\langle \widehat{f}, \widehat{h}_{1;0,k}\rangle|^2 = \sum_{h\in H} \frac{1}{2\pi} \int_{-\pi}^{\pi} |[\widehat{f}, \widehat{h}](\xi)|^2 d\xi$$
$$= \sum_{h\in H} \frac{1}{2\pi} \int_{-\pi}^{\pi} |\widehat{b_h}(\xi)|^2 |[\widehat{f}, \widehat{\phi}](\xi)|^2 d\xi = \frac{1}{2\pi} \int_{-\pi}^{\pi} |\widehat{\theta}(\xi)|^2 |[\widehat{f}, \widehat{\phi}](\xi)|^2 d\xi$$
$$= \frac{1}{2\pi} \int_{-\pi}^{\pi} |[\widehat{f}, \widehat{\eta}](\xi)|^2 d\xi = \sum_{k\in\mathbb{Z}} |\langle f, \eta(\cdot - k)\rangle|^2.$$

This completes the proof. □

For a homogeneous framelet having the refinable structure, the following result shows that it is necessarily and sufficiently linked to a nonhomogeneous framelet having the refinable structure.

Theorem 4.5.12 *Let $\Psi \subseteq L_2(\mathbb{R})$ such that $\sum_{\psi\in\Psi} \|\psi\|^2_{L_2(\mathbb{R})} < \infty$. Suppose that there exists $\phi \in L_2(\mathbb{R})$ satisfying*

$$\widehat{\phi}(2\xi) = \widehat{a}(\xi)\widehat{\phi}(\xi), \quad \widehat{\psi}(2\xi) = \widehat{b_\psi}(\xi)\widehat{\phi}(\xi), \qquad a.e.\, \xi \in \mathbb{R}, \psi \in \Psi$$

for some 2π-periodic measurable functions $\widehat{a}$ and $\widehat{b_\psi}, \psi \in \Psi$. Then Ψ is a homogeneous framelet in $L_2(\mathbb{R})$ if and only if $\{\eta; \Psi\}$ is a framelet in $L_2(\mathbb{R})$ having the same lower and upper frame bounds, where

$$\widehat{\eta}(\xi) := \widehat{\theta}(\xi)\widehat{\phi}(\xi) \quad \textit{with}$$
$$\widehat{\theta}(\xi) := \chi_{\operatorname{supp}([\widehat{\phi}, \widehat{\phi}])}(\xi) \sqrt{\sum_{\psi\in\Psi} \sum_{n=0}^{\infty} |\widehat{b_\psi}(2^n\xi)|^2 \prod_{j=0}^{n-1} |\widehat{a}(2^j\xi)|^2} \tag{4.5.12}$$

with the convention $\prod_{j=0}^{-1} := 1$.

Proof The sufficiency part ($\Leftarrow$) follows directly from Theorem 4.3.3.

Necessity ($\Rightarrow$). Suppose that Ψ is a homogeneous framelet in $L_2(\mathbb{R})$. By Theorem 4.3.4, $\{\Phi; \Psi\}$ is a framelet in $L_2(\mathbb{R})$ with $\Phi := \{2^{-j}\psi(2^{-j}\cdot) : \psi \in \Psi, j \in \mathbb{N}\}$. Note that $\sum_{h\in\Phi} \|h\|^2_{L_2(\mathbb{R})} = \sum_{\psi\in\Psi} \|\psi\|^2_{L_2(\mathbb{R})} < \infty$ and for all $j \in \mathbb{N}$ and $\psi \in \Psi$,

$$2^{-j}\widehat{\psi(2^{-j}\cdot)}(\xi) = \widehat{\psi}(2^j\xi) = \widehat{b_\psi}(2^{j-1}\xi)\widehat{\phi}(2^{j-1}\xi) = \widehat{b_\psi}(2^{j-1}\xi)\widehat{a}(2^{j-2}\xi)\cdots\widehat{a}(\xi)\widehat{\phi}(\xi).$$

Now it follows directly from Lemma 4.5.11 that $\eta \in L_2(\mathbb{R})$ and

$$\sum_{h\in\Phi}\sum_{k\in\mathbb{Z}} |\langle f, h(\cdot - k)\rangle|^2 = \sum_{k\in\mathbb{Z}} |\langle f, \eta(\cdot - k)\rangle|^2, \qquad \forall\, f \in \check{\mathscr{D}}(\mathbb{R}).$$

Applying Theorem 4.2.4 and noting that $\check{\mathscr{D}}(\mathbb{R})$ is dense in $L_2(\mathbb{R})$, we conclude that $\mathsf{AS}_0(\eta; \Psi)$ is a frame for $L_2(\mathbb{R})$, that is, $\{\eta; \Psi\}$ is a framelet in $L_2(\mathbb{R})$. □

As a direct consequence of Theorems 4.3.8 and 4.5.12, the following result characterizes a homogeneous tight framelet having the refinable structure.

Corollary 4.5.13 *Let $\phi, \psi^1, \ldots, \psi^s \in L_2(\mathbb{R})$ such that*

$$\widehat{\phi}(2\xi) = \widehat{a}(\xi)\widehat{\phi}(\xi), \quad \widehat{\psi^\ell}(2\xi) = \widehat{b_\ell}(\xi)\widehat{\phi}(\xi), \qquad a.e.\ \xi \in \mathbb{R}, \ell = 1, \ldots, s \tag{4.5.13}$$

for some 2π-periodic measurable functions $\widehat{a}, \widehat{b_1}, \ldots, \widehat{b_s}$. Then the following statements are equivalent:

(1) $\{\psi^1, \ldots, \psi^s\}$ is a homogeneous tight framelet in $L_2(\mathbb{R})$.
(2) $\{\eta; \psi^1, \ldots, \psi^s\}$ is a tight framelet in $L_2(\mathbb{R})$, where η is defined in (4.5.12).
(3) $\lim_{j\to\infty}\langle \widehat{\Theta}(2^{-j}\cdot)|\widehat{\phi}(2^{-j}\cdot)|^2, \mathbf{h}\rangle = \langle 1, \mathbf{h}\rangle$ *for all $\mathbf{h} \in \mathscr{D}(\mathbb{R})$ and $\{\widehat{a}; \widehat{b_1}, \ldots, \widehat{b_s}\}_{\widehat{\Theta}}$ is a* (frequency-based) generalized tight framelet filter bank, *that is,*

$$\widehat{\Theta}(2\xi)|\widehat{a}(\xi)|^2 + \sum_{\ell=1}^{s} |\widehat{b_\ell}(\xi)|^2 = \widehat{\Theta}(\xi), \quad a.e.\ \xi \in \sigma_{\widehat{\phi}},$$

$$\widehat{\Theta}(2\xi)\widehat{a}(\xi)\overline{\widehat{a}(\xi+\pi)} + \sum_{\ell=1}^{s} \widehat{b_\ell}(\xi)\overline{\widehat{b_\ell}(\xi+\pi)} = 0, \quad a.e.\ \xi \in \sigma_{\widehat{\phi}} \cap (\sigma_{\widehat{\phi}} - \pi),$$

where $\sigma_{\widehat{\phi}} := \operatorname{supp}([\widehat{\phi}, \widehat{\phi}])$, $\widehat{\Theta}(\xi) := |\widehat{\theta}(\xi)|^2$, *and $\widehat{\theta}$ is defined in* (4.5.12).

Proof (1) $\iff$ (2) follows from Theorem 4.5.12 and (2) $\iff$ (3) has been proved in Theorems 4.3.8 and 4.1.10. □

The following result generalizes Lemma 4.5.11.

Lemma 4.5.14 *Let $\phi, \tilde{\phi} \in L_2(\mathbb{R})$ and $H, \tilde{H} \subseteq L_2(\mathbb{R})$ such that $\sum_{h\in H}(\|h\|^2_{L_2(\mathbb{R})} + \|\tilde{h}\|^2_{L_2(\mathbb{R})}) < \infty$ and $\widehat{h}(\xi) = \widehat{b_h}(\xi)\widehat{\phi}(\xi)$, $\widehat{\tilde{h}}(\xi) = \widehat{b_{\tilde{h}}}(\xi)\widehat{\tilde{\phi}}(\xi)$ for a.e. $\xi \in \mathbb{R}$ for some*

2π-periodic measurable functions $\widehat{b_h}$ and $\widehat{b_{\tilde h}}$, $h \in H$. Define $\widehat{\theta}, \widehat{\eta}$ as in (4.5.11) *and*

$$\widehat{\mathring{\eta}}(\xi) := \widehat{\mathring{\theta}}(\xi)\widehat{\tilde\phi}(\xi) \quad \textit{with}$$

$$\widehat{\mathring{\theta}}(\xi) := \begin{cases} \frac{1}{\widehat{\theta}(\xi)} \sum_{h\in H} \overline{\widehat{b_h}(\xi)}\widehat{b_{\tilde h}}(\xi), & \xi \in \operatorname{supp}(\widehat{\theta}) \cap \operatorname{supp}([\widehat{\tilde\phi}, \widehat{\tilde\phi}]), \\ 0, & \textit{otherwise}. \end{cases} \tag{4.5.14}$$

Then $\eta, \mathring{\eta} \in L_2(\mathbb{R})$, $|\widehat{\theta}(\xi)| + |\widehat{\mathring{\theta}}(\xi)| < \infty$ for almost every $\xi \in \mathbb{R}$, and

$$\sum_{h\in H}\sum_{k\in\mathbb{Z}} \langle f, \tilde h(\cdot - k)\rangle\langle h(\cdot - k), g\rangle = \sum_{k\in\mathbb{Z}} \langle f, \mathring{\eta}(\cdot - k)\rangle\langle \eta(\cdot - k), g\rangle, \qquad f, g \in \check{\mathscr{D}}(\mathbb{R})$$

with all the above series converging absolutely.

Proof Define $\widehat{\tilde\theta}$ and $\widehat{\tilde\eta}$ as in (4.5.11) with ϕ and H being replaced by $\tilde\phi$ and $\tilde H$, respectively. By Lemma 4.5.11, we have $\eta, \tilde\eta \in L_2(\mathbb{R})$ and $|\widehat{\theta}(\xi)| + |\widehat{\tilde\theta}(\xi)| < \infty$ for a.e. $\xi \in \mathbb{R}$. By the Cauchy-Schwarz inequality, for $\xi \in \operatorname{supp}(\widehat{\theta}) \cap \operatorname{supp}([\widehat{\tilde\phi}, \widehat{\tilde\phi}])$,

$$\begin{aligned} |\widehat{\mathring{\theta}}(\xi)|^2 &\leqslant \frac{1}{|\widehat{\theta}(\xi)|^2}\Big|\sum_{h\in H} \overline{\widehat{b_h}(\xi)}\widehat{b_{\tilde h}}(\xi)\Big|^2 \leqslant \frac{1}{|\widehat{\theta}(\xi)|^2}\Big(\sum_{h\in H}|\widehat{b_h}(\xi)|^2\Big)\Big(\sum_{h\in H}|\widehat{b_{\tilde h}}(\xi)|^2\Big) \\ &= \sum_{h\in H}|\widehat{b_{\tilde h}}(\xi)|^2 = |\widehat{\tilde\theta}(\xi)|^2, \end{aligned}$$

where we used the definition of $\widehat{\theta}$ in (4.5.14). Hence, we have $|\widehat{\mathring{\theta}}(\xi)| \leqslant |\widehat{\tilde\theta}(\xi)| < \infty$ for almost every $\xi \in \mathbb{R}$ and

$$\begin{aligned} \int_{\mathbb{R}} |\widehat{\mathring{\eta}}(\xi)|^2 d\xi &= \int_{\mathbb{R}} |\widehat{\mathring{\theta}}(\xi)|^2|\widehat{\tilde\phi}(\xi)|^2 d\xi \leqslant \int_{\mathbb{R}} |\widehat{\tilde\theta}(\xi)|^2|\widehat{\tilde\phi}(\xi)|^2 d\xi = \int_{\mathbb{R}} |\widehat{\tilde\eta}(\xi)|^2 d\xi \\ &= 2\pi \sum_{\tilde h\in\tilde H} \|\tilde h\|^2_{L_2(\mathbb{R})} < \infty. \end{aligned}$$

Therefore, the function $\widehat{\mathring{\eta}} \in L_2(\mathbb{R})$ and hence, $\mathring{\eta} \in L_2(\mathbb{R})$. By Lemma 4.1.1, we have

$$\begin{aligned} \sum_{h\in H}\sum_{k\in\mathbb{Z}} \langle f, \tilde h(\cdot - k)\rangle\langle h(\cdot - k), g\rangle &= \sum_{h\in H} \frac{1}{2\pi}\int_{-\pi}^{\pi} [\widehat{f}, \widehat{\tilde h}](\xi)[\widehat{h}, \widehat{g}](\xi) d\xi \\ &= \sum_{h\in H} \frac{1}{2\pi}\int_{-\pi}^{\pi} \overline{\widehat{b_{\tilde h}}(\xi)}\widehat{b_h}(\xi)[\widehat{f}, \widehat{\tilde\phi}](\xi)[\widehat{\phi}, \widehat{g}](\xi) d\xi \end{aligned}$$

$$= \frac{1}{2\pi}\int_{-\pi}^{\pi} \overline{\widehat{\mathring{\theta}}(\xi)}\widehat{\theta}(\xi)[\widehat{f},\widehat{\tilde{\phi}}](\xi)[\widehat{\phi},\widehat{g}](\xi)d\xi$$

$$= \frac{1}{2\pi}\int_{-\pi}^{\pi} [\widehat{f},\widehat{\mathring{\eta}}](\xi)[\widehat{\eta},\widehat{g}](\xi)d\xi = \sum_{k\in\mathbb{Z}}\langle f,\mathring{\eta}(\cdot-k)\rangle\langle\eta(\cdot-k),g\rangle$$

for $f,g \in \check{\mathscr{D}}(\mathbb{R})$, where we used $[\widehat{\phi},\widehat{g}] = 0$ on $\operatorname{supp}([\widehat{\phi},\widehat{\phi}])$, and $[\widehat{f},\widehat{\tilde{\phi}}] = 0$ on $\operatorname{supp}([\widehat{\tilde{\phi}},\widehat{\tilde{\phi}}])$. □

Now a homogeneous dual framelet in $L_2(\mathbb{R})$ having the refinable structure can be characterized as follows.

Theorem 4.5.15 *Let $\phi,\tilde{\phi} \in L_2(\mathbb{R})$ and $\Psi,\tilde{\Psi} \subseteq L_2(\mathbb{R})$ such that $\sum_{\psi\in\Psi}(\|\psi\|^2_{L_2(\mathbb{R})} + \|\tilde{\psi}\|^2_{L_2(\mathbb{R})}) < \infty$ and*

$$\begin{aligned} &\widehat{\phi}(2\xi) = \widehat{a}(\xi)\widehat{\phi}(\xi), \qquad \widehat{\tilde{\phi}}(2\xi) = \widehat{\tilde{a}}(\xi)\widehat{\tilde{\phi}}(\xi), \quad a.e.\,\xi \in \mathbb{R},\\ &\widehat{\psi}(2\xi) = \widehat{b_\psi}(\xi)\widehat{\phi}(\xi), \quad \widehat{\tilde{\psi}}(2\xi) = \widehat{b_{\tilde{\psi}}}(\xi)\widehat{\tilde{\phi}}(\xi), \quad a.e.\,\xi \in \mathbb{R}, \psi \in \Psi, \end{aligned} \tag{4.5.15}$$

for some 2π-periodic measurable functions $\widehat{a},\widehat{\tilde{a}}$ and $\widehat{b_\psi},\widehat{b_{\tilde{\psi}}}$, $\psi \in \Psi$. Then the following statements are equivalent:

(1) $(\tilde{\Psi},\Psi)$ is a homogeneous dual framelet in $L_2(\mathbb{R})$.

(2) $(\{\mathring{\eta};\tilde{\Psi}\},\{\eta;\Psi\})$ is a dual framelet in $L_2(\mathbb{R})$, where $\widehat{\theta},\widehat{\eta}$ are defined in (4.5.12) *and $\widehat{\mathring{\eta}}(\xi) := \widehat{\mathring{\theta}}(\xi)\widehat{\tilde{\phi}}(\xi)$, where $K := \operatorname{supp}(\widehat{\theta}) \cap \operatorname{supp}([\widehat{\tilde{\phi}},\widehat{\phi}])$ and $\widehat{\mathring{\theta}}$ is defined by*

$$\widehat{\mathring{\theta}}(\xi) := \begin{cases} \frac{1}{\overline{\widehat{\theta}(\xi)}}\sum_{\psi\in\Psi}\sum_{n=0}^{\infty}\widehat{b_{\tilde{\psi}}}(2^n\xi)\overline{\widehat{b_\psi}(2^n\xi)\prod_{j=0}^{n-1}[\widehat{\tilde{a}}(2^j\xi)\overline{\widehat{a}(2^j\xi)}]}, & \xi \in K,\\ 0, & \xi \notin K. \end{cases}$$

(3) $\mathsf{AS}_0(\mathring{\eta};\tilde{\Psi})$ and $\mathsf{AS}_0(\eta;\Psi)$ are Bessel sequences in $L_2(\mathbb{R})$, and $(\{\widehat{\mathring{\eta}};\widehat{\tilde{\Psi}}\},\{\widehat{\eta};\widehat{\Psi}\})$ is a frequency-based dual framelet.

Proof (2) $\Longleftrightarrow$ (3) has been proved in Theorem 4.3.7. (2)$\Longrightarrow$(1) has been proved in Theorem 4.3.5. We now prove (1)$\Longrightarrow$(2). By Theorem 4.3.6, $(\{\tilde{\Phi};\tilde{\Psi}\},\{\Phi;\Psi\})$ is a dual framelet in $L_2(\mathbb{R})$, where Φ and $\tilde{\Phi}$ are defined in (4.3.14). Define $\widehat{\tilde{\theta}}$ and $\widehat{\tilde{\eta}}$ as in (4.5.12) with ϕ and Ψ being replaced by $\tilde{\phi}$ and $\tilde{\Psi}$, respectively. By Theorem 4.5.12, both $\mathsf{AS}_0(\eta;\Psi)$ and $\mathsf{AS}_0(\tilde{\eta};\tilde{\Psi})$ are frames in $L_2(\mathbb{R})$ and hence Bessel sequences in $L_2(\mathbb{R})$. In particular, $\{\tilde{\eta}(\cdot-k) : k \in \mathbb{Z}\}$ and $\mathsf{AS}_0(\emptyset;\tilde{\Psi})$ are Bessel sequences in $L_2(\mathbb{R})$. We now prove that $\mathsf{AS}_0(\mathring{\eta};\tilde{\Psi})$ is also a Bessel sequence in $L_2(\mathbb{R})$. Since $|\widehat{\mathring{\theta}}(\xi)| \leqslant |\widehat{\tilde{\theta}}(\xi)|$ for almost every $\xi \in \mathbb{R}$ (see the proof of Lemma 4.5.14), we have

$$[\widehat{\mathring{\eta}},\widehat{\mathring{\eta}}](\xi) = |\widehat{\mathring{\theta}}(\xi)|^2[\widehat{\tilde{\phi}},\widehat{\tilde{\phi}}](\xi) \leqslant |\widehat{\tilde{\theta}}(\xi)|^2[\widehat{\tilde{\phi}},\widehat{\tilde{\phi}}](\xi) = [\widehat{\tilde{\eta}},\widehat{\tilde{\eta}}](\xi), \qquad a.e.\,\xi \in \mathbb{R}.$$

Since $\{\tilde{\eta}(\cdot-k) \; : \; k \in \mathbb{Z}\}$ is a Bessel sequence in $L_2(\mathbb{R})$, it follows from the above inequality and Proposition 4.4.4 that $\{\mathring{\eta}(\cdot-k) \; : \; k \in \mathbb{Z}\}$ is a Bessel sequence in $L_2(\mathbb{R})$. Since $\mathsf{AS}_0(\emptyset; \tilde{\Psi})$ is a Bessel sequence in $L_2(\mathbb{R})$, we conclude that $\mathsf{AS}(\mathring{\eta}; \tilde{\Psi})$ is a Bessel sequence in $L_2(\mathbb{R})$. Now by Lemma 4.5.14, we have

$$\sum_{h\in\Phi}\sum_{k\in\mathbb{Z}}\langle f, \tilde{h}(\cdot-k)\rangle\langle h(\cdot-k), g\rangle = \sum_{k\in\mathbb{Z}}\langle f, \mathring{\eta}(\cdot-k)\rangle\langle \eta(\cdot-k), g\rangle, \qquad \forall\, f, g \in \check{\mathscr{D}}(\mathbb{R}).$$

Applying Theorem 4.2.5, we conclude that $(\{\mathring{\eta}; \tilde{\Psi}\}, \{\eta; \Psi\})$ is a dual framelet in $L_2(\mathbb{R})$. Therefore, we proved (1)$\Longrightarrow$(2). □

4.5.5 Homogeneous Wavelets in $L_2(\mathbb{R})$ with Refinable Structure

To study and construct homogeneous wavelets in $L_2(\mathbb{R})$, similarly we link them to nonhomogeneous wavelets with the refinable structure. To do so, we first explore the connections between a (nonhomogeneous) Riesz wavelet in $L_2(\mathbb{R})$ and the refinable structure.

Theorem 4.5.16 *Let $\Phi, \Psi \subseteq L_2(\mathbb{R})$ be as in* (4.5.1). *Suppose that $\{\Phi; \Psi\}$ is a Riesz wavelet in $L_2(\mathbb{R})$. Then there exist subsets $\tilde{\Phi}, \tilde{\Psi} \subseteq L_2(\mathbb{R})$ as in* (4.5.1) *such that $(\{\tilde{\Phi}; \tilde{\Psi}\}, \{\Phi; \Psi\})$ is a biorthogonal wavelet in $L_2(\mathbb{R})$ if and only if there exist $a \in (l_2(\mathbb{Z}))^{r\times r}$ and $b \in (l_2(\mathbb{Z}))^{s\times r}$ such that* (4.5.3) *holds with ϕ and ψ as in* (4.5.2).

Proof The necessity part ($\Rightarrow$) follows directly from Theorem 4.5.1. We now prove the sufficiency part ($\Leftarrow$). Assume that (4.5.3) holds. Since $\mathsf{AS}_0(\Phi; \Psi)$ is a Riesz basis for $L_2(\mathbb{R})$, it has a unique dual Riesz basis $\mathcal{F}^{-1}(\mathsf{AS}_0(\Phi; \Psi)) := \{\mathcal{F}^{-1}(\eta) \; : \; \eta \in \mathsf{AS}_0(\Phi; \Psi)\}$, where $\mathcal{F}$ is the frame operator associated with $\mathsf{AS}_0(\Phi; \Psi)$. Define $\tilde{\Phi} := \{\tilde{\phi}^1, \dots, \tilde{\phi}^r\}$ and $\tilde{\Psi} := \{\tilde{\psi}^1, \dots, \tilde{\psi}^s\}$ by

$$\tilde{\phi}^\ell := \mathcal{F}^{-1}(\phi^\ell), \quad \ell = 1, \dots, r \qquad \text{and} \qquad \tilde{\psi}^m := \mathcal{F}^{-1}(\psi^m), \quad m = 1, \dots, s.$$

To complete the proof, we prove $\mathcal{F}^{-1}(\mathsf{AS}_0(\Phi; \Psi)) = \mathsf{AS}_0(\tilde{\Phi}; \tilde{\Psi})$ by showing

$$\tilde{\phi}^\ell_{1;k} = \mathcal{F}^{-1}(\phi^\ell_{1;k}), \quad \tilde{\psi}^m_{2^j;k} = \mathcal{F}^{-1}(\psi^m_{2^j;k}), \quad \forall\, j \in \mathbb{N}_0, k \in \mathbb{Z} \tag{4.5.16}$$

for $\ell = 1, \dots, r$ and $m = 1, \dots, s$. By the definition of $\mathcal{F}f = \sum_{\eta\in\mathsf{AS}_0(\Phi;\Psi)}\langle f, \eta\rangle\eta$, since $\mathsf{AS}_0(\Phi; \Psi)$ is invariant under integer shifts, it is trivial to check that $\mathcal{F}(f(\cdot - k)) = (\mathcal{F}f)(\cdot - k)$ for all $k \in \mathbb{Z}$. Consequently, for any $k \in \mathbb{Z}$, we have

$$\tilde{\phi}^\ell_{1;k} = \tilde{\phi}^\ell(\cdot-k) = (\mathcal{F}^{-1}\phi^\ell)(\cdot-k) = \mathcal{F}^{-1}(\phi^\ell_{1;k}),$$
$$\tilde{\psi}^m_{1;k} = \tilde{\psi}^m(\cdot-k) = (\mathcal{F}^{-1}\psi^m)(\cdot-k) = \mathcal{F}^{-1}(\psi^m_{1;k}).$$

Hence, the first identity in (4.5.16) holds.

By Proposition 4.5.7 and Theorem 4.4.11, (4.5.3) is equivalent to saying that $\mathsf{S}_{2^{-1}}(\Phi|L_2(\mathbb{R})) \subseteq \mathsf{S}(\Phi|L_2(\mathbb{R}))$ and $\mathsf{S}_{2^{-1}}(\Psi|L_2(\mathbb{R})) \subseteq \mathsf{S}(\Phi|L_2(\mathbb{R}))$. Consequently, $\mathsf{S}_{2^{-j}}(\Phi|L_2(\mathbb{R})) \cup \mathsf{S}_{2^{-j}}(\Psi|L_2(\mathbb{R})) \subseteq \mathsf{S}(\Phi|L_2(\mathbb{R}))$ for all $j \in \mathbb{N}$. Let $j \in \mathbb{N}_0, k \in \mathbb{Z}$ and $m = 1, \dots, s$. Note that $\phi^\ell_{2^{-j},k'-2^jk} \in \mathsf{S}_{2^{-j}}(\Phi|L_2(\mathbb{R})) \subseteq \mathsf{S}(\Phi|L_2(\mathbb{R}))$ for all $\ell = 1, \dots, r$ and $k' \in \mathbb{Z}$. By $\tilde{\psi}^m \perp \mathsf{S}(\Phi|L_2(\mathbb{R}))$, we deduce that $\langle \tilde{\psi}^m_{2^j;k}, \phi^\ell_{1;k'} \rangle = \langle \tilde{\psi}^m, \phi^\ell_{2^{-j};k'-2^jk} \rangle = 0$. Let $j' \in \mathbb{N}_0, k' \in \mathbb{Z}$ and $\ell = 1, \dots, s$. If $j' \geqslant j$, then

$$\langle \tilde{\psi}^m_{2^j;k}, \psi^\ell_{2^{j'};k'} \rangle = \langle \tilde{\psi}^m, \psi^\ell_{2^{j'-j};k'-2^{j'-j}k} \rangle = \boldsymbol{\delta}(j'-j)\boldsymbol{\delta}(k'-k)\boldsymbol{\delta}(m-\ell).$$

If $0 \leqslant j' < j$, then $\psi^\ell_{2^{j'-j};k'-2^{j'-j}k} \in \mathsf{S}_{2^{j'-j}}(\Psi|L_2(\mathbb{R})) \subseteq \mathsf{S}(\Phi|L_2(\mathbb{R}))$ and therefore, $\langle \tilde{\psi}^m_{2^j;k}, \psi^\ell_{2^{j'};k'} \rangle = \langle \tilde{\psi}^m, \psi^\ell_{2^{j'-j};k'-2^{j'-j}k} \rangle = 0$. By the uniqueness of $\mathcal{F}^{-1}(\psi^m_{2^j;k})$ which satisfies $\langle \mathcal{F}^{-1}(\psi^m_{2^j;k}), \psi^m_{2^j;k} \rangle = 1$ and $\langle \mathcal{F}^{-1}(\psi^m_{2^j;k}), \eta \rangle = 0 \; \forall \, \eta \in \mathsf{AS}_0(\Phi; \Psi) \backslash \{\psi^m_{2^j;k}\}$, we conclude that the second identity in (4.5.16) holds. This completes the proof. □

We now study homogeneous wavelets in $L_2(\mathbb{R})$ having the refinable structure. It is much more difficult to link homogeneous wavelets in $L_2(\mathbb{R})$ having the refinable structure with nonhomogeneous wavelets than their framelet counterparts. For $\Psi \subseteq L_2(\mathbb{R})$ and $n \in \mathbb{Z}$, we define

$$\mathscr{V}_n(\Psi) := \overline{\operatorname{span}\{\psi_{2^j;k} \; : \; j < n, k \in \mathbb{Z}, \psi \in \Psi\}}. \tag{4.5.17}$$

To study homogeneous wavelets with the refinable structure, we need the following two auxiliary results.

Lemma 4.5.17 *Let $\mathscr{V}_0$ be a shift-invariant space and define $\mathscr{V}_1 := \{f(2\cdot) \; : \; f \in \mathscr{V}_0\}$. If $\mathscr{V}_0 \subseteq \mathscr{V}_1$, then $\dim_{\mathscr{V}_1}(\xi) = \dim_{\mathscr{V}_0}(\frac{\xi}{2}) + \dim_{\mathscr{V}_0}(\frac{\xi}{2} + \pi)$ for a.e. $\xi \in \mathbb{R}$.*

Proof Let Φ be a countable subset of $L_2(\mathbb{R})$ such that $\{\varphi(\cdot - k) \; : \; k \in \mathbb{Z}, \varphi \in \Phi\}$ is a tight frame for $\mathscr{V}_0$. Then it is obvious that $\{\varphi_{2;k} \; : \; k \in \mathbb{Z}, \varphi \in \Phi\}$ is a tight frame for $\mathscr{V}_1$, or equivalently, $\{\sqrt{2}\varphi(2(\cdot - k)), \sqrt{2}\varphi(2(\cdot - k) - 1) \; : \; k \in \mathbb{Z}, \varphi \in \Phi\}$ is a tight frame for $\mathscr{V}_1$. Therefore, by the definition of the dimension function in (4.4.9),

$$\begin{aligned}\dim_{\mathscr{V}_1}(\xi) &= \sum_{\varphi \in \Phi} \Big([\widehat{\varphi_{2;0}}, \widehat{\varphi_{2;0}}](\xi) + [\widehat{\varphi_{2;1}}, \widehat{\varphi_{2;1}}](\xi) \Big) = \sum_{\varphi \in \Phi} [\widehat{\varphi}(\tfrac{\cdot}{2}), \widehat{\varphi}(\tfrac{\cdot}{2})](\xi) \\ &= \sum_{\varphi \in \Phi} \Big([\widehat{\varphi}, \widehat{\varphi}](\tfrac{\xi}{2}) + [\widehat{\varphi}, \widehat{\varphi}](\tfrac{\xi}{2} + \pi) \Big) = \dim_{\mathscr{V}_0}(\tfrac{\xi}{2}) + \dim_{\mathscr{V}_0}(\tfrac{\xi}{2} + \pi).\end{aligned}$$

This proves the claim. □

Lemma 4.5.18 *Let $\mathscr{V}_0$ be a shift-invariant subspace of $L_2(\mathbb{R})$ such that $\mathscr{V}_0 \subseteq \mathscr{V}_1$, where $\mathscr{V}_1 := \{f(2\cdot) \; : \; f \in \mathscr{V}_0\}$. Assume that there exists a finite subset $\Psi = \{\psi^1, \dots, \psi^s\}$ such that $\{\psi(\cdot - k) \; : \; k \in \mathbb{Z}, \psi \in \Psi\}$ is a Riesz sequence in $L_2(\mathbb{R})$, $\mathscr{V}_0 \cap \mathsf{S}(\Psi|L_2(\mathbb{R})) = \{0\}$, and $\mathscr{V}_0 + \mathsf{S}(\Psi|L_2(\mathbb{R})) = \mathscr{V}_1$. Then either $\int_{-\pi}^{\pi} \dim_{\mathscr{V}_0}(\xi) d\xi = \infty$ or $\int_{-\pi}^{\pi} \dim_{\mathscr{V}_0}(\xi) d\xi = 2\pi s$.*

Proof Since $\mathscr{V}_0 \cap \mathsf{S}(\Psi|L_2(\mathbb{R})) = \{0\}$ and $\mathscr{V}_0 + \mathsf{S}(\Psi|L_2(\mathbb{R})) = \mathscr{V}_1$, by Corollary 4.4.10 we have $\dim_{\mathscr{V}_0}(\xi) + \dim_{\mathsf{S}(\Psi|L_2(\mathbb{R}))}(\xi) = \dim_{\mathscr{V}_1}(\xi)$. On the other hand, since $\{\psi(\cdot - k) : k \in \mathbb{Z}, \psi \in \Psi\}$ is a Riesz sequence in $L_2(\mathbb{R})$, it follows trivially from Theorem 4.4.12 and Corollary 4.4.10 that $\dim_{\mathsf{S}(\Psi|L_2(\mathbb{R}))}(\xi) = s$ for a.e. $\xi \in \mathbb{R}$ (see the proof of Theorem 4.5.8). Hence, $\int_{-\pi}^{\pi} \dim_{\mathscr{V}_0}(\xi)d\xi + 2\pi s = \int_{-\pi}^{\pi} \dim_{\mathscr{V}_1}(\xi)d\xi$. By Lemma 4.5.17, we deduce that

$$\int_{-\pi}^{\pi} \dim_{\mathscr{V}_1}(\xi)d\xi = \int_{-\pi}^{\pi} \Big(\dim_{\mathscr{V}_0}(\tfrac{\xi}{2}) + \dim_{\mathscr{V}_0}(\tfrac{\xi}{2} + \pi)\Big)d\xi = 2\int_{-\pi}^{\pi} \dim_{\mathscr{V}_0}(\xi)d\xi.$$

If $\int_{-\pi}^{\pi} \dim_{\mathscr{V}_0}(\xi)d\xi < \infty$, the above identities imply $\int_{-\pi}^{\pi} \dim_{\mathscr{V}_0}(\xi)d\xi = 2\pi s$. □

We now link a homogeneous wavelet in $L_2(\mathbb{R})$ having the refinable structure with a nonhomogeneous wavelet in the following two results.

Theorem 4.5.19 *Let $\phi, \psi^1, \dots, \psi^s \in L_2(\mathbb{R})$ such that* (4.5.13) *holds for some 2π-periodic measurable functions $\widehat{a}, \widehat{b_1}, \dots, \widehat{b_s}$. Define η as in* (4.5.12)*. Then the following statements are equivalent:*

(1) $\mathscr{V}_0(\Psi)$ is shift-invariant and Ψ is a homogeneous Riesz wavelet in $L_2(\mathbb{R})$.
(2) $\mathscr{V}_0(\Psi) = \mathsf{S}(\eta|L_2(\mathbb{R}))$ and $\{\eta; \Psi\}$ is a Riesz wavelet in $L_2(\mathbb{R})$.
(3) There exist $\mathring{\eta} \in L_2(\mathbb{R})$ and $\tilde{\Psi} \subseteq L_2(\mathbb{R})$ such that $(\{\mathring{\eta}; \tilde{\Psi}\}, \{\eta; \Psi\})$ is a biorthogonal wavelet in $L_2(\mathbb{R})$.
(4) There exists $\tilde{\Psi} \subseteq L_2(\mathbb{R})$ such that $(\tilde{\Psi}, \Psi)$ is a homogeneous biorthogonal wavelet in $L_2(\mathbb{R})$.

Moreover, each of the above statements implies $s = 1$ and $[\widehat{\phi}, \widehat{\phi}](\xi) \neq 0$ a.e. $\xi \in \mathbb{R}$.

Proof (1)$\Longrightarrow$(2). By Theorem 4.5.12, $\{\eta; \Psi\}$ is a framelet in $L_2(\mathbb{R})$ and $\mathscr{V}_0(\Psi) = \mathsf{S}(\eta|L_2(\mathbb{R})) \subseteq \mathsf{S}(\phi|L_2(\mathbb{R}))$. By (4.5.13) and $\mathscr{V}_0(\Psi) \subseteq \mathsf{S}(\phi|L_2(\mathbb{R}))$, we conclude that $\dim_{\mathscr{V}_0(\Psi)}(\xi) \leqslant \dim_{\mathsf{S}(\phi|L_2(\mathbb{R}))}(\xi) \leqslant 1$. It is easy to check that all the conditions in Lemma 4.5.18 hold with $\mathscr{V}_0 = \mathscr{V}_0(\Psi)$. So, we must have $\int_{-\pi}^{\pi} \dim_{\mathscr{V}_0(\Psi)}(\xi)d\xi = 2\pi s$. Consequently, we have $2\pi s = \int_{-\pi}^{\pi} \dim_{\mathscr{V}_0(\Psi)}(\xi)d\xi \leqslant 2\pi$. Thus, we must have $s = 1$ and $\dim_{\mathsf{S}(\eta|L_2(\mathbb{R}))}(\xi) = \dim_{\mathscr{V}_0(\Psi)}(\xi) = 1$ for a.e. $\xi \in \mathbb{R}$. In particular, $[\widehat{\eta}, \widehat{\eta}](\xi) \neq 0$ for a.e. $\xi \in \mathbb{R}$. By $\widehat{\eta}(\xi) = \widehat{\theta}(\xi)\widehat{\phi}(\xi)$, we conclude that $[\widehat{\phi}, \widehat{\phi}](\xi) \neq 0$ for a.e. $\xi \in \mathbb{R}$.

On the other hand, since $\{\psi_{2^j;k} : j < 0, k \in \mathbb{Z}, \psi \in \Psi\}$ is a Riesz basis for $\mathscr{V}_0(\Psi)$, it is also a frame for $\mathscr{V}_0(\Psi)$. Now it follows from the identity

$$\sum_{j<0} \sum_{\psi \in \Psi} \sum_{k \in \mathbb{Z}} |\langle f, \psi_{2^j;k} \rangle|^2 = \sum_{k \in \mathbb{Z}} |\langle f, \eta(\cdot - k) \rangle|^2,$$

that $\{\eta(\cdot - k) : k \in \mathbb{Z}\}$ is a frame for $\mathscr{V}_0(\Psi) = \mathsf{S}(\eta|L_2(\mathbb{R}))$. By Theorem 4.4.5 and $[\widehat{\eta}, \widehat{\eta}](\xi) \neq 0$ for almost every $\xi \in \mathbb{R}$, we conclude that $\{\eta(\cdot - k) : k \in \mathbb{Z}\}$ is a Riesz basis for $\mathscr{V}_0(\Psi)$. Now it is easy to see that $\{\eta; \Psi\}$ is a Riesz wavelet in $L_2(\mathbb{R})$.

(2)⟹(3). By the definition of $\mathscr{V}_0(\Psi)$, it is trivial to see that $\mathscr{V}_0(\Psi) \subseteq \mathscr{V}_1 := \{f(2\cdot) \;:\; f \in \mathscr{V}_0(\Psi)\}$ and $\Psi \subseteq \mathscr{V}_1$. Since $\mathscr{V}_0(\Psi) = \mathsf{S}(\eta|L_2(\mathbb{R}))$, we see that the condition in (4.5.3) is satisfied with $\phi = \eta$. By Theorem 4.5.16, item (3) holds.

(3)⟹(4) follows directly from Corollary 4.3.14.

(4)⟹(1). We only need to prove that $\mathscr{V}_0(\Psi)$ is shift-invariant. From the representation in (4.3.13), for any integers $j' < 0$ and $k' \in \mathbb{Z}$, we have

$$\psi(2^{j'}(\cdot - k')) = \sum_{j\in\mathbb{Z}}\sum_{\psi\in\Psi}\sum_{k\in\mathbb{Z}}\langle \psi(2^{j'}(\cdot - k')), \tilde{\psi}_{2^j;k}\rangle \psi_{2^j;k}.$$

By the biorthogonality of $\mathsf{AS}(\Psi)$ and $\mathsf{AS}(\tilde{\Psi})$, we see that $\langle \psi(2^{j'}(\cdot - k')), \tilde{\psi}_{2^j;k}\rangle = \langle \psi(2^{j'}\cdot), \tilde{\psi}_{2^j;k-2^jk'}\rangle = 0$ for all $j \geqslant 0$ and $k \in \mathbb{Z}$. Hence,

$$\psi(2^{j'}(\cdot - k')) = \sum_{j<0}\sum_{\psi\in\Psi}\sum_{k\in\mathbb{Z}}\langle \psi(2^{j'}(\cdot - k')), \tilde{\psi}_{2^j;k}\rangle \psi_{2^j;k} \in \mathscr{V}_0(\Psi).$$

This proves that $\mathscr{V}_0(\Psi)$ is shift-invariant. □

The following result characterizes a homogeneous orthogonal wavelet with the refinable structure by linking it to a nonhomogeneous orthogonal wavelet.

Corollary 4.5.20 *Let $\phi, \psi^1, \dots, \psi^s \in L_2(\mathbb{R})$ such that* (4.5.13) *holds for some 2π-periodic measurable functions $\widehat{a}, \widehat{b_1}, \dots, \widehat{b_s}$. Define η as in* (4.5.12)*. Then Ψ is a homogeneous orthogonal wavelet in $L_2(\mathbb{R})$ if and only if $\{\eta; \Psi\}$ is an orthogonal wavelet in $L_2(\mathbb{R})$, which further implies $s = 1$.*

Proof If Ψ is a homogeneous orthogonal wavelet in $L_2(\mathbb{R})$, then

$$\mathscr{V}_0(\Psi) = U^{\perp} \quad \text{with} \quad U := \overline{\operatorname{span}\{\psi_{2^j;k} \;:\; k \in \mathbb{Z}, j \in \mathbb{N}_0, \psi \in \Psi\}}$$

must be shift-invariant since U is shift-invariant. By Theorem 4.5.19, $\{\eta; \Psi\}$ is a Riesz wavelet in $L_2(\mathbb{R})$. In the proof of Theorem 4.5.19 we proved that $[\widehat{\eta}, \widehat{\eta}](\xi) \neq 0$ for a.e. $\xi \in \mathbb{R}$. Since $\{\eta(\cdot - k) \;:\; k \in \mathbb{Z}\}$ is a tight frame for $\mathscr{V}_0(\Psi)$, we conclude from Theorems 4.4.5 and 4.4.6 that $[\widehat{\eta}, \widehat{\eta}] = 1$, that is, $\{\eta(\cdot - k) \;:\; k \in \mathbb{Z}\}$ is an orthonormal system in $L_2(\mathbb{R})$. Now it is not difficult to see that $\{\eta; \Psi\}$ is an orthogonal wavelet in $L_2(\mathbb{R})$. The other direction follows trivially from Corollary 4.3.14. Moreover, we must have $s = 1$ by Theorem 4.5.19. □

As a direct consequence of Theorem 4.5.19 and Corollary 4.3.14, we have

Corollary 4.5.21 *Let $\phi, \tilde{\phi} \in L_2(\mathbb{R})$ and $\Psi = \{\psi^1, \dots, \psi^s\}, \tilde{\Psi} = \{\tilde{\psi}^1, \dots, \tilde{\psi}^s\} \subseteq L_2(\mathbb{R})$ such that* (4.5.15) *holds for some 2π-periodic measurable functions $\widehat{a}, \widehat{\tilde{a}}$ and $\widehat{b_\psi}, \widehat{b_{\tilde{\psi}}}$, $\psi \in \Psi$. Define η as in* (4.5.12) *and define $\mathring{\eta}$ as in item (2) of Theorem 4.5.15. Then $(\tilde{\Psi}, \Psi)$ is a homogeneous biorthogonal wavelet in $L_2(\mathbb{R})$ if and only if $(\{\mathring{\eta}; \tilde{\Psi}\}; \{\eta; \Psi\})$ is a biorthogonal wavelet in $L_2(\mathbb{R})$, which implies $s = 1$.*

4.6 Framelets and Wavelets in Sobolev Spaces

Wavelets and framelets can be used to characterize many function spaces such as the widely used Sobolev spaces. As we shall see in this section, characterization of the Sobolev space $H^\tau(\mathbb{R})$ with $\tau \in \mathbb{R}$ by wavelets and framelets is a direct consequence of the notion of framelets in the Sobolev space $H^\tau(\mathbb{R})$.

In this section we introduce and study framelets and wavelets in Sobolev spaces. As a byproduct, a Sobolev space will be naturally characterized by framelet coefficients. At the end of this section, we shall present several examples of framelets and wavelets in Sobolev spaces constructed in the frequency domain.

For a real number $\tau \in \mathbb{R}$, we denote by $H^\tau(\mathbb{R})$ the *Sobolev space* consisting of all tempered distributions f such that $\widehat{f} \in L_2^{loc}(\mathbb{R})$ and

$$\|f\|^2_{H^\tau(\mathbb{R})} := \frac{1}{2\pi}\int_{\mathbb{R}} |\widehat{f}(\xi)|^2(1+|\xi|^2)^\tau d\xi < \infty. \tag{4.6.1}$$

Note that $H^0(\mathbb{R}) = L_2(\mathbb{R})$ and $H^\tau(\mathbb{R})$ is a Hilbert space under the inner product

$$\langle f, g\rangle_{H^\tau(\mathbb{R})} := \frac{1}{2\pi}\int_{\mathbb{R}} \widehat{f}(\xi)\overline{\widehat{g}(\xi)}(1+|\xi|^2)^\tau d\xi, \qquad f, g \in H^\tau(\mathbb{R}).$$

Moreover, the space $H^{-\tau}(\mathbb{R})$ is the dual space of $H^\tau(\mathbb{R})$ by regarding $g \in H^{-\tau}(\mathbb{R})$ as a continuous linear functional on $H^\tau(\mathbb{R})$ through $\langle f, g\rangle = \frac{1}{2\pi}\int_{\mathbb{R}} \widehat{f}(\xi)\overline{\widehat{g}(\xi)}d\xi$ for $f \in H^\tau(\mathbb{R})$. Define an operator

$$\mathcal{J}_\tau : H^\tau(\mathbb{R}) \to L_2(\mathbb{R}) \quad \text{with} \quad \widehat{\mathcal{J}_\tau f}(\xi) := \widehat{f}(\xi)(1+|\xi|^2)^{\tau/2}. \tag{4.6.2}$$

Then by (4.6.2), we see that $\mathcal{J}_\tau$ is isometric and bijective from $H^\tau(\mathbb{R})$ to $L_2(\mathbb{R})$ and

$$\langle f, h\rangle_{H^\tau(\mathbb{R})} = \langle \mathcal{J}_{2\tau} f, h\rangle = \langle f, \mathcal{J}_{2\tau} h\rangle = \langle \mathcal{J}_\tau f, \mathcal{J}_\tau h\rangle. \tag{4.6.3}$$

It is also easy to verify that $\mathcal{J}_\tau : H^t(\mathbb{R}) \to H^{t-\tau}(\mathbb{R})$ is isometric and bijective between $H^t(\mathbb{R})$ and $H^{t-\tau}(\mathbb{R})$ for every $t \in \mathbb{R}$. Moreover, $\mathcal{J}_\tau(f(\cdot - k)) = (\mathcal{J}_\tau f)(\cdot - k)$ for all $f \in H^t(\mathbb{R})$ and $k \in \mathbb{R}$.

For $J \in \mathbb{Z}$ and $\Phi, \Psi \subseteq H^\tau(\mathbb{R})$, we say that $\mathsf{AS}_J(\Phi;\Psi)$ has *stability in* $H^\tau(\mathbb{R})$ if there exist positive constants C_1 and C_2 such that for all $g \in H^{-\tau}(\mathbb{R})$,

$$C_1\|g\|^2_{H^{-\tau}(\mathbb{R})} \leqslant \sum_{\phi\in\Phi}\sum_{k\in\mathbb{Z}} 2^{-2\tau J}|\langle g, \phi_{2^J;k}\rangle|^2 + \sum_{j=J}^{\infty}\sum_{\psi\in\Psi}\sum_{k\in\mathbb{Z}} 2^{-2\tau j}|\langle g, \psi_{2^j;k}\rangle|^2 \leqslant C_2\|g\|^2_{H^{-\tau}(\mathbb{R})}. \tag{4.6.4}$$

We now show that $\mathsf{AS}_J(\Phi;\Psi)$ has stability in $H^\tau(\mathbb{R})$ if and only if it is essentially a frame for $H^\tau(\mathbb{R})$.

Proposition 4.6.1 *Let $\tau \in \mathbb{R}$ and $\Phi, \Psi \subseteq H^\tau(\mathbb{R})$. Then $\mathsf{AS}_J(\Phi;\Psi)$ has stability in $H^\tau(\mathbb{R})$ satisfying (4.6.4) if and only if $\mathsf{AS}_J^\tau(\Phi;\Psi)$ is a frame for $H^\tau(\mathbb{R})$ satisfying*

$$C_1\|f\|^2_{H^\tau(\mathbb{R})} \leqslant \sum_{\phi\in\Phi}\sum_{k\in\mathbb{Z}} |\langle f, 2^{-\tau J}\phi_{2^J;k}\rangle_{H^\tau(\mathbb{R})}|^2 + \sum_{j=J}^{\infty}\sum_{\psi\in\Psi}\sum_{k\in\mathbb{Z}} |\langle f, 2^{-\tau j}\psi_{2^j;k}\rangle_{H^\tau(\mathbb{R})}|^2 \leqslant C_2\|f\|^2_{H^\tau(\mathbb{R})} \tag{4.6.5}$$

for all $f \in H^\tau(\mathbb{R})$, where $\mathsf{AS}_J^\tau(\Phi;\Psi)$ is obtained by normalizing $\mathsf{AS}_J(\Phi;\Psi)$ in the Sobolev space $H^\tau(\mathbb{R})$ and is defined by

$$\begin{aligned}\mathsf{AS}_J^\tau(\Phi;\Psi) := \{2^{-\tau J}\phi_{2^J;k} \;:\; k\in\mathbb{Z}, \phi\in\Phi\}\\ \cup\{2^{-\tau j}\psi_{2^j;k} \;:\; j\geqslant J, k\in\mathbb{Z}, \psi\in\Psi\}.\end{aligned} \tag{4.6.6}$$

Proof Since $\mathcal{J}_{2\tau} : H^\tau(\mathbb{R}) \to H^{-\tau}(\mathbb{R})$ is isometric and bijective, it follows from (4.6.3) that (4.6.4) is equivalent to (4.6.5) by taking $g = \mathcal{J}_{2\tau}(f)$. ☐

For $\Phi, \Psi \subseteq H^\tau(\mathbb{R})$ and $\tilde{\Phi}, \tilde{\Psi} \subseteq H^{-\tau}(\mathbb{R})$, we say that $(\mathsf{AS}_J(\tilde{\Phi};\tilde{\Psi}), \mathsf{AS}_J(\Phi;\Psi))$, or more precisely $(\mathsf{AS}_J^{-\tau}(\tilde{\Phi};\tilde{\Psi}), \mathsf{AS}_J^\tau(\Phi;\Psi))$, is *a pair of dual frames for a pair of dual Sobolev spaces* $(H^{-\tau}(\mathbb{R}), H^\tau(\mathbb{R}))$ if

(1) $\mathsf{AS}_J(\Phi;\Psi)$ has stability in $H^\tau(\mathbb{R})$, i.e., $\mathsf{AS}_J^\tau(\Phi;\Psi)$ is a frame for $H^\tau(\mathbb{R})$;
(2) $\mathsf{AS}_J(\tilde{\Phi};\tilde{\Psi})$ has stability in $H^{-\tau}(\mathbb{R})$, i.e., $\mathsf{AS}_J^{-\tau}(\tilde{\Phi};\tilde{\Psi})$ is a frame for $H^{-\tau}(\mathbb{R})$;
(3) for all $f \in H^\tau(\mathbb{R})$ and $g \in H^{-\tau}(\mathbb{R})$, the following identity holds

$$\langle f, g\rangle = \sum_{\phi\in\Phi}\sum_{k\in\mathbb{Z}} \langle f, \tilde{\phi}_{2^J;k}\rangle\langle \phi_{2^J;k}, g\rangle + \sum_{j=J}^{\infty}\sum_{\psi\in\Psi}\sum_{k\in\mathbb{Z}} \langle f, \tilde{\psi}_{2^j;k}\rangle\langle \psi_{2^j;k}, g\rangle \tag{4.6.7}$$

with the series on the right-hand side converging absolutely.

Suppose that $(\mathsf{AS}_J(\tilde{\Phi};\tilde{\Psi}), \mathsf{AS}_J(\Phi;\Psi))$ is a pair of dual frames for $(H^{-\tau}(\mathbb{R}), H^\tau(\mathbb{R}))$. By (4.6.7), it is not difficult to see that we have the following representation:

$$g = \sum_{\phi\in\Phi}\sum_{k\in\mathbb{Z}} \langle g, \phi_{2^J;k}\rangle\tilde{\phi}_{2^J;k} + \sum_{j=J}^{\infty}\sum_{\psi\in\Psi}\sum_{k\in\mathbb{Z}} \langle g, \psi_{2^j;k}\rangle\tilde{\psi}_{2^j;k}, \quad g \in H^{-\tau}(\mathbb{R}) \tag{4.6.8}$$

with the series converging unconditionally in the space $H^{-\tau}(\mathbb{R})$. Since $\mathsf{AS}_J^\tau(\Phi;\Psi)$ is a frame for $H^\tau(\mathbb{R})$, the framelet coefficients of g in the representation of (4.6.8) provide an equivalent norm to $\|g\|_{H^{-\tau}(\mathbb{R})}$ through (4.6.4). Similarly,

$$f = \sum_{\phi\in\Phi}\sum_{k\in\mathbb{Z}} \langle f, \tilde{\phi}_{2^J;k}\rangle\phi_{2^J;k} + \sum_{j=J}^{\infty}\sum_{\psi\in\Psi}\sum_{k\in\mathbb{Z}} \langle f, \tilde{\psi}_{2^j;k}\rangle\psi_{2^j;k}, \qquad f \in H^\tau(\mathbb{R})$$

with the series converging unconditionally in the space $H^\tau(\mathbb{R})$ and its frame property provides an equivalent norm of $f \in H^\tau(\mathbb{R})$ by

$$C_2^{-1}\|f\|^2_{H^\tau(\mathbb{R})} \leqslant \sum_{\phi\in\Phi}\sum_{k\in\mathbb{Z}} 2^{2\tau J}|\langle f, \tilde{\phi}_{2^J;k}\rangle|^2 + \sum_{j=J}^{\infty}\sum_{\psi\in\Psi}\sum_{k\in\mathbb{Z}} 2^{2\tau j}|\langle f, \tilde{\psi}_{2^j;k}\rangle|^2 \leqslant C_1^{-1}\|f\|^2_{H^\tau(\mathbb{R})}. \tag{4.6.9}$$

Therefore, Sobolev spaces can be characterized through equivalent sequence norms of framelet coefficients in (4.6.4) and (4.6.9).

When $\tau = 0$, we have $L_2(\mathbb{R}) = H^0(\mathbb{R})$ and the definition of a pair of dual frames for a pair of dual Sobolev spaces becomes the definition of a pair of dual frames for $L_2(\mathbb{R})$. Hence, framelets and wavelets in $L_2(\mathbb{R})$ are special cases of framelets and wavelets in Sobolev spaces. In fact, most results discussed in Sects. 4.3–4.5 for the space $L_2(\mathbb{R})$ can be easily generalized to a general Sobolev space. In the following, we only discuss some of them. We shall use the notation

$$\widehat{H^\tau(\mathbb{R})} := \{\widehat{f} \,:\, f \in H^\tau(\mathbb{R})\}, \quad \|\widehat{f}\|^2_{\widehat{H^\tau(\mathbb{R})}} := \frac{1}{2\pi}\int_{\mathbb{R}} |\widehat{f}(\xi)|^2(1+|\xi|^2)^\tau d\xi.$$

Similar to Proposition 4.1.2 and Theorem 4.3.5, we have the following result.

Proposition 4.6.2 *Let $\tau \in \mathbb{R}$. Then $(\mathsf{AS}_J(\tilde{\Phi};\tilde{\Psi}), \mathsf{AS}_J(\Phi;\Psi))$ is a pair of dual frames for a pair of dual Sobolev spaces $(H^{-\tau}(\mathbb{R}), H^\tau(\mathbb{R}))$ for some integer J if and only if $(\mathsf{AS}_J(\tilde{\Phi};\tilde{\Psi}), \mathsf{AS}_J(\Phi;\Psi))$ is a pair of dual frames for $(H^{-\tau}(\mathbb{R}), H^\tau(\mathbb{R}))$ for every integer $J \in \mathbb{Z}$.*

Proof From the following straightforward inequalities

$$\|f_{2^n;0}\|^2_{H^\tau(\mathbb{R})} = \frac{2^{-n}}{2\pi}\int_{\mathbb{R}} |\widehat{f}(2^{-n}\xi)|^2(1+|\xi|^2)^\tau d\xi = \frac{2^{2\tau n}}{2\pi}\int_{\mathbb{R}} |\widehat{f}(\xi)|^2(2^{-2n}+|\xi|^2)^\tau d\xi$$

and

$$\min(1, 2^{-2n})(1+|\xi|^2) \leqslant 2^{-2n} + |\xi|^2 \leqslant \max(1, 2^{-2n})(1+|\xi|^2), \quad \forall\, n, \xi \in \mathbb{R},$$

we deduce that

$$\min(2^{2\tau n}, 1)\|f\|^2_{H^\tau(\mathbb{R})} \leqslant \|f_{2^n;0}\|^2_{H^\tau(\mathbb{R})} \leqslant \max(2^{2\tau n}, 1)\|f\|^2_{H^\tau(\mathbb{R})}, \tag{4.6.10}$$

for all $\tau, n \in \mathbb{R}$ and $f \in H^\tau(\mathbb{R})$. If $\mathsf{AS}_J^\tau(\Phi;\Psi)$ is a frame for $H^\tau(\mathbb{R})$ for some J, by scaling as in Proposition 4.1.2 and the inequalities in (4.6.10), we see that $\mathsf{AS}_J^\tau(\Phi;\Psi)$ is a frame for $H^\tau(\mathbb{R})$ for every integer J. If (4.6.7) holds for some J, by scaling it is trivial to see that it holds for every integer J. Hence, $(\mathsf{AS}_J(\tilde{\Phi};\tilde{\Psi}), \mathsf{AS}_J(\Phi;\Psi))$ is a pair of dual frames for $(H^{-\tau}(\mathbb{R}), H^\tau(\mathbb{R}))$ for every integer $J \in \mathbb{Z}$. □

Since the equality in (4.6.10) holds only if $\tau = 0$, contrary to the case of $L_2(\mathbb{R})$, we point out that the frame bounds C_1 and C_2 in (4.6.4) as well as in (4.6.9) do depend on the integer J in Proposition 4.6.2. Due to Proposition 4.6.2, we call $(\{\tilde{\Phi};\tilde{\Psi}\},\{\Phi;\Psi\})$ *a dual framelet in a pair of dual Sobolev spaces* $(H^{-\tau}(\mathbb{R}), H^{\tau}(\mathbb{R}))$ if $(\mathsf{AS}_0(\tilde{\Phi};\tilde{\Psi}), \mathsf{AS}_0(\Phi;\Psi))$ is a pair of dual frames for $(H^{-\tau}(\mathbb{R}), H^{\tau}(\mathbb{R}))$.

Similar to Theorem 4.3.7, we now characterize dual framelets in Sobolev spaces.

Theorem 4.6.3 *Let $\tau \in \mathbb{R}$ and $\Phi, \tilde{\Phi}, \Psi, \tilde{\Psi}$ be subsets of tempered distributions. Define $\boldsymbol{\Phi}, \tilde{\boldsymbol{\Phi}}, \boldsymbol{\Psi}, \tilde{\boldsymbol{\Psi}}$ as in* (4.3.15)*. Then $(\{\tilde{\Phi};\tilde{\Psi}\},\{\Phi;\Psi\})$ is a dual framelet in $(H^{-\tau}(\mathbb{R}), H^{\tau}(\mathbb{R}))$ if and only if*

(i) there exists a positive constant C such that for all $\mathbf{g} \in \mathscr{D}(\mathbb{R})$,

$$\sum_{\boldsymbol{\varphi}\in\boldsymbol{\Phi}}\sum_{k\in\mathbb{Z}}|\langle \mathbf{g}, \boldsymbol{\varphi}_{1;0,k}\rangle|^2+\sum_{j=0}^{\infty}\sum_{\boldsymbol{\psi}\in\boldsymbol{\Psi}}\sum_{k\in\mathbb{Z}}2^{-2\tau j}|\langle \mathbf{g}, \boldsymbol{\psi}_{2^{-j};0,k}\rangle|^2 \leqslant 2\pi C\|\mathbf{g}\|^2_{\widehat{H^{-\tau}(\mathbb{R})}} \tag{4.6.11}$$

and for all $\mathbf{f} \in \mathscr{D}(\mathbb{R})$,

$$\sum_{\tilde{\boldsymbol{\varphi}}\in\tilde{\boldsymbol{\Phi}}}\sum_{k\in\mathbb{Z}}|\langle \mathbf{f}, \tilde{\boldsymbol{\varphi}}_{1;0,k}\rangle|^2+\sum_{j=0}^{\infty}\sum_{\tilde{\boldsymbol{\psi}}\in\tilde{\boldsymbol{\Psi}}}\sum_{k\in\mathbb{Z}}2^{2\tau j}|\langle \mathbf{f}, \tilde{\boldsymbol{\psi}}_{2^{-j};0,k}\rangle|^2 \leqslant 2\pi C\|\mathbf{f}\|^2_{\widehat{H^{\tau}(\mathbb{R})}}; \tag{4.6.12}$$

(ii) the pair $(\{\tilde{\boldsymbol{\Phi}};\tilde{\boldsymbol{\Psi}}\},\{\boldsymbol{\Phi};\boldsymbol{\Psi}\})$ is a frequency-based dual framelet.

Proof Define $\check{\mathscr{D}}(\mathbb{R}) := \{h : \widehat{h} \in \mathscr{D}(\mathbb{R})\}$. For $f \in \check{\mathscr{D}}(\mathbb{R})$ and $g \in H^t(\mathbb{R})$ with $t \in \mathbb{R}$, the relation $\langle f, g\rangle = \frac{1}{2\pi}\langle \widehat{f}, \widehat{g}\rangle$ holds. Since $\check{\mathscr{D}}(\mathbb{R}) \subseteq H^{-\tau}(\mathbb{R}) \cap H^{\tau}(\mathbb{R})$, the necessity part ($\Rightarrow$) is evident. We now prove the sufficiency part ($\Leftarrow$). Note that (4.6.11) is equivalent to

$$\sum_{\phi\in\Phi}\sum_{k\in\mathbb{Z}}|\langle g, \phi(\cdot-k)\rangle|^2+\sum_{j=0}^{\infty}\sum_{\psi\in\Psi}\sum_{k\in\mathbb{Z}}2^{-2\tau j}|\langle g, \psi_{2^j;k}\rangle|^2 \leqslant C\|g\|^2_{H^{-\tau}(\mathbb{R})} \tag{4.6.13}$$

for all $g \in \check{\mathscr{D}}(\mathbb{R})$. Hence, for $\phi \in \Phi$, by (4.6.13) we have $|\langle g, \phi\rangle|^2 = |\langle g, \phi_{1;0}\rangle|^2 \leqslant C\|g\|^2_{H^{-\tau}(\mathbb{R})}$ for all $g \in \check{\mathscr{D}}(\mathbb{R})$. Since $\check{\mathscr{D}}(\mathbb{R})$ is dense in $H^{-\tau}(\mathbb{R})$, we see that $\|\phi\|^2_{H^{\tau}(\mathbb{R})} = \sup_{g\in\check{\mathscr{D}}(\mathbb{R}), \|g\|_{H^{-\tau}(\mathbb{R})}\leqslant 1}|\langle g, \phi\rangle|^2 \leqslant C$. Therefore, the function $\phi \in H^{\tau}(\mathbb{R})$ and $\Phi \subseteq H^{\tau}(\mathbb{R})$. Similarly we deduce from (4.6.13) that $\Psi \subseteq H^{\tau}(\mathbb{R})$. By the same argument, we can prove from (4.6.12) that $\tilde{\Phi}, \tilde{\Psi} \subseteq H^{-\tau}(\mathbb{R})$.

By Proposition 4.2.1 and (4.6.13), we deduce that $\mathsf{AS}_0^{\tau}(\Phi;\Psi)$ is a Bessel sequence in $H^{\tau}(\mathbb{R})$. Similarly, $\mathsf{AS}_0^{-\tau}(\tilde{\Phi};\tilde{\Psi})$ is a Bessel sequence in $H^{-\tau}(\mathbb{R})$. By item (ii), we see that (4.3.10) holds for all $f, g \in \check{\mathscr{D}}(\mathbb{R})$. Since $\check{\mathscr{D}}(\mathbb{R})$ is dense in both $H^{\tau}(\mathbb{R})$ and $H^{-\tau}(\mathbb{R})$, by Theorem 4.2.5, we see that $(\mathsf{AS}_0(\tilde{\Phi};\tilde{\Psi}), \mathsf{AS}_0(\Phi;\Psi))$ is a pair of dual frames for $(H^{-\tau}(\mathbb{R}), H^{\tau}(\mathbb{R}))$. This completes the proof. ☐

We completely characterized item (ii) of Theorem 4.6.3 in Theorem 4.1.6. To study the Bessel property in item (i) of Theorem 4.6.3, we generalize the *bracket product*. For $\tau \in \mathbb{R}$ and two functions $\mathbf{f}, \mathbf{g} : \mathbb{R} \to \mathbb{C}$, we define

$$[\mathbf{f}, \mathbf{g}]_\tau(\xi) := \sum_{k\in\mathbb{Z}} \mathbf{f}(\xi + 2\pi k)\overline{\mathbf{g}(\xi + 2\pi k)}(1 + |\xi + 2\pi k|^2)^\tau, \qquad \xi \in \mathbb{R}, \tag{4.6.14}$$

provided that the series converges absolutely for almost every $\xi \in \mathbb{R}$. By the Cauchy-Schwarz inequality for $l_2(\mathbb{Z})$, we have $|[\mathbf{f}, \mathbf{g}]_\tau(\xi)|^2 \leqslant [\mathbf{f}, \mathbf{f}]_\tau(\xi)[\mathbf{g}, \mathbf{g}]_\tau(\xi)$. Note that $[\widehat{f}, \widehat{g}]_\tau = [\widehat{\mathcal{J}_{2\tau} f}, \widehat{g}] = [\widehat{f}, \widehat{\mathcal{J}_{2\tau} g}] = [\widehat{\mathcal{J}_\tau f}, \widehat{\mathcal{J}_\tau g}]$. For $\phi, \tilde{\phi} \in H^\tau(\mathbb{R})$, it follows from Lemma 4.4.1 (or Lemma 4.1.1) and (4.6.3) that for all $f, g \in \check{\mathscr{D}}(\mathbb{R})$,

$$\sum_{k\in\mathbb{Z}} \langle f, \tilde{\phi}(\cdot - k)\rangle_{H^\tau(\mathbb{R})} \langle \phi(\cdot - k), g\rangle_{H^\tau(\mathbb{R})} = \frac{1}{2\pi} \int_{-\pi}^{\pi} [\widehat{f}, \widehat{\tilde{\phi}}]_\tau(\xi)[\widehat{\phi}, \widehat{g}]_\tau(\xi) d\xi. \tag{4.6.15}$$

We now study the Bessel property in item (i) of Theorem 4.6.3, that is, (4.6.13) in the time/space domain. For simplicity, we further assume that both $\Phi = \{\phi\}$ and $\Psi = \{\psi\}$ are singletons.

Proposition 4.6.4 *Let $\tau \in \mathbb{R}$ and $\phi \in H^\tau(\mathbb{R})$. For a positive constant C, the inequality $[\widehat{\phi}, \widehat{\phi}]_\tau(\xi) \leqslant C$ holds for almost every $\xi \in \mathbb{R}$ if and only if*

$$\sum_{k\in\mathbb{Z}} |\langle g, \phi(\cdot - k)\rangle|^2 \leqslant C\|g\|^2_{H^{-\tau}(\mathbb{R})}, \qquad \forall\, g \in H^{-\tau}(\mathbb{R}). \tag{4.6.16}$$

Proof Note that $\|g\|^2_{H^{-\tau}(\mathbb{R})} = \|\mathcal{J}_{-\tau} g\|^2_{L_2(\mathbb{R})}$ and $\langle g, \phi(\cdot - k)\rangle = \langle \mathcal{J}_{-\tau} g, (\mathcal{J}_\tau \phi)(\cdot - k)\rangle$. By Proposition 4.4.4, the inequality (4.6.16) holds if and only if $[\widehat{\mathcal{J}_\tau \phi}, \widehat{\mathcal{J}_\tau \phi}](\xi) \leqslant C$ for a.e. $\xi \in \mathbb{R}$. The proof is completed by noting $[\widehat{\mathcal{J}_\tau \phi}, \widehat{\mathcal{J}_\tau \phi}] = [\widehat{\phi}, \widehat{\phi}]_\tau$. □

Theorem 4.6.5 *Let $\tau \in \mathbb{R}$ and $\psi \in H^\tau(\mathbb{R})$. Suppose that $[\widehat{\psi}, \widehat{\psi}]_t \in L_\infty(\mathbb{T})$ for some $t > \tau$. If $\tau \leqslant 0$, we further assume that $|\cdot|^{-\nu}\widehat{\psi}(\cdot) \in L_\infty([-\pi, \pi])$ for some $\nu > -\tau$ (that is, there exists a positive constant C_0 such that $|\widehat{\psi}(\xi)| \leqslant C_0|\xi|^\nu$ for almost every $\xi \in [-\pi, \pi]$). Then there exists a positive constant C such that*

$$\sum_{j=0}^{\infty} \sum_{k\in\mathbb{Z}} 2^{-2\tau j} |\langle g, \psi_{2^j;k}\rangle|^2 \leqslant C\|g\|^2_{H^{-\tau}(\mathbb{R})}, \qquad \forall\, g \in H^{-\tau}(\mathbb{R}). \tag{4.6.17}$$

Proof By Proposition 4.2.1, it suffices to prove (4.6.17) for all $g \in \check{\mathscr{D}}(\mathbb{R})$. By Lemma 4.1.1, we have

$$\sum_{k\in\mathbb{Z}} |\langle g, \psi_{2^j;k}\rangle|^2 = \frac{2^j}{2\pi} \int_{-\pi}^{\pi} |[\widehat{g}(2^j\cdot), \widehat{\psi}](\xi)|^2 d\xi$$

$$\leqslant \frac{2^j}{\pi} \int_{-\pi}^{\pi} |\widehat{g}(2^j\xi)\widehat{\psi}(\xi)|^2 d\xi + \frac{2^j}{\pi} \int_{-\pi}^{\pi} \Big| \sum_{k\in\mathbb{Z}\setminus\{0\}} \widehat{g}(2^j(\xi + 2\pi k))\overline{\widehat{\psi}(\xi + 2\pi k)} \Big|^2 d\xi,$$

where we used the inequality $(y+z)^2 \leqslant 2(y^2+z^2)$ for all $y, z \in \mathbb{R}$. Note that

$$\frac{2^j}{\pi}\int_{-\pi}^{\pi} |\widehat{g}(2^j\xi)\widehat{\psi}(\xi)|^2 d\xi = \frac{1}{\pi}\int_{\mathbb{R}} |\widehat{g}(\xi)|^2 |\widehat{\psi}(2^{-j}\xi)|^2 \chi_{[-\pi,\pi]}(2^{-j}\xi) d\xi.$$

Using the Cauchy-Schwarz inequality, we have

$$\begin{aligned}
&\Big| \sum_{k\in\mathbb{Z}\setminus\{0\}} \widehat{g}(2^j(\xi+2\pi k))\overline{\widehat{\psi}(\xi+2\pi k)}\Big|^2 \\
&\leqslant \Big(\sum_{k\in\mathbb{Z}\setminus\{0\}} |\widehat{g}(2^j(\xi+2\pi k))|^2 (1+|\xi+2\pi k|^2)^{-t}\Big)\Big(\sum_{k\in\mathbb{Z}\setminus\{0\}} |\widehat{\psi}(\xi)|^2 (1+|\xi+2\pi k|^2)^{t}\Big).
\end{aligned}$$

Define $C_1 := \|[\widehat{\psi},\widehat{\psi}]_t\|_{L_\infty(\mathbb{T})} < \infty$. Then we have

$$\begin{aligned}
\sum_{k\in\mathbb{Z}} |\langle g, \psi_{2^j;k}\rangle|^2 \leqslant & \frac{1}{\pi}\int_{\mathbb{R}} |\widehat{g}(\xi)|^2 |\widehat{\psi}(2^{-j}\xi)|^2 \chi_{[-\pi,\pi]}(2^{-j}\xi) d\xi \\
& + C_1 \frac{2^j}{\pi}\int_{-\pi}^{\pi} \sum_{k\in\mathbb{Z}\setminus\{0\}} |\widehat{g}(2^j(\xi+2\pi k))|^2 (1+|\xi+2\pi k|^2)^{-t} d\xi \\
= & \frac{1}{\pi}\int_{\mathbb{R}} |\widehat{g}(\xi)|^2 |\widehat{\psi}(2^{-j}\xi)|^2 \chi_{[-\pi,\pi]}(2^{-j}\xi) d\xi \\
& + \frac{C_1}{\pi}\int_{\mathbb{R}} |\widehat{g}(\xi)|^2 (1+|2^{-j}\xi|^2)^{-t} \chi_{\mathbb{R}\setminus[-\pi,\pi]}(2^{-j}\xi) d\xi.
\end{aligned}$$

Consequently,

$$\sum_{j=0}^{\infty}\sum_{k\in\mathbb{Z}} 2^{-2\tau j} |\langle g, \psi_{2^j;k}\rangle|^2 \leqslant \frac{1}{2\pi}\int_{\mathbb{R}} |\widehat{g}(\xi)|^2 (1+|\xi|^2)^{-\tau} (2B_1(\xi) + 2C_1 B_2(\xi)) d\xi,$$

where

$$B_1(\xi) := (1+|\xi|^2)^{\tau} \sum_{j=0}^{\infty} 2^{-2\tau j} |\widehat{\psi}(2^{-j}\xi)|^2 \chi_{[-\pi,\pi]}(2^{-j}\xi),$$

$$B_2(\xi) := (1+|\xi|^2)^{\tau} \sum_{j=0}^{\infty} 2^{-2\tau j} (1+|2^{-j}\xi|^2)^{-t} \chi_{\mathbb{R}\setminus[-\pi,\pi]}(2^{-j}\xi).$$

If $B_1, B_2 \in L_\infty(\mathbb{R})$, then it follows from the above inequality that (4.6.17) holds with $C = 2\|B_1\|_{L_\infty(\mathbb{R})} + 2C_1\|B_2\|_{L_\infty(\mathbb{R})} < \infty$. We now prove $B_1, B_2 \in L_\infty(\mathbb{R})$ by considering two cases: $\tau > 0$ or $\tau \leqslant 0$.

Suppose $\tau > 0$. By the definition in (4.6.14) and $t > \tau > 0$, we have $|\widehat{\psi}(\xi)|^2 \leqslant (1+|\xi|^2)^{-t}[\widehat{\psi}, \widehat{\psi}]_t(\xi) \leqslant C_1(1+|\xi|^2)^{-t} \leqslant C_1$ for almost every $\xi \in \mathbb{R}$. Hence,

$$B_1(\xi) \leqslant C_1(1+|\xi|^2)^\tau \sum_{j=0}^{\infty} 2^{-2\tau j} \chi_{[-\pi,\pi]}(2^{-j}\xi)$$

$$\leqslant C_1(1+|\xi|^2)^\tau \sum_{j=J_\xi}^{\infty} 2^{-2\tau j} = C_1(1+|\xi|^2)^\tau \frac{2^{-2\tau J_\xi}}{1-2^{-2\tau}},$$

where $J_\xi := \max(0, \lceil \log_2(|\xi|/\pi)\rceil)$. For $|\xi| \leqslant \pi$, it follows trivially from the above inequalities and $J_\xi = 0$ that $B_1(\xi) \leqslant \frac{C_1(1+\pi^2)^\tau}{1-2^{-2\tau}}$. For $|\xi| \geqslant \pi$, since $\tau > 0$ and $J_\xi = \lceil \log_2(|\xi|/\pi)\rceil \geqslant \log_2(|\xi|/\pi)$,

$$B_1(\xi) \leqslant C_1(1+|\xi|^2)^\tau \frac{2^{-2\tau \log_2(|\xi|/\pi)}}{1-2^{-2\tau}} = \frac{C_1 \pi^{2\tau}}{1-2^{-2\tau}}(1+|\xi|^{-2})^\tau \leqslant \frac{C_1(1+\pi^2)^\tau}{1-2^{-2\tau}}.$$

Hence, $B_1 \in L_\infty(\mathbb{R})$. We now estimate B_2. Since $t > \tau > 0$ and $2^{2j} \geqslant 1$ for $j \in \mathbb{N}_0$,

$$B_2(\xi) \leqslant (1+|\xi|^2)^\tau \sum_{j=0}^{j_\xi} 2^{2(t-\tau)j}(2^{2j}+|\xi|^2)^{-t} \leqslant (1+|\xi|^2)^{\tau-t} \sum_{j=0}^{j_\xi} 2^{2(t-\tau)j},$$

where $j_\xi := \max(0, \lfloor \log_2(|\xi|/\pi)\rfloor)$. Hence, by $t-\tau > 0$ and $0 \leqslant j_\xi \leqslant \log_2(|\xi|/\pi)$,

$$B_2(\xi) \leqslant (1+|\xi|^2)^{\tau-t} \frac{2^{2(t-\tau)(j_\xi+1)}}{2^{2(t-\tau)}-1} \leqslant (1+|\xi|^2)^{\tau-t} \frac{2^{2(t-\tau)(\log_2(|\xi|/\pi)+1)}}{2^{2(t-\tau)}-1}$$

$$= \frac{2^{2(t-\tau)}}{(2^{2(t-\tau)}-1)\pi^{2(t-\tau)}}(1+|\xi|^{-2})^{\tau-t} \leqslant \frac{2^{2(t-\tau)}}{(2^{2(t-\tau)}-1)\pi^{2(t-\tau)}}.$$

Therefore, we proved $B_2 \in L_\infty(\mathbb{R})$.

Suppose $\tau \leqslant 0$. By our assumption, we have $|\widehat{\psi}(\xi)|^2 \leqslant C_0^2|\xi|^{2\nu}$ for a.e. $\xi \in [-\pi, \pi]$. Define $J_\xi := \max(0, \lceil \log_2(|\xi|/\pi)\rceil)$. By assumption $\tau+\nu > 0$ and $\tau \leqslant 0$,

$$B_1(\xi) \leqslant C_0^2(1+|\xi|^2)^\tau |\xi|^{2\nu} \sum_{j=J_\xi}^{\infty} 2^{-2(\tau+\nu)j} = C_0^2(1+|\xi|^2)^\tau |\xi|^{2\nu} \frac{2^{-2(\tau+\nu)J_\xi}}{1-2^{-2(\tau+\nu)}}$$

$$\leqslant \frac{C_0^2 \pi^{2(\tau+\nu)}}{1-2^{-2(\tau+\nu)}}(1+|\xi|^{-2})^\tau \leqslant \frac{C_0^2 \pi^{2(\tau+\nu)}}{1-2^{-2(\tau+\nu)}},$$

where we used $J_\xi \geqslant \log_2(|\xi|/\pi)$. Hence, $B_1 \in L_\infty(\mathbb{R})$. We now estimate B_2. Define $j_\xi := \max(0, \lfloor \log_2(|\xi|/\pi)\rfloor)$ as before. Then

$$B_2(\xi) \leqslant (1+|\xi|^2)^\tau \sum_{j=0}^{j_\xi} 2^{-2\tau j}(1+|2^{-j}\xi|^2)^{-t}. \tag{4.6.18}$$

If $\tau = 0$, since $t > \tau = 0$ and $2^{2j} \geqslant 1$ for all $j \in \mathbb{N}_0$, we have

$$B_2(\xi) \leqslant \sum_{j=0}^{j_\xi}(1 + |2^{-j}\xi|^2)^{-t} = \sum_{j=0}^{j_\xi} 2^{2tj}(2^{2j} + |\xi|^2)^{-t} \leqslant (1 + |\xi|^2)^{-t} \sum_{j=0}^{j_\xi} 2^{2tj}$$
$$\leqslant (1 + |\xi|^2)^{-t} \frac{2^{2t(j_\xi+1)}}{2^{2t}-1} \leqslant (1 + |\xi|^2)^{-t} \frac{2^{2t(\log_2(|\xi|/\pi)+1)}}{2^{2t}-1}$$
$$= \frac{2^{2t}}{(2^{2t}-1)\pi^{2t}}(1 + |\xi|^{-2})^{-t} \leqslant \frac{2^{2t}}{(2^{2t}-1)\pi^{2t}},$$

where we used $j_\xi \leqslant \log_2(|\xi|/\pi)$. If $\tau < 0$ and $t \geqslant 0$, then $(1 + |2^{-j}\xi|^2)^{-t} \leqslant 1$ for all $\xi \in \mathbb{R}$. By (4.6.18),

$$B_2(\xi) \leqslant (1 + |\xi|^2)^{\tau} \sum_{j=0}^{j_\xi} 2^{-2\tau j} \leqslant (1 + |\xi|^2)^{\tau} \frac{2^{-2\tau(j_\xi+1)}}{2^{-2\tau}-1}$$
$$\leqslant (1 + |\xi|^2)^{\tau} \frac{2^{-2\tau(\log_2(|\xi|/\pi)+1)}}{2^{-2\tau}-1} = \frac{2^{-2\tau}\pi^{2\tau}}{2^{-2\tau}-1}(1 + |\xi|^{-2})^{\tau} \leqslant \frac{2^{-2\tau}\pi^{2\tau}}{2^{-2\tau}-1}.$$

If $\tau < 0$ and $t < 0$, then $(1+|2^{-j}\xi|^2)^{-t} \leqslant [2\max(1, |2^{-j}\xi|^2)]^{-t} \leqslant 2^{-t}(1+|2^{-j}\xi|^{-2t})$ for all $\xi \in \mathbb{R}$. By (4.6.18) and $j_\xi \leqslant \log_2(|\xi|/\pi)$,

$$B_2(\xi) \leqslant 2^{-t}(1 + |\xi|^2)^{\tau} \sum_{j=0}^{j_\xi} 2^{-2\tau j} + 2^{-t}(1 + |\xi|^2)^{\tau}|\xi|^{-2t} \sum_{j=0}^{j_\xi} 2^{2(t-\tau)j}$$
$$\leqslant \frac{2^{-t-2\tau}\pi^{2\tau}}{2^{-2\tau}-1} + 2^{-t}(1 + |\xi|^2)^{\iota}|\xi|^{-2t} \frac{2^{2(t-\tau)(j_\xi+1)}}{2^{2(t-\tau)}-1}$$
$$\leqslant \frac{2^{-t-2\tau}\pi^{2\tau}}{2^{-2\tau}-1} + \frac{2^{t-2\tau}\pi^{2(\tau-t)}}{2^{2(t-\tau)}-1}(1 + |\xi|^{-2})^{\tau} \leqslant \frac{2^{-t-2\tau}\pi^{2\tau}}{2^{-2\tau}-1} + \frac{2^{t-2\tau}\pi^{2(\tau-t)}}{2^{2(t-\tau)}-1},$$

where we used $(1 + |\xi|^2)^{\tau} \leqslant 1$ and $j_\xi \leqslant \log_2(|\xi|/\pi)$. Therefore, we verified that $B_2 \in L_\infty(\mathbb{R})$. ☐

It follows directly from Proposition 4.6.4 (also cf. item (ii) of Lemma 4.1.1) and the above proof of Theorem 4.6.5 that

Corollary 4.6.6 *Let $\tau \in \mathbb{R}$ and $\Phi, \Psi \subseteq H^{\tau}(\mathbb{R})$. Assume that $\sum_{\phi\in\Phi}[\widehat{\phi}, \widehat{\phi}]_\tau \in L_\infty(\mathbb{R})$ and there exists $t > \tau$ such that $\sum_{\psi\in\Psi}[\widehat{\psi}, \widehat{\psi}]_t \in L_\infty(\mathbb{R})$. If $\tau \leqslant 0$, we further assume that there exists a positive number $\nu > -\tau$ such that $|\cdot|^{-2\nu} \sum_{\psi\in\Psi} |\widehat{\psi}(\cdot)|^2 \in L_\infty([-\pi, \pi])$. Then there exists a positive constant C such that the inequality (4.6.13) for the Bessel property holds.*

We now show that the moment condition on $\widehat{\psi}$ at the origin in Theorem 4.6.5 is essentially necessary for the Bessel property in (4.6.17) to hold. The following result generalizes Lemma 4.3.10 from the special case $\tau = 0$ to general $\tau \in \mathbb{R}$.

Proposition 4.6.7 *Let $\tau \in \mathbb{R}$ and $\psi \in H^\tau(\mathbb{R})$. If the inequality* (4.6.17) *holds for some positive constant C, then*

$$\sum_{j=0}^{\infty} 2^{-2\tau j}|\widehat{\psi}(2^{-j}\xi)|^2(1+|\xi|^2)^\tau \leqslant C, \quad a.e.\, \xi \in \mathbb{R}. \tag{4.6.19}$$

For a nonnegative real number $\nu \geqslant 0$, if in addition $C_1|\xi|^\nu \leqslant |\widehat{\psi}(\xi)| \leqslant C_2|\xi|^\nu$ for almost every $\xi \in [-\varepsilon, \varepsilon]$ for some positive constants C_1, C_2 and ε, then $\nu + \tau > 0$.

Proof Let $\xi_0 \in \mathbb{R}$ and let $g \in H^{-\tau}(\mathbb{R})$ such that $\widehat{g}$ vanishes outside $(\xi_0 - \pi, \xi_0 + \pi)$. For every $j \geqslant 0$, by the same proof of Lemma 4.3.10, we have

$$2\pi \sum_{k\in\mathbb{Z}} |\langle g, \psi_{2^j;k}\rangle|^2 = \frac{1}{2\pi}\sum_{k\in\mathbb{Z}} |\langle \widehat{g}, \widehat{\psi}_{2^{-j};0,k}\rangle|^2 = \int_{\mathbb{R}} |\widehat{g}(\xi)|^2|\widehat{\psi}(2^{-j}\xi)|^2 d\xi.$$

Therefore, by (4.6.17),

$$\begin{aligned}
&\frac{1}{2\pi}\int_{\mathbb{R}} |\widehat{g}(\xi)|^2(1+|\xi|^2)^{-\tau}\sum_{j=0}^{\infty} 2^{-2\tau j}|\widehat{\psi}(2^{-j}\xi)|^2(1+|\xi|^2)^\tau d\xi\\
&\quad = \sum_{j=0}^{\infty}\sum_{k\in\mathbb{Z}} 2^{-2\tau j}|\langle g, \psi_{2^j;k}\rangle|^2 \leqslant \frac{C}{2\pi}\int_{\mathbb{R}} |\widehat{g}(\xi)|^2(1+|\xi|^2)^{-\tau} d\xi.
\end{aligned}$$

Consequently, (4.6.19) must hold for almost every $\xi \in (\xi_0 - \pi, \xi_0 + \pi)$. Since ξ_0 is arbitrary, this proves (4.6.19).

Plugging the additional assumption into (4.6.19), for a.e. $\xi \in [-\varepsilon, \varepsilon]$, we have

$$\infty > C \geqslant \sum_{j=0}^{\infty} 2^{-2\tau j} C_1|2^{-j}\xi|^\nu(1+|\xi|^2)^\tau = C_1|\xi|^\nu(1+|\xi|^2)^\tau \sum_{j=0}^{\infty} 2^{-2(\nu+\tau)j},$$

from which we must have $\nu + \tau > 0$. □

For $\Phi, \Psi \subseteq H^\tau(\mathbb{R})$ and $\tilde{\Phi}, \tilde{\Psi} \subseteq H^{-\tau}(\mathbb{R})$, we say that $(\mathsf{AS}_J(\tilde{\Phi}; \tilde{\Psi}), \mathsf{AS}_J(\Phi; \Psi))$, or more precisely $(\mathsf{AS}_J^{-\tau}(\tilde{\Phi}; \tilde{\Psi}), \mathsf{AS}_J^\tau(\Phi; \Psi))$, is *a pair of biorthogonal bases for a pair of dual Sobolev spaces* $(H^{-\tau}(\mathbb{R}), H^\tau(\mathbb{R}))$ if $\mathsf{AS}_J^\tau(\Phi; \Psi)$ is a Riesz basis for $H^\tau(\mathbb{R})$, $\mathsf{AS}_J^{-\tau}(\tilde{\Phi}; \tilde{\Psi})$ is a Riesz basis for $H^{-\tau}(\mathbb{R})$, and $\mathsf{AS}_J^{-\tau}(\tilde{\Phi}; \tilde{\Psi})$ and $\mathsf{AS}_J^\tau(\Phi; \Psi)$ are biorthogonal to each other (which is equivalent to that $\mathsf{AS}_J(\tilde{\Phi}; \tilde{\Psi})$ and $\mathsf{AS}_J(\Phi; \Psi)$ are biorthogonal to each other).

For biorthogonal bases in a pair of dual Sobolev spaces, as a direct consequence of Corollary 4.2.8 and Proposition 4.6.2, we have

Theorem 4.6.8 *Let $\tau \in \mathbb{R}$. Let $\tilde{\Phi}, \tilde{\Psi} \subseteq H^{-\tau}(\mathbb{R})$ and $\Phi, \Psi \subseteq H^{\tau}(\mathbb{R})$. Then the following statements are equivalent:*

(1) $(\mathsf{AS}_J(\tilde{\Phi}; \tilde{\Psi}), \mathsf{AS}_J(\Phi; \Psi))$ *is a pair of biorthogonal bases for* $(H^{-\tau}(\mathbb{R}), H^{\tau}(\mathbb{R}))$ *for some integer* $J \in \mathbb{Z}$.
(2) $(\mathsf{AS}_J(\tilde{\Phi}; \tilde{\Psi}), \mathsf{AS}_J(\Phi; \Psi))$ *is a pair of biorthogonal bases for* $(H^{-\tau}(\mathbb{R}), H^{\tau}(\mathbb{R}))$ *for every integer* $J \in \mathbb{Z}$.
(3) $(\{\tilde{\Phi}; \tilde{\Psi}\}, \{\Phi; \Psi\})$ *is a dual framelet in* $(H^{-\tau}(\mathbb{R}), H^{\tau}(\mathbb{R}))$, *and* $\mathsf{AS}_0(\tilde{\Phi}; \tilde{\Psi})$ *and* $\mathsf{AS}_0(\Phi; \Psi)$ *are biorthogonal to each other.*

We call $(\{\tilde{\Phi}; \tilde{\Psi}\}, \{\Phi; \Psi\})$ *a biorthogonal wavelet in* $(H^{-\tau}(\mathbb{R}), H^{\tau}(\mathbb{R}))$ if the pair $(\mathsf{AS}_0(\tilde{\Phi}; \tilde{\Psi}), \mathsf{AS}_0(\Phi; \Psi))$ is a pair of biorthogonal bases for $(H^{-\tau}(\mathbb{R}), H^{\tau}(\mathbb{R}))$. Similarly, we call $\{\Phi; \Psi\}$ *a Riesz wavelet in* $H^{\tau}(\mathbb{R})$ if $\mathsf{AS}_0^{\tau}(\Phi; \Psi)$ is a Riesz basis for $H^{\tau}(\mathbb{R})$. If $\{\Phi; \Psi\}$ is a Riesz wavelet in $H^{\tau}(\mathbb{R})$, then $\mathsf{AS}_J^{\tau}(\Phi; \Psi)$ is a Riesz basis for $H^{\tau}(\mathbb{R})$ for all integers $J \in \mathbb{Z}$.

We complete this section by providing some examples of framelets and wavelets in Sobolev spaces constructed in the frequency domain.

Example 4.6.1 Define ϕ, ψ by $\widehat{\phi} := \chi_{[c-2\pi, c]}$ and $\widehat{\psi} := \chi_{[c,2c]\cup[2c-4\pi, c-2\pi]}$ with $0 < c < 2\pi$. By Corollary 4.3.9 and $[\widehat{\phi}, \widehat{\phi}] = [\widehat{\psi}, \widehat{\psi}] = 1$, we see that $\{\phi; \psi\}$ is an orthogonal wavelet in $L_2(\mathbb{R})$. When $c = \pi$, both ϕ and ψ are real-valued and ψ is called the *Shannon (homogeneous) orthogonal wavelet* in $L_2(\mathbb{R})$. Note that $\phi, \psi \in H^{\tau}(\mathbb{R})$ for every $\tau \in \mathbb{R}$. By Theorem 4.1.6 and Corollary 4.6.6, we conclude that $(\{\phi; \psi\}, \{\phi; \psi\})$ is a biorthogonal wavelet in $(H^{-\tau}(\mathbb{R}), H^{\tau}(\mathbb{R}))$ for every $\tau \in \mathbb{R}$. In particular, $\{\phi; \psi\}$ is a Riesz wavelet in $H^{\tau}(\mathbb{R})$ for every $\tau \in \mathbb{R}$.

The above construction can be modified into smooth wavelets in the Schwartz class. We start with a real-valued smooth function $\boldsymbol{\theta} : \mathbb{R} \to \mathbb{R}$ such that

$$(\boldsymbol{\theta}(x))^2 + (\boldsymbol{\theta}(-x))^2 = 1 \ \forall\, x \in \mathbb{R} \ \text{ and } \ \boldsymbol{\theta}(x) = 0, \ x < -1; \ \boldsymbol{\theta}(x) = 1, \ x > 1. \tag{4.6.20}$$

For example, such a function $\boldsymbol{\theta}$ can be constructed by

$$\boldsymbol{\theta}(x) := \frac{h(x)}{\sqrt{(h(-x))^2 + (h(x))^2}} \quad \text{with} \quad h(x) := \int_{-\infty}^{x} \chi_{[-1,1]}(t) e^{-(1+t)^{-2}-(1-t)^{-2}} dt,$$

for $x \in \mathbb{R}$. Then $h, \boldsymbol{\theta} \in \mathscr{C}^{\infty}(\mathbb{R})$ and $\boldsymbol{\theta}$ satisfies (4.6.20). Another construction of $\boldsymbol{\theta}$ is to use the polynomials $\mathsf{P}_{m,m}(x) := \sum_{j=0}^{m-1} (-1)^j \binom{-m}{j} x^j$ defined in (2.1.4) with $m \in \mathbb{N}$:

$$\boldsymbol{\theta}(x) := \begin{cases} 0, & x < -1, \\ \sin\left(\frac{\pi}{2}\left(\frac{1+x}{2}\right)^m \mathsf{P}_{m,m}\left(\frac{1-x}{2}\right)\right), & -1 \leqslant x \leqslant 1, \\ 1, & x > 1. \end{cases}$$

Using the identity $x^m \mathsf{P}_{m,m}(1-x) + (1-x)^m \mathsf{P}_{m,m}(x) = 1$ in (2.1.5), we can directly verify that (4.6.20) holds and $\boldsymbol{\theta} \in \mathscr{C}^{m-1}(\mathbb{R})$.

Using the function $\boldsymbol{\theta}$, we now modify the characteristic function $\chi_{[c_l,c_r]}$ into a smooth function. For $c_l < c_r$ and two positive numbers $\varepsilon_l, \varepsilon_r$ satisfying $\varepsilon_l + \varepsilon_r \leqslant c_r - c_l$, we define *a bump function* $\chi_{[c_l,c_r];\varepsilon_l,\varepsilon_r}$ by

$$\chi_{[c_l,c_r];\varepsilon_l,\varepsilon_r}(\xi) := \begin{cases} \boldsymbol{\theta}\left(\frac{\xi - c_l}{\varepsilon_l}\right), & \xi < c_l + \varepsilon_l, \\ 1, & c_l + \varepsilon_l \leqslant \xi \leqslant c_r - \varepsilon_r, \\ \boldsymbol{\theta}\left(\frac{c_r - \xi}{\varepsilon_r}\right), & \xi > c_r - \varepsilon_r. \end{cases} \tag{4.6.21}$$

Then $\chi_{[c_l,c_r];\varepsilon_l,\varepsilon_r}$ is supported inside $[c_l - \varepsilon_l, c_r + \varepsilon_r]$ and equals one on $[c_l + \varepsilon_l, c_r - \varepsilon_r]$. Obviously, $\lim_{\varepsilon \to 0^+} \chi_{[c_l,c_r];\varepsilon,\varepsilon}(\xi) = \chi_{[c_l,c_r]}(\xi)$ for $\xi \in \mathbb{R} \backslash \{c_l, c_r\}$. Hence, we define $\chi_{[c_l,c_r];0,0} := \chi_{[c_l,c_r]}$.

Using $\chi_{[c_l,c_r];\varepsilon_l,\varepsilon_r}$ instead of the discontinuous function $\chi_{[c_l,c_r]}$, we now modify Example 4.6.1 so that $\widehat{\phi}$ and $\widehat{\psi}$ are smooth compactly supported functions. A function f is called *bandlimited* if its Fourier transform $\widehat{f}$ has compact support.

Example 4.6.2 Let $0 < c < 2\pi$ and $0 < \varepsilon \leqslant \min(\frac{2\pi - c}{3}, \frac{c}{3})$. Define ϕ and ψ by

$$\widehat{\phi}(\xi) := \chi_{[c-2\pi,c];\varepsilon,\varepsilon}(\xi) \quad \text{and}$$
$$\widehat{\psi}(\xi) := e^{-i\xi/2} \chi_{[c,2c];\varepsilon,2\varepsilon}(\xi) + e^{i\xi/2} \chi_{[2c-4\pi,c-2\pi];2\varepsilon,\varepsilon}(\xi).$$

Then $\operatorname{supp}(\widehat{\phi}) \subseteq [c - 2\pi - \varepsilon, c + \varepsilon]$ and $\operatorname{supp}(\widehat{\psi}) \subseteq [c - \varepsilon, 2c + 2\varepsilon] \cup [2c - 4\pi - 2\varepsilon, c - 2\pi + \varepsilon]$. If $\boldsymbol{\theta} \in \mathscr{C}^\infty(\mathbb{R})$, then $\widehat{\phi}, \widehat{\psi} \in \mathscr{C}^\infty(\mathbb{R})$. By Corollary 4.3.9 and $[\widehat{\phi}, \widehat{\phi}] = [\widehat{\psi}, \widehat{\psi}] = 1$, we see that $\{\phi; \psi\}$ is an orthogonal wavelet in $L_2(\mathbb{R})$. When $c = \pi$ and $\varepsilon = \frac{\pi}{3}$, both ϕ and ψ are real-valued and ψ is called the *Meyer (homogeneous) orthogonal wavelet* in $L_2(\mathbb{R})$. Note that $\phi, \psi \in H^\tau(\mathbb{R})$ for every $\tau \in \mathbb{R}$. By Theorem 4.1.6 and Corollary 4.6.6, we conclude that $(\{\phi; \psi\}, \{\phi; \psi\})$ is a biorthogonal wavelet in $(H^{-\tau}(\mathbb{R}), H^\tau(\mathbb{R}))$ and $\{\phi; \psi\}$ is a Riesz wavelet in $H^\tau(\mathbb{R})$ for every $\tau \in \mathbb{R}$. Note that $\widehat{\phi}(2\xi) = \widehat{a}(\xi)\widehat{\phi}(\xi)$, and $\widehat{\psi}(2\xi) = \widehat{b}(\xi)\widehat{\phi}(\xi)$, where $\widehat{a}$ and $\widehat{b}$ are 2π-periodic functions given by

$$\widehat{a}(\xi) := \chi_{[\frac{c}{2}-\pi, \frac{c}{2}]; \frac{\varepsilon}{2}, \frac{\varepsilon}{2}}(\xi), \qquad \xi \in [-\pi, \pi) \quad \text{and} \quad \widehat{b}(\xi) := e^{-i\xi} \overline{\widehat{a}(\xi + \pi)}.$$

Note that $\{\widehat{a}; \widehat{b}\}$ is a (frequency-based) orthogonal wavelet filter bank.

Example 4.6.3 Let $0 < c < \pi$ and $0 < \varepsilon < \min(c, \pi - c, \frac{2\pi - c}{3})$. Define

$$\widehat{\phi}(\xi) := \chi_{[-c,c];\varepsilon,\varepsilon}(\xi) \quad \text{and} \quad \widehat{\psi^+}(\xi) := \chi_{[c,2c];\varepsilon,2\varepsilon}(\xi), \quad \widehat{\psi^-}(\xi) := \widehat{\psi^+}(-\xi).$$

By Corollary 4.3.9, $\{\phi; \psi^+, \psi^-\}$ is a tight framelet in $L_2(\mathbb{R})$. Note that $\phi, \psi^+, \psi^- \in H^\tau(\mathbb{R})$ for every $\tau \in \mathbb{R}$. By Theorem 4.1.6 and Corollary 4.6.6, we conclude that $(\{\phi; \psi^+, \psi^-\}, \{\phi; \psi^+, \psi^-\})$ is a dual framelet in $(H^{-\tau}(\mathbb{R}), H^{\tau}(\mathbb{R}))$ for every $\tau \in \mathbb{R}$.

Finally, we provide an example of homogeneous orthogonal wavelets.

Example 4.6.4 Let $n \in \mathbb{N}$ and $K := [\frac{2^n\pi}{2^{n+1}-1}, \pi] \cup [2^n\pi, \frac{2^{2n+1}\pi}{2^{n+1}-1}]$. Define ψ by $\widehat{\psi} := \chi_{K\cup(-K)}$. By Corollary 4.3.12 and $[\widehat{\psi}, \widehat{\psi}] = 1$, ψ is a homogeneous orthogonal wavelet in $L_2(\mathbb{R})$.

4.7 Approximation by Dual Framelets and Quasi-Projection Operators*

Framelets and wavelets can provide an efficient approximation scheme for functions in a Sobolev space. In this section we investigate approximation property of dual framelets and quasi-projection operators in Sobolev spaces.

Let $(\{\tilde{\Phi}; \tilde{\Psi}\}, \{\Phi; \Psi\})$ be a dual framelet in $(H^{-\tau}(\mathbb{R}), H^{\tau}(\mathbb{R}))$. For $n \in \mathbb{N}$, the *approximation operator* $\mathcal{A}_n : H^{\tau}(\mathbb{R}) \to H^{\tau}(\mathbb{R})$ is defined to be

$$\mathcal{A}_n f := \sum_{\phi\in\Phi}\sum_{k\in\mathbb{Z}} \langle f, \tilde{\phi}(\cdot - k)\rangle \phi(\cdot - k) + \sum_{j=0}^{n-1}\sum_{\psi\in\Psi}\sum_{k\in\mathbb{Z}} \langle f, \tilde{\psi}_{2^j;k}\rangle \psi_{2^j;k}. \tag{4.7.1}$$

In this section we shall investigate how well $\mathcal{A}_n f$ approximates $f \in H^{\tau}(\mathbb{R})$ as $n \to \infty$. By the cascade structure of a dual framelet in (4.3.22), we see that $\mathcal{A}_n f = \mathcal{Q}_{2^n} f$, where $\mathcal{Q}_\lambda : H^{\tau}(\mathbb{R}) \to H^{\tau}(\mathbb{R})$ is the *quasi-projection operator* defined to be

$$\mathcal{Q}_\lambda f := \sum_{\phi\in\Phi}\sum_{k\in\mathbb{Z}} \langle f, \tilde{\phi}_{\lambda;k}\rangle \phi_{\lambda;k}, \qquad \lambda \in \mathbb{R}\backslash\{0\}. \tag{4.7.2}$$

Therefore, the approximation property of a dual framelet $(\{\tilde{\Phi}; \tilde{\Psi}\}, \{\Phi; \Psi\})$ is completely determined by $\tilde{\Phi}$ and Φ. In the following, we study the approximation property of quasi-projection operators.

Theorem 4.7.1 *Let $\tau \in \mathbb{R}$. Let $\Phi \subseteq H^{\tau}(\mathbb{R})$ and $\tilde{\Phi} \subseteq H^{-\tau}(\mathbb{R})$ be finite sets. Let $\mathcal{Q}_\lambda$ be the quasi-projection operators defined in* (4.7.2). *Let $\mathrm{w}, \tilde{\mathrm{w}} : \mathbb{R} \to [0, \infty)$ be nonnegative measurable functions such that $\tilde{\mathrm{w}}(\xi) \ne 0$ for almost every $\xi \in \mathbb{R}$ and there exists a positive constant C_{w} satisfying $\mathrm{w}(\xi)/\tilde{\mathrm{w}}(\xi) \le C_{\mathrm{w}}$ for almost every $\xi \in \mathbb{R}\backslash[-\pi, \pi]$. Then for all $\lambda > 0$,*

$$\int_{\mathbb{R}} |\widehat{\mathcal{Q}_\lambda f}(\xi) - \widehat{f}(\xi)|^2 \mathrm{w}(\lambda^{-1}\xi) d\xi \le C \int_{\mathbb{R}} |\widehat{f}(\xi)|^2 \tilde{\mathrm{w}}(\lambda^{-1}\xi) d\xi, \tag{4.7.3}$$

where $C := \max(2C_1 + 4C_2, 2C_{\mathrm{w}} + 2C_3 + 4C_4)$ *with* C_1, C_2, C_3, C_4 *being positive constants such that for almost every* $\xi \in [-\pi, \pi]$,

$$\frac{\mathrm{w}(\xi)}{\tilde{\mathrm{w}}(\xi)}\Big|1 - \sum_{\phi\in\Phi} \widehat{\tilde{\phi}}(\xi)\overline{\widehat{\phi}(\xi)}\Big|^2 \leqslant C_1, \tag{4.7.4}$$

$$\sum_{n\in\mathbb{Z}\setminus\{0\}} \frac{\mathrm{w}(\xi + 2\pi n)}{\tilde{\mathrm{w}}(\xi)}\Big|\sum_{\phi\in\Phi} \widehat{\tilde{\phi}}(\xi)\overline{\widehat{\phi}(\xi + 2\pi n)}\Big|^2 \leqslant C_2, \tag{4.7.5}$$

$$\sum_{k\in\mathbb{Z}\setminus\{0\}} \frac{\mathrm{w}(\xi)}{\tilde{\mathrm{w}}(\xi + 2\pi k)}\Big|\sum_{\phi\in\Phi} \widehat{\tilde{\phi}}(\xi + 2\pi k)\overline{\widehat{\phi}(\xi)}\Big|^2 \leqslant C_3, \tag{4.7.6}$$

$$\sum_{n\in\mathbb{Z}\setminus\{0\}} \sum_{k\in\mathbb{Z}\setminus\{0\}} \frac{\mathrm{w}(\xi + 2\pi n)}{\tilde{\mathrm{w}}(\xi + 2\pi k)}\Big|\sum_{\phi\in\Phi} \widehat{\tilde{\phi}}(\xi + 2\pi k)\overline{\widehat{\phi}(\xi + 2\pi n)}\Big|^2 \leqslant C_4. \tag{4.7.7}$$

Proof By Lemma 4.1.1 and (4.7.2), we see that

$$\widehat{\mathcal{Q}_\lambda f}(\xi) = \sum_{\phi\in\Phi} [\widehat{f}(\lambda\cdot), \widehat{\tilde{\phi}}](\tfrac{\xi}{\lambda})\widehat{\phi}(\tfrac{\xi}{\lambda}) = \sum_{k\in\mathbb{Z}} \widehat{f}(\xi + 2\pi\lambda k) \sum_{\phi\in\Phi} \overline{\widehat{\tilde{\phi}}(\tfrac{\xi}{\lambda} + 2\pi k)}\widehat{\phi}(\tfrac{\xi}{\lambda}).$$

By $|y + z|^2 \leqslant 2|y|^2 + 2|z|^2$ for $y, z \in \mathbb{C}$, we have $\int_{-\lambda\pi}^{\lambda\pi} |\widehat{\mathcal{Q}_\lambda f}(\xi) - \widehat{f}(\xi)|^2 \mathrm{w}(\frac{\xi}{\lambda})d\xi \leqslant 2(I_1 + I_2)$, where

$$I_1 := \int_{-\lambda\pi}^{\lambda\pi} |\widehat{f}(\xi)|^2 \mathrm{w}(\tfrac{\xi}{\lambda})\Big|1 - \sum_{\phi\in\Phi} \widehat{\tilde{\phi}}(\tfrac{\xi}{\lambda})\overline{\widehat{\phi}(\tfrac{\xi}{\lambda})}\Big|^2 d\xi \leqslant C_1 \int_{-\lambda\pi}^{\lambda\pi} |\widehat{f}(\xi)|^2 \tilde{\mathrm{w}}(\tfrac{\xi}{\lambda})d\xi$$

and by the Cauchy-Schwarz inequality,

$$\begin{aligned}
I_2 &:= \int_{-\lambda\pi}^{\lambda\pi} \Big|\sum_{k\in\mathbb{Z}\setminus\{0\}} \widehat{f}(\xi + 2\pi\lambda k) \sum_{\phi\in\Phi} \overline{\widehat{\tilde{\phi}}(\tfrac{\xi}{\lambda} + 2\pi k)}\widehat{\phi}(\tfrac{\xi}{\lambda})\Big|^2 \mathrm{w}(\tfrac{\xi}{\lambda})d\xi \\
&\leqslant \int_{-\lambda\pi}^{\lambda\pi} \Big(\sum_{k\in\mathbb{Z}\setminus\{0\}} |\widehat{f}(\xi + 2\pi\lambda k)|^2 \tilde{\mathrm{w}}(\tfrac{\xi}{\lambda} + 2\pi k)\Big) \\
&\quad \times \Big(\sum_{k\in\mathbb{Z}\setminus\{0\}} \frac{\mathrm{w}(\frac{\xi}{\lambda})}{\tilde{\mathrm{w}}(\frac{\xi}{\lambda} + 2\pi k)}\Big|\sum_{\phi\in\Phi} \widehat{\tilde{\phi}}(\tfrac{\xi}{\lambda} + 2\pi k)\overline{\widehat{\phi}(\tfrac{\xi}{\lambda})}\Big|^2\Big)d\xi \\
&\leqslant C_3 \int_{-\lambda\pi}^{\lambda\pi} \sum_{k\in\mathbb{Z}\setminus\{0\}} |\widehat{f}(\xi + 2\pi\lambda k)|^2 \tilde{\mathrm{w}}(\tfrac{\xi}{\lambda} + 2\pi k)d\xi = C_3 \int_{\mathbb{R}\setminus[-\lambda\pi,\lambda\pi]} |\widehat{f}(\xi)|^2 \tilde{\mathrm{w}}(\tfrac{\xi}{\lambda})d\xi.
\end{aligned}$$

Similarly, we have

$$\begin{aligned}
&\int_{\mathbb{R}\setminus[-\lambda\pi,\lambda\pi]} |\widehat{\mathcal{Q}_\lambda f}(\xi) - \widehat{f}(\xi)|^2 \mathrm{w}(\tfrac{\xi}{\lambda})d\xi \\
&= \sum_{n\in\mathbb{Z}\setminus\{0\}} \int_{-\lambda\pi}^{\lambda\pi} \Big| \sum_{k\in\mathbb{Z}} \widehat{f}(\xi+2\pi\lambda k) \sum_{\phi\in\Phi} \overline{\widehat{\tilde{\phi}}(\tfrac{\xi}{\lambda}+2\pi k)}\widehat{\phi}(\tfrac{\xi}{\lambda}+2\pi n) \\
&\quad -\widehat{f}(\xi+2\pi\lambda n)\Big|^2 \mathrm{w}(\tfrac{\xi}{\lambda}+2\pi n)d\xi \\
&\leqslant 2I_3 + 4I_4 + 4I_5,
\end{aligned}$$

where the quantities I_3, I_4, I_5 with the notation $E_\lambda := \mathbb{R}\setminus[-\lambda\pi, \lambda\pi]$ are defined by

$$\begin{aligned}
I_3 &:= \sum_{n\in\mathbb{Z}\setminus\{0\}} \int_{-\lambda\pi}^{\lambda\pi} |\widehat{f}(\xi+2\pi\lambda n)|^2 \mathrm{w}(\tfrac{\xi}{\lambda}+2\pi n)d\xi = \int_{E_\lambda} |\widehat{f}(\xi)|^2 \mathrm{w}(\tfrac{\xi}{\lambda})d\xi \\
&\leqslant C_{\mathrm{w}} \int_{E_\lambda} |\widehat{f}(\xi)|^2 \tilde{\mathrm{w}}(\tfrac{\xi}{\lambda})d\xi, \\
I_4 &:= \sum_{n\in\mathbb{Z}\setminus\{0\}} \int_{-\lambda\pi}^{\lambda\pi} |\widehat{f}(\xi)|^2 \Big| \sum_{\phi\in\Phi} \widehat{\tilde{\phi}}(\tfrac{\xi}{\lambda})\overline{\widehat{\phi}(\tfrac{\xi}{\lambda}+2\pi n)}\Big|^2 \mathrm{w}(\tfrac{\xi}{\lambda}+2\pi n)d\xi \\
&= \int_{-\lambda\pi}^{\lambda\pi} |\widehat{f}(\xi)|^2 \tilde{\mathrm{w}}(\tfrac{\xi}{\lambda}) \sum_{n\in\mathbb{Z}\setminus\{0\}} \frac{\mathrm{w}(\tfrac{\xi}{\lambda}+2\pi n)}{\tilde{\mathrm{w}}(\tfrac{\xi}{\lambda})} \Big| \sum_{\phi\in\Phi} \widehat{\tilde{\phi}}(\tfrac{\xi}{\lambda})\overline{\widehat{\phi}(\tfrac{\xi}{\lambda}+2\pi n)}\Big|^2 d\xi \\
&\leqslant C_2 \int_{-\lambda\pi}^{\lambda\pi} |\widehat{f}(\xi)|^2 \tilde{\mathrm{w}}(\tfrac{\xi}{\lambda})d\xi, \\
I_5 &:= \sum_{n\in\mathbb{Z}\setminus\{0\}} \int_{-\lambda\pi}^{\lambda\pi} \Big| \sum_{k\in\mathbb{Z}\setminus\{0\}} \widehat{f}(\xi+2\pi\lambda k) \sum_{\phi\in\Phi} \overline{\widehat{\tilde{\phi}}(\tfrac{\xi}{\lambda}+2\pi k)}\widehat{\phi}(\tfrac{\xi}{\lambda}+2\pi n)\Big|^2 \mathrm{w}(\tfrac{\xi}{\lambda}+2\pi n)d\xi \\
&\leqslant \int_{-\lambda\pi}^{\lambda\pi} \Big(\sum_{k\in\mathbb{Z}\setminus\{0\}} |\widehat{f}(\xi+2\pi\lambda k)|^2 \mathrm{w}(\tfrac{\xi}{\lambda}+2\pi k)\Big) \\
&\qquad \Big(\sum_{n\in\mathbb{Z}\setminus\{0\}} \sum_{k\in\mathbb{Z}\setminus\{0\}} \frac{\mathrm{w}(\tfrac{\xi}{\lambda}+2\pi n)}{\tilde{\mathrm{w}}(\tfrac{\xi}{\lambda}+2\pi k)} \Big| \sum_{\phi\in\Phi} \widehat{\tilde{\phi}}(\tfrac{\xi}{\lambda}+2\pi k)\overline{\widehat{\phi}(\tfrac{\xi}{\lambda}+2\pi n)}\Big|^2\Big) d\xi \\
&\leqslant C_4 \int_{-\lambda\pi}^{\lambda\pi} \sum_{k\in\mathbb{Z}\setminus\{0\}} |\widehat{f}(\xi+2\pi\lambda k)|^2 \tilde{\mathrm{w}}(\tfrac{\xi}{\lambda}+2\pi k)d\xi = C_4 \int_{\mathbb{R}\setminus[-\lambda\pi,\lambda\pi]} |\widehat{f}(\xi)|^2 \tilde{\mathrm{w}}(\tfrac{\xi}{\lambda})d\xi.
\end{aligned}$$

Putting all the estimates together, we conclude that (4.7.3) holds. □

As a direct consequence of Theorem 4.7.1, for two special cases, we have the following results.

Corollary 4.7.2 *Let $\tau \in \mathbb{R}$ and $\nu \geqslant 0$. Let $\Phi \subseteq H^\tau(\mathbb{R})$ and $\tilde{\Phi} \subseteq H^{-\tau}(\mathbb{R})$ be finite sets. Suppose that $\sum_{\tilde{\phi}\in\tilde{\Phi}}[\widehat{\tilde{\phi}}, \widehat{\tilde{\phi}}]_{-\tau} \in L_\infty(\mathbb{T})$, $\sum_{\phi\in\Phi}[\widehat{\phi}, \widehat{\phi}]_\tau \in L_\infty(\mathbb{T})$, and there exist positive constants C_0 and C_2 such that for almost every $\xi \in [-\pi, \pi]$,*

$$\Big|1 - \sum_{\phi\in\Phi} \widehat{\tilde{\phi}}(\xi)\overline{\widehat{\phi}(\xi)}\Big|^2 \leqslant C_0|\xi|^{2\nu}, \tag{4.7.8}$$

$$\sum_{n\in\mathbb{Z}\setminus\{0\}} \frac{(1 + |\xi + 2\pi n|^2)^\tau}{(1 + |\xi|^2)^\tau} \Big|\sum_{\phi\in\Phi} \widehat{\tilde{\phi}}(\xi)\overline{\widehat{\phi}(\xi + 2\pi k)}\Big|^2 \leqslant C_2|\xi|^{2\nu}. \tag{4.7.9}$$

Then there exists a positive constant C such that for all $\lambda > 0$ and $f \in H^{\tau+\nu}(\mathbb{R})$,

$$\int_{\mathbb{R}} |\widehat{\mathcal{Q}_\lambda f}(\xi) - \widehat{f}(\xi)|^2 (1 + |\tfrac{\xi}{\lambda}|^2)^\tau d\xi \leqslant C\lambda^{-2\nu} \int_{\mathbb{R}} |\widehat{f}(\xi)|^2 (1 + |\tfrac{\xi}{\lambda}|^2)^\tau |\xi|^{2\nu} d\xi. \tag{4.7.10}$$

If $(\{\tilde{\Phi}; \tilde{\Psi}\}, \{\Phi; \Psi\})$ is a dual framelet in $(H^{-\tau}(\mathbb{R}), H^\tau(\mathbb{R}))$ such that (4.7.8) and (4.7.9) hold, then $(\{\tilde{\Phi}; \tilde{\Psi}\}, \{\Phi; \Psi\})$ provides approximation order ν, i.e., (4.7.10) holds with λ and $\mathcal{Q}_\lambda$ being replaced by 2^n and $\mathcal{A}_n$, respectively for all $n \in \mathbb{N}$.

Proof Take $\mathrm{w}(\xi) = (1+|\xi|^2)^\tau$ and $\tilde{\mathrm{w}}(\xi) = (1+|\xi|^2)^\tau|\xi|^{2\nu}$ in Theorem 4.7.1. By our assumption, it is straightforward to verify that all (4.7.4)–(4.7.7) in Theorem 4.7.1 are satisfied with $C_{\mathrm{w}} = \pi^{-2\nu}$, $C_1 = C_0$,

$$C_3 = \tilde{C}\|(1 + |\cdot|^2)^\tau |\widehat{\phi}(\cdot)|^2\|_{L_\infty([-\pi,\pi])}, \qquad C_4 = \tilde{C}\Big\| \sum_{\phi\in\Phi}[\widehat{\phi}, \widehat{\phi}]_\tau \Big\|_{L_\infty(\mathbb{T})}.$$

where $\tilde{C} := \pi^{-2\nu}\| \sum_{\tilde{\phi}\in\tilde{\Phi}}[\widehat{\tilde{\phi}}, \widehat{\tilde{\phi}}]_{-\tau}\|_{L_\infty(\mathbb{T})}$. Now (4.7.10) follows from Theorem 4.7.1.

If $(\{\tilde{\Phi}; \tilde{\Psi}\}, \{\Phi; \Psi\})$ is a dual framelet in $(H^{-\tau}(\mathbb{R}), H^\tau(\mathbb{R}))$ with finite sets Φ and $\tilde{\Phi}$, by Proposition 4.6.4, we have $\sum_{\tilde{\phi}\in\tilde{\Phi}}[\widehat{\tilde{\phi}}, \widehat{\tilde{\phi}}]_{-\tau} \in L_\infty(\mathbb{R})$ and $\sum_{\phi\in\Phi}[\widehat{\phi}, \widehat{\phi}]_\tau \in L_\infty(\mathbb{R})$. Now the claim follows from the fact that $\mathcal{A}_n = \mathcal{Q}_{2^n}$ for all $n \in \mathbb{N}$. ☐

Corollary 4.7.3 *Let $\tau \in \mathbb{R}$ and $\nu \geqslant 0$. Let $\Phi \subseteq H^\tau(\mathbb{R})$ and $\tilde{\Phi} \subseteq H^{-\tau}(\mathbb{R})$ be finite subsets. Suppose that $\sum_{\tilde{\phi}\in\tilde{\Phi}}[\widehat{\tilde{\phi}}, \widehat{\tilde{\phi}}]_{-\tau} \in L_\infty(\mathbb{T})$, $\sum_{\phi\in\Phi}[\widehat{\phi}, \widehat{\phi}]_\tau \in L_\infty(\mathbb{T})$, and there exist positive constants C_0 and C_2 such that for almost every $\xi \in [-\pi, \pi]$, the inequality* (4.7.8) *holds and*

$$\sum_{n\in\mathbb{Z}\setminus\{0\}} |\xi + 2\pi n|^{2\tau} \Big|\sum_{\phi\in\Phi} \widehat{\tilde{\phi}}(\xi)\overline{\widehat{\phi}(\xi + 2\pi n)}\Big|^2 \leqslant C_2|\xi|^{2\tau+2\nu}, \tag{4.7.11}$$

Then there exists a positive constant C such that for all $\lambda > 0$ and $f \in H^{\tau+\nu}(\mathbb{R})$,

$$\int_{\mathbb{R}} |\widehat{\mathcal{Q}_\lambda f}(\xi) - \widehat{f}(\xi)|^2 |\xi|^{2\tau} d\xi \leqslant C\lambda^{-2\nu} \int_{\mathbb{R}} |\widehat{f}(\xi)|^2 |\xi|^{2\tau+2\nu} d\xi. \tag{4.7.12}$$

Proof Take $\mathrm{w}(\xi) = |\xi|^{2\tau}$ and $\tilde{\mathrm{w}}(\xi) = |\xi|^{2\tau+2\nu}$ in Theorem 4.7.1. We see that all (4.7.4)–(4.7.7) in Theorem 4.7.1 are satisfied with $C_{\mathrm{w}} = \pi^{-2\nu}$, $C_1 = C_0$,

$$C_3 = \tilde{C}\Big\| |\cdot|^{2\tau} \sum_{\phi\in\Phi} |\widehat{\phi}(\cdot)|^2 \Big\|_{L_\infty([-\pi,\pi])}, \quad C_4 = \tilde{C}\max(1, (1+\pi^{-2})^{-\tau}) \Big\| \sum_{\phi\in\Phi} [\widehat{\phi}, \widehat{\phi}]_\tau \Big\|_{L_\infty(\mathbb{T})},$$

where $\tilde{C} := \pi^{-2\nu} \max(0, (1+\pi^{-2})^{\tau}) \| \sum_{\tilde{\phi}\in\tilde{\Phi}} [\widehat{\tilde{\phi}}, \widehat{\tilde{\phi}}]_{-\tau} \|_{L_\infty(\mathbb{T})}$. Now (4.7.12) follows directly from Theorem 4.7.1. □

4.8 Frequency-Based Nonstationary Dual Framelets*

The notion of affine systems $\mathsf{AS}_J(\Phi; \Psi)$ can be generalized to nonstationary affine systems by allowing an arbitrary dilation factor (instead of the particular dilation factor 2^j) and using a different set of generating functions at each scale level j. This provides an extra flexibility for the applications of framelets and wavelets in practice. The frequency-based approach introduced in Sect. 4.1 is particularly suitable for studying nonstationary or periodic framelets and wavelets which we shall address in this and next sections. In this section we only provide some basic results on nonstationary framelets.

4.8.1 Characterization of Frequency-Based Nonstationary Dual Framelets

Let us first introduce the notion of a pair of frequency-based nonstationary dual frames. Let $J \in \mathbb{Z}$ and $\{\lambda_j\}_{j=J}^\infty$ be a sequence of nonzero real numbers. Let $\boldsymbol{\Phi}$, $\tilde{\boldsymbol{\Phi}}$, $\boldsymbol{\Psi}_j$, $\tilde{\boldsymbol{\Psi}}_j, j \geqslant J$ be subsets of $L_2^{loc}(\mathbb{R})$ (or distributions). We define

$$\mathsf{FAS}_J(\boldsymbol{\Phi}; \{\boldsymbol{\Psi}_j | \lambda_j\}_{j=J}^\infty) := \{\varphi_{\lambda_J;0,k} \; : \; k \in \mathbb{Z}, \varphi \in \boldsymbol{\Phi}\} \cup \{\psi_{\lambda_j;0,k} \; : \; k \in \mathbb{Z}, j \geqslant J, \psi \in \boldsymbol{\Psi}_j\}.$$

We shall breviate $\mathsf{FAS}_J(\boldsymbol{\Phi}; \{\boldsymbol{\Psi}_j \mid \lambda_j\}_{j=J}^\infty)$ as $\mathsf{FAS}_J(\boldsymbol{\Phi}; \{\boldsymbol{\Psi}_j\}_{j=J}^\infty)$ if the underlying scaling factors $\{\lambda_j\}_{j=J}^\infty$ are understood in advance. We say that

$$(\mathsf{FAS}_J(\tilde{\boldsymbol{\Phi}}; \{\tilde{\boldsymbol{\Psi}}_j | \lambda_j\}_{j=J}^\infty), \mathsf{FAS}_J(\boldsymbol{\Phi}; \{\boldsymbol{\Psi}_j | \lambda_j\}_{j=J}^\infty)) \tag{4.8.1}$$

is *a pair of frequency-based nonstationary dual frames* if the following identity holds: for all $\mathbf{f}, \mathbf{g} \in \mathscr{D}(\mathbb{R})$,

$$\sum_{\varphi\in\Phi}\sum_{k\in\mathbb{Z}}\langle \mathbf{f}, \tilde{\varphi}_{\lambda_J;0,k}\rangle\langle \varphi_{\lambda_J;0,k}, \mathbf{g}\rangle + \sum_{j=J}^{\infty}\sum_{\psi\in\Psi_j}\sum_{k\in\mathbb{Z}}\langle \mathbf{f}, \tilde{\psi}_{\lambda_j;0,k}\rangle\langle \psi_{\lambda_j;0,k}, \mathbf{g}\rangle = 2\pi\langle \mathbf{f}, \mathbf{g}\rangle, \tag{4.8.2}$$

where the infinite series in (4.8.2) converge in the following sense:

(1) For every $\mathbf{f}, \mathbf{g} \in \mathscr{D}(\mathbb{R})$, the following series

$$\sum_{\varphi\in\Phi}\sum_{k\in\mathbb{Z}}\langle \mathbf{f}, \tilde{\varphi}_{\lambda_J;0,k}\rangle\langle \varphi_{\lambda_J;0,k}, \mathbf{g}\rangle \quad \text{and} \quad \sum_{\psi\in\Psi_j}\sum_{k\in\mathbb{Z}}\langle \mathbf{f}, \tilde{\psi}_{\lambda_j;0,k}\rangle\langle \psi_{\lambda_j;0,k}, \mathbf{g}\rangle \tag{4.8.3}$$

converge absolutely for every integer $j \geqslant J$;

(2) For every $\mathbf{f}, \mathbf{g} \in \mathscr{D}(\mathbb{R})$, the following limit exists and

$$\lim_{J'\to\infty}\Big(\sum_{\varphi\in\Phi}\sum_{k\in\mathbb{Z}}\langle \mathbf{f}, \tilde{\varphi}_{\lambda_J;0,k}\rangle\langle \varphi_{\lambda_J;0,k}, \mathbf{g}\rangle + \sum_{j=J}^{J'-1}\sum_{\psi\in\Psi_j}\sum_{k\in\mathbb{Z}}\langle \mathbf{f}, \tilde{\psi}_{\lambda_j;0,k}\rangle\langle \psi_{\lambda_j;0,k}, \mathbf{g}\rangle\Big) = 2\pi\langle \mathbf{f}, \mathbf{g}\rangle.$$

The frequency-based (stationary) affine systems considered in Sect. 4.1 correspond to $\lambda_j = 2^{-j}$ and $\Psi_j = \Psi$ for all $j \geqslant J$, that is, the generating functions remain stationary (unchanged) at all the scale levels j.

A pair of frequency-based nonstationary dual frames can be characterized by the following result.

Theorem 4.8.1 *Let $J \in \mathbb{Z}$ and $\{\lambda_j\}_{j=J}^{\infty}$ be a sequence of nonzero real numbers such that* $\lim_{j\to\infty}\lambda_j = 0$. *Let $\Phi, \tilde{\Phi}, \Psi_j, \tilde{\Psi}_j$ for $j \geqslant J$ be subsets of $L_2^{loc}(\mathbb{R})$ such that*

$$\sum_{\varphi\in\Phi}|\varphi(\cdot)|^2, \ \sum_{\tilde{\varphi}\in\tilde{\Phi}}|\tilde{\varphi}(\cdot)|^2, \ \sum_{\psi\in\Psi_j}|\psi(\cdot)|^2, \ \sum_{\tilde{\psi}\in\tilde{\Psi}_j}|\tilde{\psi}(\cdot)|^2 \in L_1^{loc}(\mathbb{R}), \quad \forall j \geqslant J. \tag{4.8.4}$$

Then the pair in (4.8.1) *is a pair of frequency-based nonstationary dual frames if and only if*

$$\lim_{J'\to\infty}\Big\langle \mathcal{I}_{\Phi}^{0}(\lambda_J\cdot) + \sum_{j=J}^{J'-1}\mathcal{I}_{\Psi_j}^{0}(\lambda_j\cdot), \mathbf{h}\Big\rangle = \langle 1, \mathbf{h}\rangle, \qquad \forall\, \mathbf{h} \in \mathscr{D}(\mathbb{R}), \tag{4.8.5}$$

$$\mathcal{I}_{\Phi}^{\lambda_J k}(\lambda_J\xi) + \sum_{j=J}^{\infty}\mathcal{I}_{\Psi_j}^{\lambda_j k}(\lambda_j\xi) = 0, \qquad a.e.\ \xi \in \mathbb{R},\ k \in \Lambda\backslash\{0\}, \tag{4.8.6}$$

(The above sum $\sum_{j=J}^{\infty}$ is in fact finite, since $\lim_{j\to\infty} \lambda_j k = 0$ for every $k \in \mathbb{R}$.) where $\Lambda := \cup_{j=J}^{\infty}[\lambda_j^{-1}\mathbb{Z}]$ and

$$\mathcal{I}_{\boldsymbol{\Phi}}^k(\xi) := \sum_{\varphi\in\Phi} \widehat{\tilde{\varphi}}(\xi)\overline{\widehat{\varphi}(\xi+2\pi k)}, \quad k \in \mathbb{Z}, \qquad \mathcal{I}_{\boldsymbol{\Phi}}^k(\xi) = 0, \quad k \in \mathbb{R}\backslash\mathbb{Z}, \tag{4.8.7}$$

$$\mathcal{I}_{\boldsymbol{\Psi}_j}^k(\xi) := \sum_{\psi\in\Psi_j} \widehat{\tilde{\psi}}(\xi)\overline{\widehat{\psi}(\xi+2\pi k)}, \quad k \in \mathbb{Z}, \qquad \mathcal{I}_{\boldsymbol{\Psi}_j}^k(\xi) = 0, \quad k \in \mathbb{R}\backslash\mathbb{Z}. \tag{4.8.8}$$

Proof Let $\mathbf{f}, \mathbf{g} \in \mathscr{D}(\mathbb{R})$. By (4.8.4) and Lemma 4.1.1, all series in (4.8.3) converge absolutely and

$$\sum_{\psi\in\Psi_j}\sum_{k\in\mathbb{Z}} \langle \mathbf{f}, \tilde{\psi}_{\lambda_j;0,k}\rangle\langle \psi_{\lambda_j;0,k}, \mathbf{g}\rangle = 2\pi \int_{\mathbb{R}} \sum_{k\in[\lambda_j^{-1}\mathbb{Z}]} \widehat{\mathbf{f}}(\xi)\overline{\widehat{\mathbf{g}}(\xi+2\pi k)\,\mathcal{I}_{\boldsymbol{\Psi}_j}^{\lambda_j k}(\lambda_j\xi)}\, d\xi. \tag{4.8.9}$$

Since $\mathbf{f}$ and $\mathbf{g}$ have compact support and $\lim_{j\to\infty} \lambda_j = 0$, we observe that there exists an integer $J_{\mathbf{f},\mathbf{g}}$ such that $\widehat{\mathbf{f}}(\xi)\overline{\widehat{\mathbf{g}}(\xi+2\pi k)} = 0$ for all $k \in [\lambda_j^{-1}\mathbb{Z}]\backslash\{0\}$ and $j \geqslant J_{\mathbf{f},\mathbf{g}}$. Therefore, for all $j \geqslant J_{\mathbf{f},\mathbf{g}}$, the only possible nonzero term in the right-hand side of (4.8.9) is $k = 0$, that is, (4.8.9) becomes

$$\sum_{\psi\in\Psi_j}\sum_{k\in\mathbb{Z}} \langle \mathbf{f}, \tilde{\psi}_{\lambda_j;0,k}\rangle\langle \psi_{\lambda_j;0,k}, \mathbf{g}\rangle = 2\pi \int_{\mathbb{R}} \widehat{\mathbf{f}}(\xi)\overline{\widehat{\mathbf{g}}(\xi)}\,\overline{\mathcal{I}_{\boldsymbol{\Psi}_j}^{0}(\lambda_j\xi)}\, d\xi.$$

Sufficiency ($\Leftarrow$). For $J' > J$, we define

$$S_J^{J'}(\mathbf{f},\mathbf{g}) := \sum_{\varphi\in\Phi}\sum_{k\in\mathbb{Z}} \langle \mathbf{f}, \tilde{\varphi}_{\lambda_J;0,k}\rangle\langle \varphi_{\lambda_J;0,k}, \mathbf{g}\rangle + \sum_{j=J}^{J'-1}\sum_{\psi\in\Psi_j}\sum_{k\in\mathbb{Z}} \langle \mathbf{f}, \tilde{\psi}_{\lambda_j;0,k}\rangle\langle \psi_{\lambda_j;0,k}, \mathbf{g}\rangle.$$

Therefore, by (4.8.9), for $J' > J$, we have

$$\overline{S_J^{J'}(\mathbf{f},\mathbf{g})} = 2\pi \int_{\mathbb{R}} \sum_{k\in\Lambda} \overline{\widehat{\mathbf{f}}(\xi)}\widehat{\mathbf{g}}(\xi+2\pi k)\Big[\mathcal{I}_{\boldsymbol{\Phi}}^{\lambda_J k}(\lambda_J\xi) + \sum_{j=J}^{J'-1}\mathcal{I}_{\boldsymbol{\Psi}_j}^{\lambda_j k}(\lambda_j\xi)\Big]\, d\xi.$$

Now by the above identity, for all $J' > \max(J, J_{\mathbf{f},\mathbf{g}})$, we deduce that

$$\begin{aligned}\overline{S_J^{J'}(\mathbf{f},\mathbf{g})} &= 2\pi \int_{\mathbb{R}} \overline{\widehat{\mathbf{f}}(\xi)}\widehat{\mathbf{g}}(\xi)\Big[\mathcal{I}_{\boldsymbol{\Phi}}^{0}(\lambda_J\xi) + \sum_{j=J}^{J'-1}\mathcal{I}_{\boldsymbol{\Psi}_j}^{0}(\lambda_j\xi)\Big]\, d\xi \\ &\quad + 2\pi \int_{\mathbb{R}} \sum_{k\in\Lambda\backslash\{0\}} \overline{\widehat{\mathbf{f}}(\xi)}\widehat{\mathbf{g}}(\xi+2\pi k)\Big[\mathcal{I}_{\boldsymbol{\Phi}}^{\lambda_J k}(\lambda_J\xi) + \sum_{j=J}^{\infty}\mathcal{I}_{\boldsymbol{\Psi}_j}^{\lambda_j k}(\lambda_j\xi)\Big]\, d\xi.\end{aligned} \tag{4.8.10}$$

Note that all the above series/summations are in fact finite by $\mathbf{f}, \mathbf{g} \in \mathscr{D}(\mathbb{R})$ and $\lim_{j\to\infty} \lambda_j = 0$. If (4.8.6) holds, then we deduce from (4.8.10) that

$$\overline{S_J^{J'}(\mathbf{f}, \mathbf{g})} = 2\pi \int_{\mathbb{R}} \overline{\mathbf{f}(\xi)}\mathbf{g}(\xi)\Big[\mathcal{I}^0_{\Phi}(\lambda_J\xi) + \sum_{j=J}^{J'-1} \mathcal{I}^0_{\Psi_j}(\lambda_j\xi)\Big]\, d\xi. \tag{4.8.11}$$

By (4.8.11) and (4.8.5), we have $\lim_{J'\to\infty} S_J^{J'}(\mathbf{f}, \mathbf{g}) = 2\pi \int_{\mathbb{R}} \mathbf{f}(\xi)\overline{\mathbf{g}(\xi)}\, d\xi = 2\pi \langle \mathbf{f}, \mathbf{g}\rangle$.

Necessity ($\Rightarrow$). The proof of the necessity part is essentially the same as that of Lemma 4.1.5. Since $\lim_{j\to\infty} \lambda_j = 0$, the set Λ is discrete and closed. For any temporarily fixed $\xi \in \mathbb{R}$ and $k_0 \in \Lambda\backslash\{0\}$, it is important to notice that $\mathrm{dist}(k_0, \Lambda\backslash\{k_0\}) > 0$. By (4.8.10), the same argument as in the proof of Lemma 4.1.5 leads to (4.8.6). Therefore, (4.8.11) must hold, from which we deduce that (4.8.5) holds. □

For the special case that $\lambda_j = 2^{-j}$ for all $j \geqslant J$ in Theorem 4.8.1, (4.8.6) becomes

$$\sum_{\varphi\in\Phi} \tilde{\varphi}(2^{-J}\xi)\overline{\varphi(2^{-J}(\xi + 2\pi 2^m k_0))}$$
$$+ \sum_{j=J}^{m} \sum_{\psi\in\Psi_j} \tilde{\psi}(2^{-j}\xi)\overline{\psi(2^{-j}(\xi + 2\pi 2^m k_0))} = 0, \quad m \geqslant J, k_0 \in \mathbb{Z}\backslash[2\mathbb{Z}] \tag{4.8.12}$$

for almost every $\xi \in \mathbb{R}$, and (4.8.5) becomes: for all $\mathbf{h} \in \mathscr{D}(\mathbb{R})$,

$$\lim_{J'\to\infty} \Big\langle \sum_{\varphi\in\Phi} \tilde{\varphi}(2^{-J}\cdot)\overline{\varphi(2^{-J}\cdot)} + \sum_{j=J}^{J'-1} \sum_{\psi\in\Psi_j} \tilde{\psi}(2^{-j}\cdot)\overline{\psi(2^{-j}\cdot)}, \mathbf{h}\Big\rangle = \langle 1, \mathbf{h}\rangle. \tag{4.8.13}$$

By a similar argument as in Theorems 4.3.7 and 4.3.8, we have the following result whose proof is left as Exercise 4.62.

Corollary 4.8.2 *Let $J \in \mathbb{Z}$ and $\{\lambda_j\}_{j=J}^{\infty}$ be a sequence of nonzero real numbers such that $\lim_{j\to\infty} \lambda_j = 0$. Let Φ, Ψ_j be subsets of distributions in $\mathscr{D}'(\mathbb{R})$ for all integers $j \geqslant J$. Then the following statements are equivalent:*

(1) $\mathsf{FAS}_J(\Phi; \{\Psi_j|\lambda_j\}_{j=J}^{\infty})$ *is a* frequency-based nonstationary tight frame *for $L_2(\mathbb{R})$, that is, $\Phi, \Psi_j \subseteq L_2(\mathbb{R})$ for all integers $j \geqslant J$ and for all $\mathbf{f} \in L_2(\mathbb{R})$,*

$$\sum_{\varphi\in\Phi} \sum_{k\in\mathbb{Z}} |\langle \mathbf{f}, \boldsymbol{\varphi}_{\lambda_J;0,k}\rangle|^2 + \sum_{j=J}^{\infty} \sum_{\psi\in\Psi_j} \sum_{k\in\mathbb{Z}} |\langle \mathbf{f}, \boldsymbol{\psi}_{\lambda_j;0,k}\rangle|^2 = 2\pi \|\mathbf{f}\|^2_{L_2(\mathbb{R})}.$$

(2) $(\mathsf{FAS}_J(\Phi; \{\Psi_j|\lambda_j\}_{j=J}^{\infty}), \mathsf{FAS}_J(\Phi; \{\Psi_j|\lambda_j\}_{j=J}^{\infty}))$ *is a pair of frequency-based nonstationary dual frames.*

(3) *There exist* $\Phi, \Psi_j \subseteq L_2(\mathbb{R})$ *for all integers* $j \geqslant J$ *such that* $\widehat{\Phi} = \boldsymbol{\Phi}$, $\widehat{\Psi_j} = \boldsymbol{\Psi}_j$ *for all* $j \geqslant J$, *and* $\mathsf{AS}_J(\Phi; \{\Psi_j|\lambda_j^{-1}\}_{j=J}^{\infty})$ is a nonstationary tight frame *for* $L_2(\mathbb{R})$:

$$\sum_{\phi\in\Phi}\sum_{k\in\mathbb{Z}} |\langle f, \phi_{\lambda_J^{-1};k}\rangle|^2 + \sum_{j=J}^{\infty}\sum_{\psi\in\Psi_j}\sum_{k\in\mathbb{Z}} |\langle f, \psi_{\lambda_j^{-1};k}\rangle|^2 = \|f\|^2_{L_2(\mathbb{R})}, \qquad (4.8.14)$$

for all $f \in L_2(\mathbb{R})$, *where*

$$\mathsf{AS}_J(\Phi; \{\Psi_j|\lambda_j^{-1}\}_{j=J}^{\infty}) := \{\phi_{\lambda_J^{-1};k} \ : \ k \in \mathbb{Z}, \phi \in \Phi\} \cup \{\psi_{\lambda_j^{-1};k} \ : \ k \in \mathbb{Z}, j \geqslant J, \psi \in \Psi_j\}.$$

4.8.2 Sequences of Frequency-Based Nonstationary Dual Framelets

We now study a sequence of pairs of frequency-based nonstationary dual frames.

Theorem 4.8.3 *Let* J_0 *be an integer and* $\{\lambda_j\}_{j=J_0}^{\infty}$ *be a sequence of nonzero real numbers such that* $\lim_{j\to\infty} \lambda_j = 0$. *For integers* $j \geqslant J_0$, *let* $\boldsymbol{\Phi}_j, \tilde{\boldsymbol{\Phi}}_j, \boldsymbol{\Psi}_j, \tilde{\boldsymbol{\Psi}}_j$ *be subsets of* $L_2^{loc}(\mathbb{R})$ *such that for all* $j \geqslant J_0$,

$$\sum_{\varphi\in\boldsymbol{\Phi}_j} |\varphi(\cdot)|^2, \ \sum_{\tilde{\varphi}\in\tilde{\boldsymbol{\Phi}}_j} |\tilde{\varphi}(\cdot)|^2, \ \sum_{\boldsymbol{\psi}\in\boldsymbol{\Psi}_j} |\boldsymbol{\psi}(\cdot)|^2, \ \sum_{\tilde{\boldsymbol{\psi}}\in\tilde{\boldsymbol{\Psi}}_j} |\tilde{\boldsymbol{\psi}}(\cdot)|^2 \in L_1^{loc}(\mathbb{R}). \qquad (4.8.15)$$

Then $(\mathsf{FAS}_J(\tilde{\boldsymbol{\Phi}}_J; \{\tilde{\boldsymbol{\Psi}}_j|\lambda_j\}_{j=J}^{\infty}), \mathsf{FAS}_J(\boldsymbol{\Phi}_J; \{\boldsymbol{\Psi}_j|\lambda_j\}_{j=J}^{\infty}))$ *is a pair of frequency-based nonstationary dual frames for every integer* $J \geqslant J_0$ *if and only if*

$$\mathcal{I}_{\boldsymbol{\Phi}_j}^{\lambda_j k}(\lambda_j\xi) + \mathcal{I}_{\boldsymbol{\Psi}_j}^{\lambda_j k}(\lambda_j\xi) = \mathcal{I}_{\boldsymbol{\Phi}_{j+1}}^{\lambda_{j+1} k}(\lambda_{j+1}\xi), \quad k \in [\lambda_j^{-1}\mathbb{Z}] \cup [\lambda_{j+1}^{-1}\mathbb{Z}] \qquad (4.8.16)$$

for almost every $\xi \in \mathbb{R}$ *for all* $j \geqslant J_0$, *and*

$$\lim_{j\to\infty} \left\langle \sum_{\varphi\in\boldsymbol{\Phi}_j} \tilde{\varphi}(\lambda_j\cdot)\overline{\varphi(\lambda_j\cdot)}, \mathbf{h}\right\rangle = \langle 1, \mathbf{h}\rangle, \qquad \forall\, \mathbf{h} \in \mathscr{D}(\mathbb{R}), \qquad (4.8.17)$$

where $\mathcal{I}_{\boldsymbol{\Psi}_j}^k$, $k \in \mathbb{R}$ *are defined in* (4.8.8) *and* $\mathcal{I}_{\boldsymbol{\Phi}_j}^k$ *is defined in* (4.8.7) *with* $\boldsymbol{\Phi} = \boldsymbol{\Phi}_j$.

Proof By the same argument as in Theorem 4.1.6, considering the difference between two consecutive scale levels, we see that

$$(\mathsf{FAS}_J(\tilde{\boldsymbol{\Phi}}_J; \{\tilde{\boldsymbol{\Psi}}_j|\lambda_j\}_{j=J}^{\infty}), \mathsf{FAS}_J(\boldsymbol{\Phi}_J; \{\boldsymbol{\Psi}_j|\lambda_j\}_{j=J}^{\infty}))$$

is a pair of frequency-based nonstationary dual frames for all integers $J \geqslant J_0$ if and only if

$$\sum_{\varphi\in\Phi_j}\sum_{k\in\mathbb{Z}}\langle \mathbf{f}, \tilde{\varphi}_{\lambda_j;0,k}\rangle\langle \varphi_{\lambda_j;0,k}, \mathbf{g}\rangle + \sum_{\psi\in\Psi_j}\sum_{k\in\mathbb{Z}}\langle \mathbf{f}, \tilde{\psi}_{\lambda_j;0,k}\rangle\langle \psi_{\lambda_j;0,k}, \mathbf{g}\rangle$$
$$= \sum_{\eta\in\Phi_{j+1}}\sum_{k\in\mathbb{Z}}\langle \mathbf{f}, \tilde{\eta}_{\lambda_{j+1};0,k}\rangle\langle \eta_{\lambda_{j+1};0,k}, \mathbf{g}\rangle, \quad \mathbf{f},\mathbf{g}\in\mathscr{D}(\mathbb{R}), j\geqslant J_0, \tag{4.8.18}$$

and

$$\lim_{j\to\infty}\sum_{\varphi\in\Phi_j}\sum_{k\in\mathbb{Z}}\langle \mathbf{f}, \tilde{\varphi}_{\lambda_j;0,k}\rangle\langle \varphi_{\lambda_j;0,k}, \mathbf{g}\rangle = 2\pi\langle \mathbf{f},\mathbf{g}\rangle, \quad \mathbf{f},\mathbf{g}\in\mathscr{D}(\mathbb{R}). \tag{4.8.19}$$

By the identity $\langle \mathbf{f}_{\lambda_1;n_1,k_1}, \mathbf{g}_{\lambda_2;n_2,k_2}\rangle = e^{i(k_2\lambda_1^{-1}\lambda_2-k_1)n_1}\langle \mathbf{f}, \mathbf{g}_{\lambda_1^{-1}\lambda_2;n_2-\lambda_1^{-1}\lambda_2 n_1,k_2-\lambda_1\lambda_2^{-1}k_1}\rangle$, we have $\langle \mathbf{f}_{\lambda_j;0,0}, \tilde{\varphi}_{\lambda_j;0,k}\rangle = \langle \mathbf{f}, \tilde{\varphi}_{1;0,k}\rangle$ and $\langle \mathbf{f}_{\lambda_j;0,0}, \tilde{\eta}_{\lambda_{j+1};0,k}\rangle = \langle \mathbf{f}, \tilde{\eta}_{\lambda_j^{-1}\lambda_{j+1};0,k}\rangle$. Hence, the inequality (4.8.18) is equivalent to

$$\sum_{\varphi\in\Phi_j}\sum_{k\in\mathbb{Z}}\langle \mathbf{f}, \tilde{\varphi}_{1;0,k}\rangle\langle \varphi_{1;0,k}, \mathbf{g}\rangle + \sum_{\psi\in\Psi_j}\sum_{k\in\mathbb{Z}}\langle \mathbf{f}, \tilde{\psi}_{1;0,k}\rangle\langle \psi_{1;0,k}, \mathbf{g}\rangle$$
$$= \sum_{\eta\in\Phi_{j+1}}\sum_{k\in\mathbb{Z}}\langle \mathbf{f}, \tilde{\eta}_{\lambda_j^{-1}\lambda_{j+1};0,k}\rangle\langle \eta_{\lambda_j^{-1}\lambda_{j+1};0,k}, \mathbf{g}\rangle, \quad \mathbf{f},\mathbf{g}\in\mathscr{D}(\mathbb{R}), j\geqslant J_0. \tag{4.8.20}$$

By Lemma 4.1.5, the identity (4.8.16) is equivalent to (4.8.20). By Lemma 4.1.4, the equation (4.8.17) is equivalent to (4.8.19). □

For the special case $\lambda_j = 2^{-j}$ for $j \geqslant J_0$ in Theorem 4.8.3, (4.8.16) becomes

$$\sum_{\varphi\in\Phi_j}\tilde{\varphi}(2\xi)\overline{\varphi(2(\xi+2\pi k))} + \sum_{\psi\in\Psi_j}\tilde{\psi}(2\xi)\overline{\psi(2(\xi+2\pi k))}$$
$$= \sum_{\eta\in\Phi_{j+1}}\tilde{\eta}(\xi)\overline{\eta(\xi+2\pi k)}, \quad \forall\, k\in\mathbb{Z}, \tag{4.8.21}$$

and

$$\sum_{\varphi\in\Phi_j}\tilde{\varphi}(\xi)\overline{\varphi(\xi+2\pi k_0)} + \sum_{\psi\in\Psi_j}\tilde{\psi}(\xi)\overline{\psi(\xi+2\pi k_0)} = 0, \quad \forall\, k_0\in\mathbb{Z}\backslash[2\mathbb{Z}] \tag{4.8.22}$$

for a.e. $\xi\in\mathbb{R}$ and for all $j\geqslant J_0$. If in addition $\Phi_j = \Phi$ and $\Psi = \Psi_j$, then (4.8.21) is just (4.1.24) and (4.8.22) becomes (4.1.25) in item (iii) of Theorem 4.1.6.

We now connect a nonstationary dual framelet filter bank with a sequence of pairs of frequency-based nonstationary dual frames. Recall that $\mathbb{N}_0 := \mathbb{N}\cup\{0\}$.

Theorem 4.8.4 *Let $\{\mathsf{d}_j\}_{j=1}^{\infty}$ be a sequence of nonzero integers and define $\lambda_0 := 1$, $\lambda_j := \prod_{n=1}^{j} \mathsf{d}_n^{-1}$ for all $j \in \mathbb{N}$. Assume that $\lim_{j\to\infty} \lambda_j = 0$. Let $a_j, \tilde{a}_j \in l_2(\mathbb{Z})$ for $j \in \mathbb{N}$ such that there exist positive real numbers $\iota, \tilde{\iota}, C_j, \tilde{C}_j$ for $j \in \mathbb{N}$ satisfying*

$$|1 - \widehat{a_j}(\lambda_j\xi)| \leqslant C_j|\xi|^{\tau}, \quad |1 - \widehat{\tilde{a}_j}(\lambda_j\xi)| \leqslant \tilde{C}_j|\xi|^{\tilde{\tau}}, \quad a.e.\ \xi \in \mathbb{R}, j \in \mathbb{N} \tag{4.8.23}$$

with $\sum_{j=1}^{\infty} C_j < \infty$ and $\sum_{j=1}^{\infty} \tilde{C}_j < \infty$. Define frequency-based nonstationary refinable functions $\boldsymbol{\varphi}^j$ and $\tilde{\boldsymbol{\varphi}}^j$ for $j \in \mathbb{N}_0$ by

$$\boldsymbol{\varphi}^j(\xi) := \prod_{k=1}^{\infty} \widehat{a_{j+k}}(\lambda_{j+k}\lambda_j^{-1}\xi), \quad \tilde{\boldsymbol{\varphi}}^j(\xi) := \prod_{k=1}^{\infty} \widehat{\tilde{a}_{j+k}}(\lambda_{j+k}\lambda_j^{-1}\xi).$$

Then all $\boldsymbol{\varphi}^j$, $\tilde{\boldsymbol{\varphi}}^j, j \in \mathbb{N}_0$ are well-defined functions in $L_{\infty}^{loc}(\mathbb{R})$ satisfying the following frequency-based nonstationary refinement equations: For $j \in \mathbb{N}$,

$$\boldsymbol{\varphi}^{j-1}(\mathsf{d}_j\xi) = \widehat{a_j}(\xi)\boldsymbol{\varphi}^j(\xi), \quad \tilde{\boldsymbol{\varphi}}^{j-1}(\mathsf{d}_j\xi) = \widehat{\tilde{a}_j}(\xi)\tilde{\boldsymbol{\varphi}}^j(\xi), \quad a.e.\ \xi \in \mathbb{R}. \tag{4.8.24}$$

Let $\theta_{j-1}, b_{j,1}, \dots, b_{j,s_{j-1}}, \tilde{\theta}_{j-1}, \tilde{b}_{j,1}, \dots, \tilde{b}_{j,s_{j-1}} \in l_2(\mathbb{Z})$ with $s_{j-1} \in \mathbb{N}_0$ and $j \in \mathbb{N}$. Define

$$\boldsymbol{\eta}^j(\xi) := \widehat{\theta_j}(\xi)\boldsymbol{\varphi}^j(\xi), \quad \tilde{\boldsymbol{\eta}}^j(\xi) := \widehat{\tilde{\theta}_j}(\xi)\tilde{\boldsymbol{\varphi}}^j(\xi), \quad j \in \mathbb{N}_0,$$

$$\boldsymbol{\psi}^{j-1,\ell}(\mathsf{d}_j\xi) := \widehat{b_{j,\ell}}(\xi)\boldsymbol{\varphi}^j(\xi), \quad \tilde{\boldsymbol{\psi}}^{j-1,\ell}(\mathsf{d}_j\xi) := \widehat{\tilde{b}_{j,\ell}}(\xi)\tilde{\boldsymbol{\varphi}}^j(\xi), \quad j \in \mathbb{N},$$

for all $\ell = 1, \dots, s_{j-1}$. Then $\boldsymbol{\eta}^j, \tilde{\boldsymbol{\eta}}^j \in L_2^{loc}(\mathbb{R})$ and $\boldsymbol{\Psi}_j := \{\boldsymbol{\psi}^{j,1}, \dots, \boldsymbol{\psi}^{j,s_j}\}$, $\tilde{\boldsymbol{\Psi}}_j := \{\tilde{\boldsymbol{\psi}}^{j,1}, \dots, \tilde{\boldsymbol{\psi}}^{j,s_j}\}$ are finite subsets of $L_2^{loc}(\mathbb{R})$ for all $j \in \mathbb{N}_0$. Moreover, the pair

$$(\mathsf{FAS}_J(\tilde{\boldsymbol{\eta}}^J; \{\tilde{\boldsymbol{\Psi}}_j \,|\, \lambda_j\}_{j=J}^{\infty}), \mathsf{FAS}_J(\boldsymbol{\eta}^J; \{\boldsymbol{\Psi}_j \,|\, \lambda_j\}_{j=J}^{\infty}))$$

is a pair of frequency-based nonstationary dual frames for every integer $J \in \mathbb{N}_0$ if and only if

$$\lim_{j\to\infty} \langle \widehat{\Theta}_j(\lambda_j\cdot), \mathbf{h}\rangle = \langle 1, \mathbf{h}\rangle \quad \forall\, \mathbf{h} \in \mathscr{D}(\mathbb{R}) \quad with \quad \widehat{\Theta}_j(\xi) := \widehat{\tilde{\theta}_j}(\xi)\overline{\widehat{\theta}_j(\xi)}$$

and $\{(\{\tilde{a}_j; \tilde{b}_{j,1}, \dots, \tilde{b}_{j,s_{j-1}}\}, \{a_j; b_{j,1}, \dots, b_{j,s_{j-1}}\})_{\Theta_{j-1}}\}_{j=1}^{\infty}$ is a generalized nonstationary dual framelet filter bank, *that is, for all $j \in \mathbb{N}$ and $\omega \in \{0, 1, \dots, |\mathsf{d}_j| - 1\}$,*

$$\widehat{\Theta_{j-1}}(\mathsf{d}_j\xi)\widehat{\tilde{a}_j}(\xi)\overline{\widehat{a_j}(\xi + \tfrac{2\pi\omega}{\mathsf{d}_j})} + \sum_{\ell=1}^{s_{j-1}} \widehat{\tilde{b}_{j,\ell}}(\xi)\overline{\widehat{b_{j,\ell}}(\xi + \tfrac{2\pi\omega}{\mathsf{d}_j})} = \boldsymbol{\delta}(\omega)\widehat{\Theta}_j(\xi),$$

for almost every $\xi \in (\mathrm{supp}([\boldsymbol{\varphi}^j, \boldsymbol{\varphi}^j]) - \frac{2\pi\omega}{\mathsf{d}_j}) \cap \mathrm{supp}([\tilde{\boldsymbol{\varphi}}^j, \tilde{\boldsymbol{\varphi}}^j])$.

Proof Since $|z| \leqslant 1 + |1-z| \leqslant e^{|1-z|}$ for all $z \in \mathbb{C}$, by (4.8.23), we deduce that

$$\Big|\prod_{k=m}^{n} \widehat{a_{j+k}}(\lambda_{j+k}\xi)\Big| \leqslant e^{\sum_{k=m}^{n}|1-\widehat{a_{j+k}}(\lambda_{j+k}\xi)|} \leqslant e^{|\xi|^\tau \sum_{k=m}^{n} C_{j+k}} \leqslant e^{C|\xi|^\tau}, \qquad (4.8.25)$$

where $C := \sum_{k=1}^{\infty} C_k < \infty$. For all $1 \leqslant m < n$, we have the following identity

$$1 - \prod_{k=m}^{n} \widehat{a_{j+k}}(\lambda_{j+k}\xi) = \sum_{k=m}^{n} \Big(1 - \widehat{a_{j+k}}(\lambda_{j+k}\xi)\Big)\Big(\prod_{\ell=k+1}^{n} \widehat{a_{j+\ell}}(\lambda_{j+\ell}\xi)\Big)$$

with the convention $\prod_{\ell=n+1}^{n} := 1$. Therefore, by (4.8.23) and (4.8.25), we have

$$\Big|\prod_{k=1}^{m} \widehat{a_{j+k}}(\lambda_{j+k}\xi) - \prod_{k=1}^{n} \widehat{a_{j+k}}(\lambda_{j+k}\xi)\Big| \leqslant e^{2C|\xi|^\tau} \sum_{k=m}^{n} |1 - \widehat{a_{j+k}}(\lambda_{j+k}\xi)| \leqslant e^{2C|\xi|^\tau} |\xi|^\tau \sum_{k=m}^{n} C_{j+k}.$$

Since $\sum_{k=1}^{\infty} C_k < \infty$, the above inequality implies the convergence of the infinite product $\prod_{k=1}^{\infty} \widehat{a_{j+k}}(\lambda_{j+k}\xi)$. Since $\boldsymbol{\varphi}^j(\lambda_j\xi) = \prod_{k=1}^{\infty} \widehat{a_{j+k}}(\lambda_{j+k}\xi)$, it follows from (4.8.25) that $\boldsymbol{\varphi}^j \in L_\infty^{loc}(\mathbb{R})$ and (4.8.24) holds. Since all $\widehat{\theta_j}, \widehat{b_{j,\ell}} \in L_2(\mathbb{T})$, it is evident that all $\boldsymbol{\varphi}^j, \boldsymbol{\psi}^{j,\ell} \in L_2^{loc}(\mathbb{R})$. The rest of the claim can be proved by a similar argument using Theorem 4.8.3 as in Theorem 4.1.9 and is left as Exercise 4.62. □

As a direct consequence of Theorem 4.8.4, similar to Theorem 4.5.4, we have

Theorem 4.8.5 *Let $\{\mathsf{d}_j\}_{j=1}^{\infty}$ be a sequence of nonzero integers and define $\lambda_0 := 1$, $\lambda_j := \prod_{n=1}^{j} \mathsf{d}_n^{-1}$ for $j \in \mathbb{N}$. Assume that $\lim_{j\to\infty} \lambda_j = 0$. Let $\phi^j, \psi^{j,1}, \dots, \psi^{j,s_j}$ be tempered distributions whose Fourier transforms lie in $L_2^{loc}(\mathbb{R})$ for all $j \in \mathbb{N}_0$. Define $\Psi_j := \{\psi^{j,1}, \dots, \psi^{j,s_j}\}$. Then $\mathsf{AS}_J(\phi^J; \{\Psi_j|\lambda_j^{-1}\}_{j=J}^{\infty})$ is a (normalized) tight frame for $L_2(\mathbb{R})$ for all $J \in \mathbb{N}_0$ (i.e., (4.8.14) holds with $\Phi = \{\phi^J\}$) if and only if*

(1) $\lim_{j\to\infty}\langle|\widehat{\phi^j}(\lambda_j\cdot)|^2, \mathbf{h}\rangle = \langle 1, \mathbf{h}\rangle$ *for all* $\mathbf{h} \in \mathscr{D}(\mathbb{R})$*;*

(2) There exist 2π-periodic measurable functions $\widehat{a_j}, \widehat{b_{j,1}}, \dots, \widehat{b_{j,s_{j-1}}}$, $j \in \mathbb{N}$ such that

$$\widehat{\phi^{j-1}}(\mathsf{d}_j\xi) = \widehat{a_j}(\xi)\widehat{\phi^j}(\xi), \quad \widehat{\psi^{j-1,\ell}}(\mathsf{d}_j\xi) = \widehat{b_{j,\ell}}(\xi)\widehat{\phi^j}(\xi) \quad a.e.\ \xi \in \mathbb{R} \qquad (4.8.26)$$

for all $\ell = 1, \dots, s_{j-1}$, and $\{a_j; b_{j,1}, \dots, b_{j,s_{j-1}}\}_{j=1}^{\infty}$ is a generalized nonstationary tight framelet filter bank, *that is, for all $j \in \mathbb{N}$ and $\omega \in \{0, 1, \dots, |\mathsf{d}_j| - 1\}$,*

$$\widehat{a_j}(\xi)\overline{\widehat{a_j}(\xi + \tfrac{2\pi\omega}{\mathsf{d}_j})} + \sum_{\ell=1}^{s_{j-1}} \widehat{b_{j,\ell}}(\xi)\overline{\widehat{b_{j,\ell}}(\xi + \tfrac{2\pi\omega}{\mathsf{d}_j})} = \boldsymbol{\delta}(\omega),$$

for almost every $\xi \in (\mathrm{supp}([\widehat{\phi^j}, \widehat{\phi^j}]) - \frac{2\pi\omega}{\mathsf{d}_j}) \cap \mathrm{supp}([\widehat{\phi^j}, \widehat{\phi^j}])$.

Proof If $\mathsf{AS}_J(\phi^J; \{\Psi_j|\lambda_j^{-1}\}_{j=J}^{\infty})$ is a tight frame for $L_2(\mathbb{R})$ for all $J \in \mathbb{N}_0$, then we have the following cascade structure: For $f \in L_2(\mathbb{R})$,

$$\sum_{k\in\mathbb{Z}} |\langle f, \phi^j_{1;k}\rangle|^2 + \sum_{\ell=1}^{s_j}\sum_{k\in\mathbb{Z}} |\langle f, \psi^{j,\ell}_{1;k}\rangle|^2 = \sum_{k\in\mathbb{Z}} |\langle f, \phi^{j+1}_{\lambda_j\lambda_{j+1}^{-1};k}\rangle|^2, \quad \forall j \in \mathbb{N}_0.$$

Note that $\lambda_j\lambda_{j+1}^{-1} = \mathsf{d}_{j+1}$. Similar to Theorem 4.5.4, we deduce from the above identity that

$$\mathsf{S}(\{\phi^j, \psi^{j,1}, \ldots, \psi^{j,s_j}\}|L_2(\mathbb{R})) = \mathsf{S}_{\mathsf{d}_{j+1}}(\phi^{j+1}|L_2(\mathbb{R})), \qquad \forall\, j \in \mathbb{N}_0.$$

Therefore, there exist 2π-periodic measurable functions $\widehat{a_j}, \widehat{b_{j,1}}, \ldots, \widehat{b_{j,s_j}}$ such that (4.8.26) holds. The claims follow from Theorem 4.8.4 with $\tilde{\theta}_j = \theta_j = \boldsymbol{\delta}$. □

4.9 Periodic Framelets and Wavelets*

As we discussed in Sect. 1.5, discrete framelet transforms for signals on bounded intervals essentially employ periodic dual framelets and wavelets. In this section, we provide a systematic study on periodic framelets and wavelets. Periodic framelets and wavelets are unavoidably linked to nonstationary affine systems and nonstationary dual framelets on the real line, which we have discussed in Sect. 4.8.

4.9.1 Frequency-Based Periodic Dual Framelets

Define $\Omega_0 := \Gamma_0 := \{0\}$. For $j \in \mathbb{N}$, we define

$$\Omega_j := 2\pi 2^{-j}\Gamma_j \quad \text{and} \quad \Gamma_j := \{1-2^{j-1}, \ldots, 2^{j-1}\}, \tag{4.9.1}$$

where Γ_j is a complete set of distinct representatives in $\mathbb{Z}/[2^j\mathbb{Z}]$. Recall that $l(\mathbb{Z})$ is the linear space of all sequences on $\mathbb{Z}$ and $l_0(\mathbb{Z})$ is the linear space of all finitely supported sequences on $\mathbb{Z}$. Throughout this section we shall use the notation $\mathbb{N}_0 := \mathbb{N} \cup \{0\}$ and $\mathbf{f}_{n,k} := e^{-ik\cdot}\mathbf{f}(\cdot - n)$ for $k, n \in \mathbb{R}$. For $\mathbf{f}, \boldsymbol{\psi} \in l(\mathbb{Z})$ and $j \in \mathbb{N}_0$, we define a 2^j-periodic sequence $[\mathbf{f}, \boldsymbol{\psi}]_{2^j\mathbb{Z}}$ by

$$[\mathbf{f}, \boldsymbol{\psi}]_{2^j\mathbb{Z}}(n) := \sum_{k\in\mathbb{Z}} \mathbf{f}(n + 2^j k)\overline{\boldsymbol{\psi}(n + 2^j k)}, \qquad n \in \mathbb{Z}. \tag{4.9.2}$$

The following result is the discrete version of Lemma 4.1.1.

Lemma 4.9.1 *Let $j \in \mathbb{N}_0$. Let $\mathbf{f}, \mathbf{g} \in l(\mathbb{Z})$ and $\boldsymbol{\Psi}, \tilde{\boldsymbol{\Psi}} \subseteq l(\mathbb{Z})$ such that*

$$\sum_{\boldsymbol{\psi}\in\boldsymbol{\Psi}} \sum_{n\in\mathbb{Z}} \sum_{k\in\mathbb{Z}} |\mathbf{f}(n)\mathbf{f}(n+2^j k)\boldsymbol{\psi}(n)\boldsymbol{\psi}(n+2^j k)| < \infty, \tag{4.9.3}$$

$$\sum_{\tilde{\boldsymbol{\psi}}\in\tilde{\boldsymbol{\Psi}}} \sum_{n\in\mathbb{Z}} \sum_{k\in\mathbb{Z}} |\mathbf{g}(n)\mathbf{g}(n+2^j k)\tilde{\boldsymbol{\psi}}(n)\tilde{\boldsymbol{\psi}}(n+2^j k)| < \infty.$$

Then $\sum_{\boldsymbol{\psi}\in\boldsymbol{\Psi}} \sum_{\omega\in\Omega_j} |\langle \mathbf{f}, \boldsymbol{\psi}_{0,\omega}\rangle|^2 < \infty$, $\sum_{\tilde{\boldsymbol{\psi}}\in\tilde{\boldsymbol{\Psi}}} \sum_{\omega\in\Omega_j} |\langle \mathbf{g}, \tilde{\boldsymbol{\psi}}_{0,\omega}\rangle|^2 < \infty$, and

$$\begin{aligned}\sum_{\boldsymbol{\psi}\in\boldsymbol{\Psi}} \sum_{\omega\in\Omega_j} \langle \mathbf{f}, \boldsymbol{\psi}_{0,\omega}\rangle \langle \tilde{\boldsymbol{\psi}}_{0,\omega}, \mathbf{g}\rangle &= 2^j \sum_{\gamma\in\Gamma_j} \sum_{\boldsymbol{\psi}\in\boldsymbol{\Psi}} [\mathbf{f}, \boldsymbol{\psi}]_{2^j\mathbb{Z}}(\gamma) [\tilde{\boldsymbol{\psi}}, \mathbf{g}]_{2^j\mathbb{Z}}(\gamma) \\ &= \sum_{n\in\mathbb{Z}} \sum_{k\in\mathbb{Z}} \mathbf{f}(n)\overline{\mathbf{g}(n+2^j k)} \sum_{\boldsymbol{\psi}\in\boldsymbol{\Psi}} 2^j \overline{\boldsymbol{\psi}(n)} \tilde{\boldsymbol{\psi}}(n+2^j k)\end{aligned} \tag{4.9.4}$$

with all the above series converging absolutely. In particular, (4.9.3) *holds if either of the following conditions is satisfied*

(i) $\mathbf{f} \in l_0(\mathbb{Z})$ *and* $\sum_{\boldsymbol{\psi}\in\boldsymbol{\Psi}} |\boldsymbol{\psi}(n)|^2 < \infty$ *for every* $n \in \mathbb{Z}$.
(ii) $\mathbf{f} \in l_2(\mathbb{Z})$ *and* $\sum_{\boldsymbol{\psi}\in\boldsymbol{\Psi}} [\boldsymbol{\psi}, \boldsymbol{\psi}]_{2^j\mathbb{Z}} := \sum_{\boldsymbol{\psi}\in\boldsymbol{\Psi}} \sum_{k\in\mathbb{Z}} |\boldsymbol{\psi}(\cdot + 2^j k)|^2 \in l_\infty(\mathbb{Z})$.

Proof Note that (4.9.3) is equivalent to $\sum_{\boldsymbol{\psi}\in\boldsymbol{\Psi}} \sum_{\gamma\in\Gamma_j} |[|\mathbf{f}|, |\boldsymbol{\psi}|]_{2^j\mathbb{Z}}(\gamma)|^2 < \infty$. Therefore, for $\omega \in \Omega_j = 2\pi 2^{-j}\Gamma_j \subseteq 2\pi 2^{-j}\mathbb{Z}$, we have $e^{i2^j n\omega} = 1$ for all $n \in \mathbb{Z}$ and

$$\langle \mathbf{f}, \boldsymbol{\psi}_{0,\omega}\rangle = \sum_{m\in\mathbb{Z}} \mathbf{f}(m)\overline{\boldsymbol{\psi}(m)} e^{im\omega} = \sum_{\gamma\in\Gamma_j} \Big(\sum_{n\in\mathbb{Z}} \mathbf{f}(\gamma+2^j n)\overline{\boldsymbol{\psi}(\gamma+2^j n)} \Big) e^{i\gamma\omega} = \sum_{\gamma\in\Gamma_j} [\mathbf{f}, \boldsymbol{\psi}]_{2^j\mathbb{Z}}(\gamma) e^{i\gamma\omega}.$$

Since $2^{-j/2}(e^{i\gamma\omega})_{\omega\in\Omega_j, \gamma\in\Gamma_j}$ is a unitary matrix, we conclude that

$$\sum_{\boldsymbol{\psi}\in\boldsymbol{\Psi}} \sum_{\omega\in\Omega_j} |\langle \mathbf{f}, \boldsymbol{\psi}_{0,\omega}\rangle|^2 = 2^j \sum_{\boldsymbol{\psi}\in\boldsymbol{\Psi}} \sum_{\gamma\in\Gamma_j} |[\mathbf{f}, \boldsymbol{\psi}]_{2^j\mathbb{Z}}(\gamma)|^2 < \infty.$$

Similarly, we have $\sum_{\tilde{\boldsymbol{\psi}}\in\tilde{\boldsymbol{\Psi}}} \sum_{\omega\in\Omega_j} |\langle \mathbf{g}, \tilde{\boldsymbol{\psi}}_{0,\omega}\rangle|^2 = 2^j \sum_{\tilde{\boldsymbol{\psi}}\in\tilde{\boldsymbol{\Psi}}} \sum_{\gamma\in\Gamma_j} |[\mathbf{g}, \tilde{\boldsymbol{\psi}}]_{2^j\mathbb{Z}}(\gamma)|^2 < \infty$. By the Cauchy-Schwarz inequality, we see that all series in (4.9.4) converge absolutely. Moreover,

$$\begin{aligned}\sum_{\boldsymbol{\psi}\in\boldsymbol{\Psi}} \sum_{\omega\in\Omega_j} \langle \mathbf{f}, \boldsymbol{\psi}_{0,\omega}\rangle \langle \tilde{\boldsymbol{\psi}}_{0,\omega}, \mathbf{g}\rangle &= 2^j \sum_{\boldsymbol{\psi}\in\boldsymbol{\Psi}} \sum_{\gamma\in\Gamma_j} [\mathbf{f}, \boldsymbol{\psi}]_{2^j\mathbb{Z}}(\gamma) [\tilde{\boldsymbol{\psi}}, \mathbf{g}]_{2^j\mathbb{Z}}(\gamma) \\ &= 2^j \sum_{\boldsymbol{\psi}\in\boldsymbol{\Psi}} \sum_{\gamma\in\Gamma_j} \Big(\sum_{k\in\mathbb{Z}} \mathbf{f}(\gamma+2^j k)\overline{\boldsymbol{\psi}(\gamma+2^j k)} \Big) [\tilde{\boldsymbol{\psi}}, \mathbf{g}]_{2^j\mathbb{Z}}(\gamma)\end{aligned}$$

$$= 2^j \sum_{\psi \in \Psi} \sum_{n \in \mathbb{Z}} \mathbf{f}(n) \overline{\psi(n)} [\tilde{\psi}, \mathbf{g}]_{2^j\mathbb{Z}}(n)$$

$$= 2^j \sum_{n \in \mathbb{Z}} \sum_{k \in \mathbb{Z}} \sum_{\psi \in \Psi} \mathbf{f}(n) \overline{\psi(n)} \tilde{\psi}(n + 2^j k) \overline{\mathbf{g}(n + 2^j k)}.$$

This proves (4.9.4). If item (i) holds, then there exists $N \in \mathbb{N}$ such that $\mathbf{f}(n) = 0$ for all $n \in \mathbb{Z} \backslash [-N, N]$ and

$$\sum_{\psi \in \Psi} \sum_{n \in \mathbb{Z}} \sum_{k \in \mathbb{Z}} |\mathbf{f}(n)\mathbf{f}(n + 2^j k)\psi(n)\psi(n + 2^j k)|$$

$$\leqslant \|\mathbf{f}\|^2_{l_\infty(\mathbb{Z})} \sum_{n=-N}^{N} \sum_{k=-2N}^{2N} \sum_{\psi \in \Psi} |\psi(n)\psi(n + 2^j k)|$$

$$\leqslant \|\mathbf{f}\|^2_{l_\infty(\mathbb{Z})} \Big(\sum_{n=-N}^{N} \sum_{k=-2N}^{2N} \sum_{\psi \in \Psi} |\psi(n)|^2 \Big)^{1/2} \Big(\sum_{n=-N}^{N} \sum_{k=-2N}^{2N} \sum_{\psi \in \Psi} |\psi(n + 2^j k)|^2 \Big)^{1/2} < \infty.$$

If item (ii) holds, then

$$\sum_{\psi \in \Psi} \sum_{n \in \mathbb{Z}} \sum_{k \in \mathbb{Z}} |\mathbf{f}(n)\mathbf{f}(n + 2^j k)\psi(n)\psi(n + 2^j k)| = \sum_{\psi \in \Psi} \sum_{\gamma \in \Gamma_j} |[|\mathbf{f}|, |\psi|]_{2^j\mathbb{Z}}(\gamma)|^2$$

$$\leqslant \sum_{\gamma \in \Gamma_j} \sum_{\psi \in \Psi} [\mathbf{f}, \mathbf{f}]_{2^j\mathbb{Z}}(\gamma)[\psi, \psi]_{2^j\mathbb{Z}}(\gamma)$$

$$\leqslant \Big\| \sum_{\psi \in \Psi} [\psi, \psi]_{2^j\mathbb{Z}} \Big\|_{l_\infty(\mathbb{Z})} \sum_{\gamma \in \Gamma_j} [\mathbf{f}, \mathbf{f}]_{2^j\mathbb{Z}}(\gamma) = \Big\| \sum_{\psi \in \Psi} [\psi, \psi]_{2^j\mathbb{Z}} \Big\|_{l_\infty(\mathbb{Z})} \|\mathbf{f}\|^2_{l_2(\mathbb{Z})} < \infty,$$

where we used the fact that $[\psi, \psi]_{2^j\mathbb{Z}}$ is 2^j-periodic. □

For $J \in \mathbb{N}_0$ and $\Phi, \tilde{\Phi}, \Psi_j, \tilde{\Psi}_j \subseteq l(\mathbb{Z})$ with $j \geqslant J$, we define *a frequency-based periodic affine system* to be

$$\mathsf{FPAS}_J(\Phi; \{\Psi_j\}_{j=J}^\infty) := \{\varphi_{0,\omega} \ : \ \varphi \in \Phi, \omega \in \Omega_J\} \cup \cup_{j=J}^\infty \{\psi_{0,\omega} \ : \ \psi \in \Psi_j, \omega \in \Omega_j\}.$$

For simplicity of discussion, throughout this section we assume that all $\Phi, \tilde{\Phi}, \Psi_j, \tilde{\Psi}_j$ are finite sets, though all the results can be easily generalized to countable subsets $\Psi \subseteq l(\mathbb{Z})$ satisfying $\sum_{\psi \in \Psi} |\psi(n)|^2 < \infty$ for all $n \in \mathbb{Z}$. We say that

$$\big(\mathsf{FPAS}_J(\tilde{\Phi}; \{\tilde{\Psi}_j\}_{j=J}^\infty), \mathsf{FPAS}_J(\Phi; \{\Psi_j\}_{j=J}^\infty)\big) \qquad (4.9.5)$$

is *a pair of frequency-based periodic dual frames* if for all $\mathbf{f}, \mathbf{g} \in l_0(\mathbb{Z})$,

$$\lim_{J'\to\infty}\Big(\sum_{\varphi\in\Phi}\sum_{\omega\in\Omega_J}\langle \mathbf{f}, \tilde{\varphi}_{0,\omega}\rangle\langle \varphi_{0,\omega}, \mathbf{g}\rangle + \sum_{j=J}^{J'}\sum_{\psi\in\Psi_j}\sum_{\omega\in\Omega_j}\langle \mathbf{f}, \tilde{\psi}_{0,\omega}\rangle\langle \psi_{0,\omega}, \mathbf{g}\rangle\Big) = \langle \mathbf{f}, \mathbf{g}\rangle.$$

A pair of frequency-based periodic dual frames can be characterized by

Theorem 4.9.2 *Let $J \in \mathbb{N}_0$ and $\Phi, \tilde{\Phi}, \Psi_j, \tilde{\Psi}_j, j \geqslant J$ be finite subsets of $l(\mathbb{Z})$. Then the pair in* (4.9.5) *is a pair of frequency-based periodic dual frames if and only if*

$$\lim_{J'\to\infty}\Big(\sum_{\varphi\in\Phi} 2^J\tilde{\varphi}(n)\overline{\varphi(n)} + \sum_{j=J}^{J'}\sum_{\psi\in\Psi_j} 2^j\tilde{\psi}(n)\overline{\psi(n)}\Big) = 1, \quad \forall\, n \in \mathbb{Z} \tag{4.9.6}$$

and for all $m \geqslant J$ and $n \in \mathbb{Z}$,

$$\sum_{\varphi\in\Phi} 2^J\tilde{\varphi}(n)\overline{\varphi(n+2^m k_0)} + \sum_{j=J}^{m}\sum_{\psi\in\Psi_j} 2^j\tilde{\psi}(n)\overline{\psi(n+2^m k_0)} = 0,\ \forall\ k_0 \in \mathbb{Z}\backslash[2\mathbb{Z}]. \tag{4.9.7}$$

Proof By Lemma 4.9.1, we have

$$\begin{aligned}\sum_{\psi\in\Psi_j}\sum_{\omega\in\Omega_j}\langle \mathbf{f}, \tilde{\psi}_{0,\omega}\rangle\langle \psi_{0,\omega}, \mathbf{g}\rangle &= \sum_{n\in\mathbb{Z}}\sum_{k\in\mathbb{Z}}\mathbf{f}(n)\overline{\mathbf{g}(n+2^j k)}\sum_{\psi\in\Psi_j} 2^j\overline{\tilde{\psi}(n)}\psi(n+2^j k)\\ &= \sum_{n\in\mathbb{Z}}\mathbf{f}(n)\overline{\mathbf{g}(n)}\sum_{\psi\in\Psi_j} 2^j\overline{\tilde{\psi}(n)}\psi(n)\\ &\quad + \sum_{n\in\mathbb{Z}}\sum_{k_0\in\mathbb{Z}\backslash[2\mathbb{Z}]}\sum_{m=j}^{\infty}\mathbf{f}(n)\overline{\mathbf{g}(n+2^m k_0)}\sum_{\psi\in\Psi_j} 2^j\overline{\tilde{\psi}(n)}\psi(n+2^m k_0).\end{aligned}$$

Therefore,

$$\begin{aligned}S_J^{J'}(\mathbf{f},\mathbf{g}) &:= \sum_{\varphi\in\Phi}\sum_{\omega\in\Omega_J}\langle \mathbf{f}, \tilde{\varphi}_{0,\omega}\rangle\langle \varphi_{0,\omega}, \mathbf{g}\rangle + \sum_{j=J}^{J'}\sum_{\psi\in\Psi_j}\sum_{\omega\in\Omega_j}\langle \mathbf{f}, \tilde{\psi}_{0,\omega}\rangle\langle \psi_{0,\omega}, \mathbf{g}\rangle\\ &= \sum_{n\in\mathbb{Z}}\mathbf{f}(n)\overline{\mathbf{g}(n)}\Big(\sum_{\varphi\in\Phi} 2^J\overline{\tilde{\varphi}(n)}\varphi(n) + \sum_{j=J}^{J'}\sum_{\psi\in\Psi_j} 2^j\overline{\tilde{\psi}(n)}\psi(n)\Big)\\ &\quad + \sum_{n\in\mathbb{Z}}\sum_{k_0\in\mathbb{Z}\backslash[2\mathbb{Z}]}\sum_{m=J}^{\infty}\mathbf{f}(n)\overline{\mathbf{g}(n+2^m k_0)}\Big(\sum_{\varphi\in\Phi} 2^J\overline{\tilde{\varphi}(n)}\varphi(n+2^m k_0)\\ &\quad + \sum_{j=J}^{\min(m,J')}\sum_{\psi\in\Psi_j} 2^j\overline{\tilde{\psi}(n)}\psi(n+2^m k_0)\Big).\end{aligned}$$

Note that for any fixed $\mathbf{f}, \mathbf{g} \in l_0(\mathbb{Z})$, there exists $N \in \mathbb{N}$ such that $\mathbf{f}(n)\overline{\mathbf{g}(n + 2^m k_0)} = 0$ for all $n \in \mathbb{Z}, k_0 \in \mathbb{Z}\backslash[2\mathbb{Z}]$ and $m \geqslant N$. We can directly verify that $\lim_{J' \to \infty} S_J^{J'}(\mathbf{f}, \mathbf{g}) = \langle \mathbf{f}, \mathbf{g} \rangle$ for all $\mathbf{f}, \mathbf{g} \in l_0(\mathbb{Z})$ if and only if (4.9.6) and (4.9.7) are satisfied. □

For $f \in L_1(\mathbb{T})$, recall that its *Fourier transform* $\widehat{f} = \{\widehat{f}(k)\}_{k \in \mathbb{Z}} \in l(\mathbb{Z})$ is defined to be $\widehat{f}(k) := \frac{1}{2\pi} \int_{-\pi}^{\pi} f(x) e^{-ikx} dx$ for $k \in \mathbb{Z}$. The Fourier transform can be naturally extended to 2π-periodic tempered distributions. As a direct consequence of Theorem 4.9.2, we have the following result on periodic tight frames.

Corollary 4.9.3 *Let $J \in \mathbb{N}_0$ and $\boldsymbol{\Phi}, \boldsymbol{\Psi}_j, j \geqslant J$ be finite subsets of $l(\mathbb{Z})$. Then the following statements are equivalent:*

(1) $\mathsf{FPAS}_J(\boldsymbol{\Phi}; \{\boldsymbol{\Psi}_j\}_{j=J}^\infty)$ *is* a frequency-based periodic tight frame *for $l_2(\mathbb{Z})$, that is, $\boldsymbol{\Phi}, \boldsymbol{\Psi}_j \subseteq l_2(\mathbb{Z})$ for all $j \geqslant J$ and for all $\mathbf{f} \in l_2(\mathbb{Z})$,*

$$\sum_{\boldsymbol{\varphi} \in \boldsymbol{\Phi}} \sum_{\omega \in \Omega_J} |\langle \mathbf{f}, \boldsymbol{\varphi}_{0,\omega} \rangle|^2 + \sum_{j=J}^{\infty} \sum_{\boldsymbol{\psi} \in \boldsymbol{\Psi}_j} \sum_{\omega \in \Omega_j} |\langle \mathbf{f}, \boldsymbol{\psi}_{0,\omega} \rangle|^2 = \|\mathbf{f}\|_{l_2(\mathbb{Z})}^2. \tag{4.9.8}$$

(2) $(\mathsf{FPAS}_J(\boldsymbol{\Phi}; \{\boldsymbol{\Psi}_j\}_{j=J}^\infty), \mathsf{FPAS}_J(\boldsymbol{\Phi}; \{\boldsymbol{\Psi}_j\}_{j=J}^\infty))$ *is a pair of frequency-based periodic dual frames.*

(3) The identity $\sum_{\boldsymbol{\varphi} \in \boldsymbol{\Phi}} 2^J |\boldsymbol{\varphi}(n)|^2 + \sum_{j=J}^{\infty} \sum_{\boldsymbol{\psi} \in \boldsymbol{\Psi}_j} 2^j |\boldsymbol{\psi}(n)|^2 = 1$ holds for all $n \in \mathbb{Z}$, and for all $m \geqslant J$ and $n \in \mathbb{Z}$,

$$\sum_{\boldsymbol{\varphi} \in \boldsymbol{\Phi}} 2^J \boldsymbol{\varphi}(n) \overline{\boldsymbol{\varphi}(n + 2^m k_0)} + \sum_{j=J}^{m} \sum_{\boldsymbol{\psi} \in \boldsymbol{\Psi}_j} 2^j \boldsymbol{\psi}(n) \overline{\boldsymbol{\psi}(n + 2^m k_0)} = 0, \ \forall \ k_0 \in \mathbb{Z}\backslash[2\mathbb{Z}].$$

(4) There exist $\Phi, \Psi_j \subseteq L_2(\mathbb{T})$ for $j \geqslant J$ such that $\widehat{\Phi} = \boldsymbol{\Phi}$ and $\widehat{\Psi_j} = \boldsymbol{\Psi}_j$ for all $j \geqslant J$ and $\mathsf{PAS}_J(\Phi; \{\Psi_j\}_{j=J}^\infty)$ *is* a (normalized) periodic tight frame *for $L_2(\mathbb{T})$:*

$$\sum_{\phi \in \Phi} \sum_{\omega \in \Omega_J} |\langle f, \phi_{\omega,0} \rangle|^2 + \sum_{j=J}^{\infty} \sum_{\psi \in \Psi_j} \sum_{\omega \in \Omega_j} |\langle f, \psi_{\omega,0} \rangle|^2 = \|f\|_{L_2(\mathbb{T})}^2, \quad \forall f \in L_2(\mathbb{T}).$$

Proof (1)$\Longrightarrow$(2) is trivial and (2) $\Longleftrightarrow$ (3) follows from Theorem 4.9.2. We now prove (2)$\Longrightarrow$(1). By Theorem 4.9.2, for every $\mathbf{f} \in l_0(\mathbb{Z})$, we have

$$\lim_{J' \to \infty} \Big(\sum_{\boldsymbol{\varphi} \in \boldsymbol{\Phi}} \sum_{\omega \in \Omega_J} |\langle \mathbf{f}, \boldsymbol{\varphi}_{0,\omega} \rangle|^2 + \sum_{j=J}^{J'} \sum_{\boldsymbol{\psi} \in \boldsymbol{\Psi}_j} \sum_{\omega \in \Omega_j} |\langle \mathbf{f}, \boldsymbol{\psi}_{0,\omega} \rangle|^2 \Big) = \|\mathbf{f}\|_{l_2(\mathbb{Z})}^2.$$

Therefore, (4.9.8) holds for all $\mathbf{f} \in l_0(\mathbb{Z})$. In particular, we have $|\langle \mathbf{f}, \boldsymbol{\psi} \rangle|^2 \leqslant \|\mathbf{f}\|_{l_2(\mathbb{Z})}^2$ for all $j \geqslant J$, $\mathbf{f} \in l_0(\mathbb{Z})$, and $\boldsymbol{\psi} \in \boldsymbol{\Psi}_j$. Since $l_0(\mathbb{Z})$ is dense in $l_2(\mathbb{Z})$, we see that $\|\boldsymbol{\psi}\|_{l_2(\mathbb{Z})}^2 = \sup_{\mathbf{f} \in l_0(\mathbb{Z}), \|\mathbf{f}\|_{l_2(\mathbb{Z})} \leqslant 1} |\langle \mathbf{f}, \boldsymbol{\psi} \rangle|^2 \leqslant 1$. Hence, $\boldsymbol{\Psi}_j \subseteq l_2(\mathbb{Z})$ for all $j \geqslant J$. Similarly, we have $\boldsymbol{\Phi} \subseteq l_2(\mathbb{Z})$. Since $l_0(\mathbb{Z})$ is dense in $l_2(\mathbb{Z})$ and (4.9.8) holds for all $\mathbf{f} \in l_0(\mathbb{Z})$, we conclude that (4.9.8) holds for all $\mathbf{f} \in l_2(\mathbb{Z})$. (1) $\iff$ (4) follows directly from Parseval's identity: $\langle f, \psi_{\omega,0}\rangle = \langle \widehat{f}, \widehat{\psi_{\omega,0}}\rangle = \langle \widehat{f}, \widehat{\psi}_{0,\omega}\rangle$. $\square$

We also have the following characterization for a sequence of frequency-based periodic dual frames.

Theorem 4.9.4 *Let $\boldsymbol{\Phi}_j, \tilde{\boldsymbol{\Phi}}_j, \boldsymbol{\Psi}_j, \tilde{\boldsymbol{\Psi}}_j$ be finite subsets of $l(\mathbb{Z})$ for all $j \in \mathbb{N}_0$. Then*

$$\left(\mathsf{FPAS}_J(\tilde{\boldsymbol{\Phi}}_J; \{\tilde{\boldsymbol{\Psi}}_j\}_{j=J}^{\infty}), \mathsf{FPAS}_J(\boldsymbol{\Phi}_J; \{\boldsymbol{\Psi}_j\}_{j=J}^{\infty})\right) \tag{4.9.9}$$

is a pair of frequency-based periodic dual frames for every $J \in \mathbb{N}_0$ if and only if

$$\lim_{j\to\infty} \sum_{\varphi\in\boldsymbol{\Phi}_j} 2^j \tilde{\varphi}(n)\overline{\varphi(n)} = 1, \qquad \forall\, n \in \mathbb{Z} \tag{4.9.10}$$

and for every $j \in \mathbb{N}_0$ and $n \in \mathbb{Z}$,

$$\begin{aligned}\sum_{\varphi\in\boldsymbol{\Phi}_j} \tilde{\varphi}(n)\overline{\varphi(n+2^{j+1}k)} &+ \sum_{\psi\in\boldsymbol{\Psi}_j} \tilde{\psi}(n)\overline{\psi(n+2^{j+1}k)} \\ &= 2\sum_{\eta\in\boldsymbol{\Phi}_{j+1}} \tilde{\eta}(n)\overline{\eta(n+2^{j+1}k)}, \quad \forall\, k \in \mathbb{Z},\end{aligned} \tag{4.9.11}$$

$$\sum_{\varphi\in\boldsymbol{\Phi}_j} \tilde{\varphi}(n)\overline{\varphi(n+2^{j}k)} + \sum_{\psi\in\boldsymbol{\Psi}_j} \tilde{\psi}(n)\overline{\psi(n+2^{j}k)} = 0, \quad \forall\, k \in \mathbb{Z}\backslash[2\mathbb{Z}]. \tag{4.9.12}$$

Proof Considering the difference between two consecutive levels of the sequence of pairs in (4.9.9), we see that the pair in (4.9.9) is a pair of frequency-based periodic dual frames for every $J \in \mathbb{N}_0$ if and only if for all $\mathbf{f}, \mathbf{g} \in l_0(\mathbb{Z})$ and $j \in \mathbb{N}_0$,

$$\begin{aligned}\sum_{\varphi\in\boldsymbol{\Phi}_j}\sum_{\omega\in\Omega_j} \langle \mathbf{f}, \tilde{\boldsymbol{\varphi}}_{0,\omega}\rangle\langle \boldsymbol{\varphi}_{0,\omega}, \mathbf{g}\rangle &+ \sum_{\psi\in\boldsymbol{\Psi}_j}\sum_{\omega\in\Omega_j} \langle \mathbf{f}, \tilde{\boldsymbol{\psi}}_{0,\omega}\rangle\langle \boldsymbol{\psi}_{0,\omega}, \mathbf{g}\rangle \\ &= \sum_{\eta\in\boldsymbol{\Phi}_{j+1}}\sum_{\omega\in\Omega_{j+1}} \langle \mathbf{f}, \tilde{\boldsymbol{\eta}}_{0,\omega}\rangle\langle \boldsymbol{\eta}_{0,\omega}, \mathbf{g}\rangle,\end{aligned} \tag{4.9.13}$$

for all $\mathbf{f}, \mathbf{g} \in l_0(\mathbb{Z})$, and

$$\lim_{j\to\infty} \sum_{\varphi\in\boldsymbol{\Phi}_j}\sum_{\omega\in\Omega_j} \langle \mathbf{f}, \tilde{\boldsymbol{\varphi}}_{0,\omega}\rangle\langle \boldsymbol{\varphi}_{0,\omega}, \mathbf{g}\rangle = \langle \mathbf{f}, \mathbf{g}\rangle, \qquad \forall\, \mathbf{f}, \mathbf{g} \in l_0(\mathbb{Z}). \tag{4.9.14}$$

By Lemma 4.9.1, (4.9.14) is equivalent to (4.9.10). Similarly, (4.9.13) is equivalent to (4.9.11) and (4.9.12). $\square$

We now study the connection between a frequency-based periodic dual frame and a periodic dual framelet filter bank.

Theorem 4.9.5 *For $j \in \mathbb{N}_0$, let $\varphi^j, \tilde{\varphi}^j, \mathbf{a}_j, \tilde{\mathbf{a}}_j \in l(\mathbb{Z})$ such that both $\mathbf{a}_j$ and $\tilde{\mathbf{a}}_j$ are 2^j-periodic sequences such that $\mathbf{a}_j(\cdot + 2^j) = \mathbf{a}_j$ and $\tilde{\mathbf{a}}_j(\cdot + 2^j) = \tilde{\mathbf{a}}_j$, and*

$$\varphi^{j-1}(n) = \mathbf{a}_j(n)\varphi^j(n) \quad and \quad \tilde{\varphi}^{j-1}(n) = \tilde{\mathbf{a}}_j(n)\tilde{\varphi}^j(n), \quad \forall\, n \in \mathbb{Z}, j \in \mathbb{N}.$$

For $j \in \mathbb{N}$, let $\boldsymbol{\theta}_j, \mathbf{b}_{j,1}, \ldots, \mathbf{b}_{j,s_{j-1}}, \tilde{\boldsymbol{\theta}}_j, \tilde{\mathbf{b}}_{j,1}, \ldots, \tilde{\mathbf{b}}_{j,s_{j-1}} \in l(\mathbb{Z})$ with $s_{j-1} \in \mathbb{N}_0$ such that all of them are 2^j-periodic sequences. For $j \in \mathbb{N}_0$, define $\boldsymbol{\Psi}_j := \{\boldsymbol{\psi}^{j,1}, \ldots, \boldsymbol{\psi}^{j,s_j}\}$ and $\tilde{\boldsymbol{\Psi}}_j := \{\tilde{\boldsymbol{\psi}}^{j,1}, \ldots, \tilde{\boldsymbol{\psi}}^{j,s_j}\}$, where for $j \in \mathbb{N}_0$ and $\ell = 1, \ldots, s_{j-1}$,

$$\boldsymbol{\eta}^j(n) := \boldsymbol{\theta}_j(n)\varphi^j(n), \qquad \tilde{\boldsymbol{\eta}}^j(n) := \tilde{\boldsymbol{\theta}}_j(n)\tilde{\varphi}^j(n), \quad n \in \mathbb{Z}, \tag{4.9.15}$$

$$\boldsymbol{\psi}^{j-1,\ell}(n) := \mathbf{b}_{j,\ell}(n)\varphi^j(n), \quad \tilde{\boldsymbol{\psi}}^{j-1,\ell}(n) := \tilde{\mathbf{b}}_{j,\ell}(n)\tilde{\varphi}^j(n), \quad n \in \mathbb{Z}. \tag{4.9.16}$$

Then $(\mathsf{FPAS}_J(\tilde{\boldsymbol{\eta}}^J; \{\tilde{\boldsymbol{\Psi}}_j\}_{j=J}^{\infty}), \mathsf{FPAS}_J(\boldsymbol{\eta}^J; \{\boldsymbol{\Psi}_j\}_{j=J}^{\infty}))$ is a pair of frequency-based periodic dual frames for every integer $J \in \mathbb{N}_0$ if and only if

$$\lim_{j\to\infty} 2^j \boldsymbol{\Theta}_j(n)\tilde{\varphi}^j(n)\overline{\varphi^j(n)} = 1, \qquad \forall\, n \in \mathbb{Z}, \tag{4.9.17}$$

where $\boldsymbol{\Theta}_j(n) := \tilde{\boldsymbol{\theta}}_j(n)\overline{\boldsymbol{\theta}_j(n)}$ for $n \in \mathbb{Z}$ and $j \in \mathbb{N}_0$, and

$$\left\{(\{\tilde{\mathbf{a}}_j; \tilde{\mathbf{b}}_{j,1}, \ldots, \tilde{\mathbf{b}}_{j,s_{j-1}}\}, \{\mathbf{a}_j; \mathbf{b}_{j,1}, \ldots, \mathbf{b}_{j,s_{j-1}}\})_{\boldsymbol{\Theta}_{j-1}}\right\}_{j=1}^{\infty}$$

is a frequency-based generalized periodic dual framelet filter bank, *that is,*

$$\begin{gathered}\boldsymbol{\Theta}_{j-1}(n)\tilde{\mathbf{a}}_j(n)\overline{\mathbf{a}_j(n)} + \sum_{\ell=1}^{s_{j-1}} \tilde{\mathbf{b}}_{j,\ell}(n)\overline{\mathbf{b}_{j,\ell}(n)} = \boldsymbol{\Theta}_j(n), \\ \forall\, n \in \operatorname{supp}([\tilde{\varphi}^j, \tilde{\varphi}^j]_{2^j\mathbb{Z}}) \cap \operatorname{supp}([\varphi^j, \varphi^j]_{2^j\mathbb{Z}}),\end{gathered} \tag{4.9.18}$$

$$\begin{gathered}\boldsymbol{\Theta}_{j-1}(n)\tilde{\mathbf{a}}_j(n)\overline{\mathbf{a}_j(n + 2^{j-1})} + \sum_{\ell=1}^{s_{j-1}} \tilde{\mathbf{b}}_{j,\ell}(n)\overline{\mathbf{b}_{j,\ell}(n + 2^{j-1})} = 0, \\ \forall\, n \in \operatorname{supp}([\tilde{\varphi}^j, \tilde{\varphi}^j]_{2^j\mathbb{Z}}) \cap \operatorname{supp}([\varphi^j, \varphi^j]_{2^j\mathbb{Z}} - 2^{j-1}).\end{gathered} \tag{4.9.19}$$

Proof We use Theorem 4.9.4 to prove the claim. Let $\boldsymbol{\Phi}_j := \{\boldsymbol{\eta}^j\}$ and $\tilde{\boldsymbol{\Phi}}_j := \{\tilde{\boldsymbol{\eta}}^j\}$. Then (4.9.10) is equivalent to (4.9.17). By (4.9.15) and (4.9.16), we have

$$\tilde{\boldsymbol{\eta}}^{j-1}(n)\overline{\boldsymbol{\eta}^{j-1}(n+k)} = \tilde{\varphi}^j(n)\overline{\varphi^j(n+k)}\tilde{\boldsymbol{\theta}}_{j-1}(n)\overline{\boldsymbol{\theta}_{j-1}(n+k)}\tilde{\mathbf{a}}_j(n)\overline{\mathbf{a}_j(n+k)},$$

$$\sum_{\boldsymbol{\psi}\in\boldsymbol{\Psi}_{j-1}} \tilde{\boldsymbol{\psi}}(n)\overline{\boldsymbol{\psi}(n+k)} = \tilde{\varphi}^j(n)\overline{\varphi^j(n+k)} \sum_{\ell=1}^{s_{j-1}} \tilde{\mathbf{b}}_{j,\ell}(n)\overline{\mathbf{b}_{j,\ell}(n+k)}$$

for all $n, k \in \mathbb{Z}$. Since all $\boldsymbol{\Theta}_j, \mathbf{a}_j, \mathbf{b}_{j,\ell}$ are 2^j-periodic, it is straightforward to verify that (4.9.18) is equivalent to (4.9.11) with j being replaced by $j-1$, and (4.9.19) is equivalent to (4.9.12) with j being replaced by $j-1$. □

4.9.2 Periodic Framelets and Wavelets in Periodic Sobolev Spaces

For $\tau \in \mathbb{R}$, the sequence space $l_2^\tau(\mathbb{Z})$ denotes the set of all sequences $v : \mathbb{Z} \to \mathbb{C}$ such that $\|v\|^2_{l_2^\tau(\mathbb{Z})} := \sum_{k\in\mathbb{Z}} |v(k)|^2(1+|k|^2)^\tau < \infty$. Note that $l_2^\tau(\mathbb{Z})$ is a Hilbert space equipped with the inner product:

$$\langle v, w\rangle_{l_2^\tau(\mathbb{Z})} := \sum_{k\in\mathbb{Z}} v(k)\overline{w(k)}(1+|k|^2)^\tau, \qquad v, w \in l_2^\tau(\mathbb{Z}).$$

For $\tau \in \mathbb{R}$, the periodic Sobolev space $H^\tau(\mathbb{T})$ is defined to be the set of all 2π-periodic tempered distributions f on $\mathbb{R}$ such that $\widehat{f} = \{\widehat{f}(k)\}_{k\in\mathbb{Z}} \in l(\mathbb{Z})$ and

$$\|f\|^2_{H^\tau(\mathbb{T})} := \|\widehat{f}\|^2_{l_2^\tau(\mathbb{Z})} = \sum_{k\in\mathbb{Z}} |\widehat{f}(k)|^2(1+|k|^2)^\tau < \infty. \tag{4.9.20}$$

In other words, by (4.9.20), we see that $\widehat{H^\tau(\mathbb{T})} = l_2^\tau(\mathbb{Z})$. Note that $H^0(\mathbb{T}) = L_2(\mathbb{T})$ and $H^\tau(\mathbb{T})$ is a Hilbert space equipped with the inner product

$$\langle f, g\rangle_{H^\tau(\mathbb{T})} := \langle \widehat{f}, \widehat{g}\rangle_{l_2^\tau(\mathbb{Z})} = \sum_{k\in\mathbb{Z}} \widehat{f}(k)\overline{\widehat{g}(k)}(1+|k|^2)^\tau, \qquad f, g \in H^\tau(\mathbb{T}).$$

Also, $H^{-\tau}(\mathbb{T})$ is a dual space of $H^\tau(\mathbb{T})$ under the pairing

$$\langle f, g\rangle := \frac{1}{2\pi}\int_{-\pi}^{\pi} f(x)\overline{g(x)}dx = \langle \widehat{f}, \widehat{g}\rangle := \sum_{k\in\mathbb{Z}} \widehat{f}(k)\overline{\widehat{g}(k)}, \qquad f \in H^\tau(\mathbb{T}), g \in H^{-\tau}(\mathbb{T}).$$

For $\Phi, \Psi \subseteq H^\tau(\mathbb{T})$ with $j \geqslant J$, we define a properly normalized periodic affine system in the periodic Sobolev space $H^\tau(\mathbb{T})$ by

$$\mathsf{PAS}_J^\tau(\Phi; \{\Psi_j\}_{j=J}^\infty) := \{2^{-\tau J}\phi_{\omega,0} \;:\; \phi \in \Phi, \omega \in \Omega_J\} \cup \{2^{-\tau j}\psi_{\omega,0} \;:\; j \geqslant J, \psi \in \Psi_j, \omega \in \Omega_j\}.$$

For simplicity, we also define $\mathsf{PAS}_J(\Phi; \{\Psi_j\}_{j=J}^\infty) := \mathsf{PAS}_J^0(\Phi; \{\Psi_j\}_{j=J}^\infty)$.

Let $\Phi, \Psi_j \subseteq H^\tau(\mathbb{T})$ and $\tilde{\Phi}, \tilde{\Psi}_j \subseteq H^{-\tau}(\mathbb{T})$ for $j \geqslant J \geqslant 0$. We say that

$$(\mathsf{PAS}_J^{-\tau}(\tilde{\Phi}; \{\tilde{\Psi}_j\}_{j=J}^\infty), \mathsf{PAS}_J^\tau(\Phi; \{\Psi_j\}_{j=J}^\infty)) \tag{4.9.21}$$

is *a pair of periodic dual frames in a pair of dual periodic Sobolev spaces* $(H^{-\tau}(\mathbb{T}), H^{\tau}(\mathbb{T}))$ if

(1) $\mathsf{PAS}_J^{\tau}(\Phi;\{\Psi_j\}_{j=J}^{\infty})$ is a frame for $H^{\iota}(\mathbb{T})$: there exist $C_1>0$ and $C_2>0$ such that

$$C_1\|f\|_{H^{\tau}(\mathbb{T})}^2 \leqslant \sum_{\eta\in \mathsf{PAS}_J^{\tau}(\Phi;\{\Psi_j\}_{j=J}^{\infty})} |\langle f,\eta\rangle_{H^{\tau}(\mathbb{T})}|^2 \leqslant C_2\|f\|_{H^{\tau}(\mathbb{T})}^2, \qquad \forall f\in H^{\tau}(\mathbb{T}).$$

(2) $\mathsf{PAS}_J^{-\tau}(\tilde{\Phi};\{\tilde{\Psi}_j\}_{j=J}^{\infty})$ is a frame for $H^{-\tau}(\mathbb{T})$.
(3) The following identity holds: For all $f\in H^{\tau}(\mathbb{T})$ and $g\in H^{-\tau}(\mathbb{T})$,

$$\langle f,g\rangle=\sum_{\phi\in\Phi}\sum_{\omega\in\Omega_J}\langle f,\tilde{\phi}_{\omega,0}\rangle\langle\phi_{\omega,0},g\rangle+\sum_{j=J}^{\infty}\sum_{\psi\in\Psi_j}\sum_{\omega\in\Omega_j}\langle f,\tilde{\psi}_{\omega,0}\rangle\langle\psi_{\omega,0},g\rangle$$

with the series on the right-hand side converging absolutely.

As a direct consequence of the above notion, we have

$$f=\sum_{\phi\in\Phi}\sum_{\omega\in\Omega_J}\langle f,\tilde{\phi}_{\omega,0}\rangle\phi_{\omega,0}+\sum_{j=J}^{\infty}\sum_{\psi\in\Psi_j}\sum_{\omega\in\Omega_j}\langle f,\tilde{\psi}_{\omega,0}\rangle\psi_{\omega,0}, \qquad f\in H^{\tau}(\mathbb{T})$$

with the series on the right-hand side converging unconditionally in $H^{\tau}(\mathbb{T})$, and

$$g=\sum_{\phi\in\Phi}\sum_{\omega\in\Omega_J}\langle g,\phi_{\omega,0}\rangle\tilde{\phi}_{\omega,0}+\sum_{j=J}^{\infty}\sum_{\psi\in\Psi_j}\sum_{\omega\in\Omega_j}\langle g,\psi_{\omega,0}\rangle\tilde{\psi}_{\omega,0}, \qquad g\in H^{-\tau}(\mathbb{T})$$

with the series on the right-hand side converging unconditionally in $H^{-\tau}(\mathbb{T})$.

Similar to Theorem 4.6.3 we have the following result:

Theorem 4.9.6 *Let $\tau\in\mathbb{R}$ and $\Phi,\tilde{\Phi},\Psi_j,\tilde{\Psi}_j$ with $j\geqslant J\geqslant 0$ be subsets of 2π-periodic tempered distributions. Define $\boldsymbol{\Phi}:=\widehat{\Phi}, \tilde{\boldsymbol{\Phi}}:=\widehat{\tilde{\Phi}}, \boldsymbol{\Psi}_j:=\widehat{\Psi}_j, \tilde{\boldsymbol{\Psi}}_j:=\widehat{\tilde{\Psi}}_j$ for $j\geqslant J$. Then $(\mathsf{PAS}_J^{-\tau}(\tilde{\Phi};\{\tilde{\Psi}_j\}_{j=J}^{\infty}), \mathsf{PAS}_J^{\tau}(\Phi;\{\Psi_j\}_{j=J}^{\infty}))$ is a pair of periodic dual frames in $(H^{-\tau}(\mathbb{T}),H^{\tau}(\mathbb{T}))$ if and only if*

(i) there exists a positive constant C such that for all $\mathbf{g}\in l_0(\mathbb{Z})$,

$$\sum_{\varphi\in\boldsymbol{\Phi}}\sum_{\omega\in\Omega_J}2^{-2\tau J}|\langle\mathbf{g},\boldsymbol{\varphi}_{0,\omega}\rangle|^2+\sum_{j=J}^{\infty}\sum_{\boldsymbol{\psi}\in\boldsymbol{\Psi}_j}\sum_{\omega\in\Omega_j}2^{-2\tau j}|\langle\mathbf{g},\boldsymbol{\psi}_{0,\omega}\rangle|^2\leqslant C\|\mathbf{g}\|_{l_2^{-\tau}(\mathbb{Z})}^2$$

and for all $\mathbf{f}\in l_0(\mathbb{Z})$,

$$\sum_{\tilde{\varphi}\in\tilde{\boldsymbol{\Phi}}}\sum_{\omega\in\Omega_J}2^{2\tau J}|\langle\mathbf{f},\tilde{\boldsymbol{\varphi}}_{0,\omega}\rangle|^2+\sum_{j=J}^{\infty}\sum_{\tilde{\boldsymbol{\psi}}\in\tilde{\boldsymbol{\Psi}}}\sum_{\omega\in\Omega_j}2^{2\tau j}|\langle\mathbf{f},\tilde{\boldsymbol{\psi}}_{0,\omega}\rangle|^2\leqslant C\|\mathbf{f}\|_{l_2^{\tau}(\mathbb{Z})}^2;$$

(ii) $(\mathsf{FPAS}_J(\tilde{\boldsymbol{\Phi}};\{\tilde{\boldsymbol{\Psi}}_j\}_{j=J}^{\infty}), \mathsf{FPAS}_J(\boldsymbol{\Phi};\{\boldsymbol{\Psi}_j\}_{j=J}^{\infty}))$ *is a frequency-based periodic dual frame.*

Let $\Phi, \Psi_j \subseteq H^{\tau}(\mathbb{T})$ and $\tilde{\Phi}, \tilde{\Psi}_j \subseteq H^{-\tau}(\mathbb{T})$ for $j \geqslant J \geqslant 0$. We say that the pair in (4.9.21) is *a pair of periodic biorthogonal bases in* $(H^{-\tau}(\mathbb{T}), H^{\tau}(\mathbb{T}))$ if

(1) $\mathsf{PAS}_J^{\tau}(\Phi;\{\Psi_j\}_{j=J}^{\infty})$ is a Riesz basis for $H^{\tau}(\mathbb{T})$.
(2) $\mathsf{PAS}_J^{-\tau}(\tilde{\Phi};\{\tilde{\Psi}_j\}_{j=J}^{\infty})$ is a Riesz basis for $H^{-\tau}(\mathbb{T})$.
(3) $\mathsf{PAS}_J^{-\tau}(\tilde{\Phi};\{\tilde{\Psi}_j\}_{j=J}^{\infty})$ and $\mathsf{PAS}_J^{\tau}(\Phi;\{\Psi_j\}_{j=J}^{\infty}))$ are biorthogonal to each other.

The pair in (4.9.21) is a pair of periodic biorthogonal bases in $(H^{-\tau}(\mathbb{T}), H^{\tau}(\mathbb{T}))$ if and only if it is a pair of periodic dual frames in $(H^{-\tau}(\mathbb{T}), H^{\tau}(\mathbb{T}))$ and the above item (3) holds.

4.9.3 Periodic Dual Framelets and Wavelets by Periodization

One way of constructing periodic dual framelets on $\mathbb{T}$ is to apply the periodization operator to dual framelets on $\mathbb{R}$. For a continuous function $\mathbf{f} \in \mathscr{C}(\mathbb{R})$, we define

$$\mathbf{f}|_{2\pi\mathbb{Z}} := \{\mathbf{f}(2\pi k)\}_{k\in\mathbb{Z}} \in l(\mathbb{Z}),$$

which is the *periodization operator* in the frequency domain.

Theorem 4.9.7 *Let* $\boldsymbol{\Phi}, \tilde{\boldsymbol{\Phi}}, \boldsymbol{\Psi}_j, \tilde{\boldsymbol{\Psi}}_j$ *for* $j \geqslant J \geqslant 0$ *be finite subsets of* $\mathscr{C}(\mathbb{R})$. *Define*

$$\mathring{\boldsymbol{\Phi}} := \{2^{-\frac{J}{2}}\boldsymbol{\varphi}(2^{-J}\cdot)|_{2\pi\mathbb{Z}} \;:\; \boldsymbol{\varphi} \in \boldsymbol{\Phi}\}, \; \mathring{\boldsymbol{\Psi}}_j := \{2^{-\frac{j}{2}}\boldsymbol{\psi}(2^{-j}\cdot)|_{2\pi\mathbb{Z}} \;:\; \boldsymbol{\psi} \in \boldsymbol{\Psi}_j\}, \tag{4.9.22}$$

$$\mathring{\tilde{\boldsymbol{\Phi}}} := \{2^{-\frac{J}{2}}\tilde{\boldsymbol{\varphi}}(2^{-J}\cdot)|_{2\pi\mathbb{Z}} \;:\; \tilde{\boldsymbol{\varphi}} \in \tilde{\boldsymbol{\Phi}}\}, \; \mathring{\tilde{\boldsymbol{\Psi}}}_j := \{2^{-\frac{j}{2}}\tilde{\boldsymbol{\psi}}(2^{-j}\cdot)|_{2\pi\mathbb{Z}} \;:\; \tilde{\boldsymbol{\psi}} \in \tilde{\boldsymbol{\Psi}}_j\}$$

for all $j \geqslant J$. *Suppose that*

$$\text{for some } C>0 \text{ and } \tau>0, \sum_{\boldsymbol{\psi}\in\boldsymbol{\Psi}_j} |\tilde{\boldsymbol{\psi}}(\xi)\boldsymbol{\psi}(\xi)| \leqslant C|\xi|^{\tau}, \;\; \forall\, j \geqslant J, \xi \in [-\pi,\pi]. \tag{4.9.23}$$

If $(\mathsf{FAS}_J(\tilde{\boldsymbol{\Phi}};\{\tilde{\boldsymbol{\Psi}}_j\,|\,2^{-j}\}_{j=J}^{\infty}), \mathsf{FAS}_J(\boldsymbol{\Phi};\{\boldsymbol{\Psi}_j\,|\,2^{-j}\}_{j=J}^{\infty}))$ *is a pair of frequency-based (nonstationary) dual frames, then* $(\mathsf{FPAS}_J(\mathring{\tilde{\boldsymbol{\Phi}}};\{\mathring{\tilde{\boldsymbol{\Psi}}}_j\}_{j=J}^{\infty}), \mathsf{FPAS}_J(\mathring{\boldsymbol{\Phi}};\{\mathring{\boldsymbol{\Psi}}_j\}_{j=J}^{\infty}))$ *is a pair of frequency-based periodic dual frames.*

Proof By Theorem 4.8.1 with $\lambda_j = 2^{-j}$, the identities (4.8.12) and (4.8.13) must hold for almost every $\xi \in \mathbb{R}$. Since all involved functions in (4.8.12) are continuous, (4.8.12) must hold for all $\xi \in \mathbb{R}$. Plugging $\xi = 2\pi n$ with $n \in \mathbb{Z}$ into (4.8.12), we conclude that for all $n \in \mathbb{Z}, m \geqslant J$ and $k_0 \in \mathbb{Z}\backslash[2\mathbb{Z}]$,

$$\sum_{\mathring{\varphi} \in \mathring{\Phi}} 2^J \tilde{\mathring{\varphi}}(n)\overline{\mathring{\varphi}(n+2^m k_0)} + \sum_{j=J}^{m} \sum_{\mathring{\psi} \in \mathring{\Psi}_j} 2^j \tilde{\mathring{\psi}}(n)\overline{\mathring{\psi}(n+2^m k_0)}$$

$$= \sum_{\varphi \in \Phi} \tilde{\varphi}(2^{-J}2\pi n)\overline{\varphi(2^{-J}(2\pi n + 2\pi 2^m k_0))}$$

$$+ \sum_{j=J}^{m} \sum_{\psi \in \Psi_j} \tilde{\psi}(2^{-j}2\pi n)\overline{\psi(2^{-j}(2\pi n + 2\pi 2^m k_0))} = 0.$$

Therefore, (4.9.7) holds with $\Phi = \mathring{\Phi}$ and $\Psi_j = \mathring{\Psi}_j$. By our assumption in (4.9.23),

$$H(\xi) := \lim_{J' \to \infty} \sum_{\varphi \in \Phi} \tilde{\varphi}(2^{-J}\xi)\overline{\varphi(2^{-J}\xi)} + \sum_{j=J}^{J'} \sum_{\psi \in \Psi_j} \tilde{\psi}(2^{-j}\xi)\overline{\psi(2^{-j}\xi)}$$

is continuous since the above right-hand side converges uniformly on every compact set by (4.9.23). Hence, by (4.8.13), we have $H(\xi) = 1$ for all $\xi \in \mathbb{R}$. In particular,

$$\lim_{J' \to \infty} \Big(\sum_{\mathring{\varphi} \in \mathring{\Phi}} 2^J \tilde{\mathring{\varphi}}(n)\overline{\mathring{\varphi}(n)} + \sum_{j=J}^{J'} \sum_{\mathring{\psi} \in \mathring{\Psi}_j} 2^j \tilde{\mathring{\psi}}(n)\overline{\mathring{\psi}(n)} \Big) =$$

$$\lim_{J' \to \infty} \Big(\sum_{\varphi \in \Phi} \tilde{\varphi}(2^{-J}2\pi n)\overline{\varphi(2^{-J}2\pi n)} + \sum_{j=J}^{J'} \sum_{\psi \in \Psi_j} \tilde{\psi}(2^{-j}2\pi n)\overline{\psi(2^{-j}2\pi n)} \Big) = H(2\pi n) = 1$$

for every $n \in \mathbb{Z}$. By Theorem 4.9.2, ($\mathsf{FPAS}_J(\tilde{\mathring{\Phi}}; \{\tilde{\mathring{\Psi}}_j\}_{j=J}^{\infty})$, $\mathsf{FPAS}_J(\mathring{\Phi}; \{\mathring{\Psi}_j\}_{j=J}^{\infty})$) is a pair of frequency-based periodic dual frames. □

As a direct consequence of Theorem 4.9.7, we have

Corollary 4.9.8 *Let $J \in \mathbb{N}_0$ and Φ, Ψ_j with $j \geqslant J$ be finite subsets of $\mathscr{C}(\mathbb{R})$. Define $\mathring{\Phi}$ and $\mathring{\Psi}_j$ for $j \geqslant J$ as in (4.9.22). Suppose that (4.9.23) is satisfied with $\tilde{\Psi}_j = \Psi_j$. If $\mathsf{FAS}_J(\Phi; \{\Psi_j \,|\, 2^{-j}\}_{j=J}^{\infty})$ is a frequency-based tight frame for $L_2(\mathbb{R})$, then $\mathsf{FPAS}_J(\mathring{\Phi}; \{\mathring{\Psi}_j\}_{j=J}^{\infty})$ is a frequency-based periodic tight frame.*

Similarly, we have the following result.

Theorem 4.9.9 *Let $\Phi_j, \Psi_j, \tilde{\Phi}_j, \tilde{\Psi}_j, j \in \mathbb{N}_0$ be finite subsets of $\mathscr{C}(\mathbb{R})$. Suppose that*

$$\lim_{j \to \infty} \sum_{\varphi \in \Phi_j} \tilde{\varphi}(2^{-j}2\pi n)\overline{\varphi(2^{-j}2\pi n)} = 1 \qquad \forall\, n \in \mathbb{Z}. \tag{4.9.24}$$

If $(\mathsf{FAS}_J(\tilde{\Phi}_J;\{\tilde{\Psi}_j\,|\,2^{-j}\}_{j=J}^{\infty}),\mathsf{FAS}_J(\Phi_J;\{\Psi_j\,|\,2^{-j}\}_{j=J}^{\infty}))$ *is a pair of frequency-based dual frames for every* $J\in\mathbb{N}_0$*, then* $(\mathsf{FPAS}_J(\mathring{\tilde{\Phi}}_J;\{\mathring{\tilde{\Psi}}_j\}_{j=J}^{\infty}),\mathsf{FAS}_J(\mathring{\Phi}_J;\{\mathring{\Psi}_j\}_{j=J}^{\infty}))$ *is a pair of frequency-based periodic dual frames for every* $J\in\mathbb{N}_0$*, where*

$$\mathring{\Phi}_j:=\{2^{-j/2}\varphi(2^{-j}\cdot)|_{2\pi\mathbb{Z}}\;:\;\varphi\in\Phi_j\},\quad \mathring{\Psi}_j:=\{2^{-j/2}\psi(2^{-j}\cdot)|_{2\pi\mathbb{Z}}\;:\;\psi\in\Psi_j\},$$

$$\mathring{\tilde{\Phi}}_j:=\{2^{-j/2}\tilde{\varphi}(2^{-j}\cdot)|_{2\pi\mathbb{Z}}\;:\;\tilde{\varphi}\in\tilde{\Phi}_j\},\quad \mathring{\tilde{\Psi}}_j:=\{2^{-j/2}\tilde{\psi}(2^{-j}\cdot)|_{2\pi\mathbb{Z}}\;:\;\tilde{\psi}\in\tilde{\Psi}_j\}.$$

If in addition $\Phi_j=\{\varphi^{j,1},\dots,\varphi^{j,r_j}\}$ *and* $\tilde{\Phi}_j=\{\tilde{\varphi}^{j,1},\dots,\tilde{\varphi}^{j,r_j}\}$ *for all* $j\in\mathbb{N}_0$ *such that* $\varphi^j:=(\varphi^{j,1},\dots,\varphi^{j,r_j})^{\mathsf{T}}$ *and* $\tilde{\varphi}^j:=(\tilde{\varphi}^{j,1},\dots,\tilde{\varphi}^{j,r_j})^{\mathsf{T}}$ *satisfy the following nonstationary refinement equations:*

$$\varphi^{j-1}(2\xi)=\mathbf{a}_j(\xi)\varphi^j(\xi),\quad \tilde{\varphi}^{j-1}(2\xi)=\tilde{\mathbf{a}}_j(\xi)\tilde{\varphi}^j(\xi),\quad \xi\in\mathbb{R},j\in\mathbb{N}\tag{4.9.25}$$

for some $r_{j-1}\times r_j$ *matrices* $\mathbf{a}_j,\tilde{\mathbf{a}}_j$ *of* 2π*-periodic continuous functions, then* $\mathring{\varphi}^j=(\mathring{\varphi}^{j,1},\dots,\mathring{\varphi}^{j,r_j})^{\mathsf{T}}$ *and* $\mathring{\tilde{\varphi}}^j:=(\mathring{\tilde{\varphi}}^{j,1},\dots,\mathring{\tilde{\varphi}}^{j,r_j})^{\mathsf{T}}$ *also satisfy*

$$\mathring{\varphi}^{j-1}(n)=\mathring{\mathbf{a}}_j(n)\mathring{\varphi}^j(n),\quad \mathring{\tilde{\varphi}}^{j-1}(n)=\mathring{\tilde{\mathbf{a}}}_j(n)\mathring{\tilde{\varphi}}^j(n),\quad \forall\, n\in\mathbb{Z},j\in\mathbb{N},\tag{4.9.26}$$

where $\mathring{\mathbf{a}}_j$ *and* $\mathring{\tilde{\mathbf{a}}}_j$ *are* $r_{j-1}\times r_j$ *matrices of* 2^j*-periodic sequences defined by*

$$\mathring{\mathbf{a}}_j(n):=\sqrt{2}\mathbf{a}_j(2^{-j}2\pi n),\quad \mathring{\tilde{\mathbf{a}}}_j(n):=\sqrt{2}\tilde{\mathbf{a}}_j(2^{-j}2\pi n),\quad n\in\mathbb{Z}.$$

Proof By Theorem 4.8.3, the identities (4.8.21) and (4.8.22) must hold. Plugging $\xi=2\pi n$ into (4.8.21) and (4.8.22), we obtain (4.9.11) and (4.9.12). By our assumption in (4.9.24), (4.9.10) holds with $\Phi_j=\mathring{\Phi}_j$. Now the claim follows from Theorem 4.9.4. By our assumption in (4.9.25), we have $\varphi^{j-1}(2^{1-j}\xi)=\mathbf{a}_j(2^{-j}\xi)\varphi^j(2^{-j}\xi)$ for all $\xi\in\mathbb{R}$. Therefore, by $\mathring{\varphi}^j(n)=2^{-j/2}\varphi^j(2^{-j}2\pi n)$ for all $n\in\mathbb{Z}$, we have

$$\begin{aligned}\mathring{\varphi}^{j-1}(n)&=2^{(1-j)/2}\varphi^{j-1}(2^{1-j}2\pi n)=\mathbf{a}_j(2^{-j}2\pi n)2^{(1-j)/2}\varphi^j(2^{-j}2\pi n)\\&=\sqrt{2}\mathbf{a}_j(2^{-j}2\pi n)\mathring{\varphi}^j(n)=\mathring{\mathbf{a}}_j(n)\mathring{\varphi}^j(n).\end{aligned}$$

This verifies (4.9.26). All other claims can be directly verified. □

We now study the periodization operator in the time/space domain.

Lemma 4.9.10 *Define the* periodization operator $\mathscr{P}:L_1(\mathbb{R})\to L_1(\mathbb{T})$ *by*

$$\mathscr{P}f(x):=f^{per}(x):=\sum_{k\in\mathbb{Z}}f\big(\tfrac{x}{2\pi}-k\big),\qquad x\in\mathbb{R},\ f\in L_1(\mathbb{R}).\tag{4.9.27}$$

Then $\mathscr{P}f \in L_1(\mathbb{T})$ *and* $\widehat{\mathscr{P}f}(k) = \widehat{f}(2\pi k)$ $\forall\, k \in \mathbb{Z}$, *i.e.,* $\widehat{\mathscr{P}f} = \widehat{f}|_{2\pi\mathbb{Z}}$. *Moreover,*

$$\mathscr{P}(f_{\lambda;t}) = \big(\mathscr{P}(f_{\lambda;0})\big)_{\lambda^{-1}2\pi t,0}, \qquad \forall\, \lambda \in \mathbb{R}\backslash\{0\}, t \in \mathbb{R}. \tag{4.9.28}$$

Proof Since $\int_{-\pi}^{\pi} \sum_{k\in\mathbb{Z}} |f(\frac{x}{2\pi} - k)|dx = \int_{\mathbb{R}} |f(\frac{x}{2\pi})|dx = 2\pi \int_{\mathbb{R}} |f(x)|dx < \infty$, we see that $\mathscr{P}f$ is a well-defined function in $L_1(\mathbb{T})$. By calculation, for $k \in \mathbb{Z}$, we have

$$\widehat{\mathscr{P}f}(k) := \frac{1}{2\pi}\int_{-\pi}^{\pi} \mathscr{P}f(x)e^{-ikx}dx = \frac{1}{2\pi}\int_{\mathbb{R}} f(\tfrac{x}{2\pi})e^{-ikx}dx = \widehat{f}(2\pi k).$$

The identity in (4.9.28) can be directly checked. □

By $\mathcal{L}_2(\mathbb{R})$ we denote the linear space of all measurable functions f on $\mathbb{R}$ such that $\sum_{k\in\mathbb{Z}} |f(\frac{\cdot}{2\pi} - k)| \in L_2(\mathbb{T})$. It is evident that $\mathcal{L}_2(\mathbb{R}) \subseteq L_1(\mathbb{R}) \cap L_2(\mathbb{R})$.

Theorem 4.9.11 *Suppose that* $\Phi, \Psi, \tilde{\Phi}, \tilde{\Psi}$ *in* (4.5.1) *are finite subsets of* $\mathcal{L}_2(\mathbb{R})$ *such that* $(\{\tilde{\Phi}; \tilde{\Psi}\}, \{\Phi; \Psi\})$ *is a dual framelet in* $L_2(\mathbb{R})$. *For* $j \in \mathbb{N}_0$, *define*

$$\mathring{\Phi}_j := \{\mathscr{P}(\phi_{2^j;0}) \;:\; \phi \in \Phi\}, \quad \mathring{\Psi}_j := \{\mathscr{P}(\psi_{2^j;0}) \;:\; \psi \in \Psi\}, \tag{4.9.29}$$

$$\mathring{\tilde{\Phi}}_j := \{\mathscr{P}(\tilde{\phi}_{2^j;0}) \;:\; \tilde{\phi} \in \tilde{\Phi}\}, \quad \mathring{\tilde{\Psi}}_j := \{\mathscr{P}(\tilde{\psi}_{2^j;0}) \;:\; \tilde{\psi} \in \tilde{\Psi}\}. \tag{4.9.30}$$

If the two systems $\mathsf{PAS}_J(\mathring{\tilde{\Phi}}_J; \{\mathring{\tilde{\Psi}}_j\}_{j=J}^{\infty})$ *and* $\mathsf{PAS}_J(\mathring{\Phi}_J; \{\mathring{\Psi}_j\}_{j=J}^{\infty})$ *are Bessel sequences in* $L_2(\mathbb{T})$ *for all* $J \in \mathbb{N}_0$, *then the pair* $\big(\mathsf{PAS}_J(\mathring{\tilde{\Phi}}_J; \{\mathring{\tilde{\Psi}}_j\}_{j=J}^{\infty}), \mathsf{PAS}_J(\mathring{\Phi}_J; \{\mathring{\Psi}_j\}_{j=J}^{\infty})\big)$ *is a pair of periodic dual frames for* $L_2(\mathbb{T})$ *for all* $J \in \mathbb{N}_0$.

Proof By $\mathcal{L}_2(\mathbb{R}) \subseteq L_1(\mathbb{R}) \cap L_2(\mathbb{R})$, we see that all functions in $\widehat{\Phi}, \widehat{\Psi}, \widehat{\tilde{\Phi}}, \widehat{\tilde{\Psi}}$ are continuous. By Theorem 4.1.6 and Lemma 4.1.3, it follows from (4.1.10) and Lemma 4.1.4 that

$$\lim_{j\to\infty} \sum_{\phi\in\Phi} \widehat{\tilde{\phi}}(2^{-j}2\pi n)\overline{\widehat{\phi}(2^{-j}2\pi n)} = \sum_{\phi\in\Phi} \widehat{\tilde{\phi}}(0)\overline{\widehat{\phi}(0)} = 1, \qquad \forall\, n \in \mathbb{Z}.$$

Now the claim follows directly from Theorem 4.9.9. □

Corollary 4.9.12 *Suppose that* Φ, Ψ *are finite subsets of* $\mathcal{L}_2(\mathbb{R})$ *such that* $\{\Phi; \Psi\}$ *is a tight framelet in* $L_2(\mathbb{R})$. *For* $j \in \mathbb{N}_0$, *define* $\mathring{\Phi}_j$ *and* $\mathring{\Psi}_j$ *as in* (4.9.29). *Then* $\mathsf{PAS}_J(\mathring{\Phi}_J; \{\mathring{\Psi}_j\}_{j=J}^{\infty}))$ *is a periodic tight frame for* $L_2(\mathbb{T})$ *for all* $J \in \mathbb{N}_0$.

Theorem 4.9.13 *Suppose that* $\Phi, \Psi, \tilde{\Phi}, \tilde{\Psi}$ *in* (4.5.1) *are finite subsets of* $\mathcal{L}_2(\mathbb{R})$ *such that* $(\{\tilde{\Phi}; \tilde{\Psi}\}, \{\Phi; \Psi\})$ *is a biorthogonal wavelet in* $L_2(\mathbb{R})$. *For* $j \in \mathbb{N}_0$, *define* $\mathring{\Phi}_j, \mathring{\Psi}_j$ *as in* (4.9.29) *and* $\mathring{\tilde{\Phi}}_j, \mathring{\tilde{\Psi}}_j$ *as in* (4.9.30). *If* $\mathsf{PAS}_J(\mathring{\tilde{\Phi}}_J; \{\mathring{\tilde{\Psi}}_j\}_{j=J}^{\infty})$ *and* $\mathsf{PAS}_J(\mathring{\Phi}_J; \{\mathring{\Psi}_j\}_{j=J}^{\infty})$ *are Bessel sequences in* $L_2(\mathbb{T})$ *for all* $J \in \mathbb{N}_0$, *then for all* $J \in \mathbb{N}_0$, $\big(\mathsf{PAS}_J(\mathring{\tilde{\Phi}}_J; \{\mathring{\tilde{\Psi}}_j\}_{j=J}^{\infty}), \mathsf{PAS}_J(\mathring{\Phi}_J; \{\mathring{\Psi}_j\}_{j=J}^{\infty})\big)$ *is a pair of periodic biorthogonal bases for* $L_2(\mathbb{T})$.

Proof By Theorem 4.9.11, the pair $\big(\mathsf{PAS}_J(\tilde{\mathring{\Phi}}_J; \{\tilde{\mathring{\Psi}}_j\}_{j=J}^\infty), \mathsf{PAS}_J(\mathring{\Phi}_J; \{\mathring{\Psi}_j\}_{j=J}^\infty)\big)$ is a pair of periodic dual frames for $L_2(\mathbb{T})$ for all $J \in \mathbb{N}_0$. To complete the proof, we show that $\mathsf{PAS}_J(\tilde{\mathring{\Phi}}_J; \{\tilde{\mathring{\Psi}}_j\}_{j=J}^\infty)$ and $\mathsf{PAS}_J(\mathring{\Phi}_J; \{\mathring{\Psi}_j\}_{j=J}^\infty)$ are biorthogonal in $L_2(\mathbb{T})$.

For $j, j' \in \mathbb{N}_0$, $\omega \in \Omega_j$, $\omega' \in \Omega_{j'}$ and $f, g \in \mathcal{L}_2(\mathbb{R})$, by (4.9.28), we have

$$\langle (\mathscr{P}(f_{2^j;0}))_{\omega,0}, (\mathscr{P}(g_{2^{j'};0}))_{\omega',0}\rangle = \langle \mathscr{P}(f_{2^j;2^j\omega/(2\pi)}), \mathscr{P}(g_{2^{j'};2^{j'}\omega'/(2\pi)})\rangle.$$

Let $m := \frac{2^j k}{2\pi}$ and $m' := \frac{2^{j'} k'}{2\pi}$. By the definition of the set Ω_j in (4.9.1), we see that $m, m' \in \mathbb{Z}$. Hence, by the definition of the periodization operator, we conclude that

$$\begin{aligned}
&\langle (\mathscr{P}(f_{2^j;0}))_{\omega,0}, (\mathscr{P}(g_{2^{j'};0}))_{\omega',0}\rangle_{L_2(\mathbb{T})} = \langle \mathscr{P}(f_{2^j;m}), \mathscr{P}(g_{2^{j'};m'})\rangle_{L_2(\mathbb{T})}\\
&= \sum_{n\in\mathbb{Z}} \frac{1}{2\pi}\int_{-\pi}^{\pi} \mathscr{P}(f_{2^j;m})(x)\overline{g_{2^{j'};m'}(\tfrac{x}{2\pi}-n)}dx = \frac{1}{2\pi}\int_{\mathbb{R}} \mathscr{P}(f_{2^j;m})(x)\overline{g_{2^{j'};m'}(\tfrac{x}{2\pi})}dx\\
&= \sum_{n\in\mathbb{Z}} \frac{1}{2\pi}\int_{\mathbb{R}} f_{2^j;m}(\tfrac{x}{2\pi}-n)\overline{g_{2^{j'};m'}(\tfrac{x}{2\pi})}dx = \sum_{n\in\mathbb{Z}}\int_{\mathbb{R}} f_{2^j;m+2^jn}(x)\overline{g_{2^{j'};m'}(x)}dx.
\end{aligned}$$

That is, for all $j, j' \in \mathbb{N}_0$, $\omega \in 2^{-j}2\pi\mathbb{Z}$, $\omega' \in 2^{-j'}2\pi\mathbb{Z}$, and $f, g \in \mathcal{L}_2(\mathbb{R})$,

$$\langle (\mathscr{P}(f_{2^j;0}))_{\omega,0}, (\mathscr{P}(g_{2^{j'};0}))_{\omega',0}\rangle_{L_2(\mathbb{T})} = \sum_{n\in\mathbb{Z}} \langle f_{2^j;2^j\omega/(2\pi)+2^jn}, g_{2^{j'};2^{j'}\omega'/(2\pi)}\rangle_{L_2(\mathbb{R})}.$$

Since $\mathsf{AS}_J(\tilde{\Phi};\tilde{\Psi})$ and $\mathsf{AS}_J(\Phi;\Psi)$ are biorthogonal to each other in $L_2(\mathbb{R})$, using the above identity, we can trivially see that $\mathsf{PAS}_J(\tilde{\mathring{\Phi}}_J; \{\tilde{\mathring{\Psi}}_j\}_{j=J}^\infty)$ and $\mathsf{PAS}_J(\mathring{\Phi}_J; \{\mathring{\Psi}_j\}_{j=J}^\infty)$ are biorthogonal to each other in $L_2(\mathbb{T})$. □

It follows directly from Corollary 4.9.12 and Theorem 4.9.13 that

Corollary 4.9.14 *Suppose that Φ, Ψ are finite subsets of $\mathcal{L}_2(\mathbb{R})$ such that $\{\Phi;\Psi\}$ is an orthogonal wavelet in $L_2(\mathbb{R})$. For $j \in \mathbb{N}_0$, define $\mathring{\Phi}_j$ and $\mathring{\Psi}_j$ as in (4.9.29). Then $\mathsf{PAS}_J(\mathring{\Phi}_J; \{\mathring{\Psi}_j\}_{j=J}^\infty)$ is a periodic orthonormal basis for $L_2(\mathbb{T})$ for all $J \in \mathbb{N}_0$.*

4.10 Exercises

4.1. For a measurable function $f : \mathbb{R} \to \mathbb{C}$, its support $\operatorname{supp}(f)$ using closed sets is

$$\operatorname{supp}(f) := \cap\{K \,:\, K \subseteq \mathbb{R} \text{ is closed and } f(x) = 0 \text{ for a.e. } x \in \mathbb{R}\backslash K\}.$$

Prove that $f(x) = 0$ for a.e. $x \in \mathbb{R}\backslash \operatorname{supp}(f)$. *Hint:* Any open set in $\mathbb{R}$ can be written as a countable disjoint union of open intervals.

4.2. If both f and g are measurable functions such that $f(x) = g(x)$ for a.e. $x \in \mathbb{R}$, prove that $\operatorname{supp}(f) = \operatorname{supp}(g)$ with $\operatorname{supp}(f)$ and $\operatorname{supp}(g)$ being defined above.

4.3. Provide an example of a continuous function to show that it is generally not true that $f(x) \neq 0$ for almost every $x \in \operatorname{supp}(f)$. *Hint:* use a generalized Cantor set $C \subseteq [0, 1]$ such that the Lebesgue measure of C is strictly between 0 and 1 to build a continuous function $f : [0, 1] \to \mathbb{R}$ such that $f(x) > 0$ for all $x \in [0, 1]\backslash C$ and $f(x) = 0$ for all $x \in C$. Note that $\operatorname{supp}(f) = [0, 1]$ which is the closure of $[0, 1]\backslash C$.

4.4. Prove Corollary 4.1.7.

4.5. Let $a \in l_0(\mathbb{Z})$ with $\widehat{a}(0) = 1$. Define $\varphi^a(\xi) := \prod_{j=1}^{\infty} \widehat{a}(2^{-j}\xi)$. By Lemma 4.1.8, $\varphi^a = \widehat{\phi^a}$ for a tempered distribution ϕ^a. Define $\{\phi_n\}_{n\in\mathbb{N}}$ by $\widehat{\phi_n}(\xi) := \prod_{j=1}^{n} \widehat{a}(2^{-j}\xi)$.

a. Prove that $\widehat{\phi_n}$ converges to $\widehat{\phi^a}$ in the sense of distributions. *Hint:* use (4.1.33).
b. Prove that ϕ_n converges to ϕ^a in the sense of distributions.
c. Prove $\operatorname{supp}(\phi_n) \subseteq 2^{-1}\operatorname{fsupp}(a) + \cdots + 2^{-n}\operatorname{fsupp}(a)$ and $\operatorname{supp}(\phi^a) \subseteq \operatorname{fsupp}(a)$.
d. Prove that all $\widehat{\phi_n}$ and $\widehat{\phi^a}$ can be extended into analytic functions in $\mathbb{C}$.

4.6. Let $\{h_k\}_{k\in\Lambda}$ be a countable sequence in a Hilbert space $\mathcal{K}$. Define a closed subspace $\mathcal{H} := \overline{\operatorname{span}\{h_k : k \in \Lambda\}}$ of $\mathcal{K}$. If $\{h_k\}_{k\in\Lambda}$ is a Bessel sequence in $\mathcal{K}$, prove that $\{\mathcal{F}h : h \in \mathcal{K}\} = \{\mathcal{F}h : h \in \mathcal{H}\}$ and $\{\mathcal{F}h : h \in \mathcal{K}\}$ is dense in $\mathcal{H}$, where $\mathcal{F}$ is the frame operator associated with $\{h_k\}_{k\in\Lambda}$. If in addition $\{\mathcal{F}h : h \in \mathcal{H}\} = \mathcal{H}$, prove that $\{h_k\}_{k\in\Lambda}$ is a frame for $\mathcal{H}$.

4.7. If $\{f_k\}_{k\in\Lambda}$ and $\{h_k\}_{k\in\Lambda}$ are Bessel sequences in a Hilbert space $\mathcal{H}$, prove that $\{c_k f_k + d_k h_k\}_{k\in\Lambda}$ is a Bessel sequence in $\mathcal{H}$ for every $\{c_k\}_{k\in\Lambda}, \{d_k\}_{k\in\Lambda} \in l_\infty(\Lambda)$.

4.8. Let $BS(\mathcal{H}, \Lambda)$ denote the set of all Bessel sequences $\{h_k\}_{k\in\Lambda}$ in a Hilbert space $\mathcal{H}$. Let $B(\mathcal{H}, l_2(\Lambda))$ denote the set of all bounded linear operators from $\mathcal{H}$ to $l_2(\Lambda)$. Prove that the map $BS(\mathcal{H}, \Lambda) \to B(\mathcal{H}, l_2(\Lambda))$ sending a Bessel sequence to its analysis operator $\mathcal{W}$ is bijective.

4.9. Suppose that $\{h_k\}_{k\in\Lambda}$ is a Bessel sequence in a Hilbert space $\mathcal{H}$ such that for a positive constant C, $\sum_{k\in\Lambda} |\langle h, h_k\rangle|^2 \leqslant C\|h\|^2$ for all $h \in \mathcal{H}$. Prove that (i) $\|h_k\|^2 \leqslant C$ for all $k \in \Lambda$; (ii) If Λ is an infinite index set, then $\{h_0 + h_k\}_{k\in\mathbb{N}}$ cannot be a Bessel sequence in $\mathcal{H}$ for any $h_0 \in \mathcal{H}$ with $\|h_0\| \neq 0$.

4.10. Suppose that $\{h_k\}_{k\in\Lambda}$ is a frame in a Hilbert space $\mathcal{H}$ with the lower and upper frame bounds C_1 and C_2, respectively. Assume $\|h_k\| \neq 0$ for all $k \in \Lambda$. Is it true that there always exists some $k \in \Lambda$ such that $\|h_k\|^2 \geqslant C_1$? *Hint:* Let $\{e_k\}_{k\in\mathbb{N}}$ be an orthonormal basis of $\mathcal{H}$. Prove that $\{2^{-j/2}e_k : j, k \in \mathbb{N}\}$ is a tight frame for $\mathcal{H}$.

4.11. Let $T : \mathcal{H} \to \mathcal{H}$ be a bounded linear operator such that $\langle Tf, g\rangle = \langle f, Tg\rangle$ for all $f, g \in \mathcal{H}$ (i.e., $T^\star = T$). If there exist positive constants C_1 and C_2 such that $C_1\langle h, h\rangle \leqslant \langle Th, h\rangle \leqslant C_2\langle h, h\rangle$ for all $h \in \mathcal{H}$. Prove that

T is bijective, $C_2^{-1}\langle h, h\rangle \leqslant \langle T^{-1}h, h\rangle \leqslant C_1^{-1}\langle h, h\rangle$ for all $h \in \mathcal{H}$, and $T^{-1} = \frac{2}{C_1+C_2}\sum_{j=0}^{\infty}\left(1 - \frac{2T}{C_1+C_2}\right)^j$.

4.12. Let $\{h_k\}_{k\in\Lambda}$ be a Riesz sequence in a Hilbert space $\mathcal{H}$ with the lower and upper Riesz bounds C_3 and C_4, respectively. Define $\mathcal{V}c := \sum_{k\in\Lambda} c_k h_k$ for $c \in l_2(\Lambda)$.

a. Prove that $\mathcal{V}(l_2(\Lambda))$ is closed and $C_3 \leqslant \|h_k\|^2 \leqslant C_4$ for all $k \in \Lambda$.
b. If $C_3 = C_4$, prove that $\{\frac{1}{\sqrt{C_3}} h_k\}_{k\in\Lambda}$ is an orthonormal basis of $\mathcal{H}$.

4.13. If $\{h_k\}_{k\in\Lambda}$ is a frame in a Hilbert space $\mathcal{H}$ with lower and upper frame bounds C_1 and C_2, respectively, prove that $C_1 \sum_{k\in\Lambda} |c_k|^2 \leqslant \|\sum_{k\in\Lambda} c_k h_k\|^2 \leqslant C_2 \sum_{k\in\Lambda} |c_k|^2$ for all $\{c_k\}_{k\in\Lambda} \in \operatorname{ran}(\mathcal{W})$. *Hint:* Use Theorem 4.2.4.

4.14. Let $\{h_k\}_{k\in\Lambda}$ be a frame in a Hilbert space $\mathcal{H}$ with lower and upper frame bounds C_1 and C_2, respectively. Let $\{\eta_j\}_{j\in\Lambda_0}$ be an orthonormal basis of $\ker(\mathcal{V})$, where $\ker(\mathcal{V}) := \{c \in l_2(\Lambda) : \sum_{k\in\Lambda} c_k h_k = 0\}$. Define $g_k := (\eta_j(k))_{j\in\Lambda_0}$ for $k \in \Lambda$.

a. Prove $\|g_k\|_{l_2(\Lambda_0)} \leqslant 1$ for all $k \in \Lambda$. Define $\mathcal{H}_0 := \overline{\operatorname{span}\{g_k\}_{k\in\Lambda}}^{\|\cdot\|_{l_2(\Lambda_0)}} \subset l_2(\Lambda_0)$.
b. For $C_1 \leqslant \lambda \leqslant C_2$, prove that $\{h_k \oplus (\lambda g_k)\}_{k\in\Lambda}$ is a Riesz basis of $\mathcal{H} \oplus \mathcal{H}_0$ satisfying $C_1 \sum_{k\in\Lambda} |c_k|^2 \leqslant \|\sum_{k\in\Lambda} c_k(h_k \oplus \lambda g_k)\|^2_{\mathcal{H}\oplus\mathcal{H}_0} \leqslant C_2 \sum_{k\in\Lambda} |c_k|^2$ for all $c \in l_2(\Lambda)$, where $\oplus$ stands for the orthogonal direct sum.

4.15. Let $\mathcal{H}$ be a Hilbert space and $\mathcal{H}_0$ be a finite-dimensional Hilbert space. Prove that the orthogonal projection operator $P : \mathcal{H} \oplus \mathcal{H}_0 \to \mathcal{H}$ maps every closed subset S in $\mathcal{H} \oplus \mathcal{H}_0$ into a closed subset of $\mathcal{H}$. *Hint:* Every bounded closed set in $\mathcal{H}_0$ is compact. For each $f \in \mathcal{H}$, first show that $\inf_{(f,g)\in S} \|g\|_{\mathcal{H}_0}$ is attainable.

4.16. Let $\{h_k\}_{k\in\Lambda}$ be a frame in a Hilbert space $\mathcal{H}$. For every finite subset $\Lambda_0 \subset \Lambda$, prove that $\{h_k\}_{k\in\Lambda\setminus\Lambda_0}$ is a frame sequence (i.e., a frame in its closed linear span). *Hint:* use item (5) of Theorem 4.2.4 and Exercise 4.15.

4.17. Let $\mathcal{H} = \mathbb{R}^d$ or $\mathbb{C}^d$. Let $h_1, \dots, h_n \in \mathcal{H}$ and let C be the smallest constant such that $\sum_{k=1}^n |\langle h, h_k\rangle|^2 \leqslant C\|h\|^2$ for all $h \in \mathcal{H}$. Prove that $C \leqslant \sum_{k=1}^n \|h_k\|^2 \leqslant dC$.

4.18. Let $\mathcal{H} = \mathbb{R}^d$ or $\mathbb{C}^d$ and $h_1, \dots, h_n \in \mathcal{H}$. Define $F := [h_1, \dots, h_n]$ as a $d \times n$ matrix.

a. Prove that $\{h_1, \dots, h_n\}$ is a frame for $\mathcal{H}$ if and only if F is of full rank.
b. If $\{h_1, \dots, h_n\}$ is a frame for $\mathcal{H}$, describe all its dual frames in $\mathcal{H}$.
c. $\{h_1, \dots, h_n\}$ is a (normalized) tight frame for $\mathcal{H}$ if and only if $F^\star F = I_d$.
d. If $\{h_1, \dots, h_n\}$ is a tight frame for $\mathcal{H}$, prove that $\sum_{k=1}^n \|h_k\|^2 = d$. If in addition $\|h_1\| = \dots = \|h_n\|$, further deduce that $\|h_1\|^2 = \dots = \|h_n\|^2 = \frac{d}{n}$.

4.19. Define $h_1 := \sqrt{\frac{2}{3}}(1,0)^{\mathsf{T}}$, $h_2 := \sqrt{\frac{2}{3}}(-\frac{1}{2}, \frac{\sqrt{3}}{2})^{\mathsf{T}}$, and $h_2 := \sqrt{\frac{2}{3}}(-\frac{1}{2}, -\frac{\sqrt{3}}{2})^{\mathsf{T}}$.

a. Prove that $\{h_1, h_2, h_3\}$ forms a (normalized) tight frame for $\mathbb{R}^2$.
b. Find all the dual frames of $\{h_1, h_2, h_3\}$ in $\mathbb{R}^2$.
c. Generally, let $n \in \mathbb{N}$ with $n \geqslant 2$ and $\theta \in \mathbb{R}$. For $k = 0, \ldots, n-1$, define $h_{k+1} := \sqrt{\frac{2}{n}}\big(\cos(\theta + 2\pi k/n), \sin(\theta + 2\pi k/n)\big)^{\mathsf{T}}$. Prove that $\{h_1, \ldots, h_n\}$ is a (normalized) tight frame for $\mathbb{R}^2$.

4.20. Let $\{h_k\}_{k\in\Lambda}$ in a Hilbert space $\mathcal{H}$. We say that $\{h_k\}_{k\in\Lambda}$ is complete in $\mathcal{H}$ if $\langle h, h_k\rangle = 0$ for all $k \in \Lambda$ implies $h = 0$. Prove that $\{h_k\}_{k\in\Lambda}$ is complete in $\mathcal{H}$ if and only if $\operatorname{span}\{h_k\}_{k\in\Lambda}$ is dense in $\mathcal{H}$.

4.21. Let $\{h_k\}_{k\in\Lambda}$ in a Hilbert space $\mathcal{H}$. Prove that

a. $\{h_k\}_{k\in\Lambda}$ has a biorthogonal sequence $\{\tilde{h}_k\}_{k\in\Lambda}$ satisfying $\langle \tilde{h}_j, h_k\rangle = \delta_{j,k}$ for all $j, k \in \Lambda$ if and only if $\{h_k\}_{k\in\Lambda}$ is minimal (see item (7) of Theorem 4.2.7).
b. $\{h_k\}_{k\in\Lambda}$ has a unique biorthogonal sequence $\{\tilde{h}_k\}_{k\in\Lambda}$ if and only if $\{h_k\}_{k\in\Lambda}$ is minimal and complete.

4.22. Let $\{e_k\}_{k\in\mathbb{N}}$ be an orthonormal basis for a Hilbert space $\mathcal{H}$ and $h_k := \frac{1}{k}e_k$, $k \in \mathbb{N}$.

a. Prove that $\{h_k\}_{k\in\mathbb{N}}$ is a Bessel sequence, $\operatorname{span}\{h_k : k \in \mathbb{N}\}$ is dense in $\mathcal{H}$, but $\{h_k\}_{k\in\mathbb{N}}$ is not a frame sequence in $\mathcal{H}$. [In fact, $\mathcal{FH} \subsetneq \mathcal{H}$.]
b. $\{h_k\}_{k\in\mathbb{N}}$ has a unique biorthogonal sequence $\{\tilde{h}_k\}_{k\in\mathbb{N}}$ given by $\tilde{h}_k := ke_k$.
c. Prove that $\langle f, g\rangle = \sum_{k\in\mathbb{N}}\langle f, \tilde{h}_k\rangle\langle h_k, g\rangle$ holds for all $f, g \in \mathcal{H}$ but $\{\tilde{h}_k\}_{k\in\Lambda}$ is not a Bessel sequence in $\mathcal{H}$.

4.23. Let $\{e_k\}_{k\in\mathbb{N}}$ be an orthonormal basis for a Hilbert space $\mathcal{H}$.

a. Prove that $\{h_k\}_{k\in\mathbb{N}}$ is complete and minimal, where $h_k := e_k + e_{k+1}$ for $k \in \mathbb{N}$.
b. $\{\tilde{h}_k\}_{k\in\mathbb{N}}$ is biorthogonal to $\{h_k\}_{k\in\mathbb{N}}$, where $\tilde{h}_k := \sum_{j=1}^{k}(-1)^{j+k}e_j$, $k \in \mathbb{N}$.
c. $\{h_k\}_{k\in\mathbb{N}}$ is a Bessel sequence, but not a frame for $\mathcal{H}$.
d. There does not exist a sequence $\{c_k\}_{k\in\mathbb{N}}$ in $\mathbb{C}$ such that $\lim_{n\to\infty}\sum_{k=1}^{n} c_k h_k = e_1$ in $\mathcal{H}$. *Hint:* Use proof by contradiction and $\langle e_1, e_j\rangle = \lim_{n\to\infty}\sum_{k=1}^{n} c_k\langle h_k, e_j\rangle$ for all $j \in \mathbb{N}$ to uniquely determine such complex numbers c_k, $k \in \mathbb{N}$.

4.24. Let $\{h_k\}_{k\in\Lambda}$ be a frame in a Hilbert space $\mathcal{H}$ with the lower and upper frame bounds C_1 and C_2, respectively. For $\{f_k\}_{k\in\Lambda}$ in $\mathcal{H}$ such that $\sum_{k\in\Lambda}\|f_k\|^2 < C_1$, prove that $\{f_k + h_k\}_{k\in\Lambda}$ must be a frame in $\mathcal{H}$.

4.25. Let $\{h_k\}_{k\in\Lambda}$ be a Riesz basis for a Hilbert space $\mathcal{H}$ with the lower and upper Riesz bounds C_3 and C_4, respectively. For $\{f_k\}_{k\in\Lambda}$ in $\mathcal{H}$ such that $\sum_{k\in\Lambda}\|f_k\|^2 < C_3$, prove that $\{f_k + h_k\}_{k\in\Lambda}$ must be a Riesz basis for $\mathcal{H}$.

4.26. Prove Corollaries 4.2.8 and 4.2.9.

4.27. For a finite subset Φ of $L_2(\mathbb{R})$, prove $\sup_{j\in\mathbb{Z}} \operatorname{len}(\mathrm{S}(\Phi(2^j\cdot)|L_2(\mathbb{R}))) = \dim(\operatorname{span}\Phi)$.

4.28. Let $r \in \mathbb{N}$ and $\Phi_j, j \in \mathbb{Z}$ be subsets of $L_2(\mathbb{R})$ such that each Φ_j has no more than r elements. Prove $\dim(\cap_{j\in\mathbb{Z}} \mathrm{S}_{2^j}(\Phi_j|L_2(\mathbb{R}))) \leqslant r$.

4.29. For any square integrable function $f \in L_2(\mathbb{R})$ with $\|f\|_{L_2(\mathbb{R})} \ne 0$, prove $\dim(\overline{\operatorname{span}\{f(2^j\cdot) : j \in \mathbb{Z}\}}) = \operatorname{len}(\mathrm{S}(\{f(2^j\cdot) : j \in \mathbb{Z}\}|L_2(\mathbb{R})) = \infty$.

4.30. Let $\mathscr{V} \subseteq L_2(\mathbb{R})$ be shift-invariant. Prove $\rho^2(f, \mathrm{S}(\mathcal{P}_{\mathscr{V}} g|L_2(\mathbb{R}))) \leqslant \rho^2(f, \mathscr{V}) + \rho^2(f, \mathrm{S}(g|L_2(\mathbb{R})))$ for all $f, g \in L_2(\mathbb{R})$, where $\rho^2(f, \mathscr{V}) := \sup_{h\in\mathscr{V}} \|f - h\|^2_{L_2(\mathbb{R})}$ and $\mathcal{P}_{\mathscr{V}} : L_2(\mathbb{R}) \to \mathscr{V}$ is the orthogonal projection operator.

4.31. Let $\Psi = \{\psi^1, \dots, \psi^s\}$ and $\tilde{\Psi} = \{\tilde{\psi}^1, \dots, \tilde{\psi}^s\}$ in $L_2(\mathbb{R})$ such that $(\tilde{\Psi}, \Psi)$ is a homogeneous dual framelet in $L_2(\mathbb{R})$. For $n \in \mathbb{Z}$, define $\mathcal{A}_n : L_2(\mathbb{R}) \to \mathscr{V}_n(\Psi)$ by $\mathcal{A}_n f = \sum_{j<n} \sum_{\ell=1}^{s} \sum_{k\in\mathbb{Z}} \langle f, \tilde{\psi}_{2^j;k}\rangle \psi_{2^j;k}$ for $f \in L_2(\mathbb{R})$. Prove that $\mathcal{A}_n$ is well-defined and bounded, and $\widehat{\mathcal{A}_n f}(\xi) = \sum_{j<n} \left(\sum_{\psi\in\Psi} [\widehat{f}(2^j\cdot), \widehat{\tilde{\psi}}](2^{-j}\xi)\widehat{\psi}(2^{-j}\xi) \right)$ with the series $\sum_{j<n}$ absolutely converging in $L_2(\mathbb{R})$.

4.32. Let $\Psi, \tilde{\Psi}$ be finite subsets of $L_2(\mathbb{R})$ such that $(\tilde{\Psi}, \Psi)$ is a homogeneous biorthogonal wavelet in $L_2(\mathbb{R})$. Define $\mathscr{V}_0(\Psi) := \mathrm{S}(\{\psi(2^{-j}\cdot) : \psi \in \Psi, j \in \mathbb{N}\}|L_2(\mathbb{R}))$. Prove $\dim_{\mathscr{V}_0(\Psi)}(\xi) = 2\pi \sum_{\psi\in\Psi} [\widehat{\tilde{\psi}}(2^{-j}\cdot), \widehat{\psi}(2^{-j}\cdot)](\xi) = \dim_{\mathscr{V}_0(\tilde{\Psi})}(\xi)$ a.e. $\xi \in \mathbb{R}$.

4.33. For $\psi \in L_2(\mathbb{R})$ such that ψ is a homogeneous orthogonal wavelet in $L_2(\mathbb{R})$ and $\sum_{k\in\mathbb{Z}} |\widehat{\psi}(2\xi + 4\pi k)|^2 \ne 0$ a.e. $\xi \in \mathbb{R}$, prove that $\mathbb{R} \setminus \operatorname{supp}(\widehat{\psi})$ has measure zero.

4.34. Prove Theorem 4.4.13.

4.35. Let $\widehat{\phi} = \chi_{[0,2\pi)}$. Define a 2π-periodic function $\mathbf{a}$ by $\mathbf{a}(\xi) := \chi_{[0,\pi)}(\xi)$ for $\xi \in [0, 2\pi)$. Prove that (a) $\widehat{\phi}(2\xi) = \mathbf{a}(\xi)\widehat{\phi}(\xi)$ for all $\xi \in \mathbb{R}$ and $\{\phi(\cdot - k) : k \in \mathbb{Z}\}$ is an orthonormal basis of $\mathscr{V}_0 = \mathrm{S}(\phi \,|\, L_2(\mathbb{R}))$ (i.e., items (ii) and (iii) of Theorem 4.5.8 with $r = 1$ are satisfied). (b) ϕ does not generate a multiresolution analysis of $L_2(\mathbb{R})$ by showing that item (i) of Theorem 4.5.8 fails with $r = 1$.

4.36. Let $\phi \in L_2(\mathbb{R})$ and $v \in l_2(\mathbb{Z})$. Prove that $v * \phi := \sum_{k\in\mathbb{Z}} v(k)\phi(\cdot - k)$ is a well-defined measurable function in $L_2^{loc}(\mathbb{R})$.

4.37. Let $\phi \in L_2(\mathbb{R})$ and $v \in l_2(\mathbb{Z})$. Prove that $v * \phi = 0$ (i.e., $v \perp \{\phi(x - k)\}_{k\in\mathbb{Z}}$ in $l_2(\mathbb{Z})$ for almost every $x \in \mathbb{R}$) if and only if $\widehat{v}(\xi)\widehat{\phi}(\xi) = 0$ a.e. $\xi \in \mathbb{R}$.

4.38. For a measurable function ϕ such that $\int_{\mathbb{R}} |\phi(x)|^2 dx < \infty$, prove that (a) there exists a set E of measure zero such that $\{\phi(x-k)\}_{k\in\mathbb{Z}} \in l_2(\phi)$ for all $x \in \mathbb{R}\backslash E$. (b) For any set E of measure zero such that $\{\phi(x - k)\}_{k\in\mathbb{Z}} \in l_2(\phi)$ for all $x \in \mathbb{R}\backslash E$, the closed linear span of $\{\phi(x - k)\}_{k\in\mathbb{Z}}, x \in \mathbb{R}\backslash E$ is $l_2(\phi)$.

4.39. Let $\phi \in L_2(\mathbb{R})$. Prove that by properly modifying the measurable function ϕ on a set of measure 0, $\{\phi(x-k)\}_{k\in\mathbb{Z}} \in l_2(\phi)$ for all $x \in \mathbb{R}$ and $\mathrm{S}(\phi \,|\, L_2(\mathbb{R}))$ is a reproducing kernel Hilbert space.

4.40. Let $E \subset \mathbb{R}$ be Lebesgue measurable and $A : E \to \mathbb{C}^{r\times s}$ be an $r \times s$ matrix of measurable functions on E. If the rank of $A(\xi)$ is a constant integer n for all $\xi \in E$, prove that there exists an $s \times (s - n)$ matrix $V : E \to \mathbb{C}^{s\times(s-n)}$ of measurable functions on E such that $A(\xi)V(\xi) = 0$ and $\overline{V(\xi)}^{\mathsf{T}} V(\xi) = I_{s-n}$ for all $\xi \in E$.

4.41. Let $A : \mathbb{R} \to \mathbb{C}^{r\times r}$ be an $r \times r$ matrix of Lebesgue measurable functions on $\mathbb{R}$ such that $\overline{A(\xi)}^{\mathsf{T}} = A(\xi)$ for all $\xi \in \mathbb{R}^d$. Let $\lambda_j(\xi)$ be the jth largest eigenvalue of the matrix $A(\xi)$ for $j = 1, \dots, r$ with $\lambda_1(\xi) \geqslant \lambda_2(\xi) \geqslant \cdots \geqslant \lambda_r(\xi)$. Prove that all the eigenvalue functions $\lambda_1, \dots, \lambda_r$ are measurable and there exists an $r \times r$ matrix U of measurable functions on $\mathbb{R}$ such that $\overline{U(\xi)}^{\mathsf{T}} U(\xi) = I_r$ and $A(\xi) = U(\xi)\operatorname{diag}(\lambda_1(\xi), \dots, \lambda_r(\xi))\overline{U(\xi)}^{\mathsf{T}}$.

4.42. Let $A : \mathbb{R} \to \mathbb{C}^{r\times s}$ be an $r \times s$ matrix of measurable functions on $\mathbb{R}$. Let $\sigma_j(\xi)$ be the jth largest singular value of $A(\xi)$ with $\sigma_1(\xi) \geqslant \sigma_2(\xi) \geqslant \cdots \geqslant \sigma_{\min(r,s)}(\xi) \geqslant 0$. Prove that all $\sigma_1, \dots, \sigma_{\min(r,s)}$ are nonnegative measurable functions and there exist an $r \times r$ matrix U and an $s \times s$ unitary matrix V of measurable functions on $\mathbb{R}$ such that $\overline{U(\xi)}^{\mathsf{T}} U(\xi) = I_r$, $\overline{V(\xi)}^{\mathsf{T}} V(\xi) = I_s$, and the $r \times s$ matrix $\overline{U(\xi)}^{\mathsf{T}} A(\xi)V(\xi)$ is an $r \times s$ diagonal matrix with the first $\min(r, s)$ diagonal entries being $\sigma_1(\xi), \dots, \sigma_{\min(r,s)}(\xi)$ and with all its other entries being zero.

4.43. Let H be a countable subset of $L_2(\mathbb{R})$ such that $\{h(\cdot - k) \ : \ k \in \mathbb{Z}, h \in H\}$ is a Bessel sequence in $L_2(\mathbb{R})$. If $r := \operatorname{len}(\mathrm{S}(H \,|\, L_2(\mathbb{R}))) < \infty$, prove that there exists $\eta^1, \dots, \eta^r \in \mathrm{S}(H \,|\, L_2(\mathbb{R}))$ such that $\mathrm{S}(\{\eta^1, \dots, \eta^r\} \,|\, L_2(\mathbb{R})) = \mathrm{S}(H \,|\, L_2(\mathbb{R}))$ and $\sum_{\ell=1}^{r} \sum_{k\in\mathbb{Z}^d} |\langle f, \eta^\ell(\cdot - k)\rangle|^2 = \sum_{h\in H} \sum_{k\in\mathbb{Z}^d} |\langle f, h(\cdot - k)\rangle|^2$ for all $f \in L_2(\mathbb{R}^d)$.

4.44. For an $m \times n$ matrix M, recall that its pseudoinverse matrix M^+ is the unique $n \times m$ matrix satisfying $MM^+M = M$, $M^+MM^+ = M^+$, $(MM^+)^\star = MM^+$, and $(M^+M)^\star = M^+M$. Let $M = U\Sigma V^\star$ be the singular value decomposition of M such that $UU^\star = I_m$, $VV^\star = I_n$ and the $m \times n$ matrix Σ is diagonal. Prove that $M^+ = V\Sigma^+U^\star$, where Σ^+ is the transpose of Σ but with all its nonzero diagonal entries being replaced by their reciprocal.

4.45. Let $\Phi = \{\phi^1, \dots, \phi^r\}$ be a finite subset of $L_2(\mathbb{R})$ such that $\{f(\cdot - k) \ : \ k \in \mathbb{Z}, f \in \Phi\}$ is a frame for $\mathrm{S}(\Phi|L_2(\mathbb{R}))$.

a. Prove that $\mathcal{F}(h(\cdot - k)) = (\mathcal{F}h)(\cdot - k)$ for all $h \in \mathrm{S}(\Phi|L_2(\mathbb{R}))$ and show that its canonical dual frame is given by $\{(\mathcal{F}^{-1}f)(\cdot - k) \ : \ f \in \Phi\}$, where $\mathcal{F}$ is the associated frame operator of $\{f(\cdot - k) \ : \ k \in \mathbb{Z}, f \in \Phi\}$.

b. Write $\phi := (\phi^1, \dots, \phi^r)^{\mathsf{T}}$ and $\eta := (\mathcal{F}^{-1}(\phi^1), \dots, \mathcal{F}^{-1}(\phi^r))^{\mathsf{T}}$. Prove that $\widehat{\eta}(\xi) = [\widehat{\phi}, \widehat{\phi}]^+(\xi)\widehat{\phi}(\xi)$ for almost every $\xi \in \mathbb{R}$, where $[\widehat{\phi}, \widehat{\phi}]^+(\xi)$ is the unique pseudoinverse matrix of the $r \times r$ matrix $[\widehat{\phi}, \widehat{\phi}](\xi)$.

4.46. Let Φ be a finite subset of $L_2(\mathbb{R})$ and $\alpha, \beta \in \mathbb{R}$. Define a Gabor system by $G := \{f_{1;\alpha k,\beta m} := e^{-i\beta m\cdot} f(\cdot - \alpha k) \ : \ k, m \in \mathbb{Z}, f \in \Phi\}$. Suppose that G is a frame for $L_2(\mathbb{R})$. Let $\mathcal{F}$ be its associated frame operator. Prove that $\mathcal{F}(h_{1;\alpha k,\beta m}) = \mathcal{F}(h)_{1;\alpha k,\beta m}$ for all $h \in L_2(\mathbb{R})$ and show that the canonical dual frame of G is given by $\{(\mathcal{F}^{-1}f)_{1;\alpha k,\beta m} := e^{-i\beta m\cdot}(\mathcal{F}^{-1}f)(\cdot - \alpha k) \ : \ k, m \in \mathbb{Z}, f \in \Phi\}$.

4.47. Use the definition of a homogeneous dual framelet to directly prove (4.3.23).

4.48. Prove that (4.3.23) is equivalent to the identity in (4.3.18).

4.49. Prove the inequality in (4.6.10).

4.50. For $f, \phi \in H^\tau(\mathbb{R})$, prove $\sum_{k\in\mathbb{Z}} |\langle f, \phi(\cdot - k)\rangle_{H^\tau(\mathbb{R})}|^2 = \frac{1}{2\pi}\int_{-\pi}^{\pi} |[\widehat{f}, \widehat{\phi}]_\tau(\xi)|^2 d\xi$.

4.51. For $\Phi \subseteq H^\tau(\mathbb{R})$, $\mathrm{S}(\Phi|H^\tau(\mathbb{R}))$ denotes the closed linear span of $\{\phi(\cdot - k)\}_{k\in\mathbb{Z},\phi\in\Phi}$ in $H^\tau(\mathbb{R})$. For $\phi \in H^\tau(\mathbb{R})$, prove that the following statements are equivalent:

a. $\{\phi(\cdot - k) : k \in \mathbb{Z}\}$ is a frame for $\mathrm{S}(\phi|H^\tau(\mathbb{R}))$ satisfying

$$C_1\|f\|^2_{H^\tau(\mathbb{R})} \leqslant \sum_{k\in\mathbb{Z}} |\langle f, \phi(\cdot - k)\rangle_{H^\tau(\mathbb{R})}|^2 \leqslant C_2\|f\|^2_{H^\tau(\mathbb{R})}, \quad \forall f \in \mathrm{S}(\phi|H^\tau(\mathbb{R})).$$

b. $C_1 \leqslant [\widehat{\phi}, \widehat{\phi}]_\tau(\xi) \leqslant C_2$ for almost every $\xi \in \mathrm{supp}([\widehat{\phi}, \widehat{\phi}]_\tau)$.

c. $\mathcal{V} : l_2^\tau(\phi) \to \mathrm{S}(\phi|H^\tau(\mathbb{R}))$, $\{v(k)\}_{k\in\mathbb{Z}} \mapsto \sum_{k\in\mathbb{Z}} v(k)\phi(\cdot - k)$ is a well-defined bounded bijective linear operator with a bounded inverse, where $l_2^\tau(\phi) := \{v \in l_2^\tau(\mathbb{Z}) : \mathrm{supp}\,\widehat{v} \subseteq \mathrm{supp}([\widehat{\phi}, \widehat{\phi}]_\tau)\}$ is a subspace of the sequence Sobolev space $l_2^\tau(\mathbb{Z})$.

d. $\mathcal{W} : \mathrm{S}(\phi|H^\tau(\mathbb{R})) \to l_2^\tau(\phi), f \mapsto \{\langle f, \phi(\cdot - k)\rangle_{H^\tau(\mathbb{R})}\}_{k\in\mathbb{Z}}$ is a well-defined bounded bijective linear operator with a bounded inverse.

4.52. Let $\tau \in \mathbb{R}$ and $\phi \in H^\tau(\mathbb{R})$. Prove that the following statements are equivalent.

a. $\{\phi(\cdot - k) : k \in \mathbb{Z}\}$ is a Riesz basis for $\mathrm{S}(\phi|H^\tau(\mathbb{R}))$ satisfying

$$C_1\sum_{k\in\mathbb{Z}} |v(k)|^2 \leqslant \Big\|\sum_{k\in\mathbb{Z}} v(k)\phi(\cdot - k)\Big\|^2_{H^\tau(\mathbb{R})} \leqslant C_2\sum_{k\in\mathbb{Z}} |v(k)|^2, \quad \forall v \in l_0(\mathbb{Z}).$$

b. $C_1 \leqslant [\widehat{\phi}, \widehat{\phi}]_\tau(\xi) \leqslant C_2$ for almost every $\xi \in \mathbb{R}$.

c. $\mathcal{V} : l_2^\tau(\mathbb{Z}) \to \mathrm{S}(\phi|H^\tau(\mathbb{R}))$, $\{v(k)\}_{k\in\mathbb{Z}} \mapsto \sum_{k\in\mathbb{Z}} v(k)\phi(\cdot - k)$ is a well-defined bounded bijective operator with a bounded inverse.

d. $\mathcal{W} : \mathrm{S}(\phi|H^\tau(\mathbb{R})) \to l_2^\tau(\mathbb{Z}), f \mapsto \{\langle f, \phi(\cdot - k)\rangle_{H^\tau(\mathbb{R})}\}_{k\in\mathbb{Z}}$ is a well-defined bounded bijective linear operator with a bounded inverse.

4.53. If $\sum_{j<0}\sum_{\psi\in\Psi}(|\widehat{\psi}(2^j\cdot)|^2 + |\widehat{\tilde{\psi}}(2^j\cdot)|^2) \in L_1^{loc}(\mathbb{R})$, prove that for all $f, g \in \check{\mathscr{D}}(\mathbb{R})$,

$$\sum_{j<0}\sum_{\psi\in\Psi}\sum_{k\in\mathbb{Z}}\langle f, \tilde{\psi}_{2^j;k}\rangle\langle\psi_{2^j;k}, g\rangle = \sum_{\psi\in\Psi}\sum_{j=1}^{\infty}\sum_{k\in\mathbb{Z}}\langle f, 2^j\tilde{\psi}(2^j(\cdot - k))\rangle\langle 2^j\psi(2^j(\cdot - k)), g\rangle.$$

$\Longleftrightarrow$ (4.3.22) holds for all $f, g \in \check{\mathscr{D}}(\mathbb{R}) \Longleftrightarrow$ (4.3.18).

4.54. Let $\Psi \subseteq L_2(\mathbb{R})$ such that $\sum_{\psi\in\Psi}\|\psi\|^2_{L_2(\mathbb{R})} < \infty$ and $\widehat{\psi} = \chi_{E_\psi}$ with some measurable sets E_ψ for $\psi \in \Psi$. Prove that Ψ is a homogeneous tight framelet in $L_2(\mathbb{R})$ if and only if (i) up to a null set, $\mathbb{R}$ is the disjoint union of $\{2^jE_\psi\}_{j\in\mathbb{Z},\psi\in\Psi}$ and (ii) $E_\psi \cap (E_\psi + 2\pi k)$ has measure zero for all $k \in \mathbb{Z}\backslash\{0\}$

and $\psi \in \Psi$. Prove that Ψ is a homogeneous orthogonal wavelet in $L_2(\mathbb{R})$ if and only if item (ii) is replaced by (ii') up to a null set, $\mathbb{R}$ is the disjoint union of $\{E_\psi + 2\pi k\}_{k\in\mathbb{Z}}$ for every $\psi \in \Psi$.

4.55. Define $\psi \in L_2(\mathbb{R})$ by $\widehat{\psi} := \chi_{E\cup(-E)}$ with $E := [\frac{2^m\pi}{2^{m+1}-1}, \pi] \cup [2^m\pi, \frac{2^{2m+1}\pi}{2^{m+1}-1}]$ and $m \in \mathbb{N}$. Prove that ψ is a homogeneous orthogonal wavelet in $L_2(\mathbb{R})$ and show that $\mathscr{V}_0(\psi)$ defined in (4.5.17) is shift-invariant with $\operatorname{len}(\mathscr{V}_0(\psi)) \geqslant 2$ for $m \geqslant 2$.

4.56. For a homogeneous tight framelet Ψ in $L_2(\mathbb{R})$, define $\Phi := \{2^{-j}\psi(2^{-j}\cdot) : j \in \mathbb{N}, \psi \in \Psi\}$. Prove that $\mathscr{V}_0(\Psi) = \mathsf{S}(\Phi|L_2(\mathbb{R}))$ and $\mathsf{S}(\Phi \cup \Psi|L_2(\mathbb{R})) = \mathsf{S}_2(\Phi|L_2(\mathbb{R}))$, where $\mathscr{V}_0(\Psi) := \overline{\operatorname{span}\{\psi_{2^j;k} : j < 0, k \in \mathbb{Z}, \psi \in \Psi\}}$ is defined in (4.5.17).

4.57. Let $\Psi \subseteq L_2(\mathbb{R})$ such that Ψ is a homogeneous Riesz wavelet in $L_2(\mathbb{R})$. Prove that $\mathscr{V}_0(\Psi) := \overline{\operatorname{span}\{\psi_{2^j;k} : j < 0, k \in \mathbb{Z}, \psi \in \Psi\}}$ is shift-invariant if and only if there exists $\tilde{\Psi}$ such that $(\tilde{\Psi}, \Psi)$ is a homogeneous biorthogonal wavelet in $L_2(\mathbb{R})$.

4.58. Let $\psi \in L_2(\mathbb{R})$ be a homogeneous orthogonal wavelet, i.e., $\mathsf{AS}(\psi)$ is an orthonormal basis for $L_2(\mathbb{R})$. Define $\eta := \psi + \varepsilon\sqrt{2}\psi(2\cdot)$ with $0 < \varepsilon < 1$. Prove that $\mathsf{AS}(\eta)$ is a Riesz basis for $L_2(\mathbb{R})$, but there does not exist $\tilde{\eta} \in L_2(\mathbb{R})$ such that $(\tilde{\eta}, \eta)$ is a homogeneous biorthogonal wavelet in $L_2(\mathbb{R})$.

4.59. Let $\lambda_j = 2^{-j}$ in Theorem 4.8.1 for all $j \geqslant J$. Prove that (4.8.6) becomes (4.8.12), while (4.8.5) becomes (4.8.13).

4.60. If $(\{\tilde{\Phi}; \tilde{\Psi}\}, \{\Phi; \Psi\})$ is a dual framelet in $L_2(\mathbb{R})$ such that $\sum_{\phi\in\Phi}(\|\phi\|^2_{L_2(\mathbb{R})} + \|\tilde{\phi}\|^2_{L_2(\mathbb{R})}) < \infty$ and $\sum_{\psi\in\Psi}(\|\psi\|^2_{L_2(\mathbb{R})} + \|\tilde{\psi}\|^2_{L_2(\mathbb{R})}) < \infty$, prove that it is impossible for any of $\Phi, \tilde{\Phi}, \Psi, \tilde{\Psi}$ to be either the empty set $\emptyset$ or the zero element $\{0\}$.

4.61. If $\{\Phi; \Psi\}$ is a framelet in $L_2(\mathbb{R})$ with $\sum_{\phi\in\Phi} \|\phi\|^2_{L_2(\mathbb{R})} < \infty$ and $\sum_{\psi\in\Psi} \|\psi\|^2_{L_2(\mathbb{R})} < \infty$, prove that both Φ and Ψ cannot be $\emptyset$ or $\{0\}$.

4.62. Prove Corollary 4.8.2 and Theorem 4.8.4.

4.63. For $j \in \mathbb{N}$, let $a_j, b_{j,1}, \ldots, b_{j,s_j-1} \in l_0(\mathbb{Z})$ with $\widehat{a_j}(0) = 1$ and $\sum_{j=1}^\infty 2^{-j} \operatorname{len}(a_j) < \infty$.

a. Prove $\sum_{j=1}^\infty \|(1 - \widehat{a_j}(2^{-j}\xi))/\xi\|_{L_\infty(\mathbb{R})} < \infty$. Then use Theorem 4.8.4 to show that $\widehat{\phi^j}(\xi) := \prod_{k=1}^\infty \widehat{a_{j+k}}(2^{-k}\xi)$ and $\widehat{\psi^{j-1,\ell}}(\xi) := \widehat{b_{j,\ell}}(\xi/2)\widehat{\phi^j}(\xi/2)$ are well-defined functions in $L_2^{loc}(\mathbb{R})$ for all $j \in \mathbb{N}$. Define $\Psi_j := \{\psi^{j,1}, \ldots, \psi^{j,s_j-1}\}$.

b. Prove that all $\phi^j, \psi^{j,\ell}$ are compactly supported tempered distributions.

c. If $\{a_j; b_{j,1}, \ldots, b_{j,s_j-1}\}$ are tight framelet filter banks for all $j \in \mathbb{N}$, prove that $AS_J(\phi^j; \{\Psi_j|2^{-j}\}_{j=J}^\infty)$ are (normalized) tight frames for $L_2(\mathbb{R})$ for all $J \in \mathbb{N}_0$.

4.64. Let $\tau \in \mathbb{R}$ and $\psi \in H^\tau(\mathbb{R})$ such that ψ satisfies all the conditions in Theorem 4.6.5. Define $\mathring{\Psi}_j := \{\mathscr{P}(\psi_{2^j;0})\}$. Prove that $\widehat{\mathring{\Psi}_j} = \{2^{-j/2}\widehat{\psi}(2^{-j}2\pi k)\}_{k\in\mathbb{Z}}$

and there exists a positive constant C such that for all $g \in H^{-\tau}(\mathbb{T})$,

$$\sum_{j=0}^{\infty} \sum_{\psi \in \mathring{\Psi}_j} \sum_{\omega \in \Omega_j} 2^{-2\tau j} |\langle g, \psi_{\omega,0} \rangle|^2 = \sum_{j=0}^{\infty} \sum_{\widehat{\psi} \in \widehat{\mathring{\Psi}}_j} \sum_{\omega \in \Omega_j} 2^{-2\tau j} |\langle \widehat{g}, \boldsymbol{\psi}_{0,\omega} \rangle|^2 \leqslant C \|g\|^2_{H^{-\tau}(\mathbb{T})}.$$

4.65. Let $\widehat{a}$ be a 2π-periodic continuous function on $\mathbb{R}$ such that $\widehat{a}(0) = 1$. Assume that $\widehat{\phi}(\xi) := \prod_{j=1}^{\infty} \widehat{a}(2^{-j}\xi), \xi \in \mathbb{R}$ is well defined with the series $\{\prod_{j=1}^{n} \widehat{a}(2^{-j}\xi)\}_{n\in\mathbb{N}}$ converging uniformly for ξ on every bounded interval (e.g., this condition is satisfied if the inequality (4.1.30) holds). Note that $\widehat{\phi}$ is a continuous function. Prove that the following statements are equivalent:

a. $\sum_{k\in\mathbb{Z}} |\widehat{\phi}(\xi + 2\pi k)|^2 > 0$ for all $\xi \in \mathbb{R}$.
b. There exists a constant $C > 0$ such that $\sum_{k\in\mathbb{Z}} |\widehat{\phi}(\xi + 2\pi k)|^2 \geqslant C$ for all $\xi \in \mathbb{R}$.
c. There exists a compact set K of $\mathbb{R}$ such that $\widehat{\phi}(\xi) \neq 0$ for all $\xi \in K$, $(-\varepsilon, \varepsilon) \subset K$ for some $\varepsilon > 0$, and $\sum_{n\in\mathbb{Z}} \chi_K(\xi + 2\pi n) = 1$ a.e. $\xi \in \mathbb{R}$.
d. There exist a positive constant C and a compact set K of $\mathbb{R}$ such that $|\widehat{\phi}(\xi)| \geqslant C$ for all $\xi \in K$, $(-\varepsilon, \varepsilon) \subset K$ for some $\varepsilon > 0$, $\sum_{n\in\mathbb{Z}} \chi_K(\xi + 2\pi n) = 1$ a.e. $\xi \in \mathbb{R}$.
e. (Cohen's criteria) There exists a compact set K of $\mathbb{R}$ such that $(-\varepsilon, \varepsilon) \subset K$ for some $\varepsilon > 0$, $\sum_{n\in\mathbb{Z}} \chi_K(\xi + 2\pi n) = 1$ a.e. $\xi \in \mathbb{R}$, and $\widehat{a}(2^{-j}\xi) \neq 0$ for all $j \in \mathbb{N}, \xi \in K$.

4.66. Let $\widehat{a}$ be a 2π-periodic measurable function on $\mathbb{R}$ such that $\widehat{\phi}(\xi) := \prod_{j=1}^{\infty} \widehat{a}(2^{-j}\xi)$ is well defined for almost every $\xi \in \mathbb{R}$. If $|\widehat{a}(\xi)|^2 + |\widehat{a}(\xi+\pi)|^2 = 1$ a.e. $\xi \in \mathbb{R}$ and the fourth condition (i.e., item d.) of Exercise 4.65 is satisfied, prove that $\{\phi(\cdot - k) \ : \ k \in \mathbb{Z}\}$ is an orthonormal system in $L_2(\mathbb{R})$.

Chapter 5
Analysis of Refinable Vector Functions

As we discussed in Chap. 4, framelets and wavelets are often derived from a refinable vector function $\phi = (\phi_1, \dots, \phi_r)^\mathsf{T}$ satisfying the following refinement equation

$$\widehat{\phi}(2\xi) = \widehat{a}(\xi)\widehat{\phi}(\xi), \qquad a.e.\ \xi \in \mathbb{R},$$

where $\widehat{a}$ is an $r \times r$ matrix of 2π-periodic Lebesgue measurable functions. On the other hand, stability of affine systems on $\mathbb{R}$ and discrete affine systems on $\mathbb{Z}$ are closely related to properties of refinable functions. The main goal of this chapter is to investigate properties of general refinable vector functions.

We first study the distributional solutions, called refinable (vector) functions or distributions, to a refinement equation in both frequency domain and space domain. Secondly, we address linear independence and stability of integer shifts of functions. Thirdly, we study approximation properties of quasi-projection operators in $L_p(\mathbb{R})$ and approximation orders of shift-invariant subspaces in $L_p(\mathbb{R})$. Then we characterize convergence of a cascade algorithm and smoothness of a refinable function in Sobolev spaces $W_p^m(\mathbb{R})$ with $1 \leqslant p \leqslant \infty$. Finally, we provide sharp error estimate for cascade algorithms and refinable functions with perturbed filters/masks.

A norm $\|\cdot\|$ on $r \times r$ matrices is *submultiplicative* if $\|EF\| \leqslant \|E\|\|F\|$ for all $r \times r$ matrices E and F. For example, the operator norm $\|E\| := \sup_{\|x\| \leqslant 1} \|Ex\|$ is submultiplicative, where $\|\cdot\|$ is a norm on $\mathbb{C}^r$. Throughout this chapter $\|\cdot\|$ stands for a submultiplicative matrix norm. For $r, s \in \mathbb{N}$ and a Banach space $\mathscr{B}$, we define $\mathscr{B}^{r\times s}$ to be the Banach space of all $r \times s$ matrices $(b_{j,k})_{1\leqslant j\leqslant r, 1\leqslant k\leqslant s}$ (j refers to row and k refers to column) of elements in $\mathscr{B}$ with the norm:

$$\|(b_{j,k})_{1\leqslant j\leqslant r,1\leqslant k\leqslant s}\|_{\mathscr{B}^{r\times s}} := \|(\|b_{j,k}\|_{\mathscr{B}})_{1\leqslant j\leqslant r,1\leqslant k\leqslant s}\|_{\mathbb{C}^{r\times s}},$$

where $\|\cdot\|_{\mathbb{C}^{r\times s}}$ is a norm on $\mathbb{C}^{r\times s}$. In particular, we define $\mathscr{B}^r := \mathscr{B}^{r\times 1}$. Note that all norms on $\mathbb{C}^{r\times s}$ are equivalent. When $\mathscr{B}$ is a Banach space related to l_p or L_p, we

B. Han, *Framelets and Wavelets*, Applied and Numerical Harmonic Analysis,
https://doi.org/10.1007/978-3-319-68530-4_5

often use the l_p-norm on $\mathbb{C}^{r\times s}$:

$$\|(c_{j,k})_{1\leqslant j\leqslant r,1\leqslant k\leqslant s}\|_{l_p} := \Big(\sum_{j=1}^{r}\sum_{k=1}^{s}|c_{j,k}|^p\Big)^{1/p} \tag{5.0.1}$$

for $1 \leqslant p < \infty$ and $\|(c_{j,k})_{1\leqslant j\leqslant r,1\leqslant k\leqslant s}\|_{l_\infty} := \max\{|c_{j,k}| \; : \; 1 \leqslant j \leqslant r, 1 \leqslant k \leqslant s\}$ for $p = \infty$.

For $a = \{a(k)\}_{k\in\mathbb{Z}} \in (l_2(\mathbb{Z}))^{r\times r}$ with each $a(k)$ being an $r \times r$ matrix, recall that $\widehat{a}(\xi) := \sum_{k\in\mathbb{Z}} a(k)e^{-ik\xi}$, $\xi \in \mathbb{R}$ and $\widehat{a} \in (L_2(\mathbb{T}))^{r\times r}$. By $\boldsymbol{\delta}$ we denote the Kronecker/Dirac sequence such that $\boldsymbol{\delta}(0) = 1$ and $\boldsymbol{\delta}(k) = 0$ for all $k \in \mathbb{Z}\backslash\{0\}$. We shall also use $\boldsymbol{\delta}$ to denote the Dirac distribution on $\mathbb{R}$.

For a function or distribution f on $\mathbb{R}$, throughout this book f' refers to the classical derivative (that is, $f'(x) := \lim_{h\to 0}\frac{f(x+h)-f(x)}{h}$) and Df refers to the distributional derivative (that is, $\langle Df; \psi\rangle := -\langle f; \psi'\rangle$ for all $\psi \in \mathscr{D}(\mathbb{R})$). Similarly, $f^{(j)}$ means the jth classical derivative and $D^j f$ is the jth distributional derivative for $j \in \mathbb{N}_0$.

5.1 Distributional Solutions to Vector Refinement Equations

As we have seen in Sect. 4.5, wavelets and framelets are derived from refinable vector functions through the refinable structure:

$$\widehat{\phi}(2\xi) = \widehat{a}(\xi)\widehat{\phi}(\xi), \qquad \widehat{\psi}(2\xi) = \widehat{b}(\xi)\widehat{\phi}(\xi), \qquad \xi \in \mathbb{R}.$$

Note that the first equation is called the frequency-based refinement equation and can be equivalently expressed as $\phi = 2\sum_{k\in\mathbb{Z}} a(k)\phi(2\cdot -k)$ using the Fourier transform. One often first constructs a wavelet or framelet filter bank $\{a; b\}$ as we have discussed in Chaps. 2 and 3. To obtain wavelets and framelets on the real line, it is critical to establish the link between a low-pass filter a with a refinable vector function/distribution ϕ by solving the refinement equation $\phi = 2\sum_{k\in\mathbb{Z}} a(k)\phi(2\cdot -k)$. Consequently, the existence of solutions (in particular, tempered distributional solutions) to a refinement equation is one of the fundamental problems in wavelet theory.

In this section we study distributional solutions to a vector refinement equation. If one is only interested in the scalar case $r = 1$, see Lemma 4.1.8 and this section can be safely skipped.

Lemma 5.1.1 *Let $a \in (l_2(\mathbb{Z}))^{r\times r}$ such that there exist nonnegative constants C_0 and τ satisfying*

$$\|\widehat{a}(\xi)\| \leqslant \|\widehat{a}(0)\| + C_0|\xi|^\tau, \qquad \forall\, \xi \in [-\pi, \pi]. \tag{5.1.1}$$

Then for all $\xi \in \mathbb{R}$ *and* $n \in \mathbb{N}$,

$$\left\| \prod_{j=1}^{n} \widehat{a}(2^{-j}\xi) \right\| \leqslant \begin{cases} (\|\widehat{a}(0)\| + C_0 2^{-(n+1)\tau/2}|\xi|^{\tau})^n, & \text{if } \widehat{a}(0) = 0 \text{ or } \tau = 0, \\ \|\widehat{a}(0)\|^n C(\xi), & \text{otherwise}, \end{cases} \tag{5.1.2}$$

where

$$C(\xi) := e^{\frac{C_0}{\rho(1-2^{-\tau})}} \max\left(1, |\xi|^{\log_2(\rho^{-1}\| \|\hat{a}\| \|_{L_\infty(\mathbb{R})})}\right) \quad \textit{with} \quad \rho := \|\hat{a}(0)\|. \tag{5.1.3}$$

Proof Since $\widehat{a}$ is 2π-periodic, the inequality (5.1.1) must hold for all $\xi \in \mathbb{R}$. If $\widehat{a}(0) = 0$, by (5.1.1), we have

$$\left\| \prod_{j=1}^{n} \widehat{a}(2^{-j}\xi) \right\| \leqslant \prod_{j=1}^{n} \|\widehat{a}(2^{-j}\xi)\| \leqslant \prod_{j=1}^{n} (C_0|2^{-j}\xi|^{\tau}) = (C_0 2^{-(n+1)\tau/2}|\xi|^{\tau})^n.$$

If $\tau = 0$, the above argument also shows that $\left\| \prod_{j=1}^{n} \widehat{a}(2^{-j}\xi) \right\| \leqslant (\|\widehat{a}(0)\| + C_0)^n$.

If $\rho := \|\widehat{a}(0)\| > 0$ and $\tau > 0$, using the inequality $|z| \leqslant 1 + |1 - z| \leqslant e^{|1-z|}$ for all $z \in \mathbb{C}$, for $|\xi| \leqslant 2$, we have

$$\begin{aligned} \left\| \prod_{j=1}^{n} \widehat{a}(2^{-j}\xi) \right\| &\leqslant \prod_{j=1}^{n} \|\widehat{a}(2^{-j}\xi)\| \leqslant \prod_{j=1}^{n} (\rho + C_0|2^{-j}\xi|^{\tau}) = \rho^n \prod_{j=1}^{n} (1 + \tfrac{C_0}{\rho}|2^{-j}\xi|^{\tau}) \\ &\leqslant \rho^n e^{\frac{C_0}{\rho}\sum_{j=1}^{n} 2^{-\tau j}|\xi|^{\tau}} \leqslant \rho^n e^{\frac{C_0}{\rho}|\xi/2|^{\tau}\sum_{j=0}^{\infty} 2^{-\tau j}} \\ &\leqslant \rho^n e^{\frac{C_0}{\rho}\frac{1}{1-2^{-\tau}}} = \rho^n C(0). \end{aligned}$$

For $2^m < |\xi| \leqslant 2^{m+1}$ with $m \in \mathbb{N}$, noting that $|2^{-m}\xi| \leqslant 2$ and $m < \log_2|\xi|$, we deduce

$$\begin{aligned} \left\| \prod_{j=1}^{n} \widehat{a}(2^{-j}\xi) \right\| &= \prod_{j=1}^{\min(m,n)} \|\widehat{a}(2^{-j}\xi)\| \prod_{j=1}^{\max(0,n-m)} \|\widehat{a}(2^{-j}2^{-m}\xi)\| \\ &\leqslant \rho^n C(0)(\rho^{-1}\| \|\hat{a}\| \|_{L_\infty(\mathbb{R})})^{\min(m,n)} \leqslant \rho^n C(0)(\rho^{-1}\| \|\hat{a}\| \|_{L_\infty(\mathbb{R})})^m \\ &\leqslant \rho^n C(0)(\rho^{-1}\| \|\hat{a}\| \|_{L_\infty(\mathbb{R})})^{\log_2|\xi|} = \rho^n C(0)|\xi|^{\log_2(\rho^{-1}\| \|\hat{a}\| \|_{L_\infty(\mathbb{R})})}. \end{aligned}$$

Therefore, we proved (5.1.2). □

If $\|\widehat{a}(0)\| < 1$ and $\tau > 0$, then it follows from (5.1.2) that $\lim_{n\to\infty} \prod_{j=1}^{n} \widehat{a}(2^{-j}\xi) = 0$ with the limit converging uniformly for ξ on every bounded interval.

We have the following result on the existence of tempered distributional solutions to a general refinement equation (not necessarily having a finitely supported matrix-valued filter/mask).

Theorem 5.1.2 *Let $a \in (l_2(\mathbb{Z}))^{r\times r}$ such that $\widehat{a}(0) \neq 0$. Assume that there exist nonnegative constants C_0, C_a, τ, τ_a and an $r \times 1$ vector u of compactly supported tempered distributions (Note that the Fourier transform $\widehat{u}$ is a continuous function, e.g., $u \in (l_0(\mathbb{Z}))^{r\times 1}$), satisfying* (5.1.1) *and*

$$\tau_a > \log_2(\|\widehat{a}(0)\| + C_0\boldsymbol{\delta}(\tau)) \quad \text{and} \quad \|\widehat{a}(\xi)\widehat{u}(\xi) - \widehat{u}(2\xi)\| \leqslant C_a|\xi|^{\tau_a}, \quad \forall\, \xi \in \mathbb{R}. \tag{5.1.4}$$

Define $\boldsymbol{\varphi}_0 := \widehat{u}$ and the vector functions $\boldsymbol{\varphi}_n, n \in \mathbb{N}$ by

$$\boldsymbol{\varphi}_n(\xi) := \Big(\prod_{j=1}^{n} \widehat{a}(2^{-j}\xi)\Big)\widehat{u}(2^{-n}\xi), \quad \xi \in \mathbb{R}. \tag{5.1.5}$$

Then the following claims hold:

(i) $\boldsymbol{\varphi}(\xi) := \lim_{n\to\infty} \boldsymbol{\varphi}_n(\xi)$ *is a well-defined vector function with the limit uniformly converging for ξ in every bounded interval such that*

$$\|\boldsymbol{\varphi}_n(\xi) - \boldsymbol{\varphi}_m(\xi)\| \leqslant \frac{(\tilde{\rho}2^{-\tau_a})^m}{2^{\tau_a} - \tilde{\rho}} C_a C(\xi)|\xi|^{\tau_a}, \quad \forall\, \xi \in \mathbb{R}, 0 \leqslant m < n \leqslant \infty, \tag{5.1.6}$$

where $\tilde{\rho} := \|\widehat{a}(0)\| + C_0\boldsymbol{\delta}(\tau)$ and $C(\xi)$ is defined in (5.1.3) *or $C(\xi) = 1$ if $\tau = 0$.*

(ii) All $\boldsymbol{\varphi}_n$ are tempered distributions and $\lim_{n\to\infty} \boldsymbol{\varphi}_n = \boldsymbol{\varphi}$ *in the sense of tempered distributions.*

(iii) $\boldsymbol{\varphi}$ is a tempered distribution satisfying $\boldsymbol{\varphi}(2\xi) = \widehat{a}(\xi)\boldsymbol{\varphi}(\xi)$ and $\boldsymbol{\varphi}(\xi) = \widehat{u}(\xi) + \mathcal{O}(|\xi|^{\tau_a})$ as $\xi \to 0$.

(iv) There exists a vector tempered distribution ϕ such that $\widehat{\phi} = \boldsymbol{\varphi}$, $\widehat{\phi}(\xi) = \widehat{u}(\xi) + \mathcal{O}(|\xi|^{\tau_a})$ as $\xi \to 0$ and ϕ satisfies the time-domain refinement equation:

$$\phi = 2\sum_{k\in\mathbb{Z}} a(k)\phi(2\cdot -k) \tag{5.1.7}$$

with the series converging in the sense of tempered distributions.

(v) There exist vectors of tempered distributions $\phi_n, n \in \mathbb{N}$ such that $\widehat{\phi_n} = \boldsymbol{\varphi}_n$ and $\lim_{n\to\infty} \phi_n = \phi$ *in the sense of tempered distributions.*

Moreover, if in addition $a \in (l_0(\mathbb{Z}))^{r\times r}$ is finitely supported, then the vector function ϕ in item (iv) is supported inside fsupp(a).

Proof Define $e(\xi) := \widehat{a}(\xi)\widehat{u}(\xi) - \widehat{u}(2\xi)$. Then $\widehat{a}(\xi)\widehat{u}(\xi) = \widehat{u}(2\xi) + e(\xi)$ and

$$\varphi_n(\xi) = \Big(\prod_{j=1}^{n-1}\widehat{a}(2^{-j}\xi)\Big)\widehat{a}(2^{-n}\xi)\widehat{u}(2^{-n}\xi) = \varphi_{n-1}(\xi) + \Big(\prod_{j=1}^{n-1}\widehat{a}(2^{-j}\xi)\Big)e(2^{-n}\xi).$$

Therefore, for $0 \leqslant m < n < \infty$, we have

$$\varphi_n(\xi) - \varphi_m(\xi) = \sum_{k=m+1}^{n}\Big(\prod_{j=1}^{k-1}\widehat{a}(2^{-j}\xi)\Big)e(2^{-k}\xi). \tag{5.1.8}$$

By Lemma 5.1.1, we deduce from (5.1.4) and (5.1.8) that $0 < \tilde{\rho}2^{-\tau_a} < 1$ and

$$\begin{aligned}\|\varphi_n(\xi) - \varphi_m(\xi)\| &\leqslant \sum_{k=m+1}^{n} \tilde{\rho}^{k-1}C(\xi)C_a|2^{-k}\xi|^{\tau_a}\\ &\leqslant 2^{-\tau_a}C_aC(\xi)|\xi|^{\tau_a}\sum_{k=m}^{\infty}(\tilde{\rho}2^{-\tau_a})^k = \frac{(\tilde{\rho}2^{-\tau_a})^m}{2^{\tau_a}-\tilde{\rho}}C_aC(\xi)|\xi|^{\tau_a}.\end{aligned}$$

This proves (5.1.6). Now it follows directly from (5.1.6) that item (i) holds.

Setting $m = 0$ in (5.1.6) and defining $\varphi_\infty = \varphi$, by Lemma A.6.2, we see that φ and $\varphi_n, n \in \mathbb{N}$ are tempered distributions. We now prove that $\lim_{n\to\infty}\varphi_n = \varphi$ in the sense of tempered distributions. That is, we have to prove that for every $h \in \mathscr{S}(\mathbb{R})$, $\lim_{n\to\infty}\langle\varphi_n, h\rangle = \langle\varphi, h\rangle$. Since $h \in \mathscr{S}(\mathbb{R})$, h decays faster than any polynomial and hence it is trivial to see that $\int_{\mathbb{R}} C(\xi)|\xi|^{\tau_a}|h(\xi)|d\xi < \infty$. By item (i), we have $\lim_{n\to\infty}\varphi_n(\xi) = \varphi(\xi)$ for every $\xi \in \mathbb{R}$. Since $|\varphi_n(\xi)h(\xi)| \leqslant (\frac{C_a}{2^{\tau_a}-\tilde{\rho}}C(\xi)|\xi|^{\tau_a} + \|\widehat{u}(\xi)\|)|h(\xi)|$ for all $\xi \in \mathbb{R}$ and $n \in \mathbb{N}$, by the Dominated Convergence Theorem, we conclude that

$$\begin{aligned}\lim_{n\to\infty}\langle\varphi_n, h\rangle &= \lim_{n\to\infty}\int_{\mathbb{R}}\varphi_n(\xi)\overline{h(\xi)}d\xi = \int_{\mathbb{R}}\lim_{n\to\infty}\varphi_n(\xi)\overline{h(\xi)}d\xi\\ &= \int_{\mathbb{R}}\varphi(\xi)\overline{h(\xi)}d\xi = \langle\varphi, h\rangle.\end{aligned}$$

This proves that $\lim_{n\to\infty}\varphi_n = \varphi$ in the sense of tempered distributions. Hence, item (ii) holds.

By the definition of φ_n in (5.1.5), we have $\varphi_n(2\xi) = \widehat{a}(\xi)\varphi_{n-1}(\xi)$. By item (i), we see that $\varphi(2\xi) = \lim_{n\to\infty}\varphi_n(2\xi) = \widehat{a}(\xi)\lim_{n\to\infty}\varphi_{n-1}(\xi) = \widehat{a}(\xi)\varphi(\xi)$. Setting $m = 0$ and $n = \infty$ in (5.1.6), since $\varphi_0 = \widehat{u}$, we conclude that $\|\widehat{u}(\xi) - \varphi(\xi)\| = \mathscr{O}(|\xi|^{\tau_a})$ as $\xi \to 0$. This proves item (iii).

Items (iv) and (v) follows directly from items (iii) and (ii), respectively.

Suppose that $a \in (l_0(\mathbb{Z}))^{r\times r}$ is finitely supported. By the definition of ϕ_n, we have

$$\phi_n = 2\sum_{k\in\mathbb{Z}} a(k)\phi_{n-1}(2\cdot -k), \qquad n \in \mathbb{N}.$$

By induction on n, we can directly deduce from the above identity that $\operatorname{supp}(\phi_n) \subseteq 2^{-1}[\operatorname{supp}(\phi_{n-1}) + \operatorname{fsupp}(a)]$. Note that $\operatorname{supp}(\phi_0) = \operatorname{supp}(u)$. Now we see that ϕ_n is compactly supported and

$$\operatorname{supp}(\phi_n) \subseteq 2^{-n}\operatorname{supp}(u)+2^{-1}\operatorname{fsupp}(a)+\cdots+2^{-n}\operatorname{fsupp}(a) \subseteq 2^{-n}\operatorname{supp}(u)+\operatorname{fsupp}(a).$$

Since ϕ_n converges to ϕ in the sense of tempered distributions, we conclude that $\operatorname{supp}(\phi) \subseteq \operatorname{fsupp}(a)$. □

Lemma 4.1.8 in Chap. 4 is a direct consequence of Theorem 5.1.2, since both (5.1.1) and (5.1.4) are obviously satisfied with $\widehat{u} = 1$, $C_a = C_0$ and $\tau_a = \tau$.

We now study all the compactly supported distributional solutions to a vector refinement equation with a finitely supported matrix-valued filter/mask.

Theorem 5.1.3 *For $a \in (l_0(\mathbb{Z}))^{r\times r}$, define $DS(a)$ to be the set of all compactly supported vector distributional solutions to* (5.1.7) *and define*

$$E(a) = \{\{\widehat{u}^{(j)}(0)\}_{j=0}^{J_a-1} \; : \; \widehat{a}(\xi)\widehat{u}(\xi) = \widehat{u}(2\xi) + \mathscr{O}(|\xi|^{J_a}), \quad \xi \to 0, u \in (l_0(\mathbb{Z}))^{r\times 1}\},$$

where J_a is the smallest positive integer such that $J_a > \log_2 \|\widehat{a}(0)\|$. Then

(1) the mapping $DS(a) \to E(a)$ with $\phi \mapsto \{\widehat{\phi}^{(j)}(0)\}_{j=0}^{J_a-1}$ is a well-defined bijective linear mapping between the two linear spaces.

(2) $\dim(E(a)) \leqslant M_a := \sum_{j=0}^{J_a-1} m_j$, where m_j is the geometric multiplicity of the eigenvalue 2^j of $\widehat{a}(0)$, that is, $m_j := \dim(\{\vec{v} \in \mathbb{C}^r \; : \; \widehat{a}(0)\vec{v} = 2^j\vec{v}\})$.

(3) $\dim(E(a)) \geqslant 1 \iff M_a \geqslant 1$ (i.e., 2^j is an eigenvalue of $\widehat{a}(0)$ for some $j \in \mathbb{N}_0$).

Proof Let $\phi \in \mathrm{DS}(a)\backslash\{0\}$. Since $\widehat{\phi}$ is analytic and $\widehat{\phi}(2\xi) = \widehat{a}(\xi)\widehat{\phi}(\xi)$, we see that the mapping in item (1) is well defined. Since $\widehat{\phi}$ is nontrivial, there exists a nonnegative integer j such that $\widehat{\phi}^{(j)}(0) \neq 0$ and $\widehat{\phi}(\xi) = \widehat{\phi}^{(j)}(0)\frac{\xi^j}{j!} + \mathscr{O}(|\xi|^{j+1})$ as $\xi \to 0$. By $\widehat{\phi}(2\xi) = \widehat{a}(\xi)\widehat{\phi}(\xi)$, we have

$$2^j\widehat{\phi}^{(j)}(0)\frac{\xi^j}{j!} = \widehat{a}(\xi)\widehat{\phi}^{(j)}(0)\frac{\xi^j}{j!} + \mathscr{O}(|\xi|^{j+1}) = \widehat{a}(0)\widehat{\phi}^{(j)}(0)\frac{\xi^j}{j!} + \mathscr{O}(|\xi|^{j+1}), \qquad \xi \to 0.$$

That is, for every $\phi \in \mathrm{DS}(a)\backslash\{0\}$, we must have

$$\widehat{a}(0)\widehat{\phi}^{(j)}(0) = 2^j\widehat{\phi}^{(j)}(0) \quad \text{and} \quad \widehat{\phi}^{(j)}(0) \neq 0 \qquad \text{for some } j \in \mathbb{N}_0. \qquad (5.1.9)$$

Note that (5.1.9) implies $\|\widehat{a}(0)\| \geqslant 2^j$. Hence, we have $j < J_a$ by$J_a > \log_2 \|\widehat{a}(0)\|$. This shows that if $\phi \in \mathrm{DS}(a)\backslash\{0\}$, then $\{\widehat{\phi}^{(j)}(0)\}_{j=0}^{J_a-1} \neq 0$ in $E(a)$. Thus, the mapping in item (1) must be one-to-one.

We now prove that the mapping in item (1) is onto. Since $a \in (l_0(\mathbb{Z}))^{r\times r}$, there exists a positive constant C_0 such that $\|\widehat{a}(\xi) - \widehat{a}(0)\| \leqslant C_0|\xi|$ for all $\xi \in [-\pi, \pi]$. Hence, (5.1.1) is satisfied with $\tau = 1$. For $\{\widehat{u}^{(j)}(0)\}_{j=0}^{J_a-1} \in E(a)$, since $u \in (l_0(\mathbb{Z}))^{r\times 1}$, by the definition of $E(a)$ and $J_a > \log_2 \|\widehat{a}(0)\|$, there exists a positive constant C_a such that (5.1.4) is satisfied with $\tau_a = J_a$ and $\tau = 1$. Therefore, by item (iv) of Theorem 5.1.2, there exists a compactly supported distributional solution ϕ to (5.1.7) and $\widehat{\phi}(\xi) = \widehat{u}(\xi) + \mathscr{O}(|\xi|^{J_a})$ as $\xi \to 0$. Therefore, the mapping in item (1) is onto. This proves item (1).

We prove item (2). For $j \in \mathbb{N}_0$, define $E_j := \{\vec{v} \in \mathbb{C}^r : \widehat{a}(0)\vec{v} = 2^j\vec{v}\}$ and

$$E_j(a) := \{\{\widehat{u}^{(k)}(0)\}_{k=0}^{J_a-1} \in E(a) \ : \ \widehat{u}^{(\ell)}(0) = 0, \forall\, \ell = 0, \dots, j-1\}.$$

For $\{\widehat{u}^{(k)}(0)\}_{k=0}^{J_a-1} \in E_j(a)/E_{j+1}(a)$, since $\widehat{a}(\xi)\widehat{u}(\xi) = \widehat{u}(2\xi) + \mathscr{O}(|\xi|^{J_a})$ as $\xi \to 0$, by a similar argument for (5.1.9), we must have $\widehat{a}(0)\widehat{u}^{(j)}(0) = 2^j\widehat{u}^{(j)}(0)$. Therefore, $\widehat{u}^{(j)}(0) \in E_j$. Hence, the mapping $E_j(a)/E_{j+1}(a) \to E_j$ with $\{\widehat{u}^{(k)}(0)\}_{k=0}^{J_a-1} \mapsto \widehat{u}^{(j)}(0)$ is well defined and one-to-one. Consequently, by $E_{J_a}(a) = \{0\}$, we have

$$\dim(E(a)) = \dim(E_0(a)) = \sum_{j=0}^{J_a-1} \dim(E_j(a)/E_{j+1}(a)) \leqslant \sum_{j=0}^{J_a-1} \dim(E_j) = \sum_{j=0}^{J_a-1} m_j = M_a.$$

This proves item (2).

If $\dim(E(a)) \geqslant 1$, by item (2), we must have $M_a \geqslant \dim(E(a)) \geqslant 1$. Conversely, assume that $M_a \geqslant 1$. Then there exists a largest possible nonnegative integer j such that $j < J_a$ and 2^j is an eigenvalue of $\widehat{a}(0)$, that is $\widehat{a}(0)\vec{v} = 2^j\vec{v}$ for some $\vec{v} \in \mathbb{C}^r\backslash\{0\}$. Define $v_k := 0$ for all $k = 0, \dots, j-1$, $v_j := \vec{v}$, and

$$v_k := (2^k I_r - \widehat{a}(0))^{-1} \sum_{\ell=0}^{k-1} \frac{k!}{\ell!(k-\ell)!} \widehat{a}^{(k-\ell)}(0) v_\ell, \qquad k = j+1, \dots, J_a - 1,$$

where we used the fact that $\det(2^k I_r - \widehat{a}(0)) \neq 0$ for all $k > j$ by our choice of the integer j and $J_a > \log_2 \|\widehat{a}(0)\|$. Then there exists $u \in (l_0(\mathbb{Z}))^r$ such that $\widehat{u}^{(k)}(0) = v_k$ for all $k = 0, \dots, J_a - 1$. That is, $\widehat{a}(\xi)\widehat{u}(\xi) = \widehat{u}(2\xi) + \mathscr{O}(|\xi|^{J_a})$ as $\xi \to 0$. Hence, $\{\widehat{u}^{(k)}(0)\}_{k=0}^{J_a-1} \in E(a)\backslash\{0\}$ by $\widehat{u}^{(j)}(0) = \vec{v} \neq 0$. Therefore, $\dim(E(a)) \geqslant 1$. This proves item (3). □

As shown by Exercise 5.9, the inequality $\dim(E(a)) < M_a$ in item (2) of Theorem 5.1.3 can happen. For a filter $a \in (l_0(\mathbb{Z}))^{r\times r}$ satisfying

$$1 \text{ is a simple eigenvalue of } \widehat{a}(0) \text{ and } \det(2^j I_r - \widehat{a}(0)) \neq 0 \text{ for all } j \in \mathbb{N}, \tag{5.1.10}$$

by Theorem 5.1.3, we have $\dim(\mathrm{DS}(a)) = \dim(E(a)) = 1$, and up to a multiplicative constant, there exists a unique compactly supported vector distribution ϕ with $\widehat{\phi}(0) \neq 0$ satisfying the refinement equation $\phi = 2\sum_{k\in\mathbb{Z}} a(k)\phi(2\cdot -k)$. See Sect. 6.2.3 about how to plot a refinable vector function ϕ. Moreover, up to a multiplicative constant, the values $\widehat{\phi}^{(j)}(0)$ for $j \in \mathbb{N}_0$ are uniquely determined by

$$\widehat{a}(0)\widehat{\phi}(0) = \widehat{\phi}(0),$$
$$\widehat{\phi}^{(j)}(0) = [2^j I_r - \widehat{a}(0)]^{-1} \sum_{k=0}^{j-1} \frac{j!}{k!(j-k)!} \widehat{a}^{(j-k)}(0)\widehat{\phi}^{(k)}(0), \quad j \in \mathbb{N}. \tag{5.1.11}$$

The natural condition in (5.1.10) is often necessary in the classical theory of multiwavelets (i.e., $r > 1$). This natural condition in (5.1.10) simplifies the treatment of the classical theory of multiwavelets. However, as we shall see in this book, the condition (5.1.10) may fail in the setting of multiframelets and this adds a layer of complexity in the study of multiframelets.

5.2 Linear Independence of Integer Shifts of Compactly Supported Functions

By the definition of an affine system, the set $\{\phi(\cdot - k) : k \in \mathbb{Z}, \phi \in \Phi\}$ is part of an affine system $\mathsf{AS}_0(\Phi; \Psi)$. If $\mathsf{AS}_0(\Phi; \Psi)$ is a Riesz basis for $L_2(\mathbb{R})$, then by item (6) of Theorem 4.2.7, the elements in $\mathsf{AS}_0(\Phi; \Psi)$ must be ℓ_2-linearly independent; in particular, the integer shifts $\phi(\cdot - k)$, $k \in \mathbb{Z}$ and $\phi \in \Phi$ must be linearly independent. In this section, we introduce and study a stronger notion of linear independence for integer shifts of a finite set of compactly supported distributions.

Recall that $l(\mathbb{Z})$ denotes the space of all sequences $v = \{v(k)\}_{k\in\mathbb{Z}} : \mathbb{Z} \to \mathbb{C}$. In particular, by $\boldsymbol{\delta}$ we denote the Kronecker/Dirac sequence on $\mathbb{Z}$ such that $\boldsymbol{\delta}(0) = 1$ and $\boldsymbol{\delta}(k) = 0$ for all $k \in \mathbb{Z}\backslash\{0\}$. For $v = \{v(k)\}_{k\in\mathbb{Z}} \in l(\mathbb{Z})$ and a compactly supported distribution ϕ on $\mathbb{R}$, we define

$$v * \phi := \sum_{k\in\mathbb{Z}} v(k)\phi(\cdot - k) \tag{5.2.1}$$

with the series converging in the sense of distributions. Since ϕ has compact support, the convolution $v * \phi$ is a well-defined distribution on $\mathbb{R}$. Let $\phi_1, \dots, \phi_r$ be compactly supported distributions on $\mathbb{R}$. We say that the integer shifts of $\phi_1, \dots, \phi_r$ (or $(\phi_1, \dots, \phi_r)^\mathsf{T}$) are *(globally) linearly independent* if

$$\sum_{\ell=1}^{r} \sum_{k\in\mathbb{Z}} v_\ell(k)\phi_\ell(\cdot - k) = 0 \tag{5.2.2}$$

for some sequences $v_1, \ldots, v_r \in l(\mathbb{Z})$, then we must have $v_1(k) = \cdots = v_r(k) = 0$ for all $k \in \mathbb{Z}$. Define

$$\mathsf{S}(\phi_1, \ldots, \phi_r) := \{v_1 * \phi_1 + \cdots + v_r * \phi_r \;:\; v_1, \ldots, v_r \in l(\mathbb{Z})\}. \tag{5.2.3}$$

Then the integer shifts of $\phi_1, \ldots, \phi_r$ are linearly independent if and only if the mapping $(l(\mathbb{Z}))^{1\times r} \to \mathsf{S}(\phi_1, \ldots, \phi_r)$ with $(v_1, \ldots, v_r) \mapsto v_1 * \phi_1 + \cdots + v_r * \phi_r$ is a bijection.

5.2.1 *Characterization of Linear Independence*

The following result characterizes linear independence of integer shifts of compactly supported distributions.

Theorem 5.2.1 *Let $\phi_1, \ldots, \phi_r$ be compactly supported distributions on $\mathbb{R}$. The following statements are equivalent:*

(i) The integer shifts of $\phi_1, \ldots, \phi_r$ are linearly independent.
(ii) $\{\widehat{\phi_\ell}(z + 2\pi k)\}_{k\in\mathbb{Z}}, \ell = 1, \ldots, r$ are linearly independent for all $z \in \mathbb{C}$, that is, there do not exist $\zeta \in \mathbb{C}$ and $c_1, \ldots, c_r \in \mathbb{C}$ such that $|c_1| + \cdots + |c_r| \neq 0$ and

$$\sum_{\ell=1}^{r} c_\ell \widehat{\phi_\ell}(\zeta + 2\pi k) = 0, \qquad \forall\, k \in \mathbb{Z}. \tag{5.2.4}$$

(iii) There exist compactly supported $\mathscr{C}^\infty(\mathbb{R})$ functions $\tilde{\phi}_1, \ldots, \tilde{\phi}_r \in \mathscr{D}(\mathbb{R})$ such that

$$\langle \tilde{\phi}_m, \phi_\ell(\cdot - k)\rangle = \boldsymbol{\delta}(m - \ell)\boldsymbol{\delta}(k), \qquad \forall\, k \in \mathbb{Z}, \ell, m = 1, \ldots, r. \tag{5.2.5}$$

Proof Recall that the Poisson summation formula (see Theorem A.6.5):

$$\sum_{k\in\mathbb{Z}} f(x - k) = \sum_{k\in\mathbb{Z}} \widehat{f}(2\pi k) e^{i2\pi kx} \tag{5.2.6}$$

holds for any compactly supported distribution f on $\mathbb{R}$. For $\zeta \in \mathbb{C}$ and $f(x) = \eta(x)e^{-i\zeta x}$, since $\widehat{f}(z) = \widehat{\eta}(\zeta + z)$, the above Poisson summation formula can be written as

$$\sum_{k\in\mathbb{Z}} \eta(x - k)e^{-i\zeta(x-k)} = \sum_{k\in\mathbb{Z}} \widehat{\eta}(\zeta + 2\pi k)e^{i2\pi kx} \tag{5.2.7}$$

for every compactly supported distribution η on $\mathbb{R}$ and $\zeta \in \mathbb{C}$ with the above series converging in the sense of distributions. We use proof by contradiction to prove (i)$\Longrightarrow$(ii). Suppose that (ii) fails. Then there exist $\zeta \in \mathbb{C}$ and $c_1, \dots, c_r \in \mathbb{C}$ such that $|c_1| + \cdots + |c_r| \neq 0$ and (5.2.4) holds. Define $\eta := \sum_{\ell=1}^r c_\ell \phi_\ell$. Then η is a compactly supported distribution on $\mathbb{R}$. It follows from (5.2.4) that $\widehat{\eta}(\zeta + 2\pi k) = 0$ for all $k \in \mathbb{Z}$. Now by the Poisson summation formula in (5.2.7), we have

$$e^{-i\zeta x} \sum_{k\in\mathbb{Z}} \sum_{\ell=1}^{r} c_\ell e^{i\zeta k} \phi_\ell(x-k) = \sum_{k\in\mathbb{Z}} \eta(x-k) e^{-i\zeta(x-k)} = \sum_{k\in\mathbb{Z}} \widehat{\eta}(\zeta + 2\pi k) e^{i2\pi k x} = 0.$$

Since $e^{-i\zeta x} \neq 0$, defining $v_\ell(k) := c_\ell e^{i\zeta k}$ for $k \in \mathbb{Z}$ and $\ell = 1, \dots, r$, we see that (5.2.2) holds. This is a contradiction to item (i), since $v_1, \dots, v_r$ are not all identically zero. Thus, we proved (i)$\Longrightarrow$(ii).

If both (5.2.2) and item (iii) hold, then

$$v_m(n) = \Big\langle \sum_{\ell=1}^{r} \sum_{k\in\mathbb{Z}} v_\ell(k) \phi_\ell(\cdot - k), \tilde{\phi}_m(\cdot - n) \Big\rangle = 0, \qquad \forall\, n \in \mathbb{Z}, m = 1, \dots, r.$$

Hence, all $v_1, \dots, v_r$ must be identically zero. Therefore, we proved (iii)$\Longrightarrow$(i).

We now prove (ii)$\Longrightarrow$(iii) using induction on $r \in \mathbb{N}_0$. The claim is obviously true for $r = 0$, since the statements are empty. Suppose that (ii)$\Longrightarrow$(iii) holds for $r-1$. We now prove the claim for $r \geqslant 1$. Define $L : \mathscr{D}(\mathbb{R}) \to \mathbb{C}[z, z^{-1}]$ by

$$L(h)(z) := \sum_{k\in\mathbb{Z}} \langle \phi_r(\cdot - k), h \rangle z^k, \qquad z \in \mathbb{C}\backslash\{0\}, h \in \mathscr{D}(\mathbb{R}).$$

Since both h and ϕ_r are compactly supported, then $L(h)$ is a well-defined Laurent polynomial in the Laurent polynomial ring $\mathbb{C}[z, z^{-1}]$. Define

$$K := \{h \in \mathscr{D}(\mathbb{R}) \ :\ \langle \phi_\ell(\cdot - k), h \rangle = 0, \quad \forall\, k \in \mathbb{Z}, \ell = 1, \dots, r-1\}.$$

We now prove that $L(K)$ is an ideal in $\mathbb{C}[z, z^{-1}]$. Clearly, K is shift-invariant and $L(K)$ is a linear subspace of $\mathbb{C}[z, z^{-1}]$. Let $h \in K$ and $\mathsf{p}(z) = \sum_{n\in\mathbb{Z}} p_n z^n \in \mathbb{C}[z, z^{-1}]$ be a Laurent polynomial. Define $g := \sum_{n\in\mathbb{Z}} \overline{p_n} h(\cdot - n)$. Since $\{p_n\}_{n\in\mathbb{Z}} \in l_0(\mathbb{Z})$ and $h \in K$, we see that $g \in K$. Note that

$$\begin{aligned} L(g)(z) &= \sum_{k\in\mathbb{Z}} \langle \phi_r(\cdot - k), g \rangle z^k = \sum_{n\in\mathbb{Z}} \sum_{k\in\mathbb{Z}} p_n \langle \phi_r(\cdot - k), h(\cdot - n) \rangle z^k \\ &= \sum_{n\in\mathbb{Z}} \sum_{k\in\mathbb{Z}} p_n \langle \phi_r(\cdot - k + n), h \rangle z^{k-n} z^n = \sum_{n\in\mathbb{Z}} p_n z^n L(h)(z) = \mathsf{p}(z) L(h)(z). \end{aligned}$$

This shows $\mathsf{p}L(h) = L(g) \in L(K)$. Hence, $L(K)$ is an ideal in $\mathbb{C}[z, z^{-1}]$. By long division, there exists a Laurent polynomial $\mathsf{q} \in L(K)$ such that $\mathsf{q}\mathbb{C}[z, z^{-1}] = L(K)$ (in fact, we can choose a Laurent polynomial $\mathsf{q} \in L(K)$ such that q is not identically zero and $\mathrm{len}(\mathsf{q})$ is the smallest, i.e., the coefficient sequence of q has the shortest possible length).

We now prove that q must be a nonzero monomial. Otherwise, there exists $\zeta \in \mathbb{C}$ such that $\mathsf{q}(e^{i\zeta}) = 0$. Since $L(K) = \mathsf{q}\mathbb{C}[z, z^{-1}]$, we conclude that $L(h)(e^{i\zeta}) = 0$ for every $h \in K$. By induction hypothesis, there exist $\tilde{h}_1, \dots, \tilde{h}_{r-1} \in \mathscr{D}(\mathbb{R})$ such that

$$\langle \tilde{h}_m, \phi_\ell(\cdot - k)\rangle = \delta(m-\ell)\delta(k), \qquad \forall\, k \in \mathbb{Z} \quad \text{and} \quad \ell, m = 1, \dots, r-1. \tag{5.2.8}$$

For $h \in \mathscr{D}(\mathbb{R})$, we define a projection operator

$$Ph := h - \sum_{\ell=1}^{r-1}\sum_{n\in\mathbb{Z}} \langle h, \phi_\ell(\cdot - n)\rangle \tilde{h}_\ell(\cdot - n).$$

By (5.2.8), it is trivial to directly check that $Ph \in K$ and

$$\begin{aligned}
L(Ph)(z) &= \sum_{k\in\mathbb{Z}} \langle \phi_r(\cdot - k), h\rangle z^k - \sum_{\ell=1}^{r-1}\sum_{n\in\mathbb{Z}}\sum_{k\in\mathbb{Z}} \overline{\langle h, \phi_\ell(\cdot - n)\rangle}\langle \phi_r(\cdot - k), \tilde{h}_\ell(\cdot - n)\rangle z^k\\
&= \sum_{k\in\mathbb{Z}} \langle \phi_r(\cdot - k), h\rangle z^k - \sum_{\ell=1}^{r-1}\sum_{k\in\mathbb{Z}}\sum_{n\in\mathbb{Z}} \langle \phi_\ell(\cdot - k), h\rangle\langle \phi_r(\cdot - n), \tilde{h}_\ell(\cdot - k)\rangle z^n\\
&= \sum_{k\in\mathbb{Z}} \langle \phi_r(\cdot - k), h\rangle z^k - \sum_{\ell=1}^{r-1}\sum_{k\in\mathbb{Z}} \langle \phi_\ell(\cdot - k), h\rangle z^k \Big(\sum_{n\in\mathbb{Z}} \langle \phi_r(\cdot - n), \tilde{h}_\ell(\cdot - k)\rangle z^{n-k}\Big).
\end{aligned}$$

Setting $z = e^{i\zeta}$ in the above identity and defining $c_r := 1$ and $c_\ell := -\sum_{n\in\mathbb{Z}} \langle \phi_r(\cdot-n), \tilde{h}_\ell\rangle e^{i\zeta n} = -L(\tilde{h}_\ell)(e^{i\zeta}) \in \mathbb{C}$ for $\ell = 1, \dots, r-1$, we conclude that

$$\Big\langle \sum_{\ell=1}^{r}\sum_{k\in\mathbb{Z}} c_\ell e^{i\zeta k}\phi_\ell(\cdot - k), h\Big\rangle = L(Ph)(e^{i\zeta}) = 0 \qquad \forall\, h \in \mathscr{D}(\mathbb{R}).$$

That is, setting $\eta := \sum_{\ell=1}^{r} c_\ell \phi_\ell$, we proved

$$\sum_{k\in\mathbb{Z}} \eta(x - k)e^{-i\zeta(x-k)} = e^{-i\zeta x}\sum_{\ell=1}^{r}\sum_{k\in\mathbb{Z}} c_\ell e^{i\zeta k}\phi_\ell(\cdot - k) = 0.$$

By the Poisson summation formula in (5.2.7), we deduce from the above identity that $\sum_{\ell=1}^{r} c_\ell \widehat{\phi_\ell}(\zeta + 2\pi k) = \widehat{\eta}(\zeta + 2\pi k) = 0$ for all $k \in \mathbb{Z}$ (see Exercises 5.14–5.18). Since $c_r = 1 \neq 0$, this is a contradiction to item (ii). Therefore, q must be a nonzero monomial and consequently $L(K) = \mathrm{q}\mathbb{C}[z, z^{-1}] = \mathbb{C}[z, z^{-1}]$. Hence, there exists $\tilde{\phi}_r \in K$ such that $L(\tilde{\phi}_r) = 1$. Since $\tilde{\phi}_r \in K$ and $1 = L(\tilde{\phi}_r) = \sum_{n\in\mathbb{Z}} \langle \phi_r(\cdot - n), \tilde{\phi}_r\rangle z^n$, this implies that $\langle \tilde{\phi}_r, \phi_\ell(\cdot - k)\rangle = \boldsymbol{\delta}(\ell - r)\boldsymbol{\delta}(k)$ for all $k \in \mathbb{Z}$ and $\ell = 1, \dots, r$. Now define

$$\tilde{\phi}_\ell := \tilde{h}_\ell - \sum_{n\in\mathbb{Z}} \langle \tilde{h}_\ell, \phi_r(\cdot - n)\rangle \tilde{\phi}_r(\cdot - n), \qquad \ell = 1, \dots, r-1.$$

It is trivial to deduce from (5.2.8) and the above identity that (5.2.5) is satisfied. Since $\tilde{\phi}_1, \dots, \tilde{\phi}_r \in \mathscr{D}(\mathbb{R})$, the claim holds for r. Now by induction, we complete the proof of (ii)$\Longrightarrow$(iii). □

Let $\phi = (\phi_1, \dots, \phi_r)^{\mathsf{T}}$. Then item (ii) of Theorem 5.2.1 simply means $\mathrm{span}\{\widehat{\phi}(z + 2\pi k) \;:\; k \in \mathbb{Z}\} = \mathbb{C}^r$ for all $z \in \mathbb{C}$.

5.2.2 *Linearly Independent Generators of Shift-Invariant Spaces*

For any compactly supported distributions $\phi_1, \dots, \phi_r$ on the real line $\mathbb{R}$, in this section we show that the shift-invariant space $\mathrm{S}(\phi_1, \dots, \phi_r)$ always has a finite set of compactly supported generators whose integer shifts are linearly independent.

For a compactly supported distribution ϕ on $\mathbb{R}$, we define $\mathrm{fsupp}(\phi) := [m, n]$ and $\mathrm{len}(\phi) := n - m$, where $m, n \in \mathbb{Z}$ such that ϕ vanishes outside $[m, n]$ with m being the largest integer and n being the smallest integer. If ϕ is identically zero, we simply define $\mathrm{fsupp}(0) := \emptyset$ and $\mathrm{len}(0) := -\infty$.

Lemma 5.2.2 *Let ϕ and η be compactly supported distributions on $\mathbb{R}$. If $\phi = u * \eta$ for some $u \in l_0(\mathbb{Z})\backslash\{0\}$, then $\mathrm{S}(\phi) = \mathrm{S}(\eta)$.*

Proof For $v \in l(\mathbb{Z})$, since $u \in l_0(\mathbb{Z})$, we see that $v * u$ is a well-defined sequence in $l(\mathbb{Z})$. Since $v * \phi = v * (u * \eta) = (v * u) * \eta \in \mathrm{S}(\eta)$, we conclude $\mathrm{S}(\phi) \subseteq \mathrm{S}(\eta)$.

Since $\phi = u * \eta$ with $u \in l_0(\mathbb{Z})\backslash\{0\}$, it is not difficult to see that $\eta = \sum_{k=M}^{\infty} u_1(k)\phi(\cdot - k) = \sum_{k=-\infty}^{N} u_2(k)\phi(\cdot - k)$ for some $M, N \in \mathbb{Z}$ and $u_1, u_2 \in l(\mathbb{Z})$ (see Exercise 5.20). For $v \in l(\mathbb{Z})$, we have

$$\begin{aligned}\sum_{n=0}^{\infty} v(n)\eta(\cdot - n) &= \sum_{n=0}^{\infty} v(n) \sum_{k=M+n}^{\infty} u_1(k-n)\phi(\cdot - k)\\ &= \sum_{k=M}^{\infty} \phi(\cdot - k)\left(\sum_{n=0}^{k-M} v(n)u_1(k-n)\right) \in \mathrm{S}(\phi)\end{aligned}$$

and

$$\sum_{n=-\infty}^{-1} v(n)\eta(\cdot - n) = \sum_{n=-\infty}^{-1} v(n) \sum_{k=-\infty}^{N+n} u_2(k-n)\phi(\cdot - k)$$
$$= \sum_{k=-\infty}^{N-1} \phi(\cdot - k) \left(\sum_{n=k-N}^{-1} v(n)u_2(k-n) \right) \in \mathrm{S}(\phi).$$

Thus, we proved $v * \eta \in \mathrm{S}(\phi)$. Therefore, we conclude that $\mathrm{S}(\eta) \subseteq \mathrm{S}(\phi)$. □

Theorem 5.2.3 *Let ϕ be a compactly supported distribution on $\mathbb{R}$ such that ϕ is not identically zero. Then there exists a compactly supported distribution η on $\mathbb{R}$ such that $\mathrm{S}(\phi) = \mathrm{S}(\eta)$, $\phi = u * \eta$ for some $u \in l_0(\mathbb{Z})$, and the integer shifts of η are linearly independent (In fact, all such η are determined by the property $\mathrm{S}(\eta) = \mathrm{S}(\phi)$ with the smallest* $\mathrm{len}(\eta)$*). If in addition $\phi \in L_p(\mathbb{R})$ for some $1 \leqslant p \leqslant \infty$ (or $\phi \in \mathscr{C}(\mathbb{R})$), then $\eta \in L_p(\mathbb{R})$(or $\eta \in \mathscr{C}(\mathbb{R})$).*

Proof If the integer shifts of ϕ are linearly independent, we are done by taking $\eta = \phi$. Otherwise, by Theorem 5.2.1, there exists $\zeta_1 \in \mathbb{C}$ such that $\widehat{\phi}(\zeta_1 + 2\pi k) = 0$ for all $k \in \mathbb{Z}$. By the Poisson summation formula in (5.2.7), we have $\sum_{k\in\mathbb{Z}} e^{i\zeta_1 k}\phi(\cdot - k) = 0$. Define $\eta := \sum_{k=0}^{\infty} e^{i\zeta_1 k}\phi(\cdot - k)$, from which we deduce that $\phi = \eta - e^{i\zeta_1}\eta(\cdot - 1)$ (i.e., $\widehat{\phi}(\xi) = \widehat{\eta}(\xi)(1 - e^{-i(\xi-\zeta_1)})$). Since $\mathrm{fsupp}(\phi) = [m, n]$ for some $m, n \in \mathbb{Z}$, we must have $\mathrm{fsupp}(\eta) \subseteq [m, \infty)$. On the other hand, since $\eta = -\sum_{k=-\infty}^{-1} e^{i\zeta_1 k}\phi(\cdot - k)$, we see that $\mathrm{fsupp}(\eta) \subseteq (-\infty, n-1]$. Hence, we must have $\mathrm{fsupp}(\eta) \subseteq [m, n-1]$ and $\mathrm{len}(\eta) \leqslant n - m - 1 \leqslant \mathrm{len}(\phi) - 1$. We can continue this procedure if the integer shifts of η are not linearly independent. Since $\mathrm{len}(\phi) < \infty$, there exist $\zeta_1, \ldots, \zeta_n \in \mathbb{C}$ and a compactly supported distribution η such that

$$\widehat{\phi}(\xi) = \widehat{\eta}(\xi) \prod_{j=1}^{n} (1 - e^{-i(\xi - \zeta_j)}), \tag{5.2.9}$$

$\mathrm{len}(\eta) \leqslant \mathrm{len}(\phi) - n$ and the integer shifts of η are linearly independent. Note that (5.2.9) is equivalent to $\phi = u * \eta$, where $u \in l_0(\mathbb{Z})$ is defined by $\widehat{u}(\xi) := \prod_{j=1}^{n}(1 - e^{-i(\xi-\zeta_j)})$. Since ϕ is not identically zero, $u \neq 0$ and by Lemma 5.2.2, $\mathrm{S}(\phi) = \mathrm{S}(\eta)$.

By $\phi = u * \eta$ with $u \in l_0(\mathbb{Z})\backslash\{0\}$, we have $\eta = \sum_{k=M}^{\infty} v(k)\phi(\cdot - k)$ for some $M \in \mathbb{Z}$ and $v \in l(\mathbb{Z})$ (see Exercise 5.20). If $\phi \in L_p(\mathbb{R})$, since $\eta = \sum_{k=M}^{\infty} v(k)\phi(\cdot - k)$ is compactly supported, we get $\eta \in L_p(\mathbb{R})$. □

We now show that every shift-invariant space $\mathrm{S}(\phi_1, \ldots, \phi_r)$ with compactly supported distributions $\phi_1, \ldots, \phi_r$ must have a set of linearly independent generators.

Theorem 5.2.4 *Let $\phi_1, \ldots, \phi_r$ be compactly supported distributions on $\mathbb{R}$ such that not all of them are identically zero. Suppose that $\{\eta_1, \ldots, \eta_t\}$ be a finite set (including the empty set with $t = 0$) of compactly supported distributions such that the integer shifts of $\eta_1, \ldots, \eta_t$ are linearly independent and $\mathrm{S}(\eta_1, \ldots, \eta_t) \subseteq$*

$\mathsf{S}(\phi_1, \ldots \phi_r)$. *Then there exist compactly supported distributions* $\eta_{t+1}, \ldots, \eta_s$ *such that*

(i) $\mathsf{S}(\phi_1, \ldots, \phi_r) = \mathsf{S}(\eta_1, \ldots, \eta_s)$ *and the integer shifts of* $\eta_1, \ldots, \eta_s$ *are linearly independent;*

(ii) *every* ϕ_ℓ, $\ell = 1, \ldots, r$, *is a finite linear combination of* $\eta_1(\cdot - k), \ldots, \eta_s(\cdot - k), k \in \mathbb{Z}$, *that is, there exists* $u \in (l_0(\mathbb{Z}))^{r\times s}$ *such that* $\phi = u * \eta$, *where* $\phi = (\phi_1, \ldots, \phi_r)^\mathsf{T}$ *and* $\eta = (\eta_1, \ldots, \eta_s)^\mathsf{T}$*;*

(iii) $0 \leqslant t \leqslant s \leqslant r$ *such that* s *is uniquely determined by the space* $\mathsf{S}(\phi_1, \ldots, \phi_r)$.

Moreover, span$\{\widehat{\phi}(2\pi k) : k \in \mathbb{Z}\} = \mathbb{C}^r$ *if and only if* $s = r$ *and* $\det(\mathsf{u}(1)) \neq 0$, *where* $\mathsf{u}(z) := \sum_{k\in\mathbb{Z}} u(k)z^k$. *If in addition all* $\phi_1, \ldots, \phi_r \in L_p(\mathbb{R})$ *(or* $\mathscr{C}(\mathbb{R})$*), then all* $\eta_1, \ldots, \eta_s \in L_p(\mathbb{R})$ *(or* $\mathscr{C}(\mathbb{R})$*).*

Proof We use induction on $j = 0, \ldots, r$ to prove the following statement: There exist compactly supported distributions $\eta_{t+1}, \ldots, \eta_{s_j}$ satisfying

(1) $t \leqslant s_j \leqslant t + j$ and $\mathsf{S}(\phi_1, \ldots, \phi_j) \subseteq \mathsf{S}(\eta_1, \ldots, \eta_{s_j}) \subseteq \mathsf{S}(\phi_1, \ldots, \phi_r)$;
(2) Every $\phi_\ell, \ell = 1, \ldots, j$ is a finite linear combination of $\eta_1, \ldots, \eta_{s_j}$;
(3) The integer shifts of $\eta_1, \ldots, \eta_{s_j}$ are linearly independent.

The claim for $j = 0$ follows directly from our assumption, since $\mathsf{S}(\phi_1, \ldots, \phi_0) = \emptyset$ by definition. Assume that the claim holds for some $j \geqslant 0$. We now prove the claim for $j + 1$. By Theorem 5.2.1, there exist $\tilde{\eta}_1, \ldots, \tilde{\eta}_{s_j} \in \mathscr{D}(\mathbb{R})$ such that

$$\langle \tilde{\eta}_m, \eta_\ell(\cdot - k)\rangle = \boldsymbol{\delta}(m - \ell)\boldsymbol{\delta}(k), \qquad \forall\, k \in \mathbb{Z}, \ell, m = 1, \ldots, s_j. \tag{5.2.10}$$

Define

$$\varphi := \phi_{j+1} - \sum_{\ell=1}^{s_j} \sum_{k\in\mathbb{Z}} \langle \phi_{j+1}, \tilde{\eta}_\ell(\cdot - k)\rangle \eta_\ell(\cdot - k).$$

Then φ is a compactly supported distribution such that

$$\langle \varphi, \tilde{\eta}_\ell(\cdot - k)\rangle = 0, \qquad \forall\, k \in \mathbb{Z}, \ell = 1, \ldots, s_j. \tag{5.2.11}$$

If φ is identically zero, then the claim holds by taking $s_{j+1} = s_j$.

Suppose that φ is not identically zero. By Theorem 5.2.3, there exists a compactly supported distribution g such that $\mathsf{S}(\varphi) = \mathsf{S}(g)$, $\varphi = u * g$ for some $u \in l_0(\mathbb{Z})$, and the integer shifts of g are linearly independent. Define $s_{j+1} := s_j + 1$ and $\eta_{s_{j+1}} := g$. By the definition of φ, now it is straightforward to see that items (1) and (2) hold with j being replaced by $j + 1$. We now prove that the integer shifts of $\eta_1, \ldots, \eta_{s_j}, g$ must be linearly independent. Suppose that there exist $v_1, \ldots, v_{s_j}, v \in l(\mathbb{Z})$ such that $f := v_1 * \eta_1 + \cdots + v_{s_j} * \eta_{s_j} + v * g = 0$. Noting that $v * g \in \mathsf{S}(g) = \mathsf{S}(\varphi)$, by (5.2.10) and (5.2.11), we have $0 = \langle f, \tilde{\eta}_m(\cdot - n)\rangle = v_m(n)$ for $n \in \mathbb{Z}$ and $m = 1, \ldots, s_j$. Hence, we must have $f = v * g = 0$. Since the integer shifts of g are linearly independent, we conclude that $v = 0$. Therefore, we verified that the integer shifts of $\eta_1, \ldots, \eta_{s_j}, g$ are linearly independent. Thus, the claim holds for $j+1$. The proof is now completed by induction.

Taking $j = r$ and $s = s_r$ in the above claim, we proved items (i) and (ii). Now we prove item (iii). If we start with $t = 0$, by our proved claim, there exist compactly supported distributions $h_1, \ldots, h_m$ with $m \leqslant r$ such that the integer shifts of $h_1, \ldots, h_m$ are linearly independent and $\mathsf{S}(h_1, \ldots, h_m) = \mathsf{S}(\phi_1, \ldots, \phi_r)$. Note that $\mathsf{S}(\eta_1, \ldots, \eta_s) = \mathsf{S}(\phi_1, \ldots, \phi_r) = \mathsf{S}(h_1, \ldots, h_m)$ and the integer shifts of $\eta_1, \ldots, \eta_s$ are linearly independent. If $m < s$, by item (iii) of Theorem 5.2.1 (see Exercise 5.23), there exists $u \in (l_0(\mathbb{Z}))^{s\times m}$ such that $\eta = u * h$, where $h := (h_1, \ldots, h_m)^\mathsf{T}$ and $\eta := (\eta_1, \ldots, \eta_s)^\mathsf{T}$. In particular, we have

$$\widehat{\eta}(2\pi k) = \widehat{u}(0)\widehat{h}(2\pi k), \qquad \forall\, k \in \mathbb{Z}.$$

Since $m < s$ and $\widehat{u}(0)$ is an $s \times m$ matrix, there exists a nonzero vector $\vec{v} \in \mathbb{C}^{1\times s}$ such that $\vec{v}\widehat{u}(0) = 0$. Consequently, $\vec{v}\widehat{\eta}(2\pi k) = \vec{v}\widehat{u}(0)\widehat{h}(2\pi k) = 0$ for all $k \in \mathbb{Z}$. By Theorem 5.2.1, this is a contradiction to the fact that the integer shifts of $\eta_1, \ldots, \eta_s$ are linearly independent. Hence, we must have $m \geqslant s$. Exchanging the roles of η and h, we must also have $s \geqslant m$. Hence, $s = m$. This proves item (iii) by $m \leqslant r$.

Note that $\phi = u * \eta$ can be equivalently rewritten as $\widehat{\phi}(z) = \mathsf{u}(e^{-iz})\widehat{\eta}(z)$ for all $z \in \mathbb{C}$. Therefore, $\widehat{\phi}(2\pi k) = \mathsf{u}(1)\widehat{\eta}(2\pi k)$ for all $k \in \mathbb{Z}$. Since $\operatorname{span}\{\widehat{\eta}(2\pi k) \,:\, k \in \mathbb{Z}\} = \mathbb{C}^r$ and $s \leqslant r$, it is trivial to see that $\operatorname{span}\{\widehat{\phi}(2\pi k) \,:\, k \in \mathbb{Z}\} = \mathbb{C}^r$ if and only if $s = r$ and $\det(\mathsf{u}(1)) \neq 0$. □

5.2.3 *Linear Independence of Refinable Vector Functions*

In this subsection, we now study linear independence of compactly supported refinable vector functions.

Lemma 5.2.5 *Let $\phi_1, \ldots, \phi_r$ be compactly supported distributions on $\mathbb{R}$ such that the integer shifts of $\phi_1, \ldots, \phi_r$ are linearly independent. Then*

$$\mathsf{S}(\phi_1, \ldots, \phi_r) \subseteq \mathsf{S}(\phi_1(2\cdot), \ldots, \phi_r(2\cdot), \phi_1(2\cdot-1), \ldots, \phi_r(2\cdot-1)) \tag{5.2.12}$$

if and only if there exists a filter $a \in (l_0(\mathbb{Z}))^{r\times r}$ such that ϕ satisfies the refinement equation $\phi = 2\sum_{k\in\mathbb{Z}} a(k)\phi(2\cdot-k)$, where $\phi = (\phi_1, \ldots, \phi_r)^\mathsf{T}$.

Proof The sufficiency part ($\Leftarrow$) is obvious. For the necessity part ($\Rightarrow$), by (5.2.12), there exists a sequence $a : \mathbb{Z} \to \mathbb{C}^{r\times r}$ such that $\phi = 2\sum_{k\in\mathbb{Z}} a(k)\phi(2\cdot-k)$. Since the integer shifts of $\phi_1, \ldots, \phi_r$ are linearly independent, by Theorem 5.2.1, there exist $\tilde{\phi}_1, \ldots, \tilde{\phi}_r \in \mathscr{D}(\mathbb{R})$ such that (5.2.5) holds. Define $\tilde{\phi} := (\tilde{\phi}_1, \ldots, \tilde{\phi}_r)^\mathsf{T}$. Then by $\phi = 2\sum_{k\in\mathbb{Z}} a(k)\phi(2\cdot-k)$ we see that for all $n \in \mathbb{Z}$,

$$a(n) = \Big\langle 2\sum_{k\in\mathbb{Z}} a(k)\phi(2\cdot-k), \tilde{\phi}(2\cdot-n)\Big\rangle = \langle \phi, \tilde{\phi}(2\cdot-n)\rangle := \int_{\mathbb{R}} \phi(x)\overline{\tilde{\phi}(2x-n)}^\mathsf{T}\,dx.$$

Since both ϕ and $\tilde{\phi}$ have compact support, we conclude that $a \in (l_0(\mathbb{Z}))^{r\times r}$. □

To study linear independence of a refinable vector function with a finitely supported filter, we need the following result on the structure of particular zeros of a Laurent polynomial.

Lemma 5.2.6 *Let* a *be a Laurent polynomial with* $|\mathsf{a}(1)|+|\mathsf{a}(-1)| \neq 0$. *Then there exist Laurent polynomials* u *and* $\mathring{\mathsf{a}}$ *such that* $\mathsf{u}(1) \neq 0$, u *is not a monomial, and*

$$\mathsf{a}(z)\mathsf{u}(z) = \mathsf{u}(z^2)\mathring{\mathsf{a}}(z) \qquad \forall\, z \in \mathbb{C}\backslash\{0\} \tag{5.2.13}$$

if and only if either $\mathsf{a}(z_0) = \mathsf{a}(-z_0) = 0$ *for some* $z_0 \in \mathbb{C}\backslash\{0\}$ *or* $\prod_{j=0}^{m-1}(z+\zeta^{2^j})$ *is a factor of* $\mathsf{a}(z)$ *for some* $m \in \mathbb{N}$ *and* $\zeta \in \mathbb{T}\backslash\{1\}$ *such that* $\zeta^{2^m} = \zeta$, *where* $\mathbb{T} := \{z \in \mathbb{C} \,:\, |z| = 1\}$.

Proof Sufficiency ($\Leftarrow$). Suppose $\mathsf{a}(z_0) = \mathsf{a}(-z_0) = 0$ for some $z_0 \in \mathbb{C}\backslash\{0\}$. Then $(z^2 - z_0^2) \mid \mathsf{a}(z)$. Define $\mathsf{u}(z) := z - z_0^2$ and $\mathring{\mathsf{a}}(z) := \frac{\mathsf{a}(z)}{z^2-z_0^2}\mathsf{u}(z)$. Then $\mathring{\mathsf{a}}$ is a Laurent polynomial and (5.2.13) holds. Since $|\mathsf{a}(1)| + |\mathsf{a}(-1)| \neq 0$, we must have $z_0^2 \neq 1$ and $\mathsf{u}(1) \neq 0$.

Suppose that $\prod_{j=0}^{m-1}(z+\zeta^{2^j})$ is a factor of $\mathsf{a}(z)$ for some $m \in \mathbb{N}$ and $\zeta^{2^m} = \zeta$ with $\zeta \neq 1$. Define $\mathsf{u}(z) := \prod_{j=1}^{m}(z-\zeta^{2^j})$ and

$$\mathring{\mathsf{a}}(z) := \frac{\mathsf{a}(z)\mathsf{u}(z)}{\mathsf{u}(z^2)} = \frac{\mathsf{a}(z)}{\mathsf{u}(z^2)/\mathsf{u}(z)} = \frac{\mathsf{a}(z)}{\prod_{j=0}^{m-1}(z+\zeta^{2^j})},$$

where we used $\mathsf{u}(z^2) = \prod_{j=1}^{m}(z^2-\zeta^{2^j}) = \left(\prod_{j=0}^{m-1}(z-\zeta^{2^j})\right)\left(\prod_{j=0}^{m-1}(z+\zeta^{2^j})\right) = \mathsf{u}(z)\prod_{j=0}^{m-1}(z+\zeta^{2^j})$ and $\mathsf{u}(z) = \prod_{j=0}^{m-1}(z-\zeta^{2^j})$ by $\zeta^{2^m} = \zeta$. Hence, $\mathring{\mathsf{a}}$ is a well-defined Laurent polynomial. If $\zeta^{2^j} = 1$ for some $j = 0, \ldots, m-1$, then $1 = (\zeta^{2^j})^{2^{m-j}} = \zeta^{2^m} = \zeta$, a contradiction to $\zeta \neq 1$. Hence, we must have $\mathsf{u}(1) \neq 0$. This completes the proof of sufficiency.

Necessity ($\Rightarrow$). Since u is not a monomial, there exists $\zeta \in \mathbb{C}\backslash\{0\}$ such that $\mathsf{u}(\zeta) = 0$. Since $\mathsf{u}(1) \neq 0$, we must have $\zeta \neq 1$. Suppose that $|\mathsf{a}(z)| + |\mathsf{a}(-z)| \neq 0$ for all $z \in \mathbb{C}\backslash\{0\}$. We now prove that there must exist a positive integer m such that $\zeta^{2^m} = \zeta$ and $\mathsf{a}(z)$ contains the factor $\prod_{j=0}^{m-1}(z+\zeta^{2^j})$. For a nonzero complex number c, by $c^{1/2}$ we denote a properly chosen complex number such that $(c^{1/2})^2 = c$. Note that the two points $\zeta^{1/2}$ and $-\zeta^{1/2}$ are roots of $\mathsf{u}(z^2)$ by $\mathsf{u}(\zeta) = 0$. Since either $\mathsf{a}(\zeta^{1/2}) \neq 0$ or $\mathsf{a}(-\zeta^{1/2}) \neq 0$, by (5.2.13), either $\mathsf{u}(\zeta^{1/2}) = 0$ or $\mathsf{u}(-\zeta^{1/2}) = 0$. For simplicity, we define $\zeta^{1/2}$ in such a way that $\mathsf{u}(\zeta^{1/2}) = 0$. Continuing this argument, we see that $\mathsf{u}(\zeta^{2^{-n}}) = 0$ for all $n \in \mathbb{N}_0$. Since u has only finitely many roots, we must have $\zeta^{2^{-m}} = \zeta$ for some $m \in \mathbb{N}$. Hence, $\zeta^{2^m} = \zeta$ and all $\zeta, \zeta^2, \ldots, \zeta^{2^{m-1}}$ are roots of $\mathsf{u}(z)$. Therefore, $\mathsf{u}(z) = \mathsf{u}_1(z)\prod_{j=0}^{m-1}(z-\zeta^{2^j})$ for some Laurent polynomial u_1. If u_1 is not a monomial, then $\mathsf{u}_1(1) \neq 0$ and it follows from (5.2.13) that

$$\mathsf{a}(z)\mathsf{u}_1(z) = \mathsf{u}_1(z^2)\prod_{j=0}^{m-1}(z+\zeta^{2^j})\mathring{\mathsf{a}}(z). \tag{5.2.14}$$

Therefore, we can repeat the above argument by replacing u with u_1. Consequently, we can assume that u_1 is a monomial and it follows from (5.2.14) that $\prod_{j=0}^{m-1}(z+\zeta^{2^j})$ is a factor of $\mathsf{a}(z)$. □

We now investigate the linear independence of a refinable vector function through its refinement filter.

Theorem 5.2.7 *Let $\phi_1, \ldots, \phi_r$ be compactly supported distributions on $\mathbb{R}$ such that $\phi = 2\sum_{k\in\mathbb{Z}} a(k)\phi(2\cdot -k)$ for some $a \in (l_0(\mathbb{Z}))^{r\times r}$, where $\phi := (\phi_1, \ldots, \phi_r)^{\mathsf{T}}$. Assume that span$\{\widehat{\phi}(2\pi k) \;:\; k \in \mathbb{Z}\} = \mathbb{C}^r$.*

(i) The integer shifts of ϕ are linearly dependent if and only if there exist $u, \mathring{a} \in (l_0(\mathbb{Z}))^{r\times r}$ such that $\det(\mathsf{u}(1)) \neq 0$, $\det(\mathsf{u})$ is not a monomial, and (5.2.13) holds.

(ii) If the integer shifts of ϕ are linearly dependent, then one of the following must hold:

(a) $\det(\mathsf{a}(z_0)) = \det(\mathsf{a}(-z_0)) = 0$ for some $z_0 \in \mathbb{C}\backslash\{0\}$;

(b) $\prod_{j=1}^{m-1}(z+\zeta^{2^j})$ is a factor of $\det(\mathsf{a}(z))$ for some $m \in \mathbb{N}$ and some $\zeta \in \mathbb{T}\backslash\{1\}$ satisfying $\zeta^{2^m} = \zeta$.

Proof Necessity part (⇒) of item (i). By Theorem 5.2.4, there exist compactly supported distributions $\eta_1, \ldots, \eta_r$ such that items (i) and (ii) of Theorem 5.2.4 are satisfied with $s = r$. By $\phi = u * \eta$ with $u \in (l_0(\mathbb{Z}))^{r\times r}$, we have $\widehat{\phi}(z) = \mathsf{u}(e^{-iz})\widehat{\eta}(z)$. Since $\phi = 2\sum_{k\in\mathbb{Z}} a(k)\phi(2\cdot -k)$ and $\mathsf{S}(\phi_1, \ldots, \phi_r) = \mathsf{S}(\eta_1, \ldots, \eta_r)$, by Lemma 5.2.5, there exists $\mathring{a} \in (l_0(\mathbb{Z}))^{r\times r}$ such that $\eta = 2\sum_{k\in\mathbb{Z}} \mathring{a}(k)\eta(2\cdot -k)$, that is, $\widehat{\eta}(2z) = \mathring{\mathsf{a}}(e^{-iz})\widehat{\eta}(z)$. Therefore, we have

$$\mathsf{a}(e^{-iz})\mathsf{u}(e^{-iz})\widehat{\eta}(z) = \mathsf{a}(e^{-iz})\widehat{\phi}(z) = \widehat{\phi}(2z) = \mathsf{u}(e^{-i2z})\widehat{\eta}(2z) = \mathsf{u}(e^{-i2z})\mathring{\mathsf{a}}(e^{-iz})\widehat{\eta}(z).$$

Since span$\{\widehat{\eta}(z+2\pi k) \;:\; k \in \mathbb{Z}\} = \mathbb{C}^r$, we must have $\mathsf{a}(e^{-iz})\mathsf{u}(e^{-iz}) = \mathsf{u}(e^{-i2z})\mathring{\mathsf{a}}(e^{-iz})$, in other words, (5.2.13) holds. Since span$\{\widehat{\phi}(2\pi k) \;:\; k \in \mathbb{Z}\} = \mathbb{C}^r$, $\det(\mathsf{u}(1)) \neq 0$ follows directly from Theorem 5.2.4.

If $\det(\mathsf{u})$ is a monomial, then it is trivial to see from $\widehat{\phi}(z) = \mathsf{u}(e^{-iz})\widehat{\eta}(z)$ that the integer shifts of ϕ must be linearly independent. Thus, the determinant $\det(\mathsf{u})$ cannot be a monomial and we proved the necessity part of item (i).

Since the integer shifts of η are linearly independent and $\widehat{\eta}(2z) = \mathring{\mathsf{a}}(e^{-iz})\widehat{\eta}(z)$, we see that $\det(\mathring{\mathsf{a}})$ is not identically zero (see Exercise 5.29). Now it follows from (5.2.13) that

$$\det(\mathsf{a}(z))\det(\mathsf{u}(z)) = \det(\mathsf{u}(z^2))\det(\mathring{\mathsf{a}}(z)).$$

Since both $\det(\mathsf{u})$ and $\det(\mathring{\mathsf{a}})$ are not identically zero, the determinant $\det(\mathsf{a})$ cannot be identically zero. Now item (ii) follows from Lemma 5.2.6.

To complete the proof of item (i), it suffices to prove the sufficiency part of item (i). Since $\det(\mathsf{u}(1)) \neq 0$, $\det(\mathsf{u}(z))$ is not identically zero. Therefore, $\mathsf{a}(z) = \mathsf{u}(z^2)\mathring{\mathsf{a}}(z)\mathsf{u}(z)^{-1}$ for all $z \in \mathbb{C}$ such that $\det(\mathsf{u}(z)) \neq 0$. Since $\det(\mathsf{u})$ is not a monomial, there must exist $\zeta \in \mathbb{C}$ such that $\det(\mathsf{u}(e^{-i\zeta})) = 0$. Hence, there exists a $1 \times r$ nontrivial vector c of complex numbers such that $c\mathsf{u}(e^{-i\zeta}) = 0$. By $\widehat{\phi}(z) = \mathsf{a}(e^{-i2^{-1}z})\widehat{\phi}(2^{-1}z)$, we deduce that

$$\begin{aligned}\widehat{\phi}(z) &= \mathsf{u}(e^{-iz})\mathring{\mathsf{a}}(e^{-i2^{-1}z})\big(\mathsf{u}(e^{-i2^{-1}z})\big)^{-1}\widehat{\phi}(2^{-1}z)\\ &= \mathsf{u}(e^{-iz})\Big(\prod_{j=1}^{n}\mathring{\mathsf{a}}(e^{-i2^{-j}z})\Big)\big(\mathsf{u}(e^{-i2^{-n}z})\big)^{-1}\widehat{\phi}(2^{-n}z).\end{aligned}$$

For every $k \in \mathbb{Z}$, since $\det(\mathsf{u}(1)) \neq 0$ and $\det(\mathsf{u})$ is a continuous function, there always exists sufficiently large $n \in \mathbb{N}$ (depending on k) such that $\det(\mathsf{u}(e^{-i2^{-n}(\zeta+2\pi k)})) \neq 0$. Plugging $z = \zeta + 2\pi k$ into the above identity, by $c\mathsf{u}(e^{-i\zeta}) = 0$, we conclude

$$\begin{aligned}&c\widehat{\phi}(\zeta + 2\pi k)\\ &= c\mathsf{u}(e^{-i\zeta})\Big(\prod_{j=1}^{n}\mathring{\mathsf{a}}(e^{-i2^{-j}(\zeta+2\pi k)})\Big)\big(\mathsf{u}(e^{-i2^{-n}(\zeta+2\pi k)})\big)^{-1}\widehat{\phi}(2^{-n}(\zeta + 2\pi k)) = 0.\end{aligned}$$

Therefore, by Theorem 5.2.1, the integer shifts of ϕ must be linearly dependent. □

5.3 Stability of Integer Shifts of Functions in $L_p(\mathbb{R})$

We characterized in Theorem 4.4.12 when the integer shifts of a finite number of functions in $L_2(\mathbb{R})$ form a Riesz sequence in $L_2(\mathbb{R})$ (i.e., a Riesz basis in their generated shift-invariant subspace of $L_2(\mathbb{R})$). To study wavelets and framelets in $L_p(\mathbb{R})$ spaces with $1 \leqslant p \leqslant \infty$, in this section we characterize stability of integer shifts of a finite number of functions in the Banach space $L_p(\mathbb{R})$. The generalization of Theorem 4.4.12 from $p = 2$ to the general case $1 \leqslant p \leqslant \infty$ is often much more involved.

For $1 \leqslant p \leqslant \infty$, we say that the integer shifts of $\phi_1, \ldots, \phi_r \in L_p(\mathbb{R})$ (or $(\phi_1, \ldots, \phi_r)^{\mathsf{T}}$) are *stable* in $L_p(\mathbb{R})$ if there exist positive constants C_1 and C_2 such that

$$C_1 \sum_{\ell=1}^{r} \|v_\ell\|_{l_p(\mathbb{Z})} \leqslant \Big\| \sum_{\ell=1}^{r}\sum_{k\in\mathbb{Z}} v_\ell(k)\phi_\ell(\cdot - k)\Big\|_{L_p(\mathbb{R})} \leqslant C_2 \sum_{\ell=1}^{r} \|v_\ell\|_{l_p(\mathbb{Z})}, \quad \forall\, v_1, \ldots, v_r \in l_p(\mathbb{Z}). \tag{5.3.1}$$

By the standard density argument, (5.3.1) holds for all $v_1, \ldots, v_r \in l_p(\mathbb{Z})$ if and only if it holds for all $v_1, \ldots, v_r \in l_0(\mathbb{Z})$. For $1 \leqslant p \leqslant \infty$, by $\mathcal{L}_p(\mathbb{R})$ we denote the linear space of all measurable functions f on $\mathbb{R}$ such that

$$\|f\|_{\mathcal{L}_p(\mathbb{R})} := \Big\| \sum_{k\in\mathbb{Z}} |f(\cdot - k)| \Big\|_{L_p([0,1])} < \infty. \tag{5.3.2}$$

Note that $\mathcal{L}_p(\mathbb{R}) \subseteq L_p(\mathbb{R})$ and $\mathcal{L}_q(\mathbb{R}) \subseteq \mathcal{L}_p(\mathbb{R})$ for all $1 \leqslant p \leqslant q \leqslant \infty$.

We need a few auxiliary results in order to study stability of integer shifts of a finite number of functions in $L_p(\mathbb{R})$.

Lemma 5.3.1 *For* $1 \leqslant p \leqslant \infty$ *and* $\phi \in \mathcal{L}_p(\mathbb{R})$,

$$\|v * \phi\|_{\mathcal{L}_p(\mathbb{R})} = \Big\| \sum_{k\in\mathbb{Z}} v(k)\phi(\cdot - k) \Big\|_{\mathcal{L}_p(\mathbb{R})} \leqslant \|\phi\|_{\mathcal{L}_p(\mathbb{R})} \|v\|_{l_1(\mathbb{Z})}, \ \forall\, v \in l_1(\mathbb{Z}), \tag{5.3.3}$$

$$\|u * \phi\|_{L_p(\mathbb{R})} = \Big\| \sum_{k\in\mathbb{Z}} u(k)\phi(\cdot - k) \Big\|_{L_p(\mathbb{R})} \leqslant \|\phi\|_{\mathcal{L}_p(\mathbb{R})} \|u\|_{l_p(\mathbb{Z})}, \quad \forall\, u \in l_p(\mathbb{Z}). \tag{5.3.4}$$

Proof By the triangle inequality for $\|\cdot\|_{\mathcal{L}_p(\mathbb{R})}$, (5.3.3) follows from $\| \sum_{k\in\mathbb{Z}} v(k) \phi(\cdot - k)\|_{\mathcal{L}_p(\mathbb{R})} \leqslant \sum_{k\in\mathbb{Z}} |v(k)| \|\phi(\cdot - k)\|_{\mathcal{L}_p(\mathbb{R})} = \|v\|_{l_1(\mathbb{Z})} \|\phi\|_{\mathcal{L}_p(\mathbb{R})}$. (5.3.4) follows from

$$\begin{aligned}
\Big\| \sum_{k\in\mathbb{Z}} u(k)\phi(\cdot - k) \Big\|_{L_p(\mathbb{R})} &= \Big\| \Big\{ \Big\| \sum_{k\in\mathbb{Z}} u(k+n)\phi(\cdot - k) \Big\|_{L_p([0,1])} \Big\}_{n\in\mathbb{Z}} \Big\|_{l_p(\mathbb{Z})} \\
&= \Big\| \Big\| \Big\{ \sum_{k\in\mathbb{Z}} u(k+n)\phi(\cdot - k) \Big\}_{n\in\mathbb{Z}} \Big\|_{l_p(\mathbb{Z})} \Big\|_{L_p([0,1])} \\
&\leqslant \Big\| \sum_{k\in\mathbb{Z}} \|\{u(k+n)\}_{n\in\mathbb{Z}}\|_{l_p(\mathbb{Z})} |\phi(\cdot - k)| \Big\|_{L_p([0,1])} = \|u\|_{l_p(\mathbb{Z})} \|\phi\|_{\mathcal{L}_p(\mathbb{R})},
\end{aligned}$$

where we used $\|\{\sum_{k\in\mathbb{Z}} u(k+n)\phi(x-k)\}_{n\in\mathbb{Z}}\|_{l_p(\mathbb{Z})} \leqslant \sum_{k\in\mathbb{Z}} \|\{u(k+n)\}_{n\in\mathbb{Z}}\|_{l_p(\mathbb{Z})} |\phi(x-k)|$ by the triangle inequality for $\|\cdot\|_{l_p(\mathbb{Z})}$. ☐

Corollary 5.3.2 *Let* $1 \leqslant p, p' \leqslant \infty$ *such that* $\frac{1}{p} + \frac{1}{p'} = 1$. *For* $\phi \in \mathcal{L}_p(\mathbb{R})$,

$$\|\{\langle f, \phi(\cdot - k)\rangle\}_{k\in\mathbb{Z}}\|_{l_{p'}(\mathbb{Z})} \leqslant \|\phi\|_{\mathcal{L}_p(\mathbb{R})} \|f\|_{L_{p'}(\mathbb{R})}, \qquad \forall\, f \in L_{p'}(\mathbb{R}).$$

Proof By calculation, for $f \in L_{p'}(\mathbb{R})$, we have

$$\begin{aligned}
\|\{\langle f, \phi(\cdot - k)\rangle\}_{k\in\mathbb{Z}}\|_{l_{p'}(\mathbb{Z})} &= \sup \Big\{ \Big| \sum_{k\in\mathbb{Z}} \langle f, \phi(\cdot - k)\rangle \overline{u(k)} \Big| \ : \ u \in l_0(\mathbb{Z}), \|u\|_{l_p(\mathbb{Z})} \leqslant 1 \Big\} \\
&= \sup \Big\{ \Big| \Big\langle f, \sum_{k\in\mathbb{Z}} u(k)\phi(\cdot - k) \Big\rangle \Big| \ : \ u \in l_0(\mathbb{Z}), \|u\|_{l_p(\mathbb{Z})} \leqslant 1 \Big\}
\end{aligned}$$

$$\leqslant \sup\Big\{\|f\|_{L_{p'}(\mathbb{R})}\Big\|\sum_{k\in\mathbb{Z}} u(k)\phi(\cdot-k)\Big\|_{L_p(\mathbb{R})} \;:\; u\in l_0(\mathbb{Z}), \|u\|_{l_p(\mathbb{Z})}\leqslant 1\Big\}$$
$$\leqslant \sup\{\|f\|_{L_{p'}(\mathbb{R})}\|\phi\|_{\mathcal{L}_p(\mathbb{R})}\|u\|_{l_p(\mathbb{Z})} \;:\; u\in l_0(\mathbb{Z}), \|u\|_{l_p(\mathbb{Z})}\leqslant 1\}$$
$$= \|\phi\|_{\mathcal{L}_p(\mathbb{R})}\|f\|_{L_{p'}(\mathbb{R})}.$$

This completes the proof. □

Lemma 5.3.3 *Let $A(\mathbb{T})$ denote the linear space of all 2π-periodic measurable functions $h \in L_1(\mathbb{T})$ such that $\|h\|_{A(\mathbb{T})} := \sum_{k\in\mathbb{Z}} |\widehat{h}(k)| < \infty$. Then*

$$\|[\widehat{f},\widehat{g}]\|_{A(\mathbb{T})} \leqslant \|f\|_{\mathcal{L}_2(\mathbb{R})}\|g\|_{\mathcal{L}_2(\mathbb{R})}, \qquad \forall f, g \in \mathcal{L}_2(\mathbb{R}), \tag{5.3.5}$$

where $[\widehat{f},\widehat{g}](\xi) := \sum_{k\in\mathbb{Z}} \widehat{f}(\xi+2\pi k)\overline{\widehat{g}(\xi+2\pi k)}$, and

$$\|f * g\|_{\mathcal{L}_\infty(\mathbb{R})} \leqslant \|f\|_{L_1(\mathbb{R})}\|g\|_{\mathcal{L}_\infty(\mathbb{R})}, \qquad \forall f \in L_1(\mathbb{R}), g \in \mathcal{L}_\infty(\mathbb{R}). \tag{5.3.6}$$

Proof By Lemma 4.4.1, the Fourier series of $[\widehat{f},\widehat{g}]$ is $\sum_{k\in\mathbb{Z}}\langle f, g(\cdot-k)\rangle e^{-ik\xi}$. Hence,

$$\|[\widehat{f},\widehat{g}]\|_{A(\mathbb{T})} = \sum_{k\in\mathbb{Z}} |\langle f, g(\cdot-k)\rangle| \leqslant \sum_{k\in\mathbb{Z}} \int_{\mathbb{R}} |f(x)g(x-k)|dx = \int_{\mathbb{R}} |f(x)| \sum_{k\in\mathbb{Z}} |g(x-k)|dx$$
$$= \int_0^1 \Big(\sum_{n\in\mathbb{Z}} |f(x-n)|\Big)\Big(\sum_{k\in\mathbb{Z}} |g(x-k)|\Big)dx \leqslant \|f\|_{\mathcal{L}_2(\mathbb{R})}\|g\|_{\mathcal{L}_2(\mathbb{R})},$$

where we used the Cauchy-Schwarz inequality in the last inequality. This proves (5.3.5). By calculation,

$$\|f * g\|_{\mathcal{L}_\infty(\mathbb{R})} \leqslant \Big\| \int_{\mathbb{R}} |f(t)| \sum_{k\in\mathbb{Z}} |g(\cdot-k-t)|dt \Big\|_{L_\infty([0,1])} \leqslant \|f\|_{L_1(\mathbb{R})}\|g\|_{\mathcal{L}_\infty(\mathbb{R})}.$$

This proves (5.3.6). □

The following result characterizes stability of integer shifts of functions in $L_p(\mathbb{R})$.

Theorem 5.3.4 *Let $1 \leqslant p \leqslant \infty$ and $\phi_1, \dots, \phi_r \in \mathcal{L}_p(\mathbb{R})$ such that not all $\phi_1, \dots, \phi_r$ are identically zero. Then the following statements are equivalent:*

(i) The integer shifts of $\phi_1, \dots, \phi_r$ are stable in $L_p(\mathbb{R})$.
(ii) $\{\widehat{\phi_\ell}(\zeta + 2\pi k)\}_{k\in\mathbb{Z}}, \ell = 1, \dots, r$ are linearly independent for all $\zeta \in \mathbb{R}$, i.e., there do not exist $\zeta \in \mathbb{R}$ and $c_1, \dots, c_r \in \mathbb{C}$ such that $|c_1| + \cdots + |c_r| \neq 0$ and (5.2.4) holds.
(iii) There exist functions $\tilde{\phi}_1, \dots, \tilde{\phi}_r \in \mathscr{C}^\infty(\mathbb{R}) \cap \mathcal{L}_\infty(\mathbb{R})$ such that (5.2.5) holds.

Proof (iii)⟹(i). By Lemma 5.3.1, we see that the right-hand side inequality in (5.3.1) holds with $C_2 = \max(\|\phi_1\|_{\mathcal{L}_p(\mathbb{R})}, \dots, \|\phi_r\|_{\mathcal{L}_p(\mathbb{R})})$. Define $f :=$

$\sum_{\ell=1}^{r}\sum_{k\in\mathbb{Z}} v_\ell(k)\,\phi_\ell(\cdot-k)$ for $v_1,\dots,v_r \in l_0(\mathbb{Z})$. By (5.2.5), we have

$$v_\ell(k) = \langle f, \tilde\phi_\ell(\cdot - k)\rangle, \qquad \ell = 1,\dots,r \quad\text{and}\quad k\in\mathbb{Z}.$$

Let $1 \leqslant p' \leqslant \infty$ such that $\frac{1}{p}+\frac{1}{p'} = 1$. Since $\tilde\phi_\ell \in \mathcal{L}_\infty(\mathbb{R}) \subseteq \mathcal{L}_{p'}(\mathbb{R})$, by Corollary 5.3.2, we have

$$\sum_{\ell=1}^{r}\|v_\ell\|_{l_p(\mathbb{Z})} = \sum_{\ell=1}^{r}\|\{\langle f, \tilde\phi_\ell(\cdot-k)\rangle\}_{k\in\mathbb{Z}}\|_{l_p(\mathbb{Z})} \leqslant \|f\|_{L_p(\mathbb{R})}\sum_{\ell=1}^{r}\|\tilde\phi_\ell\|_{\mathcal{L}_{p'}(\mathbb{R})}.$$

Since $0 < \sum_{\ell=1}^{r}\|\tilde\phi_\ell\|_{\mathcal{L}_{p'}(\mathbb{R})} < \infty$, we see that the left-hand side inequality in (5.3.1) holds with $C_1 = (\sum_{\ell=1}^{r}\|\tilde\phi_\ell\|_{\mathcal{L}_{p'}(\mathbb{R})})^{-1}$ for all $v_1,\dots,v_r \in l_0(\mathbb{Z})$. By the standard density argument, we see that the left-hand side inequality holds for all $v_1,\dots,v_r \in l_p(\mathbb{Z})$. This proves (iii)$\Longrightarrow$(i).

We prove (i)$\Longrightarrow$(ii) using proof by contradiction. Suppose that item (ii) fails. Then there exists $\zeta\in\mathbb{R}$ and $c_1,\dots,c_r\in\mathbb{C}$ such that $|c_1|+\dots+|c_r|\neq 0$ and (5.2.4) holds. Define $f(x) := (c_1\phi_1(x)+\dots+c_r\phi_r(x))e^{-i\zeta x}$. Then (5.2.4) implies that $\widehat{f}(2\pi k) = 0$ for all $k\in\mathbb{Z}$, which is equivalent to

$$\sum_{k\in\mathbb{Z}} f(\cdot-k) = 0 \tag{5.3.7}$$

with the series converging absolutely by $f \in \mathcal{L}_p(\mathbb{R})$. If $p = \infty$, (5.3.7) is a contradiction to item (i). So, we consider the case $1 \leqslant p < \infty$. For $n\in\mathbb{N}$, we define finitely supported sequences $u_n \in l_0(\mathbb{Z})$ and a compactly supported function g by

$$u_n(k) = 1, \qquad k\in[-n,n]\cap\mathbb{Z} \quad\text{and}\quad u_n(k) = 0, \qquad \forall\, k\in\mathbb{Z}\backslash[-n,n],$$
$$g := f - h_n + \tilde h_n, \quad h_n := f\chi_{\mathbb{R}\backslash[-N_n,N_n]}, \quad \tilde h_n := \sum_{k\in\mathbb{Z}} h_n(\cdot-k)\chi_{[0,1)}$$

with $N_n := \lfloor (2n+1)^{\frac{1}{2p}}\rfloor$. Note that g is supported inside $[-N_n, N_n]$ and $\|\tilde h_n\|_{\mathcal{L}_p(\mathbb{R})} \leqslant \|h_n\|_{\mathcal{L}_p(\mathbb{R})}$. By Lemma 5.3.1, since $f - g = h_n - \tilde h_n$, we have

$$\begin{aligned}\|u_n * f - u_n * g\|_{L_p(\mathbb{R})} &\leqslant \|u_n\|_{l_p(\mathbb{Z})}\|f-g\|_{\mathcal{L}_p(\mathbb{R})}\\ &= \|u_n\|_{l_p(\mathbb{Z})}\|h_n - \tilde h_n\|_{\mathcal{L}_p(\mathbb{R})} \leqslant 2\|h_n\|_{\mathcal{L}_p(\mathbb{R})}\|u_n\|_{l_p(\mathbb{Z})}.\end{aligned}$$

On the other hand, it follows from (5.3.7) that $\sum_{k\in\mathbb{Z}} g(\cdot-k) = 0$. Define

$$\psi := \sum_{k=0}^{\infty} g(\cdot-k) = -\sum_{k=-\infty}^{-1} g(\cdot-k).$$

Since g is supported inside $[-N_n, N_n]$, from the definition of ψ, we see that ψ must be supported inside $[-N_n, N_n - 1]$ and $g = \psi - \psi(\cdot - 1)$. Consequently, we have

$$\psi(x) = \sum_{k=0}^{2N_n-1} g(x-k), \qquad \forall\, x \in [-N_n, N_n - 1].$$

Thus, by $g = f\chi_{[-N_n,N_n]} + \tilde{h}_n$ and $\|\tilde{h}_n\|_{L_p(\mathbb{R})} \leqslant \|\tilde{h}_n\|_{\mathcal{L}_p(\mathbb{R})} \leqslant \|h_n\|_{\mathcal{L}_p(\mathbb{R})}$, we have

$$\|\psi\|_{L_p(\mathbb{R})} \leqslant 2N_n\|g\|_{L_p(\mathbb{R})} = 2N_n\|f\chi_{[-N_n,N_n]} + \tilde{h}_n\|_{L_p(\mathbb{R})} \leqslant 2N_n(\|f\|_{L_p(\mathbb{R})} + \|h_n\|_{\mathcal{L}_p(\mathbb{R})}).$$

Note that $u_n * g = u_n * (\psi - \psi(\cdot - 1)) = \psi(\cdot + n) - \psi(\cdot - n - 1)$. Hence, we have

$$\|u_n * g\|_{L_p(\mathbb{R})} \leqslant 2\|\psi\|_{L_p(\mathbb{R})} \leqslant 4N_n(\|f\|_{L_p(\mathbb{R})} + \|h_n\|_{\mathcal{L}_p(\mathbb{R})}).$$

In conclusion, noting that $\|u_n\|_{l_p(\mathbb{Z})} = (2n+1)^{1/p}$, we proved

$$\begin{aligned}\frac{\|u_n * f\|_{L_p(\mathbb{R})}}{\|u_n\|_{l_p(\mathbb{Z})}} &\leqslant \frac{\|u_n * f - u_n * g\|_{L_p(\mathbb{R})}}{\|u_n\|_{l_p(\mathbb{Z})}} + \frac{\|u_n * g\|_{L_p(\mathbb{R})}}{\|u_n\|_{l_p(\mathbb{Z})}}\\ &\leqslant 2\|h_n\|_{\mathcal{L}_p(\mathbb{R})} + \frac{4N_n}{(2n+1)^{1/p}}(\|f\|_{L_p(\mathbb{R})} + \|h_n\|_{\mathcal{L}_p(\mathbb{R})}).\end{aligned}$$

Since $f \in \mathcal{L}_p(\mathbb{R})$, we see that $\lim_{n\to\infty} \|h_n\|_{\mathcal{L}_p(\mathbb{R})} = 0$. Since $\lim_{n\to\infty} \frac{4N_n}{(2n+1)^{1/p}} = 0$, the above inequalities show that $\lim_{n\to\infty} \frac{\|u_n * f\|_{L_p(\mathbb{R})}}{\|u_n\|_{l_p(\mathbb{Z})}} = 0$, which is a contradiction to item (i). Hence, (i)$\Longrightarrow$(ii) must be true.

(ii)$\Longrightarrow$(iii). Define $G(x) := \pi^{-1/2}e^{-x^2}$ and $\varphi_\ell := \phi_\ell * G$ for $\ell = 1, \dots, r$. Since $G \in \mathcal{L}_\infty(\mathbb{R})$, by Lemma 5.3.3, we have $\varphi_\ell \in \mathcal{L}_\infty(\mathbb{R}) \subseteq \mathcal{L}_2(\mathbb{R})$. Since $\widehat{G}(\xi) = e^{-\xi^2/4} \neq 0$ and $\widehat{\varphi_\ell}(\xi) = \widehat{\phi_\ell}(\xi)\widehat{G}(\xi)$, we see that item (ii) still holds if ϕ_ℓ is replaced by φ_ℓ, $\ell = 1, \dots, r$. Define $\varphi := (\varphi_1, \dots, \varphi_r)^{\mathsf{T}}$. Then $\det[\widehat{\varphi}, \widehat{\varphi}](\xi) > 0$ for all $\xi \in \mathbb{R}$. Define

$$(\widehat{\eta_1}(\xi), \dots, \widehat{\eta_r}(\xi))^{\mathsf{T}} := \widehat{\eta}(\xi) := ([\widehat{\varphi}, \widehat{\varphi}](\xi))^{-1}\widehat{\varphi}(\xi).$$

By Lemma 5.3.3, $[\widehat{\varphi_m}, \widehat{\varphi_\ell}] \in A(\mathbb{T})$ for $1 \leqslant \ell, m \leqslant r$. Thus, by Wiener's lemma in Theorem A.3.7, we see that all the entries of $([\widehat{\varphi}, \widehat{\varphi}](\xi))^{-1}$ are from $A(\mathbb{T})$. Now by (5.3.3), we see that $\eta_\ell \in \mathcal{L}_\infty(\mathbb{R})$ for all $\ell = 1, \dots, r$. Define $\tilde{\phi}_\ell := \eta_\ell * G$, $\ell = 1, \dots, r$. Note that $G(-\cdot) = G$ and G is real-valued. By the definition of $\tilde{\phi}_m$ and η_ℓ, we also have

$$\begin{aligned}\langle \tilde{\phi}_m, \phi_\ell(\cdot - k)\rangle &= \langle \eta_m * G, \phi_\ell(\cdot - k)\rangle = \langle \eta_m, (\phi_\ell * G)(\cdot - k)\rangle\\ &= \langle \eta_m, \varphi_\ell(\cdot - k)\rangle = \boldsymbol{\delta}(m-\ell)\boldsymbol{\delta}(k), \qquad k \in \mathbb{Z}.\end{aligned}$$

Hence, (5.2.5) holds. Since $G \in \mathscr{C}^\infty(\mathbb{R}) \cap \mathcal{L}_\infty(\mathbb{R})$, by Lemma 5.3.3, we conclude $\tilde{\phi}_\ell \in \mathscr{C}^\infty(\mathbb{R}) \cap \mathcal{L}_\infty(\mathbb{R})$. This proves (ii)$\Longrightarrow$(iii). □

As a direct consequence of Theorems 5.2.1 and 5.3.4, we have

Corollary 5.3.5 *Let $1 \leqslant p \leqslant \infty$ and $\phi_1, \dots, \phi_r \in L_p(\mathbb{R})$ be compactly supported functions. If the integer shifts of $\phi_1, \dots, \phi_r$ are linearly independent, then the integer shifts of $\phi_1, \dots, \phi_r$ are stable in $L_p(\mathbb{R})$.*

Proof For $\ell = 1, \dots, r$, since ϕ_ℓ is a compactly supported function in $L_p(\mathbb{R})$, it is trivial to see that $\phi_\ell \in \mathcal{L}_p(\mathbb{R})$. Now the claim follows directly from Theorems 5.2.1 and 5.3.4. □

For $\phi = (\phi_1, \dots, \phi_r)^\mathsf{T}$ with $\phi_1, \dots, \phi_r$ being compactly supported distributions on $\mathbb{R}$, we say that the integer shifts of ϕ are *stable* if

$$\operatorname{span}\{\widehat{\phi}(\xi + 2\pi k) \; : \; k \in \mathbb{Z}\} = \mathbb{C}^r, \qquad \forall\, \xi \in \mathbb{R}. \tag{5.3.8}$$

Using Theorem 5.2.1, we have the following result on stability of integer shifts of compactly supported distributions.

Theorem 5.3.6 *Let $\phi = (\phi_1, \dots, \phi_r)^\mathsf{T}$ with $\phi_1, \dots, \phi_r$ being compactly supported distributions on $\mathbb{R}$. The following statements are equivalent:*

(1) For any polynomial sequence $v \in (\mathbb{P})^{1\times r}$ satisfying $\sum_{k\in\mathbb{Z}} v(k)\phi(\cdot - k) = 0$, we must have $v(k) = 0$ for all $k \in \mathbb{Z}$.
(2) For any $v \in (l_\infty(\mathbb{Z}))^{1\times r}$ satisfying $\sum_{k\in\mathbb{Z}} v(k)\phi(\cdot - k) = 0$, we must have $v(k) = 0$ for all $k \in \mathbb{Z}$.
(3) The integer shifts of ϕ are stable, i.e., span$\{\widehat{\phi}(\xi + 2\pi k) \; : \; k \in \mathbb{Z}\} = \mathbb{C}^r$ for all $\xi \in \mathbb{R}$.
(4) There exist Schwartz functions $\tilde{\phi}_1, \dots, \tilde{\phi}_r \in \mathscr{S}(\mathbb{R})$ such that the biorthogonality relation in (5.2.5) holds and each $\tilde{\phi}_\ell$, $\ell = 1, \dots, r$ is a finite linear combination of elements $\sum_{k\in\mathbb{Z}} w(k)h(\cdot - k)$ for some $h \in \mathscr{D}(\mathbb{R})$ and $w \in l_1(\mathbb{Z})$ with exponential decay (i.e., there exists $\tau > 0$ such that $\sup_{k\in\mathbb{Z}} |w(k)|e^{-\tau|k|} < \infty$).
(5) There exist Schwartz functions $\tilde{\phi}_1, \dots, \tilde{\phi}_r \in \mathscr{S}(\mathbb{R})$ such that the biorthogonality relation in (5.2.5) holds.

Proof (1)$\Longrightarrow$(2) is trivial. Using proof by contradiction and the same argument as in the proof of (i)$\Longrightarrow$(ii) in Theorem 5.2.1, we conclude that (2)$\Longrightarrow$(3). If (3) holds, then span$\{\widehat{\phi}(2\pi k) \; : \; k \in \mathbb{Z}\} = \mathbb{C}^r$. By Theorem 5.2.4, there exist compactly supported distributions $\eta_1, \dots, \eta_r$ and a sequence $u \in (l_0(\mathbb{Z}))^{r\times r}$ such that $\phi = u * \eta$ and the integer shifts of η are linearly independent, where $\eta := (\eta_1, \dots, \eta_r)^\mathsf{T}$. Because the integer shifts of ϕ are stable and $\widehat{\phi}(\xi) = \widehat{u}(\xi)\widehat{\eta}(\xi)$, we must have $\det(\widehat{u}(\xi)) \neq 0$ for all $\xi \in \mathbb{R}$. Since the integer shifts of η are linearly independent, by Theorem 5.2.1, there exists $\tilde{\eta} := (\tilde{\eta}_1, \dots, \tilde{\eta}_r)^\mathsf{T}$ with $\tilde{\eta}_1, \dots, \tilde{\eta}_r \in \mathscr{D}(\mathbb{R})$ such that $\langle \tilde{\eta}, \eta(\cdot - k)\rangle = \boldsymbol{\delta}(k)I_r$ for all $k \in \mathbb{Z}$. Define $\widehat{v}(\xi) := (\overline{\widehat{u}(\xi)}^\mathsf{T})^{-1}$ and

$(\widehat{\tilde{\phi}_1}(\xi), \ldots, \widehat{\tilde{\phi}_r}(\xi))^\mathsf{T} := \widehat{v}(\xi)\widehat{\eta}(\xi)$. Since $u \in (l_0(\mathbb{Z}))^{r\times r}$ and $\det(\widehat{u}(\xi)) \neq 0$ for all $\xi \in \mathbb{R}$, the sequence in every entry of v must have exponential decay. Now one can directly check that $\tilde{\phi}_1, \ldots, \tilde{\phi}_r$ satisfy all the conditions in item (4).

(4)$\Longrightarrow$(5) is trivial. To prove (5)$\Longrightarrow$(1), by the biorthogonality relation in (5.2.5), it suffices to prove that for every compactly supported distribution f and $\psi \in \mathscr{S}(\mathbb{R})$, the sequence $\{\langle f; \psi(\cdot - k)\rangle\}_{k\in\mathbb{Z}}$ must have faster decay than any polynomial decay. Since f has compact support, there exists $h \in \mathscr{D}(\mathbb{R})$ such that h takes value one in the neighborhood of the support of f. Consequently, $\langle f, \psi(\cdot - k)\rangle = \langle fh, \psi(\cdot - k)\rangle = \langle f, \overline{h}\psi(\cdot - k)\rangle$. Since f is a distribution and $h \in \mathscr{D}(\mathbb{R})$, it follows directly from item (ii) of Theorem A.6.1 that the sequence $\{\langle f, \overline{h}\psi(\cdot - k)\rangle\}_{k\in\mathbb{Z}}$ has faster decay than any polynomial decay. This completes the proof. □

Similar to Theorem 5.2.7 and using the definition of stability in (5.3.8), we have the following result on stability of refinable vector functions, whose proof is left as Exercise 5.32.

Theorem 5.3.7 *Let $\phi := (\phi_1, \ldots, \phi_r)^\mathsf{T}$ be an $r \times 1$ vector of compactly supported distributions such that $\widehat{\phi}(0) \neq 0$ and $\phi = 2\sum_{k\in\mathbb{Z}} a(k)\phi(2\cdot -k)$ for some $a \in (l_0(\mathbb{Z}))^{r\times r}$. Assume that $\operatorname{span}\{\widehat{\phi}(2\pi k) : k \in \mathbb{Z}\} = \mathbb{C}^r$.*

(i) The integer shifts of $\phi_1, \ldots, \phi_r$ are not stable if and only if there exist $u, \mathring{a} \in (l_0(\mathbb{Z}))^{r\times r}$ such that $\det(\mathsf{u}(1)) \neq 0$, $\det(\mathsf{u}(\zeta)) = 0$ for some $\zeta \in \mathbb{T}$, and (5.2.13) holds.

(ii) If the integer shifts of $\phi_1, \ldots, \phi_r$ are not stable, then one of the following must hold:

(a) $\det(\mathsf{a}(\xi_0)) = \det(\mathsf{a}(-\xi_0)) = 0$ for some $\xi_0 \in \mathbb{T}$;

(b) $\prod_{j=1}^{m-1}(z+\zeta^{2^j})$ is a factor of $\det(\mathsf{a}(z))$ for some $m \in \mathbb{N}$ and some $\zeta \in \mathbb{T}\backslash\{1\}$ satisfying $\zeta^{2^m} = \zeta$.

Corollary 5.3.8 *Let $a \in l_0(\mathbb{Z})$ with $\widehat{a}(0) = 1$. Define a compactly supported distribution ϕ by $\widehat{\phi}(\xi) := \prod_{j=1}^{\infty}\widehat{a}(2^{-j}\xi), \xi \in \mathbb{R}$. Then the following statements are equivalent:*

(i) The integer shifts of ϕ are linearly independent (or stable).

(ii) There do not exist $u, \mathring{a} \in l_0(\mathbb{Z})$ such that $\mathsf{u}(1) \neq 0$, $\mathsf{u}(\zeta) = 0$ for some $\zeta \in \mathbb{C}\backslash\{0\}$ (or $\mathsf{u}(\zeta) = 0$ for some $\zeta \in \mathbb{T}$), and (5.2.13) holds.

(iii) The following two conditions are satisfied:

(a) $\mathsf{a}(z)$ and $\mathsf{a}(-z)$ do not have common zeros on $\mathbb{C}\backslash\{0\}$ (or on $\mathbb{T}$);

(b) $\prod_{j=1}^{m-1}(z+\zeta^{2^j})$ is not a factor of $\mathsf{a}(z)$ for any $m \in \mathbb{N}$ and any $\zeta \in \mathbb{T}\backslash\{1\}$ satisfying $\zeta^{2^m} = \zeta$.

Proof (ii) $\Longleftrightarrow$ (iii) has been proved in Lemma 5.2.6. The proof of (i) $\Longleftrightarrow$ (ii) appeared in the proof of Theorem 5.2.7. □

5.4 Approximation Using Quasi-Projection Operators in $L_p(\mathbb{R})$

As we have seen in Sect. 4.7, the approximation property of a dual framelet in the space $L_2(\mathbb{R})$ is completely determined by the quasi-projection operator $\mathcal{Q}_\lambda$ in (4.7.2) using scaled shift-invariant spaces. Since a shift-invariant space is part of an affine system, it is of fundamental importance in both wavelet theory and approximation theory to study the approximation properties of scaled shift-invariant spaces in the $L_p(\mathbb{R})$ spaces with $1 \leqslant p \leqslant \infty$. In this section we study the approximation properties of functions in an L_p Sobolev space using quasi-projection operators in $L_p(\mathbb{R})$, which frequently appear in approximation theory and wavelet analysis.

For a function $f : \mathbb{R} \to \mathbb{C}$, we say that f is *absolutely continuous* (on $\mathbb{R}$) if f is absolutely continuous on every bounded open interval, that is, for every given bounded open interval I and any given $\varepsilon > 0$, there exists $\delta > 0$ such that for all non-overlapping open intervals $(c_j, d_j), j = 1, \dots, J$ which are contained inside I and satisfy $\sum_{j=1}^J |d_j - c_j| < \delta$, then inequality $\sum_{j=1}^J |f(c_j) - f(d_j)| < \varepsilon$ holds.

For $m \in \mathbb{N}_0$ and $1 \leqslant p \leqslant \infty$, the ($L_p$) *Sobolev space* $W_p^m(\mathbb{R})$ consists of all functions $f \in L_p(\mathbb{R})$ such that $f, f', \dots, f^{(m-1)}$ are absolutely continuous functions in $L_p(\mathbb{R})$, $f^{(m)} \in L_p(\mathbb{R})$, and

$$\|f\|_{W_p^m(\mathbb{R})} := \|f\|_{L_p(\mathbb{R})} + \cdots + \|f^{(m)}\|_{L_p(\mathbb{R})} < \infty. \tag{5.4.1}$$

In particular, we have $W_p^0(\mathbb{R}) = L_p(\mathbb{R})$. Recall that $f^{(j)}$ is the jth classical derivative while $D^j f$ is the jth distributional derivative. Let $f \in W_p^m(\mathbb{R})$. Since $f^{(m-1)}$ is absolutely continuous, we have $f^{(m-1)}(x) = f^{(m-1)}(x_0) + \int_{x_0}^x f^{(m)}(y)dy$. By $f, \dots, f^{(m)} \in L_p(\mathbb{R})$ with $1 \leqslant p \leqslant \infty$, all $f, \dots, f^{(m)}$ can be regarded as distributions and $f^{(j)} = D^j f$ in the sense of distributions for $j = 0, \dots, m$. We shall show in Proposition 5.5.16 that $f \in W_p^m(\mathbb{R})$ if and only if $f, Df, \dots, D^m f \in L_p(\mathbb{R})$.

Next we discuss how to measure smoothness of a function using modulus of smoothness. For a sequence $v \in l(\mathbb{Z})$ and a function f on $\mathbb{R}$, we define

$$\begin{aligned} \nabla_k v &:= (\mathrm{Id} - [\![k, 0]\!])v = v - v(\cdot - k), \qquad k \in \mathbb{Z}, \\ \nabla_t f &:= (\mathrm{Id} - [\![t, 0]\!])f = f - f(\cdot - t), \qquad t \in \mathbb{R}, \end{aligned} \tag{5.4.2}$$

where $[\![t, 0]\!]f := f(\cdot - t)$. In particular, we define $\nabla := \nabla_1$. For $m \in \mathbb{N}$, the *mth modulus of smoothness* of $f \in L_p(\mathbb{R})$ is defined to be

$$\omega_m(f, \lambda)_p := \sup_{|t| \leqslant \lambda} \|\nabla_t^m f\|_{L_p(\mathbb{R})}, \qquad \lambda \geqslant 0. \tag{5.4.3}$$

Define $\nabla_t^0 f := f$ and $\omega_0(f,\lambda)_p := \|f\|_{L_p(\mathbb{R})}$. By the definition of modulus of smoothness in (5.4.3) and $\nabla_t^n = (\mathrm{Id} - [\![t,0]\!])^n = \sum_{j=0}^n \binom{n}{j}(-1)^j [\![jt,0]\!]$, we have

$$\omega_m(f,\lambda t)_p \leqslant \lceil t \rceil^m \omega_m(f,\lambda)_p \quad \text{and} \quad \omega_{m+n}(f,\lambda)_p \leqslant 2^n \omega_m(f,\lambda)_p$$

for all $\lambda, t > 0$ and $m, n \in \mathbb{N}_0$. For an absolutely continuous function f, we have $\nabla_t f(x) = \int_{-t}^0 f'(x+y)dy$. Applying Minkowski's inequality, we have $\|\nabla_t f\|_{L_p(\mathbb{R})} \leqslant |t|\|f'\|_{L_p(\mathbb{R})}$. Thus, for all $m, n \in \mathbb{N}_0$, we deduce $\|\nabla_t^{m+n} f\|_{L_p(\mathbb{R})} \leqslant |t|^m \|\nabla_t^n f^{(m)}\|_{L_p(\mathbb{R})}$ and

$$\omega_m(f,\lambda)_p \leqslant \lambda^m \|f^{(m)}\|_{L_p(\mathbb{R})}, \quad \omega_{m+n}(f,\lambda)_p \leqslant \lambda^m \omega_n(f^{(m)},\lambda)_p, \tag{5.4.4}$$

for all $f \in W_p^m(\mathbb{R})$ and $\lambda > 0$. For $m \in \mathbb{N}$, the *(cardinal) B-spline function* B_m *of order* m is defined to be

$$B_1 := \chi_{(0,1]} \quad \text{and} \quad B_m := B_{m-1} * B_1 = \int_0^1 B_{m-1}(\cdot - y)dy. \tag{5.4.5}$$

See Proposition 6.1.1 for some basic properties of B-spline functions.

To study the approximation property using quasi-projection operators, we need the following technical result by smoothing a general function in $L_p(\mathbb{R})$ into an infinitely differentiable function.

Lemma 5.4.1 *Let* ψ_0 *be a nonnegative function in* $\mathscr{C}^\infty(\mathbb{R})$ *such that* ψ_0 *vanishes outside* $[0,1]$ *and* $\int_{\mathbb{R}} \psi_0(x)dx = 1$. *Define* $\psi := \psi_0 * B_m$ *and*

$$f_\lambda(x) := \int_{\mathbb{R}} (f - \nabla_t^m f)(x)\lambda\psi(\lambda t)dt, \qquad x \in \mathbb{R}, \lambda > 0, f \in L_p(\mathbb{R}). \tag{5.4.6}$$

Then $f_\lambda \in \mathscr{C}^\infty(\mathbb{R}) \cap (\cap_{n=1}^\infty W_p^n(\mathbb{R}))$ *and*

$$\begin{aligned} \|f - f_\lambda\|_{L_p(\mathbb{R})} &\leqslant (m+1)^m \omega_m(f,\lambda^{-1})_p, \\ \|f_\lambda^{(m)}\|_{L_p(\mathbb{R})} &\leqslant 2^m \|\psi_0\|_{L_{p'}(\mathbb{R})}\lambda^m \omega_m(f,\lambda^{-1})_p, \end{aligned} \tag{5.4.7}$$

where $1 \leqslant p, p' \leqslant \infty$ *such that* $\frac{1}{p} + \frac{1}{p'} = 1$.

Proof Note that ψ is a nonnegative function in $\mathscr{C}^\infty(\mathbb{R})$ such that ψ vanishes outside $[0, m+1]$ and $\int_{\mathbb{R}} \psi(x)dx = 1$. Since

$$\nabla_t^m f = (\mathrm{Id} - [\![t,0]\!])^m f = \sum_{j=0}^m (-1)^j \frac{m!}{j!(m-j)!}[\![t,0]\!]^j f = \sum_{j=0}^m (-1)^j \frac{m!}{j!(m-j)!} f(\cdot - jt),$$

we have

$$f_\lambda = \sum_{j=1}^{m}(-1)^{j+1}\frac{m!}{j!(m-j)!}\int_{\mathbb{R}} f(\cdot - jt)\lambda\psi(\lambda t)dt$$
$$= \sum_{j=1}^{m}(-1)^{j+1}\frac{m!}{j!(m-j)!}\int_{\mathbb{R}} f(\cdot - \lambda^{-1}jt)\psi(t)dt.$$

That is, we have

$$f_\lambda = \sum_{j=1}^{m}(-1)^{j+1}\frac{m!}{j!(m-j)!}g_{j,\lambda} \quad \text{with} \quad g_{j,\lambda} := (f(\lambda^{-1}j\cdot) * \psi)(\lambda j^{-1}\cdot). \tag{5.4.8}$$

Since $\psi \in \mathscr{C}^\infty(\mathbb{R})$ has compact support, we have $g_{j,\lambda} \in \mathscr{C}^\infty(\mathbb{R}) \cap (\cap_{n=1}^\infty W_p^n(\mathbb{R}))$ and hence $f_\lambda \in \mathscr{C}^\infty(\mathbb{R}) \cap (\cap_{n=1}^\infty W_p^n(\mathbb{R}))$. By (5.4.6) and $\mathrm{supp}(\psi) \subseteq [0, m+1]$, we have

$$\begin{aligned}\|f - f_\lambda\|_{L_p(\mathbb{R})} &= \left\| \int_{\mathbb{R}} \nabla_t^m f(\cdot)\lambda\psi(\lambda t)dt \right\|_{L_p(\mathbb{R})} \leqslant \int_{\mathbb{R}} \|\nabla_t^m f\|_{L_p(\mathbb{R})}\lambda\psi(\lambda t)dt \\ &= \int_{\mathbb{R}} \|\nabla_{\lambda^{-1}t}^m f\|_{L_p(\mathbb{R})}\psi(t)dt \leqslant \int_0^{m+1} \omega_m(f, \lambda^{-1}t)_p\psi(t)dt \\ &\leqslant \omega_m(f, \lambda^{-1}(m+1))_p \leqslant (m+1)^m\omega_m(f, \lambda^{-1})_p.\end{aligned}$$

Note that $\psi^{(m)} = \psi_0 * B_m^{(m)} = \nabla^m\psi_0$ by $B_m^{(m-1)} = \nabla^{m-1}B_1$ and $DB_1 = \nabla\delta$. We deduce

$$\begin{aligned}g_{j,\lambda}^{(m)} &= \lambda^m j^{-m}(f(\lambda^{-1}j\cdot) * \psi^{(m)})(\lambda j^{-1}\cdot) = \lambda^m j^{-m}(f(\lambda^{-1}j\cdot) * (\nabla^m\psi_0))(\lambda j^{-1}\cdot) \\ &= \lambda^m j^{-m}((\nabla_{\lambda^{-1}j}^m f)(\lambda^{-1}j\cdot) * \psi_0)(\lambda j^{-1}\cdot).\end{aligned}$$

Since ψ_0 is supported inside $[0, 1]$, by Hölder's inequality, we have $|[(\nabla_{\lambda^{-1}j}^m f)(\lambda^{-1}j\cdot) * \psi_0](x)| \leqslant \|\psi_0\|_{L_{p'}(\mathbb{R})}\left(\int_0^1 |(\nabla_{\lambda^{-1}j}^m f)(\lambda^{-1}j(x-y))|^p dy\right)^{1/p}$. Thus,

$$\begin{aligned}\|g_{j,\lambda}^{(m)}\|_{L_p(\mathbb{R})} &= \lambda^{m-1/p}j^{-m+1/p}\|(\nabla_{\lambda^{-1}j}^m f)(\lambda^{-1}j\cdot) * \psi_0\|_{L_p(\mathbb{R})} \\ &\leqslant \lambda^{m-1/p}j^{-m+1/p}\|\psi_0\|_{L_{p'}(\mathbb{R})}\|(\nabla_{\lambda^{-1}j}^m f)(\lambda^{-1}j\cdot)\|_{L_p(\mathbb{R})} \\ &= \lambda^m j^{-m}\|\psi_0\|_{L_{p'}(\mathbb{R})}\|\nabla_{\lambda^{-1}j}^m f\|_{L_p(\mathbb{R})} \\ &\leqslant \lambda^m j^{-m}\|\psi_0\|_{L_{p'}(\mathbb{R})}\omega_m(f, \lambda^{-1}j)_p \\ &\leqslant \lambda^m\|\psi_0\|_{L_{p'}(\mathbb{R})}\omega_m(f, \lambda^{-1})_p.\end{aligned}$$

Using (5.4.8), we conclude that

$$\|f_\lambda^{(m)}\|_{L_p(\mathbb{R})} \leqslant \sum_{j=1}^{m} \frac{m!}{j!(m-j)!}\|g_{j,\lambda}^{(m)}\|_{L_p(\mathbb{R})} \leqslant (2^m-1)\|\psi_0\|_{L_{p'}(\mathbb{R})}\lambda^m \omega_m(f,\lambda^{-1})_p.$$

This completes the proof of (5.4.7), □

Let $1 \leqslant p, p' \leqslant \infty$ such that $\frac{1}{p}+\frac{1}{p'}=1$. Let $\phi_1,\dots,\phi_r \in \mathcal{L}_p(\mathbb{R})$ and $\tilde{\phi}_1,\dots,\tilde{\phi}_r \in \mathcal{L}_{p'}(\mathbb{R})$ (see (5.3.2) for the definition of $\mathcal{L}_p(\mathbb{R})$). For $\lambda > 0$, we define the *quasi-projection operators* $\mathcal{Q}_\lambda : L_p(\mathbb{R}) \to L_p(\mathbb{R})$ as follows: For $f \in L_p(\mathbb{R})$,

$$\mathcal{Q}_\lambda f := \sum_{\ell=1}^{r}\sum_{k\in\mathbb{Z}} \langle f, \lambda\tilde{\phi}_\ell(\lambda\cdot -k)\rangle \phi_\ell(\lambda\cdot -k) \quad \text{and} \quad \mathcal{Q} := \mathcal{Q}_1. \tag{5.4.9}$$

Then $\mathcal{Q}_\lambda f = [\mathcal{Q}(f(\lambda^{-1}\cdot))](\lambda\cdot)$. Moreover, by Lemma 5.3.1 and Corollary 5.3.2, all quasi-projection operators $\mathcal{Q}_\lambda$ are bounded linear operators since

$$\|\mathcal{Q}_\lambda\| = \|\mathcal{Q}\| \leqslant \sum_{\ell=1}^{r} \|\tilde{\phi}_\ell\|_{\mathcal{L}_{p'}(\mathbb{R})}\|\phi_\ell\|_{\mathcal{L}_p(\mathbb{R})} < \infty. \tag{5.4.10}$$

Recall that $\mathbb{P}_{m-1}$ is the space of all polynomials having degree $< m$. We now study the approximation property of quasi-projection operators $\mathcal{Q}_\lambda$ in $L_p(\mathbb{R})$.

Theorem 5.4.2 *Let $1 \leqslant p, p' \leqslant \infty$ such that $\frac{1}{p}+\frac{1}{p'}=1$. Let $\phi_1,\dots,\phi_r$ be compactly supported functions in $L_p(\mathbb{R})$ and $\tilde{\phi}_1,\dots,\tilde{\phi}_r$ be compactly supported functions in $L_{p'}(\mathbb{R})$. Let $\mathcal{Q}_\lambda$ and $\mathcal{Q}$ be the quasi-projection operators defined in (5.4.9). Let $c \in \{0,1\}$. If $\mathcal{Q}\mathsf{p} = c\mathsf{p}$ for all $\mathsf{p} \in \mathbb{P}_{m-1}$, then there is a positive constant C such that*

$$\|\mathcal{Q}_\lambda f - cf\|_{L_p(\mathbb{R})} \leqslant C\omega_m(f,\lambda^{-1})_p, \qquad \forall f \in L_p(\mathbb{R}), \lambda > 0. \tag{5.4.11}$$

In particular, for all $\lambda > 0$,

$$\|\mathcal{Q}_\lambda f - cf\|_{L_p(\mathbb{R})} \leqslant C\lambda^{-j}\omega_{m-j}(f^{(j)},\lambda^{-1})_p, \quad j=0,\dots,m, f \in W_p^j(\mathbb{R}). \tag{5.4.12}$$

If all $\phi_1,\dots,\phi_r,\tilde{\phi}_1,\dots,\tilde{\phi}_r$ vanish outside $[-N,N]$ for some $N \in \mathbb{N}_0$ and ψ_0 is as in Lemma 5.4.1, then we can take

$$C = \Big(c+\sum_{\ell=1}^{r}\|\tilde{\phi}_\ell\|_{\mathcal{L}_{p'}(\mathbb{R})}\|\phi_\ell\|_{\mathcal{L}_p(\mathbb{R})}\Big)\Big((m+1)^m+\frac{2^m}{m!}((2N)^m+(2N+1)^m)\|\psi_0\|_{L_{p'}(\mathbb{R})}\Big).$$

Proof Define f_λ as in (5.4.6) of Lemma 5.4.1. Then $f_\lambda \in \mathscr{C}^\infty(\mathbb{R}) \cap (\cap_{n=1}^\infty W_p^n(\mathbb{R}))$, the inequalities in (5.4.7) hold and

$$\mathcal{Q}_\lambda f - cf = \mathcal{Q}_\lambda(f - f_\lambda) + (\mathcal{Q}_\lambda f_\lambda - cf_\lambda) + c(f_\lambda - f).$$

By (5.4.7) and (5.4.10), the above identity implies

$$\begin{aligned} \|\mathcal{Q}_\lambda f - cf\|_{L_p(\mathbb{R})} &\leqslant (c + \|\mathcal{Q}_\lambda\|)\|f - f_\lambda\|_{L_p(\mathbb{R})} + \|\mathcal{Q}_\lambda f_\lambda - cf_\lambda\|_{L_p(\mathbb{R})} \\ &\leqslant C_1\omega_m(f, \lambda^{-1})_p + \|\mathcal{Q}_\lambda f_\lambda - cf_\lambda\|_{L_p(\mathbb{R})}, \end{aligned} \tag{5.4.13}$$

where $C_1 := (m+1)^m(c + \sum_{\ell=1}^r \|\tilde{\phi}_\ell\|_{\mathcal{L}_{p'}(\mathbb{R})}\|\phi_\ell\|_{\mathcal{L}_p(\mathbb{R})}) < \infty$.

We now study $\|\mathcal{Q}_\lambda f_\lambda - cf_\lambda\|_{L_p(\mathbb{R})}$ by estimating $\|\mathcal{Q}_\lambda f_\lambda - cf_\lambda\|_{L_p(I_k)}$ with $I_k := [k/\lambda, (k+1)/\lambda]$ for $k \in \mathbb{Z}$. For $g \in W_1^m(\mathbb{R})$, we have the following Taylor expansion at a point x_0: for all $x \in \mathbb{R}$ and $m \in \mathbb{N}$,

$$g(x) = \sum_{j=0}^{m-1} g^{(j)}(x_0)\frac{(x-x_0)^j}{j!} + \int_0^{x-x_0} g^{(m)}(x-y)\frac{y^{m-1}}{(m-1)!}dy. \tag{5.4.14}$$

For $k \in \mathbb{Z}$, we take $\mathsf{q}_{k/\lambda}$ to be the $(m-1)$th-degree Taylor polynomial of f_λ at the point k/λ as follows:

$$\mathsf{q}_{k/\lambda}(x) := \sum_{j=0}^{m-1} f_\lambda^{(j)}(k/\lambda)\frac{(x-k/\lambda)^j}{j!}, \qquad x \in \mathbb{R}.$$

Since all $\phi_1, \dots, \phi_r$ and $\tilde{\phi}_1, \dots, \tilde{\phi}_r$ vanish outside $[-N, N]$ for some $N \in \mathbb{N}_0$, we have

$$\tilde{\phi}_\ell(\lambda y - n)\phi_\ell(\lambda x - n) = 0 \qquad \forall\, n \in \mathbb{Z},\ |y-x| > 2N/\lambda,\ \ell = 1, \dots, r.$$

Then for all $x \in I_k := [k/\lambda, (k+1)/\lambda]$,

$$\{y \in \mathbb{R} \;:\; |y-x| \leqslant 2N/\lambda\} \subseteq [(k-2N)/\lambda, (k+2N+1)/\lambda] =: I_{k,N}.$$

Thus, for every $x \in I_k$, we have

$$\begin{aligned} &|\langle f_\lambda - \mathsf{q}_{k/\lambda}, \lambda\tilde{\phi}_\ell(\lambda \cdot - n)\rangle\phi_\ell(\lambda x - n)| \\ &\quad\leqslant \int_{\mathbb{R}} |f_\lambda(y) - \mathsf{q}_{k/\lambda}(y)| \cdot \lambda|\tilde{\phi}_\ell(\lambda y - n)\phi_\ell(\lambda x - n)|dy \\ &\quad= \int_{|y-x|\leqslant 2N/\lambda} |f_\lambda(y) - \mathsf{q}_{k/\lambda}(y)| \cdot \lambda|\tilde{\phi}_\ell(\lambda y - n)|dy|\phi_\ell(\lambda x - n)| \end{aligned}$$

$$\leqslant \|\lambda\tilde{\phi}_\ell(\lambda\cdot -n)\|_{L_{p'}(\mathbb{R})}\|f_\lambda - \mathsf{q}_{k/\lambda}\|_{L_p(I_{k,N})}|\phi_\ell(\lambda x-n)|$$
$$= \lambda^{1/p}\|\tilde{\phi}_\ell\|_{L_{p'}(\mathbb{R})}\|f_\lambda - \mathsf{q}_{k/\lambda}\|_{L_p(I_{k,N})}|\phi_\ell(\lambda x-n)|.$$

By assumption $\mathcal{Q}\mathsf{p} = c\mathsf{p}$ for all $\mathsf{p} \in \mathbb{P}_{m-1}$, we have $\mathcal{Q}_\lambda \mathsf{q}_{k/\lambda} = c\mathsf{q}_{k/\lambda}$. Thus, for $x \in I_k$,

$$\begin{aligned}|\mathcal{Q}_\lambda f_\lambda(x) - c\mathsf{q}_{k/\lambda}(x)| &= |\mathcal{Q}_\lambda(f_\lambda - \mathsf{q}_{k/\lambda})(x)| \\ &= \Big|\sum_{\ell=1}^{r}\sum_{n\in\mathbb{Z}}\langle f_\lambda - \mathsf{q}_{k/\lambda}, \lambda\tilde{\phi}_\ell(\lambda\cdot -n)\rangle\phi_\ell(\lambda x-n)\Big| \\ &\leqslant \sum_{\ell=1}^{r}\lambda^{1/p}\|\tilde{\phi}_\ell\|_{L_{p'}(\mathbb{R})}\|f_\lambda - \mathsf{q}_{k/\lambda}\|_{L_p(I_{k,N})}\sum_{n\in\mathbb{Z}}|\phi_\ell(\lambda x-n)|.\end{aligned}$$

Therefore,

$$\begin{aligned}\|\mathcal{Q}_\lambda f_\lambda - c\mathsf{q}_{k/\lambda}\|_{L_p(I_k)} &\leqslant \sum_{\ell=1}^{r}\lambda^{1/p}\|\tilde{\phi}_\ell\|_{L_{p'}(\mathbb{R})}\|f_\lambda - \mathsf{q}_{k/\lambda}\|_{L_p(I_{k,N})}\Big\|\sum_{n\in\mathbb{Z}}|\phi_\ell(\lambda\cdot -n)|\Big\|_{L_p(I_k)} \\ &= \|f_\lambda - \mathsf{q}_{k/\lambda}\|_{L_p(I_{k,N})}\sum_{\ell=1}^{r}\|\tilde{\phi}_\ell\|_{L_{p'}(\mathbb{R})}\|\phi_\ell\|_{\mathcal{L}_p(\mathbb{R})}.\end{aligned}$$

Consequently, we proved

$$\begin{aligned}\|\mathcal{Q}_\lambda f_\lambda - cf_\lambda\|_{L_p(I_k)} &\leqslant \|\mathcal{Q}_\lambda f_\lambda - c\mathsf{q}_{k/\lambda}\|_{L_p(I_k)} + c\|f_\lambda - \mathsf{q}_{k/\lambda}\|_{L_p(I_k)} \\ &\leqslant C_2\|f_\lambda - \mathsf{q}_{k/\lambda}\|_{L_p(I_{k,N})},\end{aligned} \tag{5.4.15}$$

where $C_2 := c + \sum_{\ell=1}^{r}\|\tilde{\phi}_\ell\|_{L_{p'}(\mathbb{R})}\|\phi_\ell\|_{\mathcal{L}_p(\mathbb{R})} < \infty$. We now estimate $\|f_\lambda - \mathsf{q}_{k/\lambda}\|_{L_p(I_{k,N})}$. By (5.4.14) with $g = f_\lambda$ and the definition of $\mathsf{q}_{k/\lambda}$, applying Minkowski's inequality, we have

$$\begin{aligned}\|f_\lambda - \mathsf{q}_{k/\lambda}\|_{L_p(I_{k,N})} &= \Big\|\int_0^{\cdot-k/\lambda} f_\lambda^{(m)}(\cdot - y)\frac{y^{m-1}}{(m-1)!}dy\Big\|_{L_p(I_{k,N})} \\ &\leqslant \Big\|\int_{-2N/\lambda}^{(2N+1)/\lambda} |f_\lambda^{(m)}(\cdot - y)|\frac{|y|^{m-1}}{(m-1)!}dy\Big\|_{L_p(I_{k,N})} \\ &\leqslant \int_{-2N/\lambda}^{(2N+1)/\lambda} \|f_\lambda^{(m)}(\cdot - y)\|_{L_p(I_{k,N})}\frac{|y|^{m-1}}{(m-1)!}dy \\ &\leqslant \|f_\lambda^{(m)}\|_{L_p(\tilde{I}_{k,N})}\int_{-2N/\lambda}^{(2N+1)/\lambda}\frac{|y|^{m-1}}{(m-1)!}dy \\ &= \frac{(2N)^m + (2N+1)^m}{m!}\lambda^{-m}\|f_\lambda^{(m)}\|_{L_p(\tilde{I}_{k,N})},\end{aligned}$$

where $\tilde{I}_{k,N} := [(k-4N-1)/\lambda, (k+4N+1)/\lambda]$. It follows from (5.4.15) that

$$\|\mathcal{Q}_\lambda f_\lambda - cf_\lambda\|_{L_p(I_k)} \leqslant C_3\lambda^{-m}\|f_\lambda^{(m)}\|_{L_p(\tilde{I}_{k,N})},$$

where $C_3 := C_2\frac{(2N)^m+(2N+1)^m}{m!} < \infty$. For $1 \leqslant p < \infty$, we have

$$\begin{aligned}\|\mathcal{Q}_\lambda f_\lambda - cf_\lambda\|_{L_p(\mathbb{R})} &= \Big(\sum_{k\in\mathbb{Z}}\|\mathcal{Q}_\lambda f_\lambda - cf_\lambda\|^p_{L_p(I_k)}\Big)^{\frac{1}{p}}\\ &\leqslant C_3\lambda^{-m}\Big(\sum_{k\in\mathbb{Z}}\|f_\lambda^{(m)}\|^p_{L_p(\tilde{I}_{k,N})}\Big)^{\frac{1}{p}} = C_3\lambda^{-m}(8N+2)^{\frac{1}{p}}\|f_\lambda^{(m)}\|_{L_p(\mathbb{R})}.\end{aligned}$$

It is straightforward to check that the above inequality also holds for $p=\infty$. Now it follows from (5.4.7) and (5.4.13) that (5.4.11) holds with $C = C_1 + C_3(8N+2)^{\frac{1}{p}}2^m\|\psi_0\|_{L_{p'}(\mathbb{R})} < \infty$.

By (5.4.4), the inequality (5.4.12) follows directly from (5.4.11). □

5.5 Accuracy and Approximation Orders of Shift-Invariant Spaces

As we have seen in Theorem 5.4.2, the approximation property of the quasi-projection operator $\mathcal{Q}$ is closely related to their ability to preserve polynomials in certain polynomial space: $\mathcal{Q}\mathsf{p} = \mathsf{p}$ for all $\mathsf{p} \in \mathbb{P}_{m-1}$. In particular, we have $\mathcal{Q}\mathbb{P}_{m-1} = \mathbb{P}_{m-1}$. By the definition of the quasi-projection operator $\mathcal{Q}$ in (5.4.9), this polynomial preservation property necessarily implies $\mathbb{P}_{m-1} = \mathcal{Q}\mathbb{P}_{m-1} \subseteq \mathsf{S}(\phi_1,\dots,\phi_r)$. Hence, it is natural to study when $\mathbb{P}_{m-1} \subseteq \mathsf{S}(\phi_1,\dots,\phi_r)$, with the largest possible integer m called the *accuracy order* of the shift-invariant space $\mathsf{S}(\phi_1,\dots,\phi_r)$.

In this section we characterize accuracy order and approximation order of shift-invariant subspaces in $L_p(\mathbb{R})$ with $1 \leqslant p \leqslant \infty$. We also study approximation by quasi-interpolation operators in $L_p(\mathbb{R})$.

5.5.1 Accuracy Order of Shift-Invariant Spaces

Since a shift-invariant space is obtained by the convolution of all sequences with its generating functions, we first study when the convolution of a polynomial sequence with a compactly supported distribution is still a polynomial.

Theorem 5.5.1 *Let φ be a compactly supported distribution on $\mathbb{R}$ and p be a polynomial. Then $\mathsf{p}|_{\mathbb{Z}} * \varphi := \sum_{k\in\mathbb{Z}} \mathsf{p}(k)\varphi(\cdot - k)$ is a polynomial if and only if*

$$\mathsf{p}^{(j)}(-i\tfrac{d}{d\xi})\widehat{\varphi}(2\pi k) := [\mathsf{p}^{(j)}(-i\tfrac{d}{d\xi})\widehat{\varphi}(\xi)]|_{\xi=2\pi k} = 0, \quad \forall\, k \in \mathbb{Z}\backslash\{0\}, j \in \mathbb{N}_0. \tag{5.5.1}$$

Moreover, if (5.5.1) *holds, then* $\mathsf{p}|_{\mathbb{Z}} * \varphi = \int_{\mathbb{R}} \mathsf{p}(\cdot - y)\varphi(y)dy = \mathsf{p} * \varphi$ *and*

$$(\mathsf{p} * \varphi)(x) = \sum_{j=0}^{\infty} [\mathsf{p}^{(j)}(-i\tfrac{d}{d\xi})\widehat{\varphi}](0)\frac{x^j}{j!} = [\mathsf{p}(x - i\tfrac{d}{d\xi})\widehat{\varphi}(\xi)]|_{\xi=0} = (\mathsf{p} * u_\varphi)(x), \tag{5.5.2}$$

where $u_\varphi \in l_0(\mathbb{Z})$ *satisfies* $\widehat{u_\varphi}(\xi) = \widehat{\varphi}(\xi) + \mathscr{O}(|\xi|^{\deg(\mathsf{p})+1})$ *as* $\xi \to 0$.

Proof For $n \in \mathbb{N}_0$, plugging $f(x) = x^n\varphi(x)$ into the Poisson summation formula in (5.2.6), noting that $\widehat{f}(\xi) = i^n\widehat{\varphi}^{(n)}(\xi)$, we have

$$\sum_{k\in\mathbb{Z}} (x-k)^n\varphi(x-k) = \sum_{k\in\mathbb{Z}} i^n\widehat{\varphi}^{(n)}(2\pi k)e^{i2\pi kx}$$

with both series converging in the sense of distributions. By the Taylor expansion of p at the point x, we have $\mathsf{p}(y) = \sum_{n=0}^{\infty} \frac{(-1)^n}{n!}\mathsf{p}^{(n)}(x)(x-y)^n$ and therefore,

$$\begin{aligned}\sum_{k\in\mathbb{Z}} \mathsf{p}(k)\varphi(x-k) &= \sum_{n=0}^{\infty} \frac{(-1)^n}{n!}\mathsf{p}^{(n)}(x)\sum_{k\in\mathbb{Z}}(x-k)^n\varphi(x-k)\\ &= \sum_{n=0}^{\infty} \frac{(-i)^n}{n!}\mathsf{p}^{(n)}(x)\sum_{k\in\mathbb{Z}}\widehat{\varphi}^{(n)}(2\pi k)e^{i2\pi kx}.\end{aligned}$$

Therefore, we proved that for any polynomial p and any compactly supported distribution φ,

$$(\mathsf{p}|_{\mathbb{Z}} * \varphi)(x) := \sum_{k\in\mathbb{Z}} \mathsf{p}(k)\varphi(x-k) = \sum_{k\in\mathbb{Z}}\sum_{n=0}^{\infty} \frac{(-i)^n}{n!}\widehat{\varphi}^{(n)}(2\pi k)\mathsf{p}^{(n)}(x)e^{i2\pi kx} \tag{5.5.3}$$

in the sense of distributions. For a function $\psi \in \mathscr{D}(\mathbb{R})$, we have (see Exercise 5.11)

$$\int_{\mathbb{R}} \mathsf{p}^{(n)}(x)e^{i2\pi kx}\psi(x)dx = \widehat{[\mathsf{p}^{(n)}\psi]}(-2\pi k) = [\mathsf{p}^{(n)}(i\tfrac{d}{d\xi})\widehat{\psi}(\xi)]|_{\xi=-2\pi k}.$$

Therefore, the identity in (5.5.3) simply means

$$\langle \mathsf{p}|_{\mathbb{Z}} * \varphi, \overline{\psi}\rangle = \sum_{k\in\mathbb{Z}}\sum_{n=0}^{\infty} \frac{(-i)^n}{n!}\widehat{\varphi}^{(n)}(2\pi k)[\mathsf{p}^{(n)}(i\tfrac{d}{d\xi})\widehat{\psi}](-2\pi k)$$
$$= \sum_{k\in\mathbb{Z}}\sum_{j=0}^{\infty} \frac{i^j}{j!}\widehat{\psi}^{(j)}(-2\pi k)\sum_{n=0}^{\infty}\frac{(-i)^n}{n!}\mathsf{p}^{(j+n)}(0)\widehat{\varphi}^{(n)}(2\pi k).$$

Rewriting the last summation in the above identity, we conclude that

$$\langle \mathsf{p}|_{\mathbb{Z}} * \varphi, \overline{\psi}\rangle = \sum_{k\in\mathbb{Z}}\sum_{j=0}^{\infty} \frac{i^j}{j!}\widehat{\psi}^{(j)}(-2\pi k)[\mathsf{p}^{(j)}(-i\tfrac{d}{d\xi})\widehat{\varphi}](2\pi k), \quad \forall\, \psi \in \mathscr{D}(\mathbb{R}). \tag{5.5.4}$$

For a polynomial q, we have

$$\begin{aligned}\langle \mathsf{q}, \overline{\psi}\rangle &= \int_{\mathbb{R}} \mathsf{q}(x)\psi(x)dx = \sum_{j=0}^{\infty}\frac{1}{j!}\mathsf{q}^{(j)}(0)\int_{\mathbb{R}} x^j\psi(x)dx \\ &= \sum_{j=0}^{\infty}\frac{i^j}{j!}\mathsf{q}^{(j)}(0)\widehat{\psi}^{(j)}(0) = \mathsf{q}(i\tfrac{d}{d\xi})\widehat{\psi}(0).\end{aligned} \tag{5.5.5}$$

If $\mathsf{p}|_{\mathbb{Z}} * \varphi$ is a polynomial q, comparing the coefficients of $\widehat{\psi}^{(j)}(2\pi k)$ in (5.5.4) with (5.5.5) (see Exercises 5.14–5.18 for details), we conclude that (5.5.1) must hold and $\mathsf{q}^{(j)}(0) = [\mathsf{p}^{(j)}(-i\frac{d}{d\xi})\widehat{\varphi}](0)$ for all $j \in \mathbb{N}_0$, which is just (5.5.2), where we also used (1.2.5) for the last identity in (5.5.2).

If the condition in (5.5.1) is satisfied, it is trivial to deduce from (5.5.4) that

$$\langle \mathsf{p}|_{\mathbb{Z}} * \varphi, \overline{\psi}\rangle = \sum_{j=0}^{\infty}\frac{i^j}{j!}\widehat{\psi}^{(j)}(0)[\mathsf{p}^{(j)}(-i\tfrac{d}{d\xi})\widehat{\varphi}](0) = \sum_{j=0}^{\infty}\frac{1}{j!}\langle x^j, \overline{\psi(x)}\rangle[\mathsf{p}^{(j)}(-i\tfrac{d}{d\xi})\widehat{\varphi}](0)$$
$$= \Big\langle \sum_{j=0}^{\infty}[\mathsf{p}^{(j)}(-\tfrac{d}{d\xi})\widehat{\varphi}](0)\tfrac{x^j}{j!}, \overline{\psi(x)}\Big\rangle.$$

Hence, $\mathsf{p} * \varphi = [\mathsf{p}(\cdot - i\frac{d}{d\xi})\widehat{\varphi}(\xi)]|_{\xi=0}$ is a polynomial.

Since $\mathsf{p} \in \mathscr{C}^{\infty}(\mathbb{R})$ and φ is compactly supported, $\int_{\mathbb{R}} \mathsf{p}(x-y)\varphi(y)dy$ is well defined. Noting that $\mathsf{p}(x-y) = \sum_{j=0}^{\infty}\frac{\mathsf{p}^{(j)}(x)}{j!}(-y)^j$, we have

$$\int_{\mathbb{R}} \mathsf{p}(x-y)\varphi(y)dy = \sum_{j=0}^{\infty}\frac{\mathsf{p}^{(j)}(x)}{j!}\int_{\mathbb{R}}\varphi(y)(-y)^j dy$$
$$= \sum_{j=0}^{\infty}\frac{\mathsf{p}^{(j)}(x)}{j!}(-i)^j\widehat{\varphi}^{(j)}(0) = [\mathsf{p}(x - i\tfrac{d}{d\xi})\widehat{\varphi}(\xi)]|_{\xi=0}.$$

By Lemma 1.2.1, this proves $\mathsf{p}|_{\mathbb{Z}} * \varphi = \int_{\mathbb{R}} \mathsf{p}(\cdot - y)\varphi(y)dy = \mathsf{p} * \varphi = \mathsf{p} * u_\varphi$. □

In fact, the convolution $\mathsf{p} * \varphi$ is a tempered distribution and (5.5.3) also holds in the sense of tempered distributions. Consequently, (5.5.4) holds for all $\psi \in \mathscr{S}(\mathbb{R})$.

We now characterize when a quasi-projection operator $\mathcal{Q}$ can preserve every polynomial from a polynomial space $\mathbb{P}_{m-1}$.

Proposition 5.5.2 *Let $\phi_1, \ldots, \phi_r, \tilde{\phi}_1, \ldots, \tilde{\phi}_r$ be compactly supported distributions on $\mathbb{R}$. Let a quasi-projection operator $\mathcal{Q}$ be defined in (5.4.9). Define $\phi := (\phi_1, \ldots, \phi_r)^\mathsf{T}$ and $\tilde{\phi} := (\tilde{\phi}_1, \ldots, \tilde{\phi}_r)^\mathsf{T}$. For $m \in \mathbb{N}$, $\mathcal{Q}\mathsf{p} = \mathsf{p}$ for all $\mathsf{p} \in \mathbb{P}_{m-1}$ if and only if*

$$\overline{\widehat{\tilde{\phi}}(\xi)}^\mathsf{T} \widehat{\phi}(\xi + 2\pi k) = \boldsymbol{\delta}(k) + \mathscr{O}(|\xi|^m), \qquad \xi \to 0 \quad \text{for all} \quad k \in \mathbb{Z}. \tag{5.5.6}$$

Moreover, the identity $\mathcal{Q}\mathbb{P}_{m-1} = \mathbb{P}_{m-1}$ holds if and only if $\overline{\widehat{\tilde{\phi}}(\xi)}^\mathsf{T} \widehat{\phi}(\xi + 2\pi k) = \mathscr{O}(|\xi|^m)$ as $\xi \to 0$ for all $k \in \mathbb{Z}\backslash\{0\}$ and $\overline{\widehat{\tilde{\phi}}(0)}^\mathsf{T} \widehat{\phi}(0) \neq 0$.

Proof Let $\mathsf{p} \in \mathbb{P}_{m-1}$. By (5.5.5) (also see Exercise 5.11), we have

$$\langle \mathsf{p}, \tilde{\phi}(\cdot - k)\rangle = \langle \mathsf{p}(\cdot + k), \tilde{\phi}\rangle = [\mathsf{p}(i\tfrac{d}{d\xi} + k)\overline{\widehat{\tilde{\phi}}(\xi)^\mathsf{T}}]|_{\xi=0} = \sum_{j=0}^{\infty} \frac{(-i)^j}{j!} \mathsf{p}^{(j)}(k) \overline{\widehat{\tilde{\phi}}^{(j)}(0)}^\mathsf{T}.$$

Let $u_\phi \in (l_0(\mathbb{Z}))^{1\times r}$ such that $\overline{\widehat{u_\phi}(\xi)}^\mathsf{T} = \widehat{\tilde{\phi}}(\xi) + \mathscr{O}(|\xi|^m)$ as $\xi \to 0$. By (1.2.5) in Lemma 1.2.1, we conclude that

$$\langle \mathsf{p}, \tilde{\phi}(\cdot - k)\rangle = \mathsf{p} * u_\phi, \quad \forall\, \mathsf{p} \in \mathbb{P}_{m-1}.$$

Consequently, for $\mathsf{p} \in \mathbb{P}_{m-1}$,

$$\mathcal{Q}\mathsf{p} = \sum_{k\in\mathbb{Z}} \langle \mathsf{p}, \tilde{\phi}(\cdot - k)\rangle \phi(\cdot - k) = \sum_{k\in\mathbb{Z}} (\mathsf{p} * u_\phi)(k)\phi(\cdot - k) = (\mathsf{p} * u_\phi) * \phi = \mathsf{p} * \varphi,$$

where $\varphi := u_\phi * \phi$. Since $\widehat{\varphi}(\xi + 2\pi k) = \widehat{u_\phi}(\xi)\widehat{\phi}(\xi + 2\pi k)$ for all $k \in \mathbb{Z}$, it follows directly from Theorem 5.5.1 that $\mathsf{p} * \varphi = \mathsf{p}$ for all $\mathsf{p} \in \mathbb{P}_{m-1}$ if and only if

$$\widehat{\varphi}(\xi + 2\pi k) = \boldsymbol{\delta}(k) + \mathscr{O}(|\xi|^m), \qquad \xi \to 0 \quad \text{for all } k \in \mathbb{Z}. \tag{5.5.7}$$

Similarly, by $\mathcal{Q}\mathsf{p} = \mathsf{p} * \varphi$ for $\mathsf{p} \in \mathbb{P}_{m-1}$, we see that $\mathcal{Q}\mathbb{P}_{m-1} = \mathbb{P}_{m-1}$ if and only if

$$\widehat{\varphi}(0) \neq 0 \quad \text{and} \quad \widehat{\varphi}(\xi + 2\pi k) = \mathscr{O}(|\xi|^m), \qquad \xi \to 0, \quad \forall\, k \in \mathbb{Z}\backslash\{0\}. \tag{5.5.8}$$

The condition (5.5.8) is called *the Strang-Fix (or moment) condition of order m.* □

For $\phi = (\phi_1, \dots, \phi_r)^\mathsf{T}$ with all entries $\phi_1, \dots, \phi_r$ being compactly supported distributions on $\mathbb{R}$, we define $\operatorname{ao}(\phi) := \sup\{m \in \mathbb{N}_0 : \mathbb{P}_{m-1} \subseteq \mathsf{S}(\phi)\}$, called the *accuracy order* of ϕ (or $\mathsf{S}(\phi)$).

A function $\varphi \in \mathsf{S}(\phi)$ satisfying (5.5.8) (or (5.5.7)) is often called *a superfunction* for a shift-invariant space $\mathsf{S}(\phi)$ in approximation theory. As demonstrated by the following result, the accuracy order property $\mathbb{P}_{m-1} \subseteq \mathsf{S}(\phi)$ with $m := \operatorname{ao}(\phi)$ can be achieved by $\mathbb{P}_{m-1} \subseteq \mathsf{S}(\varphi)$ using a suitable single function $\varphi \in \mathsf{S}(\phi)$. This is the main reason that such φ is called a superfunction in approximation theory.

Lemma 5.5.3 *Let $\phi := (\phi_1, \dots, \phi_r)^\mathsf{T}$ with $\phi_1, \dots, \phi_r$ being compactly supported distributions on $\mathbb{R}$. Define $\mathsf{S}(\phi) := \mathsf{S}(\phi_1, \dots, \phi_r)$. For any $m \in \mathbb{N}$, the relation $\mathbb{P}_{m-1} \subseteq \mathsf{S}(\phi)$ holds if and only if there exists a compactly supported distribution $\varphi \in \mathsf{S}(\phi)$ such that* (5.5.7) *(or* (5.5.8)*) holds. If in addition span$\{\widehat{\phi}(2\pi k) : k \in \mathbb{Z}\} = \mathbb{C}^r$, we can further take $\varphi = \upsilon * \phi$ for some $\upsilon \in (l_0(\mathbb{Z}))^{1\times r}$ such that* (5.5.7) *holds.*

Proof By Theorem 5.5.1, we see that (5.5.7) is equivalent to $\mathsf{p} * \varphi = \mathsf{p}$ for all $\mathsf{p} \in \mathbb{P}_{m-1}$. Similarly, (5.5.8) is equivalent to $\mathbb{P}_{m-1} * \varphi = \mathbb{P}_{m-1}$. Since $\varphi \in \mathsf{S}(\phi)$ is compactly supported, we have $\mathsf{S}(\varphi) \subseteq \mathsf{S}(\phi)$ (see Exercise 5.25). Hence, the sufficiency part ($\Leftarrow$) is trivial by $\mathbb{P}_{m-1} \subseteq \mathsf{S}(\varphi) \subseteq \mathsf{S}(\phi)$.

Necessity ($\Rightarrow$). We first assume that the integer shifts of ϕ are linearly independent. Therefore, item (iii) of Theorem 5.2.1 holds and we define $\tilde{\phi} := (\tilde{\phi}_1, \dots, \tilde{\phi}_r)^\mathsf{T}$. For every $\mathsf{p} \in \mathbb{P}_{m-1} \subseteq \mathsf{S}(\phi)$, there exists $u_\mathsf{p} \in (l(\mathbb{Z}))^{1\times r}$ such that $\mathsf{p} = u_\mathsf{p} * \phi$. By item (iii) of Theorem 5.2.1, we have $u_\mathsf{p}(k) = \langle \mathsf{p}, \tilde{\phi}(\cdot - k)\rangle$ for all $k \in \mathbb{Z}$. That is, we have $\mathsf{p} - u_\mathsf{p} * \phi = \mathcal{Q}\mathsf{p}$, where the quasi-projection operator $\mathcal{Q}$ is defined in (5.4.9). Let $\upsilon \in (l_0(\mathbb{Z}))^{1\times r}$ such that $\overline{\widehat{\upsilon}(\xi)}^\mathsf{T} = \widehat{\tilde{\phi}}(\xi) + \mathscr{O}(|\xi|^m)$ as $\xi \to 0$. By Proposition 5.5.2, (5.5.6) must hold and hence (5.5.7) holds with $\varphi := \upsilon * \phi \in \mathsf{S}(\phi)$.

We now prove the general case. By Theorem 5.2.4, there exist compactly supported distributions $\eta_1, \dots, \eta_s$ on $\mathbb{R}$ such that items (i) and (ii) of Theorem 5.2.4 are satisfied. Define $\eta := (\eta_1, \dots, \eta_s)^\mathsf{T}$. By what has been proved, there exists $\varphi := u_\eta * \eta \in \mathsf{S}(\eta) = \mathsf{S}(\phi)$ with $u_\eta \in (l_0(\mathbb{Z}))^{1\times s}$ such that (5.5.7) holds. If in addition span$\{\widehat{\phi}(2\pi k) : k \in \mathbb{Z}\} = \mathbb{C}^r$, we further have $s = r$ and $\det(\widehat{u_\eta}(0)) \ne 0$ in Theorem 5.2.4. Since $\det(\widehat{u_\eta}(0)) \ne 0$, there exists $\tilde{u} \in (l_0(\mathbb{Z}))^{r\times r}$ such that $\widehat{\tilde{u}}(\xi) = (\widehat{u}(\xi))^{-1} + \mathscr{O}(|\xi|^m)$ as $\xi \to 0$. Define $\upsilon := u_\eta * \tilde{u}$ and $\varphi := \upsilon * \phi$. By $\widehat{\phi}(\xi) = \widehat{\upsilon}(\xi)\widehat{\eta}(\xi)$, we have

$$\begin{aligned}\widehat{\varphi}(\xi + 2\pi k) &= \widehat{u_\eta}(\xi)\widehat{\tilde{u}}(\xi)\widehat{\phi}(\xi + 2\pi k) \\ &= \widehat{u_\eta}(\xi)\widehat{\eta}(\xi + 2\pi k) + \mathscr{O}(|\xi|^m) = \boldsymbol{\delta}(k) + \mathscr{O}(|\xi|^m), \quad \xi \to 0\end{aligned}$$

for all $k \in \mathbb{Z}$. □

We now study the accuracy order of a refinable vector function/distribution ϕ by linking it to its refinement filter/mask instead of the vector function ϕ itself.

Theorem 5.5.4 *Let $a \in (l_0(\mathbb{Z}))^{r\times r}$ and ϕ be an $r \times 1$ vector of compactly supported distributions such that $\phi = 2\sum_{k\in\mathbb{Z}} a(k)\phi(2\cdot -k)$. If span$\{\widehat{\phi}(2\pi k) : k \in \mathbb{Z}\}$*

$= \operatorname{span}\{\widehat{\phi}(\pi + 2\pi k) \; : \; k \in \mathbb{Z}\} = \mathbb{C}^r$ and $\mathbb{P}_{m-1} \subseteq \mathsf{S}(\phi)$ (that is, ϕ has accuracy order m), then there exists $\upsilon \in (l_0(\mathbb{Z}))^{1\times r}$ such that $\widehat{\upsilon}(0)\widehat{\phi}(0) = 1$ and

$$\widehat{\upsilon}(2\xi)\widehat{a}(\xi) = \widehat{\upsilon}(\xi) + \mathscr{O}(|\xi|^m), \qquad \widehat{\upsilon}(2\xi)\widehat{a}(\xi + \pi) = \mathscr{O}(|\xi|^m), \;\; \xi \to 0. \qquad (5.5.9)$$

Conversely, if there exists $\upsilon \in (l_0(\mathbb{Z}))^{1\times r}$ such that $\widehat{\upsilon}(0)\widehat{\phi}(0) = 1$ and (5.5.9) is satisfied, then $\mathbb{P}_{m-1} \subseteq \mathsf{S}(\phi)$.

Proof Since $\operatorname{span}\{\widehat{\phi}(2\pi k) \; : \; k \in \mathbb{Z}\} = \mathbb{C}^r$ and $\mathbb{P}_{m-1} \subseteq \mathsf{S}(\phi)$, by Lemma 5.5.3, there exists $\upsilon \in (l_0(\mathbb{Z}))^{1\times r}$ such that $\varphi := \upsilon * \phi$ satisfies (5.5.7). Since $\widehat{\varphi}(\xi) = \widehat{\upsilon}(\xi)\widehat{\phi}(\xi)$ and $\widehat{\phi}(2\xi) = \widehat{a}(\xi)\widehat{\phi}(\xi)$, for $k \in \mathbb{Z}$,

$$\begin{aligned} \widehat{\upsilon}(2\xi)\widehat{a}(\xi)\widehat{\phi}(\xi + 2\pi k) &= \widehat{\upsilon}(2\xi)\widehat{a}(\xi + 2\pi k)\widehat{\phi}(\xi + 2\pi k) \\ &= \widehat{\upsilon}(2\xi)\widehat{\phi}(2\xi + 4\pi k) = \widehat{\varphi}(2\xi + 4\pi k). \end{aligned} \qquad (5.5.10)$$

Therefore, by (5.5.7),

$$[\widehat{\upsilon}(2\xi)\widehat{a}(\xi) - \widehat{\upsilon}(\xi)]\widehat{\phi}(\xi + 2\pi k) = \widehat{\varphi}(2\xi + 4\pi k) - \widehat{\varphi}(\xi + 2\pi k) = \mathscr{O}(|\xi|^m) \qquad (5.5.11)$$

as $\xi \to 0$. Since $\operatorname{span}\{\widehat{\phi}(2\pi k) \; : \; k \in \mathbb{Z}\} = \mathbb{C}^r$, there exist $k_1, \ldots, k_r \in \mathbb{Z}$ such that $\det(\Phi(0)) \neq 0$, where $\Phi(\xi) := (\widehat{\phi}(\xi + 2\pi k_1), \ldots, \widehat{\phi}(\xi + 2\pi k_r))$ is an $r \times r$ matrix. Since $\widehat{\phi}$ is continuous, $\Phi(\xi)$ is continuous and invertible in a neighborhood of the origin. By (5.5.11), we have $[\widehat{\upsilon}(2\xi)\widehat{a}(\xi) - \widehat{\upsilon}(\xi)]\Phi(\xi) = \mathscr{O}(|\xi|^m)$ as $\xi \to 0$, from which we conclude that the first relation in (5.5.9) holds.

Replacing ξ by $\xi + \pi$ in (5.5.10), by (5.5.7), for all $k \in \mathbb{Z}$, we have

$$\widehat{\upsilon}(2\xi)\widehat{a}(\xi + \pi)\widehat{\phi}(\xi + \pi + 2\pi k) = \widehat{\varphi}(2\xi + 2\pi + 4\pi k) = \mathscr{O}(|\xi|^m), \quad \xi \to 0.$$

Since $\operatorname{span}\{\widehat{\phi}(\pi + 2\pi k) \; : \; k \in \mathbb{Z}\} = \mathbb{C}^r$, by a similar argument, the second relation in (5.5.9) follows directly from the above identity.

We now prove the converse direction. Define $\varphi := \upsilon * \phi$. Then $\widehat{\varphi}(0) = \widehat{\upsilon}(0)\widehat{\phi}(0) = 1$. For $k \in \mathbb{Z}\backslash\{0\}$, we can write $k = 2^{n-1}k_0$ for some $n \in \mathbb{N}$ and some odd integer k_0. Consequently,

$$\begin{aligned} \widehat{\varphi}(2^n\xi + 2\pi k) &= \widehat{\upsilon}(2^n\xi)\widehat{\phi}(2^n(\xi + \pi k_0)) \\ &= \widehat{\upsilon}(2^n\xi)\widehat{a}(2^{n-1}\xi)\cdots\widehat{a}(2\xi)\widehat{a}(\xi + \pi k_0)\widehat{\phi}(\xi + \pi k_0) \\ &= \widehat{\upsilon}(2\xi)\widehat{a}(\xi + \pi)\widehat{\phi}(\xi + \pi k_0) + \mathscr{O}(|\xi|^m) = \mathscr{O}(|\xi|^m), \qquad \xi \to 0. \end{aligned}$$

Therefore, (5.5.8) holds and $\mathbb{P}_{m-1} \subseteq \mathsf{S}(\phi)$ follows directly from Lemma 5.5.3. □

If (5.5.9) holds for $\upsilon \in (l_0(\mathbb{Z}))^{1\times r}$ with $\widehat{\upsilon}(0) \neq 0$, then we say that a has *order m sum rules* with *a (moment) matching filter* υ. We define $\operatorname{sr}(a \,\|\, \upsilon)$ to be the largest such possible integer m satisfying (5.5.9). If $r = 1$ and $\widehat{a}(0) = 1$, then it is trivial to see that (5.5.9) with $\widehat{\upsilon}(0) \neq 0$ is equivalent to $\widehat{a}(\xi + \pi) = \mathscr{O}(|\xi|^m)$ as

$\xi \to 0$. Indeed, the first relation in (5.5.9) is equivalent to taking $\upsilon \in l_0(\mathbb{Z})$ such that $\widehat{\upsilon}(\xi) = 1/\widehat{\phi^a}(\xi) + \mathscr{O}(|\xi|^m)$ as $\xi \to 0$, where $\widehat{\phi^a}(\xi) := \prod_{j=1}^{\infty} \widehat{a}(2^{-j}\xi)$. Since $\widehat{\upsilon}(0) = 1$, now the second relation in (5.5.9) is equivalent to $\widehat{a}(\xi + \pi) = \mathscr{O}(|\xi|^m)$ as $\xi \to 0$.

The definition of sum rules can be rewritten using cosets of a filter. Recall that for $\gamma \in \mathbb{Z}$, the γ-coset $a^{[\gamma]}$ of a filter a is defined to be

$$\begin{aligned} &\widehat{a^{[\gamma]}}(\xi) := \sum_{k \in \mathbb{Z}} a(\gamma + 2k)e^{-ik\xi}, \quad \text{that is,} \\ &a^{[\gamma]} = a(\gamma + \cdot)\downarrow 2 = \{a(\gamma + 2k)\}_{k \in \mathbb{Z}}. \end{aligned} \tag{5.5.12}$$

Using coset sequences, we can rewrite the conditions for sum rules as follows:

Lemma 5.5.5 *Let $a \in (l_0(\mathbb{Z}))^{r\times r}$ and $\upsilon \in (l_0(\mathbb{Z}))^{1\times r}$ with $\widehat{\upsilon}(0) \ne 0$. Then a satisfies order m sum rules with the matching filter υ (that is, (5.5.9) holds) if and only if*

$$\widehat{\upsilon}(\xi)\widehat{a^{[\gamma]}}(\xi) = 2^{-1}e^{i\gamma\xi/2}\widehat{\upsilon}(\xi/2) + \mathscr{O}(|\xi|^m), \quad \xi \to 0 \quad \forall\, \gamma = 0, 1. \tag{5.5.13}$$

Proof Since $\widehat{a}(\xi) = \widehat{a^{[0]}}(2\xi) + e^{-i\xi}\widehat{a^{[1]}}(2\xi)$, the equations in (5.5.9) become

$$\begin{aligned} &\widehat{\upsilon}(2\xi)\widehat{a^{[0]}}(2\xi) + e^{-i\xi}\widehat{\upsilon}(2\xi)\widehat{a^{[1]}}(2\xi) = \widehat{\upsilon}(\xi) + \mathscr{O}(|\xi|^m), \\ &\widehat{\upsilon}(2\xi)\widehat{a^{[0]}}(2\xi) - e^{-i\xi}\widehat{\upsilon}(2\xi)\widehat{a^{[1]}}(2\xi) = \mathscr{O}(|\xi|^m), \quad \xi \to 0, \end{aligned}$$

from which it is straightforward to see that (5.5.9) is equivalent to (5.5.13). ☐

In the rest of this subsection, we show that a superfunction φ has a very special structure and is tightly linked to B-spline functions defined in (5.4.5). To do so, we need the following auxiliary result which also plays a critical role in our study of vector cascade algorithms for investigating refinable vector functions.

Lemma 5.5.6 *Let $m \in \mathbb{N}_0$ and η be a compactly supported distribution on $\mathbb{R}$ such that*

$$\widehat{\eta}(\xi + 2\pi k) = \mathscr{O}(|\xi|^{m+1}), \quad \xi \to 0, \ \forall\, k \in \mathbb{Z}. \tag{5.5.14}$$

Define $[n_-, n_+] := \operatorname{fsupp}(\eta)$. Then $\eta = \nabla^{m+1} g$, $\mathsf{S}(\eta) = \mathsf{S}(g)$, and $\operatorname{fsupp}(g) \subseteq [n_-, n_+ - m - 1]$, where g is a compactly supported distribution defined by

$$g := \sum_{k=0}^{\infty} \frac{(m+k)!}{m!k!}\eta(\cdot - k). \tag{5.5.15}$$

Proof We prove the claim by induction on m. When $m = 0$, by the Poisson summation formula, (5.5.14) is equivalent to

$$\sum_{k\in\mathbb{Z}} \eta(\cdot - k) = 0. \tag{5.5.16}$$

Define $g := \sum_{k=0}^{\infty} \eta(\cdot - k)$. Then $\eta = g - g(\cdot - 1) = \nabla g$. Since η has compact support and $g = \sum_{k=0}^{\infty} \eta(\cdot - k) = -\sum_{k=-\infty}^{-1} \eta(\cdot - k)$, we see that g also has compact support and $\text{fsupp}(g) \subseteq [n_-, n_+ - 1]$. It follows from $\eta = \nabla g = (\nabla \boldsymbol{\delta}) * g$ that $\mathtt{S}(\eta) = \mathtt{S}(g)$. Hence, the claim holds for $m = 0$.

Suppose that the claim holds for $m-1$. We now prove the claim for m. By (5.5.14) and induction hypothesis, there exists a compactly supported distribution f such that $\eta = \nabla^m f$, $\text{fsupp}(f) \subseteq [n_-, n_+ - m]$ with $f = \sum_{k=0}^{\infty} \frac{(m-1+k)!}{(m-1)!k!} \eta(\cdot - k)$. By (5.5.14) and $\widehat{\eta}(\xi) = (1 - e^{-i\xi})^m \widehat{f}(\xi)$, we deduce that $\widehat{f}(2\pi k) = 0$ for all $k \in \mathbb{Z}$. Hence, (5.5.16) holds with η being replaced by f. By what has been proved, there exists a compactly supported distribution g such that $f = \nabla g$ and $g = \sum_{n=0}^{\infty} f(\cdot - n) = \sum_{n=0}^{\infty} \sum_{k=0}^{\infty} \frac{(m-1+k)!}{(m-1)!k!} \eta(\cdot - n - k) = \sum_{k=0}^{\infty} \frac{(m+k)!}{m!k!} \eta(\cdot - k)$. Also, we have $\text{fsupp}(g) \subseteq [n_-, n_+ - m - 1]$. Hence, $\eta = \nabla^m f = \nabla^{m+1} g = (\nabla^{m+1} \boldsymbol{\delta}) * g$, from which we have $\mathtt{S}(\eta) = \mathtt{S}(g)$. The proof is completed by induction. □

The following result completely characterizes a superfunction φ.

Theorem 5.5.7 *Let $m \in \mathbb{N}$. A compactly supported distribution φ on $\mathbb{R}$ satisfies*

$$\widehat{\varphi}(\xi + 2\pi k) = \mathscr{O}(|\xi|^m), \qquad \xi \to 0, \quad \forall\, k \in \mathbb{Z}\backslash\{0\} \tag{5.5.17}$$

if and only if there exists a compactly supported distribution g such that $\varphi = B_m * g$.

Proof Sufficiency ($\Leftarrow$). By $\widehat{B_m}(\xi) = (\frac{1-e^{-i\xi}}{i\xi})^m$ and $\widehat{\varphi}(\xi) = \widehat{B_m}(\xi)\widehat{g}(\xi)$, it is trivial to directly verify that (5.5.17) holds.

Necessity ($\Rightarrow$). Consider $\eta := D^m \varphi$. Then $\widehat{\eta}(\xi) = (i\xi)^m \widehat{\varphi}(\xi)$ and it follows directly from our assumption in (5.5.17) that $\widehat{\eta}(\xi + 2\pi k) = \mathscr{O}(|\xi|^m)$ as $\xi \to 0$ for all $k \in \mathbb{Z}$. By Lemma 5.5.6, there exists a compactly supported distribution g on $\mathbb{R}$ such that $\eta = \nabla^m g$, which implies $\widehat{\eta}(\xi) = (1 - e^{-i\xi})^m \widehat{g}(\xi)$. Consequently, we have $(i\xi)^m \widehat{\varphi}(\xi) = \widehat{\eta}(\xi) = (1 - e^{-i\xi})^m \widehat{g}(\xi)$, from which we conclude that $\widehat{\varphi}(\xi) = (\frac{1-e^{-i\xi}}{i\xi})^m \widehat{g}(\xi) = \widehat{B_m}(\xi)\widehat{g}(\xi)$. That is, we proved $\varphi = B_m * g$. □

5.5.2 *Approximation Order of Shift-Invariant Subspaces of $L_p(\mathbb{R})$*

To study the approximation properties of shift-invariant subspaces of $L_p(\mathbb{R})$, we need the following two auxiliary results.

The following first result deals with the membership of a function in a shift-invariant space.

Lemma 5.5.8 *Let $\phi_1, \dots, \phi_r$ be compactly supported distributions on $\mathbb{R}$. Define $\phi := (\phi_1, \dots, \phi_r)^{\mathsf{T}}$. If f is a distribution on $\mathbb{R}$ such that for all $n \in \mathbb{N}$, $f|_{(-n,n)} \in \mathsf{S}(\phi)|_{(-n,n)}$ (which means $\langle f, \psi \rangle = \langle u_n * \phi, \psi \rangle$ $\forall \psi \in \mathscr{D}(\mathbb{R})$ with supp$(\psi) \subseteq (-n, n)$ for some $u_n \in (l(\mathbb{Z}))^{1\times r}$), then $f \in \mathsf{S}(\phi)$.*

Proof By Theorem 5.2.4, we can assume that the integer shifts of ϕ are linearly independent and therefore, item (iii) of Theorem 5.2.1 holds. Define $\tilde{\phi} := (\tilde{\phi}_1, \dots, \tilde{\phi}_r)^{\mathsf{T}}$. Then there exists a positive integer N such that all $\phi_1, \dots, \phi_r, \tilde{\phi}_1, \dots, \tilde{\phi}_r$ vanish outside $[-N, N]$. Define $u(k) := \langle f, \tilde{\phi}(\cdot - k) \rangle$ for $k \in \mathbb{Z}$. By item (iii) of Theorem 5.2.1, we have

$$u_n(k) = \langle u_n * \phi, \tilde{\phi}(\cdot - k) \rangle = \langle f, \tilde{\phi}(\cdot - k) \rangle = u(k), \qquad \forall |k| \leqslant n - N.$$

Then for $\psi \in \mathscr{D}(\mathbb{R})$ with supp$(\psi) \subseteq (-n, n)$,

$$\begin{aligned}\langle u * \phi, \psi \rangle &= \Big\langle \sum_{|k| \leqslant n+N} u(k)\phi(\cdot - k), \psi \Big\rangle = \Big\langle \sum_{|k| \leqslant n+N} u_{n+2N}(k)\phi(\cdot - k), \psi \Big\rangle \\ &= \Big\langle \sum_{k\in\mathbb{Z}} u_{n+2N}(k)\phi(\cdot - k), \psi \Big\rangle = \langle f, \psi \rangle\end{aligned}$$

where we used $u_{n+2N}(k) = u(k)$ for all $|k| \leqslant n + N$. This proves $f = u * \phi \in \mathsf{S}(\phi)$. □

The following result links the approximation property of a shift-invariant subspace of $L_p(\mathbb{R})$ with the polynomial inclusion property (i.e., the accuracy order) of a shift-invariant space.

Proposition 5.5.9 *Let $1 \leqslant p \leqslant \infty$ and $\phi = (\phi_1, \dots, \phi_r)^{\mathsf{T}}$, where $\phi_1, \dots, \phi_r$ are compactly supported functions in $L_p(\mathbb{R})$. If ϕ (or $\mathsf{S}(\phi)$) provides L_p-*density order *$m - 1$:*

$$\lim_{\lambda\to\infty} \inf_{g\in \mathsf{S}_\lambda(\phi)\cap L_p(\mathbb{R})} \lambda^{m-1} \|f - g\|_{L_p(\mathbb{R})} = 0, \qquad \forall f \in \mathscr{D}(\mathbb{R}), \tag{5.5.18}$$

then $\mathbb{P}_{m-1} \subseteq \mathsf{S}(\phi)$, that is, ϕ (or $\mathsf{S}(\phi)$) has the accuracy order at least m.

Proof Let k be the smallest nonnegative integer such that $x^n \in \mathsf{S}(\phi)$ for all $n = 0, \dots, k-1$ but $\mathsf{q}(x) := x^k \notin \mathsf{S}(\phi)$. We now prove that $k \geqslant m$. Suppose not. Then $k < m$. By Lemma 5.5.8, there exists $n \in \mathbb{N}$ such that $\mathsf{q}|_{(-n,n)} \notin \mathsf{S}(\phi)|_{(-n,n)}$. Since $\mathsf{S}(\phi)|_{(-n,n)}$ is finite-dimensional, we have

$$\varepsilon := \inf_{g\in \mathsf{S}(\phi)\cap L_p(\mathbb{R})} \|\mathsf{q} - g\|_{L_p([-n,n])} > 0.$$

Since $\mathsf{q}(x) = x^k$, there exists $f \in \mathscr{D}(\mathbb{R})$ such that $f|_{(-1,1)} = \mathsf{q}|_{(-1,1)}$. For $\lambda > 0$ and $g \in \mathsf{S}(\phi)$,

$$\begin{aligned}\|f - \lambda^{-k} g(\lambda\cdot)\|_{L_p([-1,1])} &= \|\mathsf{q} - \lambda^{-k} g(\lambda\cdot)\|_{L_p([-1,1])}\\ &= \lambda^{-1/p}\|\mathsf{q}(\lambda^{-1}\cdot) - \lambda^{-k} g\|_{L_p([-\lambda,\lambda])}\\ &= \lambda^{-k-1/p}\|\mathsf{q} - g\|_{L_p([-\lambda,\lambda])}.\end{aligned}$$

Suppose $1 \leqslant p < \infty$. For $\lambda = \ell n$ with $\ell \in \mathbb{N}$, we have

$$\|\mathsf{q} - g\|_{L_p([-\lambda,\lambda])} = \|\mathsf{q} - g\|_{L_p([-\ell n,\ell n])} = \Big(\sum_{j=0}^{\ell-1} \|\mathsf{q} - g\|^p_{L_p([(2j-\ell)n,(2j+2-\ell)n])}\Big)^{1/p}.$$

Note that

$$\begin{aligned}\|\mathsf{q} - g\|^p_{L_p([(2j-\ell)n,(2j+2-\ell)n])} &= \|\mathsf{q}(\cdot + (2j-\ell+1)n) - g(\cdot + (2j-\ell+1)n)\|_{L_p([-n,n])}\\ &= \|\mathsf{q} - g_{\mathsf{q}}\|_{L_p([-n,n])} \geqslant \varepsilon,\end{aligned}$$

since $g_{\mathsf{q}} := \mathsf{q} - \mathsf{q}(\cdot+(2j-\ell+1)n) + g(\cdot+(2j-\ell+1)n) \in \mathbb{P}_{k-1} + g(\cdot+(2j-\ell+1)n) \in \mathsf{S}(\phi)$ by $\mathbb{P}_{k-1} \subseteq \mathsf{S}(\phi)$. Thus,

$$\begin{aligned}\|f - \lambda^{-k} g(\lambda\cdot)\|_{L_p([-1,1])} &= \lambda^{-k-1/p}\|\mathsf{q} - g\|_{L_p([-\lambda,\lambda])}\\ &\geqslant \lambda^{-k-1/p}\ell^{1/p}\varepsilon = \lambda^{-k} n^{-1/p}\varepsilon.\end{aligned} \tag{5.5.19}$$

It is trivial to check that (5.5.19) also holds for $p = \infty$. Since $k < m$, (5.5.19) is a contradiction to (5.5.18). Thus, we must have $k \geqslant m$. Hence, we proved $\mathbb{P}_{m-1} \subseteq \mathbb{P}_{k-1} \subseteq \mathsf{S}(\phi)$. □

Before presenting the main result on the approximation properties of a shift-invariant space in $L_p(\mathbb{R})$, we need the following auxiliary result.

Lemma 5.5.10 *For $m \in \mathbb{N}$, $\varepsilon > 0$ and a smooth function $\mathbf{v}$ (for example, $\mathbf{v}(\xi) = \sum_{j=0}^{m} c_j\xi^j$), there exist $\eta, h \in \mathscr{D}(\mathbb{R})$ such that both η and h are supported inside $(-\varepsilon, \varepsilon)$ and $\widehat{\eta}(\xi) = \mathbf{v}(\xi) + \mathscr{O}(|\xi|^m)$ and $h(\xi) = \mathbf{v}(\xi) + \mathscr{O}(|\xi|^m)$ as $\xi \to 0$.*

Proof Let $n \in \mathbb{N}$ and $\psi \in \mathscr{D}(\mathbb{R})$ such that $\widehat{\psi}(0) = 1$. Take $u \in l_0(\mathbb{Z})$ such that $\widehat{u}(\xi) = 2^n\mathbf{v}(2^n\xi)/\widehat{\psi}(\xi) + \mathscr{O}(|\xi|^m)$ as $\xi \to 0$. Then $\eta := (u * \psi)(2^n\cdot)$ is a desired function in $\mathscr{D}(\mathbb{R})$ by letting n large enough.

Let $\psi \in \mathscr{D}(\mathbb{R})$ such that $\psi(0) \neq 0$ and ψ is supported inside $(-\varepsilon, \varepsilon)$. Let p be the mth-degree Taylor polynomial of $\mathbf{v}(\xi)/\psi(\xi)$ at $\xi = 0$. Then $h(\xi) := \psi(\xi)\mathsf{p}(\xi)$ is a desired function in $\mathscr{D}(\mathbb{R})$. □

For a superfunction φ satisfying $\widehat{\varphi}(0) \neq 0$ and $\widehat{\varphi}(\xi + 2\pi k) = \mathscr{O}(|\xi|^m)$ as $\xi \to 0$ for all $k \in \mathbb{Z}\backslash\{0\}$, using Lemma 5.5.10, in the following we see that there exists a function $\tilde{\varphi} \in \mathscr{D}(\mathbb{R})$ such that their associated quasi-projection operator $\mathcal{Q}f :=$

$\sum_{k\in\mathbb{Z}}\langle f, \tilde{\varphi}(\cdot-k)\rangle\varphi(\cdot-k)$ has the desired polynomial preservation property: $\mathcal{Q}\mathsf{p} = \mathsf{p}$ for all $\mathsf{p} \in \mathbb{P}_{m-1}$.

Proposition 5.5.11 *Let* $1 \leqslant p \leqslant \infty$ *and* $\varphi \in L_p(\mathbb{R})$ *be a compactly supported function such that* (5.5.8) *is satisfied, i.e.,* φ *is a superfunction satisfying* $\widehat{\varphi}(0) \neq 0$ *and* $\widehat{\varphi}(\xi + 2\pi k) = \mathscr{O}(|\xi|^m)$ *as* $\xi \to 0$ *for all* $k \in \mathbb{Z}\backslash\{0\}$. *Then there exist a positive constant* C *and* $\tilde{\varphi} \in \mathscr{D}(\mathbb{R})$ *such that* $\widehat{\tilde{\varphi}}(\xi) = 1/\overline{\widehat{\varphi}(\xi)} + \mathscr{O}(|\xi|^m)$ *as* $\xi \to 0$ *and for all* $\lambda > 0$,

$$\Big\| f - \sum_{k\in\mathbb{Z}} \langle f, \lambda\tilde{\varphi}(\lambda\cdot - k)\rangle \varphi(\lambda\cdot - k) \Big\|_{L_p(\mathbb{R})} \leqslant C\omega_m(f, \lambda^{-1})_p, \quad \forall f \in L_p(\mathbb{R}). \qquad (5.5.20)$$

Proof The existence of $\tilde{\varphi} \in \mathscr{D}(\mathbb{R})$ follows from Lemma 5.5.10. By (5.5.8), we conclude that $\overline{\widehat{\tilde{\varphi}}(\xi)}\widehat{\varphi}(\xi + 2\pi k) = \boldsymbol{\delta}(k) + \mathscr{O}(|\xi|^m)$ as $\xi \to 0$ for all $k \in \mathbb{Z}$. Hence $\mathcal{Q}_1\mathsf{p} = \mathsf{p}$ for all $\mathsf{p} \in \mathbb{P}_{m-1}$, where $\mathcal{Q}_\lambda f := \sum_{k\in\mathbb{Z}}\langle f, \lambda\tilde{\varphi}(\lambda\cdot - k)\rangle\varphi(\lambda\cdot - k)$. Now the claim follows directly from Theorem 5.4.2. □

Finally, we are ready to state the following main theorem which completely characterizes the approximation and accuracy orders of shift-invariant subspaces in $L_p(\mathbb{R})$.

Theorem 5.5.12 *Let* $1 \leqslant p \leqslant \infty$ *and* $\phi = (\phi_1, \ldots, \phi_r)^\mathsf{T}$, *where* $\phi_1, \ldots, \phi_r$ *are compactly supported functions in* $L_p(\mathbb{R})$. *Then the following statements are equivalent:*

(1) There exist compactly supported functions $\varphi \in \mathsf{S}(\phi) \cap L_p(\mathbb{R})$, $\tilde{\varphi} \in \mathscr{D}(\mathbb{R})$ *and a positive constant* $C > 0$ *such that the approximation property in* (5.5.20) *holds.*

(2) There exists a positive constant C *such that*

$$\inf_{g\in\mathsf{S}_\lambda(\phi)\cap L_p(\mathbb{R})} \|f - g\|_{L_p(\mathbb{R})} \leqslant C\omega_m(f, \lambda^{-1})_p, \qquad \forall f \in L_p(\mathbb{R}), \lambda > 0,$$

where $\mathsf{S}_\lambda(\phi) := \{g(\lambda\cdot) \,:\, g \in \mathsf{S}(\phi)\}$.

(3) ϕ *(or* $\mathsf{S}(\phi)$*) provides* L_p-approximation order m*: there exists* $C > 0$ *such that*

$$\inf_{g\in\mathsf{S}_\lambda(\phi)\cap L_p(\mathbb{R})} \|f - g\|_{L_p(\mathbb{R})} \leqslant C\lambda^{-m}\|f^{(m)}\|_{L_p(\mathbb{R})}, \quad \forall f \in W_p^m(\mathbb{R}), \lambda > 0.$$

(4) ϕ *(or* $\mathsf{S}(\phi)$*) provides* L_p*-density order* $m-1$, *that is,* (5.5.18) *holds.*

(5) ϕ *(or* $\mathsf{S}(\phi)$*) has accuracy order* m, *that is,* $\mathbb{P}_{m-1} \subseteq \mathsf{S}(\phi)$.

(6) There is a compactly supported function $\varphi \in \mathsf{S}(\phi)$ *satisfying* (5.5.7) *(or* (5.5.8)*).*

If in addition span$\{\widehat{\phi}(2\pi k) \,:\, k \in \mathbb{Z}\} = \mathbb{C}^r$, *then each of the above is equivalent to*

(7) There exists $\upsilon \in (l_0(\mathbb{Z}))^{1\times r}$ *such that* $\varphi := \upsilon * \phi$ *satisfies* (5.5.7) *(or* (5.5.8)*).*

Proof (1)$\Longrightarrow$(2) is trivial since $\sum_{k\in\mathbb{Z}}\langle f, \lambda\tilde{\varphi}(\lambda\cdot - k)\rangle\varphi(\lambda\cdot - k) \in \mathsf{S}_\lambda(\varphi) \cap L_p(\mathbb{R}) \subseteq \mathsf{S}_\lambda(\phi) \cap L_p(\mathbb{R})$. (2)$\Longrightarrow$(3) follows from (5.4.4). (3)$\Longrightarrow$(4) is obvious. (4)$\Longrightarrow$(5) is proved in Proposition 5.5.9. (5)$\Longrightarrow$(6) is proved in Lemma 5.5.3. Since $\phi_1, \ldots, \phi_r \in L_p(\mathbb{R})$ are compactly supported, every compactly supported function $\varphi \in \mathrm{S}(\phi)$ must be in $L_p(\mathbb{R})$. Now (6)$\Longrightarrow$(1) is proved in Proposition 5.5.11. (7)$\Longrightarrow$(6) is obvious. Under the condition $\operatorname{span}\{\widehat{\phi}(2\pi k) : k \in \mathbb{Z}\} = \mathbb{C}^r$, (5)$\Longrightarrow$(7) is proved in Lemma 5.5.3. □

5.5.3 Approximation by Quasi-Interpolation Operators in $L_p(\mathbb{R})$ *

Quasi-interpolation operators differ to quasi-projection operators in that $\tilde{\phi}_1, \ldots, \tilde{\phi}_r$ are finite linear combinations of translates and derivatives of the Dirac distribution $\boldsymbol{\delta}$. Theorem 5.4.2 for quasi-projection operators requires $\tilde{\phi}_1, \ldots, \tilde{\phi}_r \in L_{p'}(\mathbb{R})$ and hence excludes quasi-interpolation operators. We now generalize Theorem 5.4.2 by allowing $\tilde{\phi}_1, \ldots, \tilde{\phi}_r$ to be compactly supported distributions. To do so, we need the following lemma.

Lemma 5.5.13 *Let f be a compactly supported distribution on $\mathbb{R}$ such that $\widehat{f}(\xi) = \mathscr{O}(|\xi|^m)$ as $\xi \to 0$ and f vanishes outside $[c,d]$ for some $c, d \in \mathbb{R}$. Then there exists a compactly supported distribution g on $\mathbb{R}$ such that $f = D^m g$ and g vanishes outside $[c,d]$. If in addition $f \in W_p^n(\mathbb{R})$ (or $f \in \mathscr{D}(\mathbb{R})$) for some $1 \leqslant p \leqslant \infty$ and $n \in \mathbb{N}_0$, then $g \in W_p^{m+n}(\mathbb{R})$ (or $g \in \mathscr{D}(\mathbb{R})$).*

Proof Let $f_0 := f$. To find f_1 such that $f_0 = Df_1$, intuitively, we may define $f_1(x) := \int_{-\infty}^x f_0(y)dy$. Since $\widehat{f_0}(0) = 0$ and f_0 vanishes outside $[c,d]$, then f_1 will be well defined and will vanish outside $[c,d]$ provided that $f_0 \in L_1(\mathbb{R})$. Rigorously, we define f_1 as follows: For $h \in \mathscr{D}(\mathbb{R})$,

$$\langle f_1, h\rangle := -\langle f_0, H\rangle \quad \text{with} \quad H(x) := \int_{-\infty}^x h(y)dy.$$

Since f_0 is compactly supported and $H \in \mathscr{C}^\infty(\mathbb{R})$, we see that $\langle f_0, H\rangle$ and hence $\langle f_1, h\rangle$ is well defined. Now it is routine to use Theorem A.6.1 to check that f_1 is a distribution on $\mathbb{R}$. Since f_0 is supported inside $[c,d]$, we now prove that f_1 must be also supported inside $[c,d]$. Let $h \in \mathscr{D}(\mathbb{R})$ such that the support of h is either contained inside $(-\infty, c)$ or (d, ∞). If $\operatorname{supp}(h) \subseteq (-\infty, c)$, then $H(x) = C := \int_{\mathbb{R}} h(y)dy$ for all $x \geqslant c$ and consequently,

$$\langle f_1, h\rangle := -\langle f_0, C\rangle = \overline{C}\widehat{f_0}(0) = 0.$$

If $\operatorname{supp}(h) \subseteq (d, \infty)$, then $H(x) = 0$ for all $x \leqslant d$ and consequently, $\langle f_1, h\rangle := -\langle f_0, 0\rangle = 0$. This proves that $\operatorname{supp}(f_1) \subseteq [c,d]$. The claim $Df_1 = f_0$ follows from $\langle Df_1, h\rangle := -\langle f_1, h'\rangle = \langle f_0, h\rangle$ since $\int_{-\infty}^x h'(y)dy = h(x)$. By $f_0 = Df_1$, we have $\widehat{f_0}(\xi) = i\xi\widehat{f_1}(\xi)$. It follows from $\widehat{f_0}(\xi) = \mathscr{O}(|\xi|^m)$ as $\xi \to 0$ that $\widehat{f_1}(\xi) = \mathscr{O}(|\xi|^{m-1})$

as $\xi \to 0$. By what has been proved, there exists compactly supported distributions $f_2, \dots, f_{m-1}$ such that $f_{j-1} = Df_j$ for all $j = 1, \dots, m$. Take $g := f_m$. Then $D^m g = D^m f_m = D^{m-1} f_{m-1} = \cdots = f_0 = f$.

If $f \in W_p^n(\mathbb{R})$ (or $f \in \mathscr{D}(\mathbb{R})$), then we indeed have $f_j(x) = \int_{-\infty}^x f_{j-1}(y)dy$ for all $j = 1, \dots, m$. Since all f_j are compactly supported, we conclude that $g \in W_p^{m+n}(\mathbb{R})$ (or $g \in \mathscr{D}(\mathbb{R})$). □

As a consequence of Lemma 5.5.13, we have

Proposition 5.5.14 *Let f be a distribution on $\mathbb{R}$ and $m \in \mathbb{N}$. If $D^m f = 0$, then $f \in \mathbb{P}_{m-1}$.*

Proof Let $\eta \in \mathscr{D}(\mathbb{R})$ as in Lemma 5.5.10 such that $\widehat{\eta}(\xi) = 1 + \mathscr{O}(|\xi|^m)$ as $\xi \to 0$. Define a polynomial $\mathsf{q}(x) := \sum_{j=0}^{m-1} (-1)^j \langle f, \overline{\eta^{(j)}} \rangle \frac{x^j}{j!}$. For $\psi \in \mathscr{D}(\mathbb{R})$, we define $\mathring{\psi} = \sum_{j=0}^{m-1} \frac{(-i)^j}{j!} \widehat{\psi}^{(j)}(0) \eta^{(j)} \in \mathscr{D}(\mathbb{R})$. By $\widehat{\eta}(\xi) = 1 + \mathscr{O}(|\xi|^m)$ as $\xi \to 0$, we have

$$\widehat{\mathring{\psi}}(\xi) = \sum_{j=0}^{m-1} \frac{(-i)^j}{j!} \widehat{\psi}^{(j)}(0)(i\xi)^j \widehat{\eta}(\xi) = \sum_{j=0}^{m-1} \widehat{\psi}^{(j)}(0) \frac{\xi^j}{j!} + \mathscr{O}(|\xi|^m) = \widehat{\psi}(\xi) + \mathscr{O}(|\xi|^m)$$

as $\xi \to 0$. By Lemma 5.5.13, there exists $h \in \mathscr{D}(\mathbb{R})$ such that $\psi = \mathring{\psi} + h^{(m)}$. Since $D^m f = 0$, we observe $\langle f, \overline{h^{(m)}} \rangle = (-1)^m \langle D^m f, \overline{h} \rangle = 0$. Hence, by $\mathsf{q}^{(j)}(0) = (-1)^j \langle f, \overline{\eta^{(j)}} \rangle$ for $j = 0, \dots, m-1$ and by the definition of $\mathring{\psi}$,

$$\begin{aligned} \langle f, \overline{\psi} \rangle &= \langle f, \overline{\mathring{\psi}} \rangle + \langle f, \overline{h^{(m)}} \rangle \\ &= \sum_{j=0}^{m-1} \frac{(-i)^j}{j!} \langle f, \overline{\eta^{(j)}} \rangle \overline{\widehat{\psi}^{(j)}(0)} = \sum_{j=0}^{m-1} \frac{i^j}{j!} \mathsf{q}^{(j)}(0) \overline{\widehat{\psi}^{(j)}(0)} = \langle \mathsf{q}, \overline{\psi} \rangle, \end{aligned}$$

where we used (5.5.5). Thus, we proved $f = \mathsf{q} \in \mathbb{P}_{m-1}$. □

The following result links a compactly supported distribution f with its distributional derivative $Df \in L_1^{loc}(\mathbb{R})$ with an absolutely continuous function on $\mathbb{R}$.

Lemma 5.5.15 *For an absolutely continuous function f on $\mathbb{R}$, $f' \in L_1^{loc}(\mathbb{R})$ and $f' = Df$ in the sense of distributions. Conversely, if f is a distribution on $\mathbb{R}$ such that both f and Df are in $L_1^{loc}(\mathbb{R})$, then f must be absolutely continuous and $f' = Df$.*

Proof Since f is absolutely continuous, we see that f is continuous, $f' \in L_1^{loc}(\mathbb{R})$ and

$$f(x) = f(0) + g(x) \quad \text{with} \quad g(x) := \int_0^x f'(t)dt, \qquad \forall\, x \in \mathbb{R}.$$

We first prove $Dg = g'$. For $\psi \in \mathscr{D}(\mathbb{R})$, note that $g\psi$ is absolutely continuous and has compact support. Hence, $\int_{\mathbb{R}} [g\psi]'(t)dt = 0$. In other words, we have

$\int_{\mathbb{R}} g(t)\psi'(t)dt = -\int_{\mathbb{R}} g'(t)\psi(t)dt$. Thus,

$$\langle Dg; \psi\rangle = -\langle g; \psi'\rangle = -\int_{\mathbb{R}} g(t)\psi'(t)dt = \int_{\mathbb{R}} g'(t)\psi(t)dt = \langle g'; \psi\rangle.$$

This proves $Dg = g'$. Therefore, $Df = g' = f'$, where we used $g(x) = \int_0^x f'(t)dt$.

Since $Df \in L_1^{loc}(\mathbb{R})$, we can define a continuous function h on $\mathbb{R}$ by $h(x) := \int_0^x Df(t)dt$ for $x \in \mathbb{R}$. Then h is absolutely continuous. By what has been proved, we must have $Dh = h' = Df$. In other words, $D(h-f) = 0$. By Proposition 5.5.14, $f = h + C$ for some constant C. Hence, f must be absolutely continuous. □

The following result provides another way of defining the Sobolev space $W_p^m(\mathbb{R})$ using distributional derivative.

Proposition 5.5.16 *For* $1 \leqslant p \leqslant \infty$ *and* $m \in \mathbb{N}_0$, $f \in W_p^m(\mathbb{R})$ *if and only if* f, Df, ..., $D^m f \in L_p(\mathbb{R})$.

Proof Necessity ($\Rightarrow$). If $f \in W_p^m(\mathbb{R})$, then all $f, \dots, f^{(m-1)}$ are absolutely continuous and therefore, can be regarded as tempered distributions. By Lemma 5.5.15, we have $D^j f = f^{(j)} \in L_p(\mathbb{R})$ for $j = 0, \dots, m$.

Sufficiency ($\Leftarrow$). Since $f, Df, \dots, D^m f \in L_p(\mathbb{R}) \subseteq L_1^{loc}(\mathbb{R})$, by Lemma 5.5.15, all $f, Df, \dots, D^{m-1}f$ are absolutely continuous and $D^j f = f^{(j)}$ for all $j = 1, \dots, m$. This proves $f \in W_p^m(\mathbb{R})$. □

We now show that the definition $W_2^m(\mathbb{R})$ with $m \in \mathbb{N}_0$ in (5.4.1) agrees with the one for $H^\tau(\mathbb{R})$ with $\tau \in \mathbb{R}$ in (4.6.1) of Sect. 4.6 when $\tau = m$ is a nonnegative integer.

Lemma 5.5.17 *Let* $m \in \mathbb{N}_0$ *and* f *be a tempered distribution on* $\mathbb{R}$. *Then* $f \in W_2^m(\mathbb{R})$ *if and only if* $\int_{\mathbb{R}} |\widehat{f}(\xi)|^2(1 + |\xi|^2)^m d\xi < \infty$, *i.e.*, $f \in H^m(\mathbb{R})$.

Proof By Proposition 5.5.16 with $p = 2$, $f \in W_2^m(\mathbb{R})$ if and only if $f, Df, \dots, D^m f \in L_2(\mathbb{R})$. By Plancherel's Theorem in Theorem A.5.6, this is equivalent to saying that $\|\widehat{D^j f}\|_{L_2(\mathbb{R})} < \infty$ for all $j = 0, \dots, m$. Since $\widehat{D^j f}(\xi) = (i\xi)^j \widehat{f}(\xi)$, now it is easy to conclude that this requirement is equivalent to $\int_{\mathbb{R}} |\widehat{f}(\xi)|^2(1 + |\xi|^2)^m d\xi < \infty$. □

Let f be a compactly supported distribution on $\mathbb{R}$ and $m \in \mathbb{N}$. Let $\eta \in \mathscr{D}(\mathbb{R})$ such that $\widehat{\eta}(\xi) = 1 + \mathscr{O}(|\xi|^m)$ as $\xi \to 0$. Take $u_f \in l_0(\mathbb{Z})$ such that $\widehat{u_f}(\xi) = \widehat{f}(\xi) + \mathscr{O}(|\xi|^m)$ as $\xi \to 0$. Define $\eta_f := u_f * \eta \in \mathscr{D}(\mathbb{R})$. Then $\widehat{f}(\xi) = \widehat{\eta_f}(\xi) + \mathscr{O}(|\xi|^m)$ as $\xi \to 0$. By Lemma 5.5.13, there exists a compactly supported distribution g_f such that

$$f = D^m g_f + \eta_f \quad \text{such that } \eta_f \in \mathscr{D}(\mathbb{R}) \text{ and} \\ g_f \text{ is a compactly supported distribution.} \tag{5.5.21}$$

Since $\eta_f \in \mathscr{D}(\mathbb{R})$, there exists a unique $\psi_f \in \mathscr{C}^\infty(\mathbb{R})$ such that $\psi_f^{(m)} = \eta_f$ and $\psi_f(x) = 0$ when $x \to -\infty$. Therefore, $D^m[g_f + \psi_f] = f$ and we call $g_f + \psi_f$ the standard mth antiderivative of the compactly supported distribution f. If there is another distribution g such that $D^m g = f$ and $g(x) = 0$ when $x \to -\infty$, by

Proposition 5.5.14, then $g_f + \psi_f - g \in \mathbb{P}_{m-1}$. Hence, we see that $g_f + \psi_f$ is unique due to the fact that $g_f(x) + \psi_f(x) = 0$ when $x \to -\infty$. Moreover, $g_f \in L_{p'}(\mathbb{R})$ if and only if $g \in L_{p'}^{loc}(\mathbb{R})$, since $g \in g_f + \psi_f - \mathbb{P}_{m-1}$.

The following general result will be needed for studying the approximation property using quasi-interpolation operators later.

Theorem 5.5.18 *Let $1 \leqslant p, p' \leqslant \infty$ such that $\frac{1}{p} + \frac{1}{p'} = 1$. Let $\phi_1, \dots, \phi_r$ be compactly supported functions in $L_p(\mathbb{R})$ and $\tilde{\phi}_1, \dots, \tilde{\phi}_r$ be compactly supported distributions on $\mathbb{R}$. As in (5.5.21), for $\ell = 1, \dots, r$, we define compactly supported distributions $g_\ell := g_{\tilde{\phi}_\ell}$ and $\eta_\ell := \eta_{\tilde{\phi}_\ell} \in \mathscr{D}(\mathbb{R})$ such that $\tilde{\phi}_\ell = D^m g_\ell + \eta_\ell$. Let $\mathcal{Q}_\lambda$ and $\mathcal{Q}$ be the quasi-projection operators defined in (5.4.9). Let $0 \le k \le m$. If $\mathcal{Q}\mathsf{p} = \mathsf{p}$ for all $\mathsf{p} \in \mathbb{P}_{m-1}$ and if $D^k g_1, \dots, D^k g_r \in L_{p'}(\mathbb{R})$, then there exists a positive constant C such that*

$$\|\mathcal{Q}_\lambda f - f\|_{L_p(\mathbb{R})} \leqslant C\lambda^{k-m}\omega_k(f^{(m-k)}, \lambda^{-1})_p, \qquad \forall f \in W_p^{m-k}(\mathbb{R}), \lambda > 0. \tag{5.5.22}$$

Proof For $f \in W_p^{m-k}(\mathbb{R})$, $\langle f, \tilde{\phi}_\ell \rangle$ and $\mathcal{Q}_\lambda f$ are well-defined, since $f^{(m-k)} \in L_p(\mathbb{R})$, $D^k g_\ell \in L_{p'}(\mathbb{R})$, $\eta_\ell \in \mathscr{D}(\mathbb{R})$, and

$$\langle f, \tilde{\phi}_\ell \rangle := \langle f, D^m g_\ell \rangle + \langle f, \eta_\ell \rangle = (-1)^{m-k} \langle f^{(m-k)}, D^k g_\ell \rangle + \langle f, \eta_\ell \rangle.$$

By $\tilde{\phi}_\ell = D^m g_\ell + \eta_\ell$, we have $\mathcal{Q}_\lambda f = \mathcal{Q}_{\lambda,1} f + \mathcal{Q}_{\lambda,2} f$, where

$$\mathcal{Q}_{\lambda,1} f := \sum_{\ell=1}^{r} \sum_{n \in \mathbb{Z}} \langle f, \lambda D^m g_\ell(\lambda \cdot - n) \rangle \phi_\ell(\lambda \cdot - n),$$

$$\mathcal{Q}_{\lambda,2} f := \sum_{\ell=1}^{r} \sum_{n \in \mathbb{Z}} \langle f, \lambda \eta_\ell(\lambda \cdot - n) \rangle \phi_\ell(\lambda \cdot - n).$$

Since $\widehat{\tilde{\phi}_\ell}(\xi) = \widehat{\eta_\ell}(\xi) + \mathscr{O}(|\xi|^m)$ as $\xi \to 0$, it follows from our assumption $\mathcal{Q}\mathsf{p} = \mathsf{p}$ for all $\mathsf{p} \in \mathbb{P}_{m-1}$ and Proposition 5.5.2 that $\mathcal{Q}_{1,2}\mathsf{p} = \mathsf{p}$ for all $\mathsf{p} \in \mathbb{P}_{m-1}$. Since $\eta_1, \dots, \eta_r \in \mathscr{D}(\mathbb{R}) \subseteq L_{p'}(\mathbb{R})$, we conclude from Theorem 5.4.2 that there exists a positive constant C_1 such that for all $\lambda > 0$,

$$\|\mathcal{Q}_{\lambda,2} f - f\|_{L_p(\mathbb{R})} \leqslant C_1 \omega_m(f, \lambda^{-1})_p \leqslant C_1 \lambda^{k-m} \omega_k(f^{(m-k)}, \lambda^{-1})_p.$$

Note that $\mathcal{Q}_{\lambda,1} f = (-1)^{m-k} \lambda^{k-m} \tilde{\mathcal{Q}}_\lambda(f^{(m-k)})$, where

$$\tilde{\mathcal{Q}}_\lambda F := \sum_{\ell=1}^{r} \sum_{n \in \mathbb{Z}} \langle F, \lambda D^k g_\ell(\lambda \cdot - n) \rangle \phi_\ell(\lambda \cdot - n), \qquad F \in L_p(\mathbb{R}).$$

For $\mathsf{p} \in \mathbb{P}_{k-1}$, we have $\langle \mathsf{p}, D^k g_\ell \rangle = (-1)^k \langle \mathsf{p}^{(k)}, g_\ell \rangle = \langle 0, g_\ell \rangle = 0$. Hence, $\tilde{\mathcal{Q}}_1 \mathsf{p} = 0$ for all $\mathsf{p} \in \mathbb{P}_{k-1}$. Since $D^k g_1, \ldots, D^k g_r \in L_{p'}$ and $\phi_1, \ldots, \phi_r \in L_p(\mathbb{R})$, we conclude from Theorem 5.4.2 that there exists a positive constant C_2 such that $\|\tilde{\mathcal{Q}}_\lambda f\|_{L_p(\mathbb{R})} \leqslant C_2 \omega_k(f, \lambda^{-1})_p$ for all $f \in L_p(\mathbb{R})$ and $\lambda > 0$. Thus,

$$\|\mathcal{Q}_{\lambda,1} f\|_{L_p(\mathbb{R})} = \lambda^{k-m} \|\tilde{\mathcal{Q}}_\lambda (f^{(m-k)})\|_{L_p(\mathbb{R})} \leqslant C_2 \lambda^{k-m} \omega_k(f^{(m-k)}, \lambda^{-1})_p.$$

In conclusion, we see that (5.5.22) holds with $C = C_1 + C_2$. □

When $k = m$, Theorem 5.5.18 is exactly Theorem 5.4.2.

Now the approximation property using quasi-interpolation operators is a direct consequence of Theorem 5.5.18.

Theorem 5.5.19 *Let $1 \leqslant p, p' \leqslant \infty$ such that $\frac{1}{p} + \frac{1}{p'} = 1$. Let $\phi_1, \ldots, \phi_r$ be compactly supported functions in $L_p(\mathbb{R})$. Let $m, n \in \mathbb{N}$ such that $1 \leqslant n \leqslant m$. Let $\tilde{\phi}_1, \ldots, \tilde{\phi}_r$ be finite linear combinations of $D^j \boldsymbol{\delta}(\cdot - t)$, $t \in \mathbb{R}$ and $j = 0, \ldots, n-1$, where $\boldsymbol{\delta}$ is the Dirac distribution on $\mathbb{R}$. Let $\mathcal{Q}_\lambda$ and $\mathcal{Q}$ be the quasi-projection operators defined in* (5.4.9). *If $\mathcal{Q}\mathsf{p} = \mathsf{p}$ for all $\mathsf{p} \in \mathbb{P}_{m-1}$, then there exists a positive constant C such that*

$$\|\mathcal{Q}_\lambda f - f\|_{L_p(\mathbb{R})} \leqslant C\lambda^{-n} \omega_{m-n}(f^{(n)}, \lambda^{-1})_p, \qquad \forall f \in W_p^n(\mathbb{R}), \lambda > 0.$$

Proof Define $x_+ := x$ for $x \geqslant 0$ and $x_+ := 0$ for $x < 0$. Then we have $D^j \boldsymbol{\delta} = D^m(\frac{x_+^{m-j-1}}{(m-j-1)!})$ for all $j = 0, \ldots, m-1$. Let $g_\ell + \psi_\ell$ be the standard mth antiderivative of $\tilde{\phi}_\ell$. Since $\tilde{\phi}_\ell$ is a finite linear combination of $D^j \boldsymbol{\delta}(\cdot - t)$ with $t \in \mathbb{R}$ and $j = 0, \ldots, n-1$, we must have $D^{m-n}[g_\ell + \psi_\ell] \in L_\infty^{loc}(\mathbb{R}) \subseteq L_{p'}^{loc}(\mathbb{R})$, or equivalently, $D^{m-n} g_\ell \in L_{p'}(\mathbb{R})$. The claim follows from Theorem 5.5.18 with $k = m - n$. □

5.6 Convergence of Cascade Algorithms in Sobolev Spaces $W_p^m(\mathbb{R})$

In this section we study refinable vector functions which are solutions to a refinement equation in a Sobolev space $W_p^m(\mathbb{R})$. As an iterative scheme, a cascade algorithm is the major mathematical tool for investigating various properties of a refinable vector function in $W_p^m(\mathbb{R})$.

Let $a \in (l_0(\mathbb{Z}))^{r \times r}$ be a finitely supported (matrix-valued) filter. The *refinement operator* $\mathcal{R}_a$ associated with the filter a is defined to be

$$\mathcal{R}_a f := 2 \sum_{k \in \mathbb{Z}} a(k) f(2 \cdot -k) \quad \text{with} \quad f = (f_1, \ldots, f_r)^\mathsf{T}. \tag{5.6.1}$$

Then ϕ is a solution to the refinement equation $\phi = 2\sum_{k \in \mathbb{Z}} a(k)\phi(2 \cdot -k)$ if and only if ϕ is a fixed point of $\mathcal{R}_a$, that is, $\mathcal{R}_a \phi = \phi$. Such a vector function ϕ is called *a refinable (vector) function* associated with a filter or (refinement) mask a. In this section we study the convergence of a cascade algorithm $\{\mathcal{R}_a^n f\}_{n=1}^\infty$ to a refinable function ϕ in Sobolev spaces $W_p^m(\mathbb{R})$ with $m \in \mathbb{N}_0$ and $1 \leqslant p \leqslant \infty$. Note that $\widehat{\mathcal{R}_a f}(\xi) = \widehat{a}(\xi/2)\widehat{f}(\xi/2)$ and ϕ satisfies the frequency-based refinement equation $\widehat{\phi}(2\xi) = \widehat{a}(\xi)\widehat{\phi}(\xi)$, where $\widehat{a}(\xi) := \sum_{k\in\mathbb{Z}} a(k)e^{-ik\xi}$.

The study of a cascade algorithm with multiplicity $r > 1$ in this section is unavoidably more complicated than the scalar case $r = 1$. The reader, who is only interested in the scalar case $r = 1$, can directly jump to Sect. 5.6.4 (also see Theorem 7.2.4) for a self-contained treatment of a scalar cascade algorithm with $r = 1$.

5.6.1 *Initial Functions in a Vector Cascade Algorithm*

Let us first study which kind of functions can serve as an initial eligible function in an iterative cascade algorithm. To do so, we need the following auxiliary result.

Lemma 5.6.1 *Let $1 \leqslant p \leqslant \infty$ and $m \in \mathbb{N}_0$. Let $g \in W_p^m(\mathbb{R})$ and $\{g_n\}_{n\in\mathbb{N}}$ be a sequence in $W_p^m(\mathbb{R})$ such that $\lim_{n\to\infty} \|g_n - g\|_{W_p^m(\mathbb{R})} = 0$ and all $g, g_n, n \in \mathbb{N}$ are supported inside $[-N, N]$ for some $N \in \mathbb{N}$. Then*

$$\lim_{n\to\infty} \widehat{g_n}^{(j)}(0) = \widehat{g}^{(j)}(0), \qquad \lim_{n\to\infty} 2^{mn}\widehat{g_n}^{(j)}(2^n\zeta) = 0, \ \forall\, \zeta \in \mathbb{R}\backslash\{0\}, j \in \mathbb{N}_0. \tag{5.6.2}$$

Proof For $f \in L_p(\mathbb{R})$ such that f is supported inside $[-N, N]$, we have $\|f\|_{L_1(\mathbb{R})} = \|f\|_{L_1([-N,N])} \leqslant (2N)^{1-1/p}\|f\|_{L_p(\mathbb{R})}$. Hence, we can assume $p = 1$.

Define $f(x) := (-ix)^j g(x)$ and $f_n(x) := (-ix)^j g_n(x)$, $x \in \mathbb{R}$ and $n \in \mathbb{N}$. Since all g, g_n are supported inside $[-N, N]$, we see that all $f, f_n \in W_1^m(\mathbb{R})$ and $\lim_{n\to\infty} \|f_n - f\|_{W_1^m(\mathbb{R})} = 0$. Since $f^{(m)} \in L_1(\mathbb{R})$, by the Riemann-Lebesgue Lemma in item (i) of Proposition A.5.1, $\lim_{n\to\infty}\widehat{f^{(m)}}(2^n\zeta) = 0$ for $\zeta \neq 0$. Noting that $\widehat{f_n^{(m)}}(\xi) = (i\xi)^m\widehat{f_n}(\xi) = (i\xi)^m\widehat{g_n}^{(j)}(\xi)$, we conclude that

$$\begin{aligned}
2^{mn}|\widehat{g_n}^{(j)}(2^n\zeta)| &= |(i\zeta)^{-m}\widehat{f_n^{(m)}}(2^n\zeta)| \\
&\leqslant |\zeta|^{-m}(|\widehat{f_n^{(m)}}(2^n\zeta) - \widehat{f^{(m)}}(2^n\zeta)| + |\widehat{f^{(m)}}(2^n\zeta)|) \\
&\leqslant |\zeta|^{-m}(\|f_n^{(m)} - f^{(m)}\|_{L_1(\mathbb{R})} + |\widehat{f^{(m)}}(2^n\zeta)|) \to 0, \quad n \to \infty.
\end{aligned}$$

Since $|\widehat{g_n}^{(j)}(0) - \widehat{g}^{(j)}(0)| = |\widehat{f_n}(0) - \widehat{f}(0)| \leqslant \|f_n - f\|_{L_1(\mathbb{R})} \to 0$ as $n \to \infty$, we have $\lim_{n\to\infty} \widehat{g_n}^{(j)}(0) = \widehat{g}^{(j)}(0)$. This proves (5.6.2). □

We now study properties of initial functions f in a cascade algorithm $\{\mathcal{R}_a^n f\}_{n=1}^\infty$.

Proposition 5.6.2 *Let $1 \leqslant p \leqslant \infty$ and $m \in \mathbb{N}_0$. Let $a \in (l_0(\mathbb{Z}))^{r\times r}$ and f be an $r \times 1$ vector of compactly supported functions in $W_p^m(\mathbb{R})$ such that span$\{\widehat{f}(2\pi k)$:*

$k \in \mathbb{Z}\} = \mathbb{C}^r$. Suppose that $\lim_{n\to\infty} \|\mathcal{R}_a^n f - \phi\|_{(W_p^m(\mathbb{R}))^r} = 0$ for some $\phi \in (W_p^m(\mathbb{R}))^r \setminus \{0\}$. Then $\phi = 2\sum_{k\in\mathbb{Z}} a(k)\phi(2\cdot -k)$, $\widehat{\phi}(0) \neq 0$ and

$$\begin{aligned}&1 \text{ is a simple eigenvalue of } \widehat{a}(0) \text{ and all other eigenvalues of } \widehat{a}(0) \text{ are}\\ &\text{less than } 2^{-m} \text{ in modulus.}\end{aligned} \tag{5.6.3}$$

Moreover, the following statements hold:

(i) There is a unique $\upsilon \in (l_0(\mathbb{Z}))^{1\times r}$ such that $\operatorname{fsupp}(\upsilon) \subseteq [0, m]$, $\widehat{\upsilon}(0)\widehat{\phi}(0) = 1$ *and*

$$\widehat{\upsilon}(0) \neq 0, \qquad \widehat{\upsilon}(2\xi)\widehat{a}(\xi) = \widehat{\upsilon}(\xi) + \mathscr{O}(|\xi|^{m+1}), \qquad \xi \to 0. \tag{5.6.4}$$

If in addition $\operatorname{span}\{\widehat{f}(\pi + 2\pi k) : k \in \mathbb{Z}\} = \mathbb{C}^r$, then a has order $m+1$ sum rules with the matching filter υ also satisfying $\widehat{\upsilon}(2\xi)\widehat{a}(\xi+\pi) = \mathscr{O}(|\xi|^{m+1})$ as $\xi \to 0$.

(ii) The initial function f must satisfy $\mathbb{P}_m \subseteq \mathsf{S}(f)$ and

$$\widehat{\upsilon}(0)\widehat{f}(0) = 1, \quad \widehat{\upsilon}(\xi)\widehat{f}(\xi + 2\pi k) = \mathscr{O}(|\xi|^{m+1}),\ \xi \to 0,\ \forall\, k \in \mathbb{Z}\setminus\{0\}. \tag{5.6.5}$$

(iii) The vector function ϕ provides L_p-approximation order $m+1$ and

$$\widehat{\upsilon}(\xi)\widehat{\phi}(\xi + 2\pi k) = \boldsymbol{\delta}(k) + \mathscr{O}(|\xi|^{m+1}), \qquad \xi \to 0,\ \forall\, k \in \mathbb{Z}. \tag{5.6.6}$$

Proof For $n \in \mathbb{N}$, define $f_n := \mathcal{R}_a^n f$ and $a_n \in (l_0(\mathbb{Z}))^{r\times r}$ by

$$\widehat{a_n}(\xi) := \widehat{a}(2^{n-1}\xi)\widehat{a}(2^{n-2}\xi)\cdots\widehat{a}(2\xi)\widehat{a}(\xi), \qquad n \in \mathbb{N}. \tag{5.6.7}$$

Since $a \in (l_0(\mathbb{Z}))^{r\times r}$ and f has compact support, there exists $N \in \mathbb{N}$ such that the filter a and the initial function f are supported inside $[-N, N]$. Thus, all $f_n, n \in \mathbb{N}$ and ϕ are supported inside $[-N, N]$ (See Exercise 5.44). Since $\widehat{\mathcal{R}_a f}(\xi) = \widehat{a}(\xi/2)\widehat{f}(\xi/2)$, we get $\widehat{f_n}(2^n\xi) = \widehat{a_n}(\xi)\widehat{f}(\xi)$. By Lemma 5.6.1 and $\widehat{a_n}(0) = [\widehat{a}(0)]^n$, we have

$$\lim_{n\to\infty} [\widehat{a}(0)]^n\widehat{f}(0) = \lim_{n\to\infty} \widehat{a_n}(0)\widehat{f}(0) = \lim_{n\to\infty} \widehat{f_n}(0) = \widehat{\phi}(0), \tag{5.6.8}$$

$$\lim_{n\to\infty} 2^{mn}[\widehat{a}(0)]^n\widehat{f}(2\pi k) = \lim_{n\to\infty} 2^{mn}\widehat{f_n}(2^n 2\pi k) = 0, \quad \forall\, k \in \mathbb{Z}\setminus\{0\}. \tag{5.6.9}$$

Since $\operatorname{span}\{\widehat{f}(2\pi k) : k \in \mathbb{Z}\} = \mathbb{C}^r$, if $\widehat{f}(0) \in \operatorname{span}\{\widehat{f}(2\pi k) : k \in \mathbb{Z}\setminus\{0\}\}$ or $\widehat{\phi}(0) = 0$, we deduce from (5.6.8) and (5.6.9) that $\lim_{n\to\infty}[\widehat{a}(0)]^n = 0$. By Lemma 5.1.1 and $\mathcal{R}_a\phi = \phi$, $\lim_{n\to\infty}[\widehat{a}(0)]^n = 0$ implies $\widehat{\phi}(\xi) = \lim_{n\to\infty}(\prod_{j=1}^n \widehat{a}(2^{-j}\xi))\widehat{\phi}(2^{-n}\xi) = 0$, a contradiction to our assumption that ϕ is not identically zero. Thus, $\widehat{\phi}(0) \neq 0$ and there exist $k_2, \dots, k_r \in \mathbb{Z}$ such that the $r \times r$ matrix $[\widehat{f}(0), \widehat{f}(2\pi k_2), \dots, \widehat{f}(2\pi k_r)]$ is invertible. Note that

$\widehat{\phi}(0) \notin \text{span}\{\widehat{f}(2\pi k) : k \in \mathbb{Z}\backslash\{0\}\}$, since otherwise (5.6.9) implies $\widehat{\phi}(0) = [\widehat{a}(0)]^n\widehat{\phi}(0) \to 0$ as $n \to \infty$, a contradiction to $\widehat{\phi}(0) \neq 0$. Consequently, $E := [\widehat{\phi}(0), \widehat{f}(2\pi k_2), \ldots, \widehat{f}(2\pi k_r)]$ must be an invertible matrix. Now by (5.6.9) and $\widehat{a}(0)\widehat{\phi}(0) = \widehat{\phi}(0)$, we have

$$[E^{-1}\widehat{a}(0)E]^n = E^{-1}[\widehat{a}(0)]^n E = \begin{bmatrix} 1 & 2^{-mn}o(1) \\ 0 & 2^{-mn}o(1)I_{r-1} \end{bmatrix}, \qquad n \to \infty,$$

from which we conclude that (5.6.3) holds.

Using the Lebniz differentiation formula, we see that (5.6.4) is equivalent to

$$\widehat{\upsilon}(0)\widehat{a}(0) = \widehat{\upsilon}(0) \quad \text{and} \quad \widehat{\upsilon}^{(j)}(0)2^j\widehat{a}(0) + \sum_{k=0}^{j-1} \frac{2^k j!}{k!(j-k)!}\widehat{\upsilon}^{(k)}(0)\widehat{a}^{(j-k)}(0) = \widehat{\upsilon}^{(j)}(0)$$

for $j = 1, \ldots, m$. Since 1 is a simple eigenvalue of $\widehat{a}(0)$, by $\widehat{a}(0)\widehat{\phi}(0) = \widehat{\phi}(0)$ and $\widehat{\upsilon}(0)\widehat{a}(0) = \widehat{\upsilon}(0)$, we see that $\widehat{\upsilon}(0)\widehat{\phi}(0) \neq 0$ if and only if $\widehat{\upsilon}(0) \neq 0$ and $\widehat{\phi}(0) \neq 0$ (see Exercise 5.55). Since $\widehat{\phi}(0) \neq 0$, there is a unique $\widehat{\upsilon}(0)$ such that $\widehat{\upsilon}(0)\widehat{a}(0) = \widehat{\upsilon}(0)$ and $\widehat{\upsilon}(0)\widehat{\phi}(0) = 1$. From the above identity, we have the following recursive formula:

$$\widehat{\upsilon}^{(j)}(0) = \sum_{k=0}^{j-1} \frac{2^k j!}{k!(j-k)!}\widehat{\upsilon}^{(k)}(0)\widehat{a}^{(j-k)}(0)[I_r - 2^j\widehat{a}(0)]^{-1}, \quad j = 1, \ldots, m. \tag{5.6.10}$$

That is, all $\widehat{\upsilon}^{(j)}(0), j = 0, \ldots, m$ are uniquely determined. By Lemma 2.1.2, there is a unique $\upsilon \in (l_0(\mathbb{Z}))^{1\times r}$ such that υ vanishes outside $[0, m]$ and $\widehat{\upsilon}^{(j)}(0), j = 0, \ldots, m$ take the prescribed values in (5.6.10). This proves the first part of item (i).

Define $g_n := \upsilon * f_n$ and $g := \upsilon * \phi$. Then $\lim_{n\to\infty} \|g_n - g\|_{W_p^m(\mathbb{R})} = 0$. By Lemma 5.6.1,

$$\lim_{n\to\infty} 2^{mn}\widehat{g_n}^{(j)}(2^n 2\pi k) = 0 \quad \forall\, k \in \mathbb{Z}\backslash\{0\}, j \in \mathbb{N}_0. \tag{5.6.11}$$

On the other hand, by $\widehat{f_n}(\xi) = \widehat{a}(\xi/2)\widehat{f_{n-1}}(\xi/2)$ and (5.6.4), for $k \in \mathbb{Z}$,

$$\begin{aligned}\widehat{g_n}(2^n(\xi + 2\pi k)) &= \widehat{\upsilon}(2^n\xi)\widehat{f_n}(2^n(\xi + 2\pi k)) = \widehat{\upsilon}(2^n\xi)\widehat{a}(2^{n-1}\xi)\widehat{f_{n-1}}(2^{n-1}(\xi + 2\pi k)) \\ &= \widehat{\upsilon}(2^{n-1}\xi)\widehat{f_{n-1}}(2^{n-1}(\xi + 2\pi k)) + \mathscr{O}(|\xi|^{m+1}) \\ &= \cdots = \widehat{\upsilon}(\xi)\widehat{f}(\xi + 2\pi k) + \mathscr{O}(|\xi|^{m+1}), \quad \xi \to 0.\end{aligned}$$

Now it follows from the above identity and (5.6.11) that for $k \in \mathbb{Z}\backslash\{0\}$ and $j = 0, \ldots, m$,

$$[\widehat{\upsilon}\widehat{f}(\cdot + 2\pi k)]^{(j)}(0) = [\widehat{g_n}(2^n(\cdot + 2\pi k))]^{(j)}(0) = 2^{jn}\widehat{g_n}^{(j)}(2^n 2\pi k) \to 0, \quad n \to \infty.$$

By (5.6.4) and $m \geqslant 0$, we have $\widehat{\upsilon}(0)\widehat{a}(0) = \widehat{\upsilon}(0)$. Now it follows from (5.6.8) that

$$\widehat{\upsilon}(0)\widehat{f}(0) = \widehat{\upsilon}(0)[\widehat{a}(0)]^n\widehat{f}(0) = \widehat{\upsilon}(0)\widehat{f_n}(0) \to \widehat{\upsilon}(0)\widehat{\phi}(0) = 1, \quad n \to \infty.$$

This proves (5.6.5). It follows directly from (5.6.5) and Lemma 5.5.3 that $\mathbb{P}_m \subseteq \mathsf{S}(f)$.

If we take $f = \phi$ as our initial function, then $f_n := \mathcal{R}_a^n \phi = \phi$ and the above same argument for item (ii) implies that (5.6.5) holds with f being replaced by ϕ. Since $\widehat{\phi}(2\xi) = \widehat{a}(\xi)\widehat{\phi}(\xi)$, we have

$$\widehat{\upsilon}(2\xi)\widehat{\phi}(2\xi) = \widehat{\upsilon}(2\xi)\widehat{a}(\xi)\widehat{\phi}(\xi) = \widehat{\upsilon}(\xi)\widehat{\phi}(\xi) + \mathscr{O}(|\xi|^{m+1}), \quad \xi \to 0.$$

Since $\widehat{\upsilon}(0)\widehat{\phi}(0) = 1$, using the Taylor series of $\widehat{\upsilon}(\xi)\widehat{\phi}(\xi)$ at $\xi = 0$, we conclude from the above relation that $\widehat{\upsilon}(\xi)\widehat{\phi}(\xi) = 1 + \mathscr{O}(|\xi|^{m+1})$ as $\xi \to 0$. This proves (5.6.6). It follows from Theorem 5.5.12 that ϕ provides L_p-approximation order $m+1$.

We now prove the second part of item (i). Since $\lim_{n\to\infty} \|\mathcal{R}_a^n f_1 - \phi\|_{W_p^m(\mathbb{R})} = 0$ is trivially true, (5.6.5) must also hold with f being replaced by f_1. By $\widehat{f_1}(2\xi) = \widehat{a}(\xi)\widehat{f}(\xi)$, in particular, for all $k \in \mathbb{Z}$,

$$\widehat{\upsilon}(2\xi)\widehat{a}(\xi+\pi)\widehat{f}(\xi+\pi+2\pi k) = \widehat{\upsilon}(2\xi)\widehat{f_1}(2\xi+2\pi+4\pi k) = \mathscr{O}(|\xi|^{m+1}), \quad \xi \to 0.$$

By our assumption $\operatorname{span}\{\widehat{f}(\pi + 2\pi k) \; : \; k \in \mathbb{Z}\} = \mathbb{C}^r$, we can directly conclude from the above identity that $\widehat{\upsilon}(2\xi)\widehat{a}(\xi + \pi) = \mathscr{O}(|\xi|^{m+1})$ as $\xi \to 0$. □

5.6.2 *Normal Form of a Matrix Filter*

To significantly reduce the complexity for studying refinable vector functions and vector cascade algorithms, in this subsection we introduce the normal form of a matrix filter, which greatly facilitates our analysis of cascade algorithms and refinable functions.

Recall that for $u \in (l(\mathbb{Z}))^{r\times s}$, we define $u^\star \in (l(\mathbb{Z}))^{s\times r}$ to be

$$u^\star(k) := \overline{u(-k)}^{\mathsf{T}}, \qquad k \in \mathbb{Z}, \quad \text{that is,} \quad \widehat{u^\star}(\xi) := (\widehat{u}(\xi))^\star := \overline{\widehat{u}(\xi)}^{\mathsf{T}}. \qquad (5.6.12)$$

For $1 \leqslant p \leqslant \infty$, we equip $u \in (l_p(\mathbb{Z}))^{r\times s}$ with the l_p-norm:

$$\|u\|_{(l_p(\mathbb{Z}))^{r\times s}} := \left(\sum_{k\in\mathbb{Z}} \|u(k)\|_{l_p}^p\right)^{1/p}$$

with $\|\cdot\|_{l_p}$ in (5.0.1) and the usual modification for $p=\infty$. For $u \in (l_p(\mathbb{Z}))^{r\times t}$ and $v \in (l_{p'}(\mathbb{Z}))^{s\times t}$ with $\frac{1}{p}+\frac{1}{p'}=1$,

$$\langle u, v\rangle := \sum_{k\in\mathbb{Z}} u(k)\overline{v(k)}^{\mathsf{T}} = [u * v^\star](0). \tag{5.6.13}$$

By the definition in (5.6.13), we observe that $\langle u, v\rangle \in \mathbb{C}^{r\times s}$, $\|u\|^2_{(l_2(\mathbb{Z}))^{r\times s}} = \operatorname{trace}(\langle u, u\rangle)$, and $\operatorname{trace}(\langle\cdot,\cdot\rangle)$ is an inner product on $(l_2(\mathbb{Z}))^{r\times s}$.

For $m \in \mathbb{N}_0$ and $\upsilon \in (l_0(\mathbb{Z}))^{1\times r}$, we define

$$\mathscr{P}_{m,\upsilon} := \{\mathsf{p} * \upsilon \;:\; \mathsf{p} \in \mathbb{P}_m\} \subseteq (\mathbb{P}_m)^{1\times r}, \tag{5.6.14}$$

$$\mathscr{V}_{m,\upsilon} := \{u \in (l_0(\mathbb{Z}))^{1\times r} \;:\; \widehat{\upsilon}(\xi)\overline{\widehat{u}(\xi)}^{\mathsf{T}} = \mathscr{O}(|\xi|^{m+1}), \quad \xi \to 0\}, \tag{5.6.15}$$

and for $1 \leqslant p \leqslant \infty$,

$$\mathscr{F}_{m,\upsilon,p} := \{f \in (W_p^m(\mathbb{R}))^r \;:\; f \text{ has compact support and satisfies (5.6.5)}\}. \tag{5.6.16}$$

For convenience, we define

$$\mathbb{P}_{-1} := \emptyset, \quad \mathscr{P}_{-1,\upsilon} := \emptyset, \quad \text{and} \quad \mathscr{V}_{-1,\upsilon} := (l_0(\mathbb{Z}))^{1\times r}. \tag{5.6.17}$$

By definitions in (5.6.14), (5.6.15) and (5.6.17), both $\mathscr{P}_{m,\upsilon}$ and $\mathscr{V}_{m,\upsilon}$ are shift-invariant, that is, $w(\cdot - k) \in \mathscr{P}_{m,\upsilon}$ and $u(\cdot - k) \in \mathscr{V}_{m,\upsilon}$ for all $w \in \mathscr{P}_{m,\upsilon}$, $u \in \mathscr{V}_{m,\upsilon}$, and $k \in \mathbb{Z}$.

The following result investigates the special eigenvalues of a transition operator and shall be used later in our study of a vector cascade algorithm.

Lemma 5.6.3 *Let $a \in l_0(\mathbb{Z})$ such that $\widehat{a}(\xi+\pi) = \mathscr{O}(|\xi|^{m+1})$ as $\xi \to 0$ (that is, the scalar filter a satisfies order $m+1$ sum rules). Define $[\mathcal{T}_a u](n) = 2\sum_{k\in\mathbb{Z}} u(k)a(k-2n)$ for $n \in \mathbb{Z}$. Then*

$$\mathcal{T}_a\mathscr{V}_{j,\boldsymbol{\delta}} \subseteq \mathscr{V}_{j,\boldsymbol{\delta}}, \qquad \mathcal{T}_a\nabla^j\boldsymbol{\delta} - 2^{-j}\overline{\widehat{a}(0)}\nabla^j\boldsymbol{\delta} \in \mathscr{V}_{j,\boldsymbol{\delta}}, \qquad \forall\, j = 0,\ldots,m. \tag{5.6.18}$$

That is, the number $2^{-j}\overline{\widehat{a}(0)}$ is the only eigenvalue with the eigenvector $\nabla^j\boldsymbol{\delta}$ of $\mathcal{T}_a|_{\mathscr{V}_{j-1,\boldsymbol{\delta}}/\mathscr{V}_{j,\boldsymbol{\delta}}}$ for $j = 0,\ldots,m$.

Proof We first prove (5.6.18). By assumption, we can write $\overline{\widehat{a}(\xi)} = (1+e^{-i\xi})^{m+1}\widehat{b}(\xi)$ for some $b \in l_0(\mathbb{Z})$ with $\widehat{b}(0) = 2^{-1-m}\overline{\widehat{a}(0)}$. For $j \in \mathbb{N}_0$ and $u \in l_0(\mathbb{Z})$,

$$\begin{aligned}\widehat{\mathcal{T}_a\nabla^j u}(\xi) &= \widehat{\nabla^j u}(\xi/2)\overline{\widehat{a}(\xi/2)} + \widehat{\nabla^j u}(\xi/2+\pi)\overline{\widehat{a}(\xi/2+\pi)}\\ &= (1-e^{-i\xi/2})^j\widehat{u}(\xi/2)(1+e^{-i\xi/2})^{m+1}\widehat{b}(\xi/2)\end{aligned}$$

$$+ (1+e^{-i\xi/2})^j\widehat{u}(\xi/2+\pi)(1-e^{-i\xi/2})^{m+1}\widehat{b}(\xi/2+\pi)$$
$$= \widehat{\nabla^j\boldsymbol{\delta}}(\xi)\widehat{c_j}(\xi),$$

where we used $\widehat{\mathcal{T}_a v}(\xi) = \widehat{v}(\xi/2)\overline{\widehat{a}(\xi/2)} + \widehat{v}(\xi/2+\pi)\overline{\widehat{a}(\xi/2+\pi)}$ and

$$\begin{aligned}\widehat{c_j}(\xi) :=&(1+e^{-i\xi/2})^{m+1-j}\widehat{u}(\xi/2)\widehat{b}(\xi/2)\\ &+ (1-e^{-i\xi/2})^{m+1-j}\widehat{u}(\xi/2+\pi)\widehat{b}(\xi/2+\pi).\end{aligned} \tag{5.6.19}$$

For $0 \leqslant j \leqslant m+1$, since $c_j \in l_0(\mathbb{Z})$ and $\mathscr{V}_{j-1,\boldsymbol{\delta}} = \{\nabla^j u : u \in l_0(\mathbb{Z})\}$, it follows from $\mathcal{T}_a \nabla^j u = \nabla^j\boldsymbol{\delta} * c_j = \nabla^j c_j$ that $\mathcal{T}_a\mathscr{V}_{j-1,\boldsymbol{\delta}} \subseteq \mathscr{V}_{j-1,\boldsymbol{\delta}}$.

By $\widehat{b}(0) = 2^{-1-m}\overline{\widehat{a}(0)}$ and (5.6.19), we have $\widehat{c_j}(0) = 2^{-j}\overline{\widehat{a}(0)}$ with $u = \boldsymbol{\delta}$ for all $j = 0,\dots,m$. Thus, taking $u = \boldsymbol{\delta}$, we have

$$\widehat{\mathcal{T}_a\nabla^j\boldsymbol{\delta}}(\xi) = 2^{-j}\overline{\widehat{a}(0)}\widehat{\nabla^j\boldsymbol{\delta}}(\xi) + \widehat{\nabla^j\boldsymbol{\delta}}(\xi)\big(\widehat{c_j}(\xi) - 2^{-j}\overline{\widehat{a}(0)}\big).$$

Since $c_j - 2^{-j}\overline{\widehat{a}(0)}\boldsymbol{\delta} \in \mathscr{V}_{0,\boldsymbol{\delta}}$, this proves (5.6.18). □

For $U \in (l_0(\mathbb{Z}))^{r\times r}$, we say that U is *strongly invertible* if $\det(\widehat{U})$ is a nonzero monomial. In other words, $(\widehat{U})^{-1}$ is an $r \times r$ matrix of 2π-periodic trigonometric polynomials. If U is strongly invertible, define U^{-1} to be the sequence in $(l_0(\mathbb{Z}))^{r\times r}$ such that $\widehat{U^{-1}}(\xi) := (\widehat{U}(\xi))^{-1}$. Then $U * U^{-1} = U^{-1} * U = \boldsymbol{\delta} I_r$.

We now introduce the normal form of a matrix-valued filter/mask.

Theorem 5.6.4 *Let $m \in \mathbb{N}_0$ and $a \in (l_0(\mathbb{Z}))^{r\times r}$. Let $v \in (l_0(\mathbb{Z}))^{1\times r}$ with $\widehat{v}(0) \ne 0$ satisfy* (5.6.4). *Then there exists a strongly invertible sequence $U \in (l_0(\mathbb{Z}))^{r\times r}$ such that $\widehat{\mathring{v}}(\xi) := \widehat{v}(\xi)\widehat{U}(\xi)$ is equal to*

$$(1+\mathscr{O}(|\xi|), \mathscr{O}(|\xi|^{m+1}), \dots, \mathscr{O}(|\xi|^{m+1})), \qquad \xi \to 0 \tag{5.6.20}$$

and the following statements hold:

(i) If ϕ is an $r \times 1$ vector of compactly supported distributions satisfying $\widehat{\phi}(2\xi) = \widehat{a}(\xi)\widehat{\phi}(\xi)$, define

$$\widehat{\mathring{a}}(\xi) := (\widehat{U}(2\xi))^{-1}\widehat{a}(\xi)\widehat{U}(\xi) \quad \textit{and} \quad \widehat{\mathring{\phi}}(\xi) := (\widehat{U}(\xi))^{-1}\widehat{\phi}(\xi), \tag{5.6.21}$$

then $\mathring{\phi}$ is an $r \times 1$ vector of compactly supported distributions satisfying $\widehat{\mathring{\phi}}(2\xi) = \widehat{\mathring{a}}(\xi)\widehat{\mathring{\phi}}(\xi)$ with $\mathring{a} \in (l_0(\mathbb{Z}))^{r\times r}$ such that the filter $\mathring{a}$ must take the *following form:*

$$\begin{bmatrix} a_{1,1} & a_{1,2} \\ a_{2,1} & a_{2,2} \end{bmatrix} \quad \text{with} \quad \widehat{a_{1,1}}(0) = 1, \quad \widehat{a_{1,2}}(\xi) = \mathscr{O}(|\xi|^{m+1}), \quad \xi \to 0, \tag{5.6.22}$$

where the filters $a_{1,1} \in l_0(\mathbb{Z}), a_{1,2} \in (l_0(\mathbb{Z}))^{1\times(r-1)}, a_{2,1} \in (l_0(\mathbb{Z}))^{(r-1)\times 1}$ *and* $a_{2,2} \in (l_0(\mathbb{Z}))^{(r-1)\times(r-1)}$.

(ii) $\mathscr{V}_{m,\upsilon}$ *is generated by* $\mathscr{B}_{m,\upsilon}$, *i.e.,* $\mathscr{V}_{m,\upsilon} = \text{span}\{u(\cdot - k) \ : \ u \in \mathscr{B}_{m,\upsilon}, k \in \mathbb{Z}\}$, *where*

$$\begin{aligned} \mathscr{B}_{m,\upsilon} &:= \{b_1 * U^\star, b_2 * U^\star, \dots, b_r * U^\star\} \\ &\text{with} \quad b_1 := (\nabla^{m+1}\boldsymbol{\delta} e_1)^\star, b_2 := (\boldsymbol{\delta} e_2)^\star, \dots, b_r := (\boldsymbol{\delta} e_r)^\star, \end{aligned} \tag{5.6.23}$$

where $e_j \in \mathbb{R}^r$ *is the jth unit coordinate vector with the only nonzero entry* 1 *at the jth entry.*

(iii) $\mathscr{P}_{m,\upsilon} = \mathscr{P}_{m,\mathring{\upsilon}} * U^{-1}$, $\mathscr{V}_{m,\upsilon} = \mathscr{V}_{m,\mathring{\upsilon}} * U^\star$, *and the following identities hold:*

$$\begin{aligned} \mathscr{P}_{m,\upsilon} &= \{\mathsf{p} \in (\mathbb{P}_m)^{1\times r} \ : \ \langle \mathsf{p}, u\rangle = 0 \ \forall\, u \in \mathscr{V}_{m,\upsilon}\}, \\ \mathscr{V}_{m,\upsilon} &= \{u \in (l_0(\mathbb{Z}))^{1\times r} \ : \ \langle \mathsf{p}, u\rangle = 0 \ \forall\, \mathsf{p} \in \mathscr{P}_{m,\upsilon}\}. \end{aligned} \tag{5.6.24}$$

(iv) *For a compactly supported vector function* $f \in (W_p^m(\mathbb{R}))^r$, $\widehat{\upsilon}(\xi)\widehat{f}(\xi + 2\pi k) = \mathscr{O}(|\xi|^{m+1})$ *as* $\xi \to 0$ *for all* $k \in \mathbb{Z}$ *if and only if there exist compactly supported functions* $g_1, \dots, g_r \in W_p^m(\mathbb{R})$ *such that* $f = \sum_{\ell=1}^r u_\ell^\star * g_\ell = [u_1^\star, \dots, u_r^\star] * g$, *where* $\{u_1, \dots, u_r\} = \mathscr{B}_{m,\upsilon}$ *and* $g := (g_1, \dots, g_r)^\mathsf{T}$.

If the function ϕ *in item (i) satisfies the additional condition* $\widehat{\upsilon}(0)\widehat{\phi}(0) \neq 0$, *then*

(v) *for any* $n \in \mathbb{N}$, *there exists a strongly invertible sequence* $U \in (l_0(\mathbb{Z}))^{r\times r}$ *such that* $\widehat{\mathring{\upsilon}}(\xi) = \widehat{\upsilon}(\xi)\widehat{U}(\xi)$ *is equal to the expression in* (5.6.20) *and item (i) holds with the additional properties*

$$\widehat{a_{2,1}}(\xi) = \mathscr{O}(|\xi|^n), \qquad \xi \to 0$$

and $(\mathring{\phi}_1, \dots, \mathring{\phi}_r)^\mathsf{T} := \mathring{\phi}$ *satisfies*

$$\widehat{\mathring{\phi}_1}(0) \neq 0 \quad \text{and} \quad \widehat{\mathring{\phi}_\ell}(\xi) = \mathscr{O}(|\xi|^n), \quad \xi \to 0, \ \forall\, \ell = 2, \dots, r. \tag{5.6.25}$$

Proof Write $\upsilon = (\upsilon_1, \dots, \upsilon_r)$. Since $\widehat{\upsilon}(0) \neq 0$, without loss of generality, we can assume $\widehat{\upsilon_1}(0) \neq 0$; otherwise, we permute the entries in υ. Since $\widehat{\upsilon_1}(0) \neq 0$, for $\ell = 2, \dots, r$, there exists $u_\ell \in l_0(\mathbb{Z})$ such that $\widehat{u_\ell}(\xi) = \widehat{\upsilon_\ell}(\xi)/\widehat{\upsilon_1}(\xi) + \mathscr{O}(|\xi|^{m+1})$ as

$\xi \to 0$. Define $U \in (l_0(\mathbb{Z}))^{r\times r}$ by

$$\widehat{U}(\xi) := \frac{1}{\widehat{\upsilon_1}(0)} \begin{bmatrix} 1 & -\widehat{u}(\xi) \\ 0 & I_{r-1} \end{bmatrix} \quad \text{with} \quad \widehat{u} := (\widehat{u_2}, \ldots, \widehat{u_r}).$$

Then U is strongly invertible and $\widehat{\mathring{\upsilon}}(\xi) := \widehat{\upsilon}(\xi)\widehat{U}(\xi)$ must satisfy (5.6.20). Item (i) can be directly checked by using $\widehat{\mathring{\upsilon}}(2\xi)\widehat{\mathring{a}}(\xi) = \widehat{\mathring{\upsilon}}(\xi) + \mathscr{O}(|\xi|^{m+1})$ as $\xi \to 0$.

By the relation $\mathring{\upsilon} = \upsilon * U$, since U is strongly invertible, the following identities can be directly checked:

$$\mathscr{P}_{m,\upsilon} = \mathscr{P}_{m,\mathring{\upsilon}} * U^{-1}, \quad \mathscr{V}_{m,\upsilon} = \mathscr{V}_{m,\mathring{\upsilon}} * U^{\star}, \quad \mathscr{F}_{m,\upsilon,p} = U * \mathscr{F}_{m,\mathring{\upsilon},p}. \tag{5.6.26}$$

Since $\widehat{\mathring{\upsilon}}$ satisfies (5.6.20), it follows from the definition of $\mathscr{V}_{m,\mathring{\upsilon}}$ that

$$\begin{aligned}\mathscr{V}_{m,\mathring{\upsilon}} &= \{(u_1, \ldots, u_r) \in (l_0(\mathbb{Z}))^{1\times r} \;:\; \widehat{u_1}(\xi) = \mathscr{O}(|\xi|^{m+1}), \quad \xi \to 0\} \\ &= (\nabla^{m+1} l_0(\mathbb{Z})) \times (l_0(\mathbb{Z}))^{1\times(r-1)}.\end{aligned}$$

Thus $\mathscr{V}_{m,\mathring{\upsilon}} = \operatorname{span}\{b_1(\cdot - k), \ldots, b_r(\cdot - k) \;:\; k \in \mathbb{Z}\}$ and we conclude from (5.6.26) that $\mathscr{V}_{m,\upsilon}$ is generated by $\mathscr{B}_{m,\upsilon}$. So, item (ii) holds,

Since $\langle \mathsf{p} * \upsilon, u\rangle = \langle \mathsf{p}, u * \upsilon^{\star}\rangle = [\mathsf{p} * (\upsilon * u^{\star})](0)$, it follows directly from (1.2.5) that the second identity in (5.6.24) holds. By $\langle \mathsf{p}, \nabla^{m+1} u\rangle = \langle \mathsf{p} * (\nabla^{m+1}\boldsymbol{\delta})^{\star}, u\rangle = \langle (\nabla^{m+1}\mathsf{p})(\cdot + m + 1), u\rangle$, we see that

$$\{\mathsf{p} \in \mathbb{P}_m \;:\; \langle \mathsf{p}, u\rangle = 0 \;\forall\, u \in \mathscr{V}_{m,\boldsymbol{\delta}}\} = \{\mathsf{p} \in \mathbb{P}_m \;:\; \langle \mathsf{p}, \nabla^{m+1} u\rangle = 0 \quad \forall\, u \in l_0(\mathbb{Z})\} = \mathbb{P}_m.$$

Hence, the first identity in (5.6.24) holds with υ being replaced by $\mathring{\upsilon}$. The general case of the first identity in (5.6.24) follows from (5.6.26). This proves item (iii).

The sufficiency part of item (iv) is trivial, since

$$\widehat{\upsilon}(\xi)\widehat{f}(\xi + 2\pi k) = \sum_{\ell=1}^{r} \widehat{\upsilon}(\xi)\overline{\widehat{u_\ell}(\xi)}^{\mathsf{T}} \widehat{g_\ell}(\xi + 2\pi k) = \mathscr{O}(|\xi|^{m+1})$$

as $\xi \to 0$ by $u_\ell \in \mathscr{B}_{m,\upsilon}$. Define $(\eta, g_2, \ldots, g_r)^{\mathsf{T}} := U^{-1} * f$, that is, $(\widehat{\eta}, \widehat{g_2}, \ldots, \widehat{g_r})^{\mathsf{T}} = (\widehat{U}(\xi))^{-1}\widehat{f}(\xi)$. By (5.6.20), $\widehat{\upsilon}(\xi)\widehat{f}(\xi + 2\pi k) = \mathscr{O}(|\xi|^{m+1})$ as $\xi \to 0$ for all $k \in \mathbb{Z}$ if and only if (5.5.14) holds. By Lemma 5.5.6, there exists a compactly supported function g_1 such that $\nabla^{m+1} g_1 = \eta$. Since $g_1 \in \mathsf{S}(\eta)$ has compact support and $\eta \in W_p^m(\mathbb{R})$, we have $g_1 \in W_p^m(\mathbb{R})$. Now one can directly check that $f = \sum_{\ell=1}^{r} u_\ell^{\star} * g_\ell$ with $u_\ell := b_\ell * U^{\star}$. This proves item (iv).

To prove item (v), assume that item (i) holds with a desired strongly invertible sequence $U \in (l_0(\mathbb{Z}))^{r\times r}$. Since $\widehat{\mathring{\upsilon}}(0)\widehat{\mathring{\phi}}(0) = \widehat{\upsilon}(0)\widehat{\phi}(0) \ne 0$ and $\widehat{\mathring{\upsilon}}$ takes the form in (5.6.20), we see that $\widehat{\mathring{\phi}_1}(0) \ne 0$, where $(\mathring{\phi}_1, \mathring{\phi}_2)^{\mathsf{T}} := \mathring{\phi}$ with $\mathring{\phi}_2$ being an $(r-1)\times 1$

column vector. Since $\widehat{\mathring{\phi}_1}(0) \neq 0$, there exists $c \in (l_0(\mathbb{Z}))^{r-1}$ such that $\widehat{c}(\xi) = \widehat{\mathring{\phi}_2}(\xi)/\widehat{\mathring{\phi}_1}(\xi) + \mathscr{O}(|\xi|^n)$ as $\xi \to 0$. Define $W \in (l_0(\mathbb{Z}))^{r\times r}$ by $\widehat{W}(\xi) := \begin{bmatrix} 1 & 0 \\ \widehat{c}(\xi) & I_{r-1} \end{bmatrix}$ and

$$\widehat{\breve{\upsilon}}(\xi) = \widehat{\mathring{\upsilon}}(\xi)\widehat{W}(\xi), \quad \widehat{\breve{a}}(\xi) = (\widehat{W}(2\xi))^{-1}\widehat{\mathring{a}}(\xi)\widehat{W}(\xi) \quad \text{and} \quad \widehat{\breve{\phi}}(\xi) := (\widehat{W}(\xi))^{-1}\widehat{\mathring{\phi}}(\xi).$$

By the definition of W, we see that W is strongly invertible and all the claims in item (i) hold with $\mathring{\upsilon}, \mathring{a}, \mathring{\phi}$ being replaced by $\breve{\upsilon}, \breve{a}, \breve{\phi}$, respectively. By the definition of c, it is trivial that (5.6.25) holds with $\mathring{\phi}$ being replaced by $\breve{\phi}$. By calculation and the above definition of $\breve{a}_{2,1}$, $\widehat{\breve{a}_{2,1}}(\xi) = \mathscr{O}(|\xi|^n)$ as $\xi \to 0$ if and only if

$$\begin{aligned} \widehat{\breve{a}_{2,1}}(\xi) &= \widehat{\mathring{a}_{2,1}}(\xi) - \widehat{c}(2\xi)\widehat{\mathring{a}_{1,1}}(\xi) + \widehat{\mathring{a}_{2,2}}(\xi)\widehat{c}(\xi) - \widehat{c}(2\xi)\widehat{\mathring{a}_{1,2}}(\xi)\widehat{c}(\xi) \\ &= \mathscr{O}(|\xi|^n), \quad \xi \to 0. \end{aligned} \tag{5.6.27}$$

Since $\widehat{\mathring{\phi}}(2\xi) = \widehat{\mathring{a}}(\xi)\widehat{\mathring{\phi}}(\xi)$, we have $\widehat{\mathring{\phi}_1}(2\xi) = \widehat{\mathring{a}_{1,1}}(\xi)\widehat{\mathring{\phi}_1}(\xi) + \widehat{\mathring{a}_{1,2}}(\xi)\widehat{\mathring{\phi}_2}(\xi)$ and $\widehat{\mathring{\phi}_2}(2\xi) = \widehat{\mathring{a}_{2,1}}(\xi)\widehat{\mathring{\phi}_1}(\xi) + \widehat{\mathring{a}_{2,2}}(\xi)\widehat{\mathring{\phi}_2}(\xi)$. By the definition of $c \in (l_0(\mathbb{Z}))^{r-1}$, we can now directly check that (5.6.27) holds. □

A matrix filter satisfying (5.6.22) is called a *normal form* of a matrix filter a. Since U is strongly invertible, properties related to the filter a can be equivalently investigated by studying the properties of its normal form. In certain sense, a normal form reduces a general matrix filter $a \in (l_0(\mathbb{Z}))^{r\times r}$ into a scalar filter $a_{1,1} \in l_0(\mathbb{Z})$ in (5.6.22) with a particular choice of a simple matching filter $\upsilon = (\upsilon_1, 0, \dots, 0)$ and $\widehat{\upsilon_1}(0) = 1$ for the filter a.

We now look at some applications of the normal form of a matrix filter.

Theorem 5.6.5 *Let $m \in \mathbb{N}_0$ and $a \in (l_0(\mathbb{Z}))^{r\times r}$. Let $\upsilon \in (l_0(\mathbb{Z}))^{1\times r}$ with $\widehat{\upsilon}(0) \neq 0$ satisfy* (5.6.4). *Let $U \in (l_0(\mathbb{Z}))^{r\times r}$ be strongly invertible such that $\widehat{\mathring{\upsilon}}(\xi) := \widehat{\upsilon}(\xi)\widehat{U}(\xi)$ satisfies* (5.6.20). *Define $\mathring{a}$ as in* (5.6.21). *Then the following statements are equivalent:*

(1) The filter a satisfies order $m+1$ sum rules *with the matching filter υ:*

$$\begin{aligned} \widehat{\upsilon}(2\xi)\widehat{a}(\xi) &= \widehat{\upsilon}(\xi) + \mathscr{O}(|\xi|^{m+1}), \quad \xi \to 0, \\ \widehat{\upsilon}(2\xi)\widehat{a}(\xi + \pi) &= \mathscr{O}(|\xi|^{m+1}), \quad \xi \to 0. \end{aligned} \tag{5.6.28}$$

(2) The filter $\mathring{a}$ satisfies order $m+1$ sum rules with the matching filter $\mathring{\upsilon}$ satisfying (5.6.20).

(3) The filter $\mathring{a}$ takes the form in (5.6.22) *and*

$$\widehat{a_{1,1}}(\xi + \pi) = \mathscr{O}(|\xi|^{m+1}), \qquad \widehat{a_{1,2}}(\xi + \pi) = \mathscr{O}(|\xi|^{m+1}), \qquad \xi \to 0. \tag{5.6.29}$$

(4) $\mathcal{S}_a \mathscr{P}_{m,\upsilon} = \mathscr{P}_{m,\upsilon}$, *where the subdivision operator* $\mathcal{S}_a : (l(\mathbb{Z}))^{s\times r} \to (l(\mathbb{Z}))^{s\times r}$ *is defined to be*

$$[\mathcal{S}_a u](n) := 2\sum_{k\in\mathbb{Z}} u(k)a(n-2k), \qquad n \in \mathbb{Z}. \tag{5.6.30}$$

(5) $\mathcal{T}_a \mathscr{V}_{m,\upsilon} = \mathscr{V}_{m,\upsilon}$, *where the transition operator* $\mathcal{T}_a : (l(\mathbb{Z}))^{s\times r} \to (l(\mathbb{Z}))^{s\times r}$ *is defined to be*

$$[\mathcal{T}_a u](n) := 2\sum_{k\in\mathbb{Z}} u(k)\overline{a(k-2n)}^{\mathsf{T}} = 2\sum_{k\in\mathbb{Z}} u(k)a^\star(2n-k), \qquad n \in \mathbb{Z}. \tag{5.6.31}$$

Any of items (1)–(5) implies that

$$\operatorname{spec}(\mathcal{S}_a|_{\mathscr{P}_{j,\upsilon}/\mathscr{P}_{j-1,\upsilon}}) = \{2^{-j}\} = \operatorname{spec}(\mathcal{T}_a|_{\mathscr{V}_{j-1,\upsilon}/\mathscr{V}_{j,\upsilon}}), \qquad j = 0,\dots,m, \tag{5.6.32}$$

where $\operatorname{spec}(T)$ *is the multiset of all eigenvalues of* T *counting multiplicity of the eigenvalues of* T.

Proof The equivalence (1) $\iff$ (2) $\iff$ (3) can be directly checked. Note that

$$\widehat{\mathcal{S}_a u}(\xi) = 2\widehat{u}(2\xi)\widehat{a}(\xi), \quad \widehat{\mathcal{T}_a u}(\xi) = \widehat{u}(\xi/2)\overline{\widehat{a}(\xi/2)}^{\mathsf{T}} + \widehat{u}(\xi/2+\pi)\overline{\widehat{a}(\xi/2+\pi)}^{\mathsf{T}}.$$

By (5.6.21),

$$\mathcal{S}_a u = (\mathcal{S}_{\mathring{a}}(u * U)) * U^{-1}, \quad \mathcal{T}_a u = (\mathcal{T}_{\mathring{a}}(u * (U^\star)^{-1})) * U^\star.$$

Hence, it suffices to prove (3) $\iff$ (4) $\iff$ (5) for a being $\mathring{a}$. (3) $\iff$ (4) follows directly from Theorem 1.2.4 and Lemma 1.2.3. (4) $\iff$ (5) follows from (5.6.24) and the identity $\langle \mathcal{S}_a u, v\rangle = \langle u, \mathcal{T}_a v\rangle$. The identity (5.6.32) follows from Lemma 5.6.3 and Theorem 1.2.5. □

Before presenting our main result on convergence of a cascade algorithm, we need a few auxiliary results. The following result studies $\mathscr{F}_{m,\upsilon,p}$ and $\mathscr{V}_{m,\upsilon}$ using different sequences $\upsilon \in (l_0(\mathbb{Z}))^{1\times r}$.

Lemma 5.6.6 *Let* $\upsilon, \tilde{\upsilon} \in (l_0(\mathbb{Z}))^{1\times r}$ *such that* $\widehat{\upsilon}(0) \neq 0$ *and* $\widehat{\tilde{\upsilon}}(0) \neq 0$. *Then* $\mathscr{F}_{m,\upsilon,p} = \mathscr{F}_{m,\tilde{\upsilon},p}$ *if and only if there exists* $c \in l_0(\mathbb{Z})$ *such that* $\widehat{c}(0) = 1$ *and*

$$\widehat{\tilde{\upsilon}}(\xi) = \widehat{c}(\xi)\widehat{\upsilon}(\xi) + \mathscr{O}(|\xi|^{m+1}), \qquad \xi \to 0. \tag{5.6.33}$$

Similarly, $\mathscr{V}_{m,\upsilon} = \mathscr{V}_{m,\tilde{\upsilon}}$ *(or* $\mathscr{P}_{m,\upsilon} = \mathscr{P}_{m,\tilde{\upsilon}}$*) if and only if there exists* $c \in l_0(\mathbb{Z})$ *such that* (5.6.33) *holds and* $\widehat{c}(0) \neq 0$.

Proof By Theorem 5.6.4, it suffices to prove the claim for $\upsilon = (\upsilon_1, 0, \dots, 0)$ with $\widehat{\upsilon_1}(0) = 1$. For this case, $\mathscr{F}_{m,\upsilon,p}$ consists of all compactly supported vector functions

$F = (f, f_2, \dots, f_r)^\mathsf{T} \in (W_p^m(\mathbb{R}))^r$ such that

$$\widehat{f}(0) = 1 \quad \text{and} \quad \widehat{f}(\xi + 2\pi k) = \mathscr{O}(|\xi|^{m+1}), \qquad \xi \to 0, \ \forall\, k \in \mathbb{Z}\backslash\{0\}. \qquad (5.6.34)$$

Write $\tilde{\upsilon} = (\tilde{\upsilon}_1, \dots, \tilde{\upsilon}_r)$. Then $\mathscr{F}_{m,\upsilon,p} = \mathscr{F}_{m,\tilde{\upsilon},p}$ implies that for all $k \in \mathbb{Z}\backslash\{0\}$,

$$\begin{aligned}\widehat{\tilde{\upsilon}_1}(\xi)\widehat{f}(\xi + 2\pi k) + \widehat{\tilde{\upsilon}_2}(\xi)\widehat{f_2}(\xi + 2\pi k) + \cdots + \widehat{\tilde{\upsilon}_r}(\xi)\widehat{f_r}(\xi + 2\pi k)\\ = \widehat{\tilde{\upsilon}}(\xi)\widehat{F}(\xi + 2\pi k) = \mathscr{O}(|\xi|^{m+1}), \qquad \xi \to 0.\end{aligned}$$

Since $f_2, \dots, f_r$ are arbitrary functions in $W_p^m(\mathbb{R})$, by (5.6.34) and the above identity, using Lemma 5.5.10, we must have $\widehat{\tilde{\upsilon}_\ell}(\xi) = \mathscr{O}(|\xi|^{m+1})$ as $\xi \to 0$ for all $\ell = 2, \dots, r$. At the point 0, we have

$$1 = \widehat{\tilde{\upsilon}}(0)\widehat{F}(0) = \widehat{\tilde{\upsilon}_1}(0)\widehat{f}(0) + \widehat{\tilde{\upsilon}_2}(0)\widehat{f_2}(0) + \cdots + \widehat{\tilde{\upsilon}_r}(0)\widehat{f_r}(0) = \widehat{\tilde{\upsilon}_1}(0)\widehat{f}(0) = \widehat{\tilde{\upsilon}_1}(0).$$

Hence, $\widehat{\tilde{\upsilon}}(\xi) = (\widehat{\tilde{\upsilon}_1}(\xi), \mathscr{O}(|\xi|^{m+1}), \dots, \mathscr{O}(|\xi|^{m+1}))$ as $\xi \to 0$ with $\widehat{\tilde{\upsilon}_1}(0) = 1$. Take $c \in l_0(\mathbb{Z})$ such that $\widehat{c}(\xi) = \widehat{\tilde{\upsilon}_1}(\xi)/\widehat{\upsilon_1}(\xi) + \mathscr{O}(|\xi|^{m+1})$ as $\xi \to 0$. Then $\widehat{c}(0) = 1$ and (5.6.33) holds.

By the definition of $\mathscr{F}_{m,\upsilon,p}$, the converse direction (i.e., the sufficiency part) is trivial. ☐

The following two technical results will be needed later.

Lemma 5.6.7 *For $c_0 = 1$ and $c_1, \dots, c_m \in \mathbb{C}$, there always exists $u \in l_0(\mathbb{Z})$ such that $\widehat{u}^{(j)}(0) = c_j$ for $j = 0, \dots, m$ and $|1 - \widehat{u}(\xi)| < 1/2$ for all $\xi \in \mathbb{R}$.*

Proof For $j = 1, \dots, m$, let $u_j \in l_0(\mathbb{Z})$ such that $\widehat{u_j}^{(j)}(0) = c_j$ and $\widehat{u_j}^{(k)}(0) = 0$ for all $k \in \{0, 1, \dots, m\}\backslash\{j\}$. Define $\widehat{u}(\xi) := 1 + \sum_{j=1}^m n^{-j}\widehat{u_j}(n\xi)$ with $n \in \mathbb{N}$. Obviously, $\widehat{u}^{(j)}(0) = c_j$ for $j = 0, \dots, m$ and $|1 - \widehat{u}(\xi)| \leqslant \sum_{j=1}^m n^{-j}\|\widehat{u_j}\|_{L_\infty(\mathbb{R})} \to 0$ as $n \to \infty$. ☐

The following result provides the existence of compactly supported initial functions $f \in \mathscr{F}_{m,\upsilon,p} \cap (\mathscr{C}^m(\mathbb{R}))^r$ such that the integer shifts of f are stable in $L_p(\mathbb{R})$ for all $1 \leqslant p \leqslant \infty$.

Lemma 5.6.8 *Let $m \in \mathbb{N}_0$ and $\upsilon \in (l_0(\mathbb{Z}))^r$ with $\widehat{\upsilon}(0) \neq 0$. Then there exists a compactly supported vector function $f \in (\mathscr{C}^m(\mathbb{R}))^r$ such that the integer shifts of f are linearly independent (and consequently, the integer shifts of f are stable in $L_p(\mathbb{R})$) and $f \in \mathscr{F}_{m,\upsilon,p} \cap (\mathscr{C}^m(\mathbb{R}))^r$ for all $1 \leqslant p \leqslant \infty$. Furthermore, the exists a compactly supported function $\phi \in (\mathscr{C}^m(\mathbb{R}))^r$ such that* (5.6.6) *holds and the integer shifts of ϕ are stable.*

Proof Let $\varphi := B_{m+2}$ be the B-spline function of order $m + 2$ in (5.4.5). Then $\varphi \in \mathscr{C}^m(\mathbb{R})$ and by $\widehat{\varphi}(\xi) = (\frac{1-e^{-i\xi}}{i\xi})^{m+2}$, the integer shifts of φ are linearly independent, since $\{z \in \mathbb{C} : \widehat{\varphi}(z) = 0\} = \pi\mathbb{Z}\backslash\{0\}$. Moreover,

$$\widehat{\varphi}(0) = 1 \quad \text{and} \quad \widehat{\varphi}(\xi + 2\pi k) = \mathscr{O}(|\xi|^{m+1}), \quad \xi \to 0, \forall\, k \in \mathbb{Z}\backslash\{0\}. \tag{5.6.35}$$

Take $n \in \mathbb{N}_0$ such that $2^n \geqslant r$. Since the integer shifts of φ are linearly independent, the integer shifts of $\varphi(2^n \cdot -\gamma), \gamma = 0, \dots, 2^n - 1$ are linearly independent. By Theorem 5.2.4, there exist compactly supported functions $\psi_2, \dots, \psi_{2^n}$ such that the integer shifts of $\varphi, \psi_2, \dots, \psi_{2^n}$ are linearly independent and $\mathsf{S}(\varphi, \psi_2, \dots, \psi_{2^n}) = \mathsf{S}(\varphi(2^n\cdot), \varphi(2^n\cdot-1), \dots, \varphi(2^n\cdot-2^n+1))$. Since $\varphi \in \mathscr{C}^m(\mathbb{R})$, we have $\psi_2, \dots, \psi_{2^n} \in \mathscr{C}^m(\mathbb{R})$. Let $U \in (l_0(\mathbb{Z}))^{r\times r}$ be a strongly invertible sequence such that $\widehat{\mathring{\upsilon}}(\xi) := \widehat{\upsilon}(\xi)\widehat{U}(\xi)$ satisfies (5.6.20). Thus, $f := U * (\varphi, \psi_2, \dots, \psi_r)^{\mathsf{T}}$ is a desired vector function satisfying (5.6.5) and the integer shifts of f are linearly independent.

Since $\widehat{\mathring{\upsilon}}_1(0) = 1$, by Lemma 5.6.7, there exists $u \in l_0(\mathbb{Z})$ such that $\widehat{u}(\xi) = (\widehat{\mathring{\upsilon}}_1(\xi))^{-1} + \mathscr{O}(|\xi|^{m+1})$ as $\xi \to 0$ and $\widehat{u}(\xi) \neq 0$ for all $\xi \in \mathbb{R}$. Define $\phi := U * (u * \varphi, \psi_2, \dots, \psi_r)^{\mathsf{T}}$. Then (5.6.6) holds and the integer shifts of ϕ are stable (but are not linearly independent if $\widehat{u}$ is not a monomial). □

5.6.3 Convergence of a Vector Cascade Algorithm in $W_p^m(\mathbb{R})$

The following technical result plays a critical role in our study of a cascade algorithm for investigating a refinable function.

Proposition 5.6.9 *Let $1 \leqslant p \leqslant \infty$, $a \in (l_0(\mathbb{Z}))^{r\times r}$ and $u_1, \dots, u_J \in (l_0(\mathbb{Z}))^{1\times r}$. For any $\rho > 0$,*

$$\lim_{n\to\infty} \rho^n \|a_n * u_\ell^\star\|_{(l_p(\mathbb{Z}))^r} = 0, \qquad \forall\, \ell = 1, \dots, J \tag{5.6.36}$$

if and only if there exist $0 < \rho_0 < 1$ and a positive constant C such that

$$\|a_n * u_\ell^\star\|_{(l_p(\mathbb{Z}))^r} \leqslant C\rho^{-n}\rho_0^n, \qquad \forall\, n \in \mathbb{N}, \ell = 1, \dots, J, \tag{5.6.37}$$

where a_n is defined in (5.6.7) by $\widehat{a_n}(\xi) := \widehat{a}(2^{n-1}\xi)\cdots\widehat{a}(2\xi)\widehat{a}(\xi)$. Under the assumptions that 1 is an eigenvalue of $\widehat{a}(0)$ and

$$\dim(span\{\widehat{u_1^\star}(0), \dots, \widehat{u_J^\star}(0)\}) = r - 1, \quad span\{\widehat{u_1^\star}(\pi), \dots, \widehat{u_J^\star}(\pi)\} = \mathbb{C}^r, \tag{5.6.38}$$

if (5.6.36) holds with $\rho > 0$, then 1 is a simple eigenvalue of $\widehat{a}(0)$ and all other eigenvalues of $\widehat{a}(0)$ are less than $2^{1/p-1}\rho$ in modulus; in particular, if (5.6.36) holds with $\rho = 2^{m+1-1/p}$ and $m \in \mathbb{N}$, then (5.6.3) must hold and the filter a must satisfy order $m + 1$ sum rules with the matching filter υ given in (5.6.4).

Proof (5.6.37)$\Longrightarrow$(5.6.36) is trivial. (5.6.36)$\Longrightarrow$(5.6.37) will be proved in Theorem 5.7.5 using the p-norm joint spectral radius in the next section.

Since $a \in (l_0(\mathbb{Z}))^{r\times r}$ and $u_1,\dots,u_J \in (l_0(\mathbb{Z}))^{1\times r}$, there exists $N \in \mathbb{N}$ such that all $a, u_1, \dots, u_J$ vanish outside $[-N, N]$. Consequently, all $a_n * u_\ell^\star$ vanish outside $[-2^nN, 2^nN]$. Let $1 \leqslant p' \leqslant \infty$ such that $\frac{1}{p}+\frac{1}{p'} = 1$. By $\widehat{a_n}(\xi)\widehat{u_\ell^\star}(\xi) = \sum_{\beta=-2^nN}^{2^nN}[a_n * u_\ell^\star](\beta)e^{-i\beta\xi}$ and $\widehat{a_n}(0) = [\widehat{a}(0)]^n$, applying Hölder's inequality, we have

$$
\begin{aligned}
2^{(1/p-1)n}\rho^n|[\widehat{a}(0)]^n\widehat{u_\ell^\star}(0)| &= 2^{-n/p'}\rho^n\Big|\sum_{\beta=-2^nN}^{2^nN}[a_n * u_\ell^\star](\beta)\Big| \\
&\leqslant 2^{-n/p'}\rho^n(2^{n+1}N+1)^{1/p'}\left(\sum_{\beta=-2^nN}^{2^nN}|[a_n * u_\ell^\star](\beta)|^p\right)^{1/p} \\
&= (2N+2^{-n})^{1/p'}\rho^n\|a_n * u_\ell^\star\|_{(l_p(\mathbb{Z}))^r}.
\end{aligned}
$$

By (5.6.36), we conclude that

$$
\lim_{n\to\infty}[2^{(1/p-1)}\rho]^n[\widehat{a}(0)]^n\widehat{u_\ell^\star}(0) = 0, \qquad \forall\, \ell = 1,\dots,J.
$$

Since 1 is an eigenvalue of $\widehat{a}(0)$, by the first condition in (5.6.38) and a similar argument as in Proposition 5.6.2 for (5.6.3), we conclude that 1 is a simple eigenvalue of $\widehat{a}(0)$ and all other eigenvalues of $\widehat{a}(0)$ are less than $2^{1/p-1}\rho$ in modulus.

In particular, if (5.6.36) holds with $\rho = 2^{m+1-1/p}$, then (5.6.3) holds and thus there exists $\upsilon \in (l_0(\mathbb{Z}))^{1\times r}$ satisfying (5.6.4). To prove (5.6.28) for sum rules, it suffices to prove the second relation in (5.6.28). By (5.6.4), we have

$$
\begin{aligned}
&\widehat{\upsilon}(2^n\xi)\widehat{a_n}(\xi+\pi)\widehat{u_\ell^\star}(\xi+\pi) \\
&\quad = \widehat{\upsilon}(2^n\xi)\widehat{a}(2^{n-1}\xi)\cdots\widehat{a}(2\xi)\widehat{a}(\xi+\pi)\widehat{u_\ell^\star}(\xi+\pi) \\
&\quad = \widehat{\upsilon}(2\xi)\widehat{a}(\xi+\pi)\widehat{u_\ell^\star}(\xi+\pi) + \mathscr{O}(|\xi|^{m+1}), \quad \xi \to 0.
\end{aligned} \tag{5.6.39}
$$

By the Lebniz differentiation formula,

$$
\begin{aligned}
&[\widehat{\upsilon}(2^n\cdot)\widehat{a_n}(\cdot+\pi)\widehat{u_\ell^\star}(\cdot+\pi)]^{(j)}(0) \\
&\quad = \sum_{k=0}^{j}\frac{j!}{k!(j-k)!}2^{(j-k)n}\widehat{\upsilon}^{(j-k)}(0)[\widehat{a_n}\widehat{u_\ell^\star}]^{(k)}(\pi).
\end{aligned} \tag{5.6.40}
$$

Now by $\widehat{a_n}(\xi)\widehat{u_\ell^\star}(\xi) = \sum_{\beta=-2^nN}^{2^nN} [a_n * u_\ell^\star](\beta)e^{-i\beta\xi}$, we conclude that

$$\begin{aligned}
|[\widehat{a_n}\widehat{u_\ell^\star}]^{(k)}(\pi)| &\leqslant (2^nN)^k \sum_{\beta=-2^nN}^{2^nN} |[a_n * u_\ell^\star](\beta)| \\
&\leqslant (2^nN)^k(2^{n+1}N+1)^{1-1/p}\|a_n * u_\ell^\star\|_{(l_p(\mathbb{Z}))^r} \\
&\leqslant 2^{n(k+1-1/p)}N^k(2N+1)^{1-1/p}\|a_n * u_\ell^\star\|_{(l_p(\mathbb{Z}))^r}.
\end{aligned}$$

For $j = 0, \dots, m$, it follows from (5.6.40) and (5.6.36) with $\rho = 2^{m+1-1/p}$ that

$$\begin{aligned}
|[\widehat{\upsilon}(2^n\cdot)\widehat{a_n}(\cdot+\pi)\widehat{u_\ell^\star}(\cdot+\pi)]^{(j)}(0)| &\leqslant \sum_{k=0}^{j} \frac{j!}{k!(j-k)!}|\widehat{\upsilon}^{(j-k)}(0)|2^{(j-k)n}|[\widehat{a_n}\widehat{u_\ell^\star}]^{(k)}(\pi)| \\
&\leqslant \sum_{k=0}^{j} \frac{j!N^k(2N+1)^{1-1/p}}{k!(j-k)!}|\widehat{\upsilon}^{(j-k)}(0)|2^{n(j+1-1/p)}\|a_n * u_\ell^\star\|_{(l_p(\mathbb{Z}))^r} \to 0
\end{aligned}$$

as $n \to \infty$. Therefore, it follows from the above relation and (5.6.39) that

$$\widehat{\upsilon}(2\xi)\widehat{a}(\xi+\pi)\widehat{u_\ell^\star}(\xi+\pi) = \mathscr{O}(|\xi|^{m+1}), \quad \xi \to 0, \ell = 1, \dots, J.$$

Since $\operatorname{span}\{\widehat{u_1^\star}(\pi), \dots, \widehat{u_J^\star}(\pi)\} = \mathbb{C}^r$, $[\widehat{u_1^\star}(\xi+\pi), \dots, \widehat{u_J^\star}(\xi+\pi)]$ has rank r for ξ in a neighborhood of 0. We deduce from the above identity that the second relation in (5.6.28) holds. □

To study the convergence of a cascade algorithm, we now introduce a quantity $\rho_m(a, \upsilon)_p$. Let $a \in (l_0(\mathbb{Z}))^{r\times r}$ and $\upsilon \in (l_0(\mathbb{Z}))^{1\times r}$. For $m \in \mathbb{N} \cup \{-1, 0\}$, we define

$$\rho_{m+1}(a, \upsilon)_p := 2\max\left\{\limsup_{n\to\infty} \|a_n * u^\star\|_{(l_p(\mathbb{Z}))^r}^{1/n} \ : \ u \in \mathscr{B}_{m,\upsilon}\right\}, \tag{5.6.41}$$

where a_n is defined in (5.6.7) and $2^na_n = \mathcal{S}_a^n(\delta I_r)$. Since $\mathscr{V}_{m,\upsilon}$ is generated by $\mathscr{B}_{m,\upsilon}$, $\mathscr{B}_{m,\upsilon}$ in the above definition can be replaced by any (not necessarily finite) set $\mathscr{B}$ of $\mathscr{V}_{m,\upsilon}$ satisfying $\mathscr{V}_{m,\upsilon} = \operatorname{span}\{u(\cdot - k) \ : \ u \in \mathscr{B}, k \in \mathbb{Z}\}$. In the next section, we shall see that $\lim_{n\to\infty} \|a_n * u^\star\|_{(l_p(\mathbb{Z}))^r}^{1/n}$ always exists for every $u \in (l_0(\mathbb{Z}))^{1\times r}$. Also note that for $\{b_1, \dots, b_r\} = \mathscr{B}_{m,\upsilon}$,

$$\begin{aligned}
\rho_{m+1}(a, \upsilon)_p &= 2\lim_{n\to\infty} \|a_n * [b_1^\star, \dots, b_r^\star]\|_{(l_p(\mathbb{Z}))^{r\times r}}^{1/n} \\
&= \lim_{n\to\infty} \|(\mathcal{S}_a^n(\delta I_r)) * [b_1^\star, \dots, b_r^\star]\|_{(l_p(\mathbb{Z}))^{r\times r}}^{1/n}.
\end{aligned}$$

Since $\mathscr{V}_{-1,\upsilon} = (l_0(\mathbb{Z}))^{1\times r}$ and $\mathscr{B}_{-1,\upsilon} = \{\boldsymbol{\delta} e_1, \dots, \boldsymbol{\delta} e_r\}$, the quantity $\rho_0(a, \upsilon)_p$ does not depend on υ and

$$\rho_0(a, \upsilon)_p := 2 \lim_{n\to\infty} \|a_n\|^{1/n}_{(l_p(\mathbb{Z}))^{r\times r}}.$$

To introduce a fundamental quantity $\mathrm{sm}_p(a)$ for studying convergence of a cascade algorithm, we introduce two quantities n_a and m_a. If 1 is a simple eigenvalue of $\widehat{a}(0)$, we define

$$n_a := \sup\{k \in \mathbb{N} \,:\, 2^{-j} \text{ is not an eigenvalue of } \widehat{a}(0) \text{ for all } 1 \leqslant j \leqslant k-1\}. \tag{5.6.42}$$

Define $n_a := 0$ if 1 is not a simple eigenvalue of $\widehat{a}(0)$. Consequently, it follows from (5.6.10) that there exists $\upsilon \in (l_0(\mathbb{Z}))^{1\times r}$ such that

$$\widehat{\upsilon}(0) \neq 0, \qquad \widehat{\upsilon}(2\xi)\widehat{a}(\xi) = \widehat{\upsilon}(\xi) + \mathscr{O}(|\xi|^{n_a}), \qquad \xi \to 0. \tag{5.6.43}$$

Define $\mathrm{sr}(a) := m_a$ to be the largest nonnegative integer such that $m_a \leqslant n_a$ and a satisfies order m_a sum rules with the matching filter υ in (5.6.43). Now the *smoothness exponent* $\mathrm{sm}_p(a)$ of the filter a is defined to be

$$\mathrm{sm}_p(a) := \tfrac{1}{p} - \log_2 \rho_{m_a}(a, \upsilon)_p \quad \text{and} \quad \mathrm{sm}(a) := \mathrm{sm}_2(a). \tag{5.6.44}$$

As a direct consequence of Proposition 5.6.9, we have

Corollary 5.6.10 *Let $a \in (l_0(\mathbb{Z}))^{r\times r}$ such that 1 is an eigenvalue of $\widehat{a}(0)$. Let $m \in \mathbb{N}_0$ and $1 \leqslant p \leqslant \infty$. If $\mathrm{sm}_p(a) > m$ or $\rho_j(a, \mathring{\upsilon})_p < 2^{m-1/p}$ for some $j \in \mathbb{N}_0$ and $\mathring{\upsilon} \in (l_0(\mathbb{Z}))^{1\times r}$ with $\widehat{\mathring{\upsilon}}(0) \neq 0$, then (5.6.3) holds, $n_a \geqslant m_a \geqslant m+1$, and a must satisfy order $m+1$ sum rules with the matching filter $\upsilon \in (l_0(\mathbb{Z}))^{1\times r}$ in (5.6.43). In particular, if 1 is an eigenvalue of $\widehat{a}(0)$, then $\mathrm{sr}(a) \geqslant \mathrm{sm}_p(a)$ for all $1 \leqslant p \leqslant \infty$.*

Proof By definition, the condition $\mathrm{sm}_p(a) > m$ means $\rho_{m_a}(a, \upsilon)_p < 2^{m-1/p}$. Hence, it suffices to prove the claims under the assumption that $\rho_j(a, \mathring{\upsilon})_p < 2^{m-1/p}$. Let $\{b_1, \dots, b_r\} = \mathscr{B}_{j-1,\mathring{\upsilon}}$. Since $\widehat{\mathring{\upsilon}}(0) \neq 0$, it follows directly from item (ii) of Theorem 5.6.4 that both (5.6.38) and (5.6.36) with $\rho = 2^{m+1-1/p}$ are satisfied with $u_\ell = b_\ell, \ell = 1, \dots, r$ and $J = r$. Now all the claims follow directly from Proposition 5.6.9. The claim $\mathrm{sr}(a) \geqslant \mathrm{sm}_p(a)$ is now a trivial consequence. □

As the main result in this section, the following result characterizes the convergence of a cascade algorithm in $W_p^m(\mathbb{R})$, which plays a central role for studying various properties of a refinable vector function.

Theorem 5.6.11 *Let $1 \leqslant p \leqslant \infty$ and $m \in \mathbb{N}_0$. Let $a \in (l_0(\mathbb{Z}))^{r\times r}$ such that (5.1.10) is satisfied and $n_a \geqslant m+1$, where n_a is defined in (5.6.42). Then there exists a compactly supported refinable (vector) function/distribution ϕ satisfying $\phi = 2\sum_{k\in\mathbb{Z}} a(k)\phi(2\cdot -k)$ with $\widehat{\upsilon}(0)\widehat{\phi}(0) = 1$, where $\upsilon \in (l_0(\mathbb{Z}))^{1\times r}$ satisfies (5.6.43).*

The following statements are equivalent:

(1) The cascade algorithm associated with the filter a converges in $W_p^m(\mathbb{R})$*, that is, for every initial (vector) function* $f \in \mathscr{F}_{m,\upsilon,p}$*, the cascade sequence* $\{\mathcal{R}_a^n f\}_{n=1}^\infty$ *is a Cauchy sequence in* $(W_p^m(\mathbb{R}))^r$*. (In fact,* $\lim_{n\to\infty} \|\mathcal{R}_a^n f - \phi\|_{(W_p^m(\mathbb{R}))^r} = 0$*).*
(2) For one function $f \in \mathscr{F}_{m,\upsilon,p}$ *(require* $f \in (\mathscr{C}^m(\mathbb{R}))^r$ *if* $p = \infty$*) such that the integer shifts of f are stable in* $L_p(\mathbb{R})$*,* $\{\mathcal{R}_a^n f\}_{n=1}^\infty$ *is a Cauchy sequence in* $(W_p^m(\mathbb{R}))^r$*.*
(3) For one initial function $f \in \mathscr{F}_{m,\upsilon,p}$ *(require* $f \in (\mathscr{C}^m(\mathbb{R}))^r$ *if* $p = \infty$*) such that the integer shifts of f are stable in* $L_p(\mathbb{R})$*, then the refinable function* $\phi \in (W_p^m(\mathbb{R}))^r$ *(or* $\phi \in (\mathscr{C}^m(\mathbb{R}))^r$ *if* $p = \infty$*) and* $\lim_{n\to\infty} \|\mathcal{R}_a^n f - \phi\|_{(W_p^m(\mathbb{R}))^r} = 0$*.*
(4) $\lim_{n\to\infty} 2^{n(m+1-1/p)} \|a_n * u^\star\|_{(l_p(\mathbb{Z}))^r} = 0$ *for all* $u \in \mathscr{B}_{m,\upsilon}$*, where the (matrix-valued) sequences* a_n *are defined in (5.6.7) by* $\widehat{a_n}(\xi) := \widehat{a}(2^{n-1}\xi)\cdots\widehat{a}(2\xi)\widehat{a}(\xi)$*.*
(5) $\rho_{m+1}(a, \upsilon)_p < 2^{1/p-m}$*.*
(6) $\operatorname{sm}_p(a) > m$*.*
(7) There exist $J \geqslant m$ *and* $\mathring{\upsilon} \in (l_0(\mathbb{Z}))^{1\times r}$ *with* $\widehat{\mathring{\upsilon}}(0) \ne 0$ *such that the filter a has order* $J+1$ *sum rules with the matching filter* $\mathring{\upsilon}$ *satisfying* $\widehat{\mathring{\upsilon}}(0) \ne 0$ *and* $\rho_{J+1}(a, \mathring{\upsilon})_p < 2^{1/p-m}$*.*

Moreover, any of the above statements implies that a satisfies order $m+1$ *sum rules with the matching filter* υ *and there exists a positive constant C such that*

$$\|a_n * u^\star\|_{(l_p(\mathbb{Z}))^r} \leqslant C 2^{n(1/p-1-m)}, \qquad \forall\, n \in \mathbb{N}, u \in \mathscr{B}_{m-1,\upsilon}. \tag{5.6.45}$$

Proof By Theorem 5.1.3 and (5.1.10), up to a multiplicative constant, there exists a unique $r \times 1$ vector ϕ of compactly supported distributions satisfying $\widehat{\phi}(2\xi) = \widehat{a}(\xi)\widehat{\phi}(\xi)$ and $\widehat{\phi}(0) \ne 0$. Since $\widehat{\upsilon}(0)\widehat{a}(0) = \widehat{\upsilon}(0)$ and $\widehat{\upsilon}(0) \ne 0$, we must have $\widehat{\upsilon}(0)\widehat{\phi}(0) \ne 0$ (see Exercise 5.55). Therefore, without loss of generality, we can assume $\widehat{\upsilon}(0)\widehat{\phi}(0) = 1$ by multiplying a nonzero constant with ϕ.

(1)$\Longrightarrow$(2) is trivial. The existence of such an initial function f in item (2) is guaranteed by Lemma 5.6.8. If item (2) holds, then there exists $\eta \in (W_p^m(\mathbb{R}))^r$ (or $\eta \in (\mathscr{C}^m(\mathbb{R}))^r$ if $p = \infty$) such that $\lim_{n\to\infty} \|\mathcal{R}_a^n f - \eta\|_{(W_p^m(\mathbb{R}))^r} = 0$. In particular, we have $\widehat{\eta}(0) = \lim_{n\to\infty} \widehat{\mathcal{R}_a^n f}(0) = \lim_{n\to\infty} [\widehat{a}(0)]^n \widehat{f}(0)$, which implies

$$\widehat{\upsilon}(0)\widehat{\eta}(0) = \lim_{n\to\infty} \widehat{\upsilon}(0)[\widehat{a}(0)]^n \widehat{f}(0) = \widehat{\upsilon}(0)\widehat{f}(0) = 1.$$

Since we assumed that $\widehat{\upsilon}(0)\widehat{\phi}(0) = 1$ and (5.1.10) is satisfied, we must have $\widehat{\eta}(0) = \widehat{\phi}(0)$. Since $\mathcal{R}_a \eta = \eta$ and $\mathcal{R}_a \phi = \phi$, by Theorem 5.1.3 and (5.1.10), we conclude that $\eta = \phi$. Thus, (2)$\Longrightarrow$(3).

Suppose that item (3) holds. By Theorem 5.6.4, without loss of generality, we assume that a takes the normal form in (5.6.22) and υ satisfies (5.6.20). Define $f_n := \mathcal{R}_a^n f$ for $n \in \mathbb{N}$. Then $f_n = 2^n \sum_{k\in\mathbb{Z}} a_n(k) f(2^n \cdot -k)$. Consequently,

$$\nabla_{2^{-n}}^{m+1} f_n = 2^n \sum_{k\in\mathbb{Z}} [\nabla^{m+1} a_n](k) f(2^n \cdot -k), \qquad n \in \mathbb{N}. \tag{5.6.46}$$

Since the integer shifts of f are stable in $L_p(\mathbb{R})$, by (5.6.46), there exists a positive constant C_1 depending only on f such that

$$2^{n(1-1/p)}\|\nabla^{m+1}a_n\|_{(l_p(\mathbb{Z}))^{r\times r}} \leqslant C_1\|\nabla_{2^{-n}}^{m+1}f_n\|_{(L_p(\mathbb{R}))^r} \leqslant C_1\omega_{m+1}(f_n, 2^{-n})_p$$

for all $n \in \mathbb{N}$. By $\nabla^{m+1}a_n = a_n * (\nabla^{m+1}\delta I_r)$, the above inequality implies

$$2^{n(1-1/p)}\|a_n * (\nabla^{m+1}\delta)e_1\|_{(l_p(\mathbb{Z}))^{r\times r}} \leqslant C_1\omega_{m+1}(f_n, 2^{-n})_p, \qquad \forall\, n \in \mathbb{N}. \qquad (5.6.47)$$

Since (5.6.20) holds for $\widehat{\upsilon}(\xi)$, $(\varphi, f^{[2]}, \ldots, f^{[r]})^{\mathsf{T}} := f \in \mathscr{F}_{m,\upsilon,p}$ simply means that (5.6.35) holds. By Proposition 5.5.11 and (5.6.35) for each entry of f_n, there exist $v_n \in (l_0(\mathbb{Z}))^r$ and a constant $C_2 \geqslant 1$ such that C_2 depends only on φ and

$$\|g_n\|_{(L_p(\mathbb{R}))^r} \leqslant C_2\omega_{m+1}(f_n, 2^{-n})_p, \qquad \forall\, n \in \mathbb{N},$$

where

$$g_n := f_n - 2^n\sum_{k\in\mathbb{Z}} v_n(k)\varphi(2^n\cdot -k) = 2^n\sum_{k\in\mathbb{Z}}(a_n - [v_n, 0, \ldots, 0])(k)f(2^n\cdot -k).$$

Since the integer shifts of f are stable in $L_p(\mathbb{R})$,

$$2^{n(1-1/p)}\|a_n - [v_n, 0, \ldots, 0]\|_{(l_p(\mathbb{Z}))^{r\times r}} \leqslant C_1\|g_n\|_{(L_p(\mathbb{R}))^r} \leqslant C_1C_2\omega_{m+1}(f_n, 2^{-n})_p$$

for all $n \in \mathbb{N}$. In particular, the above inequality implies

$$2^{n(1-1/p)}\|a_n * b_\ell^\star\|_{(l_p(\mathbb{Z}))^r} \leqslant C_1C_2\omega_{m+1}(f_n, 2^{-n})_p, \ \forall\, n \in \mathbb{N}, \ell = 2, \ldots, r, \qquad (5.6.48)$$

where $b_\ell := (\delta e_\ell)^\star$ for $\ell = 2, \ldots, r$. Define $b_1 := (\nabla^{m+1}\delta e_1)^\star$. By (5.4.4), it follows from (5.6.47) and (5.6.48) that for $\ell = 1, \ldots, r$,

$$\begin{aligned}2^{n(m+1-1/p)}\|a_n * b_\ell^\star\|_{(l_p(\mathbb{Z}))^r} &\leqslant C_1C_2 2^{mn}\omega_{m+1}(f_n, 2^{-n})_p \leqslant C_1C_2\omega_1(f_n^{(m)}, 2^{-n})_p\\ &\leqslant C_1C_2\omega_1(f_n^{(m)} - \phi^{(m)}, 2^{-n})_p + C_1C_2\omega_1(\phi^{(m)}, 2^{-n})_p\\ &\leqslant 2C_1C_2\|f_n^{(m)} - \phi^{(m)}\|_{(L_p(\mathbb{R}))^r} + C_1C_2\omega_1(\phi^{(m)}, 2^{-n})_p.\end{aligned}$$

Since $\phi^{(m)} \in (L_p(\mathbb{R}))^r$ for $1 \leqslant p < \infty$ and $\phi^{(m)} \in \mathscr{C}(\mathbb{R})$ for $p = \infty$, we have $\lim_{n\to\infty}\omega_1(\phi^{(m)}, 2^{-n})_p = 0$. By $\lim_{n\to\infty}\|f_n - \phi\|_{(W_p^m(\mathbb{R}))^r} = 0$, we conclude that item (4) holds. Thus, we proved (3)$\Longrightarrow$(4).

(4)$\Longrightarrow$(5) follows directly from Proposition 5.6.9. We now prove (5)$\Longrightarrow$(1). Suppose that (5) holds. Since (5.6.38) is satisfied with $\{u_1, \ldots, u_s\} := \mathscr{B}_{m,\upsilon}$, by (5.1.10) and Proposition 5.6.9, the filter a satisfies order $m+1$ sum rules with the matching filter υ and (5.6.37) holds with $\rho = 2^{m+1-1/p}$ and with $\{u_1, \ldots, u_J\}$

$:= \mathscr{B}_{m,\upsilon}$ for some $0 < \rho_0 < 1$ and $C > 0$. Let $f \in \mathscr{F}_{m,\upsilon,p}$. Define $\eta := \mathcal{R}_a f - f$ and $f_n := \mathcal{R}_a^n f$ for $n \in \mathbb{N}$. Since $\widehat{\upsilon}(0)\widehat{a}(0) = \widehat{\upsilon}(0)$, we have

$$\widehat{\upsilon}(0)\widehat{\eta}(0) = \widehat{\upsilon}(0)\widehat{a}(0)\widehat{f}(0) - \widehat{\upsilon}(0)\widehat{f}(0) = \widehat{\upsilon}(0)\widehat{f}(0) - \widehat{\upsilon}(0)\widehat{f}(0) = 0. \qquad (5.6.49)$$

For $k \in \mathbb{Z}\backslash\{0\}$ such that k is even, by $f \in \mathscr{F}_{m,\upsilon,p}$, the equation (5.6.5) holds and

$$\begin{aligned}\widehat{\upsilon}(\xi)\widehat{\eta}(\xi + 2\pi k) &= \widehat{\upsilon}(\xi)\widehat{a}(\xi/2)\widehat{f}(\xi/2 + 2\pi k) - \widehat{\upsilon}(\xi)\widehat{f}(\xi + 2\pi k)\\ &= \widehat{\upsilon}(\xi/2)\widehat{f}(\xi/2 + 2\pi k) + \mathscr{O}(|\xi|^{m+1}) = \mathscr{O}(|\xi|^{m+1}),\end{aligned}$$

as $\xi \to 0$. For $k \in \mathbb{Z}\backslash\{0\}$ such that k is odd, since $\operatorname{sr}(a) \geqslant m + 1$, by (5.6.28),

$$\widehat{\upsilon}(\xi)\widehat{\eta}(\xi + 2\pi k) = \widehat{\upsilon}(\xi)\widehat{a}(\xi/2 + \pi)\widehat{f}(\xi/2 + 2\pi k) - \widehat{\upsilon}(\xi)\widehat{f}(\xi + 2\pi k) = \mathscr{O}(|\xi|^{m+1})$$

as $\xi \to 0$. Therefore, using the above two relations and (5.6.49), we proved that

$$\widehat{\upsilon}(\xi)\widehat{\eta^{(m)}}(\xi + 2\pi k) = i^m(\xi + 2\pi k)^m\widehat{\upsilon}(\xi)\widehat{\eta}(\xi + 2\pi k) = \mathscr{O}(|\xi|^{m+1}), \ \xi \to 0, \forall\, k \in \mathbb{Z}.$$

By item (iv) of Theorem 5.6.4, there exist compactly supported functions $g_1, \dots, g_r \in L_p(\mathbb{R})$ such that $\eta^{(m)} = \sum_{\ell=1}^r u_\ell^\star * g_\ell$ with $u_1, \dots, u_r \in \mathscr{B}_{m,\upsilon}$. Since $f_{n+1} - f_n = \mathcal{R}_a^n \eta$,

$$\begin{aligned}f_{n+1}^{(m)} - f_n^{(m)} &= 2^{mn}\mathcal{R}_a^n\eta^{(m)} = 2^{n(m+1)}\sum_{k\in\mathbb{Z}} a_n(k)\eta^{(m)}(2^n \cdot -k)\\ &= 2^{n(m+1)}\sum_{\ell=1}^r\sum_{k\in\mathbb{Z}}[a_n * u_\ell^\star](k)g_\ell(2^n \cdot -k).\end{aligned}$$

By Lemma 5.3.1, there exists $C_3 > 0$ depending only on $g_1, \dots, g_r$ such that

$$\|f_{n+1}^{(m)} - f_n^{(m)}\|_{(L_p(\mathbb{R}))^r} \leqslant C_3 2^{n(m+1-1/p)}\sum_{\ell=1}^r \|a_n * u_\ell^\star\|_{(l_p(\mathbb{Z}))^r}.$$

In fact, by Lemma 5.5.6, we can take $C_3 = \|f\|_{(\mathcal{L}_p(\mathbb{R}))^r}(m+1)^{\operatorname{len}(f)-m-1}$. By (5.6.37) with $\rho = 2^{1/p-m-1}$, we see that $\|f_{n+1}^{(m)} - f_n^{(m)}\|_{(L_p(\mathbb{R}))^r} \leqslant rCC_3\rho_0^n$ for all $n \in \mathbb{N}$. Since $0 < \rho_0 < 1$, this inequality implies that $\{f_n^{(m)}\}_{n=1}^\infty$ is a Cauchy sequence in $L_p(\mathbb{R})$. Note that all $f_n, n \in \mathbb{N}$ are supported inside $[N_1, N_2] := \operatorname{fsupp}(f) + \operatorname{fsupp}(a)$. It follows from the fact that $f_n^{(j-1)}(x) = \int_{N_1}^x f_n^{(j)}(y)dy$ that

$$\|f_{n+1}^{(j-1)} - f_n^{(j-1)}\|_{(L_p(\mathbb{R}))^r} \leqslant (\operatorname{len}(f) + \operatorname{len}(a))\|f_{n+1}^{(j)} - f_n^{(j)}\|_{(L_p(\mathbb{R}))^r}, \qquad (5.6.50)$$

for all $j = 1, \ldots, m$. Therefore, the cascade sequence $\{f_n^{(j)}\}_{n=1}^\infty$ is a Cauchy sequence in $L_p(\mathbb{R})$ for every $j = 0, \ldots, m$. Hence, we conclude that $\{f_n\}_{n=1}^\infty$ is a Cauchy sequence in $(W_p^m(\mathbb{R}))^r$. Thus, we proved (5)$\Longrightarrow$(1).

Let $m_a \in \mathbb{N}_0$ be as in the definition of $\mathrm{sm}_p(a)$. We now prove (5)$\Longrightarrow$(6). By Corollary 5.6.10 and item (5), we have $m_a \geqslant m+1$ and therefore, $\mathscr{B}_{m_a-1,\upsilon} \subseteq \mathscr{V}_{m_a-1,\upsilon} \subseteq \mathscr{V}_{m,\upsilon}$, from which we have $\rho_{m_a}(a,\upsilon)_p \leqslant \rho_{m+1}(a,\upsilon)_p < 2^{1/p-m}$, that is, we proved $\mathrm{sm}_p(a) > m$ and (5)$\Longrightarrow$(6).

By Corollary 5.6.10, the condition $\mathrm{sm}_p(a) > m$ implies $m_a \geqslant m+1$. (6)$\Longrightarrow$(7) is now trivial by taking $J = m_a$ and $\mathring{\upsilon} = \upsilon$. To complete the proof, we show (7)$\Longrightarrow$(2). Since $\widehat{\mathring{\upsilon}}(0) \neq 0$ and 1 is a simple eigenvalue of $\widehat{a}(0)$, without loss of generality, we assume that $\widehat{\mathring{\upsilon}}(0)\widehat{\phi}(0) = 1$. By Lemma 5.6.8, there exists a compactly supported function $f \in \mathscr{F}_{J,\mathring{\upsilon},p} \cap (\mathscr{C}^m(\mathbb{R}))^r$ such that the integer shifts of f are stable in $L_p(\mathbb{R})$ and $\widehat{\mathring{\upsilon}}(\xi)\widehat{f}(\xi) = 1 + \mathscr{O}(|\xi|^{J+1})$ as $\xi \to 0$. Define $\eta := \mathcal{R}_a f - f$. By the same proof as in (5)$\Longrightarrow$(1), there exist compactly supported functions $g_1, \ldots, g_r \in W_p^J(\mathbb{R})$ such that $\eta = \sum_{\ell=1}^r u_\ell^\star * g_\ell$, where $\{u_1, \ldots, u_r\} = \mathscr{B}_{J,\mathring{\upsilon}}$. In particular, we have $\eta^{(m)} = \sum_{\ell=1}^r u_\ell^\star * g_\ell^{(m)}$ by $J \geqslant m$. Now the same proof as in (5)$\Longrightarrow$(1) shows that $\{\mathcal{R}_a^n f\}_{n=1}^\infty$ is a Cauchy sequence in $W_p^m(\mathbb{R})$. By Corollary 5.6.10, the relation (5.6.3) holds. Consequently, by $J \geqslant m$ and $n_a \geqslant m+1$, we must have $\widehat{\mathring{\upsilon}}(\xi) = \widehat{\upsilon}(\xi) + \mathscr{O}(|\xi|^{m+1})$ as $\xi \to 0$. Thus, by $J \geqslant m$, we have $f \in \mathscr{F}_{J,\mathring{\upsilon},p} \subseteq \mathscr{F}_{m,\mathring{\upsilon},p} = \mathscr{F}_{m,\upsilon,p}$. Hence, (7)$\Longrightarrow$(2).

By the proof of (3)$\Longrightarrow$(4), we see that (5.6.47) and (5.6.48) must hold with $m+1$ being replaced by m. That is, $2^{n(1-1/p)}\|a_n * u_\ell^\star\|_{(l_p(\mathbb{Z}))^r} \leqslant C_1C_2\omega_m(f_n, 2^{-n})_p$ for all $n \in \mathbb{N}$ and $u \in \mathscr{B}_{m-1,\upsilon}$. Consequently,

$$\begin{aligned}\|a_n * u_\ell^\star\|_{(l_p(\mathbb{Z}))^r} &\leqslant C_1C_2 2^{n(1/p-1)}\omega_m(f_n, 2^{-n})_p \leqslant C_1C_2 2^{n(1/p-1-m)}\omega_0(f_n^{(m)}, 2^{-n})_p \\ &= C_1C_2 2^{n(1/p-1-m)}\|f_n^{(m)}\|_{(L_p(\mathbb{R}))^r} \leqslant C 2^{n(1/p-1-m)},\end{aligned}$$

with $C := C_1C_2 \sup_{n\in\mathbb{N}} \|f_n^{(m)}\|_{(L_p(\mathbb{R}))^r} < \infty$ by $\lim_{n\to\infty} \|f_n^{(m)} - \phi^{(m)}\|_{(L_p(\mathbb{R}))^r} = 0$. This proves (5.6.45). ☐

Note that (5.6.3) implies the natural condition (5.1.10). By Proposition 5.6.2 and Corollary 5.6.10, the conditions (5.6.3) and $\mathrm{sr}(a) \geqslant m+1$ are necessary for convergence of a cascade algorithm in Theorem 5.6.11. Hence the extra conditions (5.1.10) and $n_a \geqslant m+1$ are necessary for Theorem 5.6.11 and are separately stated to guarantee the existence of a matching filter $\upsilon \in (l_0(\mathbb{Z}))^{1\times r}$ and a vector ϕ of compactly supported distributions satisfying $\widehat{\phi}(2\xi) = \widehat{a}(\xi)\widehat{\phi}(\xi)$ and $\widehat{\upsilon}(0)\widehat{\phi}(0) = 1$.

As a special case of Theorem 5.6.11, the following result will be frequently used to study framelets and wavelets in this book.

Corollary 5.6.12 *Let $1 \leqslant p \leqslant \infty$ and $m \in \mathbb{N}_0$. Let ϕ be a compactly supported refinable vector function satisfying $\phi = 2\sum_{k\in\mathbb{Z}} a(k)\phi(2\cdot -k)$ for some $a \in (l_0(\mathbb{Z}))^{r\times r}$. If $\phi \in (W_p^m(\mathbb{R}))^r$ (require $\phi \in (\mathscr{C}^m(\mathbb{R}))^r$ if $p = \infty$) and the integer* *shifts of* ϕ *are stable in* $L_p(\mathbb{R})$, *then* $\operatorname{sr}(a) \geqslant m+1$, $\operatorname{sm}_p(a) > m$, (5.6.3) *holds, and the cascade algorithm associated with the filter a converges in* $W_p^m(\mathbb{R})$.

Proof Note that $\phi = \mathcal{R}_a\phi$. By Proposition 5.6.2 and our assumption on ϕ, we must have $\widehat{\phi}(0) \neq 0$, $\operatorname{sr}(a) \geqslant m+1$, and (5.6.3) and (5.6.6) hold. Hence, item (2) of Theorem 5.6.11 is satisfied with $f = \phi$. It follows from Theorem 5.6.11 that $\operatorname{sm}_p(a) > m$ and the cascade algorithm associated with the filter a converges in $W_p^m(\mathbb{R})$. □

Faster convergence rates of cascade algorithms can be obtained by better matching the moments of an initial function $\widehat{f}$ with those of the refinable function $\widehat{\phi}$.

Corollary 5.6.13 *Let* $1 \leqslant p \leqslant \infty$ *and* $a \in (l_0(\mathbb{Z}))^{r\times r}$ *such that* 1 *is an eigenvalue of* $\widehat{a}(0)$ *and* $\operatorname{sm}_p(a) > 0$. *Let* $\upsilon \in (l_0(\mathbb{Z}))^{1\times r}$ *as in* (5.6.43). *Let* ϕ *be a compactly supported vector function such that* $\phi = 2\sum_{k\in\mathbb{Z}} a(k)\phi(2\cdot -k)$ *with* $\widehat{\upsilon}(0)\widehat{\phi}(0) = 1$. *Define* $M \in \mathbb{N}_0$ *to be the largest nonnegative integer such that* $M < \operatorname{sm}_p(a)$. *Let* m *be an integer such that* $0 \leqslant m \leqslant M$ *and* $\varepsilon > 0$. *For any* $f \in (W_p^m(\mathbb{R}))^r$ *satisfying*

$$\begin{aligned} \widehat{\upsilon}(\xi)\widehat{f}(\xi) &= 1 + \mathscr{O}(|\xi|^{M-m+1}), && \xi \to 0, \quad \textit{and} \\ \widehat{\upsilon}(\xi)\widehat{f}(\xi + 2\pi k) &= \mathscr{O}(|\xi|^{M+1}), && \xi \to 0, \ \forall\, k \in \mathbb{Z}\backslash\{0\}, \end{aligned} \tag{5.6.51}$$

there exists a positive constant C *such that for all* $n \in \mathbb{N}$,

$$\max\left(\|\mathcal{R}_a^n f - \phi\|_{(W_p^m(\mathbb{R}))^r},\ \|\mathcal{R}_a^{n+1} f - \mathcal{R}_a^n f\|_{(W_p^m(\mathbb{R}))^r}\right) \leqslant C2^{n(m-\operatorname{sm}_p(a)+\varepsilon)}.$$

The proof of the claims is the same as the proof of (7)⟹(2) and (5)⟹(1) in Theorem 5.6.11 by noting that the function $\eta = \mathcal{R}_a f - f$ (as well as $f - \phi$) satisfies $\widehat{\upsilon}(\xi)\widehat{\eta^{(m)}}(\xi + 2\pi k) = \mathscr{O}(|\xi|^{M+1})$ as $\xi \to 0$ for all $k \in \mathbb{Z}$. The existence of such an initial function $f \in (W_p^m(\mathbb{R}))^r$ satisfying (5.6.51) is guaranteed by Lemma 5.6.8.

5.6.4 Convergence of Scalar Cascade Algorithms in $W_p^m(\mathbb{R})$

For the convenience of the reader, here we provide a self-contained short treatment for a scalar cascade algorithm with $r = 1$. For $r = 1$ and $a \in l_0(\mathbb{Z})$ with $\widehat{a}(0) = 1$, the matching filter $\upsilon \in l_0(\mathbb{Z})$ must satisfy $\widehat{\upsilon}(\xi) = 1/\widehat{\phi}(\xi) + \mathscr{O}(|\xi|^{m+1})$ as $\xi \to 0$, where $\widehat{\phi}(\xi) := \prod_{j=1}^{\infty} \widehat{a}(2^{-j}\xi)$. Then

$$\mathscr{P}_{m,\upsilon} = \mathscr{P}_m := \mathbb{P}_m, \qquad \mathscr{V}_{m,\upsilon} = \mathscr{V}_m := \nabla^{m+1} l_0(\mathbb{Z}) \tag{5.6.52}$$

and $\mathscr{F}_{m,\upsilon,p} = \mathscr{F}_{m,p}$, where

$$\mathscr{F}_{m,p} := \{f \in W_p^m(\mathbb{R}) \ :\ f \text{ is compactly supported and satisfies (5.6.54)}\}, \tag{5.6.53}$$

where

$$\widehat{f}(0) = 1 \quad \text{and} \quad \widehat{f}(\xi + 2\pi k) = \mathscr{O}(|\xi|^{m+1}), \quad \xi \to 0, \ \forall\, k \in \mathbb{Z}\backslash\{0\}. \tag{5.6.54}$$

Note that for all $J \geqslant m + 2$, the B-spline functions $B_J \in \mathscr{F}_{m,p} \cap \mathscr{C}^m(\mathbb{R})$ and the integer shifts of B_J are stable in $L_p(\mathbb{R})$. For $1 \leqslant p \leqslant \infty$, the definition of $\rho_{m+1}(a, \upsilon)_p$ in (5.6.41) and $\mathrm{sm}_p(a)$ in (5.6.44) becomes

$$\rho_{m+1}(a, \upsilon)_p = \rho_{m+1}(a)_p := 2 \limsup_{n\to\infty} \|\nabla^{m+1} a_n\|_{l_p(\mathbb{Z})}^{1/n}, \tag{5.6.55}$$

where a_n is defined by $\widehat{a_n}(\xi) := \widehat{a}(2^{n-1}\xi) \cdots \widehat{a}(2\xi)\widehat{a}(\xi)$, $\mathrm{sm}(a) := \mathrm{sm}_2(a)$, and

$$\mathrm{sm}_p(a) := \tfrac{1}{p} - \log_2 \rho_{m_a}(a)_p \qquad \text{with} \quad m_a := \mathrm{sr}(a), \tag{5.6.56}$$

where $\mathrm{sr}(a)$ is the highest order of sum rules of a, that is, m_a is the unique nonnegative integer such that $\widehat{a}(\xi) = (1 + e^{i\xi})^{m_a}\widehat{u}(\xi)$ with $u \in l_0(\mathbb{Z})$ and $\widehat{u}(\pi) \neq 0$.

Before studying scalar cascade algorithms, let us introduce two auxiliary results.

Lemma 5.6.14 *Let $a, b, u \in l_0(\mathbb{Z})$ such that $\widehat{a}(\xi) = (1 + e^{-i\xi})^m\widehat{b}(\xi)$ and all a, b, u are supported inside $(-N, N)$ with $m, N \in \mathbb{N}$. Then*

$$spec(\mathcal{T}_a) = \{\overline{\widehat{a}(0)}, 2^{-1}\overline{\widehat{a}(0)}, \ldots, 2^{1-m}\overline{\widehat{a}(0)}\} \cup spec(\mathcal{T}_b) \tag{5.6.57}$$

and

$$N^{-m}\|b_n * u\|_{l_p(\mathbb{Z})} \leqslant \|\nabla^m(a_n * u)\|_{l_p(\mathbb{Z})} \leqslant 2^m\|b_n * u\|_{l_p(\mathbb{Z})}, \quad \forall\, 1 \leqslant p \leqslant \infty, \tag{5.6.58}$$

where $\widehat{a_n}(\xi) := \widehat{a}(2^{n-1}\xi) \cdots \widehat{a}(2\xi)\widehat{a}(\xi)$ and $\widehat{b_n}(\xi) := \widehat{b}(2^{n-1}\xi) \cdots \widehat{b}(2\xi)\widehat{b}(\xi)$.

Proof Since $\nabla^m(a_n * u) = \nabla^m\boldsymbol{\delta} * (a_n * u)$, it suffices to prove (5.6.57) and (5.6.58) with $m = 1$; the general case follows by recursively applying the special case $m = 1$.

Note that $\upsilon \in \mathscr{V}_0$ if and only if $\widehat{\upsilon}(0) = 0$, i.e., $\widehat{\upsilon}(\xi) = (1 - e^{i\xi})\widehat{w}(\xi)$ for some $w \in l_0(\mathbb{Z})$. By $\widehat{a}(\xi) = (1 + e^{-i\xi})\widehat{b}(\xi)$, we have $\widehat{a}(\pi) = 0$ and $\mathcal{T}_a\mathscr{V}_0 \subseteq \mathscr{V}_0$ since

$$\begin{aligned}\widehat{\mathcal{T}_a\upsilon}(\xi) =& \overline{\widehat{a}(\xi/2)}\widehat{\upsilon}(\xi) + \overline{\widehat{a}(\xi/2 + \pi)}\widehat{\upsilon}(\xi/2 + \pi)\\ =&(1 - e^{i\xi})[\overline{\widehat{b}(\xi/2)}\widehat{w}(\xi/2) + \overline{\widehat{b}(\xi/2 + \pi)}\widehat{w}(\xi/2 + \pi)] = (1 - e^{-i\xi})\widehat{\mathcal{T}_b w}(\xi),\end{aligned}$$

which also implies $\mathrm{spec}(\mathcal{T}_a|_{\mathscr{V}_0}) = \mathrm{spec}(\mathcal{T}_b)$. By $\widehat{a}(\pi) = 0$, $\widehat{\mathcal{T}_a\boldsymbol{\delta}}(0) - \overline{\widehat{a}(0)}\widehat{\boldsymbol{\delta}}(0) = \overline{\widehat{a}(0)} + \overline{\widehat{a}(\pi)} - \overline{\widehat{a}(0)} = 0$. Hence, $\mathcal{T}_a\boldsymbol{\delta} - \overline{\widehat{a}(0)}\boldsymbol{\delta} \in \mathscr{V}_0$ and $\overline{\widehat{a}(0)}$ is the eigenvalue of $\mathcal{T}_a$ acting on the one-dimensional space $l_0(\mathbb{Z})/\mathscr{V}_0$. This proves (5.6.57) with $m = 1$.

We now prove (5.6.58) with $m = 1$. By $\widehat{a}(\xi) = (1 + e^{-i\xi})\widehat{b}(\xi)$, we have

$$\widehat{\nabla(a_n * u)}(\xi) = (1 - e^{-i\xi})\widehat{a_n}(\xi)\widehat{u}(\xi) = (1 - e^{-i2^n\xi})\widehat{b_n}(\xi)\widehat{u}(\xi). \tag{5.6.59}$$

That is, $\nabla(a_n * u) = b_n * u - (b_n * u)(\cdot - 2^n)$, from which we have

$$\|\nabla(a_n * u)\|_{l_p(\mathbb{Z})} \leqslant 2\|b_n * u\|_{l_p(\mathbb{Z})}. \tag{5.6.60}$$

Multiplying $\sum_{j=0}^{2N-1} e^{-i2^n j\xi}$ to both sides of (5.6.59), we have $(1-e^{-i2^{n+1}N\xi})\widehat{b_n}(\xi)\widehat{u}(\xi) = \sum_{j=0}^{2N-1} e^{-i2^n j\xi}\widehat{\nabla(a_n * u)}(\xi)$, which is just

$$b_n * u - (b_n * u)(\cdot - 2^{n+1}N) = \sum_{j=0}^{2N-1} (\nabla(a_n * u))(\cdot - 2^n j).$$

Since $\operatorname{fsupp}(b_n * u) \subseteq (-2^n N, 2^n N)$, it follows from the above identity that

$$2\|b_n * u\|_{l_p(\mathbb{Z})} \leqslant \sum_{j=0}^{2N-1} \|(\nabla(a_n * u))(\cdot - 2^n j)\|_{l_p(\mathbb{Z})} = 2N\|\nabla(a_n * u)\|_{l_p(\mathbb{Z})}.$$

The above inequality together with (5.6.60) proves (5.6.58) with $m = 1$. □

The following is the scalar version of Proposition 5.6.9.

Proposition 5.6.15 *Let $\rho > 0$ and $1 \leqslant p \leqslant \infty$. For $a, u \in l_0(\mathbb{Z})$, if*

$$\lim_{n\to\infty} \rho^n \|a_n * u\|_{l_p(\mathbb{Z})} = 0 \tag{5.6.61}$$

with $\widehat{a_n}(\xi) := \widehat{a}(2^{n-1}\xi)\cdots\widehat{a}(2\xi)\widehat{a}(\xi)$, then

$$\lim_{n\to\infty} \|a_n * u\|_{l_p(\mathbb{Z})}^{1/n} < \rho^{-1}. \tag{5.6.62}$$

If in addition $\widehat{a}(0) = 1$ and $\widehat{u}(\pi) \neq 0$, then (5.6.61) implies $\operatorname{sr}(a) \geqslant \lfloor 1/p + \log_2 \rho \rfloor$. In particular, if $\widehat{a}(0) = 1$, then $\operatorname{sr}(a) \geqslant \operatorname{sm}_p(a)$ for all $1 \leqslant p \leqslant \infty$.

Proof The inequality (5.6.62) follows from Theorem 5.7.5 and (5.6.61). For $p = 2$, we prove (5.6.62) without using the p-norm joint spectral radius. Define $\widehat{b}(\xi) := |\widehat{a}(\xi)|^2$ and $\widehat{v}(\xi) := |\widehat{u}(\xi)|^2$. Then

$$\begin{aligned}\|a_n * u\|_{l_2(\mathbb{Z})}^2 &= \frac{1}{2\pi}\int_{-\pi}^{\pi} |\widehat{a_n}(\xi)\widehat{u}(\xi)|^2 d\xi = \frac{1}{2\pi}\int_{-\pi}^{\pi} \widehat{v}(\xi)\widehat{b_n}(\xi) d\xi \\ &= \langle v, b_n\rangle = 2^{-n}\langle v, \mathcal{S}_b^n \boldsymbol{\delta}\rangle = 2^n\langle \mathcal{T}_b v, \boldsymbol{\delta}\rangle = 2^{-n}[\mathcal{T}_b^n v](0),\end{aligned}$$

where $\widehat{b_n}(\xi) := \widehat{b}(2^{n-1}\xi)\cdots\widehat{b}(2\xi)\widehat{b}(\xi)$ and $b_n = 2^{-n}\mathcal{S}_b^n\boldsymbol{\delta}$. Consequently, we have

$$\|\mathcal{T}_b^n v\|_{l_\infty(\mathbb{Z})} = [\mathcal{T}_b^n v](0) = 2^n\|a_n * u\|^2_{l_2(\mathbb{Z})}. \tag{5.6.63}$$

Indeed, since $\widehat{\mathcal{T}_b^n v}(\xi) \geqslant 0$ by $\widehat{b}(\xi) \geqslant 0$ and $\widehat{v}(\xi) \geqslant 0$, for $k \in \mathbb{Z}$, we have

$$|[\mathcal{T}_b^n v](k)| = \frac{1}{2\pi}\left|\int_{-\pi}^{\pi} \widehat{\mathcal{T}_b^n v}(\xi)e^{-ik\xi}d\xi\right| \leqslant \frac{1}{2\pi}\int_{-\pi}^{\pi} |\widehat{\mathcal{T}_b^n v}(\xi)|d\xi = [\mathcal{T}_b^n v](0).$$

Define $V := \operatorname{span}\{\mathcal{T}_b^n v \ : \ n \in \mathbb{N}_0\}$. Then V is the smallest invariant space of $\mathcal{T}_b$ such that $v \in V$. By the relation $\operatorname{fsupp}(\mathcal{T}_b w) \subseteq \frac{1}{2}(\operatorname{fsupp}(b) + \operatorname{fsupp}(w))$, we can easily deduce (also see Lemma 5.7.3) that $V \subseteq l([-N,N])$, where N is any integer such that $\operatorname{fsupp}(b) \cup \operatorname{fsupp}(v) \subseteq [-N,N]$. Since $\lim_{n\to\infty}\|\mathcal{T}_b^n v\|^{1/n}_{l_\infty(\mathbb{Z})} = \rho(\mathcal{T}_b|_V)$ exists, (5.6.63) implies that $\lim_{n\to\infty}\|a_n * u\|^{1/n}_{l_p(\mathbb{Z})}$ exists.

Let λ be an eigenvalue of $\mathcal{T}_b|_V$ with a nontrivial eigenvector $w = \sum_{j=0}^M c_j\mathcal{T}_b^j v \in V$ for some $c_0,\dots,c_M \in \mathbb{C}$. Then $\mathcal{T}_b^n w = \lambda^n w$ and it follows from (5.6.63) that

$$|\lambda|^n\|w\|_{l_\infty(\mathbb{Z})} = \|\mathcal{T}_b^n w\|_{l_\infty(\mathbb{Z})} \leqslant \sum_{j=0}^M |c_j|\|\mathcal{T}_b^{n+j}v\|_{l_\infty(\mathbb{Z})} \leqslant \sum_{j=0}^M |c_j|2^{n+j}\|a_{n+j} * u\|^2_{l_2(\mathbb{Z})}.$$

Multiplying $(\rho^2/2)^n$ to both sides of the above inequality, we deduce from (5.6.61) with $p = 2$ that

$$\lim_{n\to\infty}(\rho^2/2)^n|\lambda|^n\|w\|_{l_\infty(\mathbb{Z})} \leqslant \lim_{n\to\infty}\sum_{j=0}^M |c_j|(2/\rho^2)^j\rho^{2(n+j)}\|a_{n+j} * u\|^2_{l_2(\mathbb{Z})} = 0,$$

from which we must have $(\rho^2/2)|\lambda| < 1$ by $\|w\|_{l_\infty(\mathbb{Z})} \neq 0$. This proves that the spectral radius $\rho(\mathcal{T}_b|_V) < 2/\rho^2$. On the other hand, by (5.6.63), we conclude that

$$\lim_{n\to\infty}\|a_n * u\|^{1/n}_{l_2(\mathbb{Z})} = 2^{-1/2}\lim_{n\to\infty}\|\mathcal{T}_b^n v\|^{\frac{1}{2n}}_{l_\infty(\mathbb{Z})} = \sqrt{\rho(\mathcal{T}_b|_V)/2} < \rho^{-1}.$$

This proves (5.6.62) with $p = 2$.

Let $m := \lfloor 1/p + \log_2\rho\rfloor$. We prove $m_a := \operatorname{sr}(a) \geqslant m$ using proof by contradiction. Suppose $m_a \leqslant m-1$. Write $\widehat{a}(\xi) = (1+e^{-i\xi})^{m_a}\widehat{b}(\xi)$ with $b \in l_0(\mathbb{Z})$ and $\widehat{b}(\pi) \neq 0$. Since $a, b, u \in l_0(\mathbb{Z})$, we assume $\operatorname{fsupp}(a) \cup \operatorname{fsupp}(b) \cup \operatorname{fsupp}(u) \subseteq (-N,N)$. By Lemma 5.6.14 and $\operatorname{fsupp}(b_n * u) \subseteq (-2^nN, 2^nN)$, using Hölder's inequality, we have

$$\|b_n * u\|_{l_1(\mathbb{Z})} \leqslant 2^{(n+1)(1-1/p)}N^{1-1/p}\|b_n * u\|_{l_p(\mathbb{Z})} \leqslant C2^{n(1-1/p)}\|\nabla^{m_a}(a_n * u)\|_{l_p(\mathbb{Z})},$$

where $C := (\frac{N}{2})^{m_a}(2N)^{1-1/p} < \infty$. Since $\widehat{a}(0) = 1$ and $\widehat{a}(\xi) = (1+e^{-i\xi})^{m_a}\widehat{b}(\xi)$, we have $\widehat{b}(0) = 2^{-m_a}$. Note that $\|\nabla^{m_a} v\|_{l_p(\mathbb{Z})} \leqslant 2^{m_a}\|v\|_{l_p(\mathbb{Z})}$. Hence,

$$2^{-m_a(n-1)}|\widehat{b}(\pi)\widehat{u}(\pi)| = |\widehat{b_n * u}(\pi)| \leqslant \|b_n * u\|_{l_1(\mathbb{Z})} \leqslant C2^{m_a}2^{n(1-1/p)}\|a_n * u\|_{l_p(\mathbb{Z})}.$$

Sine $m_a \leqslant m-1 \leqslant 1/p - 1 + \log_2 \rho$, we have $2^{m_a+1-1/p} \leqslant \rho$. The above inequality and (5.6.61) imply

$$|\widehat{b}(\pi)\widehat{u}(\pi)| \leqslant C2^{(m_a+1-1/p)n}\|a_n * u\|_{l_p(\mathbb{Z})} \leqslant C\rho^n\|a_n * u\|_{l_p(\mathbb{Z})} \to 0, \quad n \to \infty.$$

Hence, we must have $\widehat{b}(\pi)\widehat{u}(\pi) = 0$, which is a contradiction to our assumption that $\widehat{u}(\pi) \neq 0$ and $\widehat{b}(\pi) \neq 0$. Consequently, we must have $m_a \geqslant m$. □

For the convergence of a scalar cascade algorithm in the Sobolev space $W_p^m(\mathbb{R})$, now the scalar version of Theorem 5.6.11 is as follows.

Theorem 5.6.16 *Let $1 \leqslant p \leqslant \infty$ and $m \in \mathbb{N}_0$. Let $a \in l_0(\mathbb{Z})$ with $\widehat{a}(0) = 1$. Define a compactly supported refinable distribution ϕ by $\widehat{\phi}(\xi) := \prod_{j=1}^{\infty}\widehat{a}(2^{-j}\xi)$. The following statements are equivalent:*

(1) The cascade algorithm associated with the filter a converges in $W_p^m(\mathbb{R})$, that is, for every initial function $f \in \mathscr{F}_{m,p}$, the cascade sequence $\{\mathcal{R}_a^n f\}_{n=1}^{\infty}$ is a Cauchy sequence in $W_p^m(\mathbb{R})$. (In fact, $\lim_{n\to\infty}\|\mathcal{R}_a^n f - \phi\|_{W_p^m(\mathbb{R})} = 0$).

(2) For one initial function $f \in \mathscr{F}_{m,p}$ (require $f \in \mathscr{C}^m(\mathbb{R})$ if $p = \infty$) such that the integer shifts of f are stable in $L_p(\mathbb{R})$, $\{\mathcal{R}_a^n f\}_{n=1}^{\infty}$ is a Cauchy sequence in $W_p^m(\mathbb{R})$.

(3) For one function $f \in \mathscr{F}_{m,p}$ (require $f \in \mathscr{C}^m(\mathbb{R})$ if $p = \infty$) such that the integer shifts of f are stable in $L_p(\mathbb{R})$, then the standard refinable function $\phi \in W_p^m(\mathbb{R})$ (or $\phi \in \mathscr{C}^m(\mathbb{R})$ if $p = \infty$) and $\lim_{n\to\infty}\|\mathcal{R}_a^n f - \phi\|_{W_p^m(\mathbb{R})} = 0$.

(4) $\lim_{n\to\infty} 2^{n(m+1-1/p)}\|\nabla^{m+1}a_n\|_{l_p(\mathbb{Z})} = 0$, where $\widehat{a_n}(\xi) := \widehat{a}(2^{n-1}\xi)\cdots\widehat{a}(2\xi)\widehat{a}(\xi)$.

(5) $\rho_{m+1}(a)_p < 2^{1/p-m}$.

(6) $\mathrm{sm}_p(a) > m$, which is just $\rho_{m_a}(a)_p < 2^{1/p-m}$ by (5.6.56) with $m_a := \mathrm{sr}(a)$.

Moreover, any of the above statements implies that $\mathrm{sr}(a) \geqslant m+1$ and there exists a positive constant C such that

$$\|\nabla^m a_n\|_{l_p(\mathbb{Z})} \leqslant C2^{n(1/p-1-m)}, \qquad \forall\, n \in \mathbb{N}. \tag{5.6.64}$$

Proof (1)$\Longrightarrow$(2)$\Longrightarrow$(3) is trivial. We now prove (3)$\Longrightarrow$(4). Define $f_n := \mathcal{R}_a^n f$. Then by induction we have $f_n = 2^n\sum_{k\in\mathbb{Z}} a_n(k)f(2^n\cdot -k)$, $\lim_{n\to\infty}\widehat{f_n}(\xi) = \widehat{\phi}(\xi)$, and

$$\nabla_{2^{-n}}^{m+1} f_n = 2^n \sum_{k\in\mathbb{Z}}[\nabla^{m+1}a_n](k)f(2^n\cdot -k), \qquad n \in \mathbb{N}.$$

Since the integer shifts of f are stable in $L_p(\mathbb{R})$, by the above inequalities, there exists $C_1 > 0$ depending only on f such that

$$2^{n(1-1/p)}\|\nabla^{m+1}a_n\|_{(l_p(\mathbb{Z}))^{r\times r}} \leqslant C_1\|\nabla^{m+1}_{2^{-n}}f_n\|_{(L_p(\mathbb{R}))^r} \leqslant C_1\omega_{m+1}(f_n,2^{-n})_p \qquad (5.6.65)$$

for all $n \in \mathbb{N}$. By (5.4.4), it follows from (5.6.65) that

$$\begin{aligned} 2^{n(m+1-1/p)}\|\nabla^{m+1}a_n\|_{l_p(\mathbb{Z})} &\leqslant C_1 2^{nm}\omega_{m+1}(f_n,2^{-n})_p \leqslant C_1\omega_1(f_n^{(m)},2^{-n})_p \\ &\leqslant C_1\omega_1(f_n^{(m)}-\phi^{(m)},2^{-n})_p + C_1\omega_1(\phi^{(m)},2^{-n})_p. \end{aligned}$$

Since $\phi^{(m)} \in L_p(\mathbb{R})$ for $1 \leqslant p < \infty$ and $\phi^{(m)} \in \mathscr{C}(\mathbb{R})$ for $p = \infty$, we have $\lim_{n\to\infty}\omega_1(\phi^{(m)},2^{-n})_p = 0$. By $\lim_{n\to\infty}\|f_n - \phi\|_{W_p^m(\mathbb{R})} = 0$, we conclude that item (4) holds. Thus, we proved (3)$\Longrightarrow$(4).

(4)$\Longrightarrow$(5) follows from Proposition 5.6.15. By item (5), we have $\rho_{m+1}(a) := 2\limsup\|\nabla^{m+1}a_n\|^{1/n}_{l_p(\mathbb{Z})} < 2^{1/p-m}$. Since $\nabla^{m+1}a_n = a_n * \nabla^{m+1}\boldsymbol{\delta}$, we see that (5.6.61) must hold with $u := \nabla^{m+1}\boldsymbol{\delta}$ and $\rho := 2^{m+1-1/p}$. Since $\widehat{u}(\pi) \neq 0$, it follows from Proposition 5.6.15 that $m_a := \mathrm{sr}(a) \geqslant \lfloor 1/p + \log_2\rho\rfloor = m+1$. Thus, $\rho_{m_a}(a)_p \leqslant \rho_{m+1}(a)_p < 2^{1/p-m}$ which is equivalent to $\mathrm{sm}_p(a) > m$. This proves (5)$\Longrightarrow$(6).

We now prove (5)$\Longrightarrow$(1). Let $f \in \mathscr{F}_{m,p}$. Define $f_n := \mathcal{R}_a^n f$ and $h := f_1 - f$. Then $f_{n+1} - f_n = \mathcal{R}_a^n h = 2^n\sum_{k\in\mathbb{Z}} a_n(k)h(2^n\cdot -k)$. Therefore, by $f, h, f_n \in W_p^m(\mathbb{R})$,

$$f_{n+1}^{(m)} - f_n^{(m)} = 2^{n(m+1)}\sum_{k\in\mathbb{Z}} a_n(k)h^{(m)}(2^n\cdot -k), \qquad n \in \mathbb{N}. \qquad (5.6.66)$$

By Proposition 5.6.15, we see that item (5) implies $\mathrm{sr}(a) \geqslant m+1$. Since $\mathrm{sr}(a) \geqslant m+1$ and f satisfies (5.6.54), we can directly check that

$$\widehat{h}(0) = 0 \quad \text{and} \quad \widehat{h}(\xi + 2\pi k) = \mathscr{O}(|\xi|^{m+1}), \quad \xi \to 0, \ \forall\, k \in \mathbb{Z}\backslash\{0\}. \qquad (5.6.67)$$

By $\widehat{h^{(m)}}(\xi) = (i\xi)^m\widehat{h}(\xi)$, we see that (5.5.14) holds with $\eta := h^{(m)}$. By Lemma 5.5.6, there exists a compactly supported function $g \in L_p(\mathbb{R})$ (which is explicitly given in (5.5.15)) such that $h^{(m)} = \eta = \nabla^{m+1}g$. Now we deduce from (5.6.66) that

$$f_{n+1}^{(m)} - f_n^{(m)} = 2^{n(m+1)}\sum_{k\in\mathbb{Z}}[\nabla^{m+1}a_n](k)g(2^n\cdot -k), \qquad n \in \mathbb{N}.$$

Since $g \in L_p(\mathbb{R})$ is compactly supported, we see from (5.3.4) in Lemma 5.3.1 that

$$\|f_{n+1}^{(m)} - f_n^{(m)}\|_{L_p(\mathbb{R})} \leqslant 2^{n(m+1-1/p)}\|g\|_{\mathcal{L}_p(\mathbb{R})}\|\nabla^{m+1}a_n\|_{l_p(\mathbb{Z})}.$$

By item (5), there exist $0 < \rho < 1$ and $C > 0$ such that

$$\|\nabla^{m+1} a_n\|_{l_p(\mathbb{Z})} \leqslant C\rho^n 2^{n(1/p-m-1)}, \qquad \forall\, n \in \mathbb{N}.$$

Thus, $\|f_{n+1}^{(m)} - f_n^{(m)}\|_{L_p(\mathbb{R})} \leqslant C\|g\|_{\mathcal{L}_p(\mathbb{R})}\rho^n$ for all $n \in \mathbb{N}$. This proves that $\{f_n^{(m)}\}_{n\in\mathbb{N}}$ is a Cauchy sequence in $L_p(\mathbb{R})$. Note that all f_n vanish outside $[-N, N]$ for some $N \in \mathbb{N}$ and hence, $f_n^{(j-1)}(x) = \int_{-N}^{x} f_n^{(j)}(t)dt$ for $j = 1, \dots, m$. If $\{f_n^{(j)}\}_{n\in\mathbb{N}}$ is a Cauchy sequence in $L_p(\mathbb{R})$, this relation implies that $\{f_n^{(j-1)}\}_{n\in\mathbb{N}}$ is also a Cauchy sequence in $L_p(\mathbb{R})$. Thus, $\{f_n\}_{n\in\mathbb{N}}$ is a Cauchy sequence in $W_p^m(\mathbb{R})$. Since $\lim_{n\to\infty} \widehat{f_n}(\xi) = \widehat{\phi}(\xi)$, we have $\phi \in W_p^m(\mathbb{R})$ and $\lim_{n\to\infty} \|f_n - \phi\|_{W_p^m(\mathbb{R})} = 0$. Thus, (5)$\Longrightarrow$(1).

The proof of (6)$\Longrightarrow$(2) is similar. Note that $m_a = \operatorname{sr}(a)$ and item (6) must imply $m_a \geqslant m + 1$ by $\operatorname{sr}(a) \geqslant \operatorname{sm}_p(a)$ in Proposition 5.6.15. By Lemma 5.6.7, there exists $u \in l_0(\mathbb{Z})$ such that $\widehat{u}(\xi) \neq 0$ for all $\xi \in \mathbb{R}$ and $\widehat{u}(\xi) = \widehat{\phi}(\xi)/\widehat{B_{m_a+2}}(\xi) + \mathscr{O}(|\xi|^{m_a})$ as $\xi \to 0$. Define $f = u * B_{m_a+2}$. Then f satisfies all the conditions in item (2) and

$$\widehat{f}(\xi) = \widehat{\phi}(\xi) + \mathscr{O}(|\xi|^{m_a}) \quad \text{and} \quad \widehat{f}(\xi + 2\pi k) = \mathscr{O}(|\xi|^{m_a}), \quad \xi \to 0, \forall\, k \in \mathbb{Z}\backslash\{0\}.$$

Define f_n, h as in the proof of (5)$\Longrightarrow$(1). Instead of (5.6.67), due to the above relation, we have

$$\widehat{h}(\xi + 2\pi k) = \mathscr{O}(|\xi|^{m_a}), \qquad \xi \to 0, \forall\, k \in \mathbb{Z}.$$

Consequently, $\eta := h^{(m)}$ satisfies (5.5.14) with $m + 1$ being replaced by m_a and hence $h^{(m)} = \nabla^{m_a} g$ for some compactly supported function $g \in L_p(\mathbb{R})$. By (5.6.66),

$$f_{n+1}^{(m)} - f_n^{(m)} = 2^{n(m+1)} \sum_{k\in\mathbb{Z}} [\nabla^{m_a} a_n](k) g(2^n \cdot -k), \qquad n \in \mathbb{N}.$$

By the same argument as in (5)$\Longrightarrow$(1), we conclude that $\{f_n\}_{n\in\mathbb{N}}$ is a Cauchy sequence in $W_p^m(\mathbb{R})$.

To prove (5.6.64), by item (2), we see that (5.6.65) holds with $m + 1$ being replaced by m. Hence,

$$\begin{aligned} 2^{n(m+1-1/p)}\|\nabla^m a_n\|_{l_p(\mathbb{Z})} &\leqslant C_1 2^{nm}\omega_m(f_n, 2^{-n})_p \leqslant C_1\|f_n^{(m)}\|_{L_p(\mathbb{R})} \\ &\leqslant C_1\big(\|\phi^{(m)}\|_{L_p(\mathbb{R})} + \|f_n^{(m)} - \phi^{(m)}\|_{L_p(\mathbb{R})}\big). \end{aligned}$$

Now (5.6.64) is verified by noting that $\lim_{n\to\infty} \|f_n^{(m)} - \phi^{(m)}\|_{L_p(\mathbb{R})} = 0$. □

For convergence of a scalar algorithm under the extra stability condition, the scalar version of Corollary 5.6.12 is as follows:

Corollary 5.6.17 *Let $1 \leqslant p \leqslant \infty$ and $m \in \mathbb{N}_0$. Let $a \in l_0(\mathbb{Z})$ with $\widehat{a}(0) = 1$. Define $\widehat{\phi}(\xi) := \prod_{j=1}^{\infty} \widehat{a}(2^{-j}\xi)$ for $\xi \in \mathbb{R}$. If $\phi \in W_p^m(\mathbb{R})$ (require $\phi \in \mathscr{C}^m(\mathbb{R})$ if $p = \infty$) and the integer shifts of ϕ are stable in $L_p(\mathbb{R})$, then* $\operatorname{sr}(a) \geqslant m+1$, $\operatorname{sm}_p(a) > m$, *and the cascade algorithm associated with the filter a converges in $W_p^m(\mathbb{R})$.*

Proof For $0 \leqslant j \leqslant m$, we define $g(x) := (-ix)^j\phi(x)$. Since $\phi \in W_p^m(\mathbb{R})$ has compact support, we have $g \in W_p^m(\mathbb{R}) \cap W_1^m(\mathbb{R})$ and $\widehat{g^{(j)}}(\xi) = (i\xi)^j\widehat{g}(\xi) = (i\xi)^j\widehat{\phi}^{(j)}(\xi)$. As in Lemma 5.6.1, applying the Riemann-Lebesgue Lemma, we have

$$\lim_{n\to\infty} (i2^n 2\pi k)^j\widehat{\phi}^{(j)}(2^n 2\pi k) = \lim_{n\to\infty} \widehat{g^{(j)}}(2^n 2\pi k) = 0, \quad \forall\, 0 \leqslant j \leqslant m, k \neq 0. \tag{5.6.68}$$

Note that $\phi = \mathcal{R}_a\phi$ and $\widehat{\phi}(2^n\xi) = \widehat{a_n}(\xi)\widehat{\phi}(\xi)$ with $\widehat{a_n}(\xi) = \widehat{a}(2^{n-1}\xi)\cdots\widehat{a}(2\xi)\widehat{a}(\xi)$. Therefore, by (5.6.68), for $k \in \mathbb{Z}\backslash\{0\}$, we have

$$\lim_{n\to\infty} [\widehat{a_n}(\cdot)\widehat{\phi}(\cdot + 2\pi k)]^{(j)}(2\pi k) = \lim_{n\to\infty} 2^{jn}\widehat{\phi}^{(j)}(2^n 2\pi k) = 0, \quad \forall\, 0 \leqslant j \leqslant m.$$

Since $\widehat{a_n}(2\pi k) = [\widehat{a}(0)]^n = 1$, by induction on j and the Leibniz differentiation formula, we conclude from the above identity that $\widehat{\phi}^{(j)}(2\pi k) = 0$ for all $k \in \mathbb{Z}\backslash\{0\}$ and $0 \leqslant j \leqslant m$. Take $f = \phi$ and note that $\widehat{\phi}(0) = 1$. Therefore, we conclude that $f \in \mathscr{F}_{m,p}$ satisfies (5.6.54). Hence, item (2) of Theorem 5.6.16 is satisfied. The claim now follows directly from Theorem 5.6.16. □

5.7 Express $\mathrm{sm}_p(a)$ Using the p-Norm Joint Spectral Radius

Because the convergence of a cascade algorithm in the Sobolev space $W_p^{m-1}(\mathbb{R})$ is completely characterized by the inequality $\rho_m(a, \upsilon)_p < 2^{1/p-m+1}$, it is important to study various properties of the key quantity $\rho_m(a, \upsilon)_p$ and to provide several ways of computing/estimating $\rho_m(a, \upsilon)_p$.

In this section, we shall rewrite the quantity $\rho_m(a, \upsilon)_p$ (and therefore, $\mathrm{sm}_p(a)$) using p-norm joint spectral radius in this section.

5.7.1 The p-Norm Joint Spectral Radius

Let us first introduce the p-norm joint spectral radius. Let $\mathcal{A}$ be a finite collection of $r \times r$ matrices (or operators acting on a finite-dimensional normed vector space). For $1 \leqslant p \leqslant \infty$, the *p-norm joint spectral radius* is defined to be

$$\mathrm{jsr}_p(\mathcal{A}) := \lim_{n\to\infty} \|\mathcal{A}^n\|_{l_p}^{1/n} \text{ with } \|\mathcal{A}^n\|_{l_p} := \left(\sum_{A_1\in\mathcal{A}} \cdots \sum_{A_n\in\mathcal{A}} \|A_1\cdots A_n\|^p\right)^{1/p} \tag{5.7.1}$$

with the usual modification for $p = \infty$ in (5.7.1), where $\|\cdot\|$ is a norm on $r \times r$ matrices or operators.

The following result shows that the above limit exists and is independent of the choice of the norm $\|\cdot\|$ on $r \times r$ matrices or operators.

Lemma 5.7.1 *The p-norm joint spectral radius is well defined and is independent of the choice of the norm* $\|\cdot\|$. *Moreover, if* $\|\cdot\|$ *is a submultiplicative norm (that is,* $\|AB\| \leqslant \|A\|\|B\|$ *also holds), then*

$$\mathrm{jsr}_p(\mathcal{A}) = \lim_{n\to\infty} \|\mathcal{A}^n\|_{l_p}^{1/n} = \inf_{n\in\mathbb{N}} \|\mathcal{A}^n\|_{l_p}^{1/n}. \tag{5.7.2}$$

Denote by $\#\mathcal{A}$ *the cardinality of the set* $\mathcal{A}$. *Then*

$$(\#\mathcal{A})^{1/q-1/p}\,\mathrm{jsr}_p(\mathcal{A}) \leqslant \mathrm{jsr}_q(\mathcal{A}) \leqslant \mathrm{jsr}_p(\mathcal{A}), \qquad \forall\, 1 \leqslant p \leqslant q \leqslant \infty. \tag{5.7.3}$$

Moreover, $\mathrm{jsr}_p(\mathcal{A}^\mathsf{T}) = \mathrm{jsr}_p(\mathcal{A})$, *where* $\mathcal{A}^\mathsf{T} := \{A^\mathsf{T} \,:\, A \in \mathcal{A}\}$.

Proof We first prove (5.7.2). Obviously, $\limsup_{n\to\infty} \|\mathcal{A}^n\|_{l_p}^{1/n} \geqslant \inf_{n\in\mathbb{N}} \|\mathcal{A}^n\|_{l_p}^{1/n}$. To prove (5.7.2), it suffices to prove that

$$\limsup_{n\to\infty} \|\mathcal{A}^n\|_{l_p}^{1/n} \leqslant \inf_{n\in\mathbb{N}} \|\mathcal{A}^n\|_{l_p}^{1/n} =: L. \tag{5.7.4}$$

We assume $L > 0$, otherwise all elements in $\mathcal{A}$ are zero and (5.7.2) is obvious. For any $\varepsilon > 0$, there exists $m \in \mathbb{N}$ such that $\|\mathcal{A}^m\|_{l_p} \leqslant (L+\varepsilon)^m$. Define $C := \sup_{0\leqslant j<m}(L+\varepsilon)^{-j}\|\mathcal{A}^j\|_{l_p} < \infty$. Since $\|\cdot\|$ is a submultiplicative norm, by definition, we see that

$$\|\mathcal{A}^{n_1+n_2}\|_{l_p} \leqslant \|\mathcal{A}^{n_1}\|_{l_p}\|\mathcal{A}^{n_2}\|_{l_p} \qquad \forall\, n_1, n_2 \in \mathbb{N}.$$

For any $n \in \mathbb{N}$, we can write $n = m\lfloor \frac{n}{m} \rfloor + j$ for some $j = 0, \dots, m-1$. Therefore,

$$\|\mathcal{A}^n\|_{l_p} \leqslant \|\mathcal{A}^m\|_{l_p}^{\lfloor \frac{n}{m} \rfloor}\|\mathcal{A}^j\|_{l_p} \leqslant C(L+\varepsilon)^j\|\mathcal{A}^m\|_{l_p}^{\lfloor \frac{n}{m} \rfloor} \leqslant C(L+\varepsilon)^{m\lfloor \frac{n}{m} \rfloor + j} = C(L+\varepsilon)^n.$$

Now it follows directly from the above inequality that $\limsup_{n\to\infty} \|\mathcal{A}^n\|_{l_p}^{1/n} \leqslant L+\varepsilon$. Since $\varepsilon > 0$ is arbitrary, we conclude that (5.7.4) must hold. This proves (5.7.2).

Since all norms on a finite-dimensional space are equivalent to each other, for every $n \in \mathbb{N}$, the ratios between different $\|\mathcal{A}^n\|_{l_p}$ using two different norms $\|\cdot\|$ are bounded below and above by two positive constants independent of n. Now it is straightforward to see that $\mathrm{jsr}_p(\mathcal{A})$ is always well defined and is independent of the choice of $\|\cdot\|$.

The inequalities (5.7.3) are a direct consequence of Hölder's inequality for l_p and the fact $\|\cdot\|_{l_q} \leqslant \|\cdot\|_{l_p}$ for $1 \leqslant p \leqslant q \leqslant \infty$. Note that $\|(\mathcal{A}^\mathsf{T})^n\|_{l_p} = \|\mathcal{A}^n\|_{l_p}$ with $\|\cdot\| = \|\cdot\|_{l_p}$ in (5.0.1). Therefore, $\mathrm{jsr}_p(\mathcal{A}^\mathsf{T}) = \mathrm{jsr}_p(\mathcal{A})$. □

Note that if $\mathcal{A} = \{A\}$ is a singleton, then $\mathrm{jsr}_p(\{A\}) = \rho(A)$ for all $1 \leqslant p \leqslant \infty$, where $\rho(A)$ denotes the *spectral radius* of A, that is,

$$\rho(A) := \lim_{n\to\infty} \|A^n\|^{1/n} = \max\{|\lambda| \,:\, \lambda \in \mathrm{spec}(A)\}, \tag{5.7.5}$$

where $\mathrm{spec}(A)$ is the multiset of all the eigenvalues of A counting multiplicity. From now on, we shall require $\|\cdot\|$ to be a submultiplicative norm in the definition of the p-norm joint spectral radius so that the identity in (5.7.2) holds.

The following result is useful for calculating the p-norm joint spectral radius.

Proposition 5.7.2 *Let $A_1, \ldots, A_m$ be $(r+s)\times(r+s)$ matrices such that*

$$A_\ell = \begin{bmatrix} B_\ell & D_\ell \\ 0 & C_\ell \end{bmatrix}, \qquad \ell = 1, \ldots, m,$$

where $B_1, \ldots, B_m$ are $r \times r$ matrices, $C_1, \ldots, C_m$ are $s \times s$ matrices, and $D_1, \ldots, D_m$ are $r \times s$ matrices. Then

$$\mathrm{jsr}_p(\{A_1, \ldots, A_m\}) = \max\left(\mathrm{jsr}_p(\{B_1, \ldots, B_m\}), \mathrm{jsr}_p(\{C_1, \ldots, C_m\})\right), \tag{5.7.6}$$

for all $1 \leqslant p \leqslant \infty$.

Proof Define $\mathcal{A} := \{A_1, \ldots, A_m\}$, $\mathcal{B} := \{B_1, \ldots, B_m\}$ and $\mathcal{C} := \{C_1, \ldots, C_m\}$. By calculation,

$$A_{\gamma_1} \cdots A_{\gamma_n} = \begin{bmatrix} B_{\gamma_1} \cdots B_{\gamma_n} & \sum_{j=1}^{n} \left(\prod_{k=1}^{j-1} B_{\gamma_k}\right) D_{\gamma_j} \left(\prod_{k=j+1}^{n} C_{\gamma_k}\right) \\ 0 & C_{\gamma_1} \cdots C_{\gamma_n} \end{bmatrix}, \tag{5.7.7}$$

where $\gamma_1, \ldots, \gamma_n \in \{1, \ldots, m\}$ and we used the convention that $\prod_{j=1}^{0} B_{\gamma_j} = I_r$ and $\prod_{j=n+1}^{n} C_{\gamma_j} = I_s$. Now it follows trivially from (5.7.7) that $\|\mathcal{B}^n\|_{l_p} \leqslant \|\mathcal{A}^n\|_{l_p}$ and $\|\mathcal{C}^n\|_{l_p} \leqslant \|\mathcal{A}^n\|_{l_p}$. Consequently, $\rho := \max(\mathrm{jsr}_p(\mathcal{B}), \mathrm{jsr}_p(\mathcal{C})) \leqslant \mathrm{jsr}_p(\mathcal{A})$. We now prove $\mathrm{jsr}_p(\mathcal{A}) \leqslant \rho$. By (5.7.7), we have

$$\begin{aligned} \|A_{\gamma_1} \cdots A_{\gamma_n}\| &\leqslant \|B_{\gamma_1} \cdots B_{\gamma_n}\| + \|C_{\gamma_1} \cdots C_{\gamma_n}\| + \left\| \sum_{j=1}^{n} \left(\prod_{k=1}^{j-1} B_{\gamma_k}\right) D_{\gamma_j} \left(\prod_{k=j+1}^{n} C_{\gamma_k}\right) \right\| \\ &\leqslant \|B_{\gamma_1} \cdots B_{\gamma_n}\| + \|C_{\gamma_1} \cdots C_{\gamma_n}\| + \sum_{j=1}^{n} \left\| \prod_{k=1}^{j-1} B_{\gamma_k} \right\| \|D_{\gamma_j}\| \left\| \prod_{k=j+1}^{n} C_{\gamma_k} \right\|, \end{aligned}$$

where the norms on B, C, D's are induced from the norm on A through

$$\|B\| = \left\| \begin{bmatrix} B & 0 \\ 0 & 0 \end{bmatrix} \right\|, \quad \|D\| = \left\| \begin{bmatrix} 0 & D \\ 0 & 0 \end{bmatrix} \right\|, \quad \|C\| = \left\| \begin{bmatrix} 0 & 0 \\ 0 & C \end{bmatrix} \right\|.$$

For any $\varepsilon > 0$, by $\rho := \max(\mathrm{jsr}_p(\mathcal{B}), \mathrm{jsr}_p(\mathcal{C}))$, there exists a positive constant C such that

$$\|\mathcal{B}^n\|_{l_p} \leqslant C(\rho+\varepsilon)^n, \qquad \|\mathcal{C}^n\|_{l_p} \leqslant C(\rho+\varepsilon)^n, \qquad \forall n \in \mathbb{N}. \tag{5.7.8}$$

For $p = \infty$, we deduce from the above inequalities that

$$\left\| A_{\gamma_1} \cdots A_{\gamma_n} \right\| \leqslant 2C(\rho+\varepsilon)^n + C^2(\rho+\varepsilon)^{n-1} \sum_{j=1}^{n} \left\| D_{\gamma_j} \right\| \leqslant C(n)(\rho+\varepsilon)^{n-1},$$

where $C(n) := 2C(\rho+1) + nC^2 \sup_{1\leqslant \ell \leqslant m} \|D_\ell\|$. Since $\lim_{n\to\infty}[C(n)]^{1/n} = 1$, it is now straightforward to conclude that $\mathrm{jsr}_\infty(\mathcal{A}) \leqslant \rho + \varepsilon$. Since $\varepsilon > 0$ is arbitrary, we proved $\mathrm{jsr}_\infty(\mathcal{A}) \leqslant \rho$.

We now prove $\mathrm{jsr}_p(\mathcal{A}) \leqslant \rho$ for $1 \leqslant p < \infty$. By Hölder's inequality, we observe the inequality $(|c_1| + \cdots + |c_J|)^p \leqslant J^{p-1}(|c_1|^p + \cdots + |c_J|^p)$. Hence,

$$\begin{aligned}
\left\| A_{\gamma_1} \cdots A_{\gamma_n} \right\|^p \leqslant & 3^{p-1} \left\| B_{\gamma_1} \cdots B_{\gamma_n} \right\|^p + 3^{p-1} \left\| C_{\gamma_1} \cdots C_{\gamma_n} \right\|^p \\
& + 3^{p-1} \left(\sum_{j=1}^{n} \left\| \prod_{k=1}^{j-1} B_{\gamma_k} \right\| \left\| D_{\gamma_j} \right\| \left\| \prod_{k=j+1}^{n} C_{\gamma_k} \right\| \right)^p \\
\leqslant & 3^{p-1} \left\| B_{\gamma_1} \cdots B_{\gamma_n} \right\|^p + 3^{p-1} \left\| C_{\gamma_1} \cdots C_{\gamma_n} \right\|^p \\
& + 3^{p-1} n^{p-1} \sum_{j=1}^{n} \left\| \prod_{k=1}^{j-1} B_{\gamma_k} \right\|^p \left\| D_{\gamma_j} \right\|^p \left\| \prod_{k=j+1}^{n} C_{\gamma_k} \right\|^p .
\end{aligned}$$

Now we deduce from the above inequality and (5.7.8) that

$$\begin{aligned}
\|\mathcal{A}^n\|_{l_p}^p &\leqslant 3^{p-1}\|\mathcal{B}^n\|_{l_p}^p + 3^{p-1}\|\mathcal{C}^n\|_{l_p}^p + (3n)^{p-1} \sum_{j=1}^{n} \|\mathcal{B}^{j-1}\|_{l_p}^p \left(\sum_{\ell=1}^{m} \|D_\ell\|^p \right) \|\mathcal{C}^{n-j}\|_{l_p}^p \\
&\leqslant 3^{p-1} 2C(\rho+\varepsilon)^{pn} + (3n)^{p-1} C^2 (\rho+\varepsilon)^{p(n-1)} \sum_{\ell=1}^{m} \|D_\ell\|^p \leqslant \mathring{C}(n)(\rho+\varepsilon)^{p(n-1)},
\end{aligned}$$

where $\mathring{C}(n) := 2C3^{p-1}(\rho+1)^p + (3n)^{p-1}C^2 \sum_{\ell=1}^{m} \|D_\ell\|^p$. Note that $\lim_{n\to\infty}[\mathring{C}(n)]^{1/n} = 1$. We conclude from the above inequality that $\mathrm{jsr}_p(\mathcal{A}) \leqslant \rho + \varepsilon$. Since $\varepsilon > 0$ is arbitrary, we must have $\mathrm{jsr}_p(\mathcal{A}) \leqslant \rho$. This completes the proof. □

5.7.2 Rewrite $\rho_m(a, v)_p$ Using the p-Norm Joint Spectral Radius

To link the key quantity $\rho_m(a, v)_p$ with the p-norm joint spectral radius, we first recall the definition of the transition operators and their associated variations.

For a subset K of $\mathbb{R}$, we define

$$l(K) := \{u \in l(\mathbb{Z}) \ : \ u(k) = 0, \ \forall\, k \in \mathbb{Z}\backslash K\}. \tag{5.7.9}$$

For $\gamma \in \mathbb{Z}$, we define operators $\mathcal{T}_\gamma : (l_0(\mathbb{Z}))^{1\times r} \to (l_0(\mathbb{Z}))^{1\times r}$ by

$$\begin{aligned}\mathcal{T}_\gamma u(n) :=& [\mathcal{T}_a(u(\cdot - \gamma))](n) = 2\sum_{k\in\mathbb{Z}} u(k)\overline{a(\gamma + k - 2n)}^{\mathsf{T}} \\ =& 2\sum_{k\in\mathbb{Z}} u(k)a^\star(2n - k - \gamma), \qquad n \in \mathbb{Z},\end{aligned} \tag{5.7.10}$$

where $\mathcal{T}_a$ is the transition operator defined in (5.6.31). For the convenience of computation, to avoid the complex conjugate and transpose in the definition of $\mathcal{T}_\gamma$, we often use the equivalent operators $\mathring{\mathcal{T}}_\gamma : (l_0(\mathbb{Z}))^r \to (l_0(\mathbb{Z}))^r$ with $\mathring{\mathcal{T}}_\gamma u := (\mathcal{T}_\gamma u^\star)^\star$. More explicitly,

$$[\mathring{\mathcal{T}}_\gamma u](n) := 2\sum_{k\in\mathbb{Z}} a(\gamma + 2n - k)u(k), \qquad n \in \mathbb{Z}, u \in (l_0(\mathbb{Z}))^r. \tag{5.7.11}$$

In particular, we define $\mathring{\mathcal{T}}_a := \mathring{\mathcal{T}}_0$. Since $(u(\cdot + \gamma))^\star = u^\star(\cdot - \gamma)$, we see that $\mathring{\mathcal{T}}_\gamma u = \mathring{\mathcal{T}}_a(u(\cdot + \gamma))$.

The following result provides finite-dimensional invariant spaces for the transition operators.

Lemma 5.7.3 *For $a \in (l_0(\mathbb{Z}))^{r\times r}$, then $\mathcal{T}_\gamma(l(\mathrm{fsupp}(a^\star)))^{1\times r} \subseteq (l(\mathrm{fsupp}(a^\star)))^{1\times r}$ and $\mathring{\mathcal{T}}_\gamma(l(\mathrm{fsupp}(a)))^r \subseteq (l(\mathrm{fsupp}(a)))^r$ for all $\gamma = -1, 0, 1$. Moreover, if $\mathcal{T}_a u = \lambda u$ for some $\lambda \in \mathbb{C}\backslash\{0\}$ and $u \in (l_0(\mathbb{Z}))^{1\times r}\backslash\{0\}$, then $\mathrm{fsupp}(u) \subseteq \mathrm{fsupp}(a^\star)$. Similarly, if $\mathring{\mathcal{T}}_a v = \lambda v$ for some $\lambda \in \mathbb{C}\backslash\{0\}$ and $v \in (l_0(\mathbb{Z}))^r\backslash\{0\}$, then $\mathrm{fsupp}(v) \subseteq \mathrm{fsupp}(a)$.*

Proof By the definition of $\mathcal{T}_\gamma$, if $\mathrm{fsupp}(u) \subseteq \mathrm{fsupp}(a^\star)$ and $-1 \leqslant \gamma \leqslant 1$,

$$\begin{aligned}\mathrm{fsupp}(\mathcal{T}_\gamma u) &\subseteq \tfrac{1}{2}\,\mathrm{fsupp}(a^\star) + \tfrac{1}{2}\,\mathrm{fsupp}(u) + \tfrac{1}{2}\gamma \\ &\subseteq \tfrac{1}{2}\,\mathrm{fsupp}(a^\star) + \tfrac{1}{2}\,\mathrm{fsupp}(a^\star) + \tfrac{1}{2}[-1, 1] = \mathrm{fsupp}(a^\star) + \tfrac{1}{2}[-1, 1].\end{aligned}$$

Therefore, we must have $\mathrm{fsupp}(\mathcal{T}_\gamma u) \subseteq \mathrm{fsupp}(a^\star)$ for any sequence $u \in (l_0(\mathbb{Z}))^{1\times r}$ being supported inside $\mathrm{fsupp}(a^\star)$.

By $\mathcal{T}_a u = \lambda u$ and $\lambda \neq 0$, we have $u = \lambda^{-n}\mathcal{T}_a^n u$ for all $n \in \mathbb{N}$ and therefore,

$$\begin{aligned}\operatorname{fsupp}(u) &= \operatorname{fsupp}(\mathcal{T}_a^n u) \subseteq 2^{-n}\operatorname{fsupp}(u) + (2^{-n} + \cdots + 2^{-1})\operatorname{fsupp}(a^\star)\\ &\subseteq 2^{-n}\operatorname{fsupp}(u) + \operatorname{fsupp}(a^\star).\end{aligned}$$

Taking $n \to \infty$, we conclude that $\operatorname{fsupp}(u) \subseteq \operatorname{fsupp}(a^\star)$. □

Let $[m, n] := \operatorname{fsupp}(a)$. Then $\operatorname{fsupp}(a^\star) = [-n, -m]$ and we define

$$v_{r(j+n)+\ell} := \delta(\cdot - j)e_\ell^{\mathsf{T}}, \qquad -n \leqslant j \leqslant -m,\ 1 \leqslant \ell \leqslant r,$$

where e_j is the jth coordinate unit column vector in $\mathbb{R}^r$. Then $\{v_1, \dots, v_{r(n-m+1)}\}$ is a basis of $V := (l(\operatorname{fsupp}(a^\star)))^{1\times r}$ and the matrix representations of $\mathcal{T}_\gamma|_V$ and $\mathring{\mathcal{T}}_\gamma|_V$ with $\gamma = -1, 0, 1$ under this basis is

$$\mathcal{T}_\gamma \begin{bmatrix} v_1 \\ \vdots \\ v_{r(n-m+1)} \end{bmatrix} = T_\gamma \begin{bmatrix} v_1 \\ \vdots \\ v_{r(n-m+1)} \end{bmatrix} \quad \text{and} \quad \mathring{\mathcal{T}}_\gamma \begin{bmatrix} v_1^\star \\ \vdots \\ v_{r(n-m+1)}^\star \end{bmatrix} = \mathring{T}_\gamma \begin{bmatrix} v_1^\star \\ \vdots \\ v_{r(n-m+1)}^\star \end{bmatrix}$$

where the $r(n-m+1)\times r(n-m+1)$ matrices $\mathring{T}_\gamma := \left(2a(\gamma+j-2k)^{\mathsf{T}}\right)_{-n\leqslant j,k\leqslant -m} = \overline{T_\gamma}$ and

$$\begin{aligned} T_\gamma &:= \left(2\overline{a(\gamma+j-2k)}^{\mathsf{T}}\right)_{-n\leqslant j,k\leqslant -m} \\ &= 2\begin{bmatrix} \overline{a(\gamma+n)}^{\mathsf{T}} & \overline{a(\gamma+n-2)}^{\mathsf{T}} & \overline{a(\gamma+n-4)}^{\mathsf{T}} & \cdots & \overline{a(\gamma+2m-n)}^{\mathsf{T}} \\ \overline{a(\gamma+n+1)}^{\mathsf{T}} & \overline{a(\gamma+n-1)}^{\mathsf{T}} & \overline{a(\gamma+n-3)}^{\mathsf{T}} & \cdots & \overline{a(\gamma+2m-n+1)}^{\mathsf{T}} \\ \overline{a(\gamma+n+2)}^{\mathsf{T}} & \overline{a(\gamma+n)}^{\mathsf{T}} & \overline{a(\gamma+n-2)}^{\mathsf{T}} & \cdots & \overline{a(\gamma+2m-n+2)}^{\mathsf{T}} \\ \vdots & \vdots & \vdots & \ddots & \vdots \\ \overline{a(\gamma+2n-m)}^{\mathsf{T}} & \overline{a(\gamma+2n-m-2)}^{\mathsf{T}} & \overline{a(\gamma+2n-m-4)}^{\mathsf{T}} & \cdots & \overline{a(\gamma+m)}^{\mathsf{T}} \end{bmatrix}. \end{aligned}$$

As demonstrated by the following result, the shifted transition operators allow us to link the important quantity $\rho_m(a, v)_p$ with the p-norm joint spectral radius.

Theorem 5.7.4 *Let $a \in (l_0(\mathbb{Z}))^{r\times r}$ and $a_n, n \in \mathbb{N}$ be defined in (5.6.7). Let $u \in (l_0(\mathbb{Z}))^{1\times r}$. Then*

(i) For all $\gamma_0, \dots, \gamma_{n-1} \in \{0, 1\}$ and $k \in \mathbb{Z}$,

$$\begin{aligned}&2^n[a_n * u^\star](\gamma_0 + 2\gamma_1 + \cdots + 2^{n-1}\gamma_{n-1} + 2^n k) \\ &\quad = (\mathcal{T}_{\gamma_{n-1}}\mathcal{T}_{\gamma_{n-2}}\cdots\mathcal{T}_{\gamma_1}\mathcal{T}_{\gamma_0}u)^\star(k) = (\mathring{\mathcal{T}}_{\gamma_{n-1}}\mathring{\mathcal{T}}_{\gamma_{n-2}}\cdots\mathring{\mathcal{T}}_{\gamma_1}\mathring{\mathcal{T}}_{\gamma_0}u^\star)(k)\end{aligned} \tag{5.7.12}$$

and for $1 \leqslant p \leqslant \infty$,

$$\|[\mathcal{S}_a^n(\delta I_r)] * u^\star\|_{(l_p(\mathbb{Z}))^r} = 2^n \|a_n * u^\star\|_{(l_p(\mathbb{Z}))^r}$$
$$= \left(\sum_{\gamma_0=0}^{1} \cdots \sum_{\gamma_{n-1}=0}^{1} \|\mathcal{T}_{\gamma_{n-1}} \cdots \mathcal{T}_{\gamma_0} u\|^p_{(l_p(\mathbb{Z}))^{1\times r}} \right)^{1/p} . \tag{5.7.13}$$

(ii) There exists a minimal common invariant finite-dimensional subspace of $(l_0(\mathbb{Z}))^{1\times r}$, *denoted by* $V(u)$, *of* $\mathcal{T}_0$ *and* $\mathcal{T}_1$ *such that* $u \in V(u)$, $\mathcal{T}_0 V(u) \subseteq V(u)$, *and* $\mathcal{T}_1 V(u) \subseteq V(u)$. *Moreover, there exist two positive constants* C_1 *and* C_2 *such that*

$$C_1 \|\mathcal{A}^n\|_{l_p} \leqslant \left(\sum_{\gamma_1=0}^{1} \cdots \sum_{\gamma_n=0}^{1} \|\mathcal{T}_{\gamma_n} \cdots \mathcal{T}_{\gamma_1} u\|^p_{(l_p(\mathbb{Z}))^{1\times r}} \right)^{1/p} \leqslant C_2 \|\mathcal{A}^n\|_{l_p}, \tag{5.7.14}$$

for all $n \in \mathbb{N}$, *where* $\mathcal{A} := \{A_0, A_1\}$ *with* $A_0 := \mathcal{T}_0|_{V(u)}$, $A_1 := \mathcal{T}_1|_{V(u)}$ *and any linear operator* $T : V(u) \to V(u)$ *is equipped with the operator norm* $\|T\| := \sup\{\|Tv\|_{(l_p(\mathbb{Z}))^{1\times r}} \ : \ v \in V(u), \|v\|_{(l_p(\mathbb{Z}))^{1\times r}} \leqslant 1\}$.

Proof Let $j := \gamma_0 + 2\gamma_1 + \cdots + 2^{n-1}\gamma_{n-1} + 2^n k$. By $2^n a_n = \mathcal{S}_a^n(\delta I_r)$ and $\langle \mathcal{S}_a u, v\rangle = \langle u, \mathcal{T}_a v\rangle$,

$$\begin{aligned} 2^n [a_n * u^\star](j) &= 2^n \langle a_n * u^\star, \delta(\cdot - j)\rangle = 2^n \langle a_n, \delta(\cdot - j) * u\rangle \\ &= \langle \mathcal{S}_a^n(\delta I_r), u(\cdot - j)\rangle = \langle \delta I_r, \mathcal{T}_a^n(u(\cdot - j))\rangle. \end{aligned}$$

Observing that $\mathcal{T}_a(u(\cdot - 2n)) = [\mathcal{T}_a u](\cdot - n)$ for $n \in \mathbb{Z}$, we have

$$\begin{aligned} \mathcal{T}_a^n(u(\cdot - j)) &= \mathcal{T}_a^n(u(\cdot - \gamma_0 - 2\gamma_1 - \cdots - 2^{n-1}\gamma_{n-1} - 2^n k)) \\ &= \mathcal{T}_a^{n-1}((\mathcal{T}_{\gamma_0} u)(\cdot - \gamma_1 - \cdots - 2^{n-2}\gamma_{n-1} - 2^{n-1}k)) \\ &= (\mathcal{T}_{\gamma_{n-1}} \mathcal{T}_{\gamma_{n-2}} \cdots \mathcal{T}_{\gamma_1} \mathcal{T}_{\gamma_0} u)(\cdot - k). \end{aligned}$$

Hence,

$$\begin{aligned} 2^n [a_n * u^\star](j) &= \langle \delta I_r, \mathcal{T}_a^n(u(\cdot - j))\rangle = \overline{(\mathcal{T}_a^n(u(\cdot - j)))(0)}^{\mathsf{T}} \\ &= (\mathcal{T}_{\gamma_{n-1}} \mathcal{T}_{\gamma_{n-2}} \cdots \mathcal{T}_{\gamma_1} \mathcal{T}_{\gamma_0} u)^\star(k). \end{aligned}$$

This proves (5.7.12). By (5.7.12), we have (5.7.13). Thus, item (i) is verified.

Since both a and u are finitely supported, there exists $N \in \mathbb{N}$ such that both a and u vanish outside $[-N, N]$. By Lemma 5.7.3, the space $(l([-N, N]))^{1\times r}$ is a common invariant subspace of $\mathcal{T}_0$ and $\mathcal{T}_1$. Hence, there exists a minimal/smallest common invariant finite-dimensional subspace $V(u)$ of $(l([-N,N]))^{1\times r}$ with $u \in V(u)$ such that $\mathcal{T}_0 V(u) \subseteq V(u)$ and $\mathcal{T}_1 V(u) \subseteq V(u)$. On one hand,

$$\|\mathcal{T}_{\gamma_n}\cdots\mathcal{T}_{\gamma_2}\mathcal{T}_{\gamma_1}u\|_{(l_p(\mathbb{Z}))^{1\times r}} = \|A_{\gamma_n}\cdots A_{\gamma_2}A_{\gamma_1}u\|_{(l_p(\mathbb{Z}))^{1\times r}} \leqslant |\!|\!| A_{\gamma_n}\cdots A_{\gamma_2}A_{\gamma_1} |\!|\!| \|u\|_{(l_p(\mathbb{Z}))^{1\times r}}.$$

Therefore, we can take $C_2 = \|u\|_{(l_p(\mathbb{Z}))^{1\times r}} < \infty$ in (5.7.14).

On the other hand, since $V(u)$ is the minimal common invariant finite-dimensional subspace of A_0 and A_1 containing u, there exists a positive integer m such that $V(u)$ is spanned by

$$V := \{u\} \cup \{A_{\gamma_1}\cdots A_{\gamma_j}u \ : \ \gamma_1,\ldots,\gamma_j \in \{0,1\}, j = 1,\ldots,m\}.$$

For any linear operator $T : V(u) \to V(u)$, we define

$$\|T\| := \left(\sum_{v\in V} \|Tv\|^p_{(l_p(\mathbb{Z}))^{1\times r}}\right)^{1/p}.$$

One can directly check that the above $\|\cdot\|$ is indeed a norm on all linear operators on $V(u)$. For simplicity of discussion, we define

$$\|\mathcal{A}^n u\|_{l_p} := \left(\sum_{\gamma_1=0}^{1}\cdots\sum_{\gamma_n=0}^{1} \|\mathcal{T}_{\gamma_n}\cdots\mathcal{T}_{\gamma_1}u\|^p_{(l_p(\mathbb{Z}))^{1\times r}}\right)^{1/p}.$$

Since all the norms on a finite-dimensional space are equivalent, there exists a positive constant C_3 such that $|\!|\!| T |\!|\!| \leqslant C_3\|T\|$ from which we have

$$\begin{aligned}\|\mathcal{A}^n\|_{l_p} &\leqslant C_3 \left(\sum_{v\in V}\sum_{\gamma_1=0}^{1}\cdots\sum_{\gamma_n=0}^{1} \|A_{\gamma_1}\cdots A_{\gamma_n}v\|^p_{l_p}\right)^{1/p} \\ &\leqslant C_3\Big(\sum_{j=0}^{m}\|\mathcal{A}^{n+j}u\|^p_{l_p}\Big)^{1/p} \leqslant C_3\Big(\sum_{j=0}^{m}\|\mathcal{A}^j\|^p_{l_p}\Big)^{1/p}\|\mathcal{A}^n u\|_{l_p}.\end{aligned}$$

Hence, we can take $C_1 := (\sum_{j=0}^{m}\|\mathcal{A}^j\|^p_{l_p})^{-1/p}/C_3$ in (5.7.14). This completes the proof of item (ii). □

With the help of Theorem 5.7.4, we now prove that (5.6.36) implies (5.6.37) in Proposition 5.6.9.

Theorem 5.7.5 *Let $a \in (l_0(\mathbb{Z}))^{r\times r}$ and $u \in (l_0(\mathbb{Z}))^{1\times r}$. For $\rho > 0$,*

$$\lim_{n\to\infty} \rho^n \|a_n * u^\star\|_{(l_p(\mathbb{Z}))^r} = 0 \tag{5.7.15}$$

if and only if

$$\lim_{n\to\infty} \|a_n * u^\star\|_{(l_p(\mathbb{Z}))^r}^{1/n} < \rho^{-1}. \tag{5.7.16}$$

Proof (5.7.16)$\Longrightarrow$(5.7.15) is trivial. We now prove (5.7.15)$\Longrightarrow$(5.7.16). Let $\mathcal{A} = \{\mathcal{T}_0|_{V(u)}, \mathcal{T}_1|_{V(u)}\}$. By Theorem 5.7.4, there exist positive constants C_1 and C_2 such that

$$C_1 2^{-n}\|\mathcal{A}^n\|_{l_p} \leqslant \|a_n * u^\star\|_{(l_p(\mathbb{Z}))^r} \leqslant C_2 2^{-n}\|\mathcal{A}^n\|_{l_p}, \qquad \forall n \in \mathbb{N}. \tag{5.7.17}$$

Since the p-norm joint spectral radius always exists, it follows from the above inequalities that $\lim_{n\to\infty} \|a_n * u^\star\|_{(l_p(\mathbb{Z}))^r}^{1/n} = 2^{-1}\mathrm{jsr}_p(\mathcal{A})$ exists. Now (5.7.16) is equivalent to showing that $\mathrm{jsr}_p(\mathcal{A}) < 2\rho^{-1}$. If not, then $2^{-1}\rho\,\mathrm{jsr}_p(\mathcal{A}) \geqslant 1$ and by (5.7.6),

$$\inf_{n\in\mathbb{N}} 2^{-1}\rho\|\mathcal{A}^n\|_{l_p}^{1/n} = 2^{-1}\rho\,\mathrm{jsr}_p(\mathcal{A}) \geqslant 1,$$

from which and (5.7.17) we must have

$$\rho^n\|a_n * u^\star\|_{(l_p(\mathbb{Z}))^r} \geqslant C_1 2^{-n}\rho^n\|\mathcal{A}^n\|_{l_p} \geqslant C_1 > 0,$$

a contradiction to (5.7.15). Therefore, we must have $\mathrm{jsr}_p(\mathcal{A}) < 2\rho^{-1}$ and hence (5.7.16) holds. □

Now $\rho_m(a, \upsilon)_p$ can be rewritten using the p-norm joint spectral radius.

Theorem 5.7.6 *Let $a \in (l_0(\mathbb{Z}))^{r\times r}$, $\upsilon \in (l_0(\mathbb{Z}))^{1\times r}$, and $m \in \mathbb{N} \cup \{-1, 0\}$. Let V be a finite-dimensional subspace of $\mathscr{V}_{m,\upsilon}$ such that span$\{u(\cdot - k) \ : \ u \in V\} = \mathscr{V}_{m,\upsilon}$ and V is invariant under both $\mathcal{T}_0$ and $\mathcal{T}_1$ (that is, $\mathcal{T}_0 V \subseteq V, \mathcal{T}_1 V \subseteq V$). Then*

$$\begin{gathered}\rho_{m+1}(a, \upsilon)_p = \mathrm{jsr}_p(\mathcal{A}) = \mathrm{jsr}_p(\mathring{\mathcal{A}}), \quad \forall\, 1 \leqslant p \leqslant \infty \\ \text{with} \quad \mathcal{A} = \{\mathcal{T}_0|_V, \mathcal{T}_1|_V\}, \quad \mathring{\mathcal{A}} = \{\mathring{\mathcal{T}}_0|_{V^\star}, \mathring{\mathcal{T}}_1|_{V^\star}\},\end{gathered} \tag{5.7.18}$$

where $\mathcal{T}_\gamma u := \mathcal{T}_a(u(\cdot - \gamma))$, $\mathring{\mathcal{T}}_\gamma v := \mathring{\mathcal{T}}_a(v(\cdot + \gamma))$ are defined as in (5.7.10) and (5.7.11), and $V^\star := \{u^\star \ : \ u \in V\}$. Moreover, if a satisfies order $m+1$ sum rules with the matching filter υ and $\widehat{\upsilon}(0) \neq 0$, then

$$\rho_j(a, \upsilon)_p = \max(2^{1/p-j}, \rho_{m+1}(a, \upsilon)_p), \qquad \forall j = 0, \ldots, m. \tag{5.7.19}$$

Proof Since V is finite-dimensional, take $u_1, \ldots, u_J \in V$ such that V is spanned by these elements. Since V generates $\mathscr{V}_{m,\upsilon}$, the set $\{u_1, \ldots, u_J\}$ generates $\mathscr{V}_{m,\upsilon}$. Define $V_\ell := V(u_\ell)$, $\ell = 1, \ldots, J$, where $V(u_\ell)$ is the minimal common invariant subspace of $\mathcal{T}_0$ and $\mathcal{T}_1$ containing u_ℓ. Therefore, by Theorem 5.7.4

$$\begin{aligned}\rho_{m+1}(a, \upsilon)_p &= 2\max\{\lim_{n\to\infty} \|a_n * u_\ell^\star\|_{(l_p(\mathbb{Z}))^r}^{1/n} \;:\; \ell = 1, \ldots, J\} \\ &= \max\{\operatorname{jsr}_p(\mathcal{A}|_{V_\ell}) \;:\; \ell = 1, \ldots, J\} \leqslant \operatorname{jsr}_p(\mathcal{A}),\end{aligned}$$

where $\mathcal{A}|_{V_\ell} := \{\mathcal{T}_0|_{V_\ell}, \mathcal{T}_1|_{V_\ell}\}$ and $\operatorname{jsr}_p(\mathcal{A}|_{V_\ell}) \leqslant \operatorname{jsr}_p(\mathcal{A}|_V)$ since $V_\ell \subseteq V$.

On the other hand, noting that $V_1 + \cdots + V_J = V$, by Proposition 5.7.2, we have

$$\operatorname{jsr}_p(\mathcal{A}) = \operatorname{jsr}_p(\mathcal{A}|_V) = \max(\operatorname{jsr}_p(\mathcal{A}|_{V_1}), \operatorname{jsr}_p(\mathcal{A}|_{V/V_1}))$$

and for $j = 1, \ldots, J-1$,

$$\begin{aligned}\operatorname{jsr}_p(\mathcal{A}|_{V/(V_1+\cdots+V_j)}) &= \max(\operatorname{jsr}_p(\mathcal{A}|_{(V_1+\cdots+V_j)\cap V_{j+1}}), \operatorname{jsr}_p(\mathcal{A}|_{V/(V_1+\cdots+V_{j+1})})) \\ &\leqslant \max(\operatorname{jsr}_p(\mathcal{A}|_{V_{j+1}}), \operatorname{jsr}_p(\mathcal{A}|_{V/(V_1+\cdots+V_{j+1})})).\end{aligned}$$

Consequently, we conclude that

$$\operatorname{jsr}_p(\mathcal{A}) \leqslant \max(\operatorname{jsr}_p(\mathcal{A}|_{V_1}), \ldots, \operatorname{jsr}_p(\mathcal{A}|_{V_J})).$$

This proves (5.7.18).

We now prove (5.7.19). Since a is finitely supported, there exists an integer $N \geqslant m+1$ such that a vanishes outside $[-N, N]$. By Lemma 5.7.3, $(l([-N,N]))^{1\times r}$ is invariant under both $\mathcal{T}_0$ and $\mathcal{T}_1$. Define $W_j := \mathscr{V}_{j,\upsilon} \cap (l([-N,N]))^{1\times r}$ for $j = -1, 0, \ldots, m$. Since a satisfies order $m+1$ sum rules with the matching filter υ, by item (5) of Theorem 5.6.5, W_j is a common invariant subspace of $\mathcal{T}_0$ and $\mathcal{T}_1$, and W_j generates $\mathscr{V}_{j,\upsilon}$ (by enlarging N if necessary). By (5.7.18) and Proposition 5.7.2,

$$\rho_j(a, \upsilon)_p = \operatorname{jsr}_p(\mathcal{A}|_{W_{j-1}}) = \max(\operatorname{jsr}_p(\mathcal{A}|_{W_j}), \operatorname{jsr}_p(\mathcal{A}|_{W_{j-1}/W_j}))$$

for all $j = 0, \ldots, m$. By Theorem 5.6.4, since a satisfies order $m+1$ sum rules, without loss of generality, we assume that $\upsilon = (\upsilon_1, 0, \ldots, 0)$ with $\widehat{\upsilon_1}(0) \neq 0$ and a takes the normal form in (5.6.22) and (5.6.29). Since $\upsilon = (\upsilon_1, 0, \ldots, 0)$,

$$W_{j-1} = \{(\nabla^j u_1, u_2, \ldots, u_r) \;:\; \nabla^j u_1, u_2, \ldots, u_r \in l([-N,N])\}.$$

Note that $\dim(W_{j-1}/W_j) = 1$ and W_{j-1}/W_j is spanned by $u := (\nabla^j \boldsymbol{\delta}, 0, \ldots, 0) \in W_{j-1}/W_j$. Hence, for $\gamma \in \{0,1\}$, define $u_\gamma := \nabla^j \boldsymbol{\delta}(\cdot - \gamma) \in \mathscr{V}_{j-1,\boldsymbol{\delta}}$ and

$$[\mathcal{T}_\gamma u] = \mathcal{T}_a(u(\cdot - \gamma)) = 2\sum_{k\in\mathbb{Z}} u(k)\overline{a(\gamma + k - 2\cdot)}^{\mathsf{T}} = (\mathcal{T}_{a_{1,1}} u_\gamma, \mathcal{T}_{a_{2,1}} u_\gamma).$$

Note that $(0, \mathcal{T}_{a_{2,1}} u_\gamma) \in W_j$. Since $\widehat{a_{1,1}}(\xi + \pi) = \mathscr{O}(|\xi|^{m+1})$ as $\xi \to 0$ and $\widehat{a_{1,1}}(0) = 1$, by Lemma 5.6.3 and $u_\gamma - \nabla^j \delta \in \mathscr{V}_{j,\delta}$, we have $\mathcal{T}_{a_{1,1}} \nabla^j \delta - 2^{-j} \nabla^j \delta \in \mathscr{V}_{j,\delta}$, $\mathcal{T}_{a_{1,1}}(u_\gamma - \nabla^j \delta) \in \mathscr{V}_{j,\delta}$. Therefore, $\mathcal{T}_{a_{1,1}} u_\gamma - 2^{-j} u_\gamma \in \mathscr{V}_{j,\delta}$ and

$$\mathcal{T}_\gamma u - 2^{-j} u = (\mathcal{T}_{a_{1,1}} u_\gamma - 2^{-j} u_\gamma, 0) + (2^{-j} u_\gamma - 2^{-j} \nabla^j \delta, \mathcal{T}_{a_{2,1}} u_\gamma) \in W_j.$$

Consequently, it is obvious now that $\mathrm{jsr}_p(\mathcal{A}|_{W_{j-1}/W_j}) = \mathrm{jsr}_p(\{2^{-j}, 2^{-j}\}) = 2^{1/p-j}$. Therefore, we proved

$$\rho_j(a, \upsilon)_p = \mathrm{jsr}_p(\mathcal{A}|_{W_{j-1}}) = \max(2^{1/p-j}, \mathrm{jsr}_p(\mathcal{A}|_{W_j})) = \max(2^{1/p-j}, \rho_{j+1}(a, \upsilon)_p)$$

for all $j = 0, \ldots, m$. Now (5.7.19) follows from the above identity right away. □

As a direct consequence of Hölder's inequality (or (5.7.18) and (5.7.3)), we have

$$2^{1/q-1/p} \rho_m(a, \upsilon)_p \leqslant \rho_m(a, \upsilon)_q \leqslant \rho_m(a, \upsilon)_p,$$
$$1/q - 1/p + \mathrm{sm}_p(a) \leqslant \mathrm{sm}_q(a) \leqslant \mathrm{sm}_p(a),$$

for all $1 \leqslant p \leqslant q \leqslant \infty$. The identity (5.7.19) also well explains the equivalence (5) $\Longleftrightarrow$ (6) $\Longleftrightarrow$ (7) in Theorem 5.6.11.

The following result explores the relations between $\rho_j(a, \upsilon)_p$ and the orders of sum rules.

Corollary 5.7.7 *Let $a \in (l_0(\mathbb{Z}))^{r \times r}$ and $1 \leqslant p \leqslant \infty$.*

(1) Suppose that $\mathrm{sm}_p(a) \leqslant 0$. Then $\rho_0(a, \upsilon)_p = 2^{1/p - \mathrm{sm}_p(a)}$ and all the eigenvalues of $\widehat{a}(0)$ are less than or equal to $2^{-\mathrm{sm}_p(a)}$ in modulus.

(2) Suppose that $\mathrm{sm}_p(a) > 0$ and 1 is an eigenvalue of $\widehat{a}(0)$. Then 1 is a simple eigenvalue of $\widehat{a}(0)$ and all the other eigenvalues of $\widehat{a}(0)$ are less than or equal to $2^{-\mathrm{sm}_p(a)}$ in modulus. Moreover, there exists a positive constant C such that

$$\rho_j(a, \upsilon)_p = 2^{1/p-j}, \quad \|a_n * u^\star\|_{(l_p(\mathbb{Z}))^r} \leqslant C 2^{n(1/p-1-j)}, \qquad \forall\, n \in \mathbb{N}, u \in \mathscr{B}_{j-1,\upsilon}, j = 0, \ldots, m, \tag{5.7.20}$$

and a satisfies order $m+1$ sum rules with a matching filter $\upsilon \in (l_0(\mathbb{Z}))^{1 \times r}$ satisfying (5.6.3), where m is the largest nonnegative integer satisfying $m < \mathrm{sm}_p(a)$.

Proof The identity $\rho_0(a, \upsilon)_p = 2^{1/p - \mathrm{sm}_p(a)}$ in item (1) and the first identity in (5.7.20) follow directly from (5.7.19). The second claim in item (1) and the first claim in item (2) are direct consequences of Proposition 5.6.9. The inequalities in (5.7.20) is a consequence of (5.6.45) in Theorem 5.6.11. The last claim in item (2) follows from Corollary 5.6.10. □

5.8 Smoothness of Refinable Functions and Computation of $\mathrm{sm}_p(a)$

In this section we characterize the smoothness exponent of a refinable function through the quantity $\mathrm{sm}_p(a)$ and then we study how to efficiently calculate or estimate the smoothness quantity $\mathrm{sm}_p(a)$.

5.8.1 *Characterize Smoothness Exponent of a Refinable Function*

Smoothness of a function is measured using modulus of smoothness $\omega_m(\phi,\lambda)_p$. Recall that $\omega_m(\phi,\lambda)_p := \sup_{|t|\leqslant\lambda}\|\nabla_t^m\phi\|_{L_p(\mathbb{R})}$ for $\lambda \geqslant 0$ and $m \in \mathbb{N}_0$. For $1 \leqslant p < \infty$ and a function $\phi \in L_p(\mathbb{R})$, its L_p *smoothness exponent* is defined to be

$$\mathrm{sm}_p(\phi) := \sup\{m+\tau \;:\; \sup_{\lambda>0}\lambda^{-\tau}\omega_1(f^{(m)},\lambda)_p < \infty,\;\; \phi \in W_p^m(\mathbb{R}) \text{ for some } m \in \mathbb{N}_0, \tau \geqslant 0\}. \tag{5.8.1}$$

For $p = \infty$ and $\phi \in \mathscr{C}(\mathbb{R})$, we define $\mathrm{sm}_\infty(\phi)$ as in (5.8.1) by replacing $W_p^m(\mathbb{R})$ with $\mathscr{C}^m(\mathbb{R})$. If $\phi \in L_\infty(\mathbb{R})$ but $\phi \notin \mathscr{C}(\mathbb{R})$, we simply define $\mathrm{sm}_\infty(\phi) = 0$. Note that $\phi \in L_p(\mathbb{R})$ ($\phi \in \mathscr{C}(\mathbb{R})$ if $p = \infty$) implies $\mathrm{sm}_p(\phi) \geqslant 0$. If $\phi = (\phi_1,\ldots,\phi_r)^\mathsf{T}$ is a vector function, we define $\mathrm{sm}_p(\phi) := \min_{1\leqslant\ell\leqslant r}\mathrm{sm}_p(\phi_\ell)$.

Theorem 5.8.1 *Let $1 \leqslant p \leqslant \infty$ and $a \in (l_0(\mathbb{Z}))^{r\times r}$ be a finitely supported filter on $\mathbb{Z}$. Let ϕ be an $r \times 1$ vector of compactly supported distributions satisfying $\phi = 2\sum_{k\in\mathbb{Z}} a(k)\phi(2\cdot -k)$ with $\widehat{\phi}(0) \neq 0$.*

(i) If $\mathrm{sm}_p(a) > 0$, then $\phi \in (L_p(\mathbb{R}))^r$ ($\phi \in \mathscr{C}(\mathbb{R})$ if $p = \infty$) and $\mathrm{sm}_p(\phi) \geqslant \mathrm{sm}_p(a)$.
(ii) If $\phi \in (W_p^m(\mathbb{R}))^r$ ($\phi \in (\mathscr{C}^m(\mathbb{R}))^r$ if $p = \infty$) for some $m \in \mathbb{N}_0$ and if the integer shifts of ϕ are stable in $L_p(\mathbb{R})$, then $\mathrm{sm}_p(\phi) = \mathrm{sm}_p(a) > m$.

Proof Since $\widehat{\phi}(2\xi) = \widehat{a}(\xi)\widehat{\phi}(\xi)$ and $\widehat{\phi}(0) \neq 0$, we see that 1 is an eigenvalue of $\widehat{a}(0)$. Let m be the largest nonnegative integer such that $m < \mathrm{sm}_p(a) \leqslant m+1$, where we used the assumption $\mathrm{sm}_p(a) > 0$ in item (i). By Corollary 5.6.10, (5.6.3) must hold and $n_a \geqslant m_a \geqslant m+1$. In particular, 1 is a simple eigenvalue of $\widehat{a}(0)$. Therefore, there exists $\upsilon \in (l_0(\mathbb{Z}))^{1\times r}$ satisfying (5.6.43) and $\widehat{\upsilon}(0)\widehat{\phi}(0) = 1$. Since $\mathrm{sm}_p(a) > 0$, then $\phi \in (L_p(\mathbb{R}))^r$ ($\phi \in (\mathscr{C}(\mathbb{R}))^r$ if $p = \infty$) follows directly from Theorem 5.6.11.

By Theorem 5.7.6, since $\mathrm{sm}_p(a) \leqslant m+1$ and $\rho_{m_a}(a,\upsilon)_p = 2^{1/p-\mathrm{sm}_p(a)}$,

$$\begin{aligned}\rho_{m+1}(a,\upsilon)_p &= \max(2^{1/p-(m+1)}, \rho_{m_a}(a,\upsilon)_p)\\ &= \max(2^{1/p-(m+1)}, 2^{1/p-\mathrm{sm}_p(a)}) = 2^{1/p-\mathrm{sm}_p(a)}.\end{aligned}$$

For $u_\ell \in \mathscr{B}_{m-1,\upsilon}$, by the definition of $\mathscr{V}_{m,\upsilon}$, it is trivial to see that $(\nabla u_\ell^\star)^\star \in \mathscr{V}_{m,\upsilon}$. Therefore, for any $0 < \varepsilon < \mathrm{sm}_p(a) - m$, there exists a positive constant C_1 such that

$$\|a_n * (\nabla u_\ell^\star)\|_{(l_p(\mathbb{Z}))^r} \leqslant C_1 2^{n(1/p-1-\mathrm{sm}_p(a)+\varepsilon)}, \qquad \forall\, n \in \mathbb{N}, \ell = 1, \ldots, r. \tag{5.8.2}$$

Since $\phi = \mathcal{R}_a^n \phi = 2^n[a_n * \phi](2^n \cdot)$,

$$\nabla_{2^{-n}} \phi^{(m)} = 2^n 2^{nm} \sum_{k\in\mathbb{Z}} [\nabla a_n](k) \phi^{(m)}(2^n \cdot -k) = 2^{n(m+1)}[(\nabla a_n) * \phi^{(m)}](2^n \cdot). \tag{5.8.3}$$

Since $\mathrm{sm}_p(a) > 0$, by item (3) of Theorem 5.6.11, it follows from Proposition 5.6.9 that (5.6.6) holds. Since $\phi^{(m)} \in (L_p(\mathbb{R}))^r$ and (5.6.6) holds, by item (iv) of Theorem 5.6.4, $\phi^{(m)} = \sum_{\ell=1}^r u_\ell^\star * g_\ell$ for some compactly supported functions $g_1, \ldots, g_r \in L_p(\mathbb{R})$ and $u_1, \ldots, u_r \in \mathscr{B}_{m-1,\upsilon}$. Replacing $\phi^{(m)}$ on the right-hand side of (5.8.3) by $\sum_{\ell=1}^r u_\ell^\star * g_\ell$, since $\nabla a_n = a_n * \nabla \boldsymbol{\delta}$, we have

$$\begin{aligned}\nabla_{2^{-n}} \phi^{(m)} &= 2^{n(m+1)} \sum_{\ell=1}^r [(\nabla a_n) * (u_\ell^\star * g_\ell)](2^n \cdot) \\ &= 2^{n(m+1)} \sum_{\ell=1}^r \sum_{k\in\mathbb{Z}} [a_n * (\nabla u_\ell^\star)](k) g_\ell(2^n \cdot -k).\end{aligned}$$

Since all $g_1, \ldots, g_r$ have compact support, by Lemma 5.3.1, there exists a positive constant C_2 such that

$$\|\nabla_{2^{-n}} \phi^{(m)}\|_{(L_p(\mathbb{R}))^r} \leqslant C_2 2^{n(m+1-1/p)} \sum_{\ell=1}^r \|a_n * (\nabla u_\ell^\star)\|_{(l_p(\mathbb{Z}))^{r\times r}}.$$

Therefore, from the above inequality and (5.8.2), we have

$$\|\nabla_{2^{-n}} \phi^{(m)}\|_{(L_p(\mathbb{R}))^r} \leqslant r C_1 C_2 2^{n(m-\mathrm{sm}_p(m)+\varepsilon)}, \qquad \forall\, n \in \mathbb{N}_0. \tag{5.8.4}$$

For $0 < \lambda \leqslant 1$, we have the dyadic expression $\lambda = \sum_{n=j}^\infty \gamma_n 2^{-n}$ with $\gamma_n \in \{0, 1\}$ and $\gamma_j = 1$. Note that $2^{-j} \leqslant \lambda \leqslant 2^{1-j}$. Therefore, by (5.8.4), we have

$$\begin{aligned}\|\nabla_\lambda \phi^{(m)}\|_{(L_p(\mathbb{R}))^r} &\leqslant \sum_{n=j}^\infty \gamma_n \|\nabla_{2^{-n}} \phi^{(m)}\|_{(L_p(\mathbb{R}))^r} \leqslant r C_1 C_2 \sum_{n=j}^\infty 2^{n(m-\mathrm{sm}_p(a)+\varepsilon)} \\ &= C_3 2^{-j(\mathrm{sm}_p(a)-m-\varepsilon)} \leqslant C_3 \lambda^{\mathrm{sm}_p(a)-m-\varepsilon},\end{aligned}$$

where we used $2^{-j} \leqslant \lambda$ and $C_3 := r C_1 C_2/(1 - 2^{m-\mathrm{sm}_p(a)+\varepsilon}) < \infty$ by $\mathrm{sm}_p(a) - m - \varepsilon > 0$. This proves that $\lambda^{-\tau} \omega_1(\phi^{(m)}, \lambda)_p \leqslant C_3 + 2\|\phi^{(m)}\|_{(L_p(\mathbb{R}))^r}$ for all $\lambda > 0$

with $\tau = \mathrm{sm}_p(a) - m - \varepsilon > 0$. By the definition of $\mathrm{sm}_p(\phi)$, we conclude that $\mathrm{sm}_p(\phi) \geqslant m + \tau = m + \mathrm{sm}_p(a) - m - \varepsilon = \mathrm{sm}_p(a) - \varepsilon$. Since $\varepsilon > 0$ is arbitrary, $\mathrm{sm}_p(\phi) \geqslant \mathrm{sm}_p(a)$ and item (i) is proved.

We now prove item (ii). Suppose that there exist $m \in \mathbb{N}_0$, $\tau \geqslant 0$ and $C > 0$ such that $\phi \in (W_p^m(\mathbb{R}))^r$ ($\phi \in (\mathscr{C}^m(\mathbb{R}))^r$ if $p = \infty$) and

$$\omega_1(\phi^{(m)}, \lambda)_p \leqslant C\lambda^\tau, \qquad \forall\, \lambda > 0. \tag{5.8.5}$$

Since the integer shifts of ϕ are stable in $L_p(\mathbb{R})$, as proved in Theorem 5.6.11 for (3)$\Longrightarrow$(4) with $f = \phi$, the inequalities (5.6.47) and (5.6.48) hold with $f_n = \phi$. Let $\{b_1, \dots, b_r\} = \mathscr{B}_{m,\upsilon}$. That is, there exists a positive constant C_4 such that

$$2^{n(1-1/p)} \|a_n * b_\ell^\star\|_{(l_p(\mathbb{Z}))^r} \leqslant C_4 \omega_{m+1}(\phi, 2^{-n})_p, \qquad \forall\, n \in \mathbb{N}, \ell = 1, \dots, r.$$

By (5.4.4) and (5.8.5), we have

$$\begin{aligned} 2^{n(1-1/p)} \|a_n * b_\ell^\star\|_{(l_p(\mathbb{Z}))^r} &\leqslant C_4 \omega_{m+1}(\phi, 2^{-n})_p \\ &\leqslant C_4 2^{-nm} \omega_1(\phi^{(m)}, 2^{-n})_p \leqslant C C_4 2^{-n(m+\tau)}. \end{aligned} \tag{5.8.6}$$

Since $\phi^{(m)} \in (L_p(\mathbb{R}))^r$ ($\phi^{(m)} \in (\mathscr{C}(\mathbb{R}))^r$ if $p = \infty$), we have $\lim_{n\to\infty} \omega_1(\phi^{(m)}, 2^{-n})_p = 0$. Now it follows from Proposition 5.6.9 and (5.8.6) that $m_a \geqslant m + 1$ and

$$\lim_{n\to\infty} \|a_n * b_\ell^\star\|_{(l_p(\mathbb{Z}))^r}^{1/n} < 2^{1/p-1-m} \quad \text{and} \quad \lim_{n\to\infty} \|a_n * b_\ell^\star\|_{(l_p(\mathbb{Z}))^r}^{1/n} \leqslant 2^{1/p-1-m-\tau}$$

for $\ell = 1, \dots, r$. Then $\rho_{m+1}(a, \upsilon)_p < 2^{1/p-m}$ and $\rho_{m+1}(a, \upsilon)_p \leqslant 2^{1/p-m-\tau}$. Since $m_a \geqslant m + 1$, we have $\rho_{m_a}(a, \upsilon)_p \leqslant \rho_{m+1}(a, \upsilon)_p \leqslant 2^{1/p-m-\tau}$. Thus, we proved $\mathrm{sm}_p(a) \geqslant m + \tau$. Hence, by the definition of $\mathrm{sm}_p(\phi)$, we have $\mathrm{sm}_p(a) \geqslant \mathrm{sm}_p(\phi)$. Also, we have $\rho_{m_a}(a, \upsilon)_p \leqslant \rho_{m+1}(a, \upsilon)_p < 2^{1/p-m}$ which implies $\mathrm{sm}_p(a) > m \geqslant 0$. It follows from item (i) that $\mathrm{sm}_p(\phi) \geqslant \mathrm{sm}_p(a)$ and hence $\mathrm{sm}_p(\phi) = \mathrm{sm}_p(a) > m$. □

The following result is a direct consequence of Theorem 5.8.1.

Corollary 5.8.2 *Let ϕ be a compactly supported refinable function/distribution satisfying $\phi = 2\sum_{k\in\mathbb{Z}} a(k)\phi(2\cdot -k)$ for some $a \in (l_0(\mathbb{Z}))^{r\times r}$ and ϕ is not identically zero. According to Theorem 5.2.4, there exists a compactly supported vector function $\mathring{\phi}$ such that $\mathrm{S}(\phi) = \mathrm{S}(\mathring{\phi})$, the integer shifts of $\mathring{\phi}$ are linearly independent, and $\mathring{\phi} = 2\sum_{k\in\mathbb{Z}} \mathring{a}(k)\mathring{\phi}(2\cdot -k)$ for some $\mathring{a} \in (l_0(\mathbb{Z}))^{s\times s}$ with $s \leqslant r$. If $\phi \in (L_p(\mathbb{R}))^r$, then*

$$\mathrm{sm}_p(\phi) = \mathrm{sm}_p(\mathring{\phi}) = \mathrm{sm}_p(\mathring{a}).$$

In particular, for some $m \in \mathbb{N}_0$, if $\phi \in (W_p^m(\mathbb{R}))^r$ (require $\phi \in (\mathscr{C}^m(\mathbb{R}))^r$ for $p = \infty$), then $\mathrm{sm}_p(\mathring{a}) > m$ and $\mathrm{sm}_p(\phi) = \mathrm{sm}_p(\mathring{a}) > m$.

Proof Since $\mathrm{S}(\phi) = \mathrm{S}(\mathring{\phi})$ and both ϕ and $\mathring{\phi}$ are compactly supported, by $\phi \in (L_p(\mathbb{R}))^r$, we have $\mathring{\phi} \in (L_p(\mathbb{R}))^r$ and $\mathrm{sm}_p(\phi) = \mathrm{sm}_p(\mathring{\phi})$.

Since the integer shifts of $\mathring{\phi}$ are linearly independent and $\mathring{\phi} \in (L_p(\mathbb{R}))^r$, by Corollary 5.3.5, the integer shifts of $\mathring{\phi}$ are stable in $L_p(\mathbb{R})$. Since $\mathring{\phi}$ is not identically zero, it follows from Proposition 5.6.9 with $m = 0$ that $\widehat{\mathring{\phi}}(0) \neq 0$. Now it follows from Theorem 5.8.1 that $\mathrm{sm}_p(\mathring{\phi}) = \mathrm{sm}_p(\mathring{a})$.

If $\phi \in (W_p^m(\mathbb{R}))^r$, then $\mathring{\phi} \in (W_p^m(\mathbb{R}))^r$. Since the integer shifts of $\mathring{\phi}$ are stable in $L_p(\mathbb{R})$, $n_{\mathring{a}} \geqslant m + 1$ is guaranteed by Proposition 5.6.9 and by item (2) of Theorem 5.6.11, we must have $\mathrm{sm}_p(\mathring{a}) > m$ and $\mathrm{sm}_p(\phi) = \mathrm{sm}_p(\mathring{\phi}) = \mathrm{sm}_p(\mathring{a}) > m$. ☐

The problem now is how to find such a filter $\mathring{a}$ from a given filter a. This can be done by using Lemma 5.2.6 and Theorem 5.2.7. See Algorithm 5.8.6 for finding $\mathring{a}$ from a given scalar filter a.

5.8.2 Compute $\rho_m(a, \upsilon)_p$ by Taking out Basic Factors

We now discuss how to efficiently compute $\rho_m(a, \upsilon)_p$ and therefore, $\mathrm{sm}_p(a)$. The following is the general case of Lemma 5.6.14 for matrix-valued filters.

Theorem 5.8.3 *Let $a \in (l_0(\mathbb{Z}))^{r\times r}$ and $m \in \mathbb{N}_0$. Suppose that a satisfies order m sum rules with a matching filter $\upsilon \in (l_0(\mathbb{Z}))^{1\times r}$ such that $\widehat{\upsilon}(0) \neq 0$. Let $U \in (l_0(\mathbb{Z}))^{r\times r}$ be a strongly invertible sequence such that*

$$\widehat{\mathring{\upsilon}}(\xi) := \widehat{\upsilon}(\xi)\widehat{U}(\xi) = (1 + \mathscr{O}(|\xi|), \mathscr{O}(|\xi|^m), \ldots, \mathscr{O}(|\xi|^m)), \qquad \xi \to 0. \tag{5.8.7}$$

Define

$$\widehat{\mathring{a}}(\xi) := (\widehat{U}(2\xi))^{-1}\widehat{a}(\xi)\widehat{U}(\xi), \quad \widehat{b}(\xi) := (\widehat{D_m}(2\xi))^{-1}\widehat{\mathring{a}}(\xi)\widehat{D_m}(\xi)$$

with

$$\widehat{D_m}(\xi) := \begin{bmatrix} (1 - e^{-i\xi})^m & 0 \\ 0 & I_{r-1} \end{bmatrix}. \tag{5.8.8}$$

Then $\mathring{a}$ and b are finitely supported sequences in $(l_0(\mathbb{Z}))^{r\times r}$ and

$$\rho_m(a, \upsilon)_p = \rho_m(\mathring{a}, \mathring{\upsilon})_p = \rho_0(b)_p := 2 \lim_{n\to\infty} \|b_n\|_{(l_p(\mathbb{Z}))^{r\times r}}^{1/n} \tag{5.8.9}$$

with $\widehat{b_n}(\xi) := \widehat{b}(2^{n-1}\xi)\cdots\widehat{b}(2\xi)\widehat{b}(\xi)$. *Moreover,*

$$\operatorname{spec}(\mathcal{T}_a) = \operatorname{spec}(\mathcal{T}_{\mathring{a}}) = \{1, 2^{-1}, \ldots, 2^{1-m}\} \cup \operatorname{spec}(\mathcal{T}_b). \tag{5.8.10}$$

Proof Because U is strongly invertible, by $\mathring{\upsilon} = \upsilon * U$, the condition $u \in \mathscr{V}_{m-1,\mathring{\upsilon}}$ if and only if $u * U^\star \in \mathscr{V}_{m-1,\upsilon}$. Since

$$\begin{aligned}\widehat{\mathring{a}_n}(\xi) &= \widehat{\mathring{a}}(2^{n-1}\xi)\cdots\widehat{\mathring{a}}(2\xi)\widehat{\mathring{a}}(\xi)\\ &= (\widehat{U}(2^n\xi))^{-1}\widehat{a}(2^{n-1}\xi)\cdots\widehat{a}(2\xi)\widehat{a}(\xi)\widehat{U}(\xi) = (\widehat{U}(2^n\xi))^{-1}\widehat{a_n}(\xi)\widehat{U}(\xi),\end{aligned}$$

we have $\mathring{a}_n * u^\star = U^{-1}(2^n\cdot) * a_n * (u * U^\star)^\star$. Therefore,

$$\|\mathring{a}_n * u^\star\|_{(l_p(\mathbb{Z}))^r} \leqslant \|U^{-1}\|_{(l_1(\mathbb{Z}))^{r\times r}}\|a_n * (u * U^\star)^\star\|_{(l_p(\mathbb{Z}))^r},$$

which implies $\rho_m(\mathring{a}, \mathring{\upsilon})_p \leqslant \rho_m(a, \upsilon)_p$. Conversely, $a_n * (u * U^\star)^\star = U(2^n\cdot) * (\mathring{a}_n * u^\star)$ implies $\|a_n * (u * U^\star)^\star\|_{(l_p(\mathbb{Z}))^r} \leqslant \|U\|_{(l_1(\mathbb{Z}))^{r\times r}}\|\mathring{a}_n * u^\star\|_{(l_p(\mathbb{Z}))^r}$, from which we have $\rho_m(a, \upsilon)_p \leqslant \rho_m(\mathring{a}, \mathring{\upsilon})_p$. Therefore, $\rho_m(a, \upsilon)_p = \rho_m(\mathring{a}, \mathring{\upsilon})_p$.

By (5.8.7) and Theorem 5.6.5, $\mathring{a}$ must take the normal form in (5.6.22) and (5.6.29) with $m + 1$ being replaced by m. By calculation,

$$\widehat{b}(\xi) = \begin{bmatrix}(1 + e^{-i\xi})^{-m}\widehat{\mathring{a}_{1,1}}(\xi) & (1 - e^{-i2\xi})^{-m}\widehat{\mathring{a}_{1,2}}(\xi)\\ (1 - e^{-i\xi})^{m}\widehat{\mathring{a}_{2,1}}(\xi) & \widehat{\mathring{a}_{2,2}}(\xi)\end{bmatrix}$$

must be a matrix of 2π-periodic trigonometric polynomials. So, b is finitely supported. Note that $\widehat{\mathring{a}_n}(\xi) = \widehat{D_m}(2^n\xi)\widehat{b_n}(\xi)(\widehat{D_m}(\xi))^{-1}$ and $\mathscr{B}_{m-1,\mathring{\upsilon}} = \{u_1, \ldots, u_r\}$ with $u_1 := (\nabla^m\boldsymbol{\delta} e_1)^\star$ and $u_2 := (\boldsymbol{\delta} e_2)^\star, \ldots, u_r = (\boldsymbol{\delta} e_r)^\star$. Thus,

$$\widehat{\mathring{a}_n * u_1^\star}(\xi) = \widehat{D_m}(2^n\xi)\widehat{b_n}(\xi)(\widehat{D_m}(\xi))^{-1}(1 - e^{-i\xi})^m e_1 = \widehat{D_m}(2^n\xi)\widehat{b_n}(\xi)e_1.$$

For $j = 2, \ldots, r$, by $\widehat{u_j^\star}(\xi) = e_j$, we have

$$\widehat{\mathring{a}_n * u_j^\star}(\xi) = \widehat{D_m}(2^n\xi)\widehat{b_n}(\xi)(\widehat{D_m}(\xi))^{-1}\widehat{u_j^\star}(\xi) = \widehat{D_m}(2^n\xi)\widehat{b_n}(\xi)e_j.$$

Consequently, we proved that

$$\|\mathring{a}_n * [u_1^\star, \ldots, u_r^\star]\|_{(l_p(\mathbb{Z}))^{r\times r}} = \|D_m(2^n\cdot) * b_n\|_{(l_p(\mathbb{Z}))^{r\times r}}, \qquad \forall\, n \in \mathbb{N}.$$

Therefore,

$$\rho_m(\mathring{a}, \mathring{\upsilon})_p = 2 \lim_{n\to\infty} \|D_m(2^n\cdot) * b_n\|^{1/n}_{(l_p(\mathbb{Z}))^{r\times r}}. \tag{5.8.11}$$

On one hand, we have $\|D_m(2^n\cdot) * b_n\|_{(l_p(\mathbb{Z}))^{r\times r}} \leqslant \|D_m\|_{(l_1(\mathbb{Z}))^{r\times r}}\|b_n\|_{(l_p(\mathbb{Z}))^{r\times r}}$ which implies $\rho_m(\mathring{a}, \mathring{\upsilon})_p \leqslant \rho_0(b)_p$ by (5.8.11). Since both D_m and b are finitely supported, there exists $N \in \mathbb{N}$ such that both D_m and b vanish outside $[-N, N]$. As a consequence, b_n vanishes outside $[(1-2^n)N, (2^n-1)N] \subseteq [-2^nN, 2^nN]$. Using the identity $(1-x)^m \sum_{j=0}^{\infty} \binom{m+j-1}{j} x^j = 1$ for all $|x| < 1$, we have

$$(1-e^{-i\xi})^m \widehat{w}(\xi) = 1 + \sum_{k=N+1}^{N+m} c_k e^{-ik\xi} \quad \text{with} \quad \widehat{w}(\xi) := \sum_{j=0}^{N} \binom{m+j-1}{j} e^{-ij\xi}$$

for some $c_k \in \mathbb{R}$, $k = N+1, \ldots, N+m$. Define $\widehat{W}(\xi) := \operatorname{diag}(\widehat{w}(\xi), I_{r-1})$. Since b_n vanishes outside $[-2^nN, 2^nN]$, we see that

$$b_n(k) = ((W * D_m)(2^n\cdot) * b_n)(k) = (W(2^n\cdot) * [D_m(2^n\cdot) * b_n])(k),$$

for all $k \in [-2^nN, 2^nN] \cap \mathbb{Z}$. Therefore, we have $\|b_n\|_{(l_p(\mathbb{Z}))^{r\times r}} \leqslant \|W\|_{(l_1(\mathbb{Z}))^{r\times r}}\|D_m(2^n\cdot) * b_n\|_{(l_p(\mathbb{Z}))^{r\times r}}$, from which and (5.8.11) we have

$$\rho_0(b)_p = 2 \lim_{n\to\infty} \|b_n\|^{1/n}_{(l_p(\mathbb{Z}))^{r\times r}} \leqslant 2 \lim_{n\to\infty} \|D_m(2^n\cdot) * b_n\|^{1/n}_{(l_p(\mathbb{Z}))^{r\times r}} = \rho_m(\mathring{a}, \mathring{\upsilon})_p = \rho_m(a, \upsilon)_p.$$

This completes the proof of (5.8.9).

By Theorem 5.6.5, the space $\mathscr{V}_{m-1,\mathring{\upsilon}}$ is invariant under $\mathcal{T}_{\mathring{a}}$, and consequently, we have $\operatorname{spec}(\mathcal{T}_{\mathring{a}}|_{(l_0(\mathbb{Z}))^{1\times r}/\mathscr{V}_{m-1,\mathring{\upsilon}}}) = \{1, 2^{-1}, \ldots, 2^{1-m}\}$. Therefore,

$$\begin{aligned}\operatorname{spec}(\mathcal{T}_a) = \operatorname{spec}(\mathcal{T}_{\mathring{a}}) &= \operatorname{spec}(\mathcal{T}_{\mathring{a}}|_{(l_0(\mathbb{Z}))^{1\times r}/\mathscr{V}_{m-1,\mathring{\upsilon}}}) \cup \operatorname{spec}(\mathcal{T}_{\mathring{a}}|_{\mathscr{V}_{m-1,\mathring{\upsilon}}}) \\ &= \{1, \ldots, 2^{1-m}\} \cup \operatorname{spec}(\mathcal{T}_{\mathring{a}}|_{\mathscr{V}_{m-1,\mathring{\upsilon}}}).\end{aligned}$$

Since $\mathring{\upsilon}$ takes the form in (5.8.7), we have $\mathscr{V}_{m-1,\mathring{\upsilon}} = (l_0(\mathbb{Z}))^{1\times r} * D_m^\star$. Hence, for $u \in (l_0(\mathbb{Z}))^{1\times r}$, by $\mathring{a} = D_m(2\cdot) * b * D_m^{-1}$, we have

$$\mathcal{T}_{\mathring{a}}(u * D_m^\star) = (u * D_m^\star * \mathring{a}^\star)\downarrow 2 = (u * b^\star * D_m^\star(2\cdot))\downarrow 2 = (\mathcal{T}_b u) * D_m^\star.$$

Now it is trivial to see that $\operatorname{spec}(\mathcal{T}_{\mathring{a}}|_{\mathscr{V}_{m-1,\mathring{\upsilon}}}) = \operatorname{spec}(\mathcal{T}_b)$. This proves (5.8.10). □

5.8.3 *Compute $\rho_m(a, \upsilon)_2$ and $\mathrm{sm}_2(a)$ by Spectral Radius*

The reader who is only interested in the scalar case $r = 1$ can skip this subsection, since the scalar case is treated completely in Sect. 5.8.4.

For two matrices $A = (a_{j,k})_{1\leqslant j\leqslant r, 1\leqslant k\leqslant s}$ and B, the *(right) Kronecker product* $A\otimes B$ is defined to be

$$A \otimes B := \begin{bmatrix} a_{1,1}B & a_{1,2}B & \cdots & a_{1,s}B \\ a_{2,1}B & a_{2,2}B & \cdots & a_{2,s}B \\ \vdots & \vdots & \ddots & \vdots \\ a_{r,1}B & a_{r,2}B & \cdots & a_{r,s}B \end{bmatrix}. \tag{5.8.12}$$

One can check that $(A+B)\otimes C = A\otimes C + B\otimes C$, $C\otimes(A+B) = C\otimes A + C\otimes B$, and

$$(A\otimes B)(C\otimes D) = (AC)\otimes(BD), \quad (A\otimes B)^{\mathsf{T}} = A^{\mathsf{T}}\otimes B^{\mathsf{T}}.$$

We can form a long column vector $\mathrm{vec}(A)$ by putting columns of A in order as:

$$\mathrm{vec}(A) := (a_{1,1},\ldots,a_{r,1},a_{1,2},\ldots,a_{r,2},\ldots,a_{1,s},\ldots,a_{r,s})^{\mathsf{T}}. \tag{5.8.13}$$

One can directly check that

$$\mathrm{vec}(ACB) = (B^{\mathsf{T}}\otimes A)\mathrm{vec}(C). \tag{5.8.14}$$

For $a \in (l_0(\mathbb{Z}))^{r_1\times r_2}$ and $b \in (l_0(\mathbb{Z}))^{r_3\times r_4}$, the operator $\mathcal{T}_{a,b} : (l_0(\mathbb{Z}))^{r_2\times r_3} \to (l_0(\mathbb{Z}))^{r_1\times r_4}$ is defined to be

$$\mathcal{T}_{a,b}u = 2\sum_{\ell\in\mathbb{Z}}\sum_{k\in\mathbb{Z}} a(2\cdot-k-\ell)u(\ell)b(k), \qquad u\in(l_0(\mathbb{Z}))^{r_2\times r_3}. \tag{5.8.15}$$

Using the vec operation and (5.8.14), we can easily deduce that

$$\mathrm{vec}(\mathcal{T}_{a,b}u) = 2\sum_{\ell\in\mathbb{Z}} c(2\cdot-\ell)\mathrm{vec}(u(\ell)) = \mathring{\mathcal{T}}_c(\mathrm{vec}(u)),$$

where the filter $c \in (l_0(\mathbb{Z}))^{(r_1r_4)\times(r_2r_3)}$ is defined to be

$$c := \sum_{k\in\mathbb{Z}}[b(k)]^{\mathsf{T}}\otimes a(\cdot-k), \quad \text{that is,} \quad \widehat{c}(\xi) := [\widehat{b}(\xi)]^{\mathsf{T}}\otimes\widehat{a}(\xi) \tag{5.8.16}$$

and $\mathring{\mathcal{T}}_c$ is defined in (5.7.11) with $\gamma = 0$ and a being replaced by the filter c in (5.8.16).

For a square matrix A, by $\mathrm{spec}(A)$ we denote the multiset of all eigenvalues of A counting multiplicity of the eigenvalues of A.

The following result shows that for the special case $p = 2$, the quantity $\rho_m(a, \upsilon)_2$ (and therefore, $\mathrm{sm}_2(a)$) can be efficiently computed by calculating the eigenvalues of an associated finite square matrix.

Theorem 5.8.4 *Let $a \in (l_0(\mathbb{Z}))^{r\times r}$ such that the filter a satisfies order m sum rules with a matching filter $\upsilon \in (l_0(\mathbb{Z}))^{1\times r}$ and $\widehat{\upsilon}(0) \neq 0$. Form a new sequence $c \in (l_0(\mathbb{Z}))^{r^2\times r^2}$ by*

$$c(n) := \sum_{k\in\mathbb{Z}} \overline{a(k)} \otimes a(n+k), \qquad n \in \mathbb{Z}, \qquad \textit{that is,} \quad \widehat{c}(\xi) := \overline{\widehat{a}(\xi)} \otimes \widehat{a}(\xi).$$

Define a multiset by

$$E_m := \{2^{-j} \; : \; j = 0, \dots, 2m-1\} \cup \cup_{j=0}^{m-1}\{2^{-j}\lambda, 2^{-j}\overline{\lambda} \; : \; \lambda \in \operatorname{spec}(\widehat{a}(0))\backslash\{1\}\}.$$

Then $\rho_m(a,\upsilon)_2 = \max\{\sqrt{2|\lambda|} \; : \; \lambda \in \operatorname{spec}(\mathcal{T}_{a,a^\star}|_{(l_0(\mathbb{Z}))^{r\times r}})\backslash E_m\}$. *More explicitly,*

$$\rho_m(a,\upsilon)_2 = \max\{\sqrt{2|\lambda|} \; : \; \lambda \in \operatorname{spec}((2c(k-2j))_{-\operatorname{len}(a)\leqslant j,k\leqslant \operatorname{len}(a)})\backslash E_m\}.$$

Note $\mathrm{sm}(a) := \mathrm{sm}_2(a) = \frac{1}{2} - \log_2 \rho_{m_a}(a,\upsilon)_2$ *with the sum rule order* $m_a := \mathrm{sr}(a)$.

Proof By Theorem 5.8.3, without loss of generality, we can assume that $\upsilon = (\upsilon_1, 0, \dots, 0)$ with $\widehat{\upsilon_1}(0) = 1$ and a takes the following form:

$$a = \begin{bmatrix} a_{1,1} & a_{1,2} \\ a_{2,1} & a_{2,2} \end{bmatrix} \quad \text{with} \quad \widehat{a_{1,1}}(0) = 1, \tag{5.8.17}$$
$$\widehat{a_{1,1}}(\xi) = (1+e^{-i\xi})^m \widehat{A}(\xi), \quad \widehat{a_{1,2}}(\xi) = (1-e^{-i2\xi})^m \widehat{B}(\xi),$$

where $A \in l_0(\mathbb{Z})$ and $B \in (l_0(\mathbb{Z}))^{1\times(r-1)}$.

For $j \in \mathbb{N}_0$, define $D_j \in (l_0(\mathbb{Z}))^{r\times r}$ by $\widehat{D_j}(\xi) := \operatorname{diag}((1-e^{-i\xi})^j, I_{r-1})$ and define W_j to be the linear space consisting of all elements $D_j * u * D_j^\star$, where

$$u := \begin{bmatrix} u_1 & u_2 \\ u_3 & u_4 \end{bmatrix} \quad \text{with}$$
$$u_1 \in l_0(\mathbb{Z}), u_2 \in (l_0(\mathbb{Z}))^{1\times(r-1)}, u_3 \in (l_0(\mathbb{Z}))^{r-1}, u_4 \in (l_0(\mathbb{Z}))^{(r-1)\times(r-1)}. \tag{5.8.18}$$

For $j = 0, \dots, m$, we see that $\mathcal{T}_{a,a^\star} W_j \subseteq W_j$, since for all $u \in (l_0(\mathbb{Z}))^{r\times r}$,

$$\mathcal{T}_{a,a^\star}(D_j * u * D_j^\star) = \mathcal{T}_{a*D_j,(a*D_j)^\star} u = D_j * (\mathcal{T}_{b_j,b_j^\star} u) * D_j^\star, \tag{5.8.19}$$

where by (5.8.17) $b_j \in (l_0(\mathbb{Z}))^{r\times r}$ is defined to be

$$\begin{aligned} \widehat{b_j}(\xi) :=& (\widehat{D_j}(2\xi))^{-1}\widehat{a}(\xi)\widehat{D_j}(\xi) \\ =& \begin{bmatrix} (1+e^{-i\xi})^{m-j}\widehat{A}(\xi) & (1-e^{i2\xi})^{m-j}\widehat{B}(\xi) \\ (1-e^{i\xi})^j \widehat{a_{2,1}}(\xi) & \widehat{a_{2,2}}(\xi) \end{bmatrix}. \end{aligned} \tag{5.8.20}$$

We first prove that

$$\rho_m(a, \upsilon)_2 = \sqrt{2\rho(\mathcal{T}_{a,a^\star}|_{W_m})}. \tag{5.8.21}$$

For $u \in (l_0(\mathbb{Z}))^{r\times r}$, since $\mathcal{T}_{a,a^\star} u = 2(a * u * a^\star)\downarrow 2$, we have

$$\mathcal{T}^n_{a,a^\star}(D_m * u * D^\star_m) = 2^n(a_n * D_m * u * D^\star_m * a^\star_n)\downarrow 2^n. \tag{5.8.22}$$

Therefore, for $t \in \mathbb{Z}$,

$$\begin{aligned}
&2^{-2n}\|[\mathcal{T}^n_{a,a^\star}(D_m * u * D^\star_m)](t)\|^2_{l_2} = \|(a_n * D_m * u * D^\star_m * a^\star_n)(2^n t)\|^2_{l_2}\\
&= \left\| \sum_{\ell\in\mathbb{Z}}\sum_{k\in\mathbb{Z}}(a_n * D_m)(2^n t - k - \ell)u(\ell)(a_n * D_m)^\star(k)\right\|^2_{l_2}\\
&\leqslant \left(\sum_{\ell\in \mathrm{fsupp}(u)}\sum_{k\in\mathbb{Z}}\|(a_n * D_m)(2^n t - k - \ell)\|_{l_2}\|u(\ell)\|_{l_2}\|(a_n * D_m)^\star(k)\|_{l_2}\right)^2\\
&\leqslant \left(\sum_{\ell\in \mathrm{fsupp}(u)}\|u(\ell)\|^2_{l_2}\right)\sum_{\ell\in \mathrm{fsupp}(u)}\left(\sum_{k\in\mathbb{Z}}\|(a_n * D_m)(2^n t - k - \ell)\|_{l_2}\|(a_n * D_m)(-k)\|_{l_2}\right)^2\\
&\leqslant \|u\|^2_{(l_2(\mathbb{Z}))^{r\times r}}\sum_{\ell\in \mathrm{fsupp}(u)}\left(\sum_{k\in\mathbb{Z}}\|(a_n * D_m)(2^n t - k - \ell)\|^2_{l_2}\right)\left(\sum_{k\in\mathbb{Z}}\|(a_n * D_m)(-k)\|^2_{l_2}\right)\\
&= \|u\|^2_{(l_2(\mathbb{Z}))^{r\times r}}\operatorname{len}(u)\|a_n * D_m\|^4_{(l_2(\mathbb{Z}))^{r\times r}}.
\end{aligned}$$

Since $\mathcal{T}^n_{a,a^\star}(D_m * u * D^\star_m)$ must vanish outside $[-\operatorname{len}(a), \operatorname{len}(a)]$ when n is large enough, we conclude from the above inequality that

$$\|\mathcal{T}^n_{a,a^\star}(D_m * u * D_m)\|_{(l_2(\mathbb{Z}))^{r\times r}} \leqslant C2^n\|a_n * D_m\|^2_{(l_2(\mathbb{Z}))^{r\times r}},$$

where $C = \|u\|_{(l_2(\mathbb{Z}))^{r\times r}}\sqrt{(2\operatorname{len}(a)+1)\operatorname{len}(u)}$. Hence, for all $u \in (l_0(\mathbb{Z}))^{r\times r}$,

$$\lim_{n\to\infty}\|\mathcal{T}^n_{a,a^\star}(D_m * u * D^\star_m)\|^{1/n}_{(l_2(\mathbb{Z}))^{r\times r}} \leqslant 2\lim_{n\to\infty}\|a_n * D_m\|^{2/n}_{(l_2(\mathbb{Z}))^{r\times r}} = 2^{-1}[\rho_m(a, \upsilon)_2]^2.$$

This proves $2\rho(\mathcal{T}_{a,a^\star}|_{W_m}) \leqslant [\rho_m(a, \upsilon)_2]^2$. Conversely, taking $u = \boldsymbol{\delta} I_r$ in (5.8.22), we have

$$\begin{aligned}
\|a_n * D_m\|^2_{(l_2(\mathbb{Z}))^{r\times r}} &= \operatorname{trace}(\langle a_n * D_m, a_n * D_m\rangle) = \operatorname{trace}([a_n * D_m * (a_n * D_m)^\star](0))\\
&= 2^{-n}\operatorname{trace}(\mathcal{T}^n_{a,a^\star}(D_m * D^\star_m)(0))\\
&\leqslant 2^{-n}\sqrt{r}\|\mathcal{T}^n_{a,a^\star}(D_m * D^\star_m)\|_{(l_2(\mathbb{Z}))^{r\times r}},
\end{aligned}$$

from which we deduce that

$$
\begin{aligned}
[\rho_m(a, \upsilon)_2]^2 &= 4 \lim_{n \to \infty} \|a_n * D_m\|_{(l_2(\mathbb{Z}))^{r\times r}}^{2/n} \\
&\leqslant 2 \lim_{n\to\infty} \|\mathcal{T}_{a,a^\star}^n (D_m * D_m^\star)\|_{(l_2(\mathbb{Z}))^{r\times r}}^{1/n} \leqslant 2\rho(\mathcal{T}_{a,a^\star}|_{W_m}).
\end{aligned}
$$

This proves (5.8.21). We now prove that for $j = 0, \ldots, m-1$,

$$
\begin{aligned}
\operatorname{spec}(\mathcal{T}_{a,a^\star}|_{W_j/W_{j+1}}) &= \operatorname{spec}(\mathcal{T}_{b_j,b_j^\star}|_{W_0/W_1}) \\
&= \{2^{-2j}, 2^{-1-2j}\} \cup \{2^{-j}\lambda, 2^{-j}\overline{\lambda} \ : \ \lambda \in \operatorname{spec}(\widehat{a_{2,2}}(0))\}.
\end{aligned}
\tag{5.8.23}
$$

The first identity follows directly from (5.8.19). For b_j given in (5.8.20), for simplicity, we define

$$
\widehat{d_1}(\xi) := (1+e^{-i\xi})^{m-j}\widehat{A}(\xi), \quad \widehat{d_2}(\xi) := (1-e^{i2\xi})^{m-j}\widehat{B}(\xi),
$$
$$
\widehat{d_3}(\xi) := (1-e^{i\xi})^{j}\widehat{a_{2,1}}(\xi), \quad \widehat{d_4}(\xi) = \widehat{a_{2,2}}(\xi).
$$

Note that

$$
\widehat{d_1}(0) = 2^{-j}, \quad \widehat{d_4}(0) = \widehat{a_{2,2}}(0), \quad \widehat{d_1}(\pi) = 0, \quad \widehat{d_2}(0) = \widehat{d_2}(\pi) = 0. \tag{5.8.24}
$$

For u in (5.8.18), we have

$$
\mathcal{T}_{b_j,b_j^\star} \begin{bmatrix} u_1 & u_2 \\ u_3 & u_4 \end{bmatrix} =
$$
$$
\begin{bmatrix}
\mathcal{T}_{d_1,d_1^\star}u_1+\mathcal{T}_{d_1,d_2^\star}u_2+\mathcal{T}_{d_2,d_1^\star}u_3+\mathcal{T}_{d_2,d_2^\star}u_4 & \mathcal{T}_{d_1,d_3^\star}u_1+\mathcal{T}_{d_1,d_4^\star}u_2+\mathcal{T}_{d_2,d_3^\star}u_3+\mathcal{T}_{d_2,d_4^\star}u_4 \\
\mathcal{T}_{d_3,d_1^\star}u_1+\mathcal{T}_{d_3,d_2^\star}u_2+\mathcal{T}_{d_4,d_1^\star}u_3+\mathcal{T}_{d_4,d_2^\star}u_4 & \mathcal{T}_{d_3,d_3^\star}u_1+\mathcal{T}_{d_3,d_4^\star}u_2+\mathcal{T}_{d_4,d_3^\star}u_3+\mathcal{T}_{d_4,d_4^\star}u_4
\end{bmatrix}.
$$

By Lemma 5.6.3 and $\widehat{d_2}(0) = \widehat{d_2}(\pi) = 0$, all terms $\mathcal{T}_{d_j,d_k^\star}u_n$ with either $j = 2$ or $k = 2$ belong to $\mathscr{V}_{0,\delta}$ and $\mathcal{T}_{d_2,d_2^\star}u_4 \in \mathscr{V}_{1,\delta}$. Therefore,

$$
\mathcal{T}_{b_j,b_j^\star} \begin{bmatrix} u_1 & u_2 \\ u_3 & u_4 \end{bmatrix} -
$$
$$
\begin{bmatrix}
\mathcal{T}_{d_1*d_1^\star}u_1+(\mathcal{T}_{d_1}(d_2*u_2^\star))^\star+\mathcal{T}_{d_1}(d_2*u_3) & (\mathcal{T}_{d_1}(d_3*u_1^\star+d_4*u_2^\star))^\star \\
\mathcal{T}_{d_1}(d_3*u_1+d_4*u_3) & 0
\end{bmatrix} \in W_1.
\tag{5.8.25}
$$

Let U be the space of all elements u in (5.8.18) with $u_1 \in \mathscr{V}_{0,\delta}$. Since $\widehat{d_1}(\pi) = 0$ and $\widehat{d_2}(0) = 0$, by Lemma 5.6.3 and (5.8.25), noting $W_1 \subseteq U$, we have $\mathcal{T}_{b_j,b_j^\star}U \subseteq U$. By

Lemma 5.6.3 and (5.8.24), we have $\mathcal{T}_{d_1 * d_1^\star}\boldsymbol{\delta} - 2^{-2j}\boldsymbol{\delta} \in \mathscr{V}_{0,\boldsymbol{\delta}}$ and

$$\mathcal{T}_{b_j,b_j^\star}\begin{bmatrix}\boldsymbol{\delta} & 0\\ 0 & 0\end{bmatrix} \equiv \begin{bmatrix}\mathcal{T}_{d_1*d_1^\star}\boldsymbol{\delta} & 0\\ 0 & 0\end{bmatrix} \equiv 2^{-2j}\begin{bmatrix}\boldsymbol{\delta} & 0\\ 0 & 0\end{bmatrix} \quad \text{mod } U.$$

Thus, we proved $\operatorname{spec}(W_0/U) = \{2^{-2j}\}$ since W_0/U is one-dimensional. Let V be the space of all elements u in (5.8.18) with $(u_1, u_2) \in (\mathscr{V}_{0,\boldsymbol{\delta}})^{1\times r}$ and $u_3 \in (\mathscr{V}_{0,\boldsymbol{\delta}})^{r-1}$. By Lemma 5.6.3 and (5.8.25), we have $\mathcal{T}_{b_j,b_j^\star}V \subseteq V$ and $V \subseteq U$. By (5.8.25) and (5.8.24), using Lemma 5.6.3, for $\vec{u}, \vec{v} \in \mathbb{C}^{r-1}$, we have $\mathcal{T}_{d_1}\boldsymbol{\delta} - 2^{-j}\boldsymbol{\delta} \in \mathscr{V}_{0,\boldsymbol{\delta}}$ and

$$\mathcal{T}_{b_j,b_j^\star}\begin{bmatrix}0 & \vec{v}^\star\boldsymbol{\delta}\\ \vec{u}\boldsymbol{\delta} & 0\end{bmatrix} \equiv \begin{bmatrix}0 & (\mathcal{T}_{d_1}\boldsymbol{\delta})^\star[\widehat{d_4}(0)\vec{v}]^\star\\ (\mathcal{T}_{d_1}\boldsymbol{\delta})[\widehat{d_4}(0)\vec{u}] & 0\end{bmatrix} \equiv 2^{-j}\begin{bmatrix}0 & [\widehat{d_4}(0)\vec{v}]^\star\boldsymbol{\delta}\\ [\widehat{d_4}(0)\vec{u}]\boldsymbol{\delta} & 0\end{bmatrix}$$

under mod V. Since $\widehat{d_4}(0) = \widehat{a_{2,2}}(0)$, we proved

$$\operatorname{spec}(\mathcal{T}_{b_j,b_j^\star}|_{U/V}) = \{2^{-j}\lambda, 2^{-j}\overline{\lambda} \; : \; \lambda \in \operatorname{spec}(\widehat{a_{2,2}}(0))\}.$$

For the space V/W_1, since $|\widehat{d_1 * d_1^\star}(\xi+\pi)|^2 = |\widehat{d_1}(\xi+\pi)|^2 = \mathscr{O}(|\xi|^2)$ as $\xi \to 0$, by Lemma 5.6.3, we have $\mathcal{T}_{b_1*b_1^\star}(\nabla\boldsymbol{\delta}) - 2^{-1-2j}\nabla\boldsymbol{\delta} \in \mathscr{V}_{1,\boldsymbol{\delta}}$ and

$$\mathcal{T}_{b_j,b_j^\star}\begin{bmatrix}\nabla\boldsymbol{\delta} & 0\\ 0 & 0\end{bmatrix} \equiv \begin{bmatrix}\mathcal{T}_{d_1*d_1^\star}(\nabla\boldsymbol{\delta}) & 0\\ 0 & 0\end{bmatrix} \equiv 2^{-1-2j}\begin{bmatrix}\nabla\boldsymbol{\delta} & 0\\ 0 & 0\end{bmatrix} \quad \text{mod } W_1.$$

Thus, we have $\operatorname{spec}(\mathcal{T}_{b_j,b_j^\star}|_{V/W_1}) = \{2^{-1-2j}\}$. Since $W_0/W_1 = (W_0/U) \oplus (U/V) \oplus (V/W_1)$, we conclude that (5.8.23) holds. Now it is straightforward to conclude from (5.8.23) that $\operatorname{spec}(\mathcal{T}_{a,a^\star}|_{W_0/W_m}) = E_m$.

Using the vec operation in (5.8.13) and $\operatorname{fsupp}(c) = [-\operatorname{len}(a), \operatorname{len}(a)]$, we see that the matrix representation of $T_{a,a^\star}$ on its invariant subspace $(l([-\operatorname{len}(a), \operatorname{len}(a)]))^{r\times r}$ is $(2c(j-2k)^\mathsf{T})_{-\operatorname{len}(a)\leqslant j,k\leqslant \operatorname{len}(a)}$. Also, note that $W_0 = (l_0(\mathbb{Z}))^{r\times r}$ and the fact that if $\lambda \neq 0$ is an eigenvalue of $\mathcal{T}_{a,a^\star}$ with an eigenvector $u \in (l_0(\mathbb{Z}))^{r\times r}$, then u must vanish outside $[-\operatorname{len}(a), \operatorname{len}(a)]$ by Lemma 5.7.3. Therefore,

$$\{0\} \cup \operatorname{spec}(T_{a,a^\star}|_{W_m}) = \{0\} \cup \operatorname{spec}(\mathcal{T}_{a,a^\star}|_{(l([-\operatorname{len}(a),\operatorname{len}(a)]))^{r\times r}})\backslash E_m.$$

This completes the proof. □

5.8.4 *Compute* $\operatorname{sm}_p(a)$ *and* $\operatorname{sm}_p(\phi)$ *for Scalar Filters*

As indicated by the following result, we often have more ways of efficiently computing or estimating the key quantity $\operatorname{sm}_p(a)$ for the scalar case $r = 1$.

Corollary 5.8.5 *Let $a \in l_0(\mathbb{Z})$ such that $\widehat{a}(0) = 1$ and $\widehat{a}(\xi) = (1+e^{-i\xi})^m \widehat{b}(\xi)$ for some $m \in \mathbb{N}_0$ and $b \in l_0(\mathbb{Z})$ with $\widehat{b}(\pi) \ne 0$ (i.e., $m = \mathrm{sr}(a)$). Define $c \in l_0(\mathbb{Z})$ by $\widehat{c}(\xi) := |\widehat{b}(\xi)|^2$ and b_n by $\widehat{b_n}(\xi) = \widehat{b}(2^{n-1}\xi)\cdots\widehat{b}(2\xi)\widehat{b}(\xi)$ for $n \in \mathbb{N}$. Then*

(1) $\mathrm{sm}_p(a) = \mathrm{sm}_p(b) = 1/p - \log_2(\rho_0(b)_p)$, *where* $\rho_0(b)_p = 2\lim_{n\to\infty} \|b_n\|_{l_p(\mathbb{Z})}^{1/n}$.

(2) $\mathrm{sm}_\infty(a) = \mathrm{sm}_\infty(b) \leqslant -\log_2 \lambda_b$, *where*

$$\lambda_b := \rho(\mathcal{T}_b|_{l(\mathrm{fsupp}(b^\star))}) = \max\{|\lambda| \ : \ \lambda \in \mathrm{spec}((2b(2k-j))_{j,k\in \mathrm{fsupp}(b)})\}.$$

If in addition $\widehat{b}(\xi) \geqslant 0$ for all $\xi \in \mathbb{R}$, then $\mathrm{sm}_\infty(a) = \mathrm{sm}_\infty(b) = -\log_2 \lambda_b$.

(3) $\mathrm{sm}_2(a) = \mathrm{sm}_2(b) = \frac{1}{2}\mathrm{sm}_\infty(c) = -\frac{1}{2}\log_2 \lambda_c$, *where λ_c is defined similarly as λ_b above. Moreover, the multiset* $\mathrm{spec}((2c(2k-j))_{-\mathrm{len}(b)\leqslant j,k\leqslant \mathrm{len}(b)})$ *is the same as* $\mathrm{spec}((2d(2k-j))_{-\mathrm{len}\, a\leqslant j,k\leqslant \mathrm{len}(a)})\backslash\{2^0,\dots,2^{1-2m}\}$, *where* $\widehat{d}(\xi) := |\widehat{a}(\xi)|^2$.

(4) $\rho_0(b)_p = \mathrm{jsr}_p(\{B_0, B_1\})$, *where*

$$B_0 = (2b(2k-j))_{j,k\in \mathrm{fsupp}(b)} \quad \textit{and} \quad B_1 = (2b(2k-j+1))_{j,k\in\mathrm{fsupp}(b)}.$$

(5) $\rho_0(b)_\infty = 2\inf_{n\in\mathbb{N}} \max_{0\leqslant\gamma<2^n} (\sum_{k\in\mathbb{Z}} |b_n(\gamma+2^n k)|)^{1/n}$.

(6) $\rho_0(b)_\infty \leqslant \rho(\widehat{b}) = 2\lim_{n\to\infty} \|\widehat{b_n}\|_{L_\infty(\mathbb{T})}^{1/n} \leqslant \rho_0(b)_1$, *where* $\rho(\widehat{b}) := 2\inf_{n\in\mathbb{N}} \|\widehat{b_n}\|_{L_\infty(\mathbb{T})}^{1/n}$.

(7) $\{\|\widehat{b_n^{\min}}\|_{L_\infty(\mathbb{T})}\}_{n=1}^\infty$ *decreases to $\rho(\widehat{b})$, that is,* $\|\widehat{b_n^{\min}}\|_{L_\infty(\mathbb{T})} \downarrow \rho(\widehat{b})$ *as $n\to\infty$, where*

$$\widehat{b_n^{\min}}(\xi) := 2\min_{1\leqslant j\leqslant n} |\widehat{b_j}(\xi)|^{1/j} = 2\min_{1\leqslant j\leqslant n} |\widehat{b}(2^{j-1}\xi)\cdots\widehat{b}(2\xi)\widehat{b}(\xi)|^{1/j}. \qquad (5.8.26)$$

Proof Item (1) follows directly from Lemma 5.6.14. Define $\boldsymbol{\delta}_j := \boldsymbol{\delta}(\cdot - j)$. By $\mathcal{T}_b\boldsymbol{\delta}_j = \sum_{k\in\mathbb{Z}} 2\overline{b(j-2k)}\boldsymbol{\delta}_k$ for $j \in \mathrm{fsupp}(b^\star)$, the matrix representation of the operator $\mathcal{T}_b$ under the basis $\{\boldsymbol{\delta}_{-j}\}_{j\in\mathrm{fsupp}(b)}$ of the invariant subspace $l(\mathrm{fsupp}(b^\star))$ is $\mathcal{T}_b|_{l(\mathrm{fsupp}(b^\star))} = (\overline{2b(2k-j)})_{j,k\in\mathrm{fsupp}(b)}$. Thus, $\lambda_b = \rho(\mathcal{T}_b)$. Let $v \in l(\mathrm{fsupp}(b^\star))\backslash\{0\}$ be an eigenvector of an eigenvalue λ of $\mathcal{T}_b$ with $|\lambda| = \lambda_b$. Then $\mathcal{T}_b^n v = \lambda^n v$ and

$$\begin{aligned}\lambda_b^n \|v\|_{l_\infty(\mathbb{Z})} &= \|\mathcal{T}_b^n v\|_{l_\infty(\mathbb{Z})} = \sup_{k\in\mathbb{Z}} |\langle \mathcal{T}_b^n v, \boldsymbol{\delta}(\cdot - k)\rangle| = \sup_{k\in\mathbb{Z}} |\langle v, \mathcal{S}_b^n(\boldsymbol{\delta}(\cdot-k))\rangle| \\ &\leqslant \|v\|_{l_1(\mathbb{Z})} \|\mathcal{S}_b^n \boldsymbol{\delta}\|_{l_\infty(\mathbb{Z})} = 2^n \|v\|_{l_1(\mathbb{Z})} \|b_n\|_{l_\infty(\mathbb{Z})}.\end{aligned}$$

Since $\|v\|_{l_\infty(\mathbb{Z})} \ne 0$, the above inequality yields $\lambda_b = \rho(\mathcal{T}_b) \leqslant 2\lim_{n\to\infty} \|b_n\|_{l_\infty(\mathbb{Z})}^{1/n} = \rho_0(b)_\infty$. This proves $\mathrm{sm}_\infty(u) = -\log_2 \rho_0(b)_\infty \leqslant -\log_2 \lambda_b$. If $\widehat{b}(\xi) \geqslant 0$ for all $\xi \in \mathbb{R}$, then $\widehat{b_n}(\xi) := \widehat{b}(2^{n-1}\xi)\cdots\widehat{b}(2\xi)\widehat{b}(\xi) \geqslant 0$ for all $\xi \in \mathbb{R}$. Therefore,

$$|b_n(k)| = \left|\frac{1}{2\pi}\int_{-\pi}^{\pi} \widehat{b_n}(\xi)e^{-ik\xi}d\xi\right| \leqslant \frac{1}{2\pi}\int_{-\pi}^{\pi} \widehat{b_n}(\xi)d\xi = b_n(0) = 2^{-n}\overline{\mathcal{T}_b^n\boldsymbol{\delta}(0)},$$

since $b_n(0) = \langle b_n, \boldsymbol{\delta}\rangle = 2^{-n}\langle \mathcal{S}_b\boldsymbol{\delta}, \boldsymbol{\delta}\rangle = 2^{-n}\langle \boldsymbol{\delta}, \mathcal{T}_b\boldsymbol{\delta}\rangle = \overline{\mathcal{T}_b^n\boldsymbol{\delta}(0)}$. Therefore,

$$\rho_0(b)_\infty = 2\lim_{n\to\infty}\|b_n\|_{l_\infty(\mathbb{Z})}^{1/n} \leqslant 2\lim_{n\to\infty}|\mathcal{T}_b^n\boldsymbol{\delta}(0)|^{1/n} \leqslant \rho(\mathcal{T}_b) = \lambda_b.$$

Therefore, $\rho_0(b)_\infty = \lambda_b$ and $\operatorname{sm}_\infty(a) = -\log_2\lambda_b$ if $\widehat{b}(\xi) \geqslant 0$. This proves item (2).

To prove $\operatorname{sm}_2(b) = \frac{1}{2}\operatorname{sm}_\infty(c)$ in item (3), by item (2), we have

$$\begin{aligned}\rho_0(b)_2 &= 2\lim_{n\to\infty}\|b_n\|_{l_2(\mathbb{Z})}^{1/n} = 2\lim_{n\to\infty}\left(\frac{1}{2\pi}\int_{-\pi}^{\pi}|\widehat{b_n}(\xi)|^2 d\xi\right)^{\frac{1}{2n}}\\ &= 2\lim_{n\to\infty}\left(\frac{1}{2\pi}\int_{-\pi}^{\pi}\widehat{c_n}(\xi)d\xi\right)^{\frac{1}{2n}} = 2\lim_{n\to\infty}\|c_n\|_{l_\infty(\mathbb{Z})}^{\frac{1}{2n}} = \sqrt{2\rho_0(c)_\infty}.\end{aligned}$$

Therefore, $\operatorname{sm}_2(b) = 1/2 - \log_2\rho_0(b)_2 = -\log_2\rho_0(c)_\infty = \frac{1}{2}\operatorname{sm}_\infty(c)$. Now item (3) is a direct consequence of item (1) and Lemma 5.6.14.

Since $\mathcal{T}_\gamma|_{l(\operatorname{fsupp}(b^\star))} = (2b(\gamma+2k-j))_{j,k\in\operatorname{fsupp}(b)}$ for $\gamma = 0, 1$, item (4) follows from Theorem 5.7.6. Define a submultiplicative matrix norm $\left\|(b_{j,k})_{1\leqslant j,k\leqslant n}\right\| := \sup_{1\leqslant j\leqslant n}\sum_{k=1}^n|b_{j,k}|$. Then

$$\|\{B_0, B_1\}^n\|_{l_\infty} = 2^n\max\Big\{\sum_{k\in\mathbb{Z}}|b_n(\gamma+2^n k)| \;:\; \gamma = 0,\ldots,2^n-1\Big\}.$$

Hence, item (5) follows directly from item (4). The proof of $\lim_{n\to\infty}\|\widehat{b_n}\|_{L_\infty(\mathbb{T})}^{1/n} = \inf_{n\in\mathbb{N}}\|\widehat{b_n}\|_{L_\infty(\mathbb{T})}^{1/n}$ in item (6) is the same as the proof of (5.7.2) in Lemma 5.7.1. For the first and the last inequalities of item (6), we have

$$|b_n(k)| = \left|\frac{1}{2\pi}\int_{-\pi}^{\pi}\widehat{b_n}(\xi)e^{-ik\xi}d\xi\right| \leqslant \frac{1}{2\pi}\int_{-\pi}^{\pi}\|\widehat{b_n}\|_{L_\infty(\mathbb{T})}d\xi = \|\widehat{b_n}\|_{L_\infty(\mathbb{T})} \leqslant \|b_n\|_{l_1(\mathbb{Z})}.$$

Hence, $\|b_n\|_{l_\infty(\mathbb{Z})} \leqslant \|\widehat{b_n}\|_{L_\infty(\mathbb{T})} \leqslant \|b_n\|_{l_1(\mathbb{Z})}$, from which we have item (6).

To prove item (7), since $0 \leqslant \widehat{b_{n+1}^{\min}}(\xi) \leqslant \widehat{b_n^{\min}}(\xi) \leqslant 2|\widehat{b_n}(\xi)|^{1/n}$ for all $\xi \in \mathbb{R}$ and $n \in \mathbb{N}$, we see that $\{\|\widehat{b_n^{\min}}\|_{L_\infty(\mathbb{T})}\}_{n=1}^\infty$ is a nonincreasing sequence and $\rho := 2\lim_{n\to\infty}\|\widehat{b_n^{\min}}\|_{L_\infty(\mathbb{T})} \leqslant 2\lim_{n\to\infty}\|\widehat{b_n}\|_{L_\infty(\mathbb{T})}^{1/n} = \rho(\widehat{b})$.

We now show $\rho \geqslant \rho(\widehat{b})$ by proving $\|\widehat{b_J^{\min}}\|_{L_\infty(\mathbb{T})} \geqslant \rho(\widehat{b})$ for all $J \in \mathbb{Z}$. Define $I_j := \{\xi \in (-\pi, \pi] \;:\; 2|\widehat{b_j}(\xi)|^{1/j} = \widehat{b_J^{\min}}(\xi)\}$ for $j = 1,\ldots,J$. Then $\cup_{j=1}^J I_j = (-\pi, \pi]$ and $2^j|\widehat{b_j}(\xi)| \leqslant \|\widehat{b_J^{\min}}\|_{L_\infty(\mathbb{T})}^j$ for all $\xi \in I_j$ and $j = 1,\ldots,J$. For $\xi \in \mathbb{R}$, we must have $\xi + 2\pi k_\xi \in I_j$ for some $k_\xi \in \mathbb{Z}$ and $1 \leqslant j \leqslant J$. If $n \geqslant J$, since $\widehat{b}$ is

2π-periodic,

$$2^n|\widehat{b_n}(\xi)| = 2^n|\widehat{b}(\xi + 2\pi k_\xi)\cdots\widehat{b}(2^{j-1}(\xi + 2\pi k_\xi))|\cdot|\widehat{b}(2^j\xi)\cdots\widehat{b}(2^{n-1}\xi)|$$
$$\leqslant \|\widehat{b_J^{\min}}\|^j_{L_\infty(\mathbb{T})} 2^{n-j}|\widehat{b}(2^j\xi)\cdots\widehat{b}(2^{n-1}\xi)|.$$

If $n - j \geqslant J$, we can continue this procedure with ξ being replaced $2^j\xi$. Hence,

$$2^n|\widehat{b_n}(\xi)| \leqslant \|\widehat{b_J^{\min}}\|^{J\lfloor\frac{n}{J}\rfloor}_{L_\infty(\mathbb{T})} \max\{2^j\|\widehat{b_j}\|_{L_\infty(\mathbb{R})} \;:\; j = 1,\ldots,J-1\}, \quad \forall\, \xi \in \mathbb{R}, n \in \mathbb{N}.$$

Now it is trivial to deduce that $\rho(\widehat{b}) = 2\lim_{n\to\infty}\|\widehat{b_n}\|^{1/n}_{L_\infty(\mathbb{T})} \leqslant \|\widehat{b_J^{\min}}\|_{L_\infty(\mathbb{T})}$. □

We present an example here to illustrate the calculation of the quantity $\rho(\widehat{b})$.

Example 5.8.1 Let $\widehat{b}(\xi) := \mathsf{P}_{m,n}(\sin^2(\xi/2))$ with $m, n \in \mathbb{N}$ and $n \leqslant m+1$, where $\mathsf{P}_{m,n}(x) := \sum_{j=0}^{n-1}\binom{m+j-1}{j}x^j$ is defined in (2.1.4). Then

$$\rho(\mathsf{P}_{m,n}(\sin^2(\xi/2)) = \rho(\widehat{b}) = \|\widehat{b_2^{\min}}\|_{L_\infty(\mathbb{T})} = 2\mathsf{P}_{m,n}(\tfrac{3}{4}).$$

The value $\mathsf{P}_{m,n}(\frac{3}{4})$ can be estimated similarly as in (6.4.11) of Theorem 6.4.4.

Proof Let us first prove the following inequality:

$$\mathsf{P}_{m,n}(x)\mathsf{P}_{m,n}(4x(1-x)) \leqslant (\mathsf{P}_{m,n}(\tfrac{3}{4}))^2, \qquad \forall\, x \in [\tfrac{3}{4}, 1]. \tag{5.8.27}$$

Let $z(x) := 4x(1-x)$. By the definition of $\mathsf{P}_{m,n}$, we have $\mathsf{P}_{m,n}(x) = \mathsf{P}_{m,n-1}(x) + \binom{m+n-2}{n-1}x^{n-1}$ and hence

$$\mathsf{P}_{m,n}(x)\mathsf{P}_{m,n}(z) = \mathsf{P}_{m,n-1}(x)\mathsf{P}_{m,n}(z) + \binom{m+n-2}{n-1}x^{n-1}\mathsf{P}_{m,n}(z)$$
$$= \mathsf{P}_{m,n-1}(x)\mathsf{P}_{m,n-1}(z) + \binom{m+n-2}{n-1}\left(x^{n-1}\mathsf{P}_{m,n}(z) + z^{n-1}\mathsf{P}_{m,n-1}(x)\right)$$
$$= 1 + \sum_{j=1}^{n-1}\binom{m+j-1}{j}f_{m,j}(x),$$

where

$$f_{m,j}(x) := x^j\mathsf{P}_{m,j+1}(z) + z^j\mathsf{P}_{m,j}(x) = \sum_{k=0}^{j}\binom{m+k-1}{k}x^jz^k + \sum_{k=0}^{j-1}\binom{m+k-1}{k}x^kz^j.$$

By calculation,

$$\begin{aligned}
f'_{m,j}(x) &= \sum_{k=0}^{j}\binom{m+k-1}{k} jx^{j-1}z^k + \sum_{k=1}^{j}\binom{m+k-1}{k} kx^j z^{k-1}z' \\
&\quad + \sum_{k=0}^{j-1}\binom{m+k-1}{k} jx^k z^{j-1}z' + \sum_{k=1}^{j-1}\binom{m+k-1}{k} kx^{k-1}z^j \\
&= \sum_{k=0}^{j}\binom{m+k-1}{k} jx^{j-1}z^k + \sum_{k=0}^{j-1}\binom{m+k-1}{k}(m+k)x^j z^k z' \\
&\quad + \sum_{k=0}^{j-1}\binom{m+k-1}{k} jx^k z^{j-1}z' + \sum_{k=0}^{j-2}\binom{m+k-1}{k}(m+k)x^k z^j \\
&= \sum_{k=0}^{j-1}\binom{m+k-1}{k} x^k z^k\Big[jx^{j-k-1} + (m+k)x^{j-k}z' + jz^{j-k-1}z' + (m+k)z^{j-k}\Big] \\
&= \sum_{k=0}^{j-1}\binom{m+k-1}{k} x^k z^k\Big[jx^{j-k-1}(1+xz') \\
&\qquad + jz^{j-k-1}(z'+z) + (m+k-j)x^{j-k}(z' + (z/x)^{j-k})\Big],
\end{aligned}$$

where we used $\binom{m+j-1}{j} jx^{j-1}z^j = \binom{m+j-2}{j-1}(m+j-1)x^{j-1}z^j$ to shift the $k=j$ term in the first sum to be the $k=j-1$ term in the last sum in the second identity. Since $z = 4x(1-x)$ and $z'(x) = 4-8x$, we observe that $z+z' = 4-4x-4x^2 \leqslant -\frac{5}{4}$ and $1+xz' = 1+4x-8x^2 \leqslant -\frac{1}{2}$ for all $x \in [3/4, 1]$. Moreover, by $0 \leqslant z/x = 4(1-x) \leqslant 1$ for $x \in [\frac{3}{4}, 1]$ and $k \leqslant j-1$, we have $z' + (z/x)^{j-k} \leqslant z' + z/x = 8-12x \leqslant -1$ for all $x \in [\frac{3}{4}, 1]$. Noting that $0 \leqslant k \leqslant j-1$ and $1 \leqslant j \leqslant n-1$, we have $m+k-j \geqslant m-j \geqslant m-(n-1) \geqslant 0$ by $n \leqslant m+1$. Therefore, we conclude that $f'_{m,j}(x) \leqslant 0$ for all $x \in [\frac{3}{4}, 1]$ and hence $f_{m,j}$ is decreasing on $[\frac{3}{4}, 1]$. This proves that $\mathsf{P}_{m,n}(x)\mathsf{P}_{m,n}(4x(1-x))$ is a decreasing function and achieves its maximum value at $x = \frac{3}{4}$ on $[\frac{3}{4}, 1]$. Hence, (5.8.27) holds.

Set $x := \sin^2(\xi/2)$. By definition of $\widehat{b_2^{\min}}$ in (5.8.26) and $\widehat{b}(\xi) = \mathsf{P}_{m,n}(\sin^2(\xi/2))$,

$$\|\widehat{b_2^{\min}}\|_{L_\infty(\mathbb{T})} = 2\max_{0\leqslant x\leqslant 1} \min(\mathsf{P}_{m,n}(x), [\mathsf{P}_{m,n}(x)\mathsf{P}_{m,n}(4x(1-x))]^{1/2}). \tag{5.8.28}$$

Since $\mathsf{P}_{m,n}(x)$ has nonnegative coefficients and therefore increases on $[0, 1]$, we have $\mathsf{P}_{m,n}(x) \leqslant \mathsf{P}_{m,n}(3/4)$ for all $x \in [0, 3/4]$. It now follows directly from (5.8.27) and (5.8.28) that $\|\widehat{b_2^{\min}}\|_{L_\infty(\mathbb{T})} \leqslant 2\mathsf{P}_{m,n}(\frac{3}{4})$.

Since $\widehat{b}(2^j\pi/3) = \mathsf{P}_{m,n}(3/4)$ for all $j \in \mathbb{N}$, we have $\|\widehat{b_n}\|_{L_\infty(\mathbb{T})}^{1/n} \geqslant |\widehat{b_n}(2\pi/3)|^{1/n} = \mathsf{P}_{m,n}(3/4)$ for all $n \in \mathbb{N}$. Therefore, $\widehat{b_2^{\min}}(2\pi/3) = 2\mathsf{P}_{m,n}(\frac{3}{4})$

and $\rho(\widehat{b}) = 2\lim_{n\to\infty} \|\widehat{b_n}\|_{L_\infty(\mathbb{T})}^{1/n} \geqslant 2\mathsf{P}_{m,n}(3/4)$. This shows that $\|\widehat{b_2^{\min}}\|_{L_\infty(\mathbb{T})} = 2\mathsf{P}_{m,n}(\frac{3}{4}) \geqslant \rho(\widehat{b})$. On the other hand, by item (7) of Corollary 5.8.5, we have $\rho(\widehat{b}) \leqslant \|\widehat{b_2^{\min}}\|_{L_\infty(\mathbb{T})}$. This proves that $\rho(\widehat{b}) = \|\widehat{b_2^{\min}}\|_{L_\infty(\mathbb{T})} = 2\mathsf{P}_{m,n}(\frac{3}{4})$. □

Let $a_{2m,2n}$ be the filters in (2.1.11) defined by

$$\widehat{a_{2m,2n}}(\xi) := e^{im\xi}(1+e^{-i\xi})^{2m}\widehat{b}(\xi) \quad \text{with} \quad \widehat{b}(\xi) := 2^{-2m}\mathsf{P}_{m,n}(\sin^2(\xi/2))$$

and $n \leqslant m+1$. By item (6) of Corollary 5.8.5 and Example 5.8.1, we have $\rho_0(b)_\infty \leqslant \rho(\widehat{b}) = 2^{1-2m}\mathsf{P}_{m,n}(\frac{3}{4}) \leqslant \rho_1(b)_1$. Hence,

$$\begin{aligned} 2m-1-\log_2 \mathsf{P}_{m,n}(\tfrac{3}{4}) &\leqslant -\log_2 \rho_0(b)_\infty = \mathrm{sm}_\infty(a_{2m,2n}) \leqslant \mathrm{sm}_1(a_{2m,2n}) \\ &= 1-\log_2 \rho_0(b)_1 \leqslant 2m - \log_2 \mathsf{P}_{m,n}(\tfrac{3}{4}). \end{aligned} \tag{5.8.29}$$

By exactly the same proof as in Example 5.8.1 with m being replaced by $m-1/2$, noting that we still have $m-1/2+k-j \geqslant 0$ for $1 \leqslant j \leqslant n-1$ and $1 \leqslant j \leqslant n-1$ if $n \leqslant m$, we have

Example 5.8.2 Let $\widehat{b}(\xi) := \mathsf{P}_{m-1/2,n}(\sin^2(\xi/2))$ with $m, n \in \mathbb{N}$ and $n \leqslant m$, where $\mathsf{P}_{m-1/2,n}(x) := \sum_{j=0}^{n-1} \binom{m+j-3/2}{j} x^j$. Then

$$\rho(\mathsf{P}_{m-1/2,n}(\sin^2(\xi/2))) = \rho(\widehat{b}) = \|\widehat{b_2^{\min}}\|_{L_\infty(\mathbb{T})} = 2\mathsf{P}_{m-1/2,n}(\tfrac{3}{4}).$$

Let $a_{2m-1,2n}$ be the filters in (2.1.12) given by $\widehat{a_{2m-1,2n}}(\xi) = e^{im\xi}(1+e^{-i\xi})^{2m-1}\widehat{b}(\xi)$ with $\widehat{b}(\xi) := 2^{1-2m}\mathsf{P}_{m-1/2,n}(\sin^2(\xi/2))$ and $n \leqslant m$. By item (6) of Corollary 5.8.5 and Example 5.8.2, we have $\rho_0(b)_\infty \leqslant \rho(\widehat{b}) = 2^{2-2m}\mathsf{P}_{m-1/2,n}(\frac{3}{4}) \leqslant \rho_1(b)_1$. Hence,

$$\begin{aligned} 2m-2-\log_2 \mathsf{P}_{m-1/2,n}(\tfrac{3}{4}) &\leqslant -\log_2 \rho_0(b)_\infty = \mathrm{sm}_\infty(a_{2m-1,2n}) \leqslant \mathrm{sm}_1(a_{2m-1,2n}) \\ &= 1-\log_2 \rho_0(b)_1 \leqslant 2m-1-\log_2 \mathsf{P}_{m-1/2,n}(\tfrac{3}{4}). \end{aligned}$$

We now present an algorithm to find a desired filter $\mathring{a}$ in Corollary 5.8.2 from a given scalar filter a.

Algorithm 5.8.6 *Let $a \in l_0(\mathbb{Z})$ such that $\sum_{k\in\mathbb{Z}} a(k) = 1$. Define $\mathsf{a}(z) := \sum_{k\in\mathbb{Z}} a(k)z^k$.*

(S1) Let $k = 1, m = 2$ and $a_k := a$.
(S2) Compute $\mathsf{u}_k(z^2) := \gcd(\mathsf{a}_k(z), \mathsf{a}_k(-z))$. If $\mathsf{u}_k = 1$ (no nontrivial common factor), then set $J := k$ and go to (S4). Otherwise, define $\mathsf{a}_{k+1}(z) := \mathsf{u}_k(z)\mathsf{a}_k(z)/\mathsf{u}_k(z^2)$, which must be a Laurent polynomial and $\mathrm{len}(a_{k+1}) < \mathrm{len}(a_k)$;
(S3) Increase k by 1 (that is, $k \leftarrow k+1$). Then go to (S2).

(S4) If $m > \operatorname{len}(a_k) + 1$*, set* $J := k$ *and go to (S6). Otherwise, for all* $\zeta \in \mathbb{T}\backslash\{1\}$ *satisfying* $\zeta^{2^m} = \zeta$*, if* $\prod_{j=0}^{m-1}(z + \zeta^{2^j})$ *is a factor of* $\mathsf{a}_k(z)$ *(which can be checked using Euclidean long division), then set* $\mathsf{u}_k(z) := \prod_{j=1}^{m}(z - \zeta^{2^j})$ *and define* $\mathsf{a}_{k+1}(z) := \mathsf{u}_k(z)\mathsf{a}_k(z)/\mathsf{u}_k(z^2)$*, which must be a Laurent polynomial and* $\operatorname{len}(a_{k+1}) < \operatorname{len}(a_k)$*. Increase k by 1 (*$k \leftarrow k+1$*) and continue for other roots of* $\zeta^{2^m} = \zeta$*.*

(S5) Increase m by 1 *(that is,* $m \leftarrow m+1$*) and go to (S4).*

(S6) Set $\mathsf{u}(z) := \mathsf{u}_1(z)\cdots\mathsf{u}_{J-1}(z)\mathsf{u}_J(z)$ *and* $\mathring{\mathsf{a}}(z) := \mathsf{a}(z)\mathsf{u}(z)/\mathsf{u}(z^2) = a_J(z)$*.*

Then $\mathring{a}$ *and u are desired filters in Corollary 5.8.2 satisfying*

$$\mathsf{a}(z)\mathsf{u}(z) = \mathsf{u}(z^2)\mathring{\mathsf{a}}(z), \qquad \forall\, z \in \mathbb{C}\backslash\{0\},$$

$\phi = u * \mathring{\phi}$ *and the integer shifts of* $\mathring{\phi}$ *are linearly independent, where* $\widehat{\phi}(\xi) := \prod_{j=1}^{\infty}\widehat{a}(2^{-j}\xi)$ *and* $\widehat{\mathring{\phi}}(\xi) := \prod_{j=1}^{\infty}\widehat{\mathring{a}}(2^{-j}\xi)$*. Therefore,* $\operatorname{sm}_p(\phi) = \operatorname{sm}_p(\mathring{\phi}) = \operatorname{sm}_p(\mathring{a})$ *for all* $1 \leqslant p \leqslant \infty$*.*

Let $\mathsf{u}(z) := \prod_{j=1}^{m}(z - \zeta^{2^j})$. Since $\prod_{j=0}^{m-1}(z + \zeta^{2^j}) = \mathsf{u}(z^2)/\mathsf{u}(z)$, therefore the above $\prod_{j=0}^{m-1}(z + \zeta^{2^j}) \mid \mathsf{a}_k(z)$ simply means $\mathsf{u}(z^2) \mid (\mathsf{u}(z)\mathsf{a}_k(z))$. Some examples of such special polynomials u for $m = 2, 3, 4$ are

$$z^2 + z + 1, \qquad z^3 + (\tfrac{1}{2} \pm t_1 i)z^2 + (-\tfrac{1}{2} \pm t_1 i)z - 1, \qquad z^4 + z^3 + z^2 + z + 1,$$

$$z^4 + 2z^3 + 3z^2 + 2z + 1, \qquad z^4 - (\tfrac{1}{2} \pm t_2 i)z^3 - 2z^2 + (-\tfrac{1}{2} \pm t_2 i)z + 1,$$

where $t_1 = \sin(\frac{4\pi}{7}) + \sin(\frac{2\pi}{7}) - \sin(\frac{\pi}{7})$ and $t_2 = \sin(\frac{8\pi}{15}) + \sin(\frac{4\pi}{15}) + \sin(\frac{2\pi}{15}) - \sin(\frac{\pi}{15})$.

5.9 Cascade Algorithms and Refinable Functions with Perturbed Filters*

A given filter may have irrational coefficients and therefore, is often perturbed or rounded in applications. In this section we study the error estimate of cascade algorithms and refinable functions with perturbed filters.

Lemma 5.9.1 *Let* $m \in \mathbb{N}_0$ *and* $\upsilon^a, \upsilon^b \in (l_0(\mathbb{Z}))^{1\times r}$ *such that* $\widehat{\upsilon^a}(0) \neq 0$*. Then* $\mathscr{F}_{m,\upsilon^a,p} \cap \mathscr{F}_{m,\upsilon^b,p}$ *is the empty set if and only if* $\widehat{\upsilon^b}(0) = c\widehat{\upsilon^a}(0)$ *for some* $c \in \mathbb{C}\backslash\{1\}$*.*

Proof Sufficiency ($\Leftarrow$). If not, then there exists $f \in \mathscr{F}_{m,\upsilon^a,p} \cap \mathscr{F}_{m,\upsilon^b,p}$ and consequently, we must have $1 = \widehat{\upsilon^b}(0)\widehat{f}(0) = c\widehat{\upsilon^a}(0)\widehat{f}(0) = c$, contradicting $c \neq 1$.

Necessity ($\Rightarrow$). By Theorem 5.6.4, without loss of generality, we assume $\upsilon^a = (\upsilon_1^a, 0, \ldots, 0)$ with $\widehat{\upsilon_1^a}(0) = 1$. Let $\varphi \in \mathscr{C}^m(\mathbb{R})$ be a compactly supported function satisfying (5.6.35). Define $f = \varphi(1, \vec{d}^{\mathsf{T}})^{\mathsf{T}}$ with $\vec{d} \in \mathbb{C}^{r-1}$. Then $f \in \mathscr{F}_{m,\upsilon^a,p}$ and

$$\widehat{\upsilon^b}(\xi)\widehat{f}(\xi + 2\pi k) = \mathscr{O}(|\xi|^{m+1}), \qquad \xi \to 0 \quad \forall\, k \in \mathbb{Z}\backslash\{0\}. \tag{5.9.1}$$

Write $\upsilon^b = (\upsilon_1^b, \upsilon_2^b)$ with $\upsilon_1^b \in l_0(\mathbb{Z})$ and note $\widehat{\varphi}(0) = 1$. Since $\mathscr{F}_{m,\upsilon^a,p} \cap \mathscr{F}_{m,\upsilon^b,p} = \emptyset$, we see that $f \notin \mathscr{F}_{m,\upsilon^b,p}$. Therefore, due to (5.9.1) and $f \notin \mathscr{F}_{m,\upsilon^b,p}$, we must have

$$\widehat{\upsilon_1^b}(0) + \widehat{\upsilon_2^b}(0)\vec{d} = \widehat{\upsilon^b}(0)\widehat{f}(0) \neq 1, \qquad \forall\, \vec{d} \in \mathbb{C}^{r-1}, \tag{5.9.2}$$

from which we conclude that $\widehat{\upsilon_2^b}(0) = 0$. Hence, $\widehat{\upsilon^b}(0) = c\widehat{\upsilon^a}(0)$ with $c := \widehat{\upsilon_1^b}(0) \neq 1$ by (5.9.2) and $\widehat{\upsilon_2^b}(0) = 0$. ☐

We now state the main result on cascade algorithms and refinable functions with perturbed filters.

Theorem 5.9.2 *Let $m \in \mathbb{N}_0$ and $a \in (l_0(\mathbb{Z}))^{r\times r}$ such that a vanishes outside an interval I and (5.6.3) holds. Let ϕ^a be a compactly supported refinable vector function satisfying $\phi^a = 2\sum_{k\in\mathbb{Z}} a(k)\phi^a(2\cdot -k)$ with $\widehat{\phi^a}(0) \neq 0$. Let $\upsilon^a \in (l_0(\mathbb{Z}))^{1\times r}$ be the unique sequence such that* $\operatorname{fsupp}(\upsilon^a) \subseteq [0, m]$, $\widehat{\upsilon^a}(0)\widehat{\phi^a}(0) = 1$, *and* (5.6.4) *holds with $\upsilon = \upsilon^a$. If $\rho_{m+1}(a, \upsilon^a)_p < 2^{1/p-m}$ (i.e., the cascade algorithm with the filter a converges in $W_p^m(\mathbb{R})$), then there exist constants $\varepsilon > 0$ and $C > 0$ such that*

(i) $\|\upsilon^b - \upsilon^a\|_{(l_1(\mathbb{Z}))^{1\times r}} \leqslant C\|b - a\|_{(l_1(\mathbb{Z}))^{r\times r}}$ *for all $b \in N_\varepsilon(a, m, I)$;*
(ii) $\rho_{m+1}(b, \upsilon^b)_p < 2^{1/p-m}$ *for all $b \in N_\varepsilon(a, m, I)$;*
(iii) For every initial function f from the nonempty set $\mathscr{F}_{m,\upsilon^a,p} \cap \mathscr{F}_{m,\upsilon^b,p}$,

$$\|\mathcal{R}_b^n f - \mathcal{R}_a^n f\|_{(W_p^m(\mathbb{R}))^r} \leqslant C C_f \|b - a\|_{(l_1(\mathbb{Z}))^{r\times r}}, \ \forall\, n \in \mathbb{N}, b \in N_\varepsilon(a, m, I),$$

where $C_f := \|f\|_{(\mathcal{L}_p(\mathbb{R}))^r}(m + 1)^{\operatorname{len}(f)}(\operatorname{len}(f) + \operatorname{len}(I))^m$;
(iv) $\|\phi^b - \phi^a\|_{(W_p^m(\mathbb{R}))^r} \leqslant C\|b - a\|_{(l_1(\mathbb{Z}))^{r\times r}}$ *for all $b \in N_\varepsilon(a, m, I)$, where ϕ^b is the unique compactly supported vector function satisfying $\widehat{\upsilon^b}(0)\widehat{\phi^b}(0) = 1$ and the refinement equation $\phi^b = 2\sum_{k\in\mathbb{Z}} b(k)\phi^b(2\cdot -k)$,*

where $N_\varepsilon(a, m, I)$ is the set of all filters $b \in (l_0(\mathbb{Z}))^{r\times r}$ satisfying

(1) $\|b - a\|_{(l_1(\mathbb{Z}))^{r\times r}} < \varepsilon$ *and the filter b vanishes outside the interval I;*
(2) (5.6.3) is satisfied with $\widehat{a}(0)$ being replaced by $\widehat{b}(0)$;
(3) Filter b has order $m + 1$ sum rules with the unique matching filter υ^b given by

$$\operatorname{fsupp}(\upsilon^b) \subseteq [0, m], \quad \widehat{\upsilon^b}(0)\widehat{\phi^a}(0) = 1, \quad \textit{and}$$
$$\widehat{\upsilon^b}(2\xi)\widehat{b}(\xi) = \widehat{\upsilon^b}(\xi) + \mathscr{O}(|\xi|^{m+1}), \quad \xi \to 0.$$

Proof By Theorem 5.6.4, without loss of generality, we assume $\upsilon^a = (\upsilon_1^a, 0, \ldots, 0)$ with $\widehat{\upsilon_1^a}(0) = 1$ (enlarge the interval I if necessary). Since $\rho_{m+1}(a, \upsilon^a)_p < 2^{1/p-m}$, by Corollary 5.6.10, the filter a must satisfy order $m+1$ sum rules with the matching filter υ^a and therefore, by Theorem 5.6.5, the filter a takes the normal form in (5.6.22) and (5.6.29).

We first prove item (i). Write $\phi^a = (\phi_1^a, \phi_2^a)^{\mathsf{T}}$ and $\upsilon^b = (\upsilon_1^b, \upsilon_2^b)$ with $\phi_1^a \in W_p^m(\mathbb{R})$ and $\upsilon_1^b \in l_0(\mathbb{Z})$. By $\widehat{\upsilon^a}(0)\widehat{\phi^a}(0) = 1$ and $\widehat{\upsilon^a}(0) = (1, 0, \dots, 0)$, we have $\widehat{\phi_1^a}(0) = 1$. The unique solution $\widehat{\upsilon^b}(0)$ satisfying $\widehat{\upsilon^b}(0)\widehat{\phi^a}(0) = 1$ and $\widehat{\upsilon^b}(0)\widehat{b}(0) = \widehat{\upsilon^b}(0)$ is given by

$$
\begin{aligned}
\widehat{\upsilon_1^b}(0) &= 1/\big(1 + \widehat{b_{1,2}}(0)[I_{r-1} - \widehat{b_{2,2}}(0)]^{-1}\widehat{\phi_2^a}(0)\big),\\
\widehat{\upsilon_2^b}(0) &= \widehat{\upsilon_1^b}(0)\widehat{b_{1,2}}(0)[I_{r-1} - \widehat{b_{2,2}}(0)]^{-1}.
\end{aligned}
\tag{5.9.3}
$$

Since 1 is a simple eigenvalue of $\widehat{b}(0)$, considering $\det(I_r - \widehat{b}(0))$, we have

$$
\widehat{b_{1,2}}(0)[I_{r-1} - \widehat{b_{2,2}}(0)]^{-1}\widehat{b_{2,1}}(0) = 1 - \widehat{b_{1,1}}(0). \tag{5.9.4}
$$

Note that both (5.9.3) and (5.9.4) hold when b is replaced by a. Also, when ε is sufficiently small, $I_{r-1} - \widehat{b_{2,2}}(0)$ is invertible by $\rho(\widehat{b_{2,2}}(0)) < 2^{-m}$ due to (5.6.3). Consequently, there exists a constant $C_1 > 0$ such that $\|\widehat{\upsilon^b}(0) - \widehat{\upsilon^a}(0)\|_{l_1} \leqslant C_1\|b - a\|_{(l_1(\mathbb{Z}))^{r\times r}}$ for all $b \in N_\varepsilon(a, m, I)$. Using the relation in (5.6.10) which also holds if a and υ are replaced by b and υ^b due to our assumption in item (2), we conclude that there exists a constant $C_2 > 0$ such that

$$
\sum_{j=0}^{m} \|\widehat{\upsilon^b}^{(j)}(0) - \widehat{\upsilon^a}^{(j)}(0)\|_{l_1} \leqslant C_2\|b - a\|_{(l_1(\mathbb{Z}))^{r\times r}}, \qquad \forall\, b \in N_\varepsilon(a, m, I).
$$

Since both υ^a and υ^b are supported inside $[0, m]$, item (i) must hold.

We now prove items (ii)–(iv) for $b \in N_\varepsilon(a, m, I)$ under the extra condition $\mathscr{V}_{m,\upsilon^b} = \mathscr{V}_{m,\upsilon^a}$, which is automatically true for $r = 1$ since $\mathscr{V}_{m,\upsilon^b} = \mathscr{V}_{m,\delta} = \mathscr{V}_{m,\upsilon^a}$. Define $\mathring{I} := I + [-m, m]$ and

$$
\mathring{N}_\varepsilon(a, m, \mathring{I}) := \{b \in N_\varepsilon(a, m, \mathring{I}) \;:\; \mathscr{V}_{m,\upsilon^b} = \mathscr{V}_{m,\upsilon^a}\}. \tag{5.9.5}
$$

Define $V_j := (l(-\mathring{I}))^{1\times r} \cap \mathscr{V}_{j,\upsilon^a}$ for all $j \in \mathbb{N} \cup \{-1, 0\}$. Define $A_0 u := \mathcal{T}_a u$, $A_1 u := \mathcal{T}_a u(\cdot - 1)$ and $B_0 u := \mathcal{T}_b u$, $B_1 u := \mathcal{T}_b u(\cdot - 1)$. By Lemma 5.7.3, V_j is invariant under all A_0, A_1, B_0, B_1 for every $j = -1, 0, \dots, m$. Define $\mathcal{A} := \{A_0, A_1\}$ and $\mathcal{B} := \{B_0, B_1\}$. According to Theorem 5.7.6 and Lemma 5.7.1, we have $\rho_{m+1}(a, \upsilon^a)_p = \operatorname{jsr}_p(\mathcal{A}|_{V_m}) = \inf_{n\in\mathbb{N}} \|\mathcal{A}^n|_{V_m}\|_{l_p}^{1/n}$. Since $\rho_{m+1}(a, \upsilon^a)_p < 2^{1/p-m}$, there exist $0 < \rho_0 < 1$ and $n_0 \in \mathbb{N}$ such that $\|\mathcal{A}^{n_0}|_{V_m}\|_{l_p}^{1/n_0} < \rho_0^2 2^{1/p-m}$. Taking ε small enough, we have $\|\mathcal{B}^{n_0}|_{V_m}\|_{l_p}^{1/n_0} < \rho_0^2 2^{1/p-m}$ for all $b \in \mathring{N}_\varepsilon(a, m, \mathring{I})$. By the proof of Lemma 5.7.1 for showing $\operatorname{jsr}_p(\mathcal{A}) = \inf_{n\in\mathbb{N}} \|\mathcal{A}^n\|_{l_p}^{1/n}$, there exists $C_3 > 0$ such that

$$\|\mathcal{B}^n|_{V_m}\|_{l_p} \leqslant C_3\rho_0^n 2^{n(1/p-m)}, \qquad \forall\, n \in \mathbb{N}, b \in \mathring{N}_\varepsilon(a,m,\mathring{I}). \tag{5.9.6}$$

Thus, item (ii) holds for all $b \in \mathring{N}_\varepsilon(a,m,\mathring{I})$ by $\rho_{m+1}(b,\upsilon^b)_p = \lim_{n\to\infty} \|\mathcal{B}^n|_{V_m}\|_{l_p}^{1/n} \leqslant \rho_0 2^{1/p-m} < 2^{1/p-m}$.

For $b \in \mathring{N}_\varepsilon(a,m,\mathring{I})$, since $\mathscr{V}_{m,\upsilon^b} = \mathscr{V}_{m,\upsilon^a}$, by Lemma 5.6.6, as $\xi \to 0$,

$$\widehat{\upsilon^b}(\xi) = \widehat{c}(\xi)\widehat{\upsilon^a}(\xi) + \mathscr{O}(|\xi|^{m+1}) = (\widehat{c}(\xi)\widehat{\upsilon_1^a}(\xi), 0, \dots, 0) + \mathscr{O}(|\xi|^{m+1}), \tag{5.9.7}$$

for some $c \in l_0(\mathbb{Z})$ with $\widehat{c}(0) \neq 0$. By $\widehat{\upsilon^a}(0)\widehat{\phi^a}(0) = 1 = \widehat{\upsilon^b}(0)\widehat{\phi^a}(0)$, we must have $\widehat{c}(0) = 1$. By Lemma 5.6.6 again, we conclude that $\mathscr{F}_{m,\upsilon^b,p} = \mathscr{F}_{m,\upsilon^a,p}$. Note that $\{u_1, \dots, u_r\} = \mathscr{B}_{m-1,\upsilon^a} \subseteq V_{m-1}$, where $u_1^\star = \nabla^m \boldsymbol{\delta} e_1$, $u_2^\star = \boldsymbol{\delta} e_2, \dots, u_r^\star = \boldsymbol{\delta} e_r$. For $f \in \mathscr{F}_{m,\upsilon^a,p}$, by item (iv) of Theorem 5.6.4, there exist compactly supported functions $g_1, \dots, g_r \in L_p(\mathbb{R})$ such that $f^{(m)} = \sum_{\ell=1}^r u_\ell^\star * g_\ell$. Consequently,

$$\begin{aligned}[\mathcal{R}_b^n f - \mathcal{R}_a^n f]^{(m)} &= 2^{n(m+1)} \sum_{k\in\mathbb{Z}} (b_n - a_n)(k) f^{(m)}(2^n \cdot -k)\\ &= 2^{n(m+1)} \sum_{\ell=1}^{r} \sum_{k\in\mathbb{Z}} [(b_n - a_n) * u_\ell^\star](k) g_\ell(2^n \cdot -k).\end{aligned}$$

Note that $f_1^{(m)} = \nabla^m g_1$ and $f_\ell^{(m)} = g_\ell$ for $\ell = 2, \dots, r$. Since all $g_1, \dots, g_r$ have compact support, by Lemma 5.3.3 and (5.5.15), there is a constant

$$C_f := \|(g_1, \dots, g_r)^{\mathsf{T}}\|_{(\mathcal{L}_p(\mathbb{R}))^r} \leqslant \|f\|_{(\mathcal{L}_p(\mathbb{R}))^r} (m+1)^{\operatorname{len}(f)}$$

such that

$$\begin{aligned}\|[\mathcal{R}_b^n f - \mathcal{R}_a^n f]^{(m)}\|_{(L_p(\mathbb{R}))^r} &\leqslant C_f 2^{n(m+1-1/p)} \sum_{\ell=1}^{r} \|(b_n - a_n) * u_\ell^\star\|_{(l_p(\mathbb{Z}))^r}\\ &= C_f 2^{n(m-1/p)} \sum_{\ell=1}^{r} \|(\mathcal{S}_b^n(\boldsymbol{\delta} I_r) - \mathcal{S}_a^n(\boldsymbol{\delta} I_r)) * u_\ell^\star\|_{(l_p(\mathbb{Z}))^r},\end{aligned}$$

where we used the fact that $a_n = 2^{-n}\mathcal{S}_a^n(\boldsymbol{\delta} I_r)$ and $b_n = 2^{-n}\mathcal{S}_b^n(\boldsymbol{\delta} I_r)$. Note that

$$\mathcal{S}_b^n - \mathcal{S}_a^n = \sum_{j=1}^{n} \mathcal{S}_b^{j-1} \mathcal{S}_{b-a} \mathcal{S}_a^{n-j}.$$

Hence, the above identity and Theorem 5.7.4 imply

$$\|[\mathcal{R}_b^n f - \mathcal{R}_a^n f]^{(m)}\|_{(L_p(\mathbb{R}))^r} \leqslant C_f 2^{n(m-\frac{1}{p})} \sum_{j=1}^{n}\sum_{\ell=1}^{r} \|(\mathcal{S}_b^{j-1}\mathcal{S}_{b-a}\mathcal{S}_a^{n-j}(\boldsymbol{\delta} I_r)) * u_\ell^\star\|_{(l_p(\mathbb{Z}))^r}$$

$$= C_f 2^{n(m-\frac{1}{p})} \sum_{j=1}^{n}\sum_{\ell=1}^{r} \left(\sum_{\gamma_1=0}^{1} \cdots \sum_{\gamma_n=0}^{1} \|B_{\gamma_1}\cdots B_{\gamma_{j-1}}(B_{\gamma_j} - A_{\gamma_j})A_{\gamma_{j+1}}\cdots A_{\gamma_n} u_\ell\|_{l_p}^p \right)^{\frac{1}{p}}.$$

Since $u_\ell \in \mathscr{B}_{m-1,\upsilon^a} \subseteq V_{m-1}$ and A_0, A_1 are invariant on V_{m-1}, we see that $u_\ell^j := A_{\gamma_{j+1}} \cdots A_{\gamma_n} u_\ell \in V_{m-1}$. By (5.9.7) and $\widehat{c}(0) = 1$, the filter b must also take the form in (5.6.22) and (5.6.29) with a being replaced by b. Since $[B_{\gamma_j} - A_{\gamma_j}]u_\ell^j = \mathcal{T}_{b-a}(u_\ell^j(\cdot - \gamma_j))$, now it follows from Lemma 5.6.3 that $(B_{\gamma_j} - A_{\gamma_j})u_\ell^j \in V_m$. It is easy to check that $\|B_{\gamma_j} - A_{\gamma_j}\| \leqslant C_4\|b-a\|_{(l_1(\mathbb{Z}))^{r\times r}}$ for some constant $C_4 > 0$ independent of b. Therefore,

$$\|[\mathcal{R}_b^n f - \mathcal{R}_a^n f]^{(m)}\|_{(L_p(\mathbb{R}))^r}$$
$$\leqslant 2C_f C_4 2^{n(m-1/p)}\|b-a\|_{(l_1(\mathbb{Z}))^{r\times r}} \sum_{j=1}^{n}\sum_{\ell=1}^{r} \|\mathcal{B}^{j-1}|_{V_m}\|_{l_p} \|\mathcal{A}^{n-j}u_\ell\|_{l_p}.$$

Since $\rho_{m+1}(a,\upsilon^a)_p < 2^{1/p-m}$, by Theorem 5.6.11, the inequality (5.6.45) holds. By Theorem 5.7.4, $\|\mathcal{A}^{n-j}u_\ell\|_{l_p} = 2^{(n-j)}\|a_{n-j} * u_\ell^\star\|_{(l_p(\mathbb{Z}))^r}$. Consequently, we conclude from (5.9.6), (5.6.45) and the above inequality that

$$\|[\mathcal{R}_b^n f - \mathcal{R}_a^n f]^{(m)}\|_{(L_p(\mathbb{R}))^r}$$
$$\leqslant C_5 2^{n(m-1/p)}\|b-a\|_{(l_1(\mathbb{Z}))^{r\times r}} \sum_{j=1}^{n} \rho_0^{j-1} 2^{(j-1)(1/p-m)} 2^{(n-j)(1/p-m)}$$
$$= C_5 2^{m-1/p}\|b-a\|_{(l_1(\mathbb{Z}))^{r\times r}} \sum_{j=0}^{n-1} \rho_0^j \leqslant C_6\|b-a\|_{(l_1(\mathbb{Z}))^{r\times r}},$$

where $C_5 = 2rCC_3C_4C_f$ and $C_6 = C_5 2^{m-1/p}/(1-\rho_0) < \infty$ since $0 < \rho_0 < 1$. Note that all functions $\mathcal{R}_b^n f$ and $\mathcal{R}_a^n f$ are supported inside $\mathring{I} + \operatorname{supp}(f)$. Using (5.6.50), we conclude that item (iii) holds for all $b \in \mathring{N}_\varepsilon(a, m, \mathring{I})$.

Take $f = \phi^a \in \mathscr{F}_{m,\upsilon^a,p}$ in item (iii). Since $\mathscr{V}_{m,\upsilon^b} = \mathscr{V}_{m,\upsilon^a}$ and $\widehat{\upsilon^b}(0)\widehat{\phi^a}(0) = 1$, then $\phi^a \in \mathscr{F}_{m,\upsilon^b,p}$. By item (ii) and Theorem 5.6.11,

$$\|\phi^b - \phi^a\|_{(W_p^m(\mathbb{R}))^r} \leqslant \lim_{n\to\infty} \|\mathcal{R}_b^n\phi^a - \mathcal{R}_a^n\phi^a\|_{(W_p^m(\mathbb{R}))^r} \leqslant CC_{\phi^a}\|b-a\|_{(l_1(\mathbb{Z}))^{r\times r}}$$

for all $b \in \mathring{N}_\varepsilon(a, m, \mathring{I})$. This proves item (iv) for $b \in \mathring{N}_\varepsilon(a, m, \mathring{I})$.

We now prove items (ii)–(iv) for all $b \in N_\varepsilon(a, m, I)$ using Theorem 5.6.4 without assuming $\mathscr{V}_{m,\upsilon^b} = \mathscr{V}_{m,\upsilon^a}$. For $b \in N_\varepsilon(a, m, I)$, as in the proof of Theorem 5.6.4, there exists a unique $u^b \in (l_0(\mathbb{Z}))^{1\times(r-1)}$ such that u^b vanishes outside $[0, m]$ and $\widehat{u^b}(\xi) = \widehat{\upsilon_2^b}(\xi)/\widehat{\upsilon_1^b}(\xi) + \mathscr{O}(|\xi|^{m+1})$ as $\xi \to 0$. Then there is a constant $C_7 > 0$ such that

$$\sum_{j=0}^{m} \|\widehat{u^b}^{(j)}(0)\|_{l_1} \leqslant C_7 \sum_{j=0}^{m} \|\widehat{\upsilon_2^b}^{(j)}(0)\|_{l_1} \leqslant C_7 \sum_{j=0}^{m} \|\widehat{\upsilon^b}^{(j)}(0) - \widehat{\upsilon^a}^{(j)}(0)\|_{l_1},$$

since $\upsilon_2^a = 0$. Define

$$\widehat{U^b}(\xi) := \frac{1}{\widehat{\upsilon_1^b}(0)} \begin{bmatrix} 1 & -\widehat{u^b}(\xi) \\ 0 & I_{r-1} \end{bmatrix}.$$

Since u^b is supported inside $[0, m]$, by item (ii), there exists $C_8 > 0$ depending only on m and r such that

$$\|u^b\|_{(l_1(\mathbb{Z}))^{1\times(r-1)}} \leqslant C_7 C_8 \|\upsilon^b - \upsilon^a\|_{(l_1(\mathbb{Z}))^{1\times r}} \leqslant C C_7 C_8 \|b - a\|_{(l_1(\mathbb{Z}))^{r\times r}}.$$

Since $\widehat{\upsilon_1^a}(0) = 1$, by the above inequality, there exists a constant $C_9 > 0$ independent of b such that

$$\begin{aligned} &\|U^b - I_r\boldsymbol{\delta}\|_{(l_1(\mathbb{Z}))^{r\times r}} \leqslant C_9 \|b - a\|_{(l_1(\mathbb{Z}))^{r\times r}}, \\ &\|(U^b)^{-1} - I_r\boldsymbol{\delta}\|_{(l_1(\mathbb{Z}))^{r\times r}} \leqslant C_9 \|b - a\|_{(l_1(\mathbb{Z}))^{r\times r}}. \end{aligned} \tag{5.9.8}$$

Define

$$\widehat{\mathring{b}}(\xi) := (\widehat{U^b}(2\xi))^{-1}\widehat{b}(\xi)\widehat{U^b}(\xi), \qquad \widehat{\mathring{\upsilon}^b}(\xi) := \widehat{\upsilon^b}(\xi)\widehat{U^b}(\xi),$$

$$\widehat{\mathring{\phi}^b}(\xi) := (\widehat{U^b}(\xi))^{-1}\widehat{\phi^b}(\xi).$$

Taking sufficiently small ε, since $\widehat{\mathring{\upsilon}^b}(0) = (1, 0, \ldots, 0)$ and $\operatorname{fsupp}(U^b) \subseteq [0, m]$, we have $\widehat{\mathring{\upsilon}^b}(0)\widehat{\phi^a}(0) = 1$ and $\mathring{b} \in \mathring{N}_\varepsilon(a, m, \mathring{I})$ in (5.9.5) for every $b \in N_\varepsilon(a, m, I)$. Consequently, by what has been proved for the special case $\mathscr{V}_{m,\upsilon^b} = \mathscr{V}_{m,\upsilon^a}$ and by Theorem 5.7.6, we have $\rho_{m+1}(b, \upsilon^b)_p = \rho_{m+1}(\mathring{b}, \mathring{\upsilon}^b)_p < 2^{1/p-m}$. This proves item (ii).

We now prove item (iv). By what has been proved, we have $\|\mathring{\phi}^b - \phi^a\|_{(W_p^m(\mathbb{R}))^r} \leqslant C\|\mathring{b} - a\|_{(l_1(\mathbb{Z}))^{r\times r}}$. By $\mathring{b} = (U^b)^{-1}(2\cdot) * b * U^b$ and (5.9.8), there exists a constant

$C_{10} > 0$ independent of b such that

$$\|\mathring{b} - a\|_{(l_1(\mathbb{Z}))^{r\times r}} \leqslant C_{10}\|b - a\|_{(l_1(\mathbb{Z}))^{r\times r}}. \tag{5.9.9}$$

On the other hand, by $\phi^{\mathring{b}} = (U^b)^{-1} * \phi^b$, we have $\phi^b = U^b * \phi^{\mathring{b}}$ and

$$\begin{aligned}\|\phi^b - \phi^{\mathring{b}}\|_{(W_p^m(\mathbb{R}))^r} &= \|(U^b - \delta I_r) * \phi^{\mathring{b}}\|_{(W_p^m(\mathbb{R}))^r} \leqslant \|U^b - \delta I_r\|_{(l_1(\mathbb{Z}))^{r\times r}}\|\phi^{\mathring{b}}\|_{(W_p^m(\mathbb{R}))^r}\\ &\leqslant C_9\|b - a\|_{(l_1(\mathbb{Z}))^{r\times r}}(\|\phi^a\|_{(W_p^m(\mathbb{R}))^r} + C\|\mathring{b} - a\|_{(l_1(\mathbb{Z}))^{r\times r}})\\ &\leqslant C_{11}\|b - a\|_{(l_1(\mathbb{Z}))^{r\times r}},\end{aligned}$$

where $C_{11} := C_9(\|\phi^a\|_{(W_p^m(\mathbb{R}))^r} + C\varepsilon)$. Thus, $\|\phi^b - \phi^a\|_{(W_p^m(\mathbb{R}))^r} \leqslant (CC_{\phi^a}C_{10} + C_{11})\|b - a\|_{(l_1(\mathbb{Z}))^{r\times r}}$, which proves item (iv).

The proof of item (iii) is a little bit more delicate. We first show that $\mathscr{F}_{m,\upsilon^b,p} \cap \mathscr{F}_{m,\upsilon^a,p} \neq \emptyset$. Otherwise, by Lemma 5.9.1, we must have $\widehat{\upsilon^b}(0) = c\widehat{\upsilon^a}(0)$ for some $c \in \mathbb{C}\backslash\{1\}$. By $\widehat{\upsilon^b}(0)\widehat{\phi^a}(0) = 1 = \widehat{\upsilon^a}(0)\widehat{\phi^a}(0)$, we conclude that $c = 1$, which is a contradiction to $c \in \mathbb{C}\backslash\{1\}$. Hence, the intersection $\mathscr{F}_{m,\upsilon^b,p} \cap \mathscr{F}_{m,\upsilon^a,p}$ is nonempty. For $f \in \mathscr{F}_{m,\upsilon^b,p} \cap \mathscr{F}_{m,\upsilon^a,p}$, by the relation $b = U^b(2\cdot) * \mathring{b} * (U^b)^{-1}$, we have

$$\mathcal{R}_b^n f = U^b(2^n\cdot) * (\mathcal{R}_{\mathring{b}}^n g) \qquad \text{with} \quad g := (U^b)^{-1} * f.$$

Note that $g \in \mathscr{F}_{m,\upsilon^{\mathring{b}},p} = \mathscr{F}_{m,\upsilon^a,p}$. Since $f \in \mathscr{F}_{m,\upsilon^a,p}$ and $\mathscr{V}_{m,\upsilon^{\mathring{b}}} = \mathscr{V}_{m,\upsilon^a}$, by what has been proved and (5.9.9), we have

$$\|\mathcal{R}_{\mathring{b}}^n f - \mathcal{R}_a^n f\|_{(W_p^m(\mathbb{R}))^r} \leqslant CC_f\|\mathring{b} - a\|_{(l_1(\mathbb{Z}))^{r\times r}} \leqslant CC_{10}C_f\|b - a\|_{(l_1(\mathbb{Z}))^{r\times r}}.$$

Note that

$$\mathcal{R}_b^n f - \mathcal{R}_a^n f = U^b(2^n\cdot) * (\mathcal{R}_{\mathring{b}}^n(g - f)) + (U^b(2^n\cdot) - I_r\delta) * \mathcal{R}_{\mathring{b}}^n f + (\mathcal{R}_{\mathring{b}}^n f - \mathcal{R}_a^n f).$$

Then

$$\begin{aligned}\|(U^b(2^n\cdot) - I_r\delta) * \mathcal{R}_{\mathring{b}}^n f\|_{(W_p^m(\mathbb{R}))^r} &\leqslant \|U^b - I_r\delta\|_{(l_1(\mathbb{Z}))^{r\times r}}\|\mathcal{R}_{\mathring{b}}^n f\|_{(W_p^m(\mathbb{R}))^r}\\ &\leqslant C_{12}\|b - a\|_{(l_1(\mathbb{Z}))^{r\times r}},\end{aligned}$$

where $C_{12} := C_9(\sup_{n\in\mathbb{N}}\|\mathcal{R}_a^n f\|_{(W_p^m(\mathbb{R}))^r} + CC_{10}C_f\varepsilon) < \infty$. Note $\sup_{n\in\mathbb{N}}\|\mathcal{R}_a^n f\|_{(W_p^m(\mathbb{R}))^r} \leqslant CC_f$ for some constant $C > 0$ depending only on a (see the proof of Theorem 5.6.11). On the other hand,

$$\begin{aligned}\|U^b(2^n\cdot) * (\mathcal{R}_{\mathring{b}}^n(g - f))\|_{(W_p^m(\mathbb{R}))^r} &\leqslant \|U^b\|_{(l_1(\mathbb{Z}))^{r\times r}}\|\mathcal{R}_{\mathring{b}}^n(g - f)\|_{(W_p^m(\mathbb{R}))^r}\\ &\leqslant (r + C_3\varepsilon)\|\mathcal{R}_{\mathring{b}}^n(g - f)\|_{(W_p^m(\mathbb{R}))^r},\end{aligned}$$

where by (5.9.8) $\|U^b\|_{(l_1(\mathbb{Z}))^{r\times r}} \leqslant r + C_3\varepsilon$. To complete the proof of item (iii), by (5.6.50), it suffices to estimate $\|[\mathcal{R}^n_{\overset{\circ}{b}}(g-f)]^{(m)}\|_{(L_p(\mathbb{R}))^r}$.

Let $\{u_1,\ldots,u_r\} := \mathscr{B}_{m,\upsilon^a}$. Since both $f, g \in \mathscr{F}_{m,\upsilon^a,p}$, by item (iv) of Theorem 5.6.4, we have $[f-g]^{(m)} = \sum_{\ell=1}^r D_{m+1}h$ with $D_{m+1} := [u_1^\star,\ldots,u_r^\star]$ and some compactly supported function $h \in (L_p(\mathbb{R}))^r$. Therefore,

$$\begin{aligned}\|[\mathcal{R}^n_{\overset{\circ}{b}}(g-f)]^{(m)}\|_{(L_p(\mathbb{R}))^r} &= \Big\|2^{n(m+1)}\sum_{k\in\mathbb{Z}}(\mathring{b}_n * D_{m+1})(k)h(2^n\cdot -k)\Big\|_{(L_p(\mathbb{R}))^r}\\ &\leqslant 2^{n(m+1-1/p)}\|h\|_{(\mathcal{L}_p(\mathbb{R}))^r}\|\mathring{b}_n * D_{m+1}\|_{(l_p(\mathbb{Z}))^{r\times r}}.\end{aligned}$$

By (5.5.15) in Lemma 5.5.6 and $g-f = ((U^b)^{-1} - I_r\boldsymbol{\delta}) * f$, we have

$$\begin{aligned}\|h\|_{(\mathcal{L}_p(\mathbb{R}))^r} &\leqslant (m+1)^{\operatorname{len}(g-f)}\|g-f\|_{(\mathcal{L}_p(\mathbb{R}))^r}\\ &\leqslant (m+1)^{\operatorname{len}(f)+\operatorname{len}(I)}\|(U^b)^{-1} - I_r\boldsymbol{\delta}\|_{(l_1(\mathbb{Z}))^{r\times r}}\|f\|_{(\mathcal{L}_p(\mathbb{R}))^r}\\ &\leqslant C_9\|b-a\|_{(l_1(\mathbb{Z}))^{r\times r}}(m+1)^{\operatorname{len}(f)+\operatorname{len}(I)}\|f\|_{(\mathcal{L}_p(\mathbb{R}))^r},\end{aligned}$$

where we used (5.9.8). By (5.9.6), we conclude that

$$2^n\|\mathring{b}_n * D_{m+1}\|_{(l_p(\mathbb{Z}))^{r\times r}} \leqslant C_3\rho_0^n 2^{n(1/p-m)}\|D_{m+1}\|_{(l_p(\mathbb{Z}))^{r\times r}}$$

where C_3 is the constant in (5.9.6) and $0 < \rho_0 < 1$. Putting all together, since $0 < \rho_0 < 1$, we conclude that

$$\begin{aligned}\|[\mathcal{R}^n_{\overset{\circ}{b}}(g-f)]^{(m)}\|_{(L_p(\mathbb{R}))^r} &\leqslant 2^{n(m+1-1/p)}\|h\|_{(\mathcal{L}_p(\mathbb{R}))^r}\|\mathring{b}_n * D_{m+1}\|_{(l_p(\mathbb{Z}))^{r\times r}}\\ &\leqslant C_0C_f\|b-a\|_{(l_1(\mathbb{Z}))^{r\times r}},\end{aligned}$$

where $C_0 := C_3C_9(m+1)^{\operatorname{len}(f)+\operatorname{len}(I)}\|D_{m+1}\|_{(l_p(\mathbb{Z}))^{r\times r}} < \infty$. This proves item (iii). □

5.10 Exercises

5.1. For all sequences $\upsilon \in l(\mathbb{Z})$, prove $\|\upsilon\|_{l_q(\mathbb{Z})} \leqslant \|\upsilon\|_{l_p(\mathbb{Z})}$ for all $1 \leqslant p \leqslant q \leqslant \infty$. *Hint:* first consider $p = 1$ and then convert the general case into the special case $p = 1$.

5.2. For $1 \leqslant p, q, r \leqslant \infty$ such that $\frac{1}{p} + \frac{1}{q} = \frac{1}{r} + 1$, prove Young's inequality: $\|u * \upsilon\|_{l_r(\mathbb{Z})} \leqslant \|u\|_{l_p(\mathbb{Z})}\|\upsilon\|_{l_q(\mathbb{Z})}$ for all $u \in l_p(\mathbb{Z})$ and $\upsilon \in l_q(\mathbb{Z})$.

5.3. Prove that $\|f\|_{\mathcal{L}_p(\mathbb{R})} \leqslant \|f\|_{\mathcal{L}_q(\mathbb{R})}$ and $\mathcal{L}_q(\mathbb{R}) \subseteq \mathcal{L}_p(\mathbb{R})$ for all $1 \leqslant p \leqslant q \leqslant \infty$.

5.4. Prove that $\|f\|_{L_p(\mathbb{R})} \leqslant \|f\|_{\mathcal{L}_p(\mathbb{R})}$ and $\mathcal{L}_p(\mathbb{R}) \subseteq L_p(\mathbb{R})$ for all $1 \leqslant p \leqslant \infty$.

5.5. Prove that $(\mathcal{L}_p(\mathbb{R}), \|\cdot\|_{\mathcal{L}_p(\mathbb{R})})$ is a Banach space for every $1 \leqslant p \leqslant \infty$.

5.6. Let $E(a)$ be defined in Theorem 5.1.3.

a. Prove that $\{\widehat{u}^{(j)}(0)\}_{j=0}^{J_a-1} \in E(a)$ if and only if $\{\widehat{u}^{(j)}(0)\}_{j=0}^{J_a-1}$ is a solution to $A\vec{u} = 0$, where $\vec{u} := (\widehat{u}(0)^{\mathsf{T}}, \widehat{u}'(0)^{\mathsf{T}}, \ldots, \widehat{u}^{(J_a-1)}(0)^{\mathsf{T}})^{\mathsf{T}}$ and A is an $(rJ_a) \times (rJ_a)$ block matrix given by: For $j = 0, \ldots, J_a - 1$, $[A]_{j+1,k+1} := \frac{j!}{k!(j-k)!}\widehat{a}^{(j-k)}(0)$, $0 \leqslant k < j$, $[A]_{j+1,j+1} := \widehat{a}(0) - 2^j I_r$, and $[A]_{j+1,\ell} := 0_{r\times r}$ for $\ell > j+2, \ldots, J_a$.
b. Prove that $\dim(E(a)) = rJ_a - \operatorname{rank}(A)$.
c. Let n_0 be the smallest nonnegative integer such that $\det(\widehat{a}(0) - 2^j I_r) \neq 0$ for all $j \geqslant n_0$. Prove that $rJ_a - \operatorname{rank}(A)$ is the same for all $J_a \geqslant n_0$.

5.7. If (5.6.3) holds with $m = 0$, prove that

$$\mathrm{DS}(a) = \Big\{\phi \;:\; \widehat{\phi}(\xi) = \lim_{n\to\infty}\Big(\prod_{j=1}^{n}\widehat{a}(2^{-j}\xi)\Big)\widehat{\phi}(0) \quad \text{and} \quad \widehat{a}(0)\widehat{\phi}(0) = \widehat{\phi}(0)\Big\}.$$

5.8. Let $a \in (l_0(\mathbb{Z}))^{r\times r}$. Prove that $\dim(\mathrm{DS}(a)) = 1$ and $\widehat{\phi}(0) \neq 0$ for $\phi \in \mathrm{DS}(a)\backslash\{0\}$ if and only if (5.1.10) holds.

5.9. Define $a \in (l_0(\mathbb{Z}))^{2\times 2}$ by $\widehat{a}(\xi) = \begin{bmatrix} 1 & 0 \\ 1 - e^{-i\xi} & 2 \end{bmatrix}$.

a. If $\widehat{a}(\xi)\widehat{u}(\xi) = \widehat{u}(2\xi) + \mathscr{O}(|\xi|^2)$ as $\xi \to 0$ for $u \in (l_0(\mathbb{Z}))^{2\times 1}$, prove that $\widehat{u}(0) = (0,0)^{\mathsf{T}}$ and $\widehat{u}'(0) = (0,c)^{\mathsf{T}}$ for some $c \in \mathbb{C}$.
b. Show that $M_a = 2$ and $\dim(E(a)) = 1$, where M_a and $E(a)$ are defined in Theorem 5.1.3. Hence, $\dim(E(a)) < M_a$ can happen in Theorem 5.1.3.
c. Prove that $\lim_{n\to\infty}\prod_{j=1}^{n}\widehat{a}(2^{-j}\xi)$ diverges and $\lim_{n\to\infty}\prod_{j=1}^{n}[\widehat{a}(2^{-j}\xi)/2]$ exists for all $\xi \in \mathbb{R}$.

5.10. Let $\widehat{a}$ and $\widehat{\tilde{a}}$ be 2π-periodic measurable functions such that there exist positive numbers τ and C such that $|1 - \widehat{a}(\xi)| \leqslant C|\xi|^\tau$ and $|1 - \widehat{\tilde{a}}(\xi)| \leqslant C|\xi|^\tau$ a.e. $\xi \in [-\pi, \pi]$. Define $\boldsymbol{\varphi}(\xi) := \prod_{j=1}^{\infty}\widehat{a}(2^{-j}\xi)$ and $\tilde{\boldsymbol{\varphi}}(\xi) := \prod_{j=1}^{\infty}\widehat{\tilde{a}}(2^{-j}\xi)$. Prove that there exists $C_1 > 0$ such that $|1 - \boldsymbol{\varphi}(\xi)| \leqslant C_1|\xi|^\tau$ and $|1 - \tilde{\boldsymbol{\varphi}}(\xi)| \leqslant C_1|\xi|^\tau$ a.e. $\xi \in [-\pi, \pi]$. For a measurable function $\widehat{\Theta} \in L_2(\mathbb{T})$, prove that $\lim_{j\to\infty}\langle\widehat{\Theta}(2^{-j}\cdot)\tilde{\boldsymbol{\varphi}}(2^{-j}\cdot)\boldsymbol{\varphi}(2^{-j}\cdot), \mathbf{h}\rangle = \langle 1, \mathbf{h}\rangle$ for all $\mathbf{h} \in \mathscr{D}(\mathbb{R})$ if and only if $\lim_{j\to\infty}\langle\widehat{\Theta}(2^{-j}\cdot), \mathbf{h}\rangle = \langle 1, \mathbf{h}\rangle$ for all $\mathbf{h} \in \mathscr{D}(\mathbb{R})$.

5.11. Let p be a polynomial and $\psi \in \mathscr{S}(\mathbb{R})$. Prove that $\widehat{\mathsf{p}\psi}(\xi) = \mathsf{p}(i\frac{d}{d\xi})\widehat{\psi}(\xi)$, $\widehat{\mathsf{p}(\frac{d}{dx})\psi(x)}(\xi) = \mathsf{p}(i\xi)\widehat{\psi}(\xi)$, and $\int_{\mathbb{R}}\mathsf{p}(x)\psi(x)dx = \langle \mathsf{p}, \overline{\psi}\rangle = [\mathsf{p}(i\frac{d}{d\xi})\widehat{\psi}(\xi)]|_{\xi=0}$.

5.12. Let f be a compactly supported distribution on $\mathbb{R}$. Prove that f is a tempered distribution and $\widehat{f}(z) := \int_{\mathbb{R}} f(x)e^{-izx}dx =: \langle f; e^{-iz\cdot}\rangle$ is a well-defined analytic function for $z \in \mathbb{C}$.

5.13. Let f be a compactly supported distribution on $\mathbb{R}$. For $\psi \in \mathscr{S}(\mathbb{R})$, prove $f * \psi \in \mathscr{S}(\mathbb{R})$. If $\psi \in \mathscr{D}(\mathbb{R})$, prove $f * \psi \in \mathscr{D}(\mathbb{R})$. *Hint:* use Theorem A.6.1.

5.14. Let f be a compactly supported distribution on $\mathbb{R}$. Prove that there exist constants C and m such that $|\widehat{f}(\xi)| \leqslant C(1+\xi^2)^m$ for all $\xi \in \mathbb{R}$.

5.15. Let f be a compactly supported distribution and $\psi \in \mathscr{D}(\mathbb{R})$. Prove the Poisson summation formula $\langle \sum_{k\in\mathbb{Z}} f(\cdot - k); \psi\rangle = \left\langle \sum_{k\in\mathbb{Z}} \widehat{f}(2\pi k)e^{i2\pi k\cdot}; \psi\right\rangle$ by proving that $\sum_{k\in\mathbb{Z}} g(k) = \sum_{k\in\mathbb{Z}} \widehat{g}(2\pi k)$ with $g := f * (\psi(-\cdot))$.

5.16. Let $\eta \in \mathscr{S}(\mathbb{R})$ in the Schwartz class. For $m \in \mathbb{N}_0$ and $1 \leqslant p \leqslant \infty$, prove that there exists a sequence $\{\psi_n\}_{n\in\mathbb{N}}$ in $\mathscr{D}(\mathbb{R})$ such that $\lim_{n\to\infty} \|\psi_n - \eta\|_{W_p^m(\mathbb{R})} = 0$.

5.17. Let $\eta \in \mathscr{S}(\mathbb{R})$. For any positive integer m, prove that there exists a sequence $\{\psi_n\}_{n\in\mathbb{N}}$ in $\mathscr{D}(\mathbb{R})$ such that $\lim_{n\to\infty} \sum_{k\in\mathbb{Z}}(1+k^2)^m|\widehat{\psi_n}(2\pi k) - \widehat{\eta}(2\pi k)| = 0$.

5.18. For each $n \in \mathbb{Z}$, prove that there exists $\eta \in \mathscr{S}(\mathbb{R})$ such that $\widehat{\eta}(k) = \boldsymbol{\delta}(k-n)$ for all $k \in \mathbb{Z}$ and $\widehat{\eta}$ is compactly supported.

5.19. Let $\boldsymbol{\delta}$ be the Dirac distribution. For $f \in \mathscr{S}(\mathbb{R})$, prove that $\langle f, \lambda D^j \boldsymbol{\delta}(\lambda\cdot - k)\rangle = (-1)^j \lambda^{-j} \overline{f^{(j)}(k/\lambda)}$ for all $j \in \mathbb{N}_0$, $\lambda > 0$ and $k \in \mathbb{R}$.

5.20. Let ϕ and η be compactly supported distributions on $\mathbb{R}$. Prove that

a. If $\eta = \sum_{k=M}^{\infty} u(k)\phi(\cdot - k)$ for some $u \in l(\mathbb{Z})$ with $u(M) \neq 0$, then $\phi = \sum_{k=-M}^{\infty} v(k)\eta(\cdot - k)$ for some $v \in l(\mathbb{Z})$. *hint:* use $\phi = \frac{1}{u(M)}\eta(\cdot + M) - \sum_{k=1}^{\infty} \frac{u(M+k)}{u(M)}\phi(\cdot - k)$.

b. If $\eta = \sum_{k=-\infty}^{M} u(k)\phi(\cdot - k)$ for some $u \in l(\mathbb{Z})$ with $u(M) \neq 0$, then $\phi = \sum_{k=-\infty}^{-M} v(k)\eta(\cdot - k)$ for some $v \in l(\mathbb{Z})$.

5.21. Let φ be a compactly supported distribution on $\mathbb{R}$ and p be a polynomial. Prove that both φ and $\mathsf{p} * \varphi$ are tempered distributions.

5.22. Let φ be a compactly supported distribution on $\mathbb{R}$ and p be a polynomial. Prove that (5.5.3) holds in the sense of tempered distributions, that is, (5.5.4) holds for all $\psi \in \mathscr{S}(\mathbb{R})$.

5.23. Let η be a compactly supported distribution on $\mathbb{R}$ such that the integer shifts of η are linearly independent. For every $f \in \mathtt{S}(\eta)$ such that f has compact support, prove that $f = u * \eta$ for some $u \in l_0(\mathbb{Z})$. If in addition the integer shifts of f are linearly independent, show that $f = c\phi(\cdot - k)$ for some $c \in \mathbb{C}\backslash\{0\}$ and $k \in \mathbb{Z}$.

5.24. Let ϕ and ψ be compactly supported distributions on $\mathbb{R}$. If ψ is not identically zero and $\psi \in \mathtt{S}(\phi)$, prove that $\mathtt{S}(\psi) = \mathtt{S}(\phi)$.

5.25. Let $\phi_1, \ldots, \phi_r, \psi_1, \ldots, \psi_m$ be compactly supported distributions on $\mathbb{R}$. If $\psi_\ell \in \mathtt{S}(\phi_1, \ldots, \phi_r)$ for $\ell = 1, \ldots, m$, prove that $\mathtt{S}(\psi_1, \ldots, \psi_m) \subseteq \mathtt{S}(\phi_1, \ldots, \phi_r)$.

5.26. Let $\phi_1, \ldots, \phi_r$ be compactly supported distributions on $\mathbb{R}$ which are not all identically zero. We define $\mathrm{len}(\mathtt{S}(\phi_1, \ldots, \phi_r))$ to be the smallest nonnegative integer m such that there exist compactly supported distributions $\psi_1, \ldots, \psi_m$ satisfying $\mathtt{S}(\psi_1, \ldots, \psi_m) = \mathtt{S}(\phi_1, \ldots, \phi_r)$. If $\eta_1, \ldots, \eta_s$ are compactly supported distributions such that $\mathtt{S}(\eta_1, \ldots, \eta_s) = \mathtt{S}(\phi_1, \ldots, \phi_r)$ and the integer shifts of $\eta_1, \ldots, \eta_s$ are linearly independent, prove that $\mathrm{len}(\mathtt{S}(\phi_1, \ldots, \phi_r)) = s$.

5.27. Let $\phi_1, \ldots, \phi_r, \psi_1, \ldots, \psi_m$ be compactly supported distributions on $\mathbb{R}$. If $\psi_1, \ldots, \psi_m \in \mathsf{S}(\phi_1, \ldots, \phi_r)$ and $\mathrm{len}(\mathsf{S}(\psi_1, \ldots, \psi_m)) = \mathrm{len}(\mathsf{S}(\phi_1, \ldots, \phi_r))$, prove $\mathsf{S}(\psi_1, \ldots, \psi_m) = \mathsf{S}(\phi_1, \ldots, \phi_r)$.

5.28. Let v be an $r \times s$ matrix of Laurent polynomials. Prove that there exist an $r \times r$ matrix u_1, an $s \times s$ matrix u_2, and an $r \times s$ matrix D of Laurent polynomials such that $\mathsf{v}(z) = \mathsf{u}_1(z)\mathsf{D}(z)\mathsf{u}_2(z)$ (Smith form of v), both $\det(\mathsf{u}_1)$ and $\det(\mathsf{u}_2)$ are nonzero monomials, and all the nonzero entries $\mathsf{p}_1, \ldots, \mathsf{p}_{\min(r,s)}$ of D lie on the diagonal such that $\mathsf{p}_j \mid \mathsf{p}_{j+1}$ for all $j = 1, \ldots, \min(r, s) - 1$.

5.29. Let $\eta_1, \ldots, \eta_r$ be compactly supported distributions on $\mathbb{R}$ such that the integer shifts of $\eta_1, \ldots, \eta_r$ are linearly independent and $\widehat{\eta}(2z) = \mathsf{v}(e^{-iz})\widehat{\eta}(z)$ for some $v \in (l_0(\mathbb{Z}))^{r\times r}$. Prove that $\det(\mathsf{v})$ is not identically zero. *Hint:* use Exercise 5.28.

5.30. Let $1 \leqslant p \leqslant \infty$ and $\phi_1, \ldots, \phi_r \in L_p(\mathbb{R})$ be compactly supported. Define

$$\mathsf{S}(\phi_1, \ldots, \phi_r \mid l_0(\mathbb{Z})) := \{v_1 * \phi_1 + \cdots + v_r * \phi_r \ : \ v_1, \ldots, v_r \in l_0(\mathbb{Z})\},$$

$$\mathsf{S}(\phi_1, \ldots, \phi_r \mid l_p(\mathbb{Z})) := \{v_1 * \phi_1 + \cdots v_r * \phi_r \ : \ v_1, \ldots, v_r \in l_p(\mathbb{Z})\}.$$

If the integer shifts of $\phi_1, \ldots, \phi_r$ are stable, prove that $\mathsf{S}(\phi_1, \ldots, \phi_r \mid l_p(\mathbb{Z}))$ is isomorphic to $(l_p(\mathbb{Z}))^{1\times r}$ and

$$\mathsf{S}(\phi_1, \ldots, \phi_r) \cap L_p(\mathbb{R}) = \mathsf{S}(\phi_1, \ldots, \phi_r \mid l_p(\mathbb{Z})) = \overline{\mathsf{S}(\phi_1, \ldots, \phi_r \mid l_0(\mathbb{Z}))}^{\|\cdot\|_{L_p(\mathbb{R})}}.$$

5.31. Let $\phi_1, \ldots, \phi_r$ be compactly supported distributions such that the integer shifts of $\phi_1, \ldots, \phi_r$ are linearly independent. Let $u_1, \ldots, u_r \in l(\mathbb{Z})$.

a. If $\sum_{\ell=1}^r \sum_{k\in\mathbb{Z}} u_\ell \phi_\ell(\cdot - k) = \mathsf{p}$ for some polynomial $\mathsf{p} \in \mathbb{P}$, prove that all $u_1, \ldots, u_r$ must be polynomial sequences with degrees no more than $\deg(\mathsf{p})$. *Hint:* use Theorem 5.2.1.

b. If $\sum_{\ell=1}^r \sum_{k\in\mathbb{Z}} u_\ell \phi_\ell(\cdot - k)$ agrees with a polynomial on (M, ∞) for some $M \in \mathbb{N}$, prove that there exist $N \in \mathbb{N}$ and $\mathsf{p}_1, \ldots, \mathsf{p}_r \in \mathbb{P}$ such that $u_\ell(k) = \mathsf{p}_\ell(k)$ for all $k \in \mathbb{Z} \cap (N, \infty)$ and $\ell = 1, \ldots, r$. Moreover, prove that $\sum_{\ell=1}^r \sum_{k\in\mathbb{Z}} \mathsf{p}_\ell(k)\phi_\ell(\cdot - k)$ must be a polynomial. *Hint:* use Theorem 5.2.1 and idea from Theorem 5.5.1.

c. If $\sum_{\ell=1}^r \sum_{k\in\mathbb{Z}} u_\ell \phi_\ell(\cdot - k)$ agrees with a polynomial on $(-\infty, -M)$ for some $M \in \mathbb{N}$, prove that there exists $N \in \mathbb{N}$ such that every u_ℓ must agree with a polynomial sequence on $\mathbb{Z} \cap (-\infty, -N)$ for all $\ell = 1, \ldots, r$.

5.32. Prove Theorem 5.3.7 and Corollary 5.3.8.

5.33. Prove $\|AB\|_{l_2} \leqslant \|A\|_{l_2}\|B\|_{l_2}$ for matrices A and B, where $\|\cdot\|_{l_2}$ is in (5.0.1).

5.34. For $f \in W_p^m(\mathbb{R})$, prove that $\nabla_t^m f(x) = t^m \int_{\mathbb{R}} f^{(m)}(y) B_m(\frac{x-y}{t})dy$.

5.35. Let ψ be a compactly supported function in $\mathscr{C}^\infty(\mathbb{R})$ such that $\int_{\mathbb{R}} \psi(x)dx = 1$. For $f \in L_p(\mathbb{R})$, define f_λ as in (5.4.6). Prove that $f_\lambda \in \mathscr{C}^\infty(\mathbb{R}) \cap (\cap_{n=1}^\infty W_p^n(\mathbb{R}))$

and there exists a positive constant C depending only on ψ such that

$$\|f - f_\lambda\|_{L_p(\mathbb{R})} \leqslant C\omega_m(f, \lambda^{-1})_p, \quad \|f_\lambda^{(m)}\|_{L_p(\mathbb{R})} \leqslant C\lambda^m \omega_m(f, \lambda^{-1})_p.$$

5.36. Let f be a distribution on $\mathbb{R}$ such that $f|_{(-\infty,c)} = 0$ and $f|_{(d,\infty)} \in \mathbb{P}_m$ for some $c, d \in \mathbb{R}$ and $m \in \mathbb{N}_0$. Define $D^{-1}f$ by $\langle D^{-1}f, h\rangle := -\langle f, H\rangle$, $h \in \mathscr{D}(\mathbb{R})$ with $H(x) := \int_{-\infty}^{x} h(y)dy$. Prove that $D^{-1}f$ is a well-defined distribution, the distributional derivative of $D^{-1}f$ is f, and $D^{-1}f|_{(d,\infty)} \in \mathbb{P}_{m+1}$.

5.37. Let f be a compactly supported distribution on $\mathbb{R}$. Prove that there exists $n \in \mathbb{N}_0$ such that $D^{-n}f \in L_p^{loc}(\mathbb{R})$.

5.38. Let ϕ be a compactly supported distribution on $\mathbb{R}$ and $\psi \in \mathscr{D}(\mathbb{R})$ such that $\psi = 1$ in a neighborhood of fsupp(ϕ). Define the smoothness exponent of ϕ to be $\operatorname{sm}_p(\phi) := \sup\{m - n + \tau \ : \ \sup_{\lambda>0} \lambda^{-\tau}\omega_1((D^{-n}\phi)^{(m)}, \lambda)_p < \infty, D^{-n}\phi \in W_p^m(\mathbb{R}) \text{ for some } m \in \mathbb{N}_0, \tau > 0\}$. For a compactly supported function ϕ in $L_p(\mathbb{R})$, prove that the above definition of $\operatorname{sm}_p(\phi)$ agrees with the definition in (5.8.1).

5.39. Prove that $\operatorname{sm}_p(\phi) = \sup\{m - n + \tau \ : \ \sup_{\lambda>0} \lambda^{-\tau}\omega_1((\phi * B_n)^{(m)}, \lambda)_p < \infty, \phi * B_n \in W_p^m(\mathbb{R}) \text{ for some } m \in \mathbb{N}_0, \tau > 0\}$.

5.40. Let $a \in (l_0(\mathbb{Z}))^{r\times r}$ satisfy (5.6.3) with $m = 0$. Let ϕ be the unique compactly supported refinable vector function satisfying $\phi = 2\sum_{k\in\mathbb{Z}} a(k)\phi(2\cdot -k)$ with $\widehat{\phi}(0) \neq 0$. Let $1 \leqslant p \leqslant \infty$. Prove that $\operatorname{sm}_p(\phi) \geqslant \operatorname{sm}_p(a)$ and if in addition the integer shifts of ϕ are stable, then $\operatorname{sm}_p(\phi) = \operatorname{sm}_p(a)$.

5.41. Let $\upsilon \in (l_0(\mathbb{Z}))^{1\times r}$ with $\widehat{\upsilon}(0) \neq 0$. If $r > 1$, for any $m \in \mathbb{N}$, prove that there exists a strongly invertible sequence $U \in (l_0(\mathbb{Z}))^{r\times r}$ such that $\widehat{\upsilon}(\xi)\widehat{U}(\xi) = (1, 0, \ldots, 0) + \mathscr{O}(|\xi|^m)$ as $\xi \to 0$. *Hint:* By Theorem 5.6.4, one can assume $\upsilon = (\upsilon_1, 0, \ldots, 0)$ with $\widehat{\upsilon_1}(0) = 1$. Since $r > 1$, it suffices to consider $r = 2$ and for this case, one can consider $\upsilon = (\upsilon_1, \upsilon_2)$ with $\widehat{\upsilon_1}(0) = 1$ and $\widehat{\upsilon_2}(\xi) = (1 - e^{-i\xi})^m$. Then by the fact $\gcd(\widehat{\upsilon_1}, \widehat{\upsilon_2}) = 1$, there exist $u_1, u_2 \in l_0(\mathbb{Z})$ such that $\widehat{u_1}(\xi)\widehat{\upsilon_1}(\xi) + \widehat{u_2}(\xi)\widehat{\upsilon_2}(\xi) = 1$. Now construct a strongly invertible sequence $U \in (l_0(\mathbb{Z}))^{2\times r}$ to prove the claim.

5.42. For $a \in l_0(\mathbb{Z})$ satisfying $\widehat{a}(\xi + \pi) = \mathscr{O}(|\xi|^{m+1})$ as $\xi \to 0$, prove $\mathcal{T}_a\nabla^{m+1}\boldsymbol{\delta} - \left(2^{-1-m}\overline{\widehat{a}(0)} + i^{m+1}\frac{\overline{\widehat{a}^{(m)}(\pi)}}{(m+1)!}\right)\nabla^{m+1}\boldsymbol{\delta} \in \mathscr{V}_{m+1,\boldsymbol{\delta}}$. *Hint:* The proof of Lemma 5.6.3.

5.43. Let $a \in l_0(\mathbb{Z})$ and $[m, n] := \operatorname{fsupp}(a)$ with $n-m \geqslant 1$. Prove that $\mathcal{T}_a(l(K))^{1\times r} \subseteq (l(K))^{1\times r}$ for $K = [-n, -m], [1 - n, -m], [-n, -m - 1]$ and $[1 - n, -m - 1]$.

5.44. If both $a \in (l_0(\mathbb{Z}))^{r\times r}$ and $f = (f_1, \ldots, f_r)^{\mathsf{T}}$ are supported inside $[-N, N]$ for some $N \in \mathbb{N}$, then prove that $\mathcal{R}_a f$ must vanish outside $[-N, N]$.

5.45. Let $a \in (l_0(\mathbb{Z}))^{r\times r}$ satisfy (5.6.3) with $m = 0$. Let ϕ be a compactly supported refinable function/distribution satisfying $\widehat{\phi}(2\xi) = \widehat{a}(\xi)\widehat{\phi}(\xi)$ with $\widehat{\phi}(0) \neq 0$. Then there exists a unique $\widehat{\upsilon}(0) \in \mathbb{C}^{1\times r}$ such that $\widehat{\upsilon}(0)\widehat{a}(0) = \widehat{\upsilon}(0)$ and $\widehat{\upsilon}(0)\widehat{\phi}(0) = 1$. For any $r \times 1$ vector f of compactly supported distributions satisfying $\widehat{\upsilon}(0)\widehat{f}(0) = 1$ and $\sup_{\xi\in\mathbb{R}}(1+|\xi|)^{-1}|\widehat{f}(\xi)| < \infty$, prove that $\widehat{\phi}(\xi) = \lim_{n\to\infty} \widehat{a}(2^{-1}\xi)\cdots\widehat{a}(2^{-n}\xi)\widehat{f}(2^{-n}\xi)$ with the limit uniformly converging for ξ in every bounded interval. *Hint:* By Theorem 5.6.4, without loss of generality, one can assume that $\widehat{v}(0) = (1,0,\ldots,0)$, $\widehat{\phi}(0) = (1,0,\ldots,0)^{\mathsf{T}}$ and

$$a = \begin{bmatrix} a_{1,1} & a_{1,2} \\ a_{2,1} & a_{2,2} \end{bmatrix} \quad \text{with} \quad \widehat{a_{1,1}}(0) = 1, \quad \widehat{a_{1,2}}(0) = 0, \quad \widehat{a_{2,1}}(0) = 0.$$

Write $f = (f_1,\ldots,f_r)^{\mathsf{T}} = u + g$ with $u = (f_1,0,\ldots,0)^{\mathsf{T}}$ and $g = (0,f_2,\ldots,f_r)^{\mathsf{T}}$. Apply Theorem 5.1.2 to prove that $\widehat{\phi}(\xi) = \lim_{n\to\infty} \widehat{a}(2^{-1}\xi)\cdots\widehat{a}(2^{-n}\xi)\widehat{u}(2^{-n}\xi)$. Define $\widehat{b}$ to be $\widehat{a}$ except that we replace $\widehat{a_{1,1}}$ and $\widehat{a_{1,2}}$ be zero (that is, the first row of $\widehat{b}$ is zero). Then prove that

$$\widehat{a}(2^{-1}\xi)\cdots\widehat{a}(2^{-n}\xi)\widehat{h}(2^{-n}\xi) = \widehat{b}(2^{-1}\xi)\cdots\widehat{b}(2^{-n}\xi)\widehat{h}(2^{-n}\xi).$$

Since all the eigenvalues of $\widehat{b}(0)$ are less than 1 in modulus, apply Lemma 5.1.1 to conclude $\lim_{n\to\infty} \widehat{b}(2^{-1}\xi)\cdots\widehat{b}(2^{-n}\xi) = 0$.

5.46. For $b \in l_1(\mathbb{Z})$, prove $\lim_{n\to\infty} \|b_n\|_{l_1(\mathbb{Z})}^{1/n} = \inf_{n\in\mathbb{N}} \|b_n\|_{l_1(\mathbb{Z})}^{1/n}$ and $\lim_{n\to\infty} \|\widehat{b_n}\|_{L_\infty(\mathbb{T})}^{1/n} = \inf_{n\in\mathbb{N}} \|\widehat{b_n}\|_{L_\infty(\mathbb{T})}^{1/n}$ where $\widehat{b_n}(\xi) := \widehat{b}(2^{n-1}\xi)\cdots\widehat{b}(2\xi)\widehat{b}(\xi)$.

5.47. Define a real-valued function f on $\mathbb{R}$ by $f|_{[k,k+1)} := -\frac{1}{k^2}$ for all $k \in \mathbb{N}$, $f|_{[0,1)} := \sum_{k=1}^{\infty} \frac{1}{k^2} = \frac{\pi^2}{6}$, and $f|_{(-\infty,0)} := 0$. (i) Prove that $f \in L_1(\mathbb{R})$ and $\sum_{k\in\mathbb{Z}} f(\cdot - k) = 0$. (ii) Prove that there does not exist $g \in L_1(\mathbb{R})$ such that $f = \nabla g = g - g(\cdot - 1)$. *Hint:* First prove that g must take the form $g = \sum_{k=0}^{\infty} f(\cdot - k)$.

5.48. Let $\phi = (\phi_1,\ldots,\phi_r)^{\mathsf{T}}$ be a vector of compactly supported distributions such that the integer shifts of ϕ are linearly independent. If there exists $a \in (l_0(\mathbb{Z}))^{r\times r}$ such that $\phi = 2\sum_{k\in\mathbb{Z}} a(k)\phi(2\cdot -k)$, then prove that such a filter a must be unique.

5.49. Let $\phi = (\phi_1,\ldots,\phi_r)^{\mathsf{T}}$ be a vector of functions in $\mathcal{L}_p(\mathbb{R})$ for some $1 \leqslant p \leqslant \infty$ such that the integer shifts of ϕ are stable in $L_p(\mathbb{R})$. If there exists $a \in (l_p(\mathbb{Z}))^{r\times r}$ such that $\phi = 2\sum_{k\in\mathbb{Z}} a(k)\phi(2\cdot -k)$, then prove that such a filter a must be unique.

5.50. Let $a \in (l_0(\mathbb{Z}))^{2\times 2}$ be supported inside $[-1,1]$ and be given by: For $t_0, t_1, t_2 \in \mathbb{C}$,

$$a = \left\{ \begin{bmatrix} \frac{1}{4} & t_1 \\ 0 & 0 \end{bmatrix}, \begin{bmatrix} \frac{1}{2} & 0 \\ 0 & t_0 \end{bmatrix}, \begin{bmatrix} \frac{1}{4} & t_2 \\ 0 & 0 \end{bmatrix} \right\}_{[-1,1]}.$$

a. Let $\phi := (B_2(\cdot - 1), 0)^{\mathsf{T}}$. Prove that the integer shifts of ϕ are not linearly independent and are not stable in $L_p(\mathbb{R})$ for all $1 \leqslant p \leqslant \infty$.
b. Prove that ϕ is a refinable function satisfying $\phi = 2\sum_{k\in\mathbb{Z}} a(k)\phi(2\cdot -k)$.
c. If $|t_0| < 1$, prove that (5.6.3) holds with $m = 0$ and $\mathrm{DS}(a) = \{c\phi : c \in \mathbb{C}\}$.

5.51. Let $a = \{\frac{1}{4}, 0, \frac{1}{2}, 0, \frac{1}{4}\}_{[0,4]}$ and $\phi := B_2(\cdot/2)$. Prove that $\phi = 2\sum_{k\in\mathbb{Z}} a(k)\phi(2\cdot -k)$, $\operatorname{sr}(a) = 0$, and $\operatorname{ao}(\phi) = 2$ (i.e., $\mathbb{P}_1 \in \mathsf{S}(\phi)$). Hence, the condition $\operatorname{span}\{\widehat{\phi}(2\pi k) : k \in \mathbb{Z}\} = \operatorname{span}\{\widehat{\phi}(\pi + 2\pi k) : k \in \mathbb{Z}\} = \mathbb{C}^r$ cannot be dropped in Theorem 5.5.4.

5.52. Let $b \in l_0(\mathbb{Z})$ and $k_0 \in \mathbb{Z}$. If $b(k_0 + 2k) = 0$ for all $k \in \mathbb{Z}\backslash\{0\}$ and $|b(k_0)| \geqslant \sum_{k\in\mathbb{Z}} |b(k_0 + 1 + 2k)|$, prove that $\rho_0(b)_\infty = 2|b(k_0)|$.

5.53. Let $a_2^I = \{-\frac{1}{32}, 0, \frac{9}{32}, \frac{1}{2}, \frac{9}{32}, 0, -\frac{1}{32}\}_{[-3,3]}$. Prove that $\operatorname{sr}(a_2^I) = 4$, $\operatorname{sm}_\infty(a_2^I)=2$, and $\operatorname{sm}_2(a_2^I) = -\frac{1}{2}\log_2 \frac{5+3\sqrt{17}}{512} \approx 2.44076$.

5.54. Let $\mathcal{A}$ be a finite collection of $r \times r$ matrices or linear operators on $\mathbb{C}^r$ such that not all the elements in $\mathcal{A}$ are identically zero. Define $\rho := \operatorname{jsr}_\infty(\mathcal{A})$. For arbitrary $\varepsilon > 0$, prove that $\|v\|_\varepsilon := \sup_{n\in\mathbb{N}} \sup_{A_1,\ldots,A_n\in\mathcal{A}} (\rho+\varepsilon)^{-n}\|A_1\cdots A_n v\|$ for $v \in \mathbb{C}^n$ is a norm on $\mathbb{C}^n$ and $\|A\|_\varepsilon \leqslant \rho + \varepsilon$ for all $A \in \mathcal{A}$, where $\|A\|_\varepsilon := \sup_{\|v\|\leqslant 1} \|Av\|_\varepsilon$.

5.55. Let $\lambda \in \mathbb{C}$ and E be an $r \times r$ matrix such that λ is a simple eigenvalue of E. For $\vec{u}, \vec{v} \in \mathbb{C}^r$ such that $\vec{u}^\mathsf{T} E = \lambda \vec{u}$ and $E\vec{v} = \lambda\vec{v}$, prove that if $\vec{u} \neq 0$ and $\vec{v} \neq 0$, then $\vec{u}^\mathsf{T}\vec{v} \neq 0$. *Hint:* Use $E = U\operatorname{diag}(\lambda, F)U^{-1}$ and note that λ is not an eigenvalue of F, where U is an invertible $r \times r$ matrix and F is an $(r-1) \times (r-1)$ matrix.

Chapter 6
Framelets and Wavelets Derived from Refinable Functions

Because refinable functions play a central role in the study and construction of affine systems of framelets and wavelets, in this chapter we study several classes of special refinable functions. As a consequence, we provide complete analysis of framelets and wavelets (or more precisely, multiframelets and multiwavelets) that are derived from refinable (vector-valued) functions. In particular, we investigate refinable functions having analytic expressions (such as piecewise polynomials), refinable Hermite interpolants, refinable orthogonal functions, and refinable biorthogonal functions. We characterize dual framelets and biorthogonal wavelets in Sobolev spaces that are derived from refinable functions. As a consequence, we obtain criteria for tight framelets and orthogonal wavelets in $L_2(\mathbb{R})$ that are derived from refinable functions. Then we discuss how to construct biorthogonal (matrix-valued) filters with increasing orders of sum rules. Another goal of this chapter is to study convergence of scalar cascade algorithms in weighted subspaces of $L_2(\mathbb{R})$ with infinitely supported filters and then we investigate refinable functions having exponential decay. Then we investigate the existence of a smooth compactly supported dual refinable function to a given refinable function and the local linear independence of a scalar refinable function. At the end of this chapter, we address the stability issue of discrete affine systems in the sequence space $l_2(\mathbb{Z})$ and their connections to affine systems in the function setting.

6.1 Refinable Functions Having Analytic Expressions

Framelets and wavelets having analytic expressions are of interest in both theory and applications. In this section we study refinable functions having analytic expressions. B-splines are such examples and are of importance in both wavelet theory and approximation theory.

B. Han, *Framelets and Wavelets*, Applied and Numerical Harmonic Analysis,
https://doi.org/10.1007/978-3-319-68530-4_6

6.1.1 Properties of B-Spline Functions

For $m \in \mathbb{N}$, the *(cardinal) B-spline function* B_m *of order* m is defined to be

$$B_1 := \chi_{(0,1]} \quad \text{and} \quad B_m := B_{m-1} * B_1 = \int_0^1 B_{m-1}(\cdot - t)dt. \tag{6.1.1}$$

Define $\nabla f := f - f(\cdot - 1)$, $x_+ := \max(0, x)$, $x_+^c := (x_+)^c$ for $c > 0$, and $x_+^0 := \chi_{(0,\infty)}$. Some basic properties of B-spline functions are as follows:

Proposition 6.1.1 *Let B_m be the B-spline function of order m with $m \in \mathbb{N}$. Then*

(1) $\operatorname{supp}(B_m) = [0, m]$ and $B_m(x) > 0$ for all $x \in (0, m)$.
(2) $B_m = B_m(m - \cdot)$, $B_m \in \mathscr{C}^{m-2}(\mathbb{R})$, $B_m' = \nabla B_{m-1}$, $B_m|_{(k,k+1)} \in \mathbb{P}_{m-1}$ for all $k \in \mathbb{Z}$.
(3) $\widehat{B_m}(\xi) = (\frac{1-e^{-i\xi}}{i\xi})^m$ and the integer shifts of B_m are linearly independent (and therefore, stable).
(4) B_m is refinable: $B_m = 2\sum_{k\in\mathbb{Z}} a_m^B(k)B_m(2\cdot - k)$, where $\widehat{a_m^B}(\xi) := 2^{-m}(1 + e^{-i\xi})^m$.
(5) $\operatorname{sr}(a_m^B) = m$ and $\operatorname{sm}_p(a_m^B) = \operatorname{sm}_p(B_m) = 1/p + m - 1$ for all $1 \leqslant p \leqslant \infty$.
(6) $B_m(x) = \frac{1}{(m-1)!}\nabla^m x_+^{m-1} = \frac{1}{(m-1)!}\sum_{j=0}^m (-1)^j \frac{m!}{j!(m-j)!}(x - j)_+^{m-1}$ for all $x \in \mathbb{R}$.
(7) For $m > 1$, $B_m(k) = \frac{1}{(m-1)!}\sum_{j=0}^{k-1}(-1)^j \frac{m!}{j!(m-j)!}(k - j)^{m-1}$ for all $k \in \mathbb{N}$.
(8) $[\widehat{B_m}, \widehat{B_n}](\xi) := \sum_{k\in\mathbb{Z}} \widehat{B_m}(\xi + 2\pi k)\overline{\widehat{B_n}(\xi + 2\pi k)} = \sum_{k=1-n}^{m-1} B_{m+n}(n + k)e^{-ik\xi}$.

Proof Items (1) and (2) can be directly checked. Since $\widehat{B_1}(\xi) = \frac{1-e^{-i\xi}}{i\xi}$, we have $\widehat{B_m}(\xi) = (\widehat{B_1}(\xi))^m = (\frac{1-e^{-i\xi}}{i\xi})^m$. By Theorem 5.2.1 and $\{z \in \mathbb{C} : \widehat{B_m}(z) = 0\} = 2\pi\mathbb{Z}\backslash\{0\}$, the integer shifts of B_m are linearly independent and hence stable. This proves item (3). Item (4) follows from $\widehat{B_1}(2\xi) = \widehat{a_1^B}(\xi)\widehat{B_1}(\xi)$, $\widehat{B_m}(\xi) = (\widehat{B_1}(\xi))^m$, and $\widehat{a_m^B}(\xi) = (\widehat{a_1^B}(\xi))^m$.

Since $\widehat{a_m^B}(\xi) = (1 + e^{-i\xi})^m\widehat{b}(\xi)$ with $b = 2^{-m}\boldsymbol{\delta}$, by Corollary 5.8.5, we have $\rho_0(b)_p = \rho_0(2^{-m}\boldsymbol{\delta})_p = 2^{1-m}$ and therefore, $\operatorname{sm}_p(a_m^B) = 1/p - \log_2 \rho_0(b)_p = 1/p + m - 1$. Because the integer shifts of B_m are linearly independent and therefore stable, by Theorem 5.8.1, we have $\operatorname{sm}_p(\phi) = \operatorname{sm}_p(a_m^B)$. This proves item (5).

Note that the mth distributional derivative $D^m f$ of the tempered distribution $f(x) := \frac{x_+^{m-1}}{(m-1)!}$ is the Dirac distribution $\boldsymbol{\delta}$. Taking Fourier transform, we have $1 = \widehat{\boldsymbol{\delta}}(\xi) = \widehat{D^m f}(\xi) = (i\xi)^m \widehat{f}(\xi)$. Consequently,

$$\widehat{\nabla^m f}(\xi) = (1 - e^{-i\xi})^m \widehat{f}(\xi) = \left(\frac{1 - e^{-i\xi}}{i\xi}\right)^m (i\xi)^m \widehat{f}(\xi) = \left(\frac{1 - e^{-i\xi}}{i\xi}\right)^m = \widehat{B_m}(\xi).$$

Hence, $B_m = \nabla^m f$ holds in the sense of distributions, which implies $B_m(x) = \nabla^m f(x)$ for a.e. $x \in \mathbb{R}$. Since both B_m and $\nabla^m f$ are continuous functions when $m > 1$, we must have $B_m(x) = \nabla^m f(x) = \frac{1}{(m-1)!}\nabla^m x_+^{m-1}$ for all $x \in \mathbb{R}$. One can directly check that this identity also holds for $m = 1$. By the definition of $\nabla = \mathrm{Id} - [\![1, 0]\!]$,

$$\nabla^m = (\mathrm{Id} - [\![1,0]\!])^m = \sum_{j=0}^{m} (-1)^j \frac{m!}{j!(m-j)!} [\![1,0]\!]^j = \sum_{j=0}^{m} (-1)^j \frac{m!}{j!(m-j)!} [\![j,0]\!],$$

where $[\![j,0]\!]f := f(\cdot - j)$. This proves item (6) by $\nabla^m x_+^{m-1} = \sum_{j=0}^{m} (-1)^j \frac{m!}{j!(m-j)!} (x - j)_+^{m-1}$. For $m \geqslant 2$, item (6) also directly implies the following recursive relation:

$$B_m(x) = \frac{x}{m-1} B_{m-1}(x) + \frac{m-x}{m-1} B_{m-1}(x-1), \qquad x \in \mathbb{R}.$$

Item (7) follows directly from item (6). By Lemma 4.4.1 and $B_n(x) = B_n(n-x)$, since B_n is real-valued,

$$\begin{aligned}
[\widehat{B_m}, \widehat{B_n}](\xi) &= \sum_{k\in\mathbb{Z}} \langle B_m, B_n(\cdot - k)\rangle e^{-ik\xi} = \sum_{k\in\mathbb{Z}} \langle B_m, B_n(n + k - \cdot)\rangle e^{-ik\xi} \\
&= \sum_{k\in\mathbb{Z}} B_{m+n}(n+k) e^{-ik\xi} = \sum_{k=1-n}^{m-1} B_{m+n}(n+k) e^{-ik\xi}.
\end{aligned}$$

This proves item (8). □

6.1.2 Scalar Refinable Functions Having Analytic Expressions

We shall prove in this section that except B-spline functions, scalar refinable functions often do not have any analytic exprcssions. For $a \in l_0(\mathbb{Z})$, recall that $\mathrm{fsupp}(a)$ is the shortest interval such that a vanishes outside it, and $\mathrm{len}(a)$ is defined to be the length of $\mathrm{fsupp}(a)$.

To study scalar refinable functions having analytic expressions, we need the following result establishing the invertibility of the transition operator acting on its smallest invariant subspace.

Lemma 6.1.2 *Let $a = \{a(k)\}_{k\in\mathbb{Z}} \in l_0(\mathbb{Z})$ such that $\mathsf{a}(z)$ and $\mathsf{a}(-z)$ have no common zeros in $\mathbb{C}\backslash\{0\}$, where $\mathsf{a}(z) := \sum_{k\in\mathbb{Z}} a(k) z^k, z \in \mathbb{C}\backslash\{0\}$. Define $[m,n] :=$ $\mathrm{fsupp}(a)$ and assume $n > m$. Then the three square matrices $A_0 := (2a(2j - k))_{m\leqslant j,k\leqslant n-1}$, $A_1 := (2a(2j-k+1))_{m\leqslant j,k\leqslant n-1}$, and $(2a(2j-k))_{m+1\leqslant j,k\leqslant n-1}$ must be invertible.*

Proof By $\mathrm{fsupp}(a) = [m,n]$, we have $a(m)a(n) \neq 0$. Note that the first row (i.e., $j = m$) of A_0 is $(2a(m), 0, \dots, 0)$ and the last row (i.e., $j = n-1$) of A_1 is $(0, \dots, 0, 2a(n))$. Now it is not difficult to see that A_1 is invertible if and only if A_0 is invertible (and if and only if $(2a(2j-k))_{m+1\leqslant j,k\leqslant n-1}$ is invertible).

Suppose that $A_0\vec{u} = 0$ for some $\vec{u} := (u(m), \dots, u(n-1))^\mathsf{T}$ with $u(k) = 0$ for all $k \in \mathbb{Z}\backslash[m, n-1]$. Then, $(z^m, z^{m+1}, \dots, z^{n-1})A_0\vec{u} = 0$. Define $\mathsf{u}(z) := \sum_{k\in\mathbb{Z}} u(k)z^k$. Since $\mathrm{fsupp}(a) = [m, n]$ and $\mathrm{fsupp}(u) \subseteq [m, n-1]$, we have

$$0 = (z^m, z^{m+1}, \dots, z^{n-1})A_0\vec{u} = 2\sum_{j=m}^{n-1}\sum_{k=m}^{n-1} a(2j-k)u(k)z^j$$

$$= 2\sum_{j\in\mathbb{Z}}\sum_{k\in\mathbb{Z}} a(2j-k)u(k)z^j = 2\big(\mathsf{a}^{[0]}(z)\mathsf{u}^{[0]}(z) + z\mathsf{a}^{[1]}(z)\mathsf{u}^{[1]}(z)\big).$$

Since $\mathsf{a}(z)$ and $\mathsf{a}(-z)$ have no common zeros in $\mathbb{C}\backslash\{0\}$, by $\mathsf{a}(z) = \mathsf{a}^{[0]}(z^2)+z\mathsf{a}^{[1]}(z^2)$, the Laurent polynomials $\mathsf{a}^{[0]}(z)$ and $\mathsf{a}^{[1]}(z)$ have no common zeros in $\mathbb{C}\backslash\{0\}$. Now it follows from the above identity that $\mathsf{u}^{[0]}(z) = \mathsf{a}^{[1]}(z)\mathsf{q}(z)$ and $\mathsf{u}^{[1]}(z) = -z^{-1}\mathsf{a}^{[0]}(z)\mathsf{q}(z)$ for some Laurent polynomial q. Therefore,

$$\mathsf{u}(z) = \mathsf{u}^{[0]}(z^2) + z\mathsf{u}^{[1]}(z^2) = \big(\mathsf{a}^{[1]}(z^2) - z^{-1}\mathsf{a}^{[0]}(z^2)\big)\mathsf{q}(z^2) = -z^{-1}\mathsf{a}(-z)\mathsf{q}(z^2).$$

Since $\mathrm{len}(\mathsf{u}) \leqslant n-m-1 < n-m = \mathrm{len}(\mathsf{a})$, the above identity forces $\mathsf{q} = 0$. Hence, $\mathsf{u} = 0$ and $\vec{u} = 0$. This proves that A_0 is invertible. □

See Exercise 6.1 for another proof of Lemma 6.1.2.

The following result shows that a scalar refinable function that is differentiable almost everywhere with a nontrivial derivative must be absolutely continuous.

Proposition 6.1.3 *Let ϕ be a compactly supported Lebesgue measurable function such that $\int_\mathbb{R} |\phi(x)|dx < \infty$ and*

$$\phi'(x) := \lim_{h\to 0} \frac{\phi(x+h) - \phi(x)}{h} \quad \textit{exists for almost every } x \in \mathbb{R} \tag{6.1.2}$$

and $0 < \|\phi'\|_{L_1(\mathbb{R})} < \infty$. If $\phi(x) = 2\sum_{k\in\mathbb{Z}} a(k)\phi(2x-k)$ for almost every $x \in \mathbb{R}$ for some $a \in l_0(\mathbb{Z})$, then $\phi(x) = \int_{-\infty}^x \phi'(t)\,dt$ for almost every $x \in \mathbb{R}$. In particular, by modifying ϕ on a set of measure zero, ϕ must be a compactly supported absolutely continuous function on $\mathbb{R}$.

Proof Recall that ϕ' refers to the classical derivative in (6.1.2), while $D\phi$ refers to the distributional derivative. Since ϕ is differentiable almost everywhere and $\phi(x) = 2\sum_{k\in\mathbb{Z}} a(k)\phi(2x-k)$ for almost every $x \in \mathbb{R}$, we deduce (see Exercise 6.3) that

$$\phi'(x) = 4\sum_{k\in\mathbb{Z}} a(k)\phi'(2x-k), \qquad a.e.\ x \in \mathbb{R}.$$

Since we assume $\phi' \in L_1(\mathbb{R})$ with $\|\phi'\|_{L_1(\mathbb{R})} > 0$, the function ϕ' can be regarded as a nontrivial compactly supported distribution on $\mathbb{R}$. Now it follows from the above identity that $f = \phi'$ is a nontrivial compactly supported distributional solution to

$$f = 2\sum_{k\in\mathbb{Z}} 2a(k)f(2\cdot -k). \tag{6.1.3}$$

By item (3) of Theorem 5.1.3, since ϕ' is a nontrivial distributional solution to (6.1.3), we must have $4\sum_{k\in\mathbb{Z}} a(k) = 2^j$ for some $j \in \mathbb{N}_0$. Therefore, up to a multiplicative constant, the compactly supported distributional solution to (6.1.3) is unique, that is, $\dim(\mathrm{DS}(2a)) = 1$, where $\mathrm{DS}(2a)$ is the space of all compactly supported distributional solutions to the refinement equation (6.1.3).

On the other hand, since $\phi \in L_1(\mathbb{R})$, the function ϕ can be regarded as a distribution. Since $\phi(x) = 2\sum_{k\in\mathbb{Z}} a(k)\phi(2x-k)$ for a.e. $x \in \mathbb{R}$, we see that $\phi = 2\sum_{k\in\mathbb{Z}} a(k)\phi(2\cdot -k)$ holds in the sense of distributions. Now it is trivial to see that the distributional derivative $D\phi$ is also a compactly supported distributional solution to (6.1.3). Since ϕ' is nontrivial and $\dim(\mathrm{DS}(2a)) = 1$, there must exist a constant $c \in \mathbb{C}$ such that $D\phi = c\phi'$ in the sense of distributions.

Since $\phi' \in L_1(\mathbb{R})$ has compact support, we can define an absolutely continuous function $\mathring{\phi}$ on $\mathbb{R}$ by

$$\mathring{\phi}(x) := \int_{-\infty}^{x} \phi'(t)\, dt, \qquad x \in \mathbb{R}.$$

Since $\mathring{\phi}$ is absolutely continuous, $\mathring{\phi}'(x) = \phi'(x)$ for a.e. $x \in \mathbb{R}$ and $\mathring{\phi}$ is a bounded continuous function on $\mathbb{R}$ by $\phi' \in L_1(\mathbb{R})$. Therefore, $\mathring{\phi}$ can be regarded as a distribution (however, $\mathring{\phi}$ may not have compact support since $\int_{\mathbb{R}} \phi'(t)dt$ may not vanish).

Let $\psi \in \mathscr{D}(\mathbb{R})$. Since ψ and $\mathring{\phi}$ are absolutely continuous, we see that $\mathring{\phi}\psi$ is a compactly supported absolutely continuous function. Consequently, $\int_{\mathbb{R}}[\mathring{\phi}(x)\psi(x)]'dx = 0$. By $\mathring{\phi}'(x) = \phi'(x)$ for a.e. $x \in \mathbb{R}$, this implies

$$\int_{\mathbb{R}} \mathring{\phi}(x)\psi'(x)dx = -\int_{\mathbb{R}} \mathring{\phi}'(x)\psi(x)dx = -\int_{\mathbb{R}} \phi'(x)\psi(x)dx.$$

Thus, we deduce that

$$\begin{aligned}\langle D(\phi - c\mathring{\phi}); \psi\rangle &:= -\langle \phi - c\mathring{\phi}; \psi'\rangle = -\langle \phi; \psi'\rangle + \langle c\mathring{\phi}; \psi'\rangle \\ &= \langle D\phi; \psi\rangle + \int_{\mathbb{R}} c\mathring{\phi}(x)\psi'(x)dx = \langle D\phi; \psi\rangle - \int_{\mathbb{R}} c\phi'(x)\psi(x)dx \\ &= \langle D\phi; \psi\rangle - \langle c\phi'; \psi\rangle = \langle D\phi - c\phi'; \psi\rangle = 0,\end{aligned}$$

since $D\phi = c\phi'$ in the sense of distributions. Therefore, the distributional derivative $D(\phi - c\mathring{\phi}) = 0$. By Proposition 5.5.14, we must have $\phi - c\mathring{\phi} = C$ for some constant $C \in \mathbb{C}$. Since ϕ has compact support, both ϕ and ϕ' must vanish on $(-\infty, N)$ for some $N \in \mathbb{R}$. Consequently, by the definition of $\mathring{\phi}$, the function $\mathring{\phi}$ must vanish on $(-\infty, N)$. Therefore, $\phi - c\mathring{\phi}$ vanishes on $(-\infty, N)$ and consequently, we must have $C = 0$. This proves that $\phi = c\mathring{\phi}$ in the sense of distributions, from which we conclude that $\phi(x) = c\mathring{\phi}(x)$ for a.e. $x \in \mathbb{R}$. By $\mathring{\phi}'(x) = \phi'(x)$ for a.e. $x \in \mathbb{R}$ and $\|\phi'\|_{L_1(\mathbb{R})} > 0$, we further deduce that $c = 1$. This completes the proof. □

The proof of Proposition 6.1.3 actually shows that

Proposition 6.1.4 *Let ϕ be an $r \times 1$ vector of compactly supported Lebesgue measurable functions in $L_1(\mathbb{R})$ such that ϕ' exists almost everywhere and $0 < \|\phi'\|_{(L_1(\mathbb{R}))^r} < \infty$. If $\phi(x) = 2\sum_{k\in\mathbb{Z}} a(k)\phi(2x-k)$ for a.e. $x \in \mathbb{R}$ for some $a \in (l_0(\mathbb{Z}))^{r\times r}$ and if $\dim(DS(2a)) = 1$ (where $DS(2a)$ is the space of all compactly supported vector distributions f satisfying $f = 2\sum_{k\in\mathbb{Z}} 2a(k)f(2\cdot -k)$), then $\phi(x) = \int_{-\infty}^{x} \phi'(t)\,dt$ for a.e. $x \in \mathbb{R}$. In particular, by modifying ϕ on a set of measure zero, the refinable function ϕ must be an $r\times 1$ vector of compactly supported absolutely continuous functions on $\mathbb{R}$. Consequently, if $\dim(DS(2^j a)) = 1$ for all $j \in \mathbb{N}$ and if*

$$\begin{aligned} &\phi|_{(x_k,x_{k+1})} \text{ is a polynomial for all } 0 \leqslant k < n \text{ with} \\ &\qquad -\infty = x_0 < x_1 < \cdots < x_{n-1} < x_n = \infty, \end{aligned} \tag{6.1.4}$$

then there exists $m \in \mathbb{N}_0$ such that $\phi \in (\mathscr{C}^{m-1}(\mathbb{R}))^r$ and $\phi^{(m)}$ is a vector of piecewise constants.

Recall that the support $\operatorname{supp}(\phi)$ of a distribution ϕ is the smallest closed subset of $\mathbb{R}$ such that $\langle \phi; \psi\rangle = 0$ for all $\psi \in \mathscr{D}(\mathbb{R})$ satisfying $\operatorname{supp}(\psi) \subseteq \mathbb{R}\backslash \operatorname{supp}(\phi)$. Before we proceeding further, we have to characterize distributions supported on a finite number of points on $\mathbb{R}$.

Lemma 6.1.5 *Let f be a distribution on $\mathbb{R}$ such that supp$(f) = \{x_1,\ldots,x_n\}$ with $x_1 < x_2 < \cdots < x_n$. Then there exists $m \in \mathbb{N}_0$ such that $f = \sum_{\ell=1}^{n}\sum_{j=0}^{m} c_{\ell,j} (D^j\delta)(\cdot - x_\ell)$ for some $c_{\ell,j} \in \mathbb{C}$, $\ell = 1,\ldots,n$ and $j = 0,\ldots,m$.*

Proof Let $\eta := \chi_{[-1/2,1/2],1/4,1/4} \in \mathscr{D}(\mathbb{R})$ be the bump function defined in (4.6.21) such that $\eta(x) = 1$ for $x \in [-1/4, 1/4]$ and η vanishes outside $[-3/4, 3/4]$. For $\varepsilon > 0$, define $f_j := f\eta(\frac{\cdot - x_j}{\varepsilon}), j = 1,\ldots,n$. For a sufficiently small $\varepsilon > 0$, by $\operatorname{supp}(f) = \{x_1,\ldots,x_n\}$, we have $f = f_1 + \cdots + f_n$ and $\operatorname{supp}(f_j) = \{x_j\}$ for all $j = 1,\ldots,n$. As a consequence, to prove the claim, without loss of generality, we can assume that $\operatorname{supp}(f) = \{0\}$. Let $\varphi \in \mathscr{D}(\mathbb{R})$. Then φ is supported inside $[-N, N]$ for some positive integer N. By Theorem A.6.1, there exists $C > 0$ such that

$$|\langle f; h\rangle| \leqslant C\sum_{\beta=0}^{m} \|h^{(\beta)}\|_{L_\infty(\mathbb{R})} \quad \text{for all } h \in \mathscr{D}(\mathbb{R}) \text{ with support inside } [-N, N]. \tag{6.1.5}$$

Define $\psi(x) := \varphi(x) - \eta(x)\sum_{j=0}^{m} \frac{\varphi^{(j)}(0)}{j!}x^j$. Then $\psi(0) = \psi'(0) = \cdots = \psi^{(m)}(0) = 0$ and $\psi \in \mathscr{D}(\mathbb{R})$ is supported inside $[-N, N]$. Consequently, by the Taylor expansion of $\psi(x)$ at $x = 0$, there exists $C_\psi > 0$ such that for $j = 0,\ldots,m$,

$$\|\psi^{(j)}\|_{L_\infty([-\varepsilon,\varepsilon])} := \sup_{x\in[-\varepsilon,\varepsilon]} |\psi^{(j)}(x)| \leqslant C_\psi \varepsilon^{m+1-j}, \qquad \forall\, \varepsilon > 0. \tag{6.1.6}$$

It follows from the definition of ψ that

$$\langle f;\varphi\rangle=\sum_{j=0}^{m}\varphi^{(j)}(0)\langle f;x^j\eta(x)/j!\rangle+\langle f;\psi\rangle=\langle f;\psi\rangle+\sum_{j=0}^{m}c_j\varphi^{(j)}(0)$$

with $c_j:=\langle f;x^j\eta(x)/j!\rangle$. If we can prove that $\langle f;\psi\rangle=0$ is always true, then the above identity implies $f=\sum_{j=0}^{m}c_j(-1)^jD^j\boldsymbol{\delta}$ and we are done. We now use (6.1.5) to prove $\langle f;\psi\rangle=0$. Let $0<\varepsilon<1$. Since $\operatorname{supp}(f)=\{0\}$ and $\eta(x/\varepsilon)=1$ for $x\in[-\varepsilon/4,\varepsilon/4]$, we have $\langle f;\psi\rangle=\langle f;\eta(\cdot/\varepsilon)\psi\rangle$. Note that $\eta(\cdot/\varepsilon)\psi\in\mathscr{D}(\mathbb{R})$ is supported inside $[-\varepsilon,\varepsilon]\subseteq[-N,N]$. Thus, (6.1.5) holds with $h=\eta(\cdot/\varepsilon)\psi$. We now calculate $\|h^{(\beta)}\|_{L_\infty(\mathbb{R})}$. By the Leibniz differentiation formula, for $\beta=0,\ldots,m$,

$$h^{(\beta)}=[\eta(\cdot/\varepsilon)\psi]^{(\beta)}=\sum_{j=0}^{\beta}\frac{\beta!}{j!(\beta-j)!}\varepsilon^{-j}\eta^{(j)}(\cdot/\varepsilon)\psi^{(\beta-j)}.$$

Since $\eta(\cdot/\varepsilon)\psi\in\mathscr{D}(\mathbb{R})$ is supported inside $[-\varepsilon,\varepsilon]$, by (6.1.6) and the above identity,

$$\begin{aligned}\|h^{(\beta)}\|_{L_\infty(\mathbb{R})}&=\|h^{(\beta)}\|_{L_\infty([-\varepsilon,\varepsilon])}\\&\leqslant\sum_{j=0}^{\beta}\frac{\beta!}{j!(\beta-j)!}\varepsilon^{-j}\|\eta^{(j)}(\cdot/\varepsilon)\|_{L_\infty([-\varepsilon,\varepsilon])}\|\psi^{(\beta-j)}\|_{L_\infty([-\varepsilon,\varepsilon])}\\&\leqslant\sum_{j=0}^{\beta}\frac{\beta!}{j!(\beta-j)!}\varepsilon^{-j}\|\eta^{(j)}\|_{L_\infty(\mathbb{R})}C_\psi\varepsilon^{m+1+j-\beta}=\varepsilon^{m+1-\beta}C_\beta\end{aligned}$$

with $C_\beta:=C_\psi\|\eta^{(j)}\|_{L_\infty(\mathbb{R})}\sum_{j=0}^{\beta}\frac{\beta!}{j!(\beta-j)!}$. Now it follows from (6.1.5) that

$$|\langle f;\psi\rangle|=|\langle f;\eta(\cdot/\varepsilon)\psi\rangle|\leqslant C\sum_{\beta=0}^{m}C_\beta\varepsilon^{m+1-\beta}\to 0,\quad\text{as}\quad\varepsilon\to 0^+.$$

Hence, we must have $\langle f;\psi\rangle=0$ and this completes the proof. □

The following auxiliary result will be needed later and characterizes a finitely supported refinable distribution whose support is contained inside $\mathbb{Z}$.

Lemma 6.1.6 *Let ϕ be a compactly supported distribution on $\mathbb{R}$ such that $\operatorname{supp}(\phi)\subseteq\mathbb{Z}$ and $\phi=2\sum_{k\in\mathbb{Z}}a(k)\phi(2\cdot-k)$ for some $a\in l_0(\mathbb{Z})$. Then $\phi=u*(D^m\boldsymbol{\delta})$ for some $m\in\mathbb{N}_0$ and $u\in l_0(\mathbb{Z})$.*

Proof Since $\operatorname{supp}(\phi)\subseteq\mathbb{Z}$ and ϕ is compactly supported, by Lemma 6.1.5, we must have $\phi=\sum_{m=0}^{M}\phi_m$ with $\phi_m=\sum_{k\in\mathbb{Z}}u_m(k)(D^m\boldsymbol{\delta})(\cdot-k)$ for some $M\in\mathbb{N}_0$ and $u_m\in l_0(\mathbb{Z})$. Let $m=0,\ldots,M$ be fixed. For any $\psi\in\mathscr{D}(\mathbb{R})$ such that $\psi^{(j)}(k)=0$ for all $j\in\{0,\ldots,M\}\backslash\{m\}$ and $k\in\mathbb{Z}$, it follows from $\langle\phi(\cdot/2);\psi\rangle=2\sum_{k\in\mathbb{Z}}a(k)\langle\phi(\cdot-k);\psi\rangle$ that $\langle\phi_m(\cdot/2);\psi\rangle=2\sum_{k\in\mathbb{Z}}a(k)\langle\phi_m(\cdot-k);\psi\rangle$. Using Lemma 5.5.10, we conclude that $\phi_m=2\sum_{k\in\mathbb{Z}}a(k)\phi_m(2\cdot-k)$ in the sense of distributions. By the essential uniqueness of the compactly supported distributional solutions to $\phi = 2\sum_{k\in\mathbb{Z}} a(k)\phi(2\cdot - k)$, there exists $m \in \mathbb{N}_0$ such that $\phi_\ell = 0$ for all $\ell \in \{0,\ldots,M\}\backslash\{m\}$. Hence, we must have $\phi = \phi_m = \sum_{k\in\mathbb{Z}} u_m(k)(D^m\delta)(\cdot - k)$. That is, we proved $\phi = u * (D^m\delta)$ with $u = u_m \in l_0(\mathbb{Z})$. □

The following result characterizes a scalar refinable function having an analytic expression.

Theorem 6.1.7 *Let ϕ be a compactly supported distribution on $\mathbb{R}$ such that $\phi = 2\sum_{k\in\mathbb{Z}} a(k)\phi(2\cdot - k)$ for some $a \in l_0(\mathbb{Z})$. If there exists a nonempty open interval I such that $\phi|_{I+k} = f_k$ on $I + k$ for every $k \in \mathbb{Z}$ such that f_k is infinitely differentiable almost everywhere on $I + k$ and $\int_{I+k} |f_k^{(j)}(x)|dx < \infty$ for all $j \in \mathbb{N}_0$, then one of the following two cases must hold:*

(i) there exist $u \in l_0(\mathbb{Z})$ and $m \in \mathbb{N}_0$ such that $\phi = u * (D^m\delta) = \sum_{k\in\mathbb{Z}} u(k)D^m\delta(\cdot - k)$.

(ii) there exists $m \in \mathbb{N}$ such that ϕ can be identified with a compactly supported function in $W_1^{m-1}(\mathbb{R})$ (also denoted by ϕ here for simplicity) such that $\phi^{(m)}(x) = 0$ for almost every $x \in \mathbb{R}$.

Moreover, under the stronger assumption that $\phi|_{I+k} \in \mathscr{C}^\infty(I + k)$ for every $k \in \mathbb{Z}$ (this condition is satisfied if (6.1.4) *holds), then item (ii) can be replaced by*

(ii') There exist $u \in l_0(\mathbb{Z})$ and $m \in \mathbb{N}$ such that $\phi = u * B_m = \sum_{k\in\mathbb{Z}} u(k)B_m(\cdot - k)$.

Proof Without loss of generality, we can assume that $\operatorname{fsupp}(a) = [0, N]$ and $I \cap (0, 1) \neq \emptyset$. The case $N = 0$ is trivial and hence we assume $N \geqslant 1$. We first prove the claim by assuming that the integer shifts of ϕ are linearly independent. Therefore, by Theorem 5.2.7, the Laurent polynomials $\mathsf{a}(z)$ and $\mathsf{a}(-z)$ do not have common zeros. By Lemma 6.1.2, the square matrices $A_\gamma := (2a(2j - k + \gamma))_{0\leqslant j,k\leqslant N-1}$ are invertible for $\gamma = 0, 1$. Define $\Phi := (\phi(\cdot), \phi(\cdot + 1), \ldots, \phi(\cdot + N - 1))^\mathsf{T}$. By the refinement equation $\phi(\frac{\cdot}{2}) = 2\sum_{k=0}^N a(k)\phi(\cdot - k)$, we have

$$\phi(\tfrac{\cdot+\gamma}{2} + j) = 2\sum_{k\in\mathbb{Z}} a(k)\phi(\cdot + 2j - k + \gamma) = \sum_{k\in\mathbb{Z}} 2a(2j - k + \gamma)\phi(\cdot + k)$$

for $j \in \mathbb{Z}$ and $\gamma \in \{0, 1\}$. Since ϕ is supported inside $[0, N]$, we deduce from the above identity that $A_0\Phi = \Phi(\frac{\cdot}{2})$ and $A_1\Phi = \Phi(\frac{\cdot+1}{2})$ on the interval $(0, 1)$ in the sense of distributions. Since A_0 and A_1 are invertible, we have

$$\Phi = A_0^{-1}\Phi(\tfrac{\cdot}{2}), \; \Phi = A_1^{-1}\Phi(\tfrac{\cdot+1}{2}) \text{ on } (0, 1) \text{ in the sense of distributions.} \qquad (6.1.7)$$

Since $I \cap (0, 1)$ is a nonempty open set, there exist $\gamma_1, \ldots, \gamma_n \in \{0, 1\}$ such that $2^{-1}\gamma_1 + 2^{-2}\gamma_2 + \cdots + 2^{-n}\gamma_n + 2^{-n}[0, 1] \subseteq I \cap (0, 1)$. It follows from (6.1.7) that

$$\Phi = A_{\gamma_n}^{-1} \cdots A_{\gamma_1}^{-1}\Phi(2^{-n}\cdot + 2^{-n}\gamma_n + \cdots + 2^{-1}\gamma_1) \quad \text{on } (0, 1). \qquad (6.1.8)$$

This shows that each entry in $\Phi|_{(0,1)}$ can be identified on $(0,1)$ as a Lebesgue measurable function which is infinitely differentiable almost everywhere. We now define a function $\mathring{\phi}$ by $(\mathring{\phi}(x), \mathring{\phi}(x+1), \ldots, \mathring{\phi}(x+N-1))^{\mathsf{T}} := \Phi(x)$ for $x \in (0,1)$ and $\mathring{\phi}(x) = 0$ for all $x \in \mathbb{Z} \cup (\mathbb{R}\backslash[0,N])$. Then $\mathring{\phi}$ is a Lebesgue measurable function and is supported inside $[0,N]$. Considering $x \to 1^-$ and $x \to 0^+$ in (6.1.8), by our assumption that $\int_{I+k} |f_k^{(j)}(x)|dx < \infty$ for all $k \in \mathbb{Z}$, we see that

(1) $\mathring{\phi}$ is infinitely differentiable almost everywhere such that $\mathring{\phi}^{(j)} \in L_1(\mathbb{R})$ for all $j \in \mathbb{N}_0$. Therefore, all $\mathring{\phi}^{(j)}$ can be regarded as compactly supported distributions;
(2) $\mathring{\phi} = \phi$ on $\mathbb{R}\backslash\mathbb{Z}$ in the sense of distributions;
(3) $\mathring{\phi}(x) = 2\sum_{k\in\mathbb{Z}} a(k)\mathring{\phi}(2x-k)$ for a.e. $x \in \mathbb{R}$.

We can assume that ϕ is not identically zero, otherwise, item (i) trivially holds with $u = 0$.

Suppose $\|\mathring{\phi}\|_{L_1(\mathbb{R})} = 0$. By item (2), the distribution ϕ must be supported inside $[0,N] \cap \mathbb{Z}$. Hence, by Lemma 6.1.6, item (i) must hold.

Suppose $\|\mathring{\phi}\|_{L_1(\mathbb{R})} > 0$. Since $\mathring{\phi}$ can be regarded as a distribution, by item (2) and the uniqueness of distributional solutions, we have $\mathring{\phi} = \phi$. Let m be the smallest positive integer such that $\phi^{(m)}(x) = 0$ for a.e. $x \in \mathbb{R}$. By Proposition 6.1.3, we must have $\phi \in W_1^{m-1}(\mathbb{R})$. If $m = \infty$, since the integer shifts of ϕ are linearly independent, by Theorem 5.8.1, $\mathrm{sm}_1(a) = \mathrm{sm}_1(\phi) = \infty$. Therefore, the filter a satisfies sum rules of arbitrarily high orders, which is a contradiction to $a \in l_0(\mathbb{Z})$. This shows that m must be a finite positive number which can be also proved using (5.1.6) of Theorem 5.1.2. This proves item (ii).

Under the stronger assumption that $\phi|_{I+k} \in \mathscr{C}^\infty(I+k)$ for all $k \in \mathbb{Z}$, the same proof shows that item (1) can be replaced by

(1') $\mathring{\phi}|_{(k,k+1)} \in \mathscr{C}^\infty$ for all $k \in \mathbb{Z}$ and $\mathring{\phi}^{(j)} \in L_1(\mathbb{R})$ for all $j \in \mathbb{N}_0$.

Since $\phi^{(m)}|_{(k,k+1)} = 0$, we deduce from item (1') that $\phi^{(m-1)}|_{(k,k+1)}$ is a constant on $(k, k+1)$ for every $k \in \mathbb{Z}$. In other words, we must have $\phi^{(m-1)} = v * B_1$ for some $v \in l_0(\mathbb{Z})\backslash\{0\}$. Since $\phi \in W_1^{m-1}(\mathbb{R})$, the function $\phi^{(m-1)}$ agrees with the $(m-1)$th distributional derivative of ϕ. Therefore, $\widehat{\phi^{(m-1)}}(\xi) = (i\xi)^{m-1}\widehat{\phi}(\xi)$. By $\phi^{(m-1)} = v * B_1$, we have $(i\xi)^{m-1}\widehat{\phi}(\xi) = \widehat{v}(\xi)\widehat{B_1}(\xi)$. Since $\widehat{B_1}(0) = 1 \ne 0$, we must have $\widehat{v}(\xi) = (1-e^{-i\xi})^{m-1}\widehat{u}(\xi)$ for some $u \in l_0(\mathbb{Z})$. Therefore,

$$\widehat{\phi}(\xi) = (i\xi)^{1-m}\widehat{v}(\xi)\widehat{B_1}(\xi) = (\tfrac{1-e^{-i\xi}}{i\xi})^{m-1}\widehat{u}(\xi)\widehat{B_1}(\xi) = \widehat{u}(\xi)\widehat{B_m}(\xi).$$

This proves $\phi = u * B_m$ and we verified item (ii'), under the extra condition that the integer shifts of ϕ are linearly independent.

Though in general the integer shifts of ϕ may not be linearly independent, by Theorem 5.2.4, there exists a compactly supported distribution η such that $\phi = u * \eta$ and $\eta = 2\sum_{k\in\mathbb{Z}} v(k)\eta(2\cdot -k)$ for some $u, v \in l_0(\mathbb{Z})$. Since $\eta|_{I+k}$ is a finite linear combination of $\phi|_{I+n}, n \in \mathbb{Z}$ by $\eta \in \mathrm{S}(\phi)$, the corresponding condition on η instead of ϕ still holds. Now the claim follows directly from what has been proved. □

6.1.3 Refinable Functions Which Are One-Sided Analytic at Integers

In this section we study a refinable vector function which has an analytic expression (or more generally one-sided analytic at integers). As we have seen in Theorem 6.1.7, except the B-splines and finite linear combinations of the integer shifts of the Dirac distribution and their derivatives, scalar refinable functions cannot have analytic expressions in the sense of (6.1.4). As we shall see in this section, we have much more choices for refinable vector functions having analytic expressions. To do so, let us first study the eigenvalues of the transition operators and subdivision operators.

For a subset I of $\mathbb{R}$, we define $P_I : (l(\mathbb{Z}))^{1\times r} \to (l(\mathbb{Z}))^{1\times r}$ by $[P_I u](k) := u(k)$ for $k \in \mathbb{Z} \cap I$ and $[P_I u](k) := 0$ for $k \in \mathbb{Z}\backslash I$. For a complex number λ, a subset I of $\mathbb{R}$ and a positive integer n, using the short-hand notation $\lambda - \mathcal{S}_a$ for $\lambda \mathrm{Id} - \mathcal{S}_a$, we define

$$E_{\lambda,n,I} := \{u \in (l(\mathbb{Z}))^{1\times r} \; : \; P_I(\lambda - \mathcal{S}_a)^n P_I u = 0, \; \operatorname{fsupp}(u) \subseteq I\}, \tag{6.1.9}$$

where the subdivision operator $\mathcal{S}_a : (l(\mathbb{Z}))^{1\times r} \to (l(\mathbb{Z}))^{1\times r}$ is defined in (5.6.30) as $[\mathcal{S}_a u](n) := 2\sum_{k\in\mathbb{Z}} u(k)a(n-2k)$ for $n \in \mathbb{Z}$. By $\operatorname{spec}(\mathcal{S}_a)$ we denote the set of all the eigenvalues $\lambda \in \mathbb{C}$ such that $\mathcal{S}_a v = \lambda v$ for some $v \in (l(\mathbb{Z}))^{1\times r}\backslash\{0\}$.

Recall that the transition operator $\mathcal{T}_a : (l(\mathbb{Z}))^{1\times r} \to (l(\mathbb{Z}))^{1\times r}$ is defined in (5.6.31) as $[\mathcal{T}_a u](n) := 2\sum_{k\in\mathbb{Z}} u(k)\overline{a(k-2n)}^{\mathsf{T}}$ for $n \in \mathbb{Z}$. By $\operatorname{spec}(\mathcal{T}_a)$ we denote the set of all the eigenvalues $\lambda \in \mathbb{C}$ such that $\mathcal{T}_a u = \lambda u$ for some $u \in (l_0(\mathbb{Z}))^{1\times r}\backslash\{0\}$. If λ is a nonzero eigenvalue of $\mathcal{T}_a$ with a nontrivial eigenvector $u \in (l_0(\mathbb{Z}))^{1\times r}$, by $\lambda^n u = \mathcal{T}_a^n u$ and Lemma 5.7.3, we have $\operatorname{fsupp}(u) \subseteq \operatorname{fsupp}(a^\star)$. That is, $\mathcal{T}_a V \subseteq V$ with $V := (l(\operatorname{fsupp}(a^\star)))^{1\times r}$. It is easy to observe that $[P_I \mathcal{S}_a P_I]|_V$ with $I := \operatorname{fsupp}(a^\star)$ is the adjoint operator of $\mathcal{T}_a|_V$. Define $[m, n] := \operatorname{fsupp}(a)$ and

$$v_{r(j+n)+\ell} := \boldsymbol{\delta}(\cdot - j)e_\ell^{\mathsf{T}}, \qquad -n \leqslant j \leqslant -m, 1 \leqslant \ell \leqslant r,$$

where e_j is the jth coordinate unit column vector in $\mathbb{R}^r$. Then $\{v_1, \ldots, v_{r(n-m)}\}$ is a basis of V and the matrix representations of $[P_I \mathcal{S}_a P_I]|_V$ and $\mathcal{T}_a|_V$ under this basis is

$$[P_I \mathcal{S}_a P_I] \begin{bmatrix} v_1 \\ \vdots \\ v_{r(n-m)} \end{bmatrix} = S \begin{bmatrix} v_1 \\ \vdots \\ v_{r(n-m)} \end{bmatrix} \quad \text{and} \quad \mathcal{T}_a \begin{bmatrix} v_1 \\ \vdots \\ v_{r(n-m)} \end{bmatrix} = T \begin{bmatrix} v_1 \\ \vdots \\ v_{r(n-m)} \end{bmatrix},$$

where the $r(n-m) \times r(n-m)$ matrix $T := \left(2\overline{a(j-2k)}^{\mathsf{T}}\right)_{-n \leqslant j,k \leqslant -m} = \overline{S}^{\mathsf{T}}$ and

$$S := \left(2a(k-2j)\right)_{-n \leqslant j,k \leqslant -m} = 2\begin{bmatrix} a(n) & a(n+1) & \cdots & a(2n-m) \\ a(n-2) & a(n-1) & \cdots & a(2n-m-2) \\ a(n-4) & a(n-3) & \cdots & a(2m-n-4) \\ \vdots & \vdots & \ddots & \vdots \\ a(2m-n) & a(2m-n+1) & \cdots & a(m) \end{bmatrix}.$$

The following result establishes the relation of the nonzero eigenvalues of the transition operator and the subdivision operator.

Lemma 6.1.8 *Let $a \in (l_0(\mathbb{Z}))^{r\times r}$ and let I be a closed interval with integer endpoints such that* $\operatorname{fsupp}(a^\star) \subseteq I$ *(note that* $\operatorname{fsupp}(a^\star) = -\operatorname{fsupp}(a)$*). For any $\lambda \in \mathbb{C}\backslash\{0\}$ and $n \in \mathbb{N}$, the mapping $P_I : E_{\lambda,n,\mathbb{Z}} \to E_{\lambda,n,I}$ is a well-defined bijection, where $E_{\lambda,n,I}$ is defined in* (6.1.9). *In particular,* $\operatorname{spec}(\mathcal{T}_a)\backslash\{0\} = \{\overline{\lambda} \neq 0 : \lambda \in \operatorname{spec}(\mathcal{S}_a)\}$.

Proof Let $I_j := I + [-j, j]$ for $j \in \mathbb{N}_0$. Define $P_0 := P_I$ and $P_j := P_{I_j \backslash I_{j-1}}$ for all $j \in \mathbb{N}$. Then $u = \sum_{j=0}^{\infty} P_j u$. We now claim that

$$(\lambda - \mathcal{S}_a)^n u = \sum_{j=0}^{\infty}\sum_{k=0}^{\infty} P_j(\lambda - \mathcal{S}_a)^n P_k u = \sum_{j=0}^{\infty}\sum_{k=0}^{j} P_j(\lambda - \mathcal{S}_a)^n P_k u, \tag{6.1.10}$$

by proving that $P_j(\lambda - \mathcal{S}_a)^n P_k u = 0$ for all $k > j$. Since $(I_k \backslash I_{k-1}) \cap I_j = \emptyset$ for all $k > j$, we have

$$\langle P_j(\lambda - \mathcal{S}_a)^n P_k u, v\rangle = \langle P_k u, (\overline{\lambda} - \mathcal{T}_a)^n P_j v\rangle = 0, \qquad \forall\, v \in (l_0(\mathbb{Z}))^{1\times r},$$

where we used $\operatorname{fsupp}(P_k u) \subseteq I_k \backslash I_{k-1}$, $\operatorname{fsupp}(P_j v) \subseteq I_j$, and $\operatorname{fsupp}((\overline{\lambda} - \mathcal{T}_a)^n P_j v) \subseteq I_j$. Therefore, (6.1.10) holds. Since $\operatorname{fsupp}(a^\star) \subseteq I_j$, by Lemma 5.7.3, we have $\mathcal{T}_a(l(I_j))^{1\times r} \subseteq (l(I_j))^{1\times r}$. Define $W_j := (l(I_j))^{1\times r}/(l(I_{j-1}))^{1\times r}$. It is easy to check that $\mathcal{T}_a|_{W_j}$ is the adjoint operator of $P_j \mathcal{S}_a P_j|_{W_j}$. By Lemma 5.7.3, all the eigenvalues of $\mathcal{T}_a|_{W_j}$ are zero for all $j \in \mathbb{N}$. Therefore, for $\lambda \neq 0$, $\overline{\lambda} - \mathcal{T}_a$ is invertible on W_j. Since $P_j(\overline{\lambda} - \mathcal{T}_a)^n P_j = (\overline{\lambda} - \mathcal{T}_a)^n$ on W_j, $P_j(\overline{\lambda} - \mathcal{T}_a)^n P_j$ is invertible on W_j. As $P_j(\lambda - \mathcal{S}_a)^n P_j$ is the adjoint operator of $P_j(\overline{\lambda} - \mathcal{T}_a)^n P_j$, $P_j(\lambda - \mathcal{S}_a)^n P_j$ is invertible on W_j and hence, $P_j(\lambda - \mathcal{S}_a)^n P_j$ is invertible on $(l(I_j \backslash I_{j-1}))^{1\times r}$ for all $\lambda \neq 0$. Consequently, by (6.1.10), we see that $(\lambda - \mathcal{S}_a)^n u = 0$ with $\lambda \in \mathbb{C}\backslash\{0\}$ if and only if

$$P_0(\lambda - \mathcal{S}_a)^n P_0 u = 0 \quad \text{and}$$

$$P_j u = -\left[P_j(\lambda - \mathcal{S}_a)^n P_j\right]^{-1} \sum_{k=0}^{j-1} P_j(\lambda - \mathcal{S}_a)^n P_k u, \quad j \in \mathbb{N}. \tag{6.1.11}$$

If $u \in E_{\lambda,n,\mathbb{Z}}$, that is, $(\lambda - \mathcal{S}_a)^n u = 0$, then it follows from (6.1.11) that $P_0 u \in E_{\lambda,n,I}$. So, $P_I : E_{\lambda,n,\mathbb{Z}} \to E_{\lambda,n,I}$ is well defined.

Suppose that $(\lambda - \mathcal{S}_a)^n u = 0$ and $P_I u = 0$. By (6.1.11) and induction on j, we must have $P_j u = 0$ for all $j \in \mathbb{N}_0$. Consequently, $u = \sum_{j=0}^{\infty} P_j u = 0$. So, the mapping $P_I : E_{\lambda,n,\mathbb{Z}} \to E_{\lambda,n,I}$ is one-to-one.

We now prove that $P_I : E_{\lambda,n,\mathbb{Z}} \to E_{\lambda,n,I}$ is onto. Let $u_0 \in E_{\lambda,n,I}$. Then $P_0 u_0 = u_0$ and $P_0(\lambda - \mathcal{S}_a)^n P_0 u_0 = 0$. Recursively define u_j as in (6.1.11) by

$$u_j := -[P_j(\lambda - \mathcal{S}_a)^n P_j]^{-1} \sum_{k=0}^{j-1} P_j(\lambda - \mathcal{S}_a)^n P_k u_k, \qquad j \in \mathbb{N}.$$

Let $u := \sum_{j=0}^{\infty} u_j$. Then it is evident that $P_j u = u_j$ for all $j \in \mathbb{N}_0$ and (6.1.11) holds. Therefore, $u \in E_{\lambda,n,\mathbb{Z}}$ and $P_0 u = u_0$. This proves that $P_I : E_{\lambda,n,\mathbb{Z}} \to E_{\lambda,n,I}$ is onto.

By $\dim(E_{\lambda,n,\mathbb{Z}}) = \dim(E_{\lambda,n,I})$ for all $\lambda \in \mathbb{C}\backslash\{0\}$, we see that $\mathrm{spec}(\mathcal{S}_a)\backslash\{0\} = \mathrm{spec}(P_I \mathcal{S}_a P_I)\backslash\{0\}$. Since $(P_I \mathcal{T}_a P_I)^\star = P_I \mathcal{S}_a P_I$, we conclude that

$$\begin{aligned}\mathrm{spec}(\mathcal{T}_a)\backslash\{0\} &= \mathrm{spec}(\mathcal{T}_a|_{(l(I))^{1\times r}})\backslash\{0\} = \{\overline{\lambda} \neq 0 \;:\; \lambda \in \mathrm{spec}(P_I \mathcal{S}_a P_I|_{(l(I))^{1\times r}})\}\\ &= \{\overline{\lambda} \neq 0 \;:\; \lambda \in \mathrm{spec}(\mathcal{S}_a)\}.\end{aligned}$$

This completes the proof. □

We say that a function f is one-sided analytic at $x = x_0$ if there exists $\varepsilon > 0$ such that $f(x) = \sum_{j=0}^{\infty} c_j (x - x_0)^j$ for all $x \in (x_0, x_0 + \varepsilon)$ and $f(x) = \sum_{j=0}^{\infty} d_j (x - x_0)^j$ for all $x \in (x_0 - \varepsilon, x_0)$ with both series converging absolutely on $(x_0 - \varepsilon, x_0 + \varepsilon)$.

We obtain the following necessary conditions for a refinable vector function to have analytic expressions.

Theorem 6.1.9 *Let ϕ be an $r \times 1$ vector of compactly supported measurable functions in $L_1(\mathbb{R})$ such that $\phi = 2\sum_{k\in\mathbb{Z}} a(k)\phi(2\cdot -k)$ for some $a \in (l_0(\mathbb{Z}))^{r\times r}$. Suppose that there exists $\varepsilon > 0$ such that $\phi = f_k$ on $(k - \varepsilon, k) \cup (k, k + \varepsilon)$ in the sense of distributions for every $k \in \mathbb{Z}$, where f_k is a function on $(k-\varepsilon, k) \cup (k, k+\varepsilon)$ such that f_k is one-sided analytic at k (this condition is satisfied if* (6.1.4) *holds). If the integer shifts of ϕ are linearly independent, then*

(i) every nonzero eigenvalue of $\mathcal{T}_a$ in (5.6.31) *takes the form 2^{-j} for some $j \in \mathbb{N}_0$;*
(ii) $E_{2^{-j},n,\mathbb{Z}} = E_{2^{-j},1,\mathbb{Z}}$ for all $n \in \mathbb{N}$ and $j \in \mathbb{N}_0$. Consequently, the algebraic multiplicity of the eigenvalue 2^{-j} of $\mathcal{T}_a$ equals its geometric multiplicity;
(iii) $\dim(E_{2^{-j},1,\mathbb{Z}}) \leqslant 2$ for every $j \in \mathbb{N}_0$. If $\dim(E_{2^{-j},1,\mathbb{Z}}) = 2$ (i.e., 2^{-j} is a double eigenvalue of $\mathcal{T}_a$), then the filter a must satisfy order $j + 1$ sum rules and $\lim_{x\to k^+} \phi^{(j)}(x) \neq \lim_{x\to k^-} \phi^{(j)}(x)$ for some $k \in \mathbb{Z}$;
(iv) If 0 is not an eigenvalue of $\mathcal{T}_a|_{(l(\mathrm{fsupp}(a^\star)))^{1\times r}}$, then $\mathcal{T}_a|_{(l(\mathrm{fsupp}(a^\star)))^{1\times r}}$ is diagonalizable and invertible.

Proof Let $\overline{\lambda}$ be a nonzero eigenvalue of $\mathcal{T}_a$. By Lemma 6.1.8, λ is an eigenvalue of $\mathcal{S}_a$. That is, there exists $u \in E_{\lambda,1,\mathbb{Z}}\backslash\{0\}$ such that $\mathcal{S}_a u = \lambda u$. By $\phi = 2\sum_{k\in\mathbb{Z}} a(k)\phi(2\cdot -k)$, we have

$$\begin{aligned} f_u &:= \sum_{n\in\mathbb{Z}} u(n)\phi(\cdot - n) = 2\sum_{n\in\mathbb{Z}}\sum_{k\in\mathbb{Z}} u(n)a(k)\phi(2\cdot -2n-k) \\ &= \sum_{k\in\mathbb{Z}} 2\sum_{n\in\mathbb{Z}} u(n)a(k-2n)\phi(2\cdot -k) \\ &= \sum_{k\in\mathbb{Z}} [\mathcal{S}_a u](k)\phi(2\cdot -k) = \lambda\sum_{k\in\mathbb{Z}} u(k)\phi(2\cdot -k) = \lambda f_u(2\cdot). \end{aligned}$$

Since the integer shifts of ϕ are linearly independent and $u \neq 0$, we must have $f_u \neq 0$. Since ϕ is one-sided analytic, there exists $\varepsilon > 0$ such that $f_u(x) = \sum_{k=0}^{\infty} c_k x^k$ for $x \in (0,\varepsilon)$ and $f_u(x) = \sum_{k=0}^{\infty} d_k x^k$ for $x \in (-\varepsilon, 0)$. It follows from $f_u = \lambda f_u(2\cdot)$ that

$$c_k = \lambda 2^k c_k \quad \text{and} \quad d_k = \lambda 2^k d_k \qquad \forall\, k \in \mathbb{N}_0. \tag{6.1.12}$$

If $c_k = d_k = 0$ for all $k \in \mathbb{N}_0$, then $f_u = \lambda f_u(2\cdot)$ will imply $f_u = 0$, a contradiction. Hence, there exists $j \in \mathbb{N}_0$ such that $|c_j| + |d_j| \neq 0$. Consequently, (6.1.12) implies $\lambda = 2^{-j}$ and

$$f_u(x) = \begin{cases} c_j x^j, & x > 0, \\ d_j x^j, & x < 0. \end{cases} \tag{6.1.13}$$

This proves item (i).

We now prove $E_{2^{-j},n,\mathbb{Z}} = E_{2^{-j},1,\mathbb{Z}}$ by induction on n. The claim is obviously true for $n = 1$. Suppose that the claim is true for $n-1$. We now prove $E_{2^{-j},n,\mathbb{Z}} = E_{2^{-j},1,\mathbb{Z}}$. Otherwise, $E_{2^{-j},n,\mathbb{Z}} \neq E_{2^{-j},1,\mathbb{Z}}$. Since $E_{2^{-j},1,\mathbb{Z}} \subseteq E_{2^{-j},n,\mathbb{Z}}$ and $E_{2^{-j},n-1,\mathbb{Z}} = E_{2^{-j},1,\mathbb{Z}}$, there exists $v \in E_{2^{-j},n,\mathbb{Z}}\backslash E_{2^{-j},1,\mathbb{Z}}$. Define $u := (2^{-j} - \mathcal{S}_a)v$. Then $(2^{-j} - \mathcal{S}_a)^{n-1}u = (2^{-j} - \mathcal{S}_a)^n v = 0$ and therefore, $u \in E_{2^{-j},n-1,\mathbb{Z}} = E_{2^{-j},1,\mathbb{Z}}$. That is, we must have

$$\mathcal{S}_a u = 2^{-j}u \quad \text{and} \quad \mathcal{S}_a v = 2^{-j}v - u,$$

where we used $u = (2^{-j} - \mathcal{S}_a)v$ for the second identity. Define $f_u := \sum_{k\in\mathbb{Z}} u(k)\phi(\cdot - k)$ and $f_v := \sum_{k\in\mathbb{Z}} v(k)\phi(\cdot - k)$. Since $\mathcal{S}_a u = 2^{-j}u$, by our proof for item (i), (6.1.13) must hold for some $c_j, d_j \in \mathbb{C}$. By $\mathcal{S}_a v = 2^{-j}v - u$, we have

$$f_v(x) = \sum_{k\in\mathbb{Z}} v(k)\phi(x-k) = \sum_{k\in\mathbb{Z}} [\mathcal{S}_a v](k)\phi(2x-k) = 2^{-j} f_v(2x) - f_u(2x). \tag{6.1.14}$$

Since ϕ is one-sided analytic, there exists $\varepsilon > 0$ such that $f_v(x) = \sum_{k=0}^{\infty} \tilde{c}_k x^k$ for $x \in (0, \varepsilon)$ and $f_v(x) = \sum_{k=0}^{\infty} \tilde{d}_k x^k$ for $x \in (-\varepsilon, 0)$. It follows from (6.1.14) that

$$\tilde{c}_k = 2^{k-j}\tilde{c}_k - 2^j c_j \boldsymbol{\delta}(k-j) \quad \text{and} \quad \tilde{d}_k = 2^{k-j}\tilde{d}_k - 2^j d_j \boldsymbol{\delta}(k-j) \qquad \forall\, k \in \mathbb{N}_0.$$

Taking $k = j$ in the above identities, we must have $c_j = d_j = 0$. Hence, $f_u = 0$ which implies $u = 0$. Now it follows from $u = (2^{-j} - \mathcal{S}_a)v$ that $(2^{-j} - \mathcal{S}_a)v = 0$, that is, $v \in E_{2^{-j},1,\mathbb{Z}}$, a contradiction to our choice of $v \in E_{2^{-j},n,\mathbb{Z}} \backslash E_{2^{-j},1,\mathbb{Z}}$. Therefore, we must have $E_{2^{-j},n,\mathbb{Z}} = E_{2^{-j},1,\mathbb{Z}}$. By induction and Lemma 6.1.8, item (ii) holds.

Define F_j to be the space of all functions $f_u := \sum_{k\in\mathbb{Z}} u(k)\phi(\cdot - k)$ with $u \in E_{2^{-j},1,\mathbb{Z}}$ satisfying (6.1.13) for some $c_j, d_j \in \mathbb{C}$. Then we have $\dim(E_{2^{-j},1,\mathbb{Z}}) = \dim(F_j) \leqslant 2$. If $\dim(E_{2^{-j},1,\mathbb{Z}}) = 2$, then $\dim(F_j) = 2$, implying $x^j \in F_j$, i.e., there exists $u \in E_{2^{-j},1,\mathbb{Z}}$ such that $f_u(x) = \sum_{k\in\mathbb{Z}} u(k)\phi(x-k) = x^j$. Since the integer shifts of ϕ are linearly independent, we conclude from Theorem 5.5.4 that a satisfies order $j+1$ sum rules. Item (iv) follows directly from item (ii). □

Example 6.1.1 Let $a_m^B(\xi) := 2^{-m}(1 + e^{-i\xi})^m$ be the B-spline filter of order m. By Theorem 5.8.3, we have

$$\operatorname{spec}(\mathcal{T}_{a_m^B})\backslash\{0\} = \{1, \ldots, 2^{1-m}\} \cup \operatorname{spec}(\mathcal{T}_{2^{-m}\boldsymbol{\delta}}) = \{1, \ldots, 2^{2-m}, 2^{1-m}, 2^{1-m}\}.$$

Since $\operatorname{len}(a) = m$, by the above identity and item (iv) of Theorem 6.1.9, the representation matrix of the operator $\mathcal{T}_{a_m^B}$ on $l([-m, 0])$ is diagonalizable and invertible.

We shall present a few examples of refinable vector functions having analytic expressions in the next section.

6.2 Refinable Hermite Interpolants and Hermite Interpolatory Filters

Refinable vector functions having interpolation properties are of interest in theory and applications. Among them, Hermite interpolants are of particular interest in approximation theory and computational mathematics. In this section we study refinable Hermite interpolants.

For $r \in \mathbb{N}$ and a vector function $\phi = (\phi_1, \ldots, \phi_r)^\mathsf{T}$, we say that ϕ is *an order r Hermite interpolant* if all $\phi_1, \ldots, \phi_r \in \mathscr{C}^{r-1}(\mathbb{R})$ and

$$\left[\phi(k), \phi'(k), \ldots, \phi^{(r-1)}(k)\right] = \boldsymbol{\delta}(k) I_r, \quad \forall\, k \in \mathbb{Z}. \tag{6.2.1}$$

That is, $\phi_\ell^{(j)}(k) = \boldsymbol{\delta}(k)\boldsymbol{\delta}(\ell - j - 1)$ for all $k \in \mathbb{Z}, j = 0, \ldots, r-1$ and $\ell = 1, \ldots, r$. When $r = 1$, we simply say that the scalar function ϕ is interpolating, that is, ϕ is continuous and $\phi(k) = \boldsymbol{\delta}(k)$ for all $k \in \mathbb{Z}$.

6.2.1 Properties and Examples of Refinable Hermite Interpolants

We first present some basic properties of Hermite interpolants.

Lemma 6.2.1 *Let ϕ be an order r Hermite interpolant. For $f \in \mathscr{C}^{r-1}(\mathbb{R})$ and $\lambda > 0$, define a Hermite interpolation operator by*

$$\mathcal{Q}_\lambda f := \sum_{k\in\mathbb{Z}} \big(f(\lambda^{-1}k), f'(\lambda^{-1}k), \ldots, f^{(r-1)}(\lambda^{-1}k)\big)\phi(\lambda\cdot - k). \tag{6.2.2}$$

Then

(1) If $f = u * \phi$ with $u \in (l(\mathbb{Z}))^{1\times r}$, then $u(k) = \big(f(k), f'(k), \ldots, f^{(r-1)}(k)\big)$, $\forall k \in \mathbb{Z}$.
(2) The integer shifts of ϕ are linearly independent and therefore are stable in $L_p(\mathbb{R})$ for all $1 \leqslant p \leqslant \infty$.
(3) $\mathcal{Q}_\lambda f = f$ for all $f \in \mathsf{S}_\lambda(\phi) := \{g(\lambda\cdot) \ : \ g \in \mathsf{S}(\phi)\}$.
(4) $[\mathcal{Q}_\lambda f]^{(j)}(\lambda^{-1}k) = f^{(j)}(\lambda^{-1}k)$ for all $k \in \mathbb{Z}$ and $j = 0, \ldots, r-1$.
(5) If $\mathbb{P}_{m-1} \subseteq \mathsf{S}(\phi)$ with $m \geqslant r$, for $1 \leqslant p \leqslant \infty$, there exists a constant $C > 0$ such that

$$\|\mathcal{Q}_\lambda f - f\|_{L_p(\mathbb{R})} \leqslant C\lambda^{-r}\omega_{m-r}(f^{(r)}, \lambda^{-1})_p, \qquad \forall f \in W_p^r(\mathbb{R}), \lambda > 0.$$

Proof For $f = u * \phi$ with $u \in (l(\mathbb{Z}))^{1\times r}$, by (6.2.1), we have

$$\big(f(n), f'(n), \ldots, f^{(r-1)}(n)\big) = \sum_{k\in\mathbb{Z}} u(k)\big[\phi(n-k), \phi'(n-k), \ldots, \phi^{(r-1)}(n-k)\big] = u(n),$$

for all $n \in \mathbb{Z}$. Hence, item (1) holds.

If $f := u * \phi = 0$ for some $u \in (l(\mathbb{Z}))^{1\times r}$, by item (1), we must have $u(n) = \big(f(n), f'(n), \ldots, f^{(r-1)}(n)\big) = 0$ for all $n \in \mathbb{Z}$. Hence, the integer shifts of ϕ must be linearly independent. Consequently, by Corollary 5.3.5 and Theorem 5.3.4, since ϕ is a continuous compactly supported vector function, the integer shifts of ϕ are stable in $L_p(\mathbb{R})$ for all $1 \leqslant p \leqslant \infty$. This proves item (2)

Items (3) and (4) follow directly from item (1). Since $\mathbb{P}_{m-1} \subseteq \mathsf{S}(\phi)$, by item (1), we must have $\mathcal{Q}_1\mathsf{p} = \mathsf{p}$ for all $\mathsf{p} \in \mathbb{P}_{m-1}$. Now item (5) is a direct consequence of Theorem 5.5.19. □

An example of order r refinable Hermite interpolants is as follows.

Proposition 6.2.2 *For $r \in \mathbb{N}$, a vector function $\varphi = (\varphi_1, \ldots, \varphi_r)^\mathsf{T}$ is defined by:*

$$\varphi_\ell(x) := \begin{cases} (1-x)^r \frac{x^{\ell-1}}{(\ell-1)!} \sum_{j=0}^{r-\ell} \frac{(r+j-1)!}{(r-1)!j!} x^j, & x \in [0,1], \\ (1+x)^r \frac{x^{\ell-1}}{(\ell-1)!} \sum_{j=0}^{r-\ell} \frac{(r+j-1)!}{(r-1)!j!} (-x)^j, & x \in [-1,0), \\ 0, & x \in \mathbb{R}\backslash[-1,1], \end{cases} \tag{6.2.3}$$

for $\ell = 1, \dots, r$. *Then*

(i) *The vector function* φ *is an order* r *Hermite interpolant.*
(ii) *The vector function* φ *is a refinable function satisfying* $\varphi = 2\sum_{k\in\mathbb{Z}} a(k)\varphi(2\cdot -k)$, *where* $a \in (l_0(\mathbb{Z}))^{r\times r}$ *is given by:* $a(k) = 0$ *for all* $k \in \mathbb{Z}\backslash\{-1, 0, 1\}$ *and*

$$a(0) := \operatorname{diag}(2^{-1}, 2^{-2}, \dots, 2^{-r}),\ a(\pm 1) := \left[\varphi(\pm\tfrac{1}{2}), \varphi'(\pm\tfrac{1}{2}), \dots, \varphi^{(r-1)}(\pm\tfrac{1}{2})\right]a(0).$$

(iii) *The function* φ *and the filter* a *satisfy* $\varphi(-x) = \mathsf{S}\varphi(x)$ *and* $a(-k) = \mathsf{S}a(k)\mathsf{S}$ *for all* $k \in \mathbb{Z}$, *where*

$$\mathsf{S} = \operatorname{diag}((-1)^0, (-1)^1, \dots, (-1)^{r-1}).$$

Proof From the definition of φ, it is trivial to see that $\varphi^{(j)}(k) = 0$ for all $k \in \mathbb{Z}\backslash\{0\}$ and $j = 0, \dots, r-1$. Note that $(1-x)^{-r} = \sum_{j=0}^{\infty} \frac{(r+j-1)!}{(r-1)!j!} x^j$ for $|x| < 1$. Therefore,

$$(1-x)^r \sum_{j=0}^{r-\ell} \frac{(r+j-1)!}{(r-1)!j!} x^j = 1 + \mathscr{O}(x^{r-\ell+1}), \qquad x \to 0.$$

Hence, we have $\varphi_\ell(x) = \frac{x^{\ell-1}}{(\ell-1)!} + \mathscr{O}(x^r)$ as $x \to 0$, from which we deduce that $\varphi_\ell \in \mathscr{C}^{r-1}(\mathbb{R})$, $\varphi_\ell^{(\ell-1)}(0) = 1$ and $\varphi_\ell^{(j)}(0) = 0$ for all $j \in \{0, \dots, r-1\}\backslash\{\ell-1\}$. This proves item (i) by showing that φ is an order r Hermite interpolant.

Define $f(x) := \varphi(x/2)$ and $g(x) := 2\sum_{k\in\mathbb{Z}} a(k)\varphi(x-k)$. For every integer k, note that $(f-g)|_{[k,k+1]}$ is a vector of polynomials having degree at most $2r-1$. On the other hand, by the definition of the filter a and item (i), we observe that $(f-g)^{(j)}(k) = 0$ for all $k \in \mathbb{Z}$ and $j = 0, \dots, r-1$. Therefore, $(f-g)|_{[k,k+1]} = (x-k)^r(x-k-1)^r\mathsf{p}_k(x)$ for some vector polynomial p_k. Since $\deg((x-k)^r(x-k-1)^r\mathsf{p}_k(x)) \leqslant 2r-1$ and $\deg((x-k)^r(x-k-1)^r) = 2r$, we must have $\mathsf{p}_k = 0$. Hence, we proved $(f-g)|_{[k,k+1]} = 0$ for all $k \in \mathbb{Z}$. Thus, $f = g$ on the real line. This proves item (ii). Item (iii) can be directly checked. □

We now present a few examples of order r Hermite interpolants φ in Proposition 6.2.2. For $r = 1$, the function $\varphi(x) := \max(0, 1-|x|)$ is the *hat function* such that $\varphi = \frac{1}{2}\varphi(2\cdot+1) + \varphi(2\cdot) + \frac{1}{2}\varphi(2\cdot-1)$ and $\varphi(k) = \boldsymbol{\delta}(k)$ for all $k \in \mathbb{Z}$.

Example 6.2.1 For $r = 2$, the order 2 Hermite interpolant $\varphi = (\varphi_1, \varphi_2)^{\mathsf{T}}$ in Proposition 6.2.2 is the Hermite cubic splines given by

$$\varphi_1(x) = \begin{cases} (1-x)^2(1+2x), & x \in [0,1], \\ (1+x)^2(1-2x), & x \in [-1,0), \\ 0, & \text{otherwise}, \end{cases} \qquad \varphi_2(x) = \begin{cases} (1-x)^2 x, & x \in [0,1], \\ (1+x)^2 x, & x \in [-1,0), \\ 0, & \text{otherwise}. \end{cases}$$

The order 2 Hermite interpolant φ satisfies the refinement equation $\varphi = 2a(-1)\varphi(2\cdot +1) + 2a(0)\varphi(2\cdot) + 2a(1)\varphi(2\cdot -1)$ with

$$a = \left\{ \begin{bmatrix} \frac{1}{4} & \frac{3}{8} \\ -\frac{1}{16} & -\frac{1}{16} \end{bmatrix}, \begin{bmatrix} \frac{1}{2} & 0 \\ 0 & \frac{1}{4} \end{bmatrix}, \begin{bmatrix} \frac{1}{4} & -\frac{3}{8} \\ \frac{1}{16} & -\frac{1}{16} \end{bmatrix} \right\}_{[-1,1]}. \tag{6.2.4}$$

Moreover, we have $\widehat{a}(0) = \operatorname{diag}(1, 2^{-3})$, $\operatorname{spec}(\mathcal{T}_a)\backslash\{0\} = \{1, 2^{-1}, 2^{-2}, 2^{-2}, 2^{-3}, 2^{-3}\}$, and $\mathcal{T}_a|_{(l([-1,1]))^{1\times 2}}$ is diagonalizable and invertible. See Fig. 6.6 for the graphs of the order 2 Hermite interpolant $\varphi = (\varphi_1, \varphi_2)^{\mathsf{T}}$.

Example 6.2.2 For $r = 3$, the order 3 Hermite interpolant $\varphi = (\varphi_1, \varphi_2, \varphi_3)^{\mathsf{T}}$ in Proposition 6.2.2 is given by

$$\varphi_1(x) = (1-x)^3(1+3x+6x^2)\chi_{[0,1]} + (1+x)^3(1-3x+6x^2)\chi_{[-1,0)},$$

$$\varphi_2(x) = (1-x)^3 x(1+3x)\chi_{[0,1]} + (1+x)^3 x(1-3x)\chi_{[-1,0)},$$

$$\varphi_3(x) = \tfrac{1}{2}(1-x)^3 x^2 \chi_{[0,1]} + \tfrac{1}{2}(1+x)^3 x^2 \chi_{[-1,0)}.$$

The order 3 Hermite interpolant φ satisfies the refinement equation $\varphi = 2a(-1)\varphi(2\cdot +1) + 2a(0)\varphi(2\cdot) + 2a(1)\varphi(2\cdot -1)$ with

$$a = \left\{ \begin{bmatrix} \frac{1}{4} & \frac{15}{32} & 0 \\ -\frac{5}{64} & -\frac{7}{64} & \frac{3}{16} \\ \frac{1}{128} & \frac{1}{128} & -\frac{1}{32} \end{bmatrix}, \begin{bmatrix} \frac{1}{2} & 0 & 0 \\ 0 & \frac{1}{4} & 0 \\ 0 & 0 & \frac{1}{8} \end{bmatrix}, \begin{bmatrix} \frac{1}{4} & -\frac{15}{32} & 0 \\ \frac{5}{64} & -\frac{7}{64} & -\frac{3}{16} \\ \frac{1}{128} & -\frac{1}{128} & -\frac{1}{32} \end{bmatrix} \right\}_{[-1,1]}.$$

Moreover, we have $\operatorname{spec}(\widehat{a}(0)) = \{1, 2^{-4}, 2^{-5}\}$,

$$\operatorname{spec}(\mathcal{T}_a)\backslash\{0\} = \{1, 2^{-1}, 2^{-2}, 2^{-3}, 2^{-3}, 2^{-4}, 2^{-4}, 2^{-5}, 2^{-5}\},$$

and $\mathcal{T}_a|_{(l([-1,1]))^{1\times 3}}$ is diagonalizable and invertible. See Fig. 6.1 for the graphs of the order 3 Hermite interpolant $\varphi = (\varphi_1, \varphi_2, \varphi_3)^{\mathsf{T}}$.

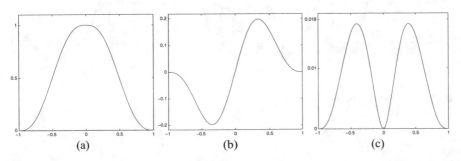

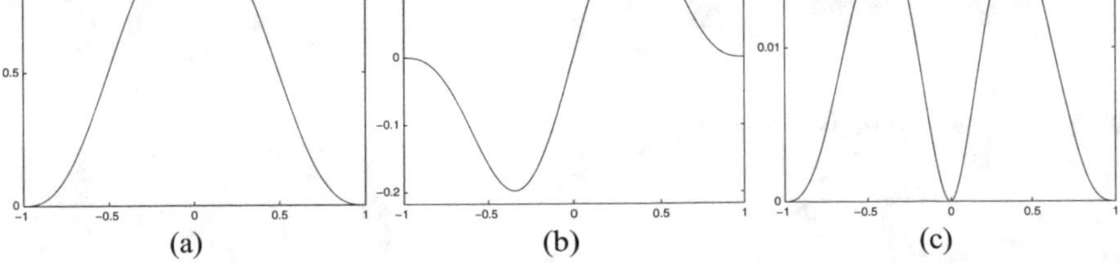

Fig. 6.1 The graphs of the order 3 spline Hermite interpolant $\varphi = (\varphi_1, \varphi_2, \varphi_3)^{\mathsf{T}}$ in Example 6.2.2. **(a)** φ_1. **(b)** φ_2. **(c)** φ_3

Example 6.2.3 For $r = 4$, the order 4 Hermite interpolant $\varphi = (\varphi_1, \varphi_2, \varphi_3, \varphi_4)^\mathsf{T}$ in Proposition 6.2.2 is given by

$$\varphi_1(x) = (1-x)^4(1+4x+10x^2+20x^3)\chi_{[0,1]} + (1+x)^4(1-4x+10x^2-20x^3)\chi_{[-1,0)},$$
$$\varphi_2(x) = (1-x)^4x(1+4x+10x^2)\chi_{[0,1]} + (1+x)^4x(1-4x+10x^2)\chi_{[-1,0)},$$
$$\varphi_3(x) = \tfrac{1}{2}(1-x)^4x^2(1+4x)\chi_{[0,1]} + \tfrac{1}{2}(1+x)^4x^2(1-4x)\chi_{[-1,0)},$$
$$\varphi_4(x) = \tfrac{1}{6}(1-x)^4x^3\chi_{[0,1]} + \tfrac{1}{6}(1+x)^4x^3\chi_{[-1,0)}.$$

The order 4 Hermite interpolant φ satisfies the refinement equation $\varphi = 2a(-1)\varphi(2\cdot +1) + 2a(0)\varphi(2\cdot) + 2a(1)\varphi(2\cdot -1)$ with $a(0) = \operatorname{diag}(1/2, 1/4, 1/8, 1/16)$ and

$$a(-1) = \begin{bmatrix} \frac{1}{4} & \frac{35}{64} & 0 & -\frac{105}{32} \\ -\frac{11}{128} & -\frac{19}{128} & \frac{15}{64} & \frac{105}{64} \\ \frac{3}{256} & \frac{1}{64} & -\frac{7}{128} & -\frac{15}{64} \\ -\frac{1}{1536} & -\frac{1}{1536} & \frac{1}{256} & \frac{3}{256} \end{bmatrix}, \quad a(1) = \begin{bmatrix} \frac{1}{4} & -\frac{35}{64} & 0 & \frac{105}{32} \\ \frac{11}{128} & -\frac{19}{128} & -\frac{15}{64} & \frac{105}{64} \\ \frac{3}{256} & -\frac{1}{64} & -\frac{7}{128} & \frac{15}{64} \\ \frac{1}{1536} & -\frac{1}{1536} & -\frac{1}{256} & \frac{3}{256} \end{bmatrix}.$$

Moreover, we have $\operatorname{spec}(\widehat{a}(0)) = \{1, 2^{-5}, 2^{-6}, 2^{-7}\}$,

$$\operatorname{spec}(\mathcal{T}_a)\backslash\{0\} = \{1, 2^{-1}, 2^{-2}, 2^{-3}, 2^{-4}, 2^{-4}, 2^{-5}, 2^{-5}, 2^{-6}, 2^{-6}, 2^{-7}, 2^{-7}\},$$

and $\mathcal{T}_a|_{(l([-1,1]))^{1\times 4}}$ is diagonalizable and invertible. See Fig. 6.2 for the graphs of the order 4 Hermite interpolant $\varphi = (\varphi_1, \varphi_2, \varphi_3, \varphi_4)^\mathsf{T}$.

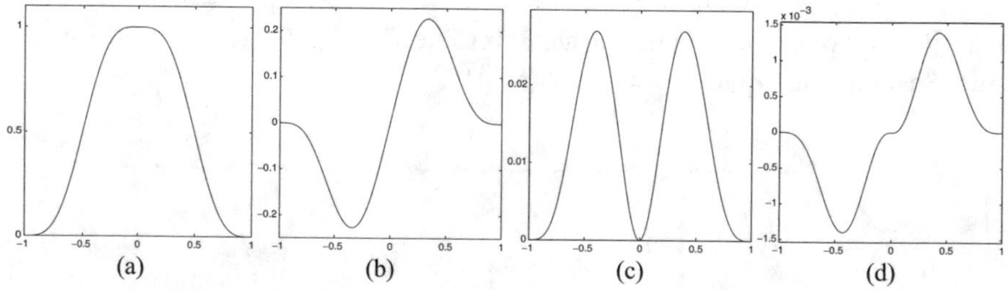

Fig. 6.2 The graphs of the order 4 spline Hermite interpolant $\varphi = (\varphi_1, \varphi_2, \varphi_3, \varphi_4)^\mathsf{T}$ in Example 6.2.3 . **(a)** φ_1. **(b)** φ_2. **(c)** φ_3. **(d)** φ_4

6.2.2 Characterization and Construction of Refinable Hermite Interpolants

The following result completely characterizes refinable Hermite interpolants.

Theorem 6.2.3 *Let $a \in (l_0(\mathbb{Z}))^{r\times r}$ and $\phi = (\phi_1, \dots, \phi_r)^\mathsf{T}$ be an $r \times 1$ vector of compactly supported distributions satisfying $\phi = 2\sum_{k\in\mathbb{Z}} a(k)\phi(2\cdot -k)$. Then ϕ is an order r Hermite interpolant if and only if*

(1) $\widehat{\phi_1}(0) = 1$;
(2) $\operatorname{sm}_\infty(a) > r-1$;
(3) 1 is a simple eigenvalue of $\widehat{a}(0)$ and all other eigenvalues of $\widehat{a}(0)$ are less than 2^{1-r} in modulus;
(4) the filter a is an order r Hermite interpolatory filter, *that is,*

$$a(0) = \operatorname{diag}(2^{-1}, \dots, 2^{-r}) \quad \text{and} \quad a(2k) = 0 \qquad \forall\, k \in \mathbb{Z}\backslash\{0\}, \tag{6.2.5}$$

and a satisfies order r sum rules with a matching filter $\upsilon \in (l_0(\mathbb{Z}))^{1\times r}$ such that

$$\widehat{\upsilon}(\xi) = (1, i\xi, \dots, (i\xi)^{r-1}) + \mathscr{O}(|\xi|^r), \quad \xi \to 0. \tag{6.2.6}$$

Moreover, if ϕ is an order r Hermite interpolant with the refinement filter a, then $\operatorname{sm}_p(\phi) = \operatorname{sm}_p(a)$ for all $1 \leqslant p \leqslant \infty$.

Proof Necessity ($\Rightarrow$). Since ϕ is an order r Hermite interpolant, we have $\phi \in (\mathscr{C}^{r-1}(\mathbb{R}))^r$ and by item (2) of Lemma 6.2.1, the integer shifts of ϕ are stable in $L_\infty(\mathbb{R})$. By Proposition 5.6.2 and Corollary 5.6.12 (also see item (2) of Theorem 5.6.11 with $p = \infty$), we see that

(i) item (3) holds, (5.6.6) holds with $m = r-1$, and $\operatorname{sm}_\infty(a) > r-1$;
(ii) The filter a must satisfy order r sum rules with a matching filter $\upsilon \in (l_0(\mathbb{Z}))^{1\times r}$ and $\widehat{\upsilon}(0)\widehat{\phi}(0) = 1$.

Note that (5.6.6) implies that $(\mathsf{p} * \upsilon) * \phi = \mathsf{p}$ for all $\mathsf{p} \in \mathbb{P}_{r-1}$. Now it follows from item (1) of Lemma 6.2.1 that

$$[\mathsf{p} * \upsilon](k) = [\mathsf{p}(k), \mathsf{p}'(k), \dots, \mathsf{p}^{(r-1)}(k)], \qquad \forall\, \mathsf{p} \in \mathbb{P}_{r-1}, k \in \mathbb{Z}. \tag{6.2.7}$$

By Theorem 1.2.1, we see that (6.2.7) holds if and only if (6.2.6) is satisfied.

Take $\mathsf{p} = 1$ in (6.2.7), we have $1 = (\mathsf{p} * \upsilon) * \phi = \sum_{k\in\mathbb{Z}} \phi_1(\cdot - k)$. Consequently, $\widehat{\phi_1}(0) = \int_\mathbb{R} \phi_1(x)dx = \int_{[0,1]}(\sum_{k\in\mathbb{Z}} \phi_1(x-k))dx = \int_{[0,1]} 1dx = 1$. Thus, item (1) holds.

By $\phi(\cdot/2) = 2\sum_{k\in\mathbb{Z}} a(k)\phi(\cdot - k) \in \mathsf{S}(\phi)$, it follows from item (1) of Lemma 6.2.1 that $2a(k) = [\phi(k/2), 2^{-1}\phi'(k/2), \dots, 2^{1-r}\phi^{(r-1)}(k/2)]$. By (6.2.1), we see that (6.2.5) holds. This proves item (4).

Sufficiency ($\Leftarrow$). Since $\operatorname{sm}_\infty(a) > r-1$ and item (3) is satisfied, by Theorem 5.6.11, the cascade algorithm associated with the filter a converges in $W_\infty^{r-1}(\mathbb{R})$. Take f to be the order r Hermite interpolant φ in Proposition 6.2.2. Since f is an order r Hermite interpolant and f is refinable, by what has been proved for necessity and Proposition 5.6.2, the equation (5.6.6) must hold with $\phi = f$ and $m = r-1$. In particular, we have $f \in \mathscr{F}_{r-1,\upsilon,\infty}$. Therefore, since $f^{(r-1)}$ is continuous, $\mathcal{R}_a^n f$ converges in $W_\infty^{r-1}(\mathbb{R})$ to some vector function $\eta \in (\mathscr{C}^{r-1}(\mathbb{R}))^r$ with $\widehat{\upsilon}(0)\widehat{\eta}(0) = 1$.

By our assumption $\widehat{\phi}_1(0) = 1$, since $\widehat{\upsilon}(0) = (1, 0, \ldots, 0)$ by (6.2.6), we have $\widehat{\upsilon}(0)\widehat{\phi}(0) = 1$. Since item (3) holds, by the uniqueness result in Theorem 5.1.3, we must have $\eta = \phi$. That is, we proved $\lim_{n\to\infty} \|f_n - \phi\|_{(W_\infty^{r-1}(\mathbb{R}))^r} = 0$ with $f_n := \mathcal{R}_a^n f$.

We claim that all f_n are order r Hermite interpolants by proving that if g is an order r Hermite interpolant, then $h := \mathcal{R}_a g$ is also an order r Hermite interpolant. By

$$h = \mathcal{R}_a g = 2\sum_{k\in\mathbb{Z}} a(k) g(2\cdot - k),$$

since g is an order r Hermite interpolant, for $j \in \mathbb{Z}$, we have

$$\begin{aligned}
&\left[h(j), h'(j), \ldots, h^{(r-1)}(j)\right] \\
&= 2\sum_{k\in\mathbb{Z}} a(k)\left[g(2j-k), g'(2j-k), \ldots, g^{(r-1)}(2j-k)\right] \operatorname{diag}(1, 2^1, \ldots, 2^{r-1}) \\
&= 2a(2j) \operatorname{diag}(1, 2^1, \ldots, 2^{r-1}).
\end{aligned}$$

Using (6.2.5), we conclude that $[h(j), h'(j), \ldots, h^{(r-1)}(j)] = \boldsymbol{\delta}(j) I_r$ for all $j \in \mathbb{Z}$. Therefore, the vector function h is an order r Hermite interpolant.

Since f is an order r Hermite interpolant, all the vector functions $f_n = \mathcal{R}_a^n f$ are order r Hermite interpolants. By $\lim_{n\to\infty} \|f_n - \phi\|_{(W_\infty^{r-1}(\mathbb{R}))^r} = 0$ and all $\phi, f_n \in (\mathscr{C}^{r-1}(\mathbb{R}))^r$, we conclude that

$$\left[\phi(j), \phi'(j), \ldots, \phi^{(r-1)}(j)\right] = \lim_{n\to\infty} \left[f_n(j), f_n'(j), \ldots, f_n^{(r-1)}(j)\right] = \boldsymbol{\delta}(j) I_r$$

for all $j \in \mathbb{Z}$. Hence, the refinable vector function ϕ is an order r Hermite interpolant.

Since the integer shifts of a Hermite interpolant ϕ are stable in $L_p(\mathbb{R})$, it follows from Theorem 5.8.1 that $\operatorname{sm}_p(\phi) = \operatorname{sm}_p(a)$ for all $1 \leqslant p \leqslant \infty$. □

By $\widehat{a}(0)\widehat{\phi}(0) = \widehat{\phi}(0)$ and $\widehat{\phi}_1(0) = 1$, we see that 1 must be an eigenvalue of $\widehat{a}(0)$. By Corollary 5.6.10, we see that items (1) and (2) of Theorem 6.2.3 imply item (3) of Theorem 6.2.3. That is, item (3) in Theorem 6.2.3 can be completely removed.

The following is a particular case of Theorem 6.2.3 for scalar refinable interpolating functions.

Corollary 6.2.4 *Let $a \in l_0(\mathbb{Z})$ with $\sum_{k\in\mathbb{Z}} a(k) = 1$. Define $\widehat{\phi}(\xi) := \prod_{j=1}^\infty \widehat{a}(2^{-j}\xi)$, $\xi \in \mathbb{R}$. Then ϕ is an interpolating function (that is, ϕ is continuous and $\phi(k) = \boldsymbol{\delta}(k)$ for all $k \in \mathbb{Z}$) if and only if $\operatorname{sm}_\infty(a) > 0$ and $a(2k) = 2^{-1}\boldsymbol{\delta}(k)$ for all $k \in \mathbb{Z}$.*

For an order r Hermite interpolatory filter, the following result shows that its matching filter has a very special structure and is useful for constructing order r Hermite interpolatory filters.

Lemma 6.2.5 *Let $a \in (l_0(\mathbb{Z}))^{r\times r}$ be an order r Hermite interpolatory filter. For an integer $m \geqslant r$, if the filter a satisfies order m sum rules with a matching filter $\upsilon \in (l_0(\mathbb{Z}))^{1\times r}$ satisfying (6.2.6), then $\widehat{\upsilon}^{(r)}(0) = \cdots = \widehat{\upsilon}^{(m-1)}(0) = 0$, more precisely,*

$$\widehat{\upsilon}(\xi) = (1, i\xi, \dots, (i\xi)^{r-1}) + \mathscr{O}(|\xi|^m), \quad \xi \to 0. \tag{6.2.8}$$

Proof By Lemma 5.5.5, the filter a satisfies order m sum rules with a matching filter υ is equivalent to (5.5.13). Since $\widehat{a^{[0]}}(\xi) = a(0) = \operatorname{diag}(2^{-1}, \dots, 2^{-r})$, the equation in (5.5.13) with $\gamma = 0$ becomes

$$\widehat{\upsilon}(2\xi)\operatorname{diag}(2^{-1}, \dots, 2^{-r}) = \widehat{\upsilon}(2\xi)\widehat{a^{[0]}}(2\xi) = 2^{-1}\widehat{\upsilon}(\xi) + \mathscr{O}(|\xi|^m), \quad \xi \to 0.$$

That is,

$$\widehat{\upsilon}(2\xi)\operatorname{diag}(2^0, \dots, 2^{1-r}) = \widehat{\upsilon}(\xi) + \mathscr{O}(|\xi|^m), \quad \xi \to 0.$$

Therefore, we have

$$\widehat{\upsilon}^{(j)}(0)[2^j \operatorname{diag}(2^0, \dots, 2^{1-r}) - I_r] = 0, \qquad \forall\, j = r, \dots, m-1,$$

from which we have $\widehat{\upsilon}^{(j)}(0) = 0$ for all $j = r, \dots, m-1$. Thus, (6.2.8) holds. □

We now construct a family of order r Hermite interpolatory filters with increasing orders of sum rules.

Theorem 6.2.6 *For $r, m \in \mathbb{N}$, there exists a unique filter $a^{H_r}_{2rm} \in (l_0(\mathbb{Z}))^{r\times r}$ such that*

(i) The matrix-valued filter $a^{H_r}_{2rm}$ is an order r Hermite interpolatory filter and $a^{H_r}_{2rm}$ vanishes outside $[1-2m, 2m-1]$;
(ii) $a^{H_r}_{2rm}$ has order 2rm sum rules with a matching filter $\upsilon \in (l_0(\mathbb{Z}))^{1\times r}$ satisfying

$$\widehat{\upsilon}(\xi) = (1, i\xi, \dots, (i\xi)^{r-1}) + \mathscr{O}(|\xi|^{2rm}), \quad \xi \to 0. \tag{6.2.9}$$

Moreover, all the filters $a^{H_r}_{2rm}$ are real-valued and have the symmetry property:

$$a^{H_r}_{2rm}(-k) = \mathsf{S}a^{H_r}_{2rm}(k)\mathsf{S}, \ \forall\, k \in \mathbb{Z} \quad \textit{with} \quad \mathsf{S} := \operatorname{diag}(1, -1, \dots, (-1)^{r-1}). \tag{6.2.10}$$

Proof Let $a = a^{H_r}_{2rm}$. Since (6.2.5) holds, we have to determine $a^{[1]}(k) = a(1+2k)$ for $k = -m, \dots, m-1$. By Lemma 5.5.5, item (ii) is equivalent to

$$\widehat{\upsilon}(\xi)e^{-i\xi/2}\widehat{a^{[1]}}(\xi) = 2^{-1}\widehat{\upsilon}(\xi/2) + \mathscr{O}(|\xi|^{2rm}), \quad \xi \to 0 \tag{6.2.11}$$

with $\upsilon \in (l_0(\mathbb{Z}))^{1\times r}$ satisfying (6.2.9). In other words, for all $\ell = 0, \dots, r-1$,

$$\sum_{j=0}^{r-1} [e^{-i\xi/2}\widehat{a^{[1]}}(\xi)]_{j+1,\ell+1}(i\xi)^j = \tfrac{1}{2}(i\xi/2)^\ell + \mathscr{O}(|\xi|^{2rm}), \qquad \xi \to 0. \tag{6.2.12}$$

Taking the Taylor expansion of $e^{-i\xi/2}\widehat{a^{[1]}}(\xi)$ at $\xi = 0$, we have

$$e^{-i\xi/2}\widehat{a^{[1]}}(\xi) = \sum_{s=0}^{2rm-1} \Big(\sum_{k=-m}^{m-1} a^{[1]}(k) \frac{(-k-\frac{1}{2})^s}{s!} \Big)(i\xi)^s + \mathscr{O}(|\xi|^{2rm}), \quad \xi \to 0.$$

Therefore, by (6.2.12), as $\xi \to 0$, we have

$$\begin{aligned}
\tfrac{1}{2}(i\xi/2)^\ell &= \sum_{s=0}^{2rm-1} \sum_{j=0}^{r-1} \sum_{k=-m}^{m-1} [a^{[1]}(k)]_{j+1,\ell+1} \frac{(-k-\frac{1}{2})^s}{s!} (i\xi)^{j+s} + \mathscr{O}(|\xi|^{2rm}) \\
&= \sum_{n=0}^{2rm+r-2} \sum_{j=0}^{\min(r-1,n)} \sum_{k=1-m}^{m} [a^{[1]}(-k)]_{j+1,\ell+1} \frac{(k-\frac{1}{2})^{n-j}}{(n-j)!} (i\xi)^n + \mathscr{O}(|\xi|^{2rm}),
\end{aligned}$$

where we used the substitution $n = s + j$ and replaced k by $-k$. Now the above equations are equivalent to: for all $\ell = 0, \dots, r-1$,

$$\sum_{j=0}^{\min(r-1,n)} \sum_{k=1-m}^{m} [a^{[1]}(-k)]_{j+1,\ell+1} \frac{(k-\frac{1}{2})^{n-j}}{(n-j)!} = 2^{-1-\ell}\boldsymbol{\delta}(n-\ell), \quad n = 0, \dots, 2rm-1.$$

Define $\mathsf{q}_n(x) := (x - 1/2)^n/n!$ for $n \in \mathbb{N}_0$. Noting that $\mathsf{q}_n^{(j)} = 0$ for all $j > n$, we can further equivalently express the above equations as: for all $\ell = 0, \dots, r-1$,

$$\sum_{j=0}^{r-1} \sum_{k=1-m}^{m} [a^{[1]}(-k)]_{j+1,\ell+1} \mathsf{q}_n^{(j)}(k) = 2^{-1-\ell}\boldsymbol{\delta}(n-\ell), \quad n = 0, \dots, 2rm-1. \tag{6.2.13}$$

For each fixed $\ell = 0, \dots, r-1$, we can rewrite the system of linear equations in (6.2.13) as the following matrix form: $A\vec{x} = \vec{b}$, where

$$A = \begin{bmatrix}
\mathsf{q}_0(1-m) & \dots & \mathsf{q}_0(m) & \cdots & \mathsf{q}_0^{(r-1)}(1-m) & \dots & \mathsf{q}_0^{(r-1)}(m) \\
\mathsf{q}_1(1-m) & \dots & \mathsf{q}_1(m) & \cdots & \mathsf{q}_1^{(r-1)}(1-m) & \dots & \mathsf{q}_1^{(r-1)}(m) \\
\vdots & \vdots & \vdots & \vdots & \vdots & \vdots & \vdots \\
\mathsf{q}_{2rm-1}(1-m) & \dots & \mathsf{q}_{2rm-1}(m) & \cdots & \mathsf{q}_{2rm-1}^{(r-1)}(1-m) & \dots & \mathsf{q}_{2rm-1}^{(r-1)}(m)
\end{bmatrix}$$

is a $(2rm) \times (2rm)$ square matrix, the vector $\vec{b} = 2^{-1-\ell}(\delta(n-\ell))_{0 \leqslant n < 2rm}$, and

$$\vec{x} = ([a^{[1]}(m-1)]_{1,\ell}, \ldots, [a^{[1]}(-m)]_{1,\ell}, \cdots, [a^{[1]}(m-1)]_{r,\ell}, \ldots, [a^{[1]}(-m)]_{r,\ell})^{\mathsf{T}}.$$

To prove that the linear system $A\vec{x} = \vec{b}$ has a unique solution, it suffices to prove that the square matrix A is invertible, that is, if $(c_0, \ldots, c_{2rm-1})A = 0$ for some $c_0, \ldots, c_{2rm-1} \in \mathbb{C}$, then we must have $c_0 = \cdots = c_{2rm-1} = 0$. Define $\mathsf{p}(x) := \sum_{n=0}^{2rm-1} c_n \mathsf{q}_n(x)$. Then $(c_0, \ldots, c_{2rm-1})A = 0$ simply means

$$\mathsf{p}^{(j)}(k) = 0 \qquad \forall\, j = 0, \ldots, r-1 \quad \text{and} \quad k = 1-m, \ldots, m.$$

Hence, $\mathsf{p}(x) = \mathsf{q}(x) \prod_{k=1-m}^{m} (x-k)^r$ for some polynomial q. But, $\deg(\mathsf{p}) < 2rm$ while $\deg(\prod_{k=1-m}^{m}(x-k)^r) = 2rm$. This forces $\mathsf{q} = 0$ and therefore, $\mathsf{p} = 0$. This proves that $c_0 = \cdots = c_{2rm-1} = 0$. Hence, A is invertible and there is a unique solution $\vec{x} = A^{-1}\vec{b}$. Since both A and $\vec{b}$ are real-valued, the solution $\vec{x}$ is also real-valued. This proves that the filter a is real-valued. It is also easy to verify that the filter $\mathring{a}(k) := \mathsf{S}a(-k)\mathsf{S}$ for $k \in \mathbb{Z}$ also satisfies all the conditions in items (i) and (ii) with a being replaced by $\mathring{a}$. By the uniqueness of the filter a, we must have $\mathring{a} = a$ and hence (6.2.10). □

We now provide an example here with $r = 2$ and $m = 2$.

Example 6.2.4 The order 2 Hermite interpolatory filter $a_8^{H_2}$ in Theorem 6.2.6 is supported inside $[-3, 3]$ and its nonzero coefficients are given by

$$a = \left\{ \begin{bmatrix} \frac{13}{1024} & \frac{5}{1024} \\ -\frac{3}{1024} & -\frac{1}{1024} \end{bmatrix}, \begin{bmatrix} \frac{243}{1024} & \frac{405}{1024} \\ -\frac{81}{1024} & -\frac{81}{1024} \end{bmatrix}, \begin{bmatrix} \frac{1}{2} & 0 \\ 0 & \frac{1}{4} \end{bmatrix}, \begin{bmatrix} \frac{243}{1024} & -\frac{405}{1024} \\ \frac{81}{1024} & -\frac{81}{1024} \end{bmatrix}, \begin{bmatrix} \frac{13}{1024} & -\frac{5}{1024} \\ \frac{3}{1024} & -\frac{1}{1024} \end{bmatrix} \right\}_{[-3,3]}.$$

Then $\operatorname{sr}(a_8^{H_2}) = 8$ and $\operatorname{sm}(a_8^{H_2}) \approx 3.394959$. See Table 6.1 for some smoothness exponents $\operatorname{sm}_2(a_{2rm}^{H_r})$ for $m = 1, \ldots, 8$. See Fig. 6.3 for the graphs of the order 2 Hermite interpolant $\phi = (\phi_1, \phi_2)^{\mathsf{T}}$ with the filter $a_8^{H_2}$.

We now show that the refinable Hermite interpolants in Proposition 6.2.2 are special cases of the family of refinable Hermite interpolants constructed in Theorem 6.2.6.

Corollary 6.2.7 *Let φ be the order r Hermite interpolant in* (6.2.3) *of Proposition 6.2.2. Then $\varphi = 2\sum_{k \in \mathbb{Z}} a_{2r}^{H_r}(k)\varphi(2 \cdot -k)$, that is, the order r Hermite interpolatory filter associated with φ is just the filter $a_{2r}^{H_r}$ in Theorem 6.2.6 with $m = 1$. Moreover, $\operatorname{sm}_p(a_{2r}^{H_r}) = \operatorname{sm}_p(\varphi) = r + 1/p$ for all $1 \leqslant p \leqslant \infty$.*

Table 6.1 The smoothness exponents $\operatorname{sm}_2(a_{2rm}^{H_r})$ for $r = 2, 3, 4$ and $m = 1, \ldots, 8$

m	1	2	3	4	5	6	7	8
$r = 2$	2.5	3.394959	3.761987	3.973156	4.134481	4.266303	4.377792	4.474377
$r = 3$	3.5	4.351051	4.918350	5.334847	5.675887	5.971379	6.235967	6.478022
$r = 4$	4.5	5.232204	5.705415	5.978377	6.162278	6.304943	6.423307	6.524627

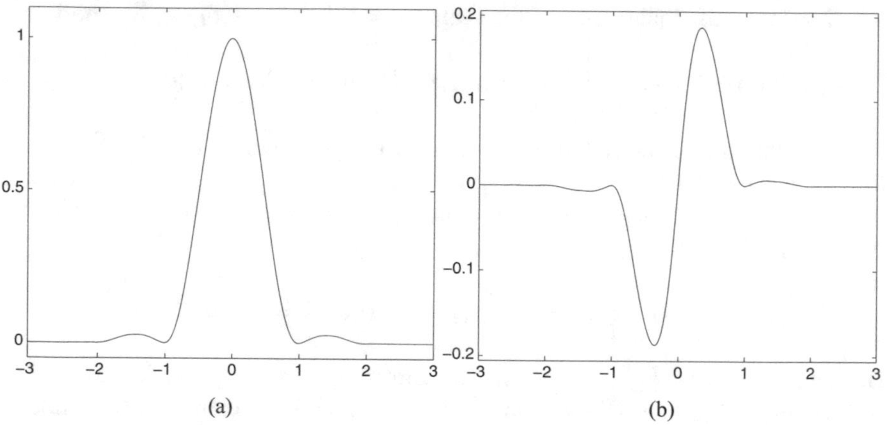

Fig. 6.3 The graphs of the order 2 Hermite interpolant $\phi = (\phi_1, \phi_2)^{\mathsf{T}}$ with the order 2 Hermite interpolatory filter $a_8^{H_2}$ in Example 6.2.4 . (**a**) ϕ_1. (**b**) ϕ_2

Proof By Proposition 6.2.2, we have $\varphi = 2\sum_{k=-1}^{1} a(k)\varphi(2\cdot-k)$. For $\mathsf{p} \in \mathbb{P}_{2r-1}$, we define $\mathsf{q} := \sum_{k\in\mathbb{Z}} \big(\mathsf{p}(k), \mathsf{p}'(k), \ldots, \mathsf{p}^{(r-1)}(k)\big)\varphi(\cdot-k)$. For each integer $k \in \mathbb{Z}$, $\mathsf{q}-\mathsf{p}$ is a polynomial on every interval $[k, k+1]$ and $[\mathsf{q}-\mathsf{p}]^{(j)}(k) = 0 = [\mathsf{q}-\mathsf{p}]^{(j)}(k+1)$ for all $j = 0, \ldots, r-1$. Since $\deg([\mathsf{q}-\mathsf{p}]|_{[k,k+1]}) < 2r$, we must have $[\mathsf{q}-\mathsf{p}]|_{[k,k+1]} = 0$. Therefore, we obtain $\mathsf{q} = \mathsf{p}$ for all $\mathsf{p} \in \mathbb{P}_{2r-1}$. Since the integer shifts of φ are stable, by Theorem 5.5.4, its filter a must satisfy order $2r$ sum rules. By the uniqueness of $a_{2r}^{H_r}$ in Theorem 6.2.6 with $m = 1$, we have $a = a_{2r}^{H_r}$.

Note that $\varphi^{(r-1)} \in \mathscr{C}(\mathbb{R})$ and $\varphi^{(r)}$ is a vector of discontinuous piecewise polynomials (since $\lim_{x\to 1^-} \frac{\varphi_\ell(x)}{(1-x)^r} > 0$). By the definition of $\mathrm{sm}_p(\varphi)$, we have $\mathrm{sm}_p(\varphi) = r + 1/p$ (See Exercise 6.7). Since the integer shifts of φ are stable in $L_p(\mathbb{R})$, by Theorem 5.8.1, we get $\mathrm{sm}_p(a_{2r}^{H_r}) = \mathrm{sm}_p(\varphi) = r + 1/p$ for all $1 \leqslant p \leqslant \infty$. □

Corollary 6.2.8 *$a_{2m}^{I} = a_{2m}^{H_1}$ for all $m \in \mathbb{N}$, where a_{2m}^{I} is defined in Theorem 2.1.3.*

Proof It is straightforward to check that a_{2m}^{I} satisfies all the conditions in Theorem 6.2.6 with $r = 1$. By the uniqueness of $a_{2m}^{H_1}$, we must have $a_{2m}^{I} = a_{2m}^{H_1}$. □

We now provide an example to analyze all order 2 refinable spline Hermite interpolants which are supported inside $[-1, 1]$ and have the symmetry property.

Theorem 6.2.9 *Let $a \in (l_0(\mathbb{Z}))^{2\times 2}$ be an order 2 Hermite interpolatory filter such that* $\mathrm{fsupp}(a) \subseteq [-1, 1]$ *and $a(-k) = \mathsf{S}a(k)\mathsf{S}$ for all $k \in \mathbb{Z}$, where $\mathsf{S} :=$* $\mathrm{diag}(1, -1)$. *Then the filter a must take the following parametric form: $a(k) = 0$ for all $k \in \mathbb{Z}\backslash\{-1, 0, 1\}$ and*

$$a = \left\{ \begin{bmatrix} \frac{1}{4} & \frac{1}{4} - 2t_2 \\ t_1 & t_2 \end{bmatrix}, \begin{bmatrix} \frac{1}{2} & 0 \\ 0 & \frac{1}{4} \end{bmatrix}, \begin{bmatrix} \frac{1}{4} & -\frac{1}{4} + 2t_2 \\ -t_1 & t_2 \end{bmatrix} \right\}_{[-1,1]}, \tag{6.2.14}$$

where $t_1, t_2 \in \mathbb{R}$. *Then* $\mathrm{sm}_p(a) \leqslant 2 + \frac{1}{p}$ *for all* $1 \leqslant p \leqslant \infty$ *(in particular,* $\mathrm{sm}_p(a) = 2+\frac{1}{p}$ *for all* $1 \leqslant p \leqslant \infty$ *if and only if* $t_1 = -\frac{1}{16}$ *and* $-\frac{1}{4} \leqslant t_2 \leqslant 0$*). Let* $\phi = (\phi_1, \phi_2)^\mathsf{T}$ *be its associated compactly supported vector distribution such that* $\widehat{\phi_1}(0) = 1$. *Then* $\phi = (\phi_1, \phi_2)^\mathsf{T}$ *is one-sided analytic at every integer and the integer shifts of* ϕ *are linearly independent if and only if one of the following three cases holds:*

(1) If $t_1 = t_2 = -1/16$*, then the filter* $a = a_4^{H_2}$ *has exactly order* 4 *sum rules,* $\mathrm{sm}_p(a) = 2 + \frac{1}{p}$ *for all* $1 \leqslant p \leqslant \infty$*, and* ϕ *is the order* 2 *(cubic spline) Hermite interpolant in Example 6.2.1. See Fig. 6.6 for the graphs of* $\phi = (\phi_1, \phi_2)^\mathsf{T}$.

(2) If $t_1 = -1/16$ *and* $t_2 = -1/8$*, then the filter a has exactly order* 3 *sum rules and* $\mathrm{sm}_p(a) = 2 + 1/p$ *for all* $1 \leqslant p \leqslant \infty$*. By Theorem 6.2.3, the refinable vector function* ϕ *is an order* 2 *Hermite interpolant and is given by*

$$\begin{aligned}\phi_1(x) &= 2(1+x)^2\chi_{[-1,-1/2]} + (1-2x^2)\chi_{(-1/2,1/2)} + 2(x-1)^2\chi_{[1/2,1]},\\ \phi_2(x) &= -\tfrac{1}{2}(x+1)^2\chi_{[-1,-1/2]} + (\tfrac{3}{2}x^2+x)\chi_{(-1/2,0)}\\ &\qquad + (x-\tfrac{3}{2}x^2)\chi_{[0,1/2)} + \tfrac{1}{2}(x-1)^2\chi_{[1/2,1]}.\end{aligned}$$

See Fig. 6.7 for the graphs of the Hermite interpolant $\phi = (\phi_1, \phi_2)^\mathsf{T}$.

(3) If $t_1 = -1/12$ *and* $t_2 = -1/4$*, then the filter a has exactly order* 2 *sum rules and* $\mathrm{sm}_p(a) = 1 + 1/p$ *for all* $1 \leqslant p \leqslant \infty$*. Though* ϕ *is not an order* 2 *Hermite interpolant, the refinable vector function* ϕ *has an analytic expression given by*

$$\begin{aligned}\phi_1(x) &= (2+3x)\chi_{[-2/3,-1/3]} + \chi_{(-1/3,1/3)} + (2-3x)\chi_{[1/3,2/3]},\\ \phi_2(x) &= -(\tfrac{2}{3}+x)\chi_{[-2/3,-1/3]} + x\chi_{(-1/3,1/3)} + (\tfrac{2}{3}-x)\chi_{[1/3,2/3]}.\end{aligned} \qquad (6.2.15)$$

See Fig. 6.8 for the graphs of $\phi = (\phi_1, \phi_2)^\mathsf{T}$.

Proof By Lemma 6.2.5, the filter a has order 3 sum rules if and only if (5.5.9) holds with $\upsilon \in (l_0(\mathbb{Z}))^{1\times 2}$ satisfying (6.2.8) for $m = 3$. By a simple calculation, we see that the filter a in (6.2.14) has order 3 sum rules if and only if $t_1 = -\frac{1}{16}$. We now show that

$$\mathrm{sm}_p(a) = \tfrac{1}{p} - \log_2 \max(|2t_2 + \tfrac{1}{4}|, \tfrac{1}{4}), \quad \forall\, 1 \leqslant p \leqslant \infty, \quad \text{if } t_1 = -\tfrac{1}{16}. \qquad (6.2.16)$$

Define

$$\widehat{U}(\xi) := \begin{bmatrix} 1 & \frac{1}{2}e^{-i\xi} - \frac{1}{2}e^{i\xi} \\ 0 & 1 \end{bmatrix}, \qquad \widehat{B}(\xi) := \begin{bmatrix} e^{i\xi} & 0 \\ 2 + 2e^{-i\xi} & e^{-i\xi} \end{bmatrix}.$$

Since (6.2.8) holds with $m = 3$, it is trivial to check that $\widehat{\upsilon}(\xi)\widehat{U}(\xi) = (1, \mathscr{O}(|\xi|^3))$ as $\xi \to 0$. Define $\widehat{b}(\xi) := (\widehat{D_3}(2\xi))^{-1}(\widehat{U}(2\xi))^{-1}\widehat{a}(\xi)\widehat{U}(\xi)\widehat{D_3}(\xi)$ as in Theorem 5.8.3, where D_3 is defined in (5.8.8) with $m = 3$. Define u by $\widehat{u}(\xi) := (\widehat{B}(2\xi))^{-1}\widehat{b}(\xi)\widehat{B}(\xi)$.

Then u is supported on $[-1, 1]$ and

$$u = \left\{ \begin{bmatrix} t_2 + \frac{1}{8} & \frac{1}{64} \\ 0 & \frac{1}{8} \end{bmatrix}, \begin{bmatrix} \frac{1}{8} + t_2 & t_2 \\ 0 & 0 \end{bmatrix}, 0 \begin{bmatrix} 0 & \frac{1}{64} \\ 0 & 0 \end{bmatrix} \right\}_{[-1,1]}.$$

By Theorem 5.8.3 and noting that both U and B are strongly invertible, we have $\rho_3(a, \upsilon)_p = \rho_0(b)_p = \rho_0(u)_p$. Since $\mathrm{fsupp}(u) = [-1, 1]$ and the space $(l([-1, 0]))^{1\times 2}$ is invariant under both $\mathcal{T}_0$ and $\mathcal{T}_1$, it follows from Theorem 5.7.6 that $\rho_0(u)_p = \mathrm{jsr}_p(\{T_0, T_1\})$ with

$$T_0 := \begin{bmatrix} 2t_2 + \frac{1}{4} & \frac{1}{32} & 0 & 0 \\ 0 & \frac{1}{4} & 0 & 0 \\ 0 & \frac{1}{32} & 2t_2 + \frac{1}{4} & t_2 \\ 0 & 0 & 0 & 0 \end{bmatrix}, \qquad T_1 := \begin{bmatrix} 2t_2 + \frac{1}{4} & t_2 & \frac{1}{4} + 2t_2 & \frac{1}{32} \\ 0 & 0 & 0 & \frac{1}{4} \\ 0 & 0 & 0 & \frac{1}{32} \\ 0 & 0 & 0 & 0 \end{bmatrix}.$$

By Proposition 5.7.2, we conclude that

$$\begin{aligned} \rho_3(a, \upsilon)_p &= \rho_0(u)_p = \mathrm{jsr}_p(\{T_0, T_1\}) \\ &= \max\left(|2t_2 + \tfrac{1}{4}|, \rho_p\left(\left\{ \begin{bmatrix} \frac{1}{4} & 0 \\ \frac{1}{32} & 2t_2 + \frac{1}{4} \end{bmatrix}, \begin{bmatrix} 0 & 0 \\ 0 & 0 \end{bmatrix} \right\} \right) \right) = \max(|2t_2 + \tfrac{1}{4}|, \tfrac{1}{4}). \end{aligned}$$

If the filter a has order 4 sum rules, then we must have $t_1 = t_2 = -1/16$ which leads to $a = a_4^{H_2}$ and therefore, (6.2.16) holds. Otherwise, the filter a has no more than 3 sum rules and we conclude that (6.2.16) holds.

Suppose that $\mathrm{sm}_p(a) > 2 + 1/p$ for some $1 \leqslant p \leqslant \infty$. By Theorem 5.6.11 or Corollary 5.6.10, we must have $\mathrm{sr}(a) \geqslant 3$. Therefore, (6.2.16) holds, which is a contradiction. Consequently, we must have $\mathrm{sm}_p(a) \leqslant 2 + 1/p$ for all $1 \leqslant p \leqslant \infty$.

When the filter a has order 3 sum rules, we must have $t_1 = -1/16$ and

$$\mathrm{spec}(\mathcal{T}_a) = \{0, 1, 1/2, 1/4, 1/4, 1/4 + 2\overline{t_2}, 1/4 + 2\overline{t_2}\}.$$

If ϕ is one-sided analytic at every integer and the integer shifts of ϕ are linearly independent, by Theorem 6.1.9, we must have $1/4 + 2\overline{t_2} = 0$ which implies $t_2 = -1/8$. This leads to item (2).

If the filter a has exactly order 2 sum rules, by calculation, we have

$$\mathrm{spec}(\mathcal{T}_a) = \{0, 1, 1/2, 1/4 + \overline{t_2 + t_3}, 1/4 + \overline{t_2 + t_3}, 1/4 + \overline{t_2 - t_3}, 1/4 + \overline{t_2 - t_3}\},$$

where $t_3 := \sqrt{1 + 16t_1 - 128t_1t_2 - 8t_2 + 16t_2}/4$. If ϕ is one-sided analytic at every integer and the integer shifts of ϕ are linearly independent, we must have $1/4 + t_2 + t_3 = 0$ and $1/4 + t_2 - t_3 = 0$. Therefore, we must have $t_1 = -1/12$ and $t_2 = -1/4$. For this case, we can directly check that a has exactly order 2 sum rules. Since $\phi \notin (\mathscr{C}^1(\mathbb{R}))^2$, ϕ is not an order 2 Hermite interpolant, but ϕ has an analytic expression given by (6.2.15). Note that ϕ is infinitely differentiable in a neighborhood of every integer and (6.2.1) still holds. One can directly conclude from (6.2.15) that the integer shifts of ϕ are linearly independent and $\operatorname{sm}_p(\phi) = 1 + 1/p$. Consequently, we have $\operatorname{sm}_p(a) = \operatorname{sm}_p(\phi) = 1 + 1/p$ for all $1 \leqslant p \leqslant \infty$. □

6.2.3 *How to Plot Refinable Vector Functions and Their Derivatives*

Let ϕ be a compactly supported vector distribution satisfying $\widehat{\phi}(0) \neq 0$ and $\phi = 2\sum_{k\in\mathbb{Z}} a(k)\phi(2\cdot -k)$ for some $a \in (l_0(\mathbb{Z}))^{r\times r}$. Suppose that $\operatorname{sm}_\infty(a) > 0$. Let $m \in \mathbb{N}_0$ such that $0 \leqslant m < \operatorname{sm}_\infty(a)$. Since 1 is an eigenvalue of $\widehat{a}(0)$ with an eigenvector $\widehat{\phi}(0)$, by Corollary 5.6.10, (5.6.3) holds and $\operatorname{sr}(a) \geqslant m+1$. Therefore, item (i) of Proposition 5.6.2 must hold, that is, there exists a unique sequence $\upsilon \in (l_0(\mathbb{Z}))^{1\times r}$ such that $\operatorname{fsupp}(\upsilon) \subseteq [0, m]$, $\widehat{\upsilon}(0)\widehat{\phi}(0) = 1$ and (5.6.4) holds. We often (re)normalize the refinable vector function ϕ satisfying $\widehat{\phi}(2\xi) = \widehat{a}(\xi)\widehat{\phi}(\xi)$ and $\widehat{\upsilon}(0)\widehat{\phi}(0) = 1$ by additionally requiring that the first nonzero entry of $\widehat{\upsilon}(0)$ should be 1 (another normalization condition $\|\widehat{\phi}(0)\|_{l_2} = 1$ is often used instead for orthogonal multiwavelets). By $\operatorname{sm}_\infty(a) > m$, Theorem 5.6.11 tells us that $\phi \in (\mathscr{C}^m(\mathbb{R}))^r$. We now discuss how to plot the vector function $\phi^{(m)}$ using the subdivision identities $\widehat{\phi}(2^n\xi) = \widehat{a_n}(\xi)\widehat{\phi}(\xi)$, where $\widehat{a_n}(\xi) := \widehat{a}(2^{n-1}\xi)\cdots\widehat{a}(2\xi)\widehat{a}(\xi)$. Note that $2^n a_n = \mathcal{S}_a^n(\delta I_r)$.

(S1) Compute $u_m \in (l_0(\mathbb{Z}))^{1\times r}$ satisfying $\mathcal{T}_a u_m = 2^{-m}u_m$ and $[\widehat{u_m}(\cdot)\overline{\widehat{\upsilon}(\cdot)}^{\mathsf{T}}]^{(m)}(0) = (-i)^m m!$ (necessarily, $\operatorname{fsupp}(u_m) \subseteq \operatorname{fsupp}(a^\star)$). For example, if ϕ is a Hermite interpolant and $0 \leqslant m < r$, then we must have $u_m = \delta e_{m+1}^{\mathsf{T}}$, where $e_{m+1} \in \mathbb{R}^r$ is the column vector with its only nonzero entry 1 at the position $m+1$.

(S2) For a desired choice $n \in \mathbb{N}$ of resolution (e.g., $6 \leqslant n \leqslant 8$), plot the graph of $\phi^{(m)}$ by putting the vector $2^{n(m+1)}[a_n * u_m^\star](k)$ at the position $2^{-n}k$ for all $k \in \mathbb{Z}$.

If $\psi := 2\sum_{k\in\mathbb{Z}} b(k)\phi(2\cdot -k)$ with $b \in (l_0(\mathbb{Z}))^{s\times r}$, then $\psi^{(m)}$ can be plotted by putting $2^{nm+1}[(\mathcal{S}_a^{n-1}b) * u_m^\star](k)$ at the point $2^{-n}k$ for all $k \in \mathbb{Z}$. Note that $\widehat{\mathcal{S}_a^{n-1}b}(\xi) = 2^{n-1}\widehat{b}(2^{n-1}\xi)\widehat{a}(2^{n-2}\xi)\cdots\widehat{a}(2\xi)\widehat{a}(\xi)$. If $b = a$, then $\psi = \phi$ and $\mathcal{S}_a^{n-1}a = 2^{n-1}a_n$.

Under the condition $\operatorname{sm}_\infty(a) > m$, in fact the eigenvector u_m in (S1) must be unique and for (S2) we further have

$$\phi^{(m)}(2^{-n}k) = 2^{n(m+1)}[a_n * u_m^\star](k), \qquad \forall\, k \in \mathbb{Z},\ n \in \mathbb{N}. \tag{6.2.17}$$

Since $\operatorname{sr}(a) \geqslant \operatorname{sm}_\infty(a) > m$, by (5.6.32) and (5.7.18) with $p = \infty$, we see $\rho(\mathcal{T}_a|_V) \leqslant \operatorname{jsr}_\infty(\{\mathcal{T}_0|_V, \mathcal{T}_1|_V\}) = \rho_{m+1}(a, \upsilon)_\infty < 2^{-m}$ with $V := \mathscr{V}_{m,\upsilon} \cap (l(\operatorname{fsupp}(a^\star)))^{1\times r}$. Hence, all $1, 2^{-1}, \ldots, 2^{-m}$ are simple eigenvalues of $\mathcal{T}_a$ while all other eigenvalues of $\mathcal{T}_a$ are less than 2^{-m} in modulus. Note that $\phi^{(m)}(n) = 2^{m+1}\sum_{k\in\mathbb{Z}} a(k)\phi^{(m)}(2n-k)$

for all $n \in \mathbb{Z}$. By (5.6.6), since 2^{-m} is a simple eigenvalue of $\mathcal{T}_a$, there exists a unique sequence u_m in (S1) and hence $u_m^\star(k) = \phi^{(m)}(k)$ for all $k \in \mathbb{Z}$. By $\phi(2^{-n}\cdot) = 2^n \sum_{k\in\mathbb{Z}} a_n(k)\phi(\cdot - k)$, i.e., $\widehat{\phi}(2^n\xi) = \widehat{a_n}(\xi)\widehat{\phi}(\xi)$, we have $\phi^{(m)}(2^{-n}k) = 2^{n(m+1)} \sum_{\ell\in\mathbb{Z}} a_n(\ell)\phi^{(m)}(\ell - k) = 2^{n(m+1)}[a_n * u_m^\star](k)$ for all $k \in \mathbb{Z}$, i.e., (6.2.17) holds.

To avoid computing an eigenvector $u_m \in (l_0(\mathbb{Z}))^{1\times r}$ of $\mathcal{T}_a$, we now discuss a more general way of plotting the vector function $\phi^{(m)}$ by replacing (S1) with (S1'):

(S1') Let $1 \leqslant \kappa \leqslant \operatorname{sr}(a) - m$. Find a sequence $u_m \in (l_0(\mathbb{Z}))^{1\times r}$ satisfying

$$\widehat{u_m}(\xi)\overline{\widehat{\upsilon}(\xi)}^{\mathsf{T}} = (-i\xi)^m + \mathscr{O}(|\xi|^{m+\kappa}), \quad \xi \to 0. \tag{6.2.18}$$

One particular choice of u_m in (S1') is $\widehat{u_m}(\xi) = (-i\xi)^m\overline{\widehat{\phi}(\xi)}^{\mathsf{T}} + \mathscr{O}(|\xi|^{m+\kappa})$ as $\xi \to 0$, which implies (6.2.18) by (5.6.6). Note that $\widehat{\phi}^{(j)}(0)$ with $j \in \mathbb{N}_0$ can be computed through (5.1.11) with $\widehat{\upsilon}(0)\widehat{\phi}(0) = 1$. If $m = 0$, then we often take

$$u_0 = \overline{\widehat{\phi}(0)}^{\mathsf{T}}\boldsymbol{\delta}, \quad \text{or equivalently} \quad u_0^\star = \widehat{\phi}(0)\boldsymbol{\delta}, \quad \text{often with} \quad \|\widehat{\phi}(0)\|_{l_2} = 1,$$

where the vector $\widehat{\phi}(0)$ is obtained by solving the linear equations induced by $\widehat{a}(0)\widehat{\phi}(0) = \widehat{\phi}(0)$. For plotting a refinable vector function ϕ associated with the refinement filter a, the above particular choice of u_0 is often sufficient.

The sequence u_m in (S1) also satisfies (6.2.18) in (S1'). Since $u_m^\star(k) = \phi^{(m)}(k)$ for all $k \in \mathbb{Z}$. Using (5.6.6) and the Poisson summation formula in (5.2.7) with $x = 0$, by $\widehat{\phi^{(m)}}(\xi) = (i\xi)^m\widehat{\phi}(\xi)$, we have

$$\widehat{\upsilon}(\xi)\widehat{u_m^\star}(\xi) = \widehat{\upsilon}(\xi)\sum_{k\in\mathbb{Z}} \phi^{(m)}(k)e^{-ik\xi} = \widehat{\upsilon}(\xi)\sum_{k\in\mathbb{Z}} \widehat{\phi^{(m)}}(\xi+2\pi k) = (i\xi)^m + \mathscr{O}(|\xi|^{\operatorname{sr}(a)})$$

as $\xi \to 0$. By $\widehat{u_m^\star}(\xi) = \overline{\widehat{u_m}(\xi)}^{\mathsf{T}}$, (6.2.18) holds with $\kappa = \operatorname{sr}(a) - m \geqslant 1$ by $\operatorname{sr}(a) \geqslant \operatorname{sm}_\infty(a) > m$. Note that in fact (5.6.6) holds with $m + 1$ being replaced by $\operatorname{sr}(a)$.

To show that the same step (S2) plots the vector function $\phi^{(m)}$, we prove

$$\lim_{n\to\infty} \|2^{n(m+1)}[a_n * u_m^\star](\cdot) - \phi^{(m)}(2^{-n}\cdot)\|_{(l_\infty(\mathbb{Z}))^r} = 0. \tag{6.2.19}$$

The larger the quantity κ in (S1') (provided that $1 \leqslant \kappa \leqslant \operatorname{sm}_\infty(a) - m + 1$), the faster the above series converges (see Corollary 5.6.13). For $m = 0$, we often just take $u_0 = \boldsymbol{\delta}\overline{\widehat{\phi}(0)}^{\mathsf{T}}$ with $\kappa = 1$. Let $\eta = \phi^{a_{2J}^I}$ be the compactly supported refinable interpolating function with the filter a_{2J}^I such that $J \geqslant m + \kappa$ and $\eta \in \mathscr{C}^{m+\kappa}(\mathbb{R})$. Define $\eta_m := u_m^\star * \eta = \sum_{k\in\mathbb{Z}} u_m^\star(k)\eta(\cdot - k)$. Let $u_0 \in (l_0(\mathbb{Z}))^{1\times r}$ be a sequence satisfying (6.2.18) with $m = 0$. Define $\eta_0 := u_0^\star * \eta$. Since $\eta \in \mathscr{C}^{m+\kappa}(\mathbb{R})$ and

$$\widehat{\eta}(\xi + 2\pi k) = \boldsymbol{\delta}(k) + \mathscr{O}(|\xi|^{m+\kappa}), \quad \xi \to 0, \ \forall k \in \mathbb{Z}, \tag{6.2.20}$$

it is trivial to check that $\eta_0 \in \mathscr{F}_{m,\upsilon,\infty}$ (in fact, $\eta_0 \in \mathscr{F}_{m+\kappa-1,\upsilon,\infty}$). By $\operatorname{sm}_\infty(a) > m$ and Theorem 5.6.11, we have $\lim_{n\to\infty} \|f_n^{(m)} - \phi^{(m)}\|_{(\mathscr{C}(\mathbb{R}))^r} = 0$, where by $\eta_0 = u_0^\star * \eta$,

$$f_n := \mathcal{R}_a^n \eta_0 = 2^n \sum_{k\in\mathbb{Z}} a_n(k)\eta_0(2^n \cdot -k) = 2^n \sum_{k\in\mathbb{Z}} [a_n * u_0^\star](k)\eta(2^n \cdot -k).$$

By $\eta_m = u_m^\star * \eta$, for all $n \in \mathbb{N}$, we define

$$g_n := 2^{n(m+1)} \sum_{k\in\mathbb{Z}} a_n(k)\eta_m(2^n \cdot -k) = 2^{n(m+1)} \sum_{k\in\mathbb{Z}} [a_n * u_m^\star](k)\eta(2^n \cdot -k).$$

We now prove $\lim_{n\to\infty} \|g_n - \phi^{(m)}\|_{(\mathscr{C}(\mathbb{R}))^r} = 0$. Since $\lim_{n\to\infty} \|f_n^{(m)} - \phi^{(m)}\|_{(\mathscr{C}(\mathbb{R}))^r} = 0$, it suffices to prove that $\lim_{n\to\infty} \|f_n^{(m)} - g_n\|_{(\mathscr{C}(\mathbb{R}))^r} = 0$. By the definition of f_n, we deduce that $f_n^{(m)} = 2^{n(m+1)} \sum_{k\in\mathbb{Z}} a_n(k)\eta_0^{(m)}(2^n \cdot -k)$. Therefore,

$$f_n^{(m)} - g_n = 2^{n(m+1)} \sum_{k\in\mathbb{Z}} a_n(k) g(2^n \cdot -k) \quad \text{with} \quad g := \eta_0^{(m)} - \eta_m.$$

Since $g \in (\mathscr{C}(\mathbb{R}))^r$ has compact support and $\widehat{\eta_0^{(m)}}(\xi) = (i\xi)^m \widehat{\eta_0}(\xi)$, by the conditions in (6.2.18) and (6.2.20), we have

$$\widehat{\upsilon}(\xi)\widehat{g}(\xi) = (i\xi)^m \widehat{\upsilon}(\xi)\widehat{u_0^\star}(\xi)\widehat{\eta}(\xi) \quad \widehat{\upsilon}(\xi)\widehat{u_m^\star}(\xi)\widehat{\eta}(\xi) = (i\xi)^m - (i\xi)^m + \mathscr{O}(|\xi|^{m+\kappa})$$

as $\xi \to 0$. We conclude from the above identity and (6.2.20) that $\widehat{\upsilon}(\xi)\widehat{g}(\xi + 2\pi k) = \mathscr{O}(|\xi|^{m+\kappa})$ as $\xi \to 0$ for all $k \in \mathbb{Z}$. Since $\kappa \geqslant 1$, by item (iv) of Theorem 5.6.4, there exists a compactly supported vector function $h \in (\mathcal{L}_\infty(\mathbb{R}))^r$ such that $g = V^\star * h$ with $V^\star := [\upsilon_1^\star, \ldots, \upsilon_r^\star]$ for some $\upsilon_1, \ldots, \upsilon_r \in \mathscr{V}_{m+\kappa-1,\upsilon} \subseteq \mathscr{V}_{m,\upsilon}$. Consequently, we have

$$f_n^{(m)} - g_n = 2^{n(m+1)} \sum_{k\in\mathbb{Z}} a_n(k)(V^\star * h)(2^n \cdot -k) = 2^{n(m+1)} \sum_{k\in\mathbb{Z}} [a_n * V^\star](k) h(2^n \cdot -k).$$

Let $C_1 := \|h\|_{(\mathcal{L}_\infty(\mathbb{R}))^r}$. By Lemma 5.3.1, we have

$$\|f_n^{(m)} - g_n\|_{(\mathscr{C}(\mathbb{R}))^r} \leqslant C_1 2^{n(m+1)} \|a_n * V^\star\|_{(l_\infty(\mathbb{Z}))^{r\times r}}.$$

By $\operatorname{sr}(a) \geqslant \operatorname{sm}_\infty(a) > m$ and (5.7.19) in Theorem 5.7.6, we have $\rho_{m+1}(a, \upsilon)_\infty = \max(2^{-m-1}, 2^{-\operatorname{sm}_\infty(a)}) < 2^{-m}$. Therefore, by $V^\star = [\upsilon_1^\star, \ldots, \upsilon_r^\star]$ with $\upsilon_1, \ldots, \upsilon_r \in \mathscr{V}_{m,\upsilon}$, for $0 < \epsilon < \min(1, \operatorname{sm}_\infty(a) - m)$, there exists a constant $C > 0$ such that $\|a_n * V^\star\|_{(l_\infty(\mathbb{Z}))^{r\times r}} \leqslant C 2^{-n\epsilon} 2^{-n(m+1)}$ for all $n \in \mathbb{N}$. Therefore, for all $n \in \mathbb{N}$,

$$\|f_n^{(m)} - g_n\|_{(\mathscr{C}(\mathbb{R}))^r} \leqslant C_1 2^{n(m+1)} \|a_n * V^\star\|_{(l_\infty(\mathbb{Z}))^{r\times r}} \leqslant C C_1 2^{-\epsilon n},$$

from which we conclude that $\lim_{n\to\infty} \|f_n^{(m)} - g_n\|_{(\mathscr{C}(\mathbb{R}))^r} = 0$. Consequently, we proved that $\lim_{n\to\infty} \|g_n - \phi^{(m)}\|_{(\mathscr{C}(\mathbb{R}))^r} = 0$. Since η is interpolating (i.e., $\eta(k) = \boldsymbol{\delta}(k)$ for all $k \in \mathbb{Z}$), we observe that $g_n(2^{-n}k) = 2^{n(m+1)}[a_n * u_m^\star](k)$ for all $k \in \mathbb{Z}$. This proves (6.2.19). Also see Theorem 7.3.1 for multivariate subdivision schemes.

One trivial choice of u_m satisfying (6.2.18) with $\kappa = 1$ is $\widehat{u_m}(\xi) := (1 - e^{i\xi})^m \widehat{u_0}(\xi)$, since $\widehat{u_0}(\xi)\overline{\widehat{v}(\xi)}^{\mathsf{T}} = 1 + \mathscr{O}(|\xi|)$ and $(1 - e^{i\xi})^m = (-i\xi)^m + \mathscr{O}(|\xi|^{m+1})$ as $\xi \to 0$. Since $u_m^\star = (\nabla^m \boldsymbol{\delta}) * u_0^\star$, we have $2^{n(m+1)} a_n * u_m^\star = 2^{n(m+1)}\nabla^m[a_n * u_0^\star]$. That is, we can use $f_n^{(m)} \approx 2^{nm}\nabla^m_{2^{-n}} f_n = 2^{n(m+1)} \sum_{k\in\mathbb{Z}} (\nabla^m[a_n * u_0^\star])(k)\eta(2^n \cdot -k)$ to plot $f_n^{(m)}$.

For a given sequence $v = (v_1, \dots, v_r) \in (l(\mathbb{Z}))^{1\times r}$, it is desirable to find a smooth function f such that $f^{(\ell)}(k) = v_{\ell+1}(k)$ for all $k \in \mathbb{Z}$ and $0 \leqslant \ell < r$. If ϕ is an order r refinable Hermite interpolant with an order r Hermite interpolatory filter $a \in (l_0(\mathbb{Z}))^{r\times r}$, then we can define a row vector function $g := v * \Phi$ with $\Phi := [\phi, \phi', \dots, \phi^{(r-1)}]$. Let f be the first entry of g. Then it follows directly from the Hermite interpolation property that $g = (f, f', \dots, f^{(r-1)})$ and $f^{(\ell)}(k) = v_{\ell+1}(k)$ for all $k \in \mathbb{Z}$ and $\ell = 0, \dots, r-1$. The above procedure for plotting the refinable vector function ϕ and its derivatives leads to a Hermite subdivision scheme for plotting the vector function g. Since ϕ is an order r Hermite interpolant, for (S1) we must have $u_m = \boldsymbol{\delta} e_{m+1}^{\mathsf{T}}$ for all $m = 0, \dots, r-1$. Then the identity (6.2.17) in (S2) leads to

$$\Phi(2^{-n}k) = 2^n a_n(k)\Lambda^{-n} = [\mathcal{S}_a^n(\boldsymbol{\delta} I_r)](k)\Lambda^{-n} \quad \text{with} \quad \Lambda = \operatorname{diag}(1, 2^{-1}, \dots, 2^{1-r})$$

for all $k \in \mathbb{Z}$ and $n \in \mathbb{N}$. Therefore,

$$g(2^{-n}k) = (v * \Phi)(2^{-n}k) = 2^n[(v \uparrow 2^n) * a_n](k)\Lambda^{-n} = [\mathcal{S}_a^n v](k)\Lambda^{-n}.$$

Define $v_0 := v$ and $v_n := [\mathcal{S}_a^n v]\Lambda^{-n} = g(2^{-n}\cdot)$ for $n \in \mathbb{N}_0$. Then the above identities become

$$v_n = [\mathcal{S}_a(\mathcal{S}_a^{n-1} v)]\Lambda^{-n} = [\mathcal{S}_a(v_{n-1}\Lambda^{n-1})]\Lambda^{-n} = \mathcal{S}_{\Lambda^{n-1}a\Lambda^{-n}} v_{n-1}.$$

The above iterative algorithm is called *the Hermite subdivision scheme* to generate the desired function/curve $f := v * \phi$. For an order r Hermite interpolatory filter a, it is trivial that the interpolation property $v_n(2k) = v_{n-1}(k)$ holds for all $k \in \mathbb{Z}$.

6.3 Compactly Supported Refinable Functions in $H^\tau(\mathbb{R})$ with $\tau \in \mathbb{R}$

To study framelets and wavelets in Sobolev spaces $H^\tau(\mathbb{R})$ with $\tau \in \mathbb{R}$, in this section we study and completely characterize compactly supported refinable (vector) functions in $H^\tau(\mathbb{R})$ with $\tau \in \mathbb{R}$. For $\tau \in \mathbb{R}$, recall that $f \in H^\tau(\mathbb{R})$ if

$$\|f\|^2_{H^\tau(\mathbb{R})} := \frac{1}{2\pi}\int_{\mathbb{R}} |\widehat{f}(\xi)|^2(1 + |\xi|^2)^\tau d\xi < \infty.$$

As we proved in Lemma 5.5.17, if $\tau = m$ for a nonnegative integer m, then the above definition agrees with the definition of $W_2^m(\mathbb{R})$ given at the beginning of Sect. 5.4 using distributional derivatives.

For a tempered distribution f on $\mathbb{R}$ such that $\widehat{f}$ is a (Lebesgue) measurable function, we define

$$\mathrm{sm}(f) := \sup\{\tau \in \mathbb{R} \; : \; f \in H^\tau(\mathbb{R})\}. \tag{6.3.1}$$

In case that the set on the right-hand side of (6.3.1) is empty, we simply define $\mathrm{sm}(f) = -\infty$. If $f = (f_1, \dots, f_r)^\mathsf{T}$, we define $\mathrm{sm}(f) := \min(\mathrm{sm}(f_1), \dots, \mathrm{sm}(f_r))$.

The following result will be needed later.

Lemma 6.3.1 *For $m \in \mathbb{N}_0$ and $0 \leqslant \tau < 1$, then $f \in H^{m+\tau}(\mathbb{R})$ implies that f, Df, ..., $D^m f \in L_2(\mathbb{R})$ and* $\lim_{\lambda\to 0^+} \lambda^{-\tau}\omega_1(D^m f, \lambda)_2 = 0$.

Proof Since $f \in H^{m+\tau}(\mathbb{R})$ implies $f \in H^m(\mathbb{R})$ by $\tau \geqslant 0$, by Proposition 5.5.16 and Lemma 5.5.17, we have $f, Df, \dots, D^m f \in L_2(\mathbb{R})$. Define $g := D^m f$. For $t \in \mathbb{R}$, by $|1 - e^{-it\xi}| = 2|\sin(t\xi/2)| \leqslant \min(|t\xi|, 2) \leqslant \min(|t\xi|, 2^{1-\tau}|t\xi|^\tau)$, we have

$$\begin{aligned} 2\pi \|g - g(\cdot - t)\|^2_{L_2(\mathbb{R})} &= \int_{\mathbb{R}} |1 - e^{-it\xi}|^2 |\widehat{g}(\xi)|^2 d\xi \\ &\leqslant |t|^{2\tau} \int_{\mathbb{R}} \min(|t\xi|^{2-2\tau}, 2^{2-2\tau}) |\xi|^{2\tau} |\widehat{g}(\xi)|^2 d\xi. \end{aligned}$$

The above inequality implies

$$(\omega_1(g, \lambda)_2)^2 \leqslant \frac{|\lambda|^{2\tau}}{2\pi} \int_{\mathbb{R}} \min(|\lambda\xi|^{2-2\tau}, 2^{2-2\tau}) |\xi|^{2\tau} |\widehat{g}(\xi)|^2 d\xi. \tag{6.3.2}$$

Since $\widehat{g}(\xi) = (i\xi)^m \widehat{f}(\xi)$ and $\int_{\mathbb{R}} |\widehat{f}(\xi)|^2 (1 + |\xi|^2)^{m+\tau} d\xi < \infty$, we have

$$\begin{aligned} \min(|\lambda\xi|^{2-2\tau}, 2^{2-2\tau}) |\xi|^{2\tau} |\widehat{g}(\xi)|^2 &\leqslant 2^{2-2\tau} |\xi|^{2\tau} |\widehat{g}(\xi)|^2 = 2^{2-2\tau} |\xi|^{2m+2\tau} |\widehat{f}(\xi)|^2 \\ &\leqslant 2^{2-2\tau} (1 + |\xi|^2)^{m+\tau} |\widehat{f}(\xi)|^2 \in L_1(\mathbb{R}). \end{aligned}$$

By the Dominated Convergence Theorem and noting that $2 - 2\tau > 0$, we have

$$\begin{aligned} &\lim_{\lambda\to 0+} \int_{\mathbb{R}} \min(|\lambda\xi|^{2-2\tau}, 2^{2-2\tau}) |\xi|^{2\tau} |\widehat{g}(\xi)|^2 d\xi \\ &\quad = \int_{\mathbb{R}} \lim_{\lambda\to 0+} \min(|\lambda\xi|^{2-2\tau}, 2^{2-2\tau}) |\xi|^{2\tau} |\widehat{g}(\xi)|^2 d\xi = 0. \end{aligned}$$

It follows from the above identity and (6.3.2) that $\lim_{\lambda\to 0+} \lambda^{-\tau}\omega_1(g, \lambda)_2 = 0$. ☐

For a compactly supported function in $H^\tau(\mathbb{R})$, we have the following result.

Lemma 6.3.2 *Let $\tau \in \mathbb{R}$ and f be a compactly supported distribution on $\mathbb{R}$. Then $f \in H^\tau(\mathbb{R})$ if and only if $[\widehat{f}, \widehat{f}]_\tau := \sum_{k\in\mathbb{Z}} |\widehat{f}(\cdot + 2\pi k)|^2(1 + |\cdot + 2\pi k|^2)^\tau \in L_\infty(\mathbb{T})$.*

Proof The sufficiency part ($\Leftarrow$) is trivial by $\|f\|_{H^\tau(\mathbb{R})} = \|[\widehat{f}, \widehat{f}]_\tau\|_{L_2(\mathbb{T})} \leqslant \|[\widehat{f}, \widehat{f}]_\tau\|_{L_\infty(\mathbb{T})}$.

Necessity ($\Rightarrow$). Since f has compact support, we can take a compactly supported function $\eta \in \mathscr{D}(\mathbb{R})$ such that η takes value $(2\pi)^{-1}$ on the support of f. Therefore, $f = 2\pi f\eta$, which implies $\widehat{f} = \widehat{f} * \widehat{\eta}$. Since $\eta \in \mathscr{D}(\mathbb{R})$, there exists $C_1 > 0$ such that $|\widehat{\eta}(\xi)| \leqslant C_1(1 + |\xi|^2)^{-1-|\tau|/2}$ for all $\xi \in \mathbb{R}$. Thus,

$$
\begin{aligned}
|\widehat{f}(\xi)|^2 = |[\widehat{f} * \widehat{\eta}](\xi)|^2 &= \left|\int_{\mathbb{R}} \widehat{f}(\zeta)\widehat{\eta}(\xi - \zeta)\, d\zeta\right|^2 \\
&\leqslant C_1^2 \left|\int_{\mathbb{R}} |\widehat{f}(\zeta)|(1 + |\xi - \zeta|^2)^{-1-|\tau|/2}\, d\zeta\right|^2 \\
&\leqslant C_1^2 \int_{\mathbb{R}} (1 + |\xi - \zeta|^2)^{-1}\, d\zeta \int_{\mathbb{R}} |\widehat{f}(\zeta)|^2(1 + |\xi - \zeta|^2)^{-1-|\tau|}\, d\zeta \\
&= \frac{C_2}{2\pi} \int_{\mathbb{R}} |\widehat{f}(\zeta)|^2(1 + |\xi - \zeta|^2)^{-1-|\tau|}\, d\zeta,
\end{aligned}
$$

where $C_2 := 2\pi C_1^2 \int_{\mathbb{R}}(1 + |\zeta|^2)^{-1}\, d\zeta < \infty$. Now we have the following estimate

$$
\begin{aligned}
[\widehat{f}, \widehat{f}]_\tau(\xi) &= \sum_{k\in\mathbb{Z}} |\widehat{f}(\xi + 2\pi k)|^2(1 + |\xi + 2\pi k|^2)^\tau \\
&\leqslant \frac{C_2}{2\pi} \sum_{k\in\mathbb{Z}} \int_{\mathbb{R}} |\widehat{f}(\zeta)|^2(1 + |\xi + 2\pi k - \zeta|^2)^{-1-|\tau|}(1 + |\xi + 2\pi k|^2)^\tau\, d\zeta \\
&= C_2 \frac{1}{2\pi} \int_{\mathbb{R}} |\widehat{f}(\zeta)|^2(1 + |\zeta|^2)^\tau A(\xi, \zeta)\, d\zeta,
\end{aligned}
$$

where

$$
A(\xi, \zeta) := \sum_{k\in\mathbb{Z}} \frac{1}{1 + |\xi + 2\pi k - \zeta|^2} \left(\frac{(1 + |\xi + 2\pi k|^2)^\tau}{(1 + |\zeta|^2)^\tau(1 + |\xi + 2\pi k - \zeta|^2)^{|\tau|}}\right).
$$

Let

$$
B(x, y) := \frac{1 + |x|^2}{(1 + |y|^2)(1 + |x - y|^2)}, \qquad x, y \in \mathbb{R}.
$$

Then

$$
2(1+|y|^2)(1+|x-y|^2) - (1+|x|^2) = 1+|x-2y|^2+2|y|^2|x-y|^2 > 0 \quad \forall\, x, y \in \mathbb{R}.
$$

It follows from the above inequality that $B(x, y) \leqslant 2$ for all $x, y \in \mathbb{R}$. Note that

$$\frac{(1+|\xi+2\pi k|^2)^\tau}{(1+|\zeta|^2)^\tau(1+|\xi+2\pi k-\zeta|^2)^{|\tau|}} = \begin{cases} [B(\xi+2\pi k,\zeta)]^\tau, & \text{if } \tau \geqslant 0, \\ [B(\zeta,\xi+2\pi k)]^{-\tau}, & \text{if } \tau < 0. \end{cases}$$

Now we can estimate $A(\xi, \zeta)$ as follows:

$$\begin{aligned} A(\xi,\zeta) &= \sum_{k\in\mathbb{Z}} \frac{1}{1+|\xi+2\pi k-\zeta|^2}\left(\frac{(1+|\xi+2\pi k|^2)^\tau}{(1+|\zeta|^2)^\tau(1+|\xi+2\pi k-\zeta|^2)^{|\tau|}}\right) \\ &\leqslant \sum_{k\in\mathbb{Z}} \frac{1}{1+|\xi+2\pi k-\zeta|^2}\Big[\max(B(\xi+2\pi k,\zeta), B(\zeta,\xi+2\pi k)\Big]^{|\tau|} \\ &\leqslant 2^{|\tau|}\sup_{x\in\mathbb{R}}\sum_{k\in\mathbb{Z}}\frac{1}{1+|x+2\pi k|^2} =: C_3 < \infty. \end{aligned}$$

Consequently, we conclude that

$$[\widehat{f},\widehat{f}]_\tau(\xi) \leqslant C_2C_3\frac{1}{2\pi}\int_{\mathbb{R}}|\widehat{f}(\zeta)|^2(1+|\zeta|^2)^\tau\, d\zeta = C_2C_3\|f\|^2_{H^\tau(\mathbb{R})} < \infty.$$

Therefore, $[\widehat{f},\widehat{f}]_\tau \in L_\infty(\mathbb{T})$. □

We now generalize Theorem 5.8.1 for the case $p = 2$ without assuming $\phi \in (L_2(\mathbb{R}))^r$. The following result completely characterizes L_2 smoothness of a refinable vector function and will play a critical role in our study of framelets and wavelets in Sobolev spaces $H^\tau(\mathbb{R})$ with $\tau \in \mathbb{R}$.

Theorem 6.3.3 *Let ϕ be an $r \times 1$ vector of compactly supported distributions such that $\widehat{\phi}(0) \neq 0$ and $\widehat{\phi}(2\xi) = \widehat{a}(\xi)\widehat{\phi}(\xi)$ for some $a \in (l_0(\mathbb{Z}))^{r\times r}$. Then* $\operatorname{sm}(\phi) \geqslant \operatorname{sm}(a)$. *If the integer shifts of ϕ are stable (that is, span$\{\widehat{\phi}(\xi+2\pi k) : k \in \mathbb{Z}\} = \mathbb{C}^r$ for every $\xi \in \mathbb{R}$), then*

(i) for any given $\tau \in \mathbb{R}$, $\phi \in (H^\tau(\mathbb{R}))^r$ if and only if $\operatorname{sm}(a) > \tau$;
(ii) $\operatorname{sm}(\phi) = \operatorname{sm}(a)$.

Proof By the definition of $\operatorname{sm}(a) = \operatorname{sm}_2(a)$, we have $\operatorname{sm}(a) = 1/2 - \log_2 \rho_m(a, \upsilon)_2$ for some $m \in \mathbb{N}_0$ and $\upsilon \in (l_0(\mathbb{Z}))^{1\times r}$. Therefore, $\rho_m(a, \upsilon)_2 = 2^{1/2-\operatorname{sm}(a)}$. Note that $\nabla^m u \in \mathscr{V}_{m-1,\upsilon}$ for all $u \in (l_0(\mathbb{Z}))^{1\times r}$. Define $\widehat{a_n}(\xi) = \widehat{a}(2^{n-1}\xi)\cdots\widehat{a}(\xi)$. Consequently, for any $\tau < \operatorname{sm}(a)$ and $0 < \varepsilon < \operatorname{sm}(a) - \tau$, by $\nabla^m a_n = a_n * (\nabla^m \delta I_r)$ and the definition of $\rho_m(a, \upsilon)_2$, there exists a positive constant C_1 such that

$$\|\nabla^m a_n\|^2_{(l_2(\mathbb{Z}))^{r\times r}} \leqslant C_1 2^{n(-1-2\tau-2\varepsilon)}, \qquad \forall\, n \in \mathbb{N}.$$

By item (i) of Theorem 5.1.2, we see that $\phi \in (H^t(\mathbb{R}))^r$ for some $t \in \mathbb{R}$. From the identity $\widehat{\phi}(2\xi) = \widehat{a}(\xi)\widehat{\phi}(\xi)$, we have $\widehat{\phi}(2^n\xi) = \widehat{a_n}(\xi)\widehat{\phi}(\xi)$ for all $n \in \mathbb{N}$. Therefore,

$$(1 - e^{-i\xi})^m (1 + |\xi|^2)^{t/2} \widehat{\phi}(2^n\xi) = \widehat{\nabla^m a_n}(\xi)(1 + |\xi|^2)^{t/2}\widehat{\phi}(\xi).$$

Since $\phi \in (H^t(\mathbb{R}))^r$, by Lemma 6.3.2 we have $[\widehat{\phi}, \widehat{\phi}]_t \in (L_\infty(\mathbb{R}))^{r\times r}$. Therefore, by the above identity, there exists a positive constant C_2 depending only on ϕ such that

$$\int_{\mathbb{R}} |1 - e^{-i\xi}|^{2m}(1 + |\xi|^2)^t \|\widehat{\phi}(2^n\xi)\|_{l_2}^2 d\xi \leqslant C_2 \|\nabla^m a_n\|^2_{(l_2(\mathbb{Z}))^{r\times r}} \leqslant C_1 C_2 2^{n(-1-2\tau-2\varepsilon)}.$$

Changing variable ξ to $2^{-n}\xi$ in the above integral, we have

$$\int_{\mathbb{R}} |1 - e^{-i2^{-n}\xi}|^{2m}(1 + |2^{-n}\xi|^2)^t \|\widehat{\phi}(\xi)\|_{l_2}^2 d\xi \leqslant C_1 C_2 2^{n(-2\tau-2\varepsilon)}.$$

Define $A_n(\xi) := (2^{-2n} + |\xi|^2)^\tau |1 - e^{-i\xi}|^{-2m}(1 + |\xi|^2)^{-t}$. Then

$$C_3 := \sup_{n\in\mathbb{N}} \sup_{2^{n-1}\leqslant|\xi|<2^n} A_n(2^{-n}\xi) = \sup_{n\in\mathbb{N}} \sup_{2^{-1}\leqslant|\xi|<1} A_n(\xi) < \infty.$$

Therefore, using the above two inequalities, we have

$$\begin{aligned}
&\int_{2^{n-1}\leqslant|\xi|<2^n} (1 + |\xi|^2)^\tau \|\widehat{\phi}(\xi)\|_{l_2}^2 d\xi \\
&= 2^{2\tau n} \int_{2^{n-1}\leqslant|\xi|<2^n} A_n(2^{-n}\xi)|1 - e^{-i2^{-n}\xi}|^{2m}(1 + |2^{-n}\xi|^2)^t \|\widehat{\phi}(\xi)\|_{l_2}^2 d\xi \\
&\leqslant C_3 2^{2\tau n} \int_{2^{n-1}\leqslant|\xi|<2^n} |1 - e^{-i2^{-n}\xi}|^{2m}(1 + |2^{-n}\xi|^2)^t \|\widehat{\phi}(\xi)\|_{l_2}^2 d\xi \\
&\leqslant C_1 C_2 C_3 2^{-2\varepsilon n}.
\end{aligned}$$

Hence, we conclude that

$$\begin{aligned}
\int_{\mathbb{R}\setminus[-1,1]} (1 + |\xi|^2)^\tau \|\widehat{\phi}(\xi)\|_{l_2}^2 d\xi &= \sum_{n=1}^\infty \int_{2^{n-1}\leqslant|\xi|<2^n} (1 + |\xi|^2)^\tau \|\widehat{\phi}(\xi)\|_{l_2}^2 d\xi \\
&\leqslant C_1 C_2 C_3 \sum_{n=1}^\infty 2^{-2\varepsilon n} < \infty.
\end{aligned}$$

Since $\widehat{\phi}$ is continuous and thus bounded on $[-1, 1]$, we conclude that $\phi \in (H^\tau(\mathbb{R}))^r$. Noting that $\tau < \operatorname{sm}(a)$ is arbitrary, we proved $\operatorname{sm}(\phi) \geqslant \operatorname{sm}(a)$.

We now prove items (i) and (ii) under the assumption that the integer shifts of ϕ are stable. If $\mathrm{sm}(a) > \tau$, then $\mathrm{sm}(\phi) \geqslant \mathrm{sm}(a) > \tau$. By the definition of $\mathrm{sm}(\phi)$, we must have $\phi \in (H^\tau(\mathbb{R}))^r$. This proves the sufficiency part ($\Leftarrow$) of item (i).

Suppose that $\phi \in (H^\tau(\mathbb{R}))^r$. By $\widehat{\phi}(2^n\xi) = \widehat{a_n}(\xi)\widehat{\phi}(\xi)$, for $m \in \mathbb{N}_0$, we have

$$(1 - e^{-i\xi})^m \widehat{\phi}(2^n\xi)(1 + |\xi|^2)^{\tau/2} = \widehat{\nabla^m a_n}(\xi)\widehat{\phi}(\xi)(1 + |\xi|^2)^{\tau/2}.$$

Since the integer shifts of ϕ are stable, there must exist $C > 0$ such that $[\widehat{\phi}, \widehat{\phi}]_\tau(\xi) \geqslant C^{-1}I_r$ for all $\xi \in \mathbb{R}$ (See Exercise 6.36). Therefore,

$$\|\nabla^m a_n\|^2_{(l_2(\mathbb{Z}))^{r\times r}} \leqslant C \int_{\mathbb{R}} |1 - e^{-i\xi}|^{2m} \|\widehat{\phi}(2^n\xi)\|^2_{l_2} (1 + |\xi|^2)^\tau d\xi, \qquad \forall\, n \in \mathbb{N}.$$

Since

$$\int_{\mathbb{R}} |1 - e^{-i\xi}|^{2m} \|\widehat{\phi}(2^n\xi)\|^2_{l_2} (1 + |\xi|^2)^\tau d\xi$$
$$= 2^{-n}2^{-2n\tau} \int_{\mathbb{R}} |1 - e^{-i2^{-n}\xi}|^{2m} \|\widehat{\phi}(\xi)\|^2_{l_2} (2^{2n} + |\xi|^2)^\tau d\xi,$$

we have

$$2^{n(1+2\tau)} \|\nabla^m a_n\|^2_{(l_2(\mathbb{Z}))^{r\times r}} \leqslant C \int_{\mathbb{R}} |1 - e^{-i2^{-n}\xi}|^{2m} \|\widehat{\phi}(\xi)\|^2_{l_2} (2^{2n} + |\xi|^2)^\tau d\xi, \qquad (6.3.3)$$

for all $n \in \mathbb{N}$. We now consider two cases: $\tau < 0$ or $\tau \geqslant 0$. Suppose that $\tau < 0$. By $|1 - e^{-i2^{-n}\xi}| \leqslant 2$, it is trivial to deduce from (6.3.3) that

$$2^{n(1+2\tau)} \|\nabla^m a_n\|^2_{(l_2(\mathbb{Z}))^{r\times r}} \leqslant C2^{2m} \int_{\mathbb{R}} \|\widehat{\phi}(\xi)\|^2_{l_2} (2^{2n} + |\xi|^2)^\tau d\xi.$$

Since $\tau < 0$, it follows from $2^{2n} + |\xi|^2 \geqslant 1 + |\xi|^2$ that $(2^{2n} + |\xi|^2)^\tau \leqslant (1 + |\xi|^2)^\tau$. Therefore, $\|\widehat{\phi}(\xi)\|^2_{l_2}(2^{2n} + |\xi|^2)^\tau \leqslant \|\widehat{\phi}(\xi)\|^2_{l_2}(1 + |\xi|^2)^\tau \in L_1(\mathbb{R})$ by $\phi \in (H^\tau(\mathbb{R}))^r$. By the Dominated Convergence Theorem,

$$\lim_{n\to\infty} 2^{n(1+2\tau)} \|\nabla^m a_n\|^2_{(l_2(\mathbb{Z}))^{r\times r}} \leqslant C2^{2m} \int_{\mathbb{R}} \|\widehat{\phi}(\xi)\|^2_{l_2} \lim_{n\to\infty} (2^{2n} + |\xi|^2)^\tau d\xi = 0,$$

where we used $\tau < 0$ and $\lim_{n\to\infty}(2^{2n} + |\xi|^2) = \infty$. In other words, we proved

$$\lim_{n\to\infty} 2^{n(1/2+\tau)} \|\nabla^m a_n\|_{(l_2(\mathbb{Z}))^{r\times r}} = 0, \qquad \forall\, n \in \mathbb{N}, m \in \mathbb{N}_0.$$

Hence, it follows from Proposition 5.6.9 and the above identity that $\rho_m(a, \upsilon)_2 \leqslant \rho_0(a, \upsilon) < 2^{1/2-\tau}$ for all $m \in \mathbb{N}_0$ and $\upsilon \in (l_0(\mathbb{Z}))^{1\times r}$. Hence, $\mathrm{sm}(a) = 1/2 - \log_2 \rho_m(a, \upsilon)_2 > \tau$. This proves the necessity part ($\Rightarrow$) of item (i) for $\tau < 0$.

Suppose that $\tau \geqslant 0$. Since the integer shifts of ϕ are stable and $\phi \in (H^\tau(\mathbb{R}))^r$, by Proposition 5.6.2 and Corollary 5.6.12 (also see Theorem 5.6.11 or Corollary 5.6.10), the filter a must satisfy order m sum rules with $m > \tau$. Assume that a takes the normal form in (5.6.22) and (5.6.29) with m being replaced by $m-1$. Then (6.3.3) can be rewritten as

$$2^{n(1+2\tau)}\|\nabla^m a_n\|^2_{(l_2(\mathbb{Z}))^{r\times r}} \leqslant C\int_{\mathbb{R}} E_n(\xi)\|\widehat{\phi}(\xi)\|^2_{l_2}(1+|\xi|^2)^\tau d\xi$$

with

$$\begin{aligned} E_n(\xi) :=& |1-e^{-i2^{-n}\xi}|^{2m}\Big(\frac{1+|2^{-n}\xi|^2}{2^{-2n}+|2^{-n}\xi|^2}\Big)^\tau \\ =& 2^{2m}\sin^{2m}(2^{-n-1}\xi)\Big(\frac{1+|2^{-n}\xi|^2}{2^{-2n}+|2^{-n}\xi|^2}\Big)^\tau. \end{aligned} \tag{6.3.4}$$

Using $|\sin(x)| \leqslant \min(|x|,1)$ and $m > \tau \geqslant 0$, we observe that

$$E_n(\xi) \leqslant \begin{cases} |2^{-n}\xi|^{2m}(\frac{2}{|2^{-n}\xi|^2})^\tau = 2^\tau|2^{-n}\xi|^{2m-2\tau} \leqslant 2^\tau, & \text{if } |2^{-n}\xi| \leqslant 1,\\ 2^{2m}(\frac{1+|2^{-n}\xi|^2}{|2^{-n}\xi|^2})^\tau = 2^{2m}(|2^{-n}\xi|^{-2}+1)^\tau \leqslant 2^{2m+\tau}, & \text{if } |2^{-n}\xi| > 1. \end{cases}$$

That is, $\sup_{n\in\mathbb{N}}\|E_n\|_{L_\infty(\mathbb{R})} < \infty$. Also, note that $\lim_{n\to\infty} E_n(\xi) = 0$ for all $\xi \in \mathbb{R}$ (Exercise 6.5). Since $\|\widehat{\phi}(\xi)\|^2_{l_2}(1+|\xi|^2)^\tau \in L_1(\mathbb{R})$ by $\phi \in (H^\tau(\mathbb{R}))^r$, by the Dominated Convergence Theorem,

$$\lim_{n\to\infty} 2^{n(1+2\tau)}\|\nabla^m a_n\|^2_{(l_2(\mathbb{Z}))^{r\times r}} \leqslant C\int_{\mathbb{R}} \lim_{n\to\infty} E_n(\xi)\|\widehat{\phi}(\xi)\|^2_{l_2}(1+|\xi|^2)^\tau d\xi = 0. \tag{6.3.5}$$

On the other hand, since the filter a takes the normal form, taking $f = \phi$, we see that (5.6.48) holds with $p = 2$ and $f_n = \phi$. That is,

$$2^{n/2}\|a_n * b_\ell^\star\|_{(l_2(\mathbb{Z}))^r} \leqslant C_1C_2\omega_{\lfloor\tau\rfloor+1}(\phi, 2^{-n})_2 \leqslant C_1C_2 2^{-n\lfloor\tau\rfloor}\omega_1(D^{\lfloor\tau\rfloor}\phi, 2^{-n})_2$$

for all $n \in \mathbb{N}$ and $\ell = 2,\dots,r$, where $b_\ell := (\boldsymbol{\delta} e_\ell)^\star$ for $\ell = 2,\dots,r$. Since $\phi \in (H^\tau(\mathbb{R}))^r$ and $0 \leqslant \tau - \lfloor\tau\rfloor < 1$, by Lemma 6.3.1, we conclude that

$$\lim_{n\to\infty} 2^{n(1/2+\tau)}\|a_n * b_\ell^\star\|_{(l_2(\mathbb{Z}))^r} = 0, \qquad \forall\, \ell = 2,\dots,r. \tag{6.3.6}$$

Define $b_1 := (\nabla^{m+1}\boldsymbol{\delta} e_1)^\star$. By (6.3.5), we see that (6.3.6) also holds for $\ell = 1$. Consequently, it follows from Proposition 5.6.9 that $\rho_m(a, \upsilon)_2 < 1/2 - \tau$, from which we conclude that $\mathrm{sm}(a) > \tau$. This proves item (i).

We already proved that $\mathrm{sm}(\phi) \geqslant \mathrm{sm}(a)$. Conversely, for any $\tau < \mathrm{sm}(\phi)$, we have $\phi \in (H^\tau(\mathbb{R}))^r$ and it follows from item (i) that $\mathrm{sm}(a) > \tau$. Consequently, $\mathrm{sm}(a) \geqslant \mathrm{sm}(\phi)$. This proves item (ii). □

As a consequence of Theorem 6.3.3, the following result generalizes Corollary 5.8.2 from integers $\tau = m$ to any real numbers τ.

Corollary 6.3.4 *Let ϕ be an $r \times 1$ vector of compactly supported distributions such that $\widehat{\phi}(0) \neq 0$ and $\widehat{\phi}(2\xi) = \widehat{a}(\xi)\widehat{\phi}(\xi)$ for some $a \in (l_0(\mathbb{Z}))^{r\times r}$. For $\tau \in \mathbb{R}$, the refinable vector distribution $\phi \in (H^\tau(\mathbb{R}))^r$ if and only if $\mathrm{sm}(\phi) > \tau$.*

Proof If $\mathrm{sm}(\phi) > \tau$, then by definition it is trivial that $\phi \in (H^\tau(\mathbb{R}))^r$. We now prove that $\phi \in (H^\tau(\mathbb{R}))^r$ implies $\mathrm{sm}(\phi) > \tau$. By Theorem 5.2.4, there exists an $s\times 1$ vector η with $1 \leqslant s \leqslant r$ of compactly supported distributions such that $\mathrm{S}(\phi) = \mathrm{S}(\eta)$ and the integer shifts of η are linearly independent (and consequently, the integer shifts of η are stable). Moreover, there exists $b \in (l_0(\mathbb{Z}))^{s\times s}$ such that $\widehat{\eta}(2\xi) = \widehat{b}(\xi)\widehat{\eta}(\xi)$. By $\mathrm{S}(\phi) = \mathrm{S}(\eta)$ and the fact that ϕ and η are compactly supported, we see that $\phi \in (H^\tau(\mathbb{R}))^r$ implies $\eta \in (H^\tau(\mathbb{R}))^s$. By item (i) of Theorem 6.3.3, we have $\mathrm{sm}(b) > \tau$. By item (ii) of Theorem 6.3.3, we conclude that $\mathrm{sm}(\phi) = \mathrm{sm}(\eta) = \mathrm{sm}(b) > \tau$. □

6.4 Framelets and Wavelets in Sobolev Spaces with Filter Banks

In this section we discuss dual framelets and biorthogonal wavelets in Sobolev spaces $H^\tau(\mathbb{R})$ such that they are derived from compactly supported refinable (vector) functions. As particular cases, we also characterize tight framelets and orthogonal wavelets in $L_2(\mathbb{R})$.

6.4.1 Dual Framelets in Sobolev Spaces and Tight Framelets in $L_2(\mathbb{R})$

The following result characterizes all compactly supported dual framelets in Sobolev spaces such that the generating functions are derived from refinable functions and filter banks. Moreover, it links a dual framelet in a pair of Sobolev spaces with a generalized (OEP-based) dual framelet filter bank.

Theorem 6.4.1 *Let $a, \tilde{a}, \theta, \tilde{\theta} \in (l_0(\mathbb{Z}))^{r\times r}$ and $b, \tilde{b} \in (l_0(\mathbb{Z}))^{s\times r}$. Let $\phi, \tilde{\phi}$ be $r \times 1$ vectors of compactly supported distributions satisfying*

$$\widehat{\phi}(2\xi) = \widehat{a}(\xi)\widehat{\phi}(\xi), \qquad \widehat{\tilde{\phi}}(2\xi) = \widehat{\tilde{a}}(\xi)\widehat{\tilde{\phi}}(\xi), \tag{6.4.1}$$

for $\xi \in \mathbb{R}$. Define $\eta, \psi, \tilde{\eta}, \tilde{\psi}$ by

$$\widehat{\eta}(\xi) := \widehat{\theta}(\xi)\widehat{\phi}(\xi), \quad \widehat{\psi}(\xi) := \widehat{b}(\xi/2)\widehat{\phi}(\xi/2),$$
$$\widehat{\tilde{\eta}}(\xi) := \widehat{\tilde{\theta}}(\xi)\widehat{\tilde{\phi}}(\xi), \quad \widehat{\tilde{\psi}}(\xi) := \widehat{\tilde{b}}(\xi/2)\widehat{\tilde{\phi}}(\xi/2), \tag{6.4.2}$$

for $\xi \in \mathbb{R}$. Define $\Theta \in (l_0(\mathbb{Z}))^{r\times r}$ by $\widehat{\Theta}(\xi) := \widehat{\tilde{\theta}}(\xi)^{\mathsf{T}}\overline{\widehat{\theta}(\xi)}$. For $\tau \in \mathbb{R}$, the pair $(\{\tilde{\eta}; \tilde{\psi}\}, \{\eta; \psi\})$ is a dual framelet in $(H^{-\tau}(\mathbb{R}), H^{\tau}(\mathbb{R}))$ if

(1) The pair $(\{\tilde{a}; \tilde{b}\}, \{a; b\})_{\Theta}$ is a generalized dual framelet filter bank, *i.e.,*

$$\widehat{\tilde{\phi}}(\xi)^{\mathsf{T}}\Big[\widehat{\tilde{a}}(\xi)^{\mathsf{T}}\widehat{\Theta}(2\xi)\overline{\widehat{a}(\xi)} + \widehat{\tilde{b}}(\xi)^{\mathsf{T}}\overline{\widehat{b}(\xi)} - \widehat{\Theta}(\xi)\Big]\overline{\widehat{\phi}(\xi + 2\pi k)} = 0,$$

$$\widehat{\tilde{\phi}}(\xi)^{\mathsf{T}}\Big[\widehat{\tilde{a}}(\xi)^{\mathsf{T}}\widehat{\Theta}(2\xi)\overline{\widehat{a}(\xi + \pi)} + \widehat{\tilde{b}}(\xi)^{\mathsf{T}}\overline{\widehat{b}(\xi + \pi)}\Big]\overline{\widehat{\phi}(\xi + 2\pi k)} = 0,$$

for all $\xi \in \mathbb{R}$ and $k \in \mathbb{Z}$;

(2) $\widehat{\tilde{\phi}}(0)^{\mathsf{T}}\widehat{\Theta}(0)\overline{\widehat{\phi}(0)} = 1$;

(3) $\phi \in (H^{\tau}(\mathbb{R}))^r$ and $\tilde{\phi} \in (H^{-\tau}(\mathbb{R}))^r$;

(4) $\widehat{\psi}(\xi) = o(|\xi|^{-\tau})$ as $\xi \to 0$ if $\tau \leqslant 0$, and $\widehat{\tilde{\psi}}(\xi) = o(|\xi|^{\tau})$ as $\xi \to 0$ if $\tau \geqslant 0$.

Conversely, if $(\{\tilde{\eta}; \tilde{\psi}\}, \{\eta; \psi\})$ is a dual framelet in $(H^{-\tau}(\mathbb{R}), H^{\tau}(\mathbb{R}))$, then items (1), (2), (4) as well as the following (3') must be satisfied, where

(3') $\eta \in (H^{\tau}(\mathbb{R}))^r$ and $\tilde{\eta} \in (H^{-\tau}(\mathbb{R}))^r$.

Under the additional condition that $\det(\widehat{\theta})$ *and* $\det(\widehat{\tilde{\theta}})$ *are not identically zero, then item (3) also holds.*

Proof Since both ϕ and $\tilde{\phi}$ have compact support and $\Theta \in (l_0(\mathbb{Z}))^{r\times r}$, it follows from Lemma 4.1.3 that item (2) is equivalent to

$$\lim_{j\to\infty} \langle \widehat{\tilde{\phi}}(2^{-j}\cdot)^{\mathsf{T}}\widehat{\Theta}(2^{-j}\cdot)\overline{\widehat{\phi}(2^{-j}\cdot)}, h\rangle = \langle 1, h\rangle, \qquad \forall\, h \in \mathscr{D}(\mathbb{R}).$$

By Theorem 4.1.10, items (1) and (2) are equivalent to that $(\{\widehat{\tilde{\eta}}; \widehat{\tilde{\psi}}\}, \{\widehat{\eta}; \widehat{\psi}\})$ is a frequency-based dual framelet. By Corollary 6.3.4, we see that item (3) is equivalent to $\operatorname{sm}(\phi) > \tau$ and $\operatorname{sm}(\tilde{\phi}) > -\tau$.

Since both ψ and $\tilde{\psi}$ have compact support, item (4) is equivalent to saying that there exists a positive integer $\nu > |\tau|$ such that $\|\widehat{\psi}(\xi)\|_{l_2} = \mathscr{O}(|\xi|^{\nu})$ as $\xi \to 0$ for $\tau \leqslant 0$ and $\|\widehat{\tilde{\psi}}(\xi)\|_{l_2} = \mathscr{O}(|\xi|^{\nu})$ as $\xi \to 0$ for $\tau \geqslant 0$.

Since item (3) implies $\operatorname{sm}(\phi) > \tau$ and $\operatorname{sm}(\tilde{\phi}) > -\tau$, it follows from Lemma 6.3.2 that for all $\tau \leqslant t < \operatorname{sm}(\phi)$ and $-\tau \leqslant \tilde{t} < \operatorname{sm}(\tilde{\phi})$, $[\widehat{\phi}, \widehat{\phi}]_t, [\widehat{\tilde{\phi}}, \widehat{\tilde{\phi}}]_{\tilde{t}} \in (L_\infty(\mathbb{T}))^{r\times r}$. Now by item (4) and Corollary 4.6.6, we conclude that $\mathsf{AS}_0(\eta; \psi)$ has stability in

$H^\tau(\mathbb{R})$ and $\mathsf{AS}_0(\tilde{\eta}; \tilde{\psi})$ has stability in $H^{-\tau}(\mathbb{R})$. By Theorem 4.6.3, we conclude that $(\{\tilde{\eta}; \tilde{\psi}\}, \{\eta; \psi\})$ is a dual framelet in $(H^{-\tau}(\mathbb{R}), H^\tau(\mathbb{R}))$.

We now prove the converse direction. Suppose that $(\{\tilde{\eta}; \tilde{\psi}\}, \{\eta; \psi\})$ is a dual framelet in $(H^{-\tau}(\mathbb{R}), H^\tau(\mathbb{R}))$. By Theorem 4.1.10, items (1) and (2) hold. Item (3') is trivial. We now prove item (4). By Proposition 4.6.7, the inequality (4.6.19) holds. Since $\widehat{\psi}$ is analytic at the origin, there exists $\nu \in \mathbb{N}_0$ such that $C_1|\xi|^\nu \leqslant |\widehat{\psi}(\xi)|_{l_2} \leqslant C_2|\xi|^\nu$ for all $\xi \in [-\varepsilon, \varepsilon]$ for some positive constants C_1, C_2 and ε. By Proposition 4.6.7, we must have $\nu + \tau > 0$. Hence, the first claim in item (4) holds if $\tau \leqslant 0$. The second claim in item (4) for $\tau \geqslant 0$ can be proved similarly.

If $\det(\widehat{\theta})$ is not identically zero, by $\widehat{\eta}(\xi) = \widehat{\theta}(\xi)\widehat{\phi}(\xi)$, we have $\mathsf{S}(\phi) = \mathsf{S}(\eta)$. Since both ϕ and η have compact support, by item (3'), then $\eta \in (H^\tau(\mathbb{R}))^r$ implies $\phi \in (H^\tau(\mathbb{R}))^r$. We can similarly prove $\tilde{\phi} \in (H^{-\tau}(\mathbb{R}))^r$ if $\det(\widehat{\tilde{\theta}})$ is not identically zero. This proves that item (3') implies item (3). □

For $\nu \geqslant 0$, we say that a function ψ has ν *vanishing moments* if $|\xi|^{-\nu}\widehat{\psi}(\xi) \in L_\infty([-\varepsilon, \varepsilon])$ for some $\varepsilon > 0$. In particular, we define

$$\operatorname{vm}(\psi) := \sup\{\nu \geqslant 0 \ : \ \psi \text{ has } \nu \text{ vanishing moments}\}. \tag{6.4.3}$$

If $\psi = (\psi_1, \ldots, \psi_s)^\mathsf{T}$, then we define $\operatorname{vm}(\psi) := \min_{1\leqslant \ell \leqslant s} \operatorname{vm}(\psi_\ell)$. Since ψ and $\tilde{\psi}$ in Theorem 6.4.1 have compact support, item (4) of Theorem 6.4.1 can be equivalently expressed as

(4') $\operatorname{vm}(\psi) > -\tau$ if $\tau \leqslant 0$, and $\operatorname{vm}(\tilde{\psi}) > \tau$ if $\tau \geqslant 0$.

Suppose that for some $\xi_1, \xi_2 \in \mathbb{R}$,

$$\operatorname{span}\{\widehat{\phi}(\xi_1 + 2\pi k) \ : \ k \in \mathbb{Z}\} = \mathbb{C}^r, \quad \operatorname{span}\{\widehat{\tilde{\phi}}(\xi_2 + 2\pi k) \ : \ k \in \mathbb{Z}\} = \mathbb{C}^r. \tag{6.4.4}$$

Since both $\widehat{\phi}$ and $\widehat{\tilde{\phi}}$ are analytic functions, we must have $\operatorname{span}\{\widehat{\phi}(\xi + 2\pi k) \ : \ k \in \mathbb{Z}\} = \mathbb{C}^r$ and $\operatorname{span}\{\widehat{\tilde{\phi}}(\xi + 2\pi k) \ : \ k \in \mathbb{Z}\} = \mathbb{C}^r$ for almost all $\xi \in \mathbb{R}$. Under the extra assumption in (6.4.4), it is trivial to deduce that $(\{\tilde{a}; \tilde{b}\}, \{a; b\})_\Theta$ is a generalized dual framelet filter bank in item (1) of Theorem 6.4.1 if and only if $(\{\tilde{a}; \tilde{b}\}, \{a; b\})_\Theta$ is a (standard) *dual framelet filter bank*, that is,

$$\begin{bmatrix} \widehat{\tilde{a}}(\xi)^\mathsf{T} & \widehat{\tilde{b}}(\xi)^\mathsf{T} \\ \widehat{\tilde{a}}(\xi + \pi)^\mathsf{T} & \widehat{\tilde{b}}(\xi + \pi)^\mathsf{T} \end{bmatrix} \begin{bmatrix} \widehat{\Theta}(2\xi) & 0 \\ 0 & I_r \end{bmatrix} \begin{bmatrix} \overline{\widehat{a}(\xi)} & \overline{\widehat{a}(\xi + \pi)} \\ \overline{\widehat{b}(\xi)} & \overline{\widehat{b}(\xi + \pi)} \end{bmatrix} = \begin{bmatrix} \widehat{\Theta}(\xi) & 0 \\ 0 & \widehat{\Theta}(\xi + \pi) \end{bmatrix}. \tag{6.4.5}$$

As a particular case of Theorem 6.4.1, we now characterize all compactly supported tight framelets in $L_2(\mathbb{R})$ which are derived from refinable functions.

Theorem 6.4.2 *Let $\theta, a \in (l_0(\mathbb{Z}))^{r\times r}$ and $b \in (l_0(\mathbb{Z}))^{s\times r}$. Let ϕ be an $r \times 1$ vector of compactly supported distributions satisfying $\widehat{\phi}(2\xi) = \widehat{a}(\xi)\widehat{\phi}(\xi)$ for all $\xi \in \mathbb{R}$.*

Define η *and* ψ *as in (6.4.2). Define* $\widehat{\Theta}(\xi) := \widehat{\theta}(\xi)^{\mathsf{T}}\overline{\widehat{\theta}(\xi)}$. *Then* $\{\eta; \psi\}$ *is a tight framelet in* $L_2(\mathbb{R})$ *if and only if*

(i) $\{a; b\}_{\Theta}$ *is a* generalized tight framelet filter bank, *that is,* $(\{a; b\}, \{a; b\})_{\Theta}$ *is a generalized dual framelet filter bank.*
(ii) $\widehat{\phi}(0)^{\mathsf{T}}\widehat{\Theta}(0)\overline{\widehat{\phi}(0)} = 1$.

If in addition $\det(\widehat{\theta})$ *is not identically zero, then items (i) and (ii) also imply* $\phi \in (L_2(\mathbb{R}))^r$.

Proof Since ϕ has compact support and $\Theta \in (l_0(\mathbb{Z}))^{r\times r}$, by Lemma 4.1.3, we see that $\widehat{\phi}(0)^{\mathsf{T}}\widehat{\theta}(0)\overline{\widehat{\phi}(0)} = 1$ if and only if

$$\lim_{j\to\infty} \langle \widehat{\phi}(2^{-j}\cdot)^{\mathsf{T}}\widehat{\Theta}(2^{-j}\cdot)\overline{\widehat{\phi}(2^{-j}\cdot)}, h\rangle = \langle 1, h\rangle, \quad \forall h \in \mathscr{D}(\mathbb{R}).$$

Sufficiency ($\Leftarrow$). By Theorem 4.1.10, the pair $(\{\widehat{\eta}; \widehat{\psi}\}, \{\widehat{\eta}; \widehat{\psi}\})$ is a frequency-based dual framelet. Since all ϕ, η, ψ have compact supports, it is trivial to see that all entries in $\widehat{\eta}, \widehat{\psi}$ belong to $L_2^{loc}(\mathbb{R})$. Therefore, by Theorem 4.3.8, we must have $\eta \in (L_2(\mathbb{R}))^r$, $\psi \in (L_2(\mathbb{R}))^s$ and $\{\eta; \psi\}$ is a tight framelet in $L_2(\mathbb{R})$.

Necessity ($\Rightarrow$). Items (i) and (ii) follow directly from Theorem 6.4.1 with $\tau = 0$.

By $\widehat{\eta}(\xi) = \widehat{\theta}(\xi)\widehat{\phi}(\xi)$, since $\det(\widehat{\theta})$ is not identically zero, we have $\mathsf{S}(\eta) = \mathsf{S}(\phi)$. Since $\eta \in (L_2(\mathbb{R}))^r$ and ϕ has compact support, we must have $\phi \in (L_2(\mathbb{R}))^r$. □

Theorem 6.4.2 facilitates the construction of tight framelets in $L_2(\mathbb{R})$ by removing the need to check the condition $\phi \in (L_2(\mathbb{R}))^r$ in advance.

6.4.2 Biorthogonal Wavelets in Sobolev Spaces and Orthogonal Wavelets in $L_2(\mathbb{R})$

We now study biorthogonal and orthogonal wavelets that are derived from refinable functions. Let us first characterize scalar orthogonal wavelets in $L_2(\mathbb{R})$.

Theorem 6.4.3 *Let* $a \in l_0(\mathbb{Z})$ *with* $\widehat{a}(0) = 1$. *Define* $\widehat{\varphi}(\xi) := \prod_{j=1}^{\infty} \widehat{a}(2^{-j}\xi)$ *and*

$$\widehat{\psi}(\xi) := \widehat{b}(\xi/2)\widehat{\varphi}(\xi/2) \quad \textit{with} \quad \widehat{b}(\xi) := e^{-i\xi}\overline{\widehat{a}(\xi+\pi)}. \tag{6.4.6}$$

Then $\{\varphi; \psi\}$ *is an orthogonal wavelet in* $L_2(\mathbb{R})$, *that is,*

$$\mathsf{AS}_0(\varphi; \psi) := \{\varphi(\cdot - k) \ : \ k \in \mathbb{Z}\} \cup \{\psi_{2^j;k} := 2^{j/2}\psi(2^j \cdot -k) \ : \ j \in \mathbb{N}_0, k \in \mathbb{Z}\}$$

is an orthonormal basis for $L_2(\mathbb{R})$ *if and only if* $\mathrm{sm}(a) > 0$ *and the filter a is an* orthogonal wavelet filter *satisfying*

$$|\widehat{a}(\xi)|^2 + |\widehat{a}(\xi+\pi)|^2 = 1. \tag{6.4.7}$$

Proof Necessity ($\Rightarrow$). Since $\{\varphi(\cdot-k) \; : \; k\in\mathbb{Z}\}$ is an orthonormal system in $L_2(\mathbb{R})$, by Lemma 4.4.1 we have $[\widehat{\varphi},\widehat{\varphi}]=1$ a.e. and by Theorem 6.3.3 (or Corollary 5.6.12), we have $\operatorname{sm}(a)>0$. It follows from $\widehat{\varphi}(2\xi)=\widehat{a}(\xi)\widehat{\varphi}(\xi)$ and $[\widehat{\varphi},\widehat{\varphi}]=1$ that

$$1=[\widehat{\varphi},\widehat{\varphi}](2\xi)=|\widehat{a}(\xi)|^2[\widehat{\varphi},\widehat{\varphi}](\xi)+|\widehat{a}(\xi+\pi)|^2[\widehat{\varphi},\widehat{\varphi}](\xi+\pi)=|\widehat{a}(\xi)|^2+|\widehat{a}(\xi+\pi)|^2.$$

Sufficiency ($\Leftarrow$). Let $f=B_1:=\chi_{(0,1]}$ and $f_n:=\mathcal{R}_a^n f$ for $n\in\mathbb{N}$. Since $\widehat{f}(0)=1$ and $\widehat{f}(2\pi k)=0$ for all $k\in\mathbb{Z}\backslash\{0\}$, we see that $f\in\mathscr{F}_{0,2}$ in (5.6.53), that is, the function f is an admissible initial function. Moreover, $\lim_{n\to\infty}\widehat{f_n}(\xi)=\widehat{\varphi}(\xi)$ for all $\xi\in\mathbb{R}$. By Theorem 5.6.16 (also see Theorem 5.6.11), the condition $\operatorname{sm}(a)>0$ implies that $\varphi\in L_2(\mathbb{R})$ and $\lim_{n\to\infty}\|f_n-\varphi\|_{L_2(\mathbb{R})}=0$. Since $[\widehat{f},\widehat{f}]=1$, by (6.4.7) and induction on $n\in\mathbb{N}$, we have $[\widehat{f_n},\widehat{f_n}]=1$, i.e., $\langle f_n,f_n(\cdot-k)\rangle=\boldsymbol{\delta}(k)$ for all $k\in\mathbb{Z}$ and $n\in\mathbb{N}$. Hence, $\langle\varphi,\varphi(\cdot-k)\rangle=\lim_{n\to\infty}\langle f_n,f_n(\cdot-k)\rangle=\boldsymbol{\delta}(k)$ for all $k\in\mathbb{Z}$. That is, $\{\varphi(\cdot-k) \; : \; k\in\mathbb{Z}\}$ is an orthonormal system in $L_2(\mathbb{R})$. By $\widehat{b}(\xi)=e^{-i\xi}\overline{\widehat{a}(\xi+\pi)}$ and (6.4.7), we can check that $\{a;b\}$ is an orthogonal wavelet filter bank satisfying

$$\begin{bmatrix}\widehat{a}(\xi) & \widehat{a}(\xi+\pi)\\ \widehat{b}(\xi) & \widehat{b}(\xi+\pi)\end{bmatrix}\overline{\begin{bmatrix}\widehat{a}(\xi) & \widehat{a}(\xi+\pi)\\ \widehat{b}(\xi) & \widehat{b}(\xi+\pi)\end{bmatrix}}^{\mathsf{T}}=I_2. \tag{6.4.8}$$

Using (6.4.8) and $[\widehat{\phi},\widehat{\phi}]=1$, we can directly check that $\mathsf{AS}_0(\varphi;\psi)$ is an orthonormal system (see Theorem 4.5.16 for details). Since $\widehat{\varphi}(0)=1$ and $\widehat{\varphi}$ is a continuous function, by (6.4.8) and Theorem 4.5.4 or Theorem 6.4.2, we see that $\mathsf{AS}_0(\varphi;\psi)$ is a tight frame for $L_2(\mathbb{R})$. Thus, $\mathsf{AS}_0(\varphi;\psi)$ is an orthonormal basis for $L_2(\mathbb{R})$. □

Applying Theorem 6.4.3, we now prove that all the Daubechies orthogonal wavelet filters a_m^D with $m\in\mathbb{N}$, defined in (2.2.4), lead to orthogonal refinable functions and orthogonal wavelets (also see Example 4.5.3). The following result also provides an asymptotic estimate of the smoothness exponents $\operatorname{sm}(a_m^D)$ for all $m\in\mathbb{N}$.

Theorem 6.4.4 *For $m\in\mathbb{N}$, let a_m^D be the Daubechies orthogonal wavelet filter defined in* (2.2.4). *Define $\widehat{\varphi}(\xi):=\prod_{j=1}^{\infty}\widehat{a_m^D}(2^{-j}\xi)$ and $\widehat{\psi}(\xi):=e^{-i\xi/2}\overline{\widehat{a_m^D}(\xi/2+\pi)}\widehat{\varphi}(\xi/2)$. Then $\{\varphi;\psi\}$ is an orthogonal wavelet in $L_2(\mathbb{R})$, $\operatorname{sm}_p(\varphi)=\operatorname{sm}_p(a_m^D)$ for all $1\leqslant p\leqslant\infty$, $|\widehat{\varphi}(\xi)|^2=1+\mathscr{O}(|\xi|^{2m})$ as $\xi\to 0$, and for all $m\in\mathbb{N}$,*

$$\tfrac{1}{2}\leqslant\tfrac{m-1}{2}\log_2\tfrac{4}{3}+\tfrac{1}{2}\leqslant\operatorname{sm}(a_m^D)\leqslant\tfrac{m-1}{2}\log_2\tfrac{4}{3}+\tfrac{1}{4}\log_2 m+\log_2\tfrac{e}{\pi^{1/4}}. \tag{6.4.9}$$

Proof Note that $|\widehat{a_m^D}(\xi)|^2=\widehat{a_{2m,2m}}(\xi):=\cos^{2m}(\xi/2)\mathsf{P}_{m,m}(\sin^2(\xi/2))$ and $|\widehat{a_m^D}(\xi)|^2+|\widehat{a_m^D}(\xi+\pi)|^2=1$. By Example 5.8.1 (and the remark afterward), we have

$$2m-1-\log_2\mathsf{P}_{m,m}(\tfrac{3}{4})\leqslant\operatorname{sm}_\infty(a_{2m,2m})\leqslant 2m-\log_2\mathsf{P}_{m,m}(\tfrac{3}{4}). \tag{6.4.10}$$

We now estimate $\mathsf{P}_{m,m}(\frac{3}{4})$ by proving

$$\frac{4\sqrt{\pi}}{e^2\sqrt{m}}3^{m-1} \leqslant \mathsf{P}_{m,m}(\tfrac{3}{4}) \leqslant 3^{m-1}, \qquad \forall\, m \in \mathbb{N}. \tag{6.4.11}$$

Note that

$$\begin{aligned}(\tfrac{3}{4})^{1-m}\mathsf{P}_{m,m}(\tfrac{3}{4}) &= \sum_{j=0}^{m-1}\binom{m+j-1}{j}\left(\frac{3}{4}\right)^{j+1-m}\\ &\leqslant \sum_{j=0}^{m-1}\binom{m+j-1}{j}\left(\frac{1}{2}\right)^{j+1-m} = 2^{1-m}\mathsf{P}_{m,m}(\tfrac{1}{2}).\end{aligned}$$

By $(1-x)^m\mathsf{P}_{m,m}(x)+x^m\mathsf{P}_{m,m}(1-x) = 1$, setting $x = 1/2$, we have $\mathsf{P}_{m,m}(\frac{1}{2}) = 2^{1-m}$. Therefore, the above inequality implies $(\frac{3}{4})^{1-m}\mathsf{P}_{m,m}(\frac{3}{4}) \leqslant 2^{1-m}\mathsf{P}_{m,m}(\frac{1}{2}) = 4^{1-m}$, from which we conclude that $\mathsf{P}_{m,m}(\frac{3}{4}) \leqslant 3^{m-1}$. On the other hand,

$$\begin{aligned}\mathsf{P}_{m,m}(\tfrac{3}{4}) &= \sum_{j=0}^{m-1}\binom{m+j-1}{j}\left(\frac{3}{4}\right)^{j} \geqslant \left(\frac{3}{4}\right)^{m-1}\sum_{j=0}^{m-1}\binom{m+j-1}{j}\\ &= \left(\frac{3}{4}\right)^{m-1}\binom{2m-1}{m} = \left(\frac{3}{4}\right)^{m-1}\frac{1}{2}\binom{2m}{m},\end{aligned}$$

where we used $\binom{m+j-1}{j} = \binom{m+j}{j} - \binom{m+j-1}{j-1}$. By the well-known *Sterling's formula*:

$$\sqrt{2\pi}n^{n+1/2}e^{-n} \leqslant n! \leqslant en^{n+1/2}e^{-n}, \qquad \forall\, n \in \mathbb{N}, \tag{6.4.12}$$

we have

$$\mathsf{P}_{m,m}(\tfrac{3}{4}) \geqslant \left(\frac{3}{4}\right)^{m-1} 2^{-1}\frac{(2m)!}{m!m!} \geqslant \left(\frac{3}{4}\right)^{m-1} 2^{-1}\frac{\sqrt{2\pi}(2m)^{2m+1/2}e^{-2m}}{e^2m^{2m+1}e^{-2m}} = \frac{4\sqrt{\pi}}{e^2\sqrt{m}}3^{m-1}.$$

This proves (6.4.11). By item (3) of Corollary 5.8.5, we have $\operatorname{sm}(a_m^D) = \frac{1}{2}\operatorname{sm}_\infty(a_{2m,2m})$. By (6.4.11), we conclude that

$$\begin{aligned}\operatorname{sm}(a_m^D) = \tfrac{1}{2}\operatorname{sm}_\infty(a_{2m,2m}) &\geqslant m - \tfrac{1}{2} - \tfrac{1}{2}\log_2 \mathsf{P}_{m,m}(\tfrac{3}{4})\\ &\geqslant m - \tfrac{1}{2} - \tfrac{m-1}{2}\log_2 3 = \tfrac{m-1}{2}\log_2\tfrac{4}{3} + \tfrac{1}{2}\end{aligned}$$

and

$$\operatorname{sm}(a_m^D) = \tfrac{1}{2}\operatorname{sm}_\infty(a_{2m,2m}) \leqslant m - \tfrac{1}{2}\log_2 \mathsf{P}_{m,m}(\tfrac{3}{4}) \leqslant \tfrac{m-1}{2}\log_2\tfrac{4}{3} + \tfrac{1}{4}\log_2 m + \log_2\tfrac{e}{\pi^{1/4}}.$$

This proves (6.4.9). In particular, by $\log_2 \frac{4}{3} > 0$ and $m \in \mathbb{N}$, the inequalities (6.4.9) imply $\operatorname{sm}(a_m^D) \geqslant \frac{1}{2} > 0$. By Theorem 6.4.3, we conclude that $\{\varphi; \psi\}$ is an orthogonal wavelet in $L_2(\mathbb{R})$. Thus, the integer shifts of φ are orthonormal and therefore, are stable. By Theorem 5.8.1, the identity $\operatorname{sm}_p(\varphi) = \operatorname{sm}_p(a_m^D)$ holds for all $1 \leqslant p \leqslant \infty$.

We now prove $|\widehat{\varphi}(\xi)|^2 = 1 + \mathscr{O}(|\xi|^{2m})$ as $\xi \to 0$. Since $|\widehat{a_m^D}(\xi)|^2 = 1 - |\widehat{a_m^D}(\xi + \pi)|^2 = 1 + \mathscr{O}(|\xi|^{2m})$ as $\xi \to 0$, by $\widehat{\varphi}(2\xi) = \widehat{a_m^D}(\xi)\widehat{\varphi}(\xi)$, we have

$$|\widehat{\varphi}(2\xi)|^2 = |\widehat{\varphi}(\xi)|^2 |\widehat{a_m^D}(\xi)|^2 = |\widehat{\varphi}(\xi)|^2 + \mathscr{O}(|\xi|^{2m}), \qquad \xi \to 0.$$

Since $\widehat{\varphi}(0) = 1$ and $|\widehat{\varphi}(\xi)|^2 = \widehat{\varphi}(\xi)\overline{\widehat{\varphi}(\xi)}$ is infinitely differentiable at the origin, using the Taylor expansion of $|\widehat{\varphi}(\xi)|^2$ at $\xi = 0$, we deduce from the above relation that $|\widehat{\varphi}(\xi)|^2 = 1 + \mathscr{O}(|\xi|^{2m})$ as $\xi \to 0$. □

Now we are ready to study biorthogonal refinable vector functions and biorthogonal multiwavelets. Let us first investigate biorthogonal refinable vector functions.

Theorem 6.4.5 *Let $\phi, \tilde{\phi}$ be $r \times 1$ vectors of compactly supported distributions satisfying* (6.4.1) *with $a, \tilde{a} \in (l_0(\mathbb{Z}))^{r\times r}$. For $\tau \in \mathbb{R}$, the pair $(\tilde{\phi}, \phi)$ is a pair of biorthogonal functions in $(H^{-\tau}(\mathbb{R}), H^{\tau}(\mathbb{R}))$, that is, $\tilde{\phi} \in (H^{-\tau}(\mathbb{R}))^r$, $\phi \in (H^{\tau}(\mathbb{R}))^r$, and $(\tilde{\phi}, \phi)$ is biorthogonal to each other:*

$$\langle \tilde{\phi}, \phi(\cdot - k)\rangle := \int_{\mathbb{R}} \tilde{\phi}(x)\overline{\phi(x-k)}^{\mathsf{T}} dx = \boldsymbol{\delta}(k) I_r, \qquad \forall\, k \in \mathbb{Z}, \tag{6.4.13}$$

if and only if

(i) *$(\tilde{a}, a)$ is a pair of biorthogonal wavelet filters:*

$$\widehat{\tilde{a}}(\xi)\overline{\widehat{a}(\xi)}^{\mathsf{T}} + \widehat{\tilde{a}}(\xi+\pi)\overline{\widehat{a}(\xi+\pi)}^{\mathsf{T}} = I_r, \qquad \xi \in \mathbb{R}; \tag{6.4.14}$$

(ii) $\operatorname{sm}(\tilde{a}) > -\tau$ *and* $\operatorname{sm}(a) > \tau$*;*

(iii) *$\widehat{\tilde{\phi}}(\xi)^{\mathsf{T}}\overline{\widehat{\phi}(\xi)} = 1 + \mathscr{O}(|\xi|^{m+1})$, where $m = \lfloor|\tau|\rfloor$ is the largest integer satisfying $m \leqslant |\tau|$;*

(iv) *For the case $\tau \geqslant 0$,* (5.6.3) *holds, $\widehat{\upsilon}(\xi)\widehat{\phi}(\xi) = 1 + \mathscr{O}(|\xi|^{m+1})$ as $\xi \to 0$, and*

$$\widehat{\tilde{\phi}}(\xi) = \overline{\widehat{\upsilon}(\xi)}^{\mathsf{T}} + \mathscr{O}(|\xi|^{m+1}), \qquad \xi \to 0, \tag{6.4.15}$$

where $\upsilon \in (l_0(\mathbb{Z}))^{1\times r}$ is given in (5.6.4) *with $\widehat{\upsilon}(0)\widehat{\phi}(0) = 1$. For the case $\tau \leqslant 0$, switch the roles of ϕ, a, υ with $\tilde{\phi}, \tilde{a}, \tilde{\upsilon}$,respectively.*

Proof Without loss of generality, we assume $\tau \geqslant 0$. Otherwise, we switch ϕ, a, υ with $\tilde{\phi}, \tilde{a}, \tilde{\upsilon}$, respectively.

Necessity ($\Rightarrow$). Note that (6.4.13) implies that the integer shifts of ϕ as well as $\tilde{\phi}$ are linearly independent and consequently are stable. By Theorem 6.3.3, we see that $\phi \in (H^{\tau}(\mathbb{R}))^r$ implies $\operatorname{sm}(a) > \tau$ and $\tilde{\phi} \in (H^{-\tau}(\mathbb{R}))^r$ implies $\operatorname{sm}(\tilde{a}) > -\tau$.

Hence, item (ii) holds. Note that (6.4.13) is equivalent to $[\widehat{\tilde{\phi}}, \widehat{\phi}] = I_r$ for almost every $\xi \in \mathbb{R}$. By $\widehat{\phi}(2\xi) = \widehat{a}(\xi)\widehat{\phi}(\xi)$ and $\widehat{\tilde{\phi}}(2\xi) = \widehat{\tilde{a}}(\xi)\widehat{\tilde{\phi}}(\xi)$ in (6.4.1), we have

$$
\begin{aligned}
I_r = [\widehat{\tilde{\phi}}, \widehat{\phi}](2\xi) &= \widehat{\tilde{a}}(\xi)[\widehat{\tilde{\phi}}, \widehat{\phi}](\xi)\overline{\widehat{a}(\xi)}^{\mathsf{T}} + \widehat{\tilde{a}}(\xi+\pi)[\widehat{\tilde{\phi}}, \widehat{\phi}](\xi+\pi)\overline{\widehat{a}(\xi+\pi)}^{\mathsf{T}} \\
&= \widehat{\tilde{a}}(\xi)\overline{\widehat{a}(\xi)}^{\mathsf{T}} + \widehat{\tilde{a}}(\xi+\pi)\overline{\widehat{a}(\xi+\pi)}^{\mathsf{T}},
\end{aligned}
$$

from which we have (6.4.14). Hence, item (i) is verified.

We now prove items (iii) and (iv). Since $\operatorname{sm}(\phi) = \operatorname{sm}(a) > \tau \geqslant 0$ and $m \leqslant \tau$, we have $\phi \in (H^m(\mathbb{R}))^r$. Since the integer shifts of ϕ are stable, by Proposition 5.6.2 and $H^m(\mathbb{R}) = W_2^m(\mathbb{R})$, we see that (5.6.3) holds and there exists $\upsilon \in (l_0(\mathbb{Z}))^{1\times r}$ satisfying $\widehat{\upsilon}(0)\widehat{\phi}(0) = 1$, (5.6.4) and (5.6.6). By Proposition 5.5.2 and (5.6.6), we have $\mathsf{p} = \mathsf{p} * \upsilon * \phi$ for all $\mathsf{p} \in \mathbb{P}_m$. Let $\mathsf{p}_j(x) := x^j$ for $j \in \mathbb{N}_0$. By (6.4.13), for $j = 0, \ldots, m$, we have

$$
i^j\widehat{\tilde{\phi}}^{(j)}(0) = \langle \tilde{\phi}, \mathsf{p}_j \rangle = \langle \tilde{\phi}, (\mathsf{p}_j * \upsilon) * \phi \rangle = \overline{\mathsf{p}_j * \upsilon(0)}^{\mathsf{T}} = i^j\overline{\widehat{\upsilon}^{(j)}(0)}^{\mathsf{T}},
$$

where we used Lemma 1.2.1 in the last identity. This proves (6.4.15). Hence, item (iv) holds. Item (iii) follows directly from (6.4.15) and (5.6.6).

Sufficiency ($\Leftarrow$). By item (iv), (5.6.3) holds and by Corollary 5.6.10, the condition $\operatorname{sm}(a) > \tau$ in item (ii) implies that a satisfies order $m+1$ sum rules with the matching filter $\upsilon \in (l_0(\mathbb{Z}))^{1\times r}$. Since $\widehat{\upsilon}(0)\widehat{\phi}(0) = 1 \neq 0$, by item (v) of Theorem 5.6.4 and item (iv), we can assume that a takes the form in (5.6.22) satisfying (5.6.28) and $\widehat{a_{2,1}}(\xi) = \mathscr{O}(|\xi|^{m+1})$ as $\xi \to 0$. Note that $\widehat{\upsilon}(\xi) = (\widehat{c}(\xi), 0, \ldots, 0) + \mathscr{O}(|\xi|^{m+1})$ as $\xi \to 0$ with $\widehat{c}(0) \neq 0$ and $c \in l_0(\mathbb{Z})$. Without loss of generality, we can further assume that $\upsilon = (c, 0, \ldots, 0)$ with $\widehat{c}(0) = 1$. Consequently, by item (v) of Theorem 5.6.4, items (iii) and (iv) imply

$$
\begin{aligned}
\widehat{\phi}(\xi) &= (1/\widehat{c}(\xi), 0, \ldots, 0)^{\mathsf{T}} + \mathscr{O}(|\xi|^{m+1}), \quad \xi \to 0, \\
\widehat{\tilde{\phi}}(\xi) &= (\overline{\widehat{c}(\xi)}, 0, \ldots, 0)^{\mathsf{T}} + \mathscr{O}(|\xi|^{m+1}), \quad \xi \to 0.
\end{aligned}
\tag{6.4.16}
$$

Let φ and ψ be as in Theorem 6.4.4 (if necessary, choose m in Theorem 6.4.4 to be sufficiently large) such that $\operatorname{sm}(\varphi) \geqslant m+1$ and $\operatorname{vm}(\psi) \geqslant m+1$. By Lemma 5.6.7 and $\widehat{c}(0) = 1 = \widehat{\varphi}(0)$, there exists $u \in l_0(\mathbb{Z})$ such that $\|1 - \widehat{u}\|_{L_\infty(\mathbb{R})} \leqslant 1/2$ and $\widehat{u}(\xi) = [\widehat{c}(\xi)\widehat{\varphi}(\xi)]^{-1} + \mathscr{O}(|\xi|^{m+1})$ as $\xi \to 0$. Define

$$
\begin{aligned}
f &:= (u * \varphi, \psi, 2^{1/2}\psi(2\cdot), \ldots, 2^{(r-2)/2}\psi(2^{r-2}\cdot))^{\mathsf{T}}, \\
\tilde{f} &:= (\tilde{f}_1, \psi, 2^{1/2}\psi(2\cdot), \ldots, 2^{(r-2)/2}\psi(2^{r-2}\cdot))^{\mathsf{T}},
\end{aligned}
$$

where $\widehat{\tilde{f}_1}(\xi) = \widehat{\varphi}(\xi)/\overline{\widehat{u}(\xi)}$. Since $\|1 - \widehat{u}\|_{L_\infty(\mathbb{R})} \leqslant 1/2$ implies $|\widehat{u}(\xi)| \geqslant 1/2$, the function $\tilde{f}_1$ is a well-defined function in $H^{m+1}(\mathbb{R})$. Since $\{\varphi; \psi\}$ is an orthogonal wavelet in $L_2(\mathbb{R})$ and $\operatorname{sm}(\varphi) \geqslant m+1$, we see that $f, \tilde{f} \in H^{m+1}(\mathbb{R})$ and $[\widehat{f}, \widehat{\tilde{f}}](\xi) = I_r$ for a.e. $\xi \in \mathbb{R}$. Moreover, by $\widehat{u}(\xi) = [\widehat{c}(\xi)\widehat{\varphi}(\xi)]^{-1} + \mathscr{O}(|\xi|^{m+1})$ and $|\widehat{\varphi}(\xi)|^2 = 1 + \mathscr{O}(|\xi|^{m+1})$ as $\xi \to 0$, we deduce from (6.4.16) that

$$\begin{aligned} \widehat{f}(\xi) &= (1/\widehat{c}(\xi), 0, \ldots, 0)^{\mathsf{T}} + \mathscr{O}(|\xi|^{m+1}) = \widehat{\phi}(\xi) + \mathscr{O}(|\xi|^{m+1}), \quad \xi \to 0,\\ \widehat{\tilde{f}}(\xi) &= (\overline{\widehat{c}(\xi)}, 0, \ldots, 0)^{\mathsf{T}} + \mathscr{O}(|\xi|^{m+1}) = \widehat{\tilde{\phi}}(\xi) + \mathscr{O}(|\xi|^{m+1}), \quad \xi \to 0. \end{aligned} \tag{6.4.17}$$

By item (iii) and $m \geqslant 0$, we trivially have $\widehat{\phi}(0) \neq 0$ and $\widehat{\tilde{\phi}}(0) \neq 0$. Therefore, by $\widehat{\phi}(0) = \widehat{a}(0)\widehat{\phi}(0)$ and $\widehat{\tilde{\phi}}(0) = \widehat{\tilde{a}}(0)\widehat{\tilde{\phi}}(0)$, we conclude that 1 is an eigenvalue of both $\widehat{a}(0)$ and $\widehat{\tilde{a}}(0)$. Applying Corollary 5.7.7 and (5.7.19) in Theorem 5.7.6 we deduce from item (ii) and $\tau \geqslant 0$ that

$$\|\widehat{a}(0)\| = 1, \quad \left\|\widehat{\tilde{a}}(0)\right\| < 2^{\tau}, \ \rho_{m+1}(a, \upsilon)_2 < 2^{1/2-\tau}, \ \rho_0(\tilde{a}, \tilde{\upsilon})_2 < 2^{1/2+\tau}. \tag{6.4.18}$$

Define $f_n := \mathcal{R}_a^n f$ and $\tilde{f}_n := \mathcal{R}_{\tilde{a}}^n \tilde{f}$ for $n \in \mathbb{N}$. Since (6.4.17) and (6.4.18) hold, applying Theorem 5.1.2, we see that $\lim_{n\to\infty} \widehat{f_n}(\xi) = \widehat{\phi}(\xi)$ and $\lim_{n\to\infty} \widehat{\tilde{f}_n}(\xi) = \widehat{\tilde{\phi}}(\xi)$ for all $\xi \in \mathbb{R}$. Define $g := \mathcal{R}_a f - f$ and $\tilde{g} := \mathcal{R}_{\tilde{a}} \tilde{f} - \tilde{f}$. By (6.4.17), we have $\widehat{g}(\xi) = \mathscr{O}(|\xi|^{m+1})$ and $\widehat{\tilde{g}}(\xi) = \mathscr{O}(|\xi|^{m+1})$ as $\xi \to 0$. Since $g, \tilde{g} \in (H^{m+1}(\mathbb{R}))^r$, by Lemma 5.5.6, there exist $h, \tilde{h} \in (H^{m+1}(\mathbb{R}))^r$ such that $g = \nabla^{m+1} h$ and $\tilde{g} = \nabla^{m+1}\tilde{h}$. Since $h \in (H^{\tau}(\mathbb{R}))^r$ has compact support, by Lemma 6.3.2, we have

$$C_1 := \left\| \sum_{k\in\mathbb{Z}} \|\widehat{h}(\cdot + 2\pi k)\|_{l_2}^2 (1 + |\cdot + 2\pi k|^2)^{\tau} \right\|_{L_\infty(\mathbb{R})} < \infty.$$

Let $\widehat{a_n}(\xi) = \widehat{a}(2^{n-1}\xi) \cdots \widehat{a}(2\xi)\widehat{a}(\xi)$. By induction on $n \in \mathbb{N}$, we have

$$f_{n+1} - f_n = \mathcal{R}_a^n g = 2^n \sum_{k\in\mathbb{Z}} a_n(k) g(2^n \cdot - k) = 2^n \sum_{k\in\mathbb{Z}} [\nabla^{m+1} a_n](k) h(2^n \cdot - k).$$

That is, $f_{n+1} - f_n = 2^n[(\nabla^{m+1} a_n) * h](2^n \cdot)$. Observe that

$$\|\eta(2^{\mu}\cdot)\|_{H^{\nu}(\mathbb{R})}^2 = 2^{(2\nu-1)\mu} \frac{1}{2\pi} \int_{\mathbb{R}} |\widehat{\eta}(\xi)|^2 (2^{-2\mu} + |\xi|^2)^{\nu} d\xi, \quad \forall\, \mu, \nu \in \mathbb{R}. \tag{6.4.19}$$

Using the above two identities, we conclude that

$$\begin{aligned} \|f_{n+1} - f_n\|_{(H^{\tau}(\mathbb{R}))^r}^2 &= 2^{2n} 2^{(2\tau-1)n} \frac{1}{2\pi} \int_{\mathbb{R}} \|\widehat{\nabla^{m+1} a_n}(\xi) \widehat{h}(\xi)\|_{l_2}^2 (2^{-2n} + |\xi|^2)^{\tau} d\xi \\ &\leqslant 2^{(2\tau+1)n} \frac{1}{2\pi} \int_{\mathbb{R}} \|\widehat{\nabla^{m+1} a_n}(\xi)\|_{l_2}^2 \|\widehat{h}(\xi)\|_{l_2}^2 (1 + |\xi|^2)^{\tau} d\xi \end{aligned}$$

$$= 2^{(2\tau+1)n}\frac{1}{2\pi}\int_{-\pi}^{\pi}\|\widehat{\nabla^{m+1}a_n}(\xi)\|_{l_2}^2\sum_{k\in\mathbb{Z}}\|\widehat{h}(\xi+2\pi k)\|_{l_2}^2(1+|\xi+2\pi k|^2)^{\tau}d\xi$$
$$\leqslant 2^{(2\tau+1)n}C_1\|\nabla^{m+1}a_n\|_{(l_2(\mathbb{Z}))^{r\times r}}^2,$$

where we used $(2^{-2n}+|\xi|^2)^{\tau}\leqslant(1+|\xi|^2)^{\tau}$ by $\tau\geqslant 0$ and $n\in\mathbb{N}$. Note that $\nabla^{m+1}\boldsymbol{\delta}\in\mathscr{V}_{m,\upsilon}$ and $\rho_{m+1}(a,\upsilon)_2<2^{1/2-\tau}$. There exist $C_2>0$ and $0<\rho_0<1$ such that

$$\|\nabla^{m+1}a_n\|_{(l_2(\mathbb{Z}))^{r\times r}}=\|a_n*\nabla^{m+1}\boldsymbol{\delta}\|_{(l_2(\mathbb{Z}))^{r\times r}}\leqslant C_2\rho_0^n2^{(-1/2-\tau)n},\qquad\forall\, n\in\mathbb{N}.$$

Thus, we obtain

$$\|f_{n+1}-f_n\|_{(H^{\tau}(\mathbb{R}))^r}^2\leqslant C_1C_2^2\rho_0^{2n},\qquad\forall\, n\in\mathbb{N}.$$

Therefore, $\{f_n\}_{n=1}^{\infty}$ is a Cauchy sequence in $H^{\tau}(\mathbb{R})$. Since $\lim_{n\to\infty}\widehat{f_n}(\xi)=\widehat{\phi}(\xi)$ for all $\xi\in\mathbb{R}$, we conclude that $\phi\in(H^{\tau}(\mathbb{R}))^r$ and $\lim_{n\to\infty}\|f_n-\phi\|_{(H^{\tau}(\mathbb{R}))^r}=0$. Similarly, we have

$$\tilde{f}_{n+1}-\tilde{f}_n=\mathcal{R}_{\tilde{a}}^n\tilde{g}=2^n\sum_{k\in\mathbb{Z}}\tilde{a}_n(k)\tilde{g}(2^n\cdot-k)=2^n\sum_{k\in\mathbb{Z}}[\nabla^{m+1}\tilde{a}_n](k)\tilde{h}(2^n\cdot-k).$$

By $\widehat{\nabla^{m+1}\tilde{a}_n}(\xi)=(1-e^{-i\xi})^{m+1}\widehat{\tilde{a}_n}(\xi)$ and (6.4.19), we deduce from the above identity that

$$\|\tilde{f}_{n+1}-\tilde{f}_n\|_{(H^{-\tau}(\mathbb{R}))^r}^2=2^{2n}2^{(-2\tau-1)n}\frac{1}{2\pi}\int_{\mathbb{R}}\|\widehat{\nabla^{m+1}\tilde{a}_n}(\xi)\widehat{\tilde{h}}(\xi)\|_{l_2}^2(2^{-2n}+|\xi|^2)^{-\tau}d\xi$$
$$\leqslant 2^{(1-2\tau)n}\frac{1}{2\pi}\int_{\mathbb{R}}\|\widehat{\tilde{a}_n}(\xi)\|_{l_2}^2\|\widehat{\tilde{h}}(\xi)\|_{l_2}^2|1-e^{-i\xi}|^{2m+2}(2^{-2n}+|\xi|^2)^{-\tau}d\xi.$$

Since $|1-e^{-i\xi}|^{2m+2}=2^{2m+2}\sin^{2m+2}(\xi/2)\leqslant|\xi|^{2m+2}$, by $\tau\geqslant 0$, we have

$$\|\widehat{\tilde{h}}(\xi)\|_{l_2}^2|1-e^{-i\xi}|^{2m+2}(2^{-2n}+|\xi|^2)^{-\tau}\leqslant\|\widehat{\tilde{h}}(\xi)\|_{l_2}^2|\xi|^{2m+2}|\xi|^{-2\tau}$$
$$=\|\widehat{\tilde{h}}(\xi)\|_{l_2}^2|\xi|^{2(m+1-\tau)}.$$

Since $m\leqslant\tau<m+1$ (by the definition of $m=\lfloor|\tau|\rfloor$) and $\tilde{h}\in(H^{m+1}(\mathbb{R}))^r$, we see that

$$C_3:=\Big\|\sum_{k\in\mathbb{Z}}\|\widehat{\tilde{h}}(\cdot+2\pi k)\|_{l_2}^2|\cdot+2\pi k|^{2(m+1-\tau)}\Big\|_{L_{\infty}(\mathbb{T})}<\infty,$$

where we used $\tau \geqslant 0$. Consequently, we conclude that

$$\|\tilde{f}_{n+1} - \tilde{f}_n\|^2_{(H^{-\tau}(\mathbb{R}))^r} \leqslant 2^{(1-2\tau)n} C_3 \|\tilde{a}_n\|^2_{(l_2(\mathbb{Z}))^{r\times r}}.$$

By $\rho_0(\tilde{a}, \tilde{\upsilon})_2 < 2^{1/2+\tau}$, there exist $C_4 > 0$ and $0 < \rho_1 < 1$ such that $\|\tilde{a}_n\|_{(l_2(\mathbb{Z}))^{r\times r}} \leqslant C_4 \rho_1^n 2^{(-1/2+\tau)n}$ for all $n \in \mathbb{N}$. Thus,

$$\|\tilde{f}_{n+1} - \tilde{f}_n\|^2_{(H^{-\tau}(\mathbb{R}))^r} \leqslant C_3 C_4^2 \rho_1^{2n}, \qquad \forall\, n \in \mathbb{N}.$$

Therefore, $\{\tilde{f}_n\}_{n=1}^{\infty}$ is a Cauchy sequence in $H^{-\tau}(\mathbb{R})$. Since $\lim_{n\to\infty} \widehat{\tilde{f}_n}(\xi) = \widehat{\tilde{\phi}}(\xi)$ for all $\xi \in \mathbb{R}$, we conclude that $\tilde{\phi} \in (H^{-\tau}(\mathbb{R}))^r$ and $\lim_{n\to\infty} \|\tilde{f}_n - \tilde{\phi}\|_{(H^{-\tau}(\mathbb{R}))^r} = 0$.

Since $[\widehat{\tilde{f}}, \widehat{f}] = I_r$ a.e., by item (i) and induction, we have $[\widehat{\tilde{f}_n}, \widehat{f_n}](\xi) = I_r$ for a.e. $\xi \in \mathbb{R}$. Since $\lim_{n\to\infty} \|f_n - \phi\|_{(H^{\tau}(\mathbb{R}))^r} = 0$ and $\lim_{n\to\infty} \|\tilde{f}_n - \tilde{\phi}\|_{(H^{-\tau}(\mathbb{R}))^r} = 0$, we conclude that $\langle \tilde{\phi}, \phi(\cdot - k)\rangle = \lim_{n\to\infty} \langle \tilde{f}_n, f_n(\cdot - k)\rangle = \boldsymbol{\delta}(k) I_r$ for all $k \in \mathbb{Z}$. $\square$

The following result completely characterizes biorthogonal multiwavelets which are derived from refinable vector functions.

Theorem 6.4.6 *Let $a, b, \tilde{a}, \tilde{b} \in (l_0(\mathbb{Z}))^{r\times r}$. Let $\phi, \tilde{\phi}$ be $r \times 1$ vectors of compactly supported distributions satisfying* (6.4.1). *Define ψ and $\tilde{\psi}$ by*

$$\widehat{\psi}(\xi) := \widehat{b}(\xi/2)\widehat{\phi}(\xi/2) \quad \text{and} \quad \widehat{\tilde{\psi}}(\xi) := \widehat{\tilde{b}}(\xi/2)\widehat{\tilde{\phi}}(\xi/2). \tag{6.4.20}$$

For $\tau \in \mathbb{R}$, the pair $(\{\tilde{\phi}; \tilde{\psi}\}, \{\phi; \psi\})$ is a biorthogonal wavelet in $(H^{-\tau}(\mathbb{R}), H^{\tau}(\mathbb{R}))$ if and only if

(I) $(\{\tilde{a}; \tilde{b}\}, \{a; b\})$ is a biorthogonal wavelet filter bank, that is,

$$\begin{bmatrix} \widehat{\tilde{a}}(\xi) & \widehat{\tilde{a}}(\xi+\pi) \\ \widehat{\tilde{b}}(\xi) & \widehat{\tilde{b}}(\xi+\pi) \end{bmatrix} \begin{bmatrix} \overline{\widehat{a}(\xi)}^{\mathsf{T}} & \overline{\widehat{b}(\xi)}^{\mathsf{T}} \\ \overline{\widehat{a}(\xi+\pi)}^{\mathsf{T}} & \overline{\widehat{b}(\xi+\pi)}^{\mathsf{T}} \end{bmatrix} = I_{2r}; \tag{6.4.21}$$

(II) $\operatorname{sm}(a) > \tau$ *and* $\operatorname{sm}(\tilde{a}) > -\tau$*;*

(III) Items (iii) and (iv) of Theorem 6.4.5 are satisfied.

(IV) $\widehat{\psi}(\xi) = o(|\xi|^{-\tau})$ as $\xi \to 0$ (i.e., $\operatorname{vm}(\psi) > -\tau$*) if $\tau \leqslant 0$, and $\widehat{\tilde{\psi}}(\xi) = o(|\xi|^{\tau})$ as $\xi \to 0$ (i.e.,* $\operatorname{vm}(\tilde{\psi}) > \tau$*) if $\tau \geqslant 0$. [i.e., item (4) of Theorem 6.4.1 holds.]*

Proof Necessity ($\Rightarrow$). If $(\{\tilde{\phi}; \tilde{\psi}\}, \{\phi; \psi\})$ is a biorthogonal wavelet in $(H^{-\tau}(\mathbb{R}), H^{\tau}(\mathbb{R}))$, then (6.4.13) trivially holds and the pair $(\{\tilde{\phi}; \tilde{\psi}\}, \{\phi; \psi\})$ is a dual framelet in $(H^{-\tau}(\mathbb{R}), H^{\tau}(\mathbb{R}))$. Consequently, items (II) and (III) follow directly from Theorem 6.4.5. Note that (6.4.13) implies the condition in (6.4.4). Now (6.4.5) and item (IV) follow directly from Theorem 6.4.1 with $\theta = \tilde{\theta} = \boldsymbol{\delta} I_r$. Since $\widehat{\Theta} = I_r$ and the matrices in (6.4.5) are square matrices, the identity (6.4.21) in item (I) follows directly from (6.4.5) with $\widehat{\Theta} = I_r$.

Sufficiency ($\Leftarrow$). Note that item (I) directly implies item (i) of Theorem 6.4.5 and item (1) of Theorem 6.4.1 with $\widehat{\Theta} = I_r$. Therefore, all the conditions in items (i)–(iv) of Theorem 6.4.5 are satisfied. By Theorem 6.4.5, we see that $\phi \in (H^\tau(\mathbb{R}))^r$, $\tilde{\phi} \in (H^{-\tau}(\mathbb{R}))^r$, and (6.4.13) holds. In particular, item (3) of Theorem 6.4.1 holds. By (6.4.13) and (6.4.21), we can easily check that $\mathsf{AS}_0(\tilde{\phi}; \tilde{\psi})$ and $\mathsf{AS}_0(\phi; \psi)$ are biorthogonal to each other (see Theorem 4.5.16 for details). From item (III) and by $m \geqslant 0$, item (iii) of Theorem 6.4.5 implies item (2) of Theorem 6.4.1. Item (IV) is just item (4) of Theorem 6.4.1. Therefore, all the conditions in items (1)–(4) of Theorem 6.4.1 are satisfied with $\theta = \tilde{\theta} = \boldsymbol{\delta} I_r$. Thus, we conclude by Theorem 6.4.1 that $(\{\tilde{\phi}; \tilde{\psi}\}, \{\phi; \psi\})$ is a dual framelet in $(H^{-\tau}(\mathbb{R}), H^\tau(\mathbb{R}))$. This proves that $(\{\tilde{\phi}; \tilde{\psi}\}, \{\phi; \psi\})$ is a biorthogonal wavelet in $(H^{-\tau}(\mathbb{R}), H^\tau(\mathbb{R}))$. □

For a matrix-valued filter $a \in (l_0(\mathbb{Z}))^{r\times r}$, the space DS$(a)$ consisting of all the compactly supported distributional solutions to $\phi = 2\sum_{k\in\mathbb{Z}} a(k)\phi(2\cdot -k)$ may have dimension greater than one. As a consequence, the technical conditions in items (iii) and (iv) of Theorem 6.4.5 cannot be avoided in general. For a filter $a \in (l_0(\mathbb{Z}))^{r\times r}$ satisfying the natural condition (5.1.10) (that is, 1 is a simple eigenvalue of $\widehat{a}(0)$ and all $2^j, j \in \mathbb{N}$ are not eigenvalues of $\widehat{a}(0)$), by Theorem 5.1.3, up to a multiplicative constant, there exists a unique nontrivial compactly supported vector distribution ϕ satisfying $\widehat{\phi}(2\xi) = \widehat{a}(\xi)\widehat{\phi}(\xi)$ with $\widehat{\phi}(0) \neq 0$.

The following result is a special case of Theorem 6.4.6 showing that Theorem 6.4.6 can be greatly simplified under the natural condition in (5.1.10).

Corollary 6.4.7 *Let $a, b, \tilde{a}, \tilde{b} \in (l_0(\mathbb{Z}))^{r\times r}$ such that* (5.1.10) *holds for both a and $\tilde{a}$. Let $\phi, \tilde{\phi}$ be $r \times 1$ vectors of compactly supported distributions satisfying the refinement equations in* (6.4.1). *Define ψ and $\tilde{\psi}$ as in* (6.4.20) *by $\widehat{\psi}(2\xi) := \widehat{b}(\xi)\widehat{\phi}(\xi)$ and $\widehat{\tilde{\psi}}(2\xi) := \widehat{\tilde{b}}(\xi)\widehat{\tilde{\phi}}(\xi)$. For $\tau \in \mathbb{R}$, the pair $(\{\tilde{\phi}; \tilde{\psi}\}, \{\phi; \psi\})$ is a biorthogonal wavelet in $(H^{-\tau}(\mathbb{R}), H^\tau(\mathbb{R}))$ if and only if*

(i) $(\{\tilde{a}; \tilde{b}\}, \{a; b\})$ *is a biorthogonal wavelet filter bank (i.e.,* (6.4.21) *holds);*
(ii) $\mathrm{sm}(a) > \tau$ *and* $\mathrm{sm}(\tilde{a}) > -\tau$*;*
(iii) $\widehat{\tilde{\phi}}(0)^{\mathsf{T}}\widehat{\phi}(0) = 1$.

Proof Let m be the largest integer such that $m \leqslant |\tau|$. By Theorem 6.4.6, it suffices to prove the sufficiency part ($\Leftarrow$) by showing that items (III) and (IV) of Theorem 6.4.6 are satisfied. Without loss of generality, we assume $\tau \geqslant 0$. Since $\mathrm{sm}(a) > \tau$ and 1 is an eigenvalue of $\widehat{a}(0)$ due to (5.1.10), by Corollary 5.6.10, the equation (5.6.3) holds and the filter a must satisfy order $m+1$ sum rules with the matching filter $\upsilon \in (l_0(\mathbb{Z}))^{1\times r}$ in (5.6.43) satisfying $\widehat{\upsilon}(0)\widehat{\phi}(0) = 1$. Since $\mathrm{sm}(a) > \tau$ and $m \leqslant \tau < m+1$, by Theorem 6.3.3, we have $\phi \in (H^m(\mathbb{R}))^r$. It follows from Proposition 5.6.2 that (5.6.6) holds. By (6.4.14) and (5.6.28) for sum rules, we have

$$\begin{aligned}\overline{\widehat{\upsilon}(2\xi)}^{\mathsf{T}} &= \widehat{\tilde{a}}(\xi)\overline{\widehat{a}(\xi)}^{\mathsf{T}}\overline{\widehat{\upsilon}(2\xi)}^{\mathsf{T}} + \widehat{\tilde{a}}(\xi+\pi)\overline{\widehat{a}(\xi+\pi)}^{\mathsf{T}}\overline{\widehat{\upsilon}(2\xi)}^{\mathsf{T}} \\ &= \widehat{\tilde{a}}(\xi)\overline{\widehat{\upsilon}(\xi)}^{\mathsf{T}} + \mathscr{O}(|\xi|^{m+1}), \quad \xi \to 0.\end{aligned}$$

By $\widehat{\tilde{\phi}}(2\xi) = \widehat{\tilde{a}}(\xi)\widehat{\tilde{\phi}}(\xi)$, since both (5.1.10) and (5.1.11) hold with a being replaced by $\tilde{a}$, there must exist $c \in \mathbb{C}$ such that $\widehat{\tilde{\phi}}(\xi) = c\overline{\widehat{\upsilon}(\xi)}^{\mathsf{T}} + \mathscr{O}(|\xi|^{m+1})$ as $\xi \to 0$.

By item (iii), we have $\overline{\widehat{\tilde{\phi}}(0)}^{\mathsf{T}}\widehat{\phi}(0) = 1 = \widehat{\upsilon}(0)\widehat{\phi}(0)$. Consequently, we must have $c = 1$. This proves (6.4.15). Consequently, it follows from (5.6.6) and (6.4.15) that item (III) of Theorem 6.4.6 is satisfied.

We now show that item (IV) of Theorem 6.4.6 is a direct consequence of (6.4.15). Note that (6.4.21) implies $\widehat{\tilde{b}}(\xi)\overline{\widehat{a}(\xi)}^{\mathsf{T}} + \widehat{\tilde{b}}(\xi+\pi)\overline{\widehat{a}(\xi+\pi)}^{\mathsf{T}} = 0$. Therefore, by (6.4.15) and (5.6.28) for sum rules, we must have

$$0 = \widehat{\tilde{b}}(\xi)\overline{\widehat{a}(\xi)}^{\mathsf{T}}\overline{\widehat{\upsilon}(2\xi)}^{\mathsf{T}} + \widehat{\tilde{b}}(\xi+\pi)\overline{\widehat{a}(\xi+\pi)}^{\mathsf{T}}\overline{\widehat{\upsilon}(2\xi)}^{\mathsf{T}} = \widehat{\tilde{b}}(\xi)\overline{\widehat{\upsilon}(\xi)}^{\mathsf{T}} + \mathscr{O}(|\xi|^{m+1})$$
$$= \widehat{\tilde{b}}(\xi)\widehat{\tilde{\phi}}(\xi) + \mathscr{O}(|\xi|^{m+1}) = \widehat{\tilde{\psi}}(2\xi) + \mathscr{O}(|\xi|^{m+1}), \qquad \xi \to 0.$$

Since $m \leqslant \tau < m+1$, we trivially deduce from the above equation that $\operatorname{vm}(\tilde{\psi}) \geqslant m+1 > \tau$. This proves the first part of item (IV) of Theorem 6.4.6. The second part of item (IV) of Theorem 6.4.6 can be proved similarly. □

The following result constructs biorthogonal wavelets with interpolating properties and shows that Theorem 6.4.6 is indeed more general than Corollary 6.4.7.

Theorem 6.4.8 *Let $a \in (l_0(\mathbb{Z}))^{r\times r}$ be an order r Hermite interpolatory filter (see item (4) of Theorem 6.2.3). Suppose $\phi = (\phi_1, \ldots, \phi_r)^{\mathsf{T}}$ is an $r \times 1$ vector of compactly supported distributions satisfying $\widehat{\phi}(2\xi) = \widehat{a}(\xi)\widehat{\phi}(\xi)$ and $\widehat{\phi_1}(0) = 1$. Define*

$$\tilde{a} = \operatorname{diag}(1, 2, \ldots, 2^{r-1})\boldsymbol{\delta} \quad \text{and} \quad \tilde{\phi} = (\boldsymbol{\delta}, -D\boldsymbol{\delta}, \ldots, (-1)^{r-1}D^{r-1}\boldsymbol{\delta})^{\mathsf{T}}, \tag{6.4.22}$$

where $\boldsymbol{\delta}$ is the Dirac sequence on $\mathbb{Z}$ or distribution on $\mathbb{R}$. Define $\widehat{\psi}(2\xi) = \widehat{b}(\xi)\widehat{\phi}(\xi)$ and $\widehat{\tilde{\psi}}(2\xi) = \widehat{\tilde{b}}(\xi)\widehat{\tilde{\phi}}(\xi)$, where

$$\widehat{b}(\xi) := e^{-i\xi}\operatorname{diag}(1, 2^{-1}, \ldots, 2^{1-r}), \qquad \widehat{\tilde{b}}(\xi) := \widehat{a}(\xi)e^{-i\xi}\overline{\widehat{a}(\xi+\pi)}^{\mathsf{T}}\widehat{\tilde{a}}(\xi)$$

If $\operatorname{sm}(a) > r - 1/2$, then $(\{\tilde{\phi}; \tilde{\psi}\}, \{\phi; \psi\})$ is a biorthogonal wavelet in $(H^{-\tau}(\mathbb{R}), H^{\tau}(\mathbb{R}))$ for all $r - 1/2 < \tau < \operatorname{sm}(a)$ and ϕ is an order r Hermite interpolant.

Proof Note that $\operatorname{sm}(a) > r - 1/2$ implies $\operatorname{sm}_\infty(a) \geqslant \operatorname{sm}(a) - 1/2 > r - 1$. Since $\widehat{a}(0)\widehat{\phi}(0) = \widehat{\phi}(0)$ with $\widehat{\phi}(0) \neq 0$, by Corollary 5.6.10 and Theorem 6.2.3, the relation (5.6.3) holds with $m = r - 1$ and $\widehat{\upsilon}(\xi)\widehat{\phi}(\xi) = 1 + \mathscr{O}(|\xi|^r)$ as $\xi \to 0$. Hence, by Theorem 6.2.3, the refinable vector function ϕ is an order r Hermite interpolant.

By direct calculation, we see that $(\{\tilde{a}; \tilde{b}\}, \{a; b\})$ is a biorthogonal wavelet filter bank and $\operatorname{sm}(\tilde{a}) = \operatorname{sm}(\operatorname{diag}(1, \ldots, 2^{r-1})\boldsymbol{\delta}) = 1/2 - r$. Thus, items (I) and (II) of Theorem 6.4.6 are verified. Let m be the largest nonnegative integer such that $m < \operatorname{sm}(a)$. Let $\upsilon \in (l_0(\mathbb{Z}))^{1\times r}$ such that (6.2.8) holds. Then

$$\widehat{\upsilon}(\xi)\widehat{\phi}(\xi) = 1 + \mathscr{O}(|\xi|^{m+1}), \qquad \xi \to 0. \tag{6.4.23}$$

Note that $\widehat{\tilde{\phi}}(\xi) = (1, -i\xi, \ldots, (-i\xi)^{r-1})^{\mathsf{T}} = \overline{\widehat{\upsilon}(\xi)}^{\mathsf{T}} + \mathscr{O}(|\xi|^{m+1})$ as $\xi \to 0$. Therefore, by (6.4.23), item (III) of Theorem 6.4.6 is satisfied. By the definition of $\tilde{a}$ and $\tilde{\phi}$ in (6.4.22), it is trivial to check that $\widehat{\tilde{\phi}}(2\xi) = \widehat{\tilde{a}}(\xi)\widehat{\tilde{\phi}}(\xi)$. Since $\tau > r-1/2 > 0$, we now check item (IV) of Theorem 6.4.6.

$$\begin{aligned}
\widehat{\tilde{\psi}}(2\xi) &= \operatorname{diag}(1, 2, \ldots, 2^{r-1}) e^{-i\xi} \overline{\widehat{a}(\xi + \pi)}^{\mathsf{T}} \operatorname{diag}(1, 2, \ldots, 2^{r-1}) \widehat{\tilde{\phi}}(\xi) \\
&= \operatorname{diag}(1, 2, \ldots, 2^{r-1}) e^{-i\xi} \overline{\widehat{a}(\xi + \pi)}^{\mathsf{T}} \operatorname{diag}(1, 2, \ldots, 2^{r-1}) \overline{\widehat{\upsilon}(\xi)}^{\mathsf{T}} + \mathscr{O}(|\xi|^{m+1}) \\
&= \operatorname{diag}(1, 2, \ldots, 2^{r-1}) e^{-i\xi} \overline{\widehat{a}(\xi + \pi)}^{\mathsf{T}} \overline{\widehat{\upsilon}(2\xi)}^{\mathsf{T}} + \mathscr{O}(|\xi|^{m+1}) \\
&= \operatorname{diag}(1, 2, \ldots, 2^{r-1}) e^{-i\xi} \mathscr{O}(|\xi|^{m+1}) + \mathscr{O}(|\xi|^{m+1}) = \mathscr{O}(|\xi|^{m+1}),
\end{aligned}$$

as $\xi \to 0$, where we used $\widehat{\phi}(\xi) = \overline{\widehat{\upsilon}(\xi)}^{\mathsf{T}} + \mathscr{O}(|\xi|^{m+1}) = \overline{(1, i\xi, \ldots, (i\xi)^{r-1})}^{\mathsf{T}} + \mathscr{O}(|\xi|^{m+1})$ as $\xi \to 0$. Thus, item (IV) of Theorem 6.4.6 is verified. Now by Theorem 6.4.6, we conclude that $(\{\tilde{\phi}; \tilde{\psi}\}, \{\phi; \psi\})$ is a biorthogonal wavelet in $(H^{-\tau}(\mathbb{R}), H^{\tau}(\mathbb{R}))$ for all $r - 1/2 < \tau < \operatorname{sm}(a)$. □

For $\tilde{a} = \operatorname{diag}(1, 2, \ldots, 2^{r-1})\boldsymbol{\delta}$, note that $\operatorname{DS}(\tilde{a}) = \{(c_1, c_2 D\boldsymbol{\delta}, \ldots, c_r D^{r-1}\boldsymbol{\delta})^{\mathsf{T}} : c_1, c_2, \ldots, c_r \in \mathbb{C}\}$ and hence, $\dim(\operatorname{DS}(\tilde{a})) = r$. Also note that 2^j is an eigenvalue of $\widehat{\tilde{a}}(0)$ for all $j = 0, \ldots, r-1$. Therefore, Corollary 6.4.7 does not apply since the natural condition in (5.1.10) fails for the filter $\tilde{a}$.

As a direct consequence of Theorem 6.4.6 with $\tau = 0$, $\tilde{a} = a$ and $\tilde{b} = b$, we have the following result, which completely characterizes an orthogonal multiwavelet derived from a refinable vector function.

Corollary 6.4.9 *Let $a, b \in (l_0(\mathbb{Z}))^{r\times r}$. Let ϕ be an $r \times 1$ vector of compactly supported distributions satisfying $\widehat{\phi}(2\xi) = \widehat{a}(\xi)\widehat{\phi}(\xi)$. Define ψ by $\widehat{\psi}(\xi) = \widehat{b}(\xi/2)\widehat{\phi}(\xi/2)$. Then $\{\phi; \psi\}$ is an orthogonal wavelet in $L_2(\mathbb{R})$ if and only if*

(1) *$\{a; b\}$ is an orthogonal wavelet filter bank, i.e.,* (6.4.21) *holds with $\tilde{a} = a$ and $\tilde{b} = b$;*
(2) $\operatorname{sm}(a) > 0$;
(3) $\|\widehat{\phi}(0)\|_{l_2}^2 = \widehat{\phi}(0)^{\mathsf{T}}\overline{\widehat{\phi}(0)} = 1$.

Proof Since $\widehat{a}(0)\widehat{\phi}(0) = \widehat{\phi}(0)$ and $\widehat{\phi}(0) \neq 0$, we see that 1 is an eigenvalue of $\widehat{a}(0)$. By $\operatorname{sm}(a) > 0$ and Corollary 5.6.10, we see that (5.1.10) holds. The claims now follow directly from Corollary 6.4.7 with $\tau = 0$. □

Applying the characterizations of orthogonal multiwavelets in Corollary 6.4.9 and tight framelets in Theorem 6.4.2, we now present two examples to illustrate the theoretical results.

Example 6.4.1 A symmetric real-valued low-pass filter a with $\operatorname{fsupp}(a) = [-1, 1]$ and $\operatorname{sr}(a) \geqslant 1$ is given by

$$a = \left\{ \begin{bmatrix} \frac{1}{4} & \frac{1}{4} \\ \frac{t}{8} & \frac{t}{8} \end{bmatrix}, \begin{bmatrix} \frac{1}{2} & 0 \\ 0 & \frac{\sqrt{8-t^2}}{4} \end{bmatrix}, \begin{bmatrix} \frac{1}{4} & -\frac{1}{4} \\ -\frac{t}{8} & \frac{t}{8} \end{bmatrix} \right\}_{[-1,1]},$$

where $t \in [-2\sqrt{2}, 2\sqrt{2}]$. Then $\{a; b\}$ is an orthogonal wavelet filter bank, where

$$b = \left\{ \begin{bmatrix} -\frac{1}{4} & -\frac{1}{4} \\ \frac{\sqrt{8-t^2}}{8} & \frac{\sqrt{8-t^2}}{8} \end{bmatrix}, \begin{bmatrix} \frac{1}{2} & 0 \\ 0 & -\frac{t}{4} \end{bmatrix}, \begin{bmatrix} -\frac{1}{4} & \frac{1}{4} \\ -\frac{\sqrt{8-t^2}}{8} & \frac{\sqrt{8-t^2}}{8} \end{bmatrix} \right\}_{[-1,1]}.$$

For $t = 2$, we have $\widehat{a}(0) = I_2$, $\dim(\operatorname{DS}(a)) = 2$, and

$$\operatorname{DS}(a) = \operatorname{span}((\chi_{(0,1]}, -\chi_{(0,1]})^{\mathsf{T}}, (\chi_{(-1,0]}, \chi_{(-1,0]})^{\mathsf{T}}).$$

Note $\operatorname{spec}(\widehat{a}(0)) = \{1, \frac{t+\sqrt{8-t^2}}{4}\}$. For $t \in [-2\sqrt{2}, 2\sqrt{2}]\backslash\{2\}$, we have $\frac{|t+\sqrt{8-t^2}|}{4} < 1$ and consequently up to a multiplicative constant there is a unique nontrivial compactly supported distribution ϕ satisfying $\widehat{\phi}(2\xi) = \widehat{a}(\xi)\widehat{\phi}(\xi)$. By Corollary 6.4.2 with $\theta = \boldsymbol{\delta} I_r$, we conclude that $\{\phi; \psi\}$ is a tight framelet in $L_2(\mathbb{R})$ for all $t \in [-2\sqrt{2}, 2\sqrt{2}]$ provided $\|\widehat{\phi}(0)\|_{l_2} = 1$, where $\widehat{\psi}(2\xi) := \widehat{b}(\xi)\widehat{\phi}(\xi)$. By Theorem 5.8.4, we have $\operatorname{sm}(a) = 0$ for $t = 0$ and $t = 2$, while numerical calculation using Theorem 5.8.4 indicates that $\operatorname{sm}(a) > 0$ for all $t \in [-2\sqrt{2}, 2\sqrt{2}]\backslash\{0, 2\}$. Hence, by Corollary 6.4.9, we see that $\{\phi; \psi\}$ is an orthogonal wavelet in $L_2(\mathbb{R})$ for all $t \in [-2\sqrt{2}, 2\sqrt{2}]\backslash\{0, 2\}$ provided $\|\widehat{\phi}(0)\|_{l_2} = 1$. For the particular case $t = 0$, it is easy to directly check that $\phi = (B_2(\cdot - 1), 0)^{\mathsf{T}}$, where B_2 is the B-spline of order 2.

Moreover, for $t = \pm\sqrt{7}$, the filter a has order 2 sum rules with a matching filter υ satisfying $\widehat{\upsilon}(0) = (1, 0)$ and $\widehat{\upsilon}'(0) = (0, \frac{i}{1-t})$. For $t = \pm\sqrt{7}$, we have $\operatorname{sr}(a) = 2$ and by Theorem 5.8.4, we obtain $\rho_2(a, \upsilon)_2 = \frac{1}{4}\sqrt{18 + 4t}$. Therefore, $\operatorname{sm}(a) = \frac{1}{2} - \log_2 \rho_2(a, \upsilon)_2 \approx 1.054582$ for $t = -\sqrt{7}$; $\operatorname{sm}(a) \approx 0.081457$ for $t = \sqrt{7}$. Hence, $\{\phi; \psi\}$ is an orthogonal wavelet in $L_2(\mathbb{R})$ with $\operatorname{vm}(\psi) = 2$. See Fig. 6.4 for the graphs of the orthogonal wavelet $\{\phi; \psi\}$ in $L_2(\mathbb{R})$ with $t = -\sqrt{7}$.

Example 6.4.2 A symmetric real-valued low-pass filter a with $\operatorname{fsupp}(a) = [-1, 2]$ and $\operatorname{sr}(a) \geqslant 1$ is given by

$$a(-1) = \begin{bmatrix} -\frac{t^4-10t^2+1}{2(t^4+6t^2+1)} & \frac{4(t^3-t)}{t^4+6t^2+1} \\ -\epsilon\frac{\sqrt{2}(t^2-\epsilon 2\sqrt{2}t-1)^2}{8(t^4+6t^2+1)} & \frac{t^4-10t^2+1}{4(t^4+6t^2+1)} \end{bmatrix}, \quad a(0) = \begin{bmatrix} -\frac{t^4-10t^2+1}{2(t^4+6t^2+1)} & 0 \\ \epsilon\frac{\sqrt{2}(t^2+\epsilon 2\sqrt{2}t-1)^2}{8(t^4+6t^2+1)} & \frac{1}{2} \end{bmatrix},$$

$$a(1) = \begin{bmatrix} 0 & 0 \\ \epsilon\frac{\sqrt{2}(t^2+\epsilon 2\sqrt{2}t-1)^2}{8(t^4+6t^2+1)} & \frac{t^4-10t^2+1}{4(t^4+6t^2+1)} \end{bmatrix}, \quad a(2) = \begin{bmatrix} 0 & 0 \\ -\epsilon\frac{\sqrt{2}(t^2-\epsilon 2\sqrt{2}t-1)^2}{8(t^4+6t^2+1)} & 0 \end{bmatrix},$$

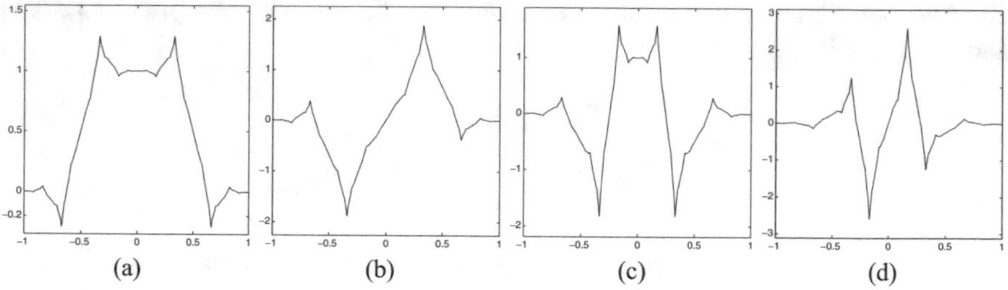

Fig. 6.4 The graphs of the orthogonal wavelet $\{\phi;\psi\}$ in $L_2(\mathbb{R})$ in Example 6.4.1 with $t=-\sqrt{7}$. (a) ϕ^1. (b) ϕ^2. (c) ψ^1. (d) ψ^2

with $\epsilon=\pm 1$ and $t\in\mathbb{R}$. Then $\{a;b\}$ is an orthogonal wavelet filter bank, where

$$b(-1)=\begin{bmatrix} -\epsilon\frac{\sqrt{2}(t^2-\epsilon 2\sqrt{2}t-1)^2}{8(t^4+6t^2+1)} & \frac{t^4-10t^2+1}{4(t^4+6t^2+1)}\\ -\epsilon\frac{(t^2-\epsilon 2\sqrt{2}t-1)^2}{4(t^4+6t^2+1)} & \frac{\sqrt{2}(t^4-10t^2+1)}{4(t^4+6t^2+1)}\end{bmatrix},\quad b(0)=\begin{bmatrix} \epsilon\frac{\sqrt{2}(t^2+\epsilon 2\sqrt{2}t-1)^2}{8(t^4+6t^2+1)} & -\frac{1}{2}\\ \epsilon\frac{(t^2+\epsilon 2\sqrt{2}t-1)^2}{4(t^4+6t^2+1)} & 0\end{bmatrix},$$

$$b(1)=\begin{bmatrix} \epsilon\frac{\sqrt{2}(t^2+\epsilon 2\sqrt{2}t-1)^2}{8(t^4+6t^2+1)} & \frac{t^4-10t^2+1}{4(t^4+6t^2+1)}\\ -\epsilon\frac{(t^2+\epsilon 2\sqrt{2}t-1)^2}{4(t^4+6t^2+1)} & -\frac{\sqrt{2}(t^4-10t^2+1)}{4(t^4+6t^2+1)}\end{bmatrix},\quad b(2)=\begin{bmatrix} -\epsilon\frac{\sqrt{2}(t^2-\epsilon 2\sqrt{2}t-1)^2}{8(t^4+6t^2+1)} & 0\\ \epsilon\frac{(t^2-\epsilon 2\sqrt{2}t-1)^2}{4(t^4+6t^2+1)} & 0\end{bmatrix},$$

with $\operatorname{fsupp}(b)=[-1,2]$. The matrix $\widehat{a}(0)$ has eigenvalues 1 and $-\frac{t^4-2t^2+1}{t^4+6t^2+1}\in[-1,0)$ taking value -1 only at $t=0$. Consequently up to a multiplicative constant there is a unique nontrivial compactly supported distribution ϕ satisfying $\widehat{\phi}(2\xi)=\widehat{a}(\xi)\widehat{\phi}(\xi)$. By Corollary 6.4.2 with $\theta=\boldsymbol{\delta}I_r$, we conclude that $\{\phi;\psi\}$ is a tight framelet in $L_2(\mathbb{R})$ for all $t\in\mathbb{R}$ provided $\|\widehat{\phi}(0)\|_{l_2}=1$, where $\widehat{\psi}(2\xi):=\widehat{b}(\xi)\widehat{\phi}(\xi)$. By Theorem 5.8.4, we have $\operatorname{sm}(a)=0$ for $t=0$, while numerical calculation using Theorem 5.8.4 indicates that $\operatorname{sm}(a)>0$ for all $t\in\mathbb{R}\backslash\{0\}$. Hence, by Corollary 6.4.9, we see that $\{\phi;\psi\}$ is an orthogonal wavelet in $L_2(\mathbb{R})$ for all $t\in\mathbb{R}\backslash\{0\}$ provided $\|\widehat{\phi}(0)\|_{l_2}=1$. If $t=\epsilon\sqrt{2}\pm\sqrt{3}$ and $\epsilon=\pm 1$, then $\operatorname{fsupp}(b)=[0,1]$ and $\operatorname{sm}(a)=0.5$; for example, if $t=\sqrt{2}+\sqrt{3}$ and $\epsilon=1$, then $\phi=(\sqrt{3}\chi_{[-2/3,-1/3]},\sqrt{3/2}\chi_{[-1/3,1/3]})^{\mathsf{T}}$. For the particular case $t=0$, it is easy to directly check that $\phi=(0,B_2(\cdot-1))^{\mathsf{T}}$, where B_2 is the B-spline of order 2.

Moreover, for $t=(\sqrt{6}+\epsilon\sqrt{2})/2$, the filter a has order 2 sum rules with a matching filter υ satisfying $\widehat{\upsilon}(0)=(2t,t^2-1)$ and $\widehat{\upsilon}'(0)=(-ti,0)$. For $t=(\sqrt{6}+\epsilon\sqrt{2})/2$, we have $\operatorname{sm}(a)=1.5$ for both $\epsilon=\pm 1$; in fact the refinable function for $\epsilon=-1$ is just $(\phi_1,-\phi_2)^{\mathsf{T}}$, where $\phi=(\phi_1,\phi_2)^{\mathsf{T}}$ is the refinable function for $\epsilon=1$. Hence, $\{\phi;\psi\}$ is an orthogonal wavelet in $L_2(\mathbb{R})$ with $\operatorname{vm}(\psi)=2$. See Fig. 6.5 for the graphs of the orthogonal wavelet $\{\phi;\psi\}$ in $L_2(\mathbb{R})$ with $t=(\sqrt{6}+\sqrt{2})/2$ and $\epsilon=1$.

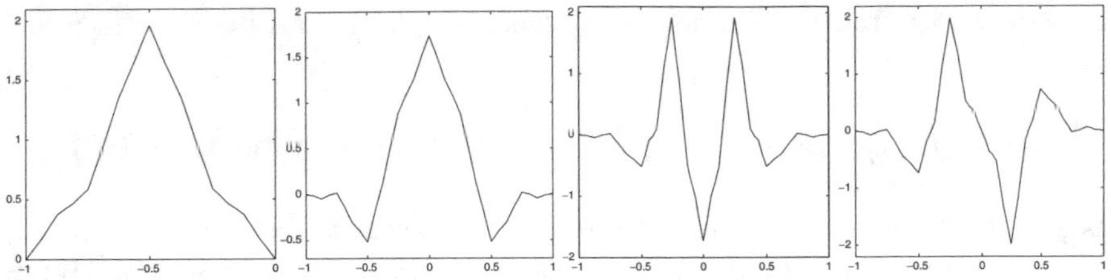

(a) (b) (c) (d)

Fig. 6.5 The graphs of the orthogonal wavelet $\{\phi; \psi\}$ in $L_2(\mathbb{R})$ in Example 6.4.2 with $t = (\sqrt{6}+\sqrt{2})/2$ and $\epsilon = 1$. **(a)** ϕ^1. **(b)** ϕ^2. **(c)** ψ^1. **(d)** ψ^2

6.5 Pairs of Biorthogonal Wavelet Filters with Increasing Orders of Sum Rules

As we have seen in Theorems 6.4.5 and 6.4.6, the key part for constructing biorthogonal wavelets is to build a pair $(\tilde{a}, a)$ of biorthogonal wavelet filters satisfying (6.4.14) and other desirable properties. The following result plays the key role for constructing a pair $(\tilde{a}, a)$ of biorthogonal wavelet filters such that $\tilde{a}$ satisfies any preassigned order of sum rules.

Theorem 6.5.1 *Let $a, \tilde{a} \in (l_0(\mathbb{Z}))^{r\times r}$ such that $(\tilde{a}, a)$ is a pair of biorthogonal wavelet filters satisfying $\widehat{\tilde{a}}(\xi)\overline{\widehat{a}(\xi)}^{\mathsf{T}} + \widehat{\tilde{a}}(\xi+\pi)\overline{\widehat{a}(\xi+\pi)}^{\mathsf{T}} = I_r$ for all $\xi \in \mathbb{R}$ (i.e., (6.4.14) holds). Suppose that 1 is a simple eigenvalue of $\widehat{\tilde{a}}(0)$ and all $2^j, j \in \mathbb{N}$ are not eigenvalues of $\widehat{\tilde{a}}(0)$ (i.e., (5.1.10) holds). If the filter $\tilde{a}$ satisfies order m sum rules with a matching filter $\tilde{\upsilon} \in (l_0(\mathbb{Z}))^{1\times r}$, then up to a multiplicative constant, all the vectors $\widehat{\tilde{\upsilon}}^{(j)}(0)$ for $j = 0, \dots, m-1$ are uniquely determined by*

$$\widehat{\tilde{\upsilon}}(\xi)\overline{\widehat{a}(\xi)}^{\mathsf{T}} = \widehat{\tilde{\upsilon}}(2\xi) + \mathscr{O}(|\xi|^m), \qquad \xi \to 0. \tag{6.5.1}$$

More precisely,

$$\begin{aligned}
&\widehat{\tilde{\upsilon}}(0)\overline{\widehat{a}(0)}^{\mathsf{T}} = \widehat{\tilde{\upsilon}}(0),\\
&\widehat{\tilde{\upsilon}}^{(j)}(0) = \sum_{\ell=0}^{j-1} \frac{j!}{\ell!(j-\ell)!}\widehat{\tilde{\upsilon}}^{(\ell)}(0)\overline{\widehat{a}^{(j-\ell)}(0)}^{\mathsf{T}}[2^j I_r - \overline{\widehat{a}(0)}^{\mathsf{T}}]^{-1}, \quad 1 \leqslant j < m.
\end{aligned} \tag{6.5.2}$$

Proof Since $\tilde{a}$ satisfies order m sum rules with a matching filter $\tilde{\upsilon} \in (l_0(\mathbb{Z}))^{1\times r}$, by the definition of sum rules, we have

$$\widehat{\tilde{\upsilon}}(2\xi)\widehat{\tilde{a}}(\xi) = \widehat{\tilde{\upsilon}}(\xi) + \mathscr{O}(|\xi|^m), \quad \widehat{\tilde{\upsilon}}(2\xi)\widehat{\tilde{a}}(\xi+\pi) = \mathscr{O}(|\xi|^m), \qquad \xi \to 0.$$

Now we deduce from the biorthogonality condition (6.4.14) and the above identities that

$$\widehat{\upsilon}(2\xi) = \widehat{\upsilon}(2\xi)\widehat{\tilde{a}}(\xi)\overline{\widehat{a}(\xi)}^{\mathsf{T}} + \widehat{\upsilon}(2\xi)\widehat{\tilde{a}}(\xi+\pi)\overline{\widehat{a}(\xi+\pi)}^{\mathsf{T}} = \widehat{\upsilon}(\xi)\overline{\widehat{a}(\xi)}^{\mathsf{T}} + \mathscr{O}(|\xi|^m)$$

as $\xi \to 0$. By our assumption on the filter a, we conclude that the above identity is equivalent to (6.5.2). Or equivalently, the identities in (5.1.11) hold and $\widehat{\upsilon}(\xi) = \overline{\widehat{\phi}(\xi)}^{\mathsf{T}} + \mathscr{O}(|\xi|^m)$ as $\xi \to 0$ if $\widehat{\phi}(2\xi) = \widehat{a}(\xi)\widehat{\phi}(\xi)$. □

For any given low-pass filter $a \in (l_0(\mathbb{Z}))^{r\times r}$, we now present a general algorithm for constructing all finitely supported dual filters $\tilde{a}$ of the given filter a such that $\tilde{a}$ has any given order of sum rules. The following algorithm is similar to the CBC (coset by coset) algorithm in Algorithm 2.6.2 but is for matrix-valued filters.

Algorithm 6.5.2 (CBC Algorithm) *Let $a \in (l_0(\mathbb{Z}))^{r\times r}$ such that 1 is a simple eigenvalue of $\widehat{a}(0)$ and all $2^j, j \in \mathbb{N}$ are not eigenvalues of $\widehat{a}(0)$ (i.e., (5.1.10) holds). Suppose that there exists a finitely supported dual filter $\mathring{a} \in (l_0(\mathbb{Z}))^{r\times r}$ of the filter a, that is, $(\mathring{a}, a)$ is a pair of biorthogonal wavelet filters satisfying $\widehat{\mathring{a}}(\xi)\overline{\widehat{a}(\xi)}^{\mathsf{T}} + \widehat{\mathring{a}}(\xi + \pi)\overline{\widehat{a}(\xi+\pi)}^{\mathsf{T}} = I_r$ for all $\xi \in \mathbb{R}$. Let $m \in \mathbb{N}$.*

(S1) Construct high-pass filters $b, \tilde{b} \in (l_0(\mathbb{Z}))^{r\times r}$ such that $(\{\mathring{a}; \tilde{b}\}, \{a; b\})$ is a biorthogonal wavelet filter bank satisfying (6.4.21) with $\tilde{a}$ being replaced by $\mathring{a}$. The existence of such high-pass filters $b, \tilde{b} \in (l_0(\mathbb{Z}))^{r\times r}$ is guaranteed by the well-known Quillen-Suslin Theorem and can be constructed from the pair $(\mathring{a}, a)$ (we do not provide details in this book about such a possible algorithm);

(S2) Define $\widehat{\tilde{a}}(\xi) := \widehat{\mathring{a}}(\xi) + \widehat{\Theta}(2\xi)\widehat{\tilde{b}}(\xi)$, where $\Theta \in (l_0(\mathbb{Z}))^{r\times r}$ is an auxiliary filter;

(S3) Calculate the vectors $\widehat{\upsilon}^{(j)}(0)$ for $j = 0, \dots, m-1$ with $\widehat{\upsilon}(0) \ne 0$ by the recursive formula in (6.5.2) (and therefore, the identity (6.5.1) holds);

(S4) Determine all the parameters in the auxiliary filter Θ by solving the following system of linear equations induced by

$$\widehat{\upsilon}(2\xi)\widehat{\Theta}(2\xi) = \widehat{\upsilon}(\xi)\overline{\widehat{b}(\xi)}^{\mathsf{T}} + \mathscr{O}(|\xi|^m), \quad \xi \to 0. \tag{6.5.3}$$

Then there always exists a solution $\Theta \in (l_0(\mathbb{Z}))^{r\times r}$ to the system of linear equations induced by (6.5.3) (we often choose Θ with the smallest len(Θ) *so that (6.5.3) has a solution). Moreover, the constructed filter $\tilde{a}$ is a dual filter of a such that $\tilde{a}$ has order m sum rules with the matching filter $\tilde{\upsilon}$, and $(\{\tilde{a}; \tilde{b}\}, \{a; \mathring{b}\})$ is a biorthogonal wavelet filter bank with $\widehat{\mathring{b}}(\xi) := \widehat{b}(\xi) - \overline{\widehat{\Theta}(\xi)}^{\mathsf{T}}\widehat{a}(\xi)$.*

Proof Directly checking (6.4.14), we see that $\tilde{a}$ is a dual filter of the filter a and $(\{\tilde{a}; \tilde{b}\}, \{a; \mathring{b}\})$ is a biorthogonal wavelet filter bank. We first prove that (6.5.3) always has a solution $\Theta \in (l_0(\mathbb{Z}))^{r\times r}$. Since $\widehat{\upsilon}(0) \ne 0$, by Theorem 5.6.4, there exist a strongly invertible sequence $U \in (l_0(\mathbb{Z}))^{r\times r}$ and $c \in l_0(\mathbb{Z})$ such that $\widehat{c}(0) = 1$ and

$$\widehat{\upsilon}(\xi) - [\widehat{c}(\xi), 0, \dots, 0]\widehat{U}(\xi) + \mathscr{O}(|\xi|^m), \qquad \xi \to 0.$$

It is now trivial to see that there exists $\mathring{\Theta} \in (l_0(\mathbb{Z}))^{r\times r}$ satisfying

$$[1, 0, \dots, 0]\widehat{\mathring{\Theta}}(2\xi) = [\widehat{c}(2\xi)]^{-1}\widehat{\upsilon}(\xi)\overline{\widehat{b}(\xi)}^{\mathsf{T}} + \mathscr{O}(|\xi|^m), \quad \xi \to 0.$$

Define $\widehat{\Theta}(\xi) := \widehat{U}(\xi)^{-1}\widehat{\mathring{\Theta}}(\xi)$. Since U is strongly invertible, we have $\Theta \in (l_0(\mathbb{Z}))^{r\times r}$ and $\widehat{\Theta}$ satisfies (6.5.3).

We now show that $\tilde{a}$ has order m sum rules. Since $(\{\mathring{a}; \tilde{b}\}, \{a; b\})$ is a biorthogonal wavelet filter bank, the identity (6.4.21) holds with $\tilde{a}$ being replaced by $\mathring{a}$, i.e.,

$$\begin{bmatrix} \overline{\widehat{a}(\xi)}^{\mathsf{T}} & \overline{\widehat{b}(\xi)}^{\mathsf{T}} \\ \overline{\widehat{a}(\xi+\pi)}^{\mathsf{T}} & \overline{\widehat{b}(\xi+\pi)}^{\mathsf{T}} \end{bmatrix} \begin{bmatrix} \widehat{\mathring{a}}(\xi) & \widehat{\mathring{a}}(\xi+\pi) \\ \widehat{\tilde{b}}(\xi) & \widehat{\tilde{b}}(\xi+\pi) \end{bmatrix} = I_{2r}.$$

By (6.5.3) and (6.5.1), it follows from $\widehat{\tilde{a}}(\xi) = \widehat{\mathring{a}}(\xi) + \widehat{\Theta}(2\xi)\widehat{\tilde{b}}(\xi)$ and $\overline{\widehat{a}(\xi)}^{\mathsf{T}}\widehat{\mathring{a}}(\xi) + \overline{\widehat{b}(\xi)}^{\mathsf{T}}\widehat{\tilde{b}}(\xi) = I_r$ (from the above identity) that as $\xi \to 0$,

$$\begin{aligned}
\widehat{\tilde{\upsilon}}(2\xi)\widehat{\tilde{a}}(\xi) &= \widehat{\tilde{\upsilon}}(2\xi)\widehat{\mathring{a}}(\xi) + \widehat{\tilde{\upsilon}}(2\xi)\widehat{\Theta}(2\xi)\widehat{\tilde{b}}(\xi) \\
&= \widehat{\tilde{\upsilon}}(2\xi)\widehat{\mathring{a}}(\xi) + \widehat{\tilde{\upsilon}}(\xi)\overline{\widehat{b}(\xi)}^{\mathsf{T}}\widehat{\tilde{b}}(\xi) + \mathscr{O}(|\xi|^m) \\
&= \widehat{\tilde{\upsilon}}(2\xi)\widehat{\mathring{a}}(\xi) + \widehat{\tilde{\upsilon}}(\xi)[I_r - \overline{\widehat{a}(\xi)}^{\mathsf{T}}\widehat{\mathring{a}}(\xi)] + \mathscr{O}(|\xi|^m) \\
&= \widehat{\tilde{\upsilon}}(\xi) + [\widehat{\tilde{\upsilon}}(2\xi) - \widehat{\tilde{\upsilon}}(\xi)\overline{\widehat{a}(\xi)}^{\mathsf{T}}]\widehat{\mathring{a}}(\xi) + \mathscr{O}(|\xi|^m) = \widehat{\tilde{\upsilon}}(\xi) + \mathscr{O}(|\xi|^m).
\end{aligned}$$

By $\overline{\widehat{a}(\xi)}^{\mathsf{T}}\widehat{\mathring{a}}(\xi+\pi) + \overline{\widehat{b}(\xi)}^{\mathsf{T}}\widehat{\tilde{b}}(\xi+\pi) = 0$ and (6.5.1), we have

$$\begin{aligned}
\widehat{\tilde{\upsilon}}(2\xi)\widehat{\tilde{a}}(\xi+\pi) &= \widehat{\tilde{\upsilon}}(2\xi)\widehat{\mathring{a}}(\xi+\pi) + \widehat{\tilde{\upsilon}}(2\xi)\widehat{\Theta}(2\xi)\widehat{\tilde{b}}(\xi+\pi) \\
&= \widehat{\tilde{\upsilon}}(2\xi)\widehat{\mathring{a}}(\xi+\pi) + \widehat{\tilde{\upsilon}}(\xi)\overline{\widehat{b}(\xi)}^{\mathsf{T}}\widehat{\tilde{b}}(\xi+\pi) + \mathscr{O}(|\xi|^m) \\
&= [\widehat{\tilde{\upsilon}}(2\xi) - \widehat{\tilde{\upsilon}}(\xi)\overline{\widehat{a}(\xi)}^{\mathsf{T}}]\widehat{\mathring{a}}(\xi+\pi) + \mathscr{O}(|\xi|^m) = \mathscr{O}(|\xi|^m).
\end{aligned}$$

This proves that $\tilde{a}$ has order m sum rules with the matching filter $\tilde{\upsilon}$. □

By Corollary 5.6.10, if $\operatorname{sm}(a) > 0$ and $\operatorname{sm}(\tilde{a}) > 0$, then the condition in (5.1.10) must hold for both a and $\tilde{a}$ and up to a multiplicative constant, there are unique vectors ϕ and $\tilde{\phi}$ of compactly supported distributions satisfying $\widehat{\phi}(2\xi) = \widehat{a}(\xi)\widehat{\phi}(\xi)$ and $\widehat{\tilde{\phi}}(2\xi) = \widehat{\tilde{a}}(\xi)\widehat{\tilde{\phi}}(\xi)$. If in addition $(\tilde{a}, a)$ is a pair of biorthogonal wavelet filters, then we must have $\widehat{\tilde{\phi}}(0)^{\mathsf{T}}\overline{\widehat{\phi}(0)} \ne 0$. Therefore, we can assume that $\widehat{\tilde{\phi}}(0)^{\mathsf{T}}\overline{\widehat{\phi}(0)} = 1$ (see Exercise 6.40). In the following examples, if not explicitly stated, the natural condition (5.1.10) is always satisfied for both a and $\tilde{a}$, and $\widehat{\tilde{\phi}}(0)^{\mathsf{T}}\overline{\widehat{\phi}(0)} = 1$.

Example 6.5.1 Let a be the order 2 Hermite interpolatory filter given in (6.2.4) such that $\operatorname{sr}(a) = 4$ and $\operatorname{sm}_p(a) = 2 + 1/p$ for all $1 \leqslant p \leqslant \infty$. By Algorithm 6.5.2, $\operatorname{diag}(1,2)\boldsymbol{\delta}$ is the unique dual filter of a supported inside $[-1,1]$ but it has no sum rules. If we define $\tilde{a}$ and $\tilde{\phi}$ as in (6.4.22) with $r = 2$ and define $\tilde{\psi}, \psi$ as in Theorem 6.4.8, then by Theorem 6.4.8, $(\{\tilde{\phi};\tilde{\psi}\},\{\phi;\psi\})$ is a biorthogonal wavelet in $(H^{-\tau}(\mathbb{R}), H^{\tau}(\mathbb{R}))$ for all $3/2 < \tau < 5/2$. Applying the CBC algorithm in Algorithm 6.5.2, we find that all the symmetric dual filter $\tilde{a}$ with $\operatorname{fsupp}(\tilde{a}) \subseteq [-2,2]$ and $\operatorname{sr}(\tilde{a}) \geqslant 2$ has the following parametric form: For $t_1, t_2 \in \mathbb{R}$,

$$\tilde{a} = \left\{ \begin{bmatrix} \frac{1}{16} - \frac{3}{8}t_1 & -\frac{1}{8}t_1 \\ \frac{15}{16}t_1 + \frac{1}{8}t_2 - \frac{7}{16} & \frac{1}{16}t_2 \end{bmatrix}, \begin{bmatrix} \frac{1}{4} & \frac{1}{4} - \frac{1}{2}t_1 \\ \frac{15}{4}t_1 - \frac{1}{4}t_2 - \frac{7}{4} & \frac{15}{4}t_1 - \frac{7}{4} \end{bmatrix}, \begin{bmatrix} \frac{3}{8} + \frac{3}{4}t_1 & 0 \\ 0 & \frac{15}{4}t_1 - \frac{1}{8}t_2 + \frac{1}{4} \end{bmatrix}, \right.$$
$$\left. \begin{bmatrix} \frac{1}{4} & -\frac{1}{4} + \frac{1}{2}t_1 \\ -\frac{15}{4}t_1 + \frac{1}{4}t_2 + \frac{7}{4} & \frac{15}{4}t_1 - \frac{7}{4} \end{bmatrix}, \begin{bmatrix} \frac{1}{16} - \frac{3}{8}t_1 & \frac{1}{8}t_1 \\ -\frac{15}{16}t_1 - \frac{1}{8}t_2 + \frac{7}{16} & \frac{1}{16}t_2 \end{bmatrix} \right\}_{[-2,2]}.$$

In particular, the sum rule $\operatorname{sr}(\tilde{a}) = 4$ if and only if $t_1 = \frac{5}{12}$ and $t_2 = \frac{9}{2}$. For this case, $\widehat{\tilde{a}}(0) = \operatorname{diag}(1, 23/16)$, $\det(\tilde{a}(2)) \ne 0$, $\operatorname{sm}(\tilde{a}) \approx -0.605049$ and by Theorem 6.4.5 $(\tilde{\phi},\phi)$ is a pair of biorthogonal functions in $(H^{-\tau}(\mathbb{R}), H^{\tau}(\mathbb{R}))$ with $-\operatorname{sm}(\tilde{a}) < \tau < 2.5$. The sum rule $\operatorname{sr}(\tilde{a}) \geqslant 3$ if and only if $t_2 = 15t_1 - \frac{7}{4}$ (calculation reveals $\operatorname{sm}(\tilde{a}) < 0$ regardless of the choice of t_1). Note that $\det(\tilde{a}(2)) = -\frac{1}{128}t_1t_2 + \frac{1}{256}t_2 - \frac{7}{128}t_1 + \frac{15}{128}t_1^2$ and $\det(\tilde{a}(2)) = 0$ if $t_2 = \frac{2t_1(7-15t_1)}{1-2t_1}$ (or $t_1 = \frac{1}{2}, t_2 = \pm\frac{\sqrt{210}}{30}$). If $t_1 = \frac{5}{16}$ and $t_2 = \frac{185}{48}$, then $\det(\tilde{a}(2)) = 0$, $\operatorname{sr}(\tilde{a}) = 2$ and $\operatorname{sm}(\tilde{a}) \approx 0.821496$. If $t_2 = \frac{2t_1(7-15t_1)}{1-2t_1}$ and $t_1 \ne \frac{1}{2}$, then $(\{\tilde{a};\tilde{b}\},\{a;b\})$ is a biorthogonal wavelet filter bank, where $b, \tilde{b}$ are given by

$$b = \left\{ \begin{bmatrix} -\frac{1}{4} & \frac{1+6t_1}{8(2t_1-1)} \\ \frac{1-t_1}{11} & \frac{1+15t_1}{11} \end{bmatrix}, \begin{bmatrix} \frac{1}{2} & 0 \\ 0 & \frac{28-60t_1}{11} \end{bmatrix}, \begin{bmatrix} -\frac{1}{4} & \frac{1+6t_1}{8(1-2t_1)} \\ \frac{t_1-1}{11} & \frac{1+15t_1}{11} \end{bmatrix} \right\}_{[-1,1]},$$

$$\tilde{b} = \left\{ \begin{bmatrix} \frac{6t_1-1}{16} & \frac{t_1}{8} \\ \frac{11(6t_1-1)}{256(2t_1-1)} & \frac{11t_1}{128(2t_1-1)} \end{bmatrix}, \begin{bmatrix} -\frac{1}{4} & \frac{2t_1-1}{4} \\ \frac{11}{64(1-2t_1)} & \frac{11}{64} \end{bmatrix}, \begin{bmatrix} \frac{5-6t_1}{8} & 0 \\ 0 & \frac{11(t_1-1)}{64(2t_1-1)} \end{bmatrix}, \right.$$
$$\left. \begin{bmatrix} -\frac{1}{4} & \frac{1-2t_1}{4} \\ \frac{11}{64(2t_1-1)} & \frac{11}{64} \end{bmatrix}, \begin{bmatrix} \frac{6t_1-1}{16} & -\frac{t_1}{8} \\ \frac{11(1-6t_1)}{256(2t_1-1)} & \frac{11t_1}{128(2t_1-1)} \end{bmatrix} \right\}_{[-2,2]},$$

with $\operatorname{fsupp}(b) = [-1,1]$ and $\operatorname{fsupp}(\tilde{b}) = [-2,2]$. See Fig. 6.6 for the graphs of the biorthogonal wavelet $(\{\tilde{\phi};\tilde{\psi}\},\{\phi;\psi\})$ in $L_2(\mathbb{R})$ with $t_1 = \frac{5}{16}$ and $t_2 = \frac{185}{48}$.

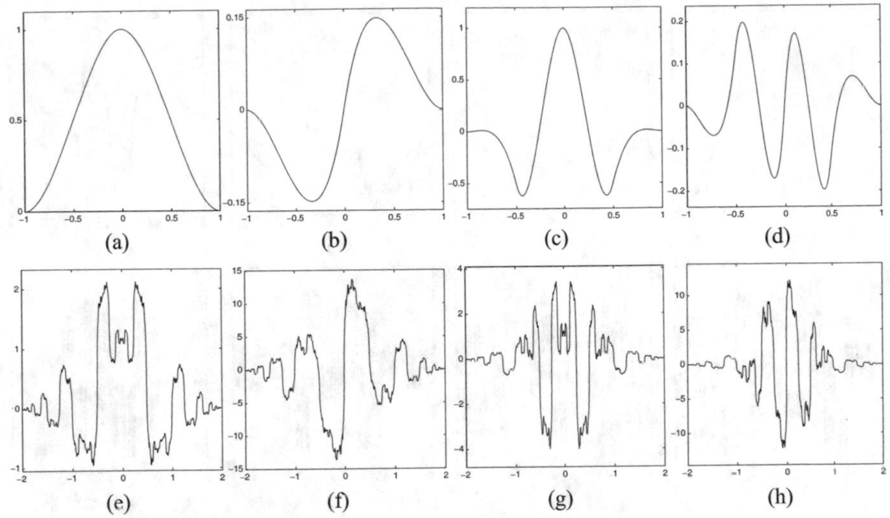

Fig. 6.6 The graphs of the biorthogonal wavelet $(\{\tilde{\phi};\tilde{\psi}\},\{\phi;\psi\})$ in $L_2(\mathbb{R})$, where $\phi=(\phi_1,\phi_2)^\mathsf{T}$ is the spline Hermite refinable interpolant in Example 6.2.1 and $\tilde{\phi}=(\tilde{\phi}_1,\tilde{\phi}_2)^\mathsf{T}$ is its dual refinable function with the dual filter $\tilde{a}$ constructed by Algorithm 6.5.2 as in Example 6.5.1 with $t_1=\frac{5}{16}$ and $t_2=\frac{185}{48}$. $\psi=(\psi_1,\psi_2)^\mathsf{T}$ is the wavelet function and $\tilde{\psi}=(\tilde{\psi}_1,\tilde{\psi}_2)^\mathsf{T}$ is the dual wavelet function. **(a)** ϕ_1. **(b)** ϕ_2. **(c)** ψ_1. **(d)** ψ_2. **(e)** $\tilde{\phi}_1$. **(f)** $\tilde{\phi}_2$. **(g)** $\tilde{\psi}_1$. **(h)** $\tilde{\psi}_2$

Example 6.5.2 Let a be the order 2 Hermite interpolatory filter given in item (2) of Theorem 6.2.9 (that is, (6.2.14) with $t_1=-1/16$ and $t_2=-1/8$) with $\operatorname{sr}(a)=3$ and $\operatorname{sm}_p(a)=2+1/p$ for all $1\leqslant p\leqslant\infty$. By Algorithm 6.5.2, all the symmetric dual filter $\tilde{a}$ with $\operatorname{fsupp}(\tilde{a})\subseteq[-1,1]$ and $\operatorname{sr}(\tilde{a})\geqslant 1$ has the following parametric form:

$$\tilde{a}=\left\{\begin{bmatrix}\frac{1}{4} & \frac{1}{8}\\ 2t_1 & t_1\end{bmatrix},\begin{bmatrix}\frac{1}{2} & 0\\ 0 & 2+2t_1\end{bmatrix},\begin{bmatrix}\frac{1}{4} & -\frac{1}{8}\\ -2t_1 & t_1\end{bmatrix}\right\}_{[-1,1]}$$

with $t_1\in\mathbb{R}$. Moreover, the pair $(\{\tilde{a};\tilde{b}\},\{a;b\})$ is a biorthogonal wavelet filter bank, where the high-pass filters b and $\tilde{b}$ are supported inside $[-1,1]$ and given by

$$b=\left\{\begin{bmatrix}-\frac{1}{4} & -\frac{1}{2}\\ -\frac{1}{16}-\frac{1}{16t_1} & -\frac{1}{8}-\frac{1}{8t_1}\end{bmatrix},\begin{bmatrix}\frac{1}{2} & 0\\ 0 & \frac{1}{4}\end{bmatrix},\begin{bmatrix}-\frac{1}{4} & \frac{1}{2}\\ \frac{1}{16}+\frac{1}{16t_1} & -\frac{1}{8}-\frac{1}{8t_1}\end{bmatrix}\right\}_{[-1,1]},$$

$$\tilde{b}=\left\{\begin{bmatrix}-\frac{1}{4} & -\frac{1}{8}\\ -2t_1 & -t_1\end{bmatrix},\begin{bmatrix}\frac{1}{2} & 0\\ 0 & -2t_1\end{bmatrix},\begin{bmatrix}-\frac{1}{4} & \frac{1}{8}\\ 2t_1 & -t_1\end{bmatrix}\right\}_{[-1,1]}.$$

If $t_1=-\frac{91}{128}$, then $\operatorname{sr}(\tilde{a})=1$ and $\operatorname{sm}(\tilde{a})\approx 0.184258$. Therefore, $(\{\tilde{\phi};\tilde{\psi}\},\{\phi;\psi\})$ is a biorthogonal wavelet in $L_2(\mathbb{R})$. Moreover, $\operatorname{sr}(\tilde{a})=2$ if and only if $t_1=-\frac{7}{8}$.

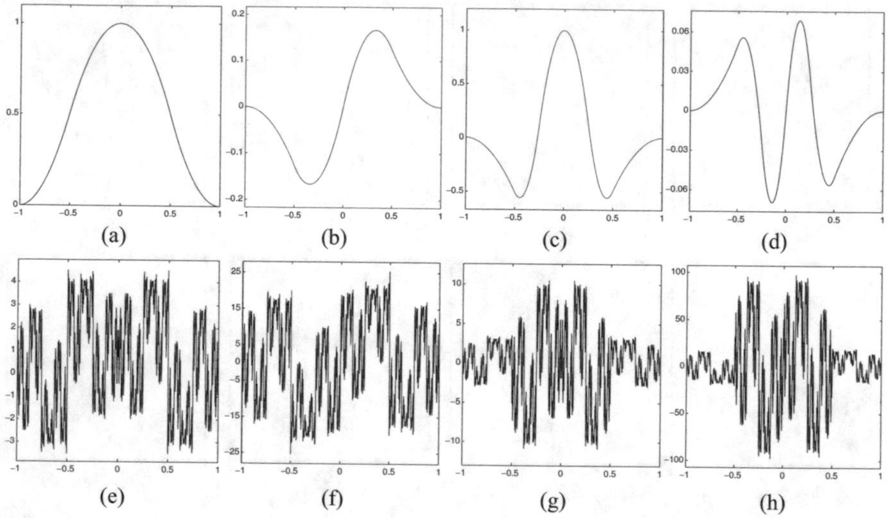

Fig. 6.7 The graphs of the biorthogonal wavelet $(\{\tilde{\phi};\tilde{\psi}\},\{\phi;\psi\})$ in $L_2(\mathbb{R})$, where $\phi=(\phi_1,\phi_2)^\mathsf{T}$ is the refinable (vector) function in item (2) of Theorem 6.2.9 and $\tilde{\phi}=(\tilde{\phi}_1,\tilde{\phi}_2)^\mathsf{T}$ is its dual refinable function with the dual filter $\tilde{a}$ constructed by Algorithm 6.5.2 as in Example 6.5.3 with $t_1=-\frac{91}{128}$ and $\operatorname{sr}(\tilde{a})=1$. $\psi=(\psi_1,\psi_2)^\mathsf{T}$ is the wavelet function and $\tilde{\psi}=(\tilde{\psi}_1,\tilde{\psi}_2)^\mathsf{T}$ is the dual wavelet function. **(a)** ϕ_1. **(b)** ϕ_2. **(c)** ψ_1. **(d)** ψ_2. **(e)** $\tilde{\phi}_1$. **(f)** $\tilde{\phi}_2$. **(g)** $\tilde{\psi}_1$. **(h)** $\tilde{\psi}_2$

For this case, by Theorem 5.8.4, we have $\rho_2(\tilde{a},\tilde{\upsilon})_2=\sqrt{37/8}$ and $\operatorname{sm}(\tilde{a})=\frac{1}{2}-\log_2\sqrt{37/8}\approx-0.604727$ and hence, $(\{\tilde{\phi};\tilde{\psi}\},\{\phi;\psi\})$ is a pair of biorthogonal wavelets in $(H^{-\tau}(\mathbb{R}),H^{\tau}(\mathbb{R}))$ with $-\operatorname{sm}(\tilde{a})<\tau<2.5$. See Fig. 6.7 for the graphs of the biorthogonal wavelet $(\{\tilde{\phi};\tilde{\psi}\},\{\phi;\psi\})$ in $L_2(\mathbb{R})$ with $t_1=-\frac{91}{128}$.

Example 6.5.3 Let a be the order 2 Hermite interpolatory filter given in item (3) of Theorem 6.2.9 (that is, (6.2.14) with $t_1=-1/12$ and $t_2=-1/4$) with $\operatorname{sr}(a)=2$ and $\operatorname{sm}_p(a)=1+1/p$ for all $1\leqslant p\leqslant\infty$. By Algorithm 6.5.2, all the symmetric dual filters $\tilde{a}$ with $\operatorname{fsupp}(\tilde{a})\subseteq[-1,1]$ and $\operatorname{sr}(\tilde{a})\geqslant 1$ have the following parametric form:

$$\tilde{a}=\left\{\begin{bmatrix}\frac{1}{4} & \frac{1}{12}\\ 3t_1 & t_1\end{bmatrix},\begin{bmatrix}\frac{1}{2} & 0\\ 0 & 2+4t_1\end{bmatrix},\begin{bmatrix}\frac{1}{4} & -\frac{1}{12}\\ -3t_1 & t_1\end{bmatrix}\right\}_{[-1,1]}$$

with $t_1\in\mathbb{R}$. Moreover, the pair $(\{\tilde{a};\tilde{b}\},\{a;b\})$ is a biorthogonal wavelet filter bank, where the high-pass filters b and $\tilde{b}$ are supported inside $[-1,1]$ and given by

$$b=\left\{\begin{bmatrix}-\frac{1}{4} & -\frac{3}{4}\\ -\frac{1}{12}-\frac{1}{24t_1} & -\frac{1}{4}-\frac{1}{8t_1}\end{bmatrix},\begin{bmatrix}\frac{1}{2} & 0\\ 0 & \frac{1}{4}\end{bmatrix},\begin{bmatrix}-\frac{1}{4} & \frac{3}{4}\\ \frac{1}{12}+\frac{1}{24t_1} & -\frac{1}{4}-\frac{1}{8t_1}\end{bmatrix}\right\}_{[-1,1]},$$

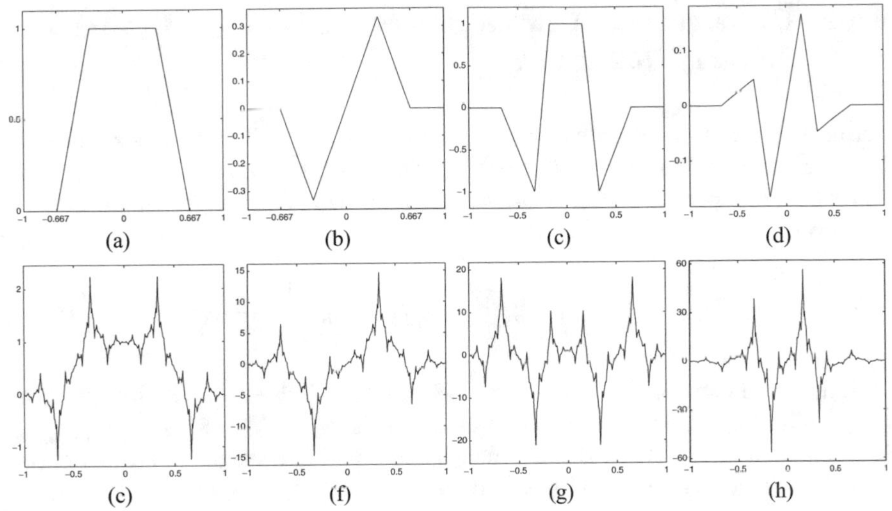

Fig. 6.8 The graphs of the biorthogonal wavelet $(\{\tilde\phi;\tilde\psi\},\{\phi;\psi\})$, where $\phi=(\phi_1,\phi_2)^{\mathsf T}$ is the refinable function in item (3) of Theorem 6.2.6 and $\tilde\phi=(\tilde\phi_1,\tilde\phi_2)^{\mathsf T}$ is its dual refinable function with the dual filter $\tilde a$ constructed by Algorithm 6.5.2 as in Theorem 6.2.9 with $t_1=-\frac{7}{16}$ and $\operatorname{sr}(\tilde a)=2$. $\psi=(\psi_1,\psi_2)^{\mathsf T}$ is the wavelet function and $\tilde\psi=(\tilde\psi_1,\tilde\psi_2)^{\mathsf T}$ is the dual wavelet function. **(a)** ϕ_1. **(b)** ϕ_2. **(c)** ψ_1. **(d)** ψ_2. **(e)** $\tilde\phi_1$. **(f)** $\tilde\phi_2$. **(g)** $\tilde\psi_1$. **(h)** $\tilde\psi_2$

$$\tilde b=\left\{\begin{bmatrix}-\frac14 & -\frac{1}{12}\\ -3t_1 & -t_1\end{bmatrix},\begin{bmatrix}\frac12 & 0\\ 0 & -4t_1\end{bmatrix},\begin{bmatrix}-\frac14 & \frac{1}{12}\\ 3t_1 & -t_1\end{bmatrix}\right\}_{[-1,1]}.$$

If $t_1=-\frac{53}{128}$, then $\operatorname{sr}(\tilde a)=1$ and $\operatorname{sm}(\tilde a)\approx 0.848293$. Therefore, $(\{\tilde\phi;\tilde\psi\},\{\phi,\psi\})$ is a biorthogonal wavelet in $L_2(\mathbb{R})$. Moreover, $\operatorname{sr}(\tilde a)=2$ if and only if $t_1=-\frac{7}{16}$. For this case, by Theorem 5.8.4, we have $\rho_2(\tilde a,\tilde\upsilon)_2=\sqrt{29/32}$ and $\operatorname{sm}(\tilde a)=\frac12-\log_2\sqrt{29/32}\approx 0.57101$. Hence, the pair $(\{\tilde\phi;\tilde\psi\},\{\phi;\psi\})$ is a biorthogonal wavelet in $L_2(\mathbb{R})$. See Fig. 6.8 for the graphs of the biorthogonal wavelet $(\{\tilde\phi;\tilde\psi\},\{\phi;\psi\})$ in $L_2(\mathbb{R})$ with $t_1=-\frac{7}{16}$.

6.6 Framelets/Wavelets with Filters of Hölder Class or Exponential Decay*

In the rest of this chapter we concentrate on scalar filters and scalar refinable functions. Though compactly supported framelets and wavelets are of importance in wavelet theory, there are many other examples of framelets and wavelets which do not have compact support, for example, the bandlimited wavelets and framelets constructed at the end of Sect. 4.6. In this section we study scalar framelets and wavelets with infinitely supported filters.

6.6.1 Convergence of Cascade Algorithms in $L_{2,\infty,0}(\mathbb{R})$ with Filters of Hölder Class

Let us first introduce weighted subspaces of $L_2(\mathbb{R})$. Recall that the bracket product in (4.1.5) is defined to be $[f, g](\xi) := \sum_{k\in\mathbb{Z}} f(\xi + 2\pi k)\overline{g(\xi + 2\pi k)}$. For $1 \leqslant p \leqslant \infty$ and $\tau \geqslant 0$, the space $L_{2,p,\tau}(\mathbb{R})$ denotes the subspace of all $f \in L_2(\mathbb{R})$ equipped with the norm:

$$\|f\|^2_{L_{2,p,\tau}(\mathbb{R})} := \left(\frac{1}{2\pi}\int_{-\pi}^{\pi} |[\widehat{e^{\tau|\cdot|}f}, \widehat{e^{\tau|\cdot|}f}](\xi)|^p d\xi\right)^{1/p} < \infty \qquad (6.6.1)$$

with the usual modification for $p = \infty$. We can also define $L_{2,p,\tau}(\mathbb{R})$ for $\tau < 0$ by using tempered distributions, but these spaces are larger than $L_2(\mathbb{R})$. Then $\|\cdot\|_{L_{2,p,\tau}(\mathbb{R})}$ is a norm and $L_{2,p,\tau}(\mathbb{R})$ is a Banach space (Exercise 6.23). We are interested in two particular cases. If $p = \infty$ and $\tau = 0$, then

$$\|f\|_{L_{2,\infty,0}(\mathbb{R})} = \sqrt{\|[\widehat{f}, \widehat{f}]\|_{L_\infty(\mathbb{T})}}, \qquad (6.6.2)$$

which is useful in the study of shift-invariant subspaces of $L_2(\mathbb{R})$ and stability of functions in $L_2(\mathbb{R})$ (for example, Proposition 4.4.4 in Sect. 4.4). If $p = 1$ and $\tau \geqslant 0$, by Plancherel's Theorem, we have

$$\|f\|^2_{L_{2,1,\tau}(\mathbb{R})} = \frac{1}{2\pi}\int_{\mathbb{R}} |\widehat{e^{\tau|\cdot|}f}(\xi)|^2 d\xi = \int_{\mathbb{R}} |f(x)|^2 e^{2\tau|x|} dx, \qquad (6.6.3)$$

which are natural subspaces to measure the exponential decay of a function in $L_2(\mathbb{R})$. In particular, $L_{2,1,0}(\mathbb{R}) = L_2(\mathbb{R})$ and all compactly supported functions $f \in L_2(\mathbb{R})$ belong to $L_{2,1,\tau}(\mathbb{R})$ for all $\tau \geqslant 0$.

To study convergence of a cascade algorithm in the spaces $L_{2,p,\tau}(\mathbb{R})$, we generalize the notion of $\operatorname{sm}(a)$ in (5.6.44) from finitely supported filters to infinitely supported filters. For 2π-periodic measurable functions $\widehat{u}$ and f, recall that the transition operator $\mathcal{T}_{\widehat{u}}$ in the frequency domain is defined to be

$$[\mathcal{T}_{\widehat{u}} f](\xi) := \overline{\widehat{u}(\xi/2)} f(\xi/2) + \overline{\widehat{u}(\xi/2 + \pi)} f(\xi/2 + \pi), \qquad \xi \in \mathbb{R}. \qquad (6.6.4)$$

For $\tau \in \mathbb{R}$, we define

$$\rho_\tau(\mathcal{T}_{|\widehat{a}|^2}) := \limsup_{n\to\infty} \|\mathcal{T}^n_{|\widehat{a}|^2}(|\sin(\cdot/2)|^\tau)\|^{1/n}_{L_\infty(\mathbb{T})}.$$

For a filter $a = \{a(k)\}_{k\in\mathbb{Z}}$, recall that $\widehat{a}(\xi) := \sum_{k\in\mathbb{Z}} a(k)e^{-ik\xi}$, $\xi \in \mathbb{R}$ and we define

$$\operatorname{sm}(a) := -\log_2 \sqrt{\rho(\mathcal{T}_{|\widehat{a}|^2})}, \qquad (6.6.5)$$

where

$$\rho(\mathcal{T}_{|\widehat{a}|^2}) := \inf\{\rho_\tau(\mathcal{T}_{|\widehat{a}|^2}) \ : \ |\widehat{a}(\cdot + \pi)|^2/|\sin(\cdot/2)|^\tau \in L_\infty(\mathbb{R}) \text{ and } \tau \geqslant 0\}.$$

If the above set is empty, then we define $\rho(\mathcal{T}_{|\widehat{a}|^2}) = \infty$. When $a \in l_0(\mathbb{Z})$ is a finitely supported filter, one can check that the above quantity $\mathrm{sm}(a)$ in (6.6.5) agrees with the definition $\mathrm{sm}(a)$ in (5.6.56) and $2\rho_{2m}(\mathcal{T}_{|\widehat{a}|^2}) = [\rho_m(a)_2]^2$ in (5.6.55) with $m := \mathrm{sr}(a)$, see Exercise 6.26 and (5.6.63).

For a filter $a \in l_1(\mathbb{Z})$, we say that a is *a filter of Hölder class* if there exist $C > 0$ and $\tau > 0$ such that

$$|\widehat{a}(x) - \widehat{a}(y)| \leqslant C|x-y|^\tau, \qquad \forall\, x, y \in \mathbb{R}. \tag{6.6.6}$$

Recall that the *standard refinable function/distribution* ϕ associated with the filter a is defined by

$$\widehat{\phi}(\xi) := \prod_{j=1}^{\infty} \widehat{a}(2^{-j}\xi), \qquad \xi \in \mathbb{R}. \tag{6.6.7}$$

If a is a filter of Hölder class defined in (6.6.6) and $\widehat{a}(0) = 1$, by Lemma 4.1.8, the function $\widehat{\phi}$ in (6.6.7) is a well-defined continuous function with $\widehat{\phi}(0) = 1$. Furthermore, Lemma 4.1.8 tells us that ϕ is a tempered distribution on $\mathbb{R}$, since $\widehat{\phi}$ is a measurable function having no more than polynomial growth.

To calculate the quantity $\mathrm{sm}(a)$, the following result generalizes Corollary 5.8.5.

Theorem 6.6.1 *Let $\tau \geqslant 0$. Suppose that $a \in l_1(\mathbb{Z})$ is a filter such that $|\widehat{a}(\xi)|^2 = |1 + e^{-i\xi}|^{2\tau}|\widehat{A}(\xi)|^2$ for all $\xi \in \mathbb{R}$, where $\widehat{A}$ is 2π-periodic and $\widehat{B}(\xi) := |\widehat{A}(\xi)|^2$ satisfies $\widehat{B}(0) \neq 0$ and $|\widehat{B}(\xi) - \widehat{B}(0)| \leqslant C|\xi|^\nu$ for all $\xi \in [-\pi, \pi]$ for some positive constants $C > 0$ and $\nu > 0$. Then*

$$\rho_{2\tau}(\mathcal{T}_{|\widehat{a}|^2}) = \lim_{n\to\infty} \|\mathcal{T}_{\widehat{B}}^n 1\|_{L_\infty(\mathbb{R})}^{1/n} = \inf_{n\in\mathbb{N}} \|\mathcal{T}_{\widehat{B}}^n 1\|_{L_\infty(\mathbb{R})}^{1/n} \quad \textit{and} \quad \rho_{2\tau}(\mathcal{T}_{|\widehat{a}|^2}) \geqslant 2^{-2\tau}|\widehat{a}(0)|^2.$$

Proof Since $\widehat{B}(0) \neq 0$, without loss of generality, we assume $\widehat{B}(0) = 1$ by considering $\widehat{a}/\sqrt{\widehat{B}(0)}$ instead of $\widehat{a}$. By $|\widehat{a}(\xi)|^2 = 2^{2\tau}|\cos(\xi/2)|^{2\tau}\widehat{B}(\xi)$, we can prove by induction on $n \in \mathbb{N}$ that

$$[\mathcal{T}_{|\widehat{a}|^2}^n(|\sin(\cdot/2)|^{2\tau})](\xi) = |\sin(\xi/2)|^{2\tau}[\mathcal{T}_{\widehat{B}}^n 1](\xi), \qquad \forall\, n \in \mathbb{N},\ \xi \in \mathbb{R}. \tag{6.6.8}$$

Consequently, by $\tau \geqslant 0$, we have $\|\mathcal{T}_{|\widehat{a}|^2}^n(|\sin(\cdot/2)|^{2\tau})\|_{L_\infty(\mathbb{T})} \leqslant \|\mathcal{T}_{\widehat{B}}^n 1\|_{L_\infty(\mathbb{T})}$ and thus,

$$\rho := \rho_{2\tau}(\mathcal{T}_{|\widehat{a}|^2}) := \limsup_{n\to\infty} \|\mathcal{T}_{|\widehat{a}|^2}^n(|\sin(\cdot/2)|^{2\tau})\|_{L_\infty(\mathbb{T})}^{1/n} \leqslant \limsup_{n\to\infty} \|\mathcal{T}_{\widehat{B}}^n 1\|_{L_\infty(\mathbb{T})}^{1/n}. \tag{6.6.9}$$

By the above definition of ρ, for any $\varepsilon > 0$, there exists $C > 0$ such that

$$\|\mathcal{T}^n_{|\widehat{a}|^2}(|\sin(\cdot/2)|^{2\tau})\|_{L_\infty(\mathbb{T})} \leqslant C(\rho+\varepsilon)^n, \qquad \forall\, n \in \mathbb{N}.$$

By $|\widehat{a}(\xi)|^2 = 2^{2\tau}|\cos(\xi/2)|^{2\tau}\widehat{B}(\xi)$, it follows from (6.6.8) and the definition of $\mathcal{T}_{\widehat{u}}$ in (6.6.4) that for $\xi \in [-\pi, \pi]$,

$$\begin{aligned}
[\mathcal{T}^n_{\widehat{B}}1](\xi) &= \frac{[\mathcal{T}^n_{|\widehat{a}|^2}(|\sin(\cdot/2)|^{2\tau})](\xi)}{|\sin(\xi/2)|^{2\tau}}\\
&= \frac{|\widehat{a}(\xi/2)|^2[\mathcal{T}^{n-1}_{|\widehat{a}|^2}(|\sin(\cdot/2)|^{2\tau})](\xi/2) + |\widehat{a}(\xi/2+\pi)|^2[\mathcal{T}^{n-1}_{|\widehat{a}|^2}(|\sin(\cdot/2)|^{2\tau})](\xi/2+\pi)}{|\sin(\xi/2)|^{2\tau}}\\
&= \widehat{B}(\xi/2)[\mathcal{T}^{n-1}_{\widehat{B}}1](\xi/2) + \frac{\widehat{B}(\xi/2+\pi)}{|\cos(\xi/4)|^{2\tau}}[\mathcal{T}^{n-1}_{|\widehat{a}|^2}(|\sin(\cdot/2)|^{2\tau})](\xi/2+\pi)\\
&\leqslant \widehat{B}(\xi/2)[\mathcal{T}^{n-1}_{\widehat{B}}1](\xi/2) + 2^\tau\|\widehat{B}\|_{L_\infty(\mathbb{T})}C(\rho+\varepsilon)^{n-1},
\end{aligned}$$

where we used

$$|\widehat{a}(\xi/2+\pi)|^2/|\sin(\xi/2)|^{2\tau} = \widehat{B}(\xi/2+\pi)/|\cos(\xi/4)|^{2\tau} \leqslant 2^\tau\|\widehat{B}\|_{L_\infty(\mathbb{T})}$$

for $\xi \in [-\pi, \pi]$. Iterating the above inequality n times, for $\xi \in [-\pi, \pi]$, we end up with

$$[\mathcal{T}^n_{\widehat{B}}1](\xi) \leqslant \prod_{j=1}^{n}\widehat{B}(2^{-j}\xi) + 2^\tau\|\widehat{B}\|_{L_\infty(\mathbb{T})}C\sum_{j=1}^{n-1}(\rho+\varepsilon)^j\prod_{k=1}^{n-j-1}\widehat{B}(2^{-k}\xi). \tag{6.6.10}$$

By our assumption on $\widehat{B}$ and $\widehat{B}(0) = 1$, it follows from Lemma 4.1.8 or Lemma 5.1.1 that there exists a positive constant C_B such that

$$0 \leqslant \prod_{j=1}^{m}\widehat{B}(2^{-j}\xi) \leqslant C_B, \qquad \forall\, \xi \in [-\pi, \pi], m \in \mathbb{N}. \tag{6.6.11}$$

It also follows from Lemma 4.1.8 or Lemma 5.1.1 that $\eta(\xi) := \prod_{j=1}^{\infty}\widehat{B}(2^{-j}\xi)$ is well defined and there exists $0 < \varepsilon_0 < \pi$ such that $1/2 \leqslant \eta(\xi) \leqslant 2$ for $\xi \in [-\varepsilon_0, \varepsilon_0]$. We now show that $\rho \geqslant 1$. By induction on $n \in \mathbb{N}$, we see that $\mathcal{T}_{|\widehat{a}|^2}(|\sin(\cdot/2)|^{2\tau})$ is a sum of 2^n terms including $|\sin(2^{-1-n}\cdot)|^{2\tau}\prod_{j=1}^{n}|\widehat{a}(2^{-j}\cdot)|^2$ as one of these terms. Hence, for $n \geqslant \log_2(\pi/\varepsilon_0)$ and $\xi \in [-\pi, \pi]$, we have $|2^{-n}\xi| \leqslant \varepsilon_0$ and by $|\widehat{a}|^2 \geqslant 0$,

$$\begin{aligned}
\|\mathcal{T}^n_{|\widehat{a}|^2}(|\sin(\cdot/2)|^{2\tau})\|_{L_\infty(\mathbb{T})} &\geqslant \Big\||\sin(2^{-1-n}\cdot)|^{2\tau}\prod_{j=1}^{n}|\widehat{a}(2^{-j}\cdot)|^2\Big\|_{L_\infty([-\pi,\pi])}\\
&= \||\sin(2^{-1-n}\cdot)|^{2\tau}\eta(\cdot)/\eta(2^{-n}\cdot)\|_{L_\infty([-\pi,\pi])} \geqslant 4^{-1}|\sin(\varepsilon_0/2)|^{2\tau} > 0,
\end{aligned}$$

from which we conclude that $\rho \geqslant 1$. Since $\widehat{a}(0) = 2^{2\tau}\widehat{B}(0) = 2^{2\tau}$, we have $\rho_{2\tau}(\mathcal{T}_{|\widehat{a}|^2}) = \rho \geqslant 1 = 2^{-2\tau}|\widehat{a}(0)|^2$. Consequently, by (6.6.10) and (6.6.11), for a.e. $\xi \in [-\pi, \pi]$ and for all $n \geqslant \log_2(\pi/\varepsilon_0)$, we have

$$\begin{aligned}[][\mathcal{T}_{\widehat{B}}^n 1](\xi) &\leqslant C_B + 2^\tau \|\widehat{B}\|_{L_\infty(\mathbb{T})} C C_B \sum_{j=1}^{n-1} (\rho+\varepsilon)^j \\ &\leqslant C_B + 2^\tau \|\widehat{B}\|_{L_\infty(\mathbb{T})} C C_B (\rho+\varepsilon)^n/(\rho+\varepsilon-1) \leqslant C_0(\rho+\varepsilon)^n,\end{aligned}$$

where we used $\rho \geqslant 1$ and $C_0 := C_B + 2^\tau \|\widehat{B}\|_{L_\infty(\mathbb{T})} C C_B/(\rho+\varepsilon-1) < \infty$. We deduce directly from the above inequality that

$$\limsup_{n\to\infty} \|\mathcal{T}_{\widehat{B}}^n 1\|_{L_\infty(\mathbb{T})}^{1/n} \leqslant \limsup_{n\to\infty} C_0^{1/n}(\rho+\varepsilon) = \rho+\varepsilon.$$

Taking $\varepsilon \to 0^+$, we have $\limsup_{n\to\infty} \|\mathcal{T}_{\widehat{B}}^n 1\|_{L_\infty(\mathbb{T})}^{1/n} \leqslant \rho$. By (6.6.9), we proved

$$\limsup_{n\to\infty} \|\mathcal{T}_{\widehat{B}}^n 1\|_{L_\infty(\mathbb{T})}^{1/n} = \rho.$$

By the definition of $\mathcal{T}_{\widehat{u}}$ in (6.6.4) and $\widehat{B} \geqslant 0$, we observe that $\|\mathcal{T}_{\widehat{B}}^{m+n} 1\|_{L_\infty(\mathbb{T})} \leqslant \|\mathcal{T}_{\widehat{B}}^m 1\|_{L_\infty(\mathbb{T})} \|\mathcal{T}_{\widehat{B}}^n 1\|_{L_\infty(\mathbb{T})}$ for all $m, n \in \mathbb{N}$. Consequently, by the same argument as in Lemma 5.7.1, we see that

$$\inf_{n\to\infty} \|\mathcal{T}_{\widehat{B}}^n 1\|_{L_\infty(\mathbb{T})}^{1/n} = \lim_{n\to\infty} \|\mathcal{T}_{\widehat{B}}^n 1\|_{L_\infty(\mathbb{T})}^{1/n} = \limsup_{n\to\infty} \|\mathcal{T}_{\widehat{B}}^n 1\|_{L_\infty(\mathbb{T})}^{1/n} = \rho.$$

This completes the proof. □

We now discuss initial functions in a cascade algorithm with an infinitely supported filter. A function $f \in L_2(\mathbb{R})$ is *admissible* with respect to a filter a if there exists a positive number $\tau > 0$ such that

$$[\widehat{a}(\cdot/2)\widehat{f}(\cdot/2) - \widehat{f}, \widehat{a}(\cdot/2)\widehat{f}(\cdot/2) - \widehat{f}]/|\sin(\cdot/2)|^\tau \in L_\infty(\mathbb{T}). \tag{6.6.12}$$

The following result shows that a compactly supported function satisfying certain moment conditions is admissible with respect to all low-pass filters.

Lemma 6.6.2 *Let $m \in \mathbb{N}$ and f be a compactly supported function in $L_2(\mathbb{R})$ such that*

$$\widehat{f}(\xi + 2\pi k) = \mathscr{O}(|\xi|^m), \qquad \xi \to 0, \quad \forall\, k \in \mathbb{Z}\backslash\{0\}. \tag{6.6.13}$$

Then there exists a positive constant C such that

$$\sum_{k\in\mathbb{Z}\backslash\{0\}} |\widehat{f}(\xi + 2\pi k)|^2 \leqslant C|\xi|^{2m}, \qquad \forall\, \xi \in [-\pi, \pi] \tag{6.6.14}$$

and f is admissible with respect to every filter a which is of Hölder class with $\widehat{a}(0) = 1$ and $\widehat{a}(\pi) = 0$.

Proof Let $u \in l_0(\mathbb{Z})$ such that $\widehat{u}(\xi) = \widehat{f}(\xi)/\widehat{B_m}(\xi) + \mathscr{O}(|\xi|^m)$ as $\xi \to 0$, where B_m is the B-spline function of order m. Define $\widehat{\eta}(\xi) := \widehat{f}(\xi) - \widehat{u}(\xi)\widehat{B_m}(\xi)$. By (6.6.13), we deduce that η is a compactly supported function in $L_2(\mathbb{R})$ and satisfies (5.5.14) with m being replaced by $m-1$. By Lemma 5.5.6, there exists a compactly supported function $g \in L_2(\mathbb{R})$ such that $\eta = \nabla^m g$. Therefore,

$$\widehat{f}(\xi) = \widehat{u}(\xi)\widehat{B_m}(\xi) + \widehat{\eta}(\xi) = \widehat{u}(\xi)\widehat{B_m}(\xi) + \widehat{\nabla^m g}(\xi) = \widehat{u}(\xi)\widehat{B_m}(\xi) + (1-e^{-i\xi})^m \widehat{g}(\xi).$$

Thus, by $\widehat{B_m}(\xi) = (\frac{1-e^{-i\xi}}{i\xi})^m$, for $\xi \in [-\pi, \pi]$, (6.6.14) follows directly from

$$\begin{aligned}
\sum_{k\in\mathbb{Z}\backslash\{0\}} |\widehat{f}(\xi+2\pi k)|^2 &\leqslant 2|\widehat{u}(\xi)|^2 \left(\sum_{k\in\mathbb{Z}\backslash\{0\}} |\widehat{B_m}(\xi+2\pi k)|^2 \right) + 2|1-e^{-i\xi}|^{2m} [\widehat{g}, \widehat{g}](\xi) \\
&\leqslant 2\|\widehat{u}\|^2_{L_\infty(\mathbb{T})} |1-e^{-i\xi}|^{2m} \left(\sum_{k\in\mathbb{Z}\backslash\{0\}} \tfrac{1}{|\xi+2\pi k|^{2m}} \right) + 2|1-e^{-i\xi}|^{2m} \|[\widehat{g}, \widehat{g}]\|_{L_\infty(\mathbb{T})} \\
&\leqslant 2^{2m} C \sin^{2m}(\xi/2) \leqslant C|\xi|^{2m},
\end{aligned}$$

where

$$C := 2\|\widehat{u}\|^2_{L_\infty(\mathbb{T})} \sup_{|\xi|\leqslant\pi} \sum_{k\in\mathbb{Z}\backslash\{0\}} \tfrac{1}{|\xi+2\pi k|^{2m}} + 2\|[\widehat{g}, \widehat{g}]\|_{L_\infty(\mathbb{T})} < \infty.$$

Define $\widehat{b}(\xi) := \widehat{a}(\xi) - 2^{-1}(1+e^{-i\xi})$ and $\widehat{g}(\xi) := 2^{-1}(1+e^{-i\xi/2})\widehat{f}(\xi/2) - \widehat{f}(\xi)$. Then g is a compactly supported function satisfying $\widehat{g}(2\pi k) = 0$ for all $k \in \mathbb{Z}$. By what has been proved, there exists $C > 0$ such that $[\widehat{g}, \widehat{g}](\xi) \leqslant C|\xi|^2$ for all $\xi \in [-\pi, \pi]$. Since a is a filter of Hölder class with $\widehat{a}(0) = 1$ and $\widehat{a}(\pi) = 0$, there exist $0 < \tau < 1$ and $C_1 \geqslant 1$ such that

$$|\widehat{b}(\xi)|^2 \leqslant C_1|\xi|^\tau, \qquad |\widehat{b}(\xi+\pi)|^2 \leqslant C_1|\xi|^\tau, \qquad \forall\, \xi \in \mathbb{R}.$$

Therefore, by $\widehat{a}(\cdot/2)\widehat{f}(\cdot/2) - \widehat{f} = \widehat{g} + \widehat{b}(\cdot/2)\widehat{f}(\cdot/2)$, we deduce that for $\xi \in [-\pi, \pi]$,

$$\begin{aligned}
&[\widehat{a}(\cdot/2)\widehat{f}(\cdot/2) - \widehat{f}, \widehat{a}(\cdot/2)\widehat{f}(\cdot/2) - \widehat{f}](\xi) \\
&\quad\leqslant 2[\widehat{g}, \widehat{g}](\xi) + 2[\widehat{b}(\cdot/2)\widehat{f}(\cdot/2), \widehat{b}(\cdot/2)\widehat{f(\cdot/2)}](\xi) \\
&\quad= 2[\widehat{g}, \widehat{g}](\xi) + 2|\widehat{b}(\cdot/2)|^2[\widehat{f}, \widehat{f}](\xi/2) + 2|\widehat{b}(\xi/2+\pi)|^2[\widehat{f}, \widehat{f}](\xi/2+\pi) \\
&\quad\leqslant 2C|\xi|^2 + 2^{2-\tau}C_1|\xi|^\tau \|[\widehat{f}, \widehat{f}]\|_{L_\infty(\mathbb{R})}.
\end{aligned}$$

Thus, (6.6.12) holds and f is admissible with respect to the filter a. □

We now characterize the convergence of a cascade algorithm $\{\mathcal{R}_a^n f\}_{n=1}^{\infty}$ in the space $L_{2,\infty,0}(\mathbb{R})$ in (6.6.2), where the refinement operator $\mathcal{R}_a f$ is defined to be $\widehat{\mathcal{R}_a f}(\xi) := \widehat{a}(\xi/2)\widehat{f}(\xi/2)$.

Theorem 6.6.3 *Let a be a filter of Hölder class with $\widehat{a}(0) = 1$. Then the following statements are equivalent:*

(i) $\widehat{a}(\pi) = 0$ and for every admissible function $f \in L_{2,\infty,0}(\mathbb{R})$ with respect to the filter a, the sequence $\{f_n\}_{n=1}^{\infty}$ with $f_n := \mathcal{R}_a^n f$ is a Cauchy sequence in $L_{2,\infty,0}(\mathbb{R})$.

(ii) For one admissible function $f \in L_{2,\infty,0}(\mathbb{R})$ with respect to the filter a such that the integer shifts of f are stable in $L_2(\mathbb{R})$, $\{f_n\}_{n=1}^{\infty}$ is a Cauchy sequence in $L_{2,\infty,0}(\mathbb{R})$.

(iii) $\rho_\tau(\mathcal{T}_{|\widehat{a}|^2}) < 1$ for all $\tau > 0$.

(iv) $\rho_\tau(\mathcal{T}_{|\widehat{a}|^2}) < 1$ and $|\widehat{a}(\cdot + \pi)|^2/|\sin(\cdot/2)|^\tau \in L_\infty(\mathbb{T})$ for at least one $\tau > 0$.

(v) $\mathrm{sm}(a) > 0$.

Let ϕ be the standard refinable function/distribution associated with the filter a satisfying (6.6.7). If $\mathrm{sm}(a) > 0$, then $\phi \in L_{2,\infty,0}(\mathbb{R})$, the coefficient sequence of $[\widehat{\phi}, \widehat{\phi}]$ is a filter of Hölder class, and

$$[\widehat{\phi}, \widehat{\phi}]_\tau := \sum_{k\in\mathbb{Z}} |\widehat{\phi}(\cdot + 2\pi k)|^2(1 + |\cdot + 2\pi k|^2)^\tau \in \mathscr{C}(\mathbb{T}), \ \forall\, 0 \leqslant \tau < \mathrm{sm}(a). \tag{6.6.15}$$

If an initial admissible function f above satisfies the extra condition $\lim_{j\to\infty} \widehat{f}(2^{-j}\xi) = 1$ for a.e. $\xi \in \mathbb{R}$, then $\lim_{n\to\infty} \|f_n - \phi\|_{L_{2,\infty,0}(\mathbb{R})} = 0$.

Proof (i)$\Longrightarrow$(ii) is trivial (the existence of such an admissible initial function f in item (ii) is guaranteed by Lemma 6.6.2, also see Exercise 6.24). Note that $\rho_\tau(\mathcal{T}_{|\widehat{a}|^2})$ is a nonincreasing function of τ. By Theorem 6.6.1 and the assumption that a is a filter of Hölder class, the claims (ii)$\Longrightarrow$(iii) and (6.6.15), as well as that the coefficient sequence of $[\widehat{\phi}, \widehat{\phi}]$ is a filter of Hölder class, have been proved by a technical argument in Theorem 2.1 of [B. Han, Refinable functions and cascade algorithms in weighted spaces with Hölder continuous masks, *SIAM J. Math. Anal.* **40** (2008), No. 1, 70–102]. (iii) implies $\widehat{a}(\pi) = 0$ (see Exercise 6.31) and hence (iii)$\Longrightarrow$(iv). (iv)$\Longrightarrow$(v) follows from the definition of $\mathrm{sm}(a)$ in (6.6.5).

We now prove (v)$\Longrightarrow$(i). Without loss of generality, for small $\tau > 0$, we can assume that both (6.6.12) and $\rho_\tau(\mathcal{T}_{|\widehat{a}|^2}) < 1$ hold. Define $\widehat{g}(\xi) := \widehat{a}(\xi/2)\widehat{f}(\xi/2) - \widehat{f}(\xi)$. (6.6.12) means $H := [\widehat{g}, \widehat{g}]/|\sin(\cdot/2)|^\tau \in L_\infty(\mathbb{T})$. By induction on $n \in \mathbb{N}$, we have $\widehat{f_n}(\xi) := \widehat{f}(2^{-n}\xi) \prod_{j=1}^n \widehat{a}(2^{-j}\xi)$ and

$$\widehat{f_{n+1}}(\xi) - \widehat{f_n}(\xi) = \widehat{g}(2^{-n}\xi) \prod_{j=1}^{n} \widehat{a}(2^{-j}\xi).$$

Therefore, we have

$$[\widehat{f_{n+1}}-\widehat{f_n},\widehat{f_{n+1}}-\widehat{f_n}](\xi) = (\mathcal{T}^n_{|\widehat{a}|^2}[\widehat{g},\widehat{g}])(\xi) \leqslant \|H\|_{L_\infty(\mathbb{T})}(\mathcal{T}^n_{|\widehat{a}|^2}(|\sin(\cdot/2)|^\tau))(\xi).$$

Since $\rho_\tau(\mathcal{T}_{|\widehat{a}|^2}) < 1$, for ρ satisfying $\rho_\tau(\mathcal{T}_{|\widehat{a}|^2}) < \rho < 1$, there exists a positive constant C such that $\|\mathcal{T}^n_{|\widehat{a}|^2}(|\sin(\cdot/2)|^\tau)\|_{L_\infty(\mathbb{T})} \leqslant C\rho^n$ for all $n \in \mathbb{N}$. Consequently, we have

$$\|f_{n+1}-f_n\|^2_{L_{2,\infty,0}(\mathbb{R})} = \|[\widehat{f_{n+1}}-\widehat{f_n},\widehat{f_{n+1}}-\widehat{f_n}]\|_{L_\infty(\mathbb{T})} \leqslant C\|H\|_{L_\infty(\mathbb{T})}\rho^n, \qquad \forall\, n \in \mathbb{N}.$$

Therefore, the sequence $\{f_n\}_{n=1}^\infty$ is a Cauchy sequence in $L_{2,\infty,0}(\mathbb{R})$. This proves (v)$\Longrightarrow$(i). □

As a direct consequence of Theorem 6.6.3, we have the following result which is useful in studying framelets and wavelets with infinite support.

Corollary 6.6.4 *Let a be a filter of Hölder class with* $\widehat{a}(0) = 1$. *Suppose that* $\varphi \in L_2(\mathbb{R})$ *satisfies* $\widehat{\varphi}(2\xi) = \widehat{a}(\xi)\widehat{\varphi}(\xi)$ *for a.e.* $\xi \in \mathbb{R}$ *and the integer shifts of* φ *are stable in* $L_2(\mathbb{R})$. *Then*

(1) $\mathrm{sm}(a) > 0$ *and* $\widehat{a}(\pi) = 0$;
(2) Its standard refinable function ϕ *(i.e.,* $\widehat{\phi}(\xi) := \prod_{j=1}^\infty \widehat{a}(2^{-j}\xi)$*) belongs to* $L_{2,\infty,0}(\mathbb{R})$ *and the integer shifts of* ϕ *are stable in* $L_2(\mathbb{R})$.

Proof Since the integer shifts of φ are stable in $L_2(\mathbb{R})$, there exists $C > 0$ such that $C^{-1} \leqslant [\widehat{\varphi},\widehat{\varphi}](\xi) \leqslant C$ for a.e. $\xi \in \mathbb{R}$. Hence $\varphi \in L_{2,\infty,0}(\mathbb{R})$. Since $\widehat{a}(\cdot/2)\widehat{\varphi}(\cdot/2) - \widehat{\varphi} = 0$, the function φ is an admissible function in $L_{2,\infty,0}(\mathbb{R})$ with respect to the filter a. Therefore, taking $f = \varphi$ in item (ii) of Theorem 6.6.3, we conclude that item (1) holds and $\phi \in L_{2,\infty,0}(\mathbb{R})$.

Since $[\widehat{\varphi},\widehat{\varphi}](\xi) \leqslant C$, we have $|\widehat{\varphi}(\xi)| \leqslant \sqrt{[\widehat{\varphi},\widehat{\varphi}](\xi)} \leqslant \sqrt{C}$ for a.e. $\xi \in \mathbb{R}$. Note that $\widehat{\phi}$ is a continuous function with $\widehat{\phi}(0) = 1$. Therefore, there exists $c > 0$ such that $|\widehat{\phi}(\xi)| \geqslant 1/2$ for all $\xi \in [-c, c]$. For any $\xi \in \mathbb{R}\backslash\{0\}$ and $n \geqslant \log_2(|\xi|/c)$, we have $|2^{-n}\xi| < c$ and therefore,

$$\begin{aligned}|\widehat{\varphi}(\xi)| &= |\widehat{\varphi}(2^{-n}\xi)|\prod_{j=1}^n |\widehat{a}(2^{-j}\xi)| \leqslant \sqrt{C}\prod_{j=1}^n |\widehat{a}(2^{-j}\xi)|\\ &\leqslant 2\sqrt{C}|\widehat{\phi}(2^{-n}\xi)|\prod_{j=1}^n |\widehat{a}(2^{-j}\xi)| = 2\sqrt{C}|\widehat{\phi}(\xi)|.\end{aligned}$$

Consequently, $[\widehat{\phi},\widehat{\phi}](\xi) \geqslant \frac{1}{4C}[\widehat{\varphi},\widehat{\varphi}](\xi) \geqslant \frac{1}{4C^2}$. Since $\phi \in L_{2,\infty,0}(\mathbb{R})$, this proves that the integer shifts of ϕ must be stable in $L_2(\mathbb{R})$. Thus, item (2) holds. □

Let $\varphi \in L_2(\mathbb{R})$ be a refinable function satisfying $\widehat{\varphi}(2\xi) = \widehat{a}(\xi)\widehat{\varphi}(\xi)$ for a.e. $\xi \in \mathbb{R}$. Let S be a measurable subset of $\mathbb{R}$ such that $[2S]\backslash S$ and $S\backslash[2S]$ have measure zero, e.g., $S = (0, \infty)$. Then $\widehat{\eta}(\xi) := \widehat{\varphi}(\xi)\chi_S(\xi)$ is also a refinable function satisfying

$\widehat{\eta}(2\xi) = \widehat{a}(\xi)\widehat{\eta}(\xi)$ for a.e. $\xi \in \mathbb{R}$. It is not necessary that the integer shifts of η are stable in $L_2(\mathbb{R})$ even if φ has this property.

6.6.2 *Biorthogonal Wavelets and Riesz Wavelets with Filters of Hölder Class*

We now characterize biorthogonal wavelets and Riesz wavelets with filters of Hölder class. The following result generalizes Theorem 6.4.5 with $r = 1$ and $\tau = 0$ to filters of Hölder class.

Corollary 6.6.5 *Let $a, \tilde{a}$ be filters of Hölder class with $\widehat{a}(0) = \widehat{\tilde{a}}(0) = 1$. Let ϕ and $\tilde{\phi}$ be the standard refinable functions/distributions associated with the filters a and $\tilde{a}$, respectively. Then the integer shifts of both ϕ and $\tilde{\phi}$ are stable in $L_2(\mathbb{R})$ and*

$$\langle \tilde{\phi}, \phi(\cdot - k)\rangle = \boldsymbol{\delta}(k), \qquad \forall\, k \in \mathbb{Z}, \tag{6.6.16}$$

if and only if $\operatorname{sm}(a) > 0, \operatorname{sm}(\tilde{a}) > 0$ *and* $\widehat{\tilde{a}}(\xi)\overline{\widehat{a}(\xi)} + \widehat{\tilde{a}}(\xi+\pi)\overline{\widehat{a}(\xi+\pi)} = 1$ *for all* $\xi \in \mathbb{R}$.

Proof Necessity ($\Rightarrow$). The inequalities $\operatorname{sm}(a) > 0$ and $\operatorname{sm}(\tilde{a}) > 0$ follow directly from Corollary 6.6.4. By $\widehat{\phi}(2\xi) = \widehat{a}(\xi)\widehat{\phi}(\xi)$ and $\widehat{\tilde{\phi}}(2\xi) = \widehat{\tilde{a}}(\xi)\widehat{\tilde{\phi}}(\xi)$, noting that (6.6.16) is equivalent to $[\widehat{\tilde{\phi}}, \widehat{\phi}] = 1$, we have

$$\begin{aligned} 1 = [\widehat{\tilde{\phi}}, \widehat{\phi}](2\xi) &= \widehat{\tilde{a}}(\xi)\overline{\widehat{a}(\xi)}[\widehat{\tilde{\phi}}, \widehat{\phi}](\xi) + \widehat{\tilde{a}}(\xi+\pi)\overline{\widehat{a}(\xi+\pi)}[\widehat{\tilde{\phi}}, \widehat{\phi}](\xi+\pi) \\ &= \widehat{\tilde{a}}(\xi)\overline{\widehat{a}(\xi)} + \widehat{\tilde{a}}(\xi+\pi)\overline{\widehat{a}(\xi+\pi)}. \end{aligned}$$

Sufficiency ($\Leftarrow$). Since $\operatorname{sm}(a) > 0$ and $\operatorname{sm}(\tilde{a}) > 0$, by Theorem 6.6.3, the cascade algorithms associated with a and $\tilde{a}$ converge in $L_{2,\infty,0}(\mathbb{R})$ and $\widehat{a}(\pi) = \widehat{\tilde{a}}(\pi) = 0$. Take $f = \chi_{[0,1]}$ and define $f_n := \mathcal{R}_a^n f$ and $\tilde{f}_n := \mathcal{R}_{\tilde{a}}^n f$. By Lemma 6.6.2, the function f is admissible with respect to both a and $\tilde{a}$ and $\lim_{j\to\infty} \widehat{f}(2^{-j}\xi) = 1$. It follows easily from $\widehat{\tilde{a}}(\xi)\overline{\widehat{a}(\xi)} + \widehat{\tilde{a}}(\xi+\pi)\overline{\widehat{a}(\xi+\pi)} = 1$ that $[\widehat{\tilde{f}_n}, \widehat{f_n}] = 1$ for all $n \in \mathbb{N}$. By $\lim_{n\to\infty} \|f_n - \phi\|_{L_{2,\infty,0}(\mathbb{R})} = 0$ and $\lim_{n\to\infty} \|\tilde{f}_n - \tilde{\phi}\|_{L_{2,\infty,0}(\mathbb{R})} = 0$, we have $[\widehat{\tilde{\phi}}, \widehat{\phi}] = 1$, i.e., (6.6.16) holds. Since $\phi, \tilde{\phi} \in L_{2,\infty,0}(\mathbb{R})$, we conclude that $\|[\widehat{\tilde{\phi}}, \widehat{\tilde{\phi}}]\|^{-1}_{L_\infty(\mathbb{T})} \leqslant [\widehat{\phi}, \widehat{\phi}](\xi) \leqslant \|[\widehat{\phi}, \widehat{\phi}]\|_{L_\infty(\mathbb{T})}$ and by Theorem 4.4.6 the integer shifts of ϕ are stable in $L_2(\mathbb{R})$. Similarly, we can prove that the integer shifts of $\tilde{\phi}$ are stable in $L_2(\mathbb{R})$. □

We are now ready to characterize scalar biorthogonal wavelets with filters of Hölder class. The following result generalizes Theorem 6.4.6 with $r = 1$ and $\tau = 0$ from finitely supported filters to filters of Hölder class.

Theorem 6.6.6 *Let $a, b, \tilde{a}, \tilde{b}$ be filters of Hölder class with $\widehat{a}(0) = \widehat{\tilde{a}}(0) = 1$. Let ϕ and $\tilde{\phi}$ be the standard refinable functions associated with the filters a and $\tilde{a}$, respectively. Define ψ and $\tilde{\psi}$ by $\widehat{\psi}(2\xi) = \widehat{b}(\xi)\widehat{\phi}(\xi)$ and $\widehat{\tilde{\psi}}(2\xi) = \widehat{\tilde{b}}(\xi)\widehat{\tilde{\phi}}(\xi)$. Then $(\{\tilde{\phi}; \tilde{\psi}\}, \{\phi; \psi\})$ is a biorthogonal wavelet in $L_2(\mathbb{R})$ if and only if*

(1) $(\{\tilde{a}; \tilde{b}\}, \{a; b\})$ *is a biorthogonal wavelet filter bank satisfying* (6.4.21)*;*
(2) $\operatorname{sm}(a) > 0$ *and* $\operatorname{sm}(\tilde{a}) > 0$.

Proof Necessity ($\Rightarrow$). By Corollary 6.6.5, item (2) holds. Item (1) can be proved by the same argument as in the proof of Theorem 6.4.6.

Sufficiency ($\Leftarrow$). By Corollary 6.6.5, the identity (6.6.16) holds. Now it follows directly from (6.6.16) and item (1) that $\mathsf{AS}_0(\{\tilde{\phi}; \tilde{\psi}\})$ and $\mathsf{AS}_0(\{\phi; \psi\})$ are biorthogonal to each other. By item (2) and Theorem 6.6.3, we see that (6.6.15) holds and $\widehat{a}(\pi) = \widehat{\tilde{a}}(\pi) = 0$. By item (1), we have $\widehat{\tilde{a}}(0)\overline{\widehat{b}(0)} + \widehat{\tilde{a}}(\pi)\overline{\widehat{b}(\pi)} = 0$, from which we deduce that $\widehat{b}(0) = 0$ and hence, $\widehat{\psi}(0) = 0$. Now it follows directly from Theorem 4.6.5 and (6.6.15) that $\mathsf{AS}_0(\{\phi; \psi\})$ is a Bessel sequence in $L_2(\mathbb{R})$. Similarly, we can prove that $\widehat{\tilde{\psi}}(0) = 0$ and $\mathsf{AS}_0(\{\tilde{\phi}; \tilde{\psi}\})$ is a Bessel sequence in $L_2(\mathbb{R})$ too. By Theorem 4.1.10 with $\widehat{\theta} = \widehat{\tilde{\theta}} = I_r$, we deduce that $(\{\widehat{\tilde{\phi}}; \widehat{\tilde{\psi}}\}, \{\widehat{\phi}; \widehat{\psi}\})$ is a frequency-based dual framelet. Therefore, we conclude that $(\{\tilde{\phi}; \tilde{\psi}\}, \{\phi; \psi\})$ is a biorthogonal wavelet in $L_2(\mathbb{R})$. □

We have the following result characterizing Riesz wavelets in $L_2(\mathbb{R})$.

Theorem 6.6.7 *Let a, b be filters of Hölder class with $\widehat{a}(0) = 1$. Let ϕ and ψ be given by*

$$\widehat{\phi}(\xi) := \prod_{j=1}^{\infty} \widehat{a}(2^{-j}\xi), \quad \widehat{\psi}(\xi) := \widehat{b}(\xi/2)\widehat{\phi}(\xi/2), \qquad \xi \in \mathbb{R}.$$

Then $\{\phi; \psi\}$ is a Riesz wavelet in $L_2(\mathbb{R})$ if and only if

(1) $\widehat{b}(0) = 0$ *and* $d(\xi) := \widehat{a}(\xi)\widehat{b}(\xi + \pi) - \widehat{a}(\xi + \pi)\widehat{b}(\xi) \neq 0$ *for all* $\xi \in \mathbb{R}$.
(2) $\operatorname{sm}(a) > 0$ *and* $\operatorname{sm}(\tilde{a}) > 0$*, where* $\widehat{\tilde{a}}(\xi) := \overline{\widehat{b}(\xi + \pi)/d(\xi)}$.

Proof Sufficiency ($\Leftarrow$). By $\widehat{a}(0) = 1$ and $\widehat{b}(0) = 0$, we have $\widehat{\tilde{a}}(0) = 1$. Define $\widehat{\tilde{b}}(\xi) := -\overline{\widehat{a}(\xi + \pi)/d(\xi)}$. Since $d(\xi) \neq 0$ and a, b are filters of Hölder class, it is evident that $\tilde{a}, \tilde{b}$ are filters of Hölder class. It is also straightforward to check that $(\{\tilde{a}; \tilde{b}\}, \{a; b\})$ is a biorthogonal wavelet filter bank. Now it follows directly from Theorem 6.6.6 that $(\{\tilde{\phi}; \tilde{\psi}\}, \{\phi; \psi\})$ is a biorthogonal wavelet in $L_2(\mathbb{R})$. In particular, the pair $\{\phi; \psi\}$ is a Riesz wavelet in $L_2(\mathbb{R})$.

Necessity ($\Rightarrow$). Since $\{\phi; \psi\}$ is a Riesz wavelet in $L_2(\mathbb{R})$ satisfying $\widehat{\phi}(2\xi) = \widehat{a}(\xi)\widehat{\phi}(\xi)$ and $\widehat{\psi}(2\xi) = \widehat{b}(\xi)\widehat{\phi}(\xi)$, by Theorems 4.5.16 and 4.5.1, there exist $\tilde{\phi}, \tilde{\psi} \in L_2(\mathbb{R})$ and $\tilde{a}, \tilde{b} \in l_2(\mathbb{Z})$ such that

(i) $(\{\tilde{\phi}; \tilde{\psi}\}, \{\phi; \psi\})$ is a biorthogonal wavelet in $L_2(\mathbb{R})$;
(ii) $\widehat{\tilde{\phi}}(2\xi) = \widehat{\tilde{a}}(\xi)\widehat{\tilde{\phi}}(\xi)$ and $\widehat{\tilde{\psi}}(2\xi) = \widehat{\tilde{b}}(\xi)\widehat{\tilde{\phi}}(\xi)$ for a.e. $\xi \in \mathbb{R}$;
(iii) $(\{\tilde{a}; \tilde{b}\}, \{a; b\})$ is a biorthogonal wavelet filter bank.

By item (iii) (that is, (6.4.21) holds with $r = 1$), we have $d(\xi)\overline{\tilde{d}(\xi)} = 1$ for a.e. $\xi \in \mathbb{R}$, where $\tilde{d}(\xi) := \widehat{\tilde{a}}(\xi)\widehat{\tilde{b}}(\xi+\pi) - \widehat{\tilde{a}}(\xi+\pi)\widehat{\tilde{b}}(\xi)$. Since the integer shifts of both $\tilde{\phi}$ and $\tilde{\psi}$ are stable in $L_2(\mathbb{R})$ because of item (i), there exists a positive constant C such that

$$C^{-1} \leqslant [\widehat{\tilde{\phi}}, \widehat{\tilde{\phi}}](\xi) \leqslant C \quad \text{and} \quad C^{-1} \leqslant [\widehat{\tilde{\psi}}, \widehat{\tilde{\psi}}](\xi) \leqslant C, \qquad a.e.\, \xi \in \mathbb{R}.$$

On the other hand, by item (ii), we have

$$\begin{aligned}
[\widehat{\tilde{\phi}}, \widehat{\tilde{\phi}}](2\xi) &= |\widehat{\tilde{a}}(\xi)|^2[\widehat{\tilde{\phi}}, \widehat{\tilde{\phi}}](\xi) + |\widehat{\tilde{a}}(\xi+\pi)|^2[\widehat{\tilde{\phi}}, \widehat{\tilde{\phi}}](\xi+\pi),\\
[\widehat{\tilde{\psi}}, \widehat{\tilde{\psi}}](2\xi) &= |\widehat{\tilde{b}}(\xi)|^2[\widehat{\tilde{\phi}}, \widehat{\tilde{\phi}}](\xi) + |\widehat{\tilde{b}}(\xi+\pi)|^2[\widehat{\tilde{\phi}}, \widehat{\tilde{\phi}}](\xi+\pi).
\end{aligned}$$

We conclude from the above inequalities and identities that $|\widehat{\tilde{a}}(\xi)| \leqslant C$ and $|\widehat{\tilde{b}}(\xi)| \leqslant C$ for a.e. $\xi \in \mathbb{R}$. Therefore, $\tilde{d} \in L_\infty(\mathbb{T})$. We deduce from $d(\xi)\overline{\tilde{d}(\xi)} = 1$ that $|d(\xi)| = |\tilde{d}(\xi)|^{-1} \geqslant \|\tilde{d}\|^{-1}_{L_\infty(\mathbb{R})} > 0$. Since d is a continuous function, we must have $d(\xi) \ne 0$ for all $\xi \in \mathbb{R}$. By (6.4.21), we must have $\widehat{\tilde{a}}(\xi) = \overline{\widehat{b}(\xi+\pi)/d(\xi)}$ and $\widehat{\tilde{b}}(\xi) = -\overline{\widehat{a}(\xi+\pi)/d(\xi)}$. Therefore, $\tilde{a}, \tilde{b}$ are filters of Hölder class. By item (i) and Corollary 6.6.4, we conclude that item (2) holds and $\widehat{a}(\pi) = \widehat{\tilde{a}}(\pi) = 0$. Thus, $\overline{\widehat{b}(0)/\widehat{d}(0)} = \widehat{\tilde{a}}(\pi) = 0$ implies $\widehat{b}(0) = 0$. This proves item (1). □

We now discuss how to effectively compute the key quantity $\rho_0(\mathcal{T}_{|\widehat{A}|^2})$ for a rational polynomial $\widehat{A}$ through approximating the infinitely supported filter A by a sequence of finitely supported filters.

Proposition 6.6.8 *Let $c \in l_0(\mathbb{Z})$ such that there exist two positive numbers $c_{\min}$ and $c_{\max}$ satisfying*

$$0 < c_{\min} \leqslant \widehat{c}(\xi) \leqslant c_{\max}, \qquad \forall\, \xi \in \mathbb{R}. \tag{6.6.17}$$

For all nonnegative integers m, n, define

$$\begin{aligned}
\widehat{c^1_{m,n}}(\xi) :=& \frac{2}{c_{\min}+c_{\max}} \sum_{j=0}^{n-1} \left(1 - \frac{2\widehat{c}(\xi)}{c_{\min}+c_{\max}}\right)^j\\
&+ \left(1 - \frac{2\widehat{c}(\xi)}{c_{\min}+c_{\max}}\right)^n \left(\frac{(1-\widehat{c}(\xi)/\widehat{c}(0))^m}{c_{\max}} + \frac{1}{\widehat{c}(0)} \sum_{\ell=0}^{m-1} \left(1 - \frac{\widehat{c}(\xi)}{\widehat{c}(0)}\right)^\ell\right)
\end{aligned} \tag{6.6.18}$$

and

$$\widehat{c^2_{m,n}}(\xi) := \tfrac{2}{c_{\min}+c_{\max}} \sum_{j=0}^{n-1} \left(1 - \tfrac{2\widehat{c}(\xi)}{c_{\min}+c_{\max}}\right)^j + \left(1 - \tfrac{2\widehat{c}(\xi)}{c_{\min}+c_{\max}}\right)^n \left(\tfrac{(1-\widehat{c}(\xi)/\widehat{c}(0))^m}{c_{\min}} + \tfrac{1}{\widehat{c}(0)} \sum_{\ell=0}^{m-1} \left(1 - \tfrac{\widehat{c}(\xi)}{\widehat{c}(0)}\right)^{\ell} \right). \tag{6.6.19}$$

Then

$$\max\left(\|\widehat{c^1_{m,n}} - \tfrac{1}{\widehat{c}}\|_{L_\infty(\mathbb{T})}, \|\widehat{c^2_{m,n}} - \tfrac{1}{\widehat{c}}\|_{L_\infty(\mathbb{T})}\right) \leqslant \max\left(\left|1 - \tfrac{c_{\max}}{\widehat{c}(0)}\right|^m, \left|1 - \tfrac{c_{\min}}{\widehat{c}(0)}\right|^m\right) \left(\tfrac{1}{c_{\min}} - \tfrac{1}{c_{\max}}\right) \left(\tfrac{c_{\max}-c_{\min}}{c_{\max}+c_{\min}}\right)^n \tag{6.6.20}$$

and

$$\widehat{c^1_{m,n}}(\xi) = \frac{1}{\widehat{c}(\xi)} + \mathscr{O}(|\xi|^m), \qquad \widehat{c^2_{m,n}}(\xi) = \frac{1}{\widehat{c}(\xi)} + \mathscr{O}(|\xi|^m), \qquad \xi \to 0. \tag{6.6.21}$$

In addition, for all even $m, n \in \mathbb{N}_0$ (when $c_{max} = \widehat{c}(0)$, m can be any nonnegative integer),

$$0 \leqslant \widehat{c^1_{m,n}}(\xi) \leqslant \frac{1}{\widehat{c}(\xi)} \leqslant \widehat{c^2_{m,n}}(\xi), \qquad \forall\, \xi \in \mathbb{R}. \tag{6.6.22}$$

Proof We observe the following basic identity:

$$\frac{1}{x} = \frac{(1-x)^n}{x} + \sum_{j=0}^{n-1} (1-x)^j, \qquad \forall\, x > 0, n \in \mathbb{N}. \tag{6.6.23}$$

Setting $x = \frac{2\widehat{c}(\xi)}{c_{\max}+c_{\min}}$ in the above identity, we have

$$\frac{1}{\widehat{c}(\xi)} = \frac{2}{c_{\max}+c_{\min}} \sum_{j=0}^{n-1} \left(1 - \frac{2\widehat{c}(\xi)}{c_{\max}+c_{\min}}\right)^j + \left(1 - \frac{2\widehat{c}(\xi)}{c_{\max}+c_{\min}}\right)^n \frac{1}{\widehat{c}(\xi)}. \tag{6.6.24}$$

Next, applying the basic identity in (6.6.23) with x and n being replaced by $\widehat{c}(\xi)/\widehat{c}(0)$ and m, we have

$$\frac{1}{\widehat{c}(\xi)} = \frac{(1-\widehat{c}(\xi)/\widehat{c}(0))^m}{\widehat{c}(\xi)} + \frac{1}{\widehat{c}(0)} \sum_{\ell=0}^{m-1} (1-\widehat{c}(\xi)/\widehat{c}(0))^{\ell}.$$

Using the above identity to replace the last fraction $\frac{1}{\widehat{c}(\xi)}$ at the end of (6.6.24), we conclude that

$$\frac{1}{\widehat{c}(\xi)} = \frac{2}{c_{\max}+c_{\min}} \sum_{j=0}^{n-1} \left(1 - \frac{2\widehat{c}(\xi)}{c_{\max}+c_{\min}}\right)^j$$

$$+\left(1-\frac{2\widehat{c}(\xi)}{c_{\max}+c_{\min}}\right)^n \left(\frac{1}{\widehat{c}(0)} \sum_{\ell=0}^{m-1}(1-\widehat{c}(\xi)/\widehat{c}(0))^\ell + \frac{(1-\widehat{c}(\xi)/\widehat{c}(0))^m}{\widehat{c}(\xi)}\right).$$

When m, n are nonnegative even integers, replacing $\widehat{c}$ in the denominator of the last fraction in the above identity by $c_{\min}$ or $c_{\max}$, we conclude from the definition of $c^1_{m,n}$ in (6.6.18) and $c^2_{m,n}$ in (6.6.19) that (6.6.21) and (6.6.22) hold, because

$$\widehat{c^2_{m,n}}(\xi) - \frac{1}{\widehat{c}(\xi)} = \left(1 - \frac{2\widehat{c}(\xi)}{c_{\max}+c_{\min}}\right)^n \left(1 - \frac{\widehat{c}(\xi)}{\widehat{c}(0)}\right)^m \left(\frac{1}{c_{\min}} - \frac{1}{\widehat{c}(\xi)}\right).$$

Similarly, the above identity holds with $c^2_{m,n}$ being replaced by $c^1_{m,n}$ and the last $c_{\min}$ being replaced by $c_{\max}$. Since

$$\left|1 - \frac{2\widehat{c}(\xi)}{c_{\max}+c_{\min}}\right| \leqslant \frac{c_{\max}-c_{\min}}{c_{\max}+c_{\min}} < 1, \qquad \forall\, \xi \in \mathbb{R},$$

the inequality (6.6.20) holds. □

If $|\widehat{A_1}(\xi)| \leqslant |\widehat{A}(\xi)| \leqslant |\widehat{A_2}(\xi)|$ for all $\xi \in \mathbb{R}$, then it is trivial to check that $\rho_0(\mathcal{T}_{|\widehat{A_1}|^2}) \leqslant \rho_0(\mathcal{T}_{|\widehat{A}|^2}) \leqslant \rho_0(\mathcal{T}_{|\widehat{A_2}|^2})$. Therefore, if $\widehat{a}(\xi) = \widehat{u}(\xi)/\widehat{c}(\xi)$ for some $u, c \in l_0(\mathbb{Z})$ such that the filter c satisfies the condition in (6.6.17). Define $\widehat{a_1}(\xi) := \widehat{u}(\xi)\widehat{c^1_{m,n}}(\xi)$ and $\widehat{a_2}(\xi) := \widehat{u}(\xi)\widehat{c^2_{m,n}}(\xi)$. For even nonnegative integers m and n,

$$|\widehat{a_1}(\xi)| \leqslant |\widehat{a}(\xi)| \leqslant |\widehat{a_2}(\xi)| \quad \text{imply} \quad \operatorname{sm}(a_2) \leqslant \operatorname{sm}(a) \leqslant \operatorname{sm}(a_1). \tag{6.6.25}$$

Since $a_1, a_2 \in l_0(\mathbb{Z})$, the quantities $\operatorname{sm}(a_1)$ and $\operatorname{sm}(a_2)$ can be efficiently computed by item (3) of Corollary 5.8.5. Then the inequalities in (6.6.25) allow us to estimate the smoothness exponent $\operatorname{sm}(a)$ for an infinitely supported filter a.

We first provide an example of Riesz wavelets in $L_2(\mathbb{R})$ derived from spline refinable functions.

Example 6.6.1 For $m \in \mathbb{N}$, let B_m be the B-spline function in (6.1.1) and a^B_m be its filter given by $\widehat{a^B_m}(\xi) := 2^{-m}(1 + e^{-i\xi})^m$. Define

$$\widehat{\psi}(2\xi) := e^{-i\xi}\overline{\widehat{a^B_m}(\xi + \pi)\widehat{B_m}(\xi)}, \quad \xi \in \mathbb{R}.$$

Then $\{B_m; \psi\}$ is a Riesz wavelet in $L_2(\mathbb{R})$.

Proof Note that $\widehat{B_m}(2\xi) = \widehat{a^B_m}(\xi)\widehat{B_m}(\xi)$ and $\widehat{\psi}(2\xi) = \widehat{b}(\xi)\widehat{B_m}(\xi)$ with $\widehat{b}(\xi) := e^{-i\xi}\overline{\widehat{a^B_m}(\xi+\pi)}$. By Theorem 6.6.7, it suffices to check items (1) and (2) of Theorem 6.6.7. Obviously, $\widehat{b}(0) = \overline{\widehat{a^B_m}(\pi)} = 0$ and $d(\xi) := \widehat{a^B_m}(\xi)\widehat{b}(\xi+\pi) - \widehat{a^B_m}(\xi+\pi)\widehat{b}(\xi) = e^{-i(\xi+\pi)}[|\widehat{a^B_m}(\xi)|^2 + |\widehat{a^B_m}(\xi+\pi)|^2] = -e^{-i\xi}[\cos^{2m}(\xi/2) + \sin^{2m}(\xi/2)] \neq 0$ for all $\xi \in \mathbb{R}$. Hence, item (1) is satisfied. By Corollary 5.8.5 or Theorem 6.6.1, it is easy to check that $\mathrm{sm}(a^B_m) = m - 1/2 > 0$. Note that

$$|\widehat{\tilde{a}}(\xi)|^2 = |\overline{\widehat{b}(\xi+\pi)/d(\xi)}|^2 = |1 + e^{-i\xi}|^{2m}|\widehat{A}(\xi)|^2$$

$$\text{with} \quad \widehat{A}(\xi) := \frac{2^{-m}}{\cos^{2m}(\xi/2) + \sin^{2m}(\xi/2)} \leqslant 2^{-1},$$

where we used $\cos^{2m}(\xi/2) + \sin^{2m}(\xi/2) = (1-x)^m + x^m \geqslant 2^{1-m}$ for all $\xi \in \mathbb{R}$ with $x := \sin^2(\xi/2)$. It follows from Theorem 6.6.1 that

$$\rho_{2m}(\mathcal{T}_{|\widehat{\tilde{a}}|^2}) = \inf_{n\in\mathbb{N}} \|\mathcal{T}^n_{|\widehat{A}|^2} 1\|^{1/n}_{L_\infty(\mathbb{R})} \leqslant \inf_{n\in\mathbb{N}} \|\mathcal{T}^n_{2^{-2}} 1\|^{1/n}_{L_\infty(\mathbb{R})} = 2^{-1}.$$

since $\mathcal{T}^n_{|c|^2} 1 = 2^n c^{2n}$ for any $c \geqslant 0$. Therefore, we have $\mathrm{sm}(\tilde{a}) = -\frac{1}{2}\log_2 \rho_{2m}(\mathcal{T}_{|\widehat{\tilde{a}}|^2}) \geqslant \frac{1}{2} > 0$. By Theorem 6.6.7, the pair $\{B_m; \psi\}$ is a Riesz wavelet in $L_2(\mathbb{R})$. □

We have the following result on Riesz wavelets in $L_2(\mathbb{R})$ satisfying the semi-orthogonality condition.

Example 6.6.2 Let a be a filter of Hölder class with $\widehat{a}(0) = 1$ and $\mathrm{sm}(a) > 0$. Let ϕ be its associated refinable function defined by $\widehat{\phi}(\xi) := \prod_{j=1}^{\infty} \widehat{a}(2^{-j}\xi)$. By Theorem 6.6.3, we have $\phi \in L_2(\mathbb{R})$, $\widehat{a}(\pi) = 0$, and the coefficient sequence of $[\widehat{\phi}, \widehat{\phi}]$ is a filter of Hölder class. Suppose that $[\widehat{\phi}, \widehat{\phi}](\xi) \neq 0$ for all $\xi \in \mathbb{R}$ (and therefore, the integer shifts of ϕ are stable in $L_2(\mathbb{R})$). Define

$$\widehat{b}(\xi) := e^{-i\xi}\overline{\widehat{a}(\xi+\pi)}[\widehat{\phi}, \widehat{\phi}](\xi+\pi), \quad \widehat{\psi}(\xi) := \widehat{b}(\xi/2)\widehat{\phi}(\xi/2).$$

We now check all the conditions in Theorem 6.6.7. Since $\widehat{a}(\pi) = 0$, we have $\widehat{b}(0) = 0$. Because the coefficient sequence of $[\widehat{\phi}, \widehat{\phi}]$ is a filter of Hölder class, the high-pass filter b is a filter of Hölder class. Since $[\widehat{\phi}, \widehat{\phi}](\xi) \neq 0$ for all $\xi \in \mathbb{R}$, we have

$$\begin{aligned} d(\xi) :=& \widehat{a}(\xi)\widehat{b}(\xi+\pi) - \widehat{a}(\xi+\pi)\widehat{b}(\xi) \\ =& -e^{-i\xi}(|\widehat{a}(\xi)|^2[\widehat{\phi}, \widehat{\phi}](\xi) + |\widehat{a}(\xi+\pi)|^2[\widehat{\phi}, \widehat{\phi}](\xi+\pi)) \\ =& -e^{-i\xi}[\widehat{\phi}, \widehat{\phi}](2\xi) \neq 0 \end{aligned}$$

for all $\xi \in \mathbb{R}$. Note that

$$\widehat{\tilde{a}}(\xi) := \frac{\overline{\widehat{b}(\xi+\pi)}}{d(\xi)} = \frac{-e^{i\xi}\widehat{a}(\xi)[\widehat{\phi},\widehat{\phi}](\xi)}{-e^{i\xi}[\widehat{\phi},\widehat{\phi}](2\xi)} = \widehat{a}(\xi)\frac{[\widehat{\phi},\widehat{\phi}](\xi)}{[\widehat{\phi},\widehat{\phi}](2\xi)}$$

and $[\widehat{\phi},\widehat{\phi}](\xi) > 0$ for all $\xi \in \mathbb{R}$. Consequently, by the definition of $\operatorname{sm}(a)$, we must have $\operatorname{sm}(\tilde{a}) = \operatorname{sm}(a) > 0$ (see Exercise 6.35). It is also straightforward to directly check that $[\widehat{\phi},\widehat{\psi}](\xi) = 0$ for all $\xi \in \mathbb{R}$. Now we conclude from Theorem 6.6.7 that $\{\phi;\psi\}$ is a Riesz wavelet in $L_2(\mathbb{R})$ with the semi-orthogonality condition: $\langle\phi,\psi(\cdot - k)\rangle = 0$ for all $k \in \mathbb{Z}$ by $[\widehat{\phi},\widehat{\psi}] = 0$.

If in addition the filter $a \in l_0(\mathbb{Z})$ is finitely supported, then $b \in l_0(\mathbb{Z})$ and both ϕ and ψ are compactly supported. By Lemma 4.4.1, we have $[\widehat{\phi},\widehat{\phi}](\xi) = \sum_{k\in\mathbb{Z}}\langle\phi,\phi(\cdot-k)\rangle e^{-ik\xi}$ which is a 2π-periodic trigonometric polynomial and can be easily computed through the identity $\mathcal{T}_{|\widehat{a}|^2}[\widehat{\phi},\widehat{\phi}] = [\widehat{\phi},\widehat{\phi}]$. That is, the 2π-periodic trigonometric polynomial $[\widehat{\phi},\widehat{\phi}]$ is the unique eigenvector of the transition operator $\mathcal{T}_{|\widehat{a}|^2}$ for the eigenvalue one under the normalization condition $[\widehat{\phi},\widehat{\phi}](0) = 1$.

6.6.3 *Refinable Functions with Filters of Exponential Decay*

We now study refinable functions and wavelets with a filter of exponential decay. For a filter $a = \{a(k)\}_{k\in\mathbb{Z}} : \mathbb{Z} \to \mathbb{C}$ and $\nu > 0$, we say that a has *exponential decay of order* ν if

$$\sup_{k\in\mathbb{Z}} |a(k)|e^{\tau|k|} < \infty \qquad \forall\, 0 \leqslant \tau < \nu. \tag{6.6.26}$$

Evidently, if a filter a has exponential decay of order $\nu > 0$ in (6.6.26), then a is a filter of Hölder class. Note that a filter a has exponential decay of order $\nu > 0$ if and only if $\widehat{a}$ is analytic inside the strip $\Gamma_\nu := \{\xi + i\zeta \,:\, \xi \in \mathbb{R}, -\nu < \zeta < \nu\}$.

We have the following result on cascade algorithms and refinable functions with exponentially decaying filters.

Theorem 6.6.9 *Let a be a filter such that $\widehat{a}(0) = 1$ and a has exponential decay of order $\nu > 0$. Then the following statements are equivalent:*

(i) $\widehat{a}(\pi) = 0$ and for every $0 < \tau < 2\nu$ and every admissible function $f \in L_{2,1,\tau}(\mathbb{R})$ with respect to the filter a, $\{f_n := \mathcal{R}_a^n f\}_{n=1}^\infty$ is a Cauchy sequence in $L_{2,1,\tau}(\mathbb{R})$.

(ii) $\widehat{a}(\pi) = 0$ and for every $0 < \tau < 2\nu$, every $1 \leqslant p \leqslant \infty$, and every admissible function $f \in L_{2,p,\tau}(\mathbb{R})$ with respect to the filter a, $\{f_n\}_{n=1}^\infty$ is a Cauchy sequence in $L_{2,p,\tau}(\mathbb{R})$.

(iii) *For some* $0 < \tau < 2\nu$, *some* $1 \leqslant p \leqslant \infty$, *and one admissible function* $f \in L_{2,p,\tau}(\mathbb{R})$ *with respect to the filter a such that the integer shifts of f are stable in* $L_2(\mathbb{R})$, $\{f_n\}_{n=1}^{\infty}$ *is a Cauchy sequence in* $L_{2,p,\tau}(\mathbb{R})$.

(iv) $\mathrm{sm}(a) > 0$.

In particular, if $\mathrm{sm}(a) > 0$ *and a has exponential decay of order* $\nu > 0$, *then its standard refinable function* ϕ *associated with the filter a must have exponential decay of order* 2ν *in* $L_2(\mathbb{R})$, *that is,*

$$\|\phi\|^2_{L_{2,1,\tau}(\mathbb{R})} = \int_{\mathbb{R}} |\phi(x)|^2 e^{2\tau|x|} dx < \infty, \qquad \forall\, 0 \leqslant \tau < 2\nu. \tag{6.6.27}$$

Proof Note that $\widehat{e^{\zeta\cdot}f}(\xi) = \widehat{f}(\xi + i\zeta)$ for $\xi, \zeta \in \mathbb{R}$. By $\|\cdot\|_{L_p(\mathbb{T})} \leqslant \|\cdot\|_{L_q(\mathbb{T})}$ for all $1 \leqslant p \leqslant q \leqslant \infty$ and (6.6.3), we have (Exercise 6.28)

$$[\widehat{f},\widehat{f}](\xi + i\zeta) := \textstyle\sum_{k\in\mathbb{Z}} |\widehat{f}(\xi + i\zeta + 2\pi k)|^2 \leqslant C_{\tau_2-\tau_1} \|f\|^2_{L_{2,1,\tau_2}(\mathbb{R})}, \\ \forall\, \xi \in \mathbb{R}, \zeta \in [-\tau_1, \tau_1], 0 \leqslant \tau_1 < \tau_2 \tag{6.6.28}$$

and for $1 \leqslant p \leqslant q \leqslant \infty$,

$$\|f\|^2_{L_{2,p,\tau_1}(\mathbb{R})} \leqslant \|f\|^2_{L_{2,q,\tau_1}(\mathbb{R})} \leqslant C_{\tau_2-\tau_1}\|f\|^2_{L_{2,1,\tau_2}(\mathbb{R})} \leqslant C_{\tau_2-\tau_1}\|f\|^2_{L_{2,p,\tau_2}(\mathbb{R})} \tag{6.6.29}$$

for all $0 \leqslant \tau_1 < \tau_2$, where

$$C_\alpha := \Big\| \sum_{k\in\mathbb{Z}} e^{-2\alpha|\cdot-k|} \Big\|_{L_\infty(\mathbb{R})} < \infty, \qquad \forall\, \alpha > 0. \tag{6.6.30}$$

Hence, $L_{2,q,\tau_2}(\mathbb{R}) \subseteq L_{2,p,\tau_1}(\mathbb{R})$ for all $1 \leqslant p, q \leqslant \infty$ and $0 \leqslant \tau_1 < \tau_2$. Now it is easy to see that (i)$\Longrightarrow$(ii)$\Longrightarrow$(iii). (iii)$\Longrightarrow$(iv) follows from Theorem 6.6.3 and $\|\cdot\|_{L_{2,\infty,0}(\mathbb{R})} \leqslant C_\tau \|\cdot\|_{L_{2,p,\tau}(\mathbb{R})}$ since $\tau > 0$.

We now prove the key part (iv)$\Longrightarrow$(i). By Theorem 6.6.3, the condition $\mathrm{sm}(a) > 0$ implies $\widehat{a}(\pi) = 0$. Therefore, we can write $\widehat{a}(\xi) = (1 + e^{-i\xi})\widehat{A}(\xi)$, where A has exponential decay of order ν. By Theorems 6.6.3 and 6.6.1, we have $\inf_{n\in\mathbb{N}} \|\mathcal{T}^n_{|\widehat{A}|^2} 1\|^{1/n}_{L_\infty(\mathbb{T})} = \rho_0(\mathcal{T}_{|\widehat{A}|^2}) = \rho_2(\mathcal{T}_{|\widehat{a}|^2}) < 1$. Therefore, for every $\rho_2(\mathcal{T}_{|\widehat{a}|^2}) < \rho < 1$, there exists $N \in \mathbb{N}$ such that $\|\mathcal{T}^N_{|\widehat{A}|^2} 1\|^{1/N}_{L_\infty(\mathbb{T})} < \rho < 1$. Since A has exponential decay of order ν, $[\mathcal{T}^N_{|\widehat{A}|^2} 1](\xi)$ is well defined for $\xi \in \Gamma_{2\nu} := \{t + i\zeta \in \mathbb{C} \,:\, t \in \mathbb{R}, -2\nu < \zeta < 2\nu\}$. Since $\widehat{A}$ is a 2π-periodic continuous function on the strip $\Gamma_{2\nu}$, there exists $\tau_1 > 0$ such that

$$\|[\mathcal{T}^N_{|\widehat{A}|^2} 1](\cdot + i\zeta)\|_{L_\infty(\mathbb{T})} \leqslant \rho^N < 1, \qquad \forall\, \zeta \in [-\tau_1, \tau_1]. \tag{6.6.31}$$

Take m to be the smallest nonnegative integer such that $2^{1-m}\nu \leqslant \tau_1$. For $n \geqslant N+m$, we can write $n = Nk + j$ with $j \in \{m, m+1, \dots, m+N-1\}$ and $k \in \mathbb{N}$. Note that $0 < \tau < 2\nu$. For every $\zeta \in [-\tau, \tau]$, by (6.6.31), we have $|2^{-j}\zeta| \leqslant 2^{-m}\tau < 2^{1-m}\nu \leqslant \tau_1$ and

$$\begin{aligned}\|[\mathcal{T}^n_{|\widehat{A}|^2} 1](\cdot + i\zeta)\|_{L_\infty(\mathbb{T})} &= \|[\mathcal{T}^j_{|\widehat{A}|^2} \mathcal{T}^{Nk}_{|\widehat{A}|^2} 1](\cdot + i\zeta)\|_{L_\infty(\mathbb{T})}\\ &\leqslant \|[\mathcal{T}^j_{|\widehat{A}|^2} 1](\cdot + i\zeta)\|_{L_\infty(\mathbb{T})} \|[\mathcal{T}^{Nk}_{|\widehat{A}|^2} 1](\cdot + i2^{-j}\zeta)\|_{L_\infty(\mathbb{T})}\\ &\leqslant \|[\mathcal{T}^j_{|\widehat{A}|^2} 1](\cdot + i\zeta)\|_{L_\infty(\mathbb{T})} \rho^{Nk} \leqslant C_1 \rho^n,\end{aligned}$$

where

$$C_1 := \sup\{\rho^{-j} \|[\mathcal{T}^j_{|\widehat{A}|^2} 1](\cdot + i\zeta)\|_{L_\infty(\mathbb{T})} \;:\; j = m, \dots, m+N-1, \zeta \in [-\tau, \tau]\} < \infty.$$

That is, we proved

$$\|[\mathcal{T}^n_{|\widehat{A}|^2} 1](\cdot + i\zeta)\|_{L_\infty(\mathbb{T})} \leqslant C_1 \rho^n, \qquad \forall\, n \in \mathbb{N}, \zeta \in [-\tau, \tau]. \tag{6.6.32}$$

We now prove that $\{f_n\}_{n=1}^\infty$ in item (i) is a Cauchy sequence in $L_{2,1,\tau}(\mathbb{R})$. Define $\widehat{g}(\xi) := \widehat{a}(\xi/2)\widehat{f}(\xi/2) - \widehat{f}(\xi)$. Since f is admissible with respect to the filter a and $f \in L_{2,1,\tau}(\mathbb{R})$, we have $g \in L_{2,1,\tau}(\mathbb{R})$ and $\widehat{g}(2\pi k) = 0$ for all $k \in \mathbb{Z}$. Define $h := \sum_{k=0}^\infty g(\cdot - k)$. Then $g = h - h(\cdot - 1)$ (that is, $\widehat{g}(\xi) = (1 - e^{-i\xi})\widehat{h}(\xi)$) and $h \in L_{2,1,\tau_0}(\mathbb{R})$ for all $0 \leqslant \tau_0 < \tau$ (see Exercise 6.29). Define $g_n := f_{n+1} - f_n$ and $\tau_0 := \tau/2$. By $\widehat{g}(\xi) = (1 - e^{-i\xi})\widehat{h}(\xi)$ and (6.6.28),

$$[\widehat{g}, \widehat{g}](\xi) = |1 - e^{-i\xi}|^2 [\widehat{h}, \widehat{h}](\xi) \leqslant C_2 |1 - e^{-i\xi}|^2, \qquad \forall\, \xi \in \Gamma_{\tau_0}, \tag{6.6.33}$$

where $C_2 := C_{\tau_0} \|h\|^2_{L_{2,1,\tau_0}(\mathbb{R})} < \infty$ with C_{τ_0} being defined in (6.6.30). By the definition $f_n := \mathcal{R}^n_a f$ and by induction on $n \in \mathbb{N}$, for $n \geqslant n_0 := 1 - \log_2(\tau_0/\nu)$, we have $\widehat{g_n}(\xi) = \widehat{g}(2^{-n}\xi) \prod_{j=1}^n \widehat{a}(2^{-j}\xi)$ for $\xi \in \Gamma_{2\nu}$ and g_n is analytic on $\Gamma_{2\nu}$, since $2^{-n}\Gamma_{2\nu} \subseteq \Gamma_{\tau_0}$ for all $n \geqslant n_0$. By induction on $n \in \mathbb{N}$, it follows from (6.6.33) that

$$[\widehat{g_n}, \widehat{g_n}](\xi) = (\mathcal{T}^n_{|\widehat{a}|^2} [\widehat{g}, \widehat{g}])(\xi) \leqslant C_2 (\mathcal{T}^n_{|\widehat{a}|^2} (|1 - e^{-i\cdot}|^2))(\xi), \qquad \xi \in \Gamma_{2\nu}. \tag{6.6.34}$$

Since $\widehat{a}(\xi) = (1 + e^{-i\xi})\widehat{A}(\xi)$, we deduce that

$$|[\mathcal{T}^n_{|\widehat{a}|^2} (|1 - e^{-i\cdot}|^2)](\xi)| = |1 - e^{-i\xi}|^2 |[\mathcal{T}^n_{|\widehat{A}|^2} 1](\xi)| \leqslant (1 + e^{2\nu})^2 |[\mathcal{T}^n_{|\widehat{A}|^2} 1](\xi)|, \quad \forall\, \xi \in \Gamma_{2\nu}.$$

Consequently, since $0 < \tau < 2\nu$, it follows from (6.6.34) and (6.6.32) that

$$[\widehat{g_n}, \widehat{g_n}](\xi) \leqslant C_2 (1 + e^{2\nu})^2 |[\mathcal{T}^n_{|\widehat{A}|^2} 1](\xi)| \leqslant C_1 C_2 (1 + e^{2\nu})^2 \rho^n \leqslant C_3 \rho^n, \quad \forall\, n \geqslant n_1, \xi \in \Gamma_\tau,$$

where $C_3 := C_1 C_2 (1 + e^{2\nu})^2 < \infty$ and $n_1 := \max(n_0, N + m)$. Note that $\widehat{g_n}(\cdot + i\zeta)$ is the Fourier transform of $e^{\zeta\cdot}(f_{n+1} - f_n)$. For $\zeta \in [-\tau, \tau]$, we have

$$2\pi \|(f_{n+1} - f_n)e^{\zeta\cdot}\|^2_{L_2(\mathbb{R})} = \|\widehat{g_n}(\cdot + i\zeta)\|^2_{L_2(\mathbb{R})}$$
$$= \|[\widehat{g_n}, \widehat{g_n}](\cdot + i\zeta)\|_{L_1(\mathbb{T})} \leqslant \|[\widehat{g_n}, \widehat{g_n}](\cdot + i\zeta)\|_{L_\infty(\mathbb{T})} \leqslant C_3 \rho^n.$$

By the definition of the space $L_{2,1,\tau}(\mathbb{R})$, it follows from the above inequalities that for all $n \geqslant n_1$,

$$\|f_{n+1} - f_n\|^2_{L_{2,1,\tau}(\mathbb{R})} = \|(f_{n+1} - f_n)e^{\tau|\cdot|}\|^2_{L_2(\mathbb{R})}$$
$$\leqslant \|(f_{n+1} - f_n)e^{\tau\cdot}\|^2_{L_2(\mathbb{R})} + \|(f_{n+1} - f_n)e^{-\tau\cdot}\|^2_{L_2(\mathbb{R})} \leqslant C_3 \rho^n / \pi.$$

This proves that $\{f_n\}_{n=1}^\infty$ is a Cauchy sequence in $L_{2,1,\tau}(\mathbb{R})$. Hence, (iv)$\Longrightarrow$(i).

We now show that ϕ has exponential decay of order 2ν. Since $\widehat{\phi}(\xi) := \prod_{j=1}^\infty \widehat{a}(2^{-j}\xi)$ and $\widehat{a}$ is analytic in the strip Γ_ν, we see that $\widehat{\phi}$ can be extended into an analytic function in $\Gamma_{2\nu}$. If $\operatorname{sm}(a) > 0$, for every $0 < \tau < 2\nu$, since $\lim_{n\to\infty} \widehat{f_n}(\xi) = \widehat{\phi}(\xi)$ (here we additionally assumed $\lim_{j\to\infty} \widehat{f}(2^{-j}\xi) = 1$ for all $\xi \in \mathbb{R}$), by what has been proved, we must have $\lim_{n\to\infty} \|f_n - \phi\|_{L_{2,1,\tau}(\mathbb{R})} = 0$ and $\phi \in L_{2,1,\tau}(\mathbb{R})$. This proves (6.6.27). □

We present an example of orthogonal wavelets with infinite support.

Example 6.6.3 Let a be a filter of Hölder class with $\widehat{a}(0) = 1$ and $\operatorname{sm}(a) > 0$. Let ϕ be its associated refinable function defined by $\widehat{\phi}(\xi) := \prod_{j=1}^\infty \widehat{a}(2^{-j}\xi)$. By Theorem 6.6.3, we have $\phi \in L_2(\mathbb{R})$, $\widehat{a}(\pi) = 0$, and the coefficient sequence of $[\widehat{\phi}, \widehat{\phi}]$ is a filter of Hölder class. Suppose that $[\widehat{\phi}, \widehat{\phi}](\xi) \neq 0$ for all $\xi \in \mathbb{R}$ (and therefore, the integer shifts of ϕ are stable in $L_2(\mathbb{R})$). Since $[\widehat{\phi}, \widehat{\phi}](\xi) > 0$ for all $\xi \in \mathbb{R}$, we define

$$\widehat{\mathring{a}}(\xi) := \widehat{a}(\xi)\frac{\sqrt{[\widehat{\phi}, \widehat{\phi}](\xi)}}{\sqrt{[\widehat{\phi}, \widehat{\phi}](2\xi)}}, \qquad \widehat{\mathring{\varphi}}(\xi) := \frac{\widehat{\phi}(\xi)}{\sqrt{[\widehat{\phi}, \widehat{\phi}](\xi)}},$$
$$\widehat{\mathring{b}}(\xi) := e^{-i\xi}\overline{\widehat{\mathring{a}}(\xi + \pi)}, \qquad \widehat{\mathring{\psi}}(\xi) := \widehat{\mathring{b}}(\xi/2)\widehat{\mathring{\varphi}}(\xi/2).$$

Since the coefficient sequence of $[\widehat{\phi}, \widehat{\phi}]$ is a filter of Hölder class and $[\widehat{\phi}, \widehat{\phi}](\xi) > 0$ for all $\xi \in \mathbb{R}$, both filters $\mathring{a}$ and $\mathring{b}$ are filters of Hölder class. Moreover, we have $\widehat{\mathring{\varphi}}(2\xi) = \widehat{\mathring{a}}(\xi)\widehat{\mathring{\varphi}}(\xi)$ and $\{\mathring{a}; \mathring{b}\}$ is an orthogonal wavelet filter bank. Note that $\operatorname{sm}(\mathring{a}) = \operatorname{sm}(a) > 0$. We conclude from Theorem 6.6.6 that $\{\mathring{\varphi}; \mathring{\psi}\}$ is an orthogonal wavelet in $L_2(\mathbb{R})$.

If in addition the filter a has exponential decay, then we conclude from Theorem 6.6.9 that the coefficient sequence of $[\widehat{\phi}, \widehat{\phi}]$ has exponential decay.

Consequently, both filters $\mathring{a}$ and $\mathring{b}$ have exponential decay. By Theorem 6.6.9 again, we see that $\{\mathring{\varphi}; \mathring{\psi}\}$ is an orthogonal wavelet in $L_2(\mathbb{R})$ with exponential decay.

6.7 Smooth Refinable Duals and Local Linear Independence of Scalar Refinable Functions*

To construct compactly supported biorthogonal wavelets in Sobolev spaces, for a given scalar refinable function ϕ having linearly independent integer shifts, it is critical to construct a refinable dual with some desirable properties. In this section we study the existence of smooth compactly supported refinable duals for a given refinable function ϕ and then investigate local linear independence of ϕ. The local linear independence plays a role in the construction of wavelets on a bounded interval.

The following result shows that for every scalar refinable function having linearly independent integer shifts, there always exist arbitrarily smooth dual refinable functions.

Theorem 6.7.1 *Let $a \in l_0(\mathbb{Z})$ such that $\widehat{a}(0) = 1$. Let ϕ be the standard refinable function/distribution associated with the filter a satisfying $\widehat{\phi}(\xi) := \prod_{j=1}^{\infty} \widehat{a}(2^{-j}\xi)$. Suppose that the integer shifts of ϕ are linearly independent. For any $m \in \mathbb{N}$, there exists $\tilde{a} \in l_0(\mathbb{Z})$ such that $\widehat{\tilde{a}}(0) = 1$, $\operatorname{sm}(\tilde{a}) \geqslant m$, $\tilde{a}$ is a dual filter of a, and $(\tilde{\phi}, \phi)$ satisfies* (6.6.16)*, where $\tilde{\phi}$ is the standard refinable function associated with the filter $\tilde{a}$.*

Proof Define $\widehat{\mathring{a}}(\xi) := 2^{-M}(1 + e^{-i\xi})^M \widehat{a}(\xi)$ with a positive integer $M \geqslant m + 1 - \operatorname{sm}(a)$. Note that $\widehat{\mathring{a}}(0) = \widehat{a}(0) = 1$. Let $\mathring{\phi}$ be the standard refinable function associated with the filter $\mathring{a}$. Then $\widehat{\mathring{\phi}}(\xi) = (\widehat{B_1}(\xi))^M \widehat{\phi}(\xi)$ for all $\xi \in \mathbb{C}$, where $\widehat{B_1}(\xi) = \frac{1-e^{-i\xi}}{i\xi}$. Observe that $\{\xi \in \mathbb{C} : \widehat{B_1}(\xi) = 0\} = 2\pi\mathbb{Z}\backslash\{0\}$ and $\widehat{\mathring{\phi}}(0) = 1 \neq 0$. By Theorem 5.2.1, the linear independence of the integer shifts of ϕ implies the linear independence of the integer shifts of $\mathring{\phi}$.

Write $\widehat{a}(\xi) = (1 + e^{-i\xi})^{\operatorname{sr}(a)} 2^M \widehat{A}(\xi)$ for some $\operatorname{sr}(a) \in \mathbb{N}_0$ and $A \in l_0(\mathbb{Z})$ with $\widehat{A}(\pi) \neq 0$. Note that

$$\widehat{\mathring{a}}(\xi) = 2^{-M}(1 + e^{-i\xi})^M \widehat{a}(\xi) = (1 + e^{-i\xi})^{\tau} \widehat{A}(\xi) \text{ with } \tau := M + \operatorname{sr}(a) \in \mathbb{N}.$$

By Corollary 5.8.5 or Theorem 6.6.1, we have $\rho_{2\tau}(\mathcal{T}_{|\widehat{\mathring{a}}|^2}) = \rho_0(\mathcal{T}_{|\widehat{A}|^2})$, from which we have

$$\begin{aligned}\operatorname{sm}(\mathring{a}) &= -\tfrac{1}{2}\log_2 \rho_{2\tau}(\mathcal{T}_{|\widehat{\mathring{a}}|^2}) = -\tfrac{1}{2}\log_2 \rho_0(\mathcal{T}_{|\widehat{A}|^2}) = M - \tfrac{1}{2}\log_2 \rho_0(\mathcal{T}_{2^{2M}|\widehat{A}|^2}) \\ &= M + \operatorname{sm}(a) \geqslant m + 1 \geqslant 2.\end{aligned}$$

Therefore, $\mathring{\phi} \in L_2(\mathbb{R})$ and $\mathrm{sm}(\mathring{\phi}) = \mathrm{sm}(\mathring{a}) = M + \mathrm{sm}(a) \geqslant 2$. Since the integer shifts of $\mathring{\phi} \in L_2(\mathbb{R})$ are linearly independent and hence stable, the function $\widehat{c}(\xi) := [\widehat{\mathring{\phi}}, \widehat{\mathring{\phi}}](\xi)$ is a 2π-periodic trigonometric polynomial and there exist positive constants $c_{\max}$ and $c_{\min}$ such that (6.6.17) holds. Now the filter u defined by $\widehat{u}(\xi) := \widehat{\mathring{a}}(\xi)\widehat{c}(\xi)/\widehat{c}(2\xi)$ is well defined. By $\widehat{\mathring{\phi}}(2\xi) = \widehat{\mathring{a}}(\xi)\widehat{\mathring{\phi}}(\xi)$, we have

$$\begin{aligned}\widehat{c}(2\xi) = [\widehat{\mathring{\phi}}, \widehat{\mathring{\phi}}](2\xi) &= |\widehat{\mathring{a}}(\xi)|^2[\widehat{\mathring{\phi}}, \widehat{\mathring{\phi}}](\xi) + |\widehat{\mathring{a}}(\xi+\pi)|^2[\widehat{\mathring{\phi}}, \widehat{\mathring{\phi}}](\xi+\pi)\\ &= |\widehat{\mathring{a}}(\xi)|^2\widehat{c}(\xi) + |\widehat{\mathring{a}}(\xi+\pi)|^2\widehat{c}(\xi+\pi),\end{aligned}$$

from which we see that u is a dual filter of $\mathring{a}$. Since $\widehat{u}(\xi) = \widehat{\mathring{a}}(\xi)\widehat{c}(\xi)/\widehat{c}(2\xi) = (1+e^{-i\xi})^{\tau}\widehat{v}(\xi)$ with $\widehat{v}(\xi) := \widehat{A}(\xi)\widehat{c}(\xi)/\widehat{c}(2\xi)$ and $\widehat{v}(\pi) \neq 0$, we get

$$\prod_{j=1}^{n} |\widehat{v}(2^{-j}\xi)|^2 = \frac{|\widehat{c}(2^{-n}\xi)|^2}{|\widehat{c}(\xi)|^2} \prod_{j=1}^{n} |\widehat{A}(2^{-j}\xi)|^2 \leqslant C \prod_{j=1}^{n} |\widehat{A}(2^{-j}\xi)|^2$$

with $C := (c_{\max}/c_{\min})^2 < \infty$. Thus, $\|\mathcal{T}^n_{|\widehat{v}|^2} 1\|_{L_\infty(\mathbb{T})} \leqslant C\|\mathcal{T}^n_{|\widehat{A}|^2} 1\|_{L_\infty(\mathbb{T})}$ for all $n \in \mathbb{N}$, from which we have $\rho_{2\tau}(\mathcal{T}_{|\widehat{u}|^2}) = \rho_0(\mathcal{T}_{|\widehat{v}|^2}) \leqslant \rho_0(\mathcal{T}_{|\widehat{A}|^2}) = 2^{-2\,\mathrm{sm}(\mathring{a})}$. It is evident that v is a filter of Hölder class and $\widehat{v}(0) \neq 0$. By Theorem 6.6.1, there exists $N \in \mathbb{N}$ such that

$$\|\mathcal{T}^N_{|\widehat{v}|^2} 1\|^{1/N}_{L_\infty(\mathbb{T})} < 2^{-2(\mathrm{sm}(\mathring{a})-1)}. \tag{6.7.1}$$

We now construct a desired finitely supported dual filter $\tilde{a}$ of a. Since $\widehat{\mathring{\phi}}(2\xi) = \mathring{\mathsf{a}}(e^{-i\xi})\widehat{\mathring{\phi}}(\xi)$ and $\{\widehat{\mathring{\phi}}(\xi+2\pi k)\}_{k\in\mathbb{Z}} \neq 0$ for all $\xi \in \mathbb{C}$, the Laurent polynomials $\mathring{\mathsf{a}}(z)$ and $\mathring{\mathsf{a}}(-z)$ must have no common zeros in $\mathbb{C}\backslash\{0\}$, where $\mathring{\mathsf{a}}(z) := \sum_{k\in\mathbb{Z}} \mathring{a}(k)z^k$. By Proposition 2.7.2, the filter a must have a finitely supported dual. Now by the CBC algorithm in Algorithm 2.6.2 or Algorithm 6.5.2, the filter a must have a finitely supported dual filter $\breve{a} \in l_0(\mathbb{Z})$ such that $\widehat{\breve{a}}(\xi) = (1+e^{-i\xi})^{\tau}\widehat{\breve{b}}(\xi)$ for some $\breve{b} \in l_0(\mathbb{Z})$. Since both u and $\breve{a}$ are dual filters of $\mathring{a}$, by $\widehat{u}(\xi) = (1+e^{-i\xi})^{\tau}\widehat{v}(\xi) = (1+e^{-i\xi})^{\tau}\widehat{A}(\xi)\widehat{c}(\xi)/\widehat{c}(2\xi)$, we can directly check that

$$\widehat{v}(\xi) = \widehat{\breve{b}}(\xi) + e^{-i\xi}\overline{\widehat{\mathring{a}}(\xi+\pi)}(1-e^{-i\xi})^{\tau}\widehat{w}(2\xi)/\widehat{c}(2\xi),$$

where $w \in l_0(\mathbb{Z})$ is defined to be

$$\widehat{w}(2\xi) := e^{i\xi}\big(\widehat{A}(\xi)\widehat{c}(\xi)\widehat{\breve{b}}(\xi+\pi) - \widehat{A}(\xi+\pi)\widehat{c}(\xi+\pi)\widehat{\breve{b}}(\xi)\big).$$

Since $c \in l_0(\mathbb{Z})$ satisfies the condition in (6.6.17), by Proposition 6.6.8, there exists a sequence $\{c_n\}_{n=1}^{\infty}$ in $l_0(\mathbb{Z})$ such that $\widehat{c_n}(0) = 1/\widehat{c}(0)$ and $\lim_{n\to\infty} \|\widehat{c_n} - \frac{1}{\widehat{c}}\|_{L_\infty(\mathbb{T})} = 0$. For $n \in \mathbb{N}$, we define

$$\widehat{v_n}(\xi) := \widehat{\mathring{b}}(\xi) + e^{-i\xi}\overline{\widehat{\mathring{a}}(\xi+\pi)(1-e^{-i\xi})^{\tau}\widehat{w}(2\xi)\widehat{c_n}(2\xi)}. \qquad (6.7.2)$$

Then all $v_n \in l_0(\mathbb{Z})$ and $\lim_{n\to\infty} \|\widehat{v_n} - \widehat{v}\|_{L_\infty(\mathbb{T})} = 0$. By (6.7.1), there exists $n_0 \in \mathbb{N}$ such that $\|\mathcal{T}^N_{|\widehat{v_n}|^2} 1\|^{1/N}_{L_\infty(\mathbb{T})} < 2^{-2(\mathrm{sm}(\mathring{a})-1)}$ for all $n \geqslant n_0$. Using Theorem 6.6.1, we conclude that

$$\rho_0(\mathcal{T}_{|\widehat{v_n}|^2}) = \inf_{j\in\mathbb{N}} \|\mathcal{T}^j_{|\widehat{v_n}|^2} 1\|^{1/j}_{L_\infty(\mathbb{T})} \leqslant \|\mathcal{T}^N_{|\widehat{v_n}|^2} 1\|^{1/N}_{L_\infty(\mathbb{T})} < 2^{-2(\mathrm{sm}(\mathring{a})-1)}, \qquad \forall\, n \geqslant n_0.$$

Define $\widehat{\tilde{u}}(\xi) := (1+e^{-i\xi})^{\tau}\widehat{v_{n_0}}(\xi)$. We conclude that $\mathrm{sm}(\tilde{u}) \geqslant -\log_2 \sqrt{\rho_0(\mathcal{T}_{|\widehat{v_{n_0}}|^2})} > \mathrm{sm}(\mathring{a}) - 1 = M + \mathrm{sm}(a) - 1 \geqslant m > 0$. Moreover, it follows directly from (6.7.2) that $\tilde{u}$ must be a finitely supported dual filter of $\mathring{a}$. Define $\widehat{\tilde{a}}(\xi) := 2^{-M}(1+e^{i\xi})^M\widehat{\tilde{u}}(\xi)$. Then $\tilde{a}$ must be a finitely supported dual filter of a and

$$\mathrm{sm}(\tilde{a}) = M + \mathrm{sm}(\tilde{u}) \geqslant 2M + \mathrm{sm}(a) - 1 \geqslant M + m > m.$$

The biorthogonality condition in (6.6.16) follows directly from Theorem 6.4.5 since $\mathrm{sm}(a) + \mathrm{sm}(\tilde{a}) \geqslant 2M + 2\,\mathrm{sm}(a) - 1 > 0$. □

For a distribution $\phi \in \mathscr{D}'(\mathbb{R})$ and an open set I, we say that $\phi = 0$ on I in the sense of distributions if $\langle \phi, \psi \rangle = 0$ for all $\psi \in \mathscr{D}(\mathbb{R})$ such that the support of ψ is contained inside I. Recall that the support, denoted by $\mathrm{supp}(\phi)$, of a distribution ϕ is the smallest closed subset of $\mathbb{R}$ such that $\phi = 0$ on $\mathbb{R}\backslash\mathrm{supp}(\phi)$. For a compactly supported distribution ϕ on $\mathbb{R}$ and a nonempty open set I, we say that *the integer shifts of ϕ are linearly independent on I* if

$$\sum_{k\in\mathbb{Z}} v(k)\phi(\cdot - k) = 0 \quad \text{on} \quad I \quad \text{for some } v \in l(\mathbb{Z}) \qquad (6.7.3)$$
$$\Longrightarrow v(k)\phi(\cdot - k) = 0 \quad \text{on} \quad I \quad \forall\, k \in \mathbb{Z}.$$

We say that the integer shifts of a compactly supported distribution ϕ are *locally linearly independent* if the integer shifts of ϕ are linearly independent on all nonempty open subsets of $\mathbb{R}$.

Evidently, the integer shifts of a compactly supported distribution ϕ are linearly independent simply means that the integer shifts of ϕ are linearly independent on $\mathbb{R}$. The following result plays an important role for constructing wavelets on bounded intervals by showing that locally linear independence is equivalent to linear independence for a compactly supported scalar (dyadic) refinable function.

Theorem 6.7.2 *Let $a \in l_0(\mathbb{Z})$ with $\widehat{a}(0) = 1$ and $\operatorname{fsupp}(a) = [0, N]$ for some $N \in \mathbb{N}_0$. Let ϕ be the standard refinable function/distribution associated with the filter a. Suppose that the integer shifts of ϕ are linearly independent. Then*

(1) For all $j \in \mathbb{N}_0$ and $M \in \mathbb{N}$, if $\sum_{k\in\mathbb{Z}} v(k)\phi(\cdot - k) = 0$ on $(0, 2^j M)$ for some $v \in l(\mathbb{Z})$, then $v(k) = 0$ for all $k = 1 - N, \dots, 2^j M - 1$.
(2) The integer shifts of ϕ are locally linearly independent.
(3) $\operatorname{supp}(\phi) = [0, N]$, that is, there are no holes within the support of ϕ.

Proof By Theorem 5.1.2, $\operatorname{supp}(\phi) \subseteq \operatorname{fsupp}(a) = [0, N]$. If $N = 0$, by $\widehat{a}(0) = 1$, then we must have $\phi = \boldsymbol{\delta}$ and all the claims obviously hold. So, we assume $N \geqslant 1$. We first prove that

$$c := \inf \operatorname{supp}(\phi) = 0 \quad \text{and} \quad d := \sup \operatorname{supp}(\phi) = N. \tag{6.7.4}$$

Note that $0 \leqslant c \leqslant d \leqslant N$. We now prove $c = 0$ using proof by contradiction. Suppose that $c > 0$. Let $c_0 := \min(2c, c + 1) > 0$. Then $0 < c < c_0$. By the definition of c in (6.7.4), there exists $\psi \in \mathscr{D}(\mathbb{R})$ such that $\operatorname{supp}(\psi) \subseteq (0, c_0)$ and $\langle \phi, \psi \rangle \neq 0$. Note that $\widehat{\phi}(2\xi) = \widehat{a}(\xi)\widehat{\phi}(\xi)$ is simply equivalent to

$$\phi(\cdot/2) = 2\sum_{k=0}^{N} a(k)\phi(\cdot - k). \tag{6.7.5}$$

Since $\phi(\cdot/2) = 0$ on $(0, c_0)$ and $\operatorname{supp}(\phi(\cdot - k)) \subseteq [c + k, N + k]$, from the above identity we deduce that

$$0 = \langle \phi(\cdot/2), \psi \rangle = 2\sum_{k=0}^{N} a(k)\langle \phi(\cdot - k), \psi \rangle = 2a(0)\langle \phi, \psi \rangle.$$

Since $\langle \phi, \psi \rangle \neq 0$, the above identity forces $a(0) = 0$, which is a contradiction to our assumption $\operatorname{fsupp}(a) = [0, N]$. Therefore, we must have $c = 0$. Similarly, we can prove $d = N$. This proves (6.7.4).

By Theorem 5.2.1 or Theorem 6.7.1, there exists a compactly supported smooth dual function $\tilde{\phi}$ of ϕ such that (6.6.16) holds. Suppose that $\operatorname{supp}(\tilde{\phi}) \subseteq [-K, K]$ with $K \in \mathbb{N}$. Then we can easily check that $\operatorname{supp}(\tilde{\phi}(\cdot - n)) \subseteq [n - K, n + K] \subseteq (0, 2^j M)$ for all $K < n < 2^j M - K$. Suppose that $f := \sum_{k\in\mathbb{Z}} v(k)\phi(\cdot - k) = 0$ on $(0, 2^j M)$. Then it follows from (6.6.16) that

$$0 = \langle f, \tilde{\phi}(\cdot - n) \rangle = \sum_{k\in\mathbb{Z}} v(k)\langle \phi(\cdot - k), \tilde{\phi}(\cdot - n) \rangle = v(n), \qquad \forall\, n = K + 1, \dots, 2^j M - K - 1. \tag{6.7.6}$$

Thus $f = f_1 + f_2$ with

$$f_1 := \sum_{k=-\infty}^{K} v(k)\phi(\cdot - k), \qquad f_2 := \sum_{k=2^jM-K}^{\infty} v(k)\phi(\cdot - k).$$

Since $\operatorname{supp}(\phi(\cdot - k)) \subseteq [k, N+k]$, we observe that $\operatorname{supp}(f_1) \subseteq (-\infty, N+K]$ and $\operatorname{supp}(f_2) \subseteq [2^jM - K, \infty)$. Let $j_0 \in \mathbb{N}$ be the smallest nonnegative integer satisfying $2^{j_0} > N + 2K$. Then $N + K < 2^j - K \leqslant 2^jM - K$ for all $j \geqslant j_0$ and $M \in \mathbb{N}$. Hence, $\operatorname{supp}(f_1)$ and $\operatorname{supp}(f_2)$ are disjoint. By $f = 0$ on $(0, 2^jM)$ and $f = f_1 + f_2$, we conclude that $f_1 = 0$ on $(0, \infty)$ and $f_2 = 0$ on $(-\infty, 2^jM)$. If $v(k) \neq 0$ for some $k = 2^jM - K, \ldots, 2^jM - 1$, then there exists a smallest integer k_0 such that $v(k_0) \neq 0$ with $2^jM - K \leqslant k_0 \leqslant 2^jM - 1$ and $f_2 = \sum_{k=k_0}^{\infty} v(k)\phi(\cdot - k)$. Since $k_0 + 1 \leqslant 2^jM$, on the interval $(-\infty, k_0 + 1)$ we have $0 = f_2 = v(k_0)\phi(\cdot - k_0)$, which forces $v(k_0) = 0$ by (6.7.4) and $\operatorname{supp}(\phi(\cdot - k_0)) \subseteq [k_0, N + k_0]$. This contradicts our assumption $v(k_0) \neq 0$. This proves that we must have $v(k) = 0$ for all $k = 2^jM - K, \ldots, 2^jM - 1$. Applying the same argument to f_1, we get $v(k) = 0$ for all $k = 1 - N, \ldots, K$. Taking into account (6.7.6), we conclude that item (1) holds for all $j \geqslant j_0$ and $M \in \mathbb{N}$.

We now prove item (1) for $j = 0, \ldots, j_0 - 1$. Suppose that item (1) holds for $j + 1$. We show that it must also hold for j. By the refinement equation in (6.7.5), we have

$$f(\cdot/2) = \sum_{k\in\mathbb{Z}} v(k)\phi(\cdot/2 - k) = \sum_{n\in\mathbb{Z}} [\mathcal{S}_a v](n)\phi(\cdot - n),$$

where $[\mathcal{S}_a v](n) = 2\sum_{j\in\mathbb{Z}} v(j)a(n - 2j)$. Since $f = 0$ on $(0, 2^jM)$, we see that $f(\cdot/2) = 0$ on $(0, 2^{j+1}M)$. By our induction hypothesis and the above identity, we conclude that $[\mathcal{S}_a v](n) = 0$ for all $n = 1 - N, \ldots, 2^{j+1}M - 1$. Since $\operatorname{fsupp}(a) = [0, N]$, we can directly deduce that

$$\mathcal{S}_a v(m - k) = 2\sum_{j=0}^{N-1} v(\lfloor m/2 \rfloor - j)a(2j - k + \operatorname{odd}(m)), \tag{6.7.7}$$

$$\forall\, m \in \mathbb{Z}, k = 0, \ldots, N-1,$$

where $\operatorname{odd}(m) := 1$ if m is an odd integer and $\operatorname{odd}(m) = 0$ otherwise (also see (2.5.10)). Since the integer shifts of ϕ are linearly independent and $\widehat{\phi}(2\xi) = \widehat{a}(\xi)\widehat{\phi}(\xi)$, as we proved in Theorem 6.7.1, the Laurent polynomials $\mathsf{a}(z)$ and $\mathsf{a}(-z)$ have no common zeros. By Lemma 6.1.2, the matrices $A_\gamma := (2a(2j - k + \gamma))_{0\leqslant j,k\leqslant N-1}$ are invertible for $\gamma = 0, 1$. Now it follows from (6.7.7) and $[\mathcal{S}_a v](n) = 0$ for all $n = 1 - N, \ldots, 2^{j+1}M - 1$ that for all $m = 0, \ldots, 2^{j+1}M - 1$,

$$\begin{aligned} 0 &= \big([\mathcal{S}_a v](m), [\mathcal{S}_a v](m-1), \ldots, [\mathcal{S}_a v](m - N + 1)\big) \\ &= \big(v(\lfloor m/2 \rfloor), v(\lfloor m/2 \rfloor - 1), \ldots, v(\lfloor m/2 \rfloor - N + 1)\big)(2a(2j - k + \operatorname{odd}(m)))_{0\leqslant j,k\leqslant N-1} \\ &= \big(v(\lfloor m/2 \rfloor), v(\lfloor m/2 \rfloor - 1), \ldots, v(\lfloor m/2 \rfloor - N + 1)\big)A_{\operatorname{odd}(m)}. \end{aligned}$$

Since $A_{\mathrm{odd}(m)}$ is invertible, we conclude that $v(\lfloor m/2\rfloor - j) = 0$ for all $j = 0, \ldots, N-1$ and $m = 0, \ldots, 2^{j+1}M - 1$, which implies $v(k) = 0$ for all $k = 1-N, \ldots, 2^jM - 1$. This proves item (1) for all $j \in \mathbb{N}_0$.

We now prove item (2). Suppose that there exist some nonempty open set $I \subseteq \mathbb{R}\backslash\mathbb{Z}$ and $v \in l(\mathbb{Z})$ such that (6.7.3) fails. Since we can translate I by integers, we can assume that I is an open interval inside $(0, 1)$ such that (6.7.3) fails. By $\mathrm{supp}(\phi(\cdot - k)) \subseteq [k, k+N]$, we see that

$$\phi(\cdot - k) = 0 \quad \text{on} \quad (0, 1), \qquad \forall\, k \in \mathbb{Z}\backslash\{1-N, \ldots, 0\}.$$

Define $\Phi := (\phi(\cdot), \phi(\cdot + 1), \ldots, \phi(\cdot + N - 1))^{\mathsf{T}}$. Then (6.7.3) fails on $I \subseteq (0, 1)$ is equivalent to

$$\vec{v}\Phi = 0 \quad \text{on} \quad I \quad \text{with} \quad \vec{v} := (v(0), \ldots, v(1-N)) \neq 0. \tag{6.7.8}$$

It has been proved in Theorem 6.1.7 that (6.1.8) holds for some $\gamma_1, \ldots, \gamma_n \in \{0, 1\}$ such that $2^{-1}\gamma_1 + 2^{-2}\gamma_2 + \cdots + 2^{-n}\gamma + 2^{-n}(0, 1) \subseteq I \cap (0, 1)$. As a consequence, $\vec{v}A_{\gamma_n} \cdots A_{\gamma_1}\Phi = 0$ on $(0, 1)$. Since both A_0 and A_1 are invertible, we can assume that (6.7.8) holds for $I = (0, 1)$. Applying item (1) with $j = 0$ and $M = 1$, we must have $v(k) = 0$ for all $k = 1-N, \ldots, 0$, that is, $\vec{v} = 0$, which is a contradiction to (6.7.8). This proves that (6.7.3) must be true for all open subsets $I \subseteq \mathbb{R}\backslash\mathbb{Z}$.

Before completing the proof of item (2), we first prove item (3). Let $I := [0, N]\backslash\,\mathrm{supp}(\phi)$. By (6.7.4), I is an open subset of $(0, N)$. By the refinement equation in (6.7.5), since $\phi(\cdot/2) = 0$ on $2I$ and therefore vanishes on $(2I)\backslash\mathbb{Z}$, by what has been proved for item (2), we must have $a(0)\phi = 0$ and $a(N)\phi(\cdot - N) = 0$ on $(2I)\backslash\mathbb{Z}$. Since $a(0)a(N) \neq 0$ and $N \geqslant 1$, this implies that $\phi = 0$ on $J\backslash\mathbb{Z}$, where $J := [2I \cup (2I - N)] \cap (0, N)$. By $I \subseteq (0, N)$, we see that $K := [(2I \mod N)\backslash\{0\}] \subseteq J$ and $|K| = 2|I|$, where $|I|$ is the Lebesgue measure of the open set I. Consequently, $\phi = 0$ on $J\backslash\mathbb{Z}$ implies $K\backslash\mathbb{Z} \subseteq I$. Thus, $2|I| = |K| = |K\backslash\mathbb{Z}| \leqslant |I|$, from which we must have $|I| = 0$. Since I is an open set, we conclude that $I = \emptyset$. This proves item (3).

We now complete the proof of item (2). Suppose that there exist some nonempty open set $I \subseteq \mathbb{R}$ and $v \in l(\mathbb{Z})$ such that $\sum_{k\in\mathbb{Z}} v(k)\phi(\cdot - k) = 0$ on I but there exists $n \in \mathbb{Z}$ such that $v(n) \neq 0$ and $\phi(\cdot - n) \neq 0$ on I. Since $\sum_{k\in\mathbb{Z}} v(k)\phi(\cdot - k) = 0$ on $I\backslash\mathbb{Z}$, by what has been proved, we must have $v(k)\phi(\cdot - k) = 0$ on $I\backslash\mathbb{Z}$ for all $k \in \mathbb{Z}$. By $v(n) \neq 0$, we must have $\phi(\cdot - n) = 0$ on $I\backslash\mathbb{Z}$ and $\phi(\cdot - n) \neq 0$ on I. Thus, $I \cap \mathbb{Z}$ cannot be empty, say $m \in I \cap \mathbb{Z}$. Since I is an open set, there exists $0 < \varepsilon < 1$ such that $(m - \varepsilon, m + \varepsilon) \subseteq I$ and $m - n \in \mathrm{supp}(\phi) = [0, N]$. Since $\phi(\cdot - n) = 0$ on $I\backslash\mathbb{Z}$, we see that $[(m - n - \varepsilon, m - n) \cup (m - n, m - n + \varepsilon)] \cap \mathrm{supp}(\phi) = \emptyset$. This is a contradiction to item (3) since $m - n \in \mathrm{supp}(\phi)$. This proves that (6.7.3) must hold for any open set I and hence item (2) holds. □

Example 6.5.3 and Fig. 6.8 show that Theorem 6.7.2 does not hold for refinable vector function $\phi = (\phi_1, \ldots, \phi_r)^{\mathsf{T}}$ with $r > 1$.

6.8 Stability of Discrete Affine Systems in the Space $l_2(\mathbb{Z})$

In this section we study connections between stability of affine systems in $L_2(\mathbb{R})$ and stability of discrete affine systems in $l_2(\mathbb{Z})$ introduced in Chap. 1.

The following result shows that the Bessel property of a discrete affine system in $l_2(\mathbb{Z})$ implies that its underlying refinable function must belong to $L_2(\mathbb{R})$.

Proposition 6.8.1 *Let $\{a; b_1, \dots, b_s\}$ be a filter bank with $a, b_1, \dots, b_s \in l_2(\mathbb{Z})$ such that*

$$\widehat{\phi}(\xi) := \lim_{J\to\infty} \prod_{j=1}^{J} \widehat{a}(2^{-j}\xi) \quad \text{exists for almost all } \xi \in \mathbb{R}. \tag{6.8.1}$$

Let $\mathcal{W}_J$ be the J-level analysis operators using the filter bank $\{a; b_1, \dots, b_s\}$ (see (1.3.4)). If there exists a positive constant C such that

$$\|\mathcal{W}_J v\|^2_{(l_2(\mathbb{Z}))^{1\times(sJ+1)}} \leqslant C\|v\|^2_{l_2(\mathbb{Z})} \qquad \forall\, v \in l_2(\mathbb{Z}), J \in \mathbb{N}, \tag{6.8.2}$$

then $\|\phi\|^2_{L_2(\mathbb{R})} \leqslant C$ and $[\widehat{\phi}, \widehat{\phi}](\xi) := \sum_{k\in\mathbb{Z}} |\widehat{\phi}(\xi + 2\pi k)|^2 \leqslant C^2$ for a.e. $\xi \in \mathbb{R}$.

Proof By $\mathcal{V}_J = \mathcal{W}_J^\star$, the inequality (6.8.2) is equivalent to

$$\|\mathcal{V}_J \vec{w}\|^2_{l_2(\mathbb{Z})} \leqslant C\|\vec{w}\|^2_{(l_2(\mathbb{Z}))^{1\times(sJ+1)}} \qquad \forall\, \vec{w} \in (l_2(\mathbb{Z}))^{1\times(sJ+1)}, J \in \mathbb{N}.$$

Consider $\vec{w} = (0, \dots, 0, v)$ with $v \in l_2(\mathbb{Z})$. Then $\mathcal{V}_J\vec{w} = 2^{-J/2}\mathcal{S}_a^J v$. Since the Fourier series of $\mathcal{S}_a^J v$ is $2^J\widehat{v}(2^J\xi)\widehat{a_J}(\xi)$ with $\widehat{a_J}(\xi) := \prod_{j=0}^{J-1}\widehat{a}(2^j\xi)$, we deduce that

$$\|\mathcal{V}_J\vec{w}\|^2_{l_2(\mathbb{Z})} = 2^{-J}\|\mathcal{S}_a^J v\|^2_{l_2(\mathbb{Z})} = \frac{1}{2\pi}\int_{-2^J\pi}^{2^J\pi} |\widehat{v}(\xi)|^2 |\widehat{a_J}(2^{-J}\xi)|^2 d\xi.$$

By $\|\mathcal{V}_J\vec{w}\|^2_{l_2(\mathbb{Z})} \leqslant C\|\vec{w}\|^2_{l_2(\mathbb{Z})} = C\|v\|^2_{l_2(\mathbb{Z})}$, the above identity yields

$$\frac{1}{2\pi}\int_{-2^J\pi}^{2^J\pi} |\widehat{v}(\xi)|^2 |\widehat{a_J}(2^{-J}\xi)|^2 d\xi = \|\mathcal{V}_J\vec{w}\|^2_{l_2(\mathbb{Z})} \leqslant C\|v\|^2_{l_2(\mathbb{Z})}. \tag{6.8.3}$$

Taking $v = \boldsymbol{\delta}$ and applying Fatou's lemma, we conclude from (6.8.3) that

$$\frac{1}{2\pi}\int_{\mathbb{R}} |\widehat{\phi}(\xi)|^2 d\xi \leqslant \liminf_{J\to\infty} \frac{1}{2\pi}\int_{-2^J\pi}^{2^J\pi} |\widehat{a_J}(2^{-J}\xi)|^2 d\xi \leqslant C,$$

where we used the assumption $\widehat{\phi}(\xi) = \lim_{J\to\infty}\widehat{a_J}(2^{-J}\xi)$ for a.e. $\xi \in \mathbb{R}$. Hence, $\widehat{\phi} \in L_2(\mathbb{R})$. By Plancherel's Theorem, $\phi \in L_2(\mathbb{R})$ and $\|\phi\|^2_{L_2(\mathbb{R})} \leqslant C$.

On the other hand, by (6.8.3), we have

$$\int_0^{2\pi} |\widehat{v}(\xi)|^2 \sum_{k=-2^{J-1}}^{2^{J-1}-1} |\widehat{a_J}(2^{-J}(\xi+2\pi k))|^2 = \int_{-2^J\pi}^{2^J\pi} |\widehat{v}(\xi)|^2 |\widehat{a_J}(2^{-J}\xi)|^2 \leqslant C \int_0^{2\pi} |\widehat{v}(\xi)|^2 d\xi,$$

for all $v \in l_2(\mathbb{Z})$, from which we conclude that for all $J \in \mathbb{N}$,

$$\sum_{k=-2^{J-1}}^{2^{J-1}-1} |\widehat{a_J}(2^{-J}(\xi + 2\pi k))|^2 \leqslant C \qquad a.e.\ \xi \in \mathbb{R}. \tag{6.8.4}$$

From the above inequality in (6.8.4), by (6.8.1), we also deduce that $|\widehat{\phi}(\xi)|^2 \leqslant C$ for a.e. $\xi \in \mathbb{R}$. Hence, by $\widehat{\phi}(\xi) = \widehat{a_J}(2^{-J}\xi)\widehat{\phi}(2^{-J}\xi)$ and (6.8.4), we obtain

$$\begin{aligned}\sum_{k=-2^{J-1}}^{2^{J-1}-1} |\widehat{\phi}(\xi + 2\pi k)|^2 &= \sum_{k=-2^{J-1}}^{2^{J-1}-1} |\widehat{a_J}(2^{-J}(\xi + 2\pi k))|^2 |\widehat{\phi}(2^{-J}(\xi + 2\pi k))|^2 \\ &\leqslant C \sum_{k=-2^{J-1}}^{2^{J-1}-1} |\widehat{a_J}(2^{-J}(\xi + 2\pi k))|^2 \leqslant C^2.\end{aligned}$$

Taking $J \to \infty$, we conclude that $[\widehat{\phi}, \widehat{\phi}](\xi) \leqslant C^2$ for a.e. $\xi \in \mathbb{R}$. □

We now prove that the Bessel property of a discrete affine system in $l_2(\mathbb{Z})$ implies the Bessel property of its underlying affine system in $L_2(\mathbb{R})$.

Theorem 6.8.2 *Let $a, b_1, \dots, b_s \in l_2(\mathbb{Z})$ such that* (6.8.1) *is satisfied. If there exists a positive constant C such that*

$$\sum_{u \in \mathsf{DAS}_J(\{a; b_1, \dots, b_s\})} |\langle v, u\rangle|^2 \leqslant C\|v\|^2_{l_2(\mathbb{Z})}, \qquad \forall\, v \in l_2(\mathbb{Z}), J \in \mathbb{N}, \tag{6.8.5}$$

where $\mathsf{DAS}_J(\{a, b_1, \dots, b_s\}), J \in \mathbb{N}$ are the discrete affine systems defined in (1.3.19)*, then the affine system $\mathsf{AS}_0(\phi; \psi^1, \dots, \psi^s)$ in* (4.3.1) *must be a Bessel sequence in $L_2(\mathbb{R})$ satisfying $\phi, \psi^1, \dots, \psi^s \in L_2(\mathbb{R})$ and*

$$\sum_{k\in\mathbb{Z}} |\langle f, \phi_{1;k}\rangle|^2 + \sum_{j=0}^{\infty}\sum_{\ell=1}^{s}\sum_{k\in\mathbb{Z}} |\langle f, \psi^{\ell}_{2^j;k}\rangle|^2 \leqslant C_1\|f\|^2_{L_2(\mathbb{R})}, \qquad \forall\, f \in L_2(\mathbb{R}), \tag{6.8.6}$$

with $C_1 := C\|[\widehat{\phi}, \widehat{\phi}]\|_{L_\infty(\mathbb{T})} < \infty$, where $\psi^1, \dots, \psi^s$ are defined to be

$$\widehat{\psi^{\ell}}(\xi) := \widehat{b_\ell}(\xi/2)\widehat{\phi}(\xi/2), \qquad \xi \in \mathbb{R}, \ell = 1, \dots, s. \tag{6.8.7}$$

Proof As we discussed in Sect. 1.3 of Chap. 1,

$$\|\mathcal{W}_J v\|^2_{(l_2(\mathbb{Z}))^{1\times(sJ+1)}} = \sum_{u\in\mathsf{DAS}_J(\{a;b_1,\dots,b_s\})} |\langle v,u\rangle|^2, \qquad \forall\, J\in\mathbb{N}.$$

As we proved in Proposition 6.8.1, the condition (6.8.5) implies that $\phi \in L_2(\mathbb{R})$ and $[\widehat{\phi},\widehat{\phi}] \in L_\infty(\mathbb{T})$. Since $b_1,\dots,b_s \in l_2(\mathbb{Z})$, by $[\widehat{\psi^\ell},\widehat{\psi^\ell}](\xi) = |\widehat{b_\ell}(\xi/2)|^2[\widehat{\phi},\widehat{\phi}](\xi/2) + |\widehat{b_\ell}(\xi/2+\pi)|^2[\widehat{\phi},\widehat{\phi}](\xi/2+\pi)$ for $\ell = 1,\dots,s$, we see that the functions $\psi^1,\dots,\psi^s$ in (6.8.7) belong to $L_2(\mathbb{R})$. Let $f \in L_2(\mathbb{R})$. Define

$$v^j(k) := \langle f,\phi_{2^j;k}\rangle, \qquad w^{\ell,j}(k) := \langle f,\psi^\ell_{2^j;k}\rangle, \qquad j\in\mathbb{Z}, k\in\mathbb{Z}, \ell=1,\dots,s. \tag{6.8.8}$$

As discussed in Sect. 4.1, we have the relations $v^{j-1}(k) = \frac{\sqrt{2}}{2}[\mathcal{T}_a v^j](k)$ and $w^{\ell,j-1}(k) = \frac{\sqrt{2}}{2}[\mathcal{T}_{b_\ell} v^j](k)$, in particular, we have

$$v^j(k) = \langle v^J, a_{J-j;k}\rangle, \quad w^{\ell,j}(k) = \langle v^J, b_{\ell,J-j;k}\rangle, \qquad k\in\mathbb{Z}, \ell=1,\dots,s, \tag{6.8.9}$$

for $j = 0,\dots,J-1$. Therefore, by our assumption in (6.8.5), we have

$$\sum_{k\in\mathbb{Z}} |\langle f,\phi_{1;k}\rangle|^2 + \sum_{j=0}^{J-1}\sum_{\ell=1}^{s}\sum_{k\in\mathbb{Z}} |\langle f,\psi^\ell_{2^j;k}\rangle|^2 = \sum_{u\in\mathsf{DAS}_J(\{a;b_1,\dots,b_s\})} |\langle v^J,u\rangle|^2 \leqslant C\|v^J\|^2_{l_2(\mathbb{Z})}.$$

Now by Proposition 4.4.4, we have

$$\begin{aligned}\|v^J\|^2_{l_2(\mathbb{Z})} &= \sum_{k\in\mathbb{Z}} |v^J(k)|^2 = \sum_{k\in\mathbb{Z}} |\langle f,\phi_{2^J;k}\rangle|^2 = \sum_{k\in\mathbb{Z}} |\langle f_{2^{-J};0},\phi(\cdot-k)\rangle|^2 \\ &\leqslant \|[\widehat{\phi},\widehat{\phi}]\|_{L_\infty(\mathbb{T})}\|f_{2^{-J};0}\|^2_{L_2(\mathbb{R})} = \|[\widehat{\phi},\widehat{\phi}]\|_{L_\infty(\mathbb{T})}\|f\|^2_{L_2(\mathbb{R})}.\end{aligned}$$

That is, we proved that for every $f \in L_2(\mathbb{R})$,

$$\sum_{k\in\mathbb{Z}} |\langle f,\phi_{1;k}\rangle|^2 + \sum_{j=0}^{J-1}\sum_{\ell=1}^{s}\sum_{k\in\mathbb{Z}} |\langle f,\psi^\ell_{2^j;k}\rangle|^2 \leqslant C\|[\widehat{\phi},\widehat{\phi}]\|_{L_\infty(\mathbb{T})}\|f\|^2_{L_2(\mathbb{R})}.$$

Taking $J\to\infty$, we conclude that (6.8.6) holds with $C_1 := C\|[\widehat{\phi},\widehat{\phi}]\|_{L_\infty(\mathbb{T})}$. □

Conversely, under the extra assumption of the Riesz basis property, we have the following result showing that the Bessel property of an affine system in $L_2(\mathbb{R})$ implies the Bessel property of its associated discrete affine system in $l_2(\mathbb{Z})$.

Theorem 6.8.3 *Let $a, b_1,\dots,b_s \in l_2(\mathbb{Z})$ such that* (6.8.1) *holds. Define $\psi^1,\dots,\psi^s$ as in* (6.8.7)*. If there exists a positive constant c such that $[\widehat{\phi},\widehat{\phi}](\xi) \geqslant c$ for a.e. $\xi\in\mathbb{R}$ and if the affine system $\mathsf{AS}_0(\phi;\psi^1,\dots,\psi^s)$ in* (4.3.1) *is a Bessel sequence* *in* $L_2(\mathbb{R})$ *satisfying* (6.8.6) *for some positive constant* C_1 *(This condition is satisfied if* $\phi \in L_2(\mathbb{R})$ *and* $a, b_1, \dots, b_s \in l_0(\mathbb{Z})$ *with* $\widehat{b_1}(0) = \cdots = \widehat{b_s}(0) = 0$*), then* (6.8.5) *holds with* $C = C_1 \| \frac{1}{[\widehat{\phi}, \widehat{\phi}]} \|_{L_\infty(\mathbb{R})} < \infty$.

Proof By Theorem 4.4.6 and $[\widehat{\phi}, \widehat{\phi}] \geqslant c > 0$ a.e., the integer shifts of ϕ are stable in $L_2(\mathbb{R})$ and $\mathcal{W}_\phi : \mathsf{S}(\phi|L_2(\mathbb{R})) \to l_2(\mathbb{Z}), f \mapsto \{\langle f, \phi(\cdot - k)\rangle\}_{k\in\mathbb{Z}}$ is a bounded and invertible linear operator. Hence, for every $v \in l_2(\mathbb{Z})$ and $J \in \mathbb{N}$, there exists $g \in \mathsf{S}(\phi|L_2(\mathbb{R}))$ such that $\mathcal{W}_\phi g = v$. Define $f := g_{2^J;0}$. Then we have $\langle f, \phi_{2^J;k}\rangle = \langle g_{2^J;0}, \phi_{2^J;k}\rangle = \langle g, \phi(\cdot - k)\rangle = v(k)$ for $k \in \mathbb{Z}$. Define v^j and $w^{\ell,j}$ as in (6.8.8). Then $v = v^J$ and (6.8.9) holds. Consequently, we have

$$\sum_{u\in \mathsf{DAS}_J(\{a;b_1,\dots,b_s\})} |\langle v, u\rangle|^2 = \sum_{k\in\mathbb{Z}} |\langle f, \phi_{1;k}\rangle|^2 + \sum_{j=0}^{J-1}\sum_{\ell=1}^{s}\sum_{k\in\mathbb{Z}} |\langle f, \psi^\ell_{2^j;k}\rangle|^2 \leqslant C_1 \|f\|^2_{L_2(\mathbb{R})}.$$

Since $\mathcal{W}_\phi$ is bounded and invertible, by item (i) of Proposition 4.4.4, we see that

$$\begin{aligned}\|v\|^2_{l_2(\mathbb{Z})} &= \sum_{k\in\mathbb{Z}} |\langle f, \phi_{2^J;k}\rangle|^2 = \sum_{k\in\mathbb{Z}} |\langle f_{2^{-J};0}, \phi(\cdot - k)\rangle|^2 \\ &\geqslant \| \tfrac{1}{[\widehat{\phi}, \widehat{\phi}]} \|^{-1}_{L_\infty(\mathbb{T})} \|f_{2^{-J};0}\|^2_{L_2(\mathbb{R})} = \| \tfrac{1}{[\widehat{\phi}, \widehat{\phi}]} \|^{-1}_{L_\infty(\mathbb{T})} \|f\|^2_{L_2(\mathbb{R})}.\end{aligned}$$

Therefore, we conclude that (6.8.5) holds with $C = C_1 \| \frac{1}{[\widehat{\phi}, \widehat{\phi}]} \|_{L_\infty(\mathbb{T})} < \infty$.

Suppose that $\phi \in L_2(\mathbb{R})$ and $a, b_1, \dots, b_s \in l_0(\mathbb{Z})$ with $\widehat{b_1}(0) = \cdots = \widehat{b_s}(0) = 0$. Then $\phi \in L_2(\mathbb{R})$ has compact support. By Theorem 6.3.3, we have $\mathrm{sm}(\phi) > 0$ and consequently by Lemma 6.3.2, $[\widehat{\phi}, \widehat{\phi}]_t \in L_\infty(\mathbb{T})$ for all $0 \leqslant t < \mathrm{sm}(\phi)$. Now it follows from Theorem 4.6.5 that $\mathsf{AS}_0(\phi; \psi^1, \dots, \psi^s)$ is a Bessel sequence in $L_2(\mathbb{R})$ satisfying (6.8.6) for some positive constant C_1. □

6.9 Exercises

6.1. Use Exercises 2.30 and 5.43 to provide another proof to Lemma 6.1.2.

6.2. Let f and g be Lebesgue measurable functions on $\mathbb{R}$. Let $E_1, E_2, E_3 \subseteq \mathbb{R}$ be measurable sets having measure zero. Suppose that $f(x) = g(x)$ for all $x \in \mathbb{R}\backslash E_1$, $f'(x)$ exists for all $x \in \mathbb{R}\backslash E_2$ and $g'(x)$ exists for all $x \in \mathbb{R}\backslash E_3$. Prove that $f'(x) = g'(x)$ for all $x \in \mathbb{R}\backslash(E_1 \cup E_2 \cup E_3)$. *Hint:* For every $x \in \mathbb{R}\backslash(E_1 \cup E_2 \cup E_3)$, there exists a sequence $\{x_n\}_{n=1}^\infty$ in $\mathbb{R}\backslash(E_1 \cup E_2 \cup E_3 \cup \{x\})$ such that $\lim_{n\to\infty} x_n = x$.

6.3. Let ϕ be a compactly supported measurable function and $E \subseteq \mathbb{R}$ be a measurable set having measure zero. Suppose that $\phi'(x)$ exists for all $x \in \mathbb{R}\backslash E$.

a. Let $g := \sum_{k\in\mathbb{Z}} u(k)\phi(\cdot - k)$ with $u \in l(\mathbb{Z})$. Prove that $g'(x)$ exists for all $x \in \mathbb{R}\backslash(\cup_{k\in\mathbb{Z}}(E + k))$.

b. If $\phi(x) = 2\sum_{k\in\mathbb{Z}} a(k)\phi(2x-k)$ for a.e. $x \in \mathbb{R}$ with $a \in l(\mathbb{Z})$, prove that $\phi'(x) = 4\sum_{k\in\mathbb{Z}} a(k)\phi'(2x-k)$ for a.e. $x \in \mathbb{R}$.

6.4. Let B_m be the B-spline function of order m with $m \in \mathbb{N}$. For convenience, we also define $B_0 := \boldsymbol{\delta}$ (the Dirac distribution). Let f be a distribution and $m \in \mathbb{N}_0$. Prove

a. $B_m * f \in \mathscr{D}'(\mathbb{R})$, where $\langle B_m * f; \psi\rangle := \langle f; \psi * B_m\rangle$ for all $\psi \in \mathscr{D}(\mathbb{R})$.
b. $[B_m * f]^{(j)} = B_{m-j} * \nabla^j f$ for $j = 1, \dots, m$, and $[B_m * f]^{(m)} = B_0 * \nabla^m f = \nabla^m f$.

6.5. Let $E_n(\xi)$ be defined in (6.3.4). If $m > \tau \geqslant 0$, prove that $\lim_{n\to\infty} E_n(\xi) = 0$.

6.6. Let f be a distribution such that $\operatorname{supp}(f) = \{0\}$. Prove that f is a finite linear combination of $\boldsymbol{\delta}, D\boldsymbol{\delta}, \dots, D^n\boldsymbol{\delta}$ for some $n \in \mathbb{N}_0$. *Hint:* use Theorem A.6.1.

6.7. Let $f : \mathbb{R} \to \mathbb{C}$ be a compactly supported function in $L_\infty(\mathbb{R})$ and $f|_{(x_k,x_{k+1})} \in \mathscr{C}^1$ for $k = 0, \dots, n-1$ with $-\infty = x_0 < x_1 < \dots < x_{n-1} < x_n = \infty$. Suppose that $\lim_{x\to x_k^+} f(x)$ and $\lim_{x\to x_k^-} f(x)$ exist for all $k = 1, \dots, n-1$ but $\lim_{x\to x_k^+} f(x) \neq \lim_{x\to x_k^-} f(x)$ for some $0 < k < n$. For $1 \leqslant p \leqslant \infty$, show that there exist $C > 0$ and $h > 0$ such that $C^{-1}\lambda^{1/p} \leqslant \omega_1(f,\lambda)_p \leqslant C\lambda^{1/p}$ for all $0 < \lambda < h$. Then use this fact to prove $\operatorname{sm}_p(f) = 1/p$ for all $1 \leqslant p \leqslant \infty$.

6.8. Let ϕ be a vector of compactly supported distributions such that ϕ satisfies $\phi = 2\sum_{k\in\mathbb{Z}} a(k)\phi(2\cdot -k)$ for some $a \in (l_0(\mathbb{Z}))^{r\times r}$. If the integer shifts of ϕ are linearly independent, prove that 0 is not an eigenvalue of $\mathcal{S}_a$, that is, if $\mathcal{S}_a v = 0$ for some $v \in (l(\mathbb{Z}))^{1\times r}$, then we must have $v(k) = 0$ for all $k \in \mathbb{Z}$.

6.9. Let ϕ be a vector of compactly supported distributions. For $j \in \mathbb{N}_0$, prove that $B_{j+1} \in \mathsf{S}(\phi)$ if and only if $x_+^j \in \mathsf{S}(\phi)$.

6.10. Let $a \in l_0(\mathbb{Z})$ and $u \in l(\mathbb{Z})$. If $\mathcal{S}_a u$ agrees with a polynomial sequence on $\mathbb{Z} \cap (M, \infty)$ for some $M \in \mathbb{N}$, prove that there exist an integer $N \in \mathbb{N}$ and a polynomial $\mathsf{p} \in \mathbb{P}$ such that $u(k) = \mathsf{p}(k)$ for all $k \in \mathbb{Z} \cap (N, \infty)$ and $\mathcal{S}_a\mathsf{p}$ must be a polynomial sequence. *Hint:* use the idea in the proof of item (1) in Theorem 1.2.4.

6.11. Let $a \in (l_0(\mathbb{Z}))^{r\times r}$ and $u \in (l(\mathbb{Z}))^{1\times r}$. If $\mathcal{S}_a u$ agrees with a vector of polynomial sequences on $\mathbb{Z} \cap (M, \infty)$ for some $M \in \mathbb{N}$, prove that there exist $N \in \mathbb{N}$ and $\mathsf{p} \in \mathbb{P}^{1\times r}$ such that $u(k) = \mathsf{p}(k)$ for all $k \in \mathbb{Z} \cap (N, \infty)$ and $\mathcal{S}_a\mathsf{p} \in \mathbb{P}^{1\times r}$.

6.12. Let ϕ be a vector of compactly supported distributions such that $\widehat{\phi}(2\xi) = \widehat{a}(\xi)\widehat{\phi}(\xi)$ for some $a \in (l_0(\mathbb{Z}))^{r\times r}$ and the integer shifts of ϕ are linearly independent. For $j \in \mathbb{N}_0$, prove that $x_+^j \in \mathsf{S}(\phi)$ if and only if $\mathcal{S}_a u = 2^{-j}u$, where $u \in (l(\mathbb{Z}))^{1\times r}$ is the unique sequence such that $x_+^j = \sum_{k\in\mathbb{Z}} u(k)\phi(x-k)$. Moreover, 2^{-j} must be an eigenvalue of $\mathcal{S}_a$ with geometric multiplicity at least 2.

6.13. Under the same assumption as in Theorem 6.1.9. Let $j \in \mathbb{N}_0$.

a. If 2^{-j} is an eigenvalue of $\mathcal{T}_a$, prove that a has order $j+1$ sum rules.
b. If 2^{-j} with $j \in \mathbb{N}_0$ is a double eigenvalue of $\mathcal{T}_a$, prove that $B_{j+1} \in \mathsf{S}(\phi)$. *Hint:* First prove $x_+^j \in \mathsf{S}(\phi)$ and then apply item (6) of Proposition 6.1.1.

c. If $B_{j+1} \in \mathrm{S}(\phi)$ with $j \in \mathbb{N}_0$, prove that 2^{-j} must be a double eigenvalue of $\mathcal{T}_a$.

6.14. Let $\varphi = (\varphi_1, \ldots, \varphi_r)^\mathsf{T}$ be the Hermite refinable function defined in Proposition 6.2.2 with $a \in (l_0(\mathbb{Z}))^{r\times r}$ being its filter. Prove

a. $B_{j+1} \in \mathrm{S}(\varphi)$ for all $j = r, \ldots, 2r-1$ and $\mathrm{S}(\varphi) = \mathrm{S}(B_{r+1}, \ldots, B_{2r})$. *Hint:* use the definition and the Hermite interpolation property of φ.
b. $\mathrm{spec}(\mathcal{T}_a)\backslash\{0\} = \{1, 2^{-1}, \ldots, 2^{1-r}\} \cup \cup_{j=r}^{2r-1}\{2^{-j}, 2^{-j}\}$ and $\mathcal{T}_a|_{(l([-1,1]))^{1\times r}}$ is diagonalizable and invertible.

6.15. Let $\upsilon \in (l_0(\mathbb{Z}))^{1\times r}$ satisfy (6.2.8) with $m \geqslant r$. Prove that

a. $\mathscr{P}_{m-1,\upsilon} = \{(\mathsf{p}, \mathsf{p}', \ldots, \mathsf{p}^{(r-1)}) \,:\, \mathsf{p} \in \mathbb{P}_{m-1}\}$.
b. $\mathscr{B}_{m-1,\upsilon} = \{\nabla^m \boldsymbol{\delta} \boldsymbol{e}_1^\mathsf{T}\} \cup \{(\nabla^{\ell-1} b_\ell)\boldsymbol{e}_1^\mathsf{T} + \boldsymbol{\delta}\boldsymbol{e}_\ell^\mathsf{T} \,:\, \ell = 2, \ldots, r\}$ generates $\mathscr{V}_{m-1,\upsilon}$, where b_ℓ satisfies $\widehat{b_\ell}(\xi) = (-1)^\ell(\frac{i\xi}{1-e^{-i\xi}})^{\ell-1} + \mathscr{O}(|\xi|^{m-\ell+1})$ as $\xi \to 0$.

6.16. Let $r, m \in \mathbb{N}$ and $\Lambda \subseteq (1 + 2\mathbb{Z})$ such that $\#\Lambda = 2m$. Prove that there exists a unique filter $a \in (l_0(\mathbb{Z}))^{r\times r}$ such that a is an order r Hermite interpolatory filter and a vanishes outside $\Lambda \cup \{0\}$, and a has order $2rm$ sum rules with a matching filter $\upsilon \in (l_0(\mathbb{Z}))^{1\times r}$ satisfying (6.2.9).

6.17. For $\phi = (\phi_1, \ldots, \phi_r)^\mathsf{T} : \mathbb{R} \to \mathbb{C}^{r\times 1}$, we say that ϕ is *interpolating* if ϕ is continuous and $\phi_\ell(\frac{\gamma}{r} + k) = \boldsymbol{\delta}(k)\boldsymbol{\delta}(\ell - 1 - \gamma)$ for all $k \in \mathbb{Z}$, $\gamma = 0, \ldots, r-1$ and $\ell = 1, \ldots, r$.

a. Prove that for every $r \in \mathbb{N}$, there exists a unique interpolating function ϕ such that ϕ vanishes outside $[-1, 1]$ and both $\phi|_{[-1,0]}$ and $\phi|_{[0,1]}$ are vectors of polynomials with degree no more than r.
b. Prove that such interpolating function ϕ is refinable with a filter $a \in (l_0(\mathbb{Z}))^{r\times r}$, $\mathrm{sr}(a) = r + 1$, and $\phi \notin \mathscr{C}^1(\mathbb{R})$.

6.18. Let $\phi = (\phi_1, \ldots, \phi_r)^\mathsf{T}$ be a vector of compactly supported distributions satisfying $\widehat{\phi}(2\xi) = \widehat{a}(\xi)\widehat{\phi}(\xi)$ with $a \in (l_0(\mathbb{Z}))^{r\times r}$. Prove that ϕ is interpolating if and only if $[1, \ldots, 1]\widehat{\phi}(0) = 1$, $\mathrm{sm}_\infty(a) > 0$, and the filter a is an interpolatory filter of type $(2, r)$, that is, $[1, \ldots, 1]\widehat{a}(0) = [1, \ldots, 1]$ and $a_{\ell,:}(2(\ell - 1) + 2k) = \boldsymbol{\delta}(k)$ for all $k \in \mathbb{Z}$ and $\ell = 1, \ldots, r$, where $a_{\ell,:}$ is a scalar sequence by taking the ℓth row of the matrix filter a, i.e., $a_{\ell,:}(rk + m - 1) := [a(k)]_{\ell,m}$ for $k \in \mathbb{Z}$ and $1 \leqslant \ell, m \leqslant r$.

6.19. Let $\phi = (\phi_1, \ldots, \phi_r)^\mathsf{T}$ be an interpolating refinable vector function satisfying $\widehat{\phi}(2\xi) = \widehat{a}(\xi)\widehat{\phi}(\xi)$ with $a \in (l_0(\mathbb{Z}))^{r\times r}$. If a satisfies order m sum rules with a matching filter $\upsilon \in (l_0(\mathbb{Z}))^{1\times r}$, prove that up to a multiplicative constant, $\widehat{\upsilon}(\xi) = (1, e^{i\xi/r}, \ldots, e^{i\xi(r-1)/r}) + \mathscr{O}(|\xi|^m)$ as $\xi \to 0$.

6.20. For every $r \in \mathbb{N}$ and $m \in \mathbb{N}$, prove that there exists a unique interpolating filter of type $(2, r)$ such that $\mathrm{fsupp}(a) \subseteq [1 - m, m]$ and a has order rm sum rules.

6.21. Let $a, b \in (l_0(\mathbb{Z}))^{2\times 2}$ with $\text{fsupp}(a) = \text{fsupp}(b) = [-1, 2]$ be given by

$$a = \left\{ \begin{bmatrix} \frac{1}{2} & 0 \\ -\frac{\sqrt{2}}{8} & \frac{1}{4} \end{bmatrix}, \begin{bmatrix} \frac{1}{2} & 0 \\ \frac{\sqrt{2}}{8} & \frac{1}{2} \end{bmatrix}, \begin{bmatrix} 0 & 0 \\ \frac{\sqrt{2}}{8} & \frac{1}{4} \end{bmatrix}, \begin{bmatrix} 0 & 0 \\ -\frac{\sqrt{2}}{8} & 0 \end{bmatrix} \right\}_{[-1,2]},$$

$$b = \left\{ \begin{bmatrix} -\frac{\sqrt{2}}{8} & \frac{1}{4} \\ -\frac{1}{4} & \frac{\sqrt{2}}{4} \end{bmatrix}, \begin{bmatrix} \frac{\sqrt{2}}{8} & -\frac{1}{2} \\ \frac{1}{4} & 0 \end{bmatrix}, \begin{bmatrix} \frac{\sqrt{2}}{8} & \frac{1}{4} \\ -\frac{1}{4} & -\frac{\sqrt{2}}{4} \end{bmatrix}, \begin{bmatrix} -\frac{\sqrt{2}}{8} & 0 \\ \frac{1}{4} & 0 \end{bmatrix} \right\}_{[-1,2]}.$$

a. Prove that $\text{sr}(a) = 2$ and $\{a; b\}$ is an orthogonal wavelet filter bank.
b. Prove that all the compactly supported refinable functions /distributions associated with the filter a are given by

$$\text{DS}(a) = \text{span}((\chi_{(-1,0]}, -\tfrac{\sqrt{2}}{2}\chi_{(-1,1]})^{\mathsf{T}}, (0, B_2(\cdot - 1))^{\mathsf{T}}).$$

6.22. Let the filters $a, b, \tilde{a}, \tilde{b} \in (l_0(\mathbb{Z}))^{2\times 2}$ be given by

$$a = \left\{ \begin{bmatrix} 0 & \frac{1}{4} \\ 0 & 0 \end{bmatrix}, \begin{bmatrix} \frac{1}{2} & \frac{1}{4} \\ 0 & \frac{1}{4} \end{bmatrix}, \begin{bmatrix} 0 & 0 \\ \frac{1}{2} & \frac{1}{4} \end{bmatrix} \right\}_{[-1,1]},$$

$$b = \left\{ \begin{bmatrix} \frac{1}{4} & -\frac{1}{2} \\ -\frac{1}{3} & \frac{1}{2} \end{bmatrix}, \begin{bmatrix} 0 & \frac{1}{2} \\ -\frac{1}{3} & \frac{1}{2} \end{bmatrix}, \begin{bmatrix} -\frac{1}{4} & 0 \\ -\frac{1}{3} & 0 \end{bmatrix} \right\}_{[-1,1]},$$

and

$$\tilde{a} = \left\{ \begin{bmatrix} \frac{1}{48} & -\frac{1}{24} \\ 0 & 0 \end{bmatrix}, \begin{bmatrix} -\frac{1}{12} & \frac{5}{24} \\ 0 & 0 \end{bmatrix}, \begin{bmatrix} \frac{19}{24} & \frac{5}{24} \\ -\frac{1}{6} & \frac{1}{3} \end{bmatrix}, \begin{bmatrix} -\frac{1}{12} & -\frac{1}{24} \\ \frac{2}{3} & \frac{1}{3} \end{bmatrix}, \begin{bmatrix} \frac{1}{48} & 0 \\ -\frac{1}{6} & 0 \end{bmatrix} \right\}_{[-2,2]},$$

$$\tilde{b} = \left\{ \begin{bmatrix} -\frac{1}{24} & \frac{1}{12} \\ \frac{1}{32} & -\frac{1}{16} \end{bmatrix}, \begin{bmatrix} \frac{1}{6} & -\frac{5}{12} \\ -\frac{1}{8} & \frac{5}{16} \end{bmatrix}, \begin{bmatrix} 0 & \frac{5}{12} \\ -\frac{5}{16} & \frac{5}{16} \end{bmatrix}, \begin{bmatrix} -\frac{1}{6} & -\frac{1}{12} \\ -\frac{1}{8} & -\frac{1}{16} \end{bmatrix}, \begin{bmatrix} \frac{1}{24} & 0 \\ \frac{1}{32} & 0 \end{bmatrix} \right\}_{[-2,2]}.$$

a. Prove that $(\{\tilde{a}; \tilde{b}\}, \{a; b\})$ is a biorthogonal wavelet filter bank.
b. Show that both a and $\tilde{a}$ have order 2 sum rules and $\text{sm}(\tilde{a}) \approx 0.36847$.
c. Prove that the refinable vector function $\phi(x) = (B_2(2x - 1), B_2(2x))^{\mathsf{T}}$ satisfies the refinement equation $\widehat{\phi}(2\xi) = \widehat{a}(\xi)\widehat{\phi}(\xi)$.

6.23. Let $\|\cdot\|_{L_{2,p,\nu}(\mathbb{R})}$ be defined as in (6.6.1) for $1 \leqslant p \leqslant \infty$ and $\nu \geqslant 0$. Prove that $\|\cdot\|_{L_{2,p,\nu}(\mathbb{R})}$ is a norm and $L_{2,p,\nu}(\mathbb{R})$ is a Banach space.

6.24. Let a be a filter of Hölder class with $\widehat{a}(0) = 1$. Define $\widehat{\phi}(\xi) := \prod_{j=1}^{\infty} \widehat{a}(2^{-j}\xi)$ for $\xi \in \mathbb{R}$. Then $\widehat{\phi}$ is a continuous function with $\widehat{\phi}(0) = 1$. Therefore, there exists $0 < \varepsilon < \pi$ such that $1/2 \leqslant |\widehat{\phi}(\xi)| \leqslant 3/2$ for all $|\xi| \leqslant \varepsilon$. Define a function η by

$$\widehat{\eta}(\xi) := \begin{cases} \widehat{\phi}(\xi), & \text{if } |\xi| \leqslant \varepsilon, \\ \widehat{\phi}(-\varepsilon)(3\pi + 2\xi)/(3\pi - 2\varepsilon), & \text{if } -3\pi/2 \leqslant \xi < -\varepsilon, \\ \widehat{\phi}(\varepsilon)(3\pi - 2\xi)/(3\pi - 2\varepsilon), & \text{if } \varepsilon < \xi \leqslant 3\pi/2, \\ 0, & \text{otherwise.} \end{cases}$$

Prove that $\widehat{\eta}$ is continuous with $\widehat{\eta}(0) = 1$, $\eta \in L_{2,\infty,0}(\mathbb{R})$, $\widehat{\eta}(\xi) - \widehat{a}(\xi/2)\widehat{\eta}(\xi/2) = 0$ for all $|\xi| \leqslant \varepsilon$ (thus η is an admissible function in $L_{2,\infty,0}(\mathbb{R})$ with respect to a).

6.25. Let $a \in l_1(\mathbb{Z})$ be a filter such that $|\widehat{a}(\cdot + \pi)|^2/|\sin(\cdot/2)|^\tau \in L_\infty(\mathbb{T})$ with $\tau > 0$. Prove that $\rho_{\tau_2}(\mathcal{T}_{|\widehat{a}|^2}) \leqslant \rho_{\tau_1}(\mathcal{T}_{|\widehat{a}|^2})$ for all $0 \leqslant \tau_1 \leqslant \tau_2 < \tau$.

6.26. For a finitely supported filter a, prove that the definition of $\operatorname{sm}(a)$ in (6.6.5) agrees with the definition $\operatorname{sm}(a)$ in (5.6.44).

6.27. Prove Corollary 6.6.5.

6.28. Prove the inequalities in (6.6.28) and (6.6.29).

6.29. Let $f \in L_{2,1,\nu}(\mathbb{R})$ with $\nu > 0$ such that $\widehat{f}(2\pi k) = 0$ for all $k \in \mathbb{Z}$. Define $h := \sum_{k=0}^{\infty} f(\cdot - k)$. Prove that $f = h - h(\cdot - 1)$ and $h \in L_{2,1,\tau}(\mathbb{R})$ for all $0 \leqslant \tau < \nu$.

6.30. Let $f \in L_{2,1,\nu}(\mathbb{R})$ for some $\nu > 0$ such that $\widehat{f}(2\pi k) = 0$ for all $k \in \mathbb{Z}\backslash\{0\}$. Prove that f is admissible (see (6.6.12)) with respect to any filter a of Hölder class satisfying $\widehat{a}(0) = 1$ and $\widehat{a}(\pi) = 0$. *Hint:* use Exercise 6.29.

6.31. Let a be a filter of Hölder class with $\widehat{a}(0)=1$. If $\liminf_{n\to\infty} \|\mathcal{T}^n_{|\widehat{a}|^2}(|\sin(\cdot/2)|^\tau) \|_{L_1(\mathbb{R})} = 0$ for some $\tau \geqslant 0$, prove that $\widehat{a}(\pi) = 0$. In particular, if $\rho_\tau(\mathcal{T}_{|\widehat{a}|^2}) < 1$ for some $\tau \geqslant 0$, then $\widehat{a}(\pi) = 0$. *Hint:* use proof by contradiction and the fact that $|1 - \prod_{j=1}^{\infty} \widehat{a}(2^{-j}\xi)| \leqslant 1/2$ in a neighborhood of the origin.

6.32. Let a be a filter such that $|\widehat{a}(\xi)| = |1 + e^{-i\xi}|^\tau |\widehat{A}(\xi)|$ for some $\tau \geqslant 0$ and some 2π-periodic trigonometric polynomial $\widehat{A}$ with $\widehat{A}(0) \neq 0$. Prove that

$$\rho_{2\tau}(\mathcal{T}_{|\widehat{a}|^2}, p) = \lim_{n\to\infty} \|\mathcal{T}^n_{|\widehat{a}|^2}(|\sin(\cdot/2)|^{2\tau})\|^{1/n}_{L_p(\mathbb{R})} = \rho_0(\mathcal{T}_{|\widehat{A}|^2}) = \inf_{n\in\mathbb{N}} \|\mathcal{T}^n_{|\widehat{A}|^2} 1\|^{1/n}_{L_\infty(\mathbb{R})}.$$

for all $1 \leqslant p \leqslant \infty$, where $\rho_{2\tau}(\mathcal{T}_{|\widehat{a}|^2}, p) := \limsup_{n\to\infty} \|\mathcal{T}^n_{|\widehat{a}|^2}(|\sin(\cdot/2)|^{2\tau})\|^{1/n}_{L_p(\mathbb{R})}$.

6.33. Let $\widehat{A}, \widehat{A_j} \in L_\infty(\mathbb{T}), j \in \mathbb{N}$ such that $\lim_{j\to\infty} \|\widehat{A_j} - \widehat{A}\|_{L_\infty(\mathbb{T})} = 0$. Prove $\limsup_{j\to\infty} \rho_0(\widehat{A_j}) \leqslant \rho_0(\widehat{A})$. If in addition $|\widehat{A}(\xi)| \leqslant |\widehat{A_j}(\xi)|$ for a.e. $\xi \in \mathbb{R}$ and for all $j \in \mathbb{N}$, then $\lim_{j\to\infty} \rho_0(\mathcal{T}_{|\widehat{A_j}|^2}) = \rho_0(\mathcal{T}_{|\widehat{A}|^2})$.

6.34. Let $b, c \in l_0(\mathbb{Z})$ be finitely supported sequences such that $\widehat{c}(\xi) \neq 0$ for all $\xi \in \mathbb{R}$. Define an infinitely supported filter a by $\widehat{a}(\xi) := \widehat{b}(\xi)/\widehat{c}(\xi)$. Prove that the filter a has exponential decay. Moreover, if $\sum_{k\in\mathbb{Z}} c(k)z^k \neq 0$ for all $z \in \{\xi + i\zeta \,:\, \xi \in \mathbb{R}, -\nu < \zeta < \nu\}$ for some $\nu > 0$, then a has exponential decay of order ν.

6.35. Let a, c be filters of Hölder class such that $\widehat{c}(\xi) \neq 0$ for all $\xi \in \mathbb{R}$. Define a filter $\mathring{a}$ by $\widehat{\mathring{a}}(\xi) := \widehat{a}(\xi)\widehat{c}(\xi)/\widehat{c}(2\xi)$. Prove that $\mathrm{sm}(\mathring{a}) = \mathrm{sm}(a)$.

6.36. Let ϕ be a vector of compactly supported distributions such that the integer shifts of ϕ are stable, that is, $\mathrm{span}\{\widehat{\phi}(\xi + 2\pi k) : k \in \mathbb{Z}\} = \mathbb{C}^r$ for all $\xi \in \mathbb{R}$. Prove that there exist a positive constant C and a positive integer N such that $\sum_{k=-N}^{N} \widehat{\phi}(\xi + 2\pi k)\overline{\widehat{\phi}(\xi + 2\pi k)}^{\mathsf{T}} \geqslant CI_r$ for all $\xi \in \mathbb{R}$ (consequently, $[\widehat{\phi}, \widehat{\phi}](\xi) \geqslant CI_r$ for all $\xi \in \mathbb{R}$). *Hint:* For every $\xi \in [-\pi, \pi]$, there exists $N_\xi \in \mathbb{N}$ such that $\mathrm{span}\{\widehat{\phi}(\xi + 2\pi k) : k = -N_\xi, \ldots, N_\xi\} = \mathbb{C}^r$, that is, $\sum_{k=-N_\xi}^{N_\xi} \widehat{\phi}(\xi + 2\pi k)\overline{\widehat{\phi}(\xi + 2\pi k)}^{\mathsf{T}} > 0$. Then use the facts that $\widehat{\phi}$ is continuous and $[-\pi, \pi]$ is a compact set to prove the claim.

6.37. Let ϕ be a compactly supported distribution on $\mathbb{R}$ such that $\mathrm{supp}(\phi)$ is a finite set and $\phi = 2\sum_{k\in\mathbb{Z}} a(k)\phi(2\cdot -k)$ for some $a \in l_0(\mathbb{Z})$. Prove that $\phi = u * (D^m\boldsymbol{\delta})$ for some $m \in \mathbb{N}_0$ and $u \in l_0(\mathbb{Z})$. *Hint:* use a similar idea as in Lemma 6.1.6.

6.38. For $n \in \mathbb{N}_0$, we define the Legendre polynomial $P_n(x) := \frac{1}{2^n n!}[(x^2-1)^n]^{(n)} = 2^n \sum_{k=0}^{n} x^k \binom{n}{k}\binom{\frac{n+k-1}{2}}{n}$. Prove $\int_{-1}^{1} P_m(x)P_n(x)dx = \frac{2}{2n+1}\boldsymbol{\delta}(m-n)$ for all $m, n \in \mathbb{N}_0$.

6.39. For $r \in \mathbb{N}$, define $\varphi := (\varphi_1, \ldots, \varphi_r)^{\mathsf{T}}$ with $\varphi_\ell(x) = \sqrt{2\ell - 1}P_{\ell-1}(2x-1)\chi_{[0,1]}$ for $\ell = 1, \ldots, r$, where $P_n, n \in \mathbb{N}_0$ are Legendre polynomials. Prove that $\langle \varphi, \varphi(\cdot - k)\rangle = \boldsymbol{\delta}(k)I_r$ for all $k \in \mathbb{Z}$ and φ satisfies the refinement equation $\varphi = 2a(0)\varphi(2\cdot) + 2a(1)\varphi(2\cdot -1)$ with $a(k) := 2^{-1}\langle \varphi(\cdot/2), \varphi(\cdot - k)\rangle$ for $k \in \{0, 1\}$.

6.40. Let $\tilde{a}, a \in (l_0(\mathbb{Z}))^{r\times r}$ such that (6.4.14) and (5.1.10) hold. Also assume that (5.1.10) holds with a being replaced by $\tilde{a}$. Let ϕ and $\tilde{\phi}$ be vectors of compactly supported distributions satisfying $\widehat{\phi}(2\xi) = \widehat{a}(\xi)\widehat{\phi}(\xi)$ and $\widehat{\tilde{\phi}}(2\xi) = \widehat{\tilde{a}}(\xi)\widehat{\tilde{\phi}}(\xi)$. If both ϕ and $\tilde{\phi}$ are not identically zero, prove that $\widehat{\tilde{\phi}}(0)^{\mathsf{T}}\widehat{\phi}(0) \neq 0$.

6.41. Let ϕ be a function in $L_1(\mathbb{R})$ or a tempered distribution. Let $c \in \mathbb{R}$ and $\epsilon \in \{-1, 1\}$. (i) (Symmetry) Prove $\phi = \epsilon\phi(c - \cdot)$ if and only if $\widehat{\phi}(\xi) = \epsilon e^{-ic\xi}\widehat{\phi}(-\xi)$. (ii) (Complex symmetry) Prove $\phi = \epsilon\overline{\phi(c - \cdot)}$ if and only if $\widehat{\phi}(\xi) = \epsilon e^{-ic\xi}\overline{\widehat{\phi}(\xi)}$.

6.42. Let $\phi = (\phi_1, \ldots, \phi_r)^{\mathsf{T}}$ be a vector of compactly supported distributions satisfying $\widehat{\phi}(2\xi) = \widehat{a}(\xi)\widehat{\phi}(\xi)$ with $a \in (l_0(\mathbb{Z}))^{r\times r}$. For $b \in (l_0(\mathbb{Z}))^{r\times r}$, define $\widehat{\psi}(2\xi) := \widehat{b}(\xi)\widehat{\phi}(\xi)$. For $c_1, \ldots, c_r, \mathring{c}_1, \ldots, \mathring{c}_r \in \mathbb{R}$ and $\epsilon_1, \ldots, \epsilon_r, \mathring{\epsilon}_1, \ldots, \mathring{\epsilon}_r \in \{-1, 1\}$, define $\mathsf{S}(\xi) := \mathrm{diag}(\epsilon_1 e^{-ic_1\xi}, \ldots, \epsilon_r e^{-ic_r\xi})$ and $\mathsf{T}(\xi) := \mathrm{diag}(\mathring{\epsilon}_1 e^{-i\mathring{c}_1\xi}, \ldots, \mathring{\epsilon}_r e^{-i\mathring{c}_r\xi})$.

a. If $\widehat{\phi}(\xi) = \mathsf{S}(\xi)\widehat{\phi}(-\xi)$ (i.e., $\phi_\ell = \epsilon_\ell\phi_\ell(c_\ell - \cdot)$ for all $\ell = 1, \ldots, r$), then $\widehat{\phi}(2\xi) = \mathsf{S}(2\xi)\widehat{a}(-\xi)\mathsf{S}^{-1}(\xi)\widehat{\phi}(\xi)$ for all $\xi \in \mathbb{R}$. If in addition $c_1, \ldots, c_r \in \mathbb{Z}$ and

$$\mathrm{span}\{\widehat{\phi}(\xi_0 + 2\pi k) : k \in \mathbb{Z}\} = \mathbb{C}^r \qquad \text{for some } \xi_0 \in \mathbb{R}, \tag{6.9.1}$$

[equivalently, $\text{len}(\mathsf{S}(\phi)) = r$.] then $\widehat{a}(\xi) = \mathsf{S}(2\xi)\widehat{a}(-\xi)\mathsf{S}^{-1}(\xi)$ for all $\xi \in \mathbb{R}$.

b. If $\widehat{a}(\xi) = \mathsf{S}(2\xi)\widehat{a}(-\xi)\mathsf{S}^{-1}(\xi)\ \forall \xi \in \mathbb{R}$, then $\mathsf{S}(2\xi)\widehat{\phi}(-2\xi) = \widehat{a}(\xi)\mathsf{S}(\xi)\widehat{\phi}(-\xi)$. If (5.1.10) holds and $\widehat{\phi}(0) \neq 0$, then $\widehat{\phi}(\xi) = \mathsf{S}(\xi)\widehat{\phi}(-\xi)$.

c. If $\widehat{\phi}(\xi) = \mathsf{S}(\xi)\widehat{\phi}(-\xi)$ and $\widehat{\psi}(\xi) = \mathsf{T}(\xi)\widehat{\psi}(-\xi)$ (i.e, $\psi_\ell = \mathring{\epsilon}_\ell \psi_\ell(\mathring{c}_\ell - \cdot)$ for all $\ell = 1, \dots, r$), then $\widehat{\psi}(2\xi) = \mathsf{T}(2\xi)\widehat{b}(-\xi)\mathsf{S}^{-1}(\xi)\widehat{\phi}(\xi)$. If in addition all $\mathring{c}_1, \dots, \mathring{c}_r \in \mathbb{Z}$ and (6.9.1) holds, then $\widehat{b}(\xi) = \mathsf{T}(2\xi)\widehat{b}(-\xi)\mathsf{S}^{-1}(\xi)$, $\forall \xi \in \mathbb{R}$.

d. If $\widehat{\phi}(\xi) = \mathsf{S}(\xi)\widehat{\phi}(-\xi)$ and $\widehat{b}(\xi) = \mathsf{T}(2\xi)\widehat{b}(-\xi)\mathsf{S}^{-1}(\xi)$, then $\widehat{\psi} = \mathsf{T}\widehat{\psi}(-\cdot)$.

e. If $\widehat{\phi}(\xi) = \mathsf{S}(\xi)\overline{\widehat{\phi}(\xi)}$ (i.e., $\phi_\ell = \epsilon_\ell \overline{\phi_\ell(c_\ell - \cdot)}$ for all $\ell = 1, \dots, r$), then $\widehat{\phi}(2\xi) = \mathsf{S}(2\xi)\overline{\widehat{a}(\xi)}\mathsf{S}^{-1}(\xi)\widehat{\phi}(\xi)$ for all $\xi \in \mathbb{R}$. If in addition $c_1, \dots, c_r \in \mathbb{Z}$ and (6.9.1) holds, then $\widehat{a}(\xi) = \mathsf{S}(2\xi)\overline{\widehat{a}(\xi)}\mathsf{S}^{-1}(\xi)$.

f. If $\widehat{a}(\xi) = \mathsf{S}(2\xi)\overline{\widehat{a}(\xi)}\mathsf{S}^{-1}(\xi)$, then $\mathsf{S}(2\xi)\overline{\widehat{\phi}(2\xi)} = \widehat{a}(\xi)\mathsf{S}(\xi)\overline{\widehat{\phi}(\xi)}$. If (5.1.10) holds and $\widehat{\phi}(0) \neq 0$, then $\widehat{\phi}(\xi) = \mathsf{S}(\xi)\overline{\widehat{\phi}(\xi)}$.

g. If $\widehat{\phi}(\xi) = \mathsf{S}(\xi)\overline{\widehat{\phi}(\xi)}$ and $\widehat{\psi}(\xi) = \mathsf{T}(\xi)\overline{\widehat{\psi}(\xi)}$ (i.e., $\psi_\ell = \mathring{\epsilon}_\ell \overline{\psi_\ell(\mathring{c}_\ell - \cdot)}$ for all $\ell = 1, \dots, r$), then $\widehat{\psi}(2\xi) = \mathsf{T}(2\xi)\overline{\widehat{b}(\xi)}\mathsf{S}^{-1}(\xi)\widehat{\phi}(\xi)$. If in addition all $\mathring{c}_1, \dots, \mathring{c}_r \in \mathbb{Z}$ and (6.9.1) holds, then $\widehat{b}(\xi) = \mathsf{T}(2\xi)\overline{\widehat{b}(\xi)}\mathsf{S}^{-1}(\xi)$ for all $\xi \in \mathbb{R}$.

h. If $\widehat{\phi}(\xi) = \mathsf{S}(\xi)\overline{\widehat{\phi}(\xi)}$ and $\widehat{b}(\xi) = \mathsf{T}(2\xi)\overline{\widehat{b}(\xi)}\mathsf{S}^{-1}(\xi)$, then $\widehat{\psi}(\xi) = \mathsf{T}(\xi)\overline{\widehat{\psi}(\xi)}$.

6.43. Let $m \in \mathbb{N}_0$ and $a \in (l_0(\mathbb{Z}))^{r\times r}$ such that 1 is a simple eigenvalue of $\widehat{a}(0)$ and all other eigenvalues of $\widehat{a}(0)$ are less than 2^{-m} in modulus. Suppose that a satisfies $m+1$ sum rules with a matching filter $\upsilon \in (l_0(\mathbb{Z}))^{1\times r}$. Let $\mathsf{S}(\xi) := \text{diag}(\epsilon_1 e^{-ic_1\xi}, \dots, \epsilon_r e^{-ic_r\xi})$ with $c_1, \dots, c_r \in \mathbb{Z}$ and $\epsilon_1, \dots, \epsilon_r \in \{-1, 1\}$. (i) If $\widehat{a}(\xi) = \mathsf{S}(2\xi)\overline{\widehat{a}(\xi)}\mathsf{S}^{-1}(\xi)$, prove $\widehat{\upsilon}(\xi) = \overline{\widehat{\upsilon}(\xi)}\mathsf{S}^{-1}(\xi) + \mathscr{O}(|\xi|^{m+1})$, $\xi \to 0$. (ii) If $\widehat{a}(\xi) = \mathsf{S}(2\xi)\widehat{a}(-\xi)\mathsf{S}^{-1}(\xi)$, prove $\widehat{\upsilon}(\xi) = \widehat{\upsilon}(-\xi)\mathsf{S}^{-1}(\xi) + \mathscr{O}(|\xi|^{m+1})$ as $\xi \to 0$.

6.44. Prove Sterling's formula: $\sqrt{2\pi}n^{n+\frac{1}{2}}e^{-n+\frac{1}{12n+1}} < n! < \sqrt{2\pi}n^{n+\frac{1}{2}}e^{-n+\frac{1}{12n}}$, for all $n \in \mathbb{N}$.

6.45. Define a 2π-periodic function by $\widehat{a_{\tau_1,\tau_2,\tau_3}}(\xi) := \frac{2^{-2\tau_1}(1+e^{-i\xi})^{2\tau_1}}{(|\sin(\xi/2)|^{2\tau_2}+|\cos(\xi/2)|^{2\tau_2})^{\tau_3}}$ with $\tau_1, \tau_2, \tau_3 \in \mathbb{R}$. If $\tau_1, \tau_2, \tau_3 > 0$ satisfying $\tau_1 > 1/4 + \max(0, (\tau_2 - 1)\tau_3/2)$, prove that $\text{sm}(a_{\tau_1,\tau_2,\tau_3}) > 0$ by showing $\rho(\mathcal{T}_{|\widehat{a_{\tau_1,\tau_2,\tau_3}}|^2}) < 1$.

6.46. Let $\phi = (\phi_1, \phi_2)^\mathsf{T}$ be the spline refinable vector function given in (6.2.15) (see item (3) of Theorem 6.2.9). Prove that the integer shifts of ϕ are linearly independent but not locally linearly independent. Therefore, Theorem 6.7.2 does not hold for $r > 1$. Also note that $\text{supp}(\phi) = [-2/3, 2/3] \subsetneq [-1, 1] = \text{fsupp}(a)$.

6.47. Let ϕ be a compactly supported Lebesgue measurable function in $L_1(\mathbb{R})$ such that $\phi'(x)$ exists for a.e. $x \in \mathbb{R}$ with $0 < \|\phi'\|_{L_1(\mathbb{R})} < \infty$. If $\phi(x) = |\mathsf{d}| \sum_{k\in\mathbb{Z}} a(k)\phi(\mathsf{d}x - k)$ for a.e. $x \in \mathbb{R}$ for some $a \in l_0(\mathbb{Z})$ and $\mathsf{d} \in \mathbb{Z}$ with $|\mathsf{d}| \geqslant 2$, prove that $\phi(x) = \int_{-\infty}^{x} \phi'(t)\,dt$ for a.e. $x \in \mathbb{R}$ and by modifying

ϕ on a set of measure zero, the function ϕ must be a compactly supported absolutely continuous function on $\mathbb{R}$.

6.48. Define $a, b \in l_0(\mathbb{Z})$ by $\widehat{a}(\xi) := e^{-i\xi}(1+e^{-i\xi}+e^{-i2\xi})(1+e^{-i\xi})/6$ and $\widehat{b}(\xi) := (1+e^{-2i\xi})/3$. Define $\widehat{\phi}(\xi) := \prod_{j=1}^{\infty} \widehat{a}(3^{-j}\xi)$ and $\widehat{\eta}(\xi) := \prod_{j=1}^{\infty} \widehat{b}(3^{-j}\xi)$. Then ϕ and η satisfy $\phi = 3\sum_{k\in\mathbb{Z}} a(k)\phi(3\cdot -k)$ and $\eta = 3\sum_{k\in\mathbb{Z}} b(k)\eta(3\cdot -k)$.

a. Prove that η is the Cantor measure μ induced by the standard ternary Cantor set on the interval $[0, 1]$.
b. The Cantor ternary function f_c on the interval $[0, 1]$ is defined by $f_c(x) := \mu([0, x))$ for $x \in [0, 1]$ so that f_c a continuous function on $[0, 1]$. Prove that $\phi(x) = f_c(2x+1)\chi_{[-1/2,0]}(x) + \chi_{(0,1/2)}(x) + f_c(2-2x)\chi_{[1/2,1]}$.
c. Prove that the integer shifts of ϕ are linearly independent by showing that $\phi(k) = \delta(k)$ for all $k \in \mathbb{Z}$. That is, ϕ is an interpolating function.

Thus, Theorem 6.7.2 does not hold for $r = 1$ and a dilation factor greater than 2.

Chapter 7
Applications of Framelets and Wavelets

In the last chapter of this book, we discuss some applications of framelets and wavelets and provide their underlying mathematics. Since many problems in applications are multidimensional, we first introduce the theory of multidimensional framelets and wavelets. Then we study subdivision schemes and their applications to curve and surface generation in computer graphics. To improve the performance of tensor product real-valued framelets and wavelets, we introduce directional tensor product complex tight framelets and explore their applications to image denoising and inpainting. Next, we discuss how to construct framelets and wavelets on the interval [0, 1] by a general method as well as their applications to numerical solutions to differential equations. Finally, we address fast multiframelet transform and its balanced approximation property.

7.1 Multidimensional Framelets and Wavelets

We briefly introduce the theory of multidimensional framelets and wavelets for problems in multiple dimensions. We also discuss how to construct multidimensional framelets and wavelets by tensor product (i.e., separable) and the projection method from one-dimensional framelets and wavelets. The main advantage of tensor product framelets and wavelets lies in that they have simple implementation and fast algorithms. Since all the results in this section can be similarly proved without essential difficulty as their one-dimensional counterpart, we left most proofs as exercises for the reader.

B. Han, *Framelets and Wavelets*, Applied and Numerical Harmonic Analysis,
https://doi.org/10.1007/978-3-319-68530-4_7

7.1.1 Multidimensional Framelet and Wavelet Filter Banks

By $l_0(\mathbb{Z}^d)$ we denote the set of all finitely supported sequences on $\mathbb{Z}^d$ and by $l(\mathbb{Z}^d)$ we denote the set of all sequences on $\mathbb{Z}^d$. We first present the fast M-framelet transform. For a filter $u = \{u(k)\}_{k\in\mathbb{Z}^d} \in l_0(\mathbb{Z}^d)$ and a $d \times d$ integer matrix M, similar to (1.1.2) and (1.1.3), we define the *subdivision operator* $\mathcal{S}_{u,\mathsf{M}} : l(\mathbb{Z}^d) \to l(\mathbb{Z}^d)$ to be

$$[\mathcal{S}_{u,\mathsf{M}} v](n) := |\det(\mathsf{M})| \sum_{k\in\mathbb{Z}^d} v(k)u(n - \mathsf{M}k), \qquad n \in \mathbb{Z}^d,\ v \in l(\mathbb{Z}^d), \tag{7.1.1}$$

and the *transition operator* $\mathcal{T}_{u,\mathsf{M}} : l(\mathbb{Z}^d) \to l(\mathbb{Z}^d)$ to be

$$[\mathcal{T}_{u,\mathsf{M}} v](n) := |\det(\mathsf{M})| \sum_{k\in\mathbb{Z}^d} v(k)\overline{u(k - \mathsf{M}n)}, \qquad n \in \mathbb{Z}^d,\ v \in l(\mathbb{Z}^d). \tag{7.1.2}$$

Let M be a $d \times d$ invertible integer matrix. Let $\tilde{a} \in l_0(\mathbb{Z}^d)$ be a low-pass filter and $\tilde{b}_1, \dots, \tilde{b}_s \in l_0(\mathbb{Z}^d)$ be high-pass filters for decomposition. For a positive integer J, *a J-level discrete M-framelet decomposition* is given by

$$v_j := |\det(\mathsf{M})|^{-1/2}\mathcal{T}_{\tilde{a},\mathsf{M}} v_{j-1}, \quad w_{\ell,j} := |\det(\mathsf{M})|^{-1/2}\mathcal{T}_{\tilde{b}_\ell,\mathsf{M}} v_{j-1}, \qquad \ell = 1, \dots, s,$$

for $j = 1, \dots, J$, where $v_0 : \mathbb{Z}^d \to \mathbb{C}$ is an input signal. Let $a \in l_0(\mathbb{Z}^d)$ be a low-pass filter and $b_1, \dots, b_s \in l_0(\mathbb{Z}^d)$ be high-pass filters for reconstruction. Now *a J-level discrete M-framelet reconstruction* is

$$\mathring{v}_{j-1} := |\det(\mathsf{M})|^{-1/2}\mathcal{S}_{a,\mathsf{M}}\mathring{v}_j + |\det(\mathsf{M})|^{-1/2}\sum_{\ell=1}^{s}\mathcal{S}_{b_\ell,\mathsf{M}}\mathring{w}_{\ell,j-1},$$

for $j = J, \dots, 1$. It is convenient to rewrite the J-level discrete M-framelet decomposition employing a filter bank $\{\tilde{a}; \tilde{b}_1, \dots, \tilde{b}_s\}$ by using *a J-level decomposition operator* $\widetilde{\mathcal{W}}_J : l(\mathbb{Z}^d) \to (l(\mathbb{Z}^d))^{1\times(sJ+1)}$ as in (1.3.4). We similarly define $\mathcal{W}_J$ if $\{a; b_1, \dots, b_s\}$ is used instead. Similarly, *a J-level discrete M-framelet reconstruction operator* $\mathcal{V}_J : (l(\mathbb{Z}^d))^{1\times(sJ+1)} \to l(\mathbb{Z}^d)$ employing a filter bank $\{a; b_1, \dots, b_s\}$ is defined as in (1.3.5). We say that a J-level *fast M-framelet transform* has the perfect reconstruction property if $\mathcal{V}_J\widetilde{\mathcal{W}}_J v_0 = v_0$ for all $v_0 \in l(\mathbb{Z}^d)$, that is, the reconstructed signal $\mathring{v}_0$ is the same as the original input signal v_0 if $\mathring{v}_J = v_J$ and $\mathring{w}_{\ell,j} = w_{\ell,j}$ for all $\ell = 1, \dots, s$ and $j = 1, \dots, J$.

For a $d \times d$ invertible integer matrix M, we define

$$\Omega_{\mathsf{M}} := [(\mathsf{M}^{\mathsf{T}})^{-1}\mathbb{Z}^d] \cap [0,1)^d \quad \text{and} \quad \Gamma_{\mathsf{M}} := [\mathsf{M}[0,1)^d] \cap \mathbb{Z}^d. \tag{7.1.3}$$

Instead of the particular choice of Ω_{M} in (7.1.3), the set Ω_{M} in this book can be chosen to be any complete set of representatives of distinct cosets of the quotient group $[(\mathsf{M}^{\mathsf{T}})^{-1}\mathbb{Z}^d]/\mathbb{Z}^d$. Similarly, the set Γ_{M} can be chosen to be any complete set of representatives of distinct cosets of the quotient group $\mathbb{Z}^d/[\mathsf{M}\mathbb{Z}^d]$.

For $u = \{u(k)\}_{k\in\mathbb{Z}^d} \in l_0(\mathbb{Z}^d)$, recall that $\widehat{u}(\xi) := \sum_{k\in\mathbb{Z}^d} u(k)e^{-ik\cdot\xi}$ for $\xi \in \mathbb{R}^d$.

Most results in Chap. 1 hold for M-framelet and M-wavelet filter banks. Similar to Theorem 1.1.1, we have

Theorem 7.1.1 *Let $a, b_1, \ldots, b_s, \tilde{a}, \tilde{b}_1, \ldots, \tilde{b}_s \in l_0(\mathbb{Z}^d)$ and M be a $d \times d$ invertible integer matrix. Then the following statements are equivalent:*

(i) The J-level fast M-framelet transform employing $(\{\tilde{a}; \tilde{b}_1, \ldots, \tilde{b}_s\}, \{a; b_1, \ldots, b_s\})$ has the perfect reconstruction property for every $J \in \mathbb{N}$, i.e., $\mathcal{V}_J\widetilde{\mathcal{W}}_J v = v$ for all $v \in l(\mathbb{Z}^d)$.

(ii) The one-level fast M-framelet transform employing $(\{\tilde{a}; \tilde{b}_1, \ldots, \tilde{b}_s\}, \{a; b_1, \ldots, b_s\})$ has the perfect reconstruction property, that is, for all $v \in l(\mathbb{Z}^d)$,

$$\mathcal{S}_{a,\mathsf{M}}\mathcal{T}_{\tilde{a},\mathsf{M}}v + \mathcal{S}_{b_1,\mathsf{M}}\mathcal{T}_{\tilde{b}_1,\mathsf{M}}v + \cdots + \mathcal{S}_{b_s,\mathsf{M}}\mathcal{T}_{\tilde{b}_s,\mathsf{M}}v = |\det(\mathsf{M})|v. \tag{7.1.4}$$

(iii) The identity (7.1.4) holds for all $v \in l_0(\mathbb{Z}^d)$.

(iv) (7.1.4) holds for the particular sequences $v = \boldsymbol{\delta}(\cdot - \gamma), \gamma \in \Gamma_{\mathsf{M}}$, more explicitly,

$$\sum_{k\in\mathbb{Z}^d} \tilde{a}(\gamma + \mathsf{M}k)\overline{a(n + \gamma + \mathsf{M}k)} + \sum_{k\in\mathbb{Z}^d}\sum_{\ell=1}^{s} \tilde{b}_\ell(\gamma + \mathsf{M}k)\overline{b_\ell(n + \gamma + \mathsf{M}k)} = |\det(\mathsf{M})|^{-1}\boldsymbol{\delta}(n)$$

for all $\gamma \in \Gamma_{\mathsf{M}}$ and $n \in \mathbb{Z}^d$.

(v) $(\{\tilde{a}; \tilde{b}_1, \ldots, \tilde{b}_s\}, \{a; b_1, \ldots, b_s\})$ is a dual M-framelet filter bank, *that is,*

$$\widehat{\tilde{a}}(\xi)\overline{\widehat{a}(\xi + 2\pi\omega)} + \sum_{\ell=1}^{s} \widehat{\tilde{b}_\ell}(\xi)\overline{\widehat{b_\ell}(\xi + 2\pi\omega)} = \boldsymbol{\delta}(\omega), \quad \forall\, \xi \in \mathbb{R}^d, \omega \in \Omega_{\mathsf{M}}, \tag{7.1.5}$$

where $\boldsymbol{\delta}$ is the Dirac sequence *such that $\boldsymbol{\delta}(0) = 1$ and $\boldsymbol{\delta}(k) = 0$ for all $k \in \mathbb{R}^d\backslash\{0\}$.*

Suppose that $(\{\tilde{a}; \tilde{b}_1, \ldots, \tilde{b}_s\}, \{a; b_1, \ldots, b_s\})$ is a dual M-framelet filter bank. If N is a $d \times d$ invertible integer matrix such that $\mathsf{M}\mathbb{Z}^d = \mathsf{N}\mathbb{Z}^d$ (which is equivalent to saying that $\mathsf{N} = \mathsf{M}E$ for a $d \times d$ integer matrix E with $|\det(E)| = 1$), then one can easily check that $(\{\tilde{a}; \tilde{b}_1, \ldots, \tilde{b}_s\}, \{a; b_1, \ldots, b_s\})$ is a dual N-framelet filter bank, due to the simple fact that $\Omega_{\mathsf{M}} = \Omega_{\mathsf{N}}$. In other words, if $(\{\tilde{a}; \tilde{b}_1, \ldots, \tilde{b}_s\}, \{a; b_1, \ldots, b_s\})$ is a dual M-framelet filter bank, then $(\{\tilde{a}; \tilde{b}_1, \ldots, \tilde{b}_s\}, \{a; b_1, \ldots, b_s\})$ is a dual $\mathsf{M}E$-framelet filter bank for any $d \times d$ integer matrix E with $|\det(E)| = 1$. This simple observation is very useful for constructing dual M-framelet filter bank, since one may use a matrix $\mathsf{M}E$ with a simple structure instead of the given matrix M by choosing a desired choice of a $d \times d$ integer matrix E with $|\det(E)| = 1$.

In particular, a dual M-framelet filter bank $(\{\tilde{a}; \tilde{b}_1, \ldots, \tilde{b}_s\}, \{a; b_1, \ldots, b_s\})$ with $s = |\det(\mathsf{M})| - 1$ satisfying (7.1.5) is called *a biorthogonal* M*-wavelet filter bank*.

Similar to Proposition 1.1.2, we have

Proposition 7.1.2 *Let $(\{\tilde{a}; \tilde{b}_1, \ldots, \tilde{b}_s\}, \{a; b_1, \ldots, b_s\})$ be a dual* M*-framelet filter bank. Define $\mathcal{V} := \mathcal{V}_1$ and $\widetilde{\mathcal{W}} := \widetilde{\mathcal{W}}_1$. Then the following statements are equivalent:*

(i) $\widetilde{\mathcal{W}}$ is onto.
(ii) $\mathcal{V}$ is one-to-one.
(iii) $\mathcal{V}\widetilde{\mathcal{W}} = \mathrm{Id}_{l(\mathbb{Z}^d)}$ and $\widetilde{\mathcal{W}}\mathcal{V} = \mathrm{Id}_{(l(\mathbb{Z}^d))^{1\times(s+1)}}$, that is, $\mathcal{V}^{-1} = \widetilde{\mathcal{W}}$ and $\widetilde{\mathcal{W}}^{-1} = \mathcal{V}$.
(iv) $s = |\det(\mathsf{M})| - 1$.

As a special case, $\{a; b_1, \ldots, b_s\}$ is called *a tight* M*-framelet filter bank* if $(\{a; b_1, \ldots, b_s\}, \{a; b_1, \ldots, b_s\})$ is a dual M-framelet filter bank, that is,

$$\widehat{a}(\xi)\overline{\widehat{a}(\xi + 2\pi\omega)} + \sum_{\ell=1}^{s} \widehat{b_\ell}(\xi)\overline{\widehat{b_\ell}(\xi + 2\pi\omega)} = \boldsymbol{\delta}(\omega), \quad \forall\, \xi \in \mathbb{R}^d, \omega \in \Omega_{\mathsf{M}}. \tag{7.1.6}$$

For $s = |\det(\mathsf{M})| - 1$, a tight M-framelet filter bank $\{a; b_1, \ldots, b_s\}$ satisfying (7.1.6) is called *an orthogonal* M*-wavelet filter bank*. Similar to Theorem 1.1.4, we have

Theorem 7.1.3 *Let $a, b_1, \ldots, b_s \in l_0(\mathbb{Z}^d)$. Then the following are equivalent:*

(i) $\|\mathcal{W}v\|^2_{(l_2(\mathbb{Z}^d))^{1\times(s+1)}} = \|v\|^2_{l_2(\mathbb{Z}^d)}$ $\forall v \in l_2(\mathbb{Z}^d)$ with $\mathcal{W} := \mathcal{W}_1$, i.e., for all $v \in l_2(\mathbb{Z}^d)$,

$$\|\mathcal{T}_{a,\mathsf{M}}v\|^2_{l_2(\mathbb{Z}^d)} + \|\mathcal{T}_{b_1,\mathsf{M}}v\|^2_{l_2(\mathbb{Z}^d)} + \cdots + \|\mathcal{T}_{b_s,\mathsf{M}}v\|^2_{l_2(\mathbb{Z}^d)} = |\det(\mathsf{M})|\|v\|^2_{l_2(\mathbb{Z}^d)}.$$

(ii) $\langle \mathcal{W}v, \mathcal{W}\tilde{v}\rangle = \langle v, \tilde{v}\rangle$ for all $v, \tilde{v} \in l_2(\mathbb{Z}^d)$.
(iii) $\mathcal{W}^\star\mathcal{W} = \mathrm{Id}_{l_2(\mathbb{Z}^d)}$, that is, $\mathcal{W}^\star\mathcal{W}v = v$ for all $v \in l_2(\mathbb{Z}^d)$.
(iv) The filter bank $\{a; b_1, \ldots, b_s\}$ is a tight M*-framelet filter bank.*

7.1.2 Multidimensional Framelets in Sobolev Spaces

For a function $f : \mathbb{R}^d \to \mathbb{C}$ and a $d \times d$ real-valued matrix U, we adopt the following notation:

$$f_{U;k,n}(x) := f_{[\![U;k,n]\!]}(x) := [\![U; k, n]\!]f(x) := |\det(U)|^{1/2} e^{-in\cdot Ux} f(Ux - k) \tag{7.1.7}$$

for $x, k, n \in \mathbb{R}^d$. In particular, we define

$$f_{U;k} := f_{U;k,0} = |\det(U)|^{1/2} f(U \cdot - k). \tag{7.1.8}$$

Let M be a $d \times d$ invertible real-valued matrix. Similar to (4.3.1), for every integer $J \in \mathbb{Z}$ and subsets Φ, Ψ of functions on $\mathbb{R}^d$, we define *a (nonhomogeneous)* M*-affine system* $\mathsf{AS}_J^{\mathsf{M}}(\Phi; \Psi)$ to be

$$\mathsf{AS}_J^{\mathsf{M}}(\Phi; \Psi) := \{\phi_{\mathsf{M}^J;k} \; : \; k \in \mathbb{Z}^d, \phi \in \Phi\} \cup \{\psi_{\mathsf{M}^j;k} \; : \; j \geqslant J, k \in \mathbb{Z}^d, \psi \in \Psi\} \tag{7.1.9}$$

and *a homogeneous* M*-affine system* $\mathsf{AS}^{\mathsf{M}}(\Psi)$ to be

$$\mathsf{AS}^{\mathsf{M}}(\Psi) := \{\psi_{\mathsf{M}^j;k} \; : \; j \in \mathbb{Z}, k \in \mathbb{Z}^d, \psi \in \Psi\}. \tag{7.1.10}$$

By the same proofs, all the claims in Sect. 4.3 hold for $\mathsf{AS}_J^{\mathsf{M}}(\Phi; \Psi)$ and $\mathsf{AS}^{\mathsf{M}}(\Psi)$. Similar to Sect. 4.6, we now introduce framelets in Sobolev spaces. For $\tau \in \mathbb{R}$, we denote by $H^\tau(\mathbb{R}^d)$ the *Sobolev space* consisting of all tempered distributions f on $\mathbb{R}^d$ such that $\widehat{f} \in L_2^{loc}(\mathbb{R}^d)$ and

$$\|f\|^2_{H^\tau(\mathbb{R}^d)} := \frac{1}{(2\pi)^d} \int_{\mathbb{R}^d} |\widehat{f}(\xi)|^2 (1 + \|\xi\|^2)^\tau d\xi < \infty.$$

Note that $H^0(\mathbb{R}^d) = L_2(\mathbb{R}^d)$ and $H^\tau(\mathbb{R}^d)$ is a Hilbert space with the inner product

$$\langle f, g \rangle_{H^\tau(\mathbb{R}^d)} := \frac{1}{(2\pi)^d} \int_{\mathbb{R}^d} \widehat{f}(\xi) \overline{\widehat{g}(\xi)} (1 + \|\xi\|^2)^\tau d\xi, \qquad f, g \in H^\tau(\mathbb{R}^d).$$

For $\Phi, \Psi \subseteq H^\tau(\mathbb{R}^d)$ and $\tilde{\Phi}, \tilde{\Psi} \subseteq H^{-\tau}(\mathbb{R}^d)$, we say that the pair $(\mathsf{AS}_J^{\mathsf{M}}(\tilde{\Phi}; \tilde{\Psi}), \mathsf{AS}_J^{\mathsf{M}}(\Phi; \Psi))$ is *a pair of dual* M*-framelets for a pair of dual Sobolev spaces* $(H^{-\tau}(\mathbb{R}^d), H^\tau(\mathbb{R}^d))$ if

(1) The affine system $\mathsf{AS}_J^{\mathsf{M}}(\Phi; \Psi)$ has *stability* in $H^\tau(\mathbb{R})$, that is, there exist positive constants C_1 and C_2 such that for all $g \in H^{-\tau}(\mathbb{R}^d)$,

$$\begin{aligned} C_1 \|g\|^2_{H^{-\tau}(\mathbb{R}^d)} \leqslant & \sum_{\phi \in \Phi} \sum_{k \in \mathbb{Z}^d} |\det(\mathsf{M})|^{-2\tau J/d} |\langle g, \phi_{\mathsf{M}^J;k} \rangle|^2 \\ & + \sum_{j=J}^{\infty} \sum_{\psi \in \Psi} \sum_{k \in \mathbb{Z}^d} |\det(\mathsf{M})|^{-2\tau j/d} |\langle g, \psi_{\mathsf{M}^j;k} \rangle|^2 \leqslant C_2 \|g\|^2_{H^{-\tau}(\mathbb{R}^d)}; \end{aligned}$$

(2) The affine system $\mathsf{AS}_J^{\mathsf{M}}(\tilde{\Phi}; \tilde{\Psi})$ has stability in $H^{-\tau}(\mathbb{R}^d)$;

(3) The following identity holds

$$\langle f, g\rangle = \sum_{\phi\in\Phi}\sum_{k\in\mathbb{Z}^d}\langle f, \tilde{\phi}_{\mathsf{M}^J;k}\rangle\langle\phi_{\mathsf{M}^J;k}, g\rangle + \sum_{j=J}^{\infty}\sum_{\psi\in\Psi}\sum_{k\in\mathbb{Z}^d}\langle f, \tilde{\psi}_{\mathsf{M}^j;k}\rangle\langle\psi_{\mathsf{M}^j;k}, g\rangle$$

for all $f \in H^\tau(\mathbb{R}^d)$ and $g \in H^{-\tau}(\mathbb{R}^d)$, with the series on the right-hand side converging absolutely.

A pair $(\{\tilde{\Phi};\tilde{\Psi}\},\{\Phi;\Psi\})$ is called *a dual* M*-framelet in* $(H^{-\tau}(\mathbb{R}^d), H^\tau(\mathbb{R}^d))$ if the pair $(\mathsf{AS}_0^{\mathsf{M}}(\tilde{\Phi};\tilde{\Psi}), \mathsf{AS}_0^{\mathsf{M}}(\Phi;\Psi))$ is a pair of dual M-framelets for a pair of dual Sobolev spaces $(H^{-\tau}(\mathbb{R}^d), H^\tau(\mathbb{R}^d))$. Similar to Theorem 4.6.3, we have

Theorem 7.1.4 *Let* M *be a* $d\times d$ *invertible real-valued matrix and define* $\mathsf{N} := (\mathsf{M}^{\mathsf{T}})^{-1}$. *Let* $\tau\in\mathbb{R}$ *and* $\Phi, \tilde{\Phi}, \Psi, \tilde{\Psi}$ *be subsets of tempered distributions on* $\mathbb{R}^d$. *Then* $(\{\tilde{\Phi};\tilde{\Psi}\},\{\Phi;\Psi\})$ *is a dual* M*-framelet in* $(H^{-\tau}(\mathbb{R}^d), H^\tau(\mathbb{R}^d))$ *if and only if*

(i) there exists a positive constant C such that

$$\sum_{\phi\in\Phi}\sum_{k\in\mathbb{Z}^d}|\langle\mathbf{g},\widehat{\phi}_{I_d;0,k}\rangle|^2 + \sum_{j=0}^{\infty}\sum_{\psi\in\Psi}\sum_{k\in\mathbb{Z}^d}|\det(\mathsf{M})|^{-2\tau j/d}|\langle\mathbf{g},\widehat{\psi}_{\mathsf{N}^j;0,k}\rangle|^2 \leqslant C\|\widehat{\mathbf{g}}\|^2_{H^{-\tau}(\mathbb{R}^d)}, \qquad \forall\, \mathbf{g}\in\mathscr{D}(\mathbb{R}^d), \tag{7.1.11}$$

where I_d *is the* $d\times d$ *identity matrix, and*

$$\sum_{\tilde{\phi}\in\tilde{\Phi}}\sum_{k\in\mathbb{Z}^d}|\langle\mathbf{f},\widehat{\tilde{\phi}}_{I_d;0,k}\rangle|^2 + \sum_{j=0}^{\infty}\sum_{\tilde{\psi}\in\tilde{\Psi}}\sum_{k\in\mathbb{Z}^d}|\det(\mathsf{M})|^{2\tau j/d}|\langle\mathbf{f},\widehat{\tilde{\psi}}_{\mathsf{N}^j;0,k}\rangle|^2 \leqslant C\|\widehat{\mathbf{f}}\|^2_{H^{\tau}(\mathbb{R}^d)}, \qquad \forall\, \mathbf{f}\in\mathscr{D}(\mathbb{R}^d); \tag{7.1.12}$$

(ii) $(\{\widehat{\tilde{\Phi}};\widehat{\tilde{\Psi}}\},\{\widehat{\Phi};\widehat{\Psi}\})$ *is a frequency-based dual* M*-framelet, that is,*

$$\lim_{J'\to\infty}\Big(\sum_{\varphi\in\widehat{\Phi}}\sum_{k\in\mathbb{Z}^d}\langle\mathbf{f},\tilde{\varphi}_{I_d;0,k}\rangle\langle\varphi_{I_d;0,k},\mathbf{g}\rangle + \sum_{j=0}^{J'-1}\sum_{\boldsymbol{\psi}\in\widehat{\Psi}}\sum_{k\in\mathbb{Z}^d}\langle\mathbf{f},\tilde{\boldsymbol{\psi}}_{\mathsf{N}^j;0,k}\rangle\langle\boldsymbol{\psi}_{\mathsf{N}^{-j};0,k},\mathbf{g}\rangle\Big) = (2\pi)^d\langle\mathbf{f},\mathbf{g}\rangle$$

for all $\mathbf{f},\mathbf{g}\in\mathscr{D}(\mathbb{R}^d)$, *where* $\sim$ *is the natural bijection between* $\{\Phi;\Psi\}$ *and* $\{\tilde{\Phi};\tilde{\Psi}\}$, *see* (4.1.3) *for details.*

A $d\times d$ matrix M is called *expansive* if all its eigenvalues are greater than one in modulus. M is called *a dilation matrix* if it is an expansive integer matrix. If M is similar to a diagonal matrix $\operatorname{diag}(\lambda_1, \ldots, \lambda_d)$ with $|\lambda_1| = \cdots = |\lambda_d|$, then M is said to be *isotropic*. For isotropic matrices, we have the following result.

Proposition 7.1.5 *A $d \times d$ matrix M is isotropic if and only if there exists a norm $\|\cdot\|_{\mathsf{M}}$ on $\mathbb{C}^d$ such that $\|\mathsf{M}x\|_{\mathsf{M}} = |\det(\mathsf{M})|^{1/d}\|x\|_{\mathsf{M}}$ for all $x \in \mathbb{C}^d$.*

Proof Necessity ($\Rightarrow$). Since M is isotropic, there is an invertible matrix E such that

$$\mathsf{M} = E^{-1}\operatorname{diag}(\lambda_1, \ldots, \lambda_d)E$$

with $|\lambda_1| = \cdots = |\lambda_d| = |\det(\mathsf{M})|^{1/d}$. For $x \in \mathbb{C}^d$, we define $\|x\|_{\mathsf{M}}^2 := \|Ex\|^2 = \bar{x}^{\mathsf{T}}\bar{E}^{\mathsf{T}}Ex$. Then $\|\mathsf{M}x\|_{\mathsf{M}} = |\det(\mathsf{M})|^{1/d}\|x\|_{\mathsf{M}}$ and $\|\cdot\|_{\mathsf{M}}$ is a norm on $\mathbb{C}^d$.

Sufficiency ($\Leftarrow$). There exists a $d \times d$ complex-valued invertible matrix E such that $E\mathsf{M}E^{-1}$ is the Jordan canonical form of M. Since $(E\mathsf{M}E^{-1})^n = E\mathsf{M}^nE^{-1}$ and $\|\mathsf{M}x\|_{\mathsf{M}} = |\det(\mathsf{M})|^{1/d}\|x\|_{\mathsf{M}}$ for all $x \in \mathbb{C}^d$, there exists a positive constant C such that for all $n \in \mathbb{N}$,

$$C^{-1}|\det(\mathsf{M})|^{n/d}\|x\|_{\mathsf{M}} \leqslant \|(E\mathsf{M}E^{-1})^n x\|_{\mathsf{M}} = \|E\mathsf{M}^nE^{-1}x\|_{\mathsf{M}} \leqslant C|\det(\mathsf{M})|^{n/d}\|x\|_{\mathsf{M}}.$$

The above inequalities force each Jordan block matrix in $E\mathsf{M}E^{-1}$ to be diagonal and its diagonal entries must have magnitude $|\det(\mathsf{M})|^{1/d}$. This proves $E\mathsf{M}E^{-1} = \operatorname{diag}(\lambda_1, \ldots, \lambda_d)$ with $|\lambda_1| = \cdots = |\lambda_d| = |\det(\mathsf{M})|^{1/d}$. Thus, M must be isotropic. □

In the following we discuss the connections between frequency-based dual framelets and filter banks. By the same proof of Theorem 4.1.9, we have

Theorem 7.1.6 *Let $\theta, a, b_1, \ldots, b_s, \tilde{\theta}, \tilde{a}, \tilde{b}_1, \ldots, \tilde{b}_s \in l_0(\mathbb{Z}^d)$ with $\widehat{a}(0) = \widehat{\tilde{a}}(0) = 1$. Let M be a $d \times d$ dilation matrix. Define the* standard M-refinable functions *ϕ and $\tilde{\phi}$ associated with a and $\tilde{a}$ by*

$$\widehat{\phi}(\xi) := \prod_{j=1}^{\infty} \widehat{a}((\mathsf{M}^{\mathsf{T}})^{-j}\xi), \qquad \widehat{\tilde{\phi}}(\xi) := \prod_{j=1}^{\infty} \widehat{\tilde{a}}((\mathsf{M}^{\mathsf{T}})^{-j}\xi), \qquad \xi \in \mathbb{R}^d. \tag{7.1.13}$$

Define $\widehat{\eta}, \widehat{\psi^1}, \ldots, \widehat{\psi^s}, \widehat{\tilde{\eta}}, \widehat{\tilde{\psi}^1}, \ldots, \widehat{\tilde{\psi}^s}$ by

$$\begin{aligned} \widehat{\eta}(\xi) &:= \widehat{\theta}(\xi)\widehat{\phi}(\xi), \quad \widehat{\psi^\ell}(\mathsf{M}^{\mathsf{T}}\xi) := \widehat{b_\ell}(\xi)\widehat{\phi}(\xi), \\ \widehat{\tilde{\eta}}(\xi) &:= \widehat{\tilde{\theta}}(\xi)\widehat{\tilde{\phi}}(\xi), \quad \widehat{\tilde{\psi}^\ell}(\mathsf{M}^{\mathsf{T}}\xi) := \widehat{\tilde{b}_\ell}(\xi)\widehat{\tilde{\phi}}(\xi), \quad \xi \in \mathbb{R}^d, \end{aligned} \tag{7.1.14}$$

for $\ell = 1, \ldots, s$. Then $(\{\widehat{\tilde{\eta}}; \widehat{\tilde{\psi}^1}, \ldots, \widehat{\tilde{\psi}^s}\}, \{\widehat{\eta}; \widehat{\psi^1}, \ldots, \widehat{\psi^s}\})$ is a frequency-based dual M-framelet if and only if

(i) $\widehat{\Theta}(0) = 1$, where $\widehat{\Theta}(\xi) := \widehat{\tilde{\theta}}(\xi)\overline{\widehat{\theta}(\xi)}$;

(ii) $(\{\tilde{a}; \tilde{b}_1, \dots, \tilde{b}_s\}, \{a; b_1, \dots, b_s\})_\Theta$ *is a dual* M*-framelet filter bank, that is,*

$$\widehat{\Theta}(\mathsf{M}^\mathsf{T}\xi)\widehat{\tilde{a}}(\xi)\overline{\widehat{a}(\xi)} + \widehat{\tilde{b}_1}(\xi)\overline{\widehat{b_1}(\xi)} + \cdots + \widehat{\tilde{b}_s}(\xi)\overline{\widehat{b_s}(\xi)} = \widehat{\Theta}(\xi), \tag{7.1.15}$$

$$\begin{aligned}\widehat{\Theta}(\mathsf{M}^\mathsf{T}\xi)\widehat{\tilde{a}}(\xi)\overline{\widehat{a}(\xi + 2\pi\omega)} + \widehat{\tilde{b}_1}(\xi)\overline{\widehat{b_1}(\xi + 2\pi\omega)} + \cdots \\ + \widehat{\tilde{b}_s}(\xi)\overline{\widehat{b_s}(\xi + 2\pi\omega)} = 0, \qquad \forall\, \xi \in \mathbb{R}^d,\ \omega \in \Omega_\mathsf{M}\backslash\{0\}.\end{aligned} \tag{7.1.16}$$

For $\tau \in \mathbb{R}$ and functions $\mathbf{f}, \mathbf{g} : \mathbb{R}^d \to \mathbb{C}$, recall that the bracket product is defined to be

$$[\mathbf{f}, \mathbf{g}]_\tau(\xi) := \sum_{k\in\mathbb{Z}^d} \mathbf{f}(\xi + 2\pi k)\overline{\mathbf{g}(\xi + 2\pi k)}(1 + \|\xi + 2\pi k\|^2)^\tau, \qquad \xi \in \mathbb{R}^d.$$

For dual framelets associated with filter banks, similar to Theorem 6.4.1, we have

Theorem 7.1.7 *Let* M *be a* $d \times d$ *isotropic dilation matrix. Let* $\theta, a, b_1, \dots, b_s, \tilde{\theta}, \tilde{a}, \tilde{b}_1, \dots, \tilde{b}_s \in l_0(\mathbb{Z}^d)$ *with* $\widehat{a}(0) = \widehat{\tilde{a}}(0) = 1$. *Let* $\phi, \tilde{\phi}$ *be defined in* (7.1.13) *and* $\eta, \psi^1, \dots, \psi^s, \tilde{\eta}, \tilde{\psi}^1, \dots, \tilde{\psi}^s$ *be defined in* (7.1.14)*. For any real number* $\tau \in \mathbb{R}$, $(\{\tilde{\eta}; \tilde{\psi}^1, \dots, \tilde{\psi}^s\}, \{\eta; \psi^1, \dots, \psi^s\})$ *is a dual* M*-framelet in a pair of Sobolev spaces* $(H^{-\tau}(\mathbb{R}^d), H^\tau(\mathbb{R}^d))$ *if and only if*

(1) $\widehat{\Theta}(0) = 1$, *where* $\widehat{\Theta}(\xi) := \widehat{\tilde{\theta}}(\xi)\overline{\widehat{\theta}(\xi)}$;
(2) $(\{\tilde{a}; \tilde{b}_1, \dots, \tilde{b}_s\}, \{a; b_1, \dots, b_s\})_\Theta$ *is a dual* M*-framelet filter bank;*
(3) $\phi \in H^\tau(\mathbb{R}^d)$ *and* $\tilde{\phi} \in H^{-\tau}(\mathbb{R}^d)$;
(4) $\widehat{\psi^\ell}(\xi) = o(\|\xi\|^{-\tau})$ *if* $\tau \leqslant 0$ *and* $\widehat{\tilde{\psi}^\ell}(\xi) = o(\|\xi\|^\tau)$ *if* $\tau \geqslant 0$ *as* $\xi \to 0$ *for all* $\ell = 1, \dots, s$.

Proof Necessity ($\Rightarrow$). Since $(\{\tilde{\eta}; \tilde{\psi}^1, \dots, \tilde{\psi}^s\}, \{\eta; \psi^1, \dots, \psi^s\})$ is a dual M-framelet in $(H^{-\tau}(\mathbb{R}^d), H^\tau(\mathbb{R}^d))$, then $(\{\widehat{\tilde{\eta}}; \widehat{\tilde{\psi}^1}, \dots, \widehat{\tilde{\psi}^s}\}, \{\widehat{\eta}; \widehat{\psi^1}, \dots, \widehat{\psi^s}\})$ must be a frequency-based dual M-framelet and consequently, items (1) and (2) follow from Theorem 7.1.6. Since $\eta \in H^\tau(\mathbb{R}^d)$ and $\widehat{\eta}(\xi) = \widehat{\theta}(\xi)\widehat{\phi}(\xi)$ with $\widehat{\theta}$ being not identically zero, noting that both η and ϕ are compactly supported, we must have $\phi \in H^\tau(\mathbb{R}^d)$ by a similar argument as in Theorem 6.4.1. Item (4) is a consequence of the Bessel properties in (7.1.11) and (7.1.12) (see the proof of item (4) in Theorem 6.4.1).

Sufficiency ($\Leftarrow$). Since $\widehat{\phi}(0) = \widehat{\tilde{\phi}}(0) = \widehat{\Theta}(0) = 1$, items (1) and (2) imply that

$$(\{\widehat{\tilde{\eta}}; \widehat{\tilde{\psi}^1}, \dots, \widehat{\tilde{\psi}^s}\}, \{\widehat{\eta}; \widehat{\psi^1}, \dots, \widehat{\psi^s}\})$$

is a frequency-based dual M-framelet. Note that $\widehat{\phi}(\mathsf{M}^\mathsf{T}\xi) = \widehat{a}(\xi)\widehat{\phi}(\xi)$. By a complicated argument using joint spectral radius, similar to Corollary 6.3.4, one can actually show that $\phi \in H^\tau(\mathbb{R}^d)$ implies $\phi \in H^t(\mathbb{R}^d)$ for some $t > \tau$. Since $\phi \in H^t(\mathbb{R}^d)$ has compact support, by the same argument as in Lemma 6.3.2, we must have $[\widehat{\phi}, \widehat{\phi}]_t \in L_\infty(\mathbb{T}^d)$. Therefore, by items (3) and (4), there exists $\nu > -\tau$ (if $\tau \leqslant 0$) such that

$$[\widehat{\psi^\ell}, \widehat{\psi^\ell}]_t \in L_\infty(\mathbb{R}^d), \qquad \|\cdot\|^{-\nu}\widehat{\psi^\ell}(\cdot) \in L_\infty([-\pi, \pi]^d) \qquad \forall\, \ell = 1, \ldots, s.$$

By the same argument as in Theorem 4.6.5 and noting that M is isotropic, we conclude that (7.1.11) holds for some positive constant C. Similarly, we can prove that (7.1.12) holds. Now the claim follows from Theorem 7.1.4. □

We say that a filter bank $\{a; b_1, \ldots, b_s\}_\Theta$ is *a tight* M*-framelet filter bank* if $(\{a; b_1, \ldots, b_s\}, \{a; b_1, \ldots, b_s\})_\Theta$ is a dual M-framelet filter bank, i.e., (7.1.15) and (7.1.16) are satisfied with $\tilde{a} = a, \tilde{b}_1 = b_1, \ldots, \tilde{b}_s = b_s$. For tight framelet filter banks, similar to Theorem 4.5.4 (without using Theorem 7.1.7) by proving frequency-based dual M-framelets, we have

Theorem 7.1.8 *Let* M *be a* $d \times d$ *dilation matrix. Let* $\theta, a, b_1, \ldots, b_s \in l_0(\mathbb{Z}^d)$ *with* $\widehat{a}(0) = 1$. *Define* ϕ *and* $\eta, \psi^1, \ldots, \psi^s$ *as in* (7.1.13) *and* (7.1.14). *Then* $\{\eta; \psi^1, \ldots, \psi^s\}$ *is a tight* M*-framelet in* $L_2(\mathbb{R}^d)$ *(that is, all the functions* $\eta, \psi^1, \ldots, \psi^s \in L_2(\mathbb{R}^d)$ *and* $\mathsf{AS}_0^{\mathsf{M}}(\{\eta; \psi^1, \ldots, \psi^s\})$ *is a normalized tight frame for* $L_2(\mathbb{R}^d)$*) if and only if* $\{a; b_1, \ldots, b_s\}_\Theta$ *is a tight* M*-framelet filter bank and* $\widehat{\Theta}(0) = 1$ *with* $\widehat{\Theta}(\xi) := |\widehat{\theta}(\xi)|^2$.

7.1.3 Framelets and Wavelets by Tensor Product and Projection Method

We now discuss how to construct multidimensional framelets and wavelets through tensor product with $\mathsf{M} = 2I_d$. For one-dimensional filters $u_1, \ldots, u_d \in l(\mathbb{Z})$, their d-dimensional tensor product filter $u_1 \otimes \cdots \otimes u_d$ is defined by

$$(u_1 \otimes \cdots \otimes u_d)(k_1, \ldots, k_d) := u_1(k_1) \cdots u_d(k_d), \qquad k_1, \ldots, k_d \in \mathbb{Z}. \qquad (7.1.17)$$

In particular, if $u_1 = \cdots = u_d = u$, then (7.1.17) is denoted by

$$\otimes^d u := u \otimes \cdots \otimes u \quad \text{with } d \text{ copies of } u. \qquad (7.1.18)$$

For one-dimensional functions $f_1, \ldots, f_d : \mathbb{R} \to \mathbb{C}$, the d-dimensional function $f_1 \otimes \cdots \otimes f_d$ is defined by

$$(f_1 \otimes \cdots \otimes f_d)(x_1, \ldots, x_d) := f_1(x_1) \cdots f_d(x_d), \qquad x_1, \ldots, x_d \in \mathbb{R}.$$

Also $\otimes^d f := f \otimes \cdots \otimes f$ with d copies of f. For two sets S and T of one-dimensional filters or functions, $S \otimes T := \{s \otimes t \,:\, s \in S, t \in T\}$.

Let $\{a^j; b^j_1, \ldots, b^j_{s_j}\}$ and $\{\tilde{a}^j; \tilde{b}^j_1, \ldots, \tilde{b}^j_{s_j}\}$ be one-dimensional filter banks for $j = 1, \ldots, d$. Let $\{\phi^j; \psi^j_1, \ldots, \psi^j_{s_j}\}$ and $\{\tilde{\phi}^j; \tilde{\psi}^j_1, \ldots, \tilde{\psi}^j_{s_j}\}$ be sets of one-dimensional functions for $j = 1, \ldots, d$. Then the following statements can be easily checked using their definitions.

(1) If $(\{\tilde{a}^j; \tilde{b}^j_1, \ldots, \tilde{b}^j_{s_j}\}, \{a^j; b^j_1, \ldots, b^j_{s_j}\})$ is a dual framelet filter bank (i.e. a dual 2-framelet filter bank) for all $j = 1, \ldots, d$, then

$$\left(\{\tilde{a}^1; \tilde{b}^1_1, \ldots, \tilde{b}^1_{s_1}\} \otimes \cdots \otimes \{\tilde{a}^d; \tilde{b}^d_1, \ldots, \tilde{b}^d_{s_d}\}, \{a^1; b^1_1, \ldots, b^1_{s_1}\} \otimes \cdots \otimes \{a^d; b^d_1, \ldots, b^d_{s_d}\}\right)$$

is a dual $2I_d$-framelet filter bank.

(2) If $\{a^j; b^j_1, \ldots, b^j_{s_j}\}$ is a tight 2-framelet filter bank for all $j = 1, \ldots, d$, then $\{a^1; b^1_1, \ldots, b^1_{s_1}\} \otimes \cdots \otimes \{a^d; b^d_1, \ldots, b^d_{s_d}\}$ is a tight $2I_d$-framelet filter bank.

(3) If $(\{\tilde{\phi}^j; \tilde{\psi}^j_1, \ldots, \tilde{\psi}^j_{s_j}\}, \{\phi^j; \psi^j_1, \ldots, \psi^j_{s_j}\})$ is a dual framelet in $L_2(\mathbb{R})$ for all $j = 1, \ldots, d$, then

$$\begin{aligned}\Big(&\{\tilde{\phi}^1; \tilde{\psi}^1_1, \ldots, \tilde{\psi}^1_{s_1}\} \otimes \cdots \otimes \{\tilde{\phi}^d; \tilde{\psi}^d_1, \ldots, \tilde{\psi}^d_{s_d}\},\\ &\{\phi^1; \psi^1_1, \ldots, \psi^1_{s_1}\} \otimes \cdots \otimes \{\phi^d; \psi^d_1, \ldots, \psi^d_{s_d}\}\Big)\end{aligned} \tag{7.1.19}$$

is a dual $2I_d$-framelet in $L_2(\mathbb{R}^d)$.

(4) If $\{\phi^j; \psi^j_1, \ldots, \psi^j_{s_j}\}$ is a tight framelet in $L_2(\mathbb{R})$ for all $j = 1, \ldots, d$, then the tensor product $\{\phi^1; \psi^1_1, \ldots, \psi^1_{s_1}\} \otimes \cdots \otimes \{\phi^d; \psi^d_1, \ldots, \psi^d_{s_d}\}$ is a tight $2I_d$-framelet in $L_2(\mathbb{R}^d)$.

(5) If $(\{\tilde{\phi}^j; \tilde{\psi}^j_1, \ldots, \tilde{\psi}^j_{s_j}\}, \{\phi^j; \psi^j_1, \ldots, \psi^j_{s_j}\})$ is a pair of biorthogonal wavelets in $L_2(\mathbb{R})$ for all $j = 1, \ldots, d$, then the pair in (7.1.19) is a pair of biorthogonal $2I_d$-wavelets in $L_2(\mathbb{R}^d)$.

(6) If $\{\phi^j; \psi^j_1, \ldots, \psi^j_{s_j}\}$ is an orthogonal wavelet in $L_2(\mathbb{R})$ for all $j = 1, \ldots, d$, then $\{\phi^1; \psi^1_1, \ldots, \psi^1_{s_1}\} \otimes \cdots \otimes \{\phi^d; \psi^d_1, \ldots, \psi^d_{s_d}\}$ is an orthogonal $2I_d$-wavelet in $L_2(\mathbb{R}^d)$.

Multidimensional framelets can be also obtained through the projection method. Let P be a $d \times n$ integer matrix. For $u \in l_0(\mathbb{Z}^n)$ and a function $f \in L_1(\mathbb{R}^n)$ so that $\widehat{f}$ is continuous, we define the *projected filter* $Pu \in l_0(\mathbb{Z}^d)$ and the *projected function* Pf on $\mathbb{R}^d$ by

$$\widehat{Pu}(\xi) := \widehat{u}(P^{\mathsf{T}}\xi) \quad \text{and} \quad \widehat{Pf}(\xi) := \widehat{f}(P^{\mathsf{T}}\xi), \qquad \xi \in \mathbb{R}^d. \tag{7.1.20}$$

Recall that $B_1 = \chi_{(0,1]}$ is the B-spline of order 1 and $a^B_1 = \{\frac{1}{2}, \frac{1}{2}\}_{[0,1]}$ is its refinement filter satisfying $\widehat{B_1}(2\xi) = \widehat{a^B_1}(\xi)\widehat{B_1}(\xi)$. In particular, the projected function $B_P := P(\otimes^n B_1)$ as defined in (7.1.20) is a piecewise polynomial called a *box spline* function with a direction matrix P. Note that the box spline function

$P(\otimes^n B_1)$ is $2I_d$-refinable with the filter $a_P := P(\otimes^n a_1^B)$, since $\widehat{B_P}(2\xi) = \widehat{a_P}(\xi)\widehat{B_P}(\xi)$ by observing

$$\widehat{a_P}(\xi) := \prod_{k\in P} \frac{1+e^{-ik\cdot\xi}}{2} \quad \text{and} \quad \widehat{B_P}(\xi) = \prod_{k\in P} \frac{1-e^{-ik\cdot\xi}}{ik\cdot\xi}, \qquad \xi \in \mathbb{R}^d, \qquad (7.1.21)$$

where $k \in P$ means that k is a column vector of P and k goes through all the columns of P once and only once. Note that $a_m^B = a_P$ and $B_m = B_P$ with $P = [1, \dots, 1]$ having m copies of ones. Some two-dimensional box splines are given by the following direction matrices:

$$P_H = \begin{bmatrix} 1 & 0 & 1 \\ 0 & 1 & 1 \end{bmatrix}, \quad P_{ZP} = \begin{bmatrix} 1 & 0 & 1 & -1 \\ 0 & 1 & 1 & 1 \end{bmatrix}, \quad P_{HH} = \begin{bmatrix} 1 & 1 & 0 & 0 & 1 & 1 \\ 0 & 0 & 1 & 1 & 1 & 1 \end{bmatrix}. \qquad (7.1.22)$$

The box spline function B_{P_H} is the hat function (or *Courant element*) taking value one at $(1,1)$ and zeros at every point in $\mathbb{Z}^2\backslash\{(1,1)\}$; $B_{P_{ZP}}$ is called the *Zwart-Powell element*; $B_{P_{HH}} = B_{P_H} * B_{P_H}$ the convolution of B_{P_H} with itself.

The following result can be directly verified.

Theorem 7.1.9 *Let filters $\Theta, a, b_1, \dots, b_s, \tilde{a}, \tilde{b}_1, \dots, \tilde{b}_s \in l_0(\mathbb{Z}^n)$ be finitely supported filters such that $(\{\tilde{a}; \tilde{b}_1, \dots, \tilde{b}_s\}, \{a; b_1, \dots, b_s\})_\Theta$ is a dual $2I_n$-framelet filter bank. Let P be a $d \times n$ integer matrix such that*

$$P^\mathsf{T}(\mathbb{Z}^d\backslash[2\mathbb{Z}^d]) \subset \mathbb{Z}^n\backslash[2\mathbb{Z}^n]. \qquad (7.1.23)$$

Then $(\{P\tilde{a}; P\tilde{b}_1, \dots, P\tilde{b}_s\}, \{Pa; Pb_1, \dots, Pb_s\})_{P\Theta}$ is a dual $2I_d$-framelet filter bank. In particular, $P(\otimes^n\{a_1^B; b_1^B\})$ is a tight $2I_d$-framelet filter bank derived from the low-pass box spline filter a_P, where $\{a_1^B; b_1^B\}$ is the Haar orthogonal wavelet filter bank with $\widehat{b_1^B}(\xi) := (1-e^{-i\xi})/2$.

Proof Since $(\{\tilde{a}; \tilde{b}_1, \dots, \tilde{b}_s\}, \{a; b_1, \dots, b_s\})_\Theta$ is a dual $2I_n$-framelet filter bank, by definition we have

$$\widehat{\Theta}(2\zeta)\widehat{\tilde{a}}(\zeta)\overline{\widehat{a}(\zeta+2\pi\eta)} + \sum_{\ell=1}^{s} \widehat{\tilde{b}_\ell}(\zeta)\overline{\widehat{b_\ell}(\zeta+2\pi\eta)} = \widehat{\Theta}(\zeta)\boldsymbol{\delta}(\eta), \quad \forall\, \zeta \in \mathbb{R}^n$$

for all $\eta \in \Omega_{2I_n}$. For $\xi \in \mathbb{R}^d$, the above identity particularly holds with $\zeta = P^\mathsf{T}\xi$. By our assumption in (7.1.23), we see that $P : [2^{-1}\mathbb{Z}^d]/\mathbb{Z}^d \to [2^{-1}\mathbb{Z}^n]/\mathbb{Z}^n$ with $k \mapsto P^\mathsf{T}k$ is injective. Consequently, for $\omega \in \Omega_\mathsf{M}$, the above identity implies

$$\widehat{\Theta}(2P^\mathsf{T}\xi)\widehat{\tilde{a}}(P^\mathsf{T}\xi)\overline{\widehat{a}(P^\mathsf{T}\xi + P^\mathsf{T}2\pi\omega)} + \sum_{\ell=1}^{s} \widehat{\tilde{b}_\ell}(P^\mathsf{T}\xi)\overline{\widehat{b_\ell}(P^\mathsf{T}\xi + P^\mathsf{T}2\pi\omega)} = \widehat{\Theta}(P^\mathsf{T}\xi)\boldsymbol{\delta}(\omega)$$

for all $\xi \in \mathbb{R}^d$ and $\omega \in \Omega_{\mathsf{M}}$. That is, we proved

$$\widehat{P\Theta}(2\xi)\widehat{P\tilde{a}}(\xi)\overline{\widehat{Pa}(\xi+2\pi\omega)}+\sum_{\ell=1}^{s}\widehat{P\tilde{b}_\ell}(\xi)\overline{\widehat{Pb_\ell}(\xi+2\pi\omega)}=\widehat{P\Theta}(\xi)\boldsymbol{\delta}(\omega), \quad \forall\, \xi \in \mathbb{R}^d$$

for all $\omega \in \Omega_{\mathsf{M}}$. Hence, we proved that $(\{P\tilde{a}; P\tilde{b}_1, \ldots, P\tilde{b}_s\}, \{Pa; Pb_1, \ldots, Pb_s\})_{P\Theta}$ is a dual $2I_d$-framelet filter bank. □

Example 7.1.1 Let P_H be given in (7.1.22). Applying Theorem 7.1.9 with $P = P_H$, we see that $\{a_{P_H}; u_1, \ldots, u_7\}$ is a tight $2I_2$-framelet filter bank, where

$$a_{P_H} = \frac{1}{8}\begin{bmatrix} 0 & 1 & 1 \\ 1 & 2 & 1 \\ 1 & 1 & 0 \end{bmatrix}, \qquad u_1 = \frac{1}{8}\begin{bmatrix} 0 & 1 & -1 \\ 1 & 0 & -1 \\ 1 & -1 & 0 \end{bmatrix}, \quad u_2 = \frac{1}{8}\begin{bmatrix} 0 & -1 & -1 \\ -1 & 0 & 1 \\ 1 & 1 & 0 \end{bmatrix},$$

$$u_3 = \frac{1}{8}\begin{bmatrix} 0 & -1 & 1 \\ -1 & 2 & -1 \\ 1 & -1 & 0 \end{bmatrix}, \quad u_4 = \frac{1}{8}\begin{bmatrix} 0 & -1 & -1 \\ 1 & 0 & -1 \\ 1 & 1 & 0 \end{bmatrix}, \quad u_5 = \frac{1}{8}\begin{bmatrix} 0 & -1 & 1 \\ 1 & -2 & 1 \\ 1 & -1 & 0 \end{bmatrix},$$

$$u_6 = \frac{1}{8}\begin{bmatrix} 0 & 1 & -1 \\ -1 & -2 & -1 \\ 1 & 1 & 0 \end{bmatrix}, \quad u_7 = \frac{1}{8}\begin{bmatrix} 0 & 1 & -1 \\ -1 & 0 & 1 \\ 1 & -1 & 0 \end{bmatrix}$$

with all filters supported on $[0,2]^2$. By Theorem 7.1.8, $\{B_{P_H}; \psi^1, \ldots, \psi^7\}$ is a tight $2I_2$-framelet in $L_2(\mathbb{R}^2)$, where $\widehat{\psi^\ell}(\xi) := \widehat{u_\ell}(\xi/2)\widehat{B_{P_H}}(\xi/2), \xi \in \mathbb{R}^2$ for $\ell = 1, \ldots, 7$.

7.2 Multidimensional Cascade Algorithms and Refinable Functions

In this section, we first investigate convergence of a cascade algorithm associated with a multidimensional scalar filter. Then we study multidimensional scalar refinable functions, their smoothness property, and biorthogonal wavelets. For their corresponding results in dimension one with $d = 1$ and $\mathsf{M} = 2$, see Sects. 5.6.4 and 5.8.4.

7.2.1 *Convergence of Cascade Algorithms in Sobolev Spaces*

We now generalize convergence of cascade algorithms in Sect. 5.6 from dimension one to every dimension but only for a scalar filter (that is, $r = 1$ in Sect. 5.6).

For $u \in l(\mathbb{Z}^d)$ and $f : \mathbb{R}^d \to \mathbb{C}$, define

$$\nabla_k u := (\mathrm{Id} - [[k, 0]])u = u - u(\cdot - k), \qquad k \in \mathbb{Z}^d,$$
$$\nabla_t f := (\mathrm{Id} - [[t, 0]])f = f - f(\cdot - t), \qquad t \in \mathbb{R}^d.$$

For $1 \leqslant j \leqslant d$, by ∂_j we denote the partial derivative with respect to the jth coordinate of $\mathbb{R}^d$. In particular, $\partial := (\partial_1, \dots, \partial_d)$. For $\mu = (\mu_1, \dots, \mu_d)^\mathsf{T} \in \mathbb{N}_0^d := (\mathbb{N} \cup \{0\})^d$, we define $|\mu| := \mu_1 + \cdots + \mu_d$, $\partial^\mu := \partial_1^{\mu_1} \cdots \partial_d^{\mu_d}$, and $\nabla^\mu := \nabla_{e_1}^{\mu_1} \cdots \nabla_{e_d}^{\mu_d}$, where $e_j = (0, \dots, 0, 1, 0, \dots, 0)^\mathsf{T} \in \mathbb{R}^d$ has its only nonzero entry 1 at the jth coordinate.

The following result is convenient for dealing with partial derivatives.

Lemma 7.2.1 *Let* M *be a* $d \times d$ *real-valued matrix and* f *be an* $n_1 \times n_2$ *matrix of smooth functions on* $\mathbb{R}^d$. *For an* $n_0 \times n_1$ *constant matrix* B *and an* $n_2 \times n_3$ *constant matrix* C,

$$[\otimes^m \partial] \otimes [Bf(\mathsf{M}\cdot)C](\cdot) = B([\otimes^m \partial] \otimes f)(\mathsf{M}\cdot)([\otimes^m \mathsf{M}] \otimes C),$$

where $\otimes$ *denotes the (right) Kronecker product on vectors and matrices (see Sect. 5.8.3 for definition).*

Proof Note that $\partial \otimes (F(\mathsf{M}\cdot)) = ((\partial \mathsf{M}) \otimes F)(\mathsf{M}\cdot)$ by

$$\begin{aligned}[\partial \otimes (F(\mathsf{M}\cdot))]_{1,i;j,k} &= [\partial]_{1,i}(F_{j,k}(\mathsf{M}\cdot)) = \sum_{\ell=1}^{d} (\partial_\ell F_{j,k})(\mathsf{M}\cdot)\mathsf{M}_{\ell,i} \\ &= ([\partial \mathsf{M}]_{1,i} F_{j,k})(\mathsf{M}\cdot) = [(\partial \mathsf{M}) \otimes F]_{1,i;j,k}(\mathsf{M}\cdot).\end{aligned}$$

By induction on m we have

$$\begin{aligned}[\otimes^m \partial] \otimes [Bf(\mathsf{M}\cdot)C](\cdot) &= ([\otimes^m (\partial \mathsf{M})] \otimes [BfC])(\mathsf{M}\cdot) = B[\otimes^m (\partial \mathsf{M}) \otimes (fC)](\mathsf{M}\cdot) \\ &= B([\otimes^m \partial] \otimes f)(\mathsf{M}\cdot)([\otimes^m \mathsf{M}] \otimes C).\end{aligned}$$

This completes the proof. □

For $a \in l_0(\mathbb{Z}^d)$, we say that a has *order* m *sum rules with respect to* M if

$$\partial^\mu \widehat{a}(2\pi\omega) = 0 \qquad \forall\, \omega \in \Omega_{\mathsf{M}} \backslash \{0\},\ \mu \in \mathbb{N}_0^d \text{ with } |\mu| < m. \tag{7.2.1}$$

We define $\mathrm{sr}(a, \mathsf{M}) := m$ with m being the largest such integer.

Let $1 \leqslant p \leqslant \infty$ and $m := \mathrm{sr}(a, \mathsf{M})$. Similar to (5.6.56), we define

$$\mathrm{sm}_p(a, \mathsf{M}) := \tfrac{d}{p} - \log_{\rho(\mathsf{M})} \rho_m(a, \mathsf{M})_p, \tag{7.2.2}$$

where $\rho(\mathsf{M})$ is the spectral radius of M (i.e., the modulus of the largest eigenvalue of M) and

$$\rho_m(a,\mathsf{M})_p := \sup\left\{\limsup_{n\to\infty} \|\nabla^\mu \mathcal{S}^n_{a,\mathsf{M}}\boldsymbol{\delta}\|^{1/n}_{l_p(\mathbb{Z}^d)} \;:\; \mu \in \mathbb{N}_0^d, |\mu| = m\right\}. \qquad (7.2.3)$$

We shall address the calculation of $\mathrm{sm}_2(a,\mathsf{M})$ and $\mathrm{sm}_\infty(a,\mathsf{M})$ in Sect. 7.2.3. Recall that the refinement operator $\mathcal{R}_{a,\mathsf{M}}$ in (5.6.1) associated with a filter a and a $d\times d$ matrix M is

$$\mathcal{R}_{a,\mathsf{M}}f := |\det(\mathsf{M})| \sum_{k\in\mathbb{Z}^d} a(k)f(\mathsf{M}\cdot - k), \qquad (7.2.4)$$

where f is a function on $\mathbb{R}^d$. Then $\widehat{\mathcal{R}_{a,\mathsf{M}}f}(\mathsf{M}^{\mathsf{T}}\xi) = \widehat{a}(\xi)\widehat{f}(\xi)$ and by induction

$$\mathcal{R}^n_{a,\mathsf{M}}f = \sum_{k\in\mathbb{Z}^d} [\mathcal{S}^n_{a,\mathsf{M}}\boldsymbol{\delta}](k)f(\mathsf{M}^n\cdot - k), \qquad n\in\mathbb{N}. \qquad (7.2.5)$$

We study the convergence of a cascade algorithm $\{\mathcal{R}^n_{a,\mathsf{M}}f\}_{n=1}^\infty$ in *Sobolev spaces* $W^m_p(\mathbb{R}^d)$. For $1\leqslant p\leqslant\infty$ and $m\in\mathbb{N}_0$, $W^m_p(\mathbb{R}^d)$ consists of all distributions f on $\mathbb{R}^d$ such that $\partial^\mu f\in L_p(\mathbb{R}^d)$ for all $\mu\in\mathbb{N}_0^d$ with $|\mu|\leqslant m$. Moreover,

$$\|f\|_{W^m_p(\mathbb{R}^d)} := \sum_{\mu\in\mathbb{N}_0^d, |\mu|\leqslant m} \|\partial^\mu f\|_{L_p(\mathbb{R}^d)} < \infty.$$

For $m\in\mathbb{N}_0$, define

$$\mathscr{V}_m := \{v\in l_0(\mathbb{Z}^d) \;:\; \widehat{v}(\xi) = \mathscr{O}(\|\xi\|^{m+1}), \xi\to 0\}.$$

One can check that $\mathscr{V}_m = \{v\in l_0(\mathbb{Z}^d) \;:\; v*\mathsf{p} = 0\,\forall\,\mathsf{p}\in\mathbb{P}_m\}$, where $\mathbb{P}_m$ denotes the set of all d-variate polynomials of total degree at most m and $v*\mathsf{p} := \sum_{k\in\mathbb{Z}^d} v(k)\mathsf{p}(\cdot - k)$.

The following result on the structure of the space $\mathscr{V}_m$ is useful in our study of multidimensional cascade algorithms.

Lemma 7.2.2 *For $v\in l_0(\mathbb{Z}^d)$, $v\in\mathscr{V}_m$ if and only if $v = \sum_{\mu\in\mathbb{N}_0^d, |\mu|=m+1} \nabla^\mu v_\mu$ for some $v_\mu\in l_0(\mathbb{Z}^d)$.*

Proof The sufficiency part is obvious since $\nabla^\mu v_\mu = (\nabla^\mu\boldsymbol{\delta}) * v_\mu$ and $|\mu| = m+1$.

Necessity ($\Rightarrow$). Using a similar idea as the long division, we can always write $v = u + \sum_{|\mu|=m+1} (\nabla^\mu\boldsymbol{\delta}) * v_\mu$ with $v_\mu\in l_0(\mathbb{Z}^d)$ and u is supported inside $\{\alpha\in\mathbb{N}_0^d \;:\; |\alpha| < m\}$. That is, using $(\nabla^\mu\boldsymbol{\delta})(\cdot - k)$ with $k\in\mathbb{Z}^d$ and $|\mu| = m+1$, we can push the support of $u = v - \sum_{|\mu|=m+1}(\nabla^\mu\boldsymbol{\delta}) * v_\mu$ into $\{\alpha\in\mathbb{N}_0^d \;:\; |\alpha| < m\}$ by a natural choice of v_μ. Then $\mathsf{p} * u = \mathsf{p} * v = 0$ for all $\mathsf{p}\in\mathbb{P}_m$ by $v\in\mathscr{V}_m$, which forces $u = 0$. ☐

The following result will be needed in our study of convergence of multidimensional cascade algorithms.

Lemma 7.2.3 *Let η be a compactly supported function in $L_p(\mathbb{R}^d)$ with $1 \leqslant p \leqslant \infty$. Then the following statements are equivalent:*

(1) η satisfies

$$\partial^\mu \widehat{\eta}(2\pi k) = 0, \qquad \forall\, k \in \mathbb{Z}^d,\, \mu \in \mathbb{N}_0^d \quad \text{with} \quad |\mu| \leqslant m \tag{7.2.6}$$

(2) $\sum_{k\in\mathbb{Z}^d} \mathsf{p}(k)\eta(\cdot - k) = 0$ for all $\mathsf{p} \in \mathbb{P}_m$.
(3) $\eta = \sum_{\mu\in\mathbb{N}_0^d, |\mu|=m+1} \nabla^\mu \eta_u$ for some compactly supported functions $\eta_\mu \in L_p(\mathbb{R}^d)$.

Proof The equivalence between items (1) and (2) is a consequence of the Poisson summation formula as we have done in Theorem 5.5.1. Item (3)$\Longrightarrow$(2) is trivial. We now prove (2)$\Longrightarrow$(3). Since η has compact support, the set $\{\eta(\cdot - k)\chi_{[0,1)^d} : k \in \mathbb{Z}^d\}$ is finite. So, pick up a basis $\eta_1, \dots, \eta_s$ from this set for the linear space generated by this set. Then η can be uniquely written as $\eta = \sum_{j=1}^s v_j * \eta_j$ for some uniquely determined $v_1, \dots, v_s \in l_0(\mathbb{Z}^d)$. If item (2) holds, then we must have $\mathsf{p} * v_j = 0$ for all $j = 1, \dots, s$ and $\mathsf{p} \in \mathbb{P}_m$. Now the claim follows from Lemma 7.2.2. □

The following result characterizes convergence of a cascade algorithm in a Sobolev space $W_p^m(\mathbb{R}^d)$ and generalizes Theorem 5.6.16 from $\mathsf{M} = 2$ to a general dilation matrix M.

Theorem 7.2.4 *Let $1 \leqslant p \leqslant \infty$, $m \in \mathbb{N}_0$ and $a \in l_0(\mathbb{Z}^d)$ with $\widehat{a}(0) = 1$. Let M be a $d \times d$ dilation matrix. We further assume that M is isotropic if $m > 0$. Let ϕ be the standard M-refinable function in (7.1.13) associated with the filter a. Then the following statements are equivalent:*

(1) The cascade algorithm associated with the filter a converges in $W_p^m(\mathbb{R})$, that is, $\{\mathcal{R}_{a,\mathsf{M}}^n f\}_{n=1}^\infty$ is a Cauchy sequence in $W_p^m(\mathbb{R}^d)$ for every compactly supported initial function $f \in W_p^m(\mathbb{R}^d)$ satisfying

$$\widehat{f}(0) = 1 \quad \text{and} \quad \partial^\mu \widehat{f}(2\pi k) = 0, \quad \forall\, k \in \mathbb{Z}^d\backslash\{0\}, \mu \in \mathbb{N}_0^d \text{ with } |\mu| \leqslant m. \tag{7.2.7}$$

In fact, $\phi \in W_p^m(\mathbb{R}^d)$ ($\phi \in \mathscr{C}^m(\mathbb{R}^d)$ if $p = \infty$) and $\lim_{n\to\infty} \|\mathcal{R}_{a,\mathsf{M}}^n f - \phi\|_{W_p^m(\mathbb{R}^d)} = 0$.

(2) For some compactly supported function $f \in W_p^m(\mathbb{R}^d)$ (require $f \in \mathscr{C}^m(\mathbb{R}^d)$ if $p = \infty$) which satisfies (7.2.7) and the integer shifts of f are stable in $L_p(\mathbb{R}^d)$, $\{\mathcal{R}_{a,\mathsf{M}}^n f\}_{n=1}^\infty$ is a Cauchy sequence in $W_p^m(\mathbb{R}^d)$.

(3) $\lim_{n\to\infty} |\det(\mathsf{M})|^{(m/d-1/p)n} \|\nabla^\mu \mathcal{S}_{a,\mathsf{M}}^n \delta\|_{l_p(\mathbb{Z}^d)}^{1/n} = 0$ for all $\mu \in \mathbb{N}_0^d$ with $|\mu| = m + 1$.

(4) $\operatorname{jsr}_p(\{\mathcal{T}_\gamma|_V : \gamma \in \Gamma_{\mathsf{M}}\}) < |\det(\mathsf{M})|^{1/p-m/d}$, *where*

$$[\mathcal{T}_\gamma v](j) := |\det(\mathsf{M})| \sum_{k\in\mathbb{Z}^d} v(k)\overline{a(\gamma + k - \mathsf{M}j)}, \qquad j \in \mathbb{Z}^d, \gamma \in \Gamma_{\mathsf{M}}$$

and V is a finite dimensional subspace of $\mathscr{V}_m$ *such that* $\mathcal{T}_\gamma V \subseteq V$ *for all* $\gamma \in \Gamma_{\mathsf{M}}$ *and* $\operatorname{span}\{v(\cdot - k) : v \in V, k \in \mathbb{Z}^d\} = \mathscr{V}_m$.

(5) $\rho_{m+1}(a, \mathsf{M})_p < |\det(\mathsf{M})|^{1/p-m/d}$.

(6) $\operatorname{sm}_p(a, \mathsf{M}) > m$.

Proof Define $f = \otimes^d B_{m+2}$, where B_{m+2} is the B-spline function of order $m + 2$. Then f is a desired function in item (2) since $f \in \mathscr{C}^m(\mathbb{R}^d)$ satisfies (7.2.7) and the integer shifts of f are stable in $L_p(\mathbb{R}^d)$. Trivially, (1)$\Longrightarrow$(2).

Suppose that item (2) holds. Define $f_n := \mathcal{R}^n_{a,\mathsf{M}} f$ for $n \in \mathbb{N}$. Then there exists a compactly supported function $f_\infty \in W^m_p(\mathbb{R}^d)$ ($f_\infty \in \mathscr{C}^m(\mathbb{R}^d)$ for $p = \infty$) such that $\lim_{n\to\infty} \|f_n - f_\infty\|_{W^m_p(\mathbb{R}^d)} = 0$. For $\mu = (\mu_1, \dots, \mu_d)^\mathsf{T} \in \mathbb{N}^d_0$, we deduce from (7.2.5) that

$$\sum_{k\in\mathbb{Z}^d} [\nabla^\mu \mathcal{S}^n_{a,\mathsf{M}} \delta](k) f(\mathsf{M}^n \cdot -k) = \nabla^{\mu,n} f_n \quad \text{with} \quad \nabla^{\mu,n} := \nabla^{\mu_1}_{\mathsf{M}^{-n}e_1} \cdots \nabla^{\mu_d}_{\mathsf{M}^{-n}e_d}.$$

Since the integer shifts of f are stable in $L_p(\mathbb{R}^d)$, there exists $C_f > 0$ depending only on f such that

$$\begin{aligned} |\det(\mathsf{M})|^{-n/p} \|\nabla^\mu \mathcal{S}^n_{a,\mathsf{M}} \delta\|_{l_p(\mathbb{Z}^d)} &\leqslant C_f \|\nabla^{\mu,n} f_n\|_{L_p(\mathbb{R}^d)} \\ &\leqslant C_f \|\nabla^{\mu,n}(f_n - f_\infty)\|_{L_p(\mathbb{R}^d)} + C_f \|\nabla^{\mu,n} f_\infty\|_{L_p(\mathbb{R}^d)}. \end{aligned}$$

Since M is isotropic when $m > 0$, there exists a positive constant C such that $\|\nabla^{\mu,n}(f_n - f_\infty)\|_{L_p(\mathbb{R}^d)} \leqslant C|\det(\mathsf{M})|^{-mn/d} \|f_n - f_\infty\|_{W^m_p(\mathbb{R}^d)}$ for every $|\mu| = m + 1$. Since $f_\infty \in W^m_p(\mathbb{R}^d)$ ($f_\infty \in \mathscr{C}^m(\mathbb{R}^d)$ for $p = \infty$) and $|\mu| = m + 1$, we must have

$$\lim_{n\to\infty} |\det(\mathsf{M})|^{mn/d} \|\nabla^{\mu,n} f_\infty\|_{L_p(\mathbb{R}^d)} = 0.$$

Combining the above three inequalities, we conclude that

$$\lim_{n\to\infty} |\det(\mathsf{M})|^{(m/d-1/p)n} \|\nabla^\mu \mathcal{S}^n_{a,\mathsf{M}} \delta\|_{l_p(\mathbb{Z}^d)} = 0, \quad \forall\, \mu \in \mathbb{N}^d_0 \quad \text{with} \quad |\mu| = m + 1.$$

Thus, (2)$\Longrightarrow$(3).

Similar to Proposition 5.6.9 and Theorem 5.6.11, using the tool of joint spectral radius, we conclude that (3) $\Longleftrightarrow$ (4) $\Longleftrightarrow$ (5) $\Longleftrightarrow$ (6). We now prove (5)$\Longrightarrow$(1). Define $\eta := \mathcal{R}_{a,\mathsf{M}} f - f$. Then by (7.2.5)

$$f_{n+1} - f_n = \mathcal{R}^n_{a,\mathsf{M}} \eta = \sum_{k\in\mathbb{Z}^d} [\mathcal{S}^n_{a,\mathsf{M}} \delta](k) \eta(\mathsf{M}^n \cdot -k).$$

Therefore, by Lemma 7.2.1

$$[\otimes^m \partial] \otimes [f_{n+1} - f_n] = \sum_{k\in\mathbb{Z}^d} [\mathcal{S}^n_{a,\mathsf{M}}\boldsymbol{\delta}](k)([\otimes^m \partial] \otimes \eta)(\mathsf{M}^n \cdot -k)(\otimes^m \mathsf{M}^n).$$

Note that item (5) implies $\operatorname{sr}(a, \mathsf{M}) \geqslant m + 1$. By (7.2.7), we can directly check that (7.2.6) holds for every component of $[\otimes^m \partial] \otimes \eta$. Therefore, by Lemma 7.2.3, $[\otimes^m \partial] \otimes \eta = \sum_{\mu\in\mathbb{N}_0^d, |\mu|=m+1} \nabla^\mu \eta_u$ for some vectors η_μ of compactly supported functions in $L_p(\mathbb{R}^d)$. Therefore, we have

$$[\otimes^m \partial] \otimes [f_{n+1} - f_n] = \sum_{\mu\in\mathbb{N}_0^d, |\mu|=m+1} \sum_{k\in\mathbb{Z}^d} [\nabla^\mu \mathcal{S}^n_{a,\mathsf{M}}\boldsymbol{\delta}](k)\eta_\mu(\mathsf{M}^n \cdot -k)(\otimes^m \mathsf{M}^n).$$

By Proposition 7.1.5, there exists a positive constant C_1 depending only on M and the vectors η_μ of compactly supported functions in $L_p(\mathbb{R}^d)$ such that

$$\begin{aligned} &\|[\otimes^m \partial] \otimes [f_{n+1} - f_n]\|_{(L_p(\mathbb{R}^d))^{1\times m^d}} \\ &\leqslant C_1 \sum_{\mu\in\mathbb{N}_0^d, |\mu|=m+1} \|\nabla^\mu \mathcal{S}^n_{a,\mathsf{M}}\boldsymbol{\delta}\|_{l_p(\mathbb{Z}^d)} |\det(\mathsf{M})|^{(m/d-1/p)n}. \end{aligned} \tag{7.2.8}$$

By our assumption in item (5), there exist $0 < \rho < 1$ and $C > 0$ such that

$$\begin{aligned} &|\det(\mathsf{M})|^{(m/d-1/p)n} \|\nabla^\mu \mathcal{S}^n_{a,\mathsf{M}}\boldsymbol{\delta}\|_{l_p(\mathbb{Z}^d)} \leqslant C\rho^n, \\ &\qquad \forall\, n \in \mathbb{N}, \mu \in \mathbb{N}_0^d \quad \text{with} \quad |\mu| = m + 1. \end{aligned} \tag{7.2.9}$$

Combining the above inequality with (7.2.8), we conclude that

$$\|[\otimes^m \partial] \otimes [f_{n+1} - f_n]\|_{(L_p(\mathbb{R}^d))^{1\times m^d}} \leqslant C_1 C \rho^n, \qquad \forall\, n \in \mathbb{N}.$$

Since all f_n are compactly supported functions with all their supports contained inside a bounded set, by $0 < \rho < 1$, this shows that $\{f_n\}_{n=1}^\infty$ is a Cauchy sequence in $W_p^m(\mathbb{R}^d)$. This proves (5)$\Longrightarrow$(1). ☐

Let $K \subseteq \mathbb{R}^d$. We define $l(K) := \{v \in l_0(\mathbb{Z}^d) \;:\; \operatorname{supp}(v) \subseteq K\}$. An example of V in item (4) of Theorem 7.2.4 is $V = \mathscr{V}_m \cap l(\sum_{j=1}^\infty \mathsf{M}^{-j}(\Gamma_\mathsf{M} - \operatorname{supp}(a)))$.

7.2.2 *Analysis of Refinable Functions and Biorthogonal Wavelets*

We now apply Theorem 7.2.4 on convergence of cascade algorithms to study multidimensional scalar refinable functions and biorthogonal wavelets.

The following result characterizes the L_p smoothness exponent for a compactly supported refinable function having stable integer shifts in $L_p(\mathbb{R}^d)$.

Corollary 7.2.5 *Let $1 \leqslant p \leqslant \infty$, $m \in \mathbb{N}_0$ and $a \in l_0(\mathbb{Z}^d)$ with $\widehat{a}(0) = 1$. Let M be a $d \times d$ dilation matrix. We further assume that M is isotropic if $m > 0$. Let ϕ be the standard M-refinable function in (7.1.13) associated with the filter a. Suppose that the integer shifts of ϕ are stable in $L_p(\mathbb{R}^d)$. Then $\phi \in W_p^m(\mathbb{R}^d)$ ($\phi \in \mathscr{C}^m(\mathbb{R}^d)$ if $p = \infty$) if and only if $\operatorname{sm}_p(a, \mathsf{M}) > m$.*

Proof Sufficiency ($\Leftarrow$). If $\operatorname{sm}_p(a, \mathsf{M}) > m$, by Theorem 7.2.4, we must have $\phi \in W_p^m(\mathbb{R}^d)$ ($\phi \in \mathscr{C}^m(\mathbb{R}^d)$ if $p = \infty$).

Necessity ($\Rightarrow$). Suppose that $\phi \in W_p^m(\mathbb{R}^d)$ ($\phi \in \mathscr{C}^m(\mathbb{R}^d)$ if $p = \infty$). It is trivial that $\mathcal{R}_{a,\mathsf{M}}\phi = \phi$ and therefore, $\{\mathcal{R}_{a,\mathsf{M}}^n\phi\}_{n=1}^\infty$ trivially converges to ϕ in $W_p^m(\mathbb{R}^d)$. We now prove that (7.2.7) holds for $f = \phi$. Since $\widehat{\phi}(\xi) = \prod_{j=1}^\infty \widehat{a}((\mathsf{M}^\mathsf{T})^{-j}\xi)$ and $\widehat{a}(0) = 1$, we have $\widehat{\phi}(0) = 1$ and $\widehat{\phi}(\mathsf{M}^\mathsf{T}\xi) = \widehat{a}(\xi)\widehat{\phi}(\xi)$. Let $\upsilon \in l_0(\mathbb{Z}^d)$ such that $\widehat{\upsilon}(\xi) = 1/\widehat{\phi}(\xi) + \mathscr{O}(\|\xi\|^{m+1})$ as $\xi \to 0$. Then $\widehat{\upsilon}(\mathsf{M}^\mathsf{T}\xi)\widehat{a}(\xi) = \widehat{\upsilon}(\xi) + \mathscr{O}(\|\xi\|^{m+1})$ as $\xi \to 0$. Define $g := \upsilon * \phi$. For $k \in \mathbb{Z}^d$, we deduce that

$$\begin{aligned}\widehat{g}((\mathsf{M}^\mathsf{T})^n(\xi + 2\pi k)) &= \widehat{\upsilon}((\mathsf{M}^\mathsf{T})^n\xi)\widehat{a}((\mathsf{M}^\mathsf{T})^{n-1}\xi)\cdots\widehat{a}(\xi)\widehat{\phi}(\xi + 2\pi k)\\ &= \widehat{\upsilon}(\xi)\widehat{\phi}(\xi + 2\pi k) + \mathscr{O}(\|\xi\|^{m+1}) = \widehat{g}(\xi + 2\pi k) + \mathscr{O}(\|\xi\|^{m+1})\end{aligned}$$

as $\xi \to 0$. That is, we proved

$$\begin{aligned}[(\otimes^j\partial) \otimes \widehat{g}](2\pi k) &= [(\otimes^j\partial) \otimes (\widehat{g}((\mathsf{M}^\mathsf{T})^n\cdot))](2\pi k)\\ &= [(\otimes^j\partial) \otimes \widehat{g}](2\pi(\mathsf{M}^\mathsf{T})^n k)(\otimes^j(\mathsf{M}^\mathsf{T})^n),\end{aligned} \tag{7.2.10}$$

for all $j = 0, \dots, m$. Define $h(x) := (\otimes^j(-ix^\mathsf{T})) \otimes g(x)$. Then $\widehat{h} = (\otimes^j\partial) \otimes \widehat{g}$. Since $\phi \in W_p^m(\mathbb{R}^d)$ has compact support and $\upsilon \in l_0(\mathbb{Z}^d)$, the function $g \in W_p^m(\mathbb{R}^d)$ must have compact support. Therefore, every component in $(\otimes^m\partial)h$ belongs to $L_1(\mathbb{R}^d)$. Noting that $\widehat{(\otimes^m\partial)h}(\xi) = (\otimes^m(i\xi^\mathsf{T})) \otimes \widehat{h}(\xi)$. For $k \in \mathbb{Z}^d\backslash\{0\}$, by the Riemann-Lebesgue lemma and $\lim_{n\to\infty} \|(\mathsf{M}^\mathsf{T})^n k\| = \infty$ (since M is expansive), we have

$$\lim_{n\to\infty} (\otimes^m(i2\pi k^\mathsf{T}\mathsf{M}^n))\widehat{h}(2\pi(\mathsf{M}^\mathsf{T})^n k) = \lim_{n\to\infty} \widehat{(\otimes^m\partial)h}(2\pi(\mathsf{M}^\mathsf{T})^n k) = 0.$$

Since $\otimes^m(i2\pi k^\mathsf{T}\mathsf{M}^n) = (\otimes^m(i2\pi k^\mathsf{T})) \otimes (\otimes^m\mathsf{M}^n)$ and M is isotropic when $m > 0$, by $k \neq 0$ and Lemma 7.2.1, the above identity implies

$$\lim_{n\to\infty} |\det(\mathsf{M})|^{mn/d}\|\widehat{h}(2\pi(\mathsf{M}^\mathsf{T})^n k)\| = 0.$$

Now we deduce from (7.2.10) and Lemma 7.2.1 that

$$[(\otimes^j\partial) \otimes \widehat{g}](2\pi k) = \lim_{n\to\infty} \widehat{h}(2\pi(\mathsf{M}^\mathsf{T})^n k)(\otimes^j(\mathsf{M}^\mathsf{T})^n) = 0,$$

for all $j = 0, \ldots, m$ and $k \in \mathbb{Z}^d \setminus \{0\}$. That is, $\partial^\mu \widehat{g}(2\pi k) = 0$ for all $|\mu| \leqslant m$ and $k \in \mathbb{Z}^d \setminus \{0\}$. Since $\widehat{g}(\xi) = \widehat{v}(\xi)\widehat{\phi}(\xi)$ and $\widehat{v}(0) = 1 \neq 0$, we must have $\partial^\mu \widehat{\phi}(2\pi k) = 0$ for all $|\mu| \leqslant m$ and $k \in \mathbb{Z}^d \setminus \{0\}$. This proves (7.2.7) with $f = \phi$. Hence, item (2) of Theorem 7.2.4 holds. Now it follows from Theorem 7.2.4 that $\mathrm{sm}_p(a, \mathsf{M}) > m$. □

A function ϕ on $\mathbb{R}^d$ is *an interpolating function* if ϕ is continuous and $\phi(k) = \boldsymbol{\delta}(k)$ for all $k \in \mathbb{Z}^d$. Another consequence of Theorem 7.2.4 is the following result on interpolating refinable functions.

Corollary 7.2.6 *Let $m \in \mathbb{N}_0$ and M be a $d \times d$ dilation matrix. Let $a \in l_0(\mathbb{Z}^d)$ with $\widehat{a}(0) = 1$ and ϕ be the standard M-refinable function in* (7.1.13) *associated with the filter a. Then ϕ is an interpolating function if and only if $\mathrm{sm}_\infty(a, \mathsf{M}) > 0$ and a is an interpolatory M-wavelet filter: $a(\mathsf{M}k) = |\det(\mathsf{M})|^{-1}\boldsymbol{\delta}(k)$ for all $k \in \mathbb{Z}^d$.*

Proof Necessity ($\Rightarrow$). If ϕ is an interpolating function, then the integer shifts of ϕ are stable in $L_\infty(\mathbb{R}^d)$, since

$$\|v\|_{l_\infty(\mathbb{Z}^d)} \leqslant \Big\| \sum_{k\in\mathbb{Z}^d} v(k)\phi(\cdot - k) \Big\|_{L_\infty(\mathbb{R}^d)} \leqslant \|v\|_{l_\infty(\mathbb{Z}^d)} \Big\| \sum_{k\in\mathbb{Z}^d} |\phi(\cdot - k)| \Big\|_{L_\infty(\mathbb{R}^d)}$$

by $v(k) = (v * \phi)(k)$ for all $k \in \mathbb{Z}^d$ and $\sum_{k\in\mathbb{Z}^d} |\phi(\cdot - k)| \in L_\infty(\mathbb{R}^d)$. It follows from Corollary 7.2.5 that $\mathrm{sm}_\infty(a, \mathsf{M}) > 0$. Note that $\phi(\mathsf{M}^{-1}\cdot) = |\det(\mathsf{M})| \sum_{k\in\mathbb{Z}^d} a(k)\phi(\cdot - k)$ and $\phi|_{\mathbb{Z}^d} = \boldsymbol{\delta}$. We trivially have $|\det(\mathsf{M})|a(j) = \phi(\mathsf{M}^{-1}j)$ for all $j \in \mathbb{Z}^d$. Since ϕ is interpolating, we conclude that $a(\mathsf{M}k) = |\det(\mathsf{M})|^{-1}\boldsymbol{\delta}(k)$ for all $k \in \mathbb{Z}^d$.

Sufficiency ($\Leftarrow$). Let $f := \otimes^d(B_2(\cdot - 1))$ and $f_n := \mathcal{R}^n_{a,\mathsf{M}} f$. Then f is a compactly supported continuous function satisfying (7.2.7) with $m = 0$. Since $\mathrm{sm}_\infty(a, \mathsf{M}) > 0$, by Theorem 7.2.4, we have $\lim_{n\to\infty} \|f_n - \phi\|_{\mathscr{C}(\mathbb{R}^d)} = 0$ and hence ϕ is continuous. Since f is interpolating and a is an interpolatory M-wavelet filter, by induction on n, we see that all f_n are interpolating. In particular, we have $\phi(k) = \lim_{n\to\infty} f_n(k) = \boldsymbol{\delta}(k)$ for all $k \in \mathbb{Z}^d$. Hence, ϕ is interpolating. □

We now use Theorem 7.2.4 to characterize multidimensional wavelets in $L_2(\mathbb{R}^d)$.

Theorem 7.2.7 *Let M be a $d \times d$ dilation matrix. Let $a, \tilde{a} \in l_0(\mathbb{Z}^d)$ with $\widehat{a}(0) = \widehat{\tilde{a}}(0) = 1$ and $\phi, \tilde{\phi}$ be their standard M-refinable functions associated with a and $\tilde{a}$, respectively. Then $\phi, \tilde{\phi} \in L_2(\mathbb{R}^d)$ and $(\tilde{\phi}, \phi)$ is biorthogonal:*

$$\langle \tilde{\phi}, \phi(\cdot - k)\rangle = \int_{\mathbb{R}^d} \tilde{\phi}(x)\overline{\phi(x - k)}dx = \boldsymbol{\delta}(k), \qquad \forall\, k \in \mathbb{Z}^d \tag{7.2.11}$$

if and only if

(1) $(\tilde{a}, a)$ *is* a pair of biorthogonal M-wavelet filters, *that is,*

$$\sum_{\omega\in\Omega_{\mathsf{M}}} \widehat{\tilde{a}}(\xi + 2\pi\omega)\overline{\widehat{a}(\xi + 2\pi\omega)} = 1, \qquad \forall\, \xi \in \mathbb{R}^d. \tag{7.2.12}$$

(2) $\mathrm{sm}_2(a, \mathsf{M}) > 0$ *and* $\mathrm{sm}_2(\tilde{a}, \mathsf{M}) > 0$.

Proof Necessity (⇒). Since $\phi, \tilde{\phi} \in L_2(\mathbb{R}^d)$ have compact support, (7.2.11) implies that the integer shifts of ϕ (and $\tilde{\phi}$) are stable in $L_2(\mathbb{R}^d)$. By Corollary 7.2.5, item (2) holds. Note that (7.2.11) is equivalent to $[\widehat{\tilde{\phi}}, \widehat{\phi}](\xi) := \sum_{k\in\mathbb{Z}^d} \widehat{\tilde{\phi}}(\xi + 2\pi k)\overline{\widehat{\phi}(\xi + 2\pi k)} = 1$. By $\widehat{\phi}(\mathsf{M}^\mathsf{T}\xi) = \widehat{a}(\xi)\widehat{\phi}(\xi)$ and $\widehat{\tilde{\phi}}(\mathsf{M}^\mathsf{T}\xi) = \widehat{\tilde{a}}(\xi)\widehat{\tilde{\phi}}(\xi)$, we deduce that

$$
\begin{aligned}
1 = [\widehat{\tilde{\phi}}, \widehat{\phi}](\xi) &= \sum_{k\in\mathbb{Z}^d} \widehat{\tilde{\phi}}(\xi + 2\pi k)\overline{\widehat{\phi}(\xi + 2\pi k)} \\
&= \sum_{k\in\mathbb{Z}^d} \widehat{\tilde{a}}((\mathsf{M}^\mathsf{T})^{-1}(\xi + 2\pi k))\overline{\widehat{a}((\mathsf{M}^\mathsf{T})^{-1}(\xi + 2\pi k))} \\
&\qquad \times \widehat{\tilde{\phi}}((\mathsf{M}^\mathsf{T})^{-1}(\xi + 2\pi k))\overline{\widehat{\phi}((\mathsf{M}^\mathsf{T})^{-1}(\xi + 2\pi k))} \\
&= \sum_{\omega\in\Omega_\mathsf{M}} \widehat{\tilde{a}}((\mathsf{M}^\mathsf{T})^{-1}\xi + 2\pi\omega)\overline{\widehat{a}((\mathsf{M}^\mathsf{T})^{-1}\xi + 2\pi\omega)}[\widehat{\tilde{\phi}}, \widehat{\phi}]((\mathsf{M}^\mathsf{T})^{-1}\xi + 2\pi\omega) \\
&= \sum_{\omega\in\Omega_\mathsf{M}} \widehat{\tilde{a}}((\mathsf{M}^\mathsf{T})^{-1}\xi + 2\pi\omega)\overline{\widehat{a}((\mathsf{M}^\mathsf{T})^{-1}\xi + 2\pi\omega)}.
\end{aligned}
$$

This proves item (1).

Sufficiency (⇐). Let $f := \chi_{[0,1)^d}$. Then $f \in L_2(\mathbb{R}^d)$ satisfies (7.2.7) with $m = 0$. Define $f_n := \mathcal{R}^n_{a,\mathsf{M}} f$ and $\tilde{f}_n := \mathcal{R}^n_{\tilde{a},\mathsf{M}} f$, $n \in \mathbb{N}$. Since $\{f(\cdot - k) \ : \ k \in \mathbb{Z}^d\}$ is an orthonormal system in $L_2(\mathbb{R}^d)$, by item (1) and induction on n, we can check that $\langle \tilde{f}_n, f_n(\cdot - k)\rangle = \boldsymbol{\delta}(k)$ for all $k \in \mathbb{Z}^d$ and $n \in \mathbb{N}$ (i.e., $[\widehat{\tilde{f}_n}, \widehat{f_n}] = 1$). By item (2) and Theorem 7.2.4, we have $\phi, \tilde{\phi} \in L_2(\mathbb{R}^d)$, $\lim_{n\to\infty} \|f_n - \phi\|_{L_2(\mathbb{R}^d)} = 0$, and $\lim_{n\to\infty} \|\tilde{f}_n - \tilde{\phi}\|_{L_2(\mathbb{R}^d)} = 0$. Consequently, $\langle \tilde{\phi}, \phi(\cdot - k)\rangle = \lim_{n\to\infty}\langle \tilde{f}_n, f_n(\cdot - k)\rangle = \boldsymbol{\delta}(k)$ for all $k \in \mathbb{Z}^d$. This proves (7.2.11). □

Multidimensional biorthogonal M-wavelets in $L_2(\mathbb{R}^d)$ are characterized by the following results.

Theorem 7.2.8 *Let* M *be a* $d \times d$ *dilation matrix. Let* $a, \tilde{a} \in l_0(\mathbb{Z}^d)$ *with* $\widehat{a}(0) = \widehat{\tilde{a}}(0) = 1$ *and* $\phi, \tilde{\phi}$ *be their standard* M*-refinable functions associated with* a *and* $\tilde{a}$*, respectively. Let* $b_1, \ldots, b_s, \tilde{b}_1, \ldots, \tilde{b}_s \in l_0(\mathbb{Z}^d)$ *with* $s := |\det(\mathsf{M})| - 1$*. Define*

$$\widehat{\psi^\ell}(\mathsf{M}^\mathsf{T}\xi) := \widehat{b_\ell}(\xi)\widehat{\phi}(\xi), \quad \widehat{\tilde{\psi}^\ell}(\mathsf{M}^\mathsf{T}\xi) := \widehat{\tilde{b}_\ell}(\xi)\widehat{\tilde{\phi}}(\xi), \quad \ell = 1, \ldots, s.$$

Then $(\{\tilde{\phi}; \tilde{\psi}^1, \ldots, \tilde{\psi}^s\}, \{\phi; \psi^1, \ldots, \psi^s\})$ *is a biorthogonal* M*-wavelet in* $L_2(\mathbb{R}^d)$*, that is,*

$$\left(\mathsf{AS}_0^\mathsf{M}(\{\tilde{\phi}; \tilde{\psi}^1, \ldots, \tilde{\psi}^s\}), \mathsf{AS}_0^\mathsf{M}(\{\phi; \psi^1, \ldots, \psi^s\})\right)$$

is a pair of biorthogonal bases in $L_2(\mathbb{R}^d)$ *if and only if*

(1) $(\{\tilde{a}; \tilde{b}_1, \ldots, \tilde{b}_s\}, \{a; b_1, \ldots, b_s\})$ *is* a biorthogonal M-wavelet filter bank*:*

$$\begin{bmatrix} \widehat{\tilde{a}}(\xi + 2\pi\omega_0) & \widehat{\tilde{b}_1}(\xi + 2\pi\omega_0) & \cdots & \widehat{\tilde{b}_s}(\xi + 2\pi\omega_0) \\ \vdots & \vdots & \ddots & \vdots \\ \widehat{\tilde{a}}(\xi + 2\pi\omega_s) & \widehat{\tilde{b}_1}(\xi + 2\pi\omega_s) & \cdots & \widehat{\tilde{b}_s}(\xi + 2\pi\omega_s) \end{bmatrix} \overline{\begin{bmatrix} \widehat{a}(\xi + 2\pi\omega_0) & \widehat{b_1}(\xi + 2\pi\omega_0) & \cdots & \widehat{b_s}(\xi + 2\pi\omega_0) \\ \vdots & \vdots & \ddots & \vdots \\ \widehat{a}(\xi + 2\pi\omega_s) & \widehat{b_1}(\xi + 2\pi\omega_s) & \cdots & \widehat{b_s}(\xi + 2\pi\omega_s) \end{bmatrix}}^{\mathsf{T}} = I_{s+1}, \tag{7.2.13}$$

where $s := |\det(\mathsf{M})| - 1$ *and* $\{\omega_0, \ldots, \omega_s\} := \Omega_{\mathsf{M}}$.

(2) $\operatorname{sm}_2(a, \mathsf{M}) > 0$ *and* $\operatorname{sm}_2(\tilde{a}, \mathsf{M}) > 0$.

Proof Necessity ($\Rightarrow$). Suppose that $(\{\tilde{\phi}; \tilde{\psi}^1, \ldots, \tilde{\psi}^s\}, \{\phi; \psi^1, \ldots, \psi^s\})$ is a biorthogonal M-wavelet in $L_2(\mathbb{R}^d)$. Define $\psi^0 := \phi$, $\tilde{\psi}^0 := \tilde{\phi}$ and $b_0 := a, \tilde{b}_0 := \tilde{a}$. Then

$$\langle \tilde{\psi}^\ell, \psi^j(\cdot - k)\rangle = \boldsymbol{\delta}(\ell - j)\boldsymbol{\delta}(k) \qquad \forall\, \ell, j = 0, \ldots, s \quad \text{and} \quad k \in \mathbb{Z}^d. \tag{7.2.14}$$

By $\widehat{\tilde{\psi}^\ell}(\mathsf{M}^\mathsf{T}\xi) = \widehat{\tilde{b}_\ell}(\xi)\widehat{\tilde{\phi}}(\xi)$ and $\widehat{\psi^j}(\mathsf{M}^\mathsf{T}\xi) = \widehat{b_j}(\xi)\widehat{\phi}(\xi)$, similar to Theorem 7.2.7, we see that item (1) must hold. Note that (7.2.11) follows directly from (7.2.14). It follows from Theorem 7.2.7 that item (2) holds.

Sufficiency ($\Leftarrow$). Note that item (1) implies (7.2.12). By Theorem 7.2.7, we see that $\phi, \tilde{\phi} \in L_2(\mathbb{R}^d)$ and (7.2.11) holds. Thus, $\psi^1, \ldots, \psi^s, \tilde{\psi}^1, \ldots, \tilde{\psi}^s \in L_2(\mathbb{R}^d)$. By (7.2.11) and (7.2.13), we can directly check that both $\mathsf{AS}_0^{\mathsf{M}}(\{\tilde{\phi}; \tilde{\psi}^1, \ldots, \tilde{\psi}^s\})$ and $\mathsf{AS}_0^{\mathsf{M}}(\{\phi; \psi^1, \ldots, \psi^s\})$ are biorthogonal to each other. By $\widehat{a}(0) = \widehat{\tilde{a}}(0) = 1$ and $s = |\det(\mathsf{M})| - 1$, we can also deduce from (7.2.13) that $\widehat{b_1}(0) = \cdots = \widehat{b_s}(0) = 0$ and $\widehat{\tilde{b}_1}(0) = \cdots = \widehat{\tilde{b}_s}(0) = 0$. Now the claim follows from Theorem 7.1.7 with $\tau = 0$. ☐

7.2.3 *Compute Smoothness Exponents* $\operatorname{sm}_2(a, \mathsf{M})$ *and* $\operatorname{sm}_\infty(a, \mathsf{M})$

In this section we first discuss how to efficiently compute the smoothness exponent $\operatorname{sm}_2(a, \mathsf{M})$. Calculating $\operatorname{sm}_p(a, \mathsf{M})$ in (7.2.2) with $p \ne 2$ is generally a difficult task. Often, one first finds a suitable basis for a finite dimensional space V in item (4) of Theorem 7.2.4 so that the representation matrices of all $\mathcal{T}_\gamma|_V, \gamma \in \Gamma_{\mathsf{M}}$ are simultaneously block lower diagonal matrices. Then calculation of the joint spectral radius $\operatorname{jsr}_p(\{\mathcal{T}_\gamma|_V \;:\; \gamma \in \Gamma_{\mathsf{M}}\})$ can be reduced by Proposition 5.7.2. Generally, computing a joint spectral radius is often expensive and difficult. However, for the special case $\widehat{a}(\xi) \geqslant 0$ for all $\xi \in \mathbb{R}^d$, $\mathrm{sm}_\infty(a, \mathsf{M})$ can be effectively computed. Let us first discuss how to compute the quantity $\mathrm{sm}_2(a, \mathsf{M})$ by calculating $\rho_m(a, \mathsf{M})_2$ in (7.2.3).

For $j \in \mathbb{N}_0$, we define $\Lambda_j := \{\mu \in \mathbb{N}_0^d \ : \ |\mu| = j\}$. For a $d \times d$ matrix M, we define a $(\#\Lambda_j) \times (\#\Lambda_j)$ matrix $S(\mathsf{M}, \Lambda_j)$ by

$$\frac{(\mathsf{M}x)^\mu}{\mu!} = \sum_{\nu \in \Lambda_j} S(\mathsf{M}, \Lambda_j)_{\mu,\nu} \frac{x^\nu}{\nu!}. \tag{7.2.15}$$

The quantity $\mathrm{sm}_2(a, \mathsf{M})$ can be efficiently computed by the following result.

Theorem 7.2.9 *Let M be a $d \times d$ dilation matrix. Let $a \in l_0(\mathbb{Z}^d)$ with $\widehat{a}(0) = 1$ such that a satisfies order m sum rules with respect to M. Define $b \in l_0(\mathbb{Z}^d)$ by $\widehat{b}(\xi) := |\widehat{a}(\xi)|^2$ and*

$$K_{b,\mathsf{M}} := \sum_{j=1}^{\infty} [\mathsf{M}^{-j} \operatorname{supp}(b)] = \Big\{ \sum_{j=1}^{\infty} \mathsf{M}^{-j} k_j \ : \ k_j \in \operatorname{supp}(b) \Big\}.$$

Then $\rho_m(a, \mathsf{M})_2$ is equal to

$$\rho_m(a, \mathsf{M})_2 = \max \Big\{ \sqrt{|\det(\mathsf{M})|}\sqrt{|t|} \ : \ t \in \operatorname{spec}\Big(|\det(\mathsf{M})| (\overline{b(j - \mathsf{M}k)})_{j,k \in K_{b,\mathsf{M}}} \Big) \backslash \{\vec{\lambda}^{-\mu} \ : \ \mu \in \mathbb{N}_0^d, |\mu| < 2m\} \Big\},$$

where $\vec{\lambda} := (\lambda_1, \ldots, \lambda_d)^{\mathsf{T}}$ with $\lambda_1, \ldots, \lambda_d$ being all the eigenvalues of M. Note that $\mathrm{sm}_2(a, \mathsf{M}) = \frac{d}{2} - \log_{\rho(\mathsf{M})} \rho_m(a, \mathsf{M})_2$ with $m := \mathrm{sr}(a, \mathsf{M})$ and $\rho(\mathsf{M})$ being the spectral radius of M.

Proof By definition of sum rules, we deduce from $\widehat{b}(\xi) = |\widehat{a}(\xi)|^2$ that b satisfies order $2m$ sum rules with respect to M. Consequently, one can check that $\mathcal{T}_{b,\mathsf{M}} \mathscr{V}_j \subseteq \mathscr{V}_j$ for all $j = 0, \ldots, 2m - 1$ and

$$\mathcal{T}_{b,\mathsf{M}}(\nabla^\mu \boldsymbol{\delta}) - \sum_{\nu \in \Lambda_j} S(\mathsf{M}^{-1}, \Lambda_j)_{\nu,\mu} (\nabla^\nu \boldsymbol{\delta}) \in \mathscr{V}_j, \quad |\mu| = j, \ j = 0, \ldots, 2m - 1.$$

which can be proved by verifying

$$\langle \nabla^\mu \boldsymbol{\delta}, \frac{(\cdot)^\eta}{\eta!} \rangle = \boldsymbol{\delta}(\mu - \eta), \qquad \langle \mathcal{T}_{b,\mathsf{M}}(\nabla^\mu \boldsymbol{\delta}), \frac{(\mathsf{M}\cdot)^\eta}{\eta!} \rangle = \boldsymbol{\delta}(\mu - \eta),$$

for all $|\eta| \leqslant |\mu| < 2m$. Since $S(AB, \Lambda_j) = S(A, \Lambda_j) S(B, \Lambda_j)$, using the Jordan canonical form of M^{-1}, we have $\operatorname{spec}(S(\mathsf{M}^{-1}, \Lambda_j)) = \{\vec{\lambda}^{-\mu} \ : \ \mu \in \Lambda_j\}$. Therefore,

$$\operatorname{spec}(\mathcal{T}_{b,\mathsf{M}}) = \operatorname{spec}(\mathcal{T}_{b,\mathsf{M}}|_{\mathscr{V}_{2m-1}}) \cup \{\vec{\lambda}^{-\mu} \ : \ \mu \in \mathbb{N}_0^d, |\mu| < 2m\}. \tag{7.2.16}$$

If $v \in l_0(\mathbb{Z}^d)$ is an eigenvector of $\mathcal{T}_{b,\mathsf{M}}$ with a nonzero eigenvalue λ, then $\lambda v = \mathcal{T}_{b,\mathsf{M}} v$, from which we have $\operatorname{supp}(v) \subseteq \mathsf{M}^{-1}\operatorname{supp}(v) + \mathsf{M}^{-1}\operatorname{supp}(b)$ and $\operatorname{supp}(v) \subseteq \mathsf{M}^{-n}\operatorname{supp}(v) + \sum_{j=1}^{n} \mathsf{M}^{-j}\operatorname{supp}(b)$. Taking $n \to \infty$, we must have $\operatorname{supp}(v) \subseteq K_{b,\mathsf{M}}$. This argument also shows that $\mathcal{T}_{b,\mathsf{M}} l(K_{b,\mathsf{M}}) \subseteq l(K_{b,\mathsf{M}})$ and $\operatorname{spec}(\mathcal{T}_{b,\mathsf{M}}|_{\mathscr{V}_j \cap l(K_{b,\mathsf{M}})}) \cup \{0\} = \operatorname{spec}(\mathcal{T}_{b,\mathsf{M}}|_{\mathscr{V}_j})$ for all $j = 0, \dots, 2m-1$. By (7.2.16), we conclude that

$$\operatorname{spec}(\mathcal{T}_{b,\mathsf{M}}|_{l(K_{b,\mathsf{M}})}) = \operatorname{spec}(\mathcal{T}_{b,\mathsf{M}}|_{\mathscr{V}_{2m-1} \cap l(K_{b,\mathsf{M}})}) \cup \{\vec{\lambda}^{-\mu} \; : \; \mu \in \mathbb{N}_0^d, |\mu| < 2m\}. \tag{7.2.17}$$

On the other hand, for $|\mu| = m$,

$$\begin{aligned}
\|\nabla^\mu \mathcal{S}_{a,\mathsf{M}}^n \boldsymbol{\delta}\|_{l_2(\mathbb{Z}^d)}^2 &= \frac{1}{(2\pi)^d} \int_{[-\pi,\pi)^d} |\widehat{\nabla^\mu \mathcal{S}_{a,\mathsf{M}}^n \boldsymbol{\delta}}(\xi)|^2 d\xi \\
&= \frac{|\det(\mathsf{M})|^{2n}}{(2\pi)^d} \int_{[-\pi,\pi)^d} |\widehat{\nabla^\mu \boldsymbol{\delta}}(\xi)|^2 \widehat{b}(\xi) \cdots \widehat{b}((\mathsf{M}^\mathsf{T})^{n-1}\xi) d\xi \\
&= |\det(\mathsf{M})|^n \langle \mathcal{S}_{b,\mathsf{M}}^n \boldsymbol{\delta}, v_\mu \rangle = |\det(\mathsf{M})|^n \langle \boldsymbol{\delta}, \mathcal{T}_{b,\mathsf{M}}^n v_\mu \rangle \\
&= |\det(\mathsf{M})|^n (\mathcal{T}_{b,\mathsf{M}}^n v_\mu)(0) \\
&\leqslant |\det(\mathsf{M})|^n \|\mathcal{T}_{b,\mathsf{M}}^n|_{\mathscr{V}_{2m-1} \cap l(K_{b,\mathsf{M}})}\| \|v_\mu\|_{l_\infty(\mathbb{Z}^d)},
\end{aligned}$$

where $\widehat{v_\mu}(\xi) := |\widehat{\nabla^\mu \boldsymbol{\delta}}(\xi)|^2$. The above inequality leads to

$$\begin{aligned}
\limsup_{n\to\infty} \|\nabla^\mu \mathcal{S}_{a,\mathsf{M}}^n \boldsymbol{\delta}\|_{l_2(\mathbb{Z}^d)}^{2/n} &\leqslant |\det(\mathsf{M})| \limsup_{n\to\infty} \|\mathcal{T}_{b,\mathsf{M}}^n|_{\mathscr{V}_{2m-1} \cap l(K_{b,\mathsf{M}})}\|^{1/n} \\
&= |\det(\mathsf{M})| \rho(\mathcal{T}_{b,\mathsf{M}}|_{\mathscr{V}_{2m-1} \cap l(K_{b,\mathsf{M}})}),
\end{aligned}$$

for all $|\mu| = m$. This proves

$$[\rho_m(a, \mathsf{M})_2]^2 \leqslant |\det(\mathsf{M})| \rho(\mathcal{T}_{b,\mathsf{M}}|_{\mathscr{V}_{2m-1} \cap l(K_{b,\mathsf{M}})}). \tag{7.2.18}$$

Conversely, any element in Λ_{2m} can be written as $\mu + \nu$ with $\mu, \nu \in \Lambda_m$. Hence, for any $j, k \in \mathbb{Z}^d$,

$$\begin{aligned}
&|[\mathcal{T}_{b,\mathsf{M}}^n([\nabla^{\mu+\nu}\boldsymbol{\delta}](\cdot - j))](k)|^2 = |\langle \mathcal{T}_{b,\mathsf{M}}^n([\nabla^{\mu+\nu}\boldsymbol{\delta}](\cdot - j)), \boldsymbol{\delta}(\cdot - k)\rangle|^2 \\
&= |\langle [\nabla^{\mu+\nu}\boldsymbol{\delta}](\cdot - j), \mathcal{S}_{b,\mathsf{M}}^n[\boldsymbol{\delta}(\cdot - k)]\rangle|^2 \\
&\leqslant \frac{|\det(\mathsf{M})|^{2n}}{(2\pi)^{2d}} \left(\int_{[-\pi,\pi)^d} |\widehat{\nabla^{\mu+\nu}\boldsymbol{\delta}}(\xi)| \widehat{b}(\xi) \cdots \widehat{b}((\mathsf{M}^\mathsf{T})^{n-1}\xi) d\xi \right)^2 \\
&\leqslant \frac{|\det(\mathsf{M})|^{2n}}{(2\pi)^{2d}} \left(\int_{[-\pi,\pi)^d} |\widehat{\nabla^{\mu}\boldsymbol{\delta}}(\xi)|^2 |\widehat{a}(\xi)|^2 \cdots |\widehat{a}((\mathsf{M}^\mathsf{T})^{n-1}\xi)|^2 d\xi \right) \\
&\qquad \times \left(\int_{[-\pi,\pi)^d} |\widehat{\nabla^{\nu}\boldsymbol{\delta}}(\xi)|^2 |\widehat{a}(\xi)|^2 \cdots |\widehat{a}((\mathsf{M}^\mathsf{T})^{n-1}\xi)|^2 d\xi \right) \\
&= |\det(\mathsf{M})|^{-2n} \|\nabla^\mu \mathcal{S}_{a,\mathsf{M}}^n \boldsymbol{\delta}\|_{l_2(\mathbb{Z}^d)}^2 \|\nabla^\nu \mathcal{S}_{a,\mathsf{M}}^n \boldsymbol{\delta}\|_{l_2(\mathbb{Z}^d)}^2,
\end{aligned}$$

where we used $\widehat{\nabla^{\mu+\nu}\boldsymbol{\delta}}(\xi) = \widehat{\nabla^{\mu}\boldsymbol{\delta}}(\xi)\overline{\widehat{\nabla^{\nu}\boldsymbol{\delta}}(\xi)}$ and $\widehat{b}(\xi) = |\widehat{a}(\xi)|^2$. Since $\mathcal{T}_{b,\mathsf{M}}^n([\nabla^{\mu+\nu}\boldsymbol{\delta}](\cdot - j))$ has support inside $K_{b,\mathsf{M}}$ for large enough n, we deduce from the above inequality that

$$\begin{aligned}\limsup_{n\to\infty}\|\mathcal{T}_{b,\mathsf{M}}^n([\nabla^{\mu+\nu}\boldsymbol{\delta}](\cdot - j))\|_{l_\infty(\mathbb{Z}^d)}^{1/n} &\leqslant |\det(\mathsf{M})|^{-1}\limsup_{n\to\infty}\|\nabla^{\mu}\mathcal{S}_{a,\mathsf{M}}^n\boldsymbol{\delta}\|_{l_2(\mathbb{Z}^d)}^{1/n}\limsup_{n\to\infty}\|\nabla^{\nu}\mathcal{S}_{a,\mathsf{M}}^n\boldsymbol{\delta}\|_{l_2(\mathbb{Z}^d)}^{1/n}\\ &\leqslant |\det(\mathsf{M})|^{-1}[\rho_m(a,\mathsf{M})_2]^2.\end{aligned}$$

By Lemma 7.2.2, $\mathscr{V}_{2m-1}$ is linearly spanned by $\nabla^{\mu}\boldsymbol{\delta}(\cdot - j), |\mu| = 2m$ and $j \in \mathbb{Z}^d$. The above inequality with (7.2.18) implies

$$[\rho_m(a,\mathsf{M})_2]^2 = |\det(\mathsf{M})|\rho(\mathcal{T}_{b,\mathsf{M}}|_{\mathscr{V}_{2m-1}\cap l(\Omega_{b,\mathsf{M}})}).$$

Note that $|\det(\mathsf{M})|\overline{(b(j-\mathsf{M}k))}_{j,k\in K_{b,\mathsf{M}}}$ is the representation matrix of $\mathcal{T}_{b,\mathsf{M}}|_{l(K_{b,\mathsf{M}})}$ under the standard basis $\{\boldsymbol{\delta}(\cdot - k) \; : \; k \in K_{b,\mathsf{M}}\}$. Now the proof is completed by the above identity and (7.2.17). $\square$

Let $K_0 \subseteq \mathbb{Z}^d$ such that $K_{b,\mathsf{M}} \subseteq K_0$. Recursively define $K_j := K_{j-1} \cap [\mathsf{M}^{-1}(K_j + \operatorname{supp}(b))]$. Then $\{K_j\}_{j=0}^{\infty}$ is a decreasing sequence of subsets of $\mathbb{Z}^d$. Therefore, $K_j = K_{j-1}$ must hold for some $j \in \mathbb{N}$ and thus $K_{b,\mathsf{M}} = K_j$.

Before discussing how to compute $\operatorname{sm}_\infty(a,\mathsf{M})$, we need the following result.

Lemma 7.2.10 *Let $a \in l_0(\mathbb{Z}^d)$ such that $\widehat{a}(\xi) \geqslant 0$ for all $\xi \in \mathbb{R}^d$. Then $\operatorname{sr}(a,\mathsf{M})$ must be an even integer.*

Proof Let $m := \operatorname{sr}(a,\mathsf{M})$ and $\omega \in \Omega_{\mathsf{M}}\backslash\{0\}$. Consider the Taylor series of $\widehat{a}(\xi)$ at $\xi = 2\pi\omega$:

$$\widehat{a}(\xi) = \sum_{|\mu|=m}\frac{\partial^{\mu}\widehat{a}(2\pi\omega)}{\mu!}(\xi - 2\pi\omega)^{\mu} + \sum_{j=m+1}^{\infty}\sum_{|\nu|=j}\frac{\partial^{\nu}\widehat{a}(2\pi\omega)}{\nu!}(\xi - 2\pi\omega)^{\nu}.$$

Since $\widehat{a}(\xi) \geqslant 0$ for all ξ near $2\pi\omega$, we must have

$$\mathsf{p}(\zeta) := \sum_{|\mu|=m}\frac{\partial^{\mu}\widehat{a}(2\pi\omega)}{\mu!}\zeta^{\mu} \geqslant 0, \qquad \forall\, \zeta \in \mathbb{R}^d.$$

Plugging $\zeta = t\xi$ into the above relation, we have

$$t^m\sum_{|\mu|=m}\frac{\partial^{\mu}\widehat{a}(2\pi\omega)}{\mu!}\xi^{\mu} = \mathsf{p}(t\xi) \geqslant 0, \qquad \forall\, t \in \mathbb{R}, \xi \in \mathbb{R}^d.$$

Suppose that m is an odd integer. Then the above inequality forces $\sum_{|\mu|=m} \frac{\partial^\mu \widehat{a}(2\pi\omega)}{\mu!} \xi^\mu = 0$ for all $\xi \in \mathbb{R}^d$, from which we must have $\partial^\mu \widehat{a}(2\pi\omega) = 0$ for all $|\mu| = m$ and $\omega \in \Omega_{\mathsf{M}} \backslash \{0\}$. But this implies $\operatorname{sr}(a, \mathsf{M}) > m$, which is a contradiction to $m = \operatorname{sr}(a, \mathsf{M})$. Hence, $\operatorname{sr}(a, \mathsf{M})$ must be an even integer. □

For a filter $a \in l_0(\mathbb{Z}^d)$ having a nonnegative symbol $\widehat{a}$, the quantity $\operatorname{sm}_\infty(a, \mathsf{M})$ can be efficiently computed by the following result.

Theorem 7.2.11 *Let* M *be a* $d \times d$ *dilation matrix and* $a \in l_0(\mathbb{Z}^d)$ *with* $\widehat{a}(0) = 1$. *Define* $m := \operatorname{sr}(a, \mathsf{M})$. *Then*

$$\rho_m(a, \mathsf{M})_\infty \geqslant \max \Big\{ |t| \ : \ t \in \operatorname{spec}(|\det(\mathsf{M})| (\overline{a(j - \mathsf{M}k)})_{j,k \in K_{a,\mathsf{M}}}) \backslash \{\vec{\lambda}^{-\mu} \ : \ \mu \in \mathbb{N}_0^d, |\mu| < m\} \Big\}, \tag{7.2.19}$$

where $\vec{\lambda} := (\lambda_1, \ldots, \lambda_d)^\mathsf{T}$ *with* $\lambda_1, \ldots, \lambda_d$ *being all the eigenvalues of* M. *If in addition* $\widehat{a}(\xi) \geqslant 0$ *for all* $\xi \in \mathbb{R}^d$, *then the inequality in* (7.2.19) *becomes an identity. Note that* $\operatorname{sm}_\infty(a, \mathsf{M}) = -\log_{\rho(\mathsf{M})} \rho_m(a, \mathsf{M})_\infty$.

Proof For $j, k \in \mathbb{Z}^d$, we have

$$\begin{aligned}
&[\mathcal{T}_{a,\mathsf{M}}^n((\nabla^\mu \boldsymbol{\delta})(\cdot - j))](k) \\
&= \langle \mathcal{T}_{a,\mathsf{M}}^n((\nabla^\mu \boldsymbol{\delta})(\cdot - j)), \boldsymbol{\delta}(\cdot - k) \rangle = \langle (\nabla^\mu \boldsymbol{\delta})(\cdot - j), \mathcal{S}_{a,\mathsf{M}}^n(\boldsymbol{\delta}(\cdot - k)) \rangle \\
&= \frac{|\det(\mathsf{M})|^n}{(2\pi)^d} \int_{[-\pi,\pi)^d} e^{-ij\cdot\xi} \widehat{\nabla^\mu \boldsymbol{\delta}}(\xi) \overline{e^{i\mathsf{M}^n k \cdot \xi} \widehat{a}(\xi) \cdots \widehat{a}((\mathsf{M}^\mathsf{T})^{n-1}\xi)} d\xi \\
&= \frac{|\det(\mathsf{M})|^n}{(2\pi)^d} \int_{[-\pi,\pi)^d} \overline{\widehat{\nabla^\mu \boldsymbol{\delta}}(\xi) \widehat{a}(\xi) \cdots \widehat{a}((\mathsf{M}^\mathsf{T})^{n-1}\xi) e^{-i(j - \mathsf{M}^n k + \mu)\cdot\xi}} d\xi \\
&= \overline{[\nabla^\mu \mathcal{S}_{a,\mathsf{M}}^n \boldsymbol{\delta}](\mathsf{M}^n k - j - \mu)}.
\end{aligned}$$

Hence, $\|\mathcal{T}_{a,\mathsf{M}}^n((\nabla^\mu \boldsymbol{\delta})(\cdot - j))\|_{l_\infty(\mathbb{Z}^d)} \leqslant \|\nabla^\mu \mathcal{S}_{a,\mathsf{M}}^n \boldsymbol{\delta}]\|_{l_\infty(\mathbb{Z}^d)}$. By $\operatorname{sr}(a, \mathsf{M}) = m$, we have $\mathcal{T}_{a,\mathsf{M}} \mathscr{V}_{m-1} \subseteq \mathscr{V}_{m-1}$. We also have $\mathcal{T}_{a,\mathsf{M}} l(K_{a,\mathsf{M}}) \subseteq l(K_{a,\mathsf{M}})$. Since $\mathscr{V}_{m-1}$ is linearly spanned by $(\nabla^\mu \boldsymbol{\delta})(\cdot - j), j \in \mathbb{Z}^d$ and $|\mu| = m$, we conclude from the above inequality that

$$\rho(\mathcal{T}_{a,\mathsf{M}}|_{\mathscr{V}_{m-1} \cap l(K_{a,\mathsf{M}})}) \leqslant \rho_m(a, \mathsf{M})_\infty.$$

Now the claim in (7.2.19) follows from (7.2.17) with b and $2m$ being replaced by a and m.

Suppose that $\widehat{a}(\xi) \geqslant 0$ for all $\xi \in \mathbb{R}^d$. By Lemma 7.2.10, m must be an even integer and hence $m = 2m_0$ for some $m_0 \in \mathbb{N}_0$. For $|\mu| = |\nu| = m_0$,

$$
\begin{aligned}
|[\nabla^{\mu+\nu} \mathcal{S}^n_{a,\mathsf{M}}\boldsymbol{\delta}](k)| &= \frac{|\det(\mathsf{M})|^n}{(2\pi)^d} \left| \int_{[-\pi,\pi)^d} \widehat{\nabla^{\mu+\nu}\boldsymbol{\delta}}(\xi)\widehat{a}(\xi)\cdots\widehat{a}((\mathsf{M}^\mathsf{T})^{n-1}\xi)e^{-ik\cdot\xi}d\xi \right| \\
&\leqslant \frac{|\det(\mathsf{M})|^n}{(2\pi)^d} \int_{[-\pi,\pi)^d} |\widehat{\nabla^{\mu}\boldsymbol{\delta}}(\xi)||\widehat{\nabla^{\nu}\boldsymbol{\delta}}(\xi)|\widehat{a}(\xi)\cdots\widehat{a}((\mathsf{M}^\mathsf{T})^{n-1}\xi)d\xi \\
&\leqslant \frac{|\det(\mathsf{M})|^n}{(2\pi)^d} \left(\int_{[-\pi,\pi)^d} |\widehat{\nabla^{\mu}\boldsymbol{\delta}}(\xi)|^2\widehat{a}(\xi)\cdots\widehat{a}((\mathsf{M}^\mathsf{T})^{n-1}\xi)d\xi \right)^{1/2} \\
&\quad \times \left(\int_{[-\pi,\pi)^d} |\widehat{\nabla^{\nu}\boldsymbol{\delta}}(\xi)|^2\widehat{a}(\xi)\cdots\widehat{a}((\mathsf{M}^\mathsf{T})^{n-1}\xi)d\xi \right)^{1/2} \\
&= \sqrt{\langle w_\mu * \mathcal{S}^n_{a,\mathsf{M}}\boldsymbol{\delta}, \boldsymbol{\delta}\rangle \langle w_\nu * \mathcal{S}^n_{a,\mathsf{M}}\boldsymbol{\delta}, \boldsymbol{\delta}\rangle} \\
&= \sqrt{\langle \boldsymbol{\delta}, \mathcal{T}^n_{a,\mathsf{M}} w_\mu\rangle \langle \boldsymbol{\delta}, \mathcal{T}^n_{a,\mathsf{M}} w_\nu\rangle} \\
&= \sqrt{[\mathcal{T}^n_{a,\mathsf{M}} w_\mu](0)[\mathcal{T}^n_{a,\mathsf{M}} w_\nu](0)} \\
&\leqslant \|\mathcal{T}^n_{a,\mathsf{M}}|_{\mathscr{V}_{m-1}\cap l(K_{a,\mathsf{M}})}\| \|w_\mu\|^{1/2}_{l_\infty(\mathbb{Z}^d)} \|w_\nu\|^{1/2}_{l_\infty(\mathbb{Z}^d)},
\end{aligned}
$$

where $\widehat{w_\mu}(\xi) := |\widehat{\nabla^{\mu}\boldsymbol{\delta}}(\xi)|^2$ and we used $w_\mu, w_\nu \in \mathscr{V}_{m-1}$. This proves $\rho_m(a,\mathsf{M})_\infty \leqslant \rho(\mathcal{T}_{a,\mathsf{M}}|_{\mathscr{V}_{m-1}\cap l(K_{a,\mathsf{M}})})$. Hence, the inequality in (7.2.19) must be an identity. □

7.3 Subdivision Schemes in Computer Graphics

Subdivision is an iterative local averaging rule in (7.1.1) to generate a smooth curve or surface from an initial control polygon or mesh in computer graphics, where a filter is often called a *mask* instead. Built on the convergence result of a cascade algorithm in Sect. 7.2, we study the convergence of subdivision schemes in $\mathscr{C}^m(\mathbb{R}^d)$. Then we provide examples of subdivision triplets for subdivision curves and surfaces in computer graphics. Finally, we illustrate subdivision schemes in computer graphics by some smooth subdivision curves.

7.3.1 Convergence of Subdivision Schemes in $\mathscr{C}^m(\mathbb{R}^d)$

A subdivision scheme is the discrete version of a cascade algorithm. The following result establishes the convergence of a subdivision scheme in $\mathscr{C}^m(\mathbb{R}^d)$ and is the theoretical foundation for applying subdivision schemes in computer graphics.

Theorem 7.3.1 *Let $m \in \mathbb{N}_0$ and M be a $d \times d$ dilation matrix. Assume that M is isotropic if $m > 0$. Let $a \in l_0(\mathbb{Z}^d)$ with $\widehat{a}(0) = 1$ and ϕ be the standard M-refinable function in* (7.1.13) *associated with the filter a. The following are equivalent:*

(1) The cascade algorithm associated with the filter a and the dilation matrix M converges in $\mathscr{C}^m(\mathbb{R}^d)$ to ϕ, that is, $\phi \in \mathscr{C}^m(\mathbb{R}^d)$ and $\lim_{n\to\infty} \|\mathcal{R}^n_{a,\mathsf{M}} f - \phi\|_{\mathscr{C}^m(\mathbb{R}^d)} = 0$ for every compactly supported function $f \in \mathscr{C}^m(\mathbb{R}^d)$ satisfying (7.2.7).

(2) $\mathrm{sm}_\infty(a, \mathsf{M}) > m$.

(3) $\phi \in \mathscr{C}^m(\mathbb{R}^d)$ and for all $\mu \in \mathbb{N}_0^d$ with $|\mu| = m$,

$$\lim_{n\to\infty} |\det(\mathsf{M})|^{|\mu|n/d} \|[\nabla^\mu \mathcal{S}^n_{a,\mathsf{M}} \boldsymbol{\delta}](\cdot) - \partial^\mu(\phi(\mathsf{M}^{-n}\cdot))\|_{l_\infty(\mathbb{Z}^d)} = 0. \tag{7.3.1}$$

(4) The subdivision scheme associated with the filter/mask a and the dilation matrix M converges in $\mathscr{C}^m(\mathbb{R}^d)$, that is, for every sequence $v \in l_\infty(\mathbb{Z}^d)$, there exists a function $g_v \in \mathscr{C}^m(\mathbb{R}^d)$ such that for all $\mu \in \mathbb{N}_0^d$ with $|\mu| \leqslant m$,

$$\lim_{n\to\infty} |\det(\mathsf{M})|^{|\mu|n/d} \|[\nabla^\mu \mathcal{S}^n_{a,\mathsf{M}} v](\cdot) - \partial^\mu(g_v(\mathsf{M}^{-n}\cdot))\|_{l_\infty(\mathbb{Z}^d)} = 0. \tag{7.3.2}$$

In fact, g_v must be $v * \phi := \sum_{k\in\mathbb{Z}^d} v(k)\phi(\cdot - k)$.

Proof (1) $\Longleftrightarrow$ (2) has been established in Theorem 7.2.4. We now prove (2)$\Longrightarrow$(3). Let $\mu = (\mu_1, \ldots, \mu_d)^\mathsf{T} \in \mathbb{N}_0^d$ with $|\mu| := \mu_1 + \cdots + \mu_d = m$. Define $f = \otimes^d B_{m+2}$. Then $f \in \mathscr{C}^m(\mathbb{R}^d)$ and (7.2.7) holds. Moreover, by item (2) of Proposition 6.1.1,

$$\partial^\mu f = \partial^\mu(\otimes^d B_{m+2}) = \nabla^\mu f_\mu \qquad \text{with} \quad f_\mu := B_{m+2-\mu_1} \otimes \cdots \otimes B_{m+2-\mu_d}.$$

Define $f_n := \mathcal{R}^n_{a,\mathsf{M}} f$, $n \in \mathbb{N}$. It follows from (7.2.5) that

$$f_{n,\mu} := \partial^\mu(f_n(\mathsf{M}^{-n}\cdot)) = \sum_{k\in\mathbb{Z}^d} [\mathcal{S}^n_{a,\mathsf{M}}\boldsymbol{\delta}](k)\partial^\mu f(\cdot - k) = \sum_{k\in\mathbb{Z}^d} [\nabla^\mu \mathcal{S}^n_{a,\mathsf{M}}\boldsymbol{\delta}](k) f_\mu(\cdot - k).$$

Define $h := \otimes^d(B_2(\cdot - 1))$ and

$$h_{n,\mu} := \sum_{k\in\mathbb{Z}^d} [\nabla^\mu \mathcal{S}^n_{a,\mathsf{M}}\boldsymbol{\delta}](k) h(\cdot - k).$$

Since $h(k) = \boldsymbol{\delta}(k)$ for all $k \in \mathbb{Z}^d$, we have $h_{n,\mu}(k) = [\nabla^\mu \mathcal{S}^n_{a,\mathsf{M}}\boldsymbol{\delta}](k)$ for all $k \in \mathbb{Z}^d$. Therefore,

$$\begin{aligned}[]
[\nabla^\mu \mathcal{S}^n_{a,\mathsf{M}}\boldsymbol{\delta}](k) - \partial^\mu(\phi(\mathsf{M}^{-n}\cdot))(k) &= [h_{n,\mu}(k) - f_{n,\mu}(k)] \\
&\quad + \partial^\mu[f_n(\mathsf{M}^{-n}\cdot) - \phi(\mathsf{M}^{-n}\cdot)](k).
\end{aligned} \tag{7.3.3}$$

Since $\operatorname{sm}_\infty(a, \mathsf{M}) > m$, by Theorem 7.2.4, $\rho_{m+1}(a, \mathsf{M})_\infty < |\det(\mathsf{M})|^{-m/d}$. Therefore, there exist $0 < \rho < 1$ and $C > 0$ such that (7.2.9) holds with $p = \infty$. Note that $h_{n,\mu} - f_{n,\mu} = \sum_{k\in\mathbb{Z}^d}[\nabla^\mu \mathcal{S}_{a,\mathsf{M}}^n \boldsymbol{\delta}](k)\eta(\cdot - k)$ with $\eta := f_\mu - h$ and $\widehat{\eta}(2\pi k) = 0$ for all $k \in \mathbb{Z}^d$. Thus, by Lemma 7.2.3, $\eta = \sum_{j=1}^d \nabla_{e_j}\eta_j$ for some compactly supported functions $\eta_j \in L_\infty(\mathbb{R}^d)$. Consequently,

$$h_{n,\mu} - f_{n,\mu} = \sum_{j=1}^{d}\sum_{k\in\mathbb{Z}^d}[\nabla^{\mu+e_j}\mathcal{S}_{a,\mathsf{M}}^n\boldsymbol{\delta}](k)\eta_j(\cdot - k).$$

Since all η_j are compactly supported functions in $L_\infty(\mathbb{R}^d)$ and $|\mu| = m$, by (7.2.9) with $p = \infty$, there exists a positive constant C_1 such that

$$\begin{aligned} &|\det(\mathsf{M})|^{mn/d}\|h_{n,\mu} - f_{n,\mu}\|_{\mathscr{C}(\mathbb{R}^d)} \\ &\quad\leqslant C_1|\det(\mathsf{M})|^{mn/d}\sum_{j=1}^{d}\|\nabla^{\mu+e_j}\mathcal{S}_{a,\mathsf{M}}^n\boldsymbol{\delta}\|_{l_\infty(\mathbb{Z}^d)} \leqslant dC_1C\rho^n. \end{aligned} \tag{7.3.4}$$

On the other hand, by Lemma 7.2.1, we have

$$(\otimes^m\partial)\otimes[f_n(\mathsf{M}^{-n}\cdot) - \phi(\mathsf{M}^{-n}\cdot)] = ((\otimes^m\partial)\otimes[f_n - \phi])(\mathsf{M}^{-n}\cdot)(\otimes^m\mathsf{M}^{-n}).$$

Consequently, by Proposition 7.1.5 and the fact that M is isotropic when $m > 0$, there exists a positive constant C_2 depending only on M such that

$$\begin{aligned} &\|(\otimes^m\partial)\otimes[f_n(\mathsf{M}^{-n}\cdot) - \phi(\mathsf{M}^{-n}\cdot)]\|_{(\mathscr{C}(\mathbb{R}^d))^{1\times d^m}} \\ &\qquad\leqslant C_2|\det(\mathsf{M})|^{-mn/d}\|(\otimes^m\partial)\otimes[f_n - \phi]\|_{(\mathscr{C}(\mathbb{R}^d))^{1\times d^m}}. \end{aligned}$$

Consequently, we conclude that

$$|\det(\mathsf{M})|^{mn/d}\|\partial^\mu(f_n(\mathsf{M}^{-n}\cdot)) - \partial^\mu(\phi(\mathsf{M}^{-n}\cdot))\|_{\mathscr{C}(\mathbb{R}^d)} \leqslant C_2d^m\|f_n - \phi\|_{\mathscr{C}^m(\mathbb{R}^d)} \to 0$$

as $n \to \infty$. Combining the above inequality with (7.3.4), we deduce from (7.3.3) that (7.3.1) holds. Thus (2)$\Longrightarrow$(3).

We now prove (3)$\Longrightarrow$(2). Let $\mu \in \mathbb{N}_0^d$ with $|\mu| = m$. Then

$$\begin{aligned} [\nabla_{e_j}\nabla^\mu\mathcal{S}_{a,\mathsf{M}}^n\boldsymbol{\delta}](k) =&[\nabla^\mu\mathcal{S}_{a,\mathsf{M}}^n\boldsymbol{\delta}](k) - [\nabla^\mu\mathcal{S}_{a,\mathsf{M}}^n\boldsymbol{\delta}](k - e_j) \\ =&\Big([\nabla^\mu\mathcal{S}_{a,\mathsf{M}}^n\boldsymbol{\delta}](k) - \partial^\mu(\phi(\mathsf{M}^{-n}\cdot))(k)\Big) \\ &- \Big([\nabla^\mu\mathcal{S}_{a,\mathsf{M}}^n\boldsymbol{\delta}](k - e_j) - \partial^\mu(\phi(\mathsf{M}^{-n}\cdot))(k - e_j)\Big) \\ &+ \Big(\partial^\mu(\phi(\mathsf{M}^{-n}\cdot))(k) - \partial^\mu(\phi(\mathsf{M}^{-n}\cdot))(k - e_j)\Big). \end{aligned}$$

This leads to

$$\begin{aligned}\|\nabla^{\mu+e_j}\mathcal{S}^n_{a,\mathsf{M}}\boldsymbol{\delta}\|_{l_\infty(\mathbb{Z}^d)} \leqslant & 2\|[\nabla^\mu \mathcal{S}^n_{a,\mathsf{M}}\boldsymbol{\delta}](\cdot) - \partial^\mu(\phi(\mathsf{M}^{-n}\cdot))\|_{l_\infty(\mathbb{Z}^d)} \\ & + \|\partial^\mu(\phi(\mathsf{M}^{-n}\cdot)) - \partial^\mu(\phi(\mathsf{M}^{-n}\cdot))(\cdot - e_j)\|_{\mathscr{C}(\mathbb{R}^d)}.\end{aligned} \tag{7.3.5}$$

Since $\phi \in \mathscr{C}^m(\mathbb{R}^d)$, by Lemma 7.2.1 we have

$$(\otimes^m \partial) \otimes (\phi(\mathsf{M}^{-n}\cdot)) = ((\otimes^m\partial)\otimes\phi)(\mathsf{M}^{-n}\cdot) \otimes (\otimes^m \mathsf{M}^{-n}).$$

Hence, by Proposition 7.1.5, since M is expansive and isotropic (when $m > 0$), we have

$$\begin{aligned}&|\det(\mathsf{M})|^{mn/d}\|(\otimes^m\partial)\otimes(\phi(\mathsf{M}^{-n}\cdot)) - (\otimes^m\partial)\otimes(\phi(\mathsf{M}^{-n}\cdot))(\cdot - e_j)\|_{(\mathscr{C}(\mathbb{R}^d))^{1\times d^m}} \\ &\quad = |\det(\mathsf{M})|^{mn/d}\left\|\Big(((\otimes^m\partial)\otimes\phi)(\mathsf{M}^{-n}\cdot) - ((\otimes^m\partial)\otimes\phi)(\mathsf{M}^{-n}\cdot - \mathsf{M}^{-n}e_j)\Big)\right. \\ &\qquad\qquad \left.\otimes(\otimes^m\mathsf{M}^{-n})\right\|_{(\mathscr{C}(\mathbb{R}^d))^{1\times d^m}} \\ &\quad \leqslant C_3\left\|((\otimes^m\partial)\otimes\phi) - ((\otimes^m\partial)\otimes\phi)(\cdot - \mathsf{M}^{-n}e_j)\right\|_{(\mathscr{C}(\mathbb{R}^d))^{1\times d^m}} \to 0\end{aligned}$$

as $n \to \infty$, where $C_3 > 0$ is a constant depending only on M. By item (3), it follows from (7.3.5) that

$$\lim_{n\to\infty} |\det(\mathsf{M})|^{mn/d}\|\nabla^{\mu+e_j}\mathcal{S}^n_{a,\mathsf{M}}\boldsymbol{\delta}\|_{l_\infty(\mathbb{Z}^d)} = 0, \qquad \forall\, |\mu| = m, j = 1,\dots,d.$$

In other words, item (3) of Theorem 7.2.4 is satisfied with $p = \infty$. Hence, it follows from Theorem 7.2.4 that $\mathrm{sm}_\infty(a,\mathsf{M}) > m$. This proves (3)$\Longrightarrow$(2).

To prove (4)$\Longrightarrow$(3), we choose $v = \boldsymbol{\delta}$. Then (7.3.2) with $\mu = 0$ forces $g_{\boldsymbol{\delta}} = \boldsymbol{\delta} * \phi = \phi$. This proves (4)$\Longrightarrow$(3). We now prove (3)$\Longrightarrow$(4). Note that we proved (3) $\Longleftrightarrow$ (2). In particular, we have $\mathrm{sm}_\infty(a,\mathsf{M}) > m$. Thus, (7.3.1) must hold for all $|\mu| \leqslant m$. For $v \in l_\infty(\mathbb{Z}^d)$, we define $g_v := v * \phi = \sum_{k\in\mathbb{Z}^d} v(k)\phi(\cdot - k)$. Since $\phi \in \mathscr{C}^m(\mathbb{R}^d)$, we have $g_v \in \mathscr{C}^m(\mathbb{R}^d)$. Note

$$[\nabla^\mu \mathcal{S}^n_{a,\mathsf{M}} v](j) = \sum_{k\in\mathbb{Z}^d} v(k)[\nabla^\mu\mathcal{S}^n_{a,\mathsf{M}}\boldsymbol{\delta}](j - \mathsf{M}^n k), \qquad j \in \mathbb{Z}^d, n\in\mathbb{N},$$

which leads to

$$\begin{aligned}&[\nabla^\mu\mathcal{S}^n_{a,\mathsf{M}}v](j) - \partial^\mu(g(\mathsf{M}^{-n}\cdot))(j) \\ &\qquad = \sum_{k\in\mathbb{Z}^d} v(k)\Big([\nabla^\mu\mathcal{S}^n_{a,\mathsf{M}}\boldsymbol{\delta}](j-\mathsf{M}^n k) - \partial^\mu(\phi(\mathsf{M}^{-n}\cdot))(j - \mathsf{M}^n k)\Big).\end{aligned}$$

Since both a and ϕ are compactly supported, the last two terms in the above identity are supported inside $\mathsf{M}^n K$ for some bounded set K for all $n \in \mathbb{N}$. Thus, there exists a positive constant C such that

$$\|[\nabla^\mu \mathcal{S}^n_{a,\mathsf{M}} v] - \partial^\mu(g(\mathsf{M}^{-n}\cdot))\|_{l_\infty(\mathbb{Z}^d)} \leqslant C\|v\|_{l_\infty(\mathbb{Z}^d)} \|[\nabla^\mu \mathcal{S}^n_{a,\mathsf{M}}\delta](\cdot) - \partial^\mu(\phi(\mathsf{M}^{-n}\cdot))\|_{l_\infty(\mathbb{Z}^d)}.$$

The claim in (7.3.2) now follows directly from the above inequality and (7.3.1). This proves (3)$\Longrightarrow$(4). □

7.3.2 Subdivision Schemes Employing Subdivision Triplets

Since a mesh modeling a geometric object in computer graphics has no natural coordinate systems as the integer lattice $\mathbb{Z}^d$ does, some extra requirements have to put on a filter/mask a and a dilation matrix M for the application of a subdivision scheme in computer graphics. We say that a finite set G of $d \times d$ integer matrices is *a symmetry group* on $\mathbb{Z}^d$ if $|\det(E)| = 1$ for every $E \in G$ and G forms a group under the matrix multiplication. To apply a subdivision scheme in computer graphics for generating smooth curves and surfaces, *a subdivision triplet* (a, M, G) is required:

(1) The real-valued mask a is *G-symmetric* with a symmetry center $c_a \in \mathbb{R}^d$, that is,

$$a(E(k - c_a) + c_a) = a(k), \qquad \forall\, k \in \mathbb{Z}^d, E \in G. \tag{7.3.6}$$

(2) The dilation matrix M is *compatible* with the symmetry group G:

$$\mathsf{M}E\mathsf{M}^{-1} \in G, \qquad \forall\, E \in G. \tag{7.3.7}$$

The following basic result holds on a subdivision triplet and links the symmetry property of a filter with the symmetry property of its underlying refinable function.

Theorem 7.3.2 *Let G be a symmetry group on $\mathbb{Z}^d$ and M be a $d \times d$ dilation matrix which is compatible with G. Let $a \in l_0(\mathbb{Z}^d)$ with $\widehat{a}(0) = 1$ such that a is G-symmetric with a symmetry center $c_a \in \mathbb{R}^d$. Let ϕ be the standard M-refinable function associated with the filter a. Then*

$$\phi(E(\cdot - c_\phi) + c_\phi) = \phi \qquad \forall\, E \in G \quad \text{with} \quad c_\phi := (\mathsf{M} - I_d)^{-1} c_a. \tag{7.3.8}$$

*If in addition a $d \times d$ matrix N is G-*equivalent *to M (i.e., $\mathsf{N} = E\mathsf{M}F$ for some $E, F \in G$), then $\phi^{\mathsf{N}} = \phi(\cdot + (\mathsf{M} - I_d)^{-1}c_a - (\mathsf{N} - I_d)^{-1}c_a)$, where $\widehat{\phi^{\mathsf{N}}}(\xi) := \prod_{j=1}^\infty \widehat{a}((\mathsf{N}^\mathsf{T})^{-j}\xi)$.*

Proof Note that (7.3.6) is equivalent to $\widehat{a}(E^\mathsf{T}\xi) = e^{i(I_d-E)c_a\cdot\xi}\widehat{a}(\xi)$ for all $E \in G$ and $\xi \in \mathbb{R}^d$. By $\widehat{\phi}(\xi) = \prod_{j=1}^\infty \widehat{a}((\mathsf{M}^\mathsf{T})^{-j}\xi)$ for $\xi \in \mathbb{R}^d$, we have

$$\begin{aligned}\widehat{\phi}(E^\mathsf{T}\xi) &= \prod_{j=1}^\infty \widehat{a}((\mathsf{M}^\mathsf{T})^{-j}E^\mathsf{T}\xi) = \prod_{j=1}^\infty \widehat{a}((\mathsf{M}^j E \mathsf{M}^{-j})^\mathsf{T}(\mathsf{M}^\mathsf{T})^{-j}\xi) \\ &= \left[\prod_{j=1}^\infty e^{i(I_d-\mathsf{M}^j E\mathsf{M}^{-j})c_a\cdot(\mathsf{M}^\mathsf{T})^{-j}\xi}\right]\left[\prod_{j=1}^\infty \widehat{a}((\mathsf{M}^\mathsf{T})^{-j}\xi)\right] \\ &= \widehat{\phi}(\xi)e^{\sum_{j=1}^\infty i(\mathsf{M}^{-j}-E\mathsf{M}^{-j})c_a\cdot\xi} = \widehat{\phi}(\xi)e^{i(I_d-E)(\mathsf{M}-I_d)^{-1}c_a\cdot\xi},\end{aligned}$$

which is equivalent to (7.3.8). If N is G-equivalent to M, then $\mathsf{M}^j\mathsf{N}^{-j} \in G$. Hence N must be a dilation matrix by $\lim_{j\to\infty}\mathsf{N}^{-j} = 0$, and

$$\begin{aligned}\widehat{\phi^\mathsf{N}}(\xi) &= \prod_{j=1}^\infty \widehat{a}((\mathsf{N}^\mathsf{T})^{-j}\xi) = \prod_{j=1}^\infty \widehat{a}((\mathsf{M}^j\mathsf{N}^{-j})^\mathsf{T}(\mathsf{M}^\mathsf{T})^{-j}\xi) = \widehat{\phi}(\xi)\prod_{j=1}^\infty e^{i(I_d-\mathsf{M}^j\mathsf{N}^{-j})c_a\cdot\xi} \\ &= \widehat{\phi}(\xi)e^{i[(\mathsf{M}-I_d)^{-1}-(\mathsf{N}-I_d)^{-1}]c_a\cdot\xi}.\end{aligned}$$

This completes the proof. □

Generally, a surface or curve is modeled by a mesh through connecting neighboring points. However, a mesh often has no natural coordinate system as $\mathbb{Z}^d$. To overcome this problem, the discrete points in $\mathbb{Z}^d$ are connected. For dimension one, the points in $\mathbb{Z}$ can be naturally connected by joining k with $k+1$ through a line segment for every $k \in \mathbb{Z}$. For dimension two, there are two basic standard meshes: The quadrilateral mesh $\mathbb{Z}^2_Q$ by connecting neighboring points through horizontal or vertical line segments, and the triangular mesh $\mathbb{Z}^2_T$ by connecting neighboring points horizontally, vertically, or along 45° degrees through line segments. See Fig. 7.1 for an illustration.

The symmetry group associated with the quadrilateral mesh $\mathbb{Z}^2_Q$ is

$$D_4 := \left\{\pm\begin{bmatrix}1&0\\0&1\end{bmatrix}, \pm\begin{bmatrix}1&0\\0&-1\end{bmatrix}, \pm\begin{bmatrix}0&1\\1&0\end{bmatrix}, \pm\begin{bmatrix}0&1\\-1&0\end{bmatrix}\right\} \qquad (7.3.9)$$

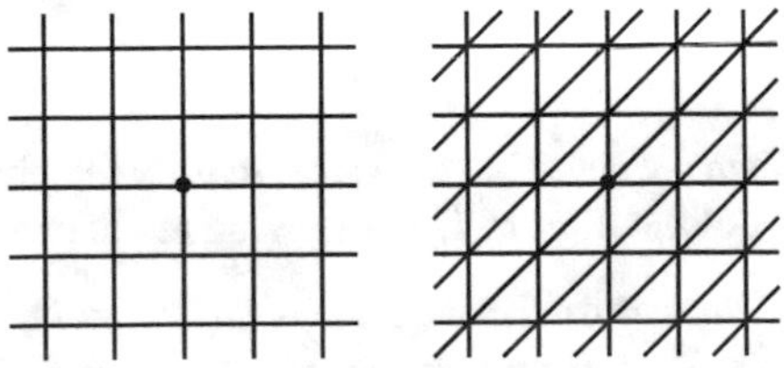

Fig. 7.1 The quadrilateral mesh $\mathbb{Z}^2_Q$ (left) and the triangular mesh $\mathbb{Z}^2_T$ (right)

and the symmetry group associated with the triangular mesh $\mathbb{Z}_T^2$ is

$$D_6 := \left\{ \pm \begin{bmatrix} 1 & 0 \\ 0 & 1 \end{bmatrix}, \pm \begin{bmatrix} 0 & -1 \\ 1 & -1 \end{bmatrix}, \pm \begin{bmatrix} -1 & 1 \\ -1 & 0 \end{bmatrix}, \pm \begin{bmatrix} 0 & 1 \\ 1 & 0 \end{bmatrix}, \pm \begin{bmatrix} 1 & -1 \\ 0 & -1 \end{bmatrix}, \pm \begin{bmatrix} -1 & 0 \\ -1 & 1 \end{bmatrix} \right\}. \tag{7.3.10}$$

Indeed, one can check that $E\mathbb{Z}_Q^2 = \mathbb{Z}_Q^2$ for all $E \in D_4$ and $F\mathbb{Z}_T^2 = \mathbb{Z}_T^2$ for all $F \in D_6$. For a triplet (a, M, G) acting on a mesh $\mathbb{Z}_G^d$, it is natural to require that $E\mathbb{Z}_G^d = \mathbb{Z}_G^d$ for all $E \in G$. Since $\mathsf{M}^{-1}\mathbb{Z}_G^d$ is the refined mesh of the coarse mesh $\mathbb{Z}_G^d$, it is also natural to require that $E\mathsf{M}^{-1}\mathbb{Z}_G^d = \mathsf{M}^{-1}\mathbb{Z}_G^d$ for all $E \in G$. This requires that M should be compatible with G. Symmetry is required on the filter/mask a to ignore the lack of a natural coordinate system on a mesh. Hence, a mask a is required to be G-symmetric for its application in computer graphics.

By solving the compatibility condition in (7.3.7), the following result tells us all the 2×2 real-valued matrices M which are compatible with either D_4 or D_6.

Proposition 7.3.3 *Let M be a 2×2 invertible real-valued matrix. Define*

$$\mathsf{M}_{\sqrt{2}} := \begin{bmatrix} 1 & 1 \\ 1 & -1 \end{bmatrix}, \qquad \mathsf{M}_{\sqrt{3}} := \begin{bmatrix} 1 & -2 \\ 2 & -1 \end{bmatrix}. \tag{7.3.11}$$

Then

(1) If M is compatible with the symmetry group D_4 in (7.3.9)*, then M must be D_4-equivalent to either cI_2 or $c\mathsf{M}_{\sqrt{2}}$ for some $c \in \mathbb{R}$.*

(2) If M is compatible with the symmetry group D_6 in (7.3.10)*, then M must be D_6-equivalent to either cI_2 or $c\mathsf{M}_{\sqrt{3}}$ for some $c \in \mathbb{R}$.*

The notion of linear-phase moments is important for almost-interpolating subdivision schemes which can preserve polynomials to certain degrees. We say that $a \in l_0(\mathbb{Z}^d)$ has *order n linear-phase moments* with phase $c \in \mathbb{R}^d$ if

$$\widehat{a}(\xi) = e^{-ic\cdot\xi} + \mathscr{O}(\|\xi\|^n), \qquad \xi \to 0. \tag{7.3.12}$$

We define $\operatorname{lpm}(a) = n$ with n being the largest such integer. For $n > 1$, the definition in (7.3.12) forces $\widehat{a}(0) = 1$ and $c = i\partial\widehat{a}(0) = \sum_{k\in\mathbb{Z}^d} a(k)k$. If in addition a is symmetric about a point $c_a \in \mathbb{R}^d$:

$$a(2c_a - k) = a(k), \qquad \forall\, k \in \mathbb{Z}^d,$$

then we must have $c = c_a$, that is, the phase c must agree with the symmetry center c_a of the filter a. The following result can be similarly proved as in Proposition 1.2.7.

Proposition 7.3.4 *Let $a \in l_0(\mathbb{Z}^d)$ and $c \in \mathbb{R}^d$. Then $\mathcal{S}_{a,\mathsf{M}}\mathsf{p} = \mathsf{p}(\mathsf{M}^{-1}(\cdot - c))$ for all $\mathsf{p} \in \mathbb{P}_{m-1}$ if and only if* $\operatorname{sr}(a, \mathsf{M}) \geqslant m$ *and* $\operatorname{lpm}(a) \geqslant m$ *with phase c.*

Let (a, M, G) be a subdivision triplet such that the associated subdivision scheme converges in $\mathscr{C}^m(\mathbb{R}^d)$. Let $v : \mathbb{Z}^d \to \mathbb{R}^r$ be a given initial coarse control mesh so that all the points $v(k), k \in \mathbb{Z}^d$ are vectors in $\mathbb{R}^r$. Then the subdivision scheme is applied to each component sequence of v independently. Hence, for simplicity of discussion, we assume $r = 1$, that is, v is a scalar sequence. By iteratively applying the subdivision scheme, one obtains a sequence $v_n := \mathcal{S}^n_{a,\mathsf{M}} v, n \in \mathbb{N}$. According to Theorem 7.3.1, one obtains a parametrization $g_{v,n}$ of the generated mesh at level n by putting the value $v_n(k)$ at the point $\mathsf{M}^{-n}(k - c_a), k \in \mathbb{Z}^d$ with respect to the reference mesh $\mathsf{M}^{-n}\mathbb{Z}^d$. If $\operatorname{lpm}(a) \geqslant j$ and $\operatorname{sr}(a, \mathsf{M}) \geqslant j$, by Proposition 7.3.4, then $g_{\mathsf{p},n}$ agrees with p on $\mathsf{M}^{-n}(k - c_a), k \in \mathbb{Z}^d$ for all polynomials $\mathsf{p} \in \mathbb{P}_{j-1}$. That is, all the polynomials up to degree $j-1$ are preserved at all levels of a subdivision scheme provided that $\operatorname{sr}(a, \mathsf{M}) \geqslant j$ and $\operatorname{lpm}(a) \geqslant j$. As $n \to \infty$, there exists a continuous function g_v on $\mathbb{R}^d$ such that g_v is the limiting surface of the finer and finer meshes $\{g_{v,n}\}_{n=1}^\infty$. A subdivision scheme with $c_a \in \mathbb{Z}^d$ is called *a primal subdivision scheme* since the value $v_n(k)$ is attached to the vertex $\mathsf{M}^{-n}(k + c_a), k \in \mathbb{Z}^d$ of the mesh $\mathsf{M}^{-n}\mathbb{Z}^d_G$; otherwise, it is called *a dual subdivision scheme*, since $v_n(k)$ is attached to the point $\mathsf{M}^{-n}k - \mathsf{M}^{-n}c_a$ which is often the center of a face of the mesh $\mathsf{M}^{-n}\mathbb{Z}^d_G$.

To efficiently compute values $\mathcal{S}_{a,\mathsf{M}}v$ on the refined reference mesh $\mathsf{M}^{-1}\mathbb{Z}^d$ from v on the coarse mesh $\mathbb{Z}^d$, we often rewrite the subdivision operator in (7.1.1) using coset masks and convolution: For $\beta, \gamma \in \mathbb{Z}^d$,

$$\begin{aligned} \mathcal{S}_{a,\mathsf{M}}v(\gamma + \mathsf{M}\beta) &= |\det(\mathsf{M})| \sum_{k\in\mathbb{Z}^d} v(k)a(\gamma + \mathsf{M}\beta - Mk) \\ &= |\det(\mathsf{M})| [v * a^{[\gamma:\mathsf{M}]}](\beta), \end{aligned} \tag{7.3.13}$$

where the *coset mask* $a^{[\gamma:\mathsf{M}]}$ of the mask a is defined to be

$$a^{[\gamma:\mathsf{M}]}(k) := a(\gamma + \mathsf{M}k), \qquad k, \gamma \in \mathbb{Z}^d. \tag{7.3.14}$$

If $\sum_{k\in\mathbb{Z}^d} a(k) = 1$ and $\operatorname{sr}(a, \mathsf{M}) \geqslant 1$, then $|\det(\mathsf{M})| \sum_{k\in\mathbb{Z}^d} a^{[\gamma:\mathsf{M}]}(k) = 1$ for all $\gamma \in \mathbb{Z}^d$. Hence, a subdivision scheme is a local averaging rule. Moreover, the identities (7.3.13) using the coset $a^{[\gamma:\mathsf{M}]}$ in (7.3.14) can be further rewritten as follows:

$$[\mathcal{S}_{a,\mathsf{M}}v](\gamma + \mathsf{M}\beta) = |\det(\mathsf{M})| [v * a^{[\gamma:\mathsf{M}]}](\beta) = \langle v(\beta + \cdot), |\det(\mathsf{M})| \overline{a^{[\gamma:\mathsf{M}]}(-\cdot)} \rangle,$$

which is attached to the point $\beta + \mathsf{M}^{-1}\gamma - \mathsf{M}^{-1}c_a$. Consequently, the filter

$$\{|\det(\mathsf{M})| \overline{a^{[\gamma:\mathsf{M}]}(-k)}\}_{k\in\mathbb{Z}^d} = \{|\det(\mathsf{M})| \overline{a(\gamma - \mathsf{M}k)}\}_{k\in\mathbb{Z}^d}, \qquad \gamma \in \Gamma_{\mathsf{M}}$$

is called the $\mathsf{M}^{-1}\gamma$*-stencil* of the mask a for computing the values $[\mathcal{S}_{a,\mathsf{M}}v](\gamma + \mathsf{M}\cdot)$ on the cosets in $\mathsf{M}^{-1}\gamma + \mathbb{Z}^d$ of the refined mesh $\mathsf{M}^{-1}\mathbb{Z}^d$. It is more convenient to use stencils for subdivision schemes in computer graphics than a filter/mask a. Due to

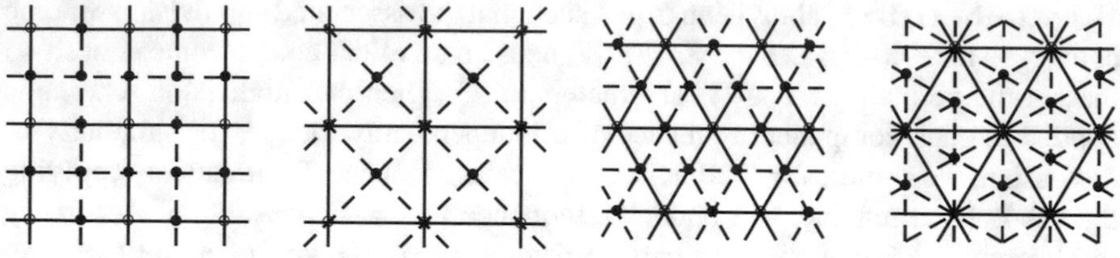

Fig. 7.2 ∘ represents vertices in the coarse mesh $\mathbb{Z}^2$ and • represents new vertices in the refinement mesh $\mathsf{M}^{-1}\mathbb{Z}^2$. The M-refinement of the reference mesh $\mathbb{Z}^2$, from left to right, are for subdivision triplets $(a, 2I_2, D_4)$, $(a, \mathsf{M}_{\sqrt{2}}, D_4)$, $(a, 2I_2, D_6)$, and $(a, \mathsf{M}_{\sqrt{3}}, D_6)$, where $\mathsf{M}_{\sqrt{2}}$ and $\mathsf{M}_{\sqrt{3}}$ are defined in (7.3.11). For better illustration purpose, we used the symmetrized triangular mesh $\{(j + k/2, \sqrt{3}k/2)^{\mathsf{T}} : j, k \in \mathbb{Z}\}$ instead of the triangular mesh $\mathbb{Z}^2_T$ in (7.1)

G-symmetry on a, if $E(\mathsf{M}^{-1}\gamma_1) \in \mathsf{M}^{-1}\gamma_2 + \mathbb{Z}^d$ for some $E \in G$, then the $(\mathsf{M}^{-1}\gamma_1)$-stencil and $(\mathsf{M}^{-1}\gamma_2)$-stencil are the same.

For dimension one and a dilation factor M, the reference coarse mesh $\mathbb{Z}$ is refined into a finer mesh $\frac{1}{\mathsf{M}}\mathbb{Z}$ by inserting new vertices at $\frac{\gamma}{\mathsf{M}} + \mathbb{Z}$ with $\gamma = 1, \dots, |\mathsf{M}| - 1$. The M-refinement of the reference coarse mesh $\mathbb{Z}^2_Q$ or $\mathbb{Z}^2_T$ is presented in Fig. 7.2.

7.3.3 1D Subdivision Triplets and Subdivision Curves

In the following, we provide a few examples of one-dimensional subdivision triplets. At the end of this subsection we provide an example of subdivision curves generated by subdivision schemes employing such subdivision triplets.

Example 7.3.1 The triplet $(a, 2, \{-1, 1\})$ is a primal subdivision triplet with

$$a = \tfrac{1}{2}\{w_3, w_2, w_1, \underline{w_0}, w_1, w_2, w_3\}_{[-3,3]},$$

where

$$w_0 = \tfrac{3+t}{4}, \quad w_1 = \tfrac{8+t}{16}, \quad w_2 = \tfrac{1-t}{8}, \quad w_3 = -\tfrac{t}{16} \quad \text{with} \quad t \in \mathbb{R}. \tag{7.3.15}$$

If $t = -\frac{1}{2}$, then $a = a^B_6(\cdot - 3)$ and $\operatorname{sr}(a, 2) = 6$, $\operatorname{lpm}(a) = 2$ and $\operatorname{sm}_p(a, 2) = 5 + 1/p$ for all $1 \leqslant p \leqslant \infty$. If $t \neq -1/2$, then $\operatorname{sr}(a, 2) = 4$. Since $\widehat{a}(\xi) = e^{i3\xi}(1 + e^{-i\xi})^4 \widehat{b}(\xi)$ with $\widehat{b}(\xi) := -\frac{t}{32} + \frac{1+t}{16}e^{-i\xi} - \frac{t}{32}e^{-i2\xi}$, by item (5) of Corollary 5.8.5, we have $\operatorname{sm}_\infty(a, 2) = 3 - \log_2(1 + t)$ provided $t > -1/2$. We only have $\operatorname{sm}_\infty(a, 2) \geqslant 3 - \log_2 |t|$ for $t \leqslant -1/2$. When $t = 0$, $a = a^B_4(\cdot - 2)$ is the centered B-spline filter of order 4 with $\operatorname{sr}(a, 2) = 4$ and $\operatorname{lpm}(a) = 2$. When $t = 1$, a is an interpolatory 2-wavelet filter with $\operatorname{sr}(a, 2) = 4$ and $\operatorname{lpm}(a) = 4$. See Fig. 7.3 for its subdivision stencils.

Fig. 7.3 The 0-stencil (left) and the $\frac{1}{2}$-stencil (right) of the primal subdivision scheme in Example 7.3.1, where $w_0, \dots, w_3$ are given in (7.3.15). It is an interpolatory 2-wavelet filter if $w_2 = \frac{1-t}{8} = 0$ (i.e. $t = 1$). Since $\mathsf{M} = 2$, each line segment (with endpoints ∘) in the coarse mesh $\mathbb{Z}$ is equally split into two line segments with one new vertex (•) in the middle

Example 7.3.2 The triplet $(a, 2, \{-1, 1\})$ is a dual subdivision triplet with

$$a = \tfrac{1}{2}\{w_2, w_1, \underline{w_0}, w_0, w_1, w_2\}_{[-2,3]},$$

where

$$w_0 = \tfrac{12+3t}{16}, \quad w_1 = \tfrac{8-3t}{32}, \quad w_2 = -\tfrac{3t}{32} \quad \text{with} \quad t \in \mathbb{R}. \tag{7.3.16}$$

If $t = -\frac{2}{3}$, then $a = a_5^B(\cdot - 2)$ and $\operatorname{sr}(a, 2) = 5$, $\operatorname{lpm}(a) = 2$ and $\operatorname{sm}_p(a, 2) = 4 + 1/p$ for all $1 \leqslant p \leqslant \infty$. Since $\widehat{a}(\xi) = e^{i2\xi}(1 + e^{-i\xi})^3 \widehat{b}(\xi)$ with $\widehat{b}(\xi) := -\frac{3t}{8} + \frac{4+3t}{32} e^{-i\xi} - \frac{3t}{8} e^{-i2\xi}$, by item (5) of Corollary 5.8.5, we have $\operatorname{sr}(a, 2) = 3$ and $\operatorname{sm}_\infty(a, 2) = 4 - \log_2(4 + 3t)$ provided $t > -2/3$. We only have $\operatorname{sm}_\infty(a, 2) \geqslant 1 - \log_2(3|t|)$ for $t \leqslant -2/3$. When $t = 0$, $a = a_3^B(\cdot - 1)$ is the shifted B-spline filter of order 3 with $\operatorname{sr}(a, 2) = 3$ and $\operatorname{lpm}(a) = 2$. When $t = 1$, $\operatorname{sr}(a, 2) = 3$ and $\operatorname{lpm}(a) = 4$. See Fig. 7.4 for its subdivision stencils.

Example 7.3.3 The triplet $(a, 3, \{-1, 1\})$ is a primal subdivision triplet with

$$a = \tfrac{1}{3}\{w_5, w_4, w_3, w_2, w_1, \underline{w_0}, w_1, w_2, w_3, w_4, w_5\}_{[-5,5]},$$

where

$$\begin{aligned} w_0 &= \tfrac{7-2t_1-8t_2}{9}, \quad w_1 = \tfrac{6-2t_1-5t_2}{9}, \quad w_2 = \tfrac{3+t_1+t_2}{9}, \\ w_3 &= \tfrac{1+t_1+4t_2}{9}, \quad w_4 = \tfrac{t_1+3t_2}{9}, \quad w_5 = \tfrac{t_2}{9} \end{aligned} \quad \text{with} \quad t_1, t_2 \in \mathbb{R}. \tag{7.3.17}$$

Fig. 7.4 The 0-stencil (left) and the $\frac{1}{2}$-stencil (right) of the dual subdivision scheme in Example 7.3.2, where w_0, w_1, w_2 are given in (7.3.16). The $\frac{1}{2}$-stencil is the same as the 0-stencil. The value $[\mathcal{S}_{a,2}v](k)$ for $k \in \mathbb{Z}$ is attached to the center $\frac{k-1}{2}$ of the line segment $[k-1, k]$ instead of the vertex $\frac{k}{2}$. Since $\mathsf{M} = 2$, each line segment is equally split into two

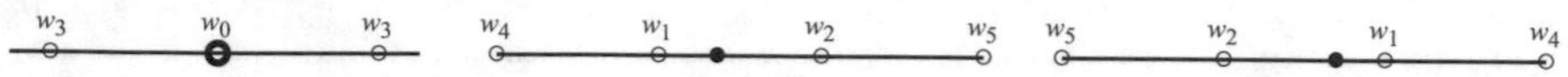

Fig. 7.5 The 0-stencil (left), the $\frac{1}{3}$-stencil (middle), and $\frac{2}{3}$-stencil of the subdivision scheme in Example 7.3.3, where $w_0, \dots, w_5$ are given in (7.3.17). Due to symmetry, $\frac{2}{3}$-stencil is the same as the $\frac{1}{3}$-stencil. It is an interpolatory 3-wavelet filter if $w_3 = \frac{1+t_1+4t_2}{9} = 0$. Since $\mathsf{M} = 3$, each line segment (with endpoints $\circ$) is equally split into three line segments with two new inserted vertices ($\bullet$) at $\frac{1}{3} + \mathbb{Z}$ and $\frac{2}{3} + \mathbb{Z}$

If $t_1 = 2/9$ and $t_2 = 1/9$, then $\operatorname{sr}(a, 3) = 5$ and $\operatorname{sm}_p(a, 3) = 4 + 1/p$ for all $1 \leqslant p \leqslant \infty$ whose 3-refinable function is the B-spline of order 5. Since $\widehat{a}(\xi) = (e^{i\xi} + 1 + e^{-i\xi})^3 \widehat{b}(\xi)$ with

$$\widehat{b}(\xi) := \tfrac{t_2}{27} e^{i2\xi} + \tfrac{t_1}{27} e^{i\xi} + \tfrac{1-2t_1-2t_2}{27} + \tfrac{t_1}{27} e^{-i\xi} + \tfrac{t_2}{27} e^{-i2\xi},$$

by a similar result to item (5) of Corollary 5.8.5, we have

$$\operatorname{sm}_\infty(a, 2) \geqslant 2 - \log_3 \max(|1 - 2t_1 - 2t_2|, |2t_1|, |2t_2|).$$

If $t_1 = 7/9$ and $t_2 = -4/9$, then a is an interpolatory 3-wavelet filter with $\operatorname{sr}(a, 3) = 4 = \operatorname{lpm}(a)$ and $\operatorname{sm}_\infty(a, 3) \geqslant \log_3 14 - 4 \approx 1.5978$. If $t_1 = 5/11$ and $t_2 = -4/11$, then a is an interpolatory 3-wavelet filter with $\operatorname{sr}(a, 3) = 3 = \operatorname{lpm}(a)$ and $\operatorname{sm}_\infty(a, 3) \geqslant 2 + \log_3(11/10) \approx 2.0867$ (Using the joint spectral radius, we in fact have $\operatorname{sm}_2(a, 3) = \log_3 11 \approx 2.18266$). See Fig. 7.5 for its subdivision stencils.

Finally we provide some subdivision curves in Fig. 7.6 generated by subdivision schemes employing the above subdivision triplets.

7.3.4 *Examples of 2D Subdivision Triplets and Stencils*

Example 7.3.4 The triplet $(a, 2I_2, D_4)$ is a primal subdivision triplet with

$$a = \frac{1}{4} \begin{bmatrix} 0 & 0 & w_7 & w_6 & w_7 & 0 & 0 \\ 0 & w_5 & w_4 & w_3 & w_4 & w_5 & 0 \\ w_7 & w_4 & w_2 & w_1 & w_2 & w_4 & w_7 \\ w_6 & w_3 & w_1 & \underline{w_0} & w_1 & w_3 & w_6 \\ w_7 & w_4 & w_2 & w_1 & w_2 & w_4 & w_7 \\ 0 & w_5 & w_4 & w_3 & w_4 & w_5 & 0 \\ 0 & 0 & w_7 & w_6 & w_7 & 0 & 0 \end{bmatrix}_{[-3,3]^2}, \qquad (7.3.18)$$

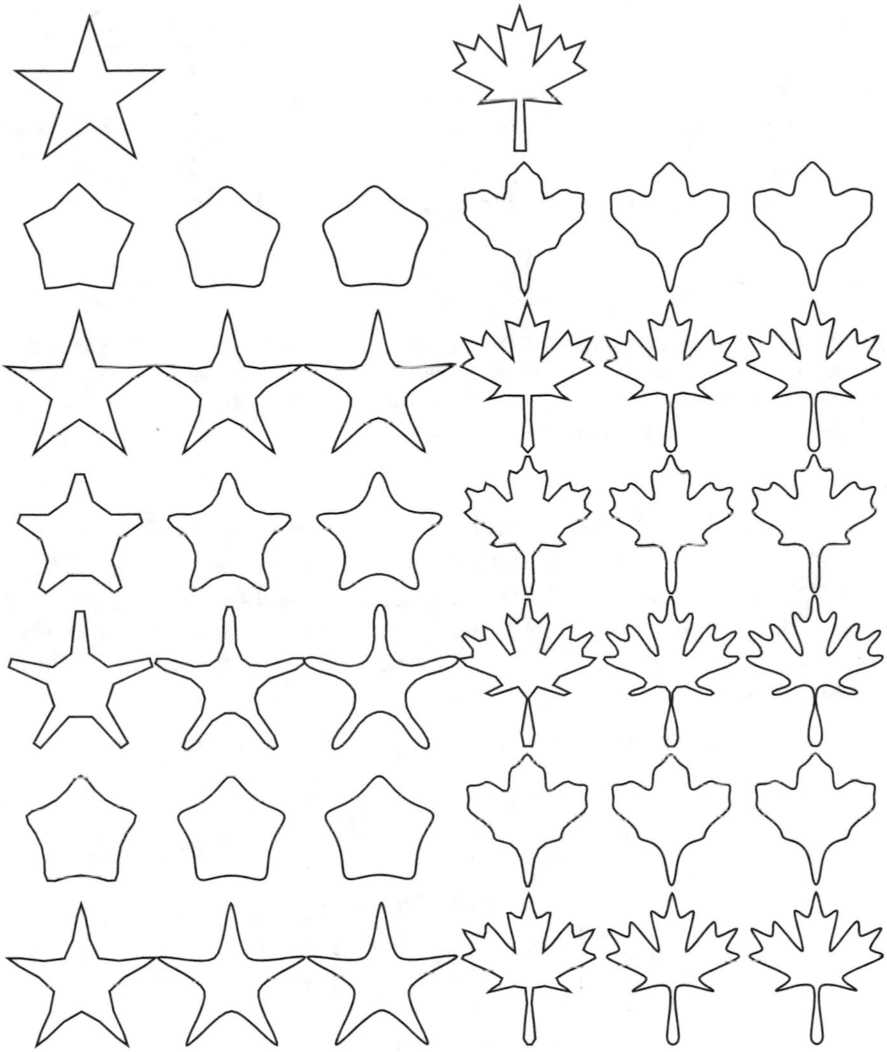

Fig. 7.6 Subdivision curves at levels 1, 2, 3 with the initial control polygons at the first row. The subdivision triplet $(a, 2, \{-1, 1\})$ in Example 7.3.1 is used with $t = -\frac{1}{2}$ (i.e., $a = a_4^B(\cdot - 2)$) for the 2nd row and with $t = 1$ (interpolatory) for the 3rd row. $(a, 2, \{-1, 1\})$ in Example 7.3.2 is used with $t = 0$ (i.e., $a = a_3^B(\cdot - 1)$, the corner cutting scheme) for the 4th row and with $t = 1$ and $\operatorname{lpm}(a) = 4$ for the 5th row. $(a, 3, \{-1, 1\})$ is used with $t_1 = \frac{2}{9}, t_2 = \frac{1}{9}$ for the 6th row and with $t_1 = \frac{5}{11}, t_2 = -\frac{4}{11}$ (interpolatory, $\operatorname{sm}_\infty(a, 3) = \log_3 11$) for the 7th row

where

$$\begin{aligned} w_0 &= \tfrac{2+t_1-4t_2}{4}, \quad & w_1 &= \tfrac{3-3t_2}{8}, \quad & w_2 &= \tfrac{2-t_2}{8}, \quad & w_3 &= \tfrac{1-t_1+2t_2}{8}, \\ w_4 &= \tfrac{1+2t_2}{16}, \quad & w_5 &= \tfrac{t_1}{16}, \quad & w_6 &= \tfrac{t_2}{8}, \quad & w_7 &= \tfrac{t_2}{16} \end{aligned} \tag{7.3.19}$$

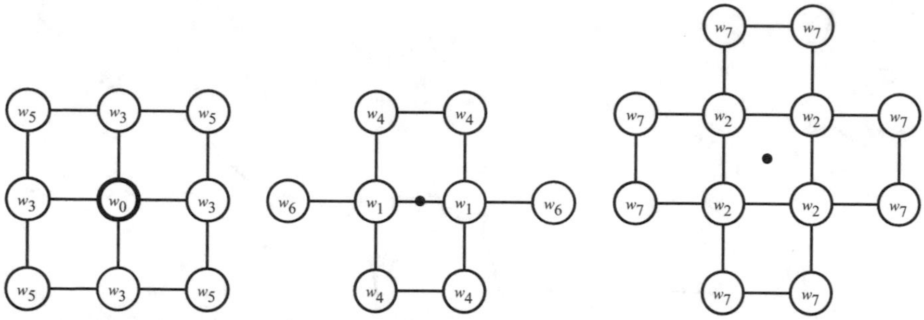

Fig. 7.7 The $(0,0)$-stencil (left), $(\frac{1}{2},0)$-stencil (middle), and $(\frac{1}{2},\frac{1}{2})$-stencil of the primal subdivision scheme in Example 7.3.4, where the weights $w_0,\dots,w_7$ are given in (7.3.19). The $(0,\frac{1}{2})$-stencil is the same as the $(\frac{1}{2},0)$-stencil. It is an interpolatory $2I_2$-wavelet filter if $w_3 = \frac{1-t_1+2t_2}{8} = 0$ and $w_5 = \frac{t_1}{16} = 0$ (i.e., $t_1 = 0$ and $t_2 = -\frac{1}{2}$)

with $t_1, t_2 \in \mathbb{R}$. Then $\operatorname{sr}(a, 2I_2) = 4$ and $\operatorname{lpm}(a) = 2$. If $t_2 = -\frac{1}{2}$, then $\operatorname{lpm}(a) = 4$. If $t_1 = 0$ and $t_2 = -\frac{1}{2}$, then a is an interpolatory $2I_2$-wavelet filter and $\operatorname{sm}_2(a, 2I_2) \approx 2.44077$ (and hence, $\operatorname{sm}_\infty(a, 2I_2) \geqslant \operatorname{sm}_2(a, 2I_2) - 1 \approx 1.44077$). If $t_1 = \frac{1}{4}$ and $t_2 = 0$, then $a = \otimes^2(a_4^B(\cdot - 2))$ is the tensor product of the centered B-spline filter of order 4 with $\operatorname{sm}_p(a, 2I_2) = 3 + 1/p$ for all $1 \leqslant p \leqslant \infty$. See Fig. 7.7 for its subdivision stencils.

Example 7.3.5 The triplet $(a, 2I_2, D_6)$ is a primal subdivision triplet with

$$a = \frac{1}{4}\begin{bmatrix} 0 & 0 & 0 & w_5 & w_4 & w_4 & w_5 \\ 0 & 0 & w_4 & w_3 & w_2 & w_3 & w_4 \\ 0 & w_4 & w_2 & w_1 & w_1 & w_2 & w_4 \\ w_5 & w_3 & w_1 & \underline{w_0} & w_1 & w_3 & w_5 \\ w_4 & w_3 & w_1 & w_1 & w_2 & w_4 & 0 \\ w_4 & w_3 & w_2 & w_3 & w_4 & 0 & 0 \\ w_5 & w_4 & w_4 & w_5 & 0 & 0 & 0 \end{bmatrix}_{[-3,3]^2}, \qquad (7.3.20)$$

where

$$w_0 = \tfrac{5-t}{8}, \quad w_1 = \tfrac{3-3t}{8}, \quad w_2 = \tfrac{1}{8}, \quad w_3 = \tfrac{1+t}{16}, \quad w_4 = \tfrac{t}{16}, \quad w_5 = 0 \qquad (7.3.21)$$

with $t \in \mathbb{R}$. Then $\operatorname{sr}(a, 2I_2) = 4$ and $\operatorname{lpm}(a) = 2$. If $t = 0$, then $a = a_{P_{HH}}(\cdot - (3,3))$ is the centered box spline filter with $\operatorname{sm}_p(a, 2I_2) = 3 + 1/p$ for all $1 \leqslant p \leqslant \infty$ and $\operatorname{lpm}(a) = 2$, where P_{HH} is defined in (7.1.22). If $t = -1$, then $\operatorname{lpm}(a) = 4$ and the mask a is an interpolatory $2I_2$-wavelet filter with $\operatorname{sm}_2(a, 2I_2) \approx 2.44077$ (and

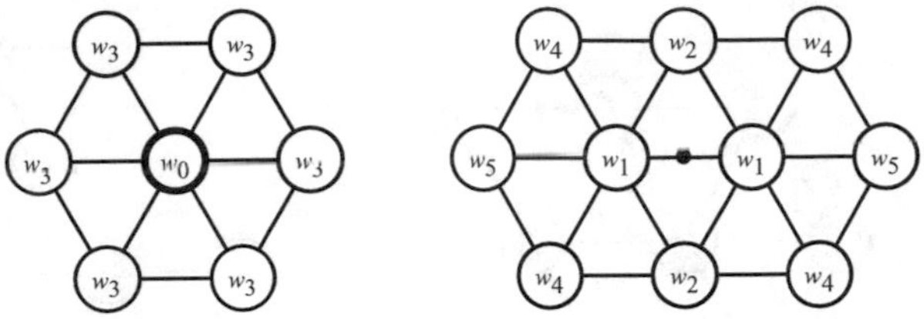

Fig. 7.8 The $(0,0)$-stencil (left) and $(\frac{1}{2},0)$-stencil (right) of the primal subdivision scheme in Example 7.3.5, where the weights $w_0,\dots,w_5$ are given in (7.3.21). The $(0,\frac{1}{2})$-stencil and $(\frac{1}{2},\frac{1}{2})$-stencil are the same as the $(\frac{1}{2},0)$-stencil. It is an interpolatory $2I_2$-wavelet filter if $w_3 = \frac{1+t}{16} = 0$

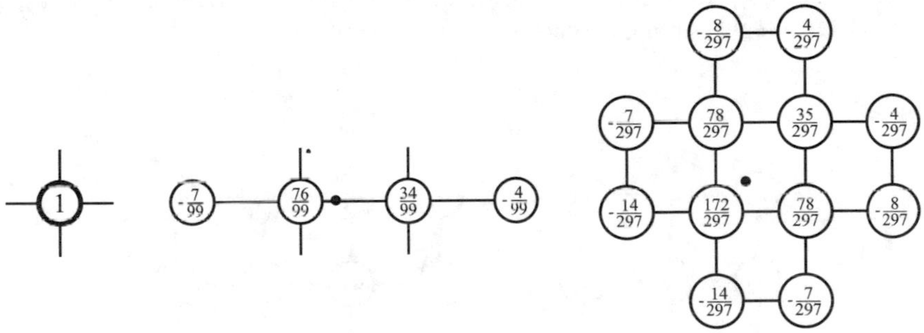

Fig. 7.9 The $(0,0)$-stencil (left), $(\frac{1}{3},0)$-stencil (middle), and $(\frac{1}{3},\frac{1}{3})$-stencil (right) of a primal subdivision triplet $(a,3I_2,D_4)$ in Example 7.3.6, where a is an interpolatory $3I_2$-wavelet filter supported on $[-5,5]^2$. The $(\frac{2}{3},0)$, $(0,\frac{1}{3})$, $(0,\frac{2}{3})$-stencils are the same as the $(\frac{1}{3},0)$-stencil, while $(\frac{1}{3},\frac{2}{3})$, $(\frac{2}{3},\frac{1}{3})$, $(\frac{2}{3},\frac{2}{3})$-stencils are the same as the $(\frac{1}{3},\frac{1}{3})$-stencil

hence, $\mathrm{sm}_\infty(a,2I_2) \geqslant \mathrm{sm}_2(a,2I_2) - 1 \approx 1.44077$). See Fig. 7.8 for its subdivision stencils.

Example 7.3.6 The triplet $(a,3I_2,D_4)$ is a primal subdivision triplet, where the interpolatory $3I_2$-wavelet filter/mask a is given through its stencils in Fig. 7.9. Then $\mathrm{sr}(a,3I_2)=3$, $\mathrm{lpm}(a)=3$ and $\mathrm{sm}_\infty(a,3I_2)=\log_3 11 \approx 2.18266$.

The triplet $(a,3I_2,D_6)$ is a primal subdivision triplet, where the interpolatory $3I_2$-wavelet filter/mask a is given through its stencils in Fig. 7.10. Then $\mathrm{sr}(a,3I_2)=3$, $\mathrm{lpm}(a)=3$ and $\mathrm{sm}_\infty(a,3I_2)=\log_3 11 \approx 2.18266$.

Example 7.3.7 The triplet $(a,\mathsf{M}_{\sqrt{2}},D_4)$ is a primal subdivision triplet, where a is given in (7.3.18) with the factor $\frac{1}{4}$ being replaced by $\frac{1}{2}$ and with $w_5=w_6=w_7=0$ and

$$w_0 = \tfrac{1}{2} + t_1 - 4t_2, \quad w_1 = \tfrac{1-2t_2}{4}, \quad w_2 = \tfrac{1}{8} - \tfrac{t_1}{2} + t_2, \quad w_3 = \tfrac{t_1}{4}, \quad w_4 = \tfrac{t_2}{4} \tag{7.3.22}$$

with $t_1,t_2 \in \mathbb{R}$. Then $\mathrm{sr}(a,\mathsf{M}_{\sqrt{2}})=4$ and $\mathrm{lpm}(a)=2$.

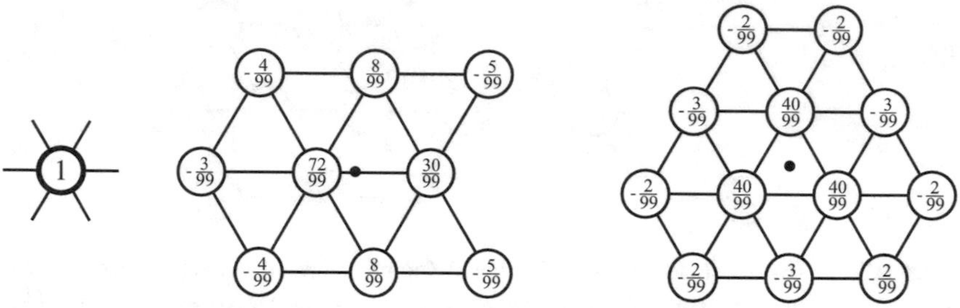

Fig. 7.10 The $(0,0)$-stencil (left), $(\frac{1}{3},0)$-stencil (middle), and $(\frac{2}{3},\frac{1}{3})$-stencil (right) of a primal subdivision triplet $(a, 3I_2, D_6)$ in Example 7.3.6, where a is an interpolatory $3I_2$-wavelet filter supported on $[-5,5]^2$. The $(\frac{2}{3},0)$, $(0,\frac{1}{3})$, $(0,\frac{2}{3})$, $(\frac{1}{3},\frac{1}{3})$, $(\frac{2}{3},\frac{2}{3})$-stencils are the same as the $(\frac{1}{3},0)$-stencil while the $(\frac{1}{3},\frac{2}{3})$-stencil is the same as the $(\frac{2}{3},\frac{1}{3})$-stencil

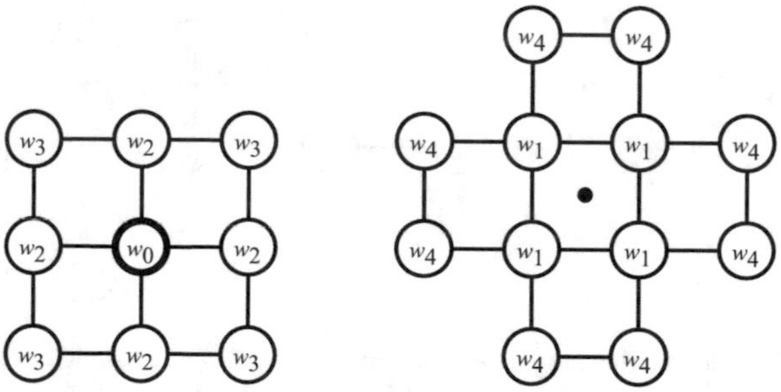

Fig. 7.11 The $(0,0)$-stencil (left) and the $(\frac{1}{2},\frac{1}{2})$-stencil (right) of the primal subdivision scheme in Example 7.3.7, where the weights $w_0,\dots,w_4$ are given in (7.3.22). It is an interpolatory $\mathsf{M}_{\sqrt{2}}$-wavelet filter if $w_2 = \frac{1}{8} - \frac{t_1}{2} + t_2 = 0$ and $w_3 = \frac{t_1}{4} = 0$ (i.e., $t_1 = 0, t_2 = -\frac{1}{8}$)

(1) $\operatorname{lpm}(a) = 4$ whenever $t_2 = -\frac{1}{8}$. If $t_1 = -\frac{17}{64}$ and $t_2 = -\frac{1}{8}$, then $\operatorname{sm}_2(a, \mathsf{M}_{\sqrt{2}}) \approx 3.760043$ and $\operatorname{sm}_\infty(a, \mathsf{M}_{\sqrt{2}}) \approx 3.555768$; If $t_1 = -\frac{1}{4}$ and $t_2 = -\frac{1}{8}$, then $\operatorname{sm}_2(a, \mathsf{M}_{\sqrt{2}}) \approx 3.720832$ and $\operatorname{sm}_\infty(a, \mathsf{M}_{\sqrt{2}}) \approx 3.331223$; If $t_1 = 0$ and $t_2 = -\frac{1}{8}$, then a is an interpolatory $\mathsf{M}_{\sqrt{2}}$-wavelet filter with $\operatorname{sm}_2(a, \mathsf{M}_{\sqrt{2}}) \approx 2.447923$ and $\operatorname{sm}_\infty(a, 2I_2) = 1.459340$. Note that $\widehat{a}(\xi) \geqslant 0$ for all $t_1 = 0, -\frac{1}{4}$ or $-\frac{17}{64}$ and $t_2 = -\frac{1}{8}$.

(2) If $t_1 = \frac{1}{8}$ and $t_2 = \frac{1}{16}$, then $\operatorname{sr}(a, \mathsf{M}_{\sqrt{2}}) = 6$ and $\operatorname{lpm}(a) = 2$ with $\operatorname{sm}_2(a, \mathsf{M}_{\sqrt{2}}) = 6$ and $\operatorname{sm}_\infty(a, \mathsf{M}_{\sqrt{2}}) = 6$. Moreover, all the coefficients of the filter a are nonnegative and $\widehat{a}(\xi) \geqslant 0$ for all $\xi \in \mathbb{R}^2$.

(3) If $t_1 = t_2 = 0$, then $a = \otimes^2(a_2^B(\cdot - 1))$ with $\operatorname{sm}_2(a, \mathsf{M}_{\sqrt{2}}) = 4$ and $\operatorname{sm}_\infty(a, \mathsf{M}_{\sqrt{2}}) = 4$. Note that $\widehat{a}(\xi) \geqslant 0$ for all $\xi \in \mathbb{R}^2$.

See Fig. 7.11 for its subdivision stencils.

An example of a dual subdivision scheme $(a, \mathsf{M}_{\sqrt{2}}, D_4)$ is given by

$$a = \begin{bmatrix} -\frac{1}{32} & 0 & 0 & -\frac{1}{32} \\ 0 & \frac{9}{32} & \frac{9}{32} & 0 \\ 0 & \frac{9}{32} & \frac{9}{32} & 0 \\ -\frac{1}{32} & 0 & 0 & -\frac{1}{32} \end{bmatrix}_{[-1,2]^2} \tag{7.3.23}$$

with $\mathrm{sr}(a, \mathsf{M}_{\sqrt{2}}) = 4$, $\mathrm{lpm}(a) = 4$ and $\mathrm{sm}_2(a, \mathsf{M}_{\sqrt{2}}) \approx 3.036544$ (and therefore, $\mathrm{sm}_\infty(a, \mathsf{M}_{\sqrt{2}}) \geqslant 2.036544$).

Example 7.3.8 The triplet $(a, \mathsf{M}_{\sqrt{3}}, D_6)$ is a primal subdivision triplet, where a is given in (7.3.20) with the factor $\frac{1}{4}$ being replaced by $\frac{1}{3}$ and

$$w_0 = \tfrac{9+4t}{9}, \quad w_1 = \tfrac{32}{81}, \quad w_2 = -\tfrac{t}{9}, \quad w_3 = -\tfrac{1}{81}, \quad w_4 = -\tfrac{2}{81}, \quad w_5 = \tfrac{t}{27} \tag{7.3.24}$$

with $t \in \mathbb{R}$. Then $\mathrm{sr}(a, \mathsf{M}_{\sqrt{3}}) = 4$ and $\mathrm{lpm}(a) = 4$. If $t = 0$, then a is an interpolatory $\mathsf{M}_{\sqrt{3}}$-wavelet filter with $\mathrm{sm}_2(a, \mathsf{M}_{\sqrt{3}}) \approx 2.529957$ and $\mathrm{sm}_\infty(a, \mathsf{M}_{\sqrt{3}}) \approx 1.560782$ (since $\widehat{a}(\xi) \geqslant 0$ for all $\xi \in \mathbb{R}^2$). If $t = -\frac{5}{9}$, then $\mathrm{sr}(a, \mathsf{M}_{\sqrt{3}}) = 5$ and $\mathrm{sm}_2(a, \mathsf{M}_{\sqrt{3}}) \approx 4.007161$ (and therefore, $\mathrm{sm}_\infty(a, \mathsf{M}_{\sqrt{3}}) \geqslant 3.007161$). If $w_3 = w_4 = w_5 = 0$ and $w_0 = \frac{2}{3}, w_1 = \frac{1}{3}, w_2 = \frac{1}{18}$, then $\mathrm{sr}(a, \mathsf{M}_{\sqrt{3}}) = 3$ and $\mathrm{lpm}(a) = 2$ with $\mathrm{sm}_2(a, \mathsf{M}_{\sqrt{3}}) \approx 2.936044$ (and hence, $\mathrm{sm}_\infty(a, \mathsf{M}_{\sqrt{3}}) \geqslant 1.936044$). See Fig. 7.12 for its subdivision stencils.

To apply two-dimensional subdivision schemes in computer graphics, special subdivision rules have to be designed for extraordinary vertices, that is, vertices having valances other than 4 for the quadrilateral mesh using D_4 or vertices having valances other than 6 for the triangular mesh using D_6. This issue will not be addressed in this book.

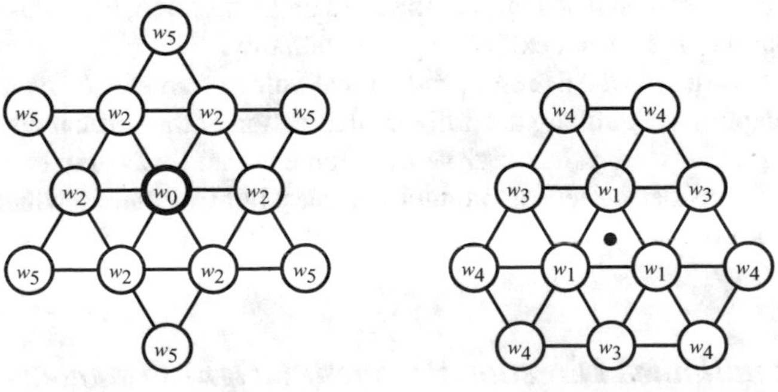

Fig. 7.12 The $(0, 0)$-stencil (left) and the $(\frac{2}{3}, \frac{1}{3})$-stencil (right) of the primal subdivision scheme in Example 7.3.8, where the weights $w_0, \ldots, w_5$ are given in (7.3.24). The $(\frac{1}{3}, \frac{2}{3})$-stencil is the same as the $(\frac{2}{3}, \frac{1}{3})$-stencil. It is an interpolatory $\mathsf{M}_{\sqrt{3}}$-wavelet filter if $t = 0$ (i.e., $w_2 = w_5 = 0$)

7.4 Directional Tensor Product Complex Tight Framelets for Image Processing

In this section we discuss applications of framelets to image processing by introducing a family of directional tensor product complex tight framelets. Such directional tensor product complex tight framelets perform much better in many applications than real-valued tensor product wavelets and framelets. In this section we concentrate on their construction and their particular applications to image processing.

Though tensor product real-valued framelets and wavelets are useful in applications, they lack directionality to capture edge-like singularities. To illustrate this point, let us look at the two-dimensional tensor product $\otimes^2\{\phi;\psi\}$ of the Haar orthogonal wavelet $\{\phi;\psi\}$ with $\phi = \chi_{[0,1]}$ and $\psi = \chi_{[0,\frac{1}{2}]} - \chi_{[\frac{1}{2},1]}$. Then $\phi \otimes \phi = \chi_{[0,1]^2}$ and

$$\phi \otimes \psi = \chi_{[0,1]\times[0,\frac{1}{2}]} - \chi_{[0,1]\times[\frac{1}{2},1]}, \quad \psi \otimes \phi = \chi_{[0,\frac{1}{2}]\times[0,1]} - \chi_{[\frac{1}{2},1]\times[0,1]},$$

$$\psi \otimes \psi = \chi_{[0,\frac{1}{2}]^2 \cup [\frac{1}{2},1]^2} - \chi_{[0,\frac{1}{2}]\times[\frac{1}{2},1]\cup[\frac{1}{2},1]\times[0,\frac{1}{2}]}.$$

We observe that $\phi \otimes \psi$ has the horizontal direction, $\psi \otimes \phi$ has the vertical direction, but $\psi \otimes \psi$ does not exhibit any directionality. The above Haar orthogonal wavelet $\{\phi;\psi\}$ has the underlying orthogonal wavelet filter bank $\{a;b\}$ with $a = \{\frac{1}{2}, \frac{1}{2}\}_{[0,1]}$ and $b = \{\frac{1}{2}, -\frac{1}{2}\}_{[0,1]}$. Then

$$a \otimes a = \begin{bmatrix} \frac{1}{4} & \frac{1}{4} \\ \frac{1}{4} & \frac{1}{4} \end{bmatrix}_{[0,1]^2}, \quad a \otimes b = \begin{bmatrix} -\frac{1}{4} & -\frac{1}{4} \\ \frac{1}{4} & \frac{1}{4} \end{bmatrix}_{[0,1]^2},$$

$$b \otimes a = \begin{bmatrix} \frac{1}{4} & -\frac{1}{4} \\ \frac{1}{4} & -\frac{1}{4} \end{bmatrix}_{[0,1]^2}, \quad b \otimes b = \begin{bmatrix} -\frac{1}{4} & \frac{1}{4} \\ \frac{1}{4} & -\frac{1}{4} \end{bmatrix}_{[0,1]^2}.$$

From above, we observe that $a \otimes b$ has the horizontal direction, $b \otimes a$ has the vertical direction, but $b \otimes b$ does not exhibit any directionality.

In short, two-dimensional tensor product real-valued framelets/wavelets can only effectively capture edges along the horizontal direction or the vertical direction. This shortcoming of tensor product real-valued framelets and wavelets can be greatly remedied by considering tensor product complex tight framelets which takes the advantages of both fast framelet transform and the discrete Fourier transform.

7.4.1 *Bandlimited Directional Complex Tight Framelets*

We now construct directional tensor product complex tight framelets in this subsection. Recall that the bump function $\chi_{[c_L,c_R];\varepsilon_L,\varepsilon_R}$ is defined in (4.6.21). Let

$s \in \mathbb{N}$ and $0 < c_1 < c_2 < \cdots < c_{s+1} := \pi$ and $\varepsilon_0, \varepsilon_1, \ldots, \varepsilon_{s+1}$ be positive real numbers satisfying

$$\varepsilon_0+\varepsilon_1 \leqslant c_1 \leqslant \tfrac{\pi}{2}-\varepsilon_1 \text{ and } \varepsilon_\ell+\varepsilon_{\ell+1} \leqslant c_{\ell+1}-c_\ell \leqslant \pi-\varepsilon_\ell-\varepsilon_{\ell+1}, \quad \forall\, \ell = 1, \ldots, s.$$

A real-valued low-pass filter a and $2s$ complex-valued high-pass filters $b_1^+, \ldots, b_s^+$, $b_1^-, \ldots, b_s^-$ are defined through their 2π-periodic Fourier series on the basic interval $(-\pi, \pi]$ as follows:

$$\widehat{a} := \chi_{[-c_1,c_1];\varepsilon_1,\varepsilon_1}, \quad \widehat{b_\ell^+} := \chi_{[c_\ell,c_{\ell+1}];\varepsilon_\ell,\varepsilon_{\ell+1}}, \quad \widehat{b_\ell^-} := \overline{\widehat{b_\ell^+}(-\cdot)}, \; \ell = 1, \ldots, s. \tag{7.4.1}$$

Then $\mathbb{C}\mathrm{TF}_{2s+1} := \{a; b_1^+, \ldots, b_s^+, b_1^-, \ldots, b_s^-\}$ is a tight framelet filter bank. Note that $\widehat{b_\ell^-} = \overline{\widehat{b_\ell^+}(-\cdot)}$ is equivalent to $b_\ell^- = \overline{b_\ell^+}$, that is, $b_\ell^-(k) = \overline{b_\ell^+(k)}$ for all $k \in \mathbb{Z}$. The tensor product complex tight framelet filter bank TP-$\mathbb{C}\mathrm{TF}_{2s+1}$ for dimension d is simply

$$\text{TP-}\mathbb{C}\mathrm{TF}_{2s+1} := \otimes^d \mathbb{C}\mathrm{TF}_{2s+1} = \otimes^d\{a; b_1^+, \ldots, b_s^+, b_1^-, \ldots, b_s^-\}.$$

We can write TP-$\mathbb{C}\mathrm{TF}_{2s+1} = \{\otimes^d a; \text{TP-}\mathbb{C}\mathrm{TF}\text{-HP}_{2s+1}\}$ with TP-$\mathbb{C}\mathrm{TF}$-$\mathrm{HP}_{2s+1} :=$ TP-$\mathbb{C}\mathrm{TF}_{2s+1} \backslash \{\otimes^d a\}$. This tensor product tight framelet filter bank TP-$\mathbb{C}\mathrm{TF}_{2s+1}$ has one real-valued low-pass filter $\otimes^d a$ and $(2s+1)^d - 1$ complex-valued high-pass filters.

To improve directionality of TP-$\mathbb{C}\mathrm{TF}_{2s+1}$, we now introduce another family of tensor product complex tight framelet filter banks TP-$\mathbb{C}\mathrm{TF}_{2s+2}$. Define filters $a, b_1^+, \ldots, b_s^+, b_1^-, \ldots, b_s^-$ as in (7.4.1). Define two auxiliary complex-valued filters a^+, a^- through their 2π-periodic Fourier series on the basic interval $(-\pi, \pi]$ by

$$\widehat{a^+} := \chi_{[0,c_1];\varepsilon_0,\varepsilon_1}, \qquad \widehat{a^-} := \overline{\widehat{a^+}(-\cdot)}. \tag{7.4.2}$$

Then $\mathbb{C}\mathrm{TF}_{2s+2} := \{a^+, a^-; b_1^+, \ldots, b_s^+, b_1^-, \ldots, b_s^-\}$, with the auxiliary filters a^+ and a^- in (7.4.2), is also a tight framelet filter bank. The tensor product complex tight framelet filter bank TP-$\mathbb{C}\mathrm{TF}_{2s+2}$ for dimension d is defined to be

$$\text{TP-}\mathbb{C}\mathrm{TF}_{2s+2} := \{\otimes^d a; \text{TP-}\mathbb{C}\mathrm{TF}\text{-HP}_{2s+2}\},$$

where TP-$\mathbb{C}\mathrm{TF}$-HP_{2s+2} consists of total $(2s+2)^d - 2^d$ complex-valued high-pass filters given by

$$\big(\otimes^d \{a^+, a^-, b_1^+, \ldots, b_s^+, b_1^-, \ldots, b_s^-\}\big) \backslash \big(\otimes^d \{a^+, a^-\}\big).$$

For applications of tensor product complex tight framelets TP-$\mathbb{C}$TF$_m$ for image and video processing, we choose the parameters as follows: For TP-$\mathbb{C}$TF$_3$,

$$c_1 = \tfrac{33}{32}, \quad c_2 = \pi, \quad \varepsilon_1 = \tfrac{69}{128}, \quad \varepsilon_2 = \tfrac{51}{512}. \tag{7.4.3}$$

For TP-$\mathbb{C}$TF$_4$,

$$c_1 = \tfrac{291}{256}, \quad c_2 = \pi, \quad \varepsilon_0 = \tfrac{35}{128}, \quad \varepsilon_1 = \tfrac{27}{64}, \quad \varepsilon_2 = \tfrac{1}{2}. \tag{7.4.4}$$

For TP-$\mathbb{C}$TF$_6$ and TP-$\mathbb{C}$TF$_5$,

$$c_1 = \tfrac{119}{128}, \; c_2 = \tfrac{\pi}{2} + \tfrac{119}{256}, \; c_3 = \pi, \; \varepsilon_0 = \tfrac{35}{128}, \; \varepsilon_1 = \tfrac{81}{128}, \; \varepsilon_2 = \tfrac{115}{256}, \; \varepsilon_3 = \tfrac{115}{256}. \tag{7.4.5}$$

See Fig. 7.13 for graphs of the one-dimensional complex tight framelet filter banks $\mathbb{C}$TF$_3$, $\mathbb{C}$TF$_4$, and $\mathbb{C}$TF$_6$ in the frequency domain.

The directionality of TP-$\mathbb{C}$TF$_m$ comes from the frequency separation property. By design, $\widehat{b_\ell^+}(\xi) \approx 0$ for $\xi \in [-\pi, 0]$. Since $\widehat{a}(0) = 1$, we can define a function $\widehat{\phi}(\xi) := \prod_{j=1}^{\infty} \widehat{a}(2^{-j}\xi)$. Hence, their associated wavelet function $\widehat{\psi}(\xi) = \widehat{b}(\xi/2)\widehat{\phi}(\xi/2)$ with $b = b_\ell^+$ or $b = b_\ell^-$ enjoys the following frequency separation property:

$$\widehat{\psi}(\xi) \approx 0, \qquad \forall\, \xi < 0 \qquad \text{or} \qquad \widehat{\psi}(\xi) \approx 0, \qquad \forall\, \xi > 0. \tag{7.4.6}$$

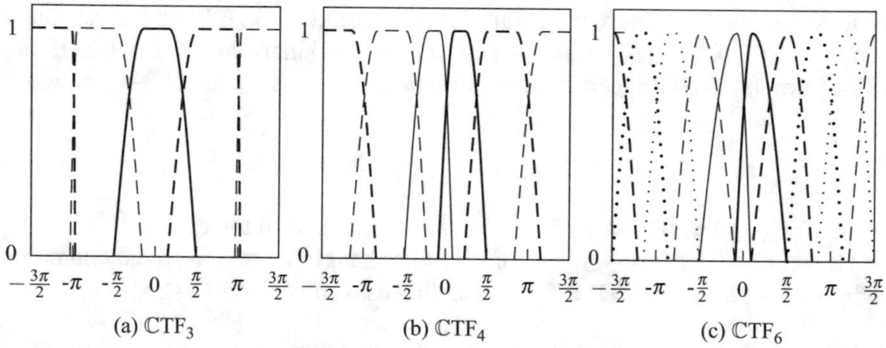

Fig. 7.13 (**a**) tight framelet filter bank $\mathbb{C}\text{TF}_3 = \{a; b_1^+, b_1^-\}$ in the frequency domain with parameters in (7.4.3). Solid line for the low-pass filter $\widehat{a}$. Dashed line for $\widehat{b_1^+}$ and thin dashed line for $\widehat{b_1^-}$. (**b**) $\mathbb{C}\text{TF}_4 = \{a^+, a^-; b_1^+, b_1^-\}$ in the frequency domain with parameters in (7.4.4). Solid line for $\widehat{a^+}$ and thin solid line for $\widehat{a^-}$. Dashed line for $\widehat{b_1^+}$ and thin dashed line for $\widehat{b_1^-}$. (**c**) $\mathbb{C}\text{TF}_6 = \{a^+, a^-; b_1^+, b_2^+, b_1^-, b_2^-\}$ in the frequency domain with parameters in (7.4.5). Solid line for $\widehat{a^+}$ and thin solid line for $\widehat{a^-}$. Dashed line for $\widehat{b_1^+}$ and thin dashed line for $\widehat{b_1^-}$. Dotted line for $\widehat{b_2^+}$ and thin dotted line for $\widehat{b_2^-}$

To see directionality, let us look at TP-$\mathbb{C}$TF$_m$ in dimension two: every wavelet function $\psi = \psi_1 \otimes \psi_2$, where ψ_1, ψ_2 are the one-dimensional wavelet functions satisfying (7.4.6). Due to the frequency separation property of b_ℓ^+ and b_ℓ^-, we see that

$$\widehat{\psi_1}(\xi) = \chi_{[\zeta_1-c_1,\zeta_1+c_1];\varepsilon_1,\varepsilon_1} \quad \text{and} \quad \widehat{\psi_2}(\xi) = \chi_{[\zeta_2-c_2,\zeta_2+c_2];\varepsilon_2,\varepsilon_2}$$

for some $\zeta_1, \zeta_2 \in \mathbb{R}$ and $c_1, c_2, \varepsilon_1, \varepsilon_2 > 0$. In other words, most energy of $\widehat{\psi}$ lies inside the rectangle $[\zeta_1 - c_1, \zeta_1 + c_1] \times [\zeta_2 - c_2, \zeta_2 + c_2]$ whose center is the point $(\zeta_1, \zeta_2)^\mathsf{T}$. Quite often $c_1 \approx c_2$. Define

$$g(\xi_1, \xi_2) := \chi_{[-c_1,c_1],\varepsilon_1,\varepsilon_1}(\xi_1)\chi_{[-c_2,c_2],\varepsilon_2,\varepsilon_2}(\xi_2), \qquad \xi_1, \xi_2 \in \mathbb{R}$$

and let f denote its inverse Fourier transform, i.e., $\widehat{f} = g$. Noting that $\overline{g(-\xi)} = g(\xi)$, we see that f is a *real-valued* tensor product function, which is almost isotropic (due to $c_1 \approx c_2$) and concentrates around the origin. Define a vector $\zeta := (\zeta_1, \zeta_2)^\mathsf{T}$ which is the mass center of the function $\widehat{\psi}$. Then $\widehat{\psi}(\xi) = g(\xi - \zeta) = \widehat{f}(\xi - \zeta)$ for all $\xi \in \mathbb{R}^2$, from which we conclude $\psi(x) = f(x)e^{i\zeta\cdot x}$, which directly leads to

$$\psi^{[r]}(x) = f(x)\cos(\zeta \cdot x), \qquad \psi^{[i]}(x) = f(x)\sin(\zeta \cdot x), \quad x \in \mathbb{R}^2,$$

where $\psi(x) = \psi^{[r]}(x) + i\psi^{[i]}(x)$ with real-valued functions $\psi^{[r]}$ and $\psi^{[i]}$. The functions $\psi^{[r]}$ and $\psi^{[i]}$ have directionality, mainly due to the directional cosine waves $\cos(\zeta \cdot x)$ and sine waves $\sin(\zeta \cdot x)$ (provided $\zeta \neq 0$). When $\|\zeta\| \neq 0$ is small, the cosine wave $\cos(\zeta \cdot x)$ and the sine wave $\sin(\zeta \cdot x)$ have low frequency (i.e., slowly oscillating waves). Thus, the elements $\psi^{[r]}$ and $\psi^{[i]}$ exhibit edge-like shapes (called *edge-like directional elements*). Such edge-like directional elements can be used to capture edge singularities. On the other hand, if $\|\zeta\| \neq 0$ is relatively large, the cosine and sine waves have high frequency (i.e., rapidly oscillating waves). Thus, the elements $\psi^{[r]}$ and $\psi^{[i]}$ exhibit texture-like shapes (called *texture-like directional elements*). See Figs. 7.14, 7.15, and 7.16 for the edge-like and texture-like directional elements of TP-$\mathbb{C}$TF$_3$, TP-$\mathbb{C}$TF$_4$ and TP-$\mathbb{C}$TF$_6$ in dimension two.

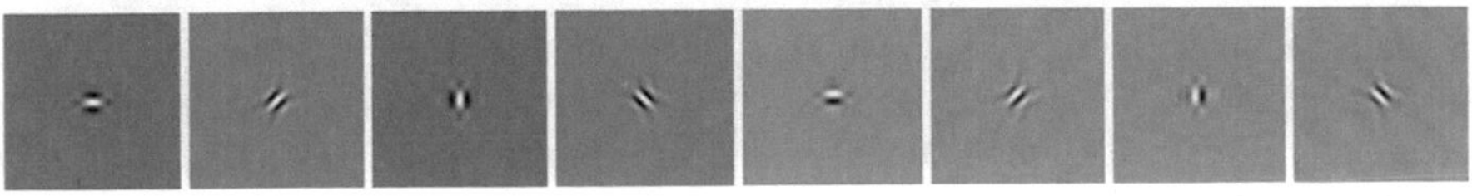

Fig. 7.14 The real part (the first four) and the imaginary part (the last four) of the generators in TP-$\mathbb{C}$TF$_3$ with parameters in (7.4.3)

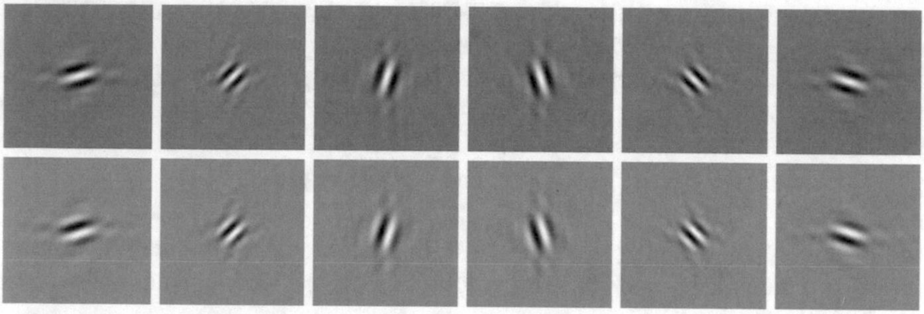

Fig. 7.15 The first row shows the real part and the second row shows the imaginary part of the generators in TP-$\mathbb{C}$TF$_4$ with parameters in (7.4.4)

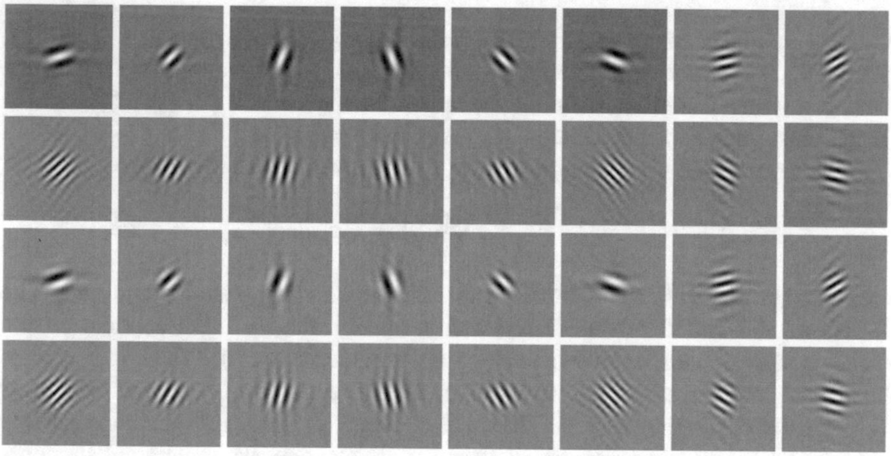

Fig. 7.16 The first two rows show the real part and the last two rows show the imaginary part of the generators in TP-$\mathbb{C}$TF$_6$ with parameters in (7.4.5)

7.4.2 Compactly Supported Directional Complex Tight Framelets

Spatially compactly supported wavelets and framelets are of particular interest in both theory and applications. In this subsection, we shall construct spatially compactly supported directional tensor product tight framelets TP-$\mathbb{C}$TF$_m$.

Let us first introduce a quantity to measure the frequency separation property for directionality. For a filter $b = \{b(k)\}_{k\in\mathbb{Z}}$ which is not identically zero, the *frequency separation* of b is measured by

$$\operatorname{Fsp}(b) := \frac{\min\{\frac{1}{\pi}\int_{-\pi}^{0} |\widehat{b}(\xi)|^2 d\xi, \frac{1}{\pi}\int_{0}^{\pi} |\widehat{b}(\xi)|^2 d\xi\}}{\frac{1}{2\pi}\int_{-\pi}^{\pi} |\widehat{b}(\xi)|^2 d\xi}. \tag{7.4.7}$$

Note that $\frac{1}{2\pi}\int_{-\pi}^{\pi}|\widehat{b}(\xi)|^2 d\xi = \|b\|_{l_2(\mathbb{Z})}^2$. It is straightforward to observe that $0 \leqslant \operatorname{Fsp}(b) \leqslant 1$. The smaller the quantity $\operatorname{Fsp}(b)$, the better frequency separation of the filter b. If b is real-valued, then $\operatorname{Fsp}(b) = 1$. It is trivial that $\operatorname{Fsp}(\overline{b}) = \operatorname{Fsp}(b)$.

The following algorithm can be used for constructing directional tensor product complex tight framelets TP-$\mathbb{C}$TF$_3$ and TP-$\mathbb{C}$TF$_4$.

Algorithm 7.4.1 *Let $a \in l_0(\mathbb{Z})$ be a real-valued low-pass filter satisfying $|\widehat{a}(\xi)|^2 + |\widehat{a}(\xi+\pi)|^2 \leqslant 1$.*

(S1) Construct a complex-valued filter a^+ by

$$\widehat{a^+}(\xi) := \widehat{a}(\xi)\widehat{u^+}(2\xi) \quad \textit{with} \quad \widehat{u^+}(\xi) := [\widehat{u_1}(\xi) + i\widehat{u_2}(\xi)]/\sqrt{2},$$

where the 2π-periodic trigonometric polynomials $\widehat{u_1}$ and $\widehat{u_2}$ are given by

$$\begin{bmatrix} \widehat{u_1}(\xi) & \widehat{u_2}(\xi) \\ \widehat{u_3}(\xi) & \widehat{u_4}(\xi) \end{bmatrix} := \begin{bmatrix} \cos(t_0) & -\sin(t_0) \\ \sin(t_0) & \cos(t_0) \end{bmatrix} \prod_{j=1}^{N} \begin{bmatrix} \cos(t_j) & -\sin(t_j) \\ e^{-i\xi}\sin(t_j) & e^{-i\xi}\cos(t_j) \end{bmatrix} \tag{7.4.8}$$

such that the real numbers $t_0, \dots, t_N \in (-\pi, \pi]$ (with properly chosen N) are found by solving the minimization problem $\min_{t_0,\dots,t_N} \int_0^{\pi} (|\widehat{a^+}(\xi+\pi)|^2 + |\widehat{a^+}(-\xi)|^2)d\xi$.

(S2) Construct an initial real-valued finitely supported tight framelet filter bank $\{a; b_1, b_2\}$ by Algorithm 3.4.1 or Algorithm 3.4.4 in Chap. 3;

(S3) Construct a complex-valued filter b^+ by $\widehat{b^+}(\xi) := \widehat{b_1}(\xi)\widehat{v_1}(2\xi) + \widehat{b_2}(\xi)\widehat{v_2}(2\xi)$ with

$$\widehat{v_1}(\xi) := [\widehat{u_1}(\xi) + i\widehat{u_2}(\xi)]/\sqrt{2}, \qquad \widehat{v_2}(\xi) := [\widehat{u_3}(\xi) + i\widehat{u_4}(\xi)]/\sqrt{2},$$

where the 2π-periodic trigonometric polynomials $\widehat{u_1}, \widehat{u_2}, \widehat{u_3}, \widehat{u_4}$ are given in (7.4.8) such that the real numbers $t_0, \dots, t_N \in (-\pi, \pi]$ (with properly chosen N) are found by solving the minimization problem $\min_{t_0,\dots,t_N} \int_0^{\pi} |\widehat{b^+}(\xi+\pi)|^2 d\xi$.

Then $\mathbb{C}$TF$_3$:= $\{a; b^+, b^-\}$ and $\mathbb{C}$TF$_4$:= $\{a^+, a^-; b^+, b^-\}$ with $a^- := \overline{a^+}$ and $b^- := \overline{b^+}$ are finitely supported tight framelet filter banks having the frequency separation property with small $\operatorname{Fsp}(a^+)$ and $\operatorname{Fsp}(b^+)$.

Applying Algorithm 7.4.1, we have the following examples of $\mathbb{C}$TF$_3$ and $\mathbb{C}$TF$_4$.

Example 7.4.1 The real-valued interpolatory low-pass filter a is given by

$$a = a_4^I = \{-\tfrac{1}{32}, 0, \tfrac{9}{32}, \underline{\tfrac{1}{2}}, \tfrac{9}{32}, 0, -\tfrac{1}{32}\}_{[-3,3]}.$$

Note that $\mathrm{sr}(a) = 4$, $\mathrm{lpm}(a) = 4$, and $\mathrm{sm}(a) \approx 2.440765$. Applying Algorithm 7.4.1 with $N = 2$, we have tight framelet filter banks $\mathbb{C}\mathrm{TF}_3 = \{a; b^+, b^-\}$ and $\mathbb{C}\mathrm{TF}_4 = \{a^+, a^-; b^+, b^-\}$ with $a^- = \overline{a^+}$ and $b^- = \overline{b^+}$, where

$$
\begin{aligned}
a^+ =&\{(-0.015745539), (0), (0.142828613 - 0.015420853866\,i), \underline{(\mathbf{0.251928624428})},\\
&(0.131640994865 + 0.1376453675\,i), (-0.017900189 + 0.24673366186\,i),\\
&(-0.0258143954 + 0.149068540345\,i), (0.018277076528\,i),\\
&(0.00111876182 - 0.00513999832\,i), (0), (-0.001142317283\,i)\}_{[-3,7]},\\
b^+ =&\{(0.000985045 + 0.000985045\,i), (0), (-0.00032895 + 0.0263894258\,i),\\
&\underline{(\mathbf{-0.01576072478 - 0.01576072478\,i})}, (-0.13971439058 + 0.055019537\,i),\\
&(0.1232412445 - 0.30425277913\,i), (0.18652437945 + 0.3281211987\,i),\\
&(-0.12314386954 - 0.03884959657\,i), (-0.022753617 - 0.06070122286\,i),\\
&(-0.0059723278 + 0.0059723278\,i), (-0.00307679 + 0.003076789\,i)\}_{[-3,7]}.
\end{aligned}
$$

Note that $\mathrm{vm}(b^+) = \mathrm{vm}(b^-) = 2$. By calculation we have $\mathrm{Fsp}(a^+) = \mathrm{Fsp}(a^-) \approx 0.2591694$ and $\mathrm{Fsp}(b^+) = \mathrm{Fsp}(b^-) \approx 0.2732070$. See Fig. 7.17 for graphs of the one-dimensional complex tight framelet filter banks $\mathbb{C}\mathrm{TF}_3$ and $\mathbb{C}\mathrm{TF}_4$.

The construction of $\mathbb{C}\mathrm{TF}_5$ and $\mathbb{C}\mathrm{TF}_6$ is similar to $\mathbb{C}\mathrm{TF}_3$ and $\mathbb{C}\mathrm{TF}_4$ in Algorithm 7.4.1 but is more complicated. Here we only provide one example of $\mathbb{C}\mathrm{TF}_5$ and $\mathbb{C}\mathrm{TF}_6$.

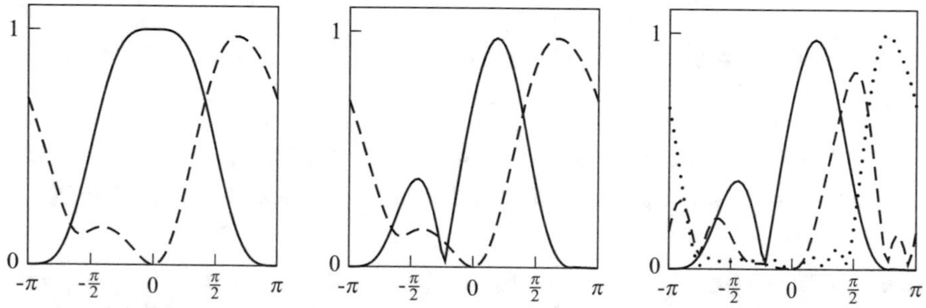

(a) $\mathbb{C}\mathrm{TF}_3$ (b) $\mathbb{C}\mathrm{TF}_4$ (c) $\mathbb{C}\mathrm{TF}_6$

Fig. 7.17 (**a**) finitely supported tight framelet filter bank $\mathbb{C}\mathrm{TF}_3 = \{a; b^+, b^-\}$ in Example 7.4.1. Solid line for the low-pass filter $|\widehat{a}|$ and dashed line for $|\widehat{b^+}|$. Note that $|\widehat{b^-}| = |\widehat{b^+}|(-\cdot)$. (**b**) $\mathbb{C}\mathrm{TF}_4 = \{a^+, a^-; b^+, b^-\}$ in Example 7.4.1. Solid line for $|\widehat{a^+}|$ and dashed line for $|\widehat{b^+}|$. (**c**) $\mathbb{C}\mathrm{TF}_6 = \{a^+, a^-; b_1^+, b_2^+, b_1^-, b_2^-\}$ in Example 7.4.2. Solid line for $|\widehat{a^+}|$, dashed line for $|\widehat{b_1^+}|$, and dotted line for $|\widehat{b_2^+}|$. Note that $|\widehat{a^-}| = |\widehat{a^+}|(-\cdot)$ and $|\widehat{b_\ell^-}| = |\widehat{b_\ell^+}|(-\cdot)$ for $\ell = 1, 2$

Example 7.4.2 Let a, a^+, a^- be the same filters as in Example 7.4.1. We have tight framelet filter banks $\mathbb{C}\mathrm{TF}_5 = \{a; b_1^+, b_2^+, b_1^-, b_2^-\}$ and $\mathbb{C}\mathrm{TF}_6 = \{a^+, a^-; b_1^+, b_2^+, b_1^-, b_2^-\}$ with $b_1^- = \overline{b_1^+}$ and $b_2^- = \overline{b_2^+}$, where

$$\begin{aligned}
b_1^+ =\{&(0.05174315731 + 0.038460624072\,i), (-0.13886192828 + 0.0109210575\,i),\\
&(-0.01905234327 - 0.09242476265\,i), (0.15045489685 - 0.094873960985\,i),\\
&\underline{(\mathbf{0.06959495342 + 0.125144925336\,i})}, (-0.10767114502 + 0.07118797344\,i),\\
&(-0.03840325863 - 0.02899490232\,i), (0.05901529201 - 0.038592301235\,i),\\
&(-0.03934841148 + 0.03235183717\,i), (0.01006951137 - 0.021158500506\,i),\\
&(0), (0.002459275717 - 0.002021989823\,i)\}_{[-4,7]},\\
b_2^+ =\{&(-0.0184562119 - 0.001240142894\,i), (0.0188601309 - 0.03852007355\,i),\\
&(0.02089541806 + 0.0190732573\,i), (-0.023209282945 + 0.06437652099\,i),\\
&\underline{(\mathbf{-0.07069213266 - 0.112865117\,i})}, (0.18919531172 + 0.09715968545\,i),\\
&(-0.21959854587 + 0.0306408377\,i), (0.16073175564 - 0.1291477234\,i),\\
&(-0.05135628985 + 0.1258753104\,i), (-0.00957992122 - 0.0474853482\,i),\\
&(0), (0.003209768116 - 0.007867206902\,i)\}_{[-4,7]}.
\end{aligned}$$

Note that $\mathrm{vm}(b_1^+) = \mathrm{vm}(b_1^-) = 2$ and $\mathrm{vm}(b_2^+) = \mathrm{vm}(b_2^-) = 2$. By calculation we have $\mathrm{Fsp}(b_1^+) = \mathrm{Fsp}(b_1^-) \approx 0.1863289$ and $\mathrm{Fsp}(b_2^+) = \mathrm{Fsp}(b_2^-) \approx 0.1960113$. See Fig. 7.17 for graphs of the complex tight framelet filter bank $\mathbb{C}\mathrm{TF}_6$.

7.4.3 *Numerical Experiments on Image Processing*

By stacking the columns of an image or video, a signal/image/video can be regarded as a column vector $\mathsf{f} = (\mathsf{f}(1), \ldots, \mathsf{f}(d))^{\mathsf{T}} \in \mathbb{R}^d$. Let $\mathsf{g} = (\mathsf{g}(1), \ldots, \mathsf{g}(d))^{\mathsf{T}}$ be an observed corrupted signal/image/video:

$$\mathsf{g}(k) = \begin{cases} \mathsf{f}(k) + \mathsf{n}(k), & \text{if } k \in \Omega, \\ \mathsf{m}(k), & \text{if } k \in \Omega^c := \{1, \ldots, d\} \backslash \Omega, \end{cases}$$

where $\Omega \subseteq \{1, \ldots, d\}$ is an observable region, $\mathsf{n}(k)$ denotes independent identically distributed (i.i.d.) Gaussian white noise with zero mean and standard deviation σ, and $\mathsf{m}(k)$ is either an unknown missing pixel or impulse noise (such as salt-and-pepper impulse noise or random-valued impulse noise). The unobservable region Ω^c is also called *an inpainting mask*. The pixels in a grayscale image or video take an integer values in $[0, 255]$. For salt-and-pepper noise, a corrupted pixel in Ω^c takes a value either 0 or 255 with equal probability. For random-valued impulse noise, a corrupted pixel in Ω^c takes an integer value from $[0, 255]$ with equal probability.

The goal of image/data restoration is to recover the true signal/image/video f from the observed corrupted image g by suppressing the Gaussian noise n in the observable region Ω while inpainting/filling the unknown/missing data/pixels in the unobservable region Ω^c. If $\Omega = \{1, \ldots, d\}$, then it is a standard denoising problem. If Ω is a known proper subset of $\{1, \ldots, d\}$, then it is a standard inpainting problem with noise. If Ω is unknown, it is a problem of removing mixed Gaussian and impulse noise. Let $\mathcal{W}$ be an $n \times d$ matrix for decomposition and $\mathcal{V}$ be a $d \times n$ matrix for reconstruction such that $\mathcal{V}\mathcal{W} = I_d$ for perfect reconstruction. If the transform is based on a tight frame, then $\mathcal{V} = \mathcal{W}^\star$ and the ratio n/d is called the *redundancy rate* of the tight frame. The main steps of any transform-based method for signal/image/video restoration are as follows:

(1) Forward transform/decomposition: transform an observed signal/image g into the transform/framelet domain $\mathsf{c} := \mathcal{W}\mathsf{g}$, where c is called a vector of frame coefficients.
(2) Thresholding frame coefficients: $\mathring{\mathsf{c}} := \eta_\lambda(\mathsf{c})$, where η is a thresholding strategy with the threshold value λ, where λ is often chosen according to the noise standard deviation σ.
(3) Backward transform/reconstruction: $\mathring{\mathsf{f}} := \mathcal{V}\mathring{\mathsf{c}}$.

Then $\mathring{\mathsf{f}}$ is an estimated signal/image/video of the true unknown signal/image f. The above procedure can be also applied iteratively with a decreasing sequence of threshold values λ. Popular choices of the thresholding operations η are the hard-thresholding η_λ^{hard} or the soft-thresholding η_λ^{soft} in (1.3.2). In this book we use the bivariate shrinkage, a variant of the soft-thresholding. For a coefficient c from the vector c of frame coefficients, the local signal variance at the frame coefficient c is estimated to be

$$\sigma_c := \sqrt{\Big(\frac{1}{\# N_c} \sum_{j \in N_c} |c_j|^2 - \sigma^2\Big)_+},$$

where $x_+ := x$ for $x > 0$ and $x_+ := 0$ otherwise, and N_c is a $[-3, 3]^2$ window centering around the coefficient c in the same band (that is, all the coefficients $c_j, j \in N_c$ are obtained from the same filter). Let c_p be the parent coefficient of c. Then the bivariate shrinkage is defined to be

$$\eta_\sigma^{bs}(c) := \eta_{\lambda_c}^{soft}(c) \qquad \text{with} \quad \lambda_c := \frac{\sigma^2}{\sigma_c} \frac{\sqrt{3}}{\sqrt{1 + |c_p/c|^2}},$$

where $\eta_{\lambda_c}^{soft}$ is the soft-thresholding in (1.3.2) with the threshold value λ_c.

The *peak signal-to-noise ratio* (PSNR) with unit decibel (dB) is often used to measure the performance:

$$\mathrm{PSNR}(\mathsf{f},\mathring{\mathsf{f}}) := 10\log_{10}\frac{255^2}{\mathrm{MSE}(\mathsf{f}-\mathring{\mathsf{f}})} \quad \text{with} \quad \mathrm{MSE}(\mathsf{f}-\mathring{\mathsf{f}}) := \frac{1}{d}\sum_{k=1}^{d}|\mathsf{f}(k)-\mathring{\mathsf{f}}(k)|^2,$$

where f is the true signal/image to be recovered and $\mathring{\mathsf{f}}$ is a restored signal/image with support $\{1,\ldots,d\}$. The larger the PSNR value, the better performance the image restoration method.

The grayscale test images and inpainting masks are given in Fig. 7.18. We assume that the standard deviation σ of Gaussian noise is known. The results for image denoising are given in Table 7.1 using the bandlimited TP-$\mathbb{C}$TF$_4$ and TP-$\mathbb{C}$TF$_6$ with parameters in (7.4.4) and (7.4.5), respectively. The corresponding results using compactly supported TP-$\mathbb{C}$TF$_4$ and TP-$\mathbb{C}$TF$_6$ in Examples 7.4.1 and 7.4.2 are also given in Table 7.1.

The results for the image inpainting problem are given in Table 7.2 with the unobservable region Ω^c (that is, the inpainting mask) being known. We use the following iterative algorithm for image inpainting: Let $\mathsf{f}_0 := \mathsf{g}$ and

$$\mathsf{g}_j := \mathsf{f}_{j-1}\chi_{\Omega^c} + \mathsf{g}\chi_{\Omega}, \qquad \mathsf{f}_j := \mathcal{W}^\star(\eta^{bs}_{\sigma_j}(\mathcal{W}\mathsf{g}_j)), \qquad j\in\mathbb{N},$$

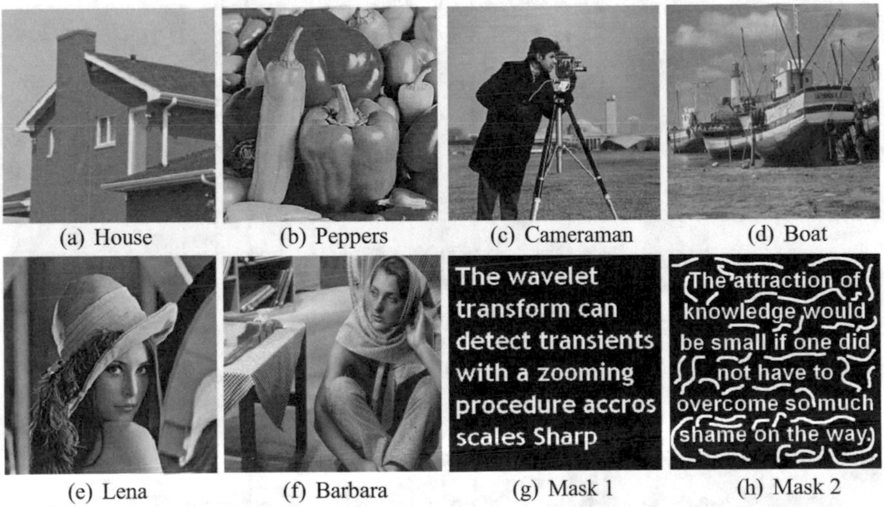

(a) House (b) Peppers (c) Cameraman (d) Boat

(e) Lena (f) Barbara (g) Mask 1 (h) Mask 2

Fig. 7.18 **(a)**–**(c)** are 256 × 256 grayscale test images and **(d)**–**(f)** are 512 × 512 grayscale test images. **(g)** is a 256 × 256 inpainting mask and **(h)** is a 512 × 512 inpainting mask

Table 7.1 PSNR values for the grayscale image denoising problem. For the two columns in each test image, the left top column is for bandlimited TP-$\mathbb{C}$TF$_4$ with parameters in (7.4.4), the right top column is for bandlimited TP-$\mathbb{C}$TF$_6$ with parameters in (7.4.5), the left bottom column is for compactly supported TP-$\mathbb{C}$TF$_4$ in Example 7.4.1 and the right bottom column is for compactly supported TP-$\mathbb{C}$TF$_6$ in Example 7.4.2. σ is the standard deviation of i.i.d. Gaussian white noise

σ	House		Peppers		Cameraman		Boat		Lena		Barbara	
5	38.56	38.93	35.82	35.88	37.28	37.45	36.53	36.92	38.12	38.37	37.42	37.84
10	34.94	35.43	31.97	32.12	32.93	33.16	33.10	33.41	35.16	35.48	33.65	34.18
20	31.89	32.31	28.77	28.98	29.07	29.37	30.03	30.26	32.33	32.57	29.97	30.54
30	30.16	30.51	26.90	27.17	27.17	27.44	28.26	28.44	30.62	30.80	27.79	28.38
50	27.93	28.12	24.53	24.82	24.91	25.21	26.12	26.25	28.46	28.54	25.21	25.71
100	24.81	24.87	21.74	21.87	22.10	22.29	23.53	23.58	25.55	25.52	22.45	22.64
5	38.42	38.65	35.87	35.88	37.26	37.34	36.53	36.82	38.05	38.30	37.05	37.56
10	34.78	35.03	31.97	32.03	32.92	33.04	33.06	33.30	35.01	35.35	33.06	33.83
20	31.77	31.93	28.71	28.77	29.08	29.20	29.98	30.10	32.11	32.40	29.30	30.21
30	30.05	30.10	26.80	26.86	27.14	27.23	28.17	28.25	30.38	30.61	27.14	28.05
50	27.80	27.77	24.38	24.44	24.85	24.93	26.01	26.06	28.21	28.34	24.68	25.43
100	24.72	24.67	21.61	21.64	22.00	22.03	23.45	23.45	25.39	25.39	22.24	22.49

Table 7.2 PSNR values for image inpainting using bandlimited TP-$\mathbb{C}$TF$_6$. Mask 1 in (g) of Fig. 7.18 is used for 256×256 images: House, Peppers, Cameraman, while Mask 2 in (h) of Fig. 7.18 is used for 512×512 images: Boat, Lena, Barbara

σ	House	Peppers	Cameraman	Boat	Lena	Barbara
0	39.73	33.23	32.77	30.76	34.17	32.65
5	36.18	31.44	31.52	29.80	32.90	31.31
10	33.85	29.73	29.97	28.77	31.74	29.84
20	31.33	27.57	27.73	27.29	30.04	27.70
30	29.64	26.10	26.25	26.20	28.82	26.24
50	27.31	24.13	24.31	24.75	27.11	24.28

Table 7.3 PSNR values for removing mixed Gaussian and impulse noise using bandlimited TP-$\mathbb{C}$TF$_6$. For the two columns in each test image, the left column is for salt-and-pepper impulse noise and the right column is for random-valued impulse noise. σ is the standard deviation of the i.i.d. Gaussian noise and $p := \#\Omega^c/d$ is the corruption percentage with the locations of Ω^c being randomly chosen from $\{1,\dots,d\}$ with uniform probability p

σ	p	House		Peppers		Cameraman		Boat		Lena		Barbara	
6	0.2	36.79	34.08	30.54	28.24	31.55	28.15	33.86	30.57	36.14	34.20	33.69	31.65
6	0.4	35.43	26.19	27.14	22.66	27.40	22.31	32.29	24.60	35.39	27.67	32.54	23.63
20	0.2	31.65	30.26	28.58	25.96	29.84	25.68	28.89	27.68	31.18	30.61	33.69	27.55
20	0.4	29.86	25.56	26.09	22.00	26.29	22.00	27.90	23.74	30.15	26.56	27.59	22.70

where $\mathcal{W}$ is the frame decomposition built from either TP-$\mathbb{C}$TF$_4$ or TP-$\mathbb{C}$TF$_6$, and σ_j is a decreasing sequence going from 512 to $\max(1, \sigma(1-\frac{p^2}{2}))$ with the *corruption percentage* $p := \#\Omega^c/d$. The algorithm stops when either $\|\mathsf{f}_j - \mathsf{f}_{j-1}\|_{l_2}$ is within a given tolerance or j reaches a fixed iteration number J (we take $J = 12$ for our numerical experiments) and the restored image is $\mathring{\mathsf{f}} := \mathsf{f}_j$.

The algorithm for removing mixed Gaussian and impulse noise is almost the same as the algorithm for image inpainting but with one added step by estimating the location of the unobservable region Ω^c by a simple technique at each iteration. The results for removing mixed Gaussian and impulse noise are given in Table 7.3. The unobservable region Ω^c is not known but we assume that the *corruption percentage* $p := \#\Omega^c/d$ is given and the locations of Ω^c is randomly selected from the whole image $\{1,\dots,d\}$ with uniform probability p. See Fig. 7.19 for an illustration of image denoising or inpainting and see Fig. 7.20 for an illustration of removing mixed Gaussian and impulse noise.

Fig. 7.19 (**a**) Noisy House with Gaussian noise $\sigma = 30$. (**b**) Denoised House with PSNR 30.15dB. (**c**) Noisy Barbara with $\sigma = 50$. (**d**) Denoised Barbara with PSNR 25.71dB. (**e**) Corrupted Peppers with inpainting Mask 1 and $\sigma = 0$. (**f**) Restored Peppers with PSNR 33.22dB. (**g**) Corrupted Boat with inpainting Mask 2 and Gaussian noise $\sigma = 20$. (**h**) Restored Boat with PSNR 27.29dB

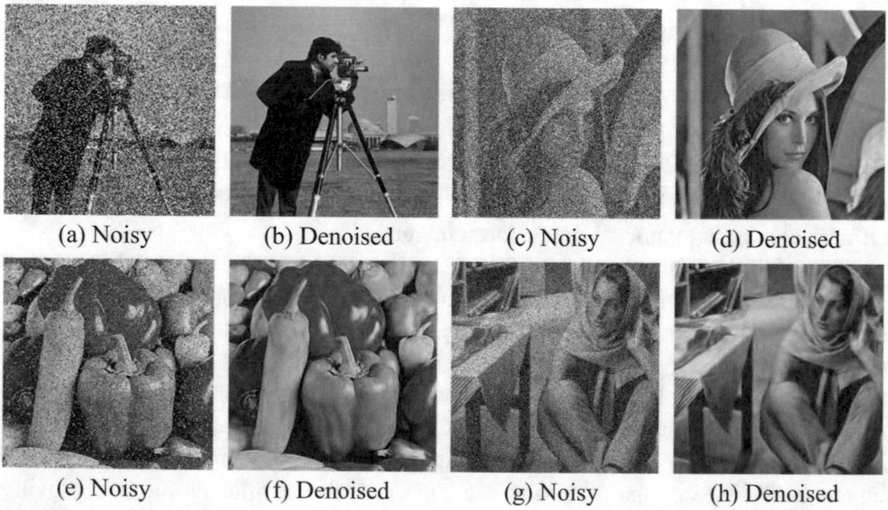

Fig. 7.20 (**a**) Noisy Cameraman with Gaussian noise $\sigma = 0$ (no Gaussian noise) and salt-and-pepper impulse noise $p = 0.3$. (**b**) Denoised Cameraman with PSNR 32.50dB. (**c**) Noisy Lena with Gaussian noise $\sigma = 15$ and salt-and-pepper impulse noise $p = 0.5$. (**d**) Denoised Lena with PSNR 30.95dB. (**e**) Noisy Peppers with Gaussian noise $\sigma = 20$ and random-valued impulse noise $p = 0.1$. (**f**) Denoised Peppers with PSNR 27.31dB. (**g**) Noisy Barbara with Gaussian noise $\sigma = 30$ and random-valued impulse noise $p = 0.2$. (**h**) Denoised Barbara with PSNR 25.93dB

7.5 Framelets/Wavelets on a Finite Interval for Numerical Algorithms*

Many problems in applications are defined in a finite interval such as $[0, 1]$ instead of the whole real line $\mathbb{R}$. This calls for framelets and wavelets on a finite interval. In this section we study how to construct framelets and wavelets on a finite interval from framelets and wavelets on the real line $\mathbb{R}$. Then we address their applications to numerical solutions of differential equations. In particular, we shall propose a general method for constructing wavelets and framelets on the interval $[0, 1]$ from wavelets and framelets with the symmetry property on the real line.

7.5.1 *Framelets on a General Domain by Restriction*

Let Φ and Ψ be two subsets of $L_2(\mathbb{R})$. Recall that an affine system $\mathsf{AS}_J(\Phi;\Psi)$ is defined in (4.3.1) for every $J \in \mathbb{Z}$. We now define the restriction of $\mathsf{AS}_J(\Phi;\Psi)$ on a measurable subset $\mathcal{I}$ of $\mathbb{R}$ to be

$$\begin{aligned}\mathsf{AS}_J(\Phi;\Psi)|_{\mathcal{I}} :=& \{f\chi_{\mathcal{I}} \ : \ f \in \mathsf{AS}_J(\Phi;\Psi)\}\\ =& \{\phi_{2^J;k}|_{\mathcal{I}} \ : \ \phi \in \Phi, k \in \mathbb{Z}\} \cup \{\psi_{2^j;k}|_{\mathcal{I}} \ : \ \psi \in \Psi, k \in \mathbb{Z}, j \geqslant J\}.\end{aligned}$$

Noting that $L_2(\mathcal{I}) \subseteq L_2(\mathbb{R})$, that is, every square integrable function f on $\mathcal{I} \subseteq \mathbb{R}$ can be naturally regarded as an element in $L_2(\mathbb{R})$, we trivially have the following result.

Proposition 7.5.1 *Let Φ and Ψ be subsets of $L_2(\mathbb{R})$. Let $J \in \mathbb{Z}$ and $\mathcal{I}$ be a measurable subset of $\mathbb{R}$.*

(i) If $\mathsf{AS}_J(\Phi;\Psi)$ is a frame for $L_2(\mathbb{R})$, that is, there exist positive constants C_1 and C_2 such that (4.3.5) *holds, then $\mathsf{AS}_J(\Phi;\Psi)|_{\mathcal{I}}$ is a frame for $L_2(\mathcal{I})$ satisfying*

$$C_1\|f\|^2_{L_2(\mathcal{I})} \leqslant \sum_{h\in \mathsf{AS}_J(\Phi;\Psi)|_{\mathcal{I}}} |\langle f, h\rangle_{L_2(\mathcal{I})}|^2 \leqslant C_2\|f\|^2_{L_2(\mathcal{I})}, \quad \forall f \in L_2(\mathcal{I}). \tag{7.5.1}$$

In particular, if $\mathsf{AS}_J(\Phi;\Psi)$ is a normalized tight frame for $L_2(\mathbb{R})$, then $\mathsf{AS}_J(\Phi;\Psi)|_{\mathcal{I}}$ is a normalized tight frame for $L_2(\mathcal{I})$ satisfying (7.5.1) *with $C_1 = C_2 = 1$.*

(ii) If $\mathsf{AS}_J(\Phi;\Psi)$ is a Bessel sequence in $L_2(\mathbb{R})$ satisfying the right-hand side inequality in (4.3.5)*, then $\mathsf{AS}_J(\Phi;\Psi)|_{\mathcal{I}}$ is a Bessel sequence in $L_2(\mathcal{I})$ satisfying the right-hand side inequality in* (7.5.1).

For a pair of dual frames on a measurable subset $\mathcal{I}$ of $\mathbb{R}$, we have

Proposition 7.5.2 *Let $\Phi, \Psi, \tilde{\Phi}, \tilde{\Psi} \subseteq L_2(\mathbb{R})$ such that $(\mathsf{AS}_J(\tilde{\Phi};\tilde{\Psi}), \mathsf{AS}_J(\Phi;\Psi))$ is a pair of dual frames for $L_2(\mathbb{R})$. Then*

(i) $(\mathsf{AS}_J(\tilde{\Phi};\tilde{\Psi})|_{\mathcal{I}}, \mathsf{AS}_J(\Phi;\Psi)|_{\mathcal{I}})$ *is a pair of dual frames for $L_2(\mathcal{I})$.*
(ii) $(\mathsf{AS}_J(\tilde{\Phi};\tilde{\Psi})|_{\mathcal{I}}\backslash\tilde{H}, \mathsf{AS}_J(\Phi;\Psi)|_{\mathcal{I}}\backslash H)$ *is a pair of dual frames for $L_2(\mathcal{I})$, where*

$$H := \{f \in \mathsf{AS}_J(\Phi;\Psi)|_{\mathcal{I}} \ : \ \|f\|_{L_2(\mathcal{I})} = 0 \ \text{ or } \ \|\tilde{f}\|_{L_2(\mathcal{I})} = 0\}$$

and $\sim$: $\mathsf{AS}_J(\Phi;\Psi) \to \mathsf{AS}_J(\tilde{\Phi};\tilde{\Psi})$ *is the default bijection between the two systems, see* (4.1.3) *for details.*

Proof By Proposition 7.5.1, all

$$\mathsf{AS}_J(\tilde{\Phi};\tilde{\Psi})|_{\mathcal{I}}, \quad \mathsf{AS}_J(\Phi;\Psi)|_{\mathcal{I}}, \quad \mathsf{AS}_J(\tilde{\Phi};\tilde{\Psi})|_{\mathcal{I}}\backslash\tilde{H}, \quad \mathsf{AS}_J(\Phi;\Psi)|_{\mathcal{I}}\backslash H$$

are Bessel sequences in $L_2(\mathcal{I})$. Since $(\mathsf{AS}_J(\tilde{\Phi};\tilde{\Psi}), \mathsf{AS}_J(\Phi;\Psi))$ is a pair of dual frames for $L_2(\mathbb{R})$, we have

$$\langle f, g\rangle = \sum_{h\in \mathsf{AS}_J(\Phi;\Psi)} \langle f, \tilde{h}\rangle\langle h, g\rangle, \qquad \forall\, f, g \in L_2(\mathbb{R})$$

with the series converging absolutely. By $L_2(\mathcal{I}) \subseteq L_2(\mathbb{R})$, the above identity still holds for all $f, g \in L_2(\mathcal{I})$. Since $\langle f, \tilde{h}\rangle\langle h, g\rangle = \langle f, \tilde{h}\chi_{\mathcal{I}}\rangle\langle h\chi_{\mathcal{I}}, g\rangle$ for all $f, g \in L_2(\mathcal{I})$, the above identity implies

$$\langle f, g\rangle_{L_2(\mathcal{I})} = \sum_{h\in \mathsf{AS}_J(\Phi;\Psi)|_{\mathcal{I}}} \langle f, \tilde{h}\rangle\langle h, g\rangle, \qquad \forall\, f, g \in L_2(\mathcal{I}). \tag{7.5.2}$$

This proves item (i). For $h \in \mathsf{AS}_J(\Phi;\Psi)$ satisfying either $\|h\chi_{\mathcal{I}}\|_{L_2(\mathbb{R})} = 0$ or $\|\tilde{h}\chi_{\mathcal{I}}\|_{L_2(\mathbb{R})} = 0$, for all $f, g \in L_2(\mathcal{I})$, we have $\langle f, \tilde{h}\chi_{\mathcal{I}}\rangle\langle h\chi_{\mathcal{I}}, g\rangle = 0$. Hence (7.5.2) still holds with $\mathsf{AS}_J(\Phi;\Psi)|_{\mathcal{I}}$ being replaced by $\mathsf{AS}_J(\Phi;\Psi)|_{\mathcal{I}}\backslash H$. This proves item (ii). □

7.5.2 Framelets and Wavelets on a Finite Interval by Symmetry

Though the constructed framelets in Propositions 7.5.1 and 7.5.2 are simple, they often have more elements than necessary in the constructed systems. To reduce redundancy rates, we now consider how to derive framelets and wavelets on $[0, 1]$ from framelets and wavelets with symmetry on the real line $\mathbb{R}$. For this purpose, we define a *folding operator* by

$$F_{c,\epsilon}(f) := \sum_{k\in\mathbb{Z}} \Big(f(\cdot - 2k) + \epsilon f(c + 2k - \cdot)\Big) \qquad \text{with} \quad \epsilon \in \{-1, 1\},\ c \in \mathbb{R}. \tag{7.5.3}$$

In other words, we first fold f at the point $\frac{c}{2}$ with symmetry if $\epsilon = 1$ or antisymmetry if $\epsilon = -1$ to obtain $g := f + \epsilon f(c - \cdot)$, and then periodize it with period 2 to have $F_{c,\epsilon}(f) = \sum_{k\in\mathbb{Z}} g(\cdot - 2k)$. If f has compact support or fast decay, then $F_{c,\epsilon}(f)$ is well-defined and

$$F_{c,\epsilon}(f)(2+\cdot) = F_{c,\epsilon}(f), \quad F_{c,\epsilon}(f)(c-\cdot) = \epsilon F_{c,\epsilon}(f). \tag{7.5.4}$$

Also observe that $F_{c,\epsilon}(f)$ on the interval $[\frac{c}{2}, \frac{c}{2}+1]$ is just obtained by repeatedly folding f at the endpoints $\frac{c}{2}$ and $\frac{c}{2}+1$ with symmetry if $\epsilon = 1$ or antisymmetry if $\epsilon = -1$. By the definition $f_{2^j;m} = 2^{j/2}f(2^j \cdot -m)$,

$$F_{c,\epsilon}(f_{2^j;m}) = \sum_{k\in\mathbb{Z}} \Big(f_{2^j;m+2^{j+1}k} + \epsilon f_{2^j;m-2^jc-2^{j+1}k}(-\cdot) \Big). \tag{7.5.5}$$

If $f(c_f - \cdot) = \epsilon_f f$ with $\epsilon_f \in \{-1,1\}$ and $c_f \in \mathbb{R}$, then

$$f_{2^j;m} = \epsilon_f f_{2^j;-m-c_f}(-\cdot) \quad \text{and} \quad F_{c,\epsilon}(f_{2^j;m}) = \epsilon\epsilon_f F_{c,c}(f_{2^j;2^jc-c_f-m}). \tag{7.5.6}$$

For compactly supported functions $f, g \in L_2(\mathbb{R})$, using (7.5.4), we deduce from the definition of $F_{c,\epsilon}(f)$ in (7.5.3) that

$$\begin{aligned}\langle f, F_{c,\epsilon}(g)\rangle &= \int_{\mathbb{R}} f(x)\overline{F_{c,\epsilon}(g)(x)}dx = \int_{\frac{c}{2}}^{\frac{c}{2}+1} F_{c,\epsilon}(f)(x)\overline{F_{c,\epsilon}(g)(x)}dx \\ &= \langle F_{c,\epsilon}(f), g\rangle.\end{aligned} \tag{7.5.7}$$

A general but simple method for constructing wavelets and framelets on a bounded interval is as follows.

Theorem 7.5.3 *Let $\Phi = \{\phi^1, \dots, \phi^r\}$, $\Psi = \{\psi^1, \dots, \psi^s\}$ and $\tilde\Phi = \{\tilde\phi^1, \dots, \tilde\phi^r\}$, $\tilde\Psi = \{\tilde\psi^1, \dots, \tilde\psi^s\}$ be finite subsets of compactly supported functions in $L_2(\mathbb{R})$ such that for $\ell = 1, \dots, r$,*

$$\phi^\ell(c^\phi_\ell - \cdot) = \epsilon^\phi_\ell \phi^\ell, \quad \tilde\phi^\ell(c^\phi_\ell - \cdot) = \epsilon^\phi_\ell \tilde\phi^\ell \quad \text{with} \quad c^\phi_\ell \in \mathbb{Z}, \epsilon^\phi_\ell \in \{-1,1\}, \tag{7.5.8}$$

and for $\ell = 1, \dots, s$,

$$\psi^\ell(c^\psi_\ell - \cdot) = \epsilon^\psi_\ell \psi^\ell, \quad \tilde\psi^\ell(c^\psi_\ell - \cdot) = \epsilon^\psi_\ell \tilde\psi^\ell \quad \text{with} \quad c^\psi_\ell \in \mathbb{Z}, \epsilon^\psi_\ell \in \{-1,1\}. \tag{7.5.9}$$

Let $\epsilon \in \{-1,1\}$ and $c \in \mathbb{Z}$. Define $\mathcal{I} := [\frac{c}{2}, \frac{c}{2}+1]$. For $j \in \mathbb{N}_0$, define

$$\begin{aligned} d^\phi_{j,\ell} &:= \lfloor 2^{j-1}c - 2^{-1}c^\phi_\ell \rfloor, \quad & d^\psi_{j,\ell} &:= \lfloor 2^{j-1}c - 2^{-1}c^\psi_\ell \rfloor, \\ o^\phi_{j,\ell} &:= \operatorname{odd}(2^jc - c^\phi_\ell), & o^\psi_{j,\ell} &:= \operatorname{odd}(2^jc - c^\psi_\ell), \end{aligned} \tag{7.5.10}$$

where odd$(m) := 1$ *if* m *is an odd integer and* odd$(m) := 0$ *if* m *is an even integer. Let* $\chi_{\mathcal{I}}$ *denote the characteristic function of the interval* $\mathcal{I}$. *For* $j \in \mathbb{N}_0$, *define*

$$\Psi_j^\ell := \begin{cases} \{F_{c,\epsilon}(\psi^\ell_{2^j;k})\chi_{\mathcal{I}} \;:\; k = d^\psi_{j,\ell}+1,\ldots,d^\psi_{j,\ell}+2^j\}, & \text{if } o^\psi_{j,\ell}=1,\\ \{F_{c,\epsilon}(\psi^\ell_{2^j;k})\chi_{\mathcal{I}} \;:\; k = d^\psi_{j,\ell}+1,\ldots,d^\psi_{j,\ell}+2^j-1\}, & \text{if } o^\psi_{j,\ell}=0, \epsilon^\psi_\ell=-\epsilon,\\ \{F_{c,\epsilon}(\psi^\ell_{2^j;k})\chi_{\mathcal{I}} \;:\; k = d^\psi_{j,\ell}+1,\ldots,d^\psi_{j,\ell}+2^j-1\} & \\ \quad \cup\{\frac{1}{\sqrt{2}}F_{c,\epsilon}(\psi^\ell_{2^j;d^\psi_{j,\ell}})\chi_{\mathcal{I}}, \frac{1}{\sqrt{2}}F_{c,\epsilon}(\psi^\ell_{2^j;d^\psi_{j,\ell}+2^j})\chi_{\mathcal{I}}\}, & \text{if } o^\psi_{j,\ell}=0,\ \epsilon^\psi_\ell=\epsilon, \end{cases} \tag{7.5.11}$$

and define $\tilde{\Psi}_j^\ell$, Φ_j^ℓ, $\tilde{\Phi}_j^\ell$ *similarly. For* $J \in \mathbb{N}_0$, *define*

$$\mathscr{B}_J := (\cup_{\ell=1}^r \Phi_J^\ell) \cup \cup_{j=J}^\infty(\cup_{\ell=1}^s \Psi_j^\ell), \quad \tilde{\mathscr{B}}_J := (\cup_{\ell=1}^r \tilde{\Phi}_J^\ell) \cup \cup_{j=J}^\infty(\cup_{\ell=1}^s \tilde{\Psi}_j^\ell). \tag{7.5.12}$$

Then for every $J \in \mathbb{N}_0$, *the following statements hold:*

(1) *If* $(\{\tilde{\Phi};\tilde{\Psi}\},\{\Phi;\Psi\})$ *is a dual framelet in* $L_2(\mathbb{R})$, *then* $(\tilde{\mathscr{B}}_J,\mathscr{B}_J)$ *is a pair of dual frames for* $L_2(\mathcal{I})$, *where* $\tilde{\mathscr{B}}_J$ *and* $\mathscr{B}_J$ *are defined in* (7.5.12).
(2) *If* $\{\Phi;\Psi\}$ *is a tight framelet in* $L_2(\mathbb{R})$, *then* $\mathscr{B}_J$ *is a (normalized) tight frame for* $L_2(\mathcal{I})$.
(3) *If* $(\{\tilde{\Phi};\tilde{\Psi}\},\{\Phi;\Psi\})$ *is a biorthogonal wavelet in* $L_2(\mathbb{R})$, *then* $(\tilde{\mathscr{B}}_J,\mathscr{B}_J)$ *is a pair of biorthogonal bases for* $L_2(\mathcal{I})$. *Moreover,* $\mathscr{B}_J$ *and* $\tilde{\mathscr{B}}_J$ *are Riesz bases for* $L_2(\mathcal{I})$.
(4) *If* $\{\Phi;\Psi\}$ *is an orthogonal wavelet in* $L_2(\mathbb{R})$, *then* $\mathscr{B}_J$ *is an orthonormal basis for* $L_2(\mathcal{I})$.

Proof Since all the elements in $\Phi, \Psi, \tilde{\Phi}, \tilde{\Psi}$ are compactly supported, there exists $N \in \mathbb{N}$ such that all these elements vanish outside $[-N,N]$ and $\mathcal{I} \subseteq [-N,N]$ as well as $|c| \leqslant N$, $\max_{\ell=1,\ldots,r}|c_\ell^\phi| \leqslant N$ and $\max_{\ell=1,\ldots,s}|c_\ell^\psi| \leqslant N$. Hence, for every $j \in \mathbb{N}_0$,

$$\begin{aligned} h_{2^j;k}(x) &= 0 \qquad \forall\, x \in [-N,N], k \in \mathbb{Z}\backslash[-N(2^j+1), N(2^j+1)],\\ h_{2^j;k}(x) &= 0 \qquad \forall\, x \in \mathbb{R}\backslash[-3N,3N], k \in [-N(2^j+1), N(2^j+1)], \end{aligned} \tag{7.5.13}$$

for $h \in \Phi \cup \Psi \cup \tilde{\Phi} \cup \tilde{\Psi}$. We first prove that if $\mathsf{AS}_J(\Phi;\Psi)$ is a Bessel sequence in $L_2(\mathbb{R})$, then $\mathscr{B}_J$ is a Bessel sequence in $L_2(\mathcal{I})$. Let $f \in L_2(\mathcal{I})$, that is, $f \in L_2(\mathbb{R})$ and f vanishes outside $\mathcal{I}$. For all $k = d^\psi_{j,\ell},\ldots,d^\psi_{j,\ell}+2^j$, by $|d^\psi_{j,\ell}| \leqslant 2^{j-1}N + 2^{-1}N$ and hence $|k| \leqslant 2^{j-1}N+2^{-1}N+2^j$, we can directly check that $\psi^\ell_{2^j;k}$ must vanish outside $[-2N-1, 2N+1]$. Indeed, for $x \notin [-2N-1,2N+1]$ and $j \in \mathbb{N}_0$,

$$|2^jx-k| \geqslant 2^j|x| - |k| \geqslant 2^j(2N+1) - 2^{j-1}N - 2^{-1}N - 2^j = 2^{j-1}3N - 2^{-1}N \geqslant N.$$

Therefore, for $f \in L_2(\mathcal{I})$, since f vanishes outside $\mathcal{I}$,

$$\sum_{h\in\Psi_j^\ell}|\langle f,h\rangle_{L_2(\mathcal{I})}|^2 \leqslant \sum_{k=d_{j,\ell}^\psi}^{d_{j,\ell}^\psi+2^j}|\langle f,F_{c,\epsilon}(\psi_{2^j;k}^\ell)\chi_{\mathcal{I}}\rangle|^2 = \sum_{k=d_{j,\ell}^\psi}^{d_{j,\ell}^\psi+2^j}|\langle F_{c,\epsilon}(f),\psi_{2^j;k}^\ell\rangle|^2$$

$$= \sum_{k=d_{j,\ell}^\psi}^{d_{j,\ell}^\psi+2^j}|\langle F_{c,\epsilon}(f)\chi_{[-2N-1,2N+1]},\psi_{2^j;k}^\ell\rangle|^2 \leqslant \sum_{k\in\mathbb{Z}}|\langle F_{c,\epsilon}(f)\chi_{[-2N-1,2N+1]},\psi_{2^j;k}^\ell\rangle|^2,$$

where we used $\langle f,F_{c,\epsilon}(\psi_{2^j;k}^\ell)\chi_{\mathcal{I}}\rangle = \langle f,F_{c,\epsilon}(\psi_{2^j;k}^\ell)\rangle$. Hence,

$$\begin{aligned}\sum_{h\in\mathscr{B}_J}|\langle f,h\rangle_{L_2(\mathcal{I})}|^2 &= \sum_{\ell=1}^{r}\sum_{h\in\Phi_J^\ell}|\langle f,h\rangle|^2 + \sum_{j=J}^{\infty}\sum_{\ell=1}^{s}\sum_{h\in\Psi_j^\ell}|\langle f,h\rangle|^2\\ &\leqslant \sum_{\ell=1}^{r}\sum_{k\in\mathbb{Z}}|\langle F_{c,\epsilon}(f)\chi_{[-2N-1,2N+1]},\phi_{2^J;k}^\ell\rangle|^2\\ &\quad+\sum_{j=J}^{\infty}\sum_{\ell=1}^{s}\sum_{k\in\mathbb{Z}}|\langle F_{c,\epsilon}(f)\chi_{[-2N-1,2N+1]},\psi_{2^j;k}^\ell\rangle|^2\\ &\leqslant C\|F_{c,\epsilon}(f)\chi_{[-2N-1,2N+1]}\|_{L_2(\mathbb{R})}^2 \leqslant C(4N+2)^2\|f\|_{L_2(\mathcal{I})}^2,\end{aligned}$$

where we used the fact that $\mathsf{AS}_J(\Phi;\Psi)$ is a Bessel sequence in $L_2(\mathbb{R})$ with an upper frame bound C. This proves that $\mathscr{B}_J$ is a Bessel sequence in $L_2(\mathcal{I})$. Similarly, $\tilde{\mathscr{B}}_J$ is a Bessel sequence in $L_2(\mathcal{I})$.

We now prove item (1). By Theorem 4.3.5, we know that $(\mathsf{AS}_J(\tilde{\Phi};\tilde{\Psi}), \mathsf{AS}_J(\Phi;\Psi))$ is a pair of dual frames in $L_2(\mathbb{R})$ for all $J \in \mathbb{Z}$. Let $f,g \in L_2(\mathcal{I})$, that is, $f,g \in L_2(\mathbb{R})$ such that both f and g vanish outside $\mathcal{I}$. Note that $F_{c,\epsilon}(g)\chi_{\mathcal{I}} = g$. Since $(\mathsf{AS}_J(\tilde{\Phi};\tilde{\Psi}), \mathsf{AS}_J(\Phi;\Psi))$ is a pair of dual frames for $L_2(\mathbb{R})$, we have

$$\begin{aligned}\langle f,g\rangle_{L_2(\mathcal{I})} &= \langle f,F_{c,\epsilon}(g)\rangle = \langle f,F_{c,\epsilon}(g)\chi_{[-3N,3N]}\rangle\\ &= \sum_{\ell=1}^{r}\sum_{k\in\mathbb{Z}}\langle f,\phi_{2^J;k}^\ell\rangle\langle\tilde{\phi}_{2^J;k}^\ell,F_{c,\epsilon}(g)\chi_{[-3N,3N]}\rangle\\ &\quad+\sum_{j=J}^{\infty}\sum_{\ell=1}^{s}\sum_{k\in\mathbb{Z}}\langle f,\psi_{2^j;k}^\ell\rangle\langle\tilde{\psi}_{2^j;k}^\ell,F_{c,\epsilon}(g)\chi_{[-3N,3N]}\rangle\\ &= \sum_{\ell=1}^{r}\sum_{k=-N(2^J+1)}^{N(2^J+1)}\langle f,\phi_{2^J;k}^\ell\rangle\langle\tilde{\phi}_{2^J;k}^\ell,F_{c,\epsilon}(g)\chi_{[-3N,3N]}\rangle\\ &\quad+\sum_{j=J}^{\infty}\sum_{\ell=1}^{s}\sum_{k=-N(2^j+1)}^{N(2^j+1)}\langle f,\psi_{2^j;k}^\ell\rangle\langle\tilde{\psi}_{2^j;k}^\ell,F_{c,\epsilon}(g)\chi_{[-3N,3N]}\rangle,\end{aligned}$$

since $\langle f, \psi^{\ell}_{2^j;k}\rangle = 0$ for all $k \notin [-N(2^j+1), N(2^j+1)]$ by the first identity in (7.5.13). Using the second identity in (7.5.13), we further have

$$\langle f, g\rangle_{L_2(\mathcal{I})} = \sum_{\ell=1}^{r} \sum_{k=-N(2^J+1)}^{N(2^J+1)} \langle f, \phi^{\ell}_{2^J;k}\rangle \langle \tilde{\phi}^{\ell}_{2^J;k}, F_{c,\epsilon}(g)\rangle$$
$$+ \sum_{j=J}^{\infty} \sum_{\ell=1}^{s} \sum_{k=-N(2^j+1)}^{N(2^j+1)} \langle f, \psi^{\ell}_{2^j;k}\rangle \langle \tilde{\psi}^{\ell}_{2^j;k}, F_{c,\epsilon}(g)\rangle$$
$$= \sum_{\ell=1}^{r} \sum_{k\in\mathbb{Z}} \langle f, \phi^{\ell}_{2^J;k}\rangle \langle F_{c,\epsilon}(\tilde{\phi}^{\ell}_{2^J;k}), g\rangle + \sum_{j=J}^{\infty} \sum_{\ell=1}^{s} \sum_{k\in\mathbb{Z}} \langle f, \psi^{\ell}_{2^j;k}\rangle \langle F_{c,\epsilon}(\tilde{\psi}^{\ell}_{2^j;k}), g\rangle.$$

By (7.5.4), we notice that $F_{c,\epsilon}(h_{2^j;m+2^{j+1}k}) = F_{c,\epsilon}(h_{2^j;m})$ for all $j, k, m \in \mathbb{Z}$. Decomposing $\mathbb{Z} = \{d^{\psi}_{j,\ell} + 1 - 2^j, \dots, d^{\psi}_{j,\ell} + 2^j\} + 2^{j+1}\mathbb{Z}$, we have

$$\sum_{k\in\mathbb{Z}} \langle f, \psi^{\ell}_{2^j;k}\rangle \langle F_{c,\epsilon}(\tilde{\psi}^{\ell}_{2^j;k}), g\rangle = \sum_{m=d^{\psi}_{j,\ell}+1-2^j}^{d^{\psi}_{j,\ell}+2^j} \sum_{k\in\mathbb{Z}} \langle f, \psi^{\ell}_{2^j;m+2^{j+1}k}\rangle \langle F_{c,\epsilon}(\tilde{\psi}^{\ell}_{2^j;m}), g\rangle. \tag{7.5.14}$$

By symmetry in (7.5.9) and the second identity in (7.5.6), we have

$$F_{c,\epsilon}(\tilde{\psi}^{\ell}_{2^j;m}) = \epsilon\epsilon^{\psi}_{\ell} F_{c,\epsilon}(\tilde{\psi}^{\ell}_{2^j;2^jc-c^{\psi}_{\ell}-m}).$$

By the definition of $d^{\psi}_{j,\ell}$ and $o^{\psi}_{j,\ell}$ in (7.5.10), we have $2d^{\psi}_{j,\ell} + o^{\psi}_{j,\ell} = 2^jc - c^{\psi}_{\ell}$. It follows from the above identity that

$$\sum_{m=d^{\psi}_{j,\ell}+1-2^j}^{d^{\psi}_{j,\ell}+o^{\psi}_{j,\ell}-1} \sum_{k\in\mathbb{Z}} \langle f, \psi^{\ell}_{2^j;m+2^{j+1}k}\rangle \langle F_{c,\epsilon}(\tilde{\psi}^{\ell}_{2^j;m}), g\rangle$$
$$= \sum_{m=d^{\psi}_{j,\ell}+1-2^j}^{d^{\psi}_{j,\ell}+o^{\psi}_{j,\ell}-1} \sum_{k\in\mathbb{Z}} \langle f, \psi^{\ell}_{2^j;m+2^{j+1}k}\rangle \langle \epsilon\epsilon^{\psi}_{\ell} F_{c,\epsilon}(\tilde{\psi}^{\ell}_{2^j;2^jc-c^{\psi}_{\ell}-m}), g\rangle$$
$$= \sum_{m=d^{\psi}_{j,\ell}+1}^{d^{\psi}_{j,\ell}+o^{\psi}_{j,\ell}+2^j-1} \sum_{k\in\mathbb{Z}} \langle f, \epsilon\epsilon^{\psi}_{\ell} \psi^{\ell}_{2^j;2^jc-c^{\psi}_{\ell}-m+2^{j+1}k}\rangle \langle F_{c,\epsilon}(\tilde{\psi}^{\ell}_{2^j;m}), g\rangle,$$

where in the last identity we changed indices by replacing $2^jc - c^{\psi}_{\ell} - m$ with the new index m. By the symmetry condition in (7.5.9) and the first identity in (7.5.6), we observe $\psi^{\ell}_{2^j;2^jc-c^{\psi}_{\ell}-m+2^{j+1}k} = \epsilon^{\psi}_{\ell}\psi^{\ell}_{2^j;m-2^jc-2^{j+1}k}(-\cdot)$ and consequently the above identity further becomes

$$\sum_{m=d^{\psi}_{j,\ell}+1-2^j}^{d^{\psi}_{j,\ell}+o^{\psi}_{j,\ell}-1}\sum_{k\in\mathbb{Z}}\langle f,\psi^{\ell}_{2^j;m+2^{j+1}k}\rangle\langle F_{c,\epsilon}(\tilde{\psi}^{\ell}_{2^j;m}),g\rangle$$
$$=\sum_{m=d^{\psi}_{j,\ell}+1}^{d^{\psi}_{j,\ell}+o^{\psi}_{j,\ell}+2^j-1}\sum_{k\in\mathbb{Z}}\langle f,\epsilon\psi^{\ell}_{2^j;m-2^jc-2^{j+1}k}(-\cdot)\rangle\langle F_{c,\epsilon}(\tilde{\psi}^{\ell}_{2^j;m}),g\rangle. \tag{7.5.15}$$

Suppose that $o^{\psi}_{j,\ell}=1$. Then it follows from (7.5.14) and (7.5.15) that

$$\sum_{k\in\mathbb{Z}}\langle f,\psi^{\ell}_{2^j;k}\rangle\langle F_{c,\epsilon}(\tilde{\psi}^{\ell}_{2^j;k}),g\rangle$$
$$=\sum_{m=d^{\psi}_{j,\ell}+1}^{d^{\psi}_{j,\ell}+2^j}\Big\langle f,\sum_{k\in\mathbb{Z}}\Big(\psi^{\ell}_{2^j;m+2^{j+1}k}+\epsilon\psi^{\ell}_{2^j;m-2^jc-2^{j+1}k}(-\cdot)\Big)\Big\rangle\langle F_{c,\epsilon}(\tilde{\psi}^{\ell}_{2^j;m}),g\rangle$$
$$=\sum_{m=d^{\psi}_{j,\ell}+1}^{d^{\psi}_{j,\ell}+2^j}\langle f,F_{c,\epsilon}(\psi^{\ell}_{2^j;m})\rangle\langle F_{c,\epsilon}(\tilde{\psi}^{\ell}_{2^j;m}),g\rangle=\sum_{h\in\Psi^{\ell}_j}\langle f,h\rangle\langle\tilde{h},g\rangle,$$

where we used (7.5.5) in the second-to-last identity and the definition of Ψ^{ℓ}_j in (7.5.11) in the last identity.

Suppose that $o^{\psi}_{j,\ell}=0$. By (7.5.9) and (7.5.5), since $2^jc-c^{\psi}_{\ell}=2d^{\psi}_{j,\ell}+o^{\psi}_{j,\ell}=2d^{\psi}_{j,\ell}$, we have $-d^{\psi}_{j,\ell}-c^{\psi}_{\ell}=d^{\psi}_{j,\ell}-2^jc$ and hence

$$\sum_{k\in\mathbb{Z}}\psi^{\ell}_{2^j;d^{\psi}_{j,\ell}+2^jq+2^{j+1}k}=\sum_{k\in\mathbb{Z}}\epsilon^{\psi}_{\ell}\psi^{\ell}_{2^j;-d^{\psi}_{j,\ell}-2^jq-2^{j+1}k-c^{\psi}_{\ell}}(-\cdot)$$
$$=\epsilon^{\psi}_{\ell}\sum_{k\in\mathbb{Z}}\psi^{\ell}_{2^j;d^{\psi}_{j,\ell}+2^jq-2^jc-2^{j+1}k}(-\cdot)$$

for all $q\in\mathbb{Z}$. Therefore,

$$\sum_{k\in\mathbb{Z}}\psi^{\ell}_{2^j;d^{\psi}_{j,\ell}+2^jq+2^{j+1}k}=\frac{1}{2}\sum_{k\in\mathbb{Z}}\Big(\psi^{\ell}_{2^j;d^{\psi}_{j,\ell}+2^jq+2^{j+1}k}+\epsilon^{\psi}_{\ell}\psi^{\ell}_{2^j;d^{\psi}_{j,\ell}+2^jq-2^jc-2^{j+1}k}(-\cdot)\Big)$$
$$=\frac{1}{2}F_{c,\epsilon^{\psi}_{\ell}}(\psi^{\ell}_{2^j;d^{\psi}_{j,\ell}+2^jq}).$$

Hence, it follows from (7.5.14) and (7.5.15) that

$$\sum_{k\in\mathbb{Z}}\langle f,\psi^{\ell}_{2^j;k}\rangle\langle F_{c,\epsilon}(\tilde{\psi}^{\ell}_{2^j;k}),g\rangle=\sum_{q=0}^{1}\Big(\sum_{k\in\mathbb{Z}}\langle f,\psi^{\ell}_{2^j;d^{\psi}_{j,\ell}+2^jq+2^{j+1}k}\rangle\Big)\langle F_{c,\epsilon}(\tilde{\psi}^{\ell}_{2^j;d^{\psi}_{j,\ell}+2^jq}),g\rangle$$

$$+\sum_{m=d^{\psi}_{j,\ell}+1}^{d^{\psi}_{j,\ell}+2^j-1}\Big\langle f,\sum_{k\in\mathbb{Z}}\Big(\psi^{\ell}_{2^j;m+2^{j+1}k}+\epsilon\psi^{\ell}_{2^j;m-2^jc-2^{j+1}k}(-\cdot)\Big)\Big\rangle\langle F_{c,\epsilon}(\tilde{\psi}^{\ell}_{2^j;m}),g\rangle$$

$$=\sum_{q=0}^{1}\langle f,\tfrac{1}{\sqrt{2}}F_{c,\epsilon^{\psi}_{\ell}}(\psi^{\ell}_{2^j;d^{\psi}_{j,\ell}+2^jq})\rangle\langle\tfrac{1}{\sqrt{2}}F_{c,\epsilon}(\tilde{\psi}^{\ell}_{2^j;d^{\psi}_{j,\ell}+2^jq}),g\rangle$$

$$+\sum_{m=d^{\psi}_{j,\ell}+1}^{d^{\psi}_{j,\ell}+2^j-1}\langle f,F_{c,\epsilon}(\psi^{\ell}_{2^j;m})\rangle\langle F_{c,\epsilon}(\tilde{\psi}^{\ell}_{2^j;m}),g\rangle=\sum_{h\in\Psi^{\ell}_j}\langle f,h\rangle\langle\tilde{h},g\rangle,$$

where in the last identity we used $F_{c,\epsilon}(\tilde{\psi}^{\ell}_{2^j;d^{\psi}_{j,\ell}})=F_{c,\epsilon}(\tilde{\psi}^{\ell}_{2^j;d^{\psi}_{j,\ell}+2^j})=0$ if $\epsilon^{\psi}_{\ell}=-\epsilon$. Indeed, by (7.5.9) and (7.5.5), taking into account $2^jc-c^{\psi}_{\ell}=2d^{\psi}_{j,\ell}$, we have

$$F_{c,\epsilon}(\tilde{\psi}^{\ell}_{2^j;d^{\psi}_{j,\ell}+2^jq})=\epsilon\epsilon^{\psi}_{\ell}F_{c,\epsilon}(\tilde{\psi}^{\ell}_{2^j;2^jc-c^{\psi}_{\ell}-d^{\psi}_{j,\ell}-2^jq})=\epsilon\epsilon^{\psi}_{\ell}F_{c,\epsilon}(\tilde{\psi}^{\ell}_{2^j;d^{\psi}_{j,\ell}+2^jq}).$$

Summarizing the above two cases, we proved that for all $f,g\in L_2(\mathcal{I})$,

$$\langle f,g\rangle_{L_2(\mathcal{I})}=\sum_{\ell=1}^{r}\sum_{k\in\mathbb{Z}}\langle f,\phi^{\ell}_{2^J;k}\rangle\langle F_{c,\epsilon}(\tilde{\phi}^{\ell}_{2^J;k}),g\rangle+\sum_{j=J}^{\infty}\sum_{\ell=1}^{s}\sum_{k\in\mathbb{Z}}\langle f,\psi^{\ell}_{2^j;k}\rangle\langle F_{c,\epsilon}(\tilde{\psi}^{\ell}_{2^j;k}),g\rangle$$

$$=\sum_{\ell=1}^{r}\sum_{h\in\Phi^{\ell}_J}\langle f,h\rangle\langle\tilde{h},g\rangle+\sum_{j=J}^{\infty}\sum_{h\in\Psi^{\mathcal{I}}_j}\langle f,h\rangle\langle\tilde{h},g\rangle.$$

This completes the proof of item (1). Item (2) is a direct consequence of item (1).

We now prove item (3). By the proved item (1), to prove item (3), it suffices to prove that $\tilde{\mathscr{B}}_J$ and $\mathscr{B}_J$ are biorthogonal to each other. By (7.5.9), it follows from (7.5.5) and (7.5.6) that

$$F_{c,\epsilon}(\psi^{\ell}_{2^j;m})=\sum_{k\in\mathbb{Z}}\Big(\psi^{\ell}_{2^j;m+2^{j+1}k}+\epsilon\epsilon^{\psi}_{\ell}\psi^{\ell}_{2^j;2^jc-c^{\psi}_{\ell}-m+2^{j+1}k}\Big).\tag{7.5.16}$$

From the above identity in (7.5.16) and (7.5.7), we have

$$\int_{\mathcal{I}}F_{c,\epsilon}(\psi^{\ell}_{2^j;m})(x)\overline{F_{c,\epsilon}(\tilde{\psi}^{p}_{2^q;n})(x)}dx=\langle F_{c,\epsilon}(\psi^{\ell}_{2^j;m}),\tilde{\psi}^{p}_{2^q;n}\rangle$$

$$=\sum_{k\in\mathbb{Z}}\langle\psi^{\ell}_{2^j;m+2^{j+1}k},\tilde{\psi}^{p}_{2^q;n}\rangle+\epsilon\epsilon^{\psi}_{\ell}\sum_{k\in\mathbb{Z}}\langle\psi^{\ell}_{2^j;2^jc-c^{\psi}_{\ell}-m+2^{j+1}k},\tilde{\psi}^{p}_{2^q;n}\rangle.\tag{7.5.17}$$

By the biorthogonality of $(\mathsf{AS}_J(\tilde{\Phi};\tilde{\Psi}), \mathsf{AS}_J(\Phi;\Psi))$ and (7.5.10), the above identity in (7.5.17) implies that $\langle f, \tilde{g}\rangle = 0$ for all $f \in \mathscr{B}_J$ and $\tilde{g} \in \tilde{\mathscr{B}}_J$ such that $\tilde{f} \neq \tilde{g}$ (here $\sim$ is the default bijection between $\mathscr{B}_J$ and $\tilde{\mathscr{B}}_J$). The identities (7.5.17) and $2d^{\psi}_{j,\ell} + o^{\psi}_{j,\ell} = 2^j c - c^{\psi}_{\ell}$ imply

$$\int_{\mathcal{I}} F_{c,\epsilon}(\psi^{\ell}_{2^j;m})(x)\overline{F_{c,\epsilon}(\tilde{\psi}^{\ell}_{2^j;m})(x)}dx = 1 + \epsilon\epsilon^{\psi}_{\ell} \sum_{k\in\mathbb{Z}} \delta(2d^{\psi}_{j,\ell} + o^{\psi}_{j,\ell} - 2m + 2^{j+1}k),$$

which equals 1 for all $m = d^{\psi}_{j,\ell} + 1, \dots, d^{\psi}_{j,\ell} + o^{\psi}_{j,\ell} + 2^j - 1$. If $o^{\psi}_{j,\ell} = 0$ and $m = d^{\psi}_{j,\ell} + 2^j q$ with $q \in \mathbb{Z}$, then the above identity yields

$$\int_{\mathcal{I}} F_{c,\epsilon}(\psi^{\ell}_{2^j;m})(x)\overline{F_{c,\epsilon}(\tilde{\psi}^{\ell}_{2^j;m})(x)}dx = 1 + \epsilon\epsilon^{\psi}_{\ell}.$$

This proves that $\tilde{\mathscr{B}}_J$ and $\mathscr{B}_J$ are biorthogonal to each other. Therefore, we proved item (3). Item (4) is a direct consequence of item (3). □

As a special case of Theorem 7.5.3, the following example is of particular interest for constructing framelets and wavelets on a bounded interval with a simple structure.

Example 7.5.1 Let $\Phi = \{\phi^1, \dots, \phi^r\}$, $\tilde{\Phi} = \{\tilde{\phi}^1, \dots, \tilde{\phi}^r\}$, $\Psi = \{\psi^1, \dots, \psi^s\}$ and $\tilde{\Psi} = \{\tilde{\psi}^1, \dots, \tilde{\psi}^s\}$ be finite subsets of compactly supported functions in $L_2(\mathbb{R})$ such that (7.5.8) and (7.5.9) are satisfied with

$$c^{\phi}_1 = \dots = c^{\phi}_r = c^{\psi}_1 = \dots = c^{\psi}_s = 0. \tag{7.5.18}$$

In addition we assume that

$$\text{all the elements in } \Phi \text{ and } \Psi \text{ vanish outside } [-1, 1]. \tag{7.5.19}$$

Take $c = 0$ in Theorem 7.5.3. Since all $d^{\phi}_{j,\ell}, d^{\psi}_{j,\ell}, o^{\phi}_{j,\ell}, o^{\psi}_{j,\ell}$ in (7.5.10) are zero, Ψ^{ℓ}_j in (7.5.11) becomes

$$\Psi^{\ell}_j = \begin{cases} \{\psi^{\ell}_{2^j;k} \; : \; k = 1, \dots, 2^j - 1\}, & \text{if } \epsilon^{\psi}_{\ell} = -\epsilon, \\ \{\psi^{\ell}_{2^j;k} \; : \; k = 1, \dots, 2^j - 1\} \cup \{\sqrt{2}\psi^{\ell}_{2^j;0}\chi_{[0,1]}, \sqrt{2}\psi^{\ell}_{2^j;2^j}\chi_{[0,1]}\}, & \text{if } \epsilon^{\psi}_{\ell} = \epsilon. \end{cases}$$

Consequently,

(i) If $\epsilon = -1$ and all the elements in Φ, Ψ are continuous, then $h(0) = h(1) = 0$ for all $h \in \mathscr{B}_J$. That is, the homogeneous Dirichlet boundary condition holds for all the elements in $\mathscr{B}_J$ on the interval $[0, 1]$.
(ii) If $\epsilon = 1$ and all the elements in Φ, Ψ are in $\mathscr{C}^1(\mathbb{R})$, then $h'(0) = h'(1) = 0$ for all $h \in \mathscr{B}_J$. That is, the homogeneous von Neumann boundary condition holds for all the elements in $\mathscr{B}_J$ on the interval $[0, 1]$.

Note that the conditions in (7.5.18) and (7.5.19) are satisfied for the orthogonal wavelet in Example 6.4.1 and all the biorthogonal wavelets in Examples 6.5.1–6.5.3. Thus, orthogonal wavelets and spline Riesz wavelets on the interval $[0, 1]$ can be easily constructed by Theorem 7.5.3.

7.5.3 Refinable Structure of Framelets and Wavelets on a Finite Interval

The refinable structure is of particular importance for numerical algorithms in applications. In this section, we discuss the refinable structure of framelets and wavelets on a finite interval constructed by Theorem 7.5.3.

For refinable vector functions with symmetry, we have (also see Exercise 6.42 for details).

Proposition 7.5.4 *Let $\phi = (\phi^1, \dots, \phi^r)^\mathsf{T}$ and $\psi = (\psi^1, \dots, \psi^s)^\mathsf{T}$ be vectors of compactly supported distributions satisfying*

$$\phi = 2\sum_{k\in\mathbb{Z}} a(k)\phi(2\cdot -k), \qquad \psi = 2\sum_{k\in\mathbb{Z}} b(k)\phi(2\cdot -k) \tag{7.5.20}$$

for some $a \in (l_0(\mathbb{Z}))^{r\times r}$ and $b \in (l_0(\mathbb{Z}))^{s\times r}$. Define

$$\mathsf{S}(\xi) := \operatorname{diag}(\epsilon_1^\phi e^{-ic_1^\phi \xi}, \dots, \epsilon_r^\phi e^{-ic_r^\phi \xi}), \quad \mathsf{T}(\xi) := \operatorname{diag}(\epsilon_1^\psi e^{-ic_1^\psi \xi}, \dots, \epsilon_s^\psi e^{-ic_s^\psi \xi}),$$

with $c_1^\phi, \dots, c_r^\phi, c_1^\psi, \dots, c_s^\psi \in \mathbb{R}$ and $\epsilon_1^\phi, \dots, \epsilon_r^\phi, \epsilon_1^\psi, \dots, \epsilon_s^\psi \in \{-1, 1\}$.

(1) If $\widehat{a}(\xi) = \mathsf{S}(2\xi)\widehat{a}(-\xi)\mathsf{S}^{-1}(\xi)$ for all $\xi \in \mathbb{R}$ and if (5.1.10) holds, then $\widehat{\phi}(\xi) = \mathsf{S}(\xi)\widehat{\phi}(-\xi)$, that is, the first identity in (7.5.8) holds.
(2) If $\widehat{\phi}(\xi) = \mathsf{S}(\xi)\widehat{\phi}(-\xi)$ and $\widehat{b}(\xi) = \mathsf{T}(2\xi)\widehat{b}(-\xi)\mathsf{S}^{-1}(\xi)$, then $\widehat{\psi}(\xi) = \mathsf{T}(\xi)\widehat{\psi}(-\xi)$, that is, the first identity in (7.5.9) holds.
(3) If $\widehat{\phi}(\xi) = \mathsf{S}(\xi)\widehat{\phi}(-\xi)$ and $\widehat{\psi}(\xi) = \mathsf{T}(\xi)\widehat{\psi}(-\xi)$ (that is, the first identities in both (7.5.8) and (7.5.9) hold) with $c_1^\phi, \dots, c_r^\phi, c_1^\psi, \dots, c_s^\psi \in \mathbb{Z}$, then there exist $(\#\vec{\Phi}_{j-1}) \times (\#\vec{\Phi}_j)$ matrices A_j and $(\#\vec{\Psi}_{j-1}) \times (\#\vec{\Phi}_j)$ matrices B_j such that

$$\vec{\Phi}_{j-1} = A_j\vec{\Phi}_j \qquad \vec{\Psi}_{j-1} = B_j\vec{\Phi}_j, \qquad j \in \mathbb{N}, \tag{7.5.21}$$

where $\vec{\Phi}_j$ and $\vec{\Psi}_j$ are column vectors by listing all the elements in $\cup_{\ell=1}^r \Phi_j^\ell$ and $\cup_{\ell=1}^s \Psi_j^\ell$, respectively. Here Ψ_j^ℓ is defined in (7.5.11) and Φ_j^ℓ is defined similarly as in (7.5.11).

Proof The equations in (7.5.20) are equivalent to $\widehat{\phi}(2\xi) = \widehat{a}(\xi)\widehat{\phi}(\xi)$ and $\widehat{\psi}(2\xi) = \widehat{b}(\xi)\widehat{\phi}(\xi)$. Define $\widehat{\eta}(\xi) := \mathsf{S}(\xi)\widehat{\phi}(-\xi)$. By our assumption $\widehat{a}(\xi)\mathsf{S}(\xi) = \mathsf{S}(2\xi)\widehat{a}(-\xi)$,

$$\widehat{a}(\xi)\widehat{\eta}(\xi) = \widehat{a}(\xi)\mathsf{S}(\xi)\widehat{\phi}(-\xi) = \mathsf{S}(2\xi)\widehat{a}(-\xi)\widehat{\phi}(-\xi) = \mathsf{S}(2\xi)\widehat{\phi}(-2\xi) = \widehat{\eta}(2\xi).$$

Therefore, η also satisfies $\eta = 2\sum_{k\in\mathbb{Z}} a(k)\eta(2\cdot -k)$ with $\widehat{\eta}(0) = \mathsf{S}(0)\widehat{\phi}(0) = \widehat{\phi}(0)$. Since (5.1.10) holds, by the uniqueness result in Theorem 5.1.3, we must have $\eta = \phi$. That is, we proved $\widehat{\phi}(\xi) = \mathsf{S}(\xi)\widehat{\phi}(-\xi)$ and hence item (1) holds.

By our assumption $\mathsf{T}(2\xi)\widehat{b}(-\xi) = \widehat{b}(\xi)\mathsf{S}(\xi)$, we have

$$\mathsf{T}(2\xi)\widehat{\psi}(-2\xi) = \mathsf{T}(2\xi)\widehat{b}(-\xi)\widehat{\phi}(-\xi) = \widehat{b}(\xi)\mathsf{S}(\xi)\widehat{\phi}(-\xi) = \widehat{b}(\xi)\widehat{\phi}(\xi) = \widehat{\psi}(2\xi).$$

This proves $\widehat{\psi}(\xi) = \mathsf{T}(\xi)\widehat{\psi}(-\xi)$. Hence, item (2) holds.

Note that if $b = a$, then $\psi = \phi$. Thus, it suffices to prove the second identity in (7.5.21). By (7.5.20), we have

$$\psi_{2^{j-1};n} = \sqrt{2}\sum_{k\in\mathbb{Z}} b(k)\phi_{2^j;k+2n}, \qquad j \in \mathbb{N},$$

which implies

$$F_{c,\epsilon}(\psi^p_{2^{j-1};n}) = \sqrt{2}\sum_{\ell=1}^{r}\sum_{k\in\mathbb{Z}}[b(k)]_{p,\ell}F_{c,\epsilon}(\phi^\ell_{2^j;k+2n}), \qquad j \in \mathbb{N}, p = 1,\ldots,s.$$

By $F_{c,\epsilon}(\phi^\ell_{2^j;m+2^{j+1}k}) = F_{c,\epsilon}(\phi^\ell_{2^j;m})$, the above identity can be rewritten as

$$\begin{aligned}
F_{c,\epsilon}(\psi^p_{2^{j-1};n}) &= \sqrt{2}\sum_{\ell=1}^{r}\sum_{m=d^\phi_{j,\ell}+1-2^j-2n}^{d^\phi_{j,\ell}+2^j-2n}\sum_{k\in\mathbb{Z}}[b(m+2^{j+1}k)]_{p,\ell}F_{c,\epsilon}(\phi^\ell_{2^j;m+2n})\\
&= \sqrt{2}\sum_{\ell=1}^{r}\sum_{m=d^\phi_{j,\ell}+1-2^j}^{d^\phi_{j,\ell}+2^j} F_{c,\epsilon}(\phi^\ell_{2^j;m})\sum_{k\in\mathbb{Z}}[b(m-2n+2^{j+1}k)]_{p,\ell}.
\end{aligned} \tag{7.5.22}$$

Due to the symmetry property in (7.5.8), it follows from (7.5.5) that $F_{c,\epsilon}(\phi^\ell_{2^j;m}) = \epsilon\epsilon^\phi_\ell F_{c,\epsilon}(\phi^\ell_{2^j;2^jc-c^\phi_\ell-m})$. Thus, by $2d^\phi_{j,\ell} + o^\phi_{j,\ell} = 2^jc - c^\phi_\ell$,

$$\begin{aligned}
&\sum_{m=d^\phi_{j,\ell}+1-2^j}^{d^\phi_{j,\ell}+o^\phi_{j,\ell}-1} F_{c,\epsilon}(\phi^\ell_{2^j;m})\sum_{k\in\mathbb{Z}}[b(m-2n+2^{j+1}k)]_{p,\ell}\\
&\quad= \sum_{m=d^\phi_{j,\ell}+1-2^j}^{d^\phi_{j,\ell}+o^\phi_{j,\ell}-1} \epsilon\epsilon^\phi_\ell F_{c,\epsilon}(\phi^\ell_{2^j;2^jc-c^\phi_\ell-m})\sum_{k\in\mathbb{Z}}[b(m-2n+2^{j+1}k)]_{p,\ell}\\
&\quad= \sum_{m=d^\phi_{j,\ell}+1}^{d^\phi_{j,\ell}+o^\phi_{j,\ell}+2^j-1} F_{c,\epsilon}(\phi^\ell_{2^j;m})\sum_{k\in\mathbb{Z}}\epsilon\epsilon^\phi_\ell[b(2^jc-c^\phi_\ell-m-2n+2^{j+1}k)]_{p,\ell}.
\end{aligned}$$

Suppose that $o^{\phi}_{j,\ell} = 1$. By $\sum_{m=d^{\phi}_{j,\ell}+1-2^j}^{d^{\phi}_{j,\ell}+2^j} = \sum_{m=d^{\phi}_{j,\ell}+1}^{d^{\phi}_{j,\ell}+2^j} + \sum_{m=d^{\phi}_{j,\ell}+1-2^j}^{d^{\phi}_{j,\ell}}$, the above identity leads to

$$\sum_{m=d^{\phi}_{j,\ell}+1-2^j}^{d^{\phi}_{j,\ell}+2^j} F_{c,\epsilon}(\phi^{\ell}_{2^j;m}) \sum_{k\in\mathbb{Z}} [b(m-2n+2^{j+1}k)]_{p,\ell} = \sum_{m=d^{\phi}_{j,\ell}+1}^{d^{\phi}_{j,\ell}+2^j} F_{c,\epsilon}(\phi^{\ell}_{2^j;m})[b^n_{j,\ell}(m)]_{p,\ell},$$

where

$$b^n_{j,\ell}(m) := \sum_{k\in\mathbb{Z}} \Big(b(m-2n+2^{j+1}k) + \epsilon\epsilon^{\phi}_{\ell} b(2^j c - c^{\phi}_{\ell} - m - 2n + 2^{j+1}k) \Big), \qquad m \in \mathbb{Z}.$$

Suppose that $o^{\phi}_{j,\ell} = 0$. By the above definition of $b^n_{j,\ell}(m)$, we have $b^n_{j,\ell}(d^{\phi}_{j,\ell} + 2^j q) = (1 + \epsilon\epsilon^{\phi}_{\ell}) \sum_{k\in\mathbb{Z}} b(d^{\phi}_{j,\ell} + 2^j q - 2n + 2^{j+1}k)$ for $q \in \mathbb{Z}$ and

$$\begin{aligned}
&\sum_{m=d^{\phi}_{j,\ell}+1-2^j}^{d^{\phi}_{j,\ell}+2^j} F_{c,\epsilon}(\phi^{\ell}_{2^j;m}) \sum_{k\in\mathbb{Z}} [b(m-2n+2^{j+1}k)]_{p,\ell} \\
&= \sum_{q=0}^{1} F_{c,\epsilon}(\phi^{\ell}_{2^j;d^{\phi}_{j,\ell}+2^j q}) \sum_{k\in\mathbb{Z}} [b(d^{\phi}_{j,\ell} + 2^j q - 2n + 2^{j+1}k)]_{p,\ell} + \sum_{m=d^{\phi}_{j,\ell}+1}^{d^{\phi}_{j,\ell}+2^j-1} F_{c,\epsilon}(\phi^{\ell}_{2^j;m})[b^n_{j,\ell}(m)]_{p,\ell} \\
&= \sum_{q=0}^{1} \tfrac{1}{\sqrt{2}} F_{c,\epsilon}(\phi^{\ell}_{2^j;d^{\phi}_{j,\ell}+2^j q}) \tfrac{1}{\sqrt{2}} [b^n_{j,\ell}(d^{\phi}_{j,\ell} + 2^j q)]_{p,\ell} + \sum_{m=d^{\phi}_{j,\ell}+1}^{d^{\phi}_{j,\ell}+2^j-1} F_{c,\epsilon}(\phi^{\ell}_{2^j;m})[b^n_{j,\ell}(m)]_{p,\ell},
\end{aligned}$$

where we used $F_{c,\epsilon}(\phi^{\ell}_{2^j;d^{\phi}_{j,\ell}+2^j q}) = 0$ for all $q \in \mathbb{Z}$ if $\epsilon^{\phi}_{\ell} = -\epsilon$.

Now it follows from (7.5.22) that there exists a $(\#\vec{\Psi}_{j-1}) \times (\#\vec{\Phi}_j)$ matrix B_j such that $\vec{\Psi}_{j-1} = B_j \vec{\Phi}_j$. This completes the proof of item (3). □

Combining Theorem 7.5.3 and Proposition 7.5.4, we have the following result.

Theorem 7.5.5 *Under the same notation as in Theorem 7.5.3, define* $\phi := (\phi^1, \ldots, \phi^r)^{\mathsf{T}}$, $\tilde{\phi} := (\tilde{\phi}^1, \ldots, \tilde{\phi}^r)^{\mathsf{T}}$, $\psi := (\psi^1, \ldots, \psi^s)^{\mathsf{T}}$, $\tilde{\psi} := (\tilde{\psi}^1, \ldots, \tilde{\psi}^s)^{\mathsf{T}}$. *Assume that* $\|\widehat{\phi}(0)\|_{l_2} = 1$ *and* $\overline{\widehat{\phi}(0)}^{\mathsf{T}} \widehat{\tilde{\phi}}(0) = 1$. *Suppose that there exist* $a, \tilde{a} \in (l_0(\mathbb{Z}))^{r\times r}$ *and* $b, \tilde{b} \in (l_0(\mathbb{Z}))^{s\times r}$ *such that*

$$\widehat{\phi}(2\xi) = \widehat{a}(\xi)\widehat{\phi}(\xi), \ \widehat{\tilde{\phi}}(2\xi) = \widehat{\tilde{a}}(\xi)\widehat{\tilde{\phi}}(\xi), \ \widehat{\psi}(2\xi) = \widehat{b}(\xi)\widehat{\phi}(\xi), \ \widehat{\tilde{\psi}}(2\xi) = \widehat{\tilde{b}}(\xi)\widehat{\tilde{\phi}}(\xi).$$

Then there exist matrices $A_j, B_j, \tilde{A}_j, \tilde{B}_j$ *such that*

$$\vec{\Phi}_{j-1} = A_j\vec{\Phi}_j, \quad \vec{\Psi}_{j-1} = B_j\vec{\Phi}_j, \quad \vec{\tilde{\Phi}}_{j-1} = \tilde{A}_j\vec{\tilde{\Phi}}_j, \quad \vec{\tilde{\Psi}}_{j-1} = \tilde{B}_j\vec{\tilde{\Phi}}_j, \tag{7.5.23}$$

for all $j \in \mathbb{N}$. *Moreover, for every* $J \in \mathbb{N}$,

(1) if all $\Phi, \Psi, \tilde{\Phi}, \tilde{\Psi}$ *are subsets of* $L_2(\mathbb{R})$ *and* $(\{\tilde{a}; \tilde{b}\}, \{a; b\})$ *is a dual framelet filter bank, then* $(\{\tilde{\Phi}; \tilde{\Psi}\}, \{\Phi; \Psi\})$ *is a dual framelet in* $L_2(\mathbb{R})$ *and* $(\tilde{\mathscr{B}}_J, \mathscr{B}_J)$ *is a pair of dual frames for* $L_2(\mathcal{I})$. *If in addition the entries of* $\vec{\Phi}_j$, *as well as the entries of* $\vec{\tilde{\Phi}}_j|_{\mathcal{I}}$, *are linearly independent, then*

$$\overline{\tilde{A}_j}^{\mathsf{T}} A_j + \overline{\tilde{B}_j}^{\mathsf{T}} B_j = \mathrm{Id}; \tag{7.5.24}$$

(2) if $\{a; b\}$ *is a tight framelet filter bank, then* $\{\Phi; \Psi\}$ *is a tight framelet in* $L_2(\mathbb{R})$ *and* $\mathscr{B}_J$ *is a normalized tight frame for* $L_2(\mathcal{I})$;

(3) if $(\{\tilde{a}; \tilde{b}\}, \{a; b\})$ *is a biorthogonal wavelet filter bank satisfying* (6.4.21) *and if* $\mathrm{sm}(a) > 0$ *and* $\mathrm{sm}(\tilde{a}) > 0$, *then* $(\{\tilde{\Phi}; \tilde{\Psi}\}, \{\Phi; \Psi\})$ *is a biorthogonal wavelet in* $L_2(\mathbb{R})$, $(\tilde{\mathscr{B}}_J, \mathscr{B}_J)$ *is a pair of biorthogonal bases for* $L_2(\mathcal{I})$, *and*

$$\text{the square matrix } \begin{bmatrix} A_j \\ B_j \end{bmatrix} \text{ has the inverse } [\overline{\tilde{A}_j}^{\mathsf{T}}, \overline{\tilde{B}_j}^{\mathsf{T}}];$$

(4) if $\{a; b\}$ *is an orthogonal wavelet filter bank and* $\mathrm{sm}(a) > 0$, *then* Ψ *is an orthogonal wavelet in* $L_2(\mathbb{R})$ *and* $\mathscr{B}_J$ *is an orthonormal basis for* $L_2(\mathcal{I})$.

Proof To prove item (1), by Theorem 6.4.1 with $\tau = 0$, we see that $(\{\tilde{\Phi}; \tilde{\Psi}\}, \{\Phi; \Psi\})$ is a dual framelet in $L_2(\mathbb{R})$. Therefore, it follows from Theorem 7.5.3 that $(\tilde{\mathscr{B}}_J, \mathscr{B}_J)$ is a pair of dual frames for $L_2(\mathcal{I})$ for every $J \in \mathbb{N}$ and

$$\langle f, \vec{\tilde{\Phi}}_{j-1}\rangle\vec{\Phi}_{j-1} + \langle f, \vec{\tilde{\Psi}}_{j-1}\rangle\vec{\Psi}_{j-1} = \langle f, \vec{\tilde{\Phi}}_j\rangle\vec{\Phi}_j, \qquad \forall f \in L_2(\mathcal{I}).$$

By (7.5.23), we deduce from the above identity that

$$\langle f, \vec{\tilde{\Phi}}_j\rangle\Big(\overline{\tilde{A}_j}^{\mathsf{T}} A_j + \overline{\tilde{B}_j}^{\mathsf{T}} B_j\Big)\vec{\Phi}_j = \langle f, \vec{\tilde{\Phi}}_j\rangle\vec{\Phi}_j, \qquad \forall f \in L_2(\mathcal{I}).$$

Since the entries in $\vec{\tilde{\Phi}}_j$ are linearly independent, the mapping $f \in L_2(\mathcal{I}) \mapsto \langle f, \vec{\tilde{\Phi}}_j\rangle \in (l_2)^{1\times \# \vec{\tilde{\Phi}}_j}$ is onto. Since the entries of $\vec{\Phi}_j$ are linearly independent, (7.5.24) follows directly from the above identity.

By Theorem 6.4.2 and the assumption in item (2), the pair $\{\Phi; \Psi\}$ is a tight framelet in $L_2(\mathbb{R})$. Now item (2) follows directly from item (1).

By Corollary 6.4.7 with $\tau = 0$ for item (3), the pair $(\{\tilde{\Phi}; \tilde{\Psi}\}, \{\Phi; \Psi\})$ is a biorthogonal wavelet in $L_2(\mathbb{R})$. Hence, by Theorem 7.5.3, the pair $(\tilde{\mathscr{B}}_J, \mathscr{B}_J)$ is a pair of biorthogonal bases for $L_2(\mathcal{I})$. Since $\vec{\tilde{\Phi}}_j$ and $\vec{\Phi}_j$ are biorthogonal to each other, the entries of $\vec{\tilde{\Phi}}_j$, as well as the entries of $\vec{\Phi}_j$, must be linearly independent. Now by item (1), the identity (7.5.24) holds.

Item (4) follows directly from Corollary 6.4.9 and item (3). □

7.5.4 Applications to Numerical Solutions of Differential Equations

Let us present a simple toy example to illustrate how to use wavelets on $\mathcal{I} := [0, 1]$ to numerically solve differential equations. Consider the classical elliptic differential equation with the homogeneous Dirichlet boundary condition:

$$-u''(x) + \alpha u(x) = f(x), \qquad x \in (0, 1) \quad \text{with} \quad u(0) = u(1) = 0,$$

where α is a nonnegative real number. Define $H_0^1(\mathcal{I}) := \{u \in L_2^1(\mathcal{I}) \ : \ u(0) = u(1) = 0\}$. A weak solution u of the above problem is to find $u \in H_0^1(\mathcal{I})$ satisfying

$$\langle u', v'\rangle_{L_2(\mathcal{I})} + \alpha\langle u, v\rangle_{L_2(\mathcal{I})} = \langle -u'', v\rangle_{L_2(\mathcal{I})} + \alpha\langle u, v\rangle_{L_2(\mathcal{I})} = \langle f, v\rangle_{L_2(\mathcal{I})},$$

for $v \in H_0^1(\mathcal{I})$. Choose n large enough (here n is related to the mesh size in certain sense) and let $\Phi_n^{\mathcal{I}} := \cup_{\ell=1}^r \Phi_n^\ell$ be constructed in Theorem 7.5.3 with the homogeneous Dirichlet boundary condition. Define $V_n^{\mathcal{I}} := \operatorname{span}\Phi_n^{\mathcal{I}}$. By Theorem 7.5.3, both $\Phi_n^{\mathcal{I}}$ and $(\cup_{\ell=1}^r \Phi_0^\ell) \cup \cup_{j=0}^{n-1} \cup_{\ell=1}^s \Psi_j^\ell$ are bases of the linear space $V_n^{\mathcal{I}}$ and $V_n^{\mathcal{I}} \subset H_0^1(\mathcal{I})$. The Galerkin scheme to solve the above problem is to find $u \in V_n^{\mathcal{I}}$ such that

$$\langle u', v'\rangle_{L_2(\mathcal{I})} + \alpha\langle u, v\rangle_{L_2(\mathcal{I})} = \langle f, v\rangle_{L_2(\mathcal{I})}, \qquad \forall\ v \in V_n^{\mathcal{I}}.$$

Define $\{v_1, \dots, v_{N_n}\} := (\cup_{\ell=1}^r \Phi_0) \cup \cup_{j=0}^{n-1} \cup_{\ell=1}^s \Psi_j^\ell$ with $N_n := \dim(V_n^{\mathcal{I}})$. Then $u \in V_n^{\mathcal{I}}$ can be represented as $u = \sum_{k=1}^{N_n} c_k v_k$ with unknown coefficients $\{c_k\}_{k=1}^{N_n}$. Now the above Galerkin scheme is equivalent to

$$\sum_{k=1}^{N_n} \big(\langle v_k', v_j'\rangle_{L_2(\mathcal{I})} + \alpha\langle v_k, v_j\rangle_{L_2(\mathcal{I})}\big)c_k = \langle f, v_j\rangle_{L_2(\mathcal{I})}, \qquad j = 1, \dots, N_n.$$

Define

$$\mathbf{A}_n := (\langle v_k', v_j'\rangle_{L_2(\mathcal{I})} + \alpha\langle v_k, v_j\rangle_{L_2(\mathcal{I})})_{1\leqslant j,k\leqslant N_n}$$

be the stiffness matrix. Define $\mathbf{x}_n := (c_1, \dots, c_{N_n})^{\mathsf{T}}$ to be the vector of unknown coefficients and $\mathbf{b}_n := (\langle f, v_1\rangle_{L_2(\mathcal{I})}, \dots, \langle f, v_{N_n}\rangle_{L_2(\mathcal{I})})^{\mathsf{T}}$. Then numerically solving the elliptic differential equation by the Galerkin scheme becomes solving the system of linear equations:

$$\mathbf{A}_n\mathbf{x}_n = \mathbf{b}_n,$$

which is often solved by iterative algorithms. There are two main advantages of using wavelets on a finite interval to numerically solve differential equations:

(i) the condition numbers $\mathrm{cond}(\mathbf{A}_n)$, which is the ratio between the largest eigenvalue and the smallest eigenvalue of $\mathbf{A}_n$, are uniformly bounded for all $n \in \mathbb{N}$. This is mainly because the Riesz basis $\mathscr{B}_0$ constructed in Theorem 7.5.3 is often not only a Riesz basis for $L_2(\mathcal{I})$ but also is a Riesz basis (after a proper normalization) for Sobolev spaces $H^\tau(\mathcal{I})$ with a wide range of $\tau \in \mathbb{R}$.
(ii) The stiffness matrix $\mathbf{A}_n$ is sparse, because the wavelet functions are compactly supported and have vanishing moments. The vector $\mathbf{b}_n$ is also sparse, because the wavelet functions possess high vanishing moments.

For simplicity, we do not further discuss the details about the above two issues in this book. We finish this section by discussing how to efficiently compute the stiffness matrix $\mathbf{A}_n$ and the vector $\mathbf{b}_n$ through refinable structure in (7.5.21). To compute $\mathbf{b}_n$, we first estimate $\{\langle f, \phi^{\ell}_{2^n;k}\rangle\}_{k\in\mathbb{Z}}$ by the following result.

Lemma 7.5.6 *Let $g \in L_2(\mathbb{R})$ be a compactly supported function. For a smooth function $f \in L_2(\mathbb{R})$,*

$$\left|\langle f, g_{2^j;k}\rangle - \sum_{\ell=0}^{m-1} 2^{-j(\ell+1/2)}\frac{\overline{c_\ell}}{\ell!}f^{(\ell)}(2^{-j}(k-\beta))\right| \leqslant C2^{-j(m+1/2)}$$

with $C := \|f^{(m)}\|_{L_\infty(2^{-j}(k+\mathrm{supp}(g)))}\frac{1}{m!}\int_{\mathbb{R}}|g(x)(x+\beta)^m|dx$, where $\beta \in \mathbb{R}$ and $c_0,\dots,c_{m-1} \in \mathbb{C}$ are given by

$$\widehat{g}(\xi) = e^{i\beta\xi}\sum_{\ell=0}^{m-1}\frac{c_\ell}{\ell!}(-i\xi)^\ell + \mathscr{O}(|\xi|^m), \qquad \xi \to 0. \tag{7.5.25}$$

In particular, if $\widehat{g}(\xi) = c_0e^{i\beta\xi} + \mathscr{O}(|\xi|^m)$, $\xi \to 0$, then $\langle f, g_{2^j;k}\rangle = \overline{c_0}2^{-j/2}f(2^{-j}(k-\beta)) + \mathscr{O}(2^{-j(m+1/2)})$ as $j \to \infty$.

Proof By $\widehat{g(\cdot-\beta)}(\xi) = e^{-i\beta\xi}\widehat{g}(\xi)$, we deduce from (7.5.25) that

$$c_\ell = i^\ell\widehat{g(\cdot-\beta)}^{(\ell)}(0) = i^\ell\int_{\mathbb{R}} g(x-\beta)(-ix)^\ell dx = \int_{\mathbb{R}} g(x)(x+\beta)^\ell dx,$$

for $\ell = 0,\dots,m-1$. Considering the Taylor expansion of f at the point $2^{-j}(k-\beta)$, we have

$$f(x) = \sum_{\ell=0}^{m-1}\frac{1}{\ell!}f^{(\ell)}(2^{-j}(k-\beta))(x-2^{-j}(k-\beta))^\ell + R_{m-1}(x)$$

with

$$R_{m-1}(x) := \int_{2^{-j}(k-\beta)}^{x} \frac{f^{(m)}(t)}{(m-1)!}(x-t)^{m-1}dt = \frac{f^{(m)}(\xi_x)}{m!}(x-2^{-j}(k-\beta))^m$$

for some ξ_x on the interval between x and $2^{-j}(k-\beta)$. Therefore,

$$\begin{aligned}\langle f, g_{2^j;k}\rangle &= \sum_{\ell=0}^{m-1} \frac{1}{\ell!} f^{(\ell)}(2^{-j}(k-\beta))\langle(\cdot - 2^{-j}(k-\beta))^{\ell}, g_{2^j;k}\rangle + \langle R_{m-1}, g_{2^j;k}\rangle \\ &= \sum_{\ell=0}^{m-1} 2^{-j(\ell+1/2)} \frac{\overline{c_\ell}}{\ell!} f^{(\ell)}(2^{-j}(k-\beta)) + \langle R_{m-1}, g_{2^j;k}\rangle\end{aligned}$$

and

$$\begin{aligned}|\langle R_{m-1}, g_{2^j;k}\rangle| &\leqslant \|f^{(m)}\|_{L_\infty(2^{-j}(k+\mathrm{supp}(g)))} \frac{1}{m!}\int_{\mathbb{R}} 2^{j/2}|g(2^j x - k)(x-2^{-j}(k-\beta))|^m dx \\ &= 2^{-j(m+1/2)}\|f^{(m)}\|_{L_\infty(2^{-j}(k+\mathrm{supp}(g)))} \frac{1}{m!}\int_{\mathbb{R}} |g(x)(x+\beta)|^m dx.\end{aligned}$$

This completes the proof. □

We often take $\beta = 0$ in (7.5.25), for which we have $c_\ell = i^\ell \widehat{g}^{(\ell)}(0)$ for $\ell = 0, \ldots, m-1$. If $g = \phi$ is a refinable vector function satisfying $\widehat{\phi}(2\xi) = \widehat{a}(\xi)\widehat{\phi}(\xi)$ and (5.6.3), then up to a multiplicative constant, all the moments $\widehat{\phi}^{(\ell)}(0), \ell = 0, \ldots, m-1$ can be uniquely computed via (5.1.11) from the filter/mask a.

By symmetry in (7.5.8), similar to (7.5.16), we have

$$F_{c,\epsilon}(\phi^{\ell}_{2^j;m}) = \sum_{k\in\mathbb{Z}} \left(\phi^{\ell}_{2^j;m+2^{j+1}k} + \epsilon\epsilon^{\phi}_{\ell}\phi^{\ell}_{2^j;2^jc-c^{\phi}_{\ell}-m+2^{j+1}k}\right). \tag{7.5.26}$$

Using the above identity, we can easily derive $\langle f, \vec{\Phi}_n\rangle$ from $\{\langle f, \phi^{\ell}_{2^n;k}\rangle\}_{k\in\mathbb{Z},\ell=1,\ldots,r}$. Now the refinable structure in (7.5.21) can be employed to compute $\mathbf{b}_n$ through the fast/discrete framelet transform.

By (7.5.26) and (7.5.7), using $\widehat{\phi}(2\xi) = \widehat{a}(\xi)\widehat{\phi}(\xi)$ and $\widehat{\psi}(2\xi) = \widehat{b}(\xi)\widehat{\phi}(\xi)$, a general stiffness matrix $\mathbf{A}_n$ can be exactly and similarly obtained by first computing

$$v_{j,n}(k) := \langle \phi^{(j)}, \phi^{(n)}(\cdot - k)\rangle := \int_{\mathbb{R}} \phi^{(j)}(x)\overline{\phi^{(n)}(x-k)}^{\mathsf{T}} dx, \qquad k \in \mathbb{Z}. \tag{7.5.27}$$

More precisely, we first compute $v_{j,k}$ by Theorem 7.5.7 to get $\langle \phi^{(j)}(2^J \cdot - k'), \phi^{(n)}(2^J \cdot - k)\rangle = 2^{-J} v_{j,n}(k-k')$ for all $k', k \in \mathbb{Z}$. Then compute the inner products $\langle \psi^{(j)}(2^p \cdot - k'), \psi^{(n)}(2^q \cdot -k)\rangle$ for $p, q = J-1, \ldots, 0$ through the fast framelet transform by recursively applying the refinable structure $\widehat{\phi}(2\xi) = \widehat{a}(\xi)\widehat{\phi}(\xi)$ and $\widehat{\psi}(2\xi) = \widehat{b}(\xi)\widehat{\phi}(\xi)$. Then (7.5.26) and (7.5.7) will be used to obtain the matrix $\mathbf{A}_n$.

Theorem 7.5.7 *Let $m \in \mathbb{N}_0$ and $\phi = (\phi^1, \ldots, \phi^r)^\mathsf{T}$ be a compactly supported vector function satisfying $\widehat{\phi}(2\xi) = \widehat{a}(\xi)\widehat{\phi}(\xi)$ for some $a \in (l_0(\mathbb{Z}))^{r\times r}$ with $\widehat{\phi}(0) \neq 0$. Suppose that $\phi \in (H^m(\mathbb{R}))^r$ (i.e., all the entries in $\phi, \phi', \ldots, \phi^{(m)}$ belong to $L_2(\mathbb{R})$) and the integer shifts of ϕ are stable in $L_2(\mathbb{R})$. Let the operator $\mathcal{T}_{a,a^\star}$ be defined in (5.8.15). For all $0 \leqslant j, n \leqslant m$, the finitely supported sequence $v_{j,n}$ in (7.5.27) is uniquely determined by*

$$\mathcal{T}_{a,a^\star} v_{j,n} = 2^{-j-n} v_{j,n} \quad \text{with} \quad \left[\widehat{\upsilon}(\xi)\widehat{v_{j,n}}(\xi)\overline{\widehat{\upsilon}(\xi)}^\mathsf{T}\right]^{(j+n)}(0) = i^{j-n}(j+n)!, \qquad (7.5.28)$$

where $\widehat{\upsilon}^{(m+1)}(0) = \cdots = \widehat{\upsilon}^{(2m)}(0) := 0$ and $\widehat{\upsilon}(0), \ldots, \widehat{\upsilon}^{(m)}(0)$ are uniquely computed by (5.6.4) with $\widehat{\upsilon}(0)\widehat{a}(0) = \widehat{\upsilon}(0)$ and $\widehat{\upsilon}(0)\widehat{\phi}(0) = 1$.

Proof By Theorem 5.8.1 and Corollary 5.8.2, we have $\operatorname{sm}(a) = \operatorname{sm}(\phi) > m$, $\operatorname{sr}(a) \geqslant m+1$, and (5.6.3) holds. Thus, it follows from Theorem 5.8.4 (with m being replaced by $m+1$) that 2^{-j-n} must be a simple eigenvalue of $\mathcal{T}_{a,a^\star}$ for all $0 \leqslant j, n \leqslant m$. By Lemma 4.4.1, we have

$$\begin{aligned}\widehat{v_{j,n}}(\xi) &= \sum_{k\in\mathbb{Z}} \langle \phi^{(j)}, \phi^{(n)}(\cdot - k)\rangle e^{-ik\xi} \\ &= [\widehat{\phi^{(j)}}, \widehat{\phi^{(n)}}](\xi) := \sum_{k\in\mathbb{Z}} \widehat{\phi^{(j)}}(\xi + 2\pi k)\overline{\widehat{\phi^{(n)}}(\xi + 2\pi k)}^\mathsf{T}.\end{aligned}$$

Since $\phi^{(j)} = 2^{j+1}\sum_{k\in\mathbb{Z}} a(k)\phi^{(j)}(2\cdot - k)$, we have $\widehat{\phi^{(j)}}(2\xi) = 2^j\widehat{a}(\xi)\widehat{\phi^{(j)}}(\xi)$. Thus

$$\begin{aligned}\widehat{v_{j,n}}(2\xi) =& [\widehat{\phi^{(j)}}, \widehat{\phi^{(n)}}](2\xi) = 2^{j+n}\widehat{a}(\xi)[\widehat{\phi^{(j)}}, \widehat{\phi^{(n)}}](\xi)\overline{\widehat{a}(\xi)}^\mathsf{T} \\ &+ 2^{j+n}\widehat{a}(\xi+\pi)[\widehat{\phi^{(j)}}, \widehat{\phi^{(n)}}](\xi+\pi)\overline{\widehat{a}(\xi+\pi)}^\mathsf{T} = 2^{j+n}\widehat{\mathcal{T}_{a,a^\star} v_{j,n}}(2\xi).\end{aligned}$$

That is, $\mathcal{T}_{a,a^\star} v_{j,n} = 2^{-j-n} v_{j,n}$. This proves that $v_{j,n}$ is an eigenvector of $\mathcal{T}_{a,a^\star}$ with the simple eigenvalue 2^{-j-n}.

By (5.6.3), we see that there are unique $\widehat{\upsilon}(0), \ldots, \widehat{\upsilon}^{(m)}(0)$ satisfying (5.6.4) with $\widehat{\upsilon}(0)\widehat{a}(0) = \widehat{\upsilon}(0)$ and $\widehat{\upsilon}(0)\widehat{\phi}(0) = 1$. That is, (5.6.4) holds with $\widehat{\upsilon}(0)\widehat{\phi}(0) = 1$. By Proposition 5.6.2, the equation (5.6.6) must hold. By $\widehat{\phi^{(\ell)}}(\xi) = (i\xi)^\ell\widehat{\phi}(\xi)$ and (5.6.6), it follows from Lemma 6.6.2 that

$$\widehat{\upsilon}(\xi)\widehat{v_{j,n}}(\xi)\overline{\widehat{\upsilon}(\xi)}^\mathsf{T} = [\widehat{\upsilon}(\xi)\widehat{\phi^{(j)}}(\xi), \widehat{\upsilon}(\xi)\widehat{\phi^{(n)}}(\xi)] = i^{j-n}\xi^{j+n} + \mathscr{O}(|\xi|^{2m+2})$$

as $\xi \to 0$. This shows that the second identity in (7.5.28) must hold for $v_{j,n}$. □

7.6 Fast Multiframelet Transform and Its Balanced Property

Though we have presented a comprehensive mathematical study on refinable vector functions and their associated multiwavelets and multiframelets, we haven't discussed yet their associated discrete multiframelet transform employing a matrix-valued filter bank. We conclude this book by discussing the discrete multiframelet transform and its basic properties using matrix-valued filters.

For $u \in (l_0(\mathbb{Z}))^{r\times s}$, recall that $u^\star(k) := \overline{u(-k)}^{\mathsf{T}}$ for all $k \in \mathbb{Z}$, i.e., $\widehat{u^\star}(\xi) := \overline{\widehat{u}(\xi)}^{\mathsf{T}}$. Let $a, \tilde{a} \in (l_0(\mathbb{Z}))^{r\times r}$ and $b, \tilde{b} \in (l_0(\mathbb{Z}))^{s\times r}$. Recall that the subdivision operator $\mathcal{S}_a$ and the transition operator $\mathcal{T}_a$ are defined in (5.6.30) and (5.6.31), respectively. For $J \in \mathbb{N}$, *the J-level discrete multiframelet decomposition* is defined by

$$\vec{v}_j := \tfrac{\sqrt{2}}{2}\mathcal{T}_{\tilde{a}}\vec{v}_{j-1}, \quad \vec{w}_j := \tfrac{\sqrt{2}}{2}\mathcal{T}_{\tilde{b}}\vec{v}_{j-1}, \qquad j = 1, \dots, J, \tag{7.6.1}$$

where $\vec{v}_0 \in (l(\mathbb{Z}))^{1\times r}$ is a (vector-valued) input signal. *The J-level discrete multiframelet reconstruction* is

$$\mathring{\vec{v}}_{j-1} := \frac{\sqrt{2}}{2}\mathcal{S}_a\mathring{\vec{v}}_j + \frac{\sqrt{2}}{2}\mathcal{S}_b\mathring{\vec{w}}_{j-1}, \qquad j = J, \dots, 1.$$

Noting that

$$\widehat{\mathcal{S}_a\vec{v}}(\xi) = 2\widehat{\vec{v}}(2\xi)\widehat{a}(\xi) \quad \text{and} \quad \widehat{\mathcal{T}_{\tilde{a}}\vec{v}}(\xi) = \widehat{\vec{v}}(\xi/2)\overline{\widehat{\tilde{a}}(\xi/2)}^{\mathsf{T}} + \widehat{\vec{v}}(\xi/2+\pi)\overline{\widehat{\tilde{a}}(\xi/2+\pi)}^{\mathsf{T}},$$

by the same argument as in Theorem 1.1.1, we have

Theorem 7.6.1 *For $a, \tilde{a} \in (l_0(\mathbb{Z}))^{r\times r}$ and $b, \tilde{b} \in (l_0(\mathbb{Z}))^{s\times r}$, the following statements are equivalent:*

(i) The J-level discrete multiframelet transform using the filter bank $(\{\tilde{a}; \tilde{b}_1, \dots, \tilde{b}_s\}, \{a; b_1, \dots, b_s\})$ has the perfect reconstruction property for every $J \in \mathbb{N}$.

(ii) Item (i) holds with $J = 1$, that is, for all $\vec{v} \in (l(\mathbb{Z}))^{1\times r}$,

$$\mathcal{S}_a\mathcal{T}_{\tilde{a}}\vec{v} + \mathcal{S}_b\mathcal{T}_{\tilde{b}}\vec{v} = 2\vec{v}. \tag{7.6.2}$$

(iii) The identity in (7.6.2) holds for all $\vec{v} \in (l_0(\mathbb{Z}))^{1\times r}$.

(iv) The identity in (7.6.2) holds for the particular sequences $\vec{v} = \boldsymbol{\delta} e_j^{\mathsf{T}}$ and $\boldsymbol{\delta}(\cdot - 1)e_j^{\mathsf{T}}$ for all $j = 1, \dots, r$, where e_j is the jth unit coordinate column vector in $\mathbb{R}^r$, i.e.,

$$\sum_{k\in\mathbb{Z}} \overline{\tilde{a}(\gamma + 2k)}^{\mathsf{T}} a(n + \gamma + 2k) + \sum_{k\in\mathbb{Z}} \overline{\tilde{b}(\gamma + 2k)}^{\mathsf{T}} b(n + \gamma + 2k) = \tfrac{1}{2}\boldsymbol{\delta}(n)I_r,$$

for all $\gamma \in \{0, 1\}$ and $n \in \mathbb{Z}$.

(v) $(\{\tilde{a};\tilde{b}\},\{a;b\})$ *is* a (matrix-valued) dual framelet filter bank*:*

$$\overline{\widehat{\tilde{a}}(\xi)}^{\mathsf{T}}\widehat{a}(\xi)+\overline{\widehat{\tilde{b}}(\xi)}^{\mathsf{T}}\widehat{b}(\xi)=I_r,\quad \overline{\widehat{\tilde{a}}(\xi)}^{\mathsf{T}}\widehat{a}(\xi+\pi)+\overline{\widehat{\tilde{b}}(\xi)}^{\mathsf{T}}\widehat{b}(\xi+\pi)=0. \qquad (7.6.3)$$

To apply a discrete multiframelet transform using matrix-valued filters for a scalar input signal, we have to convert it into a vector-valued signal first. We say that a mapping $E: l(\mathbb{Z}) \to (l(\mathbb{Z}))^{1\times r}$ is *a vector conversion operator* if it is a linear bijection. For a positive integer $r \in \mathbb{N}$, *the standard vector conversion operator* $\mathring{E}$ is naturally defined to be

$$[\mathring{E}v](k) := \big(v(rk), v(rk+1), \dots, v(rk+(r-1))\big), \qquad v \in l(\mathbb{Z}). \qquad (7.6.4)$$

Using the standard vector conversion operator, we have the following result.

Proposition 7.6.2 *Let* $\mathring{E}$ *(with* $\circ$ *on the top of E) be the standard vector conversion operator defined in* (7.6.4)*. Then*

(1) For every $m \in \mathbb{N}_0$, $\mathring{E}(\mathbb{P}_{m-1}) = \mathscr{P}_{m-1,\mathring{\upsilon}} := \{\mathsf{p} * \mathring{\upsilon} \,:\, \mathsf{p} \in \mathbb{P}_{m-1}\}$*, where*

$$\widehat{\mathring{\upsilon}}(\xi) := (1, e^{i\xi/r}, \dots, e^{i\xi(r-1)/r}). \qquad (7.6.5)$$

(2) For every $m \in \mathbb{N}_0$ *and* $\upsilon \in (l_0(\mathbb{Z}))^{1\times r}$ *with* $\widehat{\upsilon}(0) \neq 0$*, there exists a strongly invertible sequence* $U_\upsilon \in (l_0(\mathbb{Z}))^{r\times r}$ *such that* $E_\upsilon := C_{U_\upsilon} \circ \mathring{E}$ *is a vector conversion operator satisfying* $(C_{U_\upsilon} \circ \mathring{E})(\mathbb{P}_{m-1}) = \mathscr{P}_{m-1,\upsilon}$*, where* $C_{U_\upsilon} : (l(\mathbb{Z}))^{r\times r} \to (l(\mathbb{Z}))^{r\times r}$ *with* $w \mapsto \sum_{k\in\mathbb{Z}} w(\cdot - k)U_\upsilon(k)$ *is a (right) convolution operator.*

Proof By Lemma 1.2.1, for $\mathsf{p} \in \mathbb{P}_{m-1}$ we have

$$\begin{aligned}
[\mathsf{p}(r\cdot) * \mathring{\upsilon}](x) &= \sum_{j=0}^{\infty} \frac{(-i)^j}{j!} r^j \mathsf{p}^{(j)}(rx)\widehat{\mathring{\upsilon}}^{(j)}(0)\\
&= \sum_{j=0}^{m-1} \frac{(-i)^j}{j!} r^j \mathsf{p}^{(j)}(rx)\left(\delta(j), \left(\tfrac{i}{r}\right)^j, \dots, \left(\tfrac{i(r-1)}{r}\right)^j\right)\\
&= \sum_{j=0}^{m-1} \frac{1}{j!}[\mathsf{p}(r\cdot)]^{(j)}(x)\left(\delta(j), \left(\tfrac{1}{r}\right)^j, \dots, \left(\tfrac{r-1}{r}\right)^j\right)\\
&= \big(\mathsf{p}(rx), \mathsf{p}(rx+1), \dots, \mathsf{p}(rx+(r-1))\big),
\end{aligned}$$

where we used the Taylor expansion $\mathsf{p}(r(x+y)) = \sum_{j=0}^{\infty}[\mathsf{p}(r\cdot)]^{(j)}(x)\frac{y^j}{j!}$. This proves item (1).

If $r = 1$, then $\mathscr{P}_{m-1,\mathring{\upsilon}} = \mathbb{P}_{m-1} = \mathscr{P}_{m-1,\upsilon}$ and we simply take $U_\upsilon = \boldsymbol{\delta}$. If $r > 1$, by a similar argument as in Theorem 5.6.4 (see Exercise 5.41), there exists a strongly invertible sequence $U_\upsilon \in (l_0(\mathbb{Z}))^{r\times r}$ such that

$$\widehat{\mathring{\upsilon}}(\xi)\widehat{U_\upsilon}(\xi) = \widehat{\upsilon}(\xi) + \mathscr{O}(|\xi|^m), \qquad \xi \to 0. \tag{7.6.6}$$

By item (1), we have $\mathring{E}(\mathbb{P}_{m-1}) = \mathscr{P}_{m-1,\mathring{\upsilon}}$. It follows from (7.6.6) that $C_{U_\upsilon}\mathscr{P}_{m-1,\mathring{\upsilon}} = \mathscr{P}_{m-1,\upsilon}$. Hence, $E_\upsilon(\mathbb{P}_{m-1}) = \mathscr{P}_{m-1,\upsilon}$. Since U_υ is strongly invertible, C_{U_υ} must be invertible with the inverse $C_{U_\upsilon^{-1}}$. Hence, E_υ is a bijection. This proves item (2). □

For $\upsilon \in (l_0(\mathbb{Z}))^{1\times r}$ with $\widehat{\upsilon}(0) \ne 0$, if

$$\mathcal{T}_{\tilde{a}}\mathscr{P}_{m-1,\upsilon} = \mathscr{P}_{m-1,\upsilon} \quad \text{and} \quad \mathcal{T}_{\tilde{b}}\mathscr{P}_{m-1,\upsilon} = 0, \tag{7.6.7}$$

then for any input signal $v_0 \in \mathscr{P}_{m-1}$, the high-pass framelet coefficients w_j in (7.6.1) must be zero for all $j \in \mathbb{N}$. The following result characterizes (7.6.7).

Lemma 7.6.3 *Let $\tilde{a} \in (l_0(\mathbb{Z}))^{r\times r}$, $\tilde{b} \in (l_0(\mathbb{Z}))^{s\times r}$, and $\upsilon \in (l_0(\mathbb{Z}))^{1\times r}$ with $\widehat{\upsilon}(0) \ne 0$. Then*

(i) $\mathcal{T}_{\tilde{a}}\mathscr{P}_{m-1,\upsilon} = \mathscr{P}_{m-1,\upsilon}$ if and only if there exists $c \in l_0(\mathbb{Z})$ with $\widehat{c}(0) \ne 0$ such that

$$\widehat{c}(\xi)\widehat{\upsilon}(\xi)\overline{\widehat{\tilde{a}}(\xi)}^{\mathsf{T}} = \widehat{\upsilon}(2\xi) + \mathscr{O}(|\xi|^m), \qquad \xi \to 0. \tag{7.6.8}$$

(ii) $\mathcal{T}_{\tilde{b}}\mathscr{P}_{m-1,\upsilon} = 0$ if and only if

$$\widehat{\upsilon}(\xi)\overline{\widehat{\tilde{b}}(\xi)}^{\mathsf{T}} = \mathscr{O}(|\xi|^m), \qquad \xi \to 0. \tag{7.6.9}$$

Proof By the definition of the transition operator in (5.6.31), we deduce from Theorem 1.2.2 that

$$\mathcal{T}_{\tilde{a}}(\mathsf{p}*\upsilon) = \mathcal{T}_{\tilde{a}*\upsilon^\star}(\mathsf{p}) = 2[\mathsf{p}*\upsilon*\tilde{a}^\star](2\cdot) = 2\mathsf{p}(2\cdot)*u, \qquad \forall\, \mathsf{p} \in \mathbb{P}_{m-1}, \tag{7.6.10}$$

where $u \in (l_0(\mathbb{Z}))^{1\times r}$ satisfies $\widehat{u}(\xi) := \widehat{\upsilon}(\xi/2)\overline{\widehat{\tilde{a}}(\xi/2)}^{\mathsf{T}} + \mathscr{O}(|\xi|^m)$ as $\xi \to 0$. Now item (i) follows directly from Lemma 5.6.6. Item (ii) follows directly from (7.6.10) (with $\tilde{a}$ being replaced by $\tilde{b}$) and Lemma 1.2.1. □

We define $\operatorname{vm}(\tilde{b}\,|\,\upsilon) := m$, the largest nonnegative integer satisfying (7.6.9). If $(\{\tilde{a};\tilde{b}\},\{a;b\})$ is a dual framelet filter bank, then so is $(\{c\tilde{a};d\tilde{b}\},\{\overline{c^{-1}}a;\overline{d^{-1}}b\})$ for all $c,d \in \mathbb{C}\backslash\{0\}$. Therefore, by multiplying the filter $\tilde{a}$ in (7.6.8) with a nonzero number, without loss of generality, we can always assume that $\widehat{c}(0) = 1$. For $\upsilon \in (l_0(\mathbb{Z}))^{1\times r}$ with $\widehat{\upsilon}(0) \ne 0$, we say that a filter bank $\{\tilde{a};\tilde{b}\}$ has m *E_υ-balanced order* if both (7.6.8) and (7.6.9) are satisfied for some $c \in l_0(\mathbb{Z})$ with $\widehat{c}(0) = 1$. In particular, we define $\operatorname{bo}(\{\tilde{a};\tilde{b}\} \mid \upsilon) := m$ to be the largest E_υ-balanced order m. Note that $\operatorname{bo}(\{\tilde{a};\tilde{b}\} \mid \upsilon) = \operatorname{bo}(\{\tilde{a};\tilde{b}\} \mid d * \upsilon)$ for all $d \in l_0(\mathbb{Z})$ with $\widehat{d}(0) \neq 0$.

We say that a filter $a \in (l_0(\mathbb{Z}))^{r\times r}$ has *m general sum rules with a matching filter* $\upsilon \in (l_0(\mathbb{Z}))^{1\times r}$ *with* $\widehat{\upsilon}(0) \neq 0$ if

$$\widehat{\upsilon}(2\xi)\widehat{a}(\xi) = \widehat{c}(\xi)\widehat{\upsilon}(\xi) + \mathscr{O}(|\xi|^m), \quad \widehat{\upsilon}(2\xi)\widehat{a}(\xi+\pi) = \mathscr{O}(|\xi|^m), \quad \xi \to 0 \qquad (7.6.11)$$

for some $c \in l_0(\mathbb{Z})$ with $\widehat{c}(0) = 1$. We define $\operatorname{sr}(a \mid \upsilon)$ to be the largest such nonnegative integer m. Clearly, if $\widehat{c}(\xi) = 1 + \mathscr{O}(|\xi|^m)$ as $\xi \to 0$, then a has m (standard) sum rules with the matching filter υ. It is also trivial to observe that $\operatorname{sr}(a \mid \upsilon) = \operatorname{sr}(a \mid d * \upsilon)$ for all sequences $d \in l_0(\mathbb{Z})$ with $\widehat{d}(0) \neq 0$. Conversely, we have

Lemma 7.6.4 *Let* $\upsilon \in (l_0(\mathbb{Z}))^{1\times r}$ *with* $\widehat{\upsilon}(0) \neq 0$. *If a filter* $a \in (l_0(\mathbb{Z}))^{r\times r}$ *has m general sum rules with the matching filter* υ, *then there exists a sequence* $d \in l_0(\mathbb{Z})$ *with* $\widehat{d}(0) = 1$ *such that the filter a has m sum rules with the matching filter* $d * \upsilon$.

Proof Note that (7.6.11) holds for some $c \in l_0(\mathbb{Z})$ with $\widehat{c}(0) = 1$. Hence, we can define $\widehat{\eta}(\xi) = \prod_{j=1}^{\infty} \widehat{c}(2^{-j}\xi)$. Take $d \in l_0(\mathbb{Z})$ such that $\widehat{d}(\xi) = 1/\widehat{\eta}(\xi) + \mathscr{O}(|\xi|^m)$ as $\xi \to 0$. By $\widehat{\eta}(2\xi) = \widehat{c}(\xi)\widehat{\eta}(\xi)$, we have $\widehat{c}(\xi) = \widehat{d}(\xi)/\widehat{d}(2\xi) + \mathscr{O}(|\xi|)^m)$ as $\xi \to 0$. Therefore, as $\xi \to 0$,

$$\widehat{d}(2\xi)\widehat{\upsilon}(2\xi)\widehat{a}(\xi) = \widehat{d}(\xi)\widehat{\upsilon}(\xi) + \mathscr{O}(|\xi|^m), \quad \widehat{d}(2\xi)\widehat{\upsilon}(2\xi)\widehat{a}(\xi+\pi) = \mathscr{O}(|\xi|^m).$$

That is, the filter a has m sum rules with the matching filter $d * \upsilon$. □

The relation between balanced orders and general sum rules is as follows:

Theorem 7.6.5 *Let* $\tilde{a}, a \in (l_0(\mathbb{Z}))^{r\times r}$ *and* $\tilde{b}, b \in (l_0(\mathbb{Z}))^{s\times r}$. *Let* $\upsilon \in (l_0(\mathbb{Z}))^{1\times r}$ *with* $\widehat{\upsilon}(0) \neq 0$. *If* $(\{\tilde{a};\tilde{b}\},\{a;b\})$ *is a dual framelet filter bank satisfying* (7.6.3), *then* $\operatorname{sr}(a \mid \upsilon) \geqslant \operatorname{bo}(\{\tilde{a};\tilde{b}\} \mid \upsilon)$. *If in addition* $s = r$, *i.e.*, $(\{\tilde{a};\tilde{b}\},\{a;b\})$ *is a biorthogonal wavelet filter bank, then* $\operatorname{sr}(a \mid \upsilon) = \operatorname{bo}(\{\tilde{a};\tilde{b}\} \mid \upsilon)$.

Proof By definition, the filter bank $\{\tilde{a};\tilde{b}\}$ has $m := \operatorname{bo}(\{\tilde{a};\tilde{b}\} \mid \upsilon)$ E_υ-balanced order if and only if (7.6.8) and (7.6.9) hold for some $c \in l_0(\mathbb{Z})$ with $\widehat{c}(0) = 1$. By the first identity in (7.6.3), we deduce from (7.6.8) and (7.6.9) that

$$\widehat{c}(\xi)\widehat{\upsilon}(\xi) = \widehat{c}(\xi)\widehat{\upsilon}(\xi)\overline{\widehat{\tilde{a}}(\xi)}^{\mathsf{T}}\widehat{a}(\xi) + \widehat{c}(\xi)\widehat{\upsilon}(\xi)\overline{\widehat{\tilde{b}}(\xi)}^{\mathsf{T}}\widehat{b}(\xi) = \widehat{\upsilon}(2\xi)\widehat{a}(\xi) + \mathscr{O}(|\xi|^m),$$

$$0 = \widehat{c}(\xi)\widehat{\upsilon}(\xi)\overline{\widehat{\tilde{a}}(\xi)}^{\mathsf{T}}\widehat{a}(\xi+\pi) + \widehat{c}(\xi)\widehat{\upsilon}(\xi)\overline{\widehat{\tilde{b}}(\xi)}^{\mathsf{T}}\widehat{b}(\xi+\pi) = \widehat{\upsilon}(2\xi)\widehat{a}(\xi) + \mathscr{O}(|\xi|^m),$$

as $\xi \to 0$. Hence, (7.6.11) is satisfied. This proves $\operatorname{sr}(a \mid \upsilon) \geqslant \operatorname{bo}(\{\tilde{a};\tilde{b}\} \mid \upsilon)$.

Suppose that $s = r$. Then (7.6.3) is equivalent to (4.5.5). Let $m := \operatorname{sr}(a\,|\,\upsilon)$. Then (7.6.11) holds for some $c \in l_0(\mathbb{Z})$ with $\widehat{c}(0) = 1$. By (4.5.5), we have $\widehat{a}(\xi)\overline{\widehat{\tilde{a}}(\xi)}^{\mathsf{T}} + \widehat{a}(\xi+\pi)\overline{\widehat{\tilde{a}}(\xi+\pi)}^{\mathsf{T}} = I_r$. Hence, by (7.6.11),

$$\widehat{\upsilon}(2\xi) = \widehat{\upsilon}(2\xi)\widehat{a}(\xi)\overline{\widehat{\tilde{a}}(\xi)}^{\mathsf{T}} + \widehat{\upsilon}(2\xi)\widehat{a}(\xi+\pi)\overline{\widehat{\tilde{a}}(\xi+\pi)}^{\mathsf{T}} = \widehat{c}(\xi)\widehat{\upsilon}(\xi)\overline{\widehat{\tilde{a}}(\xi)}^{\mathsf{T}} + \mathscr{O}(|\xi|^m)$$

as $\xi \to 0$. By (4.5.5), we also have $\widehat{a}(\xi)\overline{\widehat{\tilde{b}}(\xi)}^{\mathsf{T}} + \widehat{a}(\xi+\pi)\overline{\widehat{\tilde{b}}(\xi+\pi)}^{\mathsf{T}} = 0$, from which and (7.6.11) we have

$$0 = \widehat{\upsilon}(2\xi)\widehat{a}(\xi)\overline{\widehat{\tilde{b}}(\xi)}^{\mathsf{T}} + \widehat{\upsilon}(2\xi)\widehat{a}(\xi+\pi)\overline{\widehat{\tilde{b}}(\xi+\pi)}^{\mathsf{T}} = \widehat{c}(\xi)\widehat{\upsilon}(\xi)\overline{\widehat{\tilde{b}}(\xi)}^{\mathsf{T}} + \mathscr{O}(|\xi|^m)$$

as $\xi \to 0$. Hence, both (7.6.8) and (7.6.9) are satisfied. This proves $\operatorname{bo}(\{\tilde{a};\tilde{b}\}\,|\,\upsilon) \geqslant m = \operatorname{sr}(a\,|\,\upsilon)$. Consequently, we must have $\operatorname{sr}(a\,|\,\upsilon) = \operatorname{bo}(\{\tilde{a};\tilde{b}\}\,|\,\upsilon)$. □

The case $\operatorname{bo}(\{\tilde{a};\tilde{b}\}\,|\,\upsilon) < \operatorname{sr}(a\,|\,\upsilon)$ can happen in Theorem 7.6.5 for a dual framelet filter bank $(\{\tilde{a};\tilde{b}\},\{a;b\})$. We have the following algorithm implementing the discrete multiframelet transform using a dual framelet filter bank $(\{\tilde{a};\tilde{b}\},\{a;b\})$.

Algorithm 7.6.6 *Let $(\{\tilde{a};\tilde{b}\},\{a;b\})$ be a dual framelet filter bank with $a, \tilde{a} \in (l_0(\mathbb{Z}))^{r\times r}$ and $b, \tilde{b} \in (l_0(\mathbb{Z}))^{s\times r}$. Suppose that a has the (largest possible) n general sum rules with a matching filter $\upsilon \in (l_0(\mathbb{Z}))^{1\times r}$ with $\widehat{\upsilon}(0) \ne 0$ (i.e., $n := \operatorname{sr}(a\,|\,\upsilon)$).*

(1) Find the E_υ-balanced order $m := \operatorname{bo}(\{\tilde{a};\tilde{b}\}\,|\upsilon)$ which is the large integer (necessarily $m \leqslant n$) satisfying (7.6.8) and (7.6.9) for some $c \in l_0(\mathbb{Z})$ with $\widehat{c}(0) = 1$.

(2) For $r = 1$, take $U_\upsilon = \boldsymbol{\delta}$. For $r > 1$, according to Theorem 5.6.4 or Exercise 5.41, construct a strongly invertible sequence $U_\upsilon \in (l_0(\mathbb{Z}))^{r\times r}$ such that (7.6.6) holds and $\widehat{U_\upsilon}^{-1}$ is an $r \times r$ matrix of 2π-periodic trigonometric polynomials. Such U_υ is called a preprocessing filter and U_υ^{-1} is called a postprocessing filter.

(3) Recursively perform the fast multiframelet transform as illustrated in Fig. 7.21.

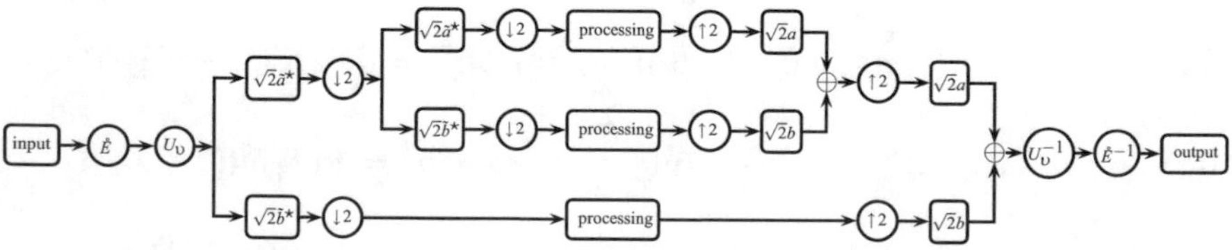

Fig. 7.21 Diagram of a two-level (balanced) discrete multiframelet transform using a dual framelet filter bank $(\{\tilde{a};\tilde{b}\},\{a;b\})$. Input and output are scalar sequences. $\mathring{E}$ is the standard vector conversion operator in (7.6.4) and $U_\upsilon \in (l_0(\mathbb{Z}))^{r\times r}$ is a strongly invertible sequence satisfying $\widehat{\mathring{\upsilon}}(\xi)\widehat{U_\upsilon}(\xi) = \widehat{\upsilon}(\xi) + \mathscr{O}(|\xi|^m)$ as $\xi \to 0$ with $m := \operatorname{bo}(\{\tilde{a};\tilde{b}\}\,|\upsilon)$, where $\mathring{\upsilon}$ is defined in (7.6.5)

To apply a fast multiframelet transform in Fig. 7.21 with $r > 1$, one has to do extra work to design a strongly invertible sequence $U_\upsilon \in (l_0(\mathbb{Z}))^{r\times r}$ satisfying $\widehat{\mathring{\upsilon}}(\xi)\widehat{U_\upsilon}(\xi) = \widehat{\upsilon}(\xi) + \mathscr{O}(|\xi|^m)$ as $\xi \to 0$ with $m := \operatorname{bo}(\{\tilde{a};\tilde{b}\} \mid \upsilon)$. If the filter bank $\{\tilde{a};\tilde{b}\}$ has m $\mathring{E}$-balanced order, then one can simply take $U_\upsilon = \boldsymbol{\delta} I_r$ so that the steps for preprocessing using U_υ and postprocessing using U_υ^{-1} can be dropped. To achieve the balanced property, the discrete multiframelet transform using a dual framelet filter bank $(\{\tilde{a};\tilde{b}\},\{a;b\})$ in Algorithm 7.6.6 is essentially implemented by using an equivalent (but balanced) dual framelet filter bank $(\{\mathring{\tilde{a}};\mathring{\tilde{b}}\},\{\mathring{a};\mathring{b}\})$ with

$$\widehat{\mathring{\tilde{a}}}(\xi) := [\overline{\widehat{U_\upsilon}(2\xi)}^{\mathsf{T}}]^{-1}\widehat{\tilde{a}}(\xi)\overline{\widehat{U_\upsilon}(\xi)}^{\mathsf{T}}, \quad \widehat{\mathring{\tilde{b}}}(\xi) := [\overline{\widehat{U_\upsilon}(2\xi)}^{\mathsf{T}}]^{-1}\widehat{\tilde{b}}(\xi)\overline{\widehat{U_\upsilon}(\xi)}^{\mathsf{T}},$$

$$\widehat{\mathring{a}}(\xi) := \widehat{U_\upsilon}(2\xi)\widehat{a}(\xi)[\widehat{U_\upsilon}(\xi)]^{-1}, \qquad \widehat{\mathring{b}}(\xi) := \widehat{U_\upsilon}(2\xi)\widehat{b}(\xi)[\widehat{U_\upsilon}(\xi)]^{-1},$$

and $\operatorname{bo}(\{\mathring{\tilde{a}};\mathring{\tilde{b}}\} \mid \mathring{\upsilon}) = \operatorname{bo}(\{\tilde{a};\tilde{b}\} \mid \upsilon)$. To achieve the perfect reconstruction property and the balanced approximation property, the linear operators $\mathring{E}, \mathring{E}^{-1}, U_\upsilon$ and U_υ^{-1} in Fig. 7.21 may be replaced by some desired linear operators $E_{\mathcal{W}} : l(\mathbb{Z}) \to (l(\mathbb{Z}))^{1\times r}$, $E_{\mathcal{V}} : (l(\mathbb{Z}))^{1\times r} \to l(\mathbb{Z})$ and $U_{\mathcal{W}}, U_{\mathcal{V}} : (l(\mathbb{Z}))^{1\times r} \to (l(\mathbb{Z}))^{1\times r}$, respectively such that $E_{\mathcal{V}}E_{\mathcal{W}} = \operatorname{Id}_{l(\mathbb{Z})}$ and $U_{\mathcal{V}}U_{\mathcal{W}} = \operatorname{Id}_{(l(\mathbb{Z}))^{1\times r}}$.

For a dual framelet filter bank, we have

Lemma 7.6.7 *Let $(\{\tilde{a};\tilde{b}\},\{a;b\})$ be a dual framelet filter bank with $\tilde{a}, a \in (l_0(\mathbb{Z}))^{r\times r}$ and $\tilde{b}, b \in (l_0(\mathbb{Z}))^{s\times r}$. Suppose that there exists an $r \times 1$ vector $\tilde{\phi}$ of compactly supported distributions satisfying $\widehat{\tilde{\phi}}(2\xi) = \widehat{\tilde{a}}(\xi)\widehat{\tilde{\phi}}(\xi)$ with $\widehat{\tilde{\phi}}(0) \ne 0$. Define $\widehat{\tilde{\psi}}(\xi) := \widehat{\tilde{b}}(\xi/2)\widehat{\tilde{\phi}}(\xi/2)$. Then*

(i) $\operatorname{vm}(\tilde{b} \mid \tilde{\upsilon}) = \operatorname{vm}(\tilde{\psi})$ *and* $\operatorname{sr}(a \mid \tilde{\upsilon}) \geqslant \operatorname{vm}(\tilde{\psi})$, *where* $\tilde{\upsilon} \in (l_0(\mathbb{Z}))^{1\times r}$ *satisfies* $\widehat{\tilde{\upsilon}}(\xi) = \overline{\widehat{\tilde{\phi}}(\xi)}^{\mathsf{T}} + \mathscr{O}(|\xi|^{m+1})$ *as* $\xi \to 0$ *with* $m := \operatorname{vm}(\tilde{\psi})$.

(ii) If $s = r$ and if 1 is a simple eigenvalue of $\widehat{\tilde{a}}(0)$ and all $2^j, j \in \mathbb{N}$ are not eigenvalues of $\widehat{\tilde{a}}(0)$, then $\operatorname{vm}(\tilde{\psi}) \geqslant \operatorname{sr}(a \mid \upsilon)$ for all $\upsilon \in (l_0(\mathbb{Z}))^{1\times r}$ with $\widehat{\upsilon}(0) \ne 0$. In particular, the identity $\operatorname{vm}(\tilde{\psi}) = \operatorname{sr}(a \mid \tilde{\upsilon})$ holds.

Proof By the definition of $\tilde{\upsilon}$, it is trivial to see that $\operatorname{vm}(\tilde{b} \mid \tilde{\upsilon}) = \operatorname{vm}(\tilde{\psi}) = m$. By the first identity in (7.6.3) and $\operatorname{vm}(\tilde{b} \mid \tilde{\upsilon}) = m$, we deduce that

$$\widehat{\tilde{\upsilon}}(\xi) = \widehat{\tilde{\upsilon}}(\xi)\overline{\widehat{\tilde{a}}(\xi)}^{\mathsf{T}}\widehat{a}(\xi) + \widehat{\tilde{\upsilon}}(\xi)\overline{\widehat{\tilde{b}}(\xi)}^{\mathsf{T}}\widehat{b}(\xi) = \widehat{\tilde{\upsilon}}(2\xi)\widehat{a}(\xi) + \mathscr{O}(|\xi|^m),$$

$$0 = \widehat{\tilde{\upsilon}}(\xi)\overline{\widehat{\tilde{a}}(\xi)}^{\mathsf{T}}\widehat{a}(\xi+\pi) + \widehat{\tilde{\upsilon}}(\xi)\overline{\widehat{\tilde{b}}(\xi)}^{\mathsf{T}}\widehat{b}(\xi+\pi) = \widehat{\tilde{\upsilon}}(2\xi)\widehat{a}(\xi) + \mathscr{O}(|\xi|^m),$$

as $\xi \to 0$. This proves $\operatorname{sr}(a \mid \tilde{\upsilon}) \geqslant m$. Hence, item (i) holds.

Let $m := \operatorname{sr}(a \,|\, \upsilon)$. By Theorem 7.6.5, we have $\operatorname{bo}(\{\tilde{a}; \tilde{b}\} \,|\, \upsilon) = m$. Hence, both (7.6.8) and (7.6.9) hold for some $c \in l_0(\mathbb{Z})$ with $\widehat{c}(0) = 1$. Choose $d \in l_0(\mathbb{Z})$ such that $\widehat{d}(\xi) = \frac{1}{\prod_{j=1}^{\infty} \widehat{c}(2^{-j}\xi)} + \mathscr{O}(|\xi|^m)$ as $\xi \to 0$. Then (7.6.8) implies

$$\widehat{d}(\xi)\widehat{\upsilon}(\xi)\overline{\widehat{\tilde{a}}(\xi)}^{\mathsf{T}} = \widehat{d}(2\xi)\widehat{\upsilon}(2\xi) + \mathscr{O}(|\xi|^m), \qquad \xi \to 0. \tag{7.6.12}$$

By our assumption on $\widehat{\tilde{a}}(0)$ and $\widehat{\tilde{\phi}}(2\xi) = \widehat{\tilde{a}}(\xi)\widehat{\tilde{\phi}}(\xi)$, we know from (5.1.11) that up to a multiplicative constant, the moments $\widehat{\tilde{\phi}}(0), \ldots, \widehat{\tilde{\phi}}^{(m-1)}(0)$ are uniquely determined. Now it follows from (7.6.12) that $\lambda\widehat{d}(\xi)\widehat{\upsilon}(\xi) = \overline{\widehat{\tilde{\phi}}(\xi)}^{\mathsf{T}} + \mathscr{O}(|\xi|^m)$ as $\xi \to 0$ for some $\lambda \in \mathbb{C}\backslash\{0\}$. By (7.6.9), we have $\operatorname{vm}(\tilde{\psi}) \geqslant m$. This proves item (ii). □

For a tight framelet filter bank, we have

Proposition 7.6.8 *Let $\{a; b\}$ be a tight framelet filter bank with $a \in (l_0(\mathbb{Z}))^{r\times r}$ and $b \in (l_0(\mathbb{Z}))^{s\times r}$. For $\upsilon \in (l_0(\mathbb{Z}))^{1\times r}$ and $\widehat{\upsilon}(0) \neq 0$,*

$$\operatorname{vm}(b \,|\, \upsilon) = \tfrac{1}{2} \operatorname{vm}\left(\widehat{\upsilon}(\xi)\overline{\widehat{\upsilon}(\xi)}^{\mathsf{T}} - \widehat{\upsilon}(\xi)\overline{\widehat{a}(\xi)}^{\mathsf{T}}\widehat{a}(\xi)\overline{\widehat{\upsilon}(\xi)}^{\mathsf{T}}\right). \tag{7.6.13}$$

Moreover, $\mathcal{T}_a\mathscr{P}_{m-1,\upsilon} = \mathscr{P}_{m-1,\upsilon}$ if and only if for some $c \in l_0(\mathbb{Z})$ with $\widehat{c}(0) \neq 0$,

$$\widehat{c}(\xi)\widehat{\upsilon}(\xi)\overline{\widehat{a}(\xi)}^{\mathsf{T}} = \widehat{\upsilon}(2\xi) + \mathscr{O}(|\xi|^m), \qquad \xi \to 0. \tag{7.6.14}$$

Consequently, $\operatorname{bo}(\{a; b\} \,|\, \upsilon) = m$ *if and only if* (7.6.14) *holds for some $c \in l_0(\mathbb{Z})$ with $\widehat{c}(0) = 1$, and $\|\widehat{\upsilon}(\xi)\overline{\widehat{a}(\xi)}^{\mathsf{T}}\|_{l_2}^2 = \|\widehat{\upsilon}(\xi)\|_{l_2}^2 + \mathscr{O}(|\xi|^{2m})$ as $\xi \to 0$. For the choice $\widehat{\upsilon}(\xi) := \overline{\widehat{\phi}(\xi)}^{\mathsf{T}}$, then* $\operatorname{bo}(\{a; b\} \,|\, \upsilon) = \operatorname{vm}(b \,|\, \upsilon) = \min(\operatorname{sr}(a \,|\, \upsilon), \frac{1}{2}\operatorname{vm}(1 - \|\widehat{\phi}\|_{l_2}^2)) = \operatorname{vm}(\psi)$, *where $\widehat{\phi}(2\xi) = \widehat{a}(\xi)\widehat{\phi}(\xi)$ and $\widehat{\psi}(2\xi) = \widehat{b}(\xi)\widehat{\phi}(\xi)$ with* $\|\widehat{\phi}(0)\|_{l_2} = 1$.

Proof By the first identity in (7.6.3), we have

$$\widehat{\upsilon}(\xi)\overline{\widehat{a}(\xi)}^{\mathsf{T}}\widehat{a}(\xi)\overline{\widehat{\upsilon}(\xi)}^{\mathsf{T}} + \|\widehat{\upsilon}(\xi)\overline{\widehat{b}(\xi)}^{\mathsf{T}}\|_{l_2}^2 = \widehat{\upsilon}(\xi)\overline{\widehat{\upsilon}(\xi)}^{\mathsf{T}},$$

from which we have (7.6.13). Now (7.6.14) follows directly from item (i) of Lemma 7.6.3. For $\widehat{\upsilon}(\xi) := \overline{\widehat{\phi}(\xi)}^{\mathsf{T}}$, (7.6.14) trivially holds for all $m \in \mathbb{N}$ with $\widehat{c} = 1$ and $\widehat{\upsilon}(\xi)\overline{\widehat{a}(\xi)}^{\mathsf{T}}\widehat{a}(\xi)\overline{\widehat{\upsilon}(\xi)}^{\mathsf{T}} = \|\widehat{\phi}(2\xi)\|_{l_2}^2$. By $\|\widehat{\phi}(0)\|_{l_2} = 1$, we have $\operatorname{vm}(\|\widehat{\phi}\|_{l_2}^2 - \|\widehat{\phi}(2\cdot)\|_{l_2}^2) = \operatorname{vm}(1 - \|\widehat{\phi}\|_{l_2}^2)$. Now the claim follows from Lemma 7.6.7. □

7.7 Exercises

7.1. Prove Theorems 7.1.1, 7.1.3, 7.1.6, 7.1.7, 7.1.8, 7.1.9, and Proposition 7.1.2.

7.2. Let $M \in \mathbb{Z}$ with $|M| \geqslant 2$ and $a \in l_0(\mathbb{Z})$. Prove that $\operatorname{sr}(a, M) \geqslant m$ if and only if $\widehat{a}(\xi) = (1 + e^{-i\xi} + \cdots + e^{-i(|M|-1)\xi})^m \widehat{b}(\xi)$ for some $b \in l_0(\mathbb{Z})$.

7.3. Let $M \in \mathbb{N}$ with $M \geqslant 2$. Let $a \in l_0(\mathbb{Z})$ such that $\widehat{a}(0) = 1$ and $\widehat{a}(\xi) = (1 + e^{-i\xi} + \cdots + e^{-i(M-1)\xi})^m \widehat{b}(\xi)$ for some $m \in \mathbb{N}_0$ and $b \in l_0(\mathbb{Z})$ with $\sum_{\gamma=1}^{M-1} |\widehat{b}(\frac{2\pi\gamma}{M})| \neq 0$ (i.e., $m = \operatorname{sr}(a, M)$). Define $c \in l_0(\mathbb{Z})$ by $\widehat{c}(\xi) := |\widehat{b}(\xi)|^2$ and b_n by $\widehat{b_n}(\xi) = \widehat{b}(M^{n-1}\xi) \cdots \widehat{b}(M\xi)\widehat{b}(\xi)$ for $n \in \mathbb{N}$. Prove that all the claims in Corollary 5.8.5 still hold for a general dilation factor M.

(1) $\operatorname{sm}_p(a, M) = \operatorname{sm}_p(b, M) = 1/p - \log_M(\rho_0(b, M)_p)$.

(2) $\operatorname{sm}_\infty(a, M) = \operatorname{sm}_\infty(b, M) \leqslant -\log_M \lambda_b$, where

$$\lambda_b := \rho(\mathcal{T}_{b,M}|_{l(\operatorname{fsupp}(b^\star))}) = \max\{|\lambda| \ : \ \lambda \in \operatorname{spec}((Mb(Mk-j))_{j,k \in \operatorname{fsupp}(b)})\}.$$

If in addition $\widehat{b}(\xi) \geqslant 0$ for all $\xi \in \mathbb{R}$, then $\operatorname{sm}_\infty(a, M) = \operatorname{sm}_\infty(b, M) = -\log_M \lambda_b$.

(3) $\operatorname{sm}_2(a, M) = \operatorname{sm}_2(b, M) = \frac{1}{2}\operatorname{sm}_\infty(c, M) = -\frac{1}{2}\log_M \lambda_c$, where λ_c is defined similarly as λ_b. Moreover, $\operatorname{spec}((Mc(Mk - j))_{-\operatorname{len}(b) \leqslant j,k \leqslant \operatorname{len}(b)})$ is the same as $\operatorname{spec}((Md(Mk - j))_{-\operatorname{len} a \leqslant j,k \leqslant \operatorname{len}(a)}) \backslash \{M^0, \ldots, M^{1-2m}\}$ with $\widehat{d}(\xi) := |\widehat{a}(\xi)|^2$.

(4) $\rho_0(b, M)_p = \operatorname{jsr}_p(\{B_0, \ldots, B_{M-1}\})$, where $B_\gamma = (Mb(\gamma + Mk - j))_{j,k \in \operatorname{fsupp}(b)}$.

(5) $\rho_0(b, M)_\infty = M \inf_{n \in \mathbb{N}} \max_{0 \leqslant \gamma < M^n} (\sum_{k \in \mathbb{Z}} |b_n(\gamma + M^n k)|)^{1/n}$.

(6) $\rho_0(b, M)_\infty \leqslant \rho(\widehat{b}, M) \leqslant \rho_0(b, M)_1$, where we define the quantity $\rho(\widehat{b}, M) := M \inf_{n \in \mathbb{N}} \|\widehat{b_n}\|_{L_\infty(\mathbb{T})}^{1/n}$.

(7) $\{\|\widehat{b_n^{\min}}\|_{L_\infty(\mathbb{T})}\}_{n=1}^\infty$ decreases to $\rho(\widehat{b}, M)$: $\|\widehat{b_n^{\min}}\|_{L_\infty(\mathbb{T})} \downarrow \rho(\widehat{b}, M)$, where

$$\widehat{b_n^{\min}}(\xi) := M \min_{1 \leqslant j \leqslant n} |\widehat{b_j}(\xi)|^{1/j} = M \min_{1 \leqslant j \leqslant n} |\widehat{b}(M^{j-1}\xi) \cdots \widehat{b}(M\xi)\widehat{b}(\xi)|^{1/j}.$$

7.4. Let $b \in l_0(\mathbb{Z})$ and $k_0 \in \mathbb{Z}$. For $M \in \mathbb{Z}$ with $|M| \geqslant 2$, if $b(k_0 + Mk) = 0$ for all $k \in \mathbb{Z}\backslash\{0\}$ and $|b(k_0)| \geqslant \sup_{1 \leqslant \gamma < M} \sum_{k \in \mathbb{Z}} |b(\gamma + k_0 + Mk)|$, prove that $\rho_0(b, M)_\infty = M|b(k_0)|$ and $\operatorname{sm}_\infty(b, M) = -1 - \log_M |b(k_0)|$.

7.5. Let $a \in l_0(\mathbb{Z})$ be given in Exercise 6.48 as $\widehat{a}(\xi) := e^{-i\xi}(1 + e^{-i\xi} + e^{-i2\xi})(1 + e^{-i\xi})/6$. Prove that a is an interpolatory 3-wavelet filter and $\operatorname{sm}_\infty(a, 3) = \log_3 2 > 0$. Hence, its associated standard 3-refinable function ϕ is a continuous interpolating function satisfying $\phi(k) = \boldsymbol{\delta}(k)$ for all $k \in \mathbb{Z}$.

7.6. Define $\widehat{a}(\xi) = (1+e^{-i\xi}+e^{-i2\xi}+e^{-i3\xi})(\frac{2-\sqrt{6}}{16}e^{i2\xi} + \frac{\sqrt{6}}{16}e^{i\xi} + \frac{\sqrt{6}}{16} + \frac{2-\sqrt{6}}{16}e^{-i\xi})$.

a. Prove that a is an interpolatory 4-wavelet and orthogonal 4-wavelet filter.

b. $\operatorname{sm}_\infty(a, 4) = -\log_4 \frac{\sqrt{6}}{4} > 0$ and $\operatorname{sm}_2(a, 4) = -\frac{1}{2}\log_4 \frac{4-\sqrt{2}}{8}$. Therefore, its associated standard 4-refinable function must be interpolating and orthogonal.

7.7. Let a be the d-dimensional Haar low-pass filter given by $\widehat{a}(\xi_1,\dots,\xi_d) := 2^{-d}\prod_{j=1}^{d}(1+e^{-i\xi_j})$. Define the high-pass filters $b_1,\dots,b_s$ with $s := 2^{d-1}(2^d-1)$ by $2^{-d}(\boldsymbol{\delta}(\cdot-\alpha)-\boldsymbol{\delta}(\cdot-\beta))$ for all $\alpha,\beta\in\{0,1\}^d$ with $\alpha\ne\beta$. Prove that $\{a;b_1,\dots,b_s\}$ is a tight $2I_d$-framelet filter bank.

7.8. Let M be a $d\times d$ invertible integer matrix. For every filter $a\in l_0(\mathbb{Z}^d)$ having nonnegative coefficients and $\operatorname{sr}(a,\mathsf{M})\geqslant 1$, prove that there always exist high-pass filters $b_1,\dots,b_s$ such that $\{a;b_1,\dots,b_s\}$ is a tight M-framelet filter bank and every high-pass filter has only two nonzero coefficients with support contained inside $\operatorname{supp}(a)$.

7.9. Prove that a $d\times d$ real-valued matrix M is isotropic if and only if there exists a norm $\|\cdot\|_{\mathsf{M}}$ on $\mathbb{R}^d$ such that $\|\mathsf{M}x\|_{\mathsf{M}} = |\det(\mathsf{M})|^{1/d}\|x\|_{\mathsf{M}}$ for all $x\in\mathbb{R}^d$. *Hint*: Apply Proposition 7.1.5 and consider $\|x+iy\| := \|x\|_{\mathsf{M}}+\|y\|_{\mathsf{M}}$ for $x,y\in\mathbb{R}^d$ instead of the norm $\|\cdot\|_{\mathsf{M}}$ on $\mathbb{C}^d$.

7.10. Let M be a $d\times d$ real-valued matrix having eigenvalues $\lambda_1,\dots,\lambda_d$ with $|\lambda_1|\leqslant\cdots\leqslant|\lambda_d|$. For any $\varepsilon>0$, prove that there exists a norm $\|\cdot\|_{\mathsf{M},\varepsilon}$ on $\mathbb{R}^d$ (or $\mathbb{C}^d$) such that $(|\lambda_1|-\varepsilon)\|x\|_{\mathsf{M},\varepsilon}\leqslant\|\mathsf{M}x\|_{\mathsf{M},\varepsilon}\leqslant(|\lambda_d|+\varepsilon)\|x\|_{\mathsf{M},\varepsilon}$.

7.11. Let M be a $d\times d$ dilation matrix and $a\in l_0(\mathbb{Z}^d)$ with $\widehat{a}(0)=1$. Prove that $\lim_{n\to\infty}\prod_{j=1}^{n}\widehat{a}((\mathsf{M}^{\mathsf{T}})^{-j}\xi)$ converges uniformly on any bounded set.

7.12. Let M be a $d\times d$ invertible integer matrix and $u\in(l_0(\mathbb{Z}^d))^{r\times r}$. For a polynomial p and $\upsilon\in(l_0(\mathbb{Z}^d))^{1\times r}$, prove that $\mathcal{T}_{u,\mathsf{M}}(\mathsf{p}*\upsilon) = |\det(\mathsf{M})|\mathsf{p}(\mathsf{M}\cdot)*\mathring{\upsilon}$, where $\mathring{\upsilon}\in(l_0(\mathbb{Z}^d))^{1\times r}$ satisfies $\widehat{\mathring{\upsilon}}(\xi) = \widehat{\upsilon}(\mathsf{M}^{-\mathsf{T}}\xi)\overline{\widehat{u}(\mathsf{M}^{-\mathsf{T}}\xi)}^{\mathsf{T}}+\mathscr{O}(\|\xi\|^{\deg(\mathsf{p})+1})$, $\xi\to 0$.

7.13. Let M be a $d\times d$ invertible integer matrix and $u\in l_0(\mathbb{Z}^d)$. For a polynomial q, prove that the following statements are equivalent:

a. $\sum_{k\in\mathbb{Z}^d}\mathsf{q}(-\mathsf{M}^{-1}\gamma-k)u(\gamma+\mathsf{M}k) = \sum_{k\in\mathbb{Z}^d}\mathsf{q}(-k)u(\mathsf{M}k)$ for all $\gamma\in\Gamma_{\mathsf{M}}$.
b. $[\mathsf{q}(-i\partial)(e^{-i\mathsf{M}^{-1}\gamma\cdot\xi}\widehat{u^{[\gamma:\mathsf{M}]}}(\xi))]|_{\xi=0} = [\mathsf{q}(-i\partial)\widehat{u^{[0:\mathsf{M}]}}(\xi)]|_{\xi=0}$ for all $\gamma\in\Gamma_{\mathsf{M}}$.
c. $[\mathsf{q}(-i\mathsf{M}^{-1}\partial)\widehat{u}(\xi)]|_{\xi=2\pi\omega}=0$ for all $\omega\in\Omega_{\mathsf{M}}\backslash\{0\}$,

where $u^{[\gamma:\mathsf{M}]} = \{u(\gamma+\mathsf{M}k)\}_{k\in\mathbb{Z}^d}$, that is, $\widehat{u^{[\gamma:\mathsf{M}]}}(\xi) := \sum_{k\in\mathbb{Z}^d}u(\gamma+\mathsf{M}k)e^{-ik\cdot\xi}$.

7.14. Let $u=\{u(\mathsf{k})\}_{\mathsf{k}\in\mathbb{Z}^d}\in l_0(\mathbb{Z}^d)$, M be a $d\times d$ invertible integer matrix, and $\mathsf{p}\in\mathbb{P}$ be a polynomial. Prove that the following statements are equivalent:

a. $\mathcal{S}_{u,\mathsf{M}}\mathsf{p}\in\mathbb{P}$, i.e., $\mathcal{S}_{u,\mathsf{M}}\mathsf{p}$ is a polynomial sequence.
b. $\sum_{\mathsf{k}\in\mathbb{Z}^d}(\partial^\mu\mathsf{p})(-\mathsf{M}^{-1}\gamma-\mathsf{k})u(\gamma+\mathsf{M}\mathsf{k}) = \sum_{\mathsf{k}\in\mathbb{Z}^d}(\partial^\mu\mathsf{p})(-\mathsf{k})u(\mathsf{M}\mathsf{k})$ for all $\mu\in\mathbb{N}_0^d$ and $\gamma\in\Gamma_{\mathsf{M}}$.
c. $[(\partial^\mu\mathsf{p})(-i\partial)(e^{-i\mathsf{M}^{-1}\gamma\cdot\xi}\widehat{u^{[\gamma:\mathsf{M}]}}(\xi))]|_{\xi=0} = [(\partial^\mu\mathsf{p})(-i\partial)\widehat{u^{[0:\mathsf{M}]}}(\xi)]|_{\xi=0}$ for all $\mu\in\mathbb{N}_0^d$ and $\gamma\in\Gamma_{\mathsf{M}}$.
d. $[(\partial^\mu\mathsf{p})(-\mathsf{M}^{-1}\gamma-i\partial)\widehat{u^{[\gamma:\mathsf{M}]}}(\xi)]|_{\xi=0} = [(\partial^\mu\mathsf{p})(-i\partial)\widehat{u^{[0:\mathsf{M}]}}(\xi)]|_{\xi=0}$ for all $\mu\in\mathbb{N}_0^d$ and $\gamma\in\Gamma_{\mathsf{M}}$.
e. $[(\partial^\mu\mathsf{p})(-i\mathsf{M}^{-1}\partial)\widehat{u}(\xi)]|_{\xi=2\pi\omega}=0$ for all $\mu\in\mathbb{N}_0^d$ and $\omega\in\Omega_{\mathsf{M}}\backslash\{\mathbf{0}\}$.

Moreover, any of the above items (1)–(5) implies $\mathcal{S}_{u,\mathsf{M}}(\mathsf{p}) = \mathsf{p}(\mathsf{M}^{-1}\cdot)*u$.

7.15. Let $u = \{u(\mathsf{k})\}_{\mathsf{k}\in\mathbb{Z}^d} \in l_0(\mathbb{Z}^d)$ and M be a $d \times d$ invertible integer matrix. For any positive integer $m \in \mathbb{N}$, prove that the following statements are equivalent:

a. $\mathcal{S}_{u,\mathsf{M}}\mathbb{P}_{m-1} \subseteq \mathbb{P}$.
b. $\mathcal{S}_{u,\mathsf{M}}\mathsf{q} \in \mathbb{P}$ for all polynomials $\mathsf{q} = (\cdot)^\mu$ with $\mu \in \mathbb{N}_0^d$ and $|\mu| = m-1$.
c. $\widehat{u}(\xi + 2\pi\omega) = \mathscr{O}(\|\xi\|^m)$ as $\xi \to 0$ for all $\omega \in \Omega_\mathsf{M}\backslash\{\mathbf{0}\}$.
d. $e^{-i\mathsf{M}^{-1}\gamma\cdot\xi}\widehat{u^{[\gamma:\mathsf{M}]}}(\xi) = \widehat{u^{[0:\mathsf{M}]}}(\xi) + \mathscr{O}(\|\xi\|^m)$ as $\xi \to 0$ for all $\gamma \in \Gamma_\mathsf{M}$.
e. $\sum_{\mathsf{k}\in\mathbb{Z}^d} u(\gamma + \mathsf{M}\mathsf{k})(\gamma + \mathsf{M}\mathsf{k})^\mu = \sum_{\mathsf{k}\in\mathbb{Z}^d} u(\mathsf{M}\mathsf{k})(\mathsf{M}\mathsf{k})^\mu$ $\forall \mu \in \mathbb{N}_0^d$ with $|\mu| < m$.

7.16. Let M be a $d\times d$ invertible integer matrix and $u \in (l_0(\mathbb{Z}^d))^{r\times r}$. For a $1 \times r$ row vector $\mathsf{p} \in \mathbb{P}^{1\times r}$ of polynomials, prove that the following are equivalent:

a. $\mathcal{S}_{u,\mathsf{M}}\mathsf{p}$ is a $1 \times r$ vector of polynomial sequences, that is, $\mathcal{S}_{u,\mathsf{M}}\mathsf{p} \in \mathbb{P}^{1\times r}$.
b. $\sum_{k\in\mathbb{Z}^d}(\partial^\mu\mathsf{p})(-\mathsf{M}^{-1}\gamma - k)u(\gamma + \mathsf{M}k) = \sum_{k\in\mathbb{Z}^d}(\partial^\mu\mathsf{p})(-k)u(\mathsf{M}k)$ for all $\gamma \in \Gamma_\mathsf{M}$ and $\mu \in \mathbb{N}_0^d$.
c. $[(\partial^\mu\mathsf{p})(-i\partial)(e^{-i\mathsf{M}^{-1}\gamma\cdot\xi}\widehat{u^{[\gamma:\mathsf{M}]}}(\xi))](0) = [(\partial^\mu\mathsf{p})(-i\partial)\widehat{u^{[0:\mathsf{M}]}}](0)$ for all $\gamma \in \Gamma_\mathsf{M}$ and $\mu \in \mathbb{N}_0^d$.
d. $[(\partial^\mu\mathsf{p})(-i\mathsf{M}^{-1}\partial)\widehat{u}](2\pi\omega) = 0$ for all $\omega \in \Omega_\mathsf{M}$ and $\mu \in \mathbb{N}_0^d$.

Moreover, if any of the above holds, then $\mathcal{S}_{u,\mathsf{M}}\mathsf{p} = \mathsf{p}(\mathsf{M}^{-1}\cdot) * u$.

7.17. Let M be a $d \times d$ invertible integer matrix and $u \in (l_0(\mathbb{Z}^d))^{r\times r}$. For $\upsilon \in (l_0(\mathbb{Z}^d))^{1\times r}$ and $m \in \mathbb{N}_0$, prove that the following are equivalent:

a. $\mathcal{S}_{u,\mathsf{M}}\mathscr{P}_{m-1,\upsilon} \in (\mathbb{P}_{m-1})^{1\times r}$ if and only if $\widehat{\upsilon}(\mathsf{M}^\mathsf{T}\xi)\widehat{u}(\xi + 2\pi\omega) = \mathscr{O}(\|\xi\|^m)$ as $\xi \to 0$ for all $\omega \in \Omega_\mathsf{M}\backslash\{0\}$, where $\mathscr{P}_{m-1,\upsilon} := \{\mathsf{p} * \upsilon : \mathsf{p} \in \mathbb{P}_{m-1}$.
b. If the above identity holds, define $\widehat{\mathring{\upsilon}}(\xi) := \widehat{\upsilon}(\mathsf{M}^\mathsf{T}\xi)\widehat{u}(\xi)$, then $\mathcal{S}_{u,\mathsf{M}}\mathscr{P}_{m-1,\upsilon} = \mathscr{P}_{m-1,\mathring{\upsilon}}$ and $\mathcal{S}_{u,\mathsf{M}}(\mathsf{p} * \upsilon) = |\det(\mathsf{M})|^{-1}\mathcal{S}_{\mathcal{S}_{u,\mathsf{M}}\upsilon,\mathsf{M}}\mathsf{p} = \mathsf{p}(\mathsf{M}^{-1}\cdot) * \mathring{\upsilon}$.

7.18. Let $u \in (l_0(\mathbb{Z}))^{r\times r}$ and $\upsilon, \mathring{\upsilon} \in (l_0(\mathbb{Z}))^{1\times r}$ with $\widehat{\mathring{\upsilon}}(0) \neq 0$. Prove that $\mathcal{S}_{u,\mathsf{M}}\mathscr{P}_{m-1,\upsilon} = \mathscr{P}_{m-1,\mathring{\upsilon}}$ if and only if $\widehat{\upsilon}(\mathsf{M}^\mathsf{T}\xi)\widehat{u}(\xi + 2\pi\omega) = \boldsymbol{\delta}(\omega)\widehat{c}(\xi)\widehat{\mathring{\upsilon}}(\xi) + \mathscr{O}(\|\xi\|^m)$, $\xi \to 0$ $\forall\, \omega \in \Omega_\mathsf{M}$ for some $2\pi\mathbb{Z}^d$-periodic trigonometric polynomial $\widehat{c}$ with $\widehat{c}(0) \neq 0$.

7.19. Let M be a $d\times d$ invertible integer matrix such that $2\mathsf{M}^{-1}$ is an integer matrix. Prove that there does not exist a real-valued orthogonal M-wavelet filter $a \in (l_0\mathbb{Z}^d)$ such that a is $\{I_d, -I_d\}$-symmetric about the center $2^{-1}\mathsf{M}c$ for some $c \in \mathbb{Z}^d$ (i.e., $a(\mathsf{M}c - k) = a(k)$ for all $k \in \mathbb{Z}^d$).

7.20. Let A, B be $d \times d$ real-valued matrices. Prove that $S(AB, \Lambda_j) = S(A, \Lambda_j)S(B, \Lambda_j)$, where $S(M, \Lambda_j)$ is defined in (7.2.15).

7.21. Let M be a $d \times d$ dilation matrix. Let G be a symmetry group which is compatible with M. If N is G-equivalent to M (i.e., $\mathsf{N} = E\mathsf{M}F$ for some $E, F \in G$), prove that N must be a dilation matrix.

7.22. Let M and N be $d\times d$ invertible integer matrices such that $\mathsf{M}\mathbb{Z}^d = \mathsf{N}\mathbb{Z}^d$. Prove that $\operatorname{sr}(a, \mathsf{M}) = \operatorname{sr}(a, \mathsf{N})$ for all $a \in l_0(\mathbb{Z}^d)$.

7.23. Let M and N be $d\times d$ invertible integer matrices such that $\mathsf{M}\mathbb{Z}^d = \mathsf{N}\mathbb{Z}^d$. Prove that if $(\{\tilde{a}; \tilde{b}_1, \ldots, \tilde{b}_s\}, \{a; b_1, \ldots, b_s\})$ is a dual M-framelet filter bank, then it is also a dual N-framelet filter bank.

7.24. For a $d \times d$ expansive integer matrix M and $a \in l_0(\mathbb{Z}^d)$ with $\widehat{a}(0) = 1$, prove that $\operatorname{sr}(a, \mathsf{M}) \geqslant \operatorname{sm}_p(a, \mathsf{M})$ for all $1 \leqslant p \leqslant \infty$.

7.25. Let M be a $d \times d$ invertible integer matrix and N be a $D \times D$ invertible integer matrix. Let P be a $d \times D$ integer matrix such that $P\mathsf{N} = \mathsf{M}P$ and $P^{\mathsf{T}}(\mathbb{Z}^d\backslash[\mathsf{M}^{\mathsf{T}}\mathbb{Z}^d]) \subseteq \mathbb{Z}^D\backslash[\mathsf{N}^{\mathsf{T}}\mathbb{Z}^D])$. D efine $\widehat{Pa}(\xi) := \widehat{a}(P^{\mathsf{T}}\xi)$ for $a \in l_0(\mathbb{Z}^D)$.

a. Prove that $\operatorname{sr}(a, \mathsf{N}) \leqslant \operatorname{sr}(Pa, \mathsf{M})$ for all $a \in (l_0(\mathbb{Z}^D))^{r\times r}$.
b. If $(\{\tilde{a}; \tilde{b}_1, \ldots, \tilde{b}_s\}, \{a; b_1, \ldots, b_s\})$ is a dual N-framelet filter bank with all filters from $l_0(\mathbb{Z}^D)$, then $(\{P\tilde{a}; P\tilde{b}_1, \ldots, P\tilde{b}_s\}, \{Pa; Pb_1, \ldots, Pb_s\})$ must be a dual M-framelet filter bank.

7.26. Let M, N be $d \times d$ and $D \times D$ invertible integer matrices, respectively. Let P be a $d\times D$ integer matrix such that $P\mathsf{N}\mathbb{Z}^D \subseteq \mathsf{M}\mathbb{Z}^d$ and $(\gamma + \mathsf{M}\mathbb{Z}^d) \cap P\mathbb{Z}^D \neq \emptyset$ for all $\gamma \in \mathbb{Z}^d$.

a. If in addition $\{k \in \mathbb{Z}^D \ : \ Pk \in \mathsf{M}\mathbb{Z}^d\} \subseteq \mathsf{N}\mathbb{Z}^D$, for every interpolatory N-wavelet filter $a \in l_0(\mathbb{Z}^D)$, prove that Pa must be an interpolatory M-wavelet filter and $\operatorname{sr}(a, \mathsf{N}) \leqslant \operatorname{sr}(Pa, \mathsf{M})$.
b. If $(\tilde{a}, a)$ is a pair of biorthogonal N-wavelet filters such that

$$\widehat{a}(P^{\mathsf{T}}\xi + 2\pi\gamma) = 0, \qquad \forall\, \omega \in [(\mathsf{N}^{\mathsf{T}})^{-1}\mathbb{Z}^D]\backslash[P^{\mathsf{T}}(\mathsf{M}^{\mathsf{T}})^{-1}\mathbb{Z}^d + \mathbb{Z}^D]$$

(a filter a satisfying this condition is called a P-projectable filter), then $(P\tilde{a}, Pa)$ must be a pair of biorthogonal M-wavelet filters.

7.27. Let $a \in l_0(\mathbb{Z}^D)$ such that $a(k_1, \ldots, k_{j-1}, c_j - k_j, k_{j+1}, \ldots, k_D) = a(k_1, \ldots, k_D)$ for all $1 \leqslant j \leqslant D$ and $k_1, \ldots, k_D \in \mathbb{Z}$ for some integers $c_1, \ldots, c_D \in \mathbb{Z}$. Define a $d \times D$ matrix P by $P = [I_d, 0]$. Prove that the filter a must be P-projectable, that is, $\widehat{a}(\xi, \pi\omega) = 0$ for all $\xi \in \mathbb{R}^d$ and $\omega \in \{0, 1\}^{D-d}\backslash\{0\}$.

7.28. Let M, N be $d \times d$ and $D \times D$ invertible integer matrices, respectively. Let P be a $d \times D$ integer matrix such that $P\mathsf{N} = \mathsf{M}P$ and $P\mathbb{Z}^D = \mathbb{Z}^d$. For every $a \in l_0(\mathbb{Z}^D)$, prove that $\operatorname{sr}(a, \mathsf{N}) \leqslant \operatorname{sr}(Pa, \mathsf{M})$ and $\operatorname{sm}_p(a, \mathsf{N}) \leqslant \operatorname{sm}_p(Pa, \mathsf{M})$ for all $1 \leqslant p \leqslant \infty$.

7.29. Let P be a $d \times D$ real-valued matrix having full rank. For any compactly supported function $f \in L_p(\mathbb{R}^D)$ with $1 \leqslant p \leqslant \infty$, prove that Pf must be a compactly supported function in $L_p(\mathbb{R}^d)$ and $\operatorname{sm}_p(f) \leqslant \operatorname{sm}_p(Pf)$, where $\widehat{Pf}(\xi) := \widehat{f}(P^{\mathsf{T}}\xi)$.

7.30. Let $a \in l_0(\mathbb{Z})$ be an interpolatory 2-wavelet filter such that $\operatorname{fsupp}(a) \subseteq [-3, 3]$. Prove that $\operatorname{sr}(a, 2) \leqslant 4$ and $\operatorname{sm}_\infty(a, 2) \leqslant 2$.

7.31. Let $a \in l_0(\mathbb{Z}^d)$ be an interpolatory $2I_d$-wavelet filter such that $\operatorname{fsupp}(a) \subseteq [-3, 3]^d$. Prove that $\operatorname{sr}(a, 2I_d) \leqslant 4$ and $\operatorname{sm}_\infty(a, 2I_d) \leqslant 2$. Thus, there is no $\mathscr{C}^2(\mathbb{R}^d)$ interpolating $2I_d$-refinable function ϕ whose mask a can be supported inside $[-3, 3]^d$.

7.32. Let M be a $d\times d$ invertible integer matrix. If $(\{\tilde{a};\tilde{b}_1,\ldots,\tilde{b}_s\},\{a;b_1,\ldots,b_s\})$ is a dual M-framelet filter bank, then $(\{E\tilde{a};E\tilde{b}_1,\ldots,E\tilde{b}_s\},\{Ea;Eb_1,\ldots,Eb_s\})$ is a dual $(E\mathsf{M})$-framelet filter bank for every $d\times d$ integer matrix E with $|\det(E)|=1$.

7.33. Let M be a $d\times d$ invertible integer matrix. Prove that (i) $\mathsf{M} = E\mathrm{diag}(\mathsf{d}_1,\ldots,\mathsf{d}_d)F$ (called the Smith normal form of M over $\mathbb{Z}$) for some $d\times d$ integer matrices E and F with $|\det(E)| = |\det(F)| = 1$ and for some $\mathsf{d}_1,\ldots,\mathsf{d}_d\in\mathbb{N}$. (ii) there exists a $d\times d$ integer matrix N such that $\mathsf{N}\mathbb{Z}^d=\mathsf{M}\mathbb{Z}^d$ and $\mathsf{N}^d=|\det(M)|I_d$.

7.34. Let M be a $d\times d$ invertible integer matrix. Define $\Omega_{\mathsf{M}} := [(\mathsf{M}^{\mathsf{T}})^{-1}\mathbb{Z}^d]\cap[0,1)^d$ (i.e., $\Omega_{\mathsf{M}}=[(\mathsf{M}^{\mathsf{T}})^{-1}\mathbb{Z}^d]/\mathbb{Z}^d$) and $\Gamma_{\mathsf{M}}:=[\mathsf{M}[0,1)^d]\cap\mathbb{Z}^d$ as in (7.1.3). Prove that $\#\Omega_{\mathsf{M}}=\#\Gamma_{\mathsf{M}}=|\det(\mathsf{M})|$. *Hint:* Use the Smith normal form of M.

7.35. Let M be a $d\times d$ expansive integer matrix with $m:=|\det(\mathsf{M})|$. For any $n\in\mathbb{N}$, prove that (i) there exists $a\in l_0(\mathbb{Z}^d)$ such that $\mathrm{sr}(a,\mathsf{M})\geqslant n$ and a is an interpolatory M-wavelet filter with $\widehat{a}(0)=1$. (ii) there exist $a,b_1,\ldots,b_{m-1}\in l_0(\mathbb{Z}^d)$ such that $\{a;b_1,\ldots,b_{m-1}\}$ is an orthogonal M-wavelet filter bank with $\mathrm{sr}(a,\mathsf{M})\geqslant n$ and $\min(\mathrm{vm}(b_1),\ldots,\mathrm{vm}(b_{m-1}))\geqslant n$.

7.36. Let M be a $d\times d$ expansive integer matrix and $\widehat{a}$ be a $2\pi\mathbb{Z}^d$-periodic continuous function on $\mathbb{R}^d$ such that $\widehat{a}(0)=1$. Assume that $\widehat{\phi}(\xi):=\prod_{j=1}^{\infty}\widehat{a}((\mathsf{M}^{\mathsf{T}})^{-j}\xi), \xi\in\mathbb{R}^d$ is well defined with the series $\{\prod_{j=1}^{n}\widehat{a}((\mathsf{M}^{\mathsf{T}})^{-j}\xi)\}_{n\in\mathbb{N}}$ converging uniformly for ξ on every compact set of $\mathbb{R}^d$ (e.g., this condition is satisfied if there exist positive constants C_0 and τ such that $|1-\widehat{a}(\xi)|\leqslant C_0|\xi|^{\tau}$ for all $\xi\in[-\pi,\pi]^d$). Prove that the following statements are equivalent:

a. There exists $C>0$ such that $\sum_{k\in\mathbb{Z}^d}|\widehat{\phi}(\xi+2\pi k)|^2\geqslant C$ for all $\xi\in\mathbb{R}^d$.
b. There exists a positive constant C and a compact set $K\subset\mathbb{R}^d$ such that $|\widehat{\phi}(\xi)|\geqslant C$ for all $\xi\in K$, $(-\varepsilon,\varepsilon)^d\subset K$ for some $\varepsilon>0$, and $\sum_{n\in\mathbb{Z}^d}\chi_K(\xi+2\pi n)=1$ a.e. $\xi\in\mathbb{R}^d$.
c. (Cohen's criteria) There exists a compact set $K\subset\mathbb{R}^d$ such that $(-\varepsilon,\varepsilon)^d\subset K$ for some $\varepsilon>0$, $\sum_{n\in\mathbb{Z}^d}\chi_K(\xi+2\pi n)=1$ a.e. $\xi\in\mathbb{R}^d$, and $\widehat{a}((\mathsf{M}^{\mathsf{T}})^{-j}\xi)\neq 0$ for all $j\in\mathbb{N}$ and $\xi\in K$.

7.37. Let M be a $d\times d$ expansive integer matrix. For $a,\tilde{a}\in l_0(\mathbb{Z}^d)$ with $\widehat{a}(0)=\widehat{\tilde{a}}(0)=1$, define $\widehat{\phi}(\xi):=\prod_{j=1}^{\infty}\widehat{a}((\mathsf{M}^{\mathsf{T}})^{-j}\xi)$ and $\widehat{\tilde{\phi}}(\xi):=\prod_{j=1}^{\infty}\widehat{\tilde{a}}((\mathsf{M}^{\mathsf{T}})^{-j}\xi)$. Suppose that the condition in item c of Exercise 7.36 is satisfied for both $\widehat{a}$ and $\widehat{\tilde{a}}$ but the associated compact sets K may be different. If both ϕ and $\tilde{\phi}$ belong to $L_2(\mathbb{R}^d)$ and $(\tilde{a},a)$ is a pair of biorthogonal M-wavelet filters, i.e., $\sum_{\omega\in\Omega_{\mathsf{M}}}\widehat{\tilde{a}}(\xi+2\pi\omega)\overline{\widehat{a}(\xi+2\pi\omega)}=1$, prove that $[\widehat{\tilde{\phi}},\widehat{\phi}](\xi)=1$ a.e. $\xi\in\mathbb{R}^d$.

7.38. For $u\in l_0(\mathbb{Z}^d)\backslash\{0\}$, define $\eta:=u*\phi$. If $\eta\in L_p(\mathbb{R}^d)$ for some $1\leqslant p\leqslant\infty$, prove that $\phi\in L_p(\mathbb{R}^d)$.

7.39. Let M be a dilation matrix and $\psi^1,\ldots,\psi^s\in L_2(\mathbb{R}^d)$. If there exists a positive constant C such that $\sum_{j=0}^{\infty}\sum_{\ell=1}^{s}\sum_{k\in\mathbb{Z}^d}|\langle f,\psi^{\ell}_{\mathsf{M}^j;k}\rangle|^2\leqslant C\|f\|^2_{L_2(\mathbb{R}^d)}$ for all $f\in L_2(\mathbb{R}^d)$, prove that $\sum_{j\in\mathbb{Z}}\sum_{\ell=1}^{s}|\widehat{\psi^{\ell}}((\mathsf{M}^{\mathsf{T}})^j\xi)|^2\leqslant C$ for almost every $\xi\in\mathbb{R}^d$. *Hint:* Check the proof of Lemma 4.3.10.

7.40. Let a be given in (7.3.23). Define

$$\widehat{b_1}(\xi_1,\xi_2) := e^{-i\xi_1}\overline{\widehat{a}(\xi_1+\pi,\xi_2+\pi)},$$
$$\widehat{b_3}(\xi_1,\xi_2) := e^{-i\xi_1}\overline{\widehat{b_2}(\xi_1+\pi,\xi_2+\pi)}, \qquad (7.7.1)$$

where

$$b_2 = \frac{1}{32}\begin{bmatrix} \sqrt{3}-2 & 0 & 0 & 2+\sqrt{3} \\ 0 & -\sqrt{3}+6 & -\sqrt{3}-6 & 0 \\ 0 & -\sqrt{3}+6 & -\sqrt{3}-6 & 0 \\ \sqrt{3}-2 & 0 & 0 & 2+\sqrt{3} \end{bmatrix}_{[-1,2]^2}.$$

Prove that $\{a; b_1, b_2, b_3\}$ is a tight $\mathsf{M}_{\sqrt{2}}$-framelet filter bank with $\mathrm{vm}(b_1) = 4$ and $\mathrm{vm}(b_2) = \mathrm{vm}(b_3) = 2$.

7.41. Let $m \in \mathbb{N}$. Prove that there exists a unique two-dimensional filter a^{2D}_{2m} such that a^{2D}_{2m} is supported inside $[1-m, m]^2 \cap \mathbb{Z}^2$, has order $2m$ sum rules with respect to $\mathsf{M}_{\sqrt{2}}$, and order $2m$ linear-phase moments with phase $(1/2, 1/2)$.

7.42. Let a^{2D}_{2m} as in Exercise 7.41. Prove $\widehat{a^{2D}_{2m}}(\xi_1,\xi_2) = \frac{1}{2}(\widehat{u}(\xi_1+\xi_2)+\widehat{u}(\xi_1-\xi_2)e^{-i\xi_2})$, where $\widehat{u}(\xi) := (\widehat{a^I_{2m}}(\xi/2) - \widehat{a^I_{2m}}(\xi/2+\pi))e^{-i\xi/2}$ and a^I_{2m} is the interpolatory filter defined in (2.1.6). In particular, show that a^{2D}_2 is given in (7.3.23).

7.43. Let a^{2D}_{2m} be given in Exercise 7.41. Define

$$\widehat{b_2}(\xi_1,\xi_2) := \tfrac{1}{2}(\widehat{v}(\xi_1+\xi_2)+\widehat{v}(\xi_1-\xi_2)e^{-i\xi_2}), \quad \widehat{v}(\xi) := 2\widehat{a^D_m}(\xi/2)\widehat{a^D_m}(\xi/2+\pi),$$

where a^D_m is the Daubechies orthogonal wavelet filter in (2.2.4). Define b_1 and b_3 as in (7.7.1) with $a = a^{2D}_{2m}$. Prove that $\{a^{2D}_{2m}; b_1, b_2, b_3\}$ is a tight $\mathsf{M}_{\sqrt{2}}$-framelet filter bank such that all b_1, b_2, b_3 have symmetry and order m vanishing moments.

7.44. Let $u \in l_0(\mathbb{Z})$ such that $|\widehat{u}(\xi)| \leqslant 1$ for all $\xi \in \mathbb{R}$. Define $v \in \mathrm{lpm}\,0$ by $|\widehat{v}(\xi)|^2 = 1 - |\widehat{u}(\xi)|^2$ and

$$\widehat{a}(\xi_1,\xi_2) := \tfrac{1}{2}(\widehat{u}(\xi_1+\xi_2)+\widehat{u}(\xi_1-\xi_2)e^{-i\xi_2}),$$
$$\widehat{b_2}(\xi_1,\xi_2) := \tfrac{1}{2}(\widehat{v}(\xi_1,\xi_2)+\widehat{v}(\xi_1-\xi_2)e^{-i\xi_2}).$$

Define b_1, b_3 in (7.7.1). Prove that $\{a; b_1, b_2, b_3\}$ is a tight $\mathsf{M}_{\sqrt{2}}$-framelet filter bank.

7.45. Let M be a $d \times d$ dilation matrix. For $u, v \in l_0(\mathbb{Z}^d)$ with $\widehat{u}(0) = \widehat{v}(0) = 1$, prove that $\mathrm{sr}(u * v, \mathsf{M}) \geqslant \mathrm{sr}(u, \mathsf{M}) + \mathrm{sr}(v, \mathsf{M})$ and

$$\mathrm{sm}_2(u * v, \mathsf{M}) \geqslant \mathrm{sm}_\infty(u * v, \mathsf{M}) \geqslant \mathrm{sm}_2(u, \mathsf{M}) + \mathrm{sm}_2(v, \mathsf{M}).$$

7.46. A one-dimensional filter $a \in l_0(\mathbb{Z})$ can be regarded as a two-dimensional filter by identifying a on $\mathbb{Z}$ with $\mathbb{Z} \times \{0\}$ in $\mathbb{Z}^2$. Let $\{a, b_1, \dots, b_s\}$ be a tight 2-framelet filter bank. Prove that $\{a; b_1, \dots, b_s\}$ is a tight $\mathsf{M}_{\sqrt{2}}$-framelet filter bank.

7.47. Regard $a \in l_0(\mathbb{Z})$ with $\widehat{a}(0) = 1$ as a two-dimensional filter as in Exercise 7.46. Prove $\mathrm{sr}(a, \mathsf{M}_{\sqrt{2}}) = \mathrm{sr}(a, 2)$. If $\mathrm{sm}_p(a, 2) \geqslant 0$, prove $\mathrm{sm}_p(a, \mathsf{M}_{\sqrt{2}}) = \mathrm{sm}_p(a, 2)$.

7.48. For $u, v \in l_0(\mathbb{Z})$ with $\widehat{u}(0) = \widehat{v}(0) = 1$, prove $\mathrm{sr}(u \otimes v, \mathsf{M}_{\sqrt{2}}) \geqslant \mathrm{sr}(u, 2) + \mathrm{sr}(v, 2)$. If $\mathrm{sm}(u, 2) \geqslant 0$ and $\mathrm{sm}(v, 2) \geqslant 0$, prove $\mathrm{sm}(u \otimes v, \mathsf{M}_{\sqrt{2}}) \geqslant \mathrm{sm}(u, 2) + \mathrm{sm}(v, 2)$.

7.49. Let $\mathsf{M}_1, \mathsf{M}_2$ be $d \times d$ invertible integer matrices and let $u_1, u_2 \in l_0(\mathbb{Z}^d)$. Prove

$$\mathcal{S}_{u_1,\mathsf{M}_1}\mathcal{S}_{u_2,\mathsf{M}_2} v = \mathcal{S}_{u_1*(u_2\uparrow \mathsf{M}_1),\mathsf{M}_1\mathsf{M}_2} v = |\det(\mathsf{M}_1\mathsf{M}_2)| u_1 * (u_2 \uparrow \mathsf{M}_1) * (v \uparrow \mathsf{M}_1\mathsf{M}_2),$$

$$\mathcal{T}_{u_2,\mathsf{M}_2}\mathcal{T}_{u_1,\mathsf{M}_1} v = \mathcal{T}_{u_1*(u_2\uparrow \mathsf{M}_1),\mathsf{M}_1\mathsf{M}_2} v = |\det(\mathsf{M}_1\mathsf{M}_2)| (u_1 * (u_2 \uparrow \mathsf{M}_1) * v) \downarrow \mathsf{M}_1\mathsf{M}_2.$$

7.50. Let $\{a; b_1, \dots, b_s\}$ be a tight M-framelet filter bank such that $\widehat{a}(0) = 1$ and $a(\mathsf{c}_a - \mathsf{k}) = \overline{a(\mathsf{k})}$ for all $\mathsf{k} \in \mathbb{Z}^d$ with $\mathsf{c}_a \in \mathbb{Z}^d$. Prove that $\mathrm{lpm}(a^\star * a) = \mathrm{lpm}(a)$ and $\min(\mathrm{vm}(b_1), \dots, \mathrm{vm}(b_s)) = \min(\mathrm{sr}(a, \mathsf{M}), \frac{1}{2}\mathrm{lpm}(a))$.

7.51. Let $a, b_1, b_2 \in l_2(\mathbb{Z})$. Suppose that $\{a; b_1, b_2\}$ is a tight framelet filter bank.

a. Prove that $|\widehat{b_1}(\xi + \pi)|^2 + |\widehat{b_2}(\xi)|^2 \geqslant A(\xi)$ a.e. $\xi \in [0, \pi]$, where

$$\begin{aligned} A(\xi) &:= 1 - \tfrac{1}{2}(|\widehat{a}(\xi)|^2 + |\widehat{a}(\xi + \pi)|^2) \\ &- \sqrt{(1 - |\widehat{a}(\xi)|^2 - |\widehat{a}(\xi + \pi)|^2) + \tfrac{1}{4}(|\widehat{a}(\xi)|^2 - |\widehat{a}(\xi + \pi)|^2)^2}. \end{aligned} \tag{7.7.2}$$

b. If b_1, b_2 have real coefficients, prove that

$$\int_0^\pi [|\widehat{b_1}(\xi+\pi)|^2 + |\widehat{b_2}(\xi)|^2] d\xi = \frac{1}{2}\int_0^\pi [2 - |\widehat{a}(\xi)|^2 - |\widehat{a}(\xi+\pi)|^2] d\xi \geqslant \frac{\pi}{2},$$

where the equal sign holds if and only if a is an orthogonal 2-wavelet filter satisfying $|\widehat{a}(\xi)|^2 + |\widehat{a}(\xi + \pi)|^2 = 1$ a.e. $\xi \in \mathbb{R}$.

7.52. Let $a \in l_2(\mathbb{Z})$ such that $|\widehat{a}(\xi)|^2 + |\widehat{a}(\xi + \pi)|^2 \leqslant 1$ a.e. $\xi \in \mathbb{R}$. Prove $0 \leqslant A(\xi) \leqslant \min(|\widehat{a}(\xi)|^2, |\widehat{a}(\xi + \pi)|^2)$, a.e. $\xi \in \mathbb{R}$, where A is defined in (7.7.2). Moreover,

a. $A(\xi) = 0$ a.e. $\xi \in [0, \pi]$ if and only if $\widehat{a}(\xi)\widehat{a}(\xi + \pi) = 0$ a.e. $\xi \in \mathbb{R}$.
b. $A(\xi) = \min(|\widehat{a}(\xi)|^2, |\widehat{a}(\xi + \pi)|^2)$ a.e. $\xi \in [0, \pi]$ if and only if $|\widehat{a}(\xi)|^2 + |\widehat{a}(\xi + \pi)|^2 = 1$ for almost every $\xi \in \mathbb{R}$ satisfying $\min(|\widehat{a}(\xi)|^2, |\widehat{a}(\xi + \pi)|^2) \neq 0$. In particular, if $|\widehat{a}(\xi)|^2 + |\widehat{a}(\xi + \pi)|^2 = 1$ a.e. $\xi \in \mathbb{R}$ (that is, a is an orthogonal wavelet filter), then $A(\xi) = \min(|\widehat{a}(\xi)|^2, |\widehat{a}(\xi + \pi)|^2)$ a.e. $\xi \in [0, \pi]$.
c. If a is the B-spline filter a_m^B of order m given by $\widehat{a_m^B}(\xi) := \cos^{2m}(\xi/2)$ with $m \in \mathbb{N}$, then $4^{-m}\sin^m(\xi) \leqslant A(\xi) \leqslant 4^{1-m}\sin^m(\xi)$ for all $\xi \in [0, \pi]$.

7.53. For a filter b such that $\|b\|_{l_2(\mathbb{Z})} \neq 0$ and b is real-valued, prove that $\mathrm{Fsp}(b) = 1$.

7.54. For any complex-valued filters $a, a^+, a^- \in l_2(\mathbb{Z})$ satisfying

$$|\widehat{a^+}(\xi)|^2 + |\widehat{a^-}(\xi)|^2 = |\widehat{a}(\xi)|^2,$$

$$\widehat{a^+}(\xi)\overline{\widehat{a^+}(\xi+\pi)} + \widehat{a^-}(\xi)\overline{\widehat{a^-}(\xi+\pi)} = \widehat{a}(\xi)\overline{\widehat{a}(\xi+\pi)}, \quad a.e.\, \xi \in \mathbb{R},$$

prove $|\widehat{a^+}(\xi+\pi)|^2 + |\widehat{a^-}(\xi)|^2 \geqslant \min(|\widehat{a}(\xi)|^2, |\widehat{a}(\xi+\pi)|^2)$ a.e. $\xi \in [0, \pi]$.

7.55. Let $v_{j,n}$ be defined in (7.5.27) with $0 \leqslant j, n \leqslant m$. If (5.6.6) holds, prove that

$$[\widehat{\upsilon}(\xi)\widehat{\upsilon_{j,n}}(\xi)\overline{\widehat{\upsilon}(\xi)}^{\mathsf{T}}]^{(j+n)}(0)$$

$$= \sum_{\ell=\min(j+n,m+1)}^{j+n} \binom{j+n}{\ell} \sum_{\mu=0}^{\min(j+n,m)} \binom{\ell}{\mu} \widehat{\upsilon}^{(\mu)}(0)\widehat{\upsilon_{j,n}}^{(\ell-\mu)}(0)\overline{\widehat{\upsilon}^{(j+n-\ell)}(0)}^{\mathsf{T}}.$$

Hence, $\widehat{\upsilon}^{(m+1)}(0), \ldots, \widehat{\upsilon}^{(2m)}(0)$ play no role and can be zero in Theorem 7.5.7.

7.56. Let P be a $d \times n$ integer matrix. Recall that the projected filter a_P is defined in (7.1.21). Prove that (7.1.23) holds if and only if $\mathrm{sr}(a, 2I_d) \geqslant 1$, that is, $\widehat{a}(2\pi\omega) = 0$ for all $\omega \in \Omega_{2I_d}\backslash\{0\}$.

Let $a, b_1, \ldots, b_s \in l_1(\mathbb{Z}^d)$ and let $\mathsf{M}, \mathsf{M}_1, \ldots, \mathsf{M}_s$ be $d \times d$ invertible integer matrices. For a given signal $v_0 \in l_\infty(\mathbb{Z}^d)$ and $J \in \mathbb{N}$, *the J-level discrete framelet decomposition* employing the filter bank $\{a\,!\,\mathsf{M}; b_1\,!\,\mathsf{M}_1, \ldots, b_s\,!\,\mathsf{M}_s\}$ is

$$v_j := |\det(\mathsf{M})|^{-1/2}\mathcal{T}_{a,\mathsf{M}}v_{j-1} \quad \text{and} \quad w_{\ell,j} := |\det(\mathsf{M}_\ell)|^{-1/2}\mathcal{T}_{b_\ell,\mathsf{M}_\ell}v_{j-1}$$

for $\ell = 1, \ldots, s$ and $j = 1, \ldots, J$. *The J-level discrete framelet reconstruction* employing the filter bank $\{a\,!\,\mathsf{M}; b_1\,!\,\mathsf{M}_1, \ldots, b_s\,!\,\mathsf{M}_s\}$ can be described by

$$\mathring{v}_{j-1} := |\det(\mathsf{M})|^{-1/2}\mathcal{S}_{a,\mathsf{M}}\mathring{v}_j + \sum_{\ell=1}^{s} |\det(\mathsf{M}_\ell)|^{-1/2}\mathcal{S}_{b_\ell,\mathsf{M}_\ell}\mathring{w}_{\ell,j}, \qquad j = J, \ldots, 1,$$

where $\mathring{v}_0$ is a reconstructed sequence on $\mathbb{Z}^d$. The *perfect reconstruction property* requires that the reconstructed sequence $\mathring{v}_0$ should be exactly the same as the original input signal v_0 if $\mathring{v}_J = v_J$ and $\mathring{w}_{\ell,j} = w_{\ell,j}$ for $j = 1, \ldots, J$ and $\ell = 1, \ldots, s$.

Define a_j and $b_{\ell,j}$ with $j \in \mathbb{N}$ and $\ell = 1, \ldots, s$ by

$$\widehat{a_j}(\xi) := \widehat{a}(\xi)\widehat{a}(\mathsf{M}^{\mathsf{T}}\xi)\cdots\widehat{a}((\mathsf{M}^{\mathsf{T}})^{j-2}\xi)\widehat{a}((\mathsf{M}^{\mathsf{T}})^{j-1}\xi),$$

$$\widehat{b_{\ell,j}}(\xi) := \widehat{a_{j-1}}(\xi)\widehat{b_\ell}((\mathsf{M}^{\mathsf{T}})^{j-1}\xi) = \widehat{a}(\xi)\widehat{a}(\mathsf{M}^{\mathsf{T}}\xi)\cdots\widehat{a}((\mathsf{M}^{\mathsf{T}})^{j-2}\xi)\widehat{b_\ell}((\mathsf{M}^{\mathsf{T}})^{j-1}\xi).$$

Define $a_1 = a$, $b_{\ell,1} = b_\ell$, and $a_0 = \boldsymbol{\delta}$. Since $a, b_1, \ldots, b_s \in l_1(\mathbb{Z}^d)$, it is straightforward to see that all $a_j, b_{\ell,j}$ are well-defined filters in $l_1(\mathbb{Z}^d) \subseteq l_2(\mathbb{Z}^d)$. For $j \in \mathbb{N}$ and $k \in \mathbb{Z}^d$, we define $a_{j;k} := |\det(\mathsf{M})|^{j/2} a_j(\cdot - \mathsf{M}^j k)$ and

$$b_{\ell,j;k} := |\det(\mathsf{M})|^{(j-1)/2} |\det(\mathsf{M}_\ell)|^{1/2} b_{\ell,j}(\cdot - \mathsf{M}^{j-1}\mathsf{M}_\ell k).$$

The *J-level discrete affine system* associated with $\{a\,!\,\mathsf{M}; b_1\,!\,\mathsf{M}_1, \ldots, b_s\,!\,\mathsf{M}_s\}$ is

$$\begin{aligned}\mathsf{DAS}_J(\{a\,!\,\mathsf{M}; b_1\,!\,\mathsf{M}_1, \ldots, b_s\,!\,\mathsf{M}_s\}) := \{a_{J;k} \,:\, k \in \mathbb{Z}^d\} \\ \cup \{b_{\ell,j;k} \,:\, k \in \mathbb{Z}^d, \ell = 1, \ldots, s, j = 1, \ldots, J\}.\end{aligned} \tag{7.7.3}$$

7.57. Prove that $v_j(k) = \langle v_0, a_{j;k} \rangle$ and $w_{\ell,j}(k) = \langle v_0, b_{\ell,j;k} \rangle$.

7.58. Let $a, b_1, \ldots, b_s \in l_1(\mathbb{Z}^d)$ and let $\mathsf{M}, \mathsf{M}_1, \ldots, \mathsf{M}_s$ be $d \times d$ invertible integer matrices. Prove that the following statemcnts are equivalent:

a. The J-level fast framelet transform employing $\{a\,!\,\mathsf{M}; b_1\,!\,\mathsf{M}_1, \ldots, b_s\,!\,\mathsf{M}_s\}$ has the perfcct reconstruction property for every $J \in \mathbb{N}$.

b. For all $v \in l_\infty(\mathbb{Z}^d)$, the filter bank $\{a\,!\,\mathsf{M}; b_1\,!\,\mathsf{M}_1, \ldots, b_s\,!\,\mathsf{M}_s\}$ satisfies

$$v = |\det(\mathsf{M})|^{-1} \mathcal{S}_{a,\mathsf{M}} \mathcal{T}_{a,\mathsf{M}} v + \sum_{\ell=1}^{s} |\det(\mathsf{M}_\ell)|^{-1} \mathcal{S}_{b_\ell,\mathsf{M}_\ell} \mathcal{T}_{b_\ell,\mathsf{M}_\ell} v. \tag{7.7.4}$$

c. (7.7.4) holds for all $v \in l_0(\mathbb{Z}^d)$.

d. The filter bank $\{a\,!\,\mathsf{M}; b_1\,!\,\mathsf{M}_1, \ldots, b_s\,!\,\mathsf{M}_s\}$ is *a tight framelet filter bank*, i.e.,

$$|\widehat{a}(\xi)|^2 + |\widehat{b_1}(\xi)|^2 + \cdots + |\widehat{b_s}(\xi)|^2 = 1, \qquad \forall\, \xi \in \mathbb{R}^d,$$

$$\chi_{\Omega_\mathsf{M}}(\omega)\widehat{a}(\xi)\overline{\widehat{a}(\xi + 2\pi\omega)} + \sum_{\ell=1}^{s} \chi_{\Omega_{\mathsf{M}_\ell}}(\omega)\widehat{b_\ell}(\xi)\overline{\widehat{b_\ell}(\xi + 2\pi\omega)} = 0,$$

for all $\xi \in \mathbb{R}^d$ and for all $\omega \in [\Omega_\mathsf{M} \cup (\cup_{\ell=1}^s \Omega_{\mathsf{M}_\ell})] \backslash \{0\}$.

7.59. Let $a, b_1, \ldots, b_s \in l_1(\mathbb{Z}^d)$ and $\mathsf{M}, \mathsf{M}_1, \ldots, \mathsf{M}_s$ be $d \times d$ invertible integer matrices. For $J \in \mathbb{N}$, define $\mathsf{DAS}_J(\{a\,!\,\mathsf{M}; b_1\,!\,\mathsf{M}_1, \ldots, b_s\,!\,\mathsf{M}_s\})$ as in (7.7.3). Prove that the following statements are equivalent:

a. $\{a\,!\,\mathsf{M}; b_1\,!\,\mathsf{M}_1, \ldots, b_s\,!\,\mathsf{M}_s\}$ is a tight framelet filter bank.

b. For all $v \in l_2(\mathbb{Z}^d)$, $v = \sum_{k\in\mathbb{Z}^d} \langle v, a_{1;k} \rangle a_{1;k} + \sum_{\ell=1}^{s} \sum_{k\in\mathbb{Z}^d} \langle v, b_{\ell,1;k} \rangle b_{\ell,1;k}$.

c. $\mathsf{DAS}_1(\{a\,!\,\mathsf{M}; b_1\,!\,\mathsf{M}_1, \ldots, b_s\,!\,\mathsf{M}_s\})$ is a (normalized) tight frame for $l_2(\mathbb{Z}^d)$:

$$\|v\|^2_{l_2(\mathbb{Z}^d)} = \sum_{k\in\mathbb{Z}^d} |\langle v, a_{1;k} \rangle|^2 + \sum_{\ell=1}^{s} \sum_{k\in\mathbb{Z}^d} |\langle v, b_{\ell,1;k} \rangle|^2, \quad \forall\, v \in l_2(\mathbb{Z}^d).$$

d. For every $j \in \mathbb{N}$ and for all $v \in l_2(\mathbb{Z}^d)$, the following identity holds:

$$\sum_{k\in\mathbb{Z}^d} \langle v, a_{j-1;k}\rangle a_{j-1;k} = \sum_{k\in\mathbb{Z}^d} \langle v, a_{j;k}\rangle a_{j;k} + \sum_{\ell=1}^{s}\sum_{k\in\mathbb{Z}^d} \langle v, b_{\ell,j;k}\rangle b_{\ell,j;k}.$$

e. For every $J \in \mathbb{N}$, the following identity holds: for all $v \in l_2(\mathbb{Z}^d)$,

$$v = \sum_{k\in\mathbb{Z}^d} \langle v, a_{J;k}\rangle a_{J;k} + \sum_{j=1}^{J}\sum_{\ell=1}^{s}\sum_{k\in\mathbb{Z}^d} \langle v, b_{\ell,j;k}\rangle b_{\ell,j;k}.$$

f. For every $J \in \mathbb{N}$, $\mathsf{DAS}_J(\{a\,!\,\mathsf{M}; b_1\,!\,\mathsf{M}_1, \ldots, b_s\,!\,\mathsf{M}_s\})$ is a (normalized) tight frame for $l_2(\mathbb{Z}^d)$, that is,

$$\|v\|^2_{l_2(\mathbb{Z}^d)} = \sum_{k\in\mathbb{Z}^d} |\langle v, a_{J;k}\rangle|^2 + \sum_{j=1}^{J}\sum_{\ell=1}^{s}\sum_{k\in\mathbb{Z}^d} |\langle v, b_{\ell,j;k}\rangle|^2, \qquad \forall\, v \in l_2(\mathbb{Z}^d).$$

7.60. Let $a, b_1, \ldots, b_s \in l_1(\mathbb{Z}^d)$ and $\mathsf{M}, \mathsf{M}_1, \ldots, \mathsf{M}_s$ be $d \times d$ invertible integer matrices. Suppose that all the eigenvalues of M are greater than one in modulus and there exist positive numbers C and τ such that $|1 - \widehat{a}(\xi)| \leqslant C\|\xi\|^\tau$ for all $\xi \in [-\pi, \pi]^d$. Define $\widehat{\phi}(\xi) := \prod_{j=1}^{\infty} \widehat{a}((\mathsf{M}^\mathsf{T})^{-j}\xi)$ and $\widehat{\psi^\ell}(\xi) := \widehat{b_\ell}(\mathsf{M}^{-\mathsf{T}}\xi)\widehat{\phi}(\mathsf{M}^{-\mathsf{T}}\xi)$, for $\ell = 1, \ldots, s$. If $\{a\,!\,\mathsf{M}; b_1\,!\,\mathsf{M}_1, \ldots, b_s\,!\,\mathsf{M}_s\}$ is a tight framelet filter bank, prove that $\{\phi\,!\,\mathsf{M}; \psi^1\,!\,\mathsf{M}_1, \ldots, \psi^s\,!\,\mathsf{M}_s\}$ is a tight framelet in $L_2(\mathbb{R}^d)$, i.e., $\phi, \psi^1, \ldots, \psi^s \in L_2(\mathbb{R}^d)$ and $\mathsf{AS}_0(\{\phi\,!\,\mathsf{M}; \psi^1\,!\,\mathsf{M}_1, \ldots, \psi^s\,!\,\mathsf{M}_s\})$ is a tight frame for $L_2(\mathbb{R}^d)$:

$$\|f\|^2_{L_2(\mathbb{R}^d)} = \sum_{k\in\mathbb{Z}^d} |\langle f, \phi(\cdot - k)\rangle|^2 + \sum_{j=0}^{\infty}\sum_{\ell=1}^{s}\sum_{k\in\mathbb{Z}^d} |\langle f, |\det(\mathsf{M}^{-1}\mathsf{M}_\ell)|^{1/2}\psi^\ell_{\mathsf{M}^j;\mathsf{M}^{-1}\mathsf{M}_\ell k}\rangle|^2,$$

for all $f \in L_2(\mathbb{R}^d)$, where

$$\begin{aligned}\mathsf{AS}_0(\{\phi\,!\,\mathsf{M}; \psi^1\,!\,\mathsf{M}_1, \ldots, \psi^s\,!\,\mathsf{M}_s\}) &:= \{\phi(\cdot - k) \;:\; k \in \mathbb{Z}^d\} \\ &\cup \{|\det(\mathsf{M}^{-1}\mathsf{M}_\ell)|^{1/2}\psi^\ell_{\mathsf{M}^j;\mathsf{M}^{-1}\mathsf{M}_\ell k} \;:\; k \in \mathbb{Z}^d, \ell = 1, \ldots, s, j \in \mathbb{N}_0\}.\end{aligned}$$

The converse also holds if in addition $\sum_{k\in\mathbb{Z}^d} |\widehat{\phi}(\xi + 2\pi k)|^2 \neq 0$ a.e. $\xi \in \mathbb{R}^d$.

Appendix A
Basics on Fourier Analysis

Before providing a self-contained brief introduction to Fourier analysis, we recall some basic facts from functional analysis and real analysis. Section A.1 on Banach spaces and Hilbert spaces is only used in Sect. 4.2 for frames and bases in Hilbert spaces. Section A.2 on $L_p(\mathbb{R})$ spaces is mainly needed for introducing Fourier analysis.

A.1 Banach Spaces and Hilbert Spaces

Let $\mathscr{X}$ be a vector space over the complex field $\mathbb{C}$ (or $\mathbb{R}$). A function $\|\cdot\| : \mathscr{X} \to [0, \infty)$ is a *norm* on $\mathscr{X}$ if for all $x, y \in \mathscr{X}$,

(i) $\|x\| \geqslant 0$, and $\|x\| = 0$ implies $x = 0$;
(ii) $\|\lambda x\| = |\lambda|\|x\|$ for all $\lambda \in \mathbb{C}$ (or $\mathbb{R}$);
(iii) $\|x + y\| \leqslant \|x\| + \|y\|$.

A space $\mathscr{X}$ equipped with a norm $\|\cdot\|$ is called a *normed linear space*. A normed liner space $(\mathscr{X}, \|\cdot\|)$ is called a *Banach space* if it is complete under the norm $\|\cdot\|$, i.e., whenever $\{x_n\}_{n\in\mathbb{N}}$ is a Cauchy sequence in $(\mathscr{X}, \|\cdot\|)$ (namely, for any $\varepsilon > 0$, there exists a positive integer N such that $\|x_n - x_m\| < \varepsilon\ \forall\ n, m \geqslant N$), then there exists $x \in \mathscr{X}$ such that $\lim_{n\to\infty} \|x_n - x\| = 0$.

By $\mathscr{B}(\mathscr{X}, \mathscr{Y})$ we denote the space of all bounded linear operators $T : \mathscr{X} \to \mathscr{Y}$, that is, T is a linear bounded operator with $\|T\| := \sup_{\|x\|\leqslant 1} \|Tx\| < \infty$. In particular, we define $\mathscr{X}^* := \mathscr{B}(\mathscr{X}, \mathbb{C})$, the dual space of $\mathscr{X}$. For $x \in \mathscr{X}$, a useful fact is that $\|x\| = \sup_{f\in\mathscr{X}^*, \|f\|\leqslant 1} |f(x)|$.

Two fundamental results on operators acting on a Banach space are as follows:

Theorem A.1.1 (The Open Mapping Theorem) *Let $\mathscr{X}$ and $\mathscr{Y}$ be Banach spaces. If $T \in \mathscr{B}(\mathscr{X}, \mathscr{Y})$ is surjective, then T is open, i.e., $T(\mathscr{O})$ is open in $\mathscr{Y}$ whenever $\mathscr{O}$ is open in $\mathscr{X}$. If T is bijective, then its inverse $T^{-1} \in \mathscr{B}(\mathscr{Y}, \mathscr{X})$.*

B. Han, *Framelets and Wavelets*, Applied and Numerical Harmonic Analysis,
https://doi.org/10.1007/978-3-319-68530-4

Theorem A.1.2 (The Uniform Boundedness Principle) *Suppose that $\mathscr{X}$ is a Banach space and $\mathscr{Y}$ is a normed space. Let $C \subseteq \mathscr{B}(\mathscr{X}, \mathscr{Y})$. If $\sup_{T\in C} \|Tx\| < \infty$ for all $x \in \mathscr{X}$, then $\sup_{T\in C} \|T\| < \infty$.*

A function $\langle \cdot, \cdot \rangle : H \times H \to \mathbb{C}$ is an *inner product* on a vector space H over $\mathbb{C}$ if for all $x, y, z \in H$,

(i) $\langle x, x\rangle \geqslant 0$, and $\langle x, x\rangle = 0$ implies $x = 0$;
(ii) $\langle x, y\rangle = \overline{\langle y, x\rangle}$;
(iii) $\langle ax + by, z\rangle = a\langle x, z\rangle + b\langle y, z\rangle$ for all $a, b \in \mathbb{C}$.

A space H equipped with an inner product $\langle \cdot, \cdot \rangle$ is called an inner product space and $\|h\| := \sqrt{\langle h, h\rangle}$ is a norm on H. An inner product space $(\mathcal{H}, \langle \cdot, \cdot \rangle)$ is a *Hilbert space* if it is complete under the norm $\|\cdot\|$.

Some basic properties on a Hilbert space is as follows:

Proposition A.1.3 *Let $(H, \langle \cdot, \cdot \rangle)$ be an inner product space. Then*

(1) Pythagorean Theorem: $\|x+y\|^2 = \|x\|^2 + \|y\|^2$, whenever $x \perp y$, i.e., $\langle x, y\rangle = 0$.
(2) The Cauchy-Schwarz Inequality: $|\langle x, y\rangle| \leqslant \sqrt{\langle x, x\rangle}\sqrt{\langle y, y\rangle} = \|x\|\|y\|$ for all $x, y \in H$, where the equal sign holds if and only if $x = \lambda y$ for some $\lambda \in \mathbb{C}$ or $y = 0$.
(3) The Parallelogram Law: $\|x+y\|^2 + \|x-y\|^2 = 2(\|x\|^2 + \|y\|^2)$ for all $x, y \in H$.

Proof Since $\langle x, y\rangle = 0$, we have $\langle y, x\rangle = 0$ and

$$\|x + y\|^2 = \langle x + y, x + y\rangle = \langle x, x\rangle + \langle y, y\rangle + \langle x, y\rangle + \langle y, x\rangle = \|x\|^2 + \|y\|^2.$$

Item (2) holds if $y = 0$. If $y \neq 0$, then $\lambda y \perp (x - \lambda y)$ with $\lambda := \langle x, y\rangle / \langle y, y\rangle$. So,

$$|\langle x, y\rangle|^2 / \|y\|^2 = \|\lambda y\|^2 \leqslant \|\lambda y\|^2 + \|x - \lambda y\|^2 = \|x\|^2,$$

which implies $|\langle x, y\rangle|^2 \leqslant \|x\|^2\|y\|^2$. The equality holds if and only if $x - \lambda y = 0$. Since

$$\|x + y\|^2 + \|x - y\|^2 = \langle x + y, x + y\rangle + \langle x - y, x - y\rangle = 2(\|x\|^2 + \|y\|^2),$$

we conclude that (3) holds. □

Theorem A.1.4 (The Riesz Representation Theorem) *For a Hilbert space $\mathcal{H}$, $\mathcal{H}^* = \mathcal{H}$, i.e., for $f \in \mathcal{H}^*$, there is a unique $y \in \mathcal{H}$ such that $f(x) = \langle x, y\rangle$ for all $x \in \mathcal{H}$.*

Proof If $f = 0$, then we take $y = 0$. Otherwise, consider $\ker(f) := \{x \in \mathcal{H} : f(x) = 0\}$. Since f is a bounded nontrivial linear functional, $\ker(f)$ is a proper closed subspace of $\mathcal{H}$. So, $(\ker(f))^{\perp} \neq \{0\}$. Pick $z \in (\ker(f))^{\perp}$ with $\|z\| = 1$. Note that $u := f(x)z - f(z)x \in \ker(f)$ by $f(u) = 0$ and so

$$0 = \langle u, z\rangle = f(x)\|z\|^2 - f(z)\langle x, z\rangle = f(x) - \langle x, \overline{f(z)}z\rangle.$$

Set $y := \overline{f(z)}z$. Then $f(x) = \langle x, y\rangle$ for all $x \in \mathcal{H}$. □

Theorem A.1.5 *Let $\mathcal{H}$ and $\mathcal{K}$ be two Hilbert spaces. Let $T : \mathcal{H} \to \mathcal{K}$ be a bounded linear operator. Then there exists a linear operator $T^\star : \mathcal{K} \to \mathcal{H}$ (called the adjoint of T) such that $\langle Tx, y\rangle_{\mathcal{K}} = \langle x, T^\star y\rangle_{\mathcal{H}}$ for all $x \in \mathcal{H}$ and $y \in \mathcal{K}$. Moreover, $\|T^\star\| = \|T\|$ and $\ker(T^\star) = (T\mathcal{H})^\perp := \{y \in \mathcal{K} : \langle y, Tx\rangle_{\mathcal{K}} = 0 \ \forall\ x \in \mathcal{H}\}$.*

Proof For $y \in \mathcal{K}$, $\ell_y : \mathcal{H} \to \mathbb{C}$ with $\ell_y(x) := \langle Tx, y\rangle_{\mathcal{K}}$ is a bounded linear functional, since $|\ell_y(x)| = |\langle Tx, y\rangle_{\mathcal{K}}| \leqslant \|Tx\|_{\mathcal{K}}\|y\|_{\mathcal{K}} \leqslant \|T\|\|x\|_{\mathcal{H}}\|y\|_{\mathcal{K}}$. Therefore, by Theorem A.1.4, there exists a unique element $T^\star y$ such that $\langle x, T^\star y\rangle_{\mathcal{H}} = \ell_y(x) = \langle Tx, y\rangle_{\mathcal{K}}$. Now it is straightforward to check that $T^\star$ is a linear operator. Note that

$$\begin{aligned}\|T^\star y\|_{\mathcal{H}} &= \sup_{x\in\mathcal{H}, \|x\|_{\mathcal{H}}\leqslant 1} |\langle x, T^\star y\rangle_{\mathcal{H}}| = \sup_{x\in\mathcal{H}, \|x\|_{\mathcal{H}}\leqslant 1} |\langle Tx, y\rangle_{\mathcal{K}}| \\ &\leqslant \sup_{x\in\mathcal{H}, \|x\|_{\mathcal{H}}\leqslant 1} \|T\|\|x\|_{\mathcal{H}}\|y\|_{\mathcal{K}} \leqslant \|T\|\|y\|_{\mathcal{K}}.\end{aligned}$$

Therefore, $\|T^\star\| \leqslant \|T\|$. By $T^{\star\star} = T$, $\|T\| = \|T^{\star\star}\| \leqslant \|T^\star\|$. Hence, $\|T^\star\| = \|T\|$.

Note that $y \in (T\mathcal{H})^\perp \iff 0 = \langle Tx, y\rangle_{\mathcal{K}} = \langle x, T^\star y\rangle_{\mathcal{H}}$ for all $x \in \mathcal{H}$, which is also equivalent to $T^\star y = 0$, that is, $y \in \ker(T^\star)$. This proves $\ker(T^\star) = (T\mathcal{H})^\perp$. □

A.2 Some Results from Real Analysis

In this section, the sets E and F always denote Lebesgue measurable subsets of the real line $\mathbb{R}$ and all involved functions are assumed to be Lebesgue measurable. The following are some basic results in real analysis.

Theorem A.2.1 (The Dominated Convergence Theorem) *Let $g \in L_1(\mathbb{R})$ and $\{f_n\}_{n\in\mathbb{N}}$ be a sequence of measurable functions such that $|f_n(x)| \leqslant g(x)$ and $\lim_{n\to\infty} f_n(x)$ exists for almost every $x \in E$. Then*

$$\int_E \lim_{n\to\infty} f_n(x)\,dx = \lim_{n\to\infty} \int_E f_n(x)\,dx. \tag{A.1}$$

Theorem A.2.2 (Monotone Convergence Theorem) *Let $\{f_n\}_{n\in\mathbb{N}}$ be a nondecreasing sequence of nonnegative measurable functions from E to $\mathbb{R}$. Then* (A.1) *holds.*

Theorem A.2.3 (Fatou's Lemma) *If $\{f_n\}_{n\in\mathbb{N}}$ is a sequence of nonnegative measurable functions from E to $\mathbb{R}$, then*

$$\int_E \liminf_{n\to\infty} f_n(x)dx \leqslant \liminf_{n\to\infty} \int_E f_n(x)dx.$$

Theorem A.2.4 (The Fubini-Tonelli Theorem) *Let $f : E\times F \to \mathbb{C}$ be a Lebesgue measurable function and let $m(x, y)$ be the two-dimensional Lebesgue measure.*

(1) (Fubini) If $f \in L_1(E \times F)$ (that is, $\int_{E\times F} |f(x, y)|dm(x, y) < \infty$), then

$$\int_{E\times F} f(x, y)dm(x, y) = \int_E \left(\int_F f(x, y)dy\right) dx = \int_F \left(\int_E f(x, y)dx\right) dy. \tag{A.2}$$

(2) (Tonelli) If f is nonnegative, then (A.2) *holds.*

By $L_p(E)$ we denote the linear space of all Lebesgue measurable functions f from E to $\mathbb{C}$ such that

$$\|f\|_{L_p(E)} := \left(\int_E |f(x)|^p\, dx\right)^{1/p} < \infty, \qquad 0 < p < \infty,$$

$$\|f\|_{L_\infty(E)} = \text{ess-sup}_{x\in E}|f(x)| := \inf\{C \geqslant 0 \ :\ \{x \in E \ :\ |f(x)| > C\} \text{ has measure } 0\} < \infty.$$

For $1 \leqslant p \leqslant \infty$, Minkowski's (or triangle) inequality holds:

$$\|f + g\|_{L_p(E)} \leqslant \|f\|_{L_p(E)} + \|g\|_{L_p(E)}$$

for all $f, g \in L_p(E)$. For $0 < p < 1$, one has

$$\|f + g\|^p_{L_p(E)} \leqslant \|f\|^p_{L_p(E)} + \|g\|^p_{L_p(E)}.$$

We say that $f \in \mathscr{C}_0(\mathbb{R})$ if f is continuous on $\mathbb{R}$ and $\lim_{|x|\to\infty} f(x) = 0$. Similarly, we say that $f \in \mathscr{C}_{uc,b}(\mathbb{R})$ if f is uniformly continuous on $\mathbb{R}$ and $\|f\|_{L_\infty(\mathbb{R})} < \infty$.

We now state two important inequalities in real analysis.

Theorem A.2.5 (Hölder's Inequality) *Let $1 \leqslant p, p' \leqslant \infty$ such that $\frac{1}{p} + \frac{1}{p'} = 1$. Then $\|fg\|_{L_1(E)} \leqslant \|f\|_{L_p(E)}\|g\|_{L_{p'}(E)}$ for all $f \in L_p(E)$ and $g \in L_{p'}(E)$. The equality holds if and only if*

(i) For $1 < p, p' < \infty$, $|f|^p/\|f\|^p_{L_p(E)} = |g|^{p'}/\|g\|^{p'}_{L_{p'}(E)}$ a.e. on E.
(ii) For $p = 1$ and $p' = \infty$, $|g(x)| = \|g\|_{L_\infty(E)}$ a.e. on the set where $f(x) = 0$. Similarly, for $p = \infty$ and $p' = 1$, $|f(x)| = \|f\|_{L_\infty(E)}$ a.e. on the set where $g(x) = 0$.

Theorem A.2.6 (Minkowski's Integral Inequality) *For* $1 \leqslant p \leqslant \infty$,

$$\left\| \int_E f(\cdot, y)dy \right\|_{L_p(F)} := \left[\int_F \left| \int_E f(x,y)dy \right|^p dx \right]^{\frac{1}{p}} \leqslant \int_E \left[\int_F |f(x,y)|^p dx \right]^{\frac{1}{p}} dy = \int_E \|f(\cdot,y)\|_{L_p(F)} dy.$$

A function $\psi = \sum_{j=1}^N c_j \chi_{[a_j,b_j]}$ with $c_j \in \mathbb{C}$ and $a_j, b_j \in \mathbb{R}$ is called a step function.

Proposition A.2.7 *The set of all step functions is dense in $L_p(\mathbb{R})$ for $0 < p < \infty$, that is, for any $f \in L_p(\mathbb{R})$ and $\varepsilon > 0$, there exists a step function ψ such that $\|f - \psi\|_{L_p(\mathbb{R})} < \varepsilon$. This is not true for $p = \infty$ by $\|1 - \psi\|_{L_\infty(\mathbb{R})} \geqslant 1$ for all step functions ψ.*

Proposition A.2.8 *Let $1 \leqslant p, p' \leqslant \infty$ such that $\frac{1}{p} + \frac{1}{p'} = 1$. Suppose that D is a dense subset of $L_{p'}(\mathbb{R})$. If $f \in L_p(\mathbb{R})$, then $\|f\|_{L_p(\mathbb{R})} = \sup_{g \in D, \|g\|_{L_{p'}(\mathbb{R})} \leqslant 1} \left| \int_{\mathbb{R}} fg \right|$.*

Theorem A.2.9 *If $f \in L_p(\mathbb{R})$ with $0 < p < \infty$, then $\lim_{t \to 0} \|f(\cdot + t) - f\|_{L_p(\mathbb{R})} = 0$.*

Proof If $f = \chi_{[a,b]}$, then it is easy to check that $\|f(\cdot+t) - f\|_{L_p(\mathbb{R})} = |2t|^{1/p}$ when t is small enough. Hence, the claim holds for $f = \chi_{[a,b]}$. Consequently, the claim holds for a step function. Let $f \in L_p(\mathbb{R})$. For any $\varepsilon > 0$, by Proposition A.2.7 there is a step function ψ such that $\|f - \psi\|_{L_p(\mathbb{R})} < \varepsilon$. Since $\lim_{t \to 0} \|\psi(\cdot + t) - \psi\|_{L_p(\mathbb{R})} = 0$, there exists $\delta > 0$ such that $\|\psi(\cdot + t) - \psi\|_{L_p(\mathbb{R})} < \varepsilon$ for all $|t| < \delta$. Thus, for $1 \leqslant p < \infty$,

$$\begin{aligned} \|f(\cdot + t) - f\|_{L_p(\mathbb{R})} &\leqslant \|f(\cdot + t) - \psi(\cdot + t)\|_{L_p(\mathbb{R})} \\ &\quad + \|\psi(\cdot + t) - \psi\|_{L_p(\mathbb{R})} + \|f - \psi\|_{L_p(\mathbb{R})} \leqslant 3\varepsilon. \end{aligned}$$

This proves $\lim_{t \to 0} \|f(\cdot + t) - f\|_{L_p(\mathbb{R})} = 0$ for $1 \leqslant p < \infty$. When $0 < p < 1$, we can prove the claim using the inequality $\|f + g\|_{L_p(\mathbb{R})}^p \leqslant \|f\|_{L_p(\mathbb{R})}^p + \|g\|_{L_p(\mathbb{R})}^p$. □

For two functions $f, g : \mathbb{R} \to \mathbb{C}$, the convolution of f and g is defined as

$$(f * g)(x) := \int_{\mathbb{R}} f(x - y) g(y)\, dy, \qquad x \in \mathbb{R}.$$

Note that $f * g = g * f$ if the above integral is well defined.

Theorem A.2.10 *For $f \in L_p(\mathbb{R})$ with $1 \leqslant p \leqslant \infty$ and $g \in L_1(\mathbb{R})$, $(f * g)(x)$ is well defined for almost every $x \in \mathbb{R}$, $f * g \in L_p(\mathbb{R})$, and $\|f * g\|_{L_p(\mathbb{R})} \leqslant \|f\|_{L_p(\mathbb{R})} \|g\|_{L_1(\mathbb{R})}$.*

Proof By Hölder's inequality, for $1 \leqslant p < \infty$, we have

$$\left| \int_{\mathbb{R}} |f(x-y)g(y)|dy \right|^p \leqslant \left(\int_{\mathbb{R}} |f(x-y)|^p |g(y)| dy \right) \left(\int_{\mathbb{R}} |g(y)| dy \right)^{p-1}.$$

Therefore, by $\int_{\mathbb{R}} |g(y)|dy = \|g\|_{L_1(\mathbb{R})}$, we have

$$\int_{\mathbb{R}} \left| \int_{\mathbb{R}} |f(x-y)g(y)|dy \right|^p dx \leqslant \|g\|_{L_1(\mathbb{R})}^{p-1} \int_{\mathbb{R}} \int_{\mathbb{R}} |f(x-y)|^p |g(y)|dydx$$
$$= \|g\|_{L_1(\mathbb{R})}^{p-1} \int_{\mathbb{R}} \int_{\mathbb{R}} |f(x-y)|^p |g(y)|dxdy = \|f\|_{L_p(\mathbb{R})}^p \|g\|_{L_1(\mathbb{R})}^p < \infty.$$

Thus, $(f * g)(x)$ is well defined for almost every $x \in \mathbb{R}$ and

$$\|f * g\|_{L_p(\mathbb{R})}^p \leqslant \int_{\mathbb{R}} \left| \int_{\mathbb{R}} |f(x-y)g(y)|dy \right|^p dx \leqslant \|f\|_{L_p(\mathbb{R})}^p \|g\|_{L_1(\mathbb{R})}^p.$$

Hence, $\|f * g\|_{L_p(\mathbb{R})} \leqslant \|f\|_{L_p(\mathbb{R})} \|g\|_{L_1(\mathbb{R})}$. When $p = \infty$, $f(\cdot)g(x - \cdot) \in L_1(\mathbb{R})$ and therefore, $(f * g)(x) = \int_{\mathbb{R}} f(y)g(x-y)dy$ is well defined for all $x \in \mathbb{R}$. Obviously $\|f * g\|_{L_\infty(\mathbb{R})} \leqslant \|f\|_{L_\infty(\mathbb{R})} \|g\|_{L_1(\mathbb{R})}$. □

For $f \in L_p(\mathbb{R})$ with $1 \leqslant p \leqslant \infty$, a point $x \in \mathbb{R}$ is called a *Lebesgue point* of f if

$$\lim_{r\to 0+} A_f(x,r) = 0, \quad \text{where} \quad A_f(x,r) := \frac{1}{2r} \int_{-r}^{r} |f(x-y) - f(x)|dy.$$

If f is continuous at point x, then x is obviously a Lebesgue point of f. *The Lebesgue Theorem* says that for $f \in L_p(\mathbb{R})$ with $1 \leqslant p \leqslant \infty$, $\lim_{r\to 0+} A_f(x,r) = 0$ a.e. $x \in \mathbb{R}$.

The following result will play a key role in our study of Fourier transform later.

Theorem A.2.11 *Let $1 \leqslant p \leqslant \infty$ and $\varphi \in L_1(\mathbb{R})$ such that $\int_{\mathbb{R}} \varphi(x)dx = 1$. Define $\varphi_\lambda := \lambda\varphi(\lambda\cdot)$ for $\lambda > 0$.*

(i) $\lim_{\lambda\to\infty} \|f * \varphi_\lambda - f\|_{L_p(\mathbb{R})} = 0$ for $f \in L_p(\mathbb{R})$ (replace $L_\infty(\mathbb{R})$ by $\mathscr{C}_{uc,b}(\mathbb{R})$).
(ii) If in addition $|\varphi(x)| \leqslant C\min(1, |x|^{-2})$ for all $x \in \mathbb{R}$ for some $C > 0$, then for $f \in L_p(\mathbb{R})$, $\lim_{\lambda\to\infty}(f * \varphi_\lambda)(x) = f(x)$ whenever x is a Lebesgue point of f. In particular, $\lim_{\lambda\to\infty}(f * \varphi_\lambda)(x) = f(x)$ for almost every $x \in \mathbb{R}$.

Proof Since $\int_{\mathbb{R}} \varphi_\lambda(x)dx = \int_{\mathbb{R}} \varphi(x)dx = 1$, we have

$$(f * \varphi_\lambda)(x) - f(x) = \int_{\mathbb{R}} [f(x-y) - f(x)]\varphi_\lambda(y)dy. \tag{A.3}$$

By Minkowski's integral inequality,

$$\|f * \varphi_\lambda - f\|_{L_p(\mathbb{R})} \leqslant \int_{\mathbb{R}} \|f(\cdot - y) - f(\cdot)\|_{L_p(\mathbb{R})} |\varphi_\lambda(y)|dy =: I.$$

By Theorem A.2.9, we observe $\lim_{y\to 0} \|f(\cdot - y) - f(\cdot)\|_{L_p(\mathbb{R})} = 0$ (where we used the assumption $f \in \mathscr{C}_{uc,b}(\mathbb{R})$ when $p = \infty$). For $\varepsilon > 0$, there exists $\delta > 0$ such that

$\|f(\cdot - y) - f(\cdot)\|_{L_p(\mathbb{R})} < \varepsilon$ for all $|y| \leqslant \delta$. Hence, we have

$$I_1 := \int_{-\delta}^{\delta} \|f(\cdot - y) - f(\cdot)\|_{L_p(\mathbb{R})} |\varphi_\lambda(y)| dy \leqslant \varepsilon \int_{\mathbb{R}} |\varphi_\lambda(y)| dy = \varepsilon \|\varphi\|_{L_1(\mathbb{R})}.$$

Since $\int_{\mathbb{R}\setminus(-\delta,\delta)} |\varphi_\lambda(y)| dy = \int_{\mathbb{R}\setminus(-\lambda\delta,\lambda\delta)} |\varphi(y)| dy \to 0$ as $\lambda \to \infty$ by $\varphi \in L_1(\mathbb{R})$,

$$I_2 := \int_{\mathbb{R}\setminus(-\delta,\delta)} \|f(\cdot - y) - f(\cdot)\|_{L_p(\mathbb{R})} |\varphi_\lambda(y)| dy \leqslant 2\|f\|_{L_p(\mathbb{R})} \int_{\mathbb{R}\setminus(-\delta,\delta)} |\varphi_\lambda(y)| dy,$$

which goes to 0 as $\lambda \to \infty$. Hence, there exists $N > 0$ such that $I_2 \leqslant \varepsilon$, $\forall\, \lambda \geqslant N$. So, $\|f * \varphi_\lambda - f\|_{L_p(\mathbb{R})} \leqslant I = I_1 + I_2 \leqslant (\|\varphi\|_{L_1(\mathbb{R})} + 1)\varepsilon\ \forall \lambda \geqslant N$. This proves item (i).

Note that $\mathbb{R} = \{|y| < \lambda^{-1}\} \cup \cup_{j=0}^{\infty} \{2^j\lambda^{-1} \leqslant |y| < 2^{j+1}\lambda^{-1}\}$. Hence,

$$\int_{|y|<\lambda^{-1}} |f(x-y) - f(x)||\varphi_\lambda(y)| dy \leqslant C\lambda \int_{-\lambda^{-1}}^{\lambda^{-1}} |f(x-y) - f(x)| dy = 2CA_f(x, \lambda^{-1}),$$

$$\int_{\{2^j/\lambda \leqslant |y| < 2^{j+1}/\lambda\}} |f(x-y) - f(x)||\varphi_\lambda(y)| dy$$

$$\leqslant C \frac{\lambda^{-1}}{(2^j\lambda^{-1})^2} \int_{|y|<2^{j+1}/\lambda} |f(x-y) - f(x)| dy = 4C2^{-j} A_f(x, 2^{j+1}\lambda^{-1}),$$

where we used assumption $|\varphi_\lambda(y)| \leqslant C\lambda \min(1, \lambda^{-2}|y|^{-2})$. By (A.3), we have

$$|(f * \varphi_\lambda)(x) - f(x)| \leqslant 2CA_f(x, \lambda^{-1}) + 8C2^{-1} \sum_{j=0}^{\infty} 2^{-j} A_f(x, 2^{j+1}\lambda^{-1})$$

$$\leqslant 8C \sum_{j=0}^{\infty} 2^{-j} A_f(x, 2^j\lambda^{-1}).$$

Since x is a Lebesgue point of f, by definition, the identity $\lim_{r\to 0+} A_f(x, r) = 0$ holds. Using Hölder's inequality, for $\frac{1}{p} + \frac{1}{p'} = 1$, we have

$$A_f(x, r) \leqslant \frac{1}{2r} \int_{-r}^{r} |f(x-y)| dy + \frac{1}{2r} \int_{-r}^{r} |f(x)| dy \leqslant (2r)^{1/p'-1} \|f\|_{L_p(\mathbb{R})} + |f(x)| < \infty.$$

Hence, there must exist a constant $C_1 > 0$ such that $A_f(x, r) \leqslant C_1$ for all $r > 0$.

Let $\varepsilon > 0$. Then there exists $N \in \mathbb{N}$ such that $\sum_{j=N+1}^{\infty} 2^{-j} < \varepsilon$. Since $A_f(x, r) \leqslant C_1$ and $\lim_{r\to 0+} A_f(x, r) = 0$, there exists $M > 0$ such that $A_f(x, 2^j\lambda^{-1}) \leqslant \varepsilon/N$ for all $\lambda \geqslant M, j = 0, \dots, N-1$. Since $\sum_{j=N+1}^{\infty} 2^{-j} A_f(x, 2^j\lambda^{-1}) \leqslant C_1 \sum_{j=N+1}^{\infty} 2^{-j} \leqslant C_1\varepsilon$,

$$|(f * \varphi_\lambda)(x) - f(x)| \leqslant 8C \sum_{j=0}^{\infty} 2^{-j} A_f(x, 2^j\lambda^{-1}) \leqslant 8C(1 + C_1)\varepsilon, \qquad \forall\, \lambda \geq M.$$

This proves the claim. □

A.3 Fourier Series

Let $\mathbb{T} := \{z \in \mathbb{C} \ : \ |z| = 1\}$ be the unit circle on the complex plane. We identify $\mathbb{T}$ with $\mathbb{R}/[2\pi\mathbb{Z}]$. For $0 < p < \infty$, by $L_p(\mathbb{T})$ we denote the linear space of all measurable functions $f : \mathbb{R} \to \mathbb{C}$ such that f is 2π-periodic: $f(x + 2\pi) = f(x)$ for all $x \in \mathbb{R}$, and

$$\|f\|_{L_p(\mathbb{T})} := \Big(\frac{1}{2\pi}\int_{\mathbb{T}} |f(x)|^p dx\Big)^{1/p} := \Big(\frac{1}{2\pi}\int_{-\pi}^{\pi} |f(x)|^p dx\Big)^{1/p} < \infty.$$

For any 2π-periodic measurable function f, $\int_{-\pi}^{\pi} |f(x)|^p dx = \int_0^{2\pi} |f(x)|^p dx$. By $L_\infty(\mathbb{T})$ we denote the linear space of all 2π-periodic measurable functions $f : \mathbb{R} \to \mathbb{C}$ such that $\|f\|_{L_\infty(\mathbb{T})} = \|f\|_{L_\infty(\mathbb{R})} < \infty$. By $\mathscr{C}(\mathbb{T})$ we denote the linear space of all 2π-periodic continuous functions $f : \mathbb{R} \to \mathbb{C}$ with $\|f\|_{\mathscr{C}(\mathbb{T})} := \sup_{x\in\mathbb{R}} |f(x)| < \infty$. Clearly, $\mathscr{C}(\mathbb{T}) \subset L_\infty(\mathbb{T})$.

For $1 \leqslant p \leqslant q \leqslant \infty$, $\|f\|_{L_p(\mathbb{T})} \leqslant \|f\|_{L_q(\mathbb{T})}$ and $\|f+g\|_{L_p(\mathbb{T})} \leqslant \|f\|_{L_p(\mathbb{T})} + \|g\|_{L_p(\mathbb{T})}$.

For $f \in L_1(\mathbb{T})$, its *Fourier series* is defined to be $\sum_{k\in\mathbb{Z}} \widehat{f}(k)e^{ikx}$, $x \in \mathbb{T}$ where the kth *Fourier coefficient* $\widehat{f}(k)$ of f is defined to be

$$\widehat{f}(k) := \frac{1}{2\pi}\int_{\mathbb{T}} f(t)e^{-ikt} dt, \qquad k \in \mathbb{Z}. \tag{A.4}$$

The *convolution* of two functions f and g in $L_1(\mathbb{T})$ is defined by

$$(f * g)(x) := \frac{1}{2\pi}\int_{\mathbb{T}} f(x - t)g(t)\, dt, \qquad x \in \mathbb{R}.$$

Theorem A.3.1 *If $f \in L_p(\mathbb{T})$ with $1 \leqslant p \leqslant \infty$ and $g \in L_1(\mathbb{T})$, then $f * g \in L_p(\mathbb{T})$, $\|f * g\|_{L_p(\mathbb{T})} \leqslant \|f\|_{L_p(\mathbb{T})}\|g\|_{L_1(\mathbb{T})}$, and $\widehat{f * g}(k) = \widehat{f}(k)\widehat{g}(k)$ for all $k \in \mathbb{Z}$.*

Proof Define $F(t,s) := f(t-s)g(s)$. Then by Tonelli's theorem,

$$\int_{-\pi}^{\pi}\int_{-\pi}^{\pi} |F(t,s)|\,dt\,ds = \int_{-\pi}^{\pi} |f(t)|\,dt \int_{-\pi}^{\pi} |g(s)|\,ds \leqslant (2\pi)^2 \|f\|_{L_p(\mathbb{T})}\|g\|_{L_1(\mathbb{T})}.$$

Since $f \in L_p(\mathbb{T})$ and $g \in L_1(\mathbb{T})$, we conclude that $F \in L_1([-\pi,\pi)^2)$. Using Minkowski's integral inequality, we have

$$\begin{aligned}\|f * g\|_{L_p(\mathbb{T})} &= \left\| \frac{1}{2\pi}\int_{-\pi}^{\pi} f(\cdot - s)g(s)\,ds \right\|_{L_p(\mathbb{T})} \leqslant \frac{1}{2\pi}\int_{-\pi}^{\pi} \|f(\cdot - s)\|_{L_p(\mathbb{T})}|g(s)|ds \\ &= \|f\|_{L_p(\mathbb{T})}\|g\|_{L_1(\mathbb{T})}.\end{aligned}$$

By $F \in L_1([-\pi,\pi)^2)$ and Fubini's Theorem, we can change the order of integration:

$$\begin{aligned}\widehat{f * g}(k) &= \frac{1}{2\pi}\int_{-\pi}^{\pi} (f * g)(t)e^{-ikt}\,dt = \frac{1}{(2\pi)^2}\int_{-\pi}^{\pi}\int_{-\pi}^{\pi} f(t-s)g(s)e^{-ikt}\,ds\,dt \\ &= \frac{1}{(2\pi)^2}\int_{-\pi}^{\pi}\int_{-\pi}^{\pi} f(t-s)e^{-ik(t-s)}g(s)e^{-iks}\,dt\,ds = \widehat{f}(k)\widehat{g}(k).\end{aligned}$$

Thus, we proved the claim. □

The *Fejér kernel* is defined to be

$$F_n(t) := \sum_{j=1-n}^{n-1} \left(1 - \frac{|j|}{n}\right) e^{ijt}, \qquad t \in \mathbb{R}, n \in \mathbb{N}.$$

Noting that $F_n(t) = \frac{1}{n}\sum_{j=0}^{n-1}\sum_{k=-j}^{j} e^{ikt}$, we have

$$F_n(t) = \frac{1}{n}\sum_{j=0}^{n-1} \frac{\sin((j+1/2)t)}{\sin(t/2)} = \frac{1}{n}\sum_{j=0}^{n-1} \frac{\cos(jt) - \cos((j+1)t)}{2\sin^2(t/2)} = \frac{\sin^2(nt/2)}{n\sin^2(t/2)}.$$

Note that $F_n(t) \geqslant 0$ for all $t \in \mathbb{R}$ and $\|F_n\|_{L_1(\mathbb{T})} = \frac{1}{2\pi}\int_{-\pi}^{\pi} F_n(t)dt = 1$. By Theorem A.3.1,

$$(f * F_n)(t) = \sum_{j=1-n}^{n-1} \left(1 - \frac{|j|}{n}\right)\widehat{f}(j)e^{ijt}.$$

Theorem A.3.2 *For $f \in L_p(\mathbb{T})$ with $1 \leqslant p \leqslant \infty$ (when $p = \infty$, replace $L_\infty(\mathbb{T})$ by $\mathscr{C}(\mathbb{T})$), then* $\lim_{n\to\infty} \|f * F_n - f\|_{L_p(\mathbb{T})} = 0$.

Proof Note that $\lim_{t\to 0}\|f(\cdot - t) - f\|_{L_p(\mathbb{T})} = 0$ (where we used the assumption $f \in \mathscr{C}(\mathbb{T})$ when $p = \infty$) and $\|f(\cdot - t)\|_{L_p(\mathbb{T})} = \|f\|_{L_p(\mathbb{T})}$. Since $\frac{1}{2\pi}\int_{-\pi}^{\pi} F_n(t)dt = 1$,

$$(f * F_n)(x) - f(x) = \frac{1}{2\pi}\int_{-\pi}^{\pi} [f(x-t) - f(x)]F_n(t)\, dt.$$

Hence, by Minkowski's integral inequality,

$$\begin{aligned}\|f * F_n - f\|_{L_p(\mathbb{T})} &= \left\| \frac{1}{2\pi}\int_{-\pi}^{\pi} [f(\cdot - t) - f(\cdot)]F_n(t)\, dt \right\|_{L_p(\mathbb{T})} \\ &\leqslant \frac{1}{2\pi}\int_{-\pi}^{\pi} \|f(\cdot - t) - f\|_{L_p(\mathbb{T})}|F_n(t)|\, dt.\end{aligned}$$

Since $\lim_{t\to 0}\|f(\cdot - t) - f\|_{L_p(\mathbb{T})} = 0$, for any given $\varepsilon > 0$, there exists $0 < \delta < \pi$ such that $\|f(\cdot - t) - f\|_{L_p(\mathbb{T})} < \varepsilon/2$ for all $t \in [-\delta, \delta]$. Therefore, by $\|F_n\|_{L_1(\mathbb{T})} = 1$,

$$\frac{1}{2\pi}\int_{-\delta}^{\delta} \|f(\cdot - t) - f(\cdot)\|_{L_p(\mathbb{T})}|F_n(t)|dt \leqslant \frac{\varepsilon}{4\pi}\int_{-\delta}^{\delta} |F_n(t)|dt \leqslant \frac{\varepsilon}{4\pi}\int_{-\pi}^{\pi} |F_n(t)|dt \leqslant \frac{\varepsilon}{2}.$$

Moreover, noting that $|F_n(t)| \leqslant \frac{1}{n\sin^2(\delta/2)}$ for all $t \in [-\pi, \pi]\backslash[-\delta, \delta]$, we have

$$\begin{aligned}&\frac{1}{2\pi}\int_{[-\pi,\pi]\backslash(-\delta,\delta)} \|f(\cdot - t) - f\|_{L_p(\mathbb{T})}|F_n(t)|\, dt \\ &\qquad \leqslant \frac{2\|f\|_{L_p(\mathbb{T})}}{2\pi}\int_{[-\pi,\pi]\backslash(-\delta,\delta)} |F_n(t)|\, dt \leqslant \frac{2\|f\|_{L_p(\mathbb{T})}}{n\sin^2(\delta/2)},\end{aligned}$$

which goes to 0 as $n \to \infty$. Therefore, there exists $N \in \mathbb{N}$ such that

$$\frac{1}{2\pi}\int_{[-\pi,\pi]\backslash(-\delta,\delta)} \|f(\cdot - t) - f\|_{L_p(\mathbb{T})}|F_n(t)|\, dt \leqslant \frac{\varepsilon}{2} \qquad \forall\, n \geqslant N.$$

Hence, we deduce that $\|f * F_n - f\|_{L_p(\mathbb{T})} \leqslant \frac{\varepsilon}{2} + \frac{\varepsilon}{2} = \varepsilon$ for all $n \geqslant N$. That is, $\lim_{n\to\infty}\|f * F_n - f\|_{L_p(\mathbb{T})} = 0$. □

Corollary A.3.3 (The Weierstrass Approximation Theorem) *The trigonometric polynomials are dense in* $\mathscr{C}(\mathbb{T})$ *and* $L_p(\mathbb{T})$ *for* $1 \leqslant p < \infty$.

Proof Since $f * F_n$ is a trigonometric polynomial, the claim follows directly from Theorem A.3.2. □

Corollary A.3.4 *If* $f, g \in L_1(\mathbb{T})$ *and* $\widehat{f}(k) = \widehat{g}(k)\ \forall k \in \mathbb{Z}$, *then* $f(t) = g(t)$ *a.e.* $t \in \mathbb{R}$.

Proof Let $h = f - g \in L_1(\mathbb{T})$. Then $\widehat{h}(k) = 0\ \forall k \in \mathbb{Z}$. Hence, $h * F_n = 0$ for all $n \in \mathbb{N}$. By Theorem A.3.2, $0 = \lim_{n\to\infty} h * F_n = h$ in $L_1(\mathbb{T})$. Hence, $h(t) = 0$ a.e. $t \in \mathbb{R}$. □

We now focus on the particular space $L_2(\mathbb{T})$, which is a Hilbert space equipped with the inner product:

$$\langle f, g\rangle := \frac{1}{2\pi}\int_{\mathbb{T}} f(t)\overline{g(t)}\,dt, \qquad f, g \in L_2(\mathbb{T}).$$

For $f, g \in L_2(\mathbb{T})$ such that $\langle f, g\rangle = 0$ (i.e., $f \perp g$), the Pythagorean Theorem says

$$\|f + g\|^2_{L_2(\mathbb{T})} = \langle f + g, f + g\rangle = \langle f, f\rangle + \langle g, g\rangle = \|f\|^2_{L_2(\mathbb{T})} + \|g\|^2_{L_2(\mathbb{T})}.$$

Theorem A.3.5 *The set $\{e^{ik\cdot} : k \in \mathbb{Z}\}$ is an orthonormal basis for $L_2(\mathbb{T})$, that is, it is an orthonormal system in $L_2(\mathbb{T})$ and $f(x) = \sum_{k\in\mathbb{Z}} \widehat{f}(k)e^{ikx}$ in $L_2(\mathbb{T})$:*

$$\lim_{n\to\infty}\Big\|\sum_{k=-n}^{n}\widehat{f}(k)e^{ik\cdot} - f\Big\|_{L_2(\mathbb{T})} = 0, \qquad \forall f \in L_2(\mathbb{T}). \tag{A.5}$$

Consequently, the following Parseval's identity holds:

$$\langle f, g\rangle = \sum_{k\in\mathbb{Z}}\widehat{f}(k)\overline{\widehat{g}(k)}, \qquad \forall f, g \in L_2(\mathbb{T}), \tag{A.6}$$

where the series on the right-hand side converges absolutely.

Proof By calculation, we have

$$\frac{1}{2\pi}\int_{\mathbb{T}} e^{ikt}\overline{e^{imt}}\,dt = \frac{1}{2\pi}\int_{-\pi}^{\pi} e^{i(k-m)t}\,dt = \begin{cases}1, & \text{if } k = m,\\ 0, & \text{if } k \neq m.\end{cases}$$

Therefore, the set $\{e^{ik\cdot} : k \in \mathbb{Z}\}$ is an orthonormal system in $L_2(\mathbb{T})$. Define $g_n := \sum_{k=-n}^{n}\widehat{f}(k)e^{ikt}$, $n \in \mathbb{N}$. Since $\langle g_n, f - g_n\rangle = 0$, by the Pythagorean Theorem, we have

$$\sum_{k=-n}^{n}|\widehat{f}(k)|^2 = \langle g_n, g_n\rangle = \|g_n\|^2_{L_2(\mathbb{T})} \leqslant \|g_n\|^2_{L_2(\mathbb{T})} + \|f - g_n\|^2_{L_2(\mathbb{T})} = \|f\|^2_{L_2(\mathbb{T})}.$$

Therefore, the following Bessel's inequality holds:

$$\sum_{k\in\mathbb{Z}}|\widehat{f}(k)|^2 \leqslant \|f\|^2_{L_2(\mathbb{T})} < \infty. \tag{A.7}$$

Note that $\|g_n - g_m\|_{L_2(\mathbb{T})}^2 = \sum_{m<|k|\leqslant n} |\widehat{f}(k)|^2$ for all $m \leqslant n$. By (A.7), we see that $\sum_{k\in\mathbb{Z}} |\widehat{f}(k)|^2$ is convergent and therefore, $\{g_n\}_{n\in\mathbb{N}}$ is a Cauchy sequence in $L_2(\mathbb{T})$. Hence, there exists $g \in L_2(\mathbb{T})$ such that $\lim_{n\to\infty} \|g_n - g\|_{L_2(\mathbb{T})} = 0$. To prove (A.5), it suffices to show that $f = g$ in $L_2(\mathbb{T})$. Since $|\widehat{g_n}(k) - \widehat{g}(k)| \leqslant \|g_n - g\|_{L_1(\mathbb{T})} \leqslant \|g_n - g\|_{L_2(\mathbb{T})} \to 0$ as $n \to \infty$, we have $\lim_{n\to\infty} |\widehat{g_n}(k) - \widehat{g}(k)| = 0$ for every $k \in \mathbb{Z}$. Note that $\widehat{g_n}(k) = \widehat{f}(k)$ whenever $|k| \leqslant n$. Hence, $\widehat{g}(k) = \lim_{n\to\infty} \widehat{g_n}(k) = \widehat{f}(k)$ for all $k \in \mathbb{Z}$. Since both f and g belong to $L_1(\mathbb{T})$ and have the same Fourier coefficients, by Corollary A.3.4, we conclude that $f = g$ in $L_2(\mathbb{R}^d)$.

For $f, g \in L_2(\mathbb{T})$, by (A.5), we have

$$\langle f, g\rangle = \lim_{n\to\infty} \Big\langle \sum_{k=-n}^{n} \widehat{f}(k)e^{ik\cdot}, \sum_{m=-n}^{n} \widehat{g}(m)e^{im\cdot}\Big\rangle = \lim_{n\to\infty} \sum_{k=-n}^{n} \widehat{f}(k)\overline{\widehat{g}(k)} = \sum_{k\in\mathbb{Z}} \widehat{f}(k)\overline{\widehat{g}(k)}.$$

The absolute convergence of the series in (A.6) follows from (A.7) and the fact that $\sum_{k\in\mathbb{Z}} |\widehat{g}(k)|^2 = \|g\|_{L_2(\mathbb{T})}^2 < \infty$. □

By $A(\mathbb{T})$ we denote the linear space of all $f \in \mathscr{C}(\mathbb{T})$ such that $\|f\|_{A(\mathbb{T})} := \sum_{k\in\mathbb{Z}} |\widehat{f}(k)| < \infty$. Note that $\|fg\|_{A(\mathbb{T})} \leqslant \|f\|_{A(\mathbb{T})}\|g\|_{A(\mathbb{T})}$ for all $f, g \in A(\mathbb{T})$.

Theorem A.3.6 *Let f be absolutely continuous on $\mathbb{T}$ and assume $f' \in L_2(\mathbb{T})$. Then $f \in A(\mathbb{T})$ and*

$$\|f\|_{A(\mathbb{T})} \leqslant \|f\|_{L_1(\mathbb{T})} + \sqrt{\frac{\pi^2}{3}}\|f'\|_{L_2(\mathbb{T})}.$$

Proof Since f is absolutely continuous, $f' \in L_1(\mathbb{T})$ and by integration by parts,

$$\widehat{f'}(k) = \frac{1}{2\pi}\int_{\mathbb{T}} f'(t)e^{-ikt}dt = \frac{1}{2\pi}\int_{\mathbb{T}} e^{-ikt}df(t) = \frac{-ik}{2\pi}\int_{T} f(t)e^{-ikt}dt = -ik\widehat{f}(k)$$

for all $k \in \mathbb{Z}$. So,

$$\|f\|_{A(\mathbb{T})} = |\widehat{f}(0)| + \sum_{k\in\mathbb{Z}\setminus\{0\}} \frac{|\widehat{f'}(k)|}{|k|} \leqslant \|f\|_{L_1(\mathbb{T})} + \Big(2\sum_{k=1}^{\infty}\frac{1}{k^2}\Big)^{1/2}\Big(\sum_{k\in\mathbb{Z}\setminus\{0\}} |\widehat{f'}(k)|^2\Big)^{1/2}$$

$$\leqslant \|f\|_{L_1(\mathbb{T})} + \sqrt{\frac{\pi^2}{3}}\|f'\|_{L_2(\mathbb{T})},$$

where we used the Bessel's inequality in (A.7) for $f' \in L_2(\mathbb{T})$. □

Theorem A.3.7 (Wiener's Lemma) *If $f \in A(\mathbb{T})$ and $f(x) \neq 0$ for all $x \in \mathbb{R}$, then $1/f \in A(\mathbb{T})$.*

Proof Since $f(x) \neq 0$ for all $x \in \mathbb{R}$, we assume $|f(x)| \geqslant 1$ for all $x \in \mathbb{R}$; otherwise, we consider $f/[\min_{x\in\mathbb{R}} |f(x)|]$. Since $f \in A(\mathbb{T})$, we can take $P(t) := \sum_{k=-N}^{N} \widehat{f}(k)e^{ikt}$ for sufficiently large N such that $\|P - f\|_{A(\mathbb{T})} \leqslant 1/3$. By $\|P - f\|_{A(\mathbb{T})} \leqslant 1/3$, we have

$|P(x) - f(x)| \leqslant 1/3$. Hence, $|P(x)| \geqslant |f(x)| - |P(x) - f(x)| \geqslant 1 - 1/3 = 2/3$. So, $\|\frac{P-f}{P}\|_{L_\infty(\mathbb{T})} \leqslant \frac{1}{2}$ and $\|P^{-n}\|_{L_\infty(\mathbb{T})} \leqslant (3/2)^n$ for all $n \in \mathbb{N}$. Consider

$$S := \sum_{n=1}^{\infty} (P - f)^{n-1} P^{-n} = \frac{1}{P} \sum_{n=1}^{\infty} \left(\frac{P - f}{P} \right)^{n-1} = \frac{1}{P} \frac{1}{1 - \frac{P-f}{P}} = \frac{1}{f}.$$

Note that $[P^{-n}]' = -nP'P^{-n-1}$. Consequently, we deduce that

$$\|[P^{-n}]'\|_{L_\infty(\mathbb{T})} \leqslant n\|P'\|_{L_\infty(\mathbb{T})} \|P^{-n-1}\|_{L_\infty(\mathbb{T})} \leqslant n\|P'\|_{L_\infty(\mathbb{T})} (3/2)^{n+1}.$$

Now by Theorem A.3.6, we have

$$\begin{aligned} \|P^{-n}\|_{A(\mathbb{T})} &\leqslant \|P^{-n}\|_{L_1(\mathbb{T})} + \sqrt{\tfrac{\pi^2}{3}} \|[P^{-n}]'\|_{L_2(\mathbb{T})} \leqslant \|P^{-n}\|_{L_\infty(\mathbb{T})} + 2\|[P^{-n}]'\|_{L_\infty(\mathbb{T})} \\ &\leqslant (3/2)^n (1 + 3n\|P'\|_{L_\infty(\mathbb{T})}) \end{aligned}$$

and $\|(P - f)^{n-1}\|_{A(\mathbb{T})} \leqslant \|P - f\|_{A(\mathbb{T})}^{n-1} \leqslant 3^{1-n}$. Consequently, we deduce that

$$\begin{aligned} \|S\|_{A(\mathbb{T})} &\leqslant \sum_{n=1}^{\infty} \|(P - f)^{n-1} P^{-n}\|_{A(\mathbb{T})} \leqslant \sum_{n=1}^{\infty} \|(P - f)^{n-1}\|_{A(\mathbb{T})} \|P^{-n}\|_{A(\mathbb{T})} \\ &\leqslant \sum_{n=1}^{\infty} \frac{3 + 9n\|P'\|_{L_\infty(\mathbb{T})}}{2^n} < \infty. \end{aligned}$$

Hence, $1/f = S \in A(\mathbb{T})$. □

A.4 Discrete Fourier Transform

For a λ-periodic function $f : \mathbb{R} \to \mathbb{C}$ with $\lambda > 0$, its Fourier series is similarly defined as $\sum_{k \in \mathbb{Z}} \widehat{f}(k) e^{it2\pi k/\lambda}$, where

$$\widehat{f}(k) := \frac{1}{\lambda} \int_0^{\lambda} f(t) e^{-it2\pi k/\lambda} \, dt, \qquad k \in \mathbb{Z}.$$

The above integral for $\widehat{f}(k)$ can be approximated by the N left-endpoint Riemann sum on the interval $[0, \lambda]$ as follows:

$$\widehat{f}(k) \approx \frac{1}{\lambda} \sum_{j=0}^{N-1} f(\tfrac{\lambda j}{N}) e^{-i\frac{\lambda j}{N} 2\pi k/\lambda} \frac{\lambda}{N} = \frac{1}{N} \sum_{j=0}^{N-1} f(\tfrac{\lambda j}{N}) e^{-i2\pi kj/N}.$$

The above approximation motivates the discrete Fourier transform. Let $\mathbb{Z}_N := \mathbb{Z}/[N\mathbb{Z}]$, an additive group. Define $\ell(\mathbb{Z}_N)$ the space of all N-periodic sequences $u : \mathbb{Z} \to \mathbb{C}$ such that $u(j+N) = u(j)$ for all $j \in \mathbb{Z}$. For $u \in \ell(\mathbb{Z}_N)$, the sequence u is uniquely determined by its N-points: $\{u(j)\}_{j=0}^{N-1}$. So, we often write $u = \{u(j)\}_{j=0}^{N-1} \in \ell(\mathbb{Z}_N)$. For $0 < p \leqslant \infty$, we also denote $\|u\|_p := (\sum_{j=0}^{N-1} |u(j)|^p)^{1/p}$.

The *N-point discrete Fourier transform (DFT)* of an N-periodic sequence $u = \{u(j)\}_{j=0}^{N-1} \in \ell(\mathbb{Z}_N)$, denoted by $\widehat{u}$, is defined by

$$\widehat{u}(k) := \sum_{j=0}^{N-1} u(j) e^{-i2\pi kj/N}, \qquad k \in \mathbb{Z}. \tag{A.8}$$

Note that $\widehat{u}$ is also an N-periodic sequence and therefore, $\widehat{u} \in \ell(\mathbb{Z}_N)$. Moreover, the N-point DFT can be written in the matrix form as

$$\begin{bmatrix} \widehat{u}(0) \\ \vdots \\ \widehat{u}(N-1) \end{bmatrix} = \mathscr{F}_N \begin{bmatrix} u(0) \\ \vdots \\ u(N-1) \end{bmatrix} \text{ with } \mathscr{F}_N := \begin{bmatrix} 1 & 1 & \cdots & 1 \\ 1 & e^{-i2\pi/N} & \cdots & e^{-i2\pi(N-1)/N} \\ \vdots & \vdots & \ddots & \vdots \\ 1 & e^{-i2\pi(N-1)/N} & \cdots & e^{-i2\pi(N-1)(N-1)/N} \end{bmatrix}.$$

It is easy to verify that the $N \times N$ Fourier matrix $\mathscr{F}_N$ satisfies $\mathscr{F}_N \overline{\mathscr{F}_N}^{\mathsf{T}} = N I_N$.

Theorem A.4.1 *Let $u = \{u(j)\}_{j=0}^{N-1} \in \ell(\mathbb{Z}_N)$ and $v = \{v(j)\}_{j=0}^{N-1} \in \ell(\mathbb{Z}_N)$. Denote their N-point DFTs by $\widehat{u}$ and $\widehat{v}$, respectively. Then*

(1) $\widehat{u}(k+N) = \widehat{u}(k)$ for all $k \in \mathbb{Z}$. That is, $\widehat{u} \in \ell(\mathbb{Z}_N)$.
(2) $\widehat{au + bv} = a\widehat{u} + b\widehat{v}$ for $a, b \in \mathbb{C}$.
(3) The N-point inverse DFT of $\widehat{u} = \{\widehat{u}(k)\}_{k=0}^{N-1}$ holds:

$$u(j) = \frac{1}{N} \sum_{k=0}^{N-1} \widehat{u}(k) e^{i2\pi kj/N}, \qquad j \in \mathbb{Z}.$$

(4) The Parseval's identity *holds: $\sum_{j=0}^{N-1} u(j)\overline{v(j)} = \frac{1}{N} \sum_{k=0}^{N-1} \widehat{u}(k)\overline{\widehat{v}(k)}$.*
(5) The convolution of u and v is defined by $[u * v](j) := \sum_{k=0}^{N-1} u(j-k)v(k)$ for $j \in \mathbb{Z}$. Then $u * v \in \ell(\mathbb{Z}_N)$ is N-periodic and $\widehat{u * v}(k) = \widehat{u}(k)\widehat{v}(k)$ for all $k \in \mathbb{Z}$.

Proof Items (1) and (2) can be verified directly. By the definition of $\widehat{u}$, we have

$$\begin{aligned} \sum_{k=0}^{N-1} \widehat{u}(k) e^{i2\pi kj/N} &= \sum_{k=0}^{N-1} \sum_{n=0}^{N-1} u(n) e^{-i2\pi kn/N} e^{i2\pi kj/N} \\ &= \sum_{n=0}^{N-1} u(n) \sum_{k=0}^{N-1} e^{i2\pi k(j-n)/N} = \sum_{n=0}^{N-1} u(n) N\delta(j-n) = Nu(j), \end{aligned}$$

where $\delta(0) = 1$ and $\delta(k) = 0$ for all $k \neq 0$. So, item (3) is verified. Item (4) can be proved by a direct calculation as follows:

$$\begin{aligned}\sum_{k=0}^{N-1} \widehat{u}(k)\overline{\widehat{v}(k)} &= \sum_{k=0}^{N-1}\sum_{j=0}^{N-1} u(j)e^{-i2\pi jk/N} \sum_{n=0}^{N-1} \overline{v(n)}e^{i2\pi nk/N} \\ &= \sum_{j=0}^{N-1}\sum_{n=0}^{N-1} u(j)\overline{v(n)} \sum_{k=0}^{N-1} e^{i2\pi k(n-j)/N} \\ &= \sum_{j=0}^{N-1}\sum_{n=0}^{N-1} u(j)\overline{v(n)} N\delta(n-j) = N\sum_{j=0}^{N-1} u(j)\overline{v(j)}.\end{aligned}$$

To prove item (5), we have

$$\begin{aligned}\widehat{u * v}(k) &= \sum_{j=0}^{N-1} [u * v](j)e^{-i2\pi kj/N} = \sum_{j=0}^{N-1}\sum_{n=0}^{N-1} u(j-n)v(n)e^{-i2\pi k(j-n+n)/N} \\ &= \sum_{n=0}^{N-1} v(n)e^{-i2\pi kn/N} \sum_{j=0}^{N-1} u(j-n)e^{-i2\pi k(j-n)/N} = \widehat{u}(k)\widehat{v}(k).\end{aligned}$$

This completes the proof. □

A.5 Fourier Transform

The *Fourier transform* of a function $f \in L_1(\mathbb{R})$ is defined to be

$$(\mathscr{F}f)(\xi) = \widehat{f}(\xi) := \int_{\mathbb{R}} f(x)e^{-ix\xi}\, dx, \qquad \xi \in \mathbb{R}. \tag{A.9}$$

By (A.9), $\widehat{(f+g)}(\xi) = \widehat{f}(\xi) + \widehat{g}(\xi)$, $\widehat{\overline{f}}(\xi) = \overline{\widehat{f}(-\xi)}$, $\widehat{f(\cdot - c)}(\xi) = e^{-ic\xi}\widehat{f}(\xi)$, and $\widehat{f(\lambda\cdot)}(\xi) = |\lambda|^{-1}\widehat{f}(\xi/\lambda)$ for $c \in \mathbb{R}$, $\lambda \in \mathbb{R}\backslash\{0\}$ and $f, g \in L_1(\mathbb{R})$.

Proposition A.5.1 *Let $f, g \in L_1(\mathbb{R})$.*

(i) (the Riemann-Lebesgue Lemma) $\widehat{f}$ is uniformly continuous and $\lim_{|\xi|\to\infty} \widehat{f}(\xi) = 0$.
(ii) If f is absolutely continuous (or differentiable) satisfying $f' \in L_1(\mathbb{R})$ and $\lim_{|x|\to\infty} f(x) = 0$, then $\widehat{f'}(\xi) = i\xi\widehat{f}(\xi)$.
(iii) If $xf(x) \in L_1(\mathbb{R})$, then $\widehat{f}$ is differentiable and $\frac{d}{d\xi}\widehat{f}(\xi) = -i\widehat{xf(x)}(\xi)$ for $\xi \in \mathbb{R}$.
(iv) $f * g \in L_1(\mathbb{R})$ and $\widehat{f * g}(\xi) = \widehat{f}(\xi)\widehat{g}(\xi)$ for all $\xi \in \mathbb{R}$.

Proof

(i) By the definition of the Fourier transform, we have

$$|\widehat{f}(\xi+t)-\widehat{f}(\xi)| = \left|\int_{\mathbb{R}} f(x)[e^{-i(\xi+t)x}-e^{-i\xi x}]\,dx\right| \leqslant \int_{\mathbb{R}^d} |f(x)(e^{-itx}-1)|\,dx.$$

Since $|f(x)(e^{-itx}-1)| \leqslant 2|f(x)| \in L_1(\mathbb{R})$, by the Dominated Convergence Theorem,

$$\lim_{t\to 0}|\widehat{f}(\xi+t)-\widehat{f}(\xi)| \leqslant \lim_{t\to 0}\int_{\mathbb{R}} |f(x)(e^{-itx}-1)|\,dx = \int_{\mathbb{R}} |f(x)| \lim_{t\to 0}|e^{-itx}-1|\,dx = 0.$$

Therefore, $\widehat{f}$ is uniformly continuous. For $\xi \neq 0$, let $\tau_\xi := \pi/\xi$. Then

$$\widehat{f}(\xi) = \int_{\mathbb{R}} f(x)e^{-i\xi x}dx = \int_{\mathbb{R}} f(x+\tau_\xi)e^{-i\xi(x+\tau_\xi)}dx = -\int_{\mathbb{R}} f(x+\tau_\xi)e^{-i\xi x}dx.$$

Hence, $\widehat{f}(\xi) = \frac{1}{2}\int_{\mathbb{R}}[f(x)-f(x+\tau_\xi)]e^{-i\xi x}dx$. Thus, $|\widehat{f}(\xi)| \leqslant \frac{1}{2}\|f-f(\cdot+\tau_\xi)\|_{L_1(\mathbb{R})} \to 0$ as $|\xi| \to \infty$ by Theorem A.2.9. Therefore, $\widehat{f} \in \mathscr{C}_0(\mathbb{R})$ and this proves item (i).

(ii) follows from integration by parts:

$$\widehat{f'}(\xi) = \int_{\mathbb{R}} f'(x)e^{-ix\xi}dx = \int_{\mathbb{R}} e^{-ix\xi}df(x) = -i\xi\int_{\mathbb{R}} f(x)e^{-ix\xi}dx = -i\xi\widehat{f}(\xi).$$

(iii) By the definition of the Fourier transform, we have

$$\frac{\widehat{f}(\xi+t)-\widehat{f}(\xi)}{t} = \frac{\int_{\mathbb{R}} f(x)[e^{-i(\xi+t)x}-e^{-i\xi x}]\,dx}{t} = \int_{\mathbb{R}} f(x)e^{-i\xi x}\left(\frac{e^{-itx}-1}{t}\right)dx.$$

Since $\left|\frac{e^{-itx}-1}{t}\right| = \left|\frac{2\sin(tx/2)}{t}\right| \leqslant |tx/t| = |x|$, we have $\left|f(x)e^{-i\xi x}\left(\frac{e^{-itx}-1}{t}\right)\right| \leqslant |xf(x)| \in L_1(\mathbb{R})$. By the Dominated Convergence Theorem,

$$\begin{aligned}\frac{d}{d\xi}\widehat{f}(\xi) &= \lim_{t\to 0}\frac{\widehat{f}(\xi+t)-\widehat{f}(\xi)}{t} = \lim_{t\to 0}\int_{\mathbb{R}} f(x)e^{-i\xi x}\left(\frac{e^{-itx}-1}{t}\right)dx \\ &= \int_{\mathbb{R}} f(x)e^{-i\xi x}\lim_{t\to 0}\left(\frac{e^{-itx}-1}{t}\right)dt = \int_{\mathbb{R}}(-ix)f(x)e^{-i\xi x}\,dx = -i\widehat{(xf(x))}(\xi).\end{aligned}$$

(iv) Since $f, g \in L_1(\mathbb{R})$, by Theorem A.2.10, $\|f*g\|_{L_1(\mathbb{R})} \leqslant \|f\|_{L_1(\mathbb{R})}\|g\|_{L_1(\mathbb{R})}$ and $f(x-y)g(y) \in L_1(\mathbb{R}^2)$. By Fubini's theorem and changing order of integration,

$$\begin{aligned}\widehat{f*g}(\xi) &= \int_{\mathbb{R}}(f*g)(x)e^{-i\xi x}dx = \int_{\mathbb{R}}\int_{\mathbb{R}} f(x-y)g(y)e^{-i\xi x}dydx \\ &= \int_{\mathbb{R}}\int_{\mathbb{R}} f(x-y)e^{-i\xi(x-y)}g(y)e^{-i\xi y}dxdy = \widehat{f}(\xi)\widehat{g}(\xi).\end{aligned}$$

This completes the proof. □

Generally, for a polynomial $\mathsf{p} \in \mathbb{P}$ and a smooth decaying function $f \in L_1(\mathbb{R})$,

$$\widehat{[\mathsf{p}(\tfrac{d}{dx})f(x)]}(\xi) = \mathsf{p}(i\xi)\widehat{f}(\xi) \quad \text{and} \quad \mathsf{p}(\tfrac{d}{d\xi})\widehat{f}(\xi) = \widehat{[\mathsf{p}(-ix)f(x)]}(\xi).$$

We now calculate the Fourier transform of the function $G(x) = \frac{1}{\sqrt{2\pi}}e^{-x^2/2}$. Since $xG(x) \in L_1(\mathbb{R})$, the function $\widehat{G}$ is differentiable and we have

$$\begin{aligned}(\widehat{G})'(\xi) &= \frac{1}{\sqrt{2\pi}}\int_{\mathbb{R}} e^{-x^2/2}e^{-i\xi x}(-ix)\,dx = \frac{i}{\sqrt{2\pi}}\int_{\mathbb{R}} e^{-i\xi x}\,de^{-x^2/2} \\ &= -\frac{\xi}{\sqrt{2\pi}}\int_{\mathbb{R}} e^{-x^2/2}e^{-i\xi x}\,dx = -\xi\widehat{G}(\xi).\end{aligned}$$

Hence, $\frac{d}{d\xi}[e^{\xi^2/2}\widehat{G}(\xi)] = 0$ and $\widehat{G}(\xi) = Ce^{-\xi^2/2}$ for some constant C. By calculation, we have $\widehat{G}(0) = \frac{1}{\sqrt{2\pi}}\int_{\mathbb{R}} e^{-x^2/2}\,dx = 1$ by $(\int_{\mathbb{R}} e^{-x^2/2}dx)^2 = \iint_{\mathbb{R}^2} e^{-(x^2+y^2)/2}dxdy = \int_0^{2\pi}\int_0^{\infty} e^{-r^2/2}rdrd\theta = 2\pi$. Thus, we conclude that $C = 1$ and $\widehat{G}(\xi) = e^{-\xi^2/2}$.

Lemma A.5.2 *For $f, g \in L_1(\mathbb{R})$, the identity $\int_{\mathbb{R}} f(y)\widehat{g}(y)dy = \int_{\mathbb{R}} \widehat{f}(\xi)g(\xi)d\xi$ holds.*

Proof Since $f, g \in L_1(\mathbb{R})$, $f(\xi)g(y) \in L_1(\mathbb{R}^2)$. By Fubini's Theorem, we have

$$\begin{aligned}\int_{\mathbb{R}} f(y)\widehat{g}(y)dy &= \int_{\mathbb{R}}\int_{\mathbb{R}} f(y)g(\xi)e^{-i\xi y}d\xi dy \\ &= \int_{\mathbb{R}}\int_{\mathbb{R}} f(y)e^{-i\xi y}g(\xi)dyd\xi = \int_{\mathbb{R}} \widehat{f}(\xi)g(\xi)d\xi.\end{aligned}$$

This completes the proof. □

The *inverse Fourier transform* of $f \in L_1(\mathbb{R})$ is defined to be

$$\mathscr{F}^{-1}(f)(x) = f^{\vee}(x) := \frac{1}{2\pi}\int_{\mathbb{R}} f(\xi)e^{i\xi x}\,d\xi, \qquad x \in \mathbb{R}.$$

Note that $\mathscr{F}^{-1}(f)(x) = \frac{1}{2\pi}\mathscr{F}(f)(-x)$.

Theorem A.5.3 *Let $G_\lambda := \lambda G(\lambda\cdot)$, where $G(x) = \frac{1}{\sqrt{2\pi}}e^{-x^2/2}$. For $f \in L_1(\mathbb{R})$,*

$$\frac{1}{2\pi}\int_{\mathbb{R}} e^{-\frac{|\xi/\lambda|^2}{2}}\widehat{f}(\xi)e^{i\xi x}d\xi = (f * G_\lambda)(x) \to f(x) \quad \text{as} \quad \lambda \to \infty$$

in both $L_1(\mathbb{R})$ norm and at every Lebesgue point x of f.

Proof Define $g_x(\xi) := \frac{1}{2\pi} e^{i\xi x} e^{-i|\xi/\lambda|^2/2} = \frac{1}{\sqrt{2\pi}} G(\xi/\lambda) e^{i\xi x} \in L_1(\mathbb{R})$. By $\widehat{G}(\xi) = e^{-\xi^2/2}$, we have $\widehat{g_x}(y) = \lambda G(\lambda(x-y)) = G_\lambda(x-y)$. By Lemma A.5.2,

$$(f * G_\lambda)(x) = \int_{\mathbb{R}} f(y) G_\lambda(x-y) dy = \int_{\mathbb{R}} f(y) \widehat{g_x}(y) dy = \int_{\mathbb{R}} \widehat{f}(\xi) g_x(\xi) d\xi.$$

The claim now follows directly from Theorem A.2.11. □

Corollary A.5.4 (The Fourier Inversion Formula) *If both f and $\widehat{f}$ are in $L_1(\mathbb{R})$, then*

$$f(x) = \mathscr{F}^{-1}(\widehat{f})(x) = \frac{1}{2\pi} \int_{\mathbb{R}} \widehat{f}(\xi) e^{i\xi x} d\xi, \qquad a.e.\ x \in \mathbb{R}.$$

Proof Note that $|e^{-|\xi/\lambda|^2/2} \widehat{f}(\xi) e^{i\xi x}| \leqslant |\widehat{f}(\xi)| \in L_1(\mathbb{R})$. By Theorem A.5.3, we have

$$f(x) = \lim_{\lambda \to \infty} \frac{1}{2\pi} \int_{\mathbb{R}} e^{-\frac{|\xi/\lambda|^2}{2}} \widehat{f}(\xi) e^{i\xi x} d\xi = \frac{1}{2\pi} \int_{\mathbb{R}} \widehat{f}(\xi) e^{i\xi x} d\xi, \qquad a.e.\, x \in \mathbb{R},$$

where we used the Dominated Convergence Theorem in the second identity. □

Theorem A.5.5 *If $f \in L_1(\mathbb{R}) \cap L_2(\mathbb{R})$, then $\widehat{f} \in L_2(\mathbb{R}) \cap \mathscr{C}_0(\mathbb{R})$ and*

$$\|f\|^2_{L_2(\mathbb{R})} = \int_{\mathbb{R}} |f(x)|^2 \, dx = \frac{1}{2\pi} \int_{\mathbb{R}} |\widehat{f}(\xi)|^2 \, d\xi = \frac{1}{2\pi} \|\widehat{f}\|^2_{L_2(\mathbb{R})}.$$

Proof Define $g := f * \overline{f(-\cdot)} = \int_{\mathbb{R}} f(\cdot + y) \overline{f(y)} dy$. Then $g \in L_1(\mathbb{R})$ with $\|g\|_{L_1(\mathbb{R})} \leqslant \|f\|^2_{L_1(\mathbb{R})}$ and $\widehat{g} = |\widehat{f}|^2 \geqslant 0$. By Theorem A.2.9,

$$|g(x) - g(0)| \leqslant \int_{\mathbb{R}} |f(x+y) - f(y)||f(y)| dy \leqslant \|f(x+\cdot) - f(\cdot)\|_{L_2(\mathbb{R})} \|f\|_{L_2(\mathbb{R})} \to 0,$$

as $x \to 0$. Thus, g is continuous at 0 and hence, 0 is a Lebesgue point of g. Since $g \in L_1(\mathbb{R})$, by Theorem A.5.3 with $x = 0$, we have $g(0) = \lim_{\lambda \to \infty} \frac{1}{2\pi} \int_{\mathbb{R}} e^{-\frac{|\xi/\lambda|^2}{2}} \widehat{g}(\xi) d\xi$. Since $\widehat{g}$ is nonnegative, by the Monotone Convergence Theorem, we conclude that

$$g(0) = \lim_{\lambda \to \infty} \frac{1}{2\pi} \int_{\mathbb{R}} e^{-\frac{|\xi/\lambda|^2}{2}} \widehat{g}(\xi) d\xi = \frac{1}{2\pi} \int_{\mathbb{R}} \lim_{\lambda \to \infty} e^{-\frac{|\xi/\lambda|^2}{2}} \widehat{g}(\xi) d\xi = \frac{1}{2\pi} \int_{\mathbb{R}} \widehat{g}(\xi) d\xi.$$

Consequently,

$$\|f\|^2_{L_2(\mathbb{R})} = \int_{\mathbb{R}} |f(x)|^2 dx = g(0) = \frac{1}{2\pi} \int_{\mathbb{R}} \widehat{g}(\xi) d\xi = \frac{1}{2\pi} \int_{\mathbb{R}} |\widehat{f}(\xi)|^2 d\xi = \frac{1}{2\pi} \|\widehat{f}\|^2_{L_2(\mathbb{R})}.$$

This completes the proof. □

For $f \in L_2(\mathbb{R})$, we define

$$\mathscr{F}(f) := \lim_{n\to\infty} \widehat{f\chi_{[-n,n]}} = \lim_{n\to\infty} \int_{-n}^{n} f(x)e^{-i\xi x}dx \quad \text{in} \quad L_2(\mathbb{R}).$$

Theorem A.5.6 (Plancherel's Theorem) *The Fourier transform $\mathscr{F} : L_2(\mathbb{R}) \to L_2(\mathbb{R})$ is well defined and $\mathscr{F}f = \widehat{f}$ for all $f \in L_1(\mathbb{R}) \cap L_2(\mathbb{R})$. Moreover,*

$$\|\widehat{f}\|^2_{L_2(\mathbb{R})} = 2\pi\|f\|^2_{L_2(\mathbb{R})} \qquad \text{and} \qquad \langle \widehat{f}, \widehat{g}\rangle = 2\pi\langle f, g\rangle, \qquad \forall\, f, g \in L_2(\mathbb{R}).$$

Similarly, we can define $\mathscr{F}^{-1}$ by $\mathscr{F}^{-1}f = f^{\vee}$ for $f \in L_1(\mathbb{R}) \cap L_2(\mathbb{R})$ and extend it to $L_2(\mathbb{R})$. Then $\mathscr{F} : L_2(\mathbb{R}) \to L_2(\mathbb{R})$ is a bijection with the inverse mapping $\mathscr{F}^{-1}$.

Proof Define $f_n := f\chi_{[-n,n]}$. It is trivial that $\lim_{n\to\infty}\|f_n - f\|_{L_2(\mathbb{R})} = 0$. Hence, $\{f_n\}_{n\in\mathbb{N}}$ is a Cauchy sequence in $L_2(\mathbb{R})$. Since $f_n \in L_1(\mathbb{R}) \cap L_2(\mathbb{R})$, by Theorem A.5.5, we see that $\{\widehat{f_n}\}_{n\in\mathbb{N}}$ must be a Cauchy sequence in $L_2(\mathbb{R})$. Therefore, there exists $\mathscr{F}(f) \in L_2(\mathbb{R})$ such that $\lim_{n\to\infty}\|\widehat{f_n} - \mathscr{F}(f)\|_{L_2(\mathbb{R})} = 0$. Hence, $\mathscr{F}(f)$ is well defined. Moreover,

$$\|\mathscr{F}(f)\|^2_{L_2(\mathbb{R})} = \lim_{n\to\infty}\|\widehat{f_n}\|^2_{L_2(\mathbb{R})} = \lim_{n\to\infty} 2\pi\|f_n\|^2_{L_2(\mathbb{R})} = 2\pi\|f\|^2_{L_2(\mathbb{R})}.$$

If in addition $f \in L_1(\mathbb{R}) \cap L_2(\mathbb{R})$, by Theorem A.5.5, we have $\widehat{f} \in L_2(\mathbb{R})$. By $f_n - f \in L_1(\mathbb{R}) \cap L_2(\mathbb{R})$, we see that

$$\lim_{n\to\infty}\|\widehat{f_n} - \widehat{f}\|^2_{L_2(\mathbb{R})} = \lim_{n\to\infty}\|\widehat{f_n - f}\|^2_{L_2(\mathbb{R})} = \lim_{n\to\infty} 2\pi\|f_n - f\|^2_{L_2(\mathbb{R})} = 0.$$

By the uniqueness of the limit of $\{\widehat{f_n}\}_{n\in\mathbb{N}}$, we must have $\mathscr{F}(f) = \widehat{f}$ in $L_2(\mathbb{R})$. □

Theorem A.5.7 (The Poisson Summation Formula) *If $f, \widehat{f} \in L_1(\mathbb{R}) \cap \mathscr{C}(\mathbb{R})$ satisfy*

$$|f(x)| + |\widehat{f}(x)| \leqslant C(1 + |x|)^{-1-\varepsilon}, \qquad \forall\, x \in \mathbb{R}, \tag{A.10}$$

for some $C, \varepsilon > 0$, then

$$\lambda \sum_{k\in\mathbb{Z}} f(x - \lambda k)e^{-i\zeta(x-\lambda k)} = \sum_{k\in\mathbb{Z}} \widehat{f}(\zeta + 2\pi k/\lambda)e^{ix2\pi k/\lambda}, \qquad \forall\, x, \zeta \in \mathbb{R}, \lambda > 0,$$

where both series converge absolutely and uniformly on $\mathbb{R}$.

Proof Define $f^{per}(x) := \lambda\sum_{k\in\mathbb{Z}} f(x - \lambda k)e^{-i\zeta(x-\lambda k)}$. Then f^{per} is λ-periodic and $f^{per} \in L_1([0, \lambda)) \cap \mathscr{C}(\mathbb{R})$ since $\int_0^{\lambda} |f^{per}(x)|dx \leqslant \lambda\|f\|_{L_1(\mathbb{R})}$ and (A.10) holds. Note

$$\widehat{f^{per}}(k) := \frac{1}{\lambda}\int_0^{\lambda} f^{per}(x)e^{-ix2\pi k/\lambda}dx = \int_{\mathbb{R}} f(x)e^{-ix(\zeta+2\pi k/\lambda)}dx = \widehat{f}(\zeta + 2\pi k/\lambda),$$

for all $k \in \mathbb{Z}$. By (A.10), we have

$$\sum_{k\in\mathbb{Z}} |\widehat{f^{per}}(k)| = \sum_{k\in\mathbb{Z}} |\widehat{f}(\zeta + 2\pi k/\lambda)| \leqslant C \sum_{k\in\mathbb{Z}} (1 + |\zeta + 2\pi k/\lambda|)^{-1-\varepsilon} < \infty.$$

Hence, $g(x) := \sum_{k\in\mathbb{Z}} \widehat{f}(\zeta + 2\pi k/\lambda) e^{ix2\pi k/\lambda}$ must be a λ-periodic continuous function having absolutely convergent Fourier series. Since $\widehat{g}(m) = \widehat{f^{per}}(m)$ for all $m \in \mathbb{Z}$ and g, f^{per} are continuous, we must have $g(x) = f^{per}(x)$ for all $x \in \mathbb{R}$. □

A.6 Distributions and Tempered Distributions

The *Schwartz class* $\mathscr{S}(\mathbb{R})$ consists of all $\mathscr{C}^\infty(\mathbb{R})$ functions φ such that

$$\rho_{\alpha,\beta}(\varphi) := \|x^\alpha \varphi^{(\beta)}(x)\|_{\mathscr{C}(\mathbb{R})} < \infty, \qquad \forall\, \alpha, \beta \in \mathbb{N}_0 := \mathbb{N} \cup \{0\}.$$

We say that $\varphi_n \to \varphi$ in $\mathscr{S}(\mathbb{R})$ as $n \to \infty$ if all $\varphi, \varphi_n \in \mathscr{S}(\mathbb{R})$ and $\lim_{n\to\infty} \rho_{\alpha,\beta}(\varphi_n - \varphi) = 0$ for all $\alpha, \beta \in \mathbb{N}_0$.

For $f, g \in \mathscr{S}(\mathbb{R})$, we define $\langle f; g\rangle := \langle f, \overline{g}\rangle = \int_{\mathbb{R}} f(x)g(x)dx$. The space $\mathscr{S}'(\mathbb{R})$ of *tempered distributions* on $\mathbb{R}$ is the dual of $\mathscr{S}(\mathbb{R})$, that is, $f \in \mathscr{S}'(\mathbb{R})$ means that f is a continuous linear functional on $\mathscr{S}(\mathbb{R})$. We say that $f_n \to f$ in $\mathscr{S}'(\mathbb{R})$ as $n \to \infty$ if all $f, f_n \in \mathscr{S}'(\mathbb{R})$ and $\lim_{n\to\infty}\langle f_n; \varphi\rangle = \langle f; \varphi\rangle$ for all $\varphi \in \mathscr{S}(\mathbb{R})$.

The (test) function space $\mathscr{D}(\mathbb{R})$ consists of all $\mathscr{C}^\infty(\mathbb{R})$ functions with compact support. We say that $\varphi_n \to \varphi$ in $\mathscr{D}(\mathbb{R})$ as $n \to \infty$ if all $\varphi, \varphi_n \in \mathscr{D}(\mathbb{R})$, all supports of φ_n are contained inside some bounded interval and $\lim_{n\to\infty} \|(\varphi_n - \varphi)^{(\beta)}\|_{\mathscr{C}(\mathbb{R})} = 0$ for all $\beta \in \mathbb{N}_0$. Note that $\mathscr{D}(\mathbb{R}) \subseteq \mathscr{S}(\mathbb{R})$ and $\varphi_n \to \varphi$ in $\mathscr{D}(\mathbb{R})$ as $n \to \infty$ implies $\varphi_n \to \varphi$ in $\mathscr{S}(\mathbb{R})$ as $n \to \infty$.

The space $\mathscr{D}'(\mathbb{R})$ of *distributions* (or generalized functions) on $\mathbb{R}$ is the dual of $\mathscr{D}(\mathbb{R})$, that is, $f \in \mathscr{D}'(\mathbb{R})$ means that f is a continuous linear functional on $\mathscr{D}(\mathbb{R})$. Note that $\mathscr{S}'(\mathbb{R}) \subseteq \mathscr{D}'(\mathbb{R})$. We say that $f_n \to f$ in $\mathscr{D}'(\mathbb{R})$ as $n \to \infty$ if all $f, f_n \in \mathscr{D}'(\mathbb{R})$ and $\lim_{n\to\infty}\langle f_n; \varphi\rangle = \langle f; \varphi\rangle$ for all $\varphi \in \mathscr{D}(\mathbb{R})$.

Theorem A.6.1

(i) A linear functional $f : \mathscr{D}(\mathbb{R}) \to \mathbb{C}$ is a distribution (that is, $f \in \mathscr{D}'(\mathbb{R})$) if and only if for every $N \in \mathbb{N}$, there exist $C > 0$ and $m \in \mathbb{N}_0$ such that

$$|\langle f; \varphi\rangle| \leqslant C \sum_{\beta=0}^{m} \|\varphi^{(\beta)}\|_{\mathscr{C}(\mathbb{R})}, \quad \forall\, \varphi \in \mathscr{D}(\mathbb{R}) \text{ with support inside } [-N, N]; \tag{A.11}$$

(ii) A linear functional $f : \mathscr{S}(\mathbb{R}) \to \mathbb{C}$ is a tempered distribution (that is, $f \in \mathscr{S}'(\mathbb{R})$) if and only if there exist $m \in \mathbb{N}_0$ and $C > 0$ such that

$$|\langle f; \varphi\rangle| \leqslant C \sum_{\alpha=0}^{m} \sum_{\beta=0}^{m} \rho_{\alpha,\beta}(\varphi) \qquad \forall\, \varphi \in \mathscr{S}(\mathbb{R}). \tag{A.12}$$

Proof

(i) Suppose that (A.11) fails. Then there exist $N \in \mathbb{N}$ and $\varphi_m \in \mathscr{D}(\mathbb{R}), m \in \mathbb{N}$ such that all φ_m are supported inside $[-N, N]$, $|\langle f; \varphi_m\rangle| = 1$, and $\sum_{\beta=0}^{m} \|\varphi_m^{(\beta)}\|_{\mathscr{C}(\mathbb{R})} \leqslant \frac{1}{m}$. Hence, $\varphi_m \to 0$ in $\mathscr{D}(\mathbb{R})$ but $|\langle f; \varphi_m\rangle| \to 1$, a contradiction to the continuity of f.

Conversely, if $\varphi_n \to \varphi$ in $\mathscr{D}(\mathbb{R})$ as $n \to \infty$, then all φ_n are supported inside $[-N, N]$ for some $N \in \mathbb{N}$. By (A.11), $\lim_{n\to\infty}\langle f; \varphi_n - \varphi\rangle = 0$. So, $f \in \mathscr{D}'(\mathbb{R})$.

(ii) Suppose that (A.12) is not true. Then there exist $\varphi_m \in \mathscr{S}(\mathbb{R}), m \in \mathbb{N}$ such that $|\langle f; \varphi_m\rangle| = 1$ and $\sum_{\alpha=0}^{m}\sum_{\beta=0}^{m} \rho_{\alpha,\beta}(\varphi_m) \leqslant \frac{1}{m}$. But this implies $\varphi_m \to 0$ in $\mathscr{S}(\mathbb{R})$ with $|\langle f; \varphi_m\rangle| \to 1$ as $n \to \infty$, a contradiction to the continuity of f.

Conversely, if $\varphi_n \to \varphi$ in $\mathscr{S}(\mathbb{R})$ as $n \to \infty$, then it is trivial to see from (A.12) that $\lim_{n\to\infty}\langle f; \varphi_n - \varphi\rangle = 0$. So, f is continuous and $f \in \mathscr{S}'(\mathbb{R})$.

□

Lemma A.6.2 *Let f be a measurable function (or a measure) on $\mathbb{R}$ such that $(1 + |\cdot|^2)^{-\tau} f(\cdot) \in L_p(\mathbb{R})$ for some $\tau \geqslant 0$ and $1 \leqslant p \leqslant \infty$ (e.g., f is a polynomial). Then f can be identified as a tempered distribution on $\mathbb{R}$ via $\langle f; h\rangle := \int_{\mathbb{R}} f(x)h(x)dx$, $h \in \mathscr{S}(\mathbb{R})$.*

Proof We use Theorem A.6.1 to prove that $f \in \mathscr{S}'(\mathbb{R})$. For $h \in \mathscr{S}(\mathbb{R})$, by Hölder's inequality with $\frac{1}{p} + \frac{1}{p'} = 1$, setting $\alpha := \lceil 1 + \tau\rceil \geqslant 1 + \tau$, we have

$$|\langle f; h\rangle| \leqslant \|(1 + |\cdot|^2)^{-\tau} f\|_{L_p} \|(1 + |\cdot|^2)^{\tau} h\|_{L_{p'}(\mathbb{R})} \leqslant C\|(1 + |\cdot|^2)^{\alpha} h\|_{\mathscr{C}(\mathbb{R})},$$

where

$$C := \|(1 + |\cdot|^2)^{-\tau} f\|_{L_p} \|(1 + |\cdot|^2)^{-(\alpha-\tau)}\|_{L_{p'}(\mathbb{R})} < \infty$$

by $\alpha - \tau \geqslant 1$. Since

$$\|(1 + |\cdot|^2)^{\alpha} h\|_{\mathscr{C}(\mathbb{R})} \leqslant 2^{2\alpha} \sum_{m=0}^{2\alpha} \|(\cdot)^m h\|_{\mathscr{C}(\mathbb{R})},$$

by Theorem A.6.1, $\langle f; \cdot\rangle$ is a continuous linear functional on $\mathscr{S}(\mathbb{R})$, that is, $f \in \mathscr{S}'(\mathbb{R})$. □

We now present some examples of distributions and tempered distributions.

(1) If $f \in L_1^{loc}(\mathbb{R})$ (or equivalently, $\int_{-n}^{n} |f| < \infty$ for all $n \in \mathbb{N}$), then $f \in \mathscr{D}'(\mathbb{R})$ through the identification $\langle f; \varphi\rangle := \int_{\mathbb{R}} f(x)\varphi(x)\, dx$ for $\varphi \in \mathscr{D}(\mathbb{R})$.
(2) The Dirac distribution $\boldsymbol{\delta} \in \mathscr{S}'(\mathbb{R})$: $\langle \boldsymbol{\delta}; \varphi\rangle := \varphi(0), \varphi \in \mathscr{S}(\mathbb{R})$. If $\phi \in L_1(\mathbb{R})$ with $\int_{\mathbb{R}} \phi(x)dx = 1$, then $\varphi_\lambda \to \boldsymbol{\delta}$ in $\mathscr{S}'(\mathbb{R})$ as $\lambda \to \infty$, where $\varphi_\lambda := \lambda\varphi(\lambda\cdot)$.
(3) Let $u = \{u(k)\}_{k\in\mathbb{Z}} \in \ell(\mathbb{Z})$. Then $u \in \mathscr{D}'(\mathbb{R})$ in the sense $\langle u; \varphi\rangle = \sum_{k\in\mathbb{Z}} u(k)\varphi(k), \varphi \in \mathscr{D}(\mathbb{R})$, i.e., $u = \sum_{k\in\mathbb{Z}} u(k)\boldsymbol{\delta}(\cdot - k) := \lim_{n\to\infty}\sum_{|k|<n} u(k)\,\boldsymbol{\delta}(\cdot - k)$ in $\mathscr{D}'(\mathbb{R})$.
(4) If $u \in \ell(\mathbb{Z})$ such that $u(\cdot)(1 + |\cdot|^2)^{-\tau} \in l_\infty(\mathbb{Z})$ for some $\tau \geqslant 0$, then $u \in \mathscr{S}'(\mathbb{R})$.

Let $f \in \mathscr{D}'(\mathbb{R})$. The support $\text{supp}(f)$ of f is the smallest closed subset of $\mathbb{R}$ such that $\langle f;\varphi\rangle = 0$ for all $\varphi \in \mathscr{D}(\mathbb{R})$ satisfying $\text{supp}(\varphi) \subseteq \mathbb{R}\setminus \text{supp}(f)$. The complex conjugate of f is defined to be $\langle \overline{f};\varphi\rangle := \overline{\langle f;\overline{\varphi}\rangle}$ for all $\varphi \in \mathscr{D}(\mathbb{R})$. If $f \in \mathscr{D}'(\mathbb{R})$, then its *distributional derivatives* $D^\beta f$, $\beta \in \mathbb{N}_0$ are also distributions, where

$$\langle D^\beta f;\varphi\rangle := (-1)^\beta \langle f;\varphi^{(\beta)}\rangle, \qquad \forall\ \varphi \in \mathscr{D}(\mathbb{R}).$$

For $\varphi_n \to \varphi$ in $\mathscr{S}(\mathbb{R})$ as $n \to \infty$, we have $\varphi_n^{(\beta)} \to \varphi^{(\beta)}$ in $\mathscr{S}(\mathbb{R})$. For $f \in \mathscr{S}'(\mathbb{R})$,

$$\langle D^\beta f;\varphi_n\rangle = (-1)^\beta \langle f;\varphi_n^{(\beta)}\rangle \to (-1)^\beta \langle f;\varphi^{(\beta)}\rangle = \langle D^\beta f;\varphi\rangle, \qquad n \to \infty.$$

Therefore, $D^\beta f \in \mathscr{S}'(\mathbb{R})$. Hence, any distributional derivative of a tempered distribution is also a tempered distribution. Moreover, for $f \in \mathscr{C}^1(\mathbb{R})$ (or f is absolutely continuous), its functional derivative $f' = Df$ in the sense of distributions.

If $f \in \mathscr{S}'(\mathbb{R})$, then its Fourier transform $\widehat{f}$ is defined to be

$$\langle \widehat{f};\varphi\rangle := \langle f;\widehat{\varphi}\rangle, \qquad \varphi \in \mathscr{S}(\mathbb{R}).$$

Theorem A.6.3

(i) The Fourier transform $\mathscr{F} : \mathscr{S}(\mathbb{R}) \to \mathscr{S}(\mathbb{R}), f \mapsto \widehat{f}$ is a homeomorphism of $\mathscr{S}(\mathbb{R})$ onto itself and $\mathscr{F}^{-1}$ is its continuous inverse.

(ii) The Fourier transform $\mathscr{F} : \mathscr{S}'(\mathbb{R}) \to \mathscr{S}'(\mathbb{R}), f \mapsto \widehat{f}$ is a homeomorphism of $\mathscr{S}'(\mathbb{R})$ onto itself and $\mathscr{F}^{-1}$ is its continuous inverse.

Proof

(i) Since $f \in \mathscr{S}(\mathbb{R})$, we have $x^\beta f(x) \in L_1(\mathbb{R})$ for all $\beta \in \mathbb{N}_0$ and therefore, by Proposition A.5.1, $\widehat{f} \in \mathscr{C}^\infty(\mathbb{R})$. Note that

$$\xi^\alpha \widehat{f}^{(\beta)}(\xi) = (-i)^\alpha \int_{\mathbb{R}} [(-ix)^\beta f(x)]^{(\alpha)} e^{-i\xi x} dx,$$

we conclude that

$$\|\xi^\alpha \widehat{f}^{(\beta)}(\xi)\|_{\mathscr{C}(\mathbb{R})} \leqslant \|[(-ix)^\beta f(x)]^{(\alpha)}\|_{L_1(\mathbb{R})} < \infty.$$

So, $\widehat{f} \in \mathscr{S}(\mathbb{R})$.

Suppose that $f_n \to f$ in $\mathscr{S}(\mathbb{R})$. Then $\lim_{n\to\infty} \rho_{\alpha,\beta}(f_n - f) = 0$. From

$$\xi^\alpha (\widehat{f_n}(\xi) - \widehat{f}(\xi))^{(\beta)} = (-i)^\alpha \int_{\mathbb{R}} (1+x^2)^{-1}(1+x^2)[(-ix)^\beta (f_n(x) - f(x))]^{(\alpha)} e^{-i\xi x} dx,$$

we have

$$\rho_{\alpha,\beta}(\widehat{f_n} - \widehat{f}) \leqslant \|(1 + |\cdot|^2)^{-1}\|_{L_1(\mathbb{R})} \|(1 + |\cdot|^2)[(-i\cdot)^\beta (f_n - f)]^{(\alpha)}\|_{\mathscr{C}(\mathbb{R})} \to 0$$

as $n \to \infty$. Hence, $\widehat{f_n} \to \widehat{f}$ in $\mathscr{S}(\mathbb{R})$. Thus, $\mathscr{F}$ is continuous. Similarly, we can prove that $\mathscr{F}^{-1}$ is continuous. Consequently, $\mathscr{F}$ must be a homeomorphism of $\mathscr{S}(\mathbb{R})$ onto itself.

(ii) We now prove that $\widehat{f} \in \mathscr{S}'(\mathbb{R})$. In fact, for $\varphi_n \to \varphi$ in $\mathscr{S}(\mathbb{R})$ as $n \to \infty$,by item (i), have $\widehat{\varphi_n} \to \widehat{\varphi}$ in $\mathscr{S}(\mathbb{R})$ as $n \to \infty$. Therefore, $\langle \widehat{f}; \varphi_n\rangle = \langle f; \widehat{\varphi_n}\rangle \to \langle f; \widehat{\varphi}\rangle = \langle \widehat{f}; \varphi\rangle$ as $n \to \infty$. Hence, $\widehat{f} \in \mathscr{S}'(\mathbb{R})$. Now item (ii) follows directly from item (i). □

We now present some examples of the Fourier transform of some tempered distributions as follows:

(1) For $f \in L_1(\mathbb{R})$, its Fourier transform $\widehat{f}^F$ defined by $\widehat{f}^F(\xi) := \int_{\mathbb{R}} f(x)e^{-i\xi x}dx$ agrees with $\widehat{f}$ in the tempered distribution sense, since by Lemma A.5.2,

$$\langle \widehat{f}^F; \varphi\rangle = \int_{\mathbb{R}} \widehat{f}^F(x)\varphi(x)dx = \int_{\mathbb{R}} f(x)\widehat{\varphi}(x)dx = \langle \widehat{f}; \varphi\rangle.$$

(2) For a polynomial $\mathsf{p} \in \mathbb{P}$, we have $\widehat{\mathsf{p}} = 2\pi\boldsymbol{\delta}\mathsf{p}(-i\frac{d}{dx})$ because

$$\begin{aligned}\langle \widehat{\mathsf{p}}; \varphi\rangle = \langle \mathsf{p}; \widehat{\varphi}\rangle &= \int_{\mathbb{R}} \mathsf{p}(\xi)\widehat{\varphi}(\xi)\, d\xi \\ &= \int_{\mathbb{R}} \widehat{[\mathsf{p}(-i\tfrac{d}{dx})\varphi(x)]}(\xi)d\xi = 2\pi[\mathsf{p}(-i\tfrac{d}{dx})\varphi(x)](0).\end{aligned}$$

(3) $\widehat{\boldsymbol{\delta}(\cdot - c)}(\xi) = e^{-ic\xi}$ by

$$\langle \widehat{\boldsymbol{\delta}(\cdot - c)}; \varphi\rangle = \langle \boldsymbol{\delta}(\cdot - c); \widehat{\varphi}\rangle = \widehat{\varphi}(c) = \int_{\mathbb{R}} \varphi(\xi)e^{-ic\xi}d\xi.$$

(4) Let $f \in L_1(\mathbb{T})$. Then $f \in \mathscr{S}'(\mathbb{R})$ and $\widehat{f} = 2\pi\{\widehat{f}(k)\}_{k\in\mathbb{Z}} \in l(\mathbb{Z})$ with $\widehat{f}(k) := \frac{1}{2\pi}\int_{\mathbb{T}} f(t)e^{-ikt}dt$. In fact,

$$\langle \widehat{f}; \varphi\rangle = \langle f; \widehat{\varphi}\rangle = \int_{\mathbb{R}} f(x)\widehat{\varphi}(x)dx = \int_{-\pi}^{\pi} f(x)\sum_{k\in\mathbb{Z}} \widehat{\varphi}(x + 2\pi k)dx.$$

By the Poisson summation formula in Theorem A.5.7 and $\widehat{\widehat{\varphi}}(x) = 2\pi\varphi(-x)$, we have

$$\sum_{k\in\mathbb{Z}} \widehat{\varphi}(x + 2\pi k) = \frac{1}{2\pi}\sum_{k\in\mathbb{Z}} \widehat{\widehat{\varphi}}(k)e^{ikx} = \sum_{k\in\mathbb{Z}} \varphi(k)e^{-ikx}.$$

Hence,

$$\langle \widehat{f}; \varphi\rangle = \int_{-\pi}^{\pi} f(x)\sum_{k\in\mathbb{Z}} \varphi(k)e^{-ikx}dx = 2\pi\sum_{k\in\mathbb{Z}} \widehat{f}(k)\varphi(k).$$

(5) Let $u \in \ell(\mathbb{Z})$ such that $u \in \mathscr{S}'(\mathbb{R})$. Then $\widehat{u} = \sum_{k\in\mathbb{Z}} u(k)e^{-ik\xi}$. Indeed,

$$\langle \widehat{u}; \varphi\rangle = \langle u; \widehat{\varphi}\rangle = \sum_{k\in\mathbb{Z}} u(k)\widehat{\varphi}(k)$$

$$= \sum_{k\in\mathbb{Z}} u(k) \int_{\mathbb{R}} \varphi(x)e^{-ikx}dx = \int_{\mathbb{R}} \varphi(x) \sum_{k\in\mathbb{Z}} u(k)e^{-ikx}dx.$$

For $f \in \mathscr{D}'(\mathbb{R})$ and $\psi \in \mathscr{D}(\mathbb{R})$, the convolution $f*\psi$ is defined to be $\langle f*\psi;\varphi\rangle := \langle f;\psi * \varphi\rangle$ for $\phi \in \mathscr{D}(\mathbb{R})$. For $f, g \in \mathscr{D}'(\mathbb{R})$, we say that $f = g$ in the sense of distributions if $\langle f;\varphi\rangle = \langle g;\varphi\rangle$ for all $\varphi \in \mathscr{D}(\mathbb{R})$.

For a compactly supported distribution f, its Fourier transform $\widehat{f}$ can be identified with the analytic function:

$$\widehat{f}(z) := \langle f; e^{-iz\cdot}\rangle = \langle f; \eta(\cdot)e^{-iz\cdot}\rangle, \qquad z \in \mathbb{C},$$

where $\eta \in \mathscr{D}(\mathbb{R})$ takes value 1 in a neighborhood of supp(f). By the continuity of f, there must exist $\tau \geqslant 0$ such that $(1 + |\cdot|^2)^{-\tau}\widehat{f} \in L_\infty(\mathbb{R})$. Moreover,

Theorem A.6.4 (Paley-Wiener's Theorem) *Let F be an entire function (i.e., F is analytic on the whole complex plane $\mathbb{C}$). Then $F|_{\mathbb{R}}$ is the Fourier transform of a tempered distribution supported inside $[-B, B]$ if and only if*

$$|F(z)| \leqslant C(1 + |z|)^N e^{B|Im(z)|} \qquad \forall\, z \in \mathbb{C} \text{ for some constants } C, N.$$

Theorem A.6.5 (The Poisson Summation Formula for Distributions) *Let f be a compactly supported distribution on $\mathbb{R}$. For $\zeta \in \mathbb{C}$ and $\lambda > 0$,*

$$\lambda \sum_{k\in\mathbb{Z}} f(\cdot - \lambda k)e^{-i\zeta(\cdot-\lambda k)} = \sum_{k\in\mathbb{Z}} \widehat{f}(\zeta + 2\pi k/\lambda)e^{i(\cdot)2\pi k/\lambda}$$

in the sense of distributions.

Proof Define

$$g(x) := \lambda\langle fe^{-i\zeta\cdot}; \varphi(\cdot + \lambda x)\rangle = \lambda[(fe^{-i\zeta\cdot}) * \varphi(-\cdot)](-\lambda x)$$

for $\varphi \in \mathscr{D}(\mathbb{R})$. Then $g \in \mathscr{D}(\mathbb{R})$, since $\widehat{g}(\xi) = \widehat{f}(\zeta + \xi/\lambda)\widehat{\varphi}(-\xi/\lambda)$ has rapid decay. By Theorem A.5.7, we have $\sum_{k\in\mathbb{Z}} g(k) = \sum_{k\in\mathbb{Z}} \widehat{g}(2\pi k)$. Note that

$$g(k) = \langle \lambda f(\cdot - \lambda k)e^{-i\zeta(\cdot-\lambda k)}; \varphi\rangle$$

and

$$\widehat{g}(2\pi k) = \widehat{f}(\zeta + 2\pi k/\lambda)\widehat{\varphi}(-2\pi k/\lambda) = \langle \widehat{f}(\zeta + 2\pi k)e^{i(\cdot)2\pi k/\lambda}; \varphi\rangle.$$

We now conclude that the identity must hold in the sense of distributions. □

Notes and Acknowledgments

The history and early developments of wavelet theory are discussed in detail in Daubechies' book [70]. Basically, the idea of wavelets originated in various forms from many areas such as the Haar basis in [120], the atomic decomposition in harmonic analysis (e.g., see [105, 179]), subband coding in engineering (e.g., see [240, 288]), and etc. The continuous wavelet transform $\mathcal{W}_\psi f$ in (4.3.26) was initially discovered by Morlet from geophysics (e.g., see [116]). A homogeneous affine system $\mathsf{AS}(\psi)$ is a direct consequence of the discretization of the continuous wavelet transform. The concept of multiresolution analysis was introduced by Mallat and Meyer (e.g., see [239, 242]). The bandlimited Meyer orthogonal wavelet in Example 4.6.2 was constructed by Meyer [242]. A family of exponentially decaying spline orthogonal wavelets was constructed in Lemarié [227] and Battle [6]. A family of compactly supported real-valued orthogonal wavelets $\psi^{a_m^D}, m \in \mathbb{N}$ was discovered in Daubechies [68], where the filters a_m^D are defined in (2.2.4). The Daubechies orthogonal low-pass filters a_m^D are closely linked to the interpolatory filters a_{2m}^I (see (2.1.6)) in Deslauriers-Dubuc [82, 90] in the study of subdivision schemes through the relation $|\widehat{a_m^D}(\xi)|^2 = \widehat{a_{2m}^I}(\xi)$. A general framework of compactly supported biorthogonal wavelets was developed in Cohen-Daubechies-Feauveau [52]. Compactly supported semi-orthogonal spline wavelets were constructed in Chui-Wang [47]. The undecimated (or stationary) wavelet transform (also called algorithme à trous) was introduced in Holschneider-Kronland-Morlet-Tchamitchian [181].

The notion of frames was introduced in Duffin-Schaeffer [92] in the setting of nonharmonic Fourier series. Homogeneous bandlimited tight framelets were constructed in Daubechies-Grossmann-Meyer [72]. Characterization of homogeneous tight framelets was obtained independently in Han [121, 123], Ron-Shen [264], and Frazier-Garrigós-Wang-Weiss [104]; further improvements were reported in Chui-Shi-Stöckler [46] and Bownik [13]. Lawton [221] observed that a finitely supported orthogonal wavelet filter bank $\{a; b\}$ with $\widehat{a}(0) = 1$ leads to a compactly supported tight framelet in $L_2(\mathbb{R})$. A general method called unitary extension

B. Han, *Framelets and Wavelets*, Applied and Numerical Harmonic Analysis,
https://doi.org/10.1007/978-3-319-68530-4

principle was introduced in Ron-Shen [264] to construct tight framelets in $L_2(\mathbb{R})$. To increase vanishing moment orders of high-pass filters, the oblique extension principle was introduced independently in Daubechies-Han-Ron-Shen [74] and Chui-He-Stöckler [39] (also see Daubechies-Han [73] and Han-Mo [157] for dual framelets).

We now provide some notes, remarks and acknowledgments of results for each chapter of the book. Any omission and incorrect comments below are the author's fault, largely due to the author's limited knowledge and personally biased viewpoints on the vast multidisciplinary area of wavelet theory and its applications. Examples in the book are computed by mathematics software `maple` and `matlab`. Figures are produced by `matlab`, `pstricks` (latex package) and `C` language.

Chapter 1

Chapter 1 systematically studies discrete framelet/wavelet transforms and their properties and implementation. The discrete approach to framelets/wavelets in Sects. 1.1–1.3 of Chap. 1 is largely the one-dimensional special version from Han [144] (published by EDP Sciences) and originated in the study of the balanced property of multiwavelets in Han [136] (published by Elsevier) and Han [139] (published by the American Mathematical Society). Most results (in particular, Lemma 1.4.5 and Theorem 1.4.7) in Sect. 1.4 first appeared in [147] (published by the American Mathematical Society). Figures 1.3 and 1.4 are from Han-Zhao [167] (published by the Society for Industrial and Applied Mathematics).

The sparsity of discrete framelet transforms was first implicitly investigated in Han [132] and later developed in Han [136, 139], on which the results in Sect. 1.2 are largely derived as special cases. The notion of linear-phase moments was first explicitly introduced in Han [140] for studying compactly supported symmetric complex orthogonal wavelets with linear-phase moments (see Sect. 2.5). The importance of linear-phase moments on symmetric orthogonal wavelets and symmetric tight framelets was discussed in Han [137, 140, 141, 143]. The usefulness of complex symmetry was first noticed in Han [143] for studying symmetric tight framelet filter banks. The notions of discrete affine systems and stability of multilevel discrete framelet transforms were introduced in Han [144]. The oblique extension principle was first introduced independently in Daubechies-Han-Ron-Shen [74] and Chui-He-Stöckler [39]. The discrete framelet transforms using OEP-based framelet filter banks were first discussed in Daubechies-Han-Ron-Shen [74]. OEP-based dual framelet filter banks were first systematically investigated in Daubechies-Han [73] for the scalar case and Han-Mo [157] for the vector case. Further developments on the oblique extension principle were discussed in Han [136, 138, 142, 145]. Most exercise problems in Chap. 1 are based on Han [136, 139, 144].

Chapter 2

Chapter 2 develops systematic algorithms for constructing almost all known (dyadic) wavelet filter banks with or without symmetry in the literature. The results in Sects. 2.4 and 2.5 are largely based on Han [140] (with permission of Springer), but the treatment here is more systematic and simplified. The CBC (coset-by-coset) algorithm for constructing biorthogonal wavelet filter banks was originated from Han [124, 126]. Algorithm 2.3.3 is similar to [141, Algorithm 2] and parts of Example 2.3.3 appeared in [141, Example 17] (published by Elsevier).

The interpolatory filters a^I_{2m} in (2.1.6) were first known in Deslauriers-Dubuc [82] in the setting of subdivision schemes. The real-valued filters $a_{m,n}$ with linear-phase moments in (2.1.11) and (2.1.12) (as well as their related filters) are called pseudo-splines in Daubechies-Han-Ron-Shen [74] (also see Selesnick [270], Dong-Shen [85, 86]). Our approach on constructing interpolatory filters and filters with linear-phase moments in Sect. 2.1 follows the approach in Han-Jia [150, Theorem 2.1], which is much more general and flexible. The convolution method for constructing interpolatory filters in Sect. 2.1.2 is a special case of Han [121, Proposition 3.4] and [123, Proposition 3.7]. The finitely supported Daubechies orthogonal filters in Sect. 2.2 was first discovered in Daubechies [68]. Further developments on real-valued orthogonal wavelets with additional properties were discussed in Daubechies [71] and Cohen-Daubechies [50]. Real-valued orthogonal wavelets with linear-phase moments are also called coiflets in Daubechies [71]. All known constructions of coiflets in the literature so far are heuristic by solving nonlinear equations; for more examples, see (in alphabetic order) Daubechies [71], Han [141], Monzón-Beylkin-Hereman [250], and references therein. It is well known in Daubechies [68, 70] (see Proposition 2.2.3) that there are no compactly supported symmetric real-valued orthogonal wavelets, except the Haar orthogonal wavelet. As noticed by two examples (i.e., $\psi^{a^S_3}$ and $\psi^{a^S_5}$) in Lawton [223], symmetry can be achieved by complex orthogonal wavelets. However, the existence of compactly supported symmetric complex orthogonal wavelets $\psi^{a^S_m}$ in (2.4.7) for all odd integers m was first established in Han [140]. The family of compactly supported symmetric complex orthogonal wavelets $\psi^{a^H_m}$ in (2.5.21) with arbitrarily increasing linear-phase moments $m \in \mathbb{N}$ was first introduced in Han [140]. See Han [137] for further investigation on complex-valued symmetric M-orthogonal wavelets with arbitrarily increasing orders of linear-phase moments. The CBC (coset by coset) algorithm in Sect. 2.6 was initially proposed in Han [124, 126] to construct scalar multivariate biorthogonal wavelets with arbitrarily high orders of vanishing moments. The CBC algorithm was further developed in Chen-Han-Riemenschneider [28] and Han [127]. For example, Han [127, Theorem 3.4] (also see Sect. 6.5) proves that if a finitely supported low-pass matrix-valued filter has one finitely supported dual filter, then it always has finitely supported dual filters having arbitrarily high orders of sum rules. The chain structure for biorthogonal wavelets in Sect. 2.7 was first introduced in Chui-Han-Zhuang [37] for a dilation factor greater than 2. The lifting scheme for scalar biorthogonal wavelets in Sect. 2.7 appeared in Sweldens [286]. For more reading on scalar biorthogonal wavelets, see (in alphabetic order) Chen-Han-Riemenschneider [28], Chui [35] (book), Chui-Han-Zhuang [37], Chui-Villiers [36] (book), Chui-Wang [47], Cohen-Daubechies [49, 51], Cohen-Daubechies-Feauveau [52], Daubechies [70], Dong-Shen [85], Han [124, 126–128], Han-Jia [151], Long-Chen [235], Mallat [240] (book), Sweldens [286], and many references therein. Many exercise problems in Chap. 2 are built on Han [137, 140, 141]. Exercises 2.42–2.45 are from Chui-Han-Zhuang [37].

Chapter 3

Chapter 3 develops general algorithms for constructing all possible OEP-based dual framelet filter banks $(\{\tilde{a};\tilde{b}_1,\tilde{b}_2\},\{a;b_1,b_2\})_\Theta$ and all possible OEP-based tight framelet filter banks $\{a;b_1,b_2\}_\Theta$ or $\{a;b_1,b_2,b_3\}_\Theta$ with or without symmetry such that the high-pass filters have short filter supports. Most results in Chap. 3 are largely built on Han [147] (published by the American Mathematical Society) and Han [143, 146] (with permission from Elsevier). In particular, Algorithms 3.2.1, 3.2.4, 3.3.6 and 3.4.1 for constructing dual or tight framelet filter banks with two high-pass filters with or without symmetry are largely from Han [147] (published by the American Mathematical Society), where Lemma 3.3.2 and Theorem 3.3.5 are also proved. Theorems 3.1.5–3.1.8, 3.3.7 and 3.6.1, Algorithm 3.6.2, Figs. 3.7 and 3.8, Examples 3.3.4 and 3.3.5 (with some modifications and improvements) first appeared in Han [143] (with permission from Elsevier). Theorems 3.1.6, 3.3.7 and 3.6.1 for the special case of real-valued filters first appeared in Han-Mo [158] (published by the Society for Industrial and Applied Mathematics). Proposition 3.5.1, Theorems 3.5.2 and 3.5.4, Algorithm 3.5.3, and parts of Fig. 3.17 and Example 3.5.3 (with some modifications) are from Han [146] (with permission from Elsevier). Algorithm 3.4.3 for 2×2 matrix-valued Fejér-Riesz Lemma uses some idea with improvements from Chui-He-Stöckler [39] (published by Elsevier).

As mentioned before, the oblique extension principle (OEP) was first introduced independently in Daubechies-Han-Ron-Shen [74] and Chui-He-Stöckler [39]. Theorem 3.3.3 and Exercises 3.21–3.23 are largely from Han-Mo [159] (published by Elsevier). Example 3.3.1 with $\Theta = \boldsymbol{\delta}$ is known in Ron-Shen [264] (published by Elsevier). Example 3.3.1 with $\Theta = \{-\frac{1}{6},\frac{4}{3},-\frac{1}{6}\}_{[-1,1]}$ essentially appeared in Chui-He-Stöckler [39] (published by Elsevier). Example 3.3.3 essentially appeared in Petukhov [255] (published by Springer). Example 3.3.2 with $\Theta = \boldsymbol{\delta}$ and the first construction in Example 3.5.3 were known in Chui-He [38] (published by Elsevier). Exercises 3.13 and 3.14 are essentially from Han [147] (published by the American Mathematical Society). Exercises 3.34–3.40 are essentially based on Han [143, 146] (published by Elsevier). For more examples and study of scalar tight framelet filter banks with or with symmetry, see (in alphabetic order) Benedetto-Li [9], Charina-Stöckler [26], Chui-He [38], Chui-He-Stöckler [39,

40], Chui-He-Stöckler-Sun [41], Daubechies-Han-Ron-Shen [74], Dong-Shen [86], Han [123, 130, 143–146], Han-Jiang-Shen-Zhuang [154], Han-Mo [158, 159], Han-Mo-Zhao [161], Han-Shen [164], Han-Zhao [167], Han-Zhao-Zhuang [168], Jiang [210], Lawton [221], Mo [248], Mo-Li [249], Petukhov [254, 255], Ron-Shen [264], Selesnick [270], Selesnick-Abdelnour [271], Shen-Li-Mo [275], and many other references therein. For scalar dual framelet filter banks with or without symmetry, see (in alphabetic order) Chui-He-Stöckler [39], Daubechies-Han [73], Daubechies-Han-Ron-Shen [74], Ehler [99], Ehler-Han [100], Han [121, 123, 136, 147], Han-Mo [157], Ron-Shen [263], and other references therein. Balanced properties of dual multiframelets have been first investigated in Han [136] and examples of dual multiframelets with balanced properties were constructed in Han [136].

Chapter 4

Chapter 4 investigates affine systems and dual framelets in the function setting on $\mathbb{R}$. The notion of frequency-based dual framelets and nonhomogeneous affine systems provides a unified framework by naturally linking many aspects of wavelet theory together. For example, frequency-based dual framelets naturally link discrete filters with nonhomogeneous affine systems on $\mathbb{R}$ without requiring the generating functions from $L_2(\mathbb{R})$, while multiwavelets/multiframelets and refinable structures are natural consequences of nonhomogeneous affine systems. Most results in Sects. 4.1, 4.3, and 4.8 of Chap. 4 are largely based or derived, often as special cases and with modifications/enhancements, from Han [138, 142] (with permission from Elsevier), where the notion of nonhomogeneous affine systems and frequency-based dual framelets were first introduced. Section 4.6 on framelets and wavelets in Sobolev spaces is a special case of Han-Shen [165] (with permission of Springer) and further developed in Han [138] (published by Elsevier).

The approximation property of dual framelets in Sect. 4.7 extends Daubechies-Han-Ron-Shen [74] and Jetter-Zhou [186]. Theorems 4.3.4 and 4.3.6 were known in Ron-Shen [263, 264] under some extra conditions, which were removed in Chui-Shi-Stöckler [46]. The characterization of homogeneous dual framelets in Theorem 4.3.11 was first appeared in Han [121, 123]. Characterization of homogeneous tight framelets in Corollary 4.3.12 was obtained independently in Han [121, 123], Ron-Shen [264], and Frazier-Garrigós-Wang-Weiss [104]; further improvements were reported in Chui-Shi-Stöckler [46] and Bownik [13]. The characterization of nonhomogeneous tight framelets or dual framelets in this book was initialized in Han [138, 142]. Lemma 4.3.10 first appeared in Chui-Shi [45]. Exercise 4.23 is from Christensen [34] (book). Exercise 4.58 appeared in Chui-Shi [45] and Daubechies [69]. Exercises 4.28 and 4.29 are largely from de Boor-DeVore-Ron [81] and Han [121]. Exercises 4.30, 4.32 and 4.33 are from Han [121]. Continuous wavelet transform in Sect. 4.3.4 was initially discovered by Morlet in geophysics (e.g., see [116]). Section 4.9 on frequency-based periodic framelets and wavelets is a natural extension of our frequency-based approach for framelets/wavelets on the real line $\mathbb{R}$. Section 4.2 on frames and bases in Hilbert spaces is classical. For more study on frames and bases in Hilbert spaces, see (in alphabetic order) Casazza-Kutyniok-Philipp [22], Christensen [33, 34] (books), Han-Kornelson-Larson [170] (book), and Heil [175]. Section 4.4 on shift-invariant subspaces and Sect. 4.5.3 on multiresolution analysis of $L_2(\mathbb{R})$ are well known in approximation theory. For further study and related topics on shift-invariant spaces, see (in alphabetic order), Aldroubi-Sun-Tang [2], Bownik [14], de Boor-DeVore-Ron [80, 81], Jetter-Plonka [184], Jetter-Zhou [185, 186], Jia [189], Ron [261] (survey article), and many references therein. For further study on multiresolution analysis, see (in alphabetic order), Chui [35] (book), Daubechies [70] (book), de Boor-DeVore-Ron [81], Mallat [239] and [240] (book), Meyer [242] (book), Jia-Shen [205], and references therein. Our proofs in Sect. 4.2 (on frames and bases in Hilbert spaces) and Sect. 4.4 (on shift-invariant spaces) of this book are elementary/simple. Cohen's criteria in item (e) of Exercise 4.65 was introduced in Cohen's doctoral thesis in [56].

Chapter 5

Chapter 5 studies refinable vector functions and their properties through convergence of vector cascade algorithms. The simple proof for Theorem 5.2.1 is a special case from Han [148] (with permission from Elsevier). Theorem 5.3.4 in Sect. 5.3 on stability of integer shifts of functions in $L_p(\mathbb{R})$ is a slightly improved special case of Jia-Micchelli [200] (published by Academic Press). Lemma 5.4.1 and Theorem 5.4.2 on approximation using quasi-projection operators are modified/enhanced special cases of Jia [192] (with permission from Elsevier). The proof of Proposition 5.5.9 follows Jia [188] (published by the American Mathematical Society).

Theorem 5.2.1 in Chap. 5 on linear independence of compactly supported functions was known in Ron [260], Ben-Artzi and Ron [7], Jia-Micchelli [201]. But the simple proof for Theorem 5.2.1 is from Han [148]. Theorems 5.2.3 and 5.2.4 were known in Lemarié [228] and Jia [189], but we presented a different proof here. Corollary 5.3.8 on stability or linear independence of scalar refinable functions was known in various forms, e.g., see Cohen-Daubechies [49], Cohen-Sun [57], and Jia-Wang [206]. Existence of distributional solutions to (vector) refinement equations in Sect. 5.1 was addressed in many papers, e.g., see Heil-Colella [176], Cavaretta-Dahmen-Micchelli [23], Jia-Jiang-Shen [197], Zhou [298], etc. The Strang-Fix condition in (5.5.8) appeared in [281]. Quasi-projection/interpolation, approximation orders and accuracy order in Sects. 5.4 and 5.5 were addressed in many papers, e.g. see (in alphabetic order) Cabrelli-Heil-Molter [18], de Boor-DeVore-Ron [80], Han [127, 132], Jetter-Plonka [184], Jetter-Zhou [185, 186], Jia [188, 190, 192], Jia-Jiang [194], Jia-Riemenschnedier-Zhou [202], Ron [261], and etc. ∞-norm joint spectral radius was introduced in Rota-Strang [266] and was further developed for wavelet theory in Daubechies-Lagarias [76–78], Jia [187], and Wang [293]. To the author's best knowledge, joint spectral radius is so far the only known tool (its main role in this book) to prove Proposition 5.6.9 for $p \neq 2$. Proposition 5.6.9 for $p = 2$ can be proved using a transition operator acting on a finite-dimensional space (see Sect. 5.8.3 for details). Section 5.7 on p-norm joint spectral radius is largely from Jia [187], Han-Jia [149], and Han [131]. Proposition 5.7.2 was known in Jia-Riemenschneider-Zhou [203]. The proof of the inequality (5.8.27) used an idea from Dong-Shen [86], where (5.8.29) was obtained. The technique for estimating (5.8.29) with $m = n$ was first presented in Daubechies [68, 70]. Convergence of vector cascade algorithms (or vector subdivision schemes) and smoothness of refinable vector functions have been studied in many papers by many researchers, to only mention a few here, see [23, 29, 51, 53, 55, 78, 96, 126, 132, 149, 150, 196, 198, 203, 204, 209, 213, 245, 246, 258, 265, 276] and references therein. Sections 5.6 and 5.9 on convergence of vector cascade algorithms and their stability under perturbation improve Han [132]. Though Sect. 5.6 appears to be technical, to the author's best knowledge, the approach in Sect. 5.6 is probably the simplest in the literature, largely thanking to the notion of the normal form of matrix-valued filters. The normal form of matrix-valued filters greatly simplifies our study of vector cascade algorithms and refinable vector functions by enabling us to employ many techniques on scalar cascade algorithms and scalar refinable functions. The normal form of matrix-valued filters was initially introduced in Han-Mo [157] and further developed in Han [132, 137, 139]. Theorem 5.8.4 for computing sm(a) with a matrix-valued filter a was known in Jia-Jiang [195]. But the simple proof to Theorem 5.8.4 improves over Han [132] by using the normal form of a matrix-valued filter. Scalar cascade algorithms and refinable functions under perturbation were addressed in Daubechies-Huang [75]. The first rigorous sharp estimate was established in Han [124] and further studied in Chen-Plonka [30] and Han [132, Section 6]. Exercise 5.9 is from Zhou [298]. Exercise 5.41 is from Han [136].

Chapter 6

Chapter 6 studies special refinable vector functions and their relations to affine systems derived from refinable functions. Results in Sect. 6.1.3 on refinable functions with analytic expression are an improved version from Han-Mo [160] (published by the American Institute of Mathematical Sciences). Section 6.2 on refinable Hermite interpolants and Hermite interpolatory filters is special cases from Han [127, 132] (published by Elsevier). Most results in Sect. 6.6 on framelets and wavelets with filters of Hölder class or exponential decay is from Han [135] (published by SIAM—the Society for Industrial and Applied Mathematics).

B-splines are widely known in approximation theory. Study of scalar refinable splines appeared in Lawton-Lee-Shen [224] and Dai-Sun-Zhang [67]. Theorem 6.2.3 on characterization of refinable Hermite interpolants is largely derived from Han [132, Corollary 5.2] and the family of Hermite interpolatory filters $a_{2rm}^{H_r}$

in Theorem 6.2.6 was first given in [127, Theorem 4.2]. Examples of Hermite interpolatory filters were constructed in Merrien [241]. For study and construction of Hermite subdivision schemes, see (in alphabetic order) Dubuc-Merrien [91], Dyn-Levin [95], Han [127, 132], Han-Mo [160], Han-Yu-Piper [166], Merrien [241], Zhou [300], and many references therein. The asymptotic smoothness estimate in (6.4.10) was first established in Daubechies [68, 70]. A special case of Theorem 6.7.2 with $\phi \in L_2(\mathbb{R})$ was known in Lemarié [228] and Meyer [243]. Theorem 6.7.1 generalizes Lemarié [229]. Proposition 6.6.8 appeared in Han-Kwon-Park [155] (published by Elsevier) and Example 6.6.1 was known in Han-Shen [163] (published by SIAM). The CBC algorithm in Sect. 6.5 is from Han [127] for constructing biorthogonal multiwavelets with arbitrarily high vanishing moments. Examples of orthogonal multiwavelets were constructed in many papers, e.g., see (in alphabetic order) Alpert [3], Cabrelli-Heil-Molter [19], Donovan-Geronimo-Hardin [88], Donovan-Geronimo-Hardin-Massopust [89](where the case $t = (\sqrt{6}+\sqrt{2})/2$ and $\epsilon = 1$ in Example 6.4.2 appeared). Goodman [111], Goodman-Lee-Tang [114], Han-Jiang [153], Plonka-Strela [257], and etc. Dual multiframelets were constructed in Han-Mo [157], Han [137], and Mo [248]. Exercises 6.17–6.19 are from Han-Kwon-Zhuang [156]. Exercises 6.24–6.33 are based on Han [135].

Chapter 7

Chapter 7 mainly studies high-dimensional scalar framelets/wavelets and discusses some applications of framelets/wavelets. Section 7.1.1 is the high dimensional version of Sect. 1.1 and is largely from Han [144] (published by EDP Sciences). Theorem 7.1.9 and Example 7.1.1 appeared in Han [145] (published by EDP Sciences). Proposition 7.1.5 is a special case of Han [142, Lemma 14] (published by Elsevier). The study of convergence of cascade algorithms in Sobolev spaces in Sect. 7.2.1 is based on the scalar case in Han [132] (published by Elsevier). Section 7.2.3 on computing smoothness exponents largely appeared in Han [131] and Han-Jia [149] (published by the Society for Industrial and Applied Mathematics). Section 7.3.1 on convergence of subdivision schemes improves Han-Jia [152] (published by the American Mathematical Society), where Example 7.3.6 is given. Theorem 7.3.2 essentially appeared in Han [133] (published by Elsevier). Proposition 7.3.3 is known in Han [129] (published by Nashboro Press). The directional complex tight framelets in Sect. 7.4 was introduced in Han-Zhao [167] (published by the Society for Industrial and Applied Mathematics).

The projection method in Theorem 7.1.9 of Chap. 7 was initially introduced in Han [128] and systematically addressed in Han [145]. Theorem 7.3.1 on the relation between subdivision schemes and cascade algorithms generalizes Han-Jia [152, Theorem 2.1]. The 4-point interpolatory subdivision scheme in Example 7.3.1 with $t = 1$ was known in Dubuc [90] and Dyn-Levin-Gregory [97]. The one-dimensional ternary interpolatory subdivision scheme in Example 7.3.3 was known in Hassan-Ivrissimitzis-Dodgson-Sabin [174] and Han-Jia [152, Theorem 3.1]. Example 7.3.4

belongs to a family of bivariate interpolatory filters constructed in Han-Jia [150]. The particular case in Example 7.3.5 with $w_3 = w_5 = 0$ is known as the butterfly scheme, which was constructed in Dyn-Gregory-Levin [93]. The two-dimensional ternary interpolatory filters in Example 7.3.6 is from Han-Jia [152]. Example 7.3.7 on quincunx filters is from Han-Jia [151]. The example in (7.3.23) appeared in Han-Jiang-Shen-Zhuang [154]. $\sqrt{3}$-subdivision schemes were studied in Jiang-Oswald-Riemenschneider [211] and several other papers. Section 7.4 on directional tensor product complex tight framelets and their applications is based on Han [144], Han-Zhao [167], Han-Mo-Zhao [161, 168], Han-Zhao-Zhuang [168], and Shen-Han-Braverman [274]. Construction of orthogonal wavelets on [0, 1] was studied in Meyer [243], Cohen-Daubechies-Vial [54], Chui-Quak [44], Han-Jiang [153], and etc. Construction of scalar biorthogonal wavelets on [0, 1] was discussed in Cohen-Daubechies-Vial [54], Dahmen-Han-Jia-Kunoth [65], Hardin-Marasovich [173] and etc. Construction of scalar wavelets on [0, 1] using symmetry was briefly mentioned in Cohen-Daubechies-Vial [54] for scalar biorthogonal wavelets with symmetry. Section 7.6 on balanced approximation property of fast multiframelet transforms is a special case in Han [137, 139]. The problem about balanced approximation property of discrete multiwavelet transforms was first noticed in Lebrun-Vetterli [226] and further studied in Chui-Jiang [43] and other papers in the function setting.

Appendix A

The simple proof of Wiener's lemma in Theorem A.3.7 is from Newman [251]. The proof of Theorem A.2.11 is from Stein-Shakarchi [279] (book). For further study on Fourier analysis and distribution theory, see Stein-Weiss [280] (book), Stein-Shakarchi [279] (book), Edwards [98] (book), and Grafakos [115] (book).

For further study and background on wavelet theory and Fourier analysis, see [1]–[302] and many references therein.

References

1. A. Aldroubi, Oblique and hierarchical multiwavelet bases. Appl. Comput. Harmon. Anal. **4**(3), 231–263 (1997)
2. A. Aldroubi, Q. Sun, W.-S. Tang, p-frames and shift invariant subspaces of L^p. J. Fourier Anal. Appl. **7**(1), 1–21 (2001)
3. B.K. Alpert, A class of bases in L^2 for the sparse representation of integral operators. SIAM J. Math. Anal. **24**(1), 246–262 (1993)
4. L.W. Baggett, P.E.T. Jorgensen, K.D. Merrill, J.A. Packer, Construction of Parseval wavelets from redundant filter systems. J. Math. Phys. **46**(8), 083502, 28 (2005)
5. L.W. Baggett, H.A. Medina, K.D. Merrill, Generalized multi-resolution analyses and a construction procedure for all wavelet sets in $\mathbf{R}^n$. J. Fourier Anal. Appl. **5**(6), 563–573 (1999)
6. G. Battle, A block spin construction of ondelettes. I. Lemarié functions. Comm. Math. Phys. **110**(4), 601–615 (1987)
7. A. Ben-Artzi, A. Ron, On the integer translates of a compactly supported function: dual bases and linear projectors. SIAM J. Math. Anal. **21**(6), 1550–1562 (1990)
8. J.J. Benedetto, Frames, sampling, and seizure prediction, in *Advances in wavelets (Hong Kong, 1997)* (Springer, Singapore, 1999), pp. 1–25
9. J.J. Benedetto, S. Li, The theory of multiresolution analysis frames and applications to filter banks. Appl. Comput. Harmon. Anal. **5**(4), 389–427 (1998)
10. M.A. Berger, Y. Wang, Bounded semigroups of matrices. Linear Algebra Appl. **166**, 21–27 (1992)
11. G. Beylkin, R.R. Coifman, V.V. Rokhlin, Fast wavelet transforms and numerical algorithms. I. Comm. Pure Appl. Math. **44**(2), 141–183 (1991)
12. T. Blu, M. Unser, Approximation error for quasi-interpolators and (multi)wavelet expansions. Appl. Comput. Harmon. Anal. **6**(2), 219–251 (1999)
13. M. Bownik, A characterization of affine dual frames in $L^2(\mathbb{R}^n)$. Appl. Comput. Harmon. Anal. **8**(2), 203–221 (2000)
14. M. Bownik, The structure of shift-invariant subspaces of $L^2(\mathbb{R}^n)$. J. Funct. Anal. **177**(2), 282–309 (2000)
15. M. Bownik, Riesz wavelets and generalized multiresolution analyses. Appl. Comput. Harmon. Anal. **14**(3), 181–194 (2003)
16. O. Bratteli, P.E.T. Jorgensen, *Wavelets Through a Looking Glass*. Applied and Numerical Harmonic Analysis (Birkhäuser Boston, Boston, MA, 2002)
17. C.S. Burrus, R.A. Gopinath, H. Guo, *Introduction to Wavelets and Wavelet Transforms: A Primer* (Prentice Hall, 1997)

B. Han, *Framelets and Wavelets*, Applied and Numerical Harmonic Analysis,
https://doi.org/10.1007/978-3-319-68530-4

18. C.A. Cabrelli, C.E. Heil, U.M. Molter, Accuracy of lattice translates of several multidimensional refinable functions. J. Approx. Theory **95**(1), 5–52 (1998)
19. C.A. Cabrelli, C.E. Heil, U.M. Molter, Self-similarity and multiwavelets in higher dimensions. Mem. Am. Math. Soc. **170**(807), viii+82 (2004)
20. J.-F. Cai, R.H. Chan, L. Shen, Z. Shen, Restoration of chopped and nodded images by framelets. SIAM J. Sci. Comput. **30**(3), 1205–1227 (2008)
21. E.J. Candès, D.L. Donoho, New tight frames of curvelets and optimal representations of objects with piecewise C^2 singularities. Comm. Pure Appl. Math. **57**(2), 219–266 (2004)
22. P.G. Casazza, G. Kutyniok, F. Philipp, Introduction to finite frame theory, in *Finite Frames*, Appl. Numer. Harmon. Anal. (Birkhäuser, New York, 2013), pp. 1–53
23. A.S. Cavaretta, W. Dahmen, C.A. Micchelli, Stationary subdivision. Mem. Am. Math. Soc. **93**(453), vi+186 (1991)
24. R.H. Chan, S.D. Riemenschneider, L. Shen, Z. Shen, Tight frame: an efficient way for high-resolution image reconstruction. Appl. Comput. Harmon. Anal. **17**(1), 91–115 (2004)
25. M. Charina, C. Conti, K. Jetter, G. Zimmermann, Scalar multivariate subdivision schemes and box splines. Comput. Aided Geom. Design **28**(5), 285–306 (2011)
26. M. Charina, J. Stöckler, Tight wavelet frames via semi-definite programming. J. Approx. Theory **162**(8), 1429–1449 (2010)
27. D.-R. Chen, On linear independence of integer translates of refinable vectors. J. Math. Anal. Appl. **222**(2), 397–410 (1998)
28. D.-R. Chen, B. Han, S.D. Riemenschneider, Construction of multivariate biorthogonal wavelets with arbitrary vanishing moments. Adv. Comput. Math. **13**(2), 131–165 (2000)
29. D.-R. Chen, R.-Q. Jia, S.D. Riemenschneider, Convergence of vector subdivision schemes in Sobolev spaces. Appl. Comput. Harmon. Anal. **12**(1), 128–149 (2002)
30. D.-R. Chen, G. Plonka, Convergence of cascade algorithms in Sobolev spaces for perturbed refinement masks. J. Approx. Theory **119**(2), 133–155 (2002)
31. D.-R. Chen, X. Zheng, Stability implies convergence of cascade algorithms in Sobolev space. J. Math. Anal. Appl. **268**(1), 41–52 (2002)
32. H.-L. Chen, *Complex Harmonic Splines, Periodic Quasi-Wavelets* (Kluwer Academic, Dordrecht, 2000)
33. O. Christensen, *An Introduction to Frames and Riesz Bases*. Applied and Numerical Harmonic Analysis (Birkhäuser Boston, Boston, MA, 2003)
34. O. Christensen, *Frames and Bases: An Introductory Course*. Applied and Numerical Harmonic Analysis (Birkhäuser Boston, Boston, MA, 2008)
35. C.K. Chui, *An Introduction to Wavelets*, volume 1 of *Wavelet Analysis and Its Applications* (Academic Press, Boston, MA, 1992)
36. C.K. Chui, J. de Villiers, *Wavelet Subdivision Methods* (CRC Press, Boca Raton, FL, 2011)
37. C.K. Chui, B. Han, X. Zhuang, A dual-chain approach for bottom-up construction of wavelet filters with any integer dilation. Appl. Comput. Harmon. Anal. **33**(2), 204–225 (2012)
38. C.K. Chui, W. He, Compactly supported tight frames associated with refinable functions. Appl. Comput. Harmon. Anal. **8**(3), 293–319 (2000)
39. C.K. Chui, W. He, J. Stöckler, Compactly supported tight and sibling frames with maximum vanishing moments. Appl. Comput. Harmon. Anal. **13**(3), 224–262 (2002)
40. C.K. Chui, W. He, J. Stöckler, Nonstationary tight wavelet frames. I. Bounded intervals. Appl. Comput. Harmon. Anal. **17**(2), 141–197 (2004)
41. C.K. Chui, W. He, J. Stöckler, Q. Sun, Compactly supported tight affine frames with integer dilations and maximum vanishing moments. Adv. Comput. Math. **18**(2–4), 159–187 (2003)
42. C.K. Chui, Q. Jiang, Surface subdivision schemes generated by refinable bivariate spline function vectors. Appl. Comput. Harmon. Anal. **15**(2), 147–162 (2003)
43. C.K. Chui, Q. Jiang, Balanced multi-wavelets in $\mathbb{R}^s$. Math. Comp. **74**(251), 1323–1344 (2005)
44. C.K. Chui, E. Quak, Wavelets on a bounded interval, in *Numerical Methods in Approximation Theory, Vol. 9 (Oberwolfach, 1991)*, volume 105 of *Internat. Ser. Numer. Math.* (Birkhäuser, Basel, 1992), pp. 53–75

45. C.K. Chui, X.L. Shi, Inequalities of Littlewood-Paley type for frames and wavelets. SIAM J. Math. Anal. **24**(1), 263–277 (1993)
46. C.K. Chui, X.L. Shi, J. Stöckler, Affine frames, quasi-affine frames, and their duals. Adv. Comput. Math. **8**(1–2), 1–17 (1998)
47. C.K. Chui, J. Wang, On compactly supported spline wavelets and a duality principle. Trans. Am. Math. Soc. **330**(2), 903–915 (1992)
48. A. Cohen, *Numerical Analysis of Wavelet Methods*, volume 32 of *Studies in Mathematics and Its Applications* (North-Holland Publishing, Amsterdam, 2003)
49. A. Cohen, I. Daubechies, A stability criterion for biorthogonal wavelet bases and their related subband coding scheme. Duke Math. J. **68**(2), 313–335 (1992)
50. A. Cohen, I. Daubechies, Orthonormal bases of compactly supported wavelets. III. Better frequency resolution. SIAM J. Math. Anal. **24**(2), 520–527 (1993)
51. A. Cohen, I. Daubechies, A new technique to estimate the regularity of refinable functions. Rev. Mat. Iberoamericana **12**(2), 527–591 (1996)
52. A. Cohen, I. Daubechies, J.-C. Feauveau, Biorthogonal bases of compactly supported wavelets. Comm. Pure Appl. Math. **45**(5), 485–560 (1992)
53. A. Cohen, I. Daubechies, G. Plonka, Regularity of refinable function vectors. J. Fourier Anal. Appl. **3**(3), 295–324 (1997)
54. A. Cohen, I. Daubechies, P. Vial, Wavelets on the interval and fast wavelet transforms. Appl. Comput. Harmon. Anal. **1**(1), 54–81 (1993)
55. A. Cohen, K. Gröchenig, L.F. Villemoes, Regularity of multivariate refinable functions. Constr. Approx. **15**(2), 241–255 (1999)
56. A. Cohen, R.D. Ryan, *Wavelets and Multiscale Signal Processing*, volume 11 of *Applied Mathematics and Mathematical Computation* (Chapman & Hall, London, 1995). Revised version of Cohen's doctoral thesis in 1992, Translated from the French by Ryan
57. A. Cohen, Q. Sun, An arithmetic characterization of the conjugate quadrature filters associated to orthonormal wavelet bases. SIAM J. Math. Anal. **24**(5), 1355–1360 (1993)
58. R.R. Coifman, D.L. Donoho, Translation invariant de-noising, in *Wavelets and Statistics* (Springer, New York, 1995), pp. 125–150
59. R.R. Coifman, M. Maggioni, Diffusion wavelets. Appl. Comput. Harmon. Anal. **21**(1), 53–94 (2006)
60. R.R. Coifman, Y. Meyer, M.V. Wickerhauser, Size properties of wavelet-packets, in *Wavelets and Their Applications*, pp. 453–470 (Jones and Bartlett, Boston, MA, 1992)
61. C. Conti, M. Cotronei, T. Sauer, Full rank interpolatory subdivision schemes: Kronecker, filters and multiresolution. J. Comput. Appl. Math. **233**(7), 1649–1659 (2010)
62. S. Dahlke, K. Gröchenig, P. Maass, A new approach to interpolating scaling functions. Appl. Anal. **72**(3–4), 485–500 (1999)
63. W. Dahmen, Stability of multiscale transformations. J. Fourier Anal. Appl. **2**(4), 341–361 (1996)
64. W. Dahmen, Multiscale and wavelet methods for operator equations, in *Multiscale Problems and Methods in Numerical Simulations*, volume 1825 of *Lecture Notes in Math.* (Springer, Berlin, 2003), pp. 31–96
65. W. Dahmen, B. Han, R.-Q. Jia, A. Kunoth, Biorthogonal multiwavelets on the interval: cubic Hermite splines. Constr. Approx. **16**(2), 221–259 (2000)
66. W. Dahmen, A. Kunoth, K. Urban, Biorthogonal spline wavelets on the interval—stability and moment conditions. Appl. Comput. Harmon. Anal. **6**(2), 132–196 (1999)
67. X.-R. Dai, Q. Sun, Z. Zhang, Compactly supported both m and n refinable distributions. East J. Approx. **6**(2), 201–209 (2000)
68. I. Daubechies, Orthonormal bases of compactly supported wavelets. Comm. Pure Appl. Math. **41**(7), 909–996 (1988)
69. I. Daubechies, The wavelet transform, time-frequency localization and signal analysis. IEEE Trans. Inform. Theory **36**(5), 961–1005 (1990)

70. I. Daubechies, *Ten Lectures on Wavelets*, volume 61 of *CBMS-NSF Regional Conference Series in Applied Mathematics* (Society for Industrial and Applied Mathematics (SIAM), Philadelphia, PA, 1992)
71. I. Daubechies, Orthonormal bases of compactly supported wavelets. II. Variations on a theme. SIAM J. Math. Anal. **24**(2), 499–519 (1993)
72. I. Daubechies, A. Grossmann, Y. Meyer, Painless nonorthogonal expansions. J. Math. Phys. **27**(5), 1271–1283 (1986)
73. I. Daubechies, B. Han, Pairs of dual wavelet frames from any two refinable functions. Constr. Approx. **20**(3), 325–352 (2004)
74. I. Daubechies, B. Han, A. Ron, Z. Shen, Framelets: MRA-based constructions of wavelet frames. Appl. Comput. Harmon. Anal. **14**(1), 1–46 (2003)
75. I. Daubechies, Y. Huang, How does truncation of the mask affect a refinable function? Constr. Approx. **11**(3), 365–380 (1995)
76. I. Daubechies, J.C. Lagarias, Two-scale difference equations. I. Existence and global regularity of solutions. SIAM J. Math. Anal. **22**(5), 1388–1410 (1991)
77. I. Daubechies, J.C. Lagarias, Sets of matrices all infinite products of which converge. Linear Algebra Appl. **161**, 227–263 (1992)
78. I. Daubechies, J.C. Lagarias, Two-scale difference equations. II. Local regularity, infinite products of matrices and fractals. SIAM J. Math. Anal. **23**(4), 1031–1079 (1992)
79. I. Daubechies, J.C. Lagarias, On the thermodynamic formalism for multifractal functions. Rev. Math. Phys. **6**(5A), 1033–1070 (1994)
80. C. de Boor, R.A. DeVore, A. Ron, Approximation from shift-invariant subspaces of $L_2(\mathbb{R}^d)$. Trans. Am. Math. Soc. **341**(2), 787–806 (1994)
81. C. de Boor, R.A. DeVore, A. Ron, The structure of finitely generated shift-invariant spaces in $L_2(\mathbf{R}^d)$. J. Funct. Anal. **119**(1), 37–78 (1994)
82. G. Deslauriers, S. Dubuc, Symmetric iterative interpolation processes. Constr. Approx. **5**(1), 49–68 (1989)
83. R.A. DeVore, Nonlinear approximation, in *Acta Numerica, 1998*, volume 7 of *Acta Numer.* (Cambridge Univ. Press, Cambridge, 1998), pp. 51–150
84. M.N. Do, M. Vetterli, Contourlets, in *Beyond Wavelets*, volume 10 of *Stud. Comput. Math.* (Academic Press/Elsevier, San Diego, CA, 2003), pp. 83–105
85. B. Dong, Z. Shen, Construction of biorthogonal wavelets from pseudo-splines. J. Approx. Theory **138**(2), 211–231 (2006)
86. B. Dong, Z. Shen, Pseudo-splines, wavelets and framelets. Appl. Comput. Harmon. Anal. **22**(1), 78–104 (2007)
87. B. Dong, Z. Shen, MRA-based wavelet frames and applications, in *Mathematics in Image Processing*, volume 19 of *IAS/Park City Math. Ser.* (Amer. Math. Soc., Providence, RI, 2013), pp. 9–158
88. G.C. Donovan, J.S. Geronimo, D.P. Hardin, Orthogonal polynomials and the construction of piecewise polynomial smooth wavelets. SIAM J. Math. Anal. **30**(5), 1029–1056 (1999)
89. G.C. Donovan, J.S. Geronimo, D.P. Hardin, P.R. Massopust, Construction of orthogonal wavelets using fractal interpolation functions. SIAM J. Math. Anal. **27**(4), 1158–1192 (1996)
90. S. Dubuc, Interpolation through an iterative scheme. J. Math. Anal. Appl. **114**(1), 185–204 (1986)
91. S. Dubuc, J.-L. Merrien, Convergent vector and Hermite subdivision schemes. Constr. Approx. **23**(1), 1–22 (2006)
92. R.J. Duffin, A.C. Schaeffer, A class of nonharmonic fourier series. Trans. Am. Math. Soc. **72**, 341–366 (1952)
93. N. Dyn, J.A. Gregory, D. Levin, A butterfly subdivision scheme for surface interpolation with tension control. ACM Trans. Graphics **9**, 160–169 (1990)
94. N. Dyn, K. Hormann, M.A. Sabin, Z. Shen, Polynomial reproduction by symmetric subdivision schemes. J. Approx. Theory **155**(1), 28–42 (2008)

95. N. Dyn, D. Levin, Analysis of Hermite-type subdivision schemes, in *Approximation Theory VIII, Vol. 2 (College Station, TX, 1995)*, volume 6 of *Ser. Approx. Decompos.* (World Sci. Publ., River Edge, NJ, 1995), pp. 117–124
96. N. Dyn, D. Levin, Subdivision schemes in geometric modelling. Acta Numer. **11**, 73–144 (2002)
97. N. Dyn, D. Levin, J.A. Gregory, A 4-point interpolatory subdivision scheme for curve design. Comput. Aided Geom. Design **4**(4), 257–268 (1987)
98. R.E. Edwards, *Fourier Series. A Modern Introduction. Vol. 1*, volume 64 of *Graduate Texts in Mathematics*, 2nd edn. (Springer, New York/Berlin, 1979)
99. M. Ehler, On multivariate compactly supported bi-frames. J. Fourier Anal. Appl. **13**(5), 511–532 (2007)
100. M. Ehler, B. Han, Wavelet bi-frames with few generators from multivariate refinable functions. Appl. Comput. Harmon. Anal. **25**(3), 407–414 (2008)
101. T. Eirola, Sobolev characterization of solutions of dilation equations. SIAM J. Math. Anal. **23**(4), 1015–1030 (1992)
102. H.G. Feichtinger, K.H. Gröchenig, Banach spaces related to integrable group representations and their atomic decompositions. I. J. Funct. Anal. **86**(2), 307–340 (1989)
103. G.B. Folland, *Real Analysis*. Pure and Applied Mathematics (New York), 2nd edn. (Wiley, New York, 1999). Modern Techniques and Their Applications, A Wiley-Interscience Publication
104. M. Frazier, G. Garrigós, K. Wang, G. Weiss, A characterization of functions that generate wavelet and related expansion. J. Fourier Anal. Appl. **3**, 883–906 (1997)
105. M. Frazier, B. Jawerth, G. Weiss, *Littlewood-Paley Theory and the Study of Function Spaces*, volume 79 of *CBMS Regional Conference Series in Mathematics* (The American Mathematical Society, Providence, RI, 1991)
106. J.-P. Gabardo, D. Han, Frames associated with measurable spaces. Adv. Comput. Math. **18**(2–4), 127–147 (2003)
107. J.S. Geronimo, D.P. Hardin, P.R. Massopust, Fractal functions and wavelet expansions based on several scaling functions. J. Approx. Theory **78**(3), 373–401 (1994)
108. S.S. Goh, B. Han, Z. Shen, Tight periodic wavelet frames and approximation orders. Appl. Comput. Harmon. Anal. **31**(2), 228–248 (2011)
109. S.S. Goh, S.L. Lee, K.-M. Teo, Multidimensional periodic multiwavelets. J. Approx. Theory **98**(1), 72–103 (1999)
110. S.S. Goh, Z.Y. Lim, Z. Shen, Symmetric and antisymmetric tight wavelet frames. Appl. Comput. Harmon. Anal. **20**(3), 411–421 (2006)
111. T.N.T. Goodman, A class of orthogonal refinable functions and wavelets. Constr. Approx. **19**(4), 525–540 (2003)
112. T.N.T. Goodman, R.-Q. Jia, D.-X. Zhou, Local linear independence of refinable vectors of functions. Proc. Roy. Soc. Edinburgh Sect. A **130**(4), 813–826 (2000)
113. T.N.T. Goodman, S.L. Lee, Wavelets of multiplicity r. Trans. Am. Math. Soc. **342**(1), 307–324 (1994)
114. T.N.T. Goodman, S.L. Lee, W.-S. Tang, Wavelets in wandering subspaces. Trans. Am. Math. Soc. **338**(2), 639–654 (1993)
115. L. Grafakos, *Classical Fourier Analysis*, volume 249 of *Graduate Texts in Mathematics*, 3rd edn. (Springer, New York, 2014)
116. A. Grossmann, J. Morlet, Decomposition of Hardy functions into square integrable wavelets of constant shape. SIAM J. Math. Anal. **15**(4), 723–736 (1984)
117. N. Guglielmi, V. Protasov, Exact computation of joint spectral characteristics of linear operators. Found. Comput. Math. **13**(1), 37–97 (2013)
118. K. Guo, G. Kutyniok, D. Labate, Sparse multidimensional representations using anisotropic dilation and shear operators, in *Wavelets and Splines: Athens 2005*. Mod. Methods Math. (Nashboro Press, Brentwood, TN, 2006), pp. 189–201
119. K. Guo, D. Labate, Optimally sparse multidimensional representation using shearlets. SIAM J. Math. Anal. **39**(1), 298–318 (2007)

120. A. Haar, Zur theorie der orthogonalen funktionen-systeme. Math. Ann. **69**, 331–371 (1910)
121. B. Han, Wavelets. Master's thesis, Institute of Mathematics, Chinese Academy of Sciences, June 1994
122. B. Han, Some applications of projection operators in wavelets. Acta Math. Sinica (N.S.) **11**(1), 105–112 (1995)
123. B. Han, On dual wavelet tight frames. Appl. Comput. Harmon. Anal. **4**(4), 380–413 (1997)
124. B. Han, *Subdivision schemes, biorthogonal wavelets and image compression*. PhD thesis, Department of Mathematical Sciences, University of Alberta, July 1998
125. B. Han, Symmetric orthonormal scaling functions and wavelets with dilation factor 4. Adv. Comput. Math. **8**(3), 221–247 (1998)
126. B. Han, Analysis and construction of optimal multivariate biorthogonal wavelets with compact support. SIAM J. Math. Anal. **31**(2), 274–304 (2000)
127. B. Han, Approximation properties and construction of Hermite interpolants and biorthogonal multiwavelets. J. Approx. Theory **110**(1), 18–53 (2001)
128. B. Han, Projectable multivariate refinable functions and biorthogonal wavelets. Appl. Comput. Harmon. Anal. **13**(1), 89–102 (2002)
129. B. Han, Classification and construction of bivariate subdivision schemes, in *Curve and Surface Fitting (Saint-Malo, 2002)*. Mod. Methods Math. (Nashboro Press, Brentwood, TN, 2003), pp. 187–197
130. B. Han, Compactly supported tight wavelet frames and orthonormal wavelets of exponential decay with a general dilation matrix. J. Comput. Appl. Math. **155**(1), 43–67 (2003)
131. B. Han, Computing the smoothness exponent of a symmetric multivariate refinable function. SIAM J. Matrix Anal. Appl. **24**(3), 693–714 (2003)
132. B. Han, Vector cascade algorithms and refinable function vectors in Sobolev spaces. J. Approx. Theory **124**(1), 44–88 (2003)
133. B. Han, Symmetric multivariate orthogonal refinable functions. Appl. Comput. Harmon. Anal. **17**(3), 277–292 (2004)
134. B. Han, Solutions in Sobolev spaces of vector refinement equations with a general dilation matrix. Adv. Comput. Math. **24**(1–4), 375–403 (2006)
135. B. Han, Refinable functions and cascade algorithms in weighted spaces with Hölder continuous masks. SIAM J. Math. Anal. **40**(1), 70–102 (2008)
136. B. Han, Dual multiwavelet frames with high balancing order and compact fast frame transform. Appl. Comput. Harmon. Anal. **26**(1), 14–42 (2009)
137. B. Han, Matrix extension with symmetry and applications to symmetric orthonormal complex M-wavelets. J. Fourier Anal. Appl. **15**(5), 684–705 (2009)
138. B. Han, Pairs of frequency-based nonhomogeneous dual wavelet frames in the distribution space. Appl. Comput. Harmon. Anal. **29**(3), 330–353 (2010)
139. B. Han, The structure of balanced multivariate biorthogonal multiwavelets and dual multiframelets. Math. Comp. **79**(270), 917–951 (2010)
140. B. Han, Symmetric orthonormal complex wavelets with masks of arbitrarily high linear-phase moments and sum rules. Adv. Comput. Math. **32**(2), 209–237 (2010)
141. B. Han, Symmetric orthogonal filters and wavelets with linear-phase moments. J. Comput. Appl. Math. **236**(4), 482–503 (2011)
142. B. Han, Nonhomogeneous wavelet systems in high dimensions. Appl. Comput. Harmon. Anal. **32**(2), 169–196 (2012)
143. B. Han, Matrix splitting with symmetry and symmetric tight framelet filter banks with two high-pass filters. Appl. Comput. Harmon. Anal. **35**(2), 200–227 (2013)
144. B. Han, Properties of discrete framelet transforms. Math. Model. Nat. Phenom. **8**(1), 18–47 (2013)
145. B. Han, The projection method for multidimensional framelet and wavelet analysis. Math. Model. Nat. Phenom. **9**(5), 83–110 (2014)
146. B. Han, Symmetric tight framelet filter banks with three high-pass filters. Appl. Comput. Harmon. Anal. **37**(1), 140–161 (2014)

147. B. Han, Algorithm for constructing symmetric dual framelet filter banks. Math. Comp. **84**(292), 767–801 (2015)
148. B. Han, On linear independence of integer shifts of compactly supported distributions. J. Approx. Theory **201**, 1–6 (2016)
149. B. Han, R.-Q. Jia, Multivariate refinement equations and convergence of subdivision schemes. SIAM J. Math. Anal. **29**(5), 1177–1199 (1998)
150. B. Han, R.-Q. Jia, Optimal interpolatory subdivision schemes in multidimensional spaces. SIAM J. Numer. Anal. **36**(1), 105–124 (1999)
151. B. Han, R.-Q. Jia, Quincunx fundamental refinable functions and quincunx biorthogonal wavelets. Math. Comp. **71**(237), 165–196 (2002)
152. B. Han, R.-Q. Jia, Optimal C^2 two-dimensional interpolatory ternary subdivision schemes with two-ring stencils. Math. Comp. **75**(255), 1287–1308 (2006)
153. B. Han, Q. Jiang, Multiwavelets on the interval. Appl. Comput. Harmon. Anal. **12**(1), 100–127 (2002)
154. B. Han, Q. Jiang, Z. Shen, X. Zhuang, Symmetric canonical quincunx tight framelets with high vanishing moments and smoothness. Math. Comp. **87**, 347–379 (2018)
155. B. Han, S.-G. Kwon, S.S. Park, Riesz multiwavelet bases. Appl. Comput. Harmon. Anal. **20**(2), 161–183 (2006)
156. B. Han, S.-G. Kwon, X. Zhuang, Generalized interpolating refinable function vectors. J. Comput. Appl. Math. **227**(2), 254–270 (2009)
157. B. Han, Q. Mo, Multiwavelet frames from refinable function vectors. Adv. Comput. Math. **18**(2–4), 211–245 (2003)
158. B. Han, Q. Mo, Splitting a matrix of Laurent polynomials with symmetry and its application to symmetric framelet filter banks. SIAM J. Matrix Anal. Appl. **26**(1), 97–124 (2004)
159. B. Han, Q. Mo, Symmetric MRA tight wavelet frames with three generators and high vanishing moments. Appl. Comput. Harmon. Anal. **18**(1), 67–93 (2005)
160. B. Han, Q. Mo, Analysis of optimal bivariate symmetric refinable Hermite interpolants. Commun. Pure Appl. Anal. **6**(3), 689–718 (2007)
161. B. Han, Q. Mo, Z. Zhao, Compactly supported tensor product complex tight framelets with directionality. SIAM J. Math. Anal. **47**(3), 2464–2494 (2015)
162. B. Han, Q. Mo, Z. Zhao, X. Zhuang, Construction and applications of compactly supported directional complex tight framelets. Preprint, 2016.
163. B. Han, Z. Shen, Wavelets with short support. SIAM J. Math. Anal. **38**(2), 530–556 (2006)
164. B. Han, Z. Shen, Compactly supported symmetric C^∞ wavelets with spectral approximation order. SIAM J. Math. Anal. **40**(3), 905–938 (2008)
165. B. Han, Z. Shen, Dual wavelet frames and Riesz bases in Sobolev spaces. Constr. Approx. **29**(3), 369–406 (2009)
166. B. Han, T.P.-Y. Yu, B. Piper, Multivariate refinable Hermite interpolant. Math. Comp. **73**(248), 1913–1935 (2004)
167. B. Han, Z. Zhao, Tensor product complex tight framelets with increasing directionality. SIAM J. Imaging Sci. **7**(2), 997–1034 (2014)
168. B. Han, Z. Zhao, X. Zhuang, Directional tensor product complex tight framelets with low redundancy. Appl. Comput. Harmon. Anal. **41**(2), 603–637 (2016)
169. B. Han, X. Zhuang, Smooth affine shear tight frames with mra structure. Appl. Comput. Harmon. Anal. **39**(2), 300–338 (2015)
170. D. Han, K. Kornelson, D.R. Larson, E. Weber, *Frames for Undergraduates*, volume 40 of *Student Mathematical Library* (American Mathematical Society, Providence, RI, 2007)
171. D. Han, D.R. Larson, Frames, bases and group representations. Mem. Am. Math. Soc. **147**(697), x+94 (2000)
172. D.P. Hardin, T.A. Hogan, Q. Sun, The matrix-valued Riesz lemma and local orthonormal bases in shift-invariant spaces. Adv. Comput. Math. **20**(4), 367–384 (2004)
173. D.P. Hardin, J.A. Marasovich, Biorthogonal multiwavelets on $[-1, 1]$. Appl. Comput. Harmon. Anal. **7**(1), 34–53 (1999)

174. M.F. Hassan, I.P. Ivrissimitzis, N.A. Dodgson, M.A. Sabin, An interpolating 4-point C^2 ternary stationary subdivision scheme. Comput. Aided Geom. Design **19**(1), 1–18 (2002)
175. C.E. Heil, What is ... a frame? Notices Am. Math. Soc. **60**(6), 748–750 (2013)
176. C.E. Heil, D. Colella, Matrix refinement equations: existence and uniqueness. J. Fourier Anal. Appl. **2**(4), 363–377 (1996)
177. C.E. Heil, G. Strang, V. Strela, Approximation by translates of refinable functions. Numer. Math. **73**(1), 75–94 (1996)
178. C.E. Heil, D.F. Walnut, Continuous and discrete wavelet transforms. SIAM Rev. **31**(4), 628–666 (1989)
179. E. Hernández, G. Weiss, *A First Course on Wavelets*. Studies in Advanced Mathematics (CRC Press, Boca Raton, FL, 1996)
180. L. Hervé, Multi-resolution analysis of multiplicity d: applications to dyadic interpolation. Appl. Comput. Harmon. Anal. **1**(4), 299–315 (1994)
181. M. Holschneider, R. Kronland-Martinet, J. Morlet, P. Tchamitchian. A real-time algorithm for signal analysis with the help of the wavelet transform, in *Wavelets (Marseille, 1987)*. Inverse Probl. Theoret. Imaging (Springer, Berlin, 1989), pp. 286–297
182. S. Jaffard, Y. Meyer, Wavelet methods for pointwise regularity and local oscillations of functions. Mem. Am. Math. Soc. **123**(587), x+110 (1996)
183. S. Jaffard, Y. Meyer, R.D. Ryan, *Wavelets*, revised edn. (Society for Industrial and Applied Mathematics (SIAM), Philadelphia, PA, 2001)
184. K. Jetter, G. Plonka, A survey on L_2-approximation orders from shift-invariant spaces, in *Multivariate Approximation and Applications* (Cambridge Univ. Press, Cambridge, 2001), pp. 73–111
185. K. Jetter, D.-X. Zhou, Order of linear approximation from shift-invariant spaces. Constr. Approx. **11**(4), 423–438 (1995)
186. K. Jetter, D.-X. Zhou, Approximation order of linear operators and finitely generated shift-invariant spaces. preprint, 1998
187. R.-Q. Jia, Subdivision schemes in L_p spaces. Adv. Comput. Math. **3**(4), 309–341 (1995)
188. R.Q. Jia, The Toeplitz theorem and its applications to approximation theory and linear PDEs. Trans. Am. Math. Soc. **347**(7), 2585–2594 (1995)
189. R.-Q. Jia, Shift-invariant spaces on the real line. Proc. Am. Math. Soc. **125**(3), 785–793 (1997)
190. R.-Q. Jia, Approximation properties of multivariate wavelets. Math. Comp. **67**(222), 647–665 (1998)
191. R.-Q. Jia, Characterization of smoothness of multivariate refinable functions in Sobolev spaces. Trans. Am. Math. Soc. **351**(10), 4089–4112 (1999)
192. R.-Q. Jia, Approximation with scaled shift-invariant spaces by means of quasi-projection operators. J. Approx. Theory **131**(1), 30–46 (2004)
193. R.-Q. Jia, Bessel sequences in Sobolev spaces. Appl. Comput. Harmon. Anal. **20**(2), 298–311 (2006)
194. R.-Q. Jia, Q. Jiang, Approximation power of refinable vectors of functions, in *Wavelet Analysis and Applications (Guangzhou, 1999)*, volume 25 of *AMS/IP Stud. Adv. Math.* (Amer. Math. Soc., Providence, RI, 2002), pp. 155–178
195. R.-Q. Jia, Q. Jiang, Spectral analysis of the transition operator and its applications to smoothness analysis of wavelets. SIAM J. Matrix Anal. Appl. **24**(4), 1071–1109 (2003)
196. R.-Q. Jia, Q. Jiang, S.L. Lee, Convergence of cascade algorithms in Sobolev spaces and integrals of wavelets. Numer. Math. **91**(3), 453–473 (2002)
197. R.-Q. Jia, Q. Jiang, Z. Shen, Distributional solutions of nonhomogeneous discrete and continuous refinement equations. SIAM J. Math. Anal. **32**(2), 420–434 (2000)
198. R.-Q. Jia, K.-S. Lau, D.-X. Zhou, L_p solutions of refinement equations. J. Fourier Anal. Appl. **7**(2), 143–167 (2001)
199. R.-Q. Jia, S. Li, Refinable functions with exponential decay: an approach via cascade algorithms. J. Fourier Anal. Appl. **17**(5), 1008–1034 (2011)

200. R.-Q. Jia, C.A. Micchelli, Using the refinement equations for the construction of pre-wavelets. II. Powers of two, in *Curves and Surfaces (Chamonix-Mont-Blanc, 1990)* (Academic Press, Boston, MA, 1991), pp. 209–246
201. R.-Q. Jia, C.A. Micchelli, On linear independence for integer translates of a finite number of functions. Proc. Edinburgh Math. Soc. (2) **36**(1), 69–85 (1993)
202. R.-Q. Jia, S.D. Riemenschneider, D.-X. Zhou, Approximation by multiple refinable functions. Can. J. Math. **49**(5), 944–962 (1997)
203. R.-Q. Jia, S.D. Riemenschneider, D.-X. Zhou, Vector subdivision schemes and multiple wavelets. Math. Comp. **67**(224), 1533–1563 (1998)
204. R.-Q. Jia, S.D. Riemenschneider, D.-X. Zhou, Smoothness of multiple refinable functions and multiple wavelets. SIAM J. Matrix Anal. Appl. **21**(1), 1–28 (1999)
205. R.-Q. Jia, Z. Shen, Multiresolution and wavelets. Proc. Edinburgh Math. Soc. (2) **37**(2), 271–300 (1994)
206. R.-Q. Jia, J. Wang, Stability and linear independence associated with wavelet decompositions. Proc. Am. Math. Soc. **117**(4), 1115–1124 (1993)
207. R.-Q. Jia, J. Wang, D.-X. Zhou, Compactly supported wavelet bases for Sobolev spaces. Appl. Comput. Harmon. Anal. **15**(3), 224–241 (2003)
208. Q. Jiang, On the regularity of matrix refinable functions. SIAM J. Math. Anal. **29**(5), 1157–1176 (1998)
209. Q. Jiang, Multivariate matrix refinable functions with arbitrary matrix dilation. Trans. Am. Math. Soc. **351**(6), 2407–2438 (1999)
210. Q. Jiang, Parameterizations of masks for tight affine frames with two symmetric/antisymmetric generators. Adv. Comput. Math. **18**(2–4), 247–268 (2003)
211. Q. Jiang, P. Oswald, S.D. Riemenschneider, $\sqrt{3}$-subdivision schemes: maximal sum rule orders. Constr. Approx. **19**(3), 437–463 (2003)
212. F. Keinert, Raising multiwavelet approximation order through lifting. SIAM J. Math. Anal. **32**(5), 1032–1049 (2001)
213. F. Keinert, *Wavelets and Multiwavelets*. Studies in Advanced Mathematics (Chapman & Hall/CRC, Boca Raton, FL, 2004)
214. A.V. Krivoshein, M.A. Skopina, Approximation by frame-like wavelet systems. Appl. Comput. Harmon. Anal. **31**(3), 410–428 (2011)
215. G. Kutyniok, W.-Q. Lim, Compactly supported shearlets are optimally sparse. J. Approx. Theory **163**(11), 1564–1589 (2011)
216. G. Kutyniok, M. Shahram, X. Zhuang, ShearLab: a rational design of a digital parabolic scaling algorithm. SIAM J. Imaging Sci. **5**(4), 1291–1332 (2012)
217. D. Labate, G. Weiss, E. Wilson, Wavelets. Notices Am. Math. Soc. **60**(1), 66–76 (2013)
218. M.-J. Lai, A. Petukhov, Method of virtual components for constructing redundant filter banks and wavelet frames. Appl. Comput. Harmon. Anal. **22**(3), 304–318 (2007)
219. M.-J. Lai, D.W. Roach, Parameterizations of univariate orthogonal wavelets with short support, in *Approximation Theory, X (St. Louis, MO, 2001)*. Innov. Appl. Math. (Vanderbilt Univ. Press, Nashville, TN, 2002), pp. 369–384
220. D. Langemann, J. Prestin, Multivariate periodic wavelet analysis. Appl. Comput. Harmon. Anal. **28**(1), 46–66 (2010)
221. W.M. Lawton, Tight frames of compactly supported affine wavelets. J. Math. Phys. **31**(8), 1898–1901 (1990)
222. W.M. Lawton, Necessary and sufficient conditions for constructing orthonormal wavelet bases. J. Math. Phys. **32**(1), 57–61 (1991)
223. W.M. Lawton, Applications of complex valued wavelet transforms to subband decomposition. IEEE Trans. Signal Process. **41**(12), 3566–3568 (1993)
224. W.M. Lawton, S.L. Lee, Z. Shen, Characterization of compactly supported refinable splines. Adv. Comput. Math. **3**(1–2), 137–145 (1995)
225. W.M. Lawton, S.L. Lee, Z. Shen, Stability and orthonormality of multivariate refinable functions. SIAM J. Math. Anal. **28**(4), 999–1014 (1997)

226. J. Lebrun, M. Vetterli, Balanced multiwavelets theory and design. IEEE Trans. Signal Process. **46**(4), 1119–1125 (1998)
227. P.G. Lemarié, Ondelettes à localisation exponentielle. J. Math. Pures Appl. (9) **67**(3), 227–236 (1988)
228. P.G. Lemarié, Fonctions à support compact dans les analyses multi-résolutions. Rev. Mat. Iberoamericana **7**(2), 157–182 (1991)
229. P.G. Lemarié-Rieusset, On the existence of compactly supported dual wavelets. Appl. Comput. Harmon. Anal. **4**(1), 117–118 (1997)
230. B. Li, L. Peng, Parametrization for balanced multifilter banks. Int. J. Wavelets Multiresolut. Inf. Process. **6**(4), 617–629 (2008)
231. S. Li, Convergence rates of vector cascade algorithms in L_p. J. Approx. Theory **137**(1), 123–142 (2005)
232. S. Li, J. Xian, Biorthogonal multiple wavelets generated by vector refinement equation. Sci. China Ser. A **50**(7), 1015–1025 (2007)
233. J.-M. Lina, M. Mayrand, Complex Daubechies wavelets. Appl. Comput. Harmon. Anal. **2**(3), 219–229 (1995)
234. R.-L. Long, *High Dimensional Wavelet Analysis* (World Publishing, Singapore, 1995) (Chinese)
235. R.-L. Long, D.-R. Chen, Biorthogonal wavelet bases on $\mathbb{R}^d$. Appl. Comput. Harmon. Anal. **2**(3), 230–242 (1995)
236. R.-L. Long, W. Chen, S. Yuan, Wavelets generated by vector multiresolution analysis. Appl. Comput. Harmon. Anal. **4**(3), 317–350 (1997)
237. R.-L. Long, Q. Mo, L^2-convergence of vector cascade algorithm. Approx. Theory Appl. (N.S.) **15**(4), 29–49 (1999)
238. J. MacArthur, K.F. Taylor, Wavelets with crystal symmetry shifts. J. Fourier Anal. Appl. **17**(6), 1109–1118 (2011)
239. S.G. Mallat, Multiresolution approximations and wavelet orthonormal bases of $L^2(\mathbf{R})$. Trans. Am. Math. Soc. **315**(1), 69–87 (1989)
240. S.G. Mallat, *A Wavelet Tour of Signal Processing*, 3rd edn. (Elsevier/Academic Press, Amsterdam, 2009)
241. J.-L. Merrien, A family of Hermite interpolants by bisection algorithms. Numer. Algorithms **2**(2), 187–200 (1992)
242. Y. Meyer, *Ondelettes et opérateurs. I, II, and III.* Actualités Mathématiques (Hermann, Paris, 1990)
243. Y. Meyer, Ondelettes sur l'intervalle. Rev. Mat. Iberoamericana **7**(2), 115–133 (1991)
244. C.A. Micchelli, H. Prautzsch, Uniform refinement of curves. Linear Algebra Appl. **114/115**, 841–870 (1989)
245. C.A. Micchelli, T. Sauer, Regularity of multiwavelets. Adv. Comput. Math. **7**(4), 455–545 (1997)
246. C.A. Micchelli, T. Sauer, On vector subdivision. Math. Z. **229**(4), 621–674 (1998)
247. C.A. Micchelli, Y. Xu, Using the matrix refinement equation for the construction of wavelets on invariant sets. Appl. Comput. Harmon. Anal. **1**(4), 391–401 (1994)
248. Q. Mo, *Compactly supported symmetry MRA wavelet frames*. Ph.D. thesis, Department of Mathematical and Statistical Sciences, University of Alberta, August 2003
249. Q. Mo, S. Li, Symmetric tight wavelet frames with rational coefficients. Appl. Comput. Harmon. Anal. **31**(2), 249–263 (2011)
250. L. Monzón, G. Beylkin, W. Hereman, Compactly supported wavelets based on almost interpolating and nearly linear phase filters (coiflets). Appl. Comput. Harmon. Anal. **7**(2), 184–210 (1999)
251. D.J. Newman, A simple proof of Wiener's $1/f$ theorem. Proc. Am. Math. Soc. **48**, 264–265 (1975)
252. I.Ya. Novikov, V.Yu. Protasov, M.A. Skopina, *Wavelet Theory*, volume 239 of *Translations of Mathematical Monographs* (American Mathematical Society, Providence, RI, 2011)

253. L. Peng, H. Wang, Construction for a class of smooth wavelet tight frames. Sci. China Ser. F **46**(6), 445–458 (2003)
254. A. Petukhov, Explicit construction of framelets. Appl. Comput. Harmon. Anal. **11**(2), 313–327 (2001)
255. A. Petukhov, Symmetric framelets. Constr. Approx. **19**(2), 309–328 (2003)
256. G. Plonka, Approximation order provided by refinable function vectors. Constr. Approx. **13**(2), 221–244 (1997)
257. G. Plonka, V. Strela, Construction of multiscaling functions with approximation and symmetry. SIAM J. Math. Anal. **29**(2), 481–510 (1998)
258. S.D. Riemenschneider, Z. Shen, Multidimensional interpolatory subdivision schemes. SIAM J. Numer. Anal. **34**(6), 2357–2381 (1997)
259. O. Rioul, Simple regularity criteria for subdivision schemes. SIAM J. Math. Anal. **23**(6), 1544–1576 (1992)
260. A. Ron, A necessary and sufficient condition for the linear independence of the integer translates of a compactly supported distribution. Constr. Approx. **5**(3), 297–308 (1989)
261. A. Ron, Introduction to shift-invariant spaces. Linear independence, in *Multivariate Approximation and Applications* (Cambridge Univ. Press, Cambridge, 2001), pp. 112–151
262. A. Ron, Z. Shen, Frames and stable bases for shift-invariant subspaces of $L_2(\mathbb{R}^d)$. Can. J. Math. **47**(5), 1051–1094 (1995)
263. A. Ron, Z. Shen, Affine systems in $L_2(\mathbb{R}^d)$. II. Dual systems. J. Fourier Anal. Appl. **3**(5), 617–637 (1997)
264. A. Ron, Z. Shen, Affine systems in $L_2(\mathbb{R}^d)$: the analysis of the analysis operator. J. Funct. Anal. **148**(2), 408–447 (1997)
265. A. Ron, Z. Shen, The Sobolev regularity of refinable functions. J. Approx. Theory **106**(2), 185–225 (2000)
266. G.-C. Rota, G. Strang, A note on the joint spectral radius. Nederl. Akad. Wetensch. Proc. Ser. A 63 Indag. Math. **22**, 379–381 (1960)
267. H.L. Royden, *Real Analysis*, 3rd edn. (Macmillan Publishing, New York, 1988)
268. A. San Antolín, R.A. Zalik, Matrix-valued wavelets and multiresolution analysis. J. Appl. Funct. Anal. **7**(1–2), 13–25 (2012)
269. I.W. Selesnick, Interpolating multiwavelet bases and the sampling theorem. IEEE Trans. Signal Process. **47**(6), 1615–1621 (1999)
270. I.W. Selesnick, Smooth wavelet tight frames with zero moments. Appl. Comput. Harmon. Anal. **10**(2), 163–181 (2001)
271. I.W. Selesnick, A. Farras Abdelnour, Symmetric wavelet tight frames with two generators. Appl. Comput. Harmon. Anal. **17**(2), 211–225 (2004)
272. I.W. Selesnick, R.G. Baraniuk, N.C. Kingsbury, The dual-tree complex wavelet transform. IEEE Signal Process. Mag. **22**(6), 123–151 (2005)
273. L. Sendur, I.W. Selesnick, Bivariate shrinkage with local variance estimation. IEEE Signal Process. Lett. **9**(12), 438–441 (2002)
274. Y. Shen, B. Han, E. Braverman, Image inpainting from partial noisy data by directional complex tight framelets. ANZIAM **58**, 247–255 (2017)
275. Y. Shen, S. Li, Q. Mo, Complex wavelets and framelets from pseudo splines. J. Fourier Anal. Appl. **16**(6), 885–900 (2010)
276. Z. Shen, Refinable function vectors. SIAM J. Math. Anal. **29**(1), 235–250 (1998)
277. Z. Shen, Wavelet frames and image restorations, in *Proceedings of the International Congress of Mathematicians. IV* (Hindustan Book Agency, New Delhi, 2010), pp. 2834–2863
278. J.-L. Starck, F. Murtagh, J.M. Fadili, *Sparse Image and Signal Processing* (Cambridge University Press, Cambridge, 2010)
279. E.M. Stein, R. Shakarchi, *Fourier Analysis*, volume 1 of *Princeton Lectures in Analysis* (Princeton University Press, Princeton, NJ, 2003). An Introduction
280. E.M. Stein, G. Weiss, *Introduction to Fourier Analysis on Euclidean Spaces* (Princeton University Press, Princeton, NJ, 1971)

281. G. Strang, G. Fix, A Fourier analysis of the finite-element method, in *Constructive Aspects of Functional Analysis* (C.I.M.E., Rome, 1973), pp. 793–840
282. G. Strang, T. Nguyen, *Wavelets and Filter Banks* (Wellesley-Cambridge Press, Wellesley, MA, 1996)
283. Q. Sun, Convergence and boundedness of cascade algorithm in Besov spaces and Triebel-Lizorkin spaces. I. Adv. Math. (China) **29**(6), 507–526 (2000)
284. Q. Sun, N. Bi, D. Huang, *An Introduction to Multiband Wavelets* (Zhejiang University Press, China, 2001). (Chinese)
285. W. Sun, X. Zhou, Irregular wavelet/Gabor frames. Appl. Comput. Harmon. Anal. **13**(1), 63–76 (2002)
286. W. Sweldens, The lifting scheme: a custom-design construction of biorthogonal wavelets. Appl. Comput. Harmon. Anal. **3**(2), 186–200 (1996)
287. M. Unser, Sampling–50 years after shannon. Proc. IEEE **88**(4), 569–587 (2000)
288. M. Vetterli, J. Kovačević, *Wavelets and Subband Coding* (Prentice Hall PTR, Englewood Cliffs, NJ, 1995)
289. L.F. Villemoes, Wavelet analysis of refinement equations. SIAM J. Math. Anal. **25**(5), 1433–1460 (1994)
290. D.F. Walnut, *An Introduction to Wavelet Analysis*. Applied and Numerical Harmonic Analysis (Birkhäuser Boston, Boston, MA, 2002)
291. G.G. Walter, X. Shen, *Wavelets and Other Orthogonal Systems*. Studies in Advanced Mathematics, 2nd edn. (Chapman & Hall/CRC, Boca Raton, FL, 2001)
292. J. Wang, Stability and linear independence associated with scaling vectors. SIAM J. Math. Anal. **29**(5), 1140–1156 (1998)
293. Y. Wang, Two-scale dilation equations and the cascade algorithm. Random Comput. Dynam. **3**(4), 289–307 (1995)
294. M.V. Wickerhauser, *Adapted Wavelet Analysis from Theory to Software* (A K Peters, Ltd., Wellesley, MA, 1994)
295. P. Wojtaszczyk, *A Mathematical Introduction to Wavelets*, volume 37 of *London Mathematical Society Student Texts* (Cambridge University Press, Cambridge, 1997)
296. R.A. Zalik, Riesz bases and multiresolution analyses. Appl. Comput. Harmon. Anal. **7**(3), 315–331 (1999)
297. Z. Zhao, *Directional tensor product complex tight framelets*. Ph.D. thesis, Department of Mathematical and Statistical Sciences, University of Alberta, August 2015
298. D.-X. Zhou, Existence of multiple refinable distributions. Michigan Math. J. **44**(2), 317–329 (1997)
299. D.-X. Zhou, The p-norm joint spectral radius for even integers. Methods Appl. Anal. **5**(1), 39–54 (1998)
300. D.-X. Zhou, Multiple refinable Hermite interpolants. J. Approx. Theory **102**(1), 46–71 (2000)
301. D.-X. Zhou, Interpolatory orthogonal multiwavelets and refinable functions. IEEE Trans. Signal Process. **50**(3), 520–527 (2002)
302. X. Zhuang, *Interpolating refinable function vectors and matrix extension with symmetry*. Ph.D. thesis, Department of Mathematical and Statistical Sciences, University of Alberta, July 2010

Index

Numbers

A

B

B. Han, *Framelets and Wavelets*, Applied and Numerical Harmonic Analysis,
https://doi.org/10.1007/978-3-319-68530-4

P

Q

R

S

T

U

V

W

Applied and Numerical Harmonic Analysis (86 volumes)

A. Saichev and W.A. Woyczyński: *Distributions in the Physical and Engineering Sciences* (ISBN 978-0-8176-3924-2)
C.E. D'Attellis and E.M. Fernandez-Berdaguer: *Wavelet Theory and Harmonic Analysis in Applied Sciences* (ISBN 978-0-8176-3953-2)
H.G. Feichtinger and T. Strohmer: *Gabor Analysis and Algorithms* (ISBN 978-0-8176-3959-4)
R. Tolimieri and M. An: *Time-Frequency Representations* (ISBN 978-0-8176-3918-1)
T.M. Peters and J.C. Williams: *The Fourier Transform in Biomedical Engineering* (ISBN 978-0-8176-3941-9)
G.T. Herman: *Geometry of Digital Spaces* (ISBN 978-0-8176-3897-9)
A. Teolis: *Computational Signal Processing with Wavelets* (ISBN 978-0-8176-3909-9)
J. Ramanathan: *Methods of Applied Fourier Analysis* (ISBN 978-0-8176-3963-1)
J.M. Cooper: *Introduction to Partial Differential Equations with MATLAB* (ISBN 978-0-8176-3967-9)
A. Procházka, N.G. Kingsbury, P.J. Payner, and J. Uhlir: *Signal Analysis and Prediction* (ISBN 978-0-8176-4042-2)
W. Bray and C. Stanojevic: *Analysis of Divergence* (ISBN 978-1-4612-7467-4)
G.T. Herman and A. Kuba: *Discrete Tomography* (ISBN 978-0-8176-4101-6)
K. Gröchenig: *Foundations of Time-Frequency Analysis* (ISBN 978-0-8176-4022-4)
L. Debnath: *Wavelet Transforms and Time-Frequency Signal Analysis* (ISBN 978-0-8176-4104-7)
J.J. Benedetto and P.J.S.G. Ferreira: *Modern Sampling Theory* (ISBN 978-0-8176-4023-1)
D.F. Walnut: *An Introduction to Wavelet Analysis* (ISBN 978-0-8176-3962-4)
A. Abbate, C. DeCusatis, and P.K. Das: *Wavelets and Subbands* (ISBN 978-0-8176-4136-8)

B. Han, *Framelets and Wavelets*, Applied and Numerical Harmonic Analysis,
https://doi.org/10.1007/978-3-319-68530-4

O. Bratteli, P. Jorgensen, and B. Treadway: *Wavelets Through a Looking Glass* (ISBN 978-0-8176-4280-80
H.G. Feichtinger and T. Strohmer: *Advances in Gabor Analysis* (ISBN 978-0-8176-4239-6)
O. Christensen: *An Introduction to Frames and Riesz Bases* (ISBN 978-0-8176-4295-2)
L. Debnath: *Wavelets and Signal Processing* (ISBN 978-0-8176-4235-8)
G. Bi and Y. Zeng: *Transforms and Fast Algorithms for Signal Analysis and Representations* (ISBN 978-0-8176-4279-2)
J.H. Davis: *Methods of Applied Mathematics with a MATLAB Overview* (ISBN 978-0-8176-4331-7)
J.J. Benedetto and A.I. Zayed: *Sampling, Wavelets, and Tomography* (ISBN 978-0-8176-4304-1)
E. Prestini: *The Evolution of Applied Harmonic Analysis* (ISBN 978-0-8176-4125-2)
L. Brandolini, L. Colzani, A. Iosevich, and G. Travaglini: *Fourier Analysis and Convexity* (ISBN 978-0-8176-3263-2)
W. Freeden and V. Michel: *Multiscale Potential Theory* (ISBN 978-0-8176-4105-4)
O. Christensen and K.L. Christensen: *Approximation Theory* (ISBN 978-0-8176-3600-5)
O. Calin and D.-C. Chang: *Geometric Mechanics on Riemannian Manifolds* (ISBN 978-0-8176-4354-6)
J.A. Hogan: *Time–Frequency and Time–Scale Methods* (ISBN 978-0-8176-4276-1)
C. Heil: *Harmonic Analysis and Applications* (ISBN 978-0-8176-3778-1)
K. Borre, D.M. Akos, N. Bertelsen, P. Rinder, and S.H. Jensen: *A Software-Defined GPS and Galileo Receiver* (ISBN 978-0-8176-4390-4)
T. Qian, M.I. Vai, and Y. Xu: *Wavelet Analysis and Applications* (ISBN 978-3-7643-7777-9)
G.T. Herman and A. Kuba: *Advances in Discrete Tomography and Its Applications* (ISBN 978-0-8176-3614-2)
M.C. Fu, R.A. Jarrow, J.-Y. Yen, and R.J. Elliott: *Advances in Mathematical Finance* (ISBN 978-0-8176-4544-1)
O. Christensen: *Frames and Bases* (ISBN 978-0-8176-4677-6)
P.E.T. Jorgensen, J.D. Merrill, and J.A. Packer: *Representations, Wavelets, and Frames* (ISBN 978-0-8176-4682-0)
M. An, A.K. Brodzik, and R. Tolimieri: *Ideal Sequence Design in Time-Frequency Space* (ISBN 978-0-8176-4737-7)
S.G. Krantz: *Explorations in Harmonic Analysis* (ISBN 978-0-8176-4668-4)
B. Luong: *Fourier Analysis on Finite Abelian Groups* (ISBN 978-0-8176-4915-9)
G.S. Chirikjian: *Stochastic Models, Information Theory, and Lie Groups, Volume 1* (ISBN 978-0-8176-4802-2)
C. Cabrelli and J.L. Torrea: *Recent Developments in Real and Harmonic Analysis* (ISBN 978-0-8176-4531-1)
M.V. Wickerhauser: *Mathematics for Multimedia* (ISBN 978-0-8176-4879-4)

B. Forster, P. Massopust, O. Christensen, K. Gröchenig, D. Labate, P. Vandergheynst, G. Weiss, and Y. Wiaux: *Four Short Courses on Harmonic Analysis* (ISBN 978-0-8176-4890-9)
O. Christensen: *Functions, Spaces, and Expansions* (ISBN 978-0-8176-4979-1)
J. Barral and S. Seuret: *Recent Developments in Fractals and Related Fields* (ISBN 978-0-8176-4887-9)
O. Calin, D.-C. Chang, and K. Furutani, and C. Iwasaki: *Heat Kernels for Elliptic and Sub-elliptic Operators* (ISBN 978-0-8176-4994-4)
C. Heil: *A Basis Theory Primer* (ISBN 978-0-8176-4686-8)
J.R. Klauder: *A Modern Approach to Functional Integration* (ISBN 978-0-8176-4790-2)
J. Cohen and A.I. Zayed: *Wavelets and Multiscale Analysis* (ISBN 978-0-8176-8094-7)
D. Joyner and J.-L. Kim: *Selected Unsolved Problems in Coding Theory* (ISBN 978-0-8176-8255-2)
G.S. Chirikjian: *Stochastic Models, Information Theory, and Lie Groups, Volume 2* (ISBN 978-0-8176-4943-2)
J.A. Hogan and J.D. Lakey: *Duration and Bandwidth Limiting* (ISBN 978-0-8176-8306-1)
G. Kutyniok and D. Labate: *Shearlets* (ISBN 978-0-8176-8315-3)
P.G. Casazza and P. Kutyniok: *Finite Frames* (ISBN 978-0-8176-8372-6)
V. Michel: *Lectures on Constructive Approximation* (ISBN 978-0-8176-8402-0)
D. Mitrea, I. Mitrea, M. Mitrea, and S. Monniaux: *Groupoid Metrization Theory* (ISBN 978-0-8176-8396-2)
T.D. Andrews, R. Balan, J.J. Benedetto, W. Czaja, and K.A. Okoudjou: *Excursions in Harmonic Analysis, Volume 1* (ISBN 978-0-8176-8375-7)
T.D. Andrews, R. Balan, J.J. Benedetto, W. Czaja, and K.A. Okoudjou: *Excursions in Harmonic Analysis, Volume 2* (ISBN 978-0-8176-8378-8)
D.V. Cruz-Uribe and A. Fiorenza: *Variable Lebesgue Spaces* (ISBN 978-3-0348-0547-6)
W. Freeden and M. Gutting: *Special Functions of Mathematical (Geo-)Physics* (ISBN 978-3-0348-0562-9)
A. Saichev and W.A. Woyczyński: *Distributions in the Physical and Engineering Sciences, Volume 2: Linear and Nonlinear Dynamics of Continuous Media* (ISBN 978-0-8176-3942-6)
S. Foucart and H. Rauhut: *A Mathematical Introduction to Compressive Sensing* (ISBN 978-0-8176-4947-0)
G. Herman and J. Frank: *Computational Methods for Three-Dimensional Microscopy Reconstruction* (ISBN 978-1-4614-9520-8)
A. Paprotny and M. Thess: *Realtime Data Mining: Self-Learning Techniques for Recommendation Engines* (ISBN 978-3-319-01320-6)
A. Zayed and G. Schmeisser: *New Perspectives on Approximation and Sampling Theory: Festschrift in Honor of Paul Butzer's 85th Birthday* (978-3-319-08800-6)
R. Balan, M. Begue, J. Benedetto, W. Czaja, and K.A Okoudjou: *Excursions in Harmonic Analysis, Volume 3* (ISBN 978-3-319-13229-7)

H. Boche, R. Calderbank, G. Kutyniok, J. Vybiral: *Compressed Sensing and its Applications* (ISBN 978-3-319-16041-2)
S. Dahlke, F. De Mari, P. Grohs, and D. Labate: *Harmonic and Applied Analysis: From Groups to Signals* (ISBN 978-3-319-18862-1)
A. Aldroubi, *New Trends in Applied Harmonic Analysis* (ISBN 978-3-319-27871-1)
M. Ruzhansky: *Methods of Fourier Analysis and Approximation Theory* (ISBN 978-3-319-27465-2)
G. Pfander: *Sampling Theory, a Renaissance* (ISBN 978-3-319-19748-7)
R. Balan, M. Begue, J. Benedetto, W. Czaja, and K.A Okoudjou: *Excursions in Harmonic Analysis, Volume 4* (ISBN 978-3-319-20187-0)
O. Christensen: *An Introduction to Frames and Riesz Bases, Second Edition* (ISBN 978-3-319-25611-5)
E. Prestini: *The Evolution of Applied Harmonic Analysis: Models of the Real World, Second Edition* (ISBN 978-1-4899-7987-2)
J.H. Davis: *Methods of Applied Mathematics with a Software Overview, Second Edition* (ISBN 978-3-319-43369-1)
M. Gilman, E. M. Smith, S. M. Tsynkov: *Transionospheric Synthetic Aperture Imaging* (ISBN 978-3-319-52125-1)
S. Chanillo, B. Franchi, G. Lu, C. Perez, E.T. Sawyer: *Harmonic Analysis, Partial Differential Equations and Applications* (ISBN 978-3-319-52741-3)
R. Balan, J. Benedetto, W. Czaja, M. Dellatorre, and K.A Okoudjou: *Excursions in Harmonic Analysis, Volume 5* (ISBN 978-3-319-54710-7)
I. Pesenson, Q.T. Le Gia, A. Mayeli, H. Mhaskar, D.X. Zhou: *Frames and Other Bases in Abstract and Function Spaces: Novel Methods in Harmonic Analysis, Volume 1* (ISBN 978-3-319-55549-2)
I. Pesenson, Q.T. Le Gia, A. Mayeli, H. Mhaskar, D.X. Zhou: *Recent Applications of Harmonic Analysis to Function Spaces, Differential Equations, and Data Science: Novel Methods in Harmonic Analysis, Volume 2* (ISBN 978-3-319-55555-3)
F. Weisz: *Convergence and Summability of Fourier Transforms and Hardy Spaces* (ISBN 978-3-319-56813-3)
C. Heil: *Metrics, Norms, Inner Products, and Operator Theory* (ISBN 978-3-319-65321-1)
S. Waldron: *An Introduction to Finite Tight Frames: Theory and Applications.* (ISBN: 978-0-8176-4814-5)
D. Joyner and C.G. Melles: *Adventures in Graph Theory: A Bridge to Advanced Mathematics.* (ISBN: 978-3-319-68381-2)
B. Han: *Framelets and Wavelets: Algorithms, Analysis, and Applications* (ISBN: 978-3-319-68529-8)

For an up-to-date list of ANHA titles, please visit **http://www.springer.com/series/4968**

Printed in the United States
By Bookmasters